Index

Contents

United States

Alabama

Albertville 51
Alexander City 55
Anniston 39
Arab 51
Athens 51
Atmore 35
Auburn 37
Bay Minette 53
Bessemer 39
Birmingham 39
Brewton 35
Butler 43
Center Point 39
Chambers County . . . 37
Clanton 57
Collinsville 49
Cullman 41
Decatur 51
Demopolis 43
Dothan 45
Enterprise 45
Eutaw 43
Evergreen 35
Florence 47
Fort Payne 49
Gadsden 39
Guntersville 51
Hayneville 57
Huntsville 51
Jackson 35
Lafayette 37
Linden 43
Livingston 43
Marion 57
McIntosh 35
Mobile 53
Monroeville 35
Montgomery 55
Oneonta 41
Opelika 37
Ozark 45
Phenix City 201
Prattville 55
Prichard 53
Russellville 47
Scottsboro 51
Selma 57
Sheffield 47
Talladega 39
Troy 59
Tuscaloosa 39
Tuscumbia 47
Tuskegee 55
Wetumpka 55

Rural service area (RSA)

AL-1 . . . . . . 41, 47, 51
AL-2 99
AL-3 93
AL-4 95
AL-5 97
AL-6 99
AL-7 99
AL-8 93

Alaska

Anchorage 89
Fairbanks 97
Juneau 65
Palmer 67
Spenard 61
Wasilla 67

Rural service area (RSA)

AK-1 63
AK-2 67
AK-3 65

Arizona

Benson 77
Bisbee 77
Casa Grande 69
Chandler 75
Flagstaff 71
Florence 69
Gila Bend 75
Glendale 75
Holbrook 81
Kingman 73
Nogales 77
Phoenix 75
Show Low 81
Sierra Vista 77
Tempe 75
Tucson 79
Willcox 77
Winslow 81
Yuma 83

Rural service area (RSA)

AZ-1 73
AZ-2 71
AZ-3 81
AZ-4 83
AZ-5 69
AZ-6 77

Arkansas

Arkadelphia 97
Arkansas City 89
Ashdown 897
Batesville 85
Benton 97
Blytheville 95
Camden 89
Clarendon 91
Clarksville 99
Clinton 85
Conway 97
De Queen 87
El Dorado 89
Eureka Springs 93
Fayetteville 93
Forrest City 91
Fort Smith 93
Harrisburg 95
Harrison 485
Heber Springs 85
Hope 897
Hot Springs 97
Jonesboro 95
Little Rock 97
Magnolia 89
Marianna 91
Marion 807
Marshall 85
Monticello 89
Morrilton 99

Mountain Home . . . 85
Mountain View 85
. . . . 99
. . . . 93
. . . . 95
. . . . 97
. . . . 99
. . . . 99
. . . . 93
. . . . 89
. . . . 97
. . . . 897
. . . . 93
Walnut Ridge 95
Warren 89
Wynne 91

Rural service area (RSA)

AR-1 . . . . . . 93, 485
AR-2 85
AR-3 85
AR-4 95
AR-5 91
AR-6 . . . . . . 85, 99
AR-7 99
AR-8 . . . . . . 93, 99
AR-9 87
AR-10 97
AR-11 897
AR-12 89

California

Anaheim 113
Anderson 119
Atascadero 131
Bakersfield 101
Barstow 121
Bishop 103
Blythe 121
Borrego Springs . . . 127
Brawley 107
Bridgeport 103
Calexico 107
Chico 105
Corning 119
Crescent City 109
Delano 101
El Centro 107
Escondido 127
Eureka 109
Fort Bragg 137
Fresno 111
Grass Valley 123
Hanford 111
Hollister 115
Independence 103
Lakeport 137
London 133
Long Beach 113
Los Angeles 113
Los Banos 115
Madera 115
Merced 115
Modesto 123
Monterey 125
Oakland 129
Oceanside 127
Oroville 105
Orland 119
Oxnard 113
Palm Springs 121
Paradise 105
Placerville 117
Red Bluff 119
Redding 119
Ridgecrest 101
Riverside 121
Sacramento 123
Salinas 125
San Andreas 133
San Bernardino . . . 121

2

Index

3

Index

Plattsburg	469
Poplar Bluff	471
Rock Port	487
Rolla	477
St. Joseph	481
St. Louis	493
Savannah	487
Sedalia	495
Shelbyville	489
Sikeston	471
Springfield	497
Stockton	467
Thayer	499
Tuscumbia	477
Warrensburg	495
Waynesville	477
West Plains	485
Rural service area (RSA)	
MO-1	487
MO-2	469
MO-3	469, 483
MO-4	469
MO-5	489
MO-6	475
MO-7	495
MO-8	493
MO-9	467, 491
MO-10	467, 495
MO-11	477
MO-12	477
MO-13	493
MO-14	467
MO-15	485
MO-16	485
MO-17	499
MO-18	471
MO-19	471

Montana

Big Timber	507
Billings	501
Bozeman	505
Butte	511
Chinook	507
Columbus	501
Glasgow	503
Glendive	503
Great Falls	505
Hamilton	511
Hardin	501
Havre	507
Helena	511
Hysham	501
Kalispell	509
Lewiston	507
Miles City	501
Missoula	511
Poison	509
Red Lodge	501
Stanford	507
Rural service area (RSA)	
MT-1	509
MT-2	507
MT-3	503
MT-4	503
MT-5	511
MT-6	511
MT-7	507
MT-8	505
MT-9	501
MT-10	501

Nebraska

Albion	521
Alliance	527
Auburn	515
Aurora	513
Beatrice	515
Bellevue	523
Blair	301

Broken Bow	525
Central City	521
Columbus	521
Deshler	515
Fremont	521
Grand Island	513
Greeley	525
Hastings	515
Holdrege	517
Kearney	517
Lexington	517
Lincoln	519
Loup City	525
McCook	517
Norfolk	521
North Platte	517
O'Neill	525
Ogallala	517
Omaha	523
Ord	525
Randolph	521
Red Cloud	515
St. Paul	525
Scottsbluff	527
Seward	513
Sidney	527
Tecumseh	515
West Point	521
York	513
Rural service area (RSA)	
NE-1	527
NE-2	525
NE-3	521
NE-4	525
NE-5	521
NE-6	517
NE-7	513
NE-8	517
NE-9	515
NE-10	515

Nevada

Boulder City	535
Carson City	537
Elko	529
Ely	531
Fallon	533
Gardnerville	537
Hawthorne	531
Henderson	535
Las Vegas	535
Laughlin	535
Lovelock	533
Reno	537
Winnemucca	533
Yerington	537
Rural service area (RSA)	
NV-1	533
NV-2	529
NV-3	537
NV-4	531
NV-5	531

New Hampshire

Claremont	541
Concord	539
Derry	399
Dover	545
Keene	541
Laconia	539
Manchester	543
Nashua	543
Ossipee	539
Portsmouth	545
Rural service area (RSA)	
NH-1	541
NH-2	539

New Jersey

Atlantic City	547
Bridgeton	557

Camden	741
Cape May Court House	547
Cumberland County	557
Flemington	555
Freehold	549
Hammonton	547
Hunterdon County	555
Long Branch	549
Mercer County	555
New Brunswick	549
Newark	589
Newton	551
Ocean County	553
Paterson	589
Salem	159
Sussex County	551
Tom's River	553
Trenton	555
Vineland	557
Rural service area (RSA)	
NJ-1	555
NJ-2	553
NJ-3	551

New Mexico

Alamogordo	843
Albuquerque	559
Artesia	569
Aztec	565
Bernalillo	559
Carlsbad	569
Clayton	567
Clovis	561
Deming	563
Farmington	565
Hobbs	569
Las Cruces	843
Las Vegas	561
Los Alamos	571
Lovington	569
Moriarty	561
Portales	561
Raton	567
Roswell	569
Santa Fe	571
Santa Rosa	561
Socorro	559
Springer	567
Truth or Consequences	843
Tucumcari	561
Rural service area (RSA)	
NM-1	565
NM-2	567
NM-3	559, 843
NM-4	561, 571
NM-5	563
NM-6	569, 843

New York

Albany	573
Albion	595
Auburn	575
Batavia	583
Bath	577
Beacon	593
Binghamton	577
Buffalo	579
Canandaigua	595
Catskill	581
Catskill Mountains	585
Corning	577
Cortland	575
Dannemora	591
Dutchess County	593
Elmira	577
Endicott	577
Fonda	573
Fredonia	583
Fulton	597
Geneseo	595
Glens Falls	573

Chickasha	681
Clinton	681
Duncan	677
Durant	679
El Reno	697
Elk City	681
Enid	683
Guthrie	699
Guymon	685
Henryetta	695
Hugo	687
Idabel	689
Kingfisher	691
Lawton	693
Marietta	679
McAlester	695
Muskogee	695
Norman	697
Nowata	701
Oklahoma City	697
Okmulgee	695
Pauls Valley	679
Pawhuska	701
Pawnee	699
Perry	699
Ponca City	699
Pryor	701
Purcell	697
Sapulpa	701
Shawnee	697
Stillwater	699
Sulphur	679
Tulsa	701
Walters	677
Watonga	691
Weatherford	681
Woodward	683

Rural service area (RSA)

OK-1	685
OK-2	683
OK-3	699
OK-4	701
OK-5	681, 691
OK-6	695
OK-7	681
OK-8	677
OK-9	679
OK-10	687, 689

Oregon

Albany	705
Ashland	709
Baker	711
Bend	703
Corvallis	705
Eugene	707
Florence	707
Grants Pass	715
Hermiston	711
Hood River	717
Klamath Falls	703
La Grande	711
McMinnville	713
Medford	709
Moro	717
Newport	705
Pendleton	711
Portland	713
Roseburg	715
Salem	713
Springfield	707
The Dalles	717
Woodburn	713

Rural service area (RSA)

OR-1	713
OR-2	717
OR-3	711
OR-4	705
OR-5	715
OR-6	703

Pennsylvania

Allentown	719
Altoona	721
Beaver Falls	743
Bedford	723
Bedford County	723
Bellefonte	747
Belvidere	719
Blair County	721
Bloomsburg	749
Bradford	725
Brookville	735
Butler	727
Carlisle	733
Chambersburg	729
Clearfield	735
Connellsville	751
Doylestown	741
Du Bois	735
Easton	719
Ebensburg	737
Ellwood City	727
Emmaus	719
Erie	731
Franklin County	729
Fulton County	729
Greensburg	743
Harrisburg	733
Hollidaysburg	721
Honesdale	745
Huntingdon	725
Indiana	735
Jim Thorpe	719
Johnstown	737
Kittanning	727
Lancaster	733
Lansdale	741
Lebanon	733
Lewistown	733
Lock Haven	747
McConnellsburg	729
Meadville	739
Middleburg	749
Monroeville	743
New Castle (system A)	727
New Castle (system B)	673
New Kensington	743
Northeast PA (region)	745
Philadelphia	741
Pittsburgh	743
Port Matilda	747
Pottstown	741
Pottsville	749
Reading	719
Ridgway	725
Scranton	745
Shamokin	749
Sharon	673
Somerset	737
State College	747
Stroudsburg	745
Sunbury	749
Titusville	739
Tunkhannock	745
Uniontown	751
Washington	743
Waynesburg	751
West Chester	741
Wilkes-Barre	745
Williamsport	753
York	733

Rural service area (RSA)

PA-1	739
PA-2	725
PA-3	747
PA-4	745
PA-5	745
PA-6	673, 727
PA-7	735
PA-8	749
PA-9	751
PA-10	723, 729
PA-11	725, 733
PA-12	733

Rhode Island

Newport	755
Providence	755
Woonsocket	755

Rural service area (RSA)

RI-1	755

South Carolina

Aiken	195
Allendale	771
Anderson	765
Bamberg	771
Barnwell	771
Beaufort	757
Bishopville	763
Camden	759
Charleston	757
Columbia	759
Conway	769
Darlington	763
Dillon	761
Edgefield	767
Florence	763
Gaffney	611
Georgetown	769
Greenville	765
Greenwood	767
Hilton Head	757
Kingstree	763
Laurens	767
Manning	763
Marion	769
Moncks Corner	757
Myrtle Beach	769
Newberry	767
Oconee County	765
Orangeburg	771
Rock Hill	611
St. George	757
St. Matthews	771
Saluda	767
Seneca	765
Spartanburg	765
Sumter	763
Walhalla	765
Walterboro	757

Rural service area (RSA)

SC-1	765
SC-2	767
SC-3	611
SC-4	759, 761, 763
SC-5	769
SC-6	763
SC-7	771
SC-8	757
SC-9	611

South Dakota

Aberdeen	773
Alexandria	779
Belle Fourche	787
Brookings	775
Clear Lake	791
Custer	783
Elk Point	789
Hayti	791
Hot Springs	777
Huron	775
Kadoka	783
Leola	773
Madison	775
Milbank	791
Mitchell	779
Mobridge	781
Mound City	781

Index

9

Plains	887	Orem	915	South Boston		937
Plainview	887	Park City	915	South Hill		943
Pleasanton	861	Price	911	Standardsville		925
Port Arthur	827	Provo	915	Staunton		931
Port Lavaca	905	Richfield	913	Virginia Beach		939
Post	863	Roosevelt	917	Warrenton		389
Quanah	903	St. George	913	Waynesboro		931
Quitman	875	Salina	913	Winchester		949
Ranger	877	Salt Lake City	915	Wise		941
Raymondville	831	Vernal	917	Woodstock		949

Rural service area (RSA) — Texas:

Rio Grande	831	Rural service area (RSA)		
Robstown	837	UT-1		915
Roby	893	UT-2		915
Rosenberg	855	UT-3		913
San Angelo	889	UT-4		913
San Antonio	891	UT-5	911, 917,	1027
San Diego	861	UT-6		913

San Marcos	823	**Vermont**			
Seminole	863	Barre			921
Sequin	891	Bennington			919
Sherman	839	Brattleboro			541
Sierra Blanca	885	Burlington			919
Sinton	837	Hyde Park			921
Snyder	893	Middlebury			919
Sonora	889	Montpelier			921
Spearman	881	Rutland			919
Stanton	829	St. Albans			921
Stephenville	833	Stowe			921

Virginia and RSA:

Sterling City	829	Rural service area (RSA)			
Stinnett	881	VT-1			921
Stratford	841	VT-2		541,	919
Sweetwater	893				
Tahoka	895	**Virginia**			
Temple	907	Amherst			935
Texarkana	897	Appomattox			935
Texas City	849	Bedford			937
Throckmorton	851	Big Stone Gap			941
Tyler	899	Blacksburg			923
Uvalde	901	Charles City			945
Van Horn	885	Charlottesville			925
Vernon	903	Chatham			929
Victoria	905	Christiansburg			923
Waco	907	Colonial Heights			945
Wellington	881	Courtland			943
Wheeler	881	Culpeper			927
Wichita Falls	909	Danville			929
Winnie	855	Deltaville			939
Winnsboro	875	Emporia			943
Woodville	857	Farmville			933

Rural service area (RSA)			Fincastle			947
TX-1		841	Franklin			943
TX-2		881	Fredericksburg			389
TX-3	887,	895	Front Royal			949
TX-4	863,	903	Gate City			941
TX-5	851,	903	Halifax			937
TX-6	847,	877	Hanover			945
TX-7	839, 875, 883, 897,	899	Harrisonburg			931
TX-8	829, 863,	893	Hillsville			923
TX-9	833, 853,	877	Lawrenceville			943
TX-10	839, 845, 899,	907	Lexington			933
TX-11	871,	899	Lovingston			925
TX-12		885	Luray			949
TX-13		885	Lynchburg			935
TX-14		885	Marion			801
TX-15	859,	907	Martinsville			937
TX-16		825	Nelson County			925
TX-17		857	New Kent			945
TX-18		901	Newport News			939
TX-19		861	Norfolk			939
TX-20	861,	905	Norton			941
TX-21		855	Pearisburg			923
			Petersburg			943
Utah			Portsmouth			939
Cedar City		913	Pulaski			923
Duchesne		917	Radford			923
Flaming Gorge area		1027	Richmond			945
Kingston		913	Roanoke			947
Logan		915	Rocky Mount			933
Manila		1027	Rustburg			935
Ogden		915	Salem			947

Virginia RSA and Washington:

Rural service area (RSA)				
VA-1				941
VA-2				801
VA-3				923
VA-4			933,	937
VA-5				933
VA-6			925,	931
VA-7			933,	937
VA-8				943
VA-9				943
VA-10			389,	949
VA-11			389,	927
VA-12			389,	939

Washington		
Aberdeen		951
Bellingham		957
Bremerton		957
Centralia		953
Chehalis		953
Chewelah		959
Clarkston		961
Cle Elum		965
Colfax		961
Dayton		961
Ellensburg		965
Everett		957
Goldendale		713
Grand Coulee		955
Grandview		965
Hoquiam		951
Kelso		953
Kennewick		965
Longview		953
Montesano		951
Morton		953
Moses Lake		955
Olympia		957
Pasco		965
Pullman		961
Richland		965
Ritzville		955
San Juan Islands		957
Seattle		957
Shelton		951
Spokane		959
Tacoma		957
Tri-Cities		965
Vancouver		713
Walla Walla		961
Waterville		963
Wenatchee		963
Yakima		965

Rural service area (RSA)			
WA-1			957
WA-2			963
WA-3			959
WA-4			951
WA-5		955,	965
WA-6			953
WA-7			713
WA-8			961

West Virginia		
Beckley		967
Bluefield		967
Buckhannon		971
Charleston		969
Charlestown		975
Clarksburg		977

Index

Canada	International	

		St. Martin	1085
		St. Vincent & Grenadines	1086
		Singapore	1086

Alberta

		All countries . . .	1059-1088			Venezuela	1087

Calgary	1031
Edmonton	1033
Fort McMurray . . .	1033
Lethbridge	1031
Lloydminster . . .	1055
Medicine Hat . . .	1031
Red Deer	1031

Anguilla	1063
Antigua	1063
Argentina	1064
Australia	1064
Bahamas	1065
Barbados	1065
Barbuda	1063

Virgin Islands (British) .	1087
Virgin Islands (U.S.) . .	1088

British Columbia

Campbell River . . .	1035
Kamloops	1035
Kelowna	1035
Prince George . . .	1035
Vancouver	1035
Victoria	1035

Bermuda	1066
Bolivia	1066
Brazil	1067
Cayman Islands . . .	1067
Costa Rica	1068
Dominican Republic .	1068
Grenada	1069
Guadeloupe	1069
Guam	1070
Guatemala	1071
Hong Kong	1071
Jamaica	1072
Martinique	1072

Manitoba

Brandon	1037
Dauphin	1037
Riverton	1037
Steinbach	1037
Winnipeg	1037

New Brunswick

Bathurst	1053
Chatham	1039
Fredericton	1039
Moncton	1039
Saint John	1039

México	1073-1081
Acapulco	1079
Aguascalientes . . .	1078
Altamira	1076
Cadereyta	1076
Cancún	1080
Celaya	1078
Chihuahua	1075
Ciudad Cuahutemoc	1075
Ciudad Juárez . .	1075
Ciudad Obregón . .	1074
Cuernavaca . . .	1081

Newfoundland

Centerville	1041
Corner Brook . . .	1041
St. John's	1041

Northwest Territories

Yellowknife	1057

Delicias	1075
Ensenada	1073
Guadalajara . . .	1077
Hermosillo	1074
Irapuato	1078
Lebu	1078
León	1078

Nova Scotia

Amherst	1043
Halifax	1043
Sydney	1043
Yarmouth	1043

Matamoros . . .	1076
Mérida	1080
Mexicali	1073
México City . . .	1081
Monterrey	1076
Morelia	1077
Nogales	1074

Ontario

Kenora	1037
London	1047
North Bay	1047
Ottawa	1047
Rimouski	1053
Sault Ste. Marie . .	1047
Thunder Bay . . .	1045
Toronto	1047

Nuevo Laredo . .	1076
Puebla	1079
Querétaro	1078
Reynosa	1076
Salamanca	1078
Saltillo	1076
San Luis Potosi . .	1078
San Luis Río Colorado	1073
Tampico	1076
Tijuana	1073
Toluca	1081
Torreón	1075
Uruapán	1077

Prince Edward Island

Charlottetown . . .	1049
Summerside . . .	1049

Québec

Alma	1051
Mont-Laurier . . .	1051
Montmagny	1053
Montréal	1051
Rivière du Loup . . .	1051
St. Georges	1053
Sept. Iles	1053
Sherbrooke	1051
Thetford Mines . . .	1051

Veracruz	1079
Zacatecas	1078
Zamora	1077
Montserrat	1082
Netherlands Antilles .	1085
New Zealand	1082
Puerto Rico (U.S.) . .	1083
St. Kitts and Nevis . .	1084
St. Lucia	1084
St. Maarten	1085

Saskatchewan

Esteven	1055
Lloydminster . . .	1055
Maple Creek . . .	1055
North Battleford . . .	1055
Prince Albert . . .	1055
Regina	1055
Saskatoon	1055

Yukon Territory

Whitehorse	1057

Cellular Telephone

Products and Publications

The service I received was great! Thanks very much.

S. Smith Chicago, Illinois

I really appreciate your prompt attention to our order.
It is a pleasure to do business with you!

S. Whittaker Columbus, Indiana

Welcome

This catalog is provided for people who have requested more information about cellular products and publications. We have examined all items carefully to insure that they are of the highest quality and have included only those that we would be happy to own ourselves.

Your call is free

Call 1–800–927–8800 to order or ask questions. Our office is open most weekdays from 9 a.m. to 5 p.m. Pacific Time (which is noon to 8 p.m. Eastern Time). You may order by voice–mail 24 hours a day.

Order conveniently

Use an American Express, Discover, MasterCard or Visa. Or ask for C.O.D. shipment. If you are with a telephone company, cellular carrier, or government agency, we will be glad to bill you instead. Most orders are shipped out the same day they are received. If you require delivery within two days, one day, or the same day, just let us know.

Postage is free if you order by mail

Of course, postage is not charged if you pick up your order in Seattle. But postage also is free if you mail your order to us with a check. That lowers our costs, and we pass the savings on to you. Information about ordering by mail is on page 33.

Expect to be satisfied with our service

We want to provide the best products, prices, and service. In fact, we will take back any item that does not meet all your expectations and send you a full refund. We are here to serve you.

U.S. cellular coverage map

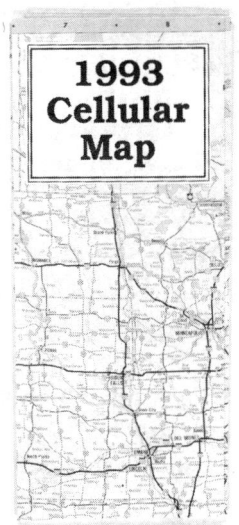

- Full–color map of all 50 states and lower Canada

- Cellular coverage areas highlighted in orange

- Shows clearly where cellular service is available

Measures 23 by 34 inches
Orders placed now ship April 1993
Folded $9. Unfolded in tube $12.

Many people have requested this map

Over the years, cellular users have asked for a national coverage map they can carry conveniently when they travel. Cellular industry personnel have requested a large coverage map they can hang on their walls. And readers of the *Cellular Travel Guide* have sought a summary map that includes all the smaller maps printed in the guide.

Perhaps this map will help you

The folded map fits easily in a glove–compartment or can be carried with other travel documents, ready when you want to check cellular coverage. The unfolded map looks great hanging on a wall, showing at a glance the areas that currently have cellular service. It is more expensive because it must be shipped in a tube. Both maps will be available in April 1993.

Choose this map if accuracy is important to you

The U.S. cellular coverage map was assembled from more than 500 city coverage maps printed in the latest *Cellular Travel Guide*. These, in turn, were created from individual system maps supplied by the cellular operating companies and from other reliable sources. Using this data, we were able to assemble the most accurate coverage map available today.

Please compare this map to our competitor's

Our map is larger. It is printed in five colors instead of two and is produced by professional cartographers. It includes roads, rivers, and geographic features, none of which are included on our competitor's. It is easy to read and has about twice as many cities. This is the map for you.

Call (800) 927–8800 to order by phone. Please see page 32 to order by mail.

Catalog

Cellular Telephone Program Handbook

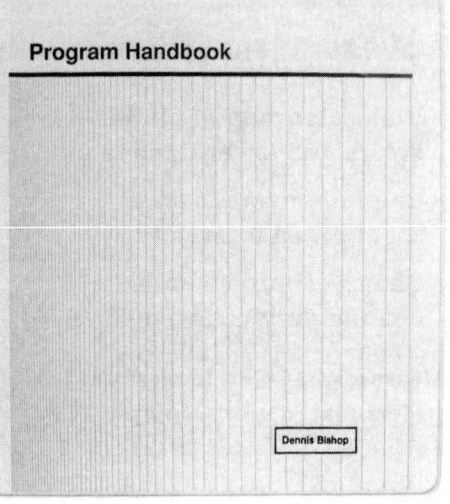

- **Tells how to program cellular telephones**

- **Complete index lists more than 360 models**

- **Specific instructions provided for each model**

396 pages in a 3–ring binder
Latest 1993 edition $149
Four quarterly updates $60

Find everything you need in one place

Open your program handbook to the index, which includes 360 models of cellular telephones. Then turn to the page for any model to find step-by-step instructions for programming the telephone's features and functions:

- telephone number
- lock code
- speakerphone

- preferred system
- system ID
- MIN option

- local use
- group ID
- other options

Save valuable time

Refer to your program handbook instead of hunting in individual manuals for the programming procedures you need. Then save time by programming the telephone right the first time. The *Cellular Telephone Program Handbook* helps you become organized to program more quickly.

Become prepared for whatever you may encounter

The handbook is designed for professional installers who have experience installing cellular telephones and want to locate complete programming procedures for different models in one place. You will learn tricks to help you handle whatever programming situations you encounter.

Choose a winner

The program handbook is written by an experienced installer who previously managed a major installation facility. He has included technical assistance numbers for all manufacturers, information about programming equipment, equipment suppliers, and system ID numbers. You will receive the latest edition of this best–selling cellular publication. Although it has doubled in pages, its price remains the same as in 1990.

Call (800) 927–8800 to order by phone.   Please see page 32 to order by mail.

DIAMONDTEL MESA 60, 60X AND 80X PORTABLES

NAM Type: E^2 PROM
Manufacturer: MITSUBISHI
Programmer: Handset
ESN Prefix: Dec 134 Hex 86
ESN/Serial Number Match: No
Handset Programmable: Yes
Number of Channels: 666

Programming Sequence: Power ON (END/FCN + 0) (simultaneously)
Hold CLR and Enter 6926232 (within 10 sec)
Enter Step Information then press SND
Press END to Write NAM Information

Step	Parameter	# of Digits	Normal
1	Mobile Number	10	XXX-XXX-XXXX
2	Security Code	4	XXXX
3	System ID	5	XXXXX
4	Local Use	1	1
5	MIN Opt	1	1
6	Initial Page Channel	3	333 or 334
7	Access Overload	2	XX
8	Preferred System	1	(0) B or (1) A
9	Group ID	2	XX
10	End-to-End Signaling	1	1
11	Roam Inhibit[1]	1	0
12	A/B Selectable[1]	1	1
13	Auto Lock[1]	1	0
14	Aux[1]	1	0

New Unlock Code: FCN + STO + 4 Digit Security Code (NAM) + New 3 Digit Unlock Code

System Select: FCN + 1 + X

X = 0	AB or BA Standard
X = 1	BA or AB Alternate
X = 2	Home
X = 3	Pref Only
X = 4	Non-Pref Only

[1]MESA 60X ONLY

Catalog

Product Operation Handbook

- How to operate different cellular telephones

- Instructions for 336 models by brand name

- Includes 210 drawings to help identify models

504 pages in a 3–ring binder
Latest 1993 edition $149
Four quarterly updates $60

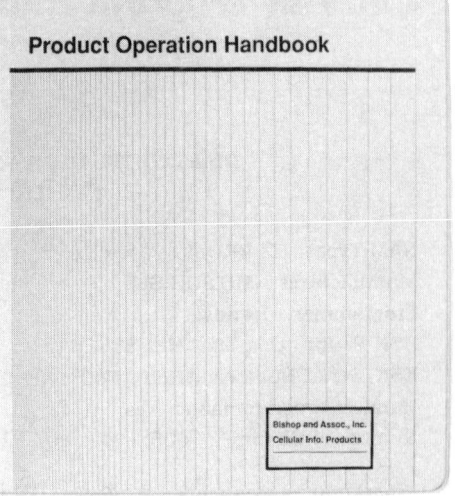

Product Operation Handbook

Bishop and Assoc., Inc.
Cellular Info. Products

Look up any cellular telephone in seconds

If you need to answer questions about different cellular telephone models, you will appreciate the convenience of this handbook. It condenses the most important instructions for every model into no–nonsense language and puts them in a convenient binder. Now when you encounter a new telephone or a customer calls with questions, all you need to know will be at your fingertips in seconds.

Better than owning 336 owner's manuals

You can find any telephone you choose in alphabetical order. For example, when you turn to the pages for the Motorola 8000X Portable, you find a line drawing of the telephone and two pages of operating instructions for its functions. "To set a new unlock code, press FCN + 6 Digit Security Code + New 3 Digit Unlock Code + STO. To display the telephone's own number, press RCL + # + RCL."

Know how to activate each telephone's functions

The following functions are included for most models: Adjusting the ringing volume, locking and unlocking the handset, storing and recalling telephone numbers from memory, viewing and resetting the call timer, selecting the A or B system, activating and deactivating the horn alert, determining the battery level, redialing the last number, using the hands-free option, and turning the silent ring option on and off.

Now you can master any cellular telephone

Your handbook makes it easy to operate any model of cellular telephone and control its functions like a professional. And to keep current, you also can order a year of updates, or use the order form in the handbook.

Call (800) 927–8800 to order by phone.   Please see page 32 to order by mail.

PANASONIC 6104EB

Stamped Transceiver Model: EF 6104EB

Handset Model: EB 1001 (Cradle)

Cosmetic Features: Illuminated Status Indicators

Display: LCD

Antenna Connector: TNC

Date Code: None

Key: MK1 with 5mm Allen Wrench

Programmer: BYTEK, CURTIS or MOTOROLA-CELNAM

AMPS Compatible: Yes

Number of Channels: 666

Marketed By: PANASONIC Distributors and Dealers

Quick Disconnect Handset: Yes

Wiring:	**Battery:**	Red	**Ground:** Green
	Horn:	Black	**Ignition:** White

Transceiver Dimensions (Inches): **Width:** 9-7/16" **Depth:** 6-5/16" **Height:** 2-3/4"

PANASONIC 6104EB

AUTOMATIC LOCK: *(NAM OPTION) Phone locks when powered "ON"*

CALL IN ABSENCE INDICATOR: END + # (simultaneously 2 sec.) + Ignition Off

CUMULATIVE CALL TIMER-READ: F + 2

DTMF TONE DURATION *(NAM OPTION)*

HANDSET VOLUME: Dial (left side of handset)

HORN ALERT: To activate: END + # (simultaneously for 2 sec.) + (press keys a second time until HORN Illuminates solid) + Ignition Off

To deactivate: Ignition On

INDIVIDUAL CALL TIMER-ACTIVATE F + 1

LAST NUMBER REDIAL: F + 0 + SND

LOCK: END + • (simultaneously for 2 sec.)

MUTE/HOLD: MUTE

PERMANENT MEMORY: *(NAM OPTION) Memory Location 00*

POWER ON/OFF: Ignition On/Off or POWER (left side of cradle)

RECALL FROM MEMORY: RCL + Memory Location (01-30) + SND

SCROLL DISPLAY-FORWARD: END (hold)

SPEAKER VOLUME: Dial (right side of handset)

SPEED DIAL: RCL + 0 + 0 + SND

STORING IN MEMORY: Telephone Number + STO + Memory Location (01-30)

SYSTEM SELECT: F + X
- X=7   Normal (AB or BA)
- X=8   Preferred (A Only or B Only)
- X=9   NonPreferred (A Only or B Only)

UNLOCK: 2, 3 or 4-Digit Unlock Code

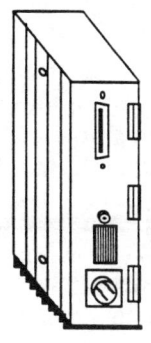

Catalog

Three–Book Cellular Package

- Three popular books at a great discount

1. Cellular Telephone Program Handbook

2. Cellular Telephone Installation Guide

3. Cellular Product Operation Handbook

All three for only $295

Own a library of books that make your life easier

If you work with cellular telephones, this three–book package will save you time and trouble. It includes everything you need to program, install, and operate all cellular telephone models. And you save $52 compared to purchasing the three books separately.

1. Program any cellular telephone you encounter

The program handbook includes step–by–step instructions for programming more than 360 models of cellular telephones. It lists the procedures for setting all features and functions, technical assistance telephone numbers for all manufacturers, information about suppliers of programming equipment, and system identification numbers.

2. Perform installations quickly and professionally

The cellular installation guide includes 182 pages of tips and techniques that help every cellular installation run smoothly. The contents include: Cellular history, how cellular operates, installation tools, installation step by step, antenna installation, wiring codes, telephone types, telephone programming, features glossary, buyer's guide, troubleshooting, tips for 10 problem vehicles, customer service skills, and assistance numbers.

3. Master the functions of all cellular models

If you work with more than one brand of cellular telephone, you'll appreciate the operation handbook. It includes basic operating instructions for 336 telephone models and 210 diagrams to identify handsets. Now when a customer calls with questions or brings a telephone in the door, you'll find the answers you need in seconds.

Call (800) 927–8800 to order by phone. Please see page 32 to order by mail.

Cellular Sales & Marketing

- **Newsletter with 12 monthly issues**
- **Promotion and sales strategy**
- **New ideas for the cellular agent**

Approximately 10 pages
Mailed to you each month
$347 annual subscription

Become serious about selling

This newsletter is full of ideas to help you attract new customers, keep them happy, and improve your business. Dealers discover promotion plans that work, and customer service managers find new ways to add customers and reduce churn. Both learn about the latest cellular telephone products before they are released to the public.

To be the best, imitate the best

You will learn how companies in other parts of the country are selling cellular. Some of their ideas have had phenomenal results. In fact, the best cellular promotions cost very little and result in tremendous media coverage. You can benefit from the experience of the best.

You will be the first to know

Changes in the cellular industry create opportunity, but you need to act while the news is fresh. *Cellular Sales & Marketing* keeps you on top of what's happening and helps you to predict where cellular is headed— while it is still early enough for you to take action.

Each issue keeps you informed

You'll find articles that focus on successful cellular dealers and operating companies. News briefs tell you what is happening in the industry. Product news covers the latest cellular equipment and accessories. And a calendar section lists upcoming conventions, seminars, and workshops. Here are just a few recent articles:

- Top seller for Centel finds best sales opportunities at community events.
- Follow–up phone call helps keep new customers after first bill arrives.
- Texas dealership tells how it signed 10,000 customers per month.

Call (800) 927–8800 to order by phone. Please see page 32 to order by mail.

Catalog

Dual Mode Cellular

- Describes the dual mode transition to digital cellular
- How dual mode will affect carriers and equipment
- How the new digital technology will work

Published 1992
280 pages
$95 hardback

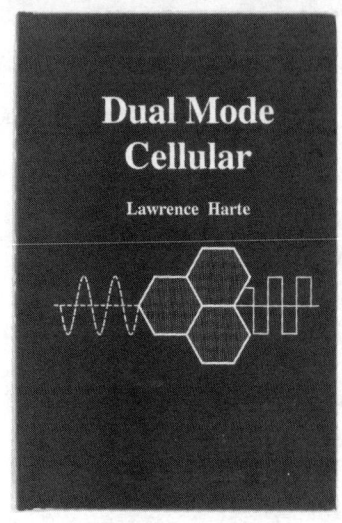

Are you ready for the transition to digital cellular?

This clearly and professionally written book will help you prepare for the coming of digital cellular, which will involve a transition period when both analog and digital service will operate simultaneously. There are three sections—first a complete overview of the existing analog cellular network and equipment, second a description of how the new dual mode technology has evolved from analog, and third considerations such as costs, patents, implementation options, and alternative technologies.

Your Dual mode questions are answered in this book

- *Analog* cellular system attributes, including mobile station signaling and call processing, as well as network design and operation.
- Dual mode *system* attributes and parameters: capacity expansion, frequency allocation, digital signal processing, and channel structure.
- Dual mode *mobile telephone* parameters with detail on both analog and digital voice channel signaling.
- Dual mode *network operation* to and from landline and mobile digital telephones, handoff, change of preferred mode, and authentication.
- Cellular system *economics,* which involve mobile telephone costs, operational costs, and dual mode marketing considerations.
- Competing *new technologies*, including frequency division multiple access, time division multiple access, and code division multiple access.

Your author is an expert on digital cellular technology

Lawrence Harte is a digital cellular development engineer for Audiovox. He has developed patents for analog and TDMA cellular equipment and has conducted presentations and seminars on digital cellular technology.

Call (800) 927–8800 to order by phone.   Please see page 32 to order by mail.

Addresses of all cellular carriers

- Includes all carriers in the *Cellular Travel Guide*
- Contact every company quickly and easily
- Choose a booklet, labels, or computer diskette

$39 printed booklet
$49 peel and stick labels
$59 computer diskette

Nearly 1000 company addresses and telephone numbers

All cellular operating companies in Canada, the Caribbean, Mexico, and the United States are listed in an easy–to–use format. And each address and telephone number has been checked twice for accuracy. Imagine how easy it will be to have all the information in one place when you own the booklet, mailing labels, or computer diskette.

Handy booklet is sorted both by company and by city

Your list includes the company name, address, and telephone number of every cellular carrier in a booklet perfect for your desk or bookshelf. Find what you need either alphabetically by company name, or alphabetically by the name of the city and state. Save time by calling the correct number the first time, and save money by not calling directory assistance.

Labels make mailings quick and easy

When you need to send a message to other cellular carriers, these labels make the job easy. Just peel them off and stick them to you mailing piece. You won't need to type addresses or convert data on a computer. This is the most convenient way to send your mailing.

Use the list over and over again if you own the diskette

Load the entire list of cellular carriers onto your computer and use your own software to make labels or use it as you please, provided you do not resell the list in any form. Please select one option from each line:

- computer   IBM–compatible or Macintosh
- disk size   3.5–inch hard diskette or 5.25–inch floppy disk
- software   DBase3, Excel, FileMaker Pro, Word, Works, WriteNow

Call (800) 927–8800 to order by phone.   Please see page 32 to order by mail.

Catalog

The Cellular Radio Handbook

- **The best reference for cellular system operation**
- **How to design, install, and operate a system**
- **Includes 42 chapters of text, 300 illustrations**

Large format
750 pages
$185 second edition

Everything you need to know about cellular systems

This reference book provides you detailed information about every aspect of operating a cellular system. You will find it well organized and easy to read, since even the technical chapters are written in the clearest English possible. Site operators, engineers, managers, technicians, and marketing personnel will need to refer to it frequently.

42 chapters cover every facet of cellular systems

What is cellular radio?
World system standards
Basic radio
Cell site selection and system design
Radio survey
Cellular radio interference
Cell plans
Units and concepts of field strength
Radio base stations
Base–station control and signaling
Cellular repeaters
Cell site antennas
Microwave and other cellular links
Base station maintenance & protection
Basic switching and trunking
Traffic engineering
More about switching and trunking
Interconnection & inter–carrier tolls

Mobile installation
Towers and masts
System installation
Equipment shelters
Budgets and cost estimates
Billing systems
Marketing
Analog & digital modulation
Noise and noise performance
Roaming
Data over cellular
Privacy methods
Rural & offshore applications
Digital GSM, CDMA, TDMA
Cordless PCN, CT2, CT3
Other mobile products
Glossary of terms
(An index is included.)

Call (800) 927–8800 to order by phone.

Please see page 32 to order by mail.

Cellular Installation Book and Videotape

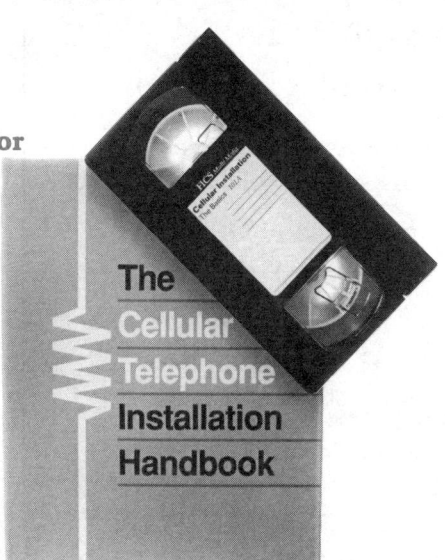

- Step–by–step instructions for installing cellular phones

- How to organize the shop, select antennas, do repairs

- Discover tips for faster and better installations

Book is 237 pages
Video is VHS and 35 minutes
Book $49, Video $49, Both $89

The Cellular Telephone Installation Handbook

Even old installers can learn new tricks

The best cellular installers have years of experience under their belts, but they keep on the lookout for new installation techniques that can save time and trouble. They'll find plenty of ideas here.

Your author has installation experience

Author Michael Losee shares tips he has learned from supervising more than 10,000 cellular installations. For example, he lists two cases when you should use a 0–db gain antenna, and four ways to unlock a cellular telephone when you don't know the unlock code. There's even advice on what to do if you drill through the gas tank.

If you install cellular, this book is for you

You'll enjoy reading this book because it is well organized and includes 110 diagrams and photographs. Here are the main topics: How to precheck a vehicle, plan installation, connect wires, install antenna, program phone, mount transceiver, test the equipment, install rural or marine, repair mistakes, and troubleshoot problems. Includes a glossary of installation terms and a list of manufacturers.

Make some popcorn and plug in the videotape

You're going to visit a Texas ranch to observe a complete cellular installation from start to finish. Along the way, you'll become acquainted with the tools of the trade, the steps to follow in completing the installation, and techniques that make the work flow smoothly. Afterward, you can see how much you've learned by taking the short quiz enclosed with the videotape. It's a lot of fun!

Call (800) 927–8800 to order by phone. Please see page 32 to order by mail.

Catalog

Simplified Cellular

- Explains how cellular works in simple terms
- Covers the phones, the network, the industry
- 136 illustrations make the concepts clear

161 pages
1993 edition
$29

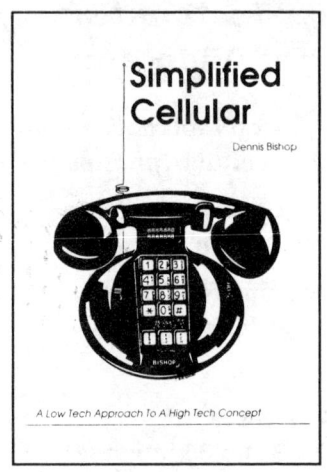

Read an introductory guide to cellular service

Written for non–technical personnel in the cellular industry, this book explains all aspects of cellular service in plain English. You will enjoy its tongue–in–cheek approach and the numerous illustrations that make cellular's high–tech concepts easy for beginners to understand.

Find answers to your questions about cellular

How does a cellular telephone work? What is involved in a mobile installation? How do you troubleshoot telephone problems? Where can a cellular telephone be used? Are all cellular portable batteries the same? How does cellular service differ from landline service?

Table of contents

How it works:	Cellular evolution
	The network
	Roaming
Telephones and telephone service:	Customer equipment
	Telephone activation
	Troubleshooting
	Equipment repair
Finding and retaining customers:	Marketing
	Customer retention
Reference:	Telephone features
	Programing terms
	Additional resources

Call (800) 927–8800 to order by phone.

Please see page 32 to order by mail.

Digital Cellular Radio

- **Tells cellular's past, present, and future**
- **How digital network will affect cellular**
- **How new digital technology will work**

448 pages
$89

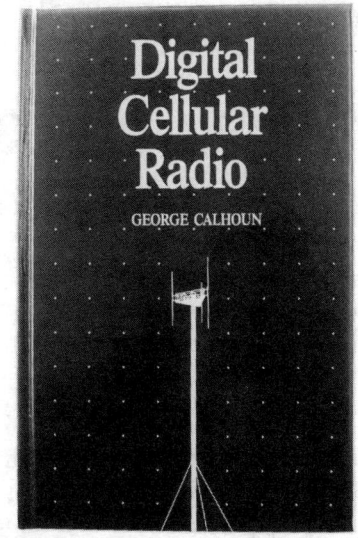

You must read this book

If your career is in the cellular industry, you cannot afford to miss this book. It is the best written and most complete text about cellular communications. Beginning with the history of mobile service and the foundations of cellular, it shows how the industry developed as it is today. Then it clearly explains how current systems and digital systems operate, and how digital technology will revolutionize the industry.

The contents are complete

Introduction: The uncertain future now facing mobile telephony. The age of FM: Mobile radio before cellular. The cellular idea. Realities of cellular: Cost, performance, efficiency, and licensing.

Basic concepts of digital communications: The emergence of digital communications. The digital vocabulary: Coding, transmission, and signal processing. Advantages of digital communications.

Design challenges for mobile telephony: Designing for frequency reuse and other special considerations of the mobile environment.

Technological alternatives for the next generation: The broad technology alternatives. Alternatives for the radio link and system architectures at the cell level and network level. Evaluating the future alternatives by cost, spectrum efficiency, and other criteria.

Managing the transition: Standards, compatibility, competition.

Own a complete reference book about cellular radio

Author George Calhoun is an authority on cellular radio, and his clearly written text is replete with explanations and diagrams you will want to refer to often. The well–organized contents are indexed.

Call (800) 927–8800 to order by phone.   Please see page 32 to order by mail.

Please see page 32 to order by mail.

Catalog

Cellular Investor

- **Industry financial news and analysis**

Approximately 10 pages
12 monthly issues for $575

Current cellular news
This is the only newsletter that focuses on financial news about the cellular industry. News reports cover sales, mergers, and takeovers of companies, plus purchases of operating licenses in specific geographic markets. Shows you the best opportunity to profit from industry growth.

Make better predictions about the future
The newsletter tells you about trends that will affect the value of cellular stocks. It projects subscriber numbers for each company and shows which stocks are under–priced based on their *private market value*.

Professionally written for cellular investors
You will find articles about the industry and individual companies, plus tables and graphs that make the information easy to use. Sample topics:
- GM poised to become a major distributor of cellular telephones.
- Celutel finds a buyer for $22.5 million in preferred stock shares.
- Southwestern Bell acquires 20% equity in Mexico's Telmex Co.

The RSA Newsletter

- **All about cellular in rural service areas**

Approximately 10 pages
$550 for 12 monthly issues

What you need to know
Keep informed about the status of all RSAs—who is acquiring them, what they are paying, and why. Learn which RSAs are the most valuable. Each issue also features in–depth articles about one U.S. state. Sample topics:

- Profile of companies that offer RSA management services.
- FCC update of tentative selectees, construction grants, licenses.
- RSAs ranked by anticipated population growth for the next five years.
- Table of recent RSA sales and table of RSA per–pop pricing trends.
- New RSA buyer arrives on the scene with venture–capital backing.

Call (800) 927–8800 to order by phone.   Please see page 32 to order by mail.

Mobile Cellular Telecommunications Systems

- **A reference for cellular system design & operation**
- **All the facts about cellular systems in one place**
- **The most popular book available on the subject**

450 pages
$76

You may be surprised

You may have thought it was impossible to find all the tools and information for designing and implementing a superior cellular system in one book. Now you've found it. It begins with an introduction to cellular systems and covers everything from cell placement to handoffs, cell splitting, switching, and data links. It is written from a technical point of view.

Find all this information in one book

- The elements of cellular radio system design
- Specifications of mobile and land stations
- Cell coverage for signal and traffic: Water, foliage, distance, etc.
- Cell–site antennas and mobile antennas
- Cochannel and non–cochannel interference reduction methods
- Frequency management and channel assignment
- Handoffs: Initiating, delaying, forced, queuing, intra/inter system
- Operational techniques and technologies
- Switching and traffic: Analog and digital, MTSO connect, etc.
- Data links and microwaves
- System evaluations; digital systems; related cellular topics

Your author is an expert on cellular systems

Dr. William Lee has both an M.S. and Ph.D. in electrical engineering from Ohio State University. A pioneer in mobile radio communications, he has worked with advanced programs for military communications systems. He has served as vice president of corporate technology and development at PacTel Cellular, and as co–chairman of CTIA's subcommittee for advanced radio techniques.

Call (800) 927–8800 to order by phone. Please see page 32 to order by mail.

Catalog

PowerStar puts an AC outlet in your car

- Powers electric accessories from your cigarette lighter
- Designed to produce clean, high–quality electricity
- Efficient enough to run a 13" color TV for 6 hours

Less than 1 pound
5 x 2.6 x 1.7 inches
Includes one–year warranty
List $149.95. Your price $129.

Imagine what you can power in your car. . .

✓ television & VCR	✓ fax	✓ shaver
✓ tape recorder	✓ stereo	✓ blender
✓ battery charger	✓ lights	✓ fan
✓ computer & printer	✓ pump	✓ power tools

Add some convenience to your life

Now you'll have power wherever you need it: Camping, picnicking, commuting, parking, even tailgate partying. When the power fails, run an extension cord into the house to run a lamp and television.

Home Power Magazine said this about PowerStar:

"The PowerStar is so simple to use we didn't need instructions...Its case is a beautiful piece of extruded, anodized aluminum...The PowerStar ran all the test loads without a problem. This included a variety of inductive loads using motors and power supplies. We had enough faith in the PowerStar to plug in over $4000 worth of computer equipment...The PowerStar offers incredible value in a midget portable inverter...The list of appliances that the PowerStar will run is endless—just about anything under 150 watts, indefinitely."   Evaluated in the February 1990 issue

Buy the highest quality and most reliable power supply

Unlike cheaper converters, PowerStar safely runs electronic equipment.
Converts 10–15 volts DC to 115 volts AC, 60 Hz modified sine wave.
400 watts peak, 200 watts for two minutes, 150 watts indefinitely.
Automatic overload shutdown/recovery; heat–based power reduction.
Note: Run engine for 10 minutes after 1 or 2 hours of 150–watt use.

Call (800) 927–8800 to order by phone.   Please see page 32 to order by mail.

Logos Automatic Call Forwarding Machine

- Receive your calls anywhere you travel in the world

- Use any phone to change the number calls are forwarded to

- Even reaches cellular phones via roamer access numbers

One–year warranty
List price $245. Your price $225.

Send your calls where you want, when you want

Logos call forwarder gives you complete control over your calls. In the standard forwarding mode, it answers a call and tells the caller: "Please hold, your call is being forwarded." Then it sends the call to whatever number you choose—you can change the number it forwards to from any other Touch–Tone telephone. If the line is busy or unanswered, the call goes back to your answering machine.

You can forward only selected calls if you wish

In the call–screening mode, *Logos* only forwards calls if the caller presses your three–digit number after hearing your answering machine message. Your message could say: "Please leave a message, or you may call me on my car telephone by dialing 888 now." The caller presses 888 to reach you or returns to the answering machine if you are out. Or for privacy, tell only your close friends the 888 code.

Logos can forward your calls to two numbers

Your answering machine could say: "Please leave a message, or in an emergency dial 855 to reach Mr. Smith's voice pager or 866 to reach Mrs. Smith at work." The caller is connected within seconds.

Logos is ready to plug in and use—and easy to control

You can use any phone to change how *Logos* forwards your calls. Just call and dial your secret ID code; then *Logos* gives you voice instructions on how to change all the options or even turn *Logos* off. Before you order *Logos*, please call your telephone company and order 3–way calling (also called conference calling). It is available in most urban areas and is necessary to use *Logos*. You also need a standard telephone wall outlet. An answering machine is optional.

Call (800) 927–8800 to order by phone.   Please see page 32 to order by mail.

Catalog

Expect the best service

Call 1–800–927–8800 toll–free if we can help you in any way. We ship most orders the same day we receive them, and everything you order is guaranteed to satisfy you, or you will be paid a full refund.

To order by telephone

• Call 1–800–927–8800 or 206–232–8800. Our office is open weekdays from 9 a.m. to 5 p.m. Pacific Time (noon to 8 p.m. Eastern Time). You may order on voice–mail 24 hours a day.

• Provide your American Express, Discover, MasterCard or Visa number and expiration date. Or request C.O.D. shipment. If you are ordering for a telephone company, cellular operating company, or government agency, you may request that a bill be sent with your order instead.

• You pay the actual cost of postage. We ship most orders by ground mail to save you money, but we can ship by priority air, second–day air, next-day air, or same–day air if you request it when you call.

To order by mail

• Please list the items you are ordering on the lines below or on a separate piece of paper. Then write your name and address below or attach your business card. Your telephone number is helpful if we have questions.

• Postage is free if you mail a check with your order. Washington state residents should add 8.2% sales tax. Canadians, please write *U.S. dollars* on your cheque or add 30% if paying in Canadian dollars.

page	item	quantity	price each	total
____	_____	_____	_____	_____
____	_____	_____	_____	_____
____	_____	_____	_____	_____
____	_____	_____	_____	_____

Postage is free if a check accompanies your order . . . .   free
Washington residents add 8.2% sales tax . . . .   _____
Amount enclosed . . . .   _____

Your telephone number: (_____)_____

Your name and address: _____

_____

_____

_____

Mail to: Communications Publishing, Box 500, Mercer Island, WA 98040

Order form

Cellular Travel Guide
Fourth Edition

This roaming guide provides all the facts you need to use your cellular telephone wherever you travel, including more than 50 countries around the world.

Use this express order form or use the regular order form on page 32.

- Your complete guide to all U.S. and Canadian cellular service.

- More than 500 maps show where you can use your telephone.

- Instructions tell how to send and receive calls easily in every city.

- Includes all new service areas, rates, and roaming agreements.

- International section covers the Caribbean, Mexico, and the world.

Reply card

Join our mailing list.

☐ Please notify me when the next edition of this guide is published.

Name
_____

Company
_____

Address
_____
_____
_____

Please write your mailing address on the left or attach your business card.

You do not need to send this postcard if you have ordered from us previously.

Number of copies	Price per copy
1 - 3	$19.90
4 - 7	17.90
8 - 15	15.90
16 - 31	13.90
32 - 63	11.90

Number ordered _____

Shipping is free if you enclose payment with your order. Otherwise it is billed at cost.

Washington residents, add 8.2% sales tax or resale card.

Canadians, send equivalent or mark cheque "U.S. funds."

For larger orders, please call (206) 232-8800.

Your name please:

_____

☐ Send to address on enclosed check.
☐ Send to address below.

_____

_____

_____

_____

Telephone: _____

Please mail this card in an envelope with your check or money order.

Communications Publishing
PO Box 500
Mercer Island, WA 98040-0500

Return address is
on opposite side

Please
apply
stamp

Mail to:

Communications Publishing
PO Box 500
Mercer Island, WA 98040-0500

Distinctive Ring Switch

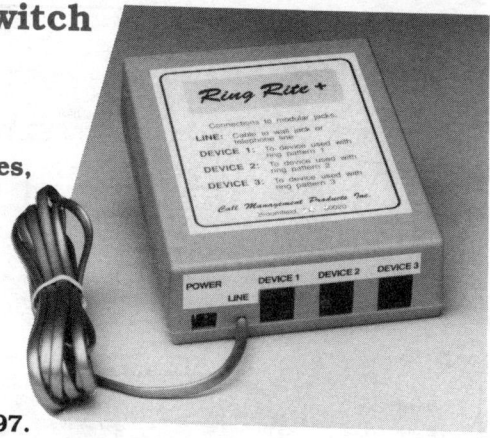

- **Give one telephone line three different numbers**

- **Works with modems, faxes, and answering machines**

- **Easy installation and automatic operation**

Only 5.5 x 4.5 x 1.6 inches
One–year warranty
List price $119. Your price $97.

Imagine three telephone numbers from a single line

Now you can have a separate number for your fax machine, modem, personal line, or business line, without paying an expensive monthly fee for each line. Distinctive ringing gives you additional telephone numbers, and each one makes your telephone ring a different way. One number will ring once, another rings quickly twice, a third rings quickly three times.

RingRite can tell the rings apart

If you use the double–ring number for your fax machine, you can send calls to your fax machine when you hear a double ring. But what if you're not in when a fax arrives? RingRite hears the double ring and routes the call to the fax machine automatically. It handles all three types of rings.

Distinctive ringing is better than normal fax switches

Some companies make switches that listen for fax or modem tones and route your calls. But they don't give you a separate telephone number for each device, and they don't transfer automatically when someone calls manually and requests your fax or modem. Distinctive ringing is better, and Ring Rite routes your distinctive ring calls automatically.

RingRite is simple to install

First call your telephone company and order distinctive ringing, also known as Custom Ringing, IdentaRing, MultiRing, Personalized Ringing, RingMaster, Ringmate, and Smart Ring. Then plug RingRite into your incoming phone jack and a power outlet. In the back of RingRite you can plug in one device for a standard ring, and two devices for two distinctive ring patterns. It's that simple. RingRite offers several special features:
- It is also works on a PBX system.
- It can be set to send devices a single normal ring if they require one.
- Barge–in protection keeps calls in progress from being interrupted.

Call (800) 927–8800 to order by phone.   Please see page 32 to order by mail.

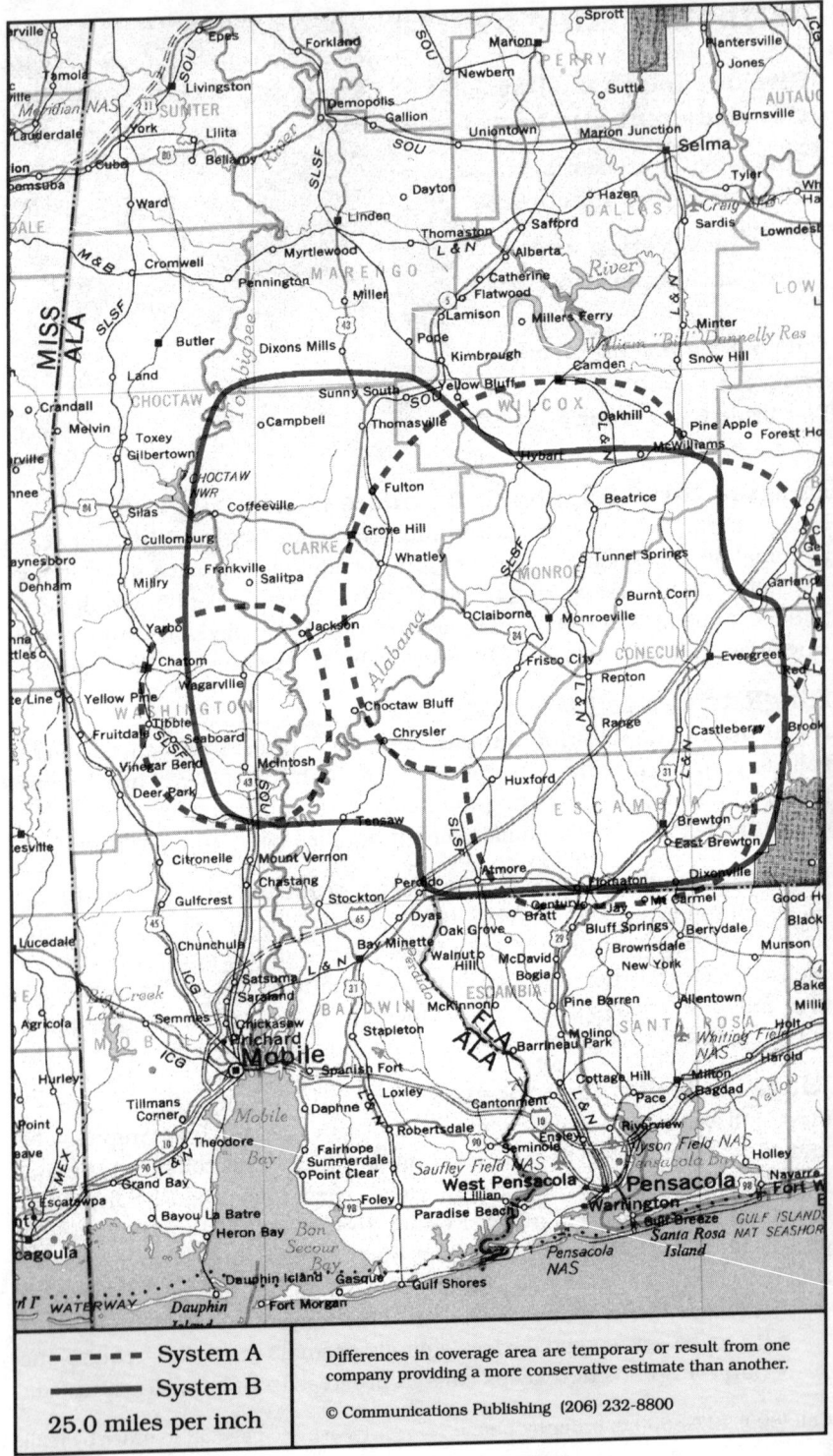

- - - - - System A
——— System B
25.0 miles per inch

Differences in coverage area are temporary or result from one
company providing a more conservative estimate than another.

© Communications Publishing (206) 232-8800

34

Atmore, Alabama

Atmore, Brewton, Evergreen, Jackson, McIntosh, Monroeville

System A	System B

System A

Cellular company

Cellular One
Licensed to: Pro-Max Communications
Managed by: Prime Cellular
522 Pike Street
Monroeville, AL 36460

Local office	800-239-8255
. . . .	205-575-2355
Corporate office Prime Cellular	201-227-1434

Sending calls

Visitor dialing instructions

Local calls	area code + 7 digits
Long distance . . .	1 + area code + 7 digits

Speed-dial numbers

Customer service	611
Emergency	911
WHOD Jackson	*945
WPIG Evergreen	*933

Receiving calls

Caller dials the roamer access number below,
hears tone, dials cellular area code and number.
. . . . 205-944-7626

Billing information

Visitor rates
Most visitors . . . . . $3 day + 85¢ minute

Visitors without roaming agreements (p. 1159)
Roamer Plus will intercept first call to bill you.

Identification numbers

City	Market	System	Billing	RSA
All served cities	312	01011	00000	AL-6

System B

Cellular company

Southeastern Cellular
210 Brookwood Road
Atmore, AL 36504

1 Henderson Street
Brewton, AL 36426

1400 College Avenue
Jackson, AL 36545

210 S Alabama Avenue
Monroeville, AL 36460

Local office	Atmore	205-368-1000
. . . .	Brewton	205-867-5004
. . . .	Jackson	205-246-7946
. . . .	Monroeville	205-575-5500
Corporate office	from Alabama	800-521-2090
. . . .		205-846-2090

Sending calls

Visitor dialing instructions

Local calls	area code + 7 digits
Long distance . . .	1 + area code + 7 digits

Speed-dial numbers

Billing information	811
Customer service	611 or 711
Emergency	911

Receiving calls

Caller dials the roamer access number below,
hears tone, dials cellular area code and number.
. . . . 205-294-7626

Billing information

Visitor rates
Most visitors . . . . . $3 day + 85¢ minute

Visitors without roaming agreements (p. 1159)
Cannot call since company takes no credit cards.

Identification numbers

City	Market	System	Billing	RSA
All served cities	312	01012	00000	AL-6

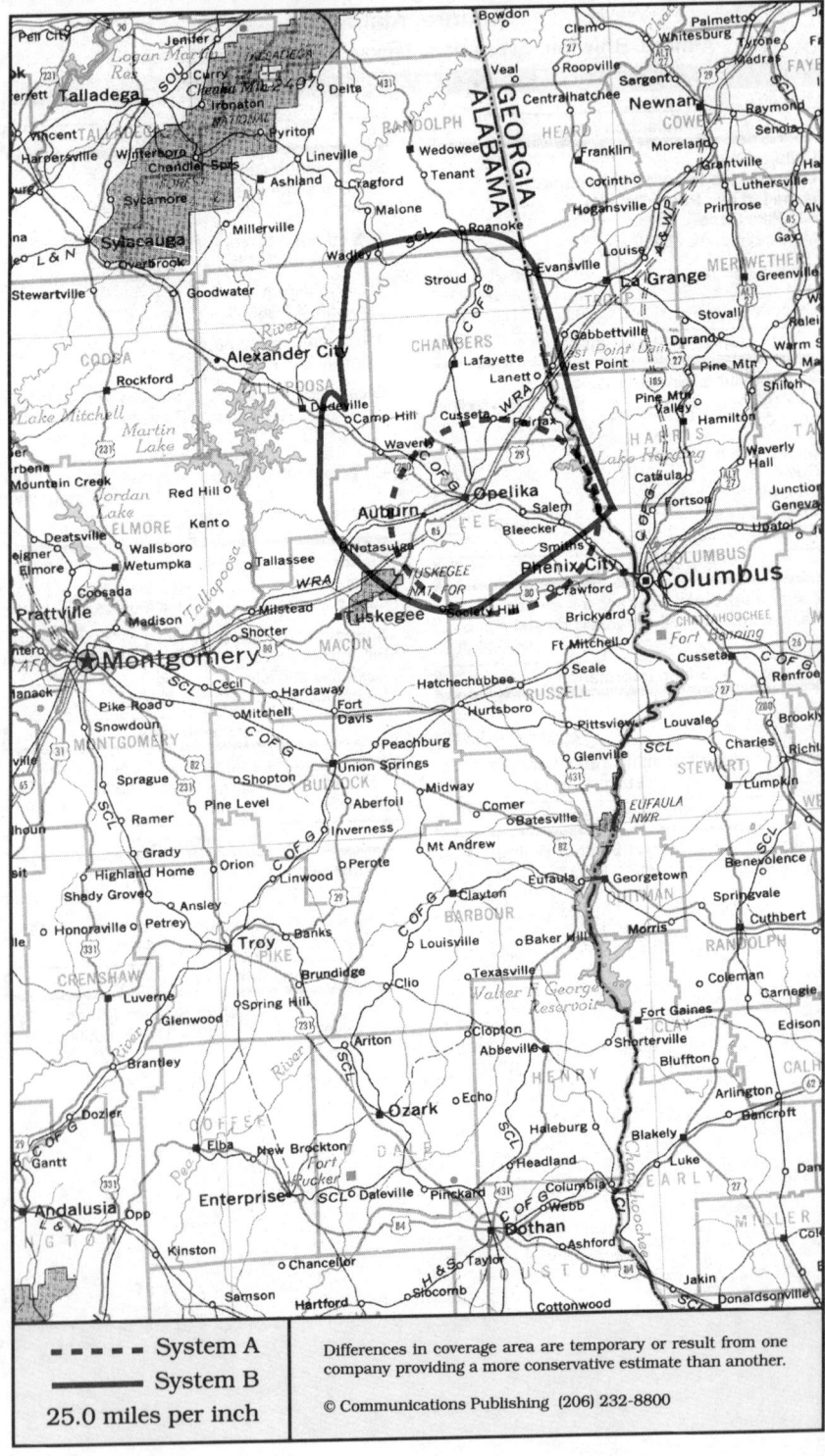

System A
System B
25.0 miles per inch

Differences in coverage area are temporary or result from one company providing a more conservative estimate than another.

© Communications Publishing (206) 232-8800

Auburn, Alabama
Auburn, Lafayette (Chambers County), Opelika

Cellular company

Cellular One
Palmer Communications
5415 Atlanta Highway
Montgomery, AL 36109

Customer service	205-260-0999
Local office	205-277-2222
Corporate office	813-433-4350

Sending calls

Visitor dialing instructions
Local calls . . . . 7 digits
Long distance . . . 1 + area code + 7 digits

Speed-dial numbers
Customer service 611
Emergency 911
Equipment problems 511
Radio station . . . . ∗102

Receiving calls

Caller dials the roamer access number below,
hears tone, dials cellular area code and number.
. . . . 205-705-7626

NationLink & RoamingAmerica (see p. 1157)
Dial ∗31 send to receive calls, ∗30 to deactivate.
Dial ∗32 send to tell caller the roamer access no.
This company's subscribers have level 6 service.

Billing information

Visitor rates
Most visitors . . . . . $3 day + 85¢ minute

Visitors without roaming agreements (p. 1159)
Call co. to use Amex, Mastercard, Visa
or Roamer Plus will intercept first call to bill you.

Identification numbers

City	Market	System	Billing	RSA
All served cities	314	00319	00000	AL-8

For full coverage area, see Columbus, GA.

Cellular company

InterCel
Interstate Cellular
421 Gilmer Avenue, PO Box 352
Lanett, AL 36863

Corporate office	800-763-2355
. . . .	205-644-2355

Sending calls

Visitor dialing instructions
Local calls . . . . . . area code + 7 digits
Long distance . . . 1 + area code + 7 digits

Speed-dial numbers
Customer service . . . . ∗611
Emergency 911

Receiving calls

Caller dials the roamer access number below,
hears tone, dials cellular area code and number.
. . . . 706-643-7626

Follow Me Roaming forwarding (see p. 1156)
Activate, dial ∗18 send. Deactivate dial ∗19 send.

Billing information

Visitor rates
Most visitors . . . . . $3 day + 85¢ minute

Visitors without roaming agreements (p. 1159)
Cannot call since company takes no credit cards.

Identification numbers

City	Market	System	Billing	RSA
Auburn	314	01902	00000	AL-8
Lafayette	311	01902	00000	AL-5

For full coverage area, see La Grange, GA.

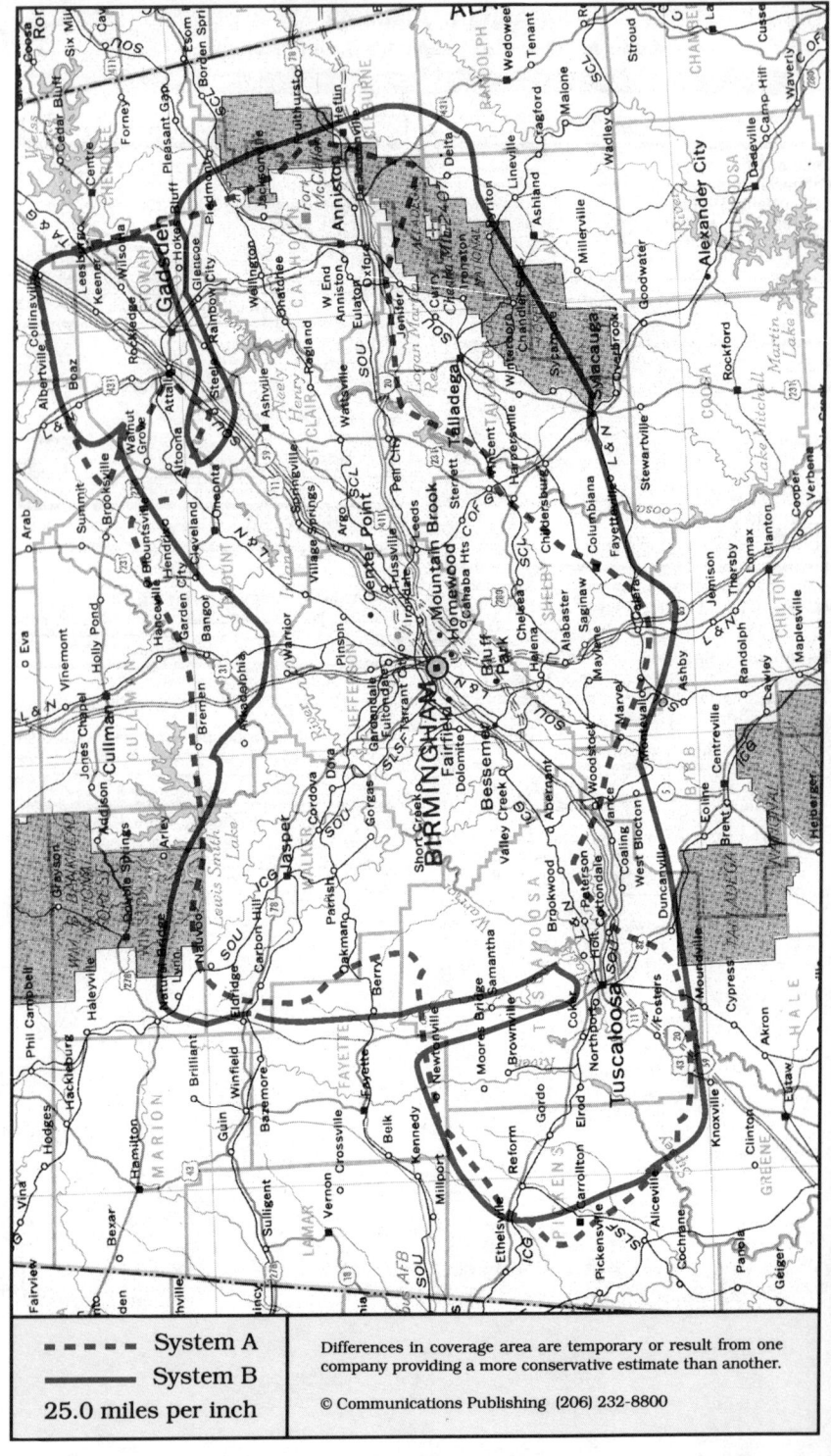

- - - - System A
————— System B
25.0 miles per inch

Birmingham, Alabama
Anniston, Birmingham, Center Point, Gadsden, Talladega, Tuscaloosa

System A	System B

Cellular company

Cellular One
Contel Cellular
100 Riverpoint Corporate Center, Suite 150
Birmingham, AL 35243

449 George Wallace Drive
Gadsden, AL 35903

Customer service		800-333-4004
. . . .		205-970-2273
Local office . .	Birmingham	205-970-2355
. . . . . .	Gadsden	205-546-2351
Corporate office		404-804-3400

Sending calls

Visitor dialing instructions
Local calls	area code + 7 digits
Long distance . . .	1 + area code + 7 digits

Speed-dial numbers
Customer service	611
Emergency	911
Traffic report	811

Receiving calls

Caller dials the roamer access number below,
hears tone, dials cellular area code and number.
Anniston	205-453-7626
Birmingham	205-936-7626
Gadsden	205-490-7626
Tuscaloosa	205-394-7626

Billing information

Visitor rates
Most visitors	$3 day + 85¢ minute

Visitors without roaming agreements (p. 1159)
Call co. to use Amex, Discover, Mastercard, Visa

Identification numbers

City	Market	System	Billing	RSA
Anniston	249	00113	30043	MSA
Birmingham	041	00113	00000	MSA
Gadsden	272	00113	30029	MSA
Tuscaloosa	222	00113	30027	MSA

Cellular company

BellSouth Mobility
Anniston sales
1030 S Quintard Avenue #B
Anniston, AL 36201-8241

Birmingham administration
100 Concourse Parkway, Suite 375
Birmingham, AL 35244-1870

Birmingham sales
3419 Colonnade Parkway, Suite 800
Birmingham, AL 35243

Customer service		800-351-2400
. . . .		205-444-9600
Local office	Anniston	205-236-8677
. . . . .	Birmingham sales	205-969-7600
.	Birmingham administration	205-444-4900
Corporate office		404-847-3600

Sending calls

Visitor dialing instructions
Local calls	area code + 7 digits
Long distance	area code + 7 digits

Speed-dial numbers
Customer service	811
Emergency	911
Roaming information	711
Technical problems	611

Receiving calls

Caller dials the roamer access number below,
hears tone, dials cellular area code and number.
Birmingham, Gadsden	205-531-7626
Heflin	205-463-8076
Knoxville	205-390-7626
Talladega	205-761-7626
Tuscaloosa	205-799-7626

Follow Me Roaming forwarding (see p. 1156)
Activate, dial ✳18 send. Deactivate dial ✳19 send.

Billing information

Visitor rates
Most visitors . . .	$2.50 day + 85¢ minute
BellSouth subscribers .	$0 day + 60¢ minute

Visitors without roaming agreements (p. 1159)
Call co. to use Amex, Mastercard, Visa

Identification numbers

City	Market	System	Billing	RSA
Anniston	249	00098	00000	MSA
Birmingham	041	00098	00000	MSA
Gadsden	272	00098	00000	MSA
Heflin, Talladega	311	00098	30352	AL-5
Tuscaloosa	222	00098	00000	MSA

For full coverage area, see Demopolis, AL.

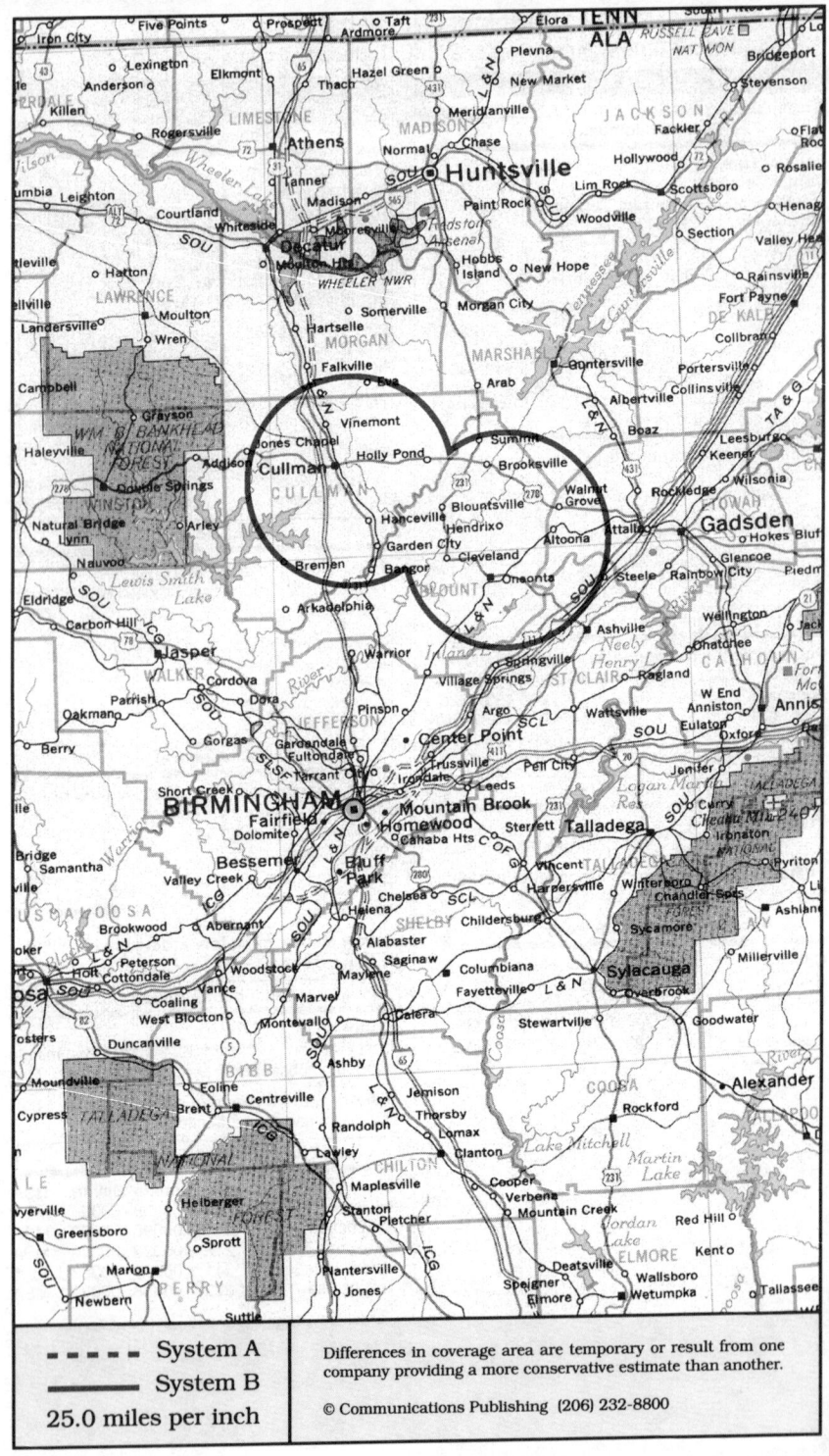

Cullman, Alabama
Cullman, Oneonta

System A	System B

Cellular company

No information about the company that will operate system A was available at press time.

Cellular company

Contel Cellular
100 Riverpoint Corporate Center, Suite 150
Birmingham, AL 35243

Customer service	800-333-4004
. . . .	205-970-2273
Local office	205-970-2355
Corporate office	404-804-3400

Sending calls

Visitor dialing instructions

Local calls	area code + 7 digits
Long distance . . .	1 + area code + 7 digits

Speed-dial numbers

Customer service	611
Emergency	911
Traffic report	811

Receiving calls

Caller dials the roamer access number below, hears tone, dials cellular area code and number.
. . . . 205-708-7626

Billing information

Visitor rates
Most visitors . . . . . $3 day + 85¢ minute

Visitors without roaming agreements (p. 1159)
Call co. to use Amex, Discover, Mastercard, Visa

Identification numbers

City	Market	System	Billing	RSA
All served cities	307	01002	30552	AL-1

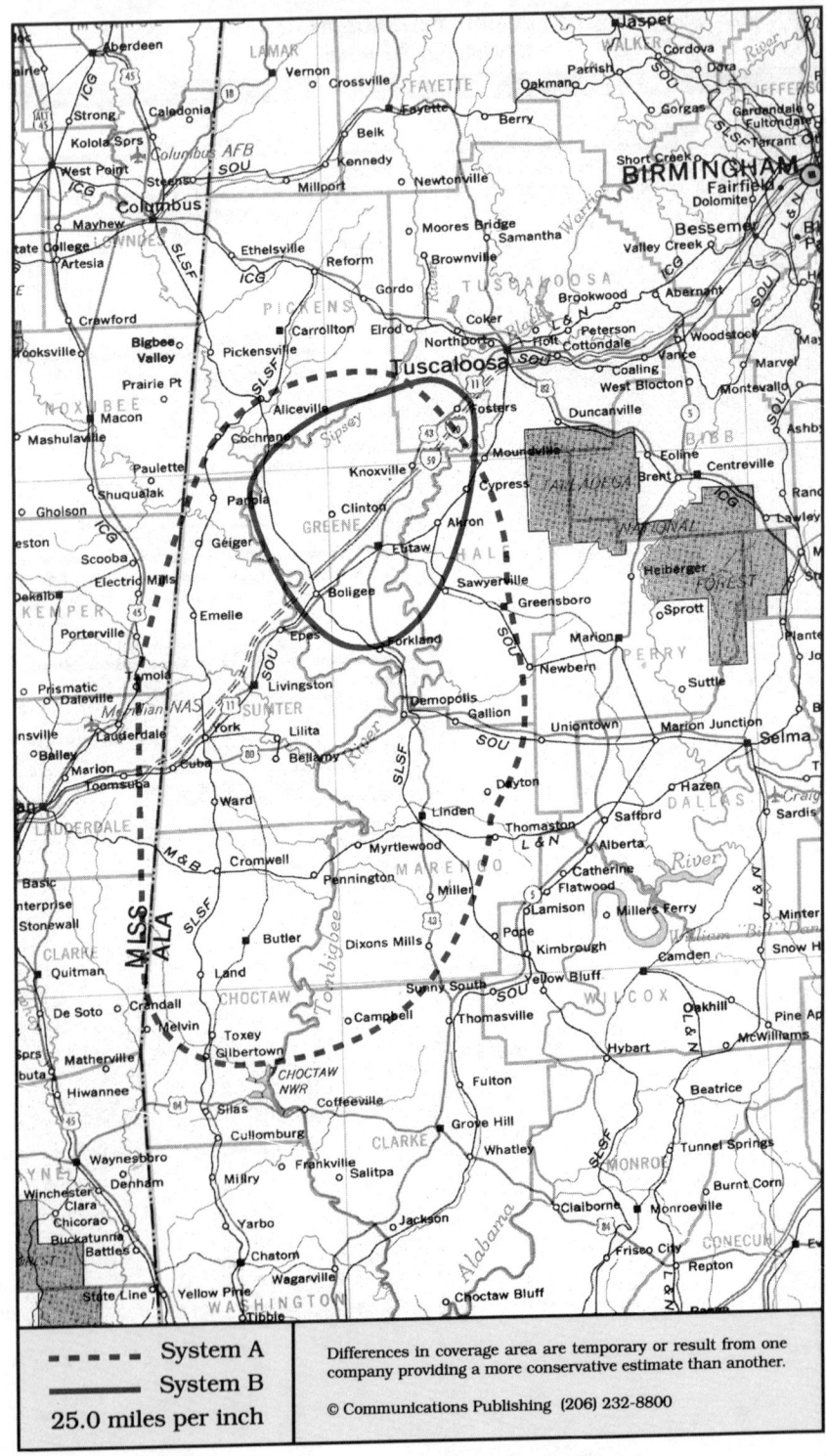

- - - - System A ———— System B 25.0 miles per inch	Differences in coverage area are temporary or result from one company providing a more conservative estimate than another. © Communications Publishing (206) 232-8800

Demopolis, Alabama
Butler, Demopolis, Eutaw, Linden, Livingston

System A	System B

System A

Cellular company

Cellular One
West Alabama Cellular Telephone
PO Box 1044
Demopolis, AL 36732

Corporate office   some areas 800-800-4002
. . . . 205-289-0063

Sending calls

Visitor dialing instructions
Local calls . . . . 1 + area code + 7 digits
Long distance . . . 1 + area code + 7 digits

Speed-dial numbers
Customer service 611
Emergency 911

Receiving calls

Caller dials the roamer access number below,
hears tone, dials cellular area code and number.
. . . . 205-216-7626

Billing information

Visitor rates
Most visitors . . . . . $3 day + 85¢ minute

Visitors without roaming agreements (p. 1159)
Call co. to use Amex, Mastercard, Visa
or Roamer Plus will intercept first call to bill you.

Identification numbers

City	Market	System	Billing	RSA
All served cities	309	01005	00000	AL-3

System B

Cellular company

BellSouth Mobility
Administration
100 Concourse Parkway, Suite 375
Birmingham, AL 35244-1870

Sales
3419 Colonnade Parkway, Suite 800
Birmingham, AL 35243

Customer service 800-351-2400
. . . . 205-444-9600
Local office . . administration 205-444-4900
. . . . sales 205-969-7600
Corporate office 404-847-3600

Sending calls

Visitor dialing instructions
Local calls . . . . . . area code + 7 digits
Long distance . . . . . area code + 7 digits

Speed-dial numbers
Customer service 811
Emergency 911
Roaming information 711
Telephone problems 611

Receiving calls

Caller dials the roamer access number below,
hears tone, dials cellular area code and number.
. . . . 205-372-7626

Follow Me Roaming forwarding (see p. 1156)
Activate, dial *18 send. Deactivate dial *19 send.

Billing information

Visitor rates
Most visitors . . . . . $3 day + 85¢ minute
BellSouth subscribers . $0 day + 60¢ minute

Visitors without roaming agreements (p. 1159)
Call co. to use Amex, Mastercard, Visa

Identification numbers

City	Market	System	Billing	RSA
All served cities	309	00098	30350	AL-3

For full coverage area, see Birmingham, AL.

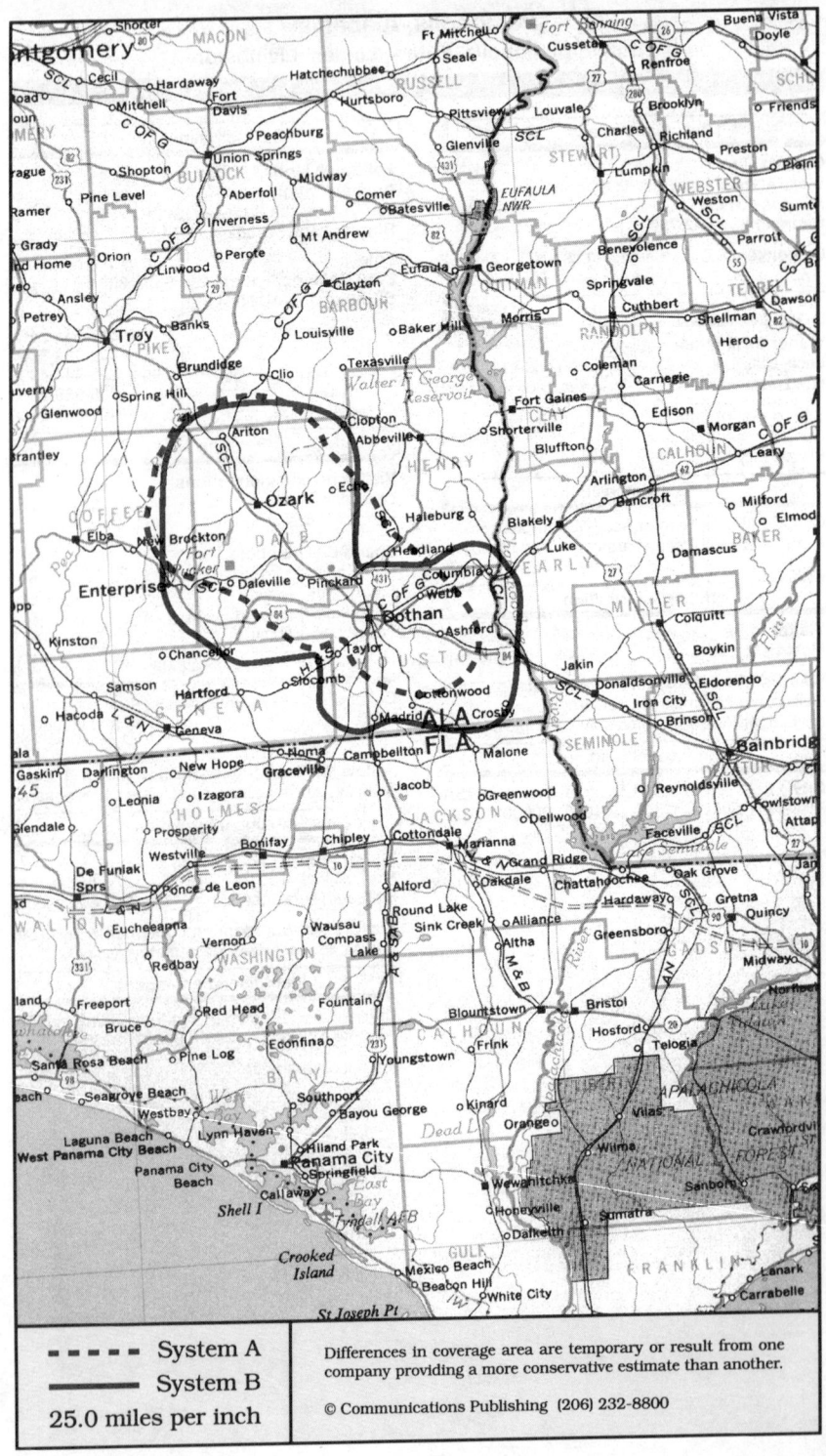

- - - - - System A
———— System B
25.0 miles per inch

Dothan, Alabama
Dothan, Enterprise, Ozark

System A	System B

Cellular company

Cellular One
Palmer Communications
3763-A Ross Clark Circle
Dothan, AL 36303

Customer service	some areas	800-756-2355
Local office	. . . . . .	205-677-2355
Corporate office	. . . . .	813-433-4350

Sending calls

Visitor dialing instructions

Local calls	area code + 7 digits
Long distance . . .	1 + area code + 7 digits

Speed-dial numbers

Customer service	611
Emergency	911
WTVY (FM radio)	*95

Receiving calls

Caller dials the roamer access number below,
hears tone, dials cellular area code and number.
. . . . 205-790-7626

NationLink & RoamingAmerica (see p. 1157)
Dial *31 send to receive calls, *30 to deactivate.
Dial *32 send to tell caller the roamer access no.
This company's subscribers have level 6 service.

Billing information

Visitor rates
Most visitors . . . . . $3 day + 85¢ minute

Visitors without roaming agreements (p. 1159)
Cannot call since company takes no credit cards.

Identification numbers

City	Market	System	Billing	RSA
All served cities	246	00329	00000	MSA

For full coverage area, see Troy, AL.

Cellular company

Centel Cellular
1530 Montgomery Highway
Dothan, AL 36303

Local office	. . . . . .	205-671-4111
Corporate office	. . . . .	312-399-2644

Sending calls

Visitor dialing instructions

Local calls	area code + 7 digits
Long distance . . .	1 + area code + 7 digits

Speed-dial numbers

Customer service	611 or 711
Directory assistance	411
Emergency	911

Receiving calls

Caller dials the roamer access number below,
hears tone, dials cellular area code and number.
. . . . 205-797-7626

Follow Me Roaming forwarding (see p. 1156)
Activate, dial *18 send. Deactivate dial *19 send.

Billing information

Visitor rates

Most visitors	$3 day + 99¢ minute
Busy or no answer	25¢ minute

Visitors without roaming agreements (p. 1159)
Cannot call since company takes no credit cards.

Identification numbers

City	Market	System	Billing	RSA
All served cities	246	00312	00000	MSA

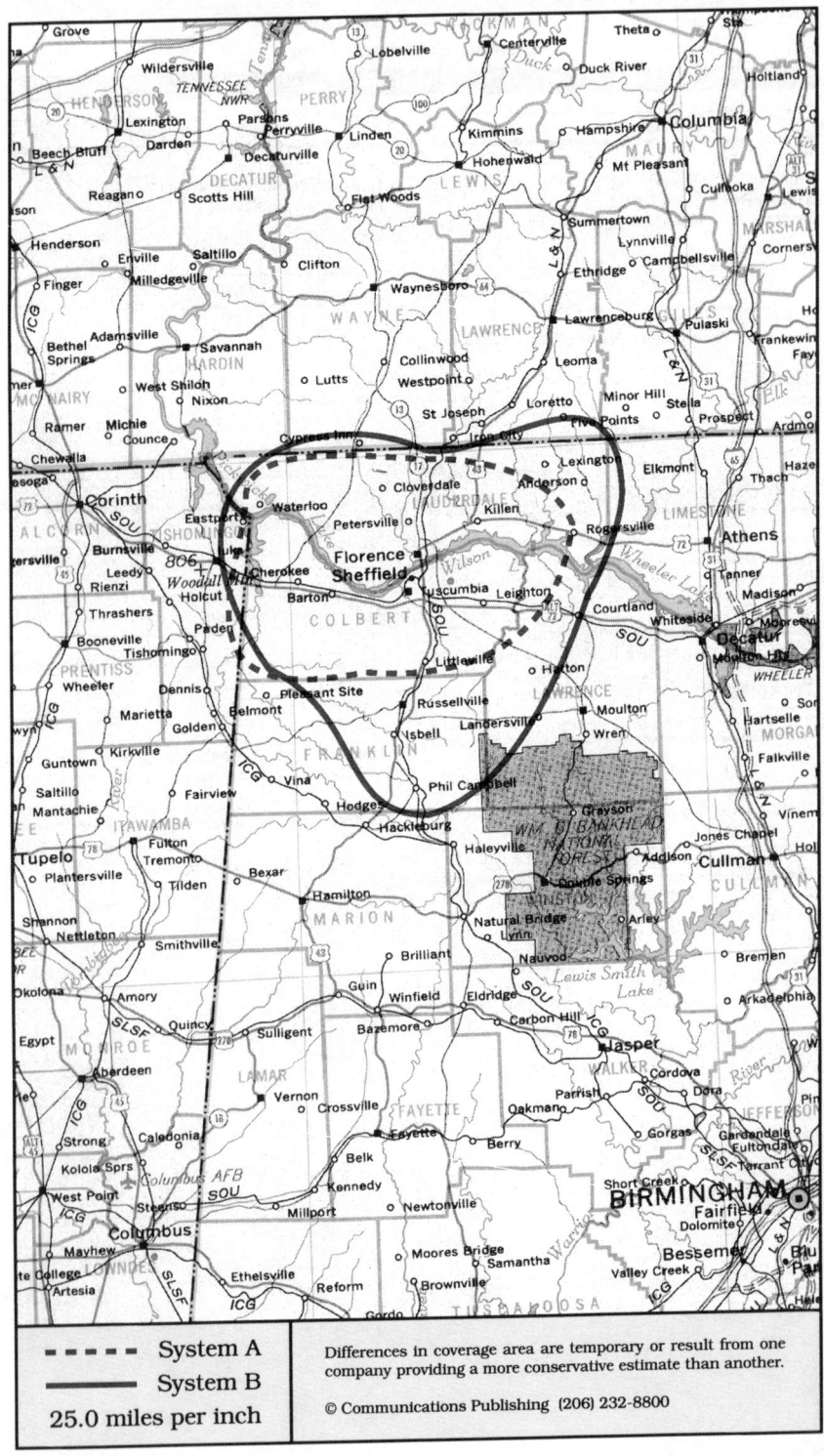

- - - - - System A	Differences in coverage area are temporary or result from one
———— System B	company providing a more conservative estimate than another.
25.0 miles per inch	© Communications Publishing (206) 232-8800

Florence, Alabama
Florence, Russellville, Sheffield, Tuscumbia

System A | System B

System A

Cellular company

Cellular One
Contel Cellular
2089 Florence Blvd
Florence, AL 35630

Customer service	800-333-4004
. . . . .	205-970-2273
Local office	205-764-9500
Corporate office	404-804-3400

Sending calls

Visitor dialing instructions
Local calls . . . . 7 digits
Long distance . . . 1 + area code + 7 digits

Speed-dial numbers
Customer service 611
Emergency 911
WERC traffic report 811

Receiving calls

Caller dials the roamer access number below,
hears tone, dials cellular area code and number.
. . . . 205-760-7626

Billing information

Visitor rates
Most visitors . . $3 per day + 85¢ per minute

Visitors without roaming agreements (p. 1159)
Call co. to use Amex, Mastercard, Visa

Identification numbers

City	Market	System	Billing	RSA
All served cities	226	00113	30025	MSA

System B

Cellular company

Shoals Cellular
Cellular Information Systems
608 Michigan Avenue
Muscle Shoals, AL 35661

Local office	205-383-5111
Corporate office	203-622-6317

Sending calls

Visitor dialing instructions
Local calls . . . . 7 digits
Long distance . . . 1 + area code + 7 digits

Speed-dial numbers
Corporate office in Connecticut 611
Customer service 611
Emergency 911
Local office in Alabama 511

Receiving calls

Caller dials the roamer access number below,
hears tone, dials cellular area code and number.
. . . . 205-370-7626

Follow Me Roaming forwarding (see p. 1156)
Activate, dial *18 send. Deactivate dial *19 send.

Billing information

Visitor rates
Most visitors . . . . . $3 day + 85¢ minute

Visitors without roaming agreements (p. 1159)
Call co. to use Amex, Mastercard, Visa

Identification numbers

City	Market	System	Billing	RSA
Florence	226	00334	00000	MSA
Russellville	307	01942	00000	AL-1

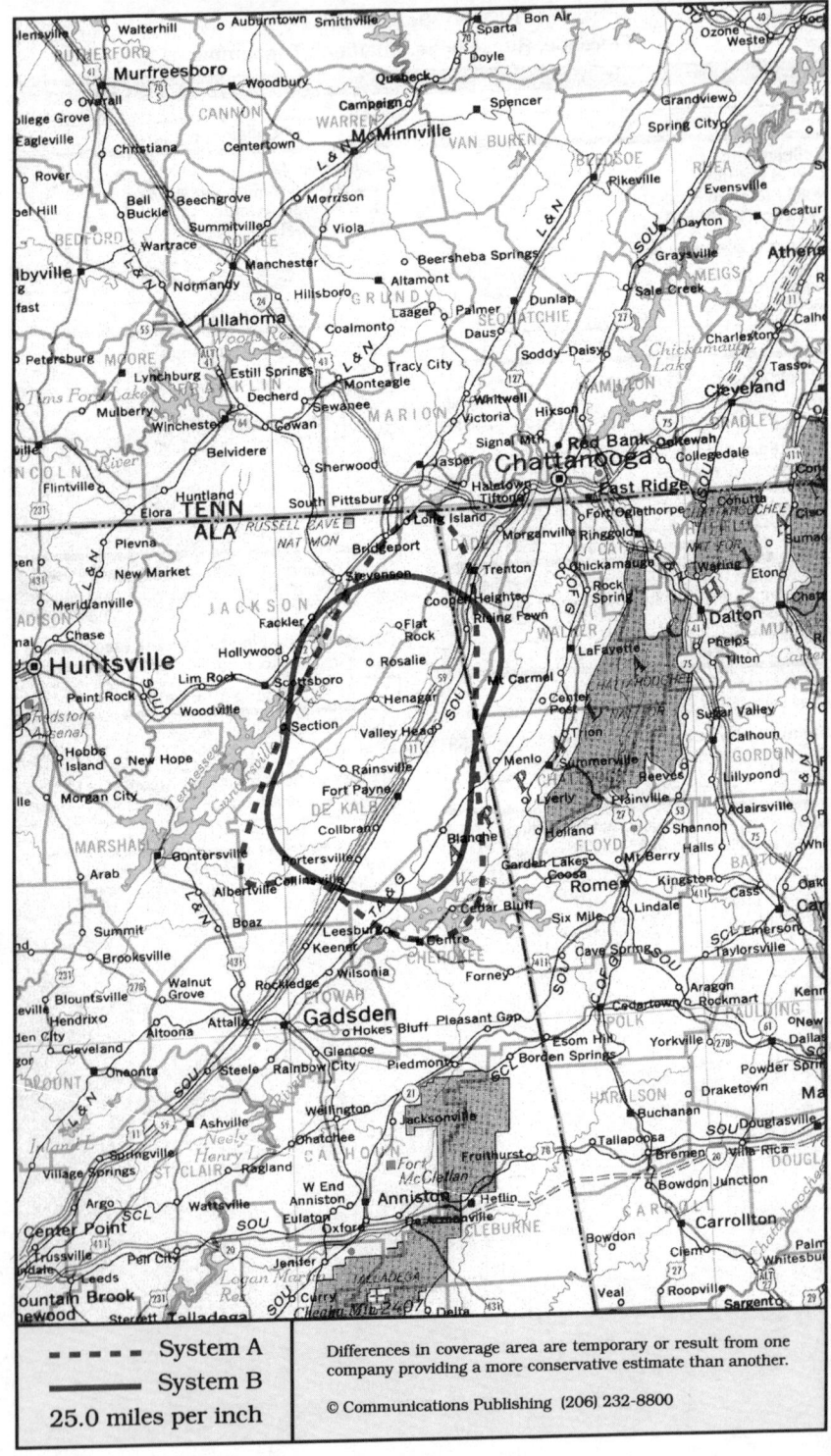

- - - - -	System A
—————	System B
25.0 miles per inch	

Differences in coverage area are temporary or result from one company providing a more conservative estimate than another.

© Communications Publishing (206) 232-8800

Fort Payne, Alabama
Collinsville, Fort Payne

System A	System B

Cellular company

Cellular One
Crowley Cellular Telecommunications
904 Gault Avenue N
Fort Payne, AL 35967

Customer service	205-830-6633
Local office	205-997-6600
Corporate office	708-843-9081

Sending calls

Visitor dialing instructions

Local calls	area code + 7 digits
Long distance . . .	1 + area code + 7 digits

Speed-dial numbers

Customer service	611
Emergency	911

Receiving calls

Caller dials the roamer access number below,
hears tone, dials cellular area code and number.
. . . . 205-656-7626

NationLink & RoamingAmerica (see p. 1157)
Dial *31 send to receive calls, *30 to deactivate.
Dial *32 send to tell caller the roamer access no.
This company's subscribers have level 8 service.

Billing information

Visitor rates
Most visitors . . . . . $3 day + 85¢ minute

Visitors without roaming agreements (p. 1159)
Call co. to use Amex, Mastercard, Visa
or Roamer Plus will intercept first call to bill you.

Identification numbers

City	Market	System	Billing	RSA
All served cities	308	00203	30361	AL-2

Cellular company

Farmers Cellular Telephone
Farmers Telephone Cooperative
PO Box 1429
Rainesville, AL 35986

Corporate office	205-638-2100

Sending calls

Visitor dialing instructions

Local calls	area code + 7 digits
Long distance . . .	1 + area code + 7 digits

Speed-dial numbers

Customer service	711
Emergency	911

Receiving calls

Caller dials the roamer access number below,
hears tone, dials cellular area code and number.
. . . . 205-638-5555

Follow Me Roaming forwarding (see p. 1156)
Activate, dial *18 send. Deactivate dial *19 send.

Billing information

Visitor rates
Most visitors . . . . . $3 day + 85¢ minute

Visitors without roaming agreements (p. 1159)
Cannot call since company takes no credit cards.

Identification numbers

City	Market	System	Billing	RSA
All served cities	308	00098	00000	AL-2

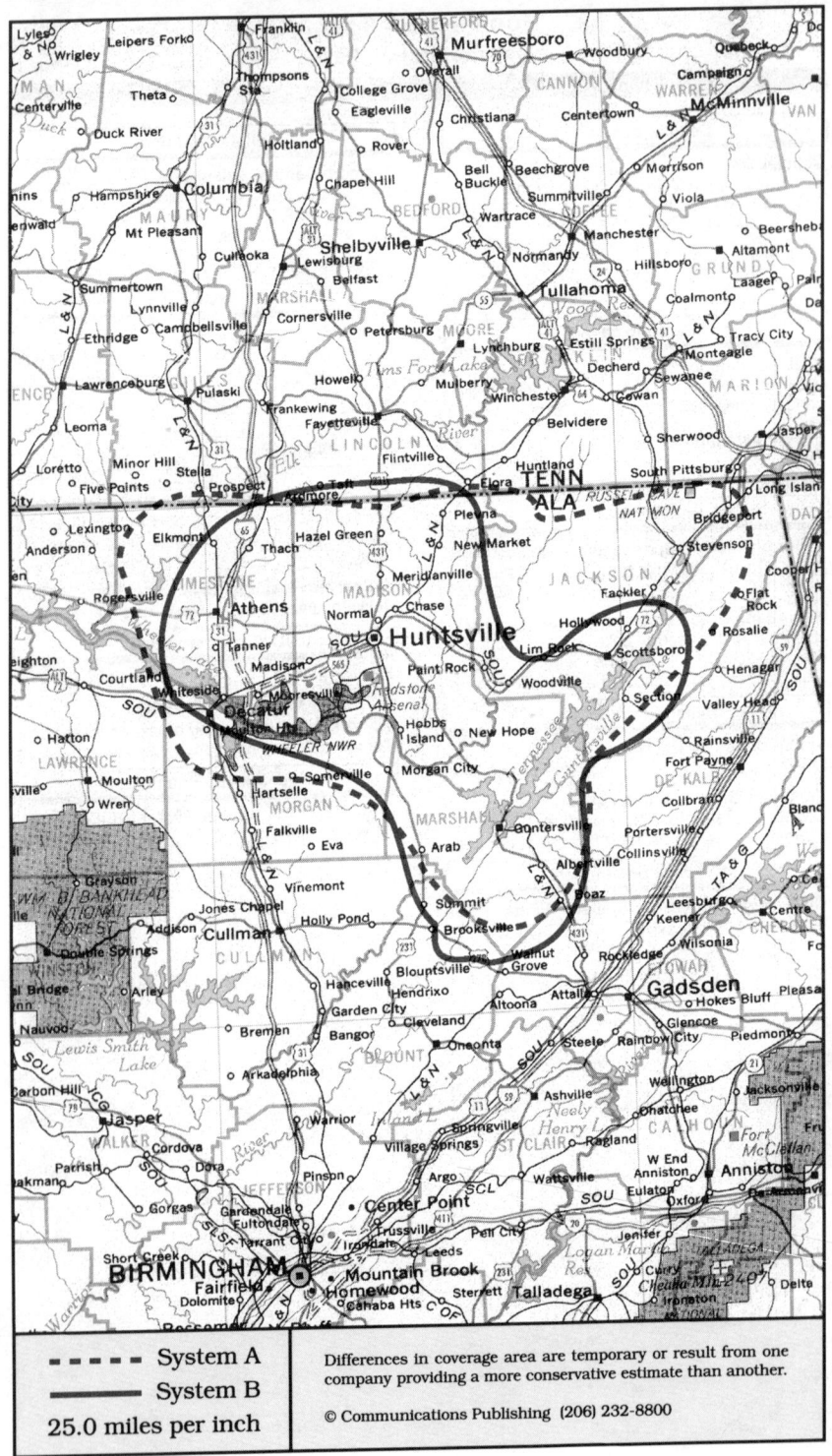

	System A
▬ ▬ ▬	System B

25.0 miles per inch

Differences in coverage area are temporary or result from one company providing a more conservative estimate than another.

© Communications Publishing (206) 232-8800

Huntsville, Alabama

Albertville, Arab, Athens, Decatur, Guntersville, Huntsville, Scottsboro

System A	System B

Cellular company

Cellular One
Crowley Cellular Telecommunications
1541 Blount Avenue
Guntersville, AL 35976

1350 Evangel Drive
Huntsville, AL 35816

Local office . .	Guntersville	205-656-6000
. . . .	Huntsville	205-830-6633
. . . . .	from Scottsboro	205-574-9001
Corporate office	. . . . .	708-843-9081

Sending calls

Visitor dialing instructions
Local calls	area code + 7 digits
Long distance . . .	1 + area code + 7 digits

Speed-dial numbers
Customer service	611
Emergency	911
Emergency road service M - F	205-508-1043
K98 fm Scottsoboro	*98
WVNN am talk	*770
WWIC am Scottsobro	*1050
WQSB fm Albertville	*111

Receiving calls

Caller dials the roamer access number below, hears tone, dials cellular area code and number.
. . . . 205-656-7626

NationLink & RoamingAmerica (see p. 1157)
Dial *31 send to receive calls, *30 to deactivate.
Dial *32 send to tell caller the roamer access no.
This company's subscribers have level 8 service.

Billing information

Visitor rates
Most visitors	$3 day + 85¢ minute

Visitors without roaming agreements (p. 1159)
Call co. to use Amex, Mastercard, Visa
or Roamer Plus will intercept first call to bill you.

Identification numbers

City	Market	System	Billing	RSA
Decatur	307	00203	00000	AL-1
Guntersville	120	00203	00000	MSA
Huntsville	120	00203	00000	MSA
Scottsboro	308	00203	00000	AL-2

Cellular company

Athens, Decatur, Huntsville, Scottsboro
BellSouth Mobility
4929 University Drive NW #G
Huntsville, AL 35816-1803

Customer service	. . . .	800-351-2400
. . . .		205-430-2800
Local office		205-837-6400
Corporate office		404-847-3600

Albertville, Arab, Guntersville
Marshall Cellular
106 S Main St, Suite 5, PO Drawer E
Arab, AL 35016

Customer service	. . . .	205-582-0291
Local office		205-586-7900

Sending calls

Visitor dialing instructions
Local calls	area code + 7 digits
Long distance	area code + 7 digits

Speed-dial numbers
Customer service	811
Emergency	911
Roaming information	711
Telephone problems	611
To report traffic problems . . .	205-651-0700

Receiving calls

Caller dials the roamer access number below, hears tone, dials cellular area code and number.
Decatur	205-340-8626
Guntersville	205-571-7626
Huntsville	205-651-7626
Scottsboro	205-574-7626

Follow Me Roaming forwarding (see p. 1156)
Activate, dial *18 send. Deactivate dial *19 send.

Billing information

Visitor rates
Most visitors	$2 day + 85¢ minute
BellSouth subscribers .	$0 day + 60¢ minute

Visitors without roaming agreements (p. 1159)
Call co. to use Amex, Mastercard, Visa

Identification numbers

City	Market	System	Billing	RSA
Decatur	307	00198	30170	AL-1
Guntersville	120	00198	00000	MSA
Huntsville	120	00198	00000	MSA
Scottsboro	308	00098	30592	AL-2

BellSouth Mobility is licensed to serve Huntsville.
Marshall Cellular is licensed to serve Marshall
County, which includes Guntersville. Both
companies operate the system jointly.

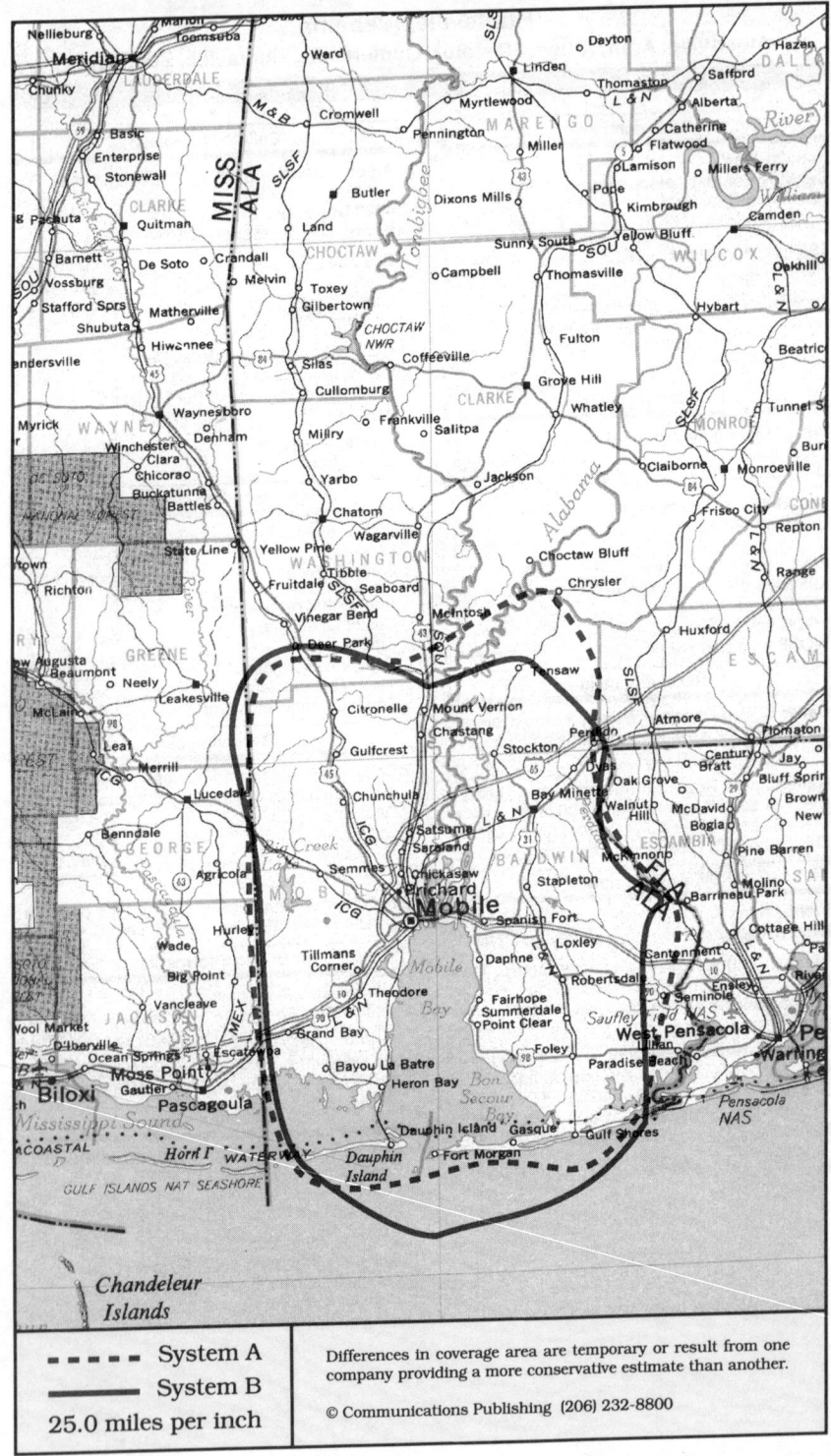

- - - - -	System A
────────	System B
25.0 miles per inch	

Differences in coverage area are temporary or result from one company providing a more conservative estimate than another.

© Communications Publishing (206) 232-8800

Mobile, Alabama
Bay Minette, Mobile, Prichard

|

System A

Cellular company

BellSouth Cellular
124 N Beltline Highway
Mobile, AL 36607

Local office	205-450-3500
Corporate office	404-604-6100

Sending calls

Visitor dialing instructions

Local calls	area code + 7 digits
Long distance . . .	1 + area code + 7 digits

Speed-dial numbers

Customer service	611
Emergency	911

Receiving calls

Caller dials the roamer access number below, hears tone, dials cellular area code and number.
. . . . 205-454-7626

Follow Me Roaming forwarding (see p. 1156)
Phone Me Anywhere forwarding (see p. 1156)
Activate, dial *18 send. Deactivate dial *19 send.

NationLink & RoamingAmerica (see p. 1157)
Caller hears your current roamer access number.
This company's subscribers have level 0 service.

Billing information

Visitor rates

Most visitors	$3 day + 95¢ minute
BellSouth subscribers .	$0 day + 60¢ minute

Visitors without roaming agreements (p. 1159)
Call co. to use Mastercard, Visa

Identification numbers

City	Market	System	Billing	RSA
All served cities	083	00081	00000	MSA

System B

Cellular company

Contel Cellular
1201 Montlimar Drive, Suite 625
Mobile, AL 36609

Customer service	800-333-4004
. . . .	205-970-2273
Local office	205-344-2414
Corporate office	404-804-3400

Sending calls

Visitor dialing instructions

Local calls	area code + 7 digits
Long distance . . .	1 + area code + 7 digits

Speed-dial numbers

Customer service	611
Directory assistance	1411
Emergency	911

Receiving calls

Caller dials the roamer access number below, hears tone, dials cellular area code and number.
. . . . 205-421-7626

Follow Me Roaming forwarding (see p. 1156)
Activate, dial *18 send. Deactivate dial *19 send.

Billing information

Visitor rates
Most visitors . . . . . $3 day + 85¢ minute

Visitors without roaming agreements (p. 1159)
Call co. to use Amex, Mastercard, Visa

Identification numbers

City	Market	System	Billing	RSA
All served cities	083	00120	00000	MSA

For full coverage area, see Pensacola, FL.

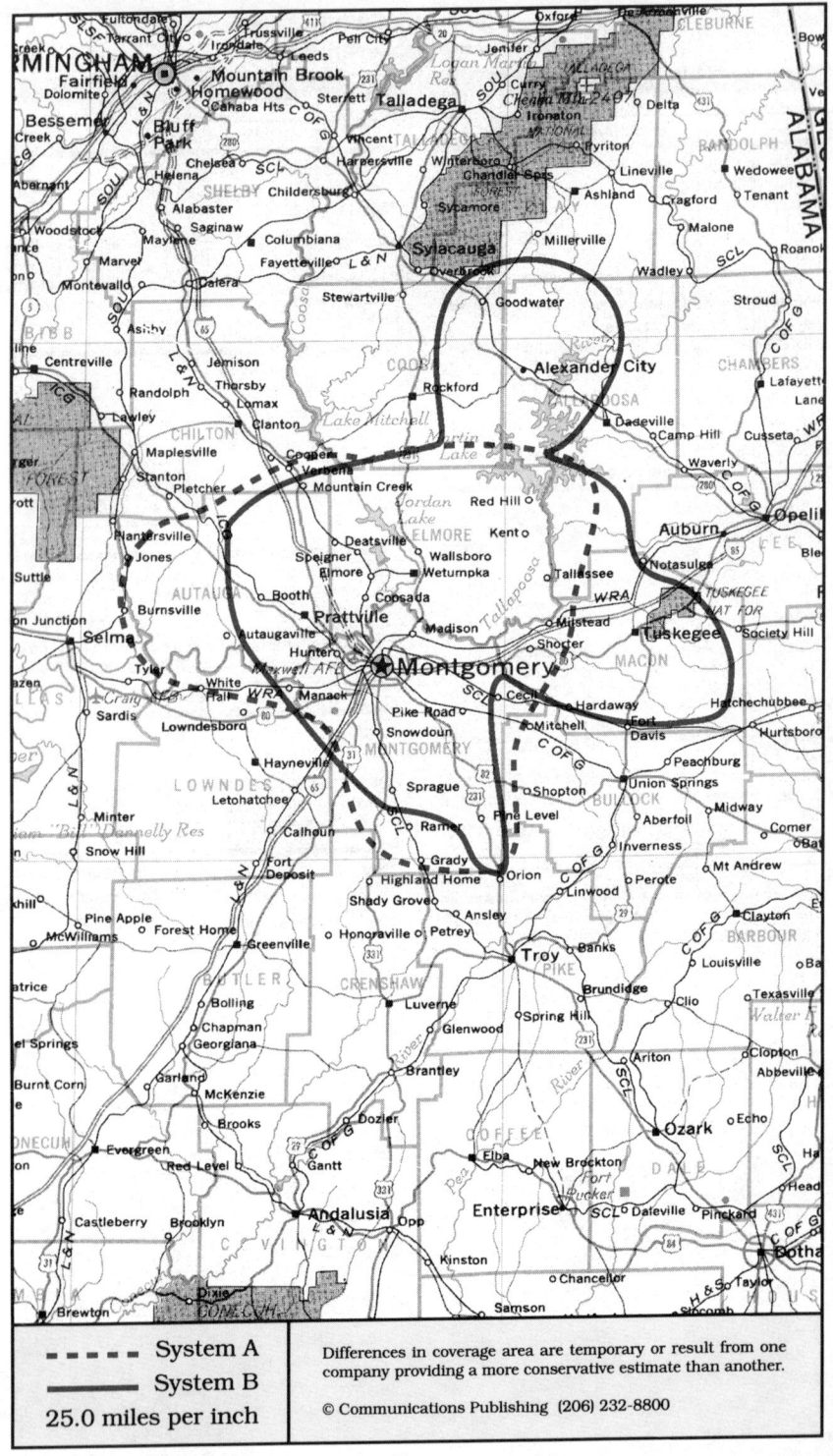

Montgomery, Alabama

Alexander City, Montgomery, Prattville, Tuskegee, Wetumpka

System A	System B

Cellular company

Cellular One
Palmer Communications
5415 Atlanta Highway
Montgomery, AL 36109

Customer service	205-260-0999
Local office	205-277-2222
Corporate office	813-433-4350

Sending calls

Visitor dialing instructions

Local calls	7 digits
Long distance	1 + area code + 7 digits

Speed-dial numbers

Customer service	611
Emergency	911
Radio station	*102

Receiving calls

Caller dials the roamer access number below,
hears tone, dials cellular area code and number.
. . . . 205-224-7626

NationLink & RoamingAmerica (see p. 1157)
Dial *31 send to receive calls, *30 to deactivate.
Dial *32 send to tell caller the roamer access no.
This company's subscribers have level 6 service.

Billing information

Visitor rates
Most visitors . . . . . $3 day + 85¢ minute

Visitors without roaming agreements (p. 1159)
Call co. to use Amex, Mastercard, Visa

Identification numbers

City	Market	System	Billing	RSA
All served cities	139	00465	00000	MSA

System A service in Tuskegee will be provided by
another company.

Cellular company

ALLTEL Mobile Communications
2880 Zelda Road
Montgomery AL 36106

Customer service	some areas	800-255-8351
Local office		205-277-8700
Corporate office		501-661-8500

Sending calls

Visitor dialing instructions

Local calls	area code + 7 digits
Long distance	1 + area code + 7 digits

Speed-dial numbers

Customer service	*611
Emergency	911
Roaming information	*811

Receiving calls

Caller dials the roamer access number below,
hears tone, dials cellular area code and number.

Alexander City	205-234-9126
Montgomery	205-399-7626
Tuskegee	205-724-7626

Follow Me Roaming forwarding (see p. 1156)
Activate, dial *18 send. Deactivate dial *19 send.

Billing information

Visitor rates
Most visitors . . . . . $3 day + 85¢ minute

Visitors without roaming agreements (p. 1159)
Call co. to use Amex, Discover, Mastercard, Visa

Identification numbers

City	Market	System	Billing	RSA
Alexander City	311	00444	30634	AL-5
Montgomery	139	00444	00000	MSA
Tuskegee	139	01016	30244	AL-8

For full coverage area, see Troy, AL.

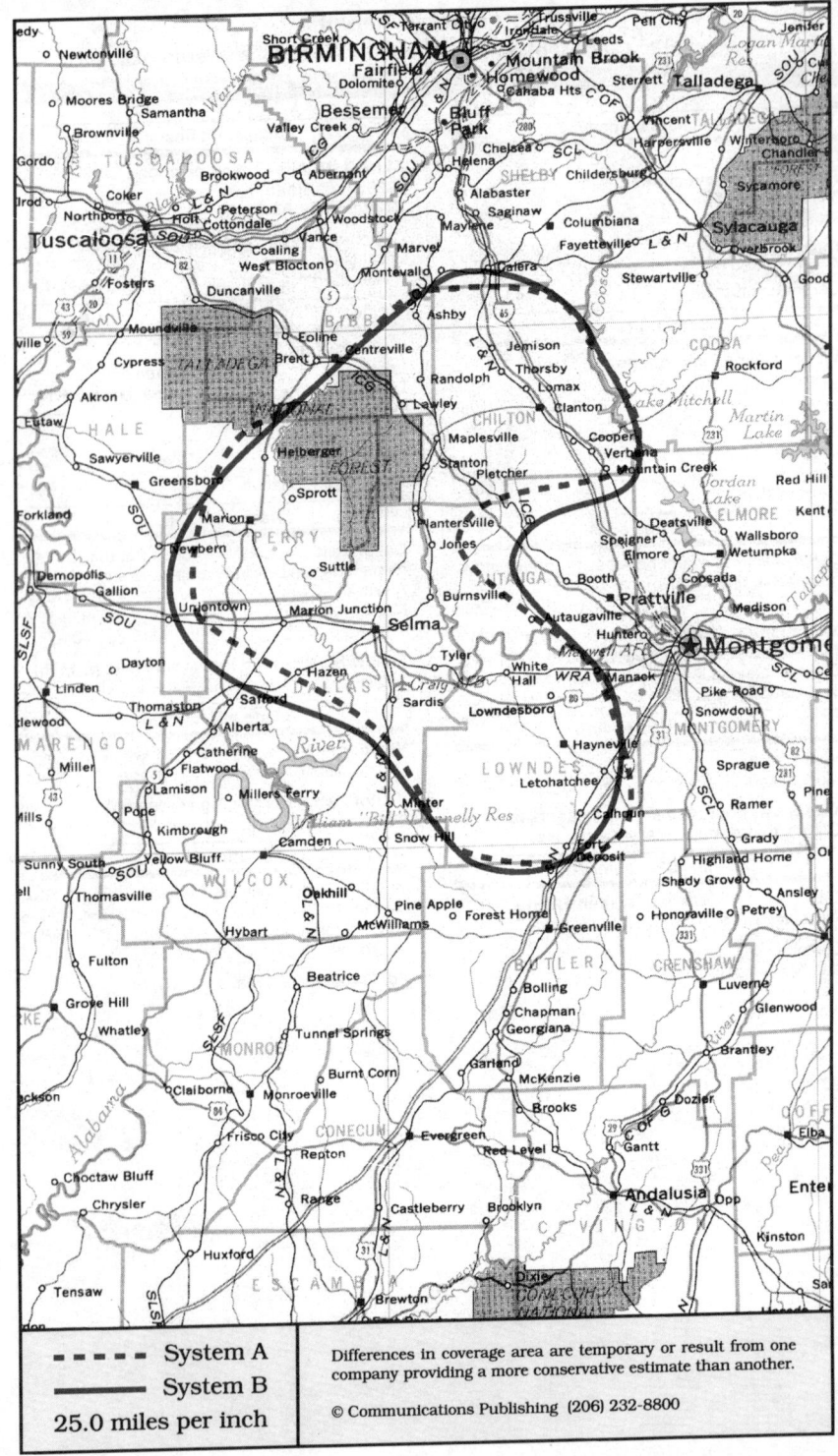

- - - - System A	Differences in coverage area are temporary or result from one
—— System B	company providing a more conservative estimate than another.
25.0 miles per inch	© Communications Publishing (206) 232-8800

Selma, Alabama
Clanton, Hayneville, Marion, Selma

System A	System B

Cellular company

Cellular One
Dominion Cellular
206 7th Street S
Clanton, AL 35045

1016 Highland Avenue
Selma, AL 36701

Local office	. . . .	Clanton	205-755-1166
	. . . . . .	Selma	205-872-9961
Corporate office	. . . .		205-755-1166

Sending calls

Visitor dialing instructions
Local calls . . . . . area code + 7 digits
Long distance . . . 1 + area code + 7 digits

Speed-dial numbers
Customer service 611
Emergency 911

Receiving calls

Caller dials the roamer access number below,
hears tone, dials cellular area code and number.
. . . . 205-419-7626

Billing information

Visitor rates
Most visitors . . . . . $3 day + 85¢ minute

Visitors without roaming agreements (p. 1159)
Roamer Plus will intercept first call to bill you.

Identification numbers

City	Market	System	Billing	RSA
All served cities	310	01007	00000	AL-4

Cellular company

Southeastern Cellular
1374 Highland Avenue
Selma, AL 36703

Local office	. . . . . .		205-875-2090
Corporate office	.	some areas	800-521-2090
	. . . .		205-846-2090

Sending calls

Visitor dialing instructions
Local calls . . . . . . area code + 7 digits
Long distance . . . 1 + area code + 7 digits

Speed-dial numbers
Billing information : 811
Customer service . . . . . 611 or 711
Emergency 911

Receiving calls

Caller dials the roamer access number below,
hears tone, dials cellular area code and number.
Clanton 205-646-8626
Selma 205-418-2626

Follow Me Roaming forwarding (see p. 1156)
Activate, dial ∗18 send. Deactivate dial ∗19 send.

Billing information

Visitor rates
Most visitors . . . . . $3 day + 85¢ minute

Visitors without roaming agreements (p. 1159)
Cannot call since company takes no credit cards.

Identification numbers

City	Market	System	Billing	RSA
All served cities	310	01008	00000	AL-4

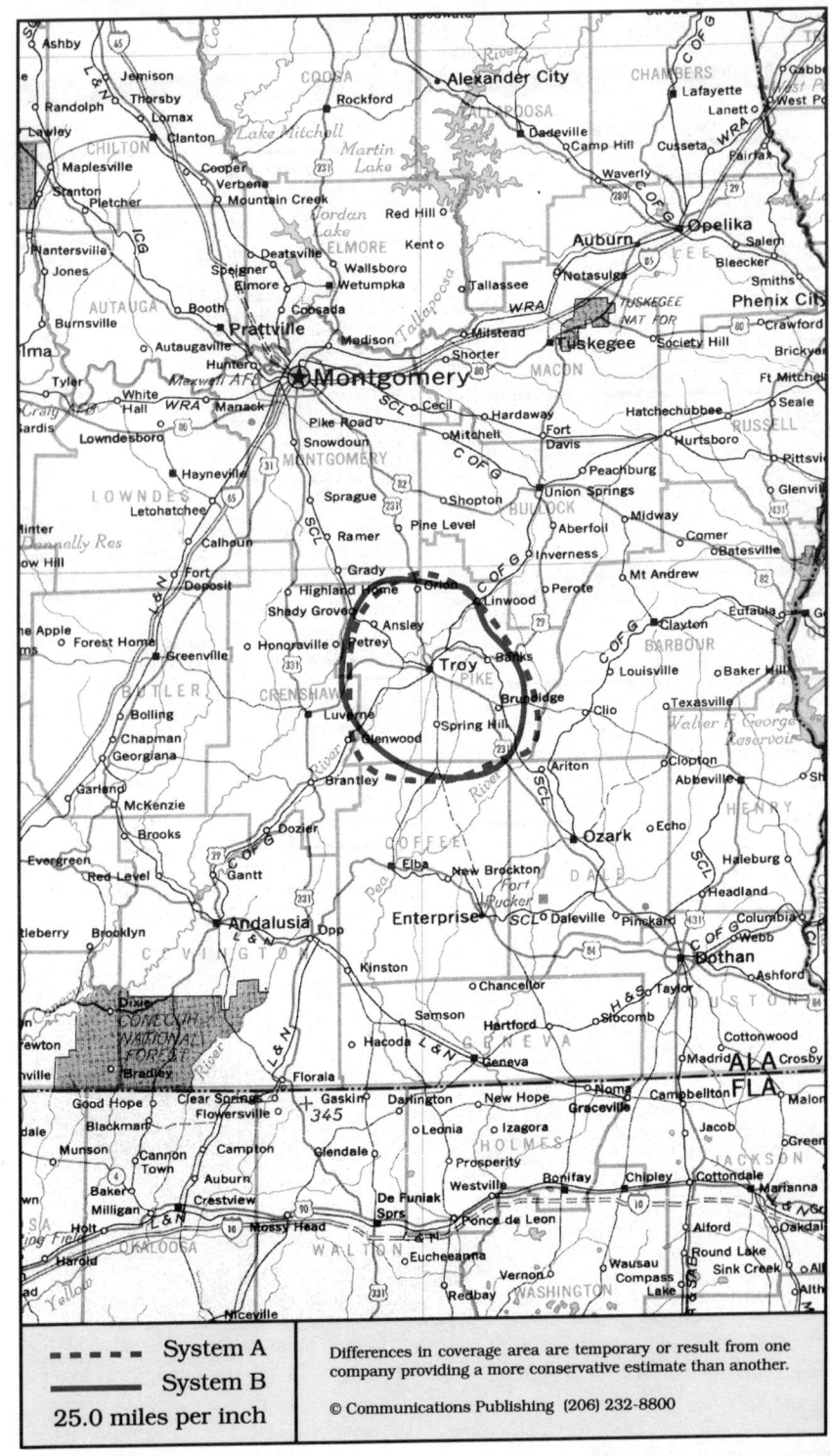

Differences in coverage area are temporary or result from one
company providing a more conservative estimate than another.

© Communications Publishing (206) 232-8800

- - - - - System A
_____ System B
25.0 miles per inch

Troy, Alabama

System A

Cellular company

Cellular One
Palmer Communications
3763-A Ross Clark Circle
Dothan, AL 36303

Customer service	some areas	800-756-2355
Local office	. . . . . .	205-677-2355
Corporate office	. . . . .	813-433-4350

Sending calls

Visitor dialing instructions

Local calls	. . . .	area code + 7 digits
Long distance	. . .	1 + area code + 7 digits

Speed-dial numbers

Customer service	. . . .	611
Emergency	. . . .	911
WTVY (FM radio)	. . . .	*95

Receiving calls

Caller dials the roamer access number below,
hears tone, dials cellular area code and number.
. . . . 205-790-7626

NationLink & RoamingAmerica (see p. 1157)
Dial *31 send to receive calls, *30 to deactivate.
Dial *32 send to tell caller the roamer access no.
This company's subscribers have level 6 service.

Billing information

Visitor rates
Most visitors . . . . . $3 day + 85¢ minute

Visitors without roaming agreements (p. 1159)
Cannot call since company takes no credit cards.

Identification numbers

City	Market	System	Billing	RSA
All served cities	313	00329	00000	AL-7

For full coverage area, see Dothan, AL.

System B

Cellular company

ALLTEL Mobile Communications
2880 Zelda Road
Montgomery AL 36106

Customer service	some areas	800-255-8351
Local office	. . . . . .	205-277-8700
Corporate office	. . . . .	501-661-8500

Sending calls

Visitor dialing instructions

Local calls	. . . . .	area code + 7 digits
Long distance	. . .	1 + area code + 7 digits

Speed-dial numbers

Customer service	. . . .	*611
Emergency	. . . .	911
Roaming information	. . . . . .	*811

Receiving calls

Caller dials the roamer access number below,
hears tone, dials cellular area code and number.
. . . . 205-670-2626

Follow Me Roaming forwarding (see p. 1156)
Activate, dial *18 send. Deactivate dial *19 send.

Billing information

Visitor rates
Most visitors . . . . . $3 day + 85¢ minute

Visitors without roaming agreements (p. 1159)
Call co. to use Amex, Discover, Mastercard, Visa

Identification numbers

City	Market	System	Billing	RSA
All served cities	313	00444	30298	AL-7

For full coverage area, see Montgomery, AL.

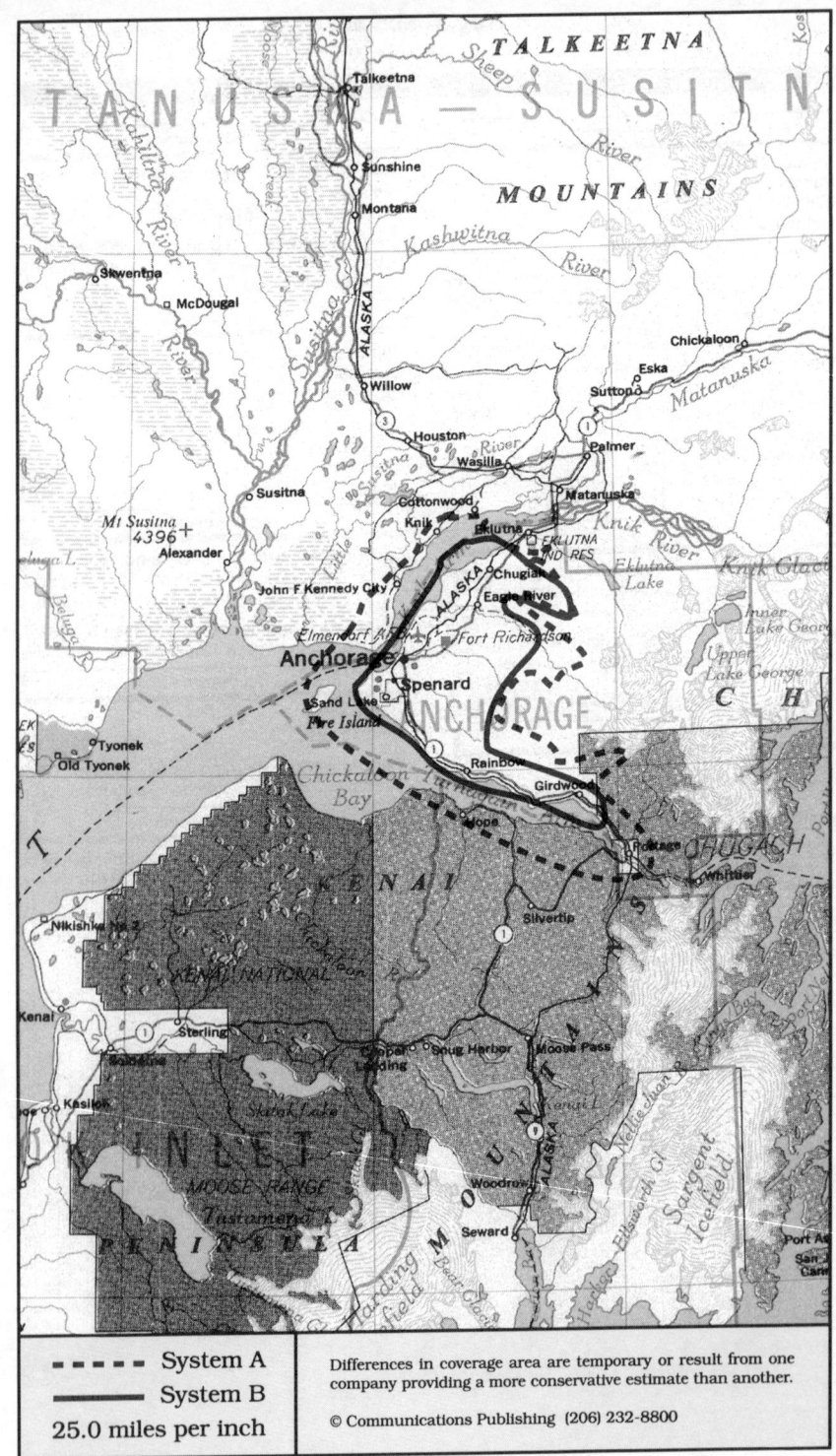

- - - -	System A
————	System B
25.0 miles per inch	

Differences in coverage area are temporary or result from one company providing a more conservative estimate than another.

© Communications Publishing (206) 232-8800

Anchorage, Alaska
Anchorage, Spenard

System A	System B

Cellular company

Cellular One
McCaw Cellular Communications
4797 Business Park Blvd, Suite 2
Anchorage, AK 99503

Local office 907-561-1122

Sending calls

Visitor dialing instructions
Local calls . . . . 7 digits
Long distance . . . . area code + 7 digits

Speed-dial numbers
Customer service 611
Directory assistance 411
Emergency 911
Make a traffic report 711

Receiving calls

Caller dials the roamer access number below,
hears tone, dials cellular area code and number.
. . . . 907-229-7626

NationLink & RoamingAmerica (see p. 1157)
Caller hears your current roamer access number.
This company's subscribers have level 0 service.

Billing information

Visitor rates
Visitors . . . . $0 day + 99¢ minute

Visitors without roaming agreements (p. 1159)
Company takes no credit cards but may bill you.

Identification numbers

City	Market	System	Billing	RSA
All served cities	187	00251	00000	MSA

For full coverage area, see Palmer, AK.

Cellular company

MACtel Cellular System
341 W Tudor Road, Suite 208
Anchorage, AK 99503

Corporate office 907-563-8000

Sending calls

Visitor dialing instructions
Local calls . . . . . . area code + 7 digits
Long distance . . . . . area code + 7 digits

Speed-dial numbers
Customer service *611
Directory assistance 411
Emergency 911
Traffic report (non-emergency) 711

Receiving calls

Caller dials the roamer access number below,
hears tone, dials cellular area code and number.
. . . . 907-244-7626

Follow Me Roaming forwarding (see p. 1156)
Activate, dial *18 send. Deactivate dial *19 send.

Billing information

Visitor rates
Visitors . . . . . . $3 day + 85¢ minute

Visitors without roaming agreements (p. 1159)
Cannot call since company takes no credit cards.

Identification numbers

City	Market	System	Billing	RSA
All served cities	187	00234	00000	MSA

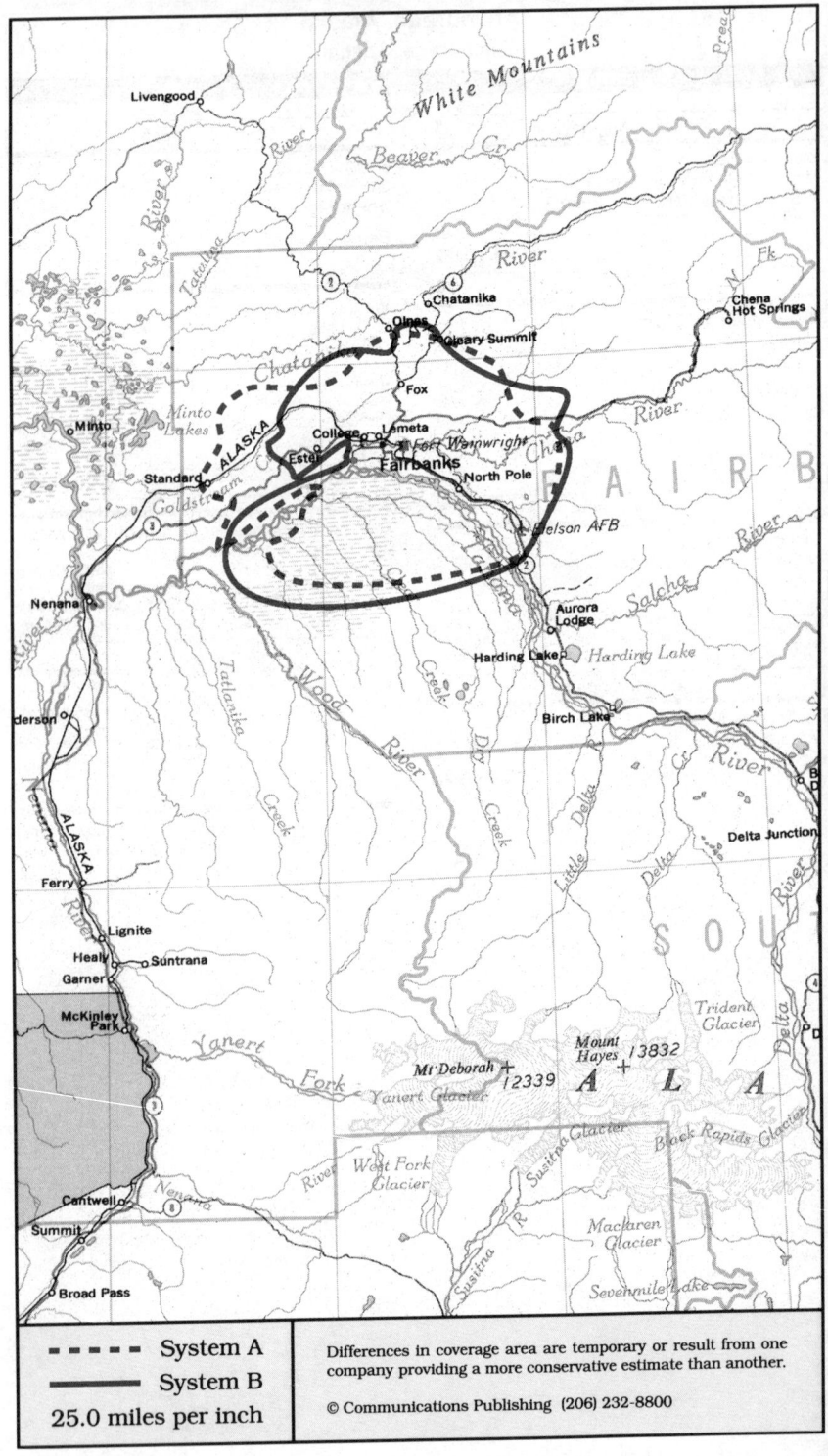

- - - - System A
——— System B
25.0 miles per inch

Differences in coverage area are temporary or result from one company providing a more conservative estimate than another.

© Communications Publishing (206) 232-8800

Fairbanks, Alaska

System A

Cellular company

Cellular One
Fairbanks Municipal Utilities System
645 5th Avenue, PO Box 72215
Fairbanks, AK 99707

Customer service (planned) . 907-459-2355
Corporate office 907-459-6000

Sending calls

Visitor dialing instructions
Local calls . . . . . area code + 7 digits
Long distance . . . 1 + area code + 7 digits

Speed-dial numbers
Customer service 611
Emergency 911

Receiving calls

Caller dials the roamer access number below,
hears tone, dials cellular area code and number.
. . . . 907-388-7626

Billing information

Visitor rates
Most visitors . . . . $2.50 day + 85¢ minute

Visitors without roaming agreements (p. 1159)
Roamer Plus will intercept first call to bill you.

Identification numbers

City	Market	System	Billing	RSA
All served cities	315	01017	00000	AK-1

System B

Cellular company

Cellulink
Pacific Telecom Cellular
200 Gaffney Road, The Alascom Building
Fairbanks, AK 99701

Customer service 800-236-9700
. . . . 414-841-1111
Local office 907-488-7788
Corporate office 414-841-1100

Sending calls

Visitor dialing instructions
Local calls . . . . . area code + 7 digits
Long distance . . . . . area code + 7 digits

Speed-dial numbers
Customer service *611
Emergency 911
Roaming information *711

Receiving calls

Caller dials the roamer access number below,
hears tone, dials cellular area code and number.
. . . . 907-322-7626

Billing information

Visitor rates
Most visitors . . . . . $3 day + 85¢ minute

Visitors without roaming agreements (p. 1159)
Call co. to use Amex

Identification numbers

City	Market	System	Billing	RSA
All served cities	315	01018	00000	AK-1

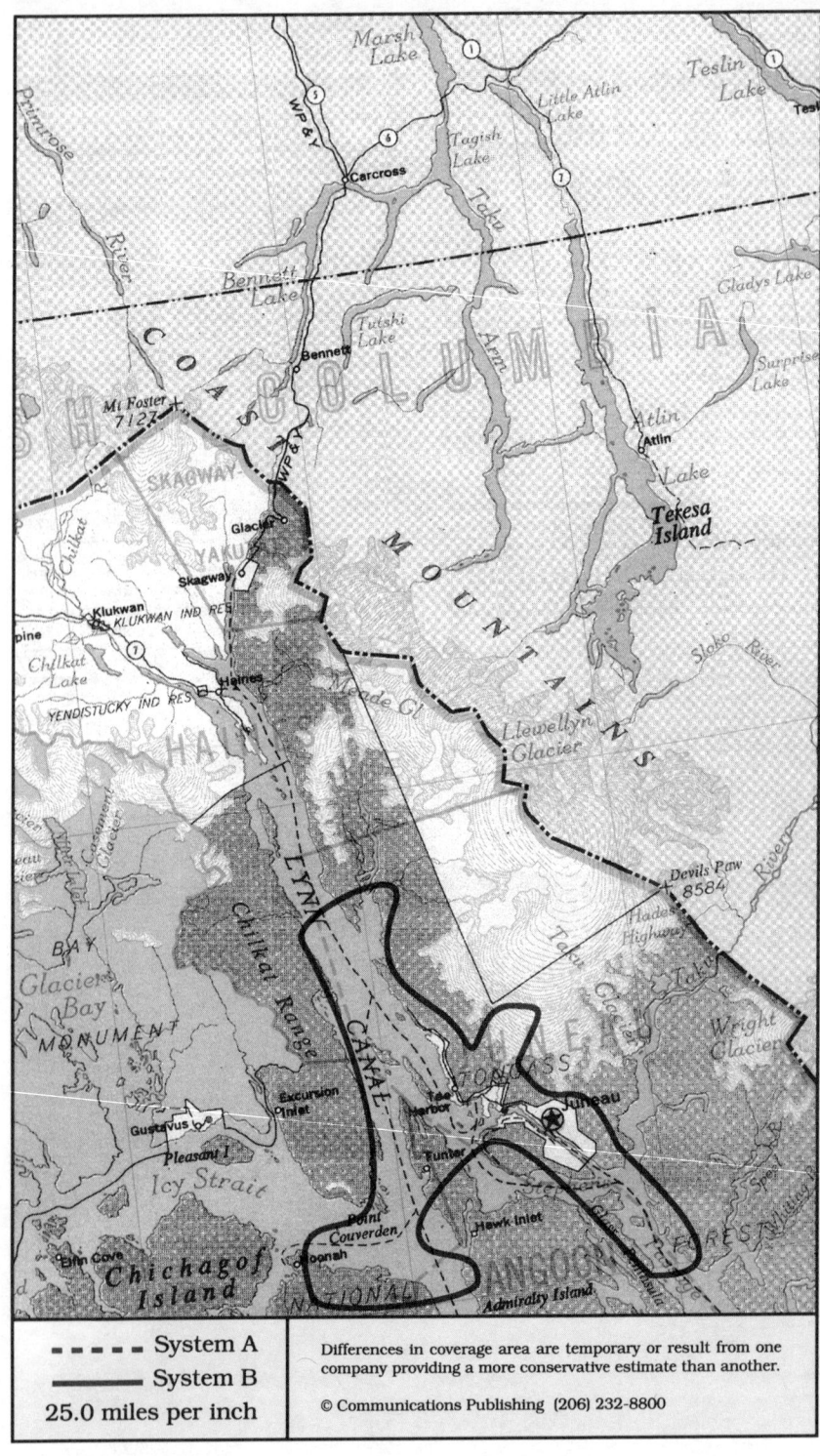

Juneau, Alaska

System A

Cellular company

No information about the company that will operate system A was available at press time.

System B

Cellular company

Cellulink
Pacific Telecom Cellular
8745 Glacier Highway, Suite 382
Juneau, AK 99801

Customer service	800-236-9700
. . . .	414-841-1111
Local office	907-586-8100
Corporate office	414-841-1100

Sending calls

Visitor dialing instructions
Local calls	area code + 7 digits
Long distance	area code + 7 digits

Speed-dial numbers
Customer service	*611
Emergency	911
Roaming information	*711

Receiving calls

Caller dials the roamer access number below, hears tone, dials cellular area code and number.
. . . . 907-321-7626

Billing information

Visitor rates
Most visitors . . . . . $3 day + 85¢ minute

Visitors without roaming agreements (p. 1159)
Call co. to use Amex

Identification numbers

City	Market	System	Billing	RSA
All served cities	317	01022	00000	AK-3

65

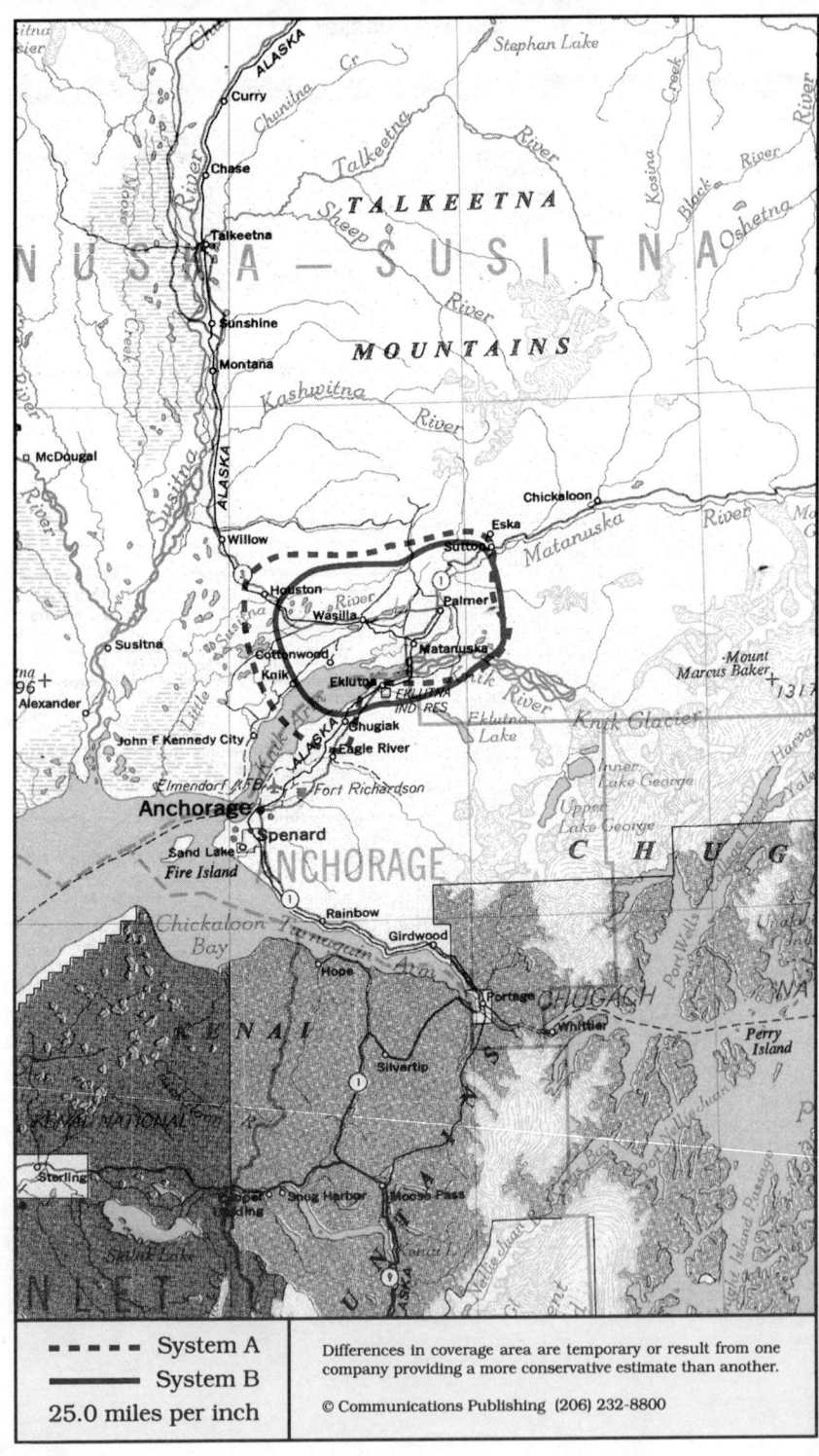

System A
System B
25.0 miles per inch

Differences in coverage area are temporary or result from one company providing a more conservative estimate than another.

© Communications Publishing (206) 232-8800

Palmer, Alaska
Palmer, Wasilla

Cellular company

Cellular One
McCaw Cellular Communications
4797 Business Park Blvd, Suite 2
Anchorage, AK 99503

Local office 907-561-1122
Corporate office 206-827-4500

Sending calls

Visitor dialing instructions
Local calls . . . . . area code + 7 digits
Long distance . . . . area code + 7 digits

Speed-dial numbers
Customer service 611
Directory assistance 411
Emergency 911
Make a traffic report 711

Receiving calls

Caller dials the roamer access number below,
hears tone, dials cellular area code and number.
. . . . 907-229-7626

Billing information

Visitor rates
Most visitors . . . . . $0 day + 99¢ minute

Visitors without roaming agreements (p. 1159)
Roamer Plus will intercept first call to bill you.

Identification numbers

City	Market	System	Billing	RSA
All served cities	316	01019	00000	AK-2

For full coverage area, see Anchorage, AK.

Cellular company

Cellular Connection
Matanuska-Kenai Cellular
701 E Parks Highway, Suite 100
Wasilla, AK 99654

Corporate office 907-373-2355

Sending calls

Visitor dialing instructions
Local calls . . . . 7 digits
Long distance . . . . . area code + 7 digits

Speed-dial numbers
Customer service *611
Emergency 911

Receiving calls

Caller dials the roamer access number below,
hears tone, dials cellular area code and number.
. . . . 907-355-7626

Billing information

Visitor rates
Most visitors . . . . . $3 day + 85¢ minute

Visitors without roaming agreements (p. 1159)
Cannot call since company takes no credit cards.

Identification numbers

City	Market	System	Billing	RSA
All served cities	316	02042	00000	AK-2

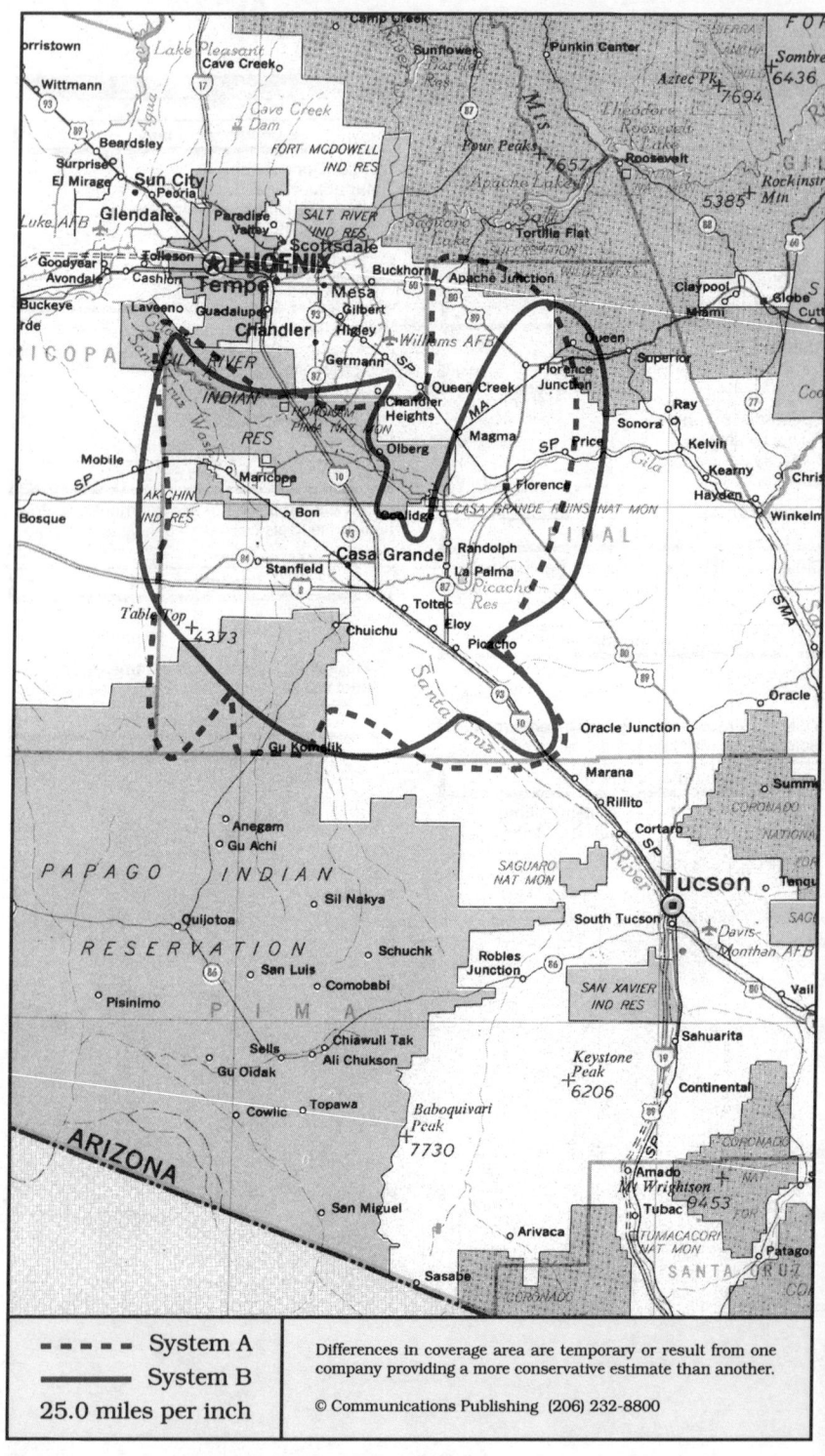

▪▪▪▪ System A	Differences in coverage area are temporary or result from one
――― System B	company providing a more conservative estimate than another.
25.0 miles per inch	© Communications Publishing (206) 232-8800

Casa Grande, Arizona
Casa Grande, Florence

System A	System B

Cellular company

United States Cellular
711 E Florence Boulevard
Casa Grande, AZ 85222

Local office	602-836-9000
Corporate office	312-399-8900

Sending calls

Visitor dialing instructions

Local calls	area code + 7 digits
Long distance . . .	1 + area code + 7 digits

Speed-dial numbers

Customer service	611
Emergency	911

Receiving calls

Caller dials the roamer access number below, hears tone, dials cellular area code and number.
. . . . 602-560-7626

Billing information

Visitor rates
Most visitors . . . . $2.50 day + 90¢ minute

Visitors without roaming agreements (p. 1159)
Call co. to use Mastercard, Visa

Identification numbers

City	Market	System	Billing	RSA
All served cities	322	01031	00000	AZ-5

Cellular company

U S West Cellular
1851 N Center
Casa Grande, AZ 85222

2201 E Camelback Road, Suite 600-B
Phoenix, AZ 85016

Customer service		800-238-7848
. . . .		206-562-2895
. . . . .	local subscribers	800-626-6611
. . . .		206-747-1771
Local office . .	Casa Grande	602-836-1200
. . . .	Phoenix	602-224-7600
Corporate office		206-747-4900

Sending calls

Visitor dialing instructions

Local calls	area code + 7 digits
Long distance	area code + 7 digits

Speed-dial numbers

Arizona highway patrol	*33
Customer service	*611
Emergency	911
Roaming information	*711

Receiving calls

Caller dials the roamer access number below, hears tone, dials cellular area code and number.
. . . . 602-705-7626

Billing information

Visitor rates
Most visitors . . . . . $0 day + 60¢ minute

Visitors without roaming agreements (p. 1159)
Call co. to use Amex, Mastercard, Visa

Identification numbers

City	Market	System	Billing	RSA
All served cities	322	01032	00000	AZ-5

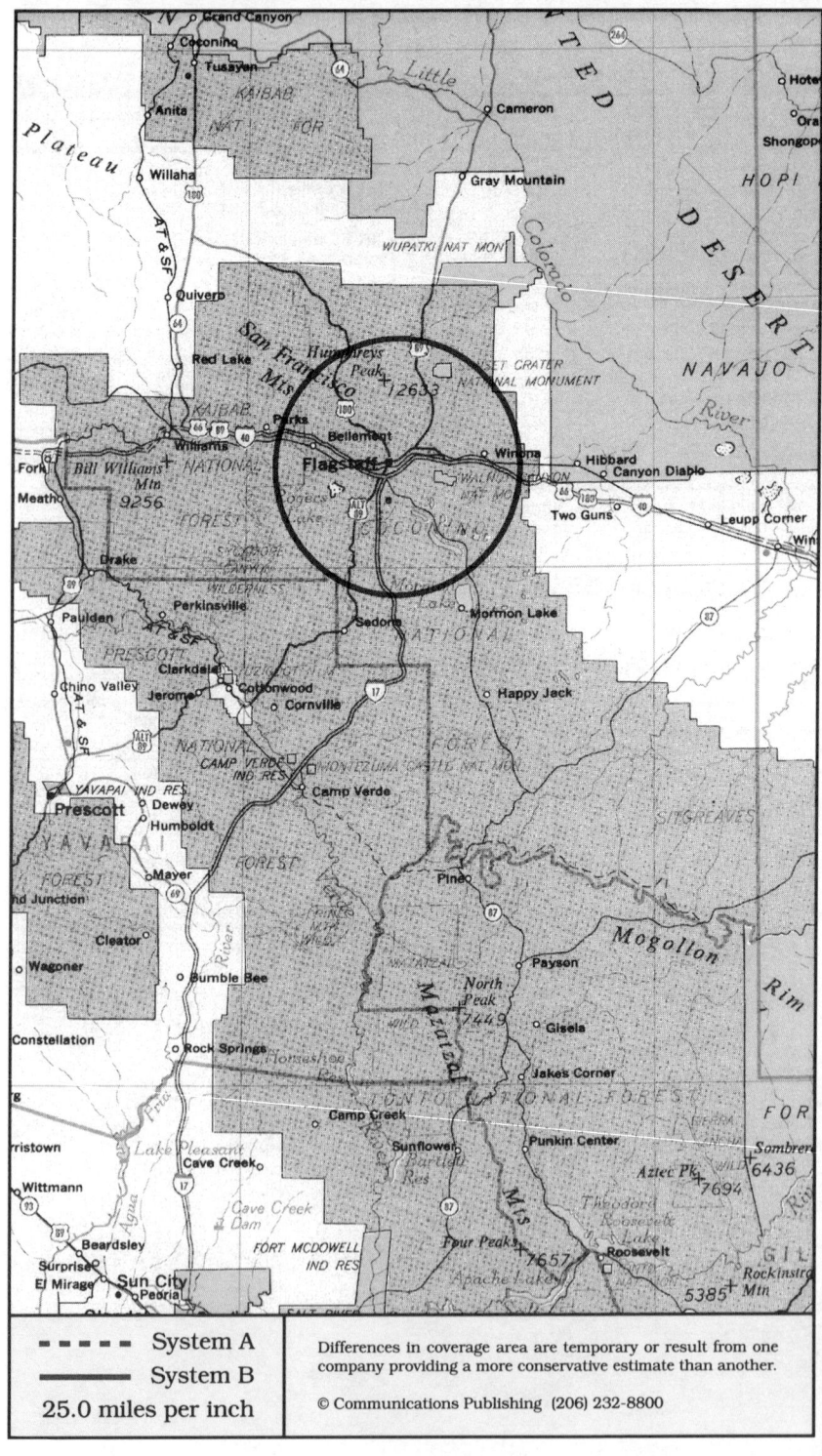

System A -----
System B ———
25.0 miles per inch

Differences in coverage area are temporary or result from one company providing a more conservative estimate than another.

© Communications Publishing (206) 232-8800

Flagstaff, Arizona

System A

Cellular company

No information about the company that will operate system A was available at press time.

System B

Cellular company

Contel Cellular
1500 E Cedar Avenue, Suite 82
Flagstaff, AZ 86004

Customer service	800-333-4004
. . . .	915-772-9222
Local office	602-779-3880
Corporate office	404-804-3400

Sending calls

Visitor dialing instructions
Local calls . . . . . . area code + 7 digits
Long distance . . . 1 + area code + 7 digits

Speed-dial numbers
Customer service 611
Emergency . . . . unavailable

Receiving calls

Caller dials the roamer access number below,
hears tone, dials cellular area code and number.
. . . . 602-699-7626

Billing information

Visitor rates
Most visitors . . . . . $2 day + 65¢ minute

Visitors without roaming agreements (p. 1159)
Call co. to use Amex, Mastercard, Visa

Identification numbers

City	Market	System	Billing	RSA
All served cities	319	01026	00000	AZ-2

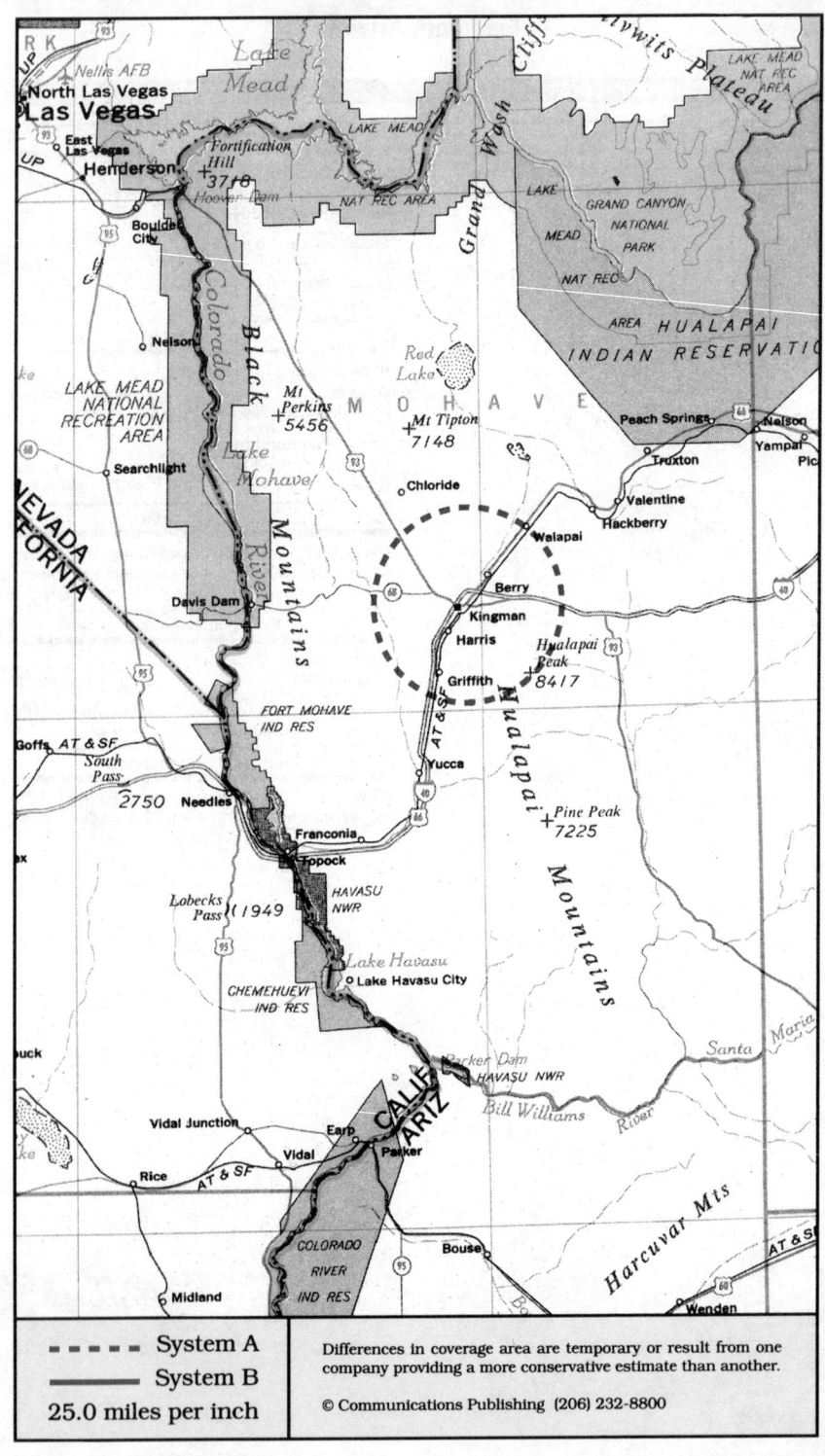

Kingman, Arizona

Cellular company

Cellular One
McCaw Cellular Communications
3763 Howard Hughes Parkway #200
Las Vegas, NV 89109-0939

Customer service	702-732-2240
Local office	702-734-1010
Corporate office	206-827-4500

Cellular company

No information about the company that will operate system B was available at press time.

Sending calls

Visitor dialing instructions

Local calls	area code + 7 digits
Long distance . . .	1 + area code + 7 digits

Speed-dial numbers

Customer service	611
Emergency	911

Receiving calls

Caller dials the roamer access number below, hears tone, dials cellular area code and number.

. . . . 702-595-7626

North American Cellular Network (see p. 1156)
All calls forward to you. Deactivate dial ∗35 send.

Billing information

Visitor rates
Most visitors . . . . . $2 day + 90¢ minute

Visitors without roaming agreements (p. 1159)
Cannot call since company takes no credit cards.

Identification numbers

City	Market	System	Billing	RSA
All served cities	318	01023	00000	AZ-1

Call switching is performed by McCaw Cellular for the licensee, Satellite Cellular.

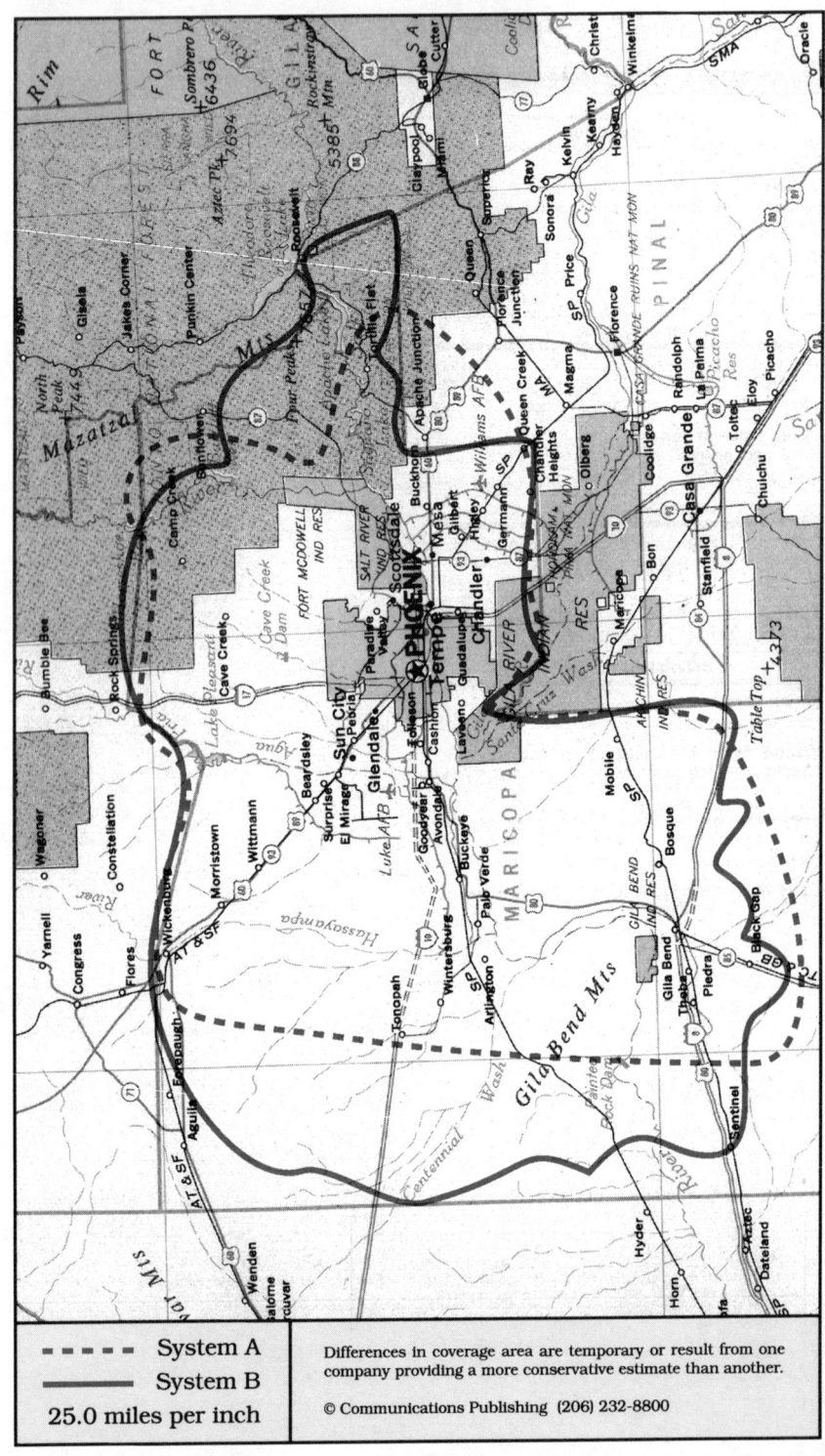

▪ ▪ ▪ ▪ System A	
—— System B	Differences in coverage area are temporary or result from one company providing a more conservative estimate than another.
25.0 miles per inch	© Communications Publishing (206) 232-8800

74

Phoenix, Arizona
Chandler, Gila Bend, Glendale, Phoenix, Tempe

System A	System B

System A

Cellular company

Bell Atlantic Mobile
528 W 21st Street, Suite 2
Tempe, AZ 85282

Customer service	800-352-7626
Local service	602-731-6100
Corporate office	908-306-7000

Sending calls

Visitor dialing instructions

Local calls	area code + 7 digits
Long distance . . .	1 + area code + 7 digits

Speed-dial numbers

Arizona highway patrol	*33
Customer service	*611
Emergency	911

Receiving calls

Caller dials the roamer access number below,
hears tone, dials cellular area code and number.
. . . . 602-390-7626

NationLink & RoamingAmerica (see p. 1157)
Dial *31 send to receive calls, *30 to deactivate.
Dial *32 send to tell caller the roamer access no.
This company's subscribers have level 8 service.

Billing information

Visitor rates

6 a.m. - 7 p.m Mon-Fri. .	$0 day + 70¢ minute
All other times	$0 day + 60¢ minute

Visitors without roaming agreements (p. 1159)
Call co. to use Mastercard, Visa
or Roamer Plus will intercept first call to bill you.

Identification numbers

City	Market	System	Billing	RSA
All served cities	026	00053	00000	MSA

System B

Cellular company

U S West Cellular
2201 E Camelback Road, Suite 600-B
Phoenix, AZ 85016

Customer service	800-238-7848
. . . .	206-562-2895
. . . . . local subscribers	800-626-6611
. . . .	206-747-1771
Local office	602-224-7600
Corporate office	206-747-4900

Sending calls

Visitor dialing instructions

Local calls	area code + 7 digits
Long distance	area code + 7 digits

Speed-dial numbers

Arizona highway patrol	*33
Customer service	*611
Emergency	911
Roaming information	*711

Receiving calls

Caller dials the roamer access number below,
hears tone, dials cellular area code and number.
. . . . 602-228-7626

Follow Me Roaming forwarding (see p. 1156)
Activate, dial *18 send. Deactivate dial *19 send.

Billing information

Visitor rates

Most visitors	$0 day + 60¢ minute

Visitors without roaming agreements (p. 1159)
Call co. to use Amex, Mastercard, Visa

Identification numbers

City	Market	System	Billing	RSA
All served cities	026	00048	00000	MSA

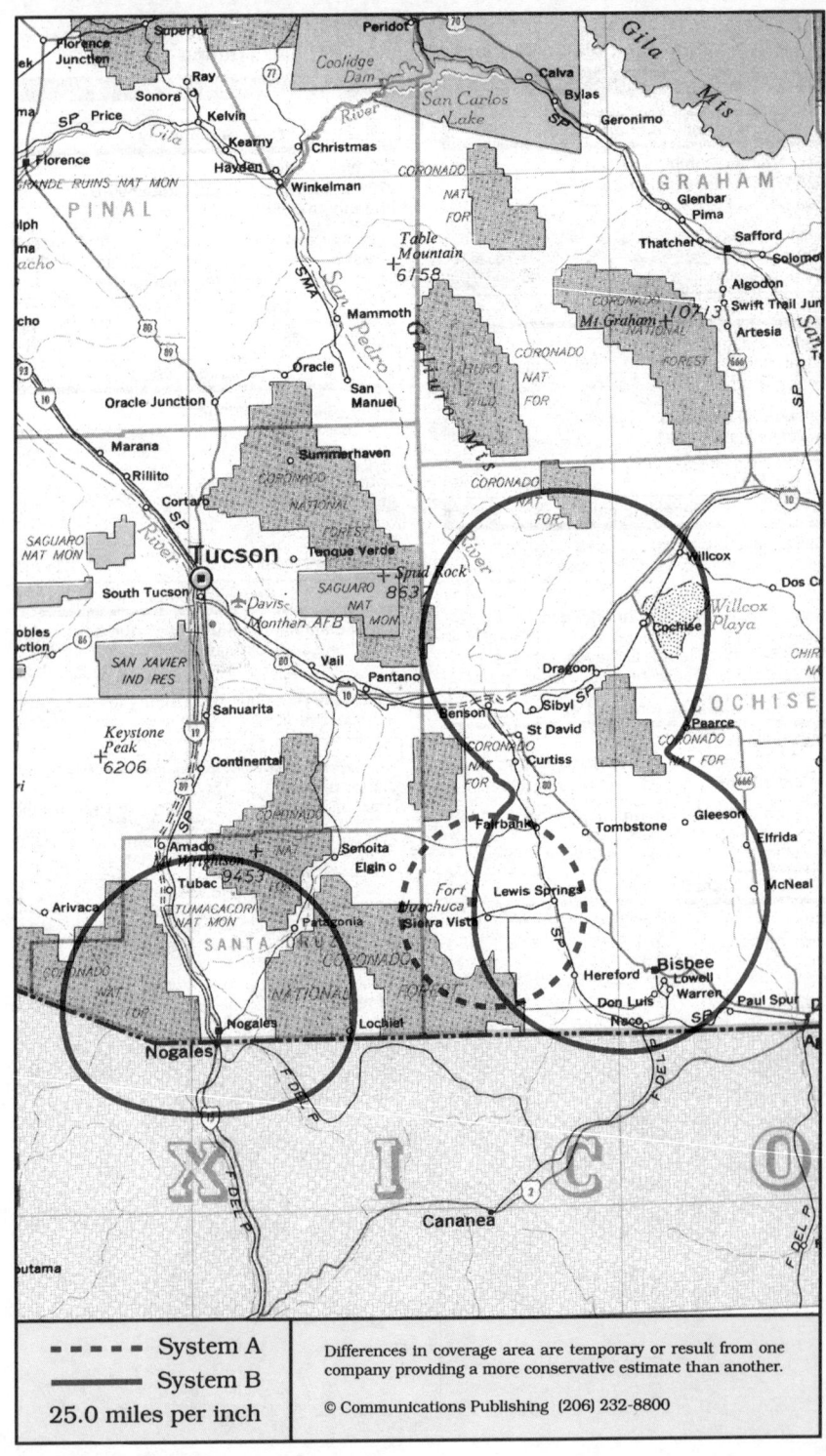

- - - - System A
——— System B
25.0 miles per inch

Differences in coverage area are temporary or result from one company providing a more conservative estimate than another.

© Communications Publishing (206) 232-8800

Sierra Vista, Arizona
Benson, Bisbee, Nogales, Sierra Vista, Willcox

System A	System B

Cellular company

Bell Atlantic Mobile
998 S Cherry
Tucson, AZ 85719

Customer service	800-352-7626
Local office	602-628-9541
Corporate office	908-306-7000

Sending calls

Visitor dialing instructions

Local calls	area code + 7 digits
Long distance . . .	1 + area code + 7 digits

Speed-dial numbers

Arizona highway patrol	*33
Customer service	*611
Emergency	911

Receiving calls

Caller dials the roamer access number below, hears tone, dials cellular area code and number.
. . . . 602-449-7626

NationLink & RoamingAmerica (see p. 1157)
Dial *31 send to receive calls, *30 to deactivate.
Dial *32 send to tell caller the roamer access no.
This company's subscribers have level 8 service.

Billing information

Visitor rates
Most visitors . . . . $2.50 day + 90¢ minute

Visitors without roaming agreements (p. 1159)
Cannot call since company takes no credit cards.

Identification numbers

City	Market	System	Billing	RSA
All served cities	323	00053	31053	AZ-6

Cellular company

Valley Telecom
Southeast Arizona Cellular
3241 N Crane Avenue
Nogales, AZ 85621

999 E Fry Boulevard, Suite 103
Sierra Vista, AZ 85635

150 W Wasson
Willcox, AZ 85643

Corporate office . .	Nogales	602-281-2222
. . (some areas)	Sierra Vista	800-834-2960
. . . .	Sierra Vista	602-458-7088
. . . (some areas)	Willcox	800-472-6497
. . . .	Willcox	602-384-2960

Sending calls

Visitor dialing instructions

Local calls	area code + 7 digits
Long distance . . .	1 + area code + 7 digits

Speed-dial numbers

Customer service	*611
Emergency	911

Receiving calls

Caller dials the roamer access number below, hears tone, dials cellular area code and number.
. . . . 602-508-7626

Billing information

Visitor rates
Most visitors . . . . . $2 day + 65¢ minute

Visitors without roaming agreements (p. 1159)
Cannot call since company takes no credit cards.

Identification numbers

City	Market	System	Billing	RSA
All served cities	323	01034	00000	AZ-6

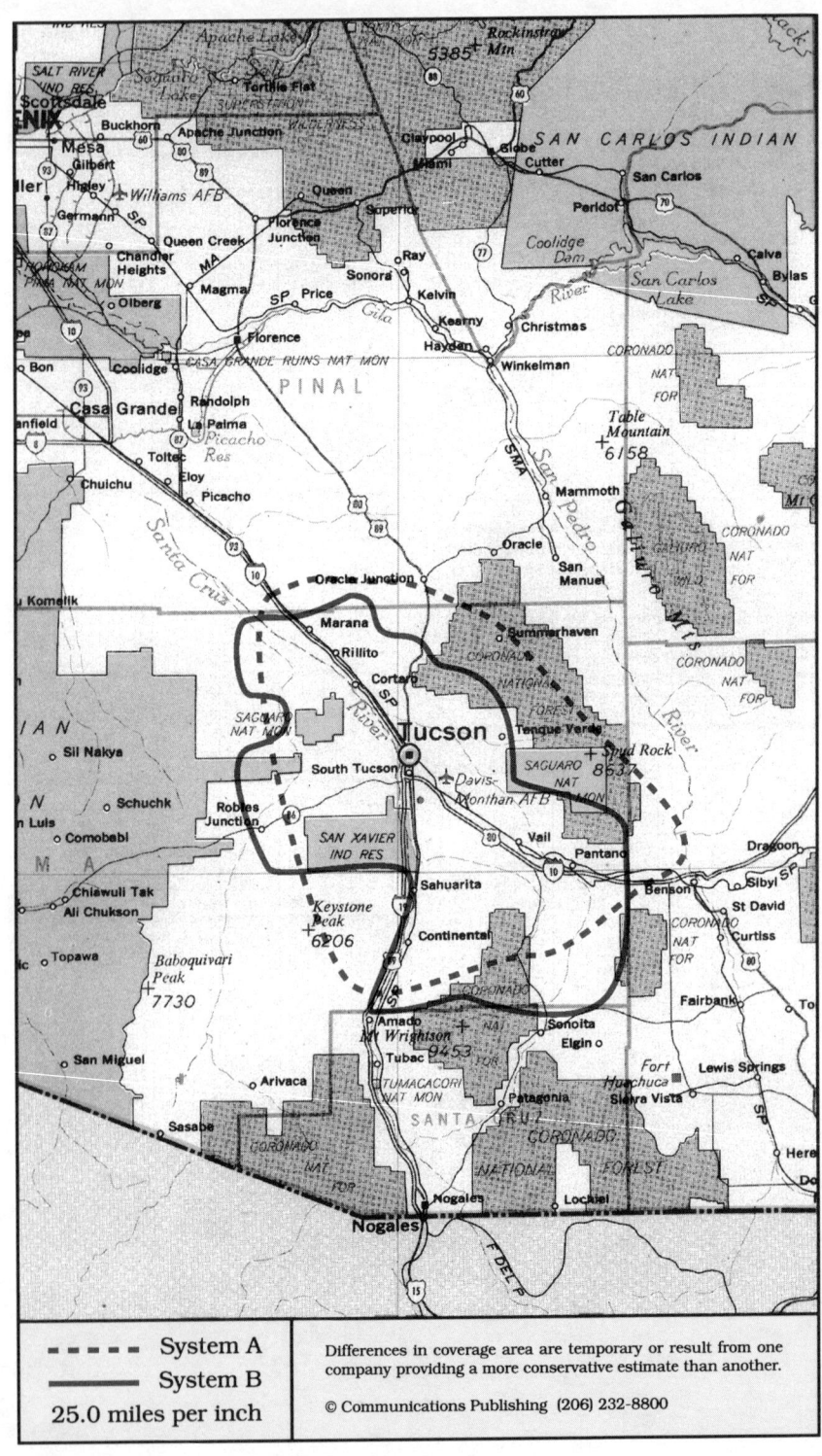

- - - - System A
——— System B

25.0 miles per inch

Differences in coverage area are temporary or result from one company providing a more conservative estimate than another.

© Communications Publishing (206) 232-8800

Tucson, Arizona

System A

Cellular company

Bell Atlantic Mobile
998 S Cherry
Tucson, AZ 85719

Customer service	800-352-7626
Local office	602-628-9541
Corporate office	908-306-7000

Sending calls

Visitor dialing instructions

Local calls	area code + 7 digits
Long distance . . .	1 + area code + 7 digits

Speed-dial numbers

Arizona highway patrol	*33
Customer service	*611
Emergency	911

Receiving calls

Caller dials the roamer access number below,
hears tone, dials cellular area code and number.
. . . . 602-449-7626

NationLink & RoamingAmerica (see p. 1157)
Dial *31 send to receive calls, *30 to deactivate.
Dial *32 send to tell caller the roamer access no.
This company's subscribers have level 8 service.

Billing information

Visitor rates

6 a.m. - 7 p.m Mon-Fri. .	$0 day + 70¢ minute
All other times	$0 day + 60¢ minute

Visitors without roaming agreements (p. 1159)
Roamer Plus will intercept first call to bill you.

Identification numbers

City	Market	System	Billing	RSA
All served cities	077	00053	30053	MSA

System B

Cellular company

U S West Cellular
3440 S Palo Verde, Suite 142
Tucson, AZ 85713

Customer service	800-238-7848
. . . .	206-562-2895
. . . . . local subscribers	800-626-6611
. . . .	206-747-1771
Local office	602-740-9100
Corporate office	206-747-4900

Sending calls

Visitor dialing instructions

Local calls	area code + 7 digits
Long distance	area code + 7 digits

Speed-dial numbers

Arizona highway patrol	*33
Customer service	*611
Emergency	911
Make a traffic report	*620
Roaming information	*711

Receiving calls

Caller dials the roamer access number below,
hears tone, dials cellular area code and number.
. . . . 602-444-7626

Follow Me Roaming forwarding (see p. 1156)
Activate, dial *18 send. Deactivate dial *19 send.

Billing information

Visitor rates
Most visitors . . . . . $0 day + 60¢ minute

Visitors without roaming agreements (p. 1159)
Call co. to use Amex, Mastercard, Visa

Identification numbers

City	Market	System	Billing	RSA
All served cities	077	00140	00000	MSA

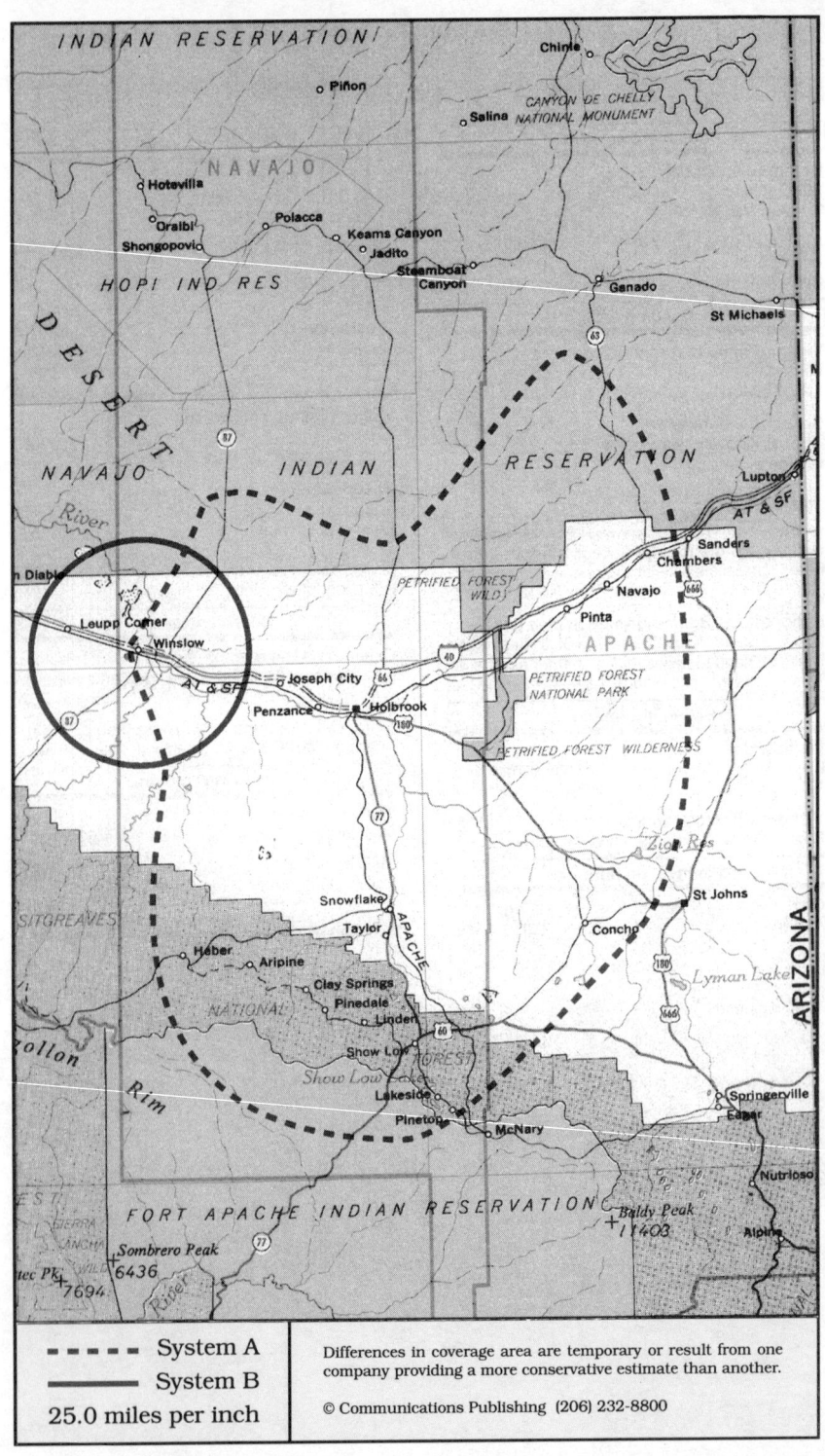

- - - - System A	Differences in coverage area are temporary or result from one company providing a more conservative estimate than another.
──── System B	
25.0 miles per inch	© Communications Publishing (206) 232-8800

Winslow, Arizona

Holbrook, Show Low, Winslow

System A	System B

Cellular company

Cellular One
Licensed to: Smith Bagley
Managed by: Sterling Cellular Management
1080 Holcomb Bridge Road, Bldg 100, Ste 200
Roswell, GA 30076

Customer service 800-552-6150
Corporate office 404-552-5030

Sending calls

Visitor dialing instructions
Local calls . . . . area code + 7 digits
Long distance . . . 1 + area code + 7 digits

Speed-dial numbers
Customer service 611
Emergency *911
Roaming information *711

Receiving calls

Caller dials the roamer access number below,
hears tone, dials cellular area code and number.
. . . . 602-521-7626

Billing information

Visitor rates
Most visitors . . . . $3.50 day + 99¢ minute
Visitors from Arizona . $3.00 day + 99¢ minute

Visitors without roaming agreements (p. 1159)
Roamer Plus will intercept first call to bill you.

Identification numbers

City	Market	System	Billing	RSA
All served cities	320	01027	00000	AZ-3

Cellular company

Century Cellunet
1701 Louisiana Avenue
Monroe, LA 71201

Customer service 800-638-9727
. . . . 318-683-3450
Corporate office 318-388-9000

Sending calls

Visitor dialing instructions
Local calls . . . . . . area code + 7 digits
Long distance . . . 1 + area code + 7 digits

Speed-dial numbers
Customer service *711
Emergency . . . . not available

Receiving calls

Caller dials the roamer access number below,
hears tone, dials cellular area code and number.
. . . . 602-587-7626

Follow Me Roaming forwarding (see p. 1156)
Activate, dial *18 send. Deactivate dial *19 send.

Billing information

Visitor rates
Most visitors . . . . . $3 day + 90¢ minute

Visitors without roaming agreements (p. 1159)
Call co. to use Amex, Mastercard, Visa

Identification numbers

City	Market	System	Billing	RSA
All served cities	320	01028	00000	AZ-3

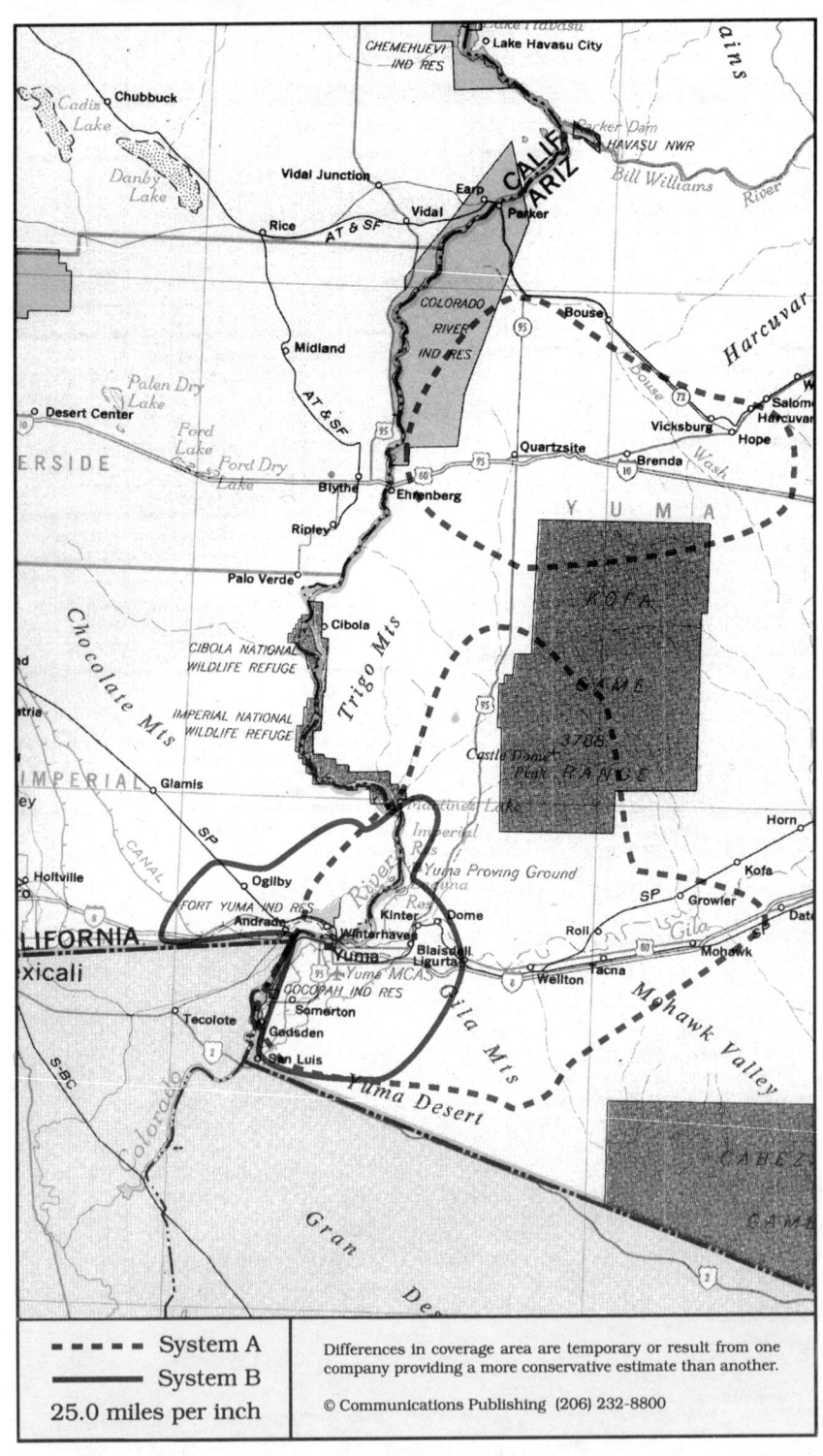

- - - - System A
———— System B
25.0 miles per inch

Differences in coverage area are temporary or result from one company providing a more conservative estimate than another.

© Communications Publishing (206) 232-8800

Yuma, Arizona

System A	System B

System A

Cellular company

Cellular One
Centennial Cellular
770 E 32nd Street, PO Box 6434
Yuma, AZ 85366-6434

Local office		602-344-5200
Corporate office	Administration	619-352-7066
. . . . . .	Management	203-972-2000

Sending calls

Visitor dialing instructions
Local calls . . . . . area code + 7 digits
Long distance . . . 0 + area code + 7 digits

Speed-dial numbers
Channel 11 television *news11
Customer service *611
Emergency 911

Receiving calls

Caller dials the roamer access number below,
hears tone, dials cellular area code and number.
. . . . 602-920-7626

NationLink & RoamingAmerica (see p. 1157)
Caller hears your current roamer access number.
This company's subscribers have level 0 service.

Billing information

Visitor rates
Most visitors . . . . . $3 day + 85¢ minute

Visitors without roaming agreements (p. 1159)
Cannot call since company takes no credit cards.

Identification numbers

City	Market	System	Billing	RSA
All served cities	321	01029	00000	AZ-4

For full coverage area, see El Centro, CA.

System B

Cellular company

Contel Cellular
1910 S 4th Avenue
Yuma, AZ 85364

Customer service	. . . .	800-333-4004
. . . .		915-772-9222
Local office		602-782-1567
Corporate office		404-804-3400

Sending calls

Visitor dialing instructions
Local calls . . . . . . area code + 7 digits
Long distance . . . 1 + area code + 7 digits

Speed-dial numbers
Customer service 611
Emergency 911
Mr. Rescue for car trouble . *HELP or *4357

Receiving calls

Caller dials the roamer access number below,
hears tone, dials cellular area code and number.
. . . . 602-580-7626

Billing information

Visitor rates
Most visitors . . . . . $2 day + 65¢ minute

Visitors without roaming agreements (p. 1159)
Call co. to use Amex, Mastercard, Visa

Identification numbers

City	Market	System	Billing	RSA
All served cities	321	01030	00000	AZ-4

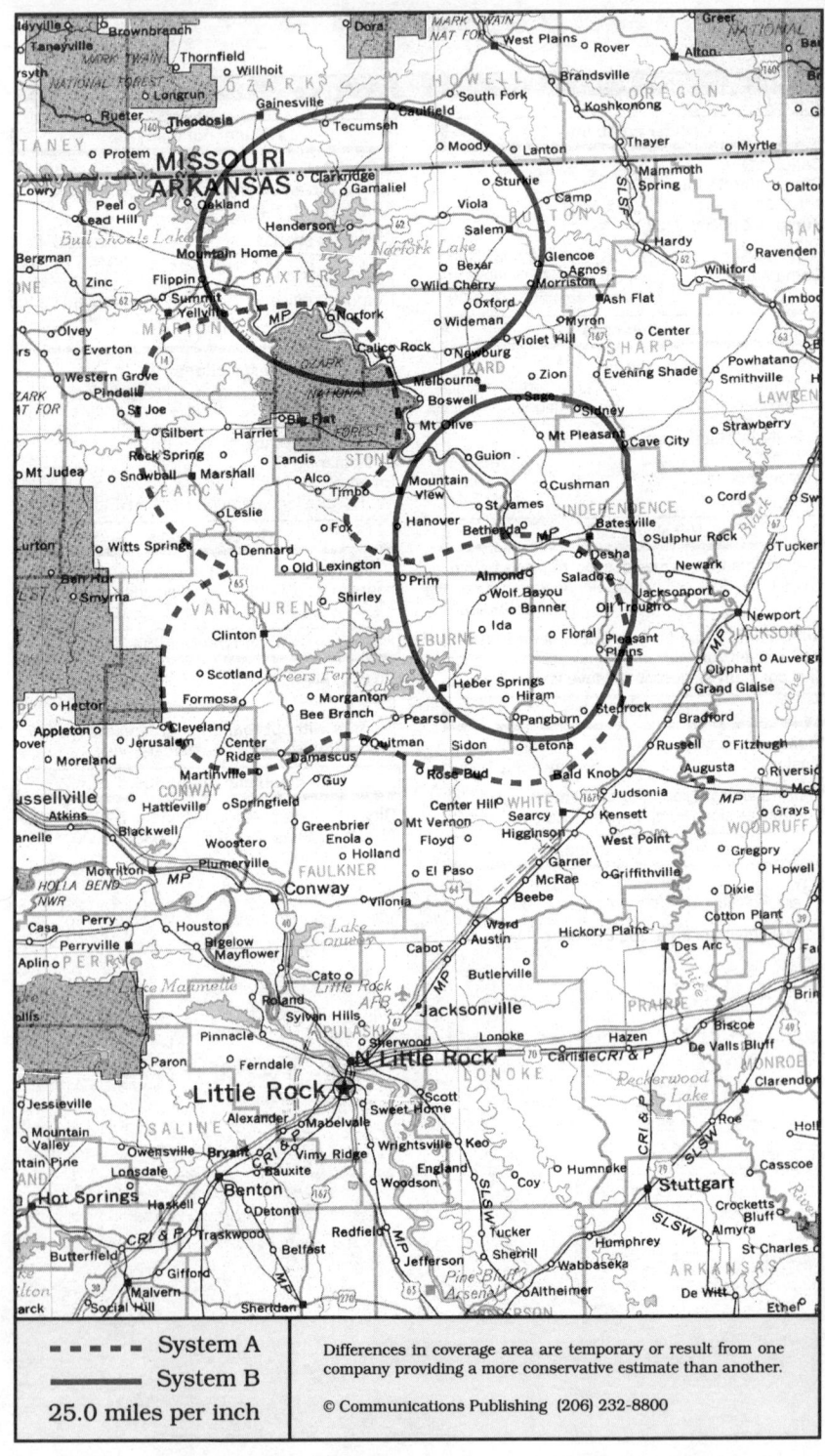

▪ ▪ ▪ ▪ ▪	System A
——————	System B
25.0 miles per inch	

Differences in coverage area are temporary or result from one company providing a more conservative estimate than another.

© Communications Publishing (206) 232-8800

Batesville, Arkansas

Batesville, Clinton, Heber Springs, Marshall, Mountain Home, Mountain View

System A	System B

Cellular company

Cellular One
Sterling Cellular Management
1080 Holcomb Bridge Road, Bldg 100, Ste 200
Roswell, GA 30076

Customer service 800-552-6150
Corporate office 404-552-5030

Cellular company

Century Cellunet
2325 Texas Boulevard, PO Box 2943
Texarkana, TX 75501

Customer service 800-638-9727
Local office 903-793-0500
Corporate office 318-388-9000

Sending calls

Visitor dialing instructions
Local calls . . . . . area code + 7 digits
Long distance . . . 1 + area code + 7 digits

Speed-dial numbers
Customer service *611
Emergency 911

Sending calls

Visitor dialing instructions
Local calls . . . . 7 digits
Long distance . . . 1 + area code + 7 digits

Speed-dial numbers
Customer service *611
Emergency 911

Receiving calls

Caller dials the roamer access number below,
hears tone, dials cellular area code and number.
. . . . 501-278-7626

Receiving calls

Caller dials the roamer access number below,
hears tone, dials cellular area code and number.
. . . . 501-698-5226

Billing information

Visitor rates
Most visitors . . . . . $3 day + 75¢ minute

Visitors without roaming agreements (p. 1159)
Roamer Plus will intercept first call to bill you.

Billing information

Visitor rates
Most visitors . . . . . $3 day + 75¢ minute

Visitors without roaming agreements (p. 1159)
Call co. to use Amex, Mastercard, Visa

Identification numbers

City	Market	System	Billing	RSA
Heber Springs	329	01047	00000	AR-6
Marshall	325	01047	00000	AR-2

Identification numbers

City	Market	System	Billing	RSA
Batesville	326	01038	00000	AR-3
Mountain Home	325	01038	00000	AR-2

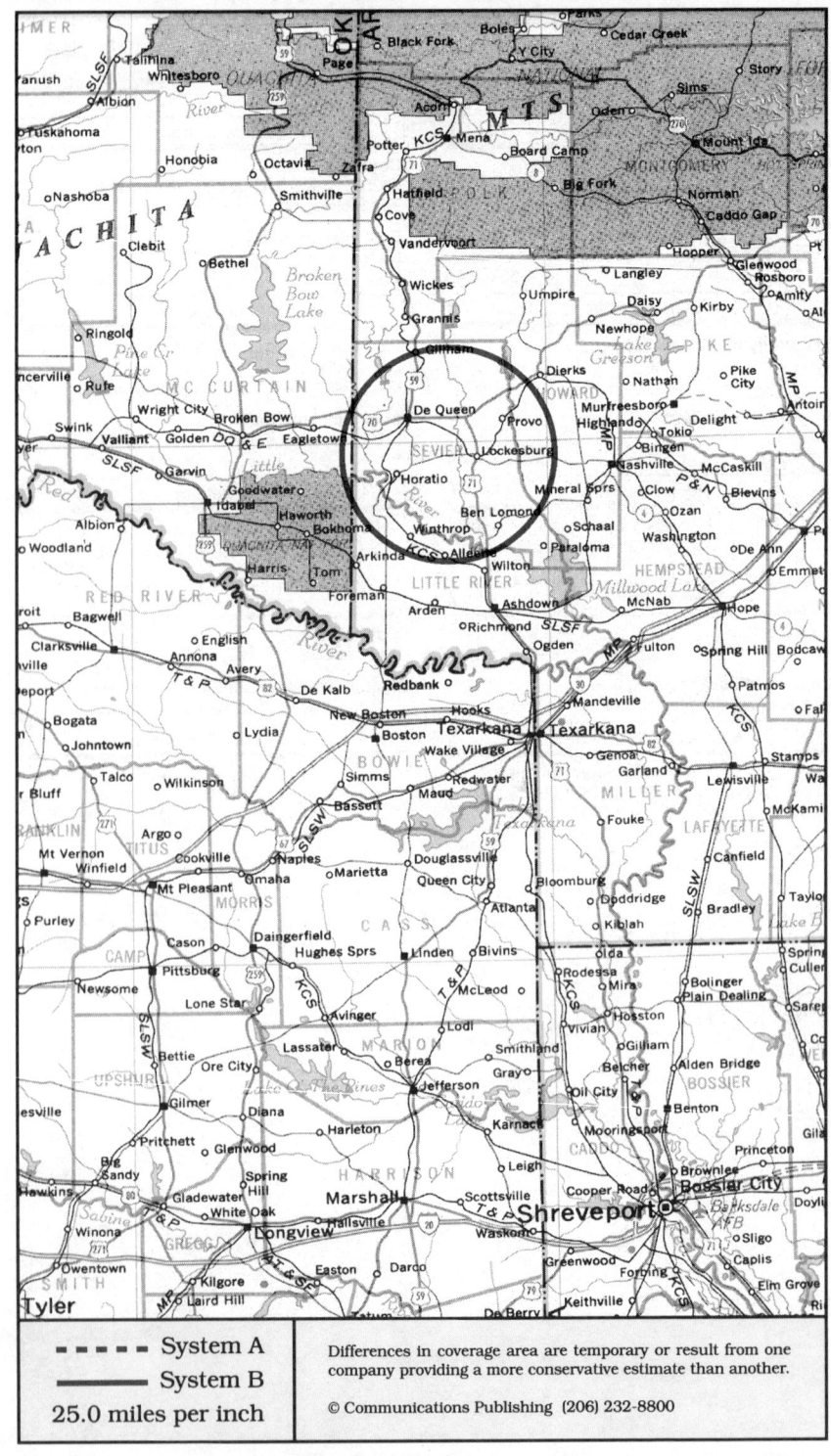

▬ ▬ ▬ System A	Differences in coverage area are temporary or result from one
▬▬▬ System B	company providing a more conservative estimate than another.
25.0 miles per inch	© Communications Publishing (206) 232-8800

De Queen, Arkansas

Cellular company

No information about the company that will operate system A was available at press time.

Cellular company

Walnut Hill Cellular
PO Box 729
Lewisville, AR 71845

Customer service	800-554-2238
Local office	501-921-4224

Sending calls

Visitor dialing instructions

Local calls	area code + 7 digits
Long distance . . .	1 + area code + 7 digits

Speed-dial numbers

Customer service	*611
Sheriff	*911

Receiving calls

Caller dials the roamer access number below, hears tone, dials cellular area code and number.
. . . . 501-584-7726

Billing information

Visitor rates
Most visitors . . . . . $3 day + 75¢ minute

Visitors without roaming agreements (p. 1159)
Cannot call since company takes no credit cards.

Identification numbers

City	Market	System	Billing	RSA
All served cities	332	01052	00000	AR-9

87

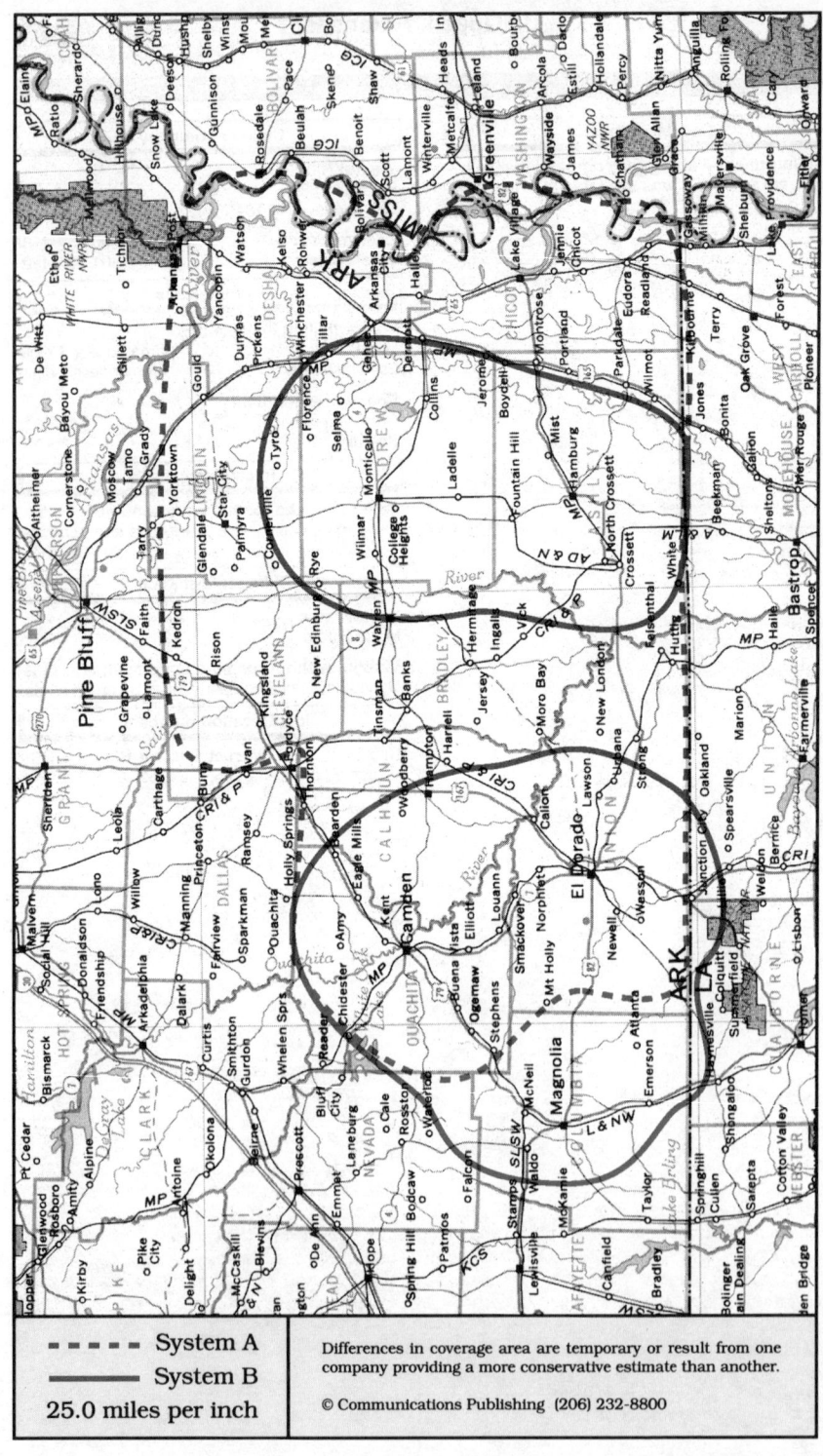

System A
System B
25.0 miles per inch

Differences in coverage area are temporary or result from one company providing a more conservative estimate than another.

© Communications Publishing (206) 232-8800

El Dorado, Arkansas

Arkansas City, Camden, El Dorado, Magnolia, Monticello, Star City, Warren

System A

Cellular company

Cellular One
Mercury Communications
211 N Washington #111
El Dorado, AR 71730

Local office	501-862-0010
Corporate office	601-948-4800

Sending calls

Visitor dialing instructions

Local calls	area code + 7 digits
Long distance	area code + 7 digits

Speed-dial numbers

Customer service	*611
Emergency	911

Receiving calls

Caller dials the roamer access number below, hears tone, dials cellular area code and number.
. . . . 501-866-7626

NationLink & RoamingAmerica (see p. 1157)
Dial *31 send to receive calls, *30 to deactivate. This company's subscribers have level 3 service.

Billing information

Visitor rates
Most visitors . . . . . . $3 day + $1 minute

Visitors without roaming agreements (p. 1159)
Call co. to use Amex, Mastercard, Visa or Roamer Plus will intercept first call to bill you.

Identification numbers

City	Market	System	Billing	RSA
All served cities	335	01057	00000	AR-12

System B

Cellular company

Century Cellunet
105 N Jefferson
El Dorado, AR 71730

Customer service	800-638-9727
Local office	501-862-3456
Corporate office	318-388-9000

Sending calls

Visitor dialing instructions

Local calls	area code + 7 digits
Long distance . . .	1 + area code + 7 digits

Speed-dial numbers

Customer service	*611
Emergency	911

Receiving calls

Caller dials the roamer access number below, hears tone, dials cellular area code and number.
. . . . 501-864-5626

Billing information

Visitor rates
Most visitors . . . . . $3 day + 75¢ minute

Visitors without roaming agreements (p. 1159)
Call co. to use Amex, Mastercard, Visa

Identification numbers

City	Market	System	Billing	RSA
All served cities	335	01058	00000	AR-12

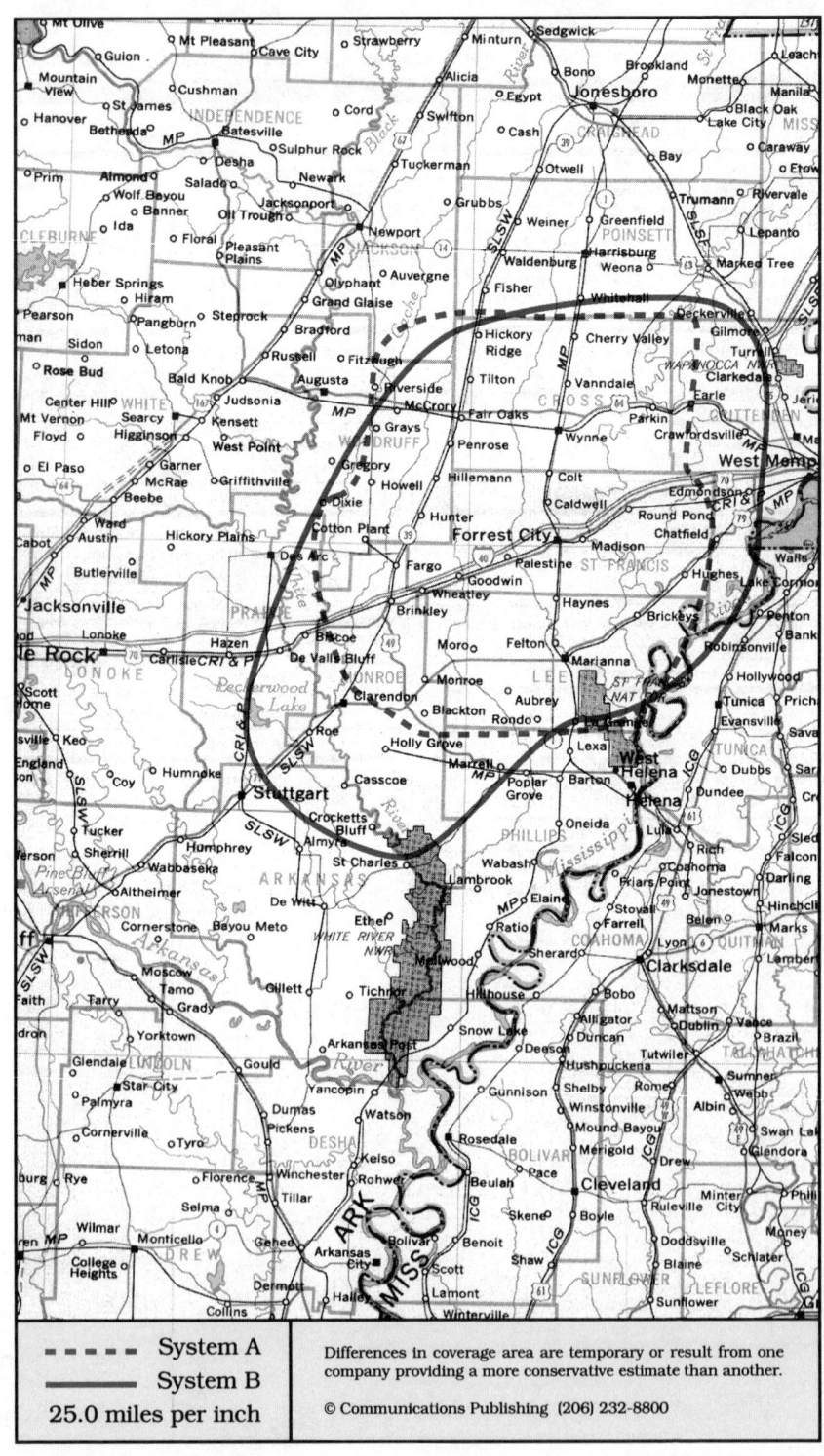

System A
System B
25.0 miles per inch

Differences in coverage area are temporary or result from one company providing a more conservative estimate than another.

© Communications Publishing (206) 232-8800

Forrest City, Arkansas
Clarendon, Forrest City, Marianna, Wynne

System A	System B

Cellular company

Cellular One
Sterling Cellular Management
1080 Holcomb Bridge Road, Bldg 100, Ste 200
Roswell, GA 30076

Customer service	800-552-6150
Corporate office	404-552-5030

Sending calls

Visitor dialing instructions

Local calls	area code + 7 digits
Long distance . . .	1 + area code + 7 digits

Speed-dial numbers

Customer service	*611
Emergency	*911
Roaming information	*711

Receiving calls

Caller dials the roamer access number below,
hears tone, dials cellular area code and number.

. . . .	501-270-7626

Billing information

Visitor rates

Most visitors	$3 day + 99¢ minute

Visitors without roaming agreements (p. 1159)
Roamer Plus will intercept first call to bill you.

Identification numbers

City	Market	System	Billing	RSA
All served cities	328	01047	00000	AR-5

Cellular company

ALLTEL Mobile Communications
2516 Cantrell Road
Little Rock, AR 72202

Customer service	some areas	800-255-8351
. . . .		501-661-5800
Local office		501-666-6688
Corporate office		501-661-8500

Sending calls

Visitor dialing instructions

Local calls	area code + 7 digits
Long distance	area code + 7 digits

Speed-dial numbers

Customer service	*611
Emergency	911
KATV news	*7
KARN radio	*92

Receiving calls

Caller dials the roamer access number below,
hears tone, dials cellular area code and number.

. . . .	501-261-8626

Follow Me Roaming forwarding (see p. 1156)
Activate, dial *18 send. Deactivate dial *19 send.

Billing information

Visitor rates

Most visitors	$3 day + 85¢ minute

Visitors without roaming agreements (p. 1159)
Cannot call since company takes no credit cards.

Identification numbers

City	Market	System	Billing	RSA
All served cities	328	01044	30502	AR-5

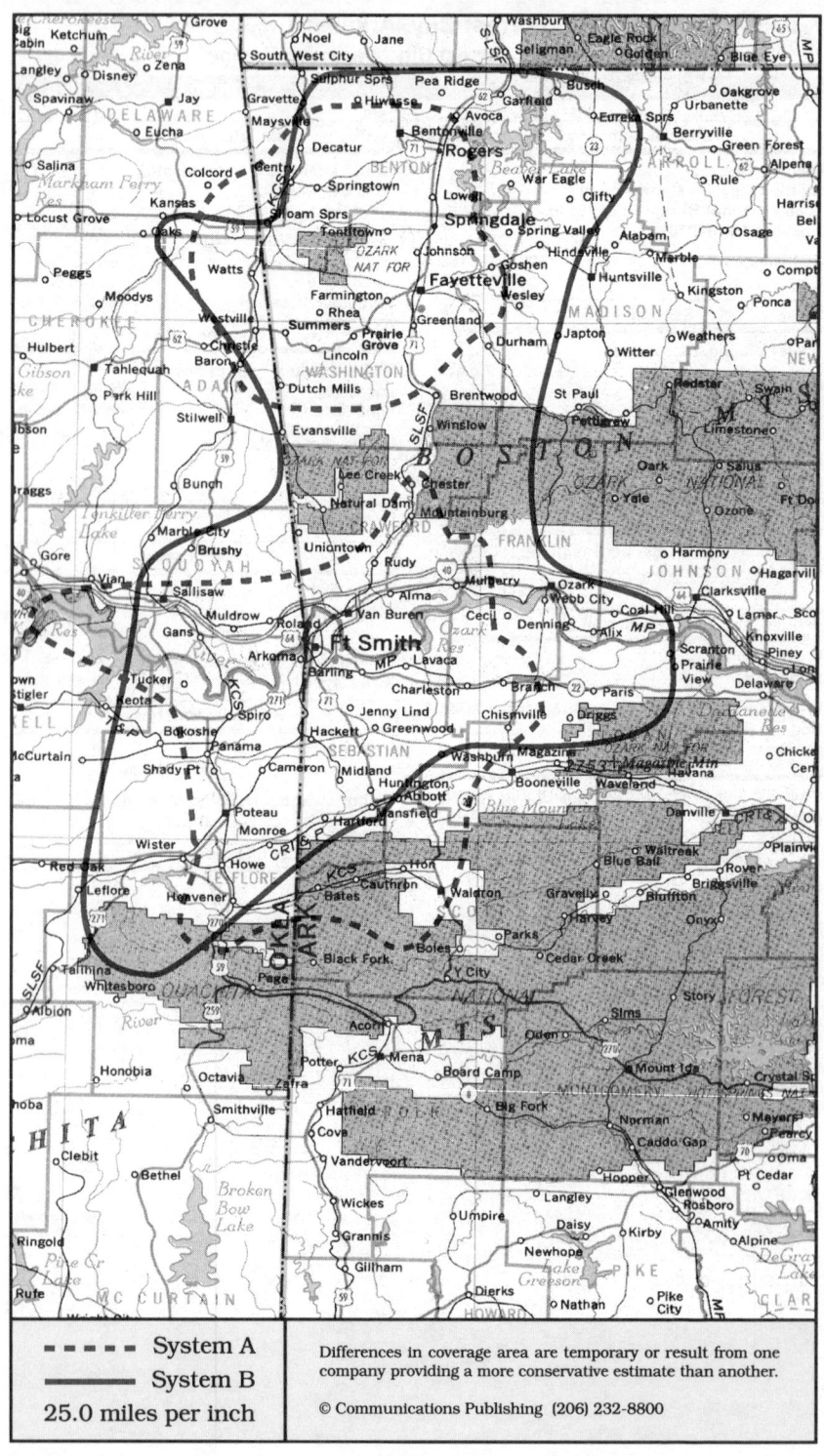

- - - -	System A
———————	System B

25.0 miles per inch

Differences in coverage area are temporary or result from one company providing a more conservative estimate than another.

© Communications Publishing (206) 232-8800

Fort Smith, Arkansas

Eureka Springs, Fayetteville, Fort Smith, Ozark (system B), Springdale, Van Buren

System A	System B

System A

Cellular company

Cellular One
McCaw Cellular Communications
3075 N College Avenue
Fayetteville, AR 72703-3417

4300 Rogers Avenue, Suite 16
Fort Smith, AR 72903

Customer service		501-530-4357
Local office . . .	Fayetteville	501-444-9100
. . . .	Fort Smith	501-783-4600
Corporate office		206-827-4500

Sending calls

Visitor dialing instructions
Local calls . . . . . area code + 7 digits
Long distance . . . 1 + area code + 7 digits

Speed-dial numbers
Customer service 611
Fayetteville office 811
Emergency 911

Receiving calls

Caller dials the roamer access number below,
hears tone, dials cellular area code and number.
Fayetteville 501-530-7626
Fort Smith 501-650-7626

NationLink & RoamingAmerica (see p. 1157)
Dial *31 send to receive calls, *30 to deactivate.
Dial *32 send to tell caller the roamer access no.
This company's subscribers have level 9 service.

North American Cellular Network (see p. 1156)
All calls forward to you. Deactivate dial *35 send.

Billing information

Visitor rates
Most visitors . . . . . $0 day + 99¢ minute

Visitors without roaming agreements (p. 1159)
Call co. to use Amex, Mastercard, Visa
or Roamer Plus will intercept first call to bill you.

Identification numbers

City	Market	System	Billing	RSA
Fayetteville	182	00607	00000	MSA
Fort Smith	165	00359	00000	MSA

System A service in Ozark, see Russellville, AR.

System B

Cellular company

ALLTEL Mobile Communications
3416 N College, Suite 4
Fayetteville, AR 72703

205 Garrison Avenue
Fort Smith, AR 72901

1415 W Walnut
Rogers, AR 72764

Customer service	some areas	800-255-8351
. . . .		501-661-5800
Local office . . .	Fayetteville	501-444-0007
. . . .	Fort Smith	501-783-0007
. . . .	Rogers	501-636-0008
Corporate office		501-661-8500

Sending calls

Visitor dialing instructions
Local calls . . . . . . area code + 7 digits
Long distance . . . 1 + area code + 7 digits

Speed-dial numbers
Customer service 611
Emergency 911

Receiving calls

Caller dials the roamer access number below,
hears tone, dials cellular area code and number.
Eureka Springs 501-253-6326
Fayetteville 501-841-7626
Fort Smith 501-651-7626

Follow Me Roaming forwarding (see p. 1156)
Activate, dial *18 send. Deactivate dial *19 send.

Billing information

Visitor rates
Most visitors . . . . . $3 day + 85¢ minute

Visitors without roaming agreements (p. 1159)
Call co. to use Amex, Discover, Mastercard, Visa

Identification numbers

City	Market	System	Billing	RSA
Eureka Springs	324	01036	00000	AR-1
Fayetteville	182	00342	00000	MSA
Fort Smith	165	00342	00000	MSA
Ozark	331	01050	00000	AR-8

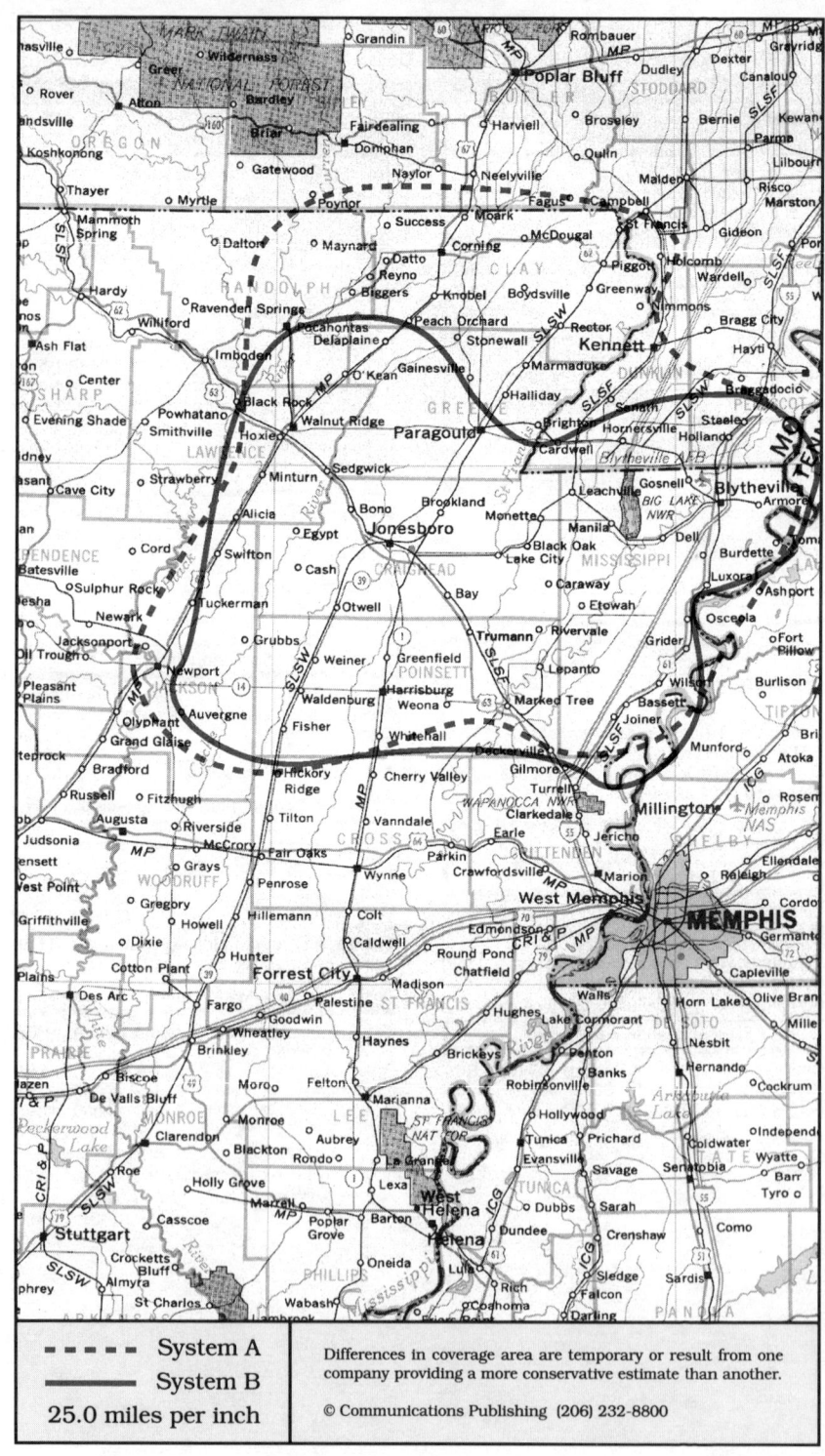

System A

System B

25.0 miles per inch

Differences in coverage area are temporary or result from one company providing a more conservative estimate than another.

© Communications Publishing (206) 232-8800

Jonesboro, Arkansas

Blytheville, Harrisburg, Jonesboro, Paragould, Walnut Ridge

System A	System B

Cellular company

Cellular One
(Managed by Quantum Communications)
2801 S Caraway Road
Jonesboro, AR 72401

Customer service .	Quantum	501-926-2273
Local office . .	Cellular One	501-935-5500
Corporate office	Quantum	612-942-7650

Sending calls

Visitor dialing instructions

Local calls	area code + 7 digits
Long distance . . .	1 + area code + 7 digits

Speed-dial numbers

Customer service	611
Emergency	911

Receiving calls

Caller dials the roamer access number below,
hears tone, dials cellular area code and number.
. . . . 501-926-7626

NationLink & RoamingAmerica (see p. 1157)
Dial ✶31 send to receive calls, ✶30 to deactivate.
This company's subscribers have level 3 service.

Billing information

Visitor rates
Most visitors . . . . . $0 day + 99¢ minute

Visitors without roaming agreements (p. 1159)
Call co. to use Amex, Mastercard, Visa
or Roamer Plus will intercept first call to bill you.

Identification numbers

City	Market	System	Billing	RSA
All served cities	327	01041	00000	AR-4

Cellular company

ALLTEL Mobile Communications
2704 S Culberhouse Road #F
Jonesboro, AR 72401

Customer service	some areas	800-255-8351
. . . .		501-661-5800
Local office		501-931-3855
Corporate office		501-661-8500

Sending calls

Visitor dialing instructions

Local calls	area code + 7 digits
Long distance	area code + 7 digits

Speed-dial numbers

Customer service	✶611
Emergency	911
KARN radio	✶92
KATV news	✶7

Receiving calls

Caller dials the roamer access number below,
hears tone, dials cellular area code and number.
. . . . 501-931-8626

Follow Me Roaming forwarding (see p. 1156)
Activate, dial ✶18 send. Deactivate dial ✶19 send.

Billing information

Visitor rates
Most visitors . . . . . $3 day + 85¢ minute

Visitors without roaming agreements (p. 1159)
Call co. to use Amex, Discover, Mastercard, Visa

Identification numbers

City	Market	System	Billing	RSA
All served cities	327	00208	30432	AR-4

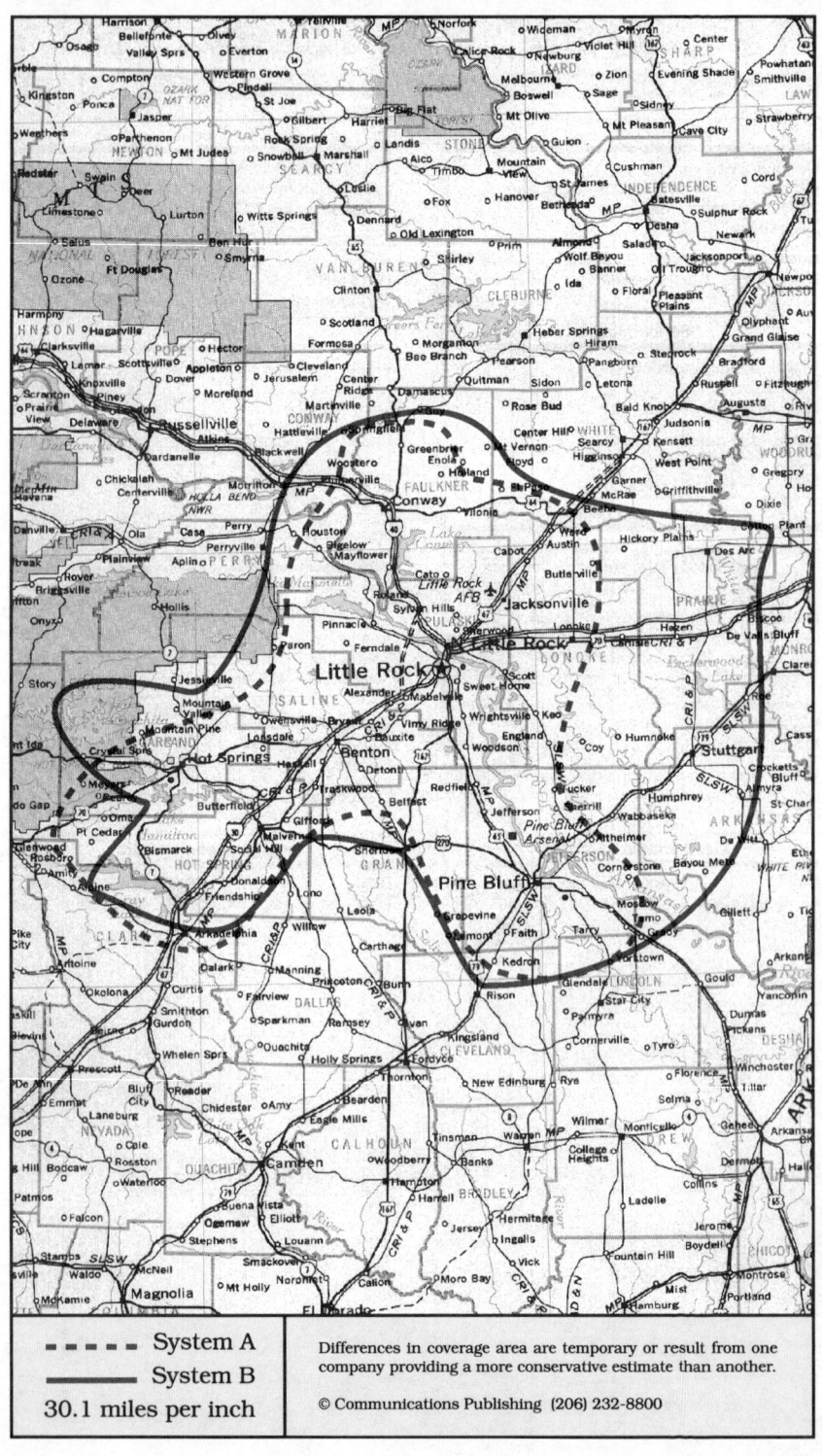

| - - - - - System A
| ————— System B
| 30.1 miles per inch

Differences in coverage area are temporary or result from one company providing a more conservative estimate than another.

© Communications Publishing (206) 232-8800

Little Rock, Arkansas
Arkadelphia, Benton, Conway, Hot Springs, Little Rock, Pine Bluff, Stuttgart

System A	System B

Cellular company

Cellular One
McCaw Cellular Communications
10500 W Markham, Suite 112
Little Rock, AR 72205

Customer service	501-221-1771
Local office	501-225-2355
Corporate office	206-827-4500

Sending calls

Visitor dialing instructions

Local calls	area code + 7 digits
Long distance . . .	1 + area code + 7 digits

Speed-dial numbers

Customer service	611
Emergency	911

Receiving calls

Caller dials the roamer access number below, hears tone, dials cellular area code and number.
. . . . 501-681-7626

NationLink & RoamingAmerica (see p. 1157)
Dial ✳31 send to receive calls, ✳30 to deactivate.
Dial ✳32 send to tell caller the roamer access no.
This company's subscribers have level 9 service.

North American Cellular Network (see p. 1156)
All calls forward to you. Deactivate dial ✳35 send.

Billing information

Visitor rates
Most visitors . . . . . $0 day + 99¢ minute

Visitors without roaming agreements (p. 1159)
Call co. to use Mastercard, Visa
or Roamer Plus will intercept first call to bill you.

Identification numbers

City	Market	System	Billing	RSA
Hot Springs	333	01053	00000	AR-10
Little Rock	092	00215	00000	MSA
Pine Bluff	291	00493	00000	MSA

Cellular company

Hot Springs, Little Rock
ALLTEL Mobile Communications
1801 Central Avenue, Suite G
Hot Springs, AR 71901

2516 Cantrell Road
Little Rock, AR 72202

Customer service	some areas	800-255-8351
. . . .		501-661-5800
Local office . . .	Hot Springs	501-624-5858
. . . .	Little Rock	501-666-6688
Corporate office		501-661-8500

Pine Bluff
Cellular Information Systems
2801 Olive Jefferson Square
Pine Bluff, AR 71603

Local office	501-536-4200
Corporate office	203-622-6317

Sending calls

Visitor dialing instructions

Local calls	area code + 7 digits
Long distance . . .	1 + area code + 7 digits

Speed-dial numbers

Customer service	✳611
Emergency	911
Make a traffic report	✳92

Receiving calls

Caller dials the roamer access number below, hears tone, dials cellular area code and number.

Hot Springs	501-622-8626
Little Rock	501-680-7626
Pine Bluff	501-550-7626

Follow Me Roaming forwarding (see p. 1156)
Activate, dial ✳18 send. Deactivate dial ✳19 send.

Billing information

Visitor rates
Most visitors . . . . . $3 day + 85¢ minute

Visitors without roaming agreements (p. 1159)
Call co. to use Amex, Discover, Mastercard, Visa

Identification numbers

City	Market	System	Billing	RSA
Hot Springs	333	00208	30212	AR-10
Little Rock	092	00208	00000	MSA
Pine Bluff	291	00208	30018	MSA

The Little Rock and Pine Bluff systems are operated jointly by the licensees.

For full coverage area, see Russellville, AR.

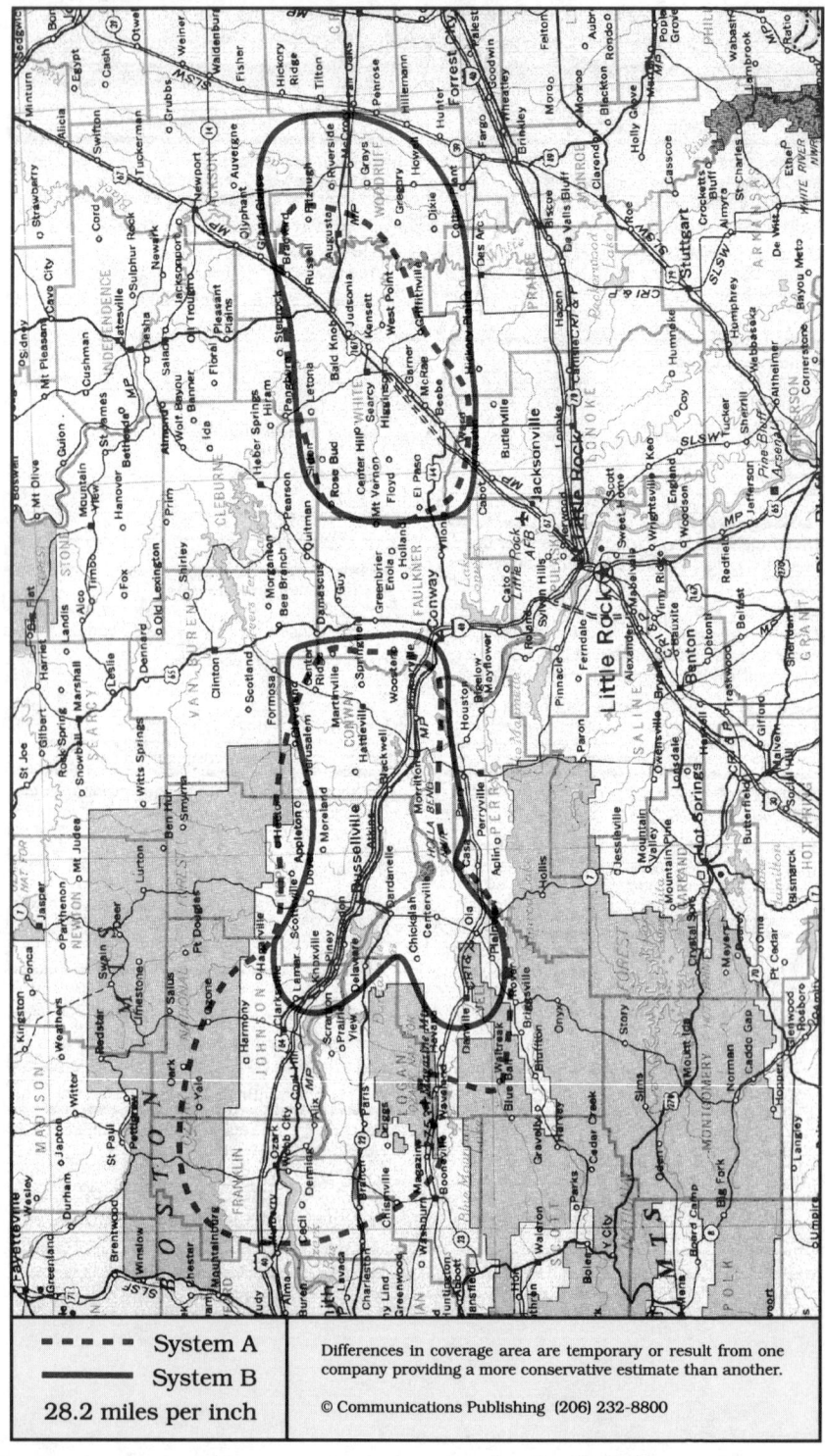

- - - - - System A
———— System B
28.2 miles per inch

Differences in coverage area are temporary or result from one company providing a more conservative estimate than another.

© Communications Publishing (206) 232-8800

Russellville, Arkansas
Clarksville, Morrilton, Ozark (system A), Russellville, Searcy

System A	System B

Cellular company

Cellular One
Sterling Cellular Management
1080 Holcomb Bridge Road, Bldg 100, Ste 200
Roswell, GA 30076

Customer service 800-552-6150
Corporate office 404-552-5030

Sending calls

Visitor dialing instructions
Local calls . . . . . area code + 7 digits
Long distance . . . 1 + area code + 7 digits

Speed-dial numbers
Customer service *611
Emergency 911

Receiving calls

Caller dials the roamer access number below,
hears tone, dials cellular area code and number.
Russellville 501-940-7626
Searcy 501-278-7626

Billing information

Visitor rates
Most visitors . . . . . $3 day + 99¢ minute

Visitors without roaming agreements (p. 1159)
Call co. to use Amex, Mastercard, Visa
or Roamer Plus will intercept first call to bill you.

Identification numbers

City	Market	System	Billing	RSA
Ozark	331	01047	00000	AR-8
Russellville	330	01047	00000	AR-7
Searcy	329	01047	00000	AR-6

Cellular company

ALLTEL Mobile Communications
2516 Cantrell Road
Little Rock, AR 72202

Customer service 800-255-8351
. . . . 501-661-5800
Local office 501-666-6688
Corporate office 501-661-8500

Sending calls

Visitor dialing instructions
Local calls . . . . . . area code + 7 digits
Long distance . . . 1 + area code + 7 digits

Speed-dial numbers
Customer service *611
Emergency 911
Make a traffic report *92

Receiving calls

Caller dials the roamer access number below,
hears tone, dials cellular area code and number.
Russellville 501-964-9626
Searcy 501-279-6626

Follow Me Roaming forwarding (see p. 1156)
Activate, dial *18 send. Deactivate dial *19 send.

Billing information

Visitor rates
Most visitors . . . . . $3 day + 85¢ minute

Visitors without roaming agreements (p. 1159)
Call co. to use Amex, Discover, Mastercard, Visa

Identification numbers

City	Market	System	Billing	RSA
Russellville	330	01048	30500	AR-7
Searcy	329	01046	30498	AR-6

For full coverage area, see Little Rock, AR.

System B service in Ozark, see Fort Smith, AR.

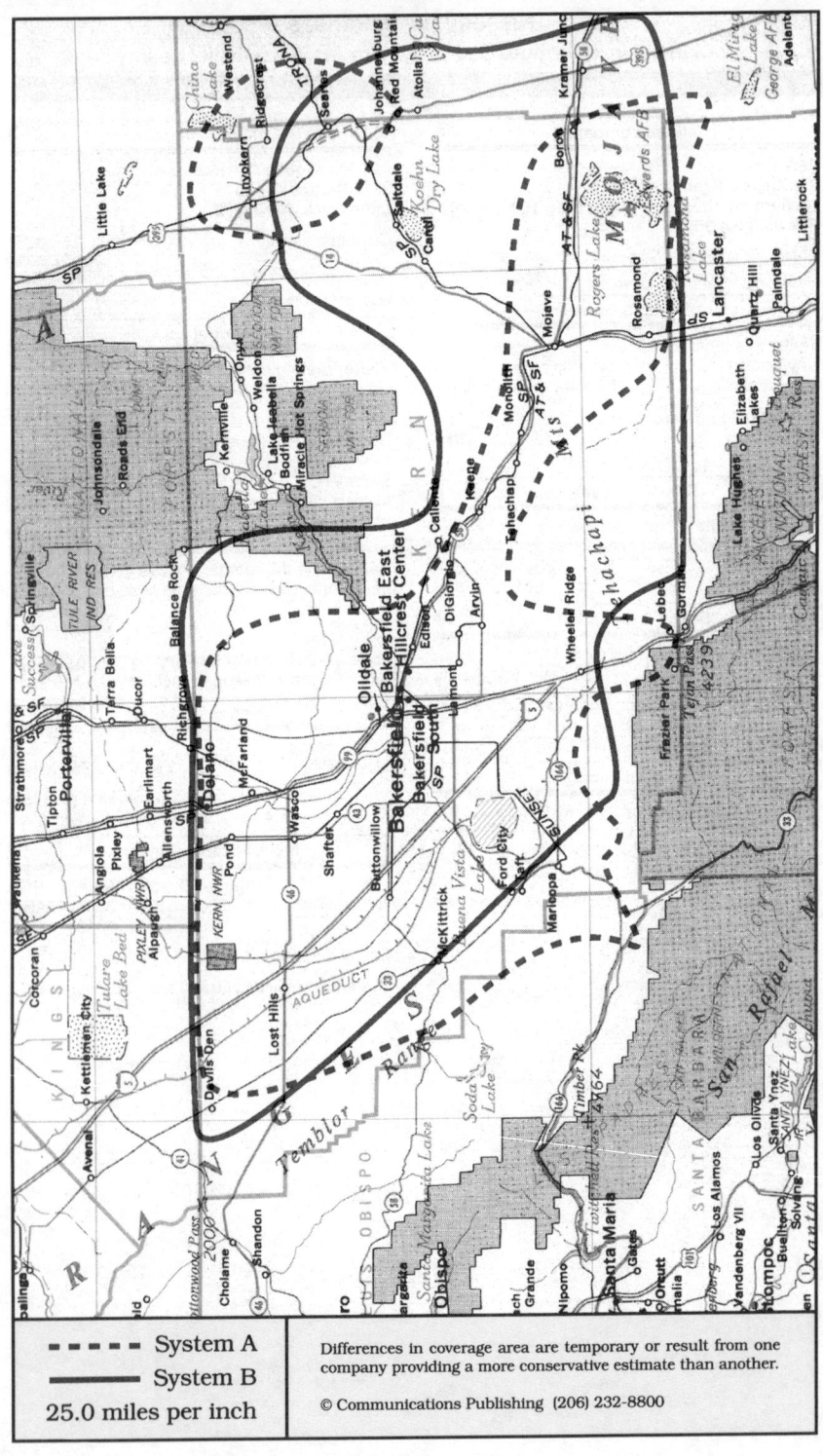

- - - - System A
———— System B
25.0 miles per inch

Differences in coverage area are temporary or result from one company providing a more conservative estimate than another.

© Communications Publishing (206) 232-8800

Bakersfield, California
Bakersfield, Delano, Ridgecrest

System A	System B

Cellular company

Bakersfield Cellular Telephone
BellSouth Cellular
4180 Truxton Avenue
Bakersfield, CA 93309

Customer service	805-327-8589
Local office	805-327-8700
Corporate office	404-604-6100

Sending calls

Visitor dialing instructions

Local calls	area code + 7 digits
Long distance . . .	1 + area code + 7 digits

Speed-dial numbers

Customer service	611
Emergency	911
News	*29

Receiving calls

Caller dials the roamer access number below,
hears tone, dials cellular area code and number.
. . . . 805-332-7626

NationLink & RoamingAmerica (see p. 1157)
Caller hears your current roamer access number.
This company's subscribers have level 0 service.

Billing information

Visitor rates
Most visitors . . . . . $2 day + 65¢ minute

Visitors without roaming agreements (p. 1159)
Call co. to use Mastercard, Visa

Identification numbers

City	Market	System	Billing	RSA
All served cities	097	00183	00000	MSA

Cellular company

Contel Cellular
1220 Oak Street, Suite M
Bakersfield, CA 93304

Customer service	800-333-4004
. . . .	805-838-2273
Local office	805-631-2355
Corporate office	404-804-3400

Sending calls

Visitor dialing instructions

Local calls	area code + 7 digits
Long distance . . .	1 + area code + 7 digits

Speed-dial numbers

Customer service	611
Directory assistance	411
Emergency	911

Receiving calls

Caller dials the roamer access number below,
hears tone, dials cellular area code and number.

Bakersfield	805-838-7626
Ridgecrest	619-382-7626

Auto-Access forwarding (see page 1156)
Activate, dial *28 send. Deactivate dial *29 send.

Follow Me Roaming forwarding (see p. 1156)
Activate, dial *18 send. Deactivate dial *19 send.

Billing information

Visitor rates
Most visitors . . . . . $2 day + 65¢ minute

Visitors without roaming agreements (p. 1159)
Call co. to use Amex, Mastercard, Visa

Identification numbers

City	Market	System	Billing	RSA
Bakersfield	097	00228	00000	MSA
Inyo County	341	01070	00000	CA-6

For full coverage area, see Fresno, CA and
Merced, CA.

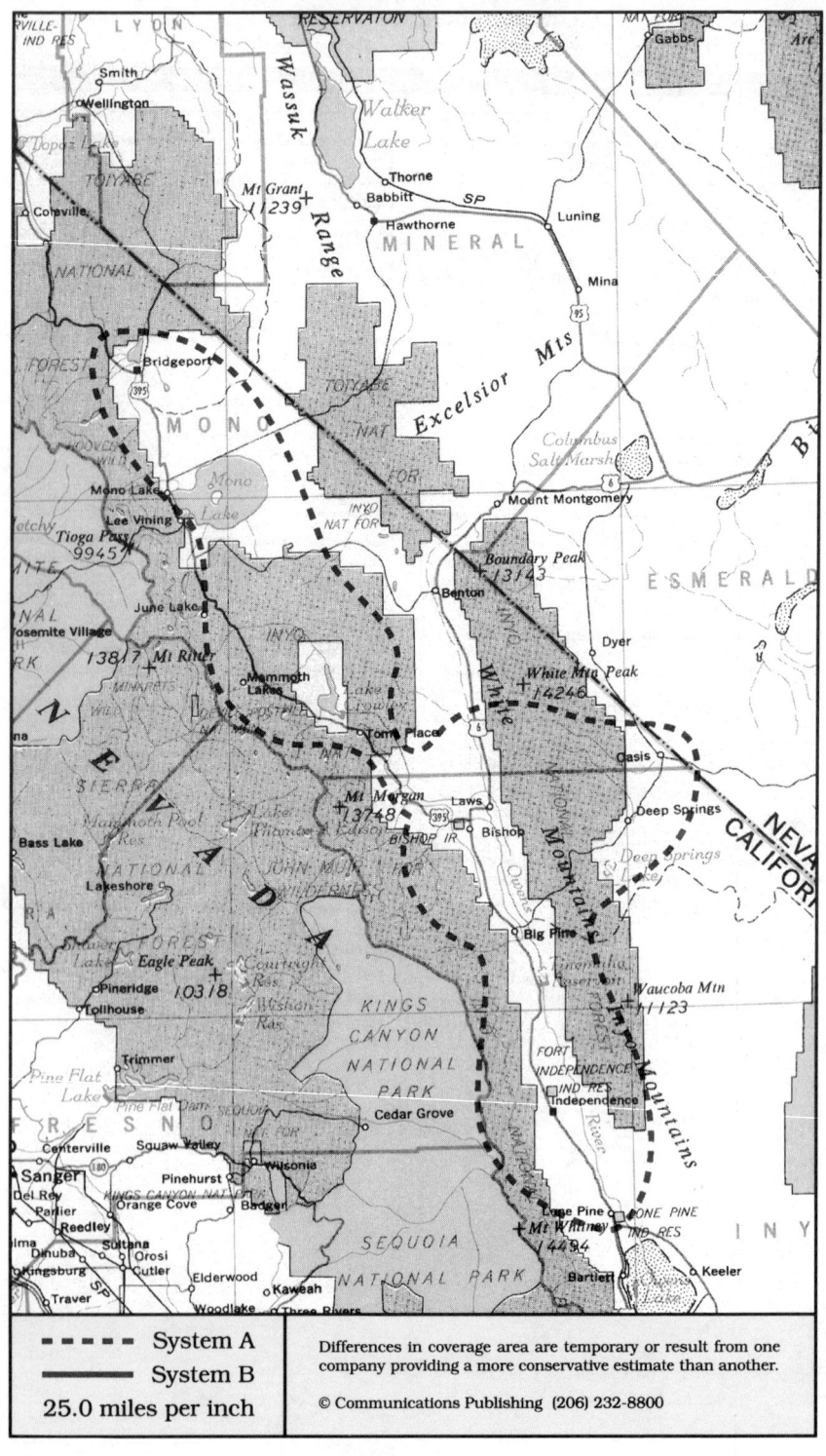

▬ ▬ ▬ System A	Differences in coverage area are temporary or result from one company providing a more conservative estimate than another.
▬▬▬▬ System B	
25.0 miles per inch	© Communications Publishing (206) 232-8800

Bishop, California
Bishop, Bridgeport, Independence

System A	System B

Cellular company

Cellular One
GenCell Management
1337 Rocking "W" Drive
Bishop, CA 93514

Customer service 800-888-7868
Local office 619-872-1001
Corporate office 707-425-8000

Sending calls

Visitor dialing instructions
Local calls . . . . . area code + 7 digits
Long distance . . . 1 + area code + 7 digits

Speed-dial numbers
Customer service 611
Emergency 911

Receiving calls

Caller dials the roamer access number below,
hears tone, dials cellular area code and number.
. . . . 619-937-7626

NationLink & RoamingAmerica (see p. 1157)
Dial *31 send to receive calls, *30 to deactivate.
This company's subscribers have level 3 service.

Billing information

Visitor rates
Most visitors . . . . . $3 day + 99¢ minute

Visitors without roaming agreements (p. 1159)
American Roaming Network intercepts call to bill.

Identification numbers

City	Market	System	Billing	RSA
All served cities	341	01069	00000	CA-6

Cellular company

Contel Cellular
5195 N Blackstone Avenue
Fresno, CA 93710-6701

Service was not operating at press time.

Customer service 800-333-4004
. . . . 209-246-2273
Local office 209-224-9222
Corporate office 404-804-3400

Sending calls

Visitor dialing instructions
Local calls . . . . . . area code + 7 digits
Long distance . . . 1 + area code + 7 digits

Speed-dial numbers
Customer service 611
Directory assistance 411
Emergency 911

Receiving calls

Caller dials the roamer access number below,
hears tone, dials cellular area code and number.
. . . . 619-382-7626

Auto-Access forwarding (see page 1156)
Activate, dial *28 send. Deactivate dial *29 send.

Billing information

Visitor rates
Most visitors . . . . . $2 day + 65¢ minute

Visitors without roaming agreements (p. 1159)
Call co. to use Amex, Mastercard, Visa

Identification numbers

City	Market	System	Billing	RSA
All served cities	341	1070	00000	CA-6

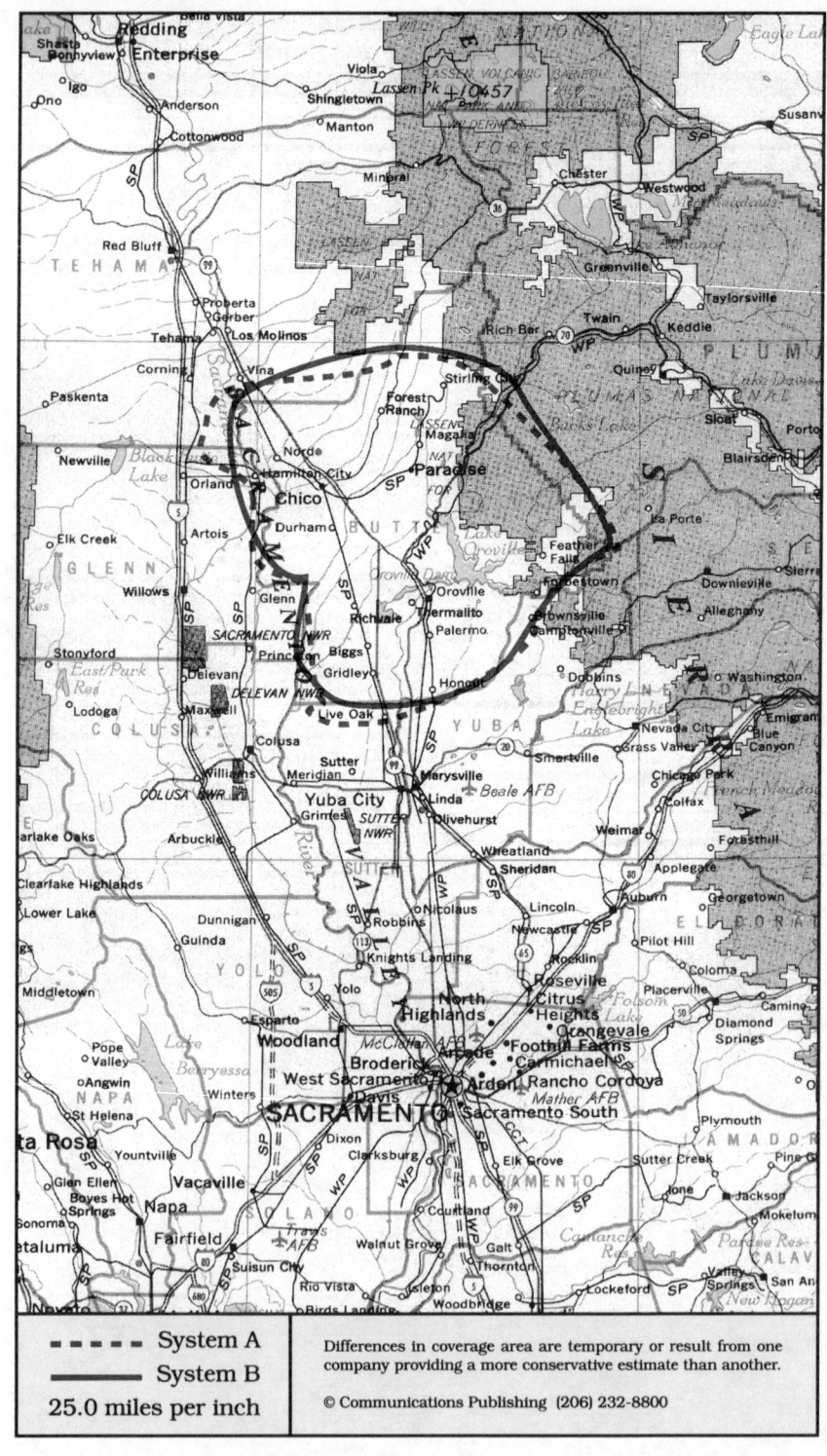

▪ ▪ ▪ ▪ System A	Differences in coverage area are temporary or result from one company providing a more conservative estimate than another.
——— System B	
25.0 miles per inch	© Communications Publishing (206) 232-8800

Chico, California
Chico, Paradise, Oroville

System A	System B

Cellular company

Cellular One
GenCell Management
1600 Mangrove Avenue, Suite P
Chico, CA 95926

Customer service	800-888-7868
Local office	916-896-1600
Corporate office	707-425-8000

Sending calls

Visitor dialing instructions

Local calls	area code + 7 digits
Long distance . . .	1 + area code + 7 digits

Speed-dial numbers

Customer service	*611
Emergency	911

Receiving calls

Caller dials the roamer access number below,
hears tone, dials cellular area code and number.
. . . . 916-521-7626

NationLink & RoamingAmerica (see p. 1157)
Dial *31 send to receive calls, *30 to deactivate.
This company's subscribers have level 5 service.

Billing information

Visitor rates
Most visitors . . . . . $3 day + 99¢ minute

Visitors without roaming agreements (p. 1159)
American Roaming Network intercepts call to bill.

Identification numbers

City	Market	System	Billing	RSA
All served cities	215	00311	00000	MSA

Cellular company

PacTel Cellular
2150 River Plaza Drive, Suite 400
Sacramento, CA 95833

Customer service	800-722-8358
. . . .	916-920-0645
Local office	916-646-3773
Corporate office	510-210-3600

Sending calls

Visitor dialing instructions

Local calls	area code + 7 digits
Long distance . . .	1 + area code + 7 digits

Speed-dial numbers

Customer service	*611
Emergency	911

Receiving calls

Caller dials the roamer access number below,
hears tone, dials cellular area code and number.
. . . . 916-520-7626

Auto-Access forwarding (see page 1156)
Activate, dial *28 send. Deactivate dial *29 send.

Follow Me Roaming forwarding (see p. 1156)
Activate, dial *18 send. Deactivate dial *19 send.

Billing information

Visitor rates

7 am - 7 pm Mon-Fri .	$0 day + 55¢ minute
All other times	$0 day + 27¢ minute

Visitors without roaming agreements (p. 1159)
Call co. to use Mastercard, Visa

Identification numbers

City	Market	System	Billing	RSA
All served cities	215	00294	00000	MSA

For full coverage area, see Sacramento, CA.

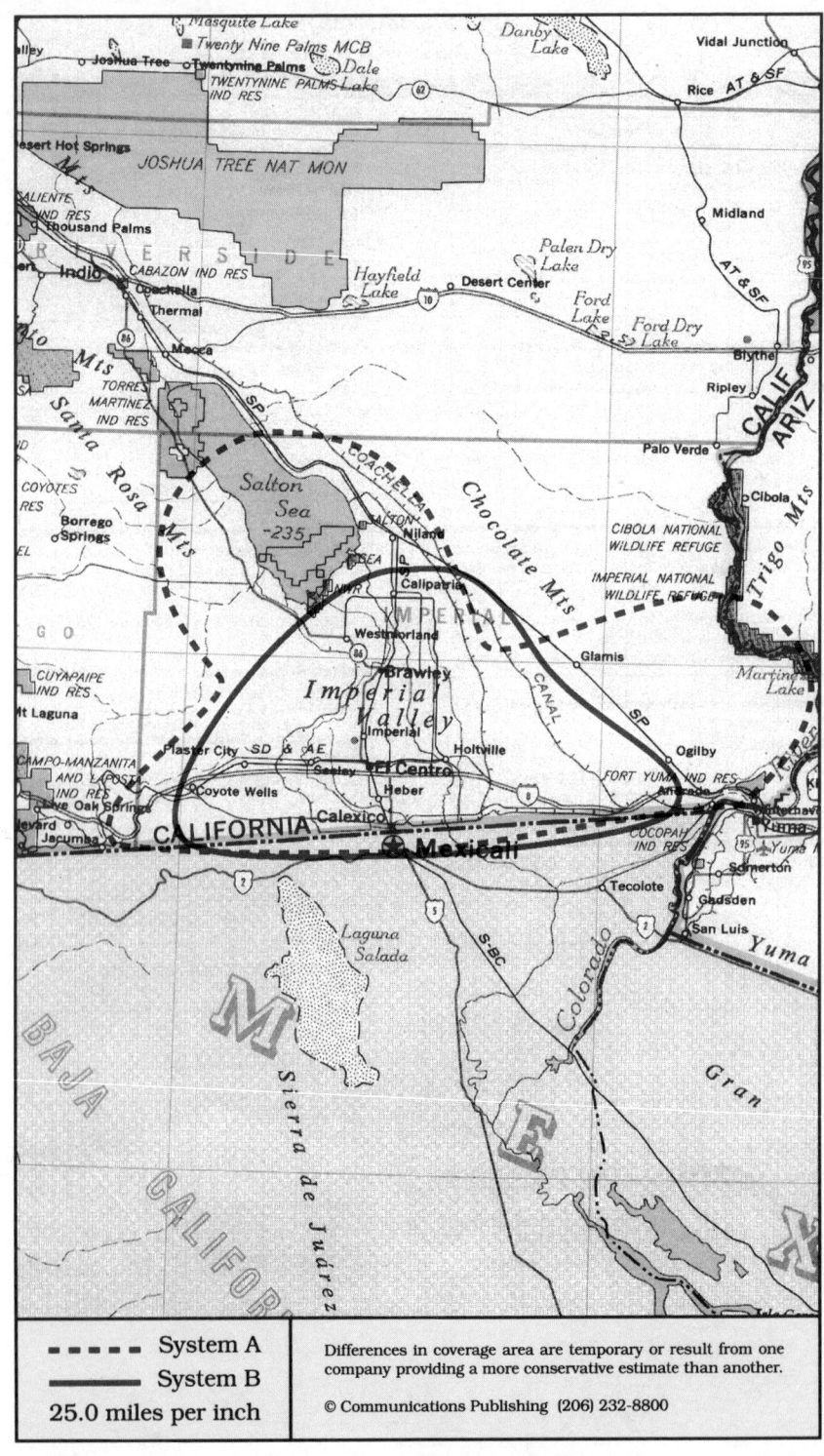

System A
System B
25.0 miles per inch

Differences in coverage area are temporary or result from one company providing a more conservative estimate than another.

© Communications Publishing (206) 232-8800

El Centro, California
Brawley, Calexico, El Centro

System A	System B

System A

Cellular company

Cellular One
Centennial Cellular
703 N La Brucherie
El Centro, CA 92243

Local office	619-337-8000
Corporate office Administration	619-352-7066
. . . . . . Management	203-972-2000

Sending calls

Visitor dialing instructions
Local calls	area code + 7 digits
Long distance . . .	0 + area code + 7 digits

Speed-dial numbers
Channel 11 television	*news11
Customer service	*611
Directory assistance	411
Emergency	911

Receiving calls

Caller dials the roamer access number below, hears tone, dials cellular area code and number.
. . . . 619-337-0915

NationLink & RoamingAmerica (see p. 1157)
Caller hears your current roamer access number. This company's subscribers have level 0 service.

Billing information

Visitor rates
Most visitors . . . . . $3 day + 85¢ minute

Visitors without roaming agreements (p. 1159)
Call co. to use Mastercard, Visa

Identification numbers

City	Market	System	Billing	RSA
All served cities	342	01029	00000	CA-7

System B

Cellular company

Contel Cellular
114 N 6th Street
El Centro, CA 92243

Customer service	800-333-4004
. . . .	915-772-9222
Local office	619-337-5507
Corporate office	404-804-3400

Sending calls

Visitor dialing instructions
Local calls	area code + 7 digits
Long distance . . .	1 + area code + 7 digits

Speed-dial numbers
Customer service	*611
Emergency	911

Receiving calls

Caller dials the roamer access number below, hears tone, dials cellular area code and number.
. . . . 619-971-7626

Auto-Access forwarding (see page 1156)
Activate, dial *28 send. Deactivate dial *29 send.

Billing information

Visitor rates
Most visitors . . . . . $2 day + 55¢ minute

Visitors without roaming agreements (p. 1159)
Call co. to use Amex, Mastercard, Visa

Identification numbers

City	Market	System	Billing	RSA
All served cities	342	01072	00000	CA-7

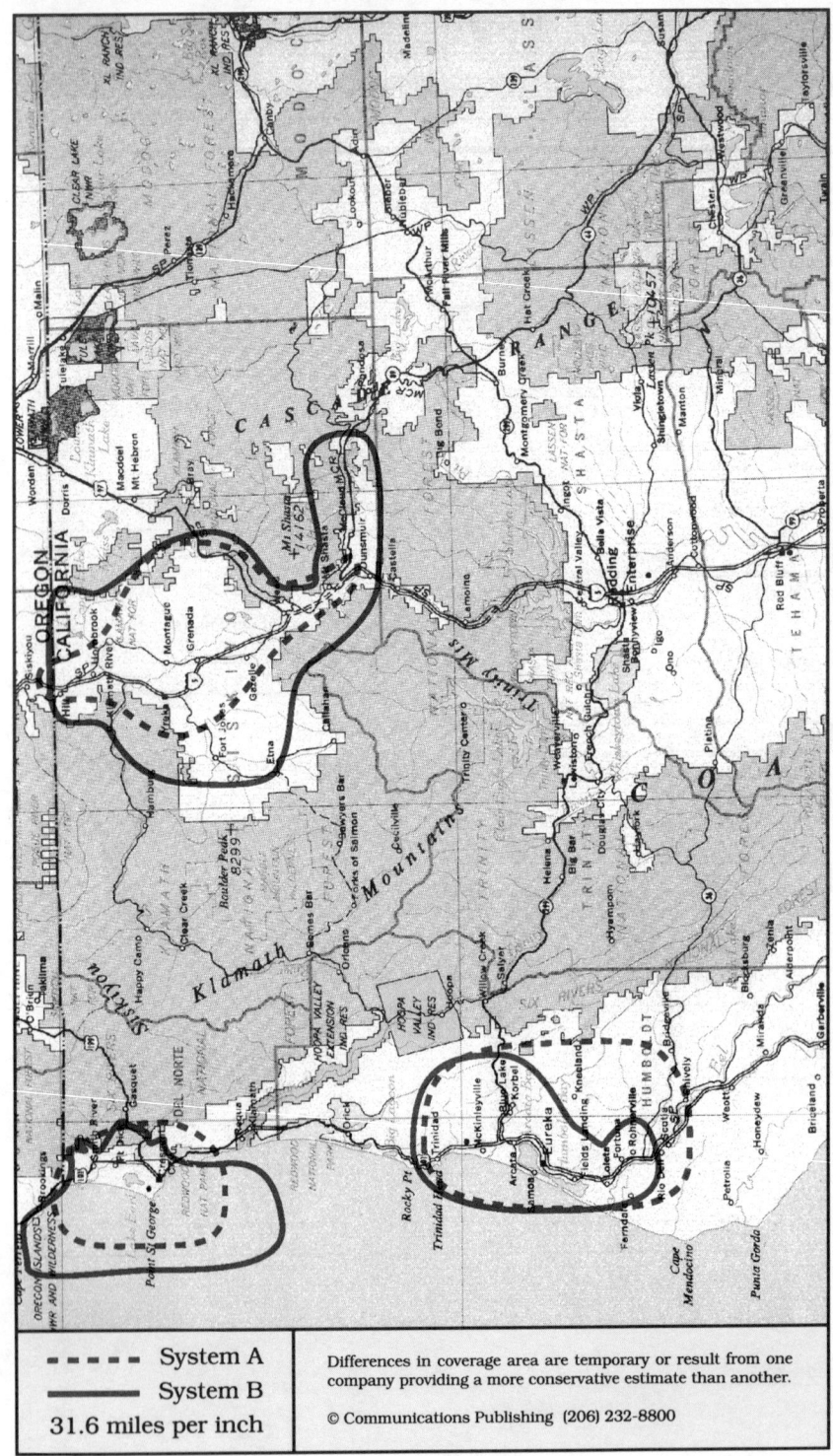

- - - - - System A
———— System B
31.6 miles per inch

Differences in coverage area are temporary or result from one company providing a more conservative estimate than another.

© Communications Publishing (206) 232-8800

Eureka, California
Crescent City, Eureka, Yreka

System A	System B

System A

Cellular company

United States Cellular
3144 Broadway, Suite C-5
Eureka, CA 95501

501 S Main
Yreka, CA 96097

Customer service	some areas	800-528-8722
. . . .		503-779-3000
Local office	Eureka	707-441-1823
. . . .	Yreka	916-842-4863
Corporate office		312-399-8900

Sending calls

Visitor dialing instructions

Local calls	7 digits
Long distance . . .	1 + area code + 7 digits

Speed-dial numbers

Customer service	611
Emergency	911

Receiving calls

Caller dials the roamer access number below,
hears tone, dials cellular area code and number.

Crescent City, Eureka . . .	707-498-7626
Yreka	916-842-0300

Billing information

Visitor rates

Most visitors	$2 day + 50¢ minute

Visitors without roaming agreements (p. 1159)
Call co. to use Amex

Identification numbers

City	Market	System	Billing	RSA
All served cities	336	01059	00000	CA-1

System B

Cellular company

Cal-One Cellular
2212 2nd Street
Eureka, CA 95501

118 Ranch Lane
Yreka, CA 96097-9514

Local office . . .	some areas	800-559-2251
. . . .	Eureka	707-444-2255
. . . .	Eureka	707-464-1200
. . . .	(some areas) Yreka	800-499-1863
. . . .	Yreka	916-842-1863
Corporate office		916-468-5222

Sending calls

Visitor dialing instructions

Local calls	area code + 7 digits
Long distance	area code + 7 digits

Speed-dial numbers

Customer service	611
Emergency	911
Road conditions (from Yreka)	*ROAD or *7623
Roaming information (from Yreka) . .	*711

Receiving calls

Caller dials the roamer access number below,
hears tone, dials cellular area code and number.

Crescent City, Eureka	707-499-7626
Yreka	916-598-7626

Follow Me Roaming forwarding (see p. 1156)
Activate, dial *18 send. Deactivate dial *19 send.

Billing information

Visitor rates

Most visitors	$2 day + 60¢ minute

Visitors without roaming agreements (p. 1159)
Call co. to use Mastercard, Visa

Identification numbers

City	Market	System	Billing	RSA
Crescent City	336	01060	00000	CA-1
Eureka	336	01060	30508	CA-1
Yreka	336	01060	30510	CA-1

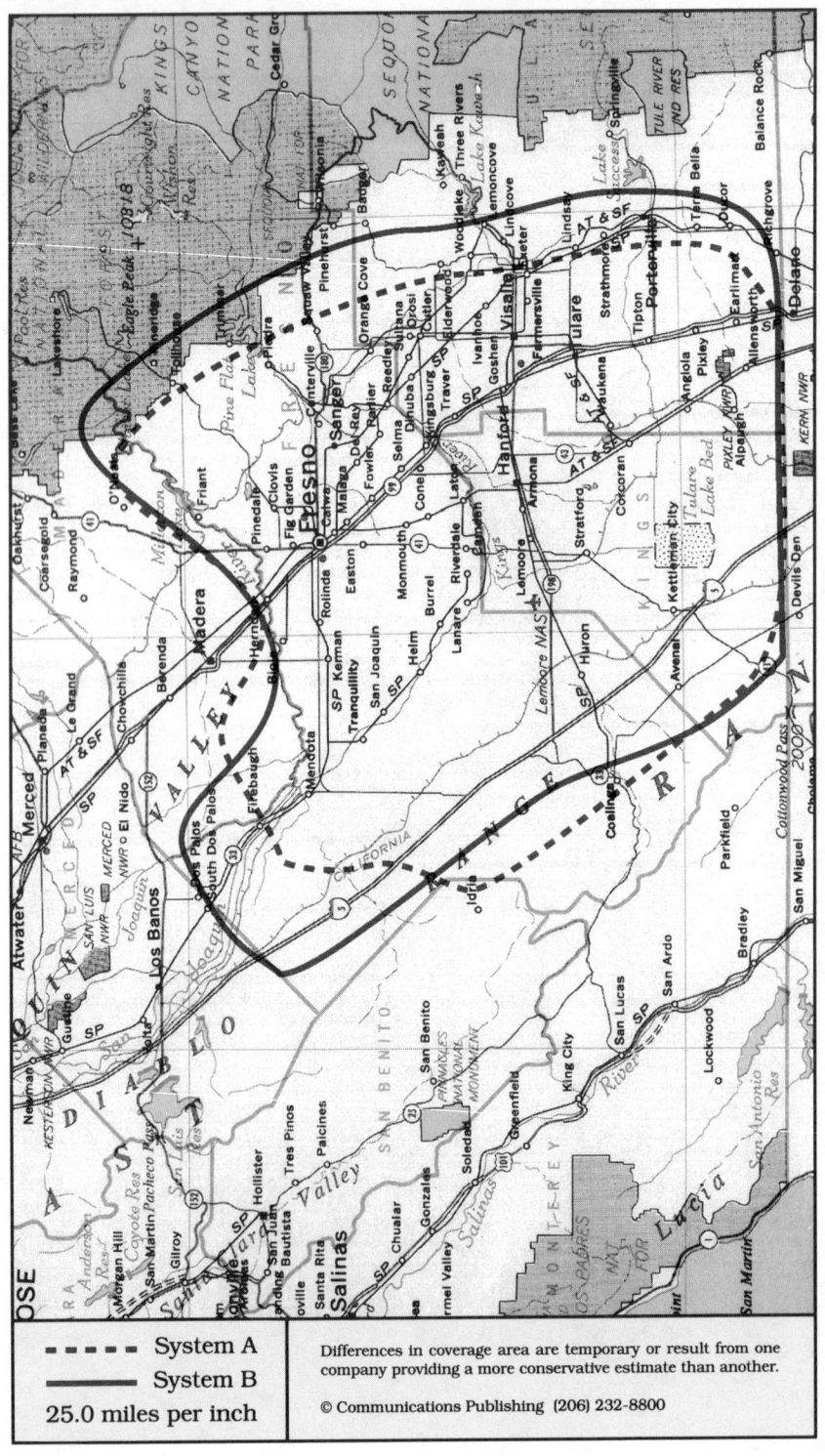

▬ ▬ ▬ ▬ ▬ System A	Differences in coverage area are temporary or result from one company providing a more conservative estimate than another.
▬▬▬▬▬ System B	
25.0 miles per inch	© Communications Publishing (206) 232-8800

Fresno, California
Fresno, Hanford, Visalia

System A	System B

Cellular company

Cellular One
McCaw Cellular Communications
5260 N Palm Avenue, Suite 120
Fresno, CA 93704

815 W Oak Street, Suite C
Visalia, CA 93291

Customer service		209-438-2468
Local office	Fresno	209-438-8888
. . . .	Visalia	209-733-5978
Corporate office		206-827-4500

Sending calls

Visitor dialing instructions
Local calls . . . . . area code + 7 digits
Long distance . . . 1 + area code + 7 digits

Speed-dial numbers
Customer service *611
Emergency 911

Receiving calls

Caller dials the roamer access number below,
hears tone, dials cellular area code and number.
Fresno 209-269-7626
Visalia 209-738-7626

NationLink & RoamingAmerica (see p. 1157)
Dial *31 send to receive calls, *30 to deactivate.
Dial *32 send to tell caller the roamer access no.
This company's subscribers have level 6 service.

North American Cellular Network (see p. 1156)
All calls forward to you. Deactivate dial *35 send.

Billing information

Visitor rates
Most visitors . . . . . $0 day + 99¢ minute

Visitors without roaming agreements (p. 1159)
Cannot call since company takes no credit cards.

Identification numbers

City	Market	System	Billing	RSA
Fresno	074	00153	00000	MSA
Hanford	347	00153	00000	CA-12
Visalia	150	00153	00000	MSA

Cellular company

Contel Cellular
5195 N Blackstone Avenue
Fresno, CA 93710-6701

3298 S Mooney Boulevard, Suite B
Visalia, CA 93277-7768

Customer service		800-333-4004
. . . .		209-246-2273
. . . .	from Visalia	209-731-2273
Local office	Fresno	209-224-9222
. . . .	Visalia	209-625-0700
Corporate office		404-804-3400

Sending calls

Visitor dialing instructions
Local calls . . . . . . area code + 7 digits
Long distance . . . 1 + area code + 7 digits

Speed-dial numbers
Customer service 611
Directory assistance 411
Emergency 911
Newspaper . . . . *BEE or *233

Receiving calls

Caller dials the roamer access number below,
hears tone, dials cellular area code and number.
Fresno 209-246-7626
Visalia 209-731-7626

Auto-Access forwarding (see page 1156)
Activate, dial *28 send. Deactivate dial *29 send.

Follow Me Roaming forwarding (see p. 1156)
Activate, dial *18 send. Deactivate dial *19 send.

Billing information

Visitor rates
Most visitors . . . . . $2 day + 65¢ minute

Visitors without roaming agreements (p. 1159)
Call co. to use Amex, Mastercard, Visa

Identification numbers

City	Market	System	Billing	RSA
Fresno	074	00162	00000	MSA
Hanford	347	00162	00000	CA-12
Visalia	150	00162	00000	MSA

For full coverage area, see Bakersfield, CA and
Merced, CA.

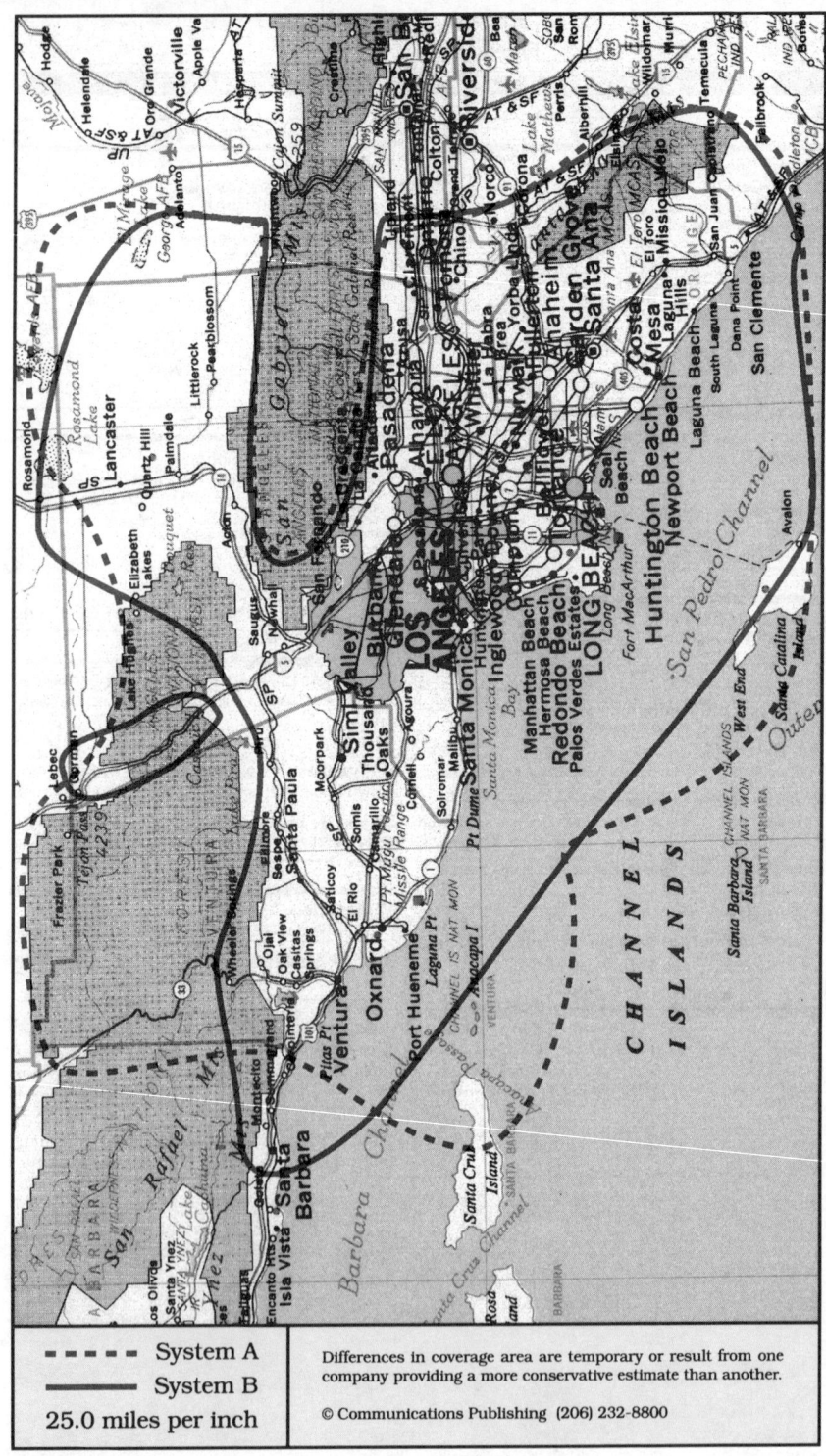

- - - - - System A	Differences in coverage area are temporary or result from one company providing a more conservative estimate than another.
——— System B	
25.0 miles per inch	© Communications Publishing (206) 232-8800

Los Angeles, California
Anaheim, Long Beach, Los Angeles, Oxnard, Santa Ana (also see Riverside)

System A	System B

Cellular company

Los Angeles
Los Angeles Cellular Telephone
17785 Center Court Drive N
Cerritos, CA 90701

Customer service	310-809-7475
Corporate office	310-924-0000

Ventura County
Cellular One
McCaw Cellular Communications
751 Daily Drive, Suite 116
Camarillo, CA 93010

Customer service	805-482-7071
Local office	805-987-0955
Corporate office	206-827-4500

Sending calls

Visitor dialing instructions
Local calls	area code + 7 digits
Long distance* . . .	0 + area code + 7 digits

(*Super Cellular and San Diego visitors use 1 +)

Speed-dial numbers
AAA road service	*222
Address directions . . .	*FIND or *3463
Customer service	611
Directory assistance	411
Emergency	911
Traffic center	*JAM or *526

Receiving calls

Caller dials the roamer access number below,
hears tone, dials cellular area code and number.
Los Angeles (city)	213-712-7626
Los Angeles (county) . . .	310-880-7626
Orange County	714-746-7626
Ventura County	805-377-7626
San Fernando, San Gabriel .	818-370-7626

NationLink & RoamingAmerica (see p. 1157)
Dial *31 send to receive calls, *30 to deactivate.
Dial *32 send to tell caller the roamer access no.
This company's subscribers have level 6 service.

Billing information

Visitor rates
Most visitors	$0 day + 99¢ minute
From San Diego system A	$0 day + 65¢ minute
From San F., Bakersfield A	$2 day + 65¢ minute
Super Cellular visitors	$0 day + 65¢ minute

Visitors without roaming agreements (p. 1159)
Cannot call since company takes no credit cards.

Identification numbers

City	Market	System	Billing	RSA
Los Angeles	002	00027	00000	MSA
Orange County	144	00027	00000	MSA
Ventura County	073	00027	30065	MSA

This region is licensed to the two companies
listed above, which operate one system jointly.

For full coverage area, see Riverside, CA.

Cellular company

PacTel Cellular
3 Park Plaza, PO Box 19651
Irvine, CA 92713-9651

Customer service	800-851-9815
. . . .	714-222-8200
Local office	714-222-7000
Corporate office	510-210-3600

Sending calls

Visitor dialing instructions
Local calls	area code + 7 digits
Long distance . . .	0 + area code + 7 digits

Speed-dial numbers
Customer service	*611
Emergency	911
Roaming information	*711

Receiving calls

Caller dials the roamer access number below,
hears tone, dials cellular area code and number.
Los Angeles	213-718-7626
Orange County	714-742-7626
Ventura County	805-657-7626
San Fernando, San Gabriel . .	818-400-7626

Auto-Access forwarding (see page 1156)
Activate, dial *28 send. Deactivate dial *29 send.

Follow Me Roaming forwarding (see p. 1156)
Activate, dial *18 send. Deactivate dial *19 send.

Billing information

Visitor rates
Most visitors	$0 day + $1.05 minute

Visitors without roaming agreements (p. 1159)
Call co. to use Amex, Mastercard, Visa

Identification numbers

City	Market	System	Billing	RSA
Los Angeles	002	00002	00000	MSA
Orange County	144	00002	00000	MSA
Ventura County	073	00002	00000	MSA

You may continue a call when traveling into San
Diego and Santa Barbara without being
disconnected.

For full coverage area, see Riverside, CA.

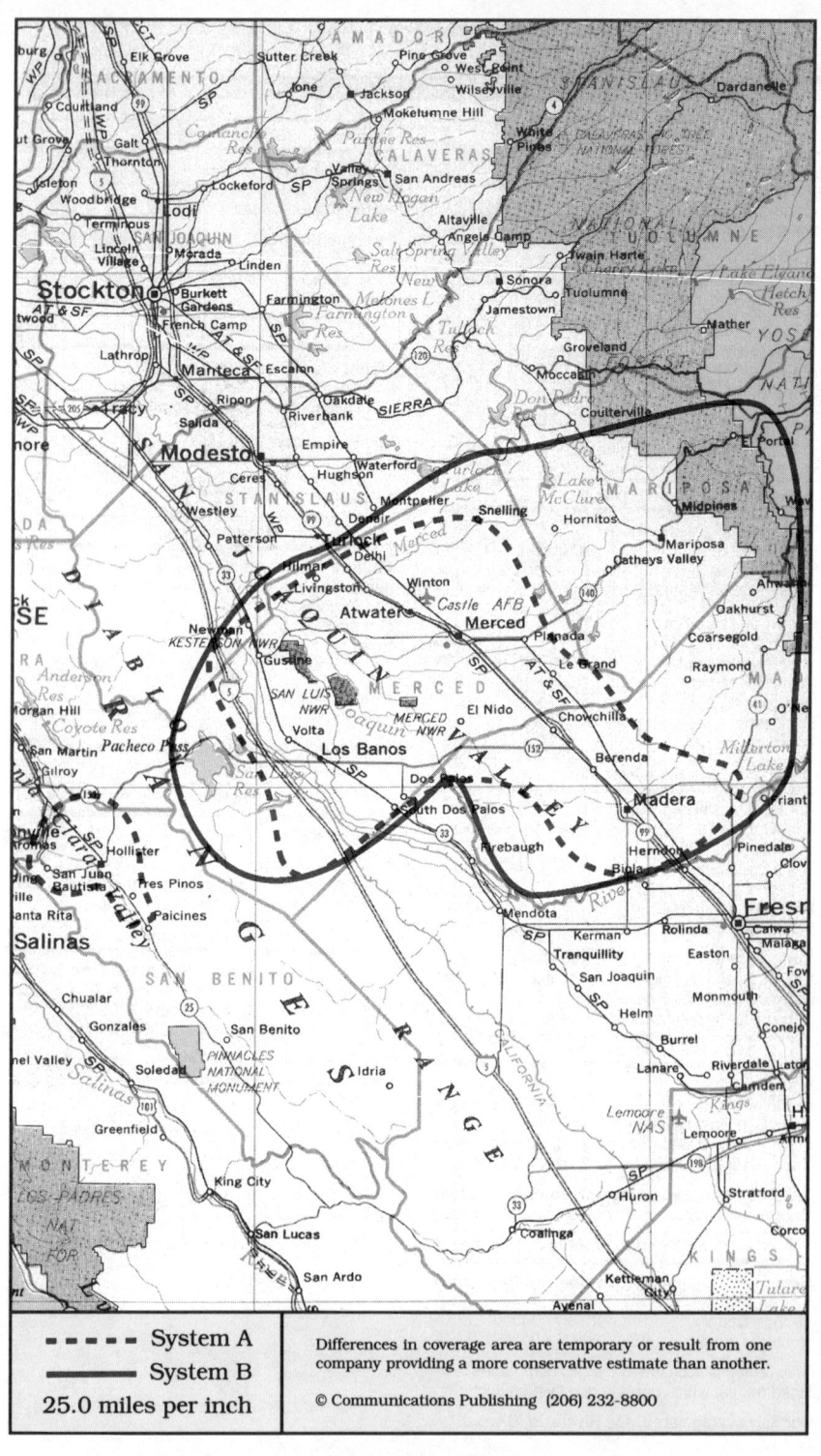

Legend	
▬ ▬ ▬ System A	
────── System B	
25.0 miles per inch	

Differences in coverage area are temporary or result from one company providing a more conservative estimate than another.

© Communications Publishing (206) 232-8800

Merced, California
Hollister, Los Banos, Madera, Merced

System A	System B

Cellular company

Cellular One
Licensed to: Cellular 2000
Managed by: Rural Cellular Management
3250 N "G" Street, Suite A
Merced, CA 95340

Local office 209-723-3100
Corporate office   Rural Cell. Mgt 707-422-2100

Sending calls

Visitor dialing instructions
Local calls . . . . . area code + 7 digits
Long distance . . . 1 + area code + 7 digits

Speed-dial numbers
Customer service 611
Directory assistance 411
Emergency 911

Receiving calls

Caller dials the roamer access number below,
hears tone, dials cellular area code and number.
. . . . 209-761-7626

NationLink & RoamingAmerica (see p. 1157)
Dial *31 send to receive calls, *30 to deactivate.
Dial *32 send to tell caller the roamer access no.
This company's subscribers have level 6 service.

North American Cellular Network (see p. 1156)
All calls forward to you. Deactivate dial *35 send.

Billing information

Visitor rates
Most visitors . . . . . $0 day + 99¢ minute

Visitors without roaming agreements (p. 1159)
Cannot call since company takes no credit cards.

Identification numbers

City	Market	System	Billing	RSA
All served cities	339	01065	30247	CA-4

Cellular company

Contel Cellular
47 W Alexander Avenue
Merced, CA 95340

Customer service 800-333-4004
. . . . 209-769-2273
Local office 209-723-8000
Corporate office 404-804-3400

Sending calls

Visitor dialing instructions
Local calls . . . . . . area code + 7 digits
Long distance . . . 1 + area code + 7 digits

Speed-dial numbers
Customer service 611
Directory assistance 411
Emergency 911

Receiving calls

Caller dials the roamer access number below,
hears tone, dials cellular area code and number.
Madera 209-974-7626
Merced 209-769-7626

Auto-Access forwarding (see page 1156)
Activate, dial *28 send. Deactivate dial *29 send.

Follow Me Roaming forwarding (see p. 1156)
Activate, dial *18 send. Deactivate dial *19 send.

Billing information

Visitor rates
Most visitors . . . . . $2 day + 65¢ minute

Visitors without roaming agreements (p. 1159)
Call co. to use Amex, Mastercard, Visa

Identification numbers

City	Market	System	Billing	RSA
Madera	339	00162	30326	CA-4
Merced	339	01066	00000	CA-4

For full coverage area, see Bakersfield, CA and
Fresno, CA.

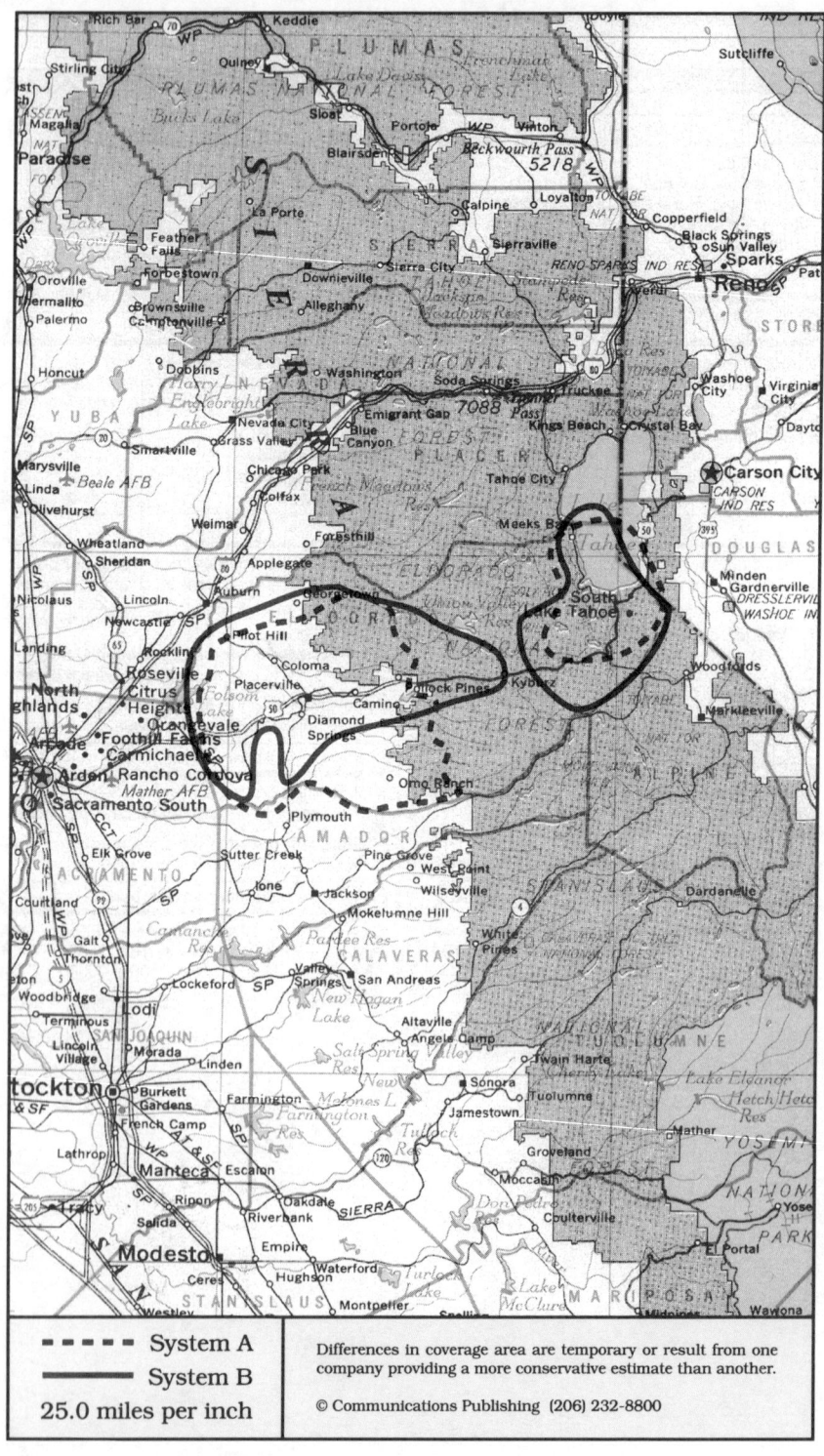

- - - - -	System A
————	System B
25.0 miles per inch	

Differences in coverage area are temporary or result from one company providing a more conservative estimate than another.

© Communications Publishing (206) 232-8800

Placerville, California
Placerville, South Lake Tahoe

|

Cellular company

Cellular One
Cellular Pacific
1166 Broadway, Suite L
Placerville, CA 95667

Corporate office   916-642-2355

Sending calls

Visitor dialing instructions
Local calls . . . . .   area code + 7 digits
Long distance . . .   1 + area code + 7 digits

Speed-dial numbers
Customer service   611
Emergency   911

Receiving calls

Caller dials the roamer access number below,
hears tone, dials cellular area code and number.
. . . .   916-425-7626

NationLink & RoamingAmerica (see p. 1157)
Dial *31 send to receive calls, *30 to deactivate.
Dial *32 send to tell caller the roamer access no.
This company's subscribers have level 6 service.

North American Cellular Network (see p. 1156)
All calls forward to you. Deactivate dial *35 send.

Billing information

Visitor rates
Most visitors . . . . . $2 day + 50¢ minute

Visitors without roaming agreements (p. 1159)
Cannot call since company takes no credit cards.

Identification numbers

City	Market	System	Billing	RSA
All served cities	346	01079	00000	CA-11

Call switching is performed by McCaw Cellular for
the licensee, Cellular Pacific.

Cellular company

Mountain Cellular
Atlantic Cellular
2849 Ray Lawyer Drive
Placerville, CA 95667

2489 Lake Tahoe Boulevard, Suite 6
South Lake Tahoe, CA 96150

Local office . . .   Placerville   800-924-3844
. . . .   Placerville   916-622-3844
. . . .   South Lake Tahoe   916-542-2242
Corporate office   401-421-7090

Sending calls

Visitor dialing instructions
Local calls . . . . . .   area code + 7 digits
Long distance . . .   1 + area code + 7 digits

Speed-dial numbers
Customer service *611
Emergency   911
Road conditions *7623
Weather *8367

Receiving calls

Caller dials the roamer access number below,
hears tone, dials cellular area code and number.
. . . .   916-957-7626

Auto-Access forwarding (see page 1156)
Activate, dial *28 send. Deactivate dial *29 send.

Follow Me Roaming forwarding (see p. 1156)
Activate, dial *18 send. Deactivate dial *19 send.

Billing information

Visitor rates
Most visitors . . . . . $3 day + 75¢ minute

Visitors without roaming agreements (p. 1159)
Cannot call since company takes no credit cards.

Identification numbers

City	Market	System	Billing	RSA
All served cities	346	01080	00000	CA-11

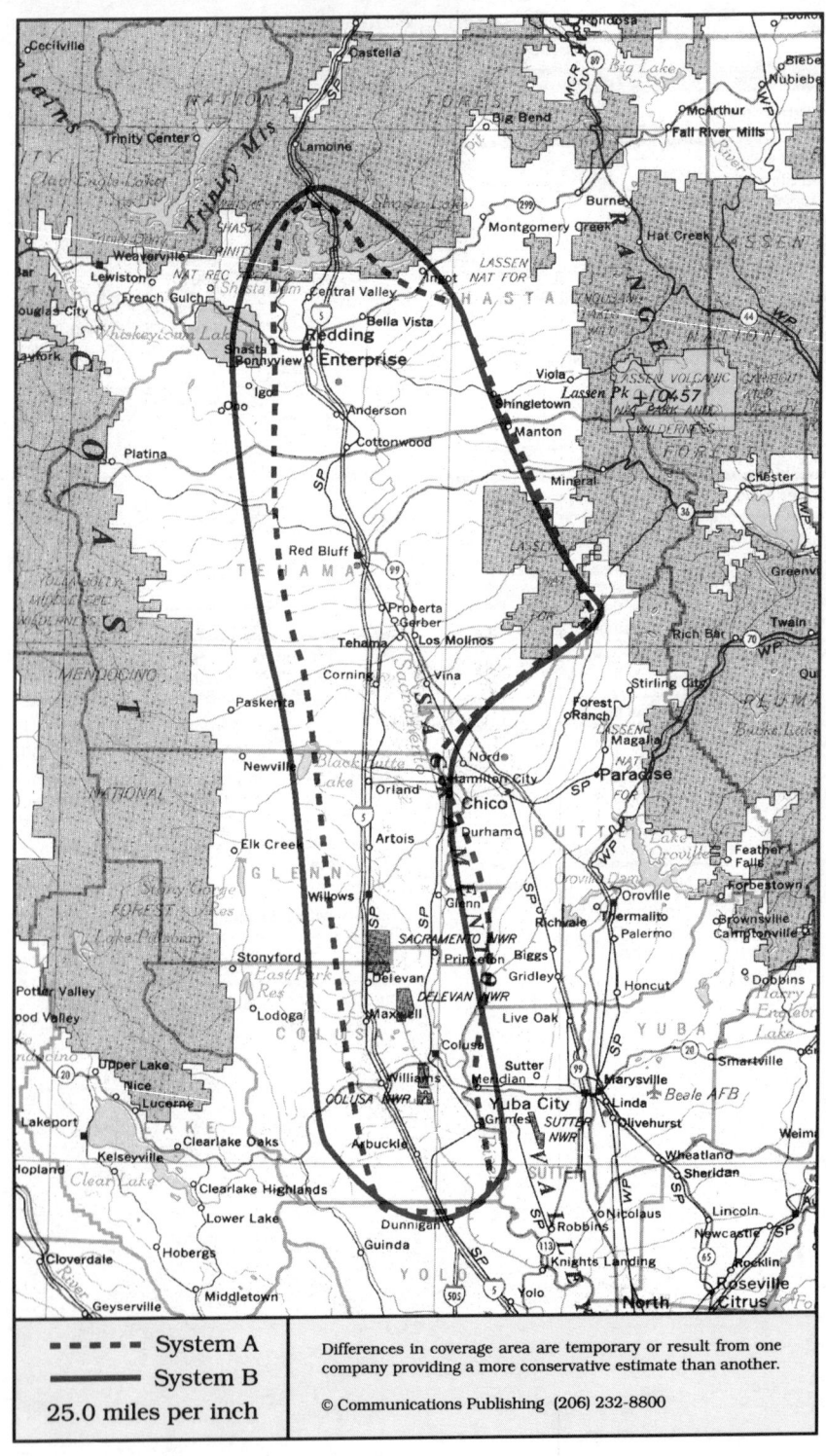

System A (dashed line)
System B (solid line)
25.0 miles per inch

Differences in coverage area are temporary or result from one company providing a more conservative estimate than another.

© Communications Publishing (206) 232-8800

Redding, California
Anderson, Corning, Orland, Red Bluff, Redding, Willows

Cellular company

Cellular One
McCaw Cellular Communications
1750 Howe Avenue, Suite 102
Sacramento, CA 95825

Customer service	800-635-9589
. . . .	916-923-2400
. . . .	209-572-1004
Local office	916-923-2222
Corporate office	206-827-4500

Sending calls

Visitor dialing instructions

Local calls	area code + 7 digits
Long distance . . .	1 + area code + 7 digits

Speed-dial numbers

Customer service	*611
Emergency	911
Road condition	*7623
Weather	*786

Receiving calls

Caller dials the roamer access number below,
hears tone, dials cellular area code and number.
. . . . 916-225-7626

NationLink & RoamingAmerica (see p. 1157)
Dial *31 send to receive calls, *30 to deactivate.
Dial *32 send to tell caller the roamer access no.
This company's subscribers have level 6 service.

North American Cellular Network (see p. 1156)
All calls forward to you. Deactivate dial *35 send.

Billing information

Visitor rates

Most visitors	$0 day + 99¢ minute

Visitors without roaming agreements (p. 1159)
Cannot call since company takes no credit cards.

North American Cellular Network (see p. 1156)
All calls forward to you. Deactivate dial *35 send.

Identification numbers

City	Market	System	Billing	RSA
Red Bluff	343	00513	00000	CA-8
Redding	254	00513	00000	MSA

For full coverage area, see Sacramento, CA.

Cellular company

PacTel Cellular
2150 River Plaza Drive, Suite 400
Sacramento, CA 95833

Customer service	800-722-8358
. . . .	916-920-0645
Local office	916-646-3773
Corporate office	510-210-3600

Sending calls

Visitor dialing instructions

Local calls	area code + 7 digits
Long distance . . .	0 + area code + 7 digits

Speed-dial numbers

Channel 3 to make a news or traffic report	*3
Customer service	*611
Emergency	911
Information services	*4636

Receiving calls

Caller dials the roamer access number below,
hears tone, dials cellular area code and number.
. . . . 916-520-7626

Auto-Access forwarding (see page 1156)
Activate, dial *28 send. Deactivate dial *29 send.

Follow Me Roaming forwarding (see p. 1156)
Activate, dial *18 send. Deactivate dial *19 send.

Billing information

Visitor rates

7 am - 7 pm Mon-Fri .	$0 day + 45¢ minute
All other times	$0 day + 15¢ minute

Visitors without roaming agreements (p. 1159)
Call co. to use Mastercard, Visa

Identification numbers

City	Market	System	Billing	RSA
Red Bluff	343	00294	00000	CA-8
Redding	254	00294	00000	MSA

For full coverage area, see Chico, CA and
Sacramento, CA.

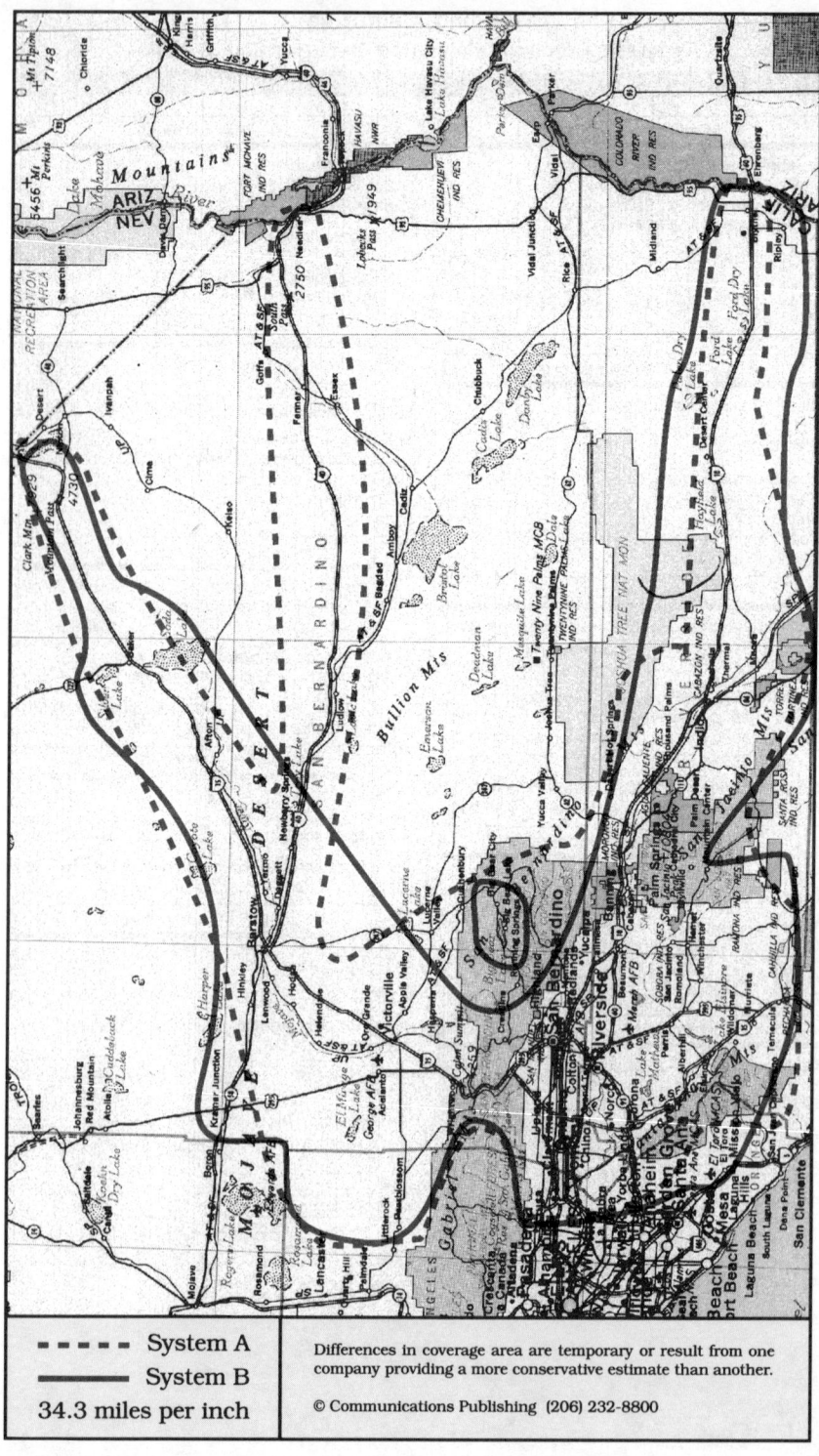

- - - - System A

———— System B

34.3 miles per inch

Differences in coverage area are temporary or result from one company providing a more conservative estimate than another.

© Communications Publishing (206) 232-8800

Riverside, California
Barstow, Blythe, Palm Springs, Riverside, San Bernardino (also see Los Angeles)

System A	System B

Cellular company

Los Angeles Cellular Telephone
17785 Center Court Drive N
Cerritos, CA 90701

Customer service	310-809-7475
Corporate office	310-924-0000

Sending calls

Visitor dialing instructions

Local calls	area code + 7 digits
Long distance . . .	1 + area code + 7 digits

Speed-dial numbers

AAA road service	*222
Address directions . . .	*FIND or *3463
Customer service	611
Directory assistance	411
Emergency	911
Traffic center	*JAM or *526

Receiving calls

Caller dials the roamer access number below,
hears tone, dials cellular area code and number.

. . . . 619-774-7626

NationLink & RoamingAmerica (see p. 1157)
Dial *31 send to receive calls, *30 to deactivate.
Dial *32 send to tell caller the roamer access no.
This company's subscribers have level 6 service.

Billing information

Visitor rates

Most visitors	$0 day + 99¢ minute

Visitors without roaming agreements (p. 1159)
Cannot call since company takes no credit cards.

Identification numbers

City	Market	System	Billing	RSA
All served cities	002	00027	00000	MSA

For full coverage area, see Los Angeles, CA.

Cellular company

PacTel Cellular
3 Park Plaza, PO Box 19651
Irvine, CA 92713-9651

Customer service	800-851-9815
. . . .	714-222-8200
Local office	714-222-7000
Corporate office	510-210-3600

Sending calls

Visitor dialing instructions

Local calls	area code + 7 digits
Long distance . . .	0 + area code + 7 digits

Speed-dial numbers

Customer service	*611
Emergency	911
Roaming information	*711

Receiving calls

Caller dials the roamer access number below,
hears tone, dials cellular area code and number.

. . . . 619-567-7626

Auto-Access forwarding (see page 1156)
Activate, dial *28 send. Deactivate dial *29 send.

Follow Me Roaming forwarding (see p. 1156)
Activate, dial *18 send. Deactivate dial *19 send.

Billing information

Visitor rates

Most visitors	$0 day + $1.05 minute

Visitors without roaming agreements (p. 1159)
Call co. to use Amex, Mastercard, Visa

Identification numbers

City	Market	System	Billing	RSA
All served cities	002	00002	00000	MSA

For full coverage area, see Los Angeles, CA.

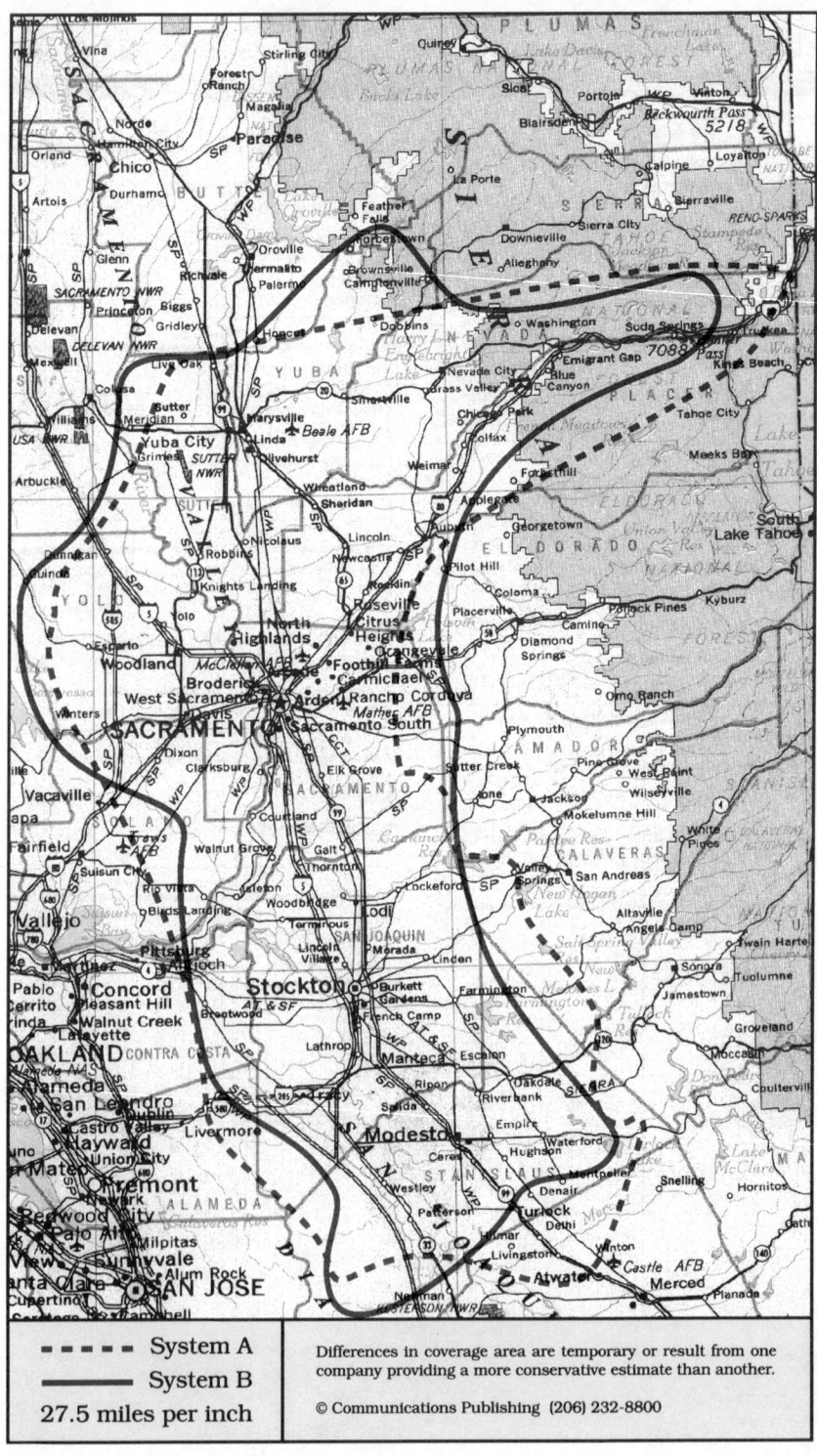

▪ ▪ ▪ ▪ ▪ **System A**	Differences in coverage area are temporary or result from one company providing a more conservative estimate than another.
——— **System B**	
27.5 miles per inch	© Communications Publishing (206) 232-8800

Sacramento, California
Grass Valley, Modesto, Sacramento, Stockton, Yuba City

System A	System B

Cellular company

Cellular One
McCaw Cellular Communications
3600 Sisk Road, Suite A-4
Modesto, CA 95356

1750 Howe Avenue, Suite 102
Sacramento, CA 95825

1128 E March Lane
Stockton, CA 95210

Customer service		800-635-9589
. . . .		916-923-2400
. . . .		209-572-1004
Local office . . .	Modesto	209-545-4475
. . . . . .	Sacramento	916-923-2222
. . . . . .	Stockton	209-476-1400
Corporate office		206-827-4500

Sending calls

Visitor dialing instructions
Local calls . . . . area code + 7 digits
Long distance . . . 1 + area code + 7 digits

Speed-dial numbers
Customer service 611
Emergency 911
Road condition *7623
Weather *786

Receiving calls

Caller dials the roamer access number below,
hears tone, dials cellular area code and number.
Modesto 209-531-7626
Sacramento, Grass Valley . . 916-425-7626
Stockton 209-481-7626
Yuba City 916-755-7626

NationLink & RoamingAmerica (see p. 1157)
Dial *31 send to receive calls, *30 to deactivate.
Dial *32 send to tell caller the roamer access no.
This company's subscribers have level 6 service.

North American Cellular Network (see p. 1156)
All calls forward to you. Deactivate dial *35 send.

Billing information

Visitor rates
Grass Valley . . . . . $2 day + 50¢ minute
Most visitors . . . . . $0 day + 99¢ minute

Visitors without roaming agreements (p. 1159)
Cannot call since company takes no credit cards.

North American Cellular Network (see p. 1156)
All calls forward to you. Deactivate dial *35 send.

Identification numbers

City	Market	System	Billing	RSA
Grass Valley	345	00129	30237	CA-10
Modesto	142	00233	00000	MSA
Sacramento	035	00129	00000	MSA
Stockton	107	00233	00000	MSA
Yuba City	274	00129	00000	MSA

McCaw serves CA-10 for licensee, Data Cellular.

For full coverage, see Placerville & Redding, CA.

Cellular company

PacTel Cellular
4231-A McHenry Avenue
Modesto, CA 95356

2150 River Plaza Drive, Suite 400
Sacramento, CA 95833

2701 E Hammer Lane, Suite 101
Stockton, CA 95210

Customer service		800-722-8358
. . . .		916-920-0645
Local office	Modesto	209-525-8607
. . . .	Sacramento	916-646-3773
. . . .	Stockton	209-952-0180
Corporate office		510-210-3600

Sending calls

Visitor dialing instructions
Local calls . . . . . . area code + 7 digits
Long distance . . . 0 + area code + 7 digits

Speed-dial numbers
Channel 3 to make a news or traffic report   *3
Customer service *611
Emergency 911
Road conditions . . . . *ROAD or *7623
Information services *4636

Receiving calls

Caller dials the roamer access number below,
hears tone, dials cellular area code and number.
Modesto 209-479-7626
Sacramento, Grass Valley . . 916-539-7626
Stockton 209-479-7626
Yuba City 916-539-7626

Auto-Access forwarding (see page 1156)
Activate, dial *28 send. Deactivate dial *29 send.

Follow Me Roaming forwarding (see p. 1156)
Activate, dial *18 send. Deactivate dial *19 send.

Billing information

Visitor rates
7 am - 7 pm Mon-Fri   . $0 day + 45¢ minute
All other times . . . . $0 day + 15¢ minute

Visitors without roaming agreements (p. 1159)
Call co. to use Mastercard, Visa

Identification numbers

City	Market	System	Billing	RSA
Grass Valley	345	00112	30180	CA-10
Modesto	142	00224	30520	MSA
Sacramento	035	00112	00000	MSA
Stockton	107	00224	00000	MSA
Yuba City	274	00112	00000	MSA

For full coverage area, see Chico, CA and
Redding, CA and Sonora, CA.

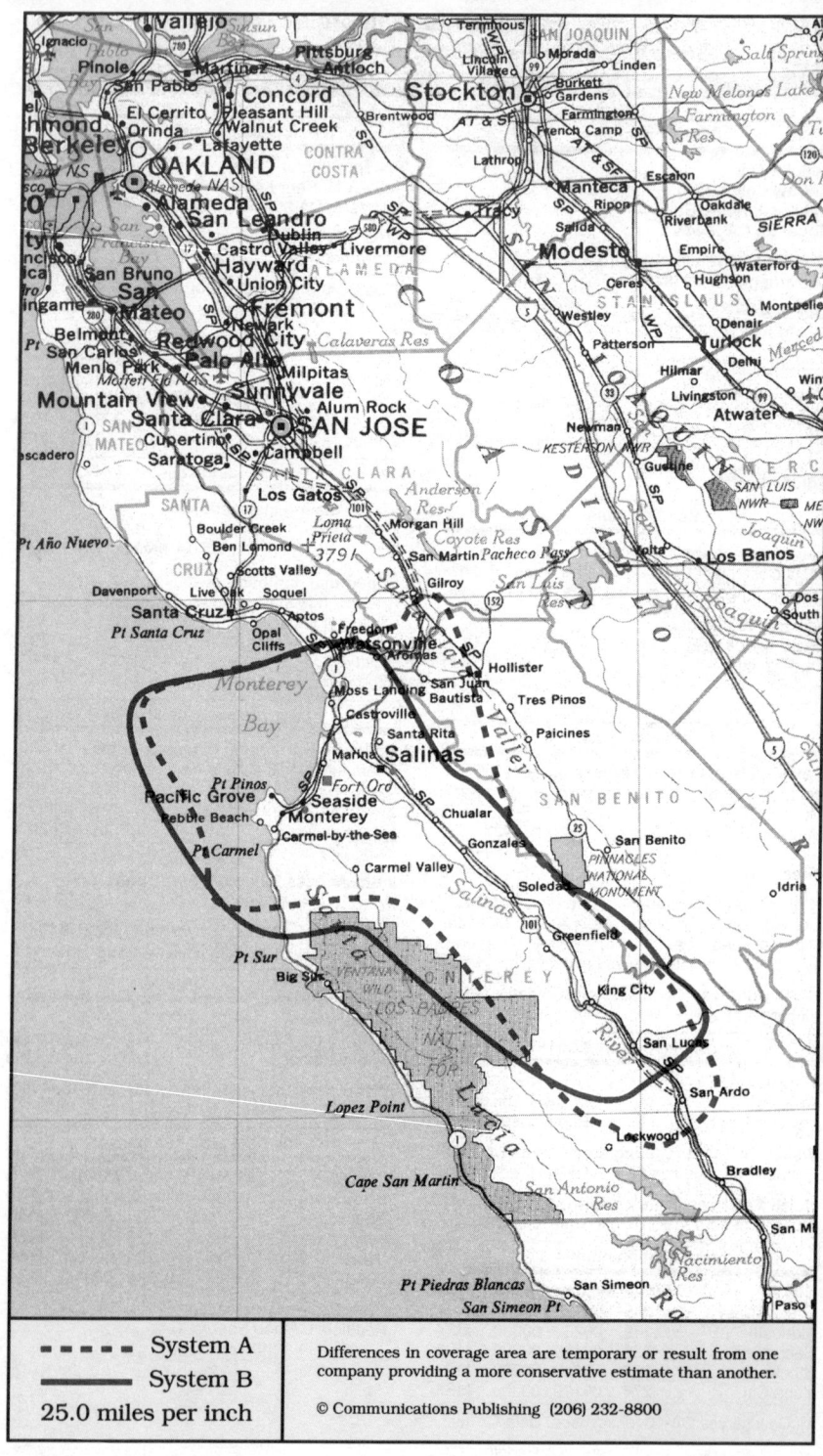

System A

System B

25.0 miles per inch

Differences in coverage area are temporary or result from one
company providing a more conservative estimate than another.

© Communications Publishing (206) 232-8800

Salinas, California
Monterey, Salinas

|

Cellular company

Cellular One
McCaw Cellular Communications
851 Delmonte Avenue
Monterey, CA 93940

Customer service 800-421-2611
Local office 408-647-8888
. . . . . . from Salinas 408-754-8888
Corporate office 206-827-4500

Sending calls

Visitor dialing instructions
Local calls . . . . area code + 7 digits
Long distance . . . 1 + area code + 7 digits

Speed-dial numbers
Customer service 611
Emergency 911

Receiving calls

Caller dials the roamer access number below,
hears tone, dials cellular area code and number.
. . . . 408-595-7626

NationLink & RoamingAmerica (see p. 1157)
Dial ∗31 send to receive calls, ∗30 to deactivate.
Dial ∗32 send to tell caller the roamer access no.
This company's subscribers have level 6 service.

North American Cellular Network (see p. 1156)
All calls forward to you. Deactivate dial ∗35 send.

Billing information

Visitor rates
Most visitors . . . . . $0 day + 99¢ minute

Visitors without roaming agreements (p. 1159)
Cannot call since company takes no credit cards.

Identification numbers

City	Market	System	Billing	RSA
All served cities	126	00527	00000	MSA

Cellular company

GTE Mobilnet
590 Brunken Avenue #J
Salinas, CA 93901

Customer service 800-366-5665
. . . . 510-416-0150
Local office 408-422-0540
Corporate office 404-391-8000

Sending calls

Visitor dialing instructions
Local calls . . . . . . area code + 7 digits
Long distance . 1 + area code + 7 digits

Speed-dial numbers
Address directions . . . . ∗627
Customer service . . . . . . ∗611
Emergency 911
Telephone problems . . . . ∗111

Receiving calls

Caller dials the roamer access number below,
hears tone, dials cellular area code and number.
. . . . 408-671-7626

Auto-Access forwarding (see page 1156)
Activate, dial ∗28 send. Deactivate dial ∗29 send.

Follow Me Roaming forwarding (see p. 1156)
Activate, dial ∗18 send. Deactivate dial ∗19 send.

Billing information

Visitor rates
Most visitors . . . . . $2 day + 50¢ minute

Visitors without roaming agreements (p. 1159)
Call co. to use Amex, Mastercard, Visa

Identification numbers

City	Market	System	Billing	RSA
All served cities	126	00040	00000	MSA

For full coverage area, see San Francisco, CA.

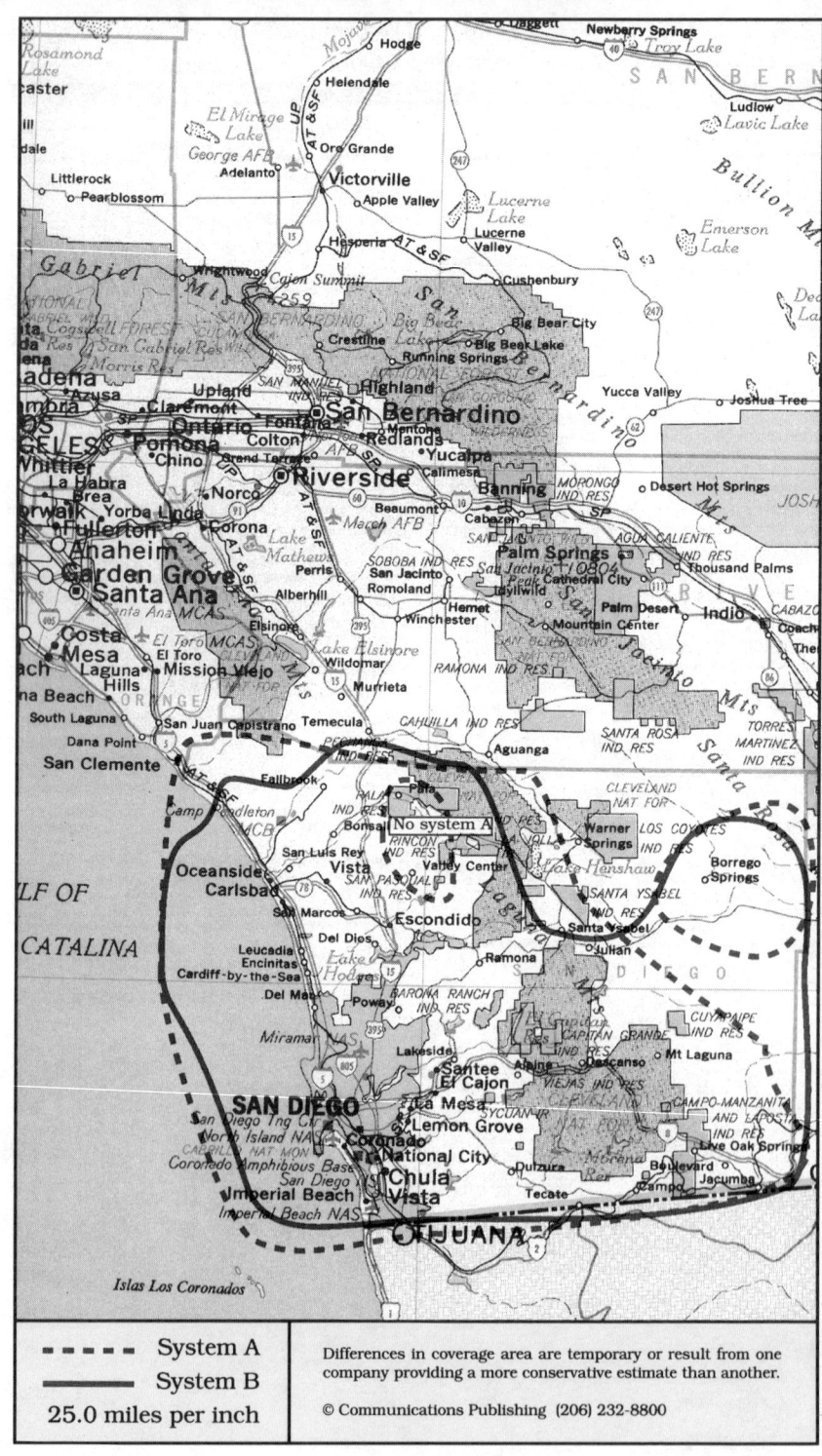

No system A

San Diego, California
Borrego Springs, Escondido, Oceanside, San Diego

System A	System B

Cellular company

U S West Cellular
7535 Metropolitan Drive
San Diego, CA 92108

Customer service	800-238-7848
. . . .	206-562-2895
. . . . . local subscribers	800-626-6611
. . . .	206-747-1771
Local office	619-291-3100
Corporate office	206-747-4900

Sending calls

Visitor dialing instructions

Local calls	area code + 7 digits
Long distance	area code + 7 digits

Speed-dial numbers

Customer service	*611
Emergency	911
Roaming information	*711

Receiving calls

Caller dials the roamer access number below,
hears tone, dials cellular area code and number.
. . . . 619-548-7626

NationLink & RoamingAmerica (see p. 1157)
Caller hears your current roamer access number.
This company's subscribers have level 0 service.

Billing information

Visitor rates
Most visitors . . . . . $0 day + 60¢ minute

Visitors without roaming agreements (p. 1159)
Call co. to use Amex, Mastercard, Visa

Identification numbers

City	Market	System	Billing	RSA
All served cities	018	00043	00000	MSA

Cellular company

PacTel Cellular
5355 Mira Sorrento Place, Suite 500
San Diego, CA 92121

Customer service	619-625-7500
Local office	619-625-7878
Corporate office	510-210-3600

Sending calls

Visitor dialing instructions

Local calls	area code + 7 digits
Long distance . . .	0 + area code + 7 digits

Speed-dial numbers

Customer service	*611
Directory assistance	411
Emergency	911
Make a traffic report	*9
Report a polluting automobile	*22

Receiving calls

Caller dials the roamer access number below,
hears tone, dials cellular area code and number.
. . . . 619-540-7626

Auto-Access forwarding (see page 1156)
Activate, dial *28 send. Deactivate dial *29 send.

Follow Me Roaming forwarding (see p. 1156)
Activate, dial *18 send. Deactivate dial *19 send.

Billing information

Visitor rates
Most visitors . . . . . $0 day + 55¢ minute

Visitors without roaming agreements (p. 1159)
Call co. to use Amex, Mastercard, Visa

Identification numbers

City	Market	System	Billing	RSA
All served cities	018	00004	00000	MSA

You may continue a call when traveling into Los
Angeles without being disconnected.

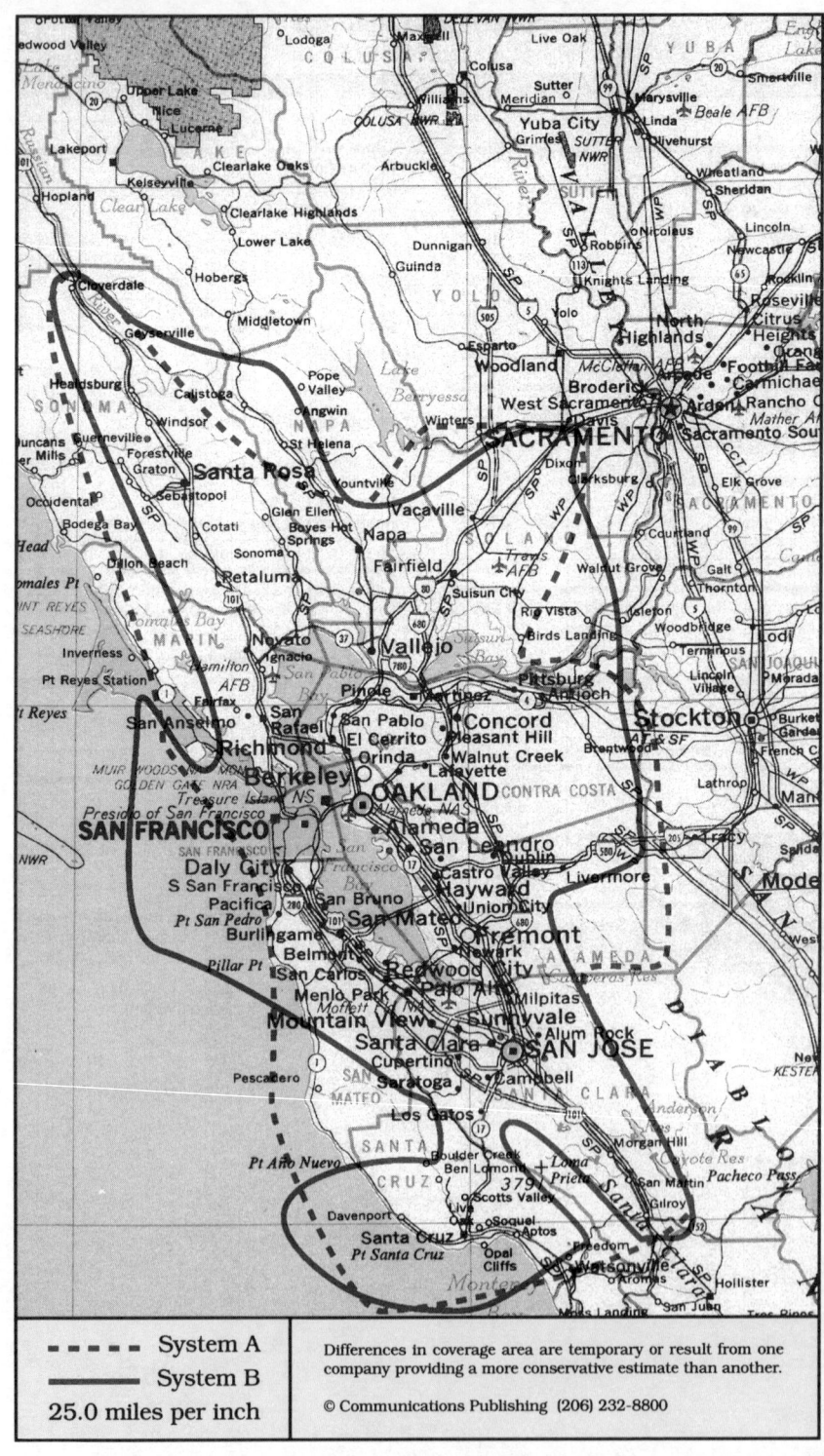

▪ ▪ ▪ ▪ System A	Differences in coverage area are temporary or result from one
—— System B	company providing a more conservative estimate than another.
25.0 miles per inch	© Communications Publishing (206) 232-8800

San Francisco, California
Oakland, San Francisco, San Jose, Santa Cruz, Santa Rosa, Vallejo

System A	System B

Cellular company

San Francisco, San Jose
Cellular One
Bay Area Cellular Telephone
651 Gateway Boulevard, Suite 1500
South San Francisco, CA 94080

Customer service	800-445-6876
Local office	415-871-9500

Santa Cruz
Cellular One
Santa Cruz Cellular Telephone
3949 Research Park Court, Suite 100
Soquel, CA 95073

Corporate office	408-464-1000

Santa Rosa, Vallejo
Cellular One
McCaw Cellular Communications
398 Tesconi Court
Santa Rosa, CA 95401

Customer service	800-762-7217
Local office (cellular phone ∗311)	707-577-8800
Corporate office	206-827-4500

Sending calls

Visitor dialing instructions

Local calls	area code + 7 digits
Long distance . . .	0 + area code + 7 digits

Speed-dial numbers

Customer service	∗611
Emergency	911
Traffic report	∗KGO81 or ∗54681

Receiving calls

Caller dials the roamer access number below,
hears tone, dials cellular area code and number.

Napa, Vallejo	707-498-7626
San Francisco	415-860-7626
San Jose	408-221-7626
Santa Rosa	707-484-7626

NationLink & RoamingAmerica (see p. 1157)
Dial ∗31 send to receive calls, ∗30 to deactivate.
Dial ∗32 send to tell caller the roamer access no.
Subscribers have level 8 service. McCaw level 6.

North American Cellular Network (see p. 1156)
All calls forward to you. Deactivate dial ∗35 send.

Billing information

Visitor rates

Most visitors	$2 day + 50¢ minute

Visitors without roaming agreements (p. 1159)
Cannot call since company takes no credit cards.

Identification numbers

City	Market	System	Billing	RSA
San Francisco	007	00031	00000	MSA
San Jose	027	00031	00000	MSA
Santa Cruz	175	00031	30017	MSA
Santa Rosa	123	00031	30015	MSA
Vallejo	111	00031	30015	MSA

Three companies operate one system jointly.

Cellular company

GTE Mobilnet
4410 Rosewood Drive
Pleasanton, CA 94588

Customer service	800-366-5665
Local office	510-416-0150
Corporate office	404-391-8000

Sending calls

Visitor dialing instructions

Local calls	area code + 7 digits
Long distance . . .	1 + area code + 7 digits

Speed-dial numbers

Address directions	∗627
Customer service	∗611
Emergency	911
Hear a traffic report ∗5101 ∗5627 ∗5726 ∗5639
Information services . . . ∗INFO or ∗4636
Make a traffic report . ∗227 or ∗483 or ∗973
Telephone problems	∗111

Receiving calls

Caller dials the roamer access number below,
hears tone, dials cellular area code and number.

Napa	707-483-7626
San Francisco	415-722-7626
San Jose	408-234-7626

Auto-Access forwarding (see page 1156)
Activate, dial ∗28 send. Deactivate dial ∗29 send.

Follow Me Roaming forwarding (see p. 1156)
Activate, dial ∗18 send. Deactivate dial ∗19 send.

Billing information

Visitor rates

Most visitors	$2 day + 50¢ minute

Visitors without roaming agreements (p. 1159)
Call co. to use Amex, Discover, Mastercard, Visa

Identification numbers

City	Market	System	Billing	RSA
San Francisco	007	00040	00000	MSA
San Jose	027	00040	00000	MSA
Santa Cruz	175	00040	00000	MSA
Santa Rosa	123	00040	00000	MSA
Vallejo	111	00040	00000	MSA

For full coverage area, see Salinas, CA and
Ukiah, CA.

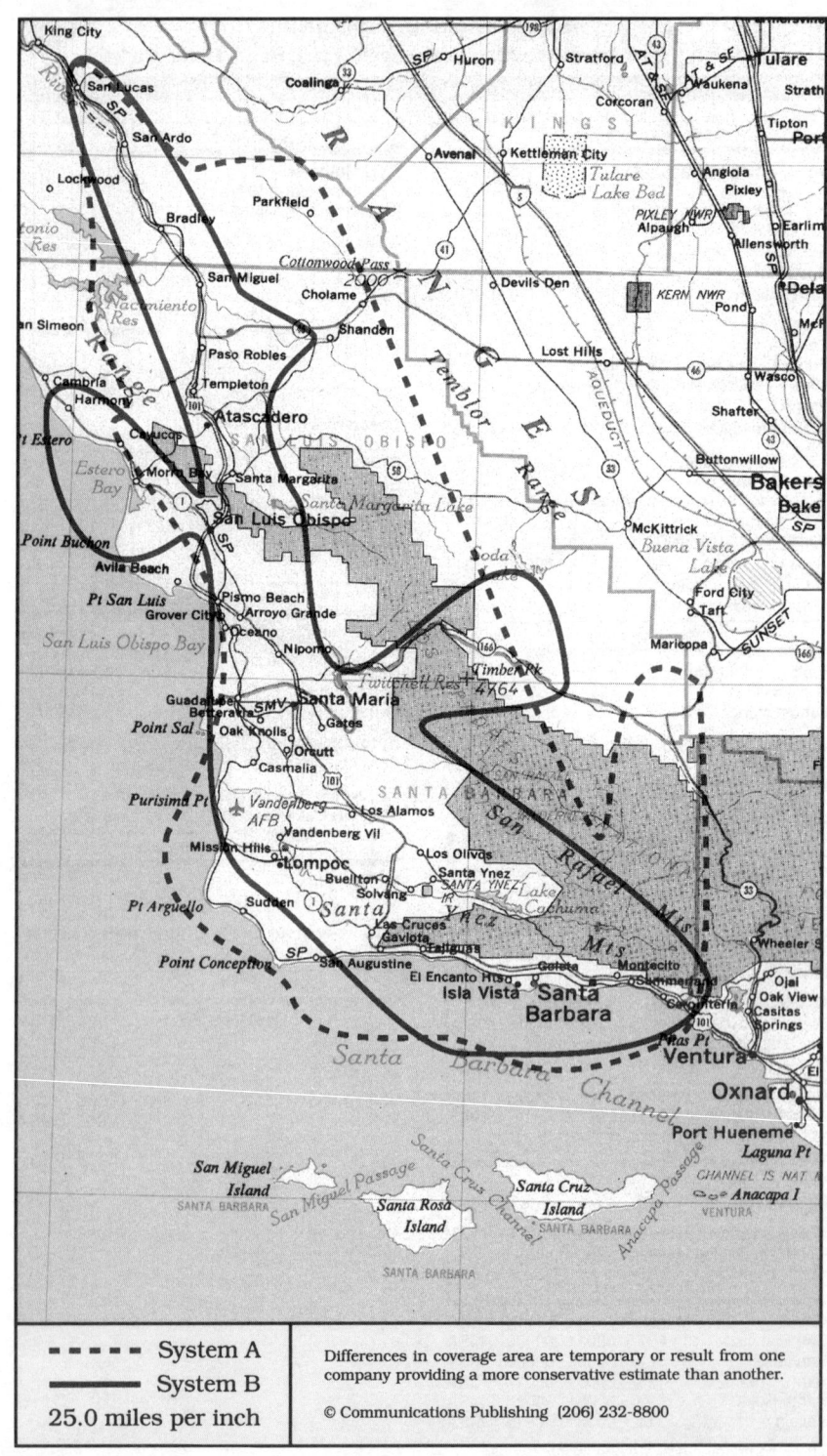

- - - - - System A ————— System B 25.0 miles per inch	Differences in coverage area are temporary or result from one company providing a more conservative estimate than another. © Communications Publishing (206) 232-8800

Santa Barbara, California
Atascadero, San Luis Obispo, Santa Barbara, Santa Maria

System A	System B

Cellular company

Cellular One
McCaw Cellular Communications
800 N Milpas Street
Santa Barbara, CA 93103

Customer service	805-962-3611
Local office	800-722-7464
. . . .	805-962-0045
Corporate office	206-827-4500

Sending calls

Visitor dialing instructions

Local calls	area code + 7 digits
Long distance . .	1 + area code + 7 digits

Speed-dial numbers

Customer service	611
Emergency	911

Receiving calls

Caller dials the roamer access number below,
hears tone, dials cellular area code and number.
. . . . 805-689-7626
(San Luis Obispo may change to 805-235-7626.)

NationLink & RoamingAmerica (see p. 1157)
Dial *31 send to receive calls, *30 to deactivate.
Dial *32 send to tell caller the roamer access no.
This company's subscribers have level 6 service.

North American Cellular Network (see p. 1156)
All calls forward to you. Deactivate dial *35 send.

Billing information

Visitor rates

Most visitors	$0 day + 99¢ minute

Visitors without roaming agreements (p. 1159)
Cannot call since company takes no credit cards.

Identification numbers

City	Market	System	Billing	RSA
San Luis Obispo	340	00531	30425	CA-5
Santa Barbara	124	00531	00000	MSA

Cellular company

GTE Mobilnet
742 Marsh Street, Suite B
San Luis Obispo, CA 93401

105 W Gutierrez Street
Santa Barbara, CA 93101

Customer service		800-366-5665
. . . .		510-416-0150
Local office .	San Luis Obispo	805-544-1565
. . . . . .	Santa Barbara	805-965-1565
Corporate office		404-391-8000

Sending calls

Visitor dialing instructions

Local calls	area code + 7 digits
Long distance . . .	1 + area code + 7 digits

Speed-dial numbers

Address directions	*627
Customer service	*611
Emergency	911
Telephone problems	*111

Receiving calls

Caller dials the roamer access number below,
hears tone, dials cellular area code and number.

San Luis Obispo	805-441-7626
Santa Barbara	805-680-7626

Auto-Access forwarding (see page 1156)
Activate, dial *28 send. Deactivate dial *29 send.

Follow Me Roaming forwarding (see p. 1156)
Activate, dial *18 send. Deactivate dial *19 send.

Billing information

Visitor rates

Most visitors	$2 day + 50¢ minute

Visitors without roaming agreements (p. 1159)
Call co. to use Amex, Mastercard, Visa

Identification numbers

City	Market	System	Billing	RSA
San Luis Obispo	340	00040	30002	CA-5
Santa Barbara	124	00040	30002	MSA

For full coverage area, see Salinas, CA.

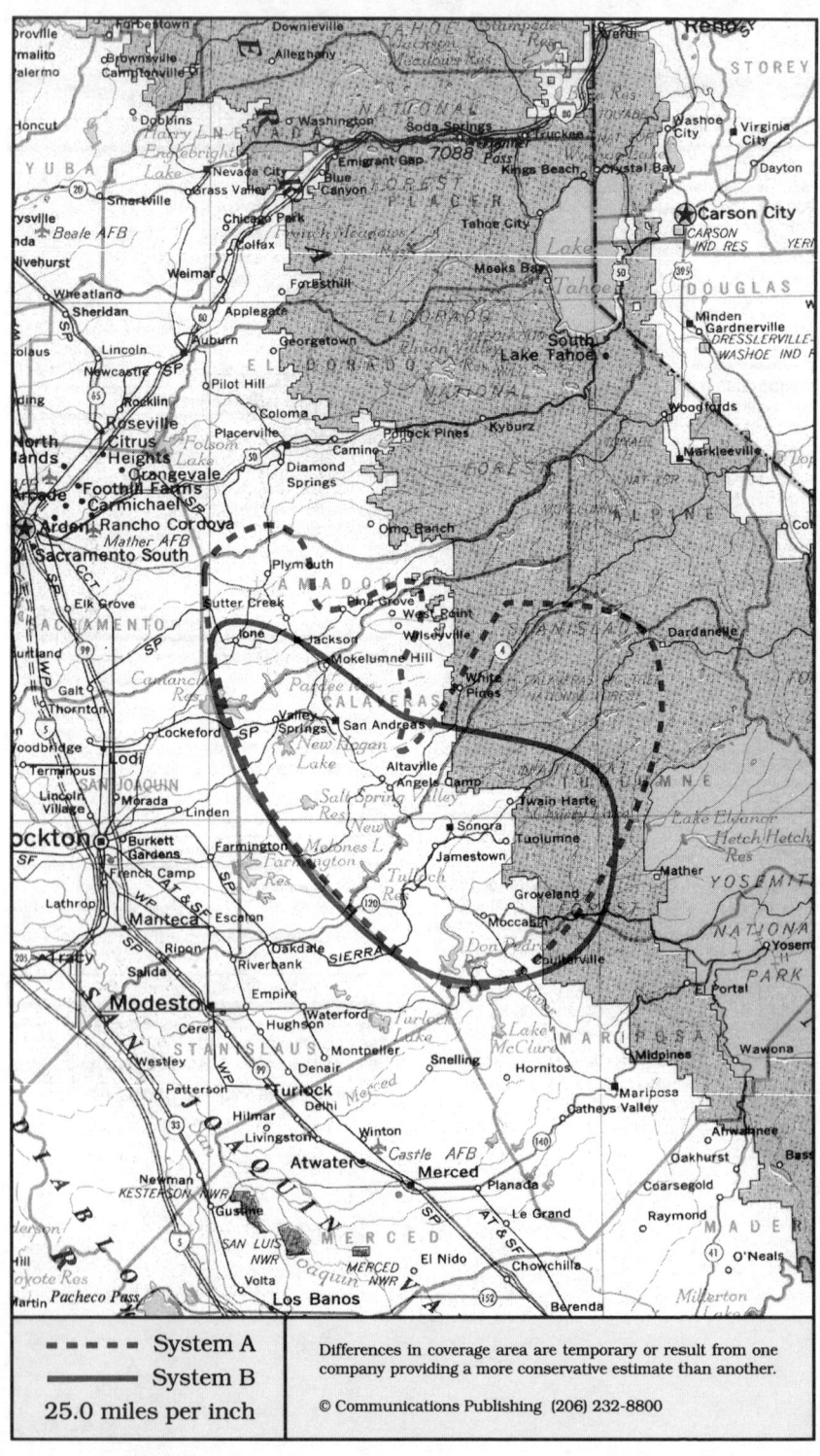

- - - - System A ———— System B 25.0 miles per inch	Differences in coverage area are temporary or result from one company providing a more conservative estimate than another. © Communications Publishing (206) 232-8800

Sonora, California
London, San Andreas, Sonora

System A	System B

System A

Cellular company

Cellular One
Licensed to: Alpine Cellular
Managed by: Rural Cellular Management
125 Peek Street, Suite C
Jackson, CA 95642

13663 Mono Way
Sonora, CA 95370

Local office	. . . .	Jackon	209-223-3223
. . . .	Sonora some areas		800-366-3663
. . . .		Sonora	209-533-3330
Corporate office	Rural Cell. Mgt		707-422-2100

Sending calls

Visitor dialing instructions
Local calls . . . . . area code + 7 digits
Long distance . . . 1 + area code + 7 digits

Speed-dial numbers
Customer service 611
Emergency 911
KVML *1450
KZSQ *92
Road conditions . . . . *ROAD or *7623

Receiving calls

Caller dials the roamer access number below,
hears tone, dials cellular area code and number.
. . . . 209-765-7626

NationLink & RoamingAmerica (see p. 1157)
Dial *31 send to receive calls, *30 to deactivate.
Dial *32 send to tell caller the roamer access no.
This company's subscribers have level 9 service.

North American Cellular Network (see p. 1156)
All calls forward to you. Deactivate dial *35 send.

Billing information

Visitor rates
Most visitors . . . . . $2 day + 50¢ minute

Visitors without roaming agreements (p. 1159)
Call co. to use Amex, Mastercard, Visa

Identification numbers

City	Market	System	Billing	RSA
All served cities	338	01063	00000	CA-3

System B

Cellular company

Licensed to: Contel Cellular
Operated by: PacTel Cellular
4231-A McHenry Avenue
Modesto, CA 95356

Customer service	. . . .		800-722-8358
. . . .			916-920-0645
Local office			209-525-8607
Corporate office	. . .	Contel	404-804-3400
. . . .		PacTel	510-210-3600

Sending calls

Visitor dialing instructions
Local calls . . . . . . area code + 7 digits
Long distance . . . 0 + area code + 7 digits

Speed-dial numbers
Customer service *611
Emergency 911

Receiving calls

Caller dials the roamer access number below,
hears tone, dials cellular area code and number.
. . . . 209-479-7626

Auto-Access forwarding (see page 1156)
Activate, dial *28 send. Deactivate dial *29 send.

Follow Me Roaming forwarding (see p. 1156)
Activate, dial *18 send. Deactivate dial *19 send.

Billing information

Visitor rates
7 am - 7 pm Mon-Fri . $0 day + 45¢ minute
All other times . . . . $0 day + 15¢ minute

Visitors without roaming agreements (p. 1159)
Call co. to use Amex, Mastercard, Visa

Identification numbers

City	Market	System	Billing	RSA
All served cities	338	00224	30182	CA-3

This area is licensed to Contel Cellular but the
system is operated by PacTel Cellular.

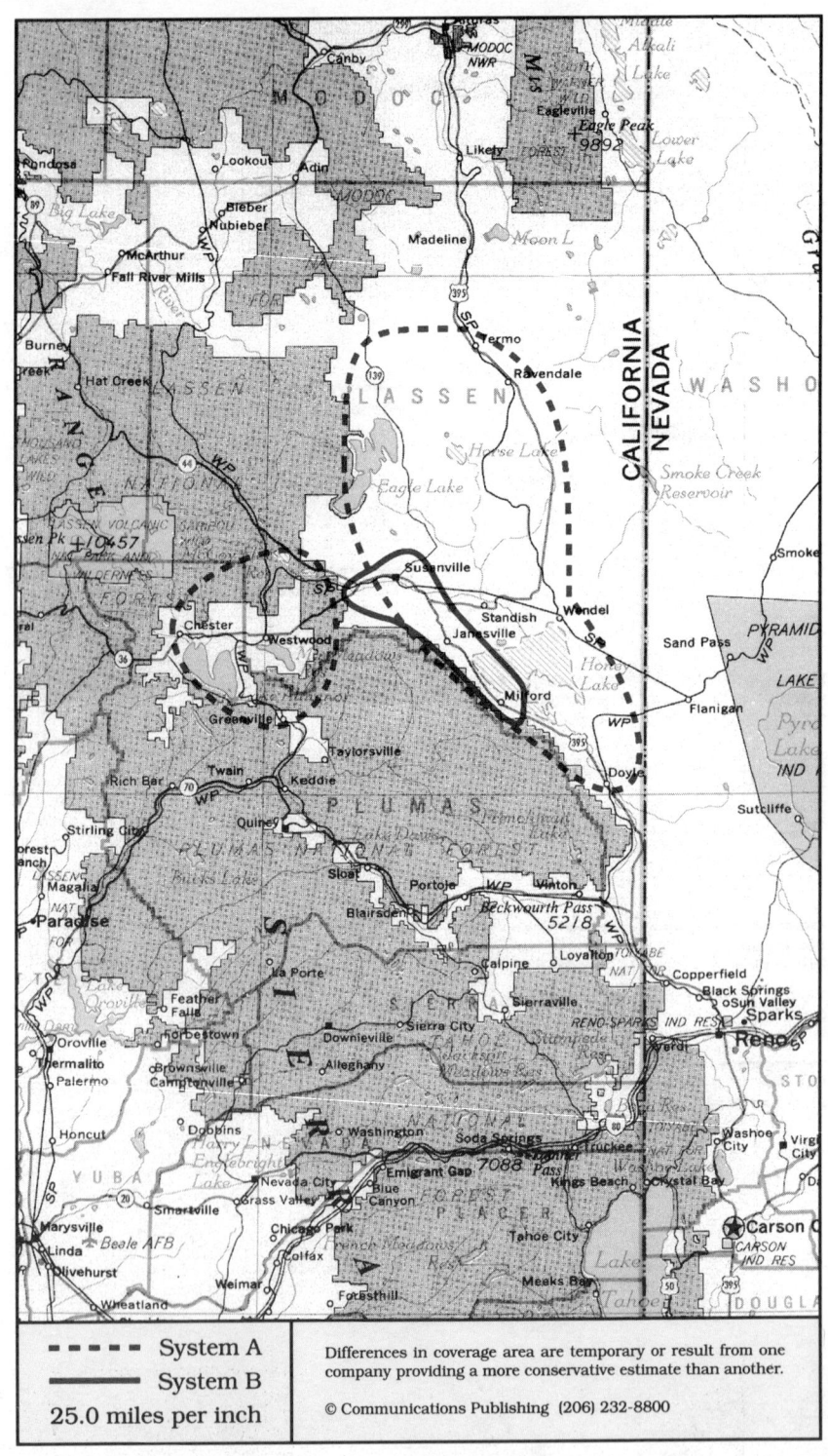

System A

System B

25.0 miles per inch

Differences in coverage area are temporary or result from one company providing a more conservative estimate than another.

© Communications Publishing (206) 232-8800

Susanville, California

System A

Cellular company

Cellular One
Managed by Quantum Communications
7901 Flying Cloud Drive, Suite 250
Eden Prairie, MN 55344

Customer service	916-251-2773
Corporate office	612-942-7650

Sending calls

Visitor dialing instructions

Local calls	area code + 7 digits
Long distance . . .	1 + area code + 7 digits

Speed-dial numbers

Customer service	611
Emergency	911

Receiving calls

Caller dials the roamer access number below,
hears tone, dials cellular area code and number.
. . . . 916-251-7626

NationLink & RoamingAmerica (see p. 1157)
Dial ∗31 send to receive calls, ∗30 to deactivate.
This company's subscribers have level 3 service.

Billing information

Visitor rates

Most visitors	$3 day + 65¢ minute

Visitors without roaming agreements (p. 1159)
Call co. to use Amex, Mastercard, Visa

Identification numbers

City	Market	System	Billing	RSA
All served cities	337	01061	00000	CA-2

System B

Cellular company

PacTel Cellular
2150 River Plaza Drive, Suite 400
Sacramento, CA 95833

Customer service	800-722-8358
. . . .	916-920-0645
Local office	916-646-3773
Corporate office	510-210-3600

Sending calls

Visitor dialing instructions

Local calls	area code + 7 digits
Long distance . . .	0 + area code + 7 digits

Speed-dial numbers

Customer service	∗611
Emergency	911

Receiving calls

Caller dials the roamer access number below,
hears tone, dials cellular area code and number.
. . . . 916-520-7626

Auto-Access forwarding (see page 1156)
Activate, dial ∗28 send. Deactivate dial ∗29 send.

Follow Me Roaming forwarding (see p. 1156)
Activate, dial ∗18 send. Deactivate dial ∗19 send.

Billing information

Visitor rates

7 am - 7 pm Mon-Fri .	$0 day + 55¢ minute
All other times	$0 day + 27¢ minute

Visitors without roaming agreements (p. 1159)
Call co. to use Diner's Club, Discover,
Mastercard, Visa

Identification numbers

City	Market	System	Billing	RSA
All served cities	337	00294	30184	CA-2

For full coverage area, see Redding, CA.

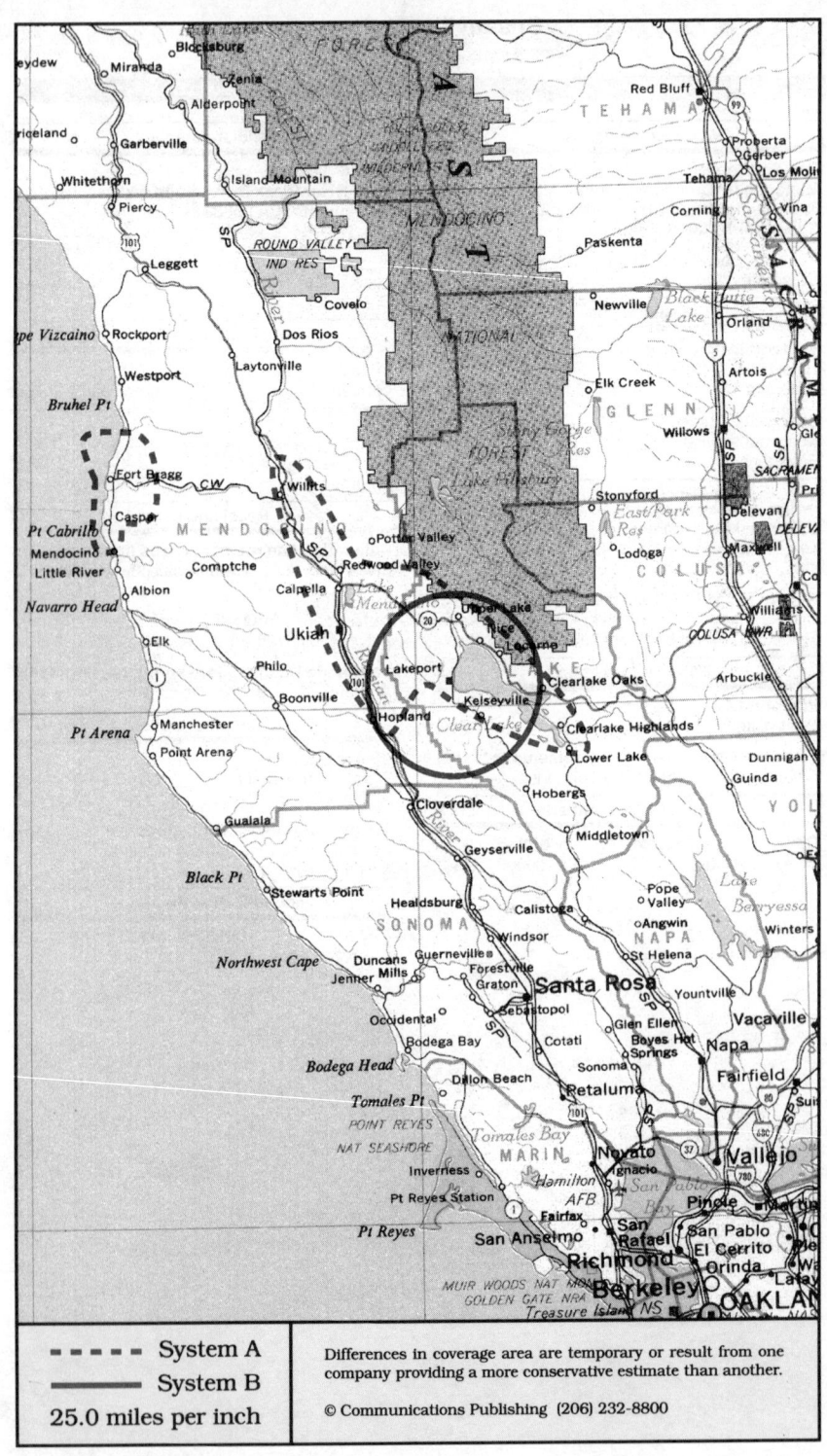

Ukiah, California
Fort Bragg, Lakeport, Ukiah

System A

Cellular company

United States Cellular
533-B S State Street
Ukiah, CA 95482

Customer service	some areas	800-528-8722
. . . .		503-779-3000
Local office	. . . . . .	707-468-0580
Corporate office	. . . . .	312-399-8900

Sending calls

Visitor dialing instructions
Local calls . . . . . area code + 7 digits
Long distance . . . 1 + area code + 7 digits

Speed-dial numbers
Customer service 611
Emergency 911

Receiving calls

Caller dials the roamer access number below,
hears tone, dials cellular area code and number.
. . . . 707-489-7626

NationLink & RoamingAmerica (see p. 1157)
Company plans to offer it in first quarter 1993.

Billing information

Visitor rates
Most visitors . . . . . $2 day + 50¢ minute

Visitors without roaming agreements (p. 1159)
Cannot call since company takes no credit cards.

Identification numbers

City	Market	System	Billing	RSA
All served cities	344	01075	00000	CA-9

System B

Cellular company

Licensed to: Contel Cellular
Operated by: GTE Mobilnet
4410 Rosewood Drive
Pleasanton, CA 94588

Customer service	. . .		800-366-5665
Local office	. . . . . .		510-416-0150
Corporate office	. . .	Contel	404-804-3400
. . . . . .		GTE Mobilnet	404-391-8000

Sending calls

Visitor dialing instructions
Local calls . . . . . area code + 7 digits
Long distance . . . 1 + area code + 7 digits

Speed-dial numbers
Customer service . . . . ✶611
Driver Guide® . . . . ✶627
Emergency 911
Information services . . . ✶INFO or ✶4636
Technical assistance . . . . ✶111

Receiving calls

Caller dials the roamer access number below,
hears tone, dials cellular area code and number.
. . . . 707-483-7626

Auto-Access forwarding (see page 1156)
Activate, dial ✶28 send. Deactivate dial ✶29 send.

Follow Me Roaming forwarding (see p. 1156)
Activate, dial ✶18 send. Deactivate dial ✶19 send.

Billing information

Visitor rates
Most visitors . . . . . $2 day + 50¢ minute

Visitors without roaming agreements (p. 1159)
Call co. to use Amex, Discover, Mastercard, Visa

Identification numbers

City	Market	System	Billing	RSA
All served cities	344	00040	00000	CA-9

For full coverage area, see San Francisco, CA.

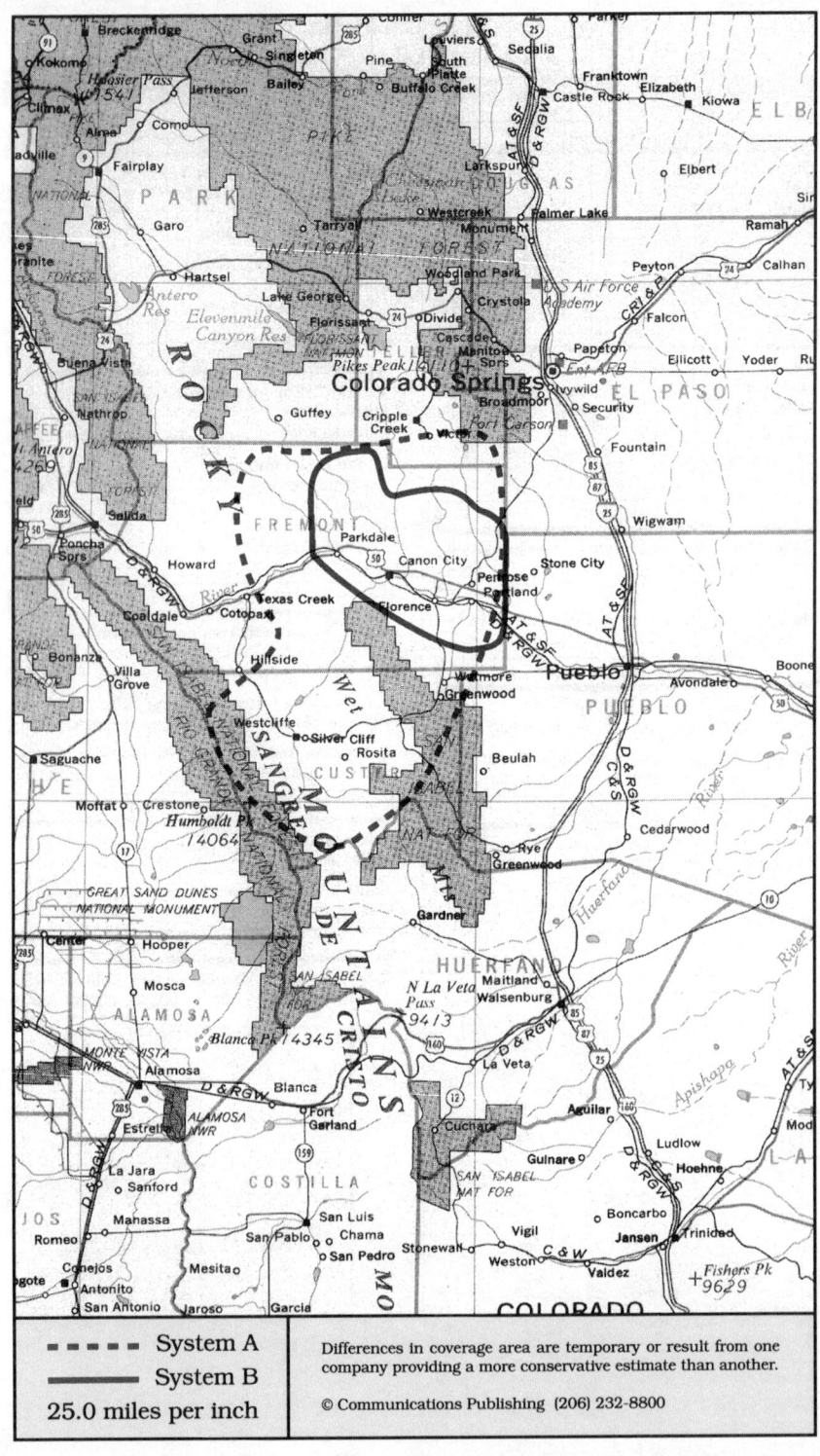

▬ ▬ ▬ ▬ ▬ **System A**	Differences in coverage area are temporary or result from one company providing a more conservative estimate than another.
▬▬▬▬▬▬▬ **System B**	
25.0 miles per inch	© Communications Publishing (206) 232-8800

Canon City, Colorado
Canon City, West Cliff

System A	System B

Cellular company

Cellular One
Celludyne II
730 Royal Gorge Boulevard
Canon City, CO 81212

Corporate office 719-269-1777

Sending calls

Visitor dialing instructions
Local calls . . . . . area code + 7 digits
Long distance . . . 1 + area code + 7 digits

Speed-dial numbers
Customer service 611
Emergency 911

Receiving calls

Caller dials the roamer access number below,
hears tone, dials cellular area code and number.
. . . . 719-269-2626

NationLink & RoamingAmerica (see p. 1157)
Caller hears your current roamer access number.
This company's subscribers have level 0 service.

North American Cellular Network (see p. 1156)
All calls forward to you. Deactivate dial ∗35 send.

Billing information

Visitor rates
Most visitors . . . . . $0 day + 99¢ minute

Visitors without roaming agreements (p. 1159)
Call co. to use Mastercard, Visa

Identification numbers

City	Market	System	Billing	RSA
All served cities	351	01089	00000	CO-4

Call switching in CO-4 is performed by McCaw Cellular for the licensee, Celludyne II.

This company reports that it offers 24-hour customer service 7 days a week and serves Custer and Freemont Counties. You may continue a call when traveling into Pueblo without being disconnected.

Cellular company

CommNet 2000
Cellular, Inc.
1702 Highway 50 W
Pueblo, CO 81008

Customer service 800-597-5528
Local office 800-729-7557
. . . . 719-429-2000
Corporate office 303-694-3234

Sending calls

Visitor dialing instructions
Local calls . . . . . . area code + 7 digits
Long distance . . . 1 + area code + 7 digits

Speed-dial numbers
Customer service 611
Emergency 911
Roaming information 711

Receiving calls

Caller dials the roamer access number below,
hears tone, dials cellular area code and number.
. . . . 719-429-7626

Billing information

Visitor rates
Most visitors . . . . . $3 day + 95¢ minute

Visitors without roaming agreements (p. 1159)
Cannot call since company takes no credit cards.

Identification numbers

City	Market	System	Billing	RSA
All served cities	351	00490	30056	CO-4

For full coverage area, see Pueblo, CO.

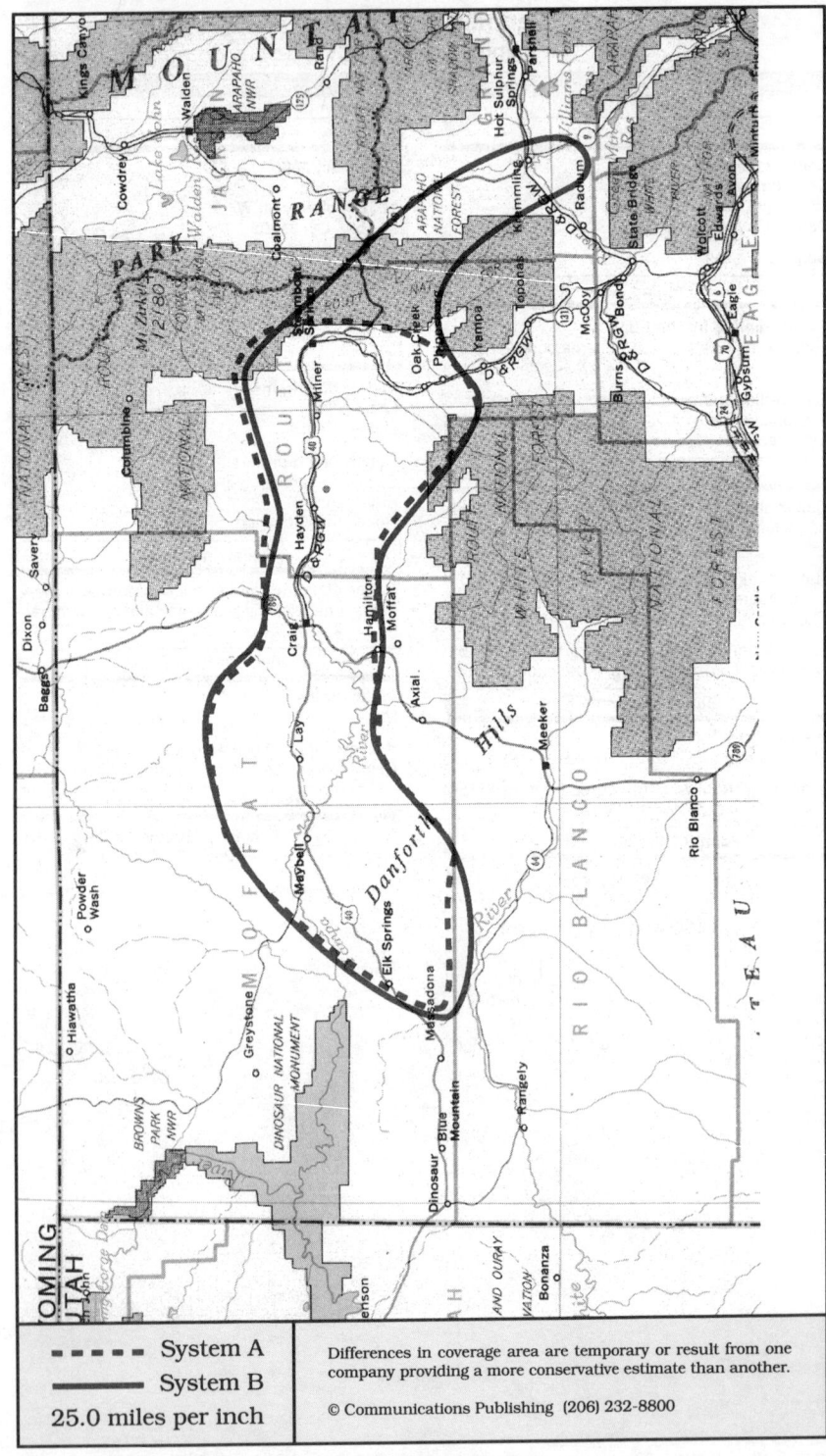

- - - - System A ———— System B 25.0 miles per inch	Differences in coverage area are temporary or result from one company providing a more conservative estimate than another. © Communications Publishing (206) 232-8800

Craig, Colorado
Craig, Steamboat Springs

Cellular company

CommNet 2000
Cellular, Inc.
1702 Highway 50 W
Pueblo, CO 81008

Customer service		800-597-5528
Local office	Pueblo	800-729-7557
. . . .	Pueblo	719-543-2000
. . . .	Craig sales agent	303-846-2000
Corporate office		303-694-3234

Sending calls

Visitor dialing instructions
Local calls . . . . . area code + 7 digits
Long distance . . . 1 + area code + 7 digits

Speed-dial numbers
Customer service 611
Emergency 911
Roaming information 711

Receiving calls

Caller dials the roamer access number below,
hears tone, dials cellular area code and number.
. . . . 303-846-7626

Billing information

Visitor rates
Most visitors . . . . . $3 day + 95¢ minute

Visitors without roaming agreements (p. 1159)
Cannot call since company takes no credit cards.

Identification numbers

City	Market	System	Billing	RSA
All served cities	348	01083	00000	CO-1

Cellular company

Union Cellular
Union Telephone
850 N State Highway 414, PO Box 160
Mountain View, WY 82939

Corporate office .	some areas	800-646-2355
. . . .		307-782-6131

Sending calls

Visitor dialing instructions
Local calls . . . . . . area code + 7 digits
Long distance . . . 1 + area code + 7 digits

Speed-dial numbers
Customer service 611
Emergency 911

Receiving calls

Caller dials the roamer access number below,
hears tone, dials cellular area code and number.
. . . . 303-734-7626

Follow Me Roaming forwarding (see p. 1156)
Activate, dial *18 send. Deactivate dial *19 send.

Billing information

Visitor rates
Most visitors . . . . . $0 day + 70¢ minute

Visitors without roaming agreements (p. 1159)
Cannot call since company takes no credit cards.

Identification numbers

City	Market	System	Billing	RSA
All served cities	348	01084	00000	CO-1

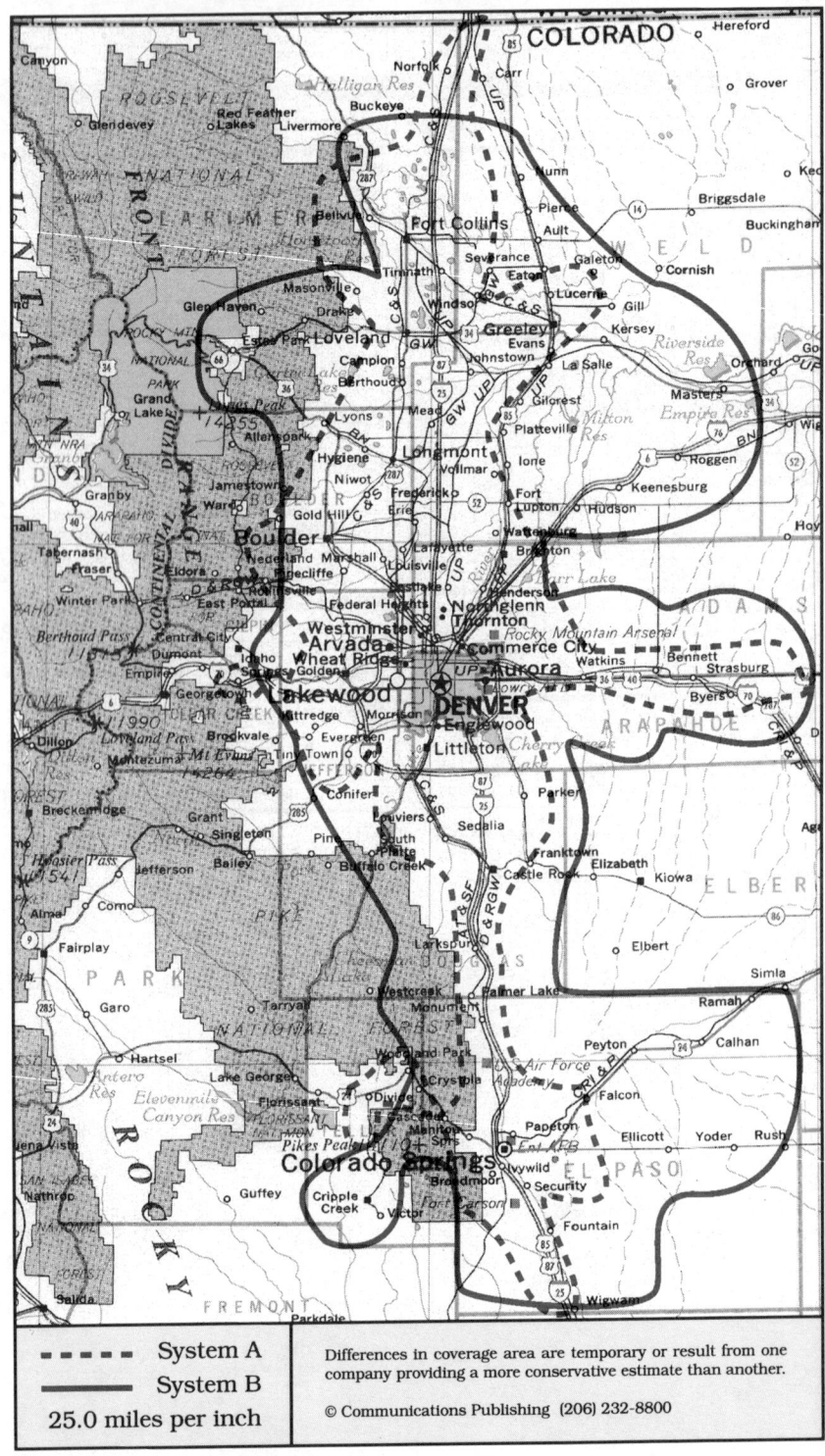

- - - - - System A
——— System B
25.0 miles per inch

Denver, Colorado
Boulder, Colorado Springs, Denver, Fort Collins, Greeley

System A	System B

Cellular company

Cellular One
McCaw Cellular Communications
1352 N Academy Boulevard
Colorado Springs, CO 80909

1001 16th Street, Suite A-125
Denver, CO 80265

Customer service	800-326-1366
. . . .	303-573-3100
. . . .	719-338-1200
Local office Colorado Springs	719-550-0100
. . . . Denver	303-537-3200
Corporate office	206-827-4500

Sending calls

Visitor dialing instructions

Local calls	area code + 7 digits
Long distance . . .	1 + area code + 7 digits

Speed-dial numbers

Customer service	611
Emergency	911

Receiving calls

Caller dials the roamer access number below,
hears tone, dials cellular area code and number.

Colorado Springs	719-338-7626
Denver (rings in all four cities)	303-888-7626
Fort Collins	303-227-7626
Greeley	303-396-7626

NationLink & RoamingAmerica (see p. 1157)
Dial *31 send to receive calls, *30 to deactivate.
Dial *32 send to tell caller the roamer access no.
This company's subscribers have level 6 service.

North American Cellular Network (see p. 1156)
All calls forward to you. Deactivate dial *35 send.

Billing information

Visitor rates

Most visitors	$0 day + 99¢ minute

Visitors without roaming agreements (p. 1159)
Call co. to use Mastercard, Visa

Identification numbers

City	Market	System	Billing	RSA
Colorado Springs	117	00045	00000	MSA
Denver, Longmont	019	00045	00000	MSA
Ft Collins, Loveland	210	00045	00000	MSA
Greeley	243	00045	00000	MSA

For full coverage area, see Pueblo, CO.

Cellular company

U S West Cellular	Local office
775 N Academy Boulevard	
Colorado Springs, CO 80909	719-637-5500
2000 S Colorado Blvd, Fl. 3, NCR	
Denver, CO 80222	303-782-1800
2422 S Trenton Way, Suite A	
Denver, CO 80231	303-368-8778
4637 S Mason Street, Suite A2	
Fort Collins, CO 80525	303-223-1113
800 8th Avenue, Suite 317	
Greeley, CO 80631	303-356-1466
8725 Sheridan Boulevard	
Westminster, CO 80003	303-426-0414
Customer service	800-238-7848
. . . .	206-562-2895
. . . . local subscribers	800-626-6611
. . . .	206-747-1771
Corporate office	206-747-4900

Sending calls

Visitor dialing instructions

Local calls	area code + 7 digits
Long distance	area code + 7 digits

Speed-dial numbers

Channel 4 television in Denver	#4
Customer service	*611
Eagle 96 radio (Fort Collins)	#96
Emergency	911
Make a traffic report (Colorado Springs)	#8811
Report a drunk driver (Denver) . . .	*384
Report traffic problem (Denver)	#TIP or #847
Roaming information	*711
Traffic report (Denver)	#JAM or #526
Tri 102 radio (Fort Collins)	#102

Receiving calls

Caller dials the roamer access number below,
hears tone, dials cellular area code and number.

Colorado Springs	719-661-7626
Denver	303-877-7626
Fort Collins	303-222-7626
Greeley	303-381-7626
Longmont	303-775-7626
Loveland	303-679-7626

Follow Me Roaming forwarding (see p. 1156)
Activate, dial *18 send. Deactivate dial *19 send.

Billing information

Visitor rates

Most visitors	$3 day + $1.25 minute

Visitors without roaming agreements (p. 1159)
Call co. to use Amex, Mastercard, Visa

Identification numbers

City	Market	System	Billing	RSA
Colorado Springs	117	00180	00000	MSA
Denver, Longmont	019	00058	00000	MSA
Ft Collins, Loveland	210	00058	00000	MSA
Greeley	243	00058	00000	MSA

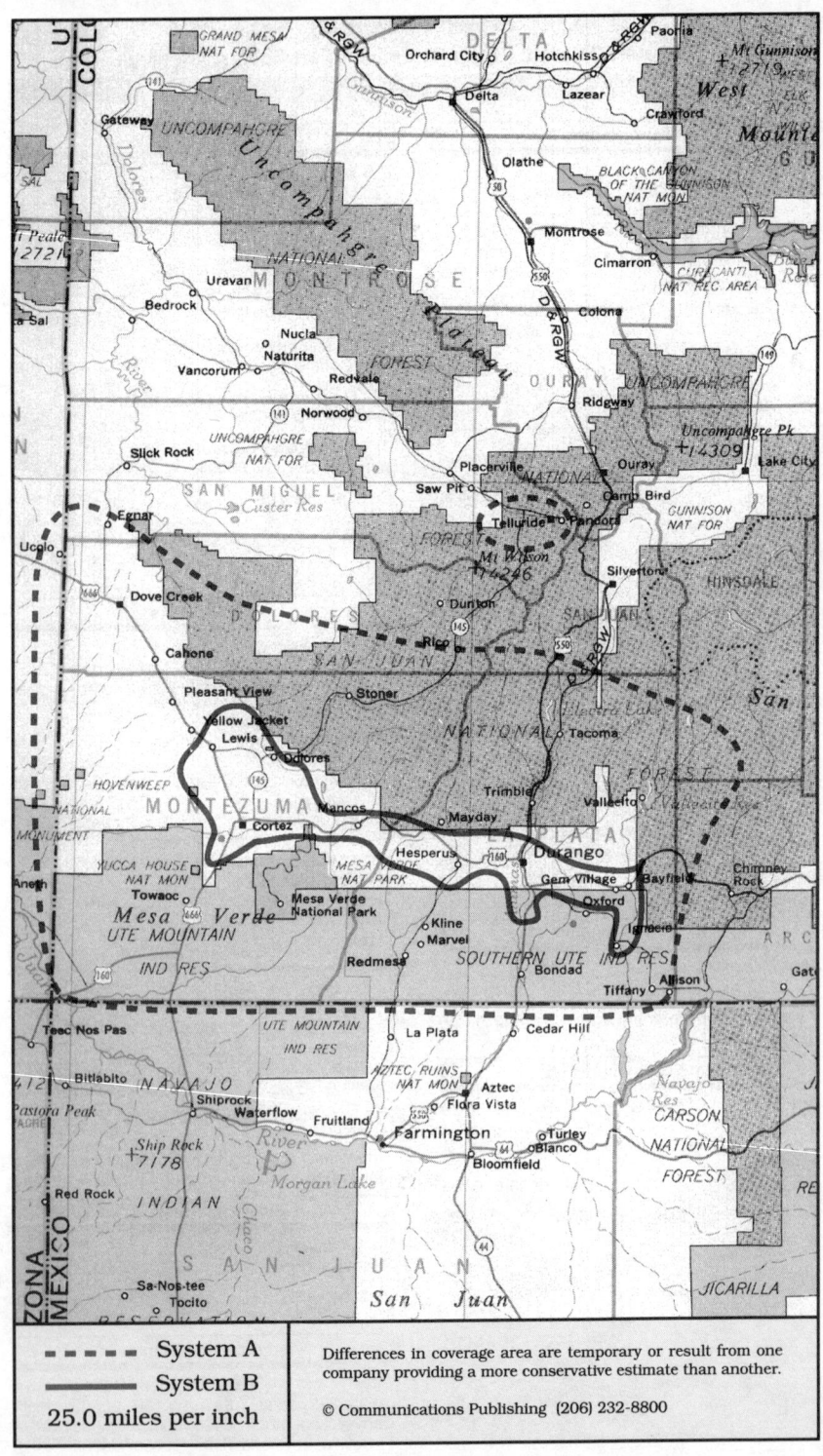

▬ ▬ ▬ System A	Differences in coverage area are temporary or result from one company providing a more conservative estimate than another.
▬▬▬ System B	
25.0 miles per inch	© Communications Publishing (206) 232-8800

Durango, Colorado
Cortez, Durango, Telluride

Cellular company

Liberty Cellular
Unitel
178 Bodo Drive, Suite D
Durango, CO 81301

Local office	303-247-3111
Corporate office	609-646-9400

Sending calls

Visitor dialing instructions

Local calls	1 + area code + 7 digits
Long distance . . .	1 + area code + 7 digits

Speed-dial numbers

Customer service	611
Emergency	911

Receiving calls

Caller dials the roamer access number below,
hears tone, dials cellular area code and number.
. . . . 303-749-7626

Billing information

Visitor rates
Most visitors . . . . . $3 day + 99¢ minute

Visitors without roaming agreements (p. 1159)
Call co. to use Mastercard, Visa

Identification numbers

City	Market	System	Billing	RSA
All served cities	353	01093	00000	CO-6

Cellular company

CommNet 2000
Cellular, Inc.
1702 Highway 50 W
Pueblo, CO 81008

Customer service	800-597-5531
Local office	800-729-7557
. . . .	719-543-2000
Corporate office	303-694-3234

Sending calls

Visitor dialing instructions

Local calls	area code + 7 digits
Long distance . . .	1 + area code + 7 digits

Speed-dial numbers

Customer service	611
Emergency	911
Roaming information	711

Receiving calls

Caller dials the roamer access number below,
hears tone, dials cellular area code and number.

Cortez	303-560-7626
Durango (planned number) . .	303-946-7626

Billing information

Visitor rates
Most visitors . . . . . $3 day + 95¢ minute

Visitors without roaming agreements (p. 1159)
Cannot call since company takes no credit cards.

Identification numbers

City	Market	System	Billing	RSA
All served cities	353	01094	00000	CO-6

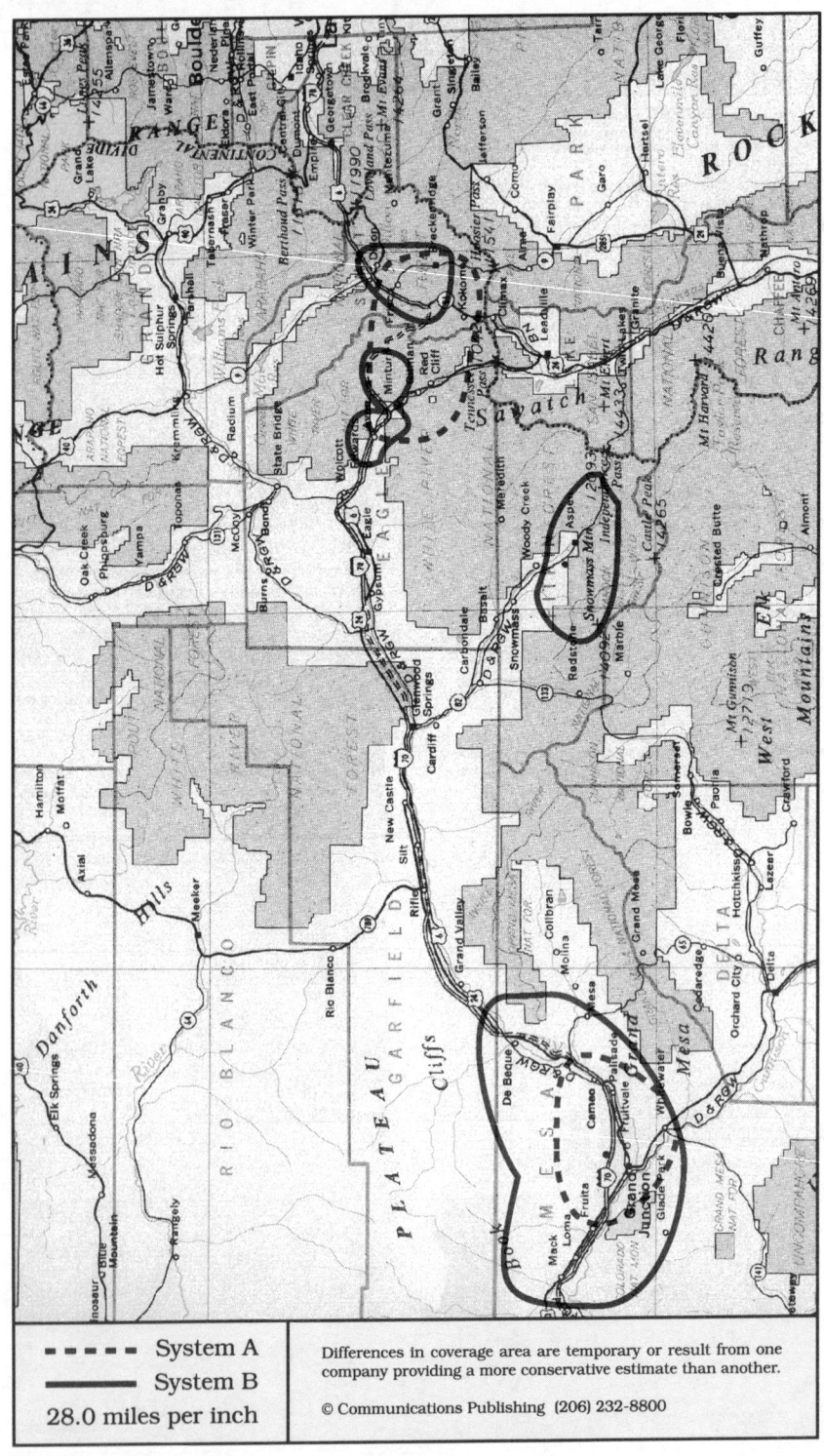

▬ ▬ ▬ ▬ System A	
▬▬▬▬▬ System B	Differences in coverage area are temporary or result from one company providing a more conservative estimate than another.
28.0 miles per inch	© Communications Publishing (206) 232-8800

Grand Junction, Colorado

Aspen, Avon, Dillon, Frisco, Grand Junction, Idaho Springs, Vail

System A	System B

Cellular company

Cellular One
McCaw Cellular Communications
PO Box 1056
Dillon, CO 80435

1048 Independent Avenue, Suite A-210
Grand Junction, CO 81505-6121

Customer service	800-326-1366
. . . . from Frisco	303-389-1200
. . . . . from Grand Junction	303-260-1200
. . . . From Vail	303-390-1200
Local office . . Dillon/Frisco	303-668-3111
. . . . . . Grand Junction	303-260-2000
Corporate office	206-827-4500

Sending calls

Visitor dialing instructions

Local calls	area code + 7 digits
Long distance	1 + area code + 7 digits

Speed-dial numbers

Customer service	611
Emergency	911

Receiving calls

Caller dials the roamer access number below,
hears tone, dials cellular area code and number.

Aspen (planned number) . .	303-948-7626
Dillon, Grand Junction, Idaho S.	303-260-7626
Frisco	303-389-7626
Vail	303-390-7626

NationLink & RoamingAmerica (see p. 1157)
Dial *31 send to receive calls, *30 to deactivate.
Dial *32 send to tell caller the roamer access no.
This company's subscribers have level 6 service.

North American Cellular Network (see p. 1156)
All calls forward to you. Deactivate dial *35 send.

Billing information

Visitor rates

Most visitors	$0 day + 99¢ minute

Visitors without roaming agreements (p. 1159)
Call co. to use Mastercard, Visa

Identification numbers

City	Market	System	Billing	RSA
All served cities	350	01087	00000	CO-3

Cellular company

U S West Cellular
225 N 5th Street, Suite 115
Grand Junction, CO 81501

953 S Frontage Road W, Suite 105
Vail, CO 81657

Customer service	800-238-7848
. . . .	206-562-2895
. . . . . local subscribers	800-626-6611
. . . .	206-747-1771
Local office . . Grand Junction	303-241-2233
. . . . Vail	303-479-9567
Corporate office	206-747-4900

Sending calls

Visitor dialing instructions

Local calls	area code + 7 digits
Long distance	area code + 7 digits

Speed-dial numbers

Customer service	*611
Emergency	911
Roaming information	*711

Receiving calls

Caller dials the roamer access number below,
hears tone, dials cellular area code and number.

Aspen	303-379-7626
Dillon	303-485-7626
Grand Junction	303-250-7626
Vail	303-471-7626

Billing information

Visitor rates

Most visitors	$3 day + 99¢ minute

Visitors without roaming agreements (p. 1159)
Call co. to use Amex, Mastercard, Visa

Identification numbers

City	Market	System	Billing	RSA
All served cities	350	00058	00000	CO-3

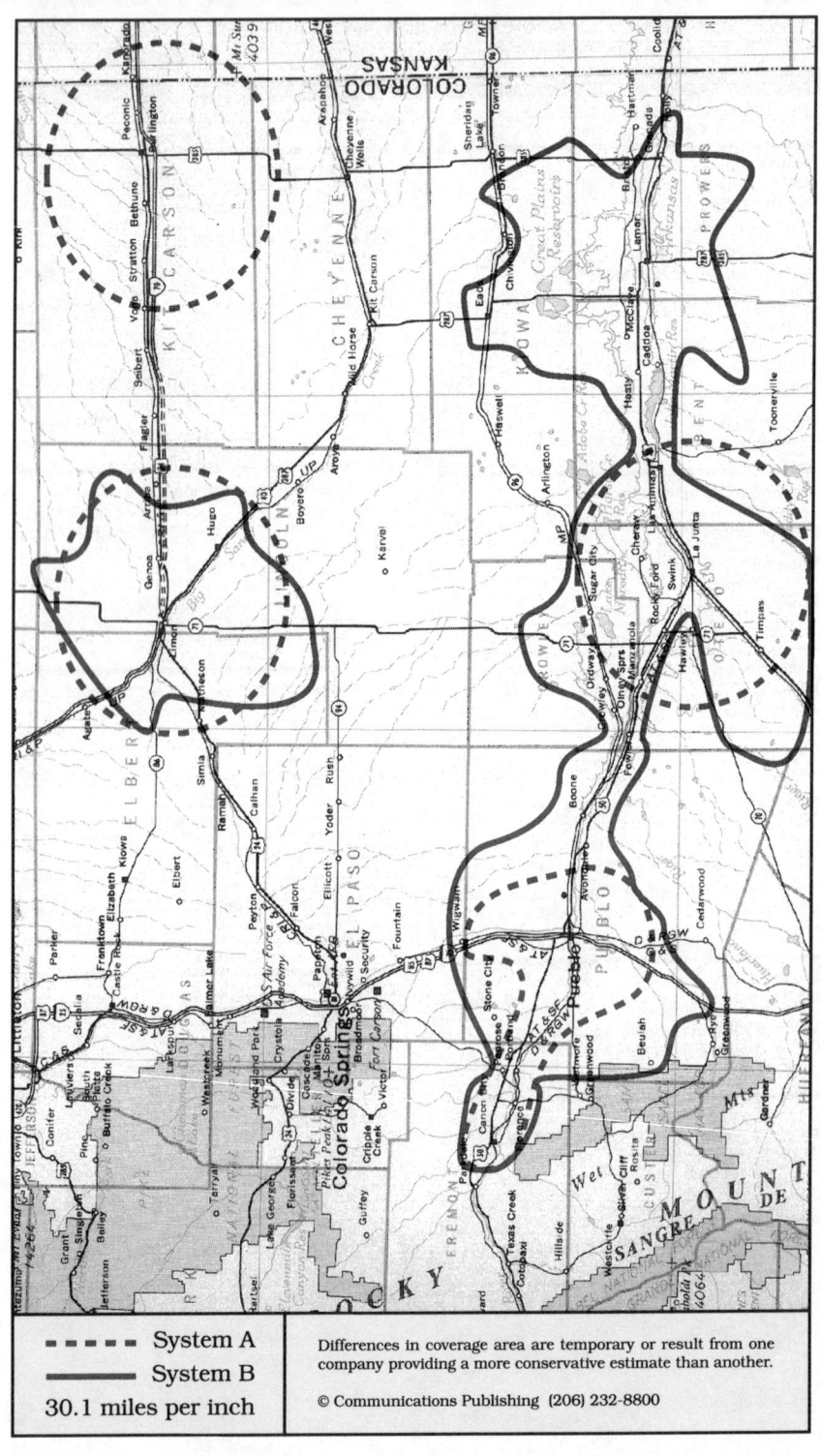

System A
System B
30.1 miles per inch

Differences in coverage area are temporary or result from one company providing a more conservative estimate than another.

© Communications Publishing (206) 232-8800

Pueblo, Colorado
Burlington, La Junta, Lamar, Limon, Pueblo

System A	System B

Cellular company

Burlington and Limon (CO-5)
Cellular One
PO Box 51404
Indianapolis, IN 46251

Corporate office	317-637-5084

La Junta (CO-8)
Cellular One
Pacific Northwest Cellular
11400 SE 8th Street, Suite 445
Bellevue, WA 98004-6431

Customer service	800-635-0304
Corporate office	206-635-0300

Pueblo
Cellular One
McCaw Cellular Comms.
301 N Main, Suite 303
Pueblo, CO 81003

Customer service	800-326-1366
. . . .	719-568-1200
Local office	719-543-7032
Corporate office	206-827-4500

Sending calls

Visitor dialing instructions
Local calls . . . . . area code + 7 digits
Long distance . . . 1 + area code + 7 digits

Speed-dial numbers
Customer service 611
Emergency 911

Receiving calls

Caller dials the roamer access number below,
hears tone, dials cellular area code and number.

Burlington	719-346-4000
La Junta	303-888-7626
Limon	719-775-3000
Pueblo	719-568-7626

NationLink & RoamingAmerica (see p. 1157)
Dial *31 send to receive calls, *30 to deactivate.
Dial *32 send to tell caller the roamer access no.
This company's subscribers have level 6 service.

North American Cellular Network (see p. 1156)
All calls forward to you. Deactivate dial *35 send.

Billing information

Visitor rates
Most visitors . . . . . $0 day + 99¢ minute

Visitors without roaming agreements (p. 1159)
Call co. to use Mastercard, Visa

Identification numbers

City	Market	System	Billing	RSA
Burlington, Limon	352	01091	00000	CO-5
La Junta	355	01099	00000	CO-8
Pueblo	241	00045	00000	MSA

Calls may continue from Pueblo to Canon City.

For full coverage area, see Denver, CO.

Cellular company

CommNet 2000
Cellular, Inc.
1702 Highway 50 W
Pueblo, CO 81008

Customer service	800-597-5528
Local office	800-729-7557
. . . .	719-543-2000
Corporate office	303-694-3234

Sending calls

Visitor dialing instructions
Local calls . . . . . area code + 7 digits
Long distance . . . 1 + area code + 7 digits

Speed-dial numbers
Customer service 611
Emergency 911
Roaming information 711

Receiving calls

Caller dials the roamer access number below,
hears tone, dials cellular area code and number.

Alamosa	719-588-7626
Fort Morgan	303-380-7626
La Junta	719-469-7626
Lamar	719-688-7626
Limon	719-740-7626
Pueblo	719-250-7626

Billing information

Visitor rates
Most visitors . . . . . $3 day + 95¢ minute

Visitors without roaming agreements (p. 1159)
Cannot call since company takes no credit cards.

Identification numbers

City	Market	System	Billing	RSA
La Junta	355	00490	30388	CO-8
Lamar	355	00490	30060	CO-8
Limon	352	00490	30138	CO-5
Pueblo	241	00490	00000	MSA

For full coverage area, see Canon City, CO.

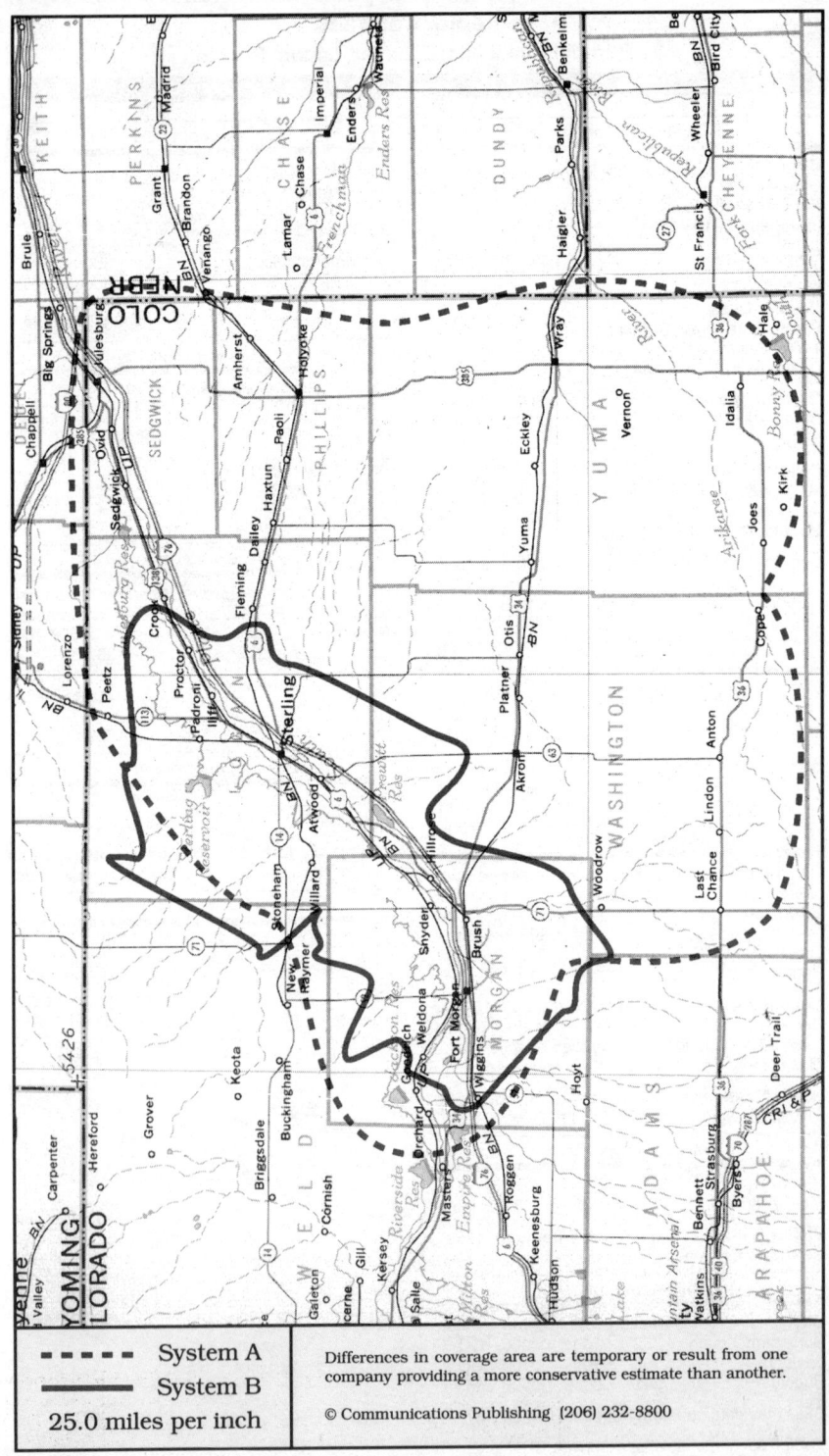

▪ ▪ ▪ ▪ System A	Differences in coverage area are temporary or result from one
——— System B	company providing a more conservative estimate than another.
25.0 miles per inch	© Communications Publishing (206) 232-8800

Sterling, Colorado
Fort Morgan, Julesburg, Sterling, Wray

System A	System B

Cellular company

Cellular One of Northeast Colorado
1220 W Platte Avenue
Fort Morgan, CO 80701-2949

Corporate office 303-867-6767

Sending calls

Visitor dialing instructions
Local calls . . . . . area code + 7 digits
Long distance . . . 1 + area code + 7 digits

Speed-dial numbers
Customer service 611
Emergency 911

Receiving calls

Caller dials the roamer access number below,
hears tone, dials cellular area code and number.
Fort Morgan 303-768-7626
Sterling 303-520-7626
Yuma 303-630-7626

Billing information

Visitor rates
Most visitors . . . . $2.50 day + 75¢ minute

Visitors without roaming agreements (p. 1159)
Call co. to use Mastercard, Visa
or Roamer Plus will intercept first call to bill you.

Identification numbers

City	Market	System	Billing	RSA
All served cities	349	01085	00000	CO-2

Cellular company

CommNet 2000
Cellular, Inc.
1702 Highway 50 W
Pueblo, CO 81008

Customer service 800-597-5528
Local office 800-729-7557
. . . . 719-543-2000
Corporate office 303-694-3234

Sending calls

Visitor dialing instructions
Local calls . . . . . . area code + 7 digits
Long distance . . . 1 + area code + 7 digits

Speed-dial numbers
Customer service 611
Emergency 911
Roaming information 711

Receiving calls

Caller dials the roamer access number below,
hears tone, dials cellular area code and number.
Fort Morgan 303-380-7626
Sterling 303-580-7626

Billing information

Visitor rates
Most visitors . . . . . $3 day + 95¢ minute

Visitors without roaming agreements (p. 1159)
Cannot call since company takes no credit cards.

Identification numbers

City	Market	System	Billing	RSA
Ft Morgan	349	00490	30256	CO-2
Sterling	349	00490	30384	CO-2

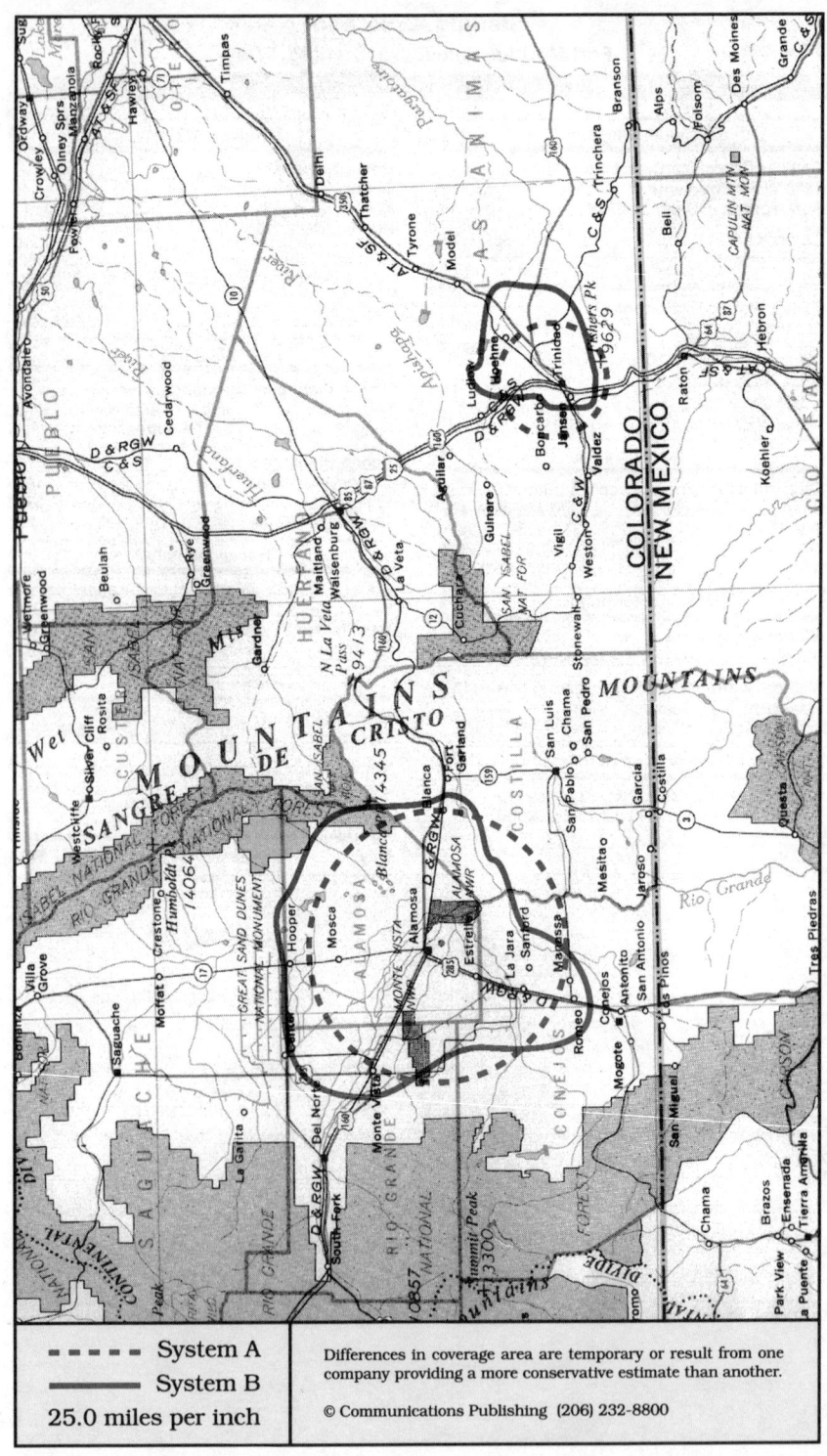

▪ ▪ ▪ ▪ ▪ System A	
▬▬▬▬ System B	Differences in coverage area are temporary or result from one company providing a more conservative estimate than another.
25.0 miles per inch	© Communications Publishing (206) 232-8800

Trinidad, Colorado
Alamosa, Trinidad

System A	System B

Cellular company

Cellular One
Pacific Northwest Cellular
11400 SE 8th Street, Suite 445
Bellevue, WA 98004-6431

Customer service	800-635-0304
Corporate office	206-635-0300

Sending calls

Visitor dialing instructions

Local calls	area code + 7 digits
Long distance . . .	1 + area code + 7 digits

Speed-dial numbers

Customer service	611
Emergency	911

Receiving calls

Caller dials the roamer access number below,
hears tone, dials cellular area code and number.

Alamosa	719-568-7626
Trinidad	303-888-7626

NationLink & RoamingAmerica (see p. 1157)
Dial *31 send to receive calls, *30 to deactivate.
Dial *32 send to tell caller the roamer access no.
This company's subscribers have level 6 service.

North American Cellular Network (see p. 1156)
All calls forward to you. Deactivate dial *35 send.

Billing information

Visitor rates

Most visitors	$0 day + 99¢ minute

Visitors without roaming agreements (p. 1159)
Call co. to use Mastercard, Visa

Identification numbers

City	Market	System	Billing	RSA
Alamosa	354	00045	30347	CO-7
Trinidad	356	00045	30291	CO-9

Cellular company

CommNet 2000
Cellular, Inc.
1702 Highway 50 W
Pueblo, CO 81008

Customer service	800-597-5528
Local office	800-729-7557
. . . .	719-543-2000
Corporate office	303-694-3234

Sending calls

Visitor dialing instructions

Local calls	area code + 7 digits
Long distance . . .	1 + area code + 7 digits

Speed-dial numbers

Customer service	611
Emergency	911
Roaming information	711

Receiving calls

Caller dials the roamer access number below,
hears tone, dials cellular area code and number.

Alamosa	719-588-7626
Trinidad	719-680-7626

Billing information

Visitor rates

Most visitors	$3 day + 95¢ minute

Visitors without roaming agreements (p. 1159)
Cannot call since company takes no credit cards.

Identification numbers

City	Market	System	Billing	RSA
Alamosa	354	00490	30058	CO-7
Trinidad	356	00490	30140	CO-9

- - - - System A	
——— System B	
25.0 miles per inch	

Differences in coverage area are temporary or result from one company providing a more conservative estimate than another.

© Communications Publishing (206) 232-8800

Bridgeport, Connecticut
Bridgeport, Springfield MA, Stamford, Willimantic

System A	System B

Cellular company

Bell Atlantic Mobile
50 Rockland Road
Norwalk, CT 06854

20 Alexander Drive, PO Box 5029
Wallingford, CT 06492

482 Pigeonhill Road
Windsor, CT 06095

1123-B Riverdale Road
West Springfield, CT 01089

Customer service		800-852-3630
Local office	Norwalk	203-852-9292
. . . .	Wallingford	203-269-8858
. . . .	Windsor	203-688-3010
. . . . .	West Springfield	413-781-6000
Corporate office		908-306-7000

Sending calls

Visitor dialing instructions

Local calls	area code + 7 digits
Long distance	area code + 7 digits

Speed-dial numbers

Customer service	*611
Emergency	911
Massachusetts state police . . .	*677
Roaming information	*711
Windsor office	*811

Receiving calls

Caller dials the roamer access number below,
hears tone, dials cellular area code and number.

Northern Connecticut	203-930-7626
Southern Connecticut . . .	203-856-7626
Springfield & Greenfield, MA .	413-531-7626

NationLink & RoamingAmerica (see p. 1157)
Caller hears your current roamer access number.
This company's subscribers have level 0 service.

Billing information

Visitor rates

Most visitors	$3 day + 99¢ minute

Visitors without roaming agreements (p. 1159)
Roamer Plus will intercept first call to bill you.

Identification numbers

City	Market	System	Billing	RSA
Bridgeport	042	00119	00000	MSA
Fairfield County	042	00119	30019	MSA
Greenfield, MA	470	01327	00000	MA-1
Hartford	032	00119	31119	MSA
New Haven County	049	00119	36119	MSA
New London	154	00119	31049	MSA
Springfield, MA	063	00119	33119	MSA
Willimantic	358	01103	00000	CT-2

For full coverage area, see Pittsfield, MA and
Providence, RI.

Cellular company

SNET Cellular
Springwich Cellular Limited Partnership
555 Long Wharf Drive, Floor 8
New Haven, CT 06511

Customer service	203-553-7633
Corporate office	203-553-7600

Sending calls

Visitor dialing instructions

Local calls	area code + 7 digits
Long distance	area code + 7 digits

Speed-dial numbers

Customer service	*711
Emergency	911
Lottery results	*946
Massachusetts state police .	*MSP or *677
Off-track betting	*682
Poison control	*XXX or *999
Roamer access	*7626
SNET Cellular information .	*INFO or *4636
State information office .	*CONN or *2666
Time of day	*TIME or *8463
Traffic and weather	*123
U.S. Coast Guard . . .	*USCG or *8724

Receiving calls

Caller dials the roamer access number below,
hears tone, dials cellular area code and number.

Connecticut	203-631-0000
Greenfield, MA	413-773-0500
Springfield, MA	413-539-7626

Follow Me Roaming forwarding (see p. 1156)
Activate, dial *18 send. Deactivate dial *19 send.

Billing information

Visitor rates

Most visitors	$3 day + $1 minute

Visitors without roaming agreements (p. 1159)
Call co. to use Mastercard, Visa
or Roamer Plus will intercept first call to bill you.

Identification numbers

City	Market	System	Billing	RSA
Bridgeport	042	00088	00000	MSA
Greenfield, MA	470	01328	00000	MA-1
New Haven	049	00088	00000	MSA
New London	154	00088	00000	MSA
Springfield, MA	063	00088	30006	MSA
Willimantic	358	00088	00000	CT-2

For full coverage area, see Torrington, CT.

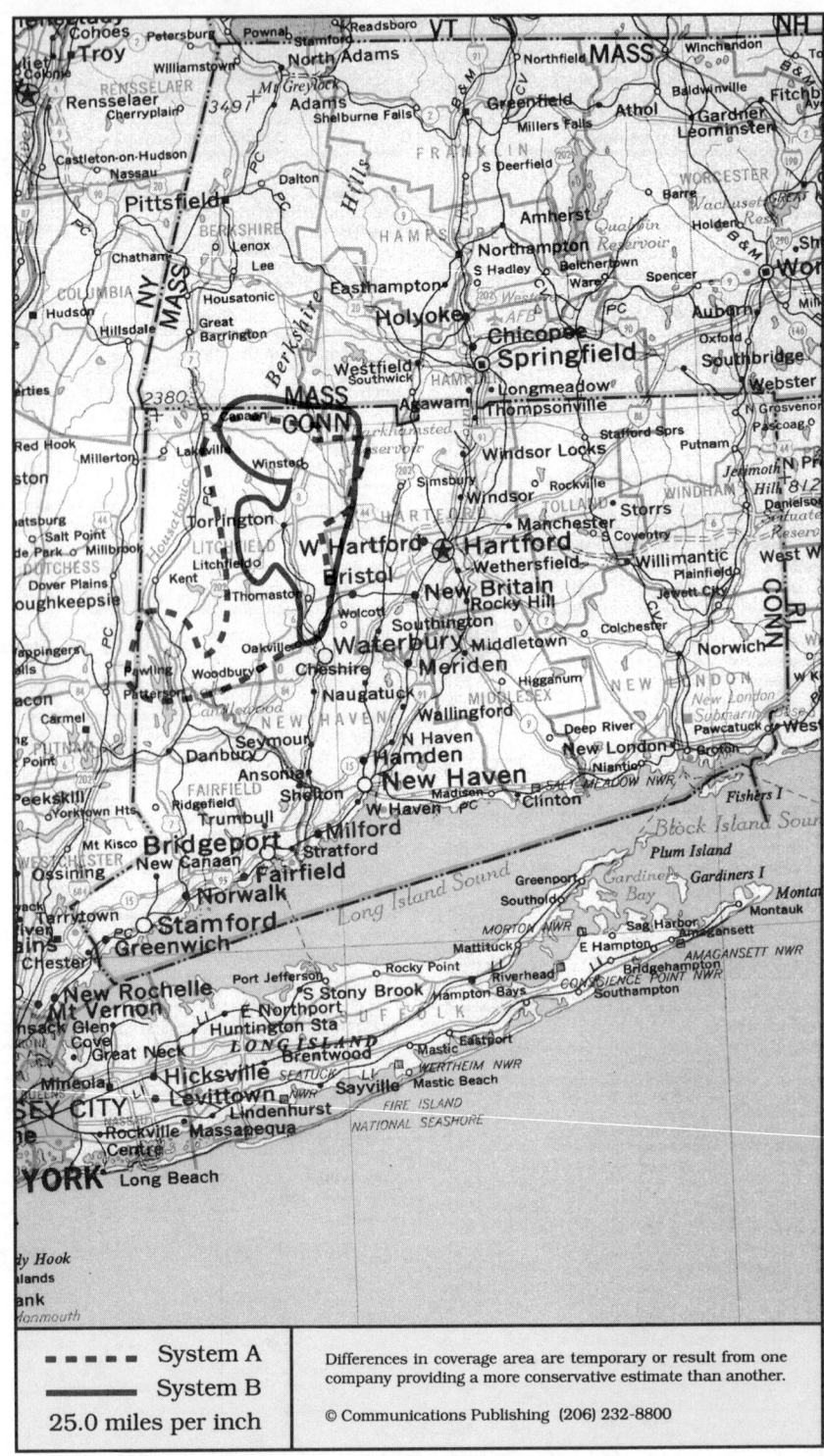

- - - - - System A	
———— System B	Differences in coverage area are temporary or result from one company providing a more conservative estimate than another.
25.0 miles per inch	© Communications Publishing (206) 232-8800

Torrington, Connecticut
Litchfield, Torrington

System A	System B

Cellular company

Cellular One
Litchfield County Cellular
777 E Main Street
Torrington, CT 06790

Corporate office 203-489-9999

Sending calls

Visitor dialing instructions
Local calls . . . . area code + 7 digits
Long distance . . . . area code + 7 digits

Speed-dial numbers
Customer service 611
Emergency 911
Roaming information 711

Receiving calls

Caller dials the roamer access number below,
hears tone, dials cellular area code and number.
. . . . 203-480-7626

Billing information

Visitor rates
Most visitors . . . . . $3 day + 99¢ minute

Visitors without roaming agreements (p. 1159)
Call co. to use Amex, Discover, Mastercard, Visa
or Roamer Plus will intercept first call to bill you.

Identification numbers

City	Market	System	Billing	RSA
All served cities	357	01101	00000	CT-1

Cellular company

SNET Cellular
Springwich Cellular Limited Partnership
555 Long Wharf Drive, Floor 8
New Haven, CT 06511

Customer service 203-553-7633
Corporate office 203-553-7600

Sending calls

Visitor dialing instructions
Local calls . . . . . . area code + 7 digits
Long distance . . . . . area code + 7 digits

Speed-dial numbers
Customer service . . . . ∗711
Emergency 911
Lottery results . . . . ∗946
Massachusetts state police . ∗MSP or ∗677
Off-track betting . . . . ∗682
Poison control . . . . . ∗XXX or ∗999
Roamer access . . . . ∗7626
SNET Cellular information . ∗INFO or ∗4636
State information office . ∗CONN or ∗2666
Time of day . . . . . . ∗TIME or ∗8463
Traffic and weather . . . . ∗123
U.S. Coast Guard . . . ∗USCG or ∗8724

Receiving calls

Caller dials the roamer access number below,
hears tone, dials cellular area code and number.
. . . . 203-631-0000

Follow Me Roaming forwarding (see p. 1156)
Activate, dial ∗18 send. Deactivate dial ∗19 send.

Billing information

Visitor rates
Most visitors . . . . . $3 day + $1 minute

Visitors without roaming agreements (p. 1159)
Call co. to use Mastercard, Visa
or Roamer Plus will intercept first call to bill you.

Identification numbers

City	Market	System	Billing	RSA
All served cities	357	00088	00000	CT-1

For full coverage area, see Bridgeport, CT.

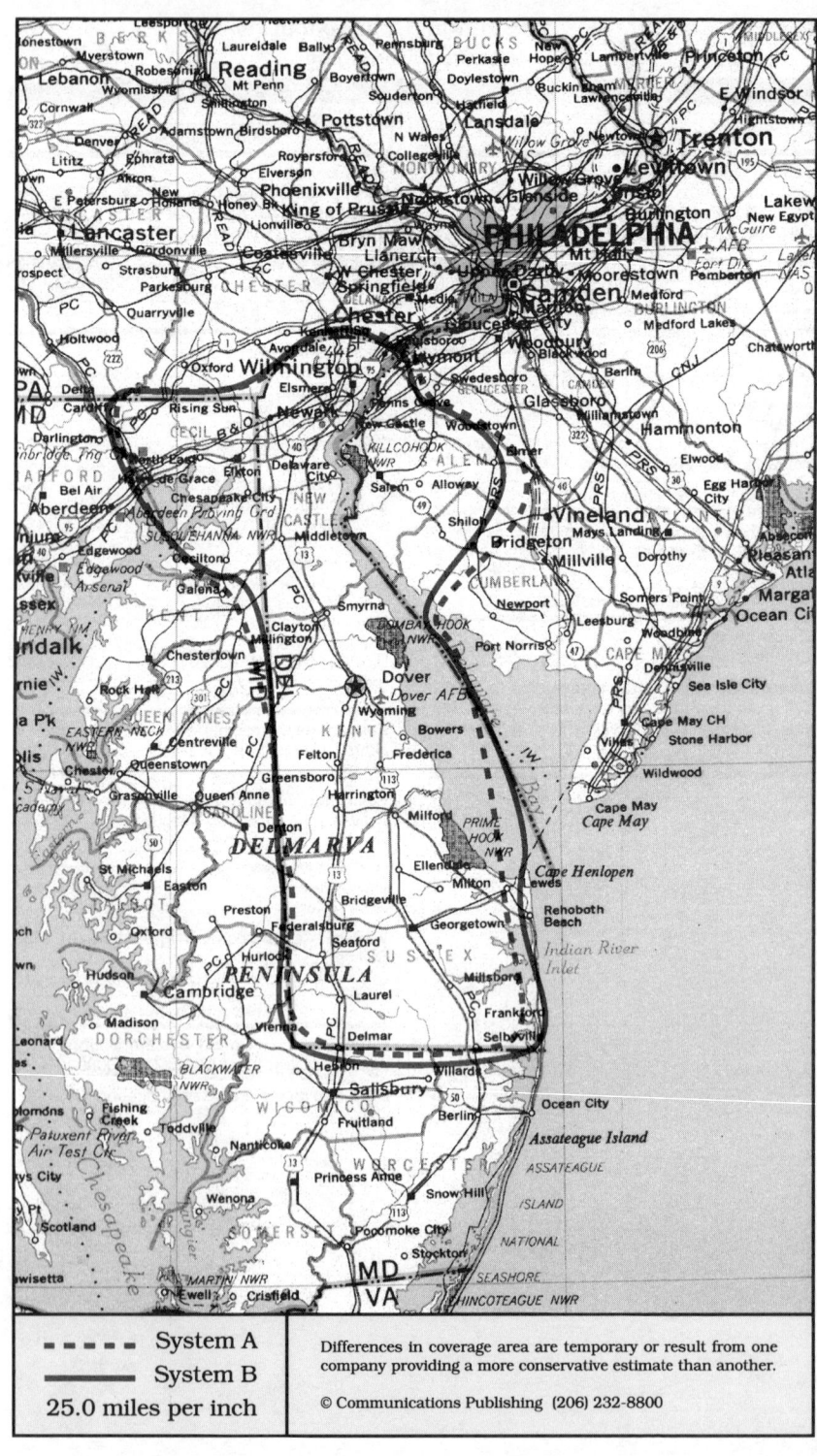

- - - - -	System A
————————	System B
25.0 miles per inch	

Differences in coverage area are temporary or result from one company providing a more conservative estimate than another.

© Communications Publishing (206) 232-8800

Wilmington, Delaware

Dover, Elkton, Salem NJ, Wilmington

Cellular company

Comcast Cellular One
18 Boulden Circle, Suite 24
New Castle, DE 19720-3494

Customer service	in DE, NJ, PA	800-233-4140
Local office	. . . . . .	302-328-7900
Corporate office	. . . . .	215-975-5000

Sending calls

Visitor dialing instructions

Local calls	. . . . .	area code + 7 digits
Long distance	. . .	area code + 7 digits

Speed-dial numbers

Coast Guard (emergency) . . .	*CG or *24
Customer service	611
Emergency	911
Hear a traffic report	*018723
Weather	*019328

Receiving calls

Caller dials the roamer access number below,
hears tone, dials cellular area code and number.
. . . . 302-740-7626

NationLink & RoamingAmerica (see p. 1157)
Caller hears your current roamer access number.
This company's subscribers have level 0 service.

North American Cellular Network (see p. 1156)
All calls forward to you. Deactivate dial *35 send.

Billing information

Visitor rates
Most visitors . . . . . $3 day + 99¢ minute

Visitors without roaming agreements (p. 1159)
Roamer Plus will intercept first call to bill you.

Identification numbers

City	Market	System	Billing	RSA
Dover	359	00123	30279	DE-1
Wilmington	069	00123	00000	MSA

Cellular company

Bell Atlantic Mobile
502 First State Boulevard
Newport, DE 19804

Customer service	. . . .	800-922-0204
Local office	. . . . . .	302-633-1000
Corporate office	. . . . .	908-306-7000

Sending calls

Visitor dialing instructions

Local calls	. . . . . .	area code + 7 digits
Long distance	. . . . .	area code + 7 digits

Speed-dial numbers

Customer service	*611
Emergency	911
Sports channel	*610
Traffic report	*WAFL or *9235

Receiving calls

Caller dials the roamer access number below,
hears tone, dials cellular area code and number.

Dover	302-270-7626
Wilmington	302-530-7626

Follow Me Roaming forwarding (see p. 1156)
Activate, dial *18 send. Deactivate dial *19 send.

Billing information

Visitor rates
Most visitors . . . . . $3 day + $1 minute

Visitors without roaming agreements (p. 1159)
Roamer Plus will intercept first call to bill you.

Identification numbers

City	Market	System	Billing	RSA
Dover	359	00008	30078	DE-1
Wilmington	069	00008	30364	MSA

For full coverage area, see Philadelphia, PA.

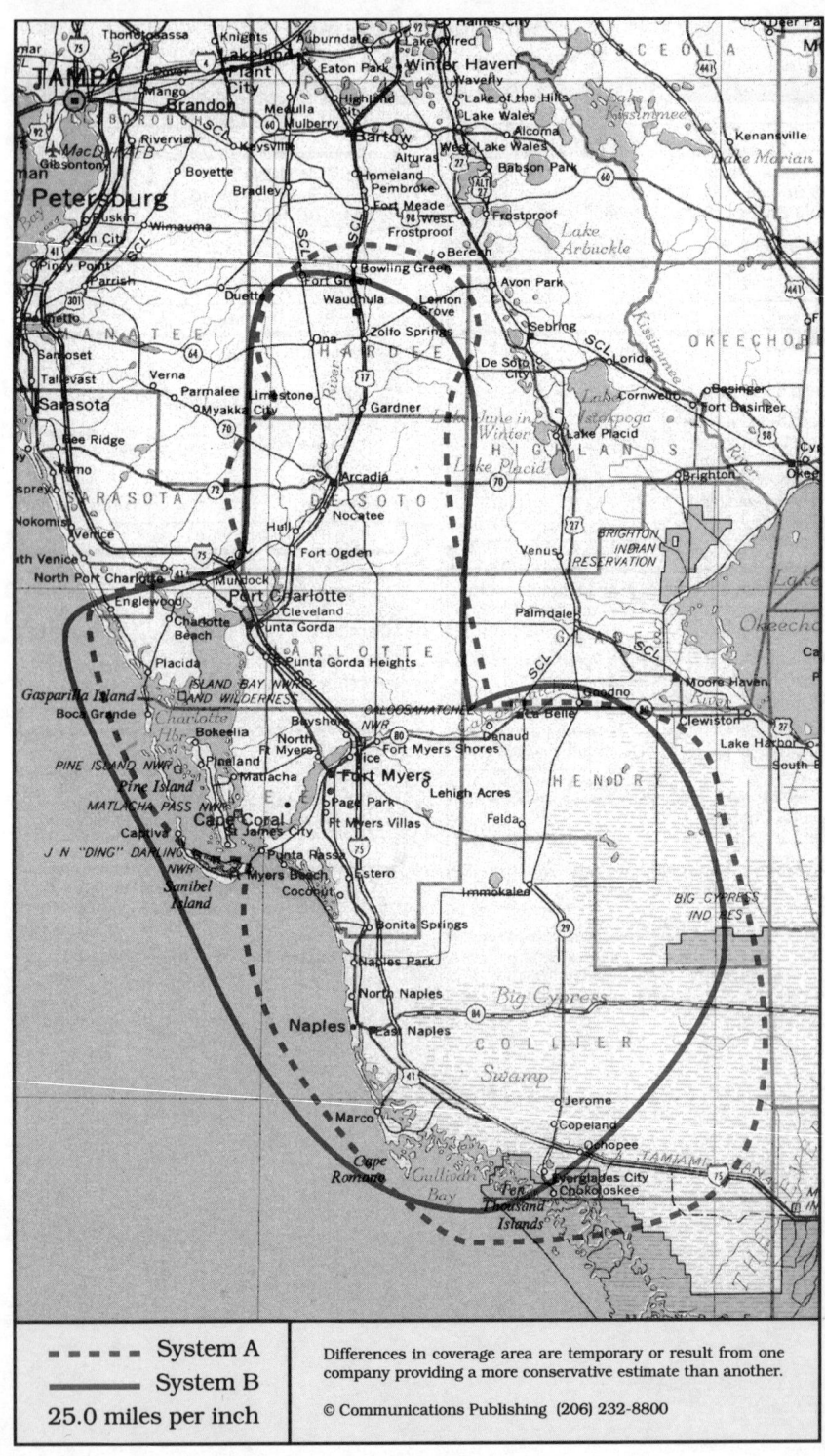

Legend	
- - - -	System A
——	System B
25.0 miles per inch	

Differences in coverage area are temporary or result from one company providing a more conservative estimate than another.

© Communications Publishing (206) 232-8800

Fort Myers, Florida
Arcadia, Fort Myers, Naples, Port Charlotte

System A	System B

Cellular company

Cellular One
Palmer Communications
11500 S Cleveland Avenue
Fort Myers, FL 33907

Cellular One of Southwest Florida
5048 Tamiami Trail N
Naples, FL 33940

Cellular One of Southwest Florida
2195-B Tamiami Trail
Port Charlotte, FL 33948

Customer service	Fort Myers	813-936-4534
. . . . .	Naples	813-263-2355
Local office	Fort Myers	813-936-2355
. . . . .	Naples	813-263-2355
. . . . .	Port Charlotte	813-255-0077
Corporate office	Fort Myers	813-433-4350
. . . . .	Naples	813-263-2355

Sending calls

Visitor dialing instructions
Local calls . . . . . area code + 7 digits
Long distance . . . 1 + area code + 7 digits

Speed-dial numbers
Customer service 611
Directory assistance 411
Emergency 911

Receiving calls

Caller dials the roamer access number below,
hears tone, dials cellular area code and number.
Fort Myers 813-851-7626
Naples 813-860-7626
Port Charlotte 813-380-7626

NationLink & RoamingAmerica (see p. 1157)
Dial *31 send to receive calls, *30 to deactivate.
Dial *32 send to tell caller the roamer access no.
This company's subscribers have level 6 service.

North American Cellular Network (see p. 1156)
All calls forward to you. Deactivate dial *35 send.

Billing information

Visitor rates
Arcadia, Fort Myers . . $3 day + 85¢ minute
Naples, Port Charlotte . $2 day + 65¢ minute

Visitors without roaming agreements (p. 1159)
Arcadia, Fort Myers Co. takes no credit cards.
Naples, Port Charlotte Use a Mastercard or Visa

Identification numbers

City	Market	System	Billing	RSA
Fort Myers	164	00355	00000	MSA
Naples	360	00355	30127	FL-1
Port Charlotte	362	00355	30295	FL-3

Palmer Communications is licensed to serve Lee
County. Cellular One of Southwest Florida is
licensed to serve FL-1 and FL-3.

Cellular company

GTE Mobilnet
4800 S Cleveland Avenue
Fort Myers, FL 33907-1320

Customer service	. . .		800-877-5665
Local office			813-277-2000
Corporate office	. .	national	404-391-8000
. . . .		regional	813-282-6000

Sending calls

Visitor dialing instructions
Local calls . . . . . area code + 7 digits
Long distance . . . 1 + area code + 7 digits

Speed-dial numbers
Customer service *611
Emergency 911
Make a traffic report 105
News tip line 620*
Report crime information . . . . . *TIP
Roaming information *711
Talk line 95*
Telephone problems *111
Weatherline *103

Receiving calls

Caller dials the roamer access number below,
hears tone, dials cellular area code and number.
Arcadia 813-993-5626
Fort Myers 813-994-7626
Immokalee 813-657-8600
LaBelle 813-675-5400
Naples 813-591-7626
Port Charlotte 813-769-7626
Wauchula 813-773-7626

Follow Me Roaming forwarding (see p. 1156)
Activate, dial *18 send. Deactivate dial *19 send.

Billing information

Visitor rates
Most visitors . . . . . $2 day + 75¢ minute

Visitors without roaming agreements (p. 1159)
Call co. to use Amex, Mastercard, Visa

Identification numbers

City	Market	System	Billing	RSA
All other areas	164	00042	00000	MSA
Naples	360	00042	00000	FL-1
Port Charlotte	362	00042	00000	FL-3

For full coverage area, see Tampa, FL.

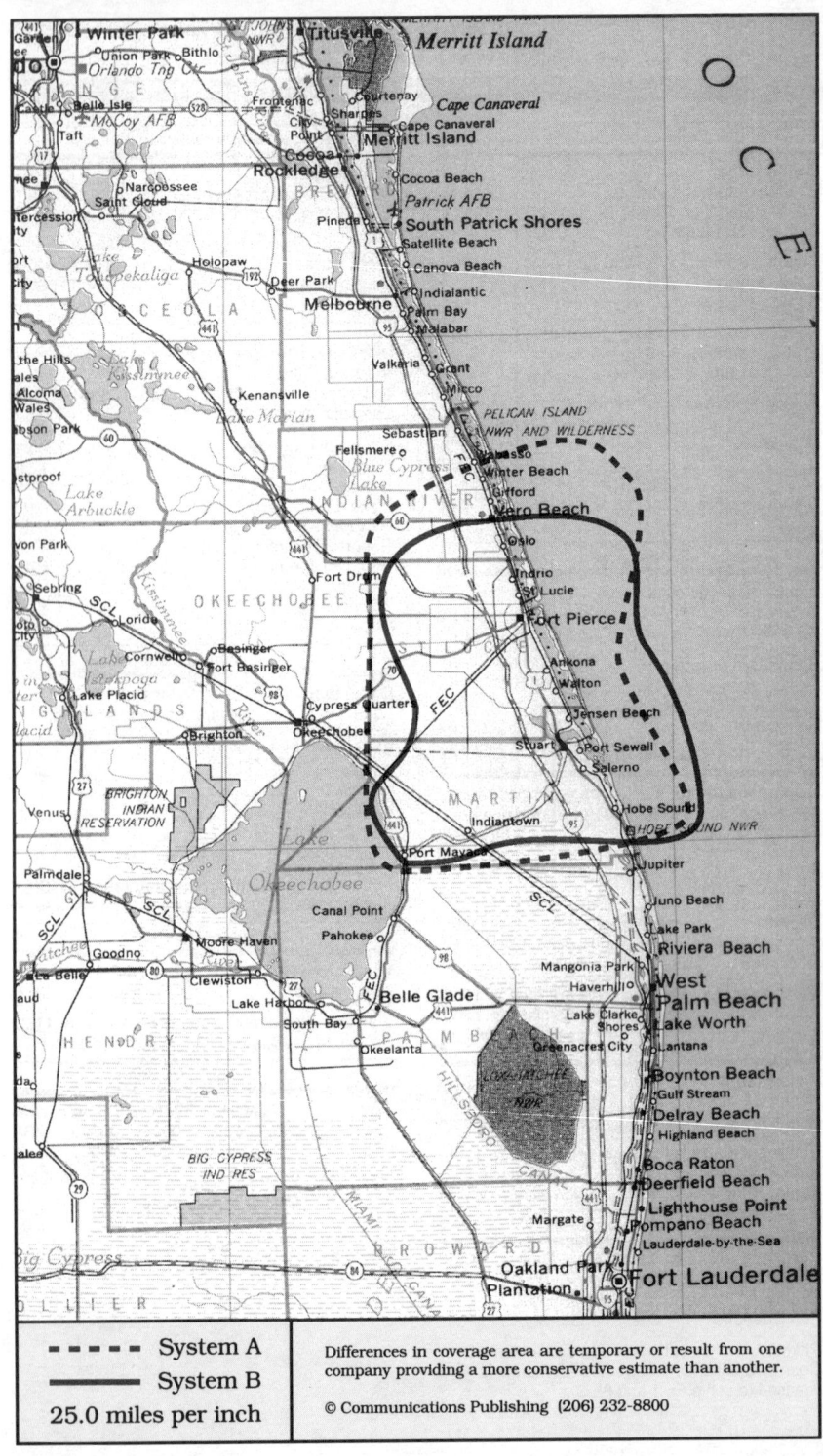

- - - - System A	Differences in coverage area are temporary or result from one company providing a more conservative estimate than another.
—— System B	
25.0 miles per inch	© Communications Publishing (206) 232-8800

Fort Pierce, Florida
Fort Pierce, Stuart, Vero Beach (system A)

System A	System B

Cellular company

Cellular One
McCaw Cellular Communications
751 N Federal Highway
Stuart, FL 34994

Customer service	800-822-3551
. . . .	407-697-9300
Local office	407-692-9365
Corporate office	206-827-4500

Sending calls

Visitor dialing instructions

Local calls	area code + 7 digits
Long distance . . .	1 + area code + 7 digits

Speed-dial numbers

Cellular One info	*INFO or *4636
Customer service	611
Emergency	911
Environmental info . .	*EARTH or *32784
Telemarketing	*767
Towing	*869
WFNG radio	*566
WINZ news and traffic	*94
WOLL news and traffic	*227
WQAM sports	*560
WTEC TV channel 12	*12

Receiving calls

Caller dials the roamer access number below,
hears tone, dials cellular area code and number.
. . . . 407-285-7626

NationLink & RoamingAmerica (see p. 1157)
Dial *31 send to receive calls, *30 to deactivate.
Dial *32 send to tell caller the roamer access no.
This company's subscribers have level 7 service.

North American Cellular Network (see p. 1156)
All calls forward to you. Deactivate dial *35 send.

Billing information

Visitor rates

Most visitors	$0 day + 99¢ minute

Visitors without roaming agreements (p. 1159)
Call co. to use Amex, Mastercard, Visa
or Roamer Plus will intercept first call to bill you.

Identification numbers

City	Market	System	Billing	RSA
Fort Pierce	208	00037	00000	MSA
Vero Beach	361	00037	30281	FL-2

For full coverage area, see Miami, FL.

Cellular company

Central Florida Cellular
United States Cellular
3221 NW Federal Highway
Jensen Beach, FL 34957-4451

Local office	407-692-3030
Corporate office	312-399-8900

Sending calls

Visitor dialing instructions

Local calls	area code + 7 digits
Long distance	area code + 7 digits

Speed-dial numbers

Customer service	611
Emergency	911
Roaming information	711

Receiving calls

Caller dials the roamer access number below,
hears tone, dials cellular area code and number.
. . . . 407-284-7626

Follow Me Roaming forwarding (see p. 1156)
Activate, dial *18 send. Deactivate dial *19 send.

Billing information

Visitor rates

Most visitors . . .	$2.50 day + 50¢ minute

Visitors without roaming agreements (p. 1159)
Call co. to use Amex, Mastercard, Visa

Identification numbers

City	Market	System	Billing	RSA
All served cities	208	00340	00000	MSA

Vero Beach is part of the Orlando, FL system.

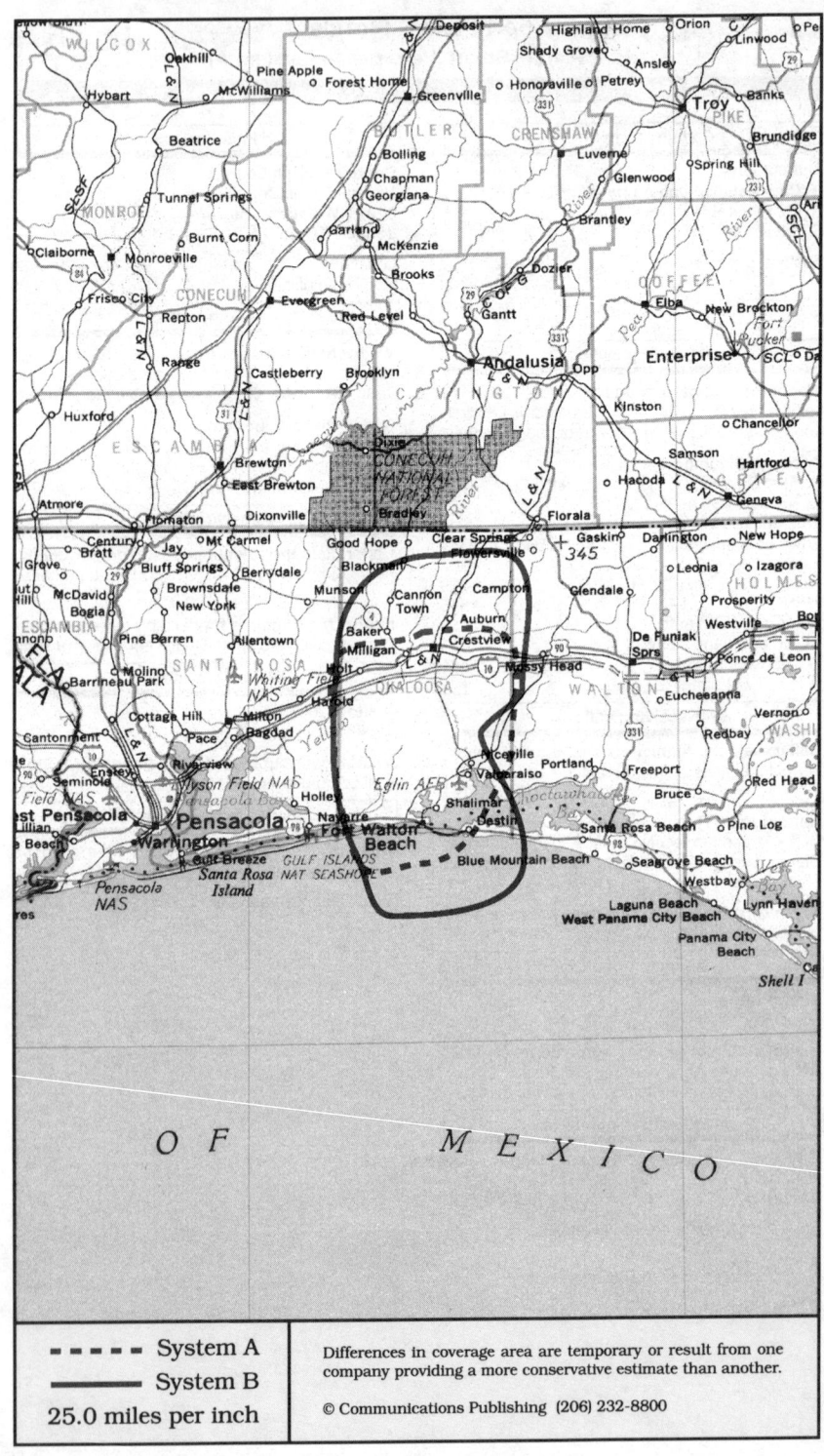

- - - - System A	Differences in coverage area are temporary or result from one company providing a more conservative estimate than another.
———— System B	
25.0 miles per inch	© Communications Publishing (206) 232-8800

Fort Walton Beach, Florida
Crestview, Fort Walton Beach

System A | System B

System A

Cellular company

Cellular One
Vanguard Cellular Systems
824 N Eglin Parkway
Fort Walton Beach, FL 32547

Customer service	904-582-4000
Local office	904-863-2355
Corporate office	919-282-3690

Sending calls

Visitor dialing instructions

Local calls	area code + 7 digits
Long distance	area code + 7 digits

Speed-dial numbers

Customer service	611
Emergency	911

Receiving calls

Caller dials the roamer access number below,
hears tone, dials cellular area code and number.
. . . . 904-582-7626

Billing information

Visitor rates
Most visitors . . . . . $3 day + 99¢ minute

Visitors without roaming agreements (p. 1159)
Call co. to use Amex, Mastercard, Visa

Identification numbers

City	Market	System	Billing	RSA
All served cities	265	00361	00000	MSA

For full coverage area, see Pensacola, FL.

System B

Cellular company

Centel Cellular
133 Beal Parkway NW
Fort Walton Beach, FL 32548

Local office	904-664-2000
Corporate office	312-399-2644

Sending calls

Visitor dialing instructions

Local calls	area code + 7 digits
Long distance . . .	1 + area code + 7 digits

Speed-dial numbers

Customer service	*611
Emergency	911

Receiving calls

Caller dials the roamer access number below,
hears tone, dials cellular area code and number.
. . . . 904-585-7626

Follow Me Roaming forwarding (see p. 1156)
Activate, dial *18 send. Deactivate dial *19 send.

Billing information

Visitor rates
Most visitors . . . . . $3 day + 99¢ minute

Visitors without roaming agreements (p. 1159)
Call co. to use Amex, Mastercard, Visa

Identification numbers

City	Market	System	Billing	RSA
All served cities	265	00344	00000	MSA

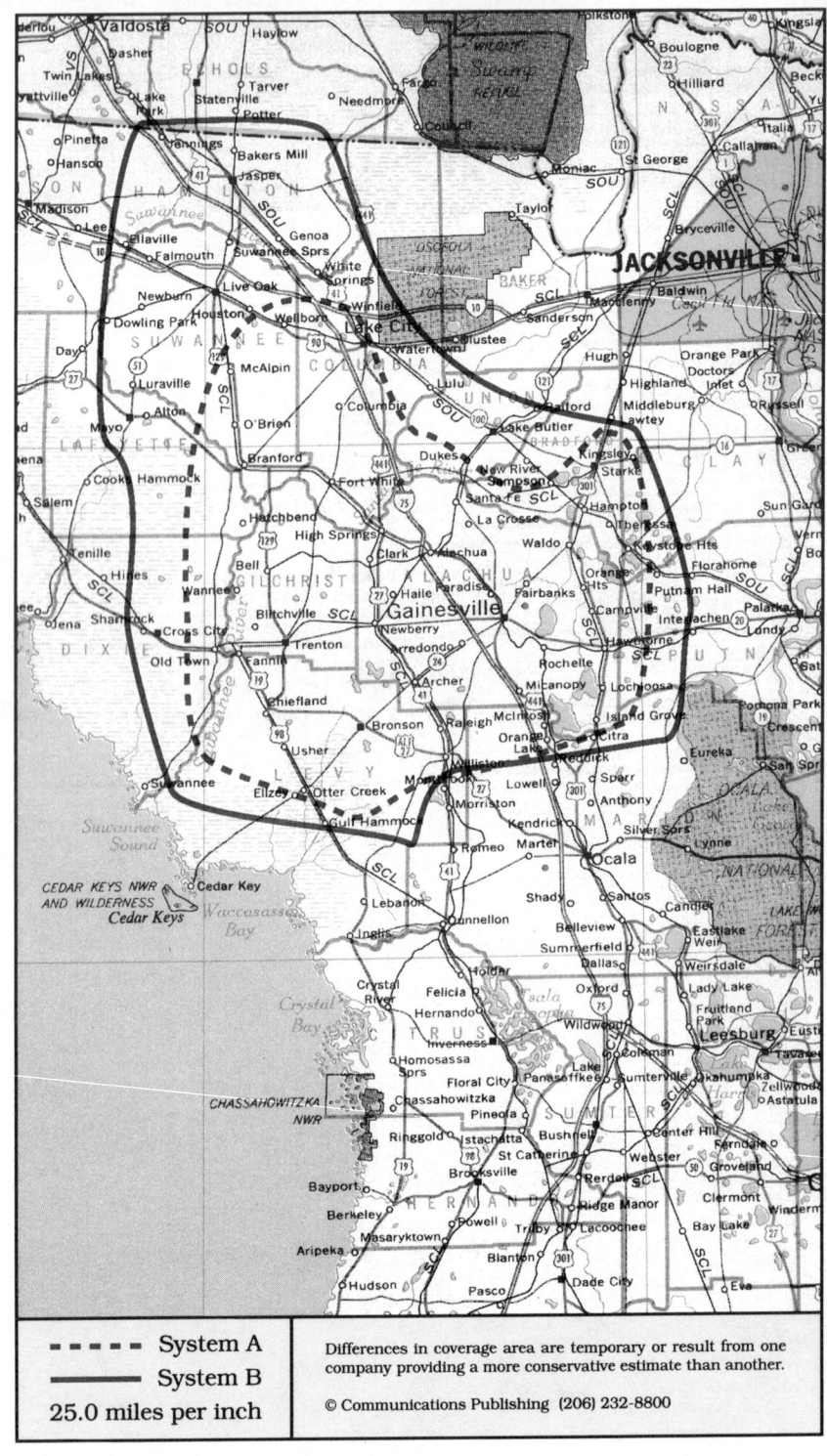

System A

System B

25.0 miles per inch

Differences in coverage area are temporary or result from one company providing a more conservative estimate than another.

© Communications Publishing (206) 232-8800

Gainesville, Florida
Chiefland, Gainesville, Lake City, Live Oak

System A	System B

Cellular company

United States Cellular
6110 NW 4th Place, Suite A
Gainesville, FL 32607

Local office   904-331-8100
Corporate office   312-399-8900

Sending calls

Visitor dialing instructions
Local calls . . . . .   area code + 7 digits
Long distance . . .   1 + area code + 7 digits

Speed-dial numbers
Customer service *611
Emergency   911

Receiving calls

Caller dials the roamer access number below,
hears tone, dials cellular area code and number.
. . . .   904-665-7626

NationLink & RoamingAmerica (see p. 1157)
Dial *31 send to receive calls, *30 to deactivate.
Dial *32 send to tell caller the roamer access no.
This company's subscribers have level 4 service.

North American Cellular Network (see p. 1156)
All calls forward to you. Deactivate dial *35 send.

Billing information

Visitor rates
Most visitors . . . . . $3 day + 99¢ minute

Visitors without roaming agreements (p. 1159)
Cannot call since company takes no credit cards.

Identification numbers

City	Market	System	Billing	RSA
Chiefland	365	01117	00000	FL-6
Gainesville	192	00365	00000	MSA

Cellular company

ALLTEL Mobile Communications
3207 SW 35th Boulevard
Gainesville, FL 32608

Customer service   some areas   800-255-8351
Local office   904-374-8500
Corporate office   501-661-8500

Sending calls

Visitor dialing instructions
Local calls . . . . . .   area code + 7 digits
Long distance . . . . .   area code + 7 digits

Speed-dial numbers
Customer service . . . . .   *611 or *711
Emergency   911

Receiving calls

Caller dials the roamer access number below,
hears tone, dials cellular area code and number.
Chiefland   904-493-3426
Gainesville   904-538-7626
Lake City   904-397-3626
Live Oak   904-362-8626

Follow Me Roaming forwarding (see p. 1156)
Activate, dial *18 send. Deactivate dial *19 send.

Billing information

Visitor rates
Most visitors . . . . . $3 day + 85¢ minute

Visitors without roaming agreements (p. 1159)
Call co. to use Amex, Discover, Mastercard, Visa

Identification numbers

City	Market	System	Billing	RSA
Chiefland	365	00348	30536	FL-6
Gainesville	192	00348	00000	MSA
Lake City, Live Oak	366	00348	30242	FL-7

For full coverage area, see Ocala, FL.

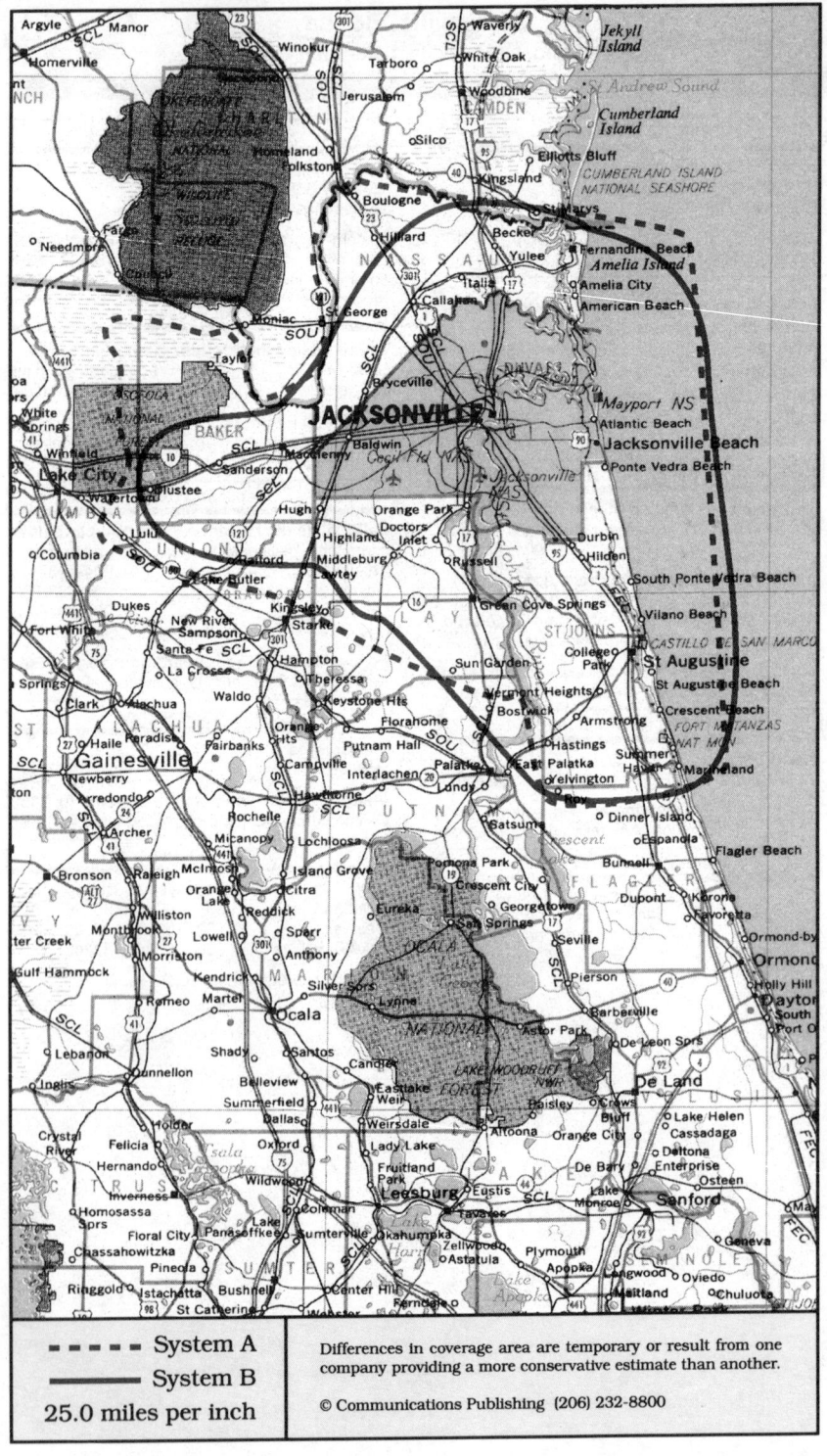

- - - System A ——— System B 25.0 miles per inch	Differences in coverage area are temporary or result from one company providing a more conservative estimate than another. © Communications Publishing (206) 232-8800

Jacksonville, Florida
Green Cove Springs, Jacksonville, MacClenny, St. Augustine

System A	System B

Cellular company

Cellular One
McCaw Cellular Communications
8081 Phillips Highway, Suite 10
Jacksonville, FL 32256

Customer service	Orlando ext.	800-822-3551
. . . .		407-865-8961
Local office		904-731-2355
Corporate office . . . . .,		206-827-4500

Sending calls

Visitor dialing instructions

Local calls	area code + 7 digits
Long distance . . .	1 + area code + 7 digits

Speed-dial numbers

Customer service	611
Emergency	911
Information services . . .	*INFO or *4636

Receiving calls

Caller dials the roamer access number below,
hears tone, dials cellular area code and number.

Jacksonville	904-631-7626
St. Augustine	904-826-7626

NationLink & RoamingAmerica (see p. 1157)
Dial *31 send to receive calls, *30 to deactivate.
Dial *32 send to tell caller the roamer access no.
This company's subscribers have level 7 service.

North American Cellular Network (see p. 1156)
All calls forward to you. Deactivate dial *35 send.

Billing information

Visitor rates

Most visitors	$0 day + 99¢ minute

Visitors without roaming agreements (p. 1159)
Call co. to use Amex, Mastercard, Visa
or Roamer Plus will intercept first call to bill you.

Identification numbers

City	Market	System	Billing	RSA
All served cities	051	00075	00000	MSA

Cellular company

BellSouth Mobility
7660 Phillips Highway #20
Jacksonville, FL 32256-6819

Customer service	800-351-2400
. . . .	407-834-2002
Local office	904-443-6800
Corporate office	404-847-3600

Sending calls

Visitor dialing instructions

Local calls	area code + 7 digits
Long distance	area code + 7 digits

Speed-dial numbers

Customer service	811
Emergency	911
Florida highway patrol . . .	904-355-9981
Roaming information	711
Telephone problems	611

Receiving calls

Caller dials the roamer access number below,
hears tone, dials cellular area code and number.

. . . .	904-635-7626

Follow Me Roaming forwarding (see p. 1156)
Activate, dial *18 send. Deactivate dial *19 send.

Billing information

Visitor rates

Most visitors	$3 day + 85¢ minute
BellSouth subscribers .	$0 day + 60¢ minute

Visitors without roaming agreements (p. 1159)
Call co. to use Amex, Mastercard, Visa

Identification numbers

City	Market	System	Billing	RSA
All served cities	051	00136	00000	MSA

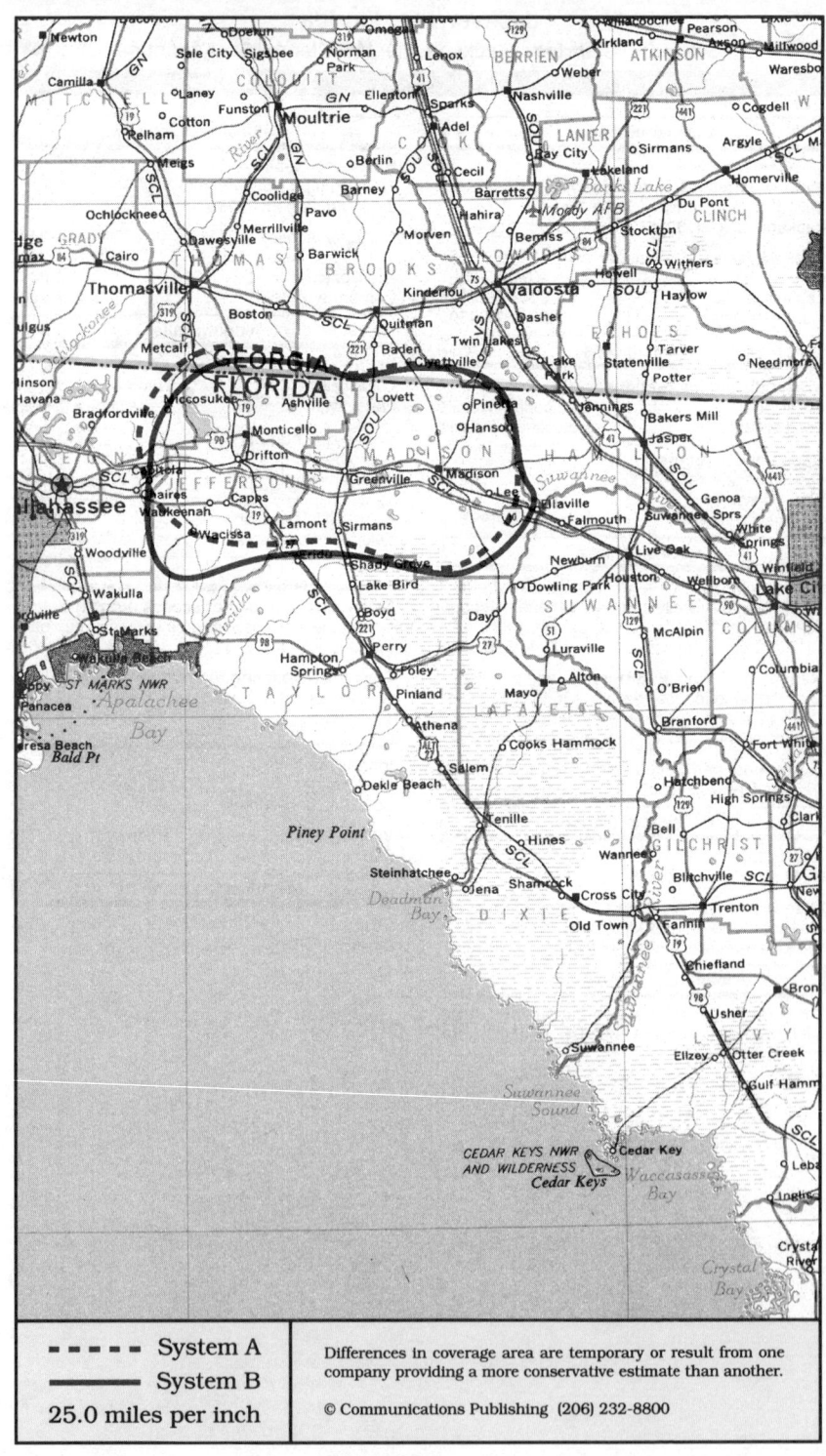

| System A |
| System B |
| 25.0 miles per inch |

Differences in coverage area are temporary or result from one company providing a more conservative estimate than another.

© Communications Publishing (206) 232-8800

Madison, Florida
Madison, Monticello

Cellular company		Cellular company	

United States Cellular
6110 NW 4th Place, Suite A
Gainesville, FL 32607

Local office	904-331-8100
Corporate office	312-399-8900

Centel Cellular
1401 Market Street
Tallahassee, FL 32312

Local office	904-847-4000
Corporate office	312-399-2644

Sending calls

Visitor dialing instructions

Local calls	area code + 7 digits
Long distance	area code + 7 digits

Speed-dial numbers

Customer service	611
Emergency	911
Roaming information	*711

Sending calls

Visitor dialing instructions

Local calls	area code + 7 digits
Long distance . . .	1 + area code + 7 digits

Speed-dial numbers

Customer service	*611
Directory assistance	411
Emergency	911

Receiving calls

Caller dials the roamer access number below,
hears tone, dials cellular area code and number.

. . . . 912-225-2876

Receiving calls

Caller dials the roamer access number below,
hears tone, dials cellular area code and number.

. . . . 904-545-7626

Billing information

Visitor rates
Most visitors . . . . . $3 day + 99¢ minute

Visitors without roaming agreements (p. 1159)
Call co. to use Mastercard, Visa

Billing information

Visitor rates
Most visitors . . . . . $3 day + 99¢ minute

Visitors without roaming agreements (p. 1159)
Cannot call since company takes no credit cards.

Identification numbers

City	Market	System	Billing	RSA
All served cities	367	01121	00000	FL-8

Identification numbers

City	Market	System	Billing	RSA
All served cities	367	01122	00000	FL-8

For full coverage area, see Tallahassee, FL.

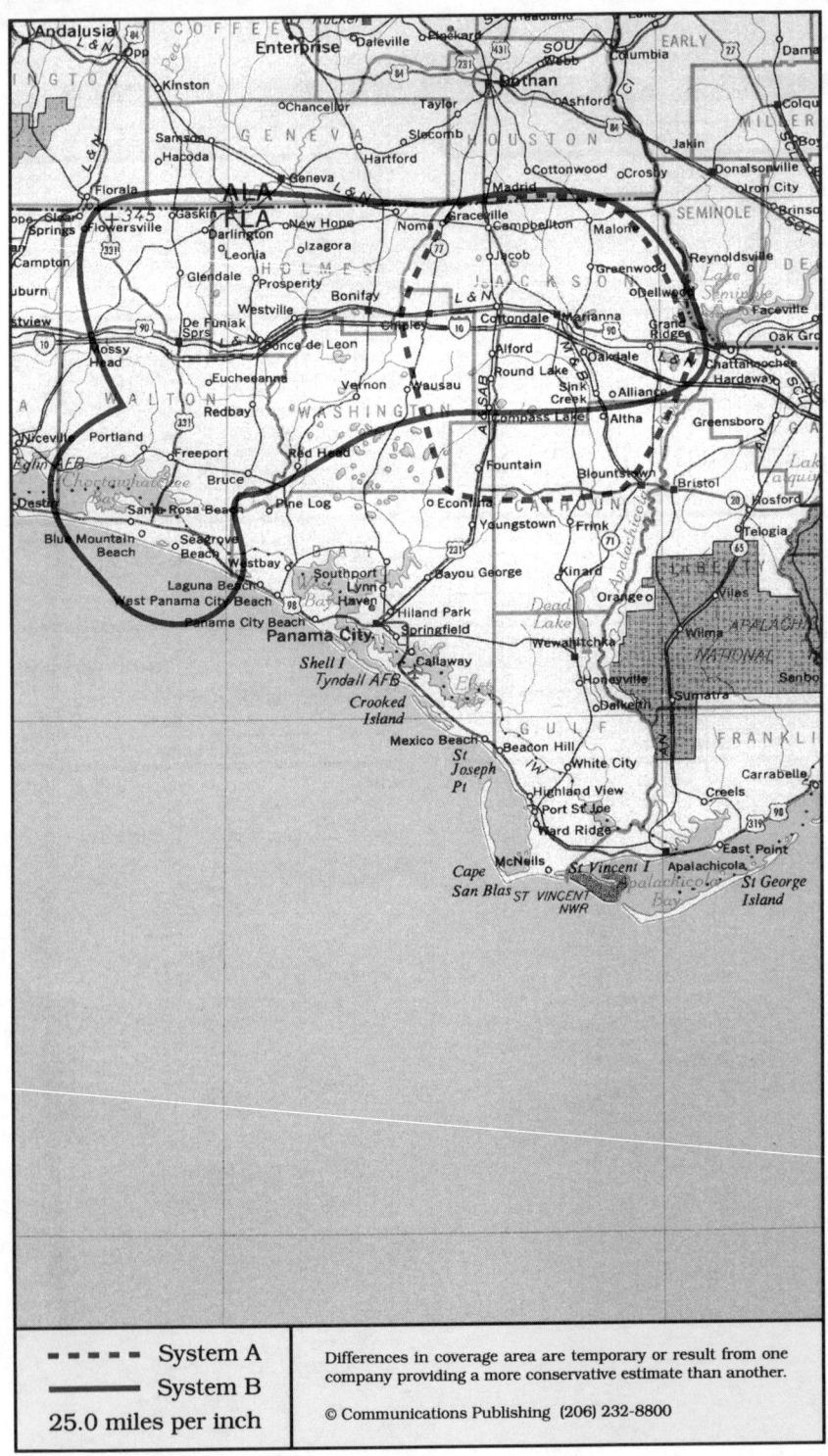

Marianna, Florida
DeFuniak Springs, Marianna

System A

Cellular company

Cellular One
Canton Cellular
4325 Lafayette Street
Marianna, FL 32446

Corporate office 904-526-7600

Sending calls

Visitor dialing instructions
Local calls . . . . . area code + 7 digits
Long distance . . . 1 + area code + 7 digits

Speed-dial numbers
Customer service *611
Emergency 911

Receiving calls

Caller dials the roamer access number below,
hears tone, dials cellular area code and number.
. . . . 904-849-7626

Billing information

Visitor rates
Most visitors . . . . . $3 day + 85¢ minute

Visitors without roaming agreements (p. 1159)
Roamer Plus will intercept first call to bill you.

Identification numbers

City	Market	System	Billing	RSA
All served cities	369	01125	00000	FL-10

System B

Cellular company

Centel Cellular
4387 Lafayette Street
Marianna, FL 32446

Local office 904-526-7700
Corporate office 312-399-2644

Sending calls

Visitor dialing instructions
Local calls . . . . . . area code + 7 digits
Long distance . . . 1 + area code + 7 digits

Speed-dial numbers
Customer service *611
Emergency 911

Receiving calls

Caller dials the roamer access number below,
hears tone, dials cellular area code and number.
DeFuniak Springs 904-585-7626
Marianna 904-526-8626

Billing information

Visitor rates
Most visitors . . . . . $3 day + 99¢ minute

Visitors without roaming agreements (p. 1159)
Cannot call since company takes no credit cards.

Identification numbers

City	Market	System	Billing	RSA
All served cities	369	01126	00000	FL-10

For full coverage area, see Panama City, FL.

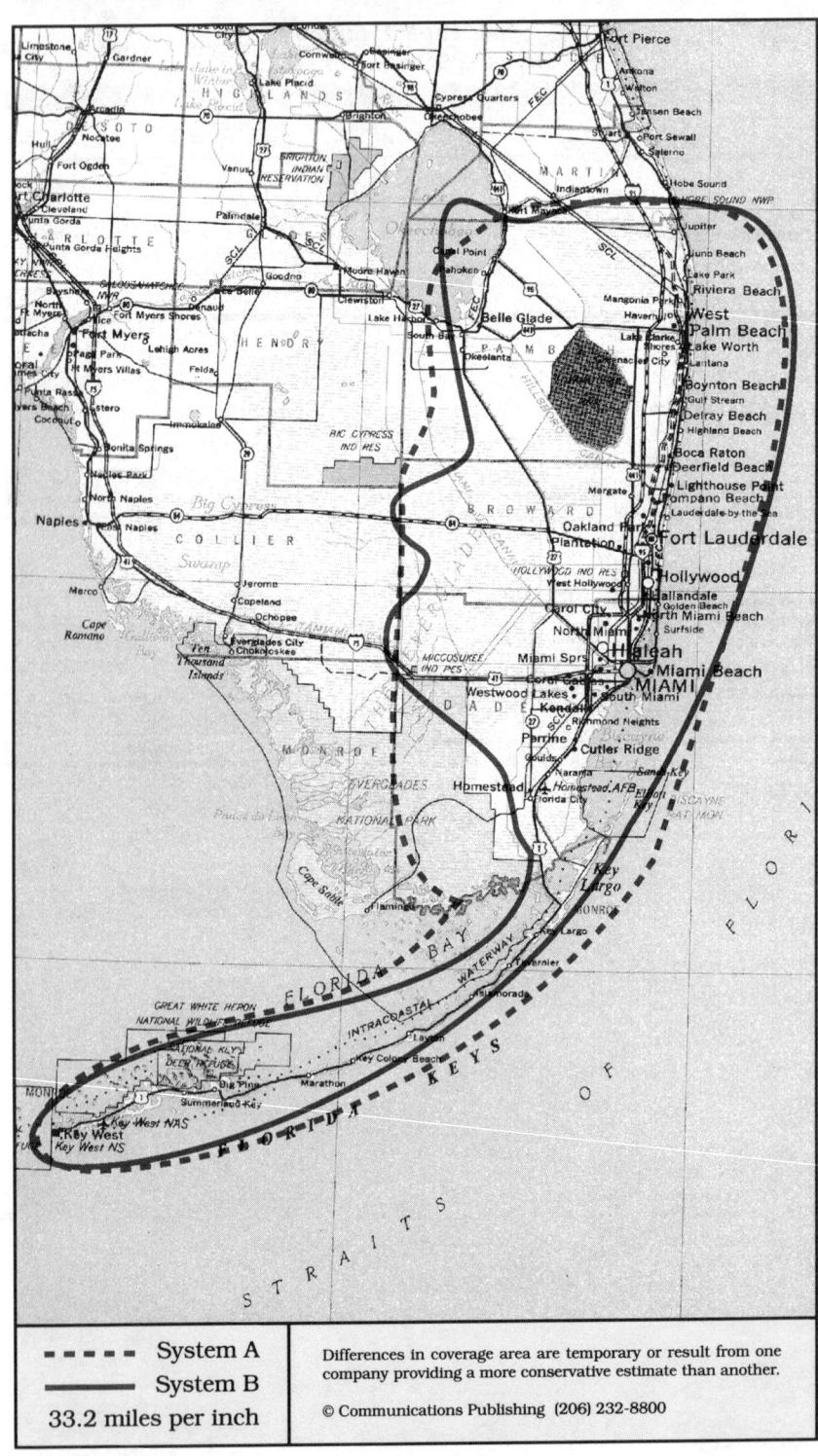

▬ ▬ ▬ ▬ System A	Differences in coverage area are temporary or result from one
▬▬▬▬▬ System B	company providing a more conservative estimate than another.
33.2 miles per inch	© Communications Publishing (206) 232-8800

Miami, Florida
Fort Lauderdale, Key West, Miami, West Palm Beach

Cellular company

Cellular One
McCaw Cellular Communications
Administration　　　　Local office
250 Australian Avenue S
West Palm Beach, FL 33401　305-833-1111

201 Alhambra Circle
Coral Gables, FL 33143　　305-444-6464

1401 W Commercial Blvd, Bay 180
Fort Lauderdale, FL 33309　305-772-4118

8215 S Dixie Highway
Miami, FL 33143　　　　305-663-8711

15540 NW 77th Court
Miami, FL 33016　　　　305-556-0022

801 Northpoint Pkwy, PO Box 24679
West Palm Beach, FL 33416　407-697-9300

1880 Okeechobee Boulevard
West Palm Beach, FL 33409　407-687-8600
Customer service . for all cities　800-822-3551
Corporate office . . . . .　　206-827-4500

Sending calls

Visitor dialing instructions
Local calls . . . . .　area code + 7 digits
Long distance . . .　1 + area code + 7 digits

Speed-dial numbers
Cellular One info . . . . . #INFO or #4636
Customer service . . . . . .　　611
Emergency . . . .　　911
Towing . . . .　　#869
WOLL is #227, WPEC TV is #12, WQAM is #560

Receiving calls

Caller dials the roamer access number below,
hears tone, dials cellular area code and number.
Boca Raton . . . .　407-479-7626
Fort Lauderdale . . . . .　305-328-7626
Key West, Miami . . . . .　305-794-7626
West Palm Beach . . . .　407-346-7626

NationLink & RoamingAmerica (see p. 1157)
Dial ∗31 send to receive calls, ∗30 to deactivate.
Dial ∗32 send to tell caller the roamer access no.
This company's subscribers have level 7 service.

North American Cellular Network (see p. 1156)
All calls forward to you. Deactivate dial ∗35 send.

Billing information

Visitor rates
Most visitors . . . . .　$0 day + 99¢ minute

Visitors without roaming agreements (p. 1159)
Call co. to use Amex, Mastercard, Visa
or Roamer Plus will intercept first call to bill you.

Identification numbers

City	Market	System	Billing	RSA
Key West	370	00037	30277	FL-11
Miami	012	00037	00000	MSA
West Palm Beach	072	00037	00000	MSA

For full coverage area, see Fort Pierce, FL.

Cellular company

BellSouth Mobility
Administration　　　　Local office
500 W Cypress Creek Road #700
Fort Lauderdale, FL 33309　305-776-3900

5805 N Andrews Way
Fort Lauderdale, FL 33309　305-776-2420

9713 Overseas Highway
Marathon, FL 33050　　305-289-0555

9700 S Dixie Highway, Suite 800
Miami, FL 33156　　　305-670-3600

1889-A Palm Beach Lakes Blvd
West Palm Beach, FL 33409　407-478-7344

Customer service　. . . .　800-351-2400
　free in S. Florida (no area code)　930-1101
Corporate office　. . . . .　404-847-3600

Sending calls

Visitor dialing instructions
Local calls . . . . . .　area code + 7 digits
Long distance . . . . .　area code + 7 digits

Speed-dial numbers
Cellular telephone problems . . . . .　611
Customer service . . . .　811
Emergency . . . .　911
Report dumping, graffiti, vandalism　.　∗ALERT
Roaming information . . . .　711
Talk shows . . . WIOD is ∗IOD, WJNO is ∗JNO
Talk shows . . . WLYF is ∗790, WPOW is ∗965
Traffic tips ∗105 or ∗475 or ∗610 or ∗697or ∗921
WCIX TV 6 news tips . . . . . . ∗TV6 or ∗886
WFTL 1400 am news talk . . . . . .　∗385
WPTV TV5 news tips . . . . ∗TV5 or ∗885

Receiving calls

Caller dials the roamer access number below,
hears tone, dials cellular area code and number.
Big Pine Key . . . .　305-872-7777
Fort Lauderdale . . . .　305-398-4450
Homestead . . . .　305-342-0015
Islamorada . . . .　305-664-7626
Key Largo . . . .　305-451-7626
Key West . . . .　305-745-4444
Miami . . . .　305-343-7626
West Palm Beach . . . . . .　407-329-1790

Follow Me Roaming forwarding (see p. 1156)
Activate, dial ∗18 send. Deactivate dial ∗19 send.

Billing information

Visitor rates
Most visitors　. . . . .　$3 day + 85¢ minute
BellSouth subscribers　.　$0 day + 60¢ minute

Visitors without roaming agreements (p. 1159)
Call co. to use Amex, Mastercard, Visa

Identification numbers

City	Market	System	Billing	RSA
Key West	370	00024	30202	FL-11
Miami	012	00024	00000	MSA
West Palm Beach	072	00024	00000	MSA

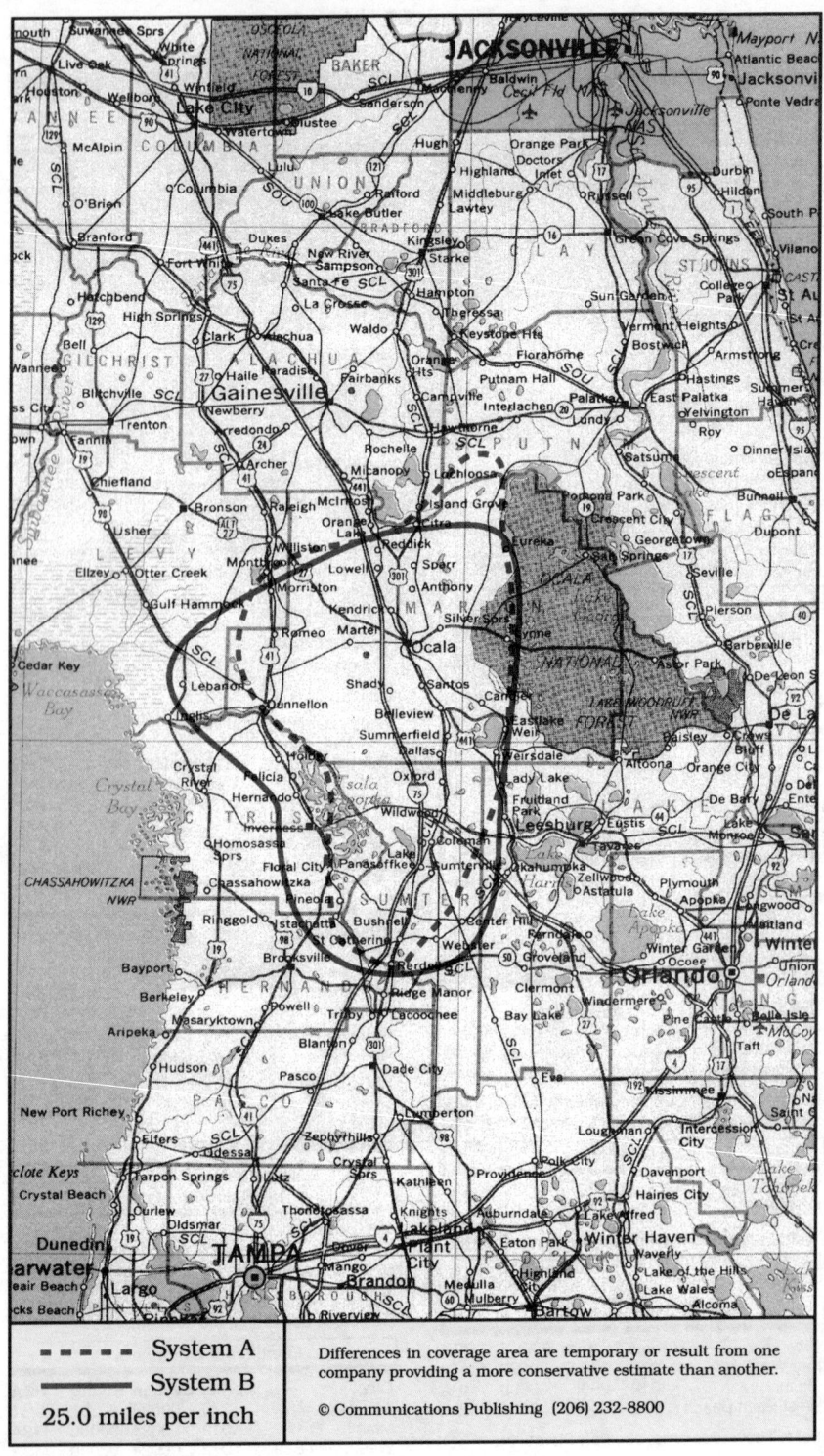

	System A
━━━━━	System B
25.0 miles per inch	

Differences in coverage area are temporary or result from one company providing a more conservative estimate than another.

© Communications Publishing (206) 232-8800

Ocala, Florida
Bushnell, Ocala, Wildwood

|

Cellular company

Cellular One
McCaw Cellular Communications
3405 SW College Road, Suite 227
Ocala, FL 32674

Customer service . .	Orlando	800-822-3551
. . . .	Orlando	407-865-8961
Local office		904-854-1999
Corporate office		206-827-4500

Sending calls

Visitor dialing instructions
Local calls . . . . . area code + 7 digits
Long distance . . . 1 + area code + 7 digits

Speed-dial numbers
Customer service *611
Emergency 911

Receiving calls

Caller dials the roamer access number below,
hears tone, dials cellular area code and number.
Marion County (Ocala) . . . 904-867-3126
Sumter County (Bushnell) . . 904-793-0626

NationLink & RoamingAmerica (see p. 1157)
Dial *31 send to receive calls, *30 to deactivate.
Dial *32 send to tell caller the roamer access no.
This company's subscribers have level 7 service.

North American Cellular Network (see p. 1156)
All calls forward to you. Deactivate dial *35 send.

Billing information

Visitor rates
Most visitors . . . . . $0 day + 99¢ minute

Visitors without roaming agreements (p. 1159)
Roamer Plus will intercept first call to bill you.

Identification numbers

City	Market	System	Billing	RSA
Bushnell	363	00175	30261	FL-4
Ocala	245	00175	30063	MSA

For full coverage area, see Orlando, FL and
Tampa, FL.

Cellular company

ALLTEL Mobile Communications
1301 SW 37th Avenue, Suite 106
Ocala, FL 32674

Customer service	some areas	800-255-8351
Local office		904-237-1100
Corporate office		501-661-8500

Sending calls

Visitor dialing instructions
Local calls . . . . . . area code + 7 digits
Long distance . . . . area code + 7 digits

Speed-dial numbers
Customer service . . . . . *611 or *711
Emergency 911
Highway patrol *47

Receiving calls

Caller dials the roamer access number below,
hears tone, dials cellular area code and number.
Bushnell, Wildwood 904-568-4426
Ocala 904-843-7626

Follow Me Roaming forwarding (see p. 1156)
Activate, dial *18 send. Deactivate dial *19 send.

Billing information

Visitor rates
Most visitors . . . . . $3 day + 85¢ minute

Visitors without roaming agreements (p. 1159)
Call co. to use Amex, Discover, Mastercard, Visa

Identification numbers

City	Market	System	Billing	RSA
Bushnell, Wildwood	363	00348	30632	FL-4
Ocala	245	00348	00000	MSA

For full coverage area, see Gainesville, FL.

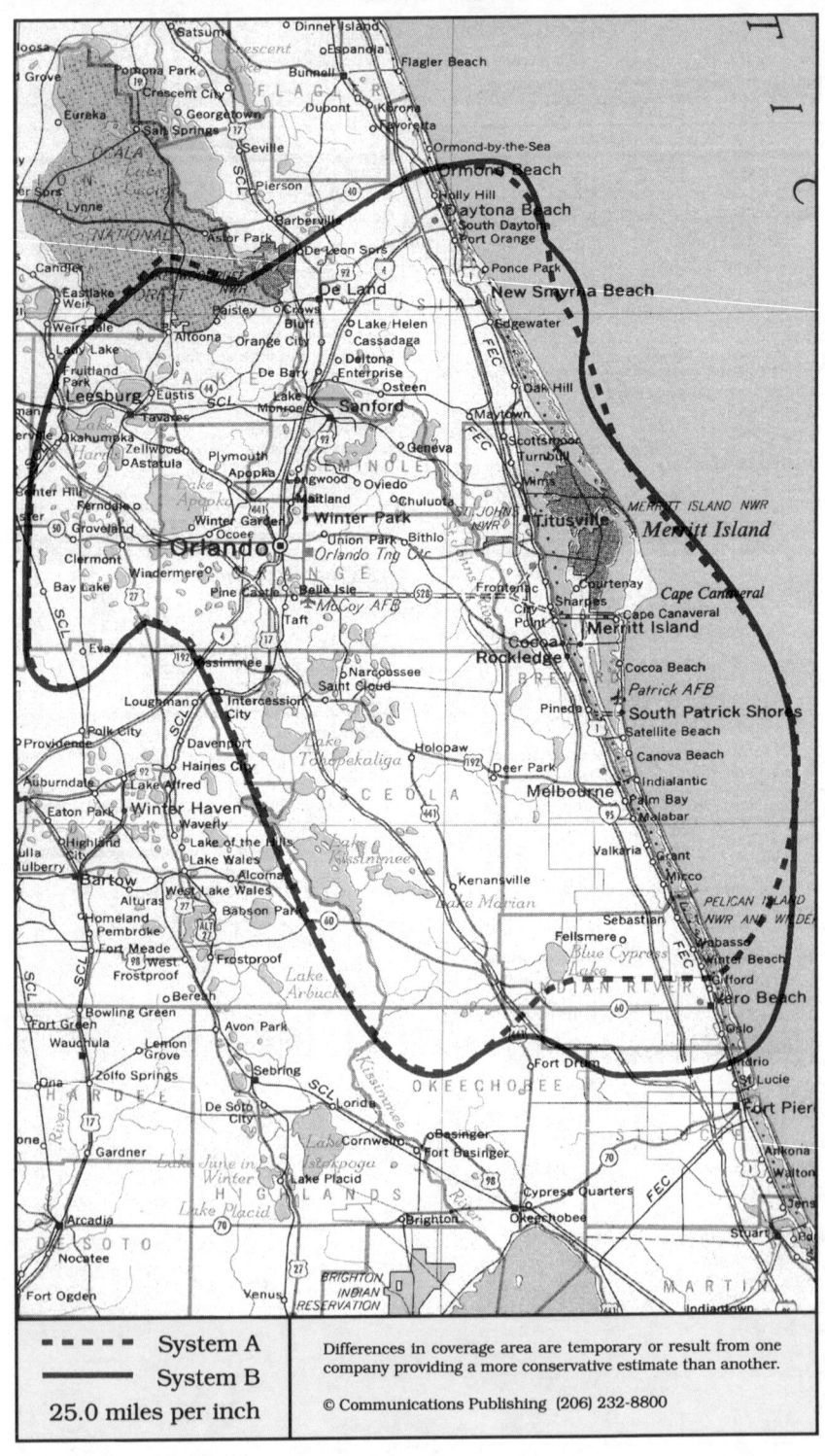

- - - - System A	Differences in coverage area are temporary or result from one
—— System B	company providing a more conservative estimate than another.
25.0 miles per inch	© Communications Publishing (206) 232-8800

Orlando, Florida
Daytona Beach, Leesburg, Melbourne, Orlando, Vero Beach (system B)

System A	System B

Cellular company

Cellular One
McCaw Cellular Communications
151 Wymore Road, Suite 1000
Altamonte Springs, FL 32714

999 3rd Street
Holly Hill, FL 32117

10601 U.S. Highway 441, Suite G-4
Leesburg, FL 34788-8873

1800 Penn Street, Suite 11
Melbourne, FL 32901

Customer service	. .	Orlando	800-822-3551
. . . .	. .		407-865-8961
Local office	Altamonte Springs		407-865-8800
. . . . . .	. .	Holly Hill	904-257-2355
. . . . . .	. .	Leesburg	904-326-8595
. . . . . .	. .	Melbourne	407-728-0099
Corporate office	. . . .		206-827-4500

Sending calls

Visitor dialing instructions
Local calls . . . . . area code + 7 digits
Long distance . . . 1 + area code + 7 digits

Speed-dial numbers
Customer service 611
Directory assistance 411
Emergency 911
Make a traffic report . . . . ∗101

Receiving calls

Caller dials the roamer access number below,
hears tone, dials cellular area code and number.
Daytona Beach 904-451-7626
Lake County (Leesburg) . . 904-360-5126
Melbourne 407-258-7626
Orlando 407-256-7626
Sanford 407-330-8626
Sebastian 407-571-7626

NationLink & RoamingAmerica (see p. 1157)
Dial ∗31 send to receive calls, ∗30 to deactivate.
Dial ∗32 send to tell caller the roamer access no.
This company's subscribers have level 7 service.

North American Cellular Network (see p. 1156)
All calls forward to you. Deactivate dial ∗35 send.

Billing information

Visitor rates
. . . . $0 day + 99¢ minute

Visitors without roaming agreements (p. 1159)
Call co. to use Amex, Mastercard, Visa
or Roamer Plus will intercept first call to bill you.

Identification numbers

City	Market	System	Billing	RSA
Daytona Beach	146	00325	00000	MSA
Leesburg	363	00175	00000	FL-4
Melbourne	137	00175	00000	MSA
Orlando	060	00175	00000	MSA
Sebastian	361	00175	30309	FL-2

Vero Beach system A is part of Fort Pierce, FL.

Cellular company

BellSouth Mobility
375 E Altamonte Drive, Suite 1600
Altamonte Springs, FL 32701

1700 Volusia Avenue, Suite 500
Daytona Beach, FL 32114

4450 W Eau Gallie Boulevard, Suite 130
Melbourne, FL 32934

Customer service	. . . .	800-351-2400
. . . .	. . . .	407-834-2002
Local office	Altamonte Springs	407-262-4000
. . . . . .	Daytona Beach	904-252-9000
. . . . . .	Melbourne	407-255-1200
Corporate office	. . . .	404-847-3600

Sending calls

Visitor dialing instructions
Local calls . . . . . area code + 7 digits
Long distance . . . . . area code + 7 digits

Speed-dial numbers
Customer service 811
Emergency 911
Hear a traffic report 123
Phone problems 611
Roamer information 711

Receiving calls

Caller dials the roamer access number below,
hears tone, dials cellular area code and number.
Daytona Beach 904-295-7626
Lake County (Leesburg) . . . 904-394-9626
Melbourne 407-543-7626
Orlando 407-222-7626
Vero Beach 407-563-6626

Follow Me Roaming forwarding (see p. 1156)
Activate, dial ∗18 send. Deactivate dial ∗19 send.

Billing information

Visitor rates
Most visitors . . . . . $3 day + 85¢ minute
BellSouth subscribers . $0 day + 60¢ minute

Visitors without roaming agreements (p. 1159)
Call co. to use Amex, Mastercard, Visa

Identification numbers

City	Market	System	Billing	RSA
Daytona Beach	146	00308	00000	MSA
Flagler Beach	364	00308	00000	FL-5
Leesburg	363	01988	30261	FL-4
Melbourne	137	00476	00000	MSA
Orlando	060	00068	00000	MSA
Vero Beach	361	01990	30452	FL-2

For full coverage area, see Flagler Beach, FL.

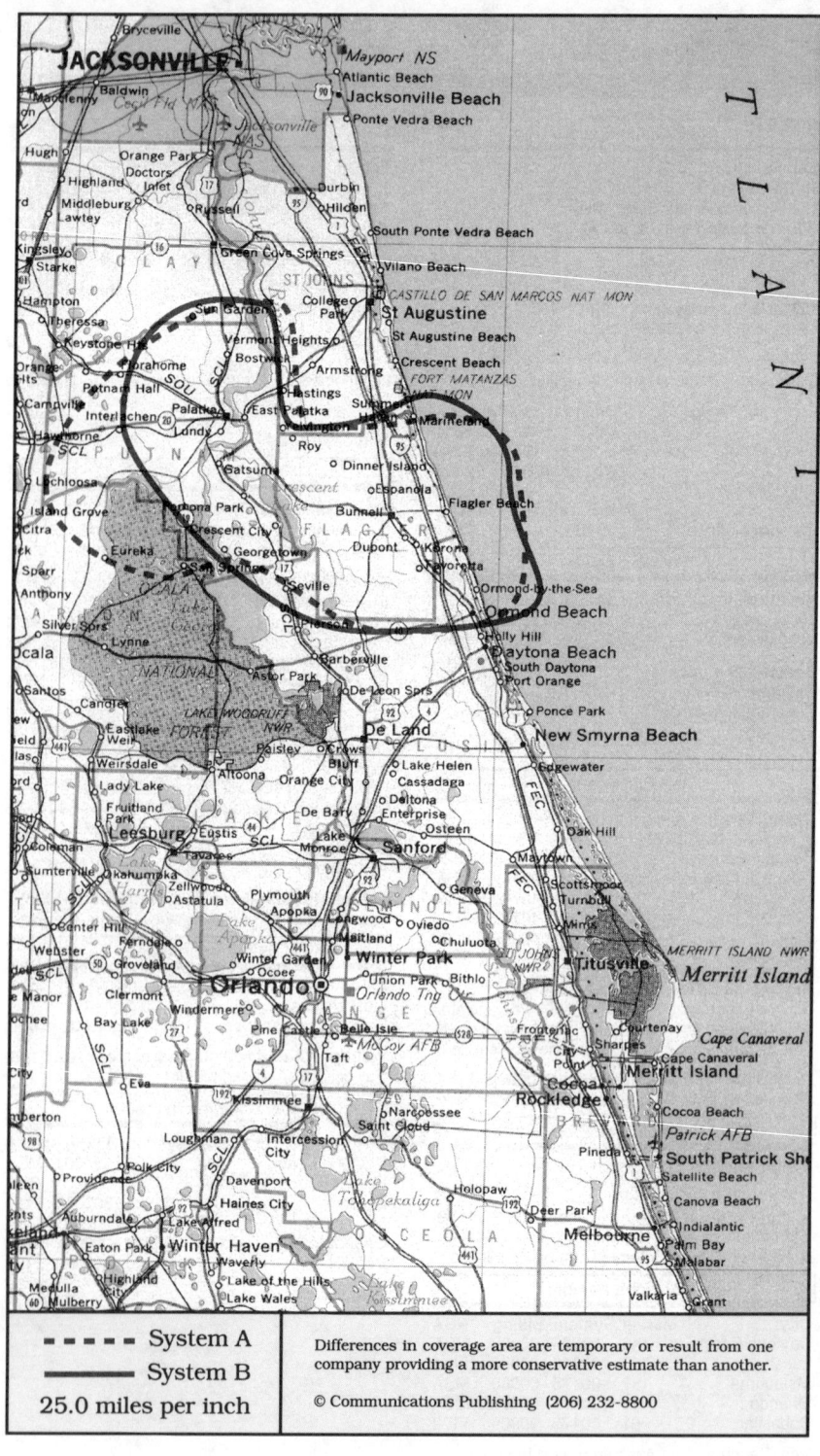

System A

System B

25.0 miles per inch

Differences in coverage area are temporary or result from one company providing a more conservative estimate than another.

© Communications Publishing (206) 232-8800

Palatka, Florida
Bunnell, Flager Beach, Ormond Beach, Palatka

System A	System B

Cellular company

Central Florida Cellular
United States Cellular
3221 NW Federal Highway
Jensen Beach, FL 34957

Local office 407-692-3030
Corporate office 312-399-8900

Sending calls

Visitor dialing instructions
Local calls . . . . . area code + 7 digits
Long distance . . . 1 + area code + 7 digits

Speed-dial numbers
Customer service *611
Emergency 911
Roaming information *711

Receiving calls

Caller dials the roamer access number below,
hears tone, dials cellular area code and number.
. . . . 904-665-7626

Billing information

Visitor rates
Most visitors . . . . . $3 day + 85¢ minute

Visitors without roaming agreements (p. 1159)
Call co. to use Amex, Discover, Mastercard, Visa

Identification numbers

City	Market	System	Billing	RSA
All served cities	364	01115	00000	FL-5

This company also will serve Palatka, FL in the future.

Cellular company

BellSouth Mobility
1700 Volusia Avenue, Suite 500
Daytona Beach, FL 32114

Customer service 800-351-2400
. . . . 407-834-2002
Local office 904-252-9000
Corporate office 404-847-3600

Sending calls

Visitor dialing instructions
Local calls . . . . . . area code + 7 digits
Long distance . . . . . area code + 7 digits

Speed-dial numbers
Customer service 811
Emergency 911
Hear a traffic report 123
Telephone problems 611
Roaming information 711

Receiving calls

Caller dials the roamer access number below,
hears tone, dials cellular area code and number.
Flagler Beach 904-439-9198
Palatka 904-329-7626

Follow Me Roaming forwarding (see p. 1156)
Activate, dial *18 send. Deactivate dial *19 send.

Billing information

Visitor rates
Most visitors . . . . . $3 day + 85¢ minute
BellSouth subscribers . $0 day + 55¢ minute

Visitors without roaming agreements (p. 1159)
Call co. to use Amex, Mastercard, Visa

Identification numbers

City	Market	System	Billing	RSA
Flagler Beach	364	00308	30338	FL-5
Palatka	364	01116	30476	FL-5

For full coverage area, see Orlando, FL.

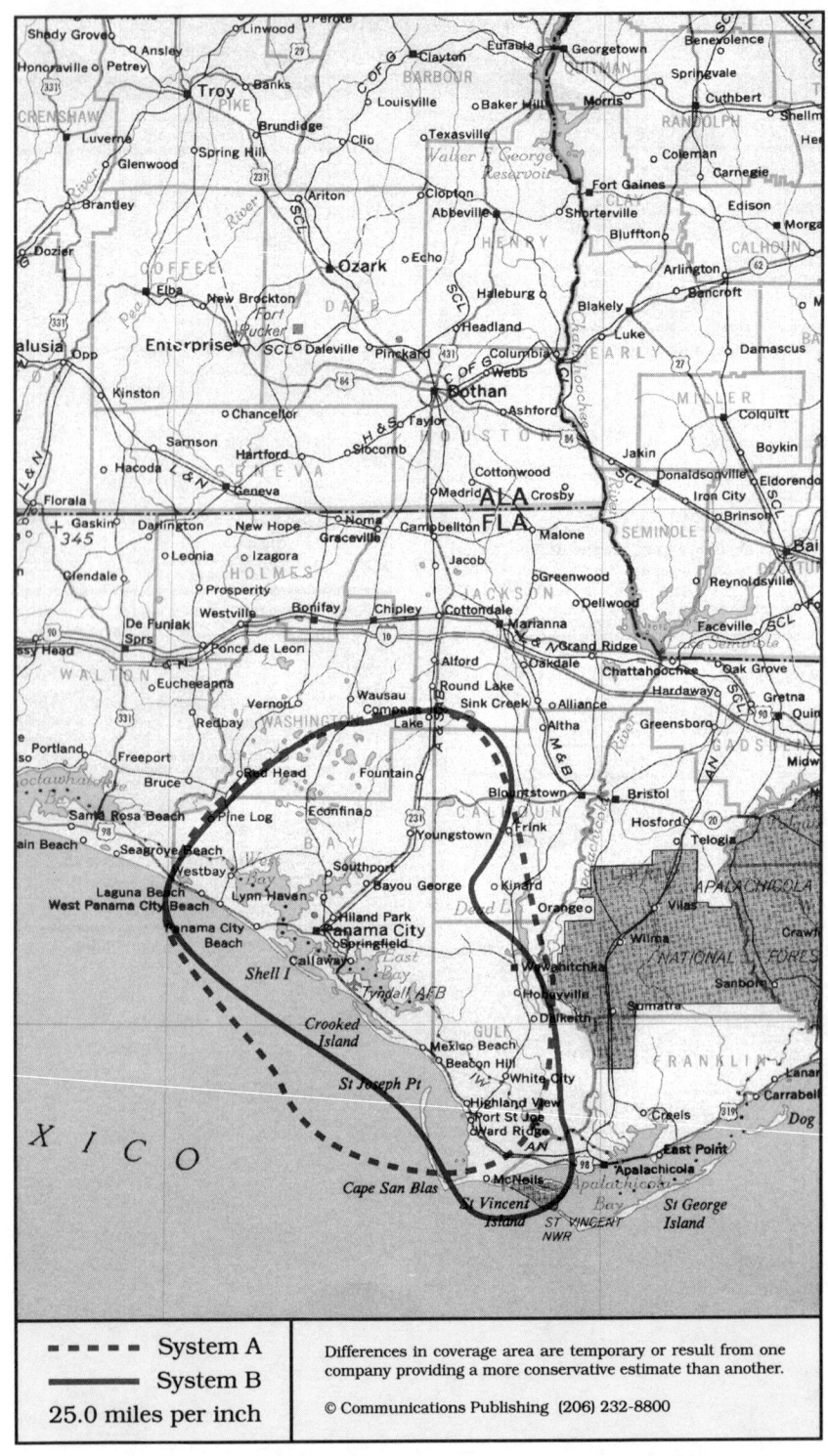

System A
System B
25.0 miles per inch

Differences in coverage area are temporary or result from one company providing a more conservative estimate than another.

© Communications Publishing (206) 232-8800

Panama City, Florida
Panama City, Port St. Joe

|

Cellular company

Cellular One
Managed by: Palmer Communications
505 W 15th Street
Panama City, FL 32401

Customer service	904-784-3507
Local office	904-769-2269
Corporate office . .	Palmer	813-433-4350

Sending calls

Visitor dialing instructions
Local calls . . . . . area code + 7 digits
Long distance . . . 1 + area code + 7 digits

Speed-dial numbers
Customer service 611
Emergency 911

Receiving calls

Caller dials the roamer access number below,
hears tone, dials cellular area code and number.
. . . . 904-866-7626

NationLink & RoamingAmerica (see p. 1157)
Dial ✳31 send to receive calls, ✳30 to deactivate.
Dial ✳32 send to tell caller the roamer access no.
This company's subscribers have level 3 service.

Billing information

Visitor rates
Most visitors . . . . . $3 day + 85¢ minute

Visitors without roaming agreements (p. 1159)
Call co. to use Amex, Discover, Mastercard, Visa

Identification numbers

City	Market	System	Billing	RSA
All served cities	283	00483	00000	MSA

Cellular company

Centel Cellular
2503 Highway 77 N
Panama City, FL 32405

112 Reid Avenue
Port St. Joe, FL 32456

Local office . .	Panama City	904-785-7000
. . . .	Port St. Joe	904-227-1000
Corporate office	312-399-2644

Sending calls

Visitor dialing instructions
Local calls . . . . area code + 7 digits
Long distance . . . . . area code + 7 digits

Speed-dial numbers
Customer service . . . . ✳611
Emergency 911

Receiving calls

Caller dials the roamer access number below,
hears tone, dials cellular area code and number.
. . . . 904-832-7626

Follow Me Roaming forwarding (see p. 1156)
Activate, dial ✳18 send. Deactivate dial ✳19 send.

Billing information

Visitor rates
Most visitors . . . . . $3 day + 99¢ minute

Visitors without roaming agreements (p. 1159)
Cannot call since company takes no credit cards.

Identification numbers

City	Market	System	Billing	RSA
Panama City	283	00462	00000	MSA
Port St. Joe	368	01124	00000	FL-9

For full coverage area, see Marianna, FL.

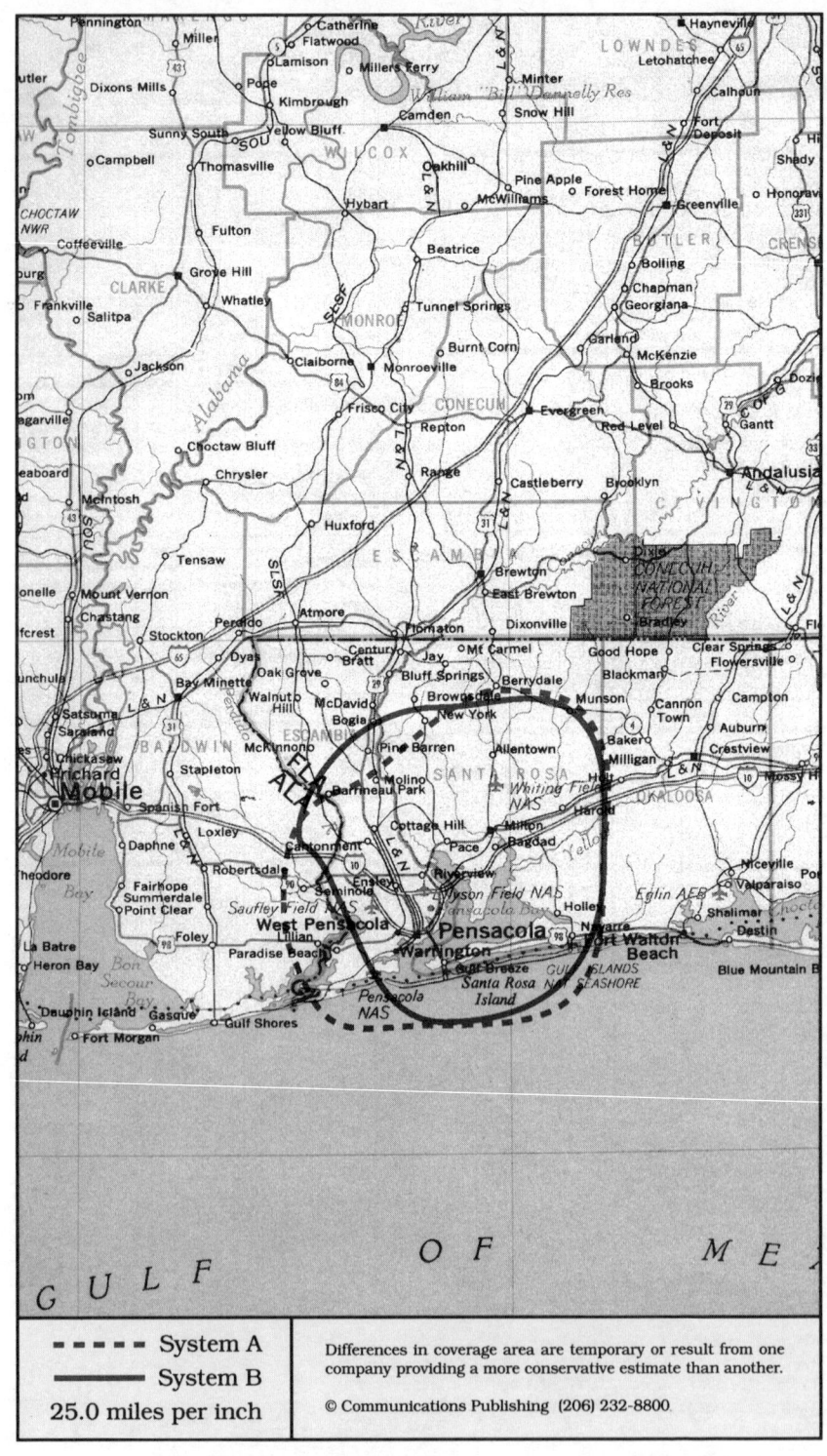

- - - - - System A —————— System B 25.0 miles per inch	Differences in coverage area are temporary or result from one company providing a more conservative estimate than another. © Communications Publishing (206) 232-8800

Pensacola, Florida
Milton, Pensacola

System A	System B

Cellular company

Cellular One
Vanguard Cellular Systems
418 E Gregory Street
Pensacola, FL 32501

Customer service	904-449-4000
Local office	904-433-7300
Corporate office	919-282-3690

Sending calls

Visitor dialing instructions

Local calls	area code + 7 digits
Long distance . . .	area code + 7 digits

Speed-dial numbers

Customer service	611
Emergency	911

Receiving calls

Caller dials the roamer access number below,
hears tone, dials cellular area code and number.
. . . . 904-449-7626

Billing information

Visitor rates
Most visitors . . . . . $3 day + 99¢ minute

Visitors without roaming agreements (p. 1159)
Call co. to use Amex, Mastercard, Visa
or Roamer Plus will intercept first call to bill you.

Identification numbers

City	Market	System	Billing	RSA
All served cities	127	00361	30021	MSA

For full coverage area, see Fort Walton Beach, FL.

Cellular company

Contel Cellular
1630 Airport Boulevard, Suite 220
Pensacola, FL 32504

Customer service	800-333-4004
. . . .	205-970-2273
Local office	904-479-8664
Corporate office	404-804-3400

Sending calls

Visitor dialing instructions

Local calls	area code + 7 digits
Long distance . . .	1 + area code + 7 digits

Speed-dial numbers

Customer service	611
Directory assistance	411
Emergency	911

Receiving calls

Caller dials the roamer access number below,
hears tone, dials cellular area code and number.
. . . . 904-572-7626

Follow Me Roaming forwarding (see p. 1156)
Activate, dial *18 send. Deactivate dial *19 send.

Billing information

Visitor rates
Most visitors . . . . . $3 day + 85¢ minute

Visitors without roaming agreements (p. 1159)
Call co. to use Amex, Mastercard, Visa

Identification numbers

City	Market	System	Billing	RSA
All served cities	127	00120	00000	MSA

For full coverage area, see Mobile, AL.

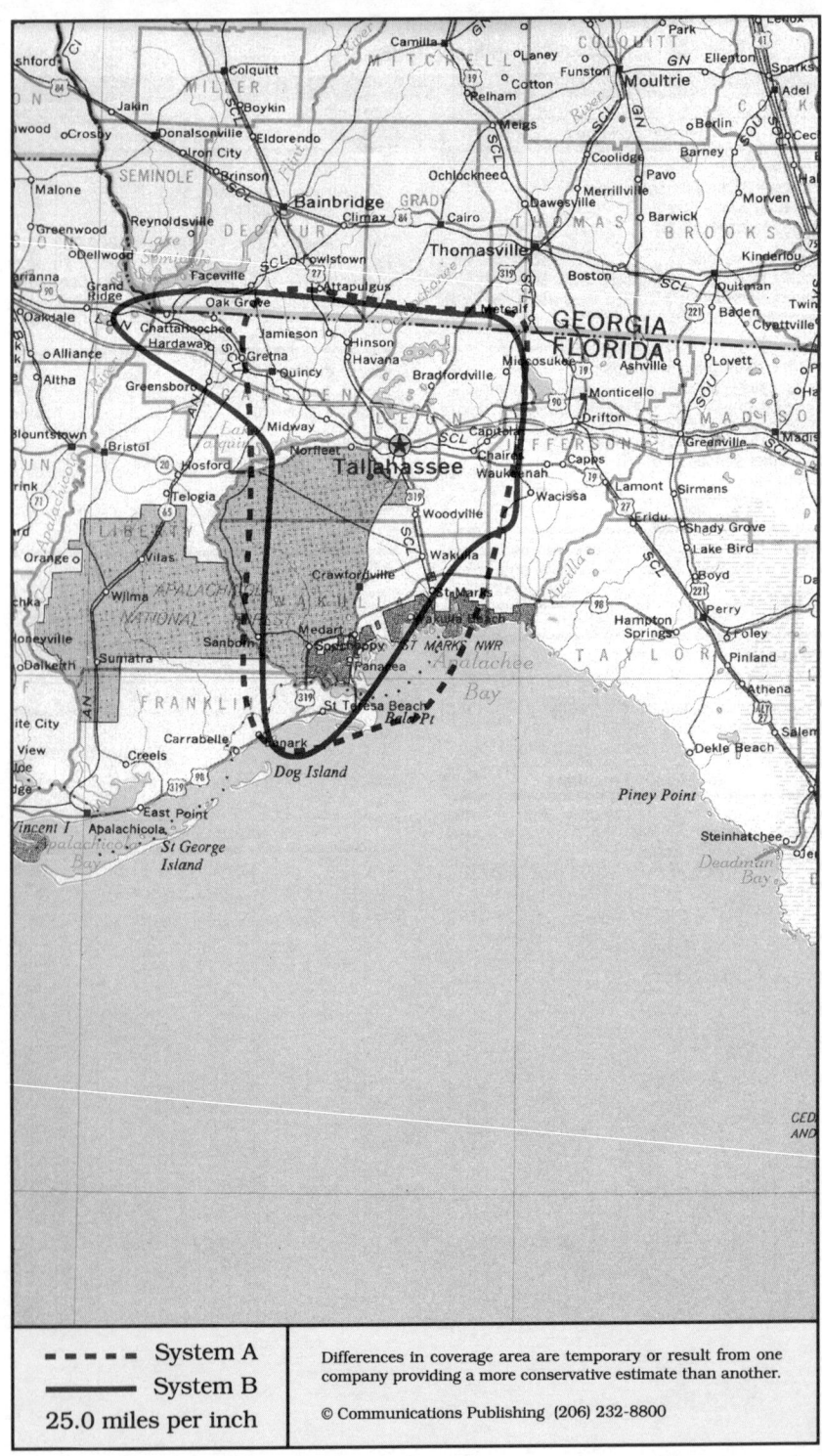

- - - - - System A	Differences in coverage area are temporary or result from one
———— System B	company providing a more conservative estimate than another.
25.0 miles per inch	© Communications Publishing (206) 232-8800

Tallahassee, Florida
Crawfordville, Quincy, Tallahassee

Cellular company

Cellular One
McCaw Cellular Communications
2735 Capital Circle NE
Tallahassee, FL 32308

Customer service . .	Orlando	800-822-3551
. . . .		407-865-8961
Local office		904-386-8999
Corporate office		206-827-4500

Sending calls

Visitor dialing instructions
Local calls . . . . area code + 7 digits
Long distance . . . 1 + area code + 7 digits

Speed-dial numbers
Customer service 611
Emergency 911

Receiving calls

Caller dials the roamer access number below,
hears tone, dials cellular area code and number.
. . . . 904-566-7626

NationLink & RoamingAmerica (see p. 1157)
Dial *31 send to receive calls, *30 to deactivate.
Dial *32 send to tell caller the roamer access no.
This company's subscribers have level 7 service.

North American Cellular Network (see p. 1156)
All calls forward to you. Deactivate dial *35 send.

Billing information

Visitor rates
Most visitors . . . . . $0 day + 99¢ minute

Visitors without roaming agreements (p. 1159)
Roamer Plus will intercept first call to bill you.

Identification numbers

City	Market	System	Billing	RSA
All served cities	168	00565	00000	MSA

Cellular company

Centel Cellular
1401 Market Street
Tallahassee, FL 32312

Local office	904-847-4000
Corporate office	312-399-2644

Sending calls

Visitor dialing instructions
Local calls . . . . . . area code + 7 digits
Long distance . . . 1 + area code + 7 digits

Speed-dial numbers
Customer service *611
Directory assistance 411
Emergency 911
WCTV news *600

Receiving calls

Caller dials the roamer access number below,
hears tone, dials cellular area code and number.
. . . . 904-545-7626

Follow Me Roaming forwarding (see p. 1156)
Activate, dial *18 send. Deactivate dial *19 send.

Billing information

Visitor rates
Most visitors . . . . . $3 day + 99¢ minute

Visitors without roaming agreements (p. 1159)
Cannot call since company takes no credit cards.

Identification numbers

City	Market	System	Billing	RSA
All served cities	168	00544	00000	MSA

For full coverage area, see Madison, FL.

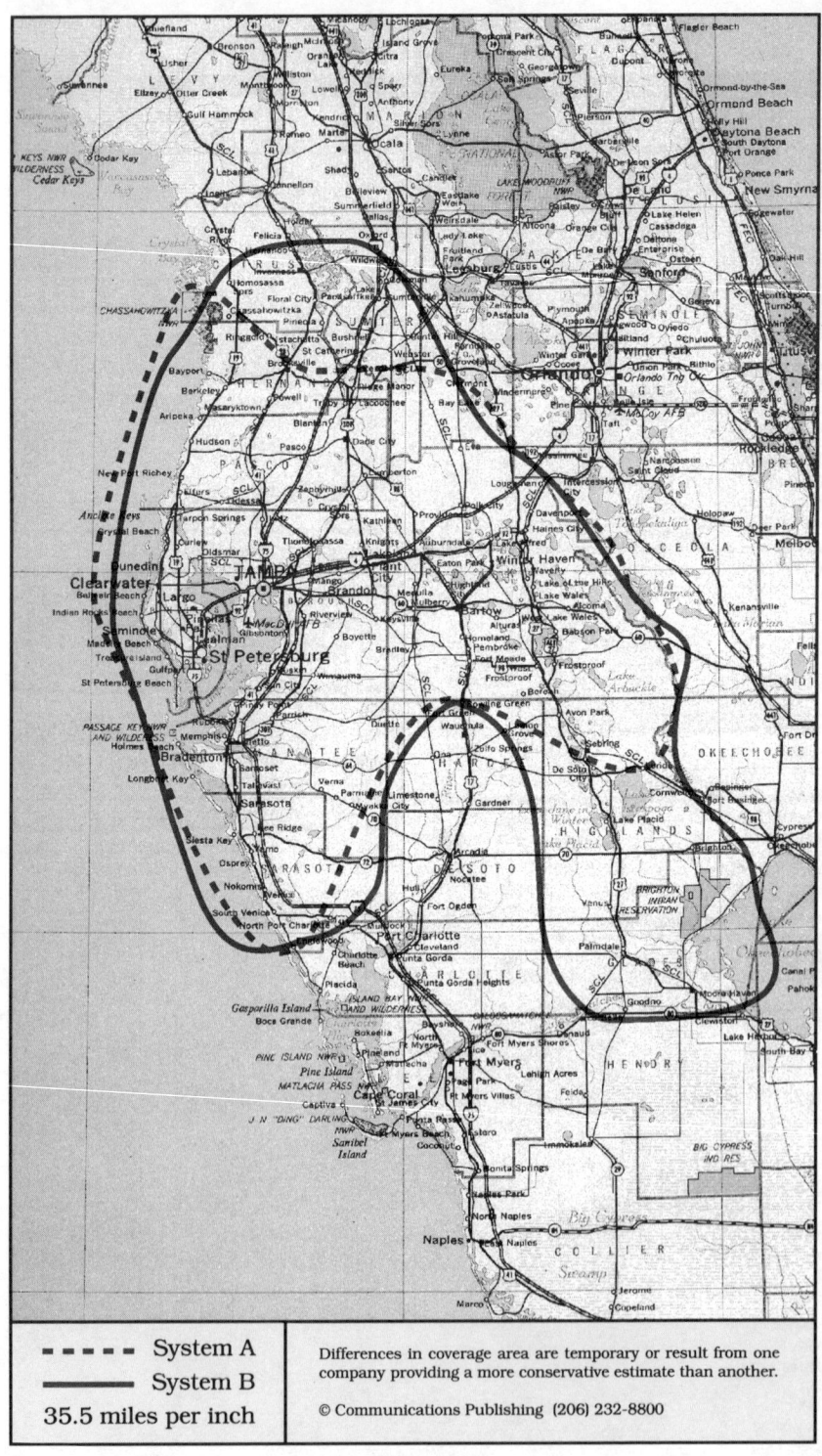

System A
System B
35.5 miles per inch

Differences in coverage area are temporary or result from one company providing a more conservative estimate than another.

© Communications Publishing (206) 232-8800

Tampa, Florida
Bradenton, Brooksville, Lakeland, St. Petersburg, Sarasota, Sebring, Tampa

System A	System B

System A

Cellular company

Cellular One
McCaw Communications   Local office
19462 Cortez Boulevard
Brooksville, FL 34601   904-544-0157

1705 34th Street N
St. Petersburg, FL 33713   813-894-9494

4832 S Tamiami Trail
Sarasota, FL 34231   813-924-2227

501 E Kennedy Blvd. #1150
Tampa, FL 33602   813-222-5555

11209 N Del Mabry Highway
Tampa, FL 33618   813-264-2644

Customer service . .	Tampa	800-822-3551
. . . .	Tampa	813-221-3400
Corporate office		206-827-4500

Sending calls

Visitor dialing instructions
Local calls . . . . . area code + 7 digits
Long distance . . . 1 + area code + 7 digits

Speed-dial numbers
Action news	*10
Customer service	611
Emergency	911
Radio station W101	9101
Television station WHVE	*1025

Receiving calls

Caller dials the roamer access number below,
hears tone, dials cellular area code and number.
Bradenton, Sarasota	813-350-7626
Citrus County	904-563-7626
Clearwater, St. Petersburg .	813-460-7626
Hernando County (Brooksville)	904-544-7626
Lakeland	813-660-7626
Sebring	813-471-4000
Tampa	813-240-7626

NationLink & RoamingAmerica (see p. 1157)
Dial *31 send to receive calls, *30 to deactivate.
Dial *32 send to tell caller the roamer access no.
This company's subscribers have level 7 service.

North American Cellular Network (see p. 1156)
All calls forward to you. Deactivate dial *35 send.

Billing information

Visitor rates
Most visitors . . . . . $0 day + 99¢ minute

Visitors without roaming agreements (p. 1159)
Call co. to use Amex, Mastercard, Visa
or Roamer Plus will intercept first call to bill you.

Identification numbers

City	Market	System	Billing	RSA
Bradenton	211	00175	00000	MSA
Brooksville	363	00175	30261	FL-4
Lakeland	114	00175	00000	MSA
Sarasota	167	00175	00000	MSA
Sebring	361	00175	30309	FL-2
Tampa	022	00175	30283	MSA

System B

Cellular company

GTE Mobilnet
600 N Westshore Boulevard, Suite 900
Tampa, FL 33609

Customer service	800-877-5665
Local office	813-282-6000
Corporate office	404-391-8000

Sending calls

Visitor dialing instructions
Local calls . . . . . area code + 7 digits
Long distance . . . 1 + area code + 7 digits

Speed-dial numbers
Customer service	*611
Emergency	911
Make a traffic report	105
News tip line	620*
Report crime information	*TIP
Roaming information	*711
Talk line	95*
Telephone problems	*111
Weatherline	*103

Receiving calls

Caller dials the roamer access number below,
hears tone, dials cellular area code and number.
Bradenton	813-745-7626
Brooksville, Dade City	904-754-7626
Clearwater	813-441-7626
Lakeland	813-647-7626
New Port Richey	813-844-7626
Sarasota	813-745-7626
Sebring	813-471-7626
Tampa	813-623-7626

Follow Me Roaming forwarding (see p. 1156)
Activate, dial *18 send. Deactivate dial *19 send.

Billing information

Visitor rates
Most visitors . . . . . $2 day + 75¢ minute

Visitors without roaming agreements (p. 1159)
Call co. to use Amex, Mastercard, Visa

Identification numbers

City	Market	System	Billing	RSA
Bradenton	211	00042	00000	MSA
Brooksville	363	00042	00000	FL-4
Lakeland	114	00042	00000	MSA
Sarasota	167	00042	00000	MSA
Sebring	361	00042	00000	FL-2
Tampa	022	00042	00000	MSA

For full coverage area, see Fort Myers, FL.

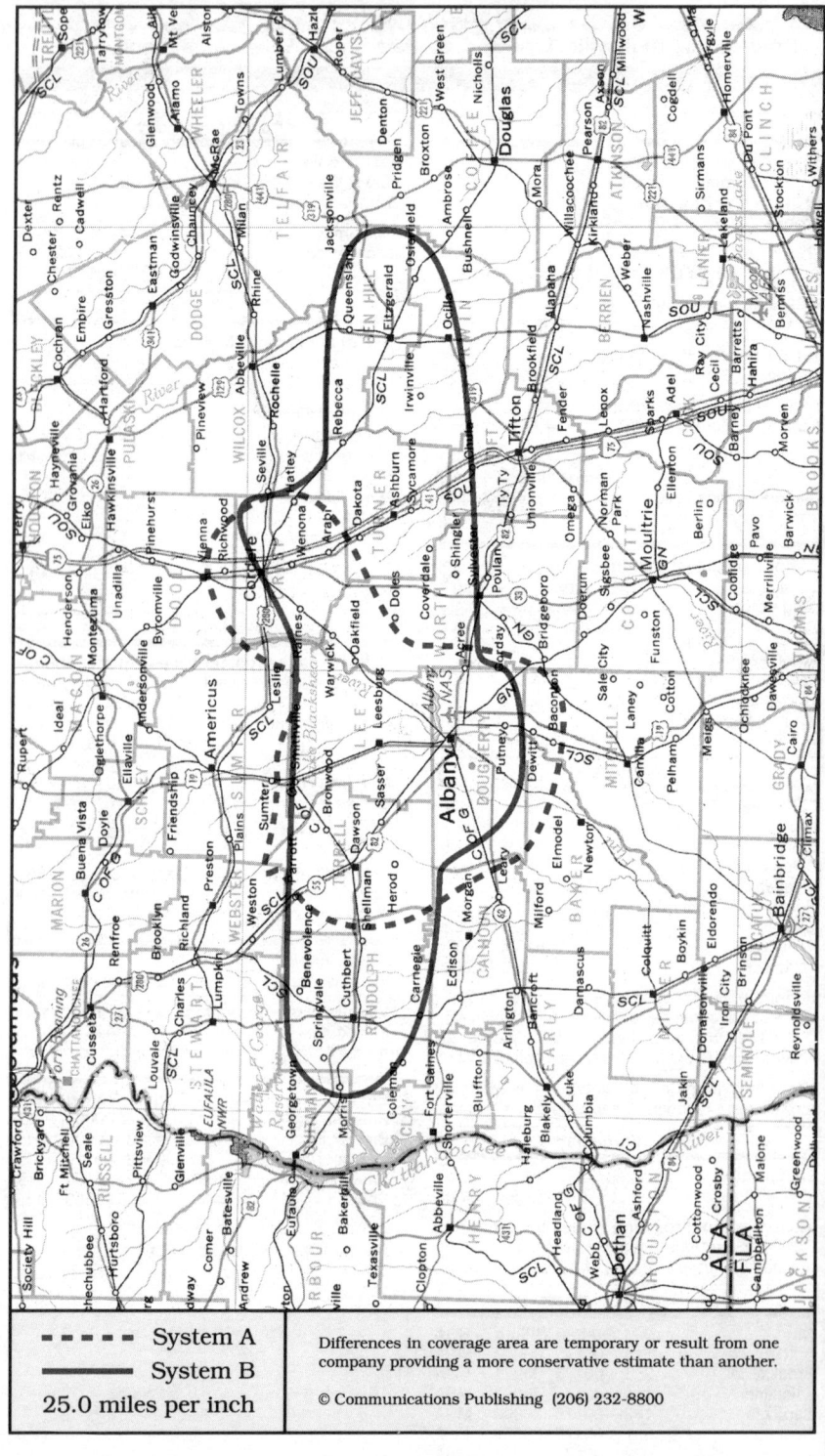

System A

System B

25.0 miles per inch

Differences in coverage area are temporary or result from one company providing a more conservative estimate than another.

© Communications Publishing (206) 232-8800

Albany, Georgia
Albany, Ashburn (system B), Cordele, Dawson, Fitzgerald

System A	System B

Cellular company

Cellular One
Palmer Communications
500 Pine Avenue
Albany, GA 31701

Customer service	800-441-2796
Local office	912-431-2355
Corporate office	813-433-4350

Sending calls

Visitor dialing instructions

Local calls	area code + 7 digits
Long distance . . .	1 + area code + 7 digits

Speed-dial numbers

Crime Stoppers	*11
Customer service	611
Emergency	911
Police	*99

Receiving calls

Caller dials the roamer access number below,
hears tone, dials cellular area code and number.
. . . . 912-886-7626

NationLink & RoamingAmerica (see p. 1157)
Dial *31 send to receive calls, *30 to deactivate.
Dial *32 send to tell caller the roamer access no.
This company's subscribers have level 6 service.

Billing information

Visitor rates
Most visitors . . . . . $3 day + 85¢ minute

Visitors without roaming agreements (p. 1159)
Call co. to use Amex, Discover, Mastercard, Visa

Identification numbers

City	Market	System	Billing	RSA
Albany	261	00241	00000	MSA
Cordele	379	01145	00000	GA-9

For full coverage area, see Columbus, GA.

Cellular company

ALLTEL Mobile Communications
2700 Dawson Road, Suite 10
Albany, GA 31707

Customer service	some areas	800-255-8351
Local office		912-888-8200
Corporate office		501-661-8500

Sending calls

Visitor dialing instructions

Local calls	area code + 7 digits
Long distance	area code + 7 digits

Speed-dial numbers

Customer service	*611
Emergency	911
Make traffic report WIKX	549

Receiving calls

Caller dials the roamer access number below,
hears tone, dials cellular area code and number.

Albany	912-881-7626
Cordele	912-276-5626

Follow Me Roaming forwarding (see p. 1156)
Activate, dial *18 send. Deactivate dial *19 send.

Billing information

Visitor rates
Most visitors . . . . . $3 day + 85¢ minute

Visitors without roaming agreements (p. 1159)
Call co. to use Amex, Discover, Mastercard, Visa

Identification numbers

City	Market	System	Billing	RSA
Albany	261	00204	00000	MSA
Ashburn	380	00204	00000	GA-10
Cordele	379	00204	30284	GA-9
Dawson	379	00204	00000	GA-9
Fitzgerald	384	00204	00000	GA-14

For full coverage area, see Valdosta, GA.

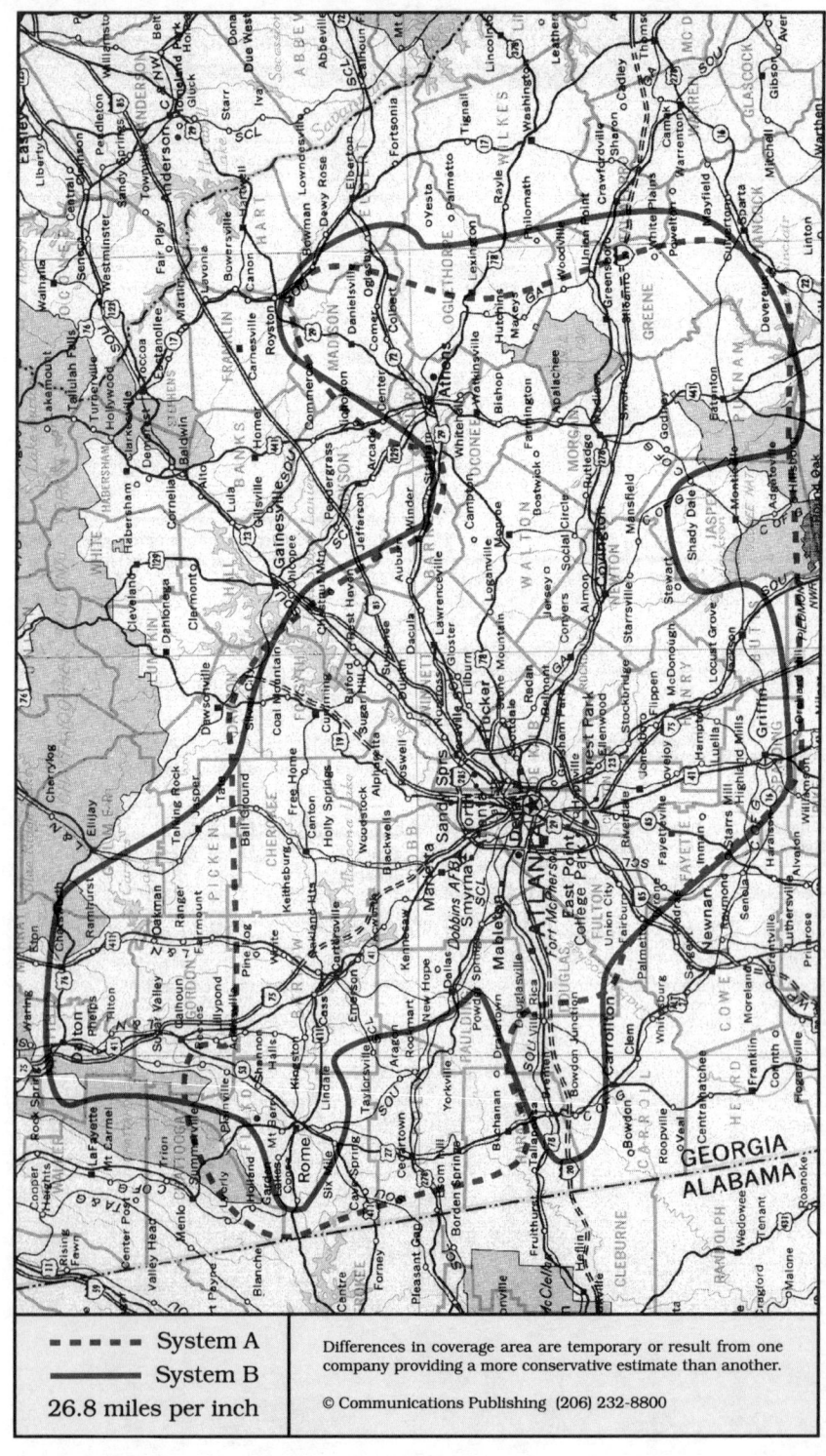

- - - - - System A
———— System B
26.8 miles per inch

Differences in coverage area are temporary or result from one company providing a more conservative estimate than another.

© Communications Publishing (206) 232-8800

Atlanta, Georgia

Athens, Atlanta, Calhoun, Cartersville, Covington, Dalton, Madison, Rome

System A	System B

Cellular company

PacTel Cellular
196 Alps Road
Athens, GA 30606

4151 Ashford Dunwoody Road NE #300
Atlanta, GA 30319-1462

Customer service	800-235-5611
. . . .	404-257-5100
Local office . . . . Athens	706-549-4665
. . . . Norcross	404-257-5000
Corporate office	510-210-3600

Sending calls

Visitor dialing instructions
Local calls	area code + 7 digits
Long distance . . .	0 + area code + 7 digits

Speed-dial numbers
Customer service	*611
Emergency	911
Make a traffic report	*640
Traffic report center	*941

Receiving calls

Caller dials the roamer access number below,
hears tone, dials cellular area code and number.
Athens	706-338-7626
Atlanta	404-558-7626
Madison	706-557-4626
Rome	706-290-7626

NationLink & RoamingAmerica (see p. 1157)
Dial *31 send to receive calls, *30 to deactivate.
Dial *32 send to tell caller the roamer access no.
This company's subscribers have level 8 service.

Billing information

Visitor rates
Most visitors . . . . . $3 day + 89¢ minute

Visitors without roaming agreements (p. 1159)
Call co. to use Amex, Mastercard, Visa
or Roamer Plus will intercept first call to bill you.

Identification numbers

City	Market	System	Billing	RSA
Athens	234	00041	00000	MSA
Atlanta	017	00041	00000	MSA
Rome	373	01133	30315	GA-3
Madison	374	01135	30321	GA-4

Cellular company

BellSouth Mobility
4805 Briarcliff Road, Suite 101
Atlanta, GA 30345

120 Interstate N Parkway E, Bldg 100, #110
Atlanta, GA 30339

114 Southfield Parkway, Building E, Suite 130
Forest Park, GA 30050

Customer service	800-351-2400
. . . .	404-847-2400
Local office . . Briarcliff Road	404-270-4000
. . . . . Interstate Pkwy	404-980-4100
. . . . . Southfield Pkwy	404-362-5700
Corporate office	404-847-3600

Sending calls

Visitor dialing instructions
Local calls	area code + 7 digits
Long distance	area code + 7 digits

Speed-dial numbers
Customer service	811
Emergency	911
Roaming information	711
Technical problems	611

Receiving calls

Caller dials the roamer access number below,
hears tone, dials cellular area code and number.
Adairsville	404-773-4626
Atlanta	404-372-7626
Athens	706-540-7626
Calhoun	404-773-4626
Cartersville	404-387-8000
Dalton	404-217-7626
Eatonton	706-485-1626
Lake Oconee	706-453-8626
Madison	404-372-7626
Rome	706-236-7626

Follow Me Roaming forwarding (see p. 1156)
Activate, dial *18 send. Deactivate dial *19 send.

Billing information

Visitor rates
Most visitors	$3 day + 85¢ minute
BellSouth subscribers .	$0 day + 60¢ minute

Visitors without roaming agreements (p. 1159)
Call co. to use Amex, Mastercard, Visa

Identification numbers

City	Market	System	Billing	RSA
Athens	234	00034	30628	MSA
Atlanta	017	00034	00000	MSA
Dalton	371	00034	30268	GA-1
Rome, Cartersville	373	00034	30272	GA-3
Madison, Eatonton	374	00034	30274	GA-4

For full coverage area, see Carrollton, GA and
Gainesville, GA and Milledgeville, GA.

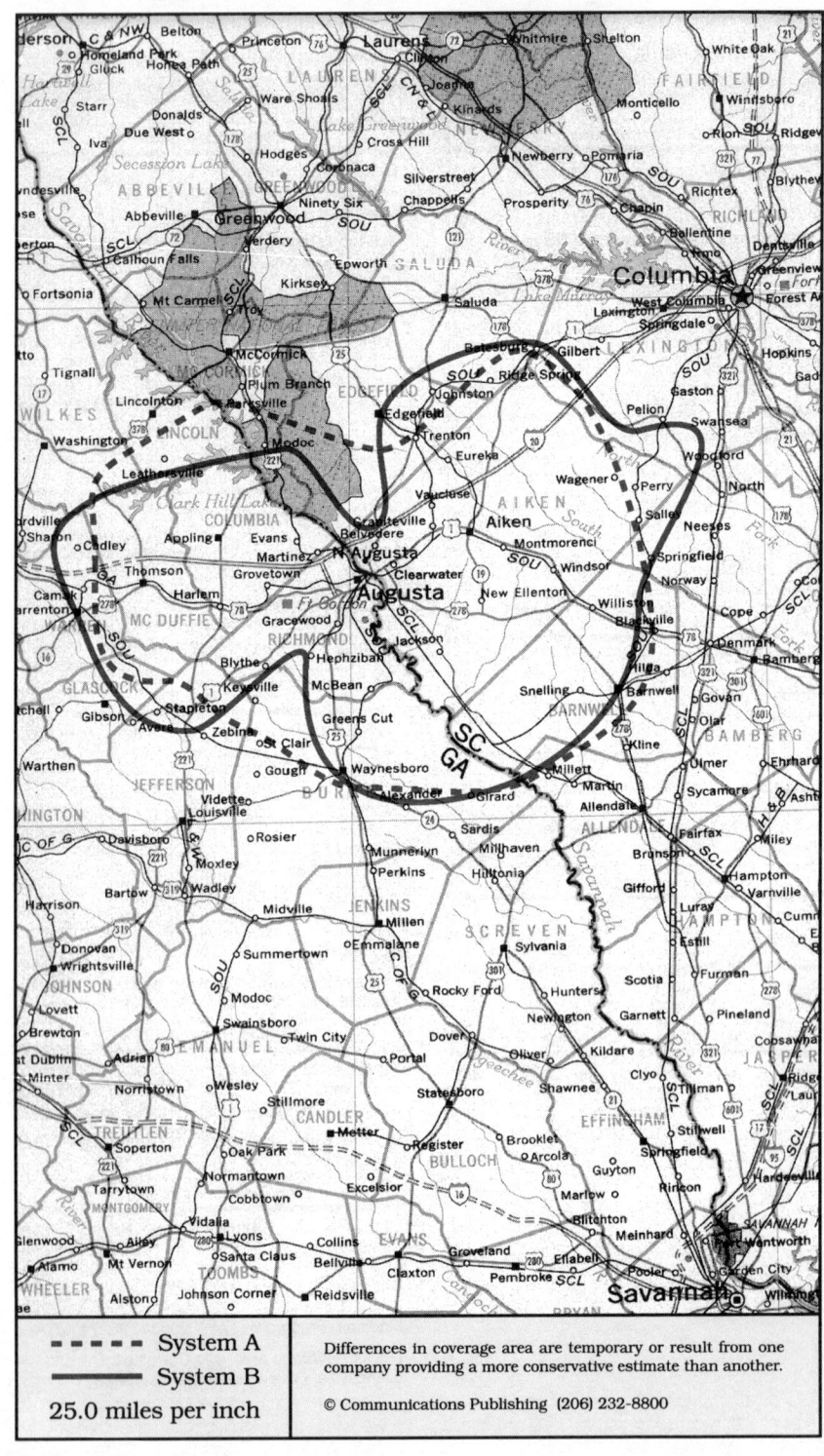

Differences in coverage area are temporary or result from one company providing a more conservative estimate than another.

© Communications Publishing (206) 232-8800

Augusta, Georgia
Aiken SC, Augusta, Thomson, Waynesboro

Cellular company

Cellular One
GTE Mobilnet
3241-C Washington Road
Augusta, GA 30907

Customer service		800-727-2444
. . . .		919-481-1181
Local office . . from Georgia		706-868-0086
. . . from South Carolina		803-640-0086
Corporate office		404-391-8000

Sending calls

Visitor dialing instructions
Local calls . . . 1 + area code + 7 digits
Long distance . . . 1 + area code + 7 digits

Speed-dial numbers
Customer service	611
Emergency	911
Make a traffic report	*2105
South Carolina highway patrol	*47

Receiving calls

Caller dials the roamer access number below,
hears tone, dials cellular area code and number.
Georgia 706-825-7626
South Carolina 803-640-7626

Follow Me Roaming forwarding (see p. 1156)
Phone Me Anywhere forwarding (see p. 1156)
Activate, dial *18 send. Deactivate dial *19 send.

NationLink & RoamingAmerica (see p. 1157)
Dial *31 send to receive calls, *30 to deactivate.
Dial *32 send to tell caller the roamer access no.
This company's subscribers have level 10 service

Billing information

Visitor rates
Most visitors . . . . . $3 day + 85¢ minute

Visitors without roaming agreements (p. 1159)
Call co. to use Amex, Mastercard, Visa

Identification numbers

City	Market	System	Billing	RSA
All served cities	108	00181	00000	MSA

Cellular company

ALLTEL Mobile Communications
2903 Washington Road
Augusta, GA 30909-2114

Customer service	some areas	800-255-8351
Local office		706-738-2355
Corporate office		501-661-8500

Sending calls

Visitor dialing instructions
Local calls . . . . . . area code + 7 digits
Long distance . . . . . area code + 7 digits

Speed-dial numbers
Customer service	*611
Emergency	911

Receiving calls

Caller dials the roamer access number below,
hears tone, dials cellular area code and number.
Georgia 706-829-7626
South Carolina 803-645-7626

Follow Me Roaming forwarding (see p. 1156)
Activate, dial *18 send. Deactivate dial *19 send.

Billing information

Visitor rates
Most visitors . . . . . $3 day + 85¢ minute

Visitors without roaming agreements (p. 1159)
Call co. to use Amex, Discover, Mastercard, Visa

Identification numbers

City	Market	System	Billing	RSA
All served cities	108	00084	00000	MSA

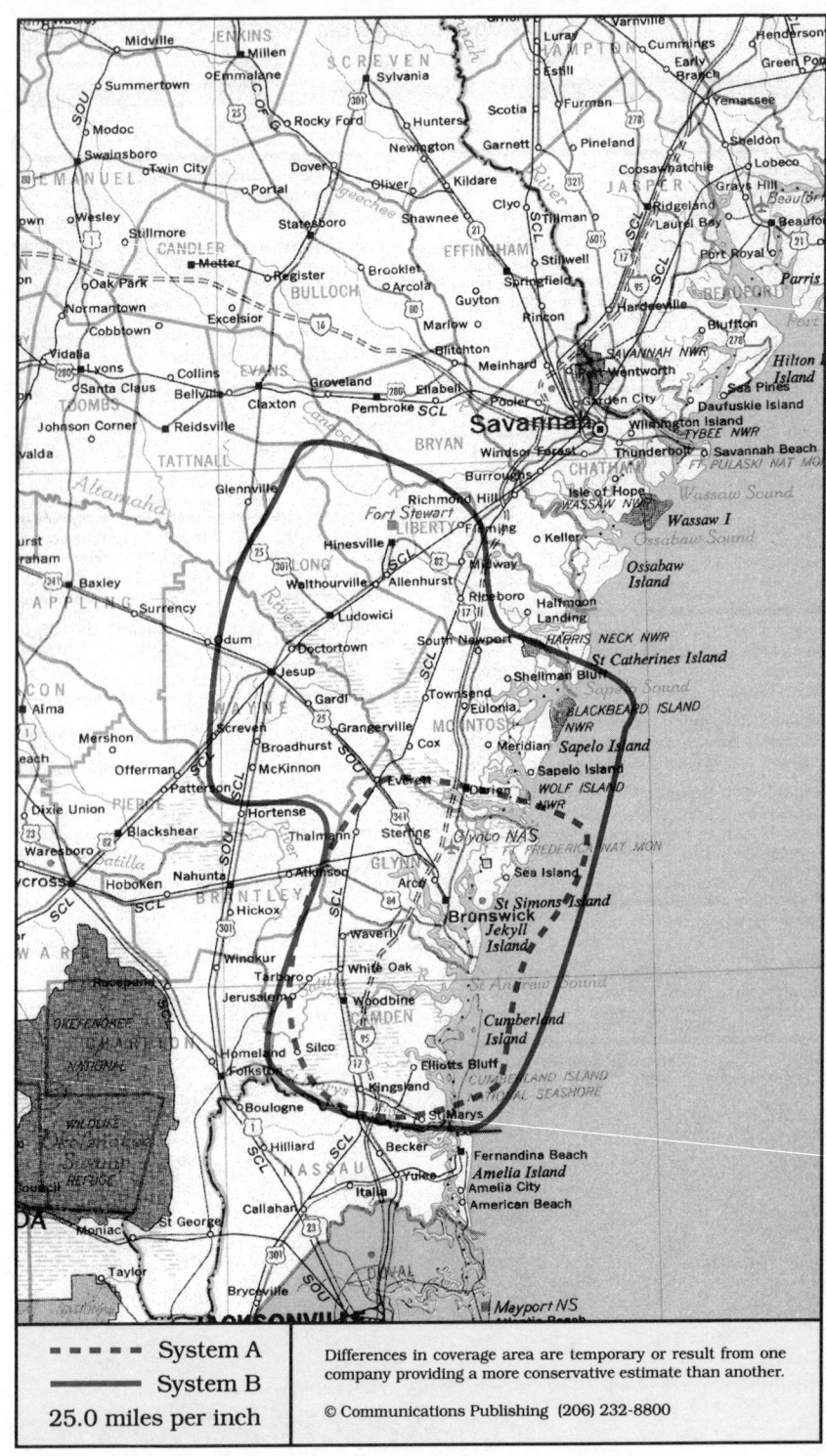

▬ ▬ ▬ ▬ System A	Differences in coverage area are temporary or result from one company providing a more conservative estimate than another.
▬▬▬▬ System B	
25.0 miles per inch	© Communications Publishing (206) 232-8800

Brunswick, Georgia
Brunswick, Hinesville, Jesup, Ludowici, Woodbine

System A	System B

Cellular company

Cellular One
Sterling Cellular Management
1080 Holcomb Bridge Road, Bldg 100, Ste 200
Roswell, GA 30076

Customer service	800-552-6150
Corporate office	404-552-5030

Sending calls

Visitor dialing instructions

Local calls	area code + 7 digits
Long distance . . .	1 + area code + 7 digits

Speed-dial numbers

Customer service	611
Emergency	911
Roaming information	711

Receiving calls

Caller dials the roamer access number below,
hears tone, dials cellular area code and number.

. . . .	912-222-7626

Billing information

Visitor rates

Most visitors	$3 day + 99¢ minute

Visitors without roaming agreements (p. 1159)
Call co. to use Amex, Mastercard, Visa
or Roamer Plus will intercept first call to bill you.

Identification numbers

City	Market	System	Billing	RSA
All served cities	382	01151	00000	GA-12

Cellular company

ALLTEL Mobile Communications
7001 Chatham Center Drive, Suite 1600
Savannah, GA 31405

Customer service	some areas	800-255-8351
Local office		800-733-7626
. . . .		912-651-6100
Corporate office		501-661-8500

Sending calls

Visitor dialing instructions

Local calls	area code + 7 digits
Long distance . . .	1 + area code + 7 digits

Speed-dial numbers

Customer service	611
Emergency	911
Hear a traffic report	*102
Weather report	*97

Receiving calls

Caller dials the roamer access number below,
hears tone, dials cellular area code and number.

. . . .	912-658-7626

Follow Me Roaming forwarding (see p. 1156)
Activate, dial *18 send. Deactivate dial *19 send.

Billing information

Visitor rates

Most visitors	$3 day + 90¢ minute

Visitors without roaming agreements (p. 1159)
Cannot call since company takes no credit cards.

Identification numbers

City	Market	System	Billing	RSA
All served cities	382	00520	30376	GA-12

For full coverage area, see Savannah, GA and
Waycross, GA.

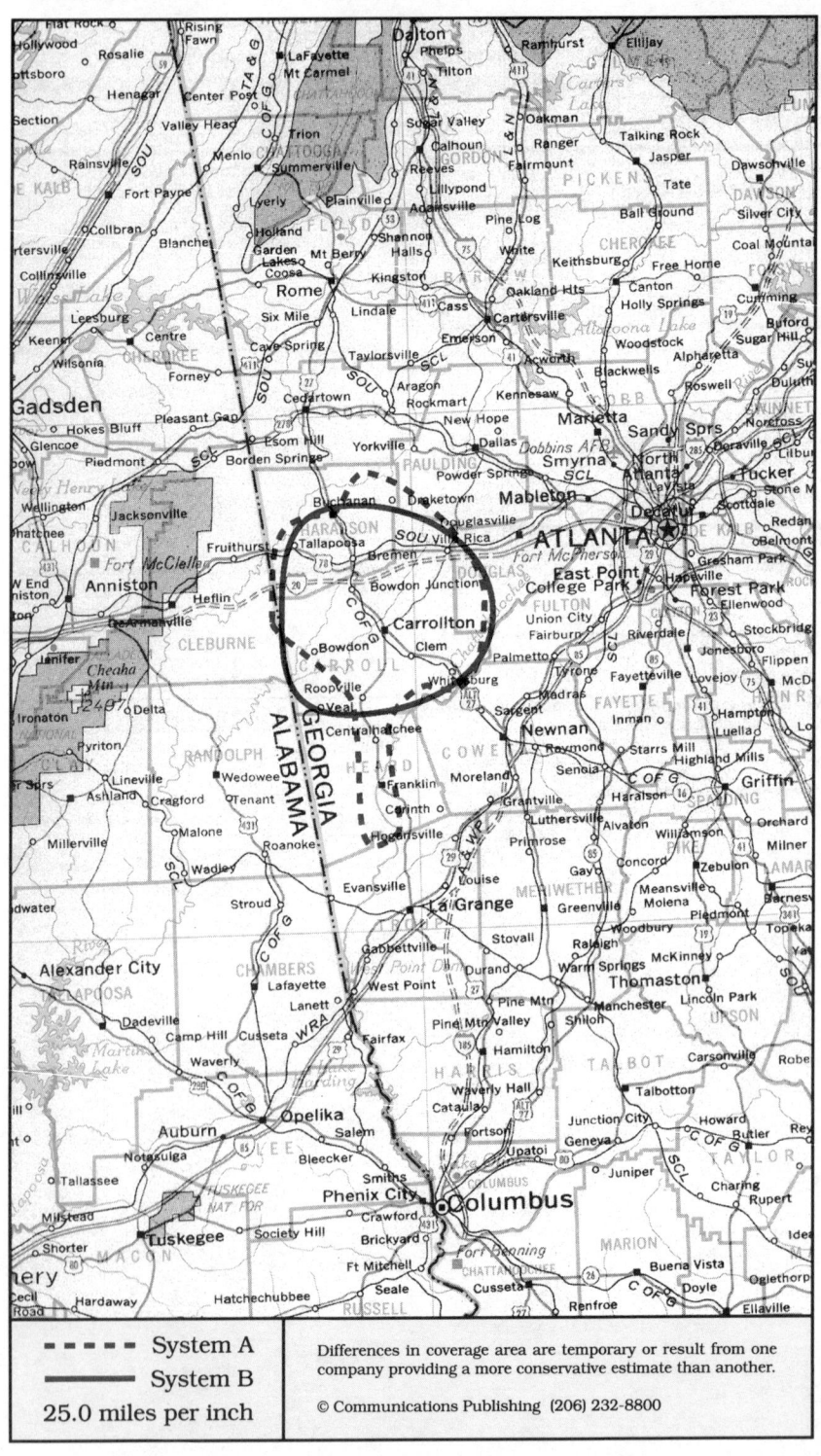

- - - - - System A
——————— System B
25.0 miles per inch

Differences in coverage area are temporary or result from one company providing a more conservative estimate than another.

© Communications Publishing (206) 232-8800

Carrollton, Georgia
Bremen, Carrollton

System A	System B

Cellular company

Cellular One
Licensed to: Blackwater Cellular
Managed by: Rural Cellular Management
26-A Bullsboro Drive
Newnan, GA 30263

Local office   404-304-1111
Corporate office   Rural Cell. Mgt   707-422-2100

Sending calls

Visitor dialing instructions
Local calls . . . . .   area code + 7 digits
Long distance . . .   1 + area code + 7 digits

Speed-dial numbers
Customer service   611
Emergency   911
Towing service . . . . .   *HELP or *4357

Receiving calls

Caller dials the roamer access number below,
hears tone, dials cellular area code and number.
. . . .   404-301-7626

NationLink & RoamingAmerica (see p. 1157)
Dial *31 send to receive calls, *30 to deactivate.
This company's subscribers have level 3 service.

Billing information

Visitor rates
Most visitors . . . . . $3 day + 85¢ minute
Visitors from Georgia   . $0 day + 79¢ minute

Visitors without roaming agreements (p. 1159)
Roamer Plus will intercept first call to bill you.

Identification numbers

City	Market	System	Billing	RSA
All served cities	375	01137	00000	GA-5

For full coverage area, see La Grange, GA.

Cellular company

BellSouth Mobility
4805 Briarcliff Road, Suite 101
Atlanta, GA 30345

Customer service   800-351-2400
. . . .   404-847-2400
Local office   404-270-4000
Corporate office   404-847-3600

Sending calls

Visitor dialing instructions
Local calls . . . . . .   area code + 7 digits
Long distance . . . . .   area code + 7 digits

Speed-dial numbers
Customer service   811
Emergency   911
Roaming information   711
Telephone problems   611

Receiving calls

Caller dials the roamer access number below,
hears tone, dials cellular area code and number.
. . . .   404-537-7626

Follow Me Roaming forwarding (see p. 1156)
Activate, dial *18 send. Deactivate dial *19 send.

Billing information

Visitor rates
Most visitors   . . . . . $3 day + 85¢ minute
BellSouth subscribers   . $0 day + 60¢ minute

Visitors without roaming agreements (p. 1159)
Call co. to use Amex, Mastercard, Visa

Identification numbers

City	Market	System	Billing	RSA
All served cities	375	00034	30276	GA-5

For full coverage area, see Atlanta, GA.

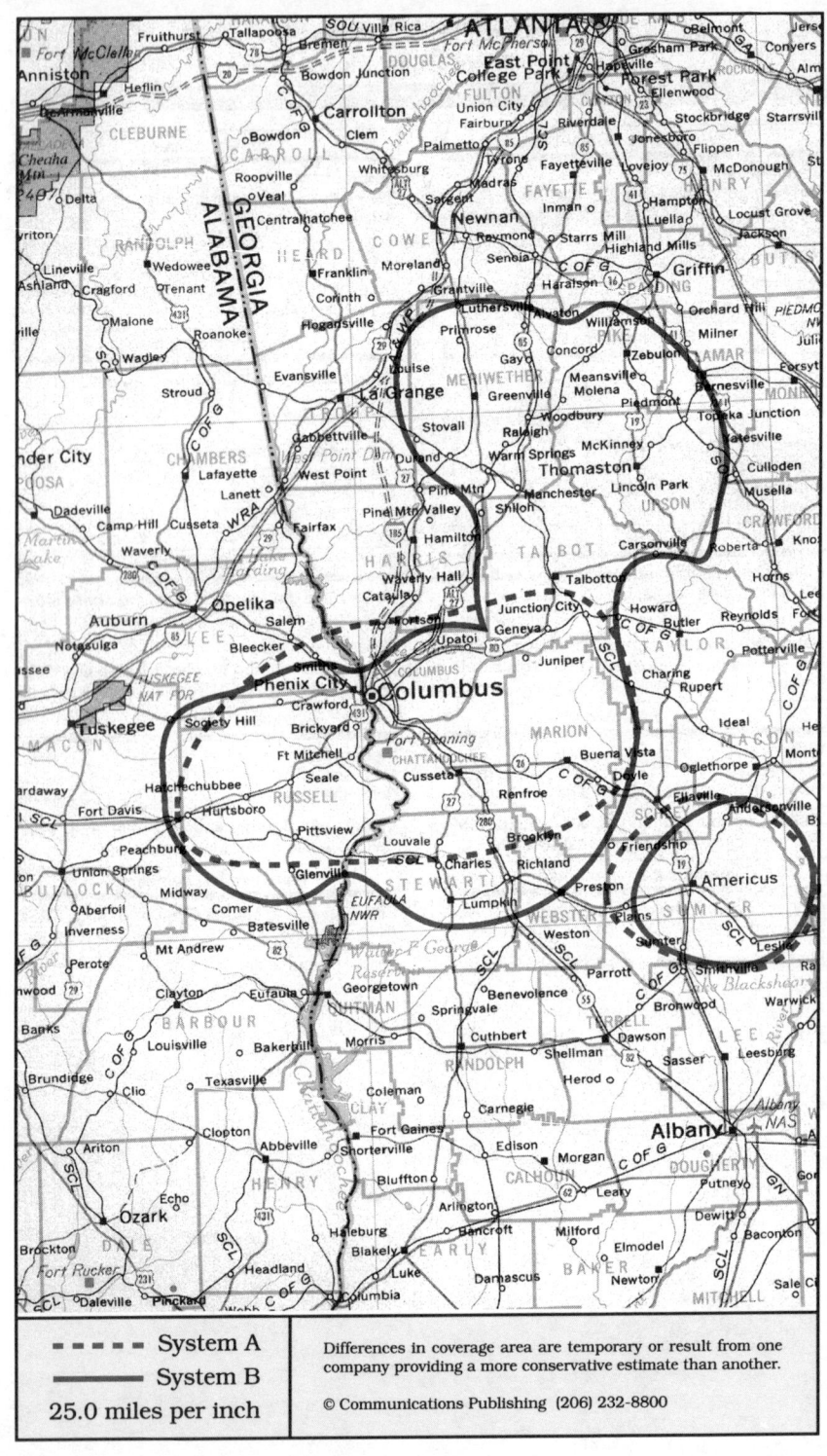

- - - - -	System A
———	System B
25.0 miles per inch	

Differences in coverage area are temporary or result from one company providing a more conservative estimate than another.

© Communications Publishing (206) 232-8800

Columbus, Georgia
Americus, Columbus, Phenix City AL, Thomaston

System A	System B

Cellular company

Cellular One
Palmer Communications
404-B Tripp Street
Americus, GA 31709

1870 Midtown Drive
Columbus, GA 31906

2308 N Expressway
Columbus, GA 31904

Customer service	706-561-1420
Local office . . . Americus	912-928-0101
. . . Columbus (Midtown)	706-596-0900
. Columbus (N Expresway)	706-322-4929
Corporate office	813-433-4350

Sending calls

Visitor dialing instructions
Local calls	area code + 7 digits
Long distance . . .	1 + area code + 7 digits

Speed-dial numbers
Columbus Hilton	*10
Customer service	*611
Emergency	911
Phenix, AL police	*99
Sunny radio	*100
WCZQ radio	*107

Receiving calls

Caller dials the roamer access number below,
hears tone, dials cellular area code and number.
Americus	912-886-7626
Columbus	706-575-7626

NationLink & RoamingAmerica (see p. 1157)
Dial *31 send to receive calls, *30 to deactivate.
Dial *32 send to tell caller the roamer access no.
This company's subscribers have level 6 service.

Billing information

Visitor rates
Most visitors . . . . . $3 day + 85¢ minute

Visitors without roaming agreements (p. 1159)
Cannot call since company takes no credit cards.

Identification numbers

City	Market	System	Billing	RSA
Americus	379	01145	00000	GA-9
Columbus	153	00319	00000	MSA

For full coverage area, see Albany, GA and
Auburn, AL.

Cellular company

Public Service Cellular
3535 Macon Road, Suite A
Columbus, GA 31907

Customer service from Georgia	800-342-7622
. . . .	706-326-1000
Local office	706-569-1556
Corporate office	912-847-4111

Sending calls

Visitor dialing instructions
Local calls	area code + 7 digits
Long distance . . .	1 + area code + 7 digits

Speed-dial numbers
Customer service	611 or 711
Emergency in Georgia	911
Phenix, AL police	*99

Receiving calls

Caller dials the roamer access number below,
hears tone, dials cellular area code and number.
. . . .	706-326-7626

Follow Me Roaming forwarding (see p. 1156)
Activate, dial *18 send. Deactivate dial *19 send.

Billing information

Visitor rates
Most visitors . . . . . $3 day + 85¢ minute

Visitors without roaming agreements (p. 1159)
Cannot call since company takes no credit cards.

Identification numbers

City	Market	System	Billing	RSA
Americus	379	00302	00000	GA-9
Columbus	153	00302	00000	MSA
Thomaston	376	00302	00000	GA-6

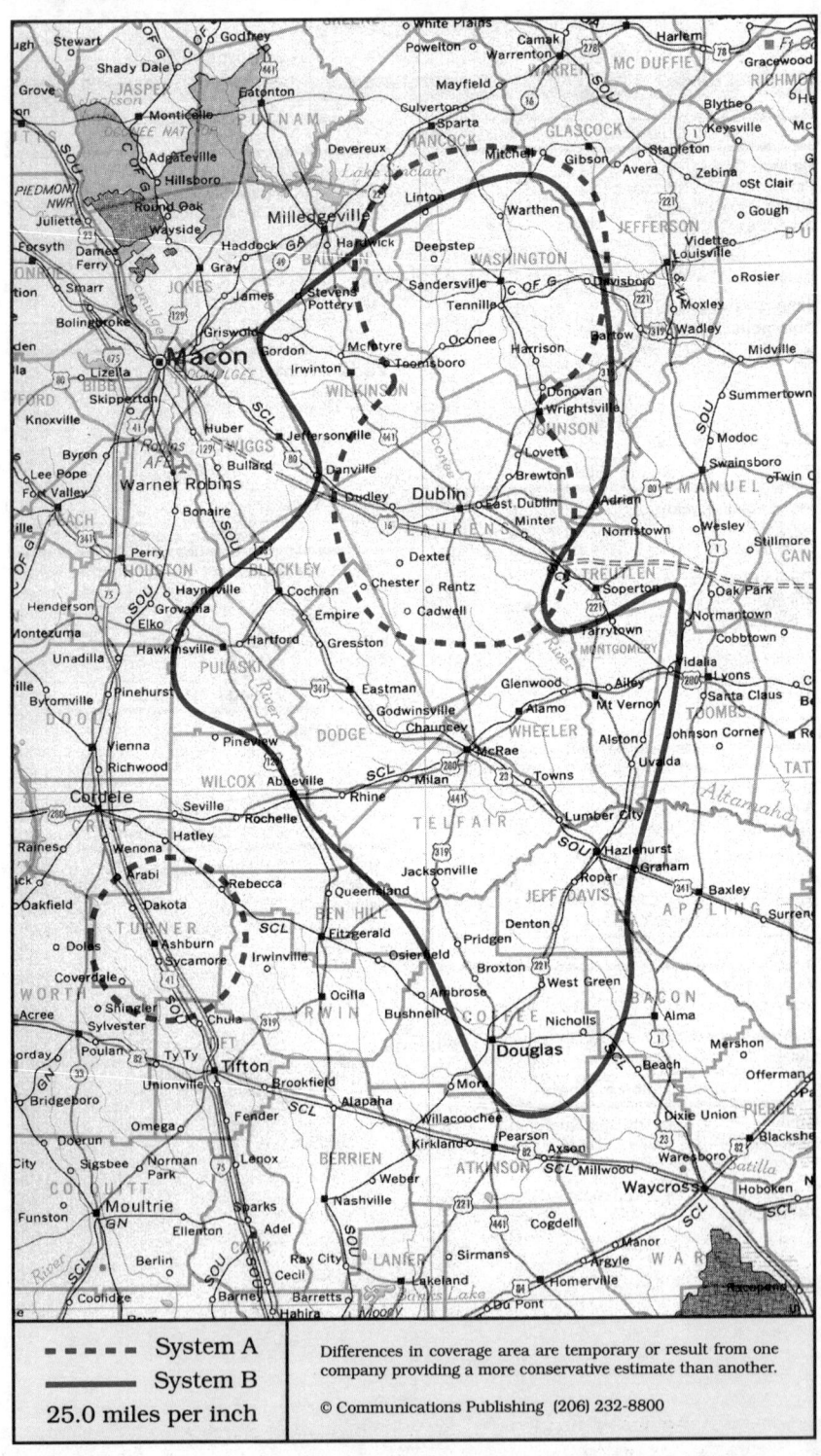

▪ ▪ ▪ System A	Differences in coverage area are temporary or result from one
──── System B	company providing a more conservative estimate than another.
25.0 miles per inch	© Communications Publishing (206) 232-8800

Dublin, Georgia
Ashburn (system A), Douglas, Dublin, Hawkinsville, Sandersville

System A	System B

Cellular company

**Cellular Georgia
Sterling Cellular Management**
1080 Holcomb Bridge Road, Bldg 100, Ste 200
Roswell, GA 30076

Customer service	800-552-6150
Corporate office	404-552-5030

Sending calls

Visitor dialing instructions

Local calls	area code + 7 digits
Long distance . . .	1 + area code + 7 digits

Speed-dial numbers

Customer service	611
Emergency	911
Roaming information	711

Receiving calls

Caller dials the roamer access number below,
hears tone, dials cellular area code and number.

Ashburn	912-424-7626
Dublin	912-277-7626

Billing information

Visitor rates

Most visitors	$3 day + 99¢ minute

Visitors without roaming agreements (p. 1159)
Roamer Plus will intercept first call to bill you.

Identification numbers

City	Market	System	Billing	RSA
Ashburn	380	01147	00000	GA-10
Dublin	377	01141	00000	GA-7

This company recently acquired GA-10 from
Quantum Communications.

This company also serves Milledgeville, GA.

Cellular company

Cellular Plus of Georgia
503 Industrial Boulevard
Dublin, GA 31021

Corporate office	from Georgia	800-362-7587
. . . .		912-272-9201

Sending calls

Visitor dialing instructions

Local calls	area code + 7 digits
Long distance . . .	1 + area code + 7 digits

Speed-dial numbers

Customer service	611
Emergency	911
Roamer access number	*511

Receiving calls

Caller dials the roamer access number below,
hears tone, dials cellular area code and number.

. . . .	912-984-7626

Follow Me Roaming forwarding (see p. 1156)
Activate, dial *18 send. Deactivate dial *19 send.

Billing information

Visitor rates

Most visitors	$3 day + 85¢ minute

Visitors without roaming agreements (p. 1159)
Cannot call since company takes no credit cards.

Identification numbers

City	Market	System	Billing	RSA
Douglas	380	01148	00000	GA-10
Dublin	377	01148	00000	GA-7

This company's subscribers can call in central
and south Georgia for $0/day + 50¢/minute.

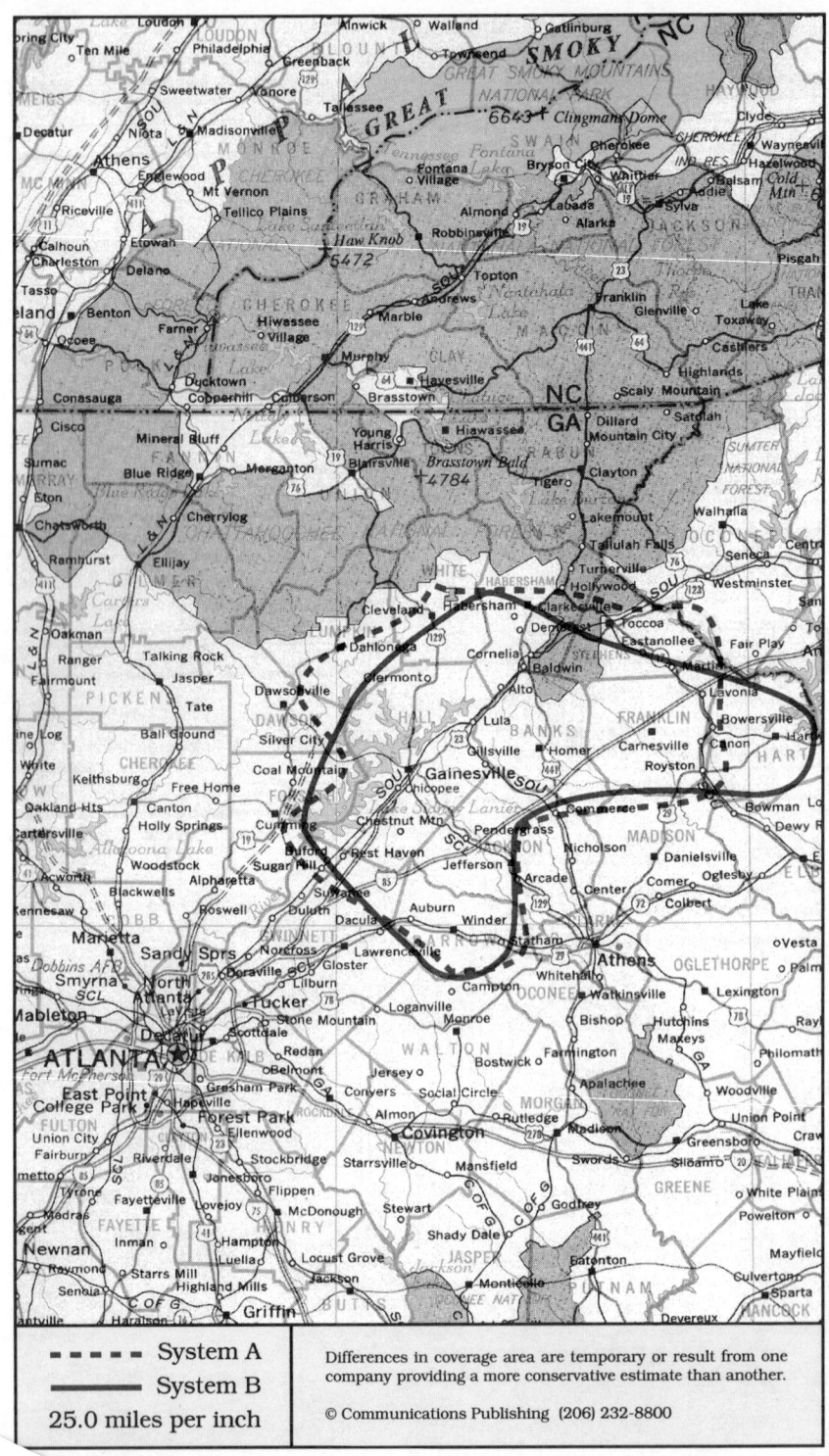

- - - - System A	Differences in coverage area are temporary or result from one company providing a more conservative estimate than another.
——— System B	
25.0 miles per inch	© Communications Publishing (206) 232-8800

Gainesville, Georgia
Bowersville, Gainesville, Winder

System A	System B

Cellular company

Cellular One
Southern Cellular
2352 Browns Bridge Road
Gainesville, GA 30504

Corporate office 404-536-5100

Sending calls

Visitor dialing instructions
Local calls . . . . . area code + 7 digits
Long distance . . . 1 + area code + 7 digits

Speed-dial numbers
Customer service . . . . ∗611
Emergency 911
Radio 103 request line . . . . . . ∗103
Radio WGGA . . . . ∗101
Radio WLAT . . . . #106

Receiving calls

Caller dials the roamer access number below,
hears tone, dials cellular area code and number.
. . . . 404-519-7626

NationLink & RoamingAmerica (see p. 1157)
Dial ∗31 send to receive calls, ∗30 to deactivate.
Dial ∗32 send to tell caller the roamer access no.
This company's subscribers have level 3 service.

Billing information

Visitor rates
Most visitors . . . . . $3 day + 85¢ minute

Visitors without roaming agreements (p. 1159)
Call co. to use Discover, Mastercard, Visa
or Roamer Plus will intercept first call to bill you.

Identification numbers

City	Market	System	Billing	RSA
All served cities	372	01131	00000	GA-2

Cellular company

BellSouth Mobility
4805 Briarcliff Road, Suite 101
Atlanta, GA 30345

Customer service	800-351-2400
. . . .	404-847-2400
Local office	404-270-4000
Corporate office	404-847-3600

Sending calls

Visitor dialing instructions
Local calls . . . . . . area code + 7 digits
Long distance . . . . . area code + 7 digits

Speed-dial numbers
Customer service 811
Emergency 911
Roaming information 711
Telephone problems 611

Receiving calls

Caller dials the roamer access number below,
hears tone, dials cellular area code and number.
. . . . 404-287-4626

Follow Me Roaming forwarding (see p. 1156)
Activate, dial ∗18 send. Deactivate dial ∗19 send.

Billing information

Visitor rates
Most visitors . . . . . $3 day + 85¢ minute
BellSouth subscribers . $0 day + 60¢ minute

Visitors without roaming agreements (p. 1159)
Call co. to use Amex, Mastercard, Visa

Identification numbers

City	Market	System	Billing	RSA
Cornelia	372	00034	30584	GA-2
Gainesville	372	00034	30270	GA-2
Hartwell	374	00034	30582	GA-4

For full coverage area, see Atlanta, GA.

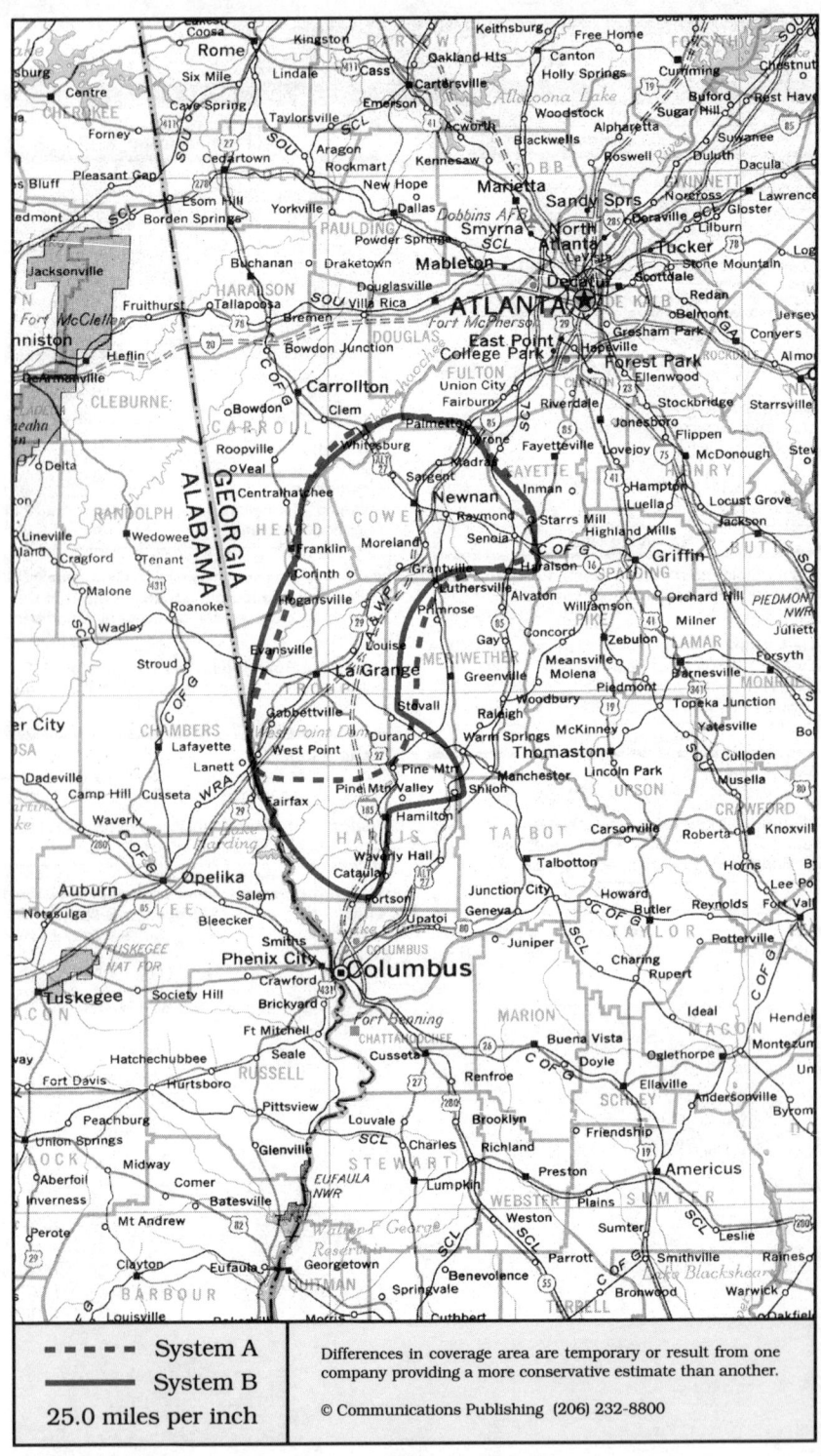

	System A
--- --- ---	System A
————	System B
25.0 miles per inch	

Differences in coverage area are temporary or result from one company providing a more conservative estimate than another.

© Communications Publishing (206) 232-8800

La Grange, Georgia
La Grange, Newnan

System A	System B

Cellular company

Cellular One
Licensed to: Blackwater Cellular
Managed by: Rural Cellular Management
26-A Bullsboro Drive
Newnan, GA 30263

Customer service	404-304-1111
Local office	706-884-1111
Corporate office Rural Cell. Mgt	707-422-2100

Sending calls

Visitor dialing instructions

Local calls	area code + 7 digits
Long distance . . .	1 + area code + 7 digits

Speed-dial numbers

Customer service	611
Emergency	911
Towing service	*HELP or *4357

Receiving calls

Caller dials the roamer access number below,
hears tone, dials cellular area code and number.
. . . . 706-301-7626

NationLink & RoamingAmerica (see p. 1157)
Dial *31 send to receive calls, *30 to deactivate.
This company's subscribers have level 3 service.

Billing information

Visitor rates

Most visitors	$3 day + 85¢ minute
Visitors from Georgia .	$0 day + 79¢ minute

Visitors without roaming agreements (p. 1159)
Roamer Plus will intercept first call to bill you.

Identification numbers

City	Market	System	Billing	RSA
All served cities	375	01137	00000	GA-5

For full coverage area, see Carrollton, GA.

Cellular company

InterCel
Interstate Cellular
421 Gilmer Avenue, PO Box 352
Lanett, AL 36863

Corporate office	800-763-2355
. . . .	404-304-2355
. . . .	706-845-2355

Sending calls

Visitor dialing instructions

Local calls	area code + 7 digits
Long distance . . .	1 + area code + 7 digits

Speed-dial numbers

Customer service	*611
Emergency	911

Receiving calls

Caller dials the roamer access number below,
hears tone, dials cellular area code and number.
. . . . 706-643-7626

Follow Me Roaming forwarding (see p. 1156)
Activate, dial *18 send. Deactivate dial *19 send.

Billing information

Visitor rates
Most visitors . . . . . $3 day + 85¢ minute

Visitors without roaming agreements (p. 1159)
Cannot call since company takes no credit cards.

Identification numbers

City	Market	System	Billing	RSA
All served cities	375	01902	00000	GA-5
Harris County	376	01902	00000	GA-6

For full coverage area, see Auburn, AL.

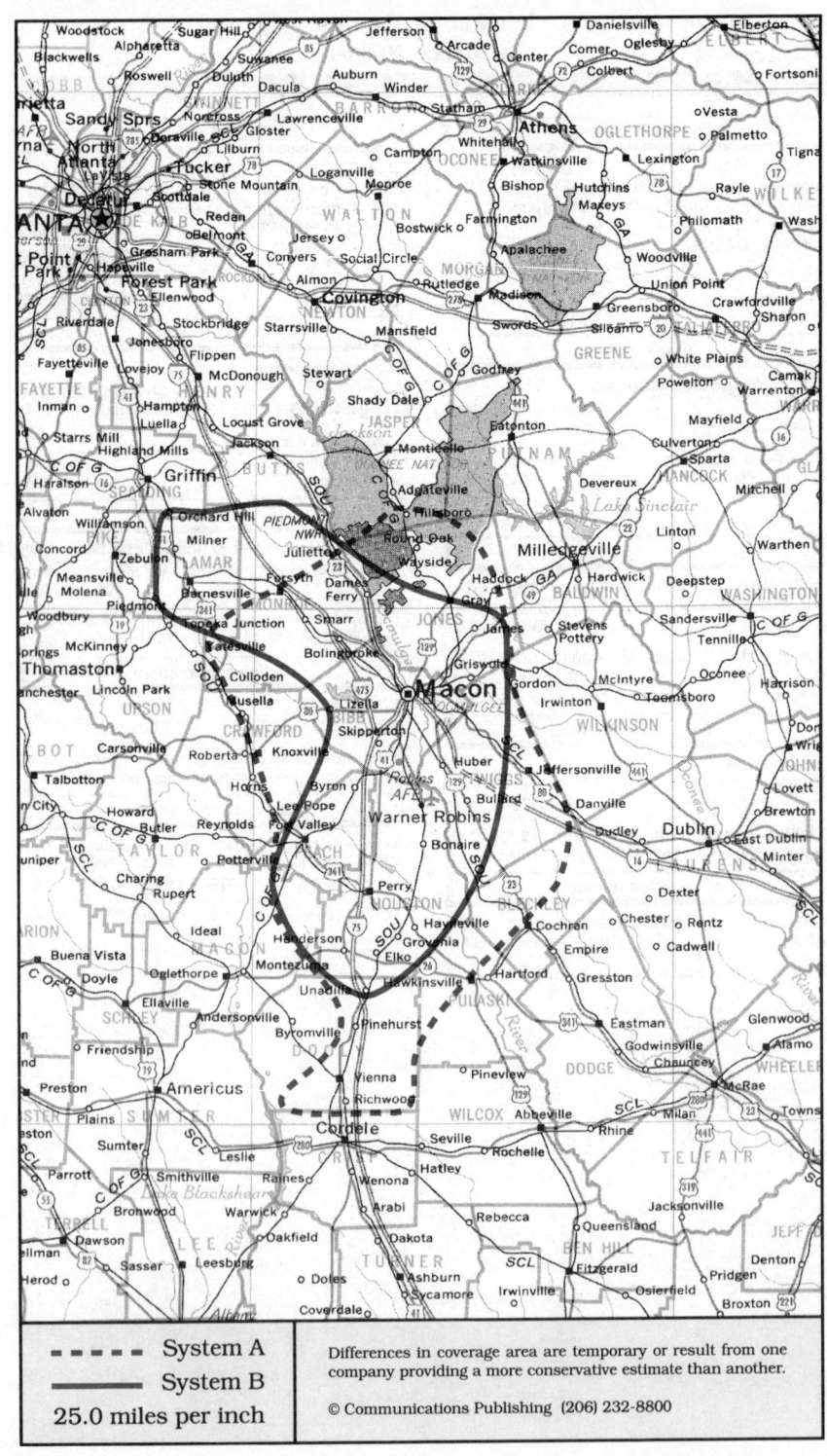

---- **System A**	Differences in coverage area are temporary or result from one company providing a more conservative estimate than another.
—— **System B**	
25.0 miles per inch	© Communications Publishing (206) 232-8800

Macon, Georgia
Forsyth, Macon, Perry, Warner Robins

System A	System B

Cellular company

Cellular One
Palmer Communications
4524 Forsyth Road, Suite 508
Macon, GA 31210

Local office	912-474-2355
Corporate office	813-433-4350

Sending calls

Visitor dialing instructions

Local calls	area code + 7 digits
Long distance . . .	1 + area code + 7 digits

Speed-dial numbers

Customer service	611
Emergency	911
WCOP	*1350
WDEN	*105
WPGA	*101

Receiving calls

Caller dials the roamer access number below,
hears tone, dials cellular area code and number.

. . . .	912-951-7626

NationLink & RoamingAmerica (see p. 1157)
Dial *31 send to receive calls, *30 to deactivate.
Dial *32 send to tell caller the roamer access no.
This company's subscribers have level 6 service.

Billing information

Visitor rates

Most visitors	$3 day + 85¢ minute

Visitors without roaming agreements (p. 1159)
Cannot call since company takes no credit cards.

Identification numbers

City	Market	System	Billing	RSA
Forsyth	376	00443	00000	GA-6
Macon	138	00443	00000	MSA

Cellular company

BellSouth Mobility
3096 Riverside Drive, Suite D
Macon, GA 32104

Customer service	800-351-2400
. . . .	404-847-2400
Local office	912-477-5066
Corporate office	404-847-3600

Sending calls

Visitor dialing instructions

Local calls	area code + 7 digits
Long distance	area code + 7 digits

Speed-dial numbers

Customer service	811
Emergency	911
Roaming information	711
Telephone problems	611
Traffic report	*123

Receiving calls

Caller dials the roamer access number below,
hears tone, dials cellular area code and number.

Forsyth	912-993-7626
Griffin	404-412-2626
Macon	912-747-7626
Warner Robins	912-542-7626

Follow Me Roaming forwarding (see p. 1156)
Activate, dial *18 send. Deactivate dial *19 send.

Billing information

Visitor rates

Most visitors	$3 day + 85¢ minute
BellSouth subscribers .	$0 day + 60¢ minute

Visitors without roaming agreements (p. 1159)
Call co. to use Amex, Mastercard, Visa

Identification numbers

City	Market	System	Billing	RSA
Forsyth	376	00426	30342	GA-6
Griffin	376	00426	30340	GA-6
Macon	138	00426	00000	MSA

Milledgeville is operated as part of this system.

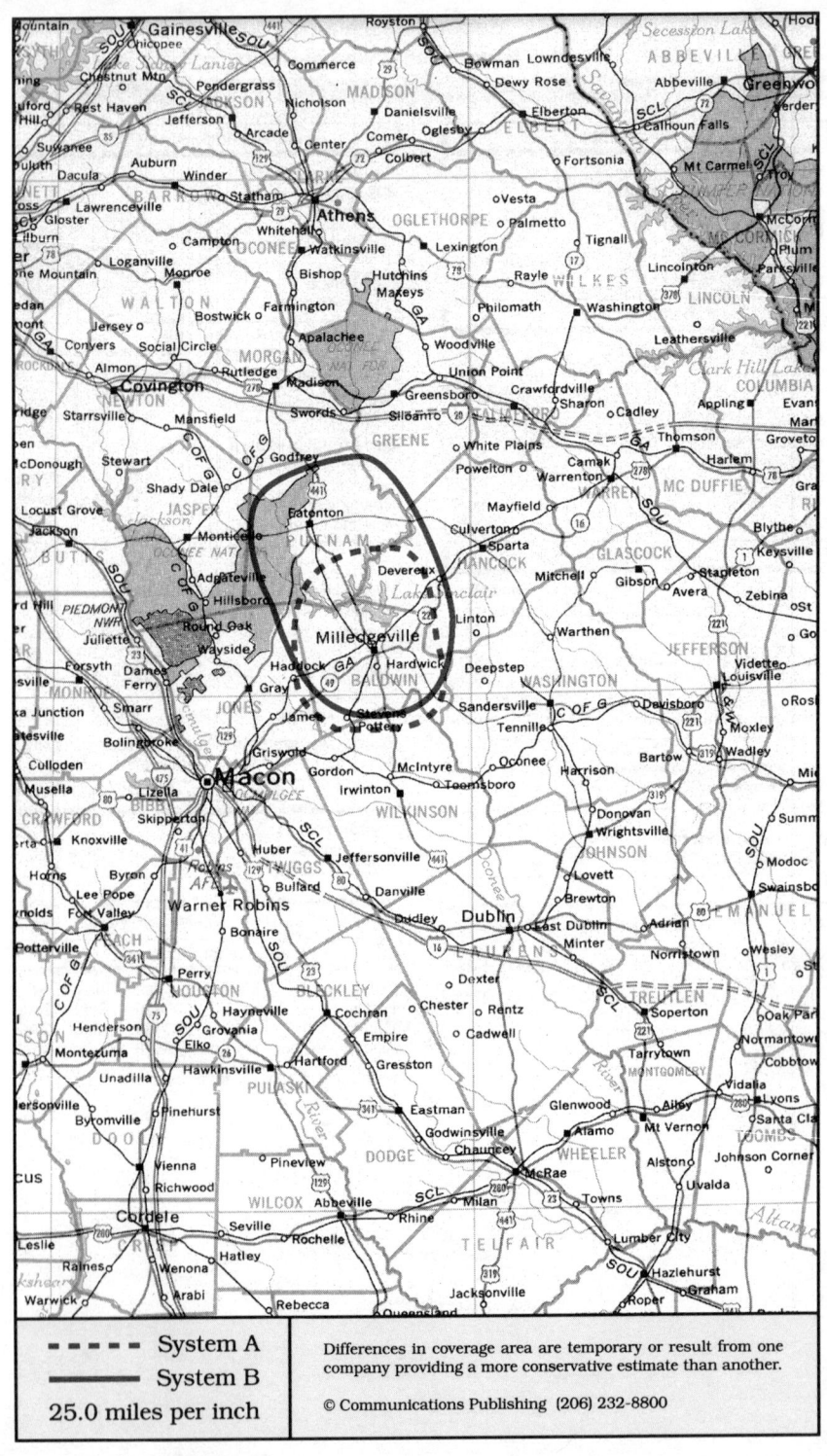

- - - - - System A	Differences in coverage area are temporary or result from one
——— System B	company providing a more conservative estimate than another.
25.0 miles per inch	© Communications Publishing (206) 232-8800

Milledgeville, Georgia
Eatonton, Milledgeville

Cellular company

Cellular Georgia
Sterling Cellular Management
1080 Holcomb Bridge Road, Bldg 100, Ste 200
Roswell, GA 30076

Customer service	800-552-6150
Corporate office	404-552-5030

Sending calls

Visitor dialing instructions

Local calls	area code + 7 digits
Long distance . . .	1 + area code + 7 digits

Speed-dial numbers

Customer service	611
Emergency	911
Roaming information	711

Receiving calls

Caller dials the roamer access number below, hears tone, dials cellular area code and number.

. . . . 912-277-7626

Billing information

Visitor rates

Most visitors . . . . . $3 day + 75¢ minute

Visitors without roaming agreements (p. 1159)
Roamer Plus will intercept first call to bill you.

Identification numbers

City	Market	System	Billing	RSA
All served cities	377	01141	00000	GA-7

This company also serves Dublin, GA.

Cellular company

BellSouth Mobility
3096 Riverside Drive, Suite D
Macon, GA 31210

Customer service	800-351-2400
. . . .	404-847-2400
Local office	912-477-5066
Corporate office	404-847-3600

Sending calls

Visitor dialing instructions

Local calls	area code + 7 digits
Long distance	area code + 7 digits

Speed-dial numbers

Customer service	811
Emergency	911
Roaming information	711
Telephone problems	611

Receiving calls

Caller dials the roamer access number below, hears tone, dials cellular area code and number.

. . . . 912-454-7626

Follow Me Roaming forwarding (see p. 1156)
Activate, dial ∗18 send. Deactivate dial ∗19 send.

Billing information

Visitor rates

Most visitors	$3 day + 85¢ minute
BellSouth subscribers .	$0 day + 60¢ minute

Visitors without roaming agreements (p. 1159)
Call co. to use Amex, Mastercard, Visa

Identification numbers

City	Market	System	Billing	RSA
All served cities	377	00426	30322	GA-7

Milledgeville is operated as part of the Macon, GA system of BellSouth Mobility.

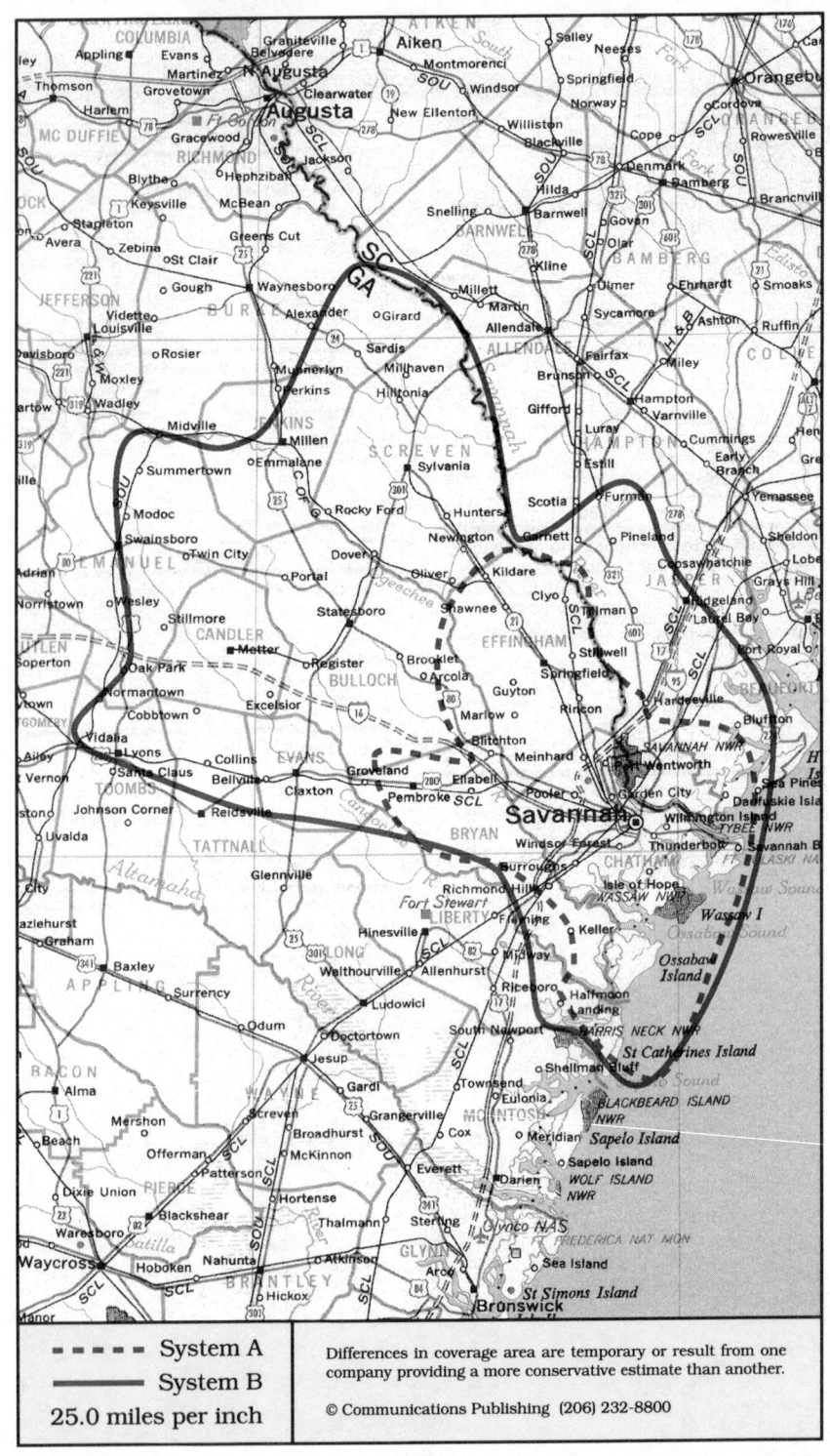

▬ ▬ ▬ ▬ System A	Differences in coverage area are temporary or result from one
▬▬▬▬▬ System B	company providing a more conservative estimate than another.
25.0 miles per inch	© Communications Publishing (206) 232-8800

Savannah, Georgia
Claxton, Metter, Millen, Savannah, Statesboro, Sylvania

System A	System B

Cellular company

Cellular One
GTE Mobilnet
401 Mall Boulevard, Suite 101-E
Savannah, GA 31406

Customer service	800-727-2444
. . . .	919-481-1181
Local office	912-352-3456
Corporate office	404-391-8000

Sending calls

Visitor dialing instructions

Local calls	7 digits
Long distance . . . 1 + area code + 7 digits	

Speed-dial numbers

Customer service	611
Emergency	911

Receiving calls

Caller dials the roamer access number below,
hears tone, dials cellular area code and number.
. . . . 912-656-7626

Follow Me Roaming forwarding (see p. 1156)
Phone Me Anywhere forwarding (see p. 1156)
Activate, dial ∗18 send. Deactivate dial ∗19 send.

NationLink & RoamingAmerica (see p. 1157)
Dial ∗31 send to receive calls, ∗30 to deactivate.
Dial ∗32 send to tell caller the roamer access no.
This company's subscribers have level 3 service.

Billing information

Visitor rates

Most visitors	$3 day + 85¢ minute

Visitors without roaming agreements (p. 1159)
Call co. to use Amex, Mastercard, Visa

Identification numbers

City	Market	System	Billing	RSA
All served cities	155	00539	00000	MSA

Cellular company

ALLTEL Mobile Communications
7001 Chatham Center Drive, Suite 1600
Savannah, GA 31405

Customer service	some areas	800-255-8351
Local office		800-733-7626
. . . .		912-651-6100
Corporate office		501-661-8500

Sending calls

Visitor dialing instructions

Local calls . . . . . . area code + 7 digits	
Long distance . . . 1 + area code + 7 digits	

Speed-dial numbers

Customer service	611
Emergency	911
Hear a traffic report	∗102
Weather report	∗97

Receiving calls

Caller dials the roamer access number below,
hears tone, dials cellular area code and number.
. . . . 912-658-7626

Follow Me Roaming forwarding (see p. 1156)
Activate, dial ∗18 send. Deactivate dial ∗19 send.

Billing information

Visitor rates

Most visitors	$3 day + 90¢ minute

Visitors without roaming agreements (p. 1159)
Cannot call since company takes no credit cards.

Identification numbers

City	Market	System	Billing	RSA
Savannah	155	00520	00000	MSA
Statesboro	378	00520	30378	GA-8

For full coverage area, see Brunswick, GA and
Waycross, GA.

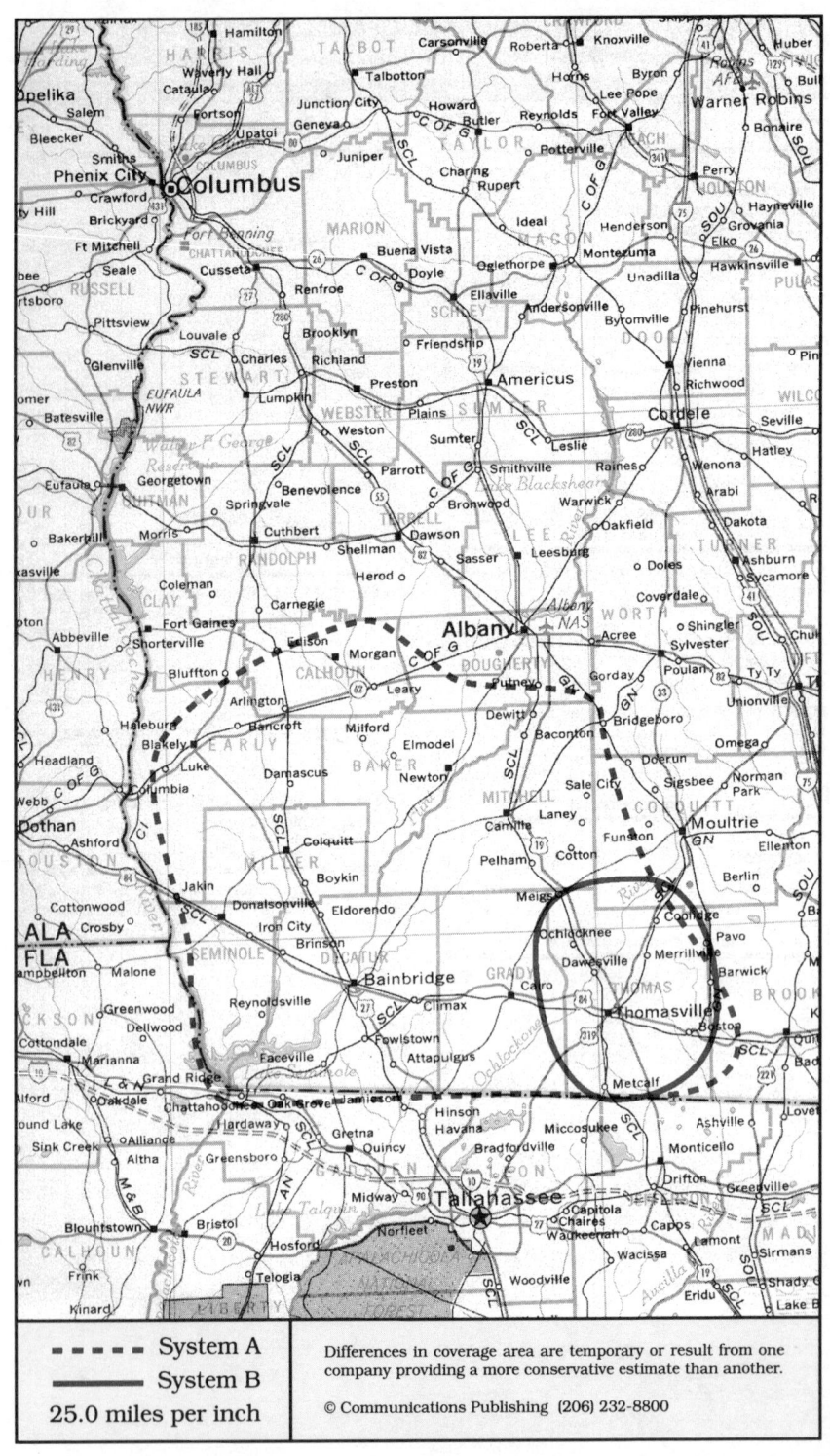

▪ ▪ ▪ ▪ ▪ System A	
▬▬▬▬ System B	
25.0 miles per inch	

Differences in coverage area are temporary or result from one company providing a more conservative estimate than another.

© Communications Publishing (206) 232-8800

Thomasville, Georgia
Bainbridge, Camilla, Donalsonville, Morgan, Thomasville

System A	System B

Cellular company

Cellular One
Mobile Communications Systems
814 S Scott Street
Bainbridge, GA 31717

Corporate office 912-243-7332

Sending calls

Visitor dialing instructions
Local calls . . . . . area code + 7 digits
Long distance . . . 1 + area code + 7 digits

Speed-dial numbers
Customer service . . . . ∗611
Emergency 911

Receiving calls

Caller dials the roamer access number below,
hears tone, dials cellular area code and number.
. . . . 912-221-7626

Billing information

Visitor rates
Most visitors . . . . . $3 day + 85¢ minute

Visitors without roaming agreements (p. 1159)
Cannot call since company takes no credit cards.

Identification numbers

City	Market	System	Billing	RSA
All served cities	383	01153	00000	GA-13

Cellular company

United States Cellular
6110 NW 4th Place, Suite A
Gainesville, FL 32607

Local office 904-331-8100
Corporate office 312-399-8900

Sending calls

Visitor dialing instructions
Local calls . . . . . . area code + 7 digits
Long distance . . . . . area code + 7 digits

Speed-dial numbers
Customer service 611
Emergency 911
Roaming information . . . . ∗711

Receiving calls

Caller dials the roamer access number below,
hears tone, dials cellular area code and number.
. . . . 912-225-2876

Billing information

Visitor rates
Most visitors . . . . . $3 day + 99¢ minute

Visitors without roaming agreements (p. 1159)
Call co. to use Mastercard, Visa

Identification numbers

City	Market	System	Billing	RSA
All served cities	383	01154	00000	GA-13

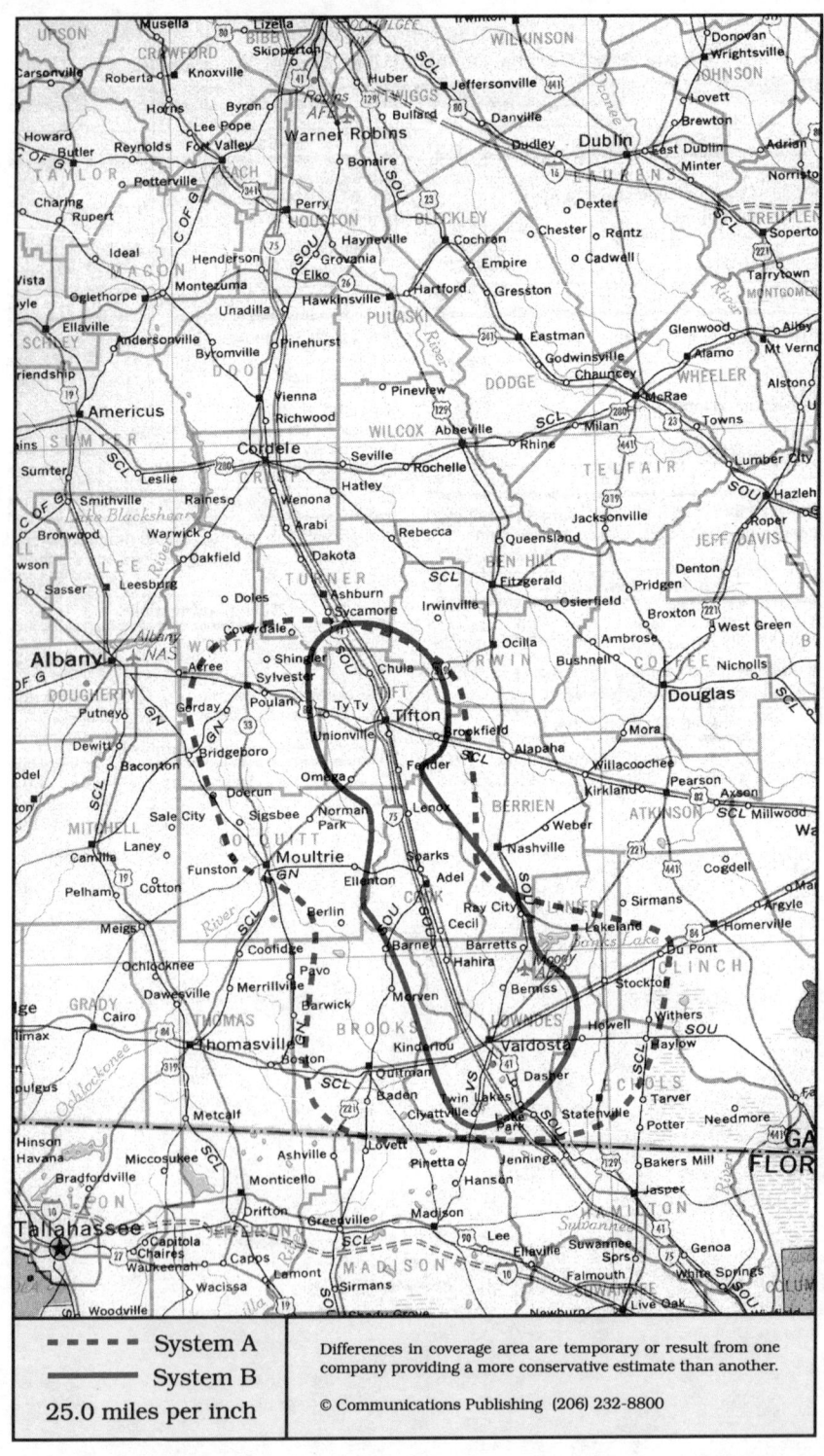

- - - - - System A	Differences in coverage area are temporary or result from one company providing a more conservative estimate than another.
——— System B	
25.0 miles per inch	© Communications Publishing (206) 232-8800

Valdosta, Georgia
Quitman, Moultrie, Tifton, Valdosta

System A	System B

Cellular company

Cellular One
GMD Limited Partnership
302-H Inner Perimeter Road
Valdosta, GA 31602

Local office	800-849-8444
. . . .	912-241-7494
Corporate office	919-321-0066

Sending calls

Visitor dialing instructions

Local calls	area code + 7 digits
Long distance . . .	1 + area code + 7 digits

Speed-dial numbers

Customer service	611
Emergency	911
Highway patrol	*HP or *47
Lowndes County sheriff	*LS or *57
Valdosta police	*VP or *87

Receiving calls

Caller dials the roamer access number below,
hears tone, dials cellular area code and number.

. . . .	912-251-7626

NationLink & RoamingAmerica (see p. 1157)
Dial *31 send to receive calls, *30 to deactivate.
This company's subscribers have level 3 service.

Billing information

Visitor rates

Most visitors	$3 day + 85¢ minute

Visitors without roaming agreements (p. 1159)
Dial 0 + 10-digit no. Operator will intercept to bill.

Identification numbers

City	Market	System	Billing	RSA
All served cities	384	01155	00000	GA-14

Cellular company

ALLTEL Mobile Communications
1707 Norman Drive
Valdosta, GA 31601

Customer service some areas	800-255-8351
Local office	912-242-5656
Corporate office	501-661-8500

Sending calls

Visitor dialing instructions

Local calls	area code + 7 digits
Long distance	area code + 7 digits

Speed-dial numbers

Customer service	*611 or *811
Emergency	911
Make traffic report WIKX	549

Receiving calls

Caller dials the roamer access number below,
hears tone, dials cellular area code and number.

Gorday	912-881-7626
Tifton	912-387-1626
Valdosta	912-560-7626

Follow Me Roaming forwarding (see p. 1156)
Activate, dial *18 send. Deactivate dial *19 send.

Billing information

Visitor rates

Most visitors	$3 day + 85¢ minute

Visitors without roaming agreements (p. 1159)
Call co. to use Amex, Discover, Mastercard, Visa

Identification numbers

City	Market	System	Billing	RSA
Tifton	384	00204	30286	GA-14
Valdosta	384	01156	30286	GA-14

For full coverage area, see Albany, GA.

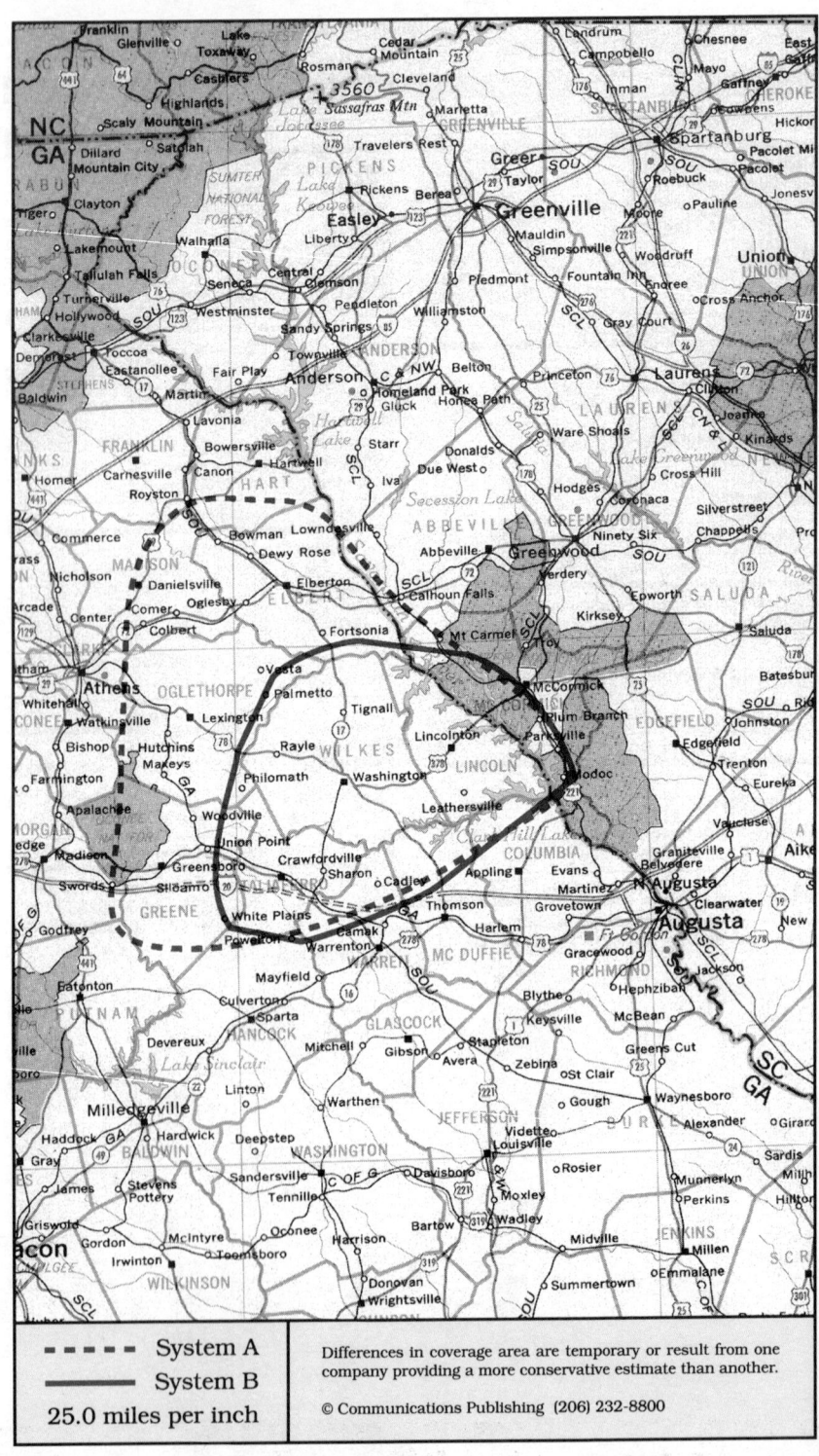

Washington, Georgia

Crawfordville, Fiberton, Greensboro, Lincolnton, Washington

System A	System B

Cellular company

PacTel Cellular
4151 Ashford Dunwoody Road NE #300
Atlanta, GA 30319-1462

Customer service	800-235-5611
. . . .	404-257-5100
Local office	404-257-5000
Corporate office	510-210-3600

Sending calls

Visitor dialing instructions

Local calls	area code + 7 digits
Long distance . . .	0 + area code + 7 digits

Speed-dial numbers

Customer service	*611
Emergency	911
Make a traffic report	*640
Traffic report center	*941

Receiving calls

Caller dials the roamer access number below,
hears tone, dials cellular area code and number.
. . . . 706-557-4626

NationLink & RoamingAmerica (see p. 1157)
Dial *31 send to receive calls, *30 to deactivate.
Dial *32 send to tell caller the roamer access no.
This company's subscribers have level 8 service.

Billing information

Visitor rates
Most visitors . . . . . $3 day + 89¢ minute

Visitors without roaming agreements (p. 1159)
Call co. to use Amex, Mastercard, Visa
or Roamer Plus will intercept first call to bill you.

Identification numbers

City	Market	System	Billing	RSA
All served cities	374	00041	30321	GA-4

For full coverage area, see Atlanta, GA.

This company reports that it will serve all of RSA
GA-4 as shown by October 1992.

Cellular company

Wilkes Cellular
PO Box 277
Washington, GA 30673

Corporate office	706-678-2355

Sending calls

Visitor dialing instructions

Local calls	area code + 7 digits
Long distance . . .	1 + area code + 7 digits

Speed-dial numbers

Customer service	611
Emergency	911

Receiving calls

Caller dials the roamer access number below,
hears tone, dials cellular area code and number.
. . . . 706-678-8626

Billing information

Visitor rates
Most visitors . . . . . $3 day + 85¢ minute

Visitors without roaming agreements (p. 1159)
Cannot call since company takes no credit cards.

Identification numbers

City	Market	System	Billing	RSA
All served cities	374	01922	00000	GA-4

Call switching is performed by Alltel Mobile for the
licensee, Wilkes Cellular.

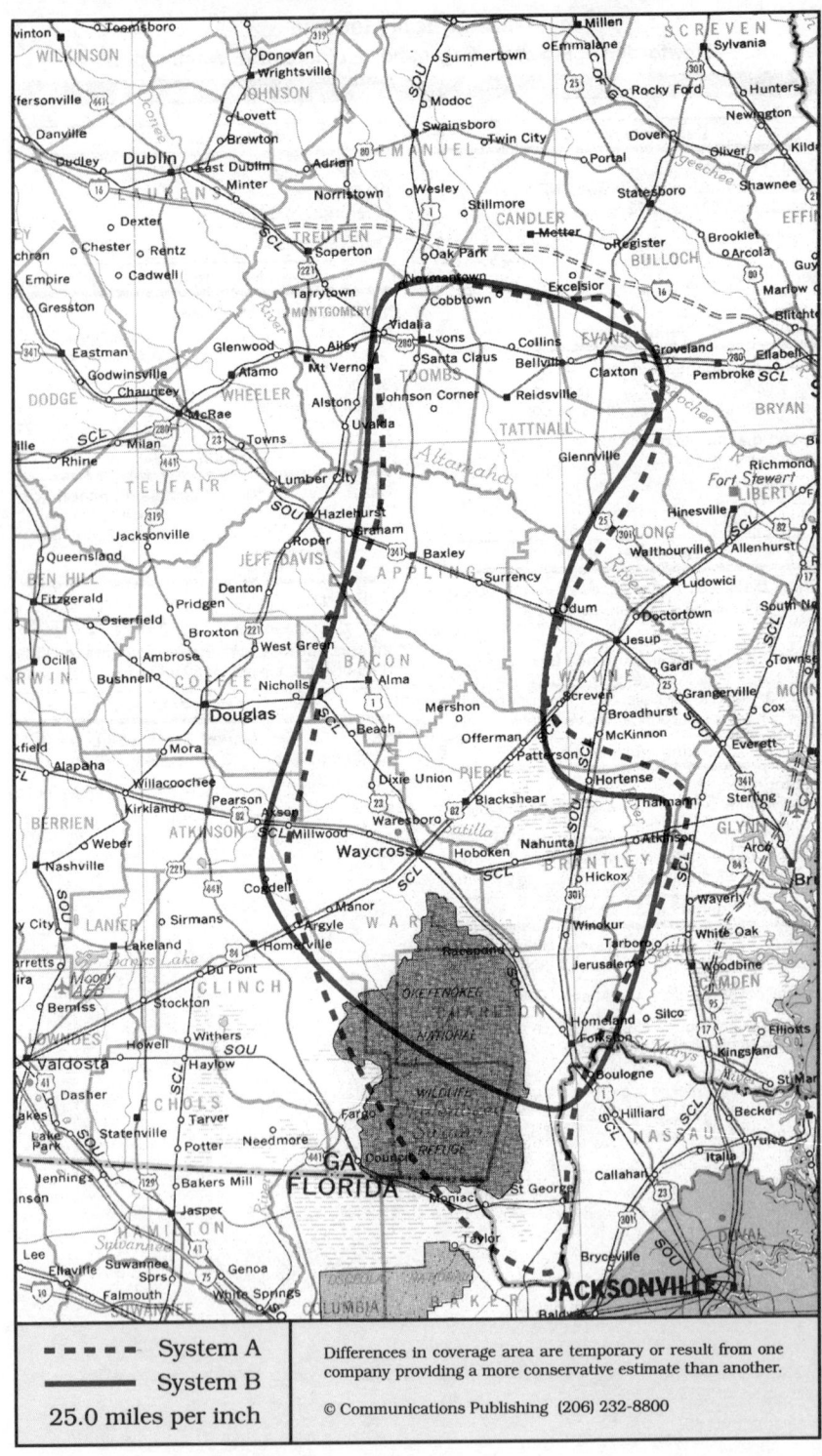

- - - -	System A
——	System B
25.0 miles per inch	

Differences in coverage area are temporary or result from one company providing a more conservative estimate than another.

© Communications Publishing (206) 232-8800

Waycross, Georgia
Folkston, Nahunta, Reidsville, Vidalia, Waycross

System A	System B

Cellular company

Cellular One
Cone Enterprises
2339 Lyons Highway, PO Box 767
Vidalia, GA 30474

PO Box 178, 721 Knight Avenue
Waycross, GA 31501

Local office . . . . Vidalia	912-538-0077
. . . . Waycross	912-285-0302
Corporate office	806-744-1661

Sending calls

Visitor dialing instructions
Local calls . . . . . area code + 7 digits
Long distance . . . 1 + area code + 7 digits

Speed-dial numbers
Customer service 611
Emergency 911

Receiving calls

Caller dials the roamer access number below,
hears tone, dials cellular area code and number.
. . . . 912-284-7626

Billing information

Visitor rates
Most visitors . . . . . $3 day + 90¢ minute

Visitors without roaming agreements (p. 1159)
Call co. to use Mastercard, Visa

Identification numbers

City	Market	System	Billing	RSA
All served cities	381	01149	00000	GA-11

Cellular company

ALLTEL Mobile Communications
7001 Chatham Center Drive, Suite 1600
Savannah, GA 31405

Customer service	some areas	800-255-8351
Local office		800-733-7626
. . . .		912-651-6100
Corporate office		501-661-8500

Sending calls

Visitor dialing instructions
Local calls . . . . . . area code + 7 digits
Long distance . . . 1 + area code + 7 digits

Speed-dial numbers
Customer service 611
Emergency 911
Hear a traffic report *102
Weather report *97

Receiving calls

Caller dials the roamer access number below,
hears tone, dials cellular area code and number.
. . . . 912-658-7626

Follow Me Roaming forwarding (see p. 1156)
Activate, dial *18 send. Deactivate dial *19 send.

Billing information

Visitor rates
Most visitors . . . . . $3 day + 90¢ minute

Visitors without roaming agreements (p. 1159)
Cannot call since company takes no credit cards.

Identification numbers

City	Market	System	Billing	RSA
All served cities	381	00520	30380	GA-11

For full coverage area, see Savannah, GA.

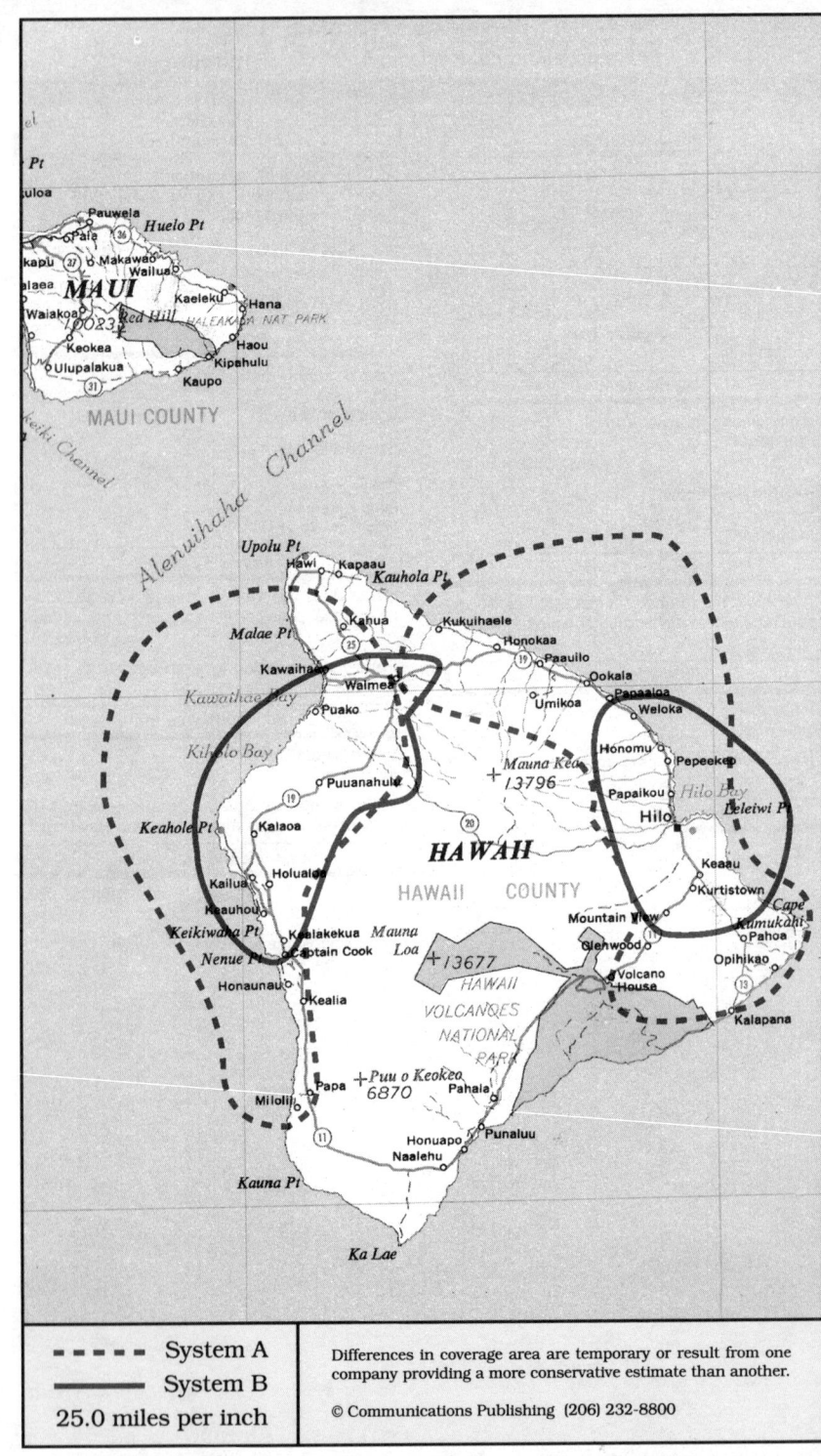

MAUI

Pauwela
Paia
Makawao
Huelo Pt
Wailua
Kaeleku
MAUI
10023
Red Hill
HALEAKALA NAT PARK
Hana
Waiakoa
Keokea
Haou
Ulupalakua
Kipahulu
Kaupo

MAUI COUNTY

Keiki Channel

Alenuihaha Channel

Upolu Pt
Hawi
Kapaau
Kauhola Pt
Kahua
Kukuihaele
Malae Pt
Honokaa
Kawaihae
Paauilo
Waimea
Ookala
Kawaihae Bay
Puako
Umikoa
Panaina
Weloka
Kiholo Bay
Puuanahulu
Mauna Kea
13796
Honomu
Pepeekeo
Keahole Pt
Kalaoa
Papaikou
Hilo
Leleiwi Pt
Kailua
HAWAII
Hilo Bay
Holualoa
HAWAII
COUNTY
Keaau
Keauhou
Kurtistown
Keikiwaha Pt
Kealakekua
Mauna
Nenue Pt
Captain Cook
Loa
Mountain View
Cape
Kumukahi
Honaunau
13677
Glenwood
Pahoa
Kealia
HAWAII
Opihikao
VOLCANOES
Volcano
House
NATIONAL
PARK
13
Papa
Puu o Keokea
6870
Pahala
Kalapana
Milolii
Honuapo
Punaluu
Naalehu
Kauna Pt

Ka Lae

- - - - - System A
————— System B
25.0 miles per inch

Differences in coverage area are temporary or result from one
company providing a more conservative estimate than another.

© Communications Publishing (206) 232-8800

222

Hilo, Hawaii
Island of Hawaii

System A	System B

Cellular company

United States Cellular
75-5660 Kopiko Street, Building A-2
Kailua Kona, HI 96740

Local office 808-326-9335
Corporate office 312-399-8900

Sending calls

Visitor dialing instructions
Local calls . . . . area code + 7 digits
Long distance . . . area code + 7 digits

Speed-dial numbers
Customer service *611
Emergency *911

Receiving calls

Caller dials the roamer access number below,
hears tone, dials cellular area code and number.
Island of Hawaii 808-936-7626

Billing information

Visitor rates
Most visitors . . . . . $5 day + 50¢ minute

Visitors without roaming agreements (p. 1159)
Call co. to use Mastercard, Visa

Identification numbers

City	Market	System	Billing	RSA
Island of Hawaii	387	01161	00000	HI-3

Cellular company

GTE Mobilnet
733 Bishop Street, Suite 1900
Honolulu, HI 96813-4010

Customer service   from Hawaii 800-635-9659
. . . .   808-522-7979
Local office   808-536-4848
Corporate office   404-391-8000

Sending calls

Visitor dialing instructions
Local calls . . . . . . area code + 7 digits
Long distance . . . 1 + area code + 7 digits

Speed-dial numbers
Customer service *611
Emergency 911
KGMB-9 news or traffic tip   *KGMB or *5462
KHON channel 2 news   .   *KHON or *5466
KIPA news or traffic tip   . . *KIPA or *5472
Make a traffic report 559
Telephone problems *111
U.S. Coast Guard . . . *USCG or *8724

Receiving calls

Caller dials the roamer access number below,
hears tone, dials cellular area code and number.
Island of Hawaii 808-987-7626

Follow Me Roaming forwarding (see p. 1156)
Activate, dial *18 send. Deactivate dial *19 send.

Billing information

Visitor rates
Most visitors . . . . . $5 day + 50¢ minute

Visitors without roaming agreements (p. 1159)
Call co. to use Amex, Discover, Mastercard, Visa

Identification numbers

City	Market	System	Billing	RSA
Island of Hawaii	387	00060	00000	HI-3

For full coverage area, see Honolulu, HI and
Lihue, HI.

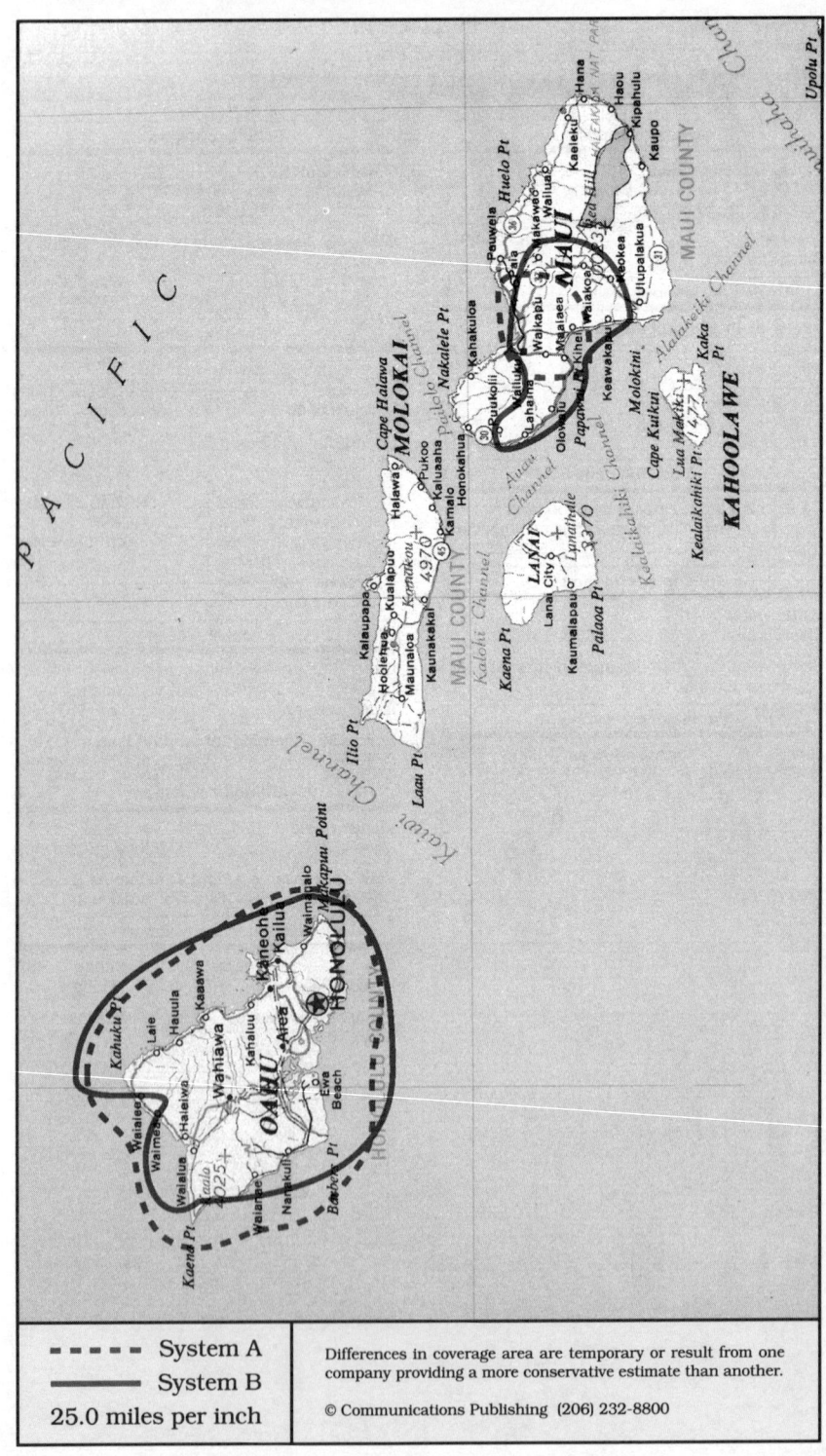

| - - - - - System A |
| ——— System B |
| 25.0 miles per inch |

Differences in coverage area are temporary or result from one company providing a more conservative estimate than another.

© Communications Publishing (206) 232-8800

Honolulu, Hawaii
Island of Maui, Island of Oahu

|

System A

Cellular company

Honolulu Cellular Telephone
Administration
500 Ala-Moana Blvd., 2 Waterfront Plaza #301
Honolulu, HI 96813

Sales
1161 Kapiolani Boulevard
Honolulu, HI 96814-2102

Customer service	from Hawaii	800-654-2355
. . . .		808-545-4755
Local office .	Kapiolani Blvd.	808-545-4765
Corporate office .	Ala-Moana	808-545-4766

Sending calls

Visitor dialing instructions
Local calls	area code + 7 digits
Long distance . . .	0 + area code + 7 digits

Speed-dial numbers
Coast Guard	*246
Customer service	*611
Emergency	911
KCCN radio	*1420
KGU radio	*76
KHVH radio	*99
KHON television	*02
KQMQ radio	*93
Puka pothole reporting	*78
Time	*846
Weather	*932

Receiving calls

Caller dials the roamer access number below,
hears tone, dials cellular area code and number.
Island of Maui	808-283-7626
Island of Oahu	808-226-7626

NationLink & RoamingAmerica (see p. 1157)
Caller hears your current roamer access number.
This company's subscribers have level 0 service.

Billing information

Visitor rates
Most visitors . . . . . $5 day + 50¢ minute

Visitors without roaming agreements (p. 1159)
Call co. to use Amex, Mastercard, Visa

Identification numbers

City	Market	System	Billing	RSA
Island of Maui	386	01159	00000	HI-2
Island of Oahu	050	00167	00000	MSA

System B

Cellular company

GTE Mobilnet
733 Bishop Street, Suite 1900
Honolulu, HI 96813-4010

Customer service	from Hawaii	800-635-9659
. . . .		808-522-7979
Local office		808-536-4848
Corporate office		404-391-8000

Sending calls

Visitor dialing instructions
Local calls	area code + 7 digits
Long distance . . .	1 + area code + 7 digits

Speed-dial numbers
Customer service	*611
Emergency	911
KGMB-9	*KGMB or *5462
KHON channel 2 news	*KHON or *5466
KNUI news or traffic tip (Mauai) . .	*900
KQMQ news or traffic tip (Oahu)	*93
KSSK news or traffic tip (Oahu) .	K59 or 559
Make a traffic report	559
Telephone problems	*111
U.S. Coast Guard . . .	*USCG or *8724

Receiving calls

Caller dials the roamer access number below,
hears tone, dials cellular area code and number.
Island of Maui	808-281-7626
Island of Oahu	808-927-7626

Follow Me Roaming forwarding (see p. 1156)
Activate, dial *18 send. Deactivate dial *19 send.

Billing information

Visitor rates
Most visitors . . . . . $5 day + 50¢ minute

Visitors without roaming agreements (p. 1159)
Call co. to use Amex, Discover, Mastercard, Visa

Identification numbers

City	Market	System	Billing	RSA
Island of Maui	386	00060	00000	HI-2
Island of Oahu	050	00060	00000	MSA

For full coverage area, see Hilo, HI and Lihue, HI.

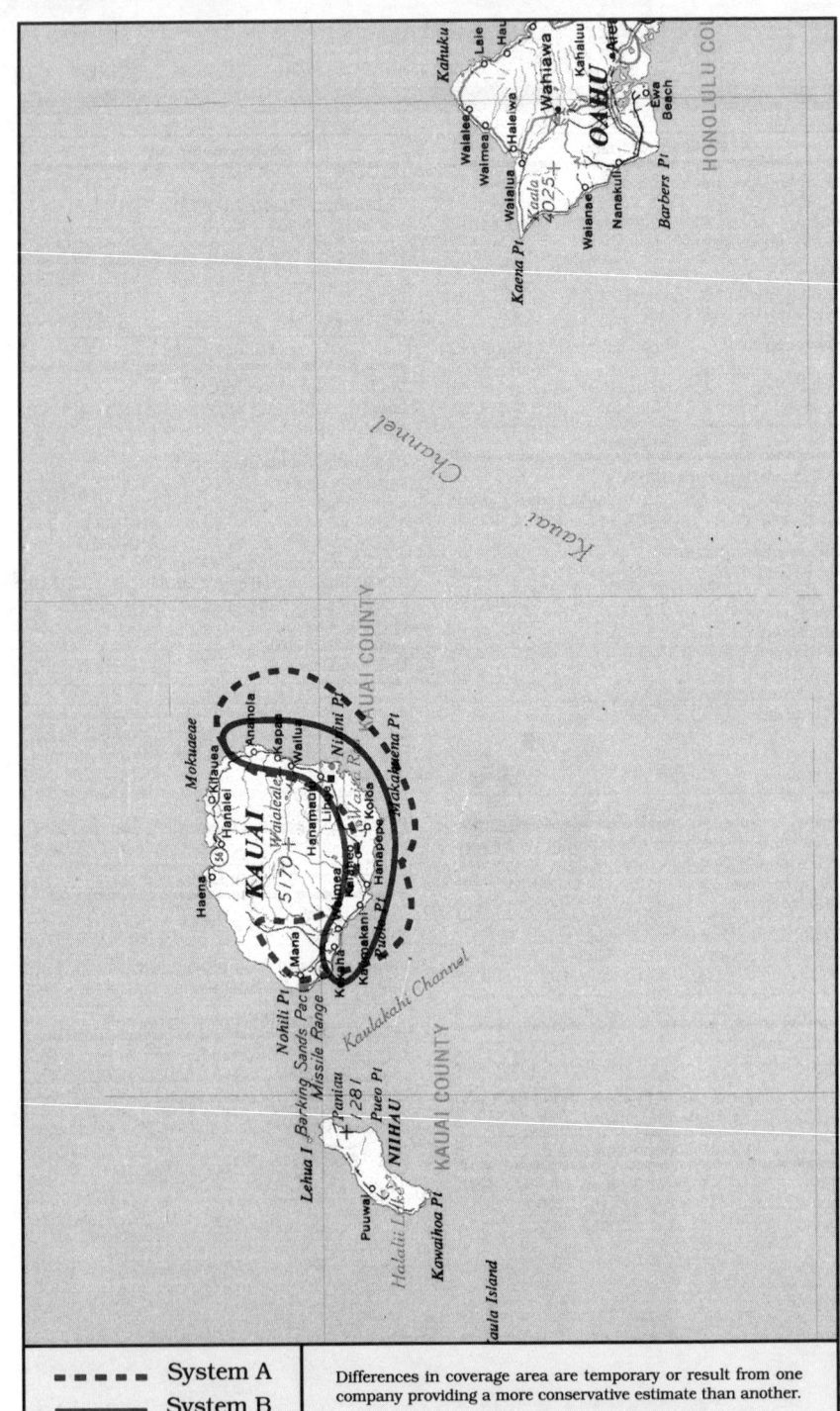

▬ ▬ ▬ ▬ System A	Differences in coverage area are temporary or result from one
▬▬▬▬ System B	company providing a more conservative estimate than another.
25.0 miles per inch	© Communications Publishing (206) 232-8800

Lihue, Hawaii
Island of Kauai

System A	System B

Cellular company

Cybertel Cellular
Ameritech Mobile Communications
3220 Kuhio Highway
Lihue, HI 96766

Local office 808-246-4626
Corporate office . . Cybertel 314-253-4993
. . . . Ameritech 708-706-7600

Sending calls

Visitor dialing instructions
Local calls . . . . . area code + 7 digits
Long distance . . . 0 + area code + 7 digits

Speed-dial numbers
Customer service 611
Emergency 911

Receiving calls

Caller dials the roamer access number below,
hears tone, dials cellular area code and number.
Island of Kauai 808-639-7626

NationLink & RoamingAmerica (see p. 1157)
Caller hears your current roamer access number.
This company's subscribers have level 0 service.

Billing information

Visitor rates
Most visitors . . . . . $5 day + 50¢ minute

Visitors without roaming agreements (p. 1159)
Call co. to use Amex, Mastercard, Visa

Identification numbers

City	Market	System	Billing	RSA
All served cities	385	01157	00000	HI-1

Cellular company

GTE Mobilnet
733 Bishop Street, Suite 1900
Honolulu, HI 96813-4010

Customer service from Hawaii 800-635-9659
. . . . 808-522-7979
Local office 808-536-4848
Corporate office 404-391-8000

Sending calls

Visitor dialing instructions
Local calls . . . . . area code + 7 digits
Long distance . . . 1 + area code + 7 digits

Speed-dial numbers
Customer service *611
Emergency 911
KGMB-9 . . . . *KGMB or *5462
KHON channel 2 news . *KHON or *5466
KUAI news or traffic tip . . *KUAI or *5824
Make a traffic report 559
Telephone problems *111
U.S. Coast Guard . . . *USCG or *8724

Receiving calls

Caller dials the roamer access number below,
hears tone, dials cellular area code and number.
Island of Kauai 808-651-7626

Follow Me Roaming forwarding (see p. 1156)
Activate, dial *18 send. Deactivate dial *19 send.

Billing information

Visitor rates
Most visitors . . . . . $5 day + 50¢ minute

Visitors without roaming agreements (p. 1159)
Call co. to use Amex, Discover, Mastercard, Visa

Identification numbers

City	Market	System	Billing	RSA
All served cities	385	00060	00000	HI-1

For full coverage area, see Hilo, HI and Honolulu, HI.

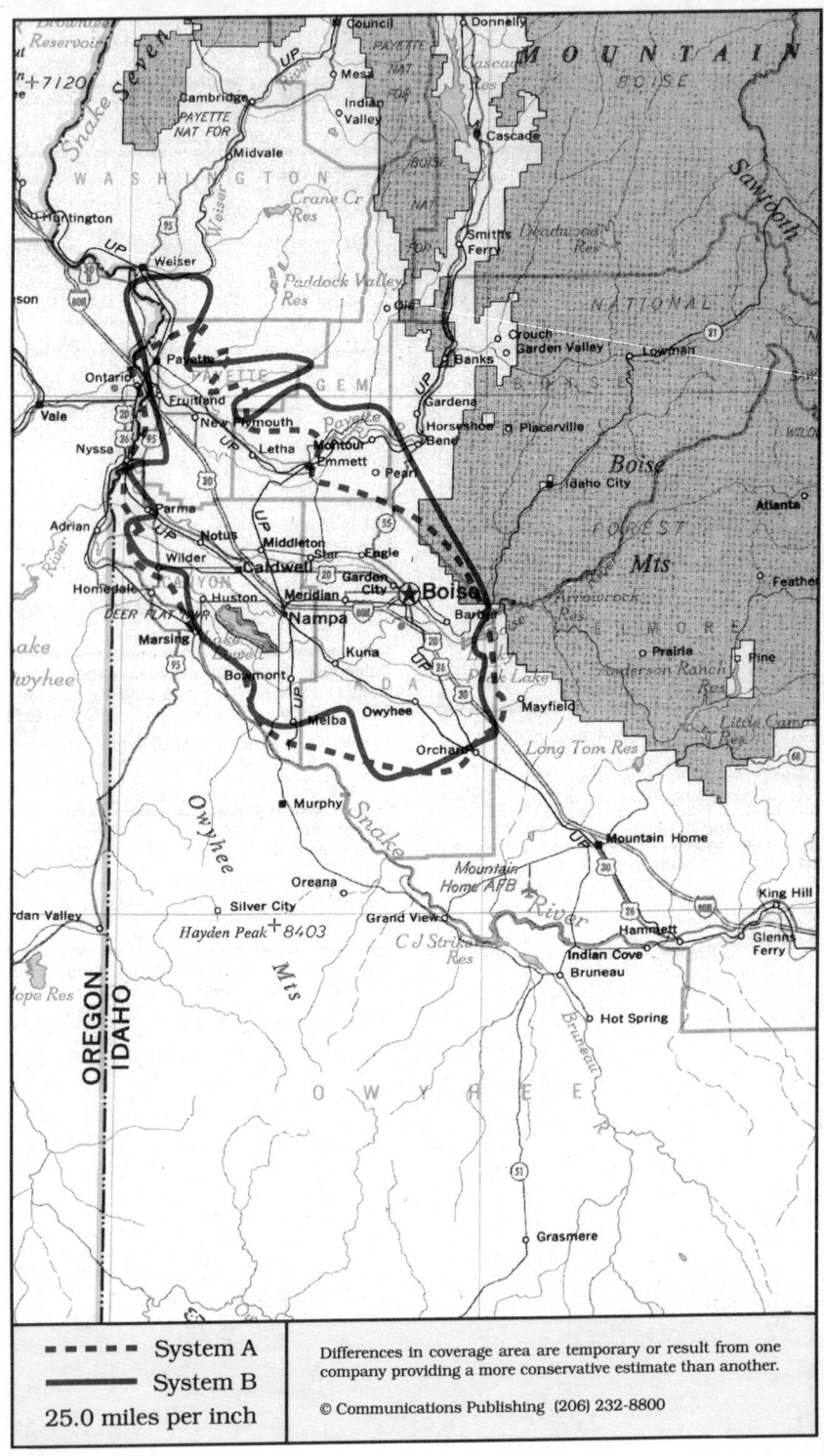

- - - - System A	Differences in coverage area are temporary or result from one company providing a more conservative estimate than another.
——— System B	
25.0 miles per inch	© Communications Publishing (206) 232-8800

Boise, Idaho
Bogus Basin, Boise, Caldwell, Nampa, Payette

System A	System B

Cellular company

Boise, Caldwell, Nampa
Cellular One
McCaw Cellular Communications
1475 Broadway
Boise, ID 83706

Customer service	208-345-2355
Local office	208-345-3100
Corporate office	206-827-4500

Payette (ID-2)
Cellular One
Pacific Northwest Cellular
11400 SE 8th Street, Suite 445
Bellevue, WA 98004-6431

Customer service	800-635-0304
. . . .	206-450-2922
Corporate office	206-635-0300

Sending calls

Visitor dialing instructions

Local calls	area code + 7 digits
Long distance . . .	1 + area code + 7 digits

Speed-dial numbers

Customer service	611
Emergency	911
Idaho State Police	*477

Receiving calls

Caller dials the roamer access number below,
hears tone, dials cellular area code and number.
. . . . 208-867-7626

NationLink & RoamingAmerica (see p. 1157)
Dial *31 send to receive calls, *30 to deactivate.
Dial *32 send to tell caller the roamer access no.
This company's subscribers have level 6 service.

North American Cellular Network (see p. 1156)
All calls forward to you. Deactivate dial *35 send.

Billing information

Visitor rates
Most visitors . . . . . . $0 day + 99¢ minute

Visitors without roaming agreements (p. 1159)
Call co. to use Amex, Discover, Mastercard, Visa

Identification numbers

City	Market	System	Billing	RSA
Boise	190	00289	00000	MSA
Nampa	391	00289	00000	ID-4
Payette	389	00289	30285	ID-2

Cellular company

U S West Cellular
7211 Franklin Road
Boise, ID 83709

Customer service	800-238-7848
. . . .	206-562-2895
. . . . local subscribers	800-626-6611
. . . .	206-747-1771
Local office	208-376-2002
Corporate office	206-747-4900

Sending calls

Visitor dialing instructions

Local calls	area code + 7 digits
Long distance	area code + 7 digits

Speed-dial numbers

Customer service	*611
Emergency	911
Roaming information	*711

Receiving calls

Caller dials the roamer access number below,
hears tone, dials cellular area code and number.

Boise	208-866-7626
Emmett	208-365-7626
Nampa	208-880-7626
Payette	208-642-7626

Follow Me Roaming forwarding (see p. 1156)
Activate, dial *18 send. Deactivate dial *19 send.

Billing information

Visitor rates
Most visitors . . . . . $3 day + 99¢ minute

Visitors without roaming agreements (p. 1159)
Call co. to use Amex, Mastercard, Visa

Identification numbers

City	Market	System	Billing	RSA
Bogus Basin	390	00272	00000	ID-3
Boise	190	00272	00000	MSA
Nampa	391	00272	00000	ID-4
Payette	389	00272	00000	ID-2

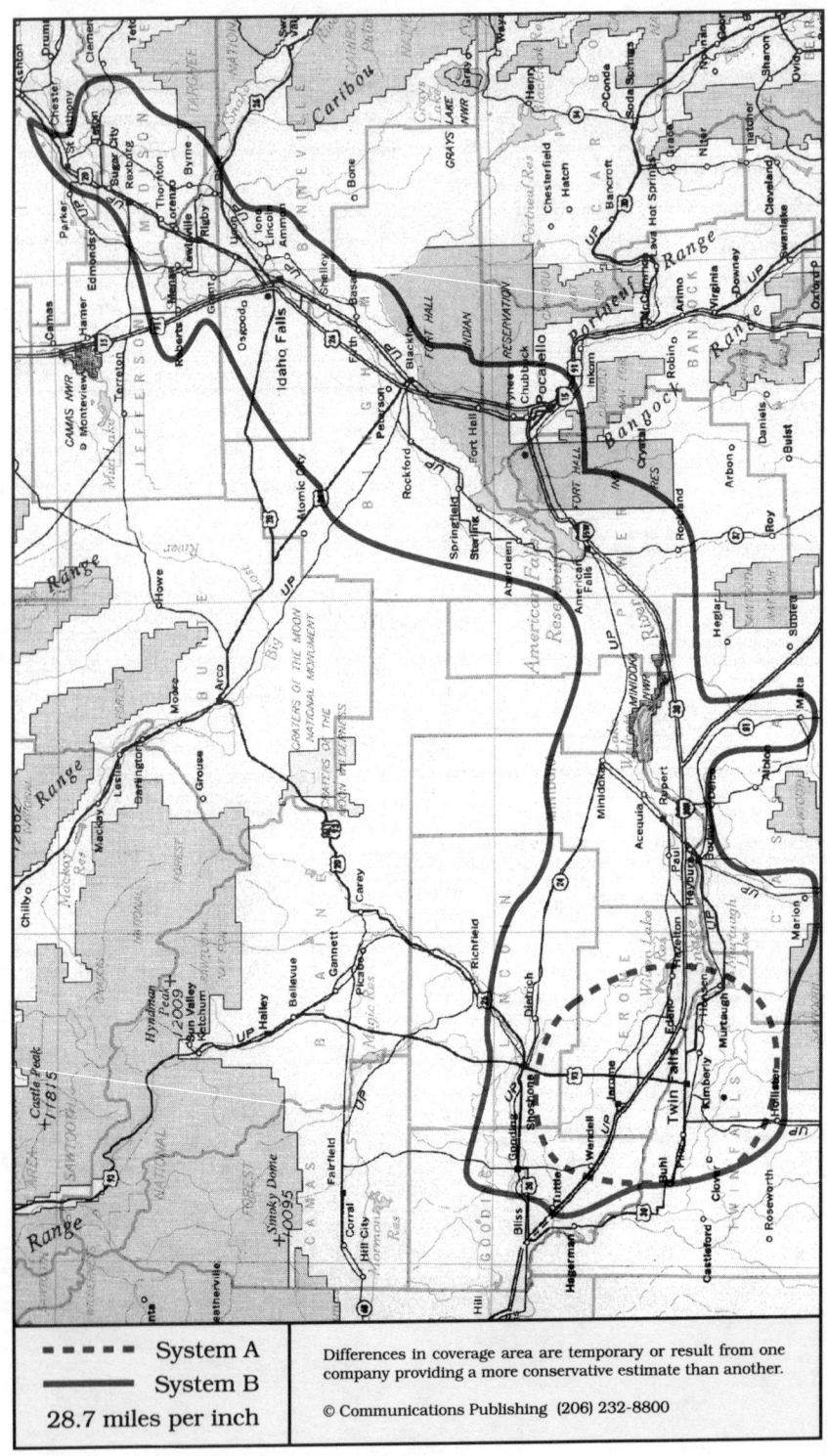

- - - - - System A
————— System B
28.7 miles per inch

Differences in coverage area are temporary or result from one company providing a more conservative estimate than another.

© Communications Publishing (206) 232-8800

Pocatello, Idaho
Gooding, Idaho Falls, Pocatello, Rexburg, Rupert, Twin Falls

System A	System B

Cellular company

Cellular One
Licensed to: Independent Cellular Telephone
Operated by: McCaw Cellular Comms
1475 Broadway
Boise, ID 83706

Customer service	208-345-2355
Local office	208-345-3100
Corporate office	206-827-4500

Sending calls

Visitor dialing instructions

Local calls	area code + 7 digits
Long distance . . .	1 + area code + 7 digits

Speed-dial numbers

Customer service	611
Emergency	911

Receiving calls

Caller dials the roamer access number below, hears tone, dials cellular area code and number.
. . . . 208-731-7626

NationLink & RoamingAmerica (see p. 1157)
Caller hears your current roamer access number. This company's subscribers have level 0 service.

North American Cellular Network (see p. 1156)
All calls forward to you. Deactivate dial *35 send.

Billing information

Visitor rates
Most visitors . . . . . $0 day + 99¢ minute

Visitors without roaming agreements (p. 1159)
Call co. to use Amex, Discover, Mastercard, Visa

Identification numbers

City	Market	System	Billing	RSA
All served cities	392	01171	00000	ID-5

Call switching is performed by McCaw Cellular for the licensee, Independent Cellular Telephone.

Cellular company

CommNet 2000
Cellular, Inc.
2052 E 17th Street
Idaho Falls, ID 83404

Customer service	800-597-5535
Local office	800-767-2007
. . . .	208-525-8039
Corporate office	303-694-3234

Sending calls

Visitor dialing instructions

Local calls	area code + 7 digits
Long distance . . .	1 + area code + 7 digits

Speed-dial numbers

Customer service	*611
Emergency	911
Roaming information	711

Receiving calls

Caller dials the roamer access number below, hears tone, dials cellular area code and number.

American Falls	208-220-7626
Blackfoot	208-680-7626
Burley	208-670-7626
Idaho Falls	208-521-7626
Pocatello	208-241-7626
Rexburg	208-351-7626
Twin Falls	208-420-7626

Billing information

Visitor rates
Most visitors . . . . . $3 day + 95¢ minute

Visitors without roaming agreements (p. 1159)
Cannot call since company takes no credit cards.

Identification numbers

City	Market	System	Billing	RSA
American Falls	393	01174	30410	ID-6
Blackfoot	393	01174	30404	ID-6
Burley	392	00174	30318	ID-5
Idaho Falls	393	00174	30408	ID-6
Pocatello	393	01174	00000	ID-6
Rexburg	393	00174	30406	ID-6
Twin Falls	392	01174	30590	ID-5

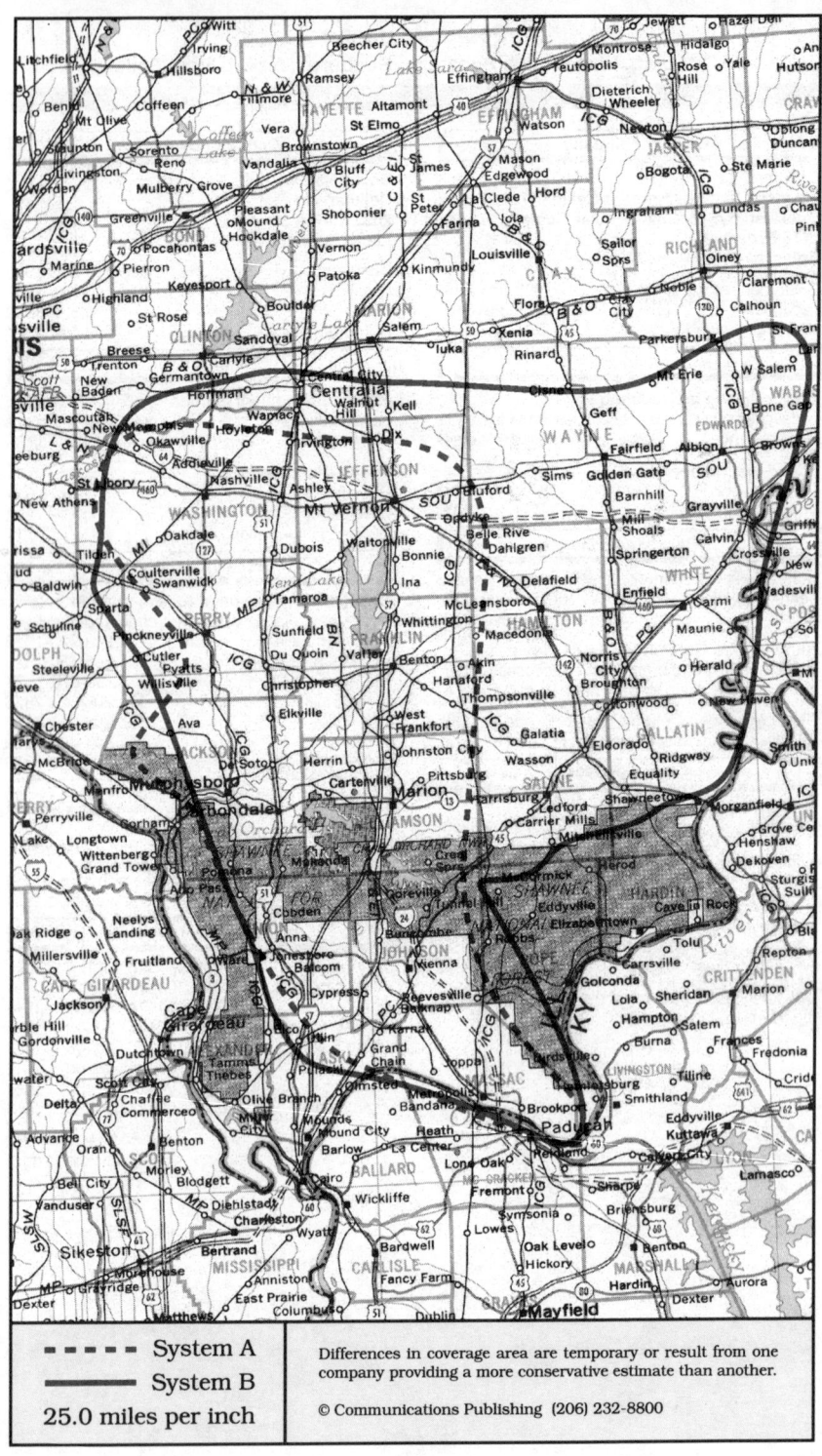

- - - - -	System A
————	System B
25.0 miles per inch	

Differences in coverage area are temporary or result from one company providing a more conservative estimate than another.

© Communications Publishing (206) 232-8800

Carbondale, Illinois

Albion, Carbondale, Centralia, Fairfield, Harrisburg, Mount Vernon, Paducah

System A	System B

Cellular company

Cellular One
HLD Cellular
Managed by: Rural Cellular Management
1376 E Main
Carbondale, IL 62901

Local office 618-549-5490
Corporate office   Rural Cell. Mgt 707-422-2100

Sending calls

Visitor dialing instructions
Local calls . . . . . area code + 7 digits
Long distance . . . 1 + area code + 7 digits

Speed-dial numbers
Customer service *611
Directory assistance 411

Receiving calls

Caller dials the roamer access number below,
hears tone, dials cellular area code and number.
. . . . 618-534-7626

NationLink & RoamingAmerica (see p. 1157)
Dial *31 send to receive calls, *30 to deactivate.
This company's subscribers have level 3 service.

Billing information

Visitor rates
Most visitors . . . . . $3 day + 99¢ minute

Visitors without roaming agreements (p. 1159)
Call co. to use Mastercard, Visa
or Roamer Plus will intercept first call to bill you.

Identification numbers

City	Market	System	Billing	RSA
All served cities	401	01189	00000	IL-8

Cellular company

Contel Cellular
300 E Main, Suite 21
Carbondale, IL 62901

Customer service 800-333-4004
. . . . 205-970-2273
Local office 618-529-2355
Corporate office 404-804-3400

Sending calls

Visitor dialing instructions
Local calls . . . . . area code + 7 digits
Long distance . . . 1 + area code + 7 digits

Speed-dial numbers
Customer service 611
Emergency 911

Receiving calls

Caller dials the roamer access number below,
hears tone, dials cellular area code and number.
Carbondale 618-525-7626
Mount Vernon 618-237-7626

Billing information

Visitor rates
Most visitors . . . . . $3 day + 85¢ minute

Visitors without roaming agreements (p. 1159)
Call co. to use Amex, Mastercard, Visa

Identification numbers

City	Market	System	Billing	RSA
Carbondale, Mt V.	401	01190	00000	IL-8
Harrisburg	402	01190	00000	IL-9

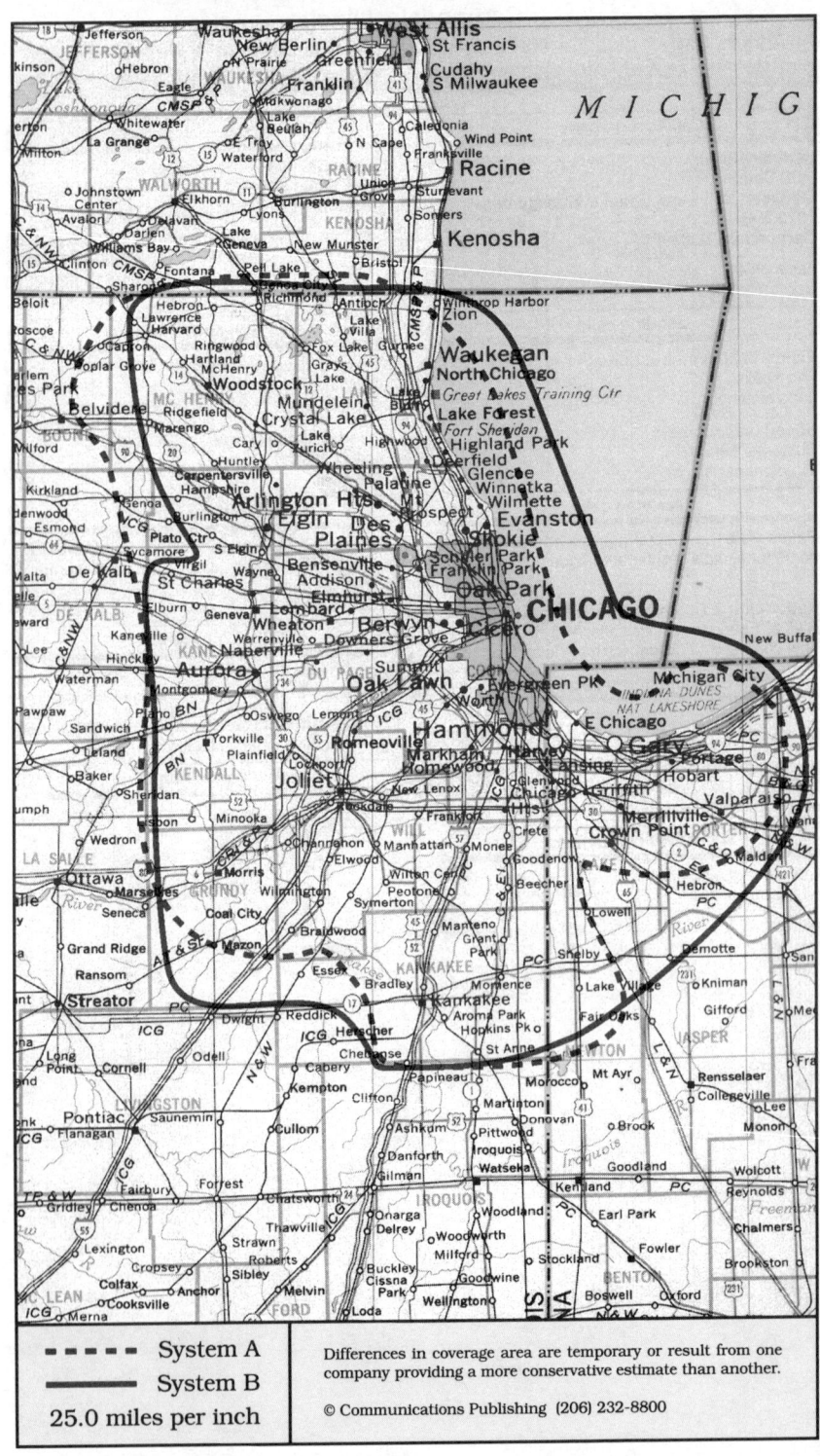

- - - - System A	Differences in coverage area are temporary or result from one company providing a more conservative estimate than another.
———— System B	
25.0 miles per inch	© Communications Publishing (206) 232-8800

Chicago, Illinois
Arlington Heights, Aurora, Chicago, Joliet, Kankakee, Gary IN, Michigan City IN

System A	System B

Cellular company

Cellular One
Southwestern Bell Mobile Systems
930 N National Parkway
Schaumburg, IL 60173

Customer service . . . direct	800-552-1551
. . . . . all departments	800-235-5663
Local office	708-762-2000
Corporate office	214-733-2000

Sending calls

Visitor dialing instructions
Local calls . . . . . area code + 7 digits
Long distance . . . 1 + area code + 7 digits

Speed-dial numbers
American Automobile Association . . . ∗222
Customer service . . . . ∗611
Emergency . . . . 0 or ∗999
Report a drunk driver . . . . . . ∗384
Roaming information . . . . ∗711
Sports report . . . ∗SPORT or ∗77678
Traffic report center . . . . ∗123

Receiving calls

Caller dials the roamer access number below,
hears tone, dials cellular area code and number.
. . . . 312-659-7626

NationLink & RoamingAmerica (see p. 1157)
Dial ∗31 send to receive calls, ∗30 to deactivate.
Dial ∗32 send to tell caller the roamer access no.
This company's subscribers have level 6 service.

Billing information

Visitor rates
Most visitors . . . . . $3 day + 85¢ minute

Visitors without roaming agreements (p. 1159)
Cannot call since company takes no credit cards.

Identification numbers

City	Market	System	Billing	RSA
Aurora	303	00001	00000	MSA
Chicago	003	00001	00000	MSA
Gary, IN	054	00001	30050	MSA
Joliet	304	00001	00000	MSA
Kankakee	273	00001	00000	MSA
La Porte, IN	394	00001	30050	IN-1

For full coverage area, see Pontiac, IL.

Cellular company

Ameritech Mobile Communications
1515 Woodfield Road, Suite 1400
Schaumburg, IL 60173

Customer service	800-221-0994
. . . .	708-706-7300
Local office	708-706-7600
Corporate office	708-706-7600

Sending calls

Visitor dialing instructions
Local calls . . . . . area code + 7 digits
Long distance . . . . area code + 7 digits

Speed-dial numbers
Customer service . . . . ∗611
Emergency . . . . 0 or ∗999
Make a traffic report 211
Roaming information . . . . . . ∗711
Smart call . . . . ∗123

Receiving calls

Caller dials the roamer access number below,
hears tone, dials cellular area code and number.
Chicago 312-550-7626
LaPorte, IN 219-873-7626

Follow Me Roaming forwarding (see p. 1156)
Activate, dial ∗18 send. Deactivate dial ∗19 send.

Billing information

Visitor rates
Most visitors . . . . . $3 day + 85¢ minute

Visitors without roaming agreements (p. 1159)
Cannot call since company takes no credit cards.

Identification numbers

City	Market	System	Billing	RSA
Aurora	303	00020	00000	MSA
Chicago	003	00020	00000	MSA
Gary, IN	054	00020	00000	MSA
Joliet	304	00020	00000	MSA
Kankakee	273	00020	00000	MSA
La Porte, IN	394	00020	01194	IN-1

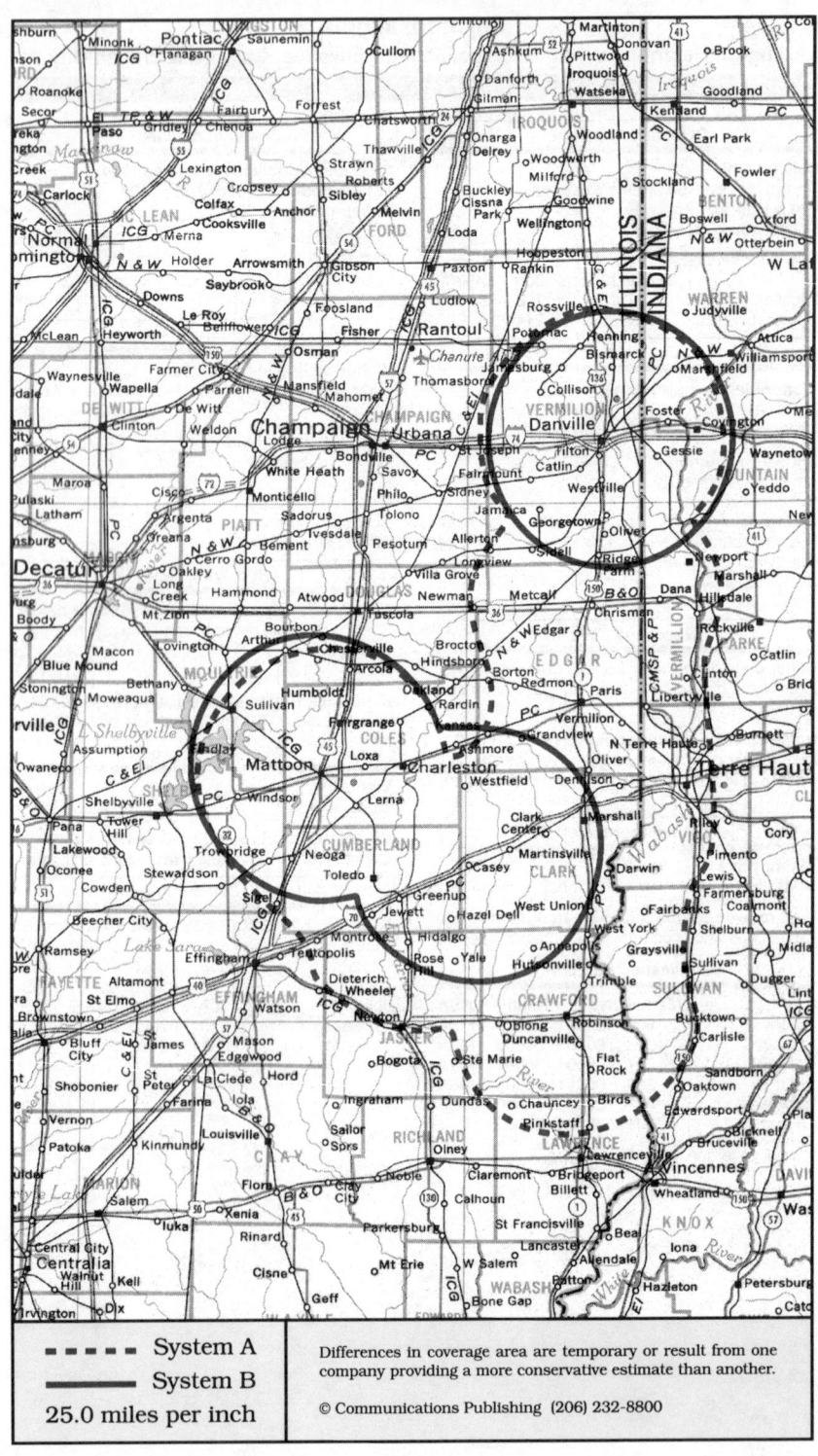

- - - - System A	Differences in coverage area are temporary or result from one company providing a more conservative estimate than another.
──── System B	
25.0 miles per inch	© Communications Publishing (206) 232-8800

Danville, Illinois
Casey, Danville, Mattoon

Cellular company

Cellular One
Licensed to: Cellular Properties
Managed by: MCMG
2 E Main Street, 28 Town Center
Danville, IL 61832

632 W Lincoln
Charleston, IL 61920

Customer service		800-735-2355
Local office . . .	Charleston	800-365-2351
. . . .	Charleston	217-345-2351
. . . .	Danville	217-442-2355
Corporate office . .	MCMG	615-791-0202

Sending calls

Visitor dialing instructions
Local calls . . . . . area code + 7 digits
Long distance . . . 1 + area code + 7 digits

Speed-dial numbers
Customer service 611
Emergency 911

Receiving calls

Caller dials the roamer access number below,
hears tone, dials cellular area code and number.
. . . . 217-497-7626

Billing information

Visitor rates
Most visitors . . . . . $3 day + 85¢ minute

Visitors without roaming agreements (p. 1159)
Roamer Plus will intercept first call to bill you.

Identification numbers

City	Market	System	Billing	RSA
All served cities	400	01187	00000	IL-7

Cellular company

Ameritech Mobile Communications
2205 W Wabash, Suite 102
Springfield, IL 62704

Customer service	800-221-0994
. . . .	708-706-7300
Local office . subject to change	217-698-7499
Corporate office	708-706-7600

Sending calls

Visitor dialing instructions
Local calls . . . . . . area code + 7 digits
Long distance . . . . . area code + 7 digits

Speed-dial numbers
Customer service . . . . ∗611
Emergency 911

Receiving calls

Caller dials the roamer access number below,
hears tone, dials cellular area code and number.
. . . . 217-474-7626

Follow Me Roaming forwarding (see p. 1156)
Activate, dial ∗18 send. Deactivate dial ∗19 send.

Billing information

Visitor rates
Most visitors . . . . . $3 day + 85¢ minute

Visitors without roaming agreements (p. 1159)
Cannot call since company takes no credit cards.

Identification numbers

City	Market	System	Billing	RSA
All served cities	400	01188	00000	IL-7

For full coverage area, see Springfield, IL.

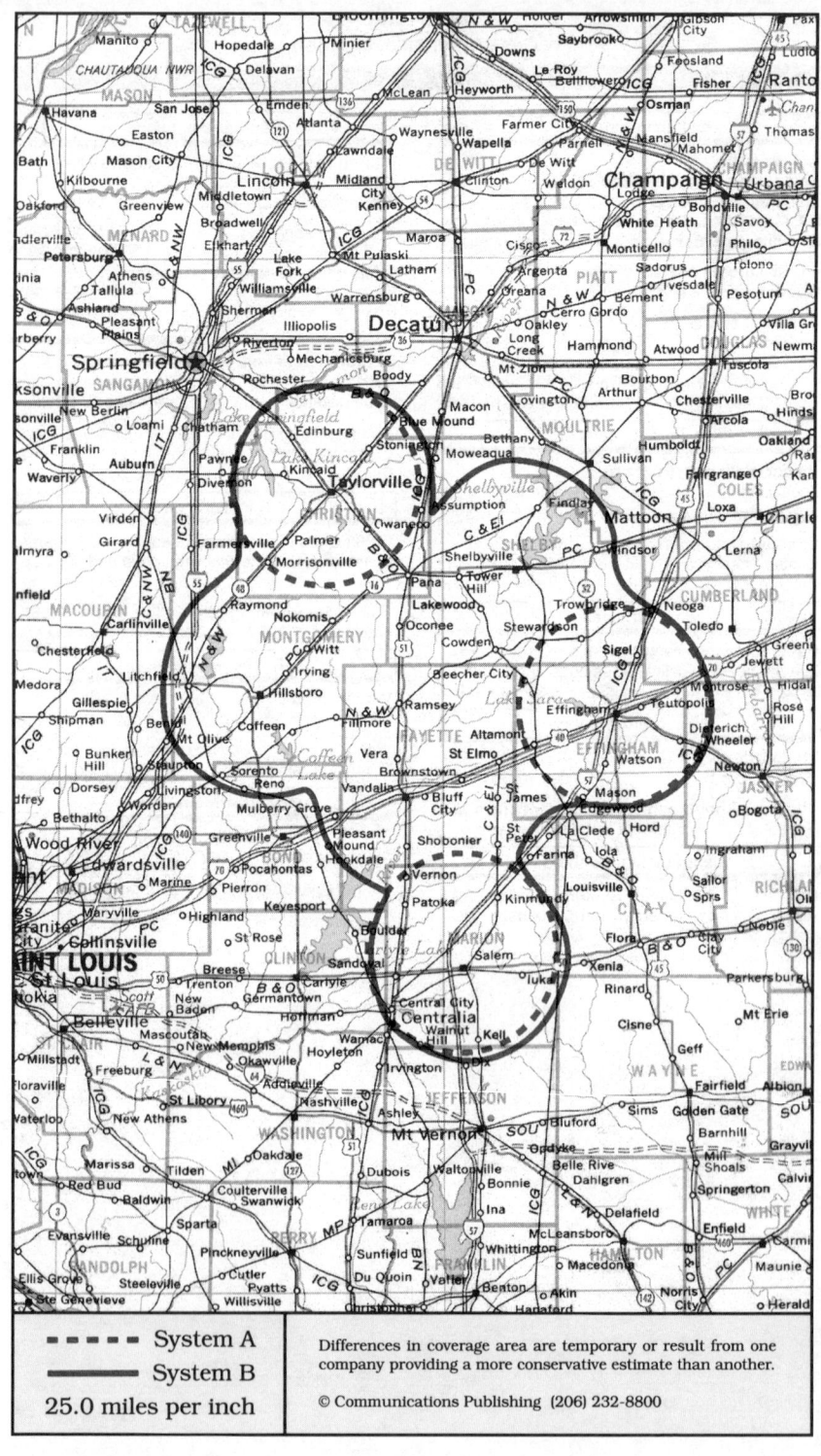

- - - - System A	Differences in coverage area are temporary or result from one company providing a more conservative estimate than another.
——— System B	
25.0 miles per inch	© Communications Publishing (206) 232-8800

238

Effingham, Illinois
Effingham, Litchfield, Salem, Taylorville, Vandalia

System A	System B

Cellular company

Cellular One
Southwestern Bell Mobile Systems
3501 E Sangamon Avenue
Springfield, IL 62707

Customer service	800-235-5663
. . . .	217-744-7393
Local office	217-744-3000
Corporate office . . Chicago	708-762-2000
. . . . Dallas	214-733-2000

Sending calls

Visitor dialing instructions

Local calls	area code + 7 digits
Long distance	area code + 7 digits

Speed-dial numbers

Customer service	*611
Emergency	911

Receiving calls

Caller dials the roamer access number below,
hears tone, dials cellular area code and number.

Effingham	217-343-7626
Salem	618-292-7626
Taylorville	217-825-7626

NationLink & RoamingAmerica (see p. 1157)
Dial *31 send to receive calls, *30 to deactivate.
Dial *32 send to tell caller the roamer access no.
This company's subscribers have level 3 service.

Billing information

Visitor rates

Most visitors	$3 day + 85¢ minute

Visitors without roaming agreements (p. 1159)
Call co. to use Amex, Discover, Mastercard, Visa

Identification numbers

City	Market	System	Billing	RSA
All served cities	399	01185	00000	IL-6

Cellular company

Ameritech Mobile Communications
2205 W Wabash, Suite 102
Springfield, IL 62704

Customer service	800-221-0994
. . . .	708-706-7300
Local office . subject to change	217-698-7499
Corporate office	708-706-7600

Sending calls

Visitor dialing instructions

Local calls	area code + 7 digits
Long distance	area code + 7 digits

Speed-dial numbers

Customer service	*611
Emergency	911

Receiving calls

Caller dials the roamer access number below,
hears tone, dials cellular area code and number.

All other cities	217-254-7626
Vandalia	618-267-7626

Billing information

Visitor rates

Most visitors	$3 day + 85¢ minute

Visitors without roaming agreements (p. 1159)
Cannot call since company takes no credit cards.

Identification numbers

City	Market	System	Billing	RSA
All served cities	399	01186	00000	IL-6

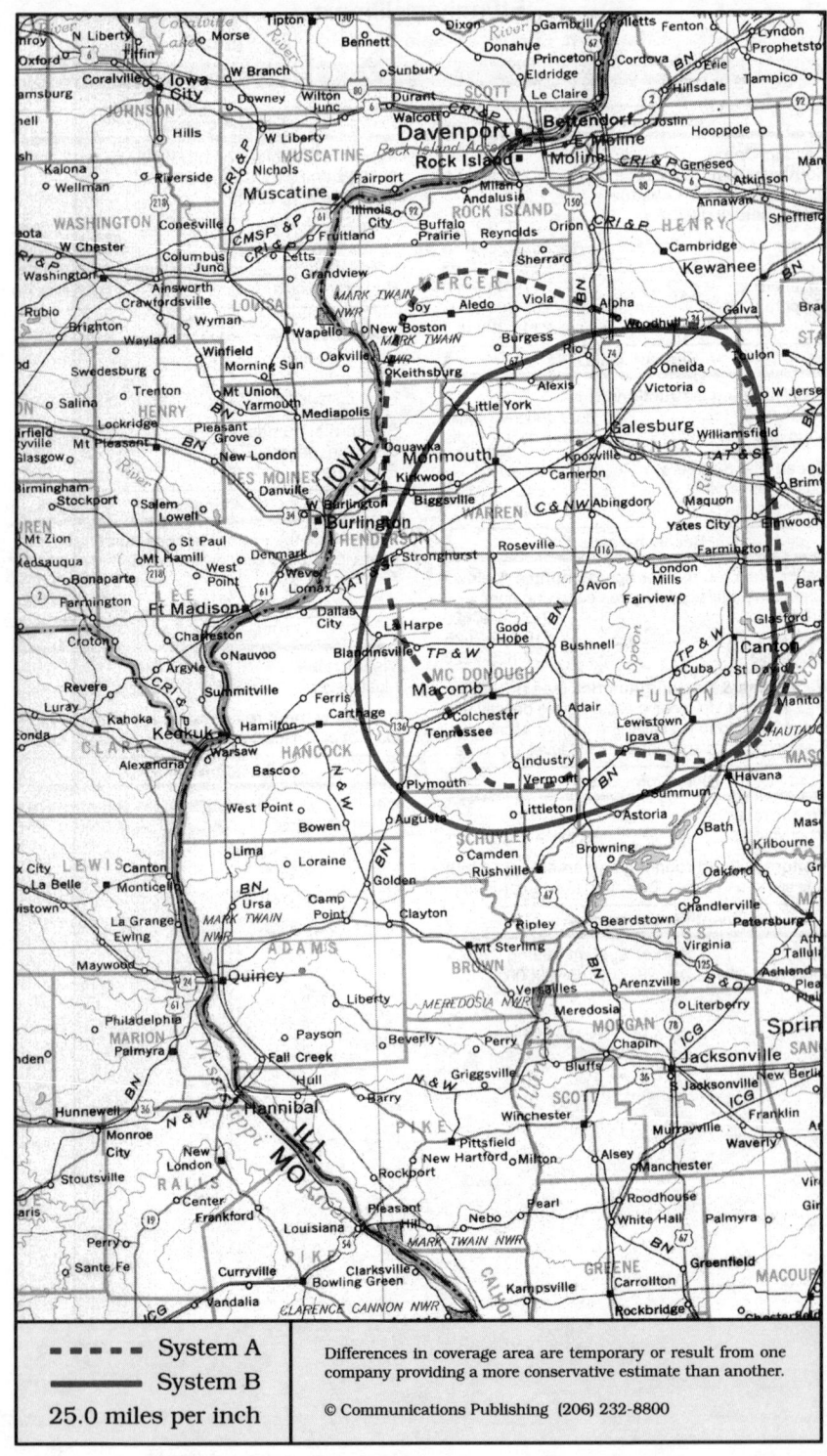

- - - - System A	Differences in coverage area are temporary or result from one company providing a more conservative estimate than another.
——— System B	
25.0 miles per inch	© Communications Publishing (206) 232-8800

240

Galesburg, Illinois
Canton, Galesburg, Macomb, Monmouth

System A	System B

Cellular company

Cellular One of Western Illinois
Dial Two
664 W Brooks Street
Galesburg, IL 61401

Corporate office 309-343-9911

Sending calls

Visitor dialing instructions
Local calls . . . . 1 + 7 digits
Long distance . . . 1 + area code + 7 digits

Speed-dial numbers
Customer service *611
Emergency 911

Receiving calls

Caller dials the roamer access number below,
hears tone, dials cellular area code and number.
. . . . 309-337-7626

Billing information

Visitor rates
Most visitors . . . . . $3 day + 99¢ minute

Visitors without roaming agreements (p. 1159)
Call co. to use Mastercard, Visa
or Roamer Plus will intercept first call to bill you.

Identification numbers

City	Market	System	Billing	RSA
All served cities	396	01179	00000	IL-3

Cellular company

Centel Cellular
888 N Henderson
Galesburg, IL 61401

Customer Service 800-423-5221
Local office 309-343-1171
Corporate office 312-399-2644

Sending calls

Visitor dialing instructions
Local calls . . . . . area code + 7 digits
Long distance . . . 1 + area code + 7 digits

Speed-dial numbers
Customer service *611
Emergency 911
Roaming information *711
WAAG *95

Receiving calls

Caller dials the roamer access number below,
hears tone, dials cellular area code and number.
. . . . 309-696-7626

Follow Me Roaming forwarding (see p. 1156)
Activate, dial *18 send. Deactivate dial *19 send.

Billing information

Visitor rates
Most visitors . . . . . $3 day + 99¢ minute

Visitors without roaming agreements (p. 1159)
Cannot call since company takes no credit cards.

Identification numbers

City	Market	System	Billing	RSA
All served cities	396	01180	00000	IL-3

For full coverage area, see Peoria, IL.

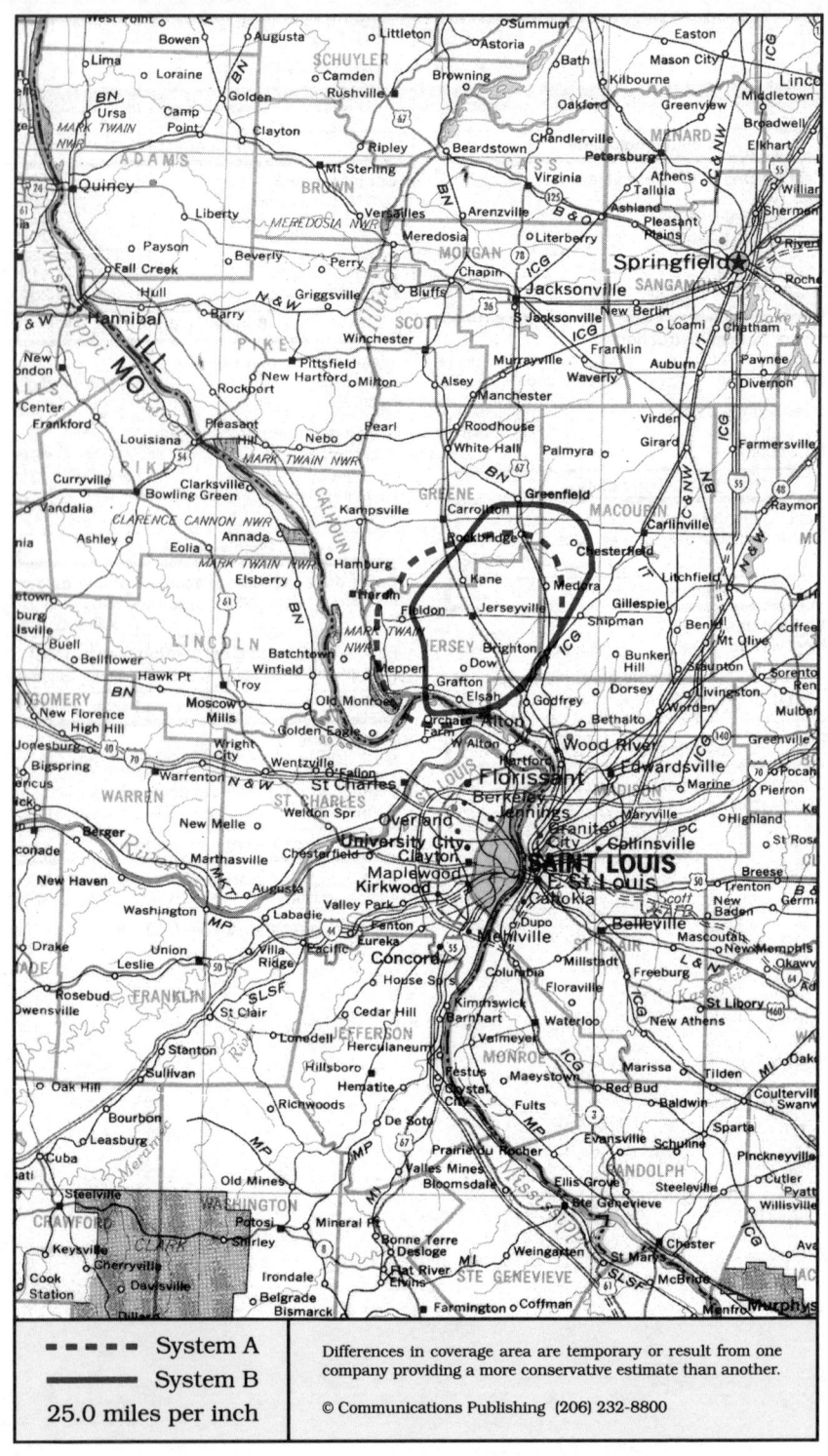

- - - - System A	Differences in coverage area are temporary or result from one company providing a more conservative estimate than another.
——— System B	
25.0 miles per inch	© Communications Publishing (206) 232-8800

Jerseyville, Illinois

System A

Cellular company

Cellular One
GenCell Management
1891 Woolner Avenue
Fairfield, CA 94533

Customer service	800-888-7868
Corporate office	707-425-8000

Sending calls

Visitor dialing instructions

Local calls	area code + 7 digits
Long distance . . .	1 + area code + 7 digits

Speed-dial numbers

Customer service	611
Emergency	911

Receiving calls

Caller dials the roamer access number below, hears tone, dials cellular area code and number.

. . . . 618-626-7626

NationLink & RoamingAmerica (see p. 1157)
Caller hears your current roamer access number. This company's subscribers have level 0 service.

Billing information

Visitor rates

Most visitors	$3 day + 99¢ minute

Visitors without roaming agreements (p. 1159)
American Roaming Network intercepts call to bill.

Identification numbers

City	Market	System	Billing	RSA
All served cities	305	00245	00000	MSA

System B

Cellular company

Licensed to: Alton Cellular (MACTEL)
Managed by: Cellular Systems International
202 Homer Adams Parkway
Godfrey, IL 62035

Local office	618-466-0199
Corporate office . . . C.S.I.	707-573-3500

Sending calls

Visitor dialing instructions

Local calls	area code + 7 digits
Long distance . . .	0 + area code + 7 digits

Speed-dial numbers

Customer service	*611
Emergency	911

Receiving calls

Caller dials the roamer access number below, hears tone, dials cellular area code and number.

. . . . 618-535-7626

Billing information

Visitor rates

Most visitors	$3 day + 85¢ minute

Visitors without roaming agreements (p. 1159)
Call co. to use Mastercard, Visa

Identification numbers

City	Market	System	Billing	RSA
All served cities	305	00586	00000	MSA

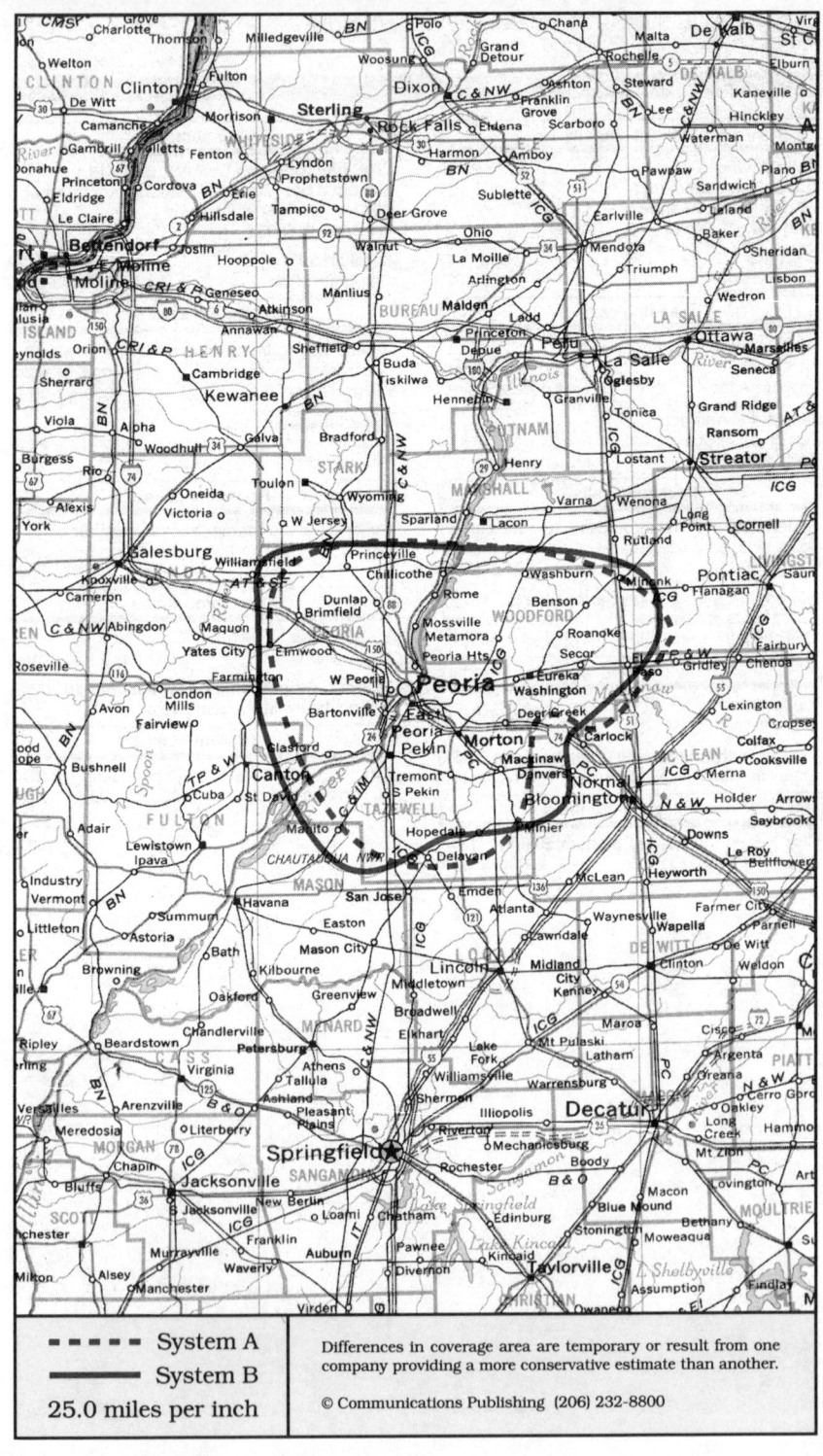

━ ━ ━ ━ System A	Differences in coverage area are temporary or result from one
━━━━━ System B	company providing a more conservative estimate than another.
25.0 miles per inch	© Communications Publishing (206) 232-8800

Peoria, Illinois
Morton, Pekin, Peoria

System A	System B

Cellular company

United States Cellular
4538 N Brandywine Drive
Peoria, IL 61614

Local office	309-685-1234
Corporate office	312-399-8900

Sending calls

Visitor dialing instructions

Local calls	area code + 7 digits
Long distance . . .	1 + area code + 7 digits

Speed-dial numbers

Customer service	611
Directory assistance	411
Emergency	911

Receiving calls

Caller dials the roamer access number below,
hears tone, dials cellular area code and number.
. . . . 309-678-7626

Billing information

Visitor rates

Most visitors	$3 day + 99¢ minute

Visitors without roaming agreements (p. 1159)
Call co. to use Amex, Mastercard, Visa
or Roamer Plus will intercept first call to bill you.

Identification numbers

City	Market	System	Billing	RSA
All served cities	103	00221	00000	MSA

Cellular company

Centel Cellular
2417 W Park 74 Drive
Peoria, IL 61615

Local office	800-423-5221
. . . .	309-693-3800
Corporate office	312-399-2644

Sending calls

Visitor dialing instructions

Local calls	area code + 7 digits
Long distance . . .	1 + area code + 7 digits

Speed-dial numbers

Customer service	*611
Emergency	911
Roaming information	*711

Receiving calls

Caller dials the roamer access number below,
hears tone, dials cellular area code and number.
. . . . 309-696-7626

Follow Me Roaming forwarding (see p. 1156)
Activate, dial *18 send. Deactivate dial *19 send.

Billing information

Visitor rates

Most visitors	$3 day + 99¢ minute

Visitors without roaming agreements (p. 1159)
Call co. to use Amex, Discover, Mastercard, Visa

Identification numbers

City	Market	System	Billing	RSA
All served cities	103	00214	00000	MSA

For full coverage area, see Galesburg, IL.

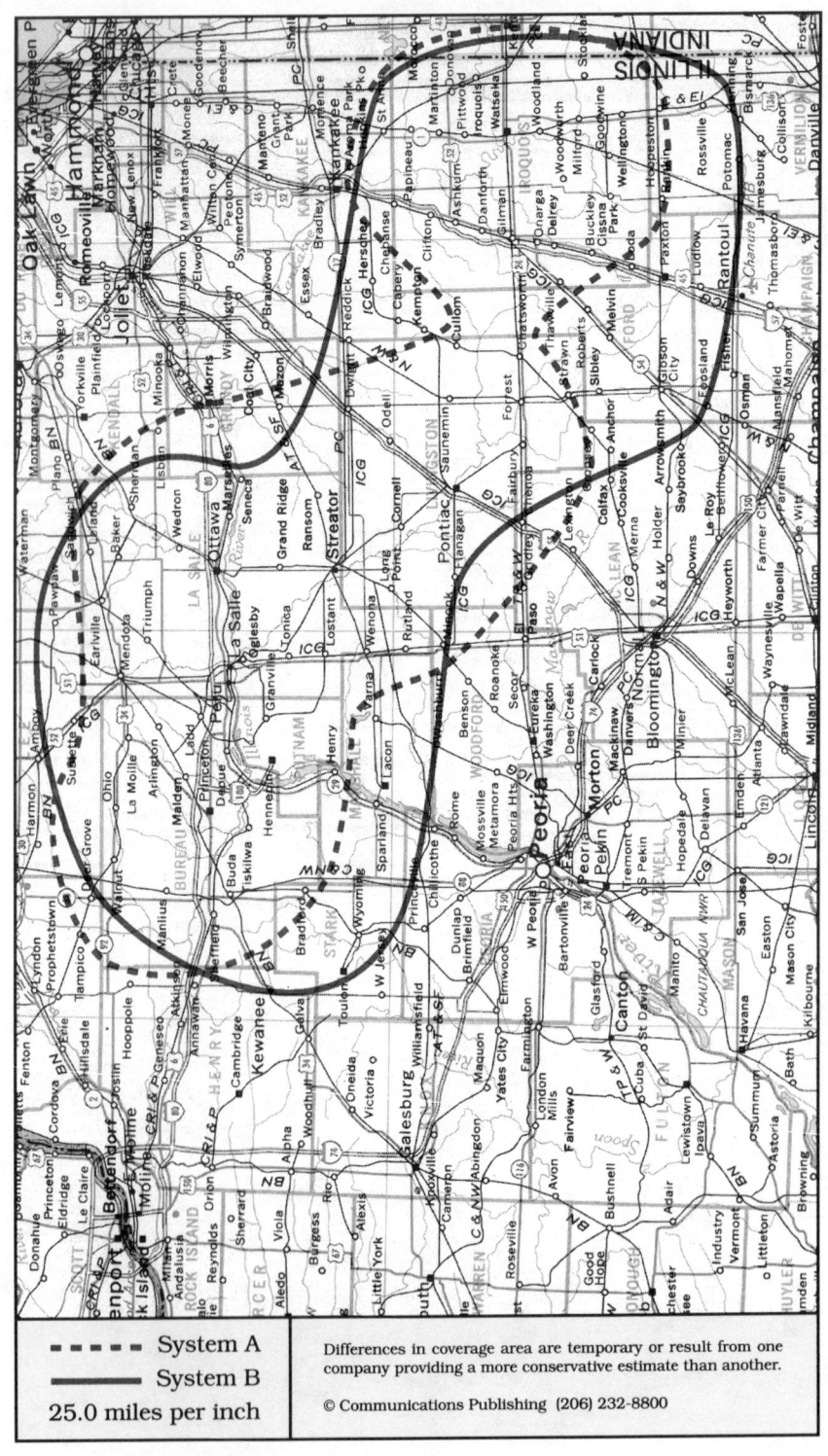

▪ ▪ ▪ ▪ ▪ System A	Differences in coverage area are temporary or result from one company providing a more conservative estimate than another.
─────── System B	
25.0 miles per inch	© Communications Publishing (206) 232-8800

Pontiac, Illinois

Henry, Lacon, La Salle, Ottawa, Pontiac, Princeton, Streator, Toulon, Watseka

System A	System B

System A

Cellular company

Cellular One
Southwestern Bell Mobile Systems
930 N National Parkway
Schaumburg, IL 60173

Customer service . . . direct	800-552-1551
. . . . all departments	800-235-5663
Local office	708-762-2000
Corporate office	214-733-2000

Sending calls

Visitor dialing instructions
Local calls . . . . . area code + 7 digits
Long distance . . . 1 + area code + 7 digits

Speed-dial numbers
American Automobile Association . . . *222
Customer service *611
Emergency . . . . 0 or *999
Report a drunk driver *384
Roaming information *711
Sports report . . . . *SPORT or *77678
Traffic report center *123

Receiving calls

Caller dials the roamer access number below,
hears tone, dials cellular area code and number.
Henry, La Salle, Ottawa, Watseka 815-228-7626
Pontiac, Streator 815-674-7626
Princeton 815-878-7626

NationLink & RoamingAmerica (see p. 1157)
Dial *31 send to receive calls, *30 to deactivate.
Dial *32 send to tell caller the roamer access no.
This company's subscribers have level 6 service.
(Princeton subscribers have level 0 service.)

Billing information

Visitor rates
Most visitors . . . . . $3 day + 85¢ minute

Visitors without roaming agreements (p. 1159)
Cannot call since company takes no credit cards.

Identification numbers

City	Market	System	Billing	RSA
La Salle, Ottawa, W.	395	00001	00000	IL-2
Pontiac, Streator	395	01177	00000	IL-2
Princeton	395	01871	00000	IL-2

For full coverage area, see Chicago, IL.

System B

Cellular company

Illinois Valley Cellular
455 Main Street
Marseilles, IL 61341

Customer service .	In Illinois	800-438-4824
. . . .		800-433-4824
Corporate office		815-795-3200

Sending calls

Visitor dialing instructions
Local calls . . . . . . area code + 7 digits
Long distance . . . 1 + area code + 7 digits

Speed-dial numbers
Customer service 611
Emergency 911

Receiving calls

Caller dials the roamer access number below,
hears tone, dials area code and number.
La Salle, Ottawa, Streator, Wat. 815-488-7626
Pontiac 815-848-7626
Princeton 815-866-7626

Follow Me Roaming forwarding (see p. 1156)
Activate, dial *18 send. Deactivate dial *19 send.

Billing information

Visitor rates
Most visitors . . . . . $3 day + 85¢ minute

Visitors without roaming agreements (p. 1159)
Call co. to use Amex, Mastercard, Visa

Identification numbers

City	Market	System	Billing	RSA
Bureau Cty, Streator	395	01178	30644	IL-2
Ottawa, Princeton	395	01178	30642	IL-2
Pontiac	395	01178	30640	IL-2

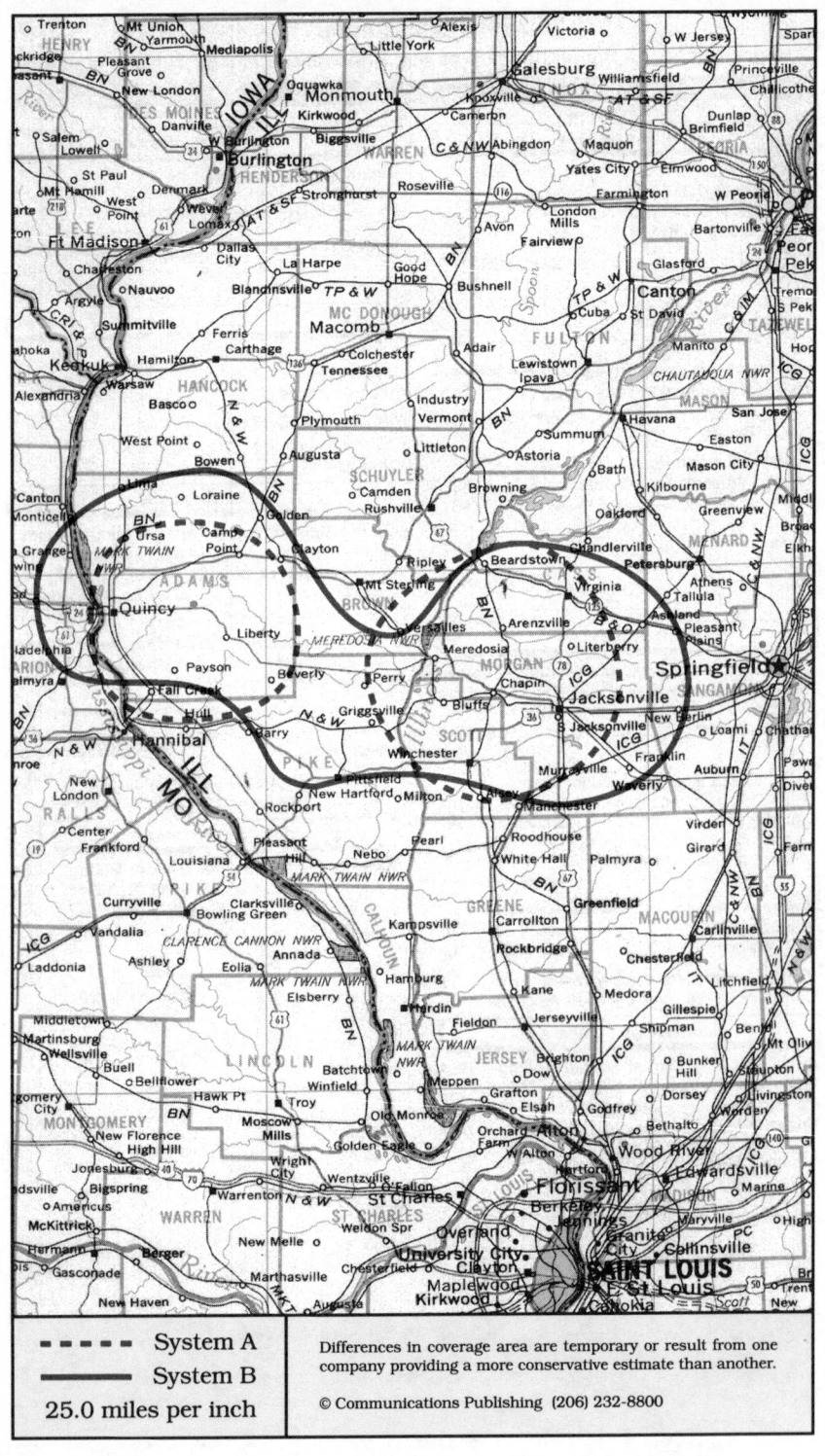

- - - - System A	Differences in coverage area are temporary or result from one company providing a more conservative estimate than another.
—— System B	
25.0 miles per inch	© Communications Publishing (206) 232-8800

Quincy, Illinois
Jacksonville, Liberty, Quincy

System A

Cellular company

Cellular One
Southwestern Bell Mobile Systems
Administration
3501 E Sangamon Avenue
Springfield, IL 62707

3502 Broadway
Quincy, IL 62301

Customer service	. . . . .		800-235-5663
. . . .			217-744-7393
Local office	. . . .	Quincy	217-224-5600
. . . . . .		Springfield	217-744-3000
Corporate office	. .	Chicago	708-762-2000
. . . . . .		Dallas	214-733-2000

Sending calls

Visitor dialing instructions
Local calls . . . . . area code + 7 digits
Long distance . . . area code + 7 digits

Speed-dial numbers
Customer service . . . . ∗611
Emergency 911

Receiving calls

Caller dials the roamer access number below,
hears tone, dials cellular area code and number.
Jacksonville 217-473-7626
Liberty, Quincy 217-257-7626

NationLink & RoamingAmerica (see p. 1157)
Dial ∗31 send to receive calls, ∗30 to deactivate.
Dial ∗32 send to tell caller the roamer access no.
This company's subscribers have level 3 service.

Billing information

Visitor rates
Most visitors . . . . . $3 day + 85¢ minute

Visitors without roaming agreements (p. 1159)
Call co. to use Amex, Discover, Mastercard, Visa

Identification numbers

City	Market	System	Billing	RSA
All served cities	397	01181	00000	IL-4

System B

Cellular company

United States Cellular
1804 Vandiver Drive
Columbia, MO 65202

Local office	. . . . . .	314-474-0400
Corporate office	. . . . .	312-399-8900

Sending calls

Visitor dialing instructions
Local calls . . . . . area code + 7 digits
Long distance . . . . . area code + 7 digits

Speed-dial numbers
Customer service 611
Emergency 911

Receiving calls

Caller dials the roamer access number below,
hears tone, dials cellular area code and number.
. . . . 217-242-7626

Follow Me Roaming forwarding (see p. 1156)
Activate, dial ∗18 send. Deactivate dial ∗19 send.

Billing information

Visitor rates
Most visitors . . . . . $3 day + 85¢ minute

Visitors without roaming agreements (p. 1159)
Cannot call since company takes no credit cards.

Identification numbers

City	Market	System	Billing	RSA
All served cities	397	01914	00000	IL-4

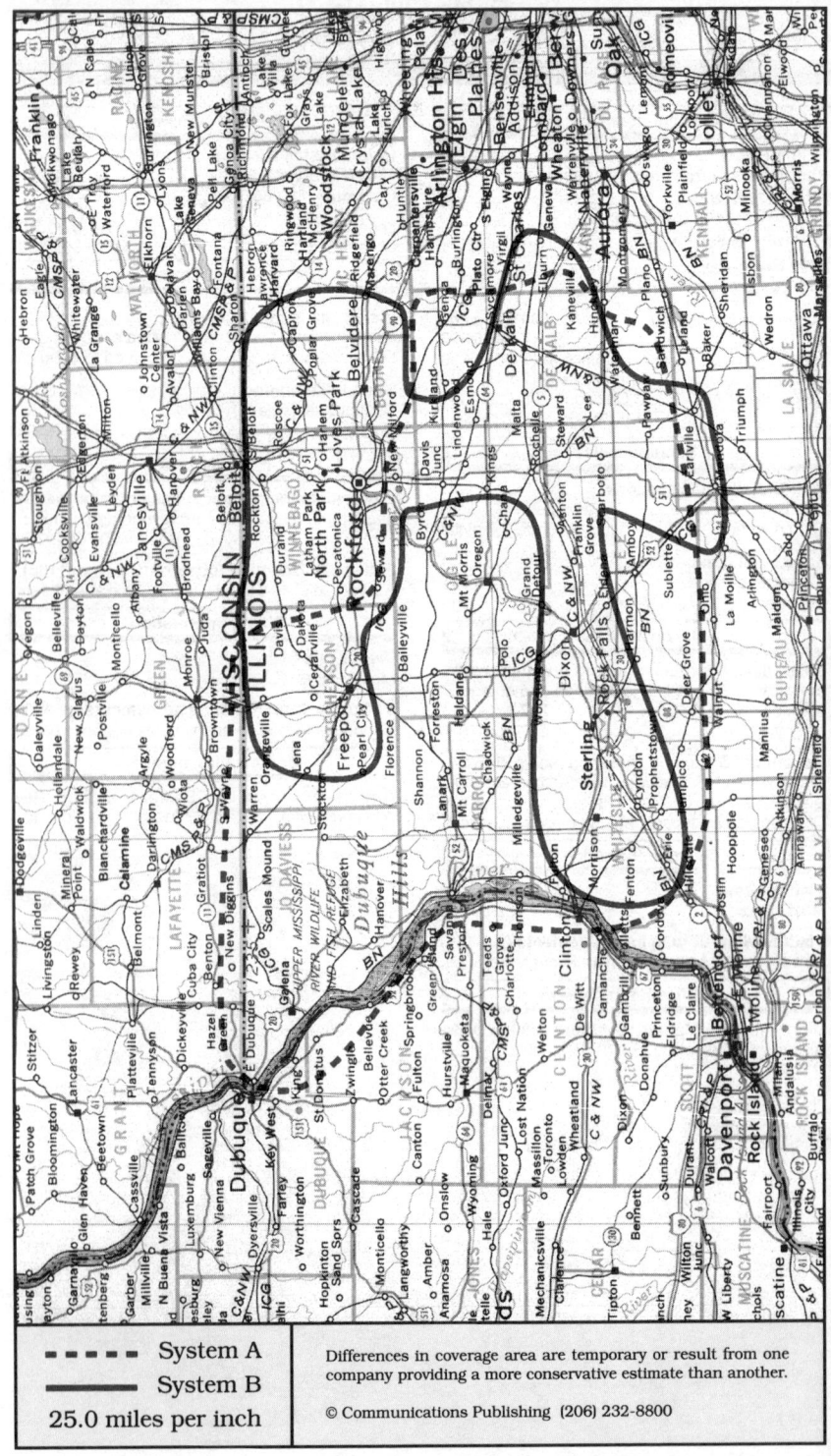

- - - - - System A
───── System B
25.0 miles per inch

Differences in coverage area are temporary or result from one company providing a more conservative estimate than another.

© Communications Publishing (206) 232-8800

Rockford, Illinois
De Kalb, Freeport, Rockford, Sterling

System A	System B

Cellular company

Cellular One
BellSouth Mobility
633 Harlem Road
Rockford, IL 61111

Customer service	800-888-5611
. . . .	815-494-2273
Local office	815-633-1111
Corporate office	404-847-3600

Sending calls

Visitor dialing instructions

Local calls	area code + 7 digits
Long distance . . .	1 + area code + 7 digits

Speed-dial numbers

Customer service	611
Emergency	911

Receiving calls

Caller dials the roamer access number below,
hears tone, dials cellular area code and number.

. . . .	815-494-7626

NationLink & RoamingAmerica (see p. 1157)
Dial ✶31 send to receive calls, ✶30 to deactivate.
Dial ✶32 send to tell caller the roamer access no.
This company's subscribers have level 6 service.

Billing information

Visitor rates

Most visitors	$3 day + 99¢ minute

Visitors without roaming agreements (p. 1159)
Cannot call since company takes no credit cards.

Identification numbers

City	Market	System	Billing	RSA
All served cities	131	00217	30033	MSA

For full coverage area, see Madison, WI.

Cellular company

Contel Cellular
6917 E State Street
Rockford, IL 61108-2692

Customer service	800-333-4004
. . . .	205-970-2273
Local office	815-395-0150
Corporate office	404-804-3400

Sending calls

Visitor dialing instructions

Local calls	area code + 7 digits
Long distance . . .	1 + area code + 7 digits

Speed-dial numbers

Customer service	611
Emergency	911

Receiving calls

Caller dials the roamer access number below,
hears tone, dials cellular area code and number.

De Kalb	815-238-7626
Freeport	815-751-7626
Rockford	815-262-7626

Follow Me Roaming forwarding (see p. 1156)
Activate, dial ✶18 send. Deactivate dial ✶19 send.

Billing information

Visitor rates

Most visitors	$3 day + 85¢ minute

Visitors without roaming agreements (p. 1159)
Call co. to use Amex, Mastercard, Visa

Identification numbers

City	Market	System	Billing	RSA
Rockford	131	00506	00000	MSA
De Kalb, Freeport	394	01176	00000	IL-1

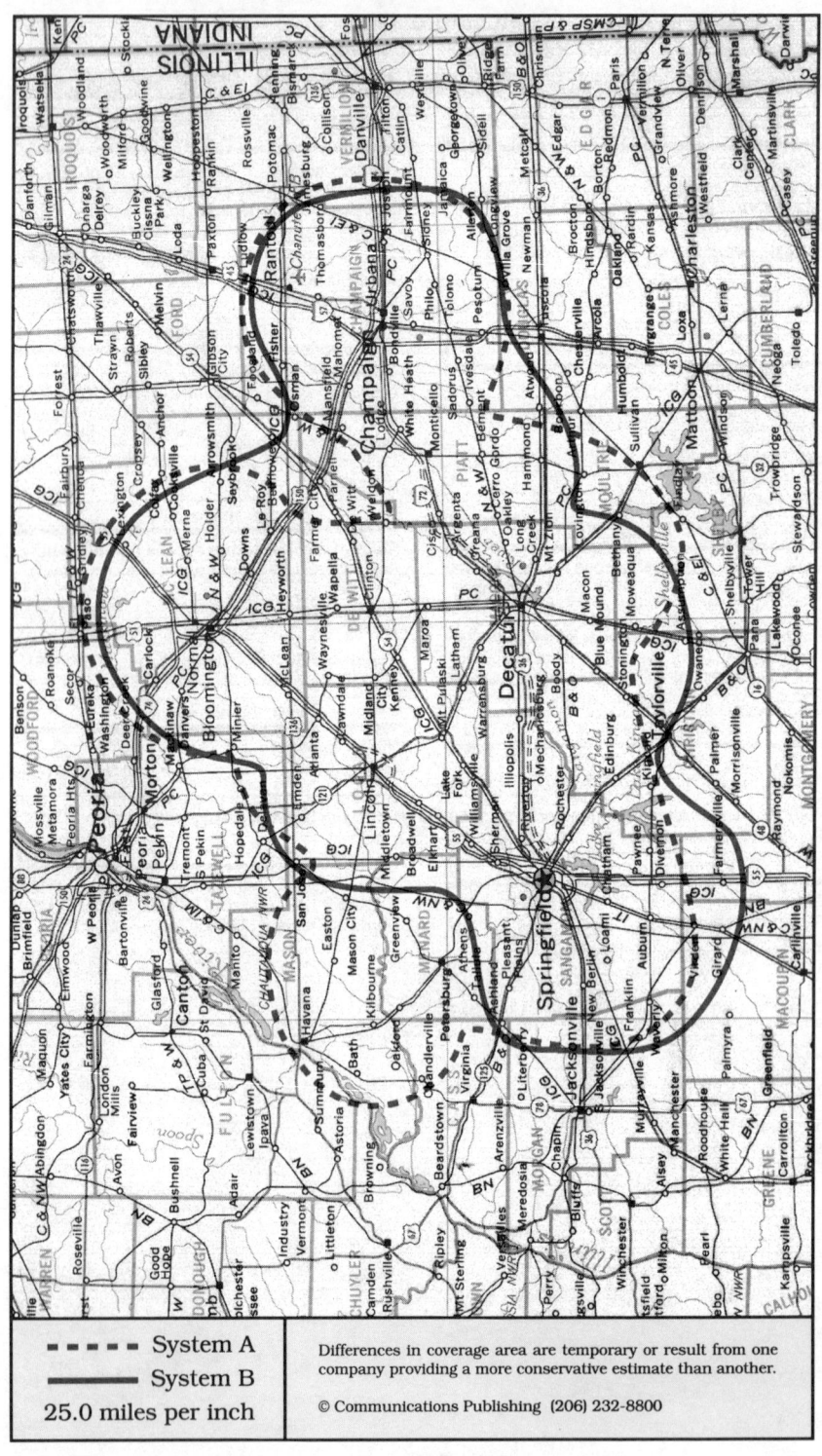

▬ ▬ ▬ ▬ System A	Differences in coverage area are temporary or result from one
▬▬▬▬▬ System B	company providing a more conservative estimate than another.
25.0 miles per inch	© Communications Publishing (206) 232-8800

252

Springfield, Illinois
Bloomington, Champaign, Decatur, Lincoln, Springfield

System A	System B

Cellular company

Cellular One
Southwestern Bell Mobile Systems
Administration
3501 E Sangamon Avenue
Springfield, IL 62707

7 Westport Court
Bloomington, IL 61704

505 E University
Champaign, IL 61820

134 N Martin Luther King Jr. Drive
Decatur, IL 62523

Customer service		800-235-5663
. . . .		217-744-7393
Local office . .	Bloomington	309-664-0290
. . . . . .	Champaign	217-351-8180
. . . . . .	Decatur	217-423-2500
. . . . . .	Springfield	708-744-3000
Corporate office . .	Chicago	708-762-2000
. . . .	Dallas	214-733-2000

Sending calls

Visitor dialing instructions
Local calls . . . . . area code + 7 digits
Long distance . . . area code + 7 digits

Speed-dial numbers
Customer service . . . . ∗611
Emergency 911

Receiving calls

Caller dials the roamer access number below,
hears tone, dials cellular area code and number.
Bloomington 309-825-7626
Champaign 217-377-7626
Decatur 217-454-7626
Lincoln 217-737-7626
Springfield 217-652-7626

NationLink & RoamingAmerica (see p. 1157)
Dial ∗31 send to receive calls, ∗30 to deactivate.
Dial ∗32 send to tell caller the roamer access no.
This company's subscribers have level 6 service.

Billing information

Visitor rates
Most visitors . . . . . $3 day + 85¢ minute

Visitors without roaming agreements (p. 1159)
Call co. to use Amex, Discover, Mastercard, Visa

Identification numbers

City	Market	System	Billing	RSA
Bloomington	250	00551	30045	MSA
Champaign	196	00551	30005	MSA
Decatur	230	00551	30003	MSA
Lincoln	398	00551	30387	IL-5
Springfield	176	00551	30001	MSA

Cellular company

Ameritech Mobile Communications
2205 W Wabash, Suite 102
Springfield, IL 62704

2905 W Springfield Avenue
Champaign, IL 61821

Customer service		800-221-0994
. . . .		708-706-7300
Local office . . .	Champaign	217-373-4500
(subject to change)	Springfield	217-698-7499
Corporate office		708-706-7600

Sending calls

Visitor dialing instructions
Local calls . . . . . area code + 7 digits
Long distance . . . 1 + area code + 7 digits

Speed-dial numbers
Customer service . . . . ∗611
Directory assistance 411
Emergency 911
Information about cellular service features ∗90
Roaming help line . . . . ∗712
Roaming information . . . . . . ∗711

Receiving calls

Caller dials the roamer access number below,
hears tone, dials cellular area code and number.
Bloomington 309-824-7626
Champaign 217-369-7626
Decatur 217-433-7626
Springfield 217-725-7626

Follow Me Roaming forwarding (see p. 1156)
Activate, dial ∗18 send. Deactivate dial ∗19 send.

Billing information

Visitor rates
Most visitors . . . . . $3 day + 85¢ minute

Visitors without roaming agreements (p. 1159)
Cannot call since company takes no credit cards.

Identification numbers

City	Market	System	Billing	RSA
Bloomington	250	00532	00000	MSA
Champaign	196	00532	00000	MSA
Decatur	230	00532	00000	MSA
Lincoln	398	00532	00000	IL-5
Springfield	176	00532	00000	MSA

For full coverage area, see Pontiac, IL.

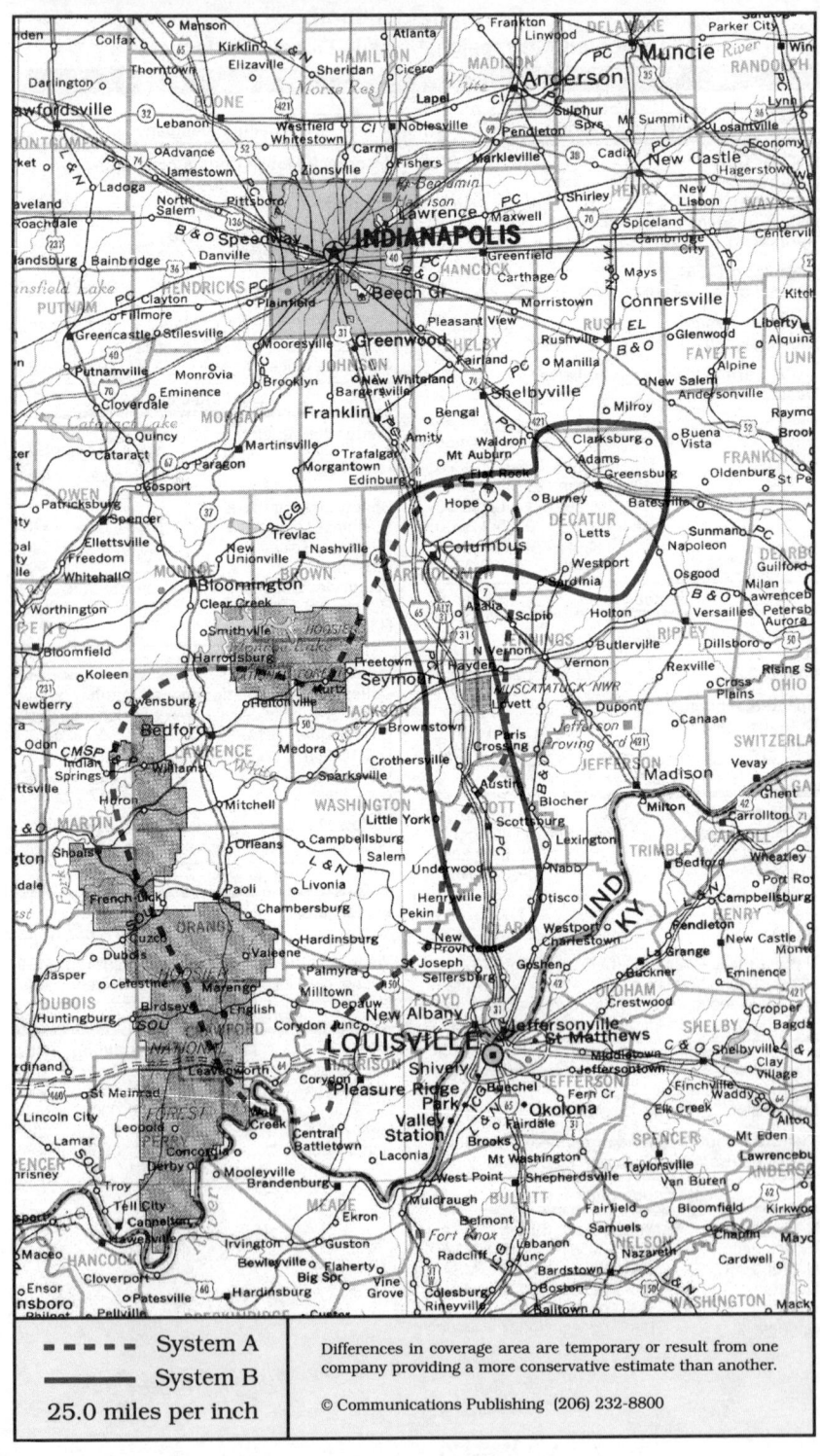

- - - - System A	Differences in coverage area are temporary or result from one company providing a more conservative estimate than another.
——— System B	
25.0 miles per inch	© Communications Publishing (206) 232-8800

Columbus, Indiana
Columbus, Scottsburg (system B), Seymour

System A

Cellular company

Cellular One
Alpha Cellular
2615 Central Avenue
Columbus, IN 47201

Customer service	800-932-0577
Corporate office	812-372-1133

Sending calls

Visitor dialing instructions

Local calls	area code + 7 digits
Long distance . . .	1 + area code + 7 digits

Speed-dial numbers

City police administrative office	*222
County police administrative office . . .	*333
Customer service	611
Emergency	911

Receiving calls

Caller dials the roamer access number below,
hears tone, dials cellular area code and number.
. . . . 812-371-7626

Billing information

Visitor rates
Most visitors . . . . . $3 day + 85¢ minute

Visitors without roaming agreements (p. 1159)
Cannot call since company takes no credit cards.

Identification numbers

City	Market	System	Billing	RSA
All served cities	410	01207	00000	IN-8

System B

Cellular company

Contel Cellular
404 Washington Street, Suite 203
Columbus, IN 47201

Customer service	800-333-4004
. . . .	502-582-2273
Local office	812-372-0555
Corporate office	404-804-3400

Sending calls

Visitor dialing instructions

Local calls	area code + 7 digits
Long distance . . .	1 + area code + 7 digits

Speed-dial numbers

Customer service	611
Emergency	911

Receiving calls

Caller dials the roamer access number below,
hears tone, dials cellular area code and number.

Columbus	812-343-7626
Scottsburg, Seymour	812-525-7626

Billing information

Visitor rates
Most visitors . . . . . $3 day + 85¢ minute

Visitors without roaming agreements (p. 1159)
Call co. to use Amex, Mastercard, Visa

Identification numbers

City	Market	System	Billing	RSA
Columbus, Seymour	410	01208	00000	IN-8
Scottsburg	411	01210	00000	IN-9

For full coverage area, see Madison, IN.

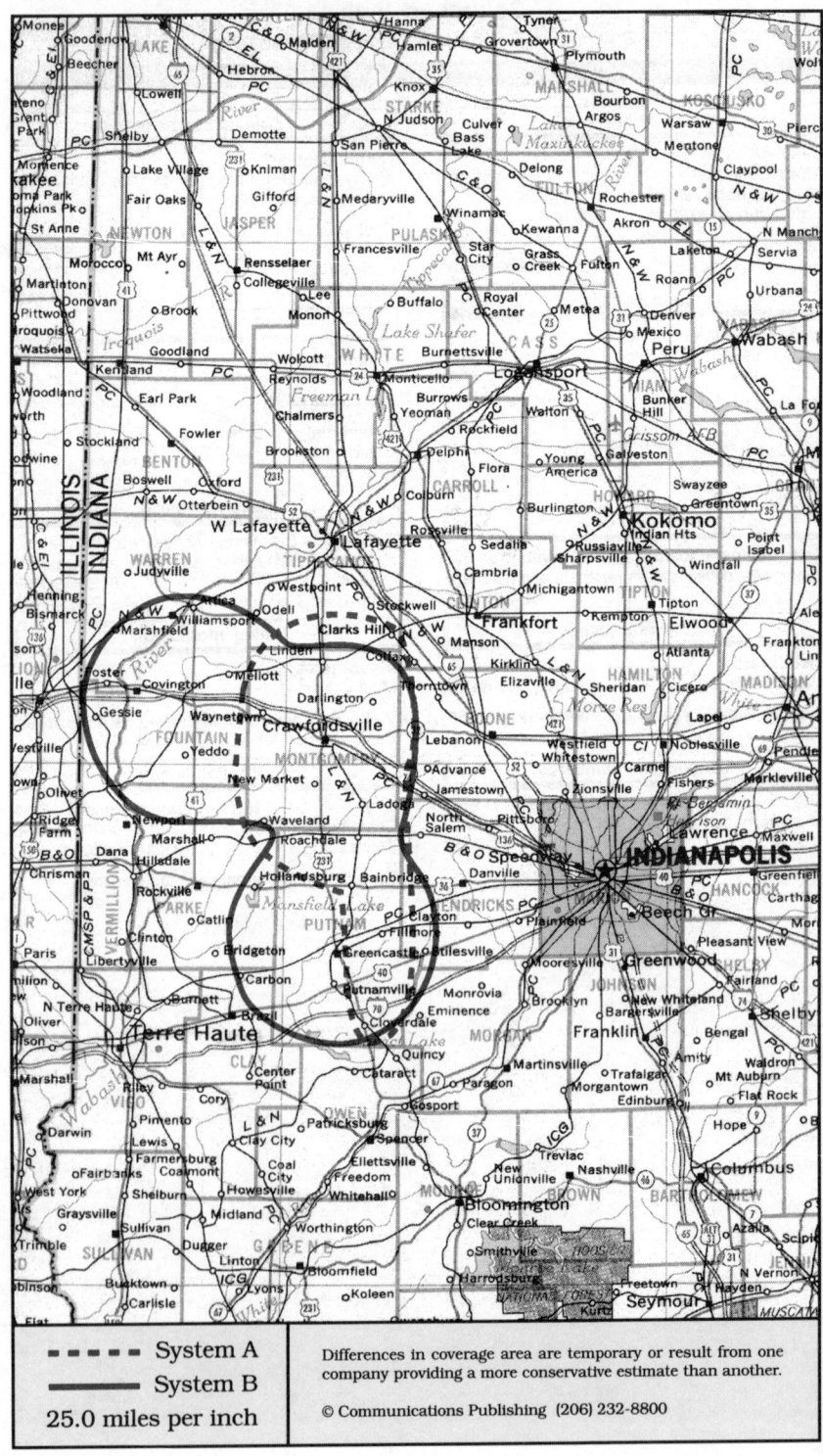

| - - - - - System A |
| ─────── System B |
| **25.0 miles per inch** |

Differences in coverage area are temporary or result from one company providing a more conservative estimate than another.

© Communications Publishing (206) 232-8800

Crawfordsville, Indiana
Covington, Crawfordsville, Greencastle

System A	System B

Cellular company

Cellular One of Indianapolis
BellSouth Cellular
2500 E 46th Street
Indianapolis, IN 46205

Customer service	317-252-5367
Local office	800-783-2351
. . . .	317-253-3300
Corporate office	404-604-6100

Sending calls

Visitor dialing instructions

Local calls	area code + 7 digits
Long distance	1 + area code + 7 digits

Speed-dial numbers

Channel 6 TV	*66
Channel 10 TV	*10
Channel 13 TV	*13
Customer service	611
Emergency	911
WENS	*971
WFBQ	947
WFMS	955
WIRE	1430
WKLR	931
WTLC	*105
WTPI	1079
WXTZ	1033
WXXP	*98
WZPL	995

Receiving calls

Caller dials the roamer access number below,
hears tone, dials cellular area code and number.
. . . . 317-443-7626

NationLink & RoamingAmerica (see p. 1157)
Dial *31 send to receive calls, *30 to deactivate.
Dial *32 send to tell caller the roamer access no.
This company's subscribers have level 6 service.

Billing information

Visitor rates
Most visitors . . . . . $3 day + 85¢ minute

Visitors without roaming agreements (p. 1159)
Call co. to use Mastercard, Visa
or Roamer Plus will intercept first call to bill you.

Identification numbers

City	Market	System	Billing	RSA
All served cities	407	00019	30377	IN-5

For full coverage area, see Indianapolis, IN.

Cellular company

United States Cellular
2509 E Market
Logansport, IN 46947

Local office	219-732-0044
Corporate office	312-399-8900

Sending calls

Visitor dialing instructions

Local calls	area code + 7 digits
Long distance	area code + 7 digits

Speed-dial numbers

Customer service	*611
Emergency	911

Receiving calls

Caller dials the roamer access number below,
hears tone, dials cellular area code and number.
. . . . 317-363-7626

Billing information

Visitor rates
Most visitors . . . . . $3 day + 85¢ minute

Visitors without roaming agreements (p. 1159)
Call co. to use Mastercard, Visa

Identification numbers

City	Market	System	Billing	RSA
All served cities	407	01202	00000	IN-5

For full coverage area, see Logansport, IN.

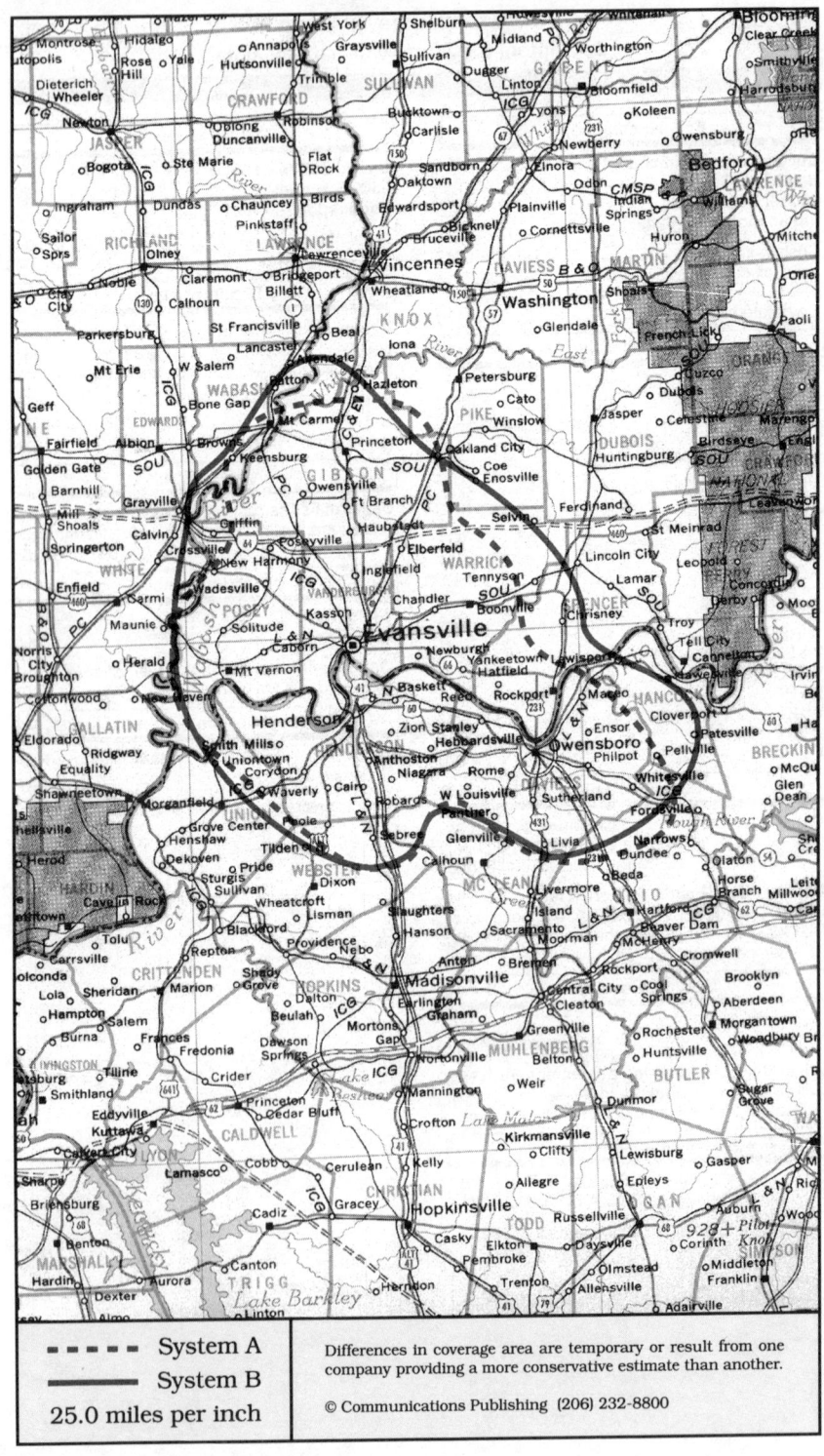

- - - - - System A	Differences in coverage area are temporary or result from one
————— System B	company providing a more conservative estimate than another.
25.0 miles per inch	© Communications Publishing (206) 232-8800

Evansville, Indiana
Evansville, Henderson KY, Owensboro KY, Princeton

System A	System B

Cellular company

United States Cellular
461 John Street
Evansville, IN 47713

3435 Frederica Street
Owensboro, KY 42301

Local office . . .	Evansville	812-464-5111
. . . .	Owensboro	502-685-5111
Corporate office		312-399-8900

Sending calls

Visitor dialing instructions

Local calls	area code + 7 digits
Long distance . . .	area code + 7 digits

Speed-dial numbers

Customer service	611
Directory assistance	411
Emergency	911
Make a traffic report . . .	812-453-1041
Roaming information	*711

Receiving calls

Caller dials the roamer access number below,
hears tone, dials cellular area code and number.

Evansville	812-453-7626
Henderson	502-831-7626
Owensboro	502-929-7626

Billing information

Visitor rates

Most visitors . . . . . $3 day + 75¢ minute

Visitors without roaming agreements (p. 1159)
Call co. to use Mastercard, Visa

Identification numbers

City	Market	System	Billing	RSA
Evansville, IN	119	00197	00000	MSA
Owensboro, KY	293	00481	00000	MSA

Cellular company

Contel Cellular
4900 Temple Avenue
Evansville, IN 47715

3221-B Frederica Street
Owensboro, KY 42301

Customer service		800-333-4004
		502-582-2273
Local office . . .	Evansville	800-333-4831
. . . .	Evansville	812-473-4484
. . . .	Owensboro	502-926-3370
Corporate office		404-804-3400

Sending calls

Visitor dialing instructions

Local calls	area code + 7 digits
Long distance . . .	1 + area code + 7 digits

Speed-dial numbers

Customer service	611
Emergency	911

Receiving calls

Caller dials the roamer access number below,
hears tone, dials cellular area code and number.

Evansville	812-455-7626
Henderson	502-823-7626
Owensboro	502-925-7626

Follow Me Roaming forwarding (see p. 1156)
Activate, dial *18 send. Deactivate dial *19 send.

Billing information

Visitor rates

Most visitors . . . . . $3 day + 85¢ minute

Visitors without roaming agreements (p. 1159)
Call co. to use Amex, Mastercard, Visa

Identification numbers

City	Market	System	Billing	RSA
Evansville, IN	119	00190	00000	MSA
Owensboro, KY	293	00190	00000	MSA

For full coverage area, see Vincennes, IN.

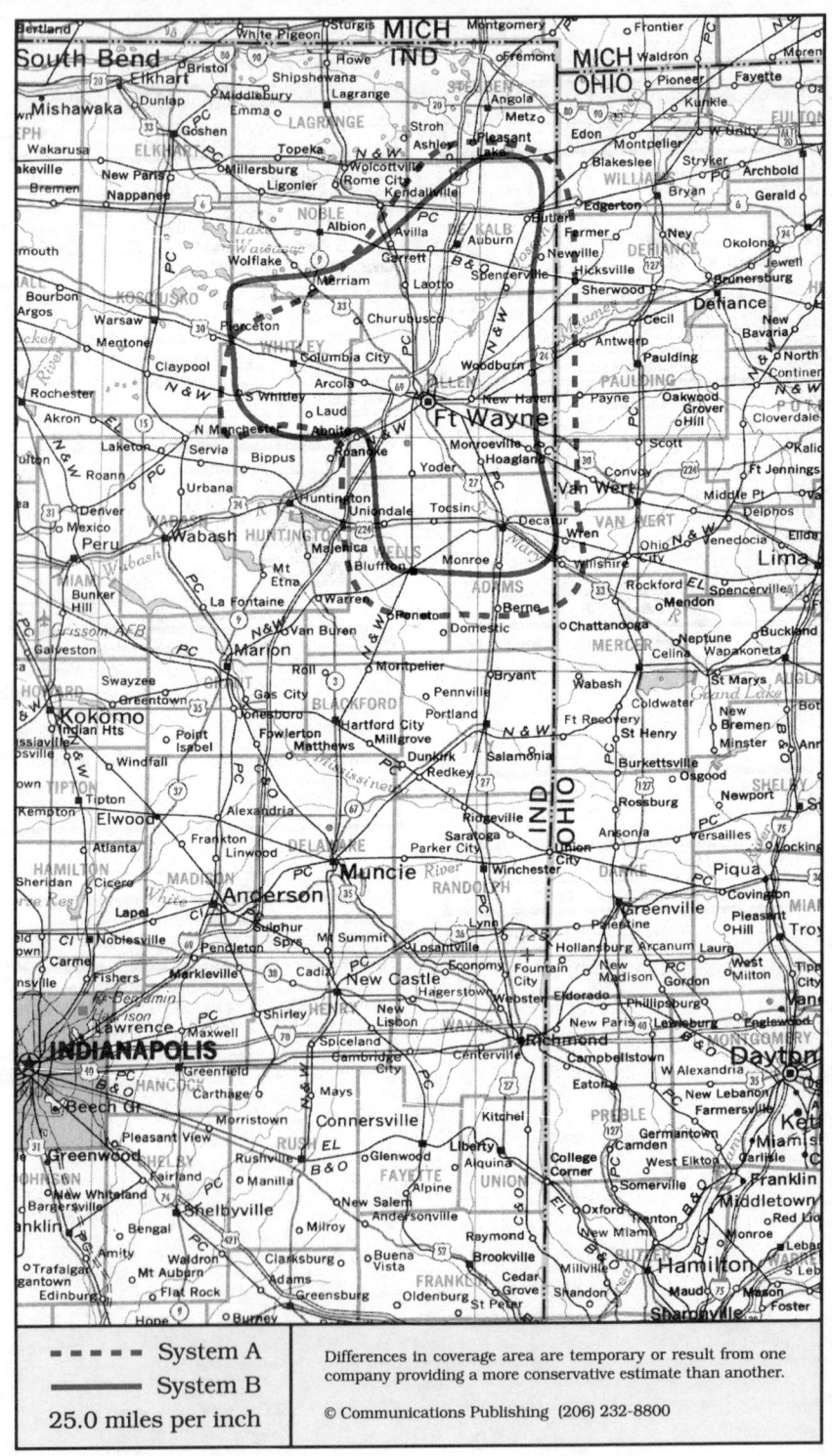

----- System A	Differences in coverage area are temporary or result from one company providing a more conservative estimate than another.
—— System B	
25.0 miles per inch	© Communications Publishing (206) 232-8800

Fort Wayne, Indiana

Cellular company

Cellular One
Centennial Cellular
421 Fernhill Avenue
Fort Wayne, IN 46805

Local office	219-484-2500
Corporate office	203-972-2000

Sending calls

Visitor dialing instructions

Local calls	area code + 7 digits
Long distance . . .	1 + area code + 7 digits

Speed-dial numbers

Customer service	*611
Emergency	911
WBYR traffic report	*989
WHJI make traffic report	*951

Receiving calls

Caller dials the roamer access number below,
hears tone, dials cellular area code and number.

. . . . 219-466-7626

NationLink & RoamingAmerica (see p. 1157)
Dial *31 send to receive calls, *30 to deactivate.
Dial *32 send to tell caller the roamer access no.
This company's subscribers have level 8 service.

Billing information

Visitor rates
Most visitors . . . . . $3 day + 95¢ minute

Visitors without roaming agreements (p. 1159)
Call co. to use Mastercard, Visa
or Roamer Plus will intercept first call to bill you.

Identification numbers

City	Market	System	Billing	RSA
All served cities	096	00199	00000	MSA

For full coverage area, see Logansport, IN.

Cellular company

GTE Mobilnet
4122 Northrop Street, Suite B-15
Fort Wayne, IN 46805

Customer service	800-669-5665
. . . .	216-642-0688
Local office	219-484-5262
Corporate office	404-391-8000

Sending calls

Visitor dialing instructions

Local calls	area code + 7 digits
Long distance . . .	1 + area code + 7 digits

Speed-dial numbers

American Automobile Association .	462-4357
Customer service	*611
Emergency	911
Roaming information	*711
Telephone problems	*111
WIBC radio	1070
WISH TV	*888

Receiving calls

Caller dials the roamer access number below,
hears tone, dials cellular area code and number.

. . . . 219-433-7626

Follow Me Roaming forwarding (see p. 1156)
Activate, dial *18 send. Deactivate dial *19 send.

Billing information

Visitor rates
Most visitors . . . . . $3 day + 85¢ minute

Visitors without roaming agreements (p. 1159)
Call co. to use Amex, Mastercard, Visa

Identification numbers

City	Market	System	Billing	RSA
All served cities	096	00080	30012	MSA

For full coverage area, see Indianapolis, IN.

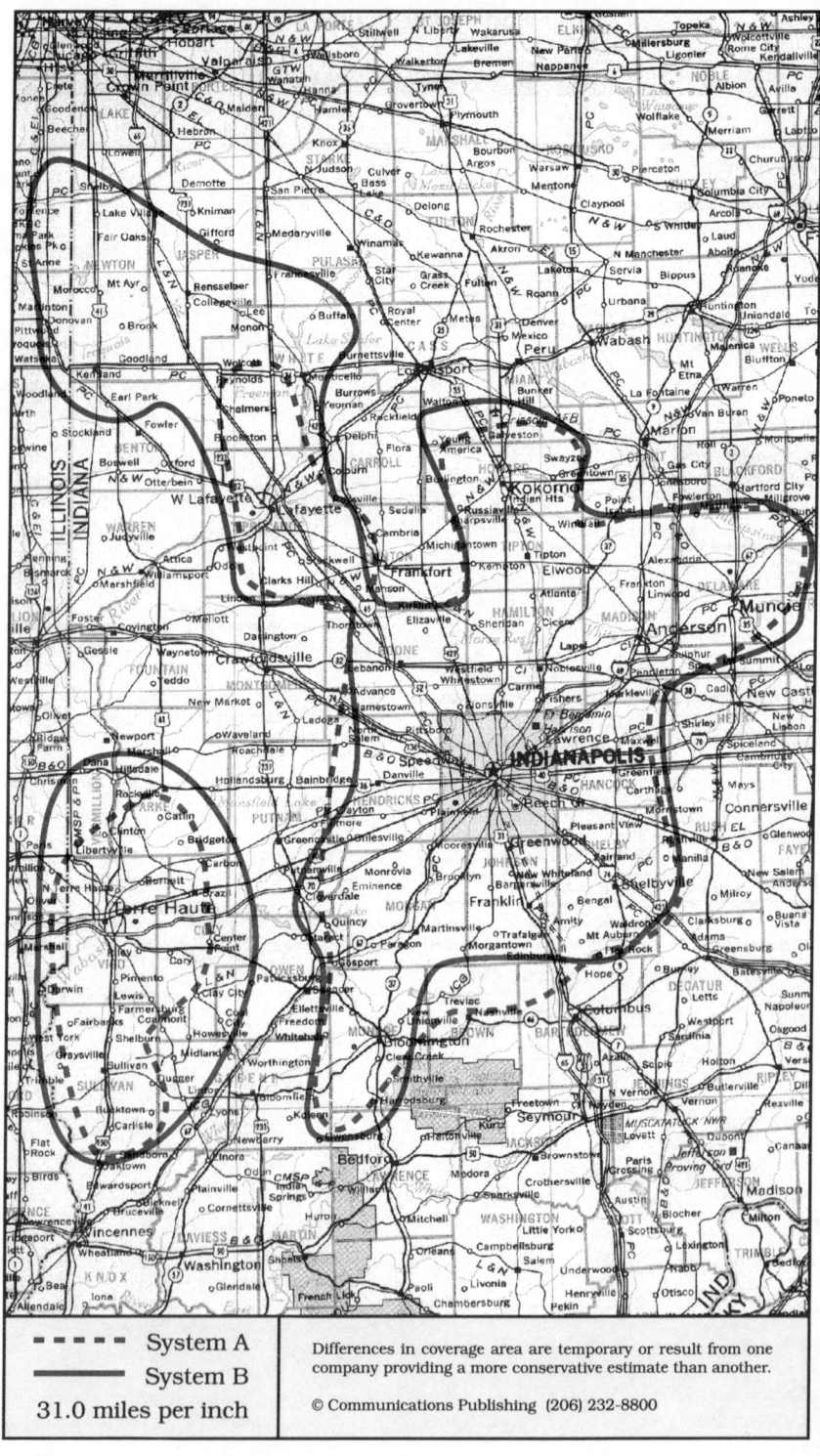

System A

System B

31.0 miles per inch

Differences in coverage area are temporary or result from one company providing a more conservative estimate than another.

© Communications Publishing (206) 232-8800

Indianapolis, Indiana

Bloomington, Indianapolis, Kokomo, Lafayette, Muncie, Rensselaer, Terre Haute

System A	System B

System A

Cellular company

Cellular One of Indianapolis
BellSouth Cellular
2500 E 46th Street
Indianapolis, IN 46205

Customer service	800-466-0067
	317-252-5367
Local office	317-253-3300
Corporate office	404-604-6100

Sending calls

Visitor dialing instructions

Local calls	area code + 7 digits
Long distance	1 + area code + 7 digits

Speed-dial numbers

Channel 6 TV	*66
Channel 10 TV	*10
Channel 13 TV	*13
Customer service	611
Emergency	911
WENS	*971
WFBQ	947
WFMS	955
WIRE	1430
WKLR	931
WTLC	*105
WTPI	1079
WXTZ	1033
WXXP	*98
WZPL	995

Receiving calls

Caller dials the roamer access number below,
hears tone, dials cellular area code and number.

Anderson	317-621-7626
Bloomington	812-327-7626
Indianapolis	317-443-7626
Kokomo	317-434-7626
Lafayette	317-426-7626
Muncie	317-749-7626
Terre Haute	812-239-7626

NationLink & RoamingAmerica (see p. 1157)
Dial *31 send to receive calls, *30 to deactivate.
Dial *32 send to tell caller the roamer access no.
This company's subscribers have level 6 service.

Billing information

Visitor rates

Most visitors	$3 day + 85¢ minute

Visitors without roaming agreements (p. 1159)
Call co. to use Mastercard, Visa

Identification numbers

City	Market	System	Billing	RSA
Anderson	217	00019	30325	MSA
Bloomington	282	00287	00000	MSA
Indianapolis	028	00019	00000	MSA
Kokomo	271	00019	30379	MSA
Lafayette	247	00415	00000	MSA
Muncie	236	00467	00000	MSA
Terre Haute	185	00567	00000	MSA

For full coverage area, see Crawfordsville, IN.

System B

Cellular company

GTE Mobilnet
11611 N Meridian, Suite 320
Carmel, IN 46032

sales office
6633 E 82nd Street
Indianapolis, IN 46250

Customer service	some areas	800-669-5665
		216-642-0688
Local office	administration	317-848-0162
	sales	317-577-2225
Corporate office		404-391-8000

Sending calls

Visitor dialing instructions

Local calls	area code + 7 digits
Long distance	1 + area code + 7 digits

Speed-dial numbers

American Automobile Association	462-4357
Customer service	*611
Emergency	911
Roaming information	*711
Telephone problems	*111
WIBC radio	1070
WISH TV	*888

Receiving calls

Caller dials the roamer access number below,
hears tone, dials cellular area code and number.

Anderson, Muncie	317-744-7626
Bloomington	812-322-7626
Indianapolis, Rensselaer	317-432-7626
Kokomo	317-438-7626
Lafayette	317-427-7626
Terre Haute	812-249-7626

Follow Me Roaming forwarding (see p. 1156)
Activate, dial *18 send. Deactivate dial *19 send.

Billing information

Visitor rates

Most visitors	$3 day + 85¢ minute

Visitors without roaming agreements (p. 1159)
Call co. to use Amex, Discover, Mastercard, Visa

Identification numbers

City	Market	System	Billing	RSA
Anderson	217	00080	00000	MSA
Bloomington	282	00080	00000	MSA
Indianapolis	028	00080	00000	MSA
Kokomo	271	00080	00000	MSA
Lafayette	247	00080	00000	MSA
Muncie	236	00080	00000	MSA
Rensselaer	403	00080	00000	IN-1
Terre Haute	185	00080	00000	MSA

For full coverage area, see Fort Wayne, IN,
Marion, IN and Richmond, IN.

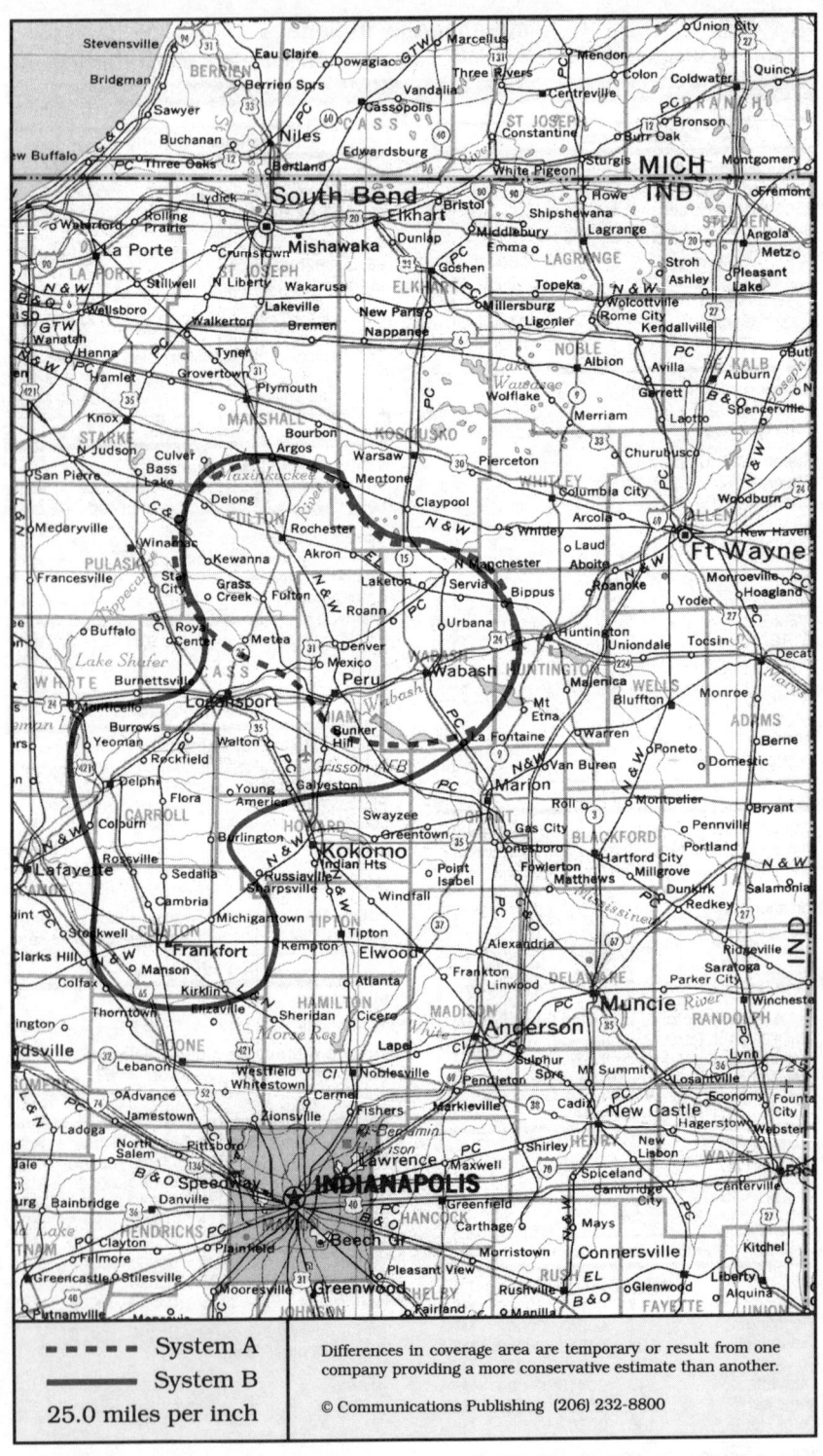

- - - - -	System A
———	System B
25.0 miles per inch	

Differences in coverage area are temporary or result from one company providing a more conservative estimate than another.

© Communications Publishing (206) 232-8800

Logansport, Indiana
Delphi, Frankfort, Logansport, Peru, Rochester, Wabash

System A	System B

Cellular company

Peru, Wabash
Cellular One
Centennial Cellular
421 Fernhill Avenue
Fort Wayne, IN 46805

Rochester
Cellular One
Centennial Cellular
850 S Marietta Street, Suite 100; PO Box 1178
South Bend, IN 46634

Local office	. .	Fort Wayne	219-484-2500
	. . . . .	South Bend	219-289-0933
Corporate office	. . . . .		203-972-2000

Sending calls

Visitor dialing instructions
Local calls	. . .	area code + 7 digits
Long distance	. . .	1 + area code + 7 digits

Speed-dial numbers
Customer service	. . . .	*611
Emergency	. . . . .	911
WBYR traffic report	. . . .	*989
WHJI make traffic report	. . . . .	*951

Receiving calls

Caller dials the roamer access number below,
hears tone, dials cellular area code and number.
Peru and Wabash	. . . . .	219-466-7626
Rochester	. . . . . .	219-298-7626

NationLink & RoamingAmerica (see p. 1157)
Dial *31 send to receive calls, *30 to deactivate.
Dial *32 send to tell caller the roamer access no.
This company's subscribers have level 8 service.

Billing information

Visitor rates
Most visitors . . . . . $3 day + 95¢ minute

Visitors without roaming agreements (p. 1159)
Call co. to use Mastercard, Visa
or Roamer Plus will intercept first call to bill you.

Identification numbers

City	Market	System	Billing	RSA
All served cities	406	01199	00000	IN-4

For full coverage area, see Fort Wayne, IN and
South Bend, IN.

Cellular company

United States Cellular
2509 E Market
Logansport, IN 46947

Local office	. . . . . .	219-732-0044
Corporate office	. . . . .	312-399-8900

Sending calls

Visitor dialing instructions
Local calls	. . . . .	area code + 7 digits
Long distance	. . . . .	area code + 7 digits

Speed-dial numbers
Customer service	. . . .	*611
Emergency	. . . .	911

Receiving calls

Caller dials the roamer access number below,
hears tone, dials cellular area code and number.
Logansport	. . . .	219-568-7626
Wabash	. . . .	219-568-7626

Billing information

Visitor rates
Most visitors . . . . . $3 day + 85¢ minute

Visitors without roaming agreements (p. 1159)
Call co. to use Mastercard, Visa

Identification numbers

City	Market	System	Billing	RSA
All served cities	406	01200	00000	IN-4

For full coverage area, see Crawfordsville, IN.

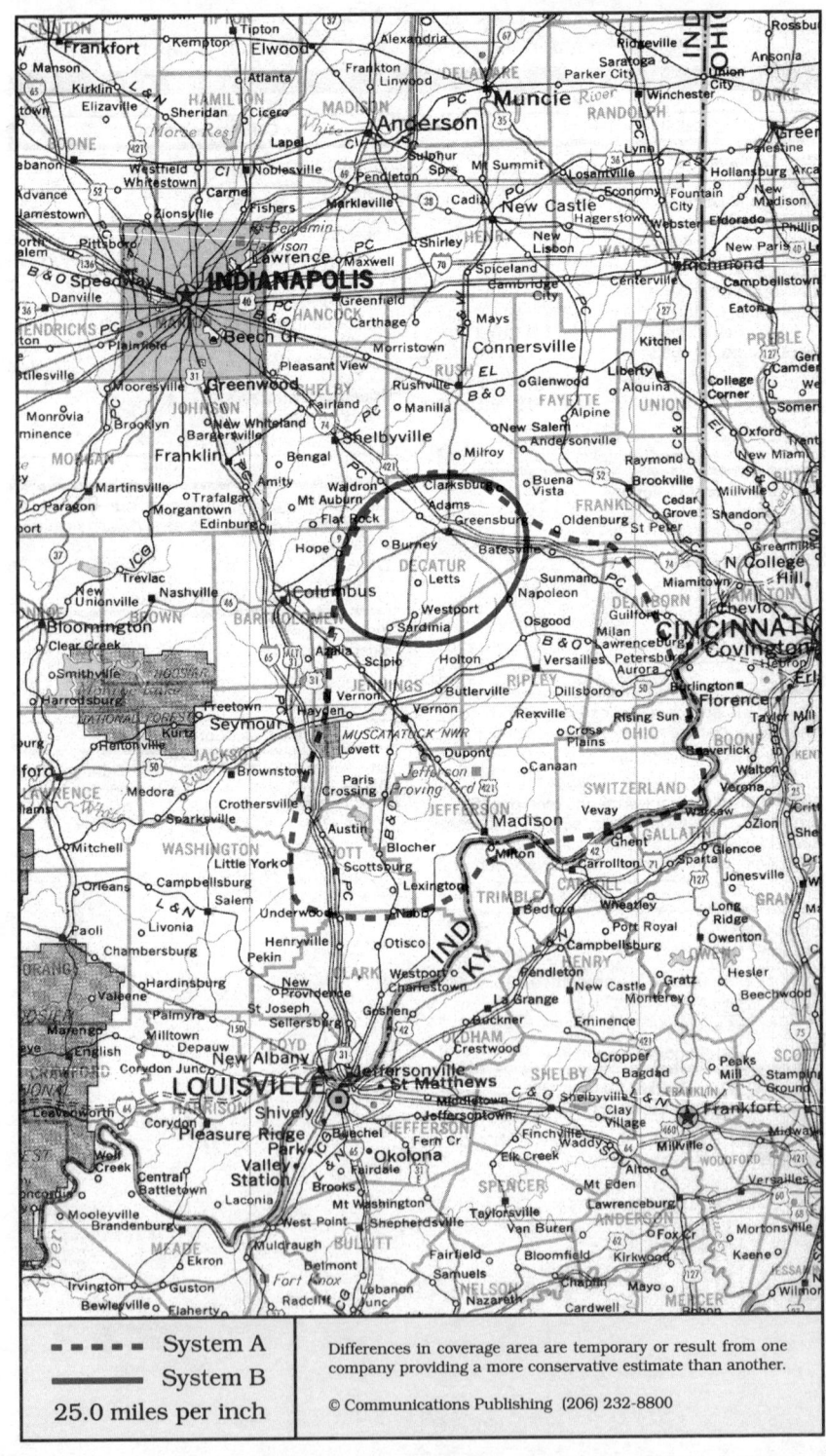

▪ ▪ ▪ ▪ ▪ System A	Differences in coverage area are temporary or result from one company providing a more conservative estimate than another.
———— System B	
25.0 miles per inch	© Communications Publishing (206) 232-8800

Madison, Indiana
Batesville, Greensburg, Madison, Scottsburg (system A)

Cellular company

Cellular One
Licensed to: SE Indiana Cellular Telephone
Managed by: CompComm
514 John Street, PO Box 347
Batesville, IN 47006

Customer service	800-999-8808
Local office	812-934-4404
Corporate office	CompComm 609-854-1000

Sending calls

Visitor dialing instructions

Local calls	area code + 7 digits
Long distance . . .	1 + area code + 7 digits

Speed-dial numbers

Customer service	611
Emergency	911

Receiving calls

Caller dials the roamer access number below, hears tone, dials cellular area code and number.

. . . . 812-662-5626

NationLink & RoamingAmerica (see p. 1157)
Dial *31 send to receive calls, *30 to deactivate. This company's subscribers have level 3 service.

Billing information

Visitor rates
Most visitors . . . . . $3 day + 85¢ minute

Visitors without roaming agreements (p. 1159)
Roamer Plus will intercept first call to bill you.

Identification numbers

City	Market	System	Billing	RSA
All served cities	411	01209	00000	IN-9

Cellular company

Contel Cellular
404 Washington Street, Suite 203
Columbus, IN 47201

Customer service	800-333-4004
. . . .	502-582-2273
Local office	812-372-0555
Corporate office	404-804-3400

Sending calls

Visitor dialing instructions

Local calls	area code + 7 digits
Long distance . . .	1 + area code + 7 digits

Speed-dial numbers

Customer service	611
Emergency	911

Receiving calls

Caller dials the roamer access number below, hears tone, dials cellular area code and number.

. . . . 812-525-7626

Billing information

Visitor rates
Most visitors . . . . . $3 day + 85¢ minute

Visitors without roaming agreements (p. 1159)
Call co. to use Amex, Mastercard, Visa

Identification numbers

City	Market	System	Billing	RSA
All served cities	411	01210	00000	IN-9

For full coverage area, see Columbus, IN.

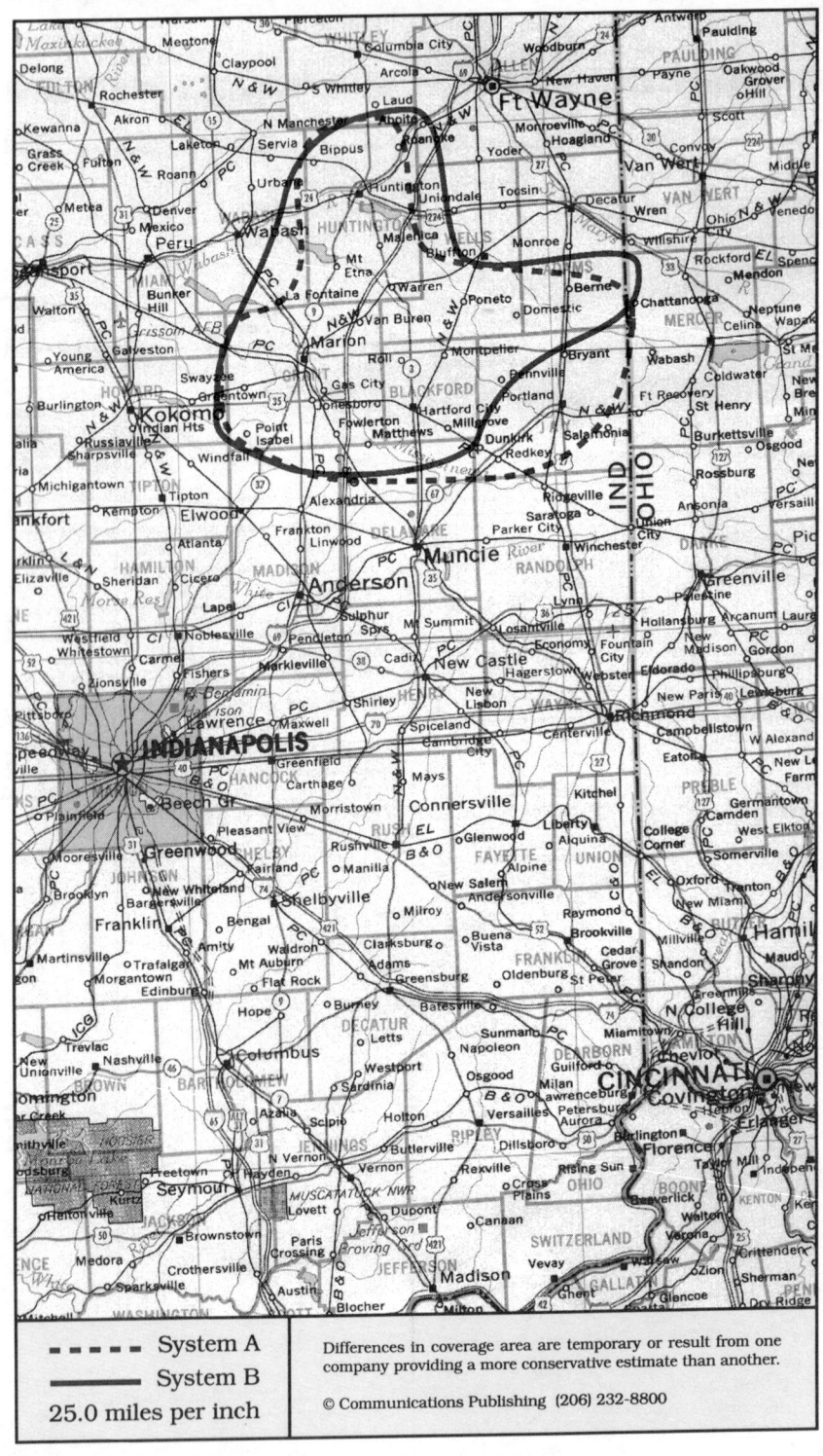

- - - - System A	Differences in coverage area are temporary or result from one company providing a more conservative estimate than another.
——— System B	
25.0 miles per inch	© Communications Publishing (206) 232-8800

Marion, Indiana

Hartford City, Huntington, Marion, Portland

Cellular company

Cellular One
Mega Comm
1305 W Johnson Street
Marion, IN 46952-2549

Corporate office 317-664-1500

Sending calls

Visitor dialing instructions
Local calls . . . . . area code + 7 digits
Long distance . . . 1 + area code + 7 digits

Speed-dial numbers
Customer service 611
Emergency 911
WBAT *1400
WCJC *993
WMRI *860

Receiving calls

Caller dials the roamer access number below,
hears tone, dials cellular area code and number.
. . . . 317-669-7626

NationLink & RoamingAmerica (see p. 1157)
Dial *31 send to receive calls, *30 to deactivate.
Dial *32 send to tell caller the roamer access no.
This company's subscribers have level 6 service.

Billing information

Visitor rates
Most visitors . . . . . $3 day + 85¢ minute

Visitors without roaming agreements (p. 1159)
Cannot call since company takes no credit cards.

Identification numbers

City	Market	System	Billing	RSA
All served cities	405	01197	00000	IN-3

Cellular company

GTE Mobilnet
11611 N Meridian, Suite 320
Carmel, IN 46032

Customer service 800-669-5665
. . . . 216-642-0688
Local office 317-848-0162
Corporate office 404-391-8000

Sending calls

Visitor dialing instructions
Local calls . . . . . area code + 7 digits
Long distance . . . 1 + area code + 7 digits

Speed-dial numbers
American Automobile Association . 462-4357
Customer service *611
Emergency 911
Roaming information *711
Telephone problems *111
WIBC radio 1070
WISH TV *888

Receiving calls

Caller dials the roamer access number below,
hears tone, dials cellular area code and number.
Huntington 219-433-7626
Marion 317-661-7626

Follow Me Roaming forwarding (see p. 1156)
Activate, dial *18 send. Deactivate dial *19 send.

Billing information

Visitor rates
Most visitors . . . . . $3 day + 85¢ minute

Visitors without roaming agreements (p. 1159)
Call co. to use Amex, Mastercard, Visa

Identification numbers

City	Market	System	Billing	RSA
All served cities	405	00080	00000	IN-3

For full coverage area, see Fort Wayne, IN and
Indianapolis, IN.

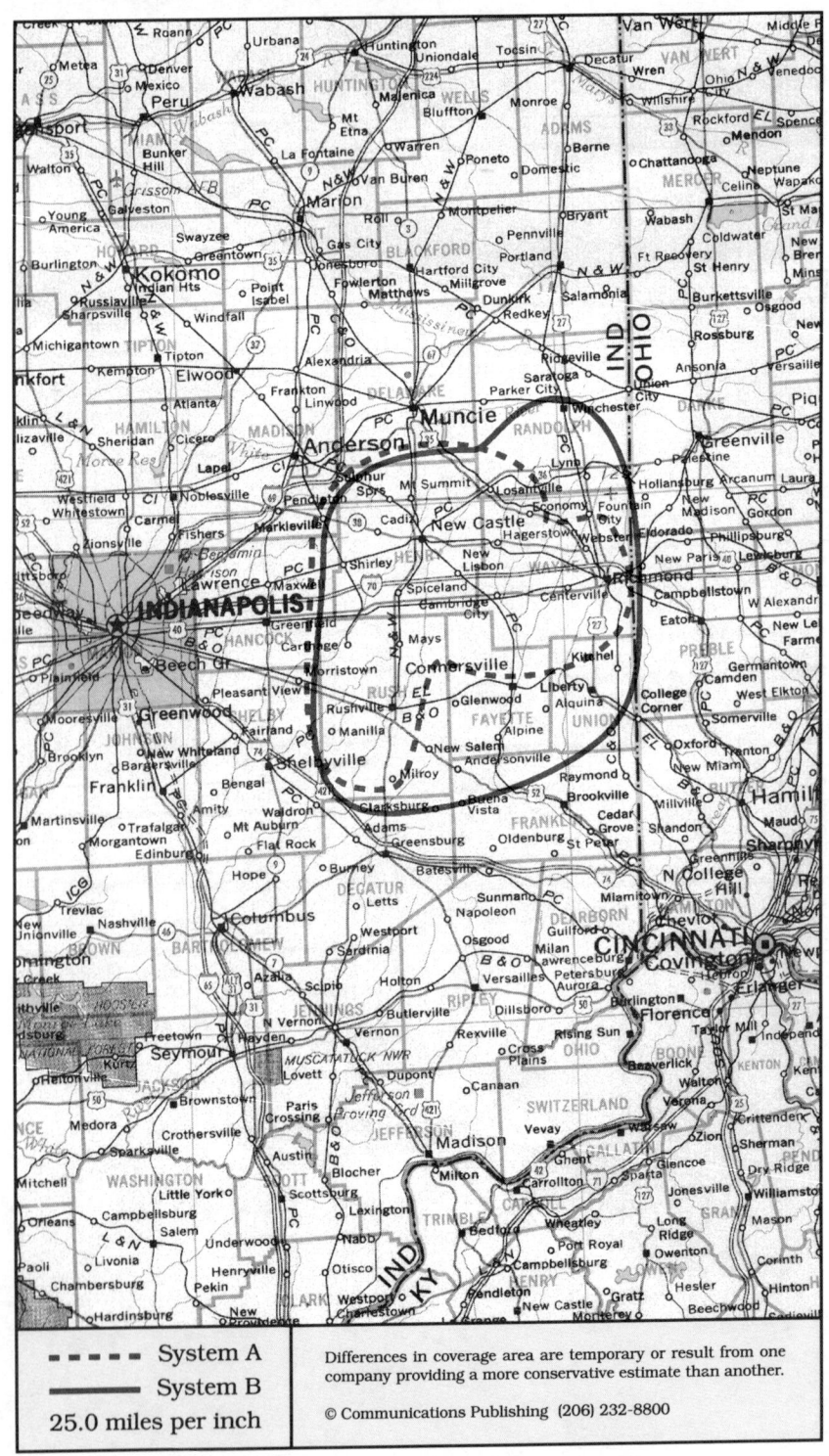

- - - -	**System A**		
————	**System B**		
25.0 miles per inch			

Differences in coverage area are temporary or result from one company providing a more conservative estimate than another.

© Communications Publishing (206) 232-8800

Richmond, Indiana
Cornersville, Liberty, New Castle, Richmond

System A	System B

System A

Cellular company

Cellular One of Richmond
New Par
8800 Governor's Hill Drive
Cincinnati, OH 45249

Local office	317-967-7111
Corporate office	614-436-4331

Sending calls

Visitor dialing instructions

Local calls	area code + 7 digits
Long distance . . .	1 + area code + 7 digits

Speed-dial numbers

Customer service	611
Emergency	911
State police	*HELP or *4357

Receiving calls

Caller dials the roamer access number below, hears tone, dials cellular area code and number.

. . . . 317-967-7626

NationLink & RoamingAmerica (see p. 1157)
Dial *31 send to receive calls, *30 to deactivate.
Dial *32 send to tell caller the roamer access no.
This company's subscribers have level 6 service.

Billing information

Visitor rates

Most visitors . . . . $2.50 day + 75¢ minute

Visitors without roaming agreements (p. 1159)
Cannot call since company takes no credit cards.

Identification numbers

City	Market	System	Billing	RSA
All served cities	408	01203	00000	IN-6

System B

Cellular company

GTE Mobilnet
11611 N Meridian, Suite 320
Carmel, IN 46032

Customer service	some areas	800-669-5665
. . . .		216-642-0688
Local office		317-848-0162
Corporate office		404-391-8000

Sending calls

Visitor dialing instructions

Local calls	area code + 7 digits
Long distance . . .	1 + area code + 7 digits

Speed-dial numbers

American Automobile Association	. 462-4357
Customer service	*611
Emergency	911
Roaming information	*711
Telephone problems	*111
WIBC radio	1070
WISH TV	*888

Receiving calls

Caller dials the roamer access number below, hears tone, dials cellular area code and number.

. . . . 317-969-7626

Follow Me Roaming forwarding (see p. 1156)
Activate, dial *18 send. Deactivate dial *19 send.

Billing information

Visitor rates

Most visitors . . . . . $3 day + 85¢ minute

Visitors without roaming agreements (p. 1159)
Call co. to use Amex, Mastercard, Visa

Identification numbers

City	Market	System	Billing	RSA
All served cities	408	00080	00000	IN-6

For full coverage area, see Indianapolis, IN.

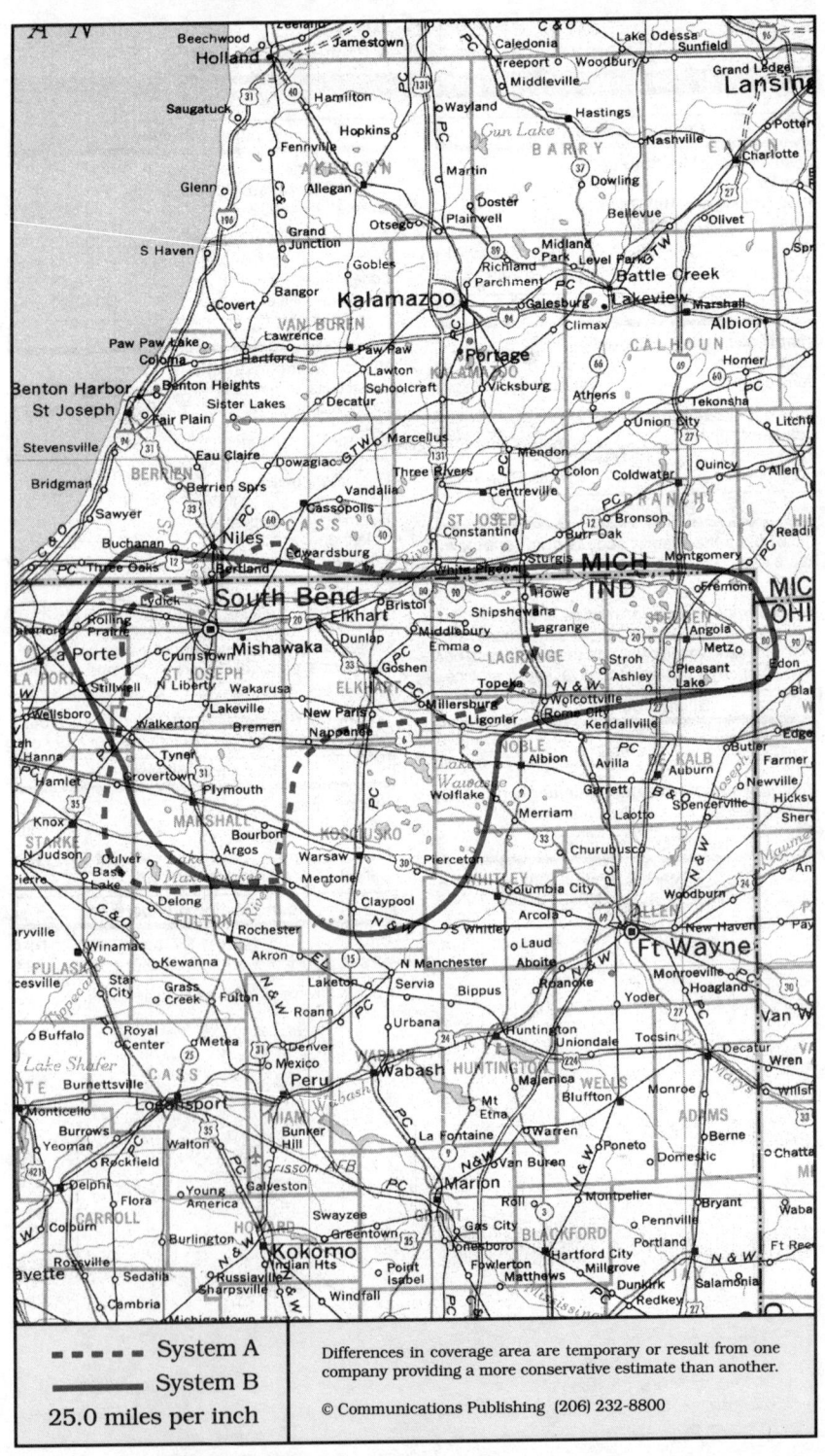

----- System A	Differences in coverage area are temporary or result from one company providing a more conservative estimate than another.
——— System B	
25.0 miles per inch	© Communications Publishing (206) 232-8800

South Bend, Indiana

Angola, Elkhart, South Bend, Warsaw

System A	System B

Cellular company

Cellular One
Cellular One
Centennial Cellular
850 S Marietta Street, Suite 100; PO Box 1178
South Bend, IN 46634

Local office 219-289-0933
Corporate office 203-972-2000

Sending calls

Visitor dialing instructions
Local calls . . . . . area code + 7 digits
Long distance . . . . area code + 7 digits

Speed-dial numbers
Customer service . . . . ∗611
Emergency 911

Receiving calls

Caller dials the roamer access number below,
hears tone, dials cellular area code and number.
. . . . 219-298-7626

NationLink & RoamingAmerica (see p. 1157)
Dial ∗31 send to receive calls, ∗30 to deactivate.
Dial ∗32 send to tell caller the roamer access no.
This company's subscribers have level 8 service.

Billing information

Visitor rates
Most visitors . . . . . $3 day + 95¢ minute

Visitors without roaming agreements (p. 1159)
Call co. to use Mastercard, Visa
or Roamer Plus will intercept first call to bill you.

Identification numbers

City	Market	System	Billing	RSA
Elkhart	223	00549	00000	MSA
South Bend	129	00549	00000	MSA

Most calls to Chicago are charged as local calls.

For full coverage area, see Logansport, IN.

Cellular company

Centel Cellular
505 W Douglas Road
Mishawaka, IN 46545

Customer service   some areas   800-548-1277
Local office   219-271-9819
Corporate office   312-399-2644

Sending calls

Visitor dialing instructions
Local calls . . . . . . area code + 7 digits
Long distance . . . 1 + area code + 7 digits

Speed-dial numbers
Customer service . . . . ∗611
Emergency 911
Make a traffic report . . . . ∗1007

Receiving calls

Caller dials the roamer access number below,
hears tone, dials cellular area code and number.
Elkhart 219-536-7626
South Bend, Warsaw 219-286-7626

Follow Me Roaming forwarding (see p. 1156)
Activate, dial ∗18 send. Deactivate dial ∗19 send.

Billing information

Visitor rates
Most visitors . . . . . $3 day + 99¢ minute

Visitors without roaming agreements (p. 1159)
Call co. to use Amex, Mastercard, Visa

Identification numbers

City	Market	System	Billing	RSA
Angola, Warsaw	404	01196	00000	IN-2
Elkhart	223	00530	00000	MSA
South Bend	129	00530	00000	MSA

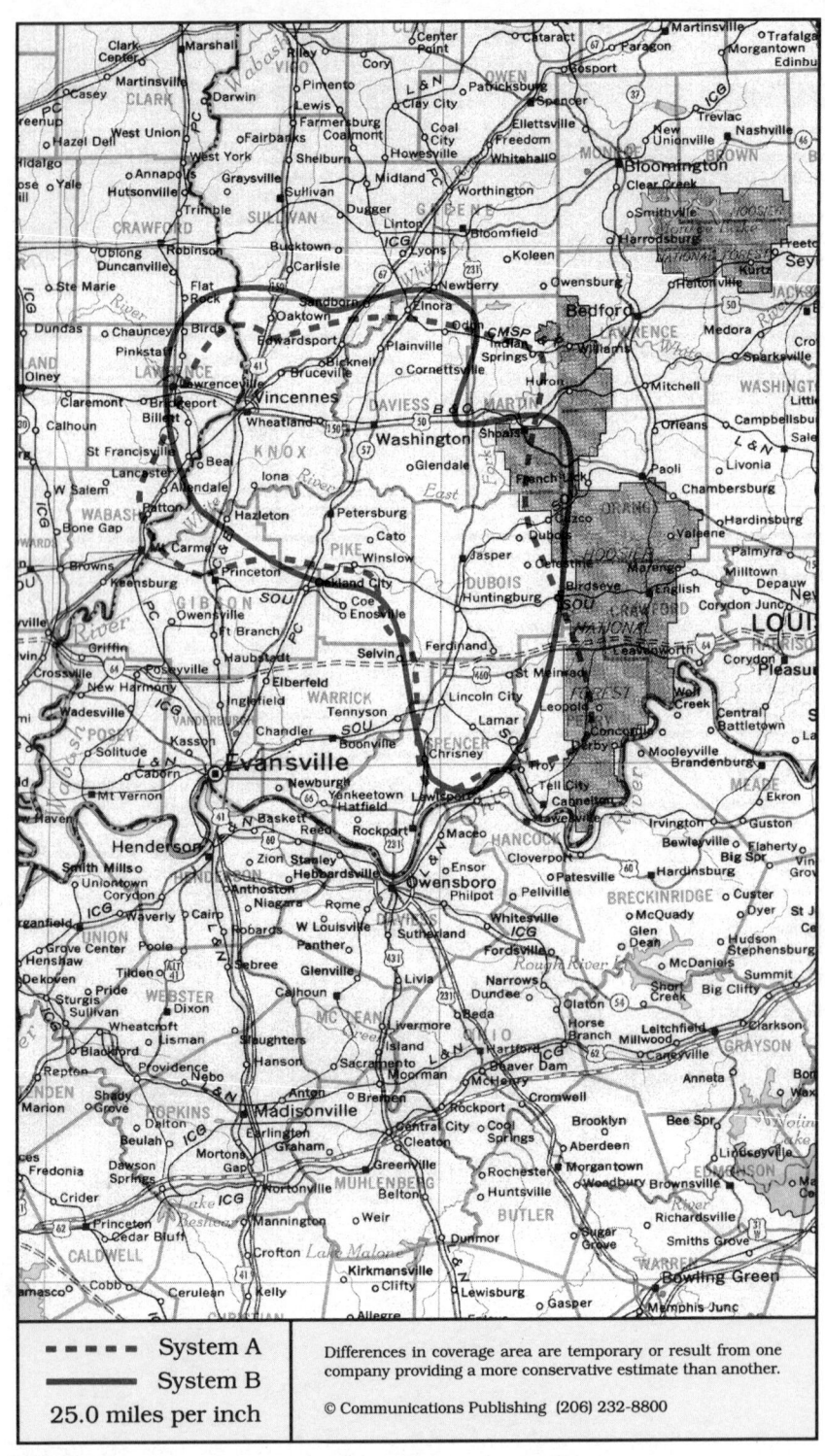

▪ ▪ ▪ ▪ ▪ System A	Differences in coverage area are temporary or result from one
——— System B	company providing a more conservative estimate than another.
25.0 miles per inch	© Communications Publishing (206) 232-8800

Vincennes, Indiana
Jasper, Shoals, Vincennes, Washington

System A	System B

Cellular company

Cellular One
Cellular of Indiana
2409 N 6th Street
Vincennes, IN 47591

Corporate office 812-886-5050

Sending calls

Visitor dialing instructions
Local calls . . . . 7 digits
Long distance . . . 1 + area code + 7 digits

Speed-dial numbers
Customer service 611
Emergency 911

Receiving calls

Caller dials the roamer access number below,
hears tone, dials cellular area code and number.
Jasper 812-630-7626
Vincennes 812-887-7626

Follow Me Roaming forwarding (see p. 1156)
Phone Me Anywhere forwarding (see p. 1156)
Activate, dial *18 send. Deactivate dial *19 send.

Billing information

Visitor rates
Most visitors . . . . $3.25 day + 80¢ minute

Visitors without roaming agreements (p. 1159)
Cannot call since company takes no credit cards.

Identification numbers

City	Market	System	Billing	RSA
All served cities	409	01205	00000	IN-7

Cellular company

Contel Cellular
4900 Temple Avenue
Evansville, IN 47715

Customer service 800-333-4004
. . . . 502-582-2273
Local office 813-473-4484
Corporate office 404-804-3400

Sending calls

Visitor dialing instructions
Local calls . . . . . . area code + 7 digits
Long distance . . . 1 + area code + 7 digits

Speed-dial numbers
Customer service 611
Emergency 911
News tip hotline (Dubois County) *101 or *104

Receiving calls

Caller dials the roamer access number below,
hears tone, dials cellular area code and number.
Jasper 812-639-7626
Vincennes 812-881-7626

Follow Me Roaming forwarding (see p. 1156)
Activate, dial *18 send. Deactivate dial *19 send.

Billing information

Visitor rates
Most visitors . . . . . $3 day + 85¢ minute

Visitors without roaming agreements (p. 1159)
Call co. to use Amex, Mastercard, Visa

Identification numbers

City	Market	System	Billing	RSA
Jasper	409	01206	00000	IN-7
Vincennes	409	00190	00000	IN-7

For full coverage area, see Evansville, IN.

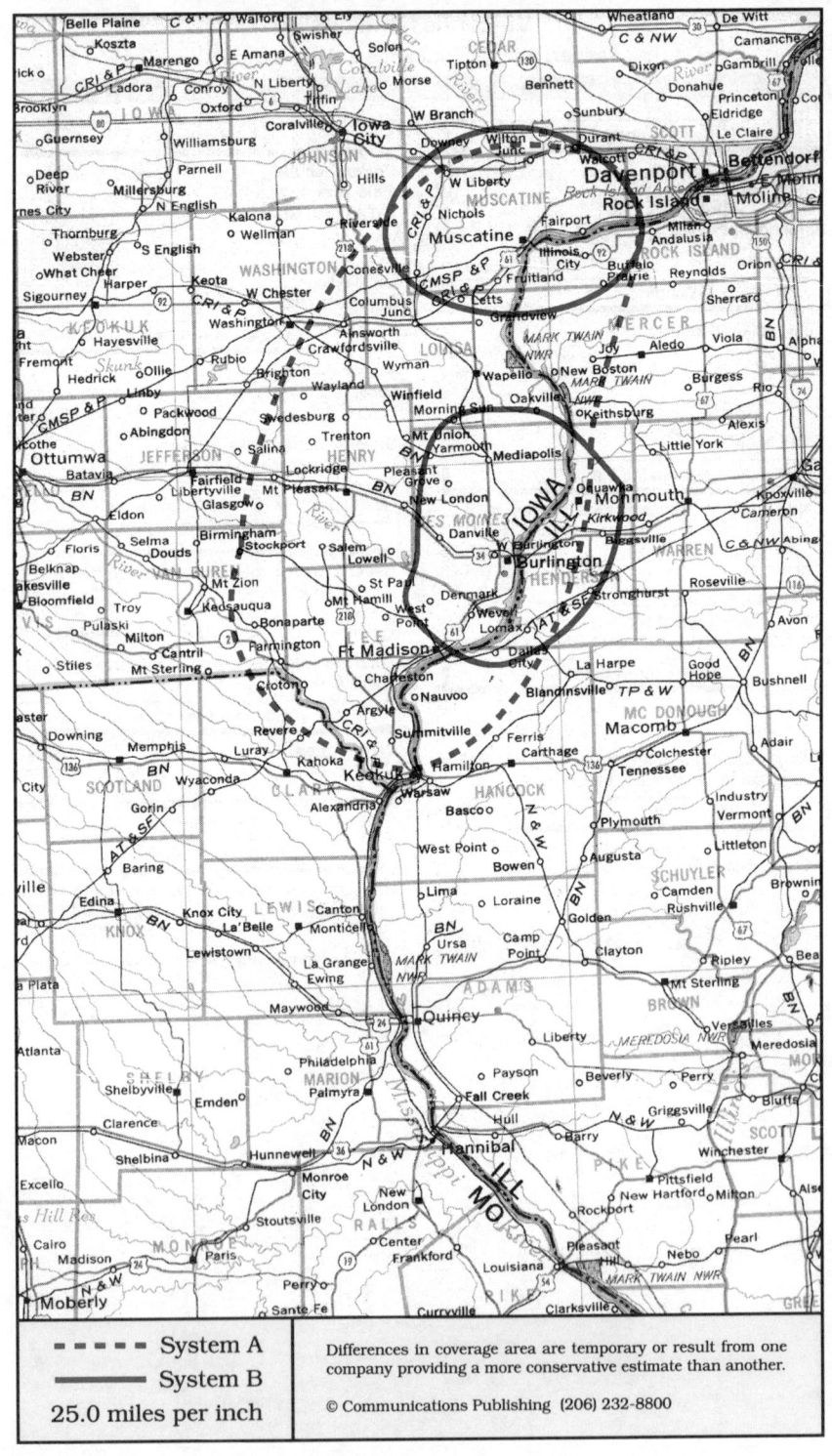

System A
System B
25.0 miles per inch

Differences in coverage area are temporary or result from one company providing a more conservative estimate than another.

© Communications Publishing (206) 232-8000

Burlington, Iowa
Burlington, Fort Madison, Mt. Pleasant, Muscatine

System A	System B

Cellular company

Cellular Plus
2010 Keokuk Street
Iowa City, IA 52240

Local office . .	some areas	800-634-7587
. . . .		319-351-5888
Corporate office	. . . . .	717-883-8832

Cellular company

Contel Cellular
2110 B Park Avenue
Muscatine, IA 52761

Customer service	. . . .	800-333-4004
. . . . .	extension 4002	404-391-8000
Local office . .	. . . .	319-263-6666
Corporate office	. . . . .	404-804-3400

Sending calls

Visitor dialing instructions

Local calls	area code + 7 digits
Long distance . . .	1 + area code + 7 digits

Speed-dial numbers

Customer service	611
Emergency	911
Iowa state patrol	*55

Sending calls

Visitor dialing instructions

Local calls	area code + 7 digits
Long distance . . .	1 + area code + 7 digits

Speed-dial numbers

Customer service	611

Receiving calls

Caller dials the roamer access number below,
hears tone, dials cellular area code and number.

. . . .	319-330-7626

Receiving calls

Caller dials the roamer access number below,
hears tone, dials cellular area code and number.

Burlington	319-750-7626
Muscatine	319-340-7626

Billing information

Visitor rates

Most visitors	$3 day + $1 minute

Visitors without roaming agreements (p. 1159)
Cannot call since company takes no credit cards.

Billing information

Visitor rates

Most visitors	$3 day + 85¢ minute

Visitors without roaming agreements (p. 1159)
Call co. to use Amex, Discover, Mastercard, Visa

Identification numbers

City	Market	System	Billing	RSA
All served cities	415	00389	30245	IA-4

For full coverage area, see Iowa City, IA and Newton, IA and Ottumwa, IA.

Identification numbers

City	Market	System	Billing	RSA
Burlington	415	00186	30324	IA-4
Muscatine	415	00186	00000	IA-4

For full coverage area, see Clinton, IA and Davenport, IA.

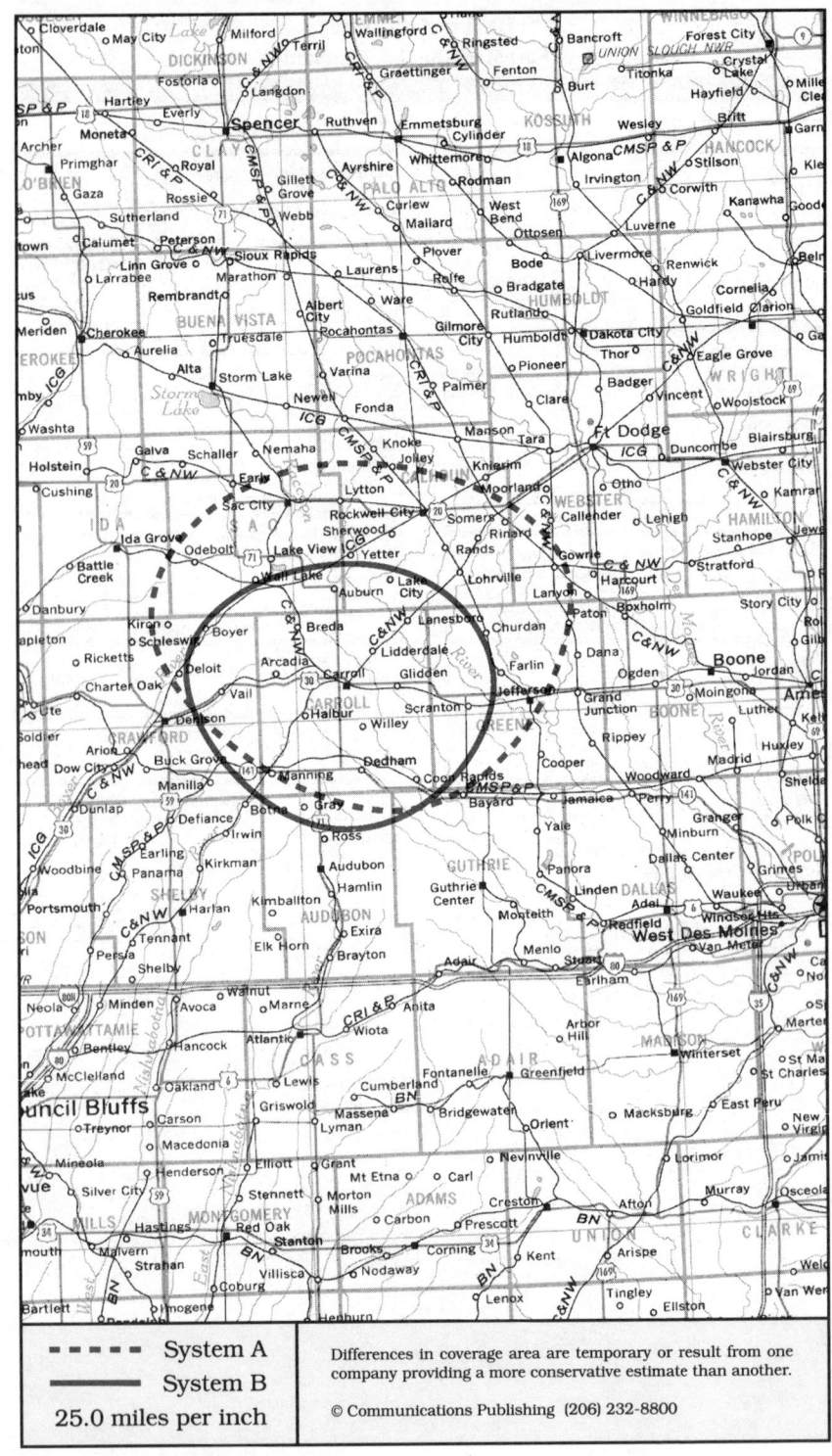

System A - - - - -
System B ———
25.0 miles per inch

Differences in coverage area are temporary or result from one
company providing a more conservative estimate than another.

© Communications Publishing (206) 232-8800

Carroll, Iowa
Carroll, Jefferson, Sac City

Cellular company

Centel Cellular
4711 Southern Hills Drive, Suite A
Sioux City, IA 51106

Local office	800-373-7545
. . . .	712-274-2494
Corporate office	312-399-2644

Sending calls

Visitor dialing instructions

Local calls	area code + 7 digits
Long distance . . .	1 + area code + 7 digits

Speed-dial numbers

Customer service	*611
Emergency	911
Iowa state patrol	*55

Receiving calls

Caller dials the roamer access number below,
hears tone, dials cellular area code and number.
. . . . 712-259-7626

Billing information

Visitor rates
Most visitors . . . . . $3 day + 99¢ minute

Visitors without roaming agreements (p. 1159)
Call co. to use Amex, Discover, Mastercard, Visa

Identification numbers

City	Market	System	Billing	RSA
All served cities	420	01227	00000	IA-9

Cellular company

United States Cellular
8475 Hickman Road
Des Moines, IA 50322

Local office	515-276-6611
Corporate office	312-399-8900

Sending calls

Visitor dialing instructions

Local calls	area code + 7 digits
Long distance	area code + 7 digits

Speed-dial numbers

Customer service	611
Emergency	911
Iowa state patrol	*55
Roaming information	*711

Receiving calls

Caller dials the roamer access number below,
hears tone, dials cellular area code and number.
. . . . 515-249-7626

Billing information

Visitor rates
Most visitors . . . . . $3 day + 99¢ minute

Visitors without roaming agreements (p. 1159)
Call co. to use Mastercard, Visa

Identification numbers

City	Market	System	Billing	RSA
All served cities	420	01228	00000	IA-9

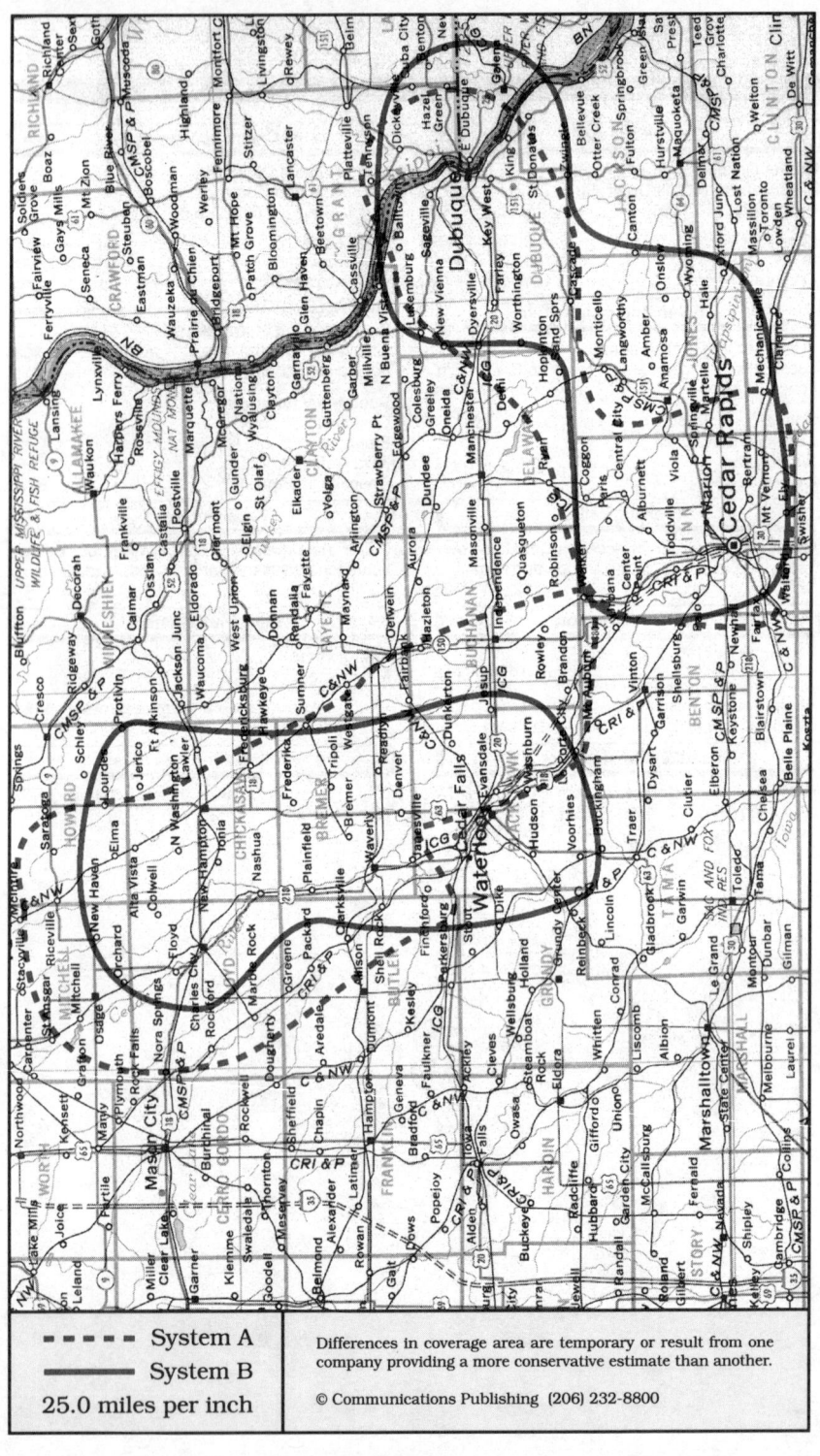

- - - - -	System A
——————	System B
25.0 miles per inch	

Differences in coverage area are temporary or result from one company providing a more conservative estimate than another.

© Communications Publishing (206) 232-8800

Cedar Rapids, Iowa
Cedar Rapids, Charles City, Dubuque, Galena IL, Waterloo

System A	System B

Cellular company

United States Cellular
300 Collins Road NE
Cedar Rapids, IA 52402

645 Century Drive
Dubuque, IA 52002

3642 University Avenue
Waterloo, IA 50701

Local office	. .	Cedar Rapids	319-350-1000
. . . . . .	. .	Dubuque	319-583-9000
. . . . . .	. .	Waterloo	319-234-4000
Corporate office	. . . .		312-399-8900

Sending calls

Visitor dialing instructions

Local calls	. . . .	area code + 7 digits
Long distance	. . . .	area code + 7 digits

Speed-dial numbers

Customer service	. . . .	611
Emergency	. . . .	911
Iowa state patrol	. . . .	*55
Make a traffic report (Cedar Rapids)	. .	*600
Roaming information	. . . .	*711

Receiving calls

Caller dials the roamer access number below,
hears tone, dials cellular area code and number.

Cedar Rapids	. . . . .	319-350-7626
Dubuque	. . . . . .	319-590-7626
Waterloo, Charles City	. . .	319-240-7626

Billing information

Visitor rates

Most visitors	. . . . .	$3 day + 99¢ minute

Visitors without roaming agreements (p. 1159)
Call co. to use Amex, Mastercard, Visa

Identification numbers

City	Market	System	Billing	RSA
Cedar Rapids	195	00303	00000	MSA
Charles City	424	01235	00000	IA-13
Dubuque	286	00331	00000	MSA
Waterloo	201	00589	00000	MSA

Cellular company

Centel Cellular
2615 11th Street SW
Cedar Rapids, IA 52404

3430 Dodge Street
Dubuque, IA 52003

131 Brookeridge Drive
Waterloo, IA 50702

Local office	. .	Cedar Rapids	800-373-5704
. . . . . .	. .	Cedar Rapids	319-366-5700
. . . . . .	. .	Dubuque	800-373-0805
. . . . . .	. .	Dubuque	319-588-0800
. . . . . .	. .	Waterloo	800-373-0405
. . . . . .	. .	Waterloo	319-236-0400
Corporate office	. . . .		312-399-2644

Sending calls

Visitor dialing instructions

Local calls	. . . . .	area code + 7 digits
Long distance	. . .	1 + area code + 7 digits

Speed-dial numbers

Customer service	. . . .	*611
Emergency	. . . .	911
KGAN news	. . . .	*2

Receiving calls

Caller dials the roamer access number below,
hears tone, dials cellular area code and number.

Cedar Rapids	. . . . . .	319-360-7626
Dubuque	. . . . . .	319-580-7626
Waterloo, Charles City	. . . .	319-230-7626

Follow Me Roaming forwarding (see p. 1156)
Activate, dial *18 send. Deactivate dial *19 send.

Billing information

Visitor rates

Most visitors	. . . .	$3 day + 99¢ minute

Visitors without roaming agreements (p. 1159)
Cannot call since company takes no credit cards.

Identification numbers

City	Market	System	Billing	RSA
Cedar Rapids	195	00286	00000	MSA
Charles City	424	01236	00000	IA-13
Dubuque	286	00286	00000	MSA
Waterloo	201	00286	00000	MSA

For full coverage area, see Iowa City, IA and
Monticello, IA.

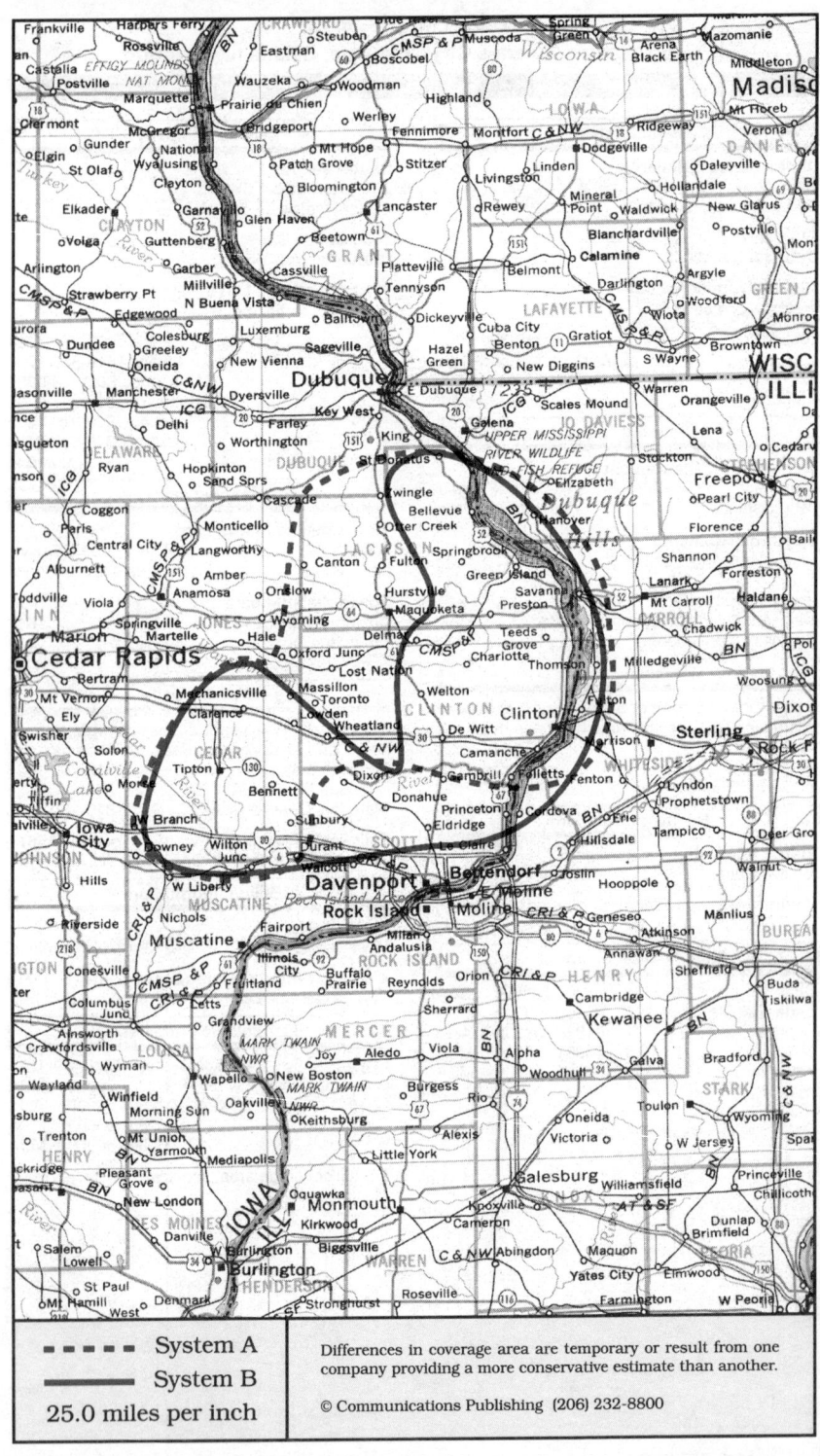

- - - - -	System A
—————	System B
25.0 miles per inch	

Differences in coverage area are temporary or result from one company providing a more conservative estimate than another.

© Communications Publishing (206) 232-8800

Clinton, Iowa
Clinton, Marquoketa, Tipton

|

System A

Cellular company

Cellular One
Iowa East Cellular Telephone
239 5th Avenue S
Clinton, IA 52732

Corporate office	800-876-2355
. . . .	319-242-3930

Sending calls

Visitor dialing instructions

Local calls	area code + 7 digits
Long distance . . .	1 + area code + 7 digits

Speed-dial numbers

Clinton police	*11
Customer service	611
Emergency	911
KCLN radio	*977

Receiving calls

Caller dials the roamer access number below,
hears tone, dials cellular area code and number.
. . . . 319-357-7626

Billing information

Visitor rates
Most visitors . . . . . $3 day + 99¢ minute

Visitors without roaming agreements (p. 1159)
Call co. to use Amex, Mastercard, Visa
or Roamer Plus will intercept first call to bill you.

Identification numbers

City	Market	System	Billing	RSA
All served cities	416	01219	00000	IA-5

For full coverage area, see Monticello, IA.

System B

Cellular company

Contel Cellular
301 N 2nd Street, Suite B
Clinton, IA 52732

Customer service	800-333-4004
. . . . . . extension 4002	404-391-8000
Local office	319-243-3663
Corporate office	404-804-3400

Sending calls

Visitor dialing instructions

Local calls	area code + 7 digits
Long distance . . .	1 + area code + 7 digits

Speed-dial numbers

Customer service	611

Receiving calls

Caller dials the roamer access number below,
hears tone, dials cellular area code and number.
. . . . 319-249-7626

Billing information

Visitor rates
Most visitors . . . . . $3 day + 85¢ minute

Visitors without roaming agreements (p. 1159)
Call co. to use Amex, Discover, Mastercard, Visa

Identification numbers

City	Market	System	Billing	RSA
All served cities	416	00186	00000	IA-5

For full coverage area, see Burlington, IA and
Davenport, IA.

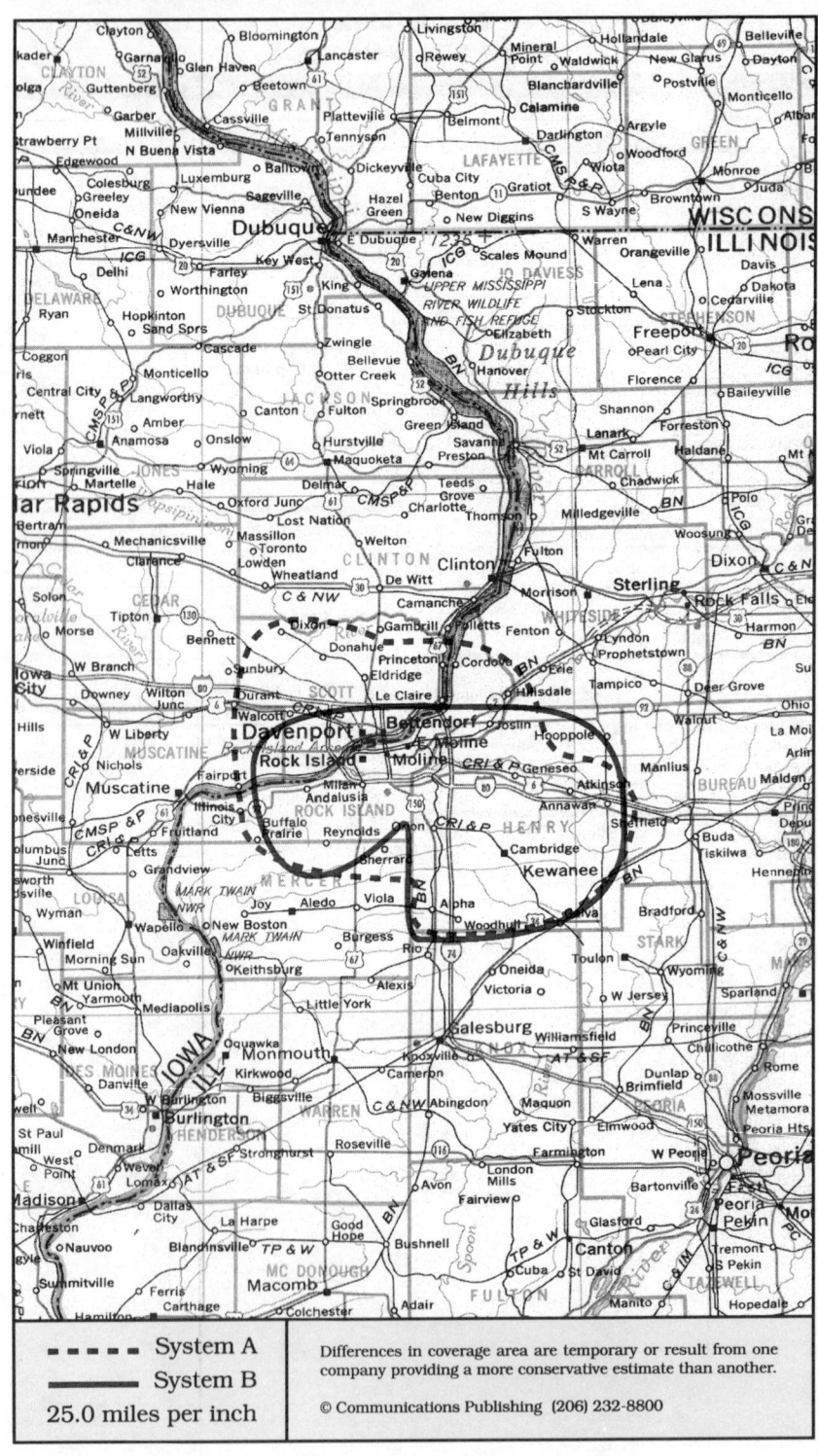

▪ ▪ ▪ ▪ ▪ **System A**	Differences in coverage area are temporary or result from one
▬▬▬▬ **System B**	company providing a more conservative estimate than another.
25.0 miles per inch	© Communications Publishing (206) 232-8800

Davenport, Iowa
Davenport, Geneseo IL, Kewanee IL, Moline IL, Rock Island IL

System A	System B

Cellular company

United States Cellular
4550 Brady
Davenport, IA 52806

Local office	319-349-8000
Corporate office	312-399-8900

Sending calls

Visitor dialing instructions

Local calls	area code + 7 digits
Long distance	area code + 7 digits

Speed-dial numbers

Customer service	611
Emergency	911
Make a traffic report	*104
Roaming information	*711

Receiving calls

Caller dials the roamer access number below, hears tone, dials cellular area code and number.
. . . . 319-349-7626

Billing information

Visitor rates
Most visitors . . . . . $3 day + 99¢ minute

Visitors without roaming agreements (p. 1159)
Call co. to use Mastercard, Visa
or Roamer Plus will intercept first call to bill you.

Identification numbers

City	Market	System	Billing	RSA
All served cities	098	00193	00000	MSA

Cellular company

Contel Cellular
5010 N Brady Street
Davenport, IA 52806-3946

Customer service	800-333-4004
. . . . . . extension 4002	404-391-8000
Local office	319-386-3441
Corporate office	404-804-3400

Sending calls

Visitor dialing instructions

Local calls	area code + 7 digits
Long distance . . .	1 + area code + 7 digits

Speed-dial numbers

Customer service	611

Receiving calls

Caller dials the roamer access number below, hears tone, dials cellular area code and number.
. . . . 319-340-7626

Follow Me Roaming forwarding (see p. 1156)
Activate, dial *18 send. Deactivate dial *19 send.

Billing information

Visitor rates
Most visitors . . . . . $3 day + 85¢ minute

Visitors without roaming agreements (p. 1159)
Call co. to use Amex, Discover, Mastercard, Visa

Identification numbers

City	Market	System	Billing	RSA
All served cities	098	00186	00000	MSA

For full coverage area, see Burlington, IA and
Clinton, IA.

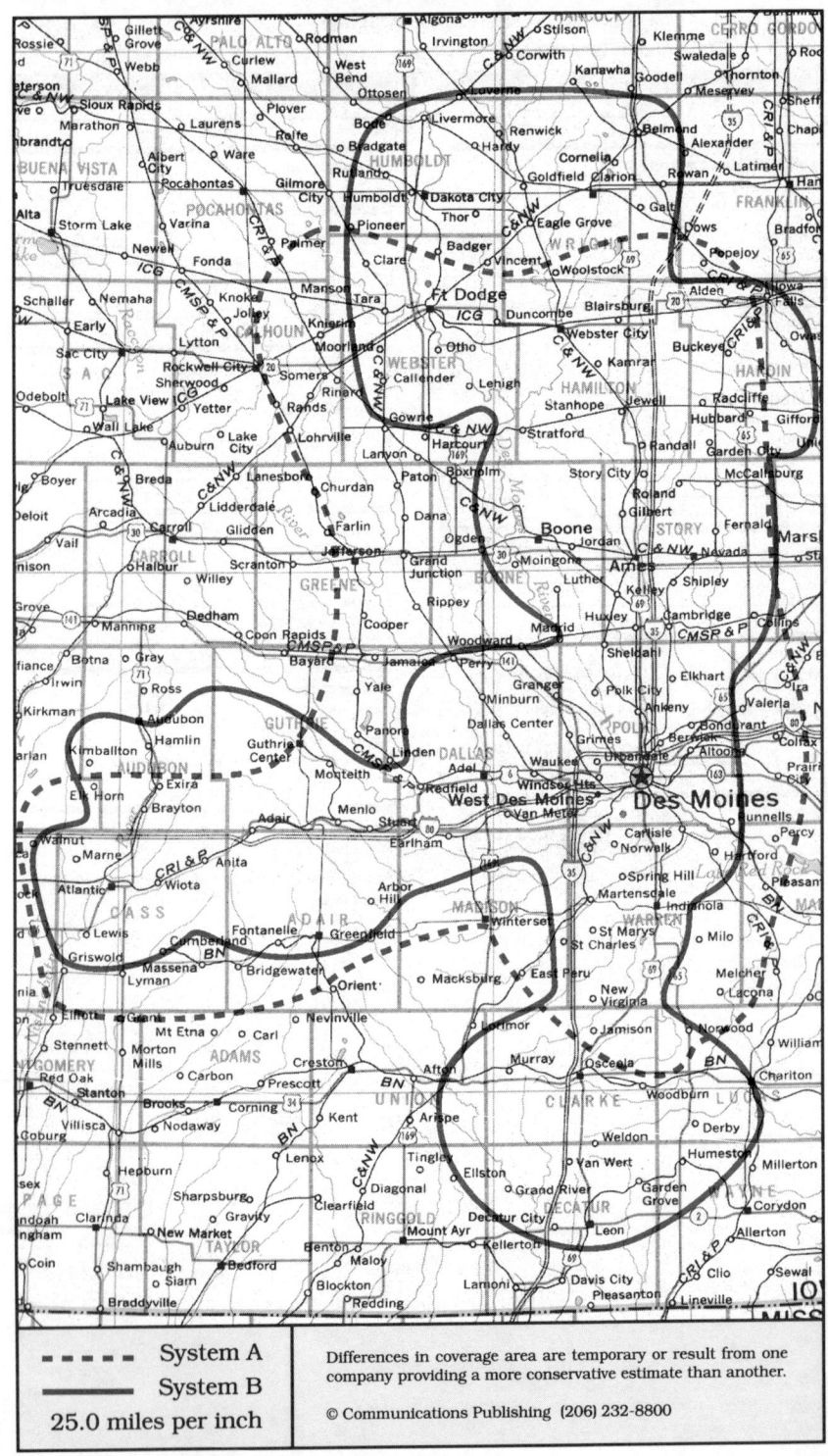

------ System A	Differences in coverage area are temporary or result from one
—— System B	company providing a more conservative estimate than another.
25.0 miles per inch	© Communications Publishing (206) 232-8800

Des Moines, Iowa
Ames, Atlantic, Des Moines, Fort Dodge, Osceola, Webster City

System A

Cellular company

United States Cellular
8475 Hickman Road
Des Moines, IA 50322-4319

Local office	515-276-6611
Corporate office	312-399-8900

Sending calls

Visitor dialing instructions

Local calls	area code + 7 digits
Long distance	area code + 7 digits

Speed-dial numbers

Customer service	611
Emergency	911
Iowa state patrol	*55
Roaming information	*711

Receiving calls

Caller dials the roamer access number below,
hears tone, dials cellular area code and number.

Atlantic	712-250-7626
Des Moines	515-249-7626
Fort Dodge	515-290-7626

Billing information

Visitor rates

Most visitors	$3 day + 99¢ minute

Visitors without roaming agreements (p. 1159)
Call co. to use Mastercard, Visa

Identification numbers

City	Market	System	Billing	RSA
Atlantic	418	01223	00000	IA-7
Des Moines	102	00195	00000	MSA
Fort Dodge	421	01229	00000	IA-10

System B

Cellular company

U S West Cellular
1200 Keo Way
Des Moines, IA 50309-1087

Customer service	800-238-7848
. . . .	206-562-2895
. . . . . local subscribers	800-626-6611
. . . .	206-747-1771
Local office	515-283-1200
Corporate office	206-747-4900

Sending calls

Visitor dialing instructions

Local calls	area code + 7 digits
Long distance	area code + 7 digits

Speed-dial numbers

Customer service	*611
Emergency	911
Iowa state patrol	#55
KGGO	*95 or *1460
Roaming information	*711

Receiving calls

Caller dials the roamer access number below,
hears tone, dials cellular area code and number.

Ames	515-239-7626
Atlantic	712-249-7626
Audubon	515-468-7626
Casey	515-740-7626
Des Moines	515-240-7626
Fort Dodge	515-570-7626
Osceola	515-340-7626
Webster City	515-835-7626

Follow Me Roaming forwarding (see p. 1156)
Activate, dial *18 send. Deactivate dial *19 send.

Billing information

Visitor rates

Most visitors	$3 day + 99¢ minute

Visitors without roaming agreements (p. 1159)
Call co. to use Amex, Mastercard, Visa

Identification numbers

City	Market	System	Billing	RSA
Ames, Ft Dodge	421	01230	00000	IA-10
Atlantic, Audubon	418	01224	00000	IA-7
Des Moines	102	00150	00000	MSA
Osceola	413	01214	00000	IA-2

For full coverage area, see Marshalltown, IA.

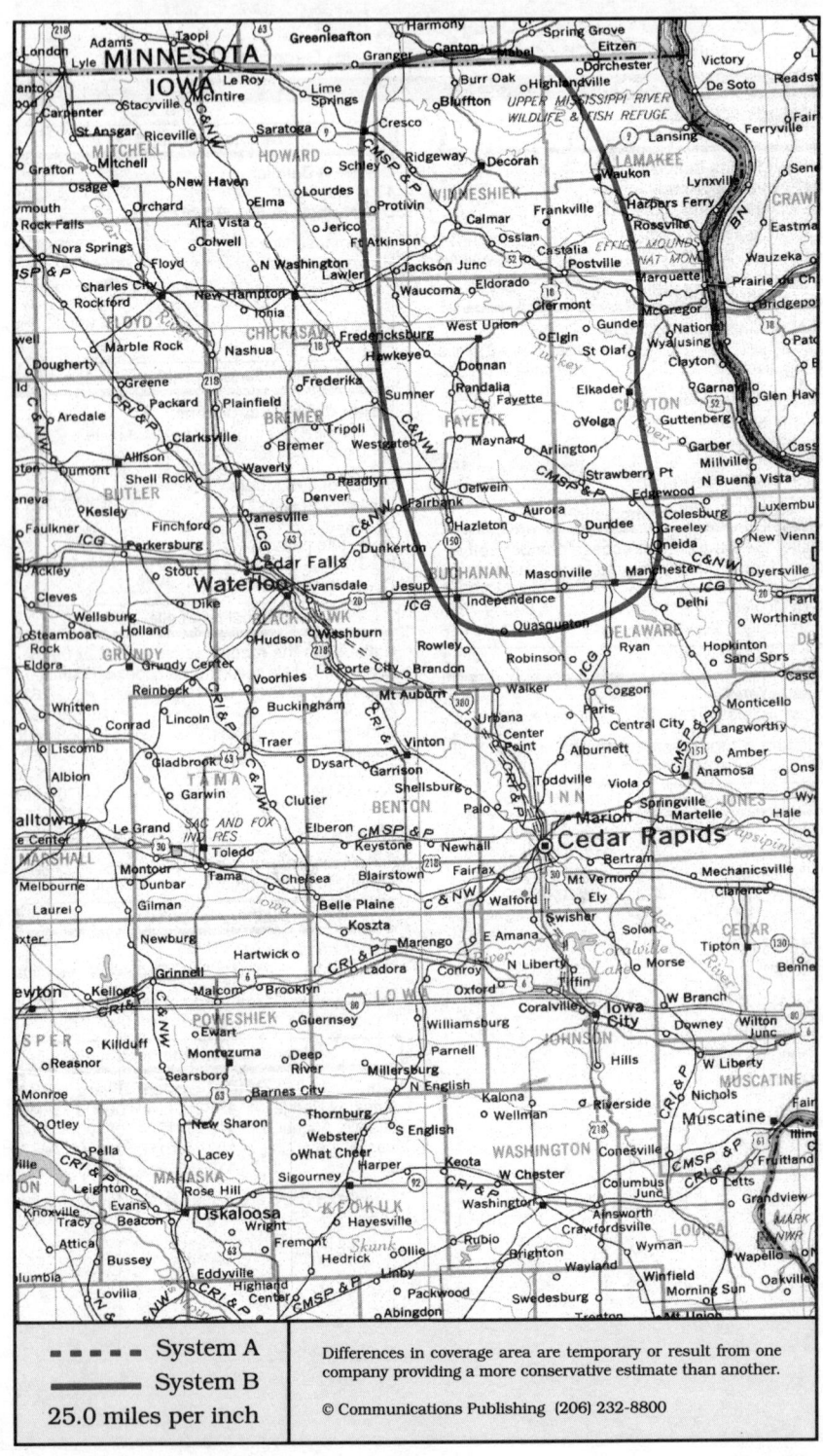

Independence, Iowa
Decorah, Independence, Manchester

System A	System B

Cellular company

No information about the company that will operate system A was available at press time.

Cellular company

United States Cellular
300 Collins Road NE
Cedar Rapids, IA 52402

Local office   319-920-1000
Corporate office   312-399-8900

Sending calls

Visitor dialing instructions
Local calls . . . . . .   area code + 7 digits
Long distance . . . . .   area code + 7 digits

Speed-dial numbers
Customer service 611
Emergency 911
Iowa state patrol . . . . ∗55
Roaming information . . . . ∗711

Receiving calls

Caller dials the roamer access number below, hears tone, dials cellular area code and number.
. . . . 319-920-7626

Billing information

Visitor rates
Most visitors . . . . . $3 day + 99¢ minute

Visitors without roaming agreements (p. 1159)
Call co. to use Amex, Mastercard, Visa

Identification numbers

City	Market	System	Billing	RSA
All served cities	423	01234	00000	IA-12

A local sales office may open in the future.

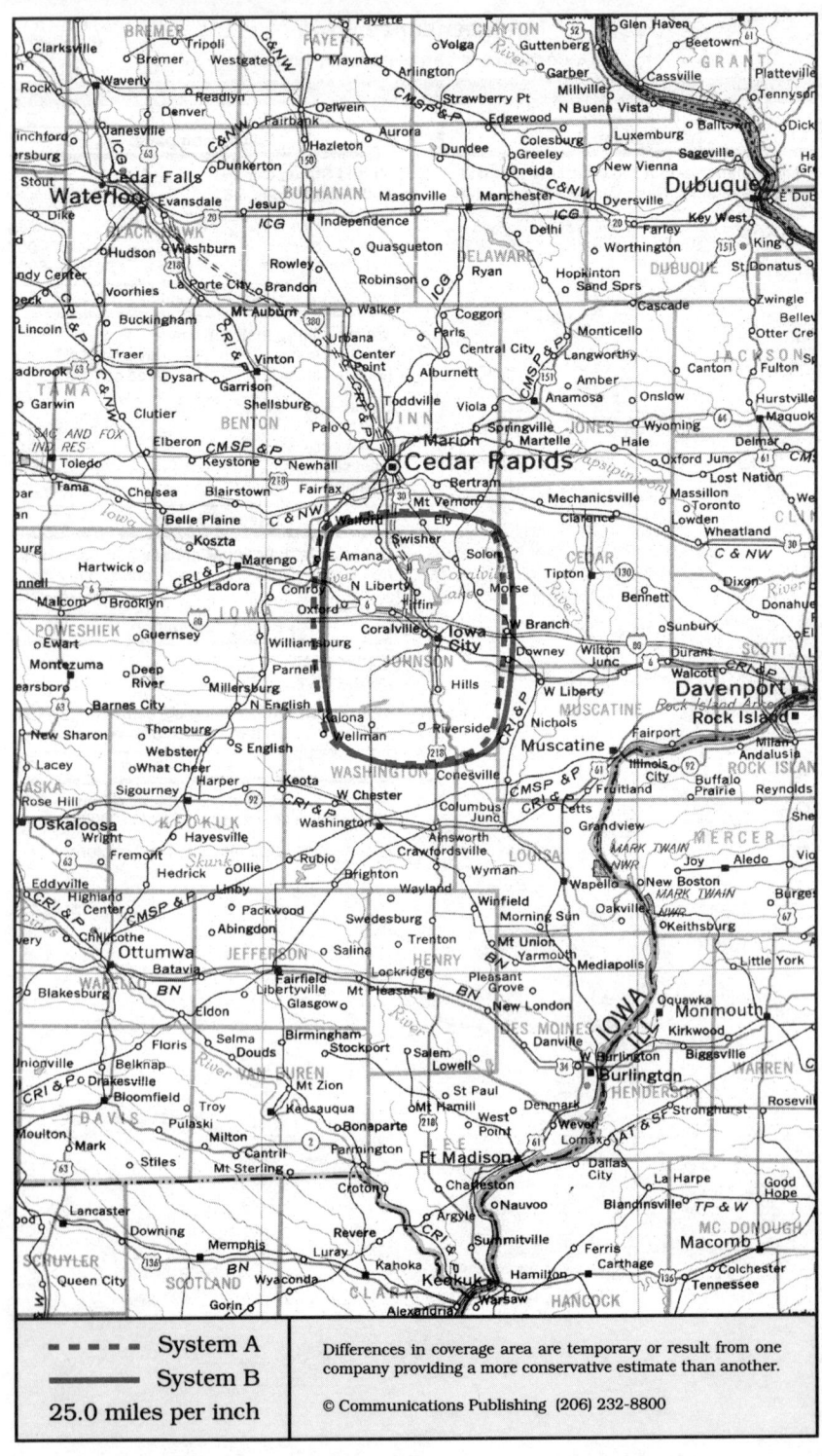

▰ ▰ ▰ ▰ ▰ **System A**	Differences in coverage area are temporary or result from one company providing a more conservative estimate than another.
────── **System B**	
25.0 miles per inch	© Communications Publishing (206) 232-8800

Iowa City, Iowa

System A	System B

System A

Cellular company

Cellular Plus
2010 Keokuk Street
Iowa City, IA 52240

Local office . .	some areas	800-634-7587
. . . .		319-351-5888
Corporate office		717-883-8832

Sending calls

Visitor dialing instructions

Local calls	area code + 7 digits
Long distance . . .	1 + area code + 7 digits

Speed-dial numbers

Customer service	611
Emergency	911
Iowa state patrol	*55

Receiving calls

Caller dials the roamer access number below, hears tone, dials cellular area code and number.
. . . . 319-330-7626

Billing information

Visitor rates

Most visitors . . . . . $3 day + $1 minute

Visitors without roaming agreements (p. 1159)
Cannot call since company takes no credit cards.

Identification numbers

City	Market	System	Billing	RSA
All served cities	296	00389	30245	MSA

For full coverage area, see Burlington, IA and Marshalltown, IA and Newton, IA and Ottumwa, IA.

System B

Cellular company

Centel Cellular
1723 2nd Street
Coralville, IA 52241

Customer service	800-373-0102
Local office	319-351-0100
Corporate office	312-399-2644

Sending calls

Visitor dialing instructions

Local calls	area code + 7 digits
Long distance . . .	1 + area code + 7 digits

Speed-dial numbers

Customer service	*611
Emergency	911
KGAN news	*2

Receiving calls

Caller dials the roamer access number below, hears tone, dials cellular area code and number.
. . . . 319-331-7626

Follow Me Roaming forwarding (see p. 1156)
Activate, dial *18 send. Deactivate dial *19 send.

Billing information

Visitor rates

Most visitors . . . . . $3 day + 99¢ minute

Visitors without roaming agreements (p. 1159)
Cannot call since company takes no credit cards.

Identification numbers

City	Market	System	Billing	RSA
All served cities	296	00286	00000	MSA

For full coverage area, see Cedar Rapids, IA and Monticello, IA.

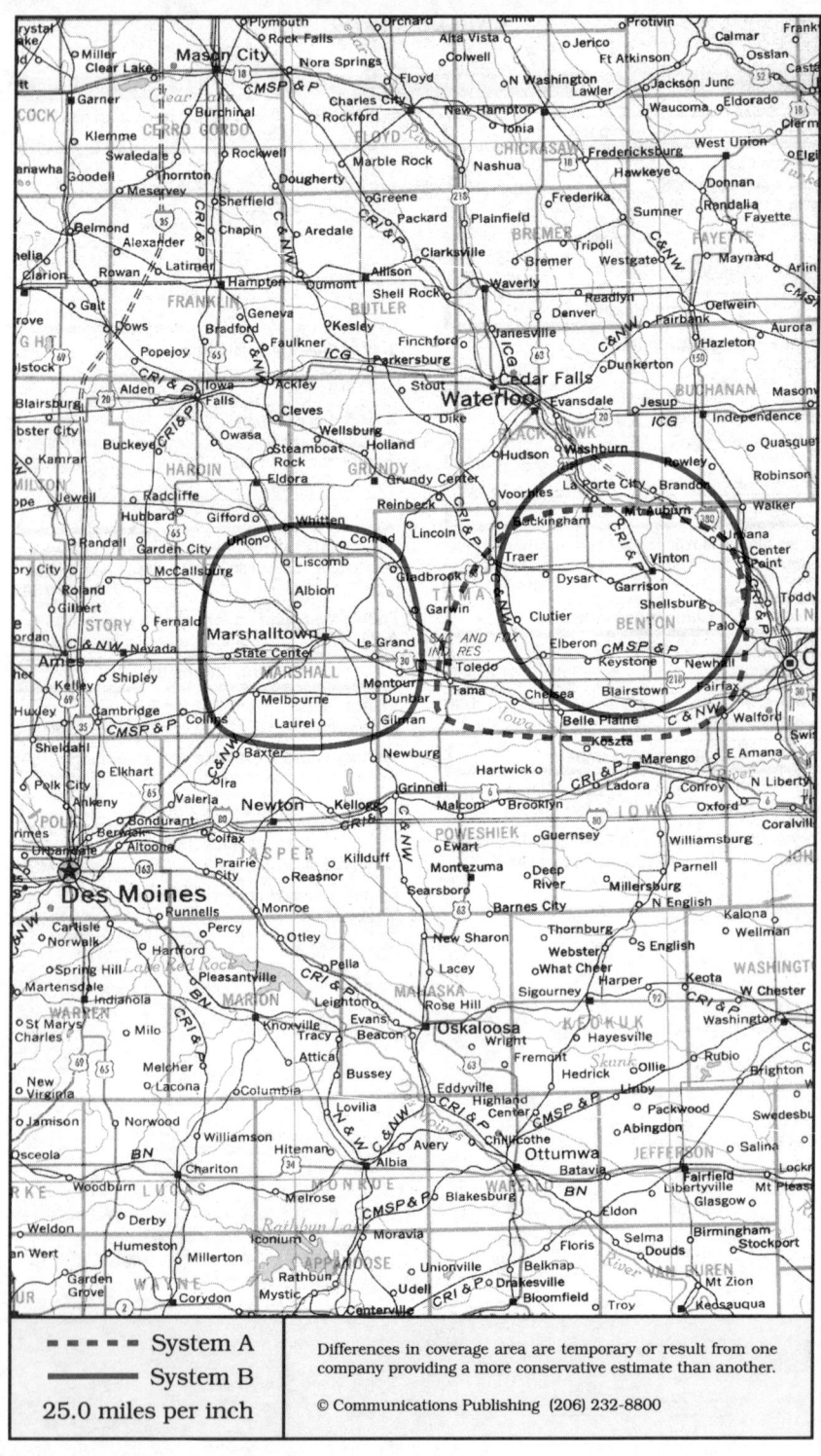

‑ ‑ ‑ ‑ ‑ System A	Differences in coverage area are temporary or result from one
───────── System B	company providing a more conservative estimate than another.
25.0 miles per inch	© Communications Publishing (206) 232-8800

Marshalltown, Iowa
Marshalltown, Toledo, Vinton

System A

Cellular company

Cellular Plus
2010 Keokuk Street
Iowa City, IA 52240

Local office . .	some areas	800-634-7587
. . . .		319-351-5888
Corporate office		717-883-8832

Sending calls

Visitor dialing instructions
Local calls	area code + 7 digits
Long distance . . .	1 + area code + 7 digits

Speed-dial numbers
Customer service	611
Emergency	911
Iowa state patrol	*55

Receiving calls

Caller dials the roamer access number below,
hears tone, dials cellular area code and number.
. . . . 319-330-7626

Billing information

Visitor rates
Most visitors . . . . . . $3 day + $1 minute

Visitors without roaming agreements (p. 1159)
Cannot call since company takes no credit cards.

Identification numbers

City	Market	System	Billing	RSA
All served cities	422	00389	30245	IA-11

This company will serve Marshalltown in the near future.

For full coverage area, see Iowa City, IA and Newton, IA.

System B

Cellular company

U S West Cellular
16 Main Street, Suite 115 (City Centre)
Marshalltown, IA 50158

Customer service		800-238-7848
. . . .		206-562-2895
. . . . .	local subscribers	800-626-6611
. . . .		206-747-1771
Local office . . .	Des Moines	515-283-1200
. . . . . .	Marshalltown	515-752-0852
Corporate office		206-747-4900

Sending calls

Visitor dialing instructions
Local calls	area code + 7 digits
Long distance	area code + 7 digits

Speed-dial numbers
Customer service	*611
Emergency	911
Iowa state patrol	#55
KGGO	*95 or *1460
Roaming information	*711

Receiving calls

Caller dials the roamer access number below,
hears tone, dials cellular area code and number.
Marshalltown, Toledo	515-750-7626
Vinton	319-360-7626

Billing information

Visitor rates
Most visitors . . . . . $3 day + 99¢ minute

Visitors without roaming agreements (p. 1159)
Call co. to use Amex, Mastercard, Visa

Identification numbers

City	Market	System	Billing	RSA
All served cities	422	01232	00000	IA-11

For full coverage area, see Des Moines, IA.

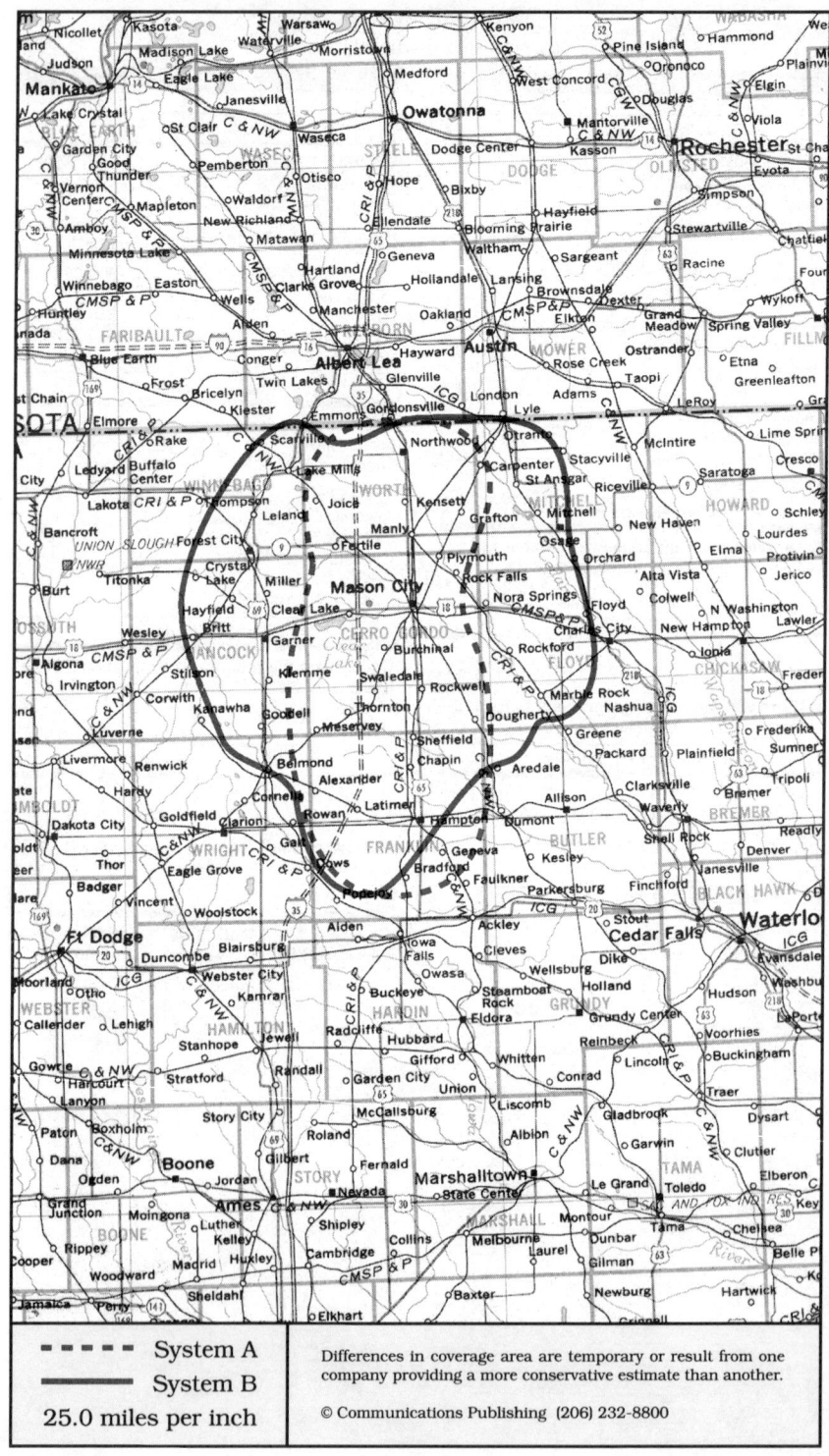

Differences in coverage area are temporary or result from one
company providing a more conservative estimate than another.

Mason City, Iowa
Hampton, Latimer, Mason City, Northwood, Osage

System A	System B

Cellular company

United States Cellular
300 Collins Road NE
Cedar Rapids, IA 52402

Local office 319-350-1000
Corporate office 312-399-8900

Sending calls

Visitor dialing instructions
Local calls . . . . . area code + 7 digits
Long distance . . . . area code + 7 digits

Speed-dial numbers
Customer service 611
Emergency 911
Iowa state patrol *55
Roaming information *711

Receiving calls

Caller dials the roamer access number below, hears tone, dials cellular area code and number.
. . . . 515-425-7626

Billing information

Visitor rates
Most visitors . . . . . $3 day + 99¢ minute

Visitors without roaming agreements (p. 1159)
Call co. to use Amex, Mastercard, Visa

Identification numbers

City	Market	System	Billing	RSA
All served cities	425	01237	00000	IA-14

Cellular company

CommNet 2000
Cellular, Inc.
1825 4th Street SW
Mason City, IA 50401

Customer service 800-597-5536
Local office 515-424-8448
Corporate office 303-694-3234

Sending calls

Visitor dialing instructions
Local calls . . . . . area code + 7 digits
Long distance . . . 1 + area code + 7 digits

Speed-dial numbers
Customer service *611
Emergency 911
Iowa state patrol . . . . #55
Roaming information 711

Receiving calls

Caller dials the roamer access number below, hears tone, dials cellular area code and number.
Latimer 515-580-7626
Mason City 515-420-7626

Billing information

Visitor rates
Most visitors . . . . . $3 day + 95¢ minute

Visitors without roaming agreements (p. 1159)
Cannot call since company takes no credit cards.

Identification numbers

City	Market	System	Billing	RSA
Latimer	425	00528	30402	IA-14
Mason City	425	00528	30314	IA-14

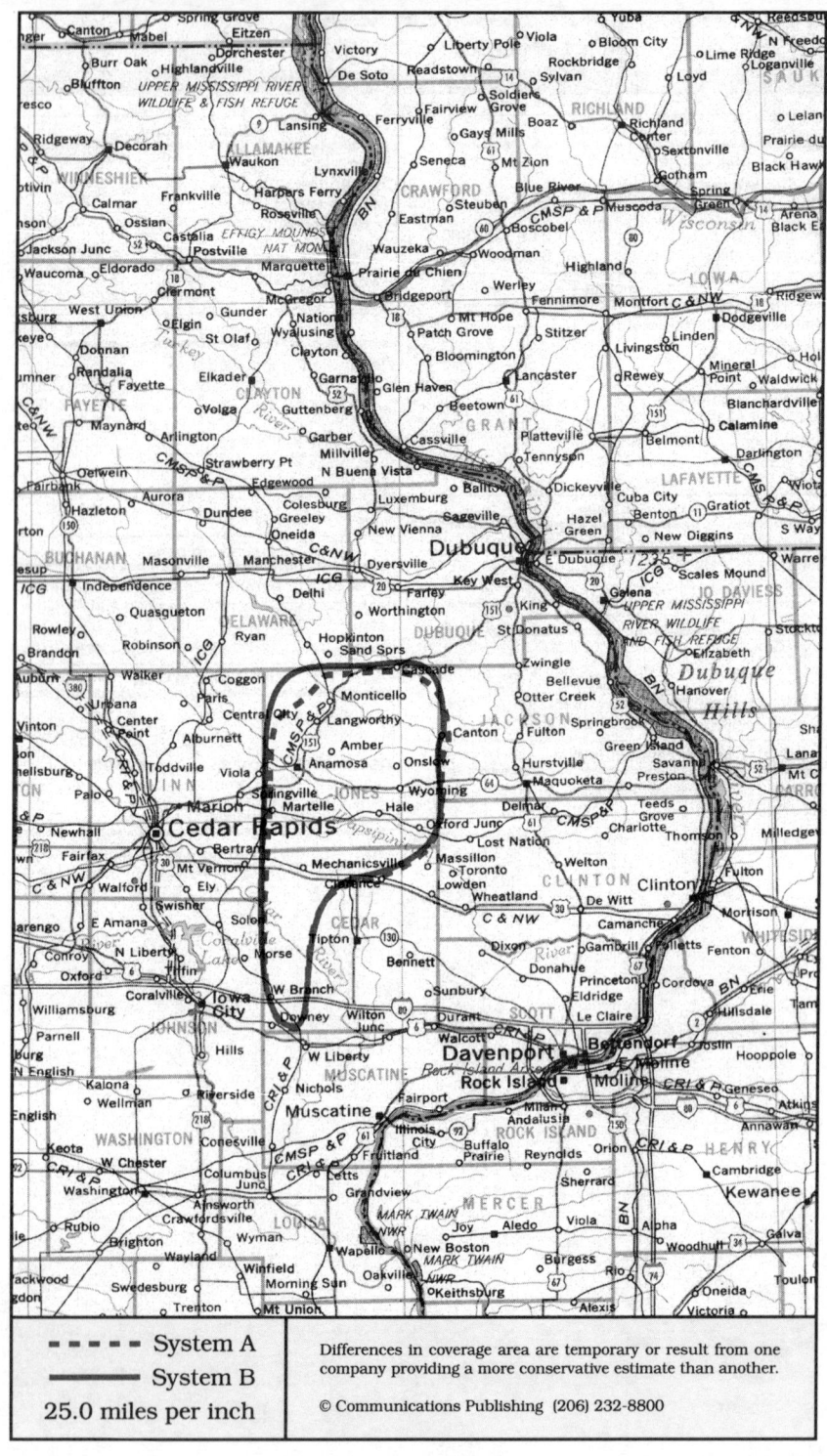

Monticello, Iowa
Anamosa, Monticello

Cellular company

Cellular One
Iowa East Cellular Telephone
239 5th Avenue S
Clinton, IA 52732

Corporate office	800-876-2355
. . . .	319-242-3930

Sending calls

Visitor dialing instructions

Local calls	area code + 7 digits
Long distance . . .	1 + area code + 7 digits

Speed-dial numbers

Customer service	611
Emergency	911

Receiving calls

Caller dials the roamer access number below,
hears tone, dials cellular area code and number.

. . . .	319-480-7626

Billing information

Visitor rates

Credit cards .	$7.50 fee + $3 day + 99¢ minute
Most visitors	$3 day + 99¢ minute

Visitors without roaming agreements (p. 1159)
Call co. to use Amex, Mastercard, Visa
or Roamer Plus will intercept first call to bill you.

Identification numbers

City	Market	System	Billing	RSA
All served cities	416	01219	00000	IA-5

For full coverage area, see Clinton, IA.

Cellular company

Centel Cellular
2615 11th Street SW
Cedar Rapids, IA 52404

Local office	800-373-5704
. . . .	319-366-5700
Corporate office	312-399-2644

Sending calls

Visitor dialing instructions

Local calls	area code + 7 digits
Long distance . . .	1 + area code + 7 digits

Speed-dial numbers

Customer service	✳611
Emergency	911
Highway patrol	✳555
KGAN news	✳2

Receiving calls

Caller dials the roamer access number below,
hears tone, dials cellular area code and number.

. . . .	319-360-7626

Billing information

Visitor rates

Most visitors	$3 day + 99¢ minute

Visitors without roaming agreements (p. 1159)
Cannot call since company takes no credit cards.

Identification numbers

City	Market	System	Billing	RSA
All served cities	416	01220	30446	IA-5

For full coverage area, see Cedar Rapids, IA and
Iowa City, IA.

Newton, Iowa

Grinnell, Knoxville, Newton, Oskaloosa, Pella, Washington, Williamsburg

System A	System B

System A

Cellular company

Cellular Plus
2010 Keokuk Street
Iowa City, IA 52240

Local office	800-634-7587
. . . .	319-351-5888
Corporate office	717-883-8832

Sending calls

Visitor dialing instructions

Local calls	area code + 7 digits
Long distance . . .	1 + area code + 7 digits

Speed-dial numbers

Customer service	611
Emergency	911
Iowa state patrol	*55

Receiving calls

Caller dials the roamer access number below,
hears tone, dials cellular area code and number.

. . . .	319-330-7626

Billing information

Visitor rates

Most visitors	$3 day + $1 minute

Visitors without roaming agreements (p. 1159)
Cannot call since company takes no credit cards.

Identification numbers

City	Market	System	Billing	RSA
All served cities	417	00389	30245	IA-6

For full coverage area, see Burlington, IA Iowa City, IA and Ottumwa, IA.

System B

Cellular company

CommNet 2000
Cellular, Inc.
1825 4th Street SW
Mason City, IA 50401

Customer service	800-597-5536
Local office	515-424-8448
Corporate office	303-694-3234

Sending calls

Visitor dialing instructions

Local calls	area code + 7 digits
Long distance . . .	1 + area code + 7 digits

Speed-dial numbers

Customer service	*611
Emergency	911
Iowa state patrol	#55
Roaming information	711

Receiving calls

Caller dials the roamer access number below,
hears tone, dials cellular area code and number.

Grinnell	515-260-7626
Knoxville	515-820-7626
Newton	515-840-7626
Oskaloosa	515-670-7626
Pella	515-820-7626
Washington	319-458-7626
Williamsburg	319-660-7626

Billing information

Visitor rates

Most visitors	$3 day + 95¢ minute

Visitors without roaming agreements (p. 1159)
Cannot call since company takes no credit cards.

Identification numbers

City	Market	System	Billing	RSA
Grinnell	417	00528	30396	IA-6
Knoxville	417	00528	30648	IA-6
Newton	417	00528	30312	IA-6
Oskaloosa	417	00528	30394	IA-6
Pella	417	00528	30390	IA-6
Washington	417	00528	30398	IA-6
Williamsburg	417	00528	30392	IA-6

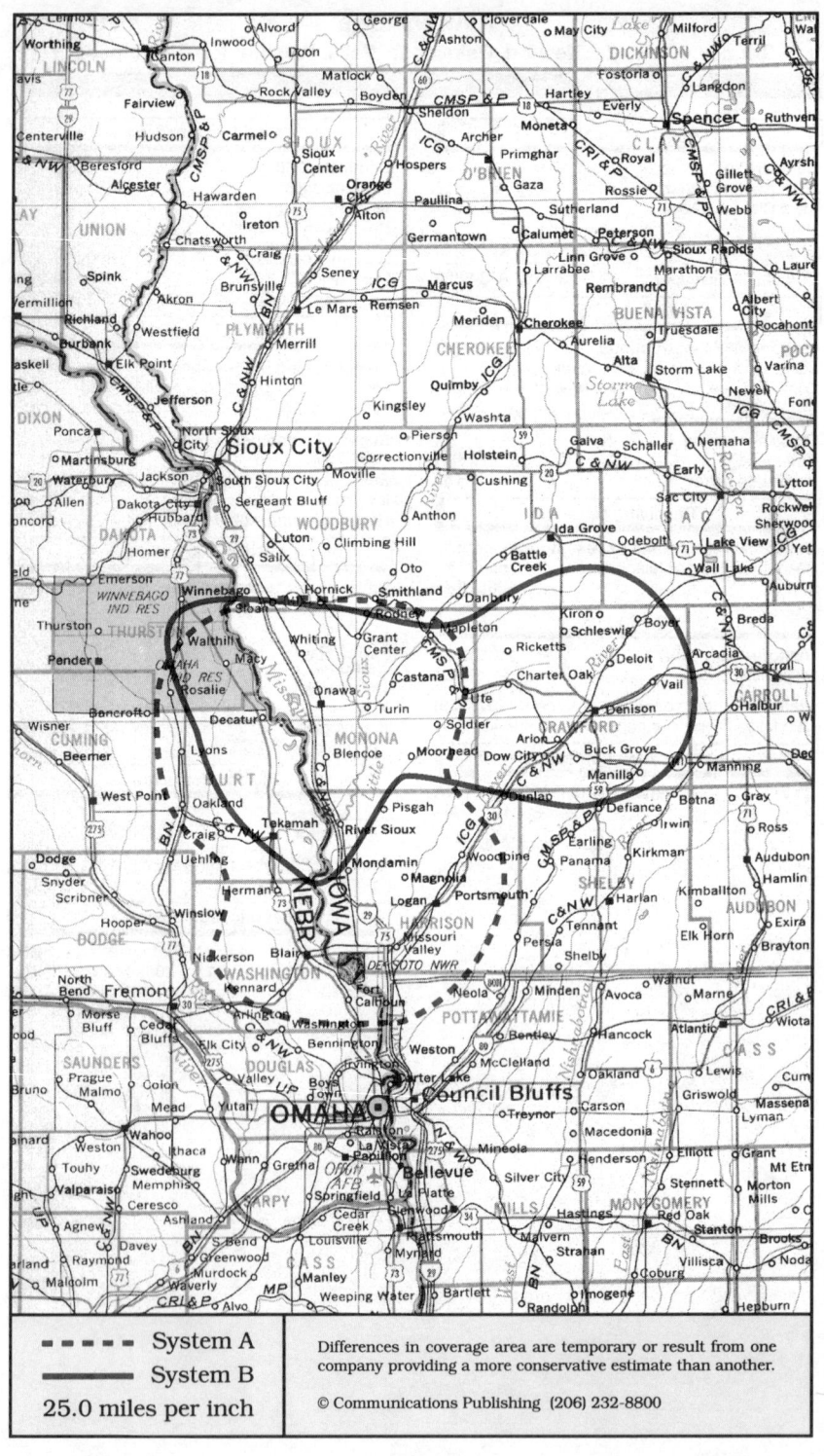

----- System A	Differences in coverage area are temporary or result from one
——— System B	company providing a more conservative estimate than another.
25.0 miles per inch	© Communications Publishing (206) 232-8800

Onawa, Iowa
Blair NE, Denison, Logan, Onawa

System A	System B

Cellular company

General Cellular
GenCell Management
1891 Woolner Avenue
Fairfield, CA 94533

Customer service	800-888-7868
Corporate office	707-425-8000

Sending calls

Visitor dialing instructions

Local calls	area code + 7 digits
Long distance . . .	1 + area code + 7 digits

Speed-dial numbers

Customer service	611
Emergency	911

Receiving calls

Caller dials the roamer access number below,
hears tone, dials cellular area code and number.
. . . . 712-644-7626

Billing information

Visitor rates
Most visitors . . . . . $3 day + 99¢ minute

Visitors without roaming agreements (p. 1159)
American Roaming Network intercepts call to bill.

Identification numbers

City	Market	System	Billing	RSA
All served cities	419	01225	00000	IA-8

Cellular company

CommNet 2000
Cellular, Inc.
124 Pierce
Sioux City, IA 51101

Customer service	800-597-5532
Local office	800-255-2083
. . . .	712-233-2083
Corporate office	303-694-3234

Sending calls

Visitor dialing instructions

Local calls	area code + 7 digits
Long distance . . .	1 + area code + 7 digits

Speed-dial numbers

Customer service	*611
Emergency	911
Iowa state patrol	#55
KMNS radio	*62
KSEZ radio	*98
Roaming information	711

Receiving calls

Caller dials the roamer access number below,
hears tone, dials cellular area code and number.

Denison	712-269-7626
Onawa	712-420-7626

Billing information

Visitor rates
Most visitors . . . . . $3 day + 95¢ minute

Visitors without roaming agreements (p. 1159)
Cannot call since company takes no credit cards.

Identification numbers

City	Market	System	Billing	RSA
Denison	419	00528	30400	IA-8
Onawa	419	00528	30062	IA-8

For full coverage area, see Sioux City, IA.

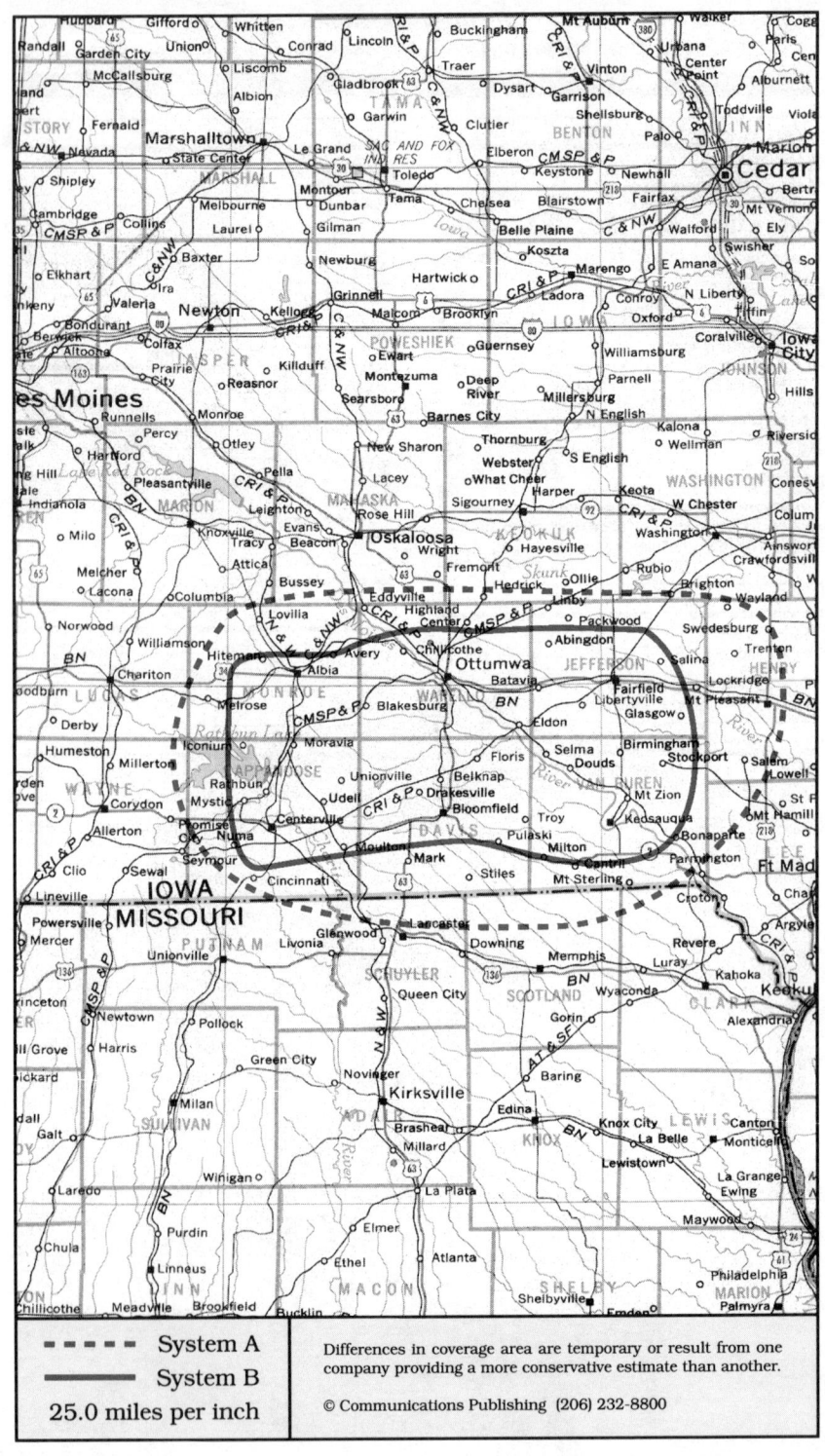

- - - - -	System A
———	System B
25.0 miles per inch	

Differences in coverage area are temporary or result from one company providing a more conservative estimate than another.

© Communications Publishing (206) 232-8800

Ottumwa, Iowa
Albia, Bloomfied, Centerville, Fairfield, Keosauqua, Ottumwa

System A	System B

Cellular company

Cellular Plus
2010 Keokuk Street
Iowa City, IA 52240

Local office . .	some areas	800-634-7587
. . . .		319-351-5888
Corporate office		717-883-8832

Sending calls

Visitor dialing instructions

Local calls	area code + 7 digits
Long distance . . .	1 + area code + 7 digits

Speed-dial numbers

Customer service	611
Emergency	911
Iowa state patrol	*55

Receiving calls

Caller dials the roamer access number below, hears tone, dials cellular area code and number.
. . . . 319-330-7626

Billing information

Visitor rates
Most visitors . . . . . . $3 day + $1 minute

Visitors without roaming agreements (p. 1159)
Cannot call since company takes no credit cards.

Identification numbers

City	Market	System	Billing	RSA
All served cities	414	00389	30245	IA-3

For full coverage area, see Burlington, IA and Iowa City, IA and Newton, IA.

Cellular company

United States Cellular
1111 Quincy Avenue, Suite 105
Ottumwa, IA 52501

Local office	515-684-8000
Corporate office	312-399-8900

Sending calls

Visitor dialing instructions

Local calls	area code + 7 digits
Long distance	area code + 7 digits

Speed-dial numbers

Customer service	611
Emergency	911
Iowa highway patrol	*55

Receiving calls

Caller dials the roamer access number below, hears tone, dials cellular area code and number.
. . . . 515-777-7626

Billing information

Visitor rates
Most visitors . . . . . $3 day + 99¢ minute

Visitors without roaming agreements (p. 1159)
Call co. to use Mastercard, Visa

Identification numbers

City	Market	System	Billing	RSA
All served cities	414	01216	00000	IA-3

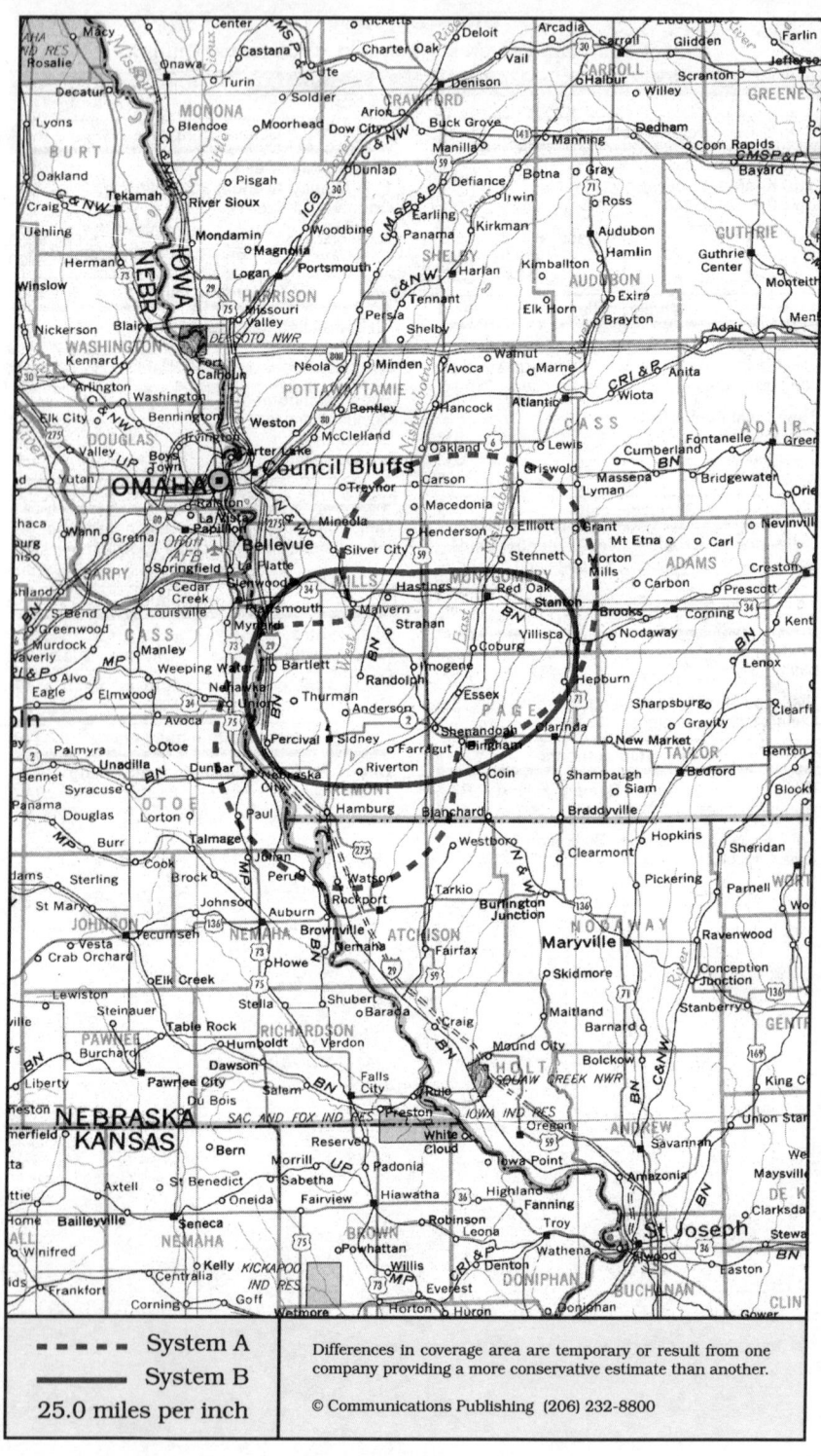

Legend	
▪▪▪▪	System A
——	System B
25.0 miles per inch	

Differences in coverage area are temporary or result from one company providing a more conservative estimate than another.

© Communications Publishing (206) 232-8800

Red Oak, Iowa
Glenwood, Red Oak, Sidney

Cellular company

United States Cellular
309 Summit Drive, PO Box 691
Maryville, MO 64468

Local office	816-562-3300
Corporate office	312-399-8900

Sending calls

Visitor dialing instructions

Local calls	area code + 7 digits
Long distance . . .	1 + area code + 7 digits

Speed-dial numbers

Customer service	*611
Emergency	911
Roaming information	*711

Receiving calls

Caller dials the roamer access number below,
hears tone, dials cellular area code and number.
. . . . 712-250-7626

Billing information

Visitor rates
Most visitors . . . . . $3 day + 85¢ minute

Visitors without roaming agreements (p. 1159)
Call co. to use Mastercard, Visa

Identification numbers

City	Market	System	Billing	RSA
All served cities	412	01211	00000	IA-1

Cellular company

Licensed to: Cellular 29 Plus
Managed by: Lincoln Telephone Cellular
605 Morton
Emerson, IA 51533

Corporate office	800-944-5526
. . . .	712-824-7231

Sending calls

Visitor dialing instructions

Local calls	area code + 7 digits
Long distance . . .	1 + area code + 7 digits

Speed-dial numbers

Customer service	611
Emergency	911

Receiving calls

Caller dials the roamer access number below,
hears tone, dials cellular area code and number.
. . . . 712-370-7626

Follow Me Roaming forwarding (see p. 1156)
Activate, dial *18 send. Deactivate dial *19 send.

Billing information

Visitor rates
Most visitors . . . . . $3 day + 99¢ minute

Visitors without roaming agreements (p. 1159)
Cannot call since company takes no credit cards.

Identification numbers

City	Market	System	Billing	RSA
All served cities	412	01212	00000	IA-1

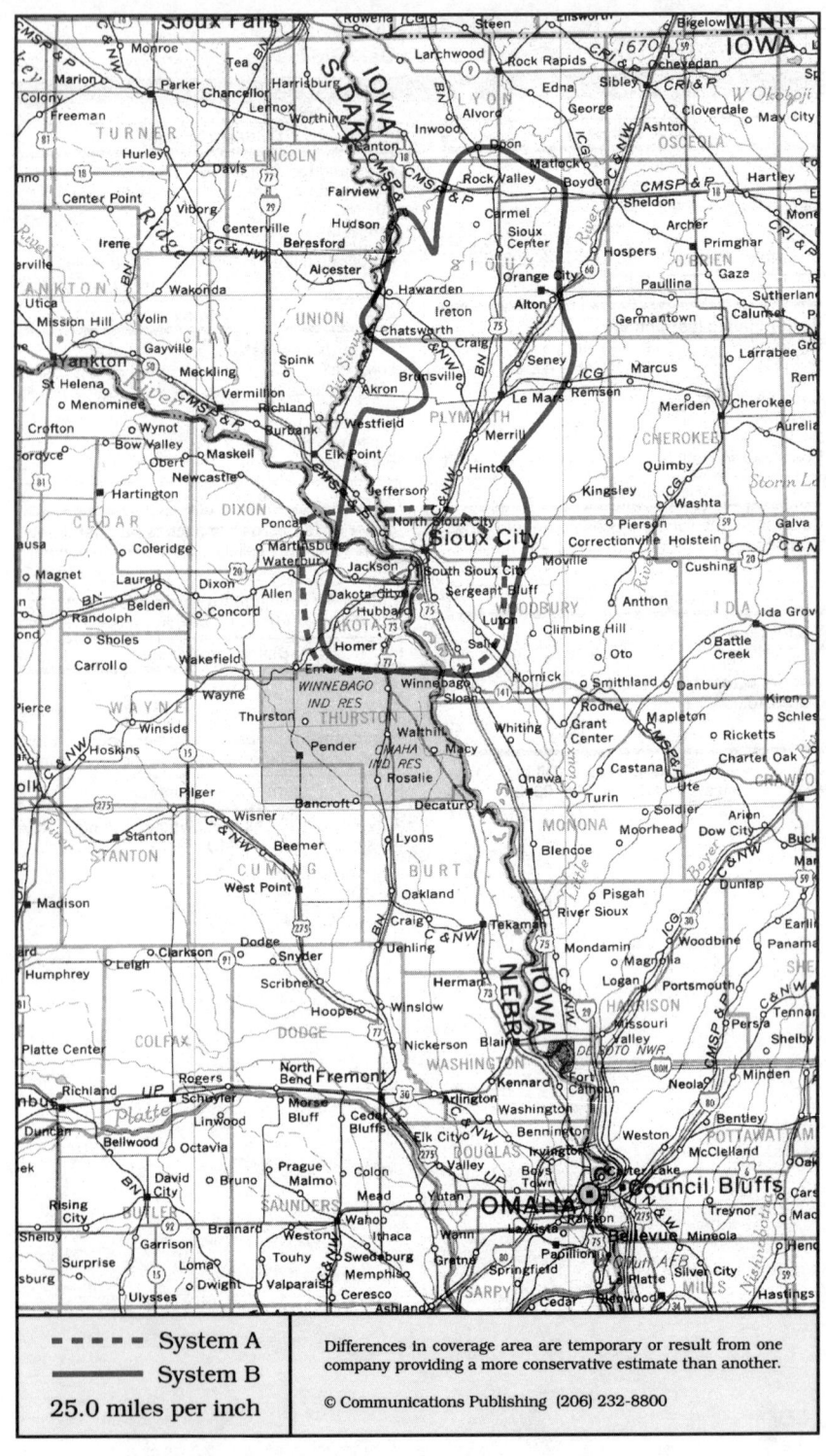

- - - - - System A	Differences in coverage area are temporary or result from one company providing a more conservative estimate than another.
———— System B	
25.0 miles per inch	© Communications Publishing (206) 232-8800

Sioux City, Iowa
Le Mars, Orange City, Sioux City

System A	System B

Cellular company

Centel Cellular
4711 Southern Hills Drive, Suite A
Sioux City, IA 51106

Local office	800-373-7545
. . . .	712-274-2494
Corporate office	312-399-2644

Sending calls

Visitor dialing instructions

Local calls	area code + 7 digits
Long distance . . .	1 + area code + 7 digits

Speed-dial numbers

Customer service	*611
Emergency	911
Iowa state patrol	*55

Receiving calls

Caller dials the roamer access number below,
hears tone, dials cellular area code and number.
. . . . 712-259-7626

NationLink & RoamingAmerica (see p. 1157)
Caller hears your current roamer access number.
This company's subscribers have level 0 service.

Billing information

Visitor rates
Most visitors . . . . . $3 day + 99¢ minute

Visitors without roaming agreements (p. 1159)
Call co. to use Amex, Discover, Mastercard, Visa

Identification numbers

City	Market	System	Billing	RSA
All served cities	253	00547	00000	MSA

Cellular company

CommNet 2000
Cellular, Inc.
124 Pierce
Sioux City, IA 51101

Customer service	800-597-5532
Local office	800-255-2083
. . . .	712-233-2083
Corporate office	303-694-3234

Sending calls

Visitor dialing instructions

Local calls	area code + 7 digits
Long distance . . .	1 + area code + 7 digits

Speed-dial numbers

Customer service	*611
Emergency	911
Iowa state patrol	#55
KMNS radio	*62
KSEZ radio	*98
Roaming information	711

Receiving calls

Caller dials the roamer access number below,
hears tone, dials cellular area code and number.

Le Mars	712-540-7626
Sioux City	712-251-7626

Billing information

Visitor rates
Most visitors . . . . . $3 day + 95¢ minute

Visitors without roaming agreements (p. 1159)
Cannot call since company takes no credit cards.

Identification numbers

City	Market	System	Billing	RSA
Le Mars	427	00528	30066	IA-16
Sioux City	253	00528	00000	MSA

For full coverage area, see Onawa, IA.

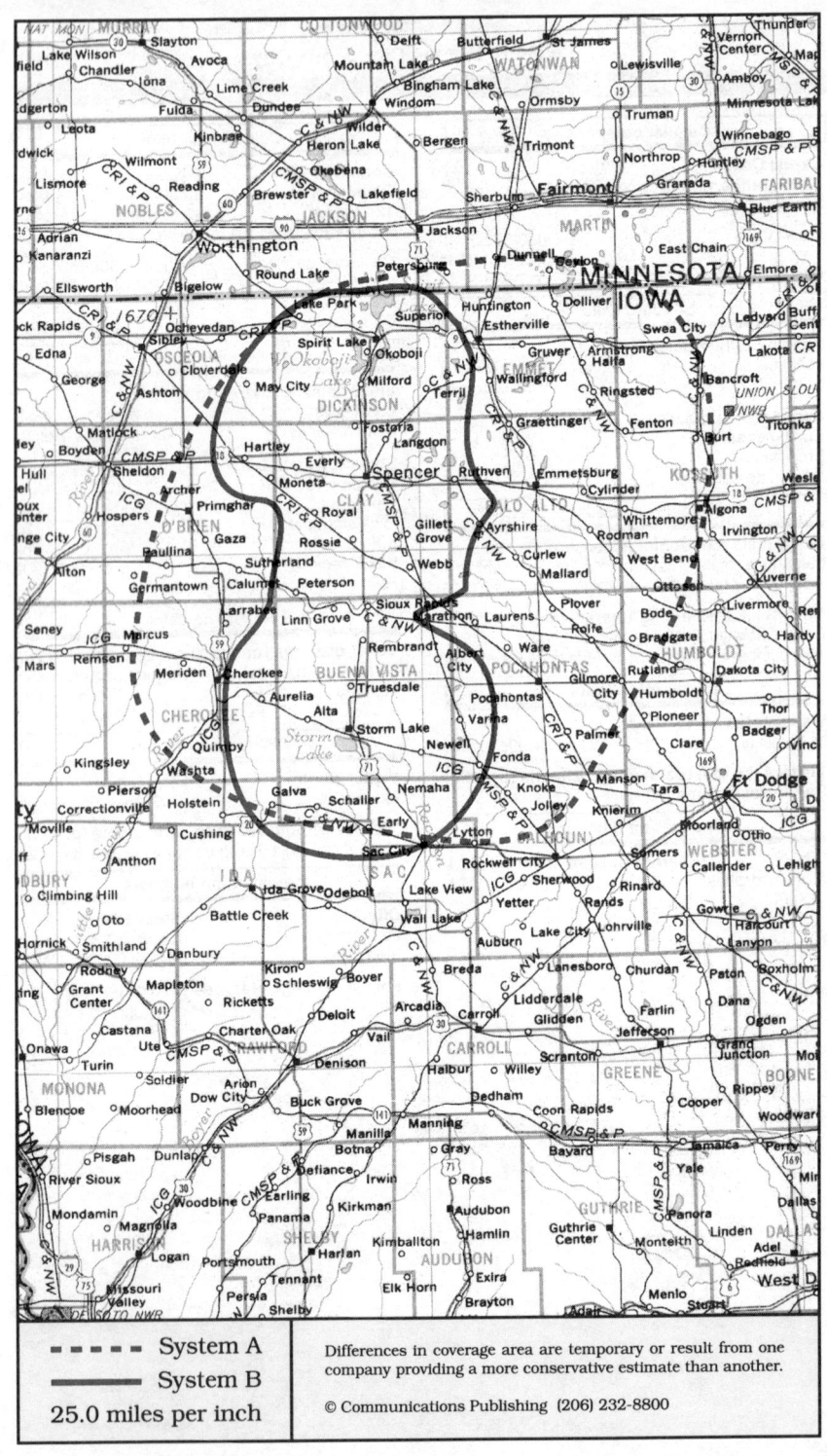

Spencer, Iowa
Cherokee, Estherville, Ida, Spencer, Spirit Lake, Storm Lake

System A	System B

Cellular company

Cellular One
Cellular Ventures
4 Grand Avenue
Spencer, IA 51031

Corporate office 712-262-4444

Sending calls

Visitor dialing instructions
Local calls . . . . 7 digits
Long distance . . . 1 + area code + 7 digits

Speed-dial numbers
Customer service . . . . ∗611
Emergency 911
Iowa state patrol . . . . #55

Receiving calls

Caller dials the roamer access number below,
hears tone, dials cellular area code and number.
. . . . 712-270-7626

Billing information

Visitor rates
Most visitors . . . . . $3 day + 85¢ minute

Visitors without roaming agreements (p. 1159)
Call co. to use Mastercard, Visa

Identification numbers

City	Market	System	Billing	RSA
All served cities	426	01239	00000	IA-15

Cellular company

CommNet 2000
Cellular, Inc.
124 Pierce Street
Sioux City, IA 51101

Customer service 800-597-5532
Local office 712-233-2083
Corporate office 303-694-3234

Sending calls

Visitor dialing instructions
Local calls . . . . . area code + 7 digits
Long distance . . . 1 + area code + 7 digits

Speed-dial numbers
Customer service . . . . ∗611
Emergency 911
Iowa state patrol . . . . #55
KMNS radio . . . . ∗62
KSEZ radio . . . . ∗98
Roaming information 711

Receiving calls

Caller dials the roamer access number below,
hears tone, dials cellular area code and number.
Spencer 712-260-7626
Storm Lake 712-299-7626

Billing information

Visitor rates
Most visitors . . . . . $3 day + 95¢ minute

Visitors without roaming agreements (p. 1159)
Cannot call since company takes no credit cards.

Identification numbers

City	Market	System	Billing	RSA
Spencer	426	00528	30064	IA-15
Storm Lake	426	00528	30650	IA-15

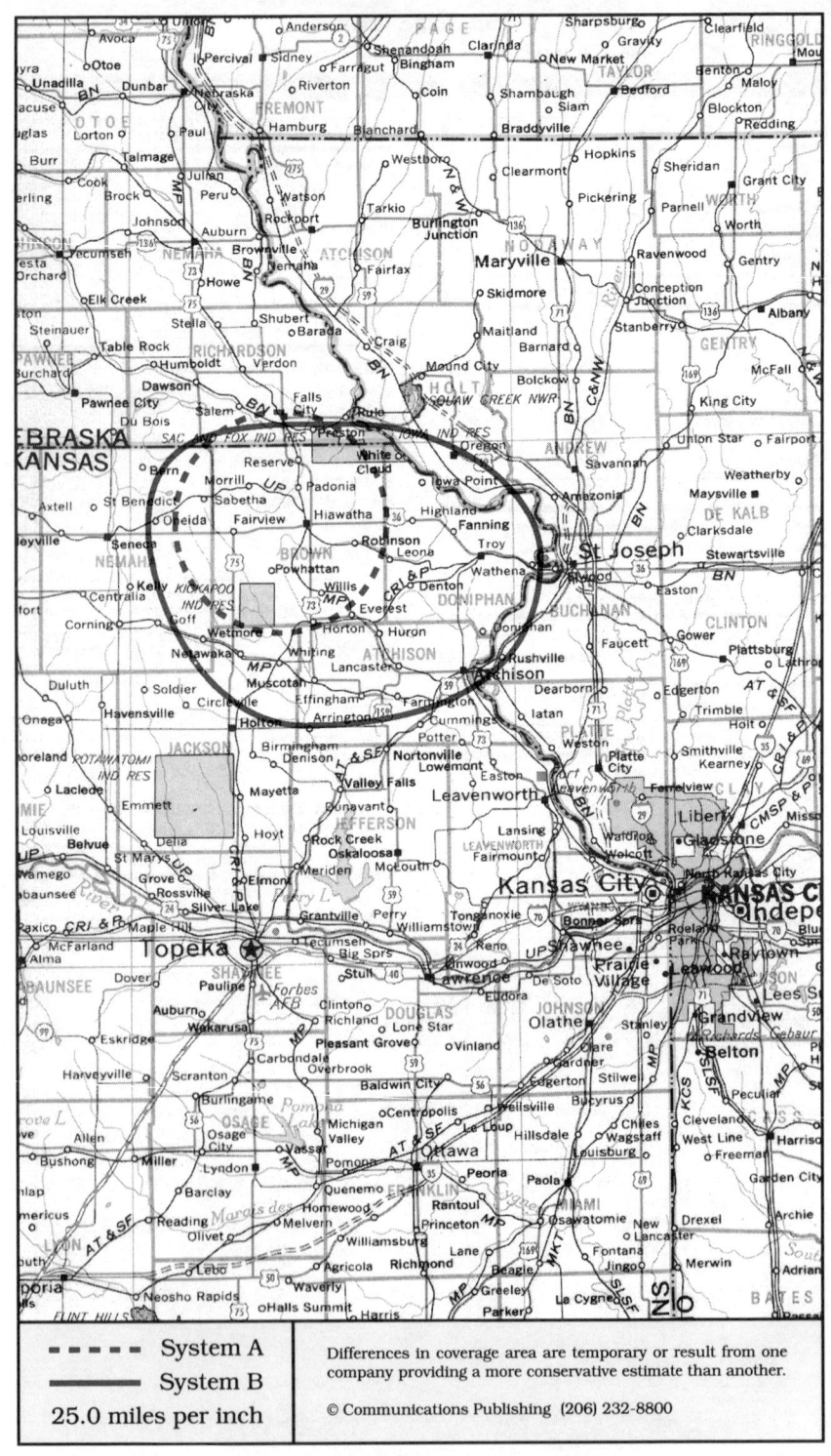

- - - - -	System A
——————	System B
25.0 miles per inch	

Differences in coverage area are temporary or result from one company providing a more conservative estimate than another.

© Communications Publishing (206) 232-8800

Atchison, Kansas
Atchison, Hiawatha, Troy, Willis

Cellular company

United States Cellular
309 Summit Drive
Maryville, MO 64468

Local office	816-562-3300
Corporate office	312-399-8900

Sending calls

Visitor dialing instructions

Local calls	area code + 7 digits
Long distance . . .	1 + area code + 7 digits

Speed-dial numbers

Customer service	*611
Emergency	911
Roaming information	*711

Receiving calls

Caller dials the roamer access number below,
hears tone, dials cellular area code and number.

. . . . 816-787-7626

Billing information

Visitor rates

Most visitors . . . . . $3 day + 85¢ minute

Visitors without roaming agreements (p. 1159)
Call co. to use Mastercard, Visa

Identification numbers

City	Market	System	Billing	RSA
All served cities	432	01251	00000	KS-5

For full coverage area, see Leavenworth, KS.

Cellular company

Kansas Cellular
Kansas Independent Cellular Networks
621 Westport Boulevard, PO Box 1607
Salina, KS 67402-1607

Customer service	800-373-5090
Corporate office	800-383-5090
. . . .	913-823-5049

Sending calls

Visitor dialing instructions

Local calls	area code + 7 digits
Long distance	area code + 7 digits

Speed-dial numbers

Customer service	*611
Roaming information	*711

Receiving calls

Caller dials the roamer access number below,
hears tone, dials cellular area code and number.

. . . . 913-658-7626

Billing information

Visitor rates

Most visitors . . . . . $3 day + 85¢ minute

Visitors without roaming agreements (p. 1159)
Call co. to use Amex, Discover, Mastercard, Visa

Identification numbers

City	Market	System	Billing	RSA
All served cities	432	01251	00000	KS-5

For full coverage area, see Emporia, KS and
Hutchinson, KS and Salina, KS.

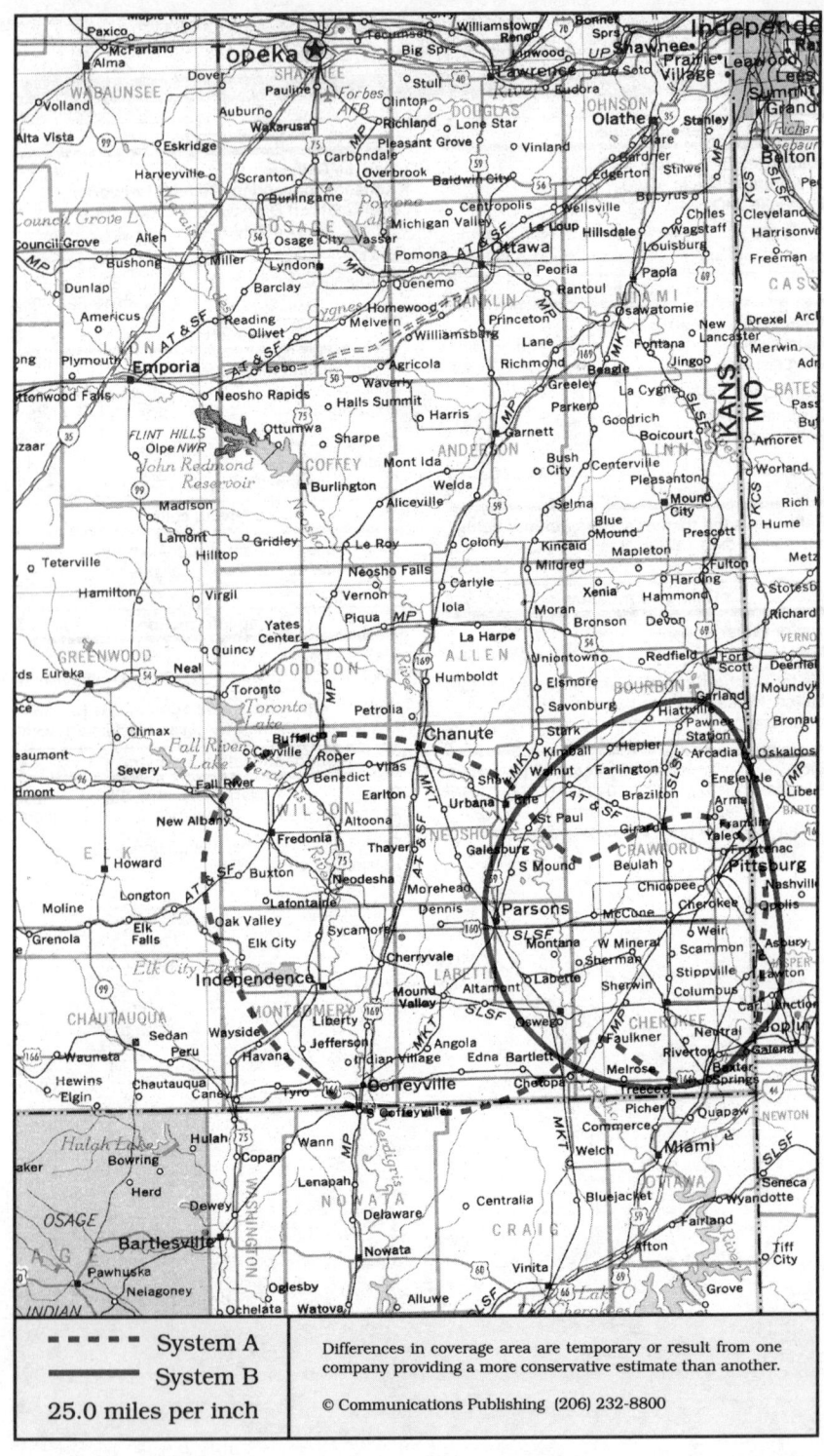

▪ ▪ ▪ ▪ System A	
——— System B	Differences in coverage area are temporary or result from one company providing a more conservative estimate than another.
25.0 miles per inch	© Communications Publishing (206) 232-8800

Coffeyville, Kansas
Coffeyville, Columbus, Independence, Parsons, Pittsburg

System A	System B

Cellular company

Kansas Cellular
Kansas Independent Cellular Networks
621 Westport Boulevard, PO Box 1607
Salina, KS 67402-1607

Customer service	800-373-5090
Corporate office	800-383-5090
. . . .	913-823-5049

Sending calls

Visitor dialing instructions

Local calls	area code + 7 digits
Long distance	area code + 7 digits

Speed-dial numbers

Customer service	*611
Roaming information	*711

Receiving calls

Caller dials the roamer access number below,
hears tone, dials cellular area code and number.
. . . . 913-658-7626

Billing information

Visitor rates
Most visitors . . . . . $3 day + 85¢ minute

Visitors without roaming agreements (p. 1159)
Call co. to use Amex, Discover, Mastercard, Visa

Identification numbers

City	Market	System	Billing	RSA
All served cities	442	01271	00000	KS-15

Cellular company

United States Cellular
1630 S Range Line Road
Joplin, MO 64804

Local office	316-249-0030
Corporate office	312-399-8900

Sending calls

Visitor dialing instructions

Local calls	area code + 7 digits
Long distance	area code + 7 digits

Speed-dial numbers

Customer service	*611 or *711
Emergency	911

Receiving calls

Caller dials the roamer access number below,
hears tone, dials cellular area code and number.
. . . . 316-249-7626

Billing information

Visitor rates
Most visitors . . . . . $3 day + 85¢ minute

Visitors without roaming agreements (p. 1159)
Call co. to use Mastercard, Visa

Identification numbers

City	Market	System	Billing	RSA
All served cities	442	01272	00000	KS-15

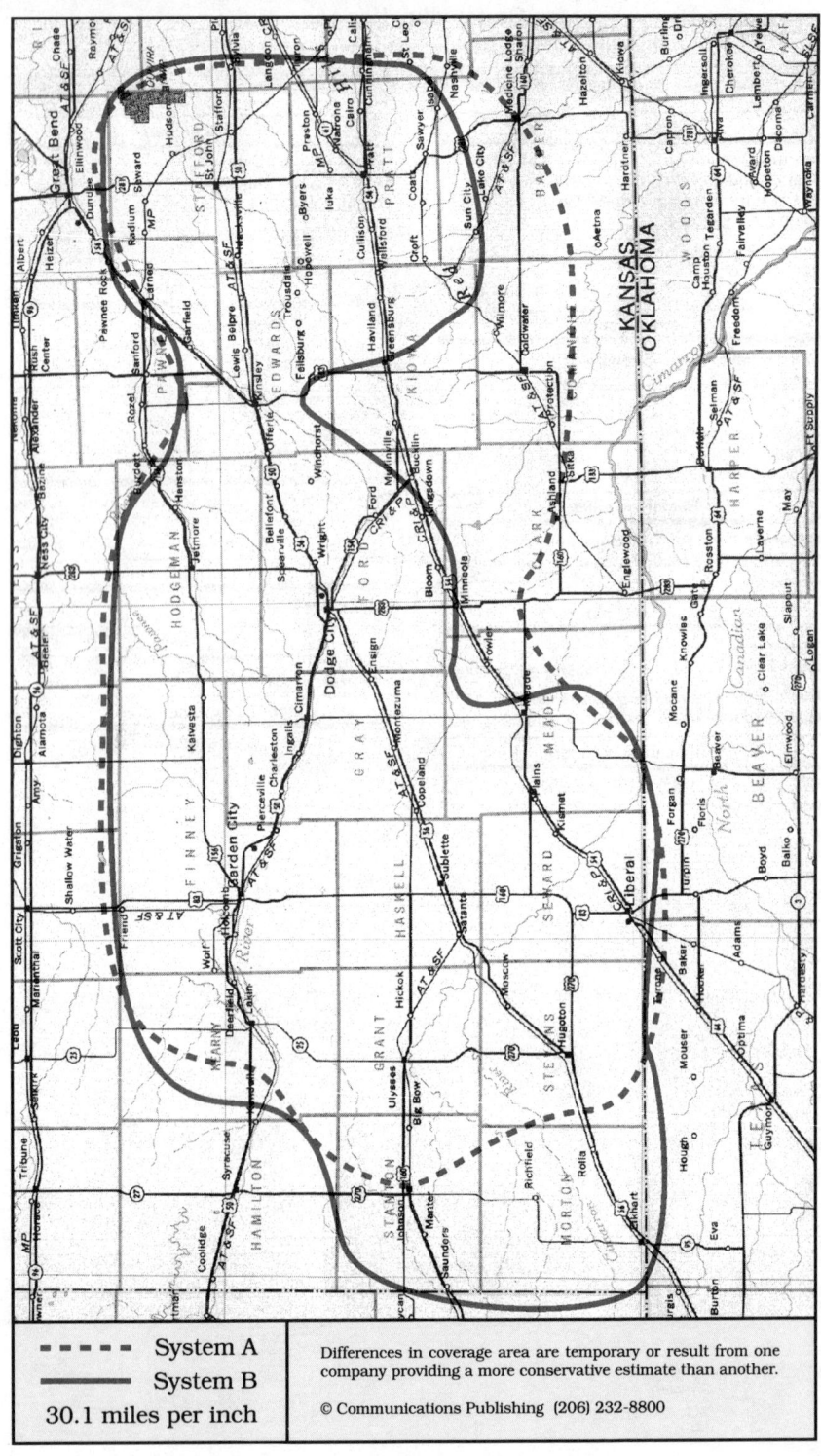

- - - - System A
───── System B
30.1 miles per inch

Differences in coverage area are temporary or result from one
company providing a more conservative estimate than another.

© Communications Publishing (206) 232-8800

Dodge City, Kansas
Dodge City, Garden City, Pratt

System A	System B

Cellular company

Cellular One
Miscellco Communications
2000 Vine
Hays, KS 67601

Customer service	800-880-2355
Local office	913-628-6111
Corporate office	601-948-1212

Cellular company

Kansas Cellular
Kansas Independent Cellular Networks
621 Westport Boulevard, PO Box 1607
Salina, KS 67402-1607

Customer service	800-373-5090
Corporate office	800-383-5090
. . . .	913-823-5049

Sending calls

Visitor dialing instructions

Local calls	area code + 7 digits
Long distance . . .	1 + area code + 7 digits

Speed-dial numbers

Customer service	611
Emergency	911

Sending calls

Visitor dialing instructions

Local calls	area code + 7 digits
Long distance	area code + 7 digits

Speed-dial numbers

Customer service	*611
Roaming information	*711

Receiving calls

Caller dials the roamer access number below,
hears tone, dials cellular area code and number.

. . . . 913-635-7626

NationLink & RoamingAmerica (see p. 1157)
Dial *31 send to receive calls, *30 to deactivate.
Dial *32 send to tell caller the roamer access no.
This company's subscribers have level 3 service.

Receiving calls

Caller dials the roamer access number below,
hears tone, dials cellular area code and number.

. . . . 913-658-7626

Follow Me Roaming forwarding (see p. 1156)
Activate, dial *18 send. Deactivate dial *19 send.

Billing information

Visitor rates

Most visitors	$3 day + 85¢ minute

Visitors without roaming agreements (p. 1159)
Call co. to use Amex, Discover, Mastercard, Visa

Billing information

Visitor rates

Most visitors	$2 day + 99¢ minute

Visitors without roaming agreements (p. 1159)
Roamer Plus will intercept first call to bill you.

Identification numbers

City	Market	System	Billing	RSA
Dodge City	439	01258	00000	KS-12
Garden City	438	01258	00000	KS-11
Pratt	440	01258	00000	KS-13

For full coverage area, see Hays, KS and
Hutchinson, KS and Salina, KS.

Identification numbers

City	Market	System	Billing	RSA
Dodge City	439	01255	00000	KS-12
Garden City	438	01255	00000	KS-11
Pratt	440	01255	00000	KS-13

For full coverage area, see Hays, KS.

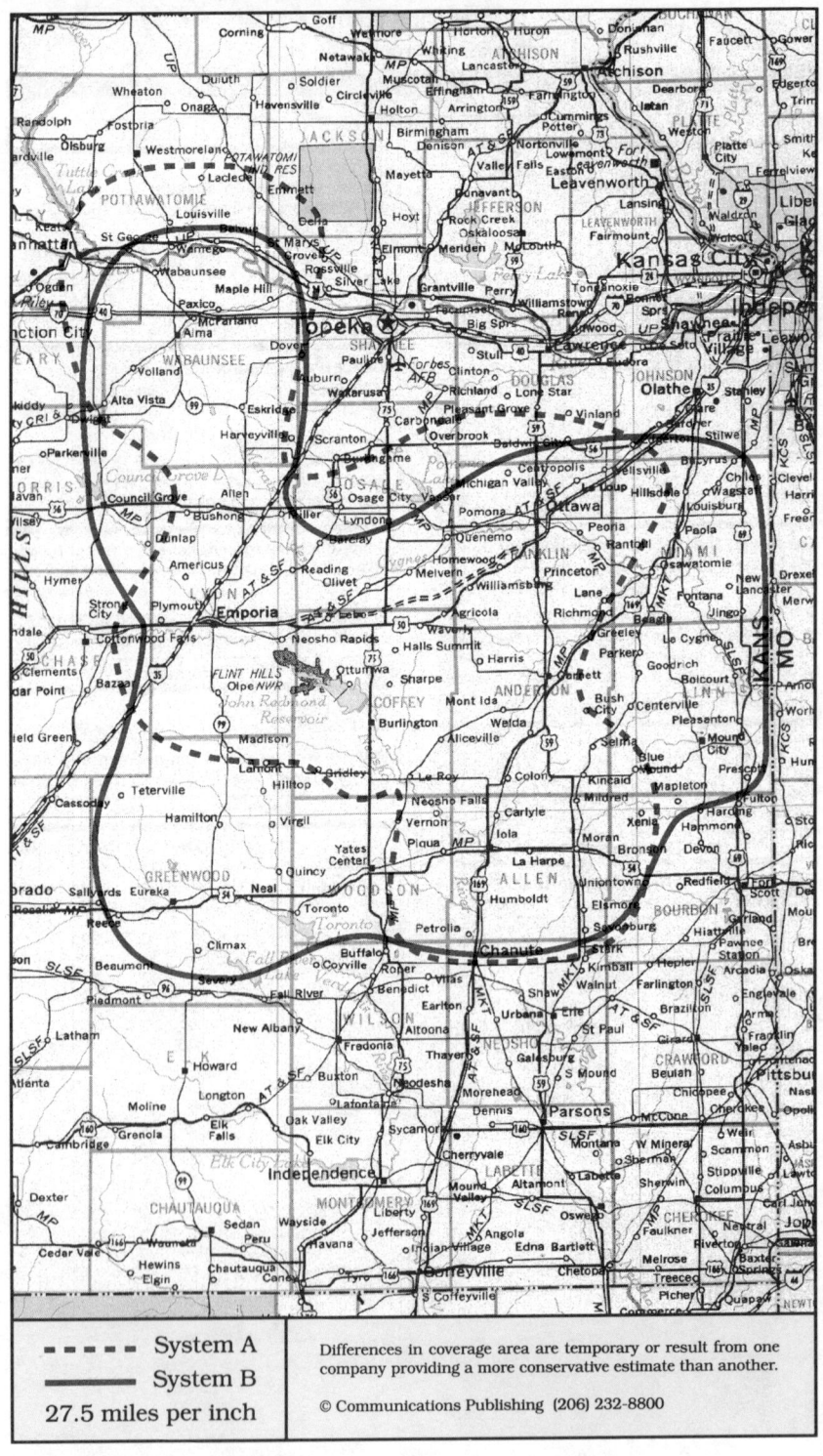

System A
System B
27.5 miles per inch

Differences in coverage area are temporary or result from one company providing a more conservative estimate than another.

© Communications Publishing (206) 232-8800

Emporia, Kansas

Alma, Burlington, Emporia, Eureka, Iola, Mound City, Ottawa

Cellular company

Cellular One of the Plains
Sterling Cellular Management
1080 Holcomb Bridge Road, Bldg 100, Ste 200
Roswell, GA 30076

Customer service	800-552-6150
Corporate office	404-552-5030

Sending calls

Visitor dialing instructions

Local calls	area code + 7 digits
Long distance . . .	1 + area code + 7 digits

Speed-dial numbers

Customer service	*611
Emergency	911
Roaming information	711

Receiving calls

Caller dials the roamer access number below,
hears tone, dials cellular area code and number.
. . . . 316-363-7626

Billing information

Visitor rates
Most visitors . . . . . $3 day + 99¢ minute

Visitors without roaming agreements (p. 1159)
Roamer Plus will intercept first call to bill you.

Identification numbers

City	Market	System	Billing	RSA
Emporia	436	01411	00000	KS-9
Iola, Ottawa	437	01411	00000	KS-10

Cellular company

Kansas Cellular
Kansas Independent Cellular Networks
621 Westport Boulevard, PO Box 1607
Salina, KS 67402-1607

Customer service	800-373-5090
Corporate office	800-383-5090
. . . .	913-823-5049

Sending calls

Visitor dialing instructions

Local calls	area code + 7 digits
Long distance	area code + 7 digits

Speed-dial numbers

Customer service	*611
Roaming information	*711

Receiving calls

Caller dials the roamer access number below,
hears tone, dials cellular area code and number.
. . . . 913-658-7626

Follow Me Roaming forwarding (see p. 1156)
Activate, dial *18 send. Deactivate dial *19 send.

Billing information

Visitor rates
Most visitors . . . . . $3 day + 85¢ minute

Visitors without roaming agreements (p. 1159)
Call co. to use Amex, Discover, Mastercard, Visa

Identification numbers

City	Market	System	Billing	RSA
Emporia	436	01258	00000	KS-9
Iola, Ottawa	437	01258	00000	KS-10

For full coverage area, see Atchison, KS and
Hutchinson, KS and Salina, KS.

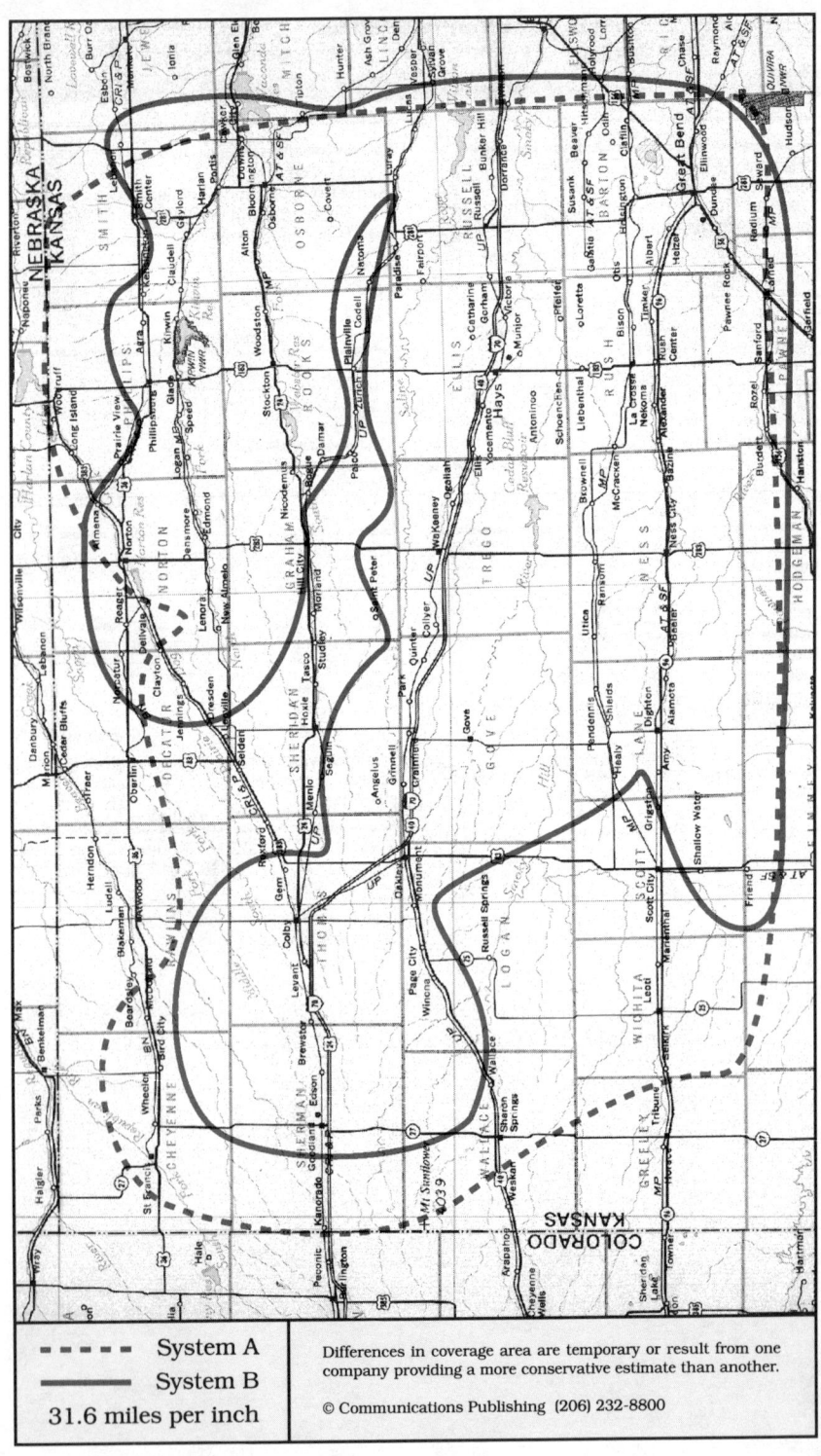

-----	System A
———	System B
31.6 miles per inch	

Differences in coverage area are temporary or result from one company providing a more conservative estimate than another.

© Communications Publishing (206) 232-8800

Hays, Kansas

Goodland, Great Bend, Hays, Oakley, Phillipsburg, Scott City

System A	System B

Cellular company

Cellular One
Miscellco Communications
2000 Vine
Hays, KS 67601

Customer service	800-880-2355
Local office	913-628-6111
Corporate office	601-948-1212

Sending calls

Visitor dialing instructions

Local calls	area code + 7 digits
Long distance . . .	1 + area code + 7 digits

Speed-dial numbers

Customer service	611
Emergency	911

Receiving calls

Caller dials the roamer access number below, hears tone, dials cellular area code and number.
. . . . 913-635-7626

NationLink & RoamingAmerica (see p. 1157)
Dial ✳31 send to receive calls, ✳30 to deactivate. Dial ✳32 send to tell caller the roamer access no. This company's subscribers have level 3 service.

Billing information

Visitor rates
Most visitors . . . . . $2 day + 99¢ minute

Visitors without roaming agreements (p. 1159)
Roamer Plus will intercept first call to bill you.

Identification numbers

City	Market	System	Billing	RSA
Goodland	428	01255	00000	KS-1
Great Bend, Hays	434	01255	00000	KS-7
Phillipsburg	429	01255	00000	KS-2
Scott City	433	01255	00000	KS-6

For full coverage area, see Dodge City, KS

Cellular company

Kansas Cellular
Kansas Independent Cellular Networks
621 Westport Boulevard, PO Box 1607
Salina, KS 67402-1607

Customer service	800-373-5090
Corporate office	800-383-5090
. . . .	913-823-5049

Sending calls

Visitor dialing instructions

Local calls	area code + 7 digits
Long distance	area code + 7 digits

Speed-dial numbers

Customer service	✳611
Roaming information	✳711

Receiving calls

Caller dials the roamer access number below, hears tone, dials cellular area code and number.
. . . . 913-658-7626

Follow Me Roaming forwarding (see p. 1156)
Activate, dial ✳18 send. Deactivate dial ✳19 send.

Billing information

Visitor rates
Most visitors . . . . . $3 day + 85¢ minute

Visitors without roaming agreements (p. 1159)
Call co. to use Amex, Discover, Mastercard, Visa

Identification numbers

City	Market	System	Billing	RSA
Goodland	428	01258	00000	KS-1
Great Bend, Hays	434	01258	00000	KS-7
Oakley	433	01258	00000	KS-6
Phillipsburg	429	01258	00000	KS-2

For full coverage area, see Dodge City, KS and Salina, KS.

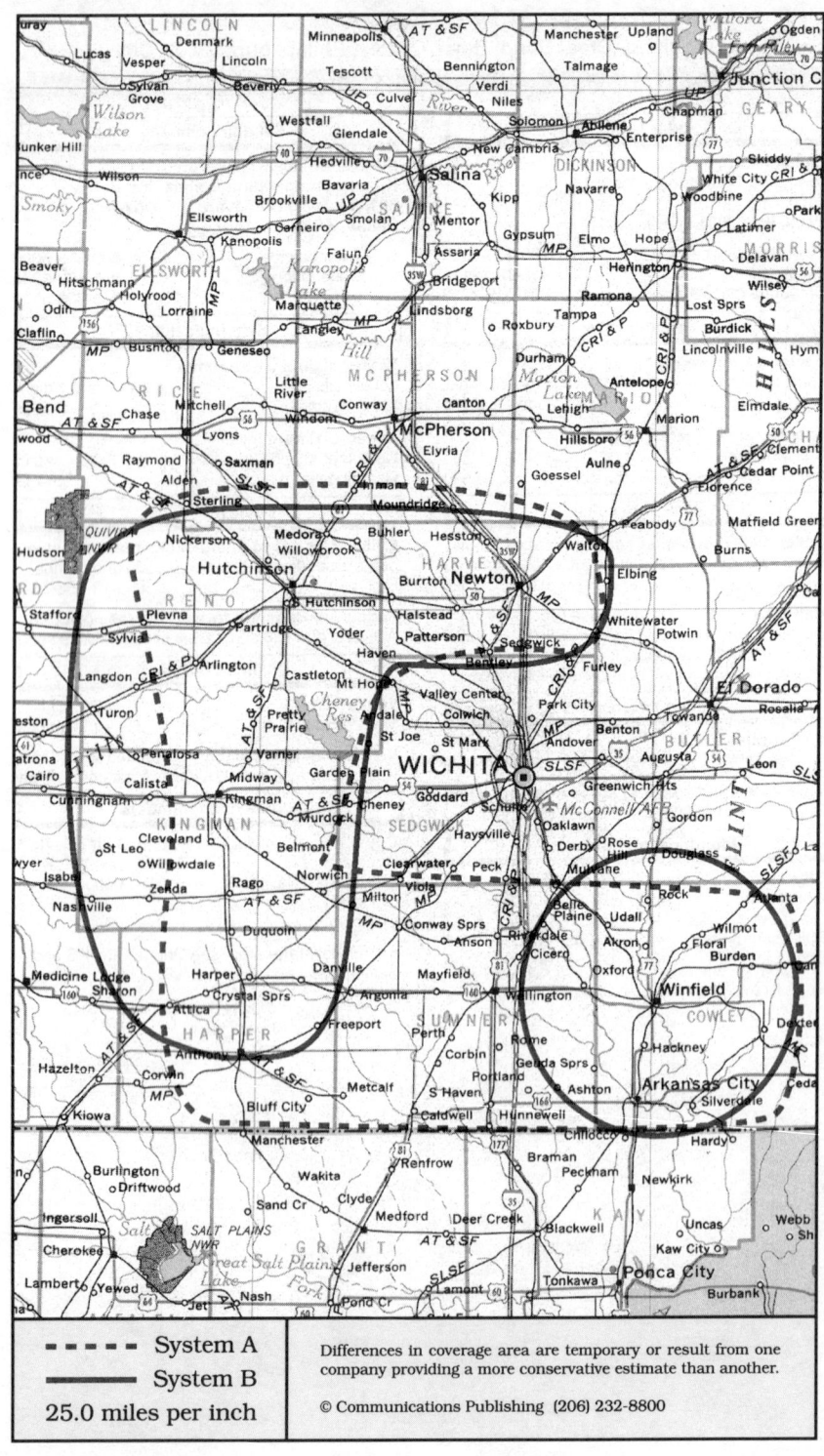

- - - - -	System A
——————	System B
25.0 miles per inch	

Differences in coverage area are temporary or result from one company providing a more conservative estimate than another.

© Communications Publishing (206) 232-8800

Hutchinson, Kansas
Anthony, Arkansas City, Hutchinson, Kingman, Wellington, Winfield

System A	System B

Cellular company

Cellular One
Bachtel Kansas 14 Limited Partnership
27 E 30th Street
Hutchinson, KS 67502

Local office	800-788-3883
. . . .	316-662-8000
Corporate office	215-972-7550

Sending calls

Visitor dialing instructions

Local calls	area code + 7 digits
Long distance . . .	1 + area code + 7 digits

Speed-dial numbers

Customer service	611
Emergency	911
Highway patrol	*47
Turnpike authority	*582

Receiving calls

Caller dials the roamer access number below,
hears tone, dials cellular area code and number.
. . . . 316-648-7626

NationLink & RoamingAmerica (see p. 1157)
Dial *31 send to receive calls, *30 to deactivate.
Dial *32 send to tell caller the roamer access no.
This company's subscribers have level 5 service.

North American Cellular Network (see p. 1156)
All calls forward to you. Deactivate dial *35 send.

Billing information

Visitor rates
Most visitors . . . . . $0 day + 99¢ minute

Visitors without roaming agreements (p. 1159)
Cannot call since company takes no credit cards.

Identification numbers

City	Market	System	Billing	RSA
All served cities	441	01269	00000	KS-14

Cellular company

Kansas Cellular
Kansas Independent Cellular Networks
621 Westport Boulevard, PO Box 1607
Salina, KS 67402-1607

Customer service	800-373-5090
Corporate office	800-383-5090
. . . .	913-823-5049

Sending calls

Visitor dialing instructions

Local calls	area code + 7 digits
Long distance	area code + 7 digits

Speed-dial numbers

Customer service	*611
Roaming information	*711

Receiving calls

Caller dials the roamer access number below,
hears tone, dials cellular area code and number.
. . . . 913-658-7626

Follow Me Roaming forwarding (see p. 1156)
Activate, dial *18 send. Deactivate dial *19 send.

Billing information

Visitor rates
Most visitors . . . . . $3 day + 85¢ minute

Visitors without roaming agreements (p. 1159)
Call co. to use Amex, Discover, Mastercard, Visa

Identification numbers

City	Market	System	Billing	RSA
All served cities	441	01258	00000	KS-14

For full coverage area, see Atchison, KS and
Dodge City, KS and Emporia, KS and Salina, KS.

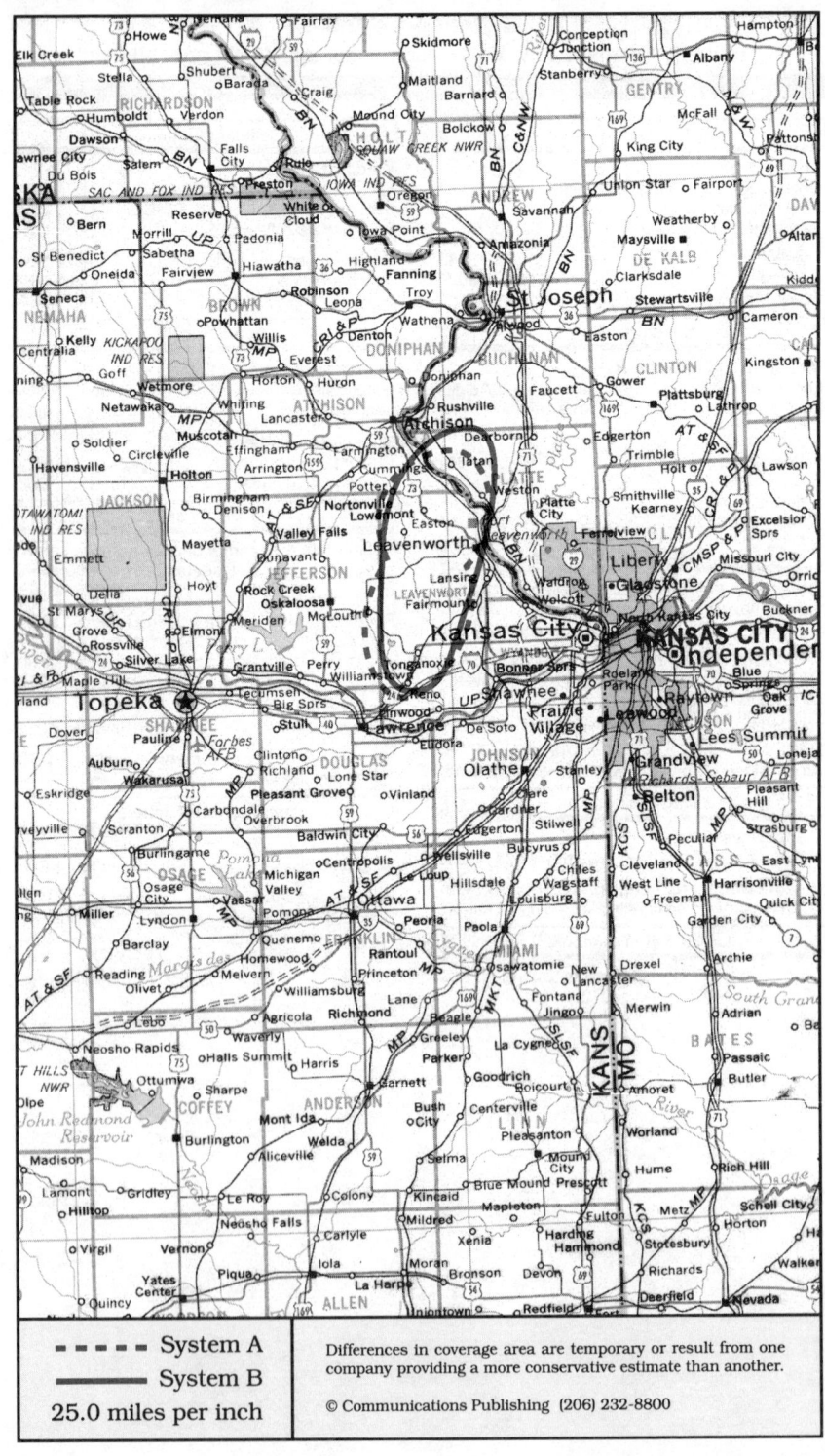

- - - - - System A
————— System B
25.0 miles per inch

Differences in coverage area are temporary or result from one company providing a more conservative estimate than another.

© Communications Publishing (206) 232-8800

Leavenworth, Kansas

System A

Cellular company

United States Cellular
309 Summit Drive
Maryville, MO 64468

Customer service   some areas   800-332-6067
Local office   816-562-3300
Corporate office   312-399-8900

Sending calls

Visitor dialing instructions
Local calls . . . . .   area code + 7 digits
Long distance . . .   1 + area code + 7 digits

Speed-dial numbers
Customer service *611
Emergency   911
Roaming information *711

Receiving calls

Caller dials the roamer access number below,
hears tone, dials cellular area code and number.
. . . .   816-787-7626

Billing information

Visitor rates
Most visitors . . . . . $3 day + 85¢ minute

Visitors without roaming agreements (p. 1159)
Call co. to use Mastercard, Visa

Identification numbers

City	Market	System	Billing	RSA
All served cities	432	01251	00000	KS-5

For full coverage area, see Atchison, KS.

System B

Cellular company

Southwestern Bell Mobile Systems
13321 W 98th Street
Lenexa, KS 66215

Customer service   some areas   800-331-0500
. . . .   913-894-1600
Local office   913-752-2300
Corporate office   214-733-2000

Sending calls

Visitor dialing instructions
Local calls . . . . . .   area code + 7 digits
Long distance . . .   1 + area code + 7 digits

Speed-dial numbers
Customer service   611
Emergency   911
Kansas Highway Patrol . . . . *HP or *47
Kansas Turnpike Authority . . *KTA or *582

Receiving calls

Caller dials the roamer access number below,
hears tone, dials cellular area code and number.
. . . .   816-223-7626

Follow Me Roaming forwarding (see p. 1156)
Activate, dial *18 send. Deactivate dial *19 send.

Billing information

Visitor rates
Most visitors . . . . . $3 day + 85¢ minute

Visitors without roaming agreements (p. 1159)
Call co. to use Amex, Discover, Mastercard, Visa

Identification numbers

City	Market	System	Billing	RSA
All served cities	432	00052	30462	KS-5

For full coverage area, see Kansas City, MO.

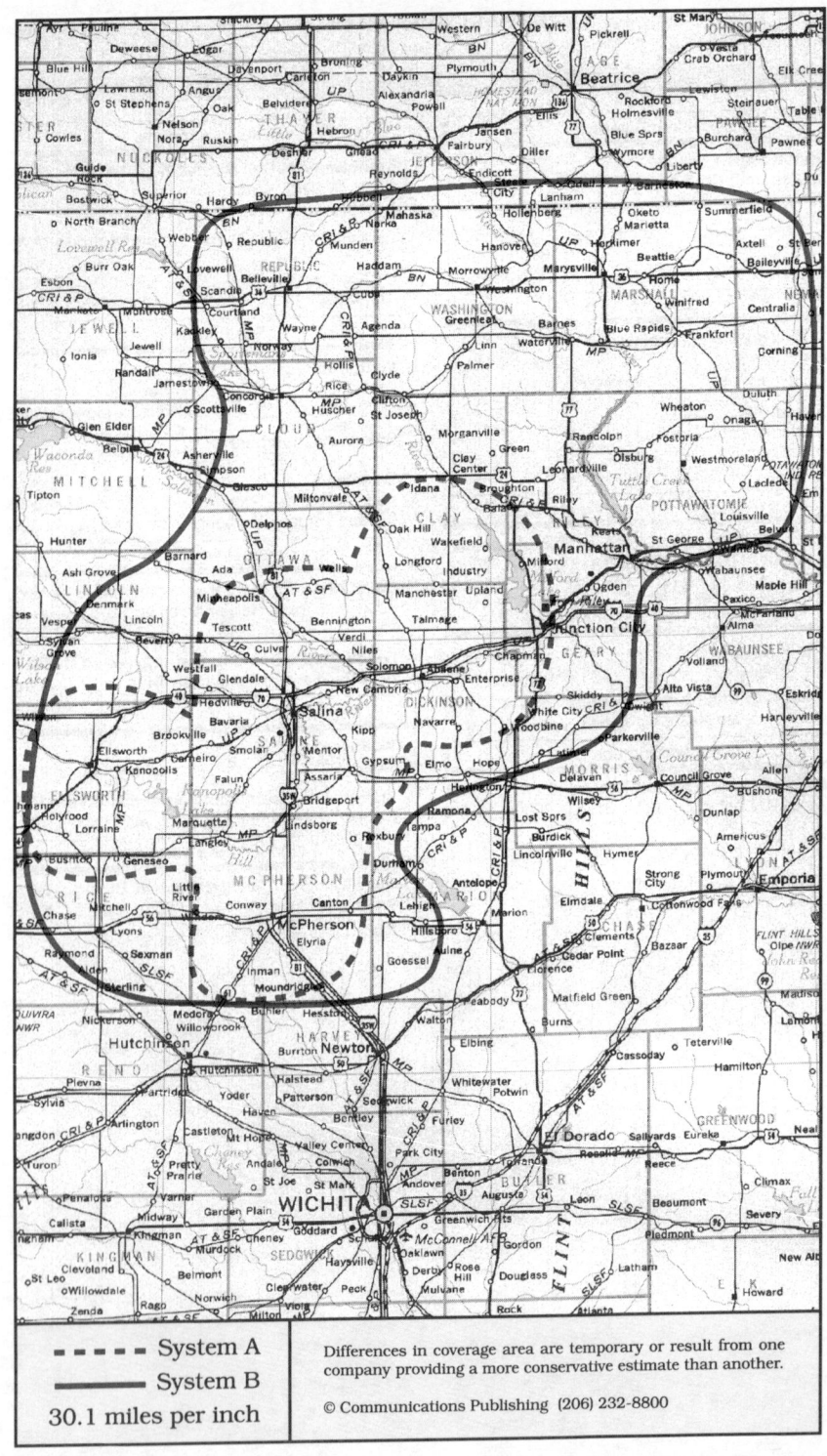

- - - - -	System A
————	System B

30.1 miles per inch

Differences in coverage area are temporary or result from one company providing a more conservative estimate than another.

© Communications Publishing (206) 232-8800

Salina, Kansas

Concordia, Junction City, Manhattan, McPherson, Salina

System A	System B

Cellular company

Cellular One
Licensed to: HBF Cellular
305 N Santa Fe
Salina, KS 67401

Corporate office 913-823-1414

Sending calls

Visitor dialing instructions
Local calls . . . . . area code + 7 digits
Long distance . . . 1 + area code + 7 digits

Speed-dial numbers
Customer service 611
Emergency 911

Receiving calls

Caller dials the roamer access number below,
hears tone, dials cellular area code and number.
. . . . 913-822-7626

NationLink & RoamingAmerica (see p. 1157)
Dial *31 send to receive calls, *30 to deactivate.
Dial *32 send to tell caller the roamer access no.
This company's subscribers have level 6 service.

Billing information

Visitor rates
Most visitors . . . . . $3 day + 85¢ minute

Visitors without roaming agreements (p. 1159)
Cannot call since company takes no credit cards.

Identification numbers

City	Market	System	Billing	RSA
All served cities	435	01257	00000	KS-8

Manhattan, KS (KS-4) will be served in 1993 by
Sterling Cellular Management at 800-552-6152.

Cellular company

Kansas Cellular
Kansas Independent Cellular Networks
621 Westport Boulevard, PO Box 1607
Salina, KS 67402-1607

Customer service 800-373-5090
Corporate office 800-383-5090
. . . . 913-823-5049

Sending calls

Visitor dialing instructions
Local calls . . . . . . area code + 7 digits
Long distance . . . . . area code + 7 digits

Speed-dial numbers
Customer service *611
Roaming information *711

Receiving calls

Caller dials the roamer access number below,
hears tone, dials cellular area code and number.
. . . . 913-658-7626

Follow Me Roaming forwarding (see p. 1156)
Activate, dial *18 send. Deactivate dial *19 send.

Billing information

Visitor rates
Most visitors . . . . . $3 day + 85¢ minute

Visitors without roaming agreements (p. 1159)
Call co. to use Amex, Discover, Mastercard, Visa

Identification numbers

City	Market	System	Billing	RSA
Concordia	430	01258	00000	KS-3
Manhattan	431	01258	00000	KS-4
Salina	435	01258	00000	KS-8

For full coverage area, see Atchison, KS and
Dodge City, KS and Emporia, KS and Hayes, KS
and Hutchinson, KS.

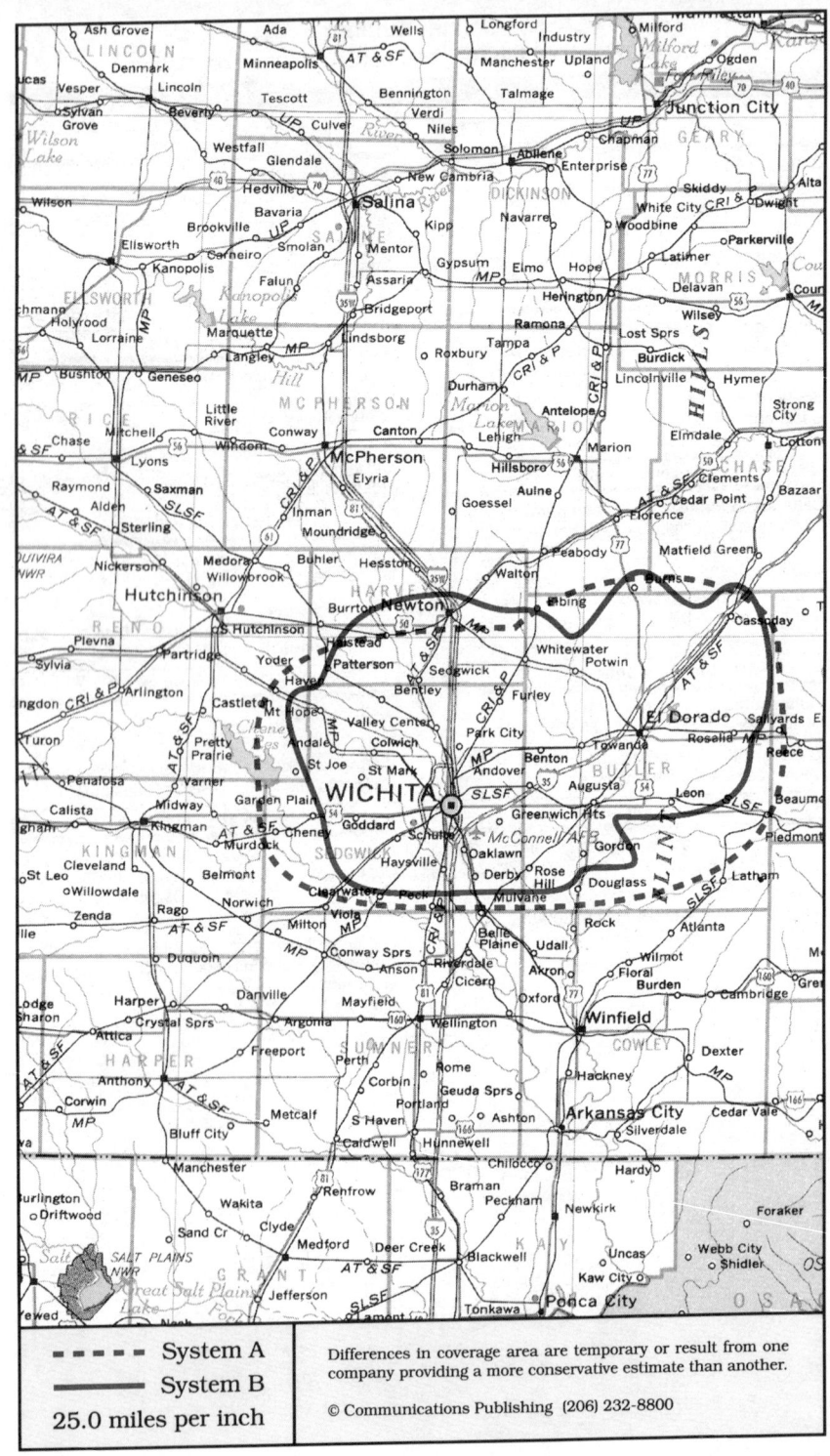

	System A
----	System B

25.0 miles per inch

Differences in coverage area are temporary or result from one company providing a more conservative estimate than another.

© Communications Publishing (206) 232-8800

Wichita, Kansas
El Dorado, Wichita

System A	System B

Cellular company

Cellular One
McCaw Cellular Communications
515 N Woodlawn
Wichita, KS 67208

Customer service	800-467-5115
. . . .	316-686-8811
Local office	316-686-2355
Corporate office	206-827-4500

Sending calls

Visitor dialing instructions

Local calls	area code + 7 digits
Long distance . . .	1 + area code + 7 digits

Speed-dial numbers

B98	*98
Customer service	611
Emergency	911
KAKE television	*10
Kansas highway patrol	*47
KFDI	*1070
Turnpike authority	*582

Receiving calls

Caller dials the roamer access number below,
hears tone, dials cellular area code and number.
. . . . 316-648-7626

North American Cellular Network (see p. 1156)
All calls forward to you. Deactivate dial *35 send.

Billing information

Visitor rates

Most visitors	$0 day + 99¢ minute

Visitors without roaming agreements (p. 1159)
Roamer Plus will intercept first call to bill you.

Identification numbers

City	Market	System	Billing	RSA
All served cities	089	00165	00000	MSA

Cellular company

Southwestern Bell Mobile Systems
3161 N Rock Road
Wichita, KS 67226-1312

Customer service	800-331-0500
. . . .	316-636-1041
Local office	316-687-2355
Corporate office	214-733-2000

Sending calls

Visitor dialing instructions

Local calls	area code + 7 digits
Long distance . . .	1 + area code + 7 digits

Speed-dial numbers

Customer service	611
Emergency	911
KNSS	*1240
KSNW	*3

Receiving calls

Caller dials the roamer access number below,
hears tone, dials cellular area code and number.
. . . . 316-651-7626

Follow Me Roaming forwarding (see p. 1156)
Activate, dial *18 send. Deactivate dial *19 send.

Billing information

Visitor rates

Most visitors	$3 day + 85¢ minute
S.B.M.S. subscribers .	$0 day + 36¢ minute

Visitors without roaming agreements (p. 1159)
Call co. to use Amex, Mastercard, Visa

Identification numbers

City	Market	System	Billing	RSA
All served cities	089	00070	00000	MSA

There is a charge for calling directory assistance.
Operator cannot bill calls to a mobile number.

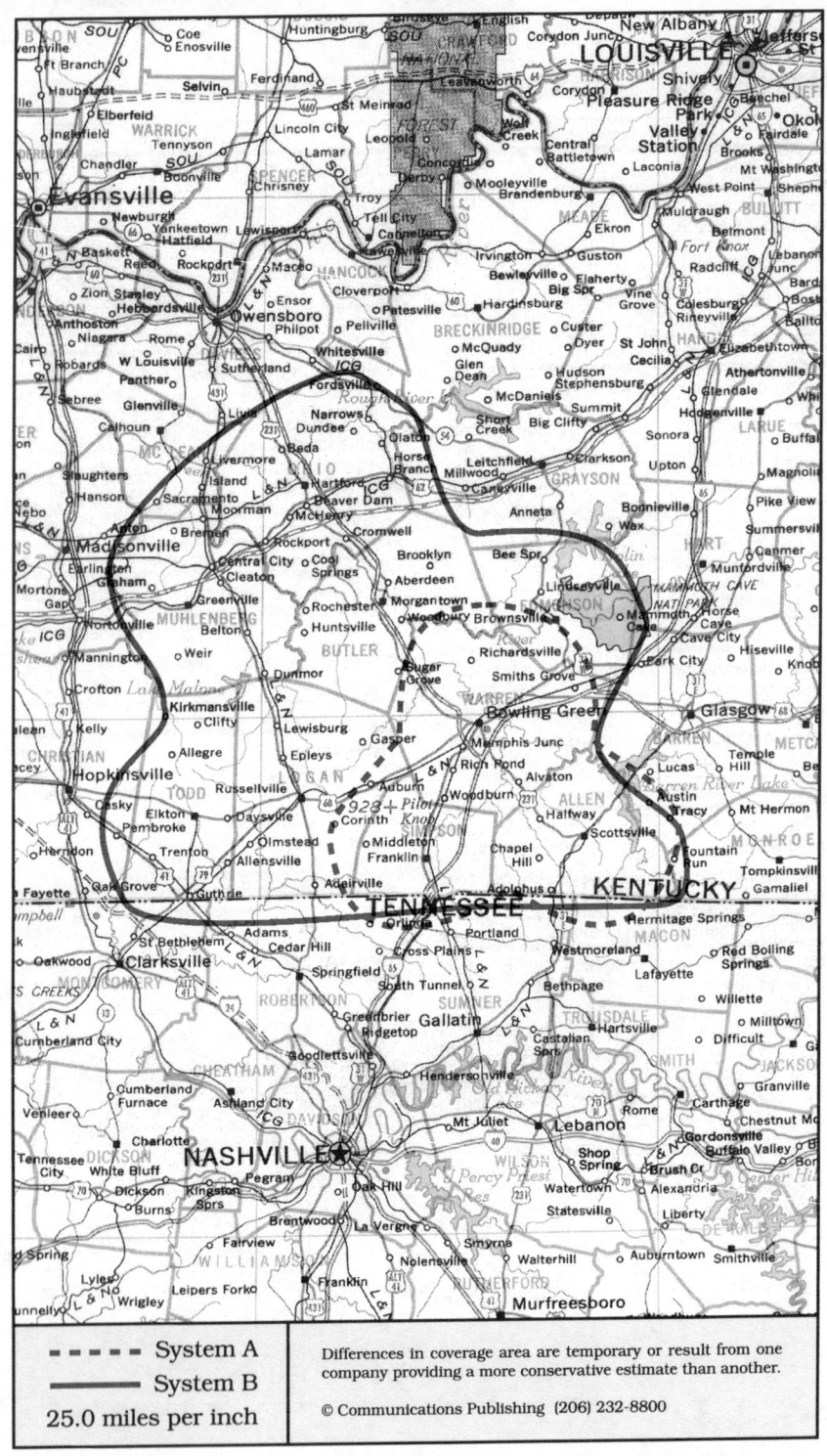

Bowling Green, Kentucky

Beaver Dam, Bowling Green, Central City, Franklin, Russellville, Scottsville

System A	System B

System A

Cellular company

Cellular One
Tsaconas Cellular
1736 U.S. 31 W Bypass
Bowling Green, KY 42101

Corporate office	800-395-8722
. . . .	502-842-5551

Sending calls

Visitor dialing instructions

Local calls	area code + 7 digits
Long distance . . .	1 + area code + 7 digits

Speed-dial numbers

Customer service	611
Emergency	911
WSBLG 107 radio	*107

Receiving calls

Caller dials the roamer access number below,
hears tone, dials cellular area code and number.
. . . . 502-792-7626

Billing information

Visitor rates
Most visitors . . . . . $2 day + 85¢ minute

Visitors without roaming agreements (p. 1159)
Cannot call since company takes no credit cards.

Identification numbers

City	Market	System	Billing	RSA
All served cities	445	01277	00000	KY-3

System B

Cellular company

Bluegrass Cellular
<u>Administration</u>
615 N Mulberry, Suite 204
Elizabethtown, KY 42701

<u>Sales office</u>
1945 Scottsville Road, Suite A3A
Bowling Green, KY 42104-5825

Customer service	in Kentucky	800-321-0310
. . . .		502-737-1876
Local office . .	Bowling Green	502-781-8999
Corporate office		502-769-0339

Sending calls

Visitor dialing instructions

Local calls	area code + 7 digits
Long distance . . .	1 + area code + 7 digits

Speed-dial numbers

Customer service	*611
Emergency	911

Receiving calls

Caller dials the roamer access number below,
hears tone, dials cellular area code and number.
. . . . 502-352-7626

Follow Me Roaming forwarding (see p. 1156)
Activate, dial *18 send. Deactivate dial *19 send.

Billing information

Visitor rates
Most visitors . . . . . $3 day + 85¢ minute

Visitors without roaming agreements (p. 1159)
Cannot call since company takes no credit cards.

Identification numbers

City	Market	System	Billing	RSA
All served cities	445	call co.	call co.	KY-3

This company asks that you call for ID numbers.

For full coverage area, see Elizabethtown, KY.

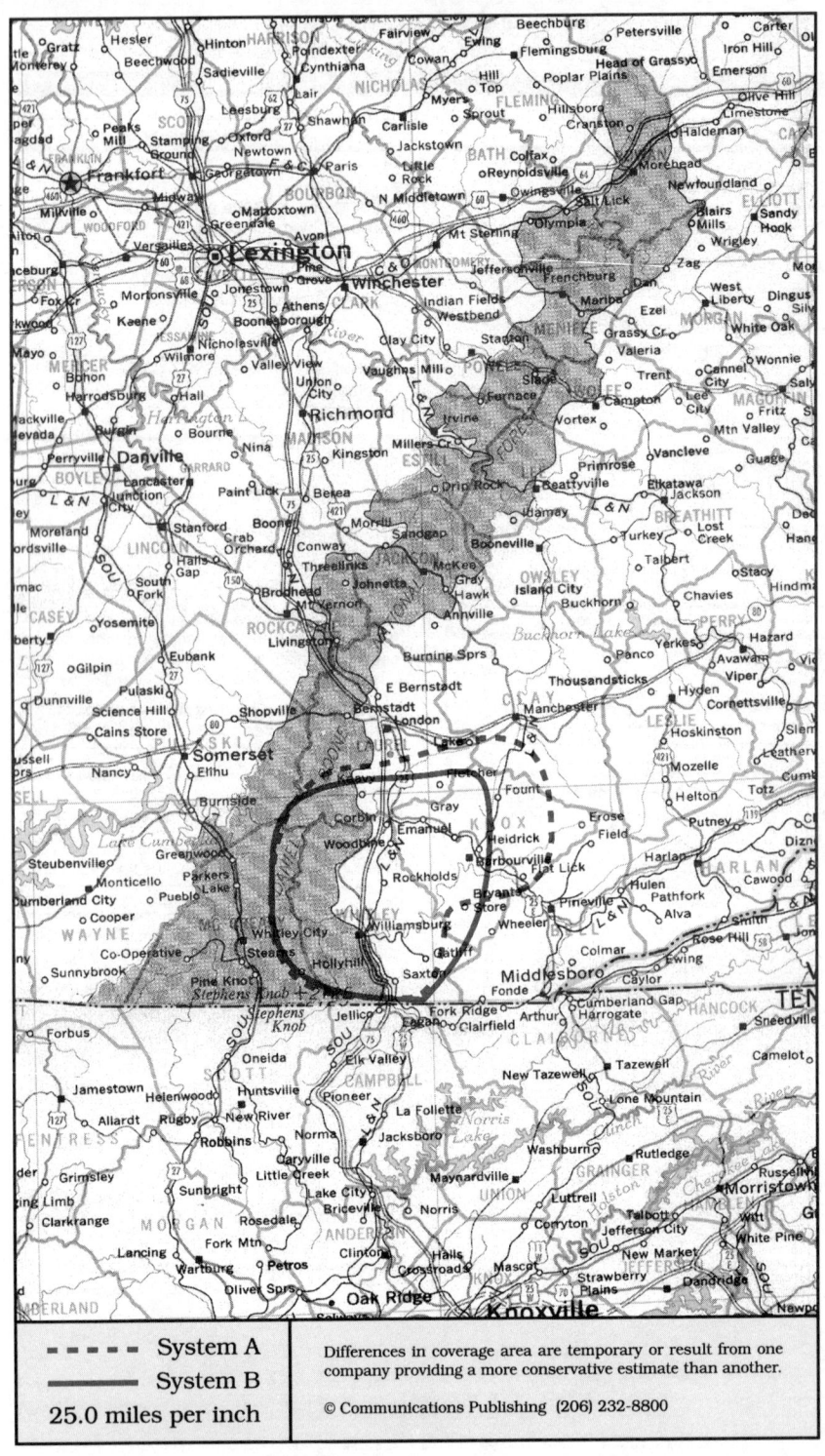

▪▪▪▪ System A	Differences in coverage area are temporary or result from one company providing a more conservative estimate than another.
▬▬▬ System B	
25.0 miles per inch	© Communications Publishing (206) 232-8800

Corbin, Kentucky
Barbourville, Corbin, Williamsburg

System A	System B

System A

Cellular company

First Kentucky Cellular
785 Cumberland Gap Parkway
Corbin, KY 40701

Corporate office 606-523-1888

Sending calls

Visitor dialing instructions
Local calls . . . . 7 digits
Long distance . . . 1 + area code + 7 digits

Speed-dial numbers
Customer service 611
Emergency 911
WCTT *107

Receiving calls

Caller dials the roamer access number below,
hears tone, dials cellular area code and number.
. . . . 606-524-7626

NationLink & RoamingAmerica (see p. 1157)
Caller hears your current roamer access number.
This company's subscribers have level 0 service.

Billing information

Visitor rates
Most visitors . . . . . $3 day + 99¢ minute

Visitors without roaming agreements (p. 1159)
Call co. to use Amex, Discover, Mastercard, Visa
or Roamer Plus will intercept first call to bill you.

Identification numbers

City	Market	System	Billing	RSA
All served cities	453	01293	00000	KY-11

System B

Cellular company

Contel Cellular
6513 Kingston Pike
Knoxville, TN 37919

Customer service 800-333-4004
. . . . 615-269-2273
Local office 615-584-2355
Corporate office 404-804-3400

Sending calls

Visitor dialing instructions
Local calls . . . . . . area code + 7 digits
Long distance . . . 1 + area code + 7 digits

Speed-dial numbers
Customer service 611
Emergency 911

Receiving calls

Caller dials the roamer access number below,
hears tone, dials cellular area code and number.
. . . . 606-549-7626

Billing information

Visitor rates
Most visitors . . . . . $3 day + 85¢ minute

Visitors without roaming agreements (p. 1159)
Call co. to use Amex, Mastercard, Visa

Identification numbers

City	Market	System	Billing	RSA
All served cities	453	01294	00000	KY-11

- - - - - System A
———— System B
25.0 miles per inch

Elizabethtown, Kentucky

Campbellsville, Elizabethtown, Glasgow, Harrodsburg, Lebanon, Radcliff

System A	System B

Cellular company

Cellular One
United Bluegrass Cellular
Danbury Cellular Telephone
111-A Reynolds Road
Glasgow, KY 42141

Sale to Horizon Cellular is pending.

Customer service	800-695-2835
. . . .	606-624-8484
Local office	502-678-2355
Corporate office	716-987-3000

Sending calls

Visitor dialing instructions

Local calls	area code + 7 digits
Long distance . . .	1 + area code + 7 digits

Speed-dial numbers

Customer service	611
Emergency	911

Receiving calls

Caller dials the roamer access number below,
hears tone, dials cellular area code and number.
. . . . 606-544-7626

NationLink & RoamingAmerica (see p. 1157)
Caller hears your current roamer access number.
This company's subscribers have level 0 service.

Billing information

Visitor rates
Most visitors . . . . . $3 day + 75¢ minute

Visitors without roaming agreements (p. 1159)
Call co. to use Mastercard, Visa
or Roamer Plus will intercept first call to bill you.

Identification numbers

City	Market	System	Billing	RSA
All served cities	447	01281	00000	KY-5

KY-4 licensed to Horizon Cellular 215-651-5900.

Cellular company

Bluegrass Cellular
Administration
615 N Mulberry, Suite 204
Elizabethtown, KY 42701

Sales offices
339 Campbellsville Bypass, Suite 132
Campbellsville, KY 42718

115 Williams
Elizabethtown, KY 42701

Customer service	in Kentucky	800-321-0310
. . . .		502-737-1876
Local office . .	Campbellsville	502-465-7005
. . . .	Elizabethtown sales	800-928-2355
. . . .	Elizabethtown sales	502-769-2731
Corporate office	administration	502-769-0339

Sending calls

Visitor dialing instructions

Local calls	area code + 7 digits
Long distance . . .	1 + area code + 7 digits

Speed-dial numbers

Customer service	*611
Emergency	911

Receiving calls

Caller dials the roamer access number below,
hears tone, dials cellular area code and number.
. . . . 502-352-7626

Follow Me Roaming forwarding (see p. 1156)
Activate, dial *18 send. Deactivate dial *19 send.

Billing information

Visitor rates
Most visitors . . . . . $3 day + 85¢ minute

Visitors without roaming agreements (p. 1159)
Cannot call since company takes no credit cards.

Identification numbers

City	Market	System	Billing	RSA
Elizabethtown	446	call co.	call co.	KY-4
Glasgow	447	call co.	call co.	KY-5

This company asks that you call for ID numbers.

For full coverage area, see Bowling Green, KY.

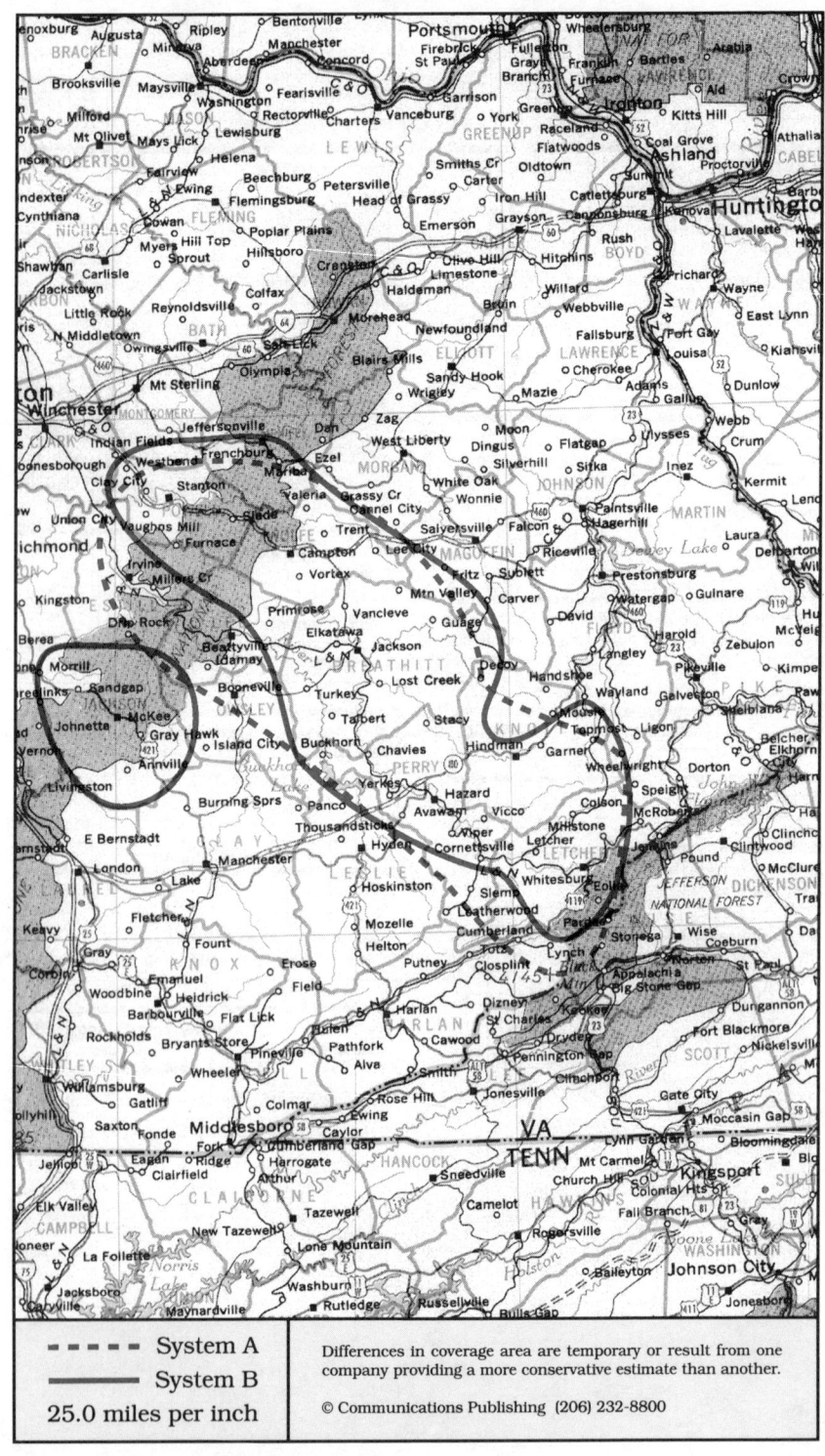

- - - - - System A	Differences in coverage area are temporary or result from one
———— System B	company providing a more conservative estimate than another.
25.0 miles per inch	© Communications Publishing (206) 232-8800

Hazard, Kentucky

Boonesville, Campton, Hazard, Jackson, Stanton, Whitesburg

System A	System B

Cellular company

Cellular One
Alpha Cellular Telephone
156 Weddington Branch Road
Pikeville, KY 41501

Corporate office 800-928-1111
. . . . 606-432-1111

Sending calls

Visitor dialing instructions
Local calls . . . . 7 digits
Long distance . . . 1 + area code + 7 digits

Speed-dial numbers
Customer service . . . . . . 611 or 711
Emergency in Pike County 911
WYMT television *57

Receiving calls

Caller dials the roamer access number below,
hears tone, dials cellular area code and number.
. . . . 606-434-7626

NationLink & RoamingAmerica (see p. 1157)
Dial *31 send to receive calls, *30 to deactivate.
Dial *32 send to tell caller the roamer access no.
This company's subscribers have level 3 service.

Billing information

Visitor rates
Most visitors . . . . . . $3 day + $1 minute

Visitors without roaming agreements (p. 1159)
Call co. to use Mastercard, Visa
or Roamer Plus will intercept first call to bill you.

Identification numbers

City	Market	System	Billing	RSA
All served cities	452	01289	00000	KY-10

For full coverage area, see Pikeville, KY.

Cellular company

Kentucky Cellular
Mountaineer Cellular General Partnership
Highway 550, PO Box 1148
Hindman, KY 41822

Corporate office 800-438-2355
. . . . 606-785-9550

Sending calls

Visitor dialing instructions
Local calls . . . . 7 digits
Long distance . . . 1 + area code + 7 digits

Speed-dial numbers
Customer service 611
Emergency 911
WKCB *107
WSKV *104

Receiving calls

Caller dials the roamer access number below,
hears tone, dials cellular area code and number.
. . . . 606-438-7626

Billing information

Visitor rates
Most visitors . . . . . $2 day + 45¢ minute

Visitors without roaming agreements (p. 1159)
Cannot call since company takes no credit cards.

Identification numbers

City	Market	System	Billing	RSA
All served cities	452	01292	30472	KY-10

Calls placed by this company's subscribers in the
area served by Appalachian Cellular of Pikeville
are charged as local calls.

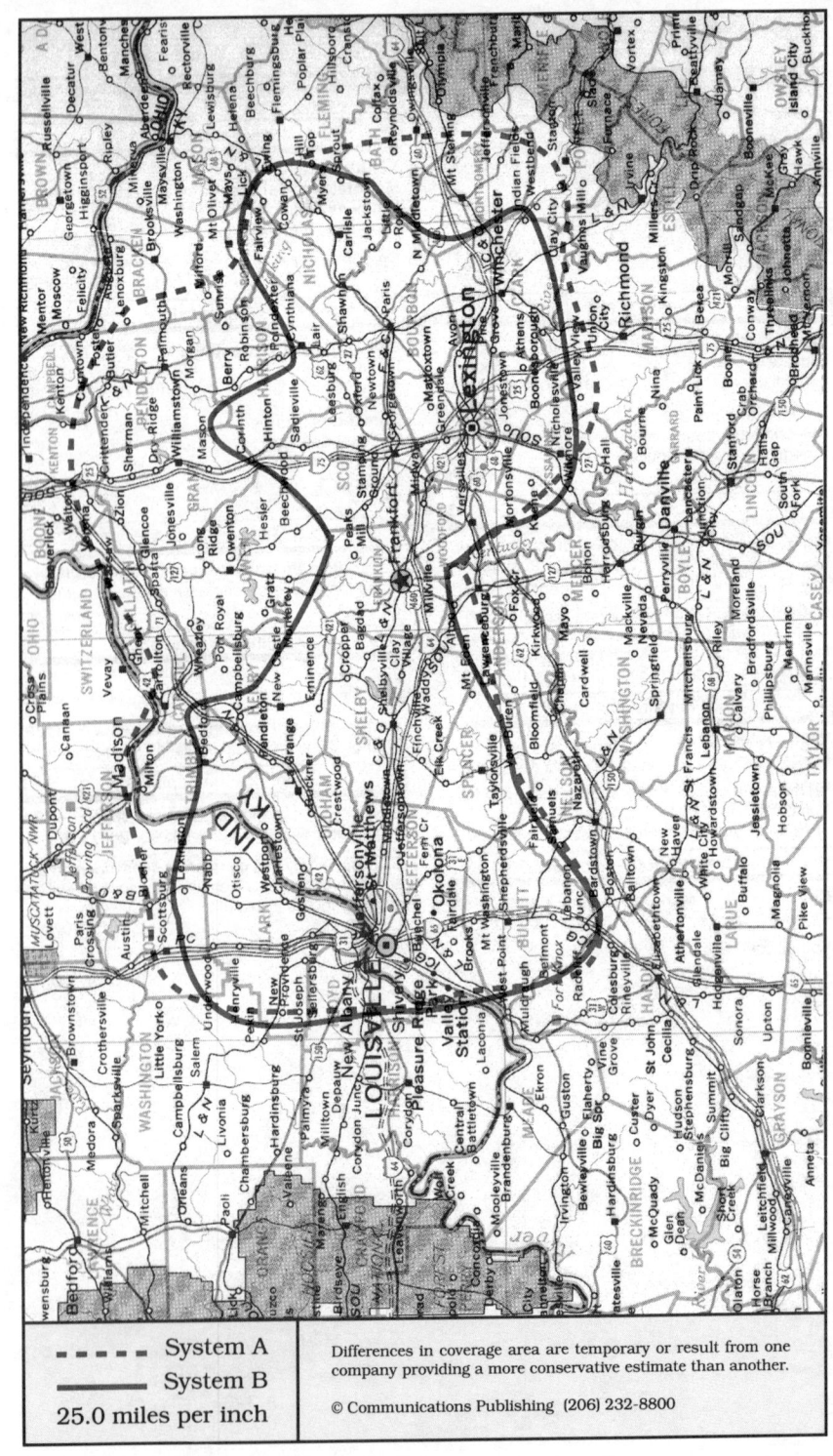

System A

System B

25.0 miles per inch

Differences in coverage area are temporary or result from one company providing a more conservative estimate than another.

© Communications Publishing (206) 232-8800

Louisville, Kentucky

Frankfort, Lexington, Louisville, New Albany IN, Owenton, Shepherdsville

System A	System B

System A

Cellular company

Cellular One
Contel Cellular
112 Mount Tabor Road, Suite 100
Lexington, KY 40517

101 S Fifth Street, Suite 2200
Louisville, KY 40202

Customer service	800-333-4004
. . . .	502-582-2273
Local office . . . Lexington	606-268-2355
. . . . . . Louisville	502-582-2355
Corporate office	404-804-3400

Sending calls

Visitor dialing instructions
Local calls . . . . . area code + 7 digits
Long distance . . . 1 + area code + 7 digits

Speed-dial numbers

Channel 27 (Lexington)	*27
Customer service	611
Emergency	911
Financial information	*5
Indiana state police	812-283-6422
Kentucky AAA	211
Make a traffic report	511
Sports	*4
State police . 502-222-0151 or 606-222-0151	
Time	*1
Weather	*2
Weekend gardener	*3
WLEX channel 18 (Lexington) . .	*18
WTVQ channel 36 (Lexington) . . .	*36
WVLK make traffic report (Lexington) . .	511

Receiving calls

Caller dials the roamer access number below,
hears tone, dials cellular area code and number.
Lexington 606-221-7626
Louisville 502-552-7626

Billing information

Visitor rates
Most visitors . . . . . $3 day + 85¢ minute

Visitors without roaming agreements (p. 1159)
Call co. to use Amex, Discover, Mastercard, Visa

Identification numbers

City	Market	System	Billing	RSA
Frankfort	449	00065	00000	KY-7
Lexington	116	00213	00000	MSA
Louisville	037	00065	00000	MSA

System B

Cellular company

BellSouth Mobility
2573 Richmond Road, Suite 360
Lexington, KY 40509

103 Whittington Parkway
Louisville, KY 40222-4980

Customer service	800-351-2400
. . . .	606-266-5153
Local office . . . . Lexington	606-268-2000
. . . . Louisville	502-329-4700
Corporate office	404-847-3600

Sending calls

Visitor dialing instructions
Local calls . . . . area code + 7 digits
Long distance . . . 1 + area code + 7 digits

Speed-dial numbers

Customer service	811
Emergency	911
Indiana state police	812-283-6422
Kentucky state police . . .	502-222-0151
Roaming information	711
Telephone problems	611

Receiving calls

Caller dials the roamer access number below,
hears tone, dials cellular area code and number.
Frankfort 502-227-0626
Lexington 606-229-7626
Louisville 502-551-7626
Shelbyville 502-633-8626

Follow Me Roaming forwarding (see p. 1156)
Activate, dial *18 send. Deactivate dial *19 send.

Billing information

Visitor rates
Most visitors . . . . . $3 day + 85¢ minute
BellSouth subscribers . $0 day + 55¢ minute

Visitors without roaming agreements (p. 1159)
Call co. to use Amex, Mastercard, Visa

Identification numbers

City	Market	System	Billing	RSA
Frankfort	449	00076	30346	KY-7
Lexington	116	00206	00000	MSA
Louisville	037	00076	00000	MSA

For full coverage area, see Richmond, KY.

| System A | - - - - - |
| System B | ——— |

25.0 miles per inch

Differences in coverage area are temporary or result from one company providing a more conservative estimate than another.

© Communications Publishing (206) 232-8800

Madisonville, Kentucky

Dixon, Kuttawa, Madisonville, Morganfield, Princeton

System A	System B

Cellular company

Cellular One
30 N Main Street
Madisonville, KY 42431

Corporate office   some areas   800-599-2355
. . . .   502-821-1111

Sending calls

Visitor dialing instructions
Local calls . . . . 1 + 7 digits
Long distance . . . 1 + area code + 7 digits

Speed-dial numbers
Customer service 611
Emergency 911

Receiving calls

Caller dials the roamer access number below,
hears tone, dials cellular area code and number.
. . . . 502-836-7626

NationLink & RoamingAmerica (see p. 1157)
Dial ✳31 send to receive calls, ✳30 to deactivate.
Dial ✳32 send to tell caller the roamer access no.
This company's subscribers have level 3 service.

Billing information

Visitor rates
Most visitors . . . . . . $3 day + $1 minute

Visitors without roaming agreements (p. 1159)
Roamer Plus will intercept first call to bill you.

Identification numbers

City	Market	System	Billing	RSA
All served cities	444	01275	00000	KY-2

Cellular company

Contel Cellular
2100 N Main Street
Madisonville, KY 42431

Customer service   800-333-4004
. . . .   502-582-2273
Local office   502-825-0350
Corporate office   404-804-3400

Sending calls

Visitor dialing instructions
Local calls . . . . . . area code + 7 digits
Long distance . . . 1 + area code + 7 digits

Speed-dial numbers
Customer service 611
Emergency 911

Receiving calls

Caller dials the roamer access number below,
hears tone, dials cellular area code and number.
. . . . 502-832-7626

Follow Me Roaming forwarding (see p. 1156)
Activate, dial ✳18 send. Deactivate dial ✳19 send.

Billing information

Visitor rates
Most visitors . . . . . $3 day + 85¢ minute

Visitors without roaming agreements (p. 1159)
Call co. to use Amex, Discover, Mastercard, Visa

Identification numbers

City	Market	System	Billing	RSA
All served cities	444	01274	00000	KY-2

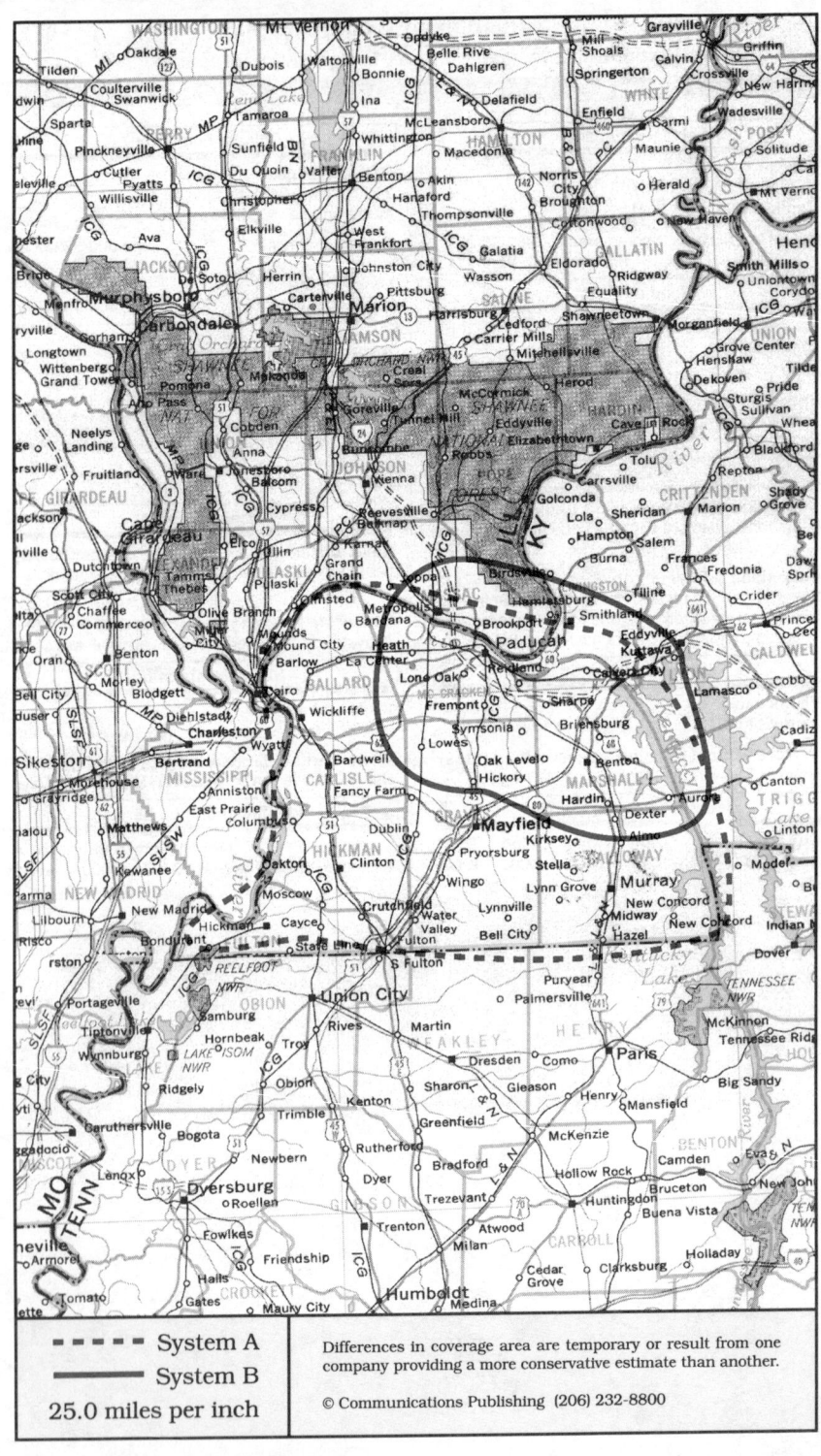

▬ ▬ ▬ ▬ System A	Differences in coverage area are temporary or result from one
▬▬▬▬ System B	company providing a more conservative estimate than another.
25.0 miles per inch	© Communications Publishing (206) 232-8800

Paducah, Kentucky
Benton, Clinton, Paducah, Mayfield, Murray, Wickliffe

System A

Cellular company

Cellular One
Licensed to: PC Cellular of Kentucky
Managed by: CompComm
1733 Kentucky Avenue
Paducah, KY 42001

Local office	800-999-2355
. . . .	502-444-0084
Corporate office CompComm	609-854-1000

Sending calls

Visitor dialing instructions
Local calls . . . . 7 digits
Long distance . . . 1 + area code + 7 digits

Speed-dial numbers
Customer service 611
Emergency 911
WKYQ *93
WKYX *57

Receiving calls

Caller dials the roamer access number below,
hears tone, dials cellular area code and number.
. . . . 502-559-7626

NationLink & RoamingAmerica (see p. 1157)
Dial *31 send to receive calls, *30 to deactivate.
This company's subscribers have level 3 service.

Billing information

Visitor rates
Most visitors . . . . . . $3 day + 85¢ minute

Visitors without roaming agreements (p. 1159)
Call co. to use Mastercard, Visa
or Roamer Plus will intercept first call to bill you.

Identification numbers

City	Market	System	Billing	RSA
All served cities	443	01273	00000	KY-1

System B

Cellular company

Contel Cellular
2320 Broadway, Suite 203
Paducah, KY 42001

Customer service	800-333-4004
. . . .	502-582-2273
Local office	502-444-9498
Corporate office	404-804-3400

Sending calls

Visitor dialing instructions
Local calls . . . . . . area code + 7 digits
Long distance . . . 1 + area code + 7 digits

Speed-dial numbers
Customer service 611
Emergency 911

Receiving calls

Caller dials the roamer access number below,
hears tone, dials cellular area code and number.
. . . . 502-832-7626

Follow Me Roaming forwarding (see p. 1156)
Activate, dial *18 send. Deactivate dial *19 send.

Billing information

Visitor rates
Most visitors . . . . . $3 day + 85¢ minute

Visitors without roaming agreements (p. 1159)
Call co. to use Amex, Mastercard, Visa

Identification numbers

City	Market	System	Billing	RSA
All served cities	443	01274	00000	KY-1

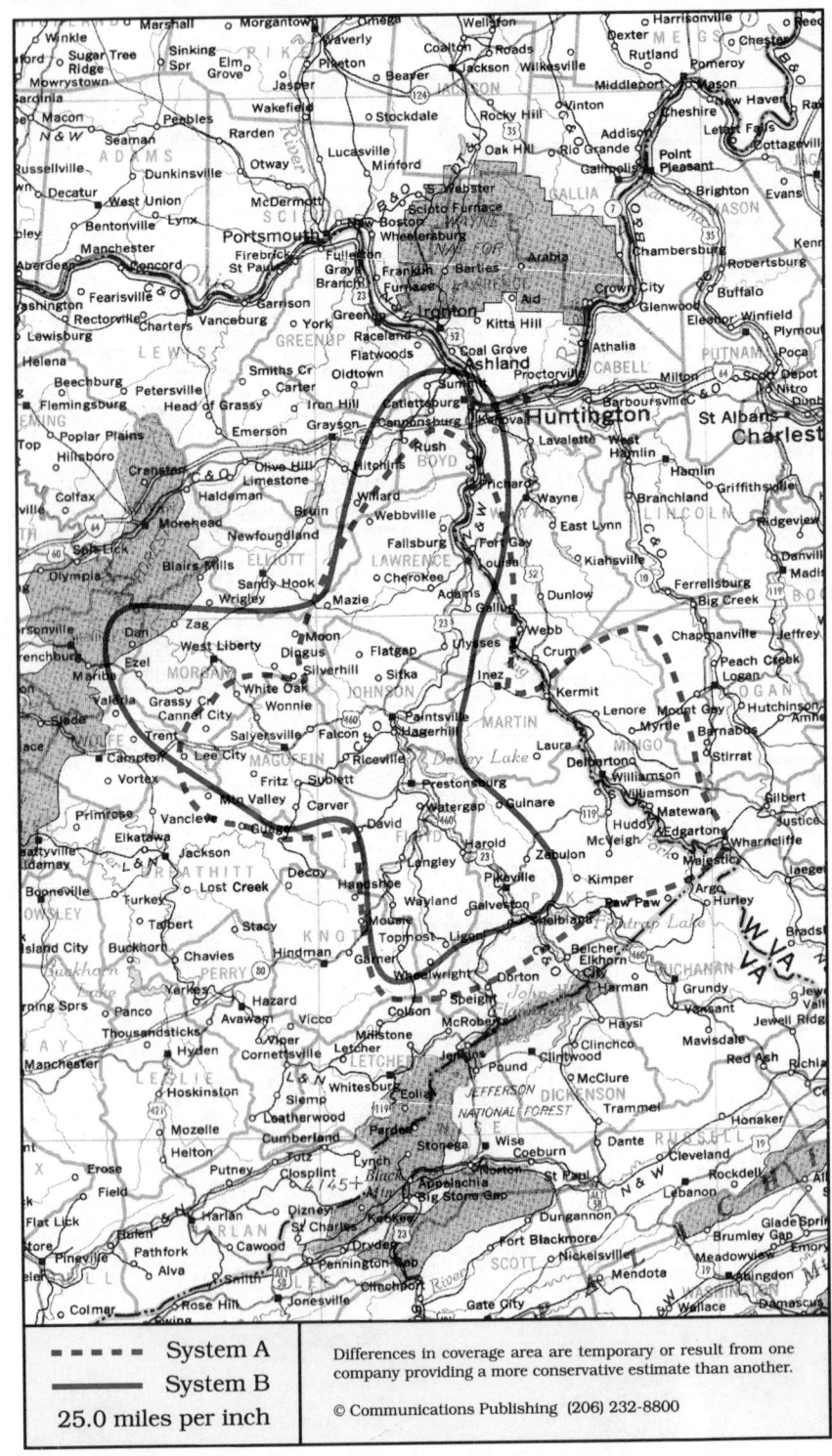

- - - -	System A
———	System B

25.0 miles per inch

Pikeville, Kentucky

Louisa, Paintsville, Pikeville, Prestonsburg, Saylersville, West Liberty

System A	System B

Cellular company

Cellular One
Alpha Cellular Telephone
156 Weddington Branch Road
Pikeville, KY 41501

Corporate office	800-928-1111
. . . .	606-432-1111

Sending calls

Visitor dialing instructions

Local calls	7 digits
Long distance . . .	1 + area code + 7 digits

Speed-dial numbers

Customer service	611 or 711
Emergency in Pike County	911
WYMT television	*57

Receiving calls

Caller dials the roamer access number below,
hears tone, dials cellular area code and number.
. . . . 606-434-7626

NationLink & RoamingAmerica (see p. 1157)
Dial *31 send to receive calls, *30 to deactivate.
This company's subscribers have level 3 service.

Billing information

Visitor rates

Most visitors	$3 day + $1 minute

Visitors without roaming agreements (p. 1159)
Call co. to use Mastercard, Visa
or Roamer Plus will intercept first call to bill you.

Identification numbers

City	Market	System	Billing	RSA
All served cities	451	01289	00000	KY-9

For full coverage area, see Hazard, KY.

Cellular company

Appalachian Cellular General Partnership
U.S. Hwy 23, One Paulsboro Row; PO Box 520
Harold, KY 41635

Corporate office	800-452-2355
. . . .	606-478-2355

Sending calls

Visitor dialing instructions

Local calls (within home area) . . .	7 digits
Local calls (outside home area) . .	1 + 7 digits
Long distance . . .	1 + area code + 7 digits

Speed-dial numbers

Customer service	*711
Emergency	911

Receiving calls

Caller dials the roamer access number below,
hears tone, dials cellular area code and number.
. . . . 606-477-7626

Billing information

Visitor rates

Most visitors	$3 day + $1 minute

Visitors without roaming agreements (p. 1159)
Cannot call since company takes no credit cards.

Identification numbers

City	Market	System	Billing	RSA
All served cities	451	1290	30470	KY-9

Calls placed by this company's subscribers in the
area served by Kentucky Cellular of Hindman are
charged as local calls.

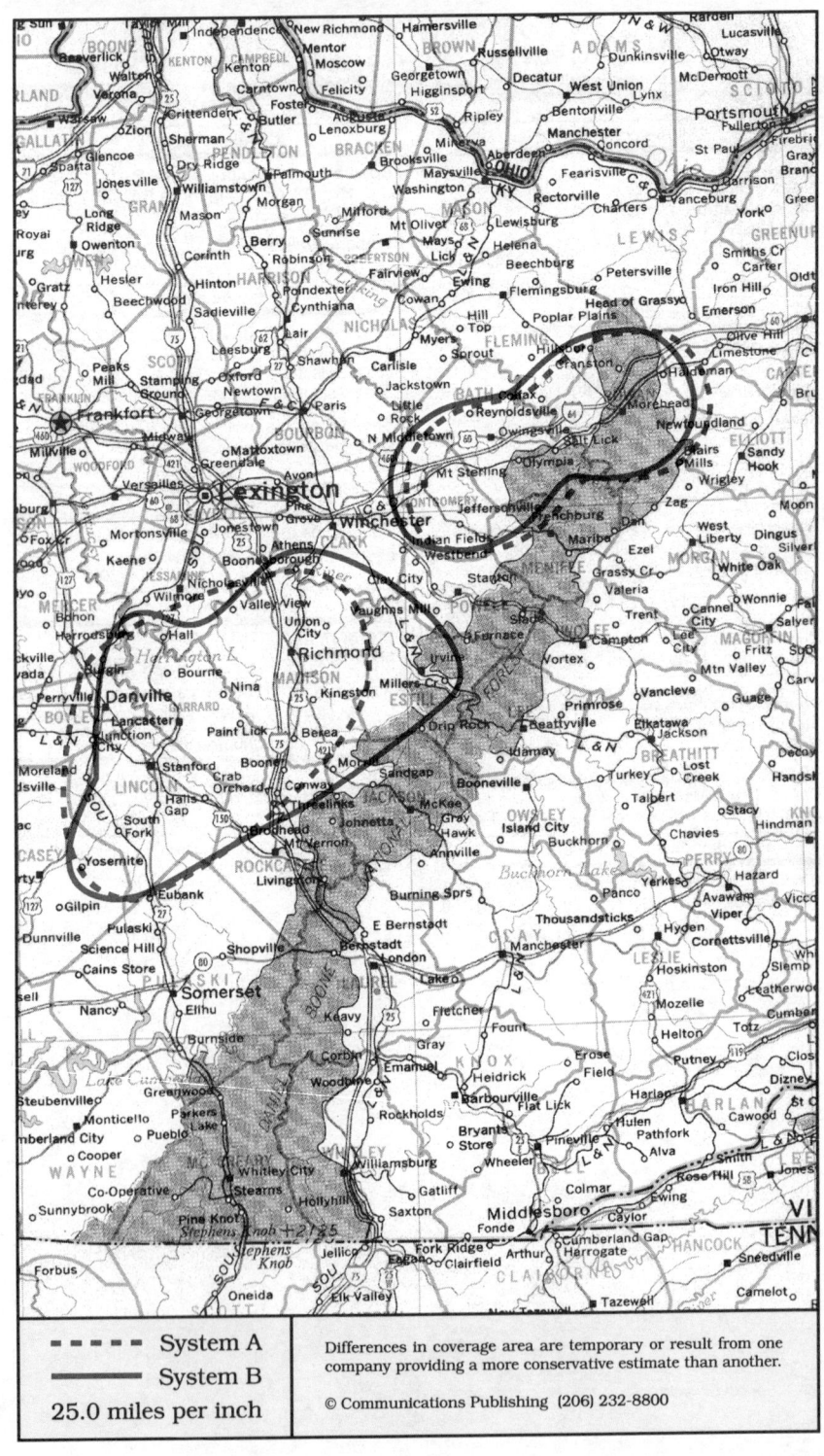

Differences in coverage area are temporary or result from one
company providing a more conservative estimate than another.

Richmond, Kentucky
Danville, Mt. Sterling, Owingsville, Richmond, Stanford

|

Cellular company

Cellular One
United Bluegrass Cellular
Danbury Cellular Telephone
124 S Keeneland Drive, Suite 3
Richmond, KY 40475

Sale to Horizon Cellular is pending.

Corporate office	some areas	800-695-2835
. . . .		606-624-8484

Sending calls

Visitor dialing instructions

Local calls	. . . . .	area code + 7 digits
Long distance	. . .	1 + area code + 7 digits

Speed-dial numbers

Customer service	611
Emergency	911

Receiving calls

Caller dials the roamer access number below,
hears tone, dials cellular area code and number.
. . . . 606-544-7626

Billing information

Visitor rates
Most visitors . . . . . $3 day + 75¢ minute

Visitors without roaming agreements (p. 1159)
Call co. to use Amex, Mastercard, Visa
or Roamer Plus will intercept first call to bill you.

Identification numbers

City	Market	System	Billing	RSA
Mt. Sterling	450	01283	00000	KY-8
Richmond	448	01283	00000	KY-6

For full coverage area, see Somerset, KY.

Cellular company

BellSouth Mobility
2573 Richmond Road, Suite 360
Lexington, KY 40509

Customer service	. . . .	800-351-2400
. . . .		606-266-5153
Local office		606-268-2000
Corporate office		404-847-3600

Sending calls

Visitor dialing instructions

Local calls	. . . . .	area code + 7 digits
Long distance	. . .	1 + area code + 7 digits

Speed-dial numbers

Customer service	811
Emergency	911
Roaming information	711
Telephone problems	611

Receiving calls

Caller dials the roamer access number below,
hears tone, dials cellular area code and number.

Mount Sterling		606-229-7626
Richmond		606-624-7626

Follow Me Roaming forwarding (see p. 1156)
Activate, dial *18 send. Deactivate dial *19 send.

Billing information

Visitor rates

Most visitors	. . .	$2.50 day + 85¢ minute
BellSouth subscribers	.	$0 day + 55¢ minute

Visitors without roaming agreements (p. 1159)
Call co. to use Amex, Mastercard, Visa
or Roamer Plus will intercept first call to bill you.

Identification numbers

City	Market	System	Billing	RSA
Mt. Sterling	450	00206	30698	KY-8
Richmond	448	00206	30334	KY-6

For full coverage area, see Lexington, KY.

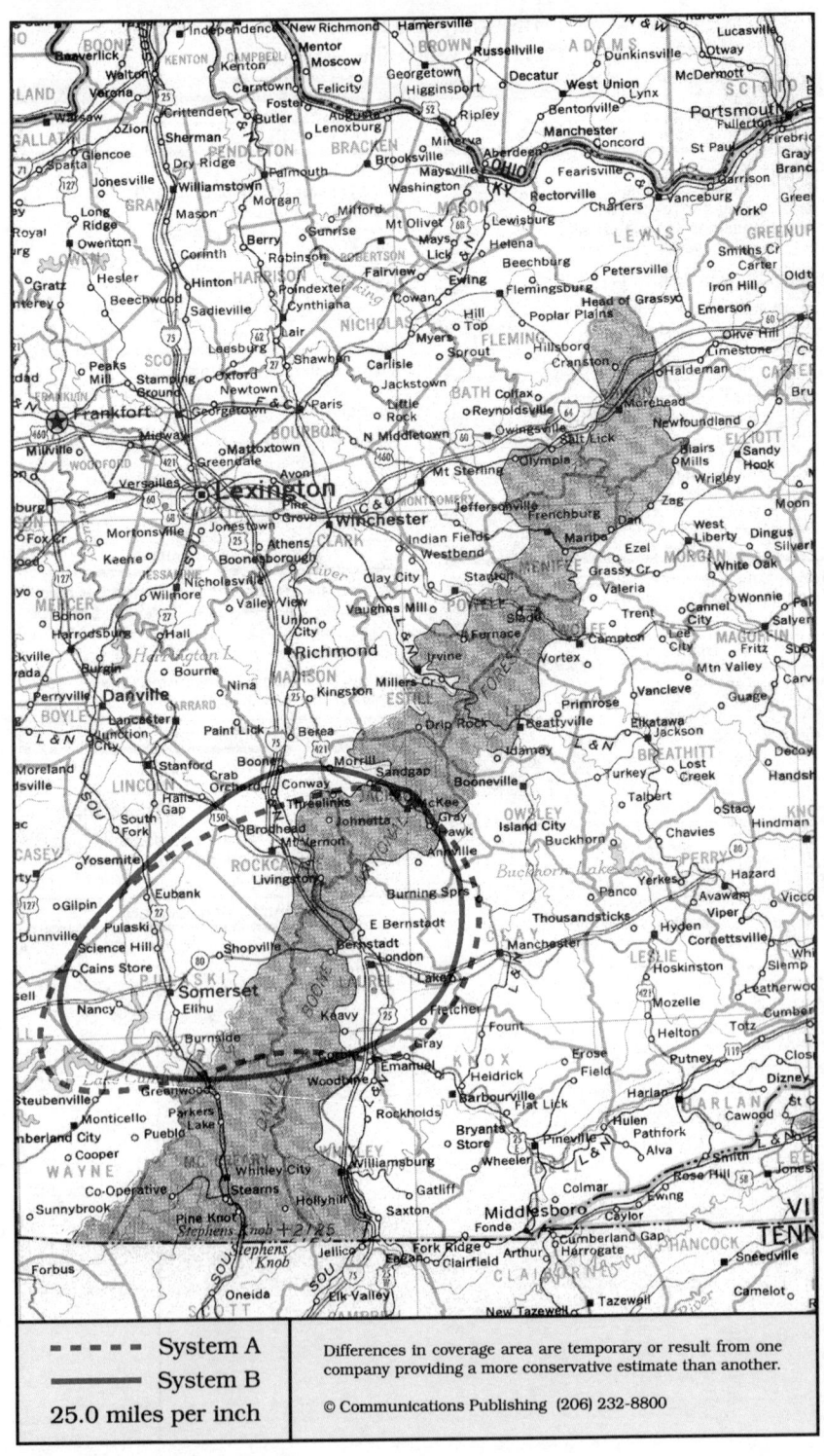

Somerset, Kentucky
London, Mt. Vernon, Somerset

Cellular company

Cellular One
United Bluegrass Cellular
Danbury Cellular Telephone
124 S Keeneland Drive, Suite 3
Richmond, KY 40475

Sale to Horizon Cellular is pending.

Corporate office   some areas 800-695-2835
. . . . 606-624-8484

Sending calls

Visitor dialing instructions
Local calls . . . . . area code + 7 digits
Long distance . . . 1 + area code + 7 digits

Speed-dial numbers
Customer service 611
Emergency 911

Receiving calls

Caller dials the roamer access number below,
hears tone, dials cellular area code and number.
. . . . 606-544-7626

Billing information

Visitor rates
Most visitors . . . . . $3 day + 75¢ minute

Visitors without roaming agreements (p. 1159)
Call co. to use Amex, Mastercard, Visa
or Roamer Plus will intercept first call to bill you.

Identification numbers

City	Market	System	Billing	RSA
All served cities	448	01283	00000	KY-6

For full coverage area, see Richmond, KY.

Cellular company

Cellular Phone of Kentucky
1535 S Main Street
London, KY 40741

Corporate office 606-878-6000

Sending calls

Visitor dialing instructions
Local calls . . . . . . area code + 7 digits
Long distance . . . 1 + area code + 7 digits

Speed-dial numbers
Customer service *611
Emergency 911
Roamers without roaming agreements . *811

Receiving calls

Caller dials the roamer access number below,
hears tone, dials cellular area code and number.
. . . . 606-877-7626

Billing information

Visitor rates
Most visitors . . . . . $3 day + 99¢ minute

Visitors without roaming agreements (p. 1159)
Call co. to use Amex, Mastercard, Visa
or Roamer Plus will intercept first call to bill you.

Identification numbers

City	Market	System	Billing	RSA
All served cities	448	01284	00000	KY-6

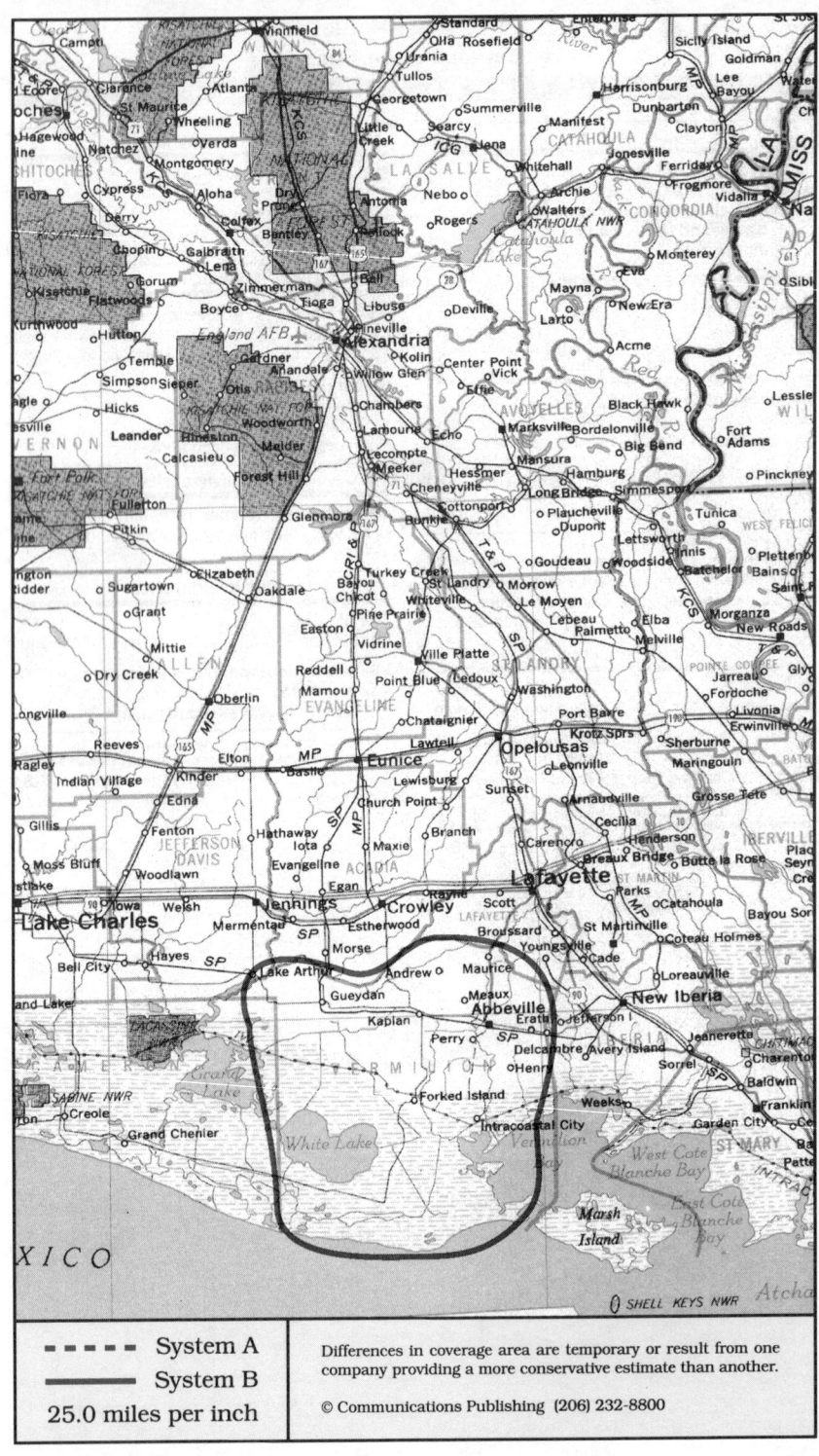

- - - - - System A
——— System B
25.0 miles per inch

Abbeville, Louisiana
Abbeville, Kaplan

System A	System B

Cellular company

No information about the company that will operate system A was available at press time.

Cellular company

Pace Communications
Kaplan Telephone
118 N Irving Avenue, PO Box 369
Kaplan, LA 70548-0369

Local office 318-643-2255
Corporate office 318-643-7171

Sending calls

Visitor dialing instructions
Local calls . . . . . . area code + 7 digits
Long distance . . . . . area code + 7 digits

Speed-dial numbers
Customer service . . . 611 or 711 or 811
Emergency 911

Receiving calls

Caller dials the roamer access number below,
hears tone, dials cellular area code and number.
. . . . 318-642-7626

Follow Me Roaming forwarding (see p. 1156)
Activate, dial *18 send. Deactivate dial *19 send.

Billing information

Visitor rates
Most visitors . . . . . $2 day + 65¢ minute

Visitors without roaming agreements (p. 1159)
Cannot call since company takes no credit cards.

Identification numbers

City	Market	System	Billing	RSA
All served cities	458	01890	00000	LA-5

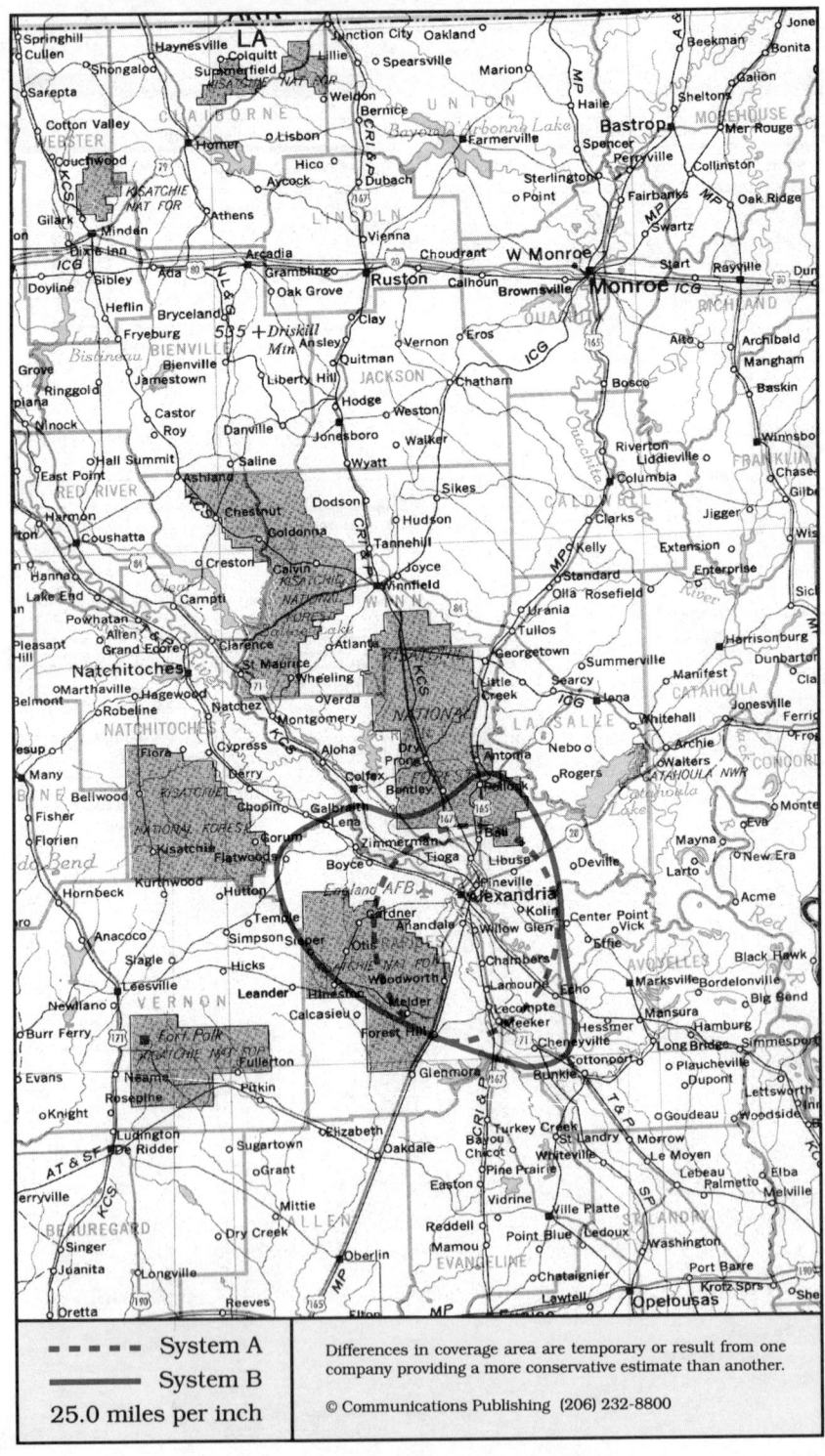

- - - - -	System A
——————	System B
25.0 miles per inch	

Differences in coverage area are temporary or result from one company providing a more conservative estimate than another.

© Communications Publishing (206) 232-8800

Alexandria, Louisiana

System A	System B

Cellular company

Cellular One
GenCell Management
1623 MacArthur Drive
Alexandria, LA 71301

Customer service	800-888-7868
Local office	318-442-5800
Corporate office	707-425-8000

Sending calls

Visitor dialing instructions

Local calls	area code + 7 digits
Long distance . . .	1 + area code + 7 digits

Speed-dial numbers

Customer service	*611
Emergency	911

Receiving calls

Caller dials the roamer access number below,
hears tone, dials cellular area code and number.
. . . . 318-447-7626

NationLink & RoamingAmerica (see p. 1157)
Dial *31 send to receive calls, *30 to deactivate.
This company's subscribers have level 3 service.

Billing information

Visitor rates
Most visitors . . . . . $3 day + 99¢ minute

Visitors without roaming agreements (page iii
American Roaming Network intercepts call to bill.

Identification numbers

City	Market	System	Billing	RSA
All served cities	205	00243	00000	MSA

Cellular company

Century Cellunet
1403-F Metro Drive
Alexandria, LA 71301

Customer service	800-638-9727
. . . .	318-683-3450
Local office	318-445-2065
Corporate office	318-388-9000

Sending calls

Visitor dialing instructions

Local calls	area code + 7 digits
Long distance . . .	1 + area code + 7 digits

Speed-dial numbers

Customer service	*611
Emergency	911

Receiving calls

Caller dials the roamer access number below,
hears tone, dials cellular area code and number.
. . . . 318-446-7626

Billing information

Visitor rates
Most visitors . . . . . $3 day + 75¢ minute

Visitors without roaming agreements (p. 1159)
Call co. to use Mastercard, Visa

Identification numbers

City	Market	System	Billing	RSA
All served cities	205	00212	00000	MSA

For full coverage area, see Natchitoches, LA.

Baton Rouge, Louisiana

Baton Rouge, Content, Donaldsonville, Livingston, Napoleonville, Scotlandville

System A	System B

Cellular company

Cellular One
10551 Coursey Boulevard, PO Box 40565
Baton Rouge, LA 70835

4696 Constitution Avenue
Baton Rouge, LA 70808

Customer service	504-291-1163
Local office . . Coursey Blvd.	504-291-5990
. . . . Constitution Ave.	504-927-2355
Corporate office	504-291-5990

Sending calls

Visitor dialing instructions

Local calls	area code + 7 digits
Long distance . . .	1 + area code + 7 digits

Speed-dial numbers

Customer service	611
Emergency	911
Louisiana state police . . .	800-525-5555
Make a traffic report	101

Receiving calls

Caller dials the roamer access number below,
hears tone, dials cellular area code and number.

Baton Rouge	504-335-7626
Gonzales	504-675-7626

NationLink & RoamingAmerica (see p. 1157)
Dial *31 send to receive calls, *30 to deactivate.
Dial *32 send to tell caller the roamer access no.
This company's subscribers have level 6 service.

Billing information

Visitor rates

Most visitors	$3 day + 85¢ minute

Visitors without roaming agreements (p. 1159)
Call co. to use Amex, Mastercard, Visa
or Roamer Plus will intercept first call to bill you.

Identification numbers

City	Market	System	Billing	RSA
All served cities	080	00085	00000	MSA

Cellular company

BellSouth Mobility
3111 S Sherwood Forest Boulevard
Baton Rouge, LA 70816

Customer service	800-351-2400
. . . .	504-296-4200
Local office	504-293-4036
Corporate office	404-847-3600

Sending calls

Visitor dialing instructions

Local calls	area code + 7 digits
Long distance	area code + 7 digits

Speed-dial numbers

Customer service	811
Emergency	911
Roaming information	711
Telephone problems	611

Receiving calls

Caller dials the roamer access number below,
hears tone, dials cellular area code and number.

. . . .	504-921-7626

Billing information

Visitor rates

Most visitors . . .	$2.50 day + 85¢ minute
BellSouth subscribers .	$0 day + 60¢ minute

Visitors without roaming agreements (p. 1159)
Call co. to use Amex, Mastercard, Visa

Identification numbers

City	Market	System	Billing	RSA
All served cities	080	00106	00000	MSA

For full coverage area, see New Orleans, LA and
Reserve, LA.

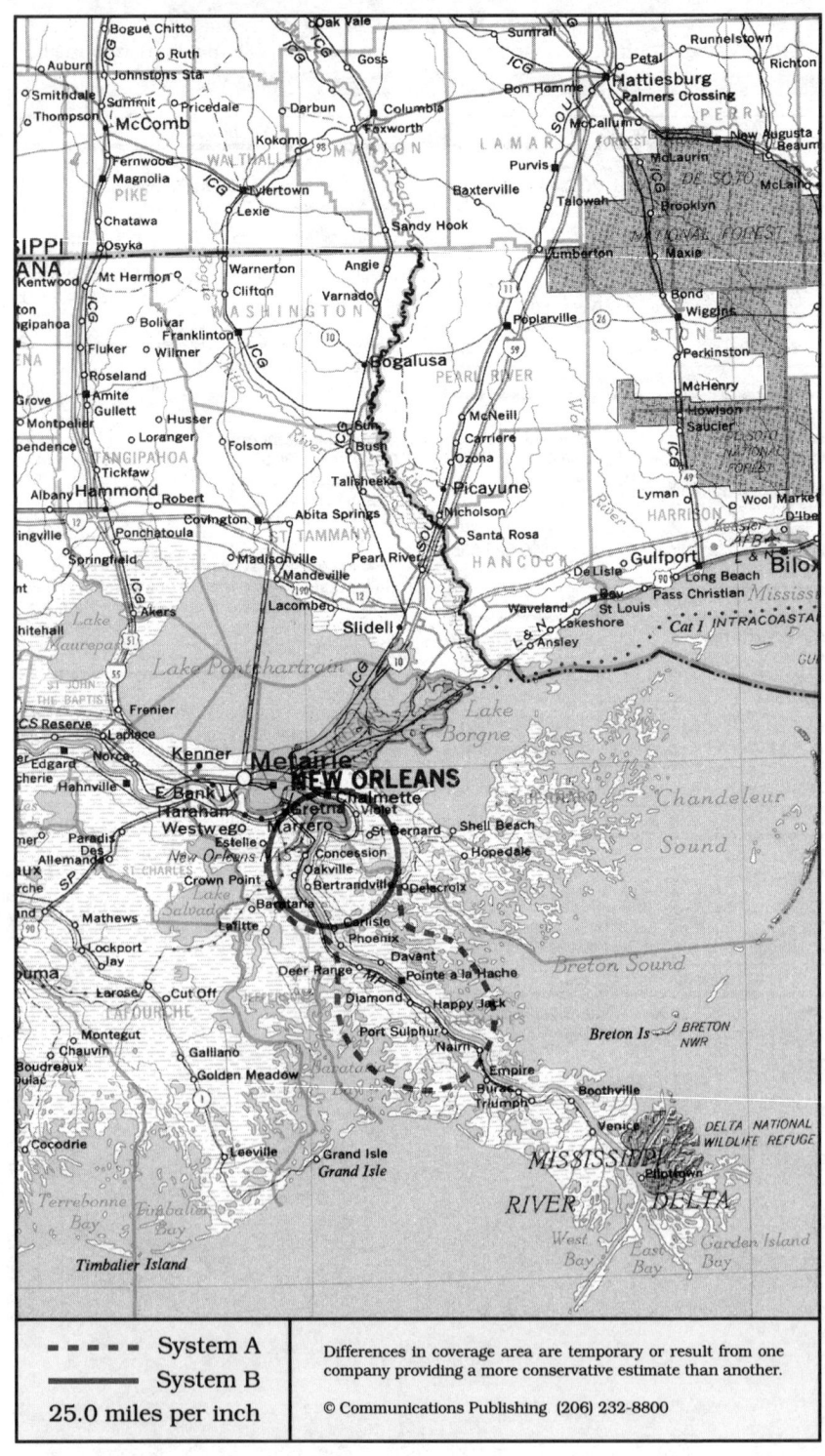

----- System A	Differences in coverage area are temporary or result from one
——— System B	company providing a more conservative estimate than another.
25.0 miles per inch	© Communications Publishing (206) 232-8800

Belle Chase, Louisiana
Belle Chase, Pointe a la Hache

Cellular company

Cellular One
Radiofone
Street address
3131 N I-10 Service Road E #400
Metairie, LA 70002-6050

Mailing address
PO Box 6228
Metairie, LA 70009

Customer service	some areas	800-452-7209
. . . .		504-837-9540
Corporate office	. . . . .	504-837-8330

Sending calls

Visitor dialing instructions
Local calls	. . . . .	area code + 7 digits
Long distance	. . .	1 + area code + 7 digits

Speed-dial numbers
Customer service	611
Emergency	911
Louisiana state police . . .	800-525-5555
Make a traffic report	101

Receiving calls

Caller dials the roamer access number below,
hears tone, dials cellular area code and number.
. . . . 504-554-7626

NationLink & RoamingAmerica (see p. 1157)
Dial *31 send to receive calls, *30 to deactivate.
Dial *32 send to tell caller the roamer access no.
This company's subscribers have level 6 service.

Billing information

Visitor rates
Most visitors . . . . . $3 day + 85¢ minute

Visitors without roaming agreements (p. 1159)
Call co. to use Amex, Mastercard, Visa

Identification numbers

City	Market	System	Billing	RSA
All served cities	462	01311	00000	LA-9

For full coverage area, see New Orleans, LA.

Cellular company

MobileTel
Administration
204 W 10th Street, PO Box 1090
Larose, LA 70373

Sales
1053 W Tunnel Boulevard
Houma, LA 70360

Customer service	. .	Larose	504-798-2355
Sales office		Houma	504-851-2355

Sending calls

Visitor dialing instructions
Local calls	. . . .	7 digits
Long distance	. . .	1 + area code + 7 digits

Speed-dial numbers
Customer service	611
Emergency	911

Receiving calls

Caller dials the roamer access number below,
hears tone, dials cellular area code and number.
. . . . 504-691-7626

Follow Me Roaming forwarding (see p. 1156)
Activate, dial *18 send. Deactivate dial *19 send.

Billing information

Visitor rates
Most visitors . . . . . $3 day + 85¢ minute

Visitors without roaming agreements (p. 1159)
Call co. to use Amex, Mastercard, Visa

Identification numbers

City	Market	System	Billing	RSA
All served cities	462	01312	00000	LA-9

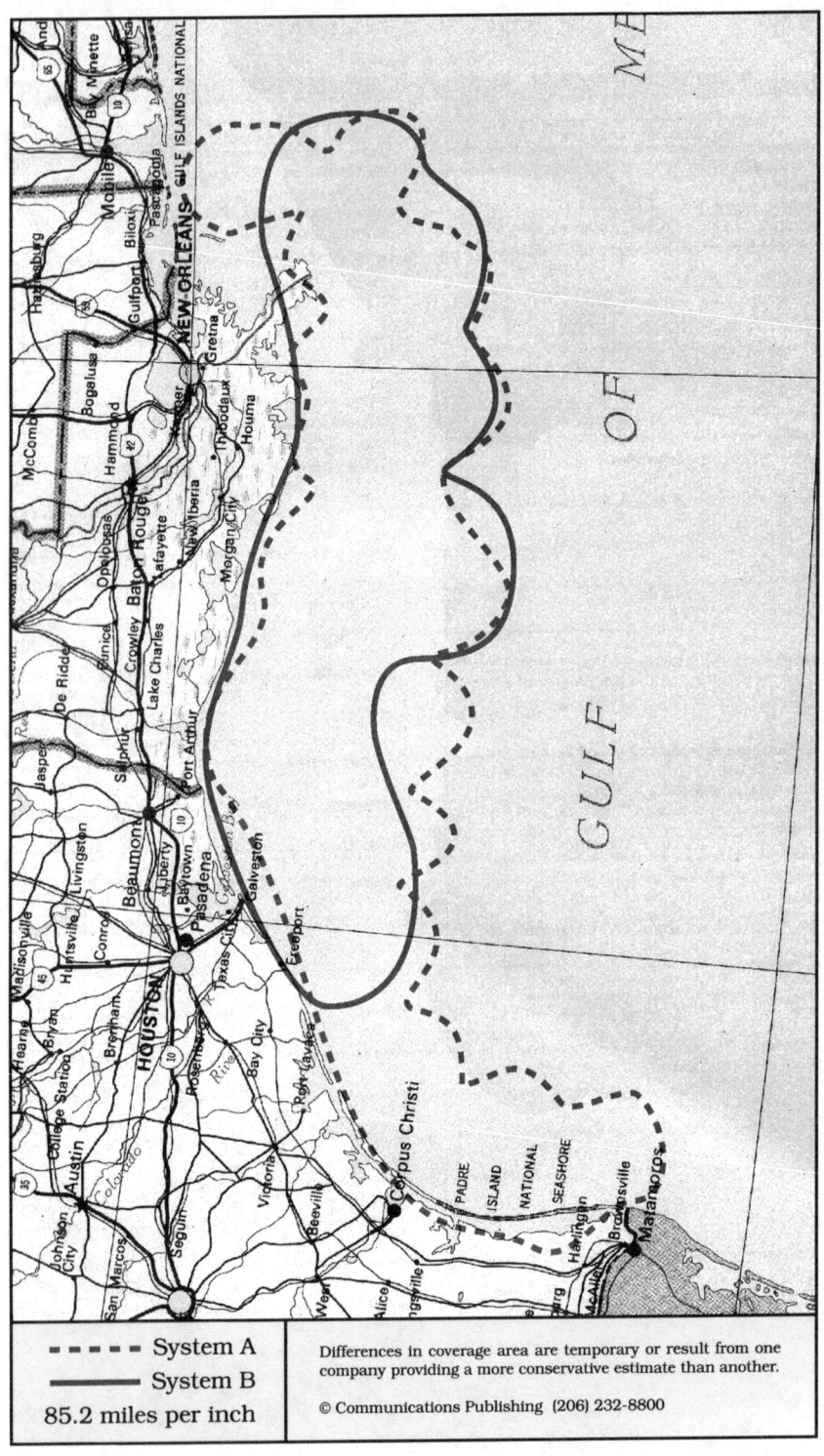

System A
System B
85.2 miles per inch

Differences in coverage area are temporary or result from one company providing a more conservative estimate than another.

© Communications Publishing (206) 232-8800

System A

Cellular company

PetroCom
Petroleum Communications
5600 NW Central Drive, Suite 100
Houston, TX 77092

3861 Ambassador Caffery Parkway, Suite 130
Lafayette, LA 70503

5901 Earhart Expressway
New Orleans, LA 70123

Local office	. . . .	Houston	713-939-1900
. . . . . .		Lafayette	318-981-9608
. . . . . .		New Orleans	800-233-8372
. . . . . .		New Orleans	504-736-9400
Corporate office	. . . . .		504-736-9400

Sending calls

Visitor dialing instructions
Local calls . . . . . area code + 7 digits
Long distance . . . 1 + area code + 7 digits

Speed-dial numbers

Apalachicola, FL to Gulfport, MS weather	211
Brownsville to Pt. O'Connor weather . .	811
Customer service	611
Directory assistance	411
Galveston, Pt. Arthur, Pt. O'Connor weather	511
Gulfport to intercoastal city water weather	311
Louisiana wildlife and fisheries hotline	*4222
U. S. Coast Guard	911
U. S. Coast Guard search & rescue	*CG or *24

Receiving calls

Caller dials the roamer access number below,
hears tone, dials cellular area code and number.
. . . . 504-736-7626

Follow Me Roaming forwarding (see p. 1156)
Phone Me Anywhere forwarding (see p. 1156)
Activate, dial *18 send. Deactivate dial *19 send.

NationLink & RoamingAmerica (see p. 1157)
Dial *31 send to receive calls, *30 to deactivate.
Dial *32 send to tell caller the roamer access no.
This company's subscribers have level 9 service.

Billing information

Visitor rates
. . . . $0 day + $2.40 minute

Visitors without roaming agreements (p. 1159)
Call this company for direct billing.

Identification numbers

City	Market	System	Billing	RSA
All served cities	306	00171	00000	MSA

Services available to Petrocom subscribers only:
For Petrocom operator service dial 711 or *11.

System B

Cellular company

Coastel Communications
RVC Services
1560 W Bay Area Boulevard, Suite 100
Friendswood, TX 77546

228 Landmark
Lafayette, LA 70506

2955 Ridgelake Drive, Suite 112
Metairie, LA 70002

Customer service	. . . .		800-822-8400
Local office . .		Friendswood	800-262-7835
. . . . . .		Friendswood	713-480-3655
. . . .		Lafayette	318-989-0444
. . . .		Metairie	504-837-6554
Corporate office	. . . . .		713-480-3655

Sending calls

Visitor dialing instructions
Local calls . . . . . area code + 7 digits
Long distance . . . 1 + area code + 7 digits

Speed-dial numbers

Corpus Christi weather	811
Customer service	611
Directory assistance	411
Galveston weather	211
New Orleans weather	311
Pensacola weather	511
Time of day	711
United States Coast Guard	911

Receiving calls

Caller dials the roamer access number below,
hears tone, dials cellular area code and number.
. . . . 318-273-7626

Billing information

Visitor rates
Most visitors . . . . . $0 day + $3 minute

Visitors without roaming agreements (p. 1159)
Call co. to use Amex, Mastercard, Visa

Identification numbers

City	Market	System	Billing	RSA
All served cities	306	00194	00000	MSA

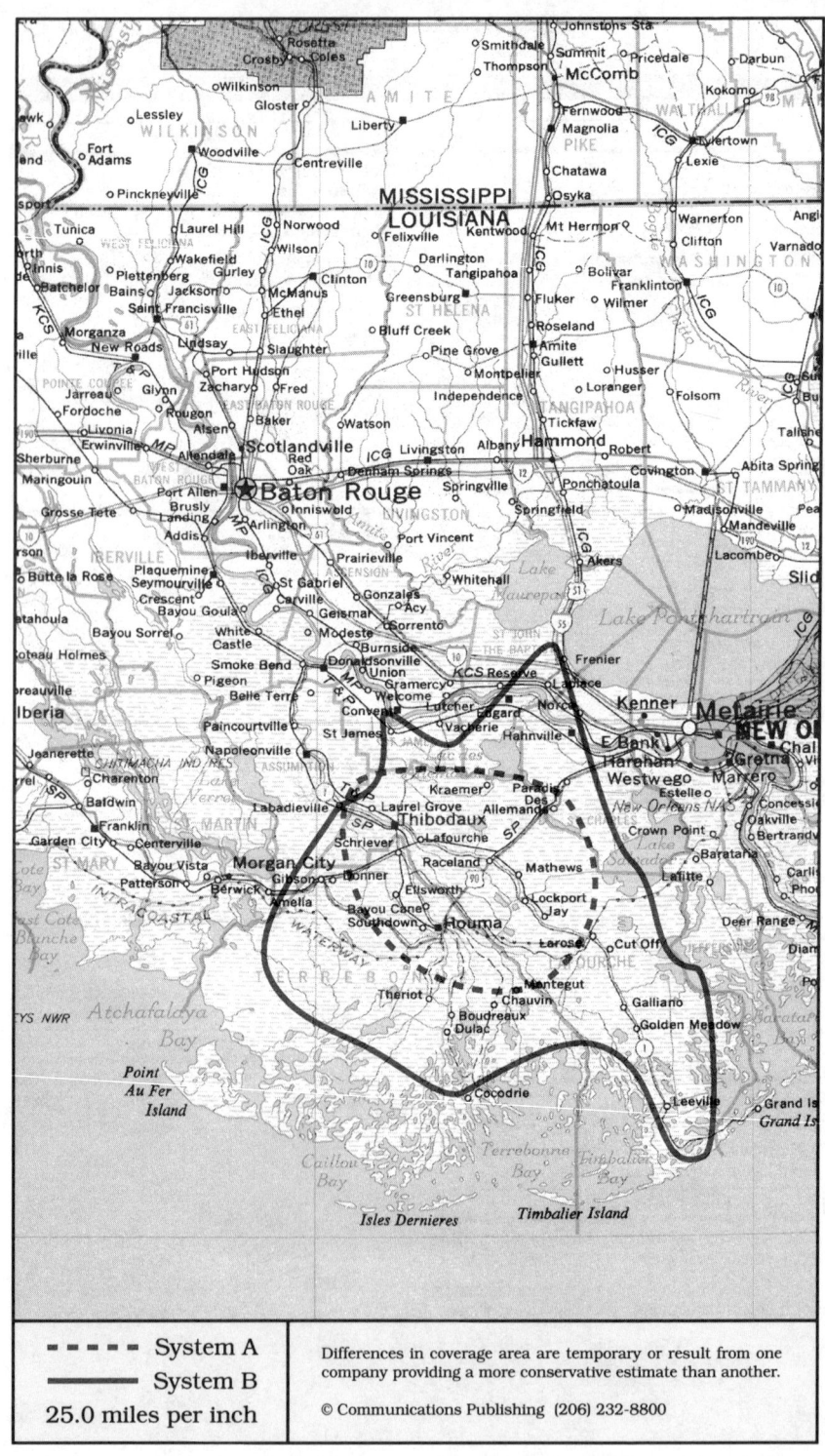

```
- - - - -  System A
─────────  System B
25.0 miles per inch
```

Differences in coverage area are temporary or result from one company providing a more conservative estimate than another.

© Communications Publishing (206) 232-8800

Houma, Louisiana
Houma, Larose, Thibodaux

System A

Cellular company

Cellular One
Radiofone
1130 Barrow Street
Houma, LA 70360

Customer service	some areas	800-452-7209
. . . .		504-837-9540
Local office		504-868-0220
Corporate office		504-837-8330

Sending calls

Visitor dialing instructions
Local calls	area code + 7 digits
Long distance . . .	1 + area code + 7 digits

Speed-dial numbers
C107 traffic	1075
Customer service	611
Emergency	911
Louisiana state police . . .	800-525-5555

Receiving calls

Caller dials the roamer access number below, hears tone, dials cellular area code and number.
. . . . 504-857-7626

NationLink & RoamingAmerica (see p. 1157)
Dial ∗31 send to receive calls, ∗30 to deactivate.
Dial ∗32 send to tell caller the roamer access no.
This company's subscribers have level 6 service.

Billing information

Visitor rates
Most visitors . . . . . $3 day + 85¢ minute

Visitors without roaming agreements (p. 1159)
Call co. to use Amex, Mastercard, Visa
or Roamer Plus will intercept first call to bill you.

Identification numbers

City	Market	System	Billing	RSA
All served cities	184	00387	00000	MSA

System B

Cellular company

MobileTel
Administration
204 W 10th Street, PO Box 1090
Larose, LA 70373

Sales
1053 W Tunnel Boulevard
Houma, LA 70360

Customer service . .	Larose	504-798-2355
Sales office	Houma	504-851-2355

Sending calls

Visitor dialing instructions
Local calls	7 digits
Long distance . . .	1 + area code + 7 digits

Speed-dial numbers
Customer service	611
Emergency	911
Roaming information	711

Receiving calls

Caller dials the roamer access number below, hears tone, dials cellular area code and number.
. . . . 504-691-7626

Follow Me Roaming forwarding (see p. 1156)
Activate, dial ∗18 send. Deactivate dial ∗19 send.

Billing information

Visitor rates
Most visitors . . . . . $3 day + 85¢ minute

Visitors without roaming agreements (p. 1159)
Call co. to use Amex, Mastercard, Visa

Identification numbers

City	Market	System	Billing	RSA
All served cities	184	00370	00000	MSA

For full coverage area, see La Place, LA.

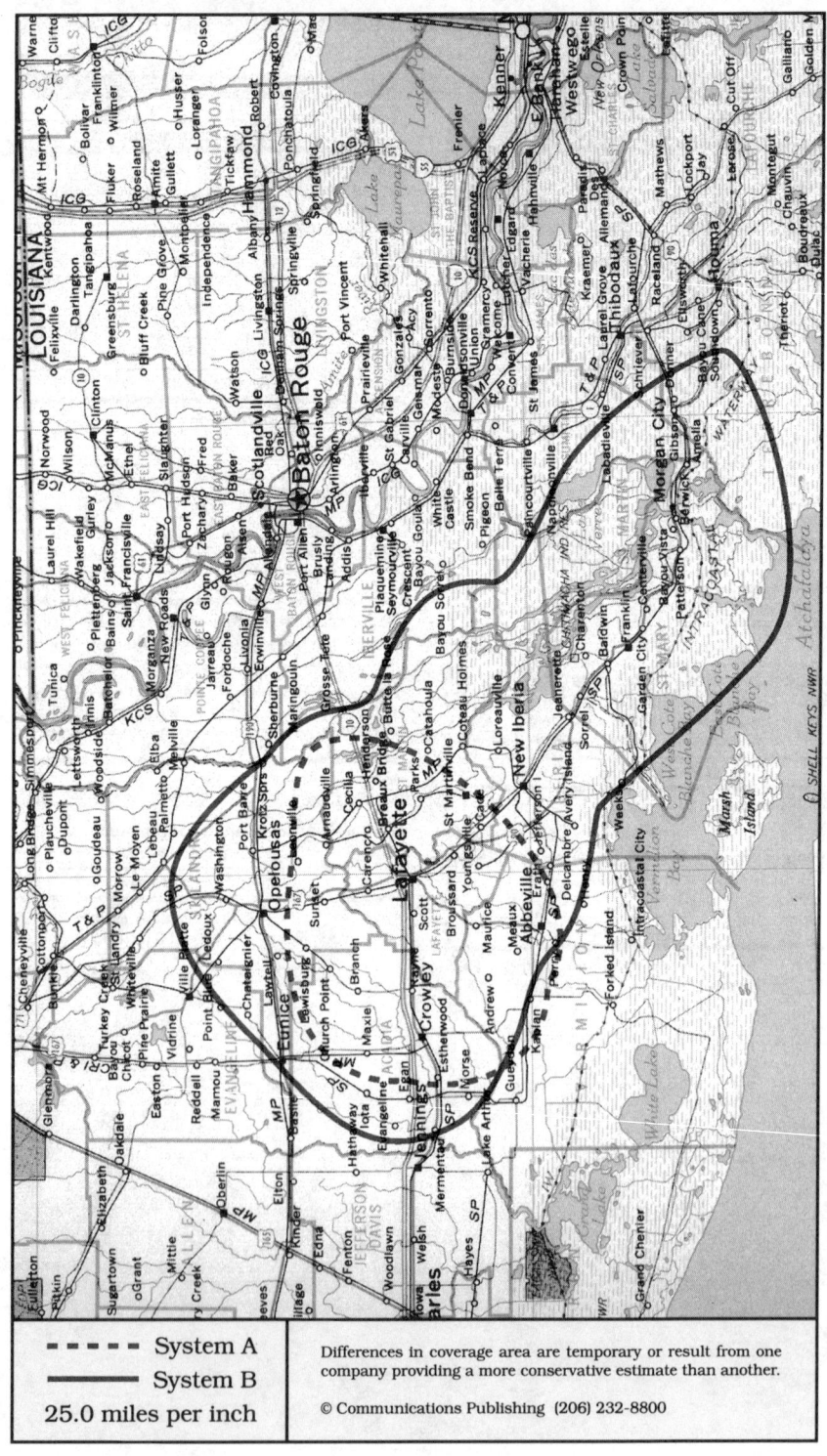

- - - - System A
——— System B

25.0 miles per inch

Differences in coverage area are temporary or result from one
company providing a more conservative estimate than another.

© Communications Publishing (206) 232-8800

Lafayette, Louisiana
Lafayette, Morgan City, New Iberia, Opelousas

Cellular company

Cellular One
McCaw Cellular Communications
5405 Johnston Street
Lafayette, LA 70503

Customer service	318-277-7611
Local office	318-984-1777
Corporate office	206-827-4500

Sending calls

Visitor dialing instructions
Local calls	area code + 7 digits
Long distance . . .	1 + area code + 7 digits

Speed-dial numbers
Customer service	611
Emergency	911
KSNB	924

Receiving calls

Caller dials the roamer access number below,
hears tone, dials cellular area code and number.
. . . . 318-277-7626

NationLink & RoamingAmerica (see p. 1157)
Dial *31 send to receive calls, *30 to deactivate.
Dial *32 send to tell caller the roamer access no.
This company's subscribers have level 9 service.

North American Cellular Network (see p. 1156)
All calls forward to you. Deactivate dial *35 send.

Billing information

Visitor rates
Lafayette	$0 day + 99¢ minute
Morgan City and New Iberia	$3 day + 99¢ minute

Visitors without roaming agreements (p. 1159)
Roamer Plus will intercept first call to bill you.

Identification numbers

City	Market	System	Billing	RSA
Lafayette	174	00431	00000	MSA
Morgan City, New I.	459	00431	30427	LA-6

McCaw Cellular reports that for Morgan City, the
new roamer access number will be 318-373-3676
and the new customer service number will be
504-372-1000. New Iberia will use the same
roamer access number and a customer service
number of 318-373-3500. These telephone
numbers were not operating at press time.

Cellular company

BellSouth Mobility
3807 Ambassador Caffery Parkway
Lafayette, LA 70503-5234

1025 Victor II Boulevard, Suite Q
Morgan City, LA 70380

Customer service	. . . .	800-351-2400
. . . .		318-988-8400
Local office	Lafayette	318-988-1891
. . . .	Morgan City	504-385-6827
Corporate office		404-847-3600

Sending calls

Visitor dialing instructions
Local calls	area code + 7 digits
Long distance . . .	1 + area code + 7 digits

Speed-dial numbers
Customer service	811
Emergency	911
Roaming information	711
Telephone problems	611

Receiving calls

Caller dials the roamer access number below,
hears tone, dials cellular area code and number.
Crowley	318-788-8426
Lafayette	318-278-7626
Morgan City	504-380-8888
New Iberia	318-373-7626
Opelousas	318-879-1526

Follow Me Roaming forwarding (see p. 1156)
Activate, dial *18 send. Deactivate dial *19 send.

Billing information

Visitor rates
Most visitors . . .	$2.50 day + 85¢ minute
BellSouth subscribers .	$0 day + 60¢ minute

Visitors without roaming agreements (p. 1159)
Call co. to use Amex, Mastercard, Visa

Identification numbers

City	Market	System	Billing	RSA
Crowley	458	00414	30048	LA-5
Lafayette	174	00414	00000	MSA
Morgan City, New I.	459	00414	30048	LA-6
Opelousas	458	00414	30048	LA-5

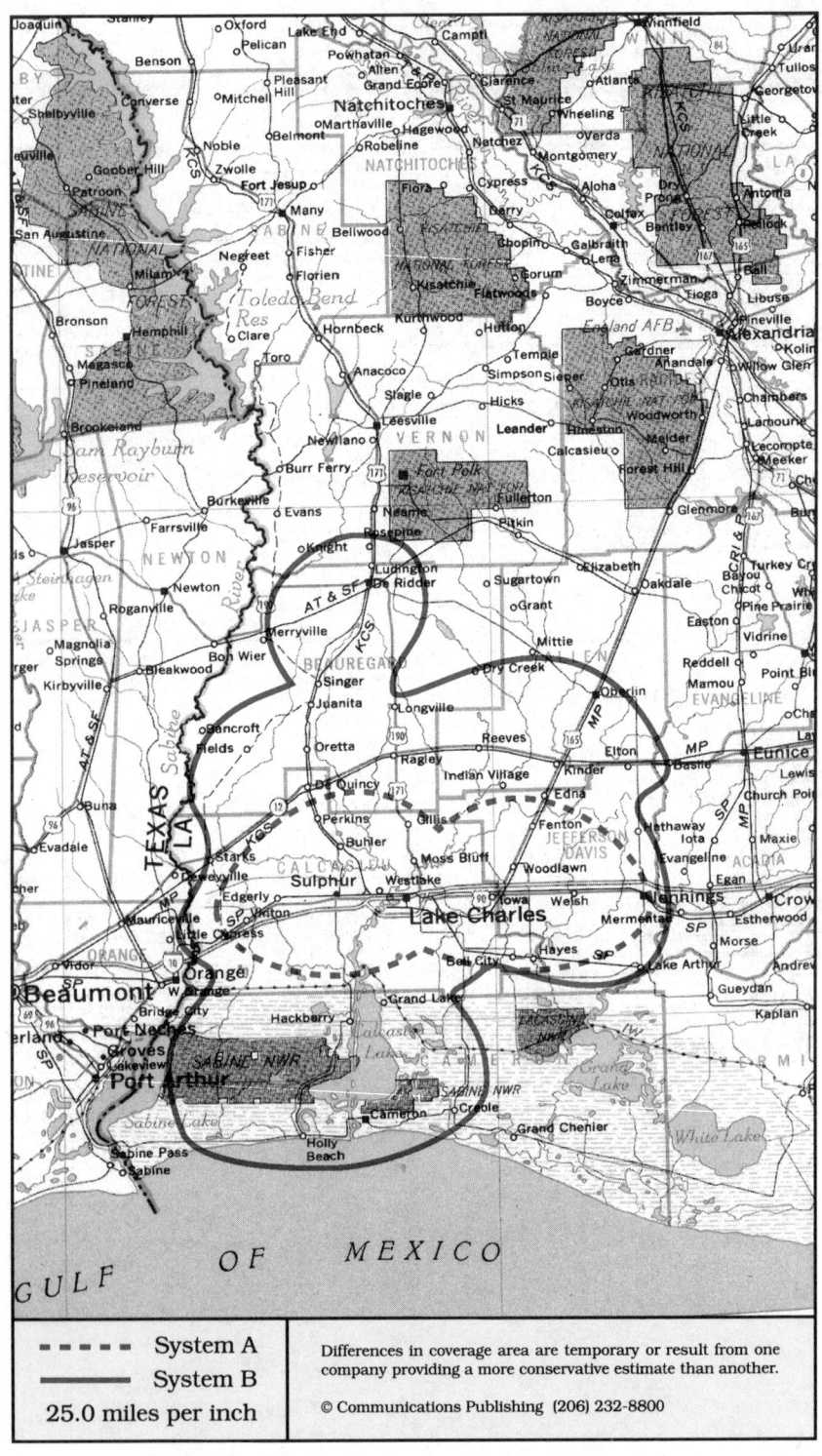

- - - - -	System A
————————	System B
25.0 miles per inch	

Differences in coverage area are temporary or result from one company providing a more conservative estimate than another.

© Communications Publishing (206) 232-8800

Lake Charles, Louisiana
Cameron, De Ridder, Jennings, Lake Charles, Sulphur

System A	System B

Cellular company

Cellular One
Lake Charles Cellular Telephone
1401 E Prien Lake Road
Lake Charles, LA 70601

Corporate office 318-475-1000

Sending calls

Visitor dialing instructions
Local calls . . . . . area code + 7 digits
Long distance . . . 1 + area code + 7 digits

Speed-dial numbers
Customer service 611
Emergency 911

Receiving calls

Caller dials the roamer access number below,
hears tone, dials cellular area code and number.
. . . . 318-438-7626

NationLink & RoamingAmerica (see p. 1157)
Dial *31 send to receive calls, *30 to deactivate.
Dial *32 send to tell caller the roamer access no.
This company's subscribers have level 6 service.

Billing information

Visitor rates
Most visitors . . . . . $3 day + 85¢ minute

Visitors without roaming agreements (p. 1159)
Cannot call since company takes no credit cards.

Identification numbers

City	Market	System	Billing	RSA
All served cities	197	00417	00000	MSA

Cellular company

Mercury Cellular and Paging
Cameron Communications
PO Box 3189
Lake Charles, LA 70602-3104

Local office 800-673-2200
. . . . 318-433-6298
Corporate office 318-583-2111

Sending calls

Visitor dialing instructions
Local calls . . . . . . area code + 7 digits
Long distance . . . 1 + area code + 7 digits

Speed-dial numbers
Customer service *611
Emergency 911

Receiving calls

Caller dials the roamer access number below,
hears tone, dials cellular area code and number.
. . . . 318-496-7626

Follow Me Roaming forwarding (see p. 1156)
Activate, dial *18 send. Deactivate dial *19 send.

Billing information

Visitor rates
Most visitors . . . $2.50 day + 85¢ minute

Visitors without roaming agreements (p. 1159)
Call co. to use Mastercard, Visa

Identification numbers

City	Market	System	Billing	RSA
De Ridder, Jennings	458	00400	00000	LA-5
Lake Charles	197	00400	00000	MSA

For full coverage area, see Leesville, LA.

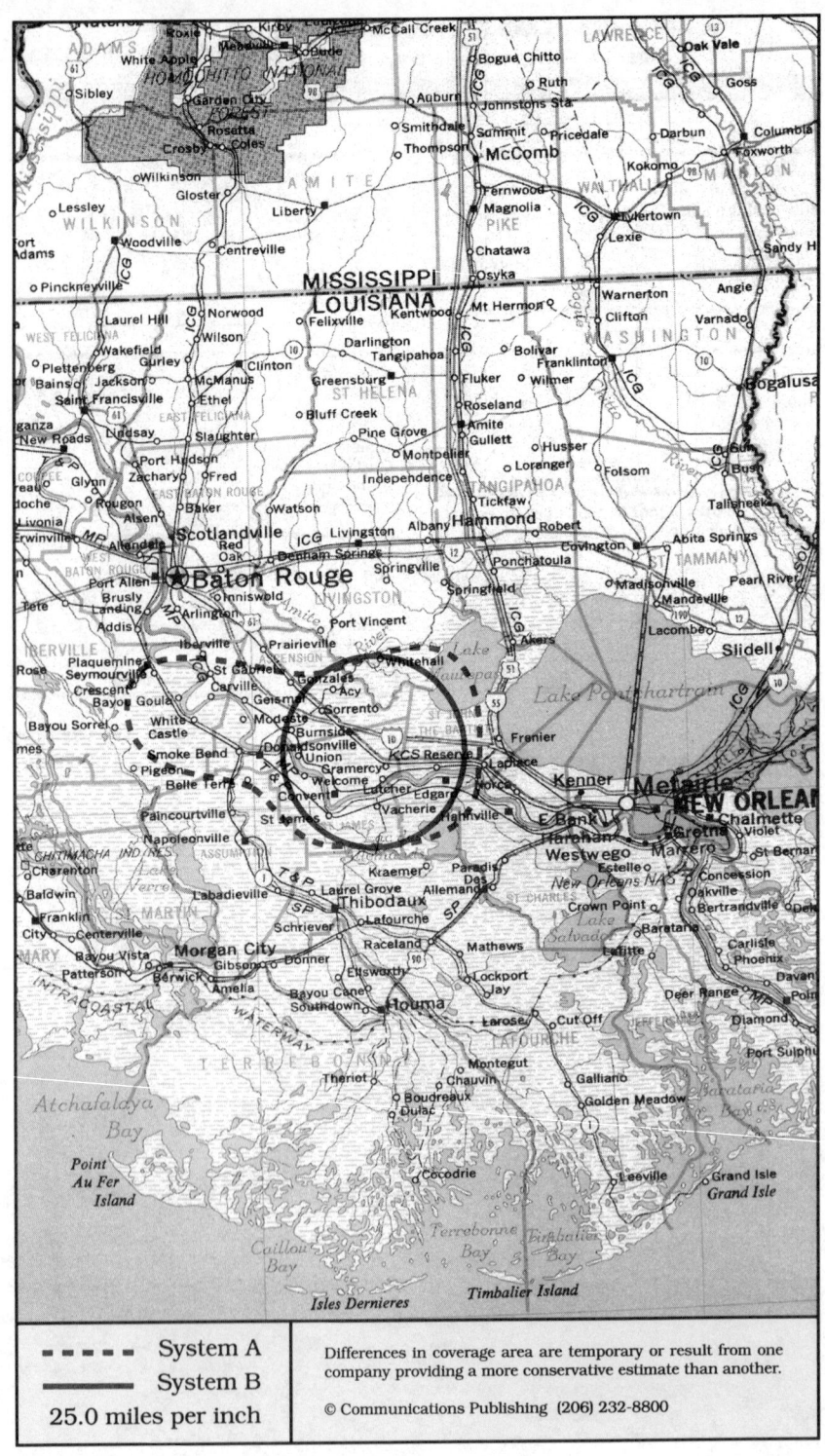

- - - - -	System A
————	System B
25.0 miles per inch	

Differences in coverage area are temporary or result from one company providing a more conservative estimate than another.

© Communications Publishing (206) 232-8800

La Place, Louisiana
Hahnville, La Place, Reserve

System A	System B

Cellular company

Cellular One of River Parishes
PriCellular
3401 Kemp Boulevard, Suite R
Wichita Falls, TX 76308

Customer service	Wichita Falls	817-691-9100
Local office	agent	504-764-7627
Corporate office		212-459-0800

Sending calls

Visitor dialing instructions

Local calls	area code + 7 digits
Long distance	area code + 7 digits

Speed-dial numbers

Customer service	611
Emergency	911

Receiving calls

Caller dials the roamer access number below,
hears tone, dials cellular area code and number.
. . . . 504-764-5626

NationLink & RoamingAmerica (see p. 1157)
Dial *31 send to receive calls, *30 to deactivate.
Dial *32 send to tell caller the roamer access no.
This company's subscribers have level 5 service.

Billing information

Visitor rates
Most visitors . . . . . $3 day + 75¢ minute

Visitors without roaming agreements (p. 1159)
Call co. to use Amex, Discover, Mastercard, Visa

Identification numbers

City	Market	System	Billing	RSA
All served cities	461	01309	00000	LA-8

Cellular company

MobileTel
Administration
204 W 10th Street, PO Box 1090
Larose, LA 70373

Sales
1053 W Tunnel Boulevard
Houma, LA 70360

Customer service . .	Larose	504-798-2355
Sales office	Houma	504-851-2355

Sending calls

Visitor dialing instructions

Local calls	7 digits
Long distance . . .	1 + area code + 7 digits

Speed-dial numbers

Customer service	611
Emergency	911

Receiving calls

Caller dials the roamer access number below,
hears tone, dials cellular area code and number.
. . . . 504-725-8555

Follow Me Roaming forwarding (see p. 1156)
Activate, dial *18 send. Deactivate dial *19 send.

Billing information

Visitor rates
Most visitors . . . . . $3 day + 85¢ minute

Visitors without roaming agreements (p. 1159)
Call co. to use Amex, Mastercard, Visa

Identification numbers

City	Market	System	Billing	RSA
All served cities	461	01310	00000	LA-8

For full coverage area, see Houma, LA.

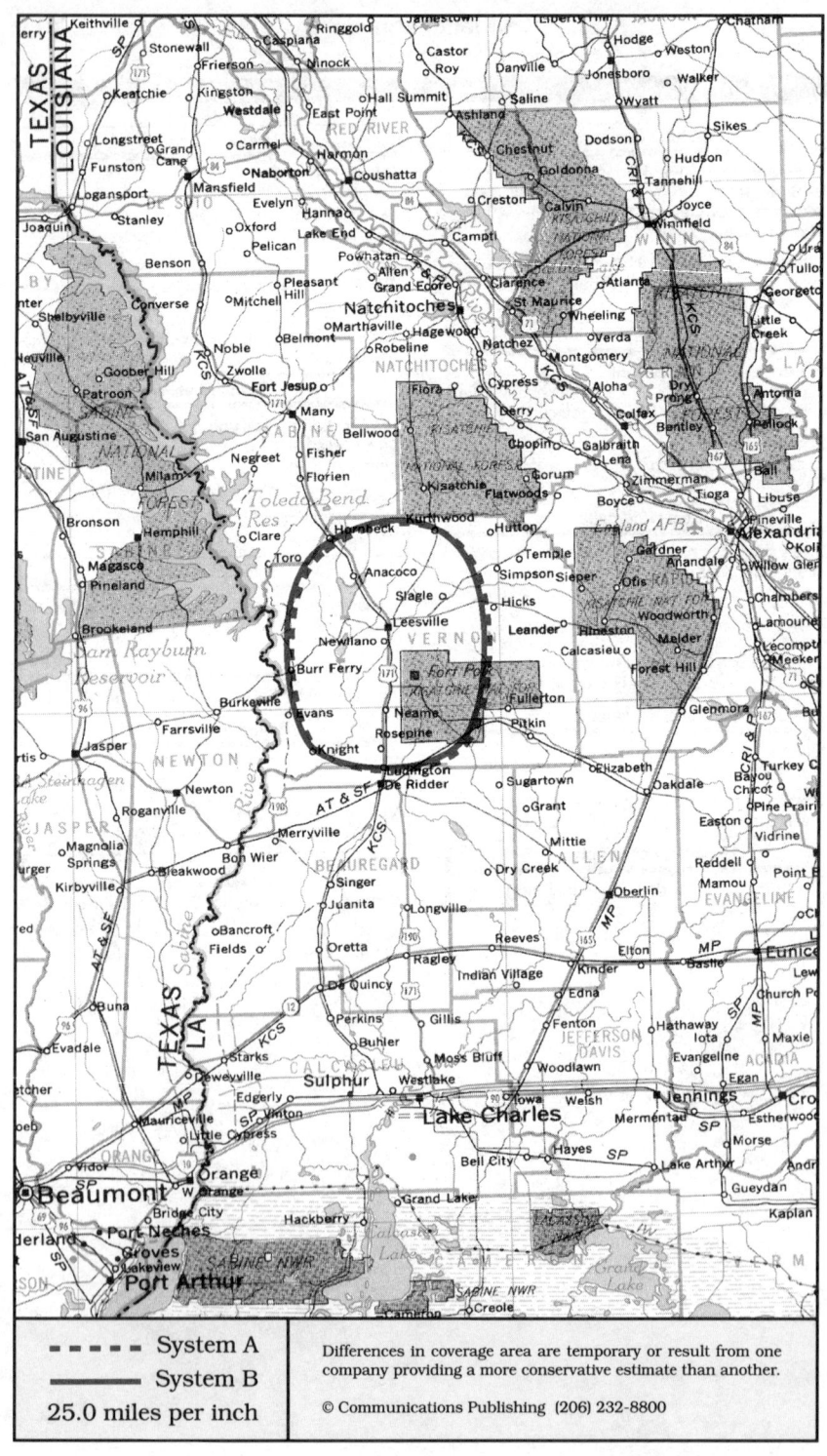

- - - - -	System A
———————	System B
25.0 miles per inch	

Differences in coverage area are temporary or result from one company providing a more conservative estimate than another.

© Communications Publishing (206) 232-8800

Leesville, Louisiana

System A

Cellular company

Cellular One
Louisiana Cellular Limited Partnership
342 Highway 1 S
Nachitoches, LA 71457

Local office	318-357-9596
. . . .	318-481-9596
Corporate office	601-445-0333

Sending calls

Visitor dialing instructions

Local calls	area code + 7 digits
Long distance . . .	1 + area code + 7 digits

Speed-dial numbers

Customer service	611
Emergency	911

Receiving calls

Caller dials the roamer access number below,
hears tone, dials cellular area code and number.
Use either number since they are interconnected.
. . . . 318-472-7626

NationLink & RoamingAmerica (see p. 1157)
Dial *31 send to receive calls, *30 to deactivate.
This company's subscribers have level 3 service.

Billing information

Visitor rates
Most visitors . . . . . $3 day + 85¢ minute

Visitors without roaming agreements (p. 1159)
Roamer Plus will intercept first call to bill you.

Identification numbers

City	Market	System	Billing	RSA
All served cities	456	01299	00000	LA-3

For full coverage area, see Natchitoches, LA.

System B

Cellular company

Mercury Cellular and Paging
Cameron Communications
PO Box 3189
Lake Charles, LA 70602-3104

Local office	800-673-2200
. . . .	318-433-6298
Corporate office	318-583-2111

Sending calls

Visitor dialing instructions

Local calls	area code + 7 digits
Long distance . . .	1 + area code + 7 digits

Speed-dial numbers

Customer service	*611
Emergency	911

Receiving calls

Caller dials the roamer access number below,
hears tone, dials cellular area code and number.
. . . . 318-496-7626

Follow Me Roaming forwarding (see p. 1156)
Activate, dial *18 send. Deactivate dial *19 send.

Billing information

Visitor rates
Most visitors . . . $2.50 day + 85¢ minute

Visitors without roaming agreements (p. 1159)
Call co. to use Mastercard, Visa

Identification numbers

City	Market	System	Billing	RSA
All served cities	456	00400	00000	LA-3

For full coverage area, see Lake Charles, LA.

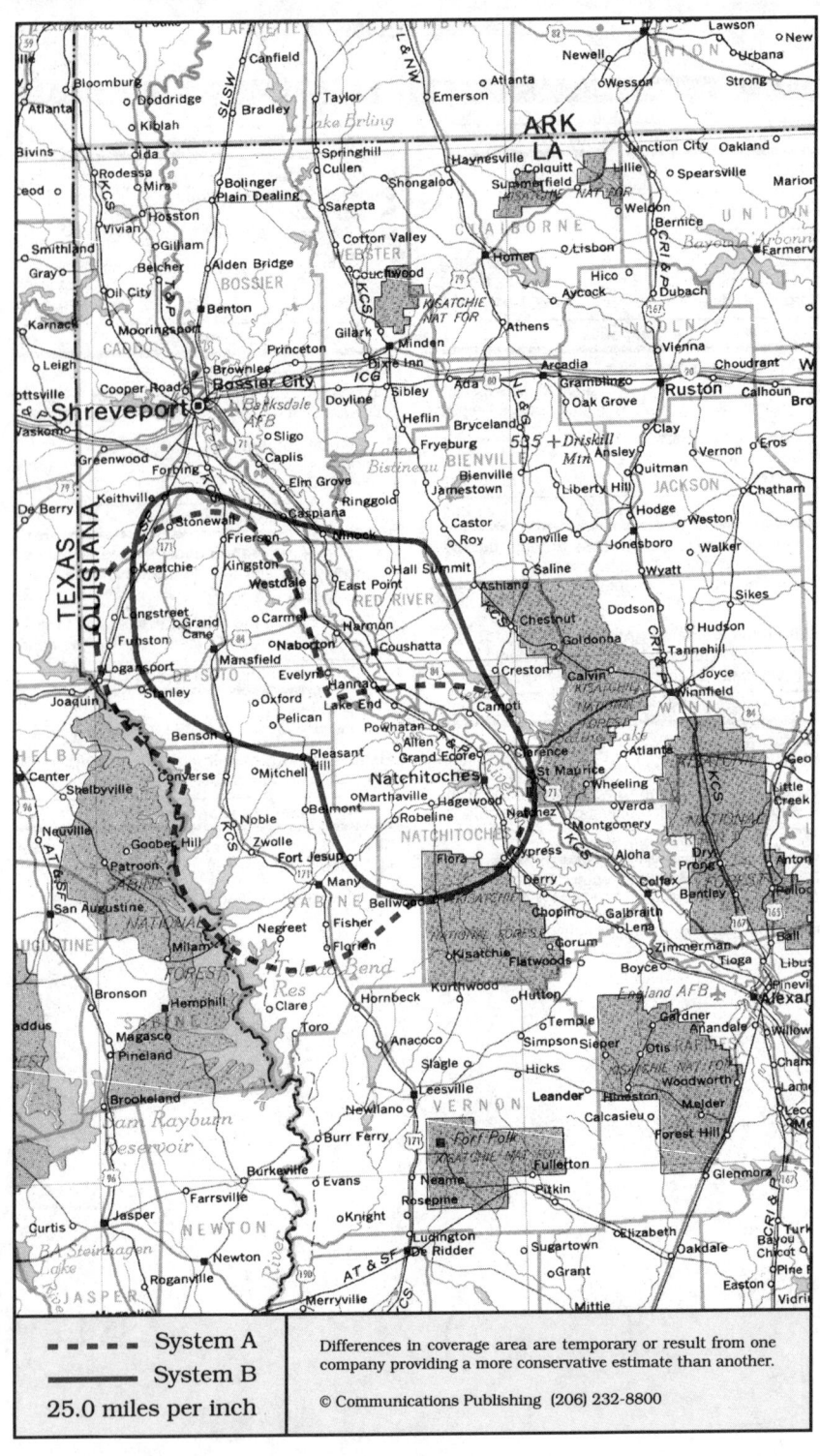

- - - - - System A	Differences in coverage area are temporary or result from one
———— System B	company providing a more conservative estimate than another.
25.0 miles per inch	© Communications Publishing (206) 232-8800

System A	System B

Cellular company

Cellular One
Louisiana Cellular Limited Partnership
342 Highway 1 S
Nachitoches, LA 71457

Local office	318-357-9596
. . . .	318-481-9596
Corporate office	601-445-0333

Sending calls

Visitor dialing instructions

Local calls	area code + 7 digits
Long distance . . .	1 + area code + 7 digits

Speed-dial numbers

Customer service	611
Emergency	911

Receiving calls

Caller dials the roamer access number below, hears tone, dials cellular area code and number. Use either number since they are interconnected.
. . . . 318-472-7626

NationLink & RoamingAmerica (see p. 1157)
Dial *31 send to receive calls, *30 to deactivate. This company's subscribers have level 5 service.

Billing information

Visitor rates
Most visitors . . . . . $3 day + 85¢ minute

Visitors without roaming agreements (p. 1159)
Roamer Plus will intercept first call to bill you.

Identification numbers

City	Market	System	Billing	RSA
All served cities	456	01299	00000	LA-3

For full coverage area, see Leesville, LA and Winnfield, LA.

Cellular company

Century Cellunet
504 College Avenue
Natchitoches, LA 71457

Customer service	800-638-9727
. . . .	318-683-3450
Local office	318-352-9574
Corporate office	318-388-9000

Sending calls

Visitor dialing instructions

Local calls	area code + 7 digits
Long distance . . .	1 + area code + 7 digits

Speed-dial numbers

Customer service	*611
Emergency	911

Receiving calls

Caller dials the roamer access number below, hears tone, dials cellular area code and number.
. . . . 318-471-7626

Follow Me Roaming forwarding (see p. 1156)
Activate, dial *18 send. Deactivate dial *19 send.

Billing information

Visitor rates
Most visitors . . . . . $3 day + 75¢ minute

Visitors without roaming agreements (p. 1159)
Call co. to use Amex, Mastercard, Visa

Identification numbers

City	Market	System	Billing	RSA
All served cities	456	01938	00000	LA-3

For full coverage area, see Alexandria, LA and Shreveport, LA.

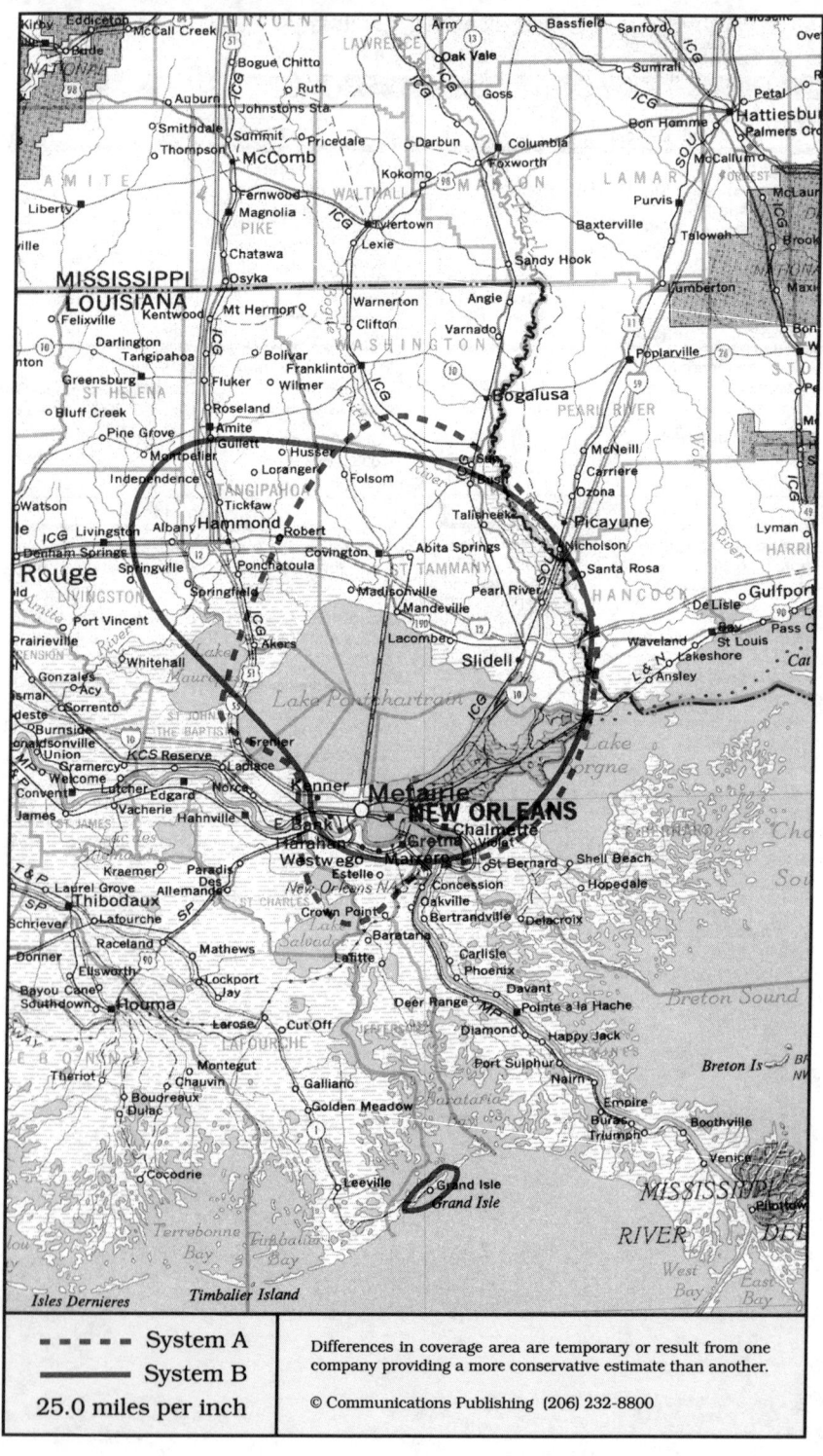

	System A
▬ ▬ ▬	System A
────────	System B
25.0 miles per inch	

Differences in coverage area are temporary or result from one company providing a more conservative estimate than another.

© Communications Publishing (206) 232-8800

New Orleans, Louisiana
Covington, Grand Isle, Hammond, Metairie, New Orleans, Slidell

System A	System B

Cellular company

Cellular One
Radiofone
Administration
PO Box 6228
Metairie, LA 70009

Sales
2701 N Causeway Boulevard
Metairie, LA 70002

Customer service	some areas	800-452-7209
. . . .		504-837-9540
Local office		504-835-1105
Corporate office		504-837-8330

Sending calls

Visitor dialing instructions
Local calls	area code + 7 digits
Long distance . . .	1 + area code + 7 digits

Speed-dial numbers
Customer service	611
Emergency	911
Louisiana state police . . .	800-525-5555
Make a traffic report	101

Receiving calls

Caller dials the roamer access number below,
hears tone, dials cellular area code and number.
. . . . 504-583-7626

NationLink & RoamingAmerica (see p. 1157)
Dial *31 send to receive calls, *30 to deactivate.
Dial *32 send to tell caller the roamer access no.
This company's subscribers have level 6 service.

Billing information

Visitor rates
Most visitors . . . . . $3 day + 85¢ minute

Visitors without roaming agreements (p. 1159)
Call co. to use Amex, Mastercard, Visa
or Roamer Plus will intercept first call to bill you.

Identification numbers

City	Market	System	Billing	RSA
All served cities	029	00057	00000	MSA

For full coverage area, see Belle Chasse, LA.

Cellular company

BellSouth Mobility
2222 Clearview Parkway
Metairie, LA 70001

Customer service	. . . .	800-351-2400
. . . .	. . . .	504-888-2110
Local office		504-883-7700
Corporate office		404-847-3600

Sending calls

Visitor dialing instructions
Local calls	area code + 7 digits
Long distance	area code + 7 digits

Speed-dial numbers
Customer service	811
Emergency	911
Roaming information	711
Telephone problems	611

Receiving calls

Caller dials the roamer access number below,
hears tone, dials cellular area code and number.
Hammond	504-543-7626
New Orleans	504-450-7626

Follow Me Roaming forwarding (see p. 1156)
Activate, dial *18 send. Deactivate dial *19 send.

Billing information

Visitor rates
Most visitors . . .	$2.50 day + 85¢ minute
BellSouth subscribers .	$0 day + 60¢ minute

Visitors without roaming agreements (p. 1159)
Call co. to use Amex, Mastercard, Visa

Identification numbers

City	Market	System	Billing	RSA
Hammond	460	00036	30050	LA-7
New Orleans	029	00036	00000	MSA

For full coverage area, see Baton Rouge, LA.

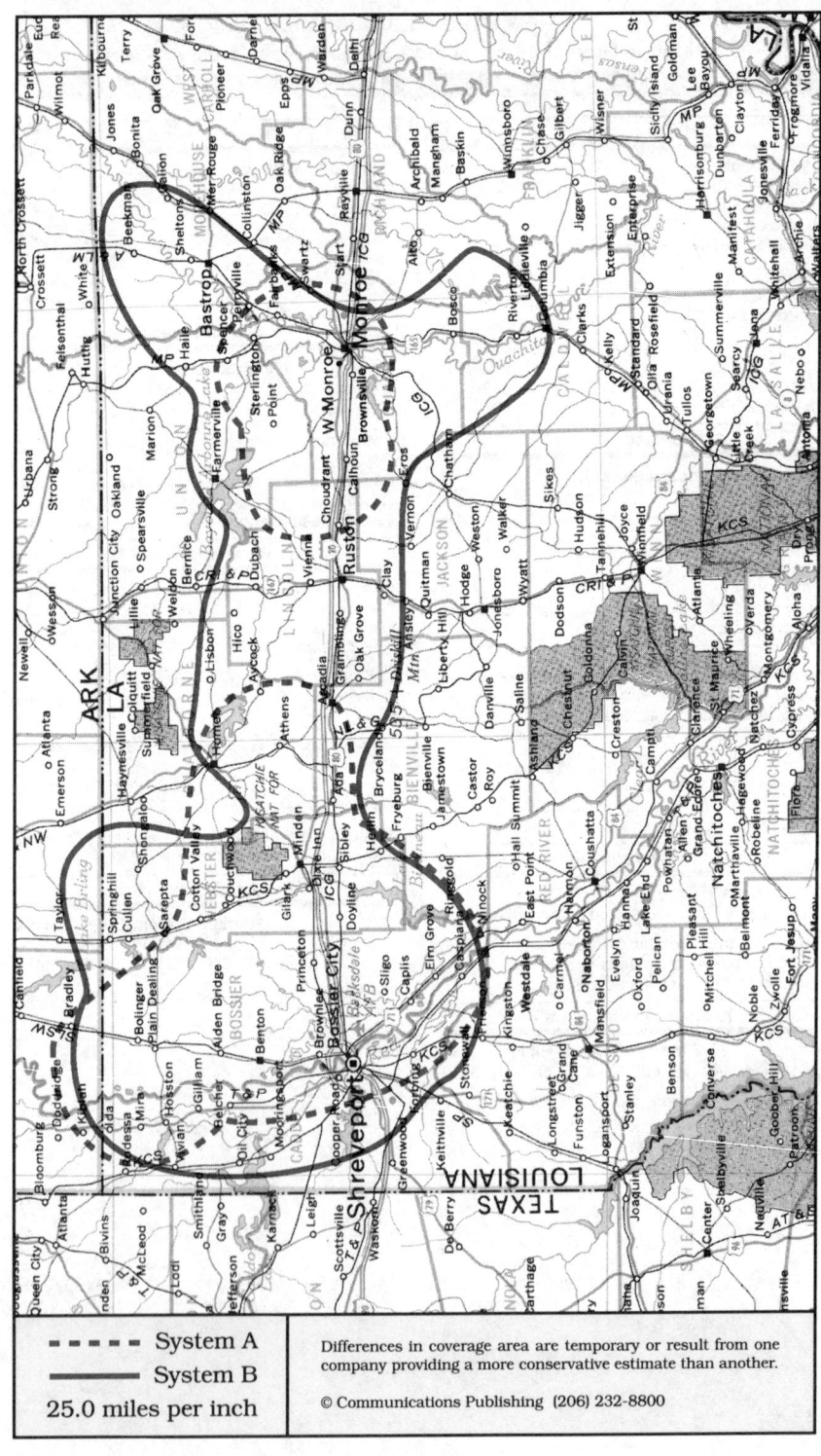

--- --- ---	System A
---	System B
25.0 miles per inch	

Differences in coverage area are temporary or result from one company providing a more conservative estimate than another.

© Communications Publishing (206) 232-8800

Shreveport, Louisiana
Bastrop, Homer, Monroe, Ruston, Shreveport

System A	System B

Cellular company

Cellular One
McCaw Cellular Communications
1130 Pecanland Road
Monroe, LA 71203

7941 Youree Drive
Shreveport, LA 71105

Customer service		800-262-3659
. . . . .	from Monroe	318-348-7500
. . . . .	from Shreveport	318-458-7500
Local office . . .	Monroe	318-348-7000
. . . . .	Shreveport	318-458-7000
Corporate office		206-827-4500

Sending calls

Visitor dialing instructions

Local calls	area code + 7 digits
Long distance . . .	1 + area code + 7 digits

Speed-dial numbers

Customer service	611
Emergency	911

Receiving calls

Caller dials the roamer access number below,
hears tone, dials cellular area code and number.

Monroe	318-348-7626
Shreveport	318-458-7626

NationLink & RoamingAmerica (see p. 1157)
Dial ∗31 send to receive calls, ∗30 to deactivate.
Dial ∗32 send to tell caller the roamer access no.
This company's subscribers have level 9 service.

North American Cellular Network (see p. 1156)
All calls forward to you. Deactivate dial ∗35 send.

Billing information

Visitor rates

Most visitors	$0 day + 99¢ minute

Visitors without roaming agreements (p. 1159)
Call co. to use Mastercard, Visa
or Roamer Plus will intercept first call to bill you.

Identification numbers

City	Market	System	Billing	RSA
Monroe	219	00463	00000	MSA
Shreveport	100	00229	00000	MSA

For full coverage area, see Longview, TX and
Texarkana, TX.

Cellular company

Century Cellunet
1701 Louisville Avenue
Monroe, LA 71201

907 N Trenton, Suite C
Ruston, LA 71270

2533 Bert Kouns Industrial Loop, Suite 121
Shreveport, LA 71118

Customer service		800-638-9727
. . . .		318-683-3450
Local office	Monroe	318-325-3600
. . . .	Ruston	318-255-8797
. . . . . .	Shreveport	318-687-8502
Corporate office		318-388-9000

Sending calls

Visitor dialing instructions

Local calls	area code + 7 digits
Long distance . . .	1 + area code + 7 digits

Speed-dial numbers

Customer service	∗611
Emergency	911

Receiving calls

Caller dials the roamer access number below,
hears tone, dials cellular area code and number.

Monroe	318-366-7626
Ruston	318-251-7626
Shreveport	318-455-7626

Follow Me Roaming forwarding (see p. 1156)
Activate, dial ∗18 send. Deactivate dial ∗19 send.

Billing information

Visitor rates

Most visitors	$3 day + 75¢ minute

Visitors without roaming agreements (p. 1159)
Call co. to use Amex, Mastercard, Visa

Identification numbers

City	Market	System	Billing	RSA
Monroe	219	00440	00000	MSA
Ruston	454	01296	00000	LA-1
Shreveport	100	00220	00000	MSA

For full coverage area, see Natchitoches, LA and
Tallulah, LA and Texarkana, TX and Winnfield,
LA.

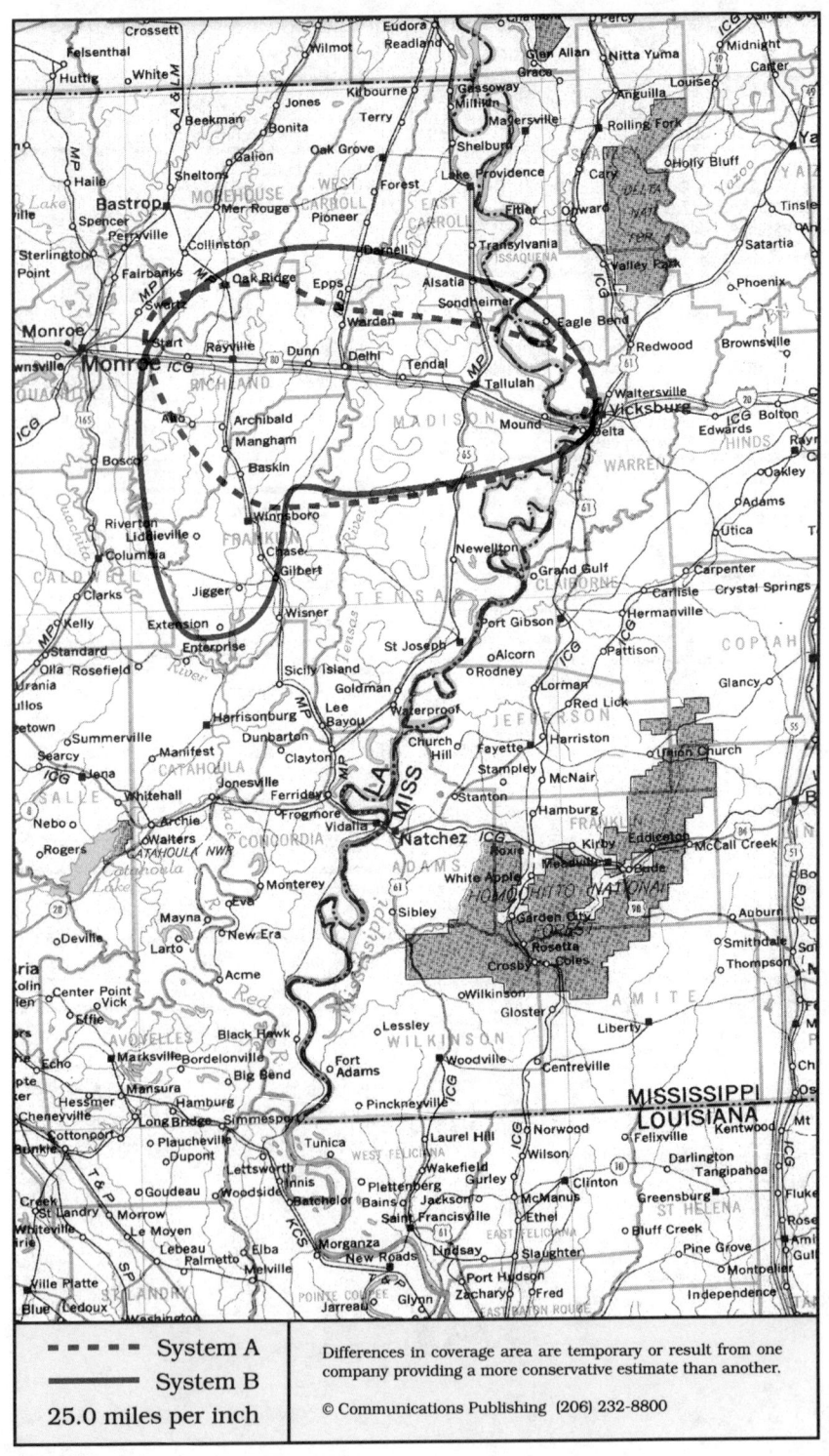

- - - -	System A
———	System B
25.0 miles per inch	

Differences in coverage area are temporary or result from one company providing a more conservative estimate than another.

© Communications Publishing (206) 232-8800

Tallulah, Louisiana
Rayville, Tallulah, Winnsboro

System A	System B

System A

Cellular company

Cellular One
Managed by: Sterling Cellular Management
904-B, 1 Broadway Square
Delhi, LA 71232

Customer service	800-552-6150
Corporate office	318-878-8151

Sending calls

Visitor dialing instructions

Local calls	area code + 7 digits
Long distance . . .	1 + area code + 7 digits

Speed-dial numbers

Customer service	611
Emergency	911

Receiving calls

Caller dials the roamer access number below, hears tone, dials cellular area code and number.
. . . . 318-729-7626

Billing information

Visitor rates

Most visitors	$3 day + 99¢ minute

Visitors without roaming agreements (p. 1159)
Call co. to use Amex
or Roamer Plus will intercept first call to bill you.

Identification numbers

City	Market	System	Billing	RSA
All served cities	455	01297	00000	LA-2

System B

Cellular company

Century Cellunet
1701 Louisville Avenue
Monroe, LA 71201

Customer service	800-638-9727
. . . .	318-683-3450
Local office	318-325-3600
Corporate office	318-388-9000

Sending calls

Visitor dialing instructions

Local calls	area code + 7 digits
Long distance . . .	1 + area code + 7 digits

Speed-dial numbers

Customer service	*611
Emergency	911

Receiving calls

Caller dials the roamer access number below, hears tone, dials cellular area code and number.
. . . . 318-366-7626

Follow Me Roaming forwarding (see p. 1156)
Activate, dial *18 send. Deactivate dial *19 send.

Billing information

Visitor rates

Most visitors	$3 day + 75¢ minute

Visitors without roaming agreements (p. 1159)
Call co. to use Amex, Mastercard, Visa

Identification numbers

City	Market	System	Billing	RSA
All served cities	455	00440	00000	LA-2

For full coverage area, see Shreveport, LA.

Winnfield, Louisiana
Columbia, Ferriday, Jena, Winnfield

System A	System B

Cellular company

Cellular One
Louisiana Cellular Limited Partnership
342 Highway 1 S
Nachitoches, LA 71457

Local office	318-357-9596
. . . .	318-481-9596
Corporate office	601-445-0333

Sending calls

Visitor dialing instructions

Local calls	area code + 7 digits
Long distance . . .	1 + area code + 7 digits

Speed-dial numbers

Customer service	611
Emergency	911

Receiving calls

Caller dials the roamer access number below, hears tone, dials cellular area code and number. Use either number since they are interconnected.
. . . . 318-472-7626

NationLink & RoamingAmerica (see p. 1157)
Dial *31 send to receive calls, *30 to deactivate. Dial *32 send to tell caller the roamer access no. This company's subscribers have level 9 service.

Billing information

Visitor rates
Most visitors . . . . . $3 day + 85¢ minute

Visitors without roaming agreements (p. 1159)
Roamer Plus will intercept first call to bill you.

Identification numbers

City	Market	System	Billing	RSA
All served cities	457	01301	00000	LA-4

For full coverage area, see Natchitoches, LA.

Cellular company

Century Cellunet
1701 Louisville Avenue
Monroe, LA 71201

Customer service	800-638-9727
. . . .	318-683-3450
Local office	318-325-3600
Corporate office	318-388-9000

Sending calls

Visitor dialing instructions

Local calls	area code + 7 digits
Long distance . . .	1 + area code + 7 digits

Speed-dial numbers

Customer service	*611
Emergency	911

Receiving calls

Caller dials the roamer access number below, hears tone, dials cellular area code and number.
. . . . 318-366-7626

Follow Me Roaming forwarding (see p. 1156)
Activate, dial *18 send. Deactivate dial *19 send.

Billing information

Visitor rates
Most visitors . . . . . $3 day + 75¢ minute

Visitors without roaming agreements (p. 1159)
Call co. to use Amex, Mastercard, Visa

Identification numbers

City	Market	System	Billing	RSA
All served cities	457	00440	30524	LA-4

For full coverage area, see Shreveport, LA and Tallulah, LA.

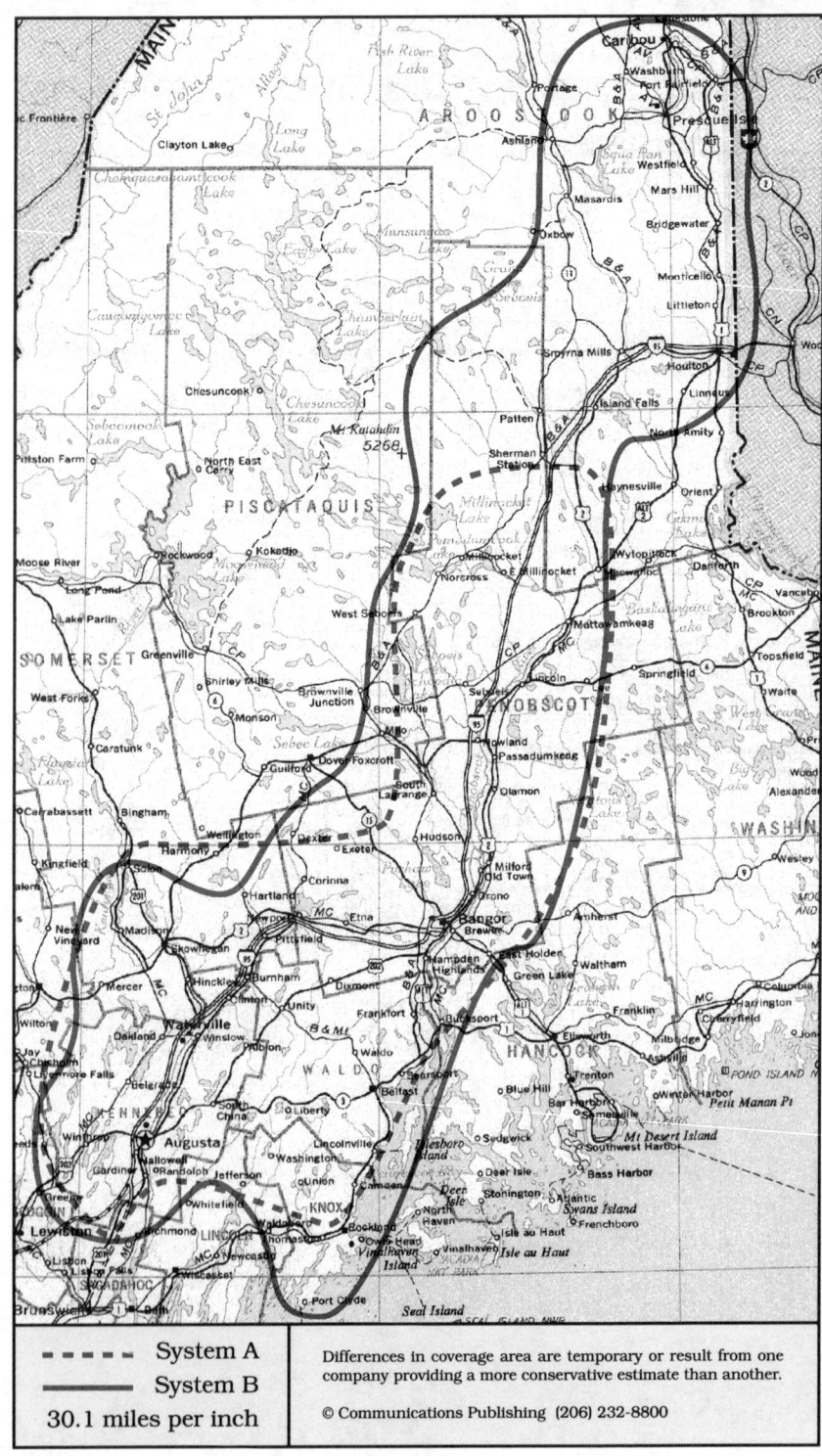

- - - -	System A
────────	System B
30.1 miles per inch	

Differences in coverage area are temporary or result from one company providing a more conservative estimate than another.

© Communications Publishing (206) 232-8800

Bangor, Maine
Augusta, Bangor, Caribou, Houlton, Presque Isle, Skowhegan, Waterville

System A	System B

Cellular company

United States Cellular
118 Mount Vernon Avenue
Augusta, ME 04330

1147 Hammond Street
Bangor, ME 04401

Customer service	some areas	800-244-0701
Local office	Augusta	207-621-0200
. . . .	Bangor	207-942-0700
Corporate office		312-399-8900

Sending calls

Visitor dialing instructions

Local calls	area code + 7 digits
Long distance	area code + 7 digits

Speed-dial numbers

Customer service	*611
Emergency	911

Receiving calls

Caller dials the roamer access number below,
hears tone, dials cellular area code and number.

Augusta, Skowhegan	. . .	207-441-7626
Bangor		207-944-7626

NationLink & RoamingAmerica (see p. 1157)
Caller hears your current roamer access number.
This company's subscribers have level 0 service.

Billing information

Visitor rates

Most visitors		$3 day + 75¢ minute

Visitors without roaming agreements (p. 1159)
Call co. to use Mastercard, Visa

Identification numbers

City	Market	System	Billing	RSA
Augusta	465	01327	00000	ME-3
Bangor	224	00271	00000	MSA
Skowhegan	464	01315	00000	ME-2

Cellular company

UNICEL
Unity Cellular Systems
269 Western Avenue
Augusta, ME 04330

1 Cumberland Place, Suite 112
Bangor, ME 04401

4 Airport Drive
Presque Isle, ME 04769

Customer service	some areas	800-244-9979
Local office	Augusta	207-621-0660
. . . .	Bangor	207-945-9979
. . . .	Presque Isle	207-764-1664
Corporate office		207-945-9979

Sending calls

Visitor dialing instructions

Local calls	area code + 7 digits
Long distance	area code + 7 digits

Speed-dial numbers

Customer service	*611
Emergency	911
Maine police	*77

Receiving calls

Caller dials the roamer access number below,
hears tone, dials cellular area code and number.

All cities except Augusta	. . .	207-745-7626
Augusta		207-242-7626

Follow Me Roaming forwarding (see p. 1156)
Activate, dial *18 send. Deactivate dial *19 send.

Billing information

Visitor rates

Most visitors		$3 day + 85¢ minute

Visitors without roaming agreements (p. 1159)
Roamer Plus will intercept first call to bill you.

Identification numbers

City	Market	System	Billing	RSA
Augusta	465	01318	00000	ME-3
Bangor	224	00254	00000	MSA
Presque Isle	464	00254	30546	ME-2

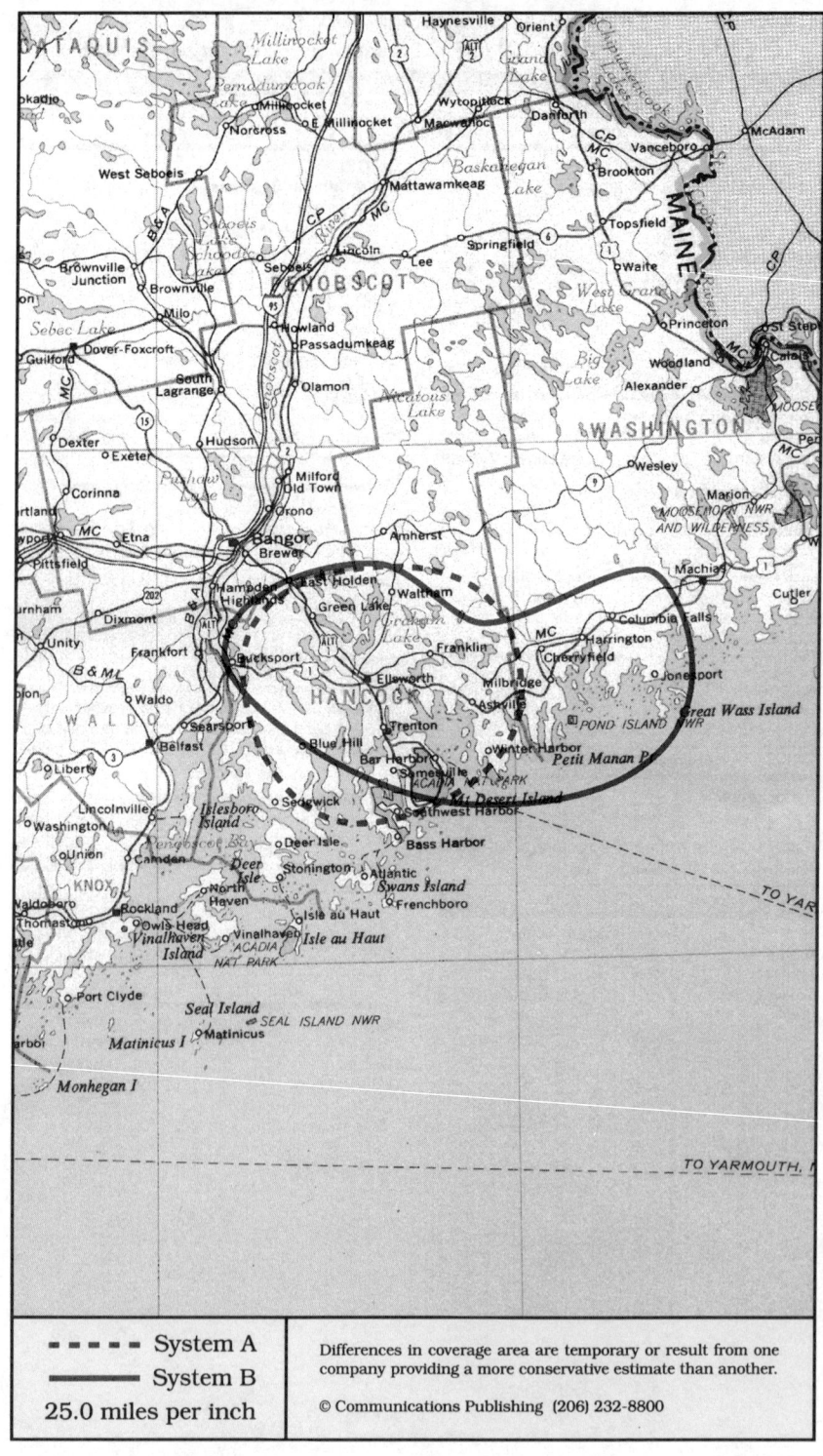

Lewiston, Maine

Cellular company

United States Cellular
118 Mount Vernon Avenue
Augusta, ME 04330

Local office	207-621-0200
Corporate office	312-399-8900

Sending calls

Visitor dialing instructions

Local calls	area code + 7 digits
Long distance	area code + 7 digits

Speed-dial numbers

Customer service	*611
Emergency	911
Roaming information	*711

Receiving calls

Caller dials the roamer access number below,
hears tone, dials cellular area code and number.
All served cities 207-754-7626

Billing information

Visitor rates
Most visitors . . . . . $3 day + 75¢ minute

Visitors without roaming agreements (p. 1159)
Call co. to use Mastercard, Visa

Identification numbers

City	Market	System	Billing	RSA
All served cities	279	00427	00000	MSA

For full coverage area, see Rumford, ME.

Cellular company

Maine Cellular
1220 Lisbon Street
Lewiston, ME 04240

Local office . . . some areas	800-564-0133
. . . .	207-782-1100
Corporate office	207-773-0800

Sending calls

Visitor dialing instructions

Local calls	area code + 7 digits
Long distance	area code + 7 digits

Speed-dial numbers

AAA road service	*222
Customer service	611
Make a traffic report	*101 or *970
Maine state police	*77

Receiving calls

Caller dials the roamer access number below,
hears tone, dials cellular area code and number.
. . . . 207-233-7626

Follow Me Roaming forwarding (see p. 1156)
Activate, dial *18 send. Deactivate dial *19 send.

Billing information

Visitor rates
Most visitors . . . . . $3 day + 90¢ minute

Visitors without roaming agreements (p. 1159)
Roamer Plus will intercept first call to bill you.

Identification numbers

City	Market	System	Billing	RSA
All served cities	279	00482	00000	MSA

For full coverage area, see Portland, ME.

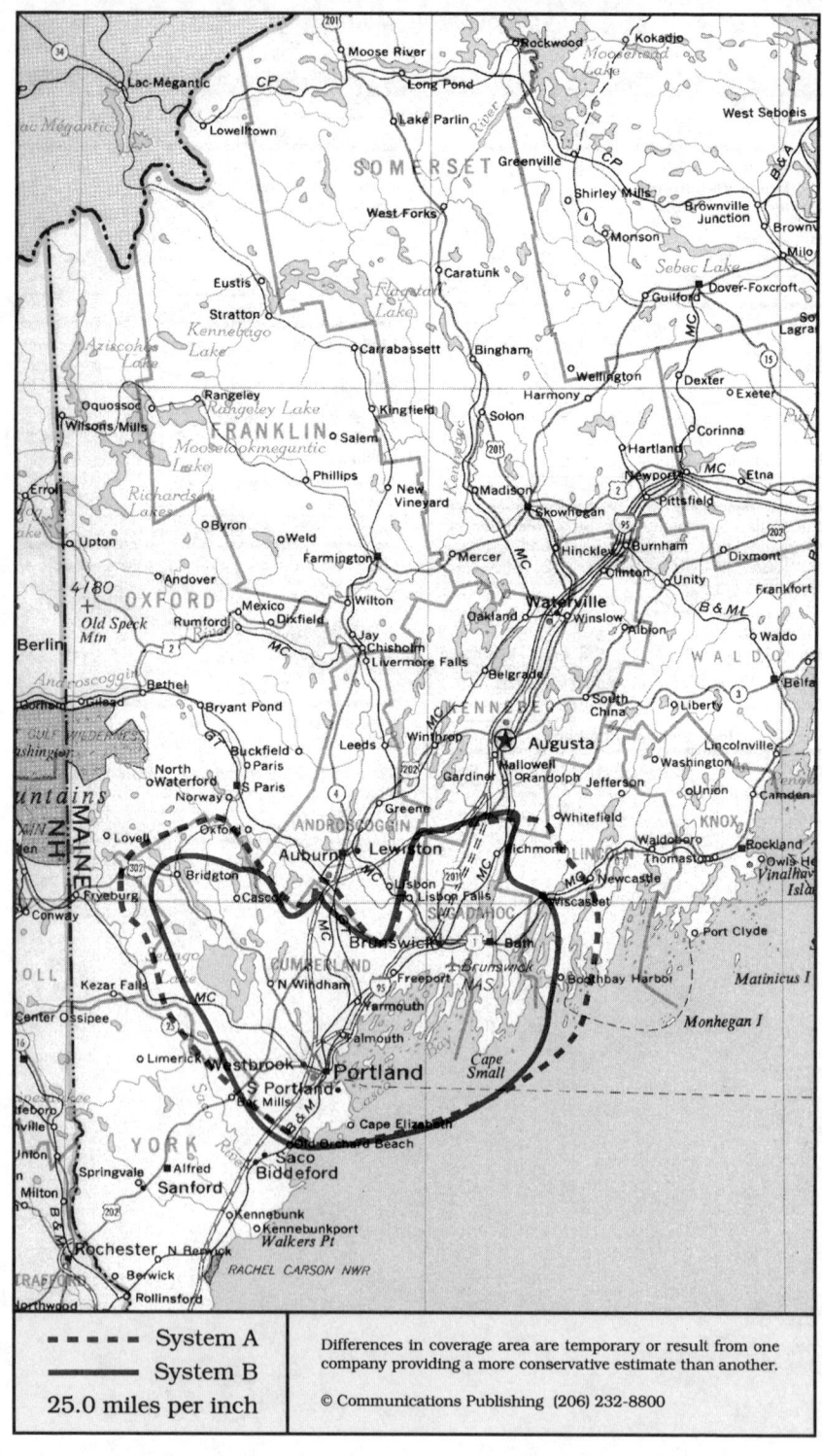

▬ ▬ ▬ ▬ System A	Differences in coverage area are temporary or result from one company providing a more conservative estimate than another.
—————— System B	
25.0 miles per inch	© Communications Publishing (206) 232-8800

Portland, Maine
Bath, Brunswick, Portland

System A	System B

Cellular company

Cellular One
Vanguard Cellular Systems
2 Thomas Drive
Westbrook, ME 04092

Customer service	207-776-2000
Local office	800-999-2369
. . . .	207-772-9805
Corporate office	919-282-3690

Sending calls

Visitor dialing instructions

Local calls	area code + 7 digits
Long distance	area code + 7 digits

Speed-dial numbers

Customer service	611
Emergency	911

Receiving calls

Caller dials the roamer access number below,
hears tone, dials cellular area code and number.
. . . . 207-776-0010

Billing information

Visitor rates
Most visitors . . . . . $3 day + 99¢ minute

Visitors without roaming agreements (p. 1159)
Call co. to use Amex, Mastercard, Visa

Identification numbers

City	Market	System	Billing	RSA
All served cities	152	00499	00000	MSA

Cellular company

Maine Cellular
190 Riverside Street Turnpike W
Portland, ME 04103

Corporate office .	some areas	800-479-0133
. . . .		207-773-0800

Sending calls

Visitor dialing instructions

Local calls	area code + 7 digits
Long distance	area code + 7 digits

Speed-dial numbers

AAA road service	*222
Coast Guard	*CG or *24
Customer service	611
Make a traffic report	*101 or *970
Maine state police	*77
National weather service	*NWS

Receiving calls

Caller dials the roamer access number below,
hears tone, dials cellular area code and number.
. . . . 207-233-7626

Follow Me Roaming forwarding (see p. 1156)
Activate, dial *18 send. Deactivate dial *19 send.

Billing information

Visitor rates
Most visitors . . . . . $3 day + 90¢ minute

Visitors without roaming agreements (p. 1159)
or Roamer Plus will intercept first call to bill you.

Identification numbers

City	Market	System	Billing	RSA
All served cities	152	00482	00000	MSA

For full coverage area, see Lewiston, ME.

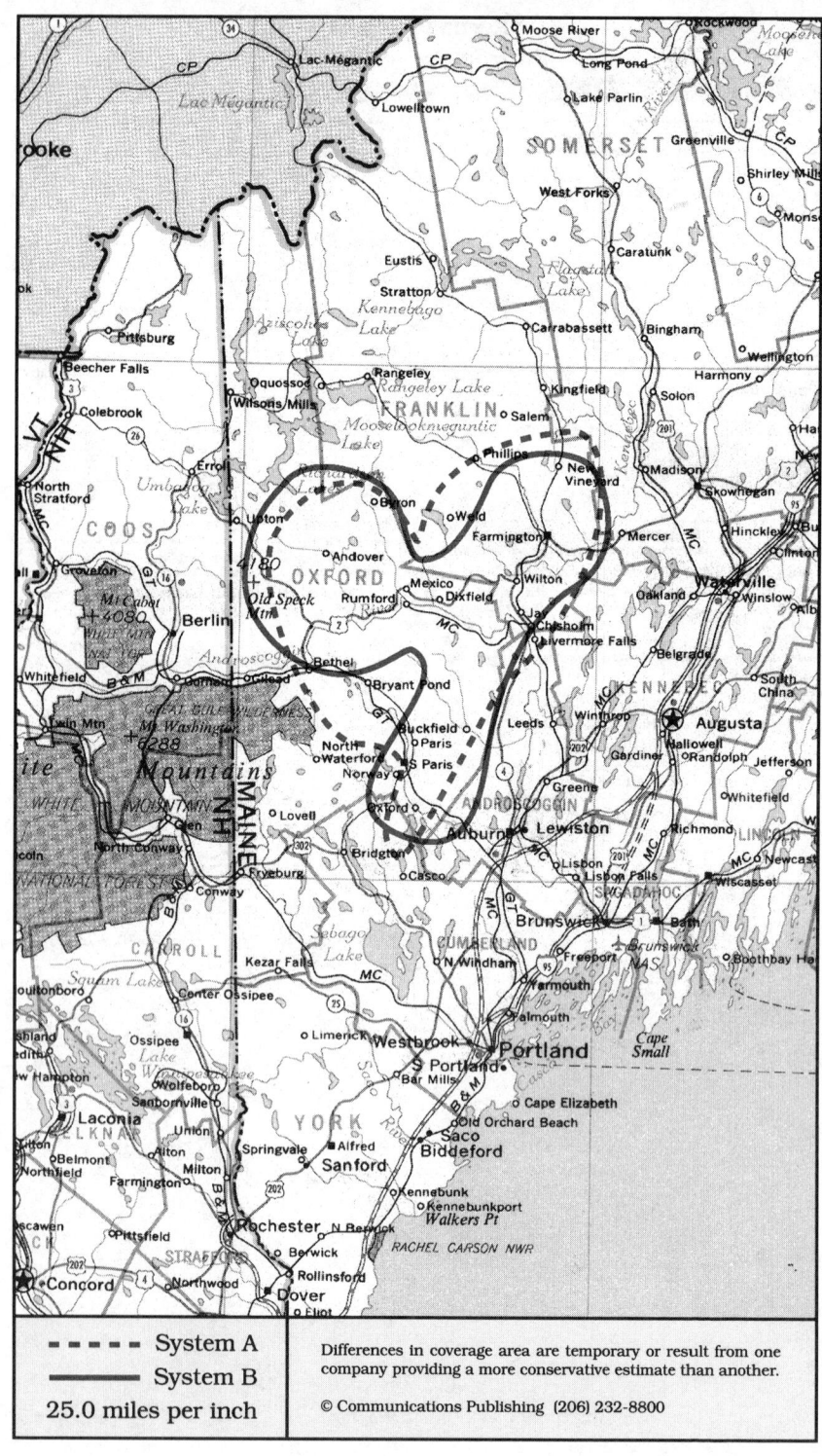

- - - - System A	Differences in coverage area are temporary or result from one company providing a more conservative estimate than another.
———— System B	
25.0 miles per inch	© Communications Publishing (206) 232-8800

Rumford, Maine
Farmington, Rumford

System A

Cellular company

United States Cellular
118 Mount Vernon Avenue
Augusta, ME 04330

Local office	207-621-0200
Corporate office	312-399-8900

Sending calls

Visitor dialing instructions

Local calls	area code + 7 digits
Long distance	area code + 7 digits

Speed-dial numbers

Customer service	*611
Emergency	911
Roaming information	*711

Receiving calls

Caller dials the roamer access number below,
hears tone, dials cellular area code and number.
. . . . 207-461-7626

Billing information

Visitor rates
Most visitors . . . . . $3 day + 75¢ minute

Visitors without roaming agreements (p. 1159)
Call co. to use Mastercard, Visa

Identification numbers

City	Market	System	Billing	RSA
All served cities	463	01313	00000	ME-1

For full coverage area, see Lewiston, ME.

System B

Cellular company

Western Maine Cellular
34 Main Street
South Paris, ME 04281

Route 25
Standish, ME 04084

Local office	in Maine	800-649-7303
. . . .		207-743-9703
Corporate office . .	Standish	207-642-9911

Sending calls

Visitor dialing instructions

Local calls	area code + 7 digits
Long distance	area code + 7 digits

Speed-dial numbers

Customer service	*611
Emergency	*77

Receiving calls

Caller dials the roamer access number below,
hears tone, dials cellular area code and number.
. . . . 207-233-7626

Follow Me Roaming forwarding (see p. 1156)
Activate, dial *18 send. Deactivate dial *19 send.

Billing information

Visitor rates
Most visitors . . . . . $3 day + 90¢ minute

Visitors without roaming agreements (p. 1159)
Roamer Plus will intercept first call to bill you.

Identification numbers

City	Market	System	Billing	RSA
All served cities	463	01314	00000	ME-1

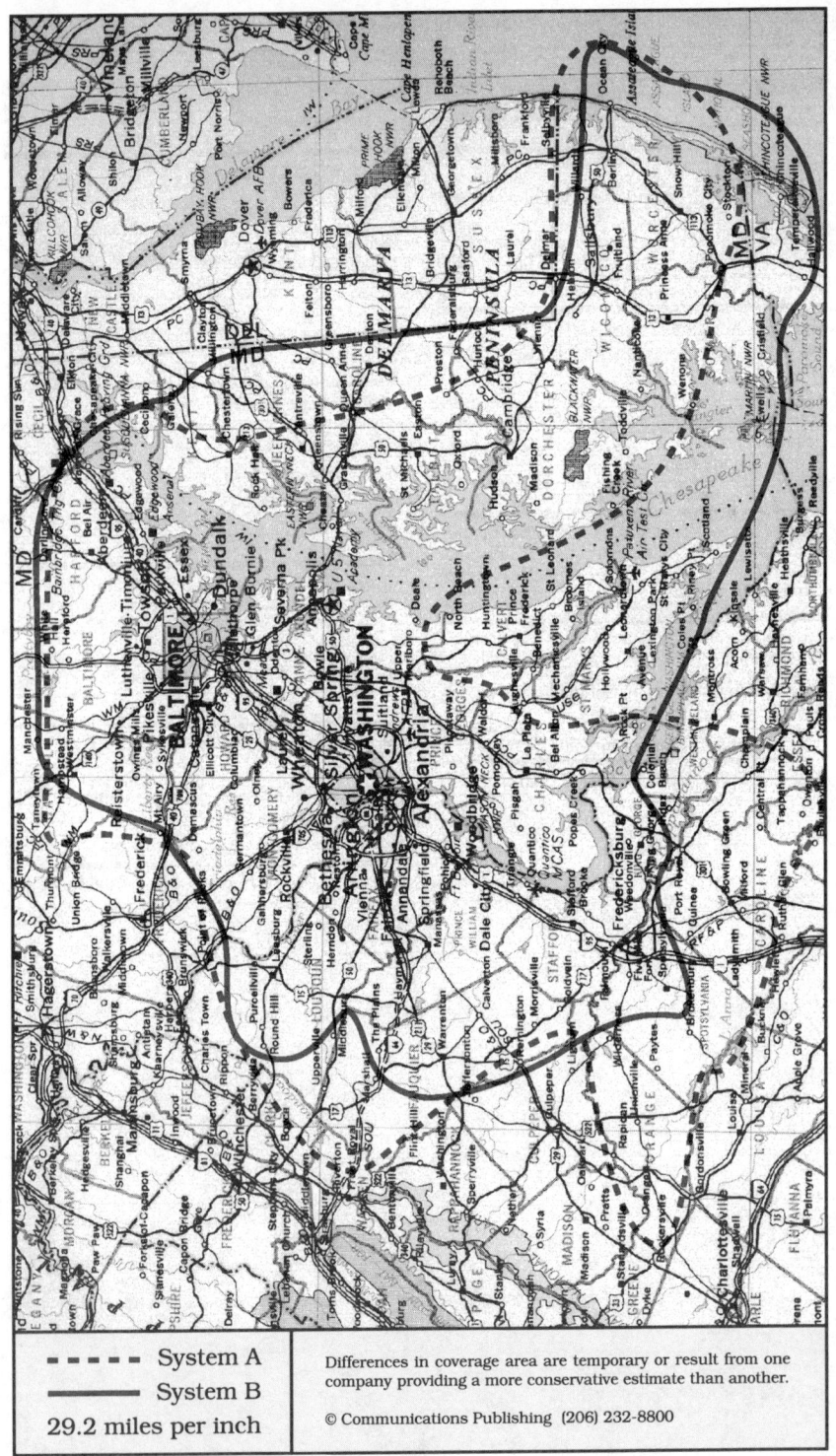

- - - - System A
——— System B
29.2 miles per inch

Differences in coverage area are temporary or result from one company providing a more conservative estimate than another.

© Communications Publishing (206) 232-8800

Baltimore, Maryland

Baltimore MD, Fredericksburg VA, Salisbury MD, Warrenton VA, Washington DC

System A	System B

System A

Cellular company

Cellular One
Southwestern Bell Mobile Systems
7855 Walker Drive, Suite 100
Greenbelt, MD 20770

Customer service	800-235-5663
. . . .	301-220-0060
Corporate office	301-220-3600

Sending calls

Visitor dialing instructions

Local calls	area code + 7 digits
Long distance	area code + 7 digits

Speed-dial numbers

Cellular information	*1394
Customer service	*611 or *811
Emergency	911
Make a traffic report in DC	*1198
Make a traffic report in MD	*1120
24 hour news	*1887

Receiving calls

Caller dials the roamer access number below,
hears tone, dials cellular area code and number.

Annapolis, MD	410-562-7626
Baltimore & Rockville, MD . .	410-440-7626
Fredericksburg, VA	703-840-7626
Salisbury, MD (Eastern Shore)	410-726-7626
Tappahannock, VA	804-450-7626
Washington, DC & Warrenton, VA	202-288-7626

NationLink & RoamingAmerica (see p. 1157)
Dial *31 send to receive calls, *30 to deactivate.
Dial *32 send to tell caller the roamer access no.
Service is not automatic and must be turned on.
This company's subscribers have level 6 service.

Billing information

Visitor rates

Most visitors	$3 day + 95¢ minute

Visitors without roaming agreements (p. 1159)
Roamer Plus will intercept first call to bill you.

Identification numbers

City	Market	System	Billing	RSA
Baltimore	014	00013	00000	MSA
Fredericksburg	691	00013	30333	VA-11
Salisbury (E. Shore)	468	00013	30345	MD-2
Tappahannock	692	00013	30343	VA-12
Warrenton	690	00013	30341	VA-10
Washington	008	00013	00000	MSA

For full coverage area, see Culpeper, VA and
Martinsburg, WV.

System B

Cellular company

Bell Atlantic Mobile
1420 Joh Avenue, Suite E
Baltimore, MD 21227

Customer service	800-922-0204
Local office	410-646-5700
Corporate office	908-306-7000

Sending calls

Visitor dialing instructions

Local calls	area code + 7 digits
Long distance	area code + 7 digits

Speed-dial numbers

Customer service	*611
Emergency	911
Make a traffic report in MD . .	301-555-9225
WBAL talk radio in MD . . .	301-555-2886
WRC talk radio in DC	202-555-9721

Receiving calls

Caller dials the roamer access number below,
hears tone, dials cellular area code and number.

Baltimore, MD	410-382-7626
Fredericksburg & Warrenton, VA	703-220-7626
Salisbury, MD (Eastern Shore) .	410-726-7626
Washington, DC	301-502-7626

Follow Me Roaming forwarding (see p. 1156)
Activate, dial *18 send. Deactivate dial *19 send.

Billing information

Visitor rates

Most visitors	$3 day + $1 minute

Visitors without roaming agreements (p. 1159)
Roamer Plus will intercept first call to bill you.

Identification numbers

City	Market	System	Billing	RSA
Baltimore	014	00018	40354	MSA
Fredericksburg	691	00018	30090	VA-11
King George	692	00018	30092	VA-12
Salisbury (E. Shore)	468	00018	30080	MD-2
Warrenton	690	00022	30088	VA-10
Washington	008	00018	00000	MSA

For full coverage area, see Frederick, MD.

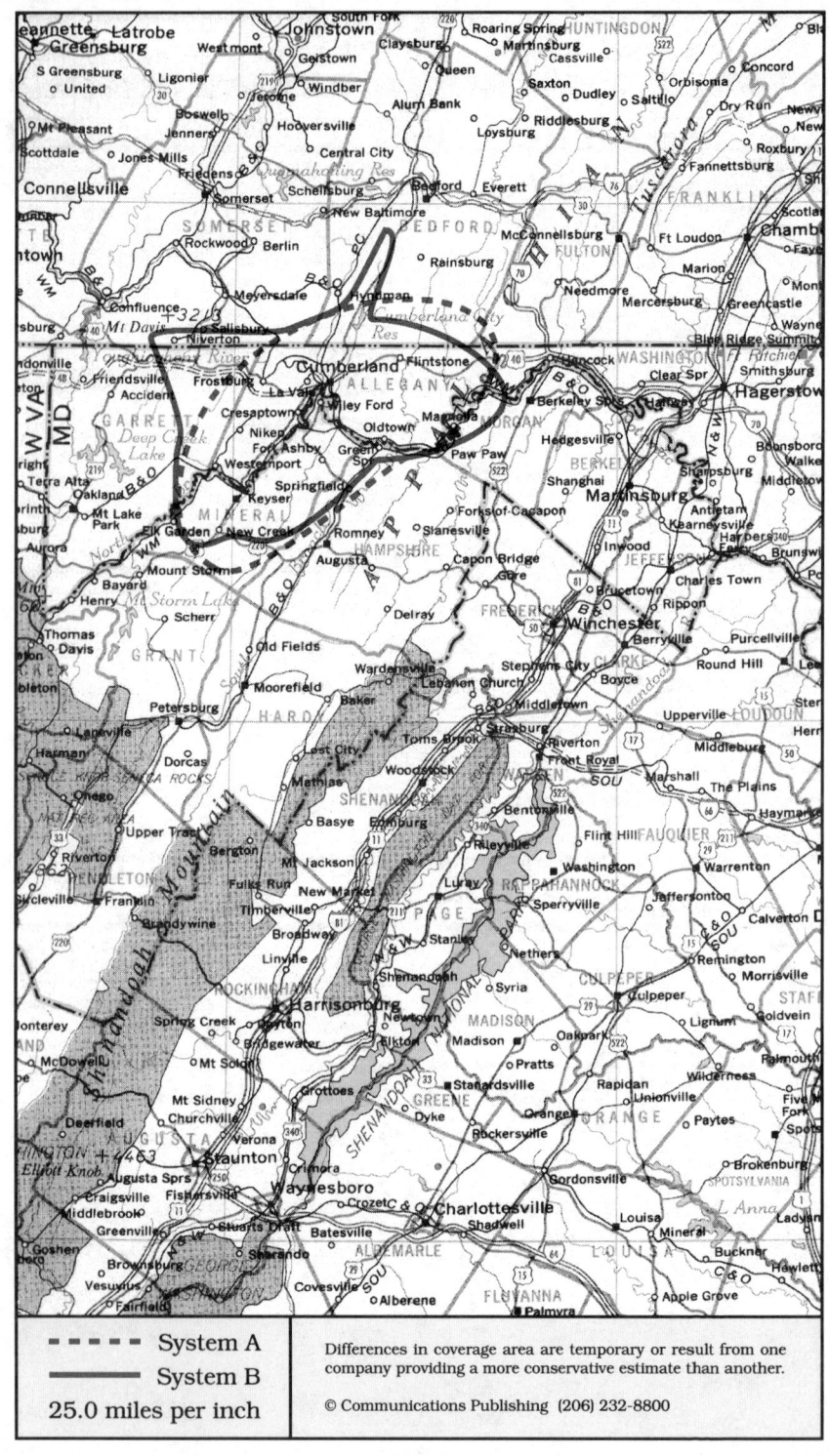

- - - - System A	Differences in coverage area are temporary or result from one
——— System B	company providing a more conservative estimate than another.
25.0 miles per inch	© Communications Publishing (206) 232-8800

Cumberland, Maryland
Cumberland, Frostburg

System A	System B

Cellular company

Cellular One
Cellular Information Systems
514 E Pleasant Valley Boulevard #1
Altoona, PA 16602

Local office	from Cumberland	301-724-2355
. . . . . .	from Altoona	814-946-4535
Corporate office	. . . . .	203-622-6317

Sending calls

Visitor dialing instructions

| Local calls . . . . . | area code + 7 digits |
| Long distance . . . | 1 + area code + 7 digits |

Speed-dial numbers

| Customer service . . . . | 611 |
| Emergency . . . . | 911 |

Receiving calls

Caller dials the roamer access number below, hears tone, dials cellular area code and number.
. . . . 301-697-7626

NationLink & RoamingAmerica (see p. 1157)
Caller hears your current roamer access number.
This company's subscribers have level 0 service.

Billing information

Visitor rates
Most visitors . . . . . $3 day + 85¢ minute

Visitors without roaming agreements (p. 1159)
Call co. to use Amex, Mastercard, Visa

Identification numbers

City	Market	System	Billing	RSA
All served cities	269	00321	00000	MSA

Cellular company

General Cellular
GenCell Management
18-20 N Centre Street
Cumberland, MD 21502

Customer service	800-888-7868
Local office	301-724-1047
Corporate office	707-425-8000

Sending calls

Visitor dialing instructions

| Local calls . . . . . . | area code + 7 digits |
| Long distance . . . | 1 + area code + 7 digits |

Speed-dial numbers

| Customer service . . . . | 611 |
| Emergency . . . . | 911 |

Receiving calls

Caller dials the roamer access number below, hears tone, dials cellular area code and number.
. . . . 301-707-7626

Billing information

Visitor rates
Most visitors . . . . . $3 day + 99¢ minute

Visitors without roaming agreements (p. 1159)
American Roaming Network intercepts call to bill.

Identification numbers

City	Market	System	Billing	RSA
All served cities	269	00304	00000	MSA

- - - - System A ———— System B 25.0 miles per inch	Differences in coverage area are temporary or result from one company providing a more conservative estimate than another. © Communications Publishing (206) 232-8800

Frederick, Maryland

System A	System B

System A

Cellular company

Cellular One of Frederick
Horizon Cellular Telephone
1170-C W Patrick Street
Frederick, MD 21702

Local office	301-846-0555
Corporate office	215-651-5900

Sending calls

Visitor dialing instructions

Local calls	area code + 7 digits
Long distance . . .	1 + area code + 7 digits

Speed-dial numbers

Customer service	611
Emergency	911
Frederick police	*FPD or *373
Report drunk drivers	*33

Receiving calls

Caller dials the roamer access number below,
hears tone, dials cellular area code and number.
. . . . 301-401-7626

NationLink & RoamingAmerica (see p. 1157)
Dial *32 send to tell caller the roamer access no.
This company's subscribers have level 3 service.

Billing information

Visitor rates

Most visitors	$3 day + 99¢ minute

Visitors without roaming agreements (p. 1159)
American Roaming Network intercepts call to bill.

Identification numbers

City	Market	System	Billing	RSA
All served cities	469	01325	00000	MD-3

System B

Cellular company

Bell Atlantic Mobile
1420 Joh Avenue, Suite E
Baltimore, MD 21227

Customer service	800-922-0204
Local office	410-646-5700
Corporate office	908-306-7000

Sending calls

Visitor dialing instructions

Local calls	area code + 7 digits
Long distance	area code + 7 digits

Speed-dial numbers

Customer service	*611
Emergency	911
Make a traffic report in MD . .	301-555-9225
WBAL talk radio in MD . . .	301-555-2886

Receiving calls

Caller dials the roamer access number below,
hears tone, dials cellular area code and number.
. . . . 301-606-7626

Follow Me Roaming forwarding (see p. 1156)
Activate, dial *18 send. Deactivate dial *19 send.

Billing information

Visitor rates

Most visitors	$3 day + $1 minute

Visitors without roaming agreements (p. 1159)
Roamer Plus will intercept first call to bill you.

Identification numbers

City	Market	System	Billing	RSA
All served cities	469	00018	30016	MD-3

For full coverage area, see Baltimore, MD.

	System A
— — — —	System A
———————	System B
25.0 miles per inch	

Differences in coverage area are temporary or result from one company providing a more conservative estimate than another.

© Communications Publishing (206) 232-8800

Hagerstown, Maryland

Cellular company

Cellular One
Vanguard Cellular Systems
580 W Northern Avenue
Hagerstown, MD 21742

Customer service	301-331-4000
Local office	301-791-2355
Corporate office	919-282-3690

Sending calls

Visitor dialing instructions

Local calls	area code + 7 digits
Long distance	area code + 7 digits

Speed-dial numbers

Customer service	611
Emergency	911
Telephone repair	811

Receiving calls

Caller dials the roamer access number below,
hears tone, dials cellular area code and number.
. . . . 301-331-7626

NationLink & RoamingAmerica (see p. 1157)
Dial *31 send to receive calls, *30 to deactivate.
Dial *32 send to tell caller the roamer access no.
This company's subscribers have level 6 service.

Billing information

Visitor rates
Most visitors . . . . . $3 day + 99¢ minute

Visitors without roaming agreements (p. 1159)
Call co. to use Amex, Mastercard, Visa

Identification numbers

City	Market	System	Billing	RSA
All served cities	257	00381	00000	MSA

Cellular company

Atlantic Cellular
1403 Dual Highway
Hagerstown, MD 21740

Local office	301-739-8300
Corporate office	401-421-7090

Sending calls

Visitor dialing instructions

Local calls	area code + 7 digits
Long distance	area code + 7 digits

Speed-dial numbers

Customer service	*611
Emergency	911
Towing	*4357 or *HELP

Receiving calls

Caller dials the roamer access number below,
hears tone, dials cellular area code and number.
. . . . 301-491-7626

Follow Me Roaming forwarding (see p. 1156)
Activate, dial *18 send. Deactivate dial *19 send.

Billing information

Visitor rates
Most visitors . . . . . $3 day + 75¢ minute

Visitors without roaming agreements (p. 1159)
Cannot call since company takes no credit cards.

Identification numbers

City	Market	System	Billing	RSA
All served cities	257	00364	00000	MSA

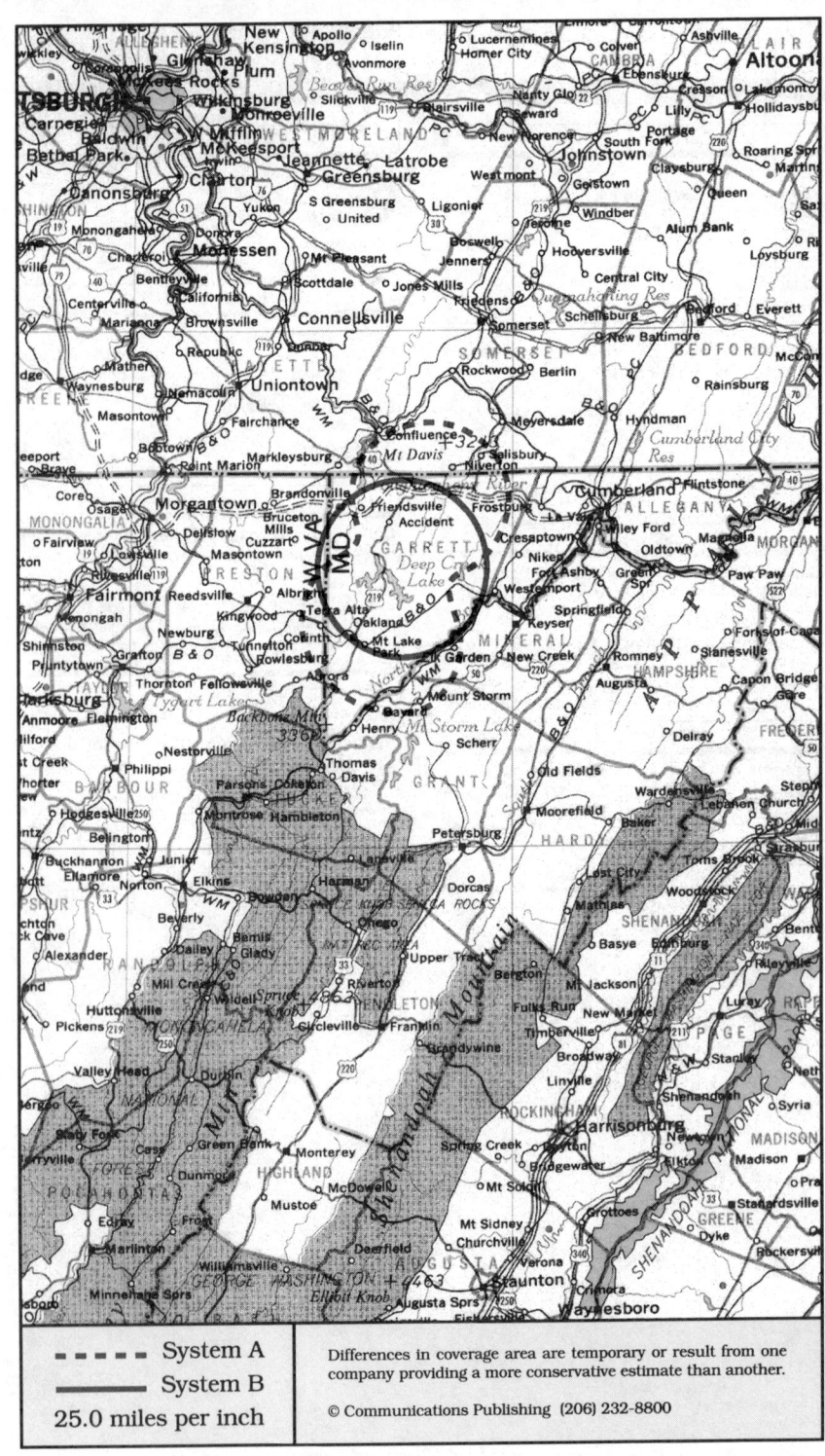

System A

System B

25.0 miles per inch

Differences in coverage area are temporary or result from one company providing a more conservative estimate than another.

© Communications Publishing (206) 232-8800

Oakland, Maryland
McHenry, Oakland

Cellular company

Cellular One of Maryland
Route 6, Box 51509 #2
Oakland, MD 21550

Corporate office 301-334-2446

Sending calls

Visitor dialing instructions
Local calls . . . . . area code + 7 digits
Long distance . . . area code + 7 digits

Speed-dial numbers
Customer service 611
Emergency 911

Receiving calls

Caller dials the roamer access number below,
hears tone, dials cellular area code and number.
. . . . 301-501-7626

Billing information

Visitor rates
Most visitors . . . . . $3 day + 99¢ minute

Visitors without roaming agreements (p. 1159)
Roamer Plus will intercept first call to bill you.

Identification numbers

City	Market	System	Billing	RSA
All served cities	467	01321	00000	MD-1

Cellular company

United States Cellular
663 Pittsburgh Street
Uniontown, PA 15401

Customer service 800-944-9066
Local office 412-437-0800
Corporate office 312-399-8900

Sending calls

Visitor dialing instructions
Local calls . . . . . . area code + 7 digits
Long distance . . . . area code + 7 digits

Speed-dial numbers
Customer service 611
Emergency 911
Roaming information *711

Receiving calls

Caller dials the roamer access number below,
hears tone, dials cellular area code and number.
. . . . 301-616-7626

Billing information

Visitor rates
Most visitors . . . . . $3 day + 99¢ minute

Visitors without roaming agreements (p. 1159)
Call co. to use Mastercard, Visa

Identification numbers

City	Market	System	Billing	RSA
All served cities	467	01322	00000	MD-1

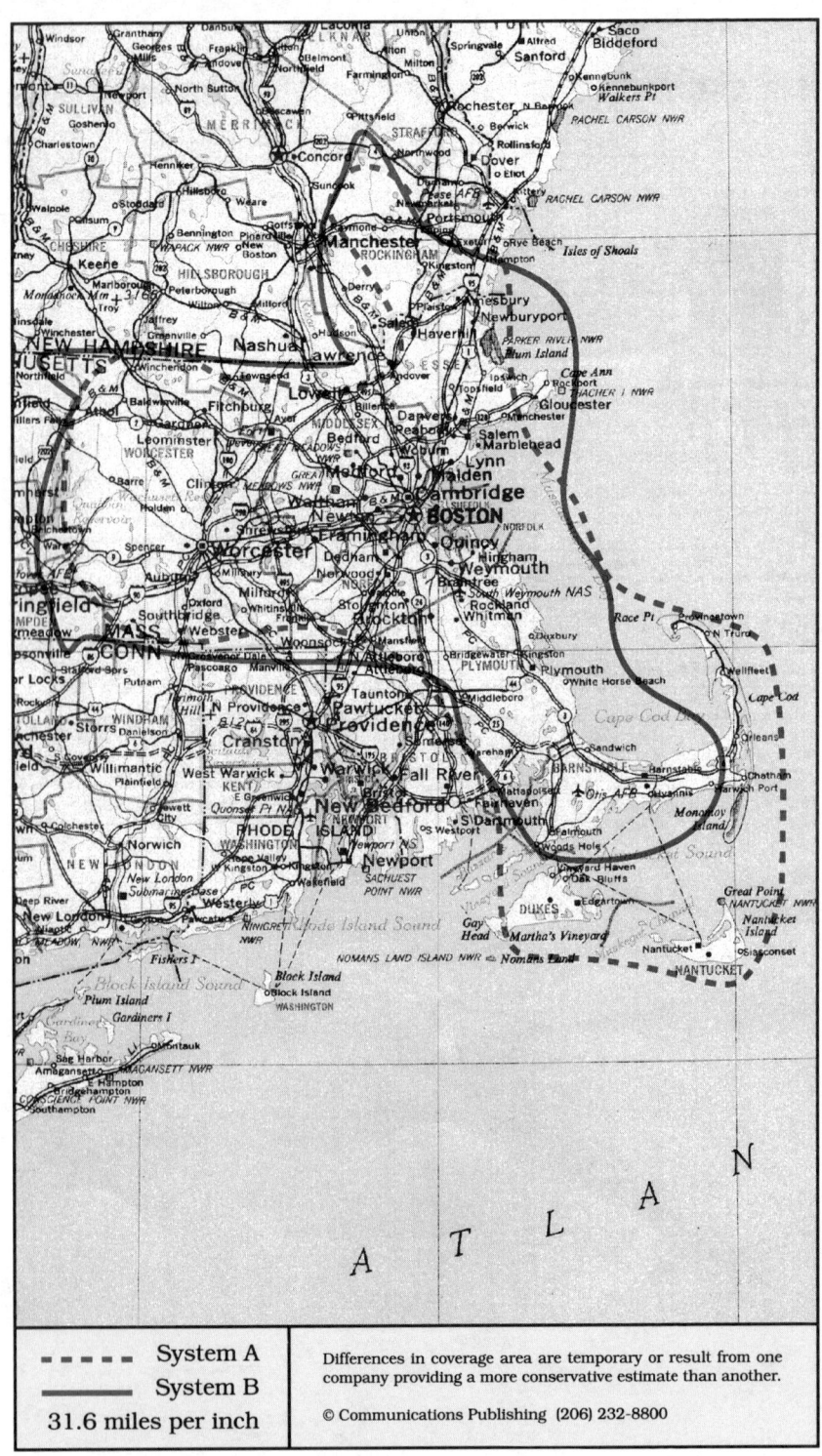

- - - - - System A	Differences in coverage area are temporary or result from one
———— System B	company providing a more conservative estimate than another.
31.6 miles per inch	© Communications Publishing (206) 232-8800

Boston, Massachusetts
Boston, Derry NH, Edgartown, Haverhill, Hyannis, Lawrence, Lowell, Worcester

System A	System B

Cellular company

Cellular One
Southwestern Bell Mobile Systems
100 Lowder Brook Drive
Westwood, MA 02090

Customer service	800-235-5663
. . . .	617-462-7000
Local office	617-462-4000
Corporate office	214-733-2000

Sending calls

Visitor dialing instructions
Local calls	area code + 7 digits
Long distance . . .	1 + area code + 7 digits

Speed-dial numbers
Customer service	*611
Emergency	*SP or 911
Hear a traffic report	1030
State police	*SP or *77
Traffic information	*STI or *784
Weather	*786

Receiving calls

Caller dials the roamer access number below,
hears tone, dials cellular area code and number.
. . . . 617-633-7626

NationLink & RoamingAmerica (see p. 1157)
Dial *31 send to receive calls, *30 to deactivate.
Dial *32 send to tell caller the roamer access no.
This company's subscribers have level 6 service.

Billing information

Visitor rates
Most visitors . . . . . $3 day + 99¢ minute

Visitors without roaming agreements (p. 1159)
Call co. to use Amex

Identification numbers

City	Market	System	Billing	RSA
Boston	006	00007	00000	MSA
Hyannis	471	00007	00000	MA-2
Worcester	055	00007	00000	MSA

Cellular company

NYNEX Mobile Communications
600 Unicorn Park Drive
Woburn, MA 01801

Customer service . . .		800-538-4747
Local office . . .	Cambridge	617-868-0300
. . . .	Hyannis	508-790-6944
. . . .	Natich	508-651-3000
. . . .	Norwell	617-982-8686
. . . .	Peabody	508-977-9020
. . . .	Woburn	617-932-1200
Corporate office		914-365-7200

Sending calls

Visitor dialing instructions
Local calls	area code + 7 digits
Long distance	area code + 7 digits

Speed-dial numbers
Celtics hotline	*33
Customer service	611
Directory assistance	411
Emergency	911
State police	*SP or *77
Stock market update	*37
Traffic report	*1
Traffic report WRKO	*68

Receiving calls

Caller dials the roamer access number below,
hears tone, dials cellular area code and number.
Boston, MA	617-571-7626
Massachusetts outside Boston .	508-776-7626
Derry, NH (Rockingham County)	603-426-7626

Follow Me Roaming forwarding (see p. 1156)
Activate, dial *18 send. Deactivate dial *19 send.

Billing information

Visitor rates
Most visitors . . . . . $3 day + 95¢ minute

Visitors without roaming agreements (p. 1159)
Call co. to use Amex, Mastercard, Visa
or Roamer Plus will intercept first call to bill you.

Identification numbers

City	Market	System	Billing	RSA
Boston	006	00028	00000	MSA
Hyannis	471	00028	30222	MA-2
Worcester	055	00028	00000	MSA

For full coverage area, see Providence, RI.

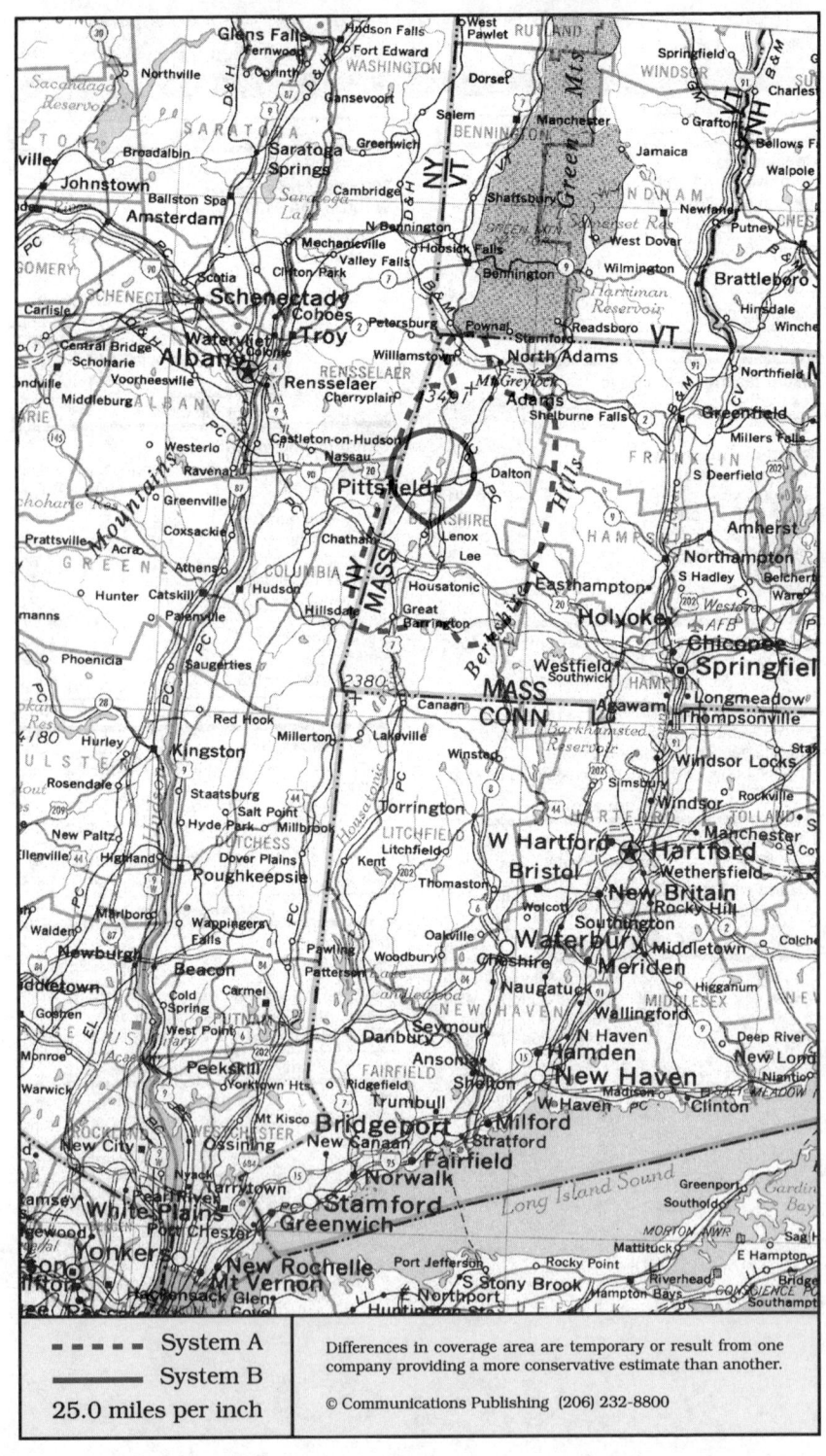

-----	System A
———	System B
25.0 miles per inch	

Differences in coverage area are temporary or result from one company providing a more conservative estimate than another.

© Communications Publishing (206) 232-8800

Pittsfield, Massachusetts

System A

Cellular company

Bell Atlantic Mobile
1123-B Riverdale Road
West Springfield, MA 01089

Customer service	. . . . .	800-852-3630
Local office	. . . . . .	413-781-6000
Corporate office	. . . . .	908-306-7000

Sending calls

Visitor dialing instructions

Local calls	. . . .	area code + 7 digits
Long distance	. . . .	area code + 7 digits

Speed-dial numbers

Customer service	*611
Emergency	911
Massachusetts state police . . .	*677
Roaming information	*711
Windsor, CT office	*811

Receiving calls

Caller dials the roamer access number below, hears tone, dials cellular area code and number.
. . . . 413-448-1000

NationLink & RoamingAmerica (see p. 1157)
Caller hears your current roamer access number. This company's subscribers have level 0 service.

Billing information

Visitor rates
Most visitors . . . . . $3 day + 99¢ minute

Visitors without roaming agreements (p. 1159)
Roamer Plus will intercept first call to bill you.

Identification numbers

City	Market	System	Billing	RSA
All served cities	213	00119	35119	MSA

For full coverage area, see Bridgeport, CT.

System B

Cellular company

NYNEX Mobile Communications
600 Unicorn Park Drive
Woburn, MA 01801

Customer service	. . . .	800-538-4747
Local office	. . . . . .	617-932-1200
Corporate office	. . . . .	914-365-7200

Sending calls

Visitor dialing instructions

Local calls	. . . . . .	area code + 7 digits
Long distance	. . . . .	area code + 7 digits

Speed-dial numbers

Customer service	611
Directory assistance	411
Emergency	911
State police	*77
Traffic report	*1

Receiving calls

Caller dials the roamer access number below, hears tone, dials cellular area code and number.
. . . . 413-446-7626

Follow Me Roaming forwarding (see p. 1156)
Activate, dial *18 send. Deactivate dial *19 send.

Billing information

Visitor rates
Most visitors . . . . . $3 day + 95¢ minute

Visitors without roaming agreements (p. 1159)
Call co. to use Amex, Mastercard, Visa or Roamer Plus will intercept first call to bill you.

Identification numbers

City	Market	System	Billing	RSA
All served cities	213	00480	00000	MSA

▄▄ ▄▄ ▄▄ System A ———— System B 25.0 miles per inch	Differences in coverage area are temporary or result from one company providing a more conservative estimate than another. © Communications Publishing (206) 232-8800

Allegan, Michigan

Cellular company

Allegan Cellular
134 Water Street
Allegan, MI 49010

Corporate office 616-673-5171

Sending calls

Visitor dialing instructions
Local calls . . . . 7 digits
Long distance . . . 1 + area code + 7 digits

Speed-dial numbers
Customer service *611
Emergency 911

Receiving calls

Caller dials the roamer access number below,
hears tone, dials cellular area code and number.
. . . . 616-650-7626

NationLink & RoamingAmerica (see p. 1157)
Dial *31 send to receive calls, *30 to deactivate.
Dial *32 send to tell caller the roamer access no.
This company's subscribers have level 5 service.

Billing information

Visitor rates
Most visitors . . . . . $3 day + 85¢ minute

Visitors without roaming agreements (p. 1159)
Roamer Plus will intercept first call to bill you.

Identification numbers

City	Market	System	Billing	RSA
All served cities	479	01345	00000	MI-8

Cellular company

Century Cellunet
2843 E Paris SE
Grand Rapids, MI 49512

Customer service 800-848-4577
. . . . 616-285-7400
Local office 616-940-0985
Corporate office 318-388-9000

Sending calls

Visitor dialing instructions
Local calls . . . . . . area code + 7 digits
Long distance . . . 1 + area code + 7 digits

Speed-dial numbers
Customer service *611
Emergency *911
Roaming information *711

Receiving calls

Caller dials the roamer access number below,
hears tone, dials cellular area code and number.
. . . . 616-240-7626

Follow Me Roaming forwarding (see p. 1156)
Activate, dial *18 send. Deactivate dial *19 send.

Billing information

Visitor rates
Most visitors . . . . . $3 day + 75¢ minute

Visitors without roaming agreements (p. 1159)
Call co. to use Amex, Discover, Mastercard, Visa

Identification numbers

City	Market	System	Billing	RSA
All served cities	479	00244	00000	MI-8

For full coverage area, see Grand Rapids, MI and
Kalamazoo, MI.

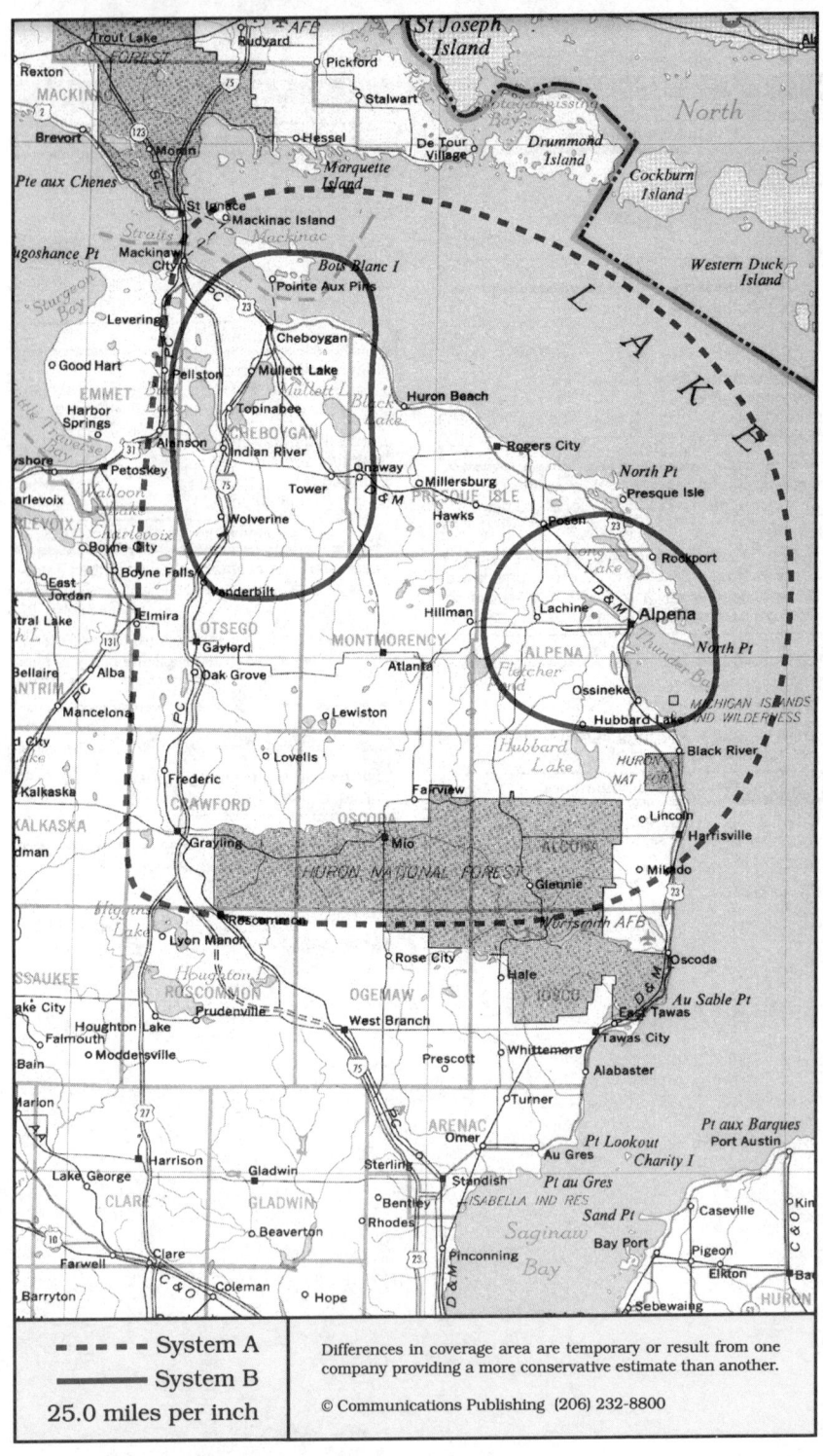

Alpena, Michigan

Alpena, Atlanta, Cheboygan, Gaylord, Grayling, Harrisville, Milo, Rogers City

System A	System B

System A

Cellular company

Cellular One
RFB Cellular
1421 W Main Street
Gaylord, MI 49735

Corporate office 517-732-0032

Sending calls

Visitor dialing instructions
Local calls . . . . . area code + 7 digits
Long distance . . . 1 + area code + 7 digits

Speed-dial numbers
Customer service *611
Emergency 911

Receiving calls

Caller dials the roamer access number below,
hears tone, dials cellular area code and number.
Cheboygan 616-290-7626
Gaylord 517-350-7626

Follow Me Roaming forwarding (see p. 1156)
Phone Me Anywhere forwarding (see p. 1156)
Activate, dial *18 send. Deactivate dial *19 send.

NationLink & RoamingAmerica (see p. 1157)
Dial *31 send to receive calls, *30 to deactivate.
Dial *32 send to tell caller the roamer access no.
This company's subscribers have level 5 service.

Billing information

Visitor rates
Most visitors . . . . . $3 day + 99¢ minute

Visitors without roaming agreements (p. 1159)
Call co. to use Amex, Mastercard, Visa

Identification numbers

City	Market	System	Billing	RSA
All served cities	475	01337	00000	MI-4

System B

Cellular company

United States Cellular
223 N Main
Cheboygan, MI 49721

Local office 616-627-9100
Corporate office 312-399-8900

Sending calls

Visitor dialing instructions
Local calls . . . . . . area code + 7 digits
Long distance . . . . . area code + 7 digits

Speed-dial numbers
Customer service 611
Emergency 911
Roaming information *711

Receiving calls

Caller dials the roamer access number below,
hears tone, dials cellular area code and number.
. . . . 616-420-7626

Billing information

Visitor rates
Most visitors . . . . . $3 day + 85¢ minute

Visitors without roaming agreements (p. 1159)
Call co. to use Mastercard, Visa

Identification numbers

City	Market	System	Billing	RSA
All served cities	475	01338	00000	MI-4

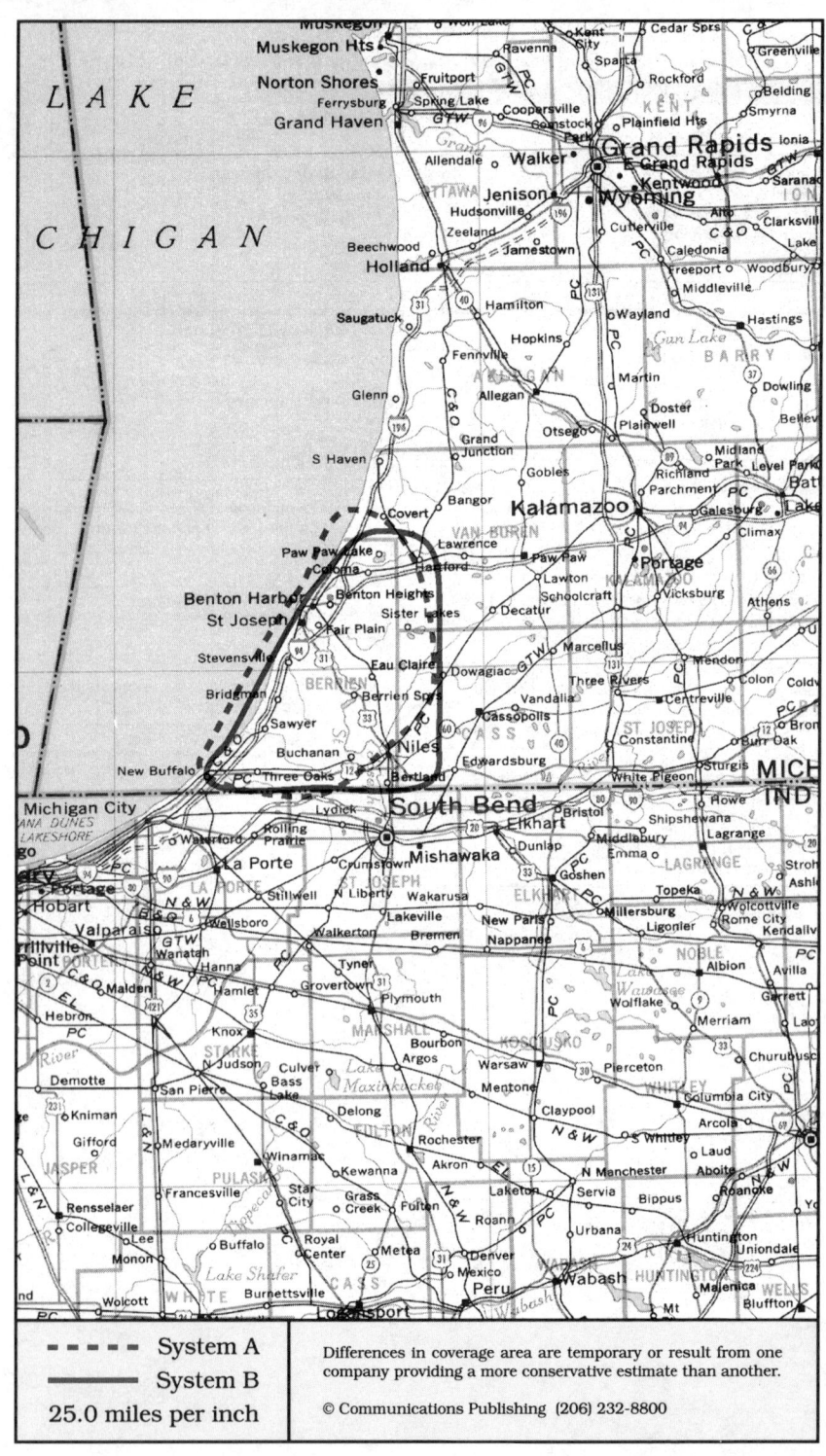

▬ ▬ ▬ ▬	System A
▬▬▬▬	System B
25.0 miles per inch	

Differences in coverage area are temporary or result from one company providing a more conservative estimate than another.

© Communications Publishing (206) 232-8800

Benton Harbor, Michigan
Benton Harbor, Niles, St. Joseph

System A	System B

Cellular company

Cellular One of Berrien County
1021 Main Street
St. Joseph, MI 49085

Customer service	616-982-9911
Corporate office	616-982-9900

Sending calls

Visitor dialing instructions

Local calls	area code + 7 digits
Long distance . . .	1 + area code + 7 digits

Speed-dial numbers

Customer service	*611
Emergency	911

Receiving calls

Caller dials the roamer access number below,
hears tone, dials cellular area code and number.
. . . . 616-926-7626

NationLink & RoamingAmerica (see p. 1157)
Dial *31 send to receive calls, *30 to deactivate.
Dial *32 send to tell caller the roamer access no.
This company's subscribers have level 6 service.

Billing information

Visitor rates
Most visitors . . . . . $3 day + 99¢ minute

Visitors without roaming agreements (p. 1159)
Roamer Plus will intercept first call to bill you.

Identification numbers

City	Market	System	Billing	RSA
All served cities	193	00277	00000	MSA

Cellular company

Century Cellunet
2018 Washington Avenue
St. Joseph, MI 49085

Customer service	800-848-4577
Local office	616-983-4999
Corporate office	318-388-9000

Sending calls

Visitor dialing instructions

Local calls	area code + 7 digits
Long distance	area code + 7 digits

Speed-dial numbers

AAA emergency	*222
Customer service	*611
Emergency	*911

Receiving calls

Caller dials the roamer access number below,
hears tone, dials cellular area code and number.
. . . . 616-921-7626

Follow Me Roaming forwarding (see p. 1156)
Activate, dial *18 send. Deactivate dial *19 send.

Billing information

Visitor rates
Most visitors . . . . . $3 day + 85¢ minute

Visitors without roaming agreements (p. 1159)
Call co. to use Amex, Discover, Mastercard, Visa

Identification numbers

City	Market	System	Billing	RSA
All served cities	193	00260	00000	MSA

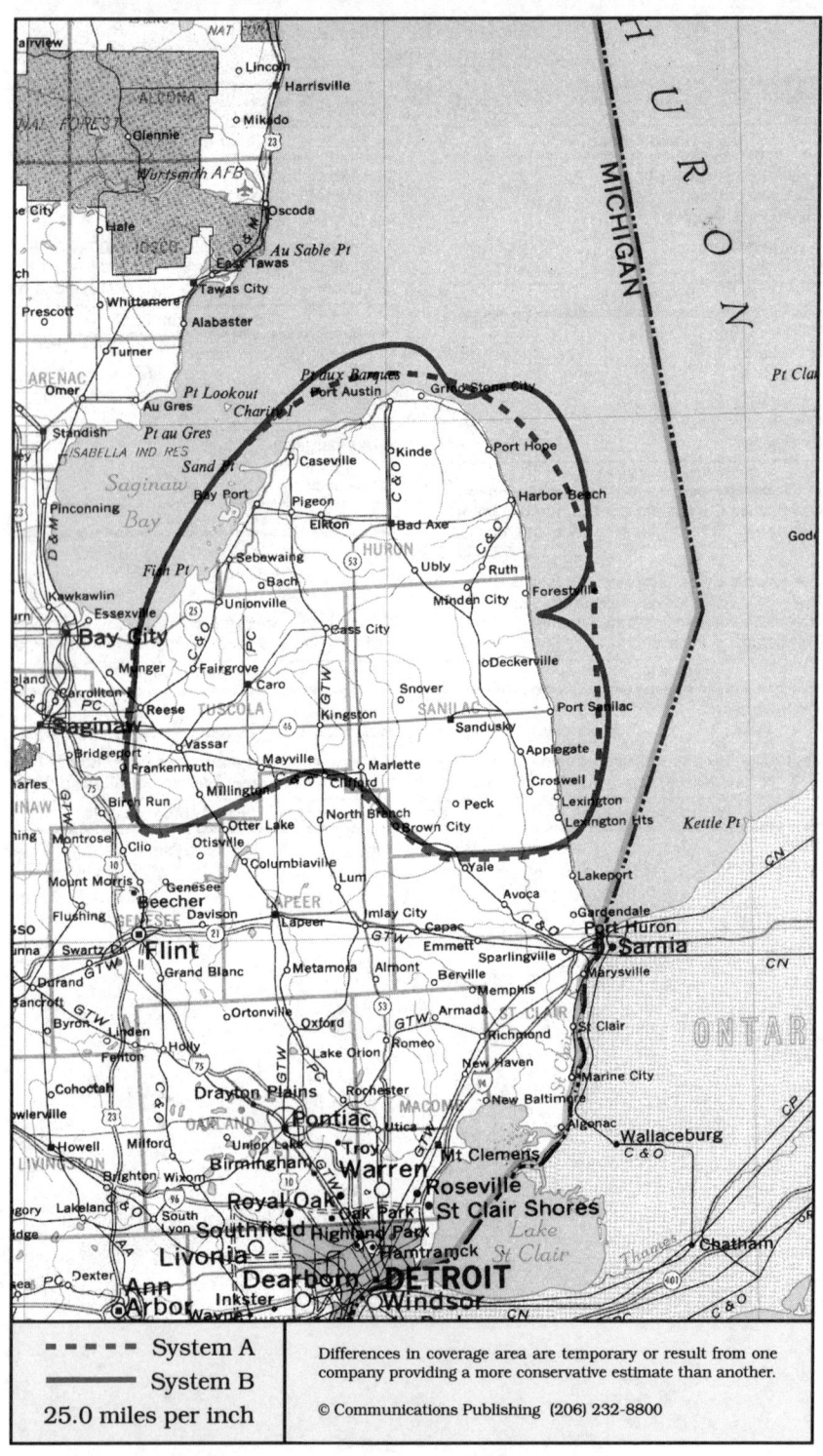

System A
System B
25.0 miles per inch

Differences in coverage area are temporary or result from one company providing a more conservative estimate than another.

© Communications Publishing (206) 232-8800

Caro, Michigan
Bad Axe, Caro, Sandusky

Cellular company

Cellular One
Lake Huron Cellular
1536 W Caro Road
Caro, MI 48723

Corporate office 800-624-8766
. . . . 517-673-1666

Sending calls

Visitor dialing instructions
Local calls . . . . . area code + 7 digits
Long distance . . . 1 + area code + 7 digits

Speed-dial numbers
Customer service 611
Emergency 911

Receiving calls

Caller dials the roamer access number below,
hears tone, dials cellular area code and number.
. . . . 517-670-7626

NationLink & RoamingAmerica (see p. 1157)
Caller hears your current roamer access number.
This company's subscribers have level 0 service.

Billing information

Visitor rates
Most visitors . . . . . $2 day + 75¢ minute

Visitors without roaming agreements (p. 1159)
Cannot call since company takes no credit cards.

Identification numbers

City	Market	System	Billing	RSA
All served cities	481	01349	00000	MI-10

Cellular company

Thumb Cellular
15 S Main Street, PO Box 650
Pigeon, MI 48755-0650

Corporate office . some areas 800-443-5057
. . . . 517-453-4333

Sending calls

Visitor dialing instructions
Local calls . . . . . area code + 7 digits
Long distance . . . 1 + area code + 7 digits

Speed-dial numbers
Customer service *611
Emergency 911

Receiving calls

Caller dials the roamer access number below,
hears tone, dials cellular area code and number.
. . . . 517-550-7626

Billing information

Visitor rates
Most visitors . . . . . $3 day + 75¢ minute

Visitors without roaming agreements (p. 1159)
Call co. to use Discover, Mastercard, Visa

Identification numbers

City	Market	System	Billing	RSA
All served cities	481	01350	00000	MI-10

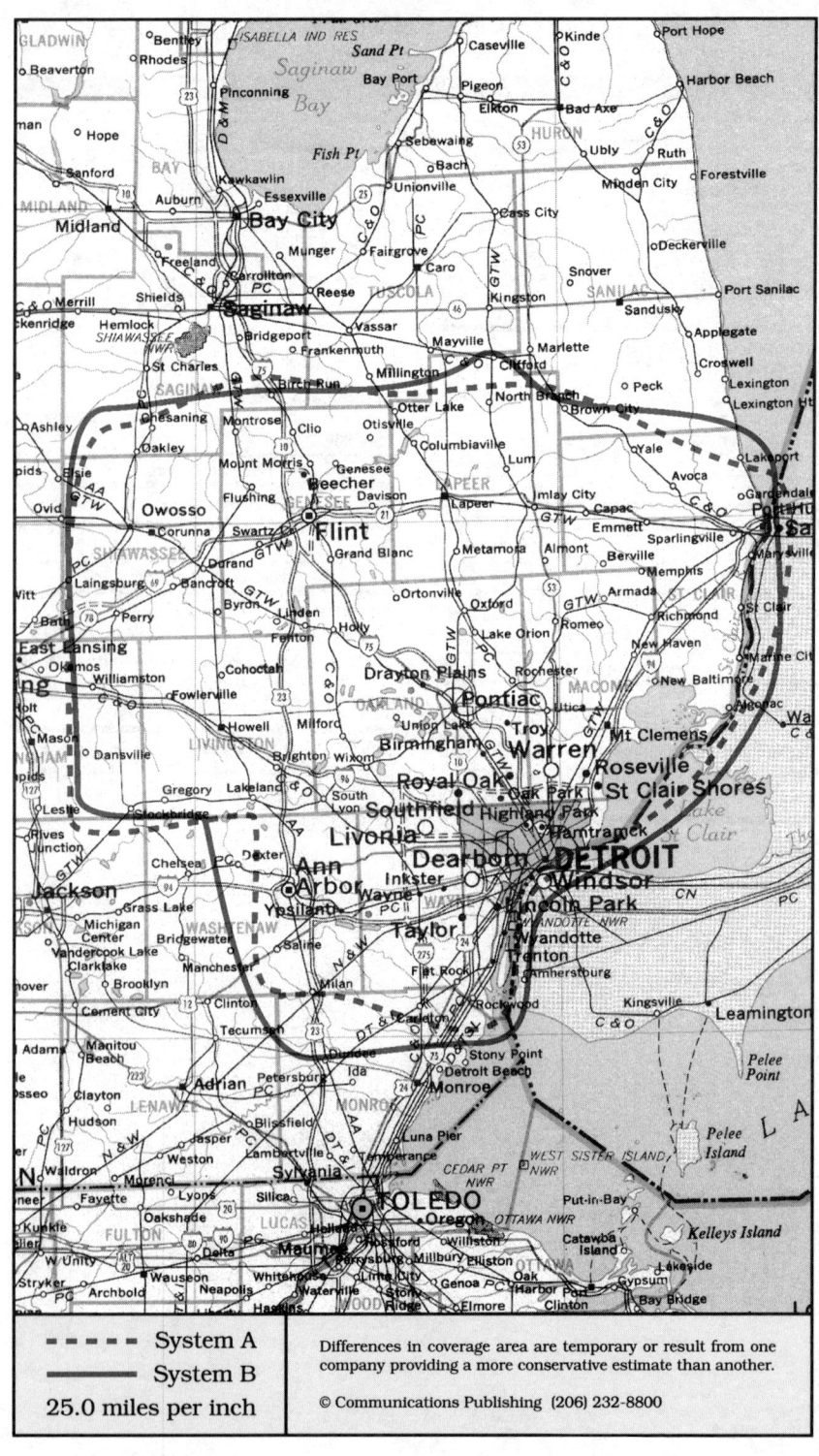

- - - - System A	Differences in coverage area are temporary or result from one company providing a more conservative estimate than another.
———— System B	
25.0 miles per inch	© Communications Publishing (206) 232-8800

Detroit, Michigan
Ann Arbor, Detroit, Flint, Owosso, Port Huron

System A	System B

Cellular company

Cellular One
New Par
31500 Northwestern Highway, Suite 300
Farmington Hills, MI 48334-2570

Customer service	800-452-3551
. . . .	313-737-5123
Local office	313-737-5100
Corporate office	614-436-4331

Sending calls

Visitor dialing instructions

Local calls	area code + 7 digits
Long distance . . .	0 + area code + 7 digits

Speed-dial numbers

Customer service	*611
Emergency	911
Make a traffic report	*995
TV 7 news	*7

Receiving calls

Caller dials the roamer access number below,
hears tone, dials cellular area code and number.
. . . . 313-938-7626

NationLink & RoamingAmerica (see p. 1157)
Dial *31 send to receive calls, *30 to deactivate.
Dial *32 send to tell caller the roamer access no.
This company's subscribers have level 6 service.

Billing information

Visitor rates
Most visitors . . . . . $3 day + 85¢ minute

Visitors without roaming agreements (p. 1159)
Roamer Plus will intercept first call to bill you.

Identification numbers

City	Market	System	Billing	RSA
Detroit	005	00021	00000	MSA
Flint	068	00021	00000	MSA
Grand Rapids	064	00021	00000	MSA

For full coverage area, see Grand Rapids, MI,
Saginaw, MI, and Toledo, OH.

Cellular company

Ameritech Mobile Communications
32255 Northwestern Highway, Suite 143
Farmington Hills, MI 48334-1573

Customer service	800-221-0994
. . . .	708-706-7300
Local office	313-737-6700
Corporate office	708-706-7600

Sending calls

Visitor dialing instructions

Local calls	area code + 7 digits
Long distance	area code + 7 digits

Speed-dial numbers

Customer service	*611
Dialing instructions	*711
Emergency	911
Roaming information	*712
TV 4 news tip hotline	*4

Receiving calls

Caller dials the roamer access number below,
hears tone, dials cellular area code and number.
. . . . 313-320-7626

Follow Me Roaming forwarding (see p. 1156)
Activate, dial *18 send. Deactivate dial *19 send.

Billing information

Visitor rates
Most visitors . . . . . $3 day + 85¢ minute

Visitors without roaming agreements (p. 1159)
Cannot call since company takes no credit cards.

Identification numbers

City	Market	System	Billing	RSA
Detroit	005	00010	00000	MSA
Flint	068	00010	00000	MSA
Grand Rapids	064	00010	00000	MSA

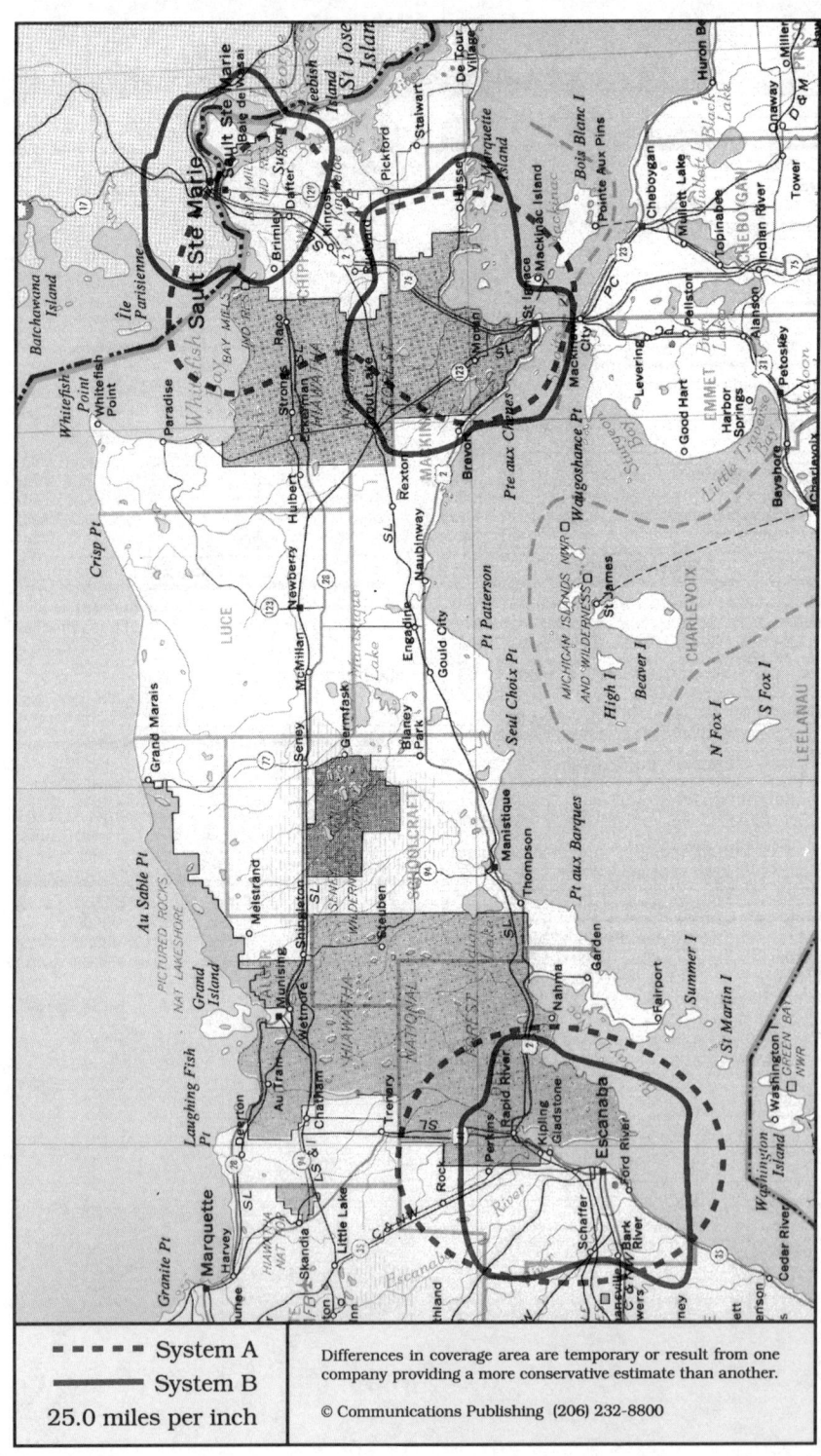

System A

System B

25.0 miles per inch

Differences in coverage area are temporary or result from one company providing a more conservative estimate than another.

© Communications Publishing (206) 232-8800

Escanaba, Michigan
Escanaba, St. Ignace, Sault Ste. Marie

System A	System B

Cellular company

Cellular One
Mackinac Cellular
1034 N Lincoln Road
Escanaba, MI 49829

Corporate office 906-786-2544

Sending calls

Visitor dialing instructions
Local calls . . . . . area code + 7 digits
Long distance . . . 1 + area code + 7 digits

Speed-dial numbers
Customer service 611
Emergency 911

Receiving calls

Caller dials the roamer access number below, hears tone, dials cellular area code and number.
. . . . 906-420-7626

Billing information

Visitor rates
Most visitors . . . . . $3 day + 99¢ minute

Visitors without roaming agreements (p. 1159)
Cannot call since company takes no credit cards.

Identification numbers

City	Market	System	Billing	RSA
All served cities	473	01333	00000	MI-2

Cellular company

Cellulink
Pacific Telecom Cellular
4600 W College Avenue
Appleton, WI 54915

Customer service 800-236-9700
. . . . 414-841-1111
Corporate office 414-841-1100

Sending calls

Visitor dialing instructions
Local calls . . . . . area code + 7 digits
Long distance . . . . . area code + 7 digits

Speed-dial numbers
Customer service *611
Emergency 911
Roaming information *711

Receiving calls

Caller dials the roamer access number below, hears tone, dials cellular area code and number.
. . . . 906-630-7626

Follow Me Roaming forwarding (see p. 1156)
Activate, dial *18 send. Deactivate dial *19 send.

Billing information

Visitor rates
Most visitors . . . . . $3 day + 85¢ minute

Visitors without roaming agreements (p. 1159)
Cannot call since company takes no credit cards.

Identification numbers

City	Market	System	Billing	RSA
All served cities	473	01332	00000	MI-2

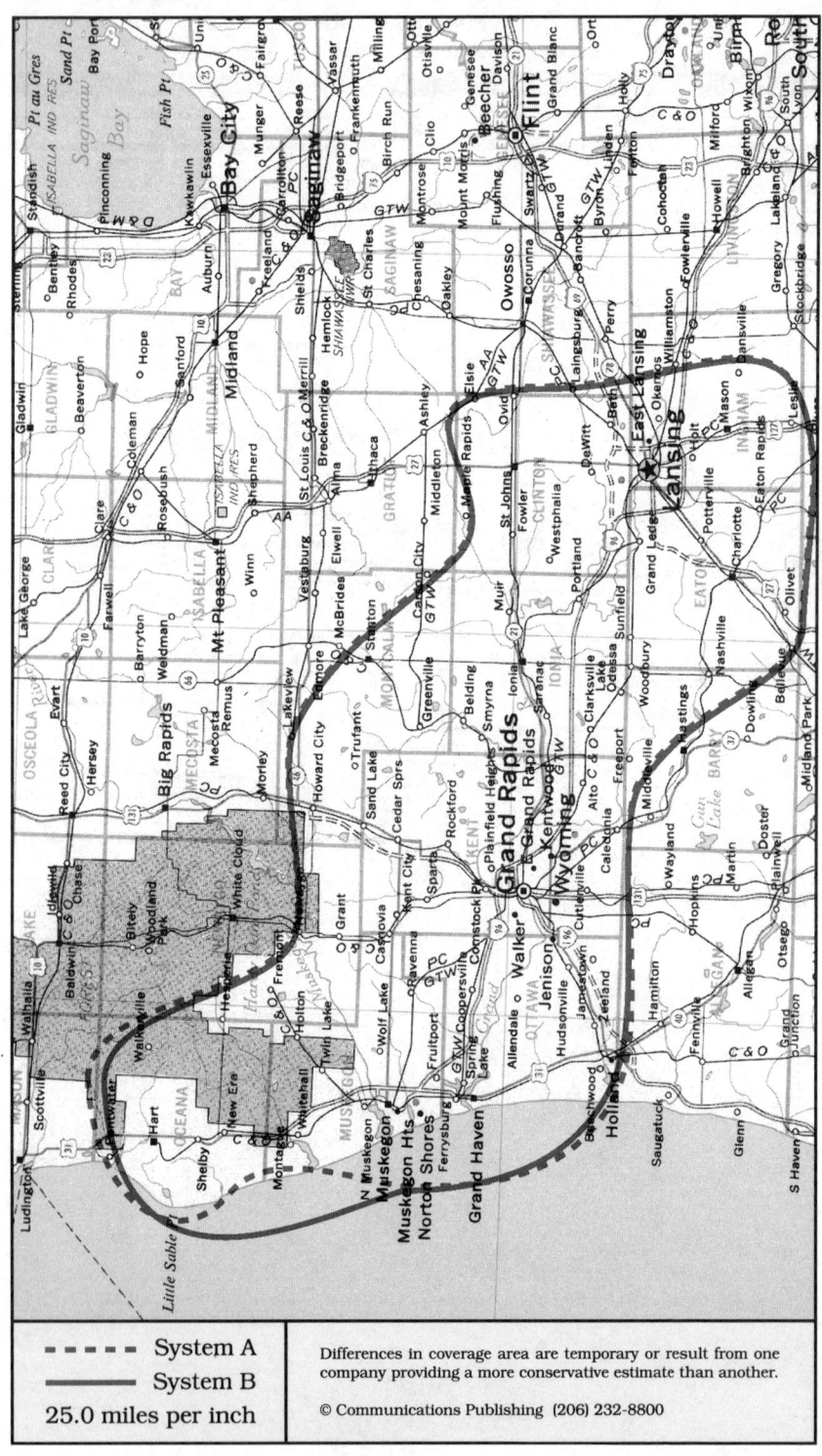

System A
System B
25.0 miles per inch

Differences in coverage area are temporary or result from one company providing a more conservative estimate than another.

© Communications Publishing (206) 232-8800

Grand Rapids, Michigan
Grand Rapids, Holland, Lansing, Muskegon

System A	System B

Cellular company

Cellular One
(affiliated with Cellular One of Detroit)
2020 Raybrook SE, Suite 306
Grand Rapids, MI 49546

3965 Okemos Road, Suite B-1
Okemos, MI 48864

Customer service	616-451-3523
. . . .	517-323-9462
Local office . . Grand Rapids	616-957-9451
. . . . Okemos (Lansing)	517-347-1801
Corporate office	313-737-5100

Sending calls

Visitor dialing instructions

Local calls	area code + 7 digits
Long distance . . .	0 + area code + 7 digits

Speed-dial numbers

Customer service	*611
Emergency	911
Make a traffic report	*995
TV 7 news	*7

Receiving calls

Caller dials the roamer access number below,
hears tone, dials cellular area code and number.

Grand Rapids, Muskegon . .	616-450-7626
Lansing	517-881-7626

NationLink & RoamingAmerica (see p. 1157)
Dial *31 send to receive calls, *30 to deactivate.
Dial *32 send to tell caller the roamer access no.
This company's subscribers have level 6 service.

Billing information

Visitor rates

Most visitors	$3 day + 85¢ minute

Visitors without roaming agreements (p. 1159)
Roamer Plus will intercept first call to bill you.

Identification numbers

City	Market	System	Billing	RSA
Grand Rapids	064	00021	00000	MSA
Lansing	078	00021	00000	MSA
Muskegon	181	00021	00000	MSA

For full coverage area, see Detroit, MI, Saginaw, MI, and Toledo, OH.

Cellular company

Century Cellunet
2843 E Paris SE
Grand Rapids, MI 49512

2339 Jolly Road
Okemos, MI 48864

Customer service	800-848-4577
. . . .	616-285-7400
Local office . . Grand Rapids	616-940-0985
. . . . . . from Muskegon	616-940-0985
. . . . Okemos (Lansing)	517-347-9000
Corporate office	318-388-9000

Sending calls

Visitor dialing instructions

Local calls	area code + 7 digits
Long distance . . .	1 + area code + 7 digits

Speed-dial numbers

Customer service	*611
Emergency	*911
Radio WODJ	*123
Roaming information	*711

Receiving calls

Caller dials the roamer access number below,
hears tone, dials cellular area code and number.

Grand Rapids, Holland . . .	616-240-7626
Lansing	517-331-7626
Muskegon	616-750-7626

Follow Me Roaming forwarding (see p. 1156)
Activate, dial *18 send. Deactivate dial *19 send.

Billing information

Visitor rates

Most visitors	$3 day + 75¢ minute

Visitors without roaming agreements (p. 1159)
Call co. to use Amex, Discover, Mastercard, Visa

Identification numbers

City	Market	System	Billing	RSA
Grand Rapids	064	00244	00000	MSA
Lansing	078	00188	00000	MSA
Muskegon	181	00448	00000	MSA

For full coverage area, see Allegan, MI, Kalamazoo, MI, Mount Pleasant, MI, Saginaw, MI, and Traverse City, MI.

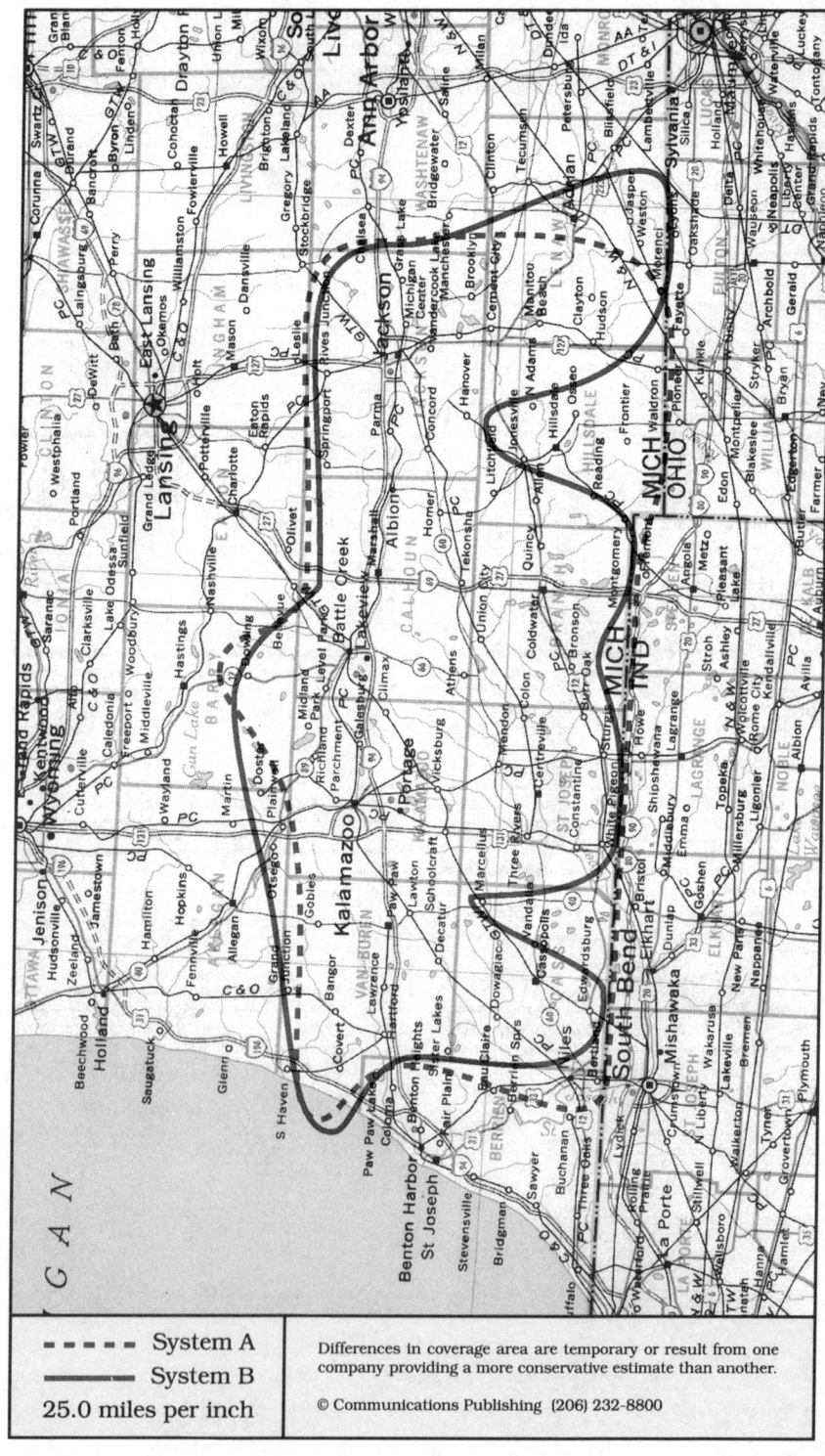

- - - - System A
——— System B
25.0 miles per inch

Differences in coverage area are temporary or result from one
company providing a more conservative estimate than another.

© Communications Publishing (206) 232-8800

Kalamazoo, Michigan
Battle Creek, Jackson, Kalamazoo, Three Rivers

System A	System B

Cellular company

Cellular One of Kalamazoo
Centennial Cellular
2505 Capital Avenue SW
Battle Creek, MI 49015

543 Mall Drive
Portage, MI 49002

Local office	. .	Battle Creek	616-965-1100
. . . .		Portage	616-580-5000
Corporate office	. . . . .		203-972-2000

Sending calls

Visitor dialing instructions

Local calls	. . . .	area code + 7 digits
Long distance	. . .	area code + 7 digits

Speed-dial numbers

Customer service	∗611
Emergency	911
WRKR	∗107

Receiving calls

Caller dials the roamer access number below,
hears tone, dials cellular area code and number.
. . . . 616-580-7626

NationLink & RoamingAmerica (see p. 1157)
Dial ∗31 send to receive calls, ∗30 to deactivate.
Dial ∗32 send to tell caller the roamer access no.
This company's subscribers have level 8 service.

Billing information

Visitor rates
Most visitors . . . . . $3 day + 85¢ minute

Visitors without roaming agreements (p. 1159)
Call co. to use Mastercard, Visa
or Roamer Plus will intercept first call to bill you.

Identification numbers

City	Market	System	Billing	MSA
Battle Creek	177	00403	00000	MSA
Jackson	207	00403	00000	MSA
Kalamazoo	132	00403	00000	MSA
Three Rivers	480	00403	00000	MI-9

Cellular company

Century Cellunet
5275 Beckley Road, Suite 4-B
Batttle Creek, MI 49015

740 Lawrence Avenue
Jackson, MI 49202

5461-C Gull Road
Kalamazoo, MI 49001

Customer service	. . . .		800-848-4577
. . . .			616-285-7400
Local office .	. .	Battle Creek	616-979-7000
. . . .		Jackson	517-783-4360
. . . .		Kalamazoo	616-342-6655
Corporate office	. . . . .		318-388-9000

Sending calls

Visitor dialing instructions

Local calls	. . . . .	area code + 7 digits
Long distance	. . .	1 + area code + 7 digits

Speed-dial numbers

AAA emergency	∗222
Customer service	∗611
Emergency	∗911
Gazette	∗45
Roaming information	∗711
WBCH	∗100
WBCK	∗930
WKHN	∗97
WLAJ	∗53
WNJC	∗953
WQLR	∗106
WQSN	∗1470

Receiving calls

Caller dials the roamer access number below,
hears tone, dials cellular area code and number.

Battle Creek	. . . .	616-967-7626
Jackson	. . . .	517-740-7626
Kalamazoo, Three Rivers	. .	616-370-7626

Follow Me Roaming forwarding (see p. 1156)
Activate, dial ∗18 send. Deactivate dial ∗19 send.

Billing information

Visitor rates
Most visitors . . . . . $3 day + 75¢ minute

Visitors without roaming agreements (p. 1159)
Call co. to use Amex, Discover, Mastercard, Visa

Identification numbers

City	Market	System	Billing	RSA
Battle Creek	177	00256	00000	MSA
Jackson	207	00374	00000	MSA
Kalamazoo	132	00386	00000	MSA
Three Rivers	480	01348	00000	MI-9

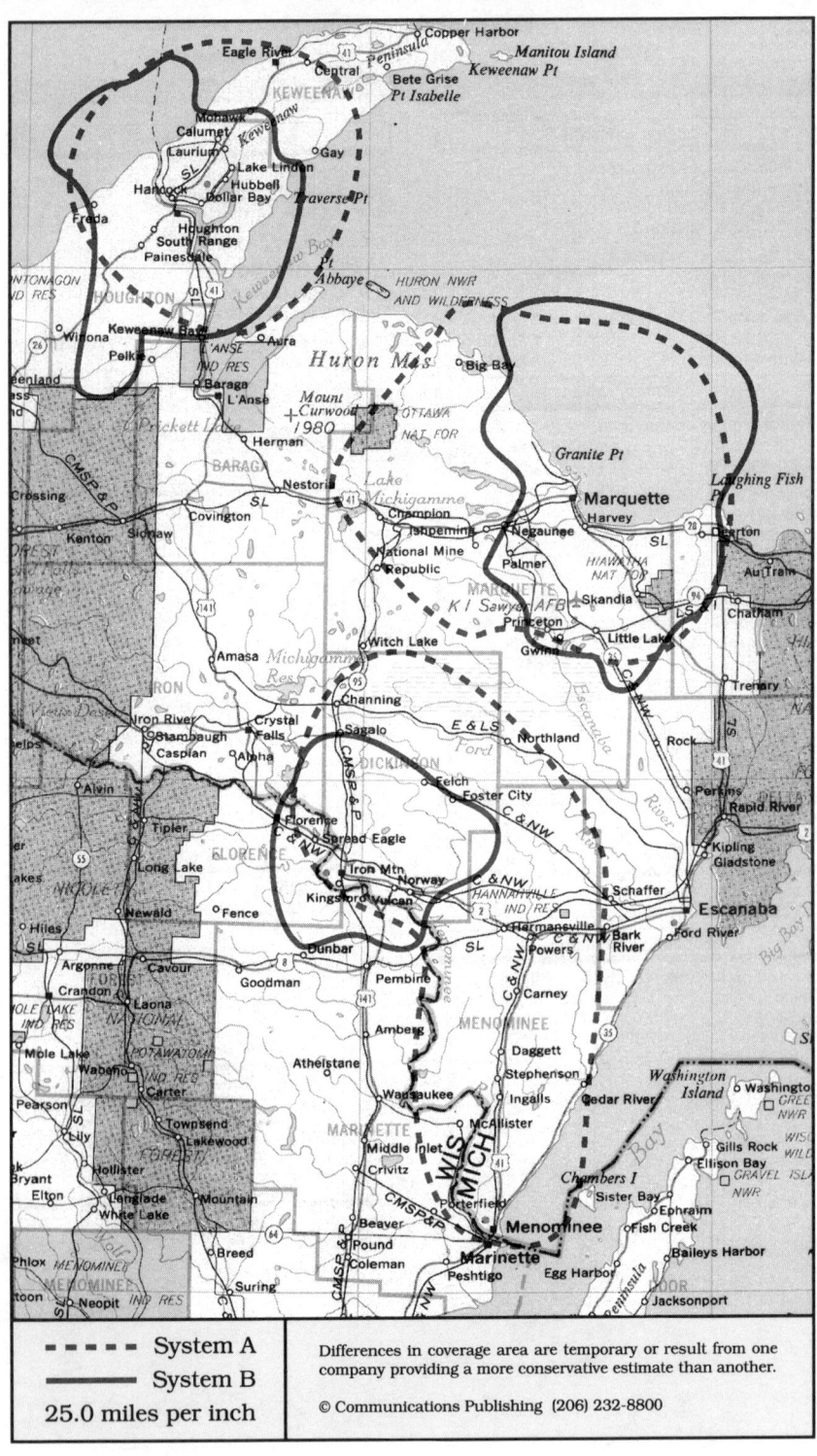

System A
System B
25.0 miles per inch

Differences in coverage area are temporary or result from one company providing a more conservative estimate than another.

© Communications Publishing (206) 232-8800

Marquette, Michigan
Houghton, Iron Mountain, Marquette, Menominee

System A	System B

Cellular company

Cellular One
Buckhead Telephone
2843 U.S. Highway 41 W
Marquette, MI 49855

Corporate office	some areas	800-491-9999
. . . .		906-228-2255

Sending calls

Visitor dialing instructions

Local calls	area code + 7 digits
Long distance . . .	1 + area code + 7 digits

Speed-dial numbers

Customer service	*611
Emergency	911

Receiving calls

Caller dials the roamer access number below, hears tone, dials cellular area code and number.
. . . .   906-360-7626

NationLink & RoamingAmerica (see p. 1157)
Caller hears your current roamer access number.
This company's subscribers have level 0 service.

Billing information

Visitor rates
Most visitors . . . . . $3 day + 99¢ minute

Visitors without roaming agreements (p. 1159)
Call co. to use Mastercard, Visa

Identification numbers

City	Market	System	Billing	RSA
All served cities	472	01331	00000	MI-1

Cellular company

Cellulink
Pacific Telecom Cellular
4600 W College Avenue
Appleton, WI 54915

Customer service	. . . .	800-236-9700
. . . .		414-841-1111
Corporate office	. . . . .	414-841-1100

Sending calls

Visitor dialing instructions

Local calls	area code + 7 digits
Long distance	area code + 7 digits

Speed-dial numbers

Customer service	*611
Emergency	911
Roaming information	*711

Receiving calls

Caller dials the roamer access number below, hears tone, dials cellular area code and number.
. . . .   906-250-7626

Billing information

Visitor rates
Most visitors . . . . . $3 day + 85¢ minute

Visitors without roaming agreements (p. 1159)
Call co. to use Mastercard, Visa

Identification numbers

City	Market	System	Billing	RSA
All served cities	472	01332	00000	MI-1

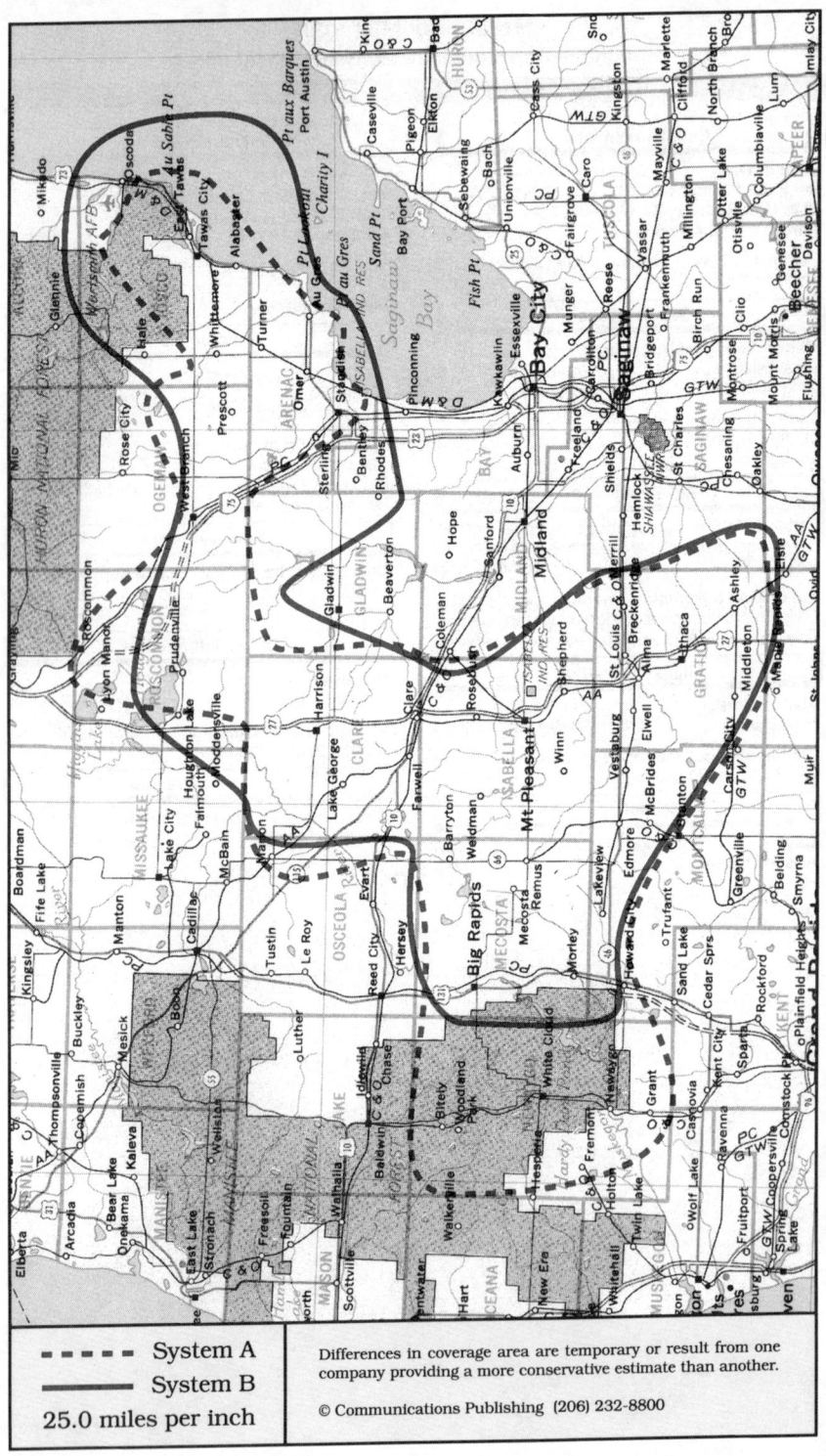

- - - - - System A	Differences in coverage area are temporary or result from one
———— System B	company providing a more conservative estimate than another.
25.0 miles per inch	© Communications Publishing (206) 232-8800

Mount Pleasant, Michigan
Alma, Big Rapids, East Tawas, Harrison, Mount Pleasant, West Branch

System A	System B

Cellular company

Cellular One
Sterling Cellular Management
2014 S Mission
Mount Pleasant, MI 48858

Customer service	800-552-6150
Local office	517-773-3310
Corporate office	404-552-5030

Sending calls

Visitor dialing instructions

Local calls	7 digits
Long distance . . .	1+ area code + 7 digits

Speed-dial numbers

Customer service	*611
Emergency	*911
Roaming information	*711

Receiving calls

Caller dials the roamer access number below,
hears tone, dials cellular area code and number.

Big Rapids	616-598-7626
Clare	517-240-7626
Mount Pleasant	517-560-7626

Billing information

Visitor rates

Most visitors	$3 day + 99¢ minute

Visitors without roaming agreements (p. 1159)
Call co. to use Amex
or Roamer Plus will intercept first call to bill you.

Identification numbers

City	Market	System	Billing	RSA
Clare, East Tawas	477	01341	00000	MI-6
Mount Pleasant	478	01341	00000	MI-7

Cellular company

Century Cellunet
2831 Bay Road
Saginaw, MI 48603

Customer service	800-848-4577
. . . .	616-285-7400
Local office	517-792-1556
Corporate office	318-388-9000

Sending calls

Visitor dialing instructions

Local calls	area code + 7 digits
Long distance	area code + 7 digits

Speed-dial numbers

AAA emergency	*222
Customer service	*611
Emergency	*911
Roaming information	*711

Receiving calls

Caller dials the roamer access number below,
hears tone, dials cellular area code and number.

Big Rapids	616-750-7626
Clare, Mount Pleasant . .	517-751-7626

Follow Me Roaming forwarding (see p. 1156)
Activate, dial *18 send. Deactivate dial *19 send.

Billing information

Visitor rates

Most visitors	$3 day + 75¢ minute

Visitors without roaming agreements (p. 1159)
Call co. to use Amex, Discover, Mastercard, Visa

Identification numbers

City	Market	System	Billing	RSA
Clare, East Tawas	477	01342	00000	MI-6
Mount Pleasant	478	00216	30344	MI-7

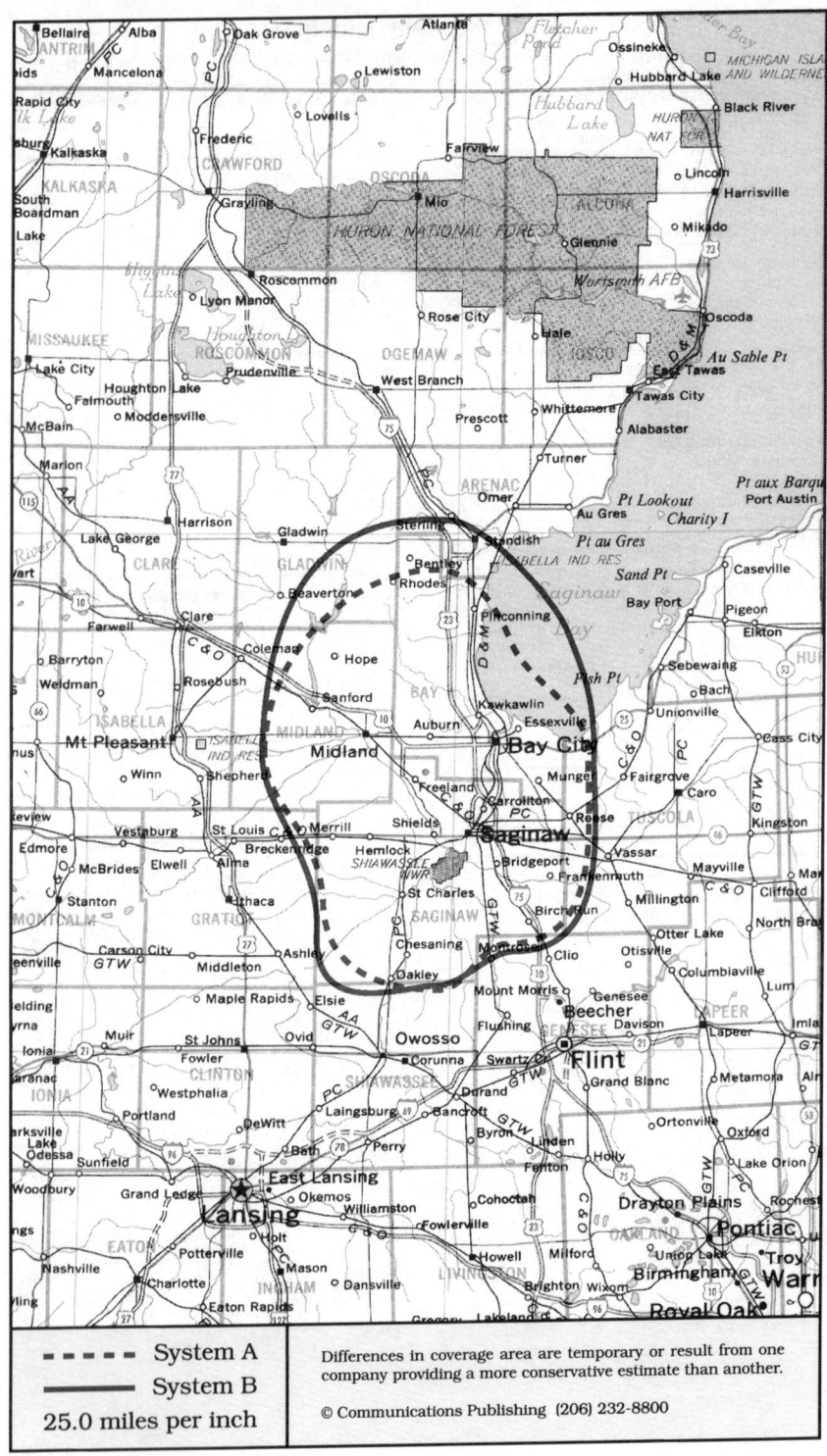

System A - - - -
System B ──────
25.0 miles per inch

Differences in coverage area are temporary or result from one company providing a more conservative estimate than another.

© Communications Publishing (206) 232-8800

Saginaw, Michigan
Bay City, Midland, Saginaw

System A

Cellular company

Cellular One
(affiliated with Cellular One of Detroit)
4901 Town Centre Road, Suite 220
Saginaw, MI 48604

Customer service		800-452-3551
. . . .		313-737-5123
Local office		517-799-1001
Corporate office . .	regional	313-737-5100
. . . .	national	614-436-4331

Sending calls

Visitor dialing instructions

Local calls	area code + 7 digits
Long distance . . .	0 + area code + 7 digits

Speed-dial numbers

Customer service	*611
Emergency	*911
WWJ	*995

Receiving calls

Caller dials the roamer access number below,
hears tone, dials cellular area code and number.
. . . . 517-798-7626

NationLink & RoamingAmerica (see p. 1157)
Dial *31 send to receive calls, *30 to deactivate.
Dial *32 send to tell caller the roamer access no.
This company's subscribers have level 6 service.

Billing information

Visitor rates
Most visitors . . . . . $3 day + 99¢ minute

Visitors without roaming agreements (p. 1159)
Roamer Plus will intercept first call to bill you.

Identification numbers

City	Market	System	Billing	RSA
All served cities	094	00021	00000	MSA

For full coverage area, see Detroit, MI.

System B

Cellular company

Century Cellunet
2831 Bay Road
Saginaw, MI 48603

Customer service	800-848-4577
. . . .	616-285-7400
Local office	517-792-1556
Corporate office	318-388-9000

Sending calls

Visitor dialing instructions

Local calls	area code + 7 digits
Long distance	area code + 7 digits

Speed-dial numbers

AAA emergency	*222
Crime stop	*27
Customer service	*611
Emergency	*911
Roaming information	*711
WHNN	*096
WJRT	*12

Receiving calls

Caller dials the roamer access number below,
hears tone, dials cellular area code and number.
. . . . 517-751-7626

Follow Me Roaming forwarding (see p. 1156)
Activate, dial *18 send. Deactivate dial *19 send.

Billing information

Visitor rates
Most visitors . . . . . $3 day + 75¢ minute

Visitors without roaming agreements (p. 1159)
Call co. to use Amex, Discover, Mastercard, Visa

Identification numbers

City	Market	System	Billing	RSA
All served cities	094	00216	00000	MSA

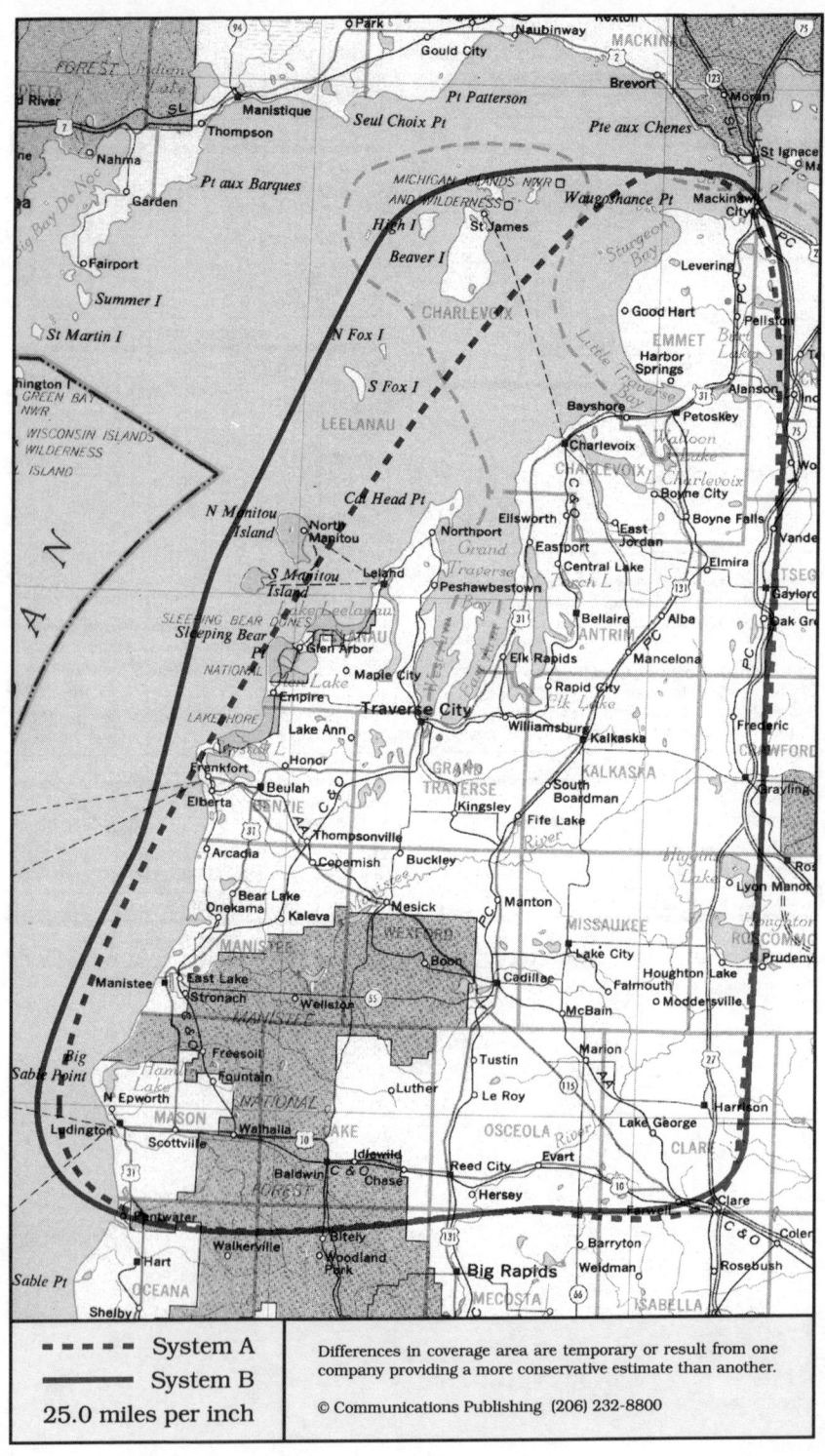

System A
System B
25.0 miles per inch

Differences in coverage area are temporary or result from one company providing a more conservative estimate than another.

© Communications Publishing (206) 232-8800

Traverse City, Michigan

Bellaire, Cadillac, Charlevoix, Lake City, Ludington, Reed City, Traverse City

System A	System B

System A

Cellular company

Cellular One
Unitel
732 Hannah Street
Traverse City, MI 49684

Local office	616-922-8787
Corporate office	609-646-9400

Sending calls

Visitor dialing instructions

Local calls	area code + 7 digits
Long distance	area code + 7 digits

Speed-dial numbers

Customer service	*611
Emergency	911

Receiving calls

Caller dials the roamer access number below,
hears tone, dials cellular area code and number.
. . . . 616-620-7626

NationLink & RoamingAmerica (see p. 1157)
Caller hears your current roamer access number.
This company's subscribers have level 0 service.

Billing information

Visitor rates
Most visitors . . . . . $3 day + 99¢ minute

Visitors without roaming agreements (p. 1159)
Call co. to use Mastercard, Visa
or Roamer Plus will intercept first call to bill you.

Identification numbers

City	Market	System	Billing	RSA
Cadillac, Ludington	476	01339	00000	MI-5
Traverse City	474	01335	00000	MI-3

System B

Cellular company

Century Cellunet
1049 Garfield Avenue
Traverse City, MI 49684

Customer service	800-848-4577
. . . .	616-285-7400
Local office	616-929-3800
Corporate office	318-388-9000

Sending calls

Visitor dialing instructions

Local calls	area code + 7 digits
Long distance	area code + 7 digits

Speed-dial numbers

AAA emergency	*222
Customer service	*611
Emergency	*911
Roaming information	*711

Receiving calls

Caller dials the roamer access number below,
hears tone, dials cellular area code and number.
. . . . 616-590-7626

Follow Me Roaming forwarding (see p. 1156)
Activate, dial *18 send. Deactivate dial *19 send.

Billing information

Visitor rates
Most visitors . . . . . $3 day + 75¢ minute

Visitors without roaming agreements (p. 1159)
Call co. to use Amex, Discover, Mastercard, Visa

Identification numbers

City	Market	System	Billing	RSA
Cadillac, Ludington	476	01336	00000	MI-5
Traverse City	474	01336	00000	MI-3

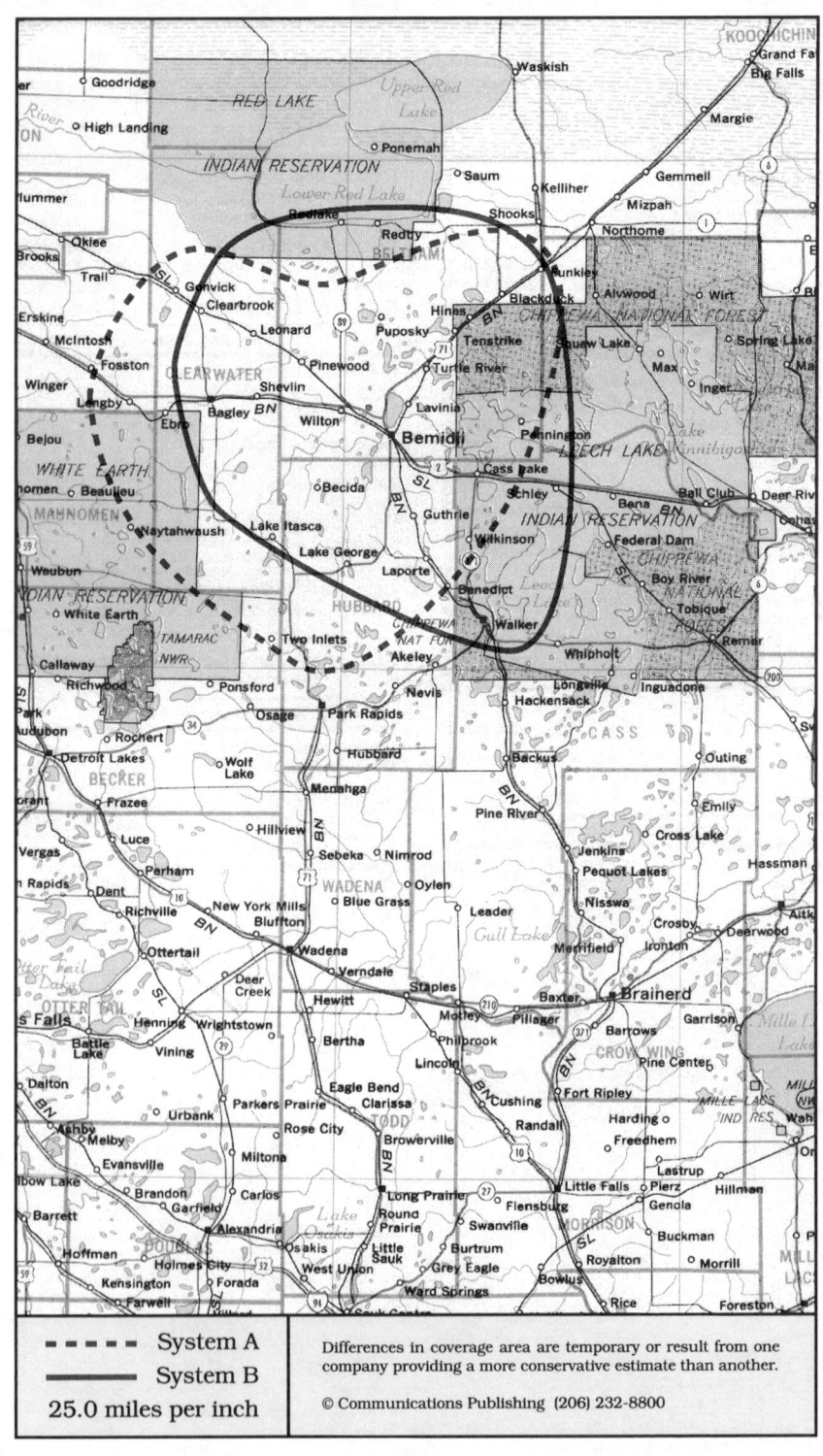

Bemidji, Minnesota
Bagley, Bemidji, Walker

Cellular company

Until June 1993
Cellular One
Quantum Communications
706 Paul Bunyan Drive NW
Bemidji, MN 56601

Customer service	800-788-2897
. . . .	218-760-2273
Local office	218-759-2355
Corporate office	612-942-7650

After June 1993
Pacific Northwest Cellular
11400 SE 8th Street #445
Bellevue, WA 98004-6431

Customer service	800-635-0304
Corporate office	206-635-0300

Sending calls

Visitor dialing instructions
Local calls . . . . . area code + 7 digits
Long distance . . . 1 + area code + 7 digits

Speed-dial numbers
Customer service 611
Emergency 911

Receiving calls

Caller dials the roamer access number below, hears tone, dials cellular area code and number.
. . . . 218-760-7626

NationLink & RoamingAmerica (see p. 1157)
Dial ∗31 send to receive calls, ∗30 to deactivate. This company's subscribers have level 3 service.

North American Cellular Network (see p. 1156)
All calls forward to you. Deactivate dial ∗35 send.

Billing information

Visitor rates
Most visitors . . . . . $3 day + 99¢ minute
McCaw Cellular visitors . $0 day + 99¢ minute
Other visitors   depends on your home company

Visitors without roaming agreements (p. 1159)
Roamer Plus will intercept first call to bill you.

Identification numbers

City	Market	System	Billing	RSA
All served cities	483	01353	00000	MN-2

Cellular company

Cellular 2000
Rural Cellular
1002 Wright Street, Suite 101
Brainerd, MN 56401

Customer service	800-450-2000
Local office	800-545-3951
. . . .	218-829-0092
Corporate office	612-762-2000

Sending calls

Visitor dialing instructions
Local calls . . . . . . area code + 7 digits
Long distance . . . . . area code + 7 digits

Speed-dial numbers
Customer service . . . . ∗611
Emergency . . . . ∗911
Roaming information . . . . ∗711

Receiving calls

Caller dials the roamer access number below, hears tone, dials cellular area code and number.
. . . . 507-330-7626

Billing information

Visitor rates
Most visitors . . . . . $3 day + 85¢ minute

Visitors without roaming agreements (p. 1159)
Cannot call since company takes no credit cards.

Identification numbers

City	Market	System	Billing	RSA
All served cities	483	01370	00000	MN-2

For full coverage area, see Grand Rapids, MN.

427

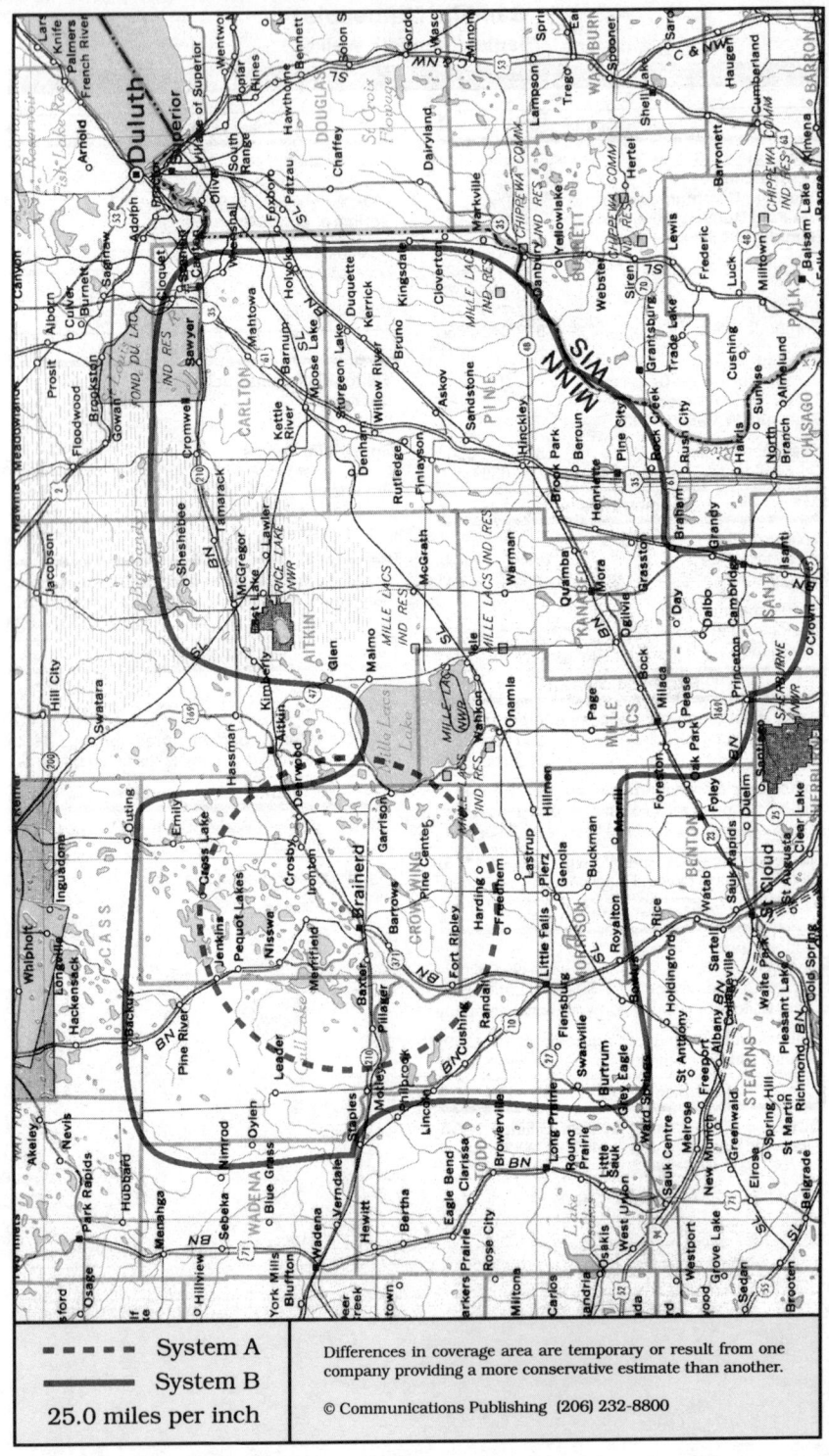

Legend	
- - - - -	System A
————	System B
25.0 miles per inch	

Differences in coverage area are temporary or result from one company providing a more conservative estimate than another.

© Communications Publishing (206) 232-8800

Brainerd, Minnesota
Brainerd, Carlton, Cambridge, Little Falls, Mora

System A	System B

Cellular company

Century Cellunet
1824 East Main Street
Onalaska, WI 54650

Local office	608-781-9000
Corporate office	318-388-9000

Sending calls

Visitor dialing instructions

Local calls	area code + 7 digits
Long distance . . .	1 + area code + 7 digits

Speed-dial numbers

Customer service	611
Emergency	911

Receiving calls

Caller dials the roamer access number below,
hears tone, dials cellular area code and number.
. . . . 218-839-7626

Billing information

Visitor rates

Most visitors	$3 day + 85¢ minute

Visitors without roaming agreements (p. 1159)
Call co. to use Amex, Mastercard, Visa

Identification numbers

City	Market	System	Billing	RSA
All served cities	487	01361	00000	MN-6

Cellular company

Cellular 2000
Rural Cellular
1002 Wright Street, Suite 101
Brainerd, MN 56401

Customer service	800-450-2000
Local office	800-545-3951
. . . .	218-829-0092
Corporate office	612-762-2000

Sending calls

Visitor dialing instructions

Local calls	area code + 7 digits
Long distance	area code + 7 digits

Speed-dial numbers

Customer service	*611
Emergency	*911
Roaming information	*711

Receiving calls

Caller dials the roamer access number below,
hears tone, dials cellular area code and number.
. . . . 507-330-7626

Billing information

Visitor rates

Most visitors	$3 day + 85¢ minute

Visitors without roaming agreements (p. 1159)
Cannot call since company takes no credit cards.

Identification numbers

City	Market	System	Billing	RSA
All served cities	487	01370	00000	MN-6

For full coverage area, see Fergus Falls, MN.

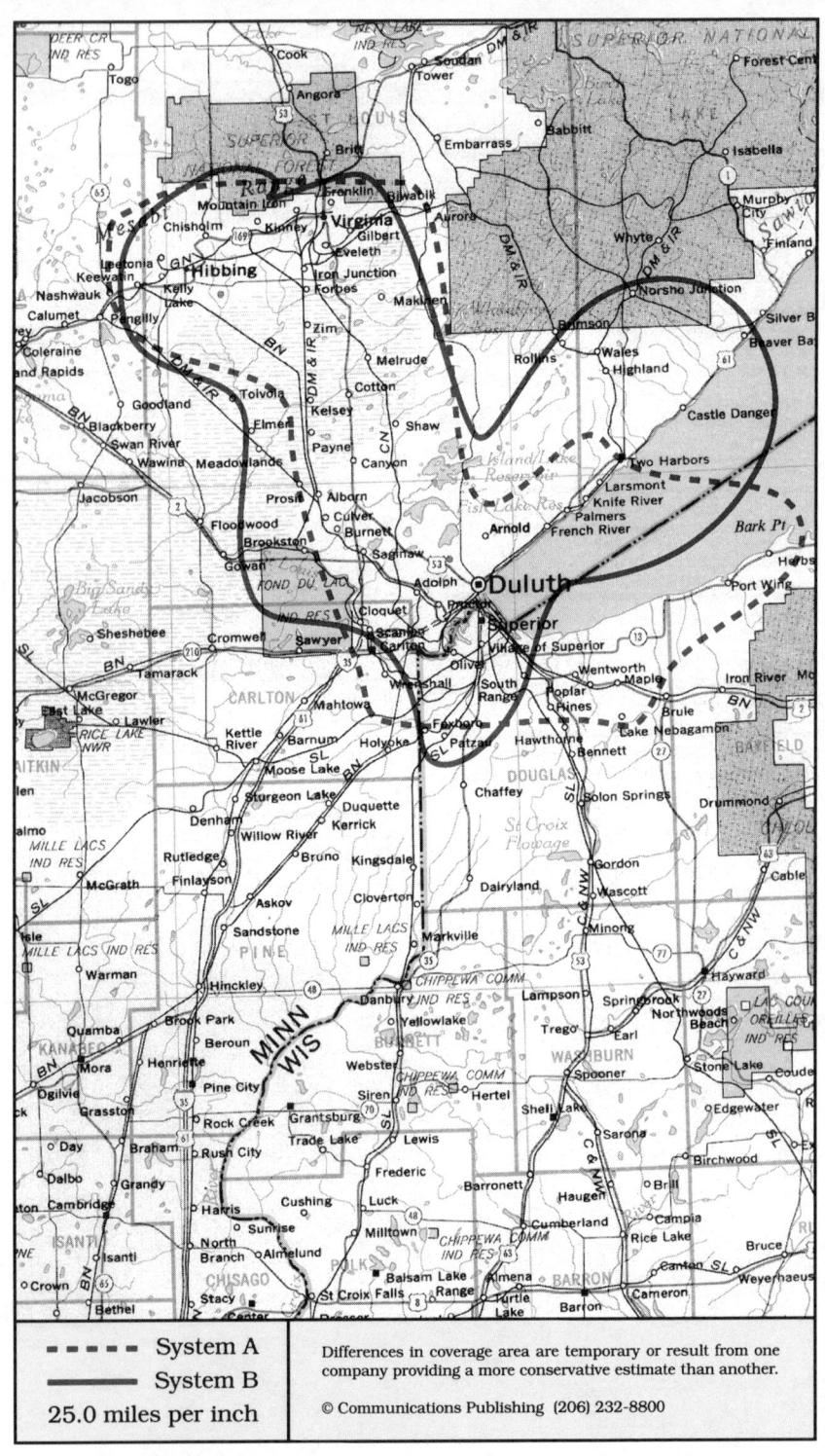

– – – – – System A	Differences in coverage area are temporary or result from one company providing a more conservative estimate than another.
———— System B	
25.0 miles per inch	© Communications Publishing (206) 232-8800

Duluth, Minnesota
Duluth, Hibbing, Two Harbors, Virginia

System A	System B

Cellular company

Cellular One of Duluth/Superior
Cellular Information Systems
224 E Central Entrance
Duluth, MN 55811

Local office	218-727-4700
Corporate office	203-622-6317

Sending calls

Visitor dialing instructions

Local calls	area code + 7 digits
Long distance . . .	1 + area code + 7 digits

Speed-dial numbers

Customer service	611
Emergency	911

Receiving calls

Caller dials the roamer access number below,
hears tone, dials cellular area code and number.
. . . . 218-348-7626

NationLink & RoamingAmerica (see p. 1157)
Dial *31 send to receive calls, *30 to deactivate.
This company's subscribers have level 3 service.

Billing information

Visitor rates
Most visitors . . . . . $3 day + 85¢ minute

Visitors without roaming agreements (p. 1159)
Call co. to use Amex, Mastercard, Visa

Identification numbers

City	Market	System	Billing	RSA
All served cities	141	00333	00000	MSA

This company is not licensed to serve MN-4.

Cellular company

U S West Cellular
5115 Burning Tree Plaza, Suite 308
Duluth, MN 55811

Customer service	800-238-7848
. . . .	206-562-2895
. . . . . local subscribers	800-626-6611
Local office	218-722-8643
Corporate office	206-747-4900

Sending calls

Visitor dialing instructions

Local calls	area code + 7 digits
Long distance . . .	1 + area code + 7 digits

Speed-dial numbers

Customer service	*611
Emergency	911
Roaming Information	*711

Receiving calls

Caller dials the roamer access number below,
hears tone, dials cellular area code and number.

Duluth	218-343-7626
Two Harbors	218-830-7626

Follow Me Roaming forwarding (see p. 1156)
Activate, dial *18 send. Deactivate dial *19 send.

Billing information

Visitor rates
Most visitors . . . . . $3 day + 99¢ minute

Visitors without roaming agreements (p. 1159)
Call co. to use Amex, Mastercard, Visa

Identification numbers

City	Market	System	Billing	RSA
Duluth	141	00316	00000	MSA
Two Harbors	485	01358	00000	MN-4

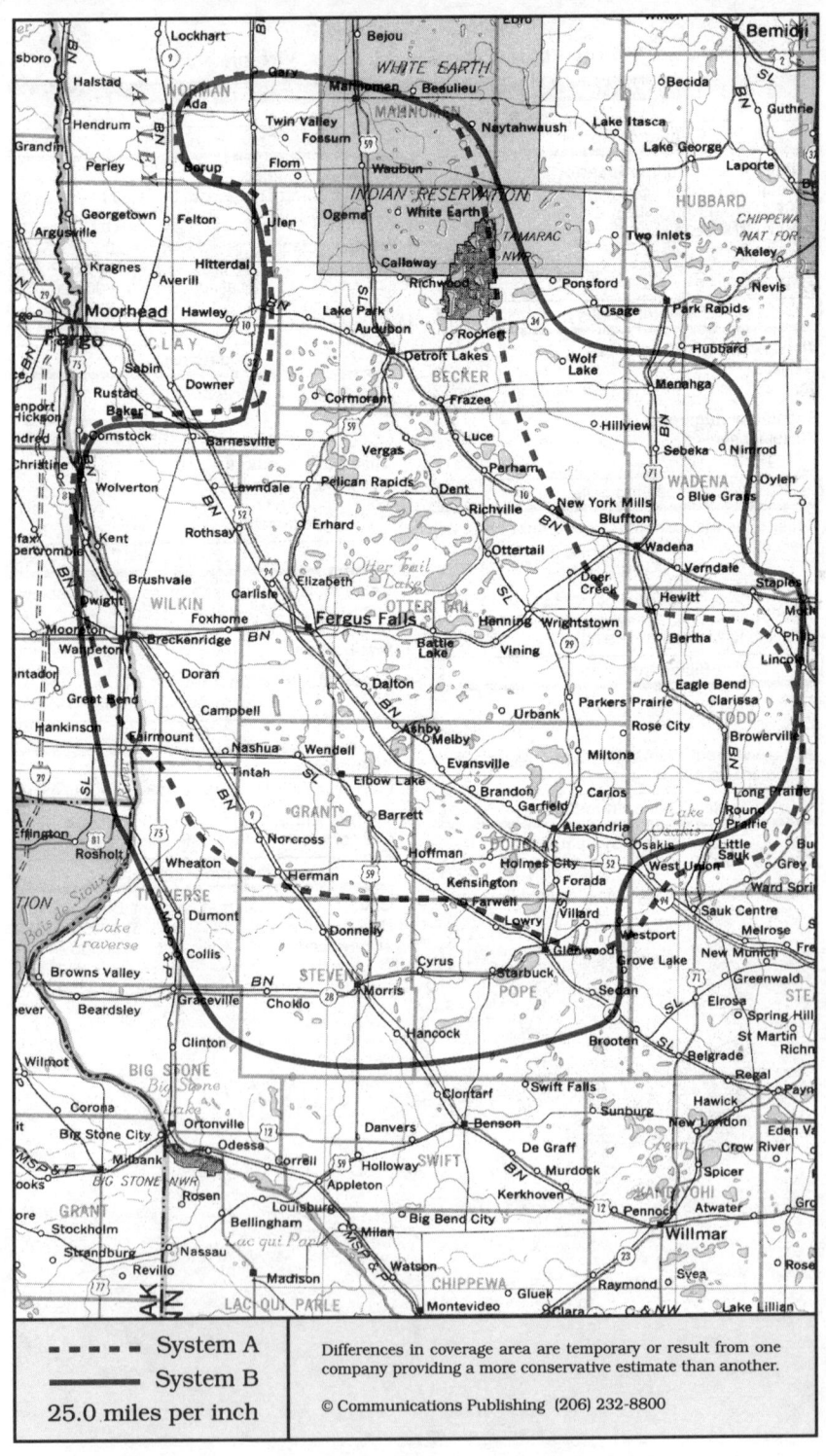

▪ ▪ ▪ ▪ ▪	System A
————	System B
25.0 miles per inch	

Differences in coverage area are temporary or result from one company providing a more conservative estimate than another.

© Communications Publishing (206) 232-8800

Fergus Falls, Minnesota
Alexandria, Breckenridge, Detroit Lakes, Fergus Falls, Long Prairie, Morris

System A	System B

Cellular company

Until June 1993
Cellular One
Quantum Communications
7901 Flying Cloud Drive, Suite 250
Eden Prairie, MN 55344

Local office	218-731-2273
Corporate office	612-942-7650

After June 1993
Pacific Northwest Cellular
11400 SE 8th Street #445
Bellevue, WA 98004-6431

Customer service	800-635-0304
Corporate office	206-635-0300

Sending calls

Visitor dialing instructions

Local calls	area code + 7 digits
Long distance . . .	1 + area code + 7 digits

Speed-dial numbers

Customer service	611
Emergency	911

Receiving calls

Caller dials the roamer access number below,
hears tone, dials cellular area code and number.
. . . . 218-731-7626

NationLink & RoamingAmerica (see p. 1157)
Dial *31 send to receive calls, *30 to deactivate.
This company's subscribers have level 3 service.

North American Cellular Network (see p. 1156)
All calls forward to you. Deactivate dial *35 send.

Billing information

Visitor rates

Most visitors	$3 day + 85¢ minute
McCaw Cellular visitors .	$0 day + 99¢ minute
Other visitors	depends on your home company

Visitors without roaming agreements (p. 1159)
Roamer Plus will intercept first call to bill you.

Identification numbers

City	Market	System	Billing	RSA
All served cities	486	01359	00000	MN-5

Cellular company

Cellular 2000
Rural Cellular
2819 Hwy 29, S. Midway Mall, PO Box 1027
Alexandria, MN 56308

Customer service	some areas	800-450-2000
Corporate office		612-762-2000

Sending calls

Visitor dialing instructions

Local calls	area code + 7 digits
Long distance	area code + 7 digits

Speed-dial numbers

Customer service	*611
Emergency	*911
Roaming information	*711

Receiving calls

Caller dials the roamer access number below,
hears tone, dials cellular area code and number.
. . . . 612-760-7626

Billing information

Visitor rates

Most visitors	$3 day + 85¢ minute

Visitors without roaming agreements (p. 1159)
Cannot call since company takes no credit cards.

Identification numbers

City	Market	System	Billing	RSA
All served cities	486	01360	00000	MN-5

For full coverage area, see Brainerd, MN.

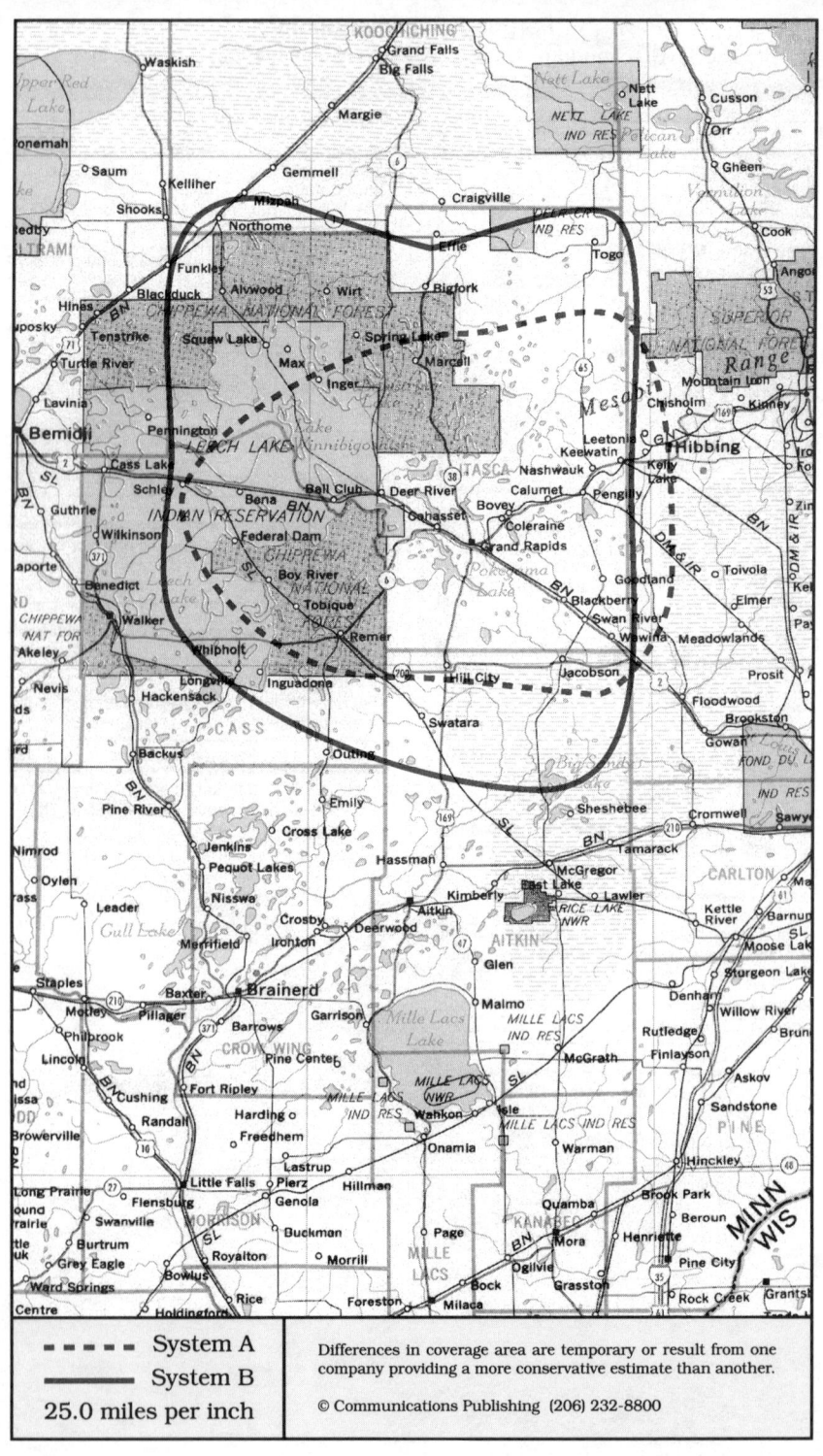

System A
System B
25.0 miles per inch

Differences in coverage area are temporary or result from one company providing a more conservative estimate than another.

© Communications Publishing (206) 232-8800

Grand Rapids, Minnesota
Cass Lake, Grand Rapids

System A	System B

System A

Cellular company

Cellular One
Palmer Communications
12800 University Drive, Suite 500
Fort Myers, FL 33907-5333

Customer service	800-441-2796
. . . .	813-433-8242
Corporate office	813-433-4350

Sending calls

Visitor dialing instructions
Local calls	area code + 7 digits
Long distance . . .	1 + area code + 7 digits

Speed-dial numbers
Customer service	611
Emergency	911

Receiving calls

Caller dials the roamer access number below, hears tone, dials cellular area code and number.
. . . . 218-259-7626

NationLink & RoamingAmerica (see p. 1157)
Caller hears your current roamer access number. This company's subscribers have level 0 service.

Billing information

Visitor rates
Most visitors . . . . . $3 day + 85¢ minute

Visitors without roaming agreements (p. 1159)
Cannot call since company takes no credit cards.

Identification numbers

City	Market	System	Billing	RSA
All served cities	484	01355	00000	MN-3

System B

Cellular company

Cellular 2000
Rural Cellular
1002 Wright Street, Suite 101
Brainerd, MN 56401

Customer service	800-450-2000
Local office	800-545-3951
. . . .	218-829-0092
Corporate office	612-762-2000

Sending calls

Visitor dialing instructions
Local calls	area code + 7 digits
Long distance	area code + 7 digits

Speed-dial numbers
Customer service	*611
Emergency	*911
Roaming information	*711

Receiving calls

Caller dials the roamer access number below, hears tone, dials cellular area code and number.
. . . . 507-330-7626

Billing information

Visitor rates
Most visitors . . . . . $3 day + 85¢ minute

Visitors without roaming agreements (p. 1159)
Cannot call since company takes no credit cards.

Identification numbers

City	Market	System	Billing	RSA
Cass Lake	487	01370	00000	MN-6
Grand Rapids	484	01370	00000	MN-3

For full coverage area, see Bemidji, MN.

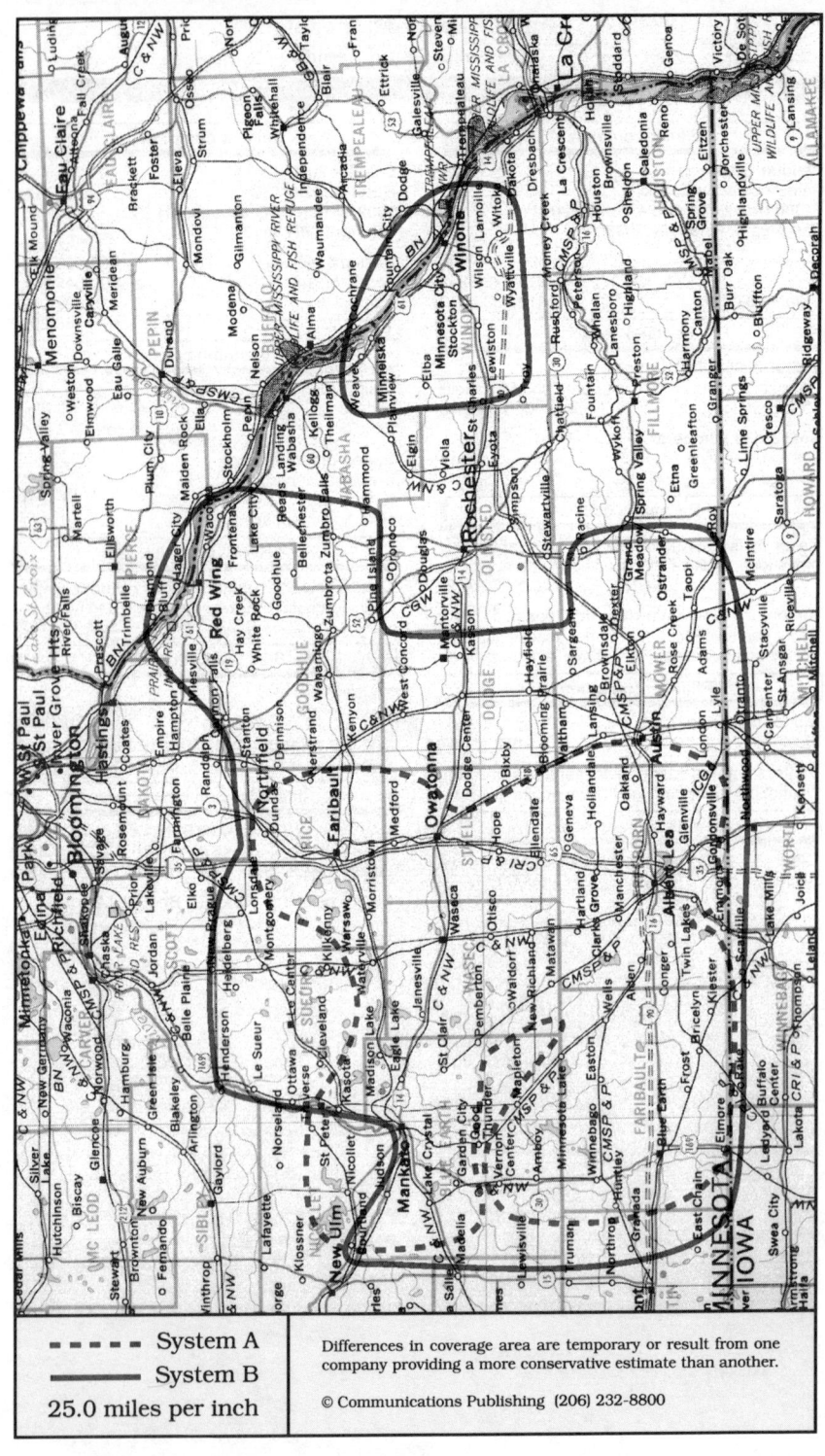

System A
System B
25.0 miles per inch

Differences in coverage area are temporary or result from one company providing a more conservative estimate than another.

© Communications Publishing (206) 232-8800

Mankato, Minnesota
Austin, Faribault, Mankato, Red Wing, Winona

System A	System B

Cellular company

Cellular One
Licensed to: Minnesota Southern Cellular Tel.
Managed by: Cooper Cellular Management
1640 Madison Avenue
Mankato, MN 56001

Local office 507-625-2355
Corporate office . . Cooper 407-425-9074

Sending calls

Visitor dialing instructions
Local calls . . . . area code + 7 digits
Long distance . . . 1 + area code + 7 digits

Speed-dial numbers
Customer service 611
Emergency 911

Receiving calls

Caller dials the roamer access number below,
hears tone, dials cellular area code and number.
. . . . 507-340-7626

Billing information

Visitor rates
Most visitors . . . . . $3 day + 85¢ minute

Visitors without roaming agreements (p. 1159)
Call co. to use Mastercard, Visa
or Roamer Plus will intercept first call to bill you.

Identification numbers

City	Market	System	Billing	RSA
All served cities	491	01369	00000	MN-10

This company is not licensed to serve MN-11.

Cellular company

Cellular 2000
Managed by: Pacific Telecom Cellular
1870 Madison Avenue, PO Box 4069
Mankato, MN 56002-4069

Local office 800-545-3950
. . . . 507-345-4990
Corporate office 414-841-1200

Sending calls

Visitor dialing instructions
Local calls . . . . . . area code + 7 digits
Long distance . . . . . area code + 7 digits

Speed-dial numbers
Customer service . . . . ∗611
Emergency . . . . ∗911
Roaming information . . . . . . ∗711

Receiving calls

Caller dials the roamer access number below,
hears tone, dials cellular area code and number.
. . . . 507-330-7626

Follow Me Roaming forwarding (see p. 1156)
Activate, dial ∗18 send. Deactivate dial ∗19 send.

Billing information

Visitor rates
Most visitors . . . . . $3 day + 99¢ minute

Visitors without roaming agreements (p. 1159)
Cannot call since company takes no credit cards.

Identification numbers

City	Market	System	Billing	RSA
Austin, Mankato	491	01370	00000	MN-10
Red Wing, Winona	492	01372	00000	MN-11

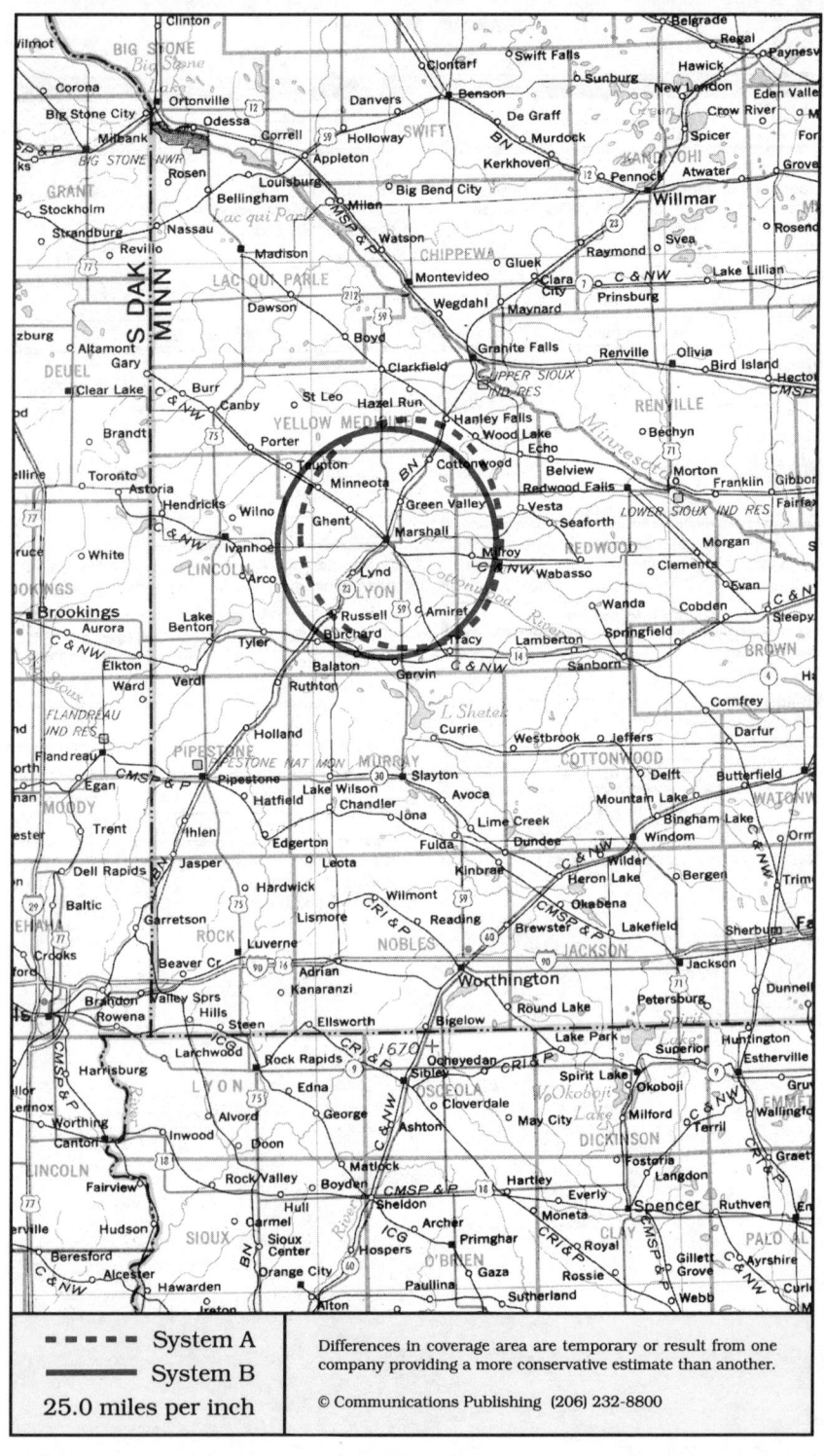

Differences in coverage area are temporary or result from one
company providing a more conservative estimate than another.

Marshall, Minnesota

Cellular company

Licensed to: MINN-8
Managed by: CompComm
900 Haddon Avenue, Suite 412
Collingswood, NJ 08108

Corporate office   CompComm 609-854-1000

Sending calls

Visitor dialing instructions
Local calls   . . . .   1 + area code + 7 digits
Long distance   . . .   1 + area code + 7 digits

Speed-dial numbers
Customer service   611
Emergency   911

Receiving calls

Caller dials the roamer access number below,
hears tone, dials cellular area code and number.
. . . .   not available

Billing information

Visitor rates
Most visitors . . . . . $3 day + 75¢ minute

Visitors without roaming agreements (p. 1159)
Roamer Plus will intercept first call to bill you.

Identification numbers

City	Market	System	Billing	RSA
All served cities	489	01365	00000	MN-8

Cellular company

Cellular 2000
Pacific Telecom Cellular
1870 Madison Avenue, PO Box 4069
Mankato, MN 56002-4069

Local office	800-545-3951
. . . .	218-829-0092
Corporate office	414-841-1200

Sending calls

Visitor dialing instructions
Local calls   . . . . . .   area code + 7 digits
Long distance . . . . .   area code + 7 digits

Speed-dial numbers
Customer service *611
Emergency *911
Roaming information *711

Receiving calls

Caller dials the roamer access number below,
hears tone, dials cellular area code and number.
. . . .   507-330-7626

Billing information

Visitor rates
Most visitors   . . . . . $3 day + 99¢ minute

Visitors without roaming agreements (p. 1159)
Cannot call since company takes no credit cards.

Identification numbers

City	Market	System	Billing	RSA
All served cities	489	01366	00000	MN-8

For full coverage area, see Willmar, MN.

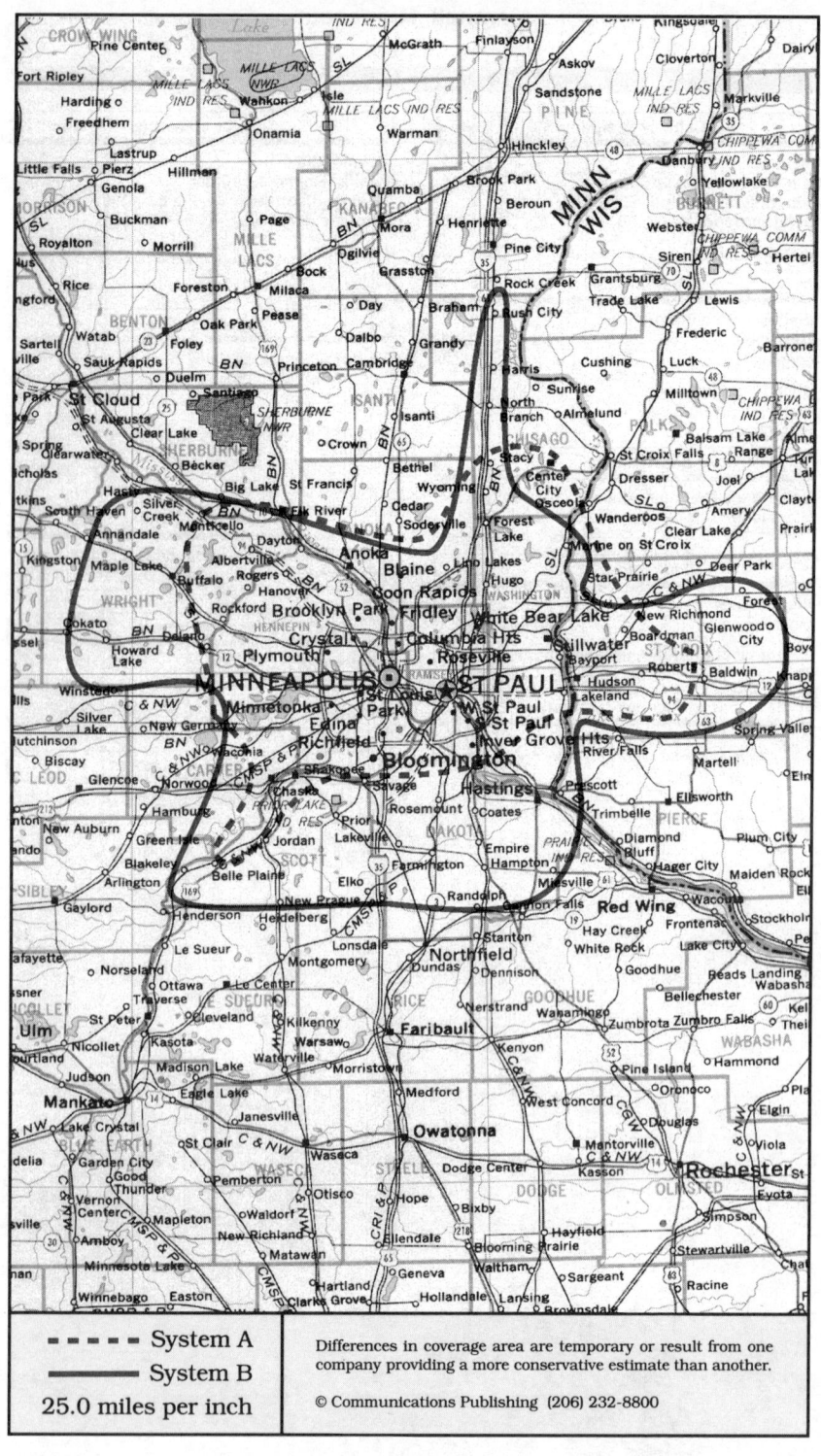

- - - - System A

———— System B

25.0 miles per inch

Differences in coverage area are temporary or result from one company providing a more conservative estimate than another.

© Communications Publishing (206) 232-8800

440

Minneapolis, Minnesota
Buffalo, Minneapolis, St. Paul, Stillwater

System A	System B

Cellular company

Cellular One
McCaw Cellular Communications
7900 Xerxes Avenue S, Suite 301
Minneapolis, MN 55431

Customer service	612-867-2273
Local office	612-831-3531
Corporate office	206-827-4500

Sending calls

Visitor dialing instructions

Local calls	area code + 7 digits
Long distance	1 + area code + 7 digits

Speed-dial numbers

Customer service	611
Emergency	911
Make a traffic report	*5795
Weather, stocks, traffic	*INFO or *4636

Receiving calls

Caller dials the roamer access number below, hears tone, dials cellular area code and number.
. . . . 612-867-7626

NationLink & RoamingAmerica (see p. 1157)
Dial *31 send to receive calls, *30 to deactivate.
Dial *32 send to tell caller the roamer access no.
This company's subscribers have level 12 service

Billing information

Visitor rates
Most visitors . . . . . $0 day + 99¢ minute

Visitors without roaming agreements (p. 1159)
Call co. to use Amex, Mastercard, Visa
or Roamer Plus will intercept first call to bill you.

Identification numbers

City	Market	System	Billing	RSA
All served cities	015	00023	00000	MSA

Cellular company

Regional office
U S West Cellular
9900 Bren Road E, Suite 306-E Opus Center
Minnetonka, MN 55343

524 S Cross Drive
Burnsville, MN 55337

929 2nd Street S
Hopkins, MN 55343

2465 N Fairview Avenue
Roseville, MN 55113

Customer service	. . . .	800-238-7848
. . . .		206-562-2895
. . . . .	local subscribers	800-626-6611
. . . .		206-747-1771
Local office	Burnsville	612-595-2600
. . . .	Hopkins	612-595-2500
. . . .	Minnetonka	612-595-2650
. . . .	Roseville	612-595-2400
Corporate office	. . . .	206-747-4900

Sending calls

Visitor dialing instructions

Local calls	area code + 7 digits
Long distance	1 + area code + 7 digits

Speed-dial numbers

Customer service	*611
Emergency	911
Make a traffic report WCCO	#830
Roaming information	*711

Receiving calls

Caller dials the roamer access number below, hears tone, dials cellular area code and number.
. . . . 612-720-7626

Follow Me Roaming forwarding (see p. 1156)
Activate, dial *18 send. Deactivate dial *19 send.

Billing information

Visitor rates
Most visitors . . . . . $3 day + 99¢ minute

Visitors without roaming agreements (p. 1159)
Call co. to use Amex, Mastercard, Visa

Identification numbers

City	Market	System	Billing	RSA
All served cities	015	00026	00000	MSA

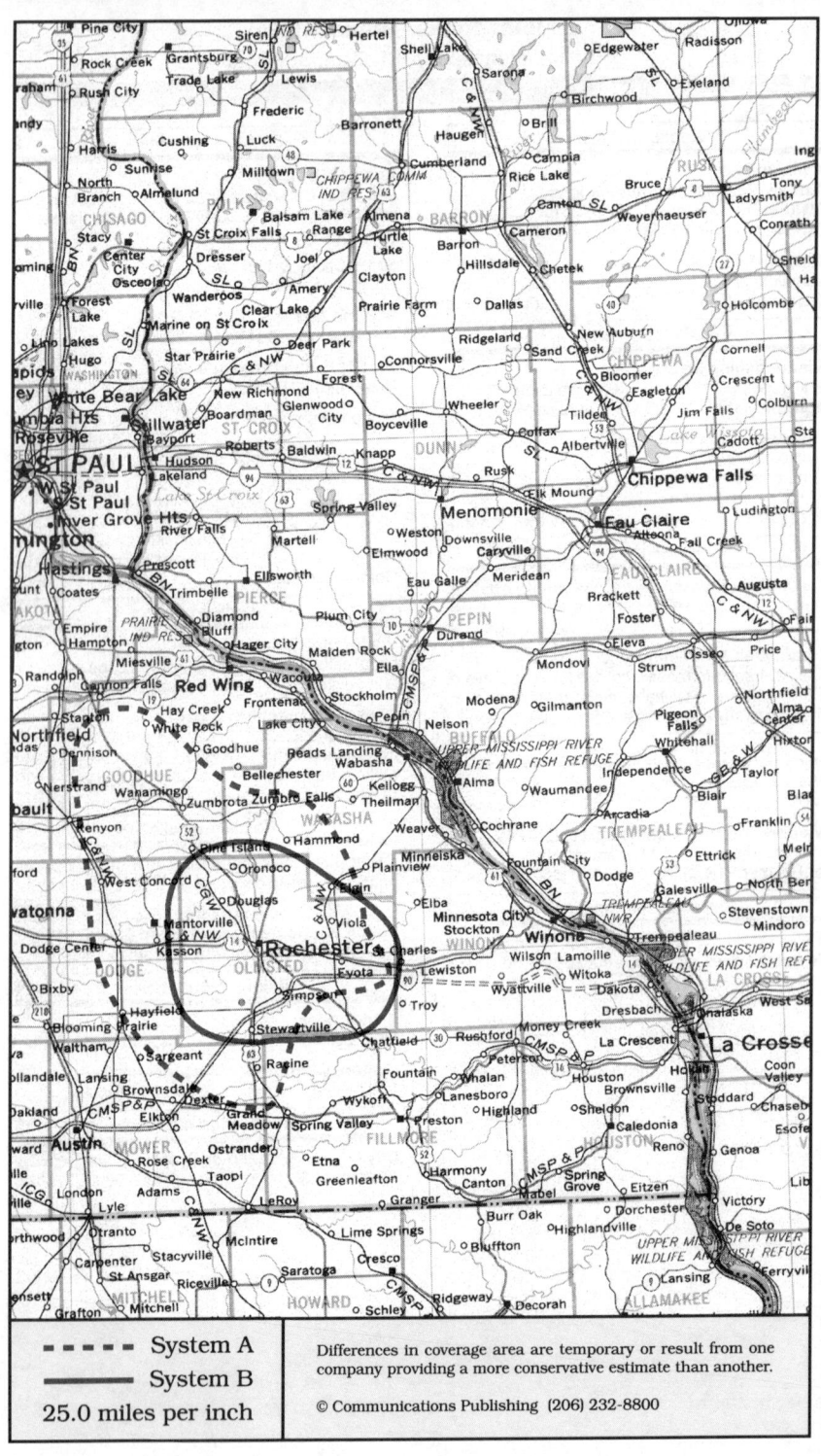

Rochester, Minnesota
Mantorville, Rochester

System A	System B

Cellular company

Cellular One
McCaw Cellular Communications
1101 N Broadway
Rochester, MN 55906

Customer service 507-254-2273
Local office 507-280-8100
Corporate office 206-827-4500

Sending calls

Visitor dialing instructions
Local calls . . . . . area code + 7 digits
Long distance . . . 1 + area code + 7 digits

Speed-dial numbers
Customer service 611
Emergency 911

Receiving calls

Caller dials the roamer access number below,
hears tone, dials cellular area code and number.
. . . . 507-254-7626

NationLink & RoamingAmerica (see p. 1157)
Dial ∗31 send to receive calls, ∗30 to deactivate.
Dial ∗32 send to tell caller the roamer access no.
This company's subscribers have level 12 service

Billing information

Visitor rates
Most visitors . . . . . $0 day + 99¢ minute

Visitors without roaming agreements (p. 1159)
Call co. to use Amex, Mastercard, Visa
or Roamer Plus will intercept first call to bill you.

Identification numbers

City	Market	System	Billing	RSA
All served cities	288	00023	30233	MSA

Cellular company

United States Cellular
1700 N Broadway, Suite 120
Rochester, MN 55906

Local office 507-288-3000
Corporate office 312-399-8900

Sending calls

Visitor dialing instructions
Local calls . . . . . . area code + 7 digits
Long distance . . . . . area code + 7 digits

Speed-dial numbers
Customer service . . . . ∗611
Emergency . . . . ∗911
Roaming information . . . . ∗711

Receiving calls

Caller dials the roamer access number below,
hears tone, dials cellular area code and number.
. . . . 507-251-7626

Billing information

Visitor rates
Most visitors . . . . . $3 day + 75¢ minute

Visitors without roaming agreements (p. 1159)
Call co. to use Mastercard, Visa

Identification numbers

City	Market	System	Billing	RSA
All served cities	288	00504	00000	MSA

System A
System B
25.0 miles per inch

Differences in coverage area are temporary or result from one company providing a more conservative estimate than another.

© Communications Publishing (206) 232-8800

St. Cloud, Minnesota
Elk River, St. Cloud, Sauk Centre

System A	System B

Cellular company

Cellular One
McCaw Cellular Communications
132 2nd Street S
Waite Park, MN 56387

Customer service	612-267-2273
Local office	612-654-0303
Corporate office	206-827-4500

Sending calls

Visitor dialing instructions

Local calls	area code + 7 digits
Long distance . . .	1 + area code + 7 digits

Speed-dial numbers

Customer service	611
Emergency	911

Receiving calls

Caller dials the roamer access number below,
hears tone, dials cellular area code and number.
. . . . 612-267-7626

Billing information

Visitor rates
Most visitors . . . . . $0 day + 99¢ minute

Visitors without roaming agreements (p. 1159)
Call co. to use Amex, Mastercard, Visa
or Roamer Plus will intercept first call to bill you.

Identification numbers

City	Market	System	Billing	RSA
All served cities	198	00023	30235	MSA

Cellular company

United States Cellular
20 32nd Avenue N
St. Cloud, MN 56303

Local office	612-252-9000
Corporate office	312-399-8900

Sending calls

Visitor dialing instructions

Local calls	area code + 7 digits
Long distance	area code + 7 digits

Speed-dial numbers

Customer service	*611
Emergency	*911

Receiving calls

Caller dials the roamer access number below,
hears tone, dials cellular area code and number.
. . . . 612-250-7626

Follow Me Roaming forwarding (see p. 1156)
Activate, dial *18 send. Deactivate dial *19 send.

Billing information

Visitor rates
Most visitors . . . . . $3 day + 99¢ minute

Visitors without roaming agreements (p. 1159)
Call co. to use Mastercard, Visa

Identification numbers

City	Market	System	Billing	RSA
All served cities	198	00534	00000	MSA

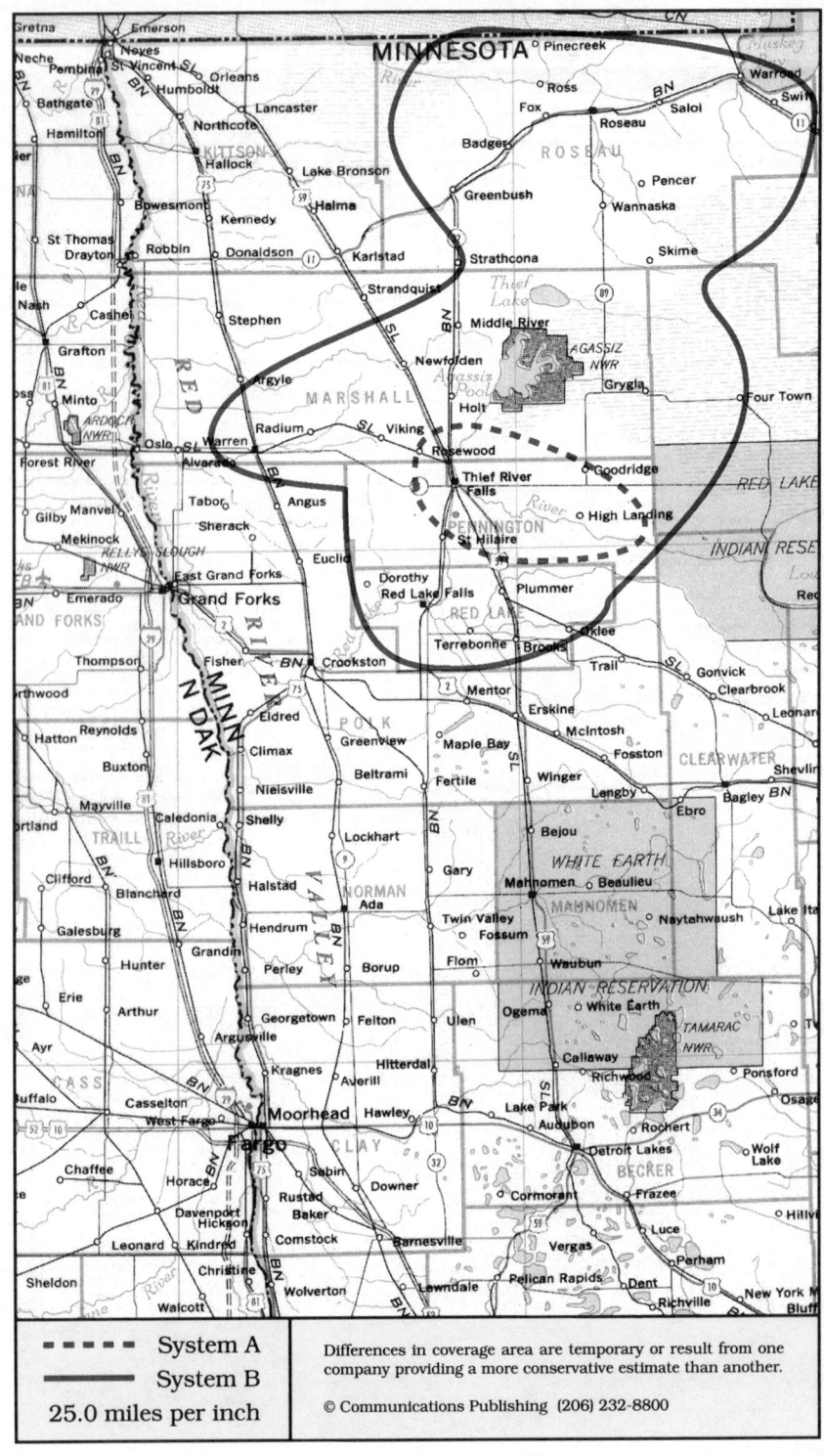

- - - - - System A	Differences in coverage area are temporary or result from one company providing a more conservative estimate than another.
───── System B	
25.0 miles per inch	© Communications Publishing (206) 232-8800

Thief River Falls, Minnesota
Red Lake Falls, Roseau, Thief River Falls, Warren

System A	System B

Cellular company

Until June 1993
Cellular One
Quantum Communications
706 Paul Bunyan Drive NW
Bemidji, MN 56601

Customer service		218-686-2273
Local office . .	some areas	800-788-2897
. . . .		218-759-2355
Corporate office		612-942-7650

After June 1993
Pacific Northwest Cellular
11400 SE 8th Street #445
Bellevue, WA 98004-6431

Customer service	800-635-0304
Corporate office	206-635-0300

Sending calls

Visitor dialing instructions

Local calls	area code + 7 digits
Long distance . . .	1 + area code + 7 digits

Speed-dial numbers

Customer service	611
Emergency	911

Receiving calls

Caller dials the roamer access number below,
hears tone, dials cellular area code and number.
. . . . 218-686-7626

NationLink & RoamingAmerica (see p. 1157)
Dial *32 send to tell caller the roamer access no.
This company's subscribers have level 3 service.

Billing information

Visitor rates

Most visitors	$3 day + 99¢ minute
McCaw Cellular visitors .	$0 day + 99¢ minute

Visitors without roaming agreements (p. 1159)
Cannot call since company takes no credit cards.

Identification numbers

City	Market	System	Billing	RSA
All served cities	482	01351	00000	MN-1

Cellular company

Cellular 2000
Rural Cellular
2819 Hwy 29, S. Midway Mall, PO Box 1027
Alexandria, MN 56308

Customer service	800-450-2000
Corporate office	612-762-2000

Sending calls

Visitor dialing instructions

Local calls	area code + 7 digits
Long distance	area code + 7 digits

Speed-dial numbers

Customer service	*611
Emergency	*911
Roaming information	*711

Receiving calls

Caller dials the roamer access number below,
hears tone, dials cellular area code and number.
. . . . 612-760-7626

Billing information

Visitor rates
Most visitors . . . . . $3 day + 85¢ minute

Visitors without roaming agreements (p. 1159)
Cannot call since company takes no credit cards.

Identification numbers

City	Market	System	Billing	RSA
All served cities	482	01360	00000	MN-1

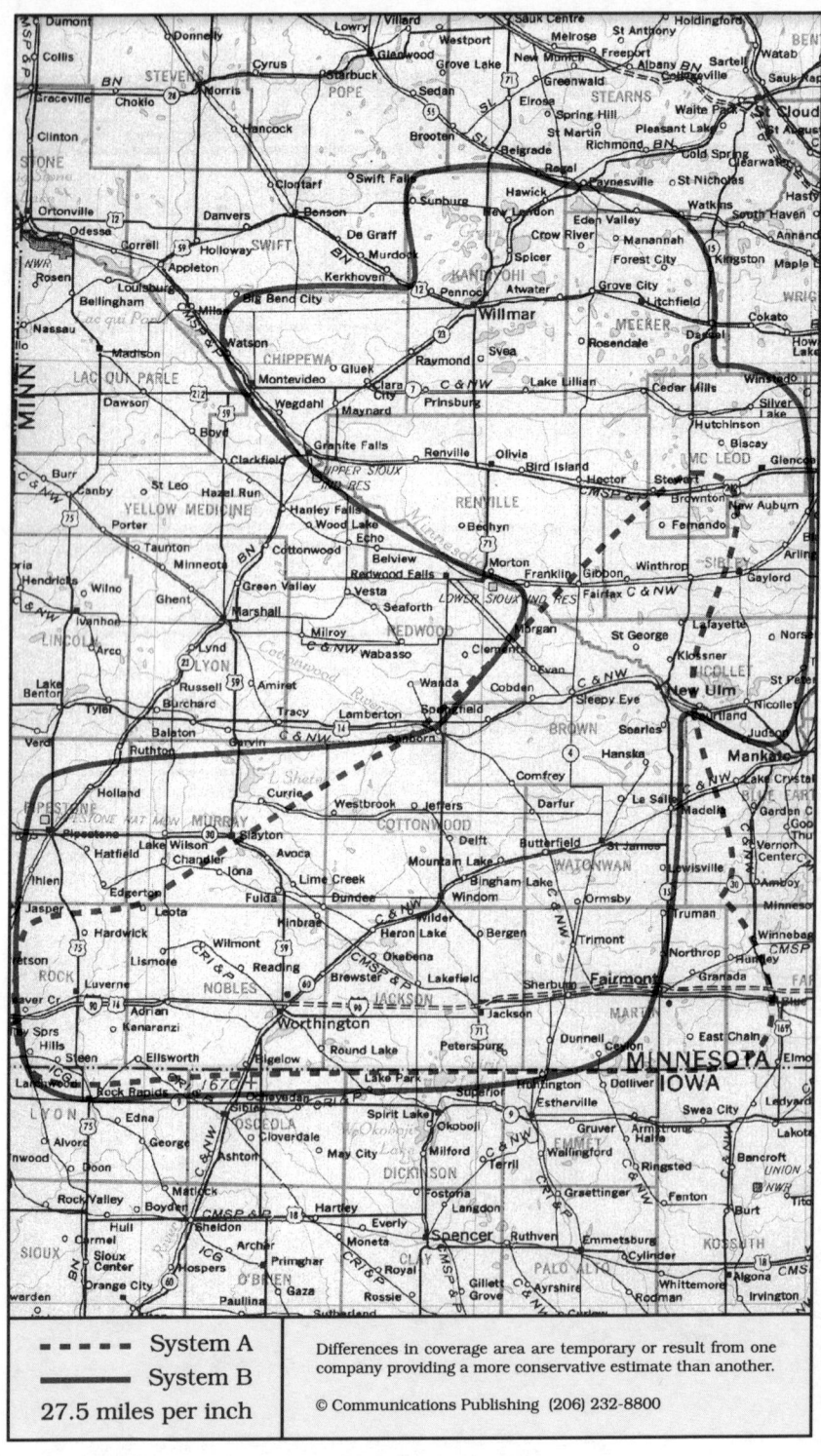

▬ ▬ ▬ ▬	System A
▬▬▬▬▬	System B
27.5 miles per inch	

Differences in coverage area are temporary or result from one company providing a more conservative estimate than another.

© Communications Publishing (206) 232-8800

Willmar, Minnesota

Fairmont, New Ulm, Willmar, Worthington

System A	System B

Cellular company

Cellular One
Greater Minnesota Cellular
1001 E Blue Earth Avenue
Fairmont, MN 56031

Customer service	605-692-6464
. . . . . from New Ulm	507-354-1700
Local office	507-235-8880
Corporate office	605-692-6464

Sending calls

Visitor dialing instructions

Local calls . . .	1 + area code + 7 digits
Long distance . . .	1 + area code + 7 digits

Speed-dial numbers

Customer service	611
Emergency	911

Receiving calls

Caller dials the roamer access number below,
hears tone, dials cellular area code and number.

Fairmont	507-230-7626
New Ulm	507-630-7626
Worthington	507-370-7626

NationLink & RoamingAmerica (see p. 1157)
Dial *31 send to receive calls, *30 to deactivate.
This company's subscribers have level 3 service.

Billing information

Visitor rates

Most visitors	$3 day + 99¢ minute

Visitors without roaming agreements (p. 1159)
Call co. to use Mastercard, Visa
or Roamer Plus will intercept first call to bill you.

Identification numbers

City	Market	System	Billing	RSA
All served cities	490	01367	00000	MN-9

Cellular company

Cellular 2000
Pacific Telecom Cellular
1870 Madison Avenue, PO Box 4069
Mankato, MN 56002-4069

Local office	800-545-3951
. . . .	218-829-0092
Corporate office	414-841-1200

Sending calls

Visitor dialing instructions

Local calls	area code + 7 digits
Long distance	area code + 7 digits

Speed-dial numbers

Customer service	*611
Emergency	*911
Roaming information	*711

Receiving calls

Caller dials the roamer access number below,
hears tone, dials cellular area code and number.

. . . .	507-330-7626

Billing information

Visitor rates

Most visitors	$3 day + 99¢ minute

Visitors without roaming agreements (p. 1159)
Cannot call since company takes no credit cards.

Identification numbers

City	Market	System	Billing	RSA
New Ulm	490	01368	00000	MN-9
Willmar,Worthington	488	01364	00000	MN-7

For full coverage area, see Marshall, MN.

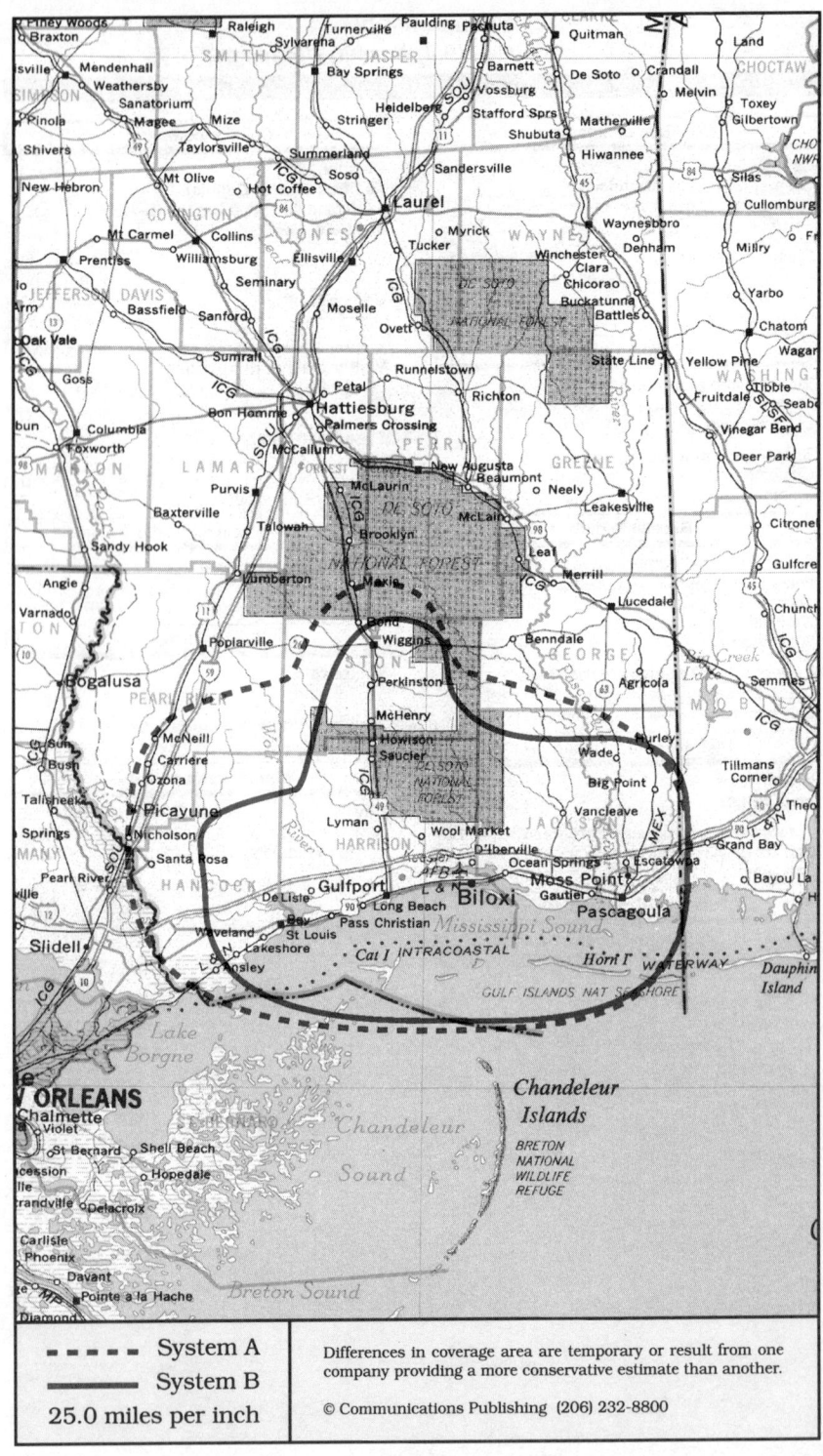

System A

System B

25.0 miles per inch

Differences in coverage area are temporary or result from one company providing a more conservative estimate than another.

© Communications Publishing (206) 232-8800

Biloxi, Mississippi
Biloxi, Gulfport, Pascagoula, Wiggins

System A	System B

Cellular company

Cellular One
Celutel
243 Beauvoir Road
Biloxi, MS 39531

Customer service 601-385-2118
Local office 601-388-1111
Corporate office 410-573-5200

Sending calls

Visitor dialing instructions
Local calls . . . . . area code + 7 digits
Long distance (in state) . . . . . 1 + 7 digits
Long distance (out state) 1 + area code + 7 digits

Speed-dial numbers
Customer service 611
Emergency 911
Highway patrol . . . . *HP or *47
News service *123

Receiving calls

Caller dials the roamer access number below,
hears tone, dials cellular area code, then
. . . . 601-380-7626

NationLink & RoamingAmerica (see p. 1157)
Dial *31 send to receive calls, *30 to deactivate.
This company's subscribers have level 3 service.

Billing information

Visitor rates
Most visitors . . . . . $3 day + 75¢ minute

Visitors without roaming agreements (p. 1159)
Call co. to use Amex, Mastercard, Visa

Identification numbers

City	Market	System	Billing	RSA
Biloxi, Gulfport	173	00281	00000	MSA
Pascagoula	252	00487	00000	MSA

Cellular company

Cellular South
Mississippi Cellular Telephone
2707 Highway 90, Suite 20
Gautier, MS 39553

9471 Three Rivers Road, 3 Rivers Place
Gulfport, MS 39503

Local office . . . . . Gautier 601-497-8400
. . . . Gulfport 601-865-0500
Corporate office 601-384-3211

Sending calls

Visitor dialing instructions
Local calls . . . . area code + 7 digits
Long distance . . . 1 + area code + 7 digits

Speed-dial numbers
Customer service 611
Emergency 911
Highway patrol . . . . *HP or *47
Roaming information *711
U.S. Coast Guard 601-762-7626

Receiving calls

Caller dials the roamer access number below,
hears tone, dials cellular area code and number.
. . . . 601-861-7626

Follow Me Roaming forwarding (see p. 1156)
Activate, dial *18 send. Deactivate dial *19 send.

Billing information

Visitor rates
Most visitors . . . $2.50 day + 75¢ minute

Visitors without roaming agreements (p. 1159)
Call co. to use Amex, Mastercard, Visa

Identification numbers

City	Market	System	Billing	RSA
Biloxi, Gulfport	173	00264	00000	MSA
Pascagoula	252	00264	00000	MSA

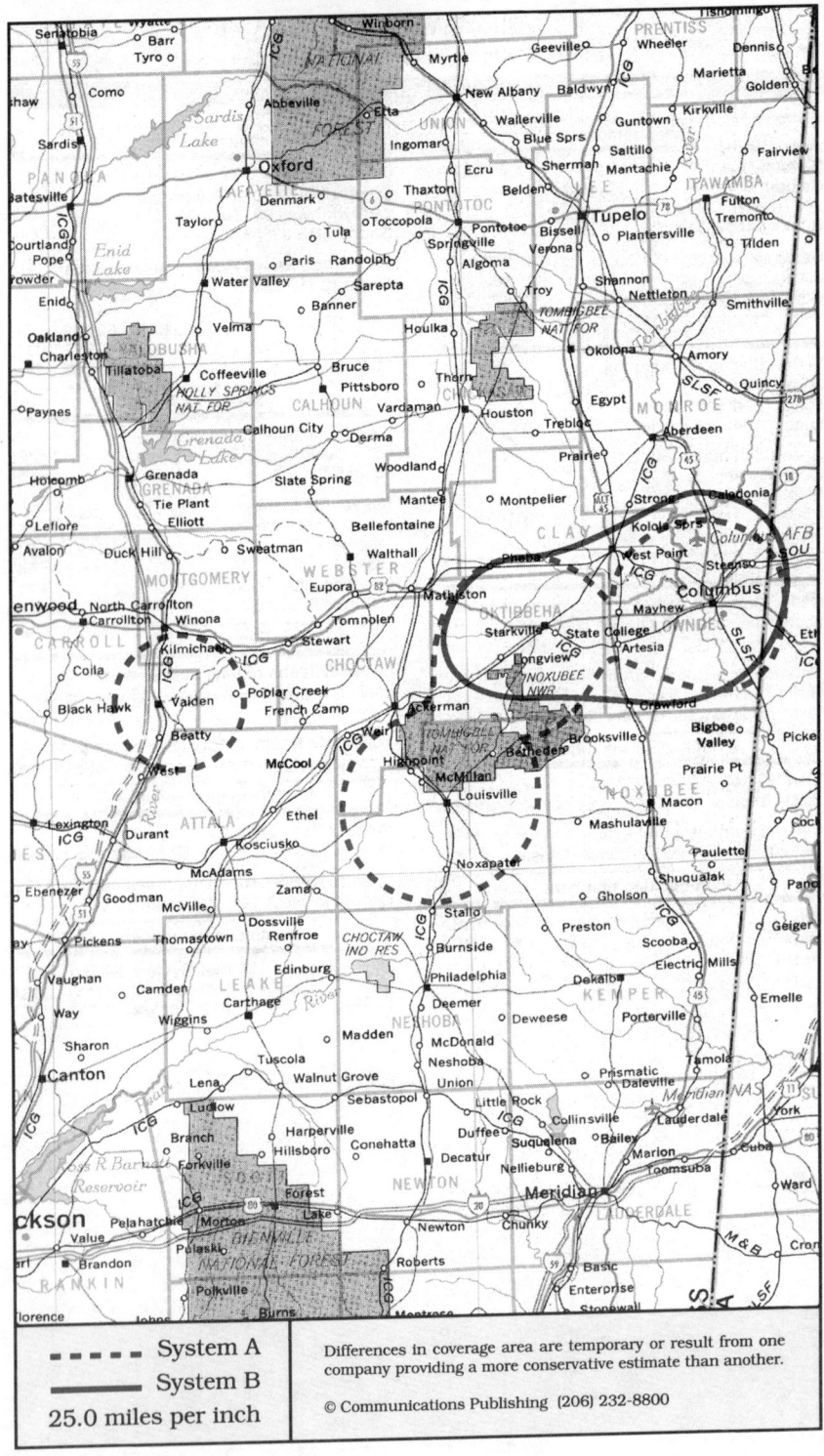

- - - - System A	Differences in coverage area are temporary or result from one
—— System B	company providing a more conservative estimate than another.
25.0 miles per inch	© Communications Publishing (206) 232-8800

Columbus, Mississippi
Columbus, Louisville, Starkville, Vaiden, Winona

System A	System B

Cellular company

Cellular One
Mercury Communications
1906-B Highway 45 N
Columbus, MS 39701

Local office	601-328-1111
. . from inside Mississippi	601-242-1111
Corporate office	601-948-4800

Sending calls

Visitor dialing instructions

Local calls	area code + 7 digits
Long distance . . .	1 + area code + 7 digits

Speed-dial numbers

Customer service	611
Emergency	911
Highway patrol	*HP or *47

Receiving calls

Caller dials the roamer access number below, hears tone, dials cellular area code and number.

. . . .   601-242-7626

NationLink & RoamingAmerica (see p. 1157)
Dial *31 send to receive calls, *30 to deactivate. This company's subscribers have level 3 service.

Billing information

Visitor rates

Most visitors	$3 day + $1 minute
In-state system A visitors.	$0 day + 50¢ minute

Visitors without roaming agreements (p. 1159)
Call co. to use Mastercard, Visa
or Roamer Plus will intercept first call to bill you.

Identification numbers

City	Market	System	Billing	RSA
All served cities	498	01383	00000	MS-6

Cellular company

Cellular South
Cellular Holding
2211 Highway 45 N, PO Box 8698
Columbus, MS 39705-8698

Customer service	800-264-2355
. . . .	601-384-3218
Local office	601-327-5700
Corporate office	601-384-3211

Sending calls

Visitor dialing instructions

Local calls	7 digits
Long distance . . .	1 + area code + 7 digits

Speed-dial numbers

Customer service	611
Emergency	911
Highway Patrol	*47
Roaming information	*711
MIX106 FM	*106
WCBI television	*TV4 or *884

Receiving calls

Caller dials the roamer access number below, hears tone, dials cellular area code and number.

Columbus	601-245-7626
Starkville	601-338-7626
West Point	601-495-7626

Follow Me Roaming forwarding (see p. 1156)
Activate, dial *18 send. Deactivate dial *19 send.

Billing information

Visitor rates

Most visitors . . .	$2 .50 day + 75¢ minute

Visitors without roaming agreements (p. 1159)
Call co. to use Amex, Mastercard, Visa

Identification numbers

City	Market	System	Billing	RSA
All served cities	498	01996	00000	MS-6

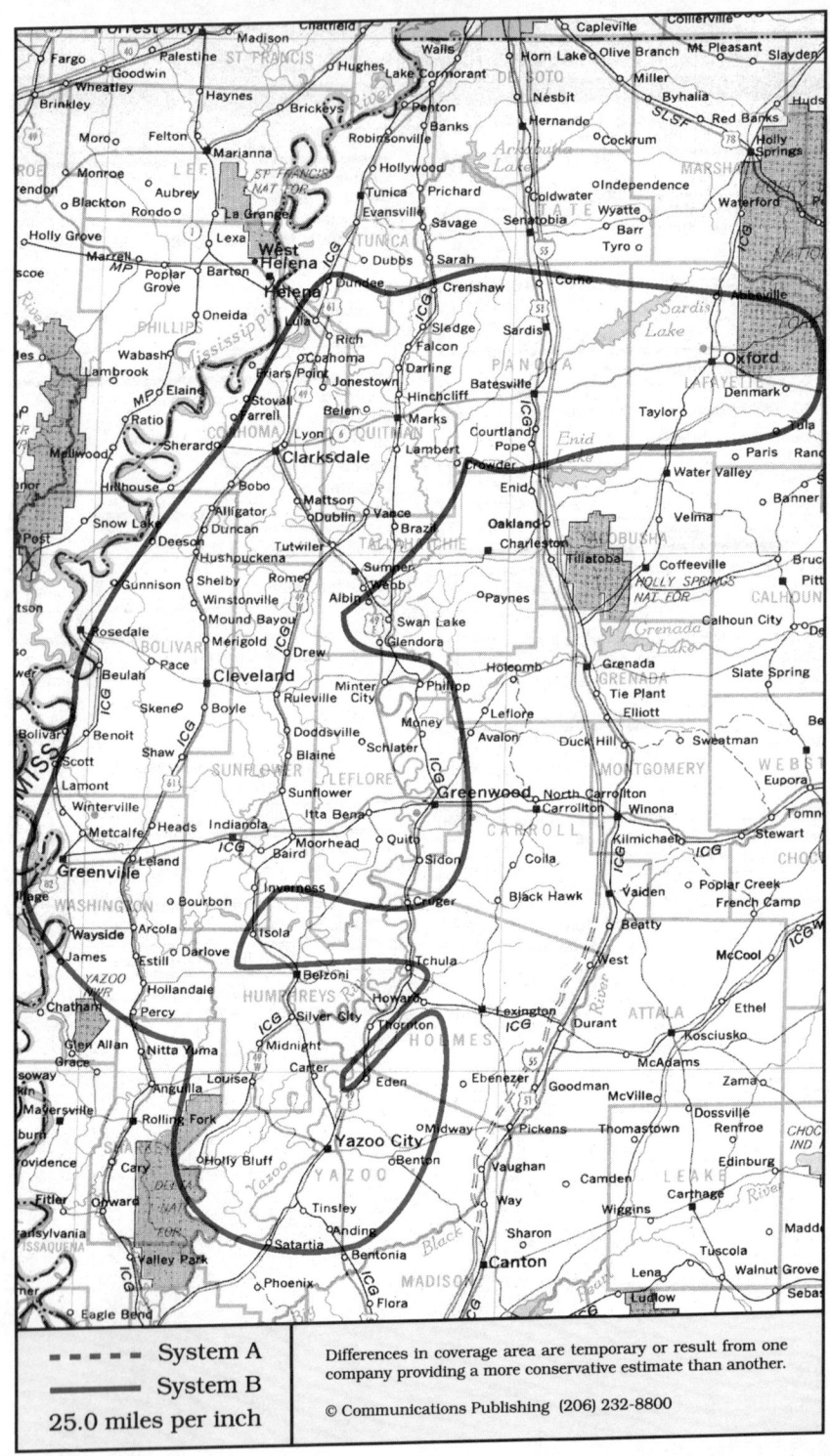

System A	System B

Cellular company

No information about the company that will operate system A was available at press time.

Cellular company

Cellular South
Cellular Holding
1515-B MS Highway 1 S
Greenville, MS 38701

902 Highway 82 W
Greenwood, MS 38930

1901-C Jackson Avenue W
Oxford, MS 38655

Customer service		800-264-2355
. . . .		601-384-3218
Local office . . .	Greenville	601-335-1010
. . . .	Greenwood	601-453-9500
. . . .	Oxford	601-234-3000
Corporate office		601-384-3211

Sending calls

Visitor dialing instructions
Local calls . . . .   7 digits
Long distance . . .   1 + area code + 7 digits

Speed-dial numbers
Customer service 611
Emergency 911
Highway Patrol *47
Roaming information *711

Receiving calls

Caller dials the roamer access number below, hears tone, dials cellular area code and number.
Batesville	601-561-7626
Clarksdale	601-624-7626
Cleveland	601-721-7626
Greenville	601-378-7626
Greenwood, Indianola . . .	601-459-7626
Louise	601-836-7626
Oxford	601-234-7626
Yazoo City	601-751-7626

Follow Me Roaming forwarding (see p. 1156)
Activate, dial *18 send. Deactivate dial *19 send.

Billing information

Visitor rates
Most visitors . . . $2 .50 day + 75¢ minute

Visitors without roaming agreements (p. 1159)
Call co. to use Amex, Mastercard, Visa

Identification numbers

City	Market	System	Billing	RSA
Clarksdale	493	01382	00000	MS-1
Greenville	497	01382	00000	MS-5
Greenwood	495	01382	00000	MS-3

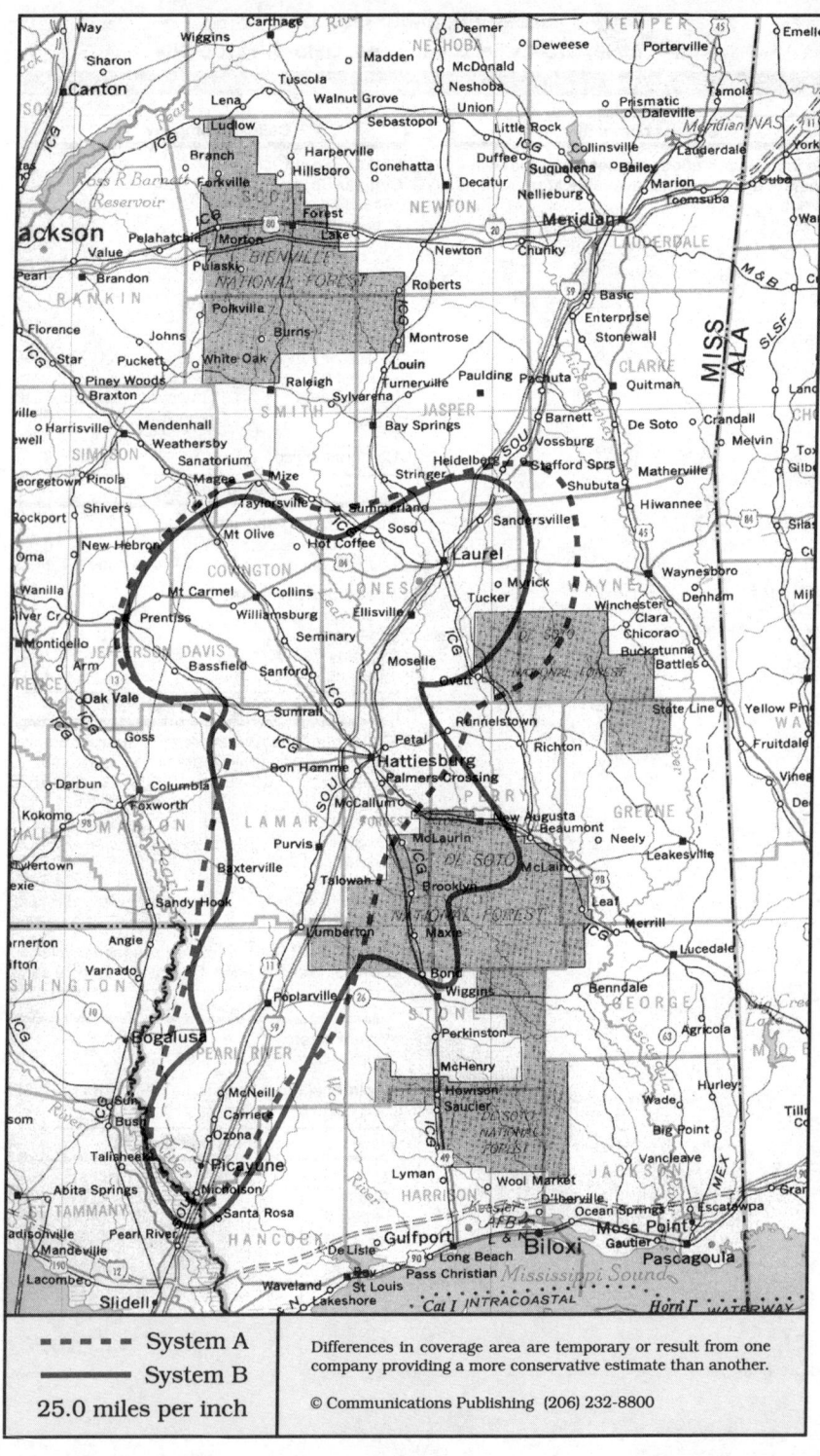

- - - - - System A
————— System B
25.0 miles per inch

Differences in coverage area are temporary or result from one company providing a more conservative estimate than another.

© Communications Publishing (206) 232-8800

Hattiesburg, Mississippi

Collins, Hattiesburg, Laurel, New Augusta, Picayune, Poplarville

System A	System B

Cellular company

Cellular One
Cellular XL Associates
806 Westover Drive
Hattiesburg, MS 39402

2257 Highway 15 N, Pinetree Shopping Center
Laurel, MS 39440

Local office . . . Hattiesburg	601-261-9300
. . . . Laurel	601-425-2211
Corporate office	601-261-9300

Sending calls

Visitor dialing instructions
Local calls . . . . area code + 7 digits
Long distance . . . . area code + 7 digits

Speed-dial numbers
Customer service 611
Emergency 911
Highway patrol *47

Receiving calls

Caller dials the roamer access number below,
hears tone, dials cellular area code and number.
Hattiesburg 601-261-7626
Laurel 601-477-7626

NationLink & RoamingAmerica (see p. 1157)
Dial *31 send to receive calls, *30 to deactivate.
This company's subscribers have level 3 service.

Billing information

Visitor rates
Most visitors . . . . . $3 day + 85¢ minute

Visitors without roaming agreements (p. 1159)
Roamer Plus will intercept first call to bill you.

Identification numbers

City	Market	System	Billing	RSA
Hattiesburg	503	01393	00000	MS-11
Laurel	502	01393	00000	MS-10

Cellular company

Cellular South
Cellular Holding
1306 Hardy Street, PO Box 15602
Hattiesburg, MS 39401

803-B N 16th Avenue
Laurel, MS 39440

Customer service	800-264-2355
. . . .	601-384-3218
Local office . . . Hattiesburg	601-582-7777
. . . . Laurel	601-425-4444
Corporate office	601-384-3211

Sending calls

Visitor dialing instructions
Local calls . . . . 7 digits
Long distance . . . 1 + area code + 7 digits

Speed-dial numbers
Customer service 611
Emergency 911
Highway Patrol *47
Roaming information *711

Receiving calls

Caller dials the roamer access number below,
hears tone, dials cellular area code and number.
Collins 601-765-7626
Hattiesburg 601-543-7626
Laurel 601-344-7626
Picayune 601-799-7626
Poplarville 601-795-7626

Follow Me Roaming forwarding (see p. 1156)
Activate, dial *18 send. Deactivate dial *19 send.

Billing information

Visitor rates
Most visitors . . . $2 .50 day + 75¢ minute

Visitors without roaming agreements (p. 1159)
Call co. to use Amex, Mastercard, Visa

Identification numbers

City	Market	System	Billing	RSA
Hattiesburg	503	01394	00000	MS-11
Laurel	502	01394	00000	MS-10

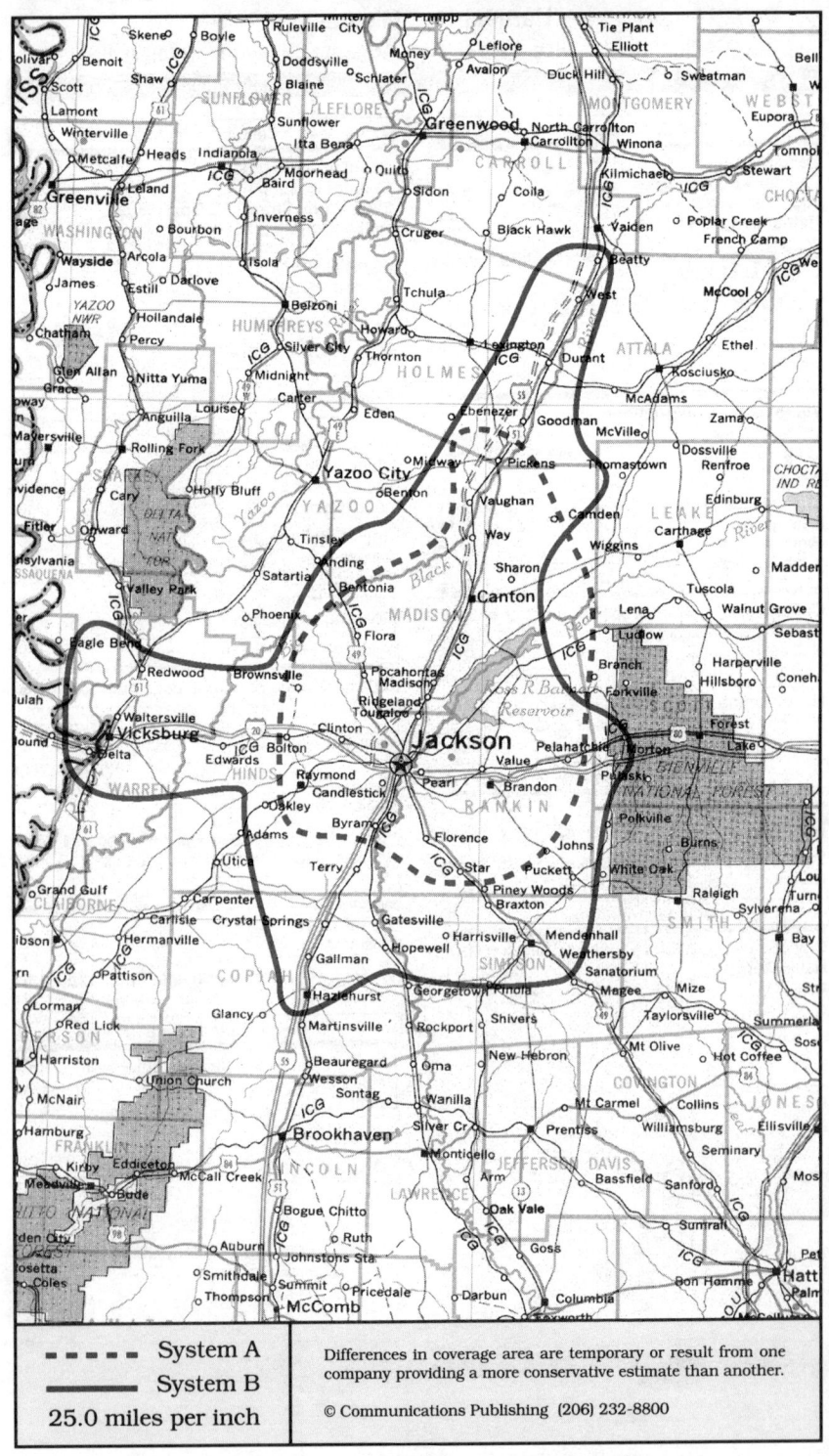

- - - - -	System A	
————	System B	
25.0 miles per inch		

Differences in coverage area are temporary or result from one company providing a more conservative estimate than another.

© Communications Publishing (206) 232-8800

Jackson, Mississippi
Canton, Crystal Springs, Hazelhurst, Jackson, Mendenhall, Vicksburg

System A	System B

Cellular company

Cellular One
Celutel
5380 I-55 N
Jackson, MS 39211

Local office	601-956-3800
Corporate office	410-573-5200

Sending calls

Visitor dialing instructions

Local calls	area code + 7 digits
Long distance	area code + 7 digits

Speed-dial numbers

Customer service	611
Emergency	911
Information hotline	*123
Make a traffic report .	*94 or *107 or *1180

Receiving calls

Caller dials the roamer access number below,
hears tone, dials cellular area code and number.
. . . . 601-331-7626

NationLink & RoamingAmerica (see p. 1157)
Caller hears your current roamer access number.
This company's subscribers have level 0 service.

Billing information

Visitor rates
Most visitors . . . . . $3 day + 99¢ minute

Visitors without roaming agreements (p. 1159)
Call co. to use Mastercard, Visa

Identification numbers

City	Market	System	Billing	RSA
All served cities	106	00205	00000	MSA

Cellular company

MCTA Cellular
ALLTEL Mobile Communications
833 E River Place
Jackson, MS 39202

Customer service	some areas	800-255-8351
. . . .		601-973-8484
Local office		601-354-1212
Corporate office		501-661-8500

Sending calls

Visitor dialing instructions

Local calls	area code + 7 digits
Long distance . . .	1 + area code + 7 digits

Speed-dial numbers

Customer service	*611
Emergency	*911
Make a traffic report	*62 or *103

Receiving calls

Caller dials the roamer access number below,
hears tone, dials cellular area code and number.

Crystal Springs, Jackson . . .	601-946-7626
Vicksburg	601-631-1626

Billing information

Visitor rates
Most visitors . . . . . $3 day + 85¢ minute

Visitors without roaming agreements (p. 1159)
Cannot call since company takes no credit cards.

Identification numbers

City	Market	System	Billing	RSA
Jackson	106	00160	00000	MSA
Crystal Springs	501	00160	30532	MS-9
Vicksburg	497	00160	30224	MS-5

This company is jointly owned by Alltel Mobile
Communications and American Cellular.
American Cellular is a subsidiary of BellSouth
Enterprises.

For full coverage area, see Meridian, MS.

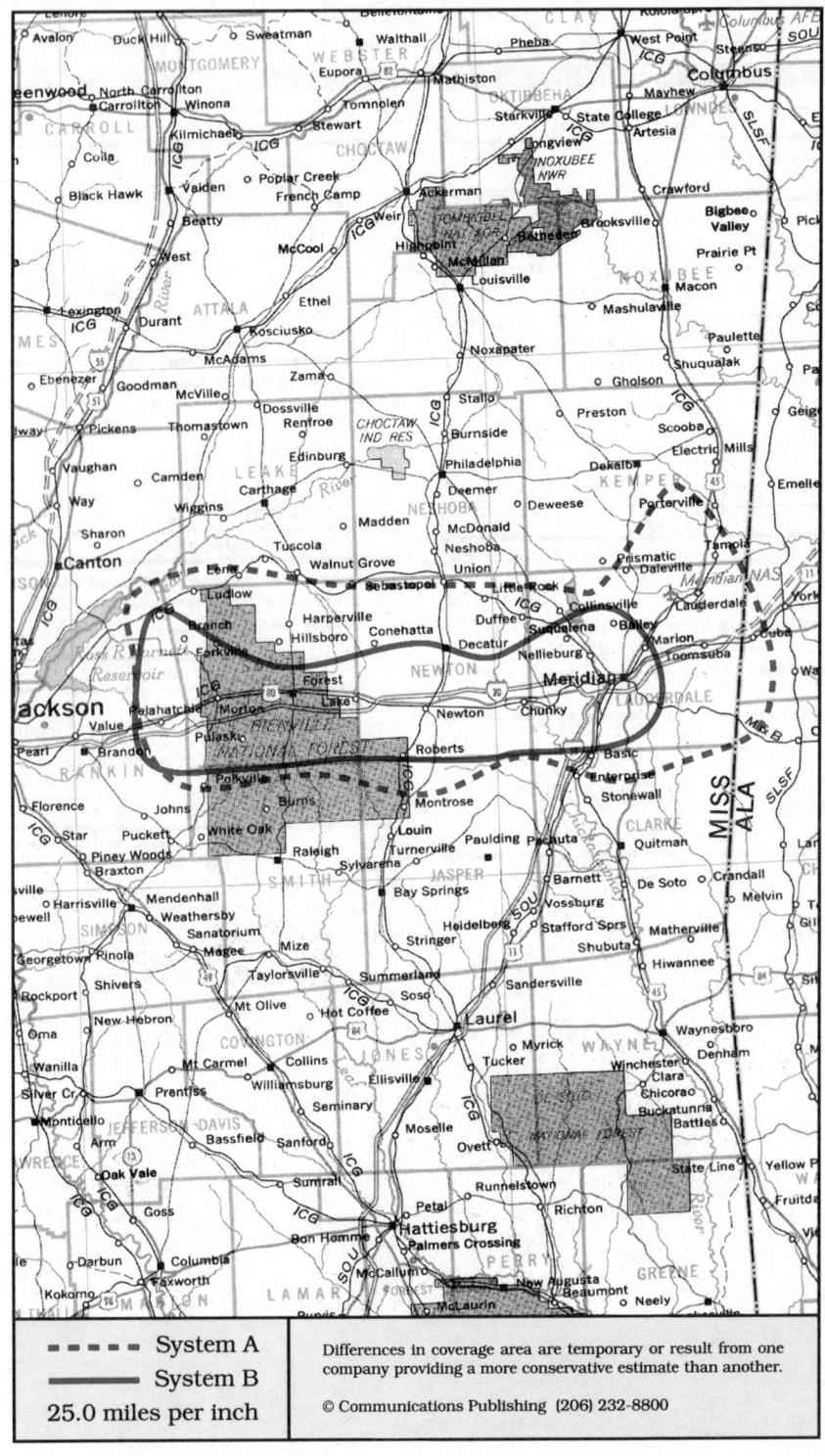

Meridian, Mississippi
Forest, Meridian

System A

System B

Cellular company

Cellular One of Meridian
Cone Enterprises
2400 15th Street, PO Box 5554
Meridian, MS 39302

Local office	601-485-2355
Corporate office	806-744-1661

Sending calls

Visitor dialing instructions

Local calls	area code + 7 digits
Long distance . . .	1 + area code + 7 digits

Speed-dial numbers

Customer service	611
Emergency	911

Receiving calls

Caller dials the roamer access number below,
hears tone, dials cellular area code and number.
. . . . 601-486-7626

Follow Me Roaming forwarding (see p. 1156)
Phone Me Anywhere forwarding (see p. 1156)
Activate, dial ✳18 send. Deactivate dial ✳19 send.

Billing information

Visitor rates

Most visitors	$3 day + 90¢ minute

Visitors without roaming agreements (p. 1159)
Call co. to use Amex, Discover, Mastercard, Visa

Identification numbers

City	Market	System	Billing	RSA
All served cities	499	01385	00000	MS-7

Cellular company

MCTA Cellular
ALLTEL Mobile Communications
833 E River Place, Suite 100
Jackson, MS 39202

Customer service	some areas	800-255-8351
. . . .		601-973-8484
Local office		601-354-1212
Corporate office		501-661-8500

Sending calls

Visitor dialing instructions

Local calls	area code + 7 digits
Long distance . . .	1 + area code + 7 digits

Speed-dial numbers

Customer service	✳611
Emergency	✳911
Make a traffic report	✳62 or ✳103

Receiving calls

Caller dials the roamer access number below,
hears tone, dials cellular area code and number.

Forest, Morton	601-946-7626
Meridian	601-486-3626

Billing information

Visitor rates

Most visitors	$3 day + 85¢ minute

Visitors without roaming agreements (p. 1159)
Cannot call since company takes no credit cards.

Identification numbers

City	Market	System	Billing	RSA
All served cities	499	00160	30534	MS-7

This company is jointly owned by Alltel Mobile
Communications and American Cellular.
American Cellular is a subsidiary of BellSouth
Enterprises.

For full coverage area, see Jackson, MS.

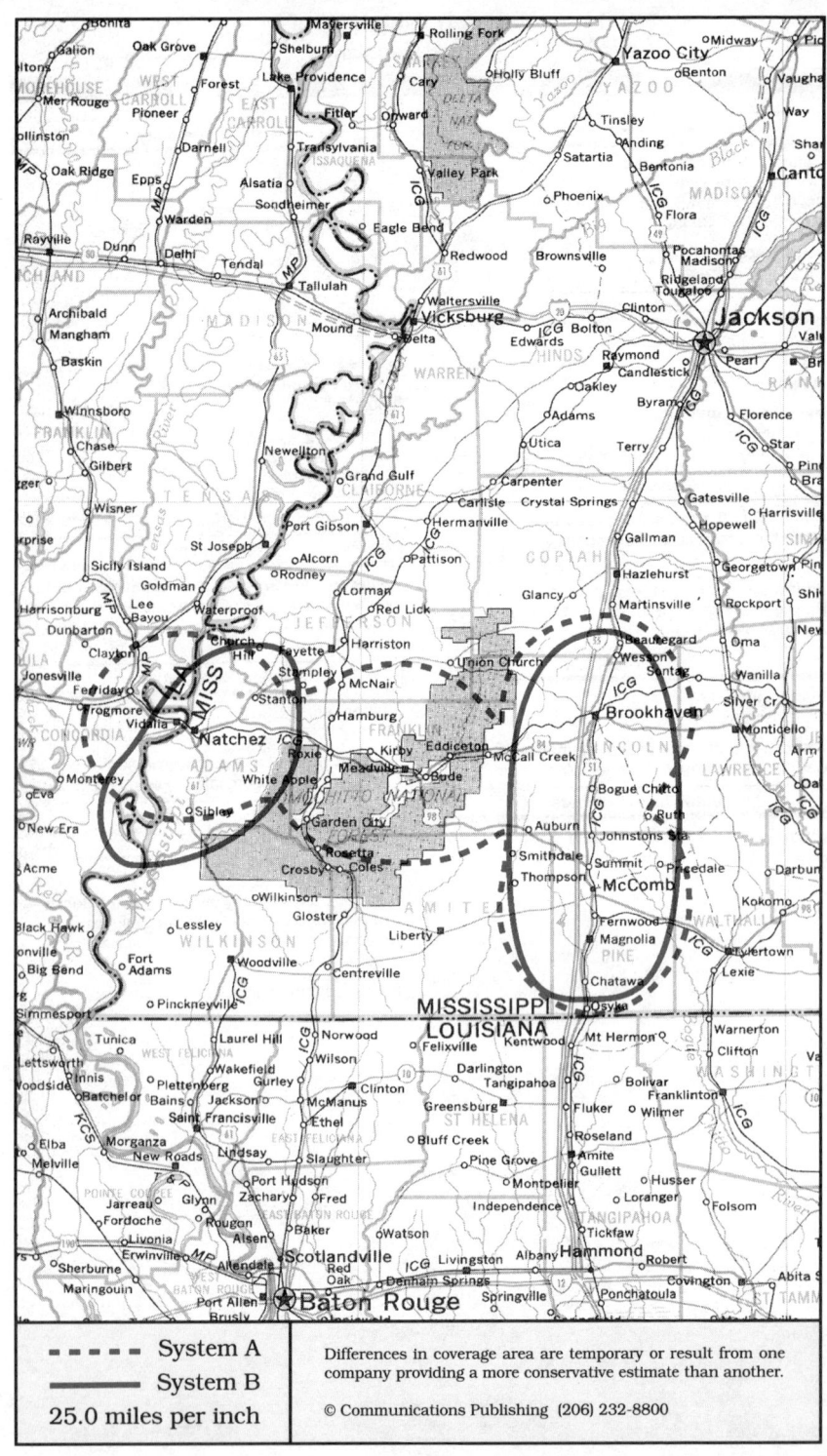

- - - - System A
——— System B
25.0 miles per inch

Natchez, Mississippi

Brookhaven, McComb, Meadville, Natchez, Vidalia LA

System A	System B

System A

Cellular company

Cellular One
Mississippi Cellular Limited Partnership
105 Lower Woodville Road, Suite 2
Natchez, MS 39120

Customer service 601-446-7266
Corporate office 601-445-0333

Sending calls

Visitor dialing instructions
Local calls . . . . . area code + 7 digits
Long distance . . . 1 + area code + 7 digits

Speed-dial numbers
Customer service 611
Emergency 911

Receiving calls

Caller dials the roamer access number below,
hears tone, dials cellular area code and number.
. . . . 601-431-7626

NationLink & RoamingAmerica (see p. 1157)
Dial ∗31 send to receive calls, ∗30 to deactivate.
Dial ∗32 send to tell caller the roamer access no.
This company's subscribers have level 9 service.

Billing information

Visitor rates
Most visitors . . . . . $3 day + 85¢ minute

Visitors without roaming agreements (p. 1159)
Roamer Plus will intercept first call to bill you.

Identification numbers

City	Market	System	Billing	RSA
All served cities	500	01387	00000	MS-8

System B

Cellular company

Cellular South
Cellular Holding
707 Brookway Boulevard
Brookhaven, MS 39601

255 Highway 61 South
Natchez, MS 39120

Customer service		800-264-2355
. . . .		601-384-3218
Local office . . .	Brookhaven	601-833-6100
. . . .	Natchez	601-445-2000
Corporate office		601-384-3211

Sending calls

Visitor dialing instructions
Local calls . . . . 7 digits
Long distance . . . 1 + area code + 7 digits

Speed-dial numbers
Customer service 611
Emergency 911
Highway Patrol . . . . ∗47
Roaming information . . . . . . ∗711

Receiving calls

Caller dials the roamer access number below,
hears tone, dials cellular area code and number.
Brookhaven 601-835-7626
McComb 601-249-7626
Natchez 601-443-7626

Follow Me Roaming forwarding (see p. 1156)
Activate, dial ∗18 send. Deactivate dial ∗19 send.

Billing information

Visitor rates
Most visitors . . . $2 .50 day + 75¢ minute

Visitors without roaming agreements (p. 1159)
Call co. to use Amex, Mastercard, Visa

Identification numbers

City	Market	System	Billing	RSA
All served cities	500	01394	00000	MS-8

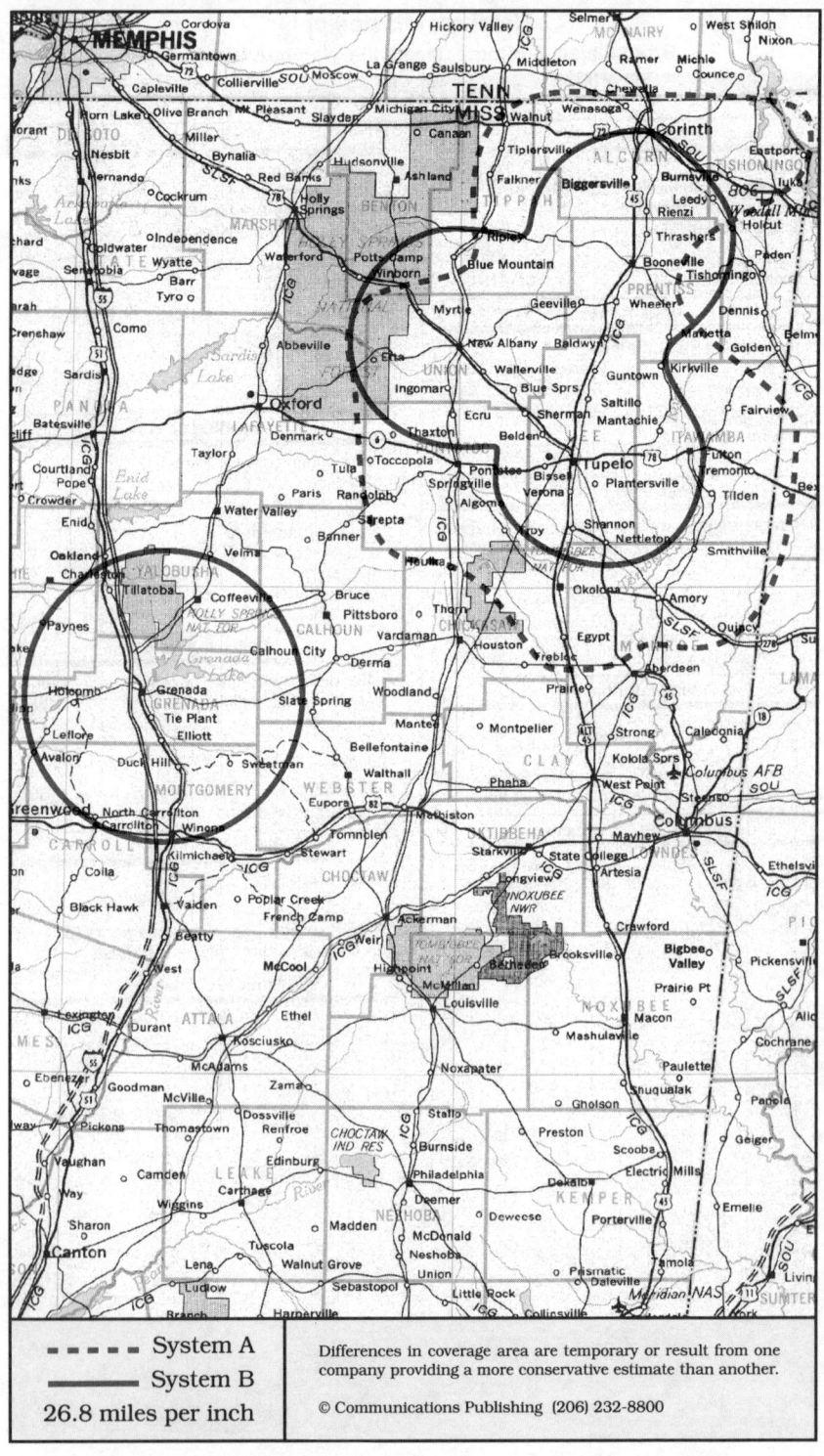

- - - - System A
———— System B
26.8 miles per inch

Differences in coverage area are temporary or result from one company providing a more conservative estimate than another.

© Communications Publishing (206) 232-8800

Tupelo, Mississippi
Booneville, Corinth, Fulton, Grenada, Okolona, Pontotoc, Tupelo

System A	System B

Cellular company

Cellular One of Northeast Mississippi
Licensed to: R & D Cellular
Managed by: Prime Cellular
852 N Gloster Street, Suite 4
Tupelo, MS 38801

Customer service		601-844-5593
Local office . .	administration	601-844-5980
. . . . . .	sales	601-844-7799
Corporate office	Prime Cellular	201-227-1434
. . . . . .	R & D Cellular	408-224-9009

Sending calls

Visitor dialing instructions
Local calls . . . . 7 digits
Long distance . . . 1 + area code + 7 digits

Speed-dial numbers
Customer service 611
Emergency 911
Roaming information 711

Receiving calls

Caller dials the roamer access number below,
hears tone, dials cellular area code and number.
. . . . 601-791-7626

Billing information

Visitor rates
Most visitors . . . . . $3 day+85¢ minute
Cellular One visitors from MS $0 day+50¢ minute

Visitors without roaming agreements (p. 1159)
Roamer Plus will intercept first call to bill you.

Identification numbers

City	Market	System	Billing	RSA
All served cities	494	01375	00000	MS-2

Cellular company

BellSouth Mobility
3849 N Gloster Street
Tupelo, MS 38801

Customer service		800-351-2400
. . . .		205-444-9600
Local office		601-841-0200
Corporate office		404-847-3600

Sending calls

Visitor dialing instructions
Local calls . . . . . . area code + 7 digits
Long distance . . . . . area code + 7 digits

Speed-dial numbers
Customer service 811
Emergency 911
Roaming information 711
Telephone problems 611

Receiving calls

Caller dials the roamer access number below,
hears tone, dials cellular area code and number.
Grenada 601-227-5499
Tupelo 601-791-1776

Follow Me Roaming forwarding (see p. 1156)
Activate, dial *18 send. Deactivate dial *19 send.

Billing information

Visitor rates
Most visitors . . . . . $3 day + 85¢ minute
BellSouth subscribers . $0 day + 60¢ minute

Visitors without roaming agreements (p. 1159)
Call co. to use Amex, Mastercard, Visa

Identification numbers

City	Market	System	Billing	RSA
Grenada	496	00062	30638	MS-4
Tupelo	494	01376	00000	MS-2

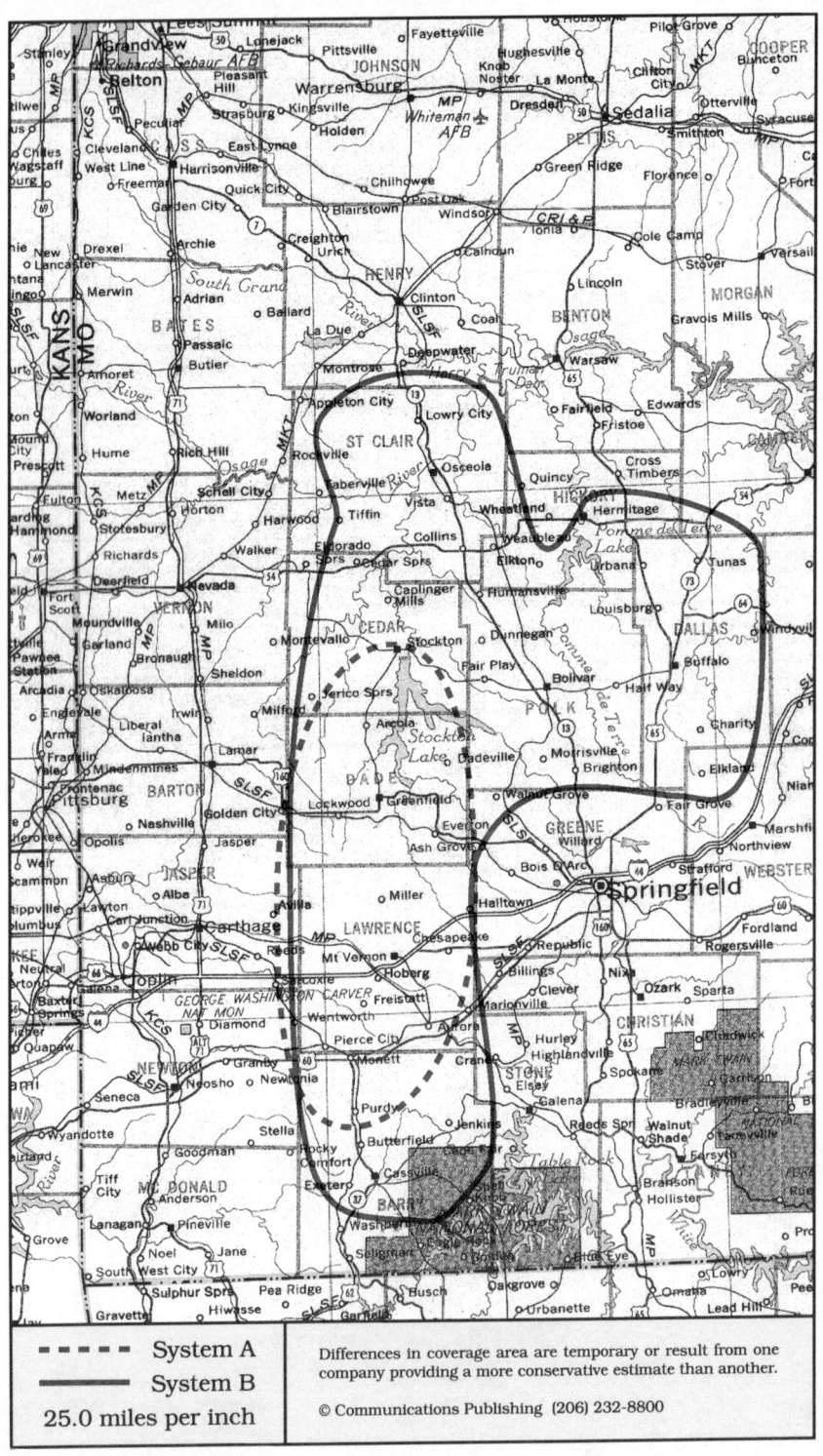

System A

System B

25.0 miles per inch

Differences in coverage area are temporary or result from one company providing a more conservative estimate than another.

© Communications Publishing (206) 232-8800

Aurora, Missouri
Aurora, Bolivar, Buffalo, Cassville, Monett, Osceola, Stockton

System A	System B

Cellular company

Cellular One of the Ozarks
100 Lincoln Street
Monett, MO 65708

Corporate office 417-235-0122

Sending calls

Visitor dialing instructions
Local calls . . . . . area code + 7 digits
Long distance . . . 1 + area code + 7 digits

Speed-dial numbers
Customer service *611
Emergency 911

Receiving calls

Caller dials the roamer access number below,
hears tone, dials cellular area code and number.
. . . . 417-861-7626

NationLink & RoamingAmerica (see p. 1157)
Dial *31 send to receive calls, *30 to deactivate.
Dial *32 send to tell caller the roamer access no.
This company's subscribers have level 9 service.

Billing information

Visitor rates
Most visitors . . . . . $3 day + 85¢ minute

Visitors without roaming agreements (p. 1159)
Cannot call since company takes no credit cards.

Identification numbers

City	Market	System	Billing	RSA
All served cities	517	01421	00000	MO-14

This company's subscribers should refer to the
roaming agreements of Crowley Cellular
Telecommunications.

Cellular company

ALLTEL Mobile Communications
3322 S Campbell, Suite N
Springfield, MO 65807

Customer service   some areas 800-255-8351
Local office 417-882-2020
Corporate office 501-661-8500

Sending calls

Visitor dialing instructions
Local calls . . . . . . area code + 7 digits
Long distance . . . . . area code + 7 digits

Speed-dial numbers
Customer service *611
Emergency 911
Highway patrol *55
Make a traffic report . . . . . *97 or *99

Receiving calls

Caller dials the roamer access number below,
hears tone, dials cellular area code and number.
Aurora 417-466-0626
Bolivar, Clinton 417-777-1626

Follow Me Roaming forwarding (see p. 1156)
Activate, dial *18 send. Deactivate dial *19 send.

Billing information

Visitor rates
Most visitors . . . . . $3 day + 85¢ minute

Visitors without roaming agreements (p. 1159)
Call co. to use Amex, Discover, Mastercard, Visa

Identification numbers

City	Market	System	Billing	RSA
Aurora, Monett	517	00546	30254	MO-14
Bolivar	513	01414	30496	MO-10
Stockton	512	00546	00000	MO-9

For full coverage area, see Lebanon, MO and
Springfield, MO.

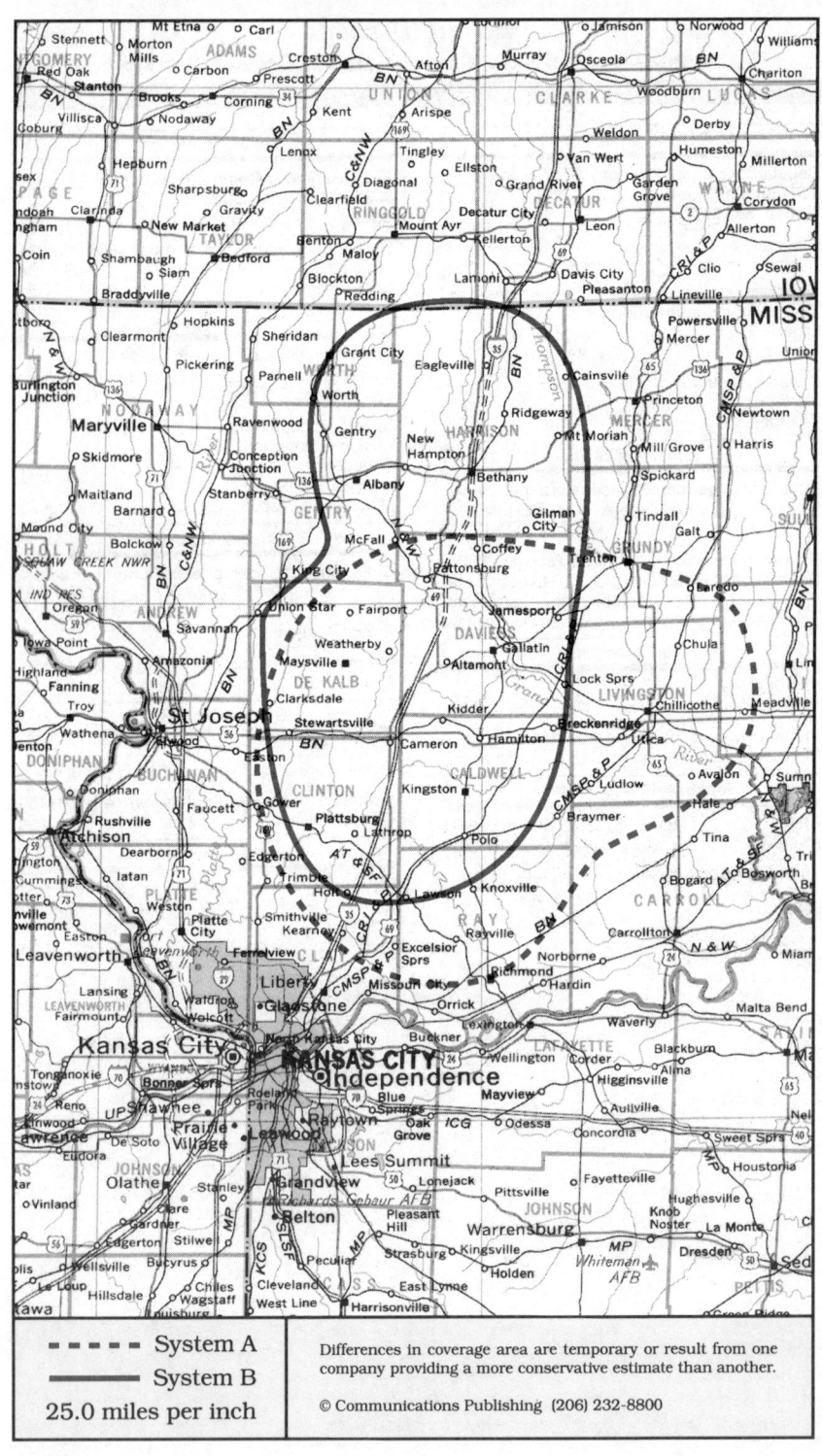

System A
System B
25.0 miles per inch

Differences in coverage area are temporary or result from one company providing a more conservative estimate than another.

© Communications Publishing (206) 232-8800

Cameron, Missouri

Albany, Bethany, Cameron, Chillicothe, Kingston, Maysville, Plattsburg

System A	System B

Cellular company

Cellular One
Licensed to: Missouri Cellular
Operated by: Quantum Communications
1125 N Morley
Moberly, MO 65270

Customer service	816-651-2273
Local office	816-263-8811
Corporate office	612-942-7650

Sending calls

Visitor dialing instructions

Local calls	area code + 7 digits
Long distance . . .	1 + area code + 7 digits

Speed-dial numbers

Customer service	611
Emergency	911
Highway Patrol	*55
Radio station 104.1	*123
Time and temperature	*711

Receiving calls

Caller dials the roamer access number below,
hears tone, dials cellular area code and number.

. . . . 816-632-7626

NationLink & RoamingAmerica (see p. 1157)
Dial *31 send to receive calls, *30 to deactivate.
This company's subscribers have level 3 service.

Billing information

Visitor rates

Most visitors . . . . . $3 day + 99¢ minute

Visitors without roaming agreements (p. 1159)
Call co. to use Amex, Mastercard, Visa
or Roamer Plus will intercept first call to bill you.

Identification numbers

City	Market	System	Billing	RSA
All served cities	507	01401	00000	MO-4

This company is not licensed to serve MO-2.

Cellular company

ALLTEL Mobile Communications
3322 S Campbell, Suite N
Springfield, MO 65807

Customer service	some areas	800-255-8351
Local office		417-882-2020
Corporate office		501-661-8500

Sending calls

Visitor dialing instructions

Local calls	area code + 7 digits
Long distance	area code + 7 digits

Speed-dial numbers

Customer service	*611
Emergency	911
Highway patrol	*55

Receiving calls

Caller dials the roamer access number below,
hears tone, dials cellular area code and number.

Bethany, Cameron	816-223-7626
Kirksville	816-626-7626

Billing information

Visitor rates

Most visitors . . . . . $3 day + 85¢ minute

Visitors without roaming agreements (p. 1159)
Call co. to use Amex, Discover, Mastercard, Visa

Identification numbers

City	Market	System	Billing	RSA
Bethany	505	1398	30492	MO-2
Cameron	507	1402	30494	MO-4
Kirksville	506	1400	30504	MO-3

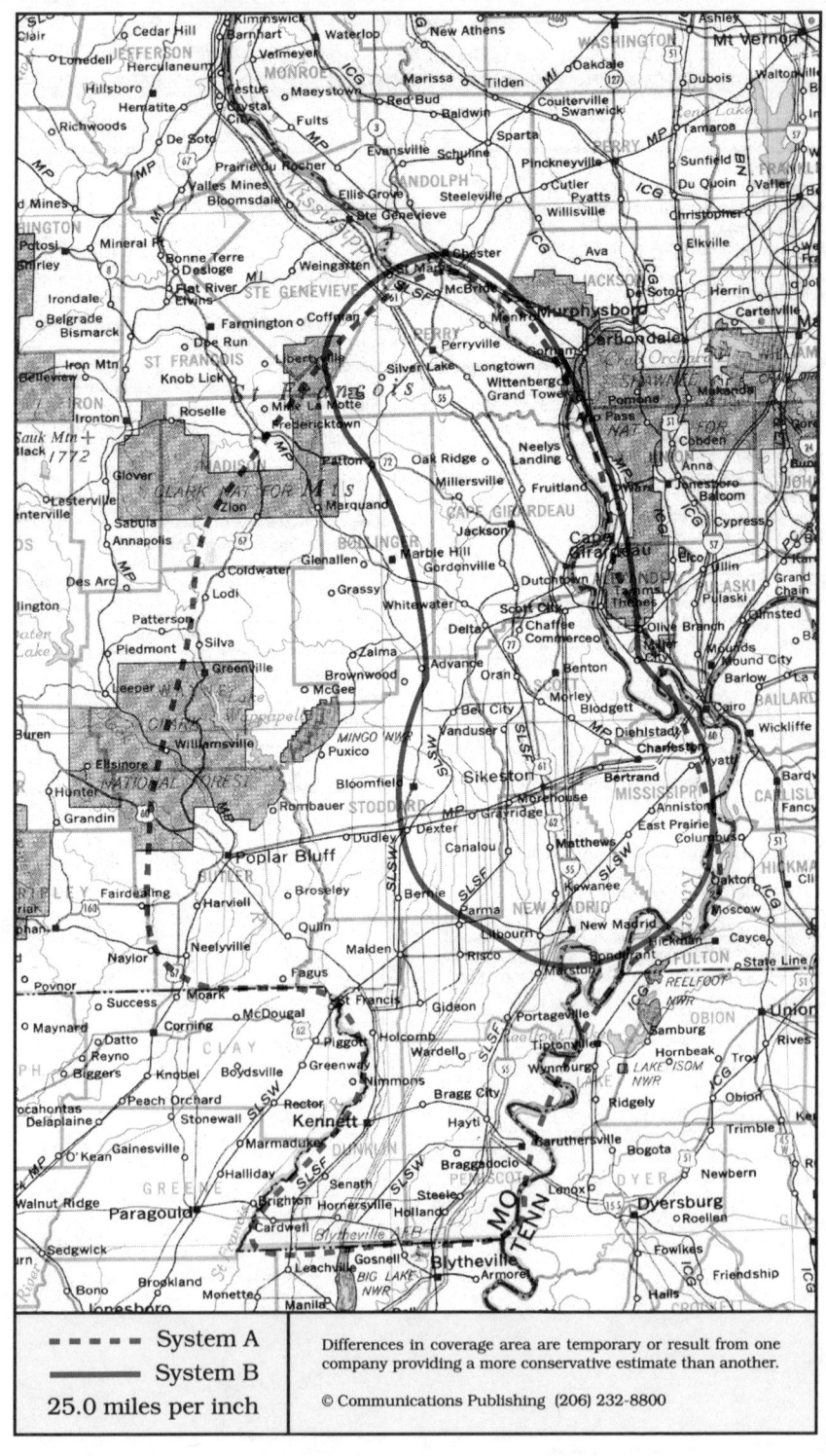

▪ ▪ ▪ ▪ ▪ **System A**	Differences in coverage area are temporary or result from one
——— **System B**	company providing a more conservative estimate than another.
25.0 miles per inch	© Communications Publishing (206) 232-8800

Cape Girardeau, Missouri
Cape Giradeau, Kennett, Poplar Bluff, Sikeston

System A	System B

Cellular company

Cybertel Cellular Telephone
500 N Kings Highway
Cape Girardeau, MO 63701

Customer service	some areas	800-662-4534
. . . .		314-444-4444
Local office		800-455-4444
. . . .		314-334-4444
Corporate office		314-423-6500

Sending calls

Visitor dialing instructions

Local calls	area code + 7 digits
Long distance . . .	1 + area code + 7 digits

Speed-dial numbers

Business update	*28
Customer service	*444 or *611
Emergency (local)	911
Hear a traffic report	*88
K103 radio	*103
KFVS news hotline	*12
KGMO radio	*1007
KMOX radio traffic report	*25
KZIM radio	*960
Make a traffic report	*66 or *666
Highway patrol	*55 or *555
News, sports, weather	*123
Police	*311
Roaming information	*211
Sports update	*78
Weather report	*98

Receiving calls

Caller dials the roamer access number below,
hears tone, dials cellular area code and number.
. . . . 314-270-7626

NationLink & RoamingAmerica (see p. 1157)
Dial *31 send to receive calls, *30 to deactivate.
Dial *32 send to tell caller the roamer access no.
This company's subscribers have level 8 service.

Billing information

Visitor rates
Most visitors . . . . . $3 day + 99¢ minute

Visitors without roaming agreements (p. 1159)
Call co. to use Amex, Mastercard, Visa

Identification numbers

City	Market	System	Billing	RSA
Cape Girardeau	521	01429	00000	MO-18
Sikeston	522	01429	00000	MO-19

For full coverage area, see St. Louis, MO.

Cellular company

Southwestern Bell Mobile Systems
13075 Manchester, Suite 100-N
St. Louis, MO 63131

Customer service	. . . .	800-331-0500
. . . .		314-821-1111
Local office		314-821-9999
Corporate office		214-733-2000

Sending calls

Visitor dialing instructions

Local calls	area code + 7 digits
Long distance . . .	1 + area code + 7 digits

Speed-dial numbers

Cellular tips	314-541-5400
Current information service	*123
Customer service	611
Emergency (Illinois)	511
Emergency (St. Louis County) . . .	311
Highway patrol (Missouri)	*55
League of Women Voters . . *VOTE or *8683
KMOX news	*1
Make a traffic report . . . . *RPT or *778
Missouri police	311
Personal traffic consultant . . *PTC or *782
Sports line	314

Receiving calls

Caller dials the roamer access number below,
hears tone, dials cellular area code and number.
. . . . 314-277-7626

Follow Me Roaming forwarding (see p. 1156)
Activate, dial *18 send. Deactivate dial *19 send.

Billing information

Visitor rates
Most visitors . . . . . $3 day + 85¢ minute

Visitors without roaming agreements (p. 1159)
Call co. to use Amex, Discover, Mastercard, Visa

Identification numbers

City	Market	System	Billing	RSA
Cape Girardeau	521	00046	30460	MO-18
Sikeston	522	00046	30522	MO-19

For full coverage area, see St. Louis, MO.

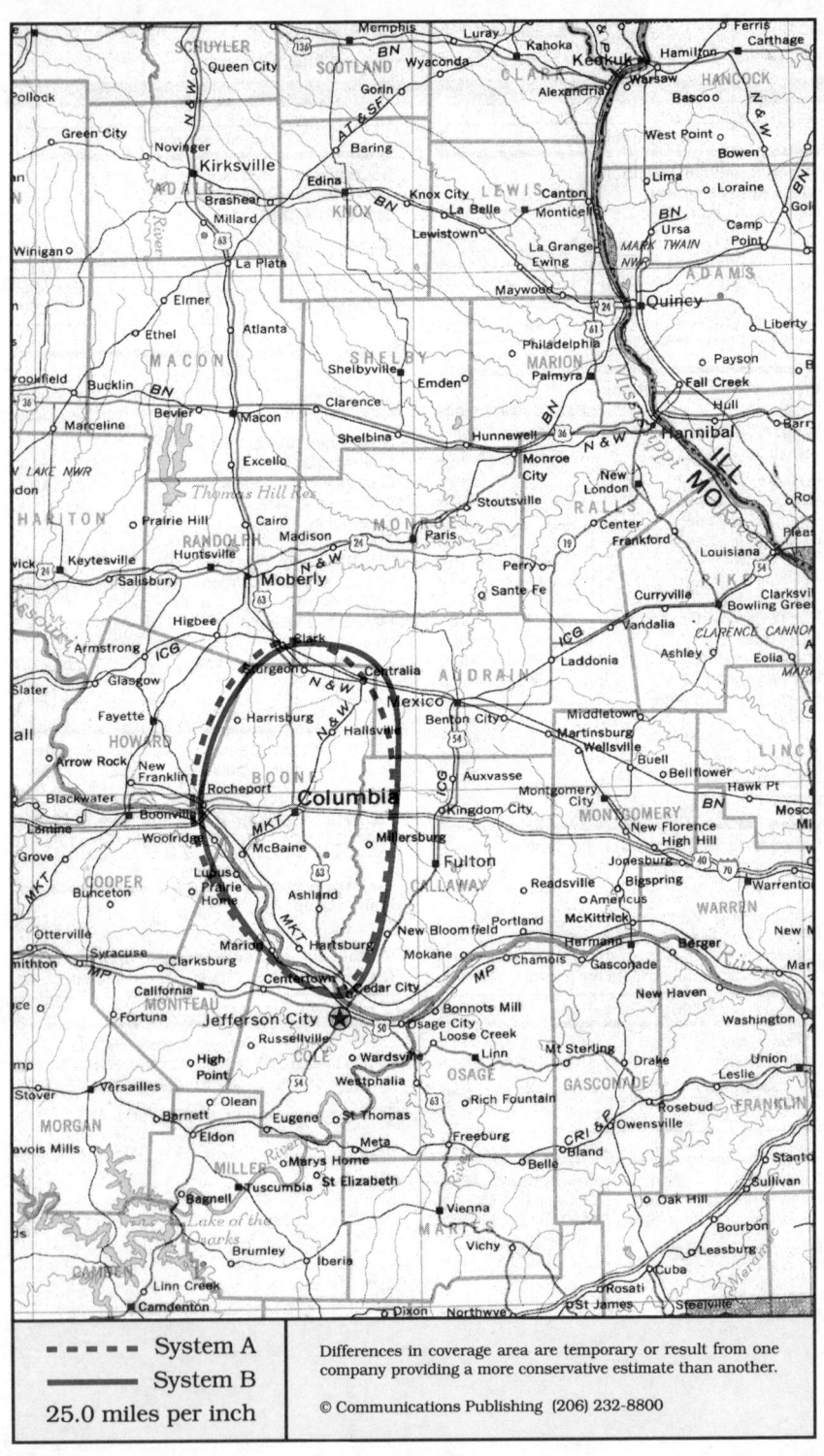

Coverage	
■ ■ ■ ■ ■	System A
—————————	System B
25.0 miles per inch	

Differences in coverage area are temporary or result from one company providing a more conservative estimate than another.

© Communications Publishing (206) 232-8800

Columbia, Missouri

Cellular company

Cellular One
Columbia Cellular Partnership
1000 W Nifong Boulevard, Bldg 5, Suite 100
Columbia, MO 65203

Corporate office 314-449-7774

Sending calls

Visitor dialing instructions
Local calls . . . . . area code + 7 digits
Long distance . . . . area code + 7 digits

Speed-dial numbers
Customer service *611
Emergency 911
Highway patrol *55

Receiving calls

Caller dials the roamer access number below,
hears tone, dials cellular area code and number.
. . . . 314-864-7626

Billing information

Visitor rates
Most visitors . . . . . $3 day + 99¢ minute

Visitors without roaming agreements (p. 1159)
Roamer Plus will intercept first call to bill you.

Identification numbers

City	Market	System	Billing	RSA
All served cities	278	00317	30115	MSA

Cellular company

United States Cellular
1804 Vandiver Drive
Columbia, MO 65202

Local office . . from Missouri 800-332-6067
. . . . 314-474-0400
Corporate office 312-399-8900

Sending calls

Visitor dialing instructions
Local calls . . . . . . area code + 7 digits
Long distance . . . . . area code + 7 digits

Speed-dial numbers
Customer service . . . . . *611 or *711
Emergency 911
Highway patrol *55

Receiving calls

Caller dials the roamer access number below,
hears tone, dials cellular area code and number.
. . . . 314-881-7626

Follow Me Roaming forwarding (see p. 1156)
Activate, dial *18 send. Deactivate dial *19 send.

Billing information

Visitor rates
Most visitors . . . . . $3 day + 75¢ minute

Visitors without roaming agreements (p. 1159)
Cannot call since company takes no credit cards.

Identification numbers

City	Market	System	Billing	RSA
All served cities	278	00298	00000	MSA

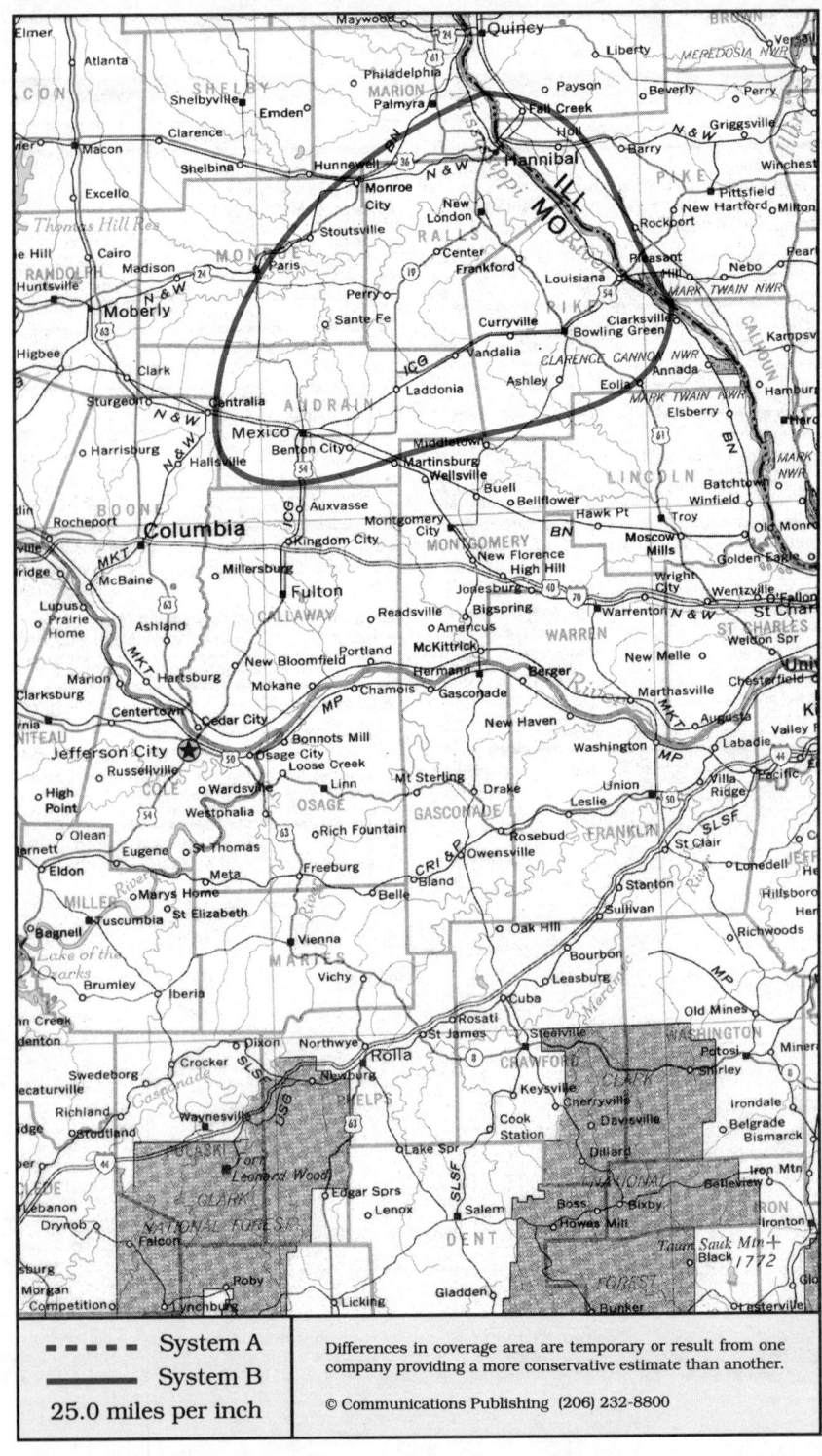

System A
System B
25.0 miles per inch

Differences in coverage area are temporary or result from one company providing a more conservative estimate than another.

© Communications Publishing (206) 232-8800

Hannibal, Missouri
Bowling Green, Hannibal, Mexico, New London

| System A | System B |

Cellular company

No information about the company that will operate system A was available at press time.

Cellular company

United States Cellular
1804 Vandiver Drive
Columbia, MO 65202

Local office . .	from Missouri	800-332-6067
. . . .		314-474-0400
Corporate office		312-399-8900

Sending calls

Visitor dialing instructions
Local calls . . . . . . area code + 7 digits
Long distance . . . . . area code + 7 digits

Speed-dial numbers
Customer service . . . . . *611 or *711
Emergency 911
Highway patrol *55

Receiving calls

Caller dials the roamer access number below,
hears tone, dials cellular area code and number.
. . . . 314-881-7626

Billing information

Visitor rates
Most visitors . . . . . $3 day + 75¢ minute

Visitors without roaming agreements (p. 1159)
Call co. to use Mastercard, Visa

Identification numbers

City	Market	System	Billing	RSA
All served cities	509	01406	00000	MO-6

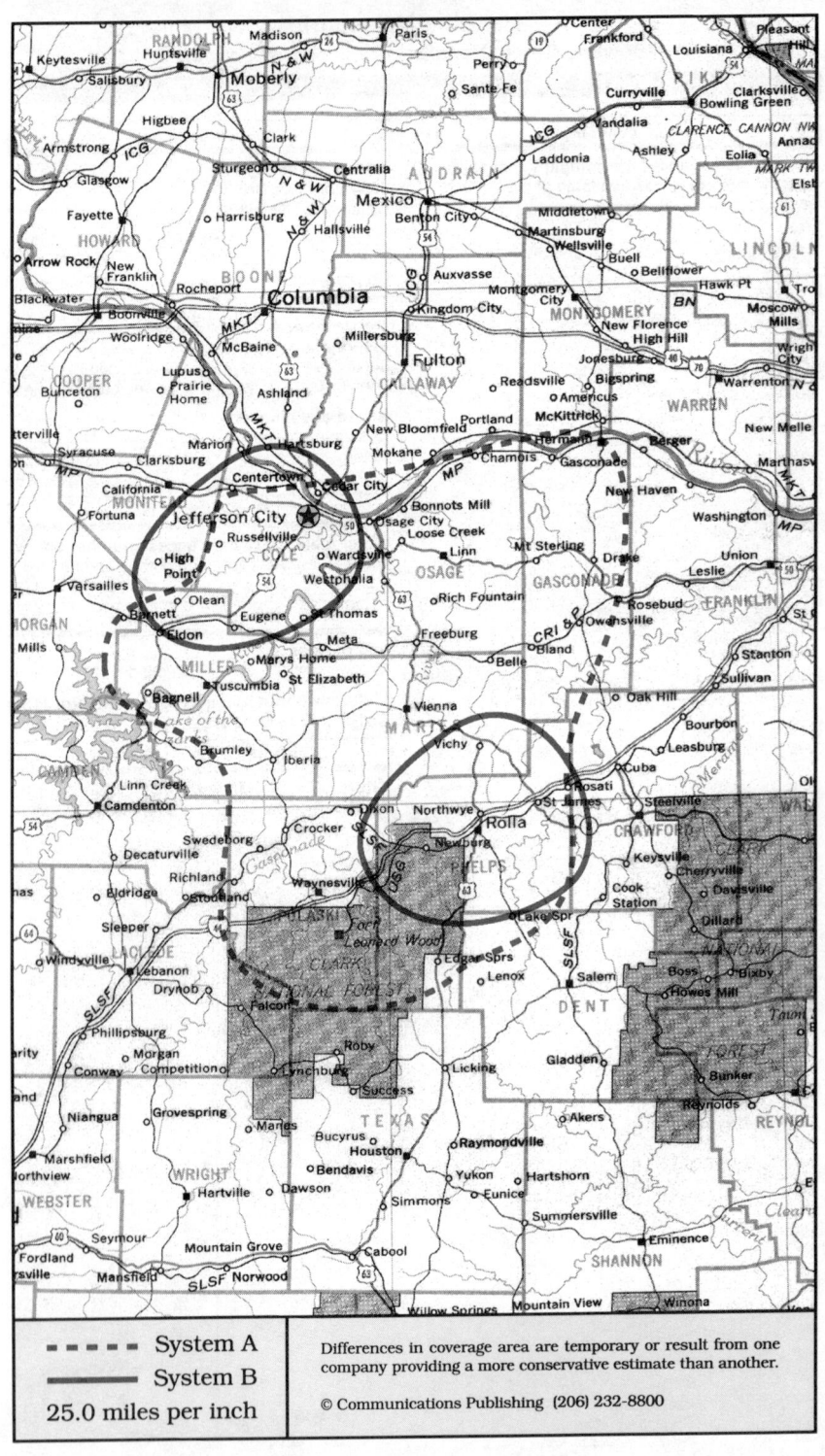

System A
System B

25.0 miles per inch

Differences in coverage area are temporary or result from one company providing a more conservative estimate than another.

© Communications Publishing (206) 232-8800

Jefferson City, Missouri
Jefferson City, Linn, Rolla, Tuscumbia, Waynesville

System A	System B

Cellular company

Missouri Cellular (Mocell)
L.F.B.
2220-B N Bishop, Highway 63, PO Box UU
Rolla, MO 65401

Corporate office 314-364-9360

Sending calls

Visitor dialing instructions
Local calls . . . . 7 digits
Long distance . . . 1 + area code + 7 digits

Speed-dial numbers
Customer service 611
Emergency 911
Missouri police . . . . ∗55

Receiving calls

Caller dials the roamer access number below,
hears tone, dials cellular area code and number.
. . . . 314-465-7626

Billing information

Visitor rates
Most visitors . . . . . $3 day + 99¢ minute

Visitors without roaming agreements (p. 1159)
Call co. to use Mastercard, Visa

Identification numbers

City	Market	System	Billing	RSA
All served cities	515	00317	30113	MO-12

Cellular company

Southwestern Bell Mobile Systems
13075 Manchester, Suite 100-N
St. Louis, MO 63131

Customer service 800-331-0500
. . . . 314-821-1111
Local office 314-821-9999
Corporate office 214-733-2000

Sending calls

Visitor dialing instructions
Local calls . . . . . . area code + 7 digits
Long distance . . . 1 + area code + 7 digits

Speed-dial numbers
Cellular tips 314-541-5400
Current information service . . . . . ∗123
Customer service 611
Highway patrol . . . . ∗55
League of Women Voters . ∗VOTE or ∗8683
Make a traffic report . . . ∗RPT or ∗778
Missouri police 311
Personal traffic consultant . . ∗PTC or ∗782
Sports line 314

Receiving calls

Caller dials the roamer access number below,
hears tone, dials cellular area code and number.
Jefferson City 314-277-7626
Rolla 314-659-7626

Follow Me Roaming forwarding (see p. 1156)
Activate, dial ∗18 send. Deactivate dial ∗19 send.

Billing information

Visitor rates
Most visitors . . . . . $3 day + 85¢ minute

Visitors without roaming agreements (p. 1159)
Call co. to use Amex, Discover, Mastercard, Visa

Identification numbers

City	Market	System	Billing	RSA
Jefferson City	514	00046	30132	MO-11
Rolla	515	00046	30226	MO-12

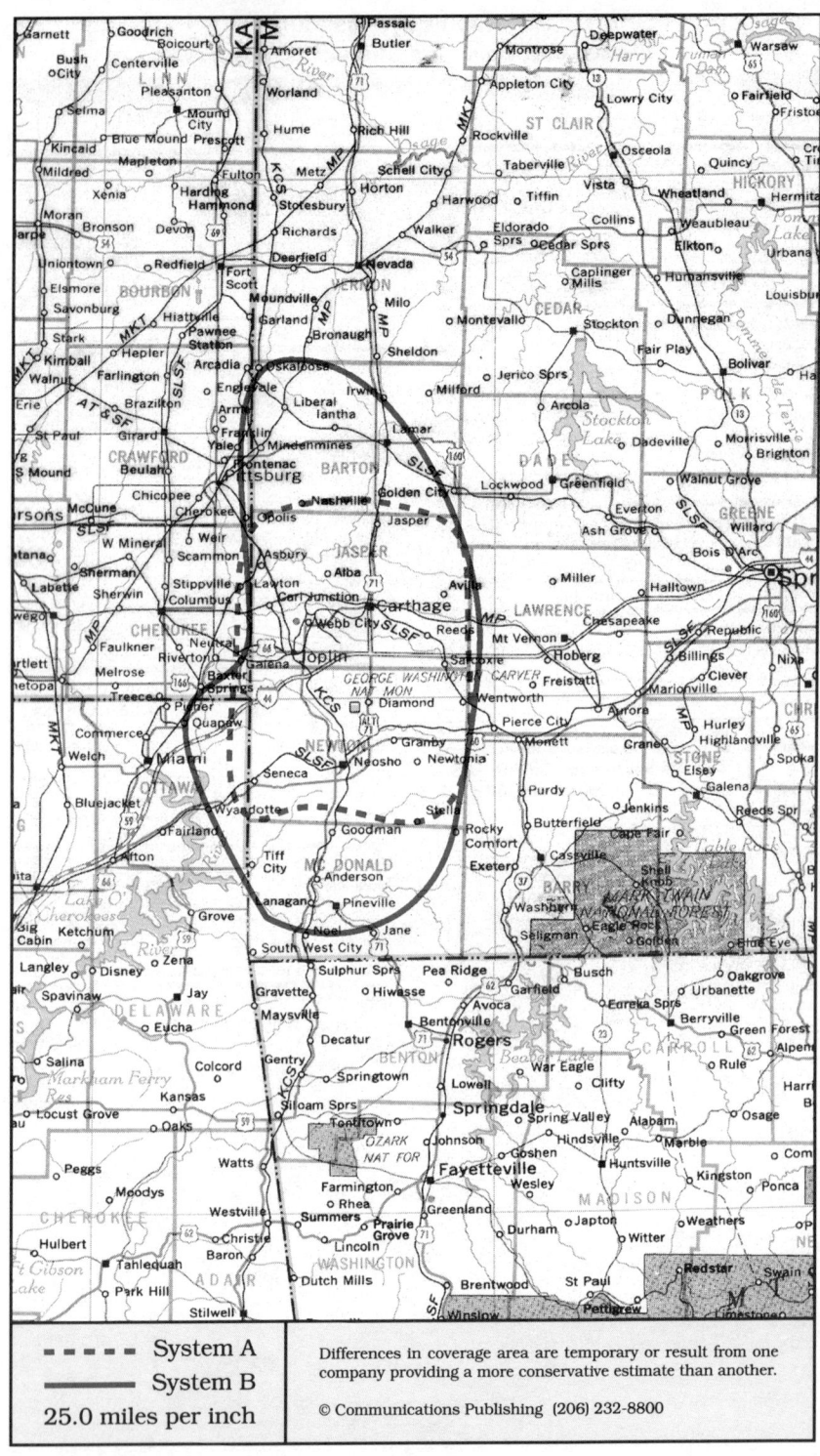

▬ ▬ ▬ System A	Differences in coverage area are temporary or result from one company providing a more conservative estimate than another.
▬▬▬▬ System B	
25.0 miles per inch	© Communications Publishing (206) 232-8800

Joplin, Missouri
Carthage, Joplin, Lamar, Neosho, Pineville

System A	System B

Cellular company

Cellular One
Crowley Cellular Telecommunications
2131 E 32nd, Suite A
Joplin, MO 64804-3019

Local office	417-782-4212
Corporate office	708-843-9081

Sending calls

Visitor dialing instructions

Local calls	area code + 7 digits
Long distance . . .	1 + area code + 7 digits

Speed-dial numbers

Customer service	*611
Emergency	911
State police	*55

Receiving calls

Caller dials the roamer access number below,
hears tone, dials cellular area code and number.

. . . . 417-437-7626

NationLink & RoamingAmerica (see p. 1157)
Dial *31 send to receive calls, *30 to deactivate.
Dial *32 send to tell caller the roamer access no.
This company's subscribers have level 6 service.

Billing information

Visitor rates

Most visitors . . . . . $3 day + 85¢ minute

Visitors without roaming agreements (p. 1159)
Call co. to use Amex, Mastercard, Visa
or Roamer Plus will intercept first call to bill you.

Identification numbers

City	Market	System	Billing	RSA
All served cities	239	00401	30069	MSA

Cellular company

United States Cellular
1630 S Range Line Road
Joplin, MO 64804

Local office	417-624-2255
Corporate office	312-399-8900

Sending calls

Visitor dialing instructions

Local calls	area code + 7 digits
Long distance	area code + 7 digits

Speed-dial numbers

Customer service	*611
Emergency	911
Highway patrol	*55
Make a traffic report . . .	417-438-4376
Roaming information	*711

Receiving calls

Caller dials the roamer access number below,
hears tone, dials cellular area code and number.

. . . . 417-438-7626

Billing information

Visitor rates

Most visitors . . . . . $3 day + 85¢ minute

Visitors without roaming agreements (p. 1159)
Call co. to use Mastercard, Visa

Identification numbers

City	Market	System	Billing	RSA
All served cities	239	00384	00000	MSA

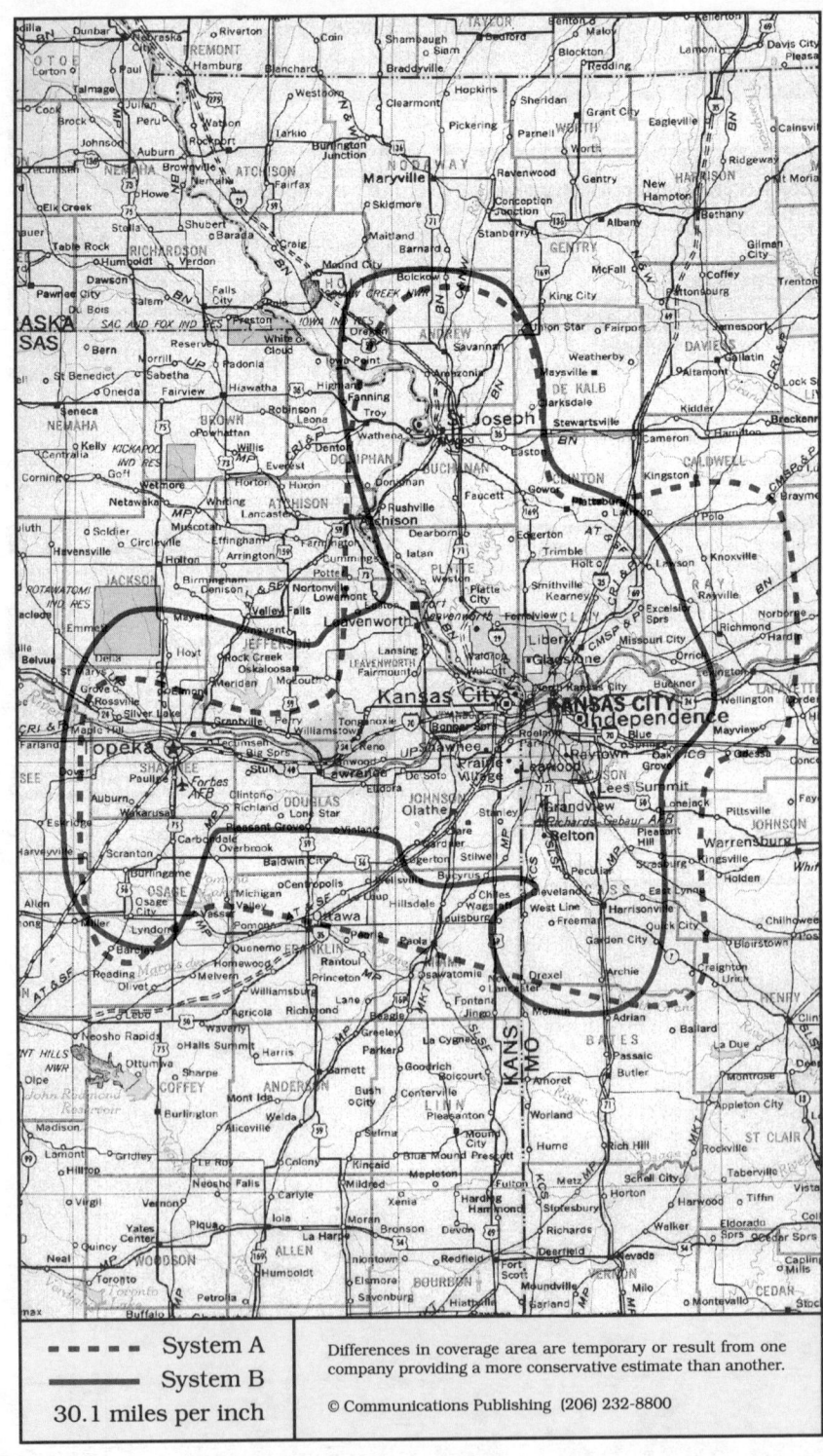

▬ ▬ ▬ ▬ System A	Differences in coverage area are temporary or result from one company providing a more conservative estimate than another.
▬▬▬▬ System B	
30.1 miles per inch	© Communications Publishing (206) 232-8800

Kansas City, Missouri
Kansas City KS & MO, Lawrence KS, Olathe KS, St. Joseph

System A	System B

Cellular company

Cellular One
McCaw Cellular Communications
1410 Kasold Drive, Suite A-6
Lawrence, KS 66049

7000 Squibb Road
Mission, KS 66202

1920 SW Wanamaker Road, Suite 130
Topeka, KS 66604-3826

Customer service	800-800-7626
. . . .	913-722-8300
. . . . . from Lawrence	913-842-0577
. . . . . from St. Joseph	816-232-6837
. . . . . from Topeka	913-234-4984
Local office . . . . Mission	913-722-8200
. . . . Lawrence	913-842-4336
. . . . Topeka	913-273-1234
Corporate office	206-827-4500

Sending calls

Visitor dialing instructions

Local calls	area code + 7 digits
Long distance . . .	1 + area code + 7 digits

Speed-dial numbers

Customer service	611
Emergency	911

Receiving calls

Caller dials the roamer access number below,
hears tone, dials cellular area code and number.

Kansas City	816-591-7626
Lawrence	913-766-7626
St. Joseph	816-351-7626
Topeka	913-221-7626

NationLink & RoamingAmerica (see p. 1157)
Dial *31 send to receive calls, *30 to deactivate.
Dial *32 send to tell caller the roamer access no.
This company's subscribers have level 9 service.

North American Cellular Network (see p. 1156)
All calls forward to you. Deactivate dial *35 send.

Billing information

Visitor rates

Most visitors	$0 day + 99¢ minute

Visitors without roaming agreements (p. 1159)
Roamer Plus will intercept first call to bill you.

Identification numbers

City	Market	System	Billing	RSA
Kansas City	024	00059	00000	MSA
Lawrence KS	301	00059	30049	MSA
St. Joseph MO	275	00059	30055	MSA
Topeka KS	179	00059	30057	MSA

Cellular company

Southwestern Bell Mobile Systems
administration
13321 W 98th Street
Lenexa, KS 66215

customer service
13228 W 99th Street
Lenexa, KS 66215

Customer service	some areas	800-331-0500
. . . .		913-894-1600
Local office . .	Independence	816-478-2355
. . . Lenexa administration		913-752-2300
. . . . . .	Lenexa sales	913-894-2355
. . . . . .	Overland Park	913-338-2355
. . . . . .	St. Joseph	816-364-6300
. . . .	Topeka	913-272-4002
Corporate office		214-733-2000

Sending calls

Visitor dialing instructions

Local calls	area code + 7 digits
Long distance . . .	1 + area code + 7 digits

Speed-dial numbers

Customer service	611
Emergency	911
Kansas Highway Patrol	*HP or *47
Kansas Turnpike Authority . .	*KTA or *582
Missouri Highway Patrol	*55
News, sports, stocks, traffic, weather .	#INFO
Recorded customer care . .	*CARE or *2273
Roaming information . .	*ROAM or *7626

Receiving calls

Caller dials the roamer access number below,
hears tone, dials cellular area code and number.

Kansas City, Lawrence . . .	816-223-7626
St. Joseph	816-387-1766
Topeka	913-231-7626

Follow Me Roaming forwarding (see p. 1156)
Activate, dial *18 send. Deactivate dial *19 send.

Billing information

Visitor rates

Most visitors	$3 day + 85¢ minute

Visitors without roaming agreements (p. 1159)
Call co. to use Amex, Discover, Mastercard, Visa

Identification numbers

City	Market	System	Billing	RSA
Kansas City	024	00052	00000	MSA
Lawrence KS	301	00052	00000	MSA
St. Joseph MO	275	00052	00000	MSA
Topeka KS	179	00052	00000	MSA

For full coverage area, see Leavenworth, KS and
Maryville, MO and Nevada, MO.

System A
System B
25.0 miles per inch

Differences in coverage area are temporary or result from one company providing a more conservative estimate than another.

© Communications Publishing (206) 232-8800

Kirksville, Missouri
Kirksville, Lancaster, Memphis

System A	System B

System A

Cellular company

United States Cellular
1111 Quincy Avenue, Suite 105
Ottumwa, IA 52501

Local office	800-967-5854
. . . .	515-684-8000
Corporate office	312-399-8900

Sending calls

Visitor dialing instructions

Local calls	area code + 7 digits
Long distance	area code + 7 digits

Speed-dial numbers

Customer service	611
Highway patrol	*55
Roaming information	*711

Receiving calls

Caller dials the roamer access number below, hears tone, dials cellular area code and number.
. . . . 816-341-7626

Billing information

Visitor rates
Most visitors . . . . . $3 day + 99¢ minute

Visitors without roaming agreements (p. 1159)
Call co. to use Mastercard, Visa

Identification numbers

City	Market	System	Billing	RSA
All served cities	506	01399	00000	MO-3

System B

Cellular company

Alltel Mobile Communications
3322 S Campbell, Suite N
Springfield, MO 65807

Customer service	some areas	800-255-8351
Local office		417-882-2020
Corporate office		501-661-8500

Sending calls

Visitor dialing instructions

Local calls	area code + 7 digits
Long distance	area code + 7 digits

Speed-dial numbers

Customer service	*611
Emergency	911
Highway patrol	*55

Receiving calls

Caller dials the roamer access number below, hears tone, dials cellular area code and number.
. . . . 816-626-7626

Billing information

Visitor rates
Most visitors . . . . . $3 day + 85¢ minute

Visitors without roaming agreements (p. 1159)
Call co. to use Amex, Discover, Mastercard, Visa

Identification numbers

City	Market	System	Billing	RSA
All served cities	506	1400	30504	MO-3

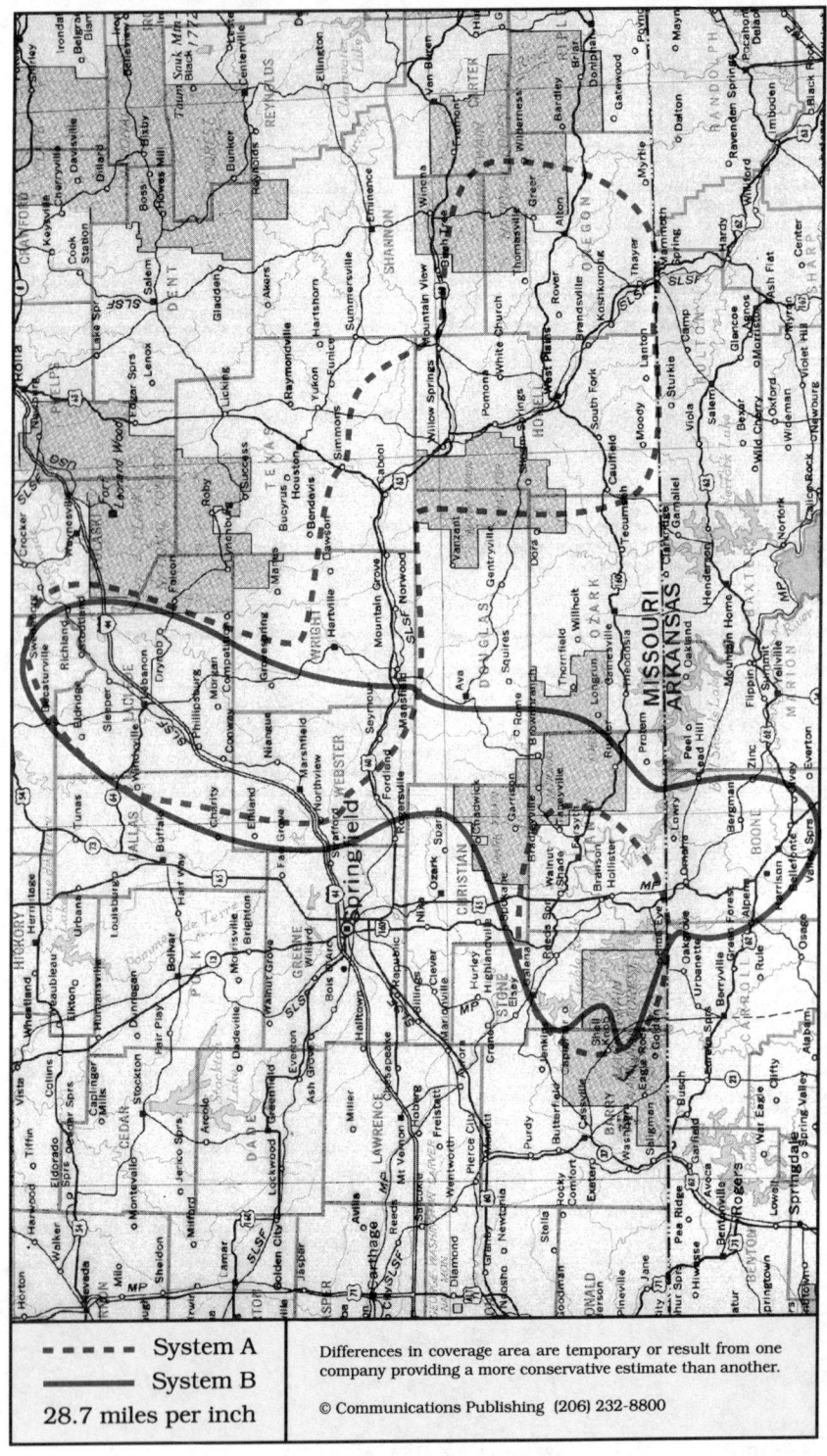

System A

System B

28.7 miles per inch

Differences in coverage area are temporary or result from one company providing a more conservative estimate than another.

© Communications Publishing (206) 232-8800

Lebanon, Missouri
Branson, Lebanon, Harrison AR, West Plains

System A	System B

Cellular company

United States Cellular
1630 S Range Line Road
Joplin, MO 64804

Local office . .	from Branson	417-331-0030
. . . . .	from Lebanon	417-531-0030
. . . .	from West Plains	417-331-4444
Corporate office		312-399-8900

Sending calls

Visitor dialing instructions

Local calls	area code + 7 digits
Long distance	area code + 7 digits

Speed-dial numbers

Customer service	*611 or *711
Emergency	911
Highway patrol	*55

Receiving calls

Caller dials the roamer access number below,
hears tone, dials cellular area code and number.

Branson, West Plains . . .	417-331-7626
Lebanon	417-531-7626

Billing information

Visitor rates

Most visitors	$3 day + 85¢ minute

Visitors without roaming agreements (p. 1159)
Call co. to use Mastercard, Visa

Identification numbers

City	Market	System	Billing	RSA
Lebanon	519	01425	00000	MO-16
West Plains	518	01423	00000	MO-15

Cellular company

ALLTEL Mobile Communications
210 S 3rd Street
Branson, MO 65616

202 Graham Street
Harrison, AR 72601

629-A W Elm
Lebanon, MO 65536

Customer service	some areas	800-255-8351
Local office	Branson	417-335-5220
. . . .	Harrison	501-743-2020
. . . .	Lebanon	417-588-4343
Corporate office		501-661-8500

Sending calls

Visitor dialing instructions

Local calls	area code + 7 digits
Long distance	area code + 7 digits

Speed-dial numbers

Customer service	*611
Emergency	911
Highway patrol	*55

Receiving calls

Caller dials the roamer access number below,
hears tone, dials cellular area code and number.

Branson	417-335-0626
Harrison	501-365-7626
Lebanon	417-588-6226

Follow Me Roaming forwarding (see p. 1156)
Activate, dial *18 send. Deactivate dial *19 send.

Billing information

Visitor rates

Most visitors	$3 day + 85¢ minute

Visitors without roaming agreements (p. 1159)
Call co. to use Amex, Discover, Mastercard, Visa

Identification numbers

City	Market	System	Billing	RSA
Branson	518	00546	30238	MO-15
Harrison	324	00546	00000	AR-1
Lebanon	519	01426	30506	MO-16

For full coverage area, see Aurora, MO and
Springfield, MO.

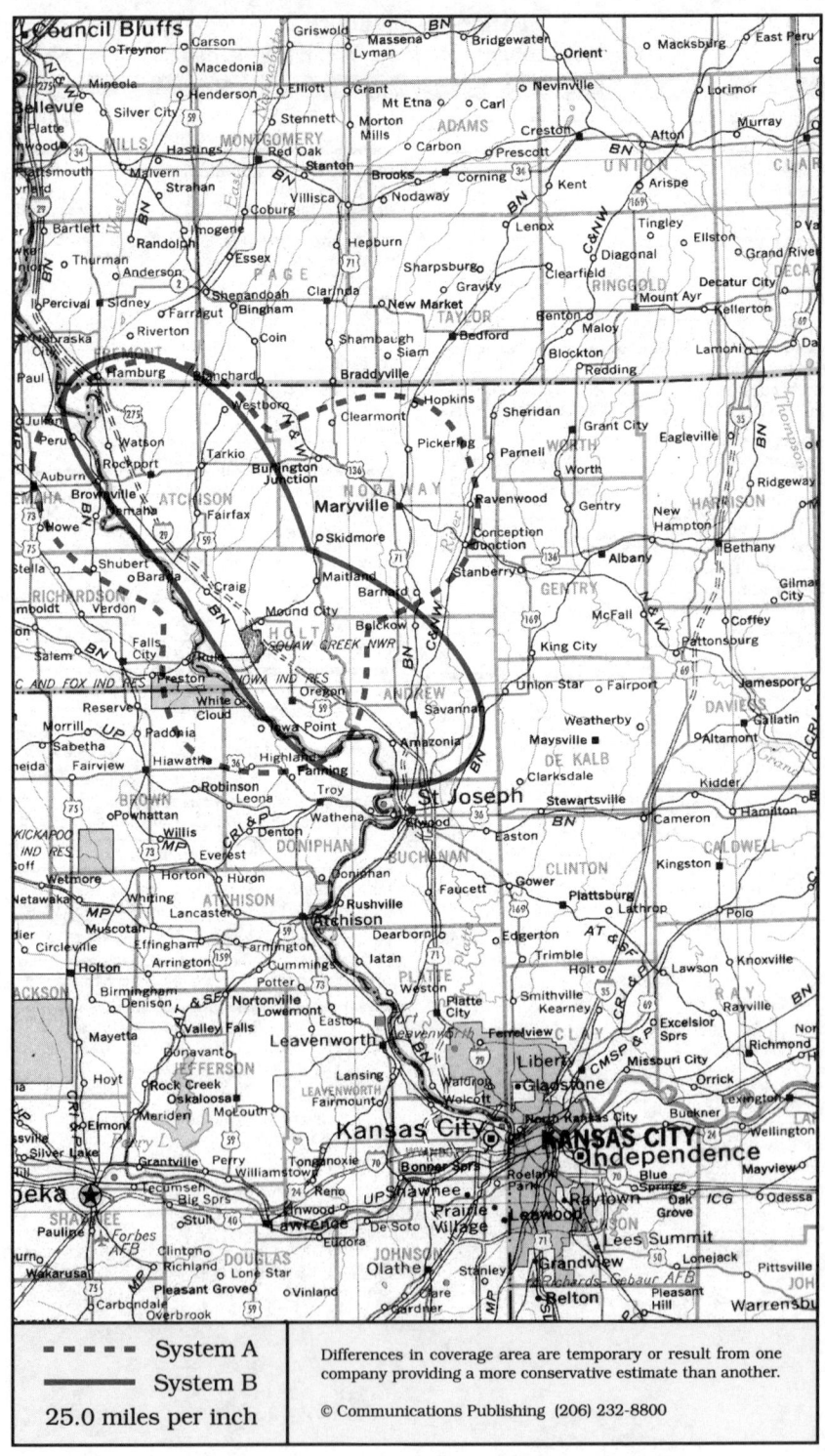

▬ ▬ ▬ ▬ System A	Differences in coverage area are temporary or result from one company providing a more conservative estimate than another.
━━━━ System B	
25.0 miles per inch	© Communications Publishing (206) 232-8800

Maryville, Missouri
Maryville, Oregon, Rock Port, Savannah

System A	System B

System A

Cellular company

United States Cellular
309 Summit Drive
Maryville, MO 64468

Local office	816-562-3300
Corporate office	312-399-8900

Sending calls

Visitor dialing instructions

Local calls	area code + 7 digits
Long distance . . .	1 + area code + 7 digits

Speed-dial numbers

Customer service	*611
Emergency	911
Roaming information	*711

Receiving calls

Caller dials the roamer access number below,
hears tone, dials cellular area code and number.
. . . . 314-881-7626

Billing information

Visitor rates
Most visitors . . . . . $3 day + 85¢ minute

Visitors without roaming agreements (p. 1159)
Call co. to use Mastercard, Visa

Identification numbers

City	Market	System	Billing	RSA
All served cities	504	01395	00000	MO-1

System B

Cellular company

Southwestern Bell Mobile Systems
3120 Karnes Road, Suite A
St. Joseph, MO 64506

Customer service	some areas	800-331-0500
. . . .		913-894-1600
Local office		816-364-6300
Corporate office		214-733-2000

Sending calls

Visitor dialing instructions

Local calls	area code + 7 digits
Long distance . . .	1 + area code + 7 digits

Speed-dial numbers

Customer service	611
Emergency	911
Highway Patrol	*HP or *47

Receiving calls

Caller dials the roamer access number below,
hears tone, dials cellular area code and number.
. . . . 816-387-1766

Follow Me Roaming forwarding (see p. 1156)
Activate, dial *18 send. Deactivate dial *19 send.

Billing information

Visitor rates
Most visitors . . . . . $3 day + 85¢ minute

Visitors without roaming agreements (p. 1159)
Call co. to use Amex, Discover, Mastercard, Visa

Identification numbers

City	Market	System	Billing	RSA
All served cities	504	00052	30366	MO-1

For full coverage area, see Kansas City, MO.

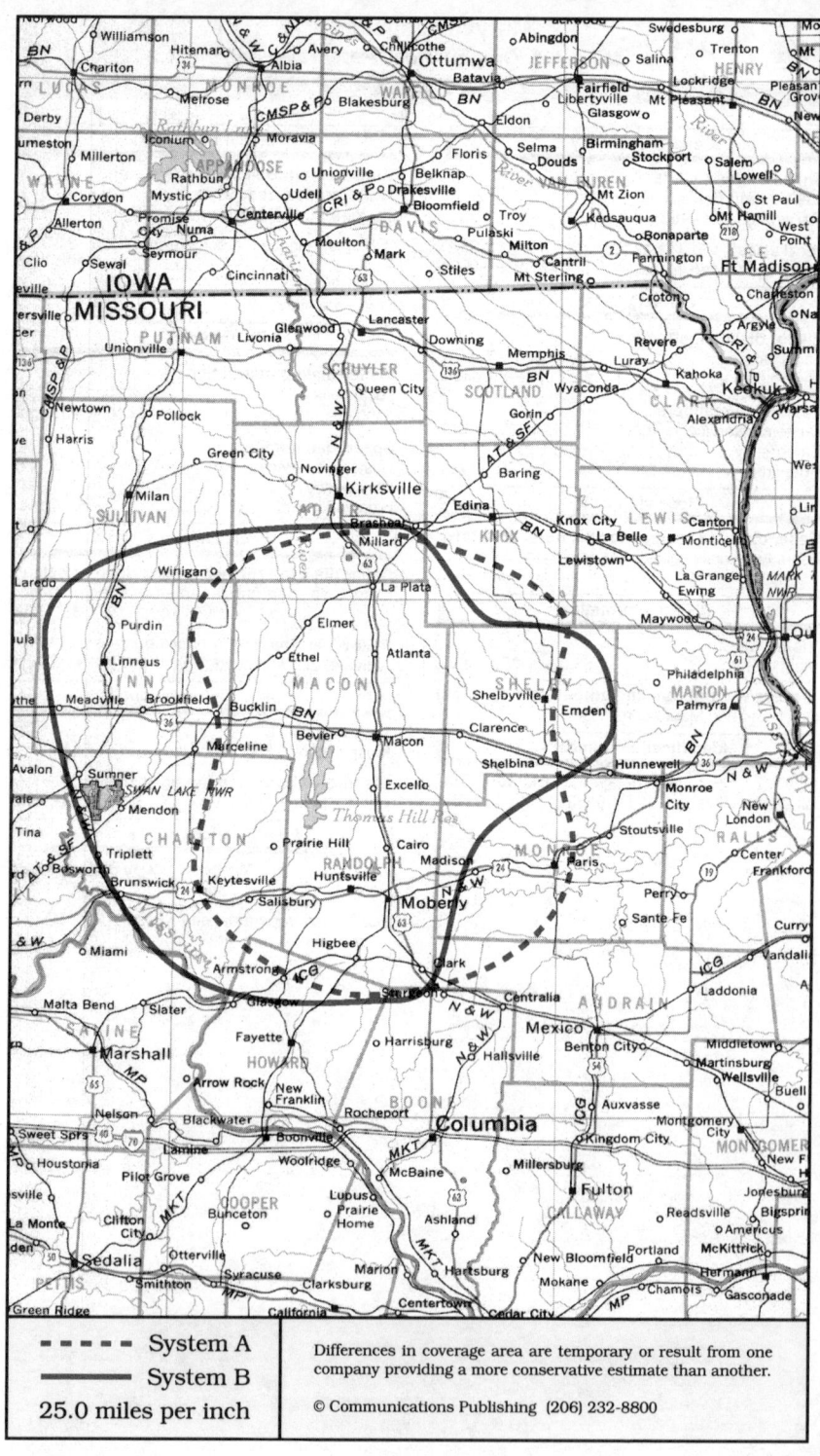

System A

System B

25.0 miles per inch

Differences in coverage area are temporary or result from one
company providing a more conservative estimate than another.

© Communications Publishing (206) 232-8800

488

Moberly, Missouri

Brookfield, Keytesville, Linneus, Macon, Moberly, Shelbyville

System A	System B

Cellular company

Cellular One
Licensed to: Missouri Cellular
Operated by: Quantum Communications
1125 N Morley
Moberly, MO 65270

Customer service	800-788-0472
. . . .	816-651-2273
Local office	816-263-8811
Corporate office	612-942-7650

Sending calls

Visitor dialing instructions

Local calls	area code + 7 digits
Long distance . . .	1 + area code + 7 digits

Speed-dial numbers

Customer service	611
Emergency	911
Highway patrol	*55
Radio station 104.1	*123
Time and temperature	*711

Receiving calls

Caller dials the roamer access number below,
hears tone, dials cellular area code and number.
. . . . 816-651-7626

NationLink & RoamingAmerica (see p. 1157)
Dial *31 send to receive calls, *30 to deactivate.
This company's subscribers have level 3 service.

Billing information

Visitor rates
Most visitors . . . . . $3 day + 99¢ minute

Visitors without roaming agreements (p. 1159)
Call co. to use Amex, Mastercard, Visa
or Roamer Plus will intercept first call to bill you.

Identification numbers

City	Market	System	Billing	RSA
All served cities	508	01403	00000	MO-5

Cellular company

Chariton Valley Cellular
Chariton Valley Telephone
630 N Morely, Suite 101
Moberly, MO 65270

Local office	816-263-2535
Corporate office	816-695-3291

Sending calls

Visitor dialing instructions

Local calls	area code + 7 digits
Long distance . . .	1 + area code + 7 digits

Speed-dial numbers

Customer service	611
Emergency	911
Highway patrol	*55
Radio station KZZT	*105

Receiving calls

Caller dials the roamer access number below,
hears tone, dials cellular area code and number.
. . . . 816-670-7626

Billing information

Visitor rates
Most visitors . . . . . $3 day + 99¢ minute

Visitors without roaming agreements (p. 1159)
Call co. to use Mastercard, Visa

Identification numbers

City	Market	System	Billing	RSA
All served cities	508	01404	00000	MO-5

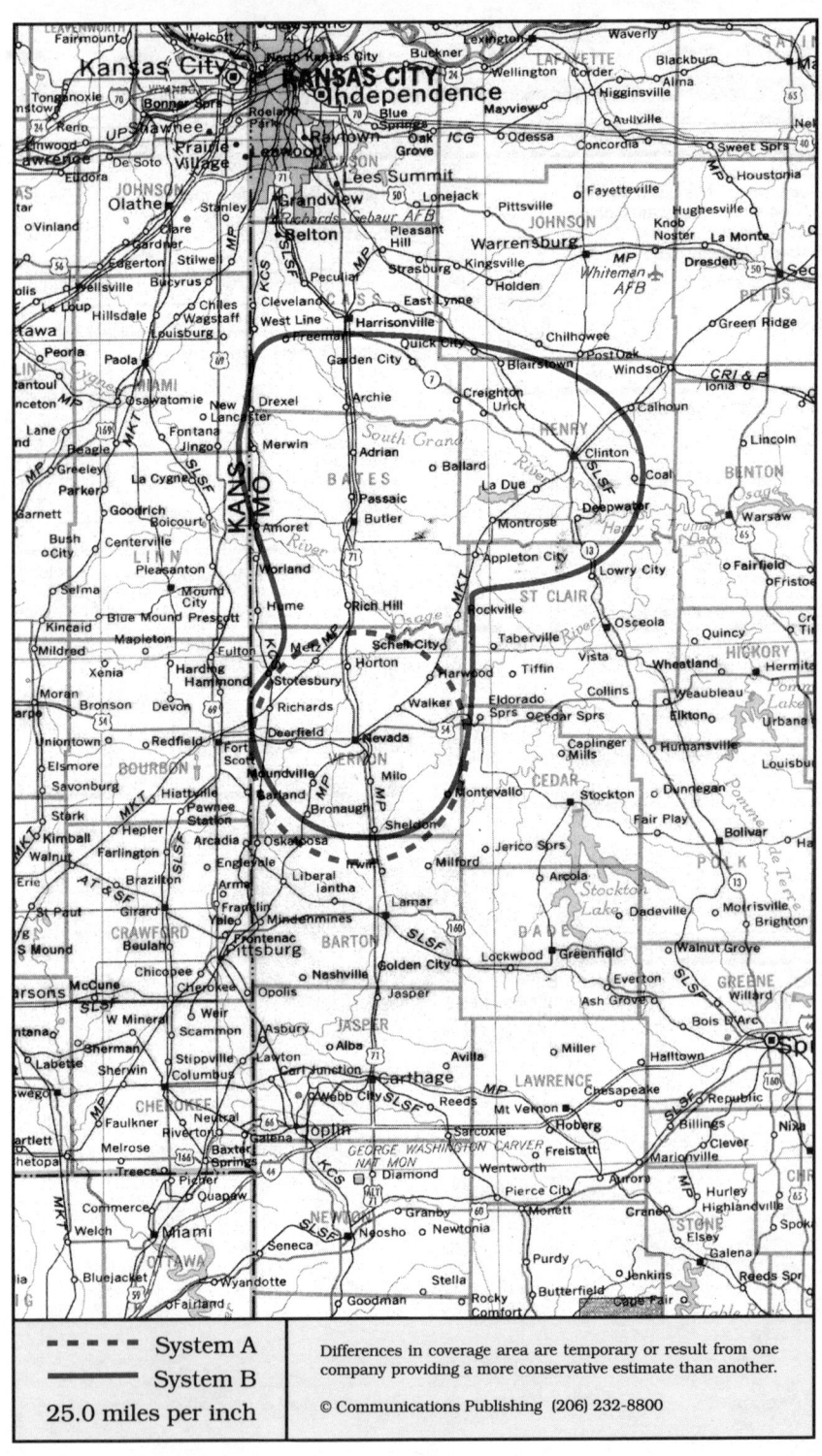

▬ ▬ ▬ ▬ System A	Differences in coverage area are temporary or result from one company providing a more conservative estimate than another.
▬▬▬▬▬ System B	
25.0 miles per inch	© Communications Publishing (206) 232-8800

Nevada, Missouri
Butler, Clinton, Nevada

System A	System B

Cellular company

Cellular One of the Plains
Sterling Cellular Management
1080 Holcomb Bridge Road, Bldg 100, Ste 200
Roswell, GA 30076

Customer service	800-552-6150
Corporate office	404-552-5030

Sending calls

Visitor dialing instructions

Local calls	7 digits
Long distance	area code + 7 digits

Speed-dial numbers

Customer service	611 or *611
Emergency	911

Receiving calls

Caller dials the roamer access number below,
hears tone, dials cellular area code and number.
. . . . 417-321-7626

Billing information

Visitor rates

Most visitors	$3 day + 99¢ minute

Visitors without roaming agreements (p. 1159)
Cannot call since company takes no credit cards.

Identification numbers

City	Market	System	Billing	RSA
All served cities	512	01411	00000	MO-9

Cellular company

Southwestern Bell Mobile Systems
13208-C E. Highway 40
Independence, MO 64055

Customer service	800-331-0500
. . . .	913-894-1600
Local office	816-478-2355
Corporate office	214-733-2000

Sending calls

Visitor dialing instructions

Local calls	area code + 7 digits
Long distance . . .	1 + area code + 7 digits

Speed-dial numbers

Customer service	611
Emergency	311
Highway Patrol	*55

Receiving calls

Caller dials the roamer access number below,
hears tone, dials cellular area code and number.
. . . . 816-223-7626

Follow Me Roaming forwarding (see p. 1156)
Activate, dial *18 send. Deactivate dial *19 send.

Billing information

Visitor rates

Most visitors	$3 day + 85¢ minute

Visitors without roaming agreements (p. 1159)
Call co. to use Amex, Discover, Mastercard, Visa

Identification numbers

City	Market	System	Billing	RSA
All served cities	512	00052	30280	MO-9

For full coverage area, see Kansas City, MO.

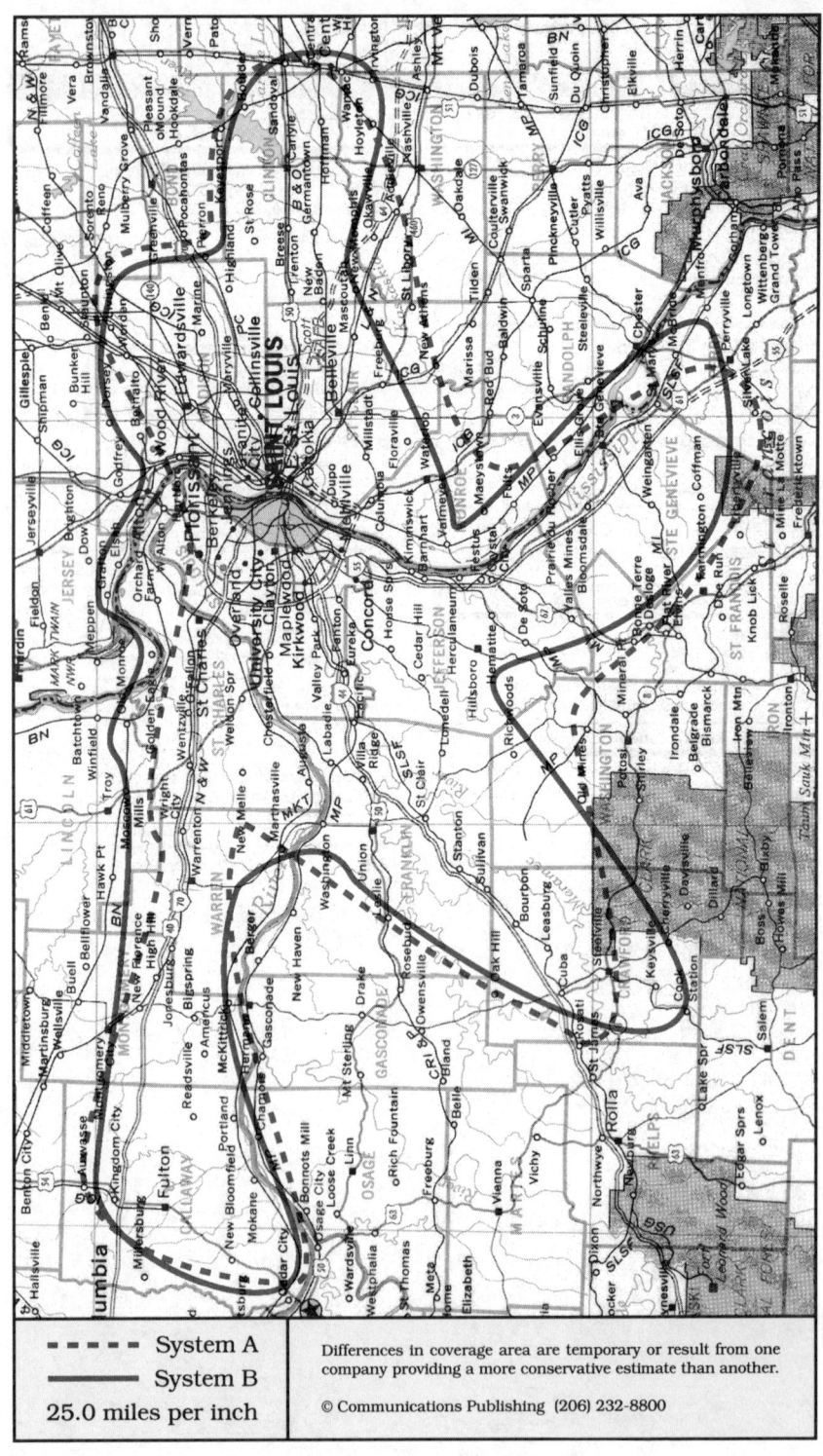

----- System A	
——— System B	
25.0 miles per inch	

Differences in coverage area are temporary or result from one company providing a more conservative estimate than another.

© Communications Publishing (206) 232-8800

St. Louis, Missouri
Carlyle IL, Farmington, Fulton, Granite City IL, St. Louis

System A	System B

Cellular company

Cybertel Cellular Telephone
1935 Beltway Drive
St. Louis, MO 63114

Customer service	some areas	800-662-4534
. . . .		314-444-4444
Corporate office	. . . .	314-423-6500

Sending calls

Visitor dialing instructions

Local calls	area code + 7 digits
Long distance . . .	1 + area code + 7 digits

Speed-dial numbers

Business update	*28
Customer service	*444 or *611
Hear a traffic report	*88
Illinois police	*511
KMOX radio traffic report	*25
Make a traffic report	*66 or *666
Missouri highway patrol . . .	*55 or *555
Missouri police	*311
Roaming information	*211
Sports update	*78
Weather report	*98

Receiving calls

Caller dials the roamer access number below,
hears tone, dials cellular area code and number.
. . . . 314-973-7626

NationLink & RoamingAmerica (see p. 1157)
Dial *31 send to receive calls, *30 to deactivate.
Dial *32 send to tell caller the roamer access no.
This company's subscribers have level 8 service.

Billing information

Visitor rates
Most visitors . . . . . $3 day + 99¢ minute

Visitors without roaming agreements (p. 1159)
Call co. to use Amex, Mastercard, Visa
or Roamer Plus will intercept first call to bill you.

Identification numbers

City	Market	System	Billing	RSA
Alton, Granite City	305	00017	00000	MSA
St. Louis	011	00017	00000	MSA
Farmington	516	00017	30273	MO-13
Fulton	511	00017	30257	MO-8

For full coverage area, see Cape Girardeau, MO.

Cellular company

Southwestern Bell Mobile Systems
13075 Manchester, Suite 100-N
St. Louis, MO 63131

Customer service	. . . .	800-331-0500
. . . .		314-821-1111
Local office	314-821-9999
Corporate office	214-733-2000

Sending calls

Visitor dialing instructions

Local calls	area code + 7 digits
Long distance . . .	1 + area code + 7 digits

Speed-dial numbers

Cellular tips	314-541-5400
Current information service	*123
Customer service	611
Emergency (Illinois)	511
Emergency (St. Louis County)	311
Highway patrol (Missouri)	*55
League of Women Voters . .	*VOTE or *8683
KMOX news	*1
Make a traffic report	*RPT or *778
Missouri police	311
Personal traffic consultant . .	*PTC or *782
Sports line	314

Receiving calls

Caller dials the roamer access number below,
hears tone, dials cellular area code and number.
. . . . 314-277-7626

Follow Me Roaming forwarding (see p. 1156)
Activate, dial *18 send. Deactivate dial *19 send.

Billing information

Visitor rates
Most visitors . . . . . $3 day + 85¢ minute

Visitors without roaming agreements (p. 1159)
Call co. to use Amex, Discover, Mastercard, Visa

Identification numbers

City	Market	System	Billing	RSA
Alton, Granite City	305	00046	00000	MSA
St. Louis	011	00046	00000	MSA
Farmington	516	01416	30422	MO-13
Fulton	511	00046	30054	MO-8

For full coverage area, see Cape Girardeau, MO.

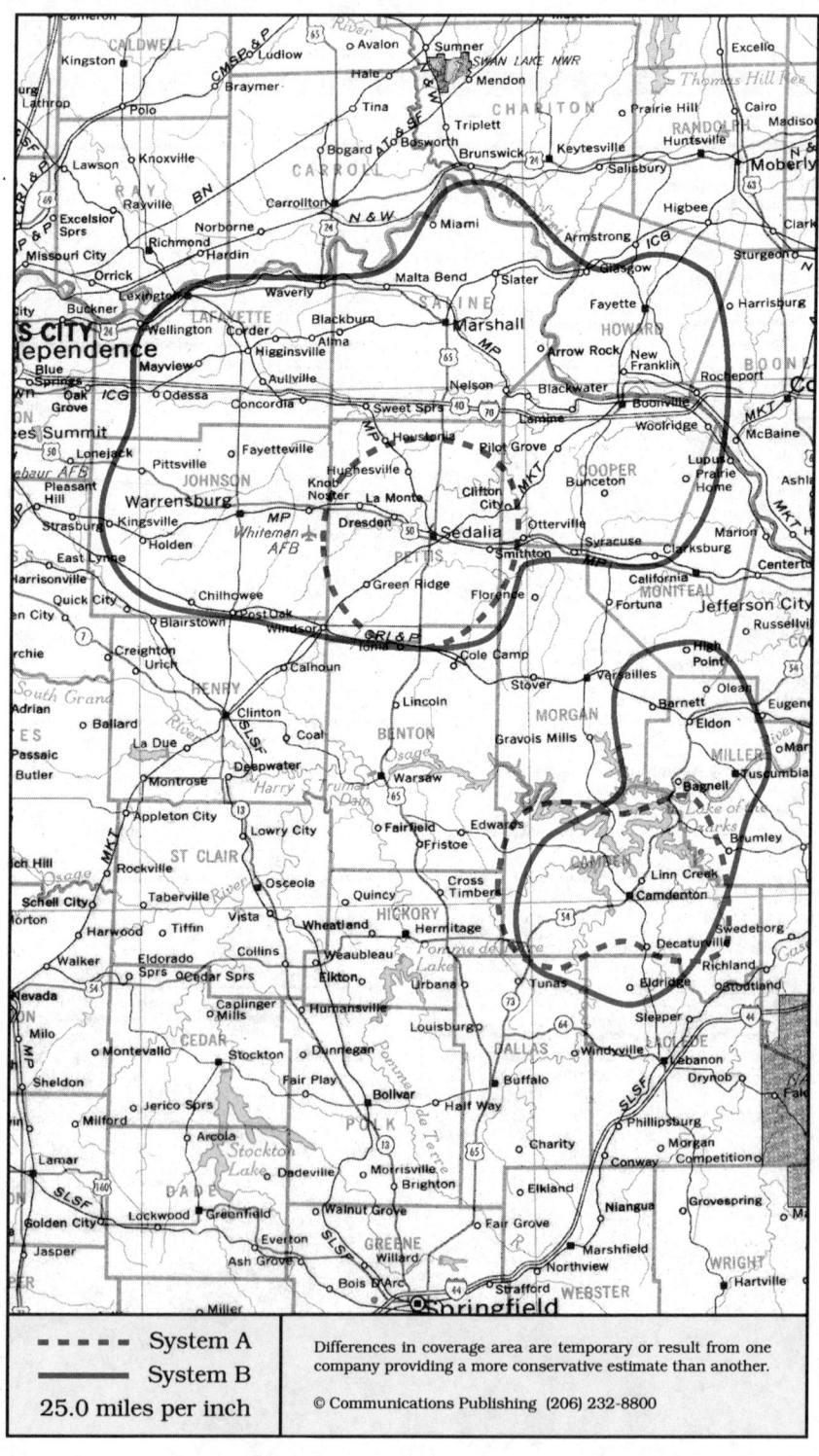

- - - - - System A	Differences in coverage area are temporary or result from one
———— System B	company providing a more conservative estimate than another.
25.0 miles per inch	© Communications Publishing (206) 232-8800

Sedalia, Missouri

Boonville, Camdenton, Lake of the Ozarks, Marshall, Sedalia, Warrensburg

System A	System B

System A

Cellular company

Cybertel Cellular Telephone
821 Thompson Boulevard
Sedalia, MO 65301-2289

Customer service	some areas	800-662-4534
. . . .		314-444-4444
Local office		800-882-9237
. . . .		816-827-4443
Corporate office		314-423-6500

Sending calls

Visitor dialing instructions
Local calls . . . . . area code + 7 digits
Long distance . . . 1 + area code + 7 digits

Speed-dial numbers
Business update . . . . ✱28
Customer service 611
Highway patrol . . . . . . ✱55 or ✱555
Police . . . . ✱311
Roaming information . . . . ✱211
Sports update . . . . ✱78
Weather report . . . . ✱98

Receiving calls

Caller dials the roamer access number below,
hears tone, dials cellular area code and number.
Camdenton 314-973-7626
Sedalia 816-281-7626

NationLink & RoamingAmerica (see p. 1157)
Dial ✱31 send to receive calls, ✱30 to deactivate.
Dial ✱32 send to tell caller the roamer access no.
This company's subscribers have level 8 service.

Billing information

Visitor rates
Most visitors . . . . . $3 day + 99¢ minute

Visitors without roaming agreements (p. 1159)
Call co. to use Amex, Mastercard, Visa

Identification numbers

City	Market	System	Billing	RSA
Camdenton	513	01413	00000	MO-10
Sedalia	510	01407	00000	MO-7

System B

Cellular company

Mid-Missouri Cellular
2113 W Broadway
Sedalia, MO 65301

Corporate office	800-242-6516
. . . .	816-620-1114

Sending calls

Visitor dialing instructions
Local calls . . . . . . area code + 7 digits
Long distance . . . 1 + area code + 7 digits

Speed-dial numbers
Customer service . . . . ✱611
Emergency 911
Highway patrol . . . . ✱55

Receiving calls

Caller dials the roamer access number below,
hears tone, dials cellular area code and number.
. . . . 816-620-7626

Follow Me Roaming forwarding (see p. 1156)
Activate, dial ✱18 send. Deactivate dial ✱19 send.

Billing information

Visitor rates
Most visitors . . . $2.50 day + 75¢ minute

Visitors without roaming agreements (p. 1159)
Call co. to use Amex, Mastercard, Visa

Identification numbers

City	Market	System	Billing	RSA
All served cities	510	01408	00000	MO-7

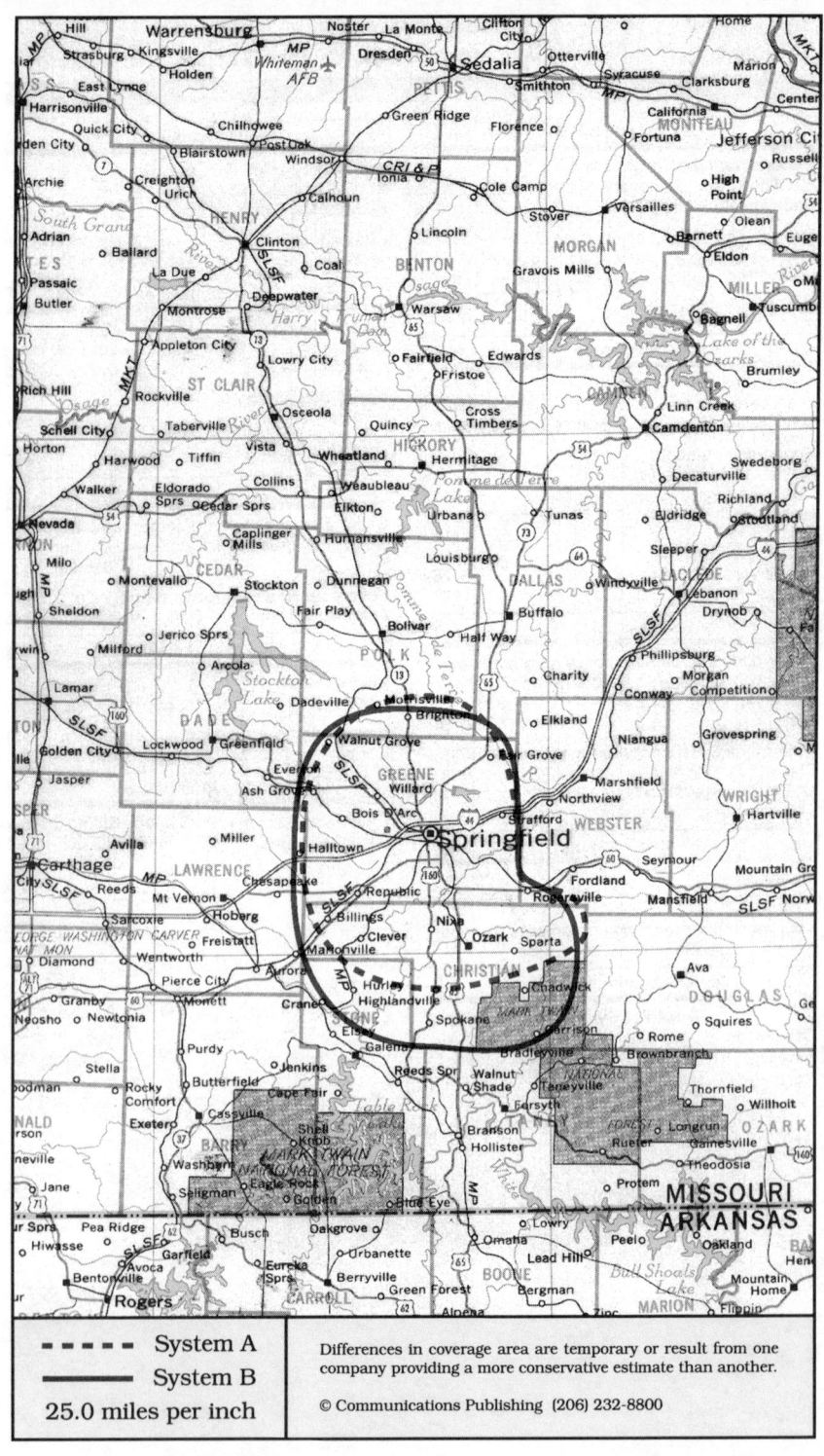

▬ ▬ ▬ ▬ System A	Differences in coverage area are temporary or result from one
▬▬▬▬▬▬ System B	company providing a more conservative estimate than another.
25.0 miles per inch	© Communications Publishing (206) 232-8800

Springfield, Missouri
Springfield, Ozark

System A	System B

Cellular company

Cellular One
Crowley Cellular Telecommunications
522 N Prince Lane
Springfield, MO 65802

Local office	417-862-6611
Corporate office	708-843-9081

Sending calls

Visitor dialing instructions

Local calls	area code + 7 digits
Long distance . . .	1 + area code + 7 digits

Speed-dial numbers

Customer service	*611
Emergency	911
State police	*55

Receiving calls

Caller dials the roamer access number below,
hears tone, dials cellular area code and number.
. . . . 417-861-7626

Billing information

Visitor rates

Most visitors	$3 day + 85¢ minute

Visitors without roaming agreements (p. 1159)
Call co. to use Amex, Mastercard, Visa
or Roamer Plus will intercept first call to bill you.

Identification numbers

City	Market	System	Billing	RSA
All served cities	163	00559	30071	MSA

Cellular company

ALLTEL Mobile Communications
3322 S Campbell, Suite N
Springfield, MO 65807

Customer service	some areas	800-255-8351
Local office		417-882-2020
Corporate office		501-661-8500

Sending calls

Visitor dialing instructions

Local calls	area code + 7 digits
Long distance	area code + 7 digits

Speed-dial numbers

Customer service	*611
Emergency	911
Highway patrol	*55
Make a traffic report	*97 or *99

Receiving calls

Caller dials the roamer access number below,
hears tone, dials cellular area code and number.
. . . . 417-839-7626

Follow Me Roaming forwarding (see p. 1156)
Activate, dial *18 send. Deactivate dial *19 send.

Billing information

Visitor rates

Most visitors	$3 day + 85¢ minute

Visitors without roaming agreements (p. 1159)
Call co. to use Amex, Discover, Mastercard, Visa

Identification numbers

City	Market	System	Billing	RSA
All served cities	163	00546	00000	MSA

For full coverage area, see Aurora, MO and
Lebanon, MO.

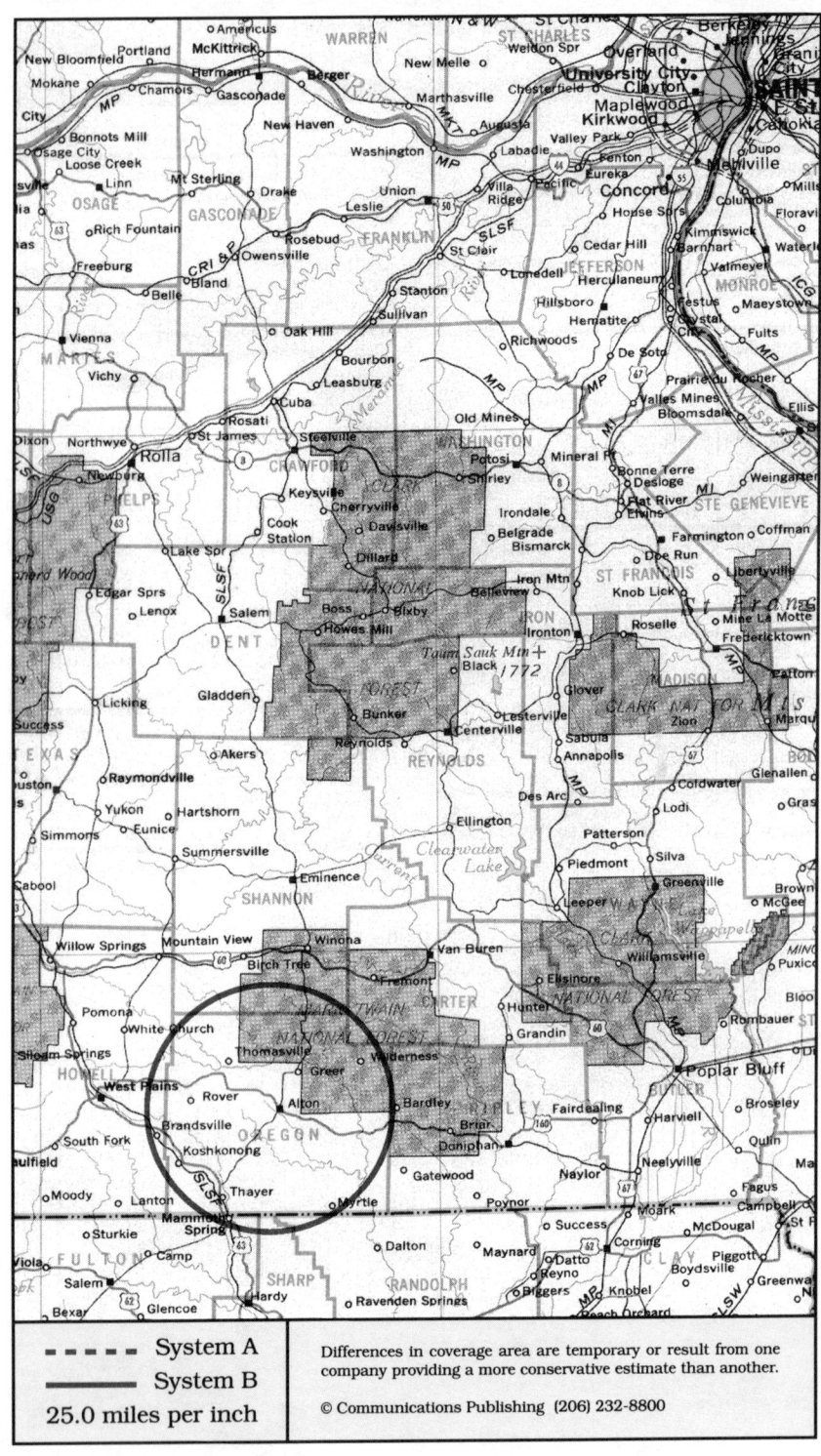

Thayer, Missouri
Alton, Thayer

System A	System B

Cellular company

No information about the company that will operate system A was available at press time.

Cellular company

United States Cellular
1630 S Range Line Road
Joplin, MO 64804

Local office	417-331-4444
Corporate office	312-399-8900

Sending calls

Visitor dialing instructions

Local calls	area code + 7 digits
Long distance	area code + 7 digits

Speed-dial numbers

Customer service	*611 or *711
Emergency	911
Highway patrol	*55

Receiving calls

Caller dials the roamer access number below, hears tone, dials cellular area code and number.
. . . . 417-331-7626

Billing information

Visitor rates
Most visitors . . . . . $3 day + 85¢ minute

Visitors without roaming agreements (p. 1159)
Call co. to use Mastercard, Visa

Identification numbers

City	Market	System	Billing	RSA
All served cities	520	01428	00000	MO-17

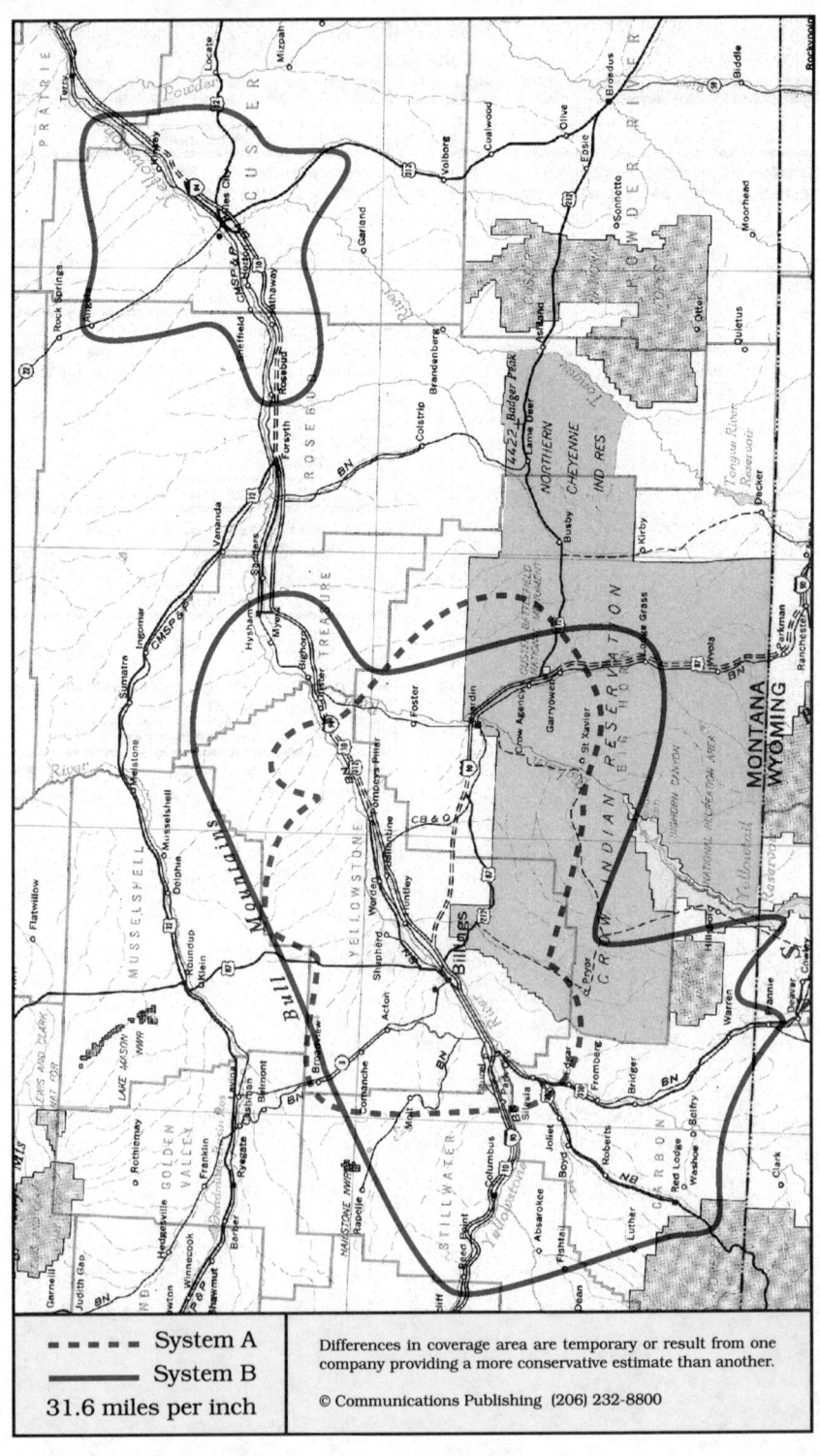

System A - - - - -
System B ———
31.6 miles per inch

Differences in coverage area are temporary or result from one company providing a more conservative estimate than another.

© Communications Publishing (206) 232-8800

Billings, Montana
Billings, Columbus, Hardin, Hysham, Miles City, Red Lodge

System A	System B

Cellular company

Cellular One
Pacific Northwest Cellular
301 S 24th Street W
Billings, MT 59102

Customer service	800-635-0304
Local office	406-652-0466
Corporate office	206-635-0300

Sending calls

Visitor dialing instructions

Local calls	area code + 7 digits
Long distance . . .	1 + area code + 7 digits

Speed-dial numbers

Customer service	611
Emergency	911

Receiving calls

Caller dials the roamer access number below,
hears tone, dials cellular area code and number.
. . . . 406-855-7626

Billing information

Visitor rates
Most visitors . . . . . $3 day + 75¢ minute

Visitors without roaming agreements (p. 1159)
Call co. to use Mastercard, Visa

Identification numbers

City	Market	System	Billing	RSA
Billings	268	00279	00000	MSA
Hardin	531	01449	00000	MT-9

Cellular company

CommNet 2000
Cellular, Inc.
1701 Grand Avenue
Billings, MT 59102

Customer service	800-597-5520
Local office	800-942-2060
. . . .	406-248-1990
Corporate office	303-694-3234

Sending calls

Visitor dialing instructions

Local calls	area code + 7 digits
Long distance . . .	1 + area code + 7 digits

Speed-dial numbers

Customer service	611
Emergency	911
Roaming information	711

Receiving calls

Caller dials the roamer access number below,
hears tone, dials cellular area code and number.

Billings	406-698-7626
Hardin	406-679-7626
Miles City	406-853-7626
Red Lodge	406-425-7626

Billing information

Visitor rates
Most visitors . . . . . $3 day + 95¢ minute

Visitors without roaming agreements (p. 1159)
Cannot call since company takes no credit cards.

Identification numbers

City	Market	System	Billing	RSA
Billings	268	00262	00000	MSA
Hardin	531	00262	30414	MT-9
Miles City	532	00262	30156	MT-10
Red Lodge	531	00262	30154	MT-9

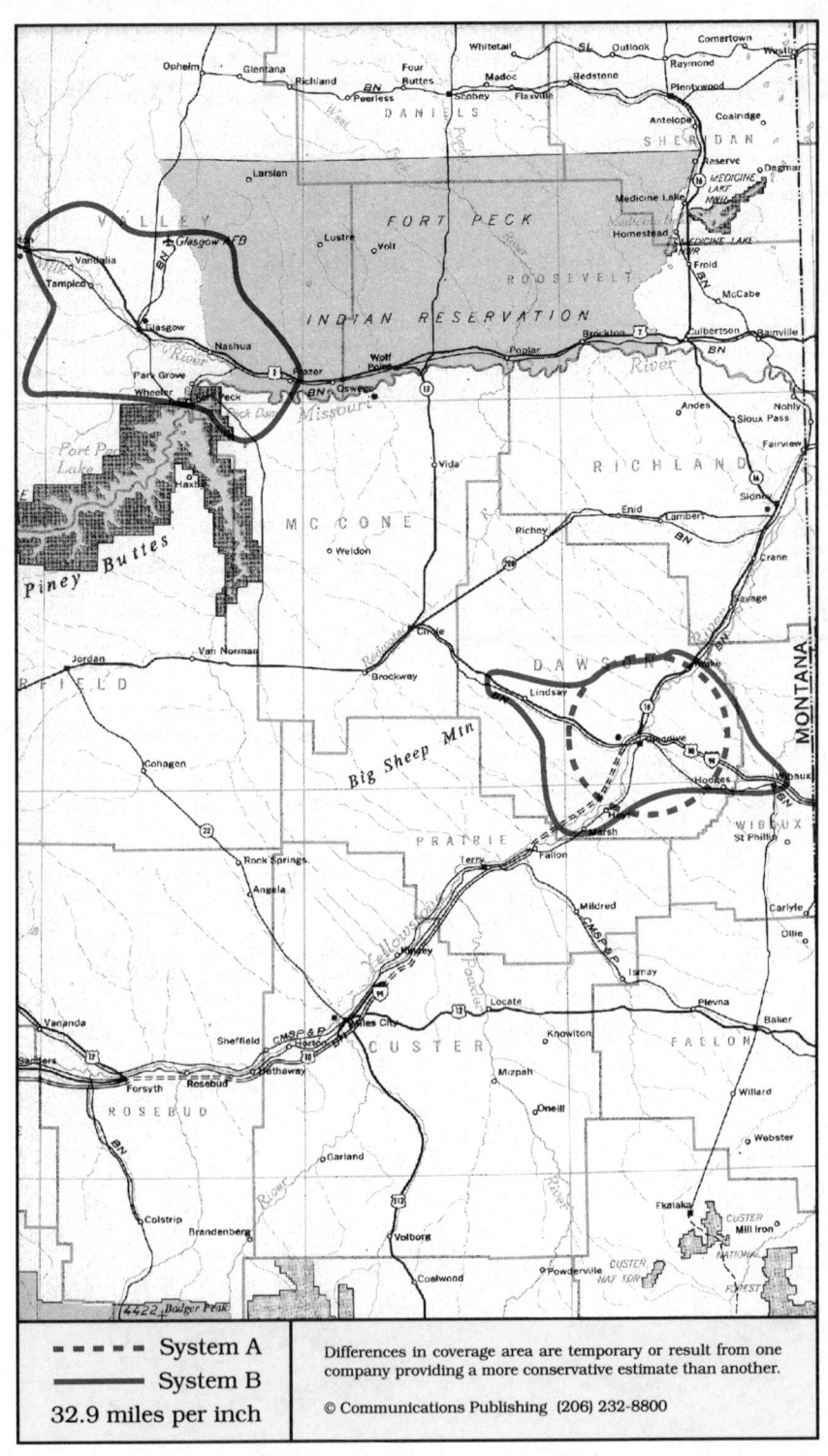

- - - - - System A	Differences in coverage area are temporary or result from one company providing a more conservative estimate than another.
———— System B	
32.9 miles per inch	© Communications Publishing (206) 232-8800

Glendive, Montana
Glasgow, Glendive

System A	System B

Cellular company

Cellular One
GenCell Management
1891 Woolner Avenue
Fairfield, CA 94533

Customer service	800-888-7868
Corporate office	707-425-8000

Sending calls

Visitor dialing instructions

Local calls	area code + 7 digits
Long distance . . .	1 + area code + 7 digits

Speed-dial numbers

Customer service	611
Emergency	911

Receiving calls

Caller dials the roamer access number below,
hears tone, dials cellular area code and number.
. . . . 406-687-6026

NationLink & RoamingAmerica (see p. 1157)
Caller hears your current roamer access number.
This company's subscribers have level 0 service.

Billing information

Visitor rates

Most visitors	$3 day + 99¢ minute

Visitors without roaming agreements (p. 1159)
American Roaming Network intercepts call to bill.

Identification numbers

City	Market	System	Billing	RSA
All served cities	526	01439	00000	MT-4

Cellular company

CommNet 2000
Cellular, Inc.
1701 Grand Avenue
Billings, MT 59102

Customer service	800-597-5520
Local office	800-942-2060
. . . .	406-248-1990
Corporate office	303-694-3234

Sending calls

Visitor dialing instructions

Local calls	area code + 7 digits
Long distance . . .	1 + area code + 7 digits

Speed-dial numbers

Customer service	611
Emergency	911
Roaming information	711

Receiving calls

Caller dials the roamer access number below,
hears tone, dials cellular area code and number.

Glasgow	406-263-7626
Glendive	406-939-7626

Billing information

Visitor rates

Most visitors	$3 day + 95¢ minute

Visitors without roaming agreements (p. 1159)
Cannot call since company takes no credit cards.

Identification numbers

City	Market	System	Billing	RSA
Glasgow	525	00262	30146	MT-3
Glendive	526	00262	30148	MT-4

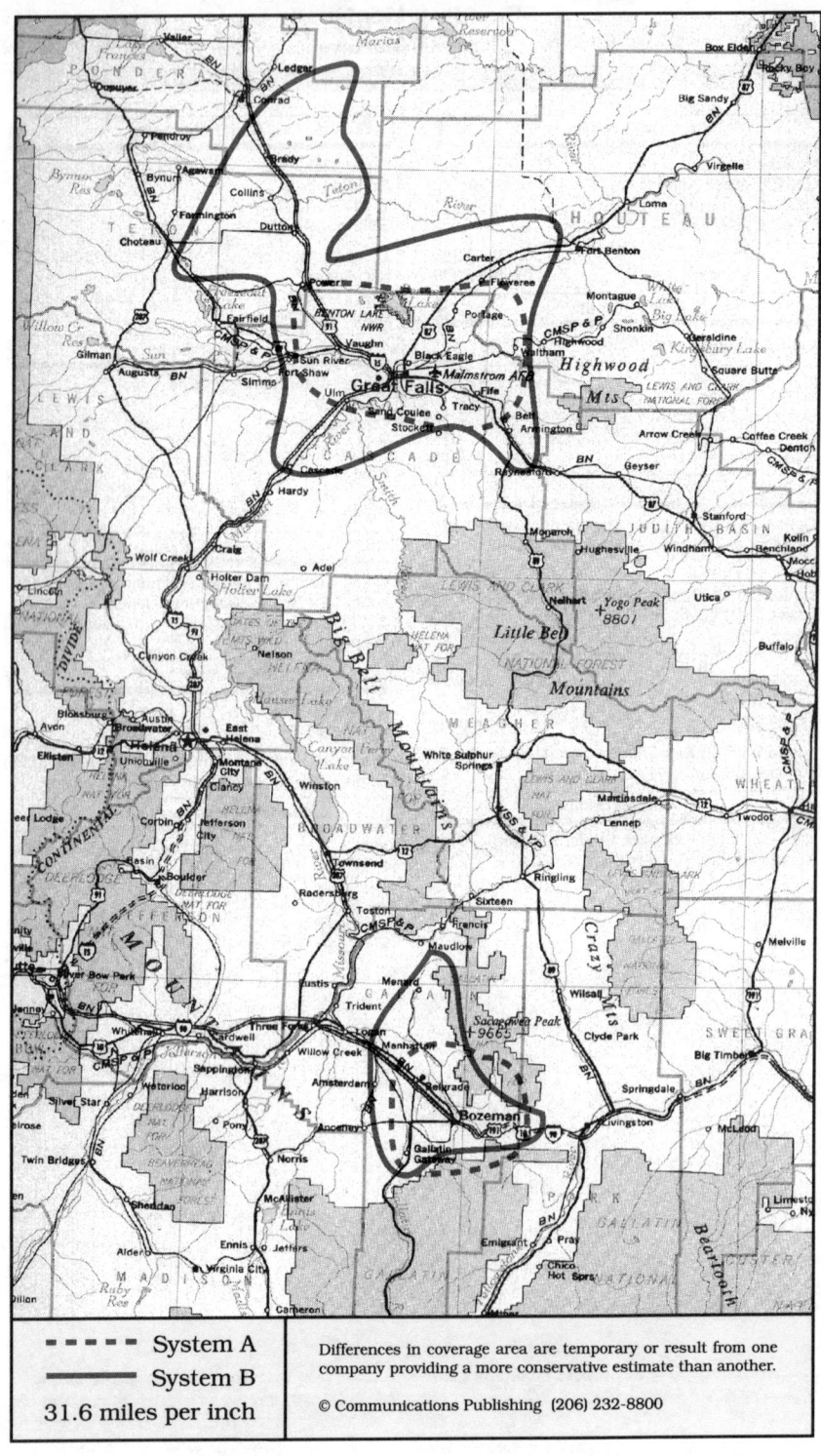

System A — - - - -
System B ————
31.6 miles per inch

Differences in coverage area are temporary or result from one company providing a more conservative estimate than another.

© Communications Publishing (206) 232-8800

System A	System B

Cellular company | ## Cellular company

Cellular One
Licensed to: Montana Cellular Telephone
Managed by: Pacific Northwest Cellular
526 1st Avenue N
Great Falls, MT 59401

Customer service	800-635-0304
. . . .	206-450-2907
Local office	406-727-2355
Corporate office	206-635-0300

CommNet 2000
Cellular, Inc.
3649 10th Avenue S
Great Falls, MT 59405

Customer service for Great Falls	800-597-5521
. . . . . . for Bozeman	800-597-5520
Local office	406-761-5422
Corporate office	303-694-3234

Sending calls

Visitor dialing instructions

Local calls	area code + 7 digits
Long distance . . .	1 + area code + 7 digits

Speed-dial numbers

Customer service	611
Emergency	911

Visitor dialing instructions

Local calls	area code + 7 digits
Long distance . . .	1 + area code + 7 digits

Speed-dial numbers

Customer service	611
Emergency	911
Roaming information	711

Receiving calls

Caller dials the roamer access number below,
hears tone, dials cellular area code and number.

Bozeman	406 -580-7626
Great Falls	406-799-7626

Caller dials the roamer access number below,
hears tone, dials cellular area code and number.

Bozeman	406-581-7626
Great Falls	406-788-7626

Billing information

Visitor rates

Most visitors	$0 day + 99¢ minute

Visitors without roaming agreements (p. 1159)
Call co. to use Mastercard, Visa

Visitor rates

Most visitors	$3 day + 95¢ minute

Visitors without roaming agreements (p. 1159)
Cannot call since company takes no credit cards.

Identification numbers

City	Market	System	Billing	RSA
Bozeman	530	01447	00000	MT-8
Great Falls	297	00373	00000	MSA

City	Market	System	Billing	RSA
Bozeman	530	00262	30152	MT-8
Great Falls	297	00262	30348	MSA

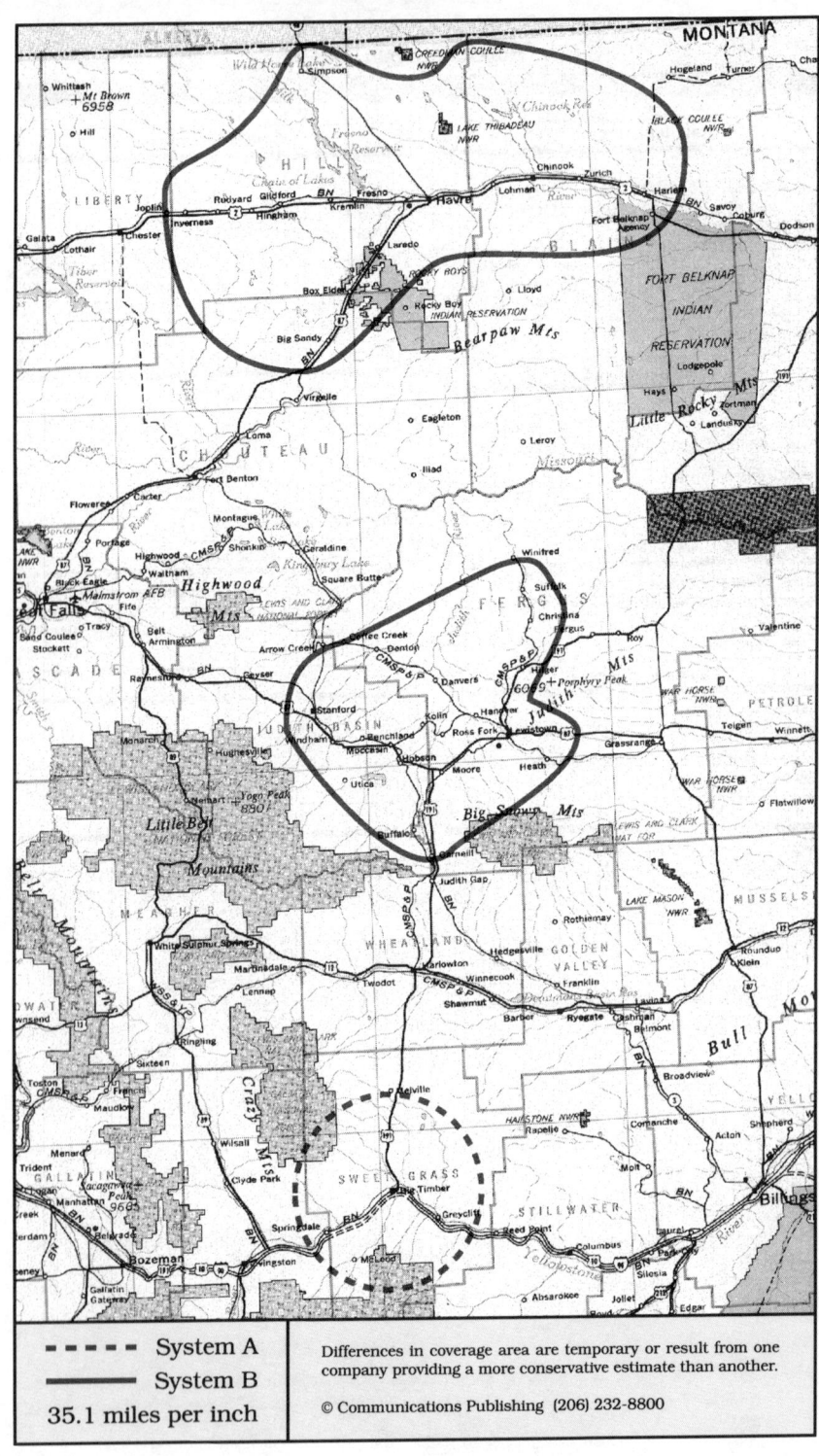

- - - - - System A
————— System B
35.1 miles per inch

Differences in coverage area are temporary or result from one company providing a more conservative estimate than another.

© Communications Publishing (206) 232-8800

Havre, Montana
Big Timber, Chinook, Havre, Lewiston, Stanford

System A	System B

Cellular company

Cellular One
GenCell Management
1891 Woolner Avenue
Fairfield, CA 94533

Customer service	800-888-7868
Corporate office	707-425-8000

Sending calls

Visitor dialing instructions

Local calls	area code + 7 digits
Long distance . . .	1 + area code + 7 digits

Speed-dial numbers

Customer service	611
Emergency	911

Receiving calls

Caller dials the roamer access number below,
hears tone, dials cellular area code and number.
. . . . 406-932-7026

NationLink & RoamingAmerica (see p. 1157)
Caller hears your current roamer access number.
This company's subscribers have level 0 service.

Billing information

Visitor rates

Most visitors	$3 day + 99¢ minute

Visitors without roaming agreements (p. 1159)
American Roaming Network intercepts call to bill.

Identification numbers

City	Market	System	Billing	RSA
All served cities	529	01445	00000	MT-7

Cellular company

CommNet 2000
Cellular, Inc.
3649 10th Avenue S
Great Falls, MT 59405

Customer service	800-597-5520
. . . . . . from Lewiston	800-597-5521
Local office	406-761-5422
Corporate office	303-694-3234

Sending calls

Visitor dialing instructions

Local calls	area code + 7 digits
Long distance . . .	1 + area code + 7 digits

Speed-dial numbers

Customer service	611
Emergency	911
Roaming information	711

Receiving calls

Caller dials the roamer access number below,
hears tone, dials cellular area code and number.

Havre	406-390-7626
Lewiston	406-366-7626

Billing information

Visitor rates

Most visitors	$3 day + 95¢ minute

Visitors without roaming agreements (p. 1159)
Cannot call since company takes no credit cards.

Identification numbers

City	Market	System	Billing	RSA
Havre	524	00262	30144	MT-2
Lewiston	529	00262	30150	MT-7

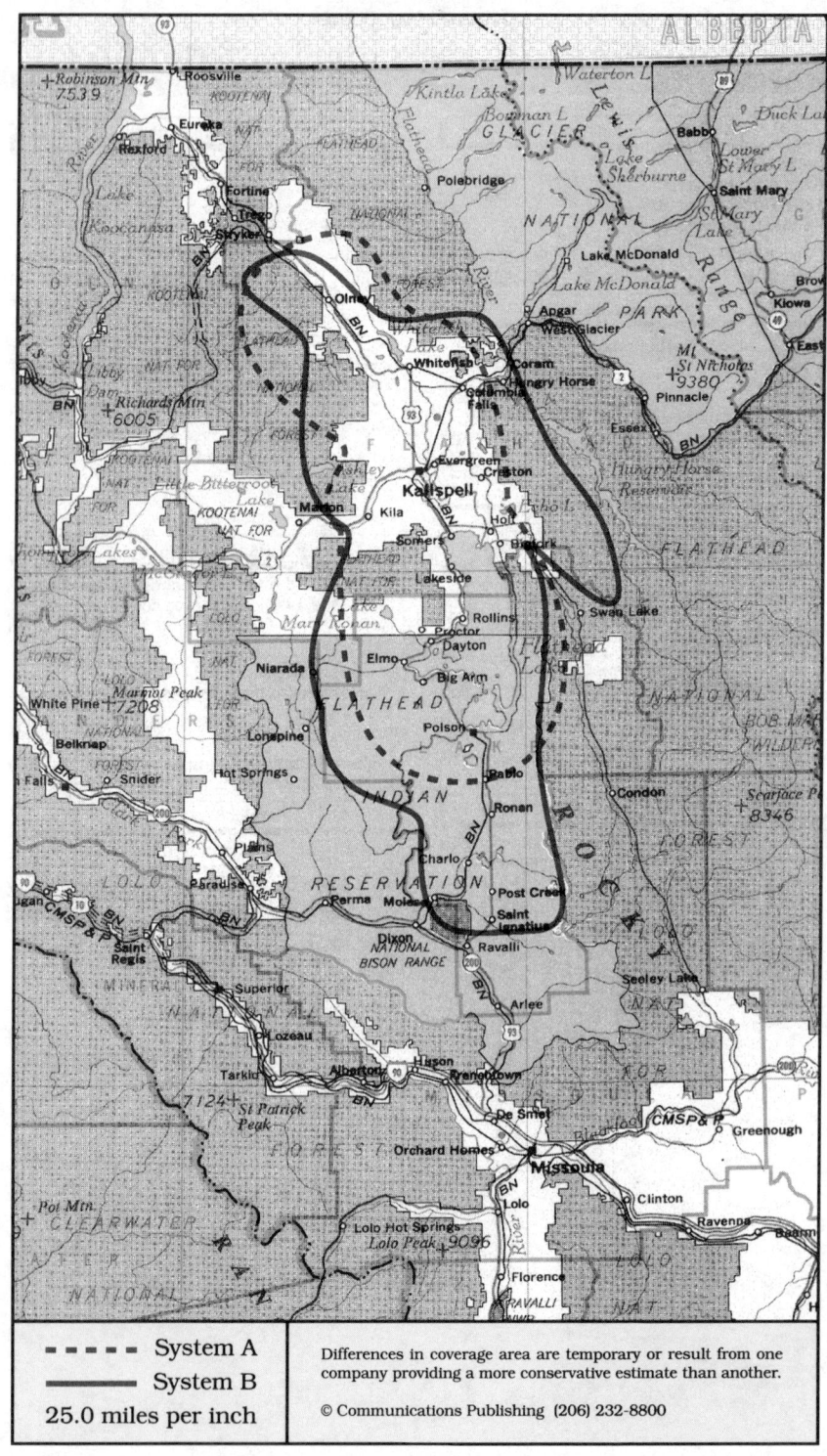

- - - - - System A
————— System B
25.0 miles per inch

Differences in coverage area are temporary or result from one company providing a more conservative estimate than another.

© Communications Publishing (206) 232-8800

Kalispell, Montana
Kalispell, Poison

System A	System B

Cellular company

Cellular One
Pacific Northwest Cellular
17 5th Street E
Kalispell, MT 59901

Customer service	800-635-0304
. . . .	406-253-2273
Local office	406-257-2355
Corporate office	206-635-0300

Sending calls

Visitor dialing instructions

Local calls	area code + 7 digits
Long distance . . .	1 + area code + 7 digits

Speed-dial numbers

Customer service	611
Emergency	911

Receiving calls

Caller dials the roamer access number below,
hears tone, dials cellular area code and number.

. . . .	406-253-7626

NationLink & RoamingAmerica (see p. 1157)
Dial *31 send to receive calls, *30 to deactivate.
This company's subscribers have level 3 service.

Billing information

Visitor rates

Most visitors	$0 day + 99¢ minute

Visitors without roaming agreements (p. 1159)
Roamer Plus will intercept first call to bill you.

Identification numbers

City	Market	System	Billing	RSA
All served cities	523	01433	00000	MT-1

Cellular company

Cellulink
Pacific Telecom Cellular
290 N Main
Kalispell, MT 59901

Customer service	800-236-9700
. . . .	414-841-1111
Local office	406-752-4422
Corporate office	414-841-1100

Sending calls

Visitor dialing instructions

Local calls	7 digits
Long distance	area code + 7 digits

Speed-dial numbers

Customer service	*611
Emergency	911
Roaming information	*711

Receiving calls

Caller dials the roamer access number below,
hears tone, dials cellular area code and number.

. . . .	406-758-7626

Billing information

Visitor rates

Most visitors	$3 day + 85¢ minute

Visitors without roaming agreements (p. 1159)
Cannot call since company takes no credit cards.

Identification numbers

City	Market	System	Billing	RSA
All served cities	523	01434	00000	MT-1

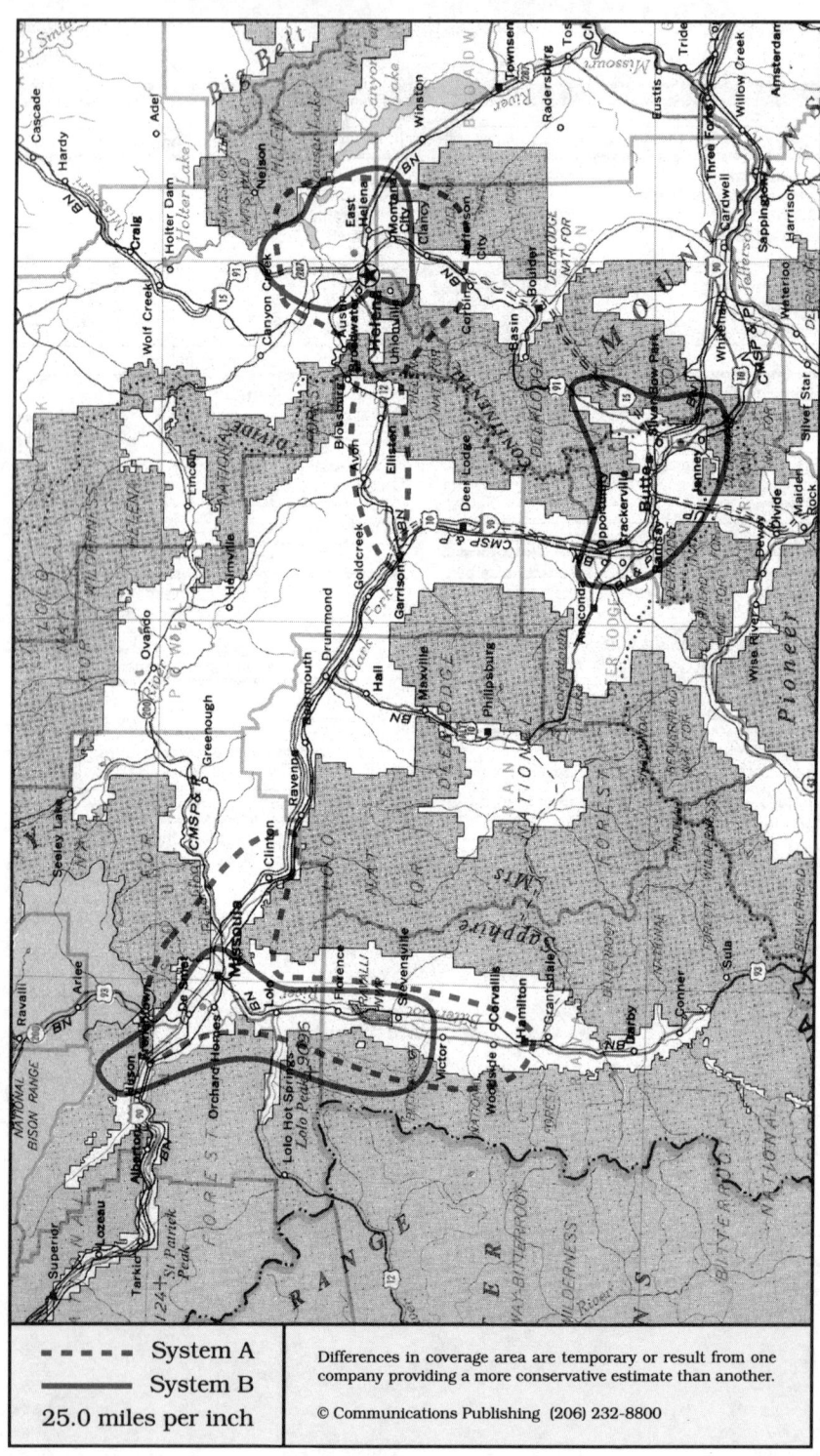

- - - - - System A
———— System B
25.0 miles per inch

Differences in coverage area are temporary or result from one company providing a more conservative estimate than another.

© Communications Publishing (206) 232-8800

Missoula, Montana
Butte, Hamilton, Helena, Missoula

System A	System B

Cellular company

Cellular One
Licensed to: Montana Cellular Telephone
Managed by: Pacific Northwest Cellular
3100 Paxson
Missoula, MT 59801

Customer service	800-635-0304
Local office	406-543-2355
Corporate office	206-635-0300

Sending calls

Visitor dialing instructions

Local calls	area code + 7 digits
Long distance . . .	1 + area code + 7 digits

Speed-dial numbers

Customer service	611
Emergency	911

Receiving calls

Caller dials the roamer access number below,
hears tone, dials cellular area code and number.
. . . .   406-240-7626

Billing information

Visitor rates
Most visitors . . . . . $0 day + 99¢ minute

Visitors without roaming agreements (p. 1159)
Call co. to use Mastercard, Visa
or Roamer Plus will intercept first call to bill you.

Identification numbers

City	Market	System	Billing	RSA
All served cities	527	01441	00000	MT-5

Cellular company

CommNet 2000
Cellular, Inc.
1530 Livingston
Missoula, MT 59801

Customer service	800-597-5520
Local office	406-542-1999
Corporate office	303-694-3234

Sending calls

Visitor dialing instructions

Local calls	area code + 7 digits
Long distance . . .	1 + area code + 7 digits

Speed-dial numbers

Customer service	611
Emergency	911
Roaming information	711

Receiving calls

Caller dials the roamer access number below,
hears tone, dials cellular area code and number.

Butte	406-490-7626
Helena	406-431-7626
Missoula	406-544-7626

Billing information

Visitor rates
Most visitors . . . . . $3 day + 95¢ minute

Visitors without roaming agreements (p. 1159)
Cannot call since company takes no credit cards.

Identification numbers

City	Market	System	Billing	RSA
Butte	528	00262	30070	MT-6
Helena	527	00262	30412	MT-5
Missoula	527	00262	30068	MT-5

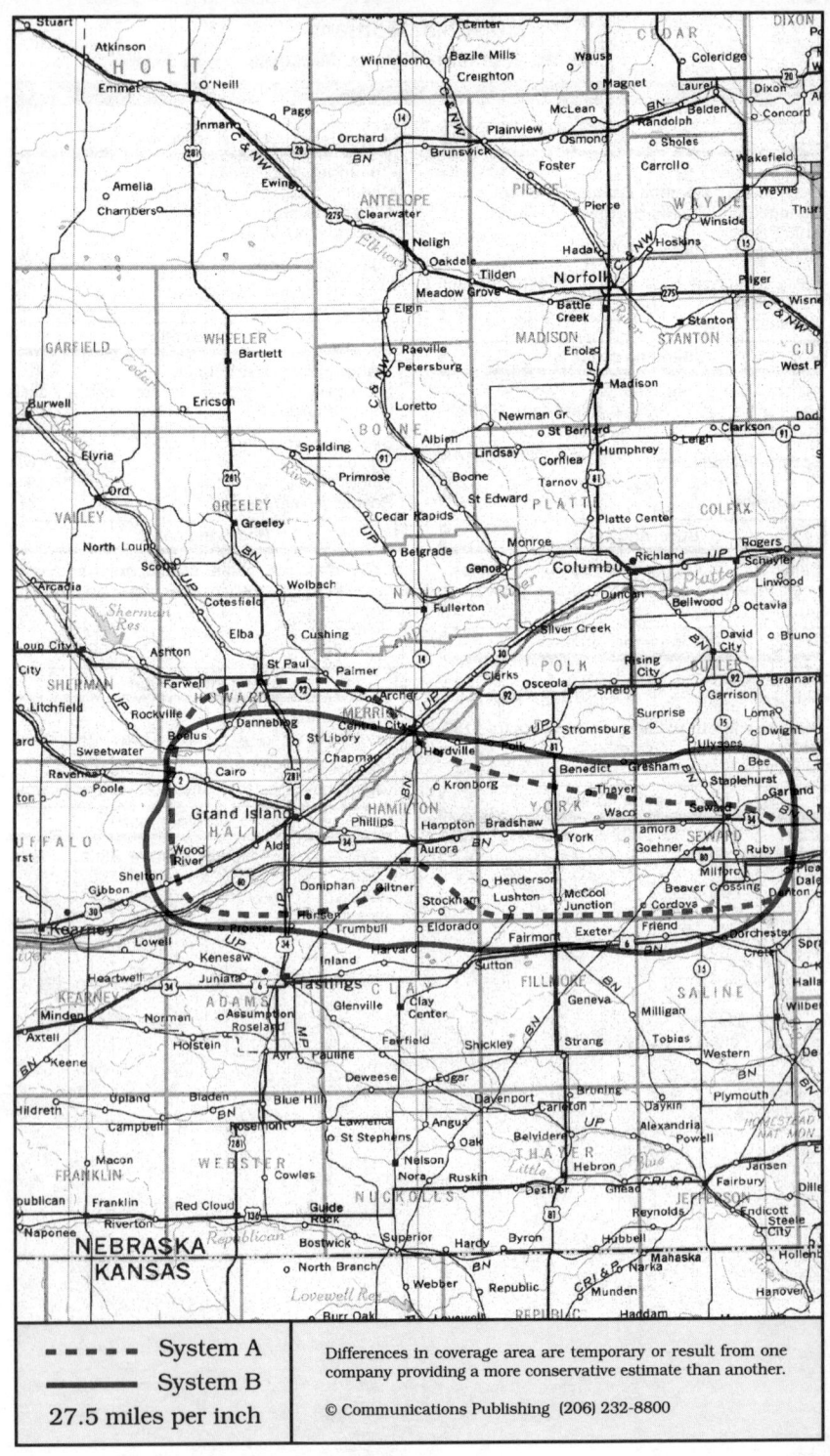

System A
System B
27.5 miles per inch

Differences in coverage area are temporary or result from one company providing a more conservative estimate than another.

© Communications Publishing (206) 232-8800

Grand Island, Nebraska
Aurora, Grand Island, Seward, York

System A	System B

Cellular company

Cellular One
GenCell Management
3537 W 13th Street, Suite 108
Grand Island, NE 68803

Local office	800-333-4847
. . . .	308-382-8600
Corporate office	707-425-8000

Sending calls

Visitor dialing instructions

Local calls	area code + 7 digits
Long distance . . .	1 + area code + 7 digits

Speed-dial numbers

Customer service	611
Emergency	911

Receiving calls

Caller dials the roamer access number below,
hears tone, dials cellular area code and number.
. . . . 308-383-7626

NationLink & RoamingAmerica (see p. 1157)
Dial *31 send to receive calls, *30 to deactivate.
This company's subscribers have level 3 service.

Billing information

Visitor rates
Most visitors . . . . . $3 day + 99¢ minute

Visitors without roaming agreements (p. 1159)
American Roaming Network intercepts call to bill.

Identification numbers

City	Market	System	Billing	RSA
All served cities	539	01465	00000	NE-7

For full coverage area, see Kearney, NE and
O'Neill, NE.

Cellular company

Nebraska Cellular
1431 N Webb Road
Grand Island, NE 68803

Corporate office	800-578-1054
. . . .	308-384-9000

Sending calls

Visitor dialing instructions

Local calls	area code + 7 digits
Long distance	area code + 7 digits

Speed-dial numbers

Customer service	*611
Emergency	911
State patrol	*55

Receiving calls

Caller dials the roamer access number below,
hears tone, dials cellular area code and number.
. . . . 308-380-7626

Billing information

Visitor rates
Most visitors . . . . . $3 day + 90¢ minute

Visitors without roaming agreements (p. 1159)
Cannot call since company takes no credit cards.

Identification numbers

City	Market	System	Billing	RSA
All served cities	539	01466	00000	NE-7

For full coverage area, see Kearney, NE and
O'Neill, NE.

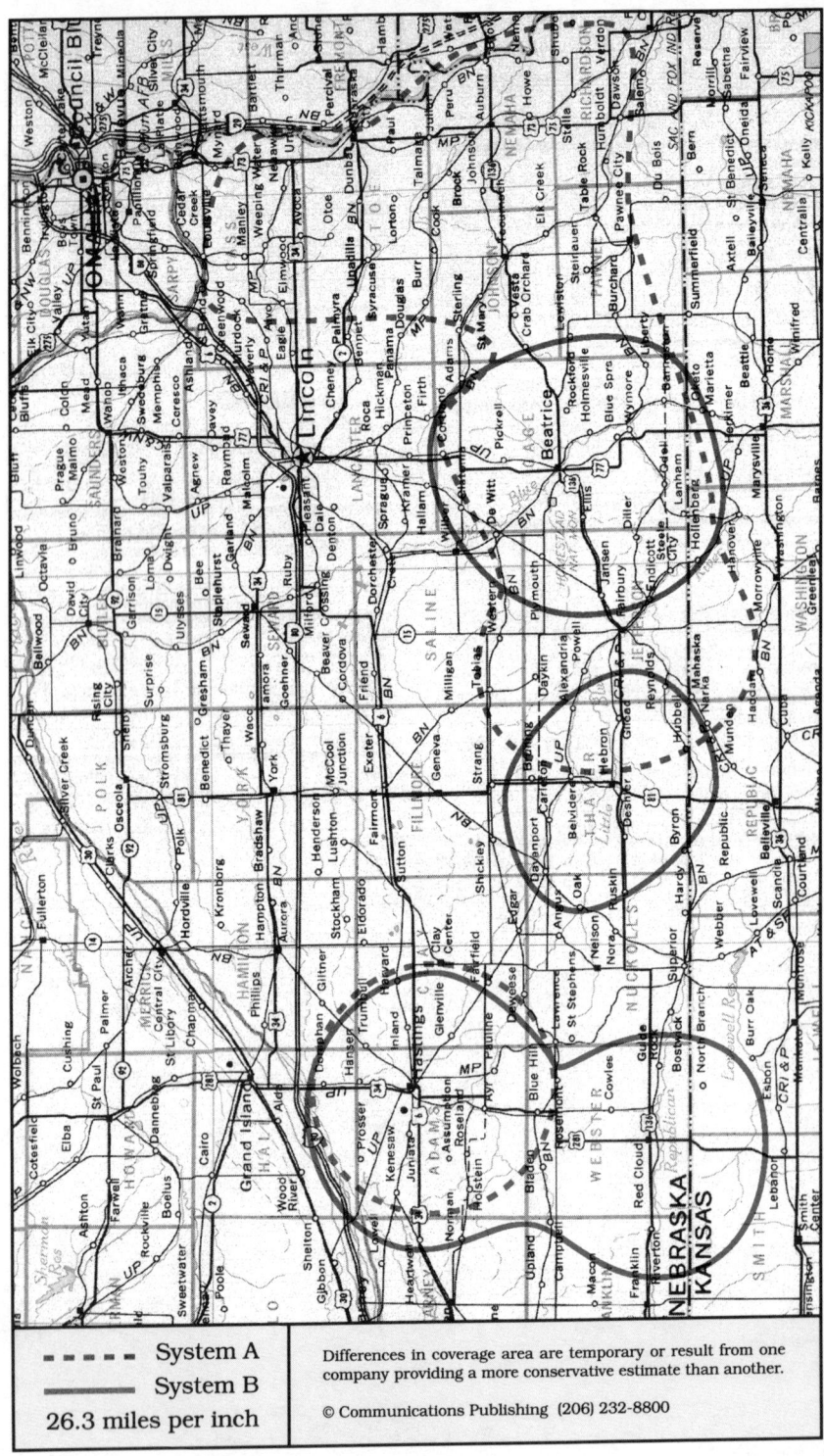

- - - - System A
———— System B
26.3 miles per inch

Differences in coverage area are temporary or result from one company providing a more conservative estimate than another.

© Communications Publishing (206) 232-8800

Hastings, Nebraska

Auburn, Beatrice, Deshler, Hastings, Red Cloud, Tecumseh

System A	System B

System A

Cellular company

X-Cell Cellular
2301 Court Street
Beatrice, NE 68310

108 N 8th Street
Nebraska City, NE 68410

Customer service	800-788-0993
Local office . . . . Beatrice	402-223-4560
. . . . . Nebraska City	402-873-4590
Corporate office	303-433-3700

Sending calls

Visitor dialing instructions

Local calls	7 digits
Long distance . . . 1 + area code + 7 digits	

Speed-dial numbers

Customer service	611
Emergency	911

Receiving calls

Caller dials the roamer access number below,
hears tone, dials cellular area code and number.
. . . . 402-223-8026

NationLink & RoamingAmerica (see p. 1157)
Caller hears your current roamer access number.
This company's subscribers have level 0 service.

Billing information

Visitor rates
Most visitors . . . . . $3 day + 75¢ minute

Visitors without roaming agreements (p. 1159)
Roamer Plus will intercept first call to bill you.

Identification numbers

City	Market	System	Billing	RSA
Beatrice	542	01471	00000	NE-10
Hastings	541	01471	00000	NE-9

System B

Cellular company

Nebraska Cellular
1431 N Webb Road
Grand Island, NE 68803

Corporate office	800-578-1054
. . . .	308-384-9000

Sending calls

Visitor dialing instructions

Local calls	area code + 7 digits
Long distance	area code + 7 digits

Speed-dial numbers

Customer service	*611
Emergency	911
State patrol	*55

Receiving calls

Caller dials the roamer access number below,
hears tone, dials cellular area code and number.
. . . . 308-380-7626

Billing information

Visitor rates
Most visitors . . . . . $3 day + 90¢ minute

Visitors without roaming agreements (p. 1159)
Cannot call since company takes no credit cards.

Identification numbers

City	Market	System	Billing	RSA
Beatrice	542	01466	00000	NE-10
Hastings	541	01466	00000	NE-9

For full coverage area, see Grand Island, NE.

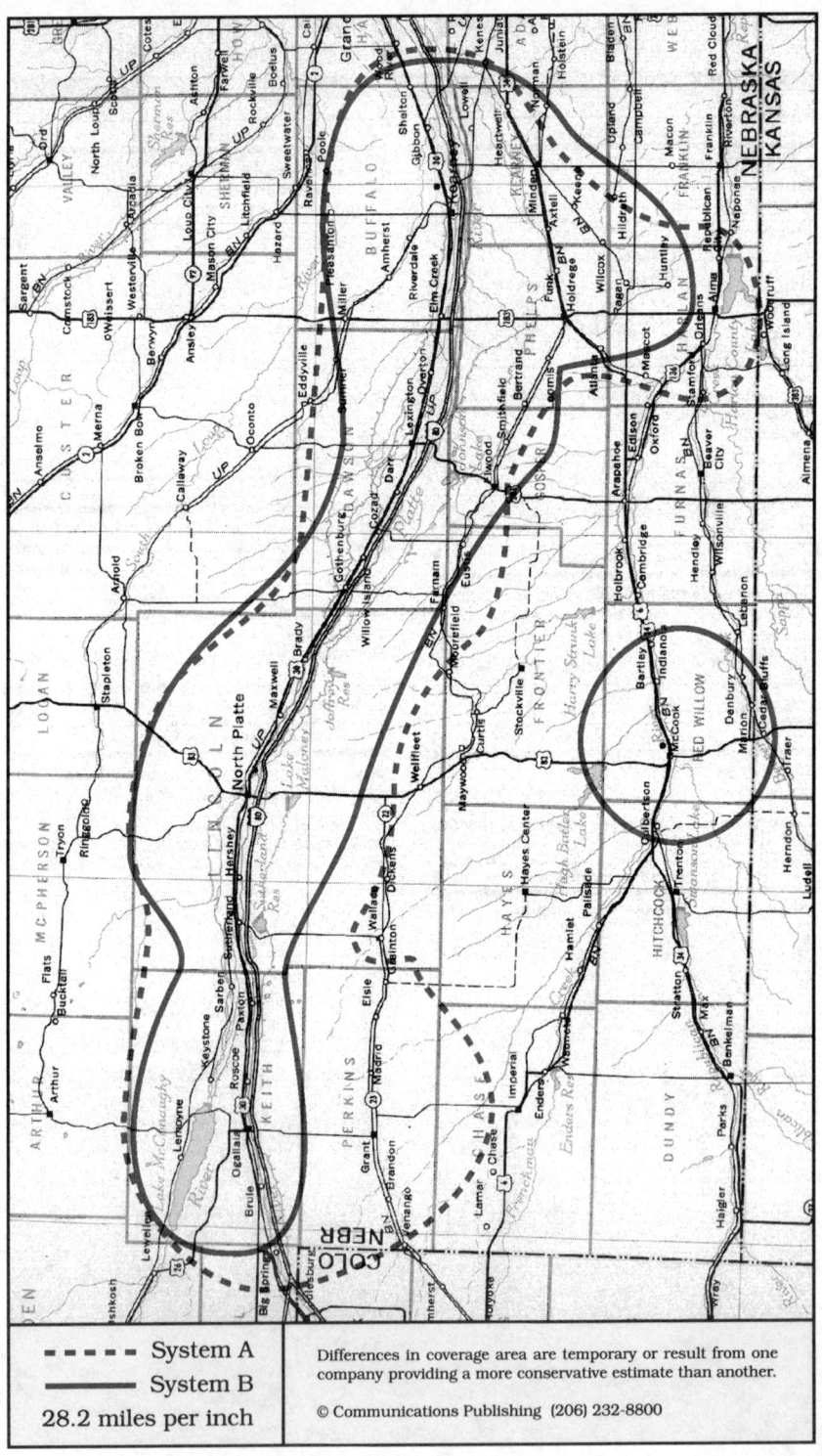

- - - - - System A
————— System B
28.2 miles per inch

Kearney, Nebraska

Holdrege, Kearney, Lexington, McCook, North Platte, Ogallala

System A	System B

System A

Cellular company

Cellular One
GenCell Management
808 E 25th Street
Kearney, NE 68847

221 S Jeffers Street #2
North Platte, NE 69101

Customer service		800-333-4847
Local office	Kearney	308-233-2000
. . . . . .	North Platte	308-539-2000
Corporate office	Cellular One	308-382-8600
. . . .	GenCell	707-425-8000

Sending calls

Visitor dialing instructions
Local calls . . . . . area code + 7 digits
Long distance . . . 1 + area code + 7 digits

Speed-dial numbers
Customer service 611
Emergency 911

Receiving calls

Caller dials the roamer access number below,
hears tone, dials cellular area code and number.
. . . . 308-233-7626

NationLink & RoamingAmerica (see p. 1157)
Dial *31 send to receive calls, *30 to deactivate.
Dial *32 send to tell caller the roamer access no.
(*32 service is only available in Holdrege NE-8.)
This company's subscribers have level 3 service.

Billing information

Visitor rates
Most visitors . . . . . $3 day + 99¢ minute

Visitors without roaming agreements (p. 1159)
American Roaming Network intercepts call to bill.

Identification numbers

City	Market	System	Billing	RSA
Holdrege	540	01465	00000	NE-8
Kearney, N. Platte	538	01465	00000	NE-6

Company plans to cover McCook by April 1992.

For full coverage area, see Grand Island, NE.

System B

Cellular company

Nebraska Cellular
1431 N Webb Road
Grand Island, NE 68803

Corporate office	800-578-1054
. . . .	308-384-9000

Sending calls

Visitor dialing instructions
Local calls . . . . . . area code + 7 digits
Long distance . . . . . area code + 7 digits

Speed-dial numbers
Customer service *611
Emergency 911
Beatrice   542   01471   00000   NE-10

Receiving calls

Caller dials the roamer access number below,
hears tone, dials cellular area code and number.
. . . . 308-380-7626

Billing information

Visitor rates
Most visitors . . . . . $3 day + 90¢ minute

Visitors without roaming agreements (p. 1159)
Cannot call since company takes no credit cards.

Identification numbers

City	Market	System	Billing	RSA
Holdrege, McCook	540	01466	00000	NE-8
Kearney, N. Platte	538	01466	00000	NE-6

For full coverage area, see Grand Island, NE.

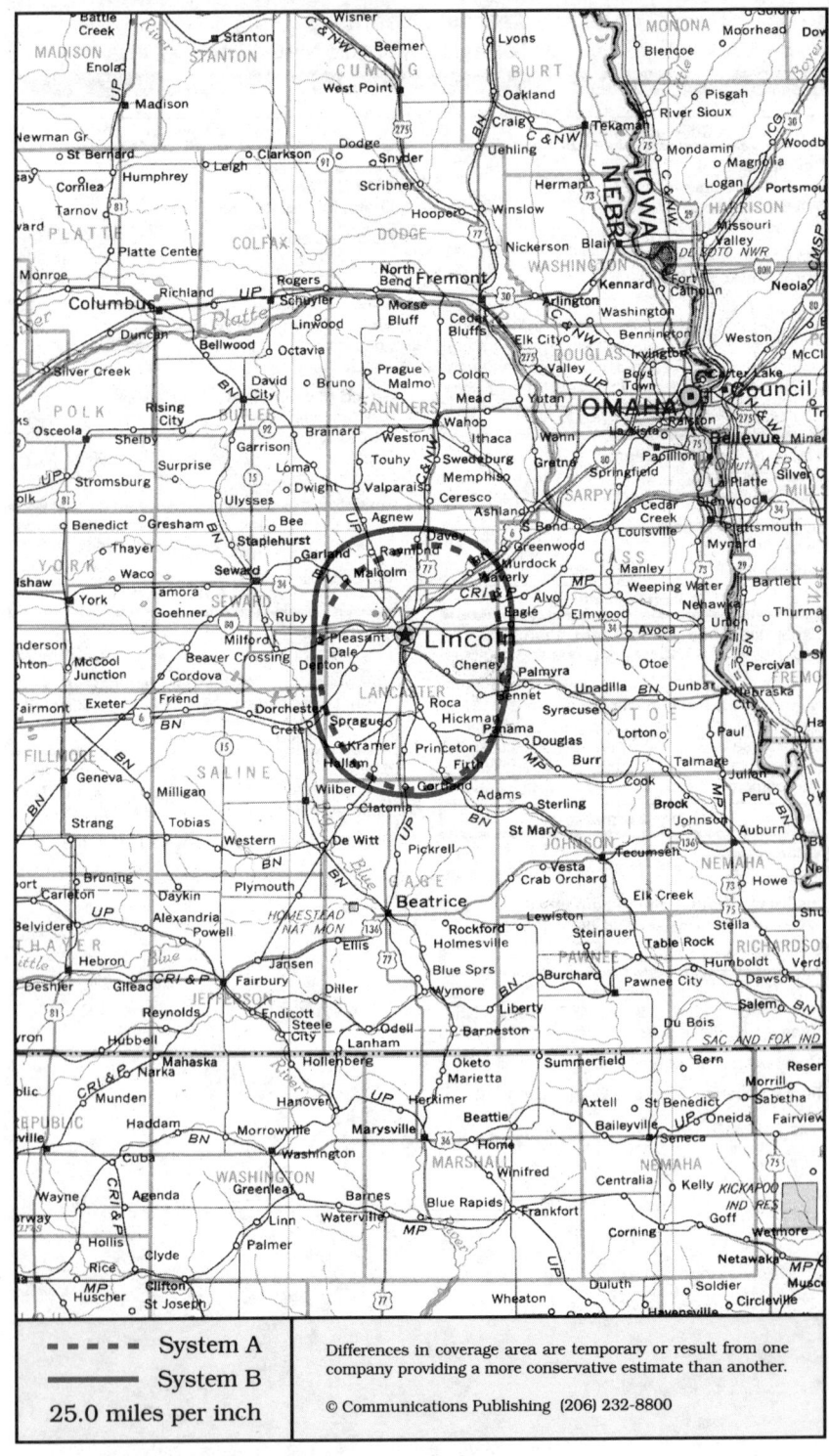

- - - - - **System A**	Differences in coverage area are temporary or result from one company providing a more conservative estimate than another.
———— **System B**	
25.0 miles per inch	© Communications Publishing (206) 232-8800

Lincoln, Nebraska

System A

Cellular company

Cellular One
Centennial Cellular
400 N 48th, Suite C07
Lincoln, NE 68504

Local office 402-466-1400
Corporate office 203-972-2000

Sending calls

Visitor dialing instructions
Local calls . . . . . area code + 7 digits
Long distance . . . 1 + area code + 7 digits

Speed-dial numbers
Customer service . . . . ✱611
Emergency 911

Receiving calls

Caller dials the roamer access number below,
hears tone, dials cellular area code and number.
. . . . 402-560-7626

NationLink & RoamingAmerica (see p. 1157)
Caller hears your current roamer access number.
This company's subscribers have level 0 service.

Billing information

Visitor rates
Most visitors . . . . . $3 day + 85¢ minute

Visitors without roaming agreements (p. 1159)
Cannot call since company takes no credit cards.

Identification numbers

City	Market	System	Billing	RSA
All served cities	172	00433	00000	MSA

System B

Cellular company

Lincoln Telephone Cellular
5745 "O" Street, PO Box 81309
Lincoln, NE 68501-1309

Corporate office 402-436-5050

Sending calls

Visitor dialing instructions
Local calls . . . . . . area code + 7 digits
Long distance . . . . area code + 7 digits

Speed-dial numbers
Customer service . . . . ✱611
Emergency 911

Receiving calls

Caller dials the roamer access number below,
hears tone, dials cellular area code and number.
. . . . 402-437-7626

Follow Me Roaming forwarding (see p. 1156)
Activate, dial ✱18 send. Deactivate dial ✱19 send.

Billing information

Visitor rates
Most visitors . . . $2.50 day + 75¢ minute

Visitors without roaming agreements (p. 1159)
Cannot call since company takes no credit cards.

Identification numbers

City	Market	System	Billing	RSA
All served cities	172	00416	00000	MSA

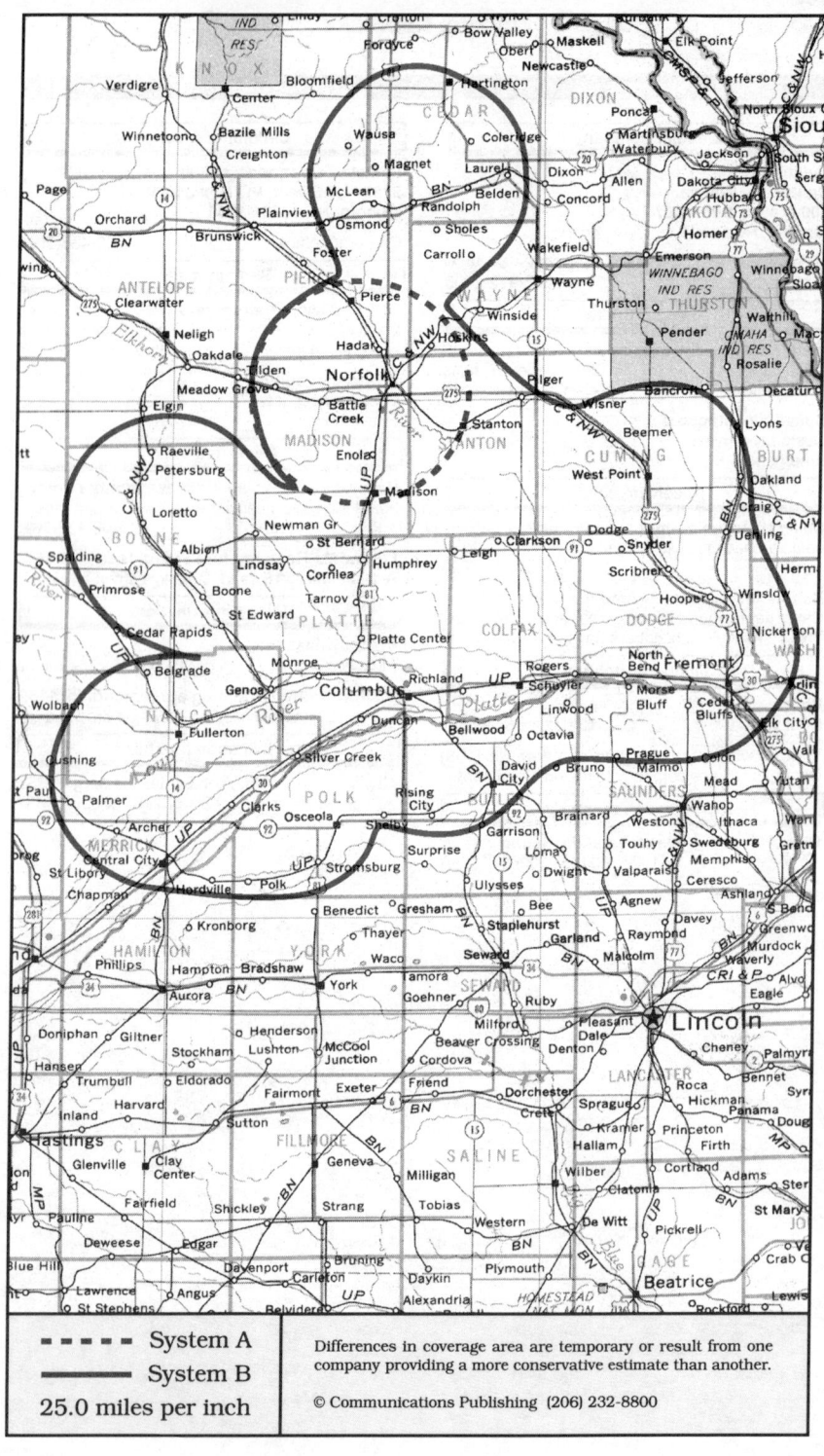

System A - - - -
System B ——————
25.0 miles per inch

Differences in coverage area are temporary or result from one company providing a more conservative estimate than another.

© Communications Publishing (206) 232-8800

Norfolk, Nebraska

Albion, Central City, Columbus, Fremont, Norfolk, Randolph, West Point

System A	System B

Cellular company

Cellular One
National Cellular
700 Omaha Avenue, Suite B
Norfolk, NE 68701

| Corporate office | some areas | 800-371-2351 |
| . . . . | | 402-371-8771 |

Sending calls

Visitor dialing instructions

| Local calls | . . . . | 7 digits |
| Long distance | . . . | 1 + area code + 7 digits |

Speed-dial numbers

| Customer service | . . . . | *611 |
| Emergency | . . . . | 911 |

Receiving calls

Caller dials the roamer access number below,
hears tone, dials cellular area code and number.

. . . . 402-640-7626

Billing information

Visitor rates

| Most visitors | . . . . . | $3 day + 99¢ minute |

Visitors without roaming agreements (p. 1159)
Call co. to use Mastercard, Visa
or Roamer Plus will intercept first call to bill you.

Identification numbers

City	Market	System	Billing	RSA
All served cities	535	01457	00000	NE-3

Cellular company

Nebraska Cellular
1431 N Webb Road
Grand Island, NE 68803

| Corporate office | . . . . . | 800-578-1054 |
| . . . . | | 308-384-9000 |

Sending calls

Visitor dialing instructions

| Local calls | . . . . . . | area code + 7 digits |
| Long distance | . . . . . | area code + 7 digits |

Speed-dial numbers

Customer service	. . . .	*611
Emergency	. . . .	911
State patrol	. . . .	*55

Receiving calls

Caller dials the roamer access number below,
hears tone, dials cellular area code and number.

. . . . 308-380-7626

Billing information

Visitor rates

| Most visitors | . . . . . | $3 day + 90¢ minute |

Visitors without roaming agreements (p. 1159)
Cannot call since company takes no credit cards.

Identification numbers

City	Market	System	Billing	RSA
Columbus, Fremont	537	01466	00000	NE-5
Norfolk, West Point	535	01466	00000	NE-3

For full coverage area, see Grand Island, NE.

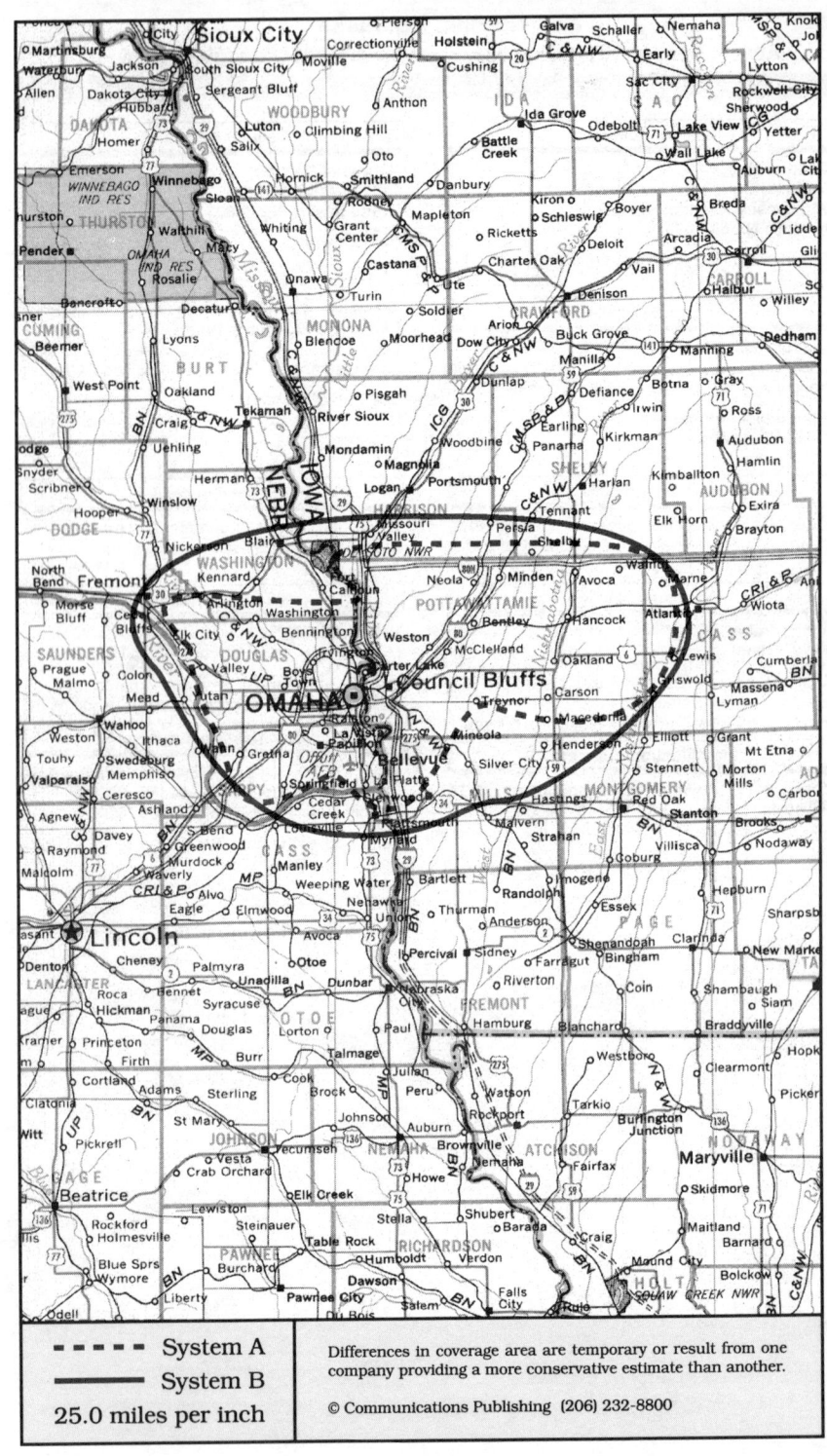

∎ ∎ ∎ ∎ System A	Differences in coverage area are temporary or result from one company providing a more conservative estimate than another.
———— System B	
25.0 miles per inch	© Communications Publishing (206) 232-8800

Omaha, Nebraska
Bellevue, Council Bluffs IA, Omaha

System A	System B

System A

Cellular company

U S West Cellular
7215 Dodge Street
Omaha, NE 68114-3611

Customer service	800-238-7848
. . . .	206-562-2895
. . . . . local subscribers	800-626-6611
. . . .	206-747-1771
Local office	402-391-9000
Corporate office	206-747-4900

Sending calls

Visitor dialing instructions
Local calls	area code + 7 digits
Long distance . . .	area code + 7 digits

Speed-dial numbers
Customer service	*611
Emergency	911
Report an accident	#333
Roaming information	*711
Traffic report WOW	#590

Receiving calls

Caller dials the roamer access number below,
hears tone, dials cellular area code and number.
. . . . 402-681-7626

Billing information

Visitor rates
Most visitors . . . . . $3 day + 99¢ minute

Visitors without roaming agreements (p. 1159)
Call co. to use Amex, Mastercard, Visa

Identification numbers

City	Market	System	Billing	RSA
All served cities	065	00137	00000	MSA

System B

Cellular company

First Cellular Omaha
Lincoln Telephone Cellular
15432 W Center Road
Omaha, NE 68144

Local office	402-330-6500
Corporate office	402-436-5050

Sending calls

Visitor dialing instructions
Local calls	area code + 7 digits
Long distance	area code + 7 digits

Speed-dial numbers
Customer service	*611
Emergency	911

Receiving calls

Caller dials the roamer access number below,
hears tone, dials cellular area code and number.
. . . . 402-677-7626

Follow Me Roaming forwarding (see p. 1156)
Activate, dial *18 send. Deactivate dial *19 send.

Billing information

Visitor rates
Most visitors . . . . . $3 day + 99¢ minute

Visitors without roaming agreements (p. 1159)
Call co. to use Amex, Discover, Mastercard, Visa

Identification numbers

City	Market	System	Billing	RSA
All served cities	065	00152	00000	MSA

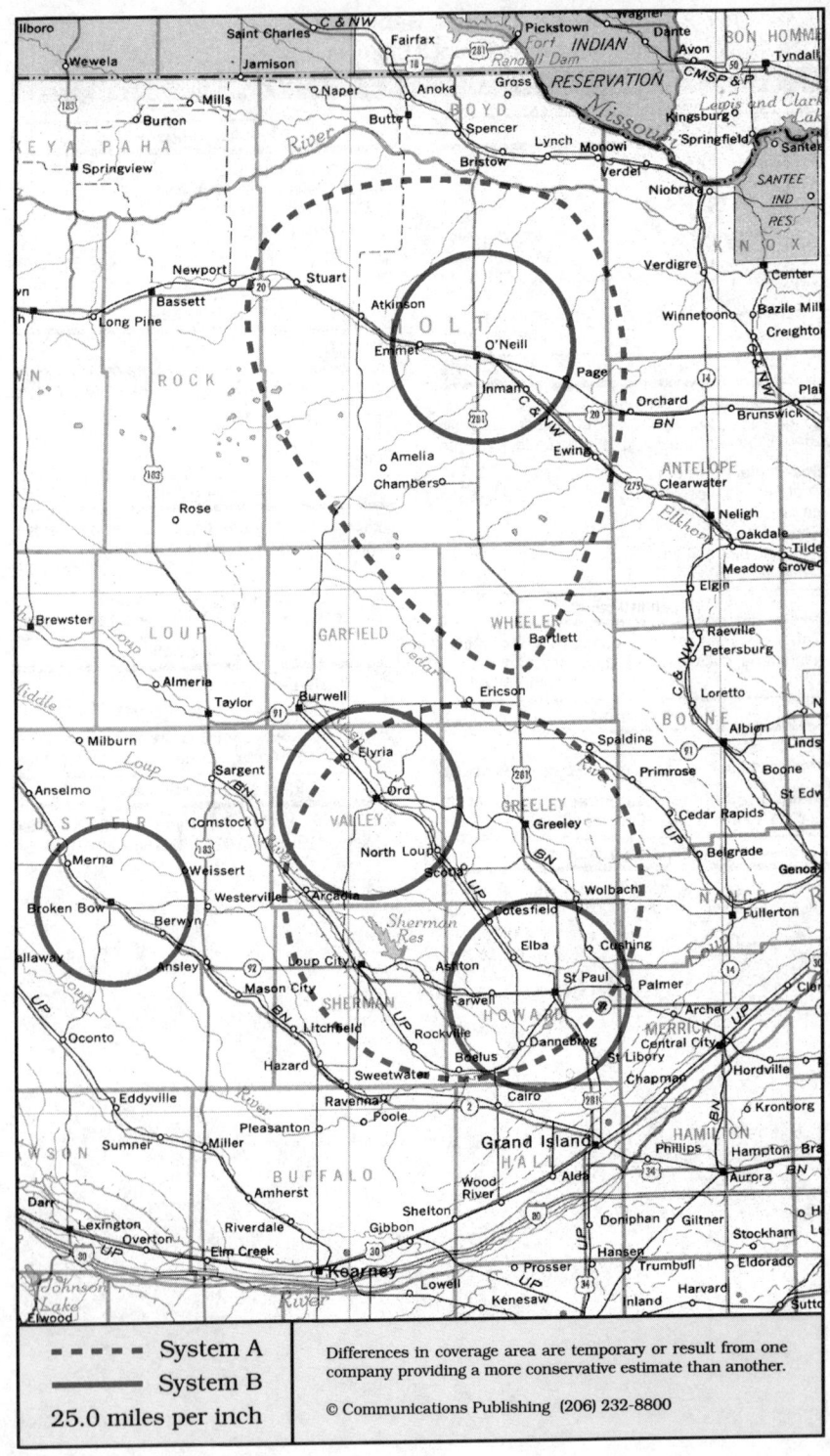

- - - - System A
———— System B
25.0 miles per inch

O'Neill, Nebraska

Broken Bow, Greeley, Loup City, O'Neill, Ord, St. Paul

System A	System B

Cellular company

Cellular One
GenCell Management
3537 W 13th Street, Suite 108
Grand Island, NE 68803

Customer service	800-333-4847
Local office	308-382-8600
Corporate office	707-425-8000

Sending calls

Visitor dialing instructions

Local calls	area code + 7 digits
Long distance . . .	1 + area code + 7 digits

Speed-dial numbers

Customer service	611
Emergency	911

Receiving calls

Caller dials the roamer access number below,
hears tone, dials cellular area code and number.
. . . . 308-383-7626

Billing information

Visitor rates

Most visitors . . . . . $3 day + 99¢ minute

Visitors without roaming agreements (p. 1159)
American Roaming Network intercepts call to bill.

Identification numbers

City	Market	System	Billing	RSA
O'Neil	534	01465	00000	NE-2
Ord	536	01465	00000	NE-4

For full coverage area, see Grand Island, NE.

Cellular company

Nebraska Cellular
1431 N Webb Road
Grand Island, NE 68803

Corporate office	800-578-1054
. . . .	308-384-9000

Sending calls

Visitor dialing instructions

Local calls	area code + 7 digits
Long distance	area code + 7 digits

Speed-dial numbers

Customer service	*611
Emergency	911
State patrol	*55

Receiving calls

Caller dials the roamer access number below,
hears tone, dials cellular area code and number.
. . . . 308-380-7626

Billing information

Visitor rates

Most visitors . . . . . $3 day + 90¢ minute

Visitors without roaming agreements (p. 1159)
Cannot call since company takes no credit cards.

Identification numbers

City	Market	System	Billing	RSA
Broken Bow, Ord	536	01466	00000	NE-4
O'Neil	534	01466	00000	NE-2

For full coverage area, see Grand Island, NE.

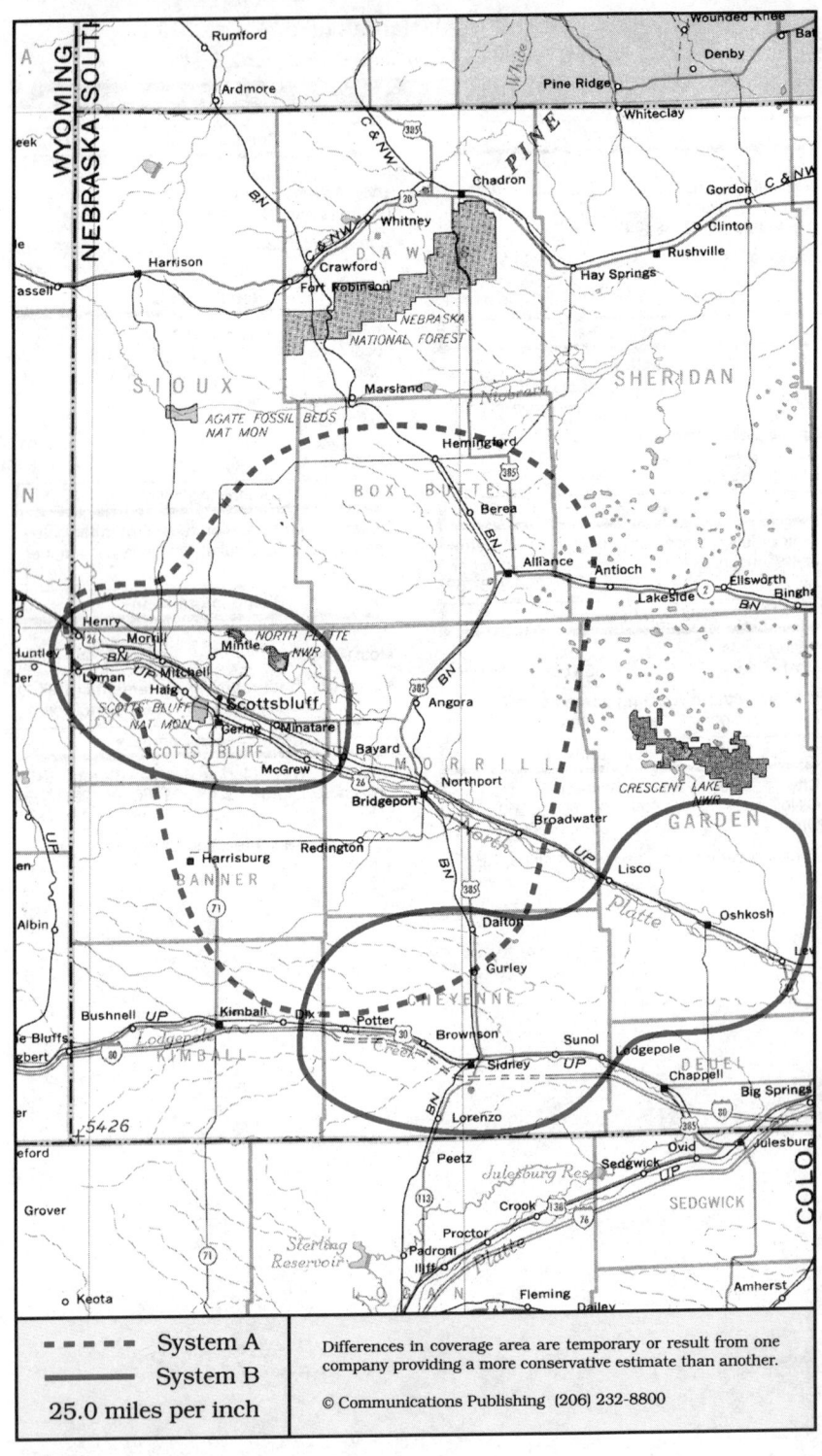

Scottsbluff, Nebraska
Alliance, Scottsbluff, Sidney

System A	System B

Cellular company

Cellular One
Gaia
2618 Avenue C
Scottsbluff, NE 69361

Corporate office 308-630-2355

Sending calls

Visitor dialing instructions
Local calls . . . . 7 digits
Long distance . . . 1 + area code + 7 digits

Speed-dial numbers
Customer service 611
Emergency 911

Receiving calls

Caller dials the roamer access number below,
hears tone, dials cellular area code and number.
. . . . 308-637-7626

Billing information

Visitor rates
Most visitors . . . . . $3 day + 85¢ minute

Visitors without roaming agreements (p. 1159)
Call co. to use Mastercard, Visa

Identification numbers

City	Market	System	Billing	RSA
All served cities	533	01453	00000	NE-1

Cellular company

Nebraska Cellular
1431 N Webb Road
Grand Island, NE 68803

Corporate office 800-578-1054
. . . . 308-384-9000

Sending calls

Visitor dialing instructions
Local calls . . . . . . area code + 7 digits
Long distance . . . . . area code + 7 digits

Speed-dial numbers
Customer service . . . . ✳611
Emergency 911
State patrol . . . . ✳55

Receiving calls

Caller dials the roamer access number below,
hears tone, dials cellular area code and number.
. . . . 308-380-7626

Billing information

Visitor rates
Most visitors . . . . . $3 day + 90¢ minute

Visitors without roaming agreements (p. 1159)
Cannot call since company takes no credit cards.

Identification numbers

City	Market	System	Billing	RSA
All served cities	533	01466	00000	NE-1

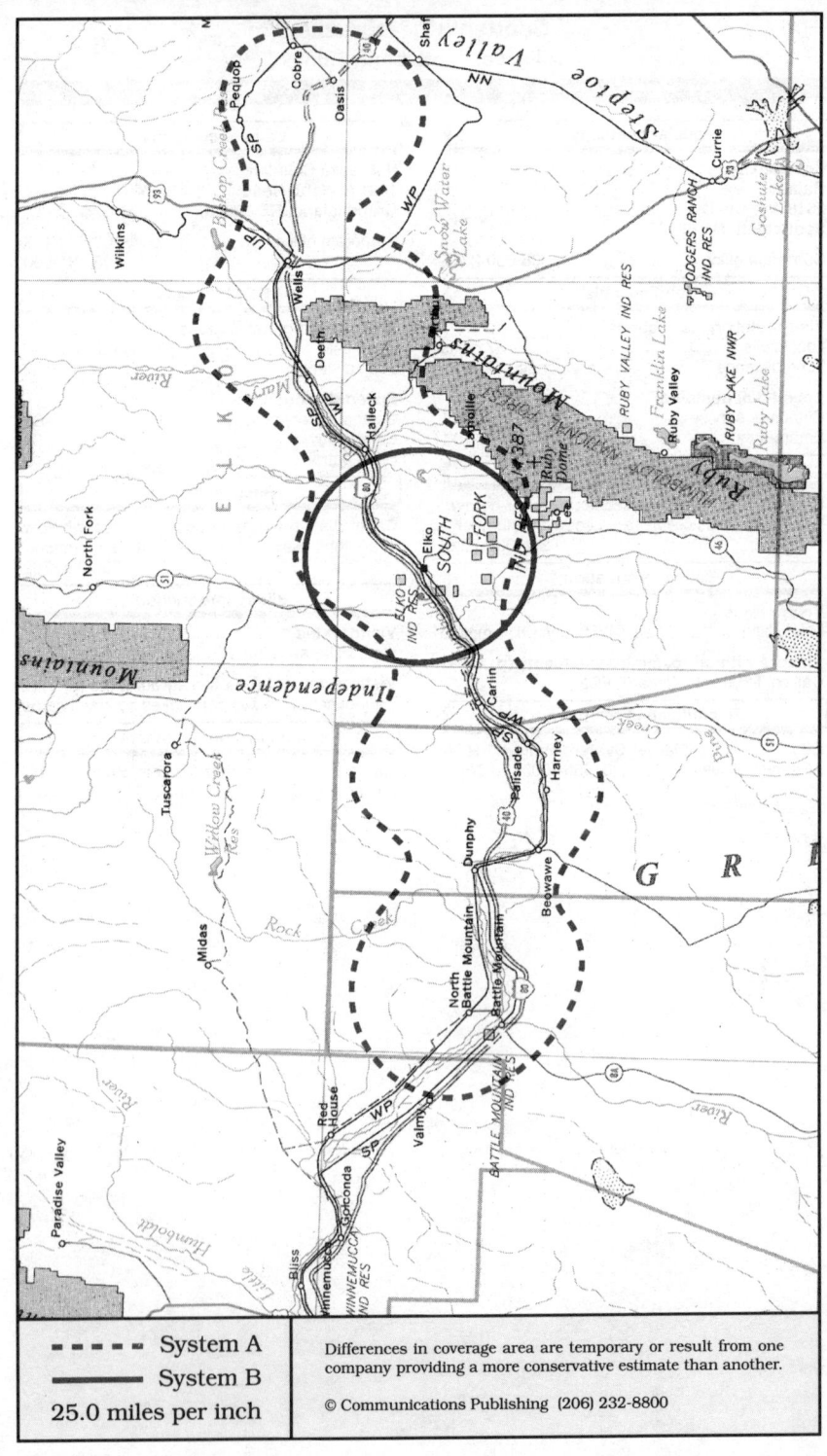

System A
System B
25.0 miles per inch

Differences in coverage area are temporary or result from one company providing a more conservative estimate than another.

© Communications Publishing (206) 232-8800

Elko, Nevada

System A

Cellular company

Cellular One
GenCell Management
2572 Idaho Street, Suite A-3
Elko, NV 89801

Customer service	800-888-7868
Local office	702-738-7290
Corporate office	707-425-8000

Sending calls

Visitor dialing instructions

Local calls	area code + 7 digits
Long distance . . .	1 + area code + 7 digits

Speed-dial numbers

Customer service	611
Emergency	911

Receiving calls

Caller dials the roamer access number below,
hears tone, dials cellular area code and number.
. . . . 702-753-2676

NationLink & RoamingAmerica (see p. 1157)
Caller hears your current roamer access number.
This company's subscribers have level 0 service.

Billing information

Visitor rates
Most visitors . . . . . $3 day + 99¢ minute

Visitors without roaming agreements (p. 1159)
American Roaming Network intercepts call to bill.

Identification numbers

City	Market	System	Billing	RSA
All served cities	544	01475	00000	NV-2

System B

Cellular company

PacTel Cellular
4016 Kietzke
Reno, NV 89502

Customer service	800-722-8358
Local office	702-829-1800
Corporate office	510-210-3600

Sending calls

Visitor dialing instructions

Local calls	area code + 7 digits
Long distance	area code + 7 digits

Speed-dial numbers

Customer service	*611
Emergency	911

Receiving calls

Caller dials the roamer access number below,
hears tone, dials cellular area code and number.
. . . . 702-753-0626

Auto-Access forwarding (see page 1156)
Activate, dial *28 send. Deactivate dial *29 send.

Follow Me Roaming forwarding (see p. 1156)
Activate, dial *18 send. Deactivate dial *19 send.

Billing information

Visitor rates

7 am - 7 pm Mon - Fri .	$0 day + 55¢ minute
Other times	$0 day + 27¢ minute

Visitors without roaming agreements (p. 1159)
Call co. to use Amex, Mastercard, Visa

Identification numbers

City	Market	System	Billing	RSA
All served cities	544	00498	00000	NV-2

This system is licensed to Alltel Mobile
Communications at 501-661-8500 and operated
by PacTel Cellular.

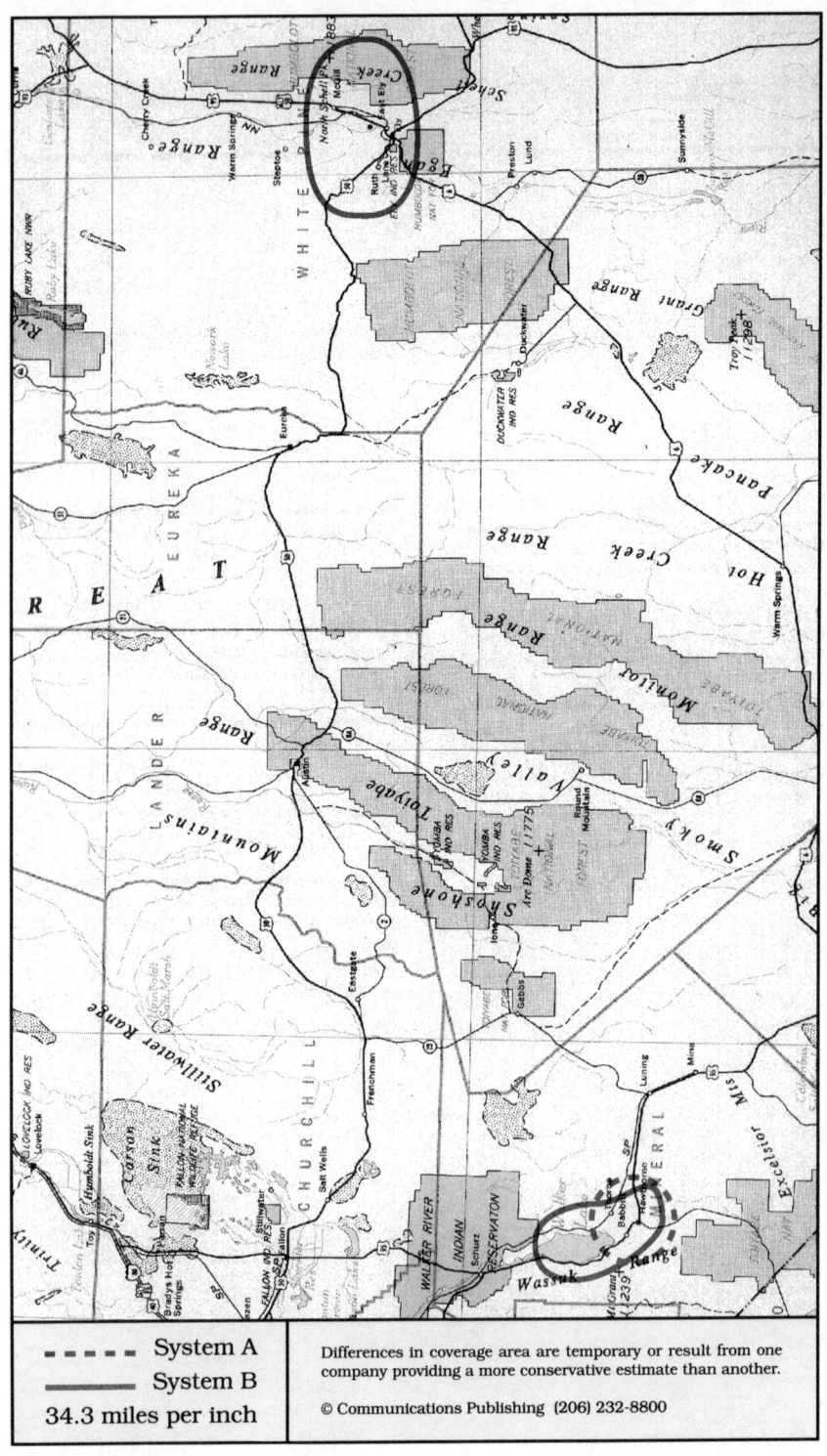

- - - - - System A	Differences in coverage area are temporary or result from one
———— System B	company providing a more conservative estimate than another.
34.3 miles per inch	© Communications Publishing (206) 232-8800

Ely, Nevada
Ely, Hawthorne

Cellular company

Cellular One
GenCell Management
2572 Idaho Street, Suite A-3
Elko, NV 89801

Customer service	800-888-7868
Local office	702-738-7290
Corporate office	707-425-8000

Sending calls

Visitor dialing instructions
Local calls	area code + 7 digits
Long distance . . .	1 + area code + 7 digits

Speed-dial numbers
Customer service	611
Emergency	911

Receiving calls

Caller dials the roamer access number below, hears tone, dials cellular area code and number.
. . . . Not available

Billing information

Visitor rates
Most visitors . . . . . $3 day + 99¢ minute

Visitors without roaming agreements (p. 1159)
American Roaming Network intercepts call to bill.

Identification numbers

City	Market	System	Billing	RSA
All served cities	546	01479	00000	NV-4

Cellular company

PacTel Cellular
4016 Kietzke Lane
Reno, NV 89502

Customer service	800-722-8358
Local office	702-829-1800
Corporate office	510-210-3600

Sending calls

Visitor dialing instructions
Local calls	area code + 7 digits
Long distance	area code + 7 digits

Speed-dial numbers
Customer service	*611
Emergency	911

Receiving calls

Caller dials the roamer access number below, hears tone, dials cellular area code and number.
. . . . 702-741-7626

Auto-Access forwarding (see page 1156)
Activate, dial *28 send. Deactivate dial *29 send.

Follow Me Roaming forwarding (see p. 1156)
Activate, dial *18 send. Deactivate dial *19 send.

Billing information

Visitor rates
7 am - 7 pm Mon - Fri .	$0 day + 55¢ minute
Other times	$0 day + 27¢ minute

Visitors without roaming agreements (p. 1159)
Call co. to use Amex, Mastercard, Visa

Identification numbers

City	Market	System	Billing	RSA
Ely	547	00498	30186	NV-5
Hawthorne	546	00498	30188	NV-4

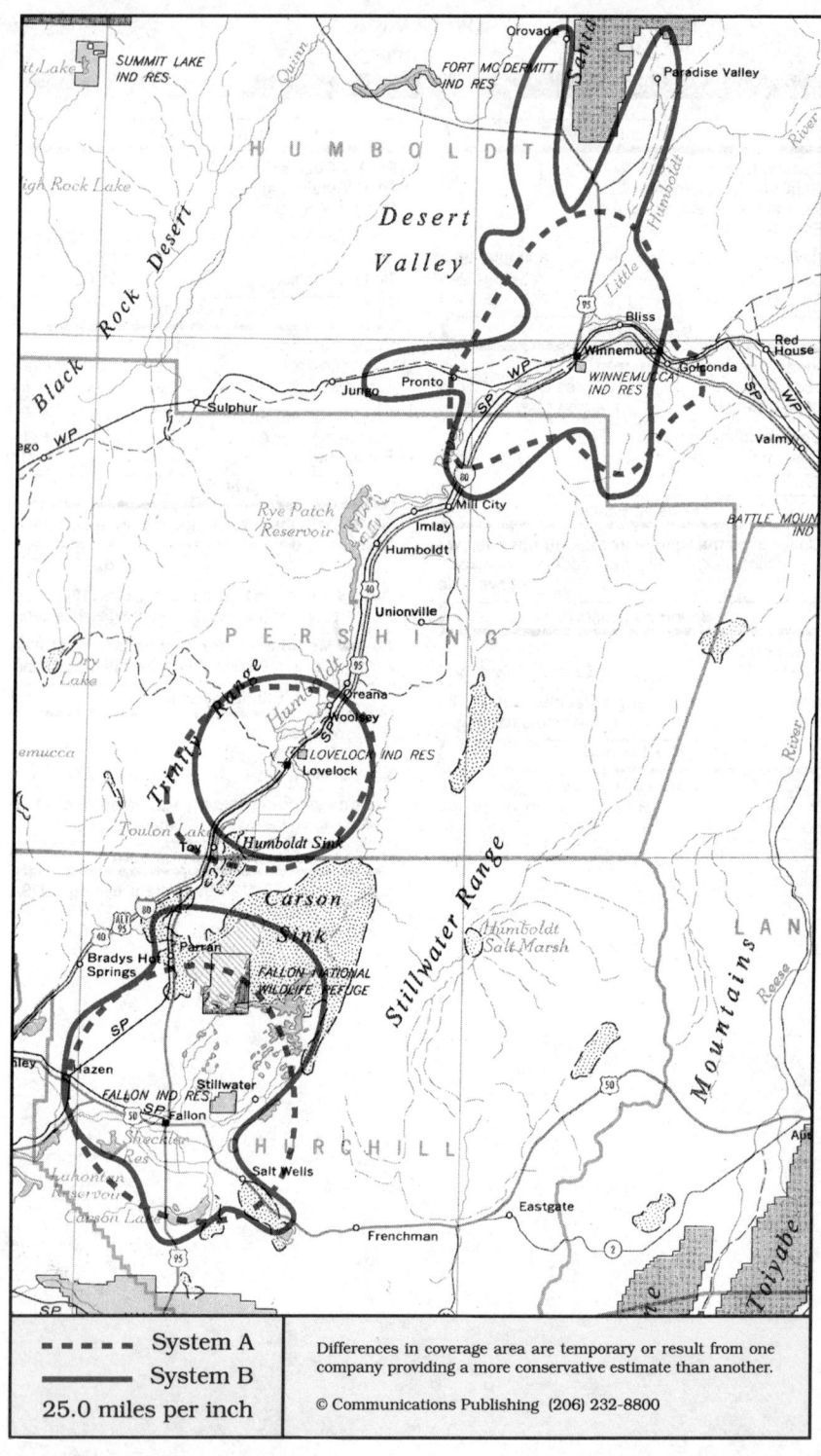

- - - - System A	Differences in coverage area are temporary or result from one company providing a more conservative estimate than another.
——— System B	
25.0 miles per inch	© Communications Publishing (206) 232-8800

Fallon, Nevada
Fallon, Lovelock, Winnemucca

System A

Cellular company

Cellular One
GenCell Management
2572 Idaho Street, Suite A-3
Elko, NV 89801

Customer service	800-888-7868
Local office	702-738-7290
Corporate office	707-425-8000

Sending calls

Visitor dialing instructions

Local calls	area code + 7 digits
Long distance . . .	1 + area code + 7 digits

Speed-dial numbers

Customer service	611
Emergency	911

Receiving calls

Caller dials the roamer access number below, hears tone, dials cellular area code and number.
. . . . 702-623-8526

NationLink & RoamingAmerica (see p. 1157)
Caller hears your current roamer access number. This company's subscribers have level 0 service.

Billing information

Visitor rates
Most visitors . . . . . $3 day + 99¢ minute

Visitors without roaming agreements (p. 1159)
American Roaming Network intercepts call to bill.

Identification numbers

City	Market	System	Billing	RSA
All served cities	543	01473	00000	NV-1

System B

Cellular company

C. C. Cellular
50 W Williams Avenue, PO Box 1450
Fallon, NV 89407-1450

Corporate office	800-992-0811
. . . .	702-426-7000

Sending calls

Visitor dialing instructions

Local calls	area code + 7 digits
Long distance	area code + 7 digits

Speed-dial numbers

Cellular information for California . . .	*INFO
Emergency	911
PacTel Cellular customer service . . .	*611
Seasonal highway reports	*ROAD or *7623

Receiving calls

Caller dials the roamer access number below, hears tone, dials cellular area code and number.
. . . . 702-427-7626

Auto-Access forwarding (see page 1156)
Activate, dial *28 send. Deactivate dial *29 send.

Billing information

Visitor rates

7 am - 7 pm Mon - Fri .	$0 day + 55¢ minute
Other times	$0 day + 27¢ minute

Visitors without roaming agreements (p. 1159)
Call co. to use Mastercard

Identification numbers

City	Market	System	Billing	RSA
All served cities	543	00498	30192	NV-1

This system is operated as part of the Reno, NV system.

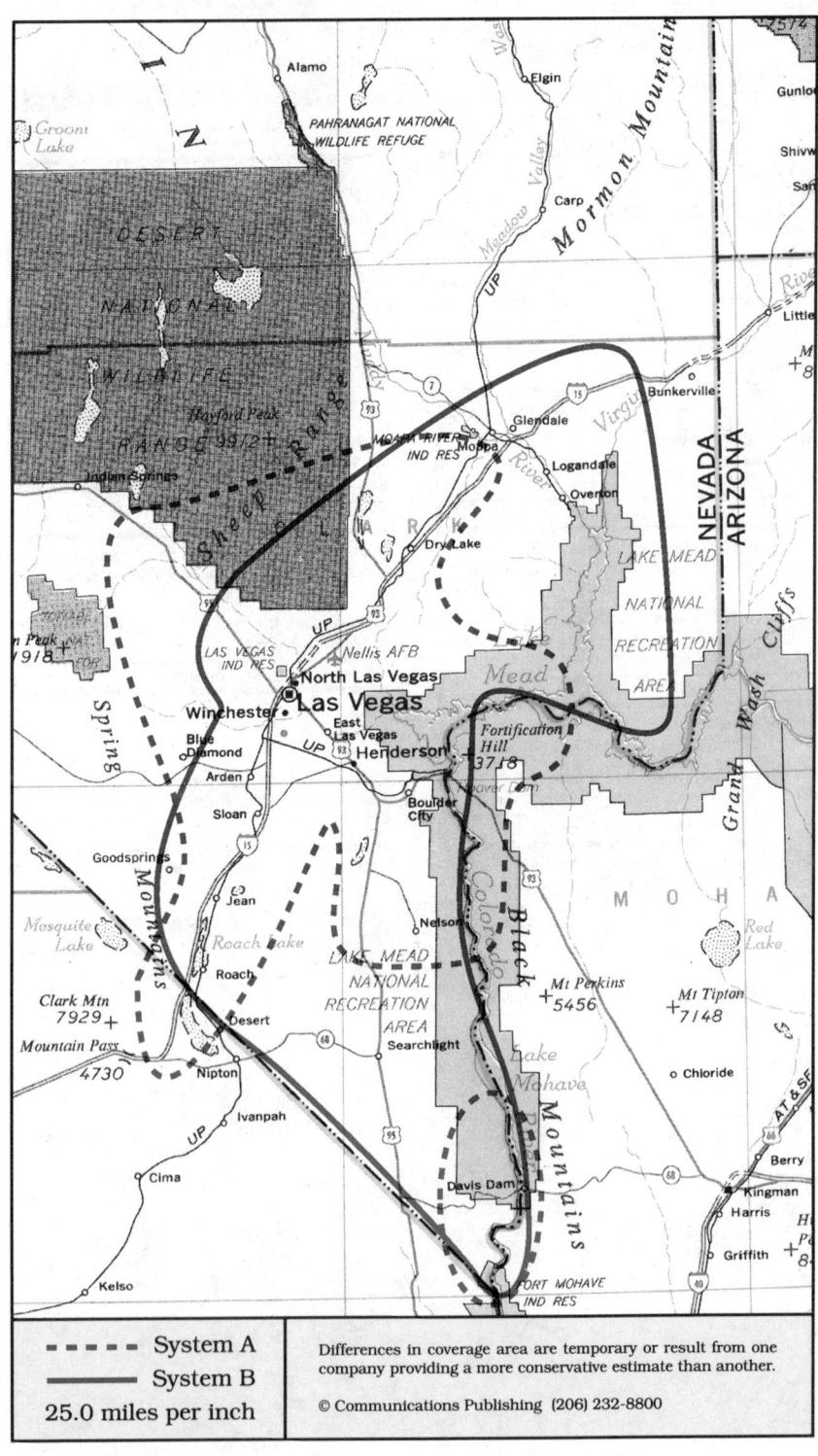

- - - - System A	Differences in coverage area are temporary or result from one company providing a more conservative estimate than another.
——— System B	
25.0 miles per inch	© Communications Publishing (206) 232-8800

Las Vegas, Nevada

Boulder City, Henderson, Las Vegas, Laughlin

System A	System B

Cellular company

Cellular One
McCaw Cellular Communications
3763 Howard Hughes Parkway #200
Las Vegas, NV 89109-0939

Customer service	702-732-2240
Local office	702-734-1010
Corporate office	206-827-4500

Sending calls

Visitor dialing instructions

Local calls	area code + 7 digits
Long distance . . .	1 + area code + 7 digits

Speed-dial numbers

Customer service	611
Emergency	911
KDWN radio	720
KKEY	931
KLAS TV-8	888
KLAV AM radio	1230
KNEW	970
KOMP radio	923
KRLB radio	575
KRLB radio	963
KVBC TV-3	883
KWNR	955

Receiving calls

Caller dials the roamer access number below,
hears tone, dials cellular area code and number.
. . . . 702-595-7626

NationLink & RoamingAmerica (see p. 1157)
Dial *31 send to receive calls, *30 to deactivate.
Dial *32 send to tell caller the roamer access no.
This company's subscribers have level 6 service.

North American Cellular Network (see p. 1156)
All calls forward to you. Deactivate dial *35 send.

Billing information

Visitor rates

Most visitors	$0 day + 99¢ minute

Visitors without roaming agreements (p. 1159)
Cannot call since company takes no credit cards.

Identification numbers

City	Market	System	Billing	RSA
All served cities	093	00211	00000	MSA

Cellular company

Centel Cellular
4022 S Industrial Road
Las Vegas, NV 89103

Customer service	800-531-2582
. . . .	702-893-8130
Local office	702-893-8100
Corporate office	312-399-2644

Sending calls

Visitor dialing instructions

Local calls	area code + 7 digits
Long distance . . .	1 + area code + 7 digits

Speed-dial numbers

Customer service	*611
Drug enforcement hotline (free) . .	388-6635
Emergency	911
Metropolitan police	702-795-3111
Nevada highway patrol . . .	702-486-4100
KDWN radio	720
KJUL FM radio	104
KLAS TV-8	888
KLAV AM radio	1230
KNEW AM radio	970
KOMP FM radio	923
KVBC TV-3	883
National park service (free) . . .	293-8932
Report drunk driving	*DUI or *384
Secret witness (free)	386-3213

Receiving calls

Caller dials the roamer access number below,
hears tone, dials cellular area code and number.
. . . . 702-379-7626

Auto-Access forwarding (see page 1156)
Activate, dial *28 send. Deactivate dial *29 send.

Follow Me Roaming forwarding (see p. 1156)
Activate, dial *18 send. Deactivate dial *19 send.

Billing information

Visitor rates

Most visitors	$3 day + 75¢ minute

Visitors without roaming agreements (p. 1159)
Cannot call since company takes no credit cards.

Identification numbers

City	Market	System	Billing	RSA
All served cities	093	00064	00000	MSA

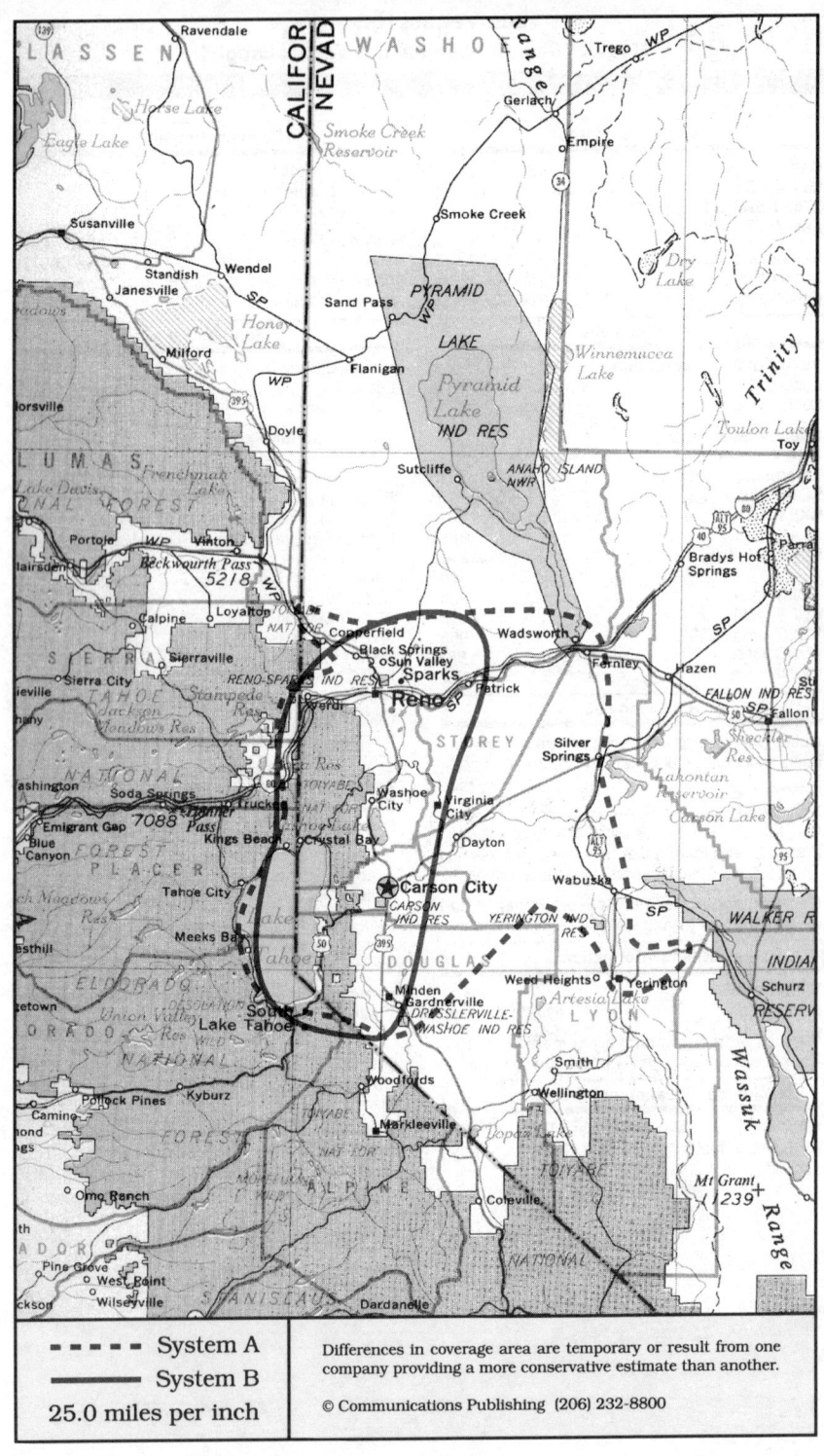

▪ ▪ ▪ ▪ System A	Differences in coverage area are temporary or result from one company providing a more conservative estimate than another.
——— System B	
25.0 miles per inch	© Communications Publishing (206) 232-8800

Reno, Nevada
Carson City, Gardnerville, Reno, Yerington

System A	System B

Cellular company

Cellular One
McCaw Cellular Communications
200 S Rock Boulevard
Reno, NV 89502

Customer service	702-858-5010
Local office	702-858-5000
Corporate office	206-827-4500

Sending calls

Visitor dialing instructions

Local calls	area code + 7 digits
Long distance . . .	1 + area code + 7 digits

Speed-dial numbers

Customer service	611
Directory assistance	411
Emergency	911

Receiving calls

Caller dials the roamer access number below, hears tone, dials cellular area code and number.
. . . . 702-742-7626

NationLink & RoamingAmerica (see p. 1157)
Dial *31 send to receive calls, *30 to deactivate. Dial *32 send to tell caller the roamer access no. This company's subscribers have level 6 service.

North American Cellular Network (see p. 1156)
All calls forward to you. Deactivate dial *35 send.

Billing information

Visitor rates
. . . . $0 day + 99¢ minute

Visitors without roaming agreements (p. 1159)
Cannot call since company takes no credit cards.

Identification numbers

City	Market	System	Billing	RSA
Carson City	545	00515	00000	NV-3
Reno	171	00515	00000	MSA

This company's subscribers pay 55¢ per minute and no daily fee in Las Vegas and Sacramento.

Cellular company

PacTel Cellular
4016 Kietzke
Reno, NV 89502

Customer service	800-722-8358
Local office	702-829-1800
Corporate office	510-210-3600

Sending calls

Visitor dialing instructions

Local calls	area code + 7 digits
Long distance	area code + 7 digits

Speed-dial numbers

Customer service	*611
Emergency	911

Receiving calls

Caller dials the roamer access number below, hears tone, dials cellular area code and number.
. . . . 702-741-7626

Auto-Access forwarding (see page 1156)
Activate, dial *28 send. Deactivate dial *29 send.

Follow Me Roaming forwarding (see p. 1156)
Activate, dial *18 send. Deactivate dial *19 send.

Billing information

Visitor rates

7 am - 7 pm Mon - Fri .	$0 day + 55¢ minute
Other times	$0 day + 27¢ minute

Visitors without roaming agreements (p. 1159)
Call co. to use Amex, Mastercard, Visa

Identification numbers

City	Market	System	Billing	RSA
Carson City	545	00498	30190	NV-3
Reno	171	00498	00000	MSA

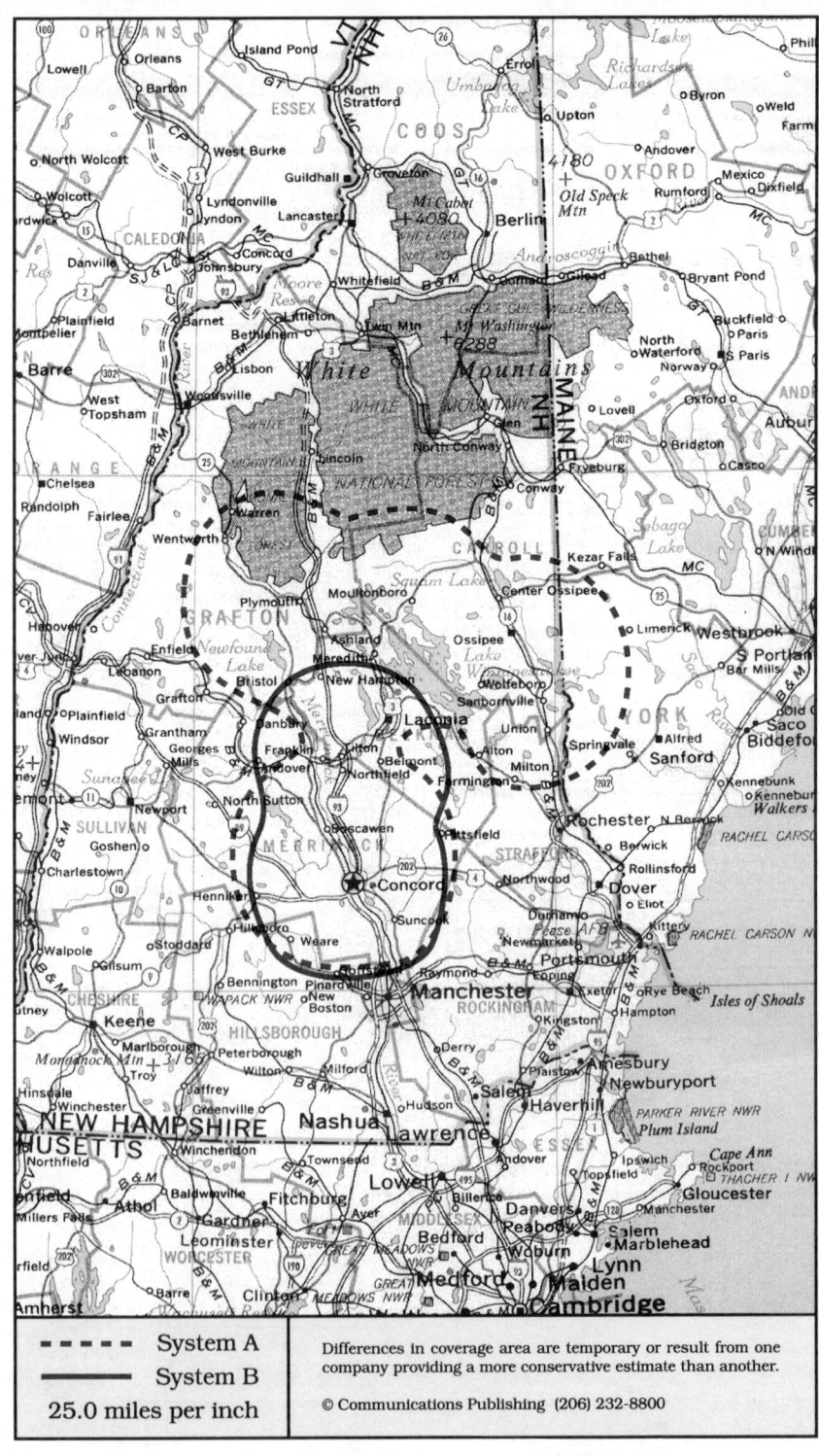

- - - - System A	Differences in coverage area are temporary or result from one
——— System B	company providing a more conservative estimate than another.
25.0 miles per inch	© Communications Publishing (206) 232-8800

Concord, New Hampshire
Concord, Laconia, Ossipee

|

Cellular company

Cellular One
Southwestern Bell Mobile Systems
100 Lowder Brook Drive
Westwood, MA 02090

Customer service	800-235-5663
. . . .	617-462-7000
Local office	617-462-4000
Corporate office	214-733-2000

Sending calls

Visitor dialing instructions
Local calls . . . . . area code + 7 digits
Long distance . . . 1 + area code + 7 digits

Speed-dial numbers
Customer service *611
State police *77

Receiving calls

Caller dials the roamer access number below,
hears tone, dials cellular area code and number.
. . . . 603-229-7626

Billing information

Visitor rates
Most visitors . . . . . $3 day + 99¢ minute

Visitors without roaming agreements (p. 1159)
Call co. to use Amex

Identification numbers

All served cities	549	01485	00000	NH-2

Cellular company

Contel Cellular
117 Manchester Street
Concord, NH 03301

Customer service	800-333-4004
. . . . . extension 4002	404-391-8000
Local office	603-224-1111
Corporate office	404-804-3400

Sending calls

Visitor dialing instructions
Local calls . . . . . . area code + 7 digits
Long distance . . . 1 + area code + 7 digits

Speed-dial numbers
Customer service 611
Emergency 911

Receiving calls

Caller dials the roamer access number below,
hears tone, dials cellular area code and number.
. . . . 603-496-7626

Follow Me Roaming forwarding (see p. 1156)
Activate, dial *18 send. Deactivate dial *19 send.

Billing information

Visitor rates
Most visitors . . . . . $3 day + 85¢ minute

Visitors without roaming agreements (p. 1159)
Call co. to use Amex, Mastercard, Visa

Identification numbers

City	Market	System	Billing	RSA
All served cities	549	01486	00000	NH-2

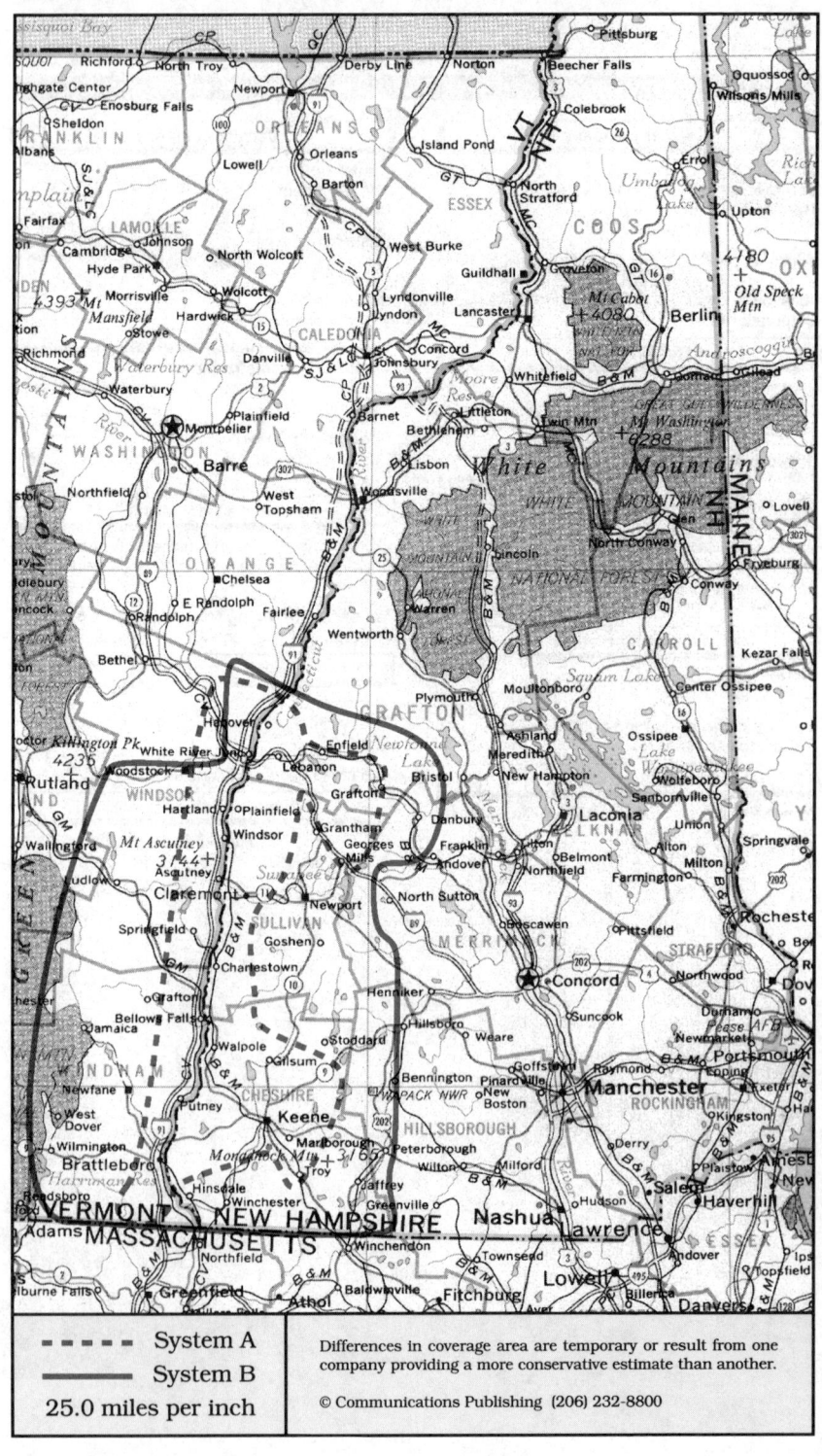

- - - - System A	Differences in coverage area are temporary or result from one company providing a more conservative estimate than another.
———— System B	
25.0 miles per inch	© Communications Publishing (206) 232-8800

Keene, New Hampshire
Claremont, Brattleboro VT, Keene

System A	System B

Cellular company

Cellular One
Atlantic Cellular
Airport Executive Plaza, Airport Road, Box 5
W Lebanon, NH 03784

Customer service	800-676-2355
. . . .	802-862-9000
Local office	603-298-9900
Corporate office	401-421-7090

Sending calls

Visitor dialing instructions

Local calls	area code + 7 digits
Long distance . . .	1 + area code + 7 digits

Speed-dial numbers

Customer service	*611
Emergency	911
Winter ski report	*SKI or *754

Receiving calls

Caller dials the roamer access number below,
hears tone, dials cellular area code and number.
. . . . 603-448-7626

NationLink & RoamingAmerica (see p. 1157)
Dial *31 send to receive calls, *30 to deactivate.
Dial *32 send to tell caller the roamer access no.
This company's subscribers have level 8 service.

Billing information

Visitor rates
Most visitors . . . . . $3 day + 99¢ minute

Visitors without roaming agreements (p. 1159)
Call co. to use Amex, Discover, Mastercard, Visa

Identification numbers

City	Market	System	Billing	RSA
Brattleboro	680	00313	30313	VT-2
Claremont, Keene	548	00313	30313	NH-1

Cellular company

United States Cellular
100 Main Street
West Lebanon, NH 03784

Customer service	800-234-8722
Local office	603-298-7111
Corporate office	312-399-8900

Sending calls

Visitor dialing instructions

Local calls	area code + 7 digits
Long distance . . .	1 + area code + 7 digits

Speed-dial numbers

Customer service	611
Emergency	911
Roaming information	*711

Receiving calls

Caller dials the roamer access number below,
hears tone, dials cellular area code and number.

Brattleboro	802-258-7626
Keene	603-358-0999
Lebanon	603-252-7626

Follow Me Roaming forwarding (see p. 1156)
Activate, dial *18 send. Deactivate dial *19 send.

Billing information

Visitor rates
Most visitors . . . . . $3 day + 85¢ minute

Visitors without roaming agreements (p. 1159)
Call co. to use Mastercard, Visa

Identification numbers

City	Market	System	Billing	RSA
Brattleboro	680	01484	00000	VT-2
Claremont, Keene	548	01484	00000	NH-1

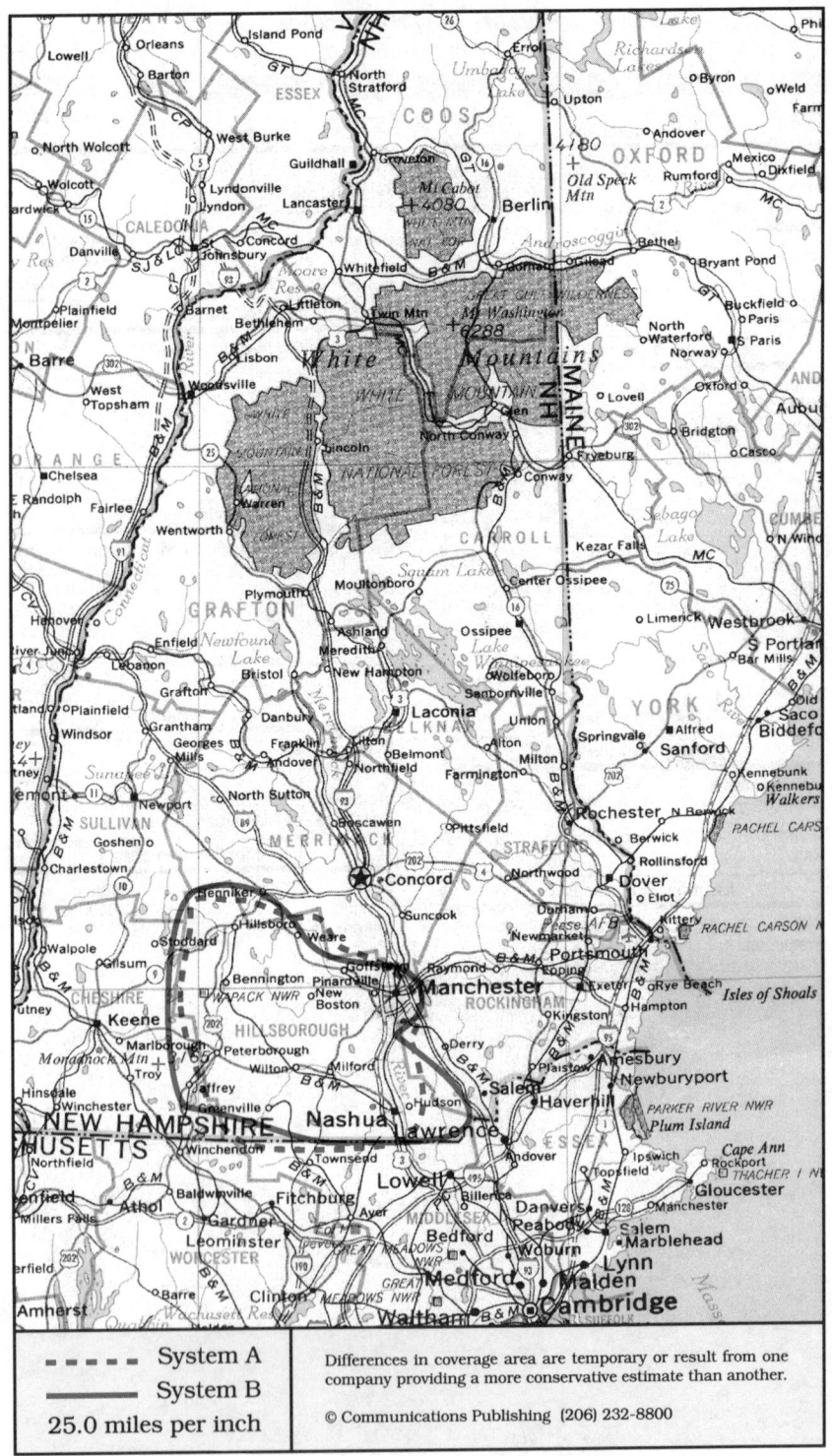

- - - - System A	Differences in coverage area are temporary or result from one company providing a more conservative estimate than another.
―――― System B	
25.0 miles per inch	© Communications Publishing (206) 232-8800

Manchester, New Hampshire
Manchester, Nashua

Cellular company

Cellular One
United States Cellular & JHP Partnership
288 Route 101
Bedford, NH 03110

Corporate office   some areas   800-662-3551
. . . .   603-471-8000

Sending calls

Visitor dialing instructions
Local calls   . . . . .   area code + 7 digits
Long distance   . . .   area code + 7 digits

Speed-dial numbers
Customer service . . . . ✶611
Directory assistance . . . . ✶411
Emergency (sheriff) 911
Medical information . . ✶NURSE or ✶68733
Mortgage interest rates . . ✶BANK or ✶2265
New Hampshire police . . . . . . ✶77
Roaming information . . . . ✶711
Time and temperature . . . . . . ✶8465
Towing by Mobile One . . ✶HELP or ✶4357
Travel & recreation . . . . ✶754
Weather (channel 50) . . . . . . ✶811
WGIR concert line . . . . ✶101
WOKQ radio . . . . ✶975
WZID traffic report . . . . ✶957

Receiving calls

Caller dials the roamer access number below,
hears tone, dials cellular area code and number.
. . . . 603-345-7626

NationLink & RoamingAmerica (see p. 1157)
Dial ✶31 send to receive calls, ✶30 to deactivate.
Dial ✶32 send to tell caller the roamer access no.
This company's subscribers have level 6 service.

Billing information

Visitor rates
Most visitors . . . . . $3 day + 85¢ minute

Visitors without roaming agreements (p. 1159)
Call co. to use Amex, Discover, Mastercard, Visa

Identification numbers

City	Market	System	Billing	RSA
All served cities	133	00445	00000	MSA

Cellular company

Contel Cellular
313 Lincoln Street
Manchester, NH 03103

Customer service   800-333-4004
. . . . . . extension 4002   404-391-8000
Local office   603-647-6916
Corporate office   404-804-3400

Sending calls

Visitor dialing instructions
Local calls . . . . . .   area code + 7 digits
Long distance . . .   1 + area code + 7 digits

Speed-dial numbers
AAA Manchester . . . . . . ✶AAA or ✶222
AAA Nashua . . . . ✶333
Automatic teller . . . . . ✶Y24 or ✶924
Customer service 611
Directory assistance . . . . . . ✶5555
Emergency 911
Films in Manchester . . . . . . ✶7777
Films in Nashua . . . . ✶8888
Hear a traffic report . . . . . . ✶9999
List of all speed dial numbers . . ✶123
Lottery winners . . . . ✶777
Manchester sales office 111
Stock quotations . . . . ✶3333
Time and temperature . . . . . ✶6666
Weather and sports . . . . . . ✶4444
WMUR television news . . . . . ✶999

Receiving calls

Caller dials the roamer access number below,
hears tone, dials cellular area code and number.
. . . . 603-493-7626

Follow Me Roaming forwarding (see p. 1156)
Activate, dial ✶18 send. Deactivate dial ✶19 send.

Billing information

Visitor rates
Most visitors . . . . . $3 day + 85¢ minute

Visitors without roaming agreements (p. 1159)
Call co. to use Amex, Mastercard, Visa

Identification numbers

City	Market	System	Billing	RSA
All served cities	133	00428	00000	MSA

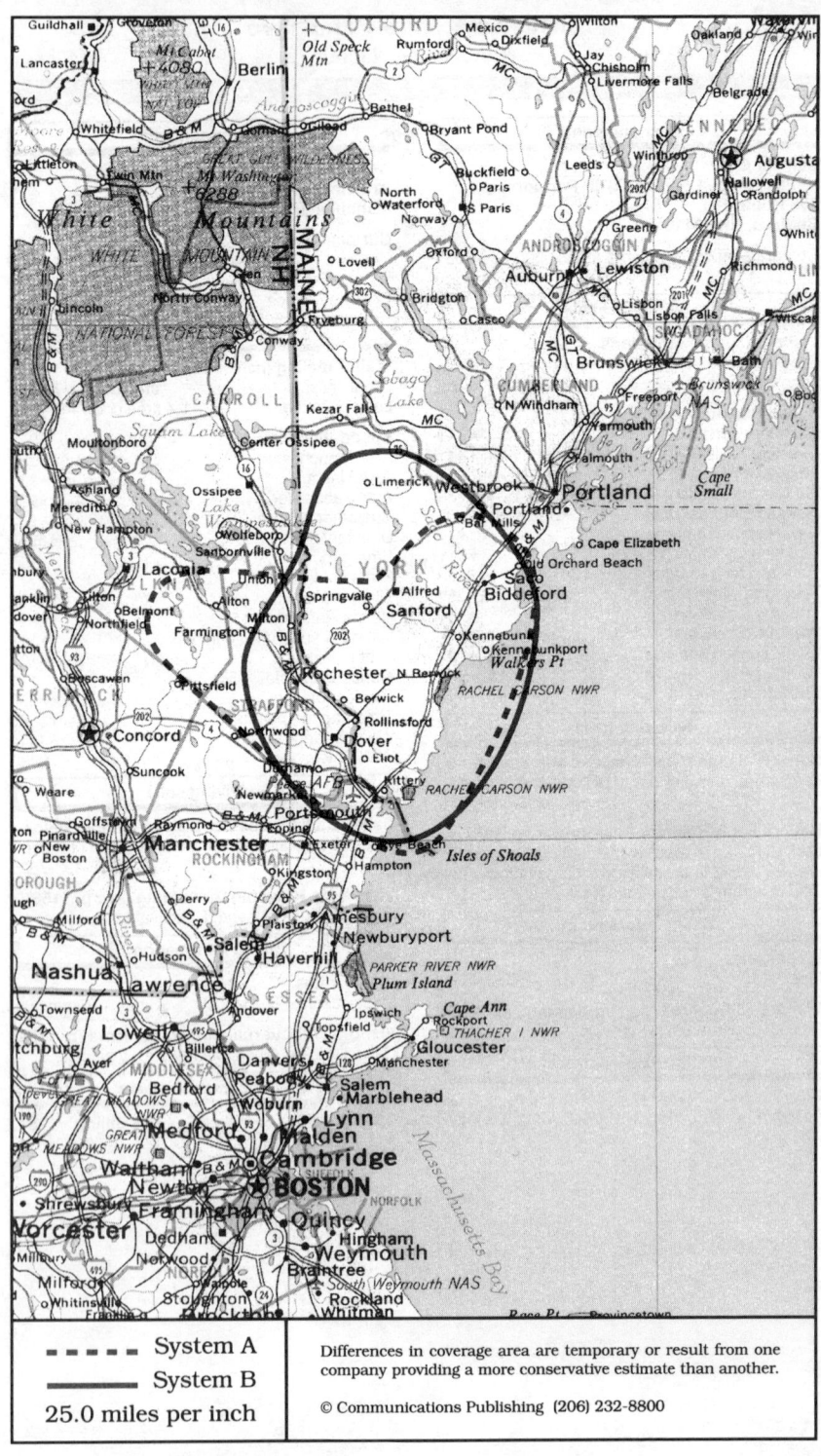

- - - - System A	Differences in coverage area are temporary or result from one company providing a more conservative estimate than another.
——— System B	
25.0 miles per inch	© Communications Publishing (206) 232-8800

Portsmouth, New Hampshire
Biddeford ME, Dover, Portsmouth

System A

Cellular company

Cellular One
Vanguard Cellular Systems
65 Main Street
Dover, NH 03820

Customer service	603-767-0900
Local office	603-742-6500
Corporate office	919-282-3690

Sending calls

Visitor dialing instructions

Local calls	area code + 7 digits
Long distance	area code + 7 digits

Speed-dial numbers

Customer service	611
Emergency	911
State police	*77

Receiving calls

Caller dials the roamer access number below,
hears tone, dials cellular area code and number.

Maine	207-284-3500
New Hampshire	603-743-7626

Billing information

Visitor rates

Most visitors	$3 day + 99¢ minute

Visitors without roaming agreements (p. 1159)
Call co. to use Amex, Mastercard, Visa

Identification numbers

City	Market	System	Billing	RSA
All served cities	156	00501	00000	MSA

System B

Cellular company

StarCellular
Saco River Cellular
323 N Street
Saco, ME 04072

Corporate office	800-346-9172
. . . .	207-283-8001

Sending calls

Visitor dialing instructions

Local calls	area code + 7 digits
Long distance	area code + 7 digits

Speed-dial numbers

AAA road service	*222
ATM locator	*924
Customer service	*611
Emergency	911
Maine police	*63
New Hampshire police	*64
News, stocks, weather . . .	*INFO or *4636
Time and weather	*321

Receiving calls

Caller dials the roamer access number below,
hears tone, dials cellular area code and number.

Maine	207-468-7626
New Hampshire	603-534-7626

Follow Me Roaming forwarding (see p. 1156)
Activate, dial *18 send. Deactivate dial *19 send.

Billing information

Visitor rates

Most visitors	$3 day + 95¢ minute

Visitors without roaming agreements (p. 1159)
Call co. to use Amex, Discover, Mastercard, Visa

Identification numbers

City	Market	System	Billing	RSA
All served cities	156	00484	00000	MSA

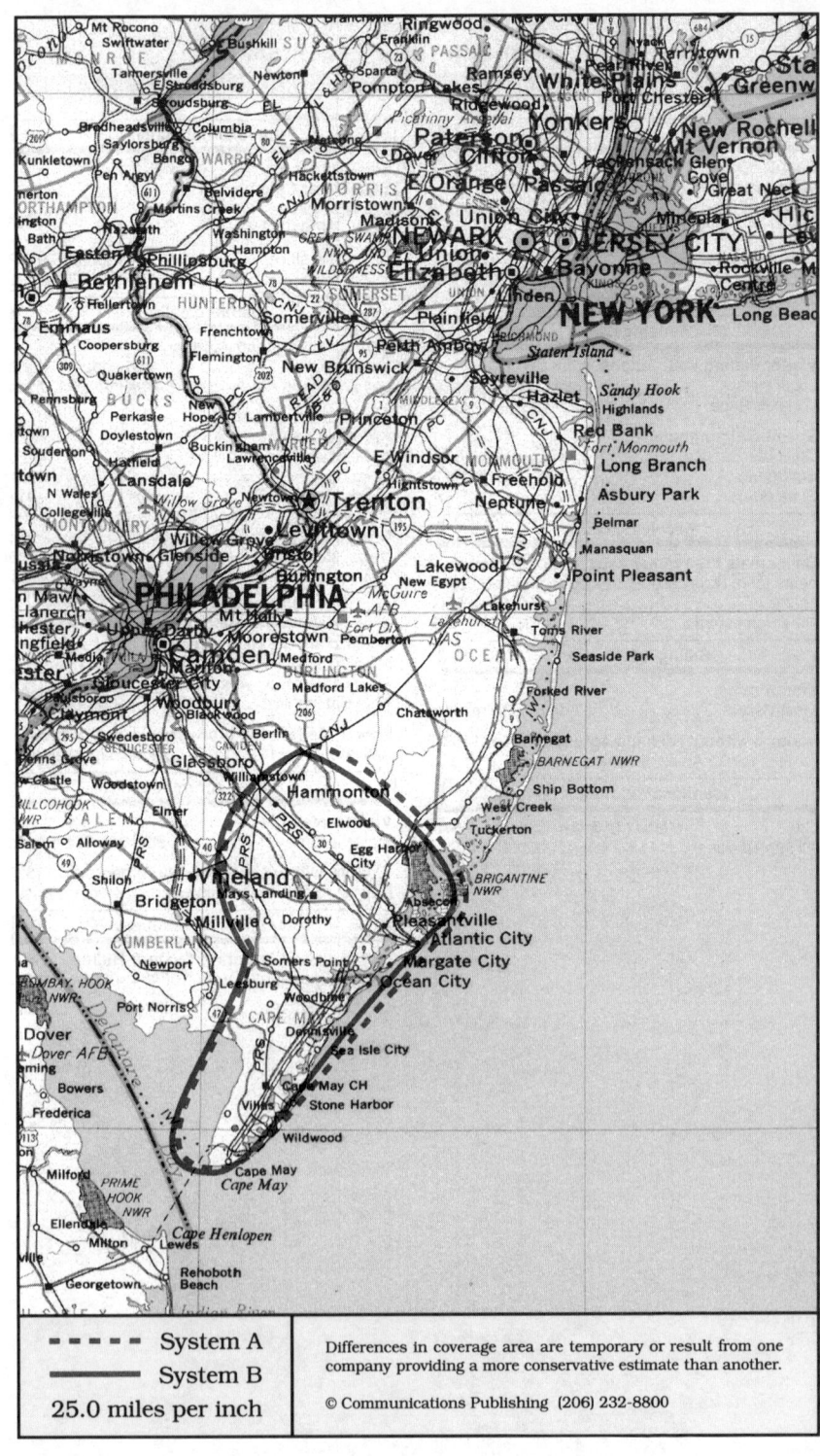

▪▪▪▪ System A	Differences in coverage area are temporary or result from one
────── System B	company providing a more conservative estimate than another.
25.0 miles per inch	© Communications Publishing (206) 232-8800

Atlantic City, New Jersey
Atlantic City, Cape May Court House, Hammonton

System A	System B

Cellular company

Comcast Cellular One
6727 Black Horse Pike
Pleasantville, NJ 08232

Customer service	some areas	800-233-4140
Local office		800-678-2351
. . . .		609-646-3700
Corporate office		215-975-5000

Sending calls

Visitor dialing instructions

Local calls	area code + 7 digits
Long distance	area code + 7 digits

Speed-dial numbers

Coast Guard (emergency) . . .	*CG or *24
Customer service	611
Emergency	911

Receiving calls

Caller dials the roamer access number below, hears tone, dials cellular area code and number.
. . . . 609-442-7626

NationLink & RoamingAmerica (see p. 1157)
Caller hears your current roamer access number. This company's subscribers have level 0 service.

North American Cellular Network (see p. 1156)
All calls forward to you. Deactivate dial *35 send.

Billing information

Visitor rates
Most visitors . . . . . $3 day + 99¢ minute

Visitors without roaming agreements (p. 1159)
Roamer Plus will intercept first call to bill you.

Identification numbers

City	Market	System	Billing	RSA
All served cities	134	00267	00000	MSA

Cellular company

Bell Atlantic Mobile
6814 Tilton Road, Unit 123-127
Pleasantville, NJ 08232

Customer service	800-922-0204
Local office	609-645-1155
Corporate office	908-306-7000

Sending calls

Visitor dialing instructions

Local calls	area code + 7 digits
Long distance	area code + 7 digits

Speed-dial numbers

Coast guard	*24
Emergency	911
Repair center	611

Receiving calls

Caller dials the roamer access number below, hears tone, dials cellular area code and number.
. . . . 609-226-7626

Follow Me Roaming forwarding (see p. 1156)
Activate, dial *18 send. Deactivate dial *19 send.

Billing information

Visitor rates
Most visitors . . . . . $3 day + $1 minute

Visitors without roaming agreements (p. 1159)
Roamer Plus will intercept first call to bill you.

Identification numbers

City	Market	System	Billing	RSA
All served cities	134	00250	00000	MSA

For full coverage area, see Tom's River, NJ.

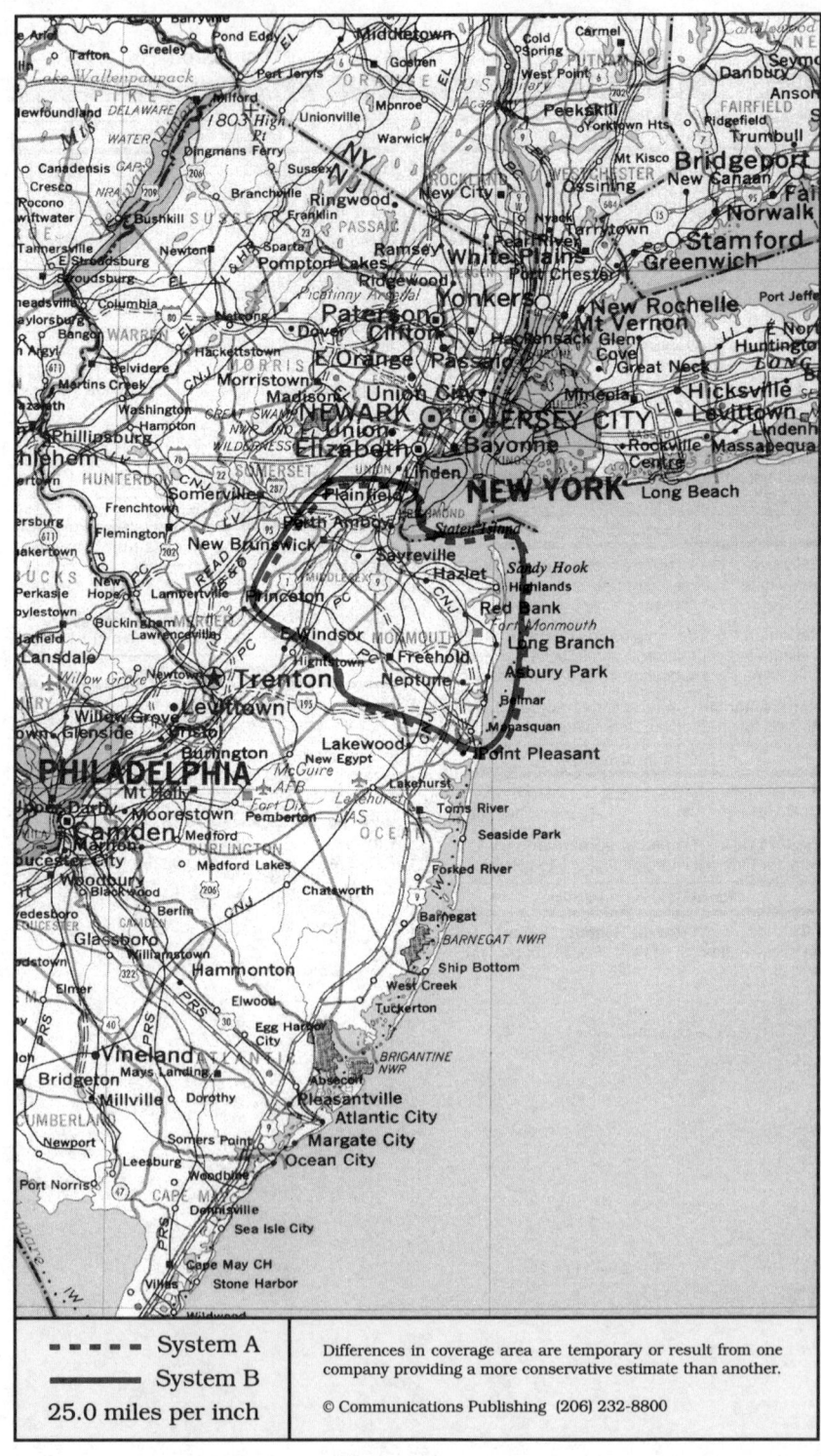

▪ ▪ ▪ ▪ ▪ System A	Differences in coverage area are temporary or result from one company providing a more conservative estimate than another.
——— System B	
25.0 miles per inch	© Communications Publishing (206) 232-8800

New Brunswick, New Jersey
Freehold, Long Branch, New Brunswick

System A	System B

Cellular company

Comcast Cellular One
485 Hwy 1 S, Bdg E, Metropolitan Corp. Plaza
Iselin, NJ 08830

Customer service	800-227-9222
. . . .	908-855-0200
Local office	908-855-0941
Corporate office	215-975-5000

Sending calls

Visitor dialing instructions

Local calls	area code + 7 digits
Long distance . . .	0 + area code + 7 digits

Speed-dial numbers

Coast Guard (emergency) . . .	∗CG or ∗24
Customer service	611
Emergency	911

Receiving calls

Caller dials the roamer access number below,
hears tone, dials cellular area code and number.

. . . . 908-610-7626

NationLink & RoamingAmerica (see p. 1157)
Caller hears your current roamer access number.
This company's subscribers have level 0 service.

North American Cellular Network (see p. 1156)
All calls forward to you. Deactivate dial ∗35 send.

Billing information

Visitor rates
Most visitors . . . . . $3 day + 99¢ minute

Visitors without roaming agreements (p. 1159)
Roamer Plus will intercept first call to bill you.

Identification numbers

City	Market	System	Billing	RSA
Long Branch	070	00173	00000	MSA
New Brunswick	062	00173	00000	MSA

For full coverage area, see Trenton, NJ.

Cellular company

NYNEX Mobile Communications
655 Route 1 & Wooding Avenue
Edison, NJ 08817

Customer service	800-227-1069
. . . .	914-365-7200
Local office	908-572-2525
Corporate office	914-365-7200

Sending calls

Visitor dialing instructions

Local calls	area code + 7 digits
Long distance	area code + 7 digits

Speed-dial numbers

Customer service	611
Emergency	911

Receiving calls

Caller dials the roamer access number below,
hears tone, dials cellular area code and number.

. . . . 201-259-7626

Billing information

Visitor rates
Most visitors . . . . . $3 day + 95¢ minute

Visitors without roaming agreements (p. 1159)
Call co. to use Amex, Mastercard, Visa
or Roamer Plus will intercept first call to bill you.

Identification numbers

City	Market	System	Billing	RSA
Long Branch	070	00022	00000	MSA
New Brunswick	062	00022	00000	MSA

For full coverage area, see New York, NY.

- - - - - System A	Differences in coverage area are temporary or result from one
————— System B	company providing a more conservative estimate than another.
25.0 miles per inch	© Communications Publishing (206) 232-8800

Newton, New Jersey
Newton (Sussex County)

System A	System B

Cellular company

Sussex Cellular
4 Union Place
Newton, NJ 07860

Corporate office 201-380-9000

Sending calls

Visitor dialing instructions
Local calls . . . . . area code + 7 digits
Long distance . . . 1 + area code + 7 digits

Speed-dial numbers
Customer service 611

Receiving calls

Caller dials the roamer access number below, hears tone, dials cellular area code and number.
. . . . 201-380-7626

Billing information

Visitor rates
Most visitors . . . . . $3 day + 99¢ minute

Visitors without roaming agreements (p. 1159)
Roamer Plus will intercept first call to bill you.

Identification numbers

City	Market	System	Billing	RSA
All served cities	552	01491	00000	NJ-3

Cellular company

Bell Atlantic Mobile
185-D Route 17 S
Paramus, NJ 07652

Customer service 800-922-0204
Local office 201-967-2355
Corporate office 908-306-7000

Sending calls

Visitor dialing instructions
Local calls . . . . . . area code + 7 digits
Long distance . . . 1 + area code + 7 digits

Speed-dial numbers
Customer service 611

Receiving calls

Caller dials the roamer access number below, hears tone, dials cellular area code and number.
. . . . 215-870-7626

Billing information

Visitor rates
Most visitors . . . . . $3 day + $1 minute

Visitors without roaming agreements (p. 1159)
Roamer Plus will intercept first call to bill you.

Identification numbers

City	Market	System	Billing	RSA
All served cities	552	00022	30086	NJ-3

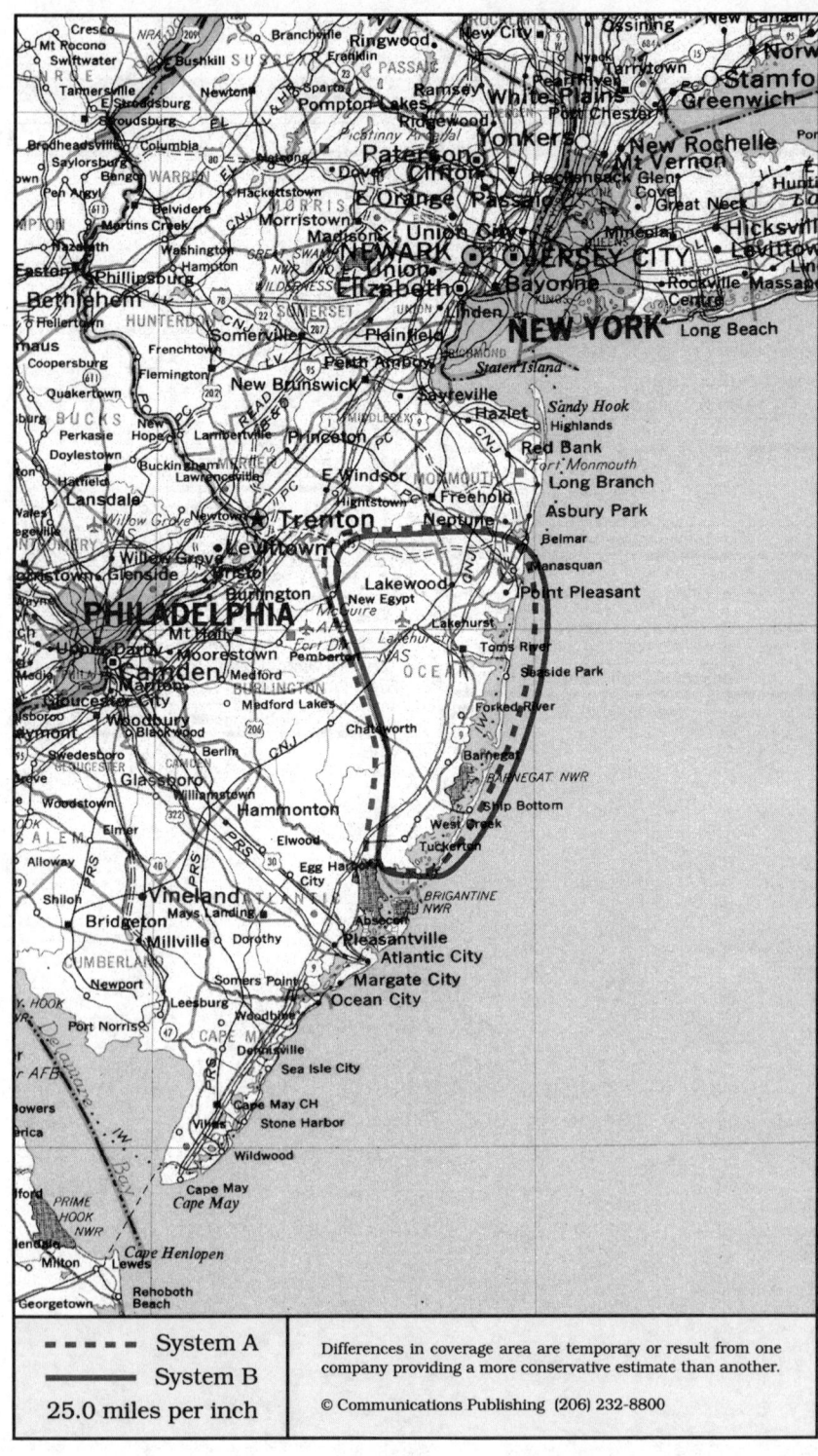

Tom's River, New Jersey
Tom's River (Ocean County)

Cellular company

Ocean County Cellular
Midland Communications
156 Route 37 E
Tom's River, NJ 08753

Corporate office	800-360-8255
. . . .	908-341-0500

Sending calls

Visitor dialing instructions

Local calls	area code + 7 digits
Long distance . . .	1 + area code + 7 digits

Speed-dial numbers

Customer service	611
Emergency	911

Receiving calls

Caller dials the roamer access number below,
hears tone, dials cellular area code and number.

. . . .	609-290-7626
. . . .	908-600-7626

NationLink & RoamingAmerica (see p. 1157)
Caller hears your current roamer access number.
This company's subscribers have level 0 service.

Billing information

Visitor rates

Most visitors	$3 day + 99¢ minute

Visitors without roaming agreements (p. 1159)
Roamer Plus will intercept first call to bill you.

Identification numbers

City	Market	System	Billing	RSA
All served cities	551	01489	00000	NJ-2

Cellular company

Bell Atlantic Mobile
6814 Tilton Road, Unit 123-127
Pleasantville, NJ 08232

Customer service	800-922-0204
Local office	609-645-1155
Corporate office	908-306-7000

Sending calls

Visitor dialing instructions

Local calls	area code + 7 digits
Long distance	area code + 7 digits

Speed-dial numbers

Coast guard	*24
Emergency	911
Repair center	611

Receiving calls

Caller dials the roamer access number below,
hears tone, dials cellular area code and number.

. . . .	908-910-7626

Follow Me Roaming forwarding (see p. 1156)
Activate, dial *18 send. Deactivate dial *19 send.

Billing information

Visitor rates

Most visitors	$3 day + $1 minute

Visitors without roaming agreements (p. 1159)
Roamer Plus will intercept first call to bill you.

Identification numbers

City	Market	System	Billing	RSA
All served cities	551	00250	30082	NJ-2

For full coverage area, see Atlantic City, NJ.

Trenton, New Jersey
Flemington (Hunterdon County), Trenton (Mercer County)

System A	System B

Cellular company

Comcast Cellular One
485 Hwy 1 S, Bdg E, Metropolitan Corp. Plaza
Iselin, NJ 08830

Customer service	800-227-9222
. . . .	908-855-0200
Local office	908-855-0941
Corporate office	215-975-5000

Sending calls

Visitor dialing instructions

Local calls	area code + 7 digits
Long distance . . .	0 + area code + 7 digits

Speed-dial numbers

Coast Guard (emergency) . . .	✶CG or ✶24
Customer service	611
Emergency	911

Receiving calls

Caller dials the roamer access number below, hears tone, dials cellular area code and number.
. . . . 908-610-7626

NationLink & RoamingAmerica (see p. 1157)
Caller hears your current roamer access number. This company's subscribers have level 0 service.

North American Cellular Network (see p. 1156)
All calls forward to you. Deactivate dial ✶35 send.

Billing information

Visitor rates

Most visitors	$3 day + 99¢ minute

Visitors without roaming agreements (p. 1159)
Roamer Plus will intercept first call to bill you.

Identification numbers

City	Market	System	Billing	RSA
Flemington	550	01487	00000	NJ-1
Trenton	121	00575	00000	MSA

For full coverage area, see New Brunswick, NJ.

Cellular company

Bell Atlantic Mobile
2990 Brunswick Pike, Route 1 N
Lawrenceville, NJ 08648

Customer service	800-922-0204
Local office	609-896-2355
Corporate office	908-306-7000

Sending calls

Visitor dialing instructions

Local calls	area code + 7 digits
Long distance . . .	1 + area code + 7 digits

Speed-dial numbers

Customer service	✶611
Emergency	911

Receiving calls

Caller dials the roamer access number below, hears tone, dials cellular area code and number.

Flemington (Hunterdon County)	908-310-7626
Trenton	609-658-7626

Follow Me Roaming forwarding (see p. 1156)
Activate, dial ✶18 send. Deactivate dial ✶19 send.

Billing information

Visitor rates

Most visitors	$3 day + $1 minute

Visitors without roaming agreements (p. 1159)
Roamer Plus will intercept first call to bill you.

Identification numbers

City	Market	System	Billing	RSA
Flemington	550	00250	30084	NJ-1
Trenton	121	00008	30362	MSA

For full coverage area, see Philadelphia, PA.

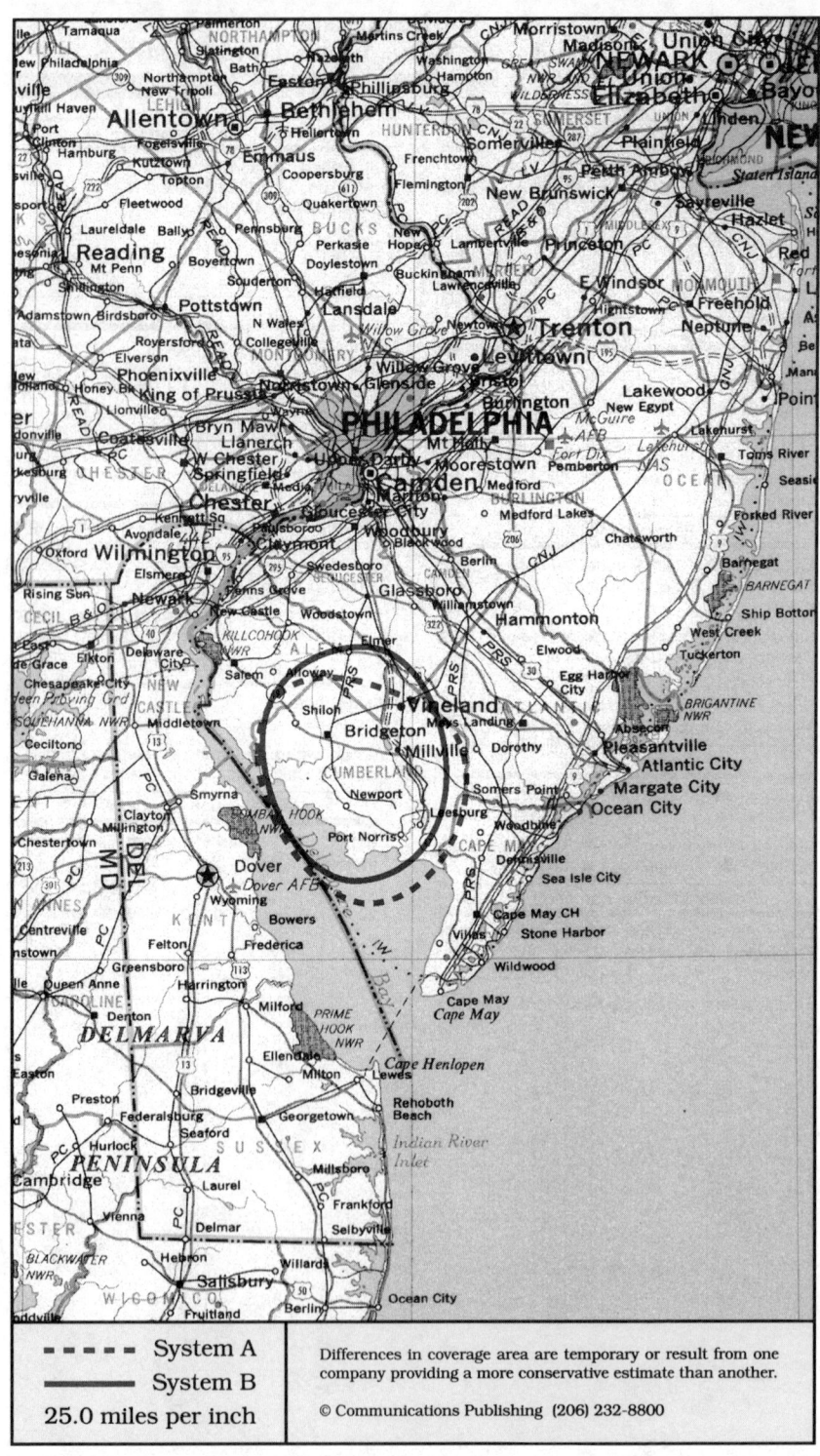

- - - - - System A	Differences in coverage area are temporary or result from one company providing a more conservative estimate than another.
───── System B	© Communications Publishing (206) 232-8800
25.0 miles per inch	

Vineland, New Jersey
Bridgeton, Vineland (Cumberland County)

System A	System B

Cellular company

United States Cellular
65 Fire Road, Suite A-2
Absecon, NJ 08201

64 N Delsea Drive
Vineland, NJ 08360

Local office	Absecon	609-272-0900
. . . .	Vineland	609-696-8900
Corporate office		312-399-8900

Sending calls

Visitor dialing instructions
Local calls	area code + 7 digits
Long distance	area code + 7 digits

Speed-dial numbers
Customer service 611

Receiving calls

Caller dials the roamer access number below,
hears tone, dials cellular area code and number.
. . . . 609-247-7626

Billing information

Visitor rates
Most visitors . . . . . $3 day + 85¢ minute

Visitors without roaming agreements (p. 1159)
Roamer Plus will intercept first call to bill you.

Identification numbers

City	Market	System	Billing	RSA
All served cities	228	00583	00000	MSA

Cellular company

Bell Atlantic Mobile
6814 Tilton Road, Unit 123-127
Pleasantville, NJ 08232

Customer service	800-922-0204
Local office	609-645-1155
Corporate office	908-306-7000

Sending calls

Visitor dialing instructions
Local calls	area code + 7 digits
Long distance	area code + 7 digits

Speed-dial numbers
Coast guard	*24
Emergency	911
Repair center	611

Receiving calls

Caller dials the roamer access number below,
hears tone, dials cellular area code and number.
. . . . 609-774-7626

Follow Me Roaming forwarding (see p. 1156)
Activate, dial *18 send. Deactivate dial *19 send.

Billing information

Visitor rates
Most visitors . . . . . $3 day + $1 minute

Visitors without roaming agreements (p. 1159)
Roamer Plus will intercept first call to bill you.

Identification numbers

City	Market	System	Billing	RSA
All served cities	228	00250	30356	MSA

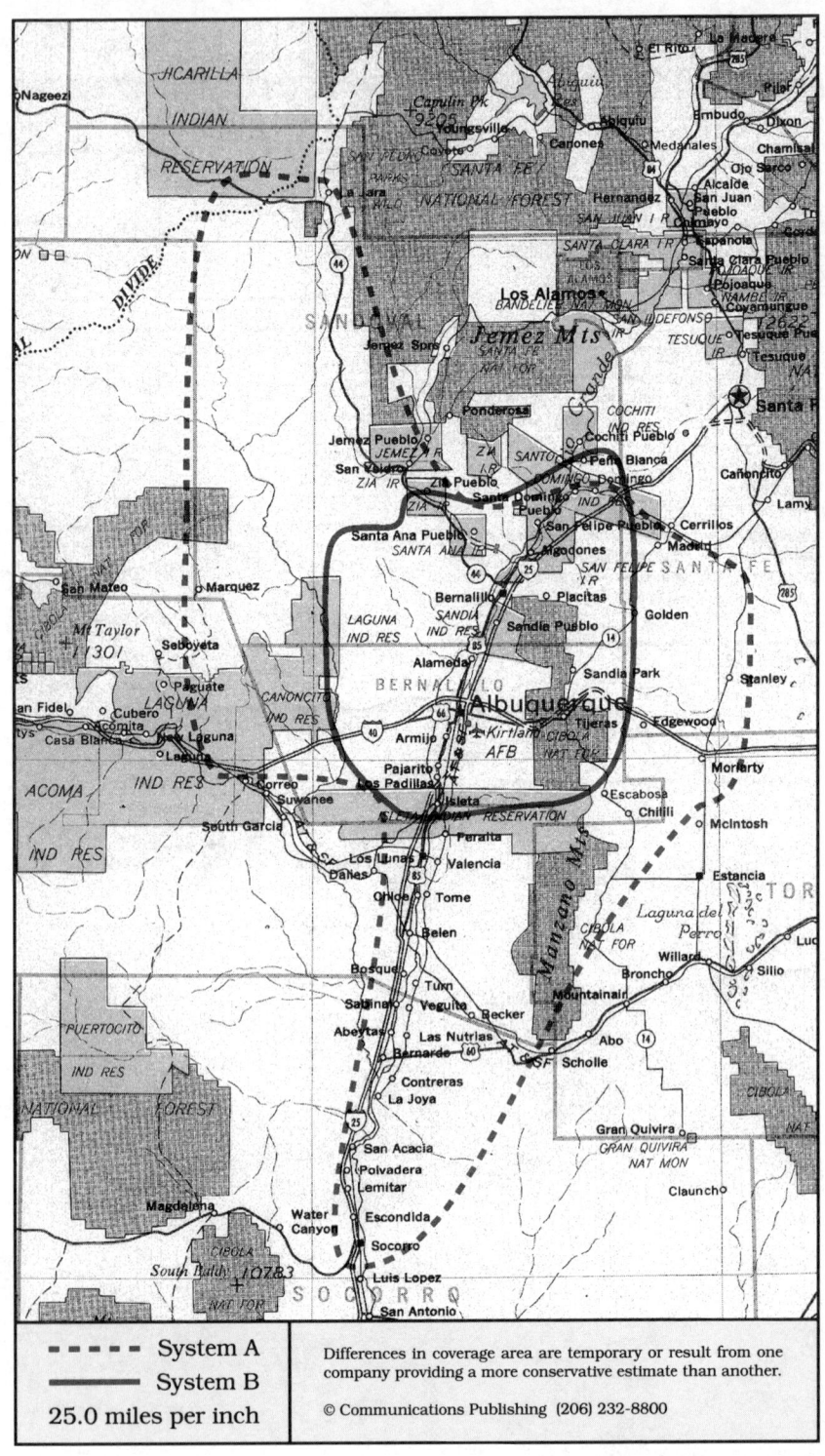

- - - - System A
———— System B
25.0 miles per inch

Differences in coverage area are temporary or result from one company providing a more conservative estimate than another.

© Communications Publishing (206) 232-8800

Albuquerque, New Mexico
Albuquerque, Bernalillo, Socorro

System A	System B

System A

Cellular company

Bell Atlantic Mobile
146 Quincy NE
Albuquerque, NM 87108

Customer service	505-262-5960
Local office	505-266-9000
Corporate office	908-306-7000

Sending calls

Visitor dialing instructions
Local calls . . . . . area code + 7 digits
Long distance . . . 1 + area code + 7 digits

Speed-dial numbers
Customer service *611
Emergency 911
Roaming information *711
Traffic watch . 515-269-1234 or 515-269-7777

Receiving calls

Caller dials the roamer access number below,
hears tone, dials cellular area code and number.
. . . . 505-269-7626

NationLink & RoamingAmerica (see p. 1157)
Dial *31 send to receive calls, *30 to deactivate.
Dial *32 send to tell caller the roamer access no.
This company's subscribers have level 8 service.

Billing information

Visitor rates
Most visitors . . . . $2.50 day + 75¢ minute

Visitors without roaming agreements (p. 1159)
Call co. to use Mastercard, Visa
or Roamer Plus will intercept first call to bill you.

Identification numbers

City	Market	System	Billing	RSA
Albuquerque	086	00079	00000	MSA
Socorro	555	00079	30076	NM-3

This company provides service in NM-3 for the
licensee.

System B

Cellular company

U S West Cellular
3301 Candelaria Road NE
Albuquerque, NM 87107

Customer service	800-238-7848
. . . .	206-562-2895
. . . . local subscribers	800-626-6611
. . . .	206-747-1771
Local office	505-881-2345
Corporate office	206-747-4900

Sending calls

Visitor dialing instructions
Local calls . . . . . area code + 7 digits
Long distance . . . . . area code + 7 digits

Speed-dial numbers
Customer service *611
Emergency 911
Roaming information *711
Traffic watch *107

Receiving calls

Caller dials the roamer access number below,
hears tone, dials cellular area code and number.
. . . . 505-263-7626

Follow Me Roaming forwarding (see p. 1156)
Activate, dial *18 send. Deactivate dial *19 send.

Billing information

Visitor rates
Most visitors . . . . . $2 day + 65¢ minute

Visitors without roaming agreements (p. 1159)
Call co. to use Amex, Mastercard, Visa

Identification numbers

City	Market	System	Billing	RSA
All served cities	086	00110	00000	MSA

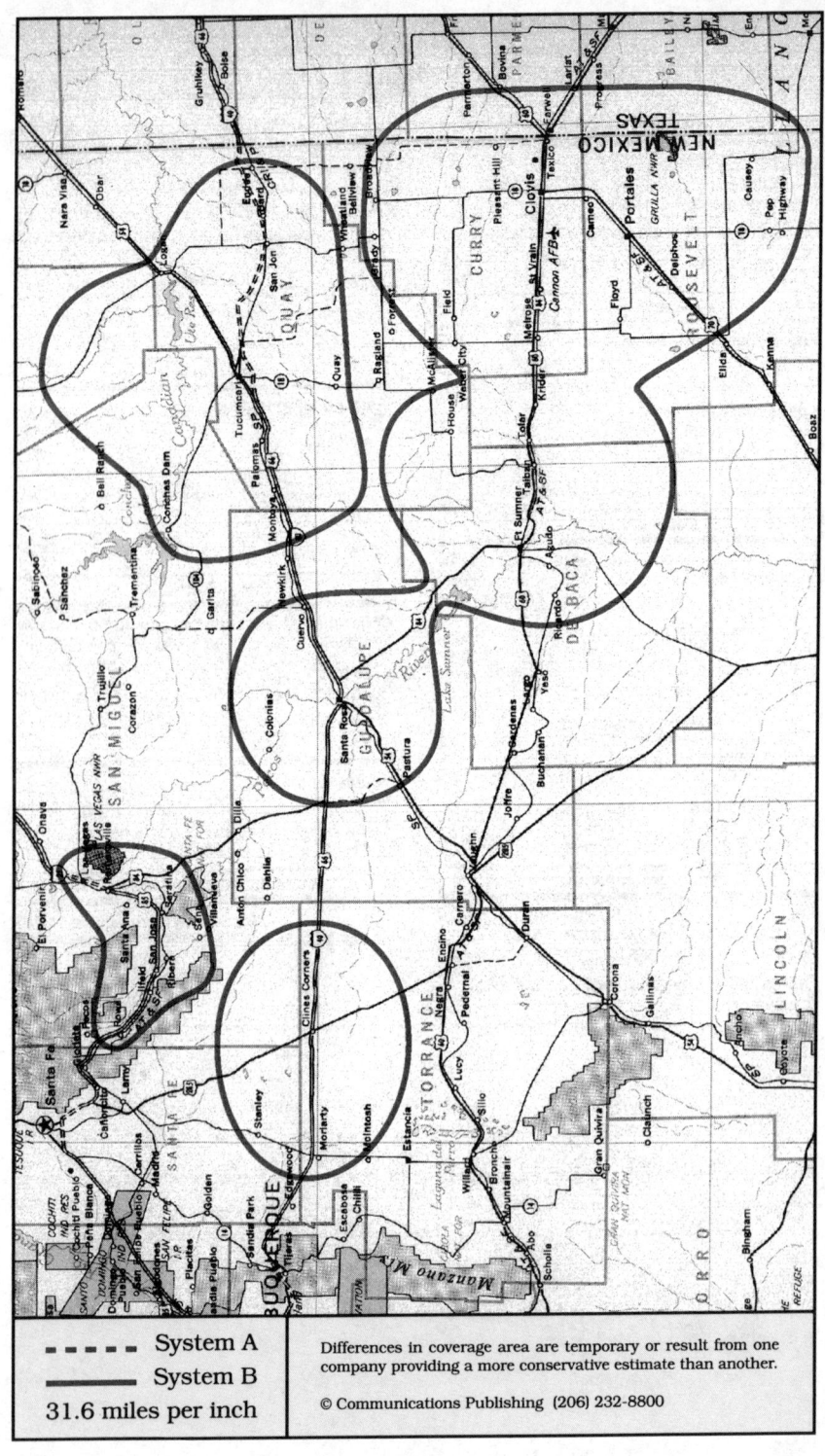

----- System A
——— System B
31.6 miles per inch

Differences in coverage area are temporary or result from one company providing a more conservative estimate than another.

© Communications Publishing (206) 232-8800

Clovis, New Mexico
Clovis, Las Vegas, Moriarty, Portales, Santa Rosa, Tucumcari

System A	System B

Cellular company

No information about the company that will operate system A was available at press time.

Cellular company

Cellular Three
7111 N Prince, PO Box 579
Clovis, NM 88102-0579

Corporate office 505-389-3333

Sending calls

Visitor dialing instructions
Local calls . . . . . . area code + 7 digits
Long distance . . . 1 + area code + 7 digits

Speed-dial numbers
Customer service 611
Emergency 911

Receiving calls

Caller dials the roamer access number below,
hears tone, dials cellular area code and number.
Except Tucumcari 505-760-7626
Tucumcari 505-487-7626

Follow Me Roaming forwarding (see p. 1156)
Activate, dial *18 send. Deactivate dial *19 send.

Billing information

Visitor rates
Most visitors . . . . . $2 day + 75¢ minute

Visitors without roaming agreements (p. 1159)
Cannot call since company takes no credit cards.

Identification numbers

City	Market	System	Billing	RSA
Except Tucumcari	556	01500	00000	NM-4
Tucumcari	556	01500	30482	NM-4

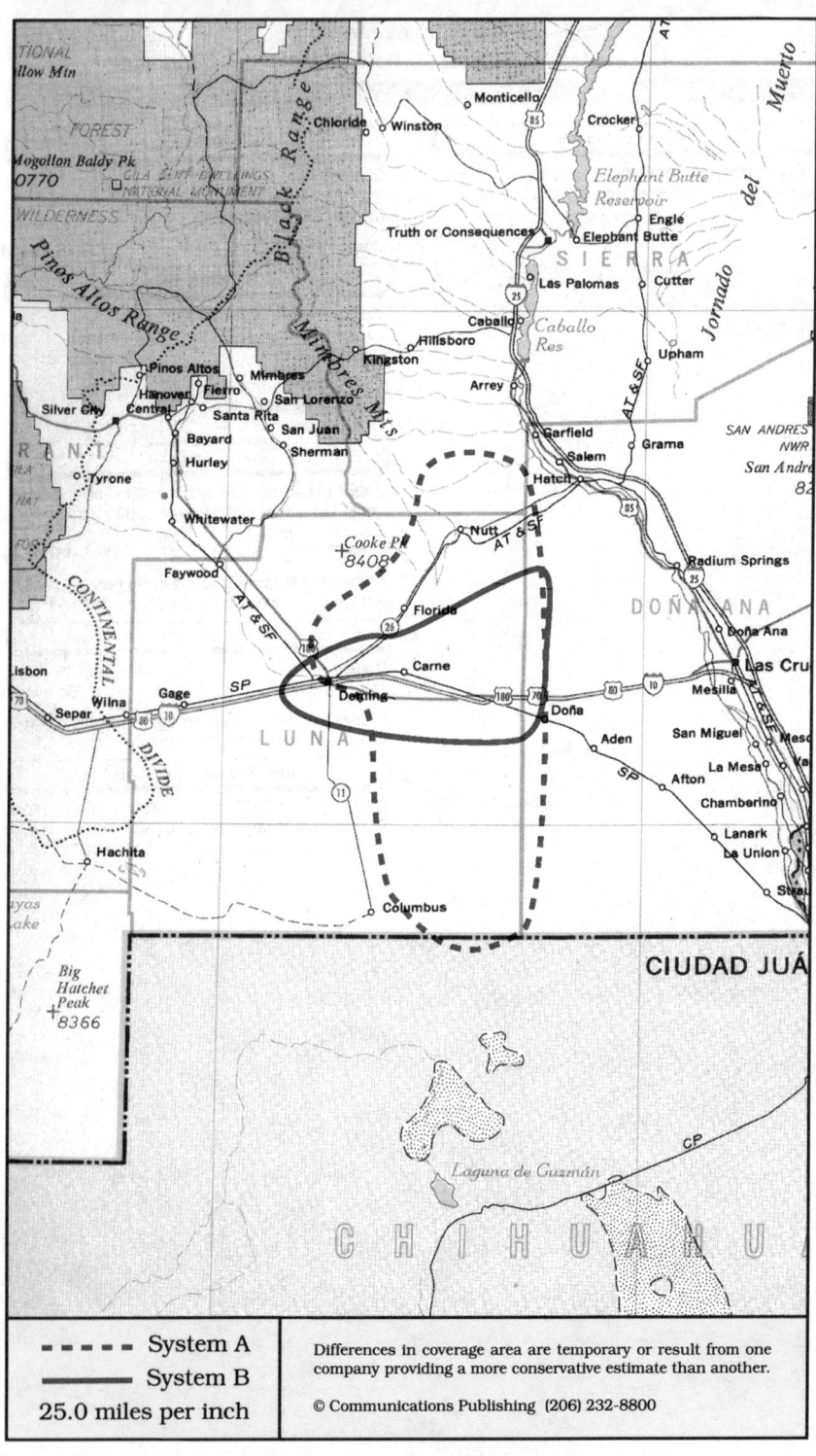

- - - - System A
——— System B
25.0 miles per inch

Differences in coverage area are temporary or result from one company providing a more conservative estimate than another.

© Communications Publishing (206) 232-8800

Deming, New Mexico

Cellular company

Centel Cellular
Rodeo Rd/County Rd, 64-A, Rte 8, Box 337-R
Santa Fe, NM 87505

Local office	505-438-3900
Corporate office	312-399-2644

Sending calls

Visitor dialing instructions

Local calls	area code + 7 digits
Long distance . . .	1 + area code + 7 digits

Speed-dial numbers

Customer service	*611

Receiving calls

Caller dials the roamer access number below,
hears tone, dials cellular area code and number.
. . . . 505-470-7626

NationLink & RoamingAmerica (see p. 1157)
Caller hears your current roamer access number.
This company's subscribers have level 0 service.

Billing information

Visitor rates

Most visitors	$3 day + 99¢ minute

Visitors without roaming agreements (p. 1159)
Cannot call since company takes no credit cards.

Identification numbers

City	Market	System	Billing	RSA
All served cities	557	01501	00000	NM-5

Cellular company

Contel Cellular
1100 S Main Street, Suite 4
Las Cruces, NM 88005

Customer service	800-333-4004
. . . .	915-772-9222
Local office	505-523-6186
Corporate office	404-804-3400

Sending calls

Visitor dialing instructions

Local calls	area code + 7 digits
Long distance . . .	1 + area code + 7 digits

Speed-dial numbers

Customer service	*611
Emergency	911

Receiving calls

Caller dials the roamer access number below,
hears tone, dials cellular area code and number.
. . . . 505-546-1626

Billing information

Visitor rates

Most visitors	$3 day + 85¢ minute

Visitors without roaming agreements (p. 1159)
Call co. to use Amex, Mastercard, Visa

Identification numbers

City	Market	System	Billing	RSA
All served cities	557	01502	00000	NM-5

For full coverage area, see El Paso, TX.

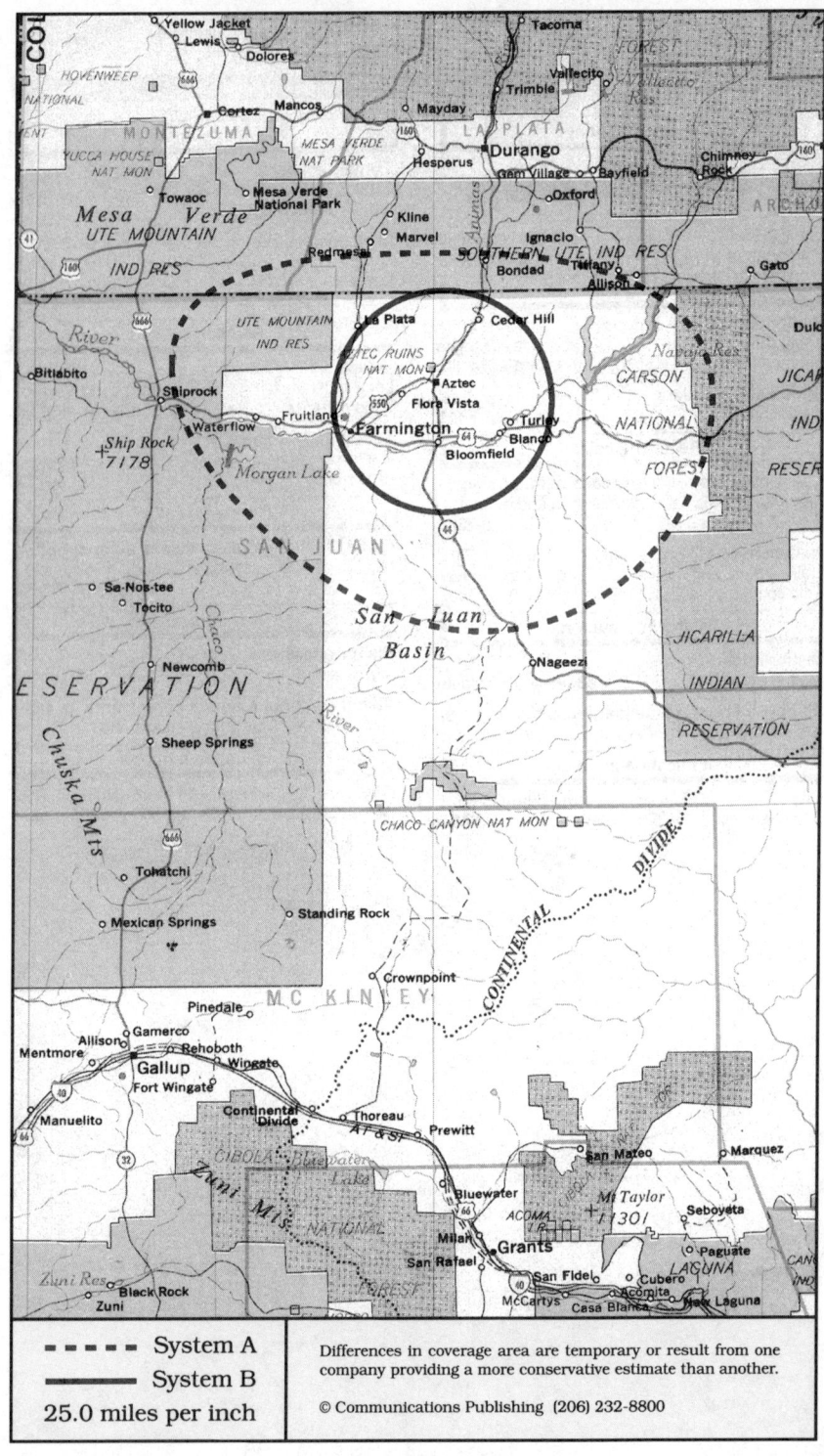

- - - - System A
———— System B
25.0 miles per inch

Differences in coverage area are temporary or result from one company providing a more conservative estimate than another.

© Communications Publishing (206) 232-8800

Farmington, New Mexico
Aztec, Farmington

System A	System B

Cellular company

Centel Cellular
5825 E Main
Farmington, NM 87402

Local office 505-326-0026
Corporate office 312-399-2644

Sending calls

Visitor dialing instructions
Local calls . . . . . area code + 7 digits
Long distance . . . 1 + area code + 7 digits

Speed-dial numbers
Customer service *611

Receiving calls

Caller dials the roamer access number below,
hears tone, dials cellular area code and number.
. . . . 505-320-7626

NationLink & RoamingAmerica (see p. 1157)
Caller hears your current roamer access number.
This company's subscribers have level 0 service.

Billing information

Visitor rates
Most visitors . . . . . $3 day + 99¢ minute

Visitors without roaming agreements (p. 1159)
Cannot call since company takes no credit cards.

Identification numbers

City	Market	System	Billing	RSA
All served cities	553	01493	00000	NM-1

Cellular company

Century Cellunet
1701 Louisville Avenue
Monroe, LA 71201

Customer service 800-638-9727
. . . . 318-683-3450
Corporate office 318-388-9000

Sending calls

Visitor dialing instructions
Local calls . . . . . . area code + 7 digits
Long distance . . . 1 + area code + 7 digits

Speed-dial numbers
Customer service *711

Receiving calls

Caller dials the roamer access number below,
hears tone, dials cellular area code and number.
. . . . 505-860-7626

Follow Me Roaming forwarding (see p. 1156)
Activate, dial *18 send. Deactivate dial *19 send.

Billing information

Visitor rates
Most visitors . . . . . $3 day + 99¢ minute

Visitors without roaming agreements (p. 1159)
Call co. to use Amex, Mastercard, Visa

Identification numbers

City	Market	System	Billing	RSA
All served cities	553	01494	00000	NM-1

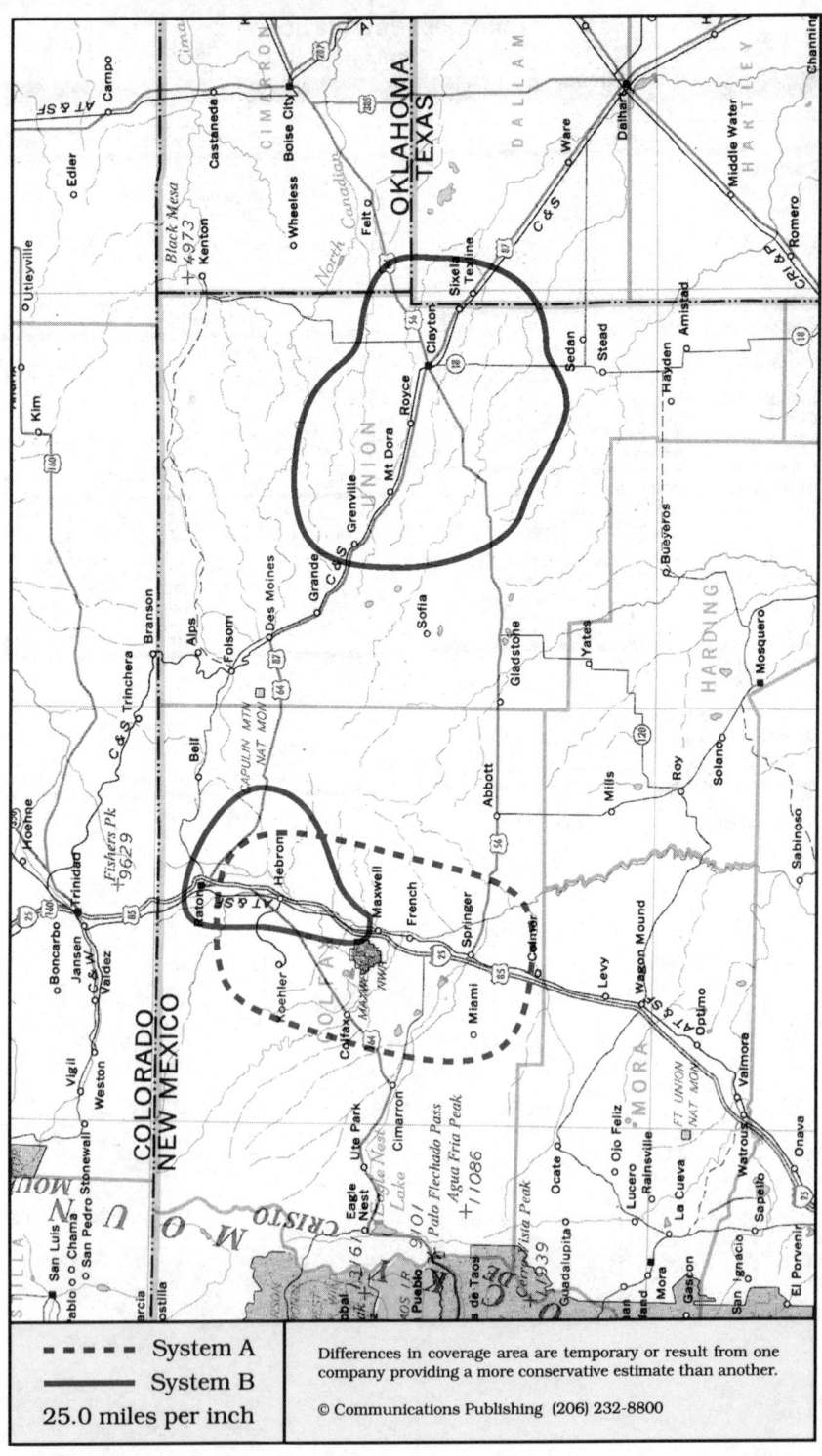

System A
System B

25.0 miles per inch

Differences in coverage area are temporary or result from one company providing a more conservative estimate than another.

© Communications Publishing (206) 232-8800

Raton, New Mexico
Clayton, Raton, Springer

Cellular company

Centel Cellular
Rodeo Rd/County Rd, 64-A, Rte 8, Box 337-R
Santa Fe, NM 87505

Local office 505-438-3900
Corporate office 312-399-2644

Sending calls

Visitor dialing instructions
Local calls . . . . . area code + 7 digits
Long distance . . . 1 + area code + 7 digits

Speed-dial numbers
Customer service *611

Receiving calls

Caller dials the roamer access number below,
hears tone, dials cellular area code and number.
. . . . 505-470-7626

NationLink & RoamingAmerica (see p. 1157)
Caller hears your current roamer access number.
This company's subscribers have level 0 service.

Billing information

Visitor rates
Most visitors . . . . . $3 day + 99¢ minute

Visitors without roaming agreements (p. 1159)
Cannot call since company takes no credit cards.

Identification numbers

City	Market	System	Billing	RSA
All served cities	554	01495	00000	NM-2

Cellular company

Cellular Three
7111 N Prince, PO Box 579
Clovis, NM 88102-0579

Corporate office 505-389-3333

Sending calls

Visitor dialing instructions
Local calls . . . . . . area code + 7 digits
Long distance . . . 1 + area code + 7 digits

Speed-dial numbers
Customer service 611
Emergency 911

Receiving calls

Caller dials the roamer access number below,
hears tone, dials cellular area code and number.
. . . . 505-374-7626

Billing information

Visitor rates
Most visitors . . . . . $2 day + 75¢ minute

Visitors without roaming agreements (p. 1159)
Cannot call since company takes no credit cards.

Identification numbers

City	Market	System	Billing	RSA
All served cities	554	01496	00000	NM-2

Call switching is performed by Cellular Three for
the licensee, which is E.N.M.R. Telephone at
(505-374-9901).

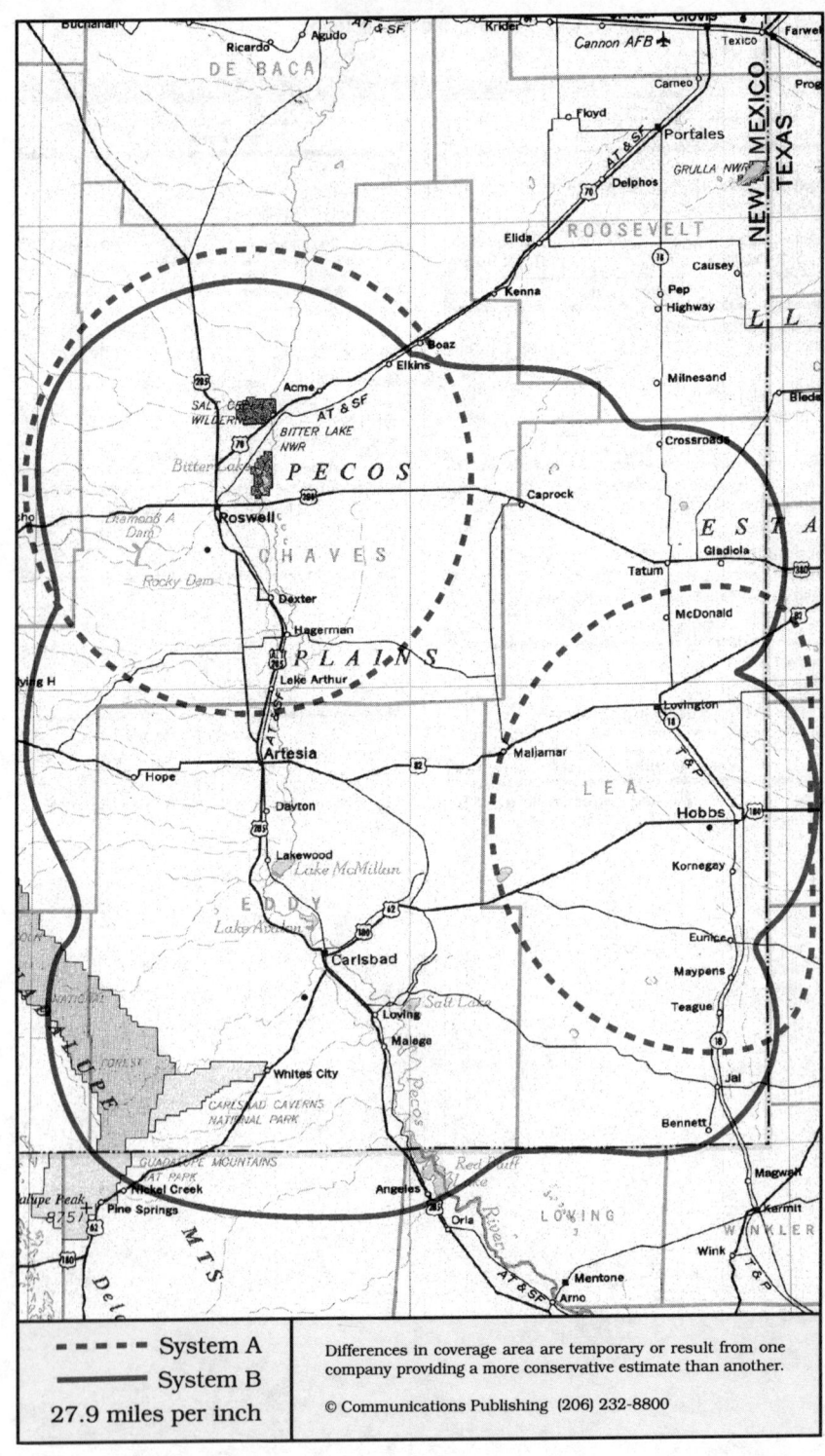

- - - - System A
——— System B
27.9 miles per inch

Differences in coverage area are temporary or result from one company providing a more conservative estimate than another.

© Communications Publishing (206) 232-8800

Roswell, New Mexico
Artesia, Carlsbad, Hobbs, Lovington, Roswell

System A

Cellular company

Cellular Information Systems
55 Holly Hill Lane, Suite 101
Greenwich, CT 06830

Local office	505-393-5077
Corporate office	203-622-6317

Sending calls

Visitor dialing instructions

Local calls	area code + 7 digits
Long distance . . .	1 + area code + 7 digits

Speed-dial numbers

Customer service	*611
Emergency	911

Receiving calls

Caller dials the roamer access number below, hears tone, dials cellular area code and number.
. . . . 505-390-7626

NationLink & RoamingAmerica (see p. 1157)
Dial *31 send to receive calls, *30 to deactivate. This company's subscribers have level 3 service.

Billing information

Visitor rates
Most visitors . . . . . $3 day + 85¢ minute

Visitors without roaming agreements (p. 1159)
Cannot call since company takes no credit cards.

Identification numbers

City	Market	System	Billing	RSA
All served cities	558	01503	00000	NM-6

System B

Cellular company

Cellular Three
7111 N Prince, PO Box 579
Clovis, NM 88102-0579

Leaco Cellular
Leaco Rural Telephone Cooperative
1500 N Love
Lovington, NM 88260

Corporate office	Cellular Three	505-389-3333
. . . .	Leaco Cellular	505-398-5352

Sending calls

Visitor dialing instructions

Local calls	area code + 7 digits
Long distance . . .	1 + area code + 7 digits

Speed-dial numbers

Customer service	611
Emergency	911

Receiving calls

Caller dials the roamer access number below, hears tone, dials cellular area code and number.

Cellular Three (Carlsbad, Roswell)	505-626-7626
Leaco Cellular (Hobbs) . . .	505-369-7626

Follow Me Roaming forwarding (see p. 1156)
Activate, dial *18 send. Deactivate dial *19 send.

Billing information

Visitor rates
Most visitors . . . . . $3 day + 75¢ minute

Visitors without roaming agreements (p. 1159)
Cannot call since company takes no credit cards.

Identification numbers

City	Market	System	Billing	RSA
Carlsbad, Roswell	558	01504	30306	NM-6
Hobbs	558	01504	30302	NM-6

Leaco Cellular is licensed to serve Lea County and portions of Eddy and Chaves as system B-3. Cellular Three is licensed to serve Chaves and Eddy counties. One system serves all counties.

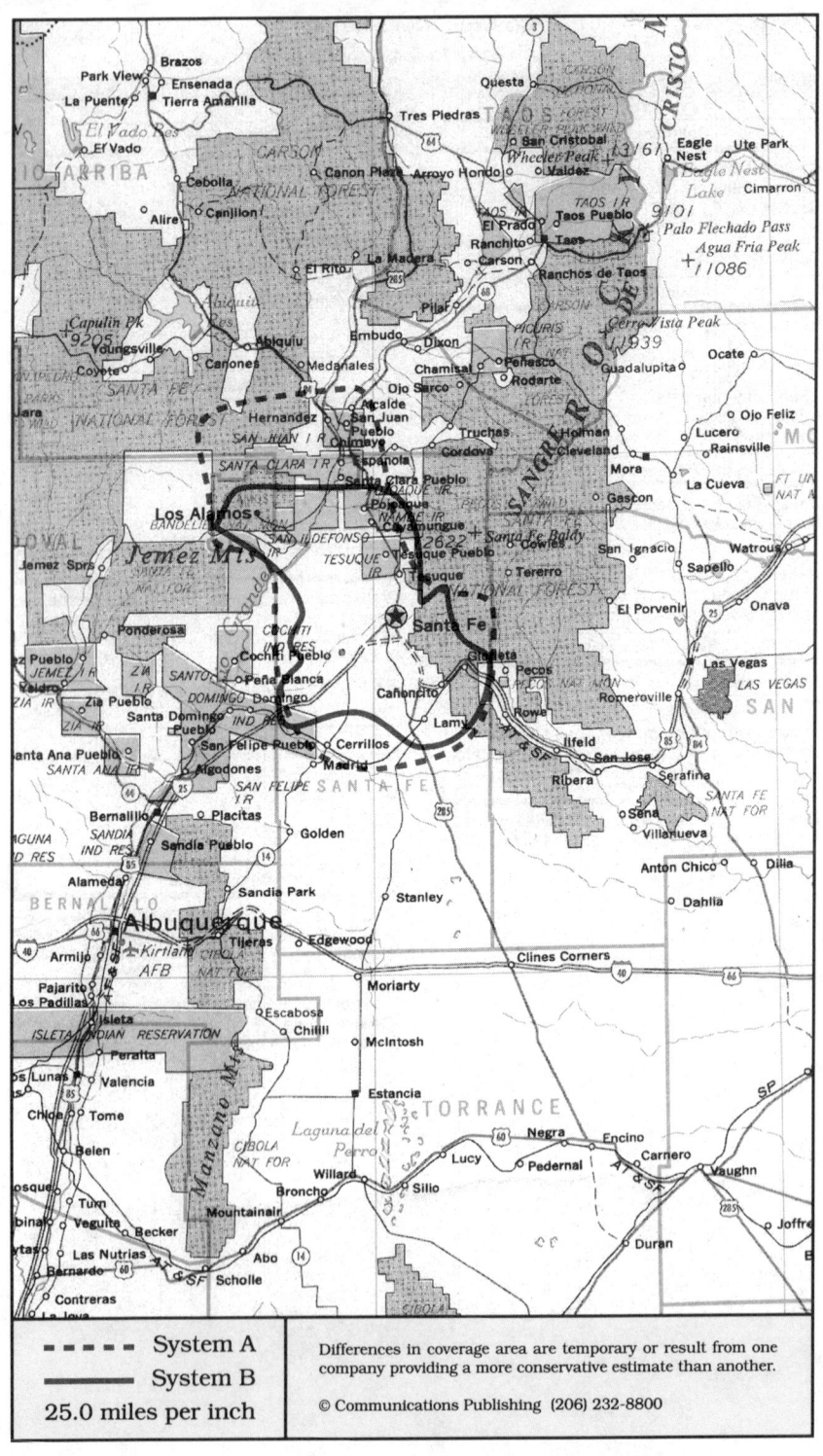

- - - - System A	Differences in coverage area are temporary or result from one company providing a more conservative estimate than another.
——— System B	
25.0 miles per inch	© Communications Publishing (206) 232-8800

Santa Fe, New Mexico
Los Alamos, Santa Fe

System A	System B

Cellular company

Centel Cellular
Rodeo Rd/County Rd, 64-A, Rte 8, Box 337-R
Santa Fe, NM 87505

Local office	505-438-3900
Corporate office	312-399-2644

Sending calls

Visitor dialing instructions

Local calls	area code + 7 digits
Long distance . . .	1 + area code + 7 digits

Speed-dial numbers

Customer service	*611

Receiving calls

Caller dials the roamer access number below, hears tone, dials cellular area code and number.
. . . . 505-470-7626

NationLink & RoamingAmerica (see p. 1157)
Caller hears your current roamer access number. This company's subscribers have level 0 service.

Billing information

Visitor rates

Most visitors	$3 day + 99¢ minute

Visitors without roaming agreements (p. 1159)
Cannot call since company takes no credit cards.

Identification numbers

City	Market	System	Billing	RSA
All served cities	556	01499	00000	NM-4

Cellular company

U S West Cellular
604 Alta Vista
Santa Fe, NM 87501

Customer service	800-238-7848
. . . .	206-562-2895
. . . . . local subscribers	800-626-6611
. . . .	206-747-1771
Local office	505-471-4400
Corporate office	206-747-4900

Sending calls

Visitor dialing instructions

Local calls	area code + 7 digits
Long distance	area code + 7 digits

Speed-dial numbers

Customer service	*611
Emergency	911
Roaming information	*711
Traffic watch	*107

Receiving calls

Caller dials the roamer access number below, hears tone, dials cellular area code and number.
. . . . 505-690-7626

Follow Me Roaming forwarding (see p. 1156)
Activate, dial *18 send. Deactivate dial *19 send.

Billing information

Visitor rates

Most visitors	$2 day + 75¢ minute

Visitors without roaming agreements (p. 1159)
Call co. to use Amex, Mastercard, Visa

Identification numbers

City	Market	System	Billing	RSA
All served cities	556	00048	30262	NM-4

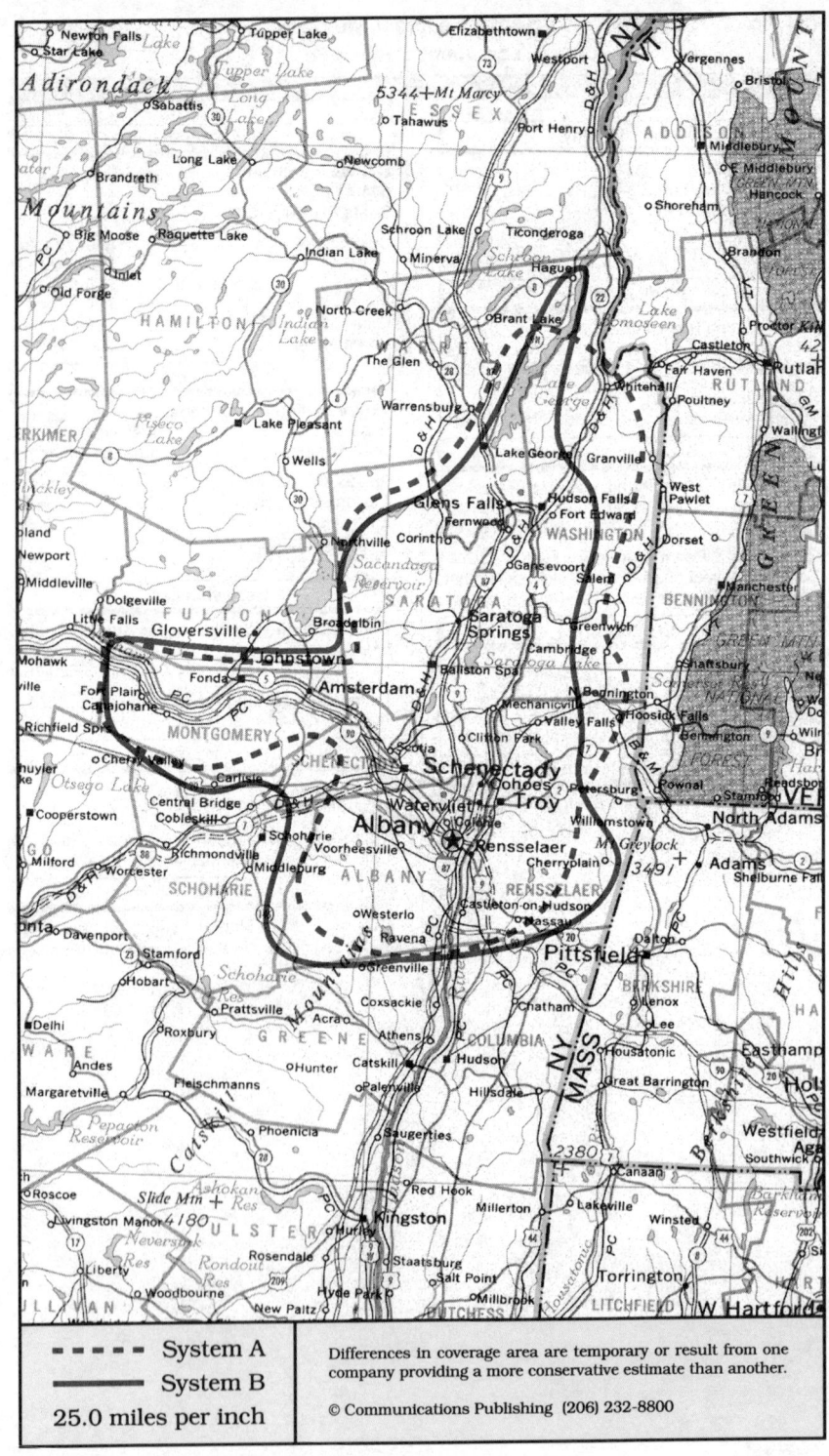

- - - -	System A
———	System B
25.0 miles per inch	

Differences in coverage area are temporary or result from one company providing a more conservative estimate than another.

© Communications Publishing (206) 232-8800

Albany, New York
Albany, Fonda, Glens Falls, Schenectady

System A	System B

Cellular company

Cellular One
Albany Telephone
Associated Communications
11 Century Hill Drive
Latham, NY 12110

Customer service	. . . . .	518-465-7300
Local office	. . . . .	518-783-3400
Corporate office	. . . .	412-281-1907

Sending calls

Visitor dialing instructions

Local calls	. . . . .	area code + 7 digits
Long distance	. . .	1 + area code + 7 digits

Speed-dial numbers

Airport information line	. .	#1AIRP or #12477
American Automobile Association	✶AAA or ✶222
CD96 radio	. . .	✶800CD96 or ✶8002396
Cellular basic information	✶CELL1 or ✶23551
Cellular One sales office	. . . . .	✶811
Channel 13 news	. . . . .	✶13
Cinema line	. . . .	#1CINE or #12463
Customer service	. . . . .	✶611
Directory assistance	. .	#1INFO or #14636
Emergency	. . . . .	✶911
Food delivery by restaurants	#1FOOD or #13663
Horoscopes	. . . .	#1HORO or #11476
Laugh line	. . . .	#1LAUG or #15284
Lottery line	. . .	#1LOTT or #15688
NewsCenter 6 news line	#1NEWS or #16397
News Center 6 news room	. . . . . . ✶6
Report a drunk driver	. .	✶DWI or ✶394
Sports line	. . . .	#1SPOR or #17767
Times union news room	✶800TIPS or ✶8008477
Traffic line	. . . .	#1TRAF or #18723
Weather line	. . . .	#1WEAT or #19328
WFLY radio	. . . . .	✶FLY or ✶359
WGY news room	. . . .	✶WGY or ✶949
WPTR radio	. . . . . .	✶1540
WPYX traffic report	.	✶PYX106 or ✶799106
WTEN news room	. . . .	✶10

Receiving calls

Caller dials the roamer access number below, hears tone, dials cellular area code and number.
. . . . 518-423-7626

NationLink & RoamingAmerica (see p. 1157)
Caller hears your current roamer access number.
This company's subscribers have level 0 service.

North American Cellular Network (see p. 1156)
All calls forward to you. Deactivate dial ✶35 send.

Billing information

Visitor rates

Most visitors	. . . . .	$3 day + 85¢ minute

Visitors without roaming agreements (p. 1159)
Roamer Plus will intercept first call to bill you.

Identification numbers

City	Market	System	Billing	RSA
Albany	044	00063	00000	MSA
Glens Falls	266	00063	00000	MSA

Cellular company

NYNEX Mobile Communications
1770 Central Avenue
Albany, NY 12205

Customer service	. . . .	800-538-4747
. . . .	617-932-1200
Local office	. . . . . .	518-452-8491
Corporate office	. . . .	914-365-7200

Sending calls

Visitor dialing instructions

Local calls	. . . . .	area code + 7 digits
Long distance	. . .	1 + area code + 7 digits

Speed-dial numbers

Customer service	. . . .	611
Directory assistance	. . . . . .	411
Emergency	. . . .	911
KLITE radio	. . . .	✶101
WGY radio	. . . .	✶81
WMVQ radio	. . . .	✶383
WNYT channel 13 television	. .	✶8813
WRGB channel 6 television	. . . . .	✶6

Receiving calls

Caller dials the roamer access number below, hears tone, dials cellular area code and number.
. . . . 518-424-7626

Follow Me Roaming forwarding (see p. 1156)
Activate, dial ✶18 send. Deactivate dial ✶19 send.

Billing information

Visitor rates

Most visitors	. . . . .	$3 day + 95¢ minute

Visitors without roaming agreements (p. 1159)
Call co. to use Amex, Mastercard, Visa
or Roamer Plus will intercept first call to bill you.

Identification numbers

City	Market	System	Billing	RSA
Albany	044	00078	00000	MSA
Glens Falls	266	00078	00000	MSA

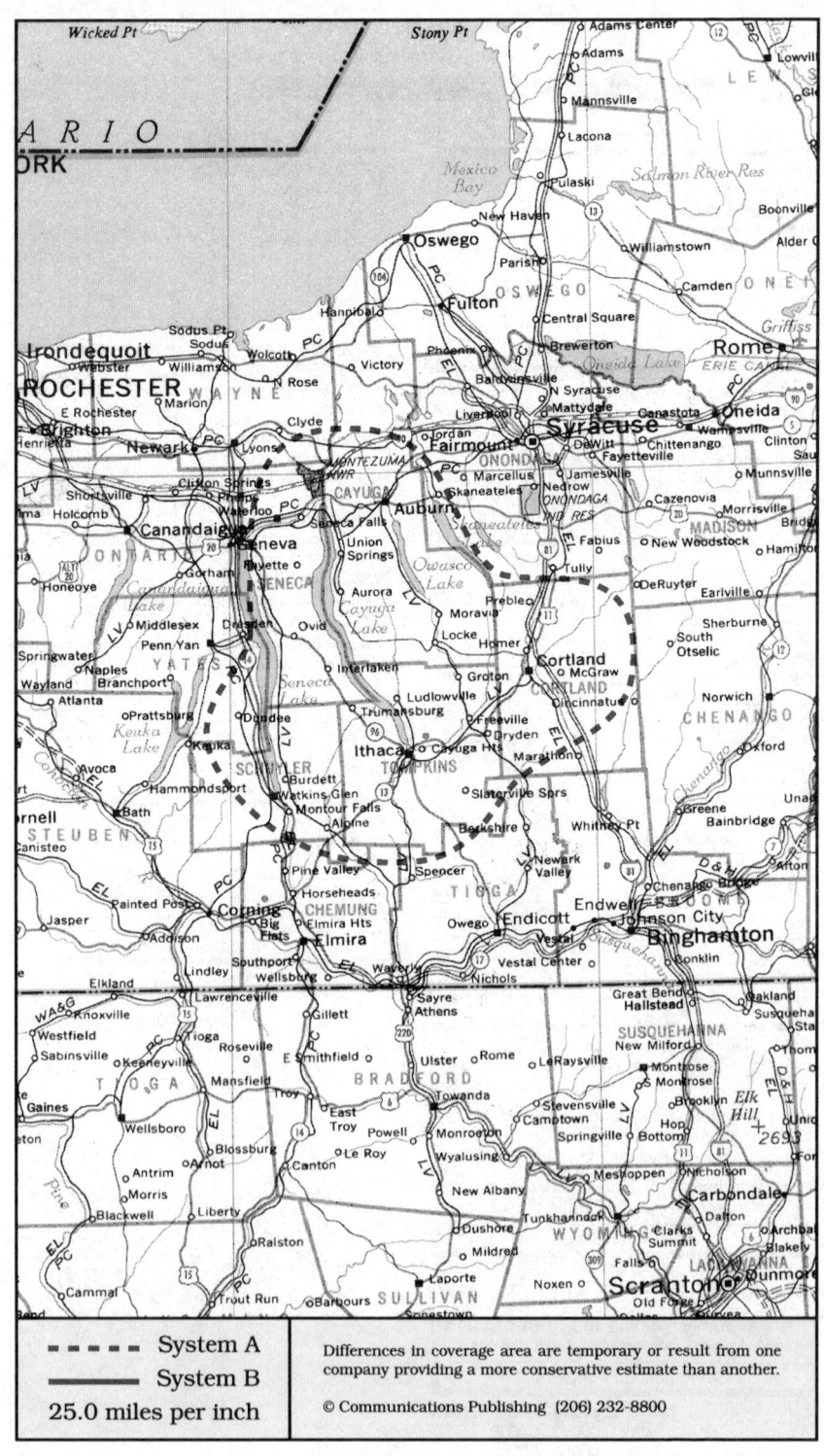

Auburn, New York
Auburn, Cortland, Ithaca, Waterloo, Watkins Glen

System A	System B

Cellular company

Cellular One
Finger Lakes Telephone
215 Fifth Street
Ithaca, NY 14850

Corporate office	some areas	800-773-2351
. . . .		607-273-0400

Sending calls

Visitor dialing instructions

Local calls	area code + 7 digits
Long distance . . .	1 + area code + 7 digits

Speed-dial numbers

Customer service	611
Emergency	911
Report drunk driver	*DWI or *394

Receiving calls

Caller dials the roamer access number below,
hears tone, dials cellular area code and number.
. . . . 607-279-7626

Billing information

Visitor rates

Most visitors	$3 day + 85¢ minute

Visitors without roaming agreements (p. 1159)
Cannot call since company takes no credit cards.

Identification numbers

City	Market	System	Billing	RSA
All served cities	562	01511	00000	NY-4

Cellular company

No information about the company that will operate system B was available at press time.

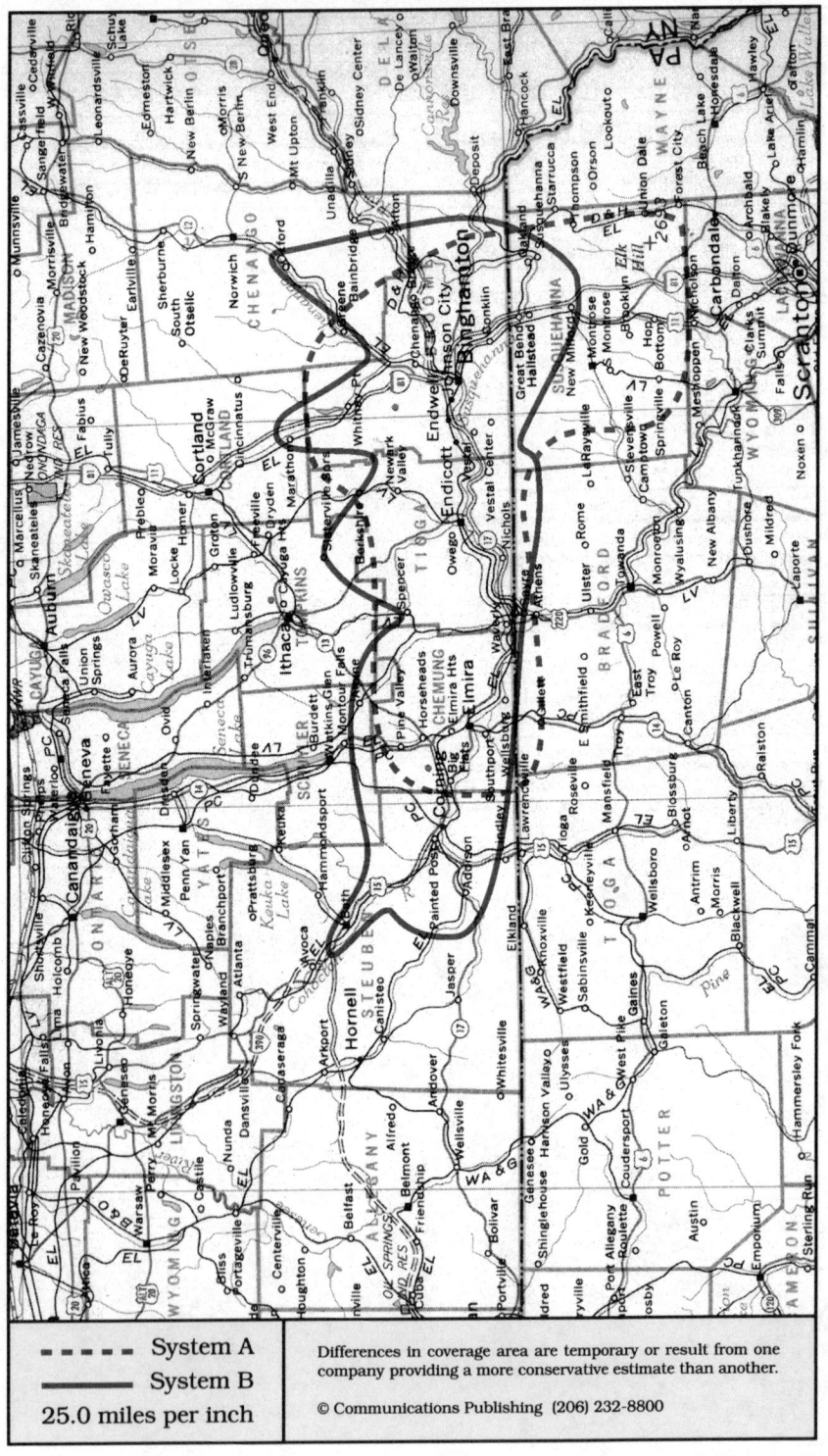

- - - - System A
———— System B
25.0 miles per inch

Differences in coverage area are temporary or result from one company providing a more conservative estimate than another.

© Communications Publishing (206) 232-8800

Binghamton, New York
Bath, Binghamton, Corning, Elmira, Endicott, Owego

System A	System B

Cellular company

Cellular One
1421 College Avenue
Elmira, NY 14901

3108 Vestal Parkway E
Vestal, NY 13850

Binghamton office is moving to Vestal. New address is above; new telephone unavailable.

Local office	Binghamton (was)	607-771-8000
. . . .	Elmira	607-737-1000
. . . .	Vestal	unavailable

Sending calls

Visitor dialing instructions

Local calls	area code + 7 digits
Long distance . . .	1 + area code + 7 digits

Speed-dial numbers

Customer service	611
Directory assistance	411
Emergency (sheriff)	911

Receiving calls

Caller dials the roamer access number below, hears tone, dials cellular area code and number.

Binghamton	607-727-7626
Elmira	607-731-7626

NationLink & RoamingAmerica (see p. 1157)
Dial ∗31 send to receive calls, ∗30 to deactivate.
Dial ∗32 send to tell caller the roamer access no.
This company's subscribers have level X service.

North American Cellular Network (see p. 1156)
All calls forward to you. Deactivate dial ∗35 send.

Billing information

Visitor rates

Most visitors	$3 day + 99¢ minute

Visitors without roaming agreements (p. 1159)
Call co. to use Amex
or Roamer Plus will intercept first call to bill you.

Identification numbers

City	Market	System	Billing	RSA
Binghamton	122	00283	30007	MSA
Elmira	284	00283	30009	MSA

Corning (NY-3) will be served in the future by Dicomm Cellular, listed under Jamestown, NY.

Cellular company

Contel Cellular
1188 Vestal Avenue
Binghamton, NY 13903

101 College Avenue
Elmira, NY 14903

Customer service	800-333-4004
. . . . . . extension 4002	404-391-8000
Local office . . . Binghamton	607-722-1322
. . . . . . Elmira	607-737-1200
Corporate office	404-804-3400

Sending calls

Visitor dialing instructions

Local calls	area code + 7 digits
Long distance . . .	1 + area code + 7 digits

Speed-dial numbers

Customer service	611
Directory assistance	411
Emergency	911
Mr. Rescue ∗HELP or ∗4357 or 1-800-447-8500	

Receiving calls

Caller dials the roamer access number below, hears tone, dials cellular area code and number.

Binghamton	607-725-7626
Corning, Elmira	607-738-7626

Follow Me Roaming forwarding (see p. 1156)
Activate, dial ∗18 send. Deactivate dial ∗19 send.

Billing information

Visitor rates

Most visitors	$3 day + 85¢ minute

Visitors without roaming agreements (p. 1159)
Call co. to use Amex, Mastercard, Visa

Identification numbers

City	Market	System	Billing	RSA
Binghamton	122	00266	00000	MSA
Corning	561	00266	30372	NY-3
Elmira	284	00266	00000	MSA

System A
System B
25.0 miles per inch

Differences in coverage area are temporary or result from one company providing a more conservative estimate than another.

© Communications Publishing (206) 232-8800

Buffalo, New York
Buffalo, Hamburg, Lockport, Niagra Falls

System A	System B

Cellular company

Cellular One
Buffalo Telephone
Associated Communications
1500 Rand Building
Buffalo, NY 14203

Customer service	716-854-0800
Local office	716-854-5076
Corporate office	412-281-1907

Sending calls

Visitor dialing instructions

Local calls	area code + 7 digits
Long distance . . .	1 + area code + 7 digits

Speed-dial numbers

Airport information line . .	#1AIRP or #12477
American Automobile Association . .	*222
Channel 7 news tip line . .	*7TV or *788
Cinema line	#1CINE or #12463
Coast Guard (Buffalo office) .	*SOS or *767
Customer service	*611
Emergency	*911
Lottery line	#1LOTT or #15688
Report a drunk driver . . .	*DWI or *394
Sales office	*811
Skiing/boating line (seasonal)	#1SKII or #17544
Talking phone book . . .	#1TALK or #18255
Thruway conditions line .	#1THRU or #18478
Time and temperature . .	#1TIME or #18463
USA customs boat reporting	#1USA or #18722
WCTR traffic and news line	*550

Receiving calls

Caller dials the roamer access number below,
hears tone, dials cellular area code and number.
. . . . 716-861-7626

NationLink & RoamingAmerica (see p. 1157)
Caller hears your current roamer access number.
This company's subscribers have level 0 service.

North American Cellular Network (see p. 1156)
All calls forward to you. Deactivate dial *35 send.

Billing information

Visitor rates
Most visitors . . . . . $3 day + 99¢ minute

Visitors without roaming agreements (p. 1159)
Roamer Plus will intercept first call to bill you.

Identification numbers

City	Market	System	Billing	RSA
All served cities	025	00003	00000	MSA

Cellular company

NYNEX Mobile Communications
2410 Walden Avenue
Buffalo, NY 14225

Customer service	800-538-4747
. . . .	617-932-1200
Local office	716-686-4300
Corporate office	914-365-7200

Sending calls

Visitor dialing instructions

Local calls	area code + 7 digits
Long distance	area code + 7 digits

Speed-dial numbers

Customer service	611
Emergency	911

Receiving calls

Caller dials the roamer access number below,
hears tone, dials cellular area code and number.
. . . . 716-863-7626

Follow Me Roaming forwarding (see p. 1156)
Activate, dial *18 send. Deactivate dial *19 send.

Billing information

Visitor rates
Most visitors . . . . . $3 day + 95¢ minute

Visitors without roaming agreements (p. 1159)
Call co. to use Amex, Mastercard, Visa
or Roamer Plus will intercept first call to bill you.

Identification numbers

City	Market	System	Billing	RSA
All served cities	025	00056	00000	MSA

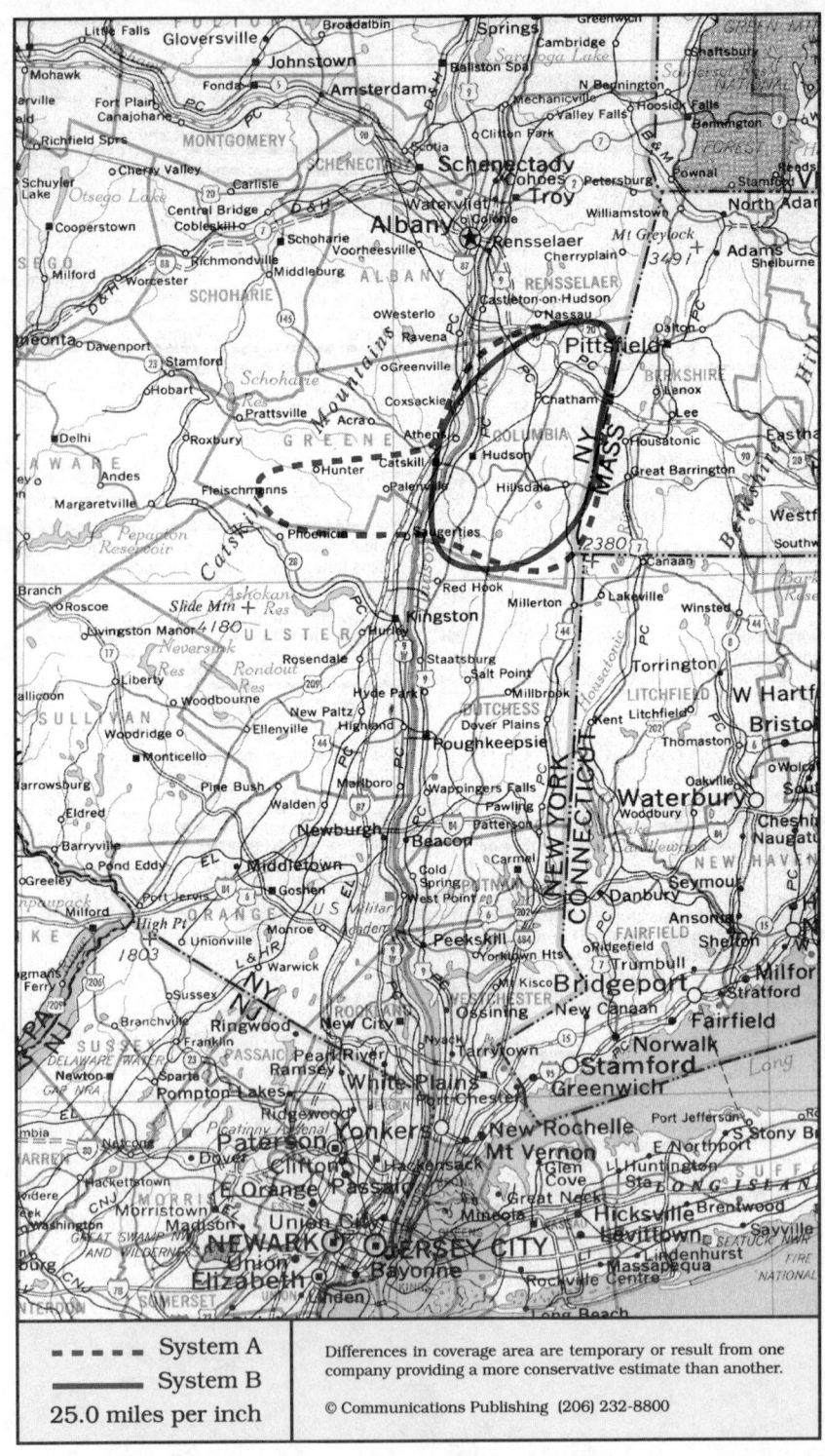

▬ ▬ ▬ ▬ System A	Differences in coverage area are temporary or result from one
——————— System B	company providing a more conservative estimate than another.
25.0 miles per inch	© Communications Publishing (206) 232-8800

Hudson, New York
Catskill, Hudson

System A

Cellular company

Cellular One
Sterling Cellular Management
300 Fairview Avenue
Hudson, NY 12534

Customer service	800-552-6150
Local office	518-828-4400
Corporate office	404-552-5030

Sending calls

Visitor dialing instructions

Local calls	area code + 7 digits
Long distance	area code + 7 digits

Speed-dial numbers

Customer service	611
Emergency	911

Receiving calls

Caller dials the roamer access number below,
hears tone, dials cellular area code and number.
. . . . 518-821-7626

Billing information

Visitor rates
Most visitors . . . . . $3 day + 99¢ minute

Visitors without roaming agreements (p. 1159)
Call co. to use Amex
or Roamer Plus will intercept first call to bill you.

Identification numbers

City	Market	System	Billing	RSA
All served cities	564	01515	00000	NY-6

System B

Cellular company

Valley Cellular
Hudson Valley RSA Limited Partnership
19 Broad Street
Kinderhook, NY 12106

Customer service .	NYNEX	800-538-4747
. . . .	NYNEX	617-932-1200
Corporate office	Valley Cellular	518-758-9557

Sending calls

Visitor dialing instructions

Local calls	area code + 7 digits
Long distance	area code + 7 digits

Speed-dial numbers

Customer service (for subscribers) . . .	811
Customer service (for visitors)	611
Emergency	911

Receiving calls

Caller dials the roamer access number below,
hears tone, dials cellular area code and number.
. . . . 518-755-7626

Billing information

Visitor rates
Most visitors . . . . . $3 day + 95¢ minute

Visitors without roaming agreements (p. 1159)
Call NYNEX to use Amex, Mastercard, Visa

Identification numbers

City	Market	System	Billing	RSA
All served cities	564	01516	00000	NY-6

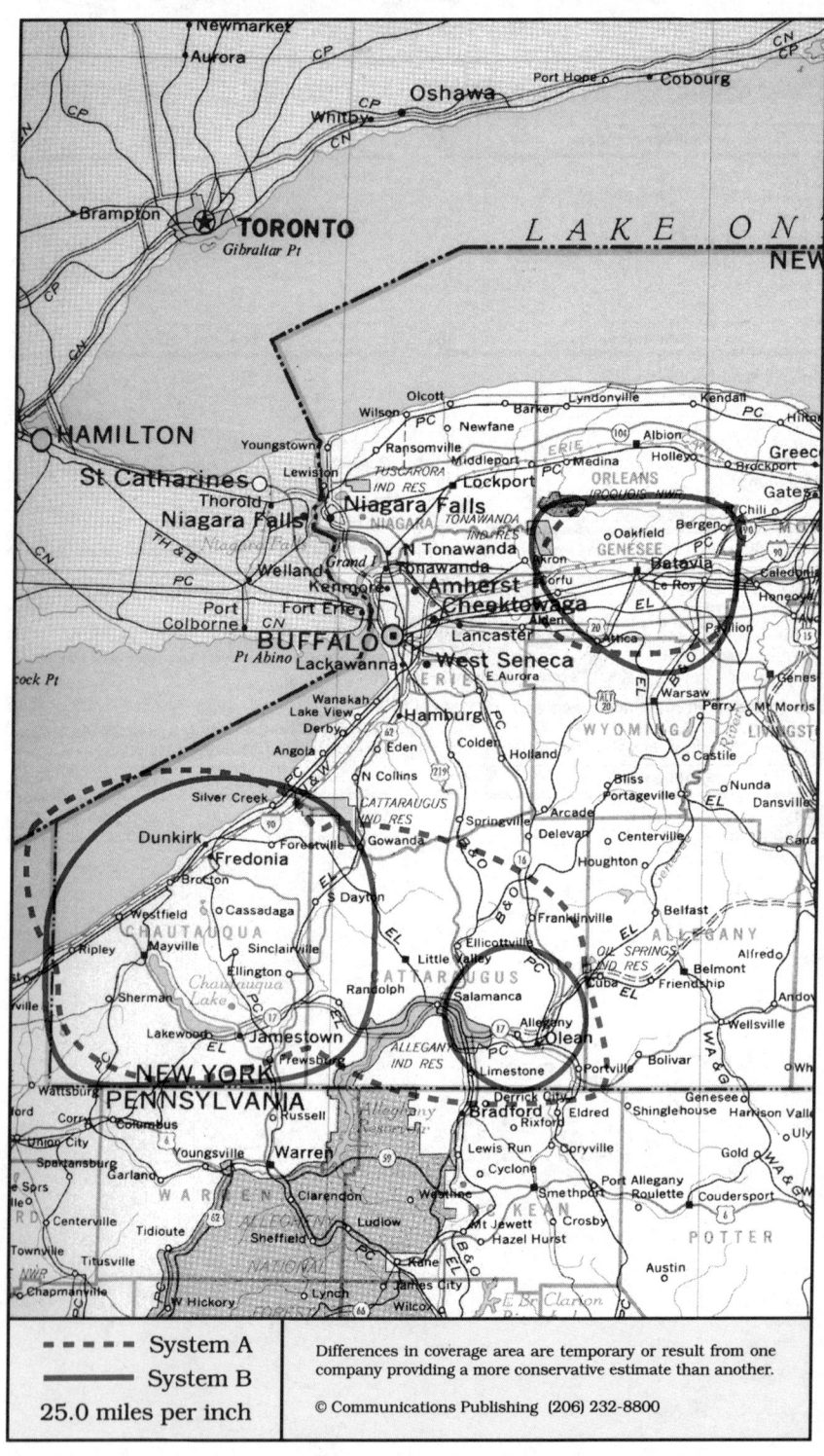

System A — — — — —
System B —————————
25.0 miles per inch

Differences in coverage area are temporary or result from one company providing a more conservative estimate than another.

© Communications Publishing (206) 232-8800

Jamestown, New York
Batavia, Fredonia, Jamestown, Mayville, Olean

System A	System B

Cellular company

Dicomm Cellular Telephone
1800 Rand Building
Buffalo, NY 14203

Corporate office	716-854-8900
. . . . . . from Batavia	716-344-6000
. . . . . . from Fredonia	716-679-5000
. . . . . . from Jamestown	716-499-5000
. . . . . . from Olean	716-378-5000

Sending calls

Visitor dialing instructions
Local calls . . . . area code + 7 digits
Long distance . . . 1 + area code + 7 digits

Speed-dial numbers
AAA roadside service *222
Customer service *611
Emergency 911
Report a drunk driver . . . *DWI or *394
WPIG radio *957

Receiving calls

Caller dials the roamer access number below,
hears tone, dials cellular area code and number.

Batavia	716-344-6626
Fredonia	716-679-5626
Jamestown	716-499-7626
Olean	716-378-7626

North American Cellular Network (see p. 1156)
All calls forward to you. Deactivate dial *35 send.

Billing information

Visitor rates
Most visitors . . . . . $0 day + 99¢ minute

Visitors without roaming agreements (p. 1159)
Call co. to use Amex, Mastercard, Visa

Identification numbers

City	Market	System	Billing	RSA
All served cities	561	01509	00000	NY-3

This company will serve Corning, NY in the future. For Corning system B, see Binghamton, NY.

Cellular company

Rochester Telephone Mobile Communications
2060 Brighton-Henrietta Town Line Road
Rochester, NY 14623

Corporate office	800-724-7862
. . . .	716-274-7000

Sending calls

Visitor dialing instructions
Local calls . . . . . . area code + 7 digits
Long distance . . . . . area code + 7 digits

Speed-dial numbers
American Automobile Association *AAA or *222
Customer service 611
Directory assistance 411
Emergency 911
Report a drunk driver . . . . *DWI or *394

Receiving calls

Caller dials the roamer access number below,
hears tone, dials cellular area code and number.
. . . . 716-721-7626

Follow Me Roaming forwarding (see p. 1156)
Activate, dial *18 send. Deactivate dial *19 send.

Billing information

Visitor rates
Most visitors . . . . . $3 day + 95¢ minute

Visitors without roaming agreements (p. 1159)
Call co. to use Mastercard, Visa

Identification numbers

City	Market	System	Billing	RSA
All served cities	561	01510	00000	NY-3

For full coverage area, see Rochester, NY.

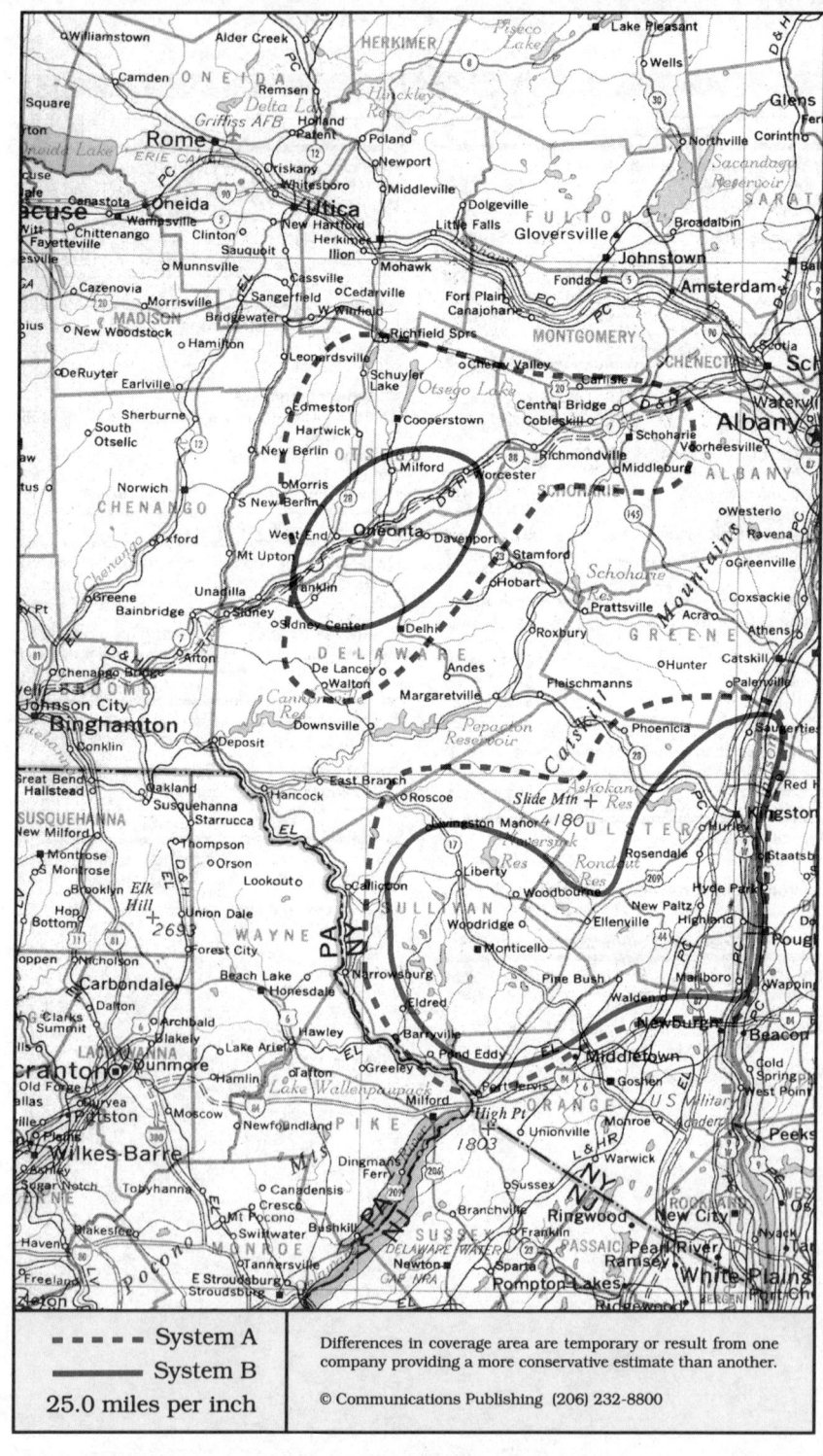

System A
System B
25.0 miles per inch

Differences in coverage area are temporary or result from one company providing a more conservative estimate than another.

© Communications Publishing (206) 232-8800

Kingston, New York
Catskill Mountains, Kingston, Monticello, Oneonta

System A	System B

Cellular company

Cellular One of Upstate New York
17 Computer Drive E
Albany, NY 12205-1109

Corporate office	518-438-2400
. . . .	518-231-1111
. . . .	607-434-1111
. . . .	914-389-1111
. . . .	914-797-1111

Sending calls

Visitor dialing instructions

Local calls	area code + 7 digits
Long distance	area code + 7 digits

Speed-dial numbers

Customer service	611
Emergency	911

Receiving calls

Caller dials the roamer access number below,
hears tone, dials cellular area code and number.

. . . . 914-389-7626

NationLink & RoamingAmerica (see p. 1157)
Caller hears your current roamer access number.
This company's subscribers have level 0 service.

Billing information

Visitor rates

Most visitors . . . . . $3 day + 99¢ minute

Visitors without roaming agreements (p. 1159)
Roamer Plus will intercept first call to bill you.

Identification numbers

City	Market	System	Billing	RSA
All served cities	563	01513	00000	NY-5

Cellular company

NYNEX Mobile Communications
99 Route 17-K
Newburgh, NY 12550

Customer service	800-538-4747
. . . .	617-932-1200
Local office	914-564-9981
Corporate office	914-365-7200

Sending calls

Visitor dialing instructions

Local calls	area code + 7 digits
Long distance	area code + 7 digits

Speed-dial numbers

Customer service	611
Emergency	911
Local office	811
Report a drunk driver	*DWI or *394

Receiving calls

Caller dials the roamer access number below,
hears tone, dials cellular area code and number.

Catskills area, Kingston . . .	914-388-7626
Oneonto	607-435-7626

Billing information

Visitor rates

Most visitors . . . . . $3 day + 95¢ minute

Visitors without roaming agreements (p. 1159)
Call co. to use Amex, Mastercard, Visa
or Roamer Plus will intercept first call to bill you.

Identification numbers

City	Market	System	Billing	RSA
All served cities	563	00404	00000	NY-5

For full coverage area, see Newburgh, NY and
New York, NY and Poughkeepsie, NY.

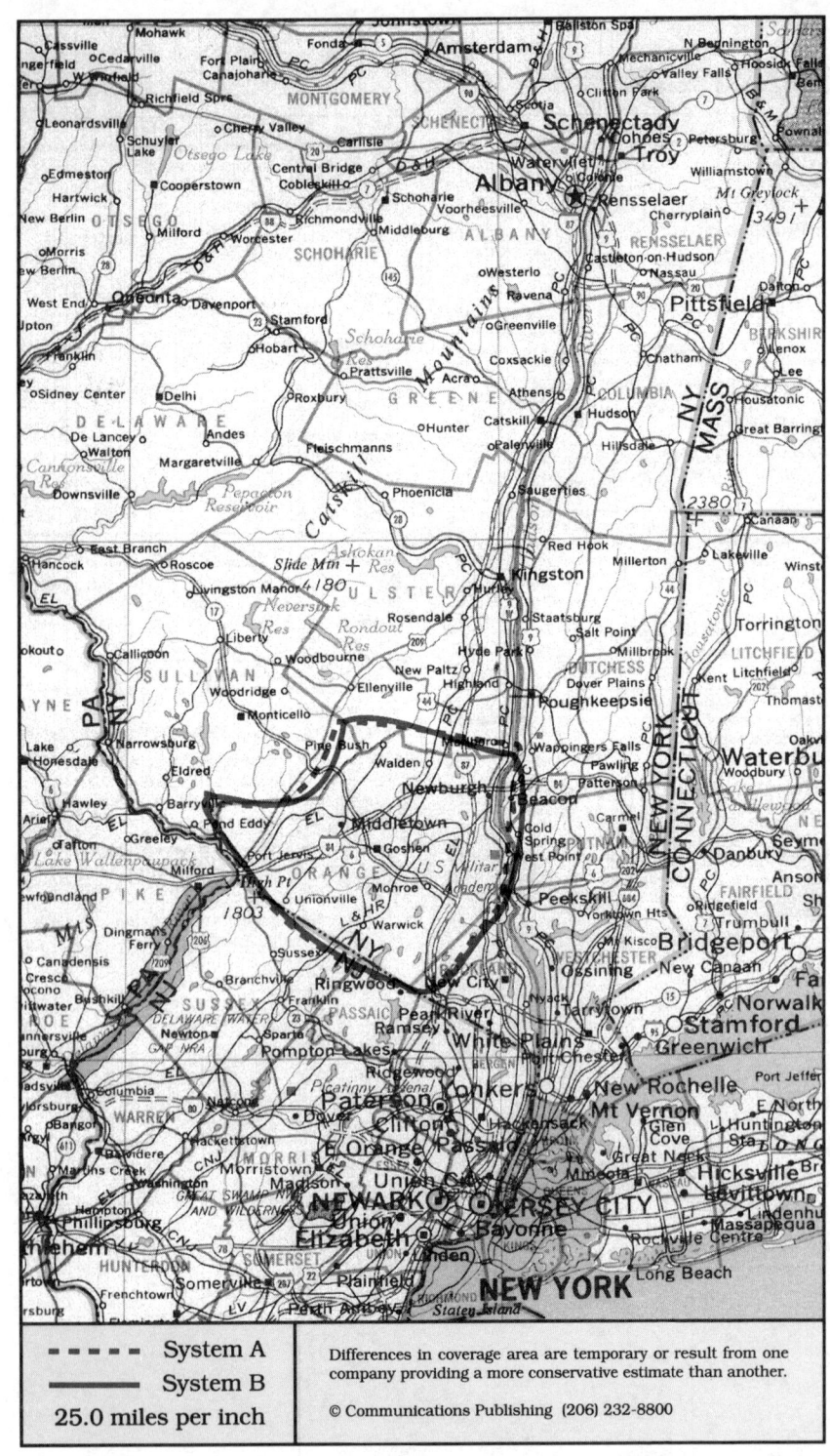

Newburgh, New York
Goshen, Middletown, Newburgh (Orange County)

System A	System B

Cellular company

Cellular One
Vanguard Cellular Systems
2 Melody Lane, Routes 6 & 17
Harriman, NY 10926

Customer service	914-542-4000
Local office	914-783-2355
Corporate office	919-282-3690

Sending calls

Visitor dialing instructions

Local calls	area code + 7 digits
Long distance	area code + 7 digits

Speed-dial numbers

Customer service	611
Emergency	911

Receiving calls

Caller dials the roamer access number below,
hears tone, dials cellular area code and number.
. . . . 914-542-7626

NationLink & RoamingAmerica (see p. 1157)
Dial *31 send to receive calls, *30 to deactivate.
Dial *32 send to tell caller the roamer access no.
This company's subscribers have level 6 service.

Billing information

Visitor rates
Most visitors . . . . . $3 day + 99¢ minute

Visitors without roaming agreements (p. 1159)
Call co. to use Amex, Mastercard, Visa

Identification numbers

City	Market	System	Billing	RSA
All served cities	151	00479	00000	MSA

Cellular company

NYNEX Mobile Communications
99 Route 17-K
Newburgh, NY 12550

Customer service	800-538-4747
. . . .	617-932-1200
Local office	914-564-9981
Corporate office	914-365-7200

Sending calls

Visitor dialing instructions

Local calls	area code + 7 digits
Long distance	area code + 7 digits

Speed-dial numbers

Customer service	611
Emergency	911
Local office	811
Report a drunk driver	*DWI or *394

Receiving calls

Caller dials the roamer access number below,
hears tone, dials cellular area code and number.
. . . . 914-344-7626

Follow Me Roaming forwarding (see p. 1156)
Activate, dial *18 send. Deactivate dial *19 send.

Billing information

Visitor rates
Most visitors . . . . . $3 day + 95¢ minute

Visitors without roaming agreements (p. 1159)
Call co. to use Amex, Mastercard, Visa
or Roamer Plus will intercept first call to bill you.

Identification numbers

City	Market	System	Billing	RSA
All served cities	151	00486	00000	MSA

For full coverage area, see Kingston, NY and
New York, NY and Poughkeepsie, NY.

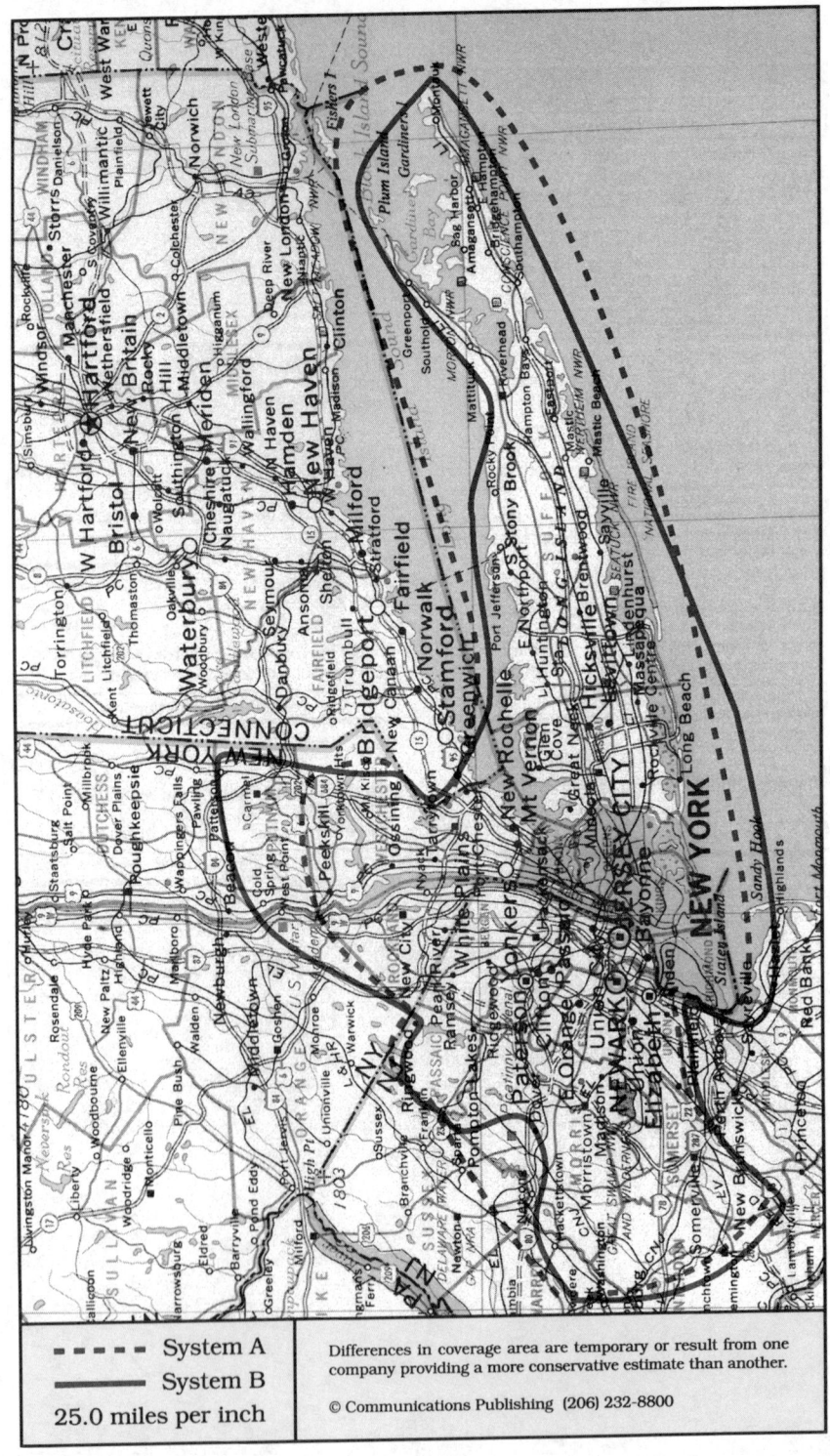

System A

System B

25.0 miles per inch

Differences in coverage area are temporary or result from one company providing a more conservative estimate than another.

© Communications Publishing (206) 232-8800

New York, New York
Long Island, Newark NJ, New City, New York, Paterson NJ, White Plains

System A	System B

Cellular company

Cellular One
McCaw Cellular Communications
15 E Midland Avenue
Paramus, NJ 07652

Customer service	800-242-7327
Local office	201-967-3000
Corporate office	206-827-4500

Sending calls

Visitor dialing instructions

Local calls	area code + 7 digits
Long distance . . .	0 + area code + 7 digits

Speed-dial numbers

Customer service	611
Emergency	911
Hear a traffic report	*22

Receiving calls

Caller dials the roamer access number below,
hears tone, dials cellular area code and number.

New Jersey	201-960-7626
New York	212-847-7626

NationLink & RoamingAmerica (see p. 1157)
NationLink & RoamingAmerica (see p. 1157)
Caller hears your current roamer access number.
This company's subscribers have level 0 service.

North American Cellular Network (see p. 1156)
North American Cellular Network (see p. 1156)
All calls forward to you. Deactivate dial *35 send.

Billing information

Visitor rates

Most visitors	$3 day + 99¢ minute

Visitors without roaming agreements (p. 1159)
Visitors without roaming agreements (p. 1159)
Roamer Plus will intercept first call to bill you.

Identification numbers

City	Market	System	Billing	RSA
All served cities	001	00025	00000	MSA

Cellular company

NYNEX Mobile Communications
2000 Corporate Drive
Orangeburg, NY 10962

Customer service	800-227-1069
Corporate office	914-365-7200

Sending calls

Visitor dialing instructions

Local calls	area code + 7 digits
Long distance	area code + 7 digits

Speed-dial numbers

Customer service	611
Emergency	911
Report a drunk driver	*DWI or *394
Traffic information	*526

Receiving calls

Caller dials the roamer access number below,
hears tone, dials cellular area code and number.

New Jersey	201-259-7626
New York	212-301-7626

Billing information

Visitor rates

Most visitors	$3 day + 95¢ minute

Visitors without roaming agreements (p. 1159)
Visitors without roaming agreements (p. 1159)
Call co. to use Amex, Mastercard, Visa
or Roamer Plus will intercept first call to bill you.

Identification numbers

City	Market	System	Billing	RSA
All served cities	001	00022	00000	MSA

For full coverage area, see Newburgh, NY and
Poughkeepsie, NY.

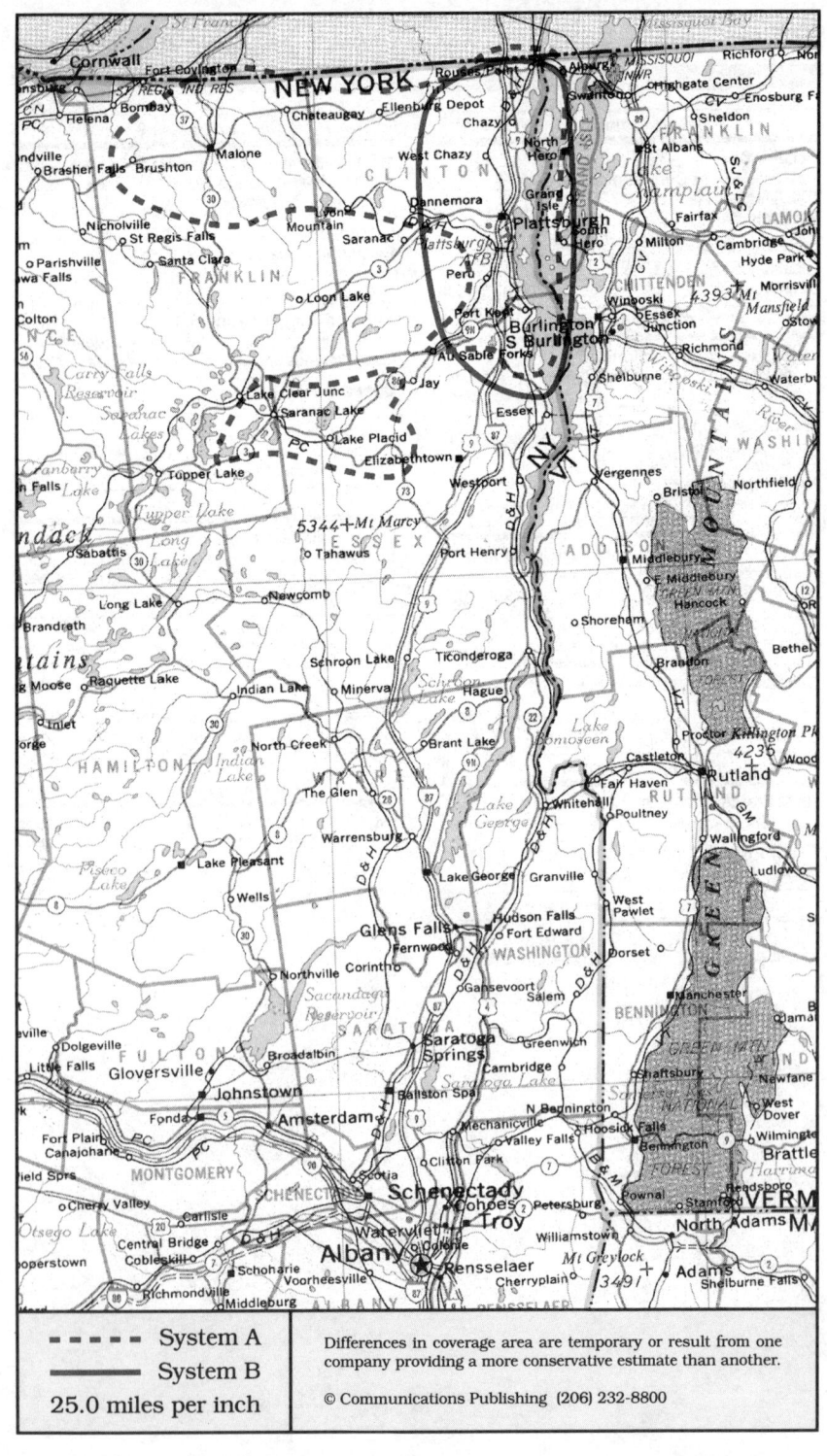

- - - - System A	Differences in coverage area are temporary or result from one company providing a more conservative estimate than another.
——— System B	
25.0 miles per inch	© Communications Publishing (206) 232-8800

Plattsburgh, New York

Dannemora, Lake Placid, Malone, Plattsburgh, Saranac Lake

System A	System B

Cellular company

Cellular One
Rural Cellular Management
332 Cornelia
Plattsburgh, NY 12901

Local office 518-561-1110
Corporate office 707-422-2100

Sending calls

Visitor dialing instructions
Local calls . . . . 7 digits
Long distance . . . 1 + area code + 7 digits

Speed-dial numbers
Customer service . . 518-561-1110 or *611
Emergency 911

Receiving calls

Caller dials the roamer access number below,
hears tone, dials cellular area code and number.
. . . . 518-572-7626

Billing information

Visitor rates
Most visitors . . . . . $3 day + 99¢ minute

Visitors without roaming agreements (p. 1159)
Roamer Plus will intercept first call to bill you.

Identification numbers

City	Market	System	Billing	RSA
All served cities	560	01507	00000	NY-2

Cellular company

Contel Cellular
65-67 S Peru Street
Plattsburgh, NY 12901

Customer service 800-333-4004
. . . . . extension 4002 404-391-8000
Local office 518-562-2355
Corporate office 404-804-3400

Sending calls

Visitor dialing instructions
Local calls . . . . . area code + 7 digits
Long distance . . . 1 + area code + 7 digits

Speed-dial numbers
Customer service 611
Directory assistance 411
Emergency 911

Receiving calls

Caller dials the roamer access number below,
hears tone, dials cellular area code and number.
. . . . 518-569-7626

Billing information

Visitor rates
Most visitors . . . . . $3 day + 85¢ minute

Visitors without roaming agreements (p. 1159)
Call co. to use Amex, Mastercard, Visa

Identification numbers

City	Market	System	Billing	RSA
All served cities	560	01508	00000	NY-2

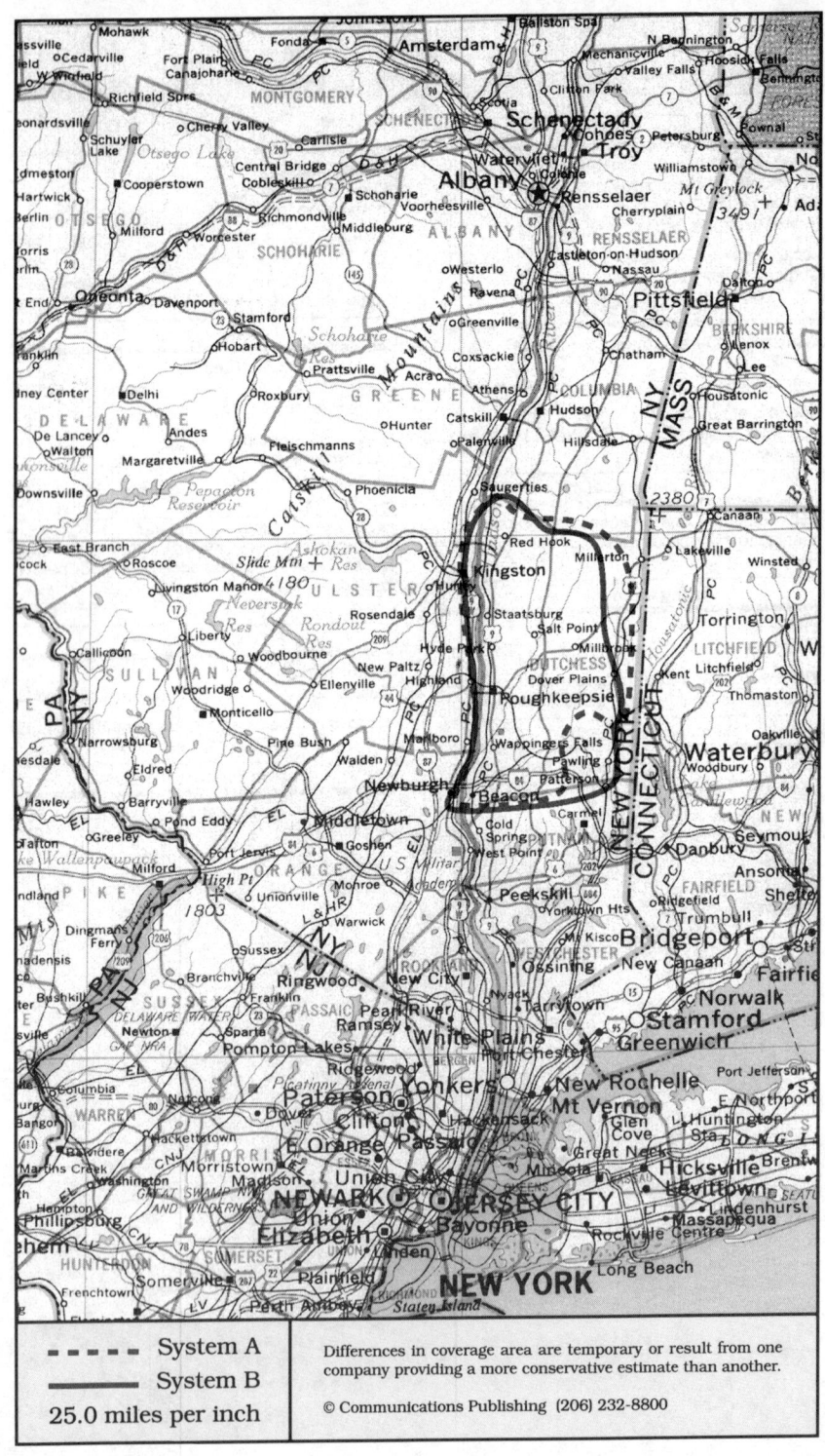

- - - - - System A	Differences in coverage area are temporary or result from one
———— System B	company providing a more conservative estimate than another.
25.0 miles per inch	© Communications Publishing (206) 232-8800

Poughkeepsie, New York
Beacon, Poughkeepsie (Dutchess County)

System A	System B

Cellular company

United States Cellular
272-A Route 9
Wappingers Falls, NY 12590

Local office	914-297-3444
Corporate office	312-399-8900

Sending calls

Visitor dialing instructions

Local calls	area code + 7 digits
Long distance	area code + 7 digits

Speed-dial numbers

Customer service	*611
Directory assistance	411
Emergency	911

Receiving calls

Caller dials the roamer access number below,
hears tone, dials cellular area code and number.
. . . . 914-453-7626

Billing information

Visitor rates
Most visitors . . . . . $3 day + 99¢ minute

Visitors without roaming agreements (p. 1159)
Call co. to use Mastercard, Visa

Identification numbers

City	Market	System	Billing	RSA
All served cities	151	00503	00000	MSA

Cellular company

NYNEX Mobile Communications
751 South Road
Poughkeepsie, NY 12601

Customer service	800-538-4747
. . . .	617-932-1200
Local office	914-298-7180
Corporate office	914-365-7200

Sending calls

Visitor dialing instructions

Local calls	area code + 7 digits
Long distance	area code + 7 digits

Speed-dial numbers

Customer service	611
Emergency	911
Local office	811
Report a drunk driver	*DWI or *394

Receiving calls

Caller dials the roamer access number below,
hears tone, dials cellular area code and number.
. . . . 914-474-7626

Follow Me Roaming forwarding (see p. 1156)
Activate, dial *18 send. Deactivate dial *19 send.

Billing information

Visitor rates
Most visitors . . . . . $3 day + 95¢ minute

Visitors without roaming agreements (p. 1159)
Call co. to use Amex, Mastercard, Visa
or Roamer Plus will intercept first call to bill you.

Identification numbers

City	Market	System	Billing	RSA
All served cities	151	00486	00000	MSA

For full coverage area, see Newburgh, NY and
New York, NY.

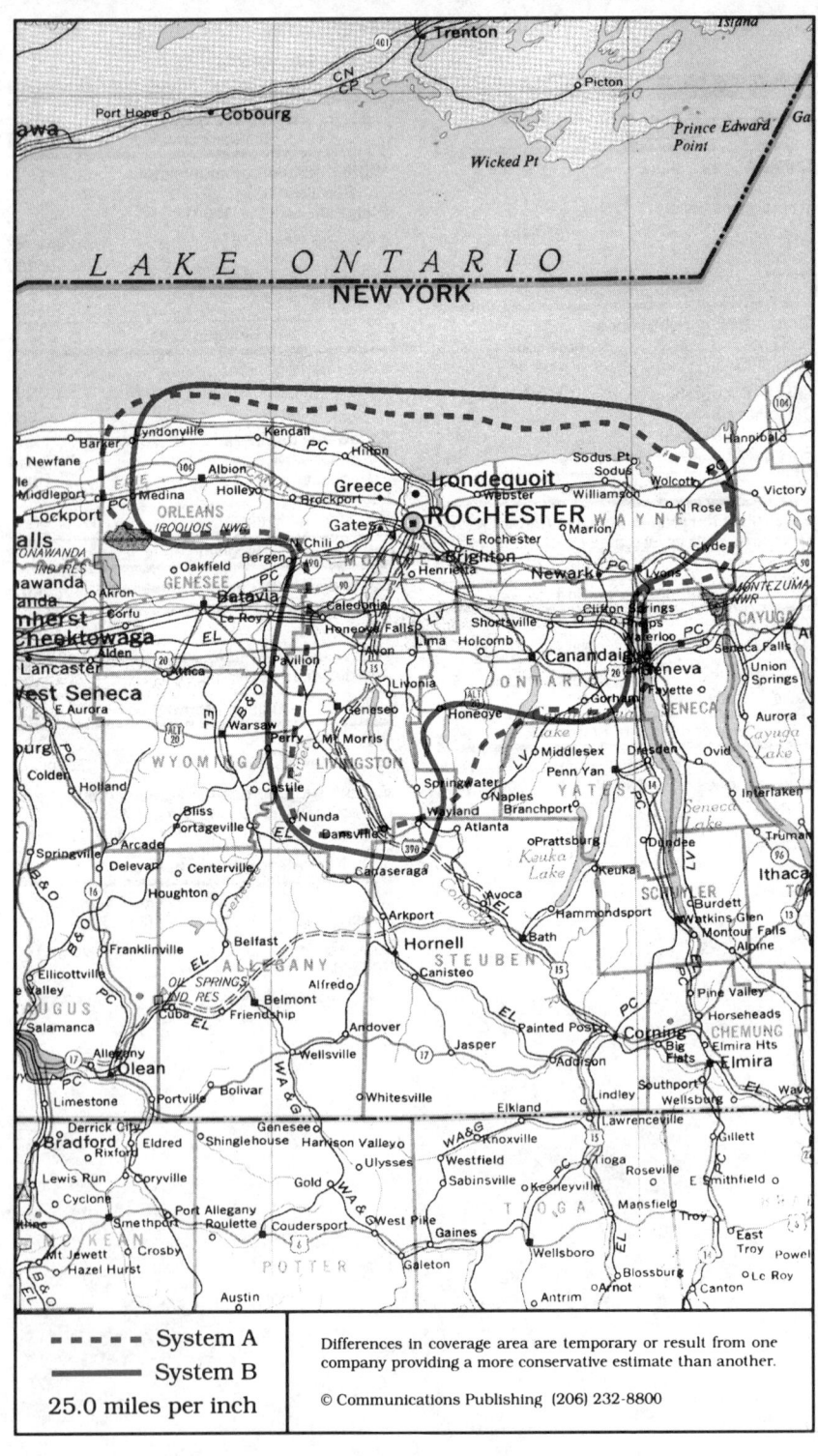

▪ ▪ ▪ ▪ ▪ System A ────── System B 25.0 miles per inch	Differences in coverage area are temporary or result from one company providing a more conservative estimate than another. © Communications Publishing (206) 232-8800

Rochester, New York
Albion, Canandaigua, Geneseo, Newark, Rochester

System A	System B

Cellular company

Cellular One
Genesee Telephone
Associated Communications
One Marine Midland Plaza, Suite 1300
Rochester, NY 14604

Customer service	some areas	800-829-4785
. . . .		716-232-6600
Local office		716-325-7000
Corporate office		412-281-1907

Sending calls

Visitor dialing instructions

Local calls	area code + 7 digits
Long distance . . .	1 + area code + 7 digits

Speed-dial numbers

American Automobile Association . . .	∗222
Customer service	∗611
Directory assistance	411
Emergency	911
Sales office	811

Receiving calls

Caller dials the roamer access number below, hears tone, dials cellular area code and number.
. . . . 716-729-7626

NationLink & RoamingAmerica (see p. 1157)
Caller hears your current roamer access number. This company's subscribers have level 0 service.

North American Cellular Network (see p. 1156)
All calls forward to you. Deactivate dial ∗35 send.

Billing information

Visitor rates
Most visitors . . . . . $3 day + 95¢ minute

Visitors without roaming agreements (p. 1159)
Roamer Plus will intercept first call to bill you.

Identification numbers

City	Market	System	Billing	RSA
All served cities	034	00117	00000	MSA

Cellular company

Rochester Telephone Mobile Communications
2060 Brighton-Henrietta Town Line Road
Rochester, NY 14623

Corporate office	800-724-7862
. . . .	716-274-7000

Sending calls

Visitor dialing instructions

Local calls	area code + 7 digits
Long distance	area code + 7 digits

Speed-dial numbers

American Automobile Association	∗AAA or ∗222
Coast Guard (emergency) .	∗SOS or ∗767
Customer service	611
Directory assistance	411
Emergency	911
Report a drunk driver	∗DWI or ∗394

Receiving calls

Caller dials the roamer access number below, hears tone, dials cellular area code and number.Caller dials the roamer access number below, hears tone, dials cellular area code and number.
. . . . 716-721-7626

Follow Me Roaming forwarding (see p. 1156)
Activate, dial ∗18 send. Deactivate dial ∗19 send.

Billing information

Visitor rates
Most visitors . . . . . $3 day + 95¢ minute

Visitors without roaming agreements (p. 1159)
Call co. to use Mastercard, Visa

Identification numbers

City	Market	System	Billing	RSA
All served cities	034	00154	00000	MSA

For full coverage area, see Jamestown, NY.

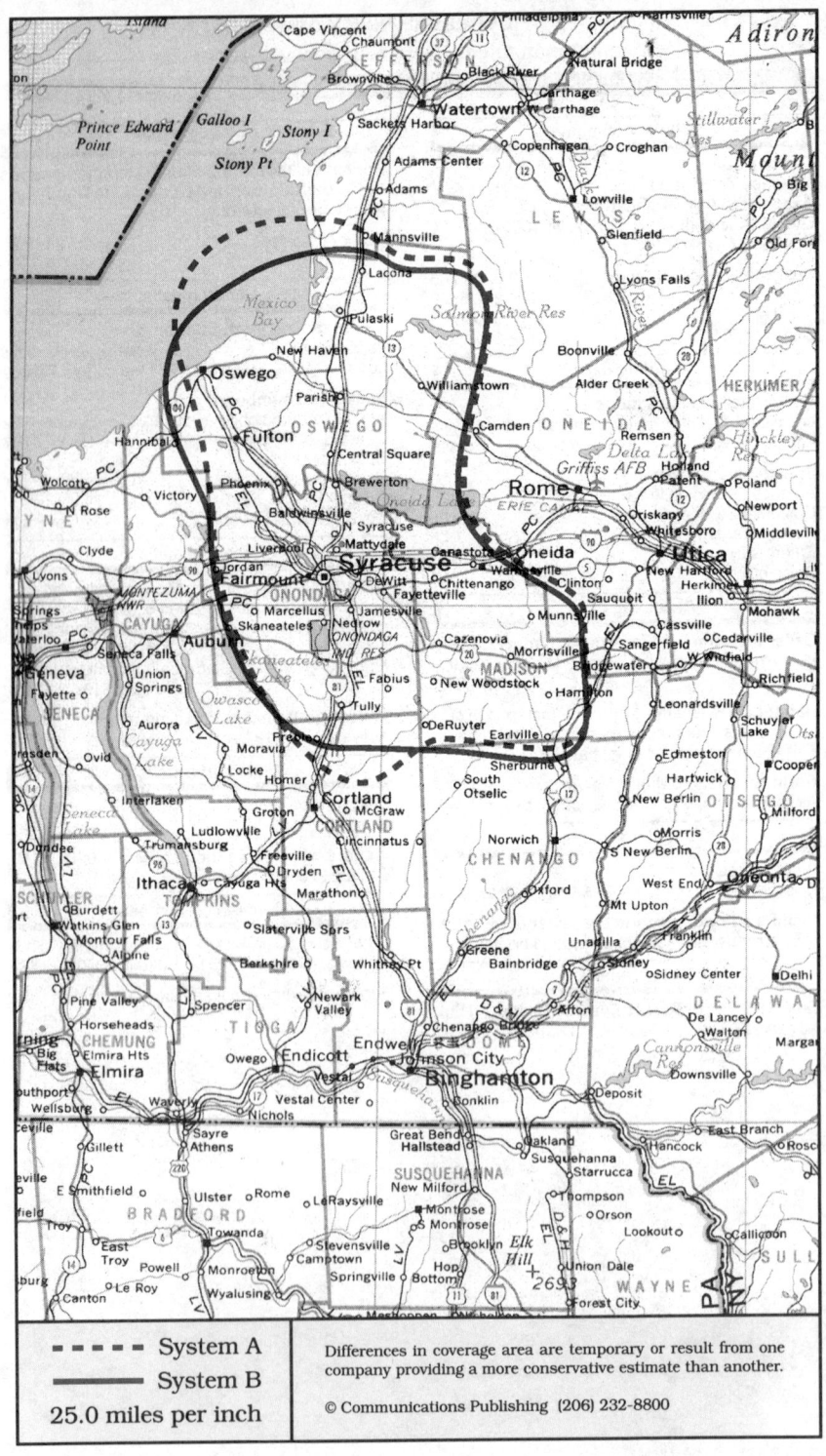

Syracuse, New York
Fulton, Oswego, Syracuse, Wampsville

System A	System B

Cellular company

Cellular One
Syracuse Telephone
2910 Erie Boulevard E
Syracuse, NY 13224

Customer service	800-541-8890
Corporate office	315-768-4400

Sending calls

Visitor dialing instructions

Local calls	area code + 7 digits
Long distance . . .	1 + area code + 7 digits

Speed-dial numbers

Customer service	*611
Emergency	911
Make a traffic report	984
Weather	786

Receiving calls

Caller dials the roamer access number below,
hears tone, dials cellular area code and number.
. . . . 315-447-7626

NationLink & RoamingAmerica (see p. 1157)
Dial *31 send to receive calls, *30 to deactivate.
Dial *32 send to tell caller the roamer access no.
This company's subscribers have level 8 service.

Billing information

Visitor rates
Most visitors . . . . . $3 day + 85¢ minute

Visitors without roaming agreements (p. 1159)
Call co. to use Amex, Discover, Mastercard, Visa
or Roamer Plus will intercept first call to bill you.

Identification numbers

City	Market	System	Billing	RSA
All served cities	053	00077	00000	MSA

Cellular company

NYNEX Mobile Communications
3433 Erie Boulevard E
DeWitt, NY 13214

Customer service	800-538-4747
. . . .	617-932-1200
Local office	315-445-9060
Corporate office	914-365-7200

Sending calls

Visitor dialing instructions

Local calls	area code + 7 digits
Long distance	area code + 7 digits

Speed-dial numbers

Customer service	611
Emergency	911
Local office	811
Make a traffic report	*570
Report drunk driving . . .	*DWI or *394
Roaming information	*711

Receiving calls

Caller dials the roamer access number below,
hears tone, dials cellular area code and number.
. . . . 315-427-7626

Follow Me Roaming forwarding (see p. 1156)
Activate, dial *18 send. Deactivate dial *19 send.

Billing information

Visitor rates
Most visitors . . . . . $3 day + 95¢ minute

Visitors without roaming agreements (p. 1159)
Call co. to use Amex, Mastercard, Visa
or Roamer Plus will intercept first call to bill you.

Identification numbers

City	Market	System	Billing	RSA
All served cities	053	00086	00000	MSA

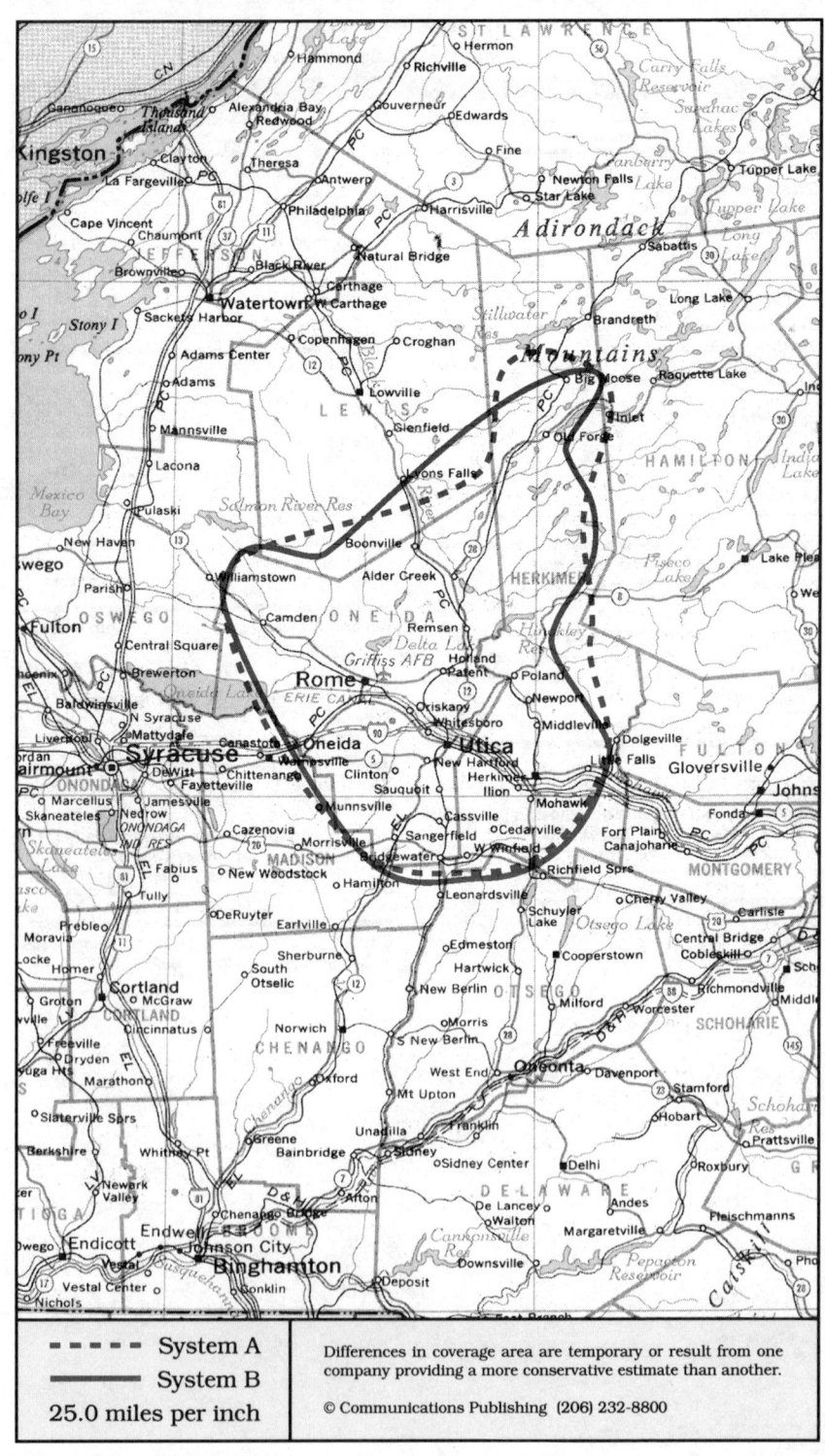

- - - - System A
——— System B
25.0 miles per inch

Differences in coverage area are temporary or result from one company providing a more conservative estimate than another.

© Communications Publishing (206) 232-8800

System A	System B

Cellular company

Cellular One
Utica Telephone
4874 Commercial Drive
New York Mills, NY 13417

Customer service	800-541-8890
. . . .	315-796-2000
Corporate office	315-768-4400

Sending calls

Visitor dialing instructions

Local calls	area code + 7 digits
Long distance . . .	1 + area code + 7 digits

Speed-dial numbers

Customer service	✳611
Emergency	911
New York Mills office	✳612
Weather	786

Receiving calls

Caller dials the roamer access number below,
hears tone, dials cellular area code and number.

. . . .	315-796-7626

Billing information

Visitor rates

Most visitors	$3 day + 85¢ minute

Visitors without roaming agreements (p. 1159)
Roamer Plus will intercept first call to bill you.

Identification numbers

City	Market	System	Billing	RSA
All served cities	115	00235	00000	MSA

Cellular company

Advantage Cellular
Rochester Telephone
258 Genesee Street, Suite 101
Utica, NY 13502

Local office	315-797-4444
Corporate office	716-274-7000

Sending calls

Visitor dialing instructions

Local calls	area code + 7 digits
Long distance	area code + 7 digits

Speed-dial numbers

Customer service	611
Emergency	911

Receiving calls

Caller dials the roamer access number below,
hears tone, dials cellular area code and number.

Rome	315-335-7626
Utica	315-794-7626

Follow Me Roaming forwarding (see p. 1156)
Activate, dial ✳18 send. Deactivate dial ✳19 send.

Billing information

Visitor rates

Most visitors	$3 day + 95¢ minute

Visitors without roaming agreements (p. 1159)
Call co. to use Mastercard, Visa

Identification numbers

City	Market	System	Billing	RSA
All served cities	115	00226	00000	MSA

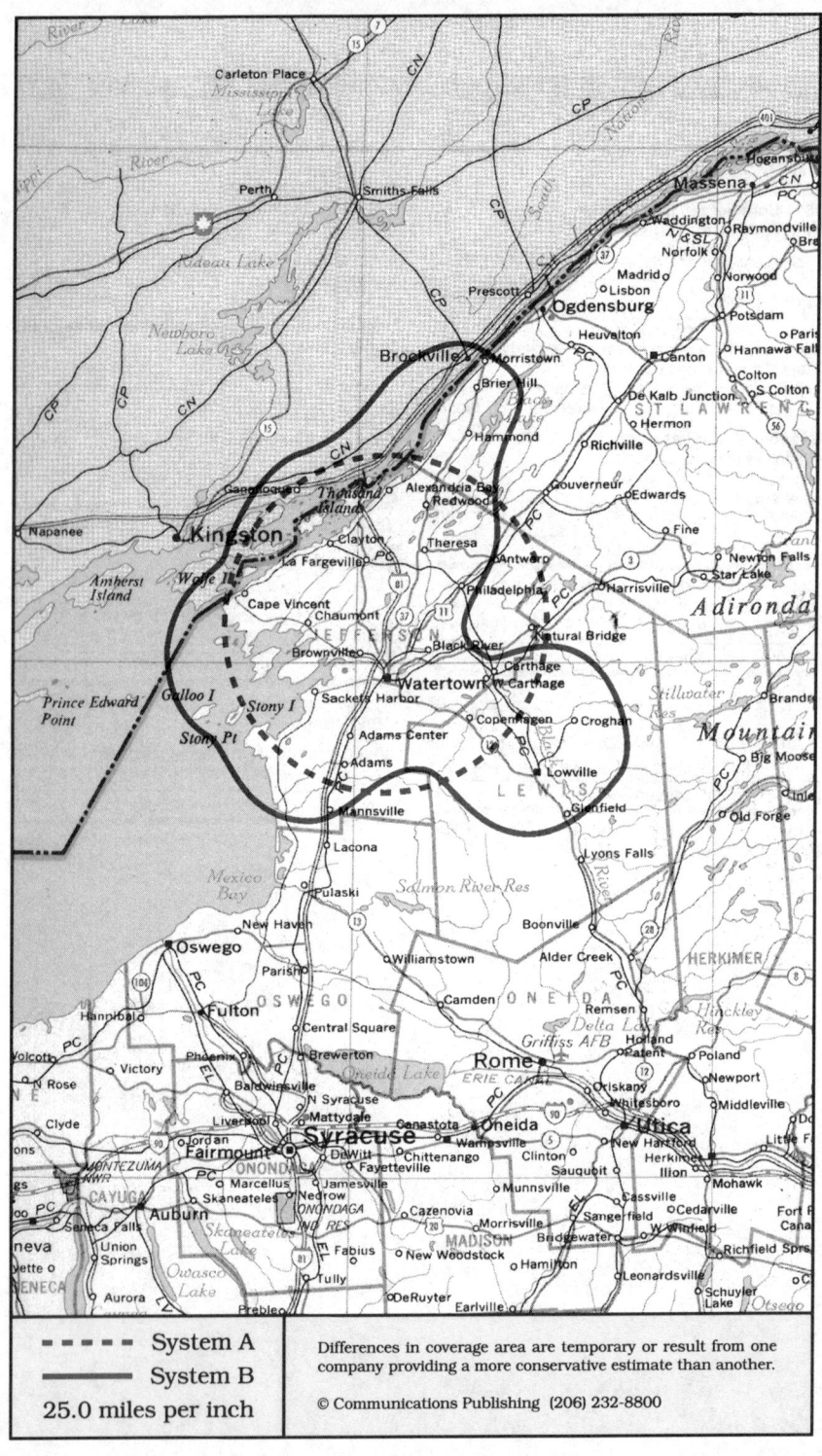

System A
System B
25.0 miles per inch

Differences in coverage area are temporary or result from one company providing a more conservative estimate than another.

© Communications Publishing (206) 232-8800

Watertown, New York
Lowville, Watertown

System A	System B

Cellular company

Cellular One
GMD Limited Partnership
Route 12 S; HC-31, Box 69
Watertown, NY 13601

Local office 800-849-8555
. . . . 315-788-3338
Corporate office 919-321-0066

Sending calls

Visitor dialing instructions
Local calls . . . . area code + 7 digits
Long distance . . . 1 + area code + 7 digits

Speed-dial numbers
Customer service 611
Emergency 911
Highway patrol . . . . . . ∗DWI or ∗394

Receiving calls

Caller dials the roamer access number below,
hears tone, dials cellular area code and number.
. . . . 315-771-7626

NationLink & RoamingAmerica (see p. 1157)
Dial ∗31 send to receive calls, ∗30 to deactivate.
This company's subscribers have level 3 service.

Billing information

Visitor rates
Most visitors . . . . . $3 day + 85¢ minute

Visitors without roaming agreements (p. 1159)
Dial 0 + 10-digit no. Operator will intercept to bill.

Identification numbers

City	Market	System	Billing	RSA
All served cities	559	01505	00000	NY-1

Cellular company

NYNEX Mobile Communications
3433 Erie Boulevard E
DeWitt, NY 13214

Customer service 800-538-4747
. . . . 617-932-1200
Local office 315-445-9060
Corporate office 914-365-7200

Sending calls

Visitor dialing instructions
Local calls . . . . . . area code + 7 digits
Long distance . . . . . area code + 7 digits

Speed-dial numbers
Customer service 611
Emergency 911
Local office 811
Make a traffic report . . . . . . ∗570
Report drunk driving . . . ∗DWI or ∗394
Roaming information . . . . . . ∗711

Receiving calls

Caller dials the roamer access number below,
hears tone, dials cellular area code and number.
. . . . 315-783-7626

Billing information

Visitor rates
Most visitors . . . . . $3 day + 95¢ minute

Visitors without roaming agreements (p. 1159)
Call co. to use Amex, Mastercard, Visa
or Roamer Plus will intercept first call to bill you.

Identification numbers

City	Market	System	Billing	RSA
All served cities	559	01506	00000	NY-1

For full coverage area, see Syracuse, NY.

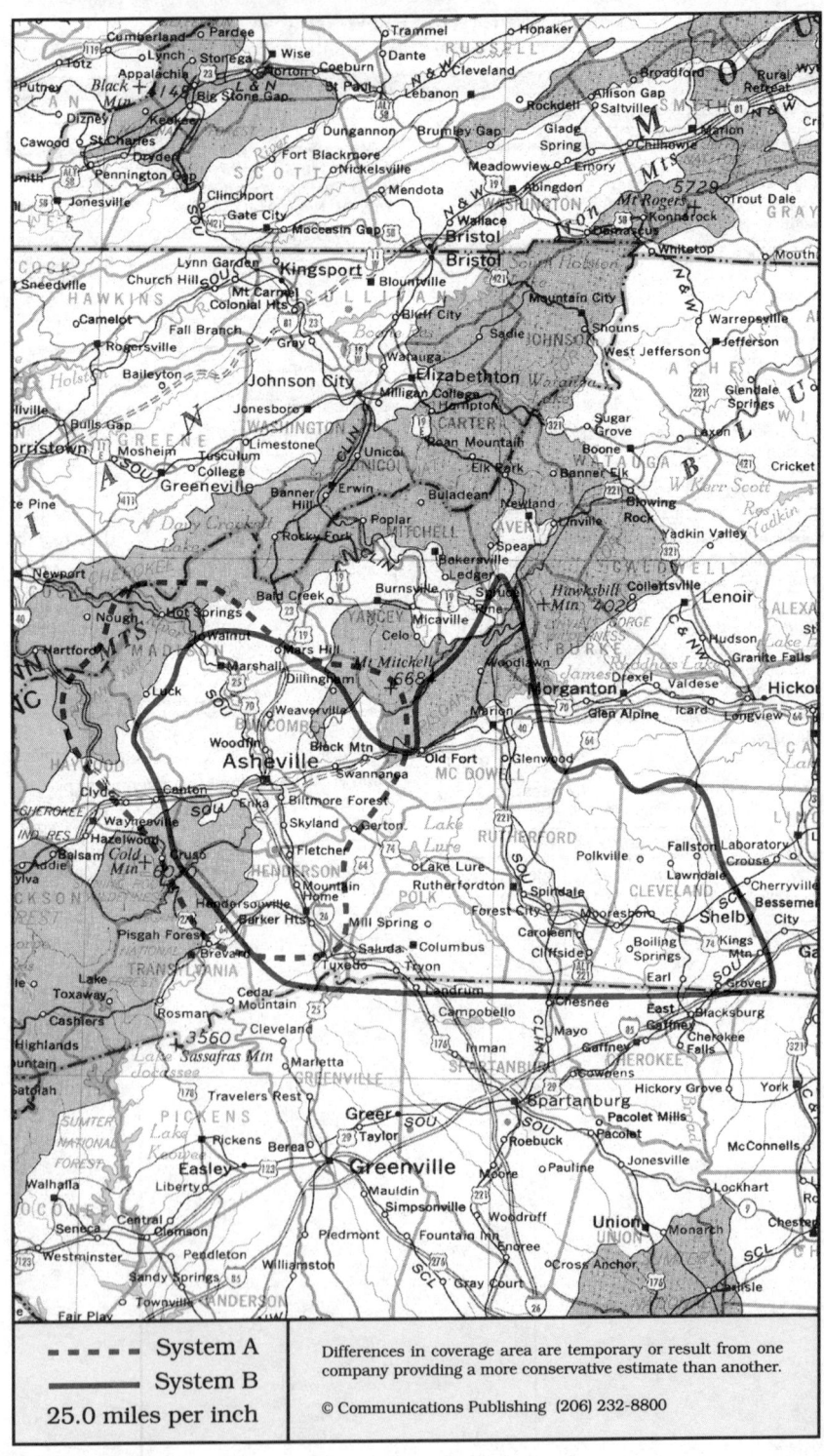

- - - - **System A**	Differences in coverage area are temporary or result from one company providing a more conservative estimate than another.
——— **System B**	
25.0 miles per inch	© Communications Publishing (206) 232-8800

Asheville, North Carolina
Asheville, Columbus, Hendersonville, Marion, Rutherfordton, Shelby

System A	System B

Cellular company

GTE Cellular Services
GTE Mobilnet
16-G Regent Park Boulevard
Asheville, NC 28806

Customer service	800-727-2444
. . . .	919-481-1181
Local office	704-251-2335
Corporate office	404-391-8000

Sending calls

Visitor dialing instructions

Local calls	7 digits
Long distance . . .	1 + area code + 7 digits

Speed-dial numbers

Customer service	611
Emergency	911

Receiving calls

Caller dials the roamer access number below,
hears tone, dials cellular area code and number.

. . . .	704-231-7626

NationLink & RoamingAmerica (see p. 1157)
Dial *31 send to receive calls, *30 to deactivate.
Dial *32 send to tell caller the roamer access no.
This company's subscribers have level 10 service

Billing information

Visitor rates

Most visitors	$3 day + 85¢ minute

Visitors without roaming agreements (p. 1159)
Call co. to use Amex, Mastercard, Visa

Identification numbers

City	Market	System	Billing	RSA
All served cities	183	00263	00000	MSA

Cellular company

United States Cellular
223 Haywood Street
Asheville, NC 28801

Local office	704-258-0000
Corporate office	312-399-8900

Sending calls

Visitor dialing instructions

Local calls	area code + 7 digits
Long distance . . .	1 + area code + 7 digits

Speed-dial numbers

Customer service	611
Emergency	911
Highway patrol	*HP or *47
Roaming information	*711

Receiving calls

Caller dials the roamer access number below,
hears tone, dials cellular area code and number.

Asheville	704-777-7626
Hendersonville	704-734-7626

Billing information

Visitor rates

Most visitors . . .	$2.50 day + 75¢ minute

Visitors without roaming agreements (p. 1159)
Call co. to use Mastercard, Visa

Identification numbers

City	Market	System	Billing	RSA
Asheville	183	00246	00000	MSA
Shelby	568	01524	00000	NC-4

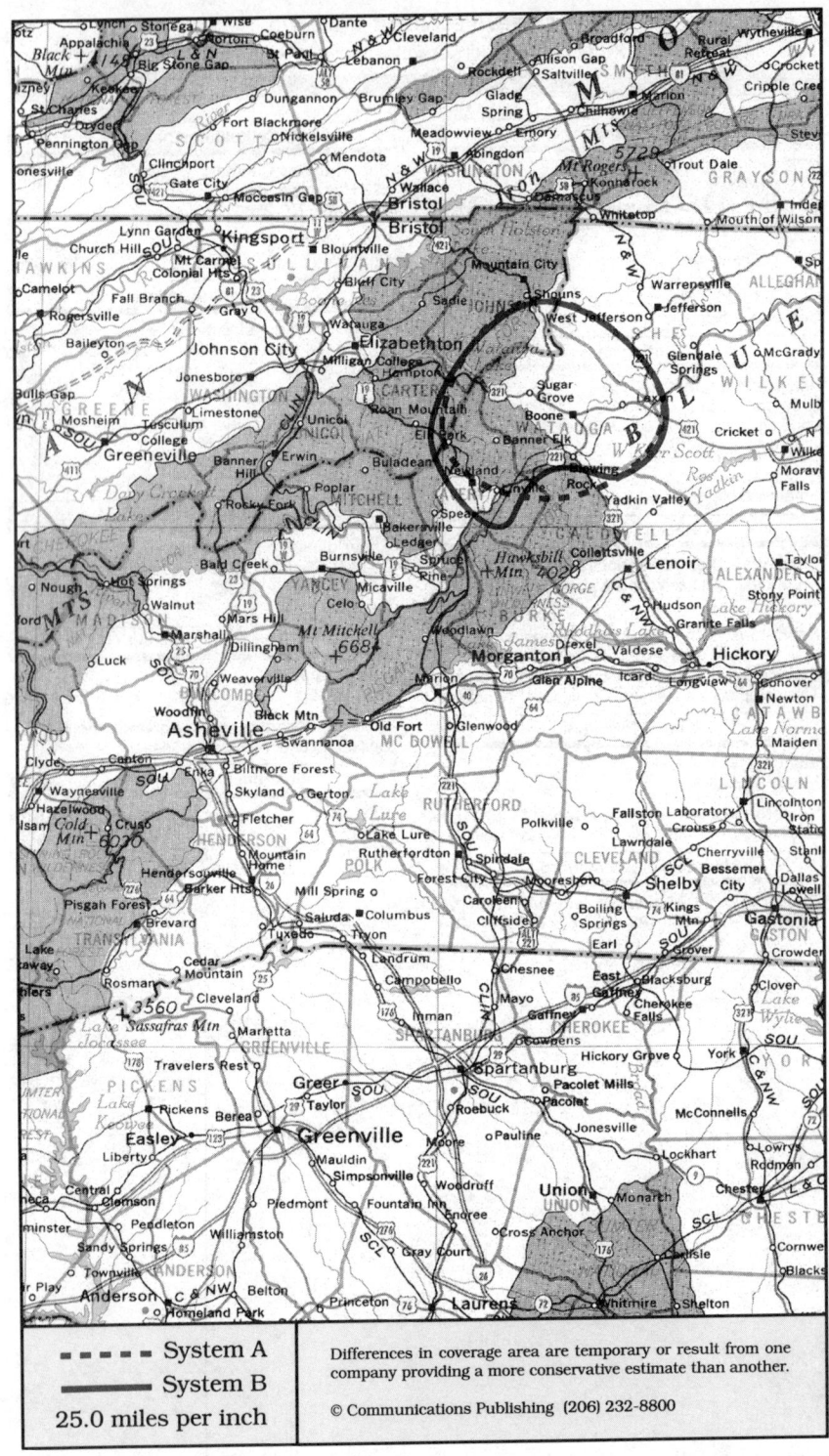

Boone, North Carolina
Boone (Avery and Watauga Counties), Newland

System A	System B

Cellular company

Cellular One
Blue Ridge Cellular Telephone
452 Harper Avenue
Lenoir, NC 28645

Corporate office 800-951-2351
. . . . 704-758-4444

Sending calls

Visitor dialing instructions
Local calls . . . . area code + 7 digits
Long distance . . . 1 + area code + 7 digits

Speed-dial numbers
Customer service *611
Emergency 911

Receiving calls

Caller dials the roamer access number below,
hears tone, dials cellular area code and number.
. . . . 704-757-7626

Follow Me Roaming forwarding (see p. 1156)
Phone Me Anywhere forwarding (see p. 1156)
Activate, dial *18 send. Deactivate dial *19 send.

Billing information

Visitor rates
Most visitors . . . . . $3 day + 99¢ minute

Visitors without roaming agreements (p. 1159)
Call co. to use Mastercard, Visa
or Roamer Plus will intercept first call to bill you.

Identification numbers

City	Market	System	Billing	RSA
All served cities	566	01519	00000	NC-2

For full coverage area, see Burnsville, NC and
Lenoir, NC.

Cellular company

Carolina West Cellular
Highway 421 W, PO Box 738
Millers Creek, NC 28651

Corporate office . some areas 800-235-5007
. . . . 919-973-5000

Sending calls

Visitor dialing instructions
Local calls . . . . . . area code + 7 digits
Long distance . . . 1 + area code + 7 digits

Speed-dial numbers
Customer service *611
Emergency 911

Receiving calls

Caller dials the roamer access number below,
hears tone, dials cellular area code and number.
. . . . 919-957-7626

Follow Me Roaming forwarding (see p. 1156)
Activate, dial *18 send. Deactivate dial *19 send.

Billing information

Visitor rates
Most visitors . . . . . $2 day + 65¢ minute

Visitors without roaming agreements (p. 1159)
Cannot call since company takes no credit cards.

Identification numbers

City	Market	System	Billing	RSA
All served cities	566	01522	00000	NC-2

For full coverage area, see Mount Airy, NC.

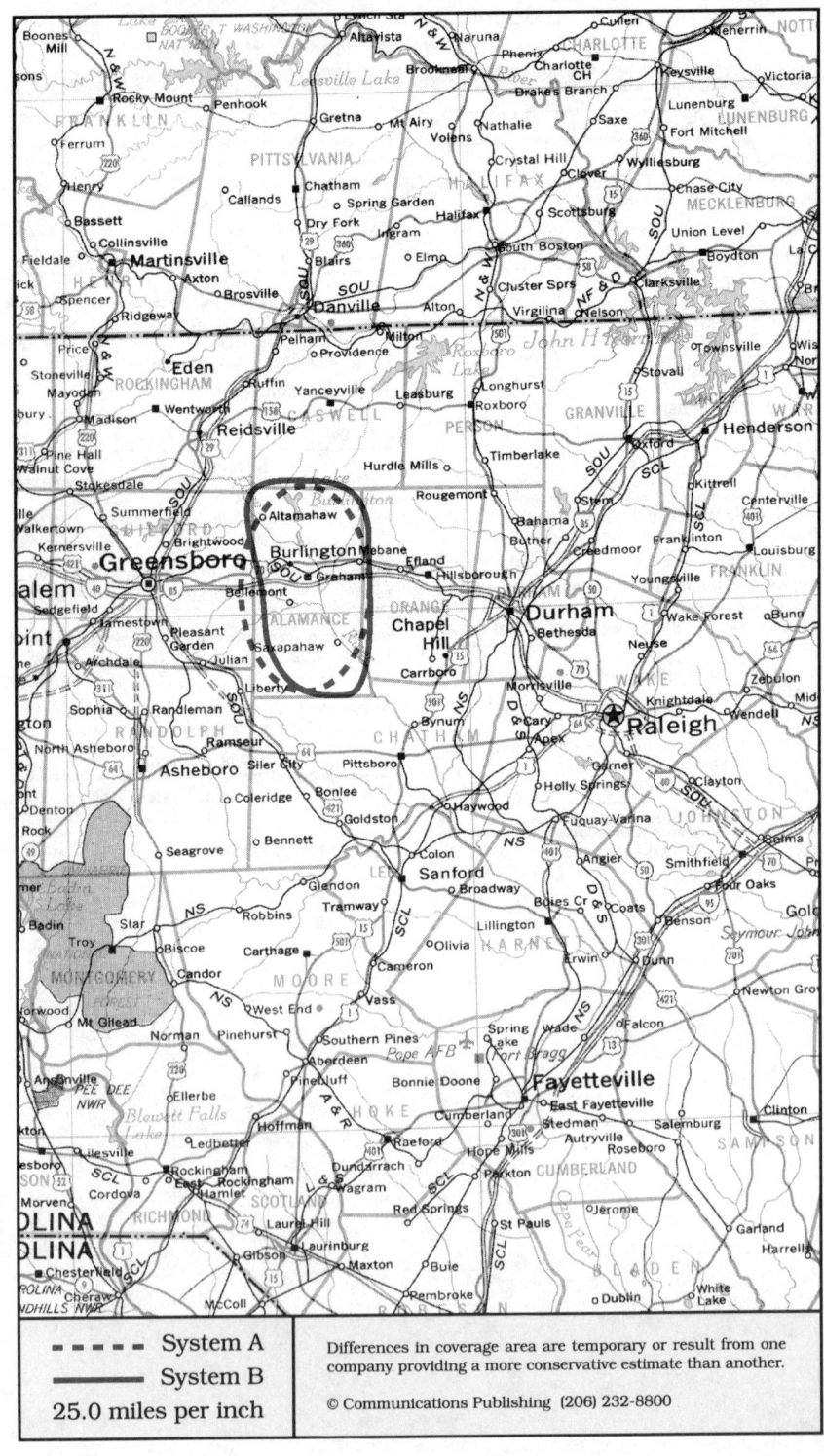

Burlington, North Carolina

System A	System B

Cellular company

Cellular One
GenCell Management
2103 Maple Avenue
Burlington, NC 27215

Customer service	800-888-7868
Local office	919-226-1500
Corporate office	707-425-8000

Sending calls

Visitor dialing instructions

Local calls	area code + 7 digits
Long distance . . .	1 + area code + 7 digits

Speed-dial numbers

Customer service	*611
Emergency	911

Receiving calls

Caller dials the roamer access number below, hears tone, dials cellular area code and number.
. . . . 919-263-7626

NationLink & RoamingAmerica (see p. 1157)
Dial *31 send to receive calls, *30 to deactivate. This company's subscribers have level 3 service.

Billing information

Visitor rates

Most visitors	$3 day + 99¢ minute

Visitors without roaming agreements (p. 1159)
American Roaming Network intercepts call to bill.

Identification numbers

City	Market	System	Billing	RSA
All served cities	280	00299	00000	MSA

Cellular company

Centel Cellular
2475 S Church Street
Burlington, NC 27215

Customer service	800-788-2260
. . . .	919-833-7494
Local office	919-228-6990
Corporate office	312-399-2644

Sending calls

Visitor dialing instructions

Local calls	area code + 7 digits
Long distance . . .	1 + area code + 7 digits

Speed-dial numbers

Customer service	*611
Emergency	911
Highway patrol	*HP or *47

Receiving calls

Caller dials the roamer access number below, hears tone, dials cellular area code and number.
. . . . 919-260-7626

Billing information

Visitor rates

Most visitors . . .	$2.50 day + 75¢ minute

Visitors without roaming agreements (p. 1159)
Call co. to use Amex, Mastercard, Visa

Identification numbers

City	Market	System	Billing	RSA
All served cities	280	00144	30052	MSA

For full coverage area, see Greensboro, NC and Raleigh, NC and Rocky Mount, NC.

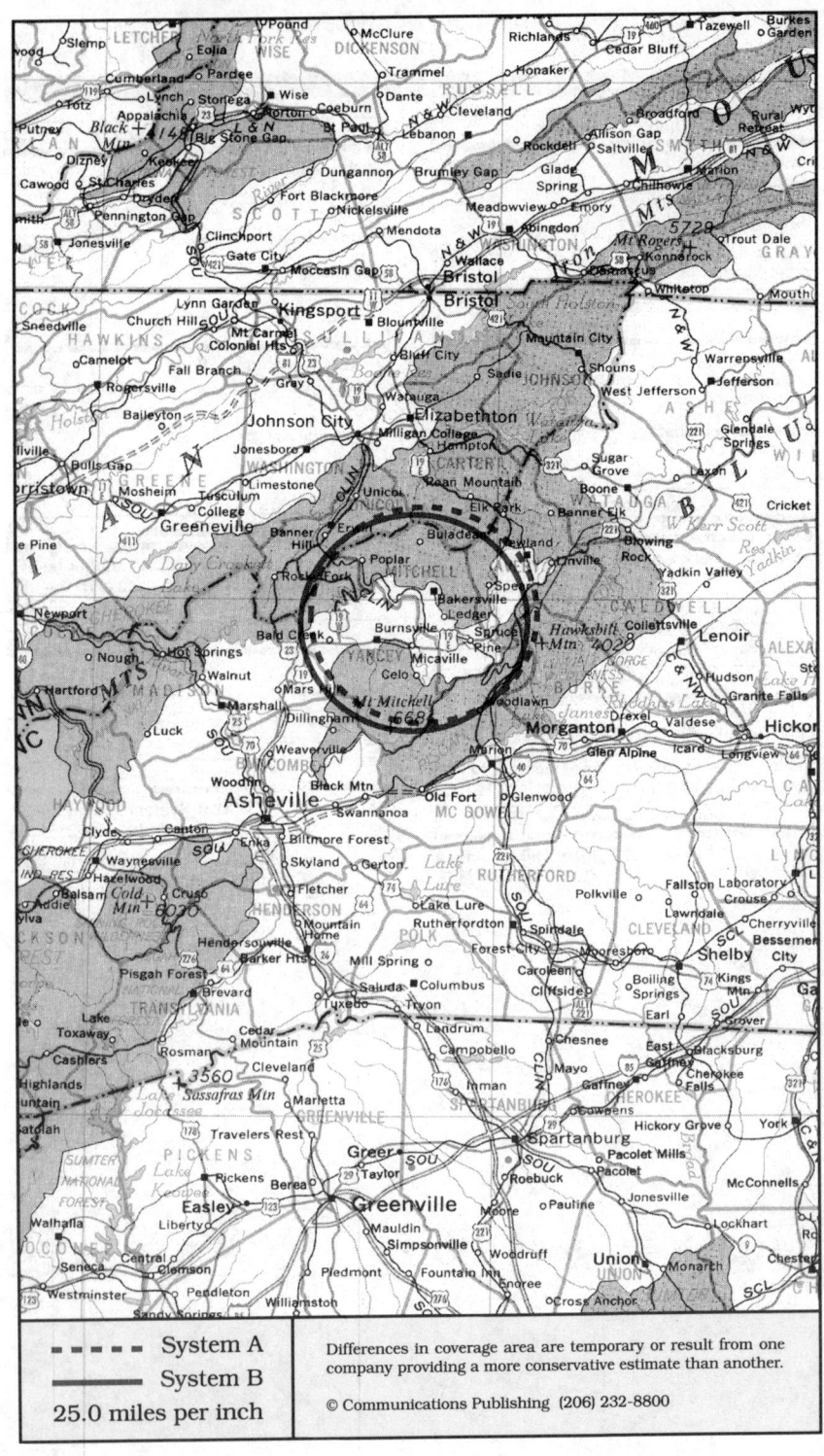

Burnsville, North Carolina
Bakersville, Burnsville (Mitchell and Yancey Counties)

System A	System B

Cellular company

Cellular One
Blue Ridge Cellular Telephone
452 Harper Avenue
Lenoir, NC 28645

Corporate office 800-951-2351
. . . . 704-758-4444

Sending calls

Visitor dialing instructions
Local calls . . . . . area code + 7 digits
Long distance . . . 1 + area code + 7 digits

Speed-dial numbers
Customer service . . . . ∗611
Emergency 911

Receiving calls

Caller dials the roamer access number below,
hears tone, dials cellular area code and number.
. . . . 704-757-7626

Follow Me Roaming forwarding (see p. 1156)
Phone Me Anywhere forwarding (see p. 1156)
Activate, dial ∗18 send. Deactivate dial ∗19 send.

Billing information

Visitor rates
Most visitors . . . . . $3 day + 99¢ minute

Visitors without roaming agreements (p. 1159)
Call co. to use Mastercard, Visa
or Roamer Plus will intercept first call to bill you.

Identification numbers

City	Market	System	Billing	RSA
All served cities	566	01519	00000	NC-2

For full coverage area, see Boone, NC.

Cellular company

United States Cellular
223 Haywood Street
Asheville, NC 28801

Local office 704-258-0000
Corporate office 312-399-8900

Sending calls

Visitor dialing instructions
Local calls . . . . . area code + 7 digits
Long distance . . . 1 + area code + 7 digits

Speed-dial numbers
Customer service 611
Emergency 911
Highway patrol . . . . ∗HP or ∗47
Roaming information . . . . . . ∗711

Receiving calls

Caller dials the roamer access number below,
hears tone, dials cellular area code and number.
. . . . 704-777-7626

Billing information

Visitor rates
Most visitors . . . $2.50 day + 75¢ minute

Visitors without roaming agreements (p. 1159)
Call co. to use Mastercard, Visa

Identification numbers

City	Market	System	Billing	RSA
All served cities	566	01524	30596	NC-2

- - - -	System A
————	System B
25.0 miles per inch	

Differences in coverage area are temporary or result from one company providing a more conservative estimate than another.

© Communications Publishing (206) 232-8800

Charlotte, North Carolina
Albemarle, Charlotte, Concord, Gaffney SC, Rock Hill SC, Salisbury

System A	System B

Cellular company

Bell Atlantic Mobile
4700 Sweden Road
Charlotte, NC 28273

Customer service	704-553-9893
. . . .	803-234-7954
Local office	704-552-5185
Corporate office	908-306-7000

Sending calls

Visitor dialing instructions

Local calls	area code + 7 digits
Long distance	area code + 7 digits

Speed-dial numbers

Channel 9 news	*09
Channel 36	*36
Customer service	611
Emergency	911
Highway patrol	*HP or *47

Receiving calls

Caller dials the roamer access number below,
hears tone, dials cellular area code and number.

Charlotte, Rock Hill	704-564-7626
Gaffney	803-270-7626
Salisbury	704-785-7626

NationLink & RoamingAmerica (see p. 1157)
Dial *31 send to receive calls, *30 to deactivate.
Dial *32 send to tell caller the roamer access no.
This company's subscribers have level 8 service.
Subscribers living in Gaffney have level 0 service.

Billing information

Visitor rates

Most visitors	$3 day + 99¢ minute

Visitors without roaming agreements (p. 1159)
Roamer Plus will intercept first call to bill you.

Identification numbers

City	Market	System	Billing	RSA
Charlotte	061	00139	00000	MSA
Gaffney	627	01641	00000	SC-3
Rock Hill	633	01653	00000	SC-9
Salisbury	579	01545	00000	NC-15

For full coverage area, see Columbia, SC and
Greenville, SC.

Cellular company

ALLTEL Mobile Communications
734 Tyvola Road
Charlotte, NC 28217

Customer service	some areas	800-255-8351
. . . .		704-529-8140
Local office		704-529-0001
Corporate office		501-661-8500

Sending calls

Visitor dialing instructions

Local calls	area code + 7 digits
Long distance	area code + 7 digits

Speed-dial numbers

Customer service	*611
Emergency	911
Highway patrol	*HP or *47
WBT radio	*928

Receiving calls

Caller dials the roamer access number below,
hears tone, dials cellular area code and number.

Albemarle	704-985-5426
Charlotte	704-534-7626
Chester	803-385-7626
Concord	704-783-7626
Gaffney	803-488-7626
Lancaster	803-283-7626
Mooresville	704-663-7626
Rock Hill	803-323-7626
Salisbury	704-639-4626

Follow Me Roaming forwarding (see p. 1156)
Activate, dial *18 send. Deactivate dial *19 send.

Billing information

Visitor rates

Most visitors	$3 day + 85¢ minute

Visitors without roaming agreements (p. 1159)
Call co. to use Amex, Discover, Mastercard, Visa

Identification numbers

City	Market	System	Billing	RSA
Charlotte	061	00114	00000	MSA
Gaffney, Chester	627	00114	30248	SC-3
Rock Hill, Lancaster	633	00014	30246	SC-9
Salisbury, Concord	579	00114	30240	NC-15

For full coverage area, see Greenville, SC and
Rockingham, NC.

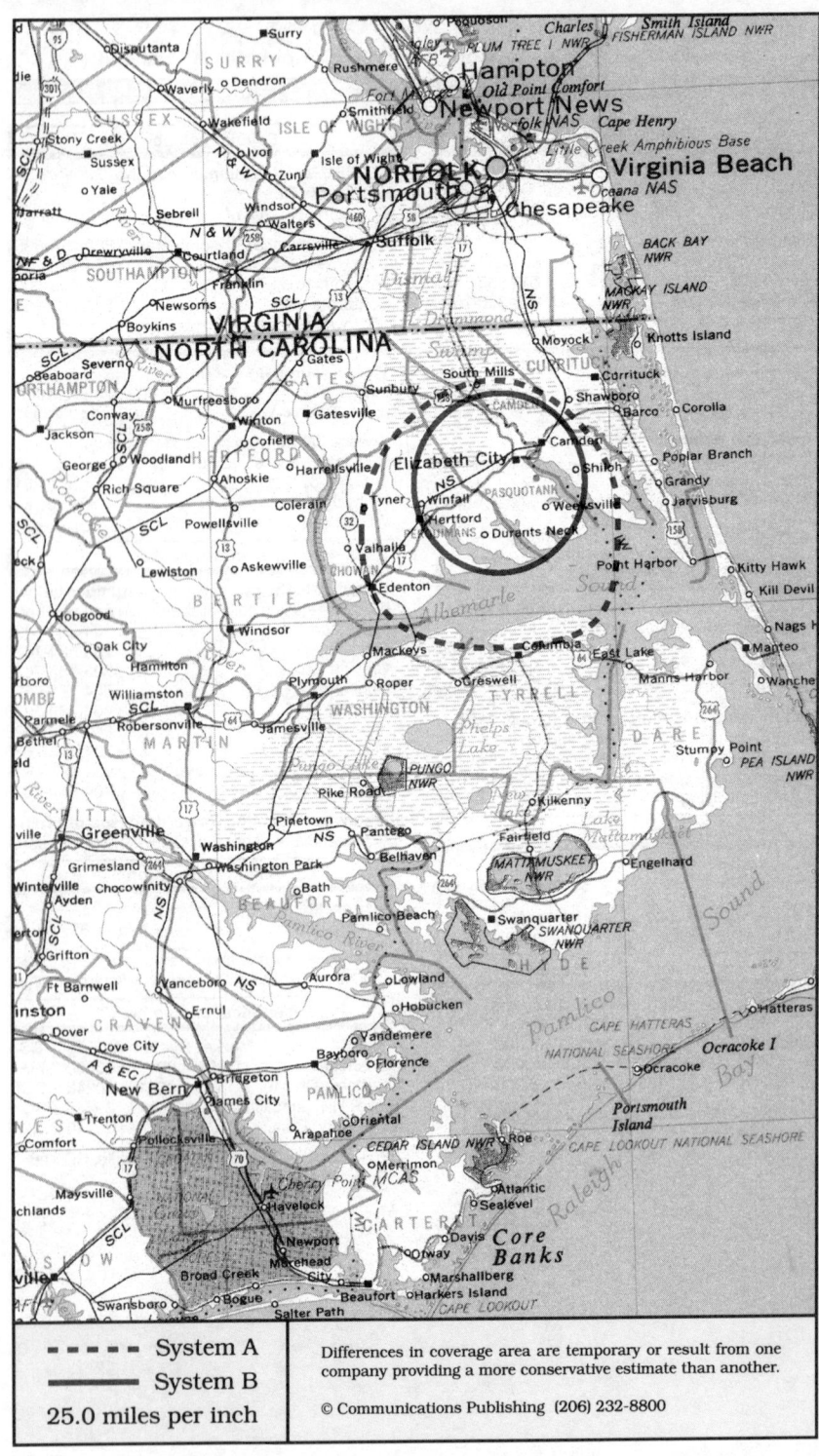

▬ ▬ ▬ System A	Differences in coverage area are temporary or result from one company providing a more conservative estimate than another.
▬▬▬▬ System B	
25.0 miles per inch	© Communications Publishing (206) 232-8800

Elizabeth City, North Carolina

System A	System B

Cellular company

United States Cellular
205 SW Greenville Boulevard, Suite 900
Greenville, NC 27834

Local office . .	some areas	800-231-2355
. . . .		919-321-2666
Corporate office		312-399-8900

Cellular company

Centel Cellular
916 N Church Street
Rocky Mount, NC 27804

Customer service	800-788-2260
. . . .	919-833-7494
Local office	919-446-1641
Corporate office	312-399-2644

Sending calls

Visitor dialing instructions

Local calls	area code + 7 digits
Long distance . . .	1 + area code + 7 digits

Speed-dial numbers

Customer service	611
Emergency	911
Highway patrol	*HP or *47
Roaming information	*711

Sending calls

Visitor dialing instructions

Local calls	area code + 7 digits
Long distance . . .	1 + area code + 7 digits

Speed-dial numbers

Customer service	611
Emergency	911
Highway patrol	*HP or *47

Receiving calls

Caller dials the roamer access number below,
hears tone, dials cellular area code and number.
. . . . 919-714-7626

Receiving calls

Caller dials the roamer access number below,
hears tone, dials cellular area code and number.
. . . . 919-331-7626

Billing information

Visitor rates
Most visitors . . . . $2.50 day + 75¢ minute

Visitors without roaming agreements (p. 1159)
Cannot call since company takes no credit cards.

Billing information

Visitor rates
Most visitors . . . $2.50 day + 75¢ minute

Visitors without roaming agreements (p. 1159)
Call co. to use Amex, Mastercard, Visa

Identification numbers

City	Market	System	Billing	RSA
All served cities	573	01533	00000	NC-9

Identification numbers

City	Market	System	Billing	RSA
All served cities	573	01534	00000	NC-9

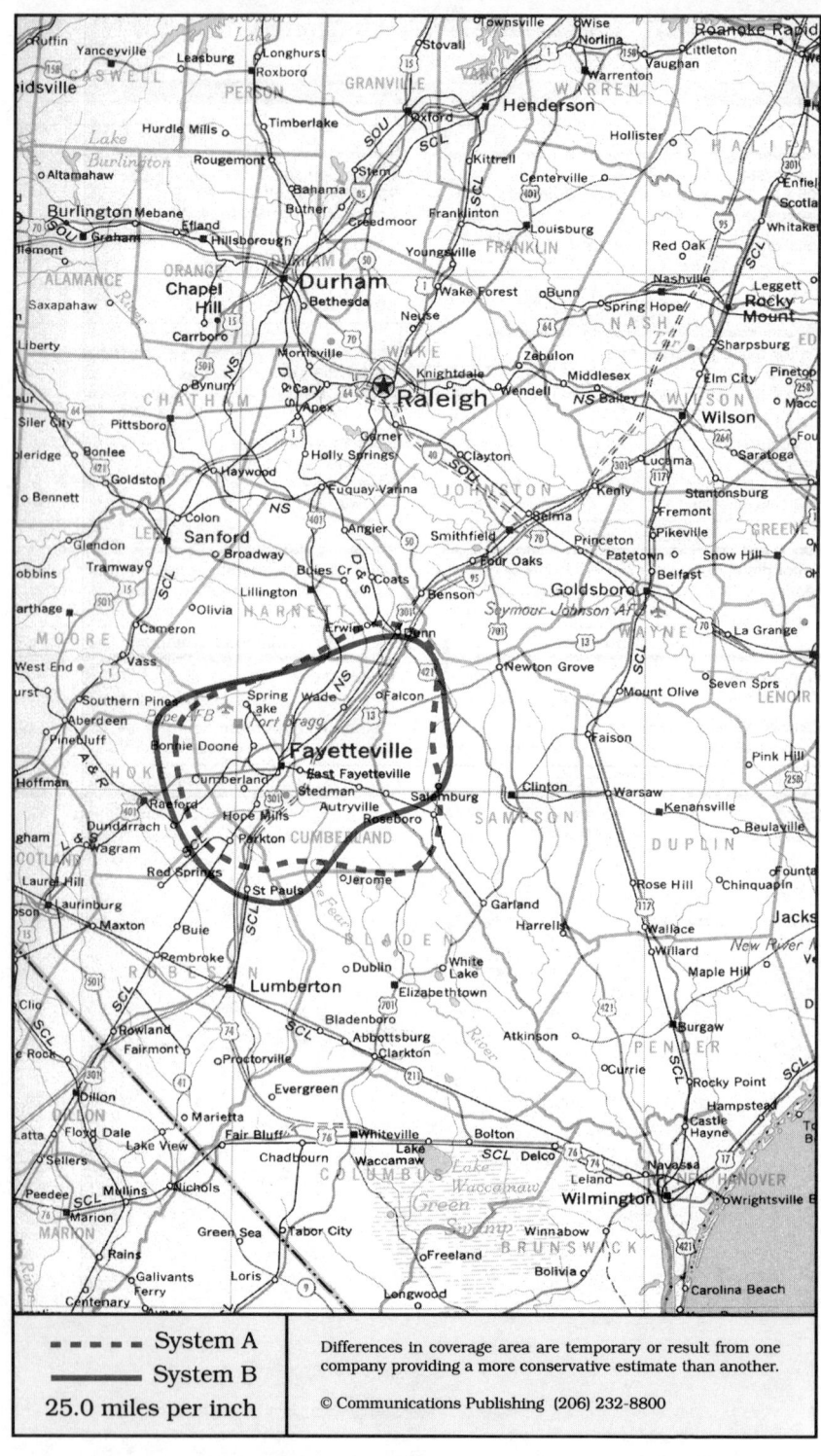

Fayetteville, North Carolina

Cellular company

Cellular One
GTE Mobilnet
3013-C Raeford Road
Fayetteville, NC 28303

Customer service	. . . . .	800-727-2444
. . . .		919-481-1181
Local office	. . . . . .	919-485-8605
Corporate office	. . . .	404-391-8000

Sending calls

Visitor dialing instructions

| Local calls | . . . . | area code + 7 digits |
| Long distance | . . . | 1 + area code + 7 digits |

Speed-dial numbers

Customer service	. . . .	611
Emergency	. . . .	911
Highway patrol	. . . . . .	*HP or *47
Local police	. . . . .	*LAW or *529
Make a traffic	. . . .	*95
Military police	. . . . .	*MP or *67
Sheriff department	. . . . . .	*1500

Receiving calls

Caller dials the roamer access number below,
hears tone, dials cellular area code and number.
. . . . 919-391-7626

Follow Me Roaming forwarding (see p. 1156)
Phone Me Anywhere forwarding (see p. 1156)
Activate, dial *18 send. Deactivate dial *19 send.

NationLink & RoamingAmerica (see p. 1157)
Dial *31 send to receive calls, *30 to deactivate.
Dial *32 send to tell caller the roamer access no.
This company's subscribers have level 3 service.

Billing information

Visitor rates
Most visitors . . . . $2.50 day + 75¢ minute

Visitors without roaming agreements (p. 1159)
Cannot call since company takes no credit cards.

Identification numbers

City	Market	System	Billing	RSA
All served cities	149	00349	00000	MSA

Cellular company

Centel Cellular
951 S McPherson Church Road, Suite 202
Fayetteville, NC 28303

Customer service	. . . .	800-788-2260
. . . .		919-833-7494
Local office	. . . .	919-484-6156
Corporate office	. . . . .	312-399-2644

Sending calls

Visitor dialing instructions

| Local calls | . . . . . . | area code + 7 digits |
| Long distance | . . . | 1 + area code + 7 digits |

Speed-dial numbers

Customer service	. . . .	611
Emergency	. . . .	911
Highway patrol	. . . .	*HP or *47

Receiving calls

Caller dials the roamer access number below,
hears tone, dials cellular area code and number.
. . . . 919-624-7626

Billing information

Visitor rates
Most visitors . . . $2.50 day + 75¢ minute

Visitors without roaming agreements (p. 1159)
Call co. to use Amex, Mastercard, Visa

Identification numbers

City	Market	System	Billing	RSA
All served cities	149	00100	00000	MSA

For full coverage area, see Goldsboro, NC.

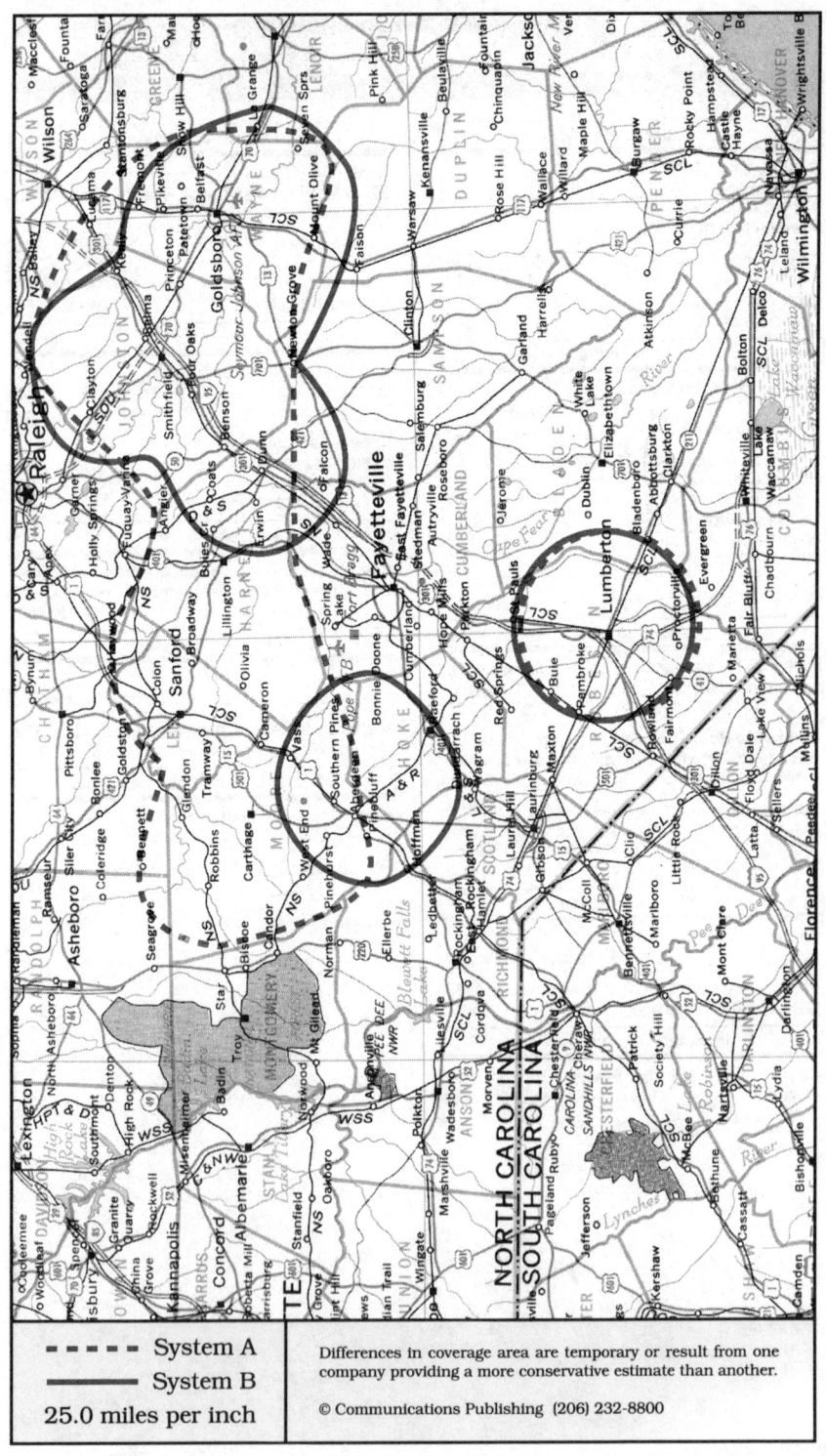

- - - - System A
———— System B
25.0 miles per inch

Differences in coverage area are temporary or result from one company providing a more conservative estimate than another.

© Communications Publishing (206) 232-8800

Goldsboro, North Carolina
Dunn, Goldsboro, Lumberton, Sanford, Smithfield, Southern Pines

System A	System B

System A

Cellular company

Cellular One
United States Cellular
410 SE Boulevard, PO Box 1130
Clinton, NC 28328

Recently acquired by United States Cellular.

Local office	919-590-3030
Corporate office	312-399-8900

Sending calls

Visitor dialing instructions

Local calls	area code + 7 digits
Long distance . . .	1 + area code + 7 digits

Speed-dial numbers

Customer service	611
Emergency	911
Roaming information	*711

Receiving calls

Caller dials the roamer access number below,
hears tone, dials cellular area code and number.

Goldsboro	919-580-7626
Lumberton	919-618-7626
Sanford	919-215-7626
Smithfield	919-915-7626

Billing information

Visitor rates

Most visitors . . . . $2.50 day + 75¢ minute

Visitors without roaming agreements (p. 1159)
Call co. to use Mastercard, Visa

Identification numbers

City	Market	System	Billing	RSA
Goldsboro, Smithfie.	574	01535	00000	NC-10
Lumberton	575	01537	00000	NC-11
Sanford, S. Pines	570	01527	00000	NC-6

For full coverage area, see Dillon, SC.

System B

Cellular company

Centel Cellular
951 S McPherson Church Road, Suite 202
Fayetteville, NC 28303

Customer service	800-788-2260
. . . .	919-833-7494
Local office	919-484-7787
Corporate office	312-399-2644

Sending calls

Visitor dialing instructions

Local calls	area code + 7 digits
Long distance . . .	1 + area code + 7 digits

Speed-dial numbers

Customer service	611
Emergency	911
Highway patrol	*HP or *47

Receiving calls

Caller dials the roamer access number below,
hears tone, dials cellular area code and number.

Goldsboro, Smithfield	919-880-7626
Laurinburg, Lumberton . . .	919-624-7626
Southern Pines	919-695-7626

Billing information

Visitor rates
Most visitors . . . $2.50 day + 75¢ minute

Visitors without roaming agreements (p. 1159)
Call co. to use Amex, Mastercard, Visa

Identification numbers

Goldsboro, Smithfie.	574	01536	00000	NC-10
Lumberton	575	01538	00000	NC-11
Southern Pines	570	01528	00000	NC-6

For full coverage area, see Fayetteville, NC.

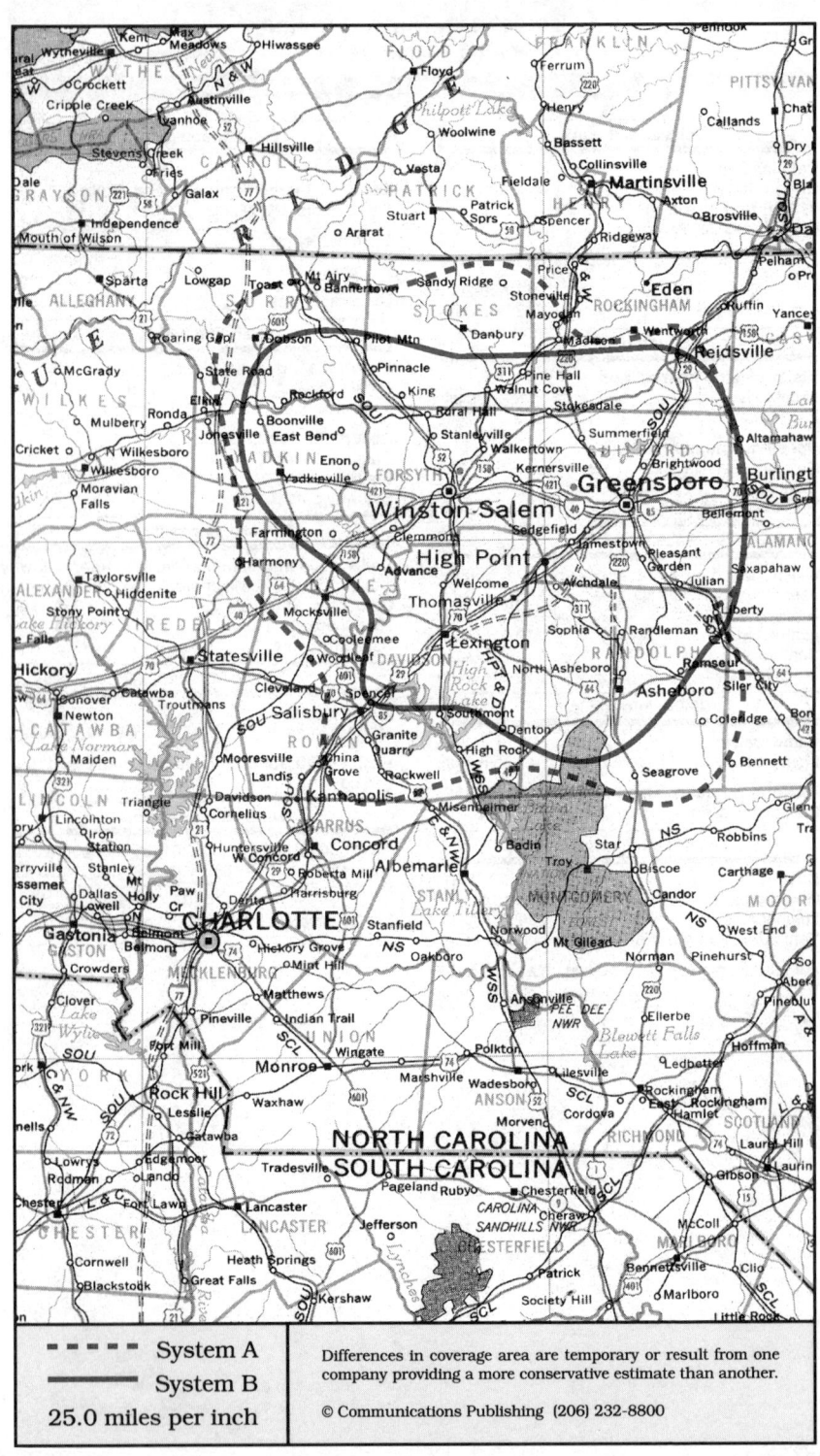

- - - - - System A	Differences in coverage area are temporary or result from one
──────── System B	company providing a more conservative estimate than another.
25.0 miles per inch	© Communications Publishing (206) 232-8800

Greensboro, North Carolina

Asheboro, Greensboro, Highpoint, Kannapolis, Mocksville, Winston-Salem

System A | System B

Cellular company

Cellular One
GTE Mobilnet
7029 Albert Pick Road, Suite 101
Greensboro, NC 27409

Customer service		800-727-2444
. . . .		919-481-1181
Local office . .	some areas	800-531-1555
. . . .		919-668-3600
Corporate office		404-391-8000

Sending calls

Visitor dialing instructions
Local calls . . . . 7 digits
Long distance . . . 1 + area code + 7 digits

Speed-dial numbers
Customer service 611
Emergency 911
Hear a traffic report *106
Highway patrol . . . . *HP or *47
Recorded information 511
WTQR radio *104

Receiving calls

Caller dials the roamer access number below,
hears tone, dials cellular area code and number.
. . . . 919-337-7626

Follow Me Roaming forwarding (see p. 1156)
Phone Me Anywhere forwarding (see p. 1156)
Activate, dial *18 send. Deactivate dial *19 send.

NationLink & RoamingAmerica (see p. 1157)
Dial *31 send to receive calls, *30 to deactivate.
Dial *32 send to tell caller the roamer access no.
This company's subscribers have level 3 service.

Billing information

Visitor rates
Most visitors . . . . $2.50 day + 75¢ minute

Visitors without roaming agreements (p. 1159)
Call co. to use Amex, Mastercard, Visa

Identification numbers

City	Market	System	Billing	RSA
All served cities	047	00095	00000	MSA

Cellular company

Centel Cellular
4003 Clifton Road
Greensboro, NC 27407

400 W Broad Avenue
High Point, NC 27262

536 Hanes Mall Boulevard, Pavilion Center
Winston-Salem, NC 27103

Customer service		800-859-8255
. . . .		919-833-7494
Local office . . .	Greensboro	919-299-3333
. . . .	High Point	919-882-0400
. . . . . .	Winston-Salem	919-760-4404
Corporate office		312-399-2644

Sending calls

Visitor dialing instructions
Local calls . . . . . . area code + 7 digits
Long distance . . . 1 + area code + 7 digits

Speed-dial numbers
Customer service *611
Emergency 911
Highway patrol . . . . *HP or *47

Receiving calls

Caller dials the roamer access number below,
hears tone, dials cellular area code and number.
Greensboro, Winston-Salem . 919-339-7626
Mocksville 919-940-7626

Billing information

Visitor rates
Most visitors . . . $2.50 day + 75¢ minute

Visitors without roaming agreements (p. 1159)
Call co. to use Amex, Mastercard, Visa

Identification numbers

City	Market	System	Billing	RSA
Greensboro, W.S.	047	00142	00000	MSA
Mocksville	579	01546	30542	NC-15

For full coverage area, see Burlington, NC.

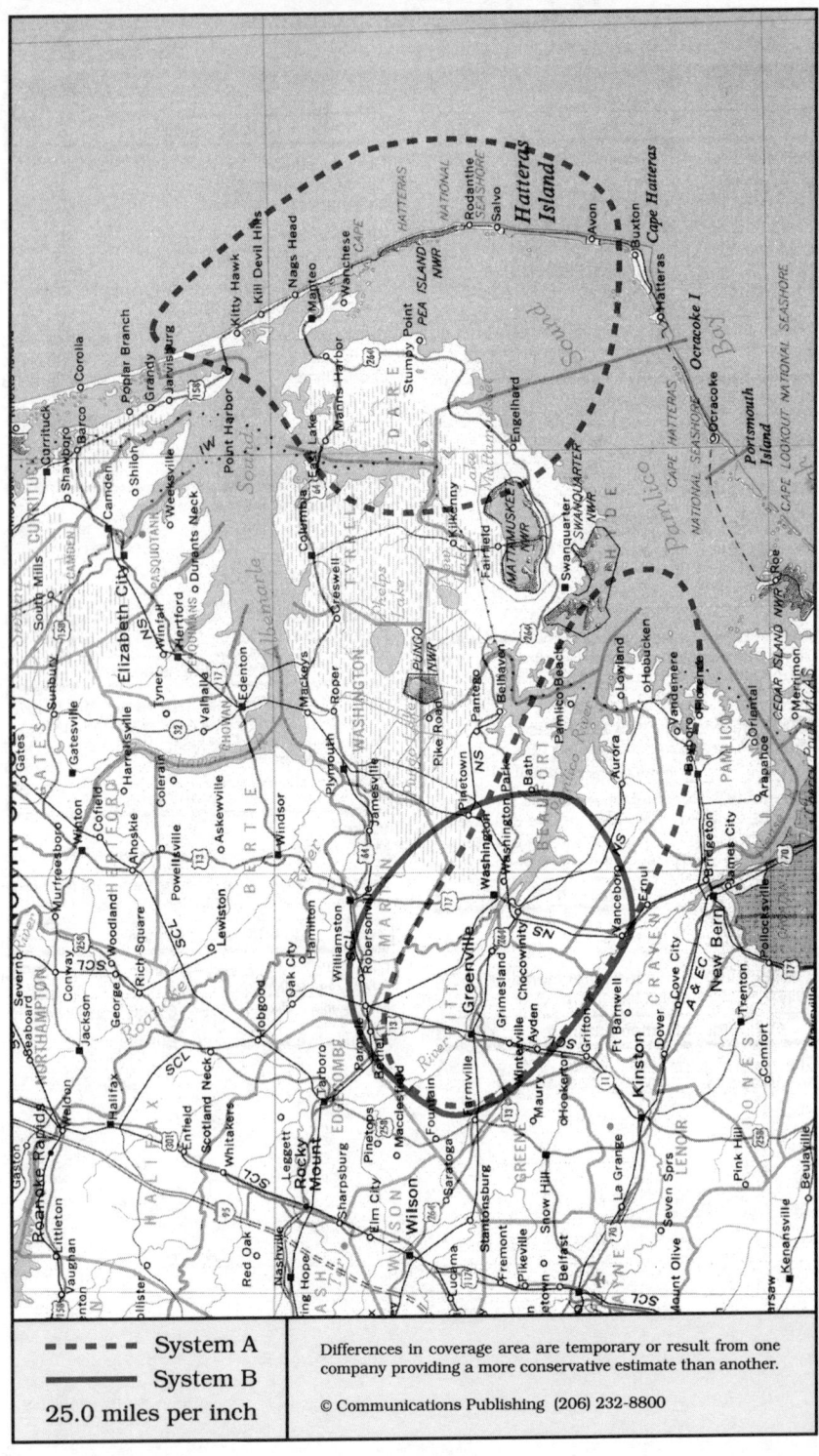

System A
System B
25.0 miles per inch

Differences in coverage area are temporary or result from one company providing a more conservative estimate than another.

© Communications Publishing (206) 232-8800

Greenville, North Carolina
Greenville, Kitty Hawk, Manteo, Washington

System A	System B

Cellular company

Cellular One
GMD Limited Partnership
333 E Arlington Boulevard
Greenville, NC 27858

Corporate office	800-849-8400
. . . .	919-321-0066

Sending calls

Visitor dialing instructions

Local calls	area code + 7 digits
Long distance . . .	1 + area code + 7 digits

Speed-dial numbers

Customer service	611
Emergency	911
Highway patrol	∗HP or ∗47
WRQR 94.3 FM	∗943

Receiving calls

Caller dials the roamer access number below, hears tone, dials cellular area code and number.

. . . .	919-916-7626

NationLink & RoamingAmerica (see p. 1157)
Dial ∗31 send to receive calls, ∗30 to deactivate. This company's subscribers have level 3 service.

Billing information

Visitor rates

Most visitors	$3 day + 85¢ minute

Visitors without roaming agreements (p. 1159)
Dial 0 + 10-digit no. Operator will intercept to bill.

Identification numbers

City	Market	System	Billing	RSA
All served cities	578	01543	00000	NC-14

Cellular company

United States Cellular
205 SW Greenville Boulevard, Suite 900
Greenville, NC 27834

Local office . . .	some areas	800-231-2355
. . . .		919-321-2666
Corporate office		312-399-8900

Sending calls

Visitor dialing instructions

Local calls	area code + 7 digits
Long distance . . .	1 + area code + 7 digits

Speed-dial numbers

Customer service	611
Emergency	911
Highway patrol	∗HP or ∗47
Roaming information	∗711

Receiving calls

Caller dials the roamer access number below, hears tone, dials cellular area code and number.

. . . .	919-714-7626

Billing information

Visitor rates

Most visitors . . .	$2.50 day + 75¢ minute

Visitors without roaming agreements (p. 1159)
Cannot call since company takes no credit cards.

Identification numbers

City	Market	System	Billing	RSA
All served cities	578	01544	00000	NC-14

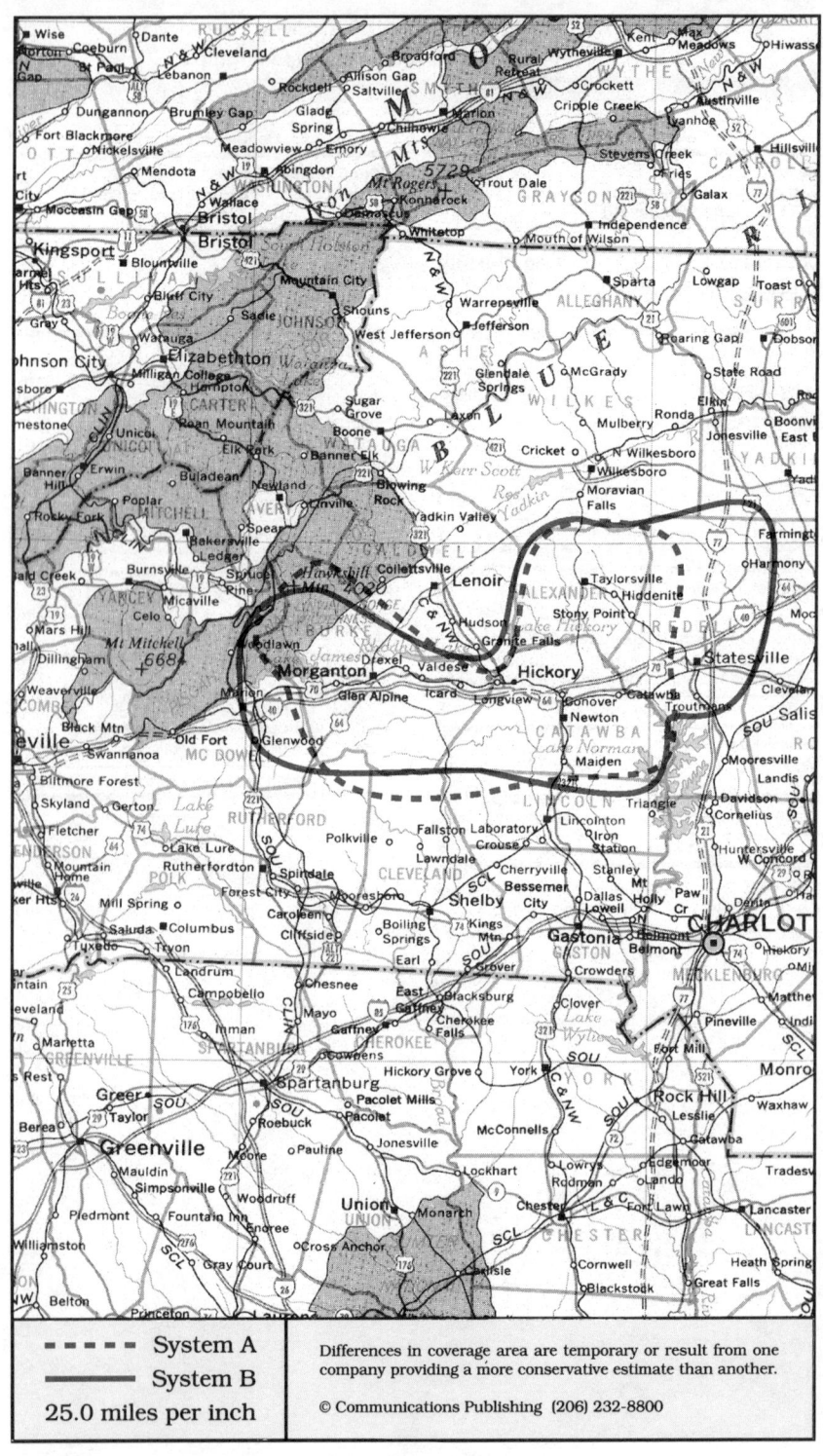

Hickory, North Carolina
Hickory, Morganton, Statesville, Taylorsville

System A	System B

Cellular company

Bell Atlantic Mobile
1260 25th Street Place SE, PO Box 2225
Hickory, NC 28603-2225

Local office 704-322-7557
Corporate office 908-306-7000

Sending calls

Visitor dialing instructions
Local calls . . . . . area code + 7 digits
Long distance . . . 1 + area code + 7 digits

Speed-dial numbers
Customer service 611
Emergency 911
Highway patrol . . . . ∗HP or ∗47

Receiving calls

Caller dials the roamer access number below,
hears tone, dials cellular area code and number.
. . . . 704-244-7626

NationLink & RoamingAmerica (see p. 1157)
Caller hears your current roamer access number.
This company's subscribers have level 0 service.

Billing information

Visitor rates
Most visitors . . . . . $3 day + 99¢ minute

Visitors without roaming agreements (p. 1159)
Roamer Plus will intercept first call to bill you.

Identification numbers

City	Market	System	Billing	RSA
All served cities	166	00385	00000	MSA

Cellular company

Centel Cellular
910 Tate Boulevard SE, Suite 108
Hickory, NC 28602

Customer service 800-788-2260
. . . . 919-833-7494
Local office 704-327-4000
Corporate office 312-399-2644

Sending calls

Visitor dialing instructions
Local calls . . . . . . area code + 7 digits
Long distance . . . 1 + area code + 7 digits

Speed-dial numbers
Customer service . . . . ∗611
Emergency 911
Highway patrol . . . . ∗HP or ∗47

Receiving calls

Caller dials the roamer access number below,
hears tone, dials cellular area code and number.
Hickory 704-381-7626
Statesville 704-871-7626

Follow Me Roaming forwarding (see p. 1156)
Activate, dial ∗18 send. Deactivate dial ∗19 send.

Billing information

Visitor rates
Most visitors . . . $2.50 day + 75¢ minute

Visitors without roaming agreements (p. 1159)
Call co. to use Amex, Mastercard, Visa

Identification numbers

City	Market	System	Billing	RSA
Hickory	166	00368	00000	MSA
Statesville	579	01546	00000	NC-15

For full coverage area, see Lenoir, NC.

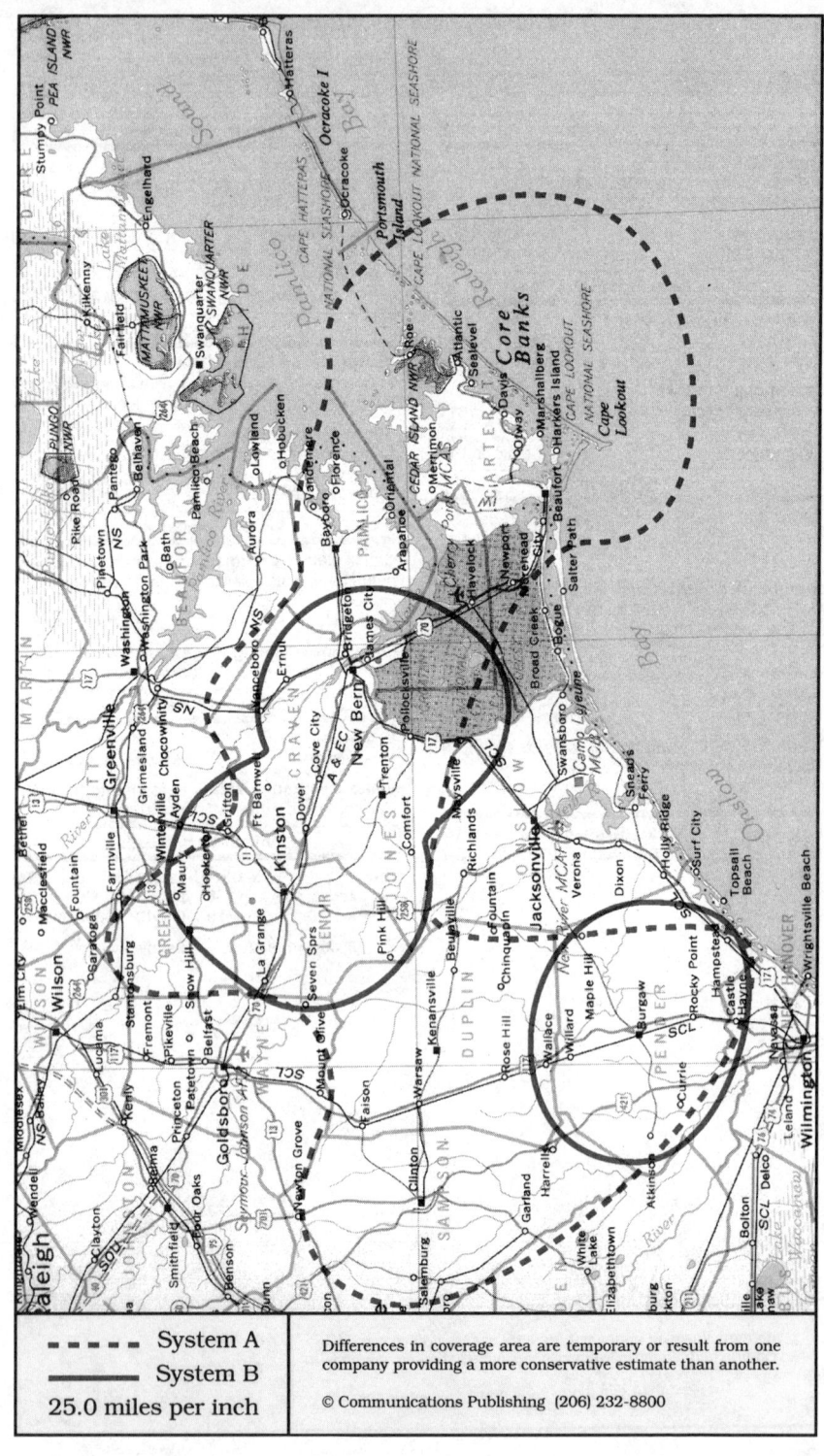

- - - - System A
———— System B
25.0 miles per inch

Differences in coverage area are temporary or result from one company providing a more conservative estimate than another.

© Communications Publishing (206) 232-8800

Kinston, North Carolina

Bayboro, Beaufort, Clinton, Kinston, Morehead City, New Bern

System A	System B

Cellular company

United States Cellular
410 SE Boulevard, PO Box 1130
Clinton, NC 28328

680-682 Stratford Boulevard
Kinston, NC 28501

Local office	. . . .	Clinton	919-590-3030
	. . . .	Kinston	919-559-5000
Corporate office	. . . .		312-399-8900

Sending calls

Visitor dialing instructions
Local calls . . . . . area code + 7 digits
Long distance . . . 1 + area code + 7 digits

Speed-dial numbers
Customer service 611
Emergency 911
Roaming information *711

Receiving calls

Caller dials the roamer access number below,
hears tone, dials cellular area code and number.
Clinton 919-590-7626
Kinston 919-559-7626

Billing information

Visitor rates
Most visitors . . . . $2.50 day + 75¢ minute

Visitors without roaming agreements (p. 1159)
Call co. to use Mastercard, Visa

Identification numbers

City	Market	System	Billing	RSA
Clinton	576	01539	00000	NC-12
Kinston	577	01541	00000	NC-13

Cellular company

Centel Cellular
2444 Commerce Road
Jacksonville, NC 28546

Customer service	. . . .	800-788-2260
	. . . .	919-833-7494
Local office	. . . .	800-877-2443
	. . . .	919-455-7787
Corporate office	. . . . .	312-399-2644

Sending calls

Visitor dialing instructions
Local calls . . . . . . area code + 7 digits
Long distance . . . 1 + area code + 7 digits

Speed-dial numbers
Customer service *611
Emergency 911
Highway patrol . . . . *HP or *47

Receiving calls

Caller dials the roamer access number below,
hears tone, dials cellular area code and number.
Clinton 919-520-7626
Kinston 919-636-7626

Billing information

Visitor rates
Most visitors . . . $2.50 day + 75¢ minute

Visitors without roaming agreements (p. 1159)
Call co. to use Amex, Mastercard, Visa

Identification numbers

City	Market	System	Billing	RSA
Clinton	576	01540	00000	NC-12
Kinston	577	01541	00000	NC-13

For full coverage area, see Fayetteville, NC and
Goldsboro, NC and Jacksonville, NC and
Wilmington, NC.

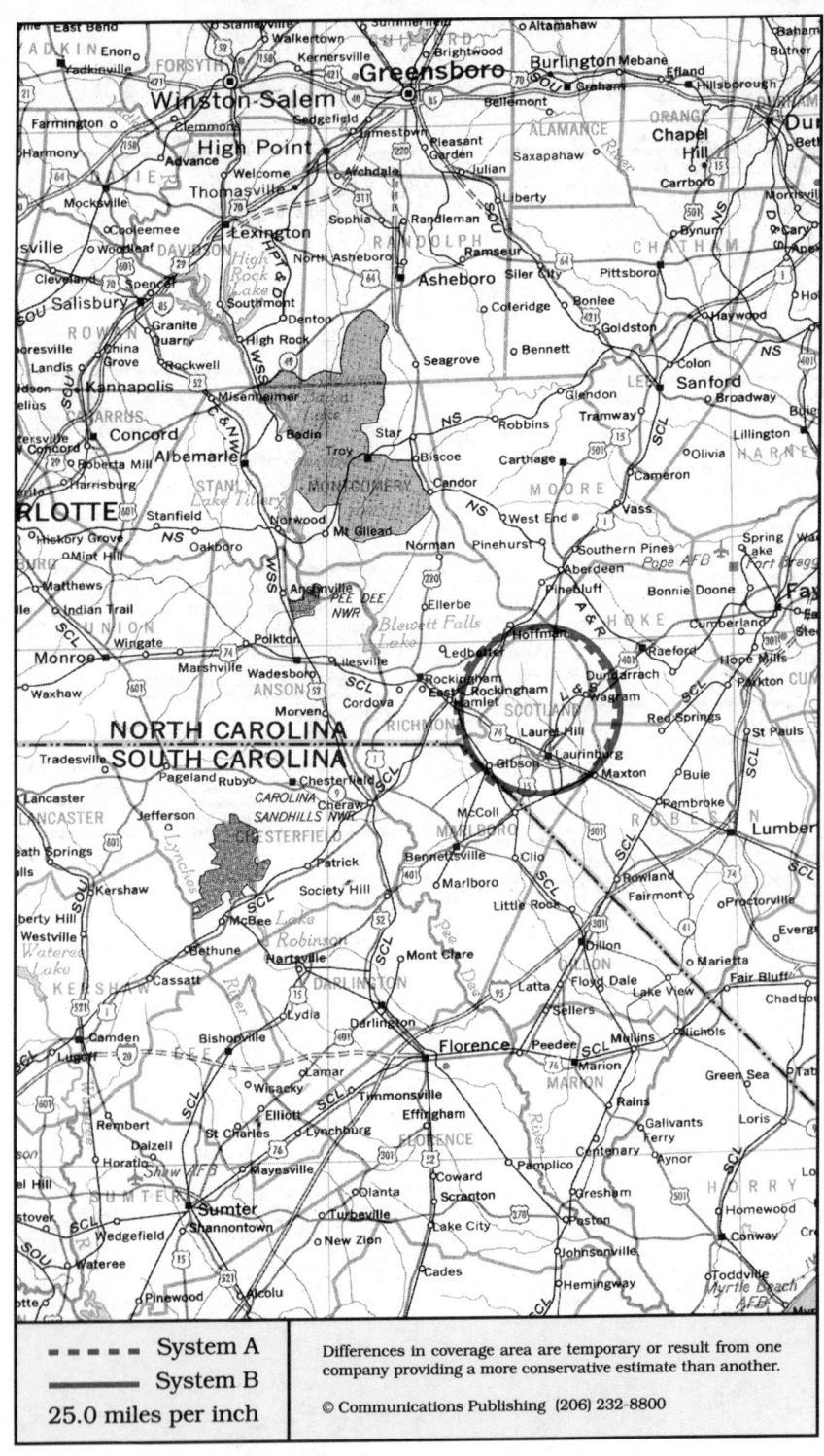

Laurinburg, North Carolina
Laurinburg (Scotland County)

System A	System B

Cellular company

Cellular One
Licensed to: SDK Enterprises
Operated by: Bell Atlantic Mobile
607 E Broad Avenue, Hwy 74; PO Box 1691
Rockingham, NC 28379

Corporate office 800-395-5845
. . . . 919-997-2323

Sending calls

Visitor dialing instructions
Local calls . . . . . area code + 7 digits
Long distance . . . 1 + area code + 7 digits

Speed-dial numbers
Customer service 611
Emergency 911
Highway patrol . . . . ✶HP or ✶47

Receiving calls

Caller dials the roamer access number below,
hears tone, dials cellular area code and number.
. . . . 704-564-7626

NationLink & RoamingAmerica (see p. 1157)
Caller hears your current roamer access number.
This company's subscribers have level 0 service.

Billing information

Visitor rates
Most visitors . . . . . $3 day + 75¢ minute

Visitors without roaming agreements (p. 1159)
Cannot call since company takes no credit cards.

Identification numbers

City	Market	System	Billing	RSA
All served cities	569	01857	00000	NC-5

This system is operated as part of the Charlotte, NC system A.

For full coverage area, see Rockingham, NC.

Cellular company

Centel Cellular
951 S McPherson Church Road, Suite 202
Fayetteville, NC 28303

Customer service 800-788-2260
. . . . 919-833-7494
Local office 919-484-6156
Corporate office 312-399-2644

Sending calls

Visitor dialing instructions
Local calls . . . . . . area code + 7 digits
Long distance . . . 1 + area code + 7 digits

Speed-dial numbers
Customer service 611
Emergency 911
Highway patrol . . . . ✶HP or ✶47

Receiving calls

Caller dials the roamer access number below,
hears tone, dials cellular area code and number.
. . . . 919-624-7626

Billing information

Visitor rates
Most visitors . . . $2.50 day + 75¢ minute

Visitors without roaming agreements (p. 1159)
Call co. to use Amex, Mastercard, Visa

Identification numbers

All served cities   569   02038   00000   NC-5

For full coverage area, see Goldsboro, NC.

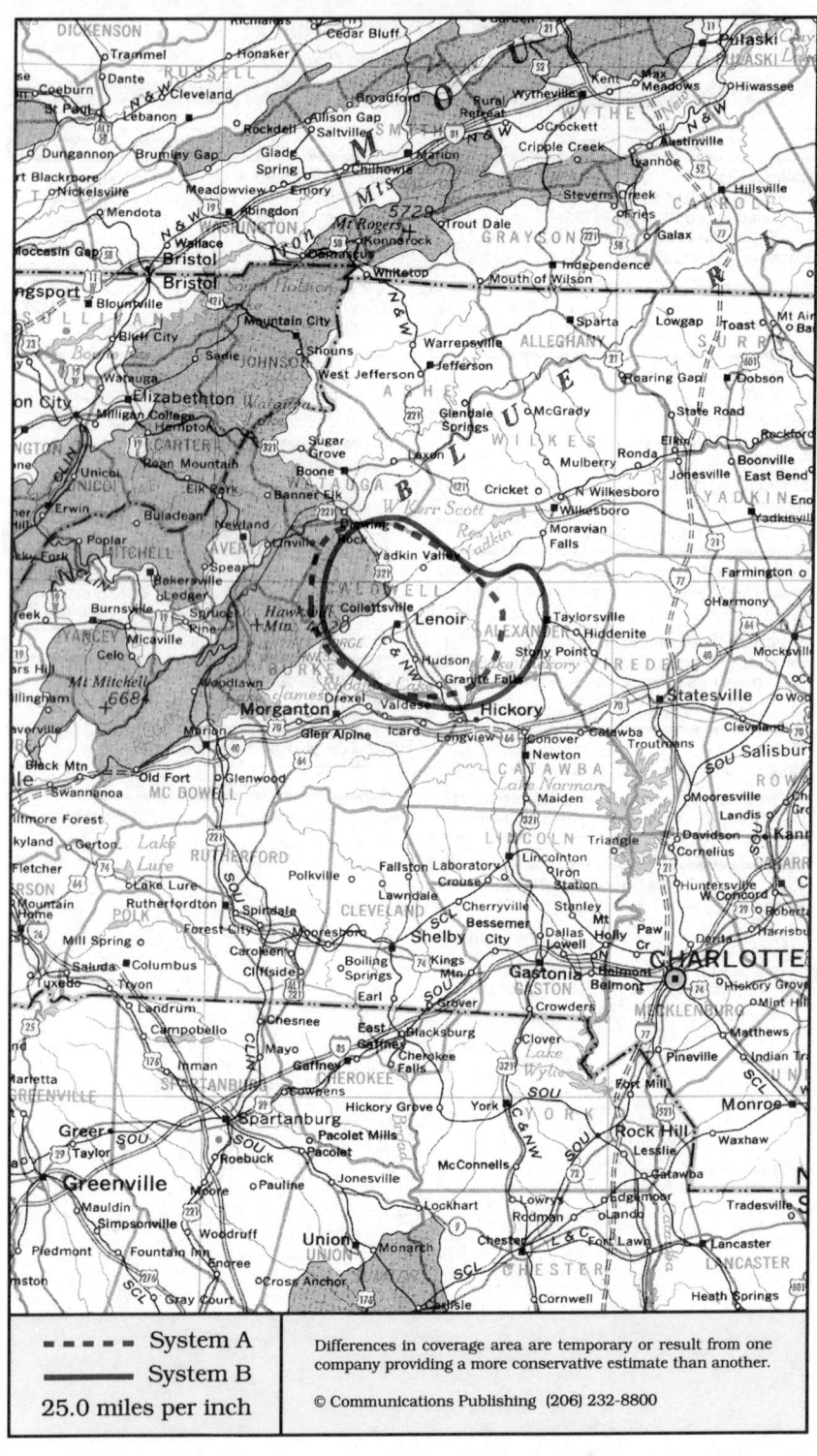

Lenoir, North Carolina
Lenoir (Caldwell County)

System A	System B

Cellular company

Cellular One
Blue Ridge Cellular Telephone
452 Harper Avenue
Lenoir, NC 28645

Corporate office 800-951-2351
. . . . 704-758-4444

Sending calls

Visitor dialing instructions
Local calls . . . . . area code + 7 digits
Long distance . . . 1 + area code + 7 digits

Speed-dial numbers
Customer service . . . . ✴611
Emergency 911

Receiving calls

Caller dials the roamer access number below,
hears tone, dials cellular area code and number.
. . . . 704-757-7626

Follow Me Roaming forwarding (see p. 1156)
Phone Me Anywhere forwarding (see p. 1156)
Activate, dial ✴18 send. Deactivate dial ✴19 send.

Billing information

Visitor rates
Most visitors . . . . . $3 day + 99¢ minute

Visitors without roaming agreements (p. 1159)
Call co. to use Mastercard, Visa
or Roamer Plus will intercept first call to bill you.

Identification numbers

City	Market	System	Billing	RSA
All served cities	566	01519	00000	NC-2

For full coverage area, see Boone, NC.

Cellular company

Centel Cellular
910 Tate Boulevard #108
Hickory, NC 28602

Customer service 800-788-2260
. . . . 919-833-7494
Local office 704-327-4000
Corporate office 312-399-2644

Sending calls

Visitor dialing instructions
Local calls . . . . . . area code + 7 digits
Long distance . . . 1 + area code + 7 digits

Speed-dial numbers
Customer service . . . . ✴611
Emergency 911
Highway patrol . . . . ✴HP or ✴47

Receiving calls

Caller dials the roamer access number below,
hears tone, dials cellular area code and number.
. . . . 704-496-7626

Billing information

Visitor rates
Most visitors . . . $2.50 day + 75¢ minute

Visitors without roaming agreements (p. 1159)
Call co. to use Amex, Mastercard, Visa

Identification numbers

City	Market	System	Billing	RSA
Lenoir	566	01992	00000	NC-2

For full coverage area, see Hickory, NC.

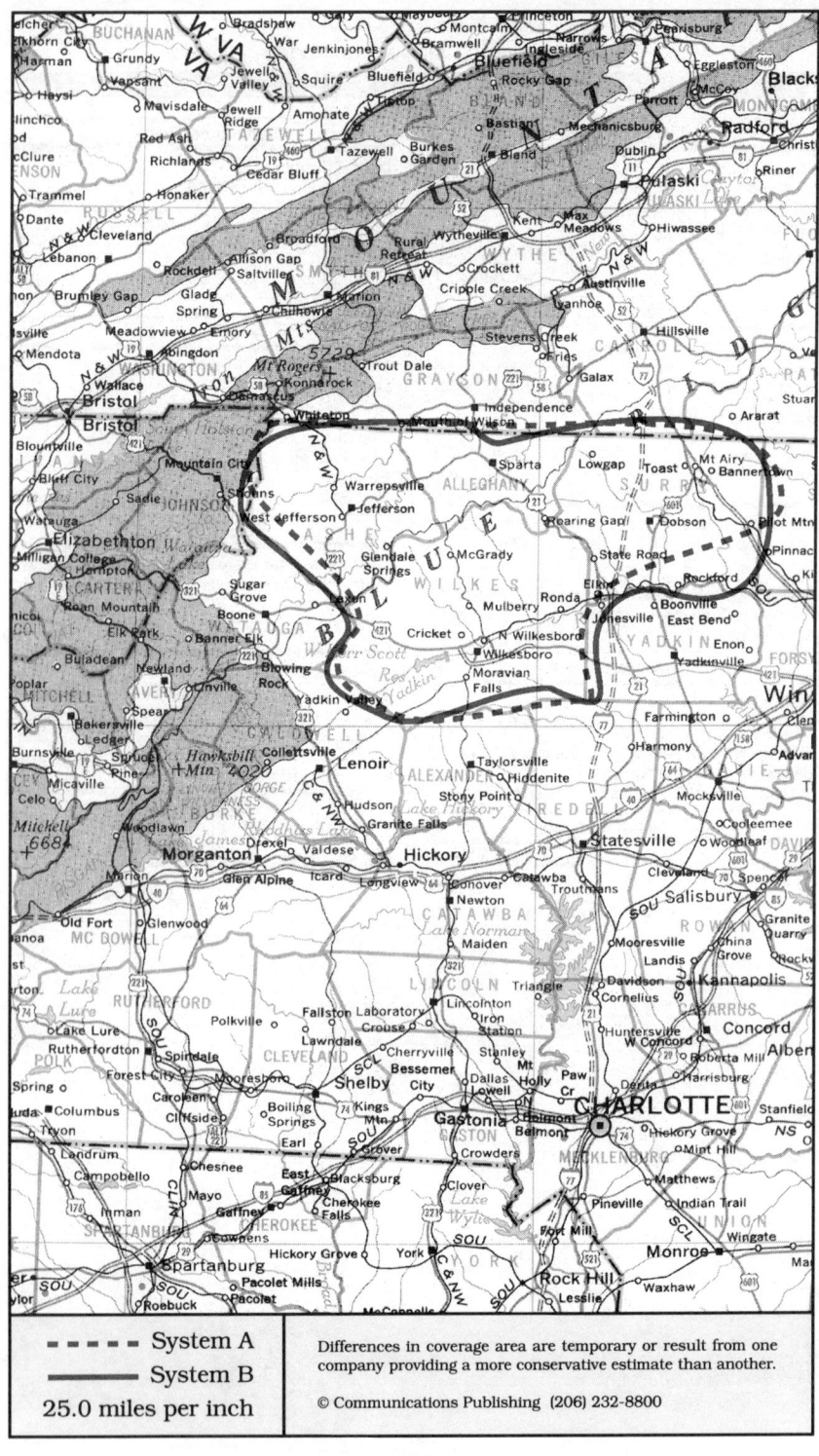

- - - - System A	Differences in coverage area are temporary or result from one
──── System B	company providing a more conservative estimate than another.
25.0 miles per inch	© Communications Publishing (206) 232-8800

Mount Airy, North Carolina
Dobson, Jefferson, Mount Airy, Wilkesboro

System A	System B

Cellular company

Cellular One
Clear Communications
K-Mart Plaza, 1418 Highway 421-B
Wilkesboro, NC 28697

Corporate office 919-838-2355

Cellular company

Carolina West Cellular
Highway 421 W, PO Box 738
Millers Creek, NC 28651

Corporate office . some areas 800-235-5007
. . . . 919-973-5000

Sending calls

Visitor dialing instructions
Local calls . . . . . area code + 7 digits
Long distance . . . 1 + area code + 7 digits

Speed-dial numbers
Customer service 611
Emergency 911

Sending calls

Visitor dialing instructions
Local calls . . . . . area code + 7 digits
Long distance . . . 1 + area code + 7 digits

Speed-dial numbers
Customer service *611
Emergency 911

Receiving calls

Caller dials the roamer access number below,
hears tone, dials cellular area code and number.
. . . . 919-921-7626

Receiving calls

Caller dials the roamer access number below,
hears tone, dials cellular area code and number.
. . . . 919-957-7626

Billing information

Visitor rates
Most visitors . . . . . $3 day + 85¢ minute

Visitors without roaming agreements (p. 1159)
Roamer Plus will intercept first call to bill you.

Billing information

Visitor rates
Most visitors . . . . . $2 day + 65¢ minute

Visitors without roaming agreements (p. 1159)
Cannot call since company takes no credit cards.

Identification numbers

City	Market	System	Billing	RSA
All served cities	567	01521	00000	NC-3

Identification numbers

City	Market	System	Billing	RSA
All served cities	567	01522	00000	NC-3

For full coverage area, see Boone, NC.

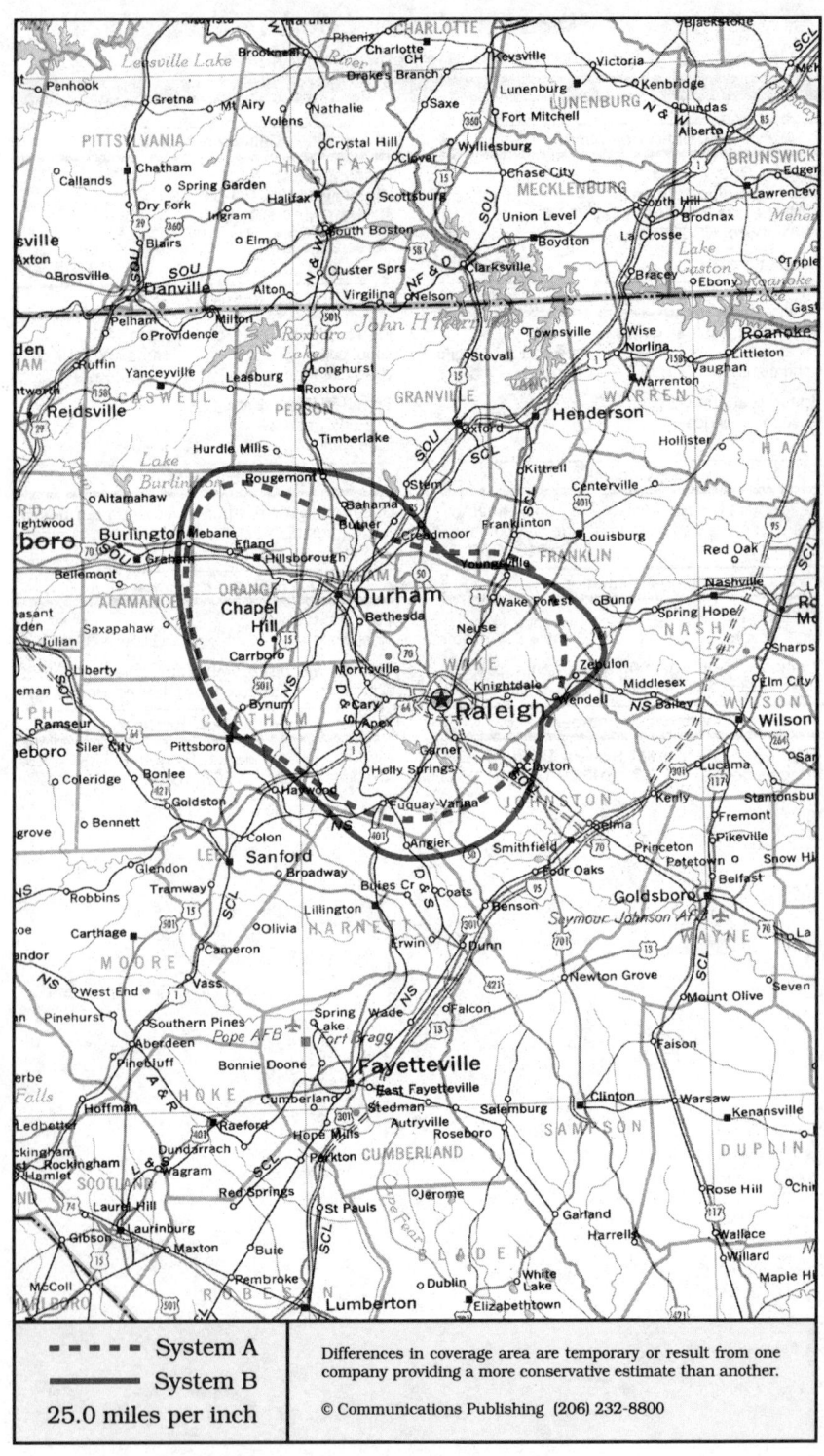

- - - - System A	Differences in coverage area are temporary or result from one company providing a more conservative estimate than another.
——— System B	
25.0 miles per inch	© Communications Publishing (206) 232-8800

Raleigh, North Carolina
Chapel Hill, Durham, Hillsborough, Raleigh

System A	System B

Cellular company

Cellular One
GTE Mobilnet
1100 Perimeter Park Drive, Suite 101
Morrisville, NC 27560

Customer service	800-727-2444
. . . .	919-481-1181
Local office	919-481-6400
Corporate office	404-391-8000

Sending calls

Visitor dialing instructions

Local calls	area code + 7 digits
Long distance . . .	1 + area code + 7 digits

Speed-dial numbers

Customer service	611
Emergency	911
Hear a traffic report	∗106
Highway patrol	∗HP or ∗47
Sports line	∗8939

Receiving calls

Caller dials the roamer access number below,
hears tone, dials cellular area code and number.
. . . . 919-740-7626

Follow Me Roaming forwarding (see p. 1156)
Phone Me Anywhere forwarding (see p. 1156)
Activate, dial ∗18 send. Deactivate dial ∗19 send.

NationLink & RoamingAmerica (see p. 1157)
Dial ∗31 send to receive calls, ∗30 to deactivate.
Dial ∗32 send to tell caller the roamer access no.
This company's subscribers have level 3 service.

Billing information

Visitor rates
Most visitors . . . . $2.50 day + 75¢ minute

Visitors without roaming agreements (p. 1159)
Call co. to use Amex, Mastercard, Visa

Identification numbers

City	Market	System	Billing	RSA
All served cities	71	00069	00000	MSA

Cellular company

Centel Cellular
557-B Pylon Drive
Raleigh, NC 27606

143 Rams Plaza
Chapel Hill, NC 27514

Customer service		800-788-2260
. . . .		919-833-7494
Local office . . .	Chapel Hill	919-933-1442
. . . .	Raleigh	919-833-7494
Corporate office		312-399-2644

Sending calls

Visitor dialing instructions

Local calls	area code + 7 digits
Long distance . . .	1 + area code + 7 digits

Speed-dial numbers

Customer service	∗611
Emergency	911
Highway patrol	∗HP or ∗47

Receiving calls

Caller dials the roamer access number below,
hears tone, dials cellular area code and number.
. . . . 919-880-7626

Billing information

Visitor rates
Most visitors . . . $2.50 day + 75¢ minute

Visitors without roaming agreements (p. 1159)
Call co. to use Amex, Mastercard, Visa

Identification numbers

City	Market	System	Billing	RSA
All served cities	71	00144	00000	MSA

For full coverage area, see Burlington, NC and
Goldsboro, NC and Rocky Mount, NC.

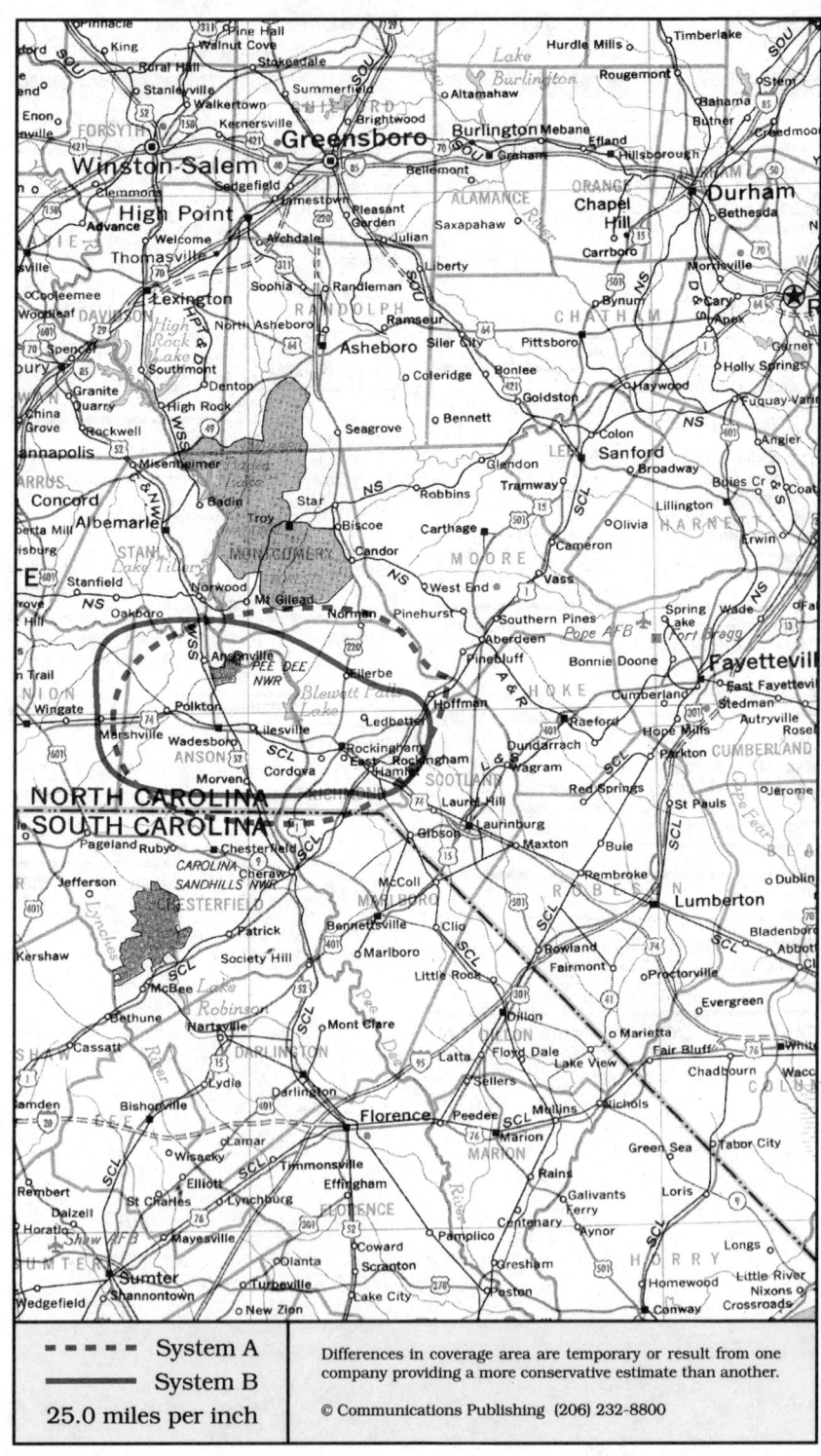

- - - - System A	Differences in coverage area are temporary or result from one
——— System B	company providing a more conservative estimate than another.
25.0 miles per inch	© Communications Publishing (206) 232-8800

Rockingham, North Carolina
Rockingham, Wadesboro

System A

Cellular company

Cellular One
Licensed to: SDK Enterprises
Operated by: Bell Atlantic Mobile
607 E Broad Avenue, Hwy 74; PO Box 1691
Rockingham, NC 28379

Corporate office	800-395-5845
. . . .	919-997-2323

Sending calls

Visitor dialing instructions

Local calls	area code + 7 digits
Long distance . . .	1 + area code + 7 digits

Speed-dial numbers

Customer service	611
Emergency	911
Highway patrol	*HP or *47

Receiving calls

Caller dials the roamer access number below,
hears tone, dials cellular area code and number.
. . . . 704-564-7626

NationLink & RoamingAmerica (see p. 1157)
Caller hears your current roamer access number.
This company's subscribers have level 0 service.

Billing information

Visitor rates
Most visitors . . . . . $3 day + 75¢ minute

Visitors without roaming agreements (p. 1159)
Cannot call since company takes no credit cards.

Identification numbers

City	Market	System	Billing	RSA
All served cities	569	01857	00000	NC-5

This system is operated as part of the Charlotte,
NC system A.

For full coverage area, see Rockingham, NC.

System B

Cellular company

ALLTEL Mobile Communications
734 Tyvola Road
Charlotte, NC 28217

Customer service	some areas	800-255-8351
. . . .		704-529-8140
Local office		704-529-0001
Corporate office		501-661-8500

Sending calls

Visitor dialing instructions

Local calls	area code + 7 digits
Long distance	area code + 7 digits

Speed-dial numbers

Customer service	*611
Emergency	911
Highway patrol	*HP or *47
WBT radio	*928

Receiving calls

Caller dials the roamer access number below,
hears tone, dials cellular area code and number.

Rockingham	919-997-1626
Wadesboro	704-694-8226

Follow Me Roaming forwarding (see p. 1156)
Activate, dial *18 send. Deactivate dial *19 send.

Billing information

Visitor rates
Most visitors . . . . . $3 day + 85¢ minute

Visitors without roaming agreements (p. 1159)
Call co. to use Amex, Discover, Mastercard, Visa

Identification numbers

City	Market	System	Billing	RSA
All served cities	569	00114	30490	NC-5

For full coverage area, see Charlotte, NC.

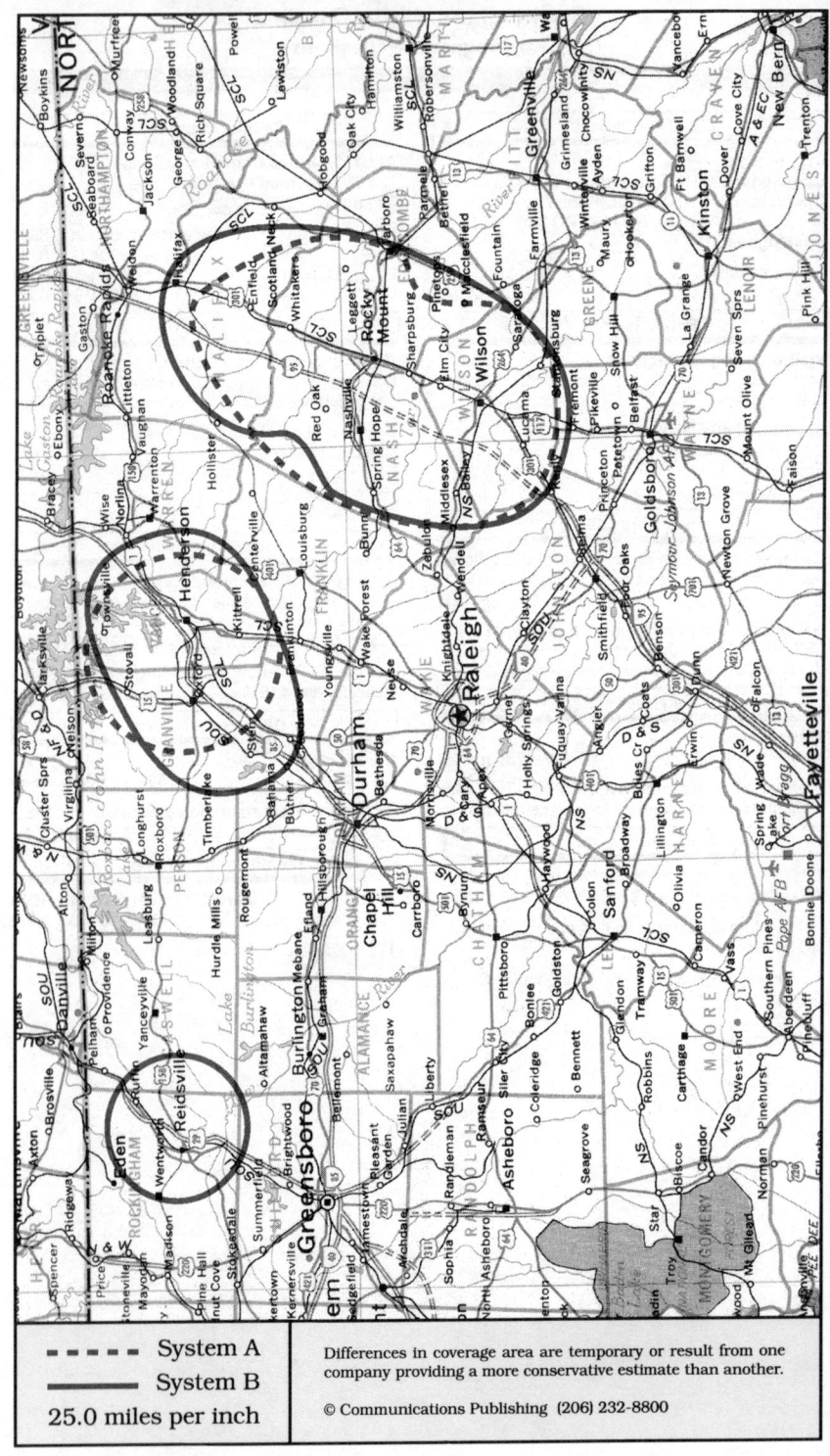

Rocky Mount, North Carolina
Henderson, Nashville, Oxford, Reidsville, Rocky Mount, Wilson

System A	System B

Cellular company

Cellular One
United States Cellular
2518 Sunset Avenue
Rocky Mount, NC 27804

Local office   919-937-1811
. . . . . . from Henderson   703-632-2000
. . . . . . from Wilson   919-321-2666
Corporate office   312-399-8900

Cellular company

Centel Cellular
916 N Church Street
Rocky Mount, NC 27804

Customer service   800-788-2260
. . . .   919-833-7494
Local office   919-446-2280
. . . . . . From Reidsville   919-299-3333
Corporate office   312-399-2644

Sending calls

Visitor dialing instructions
Local calls . . . . .   area code + 7 digits
Long distance . . .   1 + area code + 7 digits

Speed-dial numbers
Customer service   611
Emergency   911
Roaming information *711

Sending calls

Visitor dialing instructions
Local calls . . . . . .   area code + 7 digits
Long distance . . .   1 + area code + 7 digits

Speed-dial numbers
Customer service   611
Emergency   911
Highway patrol . . . .   *HP or *47

Receiving calls

Caller dials the roamer access number below,
hears tone, dials cellular area code and number.
Henderson   919-213-7626
Oxford   919-690-7626
Rocky Mount   919-813-7626
WIlson   919-236-7626

Receiving calls

Caller dials the roamer access number below,
hears tone, dials cellular area code and number.
Henderson   919-430-7626
Oxford   919-690-7626
Reidsville   919-339-7626
Rocky Mount, Wilson   919-985-7626

Billing information

Visitor rates
Most visitors . . . . $2.50 day + 75¢ minute

Visitors without roaming agreements (p. 1159)
Call co. to use Mastercard, Visa

Billing information

Visitor rates
Most visitors . . . $2.50 day + 75¢ minute

Visitors without roaming agreements (p. 1159)
Call co. to use Amex, Mastercard, Visa

Identification numbers

City	Market	System	Billing	RSA
Henderson	571	01529	00000	NC-7
Rocky Mt., Wilson	572	01531	00000	NC-8

For full coverage area, see Goldsboro, NC.

Identification numbers

Henderson	571	01530	00000	NC-7
Reidsville	571	01530	30548	NC-7
Rocky Mount	572	01532	00000	NC-8
Wilson	572	01532	30478	NC-8

For full coverage area, see Goldsboro, NC and
Greensboro, NC and Raleigh, NC.

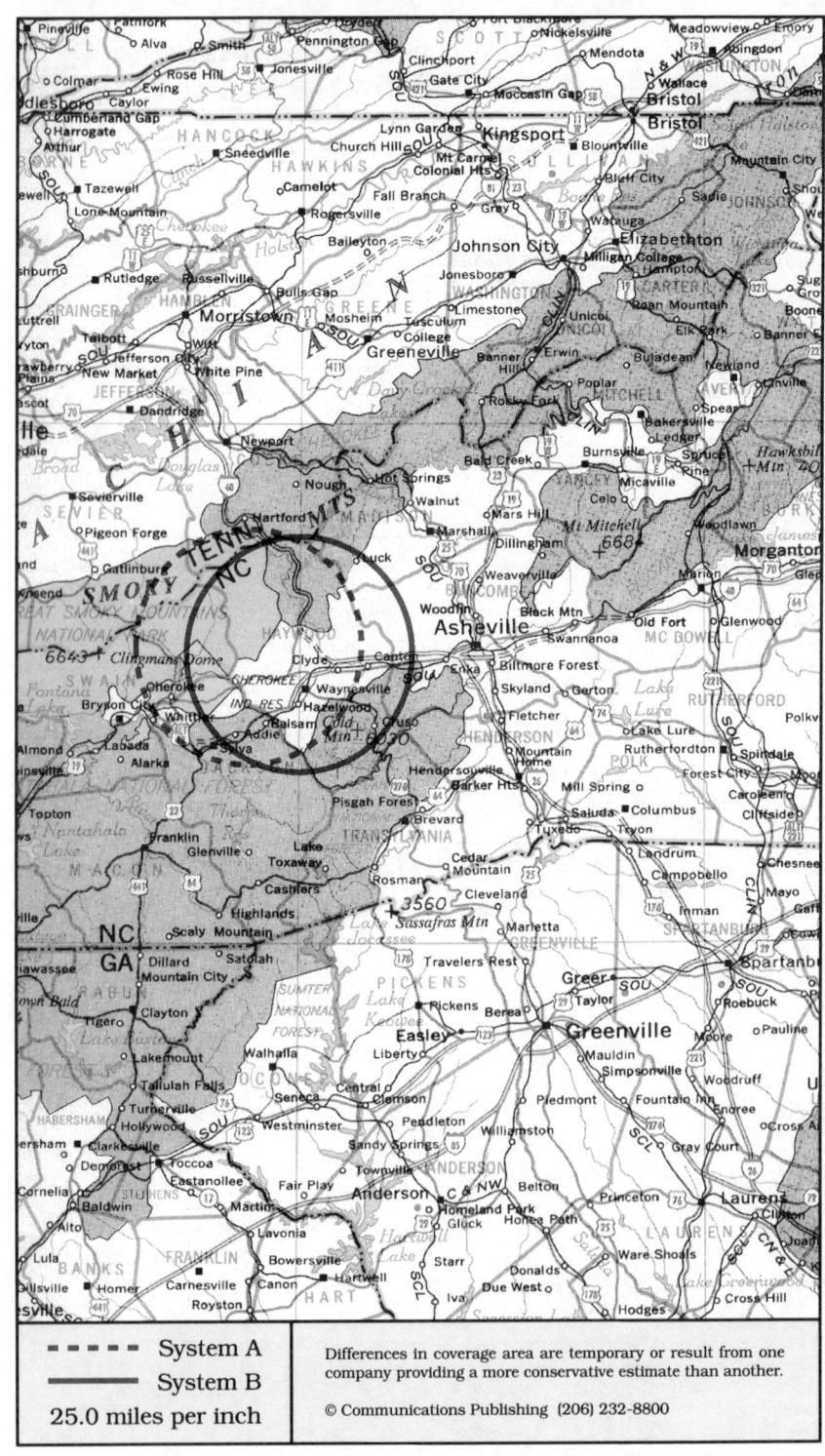

▪▪▪▪▪ System A	Differences in coverage area are temporary or result from one company providing a more conservative estimate than another.
▬▬▬ System B	
25.0 miles per inch	© Communications Publishing (206) 232-8800

Waynesville, North Carolina
Clyde, Waynesville

System A

Cellular company

Bell Atlantic Mobile
4700 Sweden Road
Charlotte, NC 28273

Customer service	704-553-9893
. . . .	803-234-7954
Local office	704-552-5185
Corporate office	908-306-7000

Sending calls

Visitor dialing instructions

Local calls	area code + 7 digits
Long distance	area code + 7 digits

Speed-dial numbers

Customer service	611
Emergency	911
Highway patrol	*HP or *47

Receiving calls

Caller dials the roamer access number below,
hears tone, dials cellular area code and number.
. . . . 803-270-7626

NationLink & RoamingAmerica (see p. 1157)
Dial *31 send to receive calls, *30 to deactivate.
This company's subscribers have level 3 service.

Billing information

Visitor rates
Most visitors . . . . . $3 day + 85¢ minute

Visitors without roaming agreements (p. 1159)
Cannot call since company takes no credit cards.

Identification numbers

City	Market	System	Billing	RSA
All served cities	565	01517	00000	NC-1

System B

Cellular company

Contel Cellular
GTE Mobile Communications
7029 Albert Pick Road, Suite 101
Greensboro, NC 27409

Customer service		800-727-2444
. . . .		919-481-1181
Local office . . .	some areas	800-531-1555
. . . .		919-668-3600
Corporate office		404-391-8000

Sending calls

Visitor dialing instructions

Local calls	7 digits
Long distance . . .	1 + area code + 7 digits

Speed-dial numbers

Customer service	611
Emergency	911
Highway patrol	*HP or *47
Recorded information	511

Receiving calls

Caller dials the roamer access number below,
hears tone, dials cellular area code and number.
. . . . 704-646-7626

Billing information

Visitor rates
Most visitors . . . . . $3 day + 85¢ minute

Visitors without roaming agreements (p. 1159)
Call co. to use Amex, Mastercard, Visa

Identification numbers

City	Market	System	Billing	RSA
All served cities	565	01518	00000	NC-1

----- System A ——— System B 25.0 miles per inch	Differences in coverage area are temporary or result from one company providing a more conservative estimate than another. © Communications Publishing (206) 232-8800

Wilmington, North Carolina
Burgaw, Jacksonville, Southport, Wilmington

System A	System B

Cellular company

Cellular One
GTE Mobilnet
320 Van Dyke Drive
Wilmington, NC 28405-3700

Customer service	800-727-2444
. . . .	919-481-1181
Local office	919-799-8898
. . . from Jacksonville	919-455-4526
Corporate office	404-391-8000

Sending calls

Visitor dialing instructions

Local calls	7 digits
Long distance . . .	1 + area code + 7 digits

Speed-dial numbers

Customer service	611
Emergency	911
Highway patrol	*HP or *47

Receiving calls

Caller dials the roamer access number below,
hears tone, dials cellular area code and number.

Jacksonville	919-340-7626
Wilmington	919-540-7626

Follow Me Roaming forwarding (see p. 1156)
Phone Me Anywhere forwarding (see p. 1156)
Activate, dial *18 send. Deactivate dial *19 send.

NationLink & RoamingAmerica (see p. 1157)
Dial *31 send to receive calls, *30 to deactivate.
Dial *32 send to tell caller the roamer access no.
This company's subscribers have level 3 service.

Billing information

Visitor rates

Most visitors	$2.50 day + 75¢ minute

Visitors without roaming agreements (p. 1159)
Call co. to use Amex, Mastercard, Visa

Identification numbers

City	Market	System	Billing	RSA
Jacksonville	258	00393	00000	MSA
Wilmington	218	00599	00000	MSA

Cellular company

Centel Cellular
2444 Commerce Road
Jacksonville, NC 28546

4512 Oleander Drive #800
Wilmington, NC 28403

Customer service		800-788-2260
. . . .		919-833-7494
Local office . . .	Jacksonville	800-877-2443
. . . .	Jacksonville	919-455-7787
. . . .	Wilmington	919-791-0800
Corporate office		312-399-2644

Sending calls

Visitor dialing instructions

Local calls	area code + 7 digits
Long distance . . .	1 + area code + 7 digits

Speed-dial numbers

Customer service	611
Emergency	911
Highway patrol	*HP or *47

Receiving calls

Caller dials the roamer access number below,
hears tone, dials cellular area code and number.

Jacksonville	919-389-7626
Wilmington	919-520-7626

Follow Me Roaming forwarding (see p. 1156)
Activate, dial *18 send. Deactivate dial *19 send.

Billing information

Visitor rates

Most visitors . . .	$2.50 day + 75¢ minute

Visitors without roaming agreements (p. 1159)
Call co. to use Amex, Mastercard, Visa

Identification numbers

City	Market	System	Billing	RSA
Jacksonville	258	00376	00000	MSA
Wilmington	218	00578	00000	MSA

For full coverage area, see Kinston, NC.

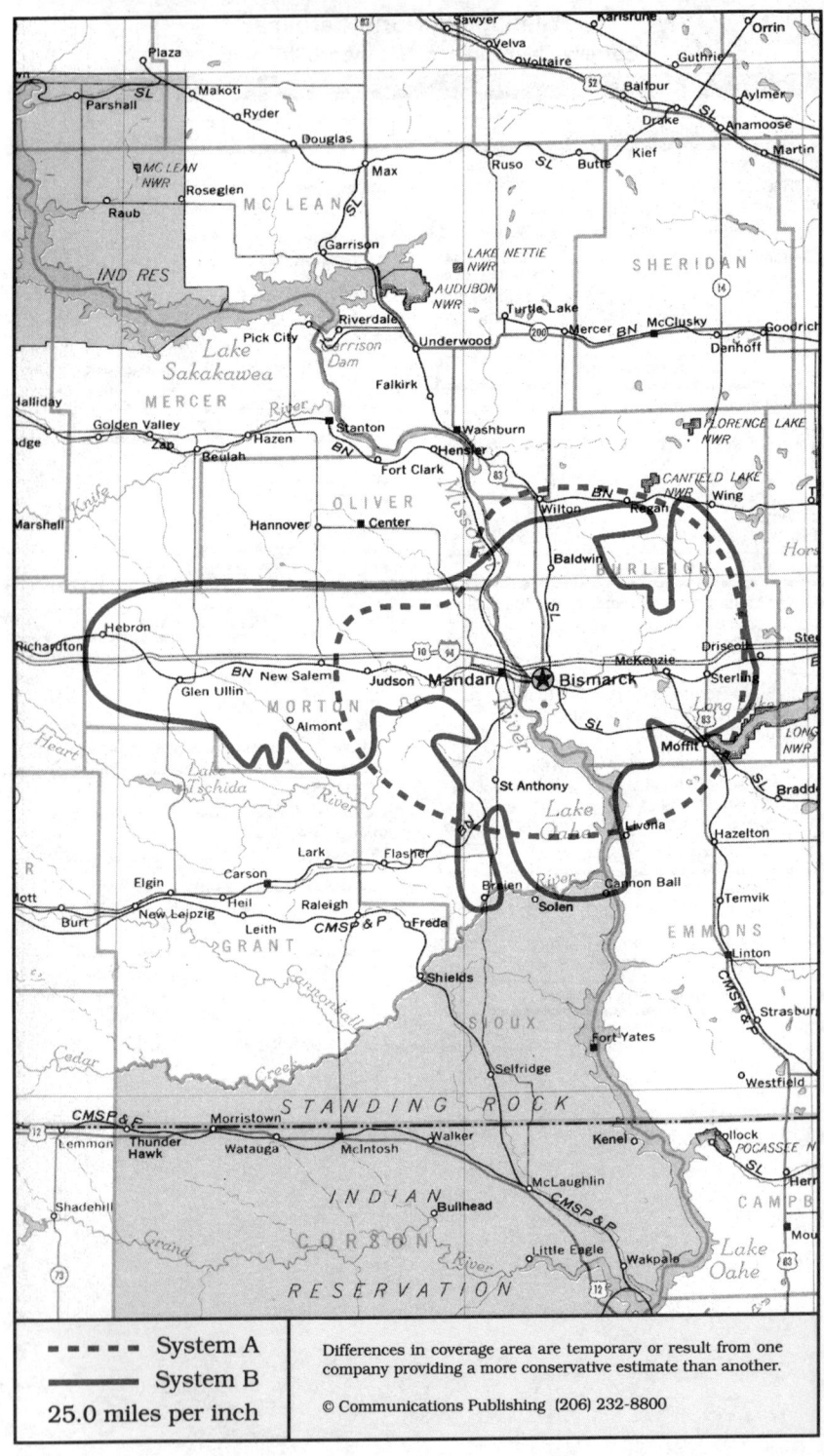

- - - - System A	Differences in coverage area are temporary or result from one
──────── System B	company providing a more conservative estimate than another.
25.0 miles per inch	© Communications Publishing (206) 232-8800

Bismarck, North Dakota

System A

Cellular company

Cellular One
Pacific Northwest Cellular
905 E Interstate Avenue
Bismarck, ND 58501

Customer service	800-635-0304
Local office	701-224-1616
Corporate office	206-635-0300

Sending calls

Visitor dialing instructions

Local calls	area code + 7 digits
Long distance . . .	1 + area code + 7 digits

Speed-dial numbers

American Automobile Association . . .	*222
Customer service	611 or 711
Emergency	911

Receiving calls

Caller dials the roamer access number below, hears tone, dials cellular area code and number.
. . . . 701-220-7626

Billing information

Visitor rates
Most visitors . . . . . $3 day + 99¢ minute

Visitors without roaming agreements (p. 1159)
Call co. to use Mastercard, Visa

Identification numbers

City	Market	System	Billing	RSA
All served cities	298	00285	00000	MSA

System B

Cellular company

U S West Cellular
904 E Divide Avenue
Bismarck, ND 58501

Customer service	800-238-7848
. . . .	206-562-2895
. . . . . local subscribers	800-626-6611
. . . .	206-747-1771
Local office	701-222-8687
Corporate office	206-747-4900

Sending calls

Visitor dialing instructions

Local calls	area code + 7 digits
Long distance . . .	area code + 7 digits

Speed-dial numbers

Customer service	*611
Emergency	911
Roaming information	*711

Receiving calls

Caller dials the roamer access number below, hears tone, dials cellular area code and number.
. . . . 701-226-7626

Billing information

Visitor rates
Most visitors . . . . . $3 day + 99¢ minute

Visitors without roaming agreements (p. 1159)
Call co. to use Amex, Mastercard, Visa

Identification numbers

City	Market	System	Billing	RSA
All served cities	298	00268	00000	MSA

For full coverage area, see Grand Forks, ND.

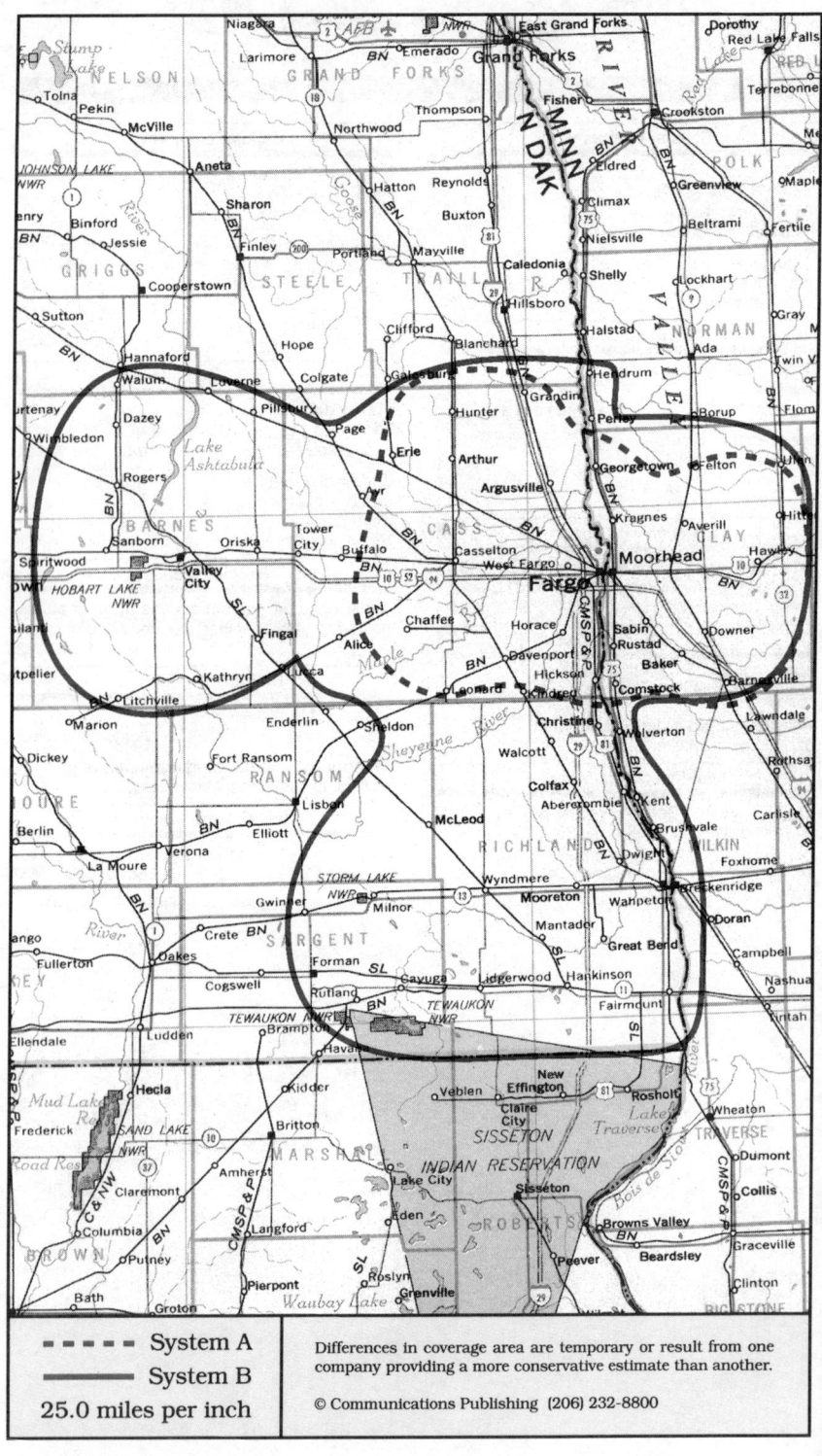

- - - - - System A
——— System B
25.0 miles per inch

Differences in coverage area are temporary or result from one company providing a more conservative estimate than another.

© Communications Publishing (206) 232-8800

Fargo, North Dakota
Fargo, Moorhead MN, Valley City, Wahpeton

System A	System B

Cellular company

Cellular One
Bachow and Associates
4417 13th Avenue SW
Fargo, ND 58103

Local office	701-281-2800
Corporate office	215-972-7550

Sending calls

Visitor dialing instructions

Local calls	area code + 7 digits
Long distance . . .	1 + area code + 7 digits

Speed-dial numbers

Customer service	611
Emergency	911

Receiving calls

Caller dials the roamer access number below,
hears tone, dials cellular area code and number.

. . . . 218-790-7626

Billing information

Visitor rates
Most visitors . . . . . $3 day + 99¢ minute

Visitors without roaming agreements (p. 1159)
Roamer Plus will intercept first call to bill you.

Identification numbers

City	Market	System	Billing	RSA
All served cities	221	347	00000	MSA

Cellular company

U S West Cellular
3301 13th Avenue SW
Fargo, ND 58103

Customer service	800-238-7848
. . . .	206-562-2895
. . . . . local subscribers	800-626-6611
. . . .	206-747-1771
Local office	701-235-5148
Corporate office	206-747-4900

Sending calls

Visitor dialing instructions

Local calls	area code + 7 digits
Long distance	area code + 7 digits

Speed-dial numbers

Customer service	*611
Emergency	911
Make a traffic report	#970
Roaming information	*711

Receiving calls

Caller dials the roamer access number below,
hears tone, dials cellular area code and number.

Fargo	701-238-7626
Valley City	701-840-7626
Wahpeton	701-640-7626

Follow Me Roaming forwarding (see p. 1156)
Activate, dial *18 send. Deactivate dial *19 send.

Billing information

Visitor rates
Most visitors . . . . . $3 day + 99¢ minute

Visitors without roaming agreements (p. 1159)
Call co. to use Amex, Mastercard, Visa

Identification numbers

City	Market	System	Billing	RSA
Fargo, Moorhead	221	00330	00000	MSA
Valley City, Wahp.	582	01552	00000	ND-3

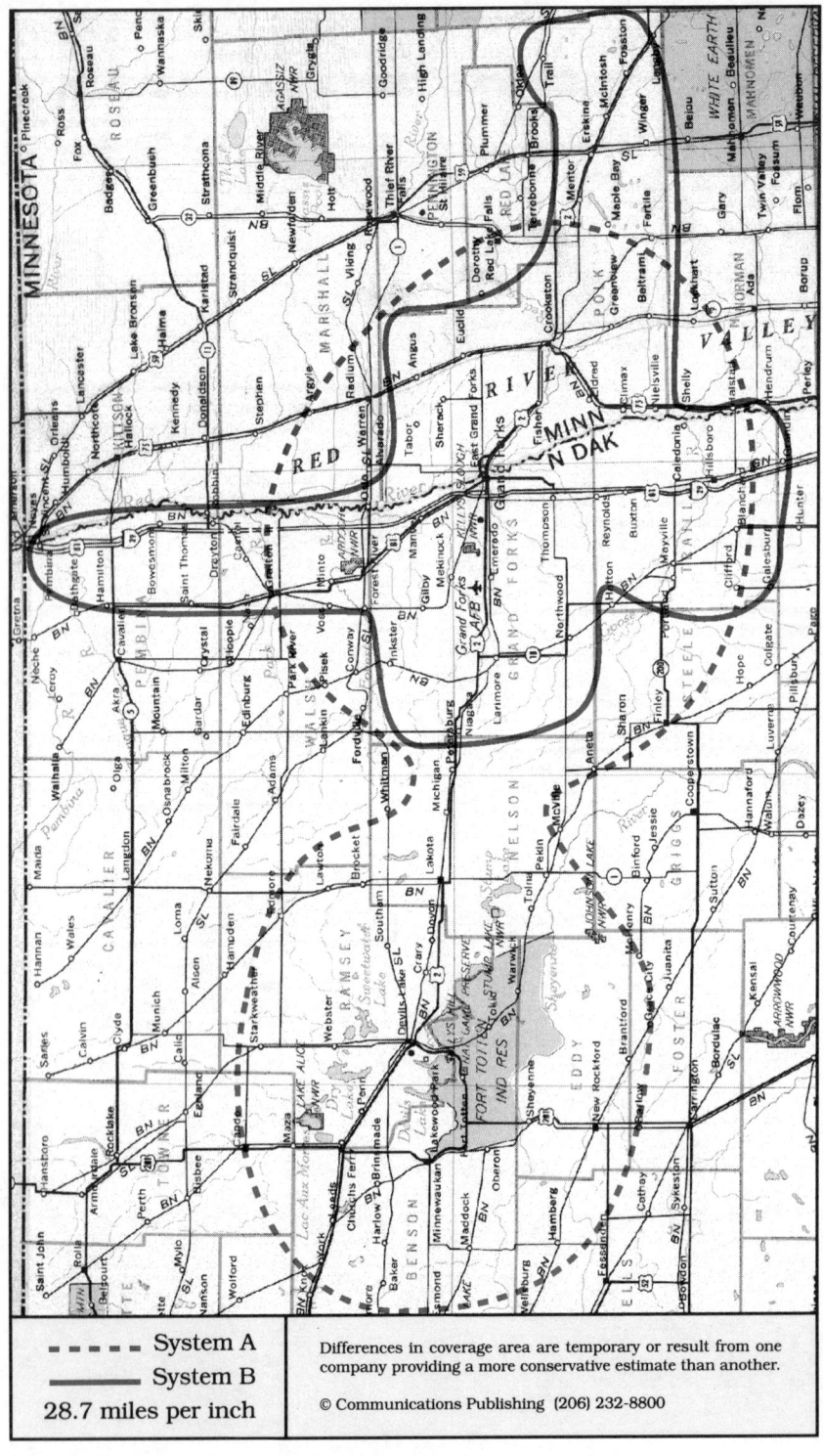

System A

System B

28.7 miles per inch

Differences in coverage area are temporary or result from one company providing a more conservative estimate than another.

© Communications Publishing (206) 232-8800

Grand Forks, North Dakota

Crookston MN, Devils Lake, Grafton, Grand Forks, Mayville, New Rockford

System A	System B

System A

Cellular company

Cellular One
Pacific Northwest Cellular
2100 S Columbia Road, Suite 116
Grand Forks, ND 58201

Customer service	800-635-0304
Local office	701-772-4201
Corporate office	206-635-0300

Sending calls

Visitor dialing instructions

Local calls	area code + 7 digits
Long distance . . .	1 + area code + 7 digits

Speed-dial numbers

Customer service	611
Emergency	911

Receiving calls

Caller dials the roamer access number below,
hears tone, dials cellular area code and number.

. . . . 218-893-7626

NationLink & RoamingAmerica (see p. 1157)
Caller hears your current roamer access number.
This company's subscribers have level 0 service.

Billing information

Visitor rates

Most visitors	$0 day + 99¢ minute

Visitors without roaming agreements (p. 1159)
Call co. to use Mastercard, Visa

Identification numbers

City	Market	System	Billing	RSA
Devils Lake/Mayville	582	00371	00000	ND-3
Grand Forks	276	00371	00000	MSA

System B

Cellular company

U S West Cellular
2600 Demers
Grand Forks, ND 58201

Customer service	800-238-7848
. . . .	206-562-2895
. . . . local subscribers	800-626-6611
. . . .	206-747-1771
Local office	701-772-4159
Corporate office	206-747-4900

Sending calls

Visitor dialing instructions

Local calls	area code + 7 digits
Long distance	area code + 7 digits

Speed-dial numbers

Customer service	*611
Emergency	911
Roaming information	*711
Traffic watch	#777

Receiving calls

Caller dials the roamer access number below,
hears tone, dials cellular area code and number.

Crookston, MN	218-280-7626
Grafton	701-520-7626
Grand Forks	218-779-7626
Mayville	701-430-7626

Billing information

Visitor rates

Most visitors	$3 day + 99¢ minute

Visitors without roaming agreements (p. 1159)
Call co. to use Amex, Mastercard, Visa

Identification numbers

City	Market	System	Billing	RSA
Grafton, Mayville	582	01552	00000	ND-3
Grand Forks, Crook.	276	00356	00000	MSA

For full coverage area, see Fargo, ND.

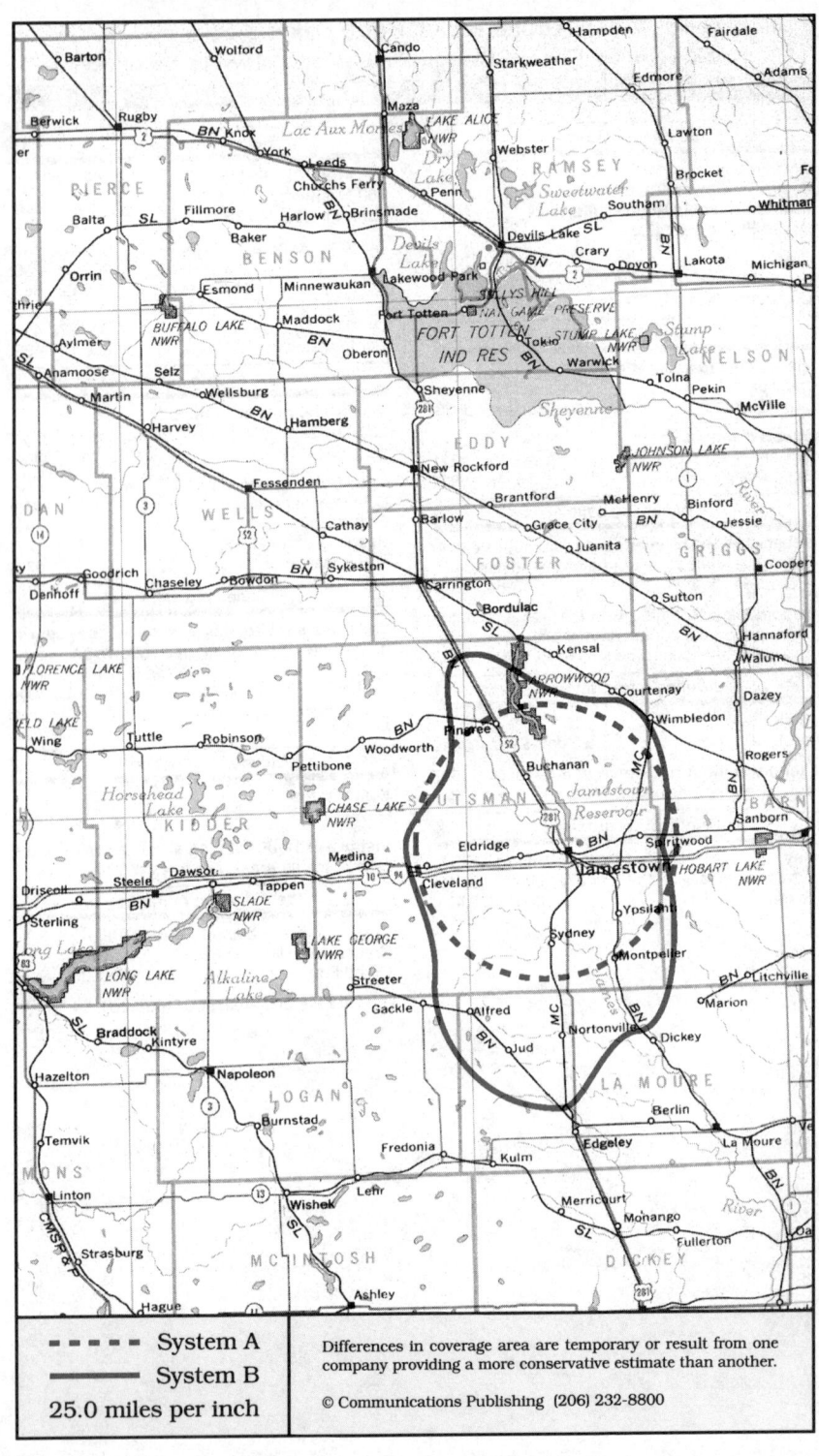

- - - - **System A**	Differences in coverage area are temporary or result from one company providing a more conservative estimate than another.
——— **System B**	
25.0 miles per inch	© Communications Publishing (206) 232-8800

Jamestown, North Dakota

Cellular company

Cellular One
Managed by: Pacific Northwest Cellular
2100 S Columbia Road, Suite 116
Grand Forks, ND 58201

Customer service	800-635-0304
Local office . from Jamestown	701-320-2273
Corporate office	206-635-0300

Sending calls

Visitor dialing instructions
Local calls	area code + 7 digits
Long distance . . .	1 + area code + 7 digits

Speed-dial numbers
Customer service	611
Emergency	911

Receiving calls

Caller dials the roamer access number below,
hears tone, dials cellular area code and number.
. . . . 701-320-7626

NationLink & RoamingAmerica (see p. 1157)
Dial *31 send to receive calls, *30 to deactivate.
This company's subscribers have level 3 service.

Billing information

Visitor rates
Most visitors . . . . . $0 day + 99¢ minute

Visitors without roaming agreements (p. 1159)
Roamer Plus will intercept first call to bill you.

Identification numbers

City	Market	System	Billing	RSA
All served cities	584	01555	00000	ND-5

Cellular company

CommNet 2000
Cellular, Inc.
1701 Grand Avenue
Billings, MT 59102

Customer service	800-597-5529
Local office	800-942-2060
. . . .	406-248-1990
Corporate office	303-694-3234

Sending calls

Visitor dialing instructions
Local calls	area code + 7 digits
Long distance . . .	1 + area code + 7 digits

Speed-dial numbers
Customer service	611
Emergency	911
Roaming information	711

Receiving calls

Caller dials the roamer access number below,
hears tone, dials cellular area code and number.
. . . . 701-269-7626

Billing information

Visitor rates
Most visitors . . . . . $3 day + 95¢ minute

Visitors without roaming agreements (p. 1159)
Cannot call since company takes no credit cards.

Identification numbers

City	Market	System	Billing	RSA
All served cities	584	01548	30194	ND-5

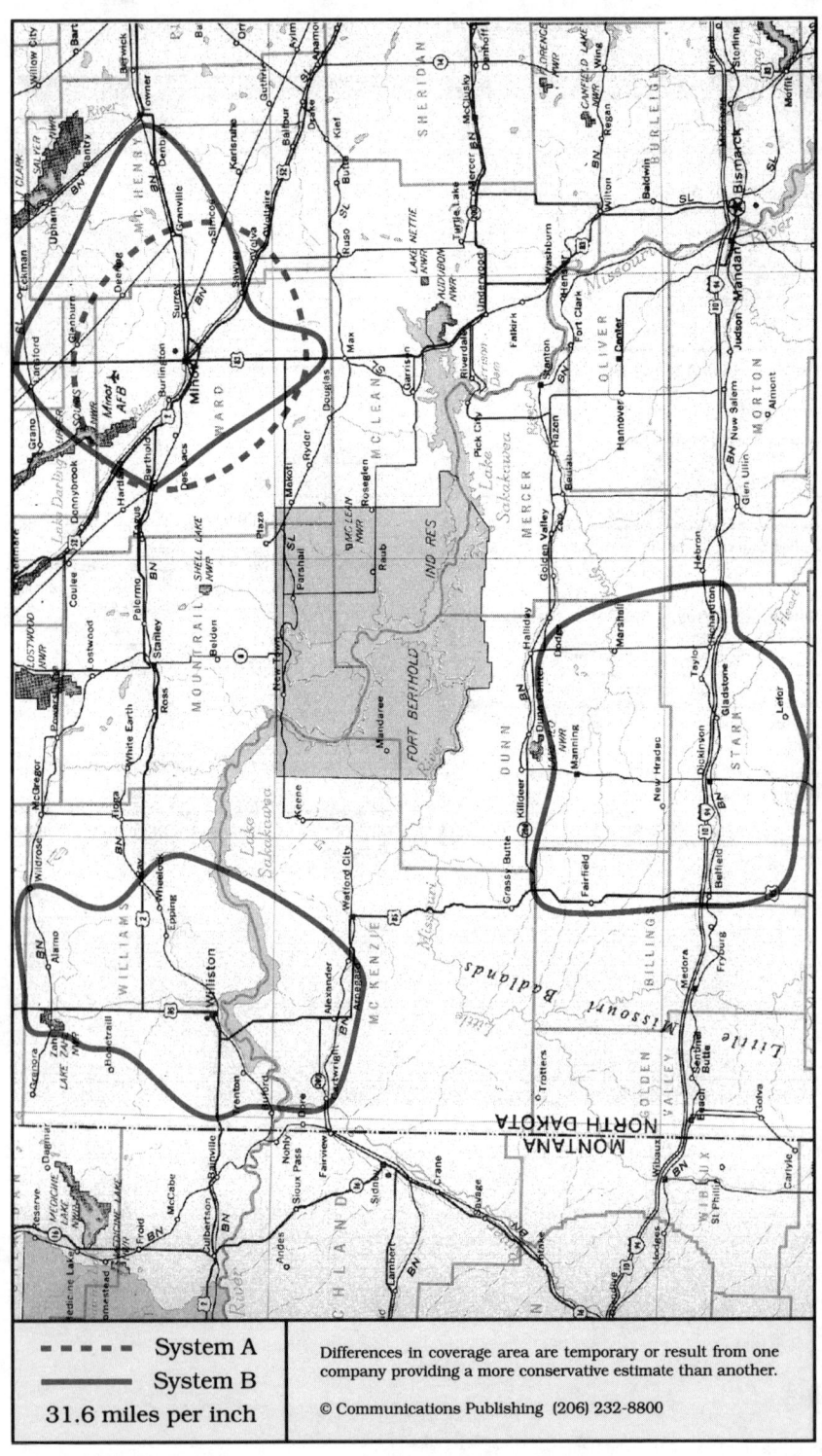

System A
System B
31.6 miles per inch

Differences in coverage area are temporary or result from one company providing a more conservative estimate than another.

© Communications Publishing (206) 232-8800

System A	System B

Cellular company

Cellular One
Pacific Northwest Cellular
11400 SE 8th Street, Suite 445
Bellevue, WA 98004-6431

This company is expected to be the licensee.

Customer service	800-635-0304
Corporate office	206-635-0300

Sending calls

Visitor dialing instructions

Local calls	area code + 7 digits
Long distance . . .	1 + area code + 7 digits

Speed-dial numbers

Customer service	611
Emergency	911

Receiving calls

Caller dials the roamer access number below,
hears tone, dials cellular area code and number.
. . . . 701-320-7626

NationLink & RoamingAmerica (see p. 1157)
Dial ∗31 send to receive calls, ∗30 to deactivate.
This company's subscribers have level 3 service.

Billing information

Visitor rates

Most visitors	$0 day + 99¢ minute

Visitors without roaming agreements (p. 1159)
Cannot call since company takes no credit cards.

Identification numbers

City	Market	System	Billing	RSA
All served cities	580	01555	00000	ND-1

Cellular company

CommNet 2000
Cellular, Inc.
1701 Grand Avenue
Billings, MT 59102

Customer service	800-597-5529
Local office	800-942-2060
. . . .	406-248-1990
Corporate office	303-694-3234

Sending calls

Visitor dialing instructions

Local calls	area code + 7 digits
Long distance . . .	1 + area code + 7 digits

Speed-dial numbers

Customer service	611
Emergency	911
Roaming information	711

Receiving calls

Caller dials the roamer access number below,
hears tone, dials cellular area code and number.

Dickinson	701-260-7626
Minot	701-240-7626
Williston	701-570-7626

Billing information

Visitor rates

Most visitors	$3 day + 95¢ minute

Visitors without roaming agreements (p. 1159)
Cannot call since company takes no credit cards.

Identification numbers

City	Market	System	Billing	RSA
Dickinson	582	01548	30160	ND-4
Granville	581	01548	30158	ND-2
Minot	580	01548	00000	ND-1
Williston	580	01548	30416	ND-1

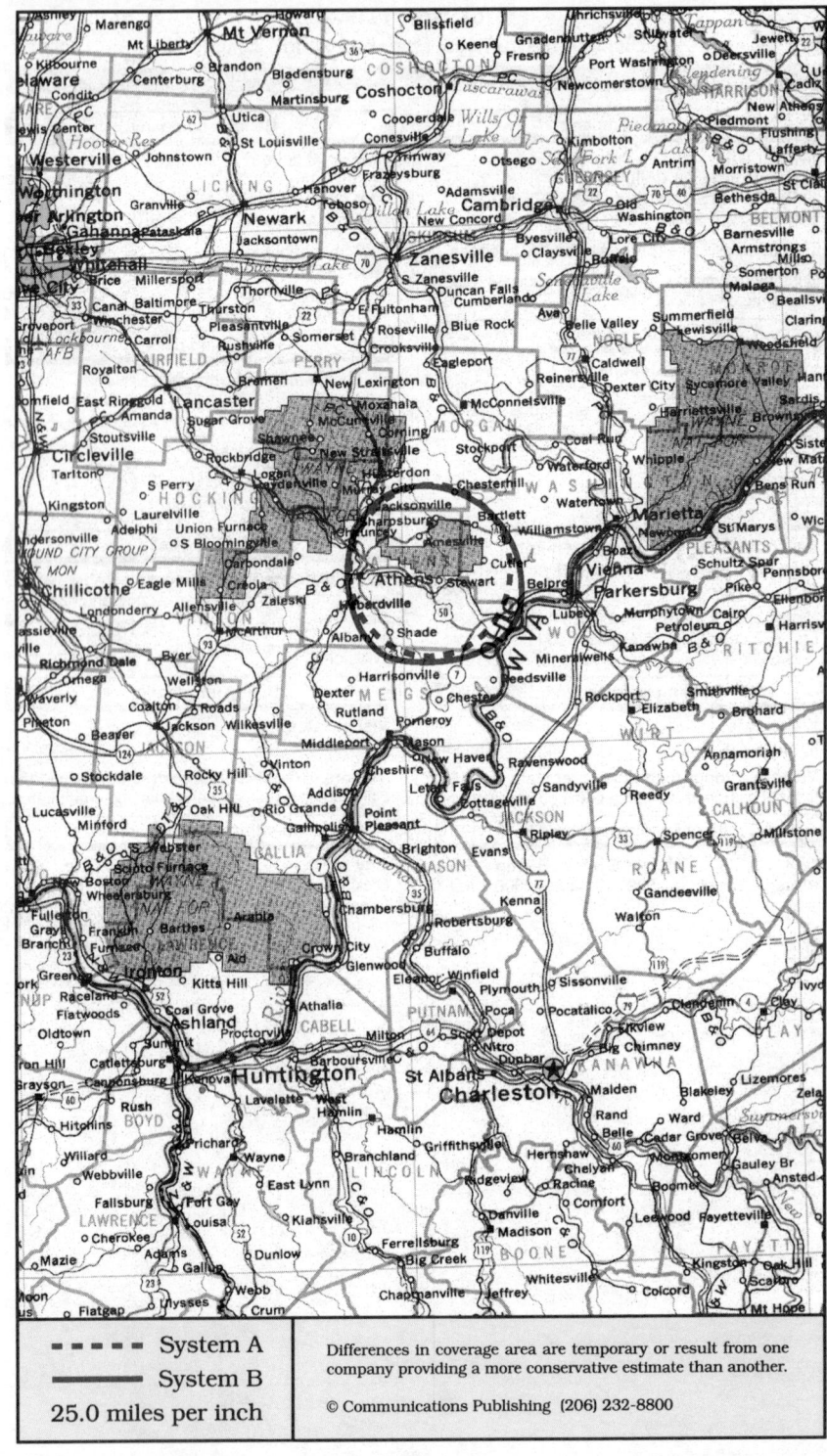

Athens, Ohio

System A

Cellular company

Cellular One
Mercury Communications
1100 E State Street
Athens, OH 45701

Local office	614-594-4800
Corporate office	601-948-4800

Sending calls

Visitor dialing instructions

Local calls	area code + 7 digits
Long distance	area code + 7 digits

Speed-dial numbers

Customer service	611
Emergency	911

Receiving calls

Caller dials the roamer access number below, hears tone, dials cellular area code and number.
. . . . 614-591-7626

NationLink & RoamingAmerica (see p. 1157)
Dial *31 send to receive calls, *30 to deactivate. This company's subscribers have level 3 service.

Billing information

Visitor rates
Most visitors . . . . . . $3 day + $1 minute

Visitors without roaming agreements (p. 1159)
Call co. to use Mastercard, Visa
or Roamer Plus will intercept first call to bill you.

Identification numbers

City	Market	System	Billing	RSA
All served cities	594	01575	00000	OH-10

System B

Cellular company

Independent Cellular Network
1015 E State Street
Athens, OH 45701

Local office	614-592-4911
Corporate office	813-489-1600

Sending calls

Visitor dialing instructions

Local calls	area code + 7 digits
Long distance	area code + 7 digits

Speed-dial numbers

Customer service	*611

Receiving calls

Caller dials the roamer access number below, hears tone, dials cellular area code and number.
. . . . 614-541-7626

Follow Me Roaming forwarding (see p. 1156)
Activate, dial *18 send. Deactivate dial *19 send.

Billing information

Visitor rates
Most visitors . . . . . $3 day + $1 minute

Visitors without roaming agreements (p. 1159)
Call co. to use Amex, Discover, Mastercard, Visa

Identification numbers

City	Market	System	Billing	RSA
All served cities	594	00032	00000	OH-10

- - - - System A	Differences in coverage area are temporary or result from one
———— System B	company providing a more conservative estimate than another.
25.0 miles per inch	© Communications Publishing (206) 232-8800

Cleveland, Ohio
Akron, Ashtabula, Canton, Carrollton, Cleveland, Elyria, Lorain

System A	System B

Cellular company

Cellular One
New Par
Administration
3 Summit Park Drive, Suite 600
Cleveland, OH 44131

764 S Broadway
Akron, OH 44311

4400 Portage Street
North Canton, OH 44720

28149 Chagrin Boulevard
Woodmere, OH 44122

Customer service		216-642-1616
. . . .		614-846-7317
Local office	Akron	216-535-0500
. . . .	North Canton	216-499-7979
. . . .	Woodmere	216-765-1444
Corporate office		614-436-4331

Sending calls

Visitor dialing instructions
Local calls . . . . . area code + 7 digits
Long distance . . . 1 + area code + 7 digits

Speed-dial numbers
Customer service	611
Driving directions	999
Emergency	911
Make a traffic report	*1041
Ohio highway patrol . . .	*HELP or *4357
Time and weather	*89

Receiving calls

Caller dials the roamer access number below,
hears tone, dials cellular area code and number.
Akron	216-697-7626
Ashtabula, Cleveland . . .	216-469-7626
Canton, Carrollton	216-495-7626

NationLink & RoamingAmerica (see p. 1157)
Dial *31 send to receive calls, *30 to deactivate.
Dial *32 send to tell caller the roamer access no.
This company's subscribers have level 6 service.

Billing information

Visitor rates
Most visitors . . . . $2.50 day + 75¢ minute

Visitors without roaming agreements (p. 1159)
Roamer Plus will intercept first call to bill you.

Identification numbers

City	Market	System	Billing	RSA
Akron	052	00015	00000	MSA
Ashtabula	587	00015	00000	OH-3
Canton	087	00015	00000	MSA
Carrollton	591	00015	00000	OH-7
Cleveland	016	00015	00000	MSA
Lorain/Elyria	136	00015	00000	MSA

Cellular company

GTE Mobilnet
6060 Rockside Woods Boulevard, Suite 431
Cleveland, OH 44131

Customer service	some areas	800-669-5665
. . . .		216-642-0688
Local office		216-642-0327
Corporate office		404-391-8000

Sending calls

Visitor dialing instructions
Local calls . . . . . . area code + 7 digits
Long distance . . . 1 + area code + 7 digits

Speed-dial numbers
Customer service	*611
Emergency	911
Telephone problems	*111

Receiving calls

Caller dials the roamer access number below,
hears tone, dials cellular area code and number.
Akron, Canton, Carrollton . .	216-388-7626
Ashtabula, Cleveland . . .	216-389-7626

Follow Me Roaming forwarding (see p. 1156)
Activate, dial *18 send. Deactivate dial *19 send.

Billing information

Visitor rates
Most visitors . . . . . $3 day + 75¢ minute

Visitors without roaming agreements (p. 1159)
Call co. to use Amex, Mastercard, Visa

Identification numbers

City	Market	System	Billing	RSA
Akron	052	00054	00000	MSA
Ashtabula	587	00054	00000	OH-3
Canton	087	00054	00000	MSA
Carrollton	591	00054	00000	OH-7
Cleveland	016	00054	00000	MSA
Lorain/Elyria	136	00054	00000	MSA

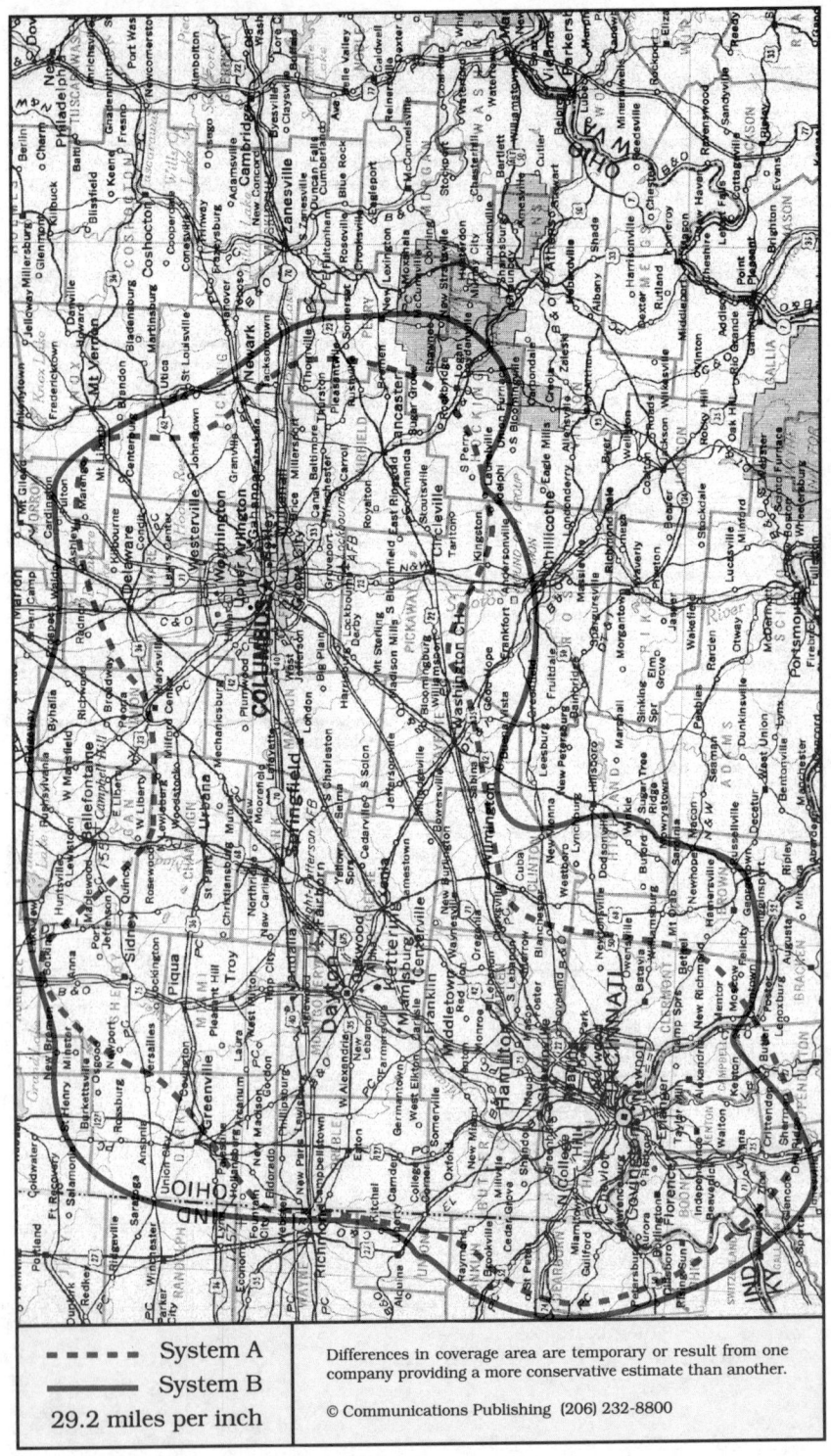

- - - - - System A
———— System B
29.2 miles per inch

Differences in coverage area are temporary or result from one company providing a more conservative estimate than another.

© Communications Publishing   (206) 232-8800

Columbus, Ohio

Cincinnati, Columbus, Dayton, Hamilton, Logan, Sidney, Springfield, Wilmington

System A	System B

Cellular company

Cellular One
New Par
350 E Wilson Bridge Road
Worthington, OH 43085

2799 Miamisburg-Centerville Road
Centerville, OH 45459

8800 Governor's Hill Drive
Cincinnati, OH 45249

900 S Patterson Boulevard
Dayton, OH 45402-2625

Customer service .	Cincinnati	513-543-6300
. . . .	Columbus	614-325-2212
Local office . . .	Cincinnati	513-543-6000
. . . .	Centerville	513-434-2355
. . . .	Dayton	513-223-9099
. . . .	Worthington	614-436-4331
Corporate office		614-436-4331

Sending calls

Visitor dialing instructions
Local calls . . . . . area code + 7 digits
Long distance . . . 1 + area code + 7 digits

Speed-dial numbers
Customer service 611
Emergency 911
Ohio highway patrol . . . ∗HELP or ∗4357
Traffic reports . . . . ∗95 or ∗105 or ∗107

Receiving calls

Caller dials the roamer access number below,
hears tone, dials cellular area code and number.

Cincinnati, Hamilton, Wilmington	513-543-7626
Columbus, Washington Court H.	614-296-7626
Dayton, Sidney, Springfield	513-477-7626
Marysville	513-935-7626

NationLink & RoamingAmerica (see p. 1157)
Dial ∗31 send to receive calls, ∗30 to deactivate.
Dial ∗32 send to tell caller the roamer access no.
This company's subscribers have level 6 service.

Billing information

Visitor rates
Most visitors . . . . $2.50 day + 75¢ minute

Visitors without roaming agreements (p. 1159)
Roamer Plus will intercept first call to bill you.

Identification numbers

City	Market	System	Billing	RSA
Cincinnati	023	00051	00000	MSA
Columbus	031	00133	00000	MSA
Dayton	040	00163	00000	MSA
Hamilton	145	00051	00000	MSA
Sidney	588	01563	00000	OH-4
Springfield	180	00163	00000	MSA
Wilmington	592	01574	00000	OH-8

Logan will be served by system A of Athens, OH.

Cellular company

Ameritech Mobile Communications
485 Metro Place S, Suite 500
Dublin, OH 43017

Customer service	800-221-0994
. . . .	708-706-7300
Local office	614-793-4500
Corporate office	708-706-7600

Sending calls

Visitor dialing instructions
Local calls . . . . . area code + 7 digits
Long distance . . . 1 + area code + 7 digits

Speed-dial numbers
Customer service . . . . ∗611
DIrectory assistance 411
Emergency 911
Recorded help line . . . . ∗90
Roaming information . . . ∗711 or ∗712
WBNS AM radio 211
WCMH news tips . . . . ∗4
WDTN . . . . ∗2
WH10 Dayton news . . . . . . ∗211
WKRC TV-12 . . . . ∗12
WNCI traffic report . . . . ∗979
WOW traffic report . . . . ∗700

Receiving calls

Caller dials the roamer access number below,
hears tone, dials cellular area code and number.

Bellefontaine	513-597-7626
Cincinnati, Hamilton	513-977-7626
Columbus, Logan	614-271-7626
Dayton, Sidney, Springfield . .	513-239-7626
Greenville, Wilmington . . .	513-238-7626
Washington Court House . . .	614-648-7626

Follow Me Roaming forwarding (see p. 1156)
Activate, dial ∗18 send. Deactivate dial ∗19 send.

Billing information

Visitor rates
Most visitors . . . . . $3 day + 85¢ minute

Visitors without roaming agreements (p. 1159)
Cannot call since company takes no credit cards.

Identification numbers

City	Market	System	Billing	RSA
Bellefontaine,Sidney	588	01564	00000	OH-4
Cincinnati	023	00014	00000	MSA
Columbus	031	00138	00000	MSA
Dayton	040	00134	00000	MSA
Hamilton	145	00014	00000	MSA
Logan	594	00138	00000	OH-10
Springfield	180	00134	00000	MSA
Wilmington	592	00138	00000	OH-8

For full coverage area, see Zanesville, OH.

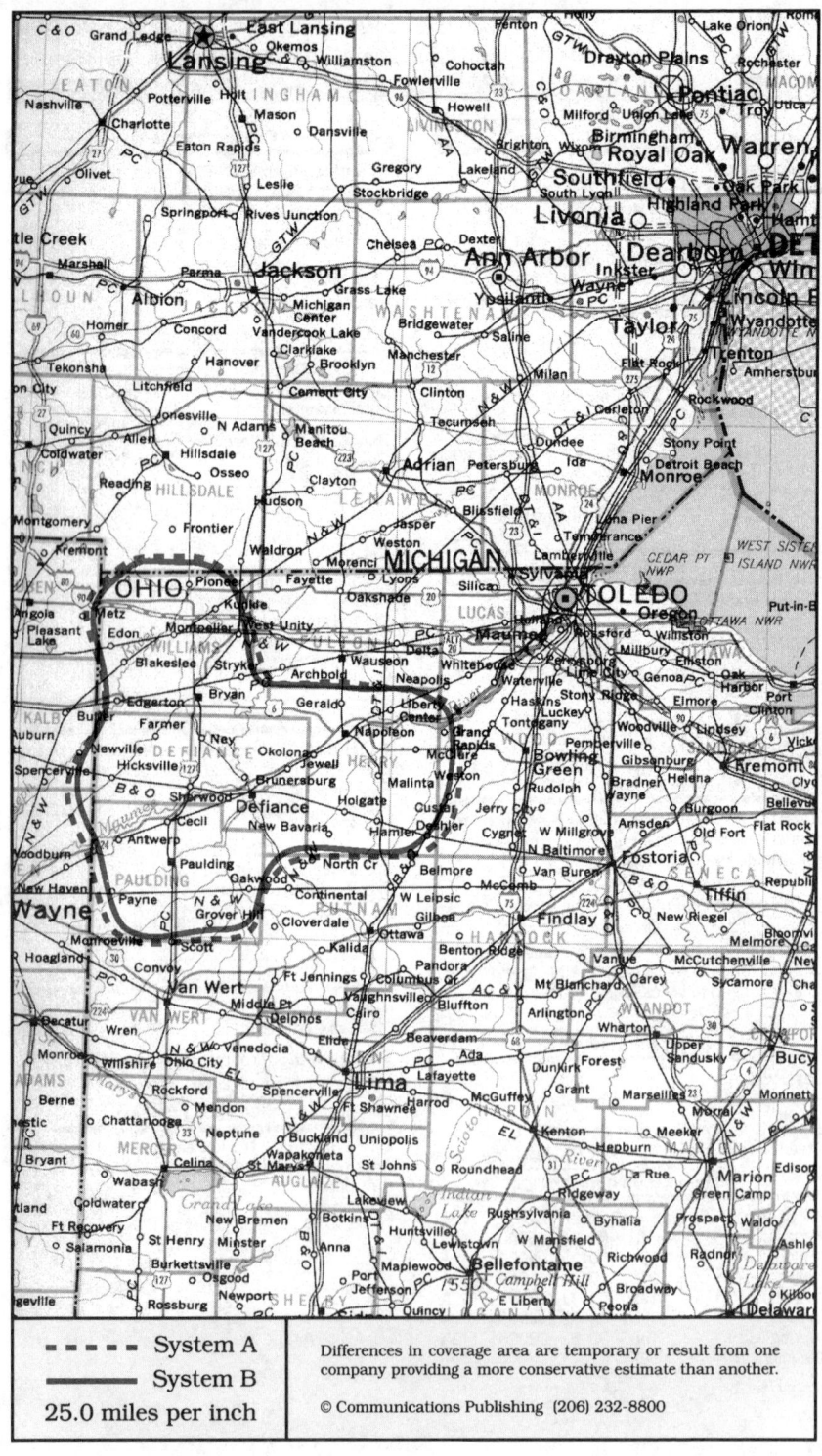

	System A
	System B
25.0 miles per inch	

Differences in coverage area are temporary or result from one company providing a more conservative estimate than another.

© Communications Publishing (206) 232-8800

Defiance, Ohio
Bryan, Defiance, Napoleon, Paulding

System A	System B

Cellular company

Liberty Cellular Phone Network
Managed by: Unitel
717 Perry Street
Defiance, OH 43512-2736

Local office   419-782-4400
Corporate office   . . .   Unitel 609-646-9400

Sending calls

Visitor dialing instructions
Local calls   . . . .   area code + 7 digits
Long distance   . . .   1 + area code + 7 digits

Speed-dial numbers
Customer service   611
Emergency   911
Farm market report   211
Time and temperature   111

Receiving calls

Caller dials the roamer access number below,
hears tone, dials cellular area code and number.
. . . .   419-769-7626

Billing information

Visitor rates
Most visitors . . . . . $3 day + 99¢ minute

Visitors without roaming agreements (p. 1159)
Call co. to use Mastercard, Visa

Identification numbers

City	Market	System	Billing	RSA
All served cities	585	01557	00000	OH-1

Cellular company

Centel Cellular
1370 Dussel Drive
Maumee, OH 43537

Local office   800-548-6019
. . . .   419-893-1077
Corporate office   312-399-2644

Sending calls

Visitor dialing instructions
Local calls . . . . . .   area code + 7 digits
Long distance . . .   1 + area code + 7 digits

Speed-dial numbers
Customer service   . . . .   ∗611
Emergency   911

Receiving calls

Caller dials the roamer access number below,
hears tone, dials cellular area code and number.
. . . .   419-262-7626

Billing information

Visitor rates
Most visitors   . . . . . $3 day + 99¢ minute

Visitors without roaming agreements (p. 1159)
Cannot call since company takes no credit cards.

Identification numbers

City	Market	System	Billing	RSA
All served cities	585	01558	00000	OH-1

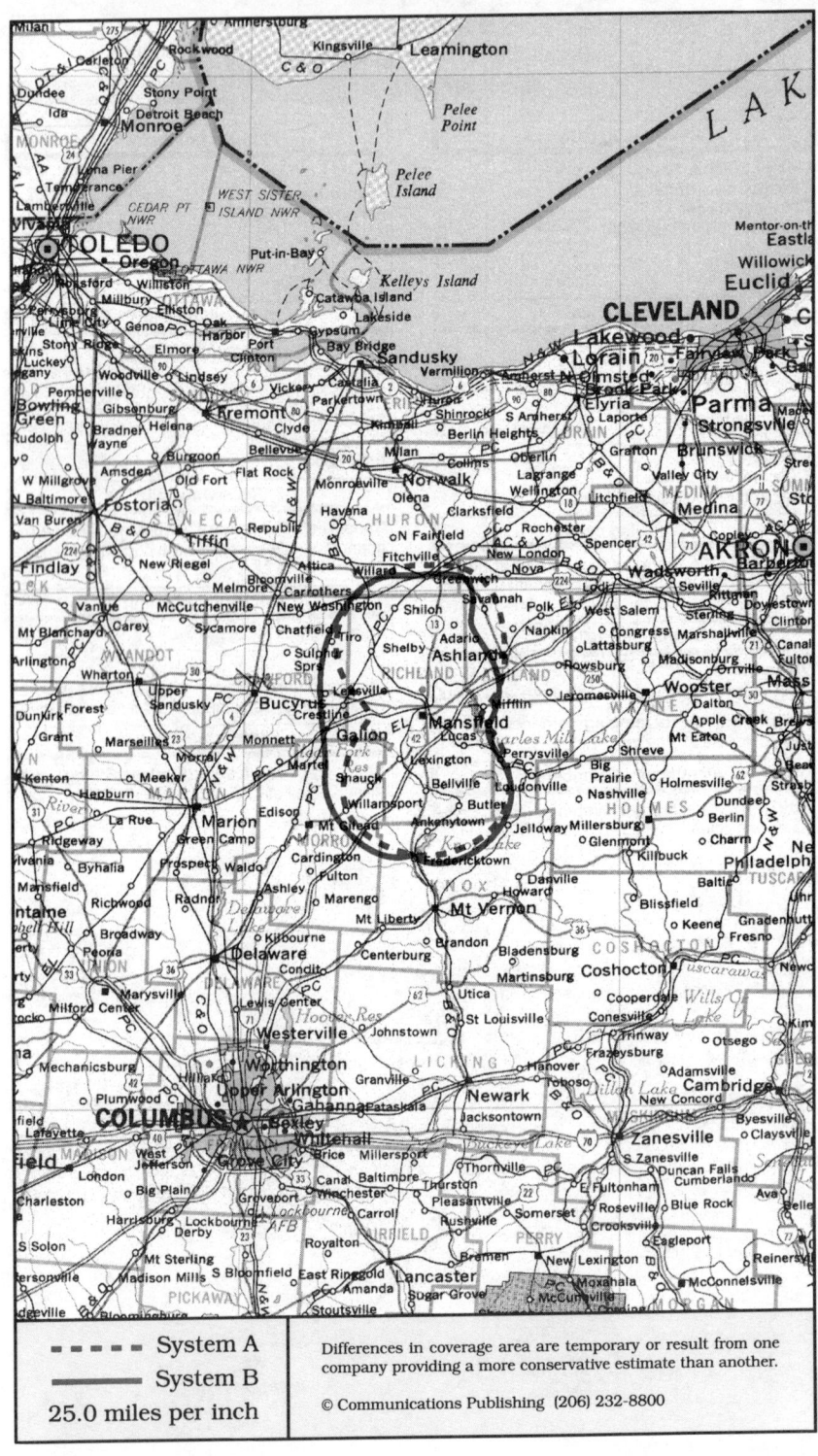

- - - - - System A
——————— System B

25.0 miles per inch

Differences in coverage area are temporary or result from one company providing a more conservative estimate than another.

© Communications Publishing (206) 232-8800

Mansfield, Ohio
Galion, Mansfield

System A

Cellular company

Cellular One
New Par
2071 W 4th Street
Mansfield, OH 44906

Customer service	419-564-5000
Local office	419-529-9000
Corporate office	614-436-4331

Sending calls

Visitor dialing instructions

Local calls	area code + 7 digits
Long distance . . .	1 + area code + 7 digits

Speed-dial numbers

Customer service	611
Emergency	911
Ohio highway patrol . . .	∗HELP or ∗4357

Receiving calls

Caller dials the roamer access number below, hears tone, dials cellular area code and number.
. . . . 419-564-7626

NationLink & RoamingAmerica (see p. 1157)
Dial ∗31 send to receive calls, ∗30 to deactivate.
Dial ∗32 send to tell caller the roamer access no.
This company's subscribers have level 6 service.

Billing information

Visitor rates
Most visitors . . . . $2.50 day + 75¢ minute

Visitors without roaming agreements (p. 1159)
Call co. to use Amex, Mastercard, Visa
or Roamer Plus will intercept first call to bill you.

Identification numbers

City	Market	System	Billing	RSA
All served cities	231	00447	00000	MSA

System B

Cellular company

Centel Cellular
270 Lexington Avenue, Suite 101
Mansfield, OH 44907

Local office	419-756-3330
Corporate office	312-399-2644

Sending calls

Visitor dialing instructions

Local calls	area code + 7 digits
Long distance . . .	1 + area code + 7 digits

Speed-dial numbers

Customer service	∗611
Emergency	911
Ohio state patrol	800-525-5555
WMAN	14

Receiving calls

Caller dials the roamer access number below, hears tone, dials cellular area code and number.
. . . . 419-543-7626

Follow Me Roaming forwarding (see p. 1156)
Activate, dial ∗18 send. Deactivate dial ∗19 send.

Billing information

Visitor rates
Most visitors . . . . . $3 day + 99¢ minute

Visitors without roaming agreements (p. 1159)
Cannot call since company takes no credit cards.

Identification numbers

City	Market	System	Billing	RSA
All served cities	231	00430	00000	MSA

For full coverage area, see Newark, OH and
Sandusky, OH.

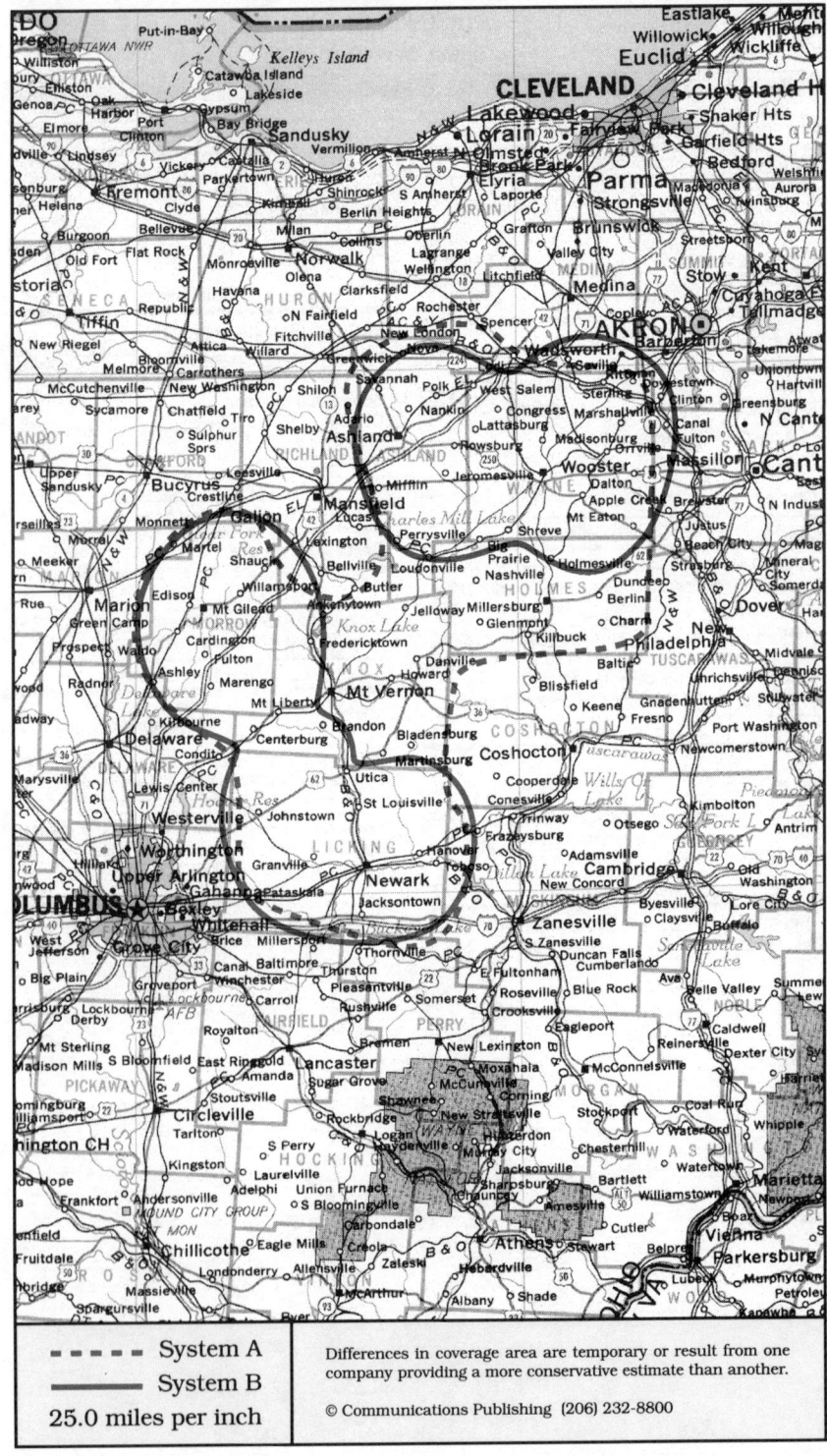

Newark, Ohio
Ashland, Mt. Gilead, Mt. Vernon, Newark, Wooster

System A	System B

Cellular company

Ohio State Cellular
Cellwave
818 Hebron Road, PO Box 2266
Heath, OH 43056

Corporate office 614-522-6446

Sending calls

Visitor dialing instructions
Local calls . . . . . area code + 7 digits
Long distance . . . 1 + area code + 7 digits

Speed-dial numbers
Customer service 611
Emergency 911

Receiving calls

Caller dials the roamer access number below,
hears tone, dials cellular area code and number.
Mt. Vernon, Newark 614-398-7626
Millersburg, Wooster 216-465-7626

NationLink & RoamingAmerica (see p. 1157)
Caller hears your current roamer access number.
This company's subscribers have level 0 service.

Billing information

Visitor rates
Most visitors . . . . . $3 day + 90¢ minute

Visitors without roaming agreements (p. 1159)
Roamer Plus will intercept first call to bill you.

Identification numbers

City	Market	System	Billing	RSA
All served cities	590	01567	00000	OH-6

Cellular company

Centel Cellular
270 Lexington Avenue, Suite 101
Mansfield, OH 44907

Local office 419-756-3330
Corporate office 312-399-2644

Sending calls

Visitor dialing instructions
Local calls . . . . . . area code + 7 digits
Long distance . . . 1 + area code + 7 digits

Speed-dial numbers
Customer service *611
Emergency 911
Ohio state patrol 800-525-5555
WMAN 14

Receiving calls

Caller dials the roamer access number below,
hears tone, dials cellular area code and number.
. . . . 419-543-7626

Billing information

Visitor rates
Most visitors . . . . . $3 day + 99¢ minute

Visitors without roaming agreements (p. 1159)
Cannot call since company takes no credit cards.

Identification numbers

City	Market	System	Billing	RSA
Ashland, Wooster	590	01568	30586	OH-6
Newark	590	01568	30676	OH-6

For full coverage area, see Mansfield, OH and
Sandusky, OH and Toledo, OH.

- - - - - System A
———— System B
25.0 miles per inch

Differences in coverage area are temporary or result from one company providing a more conservative estimate than another.

© Communications Publishing (206) 232-8800

New Philadelphia, Ohio
Cadiz, Cambridge, Dover, New Philadelphia

System A	System B

Cellular company

United States Cellular
1025 Front Avenue SW, PO Box 975
New Philadelphia, OH 44663

Local office	216-339-8805
Corporate office	312-399-8900

Sending calls

Visitor dialing instructions

Local calls	area code + 7 digits
Long distance . . .	1 + area code + 7 digits

Speed-dial numbers

Customer service	611
Emergency	911

Receiving calls

Caller dials the roamer access number below, hears tone, dials cellular area code and number.

Cambridge	614-584-7626
New Philadelphia	216-827-7626

Billing information

Visitor rates

Most visitors	$3 day + 99¢ minute

Visitors without roaming agreements (p. 1159)
Call co. to use Amex, Mastercard, Visa
or Roamer Plus will intercept first call to bill you.

Identification numbers

City	Market	System	Billing	RSA
All served cities	591	01569	00000	OH-7

For full coverage area, see Zanesville, OH.

Cellular company

Independent Cellular Network
543 W High Avenue
New Philadelphia, OH 44663

Local office	216-339-6789
Corporate office	813-489-1600

Sending calls

Visitor dialing instructions

Local calls	area code + 7 digits
Long distance . . .	1 + area code + 7 digits

Speed-dial numbers

Customer service	611
Emergency	911

Receiving calls

Caller dials the roamer access number below, hears tone, dials cellular area code and number.

. . . .	216-340-7626

Follow Me Roaming forwarding (see p. 1156)
Activate, dial ∗18 send. Deactivate dial ∗19 send.

Billing information

Visitor rates

Most visitors	$3 day + $1 minute

Visitors without roaming agreements (p. 1159)
Call co. to use Mastercard, Visa

Identification numbers

City	Market	System	Billing	RSA
All served cities	591	00032	30610	OH-7

For full coverage area, see Wheeling, WV.

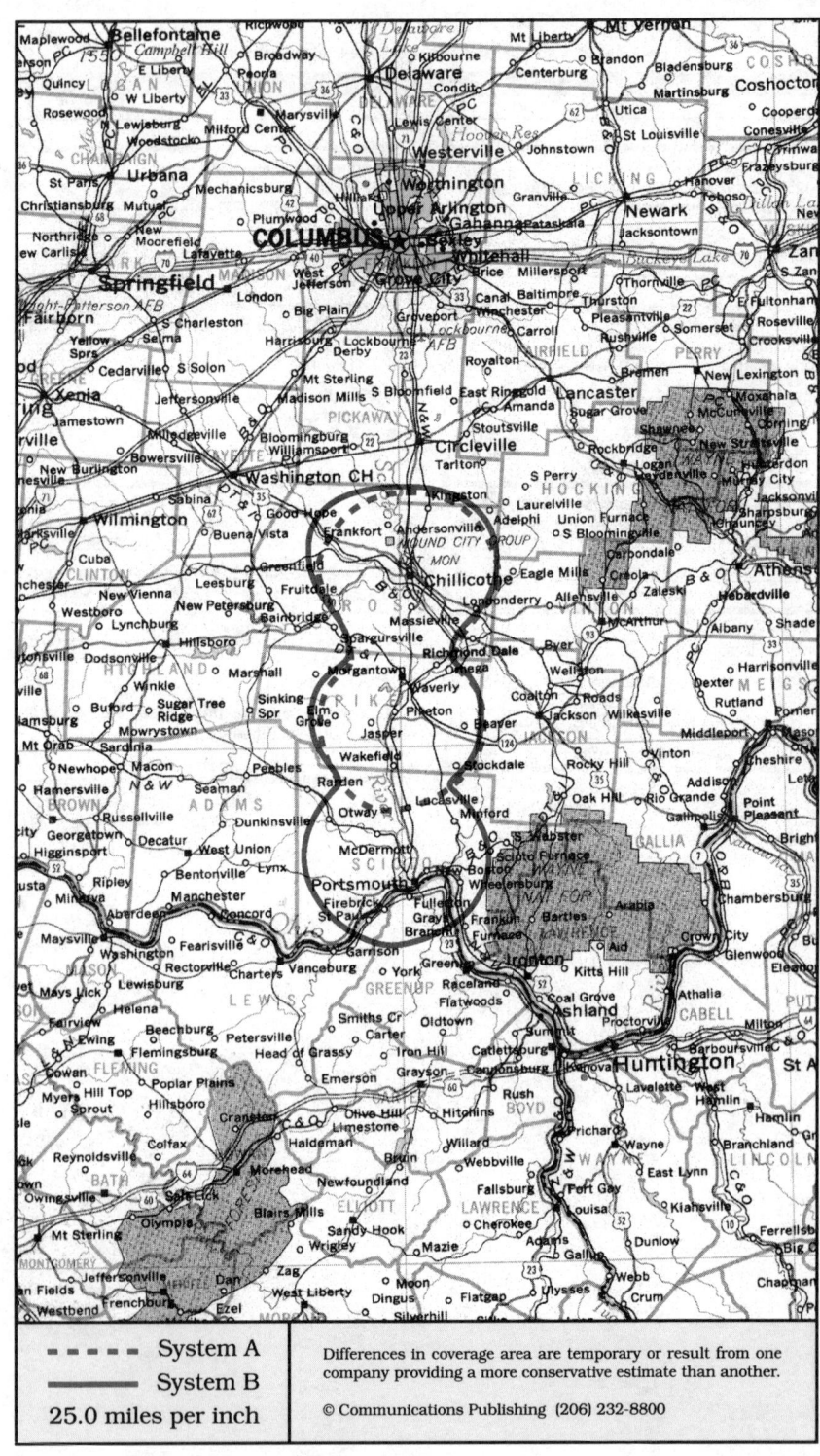

- - - - System A	Differences in coverage area are temporary or result from one company providing a more conservative estimate than another.
──────── System B	
25.0 miles per inch	© Communications Publishing (206) 232-8800

Portsmouth, Ohio
Chillicothe, Portsmouth, Waverly

System A	System B

Cellular company

Cellular One
Sterling Cellular Management
1080 Holcomb Bridge Road, Bldg 100, Ste 200
Roswell, GA 30076

Customer service 800-552-6150
Corporate office 404-552-5030

Sending calls

Visitor dialing instructions
Local calls . . . . 7 digits
Long distance . . . 1 + area code + 7 digits

Speed-dial numbers
Customer service . . . . ✳611
Emergency 911

Receiving calls

Caller dials the roamer access number below,
hears tone, dials cellular area code and number.
. . . . 614-884-7626

Billing information

Visitor rates
Most visitors . . . . . $3 day + 75¢ minute

Visitors without roaming agreements (p. 1159)
Roamer Plus will intercept first call to bill you.

Identification numbers

City	Market	System	Billing	RSA
All served cities	593	01573	00000	OH-9

Cellular company

United States Cellular
611 Chillicothe Street, PO Box 618
Portsmouth, OH 45662-0618

Local office 614-354-4111
Corporate office 312-399-8900

Sending calls

Visitor dialing instructions
Local calls . . . . . . area code + 7 digits
Long distance . . . . . area code + 7 digits

Speed-dial numbers
Customer service . . . . ✳611
Emergency 911
Roaming information . . . . ✳711

Receiving calls

Caller dials the roamer access number below,
hears tone, dials cellular area code and number.
. . . . 614-285-7626

Billing information

Visitor rates
Most visitors . . . . . $3 day + 99¢ minute

Visitors without roaming agreements (p. 1159)
Call co. to use Amex, Mastercard, Visa

Identification numbers

City	Market	System	Billing	RSA
All served cities	593	01574	00000	OH-9

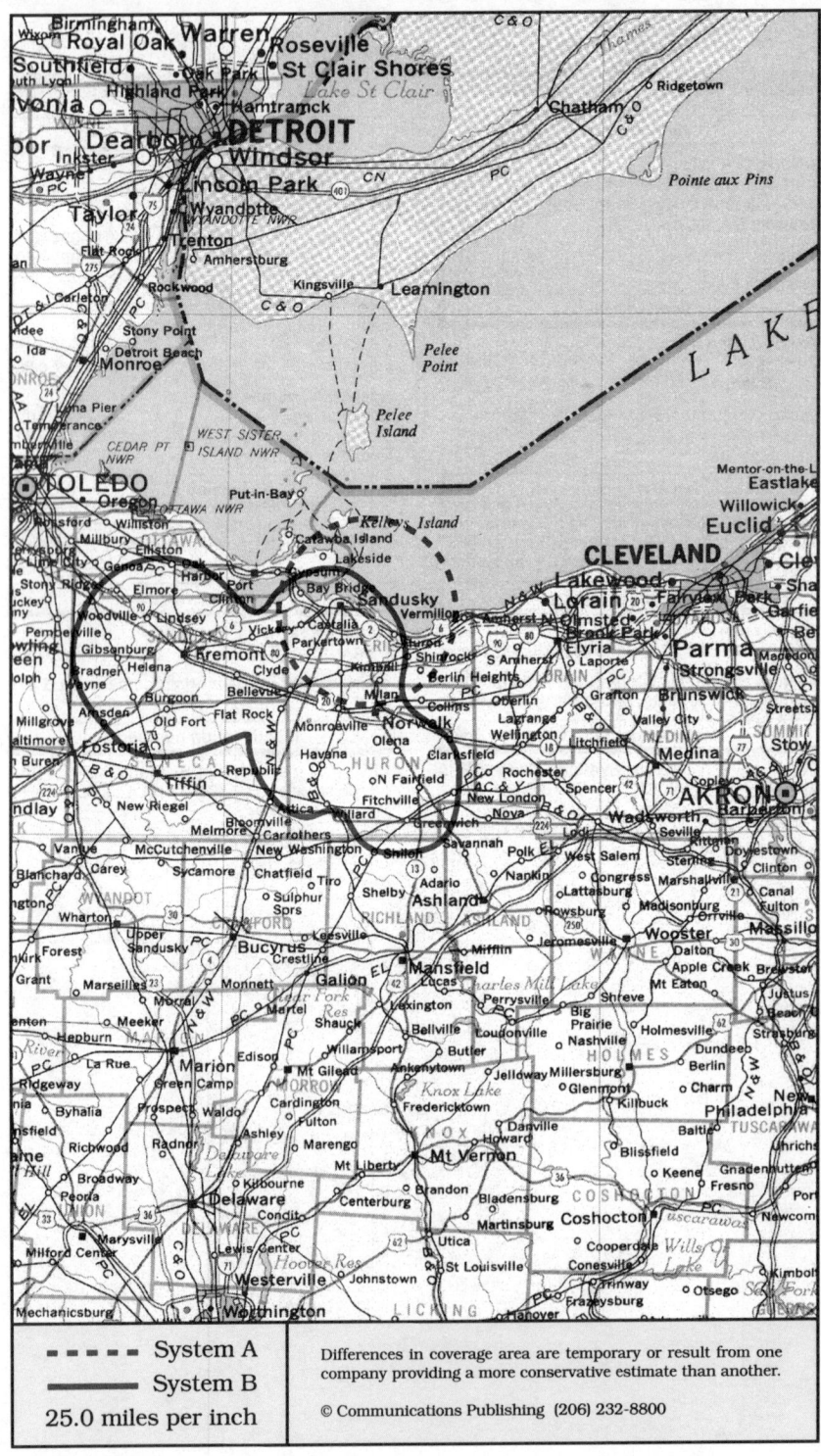

System A
System B
25.0 miles per inch

Differences in coverage area are temporary or result from one company providing a more conservative estimate than another.

© Communications Publishing (206) 232-8800

Sandusky, Ohio
Fremont, Norwalk, Sandusky

System A

Cellular company

Alpha Cellular
2615 Central Avenue
Columbus, IN 47201

Corporate office 812-372-1133

Sending calls

Visitor dialing instructions
Local calls . . . . . area code + 7 digits
Long distance . . . 1 + area code + 7 digits

Speed-dial numbers
Customer service . . . . ∗611
Emergency 911

Receiving calls

Caller dials the roamer access number below,
hears tone, dials cellular area code and number.
. . . . 419-357-7626

Billing information

Visitor rates
Most visitors . . . . . $3 day + 99¢ minute

Visitors without roaming agreements (p. 1159)
Cannot call since company takes no credit cards.

Identification numbers

City	Market	System	Billing	RSA
All served cities	586	01559	00000	OH-2

System B

Cellular company

Centel Cellular
1370 Dussel Drive
Maumee, OH 43537

Local office 800-548-6019
. . . . 419-893-1077
Corporate office 312-399-2644

Sending calls

Visitor dialing instructions
Local calls . . . . . . area code + 7 digits
Long distance . . . 1 + area code + 7 digits

Speed-dial numbers
Customer service . . . . ∗611
Emergency 911

Receiving calls

Caller dials the roamer access number below,
hears tone, dials cellular area code and number.
. . . . 419-262-7626

Billing information

Visitor rates
Most visitors . . . . . $3 day + 99¢ minute

Visitors without roaming agreements (p. 1159)
Cannot call since company takes no credit cards.

Identification numbers

City	Market	System	Billing	RSA
Fremont	586	01560	30550	OH-2
Sandusky	586	01560	00000	OH-2

For full coverage area, see Mansfield, OH and
Toledo, OH.

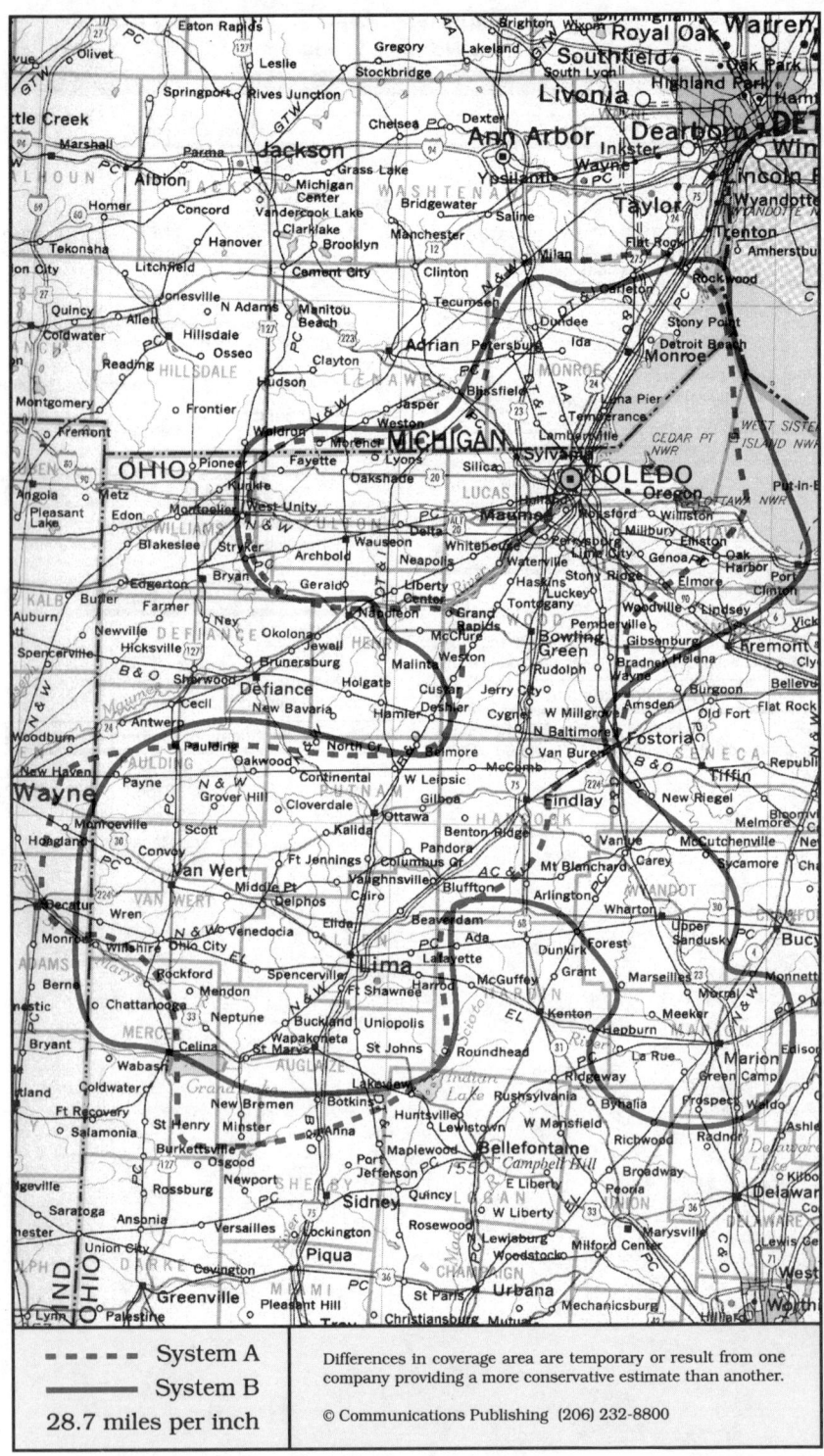

System A (dashed line)
System B (solid line)
28.7 miles per inch

Differences in coverage area are temporary or result from one company providing a more conservative estimate than another.

© Communications Publishing (206) 232-8800

Toledo, Ohio

Findlay, Lima, Marion, Monroe MI, St. Marys, Toledo, Van Wert

System A	System B

Cellular company

Cellular One
(affiliated with Cellular One of Detroit)
6710 W Central
Toledo, OH 43617

Customer service	419-243-1091
Local office	419-843-2995
Corporate office	313-737-5100

Sending calls

Visitor dialing instructions

Local calls	area code + 7 digits
Long distance . . .	0 + area code + 7 digits

Speed-dial numbers

Customer service	*611
Emergency	911
Roamer access number	511

Receiving calls

Caller dials the roamer access number below,
hears tone, dials cellular area code and number.

Lima	419-234-7626
Toledo	419-351-7626

NationLink & RoamingAmerica (see p. 1157)
Dial *31 send to receive calls, *30 to deactivate.
Dial *32 send to tell caller the roamer access no.
This company's subscribers have level 6 service.

Billing information

Visitor rates

Most visitors	$2 day + 75¢ minute

Visitors without roaming agreements (p. 1159)
Roamer Plus will intercept first call to bill you.

Identification numbers

City	Market	System	Billing	RSA
Lima	158	00021	00000	MSA
Toledo	048	00021	00000	MSA

Cellular company

Centel Cellular
1370 Dussel Drive
Maumee, OH 43537

2161 Elida Road
Lima, OH 45805-1518

Local office	Maumee	800-548-6019
. . . .		419-893-1077
. . . .	Lima	800-788-5462
. . . .		419-227-8004
Corporate office		312-399-2644

Sending calls

Visitor dialing instructions

Local calls	area code + 7 digits
Long distance . . .	1 + area code + 7 digits

Speed-dial numbers

Customer service	*611
Emergency	911

Receiving calls

Caller dials the roamer access number below,
hears tone, dials cellular area code and number.

Lima	419-235-7626
Marion	419-543-7626
Toledo, Findlay, Upper Sandusky	419-262-7626

Follow Me Roaming forwarding (see p. 1156)
Activate, dial *18 send. Deactivate dial *19 send.

Billing information

Visitor rates

Most visitors	$3 day + 99¢ minute

Visitors without roaming agreements (p. 1159)
Cannot call since company takes no credit cards.

Identification numbers

City	Market	System	Billing	RSA
Findlay	589	01566	30580	OH-5
Lima	158	00412	00000	MSA
Marion	589	01566	00000	OH-5
Toledo	048	00130	00000	MSA

For full coverage area, see Newark, OH and
Sandusky, OH.

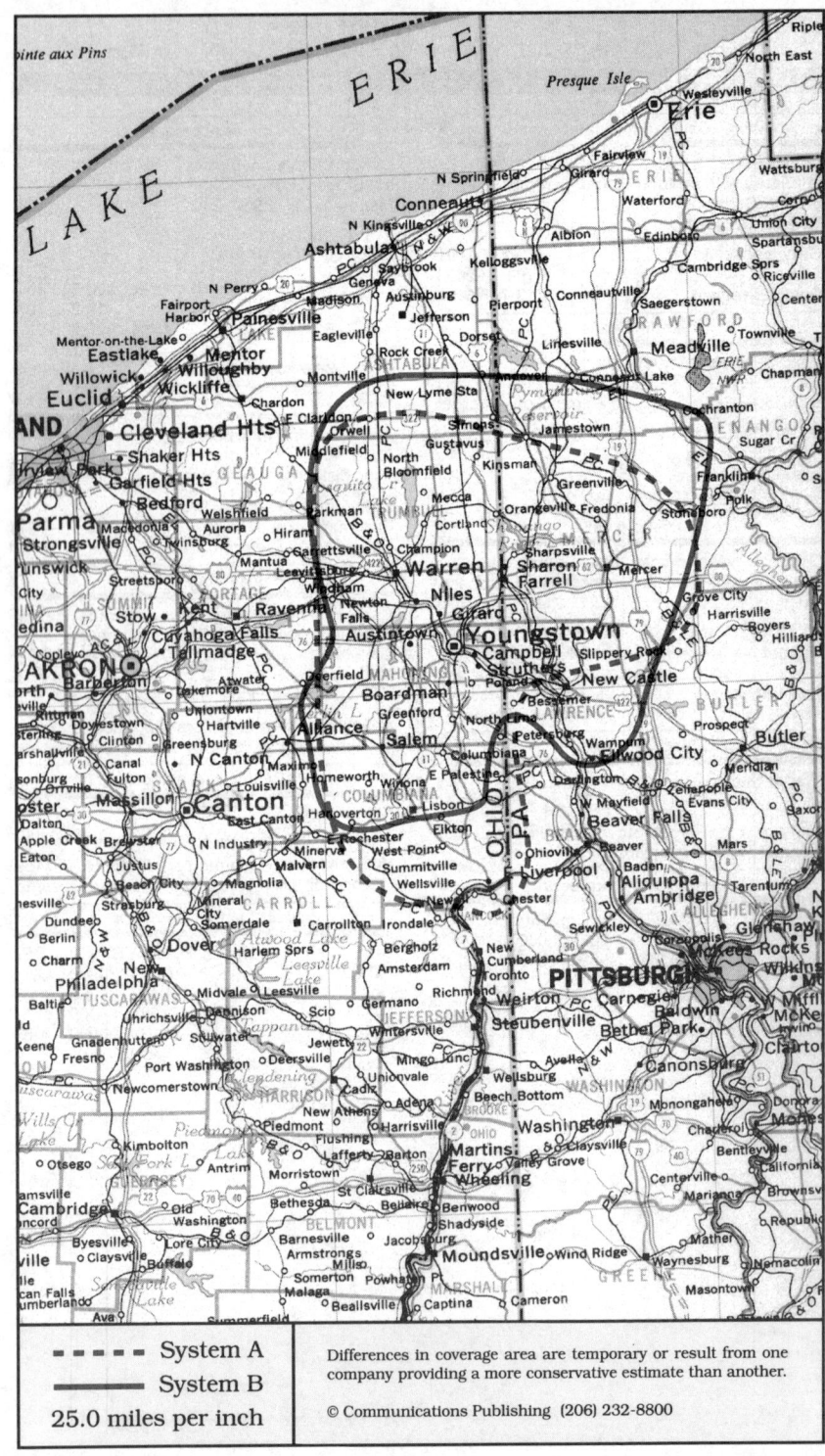

System A
System B
25.0 miles per inch

Differences in coverage area are temporary or result from one company providing a more conservative estimate than another.

© Communications Publishing (206) 232-8800

Youngstown, Ohio

Columbiana, East Liverpool, New Castle PA (B), Sharon PA, Warren, Youngstown

System A	System B

Cellular company

Wilcom Cellular
3910 South Avenue, PO Box 1685
Youngstown, OH 44501-1685

Corporate office	800-837-5505
. . . .	216-565-5000
. . . .	412-866-5000

Sending calls

Visitor dialing instructions

Local calls	area code + 7 digits
Long distance	area code + 7 digits

Speed-dial numbers

Customer service	✳611
Gold Cross Ambulance	✳11

Receiving calls

Caller dials the roamer access number below, hears tone, dials cellular area code and number.

Sharon, PA	412-866-7626
Columbiana, Warren, Youngstown	216-565-7626

NationLink & RoamingAmerica (see p. 1157)
Dial ✳31 send to receive calls, ✳30 to deactivate.
Dial ✳32 send to tell caller the roamer access no.
This company's subscribers have level 13 service

Billing information

Visitor rates

Most visitors	$3 day + 85¢ minute

Visitors without roaming agreements (p. 1159)
Roamer Plus will intercept first call to bill you.

Identification numbers

City	Market	System	Billing	RSA
East Liverpool	595	00089	30421	OH-11
Poland	066	00089	30397	MSA
Sharon	238	00089	30395	MSA
Warren,Youngstown	066	00089	00000	MSA

For full coverage area, see Butler, PA.

Cellular company

Centel Cellular
7206 Market Street
Youngstown, OH 44512

Local office	800-325-5190
. . . .	216-758-4502
Corporate office	312-399-2644

Sending calls

Visitor dialing instructions

Local calls	area code + 7 digits
Long distance . . .	1 + area code + 7 digits

Speed-dial numbers

Customer service	✳611

Receiving calls

Caller dials the roamer access number below, hears tone, dials cellular area code and number.

Sharon, PA	412-699-7626
Columbiana, Warren, Youngstown	216-727-7626

Follow Me Roaming forwarding (see p. 1156)
Activate, dial ✳18 send. Deactivate dial ✳19 send.

Billing information

Visitor rates

Most visitors	$3 day + 99¢ minute

Visitors without roaming agreements (p. 1159)
Cannot call since company takes no credit cards.

Identification numbers

City	Market	System	Billing	RSA
Columbiana	595	01578	00000	OH-11
New Castle	617	00126	00000	PA-6
Sharon	238	00126	30622	MSA
Warren,Youngstown	066	00126	00000	MSA

For full coverage area, see Meadville, PA.

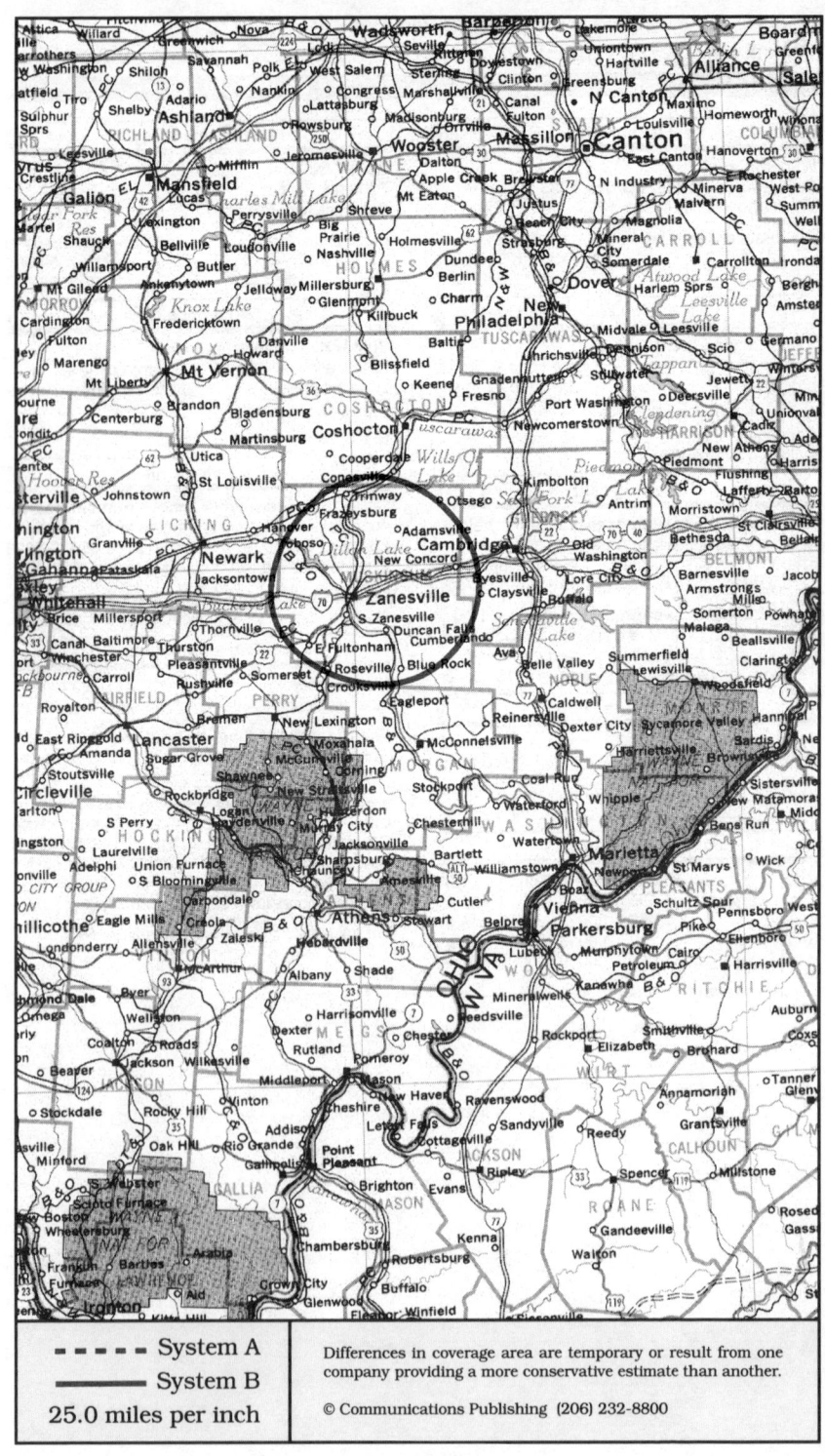

Zanesville, Ohio

System A	System B

Cellular company

United States Cellular
1025 Front Avenue SW, PO Box 975
New Philadelphia, OH 44663

Plans to begin service in 1993.

Local office	216-339-8805
Corporate office	312-399-8900

Sending calls

Visitor dialing instructions

Local calls	area code + 7 digits
Long distance . . .	1 + area code + 7 digits

Speed-dial numbers

Customer service	611
Emergency	911

Receiving calls

Caller dials the roamer access number below, hears tone, dials cellular area code and number.
. . . . not available

Billing information

Visitor rates

Most visitors	$3 day + 99¢ minute

Visitors without roaming agreements (p. 1159)
Call co. to use Amex, Mastercard, Visa
or Roamer Plus will intercept first call to bill you.

Identification numbers

City	Market	System	Billing	RSA
All served cities	591	01569	00000	OH-7

For full coverage area, see New Philadelphia, OH.

Cellular company

Ameritech Mobile Communications
485 Metro Place S, Suite 500
Dublin, OH 43017

Customer service	800-221-0994
. . . .	708-706-7300
Local office	614-793-4500
Corporate office	708-706-7600

Sending calls

Visitor dialing instructions

Local calls	area code + 7 digits
Long distance . . .	1 + area code + 7 digits

Speed-dial numbers

Customer service	*611
Emergency	911

Receiving calls

Caller dials the roamer access number below, hears tone, dials cellular area code and number.
. . . . 614-648-7626

Follow Me Roaming forwarding (see p. 1156)
Activate, dial *18 send. Deactivate dial *19 send.

Billing information

Visitor rates

Most visitors	$3 day + 85¢ minute

Visitors without roaming agreements (p. 1159)
Cannot call since company takes no credit cards.

Identification numbers

City	Market	System	Billing	RSA
All served cities	591	00138	00000	OH-7

For full coverage area, see Columbus, OH.

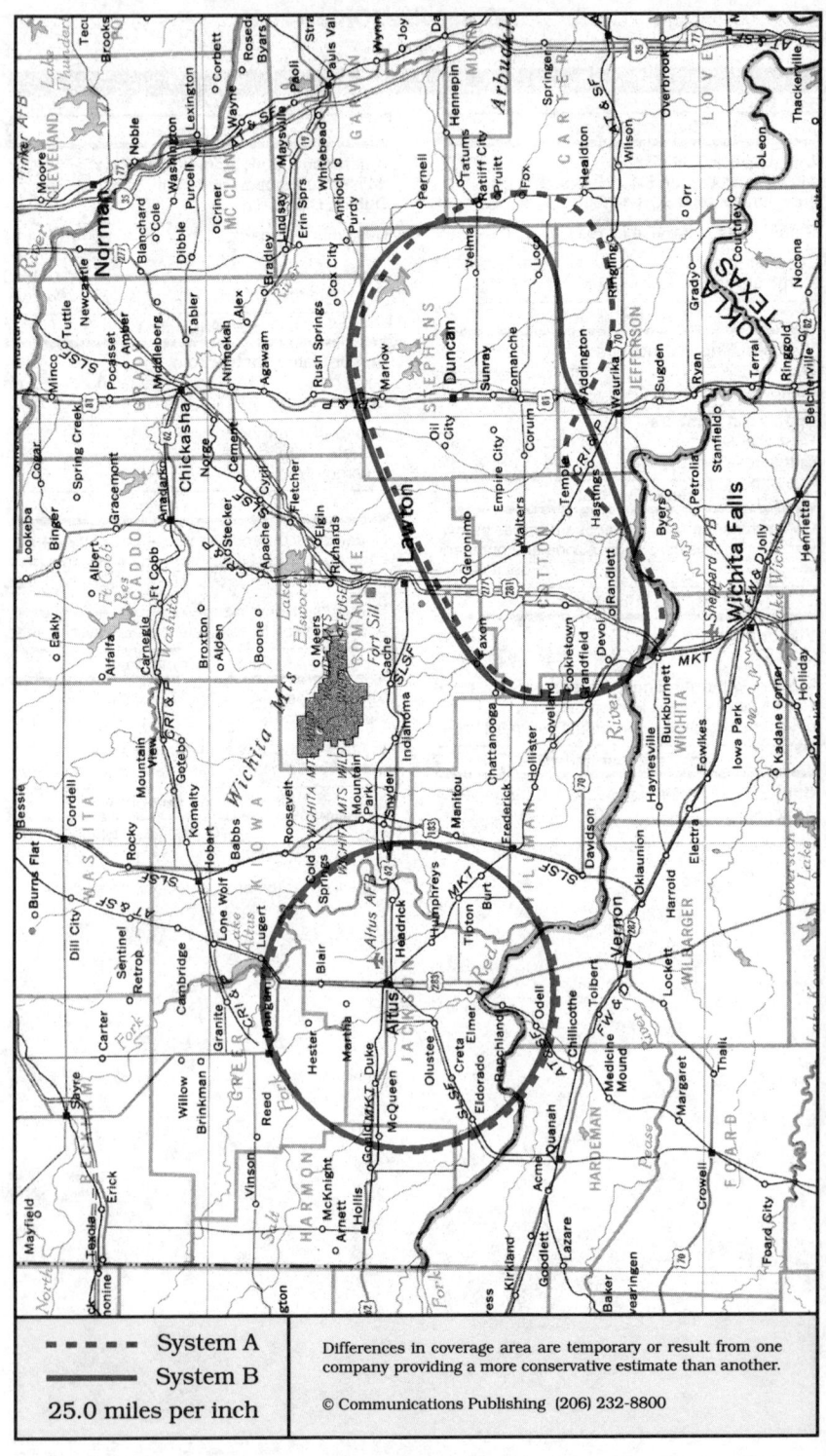

- - - - System A
——— System B
25.0 miles per inch

Differences in coverage area are temporary or result from one company providing a more conservative estimate than another.

© Communications Publishing (206) 232-8800

Altus, Oklahoma
Altus, Duncan, Walters

Cellular company

Cellular One of Southwest Oklahoma
Sooner Cellular
3000 N Main, Suite 100
Altus, OK 73521

512 N 81 Highway
Duncan, OK 73533

Local office	. . . . .	Altus	405-477-4747
Corporate office	. .	Duncan	405-252-9994

Sending calls

Visitor dialing instructions
Local calls . . . . . area code + 7 digits
Long distance . . . 1 + area code + 7 digits

Speed-dial numbers
Customer service 611
Emergency 911
Highway patrol *55

Receiving calls

Caller dials the roamer access number below,
hears tone, dials cellular area code and number.
. . . . 405-779-7626

NationLink & RoamingAmerica (see p. 1157)
Dial *31 send to receive calls, *30 to deactivate.
This company's subscribers have level 3 service.

Billing information

Visitor rates
Most visitors . . . . . $3 day + 99¢ minute

Visitors without roaming agreements (p. 1159)
Cannot call since company takes no credit cards.

Identification numbers

City	Market	System	Billing	RSA
All served cities	603	01593	00000	OK-8

For full coverage area, see Chickasha, OK.

Cellular company

United States Cellular
1001 B Avenue, Suite 130
Lawton, OK 73501

Local office	. . . .	405-355-3535
Corporate office	. . . . .	312-399-8900

Sending calls

Visitor dialing instructions
Local calls . . . . . . area code + 7 digits
Long distance . . . 1 + area code + 7 digits

Speed-dial numbers
Customer service *611
Emergency 911
Roaming information *711

Receiving calls

Caller dials the roamer access number below,
hears tone, dials cellular area code and number.
Altus 405-471-7626
Duncan 405-467-7626

Billing information

Visitor rates
Most visitors . . . . . $3 day + 75¢ minute

Visitors without roaming agreements (p. 1159)
Call co. to use Mastercard, Visa

Identification numbers

City	Market	System	Billing	RSA
All served cities	603	01594	00000	OK-8

For full coverage area, see Lawton, OK.

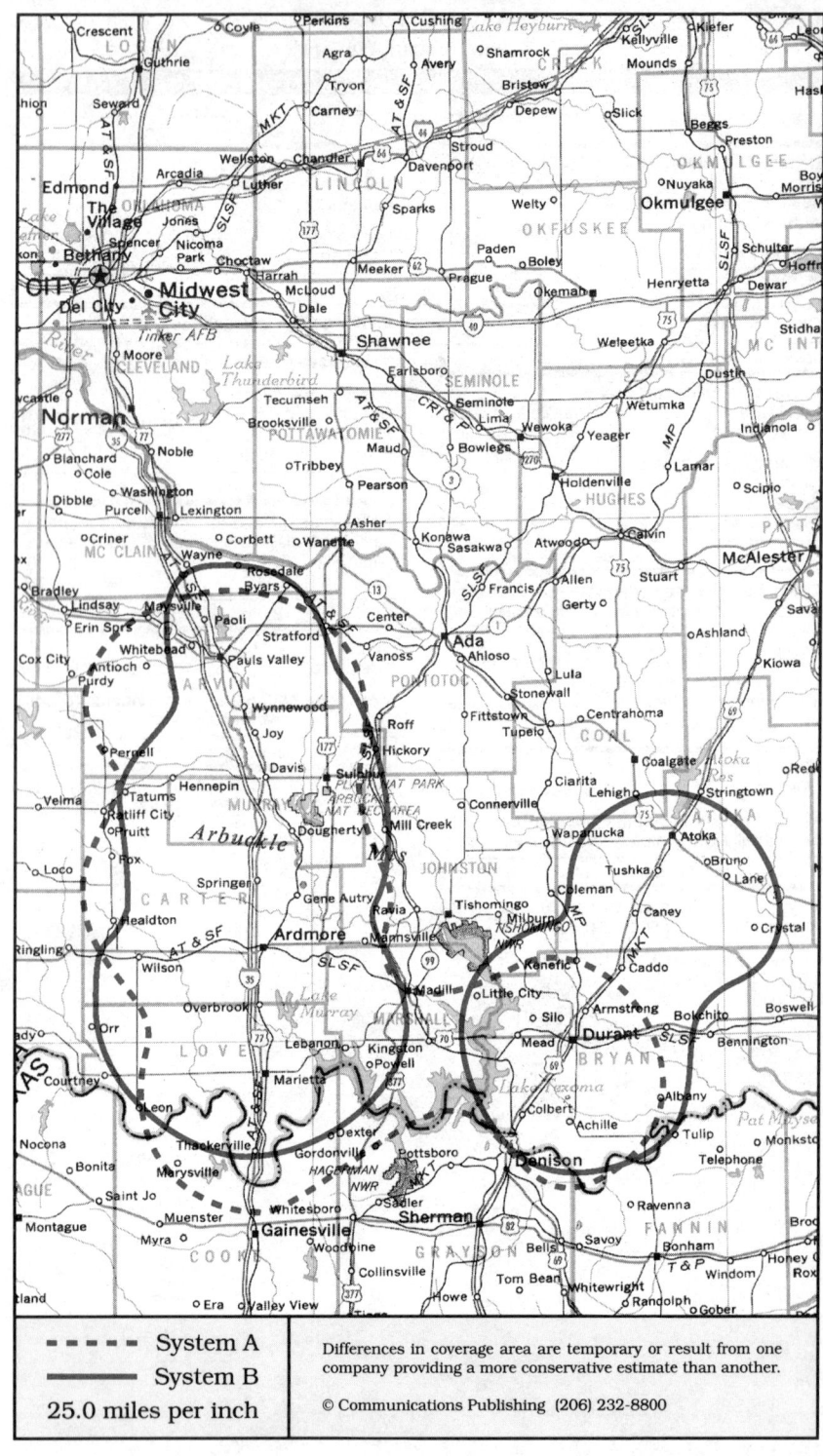

System A - - - - -
System B ——————
25.0 miles per inch

Differences in coverage area are temporary or result from one company providing a more conservative estimate than another.

© Communications Publishing (206) 232-8800

Ardmore, Oklahoma

Ardmore, Atoka, Durant, Marietta, Pauls Valley, Sulphur

System A	System B

System A

Cellular company

United States Cellular
111 N Commerce, Suite 2
Ardmore, OK 73401

Local office	405-226-3030
Corporate office	312-399-8900

Sending calls

Visitor dialing instructions

Local calls	area code + 7 digits
Long distance . . .	1 + area code + 7 digits

Speed-dial numbers

Customer service	*611
Emergency	911
Highway patrol	*55
Roaming information	*711

Receiving calls

Caller dials the roamer access number below, hears tone, dials cellular area code and number.
. . . . 405-465-7626

Billing information

Visitor rates
Most visitors . . . . . $3 day + 75¢ minute

Visitors without roaming agreements (p. 1159)
Call co. to use Mastercard, Visa

Identification numbers

City	Market	System	Billing	RSA
All served cities	604	01595	00000	OK-9

System B

Cellular company

Southwestern Bell Mobile Systems
6704 Northwest Expressway
Oklahoma City, OK 73132

Customer service	some areas	800-331-0500
. . . .		405-720-2212
Local office		405-720-0411
Corporate office		214-733-2000

Sending calls

Visitor dialing instructions

Local calls	area code + 7 digits
Long distance . . .	1 + area code + 7 digits

Speed-dial numbers

Customer service	611
Emergency	911
Roaming information	*711
Weather, sports, horoscopes	*123

Receiving calls

Caller dials the roamer access number below, hears tone, dials cellular area code and number.
. . . . 405-221-7626

Follow Me Roaming forwarding (see p. 1156)
Activate, dial *18 send. Deactivate dial *19 send.

Billing information

Visitor rates
Most visitors . . . . . $3 day + 75¢ minute

Visitors without roaming agreements (p. 1159)
Call co. to use Amex, Mastercard, Visa

Identification numbers

City	Market	System	Billing	RSA
All served cities	604	00146	30176	OK-9

For full coverage area, see Oklahoma City, OK.

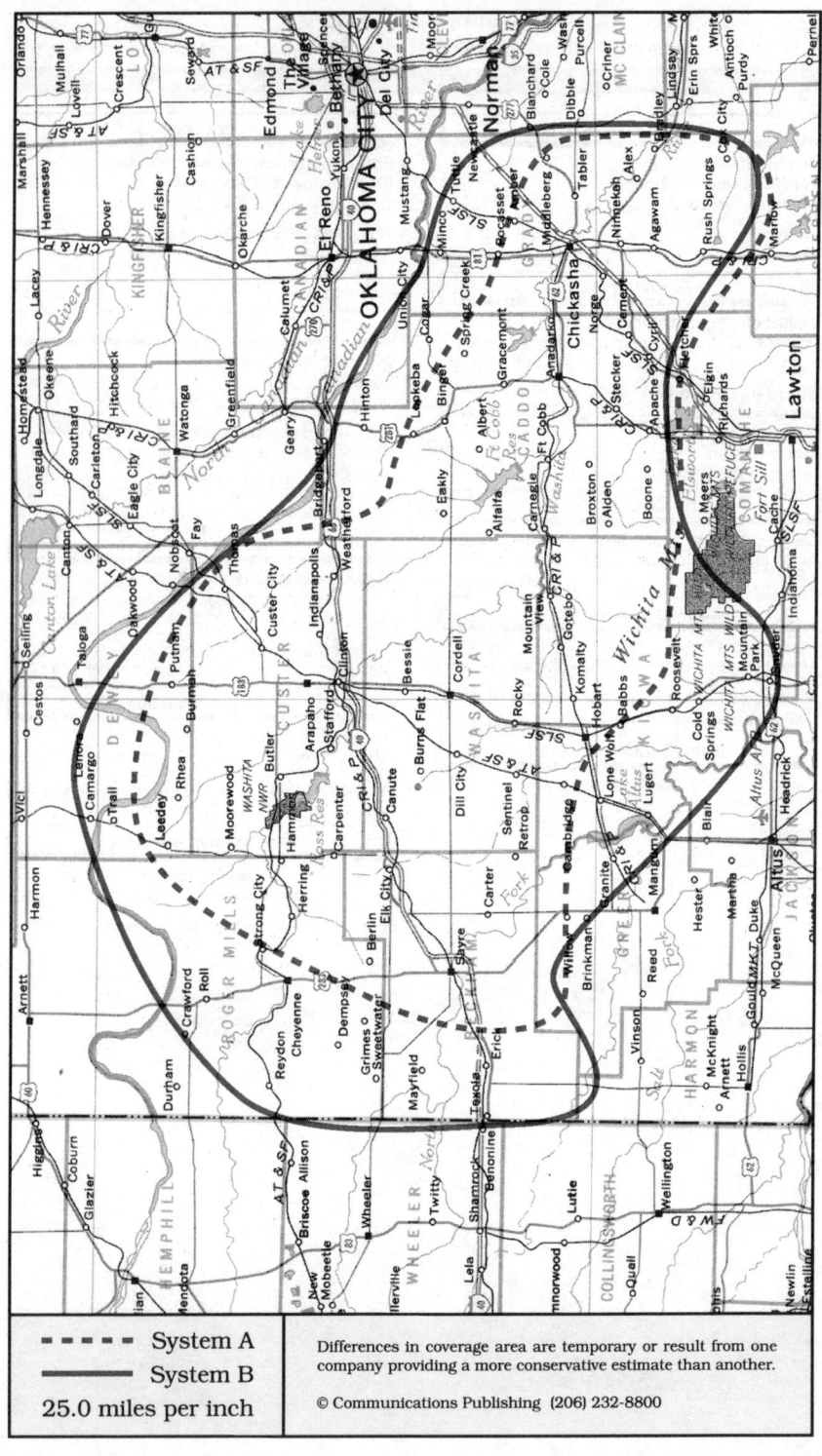

- - - - - System A
———— System B
25.0 miles per inch

Differences in coverage area are temporary or result from one company providing a more conservative estimate than another.

© Communications Publishing (206) 232-8800

Chickasha, Oklahoma
Chickasha, Clinton, Elk City, Weatherford

Cellular company

Chickasha and Elk City
Cellular One of Southwest Oklahoma
Sooner Cellular
2431 Ponderosa
Chickasha, OK 73018

100 Access Road
Elk City, OK 73648

Local office . . .	Chickasha	405-222-3232
. . . .	Elk City	405-225-2294
Corporate office		405-252-9994

Clinton and Weatherford
Cellular One of Western Oklahoma
Licensed to: Mobile Telenet
Managed by: Texoma Cellular Management
2231 Gary Boulevard
Clinton, OK 73601

Local office	405-323-4393
Corporate office	817-691-9566

Sending calls

Visitor dialing instructions

Local calls	area code + 7 digits
Long distance . . .	1 + area code + 7 digits

Speed-dial numbers

Customer service	611
Emergency	911
Highway patrol	*55

Receiving calls

Caller dials the roamer access number below,
hears tone, dials cellular area code and number.

. . . .   405-779-7626

NationLink & RoamingAmerica (see p. 1157)
Dial *31 send to receive calls, *30 to deactivate.
This company's subscribers have level 3 service.

Billing information

Visitor rates
Most visitors . . . . . $3 day + 99¢ minute

Visitors without roaming agreements (p. 1159)
Roamer Plus will intercept first call to bill you.

Identification numbers

City	Market	System	Billing	RSA
Chickasha, Elk City	602	01593	00000	OK-7
Clinton,Weatherford	600	01593	30305	OK-5

Both companies operate their systems using the
same switch.

For full coverage area, see Altus, OK.

Cellular company

Dobson Cellular Systems
1702 S 4th
Chickasha, OK 73018

2301 E Main
Weatherford, OK 73096

Customer service		800-848-4011
. . . .		405-749-9744
Local office . . .	Chickasha	405-222-0300
. . . .	Weatherford	405-772-8989
Corporate office		405-749-9744

Sending calls

Visitor dialing instructions

Local calls	area code + 7 digits
Long distance . . .	1 + area code + 7 digits

Speed-dial numbers

Customer service	611
Emergency	911

Receiving calls

Caller dials the roamer access number below,
hears tone, dials cellular area code and number.

Chickasha, Elk City	405-630-7626
Clinton, Weatherford	405-497-7626

Follow Me Roaming forwarding (see p. 1156)
Activate, dial *18 send. Deactivate dial *19 send.

Billing information

Visitor rates
Most visitors . . . . . $3 day + 75¢ minute

Visitors without roaming agreements (p. 1159)
Roamer Plus will intercept first call to bill you.

Identification numbers

City	Market	System	Billing	RSA
Chickasha, Elk City	602	01592	30518	OK-7
Clinton,Weatherford	600	01592	30305	OK-5

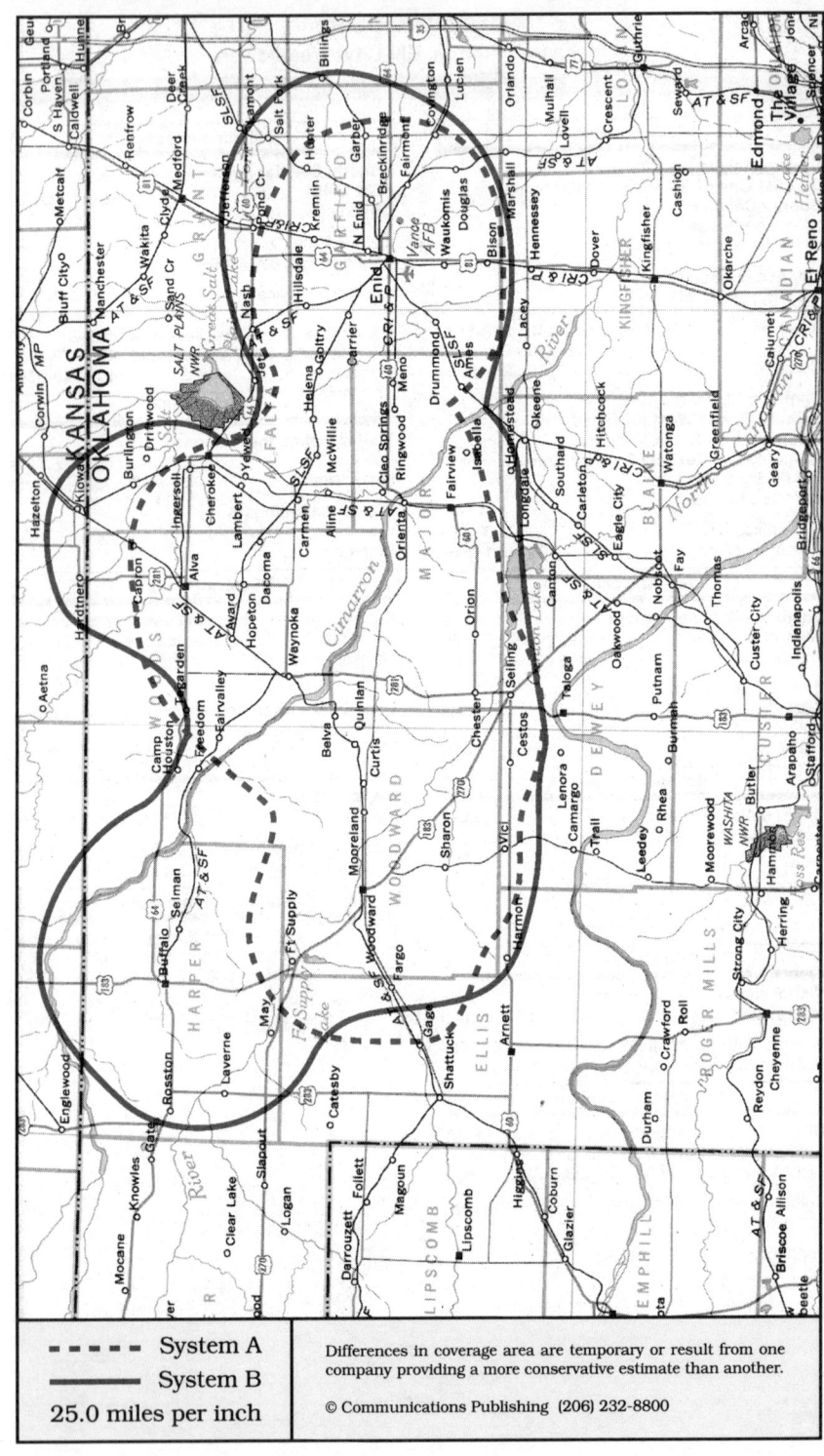

- - - - System A
———— System B
25.0 miles per inch

Differences in coverage area are temporary or result from one company providing a more conservative estimate than another.

© Communications Publishing (206) 232-8800

Enid, Oklahoma
Alva, Buffalo, Enid, Woodward

System A	System B

Cellular company

Cellular One of Enid
2312-A W Owen K.Garriott
Enid, OK 73703

Cellular One of Woodward
1125 40th Street, Suite D
Woodward, OK 73801

Customer service	800-858-1121
Local office . . . . . Enid	405-242-0141
. . . . . . Woodward	405-254-2157
Corporate office	405-751-1121

Sending calls

Visitor dialing instructions

Local calls	area code + 7 digits
Long distance . . .	1 + area code + 7 digits

Speed-dial numbers

Customer service	611
Emergency	911

Receiving calls

Caller dials the roamer access number below,
hears tone, dials cellular area code and number.
. . . . 405-548-7626

Billing information

Visitor rates
Most visitors . . . . . $3 day + 75¢ minute

Visitors without roaming agreements (p. 1159)
Roamer Plus will intercept first call to bill you.

Identification numbers

City	Market	System	Billing	RSA
Enid	302	00341	00000	MSA
Woodward	597	00341	30337	OK-2

Cellular company

Enid Cellular
Pioneer Cellular
302 W Maple, PO Box 1787
Enid, OK 73702

Pioneer Cellular
1205 Main Street, PO Box 487
Woodward, OK 73802-0487

Local office . . . . . . Enid	405-237-2355
. . . . Woodward	405-256-2355
Corporate office	405-375-2355

Sending calls

Visitor dialing instructions

Local calls	area code + 7 digits
Long distance . . .	1 + area code + 7 digits

Speed-dial numbers

Customer service	611
Emergency	911
Highway patrol	*55

Receiving calls

Caller dials the roamer access number below,
hears tone, dials cellular area code and number.

Alva	405-829-1000
Enid	405-541-7626
Woodward	405-334-7626

Follow Me Roaming forwarding (see p. 1156)
Activate, dial *18 send. Deactivate dial *19 send.

Billing information

Visitor rates
Most visitors . . . . . $3 day + 75¢ minute

Visitors without roaming agreements (p. 1159)
Call co. to use Amex, Mastercard, Visa

Identification numbers

City	Market	System	Billing	RSA
Enid	302	00324	00000	MSA
Woodward	597	01582	00000	OK-2

For full coverage area, see Kingfisher, OK.

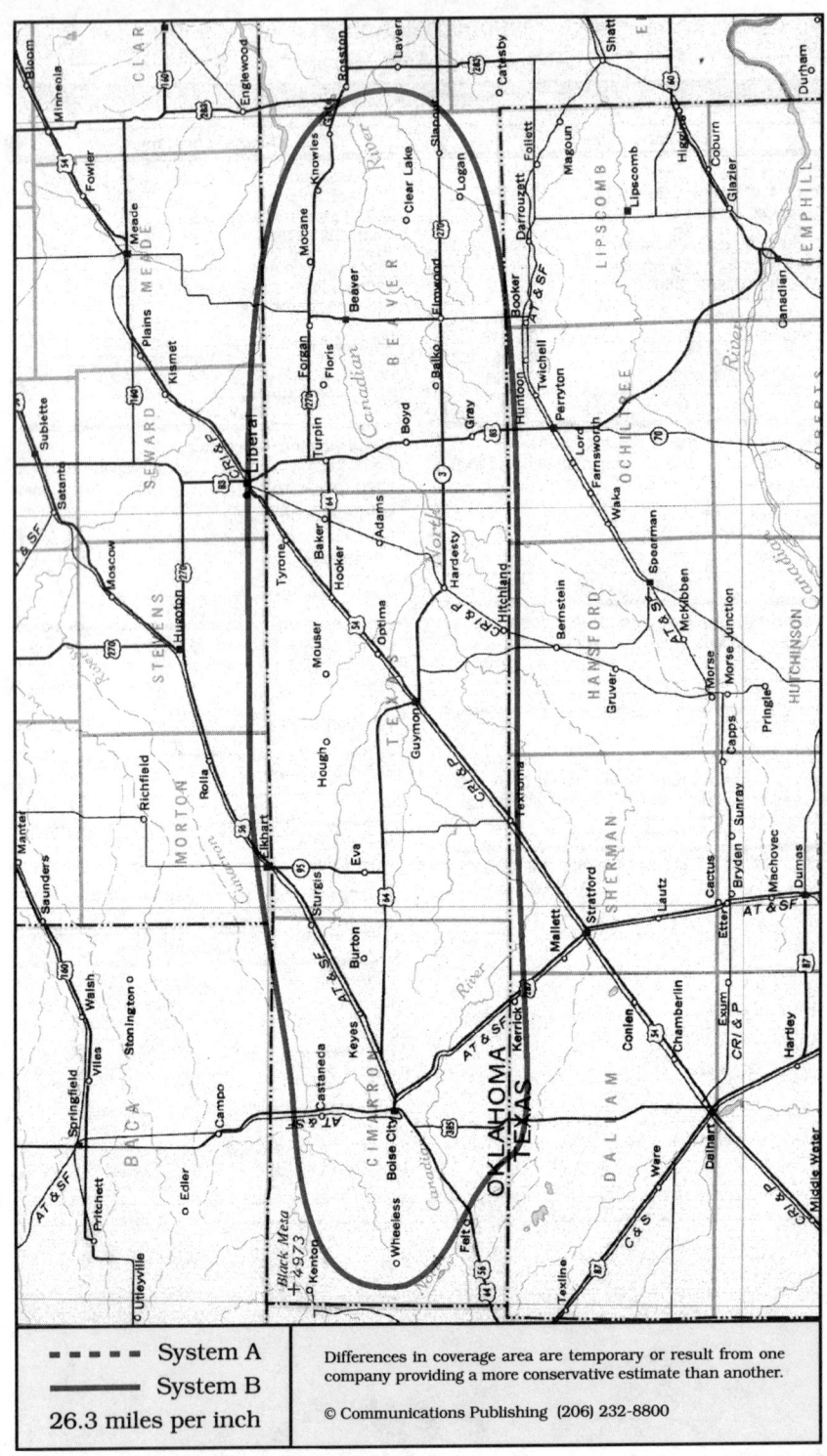

System A
System B
26.3 miles per inch

Differences in coverage area are temporary or result from one company providing a more conservative estimate than another.

© Communications Publishing (206) 232-8800

Guymon, Oklahoma
Beaver, Boise City, Guymon

System A	System B

Cellular company

No information about the company that will operate system A was available at press time.

Cellular company

PTSI Cellular
Panhandle Telecommunication Systems
603 S Main, PO Box 511
Guymon, OK 73942

Local office . . .	some areas	800-327-7525
. . . .		405-338-7525
Corporate office		405-338-2556

Sending calls

Visitor dialing instructions
Local calls . . . . . . area code + 7 digits
Long distance . . . 1 + area code + 7 digits

Speed-dial numbers
Customer service 611
Highway patrol . . . . ∗55

Receiving calls

Caller dials the roamer access number below,
hears tone, dials cellular area code and number.
. . . . 405-651-7626

Follow Me Roaming forwarding (see p. 1156)
Activate, dial ∗18 send. Deactivate dial ∗19 send.

Billing information

Visitor rates
Most visitors . . . . . $3 day + 75¢ minute

Visitors without roaming agreements (p. 1159)
Call co. to use Amex, Discover, Mastercard, Visa

Identification numbers

City	Market	System	Billing	RSA
All served cities	596	01580	00000	OK-1

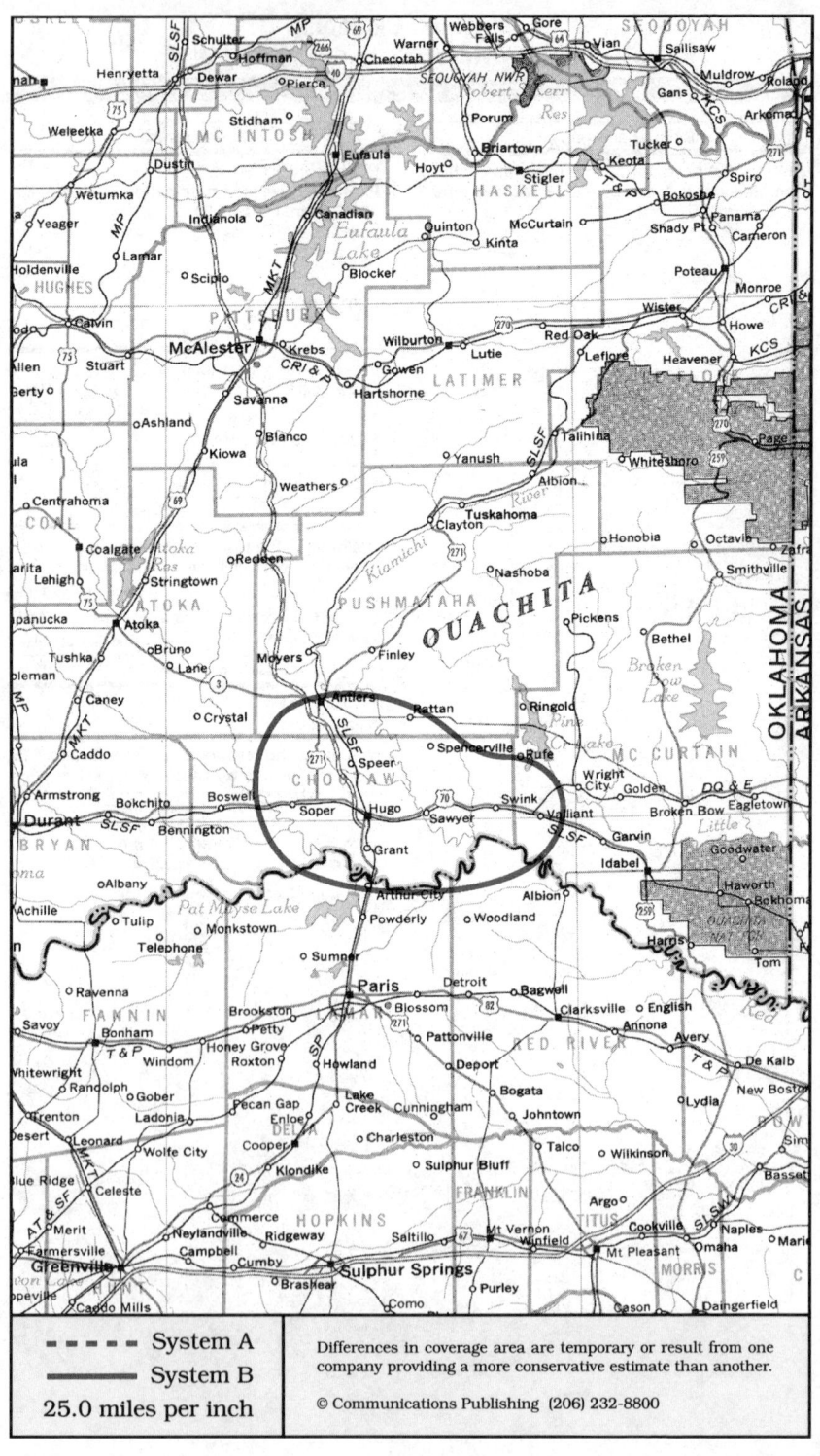

Hugo, Oklahoma
Antlers, Hugo

System A

Cellular company

United States Cellular
111 N Commerce, Suite 2
Ardmore, OK 73401

Service was not operating at press time.

Local office	405-226-3030
Corporate office	312-399-8900

Sending calls

Visitor dialing instructions

Local calls	area code + 7 digits
Long distance . . .	1 + area code + 7 digits

Speed-dial numbers

Customer service	*611
Emergency	911
Highway patrol	*55
Roaming information	*711

Receiving calls

Caller dials the roamer access number below,
hears tone, dials cellular area code and number.
. . . . 405-465-7626

Billing information

Visitor rates
Most visitors . . . . . $3 day + 75¢ minute

Visitors without roaming agreements (p. 1159)
Call co. to use Mastercard, Visa

Identification numbers

City	Market	System	Billing	RSA
All served cities	605	01597	00000	OK-10

This company also serves Idabel, OK.

System B

Cellular company

C.V. Cellular
Cross-Valliant Cellular Partnership
704 Third Avenue, PO Box 409
Warner, OK 74469-0409

Corporate office	800-324-4142
. . . .	918-463-2921

Sending calls

Visitor dialing instructions

Local calls	area code + 7 digits
Long distance . . .	1 + area code + 7 digits

Speed-dial numbers

Customer service	611
Emergency	911
State police	*55

Receiving calls

Caller dials the roamer access number below,
hears tone, dials cellular area code and number.
. . . . 405-933-7626

Billing information

Visitor rates
Most visitors . . . . . $3 day + 75¢ minute

Visitors without roaming agreements (p. 1159)
Call co. to use Amex, Mastercard, Visa

Identification numbers

City	Market	System	Billing	RSA
All served cities	605	01968	00000	OK-10

This company operates system B in one of three
portions of OK-10, the one known as B-2.

- - - - System A	Differences in coverage area are temporary or result from one company providing a more conservative estimate than another.
──── System B	
25.0 miles per inch	© Communications Publishing (206) 232-8800

Idabel, Oklahoma

System A

Cellular company

United States Cellular
111 N Commerce, Suite 2
Ardmore, OK 73401

Local office	405-226-3030
Corporate office	312-399-8900

Sending calls

Visitor dialing instructions

Local calls	area code + 7 digits
Long distance . . .	1 + area code + 7 digits

Speed-dial numbers

Customer service	*611
Emergency	911
Highway patrol	*55
Roaming information	*711

Receiving calls

Caller dials the roamer access number below, hears tone, dials cellular area code and number.
. . . . 405-465-7626

Billing information

Visitor rates

Most visitors	$3 day + 75¢ minute

Visitors without roaming agreements (p. 1159)
Call co. to use Mastercard, Visa

Identification numbers

City	Market	System	Billing	RSA
All served cities	605	01597	00000	OK-10

This service area will expand in 1993.

System B

Cellular company

Pine Cellular Phones
207 W 2nd, PO Box 548
Broken Bow, OK 74728

Corporate office	405-584-3330

Sending calls

Visitor dialing instructions

Local calls	area code + 7 digits
Long distance . . .	1 + area code + 7 digits

Speed-dial numbers

Customer service	*611
Emergency	911
Highway patrol	*55

Receiving calls

Caller dials the roamer access number below, hears tone, dials cellular area code and number.
. . . . 405-584-1130

Follow Me Roaming forwarding (see p. 1156)
Activate, dial *18 send. Deactivate dial *19 send.

Billing information

Visitor rates

Most visitors . . .	$2.50 day + 60¢ minute

Visitors without roaming agreements (p. 1159)
Cannot call since company takes no credit cards.

Identification numbers

City	Market	System	Billing	RSA
All served cities	605	01598	00000	OK-10

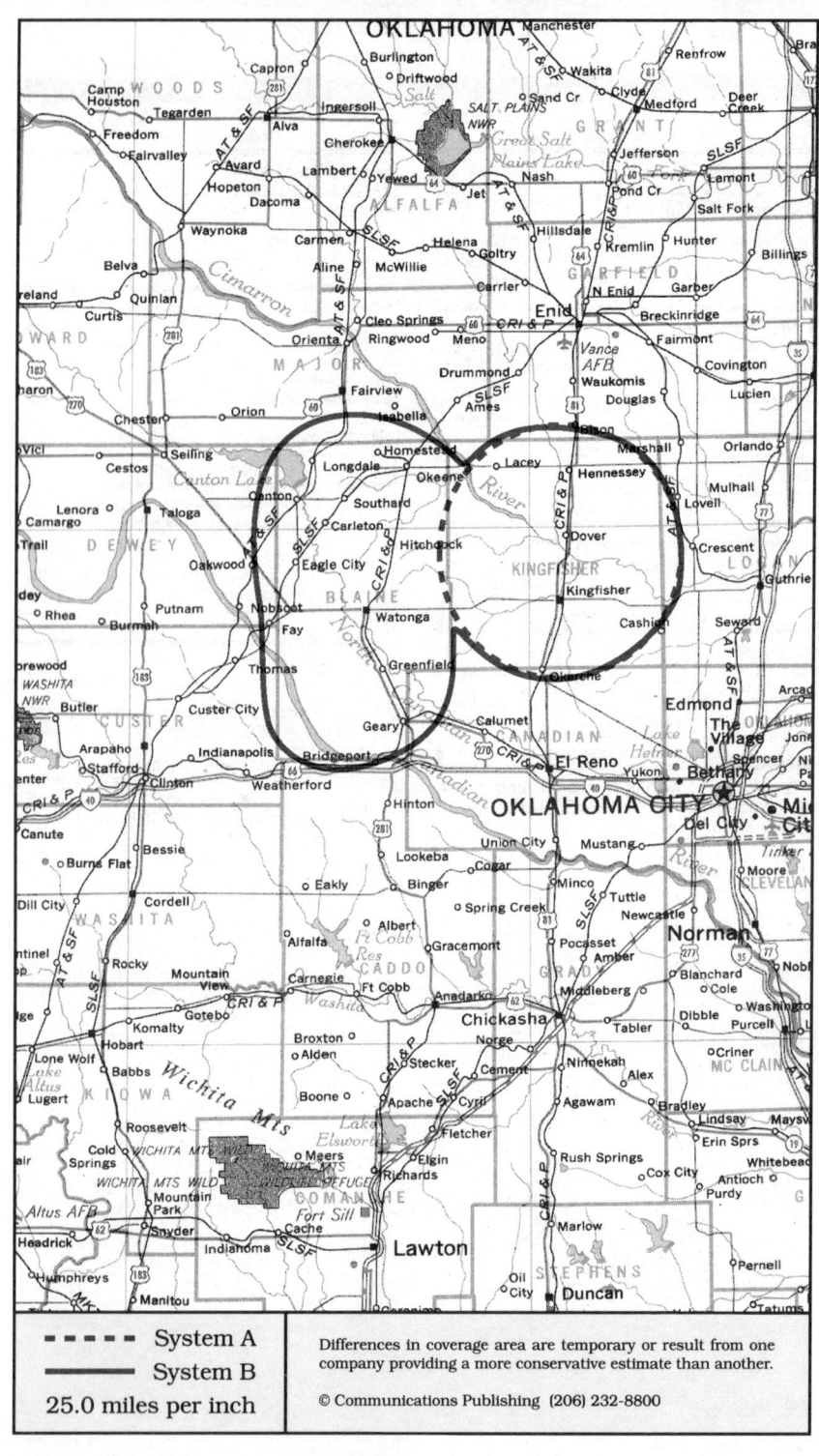

System A
System B

25.0 miles per inch

Differences in coverage area are temporary or result from one
company providing a more conservative estimate than another.

© Communications Publishing (206) 232-8800

Kingfisher, Oklahoma
Kingfisher, Watonga

System A

System B

System A

Cellular company

Cellular One of Western Oklahoma
Licensed to: Mobile Telenet
Managed by: Texoma Cellular Management
2231 Gary Boulevard, Suite B
Clinton, OK 73601

Local office	405-323-4393
Corporate office	817-691-9566

Sending calls

Visitor dialing instructions

Local calls	area code + 7 digits
Long distance . . .	1 + area code + 7 digits

Speed-dial numbers

Customer service	611
Emergency	911
Highway patrol	*55

Receiving calls

Caller dials the roamer access number below,
hears tone, dials cellular area code and number.
. . . . 405-779-7626

NationLink & RoamingAmerica (see p. 1157)
Dial *31 send to receive calls, *30 to deactivate.
This company's subscribers have level 3 service.

Billing information

Visitor rates

Most visitors	$3 day + 99¢ minute

Visitors without roaming agreements (p. 1159)
Roamer Plus will intercept first call to bill you.

Identification numbers

City	Market	System	Billing	RSA
All served cities	600	01593	30305	OK-5

System B

Cellular company

Pioneer Cellular
314 N 5th Street, PO Box 539
Kingfisher, OK 73750

Corporate office	405-375-2355

Sending calls

Visitor dialing instructions

Local calls	area code + 7 digits
Long distance . . .	1 + area code + 7 digits

Speed-dial numbers

Customer service	611
Emergency	911
Highway patrol	*55

Receiving calls

Caller dials the roamer access number below,
hears tone, dials cellular area code and number.

Kingfisher	405-368-7626
Watonga	405-623-1000

Billing information

Visitor rates

Most visitors	$3 day + 75¢ minute

Visitors without roaming agreements (p. 1159)
Call co. to use Amex, Mastercard, Visa
or Roamer Plus will intercept first call to bill you.

Identification numbers

City	Market	System	Billing	RSA
All served cities	600	01904	00000	OK-5

For full coverage area, see Enid, OK.

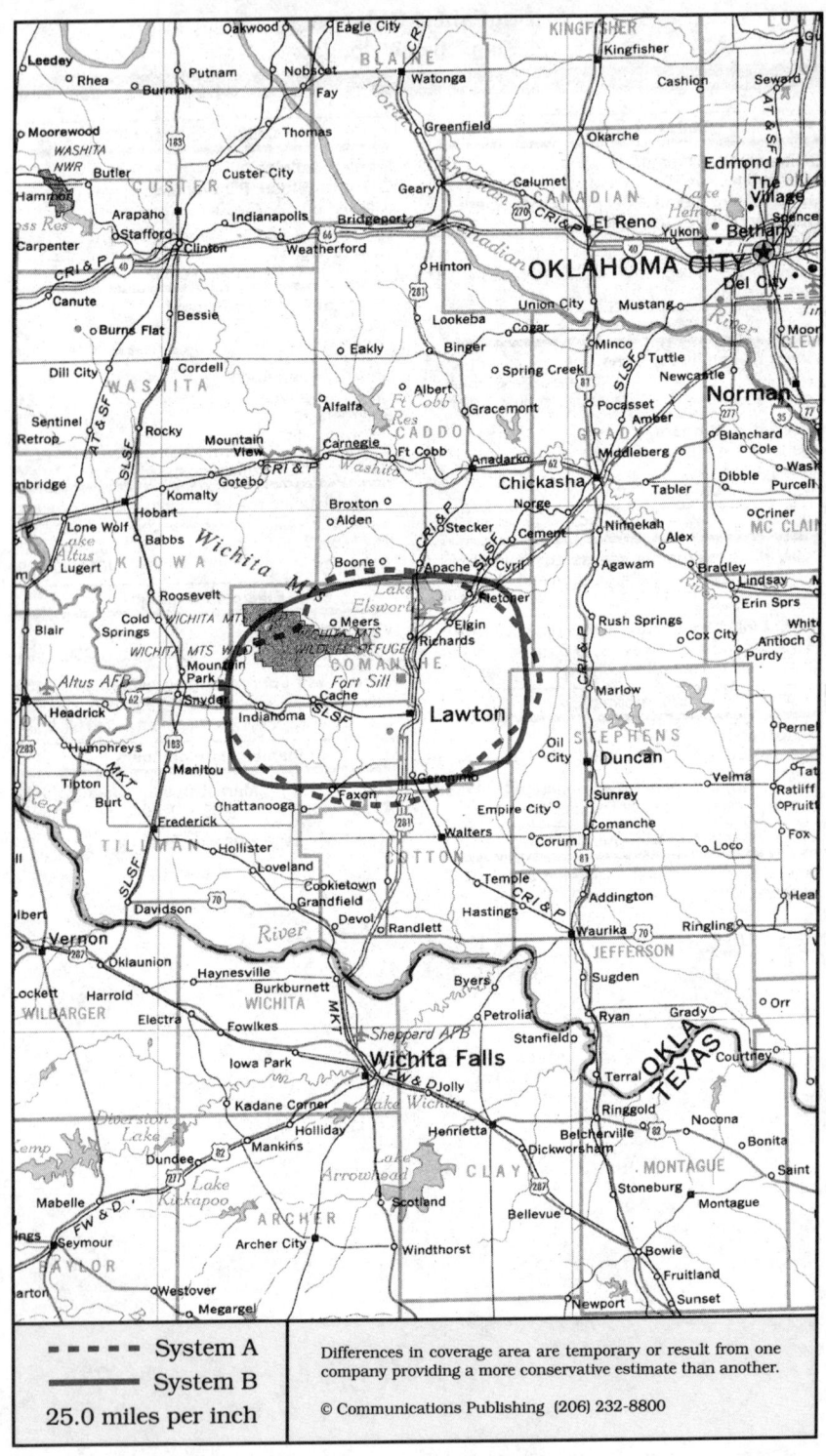

System A

System B

25.0 miles per inch

Differences in coverage area are temporary or result from one company providing a more conservative estimate than another.

© Communications Publishing (206) 232-8800

Lawton, Oklahoma

Cellular company

Cellular One
GenCell Management
1322 NW Sheridan Road
Lawton, OK 73505

Customer service	800-888-7868
Local office	405-355-3500
Corporate office . .	GenCell 707-425-8000

Sending calls

Visitor dialing instructions

Local calls	area code + 7 digits
Long distance . . .	1 + area code + 7 digits

Speed-dial numbers

Customer service (forwards to GenCell) .	*611
Emergency	911

Receiving calls

Caller dials the roamer access number below, hears tone, dials cellular area code and number.
. . . . 405-581-7626

NationLink & RoamingAmerica (see p. 1157)
Dial *31 send to receive calls, *30 to deactivate.
This company's subscribers have level 3 service.

Billing information

Visitor rates
Most visitors . . . . . $3 day + 99¢ minute

Visitors without roaming agreements (p. 1159)
American Roaming Network intercepts call to bill.

Identification numbers

City	Market	System	Billing	RSA
All served cities	260	00425	00000	MSA

Cellular company

United States Cellular
1001 B Avenue, Suite 130
Lawton, OK 73501

Local office	405-355-3535
Corporate office	312-399-8900

Sending calls

Visitor dialing instructions

Local calls	area code + 7 digits
Long distance . . .	1 + area code + 7 digits

Speed-dial numbers

Customer service	*611
Emergency	911
Roaming information	*711

Receiving calls

Caller dials the roamer access number below, hears tone, dials cellular area code and number.
. . . . 405-585-7626

Follow Me Roaming forwarding (see p. 1156)
Activate, dial *18 send. Deactivate dial *19 send.

Billing information

Visitor rates
Most visitors . . . . . $3 day + 75¢ minute

Visitors without roaming agreements (p. 1159)
Call co. to use Mastercard, Visa

Identification numbers

City	Market	System	Billing	RSA
All served cities	260	00408	00000	MSA

For full coverage area, see Altus, OK.

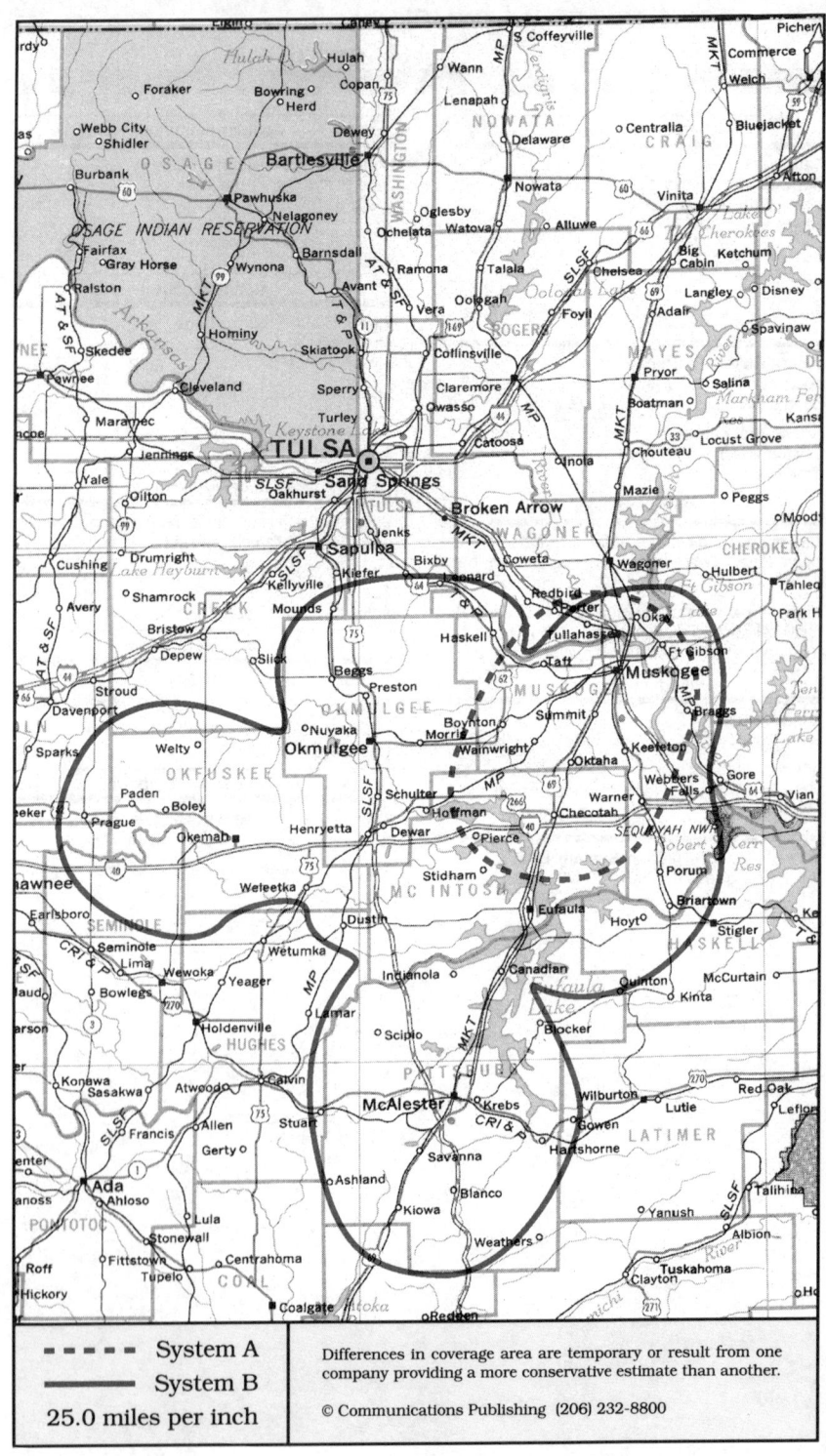

▪ ▪ ▪ ▪ System A	Differences in coverage area are temporary or result from one company providing a more conservative estimate than another.
――― System B	
25.0 miles per inch	© Communications Publishing (206) 232-8800

Muskogee, Oklahoma
Henryetta, McAlester, Muskogee, Okmulgee

System A	System B

Cellular company

Cellular One
Oklahoma Cellular
Managed by: Boston Communications Group
Operated by: McCaw Cellular Comms.
2012-A W Okmulgee Street
Muskogee, OK 74401

Customer service . .	McCaw	918-684-8484
Local office		800-682-2351
. . . .		918-682-2112
Corporate office	Boston Comms.	617-439-4141

Sending calls

Visitor dialing instructions
Local calls . . . . . area code + 7 digits
Long distance . . . 1 + area code + 7 digits

Speed-dial numbers
Customer service 611
Emergency 911
Highway patrol 918-627-0440

Receiving calls

Caller dials the roamer access number below,
hears tone, dials cellular area code and number.
. . . . 918-625-7626

North American Cellular Network (see p. 1156)
All calls forward to you. Deactivate dial *35 send.

Billing information

Visitor rates
Most visitors . . . . . $3 day + 99¢ minute

Visitors without roaming agreements (p. 1159)
Call co. to use Mastercard, Visa
or Roamer Plus will intercept first call to bill you.

Identification numbers

City	Market	System	Billing	RSA
All served cities	601	01589	00000	OK-6

This company's subscribers use the roaming
agreements of McCaw Cellular Communications.

Cellular company

OK Cellular
301 E Choctaw, PO Box 3329
McAlester, OK 74502

511 W Broadway
Muskogee, OK 74401

Local office	McAlester	918-423-4489
. . . .	Muskogee	918-687-4489
Corporate office		405-375-2355

Sending calls

Visitor dialing instructions
Local calls . . . . . . area code + 7 digits
Long distance . . . 1 + area code + 7 digits

Speed-dial numbers
Customer service 611
Emergency 911
Highway patrol *55

Receiving calls

Caller dials the roamer access number below,
hears tone, dials cellular area code and number.
Henryetta 918-758-2800
McAlester 918-421-0626
Muskogee 918-685-7626
Okmulgee 918-758-7626

Follow Me Roaming forwarding (see p. 1156)
Activate, dial *18 send. Deactivate dial *19 send.

Billing information

Visitor rates
Most visitors . . . . . $3 day + 75¢ minute

Visitors without roaming agreements (p. 1159)
Call co. to use Amex, Mastercard, Visa

Identification numbers

City	Market	System	Billing	RSA
Henryetta,McAlester	601	01590	30448	OK-6
Muskogee, Okmulg.	601	01590	00000	OK-6

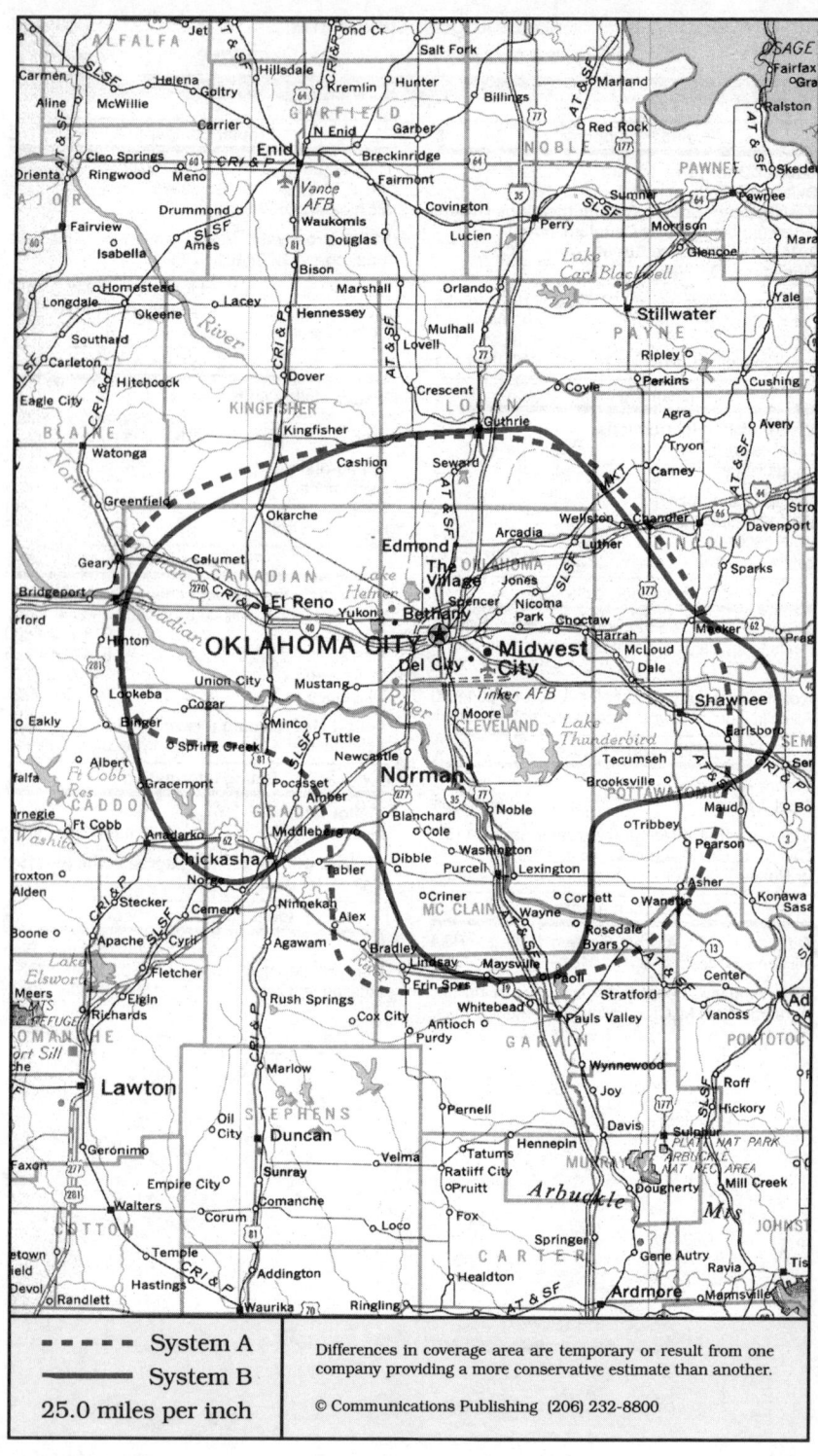

▪ ▪ ▪ ▪ ▪	System A
———	System B
25.0 miles per inch	

Differences in coverage area are temporary or result from one company providing a more conservative estimate than another.

© Communications Publishing (206) 232-8800

Oklahoma City, Oklahoma
El Reno, Norman, Oklahoma City, Purcell, Shawnee

System A	System B

Cellular company

Cellular One
McCaw Cellular Communications
5509 N Pennsylvania
Oklahoma City, OK 73112

Customer service	405-843-9665
Local office	405-843-9113
Corporate office	206-827-4500

Sending calls

Visitor dialing instructions

Local calls	area code + 7 digits
Long distance . . .	1 + area code + 7 digits

Speed-dial numbers

Customer service	611
Emergency	911

Receiving calls

Caller dials the roamer access number below,
hears tone, dials cellular area code and number.

. . . . 405-627-7626

NationLink & RoamingAmerica (see p. 1157)
Dial *31 send to receive calls, *30 to deactivate.
Dial *32 send to tell caller the roamer access no.
This company's subscribers have level 9 service.

North American Cellular Network (see p. 1156)
All calls forward to you. Deactivate dial *35 send.

Billing information

Visitor rates

Most visitors . . . . . $0 day + 99¢ minute

Visitors without roaming agreements (p. 1159)
Roamer Plus will intercept first call to bill you.

Identification numbers

City	Market	System	Billing	RSA
All served cities	045	00169	00000	MSA

Cellular company

Southwestern Bell Mobile Systems
6704 Northwest Expressway
Oklahoma City, OK 73132

Customer service	some areas	800-331-0500
. . . .		405-720-2212
Local office		405-720-0411
Corporate office		214-733-2000

Sending calls

Visitor dialing instructions

Local calls	area code + 7 digits
Long distance . . .	1 + area code + 7 digits

Speed-dial numbers

Customer service	611
Emergency	911
Roaming information	*711
Weather, sports, horoscopes	*123

Receiving calls

Caller dials the roamer access number below,
hears tone, dials cellular area code and number.

. . . . 405-630-7626

Follow Me Roaming forwarding (see p. 1156)
Activate, dial *18 send. Deactivate dial *19 send.

Billing information

Visitor rates
Most visitors . . . . . $3 day + 75¢ minute

Visitors without roaming agreements (p. 1159)
Call co. to use Amex, Mastercard, Visa

Identification numbers

City	Market	System	Billing	RSA
All served cities	045	00146	00000	MSA

For full coverage area, see Ardmore, OK and
Stillwater, OK.

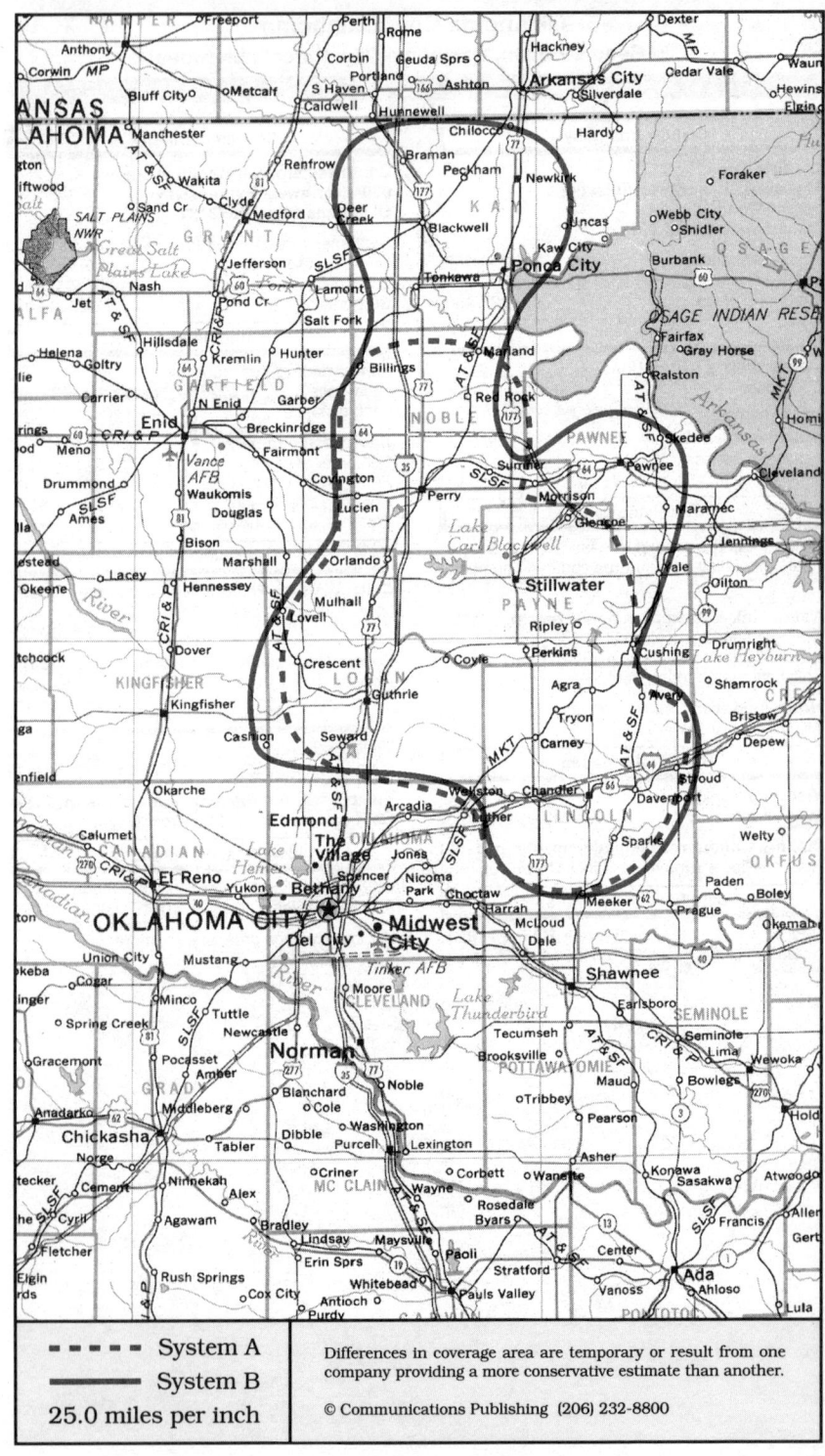

System A
System B
25.0 miles per inch

Differences in coverage area are temporary or result from one company providing a more conservative estimate than another.

© Communications Publishing (206) 232-8800

Ponca City, Oklahoma
Chandler, Guthrie, Pawnee, Perry, Ponca City, Stillwater

Cellular company

Cellular One
Stillwater Cellular
PO Box 188
Stillwater, OK 74076

Corporate office 405-377-1212

Sending calls

Visitor dialing instructions
Local calls . . . . . area code + 7 digits
Long distance . . . 1 + area code + 7 digits

Speed-dial numbers
Customer service 611
Emergency 911

Receiving calls

Caller dials the roamer access number below,
hears tone, dials cellular area code and number.
. . . . 405-747-7626

NationLink & RoamingAmerica (see p. 1157)
Dial *31 send to receive calls, *30 to deactivate.
Dial *32 send to tell caller the roamer access no.
This company's subscribers have level 3 service.

Billing information

Visitor rates
Most visitors . . . . . $3 day + 99¢ minute

Visitors without roaming agreements (p. 1159)
Roamer Plus will intercept first call to bill you.

Identification numbers

City	Market	System	Billing	RSA
All served cities	598	01583	00000	OK-3

Cellular company

Southwestern Bell Mobile Systems
6704 Northwest Expressway
Oklahoma City, OK 73132

Customer service   some areas   800-331-0500
. . . . 405-720-2212
Local office   405-720-0411
Corporate office   214-733-2000

Sending calls

Visitor dialing instructions
Local calls . . . . . . area code + 7 digits
Long distance . . . 1 + area code + 7 digits

Speed-dial numbers
Customer service 611
Emergency 911
Roaming information *711
Weather, sports, horoscopes *123

Receiving calls

Caller dials the roamer access number below,
hears tone, dials cellular area code and number.
. . . . 405-630-7626

Follow Me Roaming forwarding (see p. 1156)
Activate, dial *18 send. Deactivate dial *19 send.

Billing information

Visitor rates
Most visitors . . . . . $3 day + 75¢ minute

Visitors without roaming agreements (p. 1159)
Call co. to use Amex, Mastercard, Visa

Identification numbers

City	Market	System	Billing	RSA
All served cities	598	00146	30128	OK-3

For full coverage area, see Oklahoma City, OK.

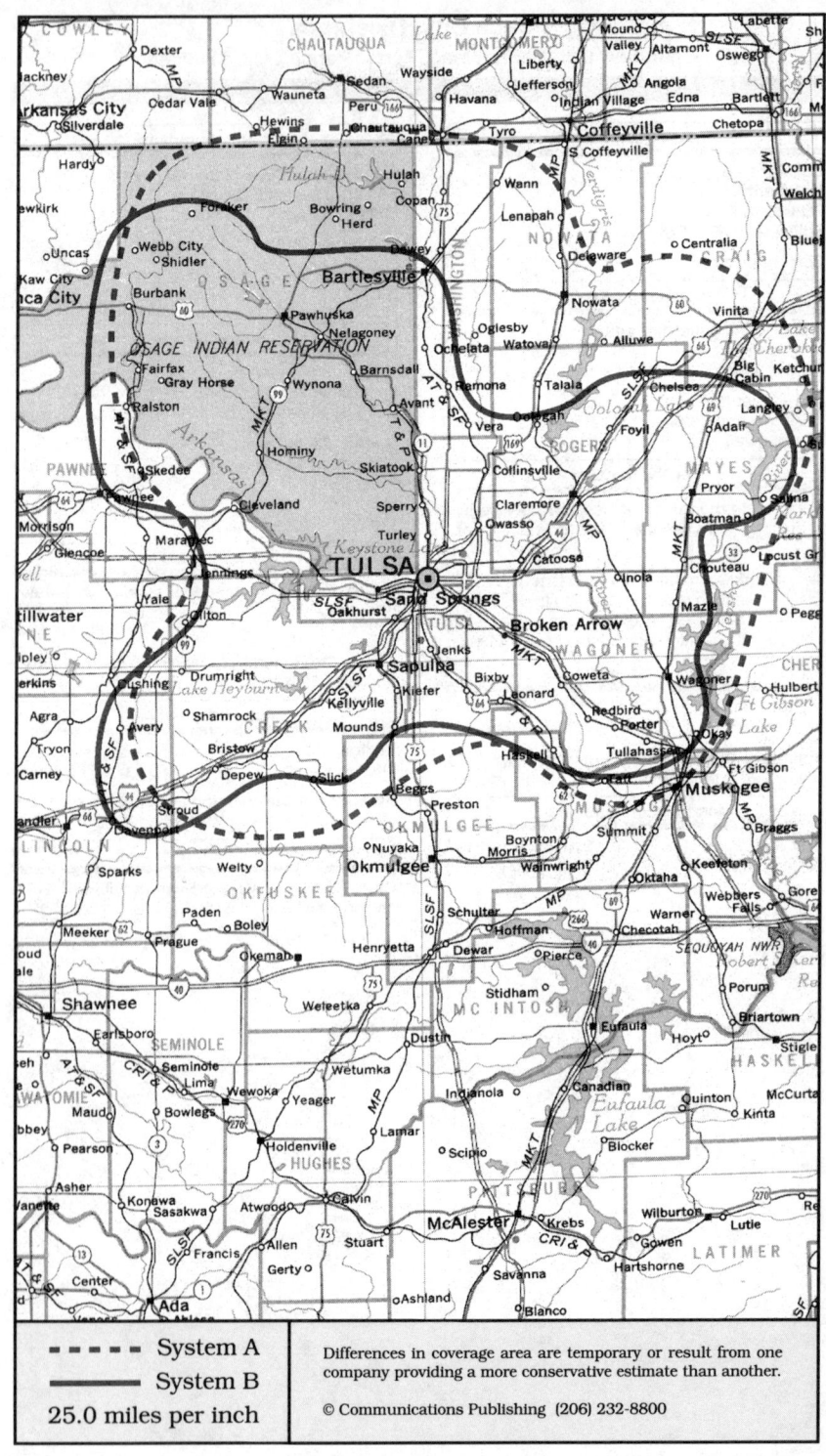

- - - - - System A	Differences in coverage area are temporary or result from one company providing a more conservative estimate than another.
——— **System B**	
25.0 miles per inch	© Communications Publishing (206) 232-8800

Tulsa, Oklahoma

Bartlesville, Broken Arrow, Nowata, Pawhuska, Pryor, Sapulpa, Tulsa

System A	System B

Cellular company

Cellular One
McCaw Cellular Communications
2325 E 71st Street
Tulsa, OK 74136

Customer service	918-625-8500
Local office	918-492-8500
Corporate office	206-827-4500

Sending calls

Visitor dialing instructions

Local calls	area code + 7 digits
Long distance . . .	1 + area code + 7 digits

Speed-dial numbers

Customer service	611
Emergency	911
Highway patrol	*55
Weather . . 918-625-RAIN or 918-625-7246	

Receiving calls

Caller dials the roamer access number below,
hears tone, dials cellular area code and number.
. . . . 918-625-7626

NationLink & RoamingAmerica (see p. 1157)
Dial *31 send to receive calls, *30 to deactivate.
Dial *32 send to tell caller the roamer access no.
This company's subscribers have level 9 service.

North American Cellular Network (see p. 1156)
All calls forward to you. Deactivate dial *35 send.

Billing information

Visitor rates

Most visitors	$0 day + 99¢ minute

Visitors without roaming agreements (p. 1159)
Call co. to use Mastercard, Visa
or Roamer Plus will intercept first call to bill you.

Identification numbers

City	Market	System	Billing	RSA
Bartlesville	599	01586	00000	OK-4
Tulsa	057	00111	00000	MSA

Cellular company

United States Cellular
6701 E 41st Street
Tulsa, OK 74145

Local office	918-665-0101
Corporate office	312-399-8900

Sending calls

Visitor dialing instructions

Local calls	area code + 7 digits
Long distance . . .	1 + area code + 7 digits

Speed-dial numbers

Customer service	611
Emergency	911
Make a traffic report	*74
Roaming information	*711
Stranded motorist line	*74

Receiving calls

Caller dials the roamer access number below,
hears tone, dials cellular area code and number.
. . . . 918-636-7626

Follow Me Roaming forwarding (see p. 1156)
Activate, dial *18 send. Deactivate dial *19 send.

Billing information

Visitor rates

Most visitors	$3 day + 75¢ minute

Visitors without roaming agreements (p. 1159)
Call co. to use Mastercard, Visa

Identification numbers

City	Market	System	Billing	RSA
Bartlesville	599	01586	00000	OK-4
Tulsa	057	00166	00000	MSA

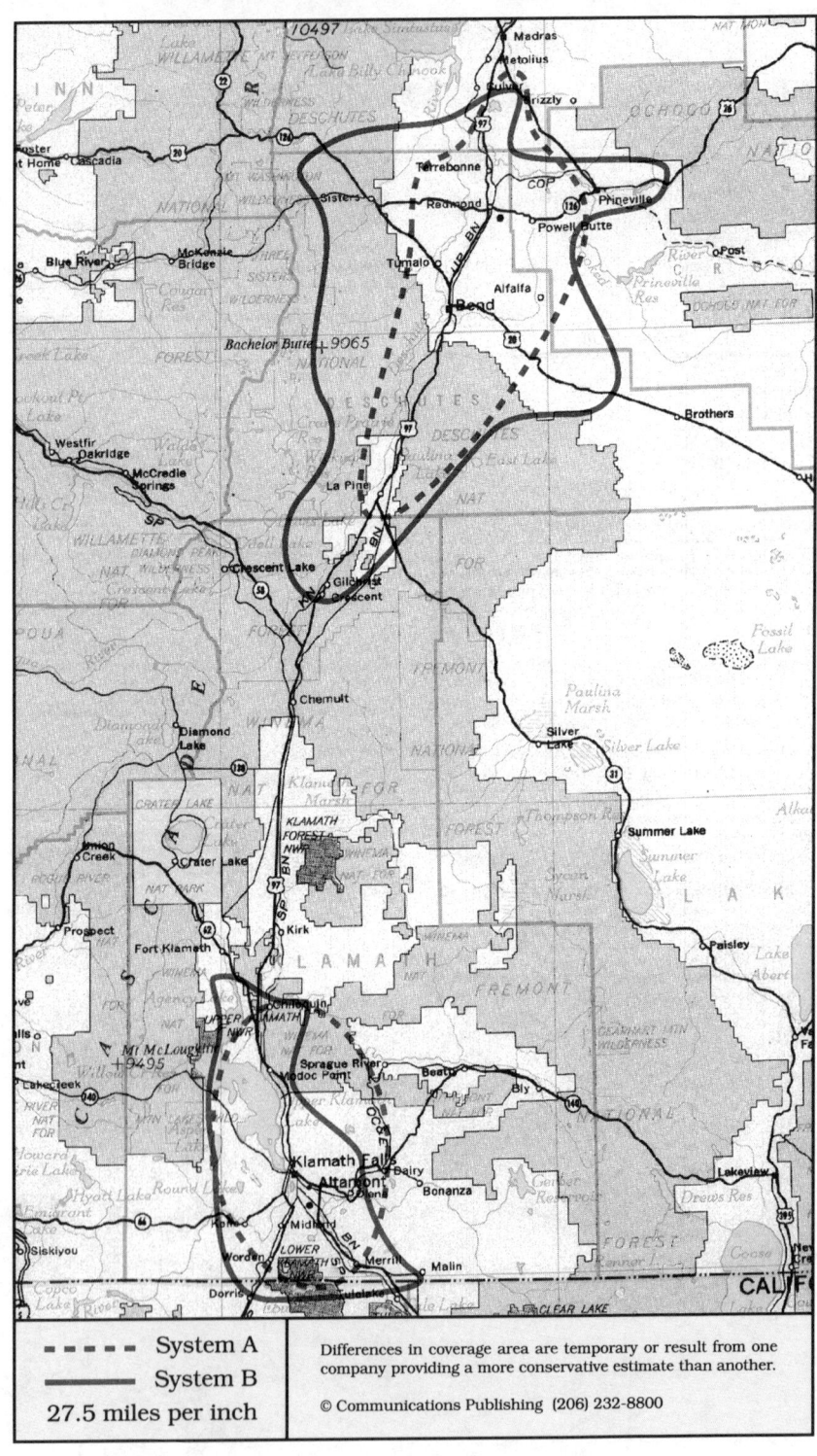

- - - - System A
———— System B
27.5 miles per inch

Differences in coverage area are temporary or result from one company providing a more conservative estimate than another.

© Communications Publishing (206) 232-8800

Bend, Oregon
Bend, Klamath Falls

System A	System B

Cellular company

Cellular One
McCaw Cellular Communications
20360 N Empire, Suite B2
Bend, OR 97701

Customer service	800-245-7626
Local office . . . . for Bend	503-385-5600
. . . . . for Klamath Falls	503-779-2451
Corporate office	206-827-4500

Sending calls

Visitor dialing instructions
Local calls	area code + 7 digits
Long distance . . .	1 + area code + 7 digits

Speed-dial numbers
Customer service	611
Emergency	911
Roadside service	*TOW or *869
Weather	*4CAST or *4278

Receiving calls

Caller dials the roamer access number below,
hears tone, dials cellular area code and number.
Bend	503-480-7626
Klamath Falls	503-891-7626

NationLink & RoamingAmerica (see p. 1157)
Dial *31 send to receive calls, *30 to deactivate.
Dial *32 send to tell caller the roamer access no.
This company's subscribers have level 6 service.

North American Cellular Network (see p. 1156)
All calls forward to you. Deactivate dial *35 send.

Billing information

Visitor rates
Most visitors	$0 day + 99¢ minute

Visitors without roaming agreements (p. 1159)
Call co. to use Discover, Mastercard, Visa

Identification numbers

City	Market	System	Billing	RSA
All served cities	611	01609	00000	OR-6

Cellular company

United States Cellular
610 Medford Center
Medford, OR 97504

Local office	503-779-3000
Corporate office	312-399-8900

Sending calls

Visitor dialing instructions
Local calls	area code + 7 digits
Long distance . . .	1 + area code + 7 digits

Speed-dial numbers
Customer service	611
Emergency	911
Towing	*869
Weather	*89

Receiving calls

Caller dials the roamer access number below,
hears tone, dials cellular area code and number.
Bend	503-420-7626
Klamath Falls	503-892-7626

Billing information

Visitor rates
Most visitors	$3 day + 85¢ minute

Visitors without roaming agreements (p. 1159)
Call co. to use Mastercard, Visa

Identification numbers

City	Market	System	Billing	RSA
All served cities	611	01610	00000	OR-6

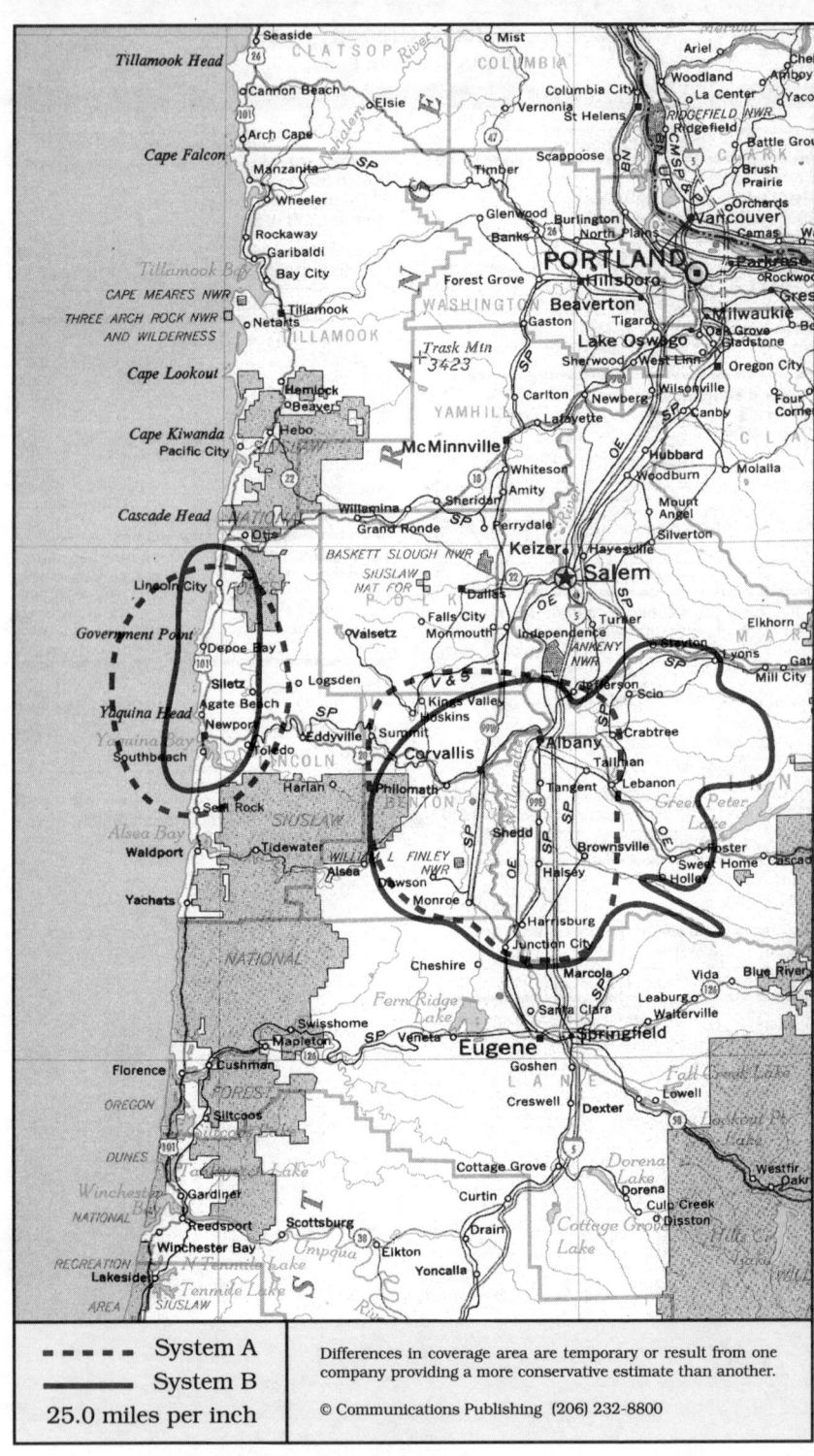

- - - - - System A	Differences in coverage area are temporary or result from one company providing a more conservative estimate than another.
——— System B	
25.0 miles per inch	© Communications Publishing (206) 232-8800

Corvallis, Oregon
Albany, Corvallis, Newport

System A	System B

Cellular company

**Cellular One
Point Communications
325 Pacific Boulevard SW
Albany, OR 97321**

Customer service	503-926-6111
. . . . . . from Newport	503-265-2900
Corporate office	503-928-2900

Sending calls

Visitor dialing instructions

Local calls	area code + 7 digits
Long distance . . .	1 + area code + 7 digits

Speed-dial numbers

Customer service	611
Emergency	911
Towing service (subscribers only)	*TOW or *869
KBCH Lincoln (subscribers only) . .	*1400
KCRF Lincoln (subscribers only) . . .	*983
KFAT Corvallis (subscribers only) . . .	*106
KNPT Newport (subscribers only) . . .	*1301
KRKT Albany (subscribers only) . .	*99
KYTE Newport (subscribers only) . .	*1027
Road conditions (subscribers only) . . .	*101
Metro One hello pages (subscribers only)	*1122

Receiving calls

Caller dials the roamer access number below,
hears tone, dials cellular area code and number.
. . . . 503-740-7626

NationLink & RoamingAmerica (see p. 1157)
Dial *31 send to receive calls, *30 to deactivate.
Dial *32 send to tell caller the roamer access no.
This company's subscribers have level 6 service.

North American Cellular Network (see p. 1156)
All calls forward to you. Deactivate dial *35 send.

Billing information

Visitor rates
Most visitors . . . . . $0 day + 99¢ minute

Visitors without roaming agreements (p. 1159)
Cannot call since company takes no credit cards.

Identification numbers

City	Market	System	Billing	RSA
All served cities	609	00061	30111	OR-4

Call switching is performed by McCaw Cellular as part of the Eugene and Portland systems for the licensee, Point Communications.

Cellular company

**U S West Cellular
2511 W 11th Avenue, Suite C
Eugene, OR 97402**

Customer service	800-238-7848
. . . .	206-562-2895
. . . . . local subscribers	800-626-6611
. . . .	206-747-1771
Local office	503-345-9596
Corporate office	206-747-4900

Sending calls

Visitor dialing instructions

Local calls	area code + 7 digits
Long distance . . .	1 + area code + 7 digits

Speed-dial numbers

Customer service	*611
Emergency	911
Roaming information	*711

Receiving calls

Caller dials the roamer access number below,
hears tone, dials cellular area code and number.
. . . . 503-990-7626

Follow Me Roaming forwarding (see p. 1156)
Activate, dial *18 send. Deactivate dial *19 send.

Billing information

Visitor rates
Most visitors . . . . . $3 day + 99¢ minute

Visitors without roaming agreements (p. 1159)
Call co. to use Amex, Mastercard, Visa

Identification numbers

City	Market	System	Billing	RSA
All served cities	609	01606	00000	OR-4

For full coverage area, see Eugene, OR.

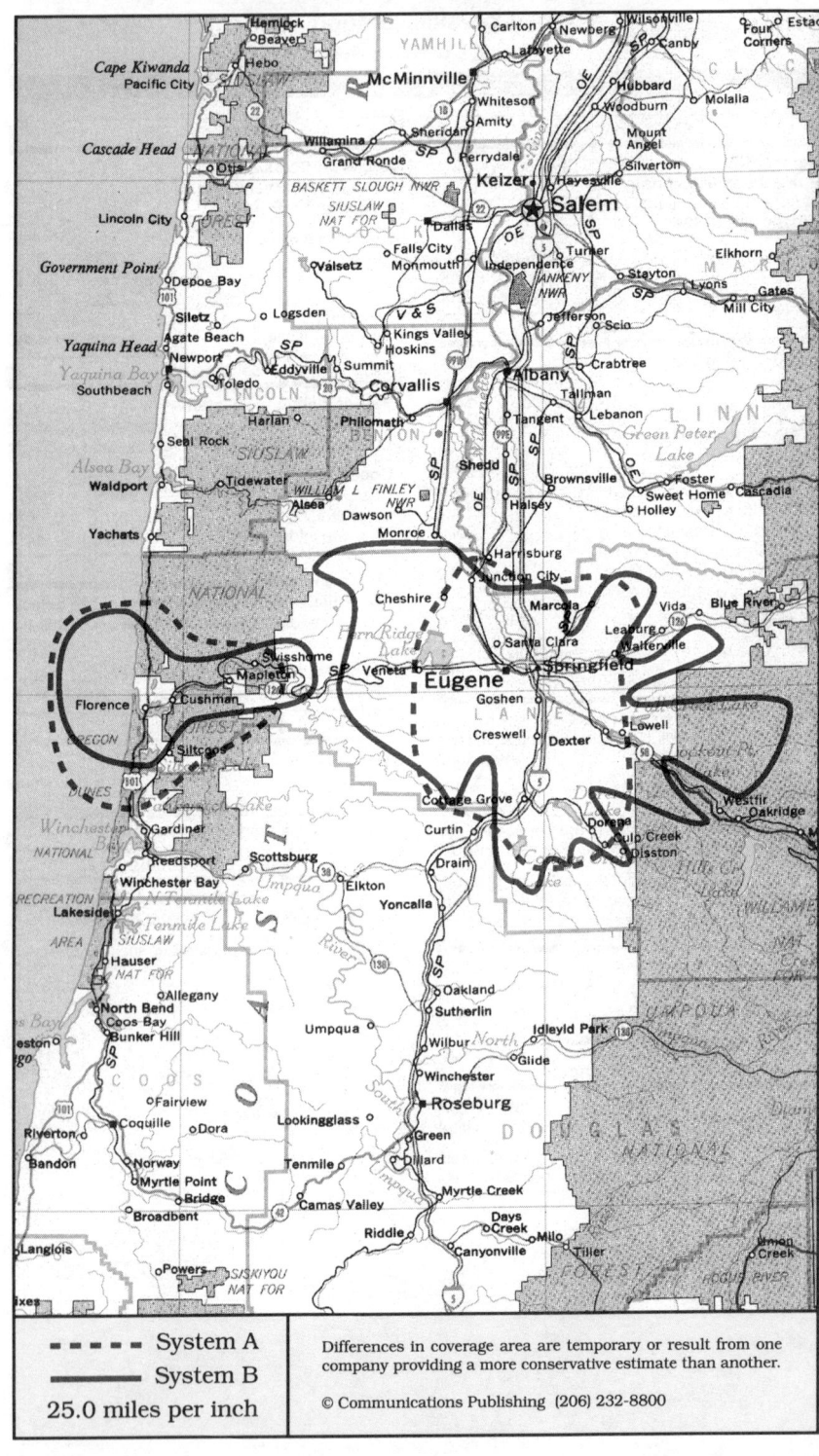

System A
System B
25.0 miles per inch

Differences in coverage area are temporary or result from one
company providing a more conservative estimate than another.

© Communications Publishing (206) 232-8800

Eugene, Oregon
Eugene, Florence, Springfield

System A	System B

Cellular company

Cellular One
McCaw Cellular Communications
1011 Valley River Way, Suite 105-B
Eugene, OR 97401

Customer service	800-245-7626
. . . .	503-345-1818
Local office	503-484-2355
Corporate office	206-827-4500

Sending calls

Visitor dialing instructions
Local calls . . . . . area code + 7 digits
Long distance . . . 1 + area code + 7 digits

Speed-dial numbers
Cellular One Connect (transfer to business) ∗555
Customer service 611
Emergency 911

Receiving calls

Caller dials the roamer access number below,
hears tone, dials cellular area code and number.
Eugene, Springfield 503-954-7626
Florence (planned for the future) 503-991-7626

NationLink & RoamingAmerica (see p. 1157)
Dial ∗31 send to receive calls, ∗30 to deactivate.
Dial ∗32 send to tell caller the roamer access no.
This company's subscribers have level 6 service.

North American Cellular Network (see p. 1156)
All calls forward to you. Deactivate dial ∗35 send.

Billing information

Visitor rates
Most visitors . . . . . $0 day + 99¢ minute

Visitors without roaming agreements (p. 1159)
Call co. to use Mastercard, Visa

Identification numbers

City	Market	System	Billing	RSA
All served cities	135	00061	00000	MSA

Cellular company

U S West Cellular
2511 W 11th Avenue, Suite C
Eugene, OR 97402

Customer service	800-238-7848
. . . .	206-562-2895
. . . . . local subscribers	800-626-6611
. . . .	206-747-1771
Local office	503-345-9596
Corporate office	206-747-4900

Sending calls

Visitor dialing instructions
Local calls . . . . . . area code + 7 digits
Long distance . . . 1 + area code + 7 digits

Speed-dial numbers
Customer service . . . . ∗611
Emergency 911
Roaming information . . . . ∗711

Receiving calls

Caller dials the roamer access number below,
hears tone, dials cellular area code and number.
. . . . 503-729-7626

Follow Me Roaming forwarding (see p. 1156)
Activate, dial ∗18 send. Deactivate dial ∗19 send.

Billing information

Visitor rates
Most visitors . . . . . $3 day + 99¢ minute

Visitors without roaming agreements (p. 1159)
Call co. to use Amex, Mastercard, Visa

Identification numbers

City	Market	System	Billing	RSA
All served cities	135	00328	00000	MSA

For full coverage area, see Corvallis, OR.

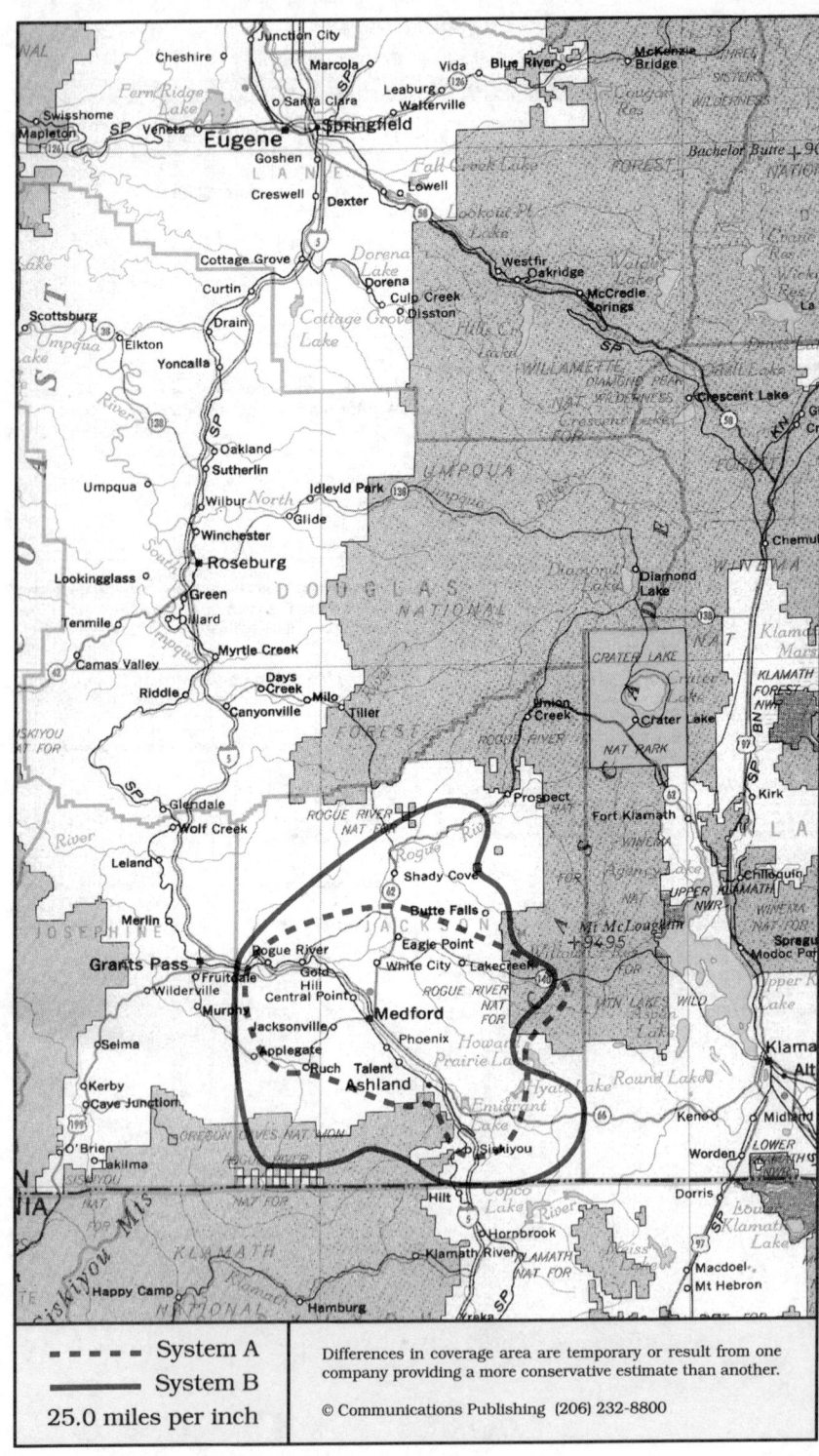

System A
System B
25.0 miles per inch

Differences in coverage area are temporary or result from one company providing a more conservative estimate than another.

© Communications Publishing (206) 232-8800

Medford, Oregon
Ashland, Medford

System A	System B

Cellular company

Cellular One
McCaw Cellular Communications
1170 Biddle Road, Suite B
Medford, OR 97504

Customer service	800-245-7626
. . . .	503-779-8990
Local office	503-779-2451
Corporate office	206-827-4500

Sending calls

Visitor dialing instructions

Local calls	area code + 7 digits
Long distance . . .	1 + area code + 7 digits

Speed-dial numbers

Customer service	611
Directory assistance	*555
Emergency	911
Roadside service	*TOW or *869
Weather conditions . .	*4CAST or *42278

Receiving calls

Caller dials the roamer access number below,
hears tone, dials cellular area code and number.

. . . .	503-944-7626

NationLink & RoamingAmerica (see p. 1157)
Dial *31 send to receive calls, *30 to deactivate.
Dial *32 send to tell caller the roamer access no.
This company's subscribers have level 6 service.

North American Cellular Network (see p. 1156)
All calls forward to you. Deactivate dial *35 send.

Billing information

Visitor rates

Most visitors	$0 day + 99¢ minute

Visitors without roaming agreements (p. 1159)
Call co. to use Mastercard, Visa

Identification numbers

City	Market	System	Billing	RSA
All served cities	229	00061	00000	MSA

Cellular company

United States Cellular
610 Medford Center
Medford, OR 97504

Local office	503-779-3000
Corporate office	312-399-8900

Sending calls

Visitor dialing instructions

Local calls	area code + 7 digits
Long distance . . .	1 + area code + 7 digits

Speed-dial numbers

Customer service	611
Emergency	911
Towing	*869
Weather	*89

Receiving calls

Caller dials the roamer access number below,
hears tone, dials cellular area code and number.

. . . .	503-821-7626

Follow Me Roaming forwarding (see p. 1156)
Activate, dial *18 send. Deactivate dial *19 send.

Billing information

Visitor rates

Most visitors	$3 day + 85¢ minute

Visitors without roaming agreements (p. 1159)
Call co. to use Mastercard, Visa

Identification numbers

City	Market	System	Billing	RSA
All served cities	229	00436	00000	MSA

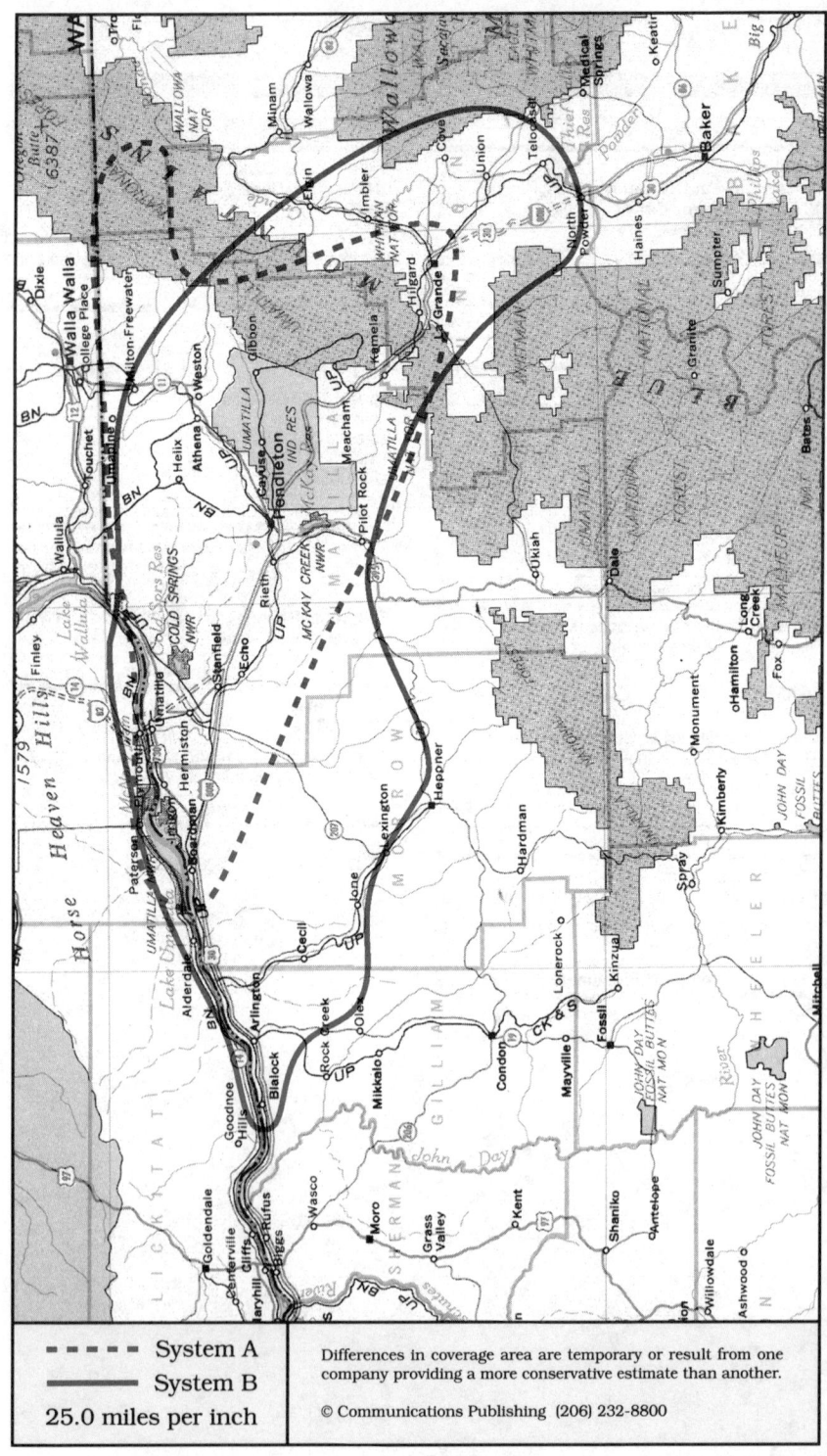

- - - - - System A	Differences in coverage area are temporary or result from one
——— System B	company providing a more conservative estimate than another.
25.0 miles per inch	© Communications Publishing (206) 232-8800

Pendleton, Oregon
Baker, Hermiston, La Grande, Pendleton

|

System A

Cellular company

Cellular One
Blue Mountain Cellular
225 SE 2nd Street
Pendleton, OR 97801

Local office 503-278-2355
Corporate office 509-529-6220

Sending calls

Visitor dialing instructions
Local calls . . . . . area code + 7 digits
Long distance . . . 1 + area code + 7 digits

Speed-dial numbers
Customer service 611
Emergency 911

Receiving calls

Caller dials the roamer access number below,
hears tone, dials cellular area code and number.
. . . . 503-969-7626

NationLink & RoamingAmerica (see p. 1157)
Dial *31 send to receive calls, *30 to deactivate.
Dial *32 send to tell caller the roamer access no.
This company's subscribers have level 6 service.

North American Cellular Network (see p. 1156)
All calls forward to you. Deactivate dial *35 send.

Billing information

Visitor rates
Most visitors . . . . . $3 day + 85¢ minute

Visitors without roaming agreements (p. 1159)
Roamer Plus will intercept first call to bill you.

Identification numbers

City	Market	System	Billing	RSA
All served cities	608	01787	30259	OR-3

System B

Cellular company

United States Cellular
370 S Main Street, Suite 101
Pendleton, OR 97801

Local office 503-278-2200
Corporate office 312-399-8900

Sending calls

Visitor dialing instructions
Local calls . . . . . . area code + 7 digits
Long distance . . . . . area code + 7 digits

Speed-dial numbers
Customer service 611
Emergency 911

Receiving calls

Caller dials the roamer access number below,
hears tone, dials cellular area code and number.
. . . . 503-379-7626

Follow Me Roaming forwarding (see p. 1156)
Activate, dial *18 send. Deactivate dial *19 send.

Billing information

Visitor rates
Most visitors . . . . . $3 day + 85¢ minute

Visitors without roaming agreements (p. 1159)
Cannot call since company takes no credit cards.

Identification numbers

City	Market	System	Billing	RSA
All served cities	608	01604	00000	OR-3

For full coverage area, see The Dalles, OR and
Yakima, WA.

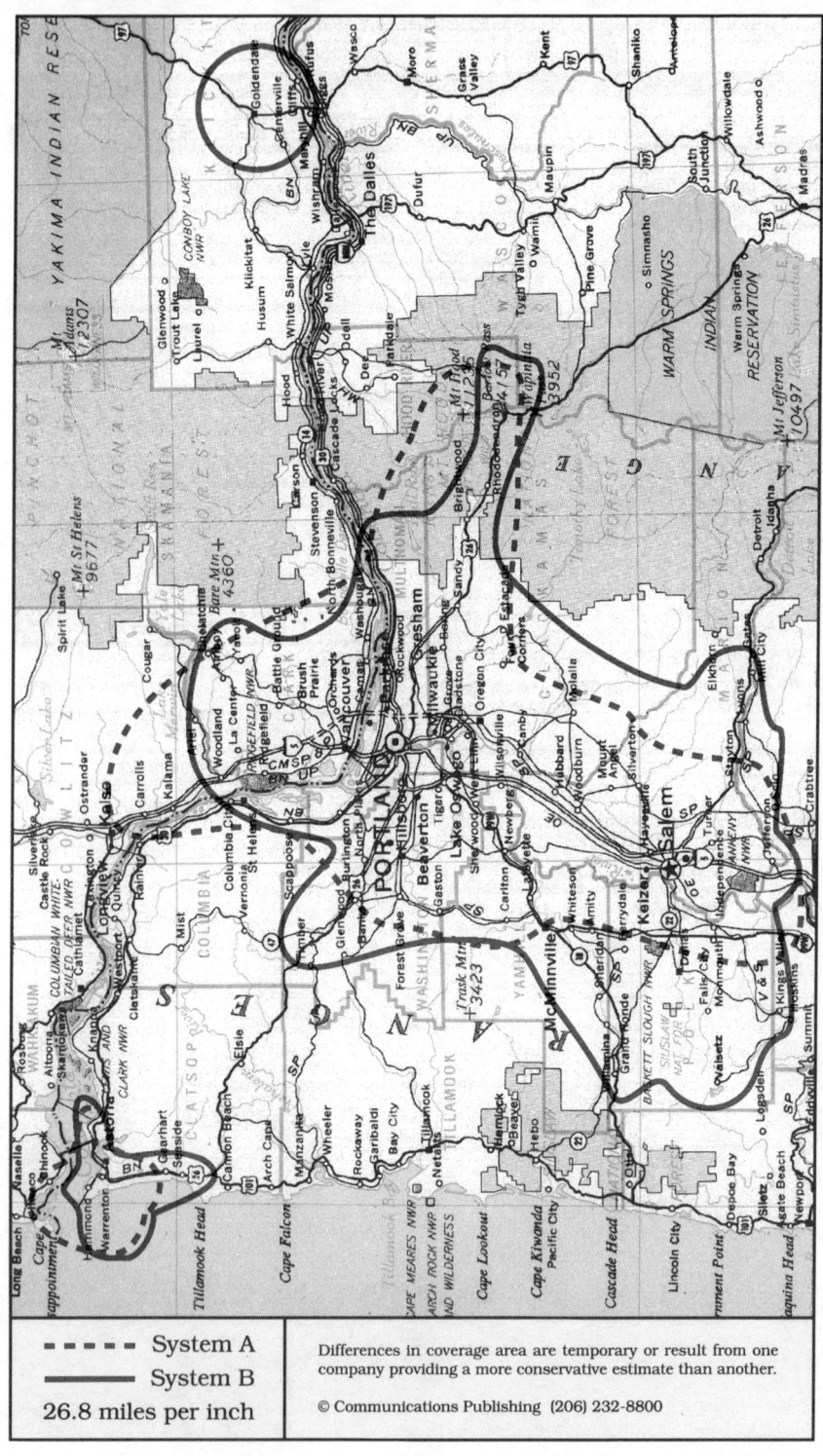

System A
System B
26.8 miles per inch

Differences in coverage area are temporary or result from one company providing a more conservative estimate than another.

© Communications Publishing (206) 232-8800

Portland, Oregon
Goldendale WA, McMinnville, Portland, Salem, Vancouver WA

System A	System B

Cellular company

Cellular One
McCaw Cellular Communications
1600 SW 4th Avenue
Portland, OR 97201

Customer service	800-245-7626
. . . . Portland	503-243-5555
. . . . Salem	503-364-3335
Local office	503-243-3333
Corporate office	206-827-4500

Sending calls

Visitor dialing instructions

Local calls	area code + 7 digits
Long distance . . .	1 + area code + 7 digits

Speed-dial numbers

Cellular One Connect (transfer to business)	*555
Customer service	611
Emergency	911
Hear a traffic report	*1190
KGW talk radio	*620
Sports line	*FAN or *326
Towing	*TOW or *869

Receiving calls

Caller dials the roamer access number below,
hears tone, dials cellular area code and number.

Astoria	503-791-7626
Beaverton	503-550-7626
Portland	503-781-7626
Salem	503-931-7626

NationLink & RoamingAmerica (see p. 1157)
Dial *31 send to receive calls, *30 to deactivate.
Dial *32 send to tell caller the roamer access no.
This company's subscribers have level 6 service.

North American Cellular Network (see p. 1156)
All calls forward to you. Deactivate dial *35 send.

Billing information

Visitor rates

Most visitors	$0 day + 99¢ minute

Visitors without roaming agreements (p. 1159)
Call co. to use Mastercard, Visa

Identification numbers

City	Market	System	Billing	RSA
Astoria	606	00061	30409	OR-1
Newberg	606	00061	30249	OR-1
Portland, Vancouver	030	00061	00000	MSA
Salem	148	00061	00000	MSA

Call switching is performed by McCaw Cellular for
the licensee, Crystal Communications.

Cellular company

GTE Mobilnet
15115 SW Sequoia Parkway, Suite 150
Portland, OR 97224

Customer service	800-366-3001
. . . .	503-620-6446
Local office	503-624-1500
Corporate office	404-391-8000

Sending calls

Visitor dialing instructions

Local calls	area code + 7 digits
Long distance . . .	1 + area code + 7 digits

Speed-dial numbers

Address directions	*627
Customer service	*611
Emergency	911
KATU TV 2 news department	*2
KEX radio	*1190
KWJJ radio station	#931
KXL make a traffic report	*750
Make a traffic report in Portland . . .	#995
Make a traffic report in Salem	*1490
Mr. Rescue	*4357
Portland events (live)	*1122
Roaming information	*711
Telephone problems	*111
Weather report	*483

Receiving calls

Caller dials the roamer access number below,
hears tone, dials cellular area code and number.

Portland	503-248-7626
Salem	503-871-7626

Follow Me Roaming forwarding (see p. 1156)
Activate, dial *18 send. Deactivate dial *19 send.

Billing information

Visitor rates

Most visitors	$3 day + 85¢ minute

Visitors without roaming agreements (p. 1159)
Call co. to use Amex, Discover, Mastercard, Visa

Identification numbers

City	Market	System	Billing	RSA
Goldendale	699	00030	00000	WA-7
McMinville	606	00030	00000	OR-1
Portland, Vancouver	030	00030	00000	MSA
Salem	148	00030	00000	MSA

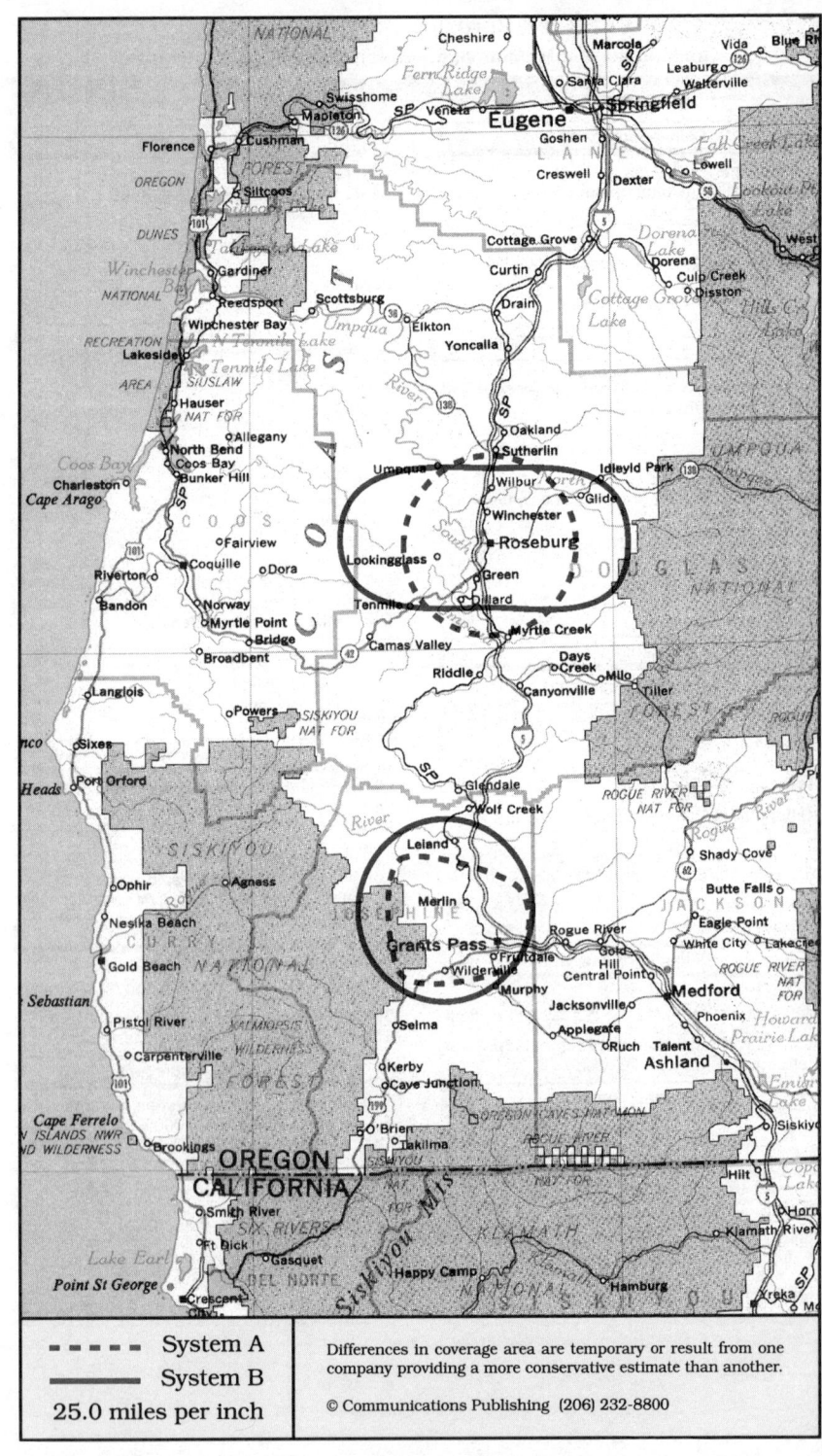

System A - - - -
System B ————

25.0 miles per inch

Differences in coverage area are temporary or result from one company providing a more conservative estimate than another.

© Communications Publishing (206) 232-8800

Roseburg, Oregon
Grants Pass, Roseburg

System A

Cellular company

United States Cellular
610 Medford Center
Medford, OR 97504

Local office		503-779-3000
. . . . . from Roseburg		503-672-0555
Corporate office		312-399-8900

Sending calls

Visitor dialing instructions
Local calls . . . . . area code + 7 digits
Long distance . . . 1 + area code + 7 digits

Speed-dial numbers
Customer service 611
Emergency 911
Towing *869
Weather *89

Receiving calls

Caller dials the roamer access number below,
hears tone, dials cellular area code and number.
. . . . 503-660-7626

Billing information

Visitor rates
Most visitors . . . . . $3 day + 85¢ minute

Visitors without roaming agreements (p. 1159)
Call co. to use Mastercard, Visa

Identification numbers

City	Market	System	Billing	RSA
All served cities	610	01607	00000	OR-5

System B

Cellular company

Contel Cellular
780 NW Garden Valley Boulevard, Suite 116
Roseburg, OR 97470

Customer service		800-333-4004
Local office		503-440-4747
Corporate office		404-804-3400

Sending calls

Visitor dialing instructions
Local calls . . . . . . area code + 7 digits
Long distance . . . 1 + area code + 7 digits

Speed-dial numbers
AAA towing . . . . *AAA or *222
Customer service 611
Emergency 911
Locksmith 333
Roadside service . . . . *HELP or *4357

Receiving calls

Caller dials the roamer access number below,
hears tone, dials cellular area code and number.
Grants Pass 503-450-7626
Myrtle Creek 503-863-9226
Roseburg 503-670-7626

Billing information

Visitor rates
Most visitors . . . . . $3 day + 85¢ minute

Visitors without roaming agreements (p. 1159)
Call co. to use Amex, Mastercard, Visa

Identification numbers

City	Market	System	Billing	RSA
All served cities	610	01608	30604	OR-5

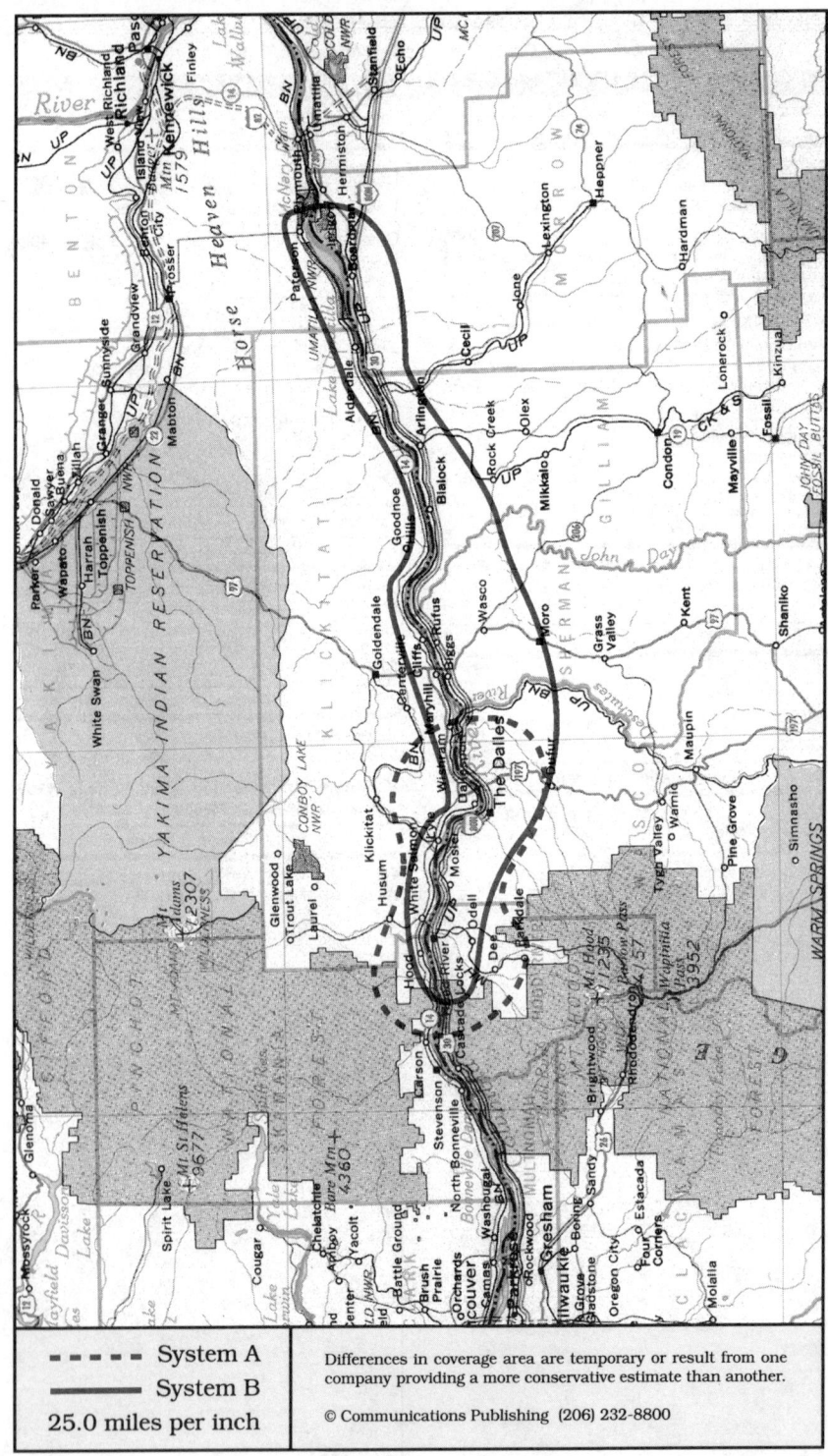

- - - - - System A	Differences in coverage area are temporary or result from one company providing a more conservative estimate than another.	
——— System B		
25.0 miles per inch	© Communications Publishing (206) 232-8800	

The Dalles, Oregon
Hood River, Moro, The Dalles

System A	System B

Cellular company

Cellular One
Pacific Northwest Cellular
11400 SE 8th Street, Suite 445
Bellevue, WA 98004-6431

Customer service	800-635-0304
Local sales agent	503-387-2196
Corporate office	206-635-0300

Sending calls

Visitor dialing instructions

Local calls	area code + 7 digits
Long distance . . .	1 + area code + 7 digits

Speed-dial numbers

Customer service	611
Emergency	911

Receiving calls

Caller dials the roamer access number below, hears tone, dials cellular area code and number.

Hood River	503-490-7626
The Dalles	503-980-7626

NationLink & RoamingAmerica (see p. 1157)
Caller hears your current roamer access number.
This company's subscribers have level 0 service.

North American Cellular Network (see p. 1156)
All calls forward to you. Deactivate dial *35 send.

Billing information

Visitor rates

Most visitors	$0 day + 99¢ minute

Visitors without roaming agreements (p. 1159)
Cannot call since company takes no credit cards.

Identification numbers

City	Market	System	Billing	RSA
All served cities	607	01601	30293	OR-2

Cellular company

United States Cellular
414 Washington, Suite 1-D
The Dalles, OR 97058

Customer service	some areas	800-543-0127
. . . .		509-248-3000
Local office		503-296-4464
Corporate office		312-399-8900

Sending calls

Visitor dialing instructions

Local calls	area code + 7 digits
Long distance	area code + 7 digits

Speed-dial numbers

Customer service	611
Emergency	911
Roaming information	*711

Receiving calls

Caller dials the roamer access number below, hears tone, dials cellular area code and number.

. . . .	503-993-7626

Billing information

Visitor rates

Most visitors	$3 day + 99¢ minute

Visitors without roaming agreements (p. 1159)
Call co. to use Mastercard, Visa

Identification numbers

City	Market	System	Billing	RSA
All served cities	607	01602	00000	OR-2

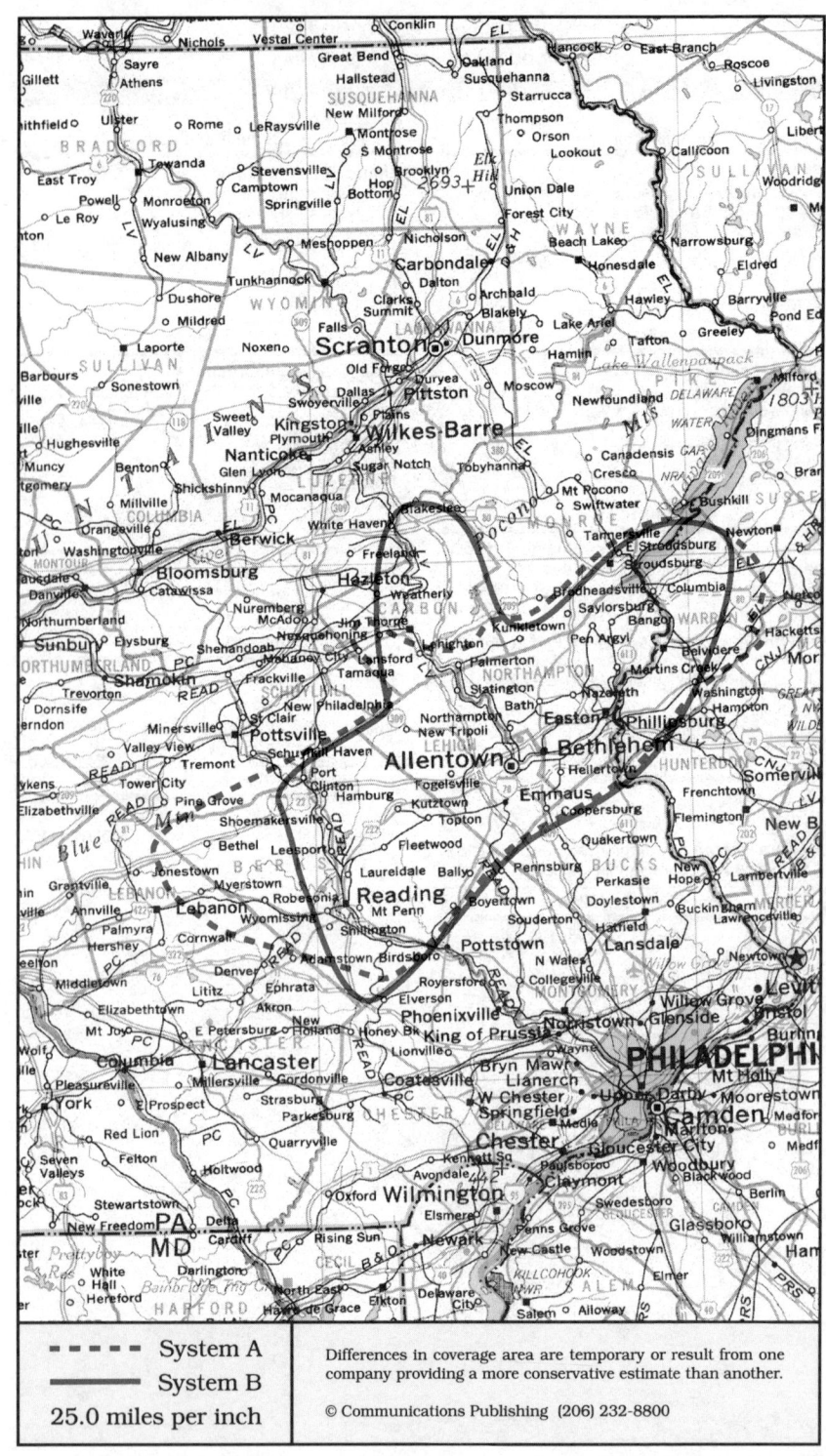

- - - - - System A	Differences in coverage area are temporary or result from one company providing a more conservative estimate than another.
———— System B	
25.0 miles per inch	© Communications Publishing (206) 232-8800

Allentown, Pennsylvania

Allentown, Belvidere, Easton, Emmaus, Jim Thorpe, Reading

System A	System B

Cellular company

Cellular One
Vanguard Cellular Systems
1906 MacArthur Road
Whitehall, PA 18052

125 Morgantown Road
Reading, PA 19611

Customer service	from Allentown	215-390-6100
. . . . .	from Reading	215-780-4100
Local office	. . . . Reading	215-373-2355
. . . . .	Whitehall	215-434-2355
Corporate office	. . . .	919-282-3690

Sending calls

Visitor dialing instructions
Local calls	. . . .	area code + 7 digits
Long distance	. . .	area code + 7 digits

Speed-dial numbers
Customer service	. . . .	611
Emergency	. . . .	911

Receiving calls

Caller dials the roamer access number below,
hears tone, dials cellular area code and number.
Allentown	. . . .	215-390-7626
Reading	. . . .	215-780-7626

NationLink & RoamingAmerica (see p. 1157)
Dial *31 send to receive calls, *30 to deactivate.
Dial *32 send to tell caller the roamer access no.
This company's subscribers have level 6 service.

Billing information

Visitor rates
Most visitors	. . . . .	$3 day + 99¢ minute

Visitors without roaming agreements (p. 1159)
Call co. to use Amex, Mastercard, Visa
or Roamer Plus will intercept first call to bill you.

Identification numbers

City	Market	System	Billing	RSA
Allentown	058	00103	00000	MSA
Reading	118	00103	30023	MSA

For full coverage area, see Scranton, PA.

Cellular company

Bell Atlantic Mobile
2415 MacArthur Road
Whitehall, PA 18052

Customer service	. . . .	800-922-0204
Local office	. . . . . .	215-432-7200
Corporate office	. . . . .	908-306-7000

Sending calls

Visitor dialing instructions
Local calls	. . . . .	area code + 7 digits
Long distance	. . . .	area code + 7 digits

Speed-dial numbers
Customer service	. . . .	*811
Emergency	. . . .	911
Repair	. . . .	*611
Sports channel	. . . .	*610
Turnpike emergency	. . . .	*11

Receiving calls

Caller dials the roamer access number below,
hears tone, dials cellular area code and number.
Allentown	. . . .	215-360-7626
Reading	. . . .	215-870-7626

Follow Me Roaming forwarding (see p. 1156)
Activate, dial *18 send. Deactivate dial *19 send.

Billing information

Visitor rates
Most visitors	. . . .	$3 day + $1 minute

Visitors without roaming agreements (p. 1159)
Roamer Plus will intercept first call to bill you.

Identification numbers

City	Market	System	Billing	RSA
Allentown	058	00008	30358	MSA
Reading	118	00008	30360	MSA

For full coverage area, see Philadelphia, PA.

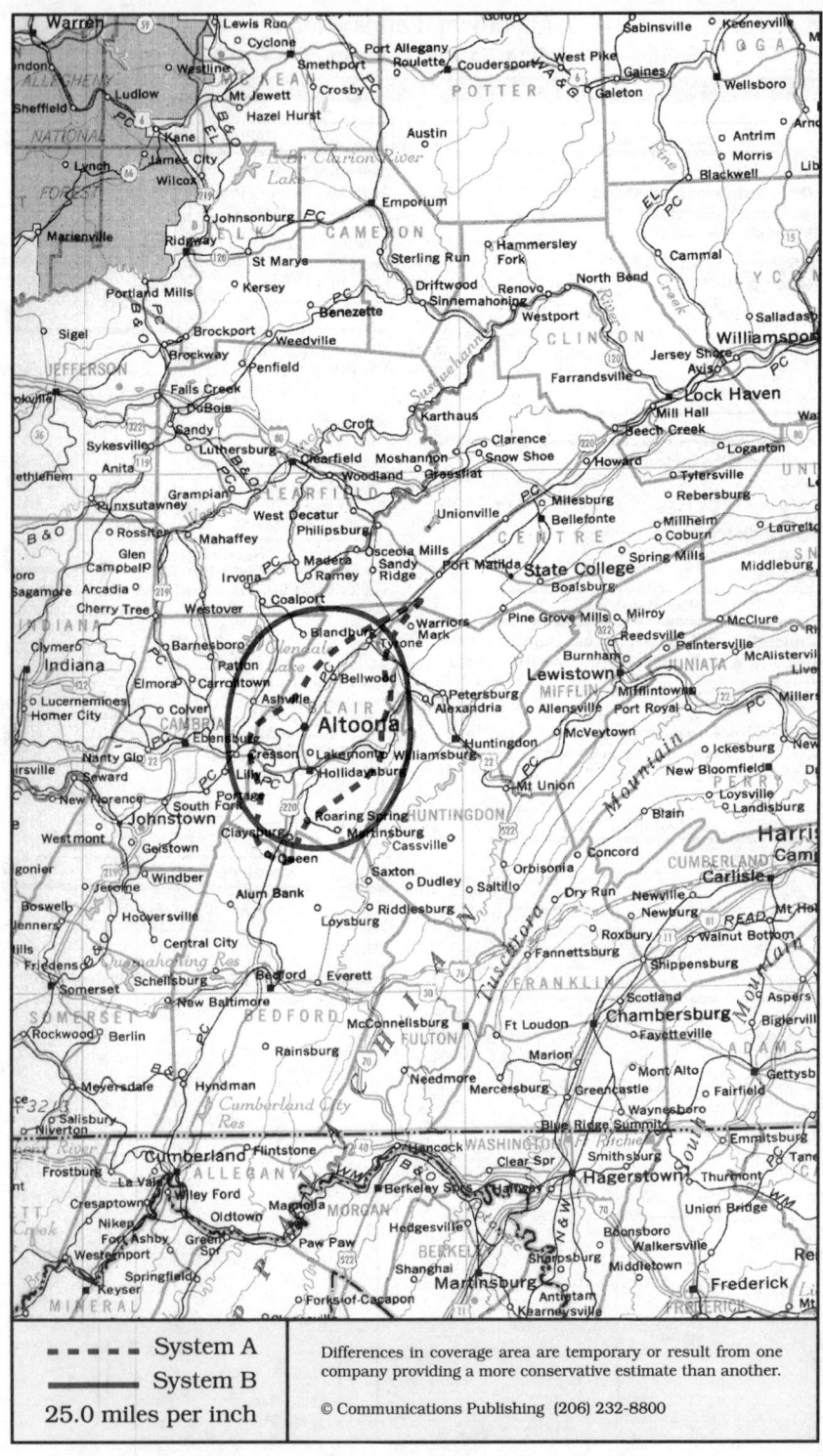

Altoona, Pennsylvania
Altoona (Blair County), Hollidaysburg

System A	System B

Cellular company

Cellular One
Cellular Information Systems
514 E Pleasant Valley Boulevard #1
Altoona, PA 16602-4802

Local office	814-946-4535
Corporate office	203-622-6317

Sending calls

Visitor dialing instructions

Local calls	area code + 7 digits
Long distance . . .	1 + area code + 7 digits

Speed-dial numbers

Customer service	611
Emergency	911
Pennsylvania AAA	412-362-1900
Turnpike emergency	*11

Receiving calls

Caller dials the roamer access number below, hears tone, dials cellular area code and number.Caller dials the roamer access number below,
hears tone, dials cellular area code and number.
. . . . 814-935-7626

Billing information

Visitor rates
Most visitors . . . . . $3 day + 85¢ minute

Visitors without roaming agreements (p. 1159)
Call co. to use Amex, Mastercard, Visa

Identification numbers

City	Market	System	Billing	RSA
All served cities	225	00247	00000	MSA

Cellular company

Independent Cellular Network
3014 Pleasant Valley Boulevard
Altoona, PA 16602

Local office	814-944-3011
Corporate office	813-489-1600

Sending calls

Visitor dialing instructions

Local calls	area code + 7 digits
Long distance . . .	1 + area code + 7 digits

Speed-dial numbers

Customer service	*611
Emergency	911

Receiving calls

Caller dials the roamer access number below, hears tone, dials cellular area code and number.
. . . . 814-931-7626

Follow Me Roaming forwarding (see p. 1156)
Activate, dial *18 send. Deactivate dial *19 send.

Billing information

Visitor rates
Most visitors . . . . . $3 day + $1 minute

Visitors without roaming agreements (p. 1159)
Cannot call since company takes no credit cards.

Identification numbers

City	Market	System	Billing	RSA
All served cities	225	00032	30020	MSA

For full coverage area, see Johnstown, PA.

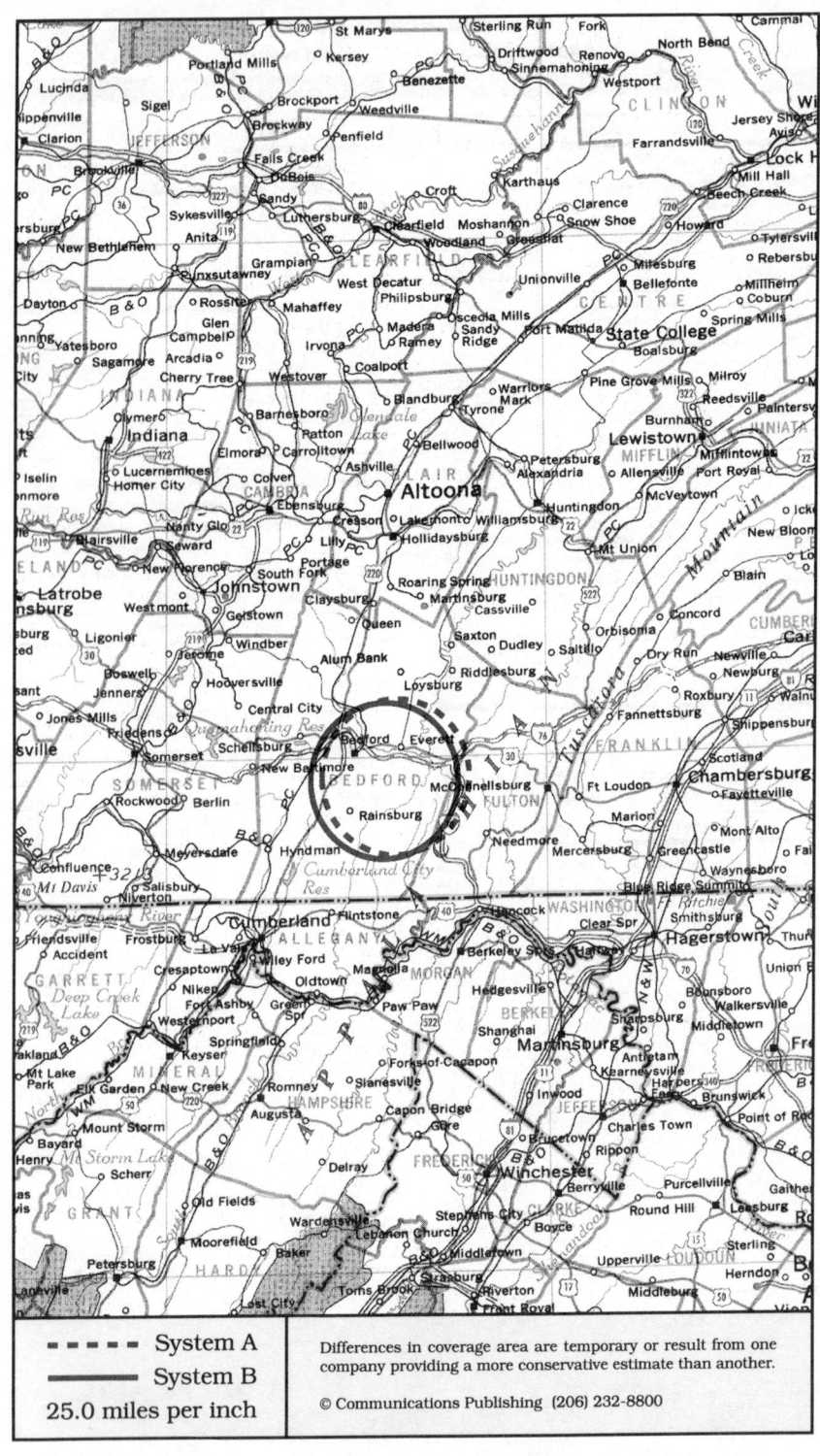

▬ ▬ ▬ ▬ System A	Differences in coverage area are temporary or result from one company providing a more conservative estimate than another.
▬▬▬▬▬ System B	
25.0 miles per inch	© Communications Publishing (206) 232-8800

Bedford, Pennsylvania
Bedford (Bedford County)

Cellular company

Cellular One
Horizon Cellular Telephone
791 Wayne Avenue
Chambersburg, PA 17201

Local office   717-264-8888
Corporate office   215-651-5900

Sending calls

Visitor dialing instructions
Local calls . . . . .   area code + 7 digits
Long distance . . .   1 + area code + 7 digits

Speed-dial numbers
Customer service   611
Emergency   911
Report drunk drivers *33
Turnpike assistance *11

Receiving calls

Caller dials the roamer access number below,
hears tone, dials cellular area code and number.
. . . .   814-977-7626

NationLink & RoamingAmerica (see p. 1157)
Dial *31 send to receive calls, *30 to deactivate.
Dial *32 send to tell caller the roamer access no.
This company's subscribers have level 3 service.

Billing information

Visitor rates
Most visitors . . . . . $3 day + 99¢ minute

Visitors without roaming agreements (p. 1159)
American Roaming Network intercepts call to bill.

Identification numbers

City	Market	System	Billing	RSA
All served cities	621	01863	00000	PA-10

For full coverage area, see Chambersburg, PA.

Cellular company

United States Cellular
663 Pittsburgh Street
Uniontown, PA 15401

Local office   800-944-9066
. . . .   412-437-0800
Corporate office   312-399-8900

Sending calls

Visitor dialing instructions
Local calls . . . . . .   area code + 7 digits
Long distance . . .   1 + area code + 7 digits

Speed-dial numbers
Customer service   611
Emergency   911

Receiving calls

Caller dials the roamer access number below,
hears tone, dials cellular area code and number.
. . . .   814-585-7626

Billing information

Visitor rates
Most visitors . . . . . $3 day + 99¢ minute

Visitors without roaming agreements (p. 1159)
Cannot call since company takes no credit cards.

Identification numbers

City	Market	System	Billing	RSA
All served cities	621	02006	00000	PA-10

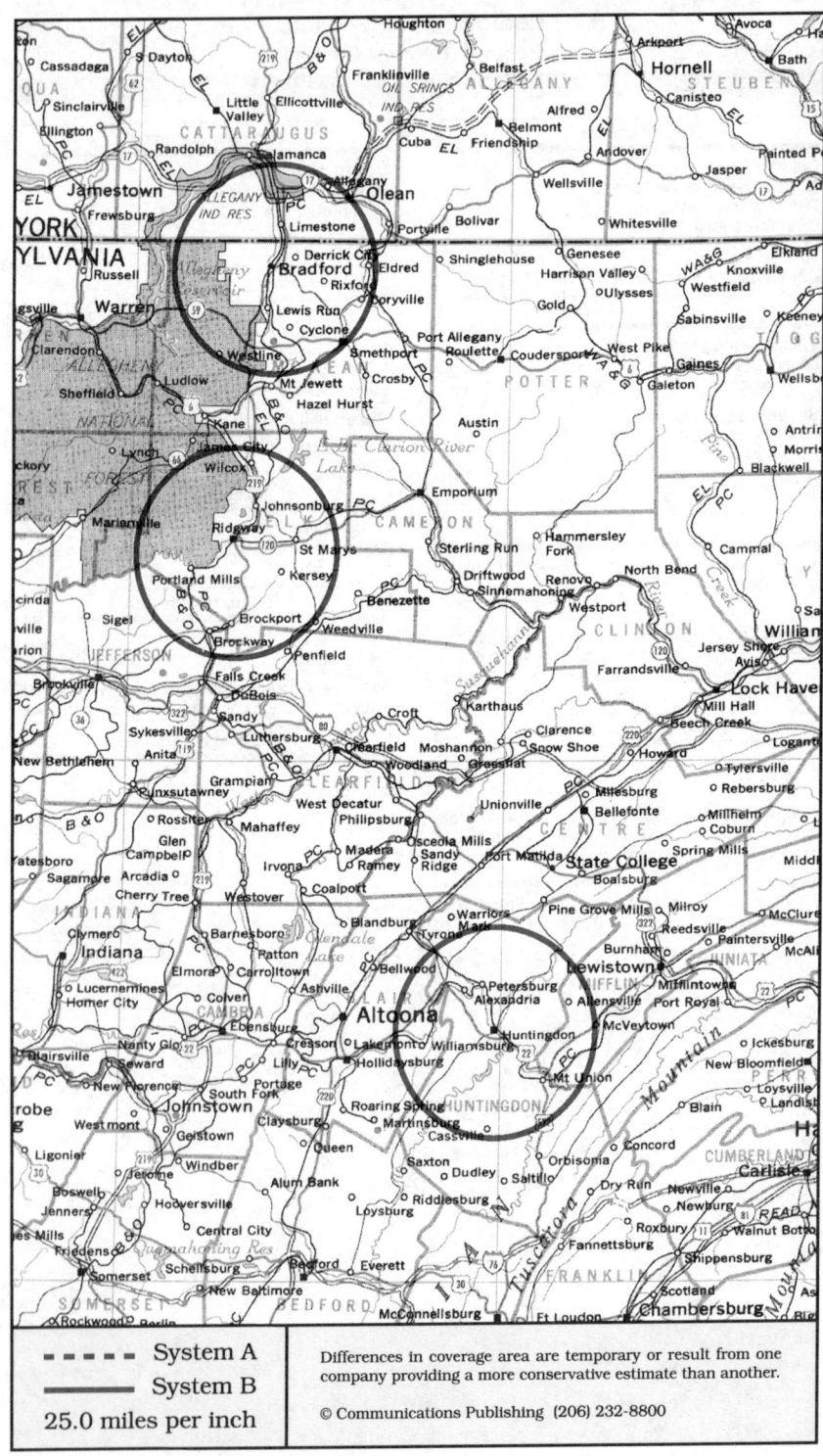

Differences in coverage area are temporary or result from one company providing a more conservative estimate than another.

Bradford, Pennsylvania
Bradford, Huntingdon, Ridgway

System A	System B

Cellular company

Cellular One
Vanguard Cellular Systems
6360 Flank Drive #800, Gateway Corp. Center
Harrisburg, PA 17112

This company is licensed to serve Bradford,
PA. Service was not operating at press time.

Customer service	717-579-4100
Local office	717-579-4004
Corporate office	919-282-3690

Sending calls

Visitor dialing instructions

Local calls	area code + 7 digits
Long distance . . .	area code + 7 digits

Speed-dial numbers

Customer service	611
Emergency	911
Information services	*411
Turnpike assistance (PA)	*11

Receiving calls

Caller dials the roamer access number below,
hears tone, dials cellular area code and number.
. . . . 717-579-7626

NationLink & RoamingAmerica (see p. 1157)
Dial *31 send to receive calls, *30 to deactivate.
Dial *32 send to tell caller the roamer access no.
This company's subscribers have level 6 service.

Billing information

Visitor rates

Most visitors	$3 day + 99¢ minute

Visitors without roaming agreements (p. 1159)
Call co. to use Amex, Mastercard, Visa
or Roamer Plus will intercept first call to bill you.

Identification numbers

City	Market	System	Billing	RSA
All served cities	622	01631	00000	PA-11

Cellular company

Bell Atlantic Mobile
6200 Babcock Boulevard
Pittsburgh, PA 15237

2895 Banksville Road
Pittsburgh, PA 15216

Customer service		800-922-0204
Local office	Babcock Boulevard	412-369-8500
. . . . .	Banksville Road	412-571-3300
Corporate office		908-306-7000

Sending calls

Visitor dialing instructions

Local calls	area code + 7 digits
Long distance . . .	1 + area code + 7 digits

Speed-dial numbers

Customer service	*611
Emergency	911
Hear a sports report	412-555-9823
Hear a traffic report	412-555-1250

Receiving calls

Caller dials the roamer access number below,
hears tone, dials cellular area code and number.

Bradford, Ridgway	814-598-7626
Huntingdon	814-599-7626

Follow Me Roaming forwarding (see p. 1156)
Activate, dial *18 send. Deactivate dial *19 send.

Billing information

Visitor rates

Most visitors	$3 day + $1 minute

Visitors without roaming agreements (p. 1159)
Roamer Plus will intercept first call to bill you.

Identification numbers

City	Market	System	Billing	RSA
Bradford, Ridgway	613	00032	30098	PA-2
Huntingdon	622	00032	30108	PA-11

Butler, Pennsylvania
Butler, Ellwood City, Kittanning, New Castle (system A)

System A	System B

Cellular company

Cellular One
Horizon Cellular
389 New Castle Road
Butler, PA 16001

Customer service	412-282-8484
Local office	412-282-8400
Corporate office	215-651-5900

Sending calls

Visitor dialing instructions

Local calls	area code + 7 digits
Long distance . . .	1 + area code + 7 digits

Speed-dial numbers

Customer service	611
Emergency	911

Receiving calls

Caller dials the roamer access number below,
hears tone, dials cellular area code and number.
. . . . 412-679-7626

NationLink & RoamingAmerica (see p. 1157)
Dial *31 send to receive calls, *30 to deactivate.
Dial *32 send to tell caller the roamer access no.
This company's subscribers have level 6 service.

North American Cellular Network (see p. 1156)
All calls forward to you. Deactivate dial *35 send.

Billing information

Visitor rates

Most visitors	$3 day + 99¢ minute

Visitors without roaming agreements (p. 1159)
Roamer Plus will intercept first call to bill you.

Identification numbers

City	Market	System	Billing	RSA
All served cities	617	00039	30303	PA-6

Call switching is performed by McCaw Cellular for
the licensee, Horizon Cellular.

For full coverage area, see Youngstown, OH

Cellular company

Bell Atlantic Mobile
6200 Babcock Boulevard
Pittsburgh, PA 15237

2895 Banksville Road
Pittsburgh, PA 15216

Customer service		800-922-0204
Local office	Babcock Boulevard	412-369-8500
. . . . .	Banksville Road	412-571-3300
Corporate office		908-306-7000

Sending calls

Visitor dialing instructions

Local calls	area code + 7 digits
Long distance . . .	1 + area code + 7 digits

Speed-dial numbers

Customer service	*611
Emergency	911
Hear a sports report	412-555-9823
Hear a traffic report	412-555-1250

Receiving calls

Caller dials the roamer access number below,
hears tone, dials cellular area code and number.
. . . . 412-629-7626

Follow Me Roaming forwarding (see p. 1156)
Activate, dial *18 send. Deactivate dial *19 send.

Billing information

Visitor rates

Most visitors	$3 day + $1 minute

Visitors without roaming agreements (p. 1159)
Roamer Plus will intercept first call to bill you.

Identification numbers

City	Market	System	Billing	RSA
All served cities	617	00032	30100	PA-6

For full coverage area, see Indiana, PA,
Pittsburgh, PA, and Youngstown, OH.

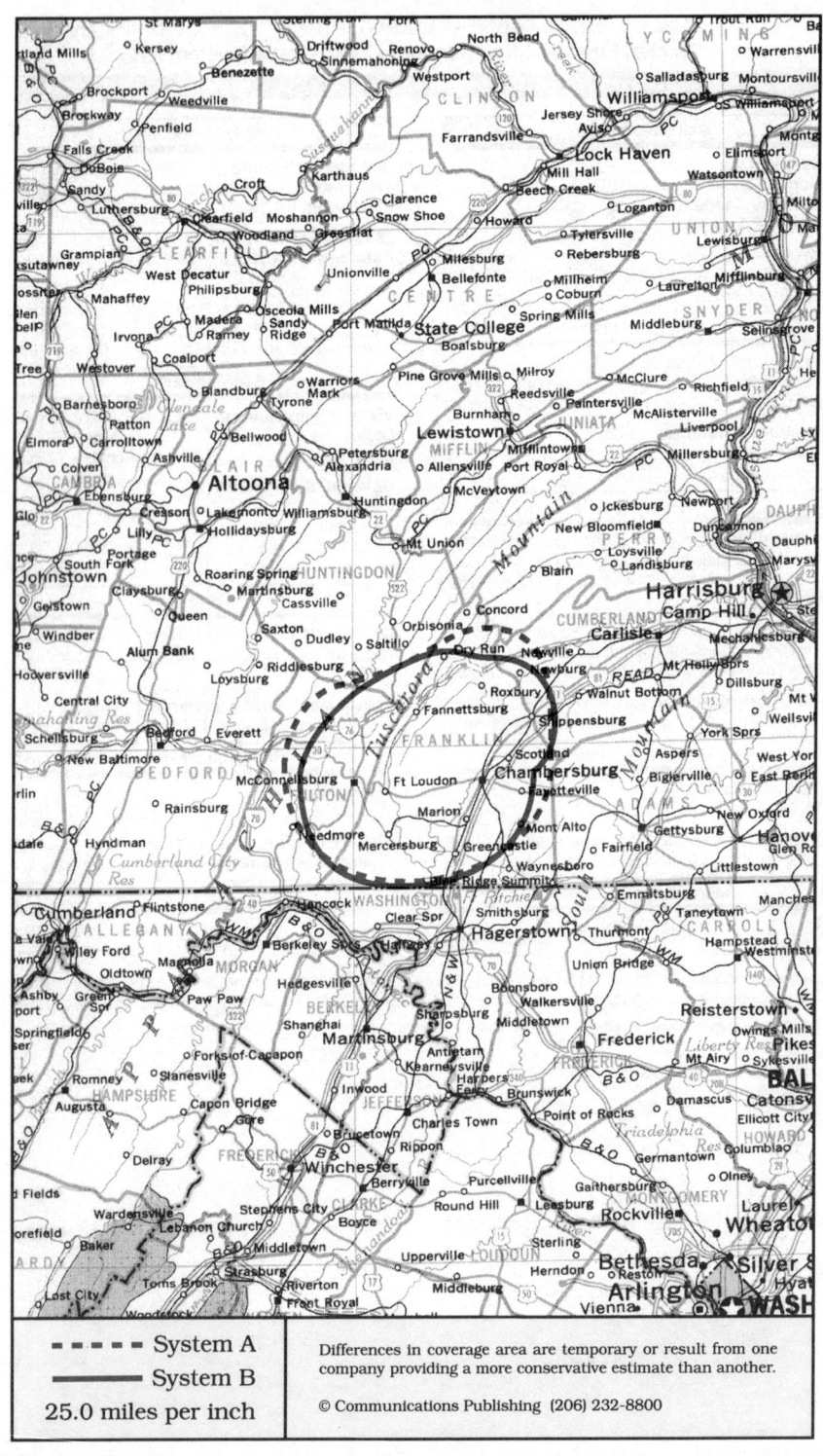

- - - - - System A
──────── System B

25.0 miles per inch

Differences in coverage area are temporary or result from one
company providing a more conservative estimate than another.

© Communications Publishing (206) 232-8800

Chambersburg, Pennsylvania
Chambersburg (Franklin and Fulton Counties), McConnellsburg

System A	System B

Cellular company

Cellular One
Horizon Cellular Telephone
791 Wayne Avenue
Chambersburg, PA 17201

Local office	717-264-8888
Corporate office	215-651-5900

Cellular company

Centel Cellular
5425 Old Jonestown Road
Harrisburg, PA 17112

Local office . . . some areas	800-848-8090
. . . .	717-545-3300
Corporate office	312-399-2644

Sending calls

Visitor dialing instructions

Local calls	area code + 7 digits
Long distance . . .	1 + area code + 7 digits

Speed-dial numbers

Customer service	611
Emergency	911
I-81 emergency assistance	*12
Report drunk drivers	*33
Turnpike assistance (PA)	*11
WCHA news tip line	*800

Sending calls

Visitor dialing instructions

Local calls	area code + 7 digits
Long distance . . .	1 + area code + 7 digits

Speed-dial numbers

Customer service	611
Emergency	911
Report a drunk driver	*DWI or *394
Roadside assistance	*949
State patrol	*12
Turnpike emergency	*11

Receiving calls

Caller dials the roamer access number below, hears tone, dials cellular area code and number.
. . . . 717-860-7626

NationLink & RoamingAmerica (see p. 1157)
Dial *31 send to receive calls, *30 to deactivate.
Dial *32 send to tell caller the roamer access no.
This company's subscribers have level 3 service.

Receiving calls

Caller dials the roamer access number below, hears tone, dials cellular area code and number.
. . . . 717-574-7626

Billing information

Visitor rates
Most visitors . . . . . $3 day + 99¢ minute

Visitors without roaming agreements (p. 1159)
Cannot call since company takes no credit cards.

Billing information

Visitor rates
Most visitors . . . . . $3 day + 99¢ minute

Visitors without roaming agreements (p. 1159)
American Roaming Network intercepts call to bill.

Identification numbers

City	Market	System	Billing	RSA
All served cities	621	01630	00000	PA-10

Identification numbers

City	Market	System	Billing	RSA
All served cities	621	01863	00000	PA-10

For full coverage area, see Bedford, PA.

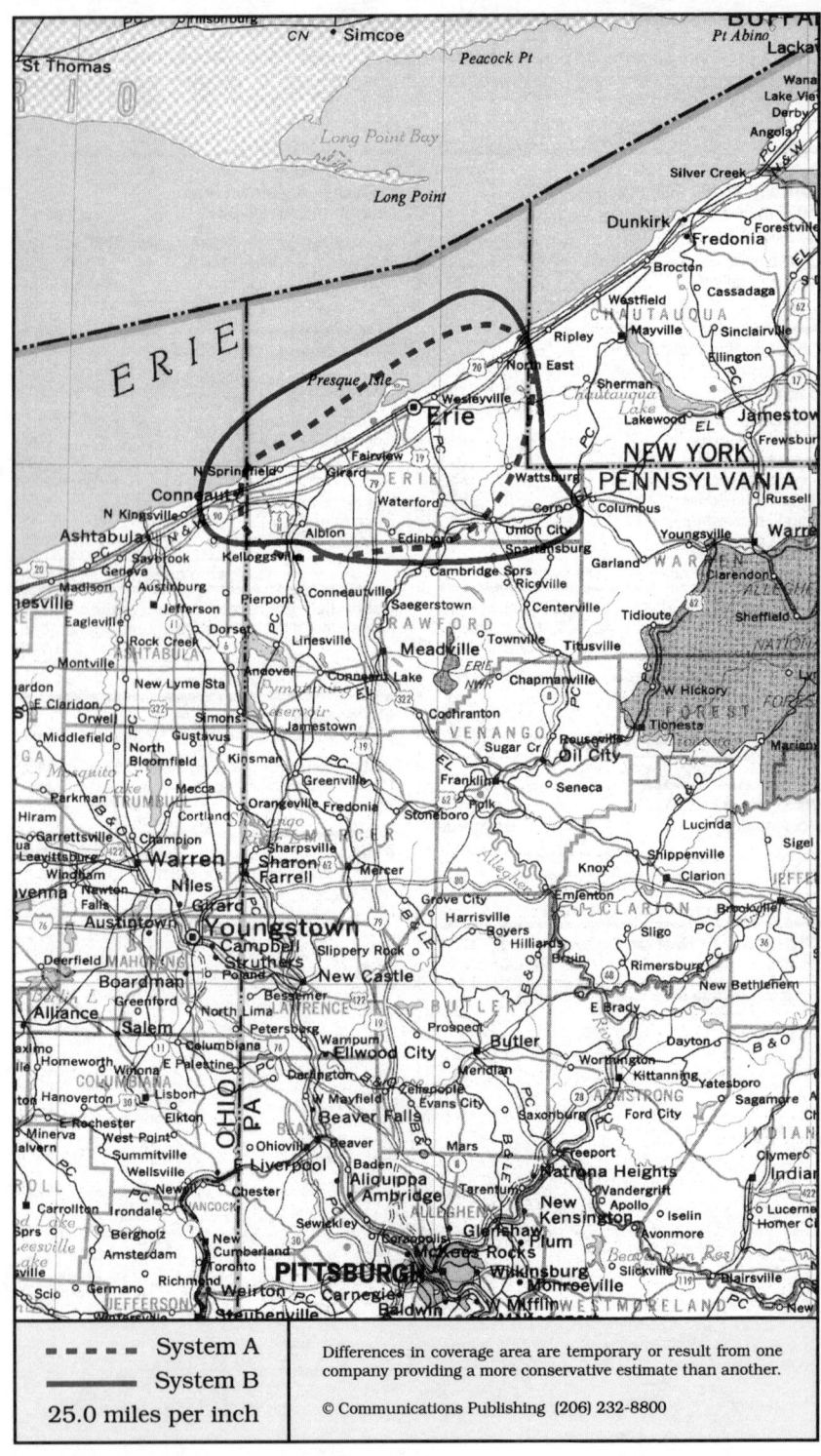

System A
System B
25.0 miles per inch

Differences in coverage area are temporary or result from one company providing a more conservative estimate than another.

© Communications Publishing (206) 232-8800

Erie, Pennsylvania
Conneaut OH, Erie

System A

Cellular company

Cellular One
McCaw Cellular Communications
4823 Peach Street
Erie, PA 16509

Customer service	800-477-3922
. . . .	814-881-0100
Local office	814-868-2355
Corporate office	206-827-4500

Sending calls

Visitor dialing instructions
Local calls . . . . area code + 7 digits
Long distance . . . 1 + area code + 7 digits

Speed-dial numbers
Customer service 611
Emergency 911

Receiving calls

Caller dials the roamer access number below,
hears tone, dials cellular area code and number.
. . . . 814-881-7626

NationLink & RoamingAmerica (see p. 1157)
Dial ∗31 send to receive calls, ∗30 to deactivate.
Dial ∗32 send to tell caller the roamer access no.
This company's subscribers have level 8 service.

North American Cellular Network (see p. 1156)
All calls forward to you. Deactivate dial ∗35 send.

Billing information

Visitor rates
Most visitors . . . . . $0 day + 99¢ minute

Visitors without roaming agreements (p. 1159)
Roamer Plus will intercept first call to bill you.

Identification numbers

City	Market	System	Billing	RSA
All served cities	130	00039	30067	MSA

System B

Cellular company

GTE Mobilnet
4318 Peach Street
Erie, PA 16509

Customer service	some areas	800-669-5665
. . . .		216-642-0688
Local office		814-866-2211
Corporate office		404-391-8000

Sending calls

Visitor dialing instructions
Local calls . . . . . . area code + 7 digits
Long distance . . . 1 + area code + 7 digits

Speed-dial numbers
Customer service . . . . ∗611
Emergency 911
Roaming information . . . . . . ∗711
Telephone problems . . . . ∗111

Receiving calls

Caller dials the roamer access number below,
hears tone, dials cellular area code and number.
. . . . 814-450-7626

Follow Me Roaming forwarding (see p. 1156)
Activate, dial ∗18 send. Deactivate dial ∗19 send.

Billing information

Visitor rates
Most visitors . . . . . $3 day + 85¢ minute

Visitors without roaming agreements (p. 1159)
Call co. to use Amex, Mastercard, Visa

Identification numbers

City	Market	System	Billing	RSA
All served cities	130	00326	00000	MSA

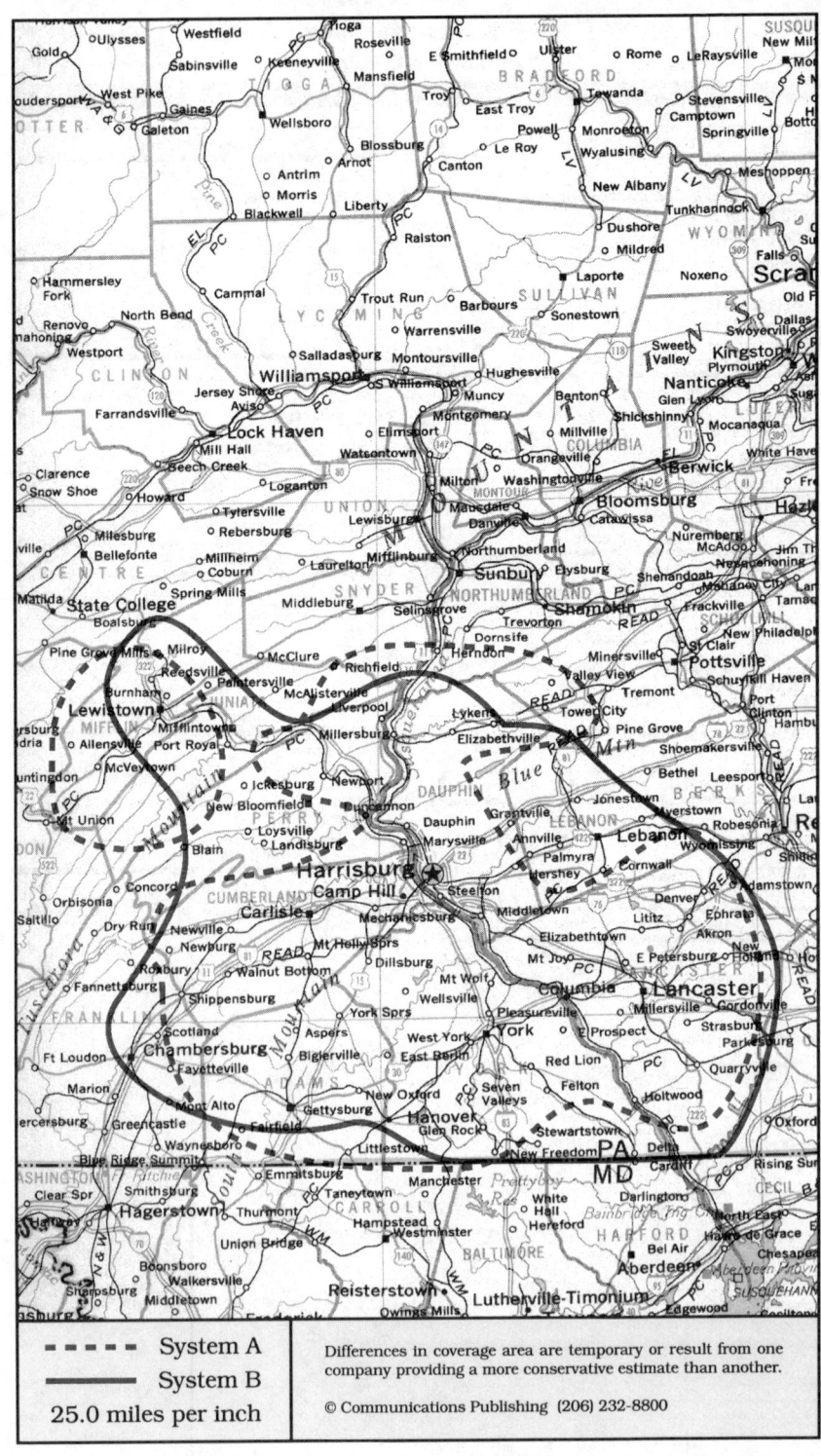

System A
System B
25.0 miles per inch

Differences in coverage area are temporary or result from one company providing a more conservative estimate than another.

© Communications Publishing (206) 232-8800

Harrisburg, Pennsylvania
Carlisle, Harrisburg, Lancaster, Lebanon, Lewistown, York

System A	System B

Cellular company

Cellular One
Vanguard Cellular Systems (regional office)
6360 Flank Drive #800, Gateway Corp. Center
Harrisburg, PA 17112

5775 Allentown Boulevard
Harrisburg, PA 17112

260 Granite Run Drive
Lancaster, PA 17601

902 N Front Street
Wormleysburg, PA 17043

1803 Mount Rose Avenue
York, PA 17403

Customer service from Harrisburg	717-579-4100
. . . . . from Lancaster	717-940-4000
. . . . . from York	717-880-4000
Local office . . . Harrisburg	717-579-2355
. . . . Lancaster	717-560-2355
. . Wormleysburg (transfer)	717-579-2355
. . . . York	717-848-2355
Corporate office . . national	919-282-3690
. . . . regional	717-579-4004

Sending calls

Visitor dialing instructions
Local calls	area code + 7 digits
Long distance . .	area code + 7 digits

Speed-dial numbers
Customer service	611
Emergency	911
Information services	*411
Report a traffic problem	*222
Turnpike emergency	*11

Receiving calls

Caller dials the roamer access number below,
hears tone, dials cellular area code and number.
Harrisburg, Lewistown . . .	717-579-7626
Lancaster	717-940-7626
York	717-880-7626

NationLink & RoamingAmerica (see p. 1157)
Dial *31 send to receive calls, *30 to deactivate.
Dial *32 send to tell caller the roamer access no.
This company's subscribers have level 6 service.

Billing information

Visitor rates
Most visitors . . . . . $3 day + 99¢ minute

Visitors without roaming agreements (p. 1159)
Call co. to use Amex, Mastercard, Visa
or Roamer Plus will intercept first call to bill you.

Identification numbers

City	Market	System	Billing	RSA
Harrisburg	084	00159	00000	MSA
Juaniata County	622	01631	00000	PA-11
Lancaster	105	00159	30011	MSA
York	099	00159	30013	MSA

Cellular company

Centel Cellular
5425 Old Jonestown Road
Harrisburg, PA 17112

3002 Hempland Road
Lancaster, PA 17601

970 Loucks Road #3A
York, PA 17404

Local office . Harrisburg (in PA)	800-848-8090
. . . . . . Harrisburg	717-545-3300
. . . . Lancaster	717-295-1600
. . . . York	717-852-0430
Corporate office	312-399-2644

Sending calls

Visitor dialing instructions
Local calls	area code + 7 digits
Long distance . . .	1 + area code + 7 digits

Speed-dial numbers
Customer service	611
Emergency	911
Report a drunk driver	*DWI or *394
Roadside assistance	*949
State patrol	*12
Turnpike emergency	*11

Receiving calls

Caller dials the roamer access number below,
hears tone, dials cellular area code and number.
. . . .   717-574-7626

Follow Me Roaming forwarding (see p. 1156)
Activate, dial *18 send. Deactivate dial *19 send.

Billing information

Visitor rates
Most visitors . . . . . $3 day + 99¢ minute

Visitors without roaming agreements (p. 1159)
Cannot call since company takes no credit cards.

Identification numbers

City	Market	System	Billing	RSA
Harrisburg	084	00096	00000	MSA
Juaniata County	622	01632	00000	PA-11
Lancaster	105	00096	00000	MSA
Lebanon	623	00096	30544	PA-12
York	099	00096	00000	MSA

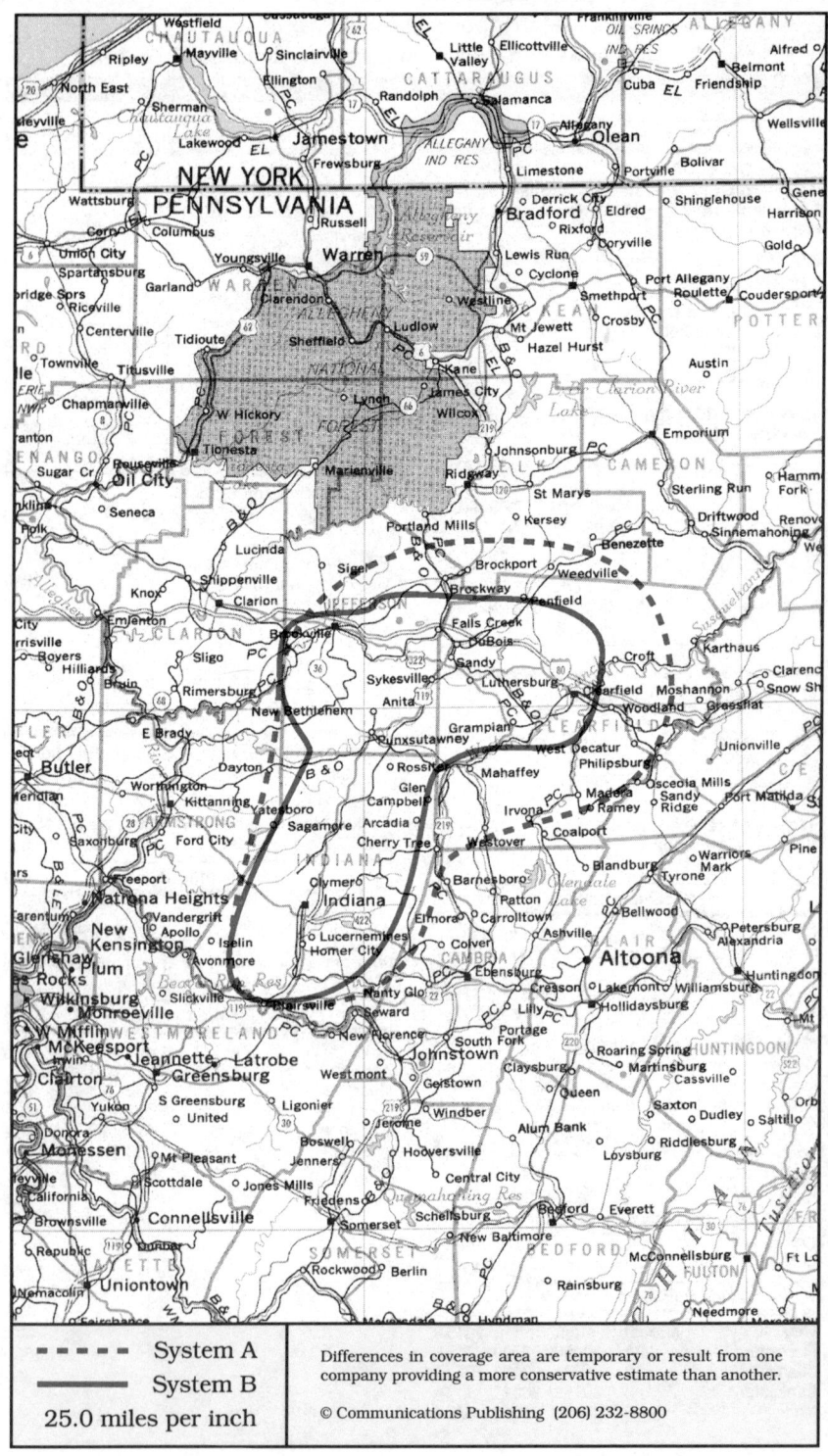

- - - - - System A
——— System B

25.0 miles per inch

Indiana, Pennsylvania
Brookville, Clearfield, Du Bois, Indiana

System A

Cellular company

Cellular One
A.M.C. Cellular Associates
1185 Wayne Avenue, PO Box 323
Indiana, PA 15701

Corporate office	800-552-8325
. . . .	412-357-8020

Sending calls

Visitor dialing instructions

Local calls	area code + 7 digits
Long distance . . .	1 + area code + 7 digits

Speed-dial numbers

Customer service	611
Emergency	911

Receiving calls

Caller dials the roamer access number below,
hears tone, dials cellular area code and number.

Du Bois	814-541-7626
Indiana	412-541-7626

NationLink & RoamingAmerica (see p. 1157)
Caller hears your current roamer access number.
This company's subscribers have level 0 service.

Billing information

Visitor rates

Most visitors	$2 day + 75¢ minute

Visitors without roaming agreements (p. 1159)
Roamer Plus will intercept first call to bill you.

Identification numbers

City	Market	System	Billing	RSA
All served cities	618	01623	00000	PA-7

System B

Cellular company

Bell Atlantic Mobile
6200 Babcock Boulevard
Pittsburgh, PA 15237

2895 Banksville Road
Pittsburgh, PA 15216

Customer service		800-922-0204
Local office	Babcock Boulevard	412-369-8500
. . . . .	Banksville Road	412-571-3300
Corporate office		908-306-7000

Sending calls

Visitor dialing instructions

Local calls	area code + 7 digits
Long distance . . .	1 + area code + 7 digits

Speed-dial numbers

Customer service	*611
Emergency	911
Hear a sports report	412-555-9823
Hear a traffic report	412-555-1250

Receiving calls

Caller dials the roamer access number below,
hears tone, dials cellular area code and number.

. . . .	814-590-7626

Follow Me Roaming forwarding (see p. 1156)
Activate, dial *18 send. Deactivate dial *19 send.

Billing information

Visitor rates

Most visitors	$3 day + $1 minute

Visitors without roaming agreements (p. 1159)
Roamer Plus will intercept first call to bill you.

Identification numbers

City	Market	System	Billing	RSA
All served cities	618	00032	30102	PA-7

For full coverage area, see Pittsburgh, PA.

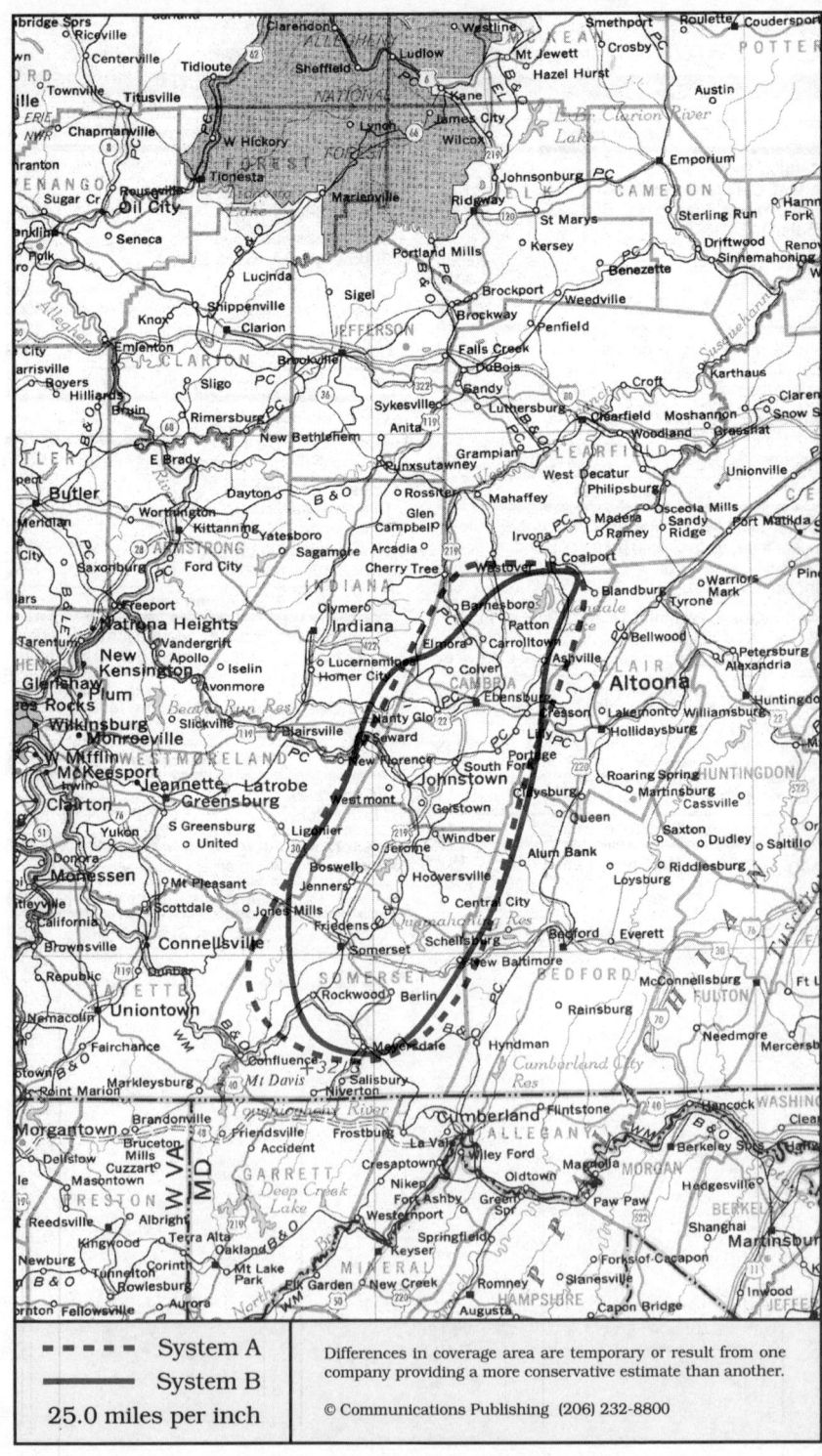

- - - - System A	Differences in coverage area are temporary or result from one
———— System B	company providing a more conservative estimate than another.
25.0 miles per inch	© Communications Publishing (206) 232-8800

Johnstown, Pennsylvania
Ebensburg, Johnstown, Somerset

System A	System B

Cellular company

Cellular One
McCaw Cellular Communications
903 Scalp Avenue, Suite 70
Johnstown, PA 15904

Customer service	800-477-3922
. . . .	814-242-0100
Local office	814-269-9500
Corporate office	206-827-4500

Sending calls

Visitor dialing instructions

Local calls	area code + 7 digits
Long distance . . .	1 + area code + 7 digits

Speed-dial numbers

Customer service	611
Emergency	911
Pennsylvania AAA	412-362-1900
Turnpike emergency	*11

Receiving calls

Caller dials the roamer access number below,
hears tone, dials cellular area code and number.
. . . . 814-242-7626

NationLink & RoamingAmerica (see p. 1157)
Dial *31 send to receive calls, *30 to deactivate.
Dial *32 send to tell caller the roamer access no.
This company's subscribers have level 8 service.

North American Cellular Network (see p. 1156)
All calls forward to you. Deactivate dial *35 send.

Billing information

Visitor rates
Most visitors . . . . . $0 day + 99¢ minute

Visitors without roaming agreements (p. 1159)
Roamer Plus will intercept first call to bill you.

Identification numbers

City	Market	System	Billing	RSA
All served cities	143	00039	30051	MSA

For full coverage area, see Pittsburgh, PA.

Cellular company

Independent Cellular Network
1405 Eisenhower Boulevard
Johnstown, PA 15904

Local office	814-266-1616
Corporate office	813-489-1600

Sending calls

Visitor dialing instructions

Local calls	area code + 7 digits
Long distance . . .	1 + area code + 7 digits

Speed-dial numbers

Customer service	*611
Emergency	911

Receiving calls

Caller dials the roamer access number below,
hears tone, dials cellular area code and number.
. . . . 814-931-7626

Follow Me Roaming forwarding (see p. 1156)
Activate, dial *18 send. Deactivate dial *19 send.

Billing information

Visitor rates
Most visitors . . . . . $3 day + $1 minute

Visitors without roaming agreements (p. 1159)
Cannot call since company takes no credit cards.

Identification numbers

City	Market	System	Billing	RSA
All served cities	143	00032	30026	MSA

For full coverage area, see Altoona, PA.

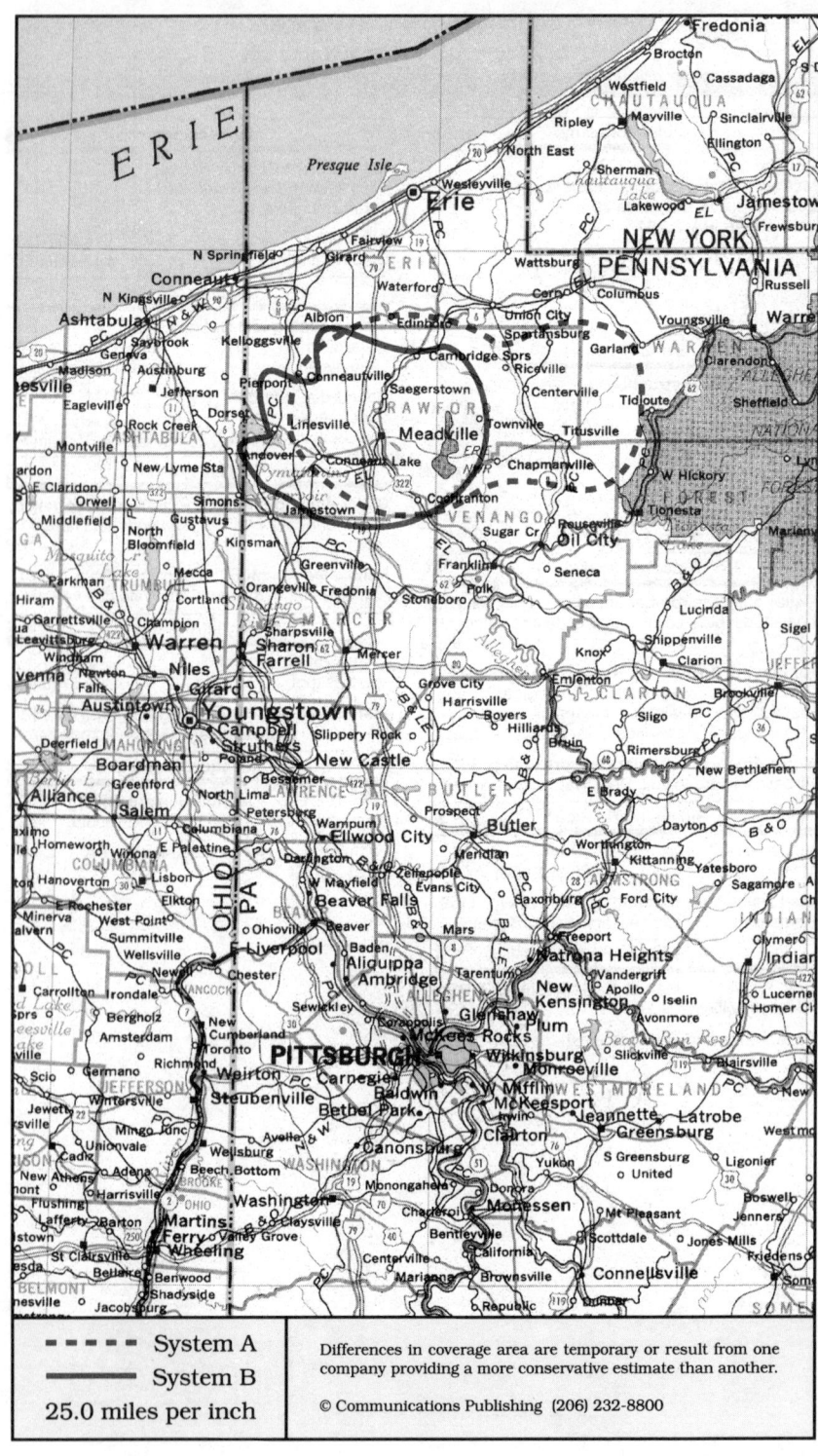

- - - - - System A	Differences in coverage area are temporary or result from one
——— System B	company providing a more conservative estimate than another.
25.0 miles per inch	© Communications Publishing (206) 232-8800

Meadville, Pennsylvania
Meadville, Titusville

Cellular company

Cellular One
1387 Conneaut Lake Road, Suite 1
Meadville, PA 16335

May be purchased by Horizon Cellular.

Local office 814-724-6666

Sending calls

Visitor dialing instructions
Local calls . . . . . area code + 7 digits
Long distance . . . 1 + area code + 7 digits

Speed-dial numbers
Customer service 611
Emergency 911

Receiving calls

Caller dials the roamer access number below,
hears tone, dials cellular area code and number.
. . . . 814-724-9222

NationLink & RoamingAmerica (see p. 1157)
Caller hears your current roamer access number.
This company's subscribers have level 0 service.

North American Cellular Network (see p. 1156)
All calls forward to you. Deactivate dial *35 send.

Billing information

Visitor rates
Most visitors . . . . . $0 day + 99¢ minute

Visitors without roaming agreements (p. 1159)
Roamer Plus will intercept first call to bill you.

Identification numbers

City	Market	System	Billing	RSA
All served cities	612	01611	00000	PA-1

Call switching is performed by McCaw Cellular as
part of the Pittsburgh, PA system for the licensee.

Cellular company

Centel Cellular
7206 Market Street
Youngstown, OH 44512

Local office 800-325-5190
. . . . 216-758-4502
Corporate office 312-399-2644

Sending calls

Visitor dialing instructions
Local calls . . . . . . area code + 7 digits
Long distance . . . 1 + area code + 7 digits

Speed-dial numbers
Customer service *611

Receiving calls

Caller dials the roamer access number below,
hears tone, dials cellular area code and number.
. . . . 412-699-7626

Billing information

Visitor rates
Most visitors . . . . . $3 day + 99¢ minute

Visitors without roaming agreements (p. 1159)
Cannot call since company takes no credit cards.

Identification numbers

City	Market	System	Billing	RSA
All served cities	612	01612	00000	PA-1

For full coverage area, see Youngstown, OH.

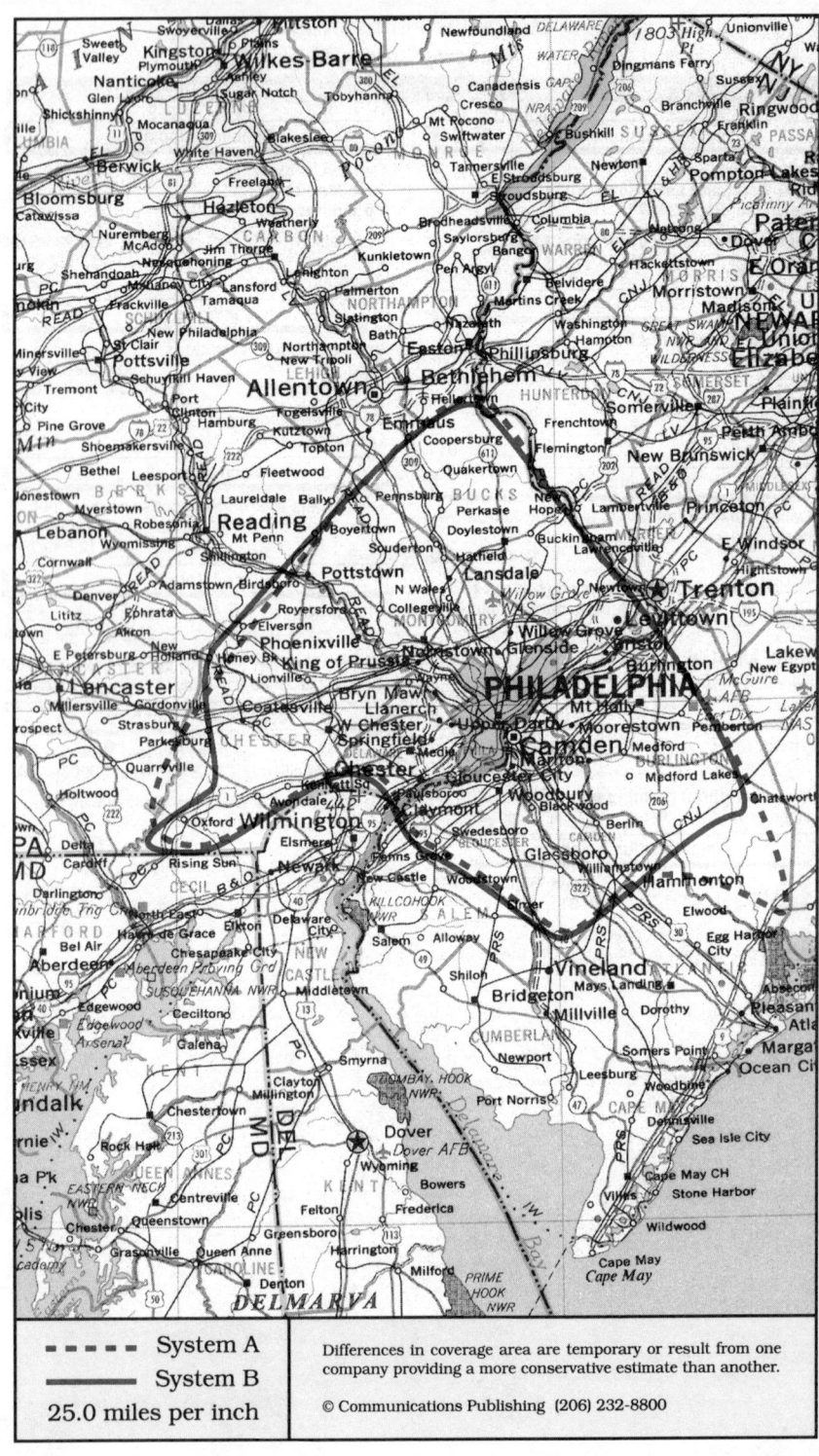

▪ ▪ ▪ ▪ ▪ System A ━━━━━ System B **25.0 miles per inch**	Differences in coverage area are temporary or result from one company providing a more conservative estimate than another. © Communications Publishing (206) 232-8800

Philadelphia, Pennsylvania

Camden NJ, Doylestown, Lansdale, Philadelphia, Pottstown, West Chester

System A	System B

Cellular company

Comcast Metrophone
480 E Swedesford Road
Wayne, PA 19087-1867

Customer service	800-234-9666
Corporate office	215-975-5000

Sending calls

Visitor dialing instructions

Local calls	area code + 7 digits
Long distance	area code + 7 digits

Speed-dial numbers

AAA road service	*222
AAA weather and road conditions	215-350-4636
Cellular information, movies, events . .	*123
Customer service	*611
Donnelly Directory, news & entertainment	*12
Emergency	*911
Instant traffic reports	*22
KYW news radio to report traffic problems	*99
KYW television	*599
Sports channel	*610
Turnpike emergency	*11

Receiving calls

Caller dials the roamer access number below, hears tone, dials cellular area code and number.
. . . . 215-350-7626

NationLink & RoamingAmerica (see p. 1157)
Caller hears your current roamer access number.
This company's subscribers have level 0 service.

North American Cellular Network (see p. 1156)
All calls forward to you. Deactivate dial *35 send.

Billing information

Visitor rates

Most visitors	$3 day + 99¢ minute

Visitors without roaming agreements (p. 1159)
Roamer Plus will intercept first call to bill you.

Identification numbers

City	Market	System	Billing	RSA
All served cities	004	00029	00000	MSA

Cellular company

Bell Atlantic Mobile
1301 Bristol Pike
Bensalem, PA 19020

1211 Route 73
Mount Laurel, PA 08054

155 N 22nd Street
Philadelphia, PA 19103

110 E Swedesford Road
Wayne, PA 19087

Customer service		800-922-0204
Local office	Bensalem	215-639-2288
. . . . . .	Mount Laurel	609-234-6020
. . . . . .	Philadelphia	215-751-9497
. .	Wayne (King of Prussia)	215-971-1600
Corporate office		908-306-7000

Sending calls

Visitor dialing instructions

Local calls	area code + 7 digits
Long distance	area code + 7 digits

Speed-dial numbers

Customer service	*611
Emergency	911
Sports channel	*610
Traffic reports and news	*1350
Turnpike emergency	*11

Receiving calls

Caller dials the roamer access number below, hears tone, dials cellular area code and number.
. . . . 215-870-7626

Follow Me Roaming forwarding (see p. 1156)
Activate, dial *18 send. Deactivate dial *19 send.

Billing information

Visitor rates

Most visitors	$3 day + $1 minute

Visitors without roaming agreements (p. 1159)
Roamer Plus will intercept first call to bill you.

Identification numbers

City	Market	System	Billing	RSA
All served cities	004	00008	00000	MSA

For full coverage area, see Allentown, PA and
Wilmington, DE.

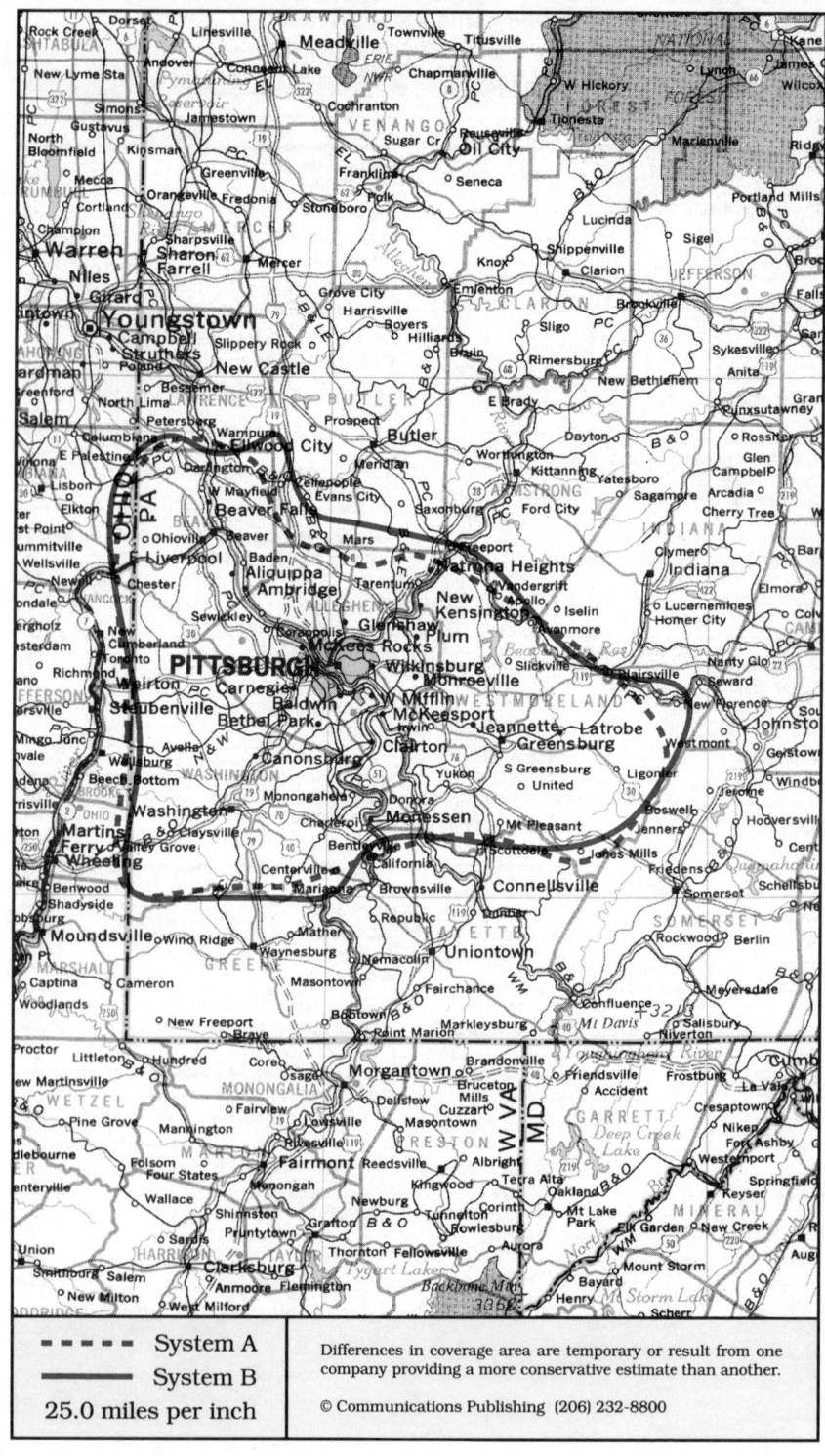

Pittsburgh, Pennsylvania

Beaver Falls, Greensburg, Monroeville, New Kensington, Pittsburgh, Washington

System A	System B

Cellular company

Cellular One
McCaw Cellular Communications
2630 Liberty Avenue
Pittsburgh, PA 15222

Customer service	800-477-3922
. . . . .	412-471-3922
Local office	412-471-4400
Corporate office	206-827-4500

Sending calls

Visitor dialing instructions

Local calls	area code + 7 digits
Long distance . . .	1 + area code + 7 digits

Speed-dial numbers

Customer service	611
Emergency	911
KDKA radio (free)	412-261-1020
Pennsylvania AAA (free) . .	412-362-1900
Turnpike emergency	*11

Receiving calls

Caller dials the roamer access number below,
hears tone, dials cellular area code and number.
. . . . 412-298-7626

NationLink & RoamingAmerica (see p. 1157)
Dial *31 send to receive calls, *30 to deactivate.
Dial *32 send to tell caller the roamer access no.
This company's subscribers have level 8 service.

North American Cellular Network (see p. 1156)
All calls forward to you. Deactivate dial *35 send.

Billing information

Visitor rates

Most visitors	$0 day + 99¢ minute

Visitors without roaming agreements (p. 1159)
Roamer Plus will intercept first call to bill you.

Identification numbers

City	Market	System	Billing	RSA
All served cities	013	00039	00000	MSA

For full coverage area, see Johnstown, PA and
Wheeling, WV.

Cellular company

Bell Atlantic Mobile
6200 Babcock Boulevard
Pittsburgh, PA 15237

2895 Banksville Road
Pittsburgh, PA 15216

Customer service		800-922-0204
Local office	Babcock Boulevard	412-369-8500
. . . . .	Banksville Road	412-571-3300
Corporate office		908-306-7000

Sending calls

Visitor dialing instructions

Local calls	area code + 7 digits
Long distance . . .	1 + area code + 7 digits

Speed-dial numbers

Customer service	*611
Emergency	911
Hear a sports report	412-555-9823
Hear a traffic report	412-555-1250

Receiving calls

Caller dials the roamer access number below,
hears tone, dials cellular area code and number.
. . . . 412-855-7626

Follow Me Roaming forwarding (see p. 1156)
Activate, dial *18 send. Deactivate dial *19 send.

Billing information

Visitor rates

Most visitors	$3 day + $1 minute

Visitors without roaming agreements (p. 1159)
Roamer Plus will intercept first call to bill you.

Identification numbers

City	Market	System	Billing	RSA
All served cities	013	00032	00000	MSA

For full coverage area, see Indiana, PA and New
Castle, PA.

▪ ▪ ▪ ▪ System A	Differences in coverage area are temporary or result from one company providing a more conservative estimate than another.
——— System B	
25.0 miles per inch	© Communications Publishing (206) 232-8800

Scranton, Pennsylvania

Honesdale (Northeast PA), Scranton, Stroudsburg, Tunkhannock, Wilkes-Barre

System A	System B

Cellular company

Cellular One
Vanguard Cellular Systems
Route 6, Siniawa Plaza
Scranton, PA 18508

1250 N 9th, Suite 4
Stroudsburg, PA 18360

277 Mundy Street
Wilkes-Barre, PA 18702

Customer service		717-650-3100
. . . . . from Honesdale		717-470-4000
Local office . . .	Scranton	717-961-2355
. . . . . .	Stroudsburg	717-424-2355
. . . . . .	Wilkes-Barre	717-825-2355
Corporate office		919-282-3690

Sending calls

Visitor dialing instructions
Local calls . . . . . area code + 7 digits
Long distance . . . area code + 7 digits

Speed-dial numbers
Customer service 611
Emergency 911

Receiving calls

Caller dials the roamer access number below,
hears tone, dials cellular area code and number.
Except Honesdale 717-650-7626
Honesdale 717-470-7626

NationLink & RoamingAmerica (see p. 1157)
Dial ∗31 send to receive calls, ∗30 to deactivate.
Dial ∗32 send to tell caller the roamer access no.
This company's subscribers have level 6 service.

Billing information

Visitor rates
Most visitors . . . . . $3 day + 99¢ minute

Visitors without roaming agreements (p. 1159)
Call co. to use Amex, Mastercard, Visa
or Roamer Plus will intercept first call to bill you.

Identification numbers

City	Market	System	Billing	RSA
Except Honesdale	056	00103	30103	MSA
Honesdale	616	01619	00000	PA-5

For full coverage area, see Allentown, PA.

Cellular company

Cellular Plus
1400 Spruce Street
Avoca, PA 18641

Customer service	some areas	800-777-7587
. . . .		717-654-7587
Corporate office		717-883-8832

Sending calls

Visitor dialing instructions
Local calls . . . . . area code + 7 digits
Long distance . . . . . area code + 7 digits

Speed-dial numbers
Customer service 611
Emergency 911

Receiving calls

Caller dials the roamer access number below,
hears tone, dials cellular area code and number.
. . . . 717-881-7626

Billing information

Visitor rates
Most visitors . . . . . $3 day + $1 minute

Visitors without roaming agreements (p. 1159)
Cannot call since company takes no credit cards.

Identification numbers

City	Market	System	Billing	RSA
Honesdale	616	00172	30288	PA-5
Scrantonx	056	00172	30288	MSA
Tunkhannock	615	00172	00000	PA-4

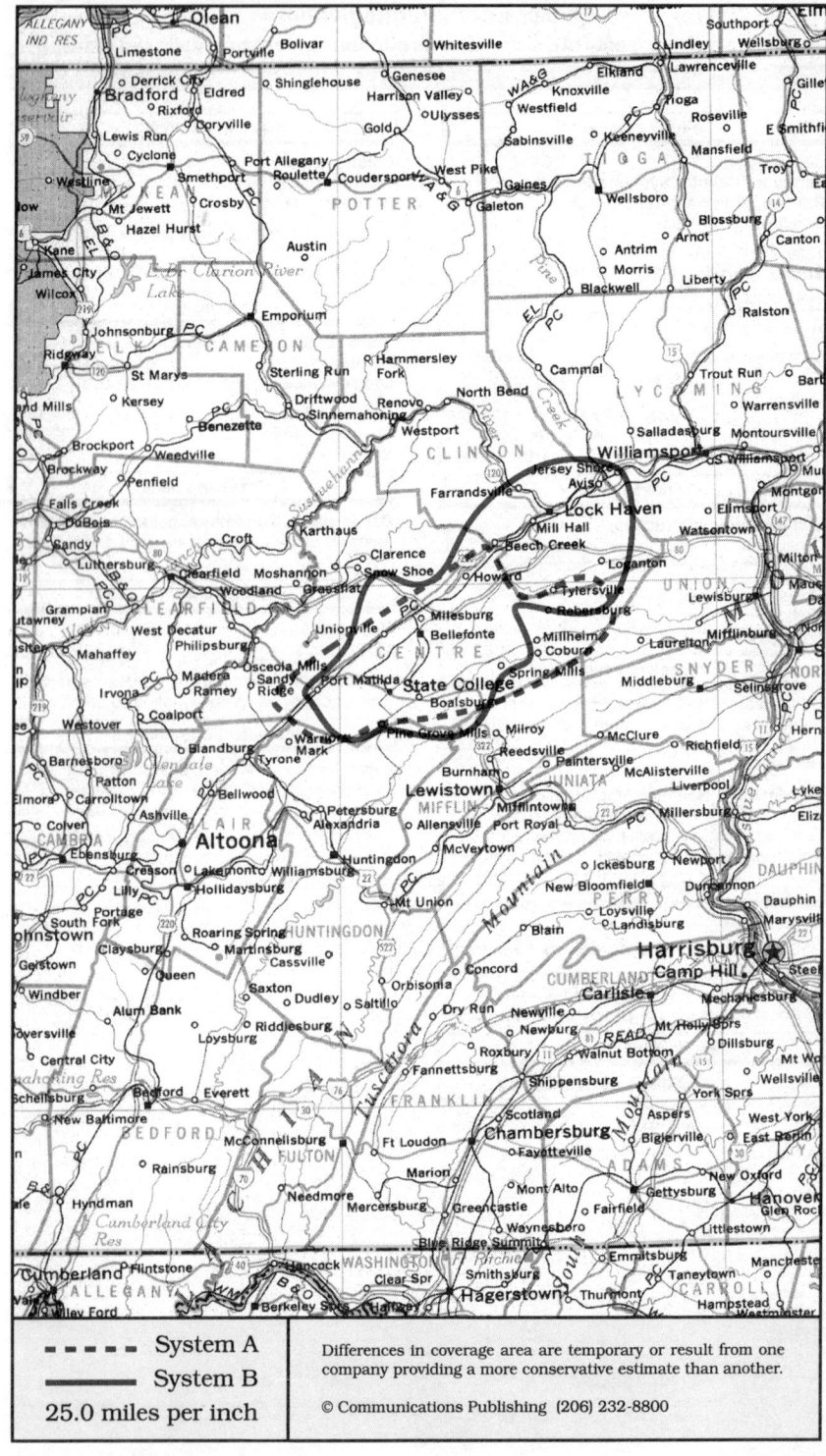

▬ ▬ ▬ ▬ System A	Differences in coverage area are temporary or result from one company providing a more conservative estimate than another.
──────── System B	
25.0 miles per inch	© Communications Publishing (206) 232-8800

State College, Pennsylvania
Bellefonte, Lock Haven, Port Matilda, State College

System A	System B

Cellular company

Cellular One
Vanguard Cellular Systems
2601 E College Avenue
State College, PA 16801

Customer service	814-880-4000
Local office	814-237-2355
Corporate office	919-282-3690

Sending calls

Visitor dialing instructions

Local calls	area code + 7 digits
Long distance	area code + 7 digits

Speed-dial numbers

Customer service	*611
Emergency	911

Receiving calls

Caller dials the roamer access number below, hears tone, dials cellular area code and number.
. . . . 814-880-7626

NationLink & RoamingAmerica (see p. 1157)
Dial *31 send to receive calls, *30 to deactivate.
Dial *32 send to tell caller the roamer access no.
This company's subscribers have level 6 service.

Billing information

Visitor rates
Most visitors . . . . . $3 day + 99¢ minute

Visitors without roaming agreements (p. 1159)
Call co. to use Amex, Mastercard, Visa

Identification numbers

City	Market	System	Billing	RSA
All served cities	259	00159	30019	MSA

This system is operated as part of Harrisburg, PA.

Service in PA-3 will be provided by Gaia, Inc.

Cellular company

Cellular Plus
2330 Commercial Boulevard, Suite 300
State College, PA 16801

Local office . . . some areas	800-877-7587
. . . .	814-231-3900
Corporate office	717-883-8832

Sending calls

Visitor dialing instructions

Local calls	area code + 7 digits
Long distance . . .	1 + area code + 7 digits

Speed-dial numbers

Customer service	611
Emergency	911

Receiving calls

Caller dials the roamer access number below, hears tone, dials cellular area code and number.
. . . . 814-571-7626

Follow Me Roaming forwarding (see p. 1156)
Activate, dial *18 send. Deactivate dial *19 send.

Billing information

Visitor rates
Most visitors . . . . . $3 day + $1 minute

Visitors without roaming agreements (p. 1159)
Call co. to use Mastercard, Visa

Identification numbers

City	Market	System	Billing	RSA
Lock Haven	614	01616	00000	PA-3
State College	259	00032	30034	MSA

This system is operated as part of Pittsburgh, PA.

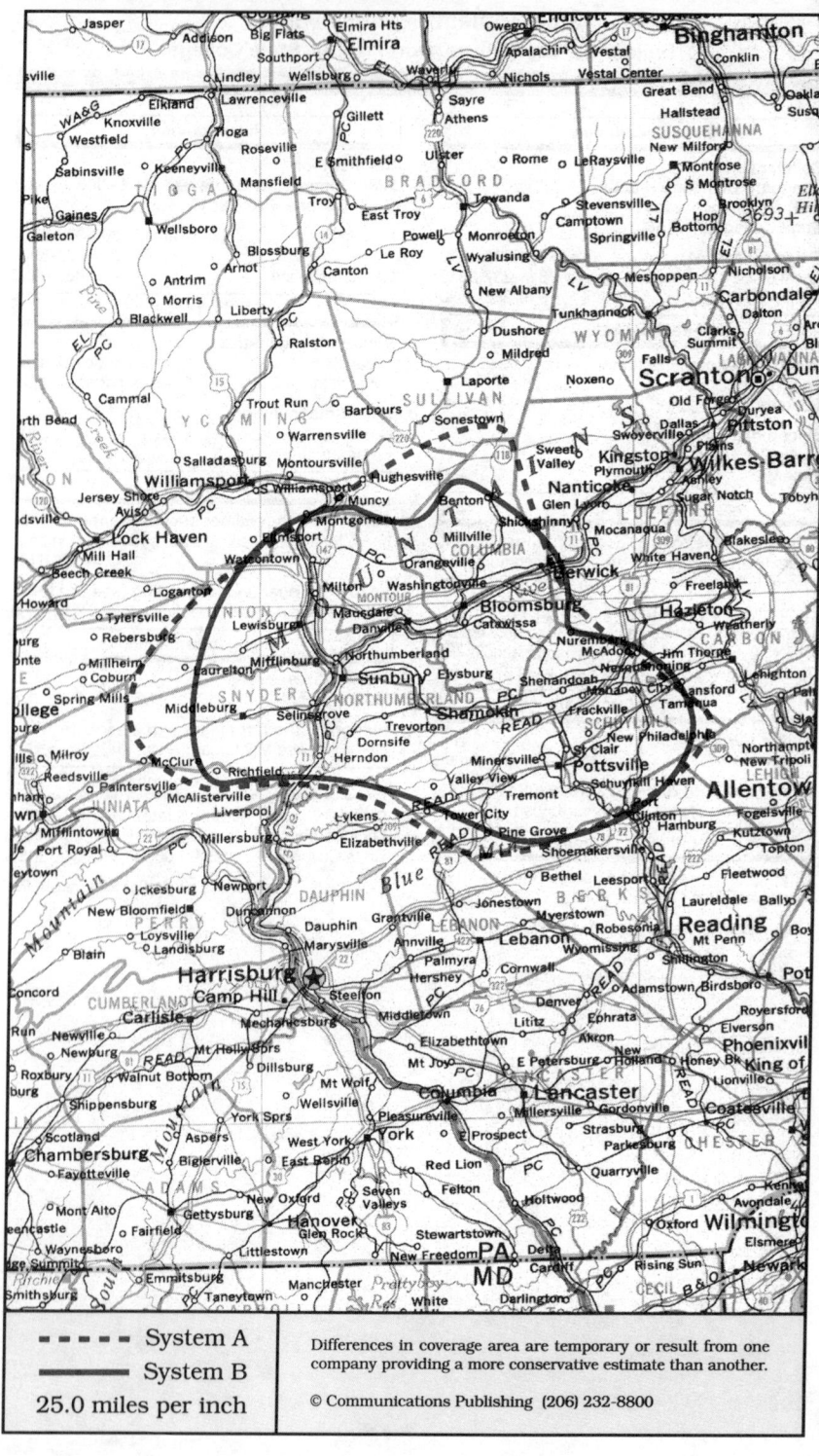

Sunbury, Pennsylvania

Bloomsburg, Middleburg, Pottsville, Shamokin, Sunbury

System A	System B

Cellular company

**Cellular One
Sunshine Cellular
Managed by: MCMG
R. R. 1, Box 3A, Route 15
Winfield, PA 17889-9799**

Local office	800-487-2355
. . . .	717-524-2351
Corporate office . .	MCMG 615-791-0202

Sending calls

Visitor dialing instructions

Local calls	area code + 7 digits
Long distance . . .	1 + area code + 7 digits

Speed-dial numbers

AAA service	*AAA or *222
Customer service	611
Emergency	911

Receiving calls

Caller dials the roamer access number below,
hears tone, dials cellular area code and number.

Either covers all cities . . .	717-850-7626
Either covers all cities . . .	717-640-7626

Billing information

Visitor rates

Most visitors	$3 day + 99¢ minute

Visitors without roaming agreements (p. 1159)
Call co. to use Amex, Mastercard, Visa
or Roamer Plus will intercept first call to bill you.

Identification numbers

City	Market	System	Billing	RSA
All served cities	619	01625	00000	PA-8

Cellular company

**United States Cellular
2062 Lycoming Creek Road
Williamsport, PA 17701-1192**

Local office	717-321-9500
Corporate office	312-399-8900

Sending calls

Visitor dialing instructions

Local calls	area code + 7 digits
Long distance	area code + 7 digits

Speed-dial numbers

Customer service	611
DIrectory assistance	411
Emergency	911

Receiving calls

Caller dials the roamer access number below,
hears tone, dials cellular area code and number.

. . . .	717-490-7626

Billing information

Visitor rates

Most visitors	$3 day + 99¢ minute

Visitors without roaming agreements (p. 1159)
Call co. to use Mastercard, Visa

Identification numbers

City	Market	System	Billing	RSA
All served cities	619	01626	00000	PA-8

For full coverage area, see Williamsport, PA.

▬ ▬ ▬ ▬ ▬ System A	Differences in coverage area are temporary or result from one company providing a more conservative estimate than another.
▬▬▬▬▬ System B	
25.0 miles per inch	© Communications Publishing (206) 232-8800

Uniontown, Pennsylvania
Connellsville, Uniontown, Waynesburg

System A	System B

System A

Cellular company

United States Cellular
663 Pittsburgh Street
Uniontown, PA 15401

Local office	800-944-9066
. . . .	412-437-0800
Corporate office	312-399-8900

Sending calls

Visitor dialing instructions

Local calls	area code + 7 digits
Long distance . . .	1 + area code + 7 digits

Speed-dial numbers

Customer service	611
Emergency	911

Receiving calls

Caller dials the roamer access number below, hears tone, dials cellular area code and number.
. . . . 412-557-7626

Billing information

Visitor rates

Most visitors . . . . . $3 day + 85¢ minute

Visitors without roaming agreements (p. 1159)
Cannot call since company takes no credit cards.

Identification numbers

City	Market	System	Billing	RSA
All served cities	620	01627	00000	PA-9

System B

Cellular company

Bell Atlantic Mobile
6200 Babcock Boulevard
Pittsburgh, PA 15237

2895 Banksville Road
Pittsburgh, PA 15216

Customer service		800-922-0204
Local office	Babcock Boulevard	412-369-8500
. . . .	Banksville Road	412-571-3300
Corporate office		908-306-7000

Sending calls

Visitor dialing instructions

Local calls	area code + 7 digits
Long distance . . .	1 + area code + 7 digits

Speed-dial numbers

Customer service	*611
Emergency	911
Hear a sports report	412-555-9823
Hear a traffic report	412-555-1250

Receiving calls

Caller dials the roamer access number below, hears tone, dials cellular area code and number.
. . . . 412-582-7626

Follow Me Roaming forwarding (see p. 1156)
Activate, dial *18 send. Deactivate dial *19 send.

Billing information

Visitor rates

Most visitors . . . . . $3 day + $1 minute

Visitors without roaming agreements (p. 1159)
Roamer Plus will intercept first call to bill you.

Identification numbers

City	Market	System	Billing	RSA
All served cities	620	00032	30104	PA-9

For full coverage area, see Pittsburgh, PA.

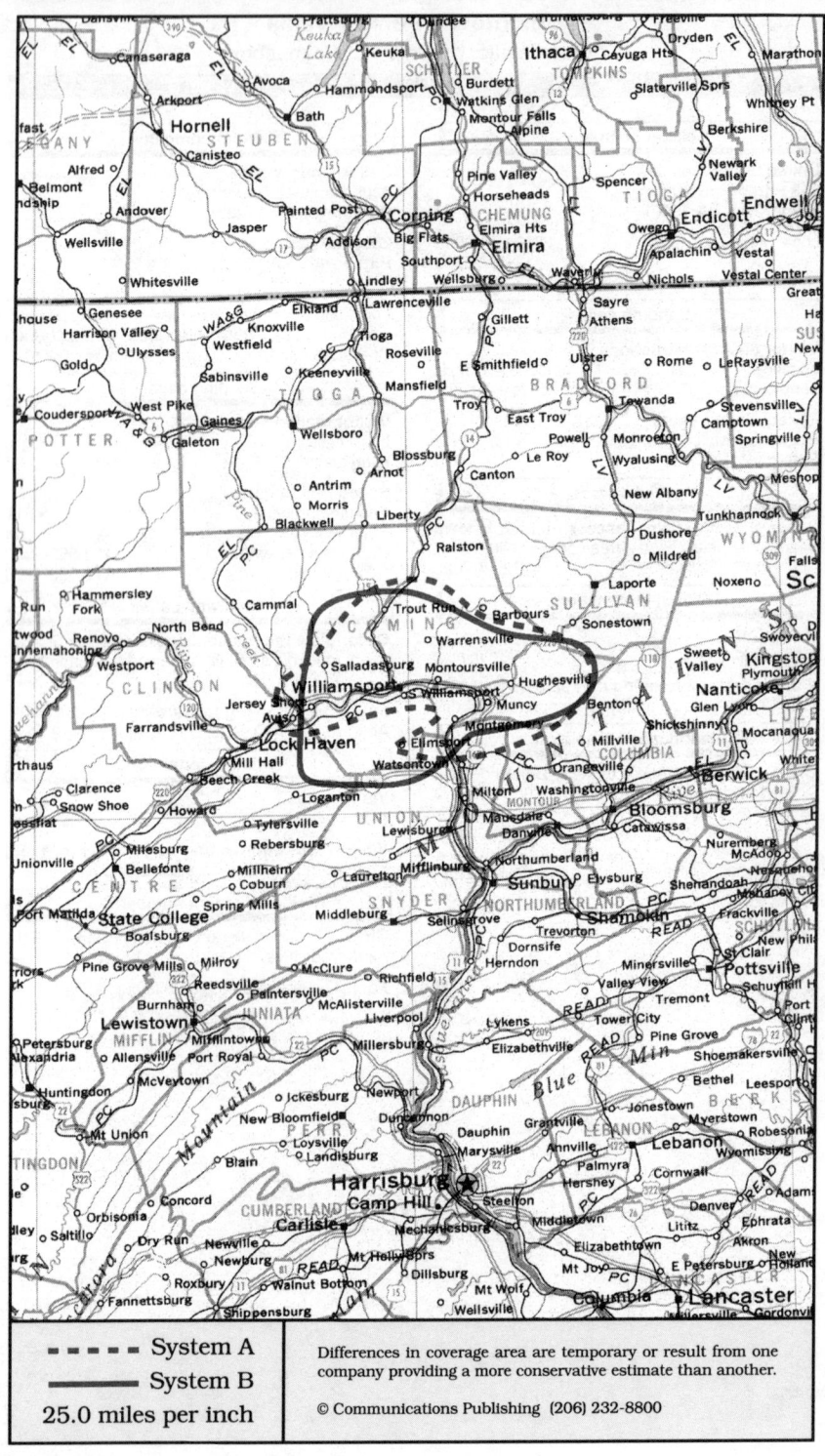

Williamsport, Pennsylvania

System A

Cellular company

Cellular One
Vanguard Cellular Systems
2495 E 3rd Street
Williamsport, PA 17701

Customer service	717-971-5100
Local office	717-326-2355
Corporate office	919-282-3690

Sending calls

Visitor dialing instructions

Local calls	area code + 7 digits
Long distance	area code + 7 digits

Speed-dial numbers

Customer service	611
Emergency	911
WYOU television	*22

Receiving calls

Caller dials the roamer access number below,
hears tone, dials cellular area code and number.
. . . . 717-971-7626

NationLink & RoamingAmerica (see p. 1157)
Dial *31 send to receive calls, *30 to deactivate.
Dial *32 send to tell caller the roamer access no.
This company's subscribers have level 6 service.

Billing information

Visitor rates
Most visitors . . . . . $3 day + 99¢ minute

Visitors without roaming agreements (p. 1159)
Call co. to use Amex, Mastercard, Visa

Identification numbers

City	Market	System	Billing	RSA
All served cities	251	00597	00000	MSA

System B

Cellular company

United States Cellular
2062 Lycoming Creek Road
Williamsport, PA 17701-1192

Local office	717-321-9500
Corporate office	312-399-8900

Sending calls

Visitor dialing instructions

Local calls	area code + 7 digits
Long distance	area code + 7 digits

Speed-dial numbers

Customer service	611
DIrectory assistance	411
Emergency	911

Receiving calls

Caller dials the roamer access number below,
hears tone, dials cellular area code and number.
. . . . 717-220-7626

Follow Me Roaming forwarding (see p. 1156)
Activate, dial *18 send. Deactivate dial *19 send.

Billing information

Visitor rates
Most visitors . . . . . $3 day + 99¢ minute

Visitors without roaming agreements (p. 1159)
Call co. to use Mastercard, Visa

Identification numbers

City	Market	System	Billing	RSA
All served cities	251	00576	00000	MSA

For full coverage area, see Sunbury, PA.

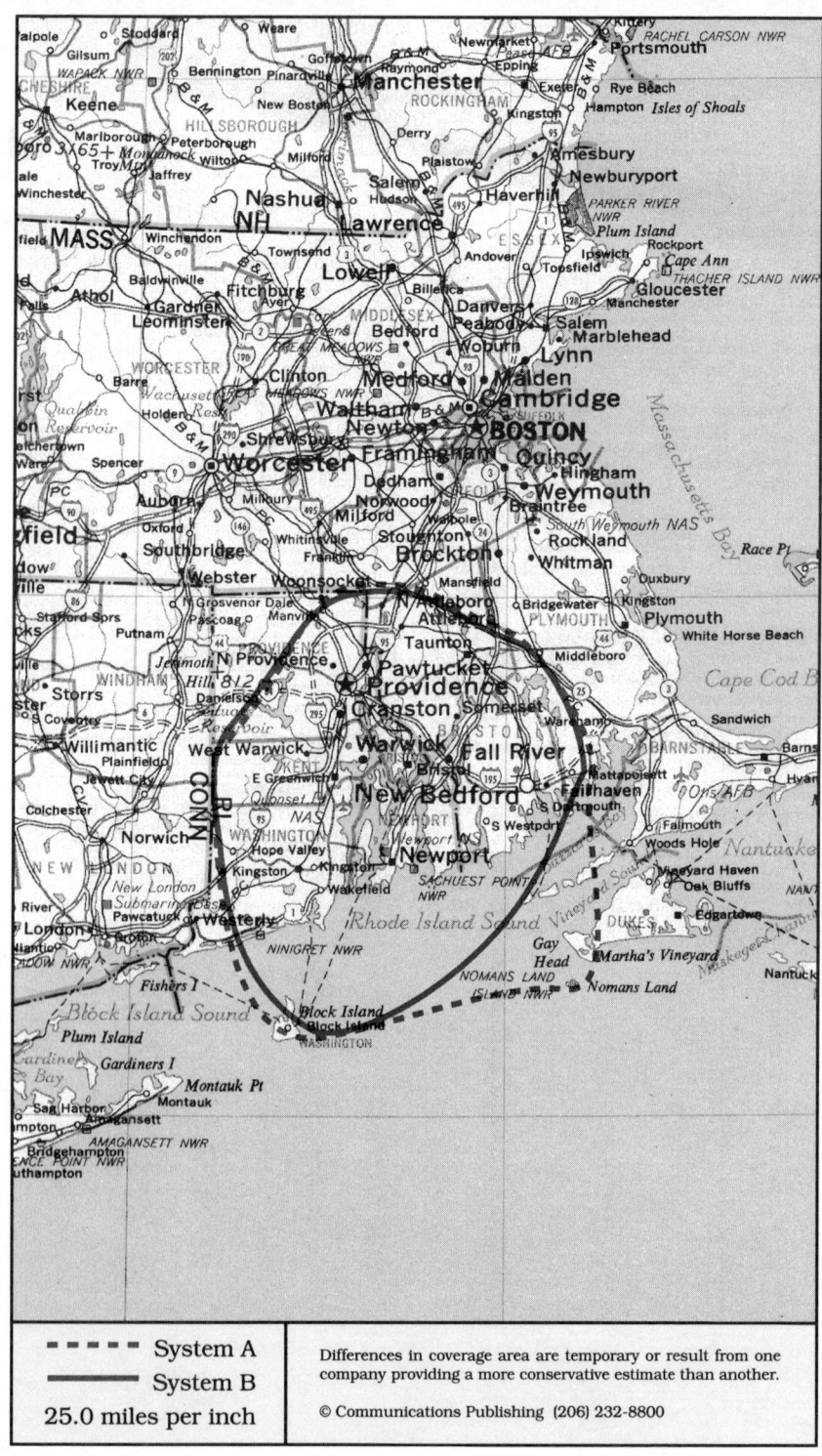

Providence, Rhode Island
New Bedford MA, Newport, Providence, Woonsocket

System A	System B

Cellular company

Bell Atlantic Mobile
1 Franklin Square
Providence, RI 02903

Customer service 800-852-3630
Local office 401-272-3800
Corporate office 908-306-7000

Sending calls

Visitor dialing instructions
Local calls . . . . . area code + 7 digits
Long distance . . . . area code + 7 digits

Speed-dial numbers
Customer service *611
Emergency 911
Information services *22
Make a traffic report *104
Roaming information *711
State police *677

Receiving calls

Caller dials the roamer access number below,
hears tone, dials cellular area code and number.
New Bedford, MA 508-971-7626
Newport, Providence 401-523-7626

NationLink & RoamingAmerica (see p. 1157)
Caller hears your current roamer access number.
This company's subscribers have level 0 service.

Billing information

Visitor rates
Most visitors . . . . . $3 day + 99¢ minute

Visitors without roaming agreements (p. 1159)
Roamer Plus will intercept first call to bill you.

Identification numbers

City	Market	System	Billing	RSA
New Bedford	076	00119	00000	MSA
Newport	624	01635	00000	RI-1
Providence	038	00119	00000	MSA

For full coverage area, see Bridgeport, CT.

Cellular company

NYNEX Mobile Communications
26 Ship Street
Providence, RI 02903

Customer service 800-538-4747
. . . . 617-932-1200
Local office 401-751-4070
Corporate office 914-365-7200

Sending calls

Visitor dialing instructions
Local calls . . . . . . area code + 7 digits
Long distance . . . . . area code + 7 digits

Speed-dial numbers
American Automobile Association . . . *82
Customer service 611
Directory assistance 411
Emergency 911
Stock market update *37
Time of day . . . . *TIME or *8463
Weather report *99

Receiving calls

Caller dials the roamer access number below,
hears tone, dials cellular area code and number.
New Bedford, MA 617-571-7626
Newport, Providence 401-529-7626

Follow Me Roaming forwarding (see p. 1156)
Activate, dial *18 send. Deactivate dial *19 send.

Billing information

Visitor rates
Most visitors . . . . . $3 day + 95¢ minute

Visitors without roaming agreements (p. 1159)
Call co. to use Amex, Mastercard, Visa
or Roamer Plus will intercept first call to bill you.

Identification numbers

City	Market	System	Billing	RSA
New Bedford	076	00028	00000	MSA
Newport	624	01636	00000	RI-1
Providence	038	00028	00000	MSA

For full coverage area, see Boston, MA.

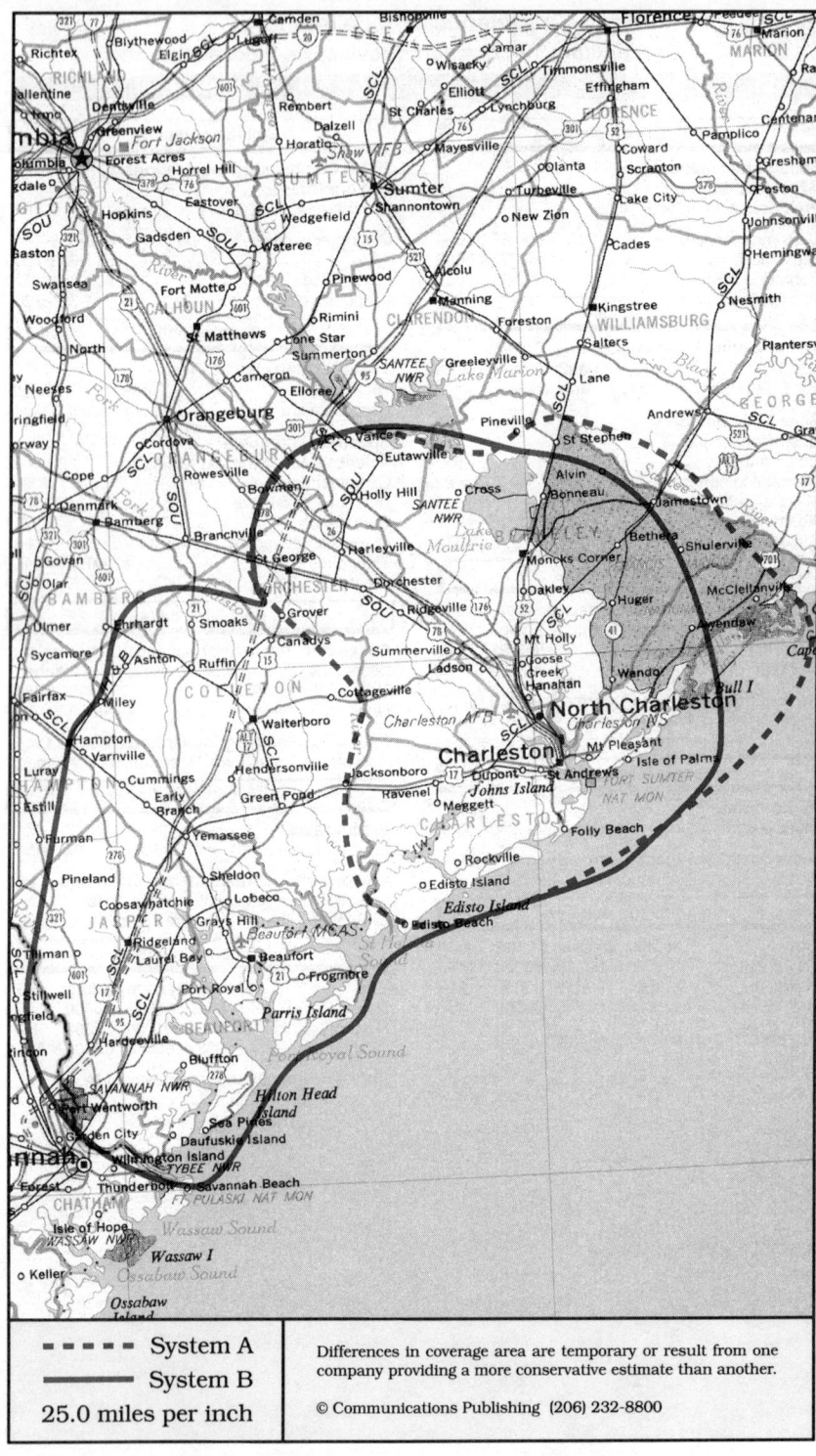

System A
System B
25.0 miles per inch

Differences in coverage area are temporary or result from one company providing a more conservative estimate than another.

© Communications Publishing (206) 232-8800

Charleston, South Carolina
Beaufort, Charleston, Hilton Head, Moncks Corner, St. George, Walterboro

System A	System B

System A

Cellular company

Cellular One
GTE Mobilnet
1941 Savage Road, Suite 300-A
West Ashley, SC 29407

Customer service	800-727-2444
. . . .	919-481-1181
Local office	803-763-6370
Corporate office	404-391-8000

Sending calls

Visitor dialing instructions

Local calls	area code + 7 digits
Long distance . . .	1 + area code + 7 digits

Speed-dial numbers

Customer service	611
Emergency	91
Highway patrol	*HP or *47
WAVE radio	*96
WDXZ radio	*104
WKCN radio	*910
WKQB radio	*7107
WXTC radio	*9982

Receiving calls

Caller dials the roamer access number below,
hears tone, dials cellular area code and number.

. . . . 803-696-7626

Follow Me Roaming forwarding (see p. 1156)
Phone Me Anywhere forwarding (see p. 1156)
Activate, dial *18 send. Deactivate dial *19 send.

NationLink & RoamingAmerica (see p. 1157)
Dial *31 send to receive calls, *30 to deactivate.
Dial *32 send to tell caller the roamer access no.
This company's subscribers have level 14 service

Billing information

Visitor rates
Most visitors . . . . . . $2 day + 50¢ minute

Visitors without roaming agreements (p. 1159)
Call co. to use Amex, Discover, Mastercard, Visa

Identification numbers

City	Market	System	Billing	RSA
All served cities	090	00127	00000	MSA

System B

Cellular company

Centel Cellular
3900 Leeds Avenue
North Charleston, SC 29405

Local office	800-745-7330
. . . .	803-744-7330
Corporate office	312-399-2644

Sending calls

Visitor dialing instructions

Local calls	area code + 7 digits
Long distance . . .	1 + area code + 7 digits

Speed-dial numbers

Coast guard	*CG or *24
Customer service	*611
Emergency	911
Highway patrol	*HP or *47

Receiving calls

Caller dials the roamer access number below,
hears tone, dials cellular area code and number.

Beaufort, Charleston	803-729-7626
Hilton Head	803-384-7626
Walterboro	803-893-7626

Follow Me Roaming forwarding (see p. 1156)
Activate, dial *18 send. Deactivate dial *19 send.

Billing information

Visitor rates
Most visitors . . . . . $3 day + 99¢ minute

Visitors without roaming agreements (p. 1159)
Cannot call since company takes no credit cards.

Identification numbers

City	Market	System	Billing	RSA
Beaufort,Walterboro	632	01652	00000	SC-8
Charleston	090	00156	00000	MSA

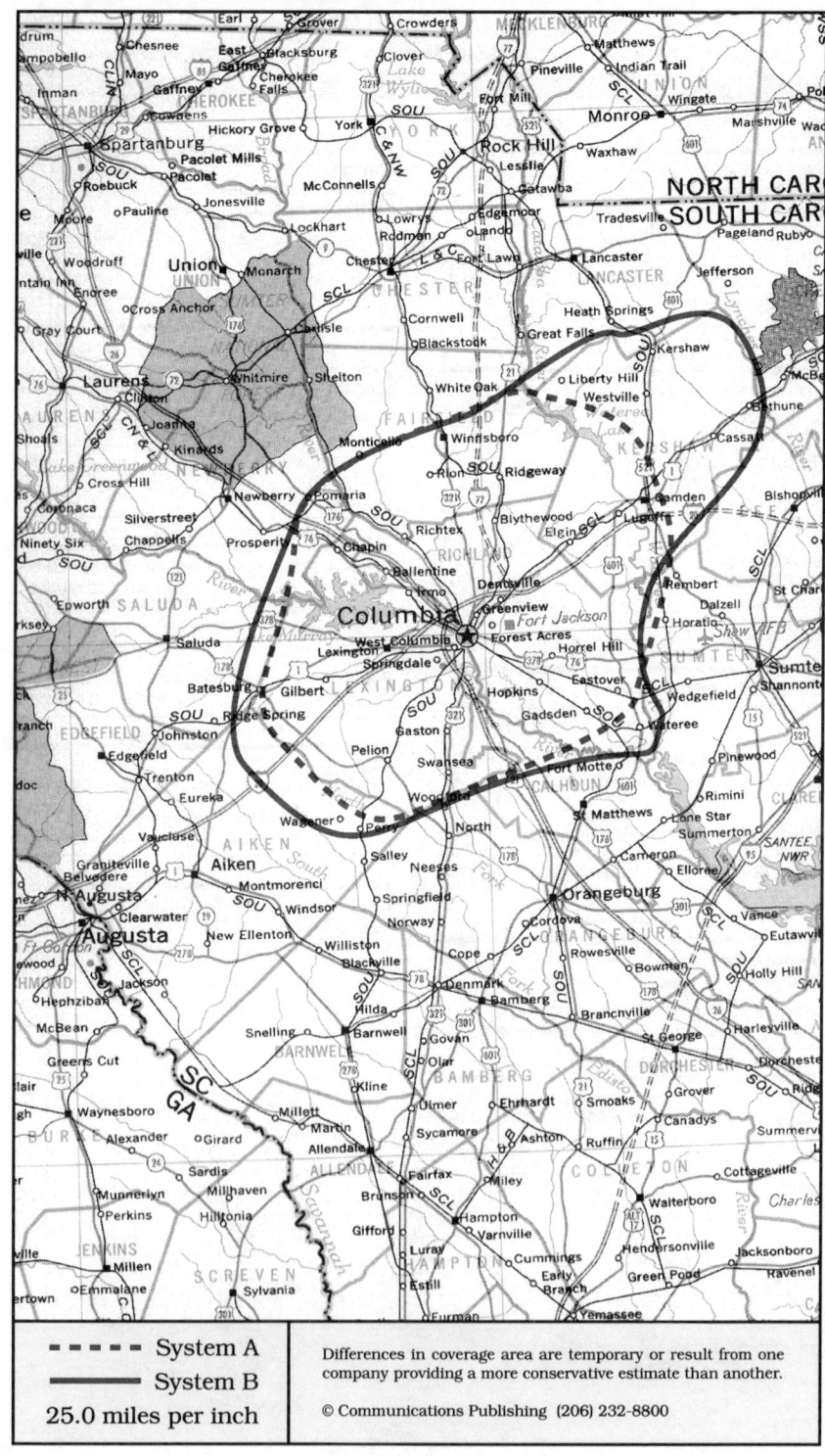

▬ ▬ ▬ System A	Differences in coverage area are temporary or result from one
▬▬▬ System B	company providing a more conservative estimate than another.
25.0 miles per inch	© Communications Publishing (206) 232-8800

Columbia, South Carolina
Camden, Columbia

System A	System B

Cellular company

Bell Atlantic Mobile
910 Gracern Road
Columbia, SC 29210

Customer service	803-731-3958
Local office	803-731-8300
Corporate office	908-306-7000

Sending calls

Visitor dialing instructions

Local calls	area code + 7 digits
Long distance . . .	1 + area code + 7 digits

Speed-dial numbers

Customer service	611
Emergency	911
Highway patrol	*HP or *47
Roaming information	711

Receiving calls

Caller dials the roamer access number below,
hears tone, dials cellular area code and number.
. . . . 803-730-7626

NationLink & RoamingAmerica (see p. 1157)
Dial *31 send to receive calls, *30 to deactivate.
Dial *32 send to tell caller the roamer access no.
This company's subscribers have level 8 service.

Billing information

Visitor rates
Most visitors . . . . $2.50 day + 75¢ minute

Visitors without roaming agreements (p. 1159)
Roamer Plus will intercept first call to bill you.

Identification numbers

City	Market	System	Billing	RSA
All served cities	095	00189	00000	MSA

For full coverage area, see Charlotte, NC.

Cellular company

BellSouth Mobility
832 Dutch Square Boulevard
Columbia, SC 29210

Customer service	800-351-2400
. . . .	803-750-8240
Local office	803-750-2200
Corporate office	404-847-3600

Sending calls

Visitor dialing instructions

Local calls	area code + 7 digits
Long distance	area code + 7 digits

Speed-dial numbers

Customer service	811
Emergency	911
Roaming information	711
Telephone problems	611

Receiving calls

Caller dials the roamer access number below,
hears tone, dials cellular area code and number.

Camden	803-424-7626
Columbia	803-331-7626

Follow Me Roaming forwarding (see p. 1156)
Activate, dial *18 send. Deactivate dial *19 send.

Billing information

Visitor rates

Most visitors	$3 day + 85¢ minute
BellSouth subscribers .	$0 day + 60¢ minute

Visitors without roaming agreements (p. 1159)
Call co. to use Amex, Mastercard, Visa

Identification numbers

City	Market	System	Billing	RSA
Camden	628	00182	30228	SC-4
Columbia	095	00182	00000	MSA

For full coverage area, see Florence, SC.

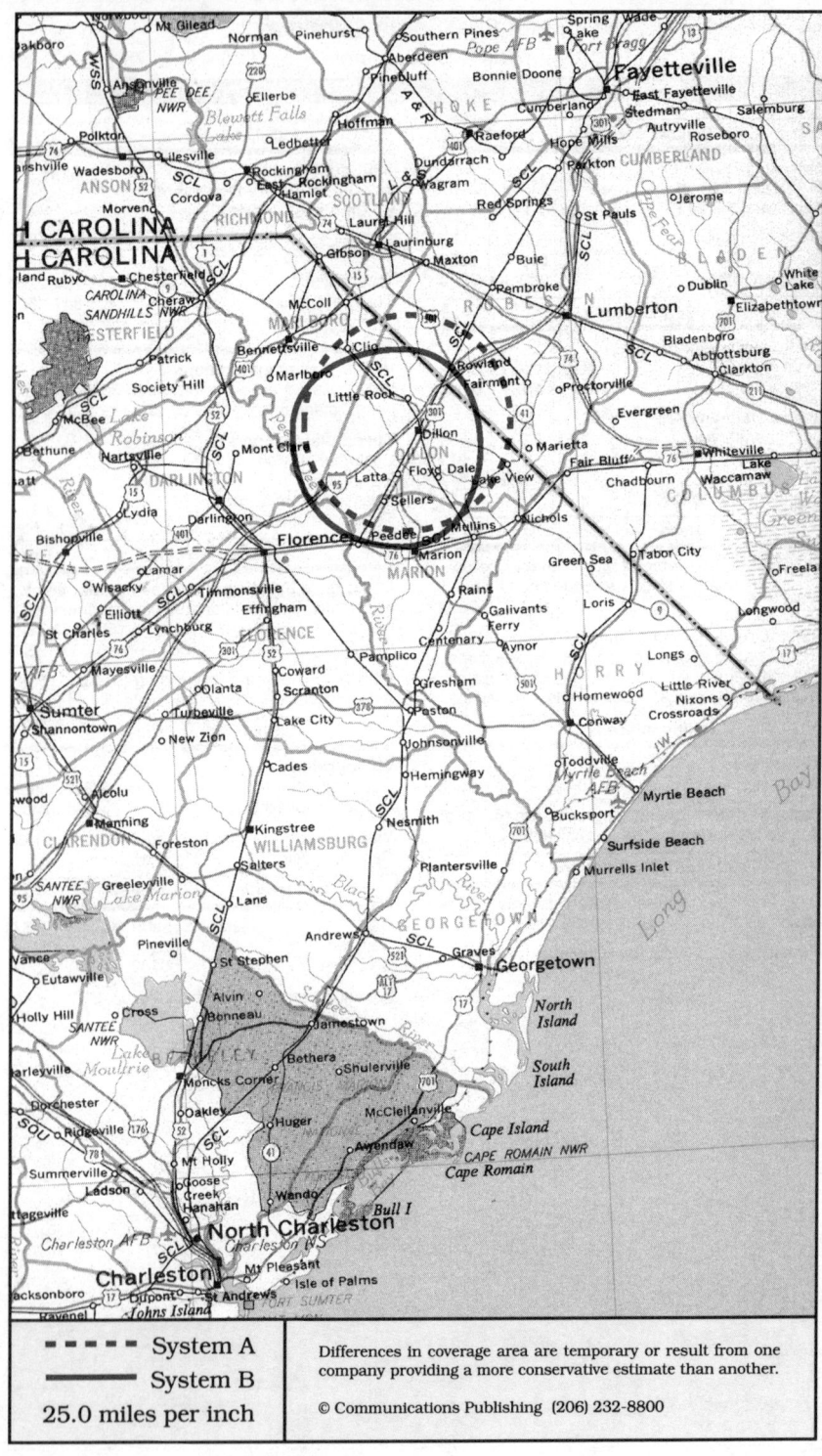

- - - - - System A	Differences in coverage area are temporary or result from one company providing a more conservative estimate than another.
——— System B	
25.0 miles per inch	© Communications Publishing (206) 232-8800

Dillon, South Carolina

System A

Cellular company

Cellular One
United States Cellular
410 SE Boulevard, PO Box 1130
Clinton, NC 28328

Recently acquired by United States Cellular.

Local office 919-590-3030
Corporate office 312-399-8900

Sending calls

Visitor dialing instructions
Local calls . . . . . area code + 7 digits
Long distance . . . 1 + area code + 7 digits

Speed-dial numbers
Customer service 611
Emergency 911
Roaming information *711

Receiving calls

Caller dials the roamer access number below,
hears tone, dials cellular area code and number.
. . . . 615-679-7626

Billing information

Visitor rates
Most visitors . . . . . $3 day + 99¢ minute

Visitors without roaming agreements (p. 1159)
Call co. to use Mastercard, Visa

Identification numbers

City	Market	System	Billing	RSA
All served cities	628	01643	00000	SC-4

For full coverage area, see Goldsboro, NC.

System B

Cellular company

BellSouth Mobility
2875 David McLeod Boulevard #A11
Florence, SC 29501-4099

Customer service 800-351-2400
. . . . 803-750-8240
Local office 803-665-2600
Corporate office 404-847-3600

Sending calls

Visitor dialing instructions
Local calls . . . . . . area code + 7 digits
Long distance . . . 1 + area code + 7 digits

Speed-dial numbers
Customer service 811
Emergency 911
Roaming information 711
Telephone problems 611

Receiving calls

Caller dials the roamer access number below,
hears tone, dials cellular area code and number.
. . . . 803-841-7626

Follow Me Roaming forwarding (see p. 1156)
Activate, dial *18 send. Deactivate dial *19 send.

Billing information

Visitor rates
Most visitors . . . . . $3 day + 85¢ minute
BellSouth subscribers . $0 day + 60¢ minute

Visitors without roaming agreements (p. 1159)
Call co. to use Amex, Mastercard, Visa

Identification numbers

City	Market	System	Billing	RSA
All served cities	628	00350	30236	SC-4

For full coverage area, see Florence, SC.

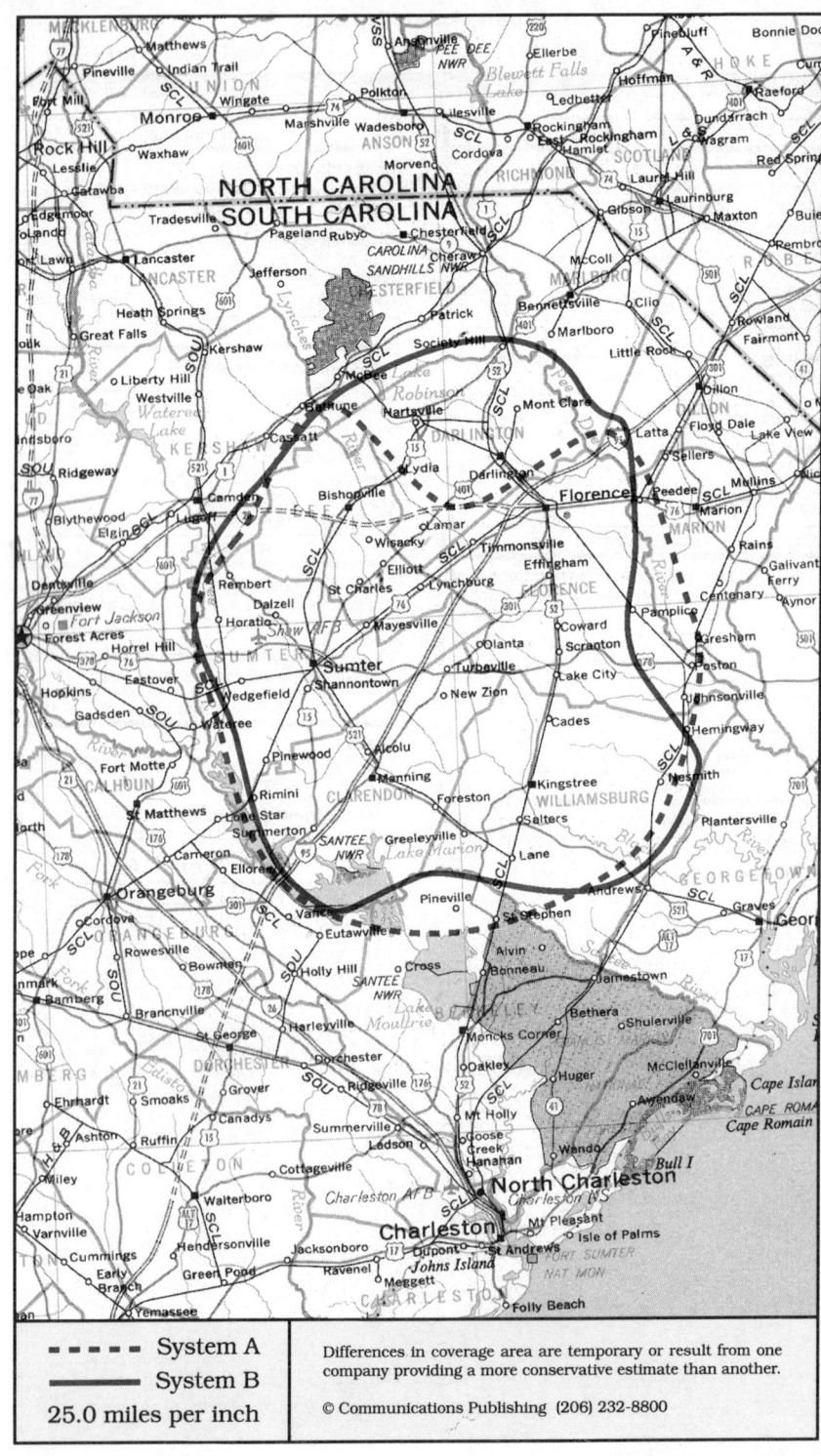

Legend	
----	System A
——	System B
25.0 miles per inch	

Differences in coverage area are temporary or result from one company providing a more conservative estimate than another.

© Communications Publishing (206) 232-8800

Florence, South Carolina

Bishopville, Darlington, Florence, Kingstree, Manning, Sumter

System A	System B

Cellular company

Cellular One
GTE Mobilnet
181 E Evans Street, BTC-105
Florence, SC 29506

Customer service	800-727-2444
. . . .	919-481-1181
Local office	803-664-2898
Corporate office	404-391-8000

Sending calls

Visitor dialing instructions

Local calls	7 digits
Long distance . . .	1 + area code + 7 digits

Speed-dial numbers

Channel 13 news room	∗8813
Customer service	611
Emergency	911
Police department	∗267
WJMX radio	∗103

Receiving calls

Caller dials the roamer access number below,
hears tone, dials cellular area code and number.

. . . . 803-687-7626

Follow Me Roaming forwarding (see p. 1156)
Phone Me Anywhere forwarding (see p. 1156)
Activate, dial ∗18 send. Deactivate dial ∗19 send.

NationLink & RoamingAmerica (see p. 1157)
Dial ∗31 send to receive calls, ∗30 to deactivate.
Dial ∗32 send to tell caller the roamer access no.
This company's subscribers have level 14 service

Billing information

Visitor rates

Most visitors	$2 day + 50¢ minute

Visitors without roaming agreements (p. 1159)
Call co. to use Amex, Mastercard, Visa

Identification numbers

City	Market	System	Billing	RSA
Florence	264	00377	00000	MSA
Manning, Sumter, K.	630	01647	00000	SC-6

Cellular company

BellSouth Mobility
2875 David McLeod Boulevard #A11
Florence, SC 29501-4099

Customer service	800-351-2400
. . . .	803-750-8240
Local office	803-665-2600
Corporate office	404-847-3600

Sending calls

Visitor dialing instructions

Local calls	area code + 7 digits
Long distance . .	1 + area code + 7 digits

Speed-dial numbers

Customer service	811
Emergency	911
Roaming information	711
Telephone problems	611

Receiving calls

Caller dials the roamer access number below,
hears tone, dials cellular area code and number.

Darlington	803-621-7626
Florence	803-678-7626
Kingstree	803-382-7626
Manning	803-473-7626
Sumter	803-491-7626
Turbeville	803-659-7626

Follow Me Roaming forwarding (see p. 1156)
Activate, dial ∗18 send. Deactivate dial ∗19 send.

Billing information

Visitor rates

Most visitors	$3 day + 85¢ minute
BellSouth subscribers .	$0 day + 60¢ minute

Visitors without roaming agreements (p. 1159)
Call co. to use Amex, Mastercard, Visa

Identification numbers

City	Market	System	Billing	RSA
Darlington	628	00350	30236	SC-4
Florence	264	00350	00000	MSA
Kingstree	630	00350	30234	SC-6
Manning, Sumter	630	00182	30230	SC-6

Manning and Sumter service areas are operated
as part of the Columbia, SC system.

For full coverage area, see Columbia, SC and
Dillon, SC.

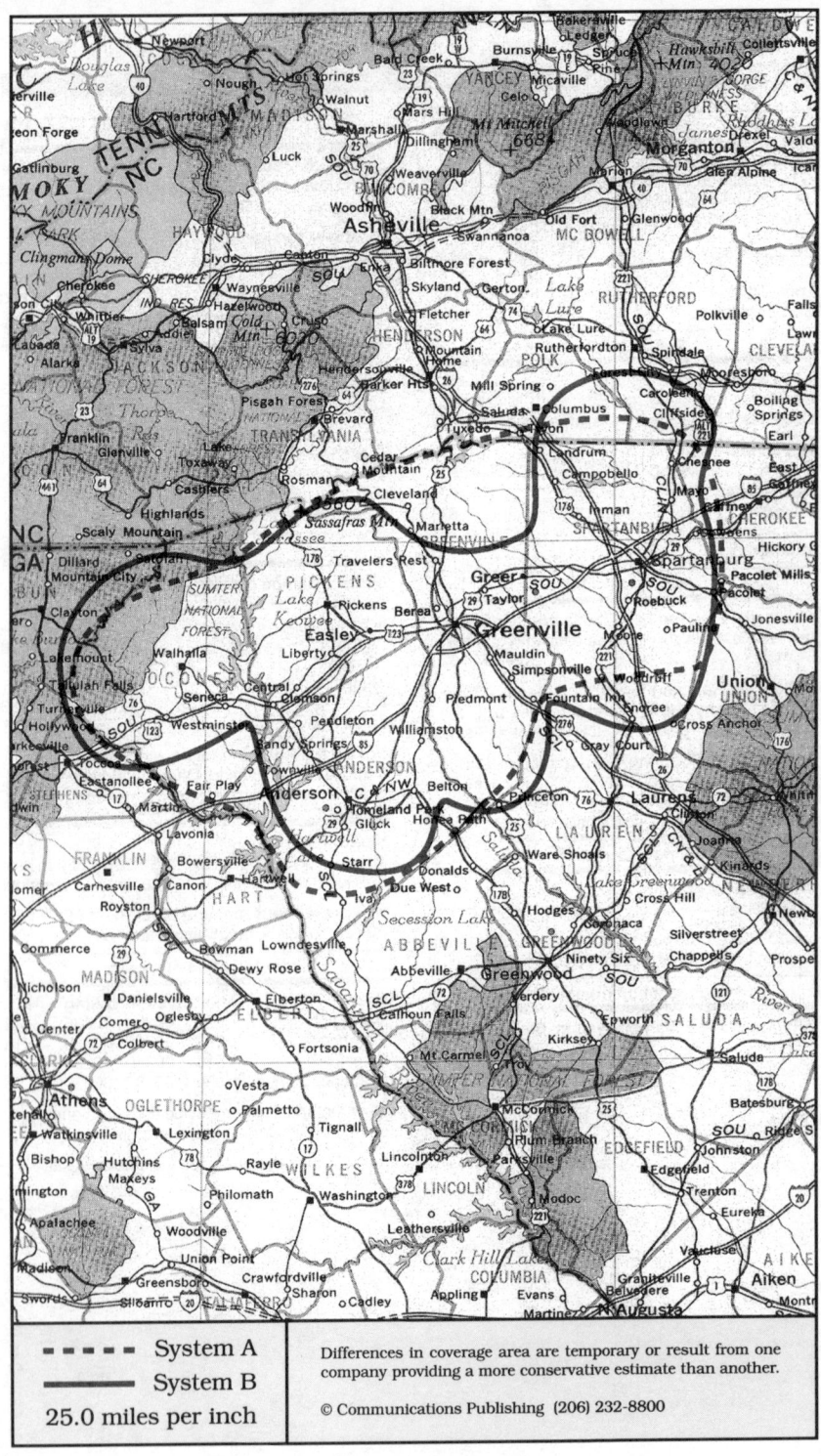

System A
System B
25.0 miles per inch

Differences in coverage area are temporary or result from one company providing a more conservative estimate than another.

© Communications Publishing (206) 232-8800

Greenville, South Carolina

Anderson, Greenville, Seneca (Oconee County), Spartanburg, Walhalla

System A	System B

Cellular company

Bell Atlantic Mobile
806 N Main Street
Anderson, SC 29621

455 Congaree Road
Greenville, SC 29607

Customer service		803-234-1400
Local office . . .	Anderson	803-225-4626
. . . .	Greenville	803-234-7954
Corporate office		908-306-7000

Sending calls

Visitor dialing instructions
Local calls . . . . . area code + 7 digits
Long distance . . . 1 + area code + 7 digits

Speed-dial numbers
Customer service	611
Directory assistance	411
Emergency	911
Highway patrol	*HP or *47
Roaming information	711

Receiving calls

Caller dials the roamer access number below,
hears tone, dials cellular area code and number.
. . . . 803-270-7626

NationLink & RoamingAmerica (see p. 1157)
Dial *31 send to receive calls, *30 to deactivate.
Dial *32 send to tell caller the roamer access no.
This company's subscribers have level 8 service.

Billing information

Visitor rates
Most visitors . . . . . $3 day + 99¢ minute

Visitors without roaming agreements (p. 1159)
Roamer Plus will intercept first call to bill you.

Identification numbers

City	Market	System	Billing	RSA
Anderson	227	00139	30939	MSA
Greenville	067	00139	30139	MSA
Seneca	625	01637	00000	SC-1

Cellular company

Centel Cellular
3470 Cinema Center
Anderson, SC 29621

704 Congaree Road
Greenville, SC 29607

Local office	Anderson	800-736-8795
. . . .	Anderson	803-226-8795
. . . .	Greenville	800-849-8736
. . . .	Greenville	803-234-6000
Corporate office		312-399-2644

Sending calls

Visitor dialing instructions
Local calls . . . . . . area code + 7 digits
Long distance . . . 1 + area code + 7 digits

Speed-dial numbers
Customer service	*611
Emergency	911

Receiving calls

Caller dials the roamer access number below,
hears tone, dials cellular area code and number.
Anderson 803-933-7626
Greenville, Seneca, Spartanburg 803-230-7626

Follow Me Roaming forwarding (see p. 1156)
Activate, dial *18 send. Deactivate dial *19 send.

Billing information

Visitor rates
Most visitors . . . . . $3 day + 99¢ minute

Visitors without roaming agreements (p. 1159)
Call co. to use Amex, Mastercard, Visa

Identification numbers

City	Market	System	Billing	RSA
Anderson	227	00116	30624	MSA
Greenville	067	00116	00000	MSA
Seneca	625	01638	00000	SC-1

For full coverage area, see Greenwood, SC.

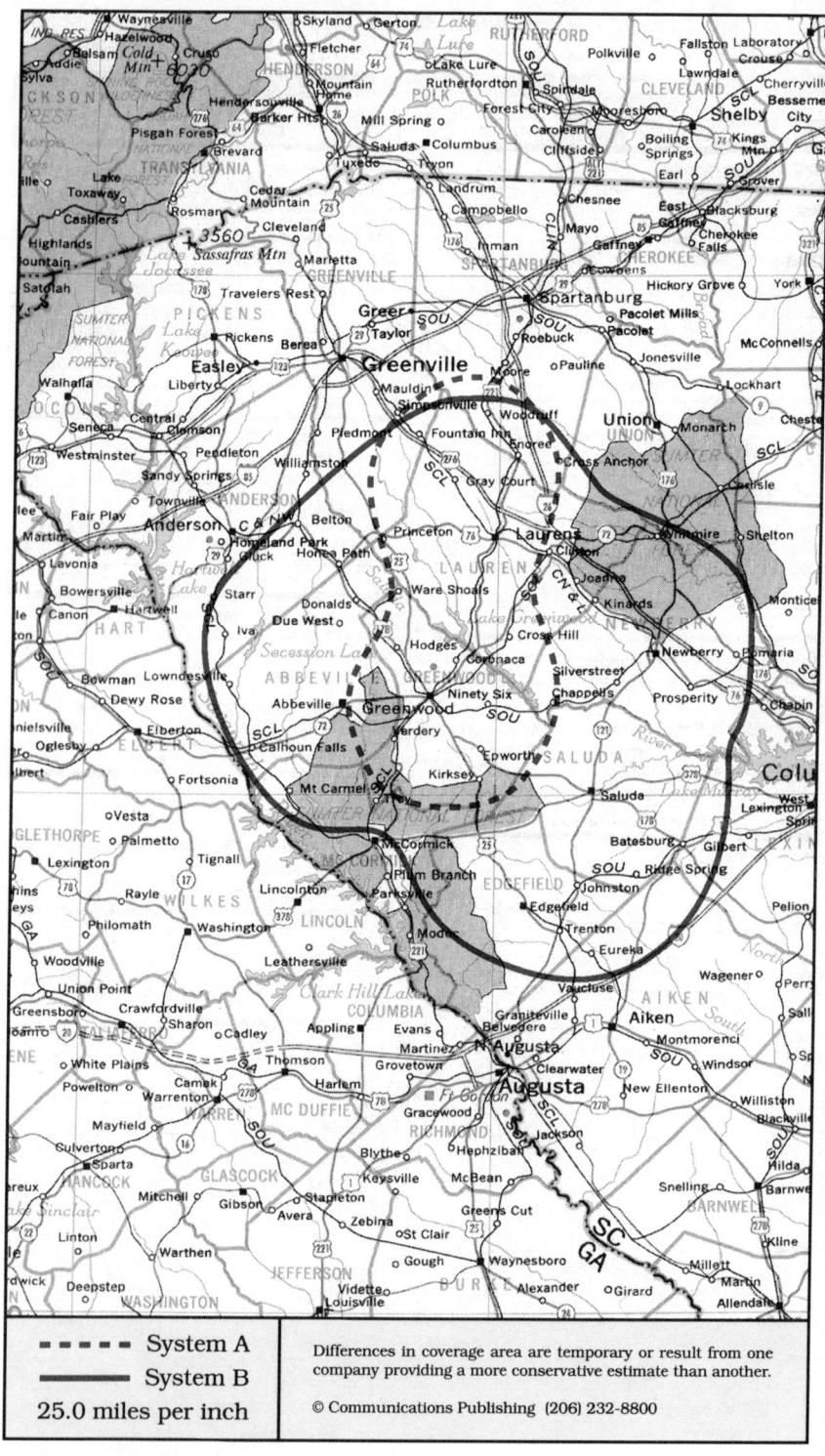

Greenwood, South Carolina
Edgefield, Greenwood, Laurens, Newberry, Saluda

System A	System B

Cellular company

Cellular One
Ally
1601 Bypass 72 NE, PO Box 3343
Greenwood, SC 29648

Local office . . some areas 800-352-3551
. . . . 803-223-4445
Corporate office 912-987-7055

Sending calls

Visitor dialing instructions
Local calls . . . . . area code + 7 digits
Long distance . . . 1 + area code + 7 digits

Speed-dial numbers
Customer service . . . . ✳611
Emergency 911

Receiving calls

Caller dials the roamer access number below,
hears tone, dials cellular area code and number.
. . . . 803-924-7626

Follow Me Roaming forwarding (see p. 1156)
Phone Me Anywhere forwarding (see p. 1156)
Activate, dial ✳18 send. Deactivate dial ✳19 send.

Billing information

Visitor rates
Most visitors . . . . . $3 day + 95¢ minute

Visitors without roaming agreements (p. 1159)
Cannot call since company takes no credit cards.

Identification numbers

City	Market	System	Billing	RSA
All served cities	626	01639	00000	SC-2

Cellular company

Centel Cellular
704 Congaree Road
Greenville, SC 29607

Local office 800-849-8736
. . . . 803-234-6000
Corporate office 312-399-2644

Sending calls

Visitor dialing instructions
Local calls . . . . . . area code + 7 digits
Long distance . . . 1 + area code + 7 digits

Speed-dial numbers
Customer service . . . . ✳611
Emergency 911

Receiving calls

Caller dials the roamer access number below,
hears tone, dials cellular area code and number.
. . . . 803-942-7626

Billing information

Visitor rates
Most visitors . . . . . $3 day + 99¢ minute

Visitors without roaming agreements (p. 1159)
Call co. to use Amex, Mastercard, Visa

Identification numbers

City	Market	System	Billing	RSA
All served cities	626	01640	00000	SC-2

For full coverage area, see Greenville, SC.

- - - - System A ———— System B 25.0 miles per inch	Differences in coverage area are temporary or result from one company providing a more conservative estimate than another. © Communications Publishing (206) 232-8800

Myrtle Beach, South Carolina
Conway, Georgetown, Marion, Myrtle Beach

System A	System B

Cellular company

Cellular One
Vanguard Cellular Systems
918 Frontage Road E
Myrtle Beach, SC 29577

Customer service	803-450-4000
Local office	800-487-6256
. . . .	803-448-5478
Corporate office	919-282-3690

Sending calls

Visitor dialing instructions

Local calls	area code + 7 digits
Long distance . . .	1 + area code + 7 digits

Speed-dial numbers

Customer service	611
Emergency	911
Highway patrol	∗HP or ∗47
U.S. Coast Guard	∗CG or ∗24

Receiving calls

Caller dials the roamer access number below,
hears tone, dials cellular area code and number.

. . . .	803-450-7626

NationLink & RoamingAmerica (see p. 1157)
Dial ∗31 send to receive calls, ∗30 to deactivate.
Dial ∗32 send to tell caller the roamer access no.
This company's subscribers have level 6 service.

Billing information

Visitor rates

Most visitors	$3 day + 99¢ minute

Visitors without roaming agreements (p. 1159)
Cannot call since company takes no credit cards.

Identification numbers

City	Market	System	Billing	RSA
All served cities	629	01645	00000	SC-5

Cellular company

BellSouth Mobility
1527 N Kings Highway #2-N
Myrtle Beach, SC 29577

Customer service	800-351-2400
. . . .	803-750-8240
Local office	803-626-5522
Corporate office	404-847-3600

Sending calls

Visitor dialing instructions

Local calls	area code + 7 digits
Long distance . . .	1 + area code + 7 digits

Speed-dial numbers

Customer service	811
Emergency	911
Roaming information	711
Telephone problems	611

Receiving calls

Caller dials the roamer access number below,
hears tone, dials cellular area code and number.

Georgetown	803-527-7626
Myrtle Beach	803-340-7626

Follow Me Roaming forwarding (see p. 1156)
Activate, dial ∗18 send. Deactivate dial ∗19 send.

Billing information

Visitor rates

Most visitors	$3 day + 85¢ minute
BellSouth subscribers .	$0 day + 60¢ minute

Visitors without roaming agreements (p. 1159)
Call co. to use Amex, Mastercard, Visa

Identification numbers

City	Market	System	Billing	RSA
All served cities	629	00350	30232	SC-5

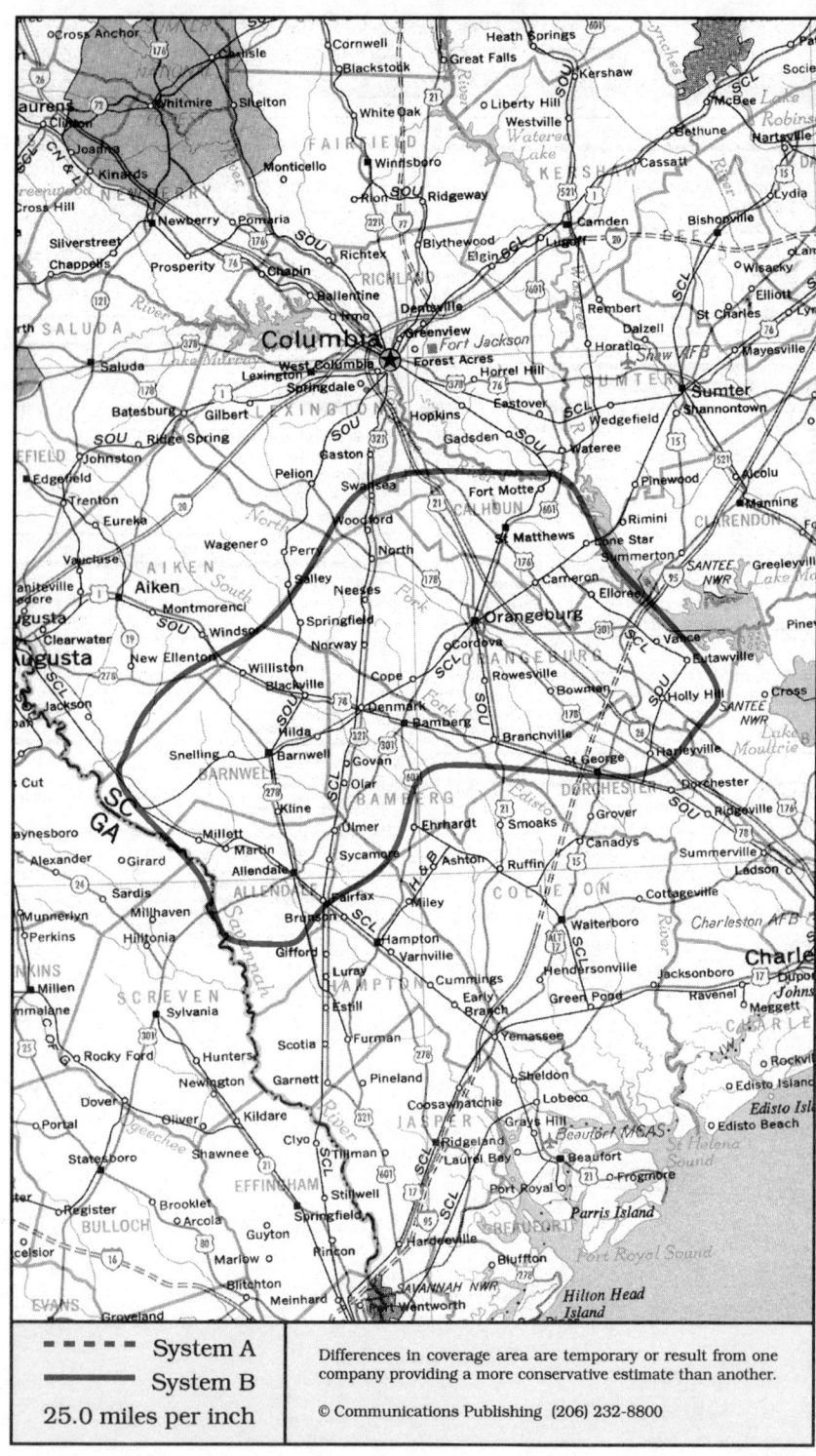

- - - - - System A

————— System B

25.0 miles per inch

Orangeburg, South Carolina
Allendale, Bamberg, Barnwell, Orangeburg, St. Matthews

System A	System B

Cellular company

No information about the company that will operate system A was available at press time.

Cellular company

ALLTEL Mobile Communications
2903 Washington Road
Augusta, GA 30909-2114

Customer service	some areas	800-255-8351
Local office		706-738-2355
Corporate office		501-661-8500

Sending calls

Visitor dialing instructions
Local calls	area code + 7 digits
Long distance	area code + 7 digits

Speed-dial numbers
Customer service	∗611
Emergency	911

Receiving calls

Caller dials the roamer access number below, hears tone, dials cellular area code and number.
Barnwell	803-259-8626
Orangeburg	803-533-8626

Billing information

Visitor rates
Most visitors . . . . . $3 day + 85¢ minute

Visitors without roaming agreements (p. 1159)
Call co. to use Amex, Discover, Mastercard, Visa

Identification numbers

City	Market	System	Billing	RSA
Barnwell	631	2022	30488	SC-7
Orangeburg	631	2022	30250	SC-7

For full coverage area, see Augusta, GA.

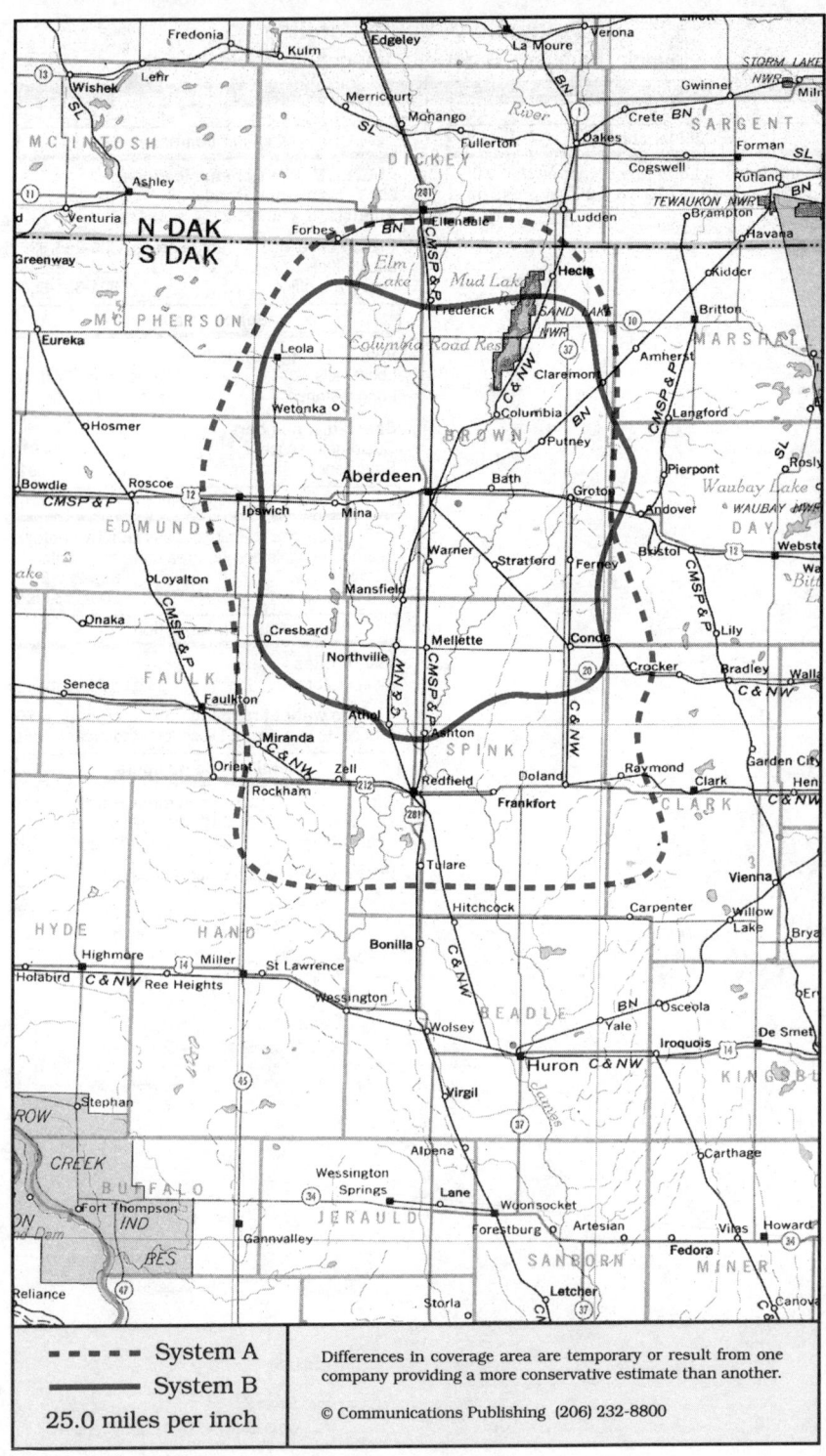

- - - -	System A
———	System B

25.0 miles per inch

Differences in coverage area are temporary or result from one company providing a more conservative estimate than another.

© Communications Publishing (206) 232-8800

Aberdeen, South Dakota
Aberdeen, Leola, Redfield

System A	System B

Cellular company

Cellular One of Aberdeen
Dakota Cellular
1923 SE 6th Avenue, Suite 109
Aberdeen, SD 57401

Corporate office 605-229-2646

Sending calls

Visitor dialing instructions
Local calls . . . . . area code + 7 digits
Long distance . . . 1 + area code + 7 digits

Speed-dial numbers
Customer service 611
Emergency 911

Receiving calls

Caller dials the roamer access number below,
hears tone, dials cellular area code and number.
Aberdeen 605-380-7626
Redfield 605-460-7626

NationLink & RoamingAmerica (see p. 1157)
Dial *31 send to receive calls, *30 to deactivate.
This company's subscribers have level 3 service.

Billing information

Visitor rates
Most visitors . . . . . $3 day + 85¢ minute

Visitors without roaming agreements (p. 1159)
Roamer Plus will intercept first call to bill you.

Identification numbers

City	Market	System	Billing	RSA
All served cities	636	01659	00000	SD-3

This service is operated as part of the Huron, SD system.

This company's subscribers temporarily use the roaming agreements of Greater South Dakota Cellular.

Cellular company

CommNet 2000
Cellular, Inc.
510 6th Street, Suite B
Brookings, SD 57006

Customer service 800-597-5532
Local office 605-692-2000
Corporate office 303-694-3234

Sending calls

Visitor dialing instructions
Local calls . . . . . area code + 7 digits
Long distance . . . 1 + area code + 7 digits

Speed-dial numbers
Customer service 611
Emergency 911
Roaming information 711

Receiving calls

Caller dials the roamer access number below,
hears tone, dials cellular area code and number.
. . . . 605-228-7626

Billing information

Visitor rates
Most visitors . . . . . $3 day + 95¢ minute

Visitors without roaming agreements (p. 1159)
Cannot call since company takes no credit cards.

Identification numbers

City	Market	System	Billing	RSA
All served cities	636	00528	30316	SD-3

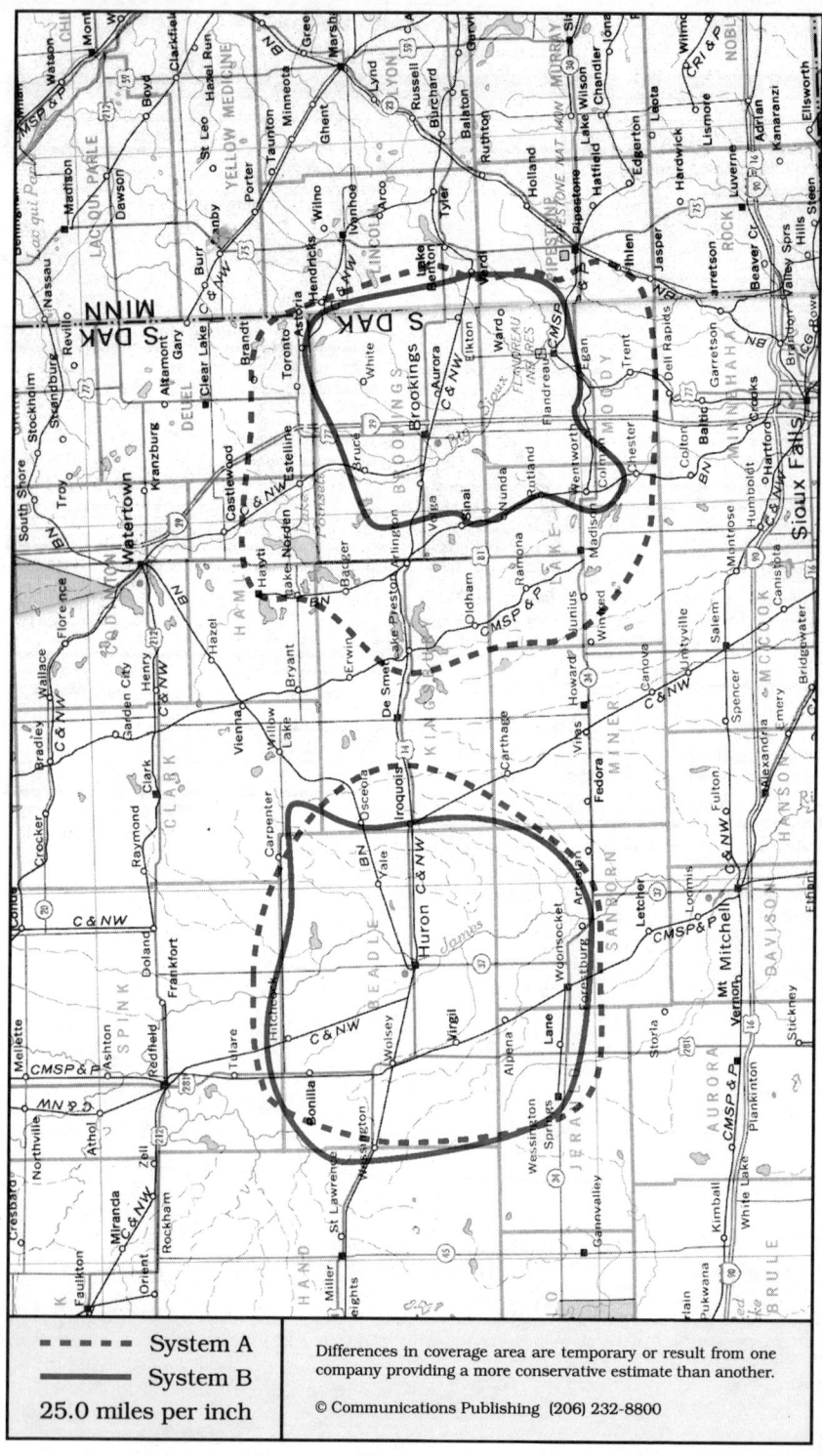

System A
System B
25.0 miles per inch

Differences in coverage area are temporary or result from one company providing a more conservative estimate than another.

© Communications Publishing (206) 232-8800

Brookings, South Dakota
Brookings, Huron, Madison, Woonsocket

System A	System B

Cellular company

Cellular One
Greater South Dakota Cellular
1819 6th Street, Suite B; PO Box 709
Brookings, SD 57006

Corporate office	some areas	800-447-4028
. . . .		605-692-6464

Sending calls

Visitor dialing instructions

Local calls	7 digits
Long distance . . .	1 + area code + 7 digits

Speed-dial numbers

Customer service	611
Emergency	911

Receiving calls

Caller dials the roamer access number below,
hears tone, dials cellular area code and number.

Brookings	605-690-7626
Colman	605-530-7626
Huron	605-350-7626
Madison	605-480-7626

NationLink & RoamingAmerica (see p. 1157)
Dial ✱31 send to receive calls, ✱30 to deactivate.
This company's subscribers have level 3 service.

Billing information

Visitor rates

Most visitors	$3 day + 85¢ minute

Visitors without roaming agreements (p. 1159)
Call co. to use Mastercard, Visa
or Roamer Plus will intercept first call to bill you.

Identification numbers

City	Market	System	Billing	RSA
All served cities	641	01669	00000	SD-8

Cellular company

CommNet 2000
Cellular, Inc.
510 6th Street, Suite B
Brookings, SD 57006

Customer service		800-597-5532
Local office		605-692-2000
Corporate office		303-694-3234

Sending calls

Visitor dialing instructions

Local calls	area code + 7 digits
Long distance . . .	1 + area code + 7 digits

Speed-dial numbers

Customer service	611
Emergency	911
Roaming information	711

Receiving calls

Caller dials the roamer access number below,
hears tone, dials cellular area code and number.

Brookings	605-695-7626
Huron	605-354-7626

Billing information

Visitor rates

Most visitors	$3 day + 95¢ minute

Visitors without roaming agreements (p. 1159)
Cannot call since company takes no credit cards.

Identification numbers

City	Market	System	Billing	RSA
Brookings	641	00528	30166	SD-8
Huron	641	00528	30418	SD-8

For full coverage area, see Mitchell, SD.

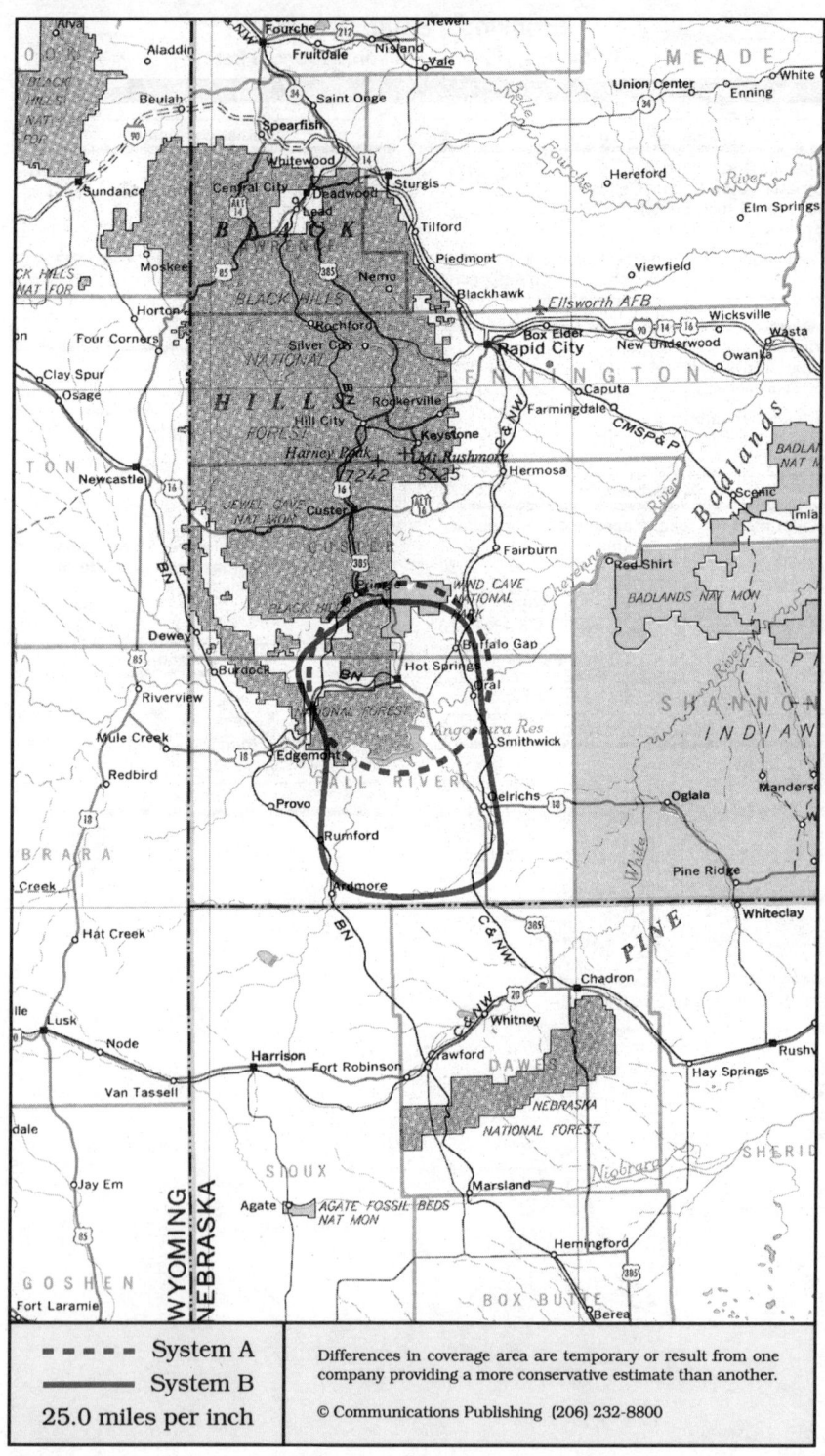

System A	- - - - -
System B	—————

25.0 miles per inch

Differences in coverage area are temporary or result from one company providing a more conservative estimate than another.

© Communications Publishing (206) 232-8800

Hot Springs, South Dakota

System A

Cellular company

Cellular One
GenCell Management
2449 W Chicago Street
Rapid City, SD 57702

Local office	605-348-8800
Corporate office	707-425-8000

Sending calls

Visitor dialing instructions

Local calls	area code + 7 digits
Long distance . . .	1 + area code + 7 digits

Speed-dial numbers

Customer service	611
Emergency	911

Receiving calls

Caller dials the roamer access number below, hears tone, dials cellular area code and number.
. . . . 605-381-7626

Billing information

Visitor rates

Most visitors	$3 day + 85¢ minute

Visitors without roaming agreements (p. 1159)
First call intercepted for a credit card or call co.

Identification numbers

City	Market	System	Billing	RSA
All served cities	638	01653	00000	SD-5

System B

Cellular company

CommNet 2000
Cellular, Inc.
510 6th Street, Suite B
Brookings, SD 57006

Customer service	800-597-5530
Local office	605-692-2000
Corporate office	303-694-3234

Sending calls

Visitor dialing instructions

Local calls	area code + 7 digits
Long distance . . .	1 + area code + 7 digits

Speed-dial numbers

Customer service	611
Emergency	911
Roaming information	711

Receiving calls

Caller dials the roamer access number below, hears tone, dials cellular area code and number.
. . . . 605-890-7626

Billing information

Visitor rates

Most visitors	$3 day + 95¢ minute

Visitors without roaming agreements (p. 1159)
Cannot call since company takes no credit cards.

Identification numbers

City	Market	System	Billing	RSA
All served cities	638	00528	30260	SD-5

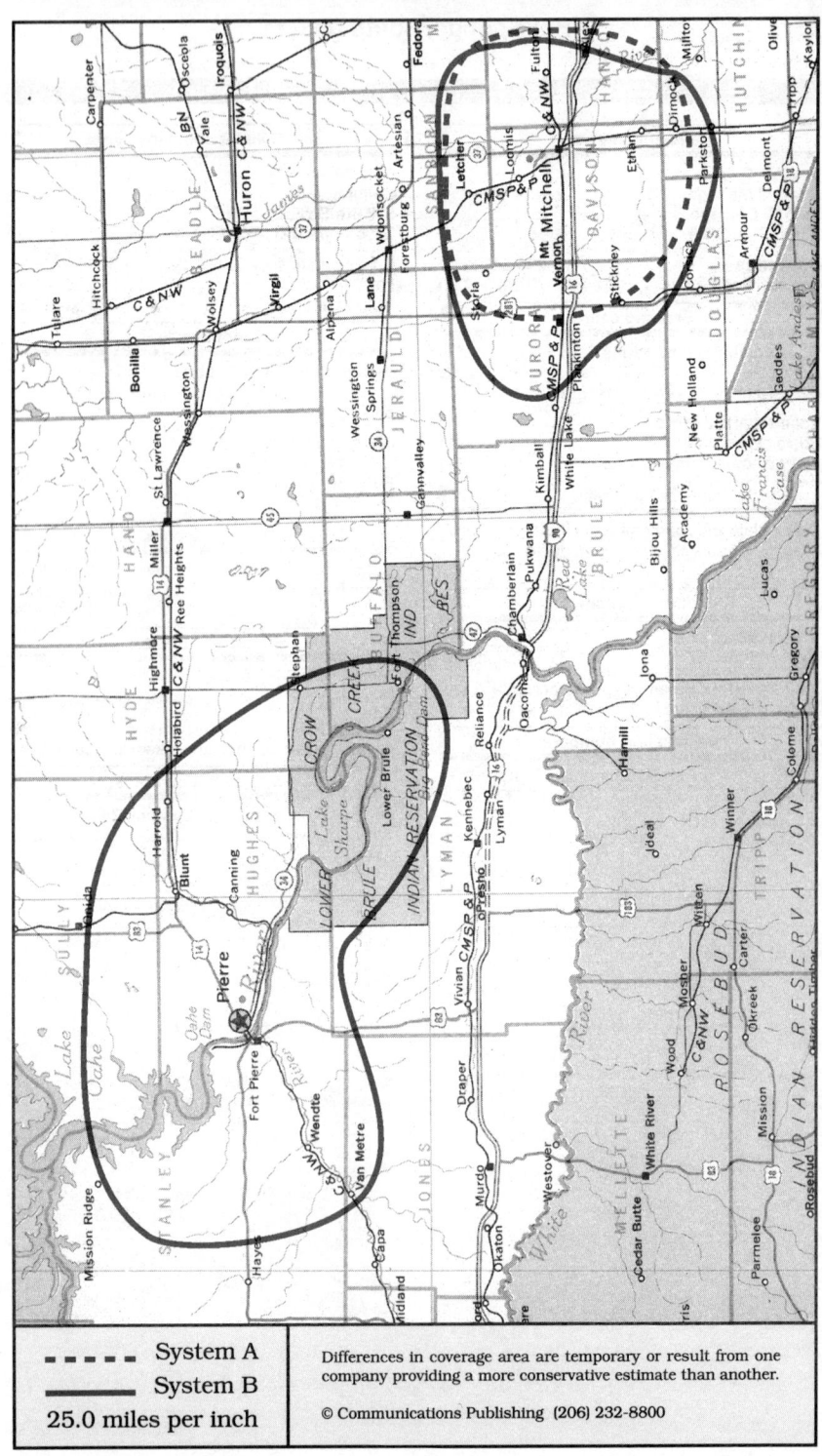

- - - - System A
───── System B
25.0 miles per inch

Differences in coverage area are temporary or result from one company providing a more conservative estimate than another.

© Communications Publishing (206) 232-8800

Mitchell, South Dakota
Alexandria, Mitchell, Pierre

System A	System B

Cellular company

Licensed to:
PriCellular
45 Rockefeller Plaza, Suite 3201
New York, NY 10020

Operated by:
Cellular One
Greater South Dakota Cellular
1819 6th Street, Suite B; PO Box 709
Brookings, SD 57006

Greater South Dakota Cellular	605-692-6464
PriCellular	212-459-0800

Sending calls

Visitor dialing instructions

Local calls	area code + 7 digits
Long distance . . .	1 + area code + 7 digits

Speed-dial numbers

Customer service	611
Emergency	911

Receiving calls

Caller dials the roamer access number below,
hears tone, dials cellular area code and number.
. . . . 605-770-7626

NationLink & RoamingAmerica (see p. 1157)
Caller hears your current roamer access number.
This company's subscribers have level 0 service.

Billing information

Visitor rates

Most visitors	$3 day + 85¢ minute

Visitors without roaming agreements (p. 1159)
Call co. to use Amex, Mastercard, Visa
or Roamer Plus will intercept first call to bill you.

Identification numbers

City	Market	System	Billing	RSA
All served cities	640	01667	00000	SD-7

Cellular company

CommNet 2000
Cellular, Inc.
510 6th Street, Suite B
Brookings, SD 57006

Customer service	800-597-5532
Local office	605-692-2000
Corporate office	303-694-3234

Sending calls

Visitor dialing instructions

Local calls	area code + 7 digits
Long distance . . .	1 + area code + 7 digits

Speed-dial numbers

Customer service	611
Emergency	911
Roaming information	711

Receiving calls

Caller dials the roamer access number below,
hears tone, dials cellular area code and number.

Mitchell	605-999-7626
Pierre	605-222-7626

Billing information

Visitor rates

Most visitors	$3 day + 95¢ minute

Visitors without roaming agreements (p. 1159)
Cannot call since company takes no credit cards.

Identification numbers

City	Market	System	Billing	RSA
Ft. Pierre	640	00528	30162	SD-6
Mitchell, Pierre	640	00528	30164	SD-7

For full coverage area, see Brookings, SD.

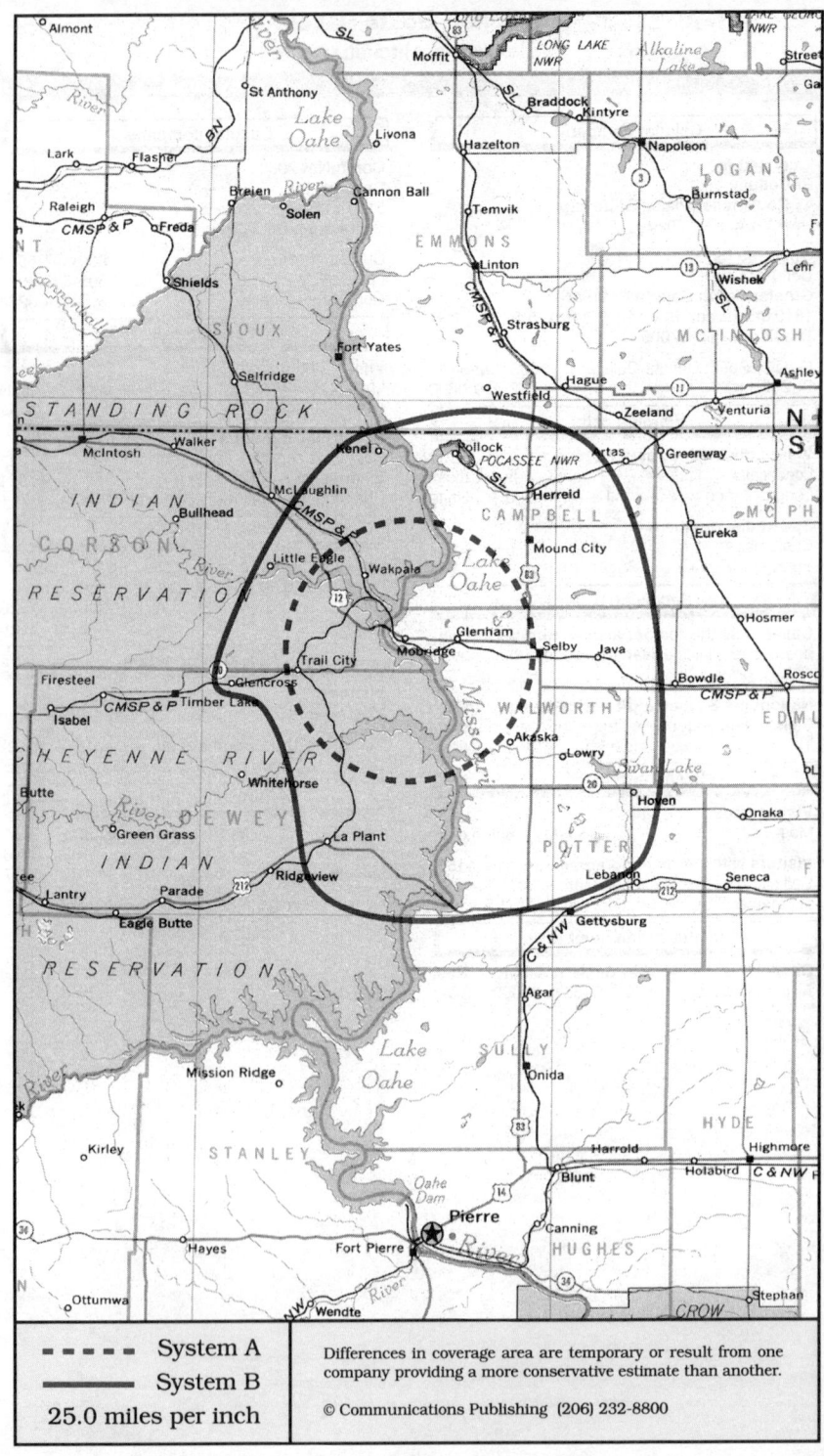

System A – – – –
System B ————
25.0 miles per inch

Differences in coverage area are temporary or result from one company providing a more conservative estimate than another.

© Communications Publishing (206) 232-8800

Mobridge, South Dakota
Mobridge, Mound City

|

System A

Cellular company

GenCell Management
1891 Woolner Avenue
Fairfield, CA 94533

Customer service 800-888-7868
Corporate office 707-425-8000

Sending calls

Visitor dialing instructions
Local calls . . . . area code + 7 digits
Long distance . . . 1 + area code + 7 digits

Speed-dial numbers
Customer service 611
Emergency 911

Receiving calls

Caller dials the roamer access number below,
hears tone, dials cellular area code and number.
. . . . Not available at press time

Billing information

Visitor rates
Most visitors . . . . . $3 day + 99¢ minute

Visitors without roaming agreements (p. 1159)
American Roaming Network intercepts call to bill.

Identification numbers

City	Market	System	Billing	RSA
All served cities	635	01657	00000	SD-2

System B

Cellular company

CommNet 2000
Cellular, Inc.
510 6th Street, Suite B
Brookings, SD 57006

Customer service 800-597-5532
Local office 605-692-2000
Corporate office 303-694-3234

Sending calls

Visitor dialing instructions
Local calls . . . . . . area code + 7 digits
Long distance . . . 1 + area code + 7 digits

Speed-dial numbers
Customer service 611
Emergency 911
Roaming information 711

Receiving calls

Caller dials the roamer access number below,
hears tone, dials cellular area code and number.
. . . . 605-848-7626

Billing information

Visitor rates
Most visitors . . . . . $3 day + 95¢ minute

Visitors without roaming agreements (p. 1159)
Cannot call since company takes no credit cards.

Identification numbers

City	Market	System	Billing	RSA
All served cities	635	00528	30252	SD-2

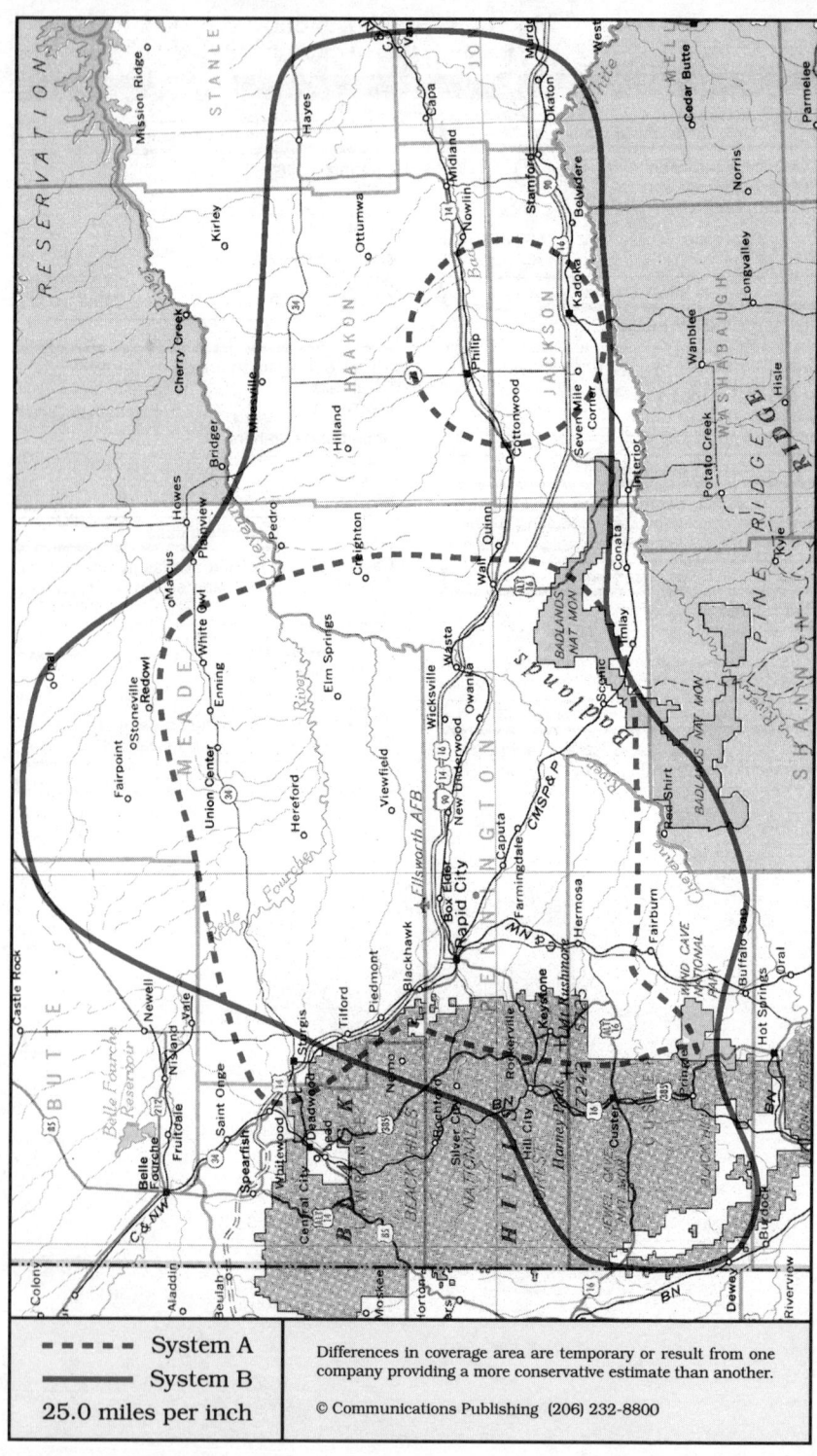

System A - - - -
System B ——
25.0 miles per inch

Differences in coverage area are temporary or result from one company providing a more conservative estimate than another.

© Communications Publishing (206) 232-8800

Rapid City, South Dakota
Custer, Kadoka, Philip, Rapid City, Sturgis

System A	System B

Cellular company

Cellular One
GenCell Management
2449 W Chicago Street
Rapid City, SD 57702

Local office	605-348-8800
Corporate office	707-425-8000

Sending calls

Visitor dialing instructions

Local calls	area code + 7 digits
Long distance . . .	1 + area code + 7 digits

Speed-dial numbers

Customer service	611
Emergency	911

Receiving calls

Caller dials the roamer access number below,
hears tone, dials cellular area code and number.
. . . . 605-381-7626

Billing information

Visitor rates

Most visitors	$3 day + 85¢ minute

Visitors without roaming agreements (p. 1159)
First call intercepted for a credit card or call co.

Identification numbers

City	Market	System	Billing	RSA
Philip	639	01655	00000	SD-6
Rapid City	289	00511	00000	MSA

Cellular company

Contel Cellular
3220 W Chicago Street
Rapid City, SD 57702

Customer service	800-333-4004
. . . . . extension 4002	404-391-8000
Local office	605-342-4000
Corporate office	404-804-3400

Sending calls

Visitor dialing instructions

Local calls	area code + 7 digits
Long distance . . .	1 + area code + 7 digits

Speed-dial numbers

Customer service	*611
Emergency	911

Receiving calls

Caller dials the roamer access number below,
hears tone, dials cellular area code and number.
. . . . 605-390-7626

Billing information

Visitor rates

Most visitors	$3 day + 85¢ minute

Visitors without roaming agreements (p. 1159)
Call co. to use Amex, Mastercard, Visa

Identification numbers

City	Market	System	Billing	RSA
Custer	638	01898	00000	SD-5
Philip	639	01864	00000	SD-6
Rapid City	289	00494	00000	MSA

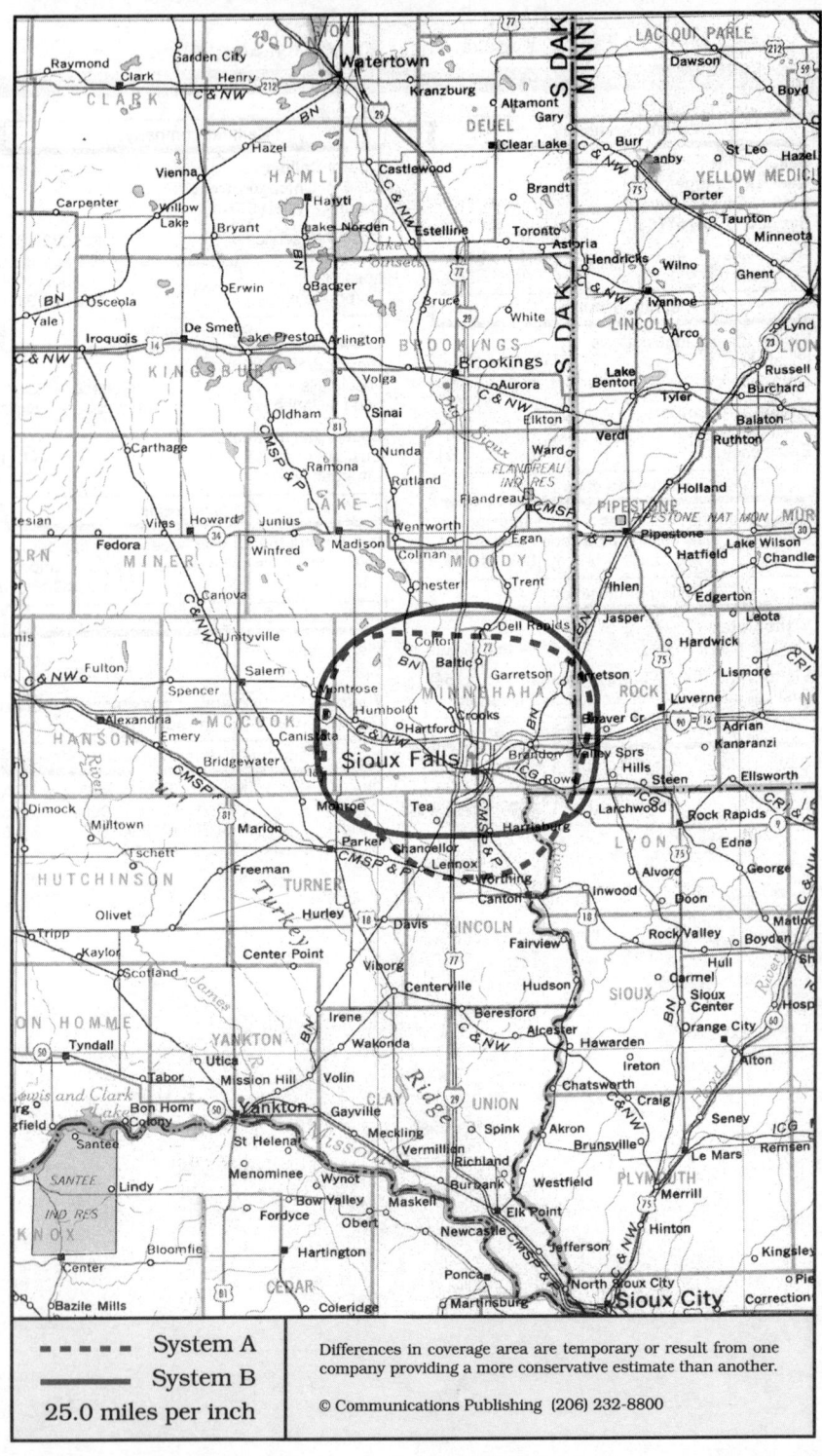

---- System A	Differences in coverage area are temporary or result from one
—— System B	company providing a more conservative estimate than another.
25.0 miles per inch	© Communications Publishing (206) 232-8800

Sioux Falls, South Dakota

System A	System B

System A

Cellular company

Cellular One
1100 W Delaware Street, PO Box 85208
Sioux Falls, SD 57118

Customer service	800-676-4414
. . . .	605-336-8504
Corporate office	605-336-0520

Sending calls

Visitor dialing instructions

Local calls	area code + 7 digits
Long distance . . .	1 + area code + 7 digits

Speed-dial numbers

Customer service	611
Emergency	911

Receiving calls

Caller dials the roamer access number below,
hears tone, dials cellular area code and number.
. . . . 605-360-7626

NationLink & RoamingAmerica (see p. 1157)
Dial ✳31 send to receive calls, ✳30 to deactivate.
Dial ✳32 send to tell caller the roamer access no.
This company's subscribers have level 3 service.

Billing information

Visitor rates
Most visitors . . . . . $3 day + 99¢ minute

Visitors without roaming agreements (p. 1159)
Roamer Plus will intercept first call to bill you.

Identification numbers

City	Market	System	Billing	RSA
All served cities	267	00555	00000	MSA

System B

Cellular company

U S West Cellular
3101 W 41st Street, Suite 111
Sioux Falls, SD 57105

Customer service	800-238-7848
. . . .	206-562-2895
. . . . . local subscribers	800-626-6611
. . . .	206-747-1771
Local office	605-336-2240
Corporate office	206-747-4900

Sending calls

Visitor dialing instructions

Local calls	area code + 7 digits
Long distance . . .	1 + area code + 7 digits

Speed-dial numbers

Customer service	✳611
Emergency	911
KELO make a traffic report	#925
Roaming information	✳711

Receiving calls

Caller dials the roamer access number below,
hears tone, dials cellular area code and number.
. . . . 605-366-7626

Follow Me Roaming forwarding (see p. 1156)
Activate, dial ✳18 send. Deactivate dial ✳19 send.

Billing information

Visitor rates
Most visitors . . . . . $3 day + 99¢ minute

Visitors without roaming agreements (p. 1159)
Call co. to use Amex, Mastercard, Visa

Identification numbers

City	Market	System	Billing	RSA
All served cities	267	00540	00000	MSA

For full coverage area, see Yankton, SD.

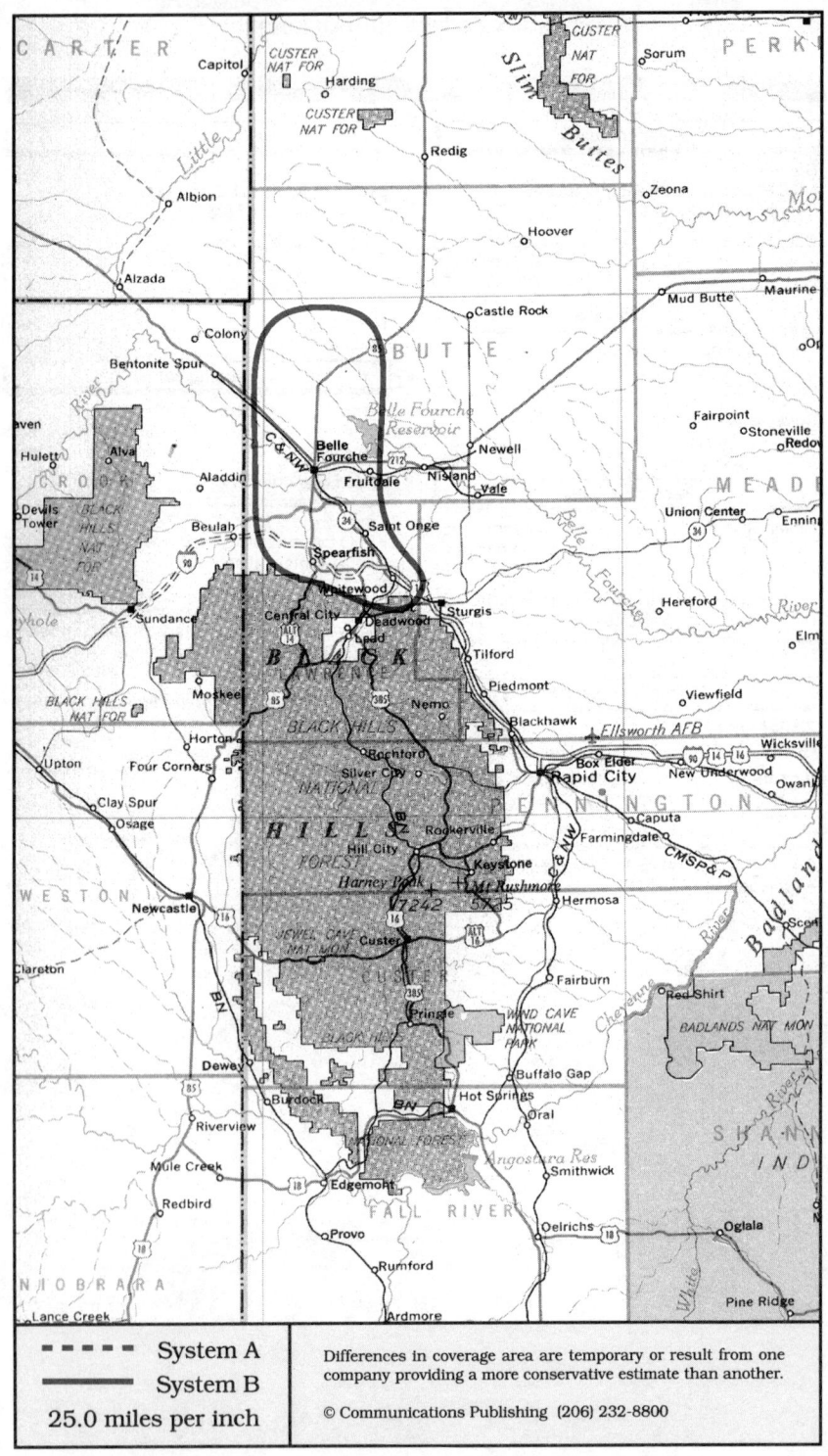

- - - - - System A ────── System B **25.0 miles per inch**	Differences in coverage area are temporary or result from one company providing a more conservative estimate than another. © Communications Publishing (206) 232-8800

Spearfish, South Dakota
Belle Fourche, Spearfish

System A	System B

Cellular company

No information about the company that will operate system A was available at press time.

Cellular company

**CommNet 2000
Cellular, Inc.
510 6th Street, Suite B
Brookings, SD 57006**

Customer service	800-597-5530
Local office	605-692-2000
Corporate office	303-694-3234

Sending calls

Visitor dialing instructions

Local calls	area code + 7 digits
Long distance . . .	1 + area code + 7 digits

Speed-dial numbers

Customer service	611
Emergency	911
Roaming information	711

Receiving calls

Caller dials the roamer access number below, hears tone, dials cellular area code and number.
. . . . 605-645-7626

Billing information

Visitor rates
Most visitors . . . . . $3 day + 95¢ minute

Visitors without roaming agreements (p. 1159)
Cannot call since company takes no credit cards.

Identification numbers

City	Market	System	Billing	RSA
All served cities	634	00528	30258	SD-1

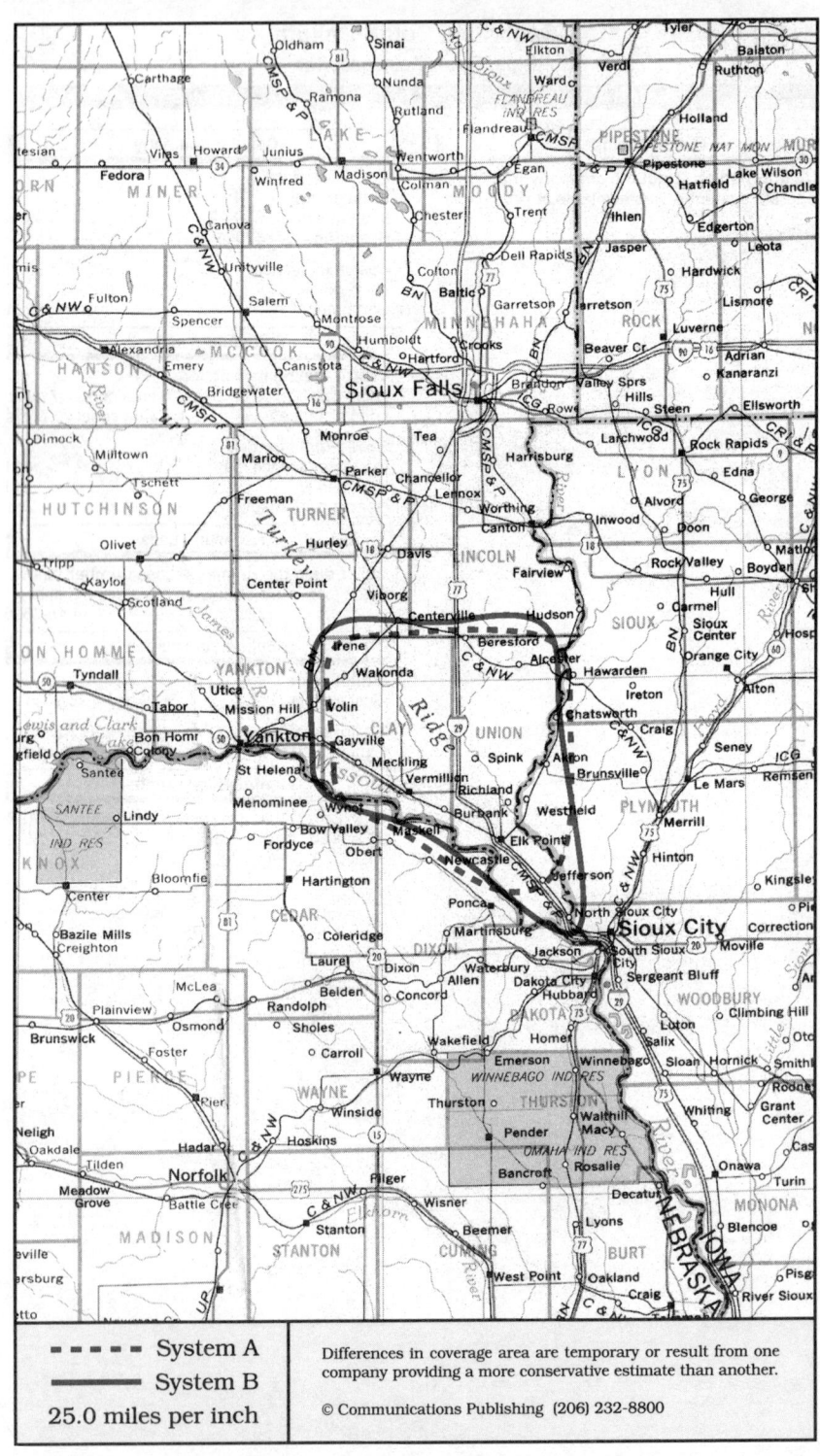

	System A
▬▬▬	System B
25.0 miles per inch	

Differences in coverage area are temporary or result from one company providing a more conservative estimate than another.

© Communications Publishing (206) 232-8800

Vermillion, South Dakota
Elk Point, Vermillion

System A	System B

Cellular company

Cellular One
GenCell Management
525 5th Street Plaza SE
Watertown, SD 57201

Customer service	800-888-7868
Local office	800-333-3414
. . . .	605-882-1111
Corporate office	707-425-8000

Sending calls

Visitor dialing instructions

Local calls	area code + 7 digits
Long distance . . .	1 + area code + 7 digits

Speed-dial numbers

Customer service	611
Emergency	911

Receiving calls

Caller dials the roamer access number below,
hears tone, dials cellular area code and number.
. . . . 605-661-7626

NationLink & RoamingAmerica (see p. 1157)
Dial *31 send to receive calls, *30 to deactivate.
This company's subscribers have level 3 service.

Billing information

Visitor rates

Most visitors	$3 day + 99¢ minute

Visitors without roaming agreements (p. 1159)
American Roaming Network intercepts call to bill.

Identification numbers

City	Market	System	Billing	RSA
All served cities	642	01671	00000	SD-9

For full coverage area, see Yankton, SD.

Cellular company

CommNet 2000
Cellular, Inc.
124 Pierce
Sioux City, IA 51101

Customer service	800-597-5532
Local office	800-255-2083
. . . .	712-233-2083
Corporate office	303-694-3234

Sending calls

Visitor dialing instructions

Local calls	area code + 7 digits
Long distance . . .	1 + area code + 7 digits

Speed-dial numbers

Customer service	611
Emergency	911
Roaming information	711

Receiving calls

Caller dials the roamer access number below,
hears tone, dials cellular area code and number.
. . . . 605-670-7626

Billing information

Visitor rates

Most visitors	$3 day + 95¢ minute

Visitors without roaming agreements (p. 1159)
Cannot call since company takes no credit cards.

Identification numbers

City	Market	System	Billing	RSA
All served cities	642	00528	30264	SD-9

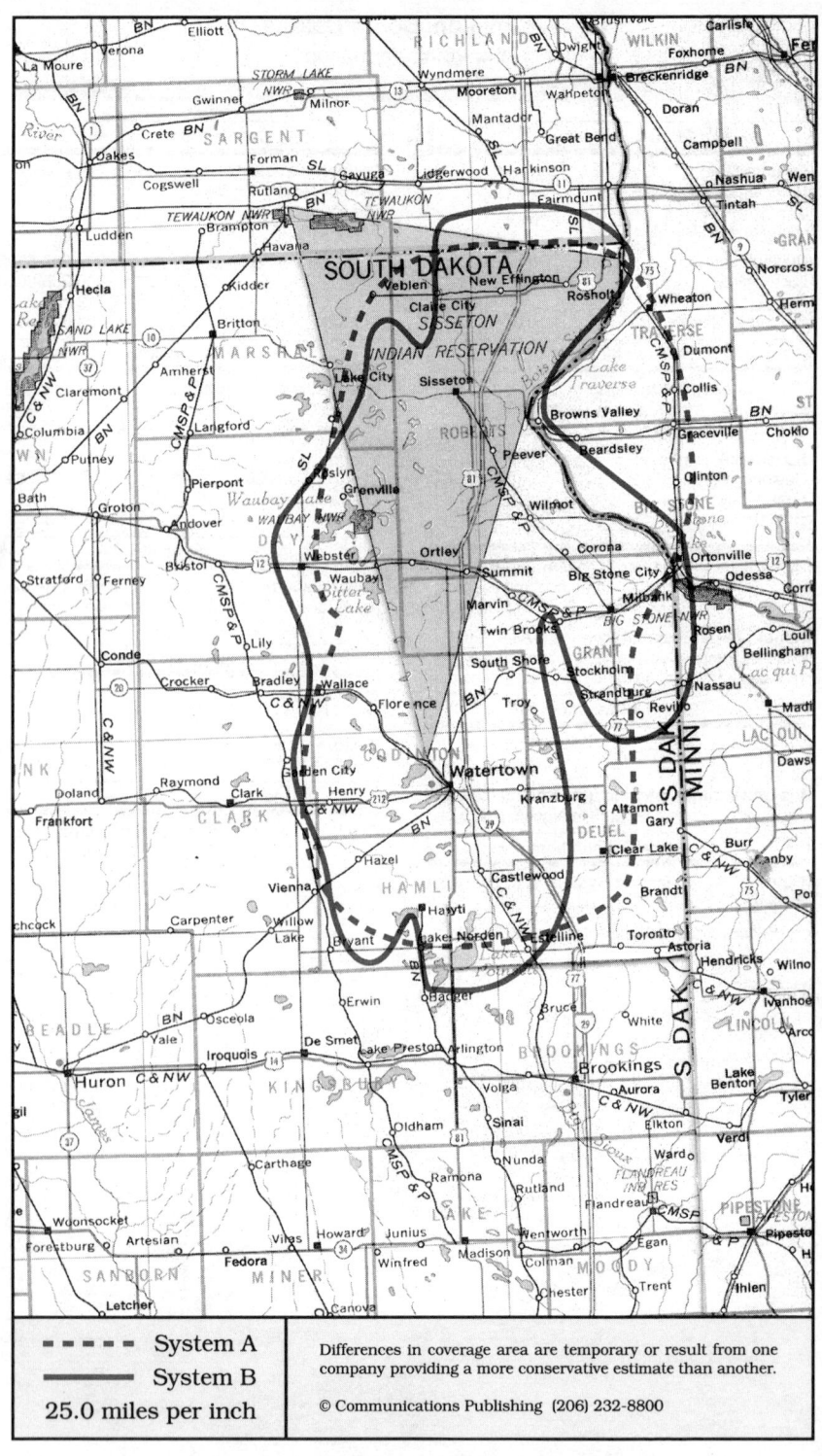

System A

System B

25.0 miles per inch

Differences in coverage area are temporary or result from one company providing a more conservative estimate than another.

© Communications Publishing (206) 232-8800

Watertown, South Dakota
Clear Lake, Hayti, Milbank, Sisseton, Watertown, Webster

System A	System B

Cellular company

Cellular One
GenCell Management
1100 Broadway, Suite 2
Yankton, SD 57078

Customer service 800-888-7868
Local office 605-665-1935
Corporate office 707-425-8000

Sending calls

Visitor dialing instructions
Local calls . . . . . area code + 7 digits
Long distance . . . 1 + area code + 7 digits

Speed-dial numbers
Customer service 611
Emergency 911

Receiving calls

Caller dials the roamer access number below,
hears tone, dials cellular area code and number.
. . . . 605-881-7626

NationLink & RoamingAmerica (see p. 1157)
Caller hears your current roamer access number.
This company's subscribers have level 0 service.

Billing information

Visitor rates
Most visitors . . . . . $3 day + 99¢ minute

Visitors without roaming agreements (p. 1159)
American Roaming Network intercepts call to bill.

Identification numbers

City	Market	System	Billing	RSA
All served cities	637	01661	00000	SD-4

Cellular company

Cellular 2000
Pacific Telecom Cellular
1870 Madison Avenue, PO Box 4069
Mankato, MN 56002-4069

Local office 800-545-3950
. . . . 507-345-4990
Corporate office 414-841-1200

Sending calls

Visitor dialing instructions
Local calls . . . . . . area code + 7 digits
Long distance . . . . . area code + 7 digits

Speed-dial numbers
Customer service . . . . ∗611
Emergency . . . . ∗911
Roaming information . . . . . . ∗711

Receiving calls

Caller dials the roamer access number below,
hears tone, dials cellular area code and number.
. . . . 507-330-7626

Billing information

Visitor rates
Most visitors . . . . . $3 day + 99¢ minute

Visitors without roaming agreements (p. 1159)
Cannot call since company takes no credit cards.

Identification numbers

City	Market	System	Billing	RSA
All served cities	637	01662	00000	SD-4

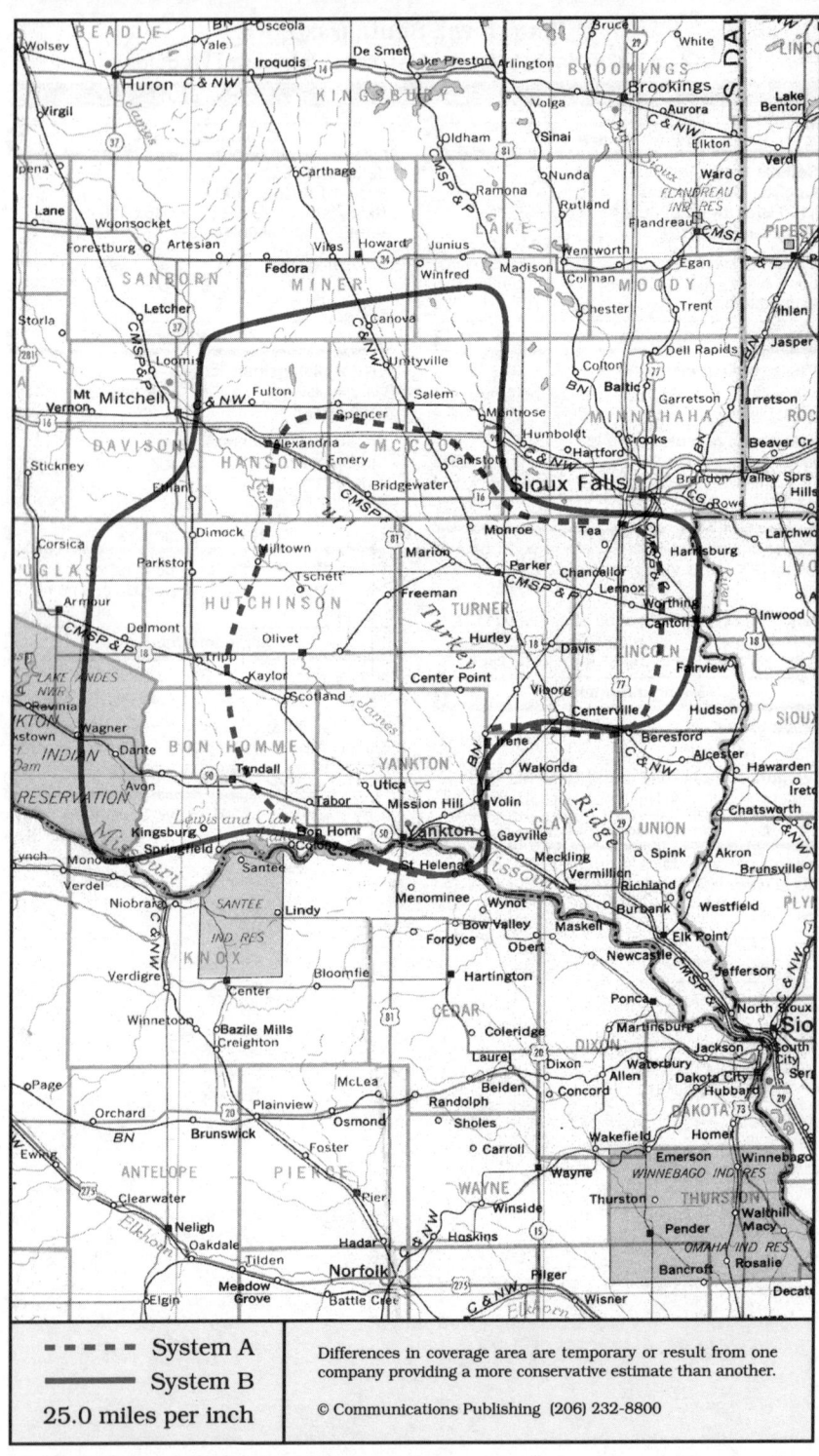

- - - - System A
————— System B
25.0 miles per inch

Differences in coverage area are temporary or result from one company providing a more conservative estimate than another.

© Communications Publishing (206) 232-8800

Yankton, South Dakota
Olivet, Parker, Salem, Tyndall, Yankton

System A	System B

Cellular company

Cellular One
GenCell Management
1100 Broadway, Suite 2
Yankton, SD 57078

Customer service	800-888-7868
Local office	605-665-1935
Corporate office	707-425-8000

Sending calls

Visitor dialing instructions

Local calls	area code + 7 digits
Long distance . . .	1 + area code + 7 digits

Speed-dial numbers

Customer service	611
Emergency	911

Receiving calls

Caller dials the roamer access number below, hears tone, dials cellular area code and number.
. . . . 605-661-7626

NationLink & RoamingAmerica (see p. 1157)
Dial *31 send to receive calls, *30 to deactivate.
This company's subscribers have level 3 service.

Billing information

Visitor rates
Most visitors . . . . . $3 day + 99¢ minute

Visitors without roaming agreements (p. 1159)
American Roaming Network intercepts call to bill.

Identification numbers

City	Market	System	Billing	RSA
All served cities	642	01671	00000	SD-9

For full coverage area, see Vermillion, SD.

Cellular company

U S West Cellular
3101 W 41st Street, Suite 111
Sioux Falls, SD 57105

Customer service	800-238-7848
. . . .	206-562-2895
. . . . . local subscribers	800-626-6611
. . . .	206-747-1771
Local office	605-336-9970
Corporate office	206-747-4900

Sending calls

Visitor dialing instructions

Local calls	area code + 7 digits
Long distance . . .	1 + area code + 7 digits

Speed-dial numbers

Customer service	*611
Emergency	911
KELO traffic report	#925
Roaming information	*711

Receiving calls

Caller dials the roamer access number below, hears tone, dials cellular area code and number.
. . . . 605-660-7626

Follow Me Roaming forwarding (see p. 1156)
Activate, dial *18 send. Deactivate dial *19 send.

Billing information

Visitor rates
Most visitors . . . . . $3 day + 99¢ minute

Visitors without roaming agreements (p. 1159)
Call co. to use Amex, Mastercard, Visa

Identification numbers

City	Market	System	Billing	RSA
All served cities	642	00540	00000	SD-9

For full coverage area, see Sioux Falls, SD.

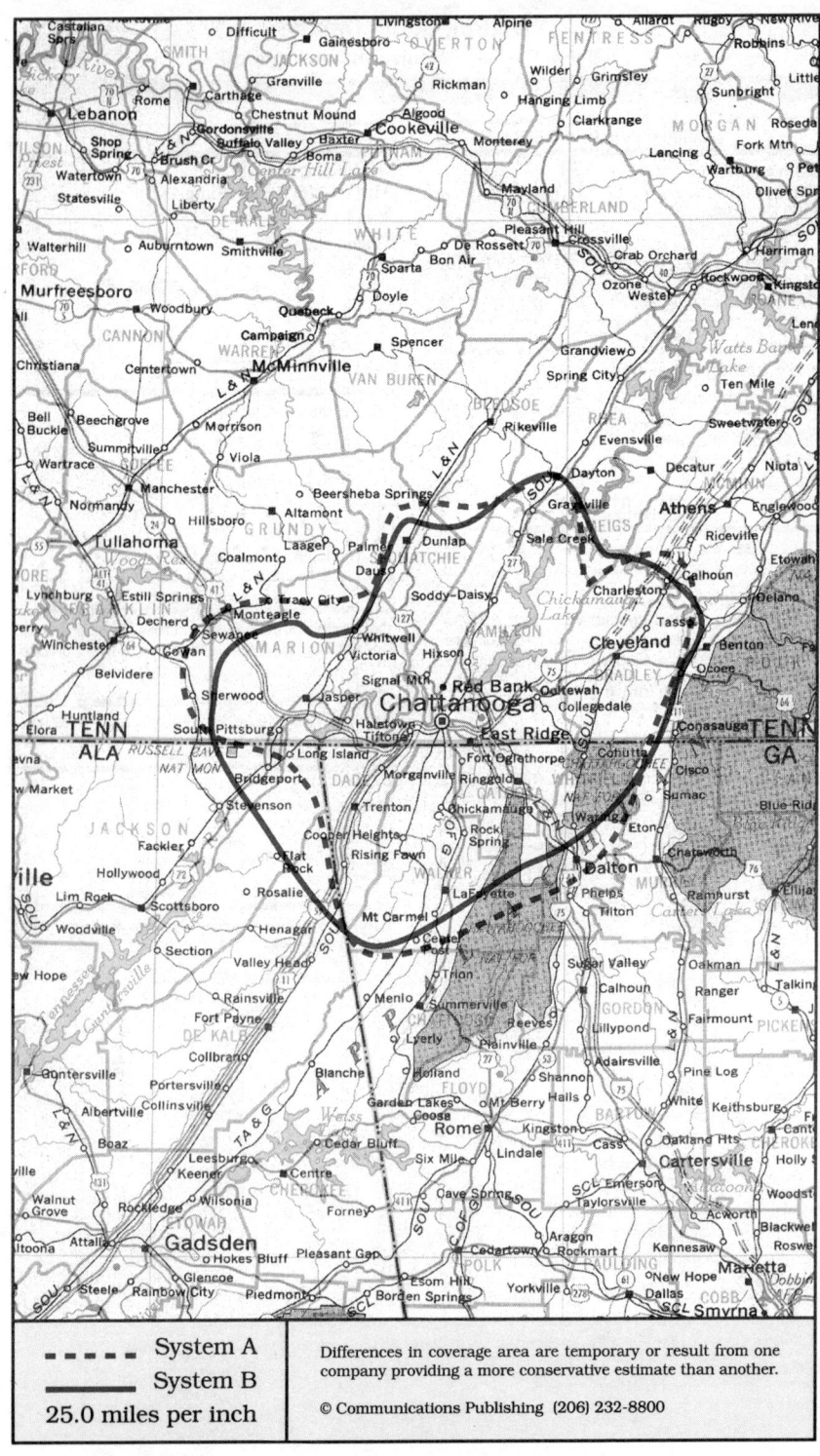

▪ ▪ ▪ System A	
──── System B	
25.0 miles per inch	

Differences in coverage area are temporary or result from one company providing a more conservative estimate than another.

© Communications Publishing (206) 232-8800

Chattanooga, Tennessee
Chattanooga, Cleveland, Dayton, Dunlap, Jasper, Lafayette GA, Trenton GA

System A	System B

Cellular company

Cellular One
Contel Cellular
5959 Shallowford Road, Suite 109
Chattanooga, TN 37421

Customer service	800-333-4004
. . . .	615-269-2273
Local office	615-892-2355
Corporate office	404-804-3400

Cellular company

BellSouth Mobility
2020 Gunbarrel Road, Suite 401
Chattanooga, TN 37421-2663

Customer service	800-351-2400
. . . .	615-855-9220
Local office	615-894-2020
Corporate office	404-847-3600

Sending calls

Visitor dialing instructions

Local calls	area code + 7 digits
Long distance . . .	1 + area code + 7 digits

Speed-dial numbers

Customer service	611
Emergency	911
Tennessee highway patrol . .	615-821-5151
Weather	*99

Sending calls

Visitor dialing instructions

Local calls	area code + 7 digits
Long distance	area code + 7 digits

Speed-dial numbers

Customer service	811
Emergency	911
Roaming information	711
Telephone problems	611
Tennessee highway patrol . .	615-821-5151

Receiving calls

Caller dials the roamer access number below,
hears tone, dials cellular area code and number.

. . . .	615-667-7626

Receiving calls

Caller dials the roamer access number below,
hears tone, dials cellular area code and number.

Chattanooga	615-488-7626
Cleveland	615-339-7626

Follow Me Roaming forwarding (see p. 1156)
Activate, dial *18 send. Deactivate dial *19 send.

Billing information

Visitor rates

Most visitors	$3 day + 60¢ minute

Visitors without roaming agreements (p. 1159)
Call co. to use Amex, Mastercard, Visa

Billing information

Visitor rates

Most visitors	$3 day + 85¢ minute
BellSouth subscribers .	$0 day + 55¢ minute

Visitors without roaming agreements (p. 1159)
Call co. to use Amex, Mastercard, Visa

Identification numbers

City	Market	System	Billing	RSA
Chattanooga	088	00161	00000	MSA
Cleveland	649	01685	00000	TN-7

For full coverage area, see Knoxville, TN.

Identification numbers

City	Market	System	Billing	RSA
Chattanooga	088	00148	00000	MSA
Cleveland	649	00148	30218	TN-7

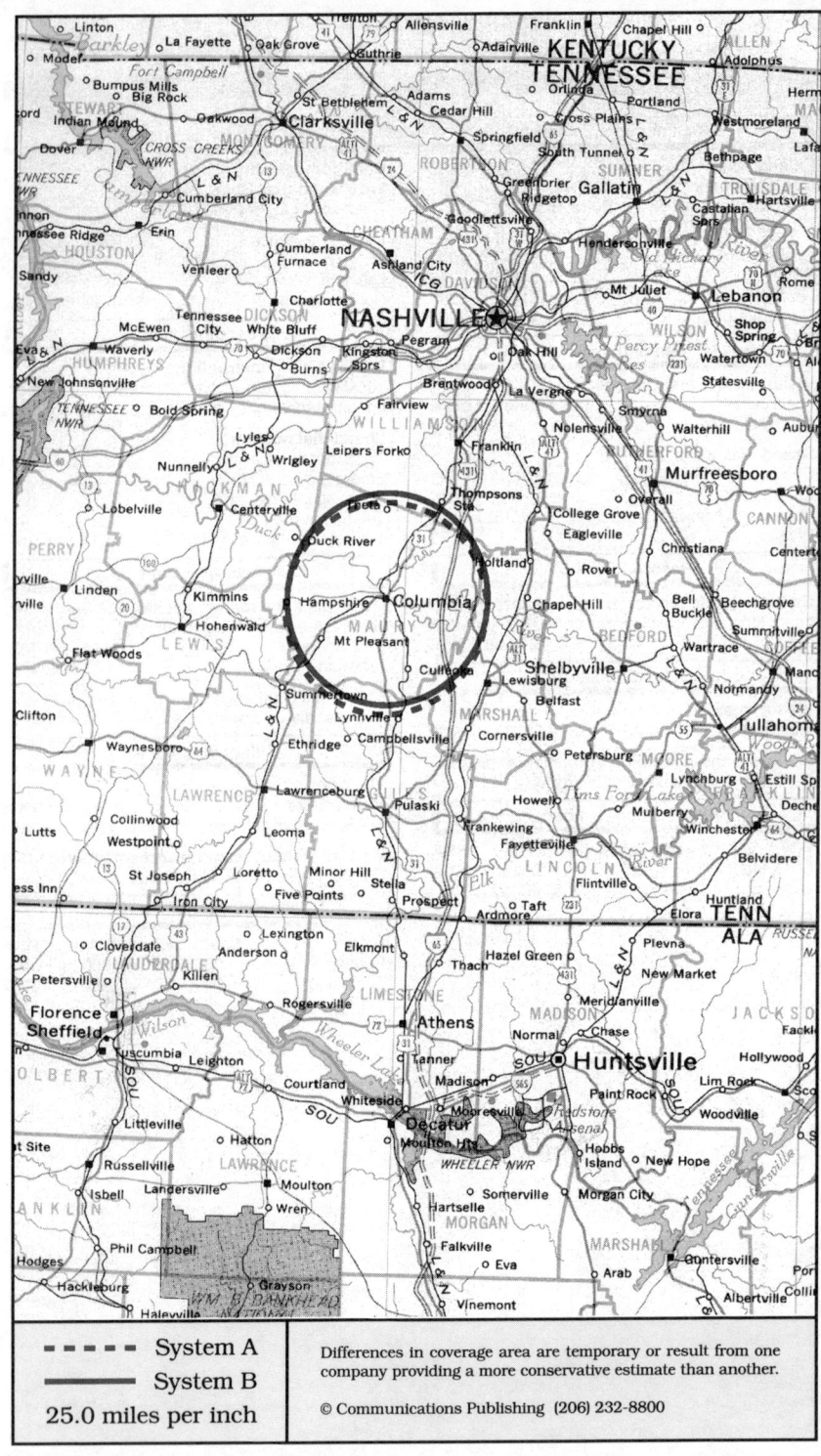

Columbia, Tennessee

|

System A

Cellular company

Cellular One
Licensed to: Ten Woodland Road
Managed by: MCMG
1262 Old Hillsboro Road
Franklin, TN 37064

Customer service	615-685-1255
Corporate office . .	MCMG 615-791-0202

Sending calls

Visitor dialing instructions

Local calls	area code + 7 digits
Long distance . . .	1 + area code + 7 digits

Speed-dial numbers

Customer service	611
Emergency	911

Receiving calls

Caller dials the roamer access number below, hears tone, dials cellular area code and number.
. . . . 615-580-7626

Billing information

Visitor rates

Most visitors	$3 day + 85¢ minute

Visitors without roaming agreements (p. 1159)
Roamer Plus will intercept first call to bill you.

Identification numbers

City	Market	System	Billing	RSA
All served cities	651	01689	00000	TN-9

This system is operated as part of the Shelbyville, TN system A.

System B

Cellular company

BellSouth Mobility
15 Century Boulevard, Suite 103
Nashville, TN 37214

Customer service	800-351-2400
. . . .	615-871-4023
Local office	615-871-2000
Corporate office	404-847-3600

Sending calls

Visitor dialing instructions

Local calls	area code + 7 digits
Long distance . . .	1 + area code + 7 digits

Speed-dial numbers

Customer service	811
Emergency	911
Roaming information	711
Telephone problems	611

Receiving calls

Caller dials the roamer access number below, hears tone, dials cellular area code and number.
. . . . 615-380-7626

Follow Me Roaming forwarding (see p. 1156)
Activate, dial *18 send. Deactivate dial *19 send.

Billing information

Visitor rates

Most visitors	$3 day + 85¢ minute
BellSouth subscribers .	$0 day + 55¢ minute

Visitors without roaming agreements (p. 1159)
Call co. to use Amex, Mastercard, Visa

Identification numbers

City	Market	System	Billing	RSA
All served cities	651	00118	30168	TN-9

For full coverage area, see Nashville, TN.

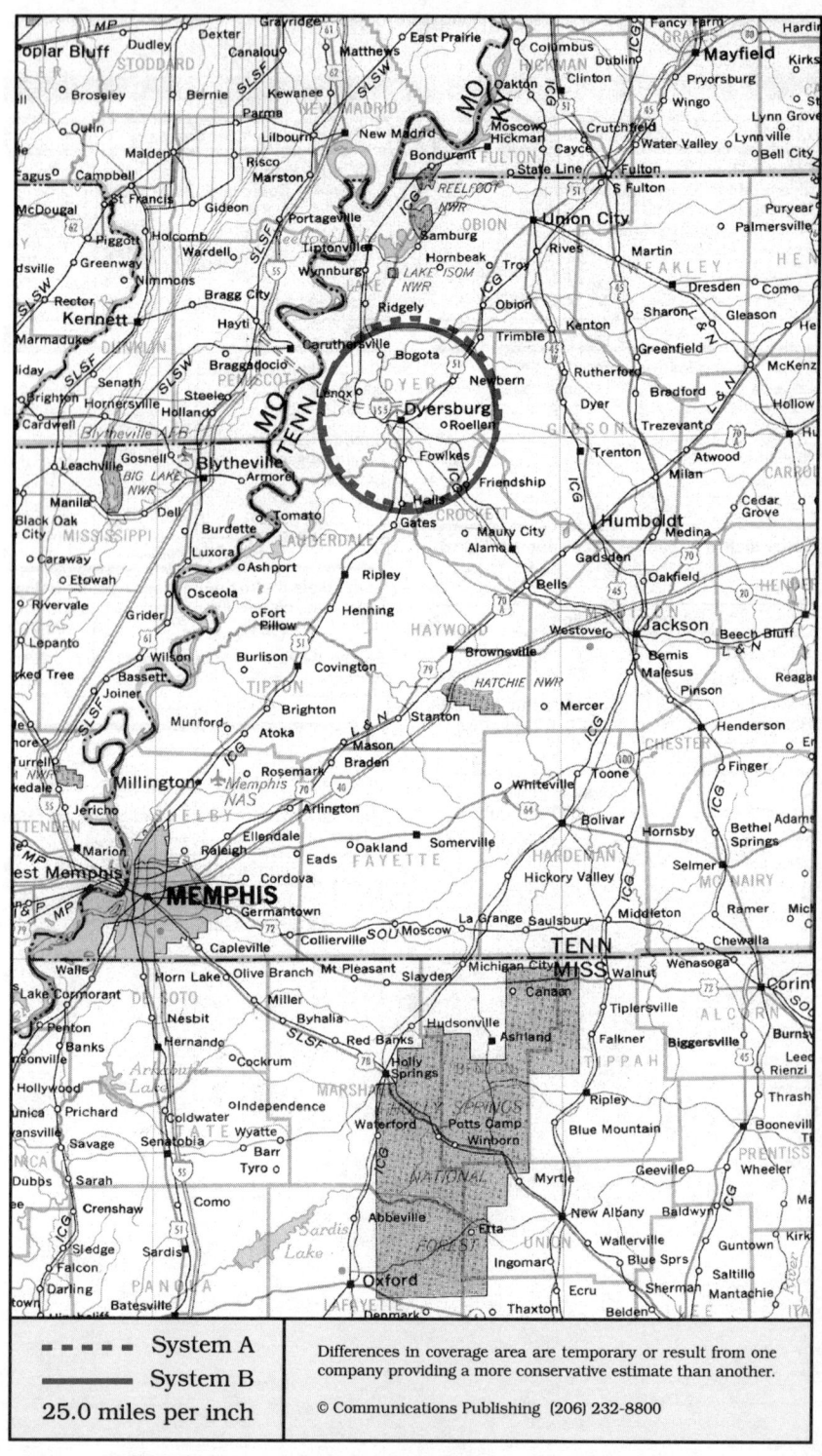

Dyersburg, Tennessee

Cellular company

Cellular One
Contel Cellular
1835 Moriah Woods Boulevard, Suite 5
Memphis, TN 38117

Customer service	800-333-4004
. . . .	615-269-2273
Local office	901-683-2355
Corporate office	404-804-3400

Sending calls

Visitor dialing instructions

Local calls	area code + 7 digits
Long distance . . .	1 + area code + 7 digits

Speed-dial numbers

Customer service	611
Directory assistance	411
Emergency	911
Make a traffic report	211
Tennessee highway patrol . .	615-821-5151

Receiving calls

Caller dials the roamer access number below,
hears tone, dials cellular area code and number.
. . . . 901-483-7626

Billing information

Visitor rates
Most visitors . . . . . $3 day + 85¢ minute

Visitors without roaming agreements (p. 1159)
Call co. to use Amex, Mastercard, Visa
Cannot call since company takes no credit cards.

Identification numbers

City	Market	System	Billing	RSA
All served cities	643	01673	00000	TN-1

Cellular company

Yorkville Telephone Cooperative
2 Nebo-Yorkville Road, PO Box 8
Yorkville, TN 38389

Corporate office	901-643-6121

Sending calls

Visitor dialing instructions

Local calls	area code + 7 digits
Long distance . . .	1 + area code + 7 digits

Speed-dial numbers

Customer service	611
Emergency	911

Receiving calls

Caller dials the roamer access number below,
hears tone, dials cellular area code and number.
. . . . 901-643-7626

Billing information

Visitor rates
Most visitors . . . . . $3 day + 85¢ minute

Visitors without roaming agreements (p. 1159)
Cannot call since company takes no credit cards.

Identification numbers

City	Market	System	Billing	RSA
All served cities	643	01998	00000	TN-1

This service is operated as part of the Memphis,
TN system B.

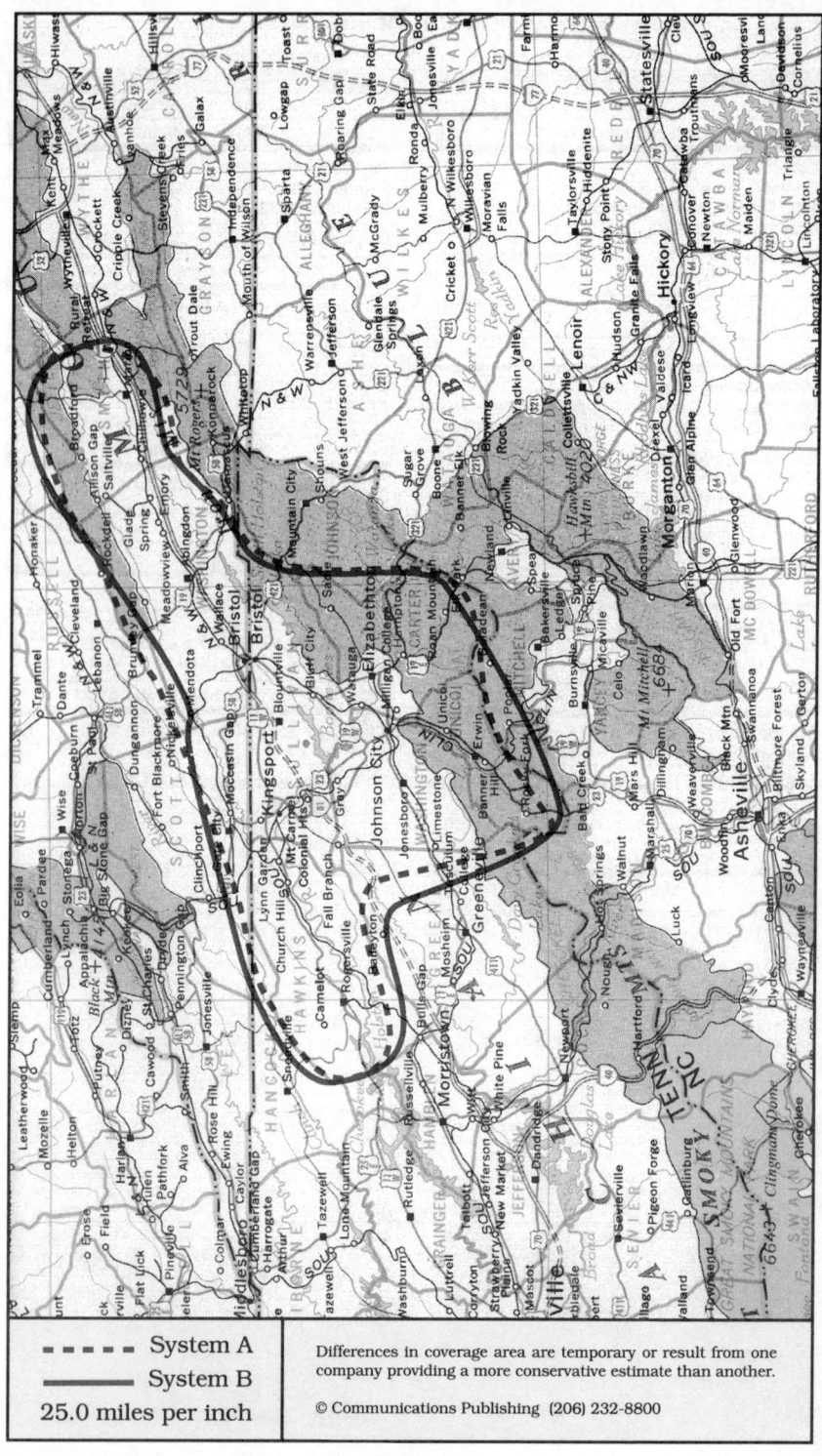

▬ ▬ ▬ System A	Differences in coverage area are temporary or result from one
▬▬▬▬ System B	company providing a more conservative estimate than another.
25.0 miles per inch	© Communications Publishing (206) 232-8800

Johnson City, Tennessee

Bristol, Johnson City, Kingsport (Tri Cities); Mountain City, Marion VA

System A

Cellular company

Cellular One
Contel Cellular
660 Eastern Star Road
Kingsport, TN 37663

Customer service	800-333-4004
. . . .	615-269-2273
Local office	615-341-2355
. . . .	615-349-4500
Corporate office	404-804-3400

Sending calls

Visitor dialing instructions

Local calls	area code + 7 digits
Long distance . . .	1 + area code + 7 digits

Speed-dial numbers

Customer service	611
Emergency	911

Receiving calls

Caller dials the roamer access number below,
hears tone, dials cellular area code and number.

Bristol, Johnson City, Kingsport	615-341-7626
Marion, VA	703-782-7626

Billing information

Visitor rates
Most visitors . . . . . $3 day + 85¢ minute

Visitors without roaming agreements (p. 1159)
Call co. to use Amex, Mastercard, Visa

Identification numbers

City	Market	System	Billing	RSA
Johnson City	085	00149	00000	MSA
Marion, VA	682	01751	00000	VA-2
Mountain City	650	01687	00000	TN-8

System B

Cellular company

Centel Cellular
664 Eastern Star Road
Kingsport, TN 37663

Local office	800-233-1985
. . . .	615-349-5000
Corporate office	312-399-2644

Sending calls

Visitor dialing instructions

Local calls	area code + 7 digits
Long distance . . .	1 + area code + 7 digits

Speed-dial numbers

Customer service	611
Emergency	911
WJHL radio	*11

Receiving calls

Caller dials the roamer access number below,
hears tone, dials cellular area code and number.

. . . . 615-335-7626

Billing information

Visitor rates
Most visitors . . . . . $3 day + 99¢ minute

Visitors without roaming agreements (p. 1159)
Cannot call since company takes no credit cards.

Identification numbers

City	Market	System	Billing	RSA
Johnson City	085	00074	00000	MSA
Marion, VA	682	01752	00000	VA-2
Mountain City	650	01688	00000	TN-8

For full coverage area, see Morristown, TN and Norton, VA.

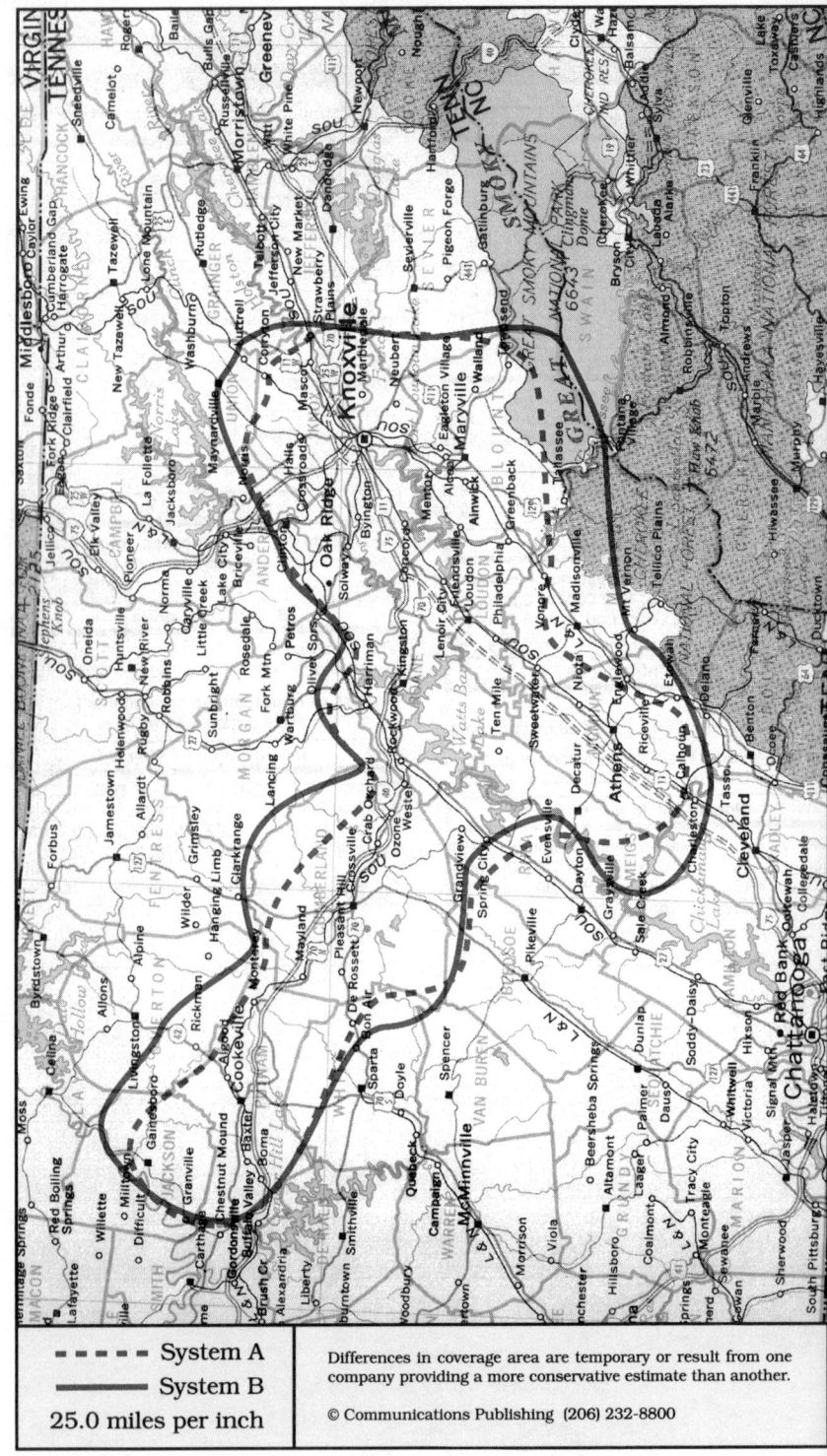

System A
System B
25.0 miles per inch

Differences in coverage area are temporary or result from one company providing a more conservative estimate than another.

© Communications Publishing (206) 232-8800

Knoxville, Tennessee

Athens, Cookeville, Kingston, Knoxville, Loudon, Maryville, Oak Ridge

System A

Cellular company

Cellular One
Contel Cellular
485 Highway 70 S
Crossville, TN 38555

6513 Kingston Pike
Knoxville, TN 37919

Customer service		800-333-4004
. . . .		615-269-2273
Local office . . .	Crossville	615-484-8840
. . . .	Knoxville	615-584-2355
Corporate office		404-804-3400

Sending calls

Visitor dialing instructions
Local calls . . . . . area code + 7 digits
Long distance . . . 1 + area code + 7 digits

Speed-dial numbers
Customer service 611
Emergency 911

Receiving calls

Caller dials the roamer access number below,
hears tone, dials cellular area code and number.
Cookeville, Crossville . . . 615-260-7626
Knoxville 615-567-7626

Billing information

Visitor rates
Most visitors . . . . . $3 day + 85¢ minute

Visitors without roaming agreements (p. 1159)
Call co. to use Amex, Mastercard, Visa

Identification numbers

City	Market	System	Billing	RSA
Athens, Lenoir City	649	01685	00000	TN-7
Cookeville	645	01677	30349	TN-3
Knoxville	079	00093	00000	MSA

For full coverage area, see Chattanooga, TN and Nashville, TN.

System B

Cellular company

United States Cellular
440 S Lowe Avenue, Suite 28
Cookeville, TN 38501

8401 Kingston Pike
Knoxville, TN 37919

Local office . . .	Cookeville	615-528-5103
. . . .	Knoxville	615-539-4600
Corporate office		312-399-8900

Sending calls

Visitor dialing instructions
Local calls . . . . . . area code + 7 digits
Long distance . . . 1 + area code + 7 digits

Speed-dial numbers
Customer service 611
Emergency 911
Make a traffic report 615-679-2277

Receiving calls

Caller dials the roamer access number below,
hears tone, dials cellular area code and number.
Athens 615-680-7626
Cookeville 615-979-7626
Knoxville 615-679-7626

Follow Me Roaming forwarding (see p. 1156)
Activate, dial ✱18 send. Deactivate dial ✱19 send.

Billing information

Visitor rates
Most visitors . . . $2.50 day + 85¢ minute

Visitors without roaming agreements (p. 1159)
Call co. to use Mastercard, Visa

Identification numbers

City	Market	System	Billing	RSA
Athens	649	01686	00000	TN-7
Cookeville	645	01678	00000	TN-3
Knoxville	079	00104	00000	MSA

For full coverage area, see Newport, TN.

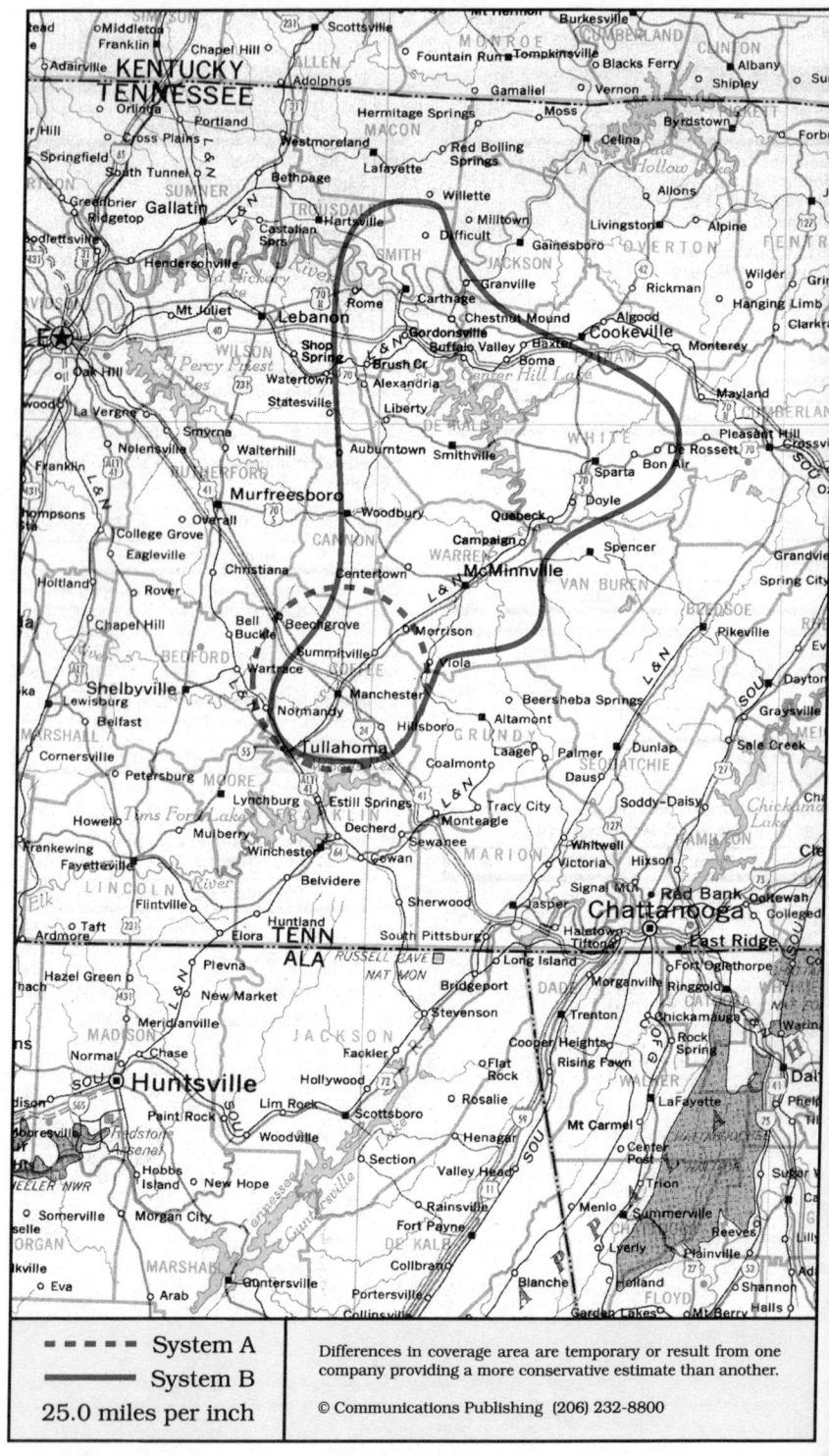

System A
System B
25.0 miles per inch

Differences in coverage area are temporary or result from one company providing a more conservative estimate than another.

© Communications Publishing (206) 232-8800

McMinnville, Tennessee
Carthage, Manchester, McMinnville, Smithville, Sparta

System A

Cellular company

Cellular One
First Tennessee Cellular
Nexus Cellular
PO Box 558
Manchester, TN 37355

Customer service	. .	McCaw	615-723-6529
Local office	. . . . .		615-728-5851
Corporate office	. .	Nexus	206-441-0334

Sending calls

Visitor dialing instructions
Local calls	area code + 7 digits
Long distance . . .	1 + area code + 7 digits

Speed-dial numbers
Customer service	611
Emergency	911

Receiving calls

Caller dials the roamer access number below, hears tone, dials cellular area code and number.
. . . . 615-723-7626

NationLink & RoamingAmerica (see p. 1157)
Caller hears your current roamer access number. This company's subscribers have level 0 service.

North American Cellular Network (see p. 1156)
All calls forward to you. Deactivate dial *35 send.

Billing information

Visitor rates
Most visitors	$0 day + 99¢ minute

Visitors without roaming agreements (p. 1159)
Call co. to use Mastercard, Visa

Identification numbers

City	Market	System	Billing	RSA
All served cities	644	01675	00000	TN-2

Call switching is performed by McCaw Cellular for the licensee, Nexus Cellular.

System B

Cellular company

Advantage Cellular Systems
DeKalb Telephone Cooperative
116 Highway 70, PO Box 457
Alexandria, TN 37012

Local office	. . . .		800-772-8645
. . . .			615-464-2355
Corporate office	.	DeKalb Tel.	615-529-2151

Sending calls

Visitor dialing instructions
Local calls	area code + 7 digits
Long distance . . .	1 + area code + 7 digits

Speed-dial numbers
Customer service	611
Emergency	911

Receiving calls

Caller dials the roamer access number below, hears tone, dials cellular area code and number.
. . . . 615-464-7626

Follow Me Roaming forwarding (see p. 1156)
Activate, dial *18 send. Deactivate dial *19 send.

Billing information

Visitor rates
Most visitors . . .	$2.50 day + 70¢ minute

Visitors without roaming agreements (p. 1159)
Call co. to use Amex, Mastercard, Visa
or Roamer Plus will intercept first call to bill you.

Identification numbers

City	Market	System	Billing	RSA
All served cities	644	01676	00000	TN-2

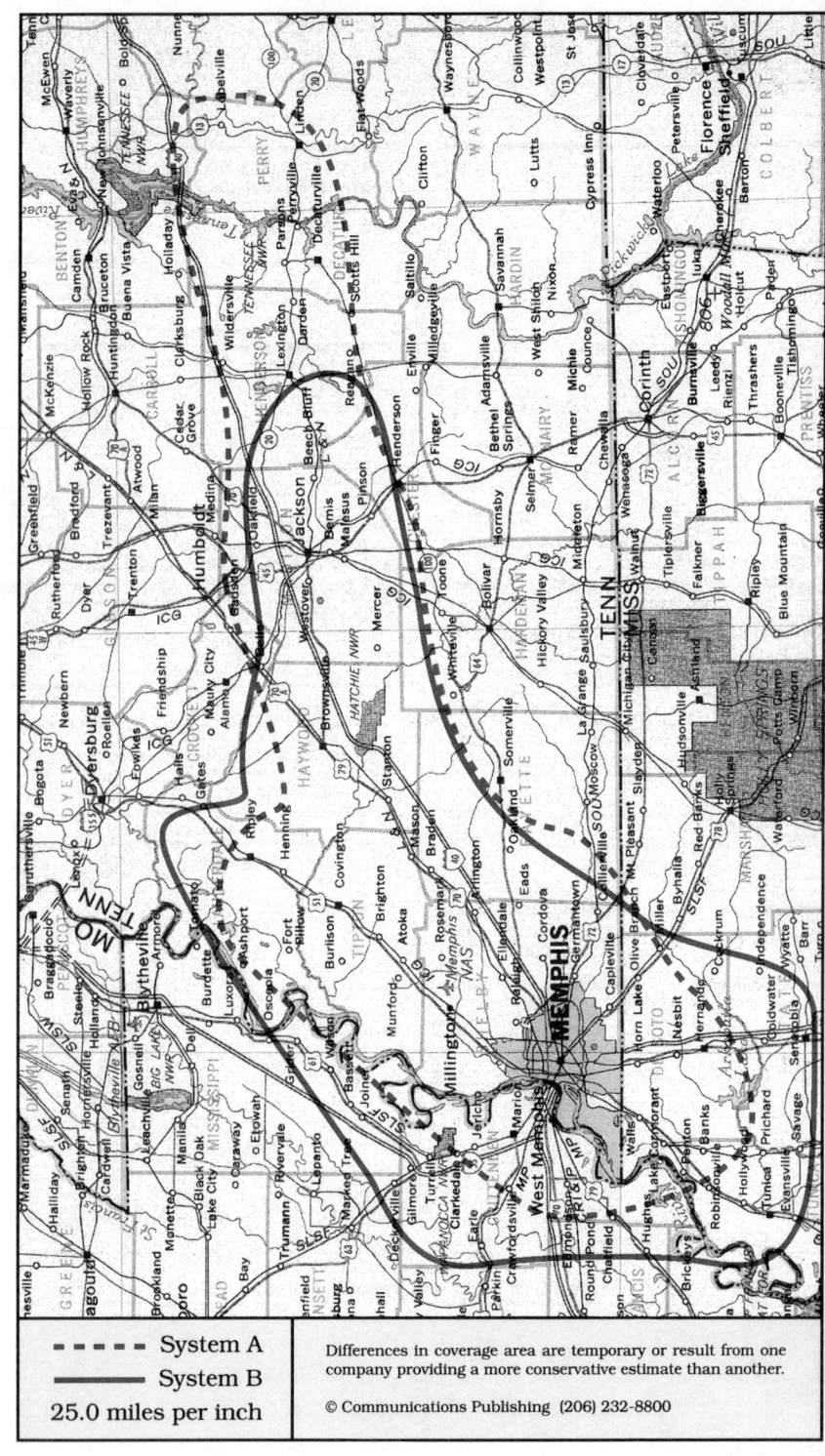

- - - - System A
———— System B
25.0 miles per inch

Differences in coverage area are temporary or result from one company providing a more conservative estimate than another.

© Communications Publishing (206) 232-8800

Memphis, Tennessee

Brownsville, Jackson, Marion AR, Memphis, Millington, Ripley, Senatobia MS

System A	System B

Cellular company

Cellular One
Contel Cellular
1835 Moriah Woods Boulevard, Suite 5
Memphis, TN 38117

Customer service	800-333-4004
. . . .	615-269-2273
Local office	901-683-2355
Corporate office	404-804-3400

Sending calls

Visitor dialing instructions

Local calls	area code + 7 digits
Long distance . . .	1 + area code + 7 digits

Speed-dial numbers

Customer service	611
Directory assistance	411
Emergency	911
Make a traffic report	211
Tennessee highway patrol . .	615-821-5151

Receiving calls

Caller dials the roamer access number below,
hears tone, dials cellular area code and number.
Use either number since they are interconnected.

Jackson	901-988-6999
Memphis	901-483-7626

Billing information

Visitor rates

Most visitors	$3 day + 85¢ minute

Visitors without roaming agreements (p. 1159)
Call co. to use Amex, Mastercard, Visa

Identification numbers

City	Market	System	Billing	RSA
Jackson	647	01681	00000	TN-5
Memphis	036	00143	00000	MSA

Cellular company

BellSouth Mobility
717 S White Station Road, Suite 1
Memphis, TN 38117

Customer service	800-351-2400
. . . .	901-680-8500
Local office	901-763-2165
Corporate office	404-847-3600

Sending calls

Visitor dialing instructions

Local calls	area code + 7 digits
Long distance	area code + 7 digits

Speed-dial numbers

Customer service	811
Emergency	911
Roaming information	711
Telephone problems	611
Tennessee highway patrol . .	615-821-5151

Receiving calls

Caller dials the roamer access number below,
hears tone, dials cellular area code and number.

Jackson	901-426-2999
Memphis	901-485-7626

Follow Me Roaming forwarding (see p. 1156)
Activate, dial *18 send. Deactivate dial *19 send.

Billing information

Visitor rates

Most visitors	$3 day + 85¢ minute
BellSouth subscribers .	$0 day + 55¢ minute

Visitors without roaming agreements (p. 1159)
Call co. to use Amex, Mastercard, Visa

Identification numbers

City	Market	System	Billing	RSA
Jackson	647	00062	30300	TN-5
Memphis	036	00062	00000	MSA

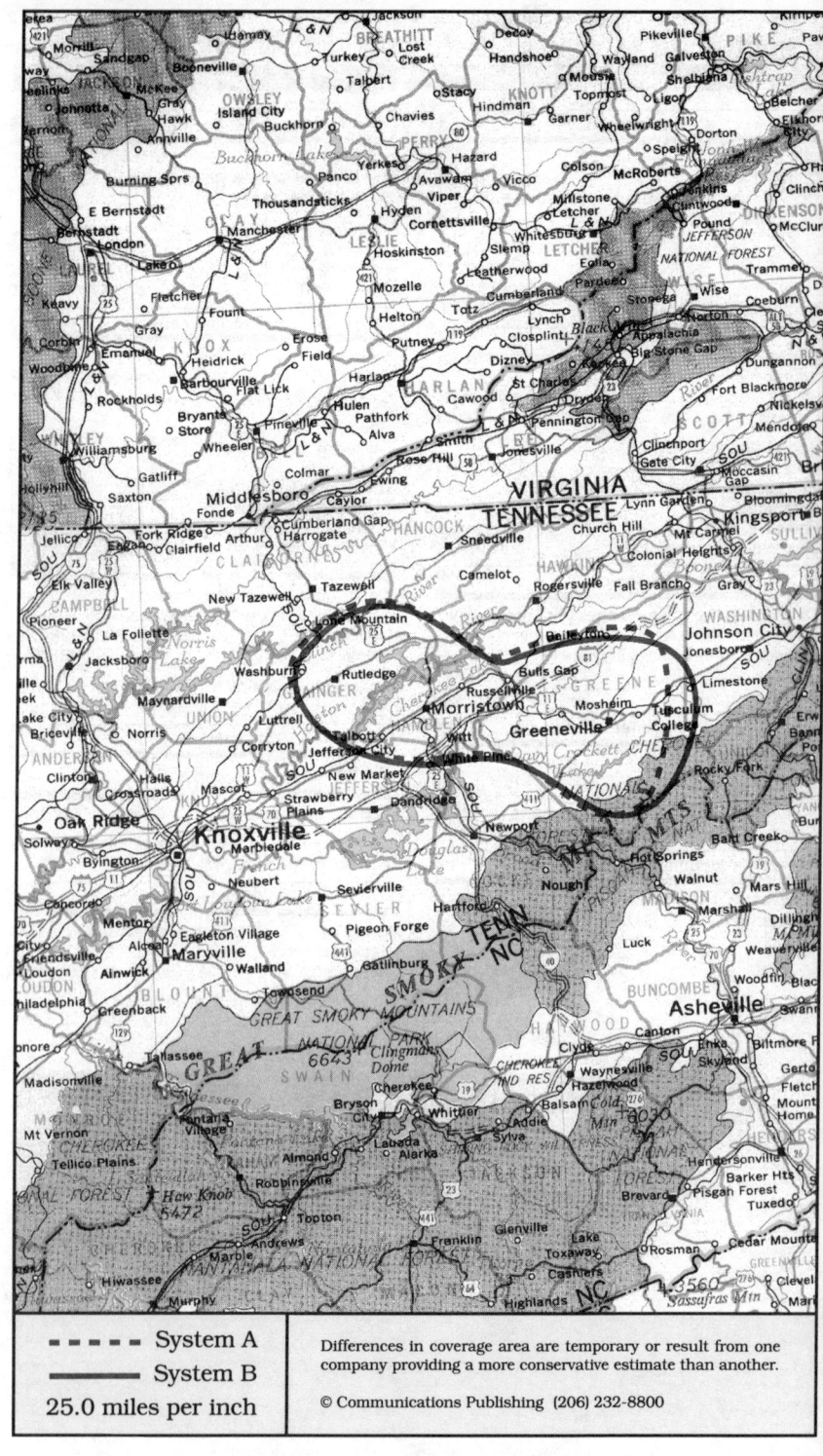

- - - - System A ——— System B 25.0 miles per inch	Differences in coverage area are temporary or result from one company providing a more conservative estimate than another. © Communications Publishing (206) 232-8800

Morristown, Tennessee

Greeneville (Greene Co.), Morristown (Hamblen Co.), Rutledge (Grainger Co.)

System A	System B

Cellular company

Cellular One
Bachtel Cellular
2217 W Andrew Johnson Highway
Morristown, TN 37814

Local office	800-487-1829
. . . .	615-581-3555
Corporate office	215-972-7550

Sending calls

Visitor dialing instructions

Local calls	area code + 7 digits
Long distance . . .	1 + area code + 7 digits

Speed-dial numbers

Customer service	611
Emergency	911

Receiving calls

Caller dials the roamer access number below,
hears tone, dials cellular area code and number.

Greeneville (Greene County) .	615-636-7626
Morristown (Hamblen County)	615-585-7626

Billing information

Visitor rates
Most visitors . . . . . $3 day + 85¢ minute

Visitors without roaming agreements (p. 1159)
Cannot call since company takes no credit cards.

Identification numbers

City	Market	System	Billing	RSA
All served cities	646	01679	00000	TN-4

Call switching and roaming agreements are
provided by Contel Cellular.

For full coverage area, see Newport, TN.

Cellular company

Centel Cellular
664 Eastern Star Road
Kingsport, TN 37663

Local office	800-233-1985
. . . .	615-349-5000
Corporate office	312-399-2644

Sending calls

Visitor dialing instructions

Local calls	area code + 7 digits
Long distance . . .	1 + area code + 7 digits

Speed-dial numbers

Customer service	611
Emergency	911
WJHL radio	★11

Receiving calls

Caller dials the roamer access number below,
hears tone, dials cellular area code and number.

. . . .	615-335-7626

Billing information

Visitor rates
Most visitors . . . . . $3 day + 99¢ minute

Visitors without roaming agreements (p. 1159)
Cannot call since company takes no credit cards.

Identification numbers

City	Market	System	Billing	RSA
All served cities	646	01896	00000	TN-4

For full coverage area, see Johnson City, TN.

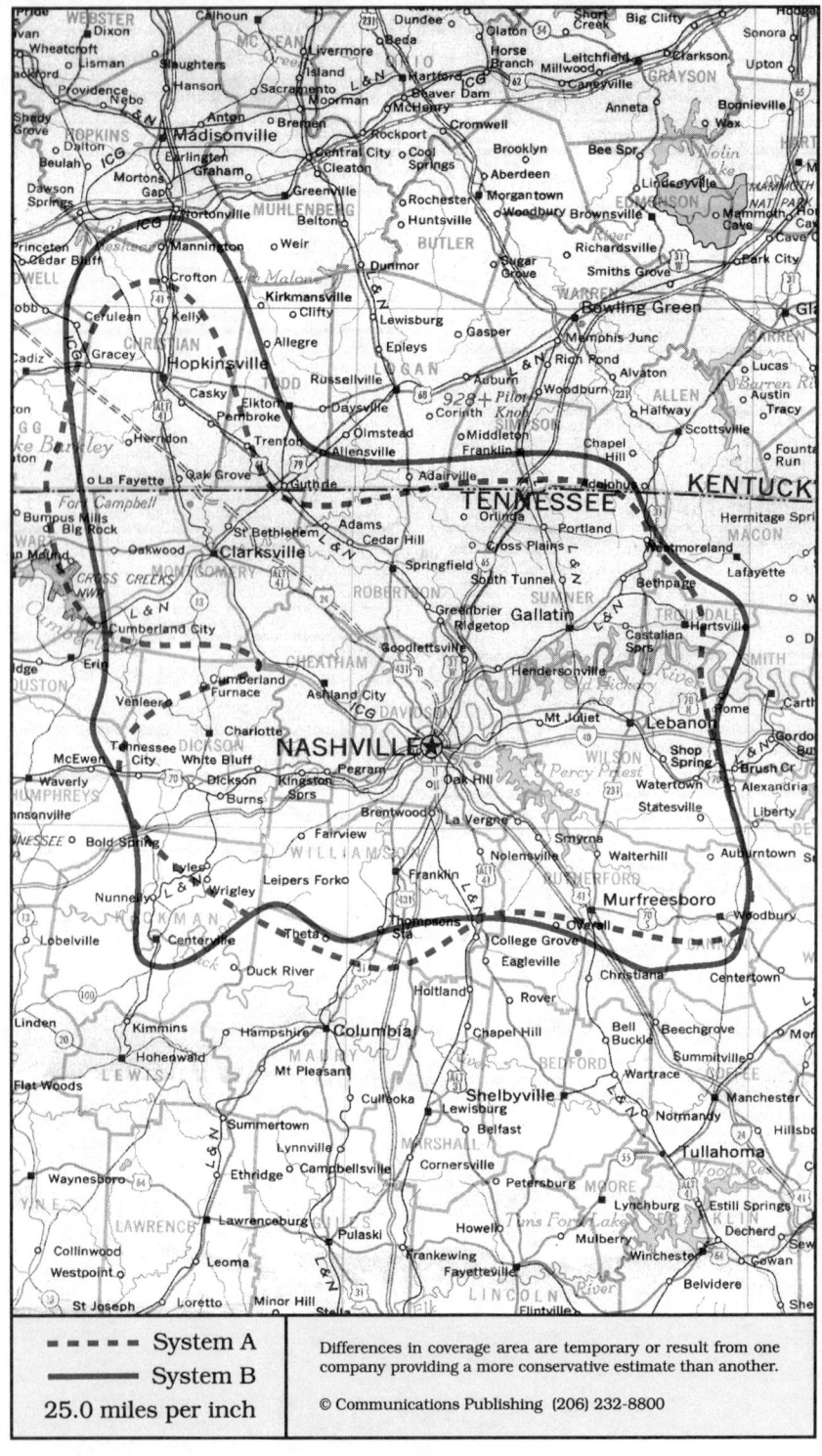

----- System A
——— System B
25.0 miles per inch

Differences in coverage area are temporary or result from one company providing a more conservative estimate than another.

© Communications Publishing (206) 232-8800

Nashville, Tennessee
Clarksville, Gallatin, Hopkinsville KY, Lebanon, Murfreesboro, Nashville

System A	System B

Cellular company

Cellular One
Contel Cellular
2804-B Guthrie Highway
Clarksville, TN 37040

624 Grassmere Park Drive, Suite 4
Nashville, TN 37211

Customer service	800-333-4004
. . . .	615-269-2273
Local office . . . Clarksville	615-648-2355
. . . . . Nashville	615-832-2355
Corporate office	404-804-3400

Sending calls

Visitor dialing instructions

Local calls	area code + 7 digits
Long distance . . .	1 + area code + 7 digits

Speed-dial numbers

Customer service	611
Emergency	911
Stock report	**S or **7
Time	**T or **8
Weather	**W or **9

Receiving calls

Caller dials the roamer access number below,
hears tone, dials cellular area code and number.

. . . .	615-351-7626

Billing information

Visitor rates

Most visitors	$3 day + 60¢ minute

Visitors without roaming agreements (p. 1159)
Call co. to use Amex, Mastercard, Visa

Identification numbers

City	Market	System	Billing	RSA
Clarksville, Hopkin.	209	00179	00000	MSA
Nashville	046	00179	00000	MSA

Cellular company

BellSouth Mobility
1850 Business Park Drive, Suite 110
Clarksville, TN 37040

15 Century Boulevard, Suite 103
Nashville, TN 37214

Customer service	800-351-2400
. . . .	615-871-4023
Local office . . . Clarksville	615-552-0006
. . . . Nashville	615-871-2000
Corporate office	404-847-3600

Sending calls

Visitor dialing instructions

Local calls	area code + 7 digits
Long distance . . .	1 + area code + 7 digits

Speed-dial numbers

Customer service	811
Emergency	911
Roaming information	711
Telephone problems	611

Receiving calls

Caller dials the roamer access number below,
hears tone, dials cellular area code and number.

Clarksville	615-553-7626
Nashville	615-943-7626

Follow Me Roaming forwarding (see p. 1156)
Activate, dial *18 send. Deactivate dial *19 send.

Billing information

Visitor rates

Most visitors . . .	$2.50 day + 85¢ minute
BellSouth subscribers .	$0 day + 55¢ minute

Visitors without roaming agreements (p. 1159)
Call co. to use Amex, Mastercard, Visa

Identification numbers

City	Market	System	Billing	RSA
Clarksville, Hopkin.	209	00296	00000	MSA
Nashville	046	00118	00000	MSA

For full coverage area, see Columbia, TN.

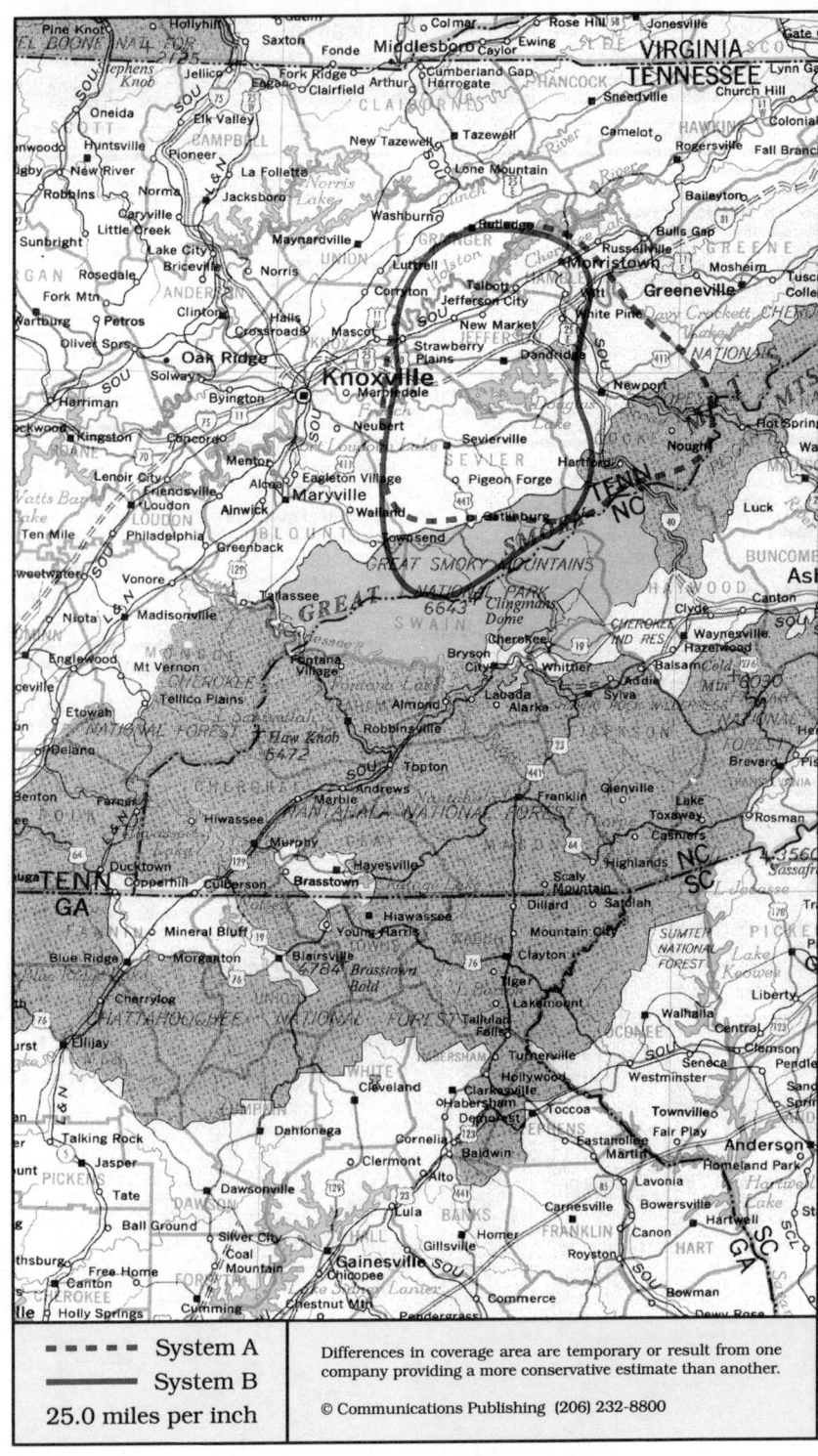

– – – – – System A	Differences in coverage area are temporary or result from one
—————— System B	company providing a more conservative estimate than another.
25.0 miles per inch	© Communications Publishing (206) 232-8800

Newport, Tennessee
Gatlinburg (Sevier Co.), Jefferson City (Jefferson Co.), Newport (Cocke Co.)

System A

Cellular company

Cellular One
Bachtel Cellular
2217 W Andrew Johnson Highway
Morristown, TN 37814

Local office	800-487-1829
. . . .	615-581-3555
Corporate office	215-972-7550

Sending calls

Visitor dialing instructions

Local calls	area code + 7 digits
Long distance . . .	1 + area code + 7 digits

Speed-dial numbers

Customer service	611
Emergency	911

Receiving calls

Caller dials the roamer access number below,
hears tone, dials cellular area code and number.

Cocke County (may change) .	615-585-7626
Jefferson and Sevier Counties	615-850-7626

Billing information

Visitor rates

Most visitors	$3 day + 85¢ minute

Visitors without roaming agreements (p. 1159)
Cannot call since company takes no credit cards.

Identification numbers

City	Market	System	Billing	RSA
All served cities	646	01679	00000	TN-4

For full coverage area, see Morristown, TN.

System B

Cellular company

United States Cellular
8401 Kingston Pike
Knoxville, TN 37919

Local office	615-539-4600
Corporate office	312-399-8900

Sending calls

Visitor dialing instructions

Local calls	area code + 7 digits
Long distance . . .	1 + area code + 7 digits

Speed-dial numbers

Customer service	611
Emergency	911
Make a traffic report	615-679-2277

Receiving calls

Caller dials the roamer access number below,
hears tone, dials cellular area code and number.

. . . .	615-680-7626

Billing information

Visitor rates

Most visitors . . .	$2.50 day + 85¢ minute

Visitors without roaming agreements (p. 1159)
Call co. to use Mastercard, Visa

Identification numbers

City	Market	System	Billing	RSA
All served cities	646	01680	00000	TN-4

For full coverage area, see Knoxville, TN.

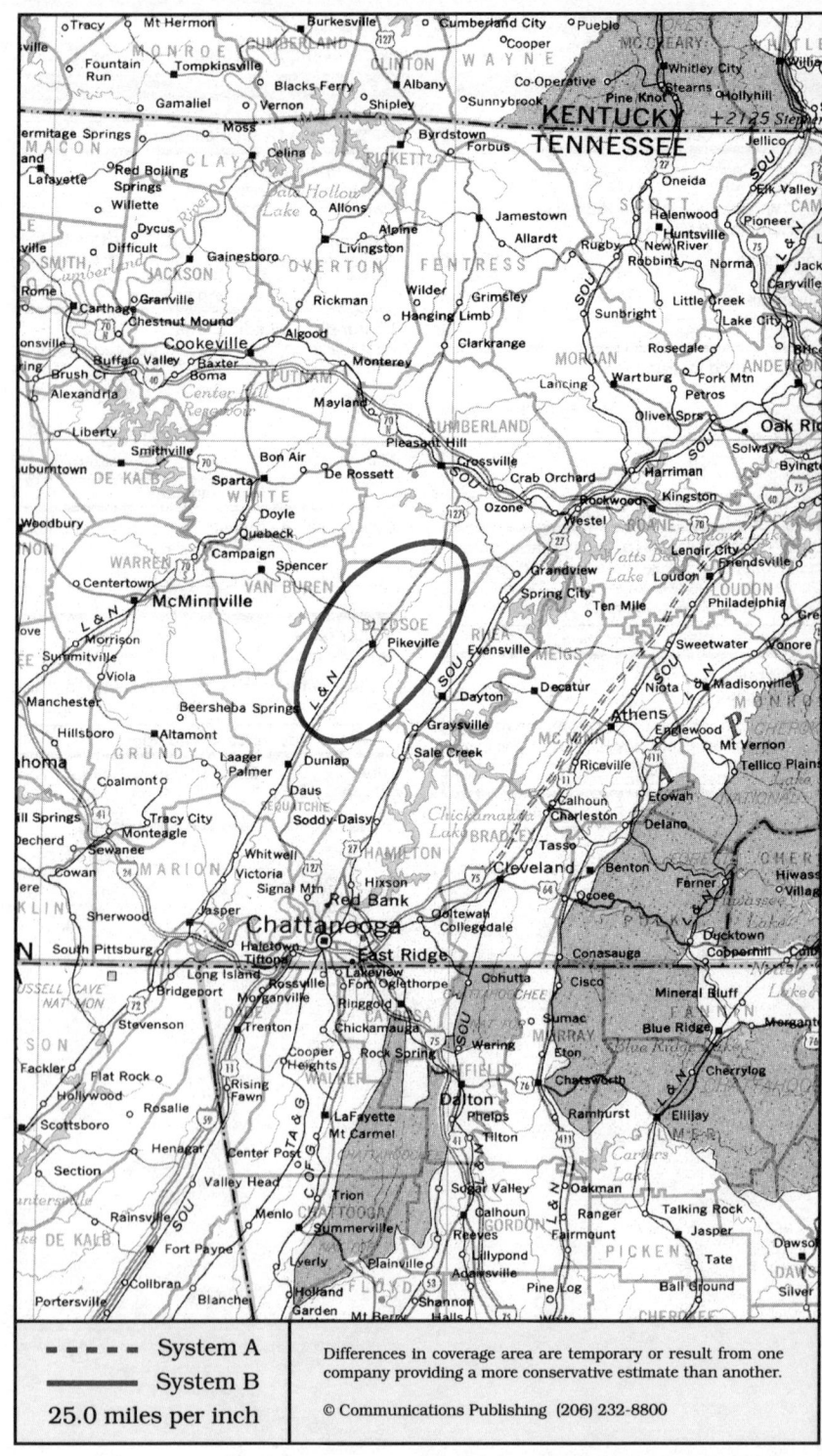

Pikeville, Tennessee

System A	System B

Cellular company

Cellular One
Contel Cellular
5959 Shallowford Road, Suite 109
Chattanooga, TN 37421

Service was not operating at press time.

Customer service	800-333-4004
. . . .	615-269-2273
Local office	615-892-2355
Corporate office	404-804-3400

Sending calls

Visitor dialing instructions

Local calls	area code + 7 digits
Long distance . . .	1 + area code + 7 digits

Speed-dial numbers

Customer service	611
Emergency	911
Tennessee highway patrol . .	615-821-5151
Weather	*99

Receiving calls

Caller dials the roamer access number below,
hears tone, dials cellular area code and number.
. . . . 615-667-7626

Billing information

Visitor rates
Most visitors . . . . . $3 day + 60¢ minute

Visitors without roaming agreements (p. 1159)
Call co. to use Amex, Mastercard, Visa

Identification numbers

City	Market	System	Billing	RSA
Pikeville	649	01685	00000	TN-7

Cellular company

Bledsoe Telephone Cooperative
203 Cumberland Avenue, PO Box 609
Pikeville, TN 37367

Customer service	615-447-2755
Corporate office	615-447-2121

Sending calls

Visitor dialing instructions

Local calls	area code + 7 digits
Long distance	area code + 7 digits

Speed-dial numbers

Customer service	611
Emergency	911

Receiving calls

Caller dials the roamer access number below,
hears tone, dials cellular area code and number.
. . . . 615-447-7626

Billing information

Visitor rates
Most visitors . . . $2.50 day + 85¢ minute

Visitors without roaming agreements (p. 1159)
Cannot call since company takes no credit cards.

Identification numbers

City	Market	System	Billing	RSA
All served cities	649	01884	00000	TN-7

- - - - System A ——— System B 25.0 miles per inch	Differences in coverage area are temporary or result from one company providing a more conservative estimate than another. © Communications Publishing (206) 232-8800

Shelbyville, Tennessee

Fayetteville, Lewisburg, Lynnburg, Pulaski, Shelbyville, Tullahoma, Winchester

System A	System B

System A

Cellular company

Cellular One of Middle Tennessee
Licensed to: Mid-Tenn Cellular Partners
Managed by: MCMG
607 Delray Street
Shelbyville, TN 37160

Customer service	615-685-1255
Local office	615-684-2355
Corporate office . . MCMG	615-791-0202

Sending calls

Visitor dialing instructions
Local calls . . . . . area code + 7 digits
Long distance . . . 1 + area code + 7 digits

Speed-dial numbers
Customer service 611
Emergency 911

Receiving calls

Caller dials the roamer access number below,
hears tone, dials cellular area code and number.
. . . . 615-580-7626

Billing information

Visitor rates
Most visitors . . . . . $3 day + 85¢ minute

Visitors without roaming agreements (p. 1159)
Roamer Plus will intercept first call to bill you.

Identification numbers

City	Market	System	Billing	RSA
All served cities	648	01683	00000	TN-6

System B

Cellular company

United States Cellular
628 Madison Street
Shelbyville, TN 37160

Local office . . . some areas	800-354-8420
. . . .	615-684-4800
Corporate office	312-399-8900

Sending calls

Visitor dialing instructions
Local calls . . . . . . area code + 7 digits
Long distance . . . 1 + area code + 7 digits

Speed-dial numbers
Customer service 611
Emergency 911
Make a traffic report 615-679-2277
Roaming information 711

Receiving calls

Caller dials the roamer access number below,
hears tone, dials cellular area code and number.
. . . . 615-659-7626

Billing information

Visitor rates
Most visitors . . . $2.50 day + 85¢ minute

Visitors without roaming agreements (p. 1159)
Call co. to use Mastercard, Visa

Identification numbers

City	Market	System	Billing	RSA
All served cities	648	01684	00000	TN-6

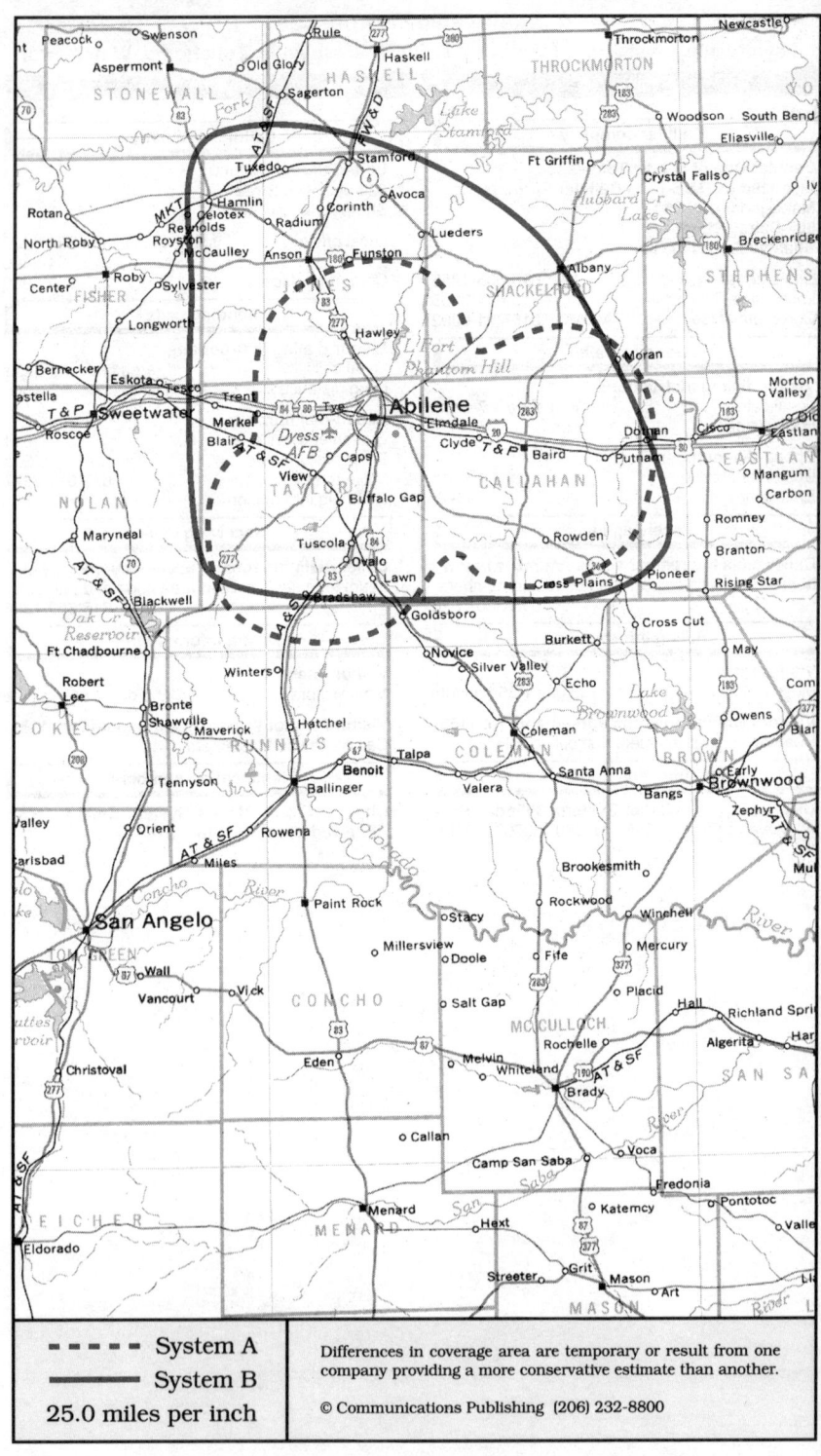

- - - - - System A	Differences in coverage area are temporary or result from one company providing a more conservative estimate than another.
———— System B	
25.0 miles per inch	© Communications Publishing (206) 232-8800

Abilene, Texas
Abilene, Anson, Baird

System A	System B

Cellular company

Cellular One
4102 Buffalo Gap Road
Abilene, TX 79605

Local office 915-695-0117
Corporate office 504-837-8330

Sending calls

Visitor dialing instructions
Local calls . . . . . area code + 7 digits
Long distance . . . 1 + area code + 7 digits

Speed-dial numbers
Customer service 611
Emergency 911

Receiving calls

Caller dials the roamer access number below,
hears tone, dials cellular area code and number.
. . . . 915-674-3333

Billing information

Visitor rates
Most visitors . . . . . $3 day + 75¢ minute

Visitors without roaming agreements (p. 1159)
Call co. to use Amex, Discover, Mastercard, Visa
or Roamer Plus will intercept first call to bill you.

Identification numbers

City	Market	System	Billing	RSA
All served cities	220	00131	00000	MSA

This company's subscribers use the roaming
agreements listed for Radiophone in the red
pages.

Cellular company

Southwestern Bell Mobile Systems
3329 Turner Drive, Suite 108-S
Abilene, TX 79606

Local office 915-698-7626
Corporate office 214-733-2000

Sending calls

Visitor dialing instructions
Local calls . . . . . . area code + 7 digits
Long distance . . . 1 + area code + 7 digits

Speed-dial numbers
Customer service 611
Emergency 911

Receiving calls

Caller dials the roamer access number below,
hears tone, dials cellular area code and number.
. . . . 915-668-7626

Follow Me Roaming forwarding (see p. 1156)
Activate, dial ✳18 send. Deactivate dial ✳19 send.

Billing information

Visitor rates
Most visitors . . . . . $3 day + 75¢ minute

Visitors without roaming agreements (p. 1159)
Call co. to use Amex, Discover, Mastercard, Visa

Identification numbers

City	Market	System	Billing	RSA
All served cities	220	00422	30008	MSA

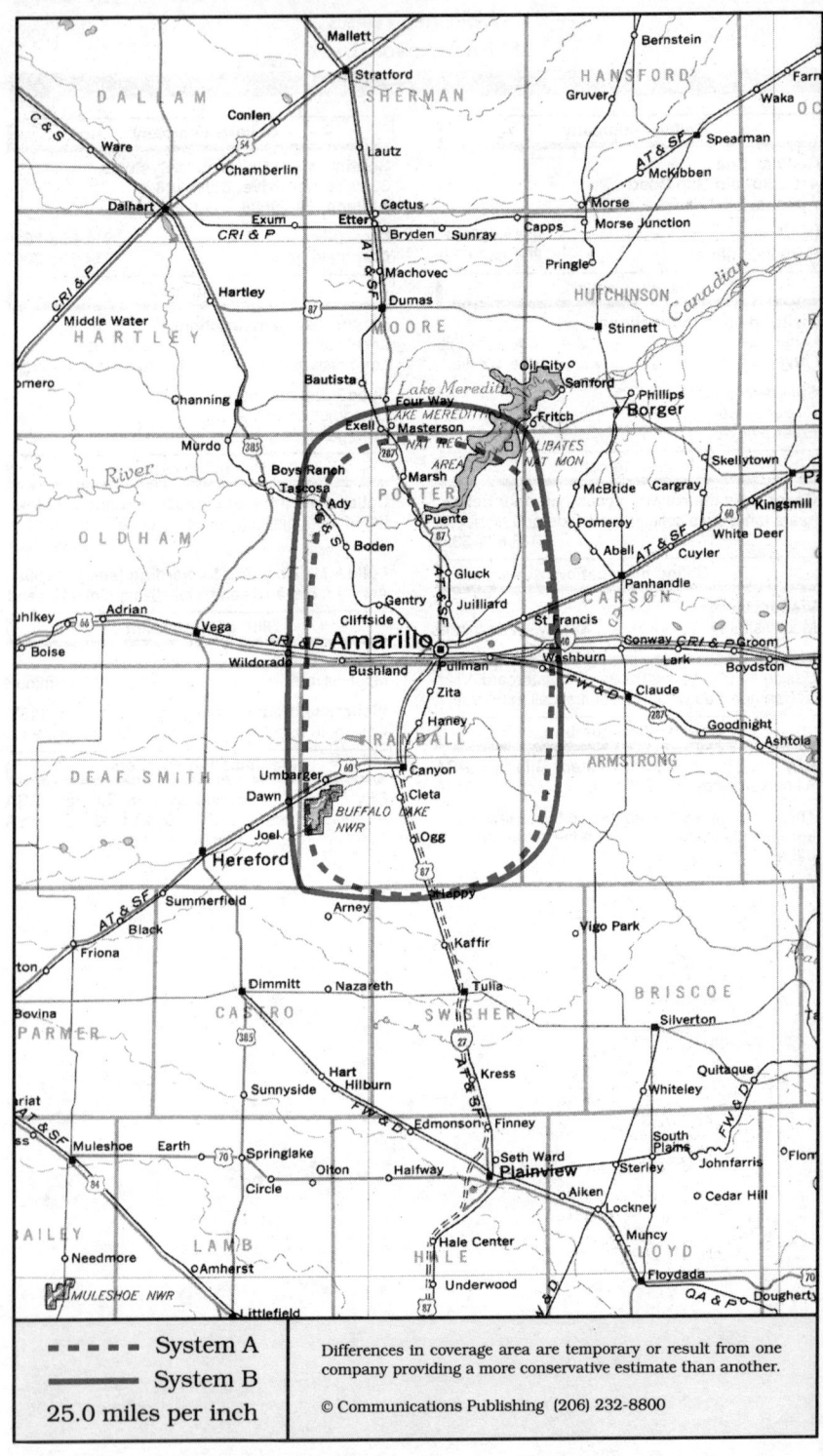

Amarillo, Texas
Amarillo, Canyon

System A	System B

Cellular company

Cellular One of Amarillo
Amarillo Cellular Telephone
105 E 9th
Amarillo, TX 79101

Corporate office 806-374-1900

Sending calls

Visitor dialing instructions
Local calls . . . . . area code + 7 digits
Long distance . . . 1 + area code + 7 digits

Speed-dial numbers
Customer service 611
Emergency 911

Receiving calls

Caller dials the roamer access number below,
hears tone, dials cellular area code and number.
. . . . 806-678-7626

NationLink & RoamingAmerica (see p. 1157)
Caller hears your current roamer access number.
This company's subscribers have level 0 service.

Billing information

Visitor rates
Most visitors . . . . . $3 day + 75¢ minute

Visitors without roaming agreements (p. 1159)
Roamer Plus will intercept first call to bill you.

Identification numbers

City	Market	System	Billing	RSA
All served cities	188	00249	00000	MSA

Cellular company

Southwestern Bell Mobile Systems
6605 Interstate 40 W, Suite B
Amarillo, TX 79106

Local office 806-353-7447
Corporate office 214-733-2000

Sending calls

Visitor dialing instructions
Local calls . . . . . . area code + 7 digits
Long distance . . . 1 + area code + 7 digits

Speed-dial numbers
Customer service 611
Emergency 911
Roaming information 711

Receiving calls

Caller dials the roamer access number below,
hears tone, dials cellular area code and number.
. . . . 806-679-7626

Follow Me Roaming forwarding (see p. 1156)
Activate, dial *18 send. Deactivate dial *19 send.

Billing information

Visitor rates
Most visitors . . . . . $2 day + 75¢ minute

Visitors without roaming agreements (p. 1159)
Call co. to use Amex, Discover, Mastercard, Visa

Identification numbers

City	Market	System	Billing	RSA
All served cities	188	00422	30010	MSA

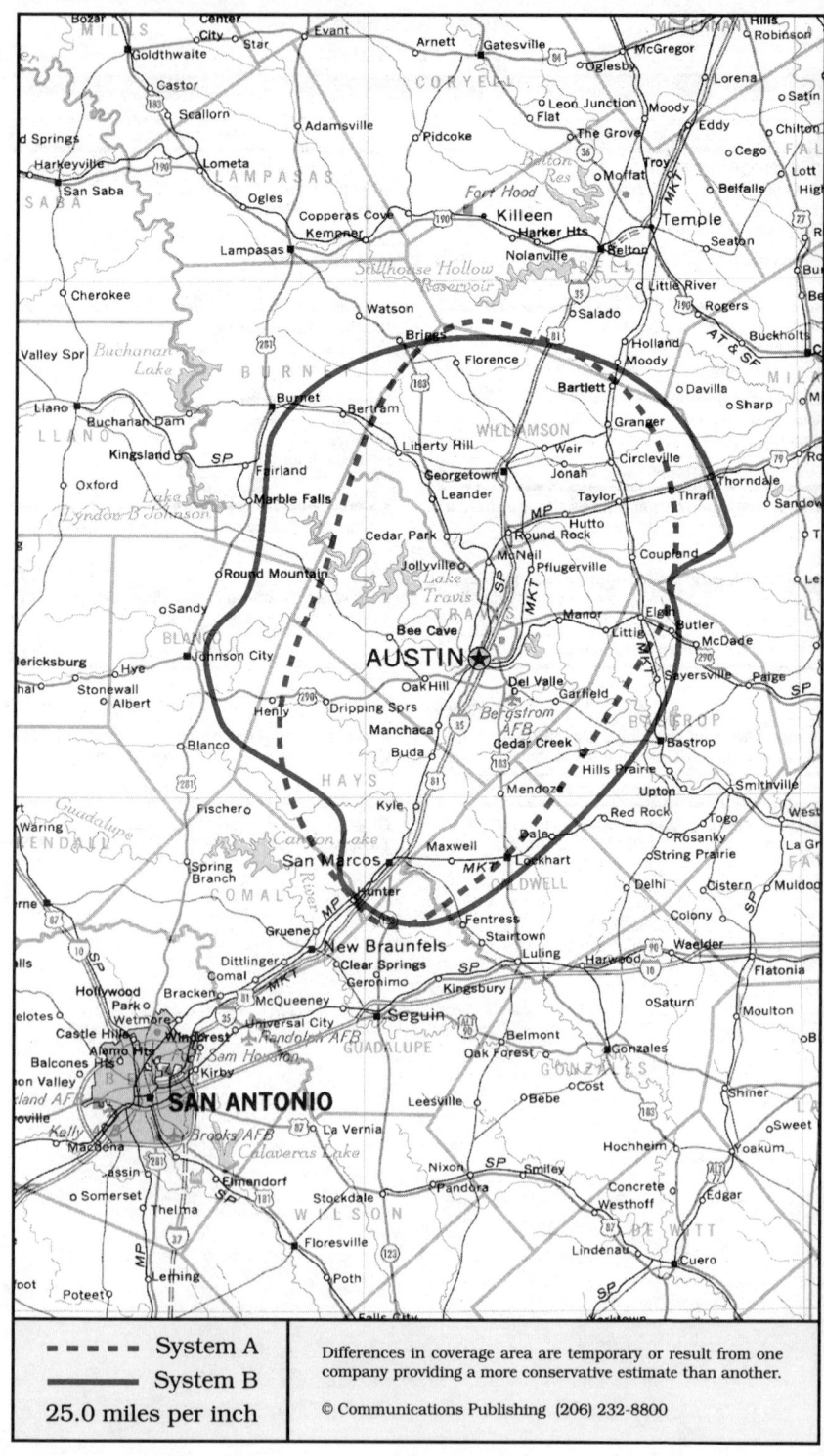

System A
System B
25.0 miles per inch

Differences in coverage area are temporary or result from one company providing a more conservative estimate than another.

© Communications Publishing (206) 232-8800

Austin, Texas
Austin, Georgetown, San Marcos

System A	System B

Cellular company

Cellular One
McCaw Cellular Communications
1120 Capital of Texas Highway, Bldg 1, #100
Austin, TX 78746

Customer service	800-262-3659
. . . .	512-750-7500
Local office	512-750-7700
Corporate office	206-827-4500

Sending calls

Visitor dialing instructions

Local calls	area code + 7 digits
Long distance . . .	1 + area code + 7 digits

Speed-dial numbers

Customer service	611
Emergency	911

Receiving calls

Caller dials the roamer access number below,
hears tone, dials cellular area code and number.
. . . . 512-422-7626

NationLink & RoamingAmerica (see p. 1157)
Dial *31 send to receive calls, *30 to deactivate.
Dial *32 send to tell caller the roamer access no.
This company's subscribers have level 9 service.

North American Cellular Network (see p. 1156)
All calls forward to you. Deactivate dial *35 send.

Billing information

Visitor rates
Most visitors . . . . . $0 day + 99¢ minute

Visitors without roaming agreements (p. 1159)
Call co. to use Amex, Discover, Mastercard, Visa
or Roamer Plus will intercept first call to bill you.

Identification numbers

City	Market	System	Billing	RSA
All served cities	075	00107	00000	MSA

For full coverage area, see San Antonio, TX and
Waco, TX.

Cellular company

GTE Mobilnet
6010 N Interstate Highway 35
Austin, TX 78752

Customer service	800-347-5665
. . . .	713-876-9144
Local office	512-458-2121
Corporate office	404-391-8000

Sending calls

Visitor dialing instructions

Local calls	area code + 7 digits
Long distance . . .	1 + area code + 7 digits

Speed-dial numbers

Customer service	*611
Emergency	911
Roaming information	*711
Telephone problems	*111

Receiving calls

Caller dials the roamer access number below,
hears tone, dials cellular area code and number.
. . . . 512-461-7626

Follow Me Roaming forwarding (see p. 1156)
Activate, dial *18 send. Deactivate dial *19 send.

Billing information

Visitor rates
Most visitors . . . . . $3 day + 85¢ minute

Visitors without roaming agreements (p. 1159)
Call co. to use Amex, Mastercard, Visa

Identification numbers

City	Market	System	Billing	RSA
All served cities	075	00164	00000	MSA

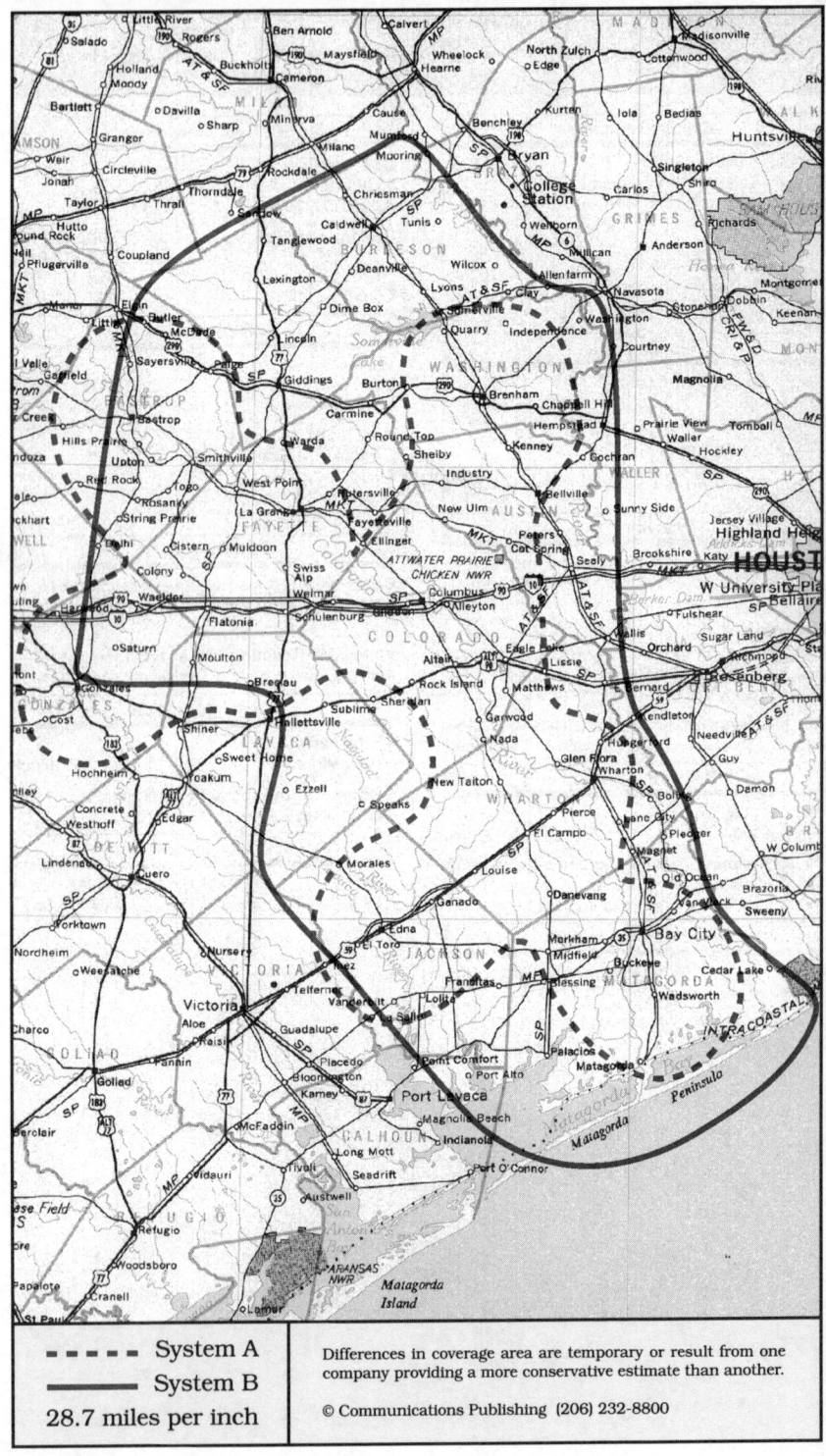

- - - - - System A
———— System B
28.7 miles per inch

Differences in coverage area are temporary or result from one company providing a more conservative estimate than another.

© Communications Publishing (206) 232-8800

Bay City, Texas

Bastrop, Bay City, Brenham, Caldwell, Edna, El Campo, Giddings, Gonzales

System A	System B

System A

Cellular company

Cellular One of Mid-South Texas
1611 N Mechanic, PO Box 1302
El Campo, TX 77437

Corporate office	some areas	800-949-2355
. . . .		409-543-1922

Sending calls

Visitor dialing instructions

Local calls	. . .	1 + area code + 7 digits
Long distance	. . .	1 + area code + 7 digits

Speed-dial numbers

Customer service	611
Emergency	911

Receiving calls

Caller dials the roamer access number below,
hears tone, dials cellular area code and number.
. . . . 409-541-2626

Billing information

Visitor rates
Most visitors . . . . . $3 day + 85¢ minute

Visitors without roaming agreements (p. 1159)
Call co. to use Amex, Discover, Mastercard, Visa
or Roamer Plus will intercept first call to bill you.

Identification numbers

City	Market	System	Billing	RSA
All served cities	667	01721	00000	TX-16

System B

Cellular company

GTE Mobilnet
100 Glenborough, Suite 800
Houston, TX 77067

Customer service	. . . .	800-347-5665
. . . .		713-876-9144
Local office	. . . . . .	713-876-5000
Corporate office	. . . .	404-391-8000

Sending calls

Visitor dialing instructions

Local calls	. . . . . .	area code + 7 digits
Long distance	. . .	1 + area code + 7 digits

Speed-dial numbers

Customer service	*611
Emergency	911
Roaming information	*711
Telephone problems	*111

Receiving calls

Caller dials the roamer access number below,
hears tone, dials cellular area code and number.
. . . . 713-824-7626

Follow Me Roaming forwarding (see p. 1156)
Activate, dial *18 send. Deactivate dial *19 send.

Billing information

Visitor rates
Most visitors . . . . . $3 day + 85¢ minute

Visitors without roaming agreements (p. 1159)
Call co. to use Amex, Mastercard, Visa

Identification numbers

City	Market	System	Billing	RSA
All served cities	667	00012	00000	TX-16

For full coverage area, see Bryan, TX and
Houston, TX and Victoria, TX.

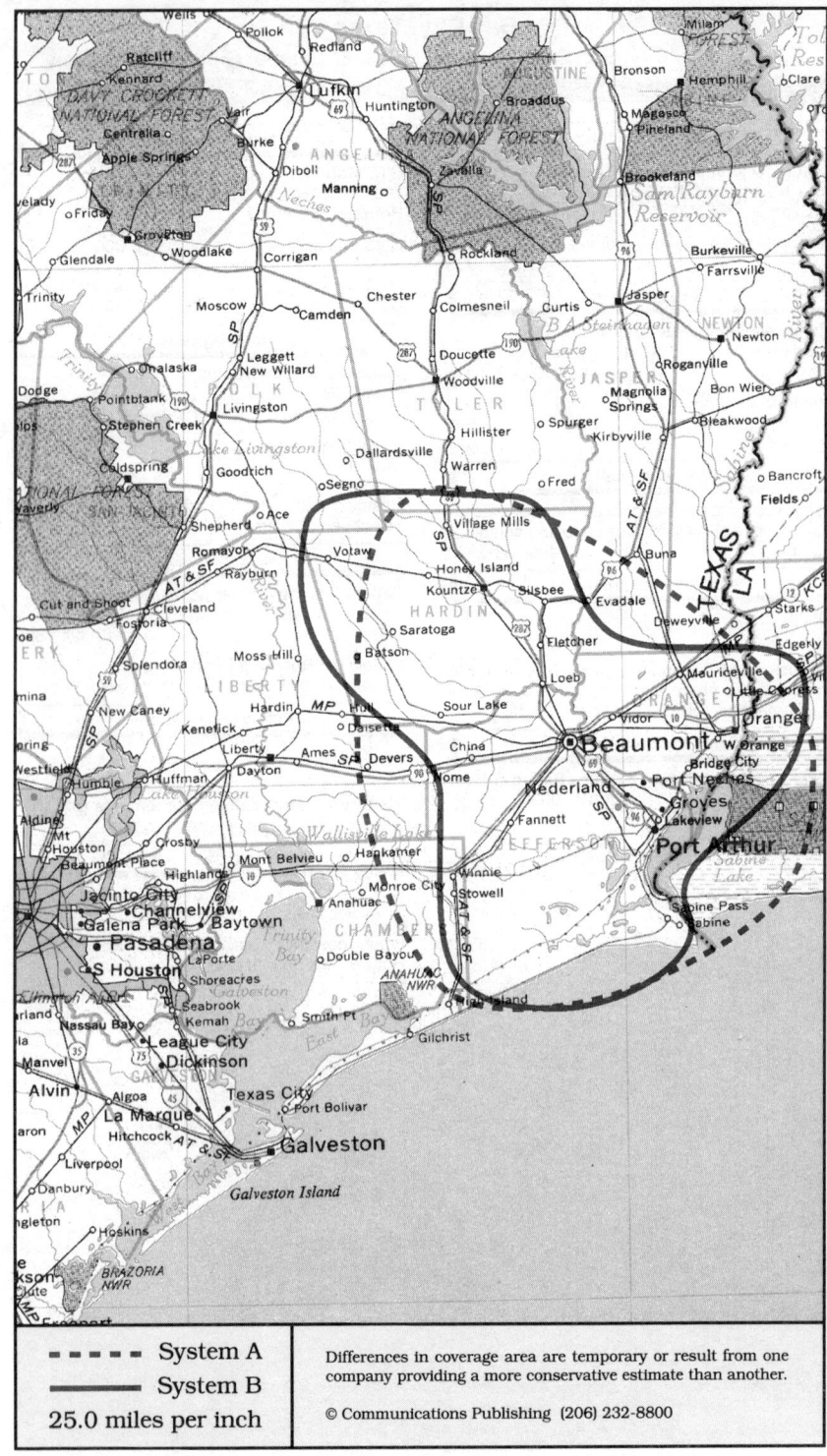

----- System A	Differences in coverage area are temporary or result from one company providing a more conservative estimate than another.
——— System B	
25.0 miles per inch	© Communications Publishing (206) 232-8800

Beaumont, Texas
Beaumont, Kountze, Port Arthur, Orange

System A	System B

System A

Cellular company

Cellular One
Centennial Cellular
4414 Dowlen Road, Suite 105
Beaumont, TX 77706

Local office	409-898-8000
Corporate office	203-972-2000

Sending calls

Visitor dialing instructions

Local calls	area code + 7 digits
Long distance . . .	1 + area code + 7 digits

Speed-dial numbers

Customer service	*611
Emergency	911

Receiving calls

Caller dials the roamer access number below,
hears tone, dials cellular area code and number.

. . . . 409-893-7626

NationLink & RoamingAmerica (see p. 1157)
Dial *31 send to receive calls, *30 to deactivate.
Dial *32 send to tell caller the roamer access no.
This company's subscribers have level 9 service.

Billing information

Visitor rates

Most visitors . . . . . $3 day + 85¢ minute

Visitors without roaming agreements (p. 1159)
Cannot call since company takes no credit cards.

Identification numbers

City	Market	System	Billing	RSA
All served cities	101	00185	00000	MSA

System B

Cellular company

GTE Mobilnet
100 Glenborough, Suite 800
Houston, TX 77067

Customer service	800-347-5665
. . . .	713-876-9144
Local office	713-876-5000
Corporate office	404-391-8000

Sending calls

Visitor dialing instructions

Local calls	area code + 7 digits
Long distance . . .	1 + area code + 7 digits

Speed-dial numbers

Customer service	*611
Emergency	911
Mr. Rescue (roadside assistance)	HELP or 4357
Roaming information	*711
Telephone problems	*111

Receiving calls

Caller dials the roamer access number below,
hears tone, dials cellular area code and number.

. . . . 409-838-7626

Follow Me Roaming forwarding (see p. 1156)
Activate, dial *18 send. Deactivate dial *19 send.

Billing information

Visitor rates
Most visitors . . . . . $3 day + 85¢ minute

Visitors without roaming agreements (p. 1159)
Call co. to use Amex, Mastercard, Visa

Identification numbers

City	Market	System	Billing	RSA
All served cities	101	00012	00000	MSA

For full coverage area, see Houston, TX and
Huntsville, TX.

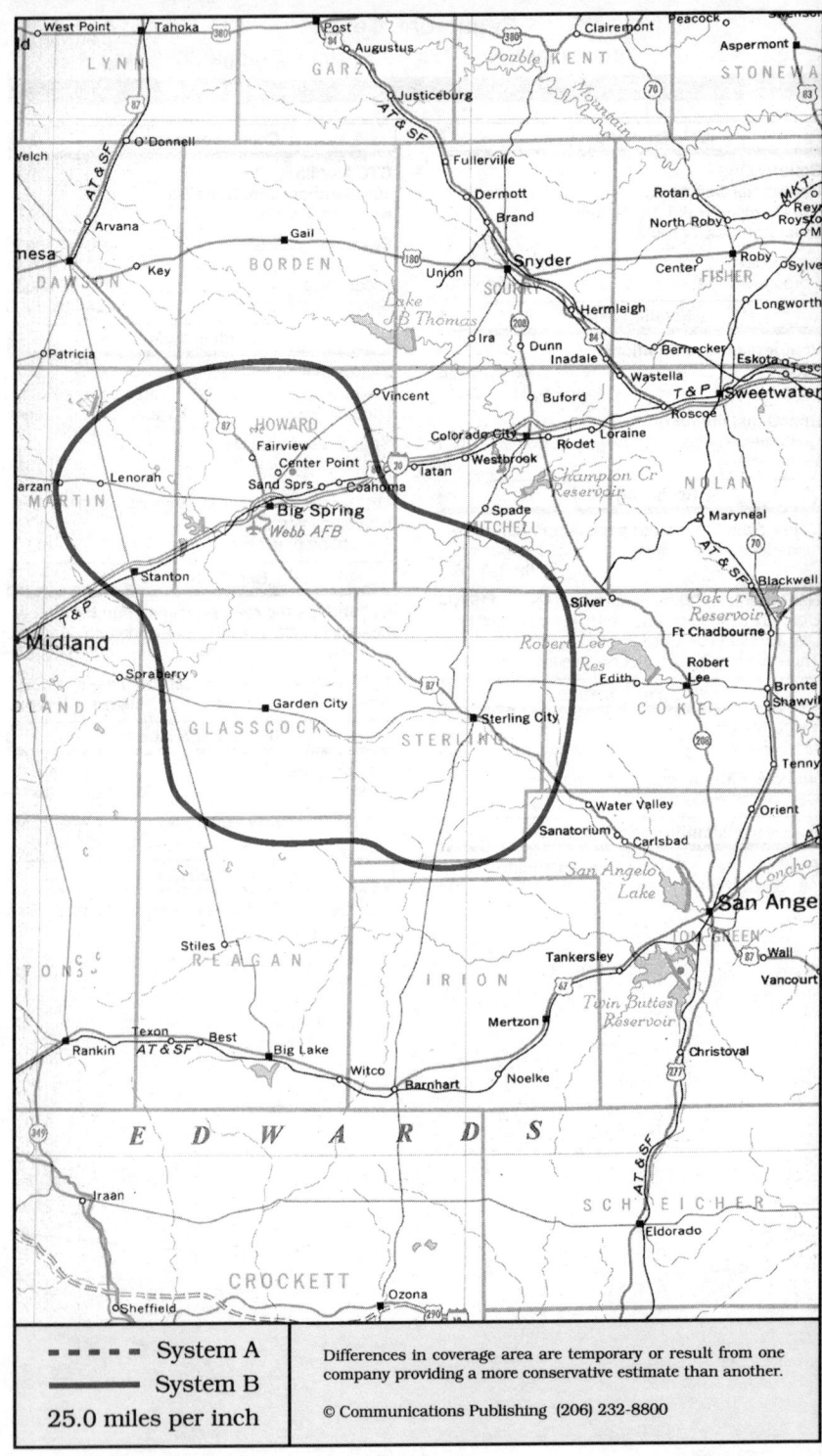

Big Spring, Texas
Big Spring, Garden City, Stanton, Sterling City

System A	System B

Cellular company

No information about the company that will operate system A was available at press time.

Cellular company

Wes-Tex Cellular
Wes-Tex Telecommunications
Wes-Tex Telephone Cooperative
W Loop 214, PO Box 1329
Stanton, TX 79782

Local office	915-756-3826
Corporate office	915-756-3393

Sending calls

Visitor dialing instructions

Local calls	area code + 7 digits
Long distance . . .	1 + area code + 7 digits

Speed-dial numbers

Customer service	611
Emergency	911

Receiving calls

Caller dials the roamer access number below, hears tone, dials cellular area code and number.
. . . . 915-270-7626

Billing information

Visitor rates
Most visitors . . . . . $2 day + 50¢ minute

Visitors without roaming agreements (p. 1159)
Cannot call since company takes no credit cards.

Identification numbers

City	Market	System	Billing	RSA
All served cities	659	01878	00000	TX-8

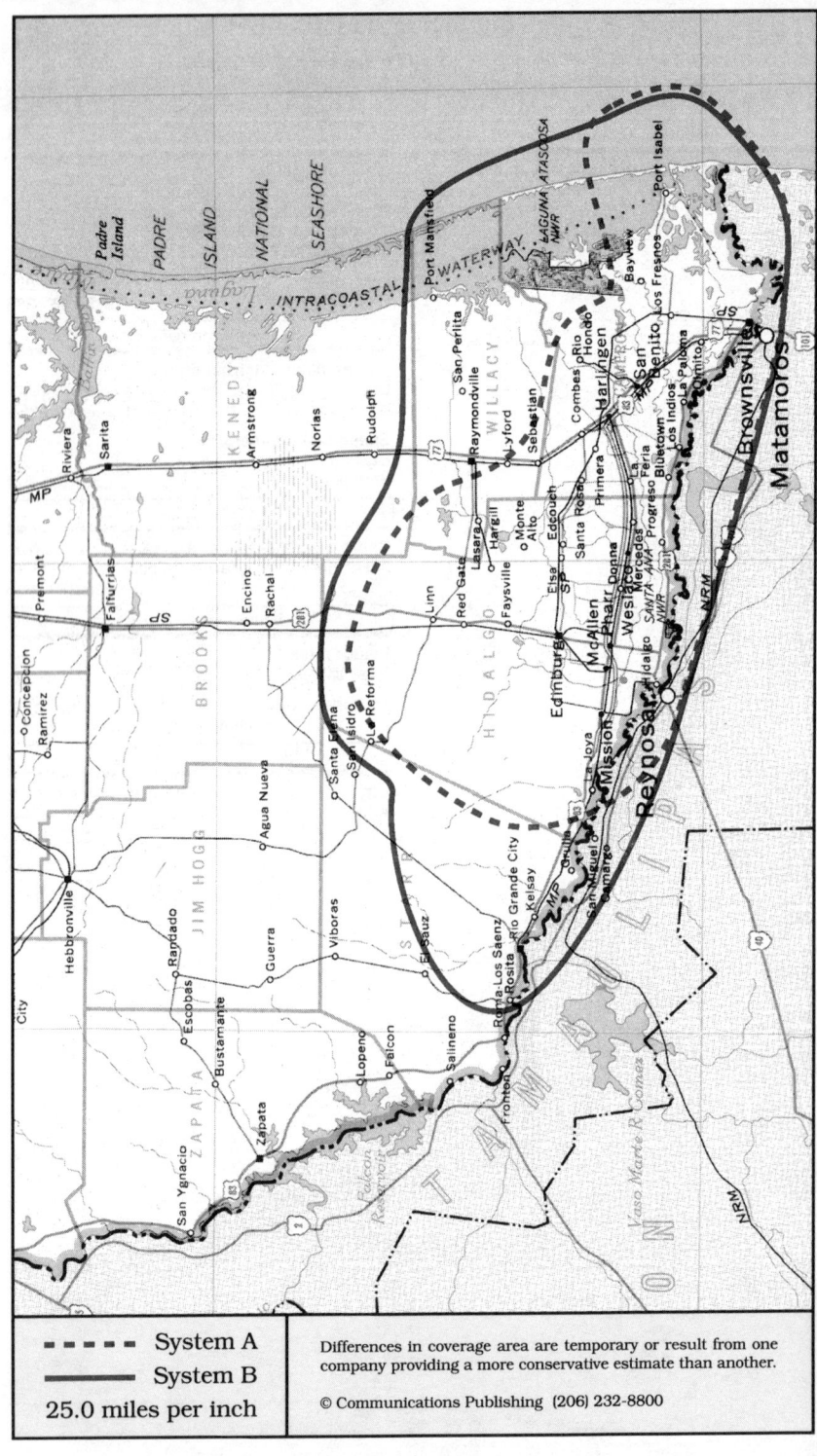

▪▪▪▪▪ System A	
——— System B	Differences in coverage area are temporary or result from one company providing a more conservative estimate than another.
25.0 miles per inch	© Communications Publishing (206) 232-8800

Brownsville, Texas

Brownsville, Edinburg, Harlingen, McAllen, Mission, Raymondville, Rio Grande

System A	System B

Cellular company

Cellular One of the Rio Grande Valley
Celutel
943 N Expressway, Suite 18
Brownsville, TX 78520

508 S 10th Street
McAllen, TX 78501

Customer service	.	Brownsville	210-546-2355
. . . . .		from Harlingen	210-425-2355
. . . . .		from McAllen	210-686-2355
Local office	. . .	Brownsville	210-544-5885
. . . . . .		McAllen	210-686-2355
Corporate office	. . . . .		410-573-5200

Sending calls

Visitor dialing instructions
Local calls . . . . . area code + 7 digits
Long distance . . . 1 + area code + 7 digits

Speed-dial numbers
Customer service 611
Emergency 911

Receiving calls

Caller dials the roamer access number below,
hears tone, dials cellular area code and number.
Brownsville 210-551-7626
Harlingen 210-291-7626
McAllen 210-638-7626

NationLink & RoamingAmerica (see p. 1157)
Dial *31 send to receive calls, *30 to deactivate.
Dial *32 send to tell caller the roamer access no.
This company's subscribers have level 3 service.

Billing information

Visitor rates
Most visitors . . . . . $3 day + 75¢ minute

Visitors without roaming agreements (p. 1159)
Call co. to use Amex, Mastercard, Visa

Identification numbers

City	Market	System	Billing	RSA
Brownsville	162	00451	00000	MSA
McAllen	128	00451	00000	MSA

Cellular company

Southwestern Bell Mobile Systems
6906 W Expressway 83
Harlingen, TX 78552

628 N McColl
McAllen, TX 78501

Local office	. . . .	Harlingen	210-428-6200
. . . .		McAllen	210-630-6060
Corporate office	. . . . .		214-733-2000

Sending calls

Visitor dialing instructions
Local calls . . . . . area code + 7 digits
Long distance . . . 1 + area code + 7 digits

Speed-dial numbers
Customer service 611
Emergency 911

Receiving calls

Caller dials the roamer access number below,
hears tone, dials cellular area code and number.
Brownsville 210-549-7626
Harlingen 210-290-7626
McAllen 210-330-7626
Weslaco 210-963-7626

Follow Me Roaming forwarding (see p. 1156)
Activate, dial *18 send. Deactivate dial *19 send.

Billing information

Visitor rates
Most visitors . . . . . $3 day + 90¢ minute

Visitors without roaming agreements (p. 1159)
Call co. to use Amex, Mastercard, Visa

Identification numbers

City	Market	System	Billing	RSA
Brownsville	162	00278	00000	MSA
McAllen	128	00278	00000	MSA

For full coverage area, see Kingsville, TX.

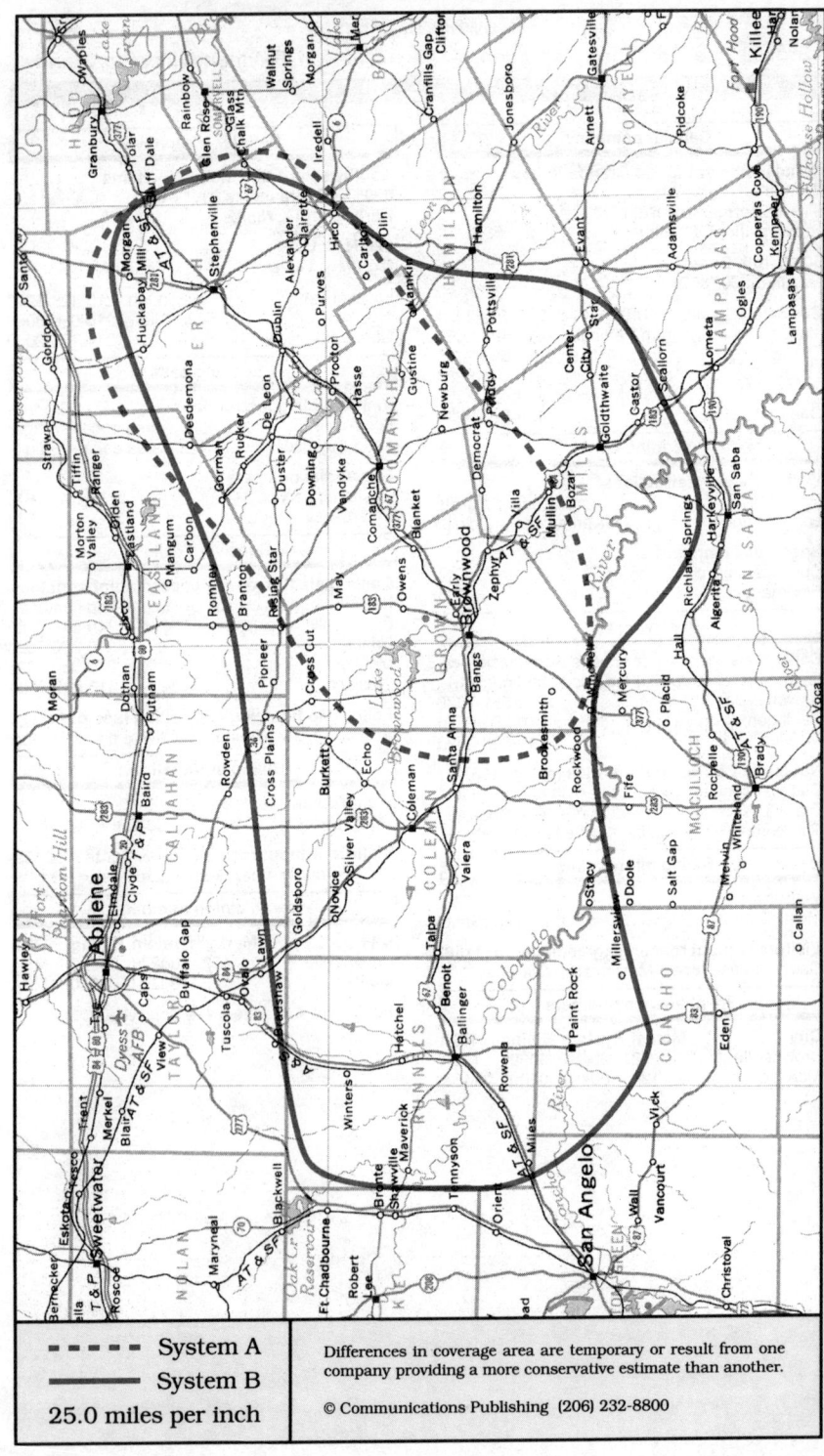

System A
System B
25.0 miles per inch

Differences in coverage area are temporary or result from one company providing a more conservative estimate than another.

© Communications Publishing (206) 232-8800

Brownwood, Texas
Ballinger, Brownwood, Coleman, Comanche, Goldthwaite, Stephenville

System A	System B

Cellular company

Cellular One
Licensed to: Lone Star Cellular
Managed by: Prime Cellular
300 N Main, Suite B
Brownwood, TX 76801

Local office	915-643-2355
Corporate office Prime Cellular	201-227-1434

Sending calls

Visitor dialing instructions

Local calls	area code + 7 digits
Long distance . . .	1 + area code + 7 digits

Speed-dial numbers

Customer service	611
Emergency	911

Receiving calls

Caller dials the roamer access number below,
hears tone, dials cellular area code and number.

Brownwood	915-642-7626
Stephenville	817-977-2222

NationLink & RoamingAmerica (see p. 1157)
Caller hears your current roamer access number.
This company's subscribers have level 0 service.

Billing information

Visitor rates

Most visitors	$3 day + 99¢ minute

Visitors without roaming agreements (p. 1159)
Roamer Plus will intercept first call to bill you.

Identification numbers

City	Market	System	Billing	RSA
All served cities	660	01707	33607	TX-9

For full coverage area, see Hillsboro, TX and
Mineral Wells, TX.

Cellular company

Licensed to: Mid-Tex Cellular
Operated by: West Central Cellular
3367 Knickerbocker Road, PO Box 991
San Angelo, TX 76904-7812

Corporate office	800-695-0150
. . . .	915-949-6799

Sending calls

Visitor dialing instructions

Local calls	area code + 7 digits
Long distance . . .	1 + area code + 7 digits

Speed-dial numbers

Customer service	611
Emergency	911

Receiving calls

Caller dials the roamer access number below,
hears tone, dials cellular area code and number.

. . . .	915-647-7626

Billing information

Visitor rates

Most visitors	$2 day + 50¢ minute

Visitors without roaming agreements (p. 1159)
Call co. to use Discover, Mastercard, Visa

Identification numbers

City	Market	System	Billing	RSA
Brownwood	660	01932	30310	TX-9
Except Brownwood	660	01932	00000	TX-9

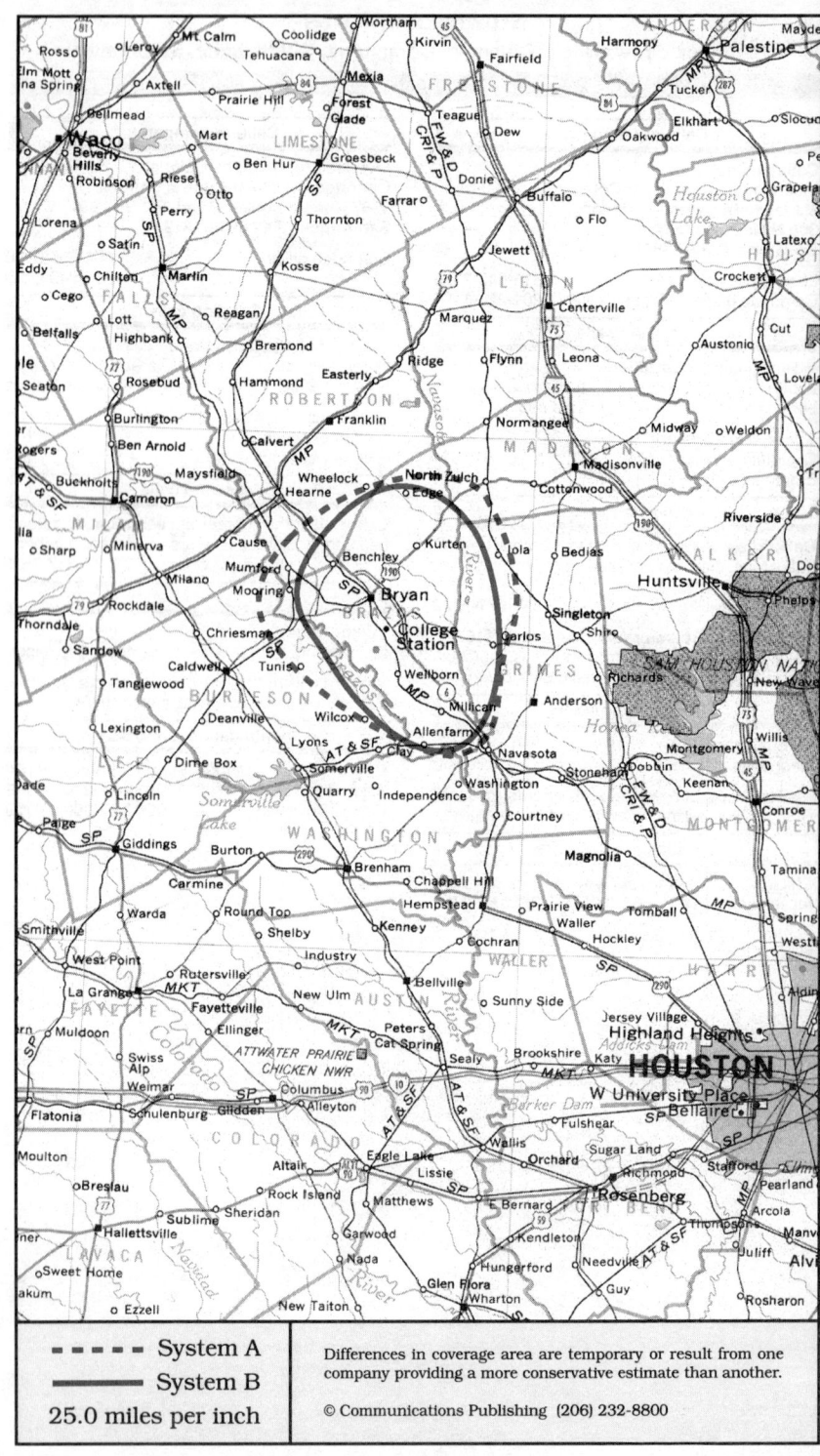

----- System A	Differences in coverage area are temporary or result from one company providing a more conservative estimate than another.
——— System B	
25.0 miles per inch	© Communications Publishing (206) 232-8800

Bryan, Texas
Bryan, College Station

System A	System B

System A

Cellular company

Cellular One
McCaw Cellular Communications
2551 Texas Avenue S, Suite A
College Station, TX 77840

Customer service	800-262-3659
. . . .	409-777-7500
Local office	409-777-7000
Corporate office	206-827-4500

Sending calls

Visitor dialing instructions

Local calls	area code + 7 digits
Long distance . . .	1 + area code + 7 digits

Speed-dial numbers

Customer service	611
Emergency	911

Receiving calls

Caller dials the roamer access number below,
hears tone, dials cellular area code and number.
. . . . 409-777-7626

NationLink & RoamingAmerica (see p. 1157)
Dial *31 send to receive calls, *30 to deactivate.
Dial *32 send to tell caller the roamer access no.
This company's subscribers have level 9 service.

North American Cellular Network (see p. 1156)
All calls forward to you. Deactivate dial *35 send.

Billing information

Visitor rates
Most visitors . . . . . $0 day + 99¢ minute

Visitors without roaming agreements (p. 1159)
Call co. to use Mastercard, Visa
or Roamer Plus will intercept first call to bill you.

Identification numbers

City	Market	System	Billing	RSA
All served cities	287	00297	00000	MSA

System B

Cellular company

GTE Mobilnet
100 Glenborough, Suite 800
Houston, TX 77067

Customer service	800-347-5665
. . . .	713-876-9144
Local office	713-876-5000
Corporate office	404-391-8000

Sending calls

Visitor dialing instructions

Local calls	area code + 7 digits
Long distance . . .	1 + area code + 7 digits

Speed-dial numbers

Customer service	*611
Emergency	911
Mr. Rescue (roadside assistance)	HELP or 4357
Roaming information	*711
Telephone problems	*111

Receiving calls

Caller dials the roamer access number below,
hears tone, dials cellular area code and number.
. . . . 409-268-7626

Follow Me Roaming forwarding (see p. 1156)
Activate, dial *18 send. Deactivate dial *19 send.

Billing information

Visitor rates
Most visitors . . . . . $3 day + 85¢ minute

Visitors without roaming agreements (p. 1159)
Call co. to use Amex, Mastercard, Visa

Identification numbers

City	Market	System	Billing	RSA
All served cities	287	00012	00000	MSA

For full coverage area, see Bay City, TX and
Huntsville, TX.

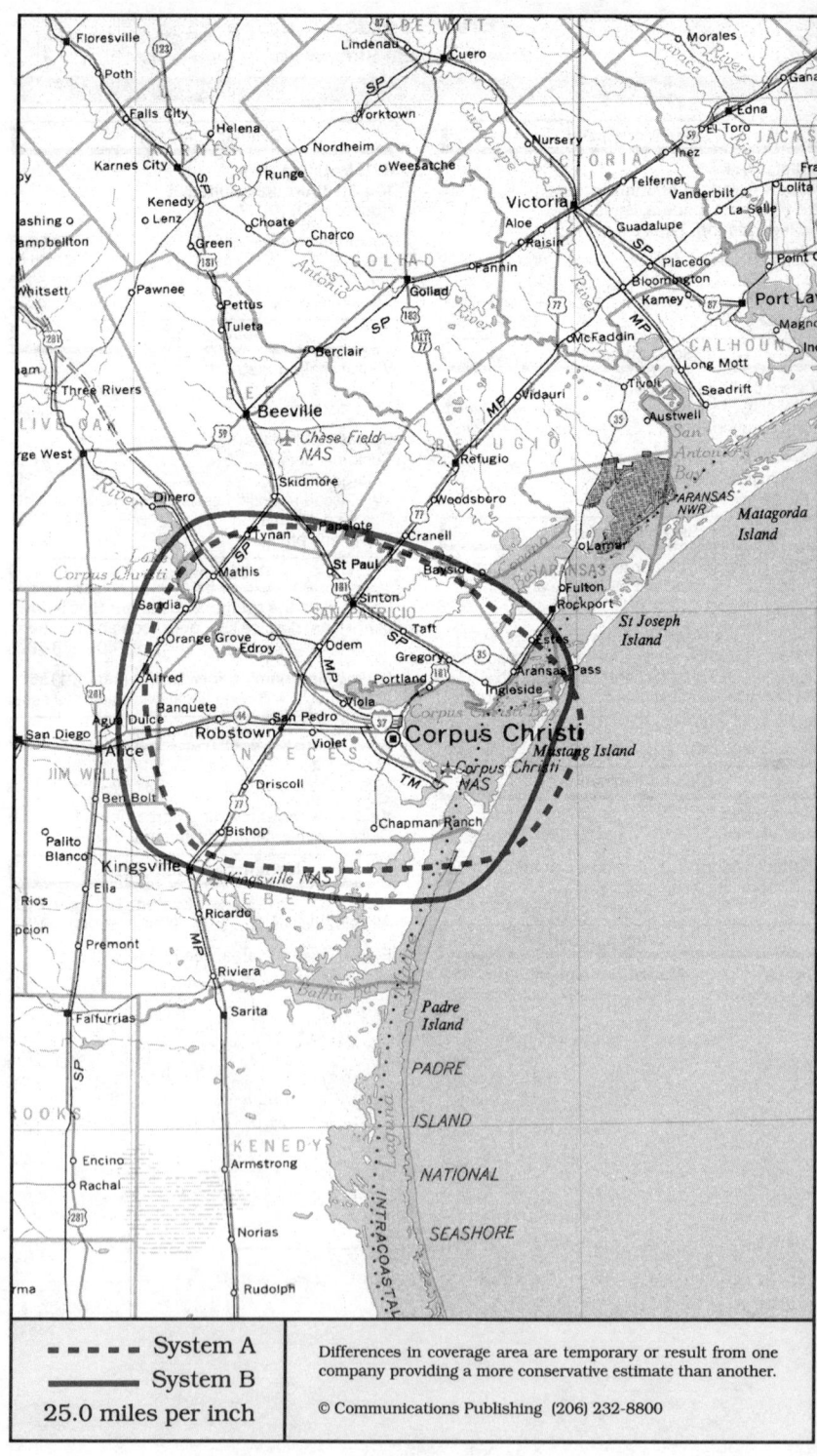

- - - - System A ——— System B 25.0 miles per inch	Differences in coverage area are temporary or result from one company providing a more conservative estimate than another. © Communications Publishing (206) 232-8800

Corpus Christi, Texas
Corpus Christi, Robstown, Sinton

System A	System B

Cellular company

Cellular One
McCaw Cellular Communications
5425 S Padre Island Drive, Suite 169
Corpus Christi, TX 78411

Customer service	800-262-3659
. . . . .	512-946-7500
Local office	512-946-7000
Corporate office	206-827-4500

Sending calls

Visitor dialing instructions

Local calls	area code + 7 digits
Long distance . . .	1 + area code + 7 digits

Speed-dial numbers

Cellular One voice mail	711
Channel 3 newsroom . . .	*TV3 or *883
Channel 6 newsroom . . .	*TV6 or *886
Corpus Christi police dept. . .	*CCPD or *2273
Customer service	611
Emergency	911
KEYS am call-in line . . .	*K99 or *5991440
K99 fm request/contest line .	*K99 or *599
Z95 fm request/contest line	*95

Receiving calls

Caller dials the roamer access number below,
hears tone, dials cellular area code and number.
. . . . 512-946-7626

NationLink & RoamingAmerica (see p. 1157)
Dial *31 send to receive calls, *30 to deactivate.
Dial *32 send to tell caller the roamer access no.
This company's subscribers have level 9 service.

North American Cellular Network (see p. 1156)
All calls forward to you. Deactivate dial *35 send.

Billing information

Visitor rates

Most visitors	$0 day + 99¢ minute

Visitors without roaming agreements (p. 1159)
Call co. to use Amex, Discover, Mastercard, Visa

Identification numbers

City	Market	System	Billing	RSA
All served cities	112	00191	00000	MSA

Cellular company

Southwestern Bell Mobile Systems
3501 S Padre Island Drive
Corpus Christi, TX 78415

Local office	512-854-5678
Corporate office	214-733-2000

Sending calls

Visitor dialing instructions

Local calls	area code + 7 digits
Long distance . . .	1 + area code + 7 digits

Speed-dial numbers

Customer service	611
Emergency	911

Receiving calls

Caller dials the roamer access number below,
hears tone, dials cellular area code and number.
. . . . 512-877-7626

Follow Me Roaming forwarding (see p. 1156)
Activate, dial *18 send. Deactivate dial *19 send.

Billing information

Visitor rates

Most visitors	$3 day + 90¢ minute

Visitors without roaming agreements (p. 1159)
Call co. to use Amex, Mastercard, Visa

Identification numbers

City	Market	System	Billing	RSA
All served cities	112	00184	00000	MSA

For full coverage area, see Kingsville, TX.

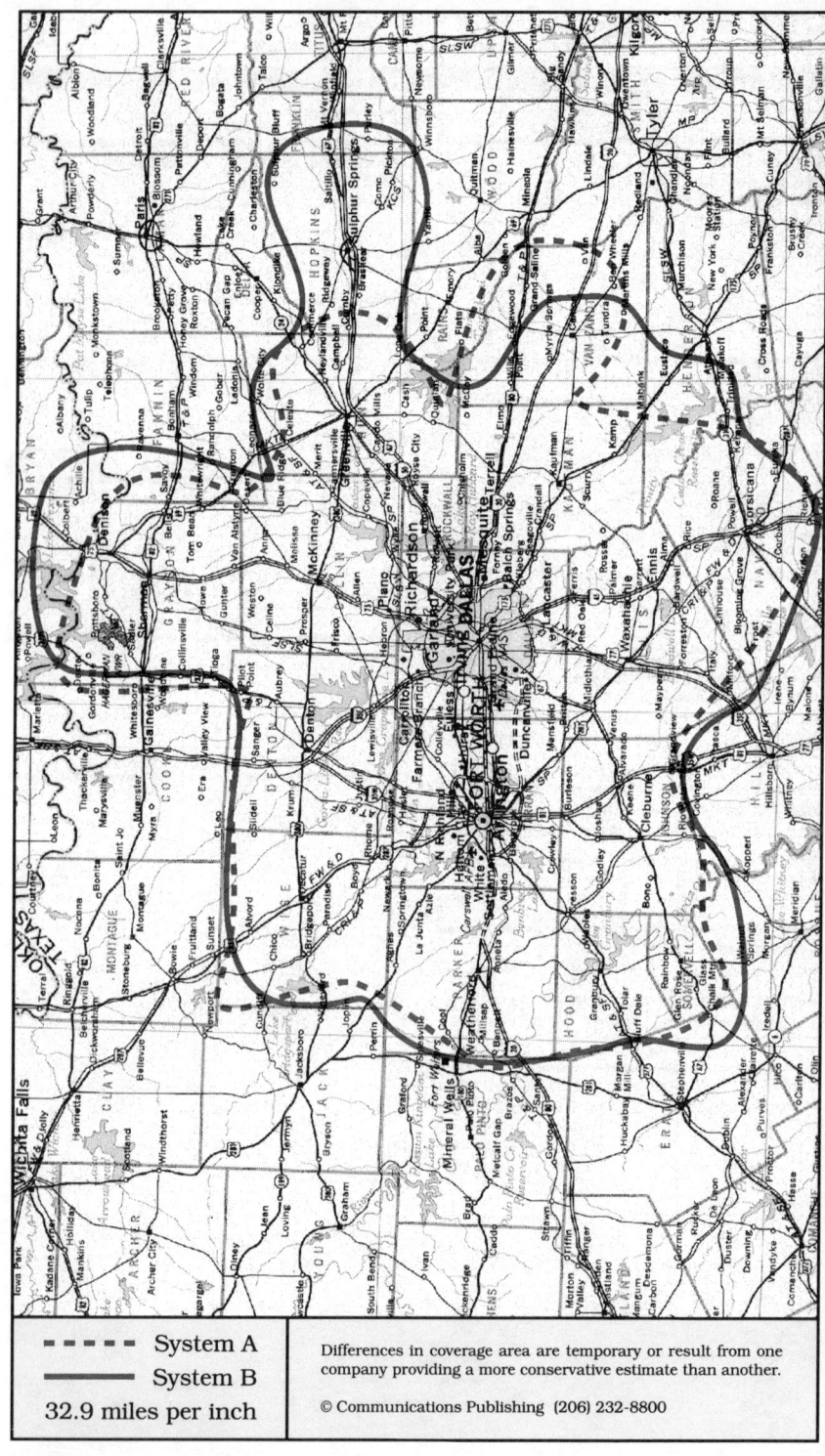

- - - - - System A
———— System B
32.9 miles per inch

Differences in coverage area are temporary or result from one company providing a more conservative estimate than another.

© Communications Publishing (206) 232-8800

Dallas, Texas

Corsicana, Dallas, Denison, Ft. Worth, Greenville, Sherman

System A	System B

Cellular company

MetroCel Cellular
McCaw Cellular Communications
17300 N Dallas Parkway, Suite 1000
Dallas, TX 75248

Customer service	800-525-6282
. . . .	214-263-4921
Local office	214-407-6100
Corporate office	206-827-4500

Sending calls

Visitor dialing instructions

Local calls	area code + 7 digits
Long distance . . .	1 + area code + 7 digits

Speed-dial numbers

Customer service	611
Emergency	911

Receiving calls

Caller dials the roamer access number below,
hears tone, dials cellular area code and number.

. . . . 214-850-7626

NationLink & RoamingAmerica (see p. 1157)
Dial *31 send to receive calls, *30 to deactivate.
Dial *32 send to tell caller the roamer access no.
This company's subscribers have level 6 service.

North American Cellular Network (see p. 1156)
All calls forward to you. Deactivate dial *35 send.

Billing information

Visitor rates

Most visitors	$0 day + 99¢ minute

Visitors without roaming agreements (p. 1159)
Roamer Plus will intercept first call to bill you.

Identification numbers

City	Market	System	Billing	RSA
Canton, Corsicana	661	01709	00000	TX-10
Dallas, Ft. Worth	009	00033	00000	MSA
Denison, Sherman	292	00545	00000	MSA
Greenville	658	01703	00000	TX-7

Call switching in TX-7 and TX-10 is performed by
McCaw Cellular for the licensee.

Cellular company

Southwestern Bell Mobile Systems
17330 Preston Road, Suite 100-A
Dallas, TX 75252

Customer service	some areas	800-331-0500
. . . .	. . . .	214-988-8484
Corporate office		214-733-2000

Sending calls

Visitor dialing instructions

Local calls	area code + 7 digits
Long distance . . .	1 + area code + 7 digits

Speed-dial numbers

Customer service	611
Emergency	911
Fast facts (sports, news, weather, etc.) .	*123
Mr. Rescue (roadside assistance)	HELP or 4357
Roaming information	*711

Receiving calls

Caller dials the roamer access number below,
hears tone, dials cellular area code and number.

. . . . 214-384-7626

Follow Me Roaming forwarding (see p. 1156)
Activate, dial *18 send. Deactivate dial *19 send.

Billing information

Visitor rates

Most visitors	$3 day + 75¢ minute

Visitors without roaming agreements (p. 1159)
Call co. to use Amex, Mastercard, Visa

Identification numbers

City	Market	System	Billing	RSA
Canton, Corsicana	661	00038	30130	TX-10
Dallas, Ft. Worth	009	00038	00000	MSA
Denison, Sherman	292	00038	00000	MSA
Greenville	658	00038	30038	TX-7

For full coverage area, see Gainesville, TX and
Mineral Wells, TX.

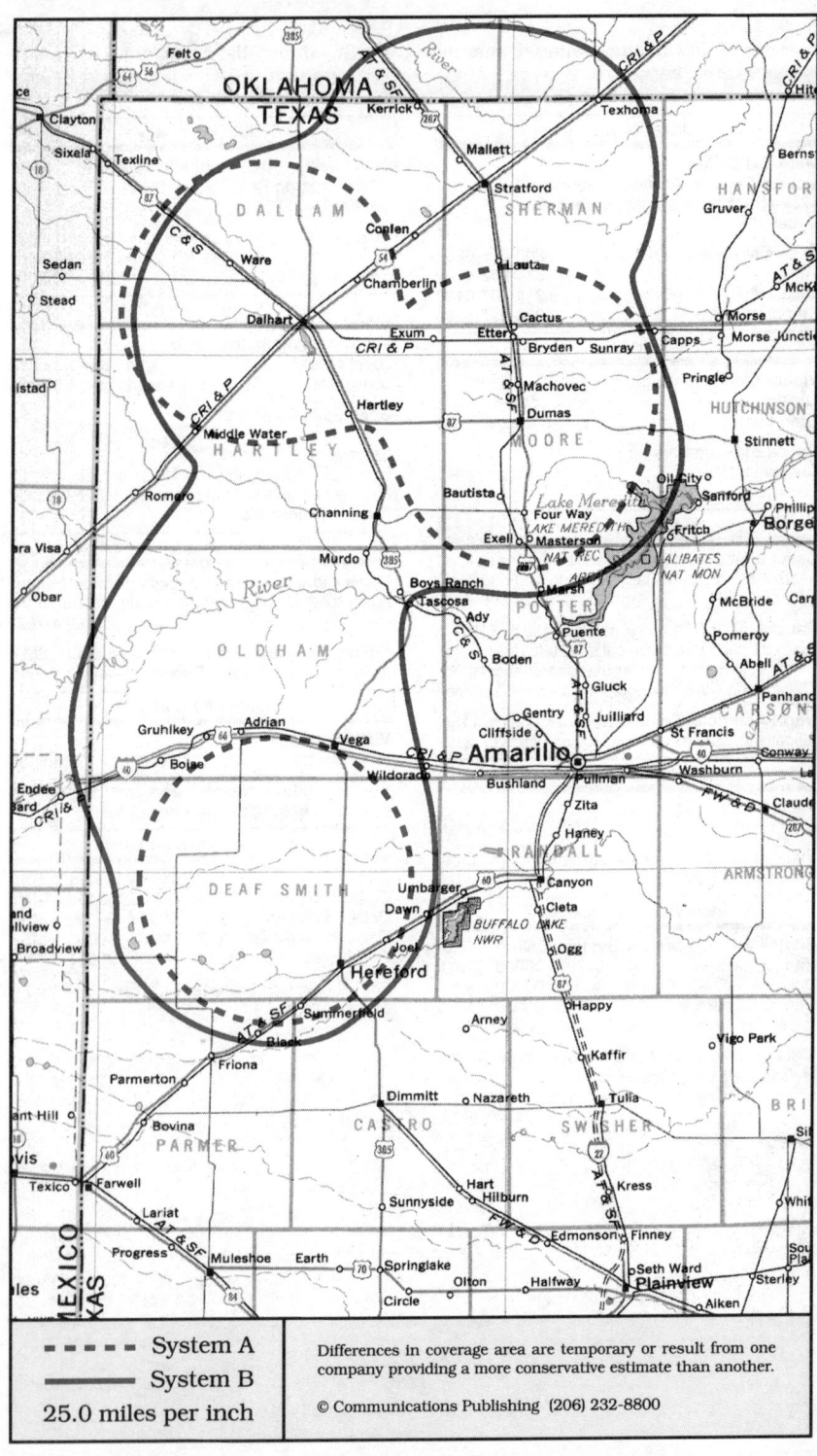

System A
System B
25.0 miles per inch

Dumas, Texas
Channing, Dalhart, Dumas, Hereford, Stratford

|

System A

Cellular company

Cellular One
Licensed to: National Cellular Partners
Operated by: Amarillo Cellular Telephone
105 E 9th
Amarillo, TX 79101

Corporate office 806-374-1900

Sending calls

Visitor dialing instructions
Local calls . . . . . area code + 7 digits
Long distance . . . 1 + area code + 7 digits

Speed-dial numbers
Customer service 611
Emergency 911

Receiving calls

Caller dials the roamer access number below,
hears tone, dials cellular area code and number.
. . . . 806-678-7626

NationLink & RoamingAmerica (see p. 1157)
Caller hears your current roamer access number.
This company's subscribers have level 0 service.

Billing information

Visitor rates
Most visitors . . . . . $3 day + 75¢ minute

Visitors without roaming agreements (p. 1159)
Roamer Plus will intercept first call to bill you.

Identification numbers

City	Market	System	Billing	RSA
All served cities	652	01691	00000	TX-1

System B

Cellular company

XIT Cellular
XIT Telecommunication & Technology
Highway 87 N, PO Box 1391
Dalhart, TX 79022

1545 S Dumas Avenue
Dumas, TX 79029

1009 W Park Avenue
Hereford, TX 79045

Local office . . . . . Dalhart 800-232-3312
. . . . Dalhart 806-384-3333
. . . . Dumas 806-935-8777
. . . . Hereford 806-364-1426
Corporate office 806-384-3311

Sending calls

Visitor dialing instructions
Local calls . . . . . . area code + 7 digits
Long distance . . . 1 + area code + 7 digits

Speed-dial numbers
Customer service 611
Emergency 911

Receiving calls

Caller dials the roamer access number below,
hears tone, dials cellular area code and number.
Dalhart 806-333-7626
Dumas 806-922-7626
Hereford 806-344-7626
Stratford 806-753-7626

Billing information

Visitor rates
Most visitors . . . . . $3 day + 75¢ minute

Visitors without roaming agreements (p. 1159)
Call co. to use Mastercard, Visa

Identification numbers

City	Market	System	Billing	RSA
All served cities	652	01692	00000	TX-1

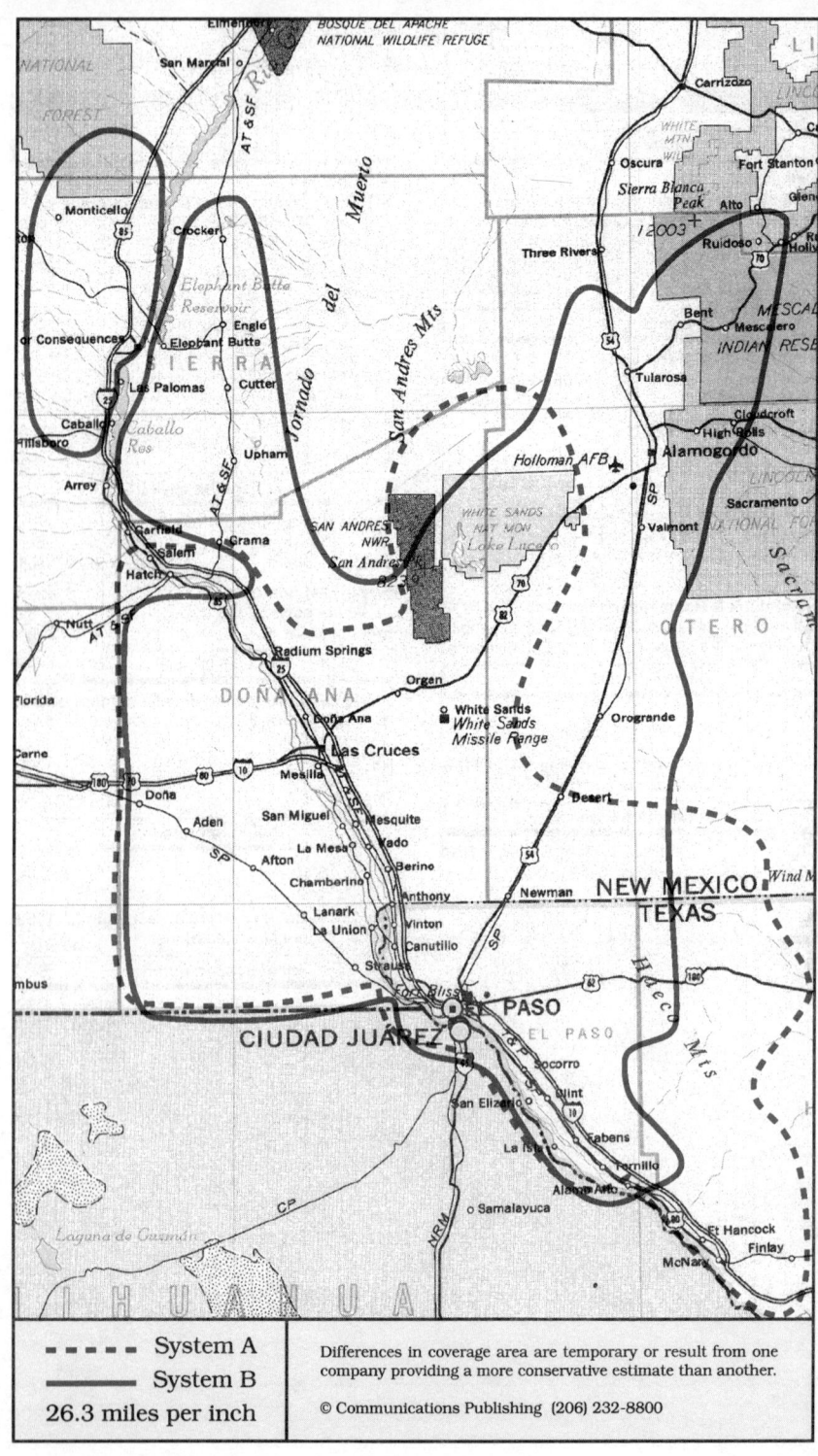

BOSQUE DEL APACHE
NATIONAL WILDLIFE REFUGE

- - - - System A
——— System B
26.3 miles per inch

Differences in coverage area are temporary or result from one company providing a more conservative estimate than another.

© Communications Publishing (206) 232-8800

El Paso, Texas

Alamogordo NM, El Paso, Las Cruces NM, Truth or Consequences NM

System A	System B

Cellular company

Bell Atlantic Mobile
423 Executive Center Boulevard
El Paso, TX 79902-1003

1810 W Amador
Las Cruces, NM 88005

Local office	. . .	El Paso	915-532-5559
	. . . . . .	Las Cruces	505-526-2233
Corporate office	. .	regional	602-948-8543
	. . . .	national	908-306-7000

Sending calls

Visitor dialing instructions
Local calls	. . . .	area code + 7 digits
Long distance	. . .	1 + area code + 7 digits

Speed-dial numbers
Customer service	*611
Emergency	911

Receiving calls

Caller dials the roamer access number below,
hears tone, dials cellular area code and number.
El Paso	915-549-7626
Las Cruces	505-642-7626

NationLink & RoamingAmerica (see p. 1157)
Dial *31 send to receive calls, *30 to deactivate.
Dial *32 send to tell caller the roamer access no.
This company's subscribers have level 8 service.

Billing information

Visitor rates
Most visitors	$3 day + 99¢ minute

Visitors without roaming agreements (p. 1159)
Roamer Plus will intercept first call to bill you.

Identification numbers

City	Market	System	Billing	RSA
El Paso	081	00097	00000	MSA
Las Cruces	285	00097	30097	MSA

Cellular company

Contel Cellular
5858 Gateway Boulevard E
El Paso, TX 79905-1923

1100 S Main Street, Suite 4
Las Cruces, NM 88005

Customer service	. . . .		800-333-4004
	. . . .		915-772-9222
Local office	El Paso	915-779-7373
	. . . .	Las Cruces	505-523-6186
Corporate office	. . . .		404-804-3400

Sending calls

Visitor dialing instructions
Local calls	. . . . .	area code + 7 digits
Long distance	. . .	1 + area code + 7 digits

Speed-dial numbers
Customer service	*611
Emergency	911

Receiving calls

Caller dials the roamer access number below,
hears tone, dials cellular area code and number.
Alamogordo	505-430-7626
El Paso	915-525-7626
Las Cruces	505-644-7626
Truth or Consequences . . .	505-740-7626

Follow Me Roaming forwarding (see p. 1156)
Activate, dial *18 send. Deactivate dial *19 send.

Billing information

Visitor rates
Most visitors	$3 day + 85¢ minute

Visitors without roaming agreements (p. 1159)
Call co. to use Amex, Mastercard, Visa

Identification numbers

City	Market	System	Billing	RSA
Alamogordo	558	01504	00000	NM-6
El Paso	081	00092	00000	MSA
Las Cruces	285	00092	00000	MSA
Truth or Consequ.	555	01498	00000	NM-3

For full coverage area, see Deming, NM.

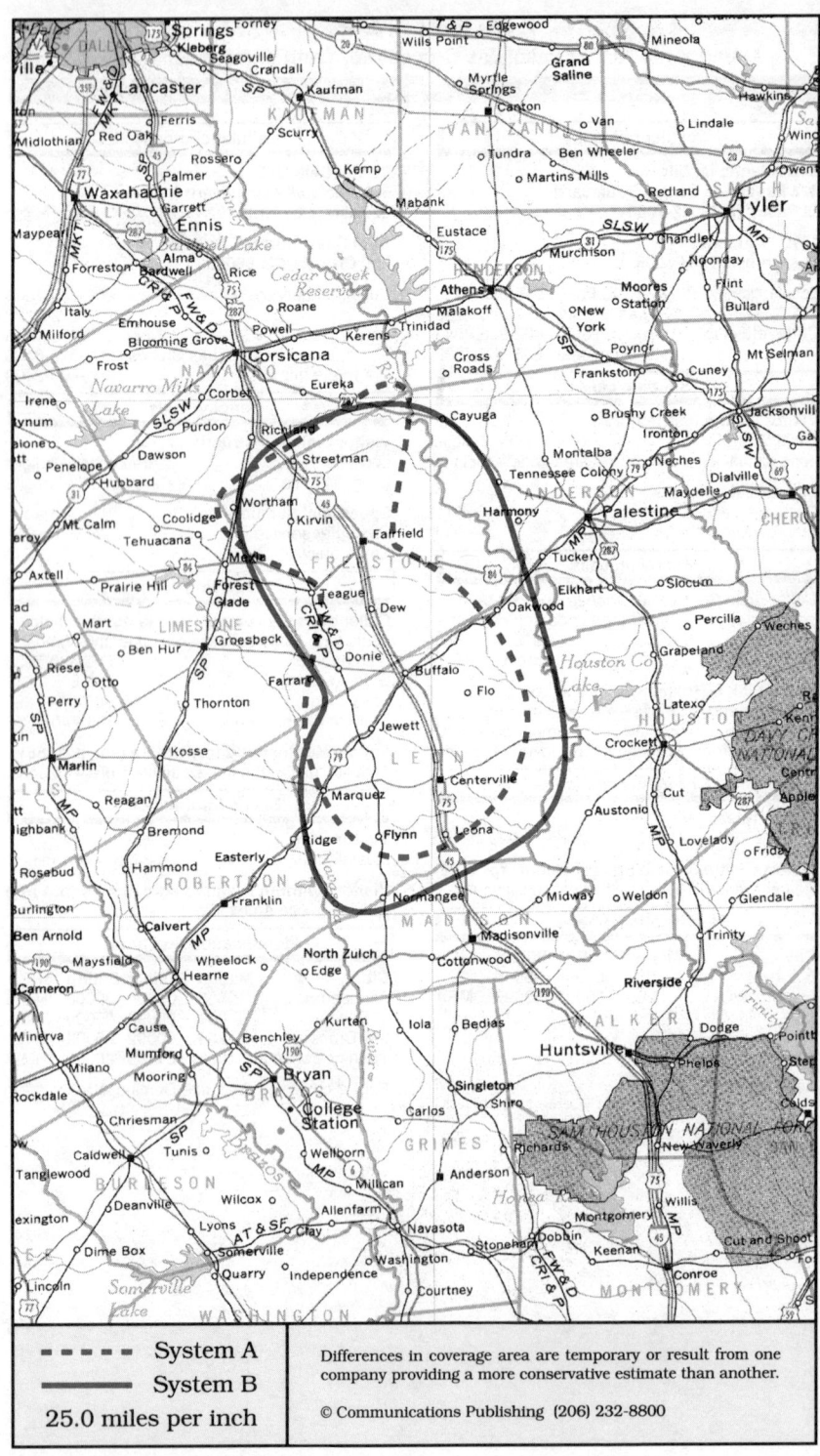

- - - - System A	Differences in coverage area are temporary or result from one
——— System B	company providing a more conservative estimate than another.
25.0 miles per inch	© Communications Publishing (206) 232-8800

Fairfield, Texas
Buffalo, Centerville, Fairfield

System A	System B

System A

Cellular company

MetroCel Cellular
McCaw Cellular Communications
17300 N Dallas Parkway, Suite 1000
Dallas, TX 75248

Customer service	800-525-6282
. . . .	214-263-4921
Local office	214-407-6100
Corporate office	206-827-4500

Sending calls

Visitor dialing instructions
Local calls . . . . . area code + 7 digits
Long distance . . . 1 + area code + 7 digits

Speed-dial numbers
Customer service 611
Emergency 911

Receiving calls

Caller dials the roamer access number below,
hears tone, dials cellular area code and number.
. . . . 214-850-7626

North American Cellular Network (see p. 1156)
All calls forward to you. Deactivate dial *35 send.

Billing information

Visitor rates
Most visitors . . . . . $0 day + 99¢ minute

Visitors without roaming agreements (p. 1159)
Cannot call since company takes no credit cards.

Identification numbers

City	Market	System	Billing	RSA
All served cities	661	01709	00000	TX-10

Call switching is performed by McCaw Cellular for the licensee.

For full coverage area, see Dallas, TX.

System B

Cellular company

GTE Mobilnet
100 Glenborough, Suite 800
Houston, TX 77067

Customer service	800-347-5665
. . . .	713-876-9144
Local office	713-876-5000
Corporate office	404-391-8000

Sending calls

Visitor dialing instructions
Local calls . . . . . . area code + 7 digits
Long distance . . . 1 + area code + 7 digits

Speed-dial numbers
Customer service *611
Emergency 911
Mr. Rescue (roadside assistance) HELP or 4357
Roaming information *711
Telephone problems *111

Receiving calls

Caller dials the roamer access number below,
hears tone, dials cellular area code and number.
. . . . 713-824-7626

Follow Me Roaming forwarding (see p. 1156)
Activate, dial *18 send. Deactivate dial *19 send.

Billing information

Visitor rates
Most visitors . . . . . $3 day + 85¢ minute

Visitors without roaming agreements (p. 1159)
Call co. to use Amex, Mastercard, Visa

Identification numbers

City	Market	System	Billing	RSA
All served cities	661	00012	30266	TX-10

For full coverage area, see Huntsville, TX.

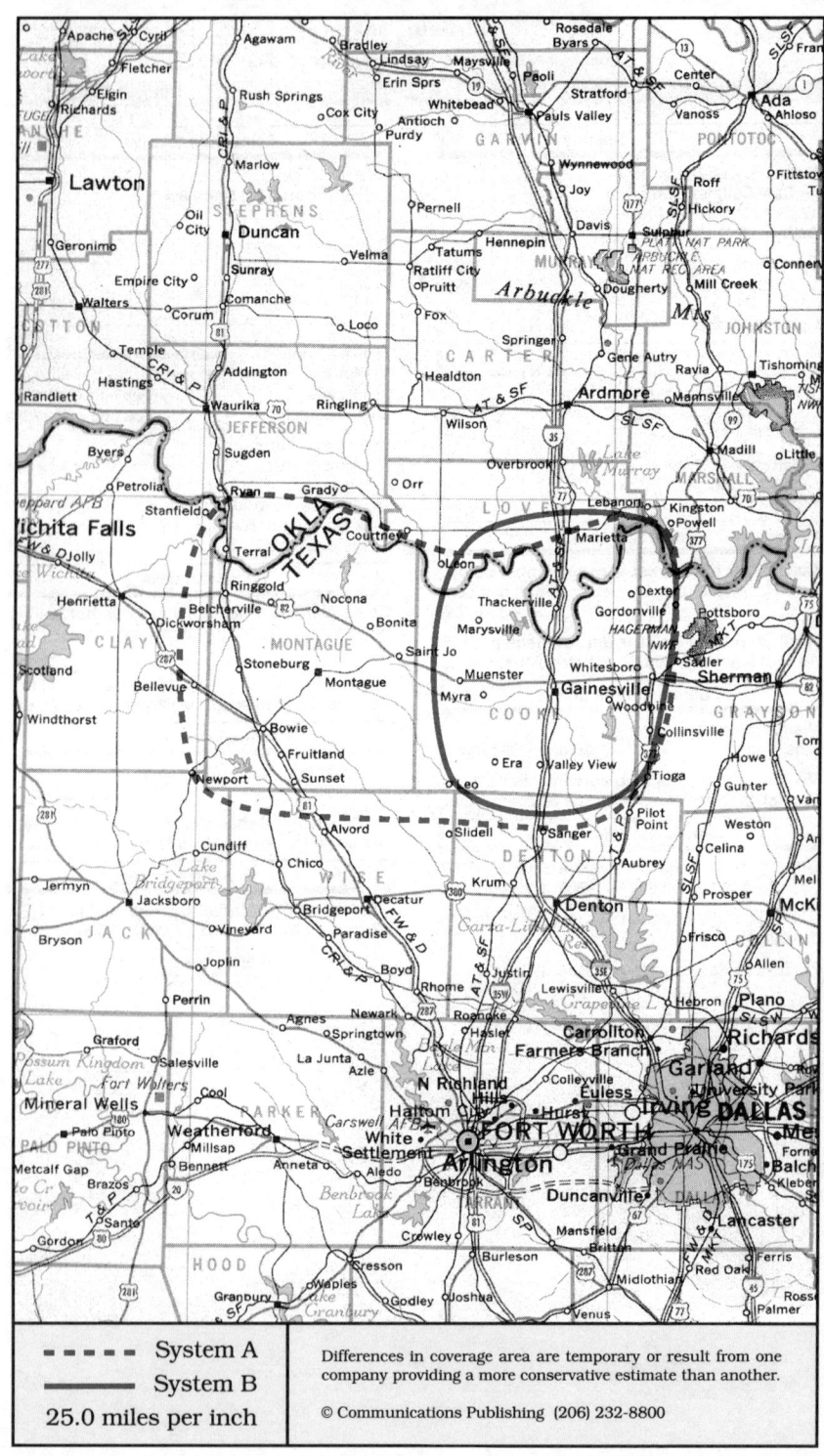

- - - - System A	Differences in coverage area are temporary or result from one company providing a more conservative estimate than another.
———— System B	
25.0 miles per inch	© Communications Publishing (206) 232-8800

Gainesville, Texas
Bowie, Gainesville, Montague, Nocona

System A	System B

Cellular company

Cellular One of Gainesville
PriCellular
1500 E Highway 82, Suite 1-A
Gainesville, TX 76240

Local office	817-665-2083
Corporate office	212-459-0800

Sending calls

Visitor dialing instructions

Local calls	area code + 7 digits
Long distance . . .	1 + area code + 7 digits

Speed-dial numbers

Customer service	*611
Emergency	911
Weather	811

Receiving calls

Caller dials the roamer access number below,
hears tone, dials cellular area code and number.
. . . . 817-727-1026

NationLink & RoamingAmerica (see p. 1157)
Dial *31 send to receive calls, *30 to deactivate.
Dial *32 send to tell caller the roamer access no.
This company's subscribers have level 5 service.

Billing information

Visitor rates
Most visitors . . . . . $3 day + 75¢ minute

Visitors without roaming agreements (p. 1159)
Call co. to use Amex, Discover, Mastercard, Visa

Identification numbers

City	Market	System	Billing	RSA
All served cities	657	00595	30287	TX-6

For full coverage area, see Wichita Falls, TX.

Cellular company

Southwestern Bell Mobile Systems
17330 Preston Road, Suite 100-A
Dallas, TX 75252

Customer service	some areas	800-331-0500
. . . .		214-988-8484
Corporate office		214-733-2000

Sending calls

Visitor dialing instructions

Local calls	area code + 7 digits
Long distance . . .	1 + area code + 7 digits

Speed-dial numbers

Customer service	611
Emergency	911
Roaming information	*711

Receiving calls

Caller dials the roamer access number below,
hears tone, dials cellular area code and number.
. . . . 817-994-7626

Follow Me Roaming forwarding (see p. 1156)
Activate, dial *18 send. Deactivate dial *19 send.

Billing information

Visitor rates
Most visitors . . . . . $3 day + 75¢ minute

Visitors without roaming agreements (p. 1159)
Call co. to use Amex, Mastercard, Visa

Identification numbers

City	Market	System	Billing	RSA
All served cities	657	00038	30036	TX-6

For full coverage area, see Dallas, TX.

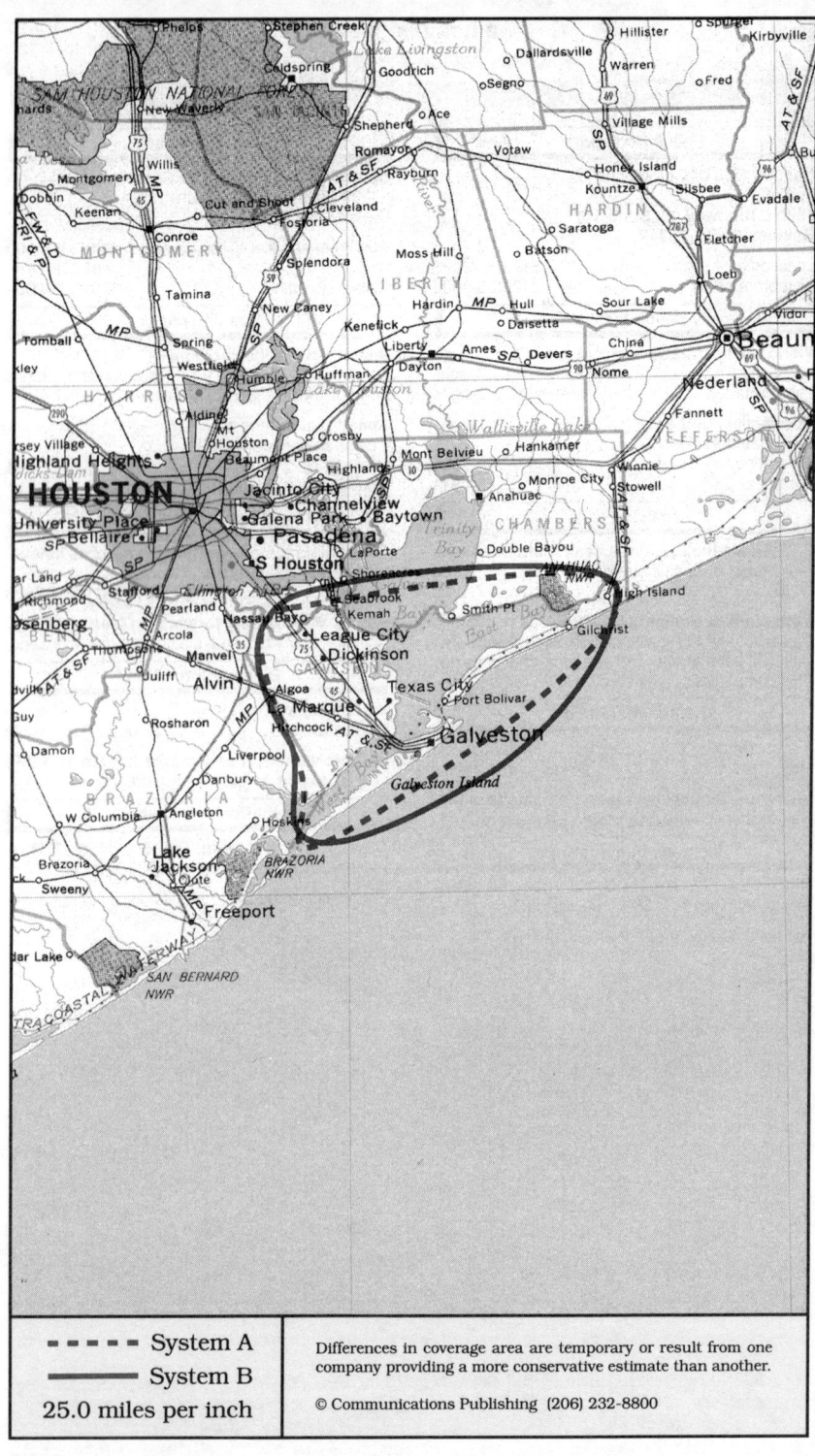

───── System A

───── System B

25.0 miles per inch

Differences in coverage area are temporary or result from one company providing a more conservative estimate than another.

© Communications Publishing (206) 232-8800

Galveston, Texas

Dickinson, Galveston, League City, La Marque, Texas City

System A	System B

Cellular company

Cellular One
Galveston Cellular Telephone
3128 Broadway
Galveston, TX 77550

Customer service	409-763-7078
Corporate office	409-763-7000

Sending calls

Visitor dialing instructions

Local calls	area code + 7 digits
Long distance . . .	1 + area code + 7 digits

Speed-dial numbers

Customer service	611
Emergency	911

Receiving calls

Caller dials the roamer access number below,
hears tone, dials cellular area code and number.
. . . . 409-771-7626

NationLink & RoamingAmerica (see p. 1157)
Dial ✳31 send to receive calls, ✳30 to deactivate.
Dial ✳32 send to tell caller the roamer access no.
This company's subscribers have level 9 service.

Billing information

Visitor rates

Most visitors	$3 day + 75¢ minute

Visitors without roaming agreements (p. 1159)
Cannot call since company takes no credit cards.

Identification numbers

City	Market	System	Billing	RSA
All served cities	170	00367	00000	MSA

Cellular company

GTE Mobilnet
100 Glenborough, Suite 800
Houston, TX 77067

Customer service	800-347-5665
. . . .	713-876-9144
Local office	713-876-5000
Corporate office	404-391-8000

Sending calls

Visitor dialing instructions

Local calls	area code + 7 digits
Long distance . . .	1 + area code + 7 digits

Speed-dial numbers

Customer service	✳611
Emergency	911
Mr. Rescue (roadside assistance)	HELP or 4357
Roaming information	✳711
Telephone problems	✳111

Receiving calls

Caller dials the roamer access number below,
hears tone, dials cellular area code and number.
. . . . 409-766-7626

Follow Me Roaming forwarding (see p. 1156)
Activate, dial ✳18 send. Deactivate dial ✳19 send.

Billing information

Visitor rates

Most visitors	$3 day + 85¢ minute

Visitors without roaming agreements (p. 1159)
Call co. to use Amex, Mastercard, Visa

Identification numbers

City	Market	System	Billing	RSA
All served cities	170	00012	00000	MSA

For full coverage area, see Houston, TX.

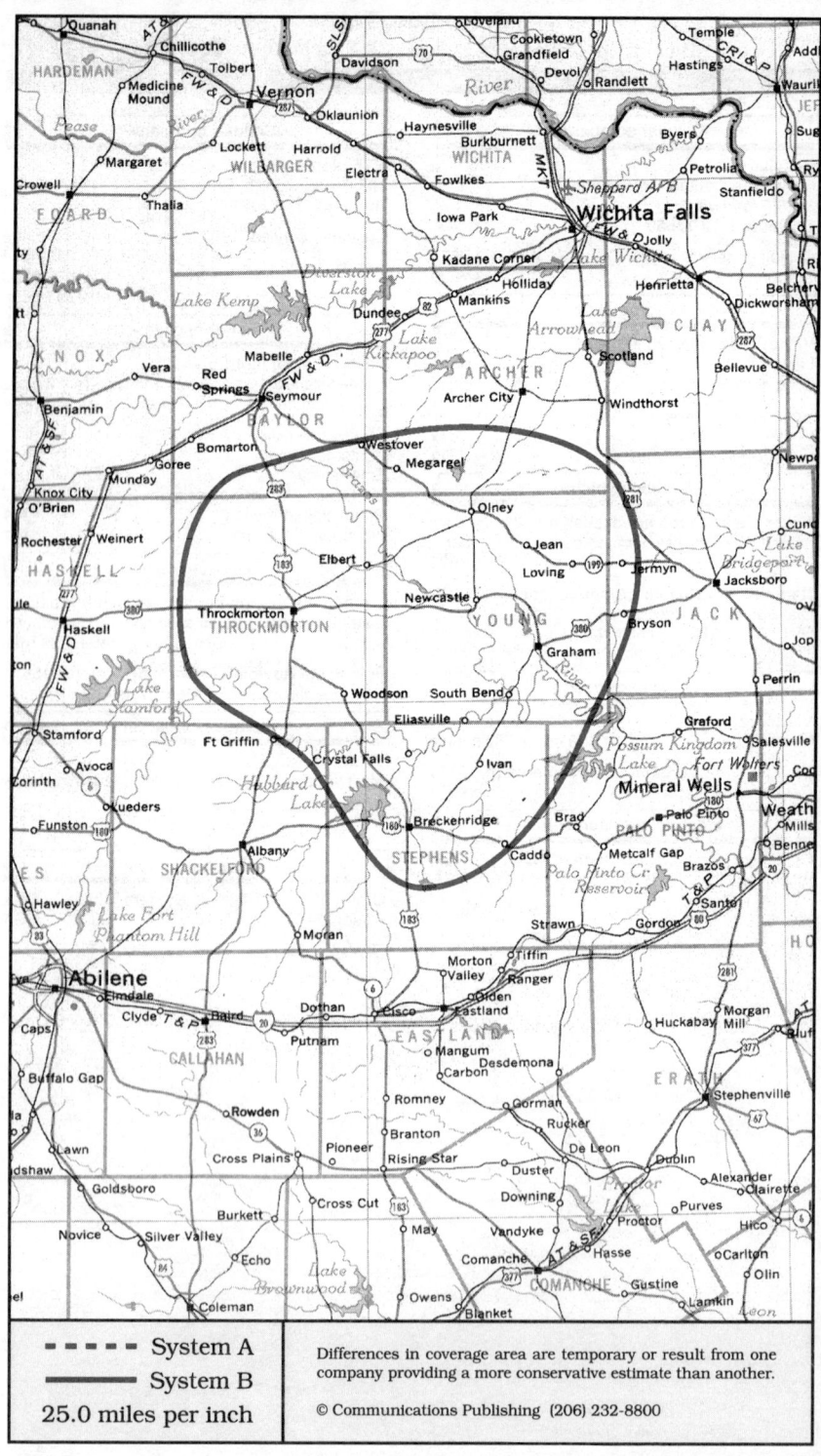

Graham, Texas
Breckenridge, Graham, Olney, Throckmorton

System A	System B

Cellular company

Cellular One of North Texas
805 Hillcrest Plaza
Vernon, TX 76384

This service was not operating at press time, and a sale of the license was pending.

Corporate office 817-553-1885

Sending calls

Visitor dialing instructions
Local calls . . . . . area code + 7 digits
Long distance . . . 1 + area code + 7 digits

Speed-dial numbers
Customer service 611
Emergency 911
Texas state police *54

Receiving calls

Caller dials the roamer access number below, hears tone, dials cellular area code and number.
. . . . 817-886-7626

NationLink & RoamingAmerica (see p. 1157)
Dial *31 send to receive calls, *30 to deactivate. Dial *32 send to tell caller the roamer access no. This company's subscribers have level 3 service.

Billing information

Visitor rates
Most visitors . . . . . $3 day + 99¢ minute

Visitors without roaming agreements (p. 1159)
Cannot call since company takes no credit cards.

Identification numbers

City	Market	System	Billing	RSA
All served cities	656	01593	30323	TX-5

The same company serves Vernon, TX.

Cellular company

Brazos Cellular Communications
209 W Elm
Olney, TX 76374

Corporate office 800-322-7430
. . . . 817-873-5100

Sending calls

Visitor dialing instructions
Local calls . . . . . . area code + 7 digits
Long distance . . . 1 + area code + 7 digits

Speed-dial numbers
Customer service 611
Emergency 911

Receiving calls

Caller dials the roamer access number below, hears tone, dials cellular area code and number.
. . . . 817-873-7626

Follow Me Roaming forwarding (see p. 1156)
Activate, dial *18 send. Deactivate dial *19 send.

Billing information

Visitor rates
Most visitors . . . . . $2 day + 75¢ minute

Visitors without roaming agreements (p. 1159)
Cannot call since company takes no credit cards.

Identification numbers

City	Market	System	Billing	RSA
All served cities	656	01976	00000	TX-5

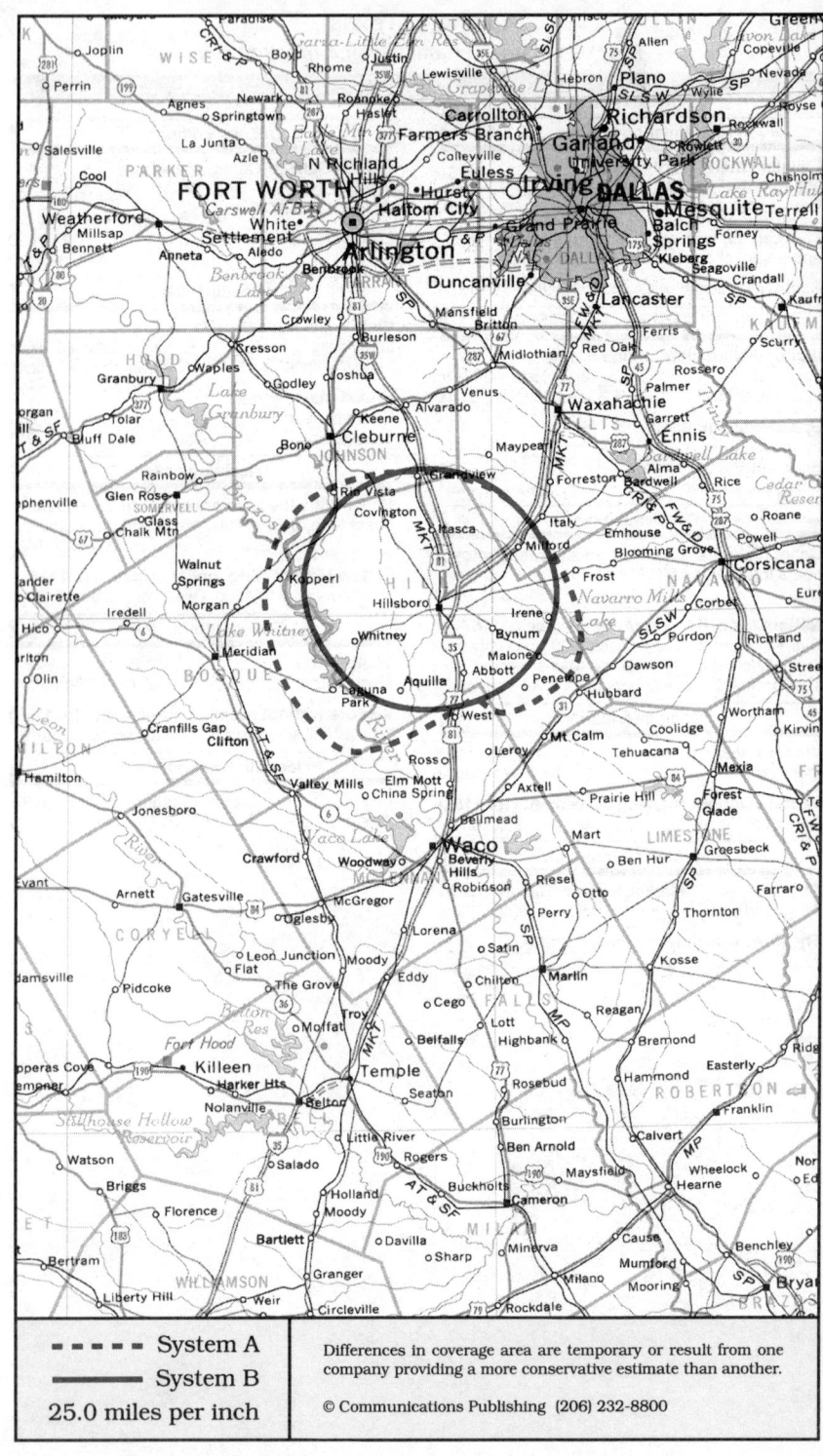

- - - - System A
———— System B
25.0 miles per inch

Differences in coverage area are temporary or result from one
company providing a more conservative estimate than another.

© Communications Publishing (206) 232-8800

Hillsboro, Texas
Hillsboro (Hill County)

|

Cellular company

Cellular One
Licensed to: Lone Star Cellular
Managed by: Prime Cellular
300 N Main, Suite B
Brownwood, TX 76801

Local office 915-643-2355
Corporate office   Prime Cellular 201-227-1434

Sending calls

Visitor dialing instructions
Local calls . . . . .   area code + 7 digits
Long distance . . .   1 + area code + 7 digits

Speed-dial numbers
Customer service   611
Emergency   911

Receiving calls

Caller dials the roamer access number below,
hears tone, dials cellular area code and number.
. . . .   817-744-7626

Billing information

Visitor rates
Most visitors . . . . . $3 day + 99¢ minute

Visitors without roaming agreements (p. 1159)
Roamer Plus will intercept first call to bill you.

Identification numbers

City	Market	System	Billing	RSA
All served cities	660	01859	30367	TX-9

This company also serves Brownwood, TX and
Mineral Wells, TX.

This service is operated as part of the Dallas, TX
system A.

Cellular company

Centel Cellular
2505 Hancock Drive
Temple, TX 76504

Local office   817-771-0077
Corporate office   312-399-2644

Sending calls

Visitor dialing instructions
Local calls . . . . .   area code + 7 digits
Long distance . . .   1 + area code + 7 digits

Speed-dial numbers
Customer service *611
Emergency   911

Receiving calls

Caller dials the roamer access number below,
hears tone, dials cellular area code and number.
. . . .   817-770-7626

Billing information

Visitor rates
Most visitors . . . . . $3 day + 99¢ minute

Visitors without roaming agreements (p. 1159)
Call co. to use Amex, Discover, Mastercard, Visa

Identification numbers

City	Market	System	Billing	RSA
All served cities	660	01934	00000	TX-9

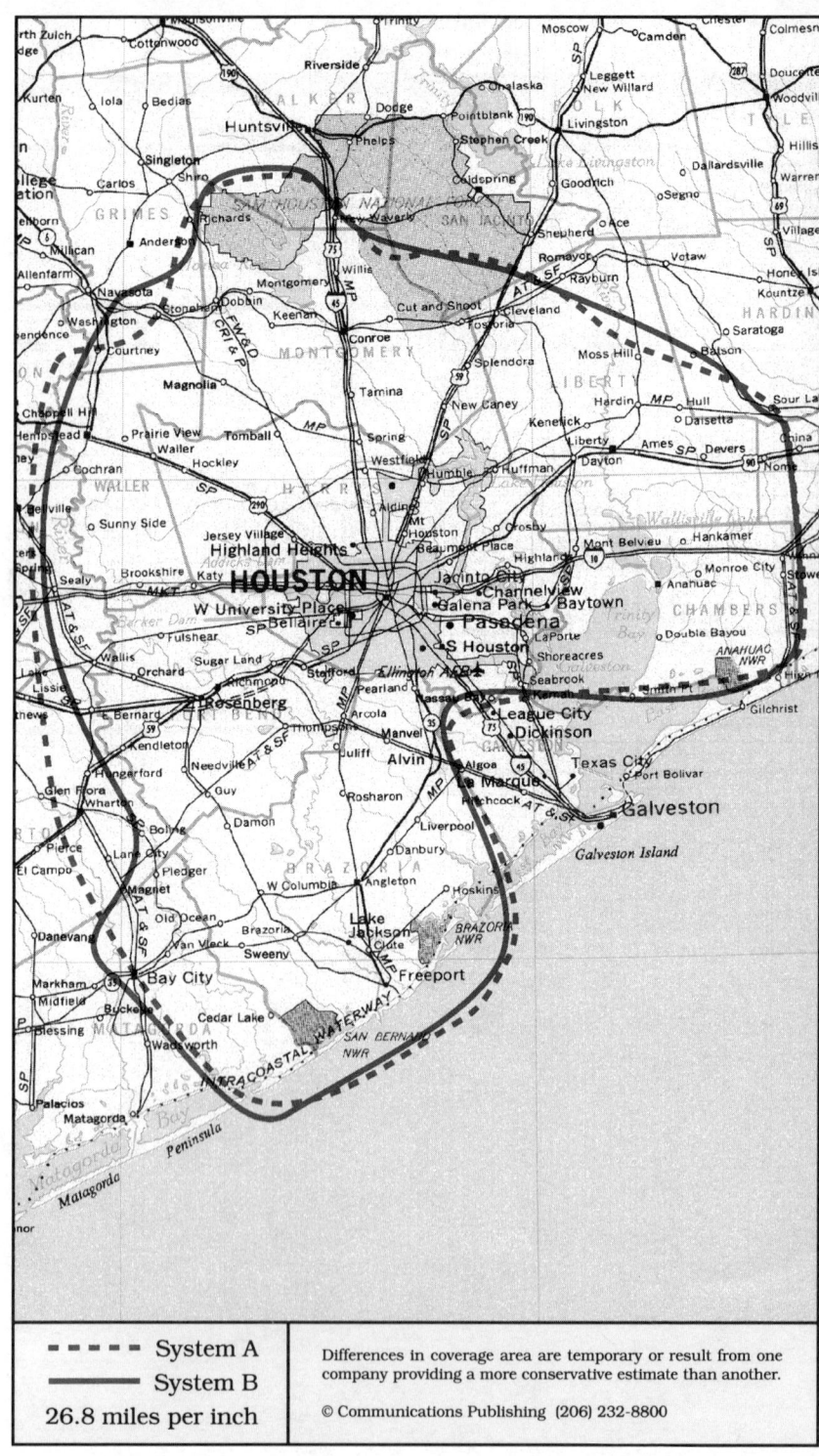

- - - - - System A ———— System B 26.8 miles per inch	Differences in coverage area are temporary or result from one company providing a more conservative estimate than another. © Communications Publishing (206) 232-8800

Houston, Texas

Baytown, Conroe, Freeport, Houston, Lake Jackson, Rosenberg, Winnie

System A	System B

Cellular company

Houston Cellular Telephone
BellSouth Cellular
One W Loop S, Suite 300
Houston, TX 77027-9009

Customer service	800-826-7626
. . . .	713-850-1211
Local office	713-850-9933
Corporate office	404-604-6100

Sending calls

Visitor dialing instructions

Local calls	area code + 7 digits
Long distance . . .	1 + area code + 7 digits

Speed-dial numbers

Auto help line of America . .	*SOS or *767
Customer service	611
Emergency	911
Express floral services . . .	*ABC or *234
Hear a traffic report	*777
Houston entertainment info	*EVENT or *38368
Make a traffic report	*700
Stock quotations weekdays .	*EFH or *334
Ticketmaster	*TIX or *849
Time of day	*TIME or *8463
Weather report	*WEA or *932

Receiving calls

Caller dials the roamer access number below,
hears tone, dials cellular area code and number.
. . . . 713-825-7626

NationLink & RoamingAmerica (see p. 1157)
Dial *31 send to receive calls, *30 to deactivate.
Dial *32 send to tell caller the roamer access no.
This company's subscribers have level 9 service.

Billing information

Visitor rates
Most visitors . . . . . $3 day + 75¢ minute

Visitors without roaming agreements (p. 1159)
Roamer Plus will intercept first call to bill you.

Identification numbers

City	Market	System	Billing	RSA
All served cities	010	00035	00000	MSA

Cellular company

GTE Mobilnet
100 Glenborough, Suite 800
Houston, TX 77067

Customer service	800-347-5665
. . . .	713-876-9144
Local office	713-876-5000
Corporate office	404-391-8000

Sending calls

Visitor dialing instructions

Local calls	area code + 7 digits
Long distance . . .	1 + area code + 7 digits

Speed-dial numbers

Customer service	*611
Emergency	911
Hear a traffic report	*123
Make a traffic report	*847
Mr. Rescue (roadside assistance)	HELP or 4357
Roaming information	*711
Telephone problems	*111

Receiving calls

Caller dials the roamer access number below,
hears tone, dials cellular area code and number.
. . . . 713-824-7626

Follow Me Roaming forwarding (see p. 1156)
Activate, dial *18 send. Deactivate dial *19 send.

Billing information

Visitor rates
Most visitors . . . . . $3 day + 85¢ minute

Visitors without roaming agreements (p. 1159)
Call co. to use Amex, Mastercard, Visa

Identification numbers

City	Market	System	Billing	RSA
Houston area	010	00012	00000	MSA
Winnie (Chambers)	672	00012	00000	TX-21

For full coverage area, see Bay City, TX and
Beaumont, TX and Huntsville, TX and Galveston,
TX.

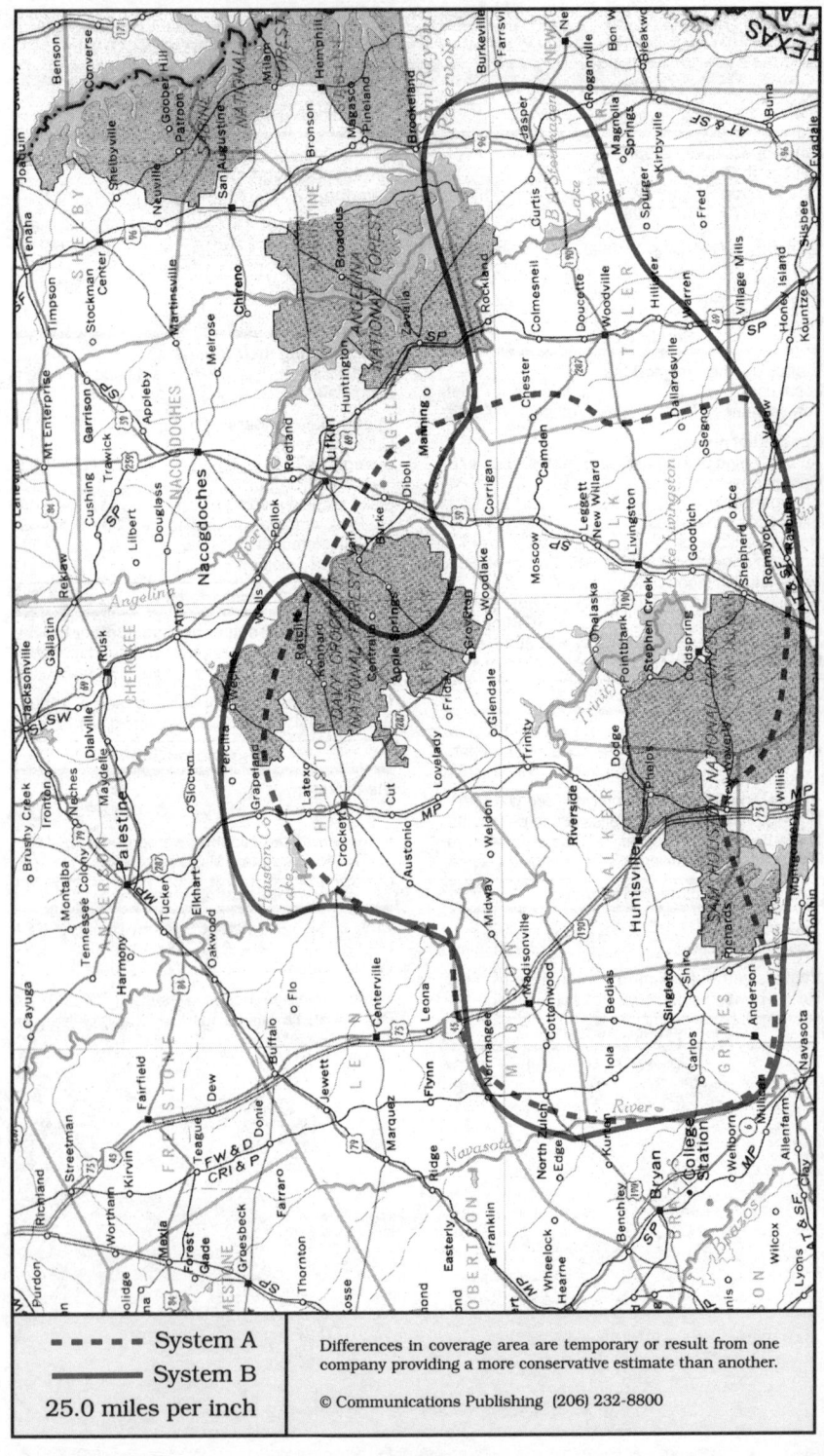

System A
System B
25.0 miles per inch

Differences in coverage area are temporary or result from one company providing a more conservative estimate than another.

© Communications Publishing (206) 232-8800

Huntsville, Texas

Crockett, Huntsville, Jasper, Livingston, Madisonville, Woodville

System A	System B

Cellular company

Cellular One
Eastex Cellular Limited Partnership
1600 Financial Plaza, Suite 760
Huntsville, TX 77340

Corporate office 409-291-3000

Sending calls

Visitor dialing instructions
Local calls . . . . . area code + 7 digits
Long distance . . . 1 + area code + 7 digits

Speed-dial numbers
Customer service 611
Emergency 911

Receiving calls

Caller dials the roamer access number below,
hears tone, dials cellular area code and number.
. . . . 409-661-7626

NationLink & RoamingAmerica (see p. 1157)
Dial *31 send to receive calls, *30 to deactivate.
Dial *32 send to tell caller the roamer access no.
This company's subscribers have level 8 service.

Billing information

Visitor rates
Most visitors . . . . . $3 day + 99¢ minute

Visitors without roaming agreements (p. 1159)
Roamer Plus will intercept first call to bill you.

Identification numbers

City	Market	System	Billing	RSA
All served cities	668	01723	00000	TX-17

Cellular company

GTE Mobilnet
100 Glenborough, Suite 800
Houston, TX 77067

Customer service 800-347-5665
. . . . 713-876-9144
Local office 713-876-5000
Corporate office 404-391-8000

Sending calls

Visitor dialing instructions
Local calls . . . . . . area code + 7 digits
Long distance . . . 1 + area code + 7 digits

Speed-dial numbers
Customer service *611
Emergency 911
Mr. Rescue (roadside assistance) HELP or 4357
Roaming information *711
Telephone problems *111

Receiving calls

Caller dials the roamer access number below,
hears tone, dials cellular area code and number.
. . . . 713-824-7626

Follow Me Roaming forwarding (see p. 1156)
Activate, dial *18 send. Deactivate dial *19 send.

Billing information

Visitor rates
Most visitors . . . . . $3 day + 85¢ minute

Visitors without roaming agreements (p. 1159)
Call co. to use Amex, Mastercard, Visa

Identification numbers

City	Market	System	Billing	RSA
All served cities	668	00012	00000	TX-17

For full coverage area, see Beaumont, TX and
Bryan, TX and Fairfield, TX and Houston, TX and
Lufkin, TX.

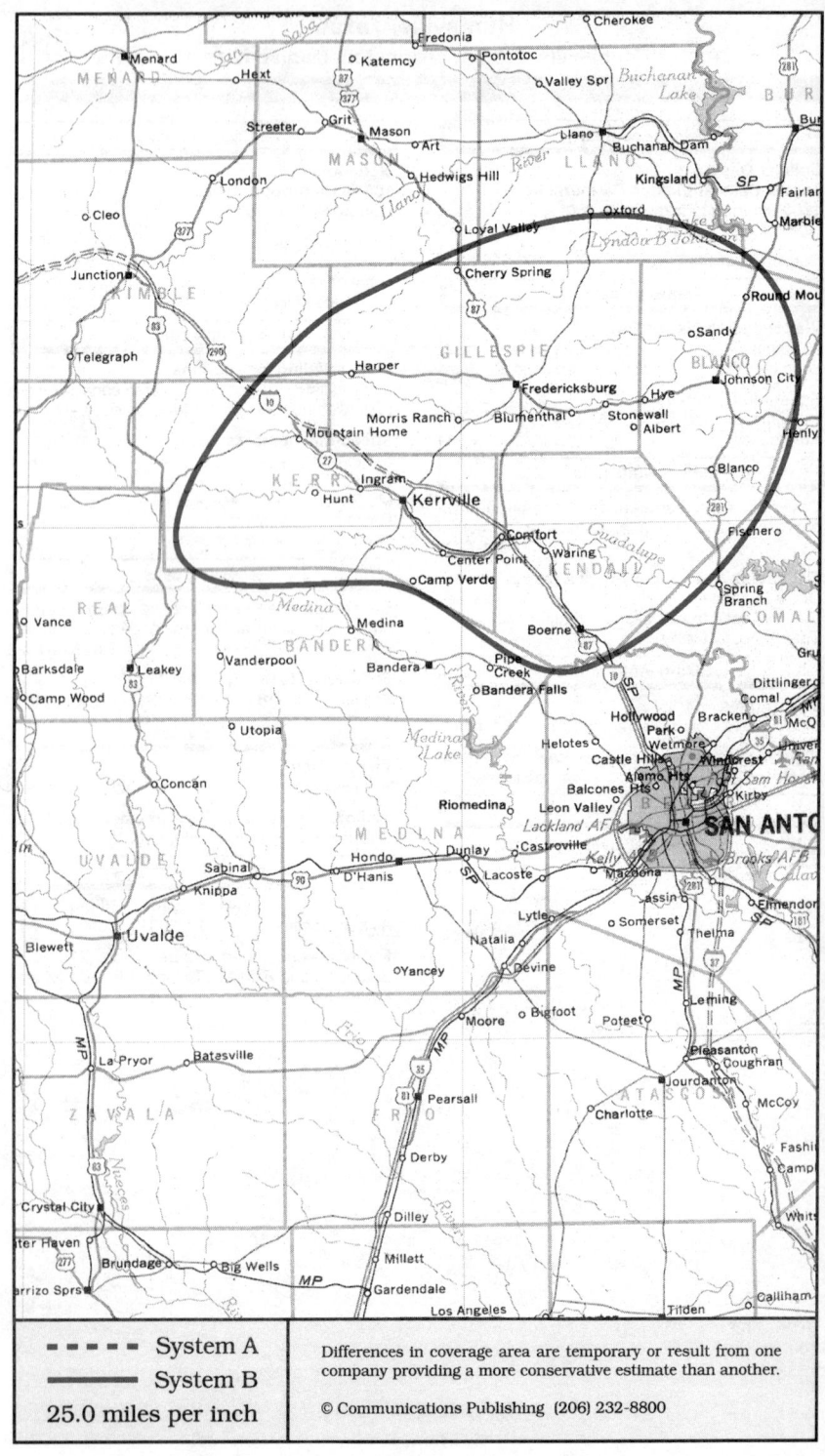

----- System A	Differences in coverage area are temporary or result from one company providing a more conservative estimate than another.
——— System B	
25.0 miles per inch	© Communications Publishing (206) 232-8800

Kerrville, Texas
Boerne, Fredericksburg, Johnson City, Kerrville

System A	System B

Cellular company

No information about the company that will operate system A was available at press time.

Cellular company

Five Star Cellular
Kerrville Telephone
955 Water Street, PO Box 1128
Kerrville, TX 78029-1128

Corporate office   512-896-1200

Sending calls

Visitor dialing instructions
Local calls . . . .   area code + 7 digits
Long distance . . .   1 + area code + 7 digits

Speed-dial numbers
Customer service   611
Emergency   911

Receiving calls

Caller dials the roamer access number below, hears tone, dials cellular area code and number.
. . . .   512-260-7626

Follow Me Roaming forwarding (see p. 1156)
Activate, dial *18 send. Deactivate dial *19 send.

Billing information

Visitor rates
Most visitors . . . . . $2 day + 75¢ minute

Visitors without roaming agreements (p. 1159)
Call co. to use Amex, Mastercard, Visa

Identification numbers

City	Market	System	Billing	RSA
All served cities	666	00122	30198	TX-15

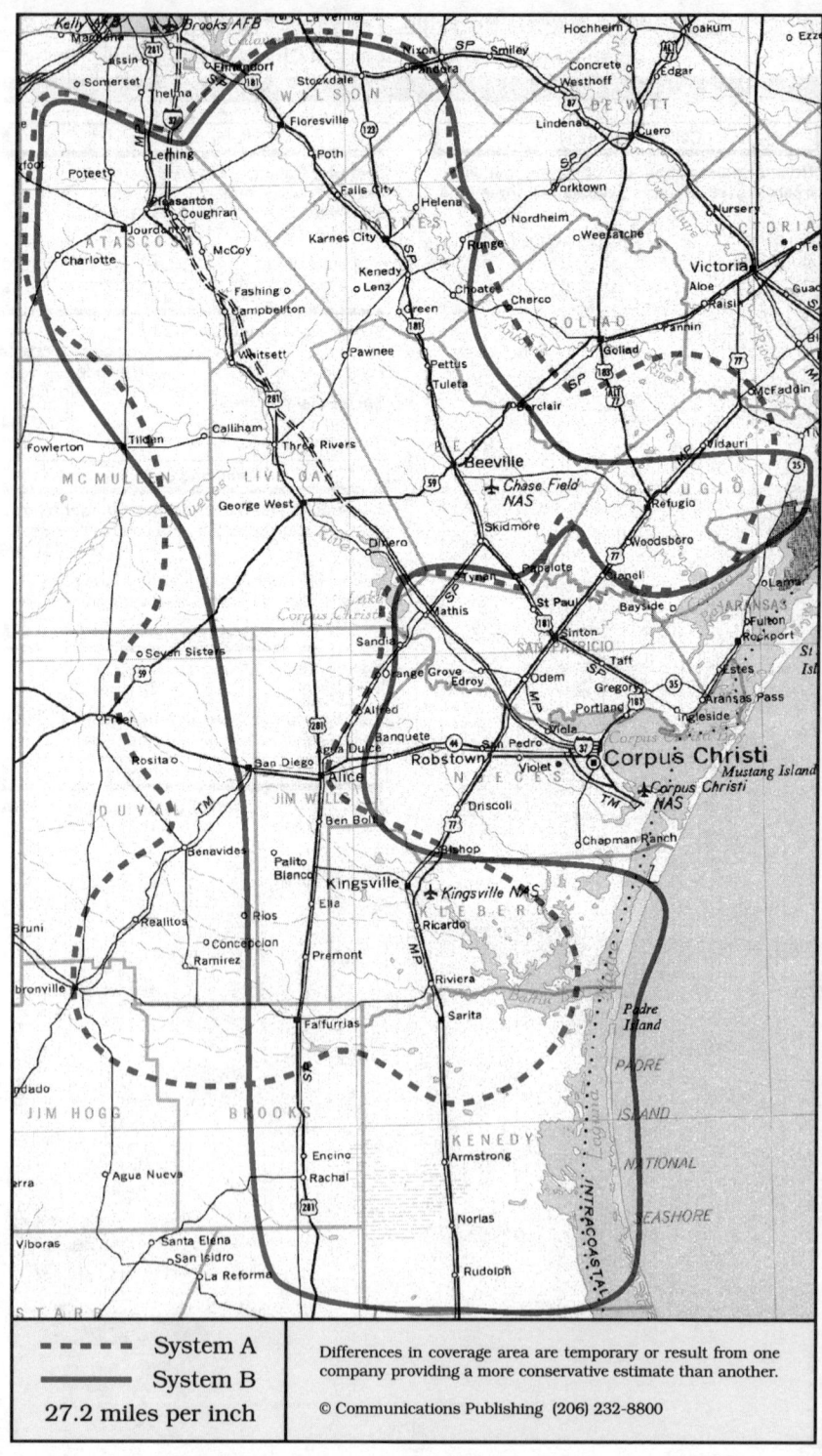

Kingsville, Texas

Alice, Beeville, Falfurrias, Floresville, Kingsville, Pleasanton, San Diego

System A	System B

System A

Cellular company

United States Cellular
112 N St. Mary's Street
Beeville, TX 78102

Local office	512-362-1000
Corporate office	312-399-8900

Sending calls

Visitor dialing instructions

Local calls	area code + 7 digits
Long distance . . .	1 + area code + 7 digits

Speed-dial numbers

Customer service	611
Emergency	911

Receiving calls

Caller dials the roamer access number below,
hears tone, dials cellular area code and number.
. . . . 512-319-7626

Billing information

Visitor rates

Most visitors	$3 day + 85¢ minute

Visitors without roaming agreements (p. 1159)
Call co. to use Amex, Mastercard, Visa

Identification numbers

City	Market	System	Billing	RSA
Beeville	671	01729	00000	TX-20
Kingsville	670	01727	00000	TX-19

For full coverage area, see Uvalde, TX.

System B

Cellular company

Southwestern Bell Mobile Systems
3501 S Padre Island Drive
Corpus Christi, TX 78415

1275 NE Loop 410
San Antonio, TX 78209

Customer service		800-333-2355
. . . .		210-841-5550
Local office . .	Corpus Christi	512-854-5678
. . . .	San Antonio	210-841-5500
Corporate office		214-733-2000

Sending calls

Visitor dialing instructions

Local calls	area code + 7 digits
Long distance . . .	1 + area code + 7 digits

Speed-dial numbers

Customer service	611
Emergency	911

Receiving calls

Caller dials the roamer access number below,
hears tone, dials cellular area code and number.
. . . . 512-877-7626

Follow Me Roaming forwarding (see p. 1156)
Activate, dial ✳18 send. Deactivate dial ✳19 send.

Billing information

Visitor rates

Most visitors	$3 day + 90¢ minute

Visitors without roaming agreements (p. 1159)
Call co. to use Amex, Mastercard, Visa

Identification numbers

City	Market	System	Billing	RSA
Beeville	671	00122	30282	TX-20
Kingsville, 3-Rivers	670	00184	30216	TX-19
Pleasanton	670	00122	30214	TX-19

For full coverage area, see Corpus Christi, TX
and San Antonio, TX and Uvalde, TX.

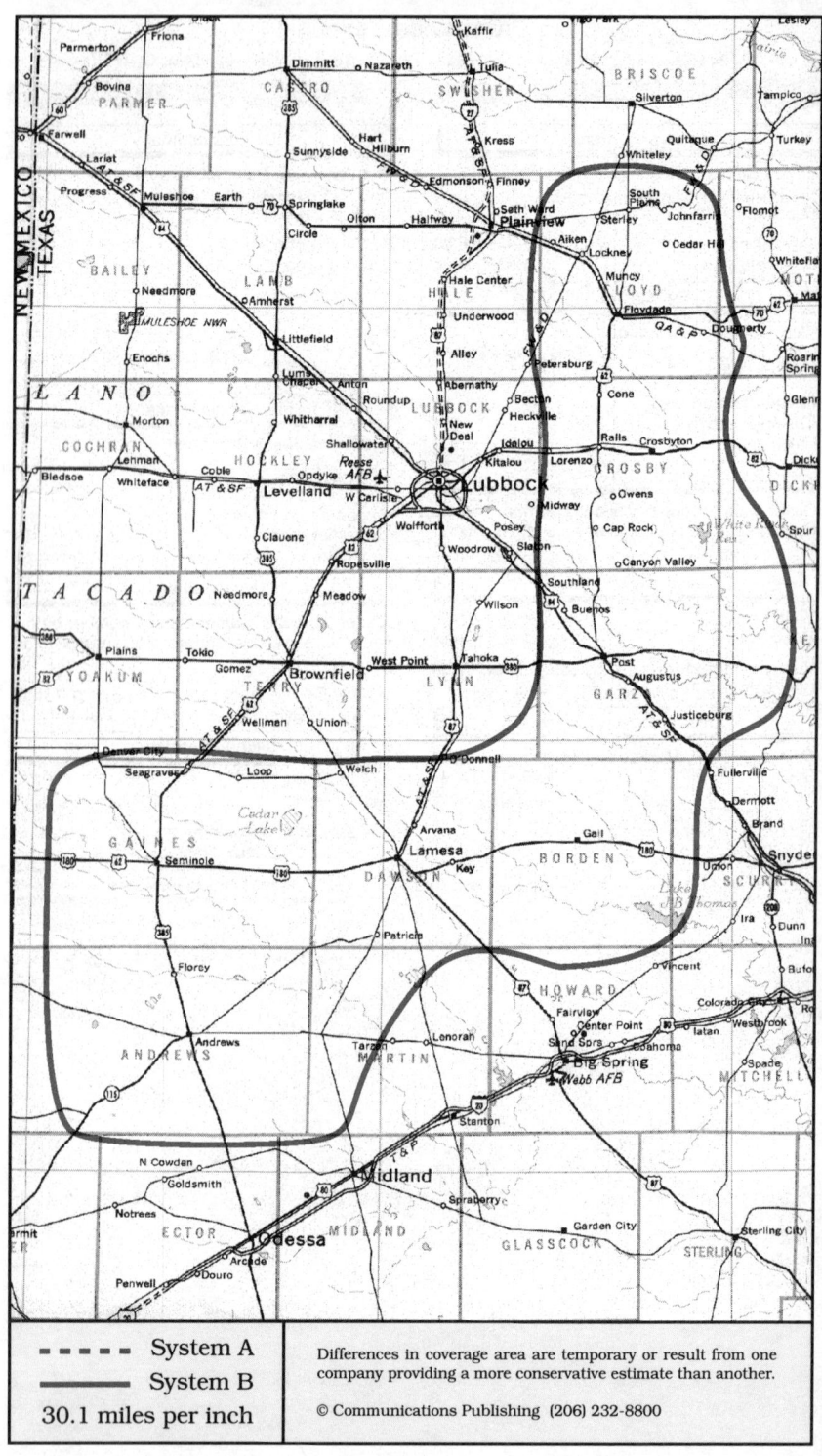

- - - - - System A
——— System B
30.1 miles per inch

Differences in coverage area are temporary or result from one company providing a more conservative estimate than another.

© Communications Publishing (206) 232-8800

Lamesa, Texas

Andrews, Crosbyton, Floydada, Gail, Lamesa, Post, Seminole

System A	System B

Cellular company

No information about the company that will operate system A was available at press time.

Cellular company

Digital Cellular of Texas
Poka Lambro Telecommunications
PO Box 53118
Lubbock, TX 79453

Physical location
11.5 miles N of Tahoka on U.S. Highway 87
Tahoka, TX 79373

Corporate office . some areas 800-662-8805
. . . . 806-924-5432

Sending calls

Visitor dialing instructions
Local calls . . . . . . area code + 7 digits
Long distance . . . 1 + area code + 7 digits

Speed-dial numbers
Customer service 611
Emergency 911

Receiving calls

Caller dials the roamer access number below,
hears tone, dials cellular area code and number.
. . . . 806-759-7626

Follow Me Roaming forwarding (see p. 1156)
Activate, dial *18 send. Deactivate dial *19 send.

Billing information

Visitor rates
Most visitors . . . . . $2 day + 65¢ minute

Visitors without roaming agreements (p. 1159)
Cannot call since company takes no credit cards.

Identification numbers

City	Market	System	Billing	RSA
Floydada	655	01706	00000	TX-4
Lamesa, Seminole	659	01706	30220	TX-8

For full coverage area, see Tahoka, TX.

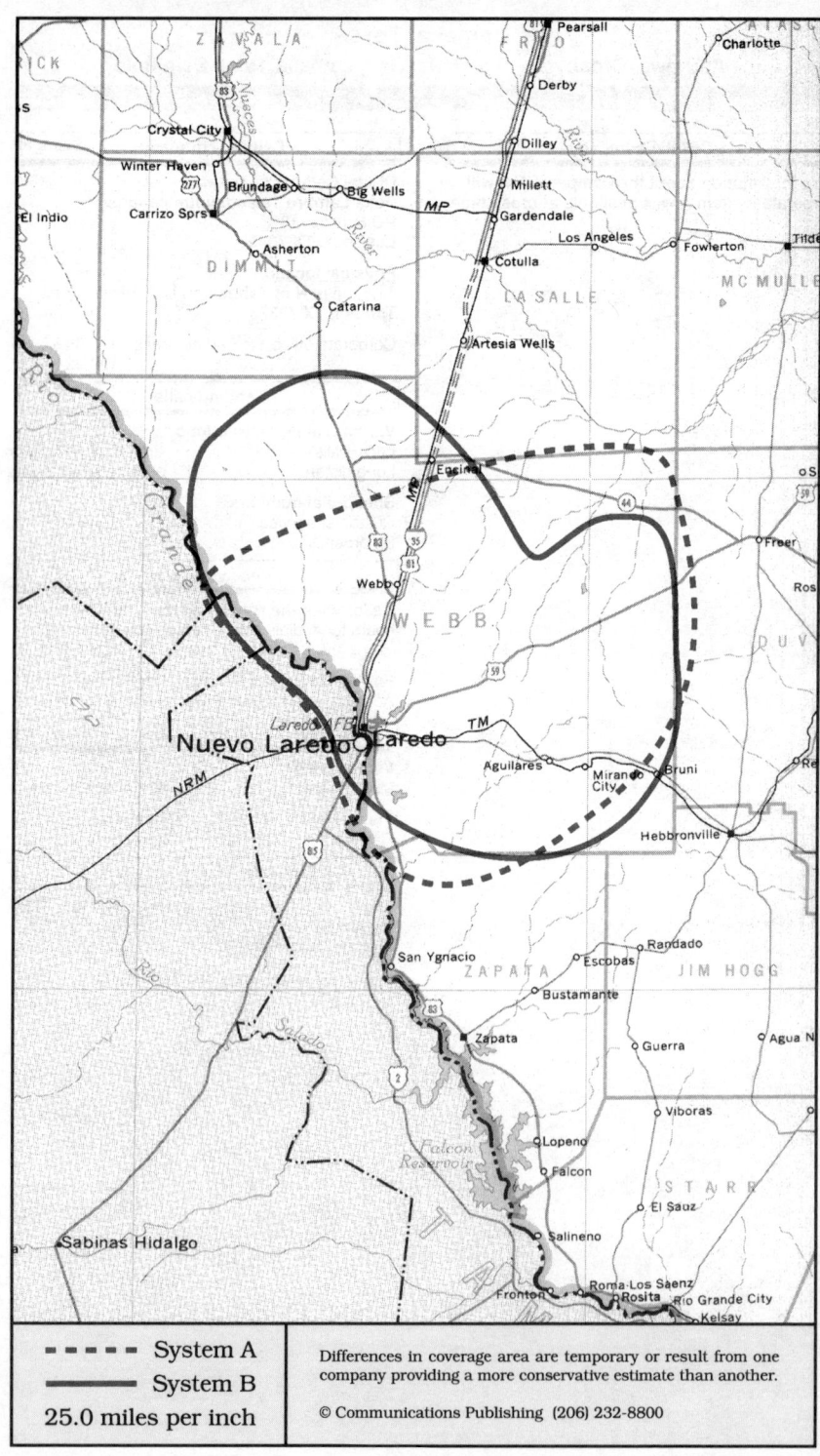

- - - - System A	Differences in coverage area are temporary or result from one
——— System B	company providing a more conservative estimate than another.
25.0 miles per inch	© Communications Publishing (206) 232-8800

System A	System B

Cellular company

United States Cellular
1510 Calle Del Norte, Suite 14
Laredo, TX 78041

Local office . .	some areas	800-334-0023
. . . .		512-727-5914
Corporate office		312-399-8900

Sending calls

Visitor dialing instructions

Local calls	area code + 7 digits
Long distance . . .	1 + area code + 7 digits

Speed-dial numbers

Customer service	611
Emergency	911

Receiving calls

Caller dials the roamer access number below,
hears tone, dials cellular area code and number.
. . . . 512-744-7626

NationLink & RoamingAmerica (see p. 1157)
Dial *31 send to receive calls, *30 to deactivate.
Dial *32 send to tell caller the roamer access no.
This company's subscribers have level 5 service.

Billing information

Visitor rates
Most visitors . . . . . $3 day + 99¢ minute

Visitors without roaming agreements (p. 1159)
Call co. to use Mastercard, Visa

Identification numbers

City	Market	System	Billing	RSA
All served cities	281	00419	00000	MSA

Cellular company

Laredo Cellular
Cellular Information Systems
4519 San Bernardo, Suite 1-C
Laredo, TX 78041

Local office	512-722-2333
Corporate office	203-622-6317

Sending calls

Visitor dialing instructions

Local calls	area code + 7 digits
Long distance . . .	1 + area code + 7 digits

Speed-dial numbers

Customer service	611
Emergency	911
Police department	HELP or 4357

Receiving calls

Caller dials the roamer access number below,
hears tone, dials cellular area code and number.
. . . . 512-763-7626

Follow Me Roaming forwarding (see p. 1156)
Activate, dial *18 send. Deactivate dial *19 send.

Billing information

Visitor rates
Most visitors . . . . . $3 day + 85¢ minute

Visitors without roaming agreements (p. 1159)
Call co. to use Amex, Mastercard, Visa

Identification numbers

City	Market	System	Billing	RSA
All served cities	281	00402	00000	MSA

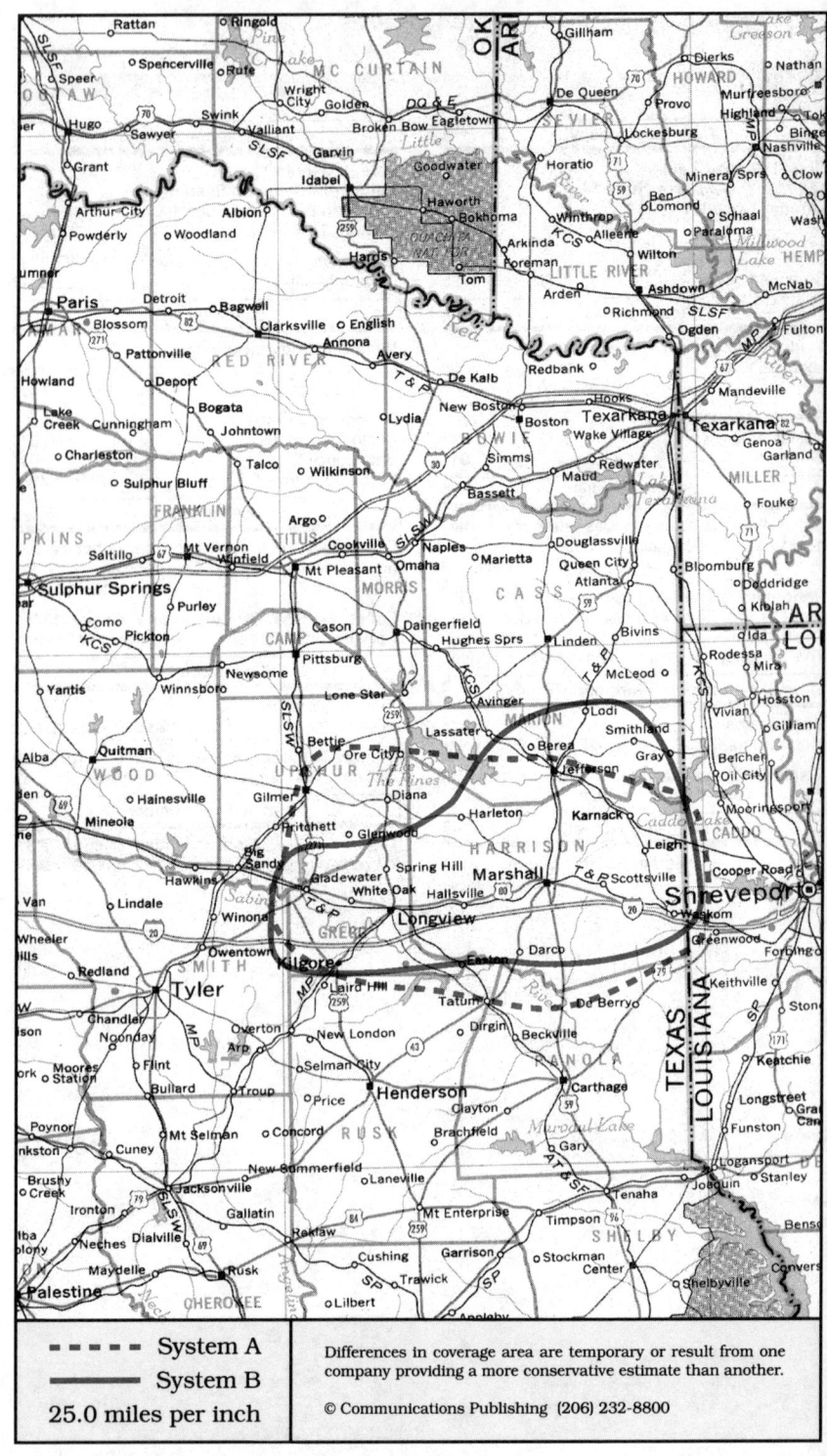

▪ ▪ ▪ ▪ ▪ System A	Differences in coverage area are temporary or result from one company providing a more conservative estimate than another.
────── System B	
25.0 miles per inch	© Communications Publishing (206) 232-8800

Longview, Texas
Jefferson, Kilgore, Longview, Marshall

System A	System B

System A

Cellular company

Cellular One
McCaw Cellular Communications
305 W Loop 281
Longview, TX 75601

Customer service	some areas	800-262-5659
. . . . . .	from Kilgore	903-985-9500
. . . . . .	from Longview	903-237-9500
. . . . . .	from Marshall	903-926-9500
Local office .	from Longview	903-237-8700
. . . . . .	from Marshall	903-926-7700
Corporate office	. . . . .	206-827-4500

Sending calls

Visitor dialing instructions

Local calls	area code + 7 digits
Long distance . . .	1 + area code + 7 digits

Speed-dial numbers

Customer service	611
Emergency	911

Receiving calls

Caller dials the roamer access number below,
hears tone, dials cellular area code and number.

Kilgore	903-985-7626
Longview	903-237-8626
Marshall	903-926-7626

NationLink & RoamingAmerica (see p. 1157)
Dial ✶31 send to receive calls, ✶30 to deactivate.
Dial ✶32 send to tell caller the roamer access no.
This company's subscribers have level 9 service.

North American Cellular Network (see p. 1156)
All calls forward to you. Deactivate dial ✶35 send.

Billing information

Visitor rates

Most visitors . . . . . $0 day + 99¢ minute

Visitors without roaming agreements (p. 1159)
Call co. to use Amex, Mastercard, Visa
or Roamer Plus will intercept first call to bill you.

Identification numbers

City	Market	System	Billing	RSA
All served cities	206	00229	00000	MSA

For full coverage area, see Shreveport, LA.

System B

Cellular company

Centel Cellular
415 Loop 281 E
Longview, TX 75601

Local office	800-657-5251
. . . .	903-663-1161
Corporate office	312-399-2644

Sending calls

Visitor dialing instructions

Local calls	area code + 7 digits
Long distance . . .	1 + area code + 7 digits

Speed-dial numbers

Customer service	✶611
Emergency	911
KLTV	✶77

Receiving calls

Caller dials the roamer access number below,
hears tone, dials cellular area code and number.

Longview	903-738-7626
Marshall	903-930-7626

Follow Me Roaming forwarding (see p. 1156)
Activate, dial ✶18 send. Deactivate dial ✶19 send.

Billing information

Visitor rates

Most visitors . . . . . $3 day + 99¢ minute

Visitors without roaming agreements (p. 1159)
Cannot call since company takes no credit cards.

Identification numbers

City	Market	System	Billing	RSA
All served cities	206	00418	00000	MSA

For full coverage area, see Tyler, TX.

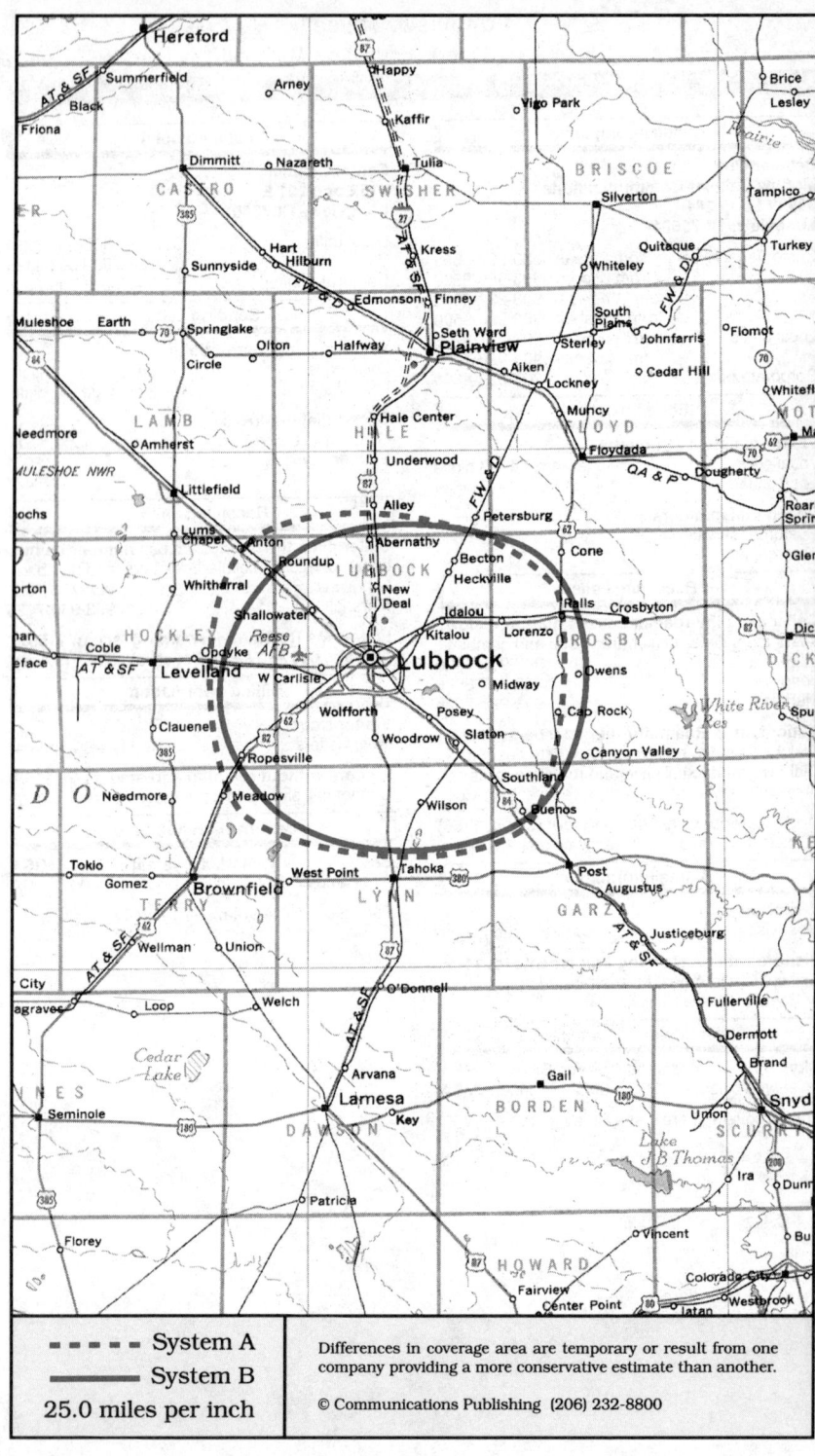

---- System A —— System B 25.0 miles per inch	Differences in coverage area are temporary or result from one company providing a more conservative estimate than another. © Communications Publishing (206) 232-8800

Lubbock, Texas

Cellular company

Cellular One of Lubbock
Cellular Information Systems
3103 34th Street
Lubbock, TX 79410-3227

Local office	806-797-2355
Corporate office	203-622-6317

Sending calls

Visitor dialing instructions

Local calls	area code + 7 digits
Long distance . . .	1 + area code + 7 digits

Speed-dial numbers

Customer service	*611
Emergency	911

Receiving calls

Caller dials the roamer access number below,
hears tone, dials cellular area code and number.
. . . . 806-777-7626

NationLink & RoamingAmerica (see p. 1157)
Dial *31 send to receive calls, *30 to deactivate.
This company's subscribers have level 3 service.

Billing information

Visitor rates
Most visitors . . . . . $3 day + 85¢ minute

Visitors without roaming agreements (p. 1159)
Call co. to use Mastercard, Visa

Identification numbers

City	Market	System	Billing	RSA
All served cities	161	00439	00000	MSA

Cellular company

Southwestern Bell Mobile Systems
5109 82nd Street, Suite 4
Lubbock, TX 79424

Local office	806-798-4300
Corporate office	214-733-2000

Sending calls

Visitor dialing instructions

Local calls	area code + 7 digits
Long distance . . .	1 + area code + 7 digits

Speed-dial numbers

Customer service	611
Emergency	911

Receiving calls

Caller dials the roamer access number below,
hears tone, dials cellular area code and number.
. . . . 806-789-7626

Follow Me Roaming forwarding (see p. 1156)
Activate, dial *18 send. Deactivate dial *19 send.

Billing information

Visitor rates
Most visitors . . . . . $2 day + 75¢ minute

Visitors without roaming agreements (p. 1159)
Call co. to use Amex, Discover, Mastercard, Visa

Identification numbers

City	Market	System	Billing	RSA
All served cities	161	00422	00000	MSA

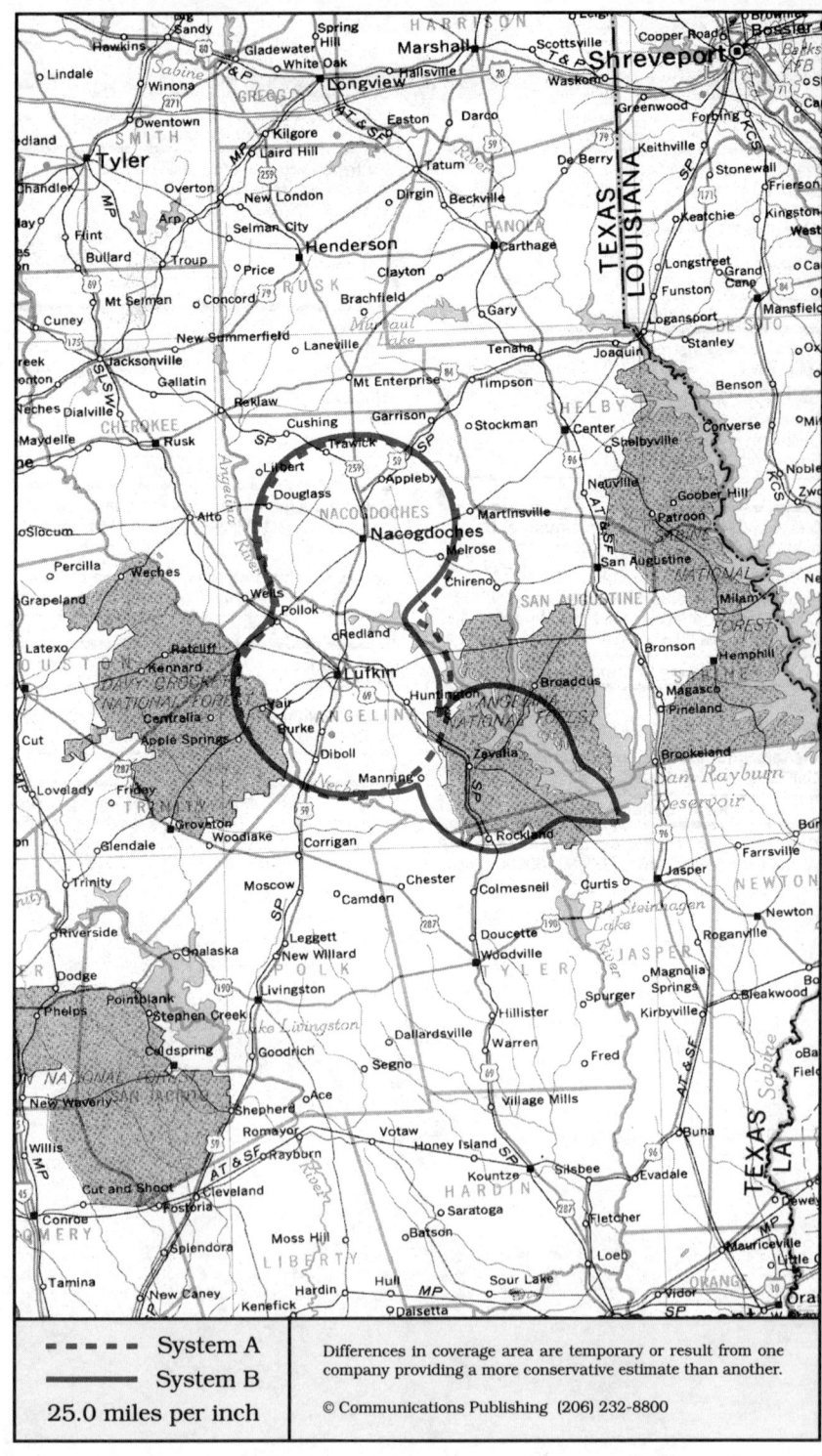

Differences in coverage area are temporary or result from one company providing a more conservative estimate than another.

© Communications Publishing (206) 232-8800

- - - - - System A
——— System B
25.0 miles per inch

Lufkin, Texas
Lufkin, Nacogdoches

System A	System B

Cellular company

United States Cellular
535 W Southwest Loop 323, Suite 208
Tyler, TX 75701

Local office	800-685-6365
. . . .	903-561-2355
Corporate office	312-399-8900

Sending calls

Visitor dialing instructions
Local calls	area code + 7 digits
Long distance	area code + 7 digits

Speed-dial numbers
Customer service	611
Emergency	911
Roaming information	*711

Receiving calls

Caller dials the roamer access number below,
hears tone, dials cellular area code and number.
. . . . 409-554-7626

NationLink & RoamingAmerica (see p. 1157)
Caller hears your current roamer access number.
This company's subscribers have level 0 service.

Billing information

Visitor rates
Most visitors . . . . . $3 day + 85¢ minute

Visitors without roaming agreements (p. 1159)
Call co. to use Mastercard, Visa

Identification numbers

City	Market	System	Billing	RSA
All served cities	662	01711	00000	TX-11

For full coverage area, see Tyler, TX.

Cellular company

GTE Mobilnet
100 Glenborough, Suite 800
Houston, TX 77067

Customer service	800-347-5665
. . . .	713-876-9144
Local office	713-876-5000
Corporate office	404-391-8000

Sending calls

Visitor dialing instructions
Local calls	area code + 7 digits
Long distance . . .	1 + area code + 7 digits

Speed-dial numbers
Customer service	*611
Emergency	911
Roaming information	*711
Telephone problems	*111

Receiving calls

Caller dials the roamer access number below,
hears tone, dials cellular area code and number.
. . . . 713-824-7626

Follow Me Roaming forwarding (see p. 1156)
Activate, dial *18 send. Deactivate dial *19 send.

Billing information

Visitor rates
Most visitors . . . . . $3 day + 85¢ minute

Visitors without roaming agreements (p. 1159)
Call co. to use Amex, Mastercard, Visa

Identification numbers

City	Market	System	Billing	RSA
All served cities	662	00012	00000	TX-11

For full coverage area, see Huntsville, TX.

871

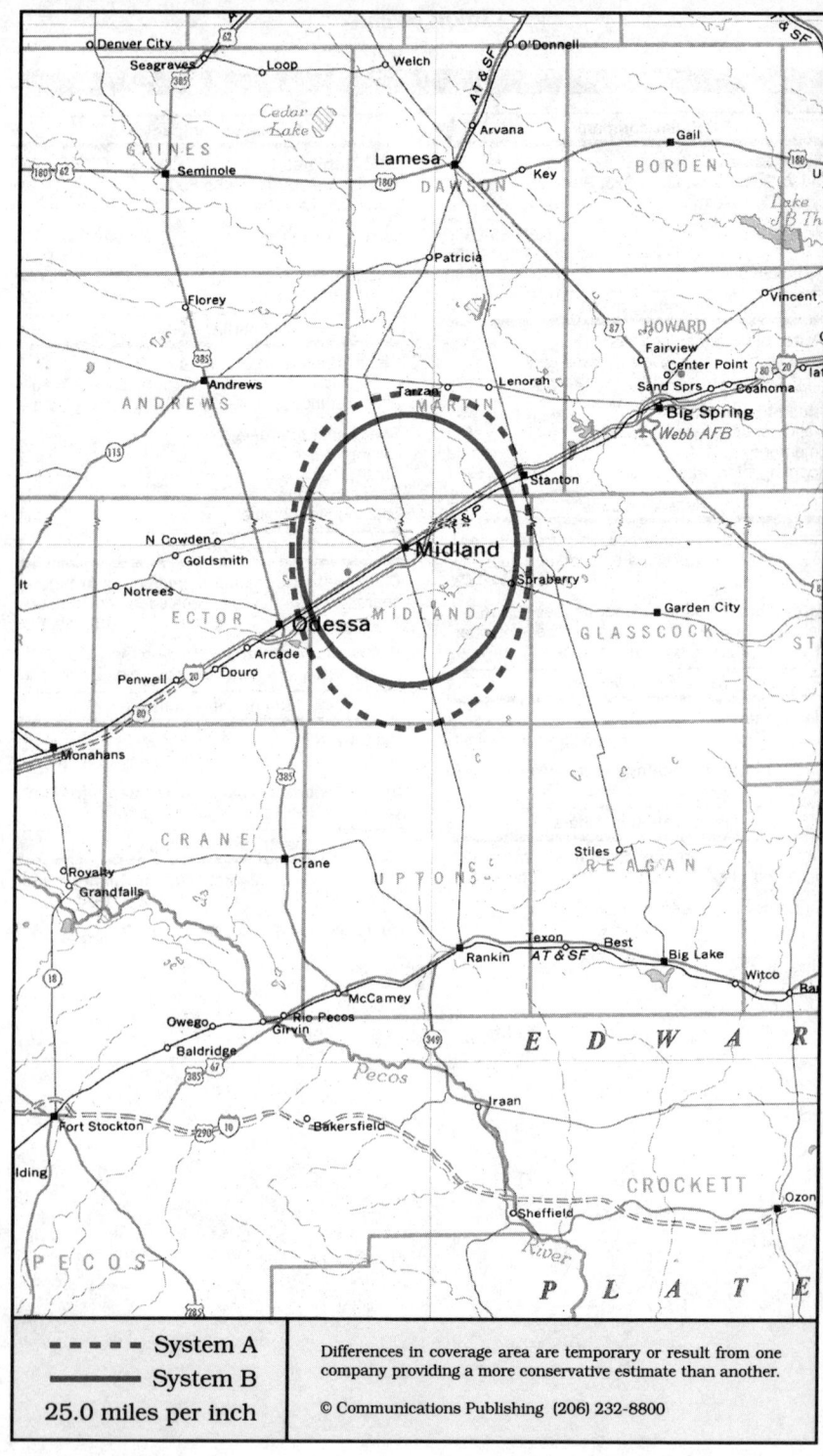

- - - - System A **———— System B** 25.0 miles per inch	Differences in coverage area are temporary or result from one company providing a more conservative estimate than another. © Communications Publishing (206) 232-8800

Midland, Texas

Cellular company

Cellular One
GenCell Management
1308 S Midkiff, Suite 203
Midland, TX 79701

Customer service	800-888-7868
Local office	915-686-0612
Corporate office	707-425-8000

Sending calls

Visitor dialing instructions

Local calls	area code + 7 digits
Long distance . . .	1 + area code + 7 digits

Speed-dial numbers

Customer service	611
Emergency	911

Receiving calls

Caller dials the roamer access number below,
hears tone, dials cellular area code and number.
. . . . 915-557-7626

NationLink & RoamingAmerica (see p. 1157)
Caller hears your current roamer access number.
This company's subscribers have level 0 service.

Billing information

Visitor rates
Most visitors . . . . . $3 day + 99¢ minute

Visitors without roaming agreements (p. 1159)
American Roaming Network intercepts call to bill.

Identification numbers

City	Market	System	Billing	RSA
All served cities	295	00459	00000	MSA

For full coverage area, see Pecos, TX.

Cellular company

Southwestern Bell Mobile Systems
3900 S County Road #1290
Midland, TX 79711

Customer service	915-563-4617
Local office	915-563-4611
Corporate office	214-733-2000

Sending calls

Visitor dialing instructions

Local calls	area code + 7 digits
Long distance . . .	1 + area code + 7 digits

Speed-dial numbers

Customer service	611
Emergency	911

Receiving calls

Caller dials the roamer access number below,
hears tone, dials cellular area code and number.
. . . . 915-559-7626

Follow Me Roaming forwarding (see p. 1156)
Activate, dial ∗18 send. Deactivate dial ∗19 send.

Billing information

Visitor rates
Most visitors . . . . . $2 day + 75¢ minute

Visitors without roaming agreements (p. 1159)
Call co. to use Amex, Discover, Mastercard, Visa

Identification numbers

City	Market	System	Billing	RSA
All served cities	295	00422	30004	MSA

For full coverage area, see Odessa, TX.

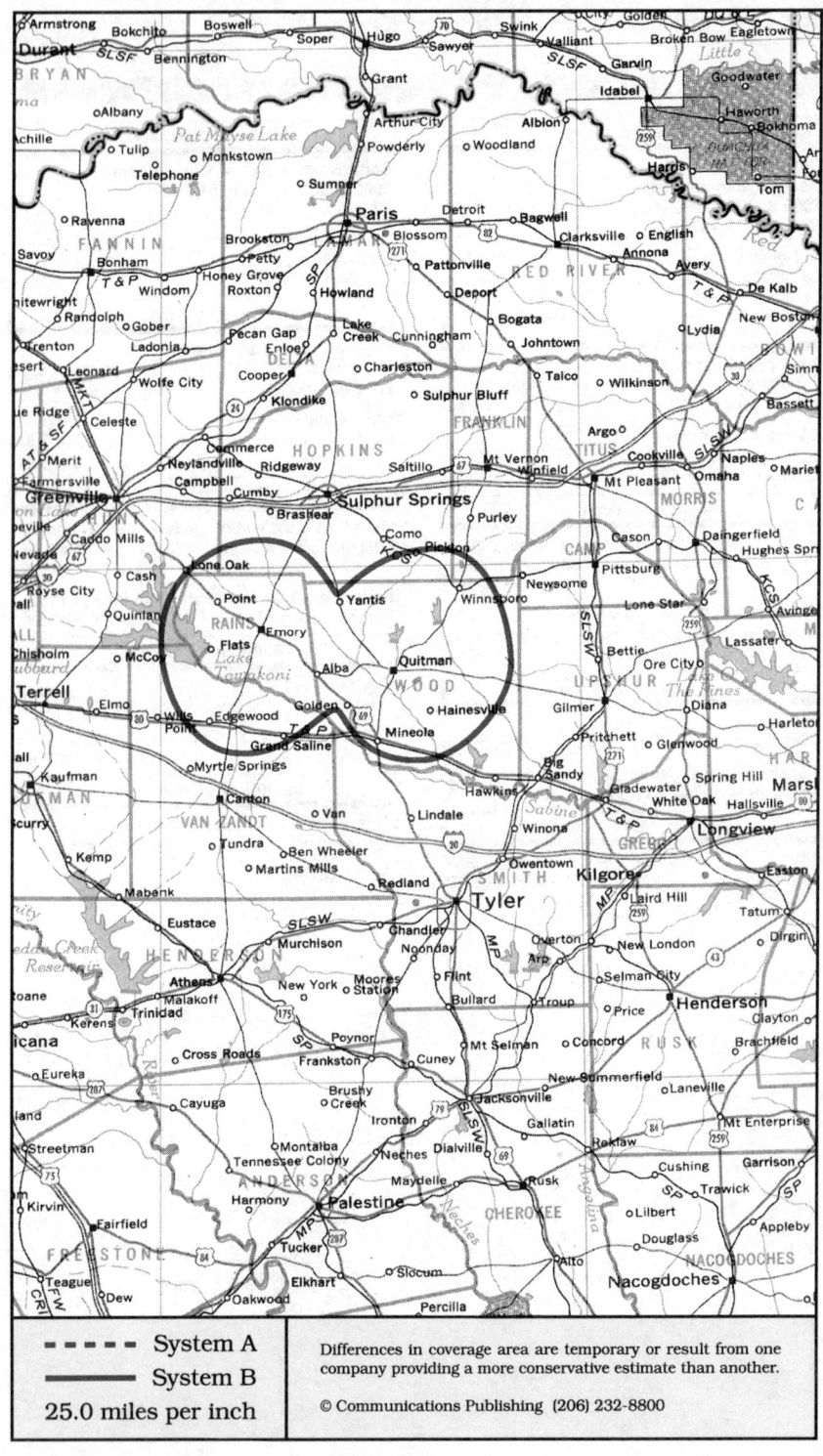

- - - - - System A	Differences in coverage area are temporary or result from one company providing a more conservative estimate than another.
——— System B	
25.0 miles per inch	© Communications Publishing (206) 232-8800

Mineola, Texas
Emory, Mineola, Quitman, Winnsboro

Cellular company

No information about the company that will operate system A was available at press time.

Cellular company

Peoples Cellular
102 N Stephens, PO Box 1206
Quitman, TX 75783

Corporate office 903-878-2197

Sending calls

Visitor dialing instructions
Local calls . . . . 1 + area code + 7 digits
Long distance . . . 1 + area code + 7 digits

Speed-dial numbers
Customer service 611
Emergency 911

Receiving calls

Caller dials the roamer access number below,
hears tone, dials cellular area code and number.
. . . . 903-850-7626

Follow Me Roaming forwarding (see p. 1156)
Activate, dial *18 send. Deactivate dial *19 send.

Billing information

Visitor rates
Most visitors . . . . . $3 day + 75¢ minute

Visitors without roaming agreements (p. 1159)
Cannot call since company takes no credit cards.

Identification numbers

City	Market	System	Billing	RSA
All served cities	658	01868	00000	TX-7

This company serves TX-7 B3, which is one of the six portions of rural area TX-7.

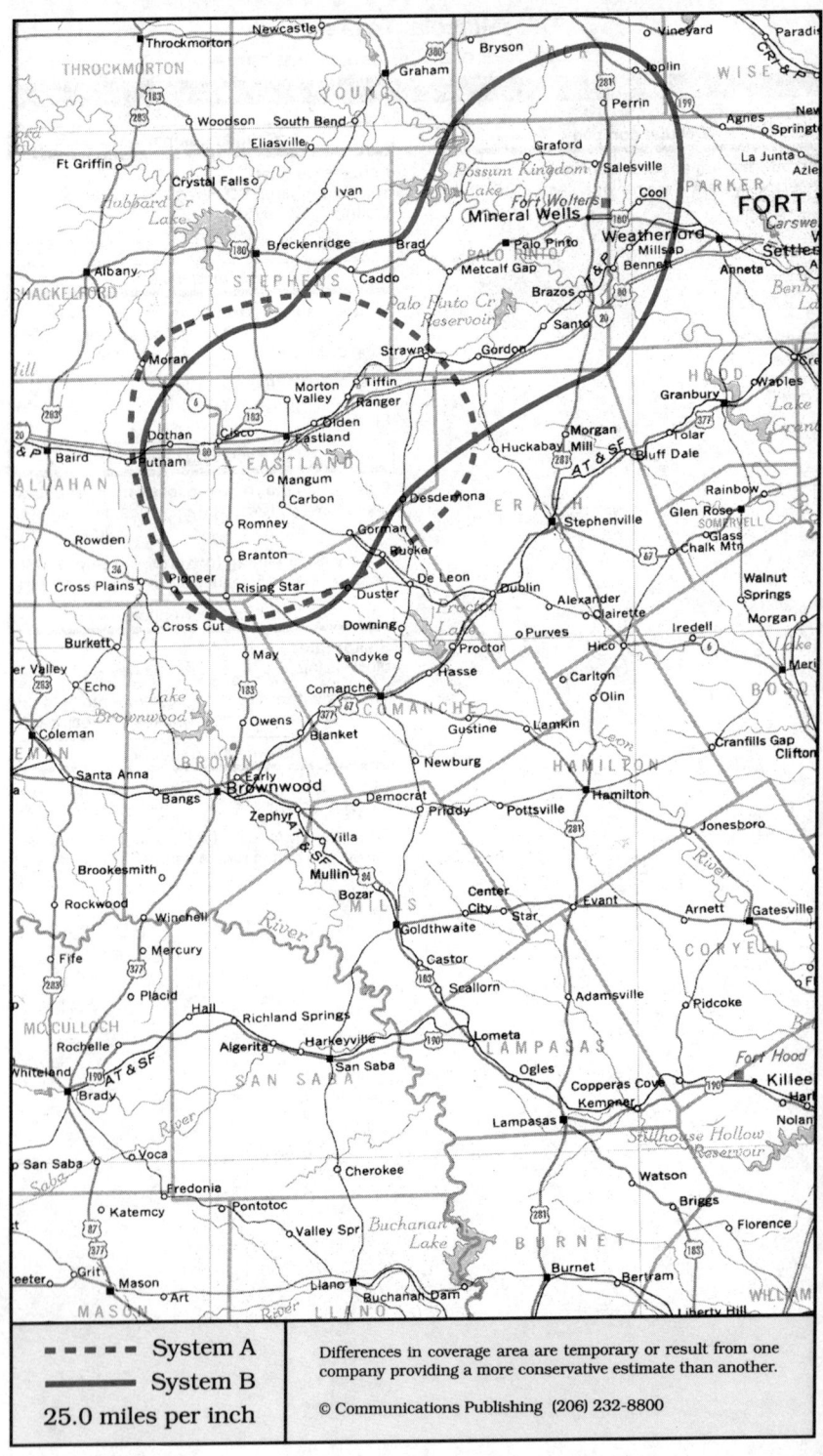

- - - - - System A ————— System B 25.0 miles per inch	Differences in coverage area are temporary or result from one company providing a more conservative estimate than another. © Communications Publishing (206) 232-8800

Mineral Wells, Texas
Cisco, Eastland, Mineral Wells, Palo Pinto, Ranger

System A	System B

Cellular company

Cellular One
Licensed to: Lone Star Cellular
Managed by: Prime Cellular
300 N Main, Suite B
Brownwood, TX 76801

Local office 915-643-2355
Corporate office   Prime Cellular 201-227-1434

Sending calls

Visitor dialing instructions
Local calls . . . . . area code + 7 digits
Long distance . . . 1 + area code + 7 digits

Speed-dial numbers
Customer service 611
Emergency 911

Receiving calls

Caller dials the roamer access number below,
hears tone, dials cellular area code and number.
. . . . 915-642-7626

NationLink & RoamingAmerica (see p. 1157)
Caller hears your current roamer access number.
This company's subscribers have level 0 service.

Billing information

Visitor rates
Most visitors . . . . . $3 day + 99¢ minute

Visitors without roaming agreements (p. 1159)
Roamer Plus will intercept first call to bill you.

Identification numbers

City	Market	System	Billing	RSA
All served cities	660	01707	33607	TX-9

For full coverage area, see Brownwood, TX.

Cellular company

Southwestern Bell Mobile Systems
17330 Preston Road, Suite 100-A
Dallas, TX 75252

Customer service   some areas 800-331-0500
. . . . 214-988-8484
Corporate office 214-733-2000

Sending calls

Visitor dialing instructions
Local calls . . . . . . area code + 7 digits
Long distance . . . 1 + area code + 7 digits

Speed-dial numbers
Customer service 611
Emergency 911
Fast facts (sports, news, weather, etc.) . *123
Mr. Rescue (roadside assistance) HELP or 4357
Roaming information *711

Receiving calls

Caller dials the roamer access number below,
hears tone, dials cellular area code and number.
. . . . 214-384-7626

Follow Me Roaming forwarding (see p. 1156)
Activate, dial *18 send. Deactivate dial *19 send.

Billing information

Visitor rates
Most visitors . . . . . $3 day + 75¢ minute

Visitors without roaming agreements (p. 1159)
Call co. to use Amex, Mastercard, Visa

Identification numbers

City	Market	System	Billing	RSA
Cisco	660	00038	30278	TX-9
Mineral Wells	657	00038	30036	TX-6

For full coverage area, see Dallas, TX.

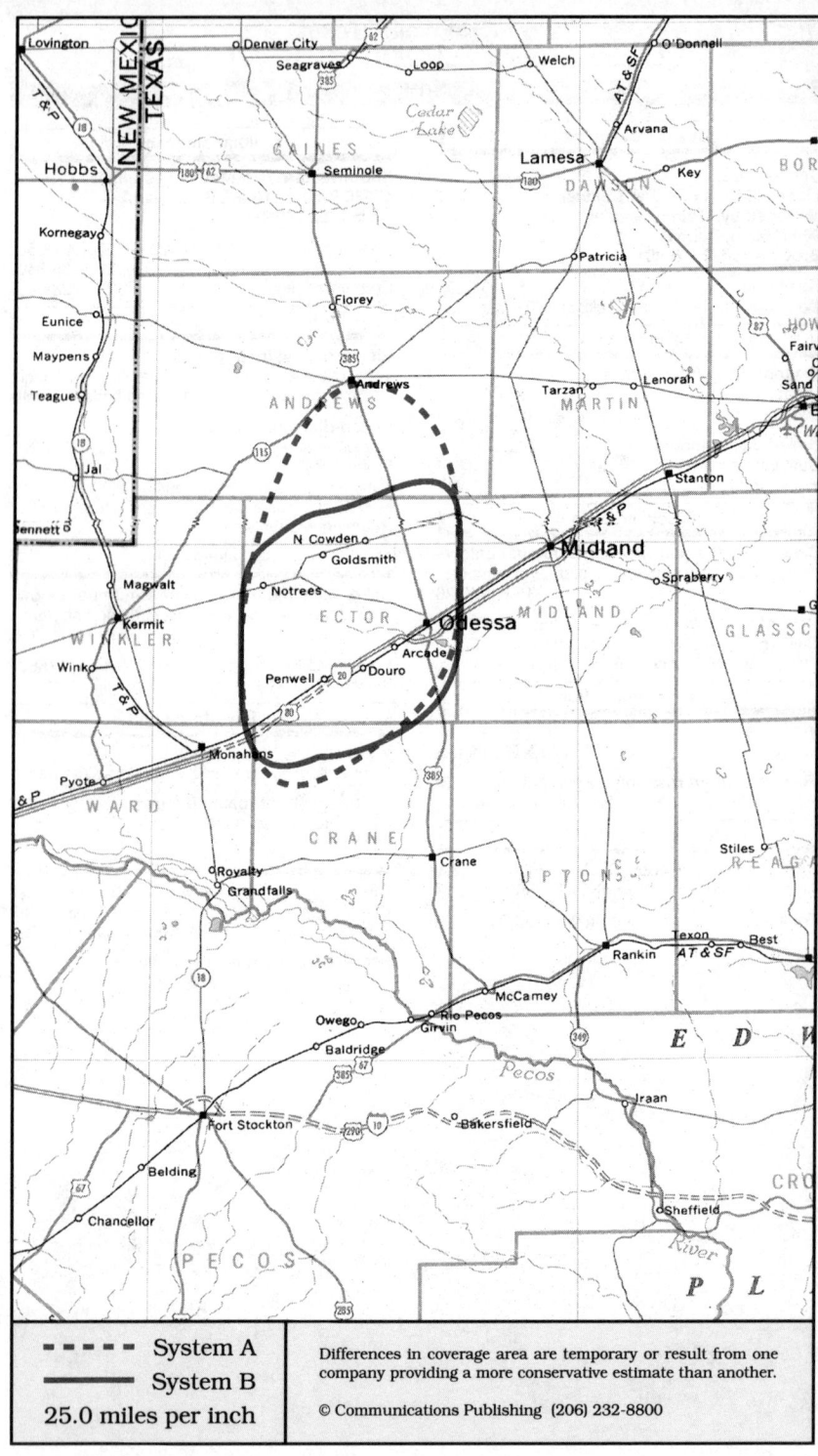

- - - - - System A	Differences in coverage area are temporary or result from one
———— System B	company providing a more conservative estimate than another.
25.0 miles per inch	© Communications Publishing (206) 232-8800

Odessa, Texas

System A

Cellular company

Cellular One of Odessa
Cellular Information Systems
3103 34th Street
Lubbock, TX 79410-3227

Local office	915-563-2355
Corporate office	203-622-6317

Sending calls

Visitor dialing instructions

Local calls	area code + 7 digits
Long distance . . .	1 + area code + 7 digits

Speed-dial numbers

Customer service	611
Emergency	911

Receiving calls

Caller dials the roamer access number below, hears tone, dials cellular area code and number.
. . . . 915-777-7626

Billing information

Visitor rates
Most visitors . . . . . $3 day + 85¢ minute

Visitors without roaming agreements (p. 1159)
Call co. to use Mastercard, Visa

Identification numbers

City	Market	System	Billing	RSA
All served cities	255	00475	00000	MSA

System B

Cellular company

Southwestern Bell Mobile Systems
3900 S County Road #1290
Midland, TX 79711

Customer service	915-563-4617
Local office	915-563-4611
Corporate office	214-733-2000

Sending calls

Visitor dialing instructions

Local calls	area code + 7 digits
Long distance . . .	1 + area code + 7 digits

Speed-dial numbers

Customer service	611
Emergency	911

Receiving calls

Caller dials the roamer access number below, hears tone, dials cellular area code and number.
. . . . 915-559-7626

Follow Me Roaming forwarding (see p. 1156)
Activate, dial *18 send. Deactivate dial *19 send.

Billing information

Visitor rates
Most visitors . . . . . $2 day + 75¢ minute

Visitors without roaming agreements (p. 1159)
Call co. to use Amex, Discover, Mastercard, Visa

Identification numbers

City	Market	System	Billing	RSA
All served cities	255	00422	00000	MSA

For full coverage area, see Midland, TX.

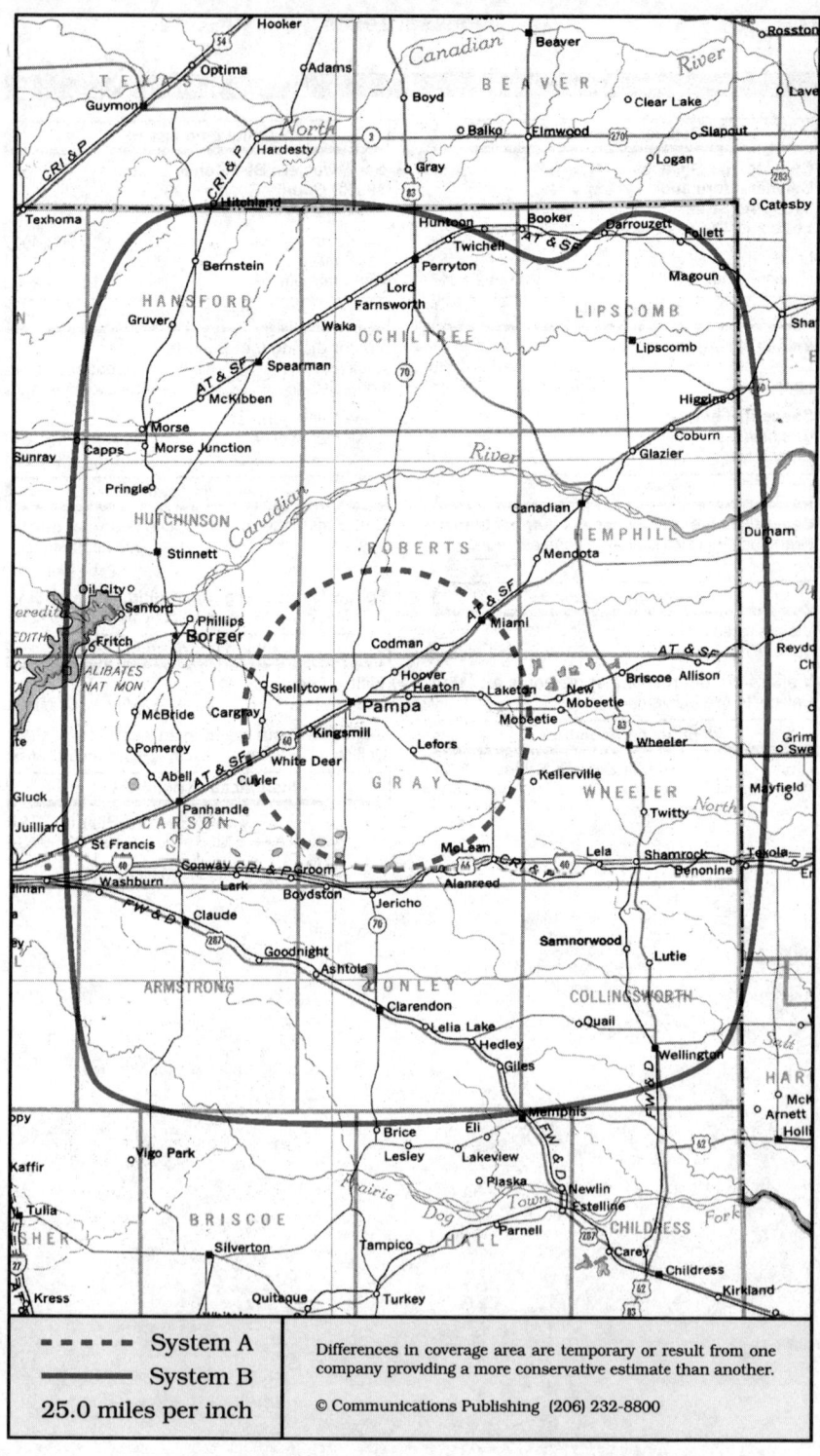

- - - - System A	Differences in coverage area are temporary or result from one company providing a more conservative estimate than another.
—— System B	
25.0 miles per inch	© Communications Publishing (206) 232-8800

Pampa, Texas

Borger, Pampa, Panhandle, Perryton, Spearman, Stinnett, Wellington, Wheeler

System A	System B

Cellular company

Century Cellunet
1916 N Hobart
Pampa, TX 79065

Customer service	800-638-9727
. . . .	318-683-3450
Local office	806-669-3435
Corporate office	318-388-9000

Sending calls

Visitor dialing instructions

Local calls	area code + 7 digits
Long distance . . .	1 + area code + 7 digits

Speed-dial numbers

Customer service	611
Emergency	911

Receiving calls

Caller dials the roamer access number below,
hears tone, dials cellular area code and number.
. . . . 806-664-7626

Billing information

Visitor rates
Most visitors . . . . . $3 day + 75¢ minute

Visitors without roaming agreements (p. 1159)
Call co. to use Amex, Mastercard, Visa

Identification numbers

City	Market	System	Billing	RSA
All served cities	653	01693	00000	TX-2

Cellular company

Dobson Cellular Systems
2131 Perryton Parkway
Pampa, TX 79065

Local office . . .	some areas	800-882-4154
. . . .		806-665-0500
Corporate office		405-749-9744

Sending calls

Visitor dialing instructions

Local calls	area code + 7 digits
Long distance . . .	1 + area code + 7 digits

Speed-dial numbers

Customer service	611
Emergency	911

Receiving calls

Caller dials the roamer access number below,
hears tone, dials cellular area code and number.
Use either number since they are interconnected.

Cheyenne, OK	405-497-7626
Borger, TX	806-898-7626

Follow Me Roaming forwarding (see p. 1156)
Activate, dial ∗18 send. Deactivate dial ∗19 send.

Billing information

Visitor rates
Most visitors . . . . . $3 day + 75¢ minute

Visitors without roaming agreements (p. 1159)
Call co. to use Amex, Mastercard, Visa

Identification numbers

City	Market	System	Billing	RSA
All served cities	653	01592	30382	TX-2

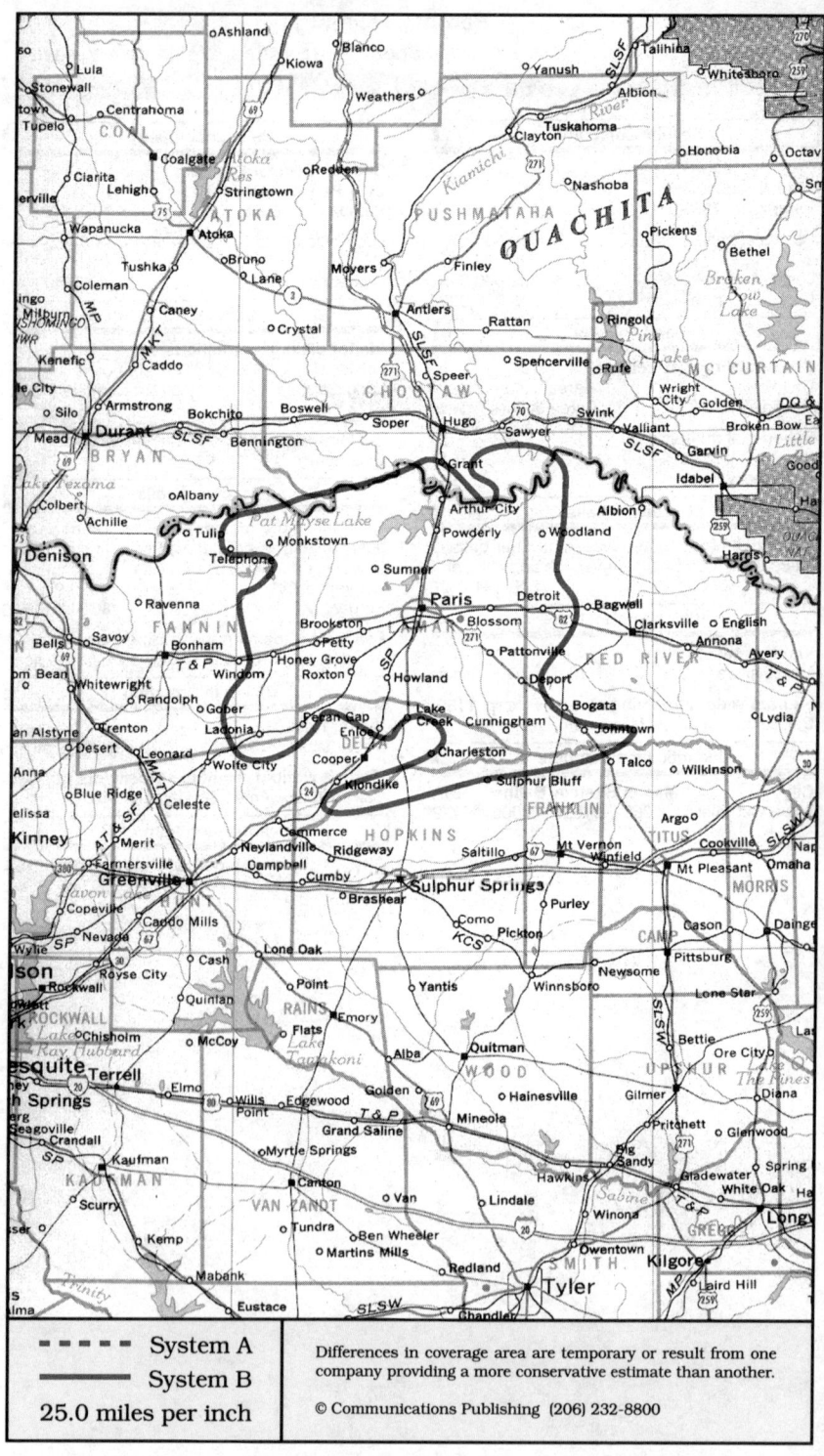

Paris, Texas
Paris (Lamar County)

System A	System B

Cellular company

No information about the company that will operate system A was available at press time.

Cellular company

Lamar County Cellular
1331 Clarksville Street
Paris, TX 75460

Customer service	800-242-4046
. . . .	318-683-3450
Corporate office	903-785-8852

Sending calls

Visitor dialing instructions
Local calls . . . . 7 digits
Long distance . . . 1 + area code + 7 digits

Speed-dial numbers
Customer service *611
Emergency 911

Receiving calls

Caller dials the roamer access number below, hears tone, dials cellular area code and number.
. . . . 903-982-7626

Billing information

Visitor rates
Most visitors . . . . . $3 day + 75¢ minute

Visitors without roaming agreements (p. 1159)
Call co. to use Amex, Mastercard, Visa

Identification numbers

City	Market	System	Billing	RSA
All served cities	658	01872	00000	TX-7

Customer service and roaming agreements are provided by Century Cellunet.

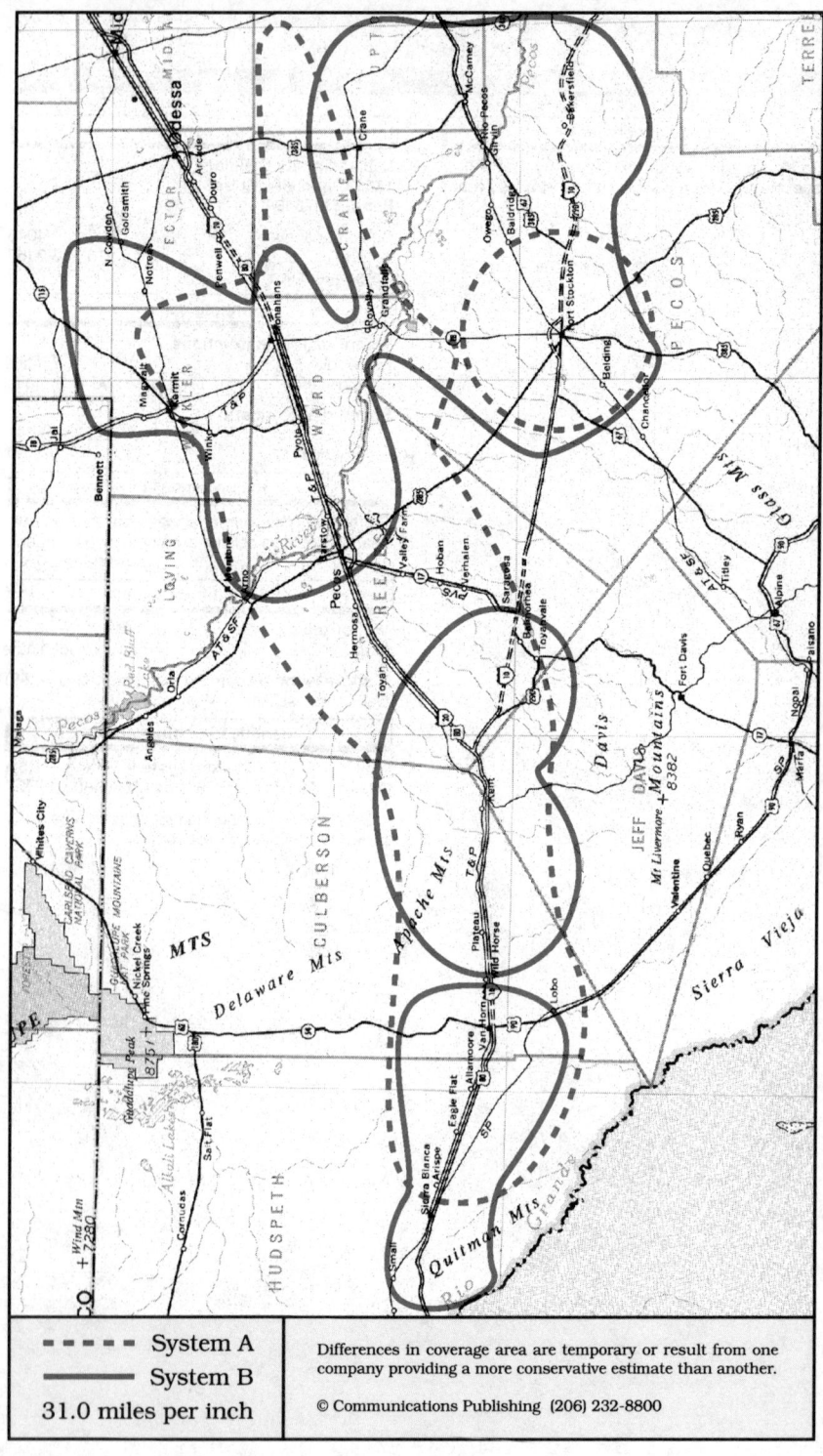

- - - - System A	Differences in coverage area are temporary or result from one company providing a more conservative estimate than another.
—— System B	
31.0 miles per inch	© Communications Publishing (206) 232-8800

Pecos, Texas

Ft. Stockton, Kermit, Monahans, Pecos, Sierra Blanca, Van Horn

System A	System B

Cellular company

Cellular One
GenCell Management
1308 SE Midkiss, Suite 203
Midland, TX 79701

Customer service	800-888-7868
Local office	915-686-0612
Corporate office	707-425-8000

Sending calls

Visitor dialing instructions

Local calls	area code + 7 digits
Long distance . . .	1 + area code + 7 digits

Speed-dial numbers

Customer service	611
Emergency	911

Receiving calls

Caller dials the roamer access number below, hears tone, dials cellular area code and number.

Ft. Stockton, Monahans . .	915-556-7626
Pecos, Van Horn	915-284-7626

NationLink & RoamingAmerica (see p. 1157)
Caller hears your current roamer access number.
This company's subscribers have level 0 service.

Billing information

Visitor rates

Most visitors	$3 day + 99¢ minute

Visitors without roaming agreements (p. 1159)
American Roaming Network intercepts call to bill.

Identification numbers

City	Market	System	Billing	RSA
Monahans	665	01717	00000	TX-14
Pecos	664	01713	00000	TX-13
Van Horn	663	01713	00000	TX-12

For full coverage area, see Midland, TX.

Cellular company

Cellular Three
7111 N Prince, PO Box 579
Clovis, NM 88102-0579

Corporate office	505-389-3333

Sending calls

Visitor dialing instructions

Local calls	area code + 7 digits
Long distance . . .	1 + area code + 7 digits

Speed-dial numbers

Customer service	611
Emergency	911

Receiving calls

Caller dials the roamer access number below, hears tone, dials cellular area code and number.

. . . .	505-760-7626

Billing information

Visitor rates

Most visitors	$3 day + 85¢ minute

Visitors without roaming agreements (p. 1159)
Cannot call since company takes no credit cards.

Identification numbers

City	Market	System	Billing	RSA
Monahans	665	01716	30514	TX-14
Pecos	664	01716	00000	TX-13
Van Horn	663	01716	00000	TX-12

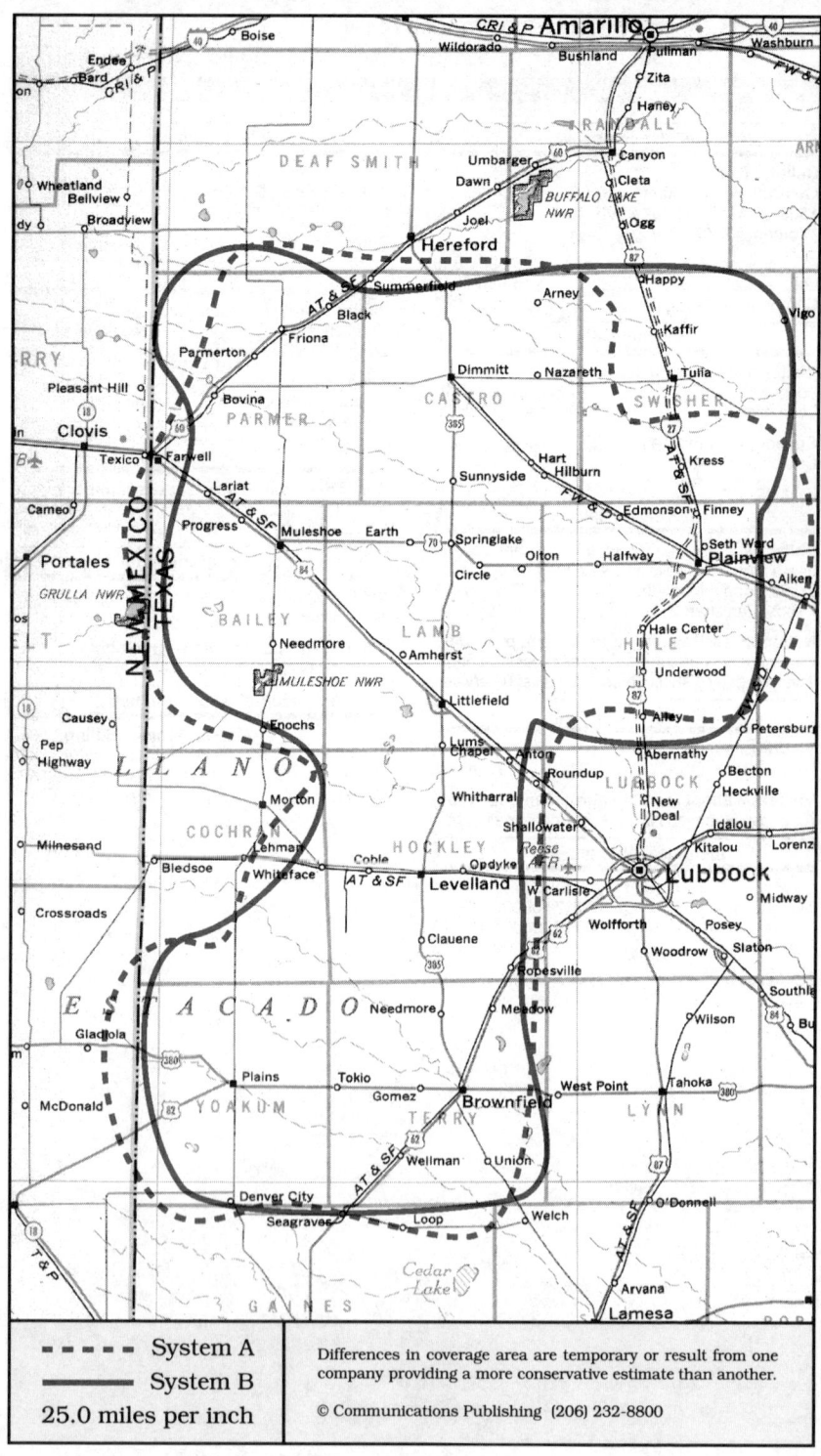

System A
System B
25.0 miles per inch

Differences in coverage area are temporary or result from one company providing a more conservative estimate than another.

© Communications Publishing (206) 232-8800

Plainview, Texas

Brownfield, Dimmitt, Farwell, Levelland, Littlefield, Muleshoe, Plains, Plainview

System A	System B

Cellular company

Liberty Cellular Phone Network
Unitel
303 Avenue H
Levelland, TX 79336

Local office	806-894-8004
Corporate office	609-646-9400

Sending calls

Visitor dialing instructions

Local calls	area code + 7 digits
Long distance . . .	1 + area code + 7 digits

Speed-dial numbers

Customer service	611
Emergency	911

Receiving calls

Caller dials the roamer access number below, hears tone, dials cellular area code and number.
. . . . 806-893-7626

NationLink & RoamingAmerica (see p. 1157)
Dial ∗31 send to receive calls, ∗30 to deactivate.
Dial ∗32 send to tell caller the roamer access no.
This company's subscribers have level 3 service.

Billing information

Visitor rates

Most visitors	$3 day + 99¢ minute

Visitors without roaming agreements (p. 1159)
Cannot call since company takes no credit cards.

Identification numbers

City	Market	System	Billing	RSA
All served cities	654	01695	00000	TX-3

For full coverage area, see Tahoka, TX.

Cellular company

Cellular Three
7111 N Prince, PO Box 579
Clovis, NM 88102-0579

Corporate office . from Texas	806-481-3333
. . . .	505-389-3333

Sending calls

Visitor dialing instructions

Local calls	area code + 7 digits
Long distance . . .	1 + area code + 7 digits

Speed-dial numbers

Customer service	611
Emergency	911

Receiving calls

Caller dials the roamer access number below, hears tone, dials cellular area code and number.

Brownfield, Levelland, Plains, Tulia	806-638-7626
Hub	806-265-7626
Plainview	806-774-7626
Summerfield	806-357-7626

Follow Me Roaming forwarding (see p. 1156)
Activate, dial ∗18 send. Deactivate dial ∗19 send.

Billing information

Visitor rates

Most visitors	$2 day + 75¢ minute

Visitors without roaming agreements (p. 1159)
Cannot call since company takes no credit cards.

Identification numbers

City	Market	System	Billing	RSA
All served cities	654	01696	00000	TX-3

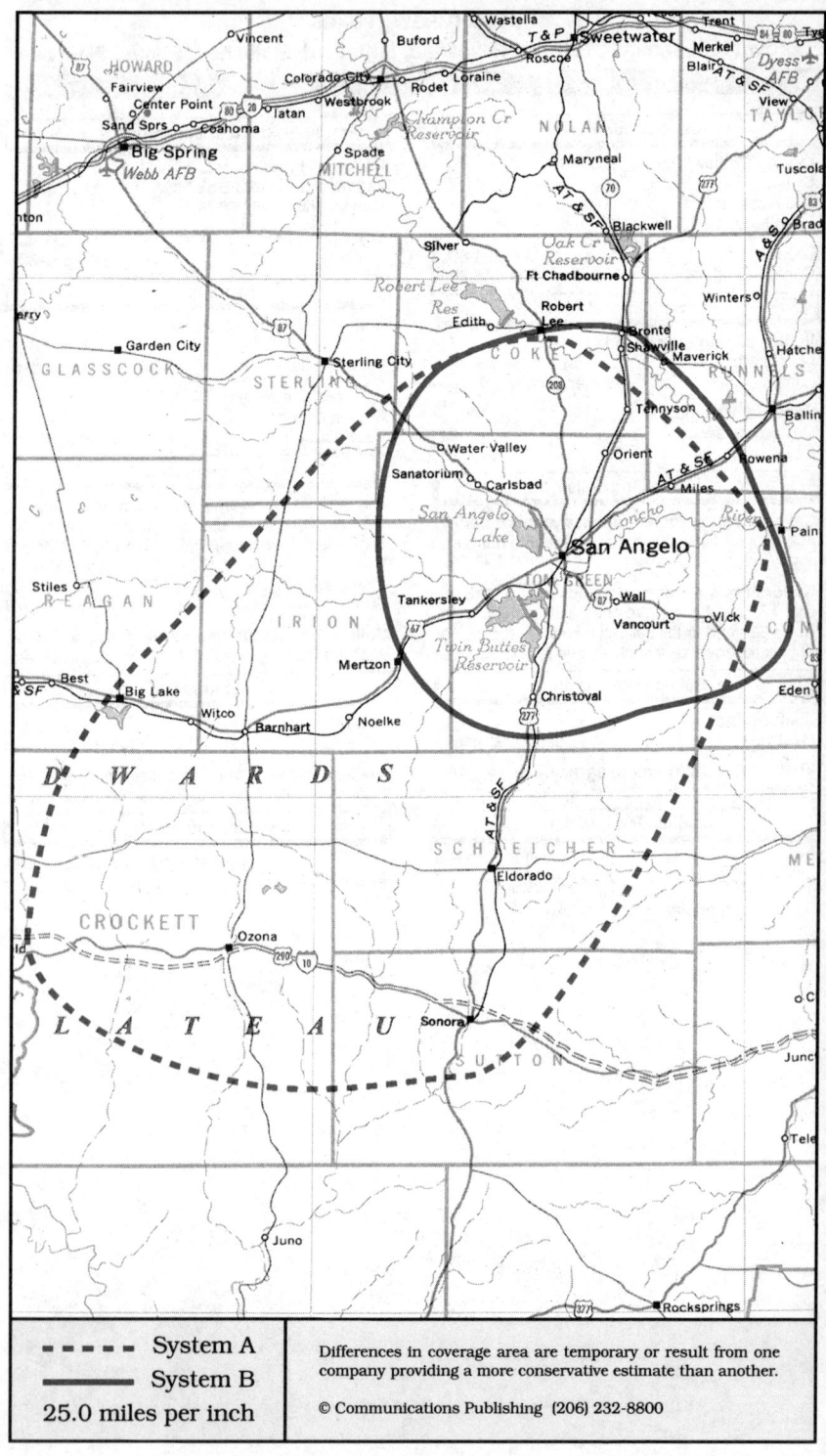

- - - - - System A
——— System B
25.0 miles per inch

San Angelo, Texas
Big Lake, Eldorado, Ozona, San Angelo, Sonora

System A	System B

Cellular company

Cellular One
GenCell Management
1723 Knickerbocker Road
San Angelo, TX 76904-5521

Customer service 800-888-7868
Local office 915-949-9900
Corporate office 707-425-8000

Sending calls

Visitor dialing instructions
Local calls . . . . . area code + 7 digits
Long distance . . . 1 + area code + 7 digits

Speed-dial numbers
Customer service *611
Emergency 911

Receiving calls

Caller dials the roamer access number below, hears tone, dials cellular area code and number.
. . . . 915-657-2657

NationLink & RoamingAmerica (see p. 1157)
Dial *31 send to receive calls, *30 to deactivate. This company's subscribers have level 3 service.

Billing information

Visitor rates
Most visitors . . . . . $3 day + 85¢ minute

Visitors without roaming agreements (p. 1159)
American Roaming Network intercepts call to bill.

Identification numbers

City	Market	System	Billing	RSA
All served cities	294	00529	00000	MSA

For full coverage area, see Ft. Stockton, TX.

Cellular company

West Central Cellular
CT Cube
3367 Knickerbocker Road
San Angelo, TX 76904-7812

Corporate office 915-944-9016

Sending calls

Visitor dialing instructions
Local calls . . . . . area code + 7 digits
Long distance . . . 1 + area code + 7 digits

Speed-dial numbers
Customer service 611
Emergency 911
Roaming information 811
24-hour assistance 7626

Receiving calls

Caller dials the roamer access number below, hears tone, dials cellular area code and number.
. . . . 915-656-7626

Follow Me Roaming forwarding (see p. 1156)
Activate, dial *18 send. Deactivate dial *19 send.

Billing information

Visitor rates
Most visitors . . . . . $3 day + 85¢ minute

Visitors without roaming agreements (p. 1159)
Cannot call since company takes no credit cards.

Identification numbers

City	Market	System	Billing	RSA
All served cities	294	00510	00000	MSA

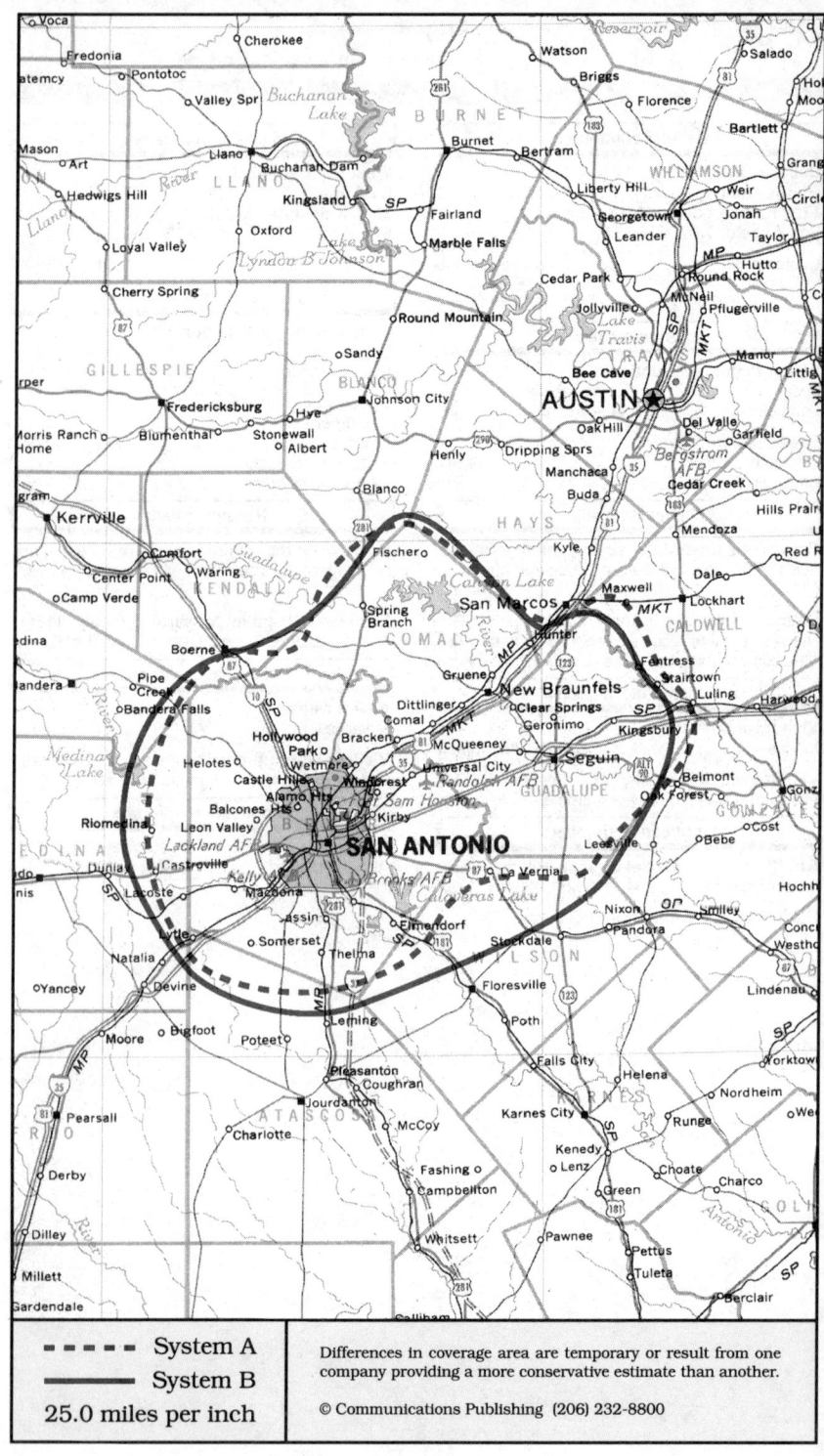

- - - - System A	Differences in coverage area are temporary or result from one company providing a more conservative estimate than another.
——— System B	
25.0 miles per inch	© Communications Publishing (206) 232-8800

San Antonio, Texas
New Braunfels, San Antonio, Sequin

System A	System B

Cellular company

Cellular One
McCaw Cellular Communications
2727 NW Loop 410
San Antonio, TX 78230

Customer service	800-262-3659
. . . .	210-861-7500
. . . .	512-861-7500
Local office	210-861-7000
Corporate office	206-827-4500

Sending calls

Visitor dialing instructions
Local calls	area code + 7 digits
Long distance . . .	1 + area code + 7 digits

Speed-dial numbers
Customer service	611
Emergency	911
KTSA talk show	*555

Receiving calls

Caller dials the roamer access number below,
hears tone, dials cellular area code and number.
. . . .	210-240-7626
. . . .	512-240-7626

NationLink & RoamingAmerica (see p. 1157)
Dial *31 send to receive calls, *30 to deactivate.
Dial *32 send to tell caller the roamer access no.
This company's subscribers have level 9 service.

North American Cellular Network (see p. 1156)
All calls forward to you. Deactivate dial *35 send.

Billing information

Visitor rates
Most visitors	$0 day + 99¢ minute

Visitors without roaming agreements (p. 1159)
Call co. to use Amex, Discover, Mastercard, Visa
or Roamer Plus will intercept first call to bill you.

Identification numbers

City	Market	System	Billing	RSA
All served cities	033	00151	00000	MSA

Cellular company

Southwestern Bell Mobile Systems
1275 NE Loop 410
San Antonio, TX 78209

Customer service		800-333-2355
. . . .		210-841-5550
Local office . . .	some areas	800-321-0611
. . . .		210-841-5500
Corporate office		214-733-2000

Sending calls

Visitor dialing instructions
Local calls	area code + 7 digits
Long distance . . .	1 + area code + 7 digits

Speed-dial numbers
Customer service	611
Emergency	911
KTSA AM 550 traffic desk	*550

Receiving calls

Caller dials the roamer access number below,
hears tone, dials cellular area code and number.
. . . .	210-260-7626

Follow Me Roaming forwarding (see p. 1156)
Activate, dial *18 send. Deactivate dial *19 send.

Billing information

Visitor rates
Most visitors	$3 day + 90¢ minute

Visitors without roaming agreements (p. 1159)
Call co. to use Amex, Mastercard, Visa

Identification numbers

City	Market	System	Billing	RSA
All served cities	033	00122	00000	MSA

For full coverage area, see Kingsville, TX.

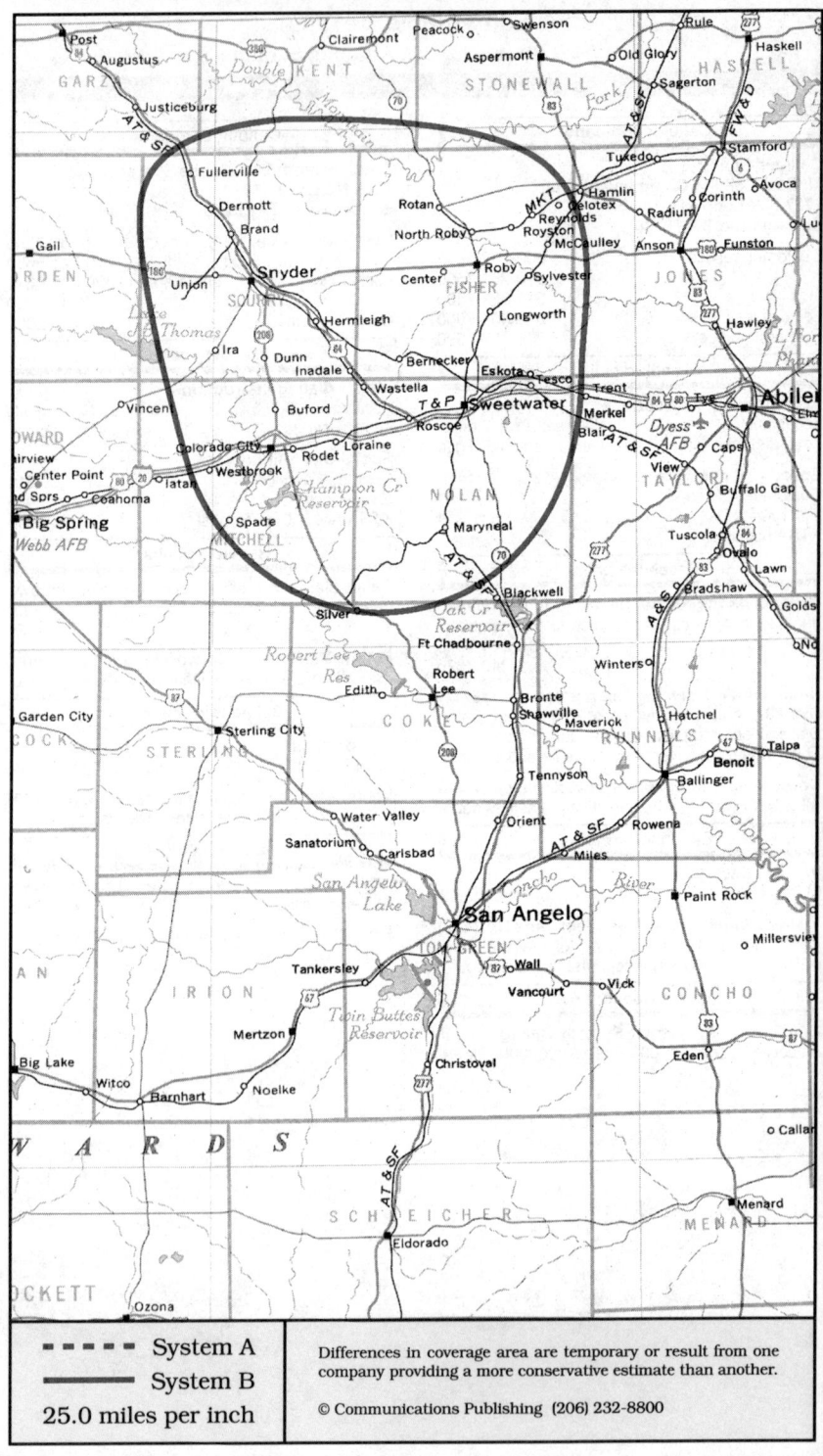

System A
System B
25.0 miles per inch

Differences in coverage area are temporary or result from one company providing a more conservative estimate than another.

© Communications Publishing (206) 232-8800

Snyder, Texas
Colorado City, Roby, Snyder, Sweetwater

System A	System B

Cellular company

No information about the company that will operate system A was available at press time.

Cellular company

Texas Cellular
Taylor Telecommunications (TTI)
N Interstate 20 Access Road, PO Box 337
Merkel, TX 79536

Customer service	some areas	800-424-7182
. . . .		915-691-1013
Local office		915-928-3200

Sending calls

Visitor dialing instructions
Local calls . . . . . . area code + 7 digits
Long distance . . . 1 + area code + 7 digits

Speed-dial numbers
Customer service 611
Emergency 911

Receiving calls

Caller dials the roamer access number below, hears tone, dials cellular area code and number.
. . . . 915-575-7626

Follow Me Roaming forwarding (see p. 1156)
Activate, dial *18 send. Deactivate dial *19 send.

Billing information

Visitor rates
Most visitors . . . . . $2 day + 75¢ minute

Visitors without roaming agreements (p. 1159)
Cannot call since company takes no credit cards.

Identification numbers

City	Market	System	Billing	RSA
All served cities	659	01876	00000	TX-8

Customer service and billing are provided by Southwestern Bell Mobile Systems.

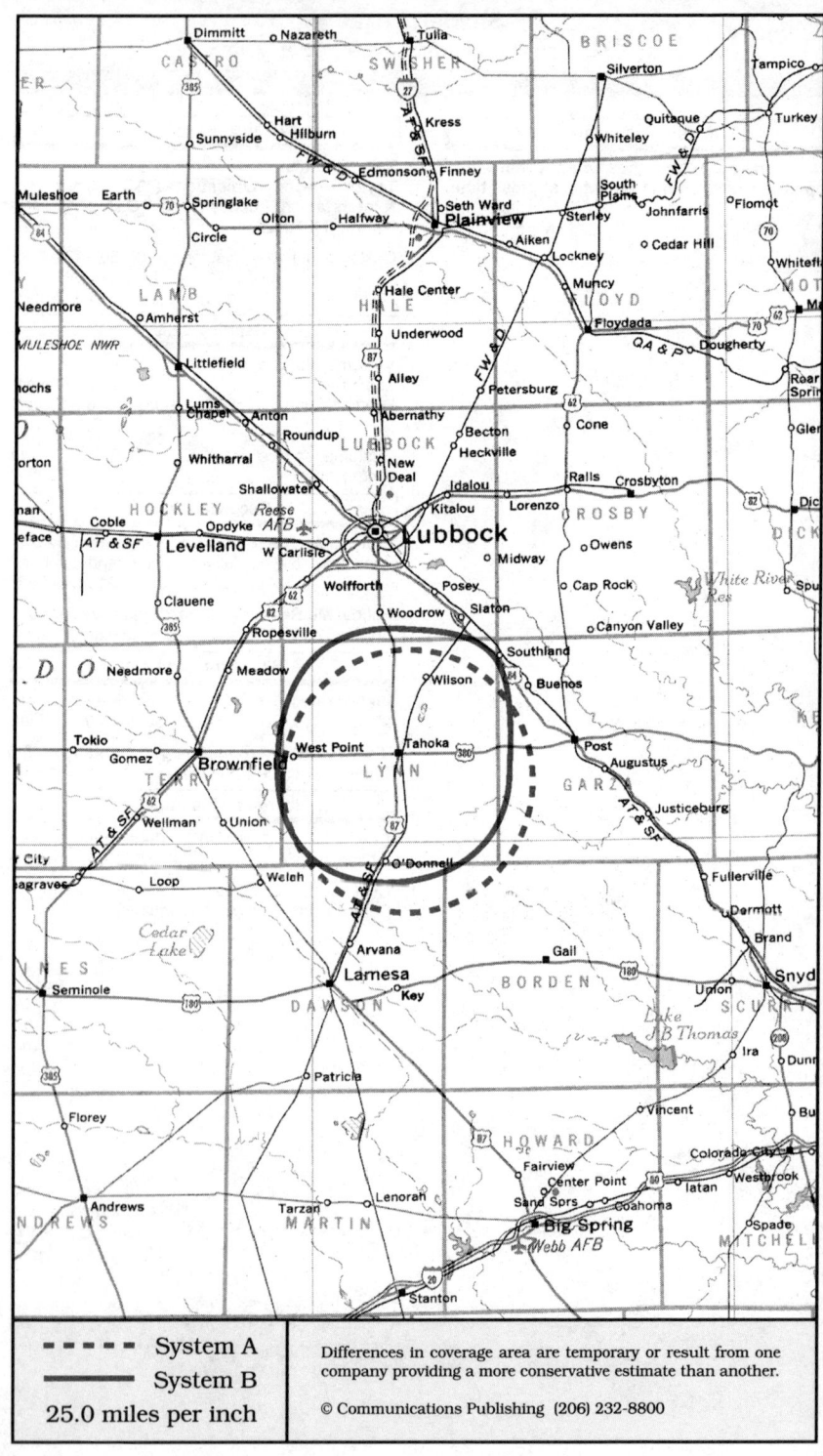

- - - - - System A
——— System B
25.0 miles per inch

System A

Cellular company

Liberty Cellular Phone Network
Unitel
303 Avenue H
Levelland, TX 79336

Local office 806-894-8004
Corporate office 609-646-9400

Sending calls

Visitor dialing instructions
Local calls . . . . . area code + 7 digits
Long distance . . . 1 + area code + 7 digits

Speed-dial numbers
Customer service 611
Emergency 911

Receiving calls

Caller dials the roamer access number below, hears tone, dials cellular area code and number.
. . . . 806-893-7626

NationLink & RoamingAmerica (see p. 1157)
Dial *31 send to receive calls, *30 to deactivate.
Dial *32 send to tell caller the roamer access no.
This company's subscribers have level 3 service.

Billing information

Visitor rates
Most visitors . . . . . $3 day + 99¢ minute

Visitors without roaming agreements (p. 1159)
Cannot call since company takes no credit cards.

Identification numbers

City	Market	System	Billing	RSA
All served cities	654	01695	00000	TX-3

For full coverage area, see Plainview, TX.

System B

Cellular company

Digital Cellular of Texas
Poka Lambro Telecommunications
PO Box 53118
Lubbock, TX 79453

Physical location
11.5 miles N of Tahoka on U.S. Highway 87
Tahoka, TX 79373

Corporate office . some areas 800-662-8805
. . . . 806-924-5432

Sending calls

Visitor dialing instructions
Local calls . . . . . . area code + 7 digits
Long distance . . . 1 + area code + 7 digits

Speed-dial numbers
Customer service 611
Emergency 911

Receiving calls

Caller dials the roamer access number below, hears tone, dials cellular area code and number.
. . . . 806-759-7626

Follow Me Roaming forwarding (see p. 1156)
Activate, dial *18 send. Deactivate dial *19 send.

Billing information

Visitor rates
Most visitors . . . . . $2 day + 65¢ minute

Visitors without roaming agreements (p. 1159)
Cannot call since company takes no credit cards.

Identification numbers

City	Market	System	Billing	RSA
All served cities	654	01706	30220	TX-3

For full coverage area, see Lamesa, TX.

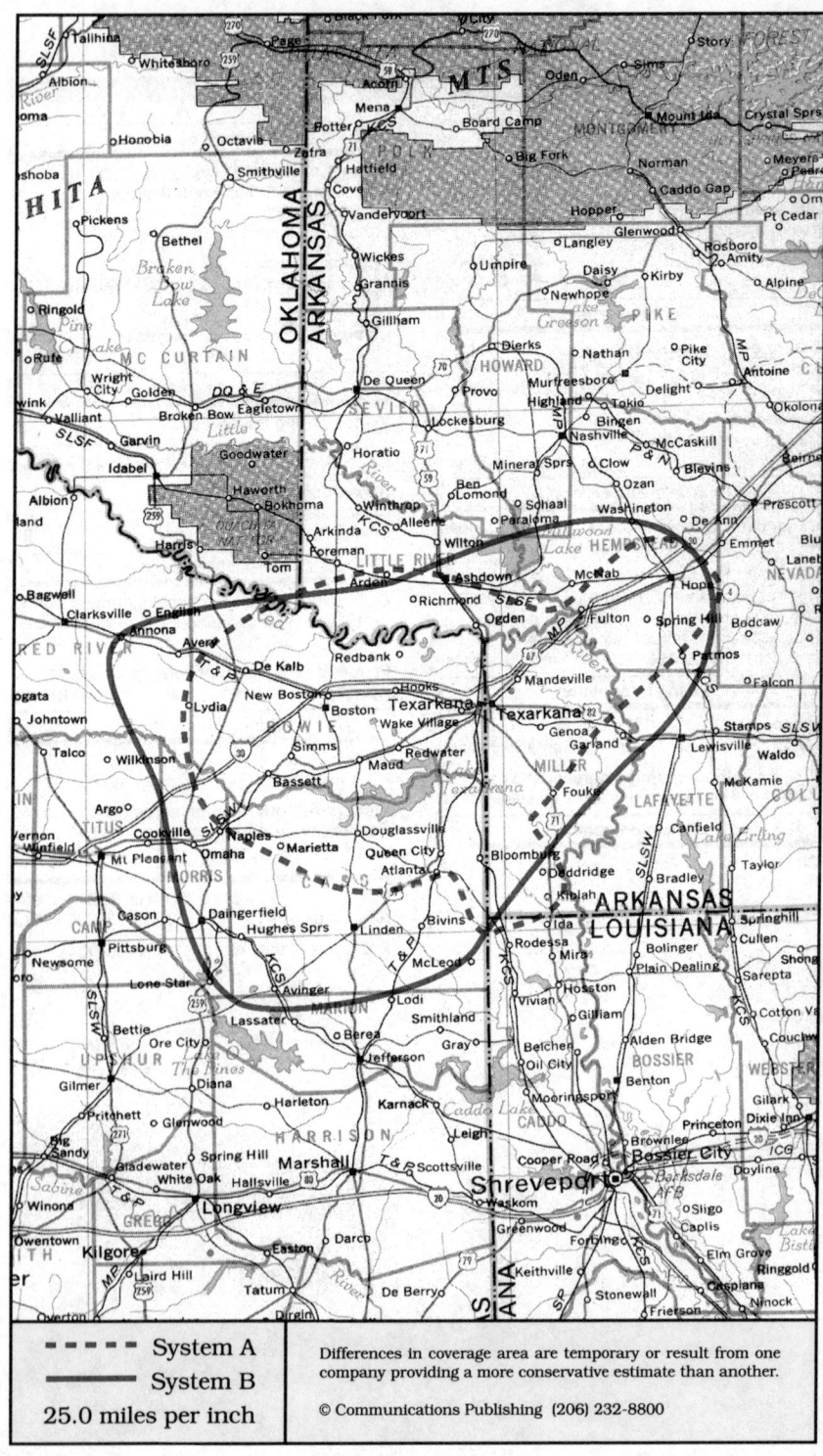

- - - - System A	Differences in coverage area are temporary or result from one company providing a more conservative estimate than another.
———— System B	
25.0 miles per inch	© Communications Publishing (206) 232-8800

Texarkana, Texas
Ashdown AR, Atlanta, Boston, Dangerfield, Hope AR, Linden, Texarkana AR & TX

System A	System B

Cellular company

Cellular One
McCaw Cellular Communications
2700 Richmond Road
Texarkana, TX 75503

Customer service	some areas	800-262-5659
. . . .		903-277-7500
Local office		903-277-7000
Corporate office		206-827-4500

Sending calls

Visitor dialing instructions
Local calls . . . . . area code + 7 digits
Long distance . . . 1 + area code + 7 digits

Speed-dial numbers
Customer service 611
Emergency 911
KTBS television in Shreveport *3
Radio 107 . . . . *107

Receiving calls

Caller dials the roamer access number below,
hears tone, dials cellular area code and number.
. . . . 903-277-7626

NationLink & RoamingAmerica (see p. 1157)
Dial *31 send to receive calls, *30 to deactivate.
Dial *32 send to tell caller the roamer access no.
This company's subscribers have level 9 service.

North American Cellular Network (see p. 1156)
All calls forward to you. Deactivate dial *35 send.

Billing information

Visitor rates
Most visitors . . . . . $0 day + 99¢ minute

Visitors without roaming agreements (p. 1159)
Call co. to use Amex, Mastercard, Visa
or Roamer Plus will intercept first call to bill you.

Identification numbers

City	Market	System	Billing	RSA
All served cities	240	00229	00000	MSA

For full coverage area, see Shreveport, LA.

Cellular company

Century Cellunet
2325 Texas Boulevard
Texarkana, TX 75501

Customer service		800-638-9727
. . . .		318-683-3450
Local office . . .	some areas	800-825-1637
. . . .		903-793-0500
Corporate office		318-388-9000

Sending calls

Visitor dialing instructions
Local calls . . . . . . area code + 7 digits
Long distance . . . 1 + area code + 7 digits

Speed-dial numbers
Customer service *611
Emergency 911

Receiving calls

Caller dials the roamer access number below,
hears tone, dials cellular area code and number.
Use either number since they are interconnected.
Atlanta 903-728-7626
Hope and Texarkana . . . 903-831-8626

Follow Me Roaming forwarding (see p. 1156)
Activate, dial *18 send. Deactivate dial *19 send.

Billing information

Visitor rates
Most visitors . . . . . $3 day + 75¢ minute

Visitors without roaming agreements (p. 1159)
Call co. to use Amex, Mastercard, Visa

Identification numbers

City	Market	System	Billing	RSA
Atlanta	658	00550	00000	TX-7
Hope, AR	334	00550	00000	AR-11
Texarkana	240	00550	00000	MSA

For full coverage area, see Shreveport, LA.

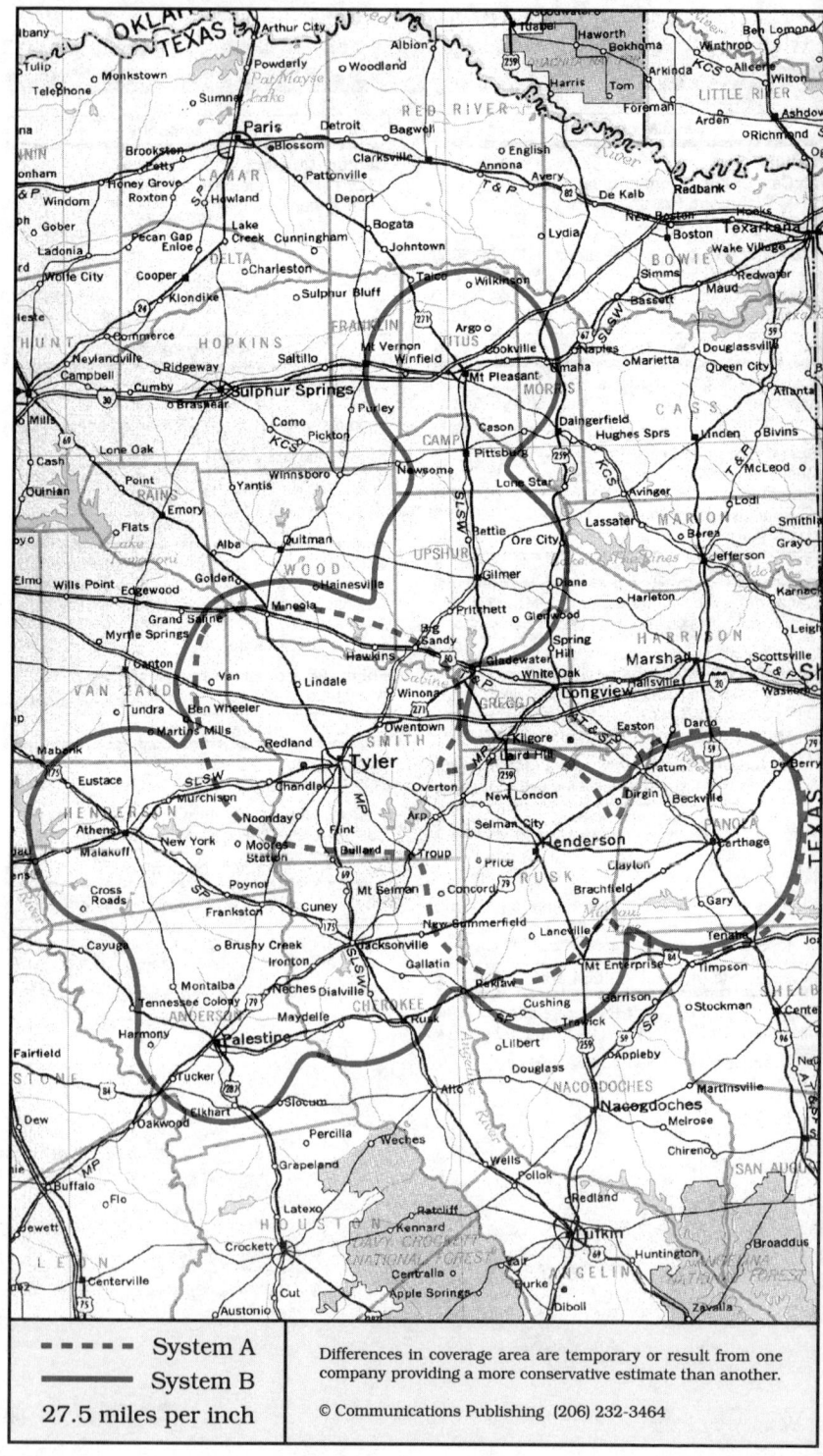

- - - - - System A	
———— System B	Differences in coverage area are temporary or result from one company providing a more conservative estimate than another.
27.5 miles per inch	© Communications Publishing (206) 232-3464

Tyler, Texas
Athens, Gilmer, Henderson, Jacksonville, Mt. Pleasant, Palestine, Tyler

System A	System B

Cellular company

United States Cellular
535 W Southwest Loop 323, Suite 208
Tyler, TX 75701

Local office 903-561-2355
Corporate office 312-399-8900

Sending calls

Visitor dialing instructions
Local calls . . . . . area code + 7 digits
Long distance . . . . area code + 7 digits

Speed-dial numbers
Customer service 611
Emergency 911
Roaming information *711

Receiving calls

Caller dials the roamer access number below,
hears tone, dials cellular area code and number.
. . . . 903-530-7626

NationLink & RoamingAmerica (see p. 1157)
Dial *31 send to receive calls, *30 to deactivate.
Dial *32 send to tell caller the roamer access no.
This company's subscribers have level 3 service.

North American Cellular Network (see p. 1156)
All calls forward to you. Deactivate dial *35 send.

Billing information

Visitor rates
Most visitors . . . . . $3 day + 75¢ minute

Visitors without roaming agreements (p. 1159)
Call co. to use Mastercard, Visa

Identification numbers

City	Market	System	Billing	RSA
All served cities	237	00579	00000	MSA

This company will serve Jacksonville in the near future.

For full coverage area, see Lufkin, TX.

Cellular company

Gilmer, Henderson, Mt. Pleasant
Centel Cellular
415 Loop 281 E
Longview, TX 75601

Local office 800-657-5251
. . . . 903-663-1161
Corporate office 312-399-2644

Athens, Jacksonvile, Tyler
Centel Cellular
212 Grande Boulevard, Suite C-100
Tyler, TX 75703

Local office 800-242-5775
. . . . 903-561-5575
Corporate office 312-399-2644

Sending calls

Visitor dialing instructions
Local calls . . . . . area code + 7 digits
Long distance . . . 1 + area code + 7 digits

Speed-dial numbers
Customer service *611
Emergency 911
KLTV *77

Receiving calls

Caller dials the roamer access number below,
hears tone, dials cellular area code and number.
Except Gilmer, Mt. Pleasant . . 903-571-7626
Gilmer, Mt. Pleasant 903-738-7626

Billing information

Visitor rates
Most visitors . . . . . $3 day + 99¢ minute

Visitors without roaming agreements (p. 1159)
Cannot call since company takes no credit cards.

Identification numbers

City	Market	System	Billing	RSA
Athens	661	01916	00000	TX-10
Gilmer	658	01870	00000	TX-7
Jacksonville	662	01712	00000	TX-11
Mt. Pleasant	658	01866	00000	TX-7
Tyler	237	00558	00000	MSA

For full coverage area, see Longview, TX.

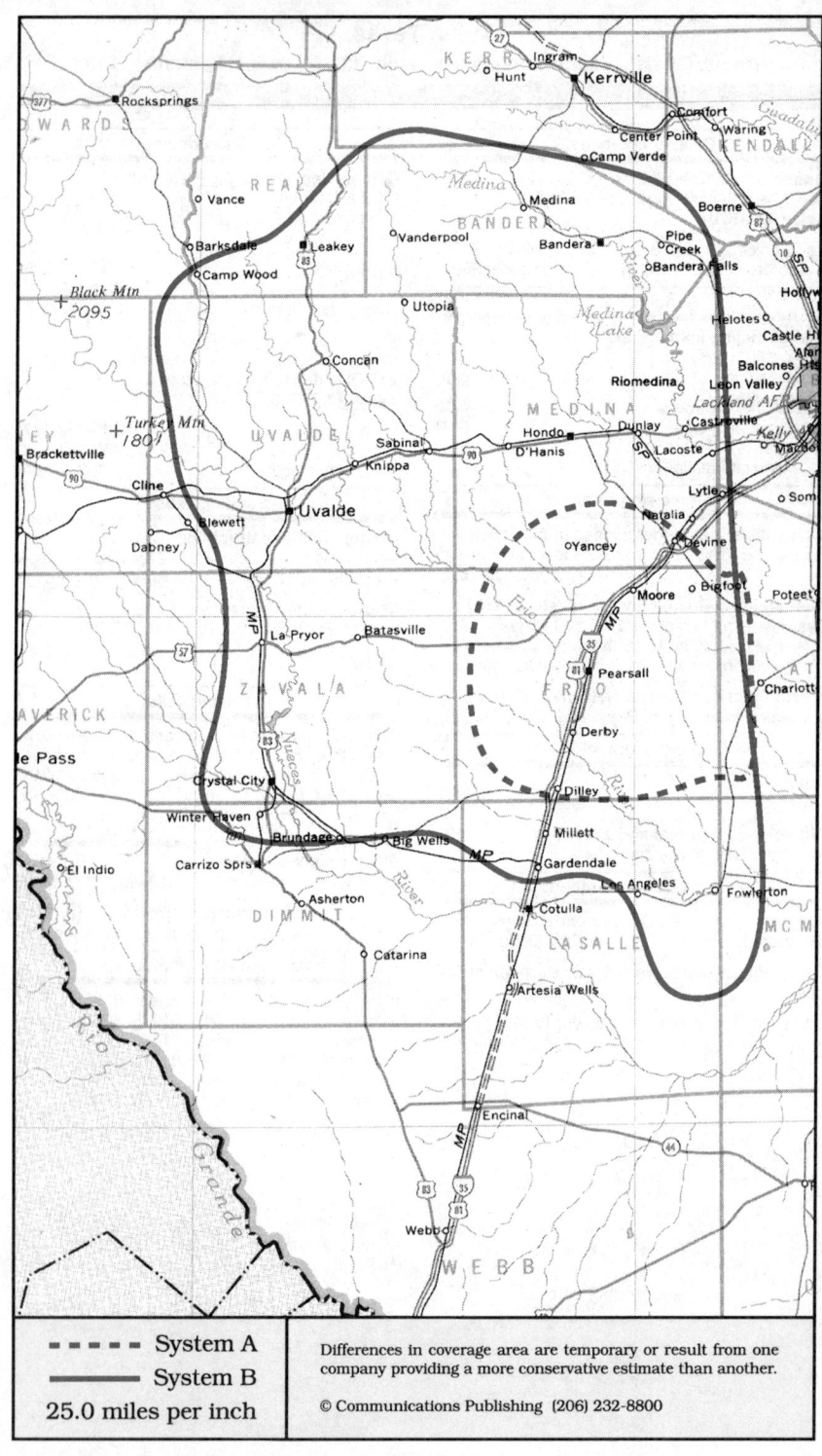

System A
System B
25.0 miles per inch

Differences in coverage area are temporary or result from one
company providing a more conservative estimate than another.

© Communications Publishing (206) 232-8800

Uvalde, Texas
Bandera, Crystal City, Hondo, Leakey, Pearsall, Uvalde

System A	System B

Cellular company

United States Cellular
112 N St. Mary's Street
Beeville, TX 78102

Local office	512-362-1000
Corporate office	312-399-8900

Sending calls

Visitor dialing instructions

Local calls	area code + 7 digits
Long distance . . .	1 + area code + 7 digits

Speed-dial numbers

Customer service	611
Emergency	911

Receiving calls

Caller dials the roamer access number below,
hears tone, dials cellular area code and number.
. . . . 512-319-7626

Billing information

Visitor rates
Most visitors . . . . . $3 day + 85¢ minute

Visitors without roaming agreements (p. 1159)
Call co. to use Amex, Mastercard, Visa

Identification numbers

City	Market	System	Billing	RSA
All served cities	669	01725	00000	TX-18

For full coverage area, see Kingsville, TX.

Cellular company

Southwestern Bell Mobile Systems
1275 NE Loop 410
San Antonio, TX 78209

Customer service	800-333-2355
. . . .	210-841-5550
Local office	210-841-5500
Corporate office	214-733-2000

Sending calls

Visitor dialing instructions

Local calls	area code + 7 digits
Long distance . . .	1 + area code + 7 digits

Speed-dial numbers

Customer service	611
Emergency	911
To reach roamer access number	511

Receiving calls

Caller dials the roamer access number below,
hears tone, dials cellular area code and number.
. . . . 210-260-7626

Follow Me Roaming forwarding (see p. 1156)
Activate, dial ∗18 send. Deactivate dial ∗19 send.

Billing information

Visitor rates
Most visitors . . . . . $3 day + 90¢ minute

Visitors without roaming agreements (p. 1159)
Call co. to use Amex, Mastercard, Visa

Identification numbers

City	Market	System	Billing	RSA
Uvalde	669	00122	30204	TX-18

For full coverage area, see Kingsville, TX and
San Antonio, TX.

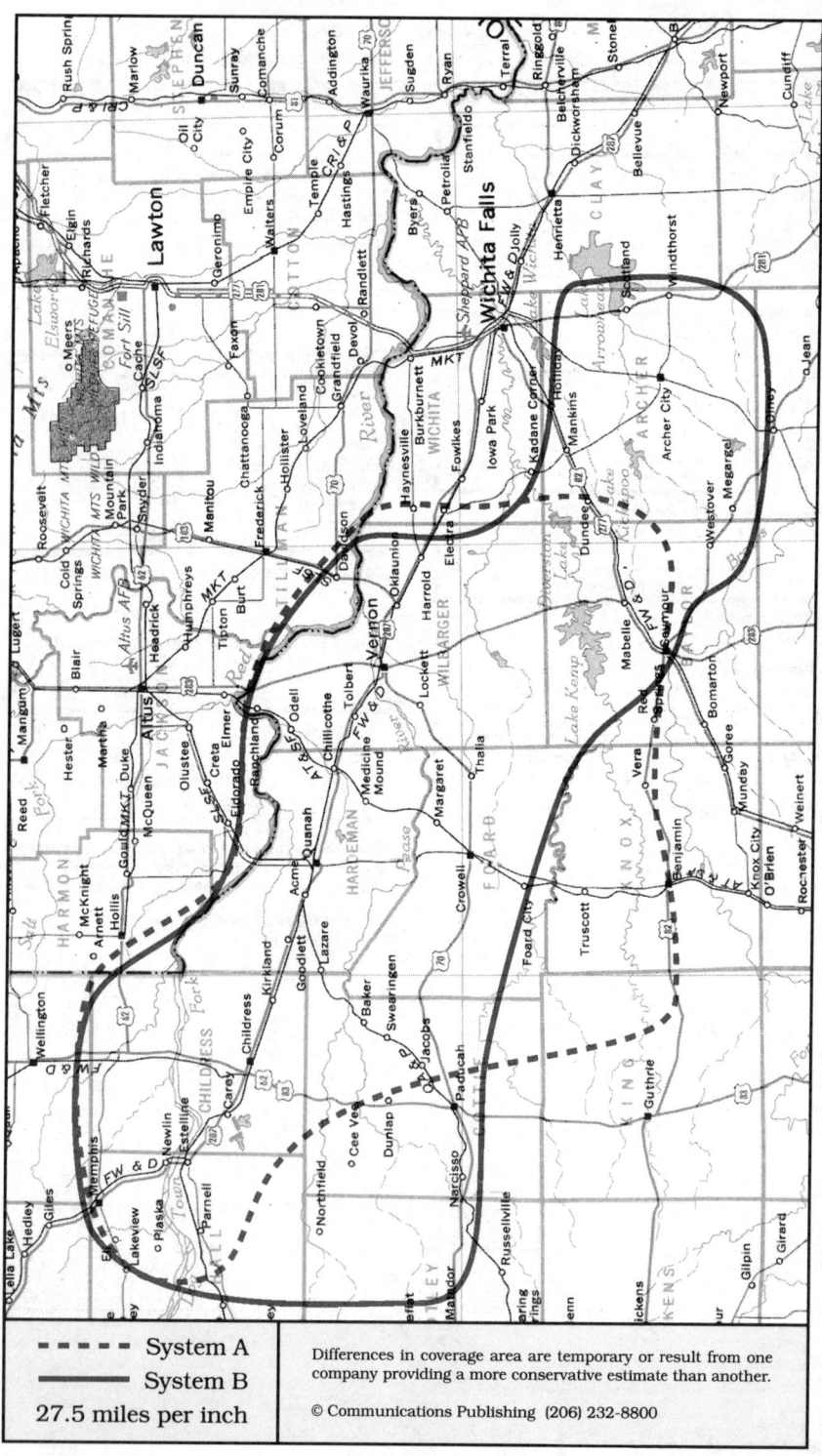

System A
System B
27.5 miles per inch

Differences in coverage area are temporary or result from one company providing a more conservative estimate than another.

© Communications Publishing (206) 232-8800

Vernon, Texas

Archer City, Childress, Memphis, Quanah, Vernon

System A	System B

Cellular company

Cellular One of North Texas
805 Hillcrest Plaza
Vernon, TX 76384

Sale of this system is pending.

Corporate office 817-553-1885

Sending calls

Visitor dialing instructions
Local calls . . . . area code + 7 digits
Long distance . . . 1 + area code + 7 digits

Speed-dial numbers
Customer service 611
Emergency 911
Texas state police *54

Receiving calls

Caller dials the roamer access number below,
hears tone, dials cellular area code and number.
. . . . 817-886-7626

Billing information

Visitor rates
Most visitors . . . . . $3 day + 99¢ minute

Visitors without roaming agreements (p. 1159)
Cannot call since company takes no credit cards.

Identification numbers

City	Market	System	Billing	RSA
All served cities	656	01699	30323	TX-5

Cellular company

United States Cellular
2301 Kell Boulevard, Suite B
Wichita Falls, TX 76308

Local office 817-761-2500
Corporate office 312-399-8900

Sending calls

Visitor dialing instructions
Local calls . . . . . area code + 7 digits
Long distance . . . . . area code + 7 digits

Speed-dial numbers
Customer service *611
Emergency 911
Roaming information *711

Receiving calls

Caller dials the roamer access number below,
hears tone, dials cellular area code and number.
. . . . 817-733-7626

Billing information

Visitor rates
Most visitors . . . . . $3 day + 75¢ minute

Visitors without roaming agreements (p. 1159)
Call co. to use Mastercard, Visa

Identification numbers

City	Market	System	Billing	RSA
Childress, Memphis	655	01698	00000	TX-4
Quanah, Vernon	656	01700	00000	TX-5

For full coverage area, see Wichita Falls, TX.

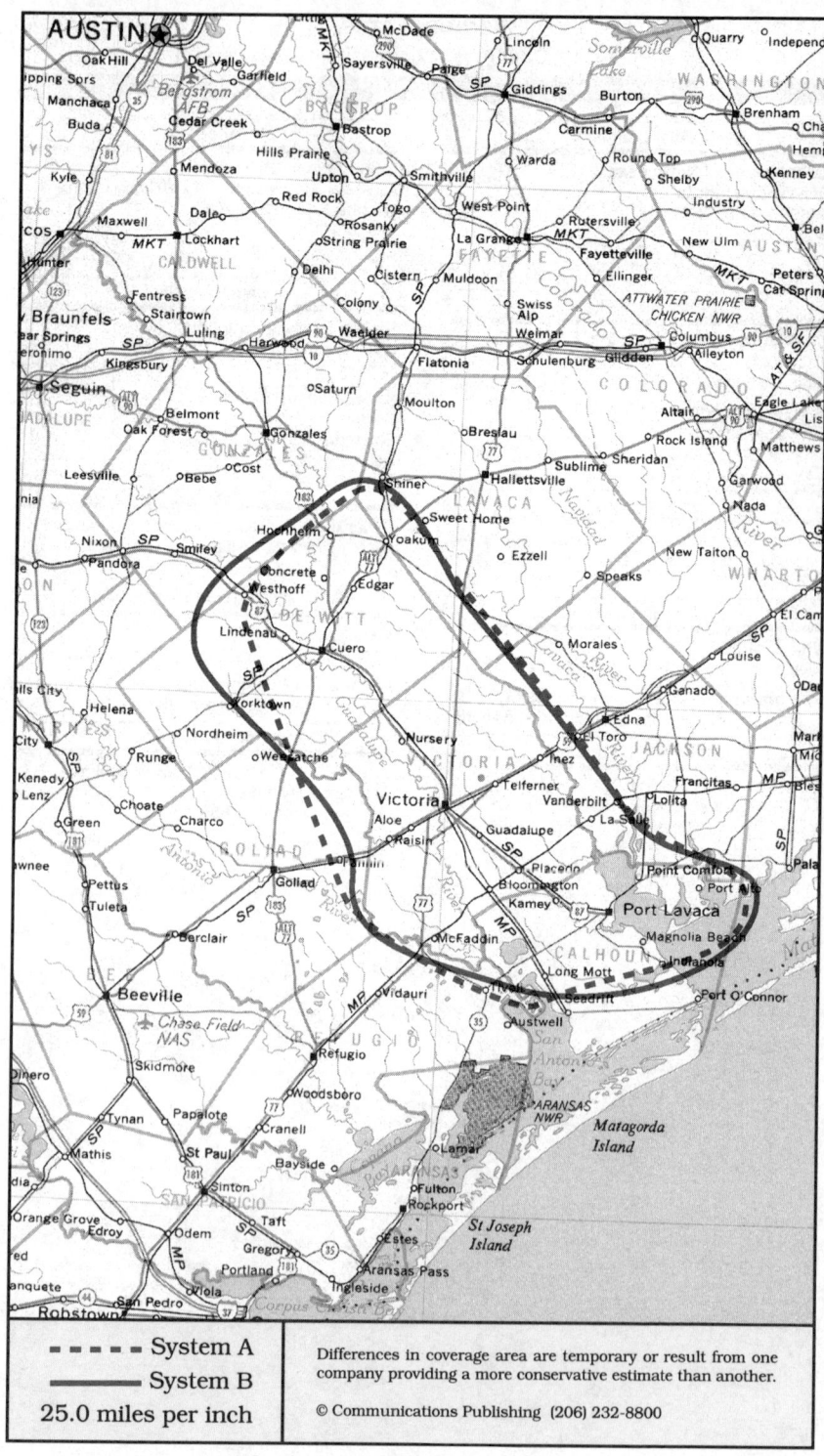

System A

System B

25.0 miles per inch

Differences in coverage area are temporary or result from one company providing a more conservative estimate than another.

© Communications Publishing (206) 232-8800

Victoria, Texas
Cuero, Port Lavaca, Victoria

System A	System B

Cellular company

Cellular One
Victoria Cellular
5606 N Navarro #204-B, PO Box 3788
Victoria, TX 77904

Corporate office 512-573-1100

Sending calls

Visitor dialing instructions
Local calls . . . . . area code + 7 digits
Long distance . . . 1 + area code + 7 digits

Speed-dial numbers
Customer service 611
Emergency 911

Receiving calls

Caller dials the roamer access number below,
hears tone, dials cellular area code and number.
. . . . 512-571-7626

NationLink & RoamingAmerica (see p. 1157)
Dial *31 send to receive calls, *30 to deactivate.
Dial *32 send to tell caller the roamer access no.
This company's subscribers have level 3 service.

Billing information

Visitor rates
Most visitors . . . . . $3 day + 99¢ minute

Visitors without roaming agreements (p. 1159)
Call co. to use Amex, Mastercard, Visa

Identification numbers

City	Market	System	Billing	RSA
All served cities	300	00581	00000	MSA

Cellular company

GTE Mobilnet
100 Glenborough, Suite 800
Houston, TX 77067

Customer service 800-347-5665
. . . . 713-876-9144
Local office 713-876-5000
Corporate office 404-391-8000

Sending calls

Visitor dialing instructions
Local calls . . . . . . area code + 7 digits
Long distance . . . 1 + area code + 7 digits

Speed-dial numbers
Customer service *611
Emergency 911
Mr. Rescue (roadside assistance) HELP or 4357
Roaming information *711
Telephone problems *111

Receiving calls

Caller dials the roamer access number below,
hears tone, dials cellular area code and number.
. . . . 512-550-7626

Follow Me Roaming forwarding (see p. 1156)
Activate, dial *18 send. Deactivate dial *19 send.

Billing information

Visitor rates
Most visitors . . . . . $3 day + 85¢ minute

Visitors without roaming agreements (p. 1159)
Call co. to use Amex, Mastercard, Visa

Identification numbers

City	Market	System	Billing	RSA
Port Lavaca	671	00012	00000	TX-20
Victoria	300	00562	00000	MSA

For full coverage area, see Bay City, TX.

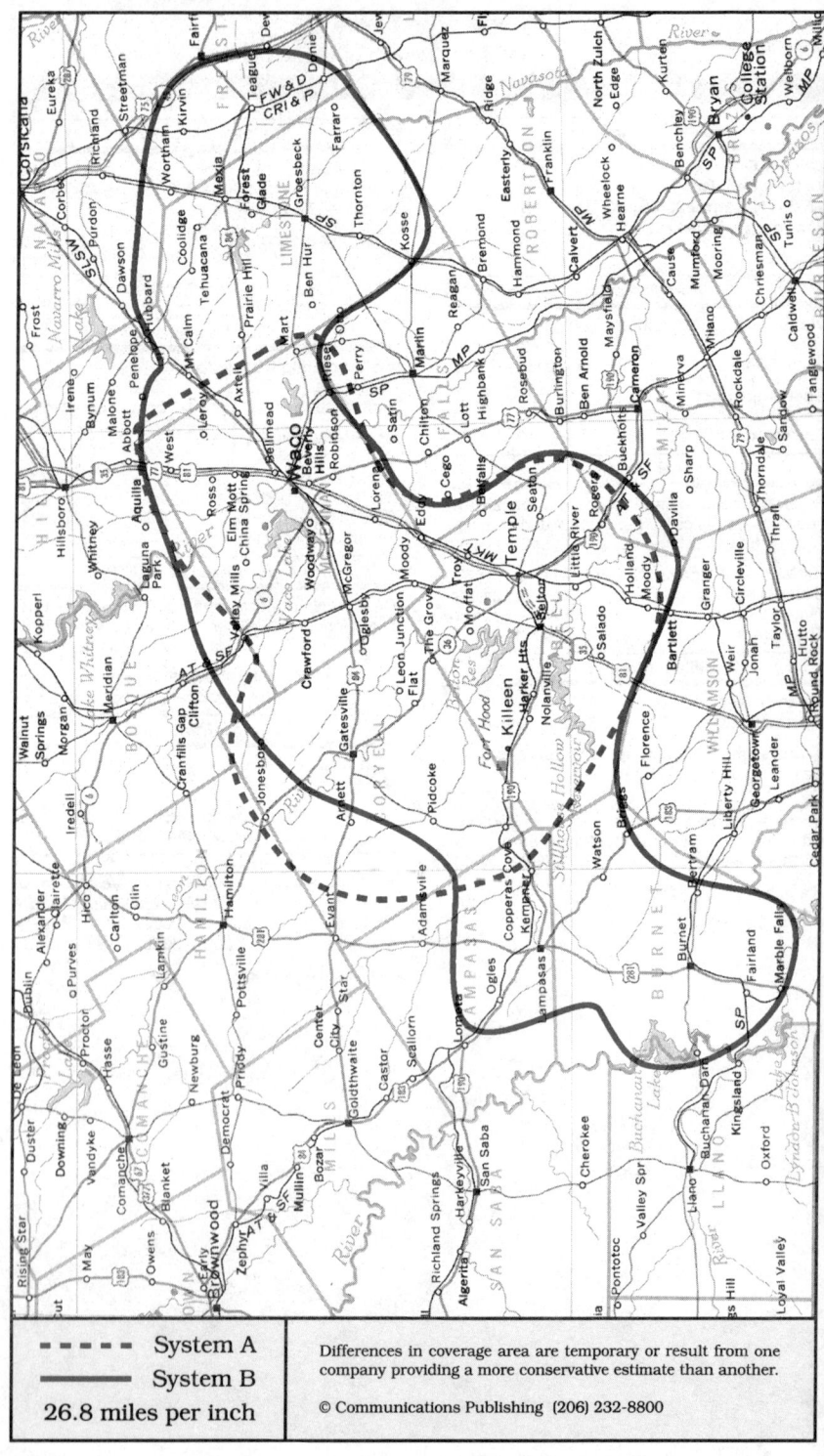

- - - - System A
———— System B
26.8 miles per inch

Differences in coverage area are temporary or result from one company providing a more conservative estimate than another.

© Communications Publishing (206) 232-8800

Waco, Texas
Cameron, Killeen, Lampasas, Mexia, Temple, Waco

System A	System B

Cellular company

Cellular One
McCaw Cellular Communications
2901 SW H.K. Dodgen Loop
Temple, TX 76502

1619-A N Valley Mills Drive
Waco, TX 76710

Customer service	800-262-3659
. . . .	512-750-7500
. . . . . . from Temple	817-760-7500
. . . . Waco	817-776-4073
Local office . . . . Temple	817-771-5474
. . from Killeen to Temple	817-526-0077
. . . . Waco	817-776-3933
Corporate office	206-827-4500

Sending calls

Visitor dialing instructions

Local calls	area code + 7 digits
Long distance . . .	1 + area code + 7 digits

Speed-dial numbers

Customer service	611
Emergency	911
Weather for Waco	817-756-5555

Receiving calls

Caller dials the roamer access number below,
hears tone, dials cellular area code and number.

Killeen	817-289-7626
Temple	817-760-7626
Waco	817-744-7626

NationLink & RoamingAmerica (see p. 1157)
Dial *31 send to receive calls, *30 to deactivate.
Dial *32 send to tell caller the roamer access no.
This company's subscribers have level 9 service.
(Waco subscribers have level 6 service.)

North American Cellular Network (see p. 1156)
All calls forward to you. Deactivate dial *35 send.

Billing information

Visitor rates

Most visitors in Killeen, T.	$0 day + 99¢ minute
Most visitors in Waco .	$3 day + 99¢ minute

Visitors without roaming agreements (p. 1159)
Call co. to use Amex, Discover, Mastercard, Visa
or Roamer Plus will intercept first call to bill you.

Identification numbers

City	Market	System	Billing	RSA
Killeen, Temple	160	00409	00000	MSA
Waco	194	00587	00000	MSA

For full coverage area, see Austin, TX.

Cellular company

Centel Cellular
2201 S W.S. Young Drive, Suite 110-C
Killeen, TX 76543

2505 Hancock Drive
Temple, TX 76504

1411 N Valley Mills Drive, Suite B
Waco, TX 76710

Customer service	800-743-3365
. . . . from Killeen	817-526-9977
. . . . from Waco	817-752-2200
Local office . . . . . Killeen	817-526-9977
. . . . Temple	817-771-0077
. . . . Waco	817-751-1346
Corporate office	312-399-2644

Sending calls

Visitor dialing instructions

Local calls	area code + 7 digits
Long distance . . .	1 + area code + 7 digits

Speed-dial numbers

Customer service	*611
Emergency	911

Receiving calls

Caller dials the roamer access number below,
hears tone, dials cellular area code and number.

. . . .	817-770-7626

Follow Me Roaming forwarding (see p. 1156)
Activate, dial *18 send. Deactivate dial *19 send.

Billing information

Visitor rates

Most visitors	$3 day + 99¢ minute

Visitors without roaming agreements (p. 1159)
Cannot call since company takes no credit cards.

Identification numbers

City	Market	System	Billing	RSA
Burnet, Lampasas	666	01720	00000	TX-15
Killeen, Temple	160	00392	00000	MSA
Cameron, Mexia	661	01920	00000	TX-10
Waco	194	00566	00000	MSA

- - - - -	System A
————————	System B

25.0 miles per inch

Differences in coverage area are temporary or result from one company providing a more conservative estimate than another.

© Communications Publishing (206) 232-8800

Wichita Falls, Texas
Henrietta, Wichita Falls

System A	System B

Cellular company

Cellular One of Wichita Falls
PriCellular
3401 Kemp Boulevard, Suite R
Wichita Falls, TX 76308

Local office	817-691-9100
Corporate office	212-459-0800

Sending calls

Visitor dialing instructions

Local calls	area code + 7 digits
Long distance . . .	1 + area code + 7 digits

Speed-dial numbers

Customer service	*611
Emergency	911
Weather	811

Receiving calls

Caller dials the roamer access number below,
hears tone, dials cellular area code and number.
. . . . 817-781-7626

NationLink & RoamingAmerica (see p. 1157)
Dial *31 send to receive calls, *30 to deactivate.
Dial *32 send to tell caller the roamer access no.
This company's subscribers have level 5 service.

Billing information

Visitor rates
Most visitors . . . . . $3 day + 75¢ minute

Visitors without roaming agreements (p. 1159)
Call co. to use Amex, Discover, Mastercard, Visa

Identification numbers

City	Market	System	Billing	RSA
All served cities	233	00595	00000	MSA

For full coverage area, see Gainesville, TX.

Cellular company

United States Cellular
2301 Kell Boulevard, Suite B
Wichita Falls, TX 76308

Local office	817-761-2500
Corporate office	312-399-8900

Sending calls

Visitor dialing instructions

Local calls	area code + 7 digits
Long distance	area code + 7 digits

Speed-dial numbers

Customer service	*611
Emergency	911
Roaming information	*711

Receiving calls

Caller dials the roamer access number below,
hears tone, dials cellular area code and number.
. . . . 817-733-7626

Follow Me Roaming forwarding (see p. 1156)
Activate, dial *18 send. Deactivate dial *19 send.

Billing information

Visitor rates
Most visitors . . . . . $3 day + 75¢ minute

Visitors without roaming agreements (p. 1159)
Call co. to use Mastercard, Visa

Identification numbers

City	Market	System	Billing	RSA
All served cities	233	00574	00000	MSA

For full coverage area, see Vernon, TX.

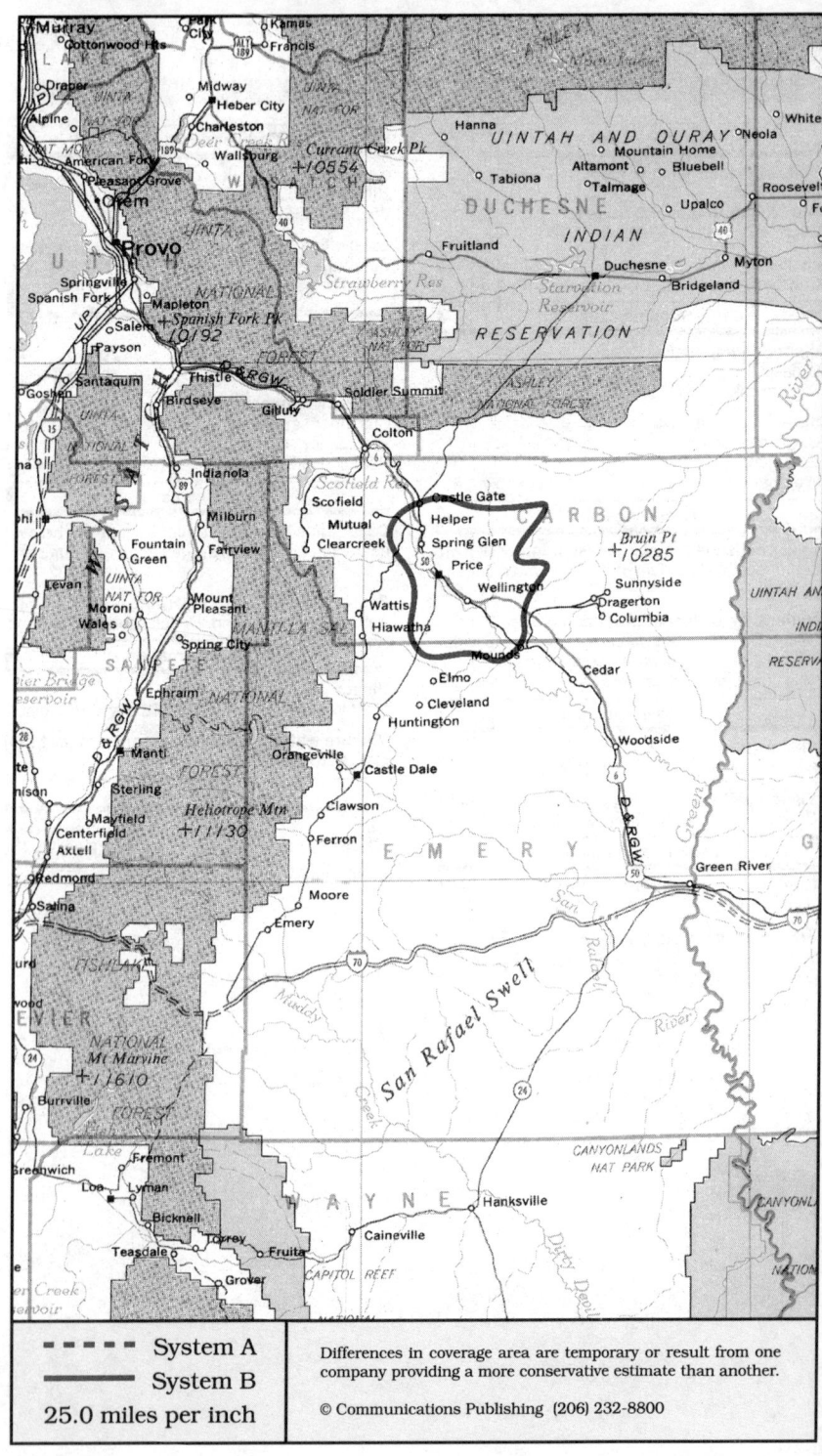

Price, Utah

Cellular company

Cellular One
American Rural Cellular
480 E 100 S
Vernal, UT 84078

This service was not operating at press time.
For nearby coverage area, see Vernal, UT.

Corporate office 801-789-4444

Sending calls

Visitor dialing instructions
Local calls . . . . . area code + 7 digits
Long distance . . . 1 + area code + 7 digits

Speed-dial numbers
Customer service 611
Emergency 911

Receiving calls

Caller dials the roamer access number below,
hears tone, dials cellular area code and number.
. . . . 801-790-7626

Billing information

Visitor rates
Most visitors . . . . . $2 day + 75¢ minute

Visitors without roaming agreements (p. 1159)
Cannot call since company takes no credit cards.

Identification numbers

City	Market	System	Billing	RSA
All served cities	677	01741	00000	UT-5

Cellular company

CommNet 2000
Cellular, Inc.
370 W St. George Boulevard
St. George, UT 84770

Customer service 800-597-5526
Local office 800-869-8062
. . . . 801-628-2002
Corporate office 303-694-3234

Sending calls

Visitor dialing instructions
Local calls . . . . . area code + 7 digits
Long distance . . . 1 + area code + 7 digits

Speed-dial numbers
Customer service 611
Emergency 911
Roaming information 711

Receiving calls

Caller dials the roamer access number below,
hears tone, dials cellular area code and number.
. . . . 801-650-7626

Billing information

Visitor rates
Most visitors . . . . . $3 day + 95¢ minute

Visitors without roaming agreements (p. 1159)
Cannot call since company takes no credit cards.

Identification numbers

City	Market	System	Billing	RSA
All served cities	677	01740	30292	UT-5

For full coverage area, see St. George, UT.

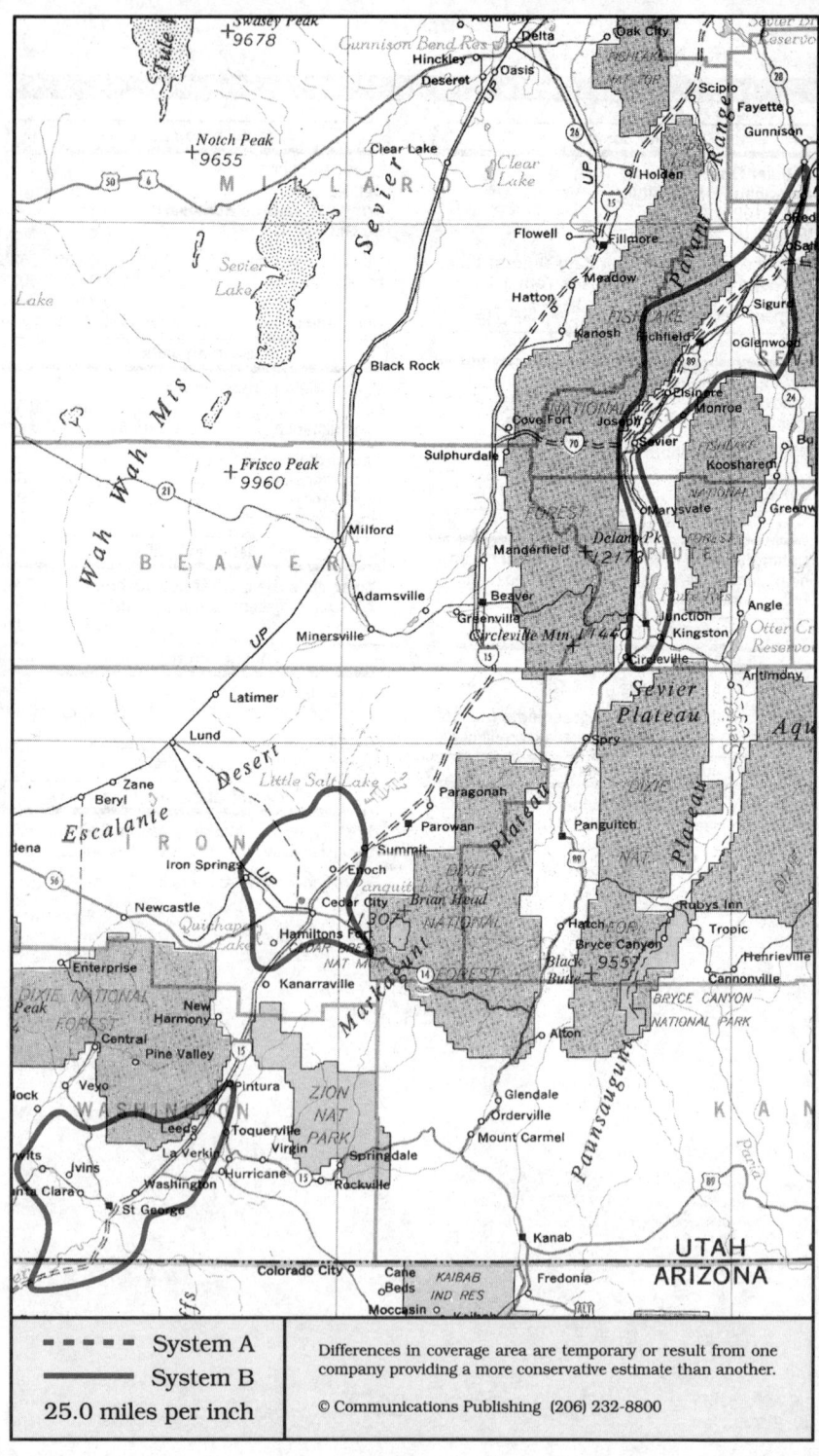

----- System A	Differences in coverage area are temporary or result from one company providing a more conservative estimate than another.
——— System B	
25.0 miles per inch	© Communications Publishing (206) 232-8800

St. George, Utah
Cedar City, Kingston, Richfield, St. George, Salina

System A	System B

Cellular company

No information about the company that will operate system A was available at press time.

Cellular company

CommNet 2000
Cellular, Inc.
370 W St. George Boulevard
St. George, UT 84770

Customer service	800-597-5527
Local office	800-869-8062
. . . .	801-628-2002
Corporate office	303-694-3234

Sending calls

Visitor dialing instructions
Local calls . . . . . . area code + 7 digits
Long distance . . . 1 + area code + 7 digits

Speed-dial numbers
Customer service 611
Emergency 911
Roaming information 711

Receiving calls

Caller dials the roamer access number below, hears tone, dials cellular area code and number.
Cedar City, Richfield 801-590-7626
Richfield 801-896-7626
St. George 801-632-7626

Billing information

Visitor rates
Most visitors . . . . . $3 day + 95¢ minute

Visitors without roaming agreements (p. 1159)
Cannot call since company takes no credit cards.

Identification numbers

City	Market	System	Billing	RSA
Cedar City	676	01740	00000	UT-4
Kingston	678	01740	30294	UT-6
Richfield	675	01740	30072	UT-3
St. George	676	01740	30420	UT-4

For full coverage area, see Price, UT.

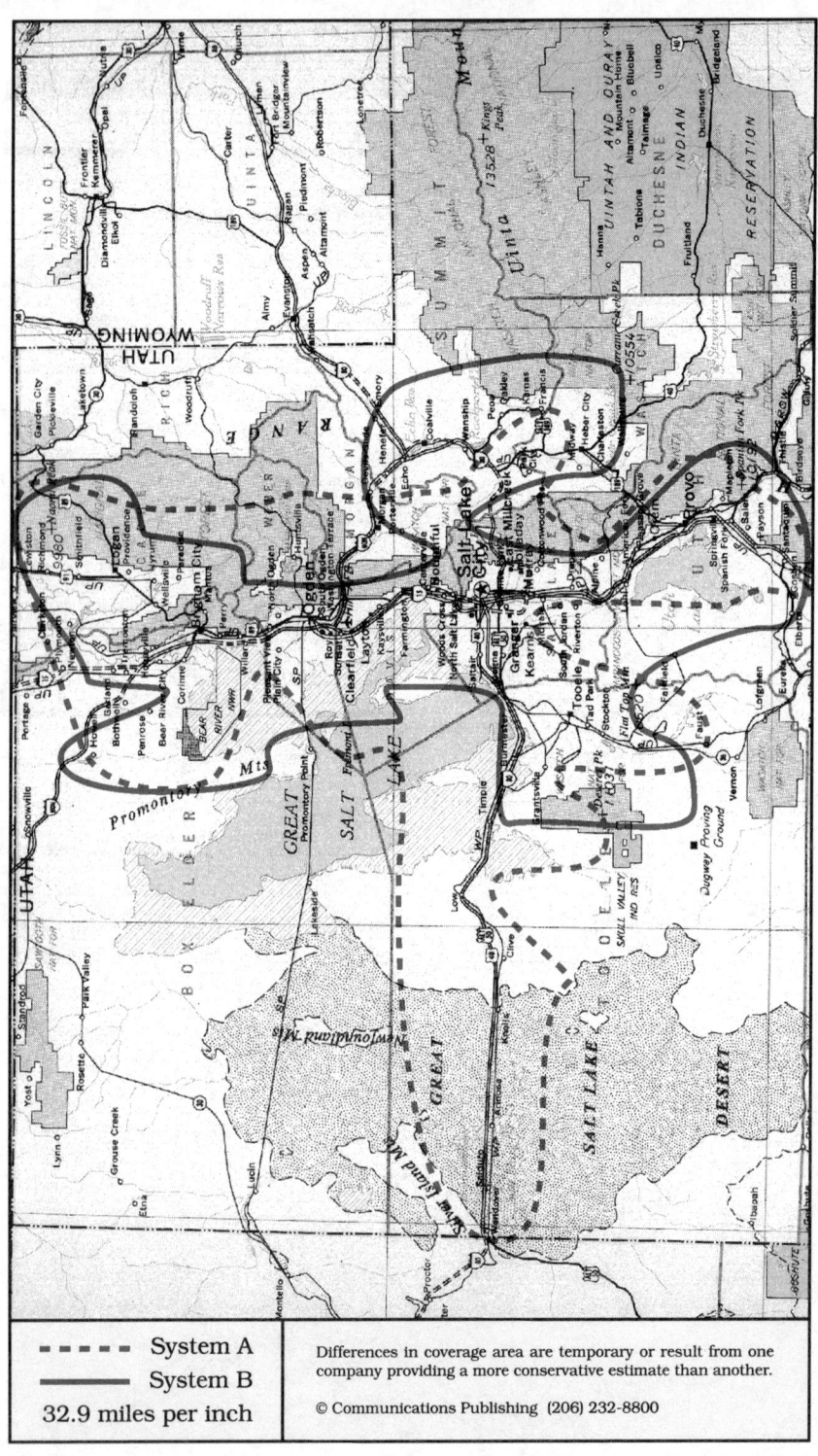

System A

System B

32.9 miles per inch

Differences in coverage area are temporary or result from one
company providing a more conservative estimate than another.

© Communications Publishing (206) 232-8800

Salt Lake City, Utah
Logan, Ogden, Orem, Park City, Provo, Salt Lake City

System A	System B

Cellular company

Cellular One
McCaw Cellular Communications
5 Triad Center, Suite 400
Salt Lake City, UT 84180

Customer service	801-575-1222
Local office	801-575-1200
Corporate office	206-827-4500

Sending calls

Visitor dialing instructions

Local calls	area code + 7 digits
Long distance . . .	1 + area code + 7 digits

Speed-dial numbers

Customer service	611
Emergency	911
Make a traffic report	211 or 311
Time and temperature	*1

Receiving calls

Caller dials the roamer access number below,
hears tone, dials cellular area code and number.

Park City, Salt Lake City . .	801-580-7626
Provo	801-376-7626

NationLink & RoamingAmerica (see p. 1157)
Dial *31 send to receive calls, *30 to deactivate.
Dial *32 send to tell caller the roamer access no.
This company's subscribers have level 6 service.

North American Cellular Network (see p. 1156)
All calls forward to you. Deactivate dial *35 send.

Billing information

Visitor rates

Most visitors	$0 day + 99¢ minute

Visitors without roaming agreements (p. 1159)
Call co. to use Amex, Mastercard, Visa

Identification numbers

City	Market	System	Billing	RSA
Brigham, Logan	673	00091	00000	UT-1
Heber City,Park City	674	00091	30363	UT-2
Orem, Provo	159	00091	00000	MSA
Salt Lake City	039	00091	00000	MSA

Call switching in UT-2 is performed by McCaw
Cellular for the licensee, Omega Cellular
Partners.

Cellular company

U S West Cellular
4021 S 700 E, Suite 100
Salt Lake City, UT 84107

Customer service	800-238-7848
. . . .	206-562-2895
. . . . . local subscribers	800-626-6611
. . . .	206-747-1771
Local office	801-269-1060
Corporate office	206-747-4900

Sending calls

Visitor dialing instructions

Local calls	area code + 7 digits
Long distance	area code + 7 digits

Speed-dial numbers

Customer service	*611
Emergency	911
Report a drunk driver	*5731
Roaming information	*711
Traffic watch	*55

Receiving calls

Caller dials the roamer access number below,
hears tone, dials cellular area code and number.

Brigham City	801-720-7626
Golden Spike	801-279-7626
Heber City, Park City	801-640-7626
Logan	801-770-7626
Provo	801-372-7626
Salt Lake City	801-573-7626

Follow Me Roaming forwarding (see p. 1156)
Activate, dial *18 send. Deactivate dial *19 send.

Billing information

Visitor rates

Most visitors	$3 day + 99¢ minute

Visitors without roaming agreements (p. 1159)
Call co. to use Amex, Mastercard, Visa

Identification numbers

City	Market	System	Billing	RSA
Brigham, Logan	673	01734	00000	UT-1
Heber City,Park City	674	01736	00000	UT-2
Orem, Provo	159	00488	00000	MSA
Salt Lake City	039	00094	00000	MSA

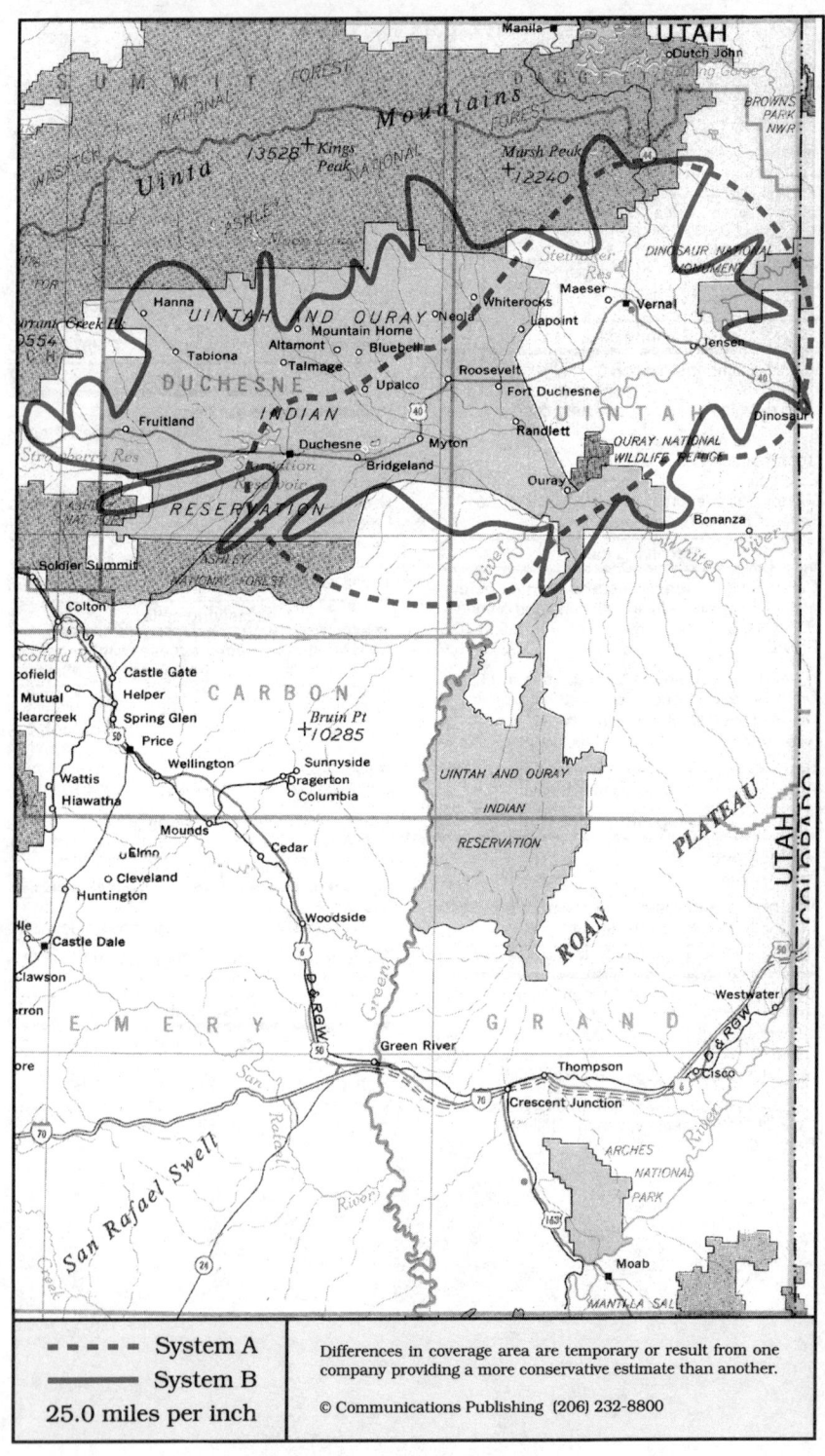

- - - - - System A	Differences in coverage area are temporary or result from one company providing a more conservative estimate than another.
———— System B	
25.0 miles per inch	© Communications Publishing (206) 232-8800

Vernal, Utah
Duchesne, Roosevelt, Vernal

System A	System B

System A

Cellular company

Cellular One
American Rural Cellular
480 E 100 S
Vernal, UT 84078

Corporate office 801-789-4444

Sending calls

Visitor dialing instructions
Local calls . . . . . area code + 7 digits
Long distance . . . 1 + area code + 7 digits

Speed-dial numbers
Customer service 611
Emergency 911

Receiving calls

Caller dials the roamer access number below,
hears tone, dials cellular area code and number.
. . . . 801-790-7626

Billing information

Visitor rates
Most visitors . . . . . $2 day + 75¢ minute

Visitors without roaming agreements (p. 1159)
Cannot call since company takes no credit cards.

Identification numbers

City	Market	System	Billing	RSA
All served cities	677	01741	00000	UT-5

System B

Cellular company

UBET Cellular
Uintah Basin Telephone Association
N Myton Bench, W Highway 40; PO Box 157
Roosevelt, UT 84066

1827 S 1500 E
Vernal, UT 84078

Local office . . . . Roosevelt 801-722-2355
. . . . Vernal 801-789-8808
Corporate office 801-646-5007

Sending calls

Visitor dialing instructions
Local calls . . . . 7 digits
Long distance . . . 1 + area code + 7 digits

Speed-dial numbers
Customer service 611
Emergency 911

Receiving calls

Caller dials the roamer access number below,
hears tone, dials cellular area code and number.
. . . . 801-823-7626

Follow Me Roaming forwarding (see p. 1156)
Activate, dial *18 send. Deactivate dial *19 send.

Billing information

Visitor rates
Most visitors . . . . . $2 day + 70¢ minute

Visitors without roaming agreements (p. 1159)
Call co. to use Mastercard, Visa

Identification numbers

City	Market	System	Billing	RSA
All served cities	677	01858	00000	UT-5

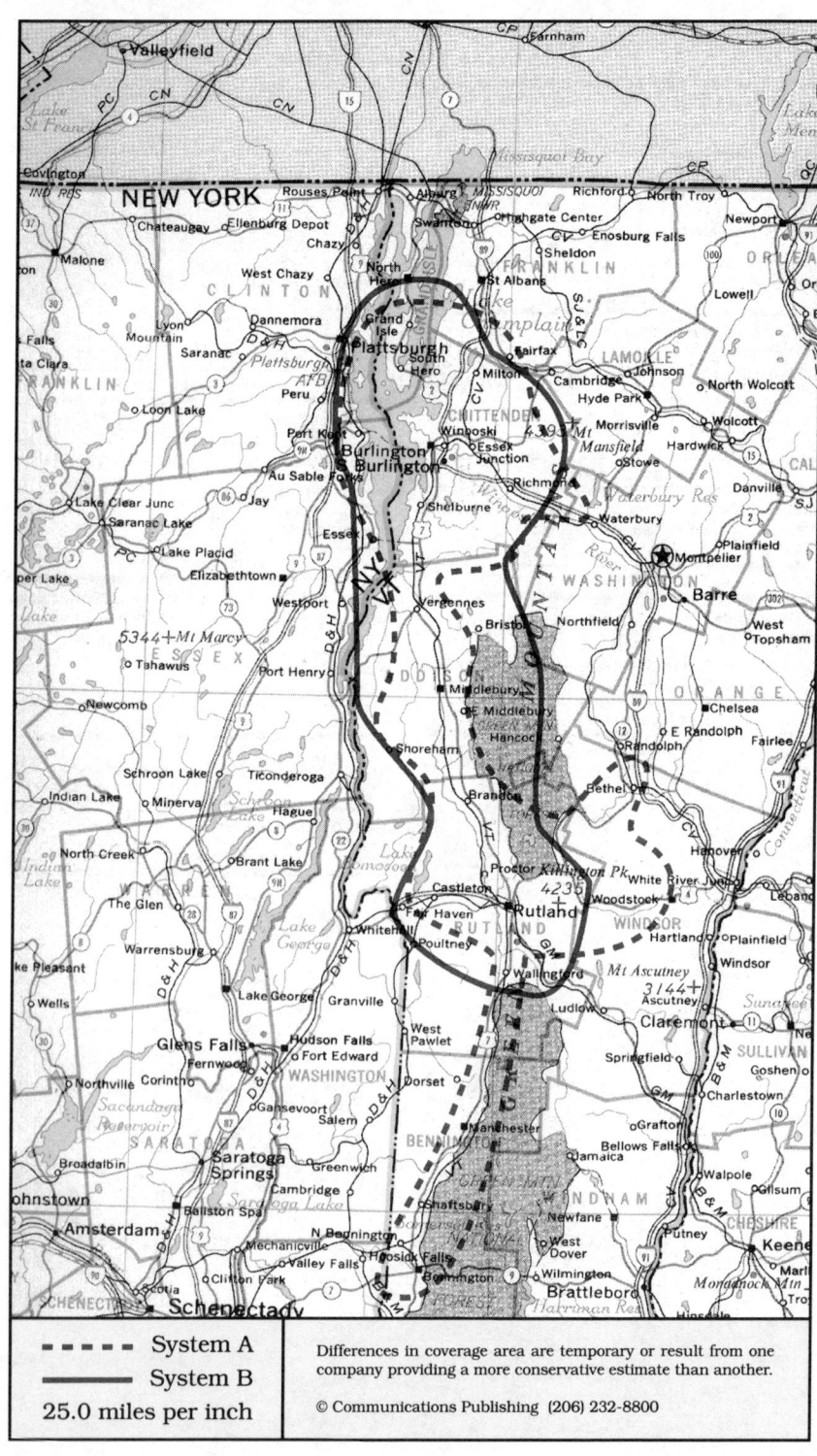

- - - - System A	Differences in coverage area are temporary or result from one company providing a more conservative estimate than another.
———— System B	
25.0 miles per inch	© Communications Publishing (206) 232-8800

Burlington, Vermont
Bennington, Burlington, Middlebury, Rutland

System A	System B

Cellular company

Cellular One
Atlantic Cellular
3 Baldwin Avenue
S Burlington, VT 05403

Local office	800-676-2355
. . . .	802-862-9000
Corporate office	401-421-7090

Sending calls

Visitor dialing instructions

Local calls	area code + 7 digits
Long distance . . .	1 + area code + 7 digits

Speed-dial numbers

Customer service	*611
Emergency	911

Receiving calls

Caller dials the roamer access number below,
hears tone, dials cellular area code and number.
. . . . 802-343-7626

NationLink & RoamingAmerica (see p. 1157)
Dial *31 send to receive calls, *30 to deactivate.
Dial *32 send to tell caller the roamer access no.
This company's subscribers have level 8 service.

Billing information

Visitor rates
Most visitors . . . . . $3 day + 99¢ minute

Visitors without roaming agreements (p. 1159)
Call co. to use Amex, Discover, Mastercard, Visa

Identification numbers

City	Market	System	Billing	RSA
Burlington	248	00313	00000	MSA
Middlebury	680	00313	00000	VT-2

Cellular company

Contel Cellular
1335 Shelburne Road
S Burlington, VT 05403

Customer service	800-333-4004
. . . . . . extension 4002	404-391-8000
Local office	802-865-3100
Corporate office	404-804-3400

Sending calls

Visitor dialing instructions

Local calls	area code + 7 digits
Long distance . . .	1 + area code + 7 digits

Speed-dial numbers

Customer service	611
Directory assistance	411
Emergency	911

Receiving calls

Caller dials the roamer access number below,
hears tone, dials cellular area code and number.
. . . . 802-238-7626

Billing information

Visitor rates
Most visitors . . . . . $3 day + 85¢ minute

Visitors without roaming agreements (p. 1159)
Call co. to use Amex, Mastercard, Visa

Identification numbers

City	Market	System	Billing	RSA
Burlington	248	00300	00000	MSA
Middlebury, Rutland	680	01748	00000	VT-2

For full coverage area, see Plattsburgh, NY.

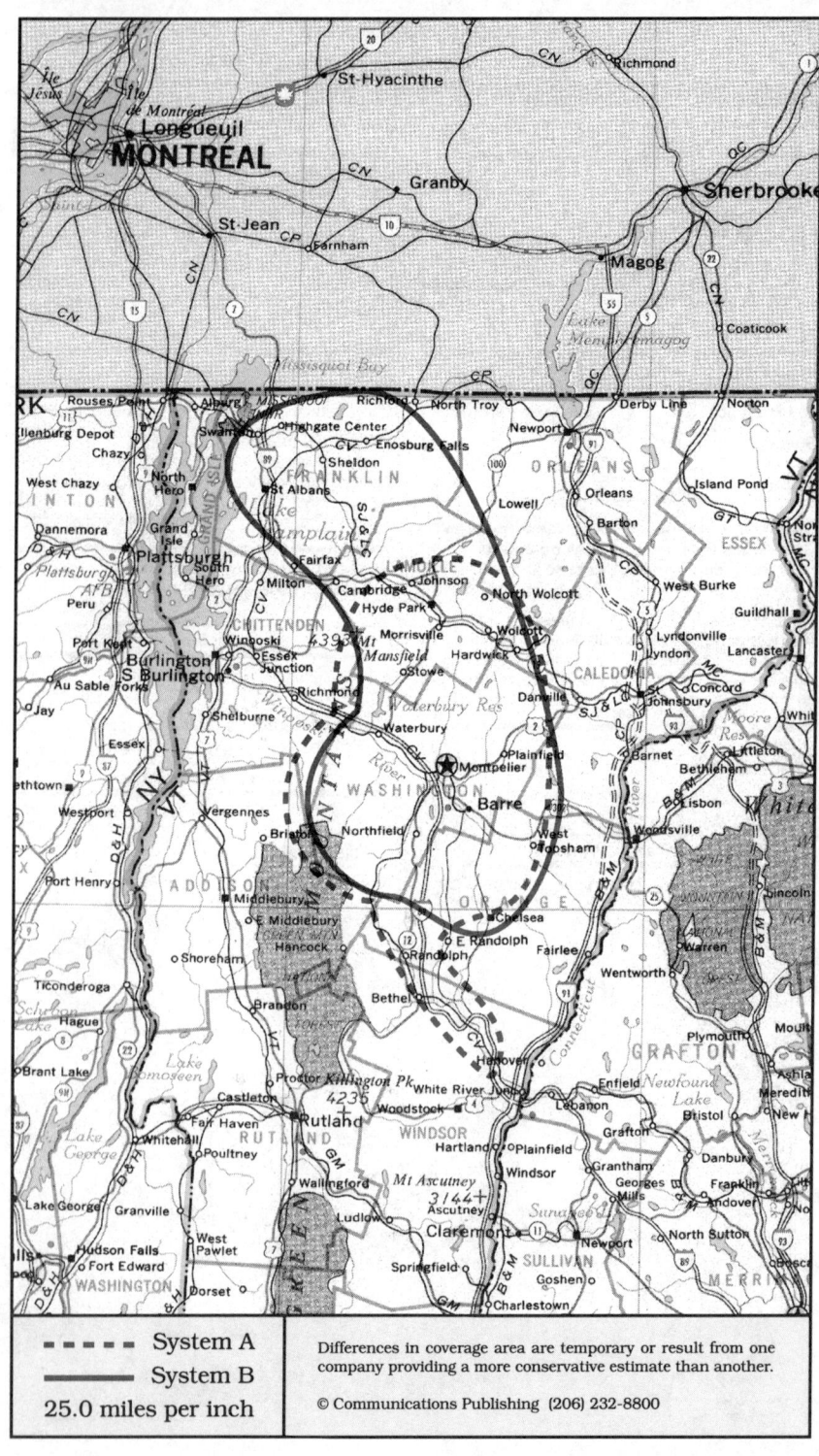

- - - - - System A
———— System B
25.0 miles per inch

Differences in coverage area are temporary or result from one company providing a more conservative estimate than another.

© Communications Publishing (206) 232-8800

Montpelier, Vermont
Barre, Hyde Park, Montpelier, St. Albans, Stowe

System A	System B

Cellular company

Cellular One
Licensed to: PC Cellular of Vermont
Managed by: CompComm
89 Main Street
Montpelier, VT 05602

Local office	802-229-2828
Corporate office CompComm	609-854-1000

Sending calls

Visitor dialing instructions

Local calls	area code + 7 digits
Long distance . . .	1 + area code + 7 digits

Speed-dial numbers

Customer service	611
Emergency	911
Vermont ski report	*SKI or *754
Vermont weather	*WX or *99
WDEV radio	*550

Receiving calls

Caller dials the roamer access number below,
hears tone, dials cellular area code and number.
. . . . 802-229-8626

NationLink & RoamingAmerica (see p. 1157)
Dial *31 send to receive calls, *30 to deactivate.
This company's subscribers have level 3 service.

Billing information

Visitor rates
Most visitors . . . . . $3 day + 75¢ minute

Visitors without roaming agreements (p. 1159)
Call co. to use Mastercard, Visa
or Roamer Plus will intercept first call to bill you.

Identification numbers

City	Market	System	Billing	RSA
All served cities	679	01745	00000	VT-1

Cellular company

Contel Cellular
81 River Street
Montpelier, VT 05602

Customer service	800-333-4004
. . . . . extension 4002	404-391-8000
Local office	802-223-3333
Corporate office	404-804-3400

Sending calls

Visitor dialing instructions

Local calls	area code + 7 digits
Long distance . . .	1 + area code + 7 digits

Speed-dial numbers

Customer service	611
Directory assistance	411
Emergency	911

Receiving calls

Caller dials the roamer access number below,
hears tone, dials cellular area code and number.
. . . . 802-238-7626

Billing information

Visitor rates
Most visitors . . . . . $3 day + 85¢ minute

Visitors without roaming agreements (p. 1159)
Call co. to use Amex, Mastercard, Visa

Identification numbers

City	Market	System	Billing	RSA
All served cities	548	01746	00000	VT-1

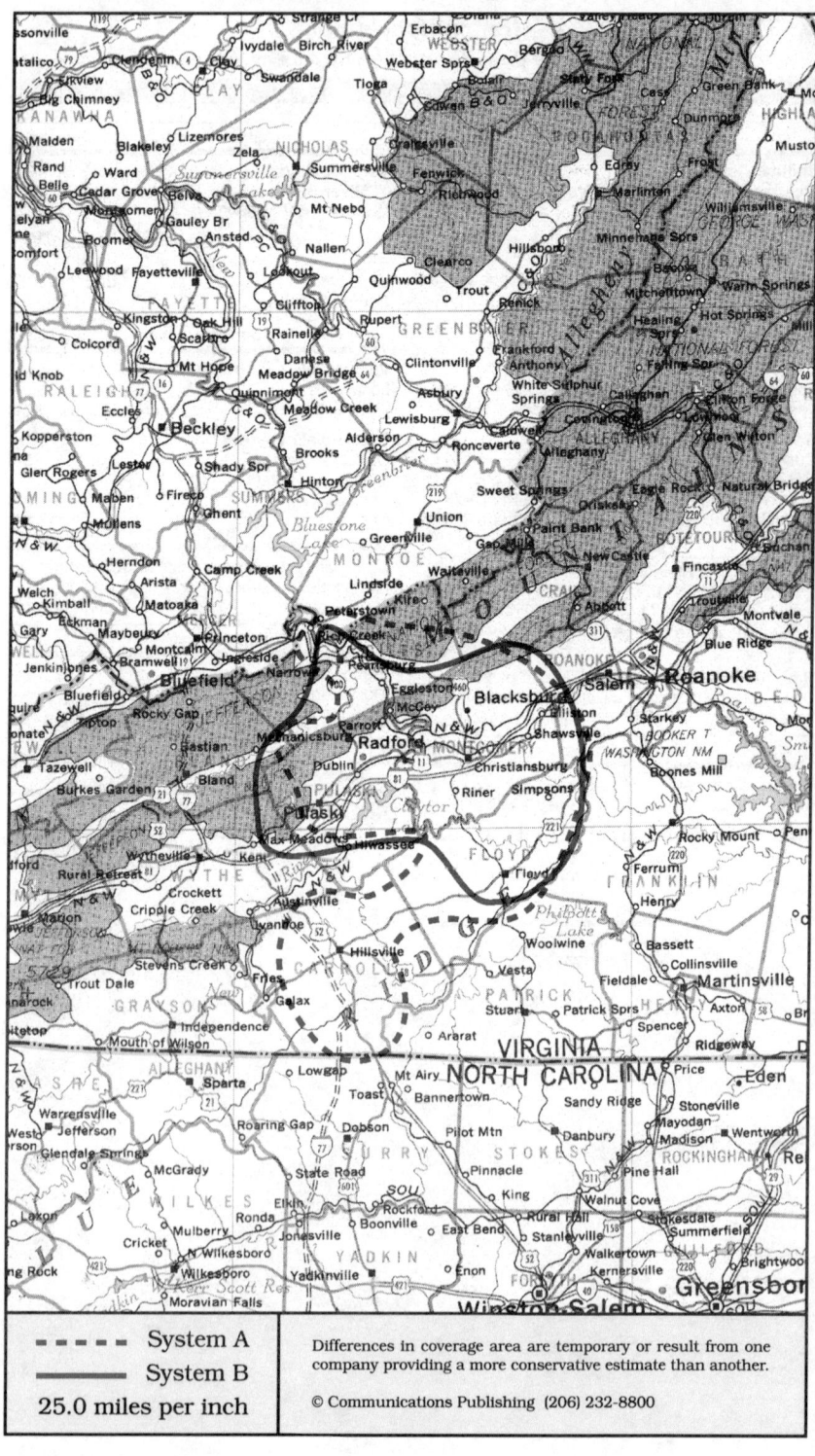

- - - - - System A	Differences in coverage area are temporary or result from one
—— System B	company providing a more conservative estimate than another.
25.0 miles per inch	© Communications Publishing (206) 232-8800

Blacksburg, Virginia
Blacksburg, Christiansburg, Hillsville, Pearisburg, Pulaski, Radford

System A	System B

Cellular company

Blue Ridge Cellular
1580 N Franklin Street, PO Box 794
Christiansburg, VA 24073

Corporate office 703-382-2656

Sending calls

Visitor dialing instructions
Local calls . . . . . area code + 7 digits
Long distance . . . 1 + area code + 7 digits

Speed-dial numbers
Customer service 611
Emergency 911

Receiving calls

Caller dials the roamer access number below,
hears tone, dials cellular area code and number.
. . . . 703-320-7626

Billing information

Visitor rates
Most visitors . . . . . $3 day + 85¢ minute

Visitors without roaming agreements (p. 1159)
Cannot call since company takes no credit cards.

Identification numbers

City	Market	System	Billing	RSA
All served cities	683	01753	00000	VA-3

Cellular company

Contel Cellular
2803 S Main Street
Blacksburg, VA 24060

Customer service 800-333-4004
. . . . 804-552-1239
Local office 703-552-3119
Corporate office 404-804-3400

Sending calls

Visitor dialing instructions
Local calls . . . . . . area code + 7 digits
Long distance . . . 1 + area code + 7 digits

Speed-dial numbers
Customer service 611
Emergency 911

Receiving calls

Caller dials the roamer access number below,
hears tone, dials cellular area code and number.
. . . . 703-230-7626

Billing information

Visitor rates
Most visitors . . . . . $3 day + $1 minute

Visitors without roaming agreements (p. 1159)
Call co. to use Amex, Mastercard, Visa

Identification numbers

City	Market	System	Billing	RSA
All served cities	683	01754	00000	VA-3

For full coverage area, see Lexington, VA and
Roanoke, VA.

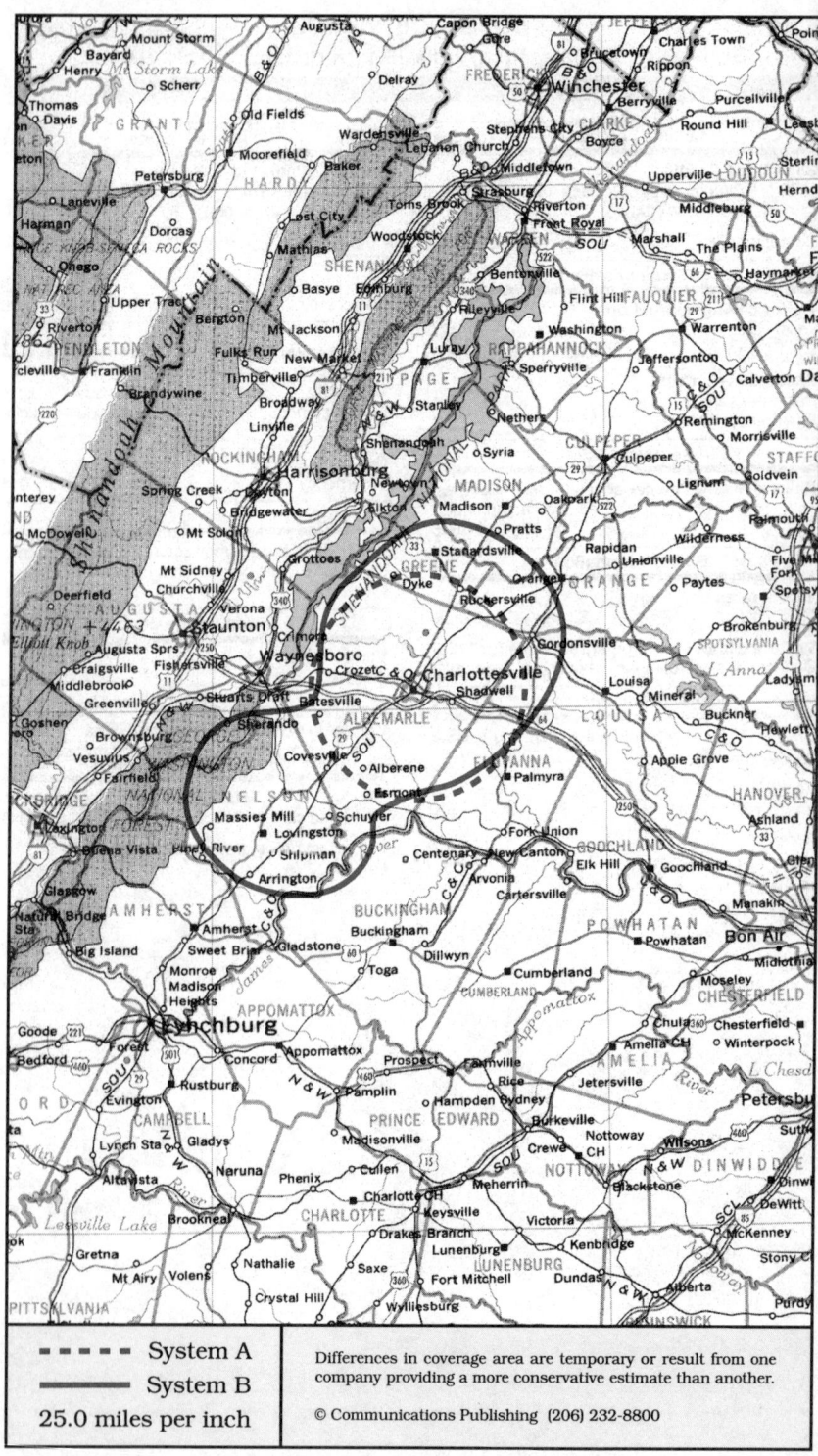

- - - - -	System A
———	System B
25.0 miles per inch	

Differences in coverage area are temporary or result from one company providing a more conservative estimate than another.

© Communications Publishing (206) 232-8800

Charlottesville, Virginia
Charlottesville, Lovingston (Nelson County), Standardsville

System A	System B

System A

Cellular company

Cellular One
Centennial Cellular
2329 Seminole Trail
Charlottesville, VA 22901

Customer service	800-763-0138
Local office	804-973-1767
Corporate office Administration	703-345-0808

Sending calls

Visitor dialing instructions

Local calls	area code + 7 digits
Long distance . . .	1 + area code + 7 digits

Speed-dial numbers

Customer service	611
Emergency	911

Receiving calls

Caller dials the roamer access number below, hears tone, dials cellular area code and number.
. . . . 804-960-7626

NationLink & RoamingAmerica (see p. 1157)
Caller hears your current roamer access number.
This company's subscribers have level 0 service.

Billing information

Visitor rates

Most visitors	$3 day + 85¢ minute

Visitors without roaming agreements (p. 1159)
Roamer Plus will intercept first call to bill you.

Identification numbers

City	Market	System	Billing	RSA
All served cities	256	00309	00000	MSA

For the system A company that will serve Nelson County in the future, see Harrisonburg, VA.

System B

Cellular company

Centel Cellular
1746 Rio Hill Center
Charlottesville, VA 22901

Local office	804-973-9100
Corporate office	312-399-2644

Sending calls

Visitor dialing instructions

Local calls	area code + 7 digits
Long distance . . .	1 + area code + 7 digits

Speed-dial numbers

Customer service	*611
Emergency	911

Receiving calls

Caller dials the roamer access number below, hears tone, dials cellular area code and number.
. . . . 804-981-7626

Billing information

Visitor rates

Most visitors	$3 day + 99¢ minute

Visitors without roaming agreements (p. 1159)
Cannot call since company takes no credit cards.

Identification numbers

City	Market	System	Billing	RSA
Charlottesville	256	00292	00000	MSA
Lovingston	686	01954	00000	VA-6

For full coverage area, see Culpeper, VA.

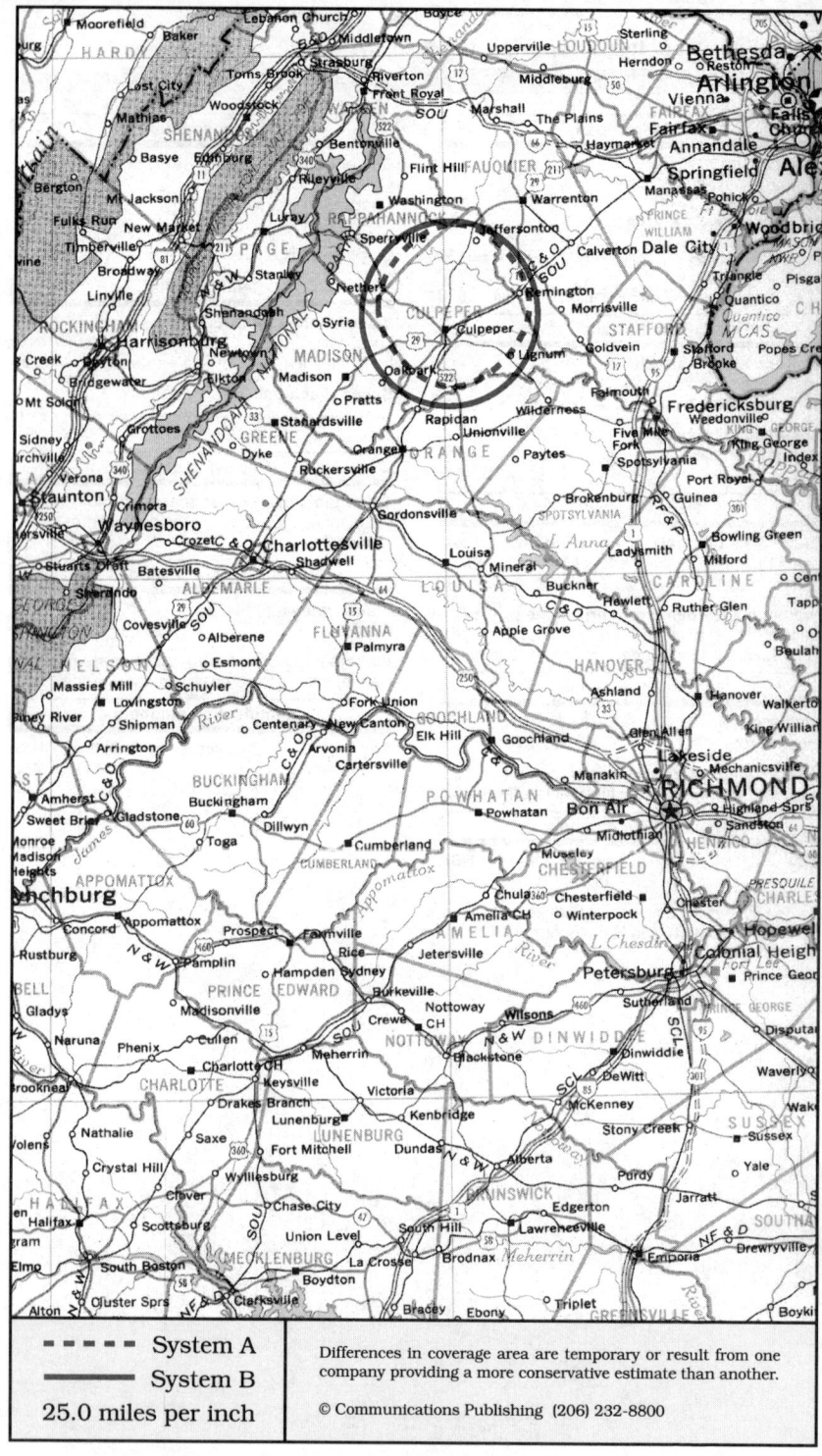

Culpeper, Virginia

Cellular company

Cellular One
Southwestern Bell Mobile Systems
7855 Walker Drive, Suite 100
Greenbelt, MD 20770

Customer service	some areas	800-235-5663
. . . .		301-220-0060
Corporate office	. . . . .	301-220-3600

Sending calls

Visitor dialing instructions
Local calls . . . . . area code + 7 digits
Long distance . . . . area code + 7 digits

Speed-dial numbers
Cellular information *1394
Customer service . . . . . *611 or *811
Emergency 911
24 hour news *1887

Receiving calls

Caller dials the roamer access number below,
hears tone, dials cellular area code and number.
. . . . 703-840-7626

NationLink & RoamingAmerica (see p. 1157)
Dial *31 send to receive calls, *30 to deactivate.
Dial *32 send to tell caller the roamer access no.
Service is not automatic and must be turned on.
This company's subscribers have level 6 service.

Billing information

Visitor rates
Most visitors . . . . . $3 day + 95¢ minute

Visitors without roaming agreements (p. 1159)
Roamer Plus will intercept first call to bill you.

Identification numbers

City	Market	System	Billing	RSA
All served cities	691	00013	30333	VA-11

For full coverage area, see Baltimore, MD.

Cellular company

Centel Cellular
1746 Rio Hill Center
Charlottesville, VA 22901

Local office	804-973-9100
Corporate office	312-399-2644

Sending calls

Visitor dialing instructions
Local calls . . . . . . area code + 7 digits
Long distance . . . 1 + area code + 7 digits

Speed-dial numbers
Customer service *611
Emergency 911

Receiving calls

Caller dials the roamer access number below,
hears tone, dials cellular area code and number.
. . . . 804-981-7626

Billing information

Visitor rates
Most visitors . . . . . $3 day + 99¢ minute

Visitors without roaming agreements (p. 1159)
Cannot call since company takes no credit cards.

Identification numbers

City	Market	System	Billing	RSA
All served cities	691	02026	00000	VA-11

For full coverage area, see Charlottesville, VA.

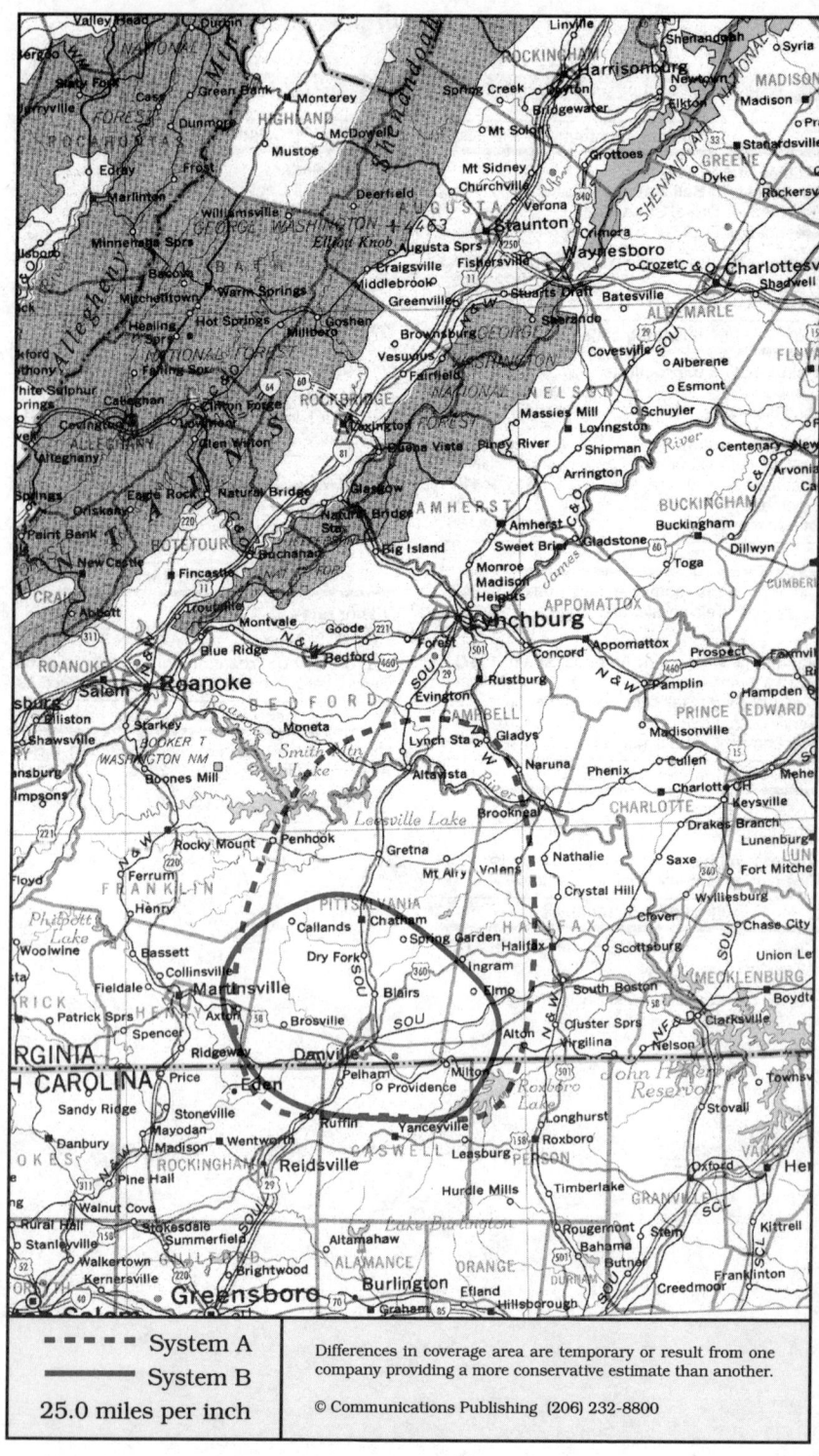

Danville, Virginia
Chatham, Danville

System A	System B

Cellular company

Cellular One
GTE Mobilnet
101 Mall Drive
Danville, VA 24540

Customer service 800-727-2444
. . . . 919-481-1181
Local office 804-791-3555
Corporate office 404-391-8000

Sending calls

Visitor dialing instructions
Local calls . . . . 7 digits
Long distance . . . 1 + area code + 7 digits

Speed-dial numbers
Customer service 611
Emergency 911
State police . . . . ∗77
WBTM . . . . ∗1330

Receiving calls

Caller dials the roamer access number below,
hears tone, dials cellular area code and number.
. . . . 804-770-7626

NationLink & RoamingAmerica (see p. 1157)
Caller hears your current roamer access number.
This company's subscribers have level 0 service.

Billing information

Visitor rates
Most visitors . . . . . $3 day + 85¢ minute

Visitors without roaming agreements (p. 1159)
Call co. to use Mastercard, Visa

Identification numbers

City	Market	System	Billing	RSA
All served cities	262	00323	00000	MSA

Cellular company

Centel Cellular
747 Piney Forest Road
Danville, VA 24540

Local office 804-791-3100
Corporate office 312-399-2644

Sending calls

Visitor dialing instructions
Local calls . . . . . . area code + 7 digits
Long distance . . . 1 + area code + 7 digits

Speed-dial numbers
Customer service . . . . ∗611
Emergency 911

Receiving calls

Caller dials the roamer access number below,
hears tone, dials cellular area code and number.
. . . . 804-250-7626

Billing information

Visitor rates
Most visitors . . . . . $3 day + 99¢ minute

Visitors without roaming agreements (p. 1159)
Cannot call since company takes no credit cards.

Identification numbers

City	Market	System	Billing	RSA
All served cities	262	00306	00000	MSA

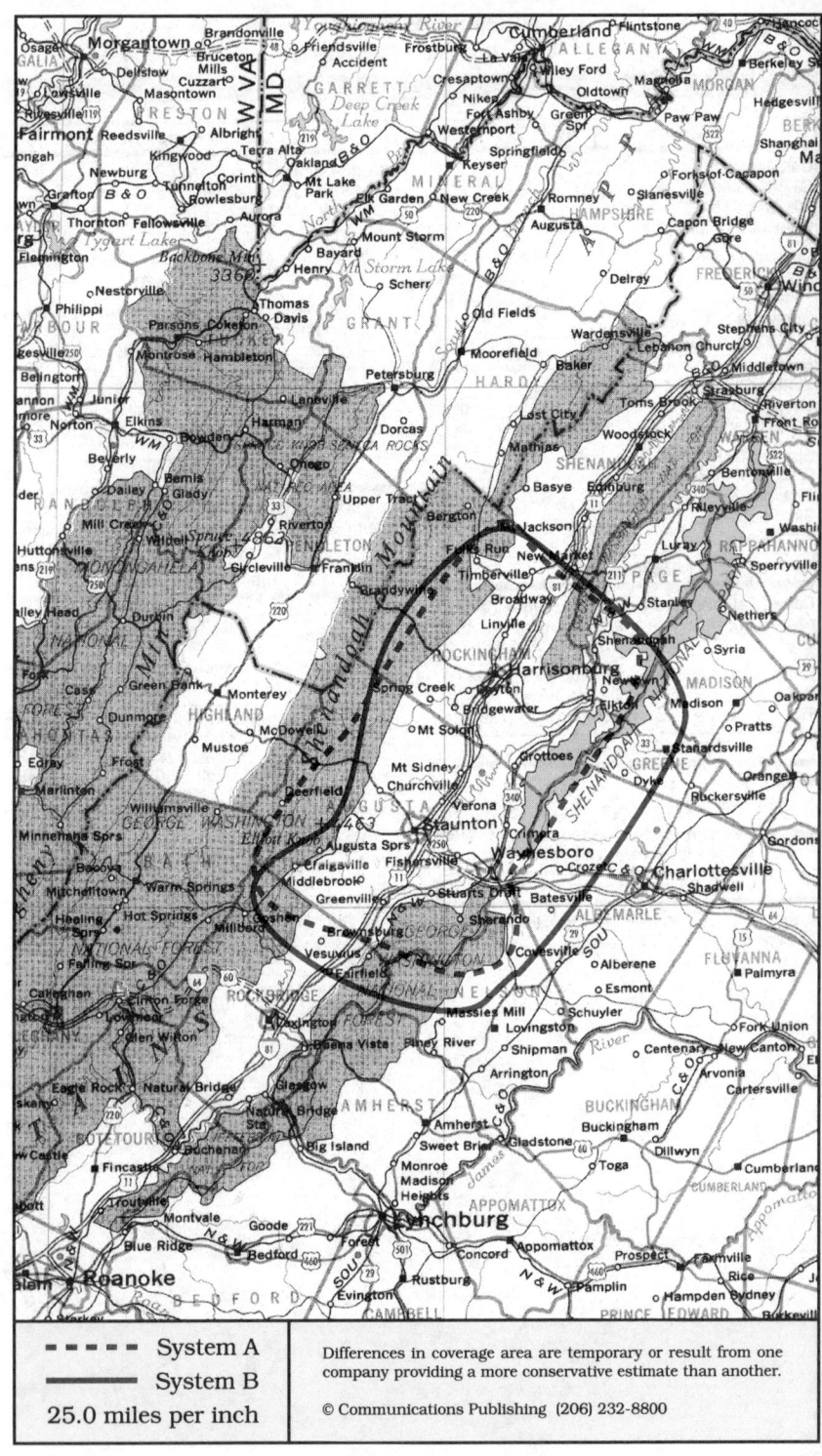

Harrisonburg, Virginia
Harrisonburg, Staunton, Waynesboro

System A	System B

Cellular company

Cellular One
Virginia Cellular
29-A N Augusta Street, PO Box 1002
Staunton, VA 24401-1002

| Corporate office | . . . . . | 800-444-4235 |
| . . . . | | 703-886-1065 |

Sending calls

Visitor dialing instructions

| Local calls | . . . . . | area code + 7 digits |
| Long distance | . . . | 1 + area code + 7 digits |

Speed-dial numbers

Customer service	. . . .	611
Emergency	. . . .	911
Highway patrol	. . . . . .	*HP or *47
Roaming information	. . . . .	711
Technical support	. . . . . .	811

Receiving calls

Caller dials the roamer access number below,
hears tone, dials cellular area number.
. . . . 703-480-7626

NationLink & RoamingAmerica (see p. 1157)
Caller hears your current roamer access number.
This company's subscribers have level 0 service.

Billing information

Visitor rates
Most visitors . . . . . $3 day + 85¢ minute

Visitors without roaming agreements (p. 1159)
Roamer Plus will intercept first call to bill you.

Identification numbers

City	Market	System	Billing	RSA
All served cities	686	01759	00000	VA-6

Cellular company

CFW Cellular
CFW Communications
2061 Evelyn Byrd Avenue #C, PO Box 899
Harrisonburg, VA 22801

1105 Greenville Avenue, PO Box 2827
Staunton, VA 24401

524 W Broad Street, PO Box 1527
Waynesboro, VA 22980

Customer service	. .	CFW	800-432-6353
Local office	. .	Harrisonburg	703-432-6353
. . . .		Staunton	703-885-5522
. . . .		Waynesboro	703-946-2355

Sending calls

Visitor dialing instructions

| Local calls | . . . . . . | area code + 7 digits |
| Long distance | . . . | 1 + area code + 7 digits |

Speed-dial numbers

| Customer service | . . . . | 611 |
| Emergency | . . . . | 911 |

Receiving calls

Caller dials the roamer access number below,
hears tone, dials cellular area code and number.
. . . . 703-470-7626

Follow Me Roaming forwarding (see p. 1156)
Activate, dial *18 send. Deactivate dial *19 send.

Billing information

Visitor rates
Most visitors . . . . . $3 day + 99¢ minute

Visitors without roaming agreements (p. 1159)
Cannot call since company takes no credit cards.

Identification numbers

City	Market	System	Billing	RSA
All served cities	686	00502	30040	VA-6

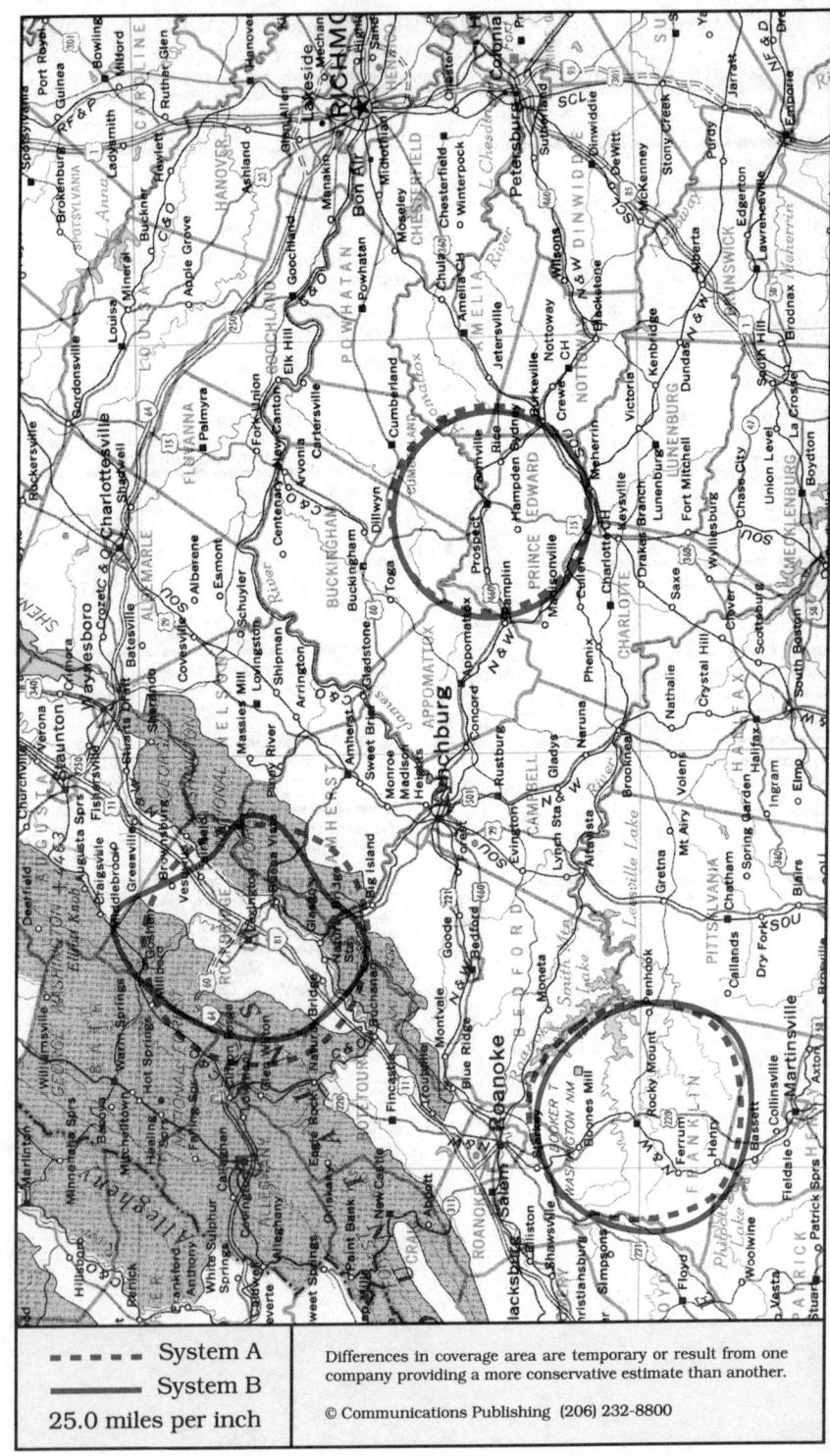

- - - - - System A
————— System B
25.0 miles per inch

Differences in coverage area are temporary or result from one company providing a more conservative estimate than another.

© Communications Publishing (206) 232-8800

Lexington, Virginia
Farmville, Lexington, Rocky Mount

System A	System B

System A

Cellular company

United States Cellular
301 Liberty Street
Martinsville, VA 24112

Local office	703-632-2000
Corporate office	312-399-8900

Sending calls

Visitor dialing instructions

Local calls	area code + 7 digits
Long distance . . .	1 + area code + 7 digits

Speed-dial numbers

Customer service	611
Emergency	911

Receiving calls

Caller dials the roamer access number below, hears tone, dials cellular area code and number.
. . . . 703-570-7626

Billing information

Visitor rates

Most visitors	$3 day + 99¢ minute

Visitors without roaming agreements (p. 1159)
Cannot call since company takes no credit cards.

Identification numbers

City	Market	System	Billing	RSA
Farmville	687	01761	00000	VA-7
Lexington	685	01757	00000	VA-5
Rocky Mount	684	01755	00000	VA-4

For full coverage area, see Martinsville, VA.

System B

Cellular company

Lexington and Rocky Mount
Contel Cellular
6 Fairway Village
Hardy, VA 24101

Farmville
Contel Cellular
3210 S Crater Road
Petersburg, VA 23805

Customer service	800-333-4004
. . . .	804-552-1239
Local office . . . . . Hardy	703-721-5300
. . . Hardy from Lexington	703-460-2009
. . . . Petersburg	804-861-6400
Corporate office	404-804-3400

Sending calls

Visitor dialing instructions

Local calls	area code + 7 digits
Long distance . . .	1 + area code + 7 digits

Speed-dial numbers

Customer service	611
Emergency	911

Receiving calls

Caller dials the roamer access number below, hears tone, dials cellular area code and number.

Farmville	804-347-7626
Lexington	703-460-7626
Rocky Mount	703-420-7626

Billing information

Visitor rates

Most visitors	$3 day + 85¢ minute

Visitors without roaming agreements (p. 1159)
Call co. to use Amex, Mastercard, Visa

Identification numbers

City	Market	System	Billing	RSA
Farmville	687	01762	00000	VA-7
Lexington	685	01758	00000	VA-5
Rocky Mount	684	01756	00000	VA-4

For full coverage area, see Blacksburg, VA and Richmond, VA and Roanoke, VA.

- - - - System A	Differences in coverage area are temporary or result from one
——— System B	company providing a more conservative estimate than another.
25.0 miles per inch	© Communications Publishing (206) 232-8800

Lynchburg, Virginia
Amherst, Appomattox, Lynchburg, Rustburg

System A	System B

Cellular company

Cellular One
Centennial Cellular
2505 Wards Road, Suite F
Lynchburg, VA 24502

Customer service	800-763-0138
Local office	804-832-0808
Corporate office Administration	703-345-0808
. . . . Management	203-972-2000

Sending calls

Visitor dialing instructions

Local calls	area code + 7 digits
Long distance . . .	1 + area code + 7 digits

Speed-dial numbers

Customer service	611
Emergency	911

Receiving calls

Caller dials the roamer access number below,
hears tone, dials cellular area code and number.
. . . . 804-660-7626

NationLink & RoamingAmerica (see p. 1157)
Caller hears your current roamer access number.
This company's subscribers have level 0 service.

Billing information

Visitor rates

Most visitors	$3 day + 85¢ minute

Visitors without roaming agreements (p. 1159)
Roamer Plus will intercept first call to bill you.

Identification numbers

City	Market	System	Billing	RSA
All served cities	203	00441	00000	MSA

Cellular company

Centel Cellular
3506 Mayflower Drive
Lynchburg, VA 24501

Local office	804-528-3500
Corporate office	312-399-2644

Sending calls

Visitor dialing instructions

Local calls	area code + 7 digits
Long distance . . .	1 + area code + 7 digits

Speed-dial numbers

Customer service	*611
Directory assistance	411
Emergency	911

Receiving calls

Caller dials the roamer access number below,
hears tone, dials cellular area code and number.
. . . . 804-841-7626

Billing information

Visitor rates

Most visitors	$3 day + 99¢ minute

Visitors without roaming agreements (p. 1159)
Cannot call since company takes no credit cards.

Identification numbers

City	Market	System	Billing	RSA
All served cities	203	00424	00000	MSA

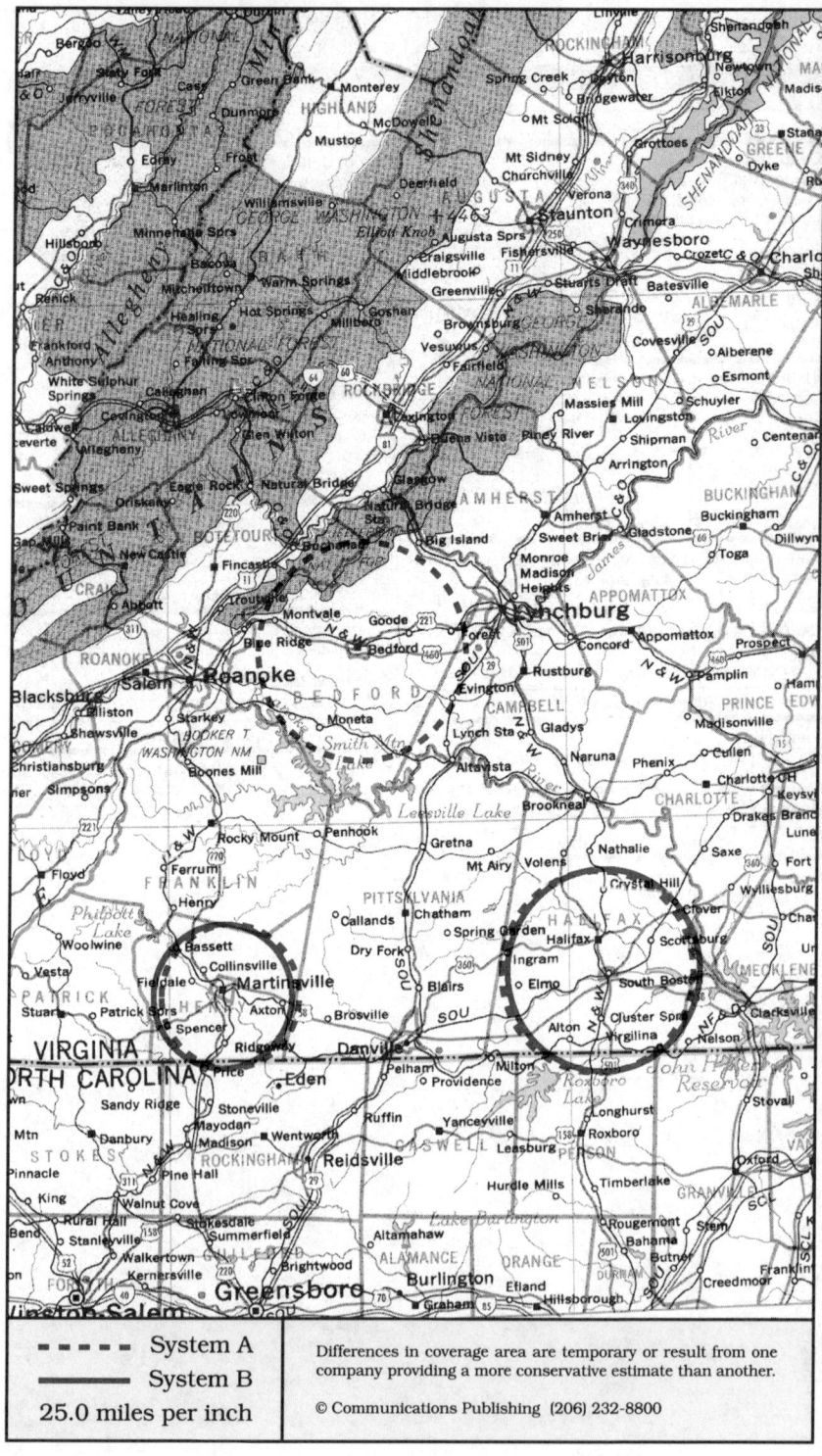

▬ ▬ ▬ ▬ System A	Differences in coverage area are temporary or result from one
▬▬▬▬▬ System B	company providing a more conservative estimate than another.
25.0 miles per inch	© Communications Publishing (206) 232-8800

Martinsville, Virginia
Bedford, Halifax, Martinsville, South Boston

System A	System B

Cellular company

United States Cellular
301 Liberty Street
Martinsville, VA 24112

Local office	703-632-2000
Corporate office	312-399-8900

Sending calls

Visitor dialing instructions

Local calls	area code + 7 digits
Long distance . . .	1 + area code + 7 digits

Speed-dial numbers

Customer service	611
Emergency	911

Receiving calls

Caller dials the roamer access number below,
hears tone, dials cellular area code and number.

Bedford, Martinsville	703-340-7626
South Boston	804-470-7626

Billing information

Visitor rates

Most visitors	$3 day + 99¢ minute

Visitors without roaming agreements (p. 1159)
Cannot call since company takes no credit cards.

Identification numbers

City	Market	System	Billing	RSA
Bedford, Martinsville	684	01755	00000	VA-4
South Boston	687	01761	00000	VA-7

For full coverage area, see Lexington, VA.

Cellular company

Centel Cellular
3506 Mayflower Drive
Lynchburg, VA 24501

Local office	804-528-3500
Corporate office	312-399-2644

Sending calls

Visitor dialing instructions

Local calls	area code + 7 digits
Long distance . . .	1 + area code + 7 digits

Speed-dial numbers

Customer service	*611
Directory assistance	411
Emergency	911

Receiving calls

Caller dials the roamer access number below,
hears tone, dials cellular area code and number.

. . . .	804-841-7626

Billing information

Visitor rates

Most visitors	$3 day + 99¢ minute

Visitors without roaming agreements (p. 1159)
Cannot call since company takes no credit cards.

Identification numbers

City	Market	System	Billing	RSA
Martinsville	684	01986	00000	VA-4
South Boston	687	01894	00000	VA-7

This company may provide service in the future in Bedford, VA.

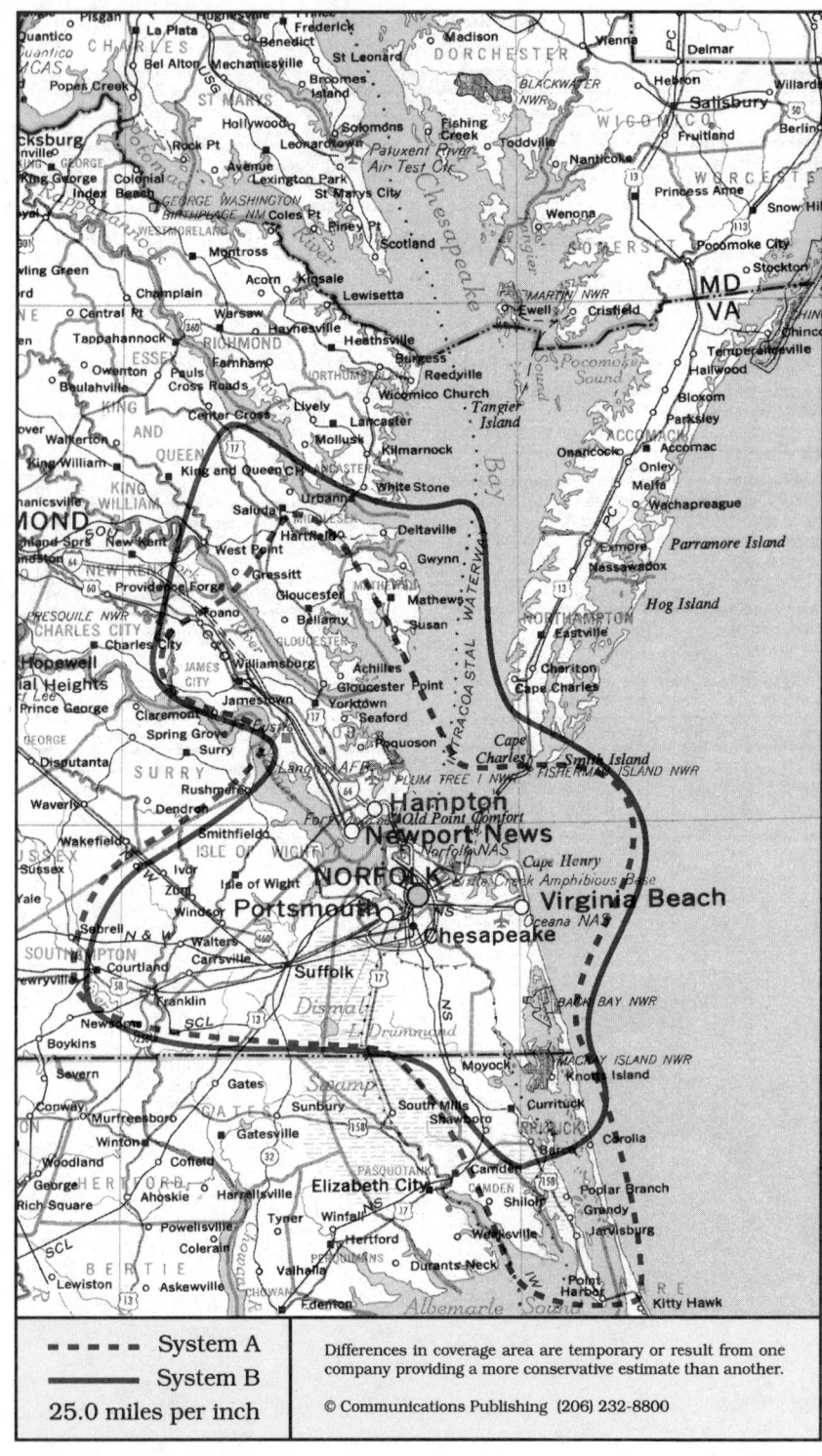

	System A
— — — —	System B
25.0 miles per inch	

Differences in coverage area are temporary or result from one company providing a more conservative estimate than another.

© Communications Publishing (206) 232-8800

Norfolk, Virginia
Deltaville, Newport News, Norfolk, Portsmouth, Virginia Beach

System A	System B

Cellular company

Centel Cellular
11751 Rocklanding Drive 11-8
Newport News, VA 23606

349 Southport Circle, Suite 101
Virginia Beach, VA 23452

Local office .	Newport News	804-873-1941
. . . . .	Virginia Beach	800-473-2355
. . . . .	Virginia Beach	804-473-9600
Corporate office	. . . . .	312-399-2644

Sending calls

Visitor dialing instructions
Local calls . . . . . area code + 7 digits
Long distance . . . 1 + area code + 7 digits

Speed-dial numbers
Customer service 611
Emergency 911

Receiving calls

Caller dials the roamer access number below,
hears tone, dials cellular area code and number.
Newport News 804-879-7626
Norfolk 804-434-7626

Billing information

Visitor rates
Most visitors . . . . . $3 day + 99¢ minute

Visitors without roaming agreements (p. 1159)
Cannot call since company takes no credit cards.

Identification numbers

City	Market	System	Billing	RSA
Newport News	104	00083	00000	MSA
Norfolk	043	00083	00000	MSA

For full coverage area, see Petersburg, VA.

Cellular company

Contel Cellular
813 Diligence Drive, Suite 119
Newport News, VA 23606

4001 Virginia Beach Boulevard, Suite 105
Virginia Beach, VA 23452-1764

Customer service	. . . .	800-333-4004
. . . .		804-552-1239
Local office . .	Newport News	804-873-3663
. . . . . .	Virginia Beach	804-498-1000
Corporate office	. . . .	404-804-3400

Sending calls

Visitor dialing instructions
Local calls . . . . . . area code + 7 digits
Long distance . . . 1 + area code + 7 digits

Speed-dial numbers
Customer service 611
Directory assistance 411
Emergency 911

Receiving calls

Caller dials the roamer access number below,
hears tone, dials cellular area code and number.
Deltaville 804-347-7626
Newport News, Norfolk . . . 804-621-7626
Greater Hampton Roads . . . 804-880-7626

Follow Me Roaming forwarding (see p. 1156)
Activate, dial ∗18 send. Deactivate dial ∗19 send.

Billing information

Visitor rates
Most visitors . . . . . $3 day + 85¢ minute

Visitors without roaming agreements (p. 1159)

Identification numbers

City	Market	System	Billing	RSA
Deltaville	692	00168	00000	VA-12
Newport News	104	00168	00000	MSA
Norfolk	043	00168	00000	MSA

For full coverage area, see Petersburg, VA.

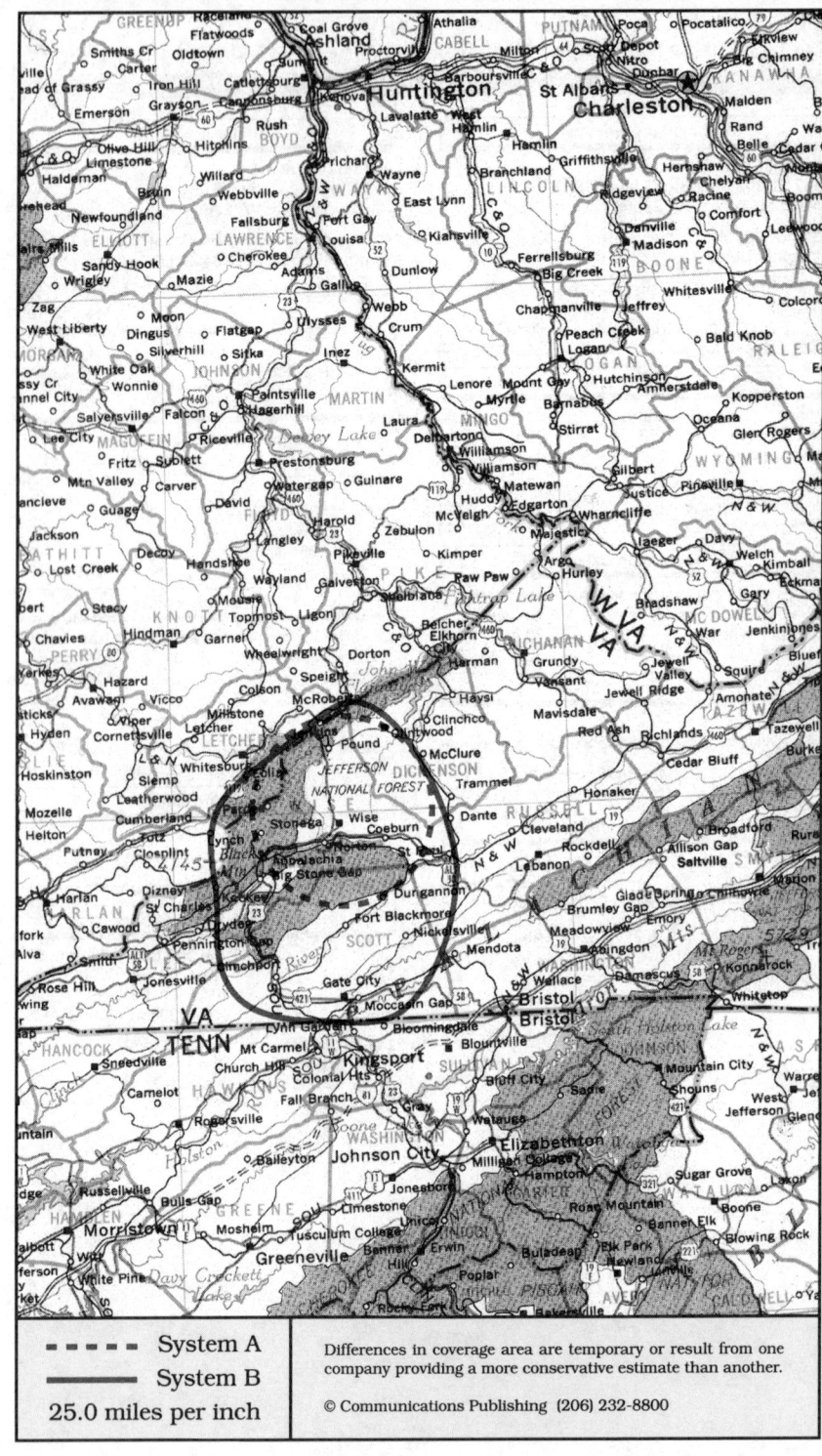

- - - - - System A	Differences in coverage area are temporary or result from one
————— System B	company providing a more conservative estimate than another.
25.0 miles per inch	© Communications Publishing (206) 232-8800

Norton, Virginia
Big Stone Gap, Gate City, Norton, Wise

System A	System B

Cellular company

Cellular One
Licensed to: SDK Enterprises
Operated by: Bell Atlantic Mobile
607 E Broad Avenue, Hwy 74; PO Box 1691
Rockingham, NC 28379

Corporate office	800-395-5845
. . . .	919-997-2323
Local sales agent in Wise . .	800-676-7085

Sending calls

Visitor dialing instructions

Local calls	area code + 7 digits
Long distance . . .	1 + area code + 7 digits

Speed-dial numbers

Customer service	611
Emergency	911
Highway patrol	∗HP or ∗47

Receiving calls

Caller dials the roamer access number below,
hears tone, dials cellular area code and number.
. . . . 704-564-7626

NationLink & RoamingAmerica (see p. 1157)
Caller hears your current roamer access number.
This company's subscribers have level 0 service.

Billing information

Visitor rates
Most visitors . . . . . $3 day + 75¢ minute

Visitors without roaming agreements (p. 1159)
Cannot call since company takes no credit cards.

Identification numbers

City	Market	System	Billing	RSA
All served cities	681	01749	00000	VA-1

This system is operated as part of the Charlotte,
NC system A.

Cellular company

Centel Cellular
664 Eastern Star Road
Kingsport, TN 37663

Local office	800-233-1985
. . . .	615-349-5000
Corporate office	312-399-2644

Sending calls

Visitor dialing instructions

Local calls	area code + 7 digits
Long distance . . .	1 + area code + 7 digits

Speed-dial numbers

Customer service	611
Emergency	911
WJHL radio	∗11

Receiving calls

Caller dials the roamer access number below,
hears tone, dials cellular area code and number.
. . . . 615-335-7626

Billing information

Visitor rates
Most visitors . . . . . $3 day + 99¢ minute

Visitors without roaming agreements (p. 1159)
Cannot call since company takes no credit cards.

Identification numbers

City	Market	System	Billing	RSA
All served cities	681	01750	00000	VA-1

For full coverage area, see Johnson City, TN.

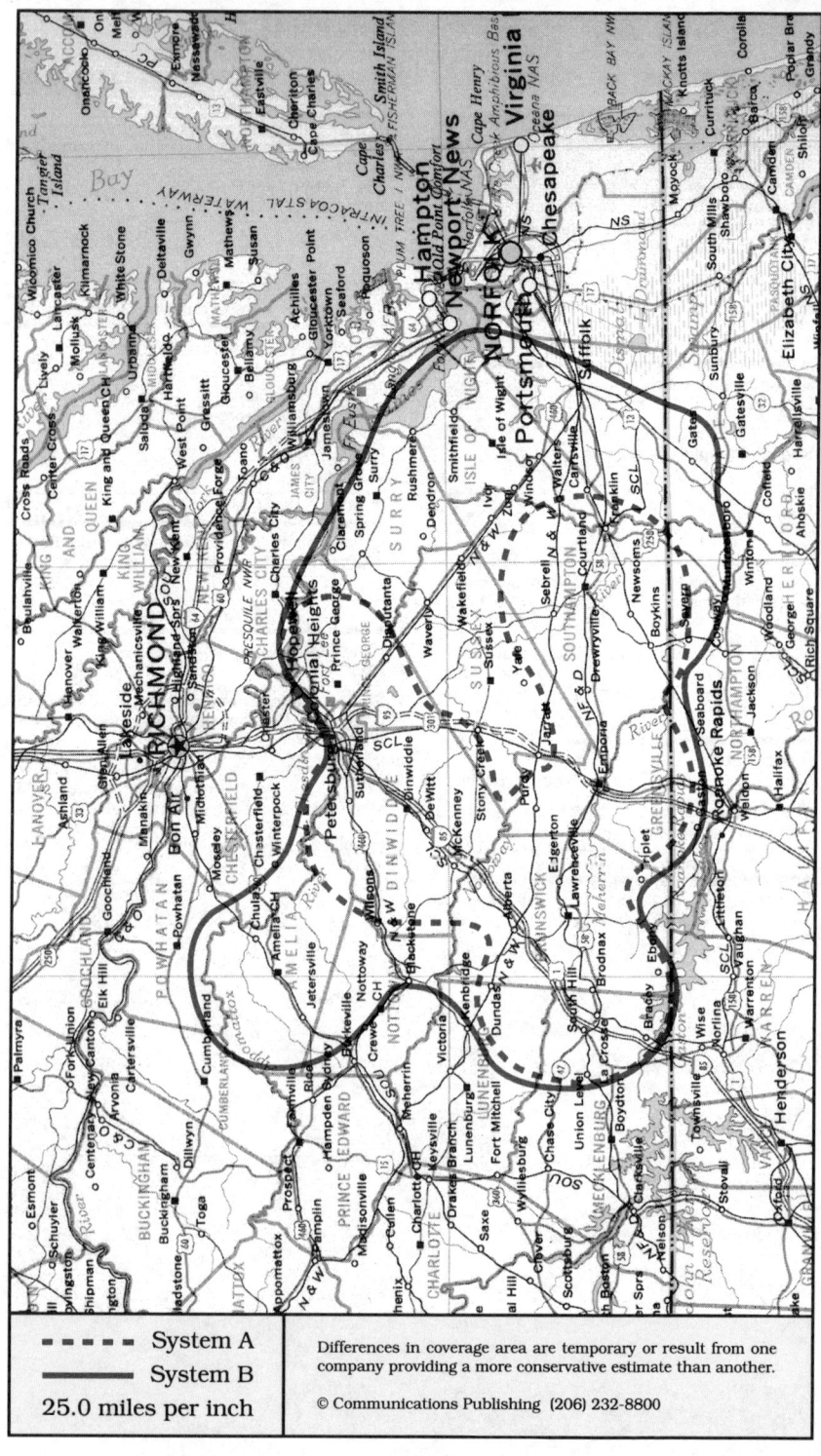

System A

System B

25.0 miles per inch

Differences in coverage area are temporary or result from one company providing a more conservative estimate than another.

© Communications Publishing (206) 232-8800

Petersburg, Virginia
Courtland, Emporia, Franklin, Lawrenceville, Petersburg, South Hill

System A	System B

Cellular company

Centel Cellular
10501 S Crater Road
Petersburg, VA 23805

Local office	800-788-1760
Corporate office	312-399-2644

Sending calls

Visitor dialing instructions

Local calls	area code + 7 digits
Long distance . . .	1 + area code + 7 digits

Speed-dial numbers

Customer service	611
Emergency	911

Receiving calls

Caller dials the roamer access number below,
hears tone, dials cellular area code and number.

. . . .	804-720-7626

Billing information

Visitor rates

Most visitors	$3 day + 99¢ minute

Visitors without roaming agreements (p. 1159)
Cannot call since company takes no credit cards.

Identification numbers

City	Market	System	Billing	RSA
Franklin	689	01765	00000	VA-9
Petersburg	235	00491	00000	MSA
South Hill	688	01763	00000	VA-8

For full coverage area, see Norfolk, VA.

Cellular company

Contel Cellular
3210 S Crater Road
Petersburg, VA 23805

Customer service	800-333-4004
. . . .	804-552-1239
Local office	804-861-6400
Corporate office	404-804-3400

Sending calls

Visitor dialing instructions

Local calls	area code + 7 digits
Long distance . . .	1 + area code + 7 digits

Speed-dial numbers

Customer service	611
Directory assistance	411
Emergency	911

Receiving calls

Caller dials the roamer access number below,
hears tone, dials cellular area code and number.

Franklin (VA-9)	804-621-7626
Petersburg, South Hill (VA-8) .	804-731-7626

Follow Me Roaming forwarding (see p. 1156)
Activate, dial *18 send. Deactivate dial *19 send.

Billing information

Visitor rates

Most visitors	$3 day + 85¢ minute

Visitors without roaming agreements (p. 1159)
Call co. to use Amex, Discover, Mastercard, Visa

Identification numbers

City	Market	System	Billing	RSA
Franklin	689	00168	00000	VA-9
Petersburg	235	00170	00000	MSA
South Hill	688	01764	00000	VA-8

For full coverage area, see Norfolk, VA.

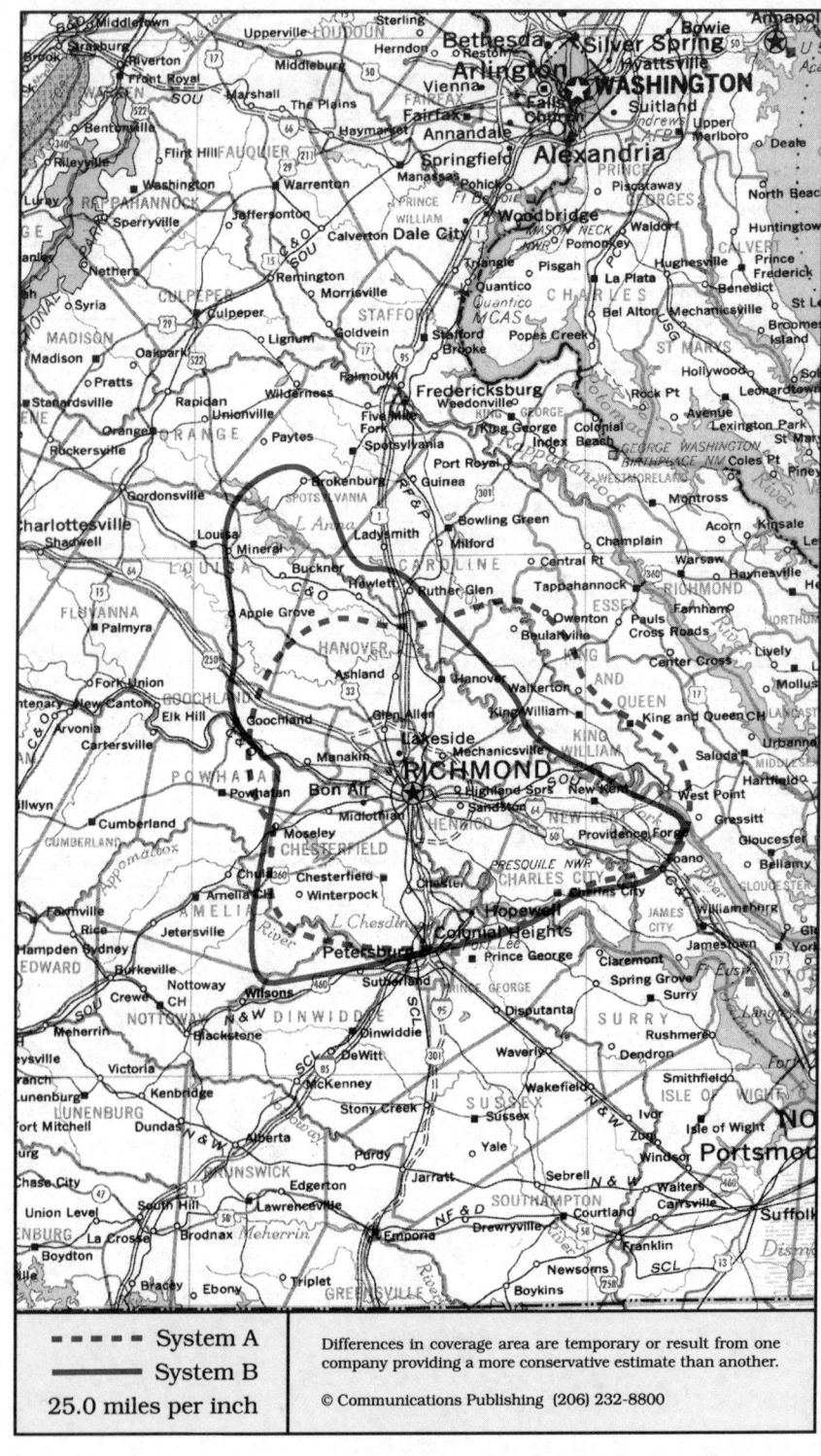

- - - - System A
——— System B
25.0 miles per inch

Differences in coverage area are temporary or result from one company providing a more conservative estimate than another.

© Communications Publishing (206) 232-8800

Richmond, Virginia

Charles City, Colonial Heights, Hanover, New Kent, Richmond

System A	System B

Cellular company

Cellular One
Richmond Cellular Telephone
9211 Arboretum Parkway, Suite 500
Richmond, VA 23236

Customer service	804-330-2456
Corporate office	804-330-7282

Sending calls

Visitor dialing instructions

Local calls	area code + 7 digits
Long distance . . .	1 + area code + 7 digits

Speed-dial numbers

Customer service	611
Directory assistance	411
Emergency	911

Receiving calls

Caller dials the roamer access number below, hears tone, dials cellular area code and number.
. . . . 804-356-7626

NationLink & RoamingAmerica (see p. 1157)
Dial *31 send to receive calls, *30 to deactivate.
Dial *32 send to tell caller the roamer access no.
This company's subscribers have level 8 service.

Billing information

Visitor rates
Most visitors . . . . . $3 day + 99¢ minute

Visitors without roaming agreements (p. 1159)
Cannot call since company takes no credit cards.

Identification numbers

City	Market	System	Billing	RSA
All served cities	059	00071	00000	MSA

Cellular company

Contel Cellular
7320 Staples Mill Road
Richmond, VA 23228

Customer service	800-333-4004
. . . .	804-552-1239
Local office	804-266-6400
Corporate office	404-804-3400

Sending calls

Visitor dialing instructions

Local calls	area code + 7 digits
Long distance . . .	1 + area code + 7 digits

Speed-dial numbers

Customer service	611
Emergency	911

Receiving calls

Caller dials the roamer access number below, hears tone, dials cellular area code and number.
. . . . 804-347-7626

Follow Me Roaming forwarding (see p. 1156)
Activate, dial *18 send. Deactivate dial *19 send.

Billing information

Visitor rates
Most visitors . . . . . $3 day + 85¢ minute

Visitors without roaming agreements (p. 1159)
Call co. to use Amex, Mastercard, Visa

Identification numbers

City	Market	System	Billing	RSA
All served cities	059	00170	00000	MSA

For full coverage area, see Norfolk, VA and Petersburg, VA.

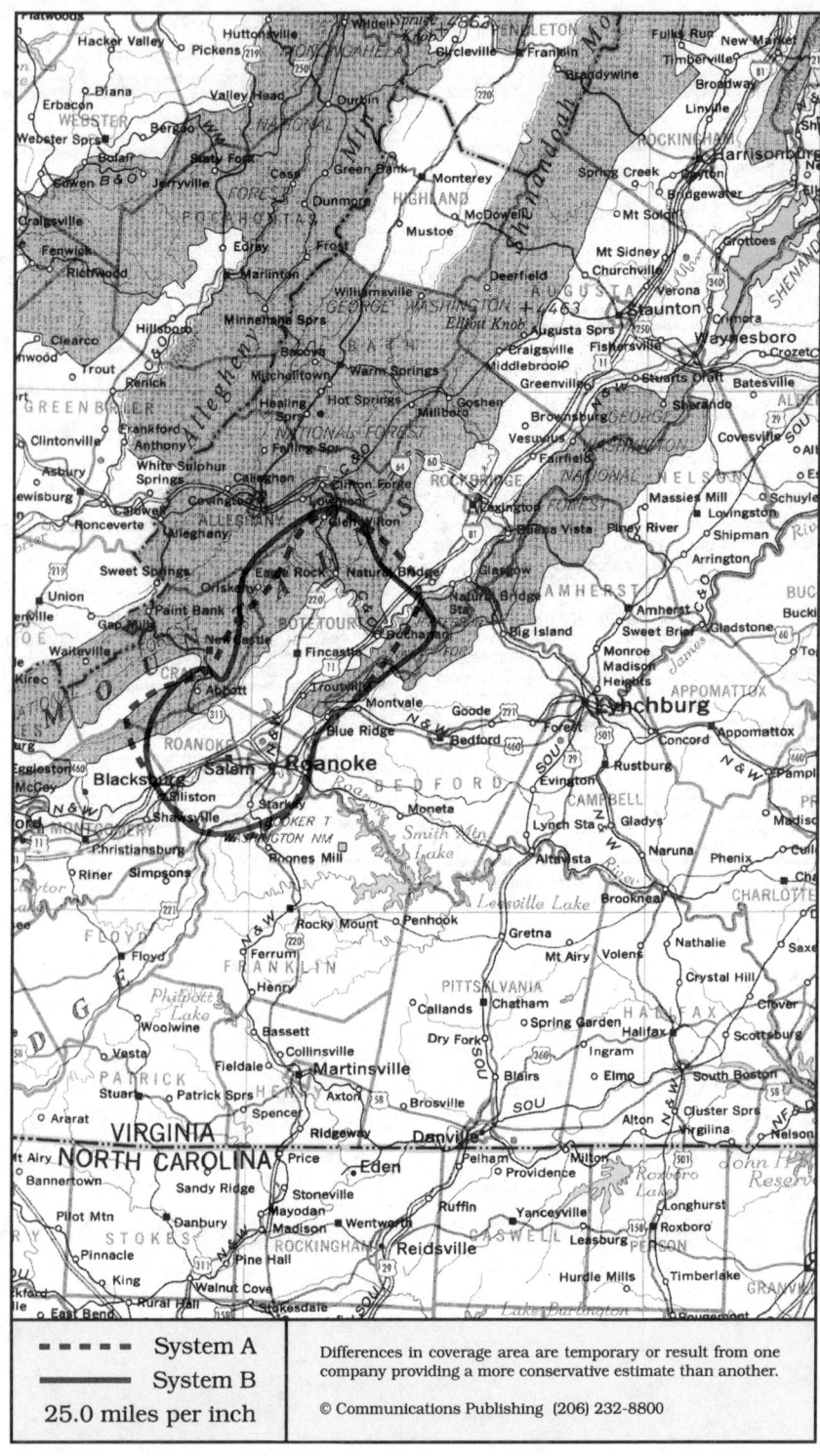

▪ ▪ ▪ ▪ ▪ System A	Differences in coverage area are temporary or result from one
──────── System B	company providing a more conservative estimate than another.
25.0 miles per inch	© Communications Publishing (206) 232-8800

Roanoke, Virginia
Fincastle, Roanoke, Salem

System A

Cellular company

Cellular One
Centennial Cellular
5228 Valleypointe Pkwy, Suite 2; PO Box 7331
Roanoke, VA 24019

Customer service	800-763-0138
Local office	703-345-0808
Corporate office Administration	703-345-0808
. . . . . . Management	203-972-2000

Sending calls

Visitor dialing instructions

Local calls	area code + 7 digits
Long distance . . .	1 + area code + 7 digits

Speed-dial numbers

Customer service	611
Emergency	911
Make a traffic report	960

Receiving calls

Caller dials the roamer access number below,
hears tone, dials cellular area code and number.
. . . .   703-580-7626

NationLink & RoamingAmerica (see p. 1157)
Caller hears your current roamer access number.
This company's subscribers have level 0 service.

Billing information

Visitor rates
Most visitors . . . . . $3 day + 85¢ minute

Visitors without roaming agreements (p. 1159)
Roamer Plus will intercept first call to bill you.

Identification numbers

City	Market	System	Billing	RSA
All served cities	157	00519	00000	MSA

System B

Cellular company

Contel Cellular
1502 Williamson Road
Roanoke, VA 24012

Customer service	800-333-4004
. . . .	804-552-1239
Local office	703-345-1093
Corporate office	404-804-3400

Sending calls

Visitor dialing instructions

Local calls	area code + 7 digits
Long distance . . .	1 + area code + 7 digits

Speed-dial numbers

Customer service	611
Emergency	911

Receiving calls

Caller dials the roamer access number below,
hears tone, dials cellular area code and number.
. . . .   703-520-7626

Follow Me Roaming forwarding (see p. 1156)
Activate, dial *18 send. Deactivate dial *19 send.

Billing information

Visitor rates
Most visitors . . . . . $3 day + 85¢ minute

Visitors without roaming agreements (p. 1159)
Call co. to use Amex, Mastercard, Visa

Identification numbers

City	Market	System	Billing	RSA
All served cities	157	00502	00000	MSA

For full coverage area, see Blacksburg, VA and
Lexington, VA.

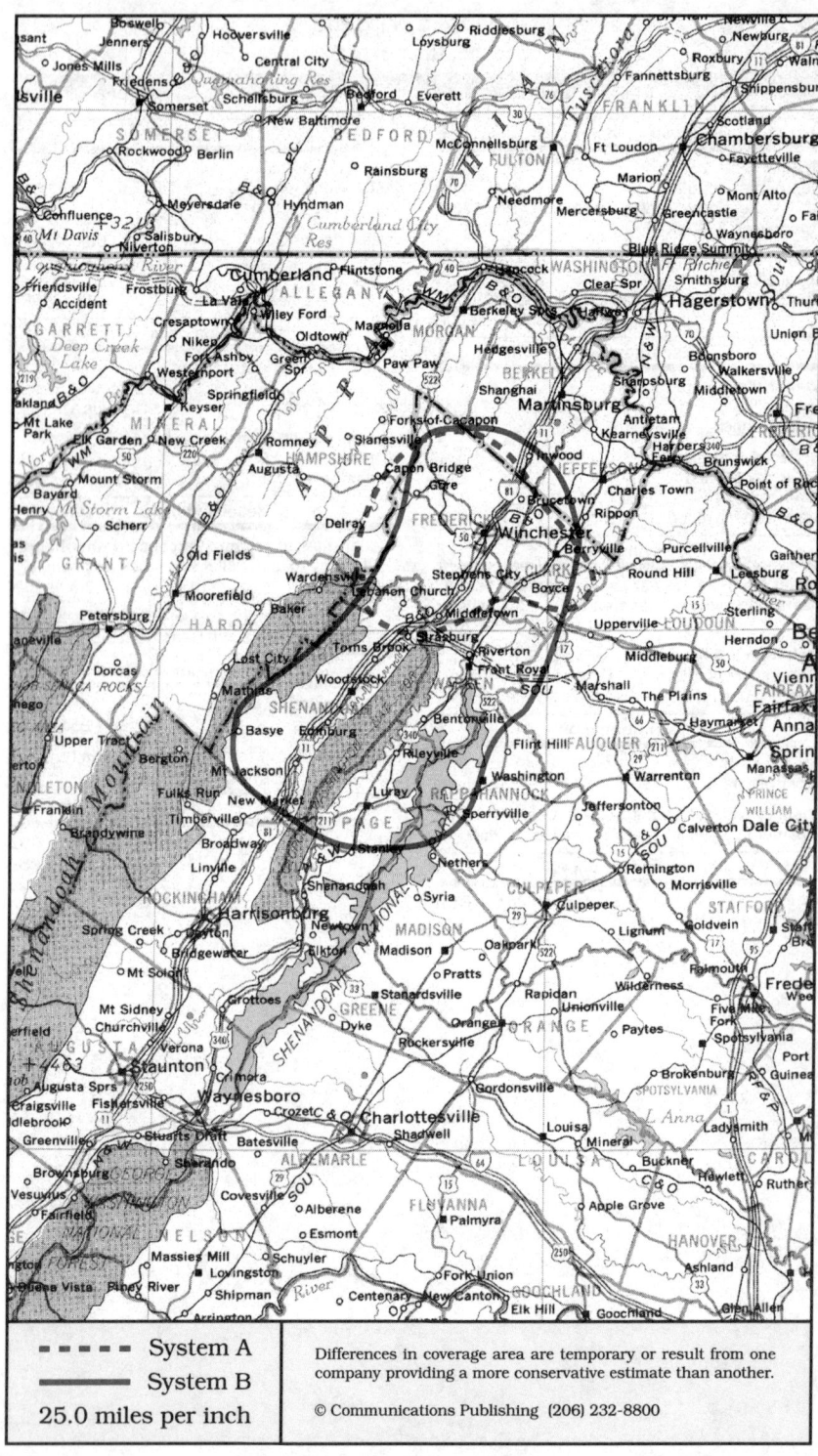

Winchester, Virginia
Front Royal, Luray, Winchester, Woodstock

System A	System B

Cellular company

Cellular One
Southwestern Bell Mobile Systems
7855 Walker Drive, Suite 100
Greenbelt, MD 20770

Customer service	some areas	800-235-5663
. . . .		301-220-0060
Corporate office	. . . . .	301-220-3600

Sending calls

Visitor dialing instructions
Local calls	area code + 7 digits
Long distance	area code + 7 digits

Speed-dial numbers
Cellular information	*1394
Customer service	*611 or *811
Emergency	911
24 hour news	*1887

Receiving calls

Caller dials the roamer access number below,
hears tone, dials cellular area code and number.
. . . . 703-678-3333

NationLink & RoamingAmerica (see p. 1157)
Dial *31 send to receive calls, *30 to deactivate.
Dial *32 send to tell caller the roamer access no.
Service is not automatic and must be turned on.
This company's subscribers have level 6 service.

Billing information

Visitor rates
Most visitors . . . . . $3 day + 95¢ minute

Visitors without roaming agreements (p. 1159)
Roamer Plus will intercept first call to bill you.

Identification numbers

City	Market	System	Billing	RSA
All served cities	690	00013	30341	VA-10

For full coverage area, see Baltimore, MD.

Cellular company

Shenandoah Cellular
Administration
124 S Main Street, PO Box 459
Edinburg, VA 22824

2604 Valley Avenue
Winchester, VA 22601

Local office		800-388-4355
. . . .		703-722-5003
Corporate office . .	Edinburg	703-984-5321

Sending calls

Visitor dialing instructions
Local calls	area code + 7 digits
Long distance . . .	1 + area code + 7 digits

Speed-dial numbers
Customer service	611
Emergency	911

Receiving calls

Caller dials the roamer access number below,
hears tone, dials cellular area code and number.
Edinburg, Luray	703-333-7626
Front Royal, Winchester . . .	703-336-7626

Follow Me Roaming forwarding (see p. 1156)
Activate, dial *18 send. Deactivate dial *19 send.

Billing information

Visitor rates
Most visitors . . . . . $3 day + $1 minute

Visitors without roaming agreements (p. 1159)
Call co. to use Amex, Mastercard, Visa
or Roamer Plus will intercept first call to bill you.

Identification numbers

City	Market	System	Billing	RSA
All served cities	690	01912	00000	VA-10

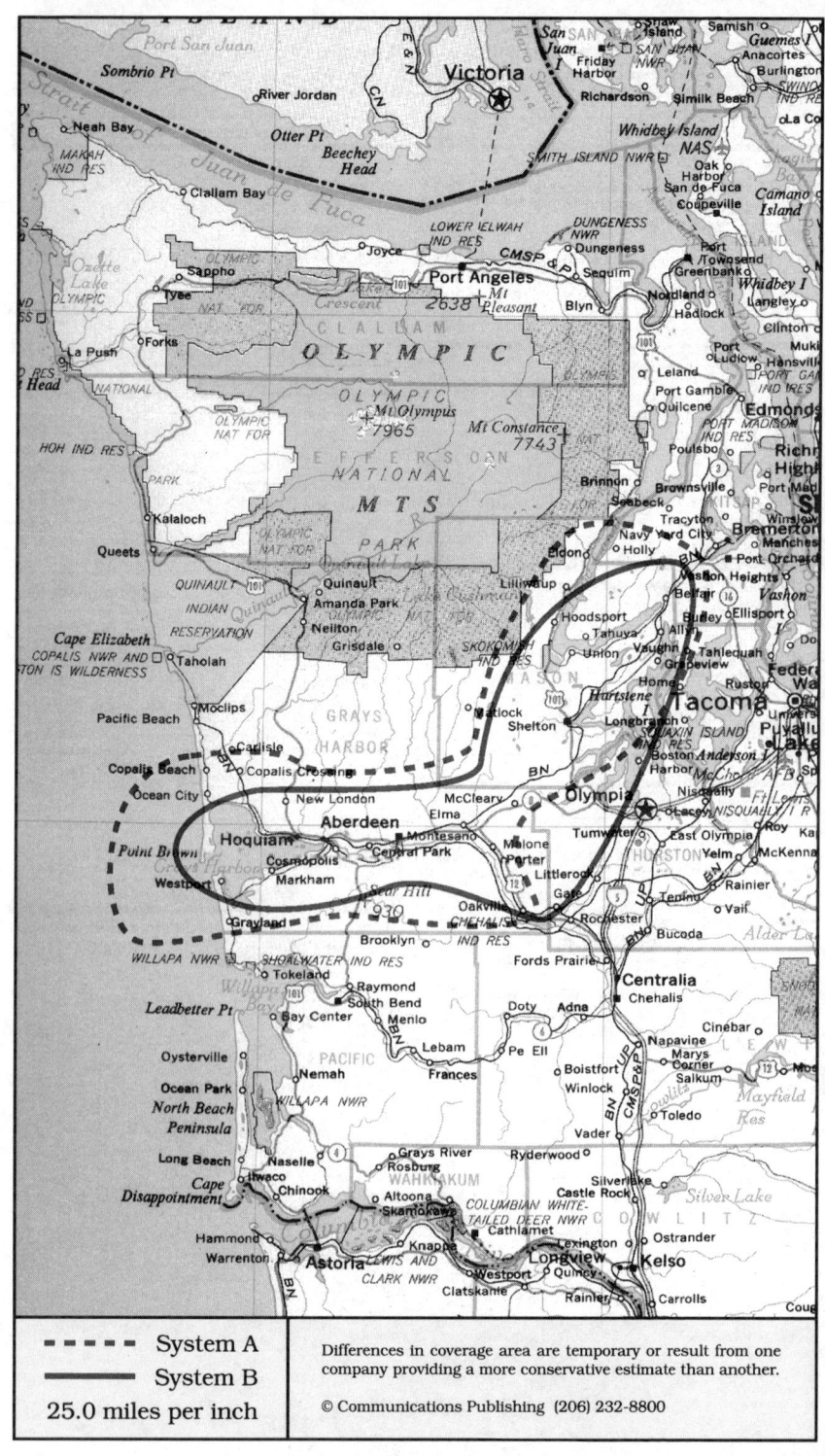

Legend	
- - - -	System A
———	System B
25.0 miles per inch	

Differences in coverage area are temporary or result from one company providing a more conservative estimate than another.

© Communications Publishing (206) 232-8800

Aberdeen, Washington
Aberdeen, Hoquiam, Montesano, Shelton

System A	System B

Cellular company

United States Cellular
2730 Simpson Avenue
Hoquiam, WA 98550-2931

Customer service	800-400-6850
Local office	206-749-7000
Corporate office	312-399-8900

Sending calls

Visitor dialing instructions

Local calls	area code + 7 digits
Long distance	area code + 7 digits

Speed-dial numbers

Customer service	611
Emergency	911
Roaming information	711

Receiving calls

Caller dials the roamer access number below, hears tone, dials cellular area code and number. Use either number since they are interconnected.

Aberdeen	206-590-7626
Elma	206-482-7626

NationLink & RoamingAmerica (see p. 1157)
Dial *31 send to receive calls, *30 to deactivate.
Dial *32 send to tell caller the roamer access no.
This company's subscribers have level 6 service.

Billing information

Visitor rates

Most visitors	$3.50 day + 85¢ minute

Visitors without roaming agreements (p. 1159)
Cannot call since company takes no credit cards.

Identification numbers

City	Market	System	Billing	RSA
All served cities	696	01779	00000	WA-4

Cellular company

U S West Cellular
4706 20th Street E, Suite A
Tacoma, WA 98424

Customer service	800-238-7848
. . . .	206-562-2895
. . . . . local subscribers	800-626-6611
. . . .	206-747-1771
Local office	206-922-0775
Corporate office	206-747-4900

Sending calls

Visitor dialing instructions

Local calls	area code + 7 digits
Long distance	area code + 7 digits

Speed-dial numbers

Customer service	*611
Emergency	911
Roaming information	*711

Receiving calls

Caller dials the roamer access number below, hears tone, dials cellular area code and number.

Aberdeen, Hoquiam	206-580-7626
Shelton	206-538-7626

Billing information

Visitor rates

Most visitors	$3 day + 99¢ minute

Visitors without roaming agreements (p. 1159)
Call co. to use Amex, Mastercard, Visa

Identification numbers

City	Market	System	Billing	RSA
All served cities	696	01780	00000	WA-4

For full coverage area, see Seattle, WA.

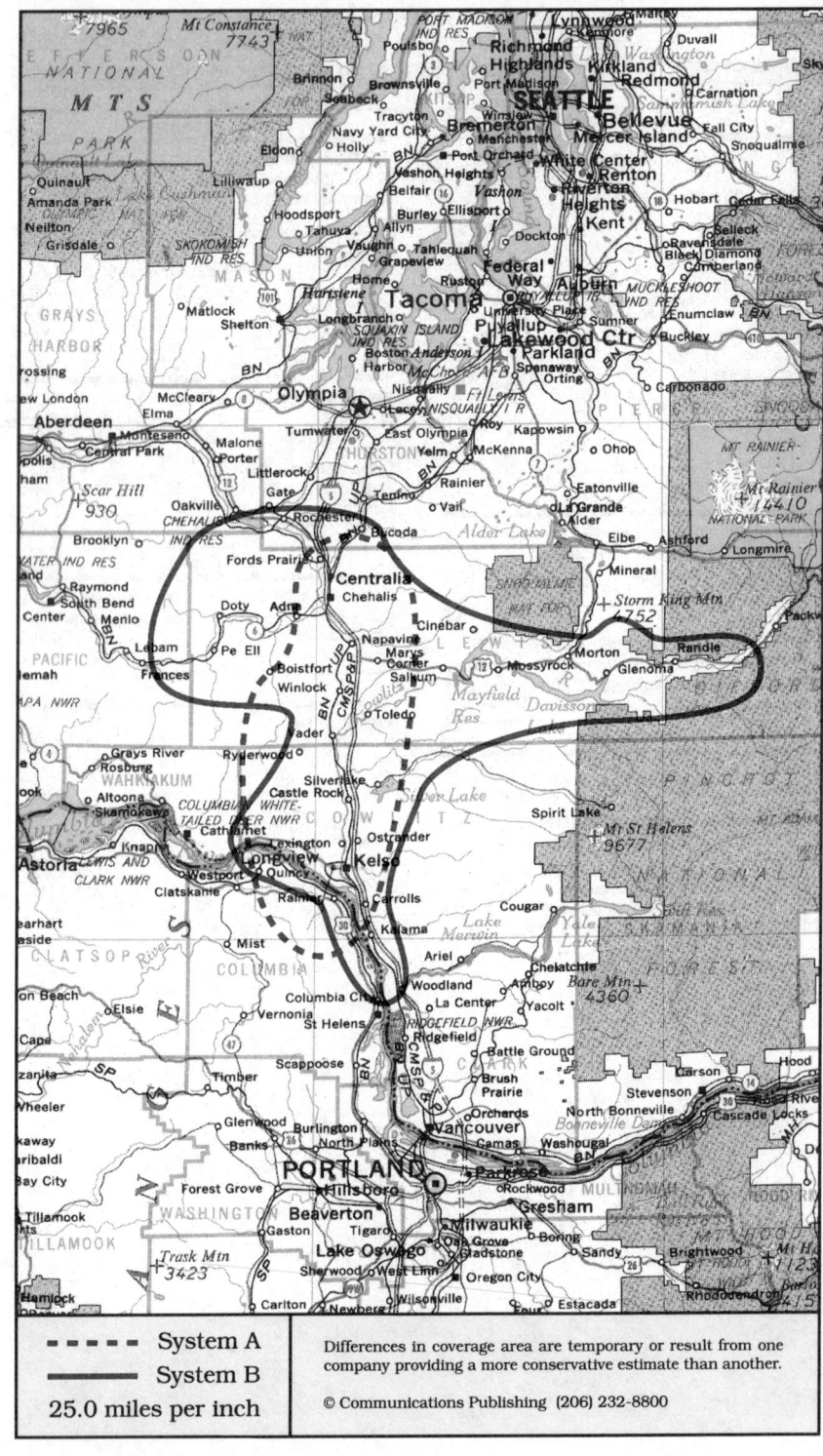

▪ ▪ ▪ ▪ ▪ System A	Differences in coverage area are temporary or result from one
━━━━━ System B	company providing a more conservative estimate than another.
25.0 miles per inch	© Communications Publishing (206) 232-8800

Longview, Washington
Centralia, Chehalis, Kelso, Longview, Morton

System A	System B

System A

Cellular company

Cellular One
McCaw Cellular Communications
3077 20th Street E, Suite B
Fife, WA 98424

1600 SW 4th Avenue
Portland, OR 97201

Customer service for the north	800-782-3551
. . . .	206-279-0500
Customer service for the south	800-245-7626
. . . .	206-423-5555
Local office . . . . Fife	206-922-4700
. . . . . . Portland	503-243-3333
Corporate office	206-827-4500

Sending calls

Visitor dialing instructions
Local calls	area code + 7 digits
Long distance . . .	1 + area code + 7 digits

Speed-dial numbers
Customer service	611
Directory assistance	411
Emergency	911
Report traffic problems	311
Washington state patrol	211

Receiving calls

Caller dials the roamer access number below,
hears tone, dials cellular area code and number.
Centralia, Chehalis	206-972-7626
Kelso, Longview	206-430-7626

North American Cellular Network (see p. 1156)
All calls forward to you. Deactivate dial *35 send.

Billing information

Visitor rates
Most visitors	$0 day + 99¢ minute

Visitors without roaming agreements (p. 1159)
Cannot call since company takes no credit cards.

Identification numbers

City	Market	System	Billing	RSA
Centralia, Chehalis	698	01783	00000	WA-6
Kelso, Longview	698	01783	30243	WA-6

For full coverage area, see Seattle, WA.

System B

Cellular company

United States Cellular
1318 Washington Way
Longview, WA 98632

Local office	800-400-6850
. . . .	206-423-9000
. . . . . from Chehalis	206-520-0000
Corporate office	312-399-8900

Sending calls

Visitor dialing instructions
Local calls	area code + 7 digits
Long distance	area code + 7 digits

Speed-dial numbers
Customer service	611
Emergency	911
Roaming information	711

Receiving calls

Caller dials the roamer access number below,
hears tone, dials cellular area code and number.
. . . .	206-749-7626

Follow Me Roaming forwarding (see p. 1156)
Activate, dial *18 send. Deactivate dial *19 send.

Billing information

Visitor rates
Most visitors . . .	$3.50 day + 85¢ minute

Visitors without roaming agreements (p. 1159)
Cannot call since company takes no credit cards.

Identification numbers

City	Market	System	Billing	RSA
All served cities	698	01784	00000	WA-6

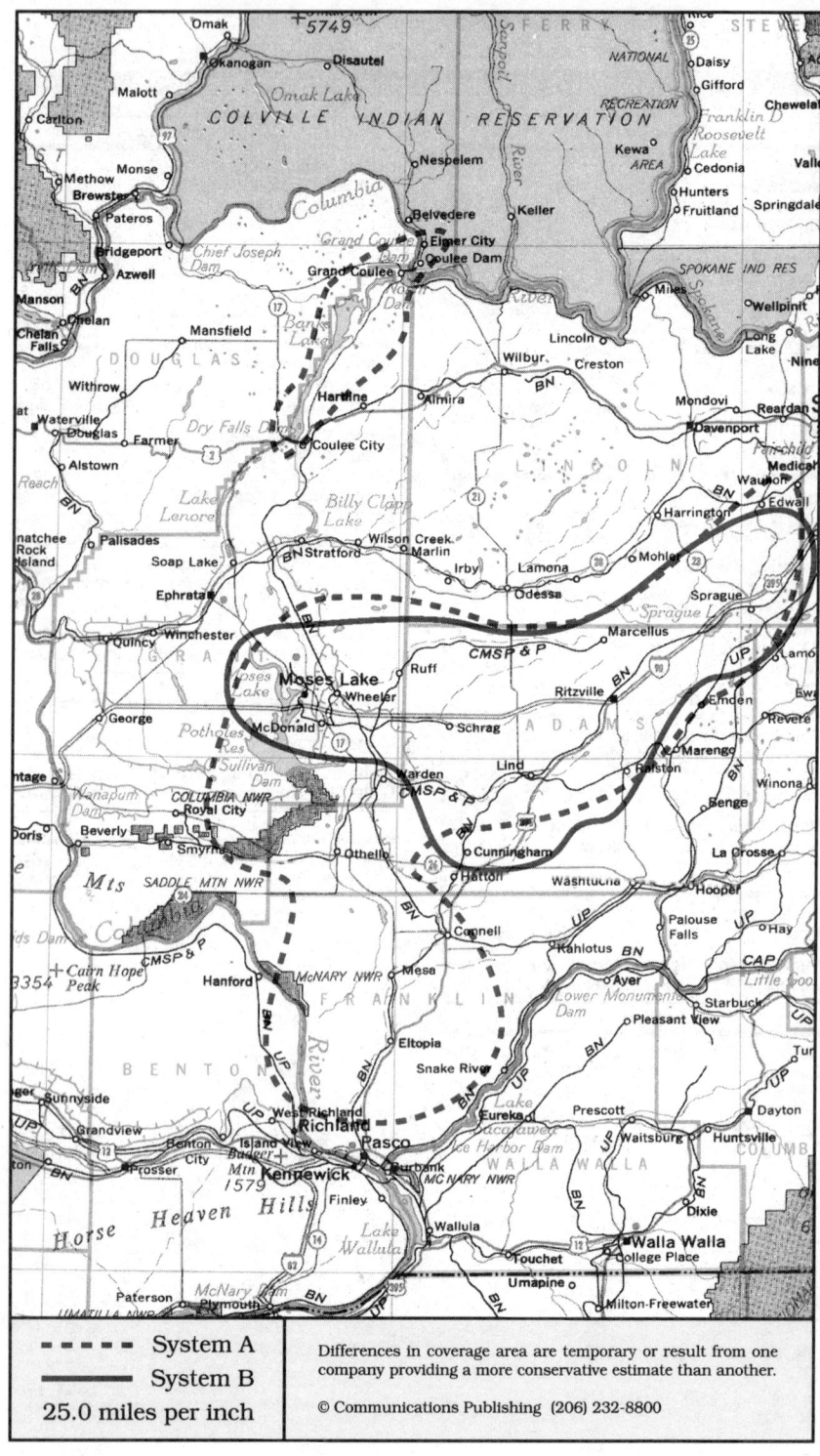

Legend	
- - - - -	System A
————	System B
25.0 miles per inch	

Differences in coverage area are temporary or result from one company providing a more conservative estimate than another.

© Communications Publishing (206) 232-8800

Moses Lake, Washington
Grand Coulee, Moses Lake, Ritzville

System A	System B

System A

Cellular company

Cellular One
McCaw Cellular Communications
308 S Balsam
Moses Lake, WA 98837

Customer service	. . . . .	800-327-6111
Local office	. . . . . .	509-765-9280
Corporate office	. . . . .	206-827-4500

Sending calls

Visitor dialing instructions

Local calls	area code + 7 digits
Long distance . . .	1 + area code + 7 digits

Speed-dial numbers

Customer service	611
Emergency	911

Receiving calls

Caller dials the roamer access number below,
hears tone, dials cellular area code and number.
. . . . 509-750-7626

North American Cellular Network (see p. 1156)
All calls forward to you. Deactivate dial *35 send.

Billing information

Visitor rates
Most visitors . . . . . $0 day + 99¢ minute

Visitors without roaming agreements (p. 1159)
Call co. to use Amex, Mastercard, Visa

Identification numbers

City	Market	System	Billing	RSA
All served cities	697	00601	30231	WA-5

For full coverage area, see Yakima, WA.

System B

Cellular company

Inland Cellular Telephone
<u>Administration</u>
9 S 2nd, PO Box 688
Roslyn, WA 98941

<u>Sales</u>
S 213 Montgomery, PO Box 221
Uniontown, WA 99179

Local office	. . .	Uniontown	800-346-1968
	. . . .	Uniontown	509-649-2211
Corporate office	.	some areas	800-346-1968
	. . . .		509-649-2500

Sending calls

Visitor dialing instructions

Local calls	area code + 7 digits
Long distance . . .	1 + area code + 7 digits

Speed-dial numbers

Customer service	*611
Emergency	911

Receiving calls

Caller dials the roamer access number below,
hears tone, dials cellular area code and number.
. . . . 509-770-7626

Billing information

Visitor rates
Most visitors . . . . . $3 day + 99¢ minute

Visitors without roaming agreements (p. 1159)
Cannot call since company takes no credit cards.

Identification numbers

City	Market	System	Billing	RSA
All served cities	697	02030	00000	WA-5

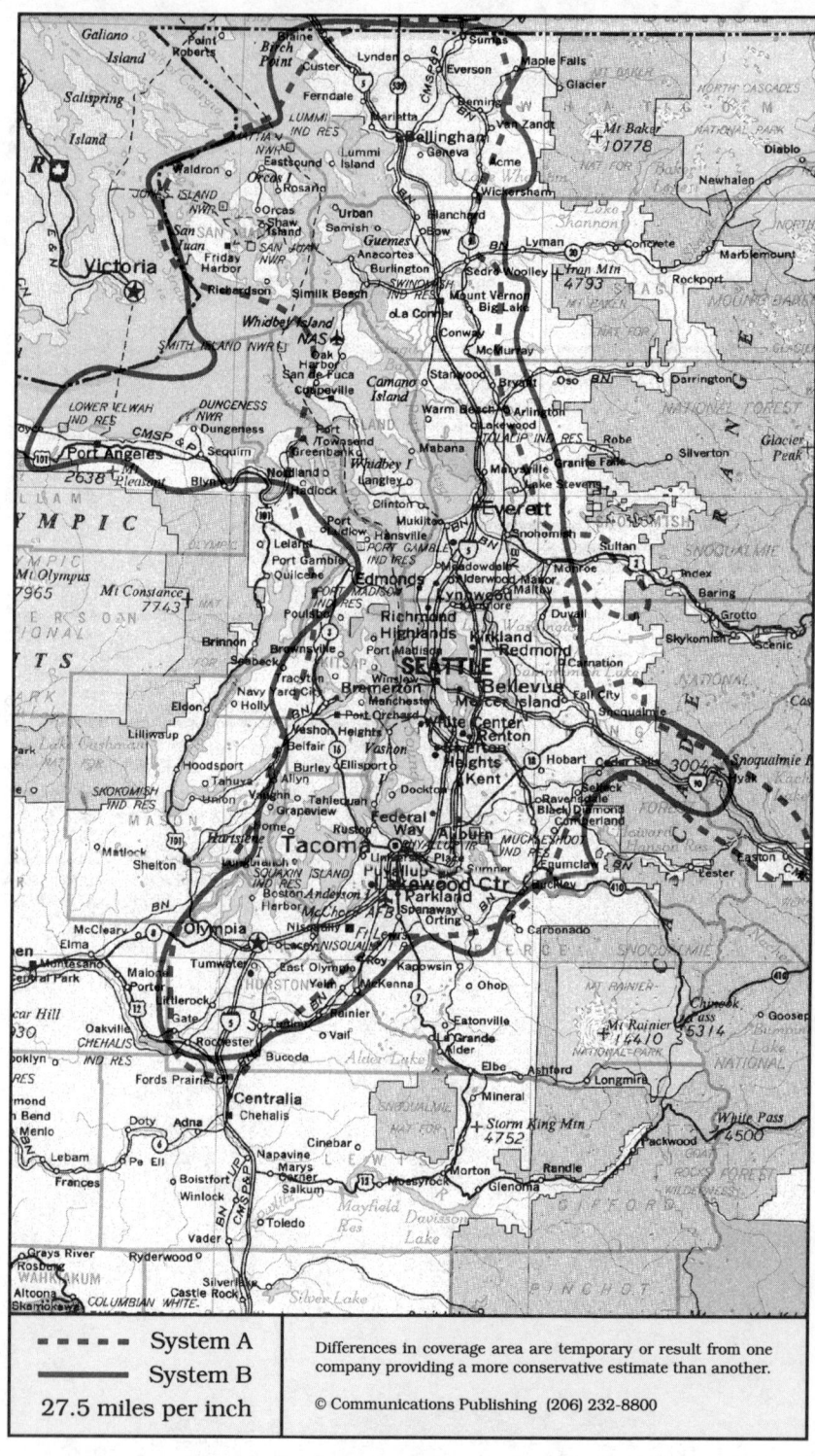

- - - - System A	Differences in coverage area are temporary or result from one
——— System B	company providing a more conservative estimate than another.
27.5 miles per inch	© Communications Publishing (206) 232-8800

Seattle, Washington

Bellingham, Bremerton, Everett, Olympia, San Juan Islands, Seattle, Tacoma

System A	System B

Cellular company

Cellular One
McCaw Cellular Communications
13221 NE 20th Street   Local office
Bellevue, WA 98005   206-644-2424

1337 Lincoln Street, Suite 2
Bellingham, WA 98226   206-734-3320

3077 20th Street E, Suite B
Fife, WA 98424   206-922-4700

617 Eastlake Avenue E
Seattle, WA 98109   206-624-5700

Customer service	800-782-3551
. . . . . from Seattle	206-389-2273
. . . . . from Tacoma	206-279-0500
Corporate office	206-827-4500

Sending calls

Visitor dialing instructions
Local calls . . . . area code + 7 digits
Long distance . . . 1 + area code + 7 digits

Speed-dial numbers
Channel 5 news	*TV-5 or *885
Coast Guard in Seattle	*CG or *24
Customer service	611
Directory assistance	411
Emergency	911
Report traffic problems	311
Washington state patrol	211

Receiving calls

Caller dials the roamer access number below,
hears tone, dials cellular area code and number.
. . . . 206-972-7626

NationLink & RoamingAmerica (see p. 1157)
Dial *31 send to receive calls, *30 to deactivate.
Dial *32 send to tell caller the roamer access no.
This company's subscribers have level 6 service.

North American Cellular Network (see p. 1156)
All calls forward to you. Deactivate dial *35 send.

Billing information

Visitor rates
Most visitors . . . . . $0 day + 99¢ minute

Visitors without roaming agreements (p. 1159)
Cannot call since company takes no credit cards.

Identification numbers

City	Market	System	Billing	RSA
Bellingham	270	00047	00000	MSA
Bremerton	212	00047	00000	MSA
Mt Vernon,San Juan	693	00047	00000	WA-1
Olympia	242	00047	00000	MSA
Seattle	020	00047	00000	MSA
Tacoma	082	00047	00000	MSA

For full coverage area, see Longview, WA.

Cellular company

U S West Cellular
11040 Main Street, Suite 100   Local office
Bellevue, WA 98004   206-455-2656

1329 King Street
Bellingham, WA 98226   206-733-2159

717 128th SW, Suite A-105
Everett, WA 98204   206-356-2656

4706 20th Street E, Suite A
Tacoma, WA 98424   206-922-0775

Customer service	800-238-7848
. . . .	206-562-2895
. . . . . local subscribers	800-626-6611
. . . .	206-747-1771
Corporate office	206-747-4900

Sending calls

Visitor dialing instructions
Local calls . . . . . area code + 7 digits
Long distance . . . . . area code + 7 digits

Speed-dial numbers
Customer service	*611
Emergency	911
Roaming information	*711

Receiving calls

Caller dials the roamer access number below,
hears tone, dials cellular area code and number.
Anacortes	206-929-7626
Bellingham	206-739-7626
Bremerton	206-731-7626
Friday Harbor (San Juan Island)	206-378-7626
Medina	206-236-1472
Olympia	206-791-7626
Seattle, Tacoma	206-947-7626

Follow Me Roaming forwarding (see p. 1156)
Activate, dial *18 send. Deactivate dial *19 send.

Billing information

Visitor rates
Most visitors . . . . . $3 day + 99¢ minute

Visitors without roaming agreements (p. 1159)
Call co. to use Amex, Mastercard, Visa

Identification numbers

City	Market	System	Billing	RSA
Bellingham	270	00258	00000	MSA
Bremerton	212	00276	00000	MSA
Mt Vernon,San Juan	693	01774	00000	WA-1
Olympia	242	00456	00000	MSA
Seattle	020	00006	00000	MSA
Tacoma	082	00006	00000	MSA

For full coverage area, see Aberdeen, WA.

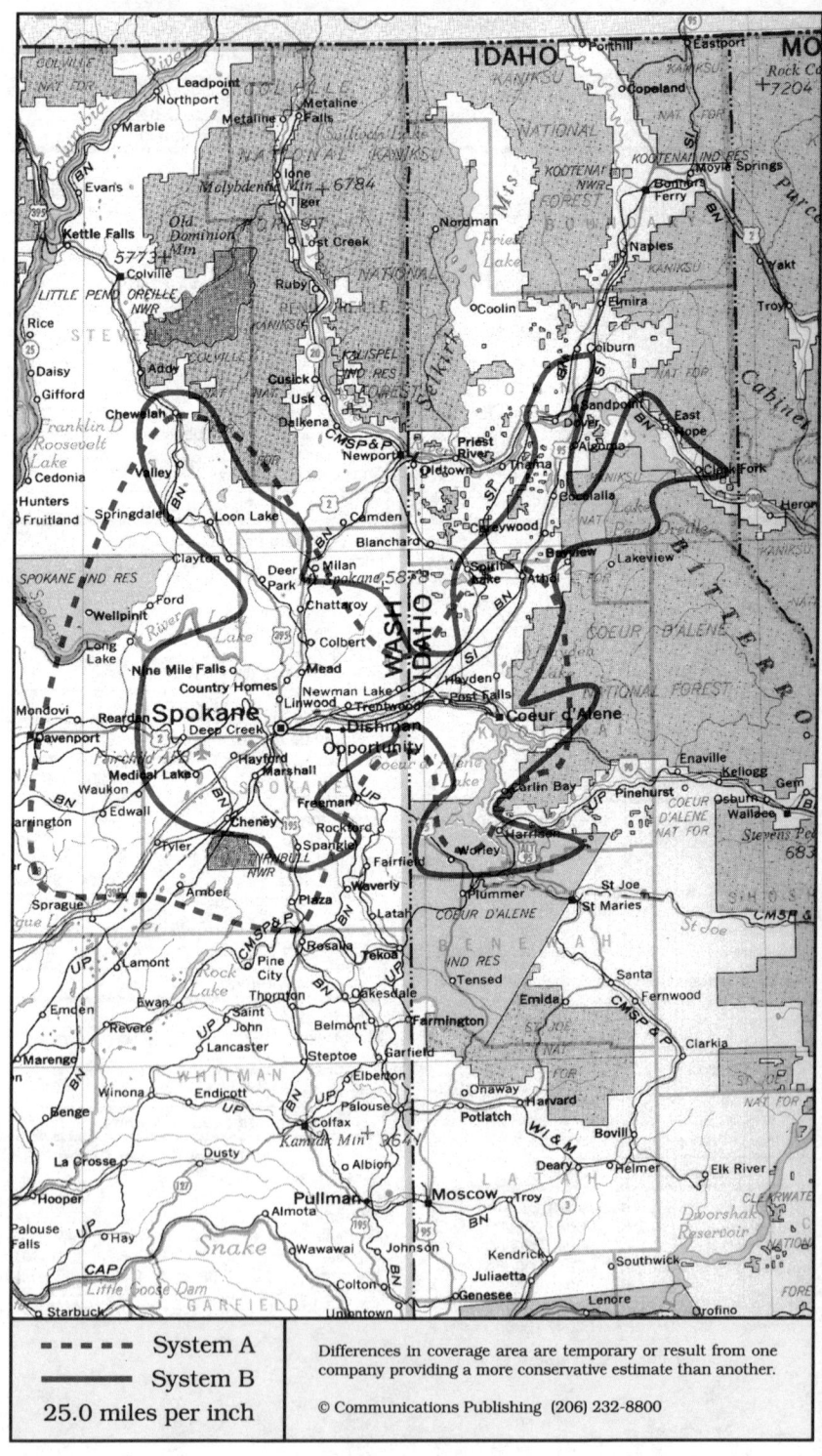

System A
System B
25.0 miles per inch

Differences in coverage area are temporary or result from one company providing a more conservative estimate than another.

© Communications Publishing (206) 232-8800

Spokane, Washington
Chewelah, Coeur d'Alene ID, Sandpoint ID, Spokane

System A	System B

Cellular company

Idaho
Cellular One
North American Cellular
W 610 Hubbard Street, Suite 131
Coeur d'Alene, ID 83814

Corporate office 208-765-3233

Washington
Cellular One
McCaw Cellular Communications
802 N Washington
Spokane, WA 99201

Customer service 800-327-6111
. . . . 509-838-2273
Local office 509-324-1000
Corporate office 206-827-4500

Sending calls

Visitor dialing instructions
Local calls . . . . . area code + 7 digits
Long distance . . . 1 + area code + 7 digits

Speed-dial numbers
Customer service 611
Emergency 911
Highway patrol in Idaho *477
KXLY news traffic *920
Q-6 television 600
Spokane sales office 811

Receiving calls

Caller dials the roamer access number below,
hears tone, dials cellular area code and number.
Coeur d'Alene 208-769-3900
Spokane 509-994-7626

NationLink & RoamingAmerica (see p. 1157)
Dial *31 send to receive calls, *30 to deactivate.
Dial *32 send to tell caller the roamer access no.
This company's subscribers have level 6 service.

North American Cellular Network (see p. 1156)
All calls forward to you. Deactivate dial *35 send.

Billing information

Visitor rates
Visitors in Idaho . . $3 day + 75¢ minute
Visitors in Washington . $0 day + 99¢ minute

Visitors without roaming agreements (p. 1159)
Call co. to use Mastercard, Visa

Identification numbers

City	Market	System	Billing	RSA
Coeur d'Alene	388	01163	00000	ID-1
Spokane	109	00231	00000	MSA

Call switching in Coeur d'Alene is performed by
McCaw Cellular for the licensee, North American
Cellular.

Cellular company

U S West Cellular
2005 Ironwood Parkway, Suite 138
Coeur d'Alene, ID 83814

1300 N Mullan Road
Spokane, WA 99206

Customer service 800-238-7848
. . . . 206-562-2895
. . . . . . local subscribers 800-626-6611
. . . . 206-747-1771
Local office . . Coeur d'Alene 208-664-4729
. . . . Spokane 509-922-2005
Corporate office 206-747-4900

Sending calls

Visitor dialing instructions
Local calls . . . . . . area code + 7 digits
Long distance . . . . . area code + 7 digits

Speed-dial numbers
Customer service *611
Emergency 911
Roaming information *711
Traffic watch in Spokane . . . . . . #920

Receiving calls

Caller dials the roamer access number below,
hears tone, dials cellular area code and number.
Chewelah 509-936-7626
Coeur d'Alene 208-661-7626
Spokane 509-993-7626

Follow Me Roaming forwarding (see p. 1156)
Activate, dial *18 send. Deactivate dial *19 send.

Billing information

Visitor rates
Most visitors . . . . . $3 day + 99¢ minute

Visitors without roaming agreements (p. 1159)
Call co. to use Amex, Mastercard, Visa

Identification numbers

City	Market	System	Billing	RSA
Chewelah	695	01778	00000	WA-3
Coeur d'Alene	388	01164	00000	ID-1
Spokane	109	00222	00000	MSA

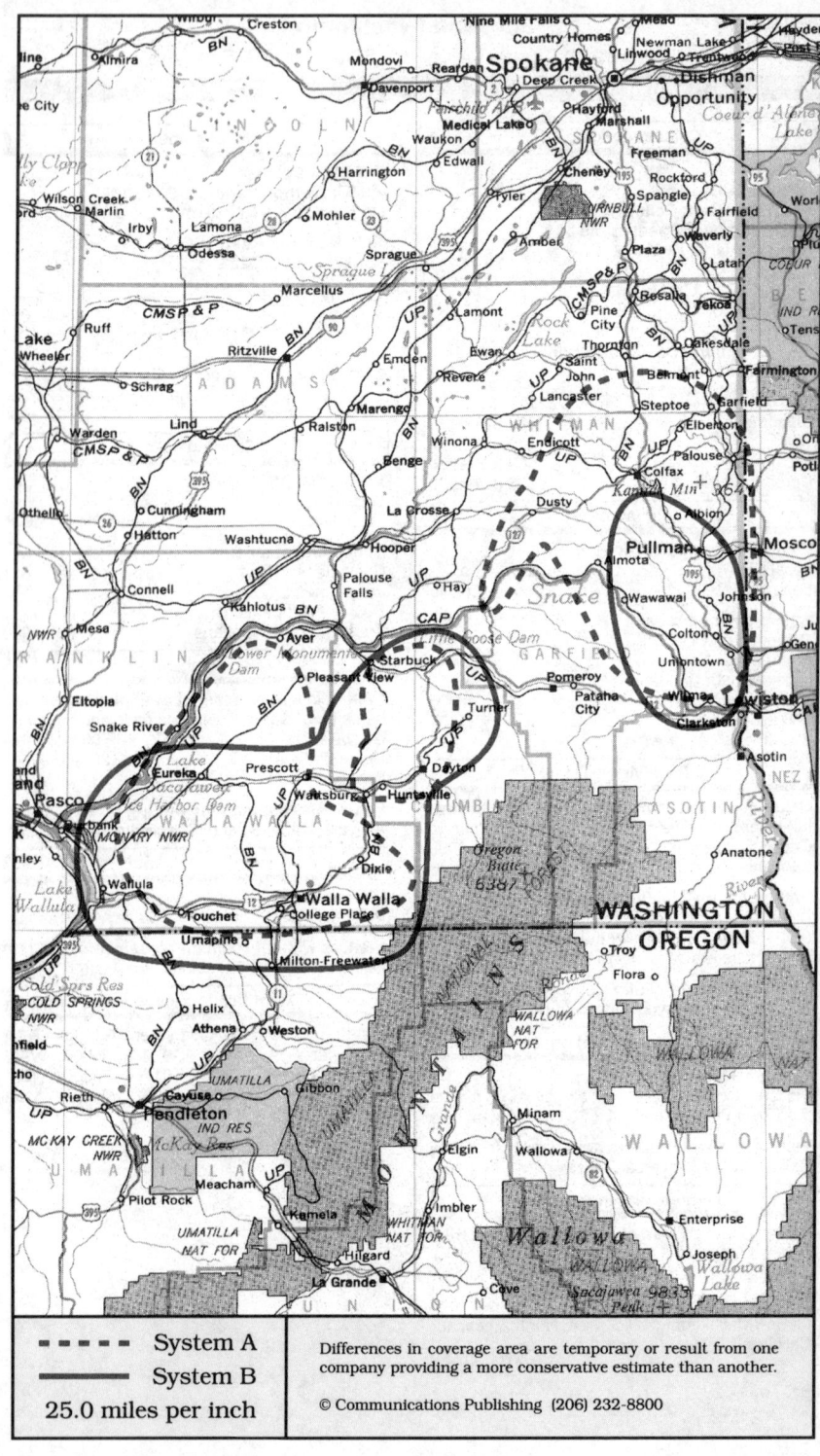

▪ ▪ ▪ ▪ ▪ **System A**	Differences in coverage area are temporary or result from one
——— **System B**	company providing a more conservative estimate than another.
25.0 miles per inch	© Communications Publishing (206) 232-8800

Walla Walla, Washington
Clarkston, Colfax, Dayton, Pullman, Walla Walla

System A	System B

Cellular company

Cellular One
Blue Mountain Cellular
1473 W Rose Street, PO Box 754
Walla Walla, WA 99362

Corporate office 509-520-5000

Sending calls

Visitor dialing instructions
Local calls . . . . . area code + 7 digits
Long distance . . . 1 + area code + 7 digits

Speed-dial numbers
Customer service 611
Emergency 911

Receiving calls

Caller dials the roamer access number below,
hears tone, dials cellular area code and number.
. . . . 509-520-7626

NationLink & RoamingAmerica (see p. 1157)
Caller hears your current roamer access number.
This company's subscribers have level 0 service.

North American Cellular Network (see p. 1156)
All calls forward to you. Deactivate dial *35 send.

Billing information

Visitor rates
Most visitors . . . . . $3 day + 85¢ minute

Visitors without roaming agreements (p. 1159)
Call co. to use Amex, Mastercard, Visa
or Roamer Plus will intercept first call to bill you.

Identification numbers

City	Market	System	Billing	RSA
All served cities	700	01787	00000	WA-8

Cellular company

Inland Cellular Telephone
S 213 Montgomery, PO Box 221
Uniontown, WA 99179

Local office	800-248-8822
. . . .	509-229-2500
Corporate office . some areas	800-346-1968
. . . .	509-649-2500

Sending calls

Visitor dialing instructions
Local calls . . . . . area code + 7 digits
Long distance . . . 1 + area code + 7 digits

Speed-dial numbers
Customer service *611
Emergency 911

Receiving calls

Caller dials the roamer access number below,
hears tone, dials cellular area code and number.
Clarkston 509-751-7626
Uniontown 509-229-7626
Walla Walla 509-629-7626

Follow Me Roaming forwarding (see p. 1156)
Activate, dial *18 send. Deactivate dial *19 send.

Billing information

Visitor rates
Most visitors . . . . . $3 day + 99¢ minute

Visitors without roaming agreements (p. 1159)
Cannot call since company takes no credit cards.

Identification numbers

City	Market	System	Billing	RSA
All served cities	700	01788	00000	WA-8

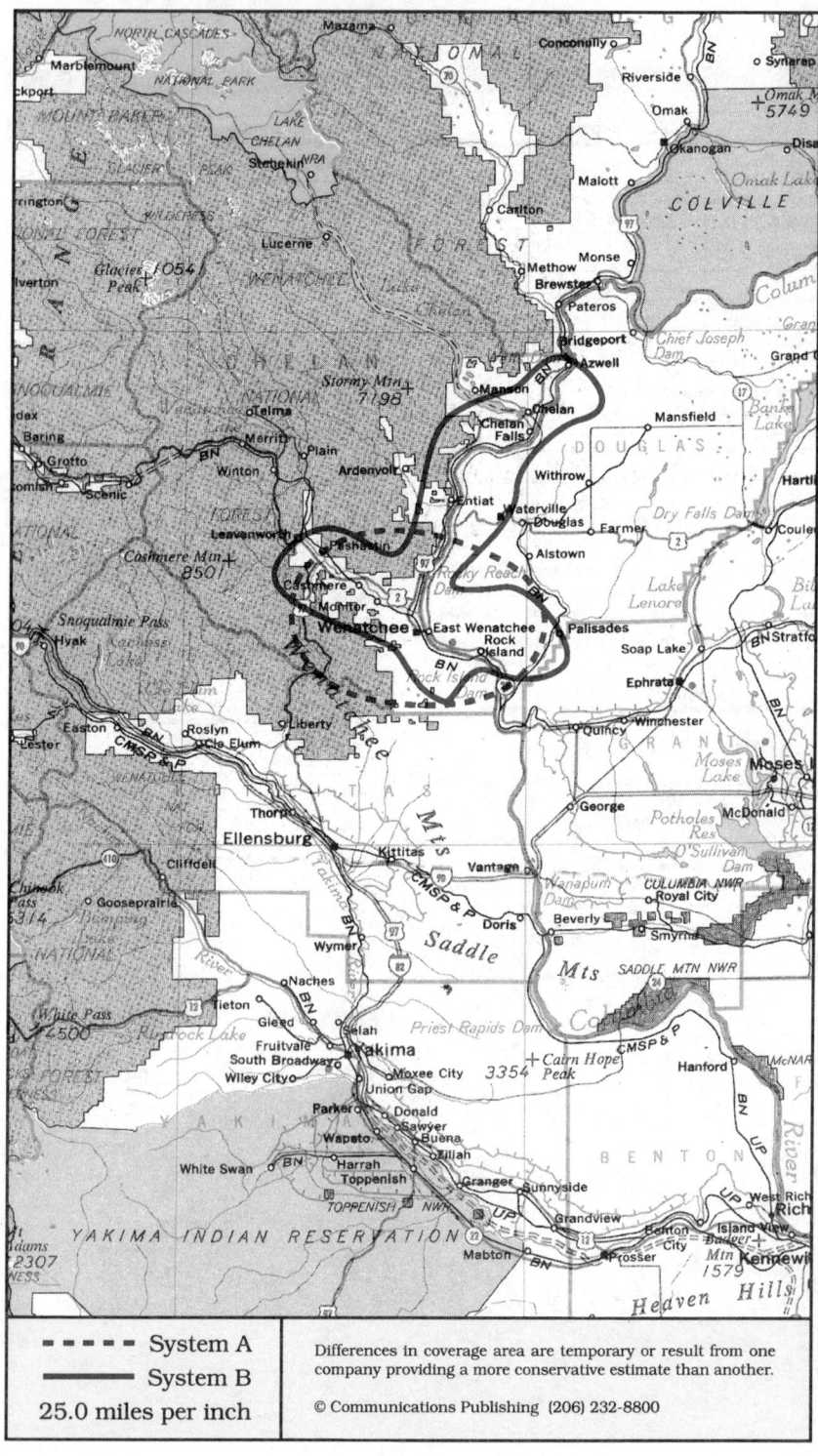

--- System A
— System B
25.0 miles per inch

Differences in coverage area are temporary or result from one company providing a more conservative estimate than another.

© Communications Publishing (206) 232-8800

Wenatchee, Washington
Waterville, Wenatchee

|

System A

Cellular company

Cellular One
Mercury Communications
235 N Mission
Wenatchee, WA 98801

Local office	509-663-1100
Corporate office	601-948-4800

Sending calls

Visitor dialing instructions

Local calls	area code + 7 digits
Long distance . . .	1 + area code + 7 digits

Speed-dial numbers

Customer service	611
Emergency	911

Receiving calls

Caller dials the roamer access number below,
hears tone, dials cellular area code and number.
. . . . 509-669-7626

NationLink & RoamingAmerica (see p. 1157)
Dial *31 send to receive calls, *30 to deactivate.
Dial *32 send to tell caller the roamer access no.
This company's subscribers have level 3 service.

Billing information

Visitor rates

Most visitors	$3 day + 85¢ minute
Some visitors	$0 day + 99¢ minute

Visitors without roaming agreements (p. 1159)
Call co. to use Amex, Mastercard, Visa

Identification numbers

City	Market	System	Billing	RSA
All served cities	694	01775	00000	WA-2

System B

Cellular company

Contel Cellular
104 S Mission
Wenatchee, WA 98801

Customer service	800-333-4004
. . . .	209-244-2980
Local office	509-664-6600
Corporate office	404-804-3400

Sending calls

Visitor dialing instructions

Local calls	area code + 7 digits
Long distance . . .	1 + area code + 7 digits

Speed-dial numbers

Customer service	611
Emergency	911
State patrol	*211

Receiving calls

Caller dials the roamer access number below,
hears tone, dials cellular area code and number.
. . . . 509-664-6226

Billing information

Visitor rates

Most visitors	$3 day + 85¢ minute

Visitors without roaming agreements (p. 1159)
Call co. to use Amex, Mastercard, Visa

Identification numbers

City	Market	System	Billing	RSA
All served cities	694	01776	00000	WA-2

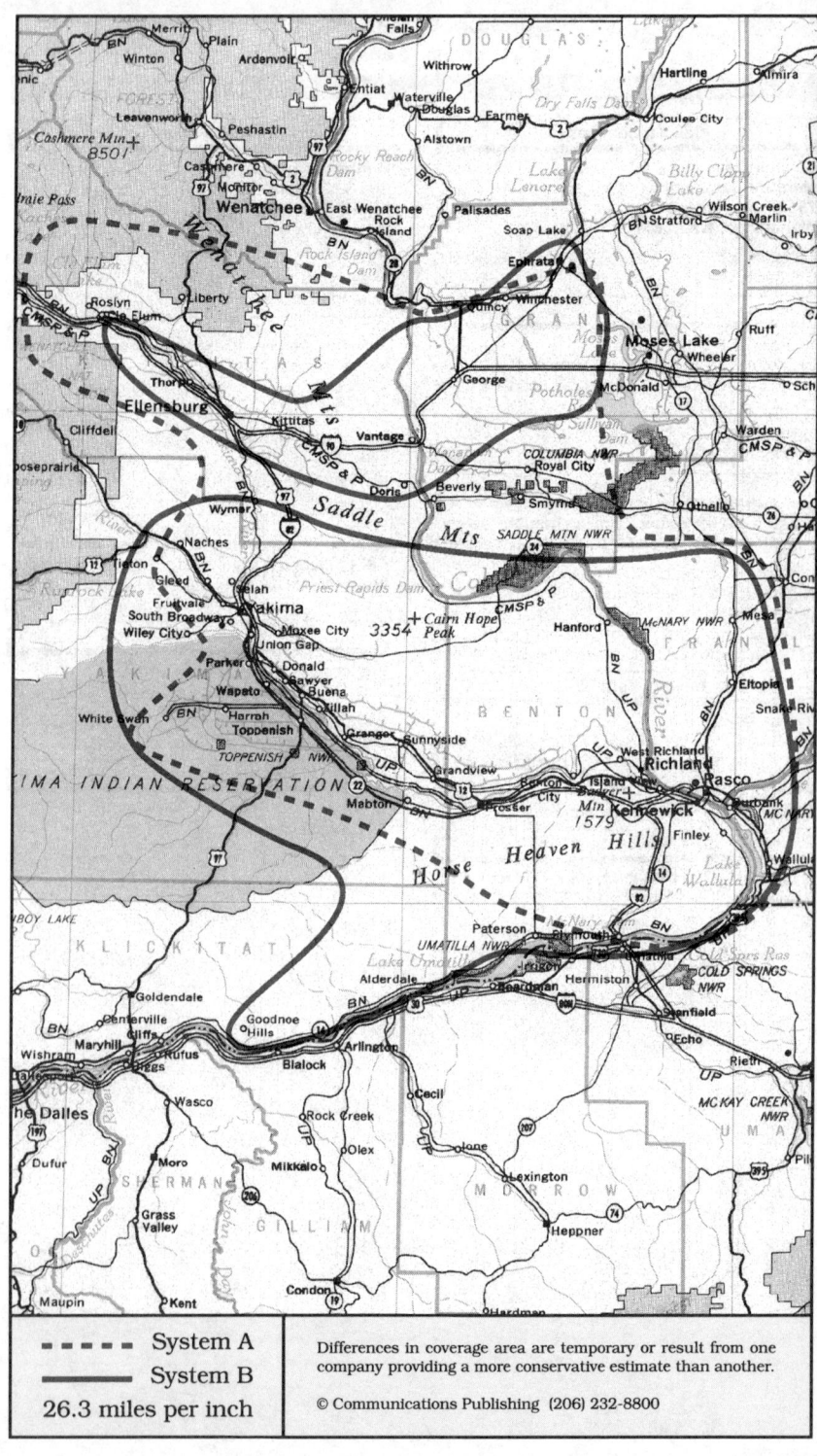

Yakima, Washington

Cle Elum, Ellensburg, Grandview, Kennewick/Pasco/Richland (Tri-Cities), Yakima

System A	System B

Cellular company

Cellular One
McCaw Cellular Communications
504 N Pine
Ellensburg, WA 98926

1103 Columbia Center Boulevard
Kennewick, WA 99336

901 W Yakima Avenue, Suite 4-B
Yakima, WA 99336

Customer service		800-327-6111
. . . .		509-838-2273
Local office . . .	Ellensburg	509-962-2663
. . . .	Kennewick	509-735-1300
. . . .	Yakima	509-454-2663
Corporate office		206-827-4500

Sending calls

Visitor dialing instructions

Local calls	area code + 7 digits
Long distance . . .	1 + area code + 7 digits

Speed-dial numbers

Customer service	611
Emergency	911

Receiving calls

Caller dials the roamer access number below,
hears tone, dials cellular area code and number.

Ellensburg	509-856-7626
Kennewick, Pasco, Richland .	509-546-7626
Ritzville	509-660-7626
Yakima	509-952-7626

NationLink & RoamingAmerica (see p. 1157)
Dial *31 send to receive calls, *30 to deactivate.
Dial *32 send to tell caller the roamer access no.
This company's subscribers have level 6 service.

North American Cellular Network (see p. 1156)
All calls forward to you. Deactivate dial *35 send.

Billing information

Visitor rates

Most visitors	$0 day + 99¢ minute

Visitors without roaming agreements (p. 1159)
Call co. to use Amex, Mastercard, Visa

Identification numbers

City	Market	System	Billing	RSA
Cle Elum,Ellensburg	697	00601	30231	WA-5
K., Pasco, Richland	214	00601	30229	MSA
Grandview, Yakima	191	00601	30227	MSA

For full coverage area, see Moses Lake, WA.

Cellular company

United States Cellular
109 E 4th
Ellensburg, WA 98926

6713 W Clearwater Avenue, Suite D
Kennewick, WA 99336

215 N 3rd Avenue, Building F
Yakima, WA 98902

Local office . . .	Ellensburg	509-925-9259
. . . .	Kennewick	509-783-3300
. . . .	Yakima	509-248-3000
Corporate office		312-399-8900

Sending calls

Visitor dialing instructions

Local calls	area code + 7 digits
Long distance . . .	area code + 7 digits

Speed-dial numbers

Customer service	611
Emergency	911
Roaming information	*711

Receiving calls

Caller dials the roamer access number below,
hears tone, dials cellular area code and number.

Ellensburg	509-929-7626
Kennewick, Pasco, Richland .	509-948-7626
Yakima	509-945-7626

Follow Me Roaming forwarding (see p. 1156)
Activate, dial *18 send. Deactivate dial *19 send.

Billing information

Visitor rates

Most visitors	$3 day + 99¢ minute

Visitors without roaming agreements (p. 1159)
Call co. to use Mastercard, Visa

Identification numbers

City	Market	System	Billing	RSA
Cle Elum,Ellensburg	697	01782	00000	WA-5
K., Pasco, Richland	214	00500	00000	MSA
Grandview, Yakima	191	00580	00000	MSA

For full coverage area, see Pendleton, OR.

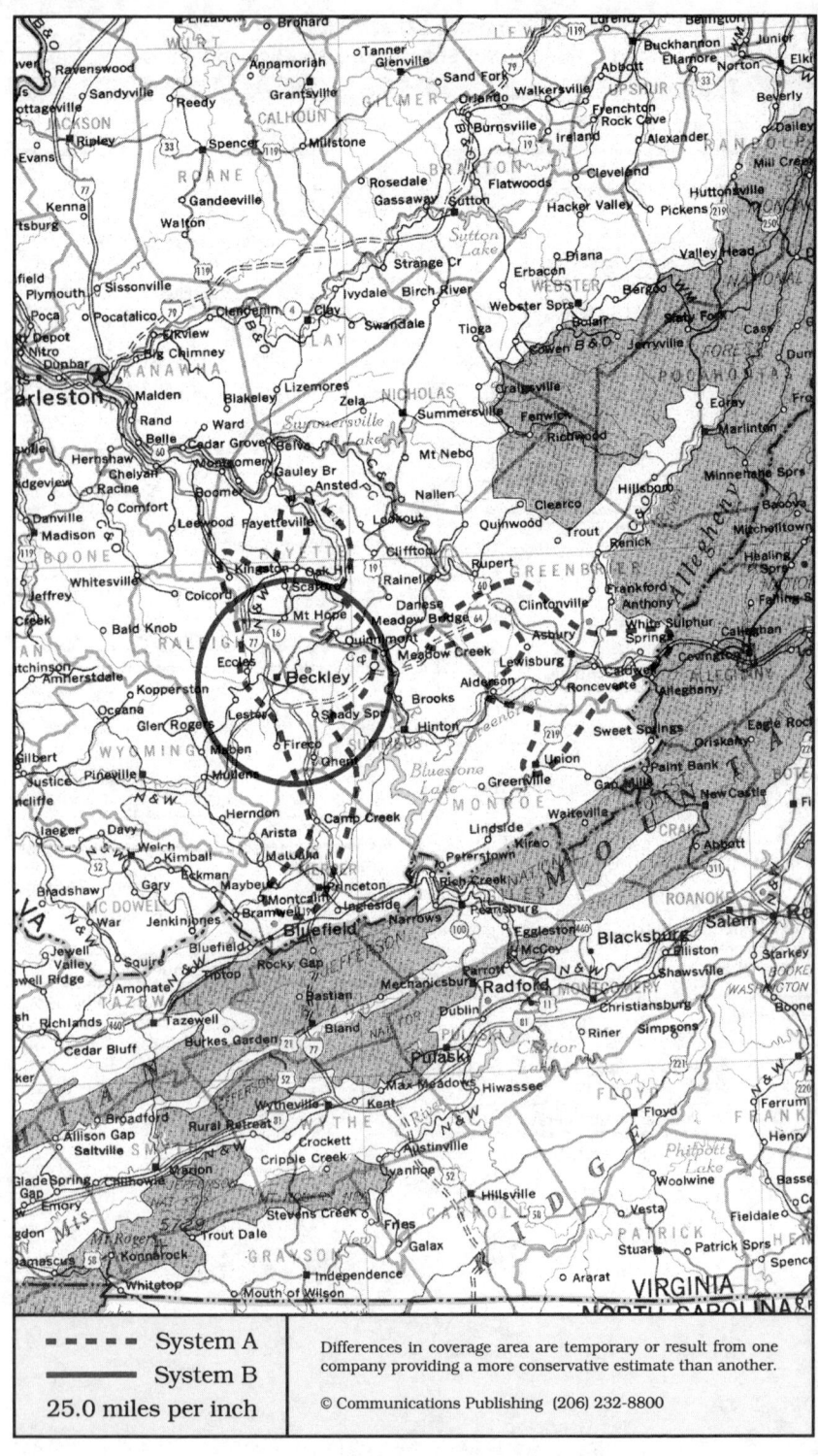

▪ ▪ ▪ ▪ **System A**	Differences in coverage area are temporary or result from one company providing a more conservative estimate than another.
——— **System B**	
25.0 miles per inch	© Communications Publishing (206) 232-8800

Beckley, West Virginia
Beckley, Bluefield, Fayetteville, Lewisburg, Princeton

System A	System B

Cellular company

Cellular One
Highland Cellular
1708 Harper Road
Beckley, WV 25801

| Corporate office | . . . . . | 800-795-2355 |
| | . . . . | 304-255-5222 |

Sending calls

Visitor dialing instructions

Local calls (inside local calling area)	.	7 digits
Local calls (outside local calling area)	1 + 7 digits
Long distance	. . .	1 + area code + 7 digits

Speed-dial numbers

| Customer service | . . . . | 611 |
| Emergency | . . . . | 911 |

Receiving calls

Caller dials the roamer access number below,
hears tone, dials cellular area code and number.

Beckley	. . . .	304-673-7626
Bluefield, Princeton	. . . .	304-320-7626
Lewisburg, White Sulphur Springs	304-647-8026

Billing information

Visitor rates

| Most visitors | . . . . . | $3 day + 95¢ minute |

Visitors without roaming agreements (p. 1159)
Roamer Plus will intercept first call to bill you.

Identification numbers

City	Market	System	Billing	RSA
All served cities	707	01801	00000	WV-7

Cellular company

ALLTEL Mobile Communications
10825 Financial Parkway #401
Little Rock, AR 72211

There is no local office, but the sales office of
Charleston, WV system B may assist you at
304-925-4000.

| Corporate office | . . . . . | 501-661-8500 |

Sending calls

Visitor dialing instructions

| Local calls | . . . . . . | area code + 7 digits |
| Long distance | . . . | 1 + area code + 7 digits |

Speed-dial numbers

| Customer service | . . . . | 611 |
| Emergency | . . . . | 911 |

Receiving calls

Caller dials the roamer access number below,
hears tone, dials cellular area code and number.

| . . . . | 304-542-7626 |

Billing information

Visitor rates

| Most visitors | . . . . | $3 day + 85¢ minute |

Visitors without roaming agreements (p. 1159)
Call co. to use Amex, Mastercard, Visa

Identification numbers

City	Market	System	Billing	RSA
All served cities	707	01802	00000	WV-7

Call switching is performed by Independent
Cellular Network for the licensee, ALLTEL Mobile
Communications.

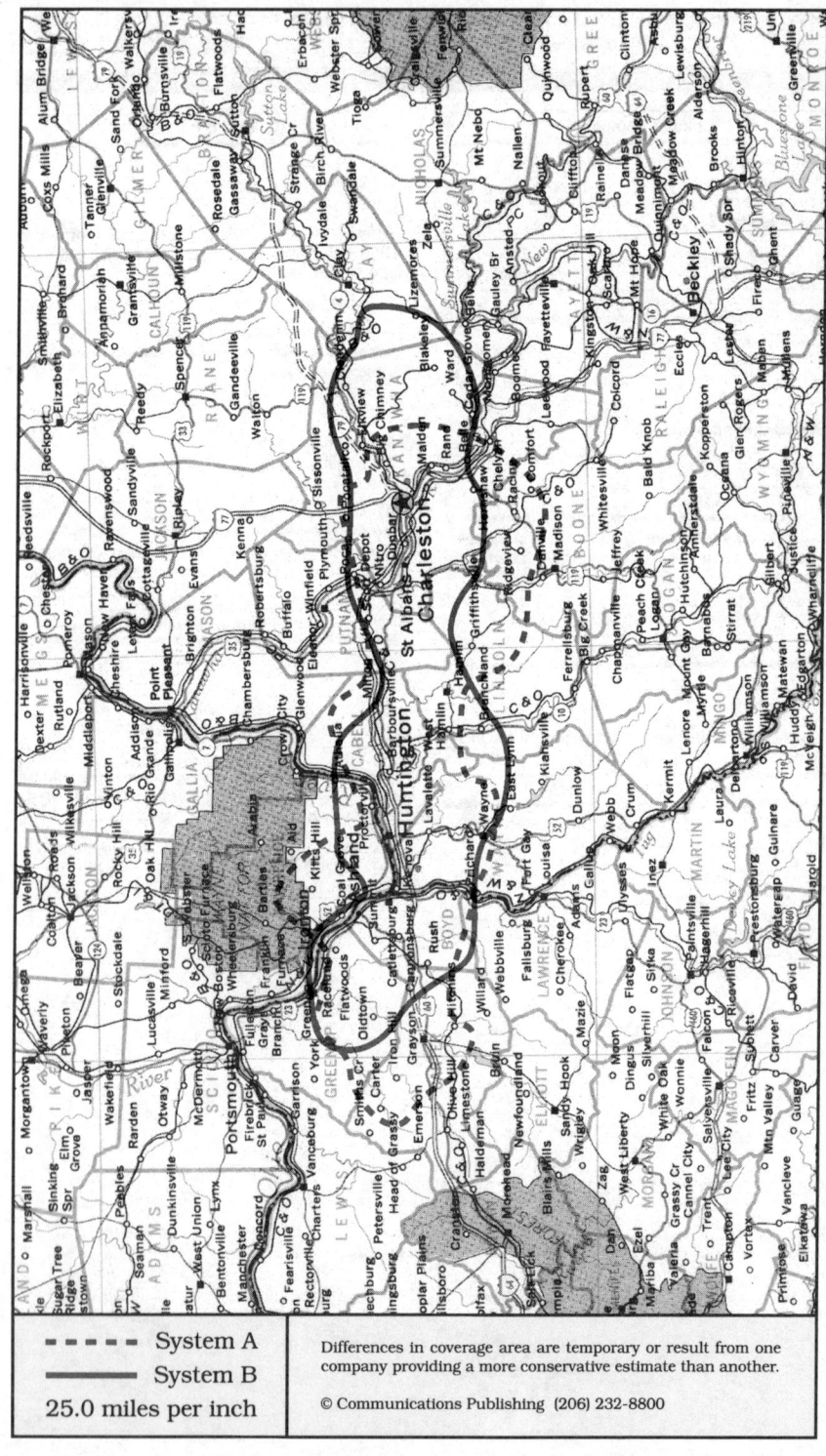

▪ ▪ ▪ ▪ ▪ **System A**	Differences in coverage area are temporary or result from one company providing a more conservative estimate than another.
━━━━━ **System B**	
25.0 miles per inch	© Communications Publishing (206) 232-8800

Charleston, West Virginia
Ashland KY, Charleston, Grayson, Huntington

System A	System B

Cellular company

Cellular One
Vanguard Cellular Systems
238 Russell Road
Ashland, KY 41101

701 Lee Street
Charleston, WV 25301

4341 Route 60 E, Building 2
Huntington, WV 25705

Customer service	. .	Ashland	606-922-4000
	. . . . . .	Charleston	304-545-4000
	. . . . . .	Huntington	304-633-4000
Local office	. . .	Ashland	606-325-2355
	. . . . . .	Charleston	304-345-2355
	. . . . . .	Huntington	304-736-2355
Corporate office	. . . .		919-282-3690

Sending calls

Visitor dialing instructions

Local calls	. . . .	area code + 7 digits
Long distance	. . .	area code + 7 digits

Speed-dial numbers

Customer service	611
Emergency	911

Receiving calls

Caller dials the roamer access number below,
hears tone, dials cellular area code and number.

Ashland	606-922-7626
Charleston	304-545-7626
Huntington	304-633-7626

Billing information

Visitor rates

Most visitors	$3 day + 99¢ minute

Visitors without roaming agreements (p. 1159)
Call co. to use Amex, Mastercard, Visa

Identification numbers

City	Market	System	Billing	RSA
Ashland	110	00307	30307	MSA
Charleston	140	00307	00000	MSA
Huntington	110	00307	30047	MSA

Cellular company

Independent Cellular Network
925 Winchester Avenue
Ashland, KY 41101

4227 MacCorkle Avenue SE
Charleston, WV 25304

3322 U.S. Highway 60 E
Huntington, WV 25705

Local office	. . . .	Ashland	606-324-4426
	. . . .	Charleston	304-925-4000
	. . . .	Huntington	304-525-4101
Corporate office	. . . . .		813-489-1600

Sending calls

Visitor dialing instructions

Local calls	. . . . . .	area code + 7 digits
Long distance	. . .	1 + area code + 7 digits

Speed-dial numbers

Customer service	*611
Emergency	911

Receiving calls

Caller dials the roamer access number below,
hears tone, dials cellular area code and number.

Ashland	606-831-7626
Charleston	304-542-7626
Huntington	304-544-7626

Follow Me Roaming forwarding (see p. 1156)
Activate, dial *18 send. Deactivate dial *19 send.

Billing information

Visitor rates

Most visitors	$3 day + $1 minute

Visitors without roaming agreements (p. 1159)
Call co. to use Amex, Mastercard, Visa

Identification numbers

City	Market	System	Billing	RSA
Ashland, Huntington	110	00032	30024	MSA
Charleston	140	00032	30022	MSA

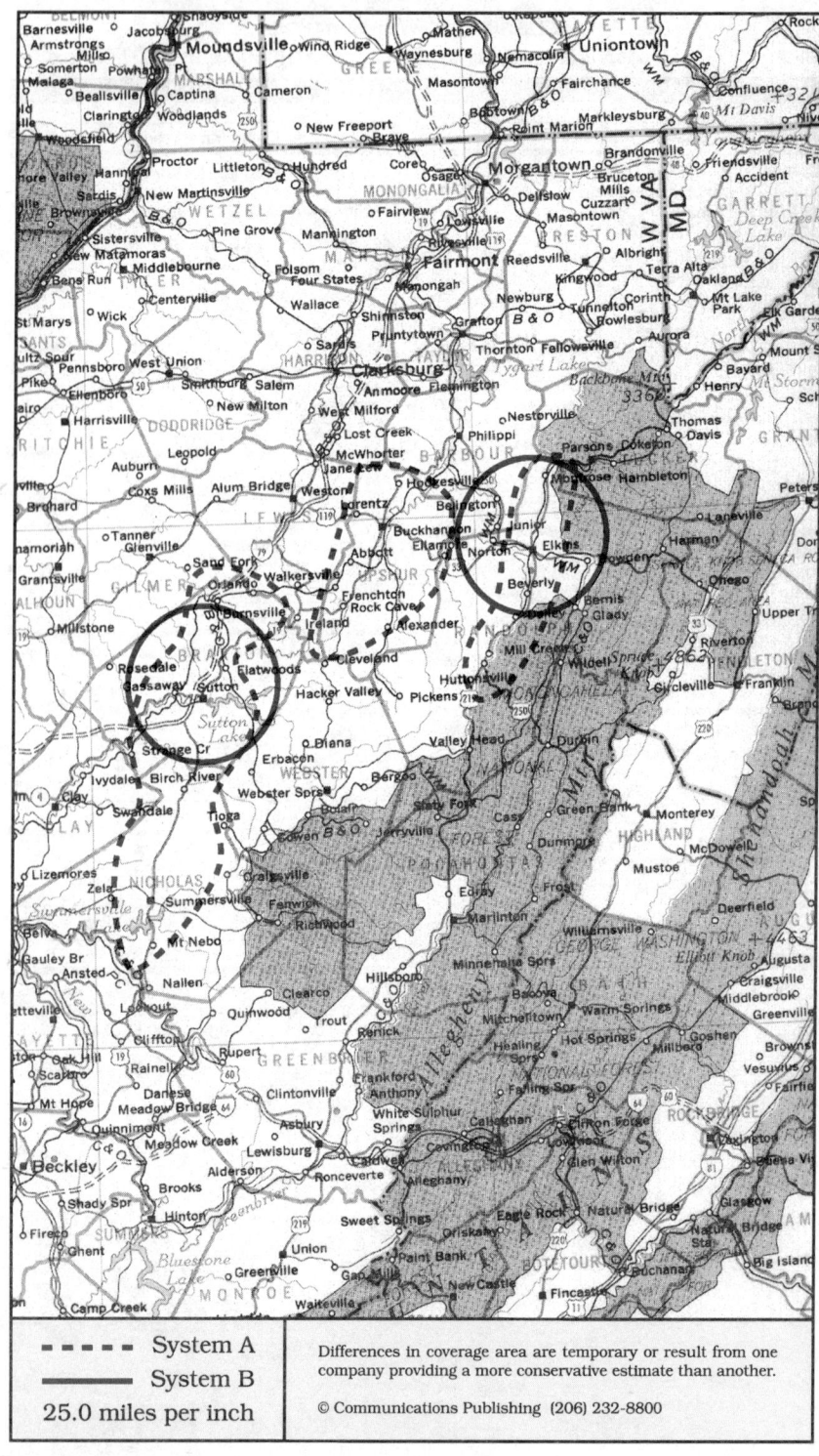

System A

System B

25.0 miles per inch

Differences in coverage area are temporary or result from one company providing a more conservative estimate than another.

© Communications Publishing (206) 232-8800

Elkins, West Virginia
Buckhannon, Elkins, Summersville, Sutton

System A	System B

Cellular company

Cellular One
Licensed to: Easterbrook Cellular
Managed by: Rural Cellular Management
1505 Harrison Avenue
Elkins, WV 26241

Local office 304-636-6400
Corporate office   Rural Cellular   707-422-2100

Sending calls

Visitor dialing instructions
Local calls . . . . . area code + 7 digits
Long distance . . . 1 + area code + 7 digits

Speed-dial numbers
Customer service *611
Emergency 911

Receiving calls

Caller dials the roamer access number below,
hears tone, dials cellular area code and number.
. . . . 304-642-7626

NationLink & RoamingAmerica (see p. 1157)
Dial *31 send to receive calls, *30 to deactivate.
This company's subscribers have level 3 service.

Billing information

Visitor rates
Most visitors . . . . . $3 day + 85¢ minute

Visitors without roaming agreements (p. 1159)
Cannot call since company takes no credit cards.

Identification numbers

City	Market	System	Billing	RSA
All served cities	705	01797	00000	WV-5

Cellular company

United States Cellular
663 Pittsburgh Street
Uniontown, PA 15401

Local office 800-944-9066
. . . . 412-437-0800
Corporate office 312-399-8900

Sending calls

Visitor dialing instructions
Local calls . . . . . . area code + 7 digits
Long distance . . . . . area code + 7 digits

Speed-dial numbers
Customer service 611
Emergency 911
Roaming information *711

Receiving calls

Caller dials the roamer access number below,
hears tone, dials cellular area code and number.
. . . . 304-678-7626

Billing information

Visitor rates
Most visitors . . . . . $3 day + 99¢ minute

Visitors without roaming agreements (p. 1159)
Cannot call since company takes no credit cards.

Identification numbers

City	Market	System	Billing	RSA
All served cities	705	01798	00000	WV-5

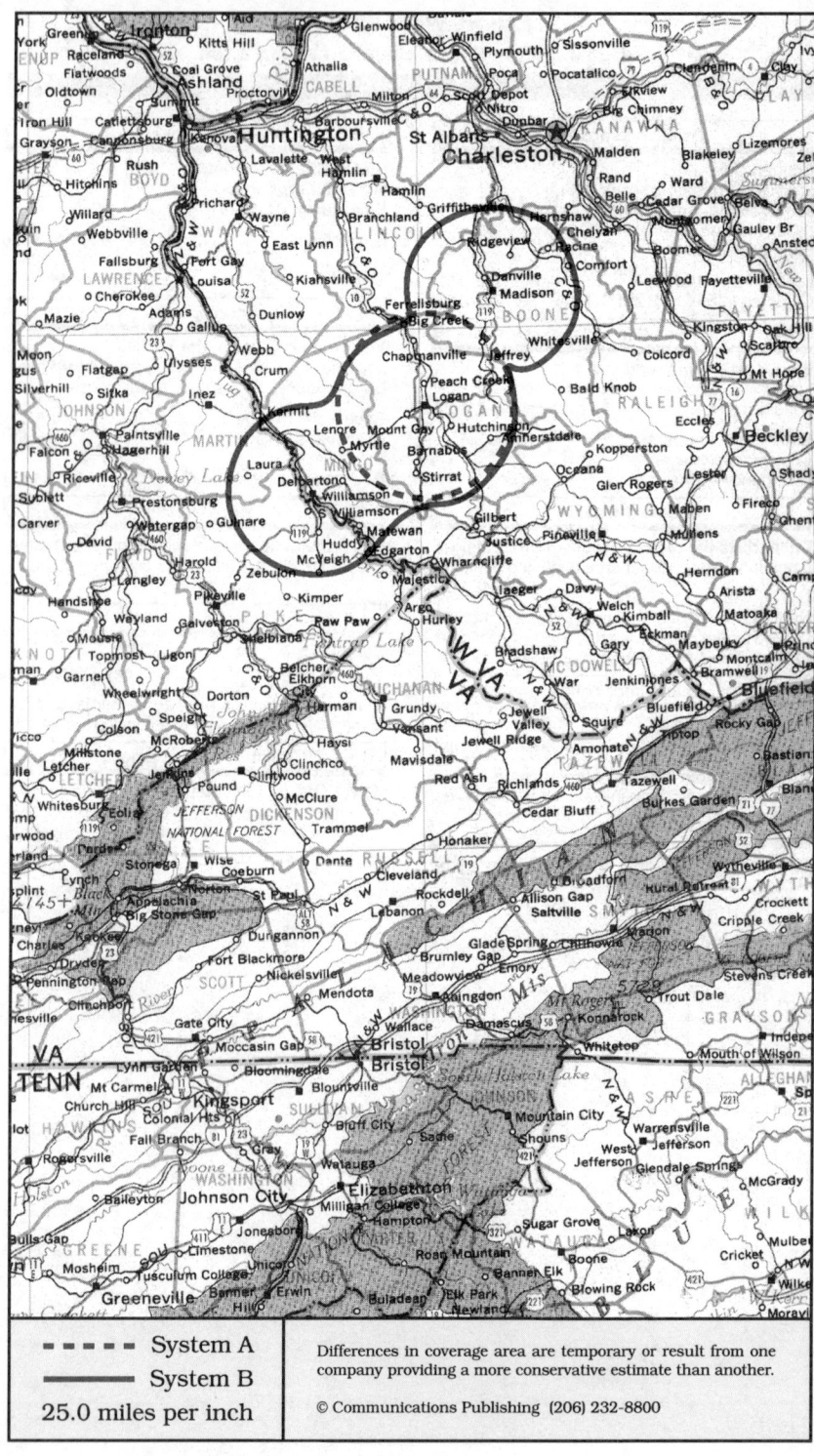

- - - - - System A
━━━━━ System B
25.0 miles per inch

Differences in coverage area are temporary or result from one company providing a more conservative estimate than another.

© Communications Publishing (206) 232-8800

Logan, West Virginia
Logan, Madison, Williamson

System A	System B

Cellular company

Cellular One
Clear Communications Group
11 White Street
Logan, WV 25601

Local office . .	some areas	800-834-1110
. . . .		304-752-1111
Corporate office		404-843-9100

Sending calls

Visitor dialing instructions
Local calls	area code + 7 digits
Long distance . .	1 + area code + 7 digits

Speed-dial numbers
Customer service	611
Emergency	911

Receiving calls

Caller dials the roamer access number below,
hears tone, dials cellular area code and number.
. . . . 304-687-7626

NationLink & RoamingAmerica (see p. 1157)
Caller hears your current roamer access number.
This company's subscribers have level 0 service.

Billing information

Visitor rates
Most visitors . . . . . . $3 day + $1 minute

Visitors without roaming agreements (p. 1159)
Call co. to use Amex, Discover, Mastercard, Visa
or Roamer Plus will intercept first call to bill you.

Identification numbers

City	Market	System	Billing	RSA
All served cities	706	01799	00000	WV-6

Cellular company

Independent Cellular Network
403 Justice Avenue
Logan, WV 25601

Local office	304-752-5200
Corporate office	813-489-1600

Sending calls

Visitor dialing instructions
Local calls	area code + 7 digits
Long distance . . .	1 + area code + 7 digits

Speed-dial numbers
Customer service	*611
Emergency	911

Receiving calls

Caller dials the roamer access number below,
hears tone, dials cellular area code and number.
. . . . 304-784-7626

Follow Me Roaming forwarding (see p. 1156)
Activate, dial *18 send. Deactivate dial *19 send.

Billing information

Visitor rates
Most visitors . . . . . $3 day + $1 minute

Visitors without roaming agreements (p. 1159)
Call co. to use Amex, Mastercard, Visa

Identification numbers

City	Market	System	Billing	RSA
All served cities	706	00032	30606	WV-6

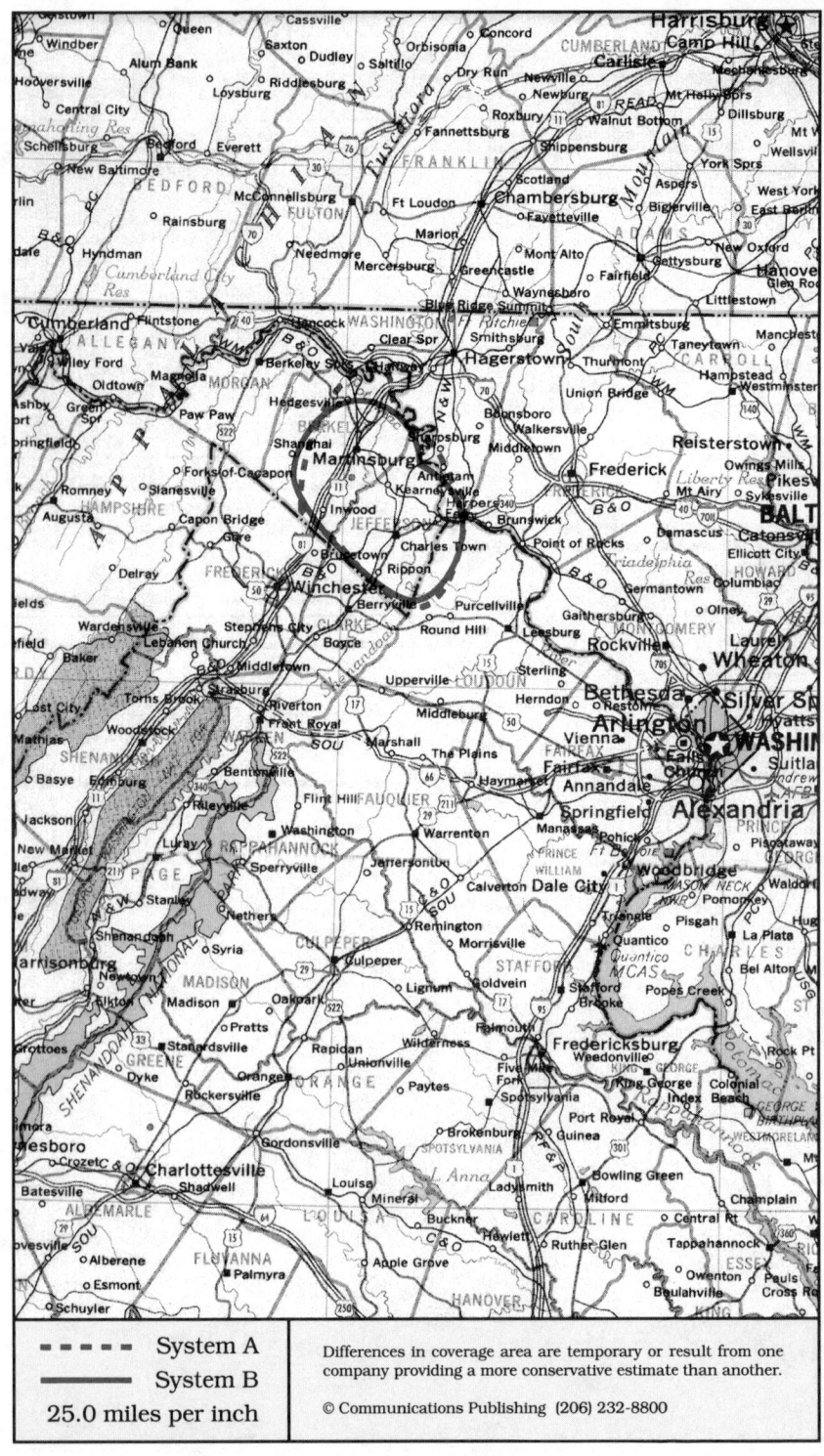

Martinsburg, West Virginia
Charlestown, Martinsburg

System A	System B

Cellular company

Cellular One
Southwestern Bell Mobile Systems
7855 Walker Drive, Suite 100
Greenbelt, MD 20770

Customer service	some areas	800-235-5663
. . . .		301-220-0060
Corporate office	. . . . .	301-220-3600

Sending calls

Visitor dialing instructions

Local calls	. . . . .	area code + 7 digits
Long distance	. . . .	area code + 7 digits

Speed-dial numbers

Cellular information	*1394
Customer service	*611 or *811
Emergency	911
24 hour news	*1887

Receiving calls

Caller dials the roamer access number below, hears tone, dials cellular area code and number.
. . . . 304-264-6666

NationLink & RoamingAmerica (see p. 1157)
Dial *31 send to receive calls, *30 to deactivate.
Dial *32 send to tell caller the roamer access no.
Service is not automatic and must be turned on.
This company's subscribers have level 6 service.

Billing information

Visitor rates
Most visitors . . . . . $3 day + 95¢ minute

Visitors without roaming agreements (p. 1159)
Cannot call since company takes no credit cards.

Identification numbers

City	Market	System	Billing	RSA
All served cities	704	00013	30339	WV-4

For full coverage area, see Baltimore, MD.

Cellular company

United States Cellular
1345 Edwin Miller Boulevard
Martinsburg, WV 25401

Local office	304-264-0400
Corporate office	312-399-8900

Sending calls

Visitor dialing instructions

Local calls	. . . . .	area code + 7 digits
Long distance	. . . . .	area code + 7 digits

Speed-dial numbers

Customer service	611
Emergency	911

Receiving calls

Caller dials the roamer access number below, hears tone, dials cellular area code and number.
. . . . 304-671-7626

Billing information

Visitor rates
Most visitors . . . . . $3 day + 99¢ minute

Visitors without roaming agreements (p. 1159)
Call co. to use Mastercard, Visa

Identification numbers

City	Market	System	Billing	RSA
All served cities	704	01796	00000	WV-4

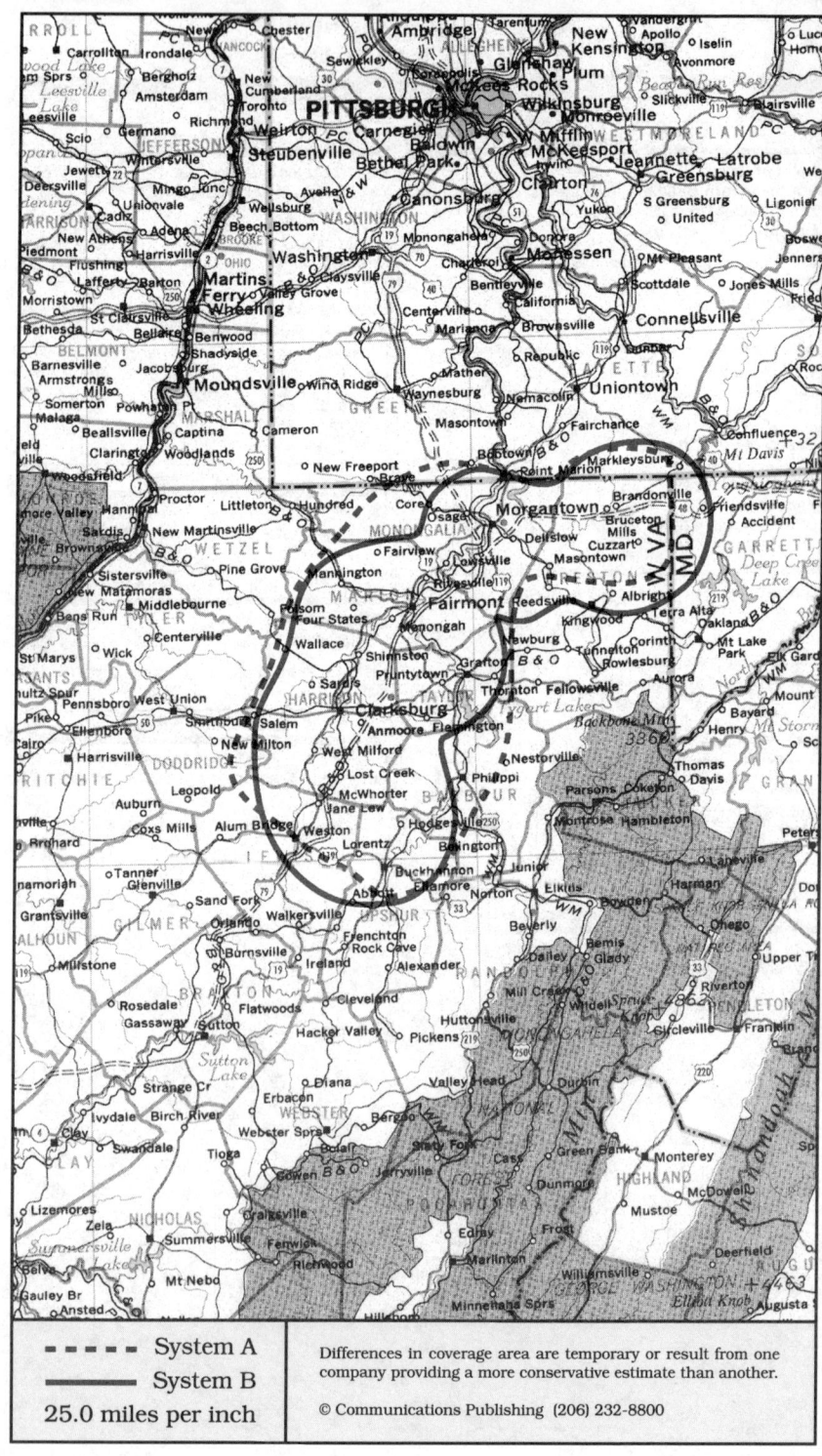

- - - - System A	Differences in coverage area are temporary or result from one company providing a more conservative estimate than another.
——— System B	
25.0 miles per inch	© Communications Publishing (206) 232-8800

Morgantown, West Virginia
Clarksburg, Fairmont, Morgantown

System A	System B

Cellular company

Mountaineer Mobile
1390 University Avenue
Morgantown, WV 26505

Corporate office   304-291-6565

Sending calls

Visitor dialing instructions
Local calls . . . . .   area code + 7 digits
Long distance . . . .   area code + 7 digits

Speed-dial numbers
Banking & financial information   BEST or 2378
Customer service *611
Emergency   911
Lakeview fitness center .   FITNESS or 3486377
Lakeview pro shop . . . .   GOLF or 4653
Lakeview resort . . .   RESORT or 737678
News tip line . . . . .   WDTV5 or 93885
Roadside assistance (AAA) . . .   AAAHELP
Roadside asssitance (not AAA) . . .   ASSIST
Sports and ticket information . .   WVU or 988
Time and weather . . . . . .   TIME or 8463

Receiving calls

Caller dials the roamer access number below,
hears tone, dials cellular area code and number.
. . . .   304-282-7626

NationLink & RoamingAmerica (see p. 1157)
Caller hears your current roamer access number.
This company's subscribers have level 0 service.

Billing information

Visitor rates
Most visitors . . . . . $3 day + 99¢ minute

Visitors without roaming agreements (p. 1159)
Cannot call since company takes no credit cards.

Identification numbers

City	Market	System	Billing	RSA
All served cities	703	01793	00000	WV-3

Cellular company

United States Cellular
1345 Edwin Miller Boulevard
Martinsburg, WV 25401

Local office   304-264-0400
Corporate office   312-399-8900

Sending calls

Visitor dialing instructions
Local calls . . . . . .   area code + 7 digits
Long distance . . . . .   area code + 7 digits

Speed-dial numbers
Customer service   611
Emergency   911
Roaming information *711

Receiving calls

Caller dials the roamer access number below,
hears tone, dials cellular area code and number.
. . . .   304-288-7626

Billing information

Visitor rates
Most visitors . . . . . $3 day + 99¢ minute

Visitors without roaming agreements (p. 1159)
Call co. to use Mastercard, Visa

Identification numbers

City	Market	System	Billing	RSA
All served cities	703	01794	00000	WV-3

- - - - - System A	Differences in coverage area are temporary or result from one
——— System B	company providing a more conservative estimate than another.
25.0 miles per inch	© Communications Publishing (206) 232-8800

Parkersburg, West Virginia
Marietta OH, Parkersburg, Vienna

System A	System B

Cellular company

Cellular One
McCaw Cellular Communications
2630 Liberty Avenue
Pittsburgh, PA 15222

Customer service	800-477-3922
. . . .	304-482-0100
Local office	412-471-4400
Corporate office	206-827-4500

Sending calls

Visitor dialing instructions

Local calls	area code + 7 digits
Long distance . . .	1 + area code + 7 digits

Speed-dial numbers

Customer service	611
Emergency	911

Receiving calls

Caller dials the roamer access number below,
hears tone, dials cellular area code and number.
. . . . 304-482-7626

NationLink & RoamingAmerica (see p. 1157)
Caller hears your current roamer access number.
This company's subscribers have level 0 service.

North American Cellular Network (see p. 1156)
All calls forward to you. Deactivate dial *35 send.

Billing information

Visitor rates
Most visitors . . . . . $0 day + 99¢ minute

Visitors without roaming agreements (p. 1159)
Cannot call since company takes no credit cards.

Identification numbers

City	Market	System	Billing	RSA
All served cities	200	00485	00000	MSA

Cellular company

Independent Cellular Network
6600 Emerson Avenue
Parkersburg, WV 26101

Local office	304-485-5600
Corporate office	813-489-1600

Sending calls

Visitor dialing instructions

Local calls	area code + 7 digits
Long distance . . .	1 + area code + 7 digits

Speed-dial numbers

Customer service	*611
Emergency	911
Time and temperature	304-422-4541

Receiving calls

Caller dials the roamer access number below,
hears tone, dials cellular area code and number.

Marietta	614-525-7626
Parkersburg	304-481-7626

Follow Me Roaming forwarding (see p. 1156)
Activate, dial *18 send. Deactivate dial *19 send.

Billing information

Visitor rates
Most visitors . . . . . $3 day + $1 minute

Visitors without roaming agreements (p. 1159)
Call co. to use Amex, Mastercard, Visa

Identification numbers

City	Market	System	Billing	RSA
All served cities	200	00032	30030	MSA

▬ ▬ ▬ ▬ System A	Differences in coverage area are temporary or result from one
▬▬▬▬▬ System B	company providing a more conservative estimate than another.
25.0 miles per inch	© Communications Publishing (206) 232-8800

Ripley, West Virginia

System A

Cellular company

Cellular One of the Plains
Sterling Cellular Management
1080 Holcomb Bridge Road, Bldg 100, Ste 200
Roswell, GA 30076

Customer service 800-552-6150
Corporate office 404-552-5030

Sending calls

Visitor dialing instructions
Local calls . . . . . area code + 7 digits
Long distance . . . area code + 7 digits

Speed-dial numbers
Customer service . . . . ∗611
Emergency 911

Receiving calls

Caller dials the roamer access number below,
hears tone, dials cellular area code and number.
. . . . 304-532-7626

Billing information

Visitor rates
Most visitors . . . . . $3 day + 99¢ minute

Visitors without roaming agreements (p. 1159)
Roamer Plus will intercept first call to bill you.

Identification numbers

City	Market	System	Billing	RSA
All served cities	701	01789	00000	WV-1

System B

Cellular company

Bell Atlantic Mobile
6200 Babcock Boulevard
Pittsburgh, PA 15237

2895 Banksville Road
Pittsburgh, PA 15216

Customer service . . . 800-922-0204
Local office Babcock Boulevard 412-369-8500
. . . . . Banksville Road 412-571-3300
Corporate office 908-306-7000

Sending calls

Visitor dialing instructions
Local calls . . . . . area code + 7 digits
Long distance . . . 1 + area code + 7 digits

Speed-dial numbers
Customer service . . . . ∗611
Emergency 911

Receiving calls

Caller dials the roamer access number below,
hears tone, dials cellular area code and number.
. . . . 304-377-7626

Follow Me Roaming forwarding (see p. 1156)
Activate, dial ∗18 send. Deactivate dial ∗19 send.

Billing information

Visitor rates
Most visitors . . . . . $3 day + $1 minute

Visitors without roaming agreements (p. 1159)
Roamer Plus will intercept first call to bill you.

Identification numbers

City	Market	System	Billing	RSA
All served cities	701	00032	30094	WV-1

This system is operated as part of Pittsburgh, PA.

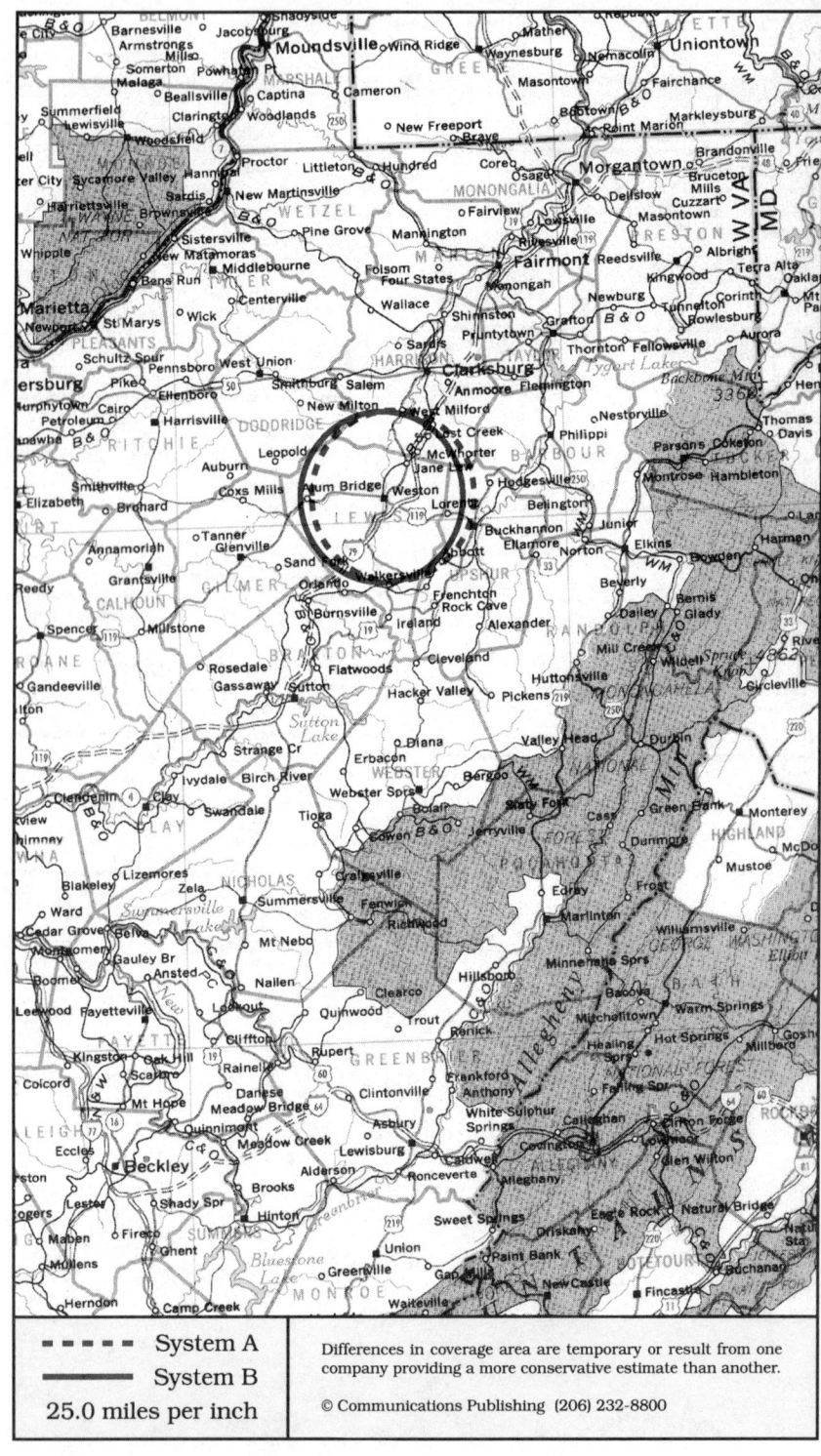

Weston, West Virginia

System A

Cellular company

United States Cellular
663 Pittsburgh Street
Uniontown, PA 15401

Local office	800-944-9066
. . . .	412-437-0800
Corporate office	312-399-8900

Sending calls

Visitor dialing instructions

Local calls	area code + 7 digits
Long distance	area code + 7 digits

Speed-dial numbers

Customer service	611
Emergency	911
Roaming information	*711

Receiving calls

Caller dials the roamer access number below,
hears tone, dials cellular area code and number.
. . . . 304-678-7626

Billing information

Visitor rates
Most visitors . . . . . $3 day + 99¢ minute

Visitors without roaming agreements (p. 1159)
Cannot call since company takes no credit cards.

Identification numbers

City	Market	System	Billing	RSA
All served cities	702	01791	00000	WV-2

System B

Cellular company

Bell Atlantic Mobile
6200 Babcock Boulevard
Pittsburgh, PA 15237

2895 Banksville Road
Pittsburgh, PA 15216

Customer service		800-922-0204
Local office	Babcock Boulevard	412-369-8500
. . . . .	Banksville Road	412-571-3300
Corporate office		908-306-7000

Sending calls

Visitor dialing instructions

Local calls	area code + 7 digits
Long distance . . .	1 + area code + 7 digits

Speed-dial numbers

Customer service	*611
Emergency	911

Receiving calls

Caller dials the roamer access number below,
hears tone, dials cellular area code and number.
. . . . 304-266-7626

Follow Me Roaming forwarding (see p. 1156)
Activate, dial *18 send. Deactivate dial *19 send.

Billing information

Visitor rates
Most visitors . . . . . $3 day + $1 minute

Visitors without roaming agreements (p. 1159)
Roamer Plus will intercept first call to bill you.

Identification numbers

City	Market	System	Billing	RSA
All served cities	702	00032	30096	WV-2

This system is operated as part of Pittsburgh, PA.

Wheeling, West Virginia
St. Clairsville OH, Steubenville OH, Weirton, Wheeling

System A	System B

Cellular company

St. Clairsville & Wheeling
Cellular One
McCaw Cellular Communications
2630 Liberty Avenue
Pittsburgh, PA 15222

Customer service		800-477-3922
. . . . from Wheeling		304-281-0100
Local office		412-471-4400
Corporate office		206-827-4500

Steubenville & Weirton
Steel Valley Cellular
PO Box 9
Steubenville, OH 43952

Corporate office	614-283-1234

Sending calls

Visitor dialing instructions

Local calls	area code + 7 digits
Long distance . . .	1 + area code + 7 digits

Speed-dial numbers

Customer service	611
Emergency	911

Receiving calls

Caller dials the roamer access number below,
hears tone, dials cellular area code and number.

St. Clairsville, Wheeling . . .	304-281-7626
Steubenville, Weirton . . .	304-670-7626

NationLink & RoamingAmerica (see p. 1157)
Only for Cellular One (St. Clairsville & Wheeling)
Dial *31 send to receive calls, *30 to deactivate.
Dial *32 send to tell caller the roamer access no.
This company's subscribers have level 8 service.

North American Cellular Network (see p. 1156)
All calls forward to you. Deactivate dial *35 send.

Billing information

Visitor rates

In St. Clairsville, Wheeling	$0 day + 99¢ minute
In Steubenville, Weirton	$3.50 day + 85¢ minute

Visitors without roaming agreements (p. 1159)
Roamer Plus will intercept first call to bill you.

Identification numbers

City	Market	System	Billing	RSA
St. Clair.,Wheeling	178	00039	30059	MSA
Steuben., Weirton	199	00039	30317	MSA

Call switching for Steubenville and Weirton is
performed by McCaw Cellular for the licensee,
Steel Valley Cellular.

For full coverage area, see Pittsburgh, PA.

Cellular company

Independent Cellular Network
51342 National Road, Suite J
St. Clairsville, OH 43950

4102 Sunset Boulevard
Steubenville, OH 43952

Local office . .	St. Clairsville	614-695-9611
. . . .	Steubenville	614-264-6696
. . . . . .	from Wheeling	304-233-5600
Corporate office		813-489-1600

Sending calls

Visitor dialing instructions

Local calls	area code + 7 digits
Long distance . . .	1 + area code + 7 digits

Speed-dial numbers

Customer service	*611
Emergency	911

Receiving calls

Caller dials the roamer access number below,
hears tone, dials cellular area code and number.

St. Clairsville, Wheeling . . .	614-391-7626
Steubenville, Weirton	614-381-7626

Follow Me Roaming forwarding (see p. 1156)
Activate, dial *18 send. Deactivate dial *19 send.

Billing information

Visitor rates

Most visitors	$3 day + $1 minute

Visitors without roaming agreements (p. 1159)
Call co. to use Amex, Mastercard, Visa

Identification numbers

City	Market	System	Billing	RSA
St. Clair.,Wheeling	178	00032	00000	MSA
Steuben., Weirton	199	00032	00000	MSA

For full coverage area, see New Philadelphia,
OH.

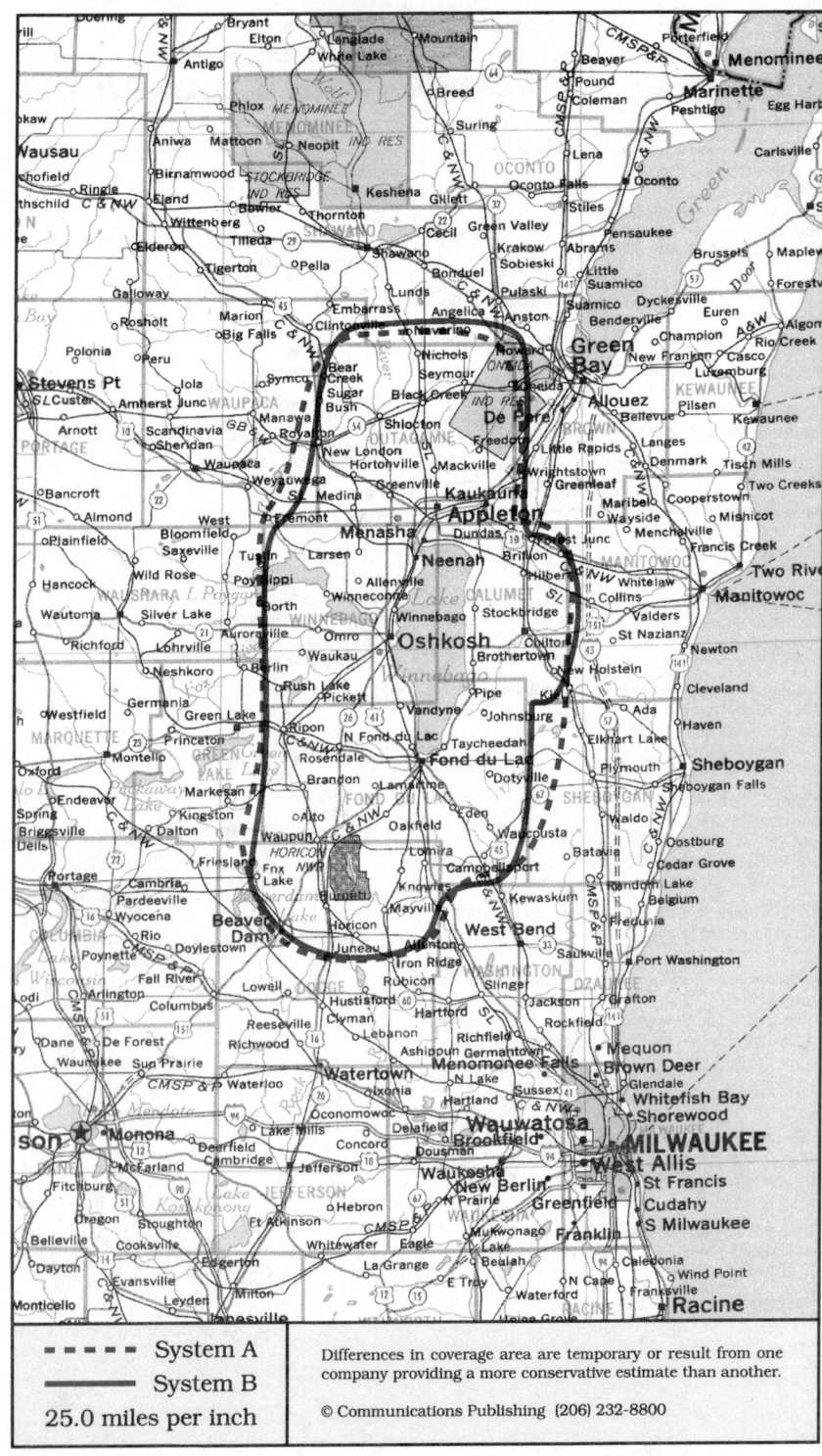

▪ ▪ ▪ ▪ ▪ System A	Differences in coverage area are temporary or result from one
──── System B	company providing a more conservative estimate than another.
25.0 miles per inch	© Communications Publishing (206) 232-8800

Appleton, Wisconsin

Appleton, Chilton, Fond du Lac, Neenah, Oshkosh

System A	System B

Cellular company

Cellular One
BellSouth Cellular
2740 W College Avenue
Appleton, WI 54914-2915

Customer service	800-888-5611
. . . .	414-738-0110
Local office	414-738-7787
Corporate office	404-604-6100

Sending calls

Visitor dialing instructions

Local calls	area code + 7 digits
Long distance . . .	1 + area code + 7 digits

Speed-dial numbers

Customer service	611
Emergency	911

Receiving calls

Caller dials the roamer access number below,
hears tone, dials cellular area code and number.

Appleton	414-428-7626
Fond du Lac	608-575-7626

NationLink & RoamingAmerica (see p. 1157)
Dial *31 send to receive calls, *30 to deactivate.
Dial *32 send to tell caller the roamer access no.
This company's subscribers have level 6 service.

Billing information

Visitor rates

Most visitors	$3 day + 99¢ minute

Visitors without roaming agreements (p. 1159)
Call co. to use Amex, Discover, Mastercard, Visa

Identification numbers

City	Market	System	Billing	RSA
Appleton, Oshkosh	125	00217	30039	MSA
Fond du Lac	716	00217	30121	WI-9

For full coverage area, see Green Bay, WI and
Madison, WI.

Cellular company

Cellulink
Pacific Telecom Cellular
4600 W College Avenue
Appleton, WI 54915

Customer service	800-236-9700
. . . .	414-841-1111
Corporate office	414-841-1100

Sending calls

Visitor dialing instructions

Local calls	area code + 7 digits
Long distance	area code + 7 digits

Speed-dial numbers

Customer service	*611
Emergency	911
Roaming information	*711

Receiving calls

Caller dials the roamer access number below,
hears tone, dials cellular area code and number.

Appleton	414-585-7626
Fond du Lac	414-579-7626

Follow Me Roaming forwarding (see p. 1156)
Activate, dial *18 send. Deactivate dial *19 send.

Billing information

Visitor rates

Most visitors	$3 day + 85¢ minute

Visitors without roaming agreements (p. 1159)
Cannot call since company takes no credit cards.

Identification numbers

City	Market	System	Billing	RSA
Appleton, Oshkosh	125	00240	00000	MSA
Fond du Lac	716	00240	00000	WI-9

For full coverage area, see Waupaca, WI.

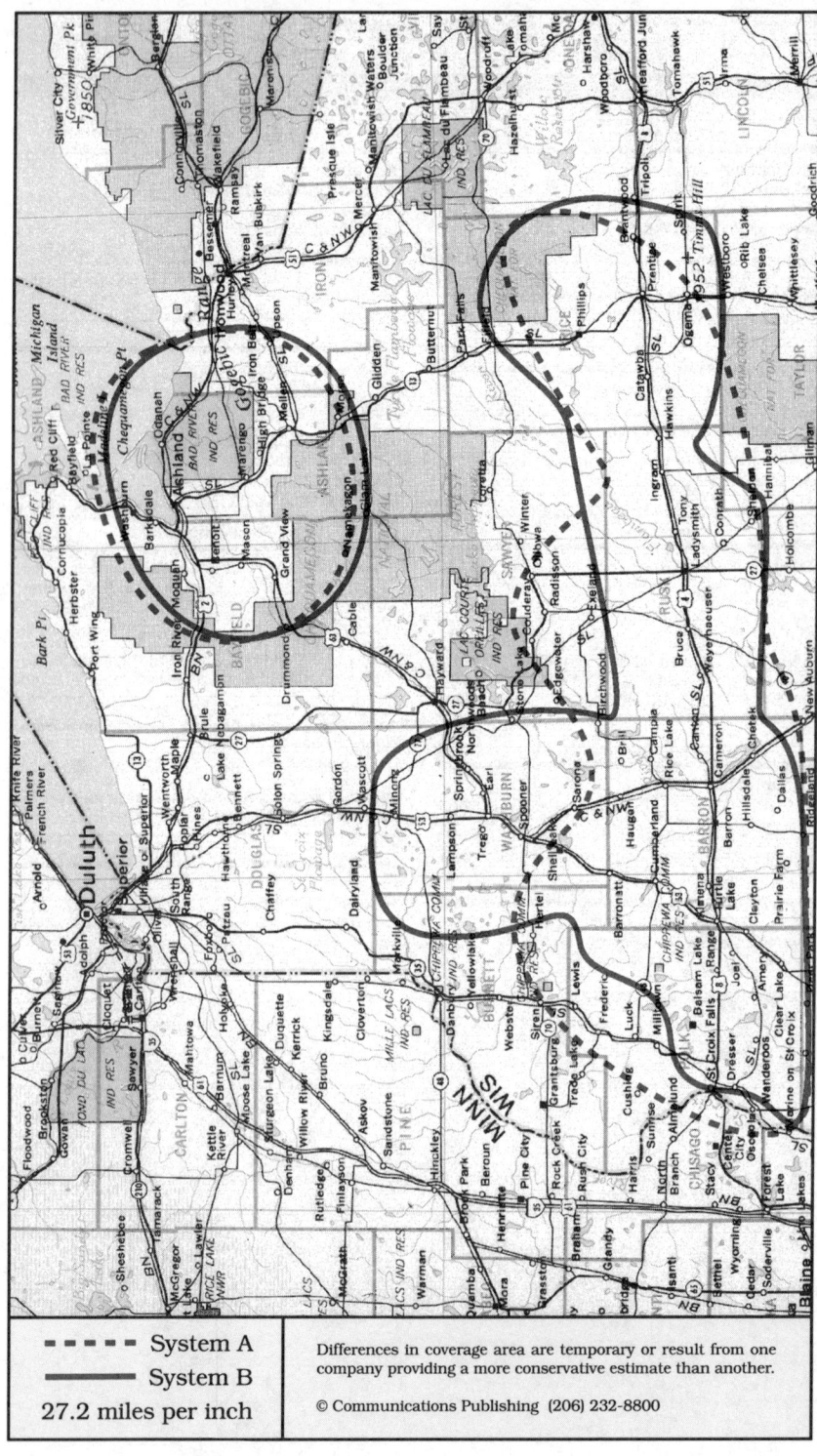

- - - - - System A
———— System B
27.2 miles per inch

Differences in coverage area are temporary or result from one company providing a more conservative estimate than another.

© Communications Publishing (206) 232-8800

Ashland, Wisconsin
Ashland, Ladysmith, Phillips, Rice Lake, Spooner

System A	System B

Cellular company

Cellular One
Cellular Information Systems
55 Holly Hill Lane #101, PO Box 4507
Greenwich, CT 06830

Corporate office 203-622-6317

Sending calls

Visitor dialing instructions
Local calls . . . . . area code + 7 digits
Long distance . . . 1 + area code + 7 digits

Speed-dial numbers
Customer service *611
Emergency 911

Receiving calls

Caller dials the roamer access number below,
hears tone, dials cellular area code and number.
Ashland 715-682-8726
Rice Lake 715-296-7626

NationLink & RoamingAmerica (see p. 1157)
Dial *31 send to receive calls, *30 to deactivate.
This company's subscribers have level 3 service.

Billing information

Visitor rates
Most visitors . . . . . $3 day + 85¢ minute

Visitors without roaming agreements (p. 1159)
Call co. to use Amex, Mastercard, Visa

Identification numbers

City	Market	System	Billing	RSA
Ashland, Ladysmith	709	01805	00000	WI-2
Rice Lake, Spooner	708	01803	00000	WI-1

For full coverage area, see Eau Claire, WI and
Menomonie, WI.

Cellular company

Cellulink
Pacific Telecom Cellular
4600 W College Avenue
Appleton, WI 54915

Customer service 800-236-9700
. . . . 414-841-1111
Corporate office 414-841-1100

Sending calls

Visitor dialing instructions
Local calls . . . . . . area code + 7 digits
Long distance . . . . . area code + 7 digits

Speed-dial numbers
Customer service *611
Emergency 911
Roaming information *711

Receiving calls

Caller dials the roamer access number below,
hears tone, dials cellular area code and number.
Rice Lake, Spooner 715-491-7626
Ashland, Ladysmith 715-492-7626

Billing information

Visitor rates
Most visitors . . . . . $3 day + 85¢ minute

Visitors without roaming agreements (p. 1159)
Cannot call since company takes no credit cards.

Identification numbers

City	Market	System	Billing	RSA
Rice Lake	708	01804	00000	WI-1
Ashland, Ladysmith	709	01806	00000	WI-2

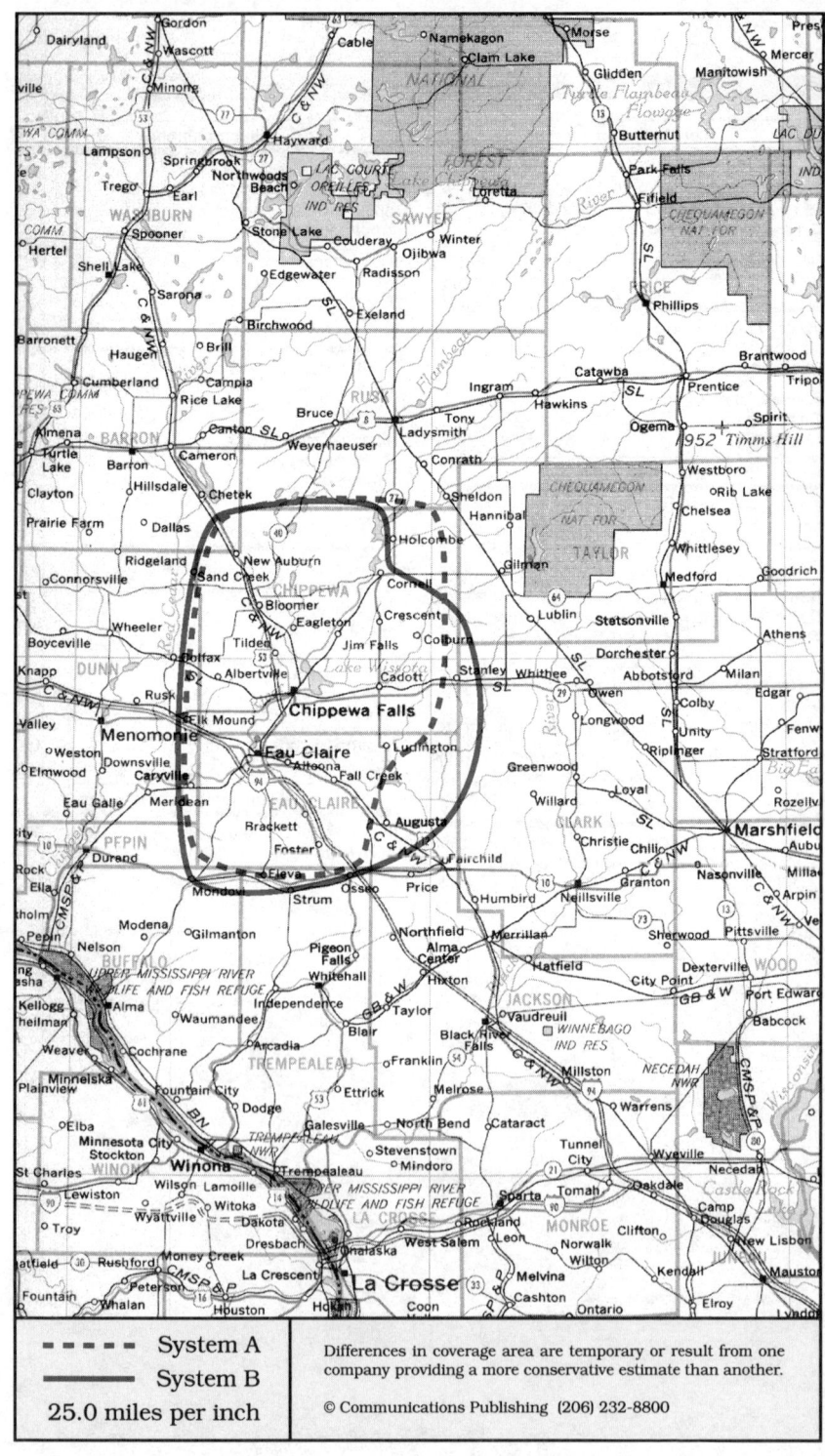

- - - - - System A	Differences in coverage area are temporary or result from one
———— System B	company providing a more conservative estimate than another.
25.0 miles per inch	© Communications Publishing (206) 232-8800

Eau Claire, Wisconsin
Chippewa Falls, Eau Claire

System A	System B

Cellular company

Cellular One
Cellular Information Systems
55 Holly Hill Lane #101, PO Box 4507
Greenwich, CT 06830

Corporate office 203-622-6317

Sending calls

Visitor dialing instructions
Local calls . . . . . area code + 7 digits
Long distance . . . 1 + area code + 7 digits

Speed-dial numbers
Customer service *611
Emergency 911

Receiving calls

Caller dials the roamer access number below,
hears tone, dials cellular area code and number.
. . . . 715-828-7626

NationLink & RoamingAmerica (see p. 1157)
Dial *31 send to receive calls, *30 to deactivate.
This company's subscribers have level 3 service.

Billing information

Visitor rates
Most visitors . . . . . $3 day + 85¢ minute

Visitors without roaming agreements (p. 1159)
Call co. to use Amex, Mastercard, Visa

Identification numbers

City	Market	System	Billing	RSA
All served cities	232	00335	00000	MSA

For full coverage area, see Ashland, WI and
Menomonie, WI.

Cellular company

Cellulink
Licensed to: Eau Claire Cellular Telephone
Managed by: Pacific Telecom Cellular
3410 Oakwood Mall Drive
Eau Claire, WI 54701

Customer service 800-597-5534
Local office 715-835-7370
Corporate office 414-841-1100

Sending calls

Visitor dialing instructions
Local calls . . . . . area code + 7 digits
Long distance . . . 1 + area code + 7 digits

Speed-dial numbers
Customer service *611
Emergency 911
Time 715-834-0123

Receiving calls

Caller dials the roamer access number below,
hears tone, dials cellular area code and number.
. . . . 715-577-7626

Follow Me Roaming forwarding (see p. 1156)
Activate, dial *18 send. Deactivate dial *19 send.

Billing information

Visitor rates
Most visitors . . . . . $3 day + 95¢ minute

Visitors without roaming agreements (p. 1159)
Call co. to use Amex, Discover, Mastercard, Visa

Identification numbers

City	Market	System	Billing	RSA
All served cities	232	00318	00000	MSA

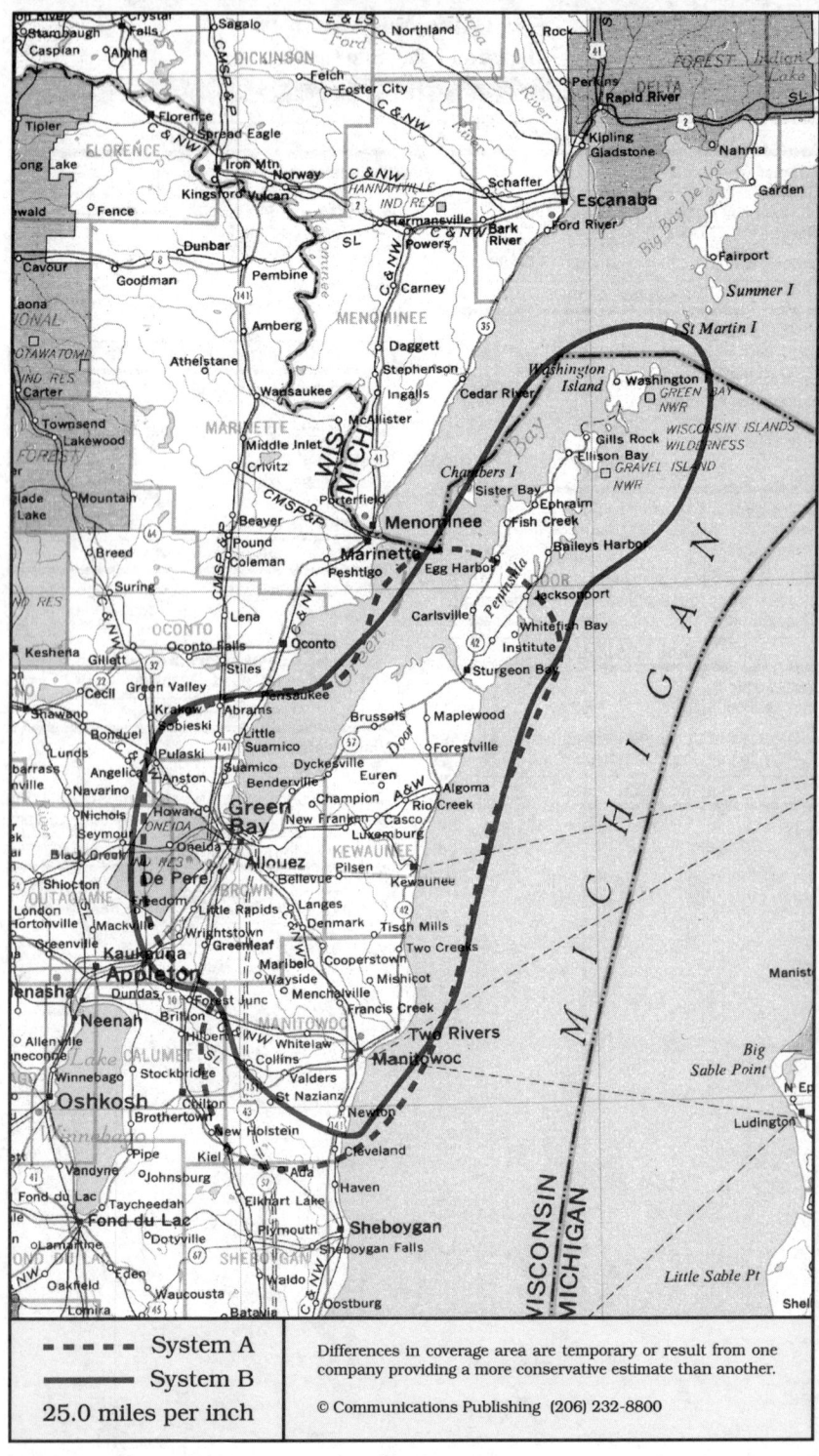

- - - - - System A	Differences in coverage area are temporary or result from one
───── System B	company providing a more conservative estimate than another.
25.0 miles per inch	© Communications Publishing (206) 232-8800

Green Bay, Wisconsin
Algoma, Ellison Bay, Green Bay, Manitowoc, Sturgeon Bay, Two Rivers

System A	System B

Cellular company

Cellular One
CellSouth Cellular
813 S Oneida Street
Green Bay, WI 54304

Customer service	. . . . .	800-888-5611
. . . . . .		414-496-2273
Local office	. . . . .	414-496-4385
Corporate office	. . . . .	404-604-6100

Sending calls

Visitor dialing instructions

Local calls	. . . . .	area code + 7 digits
Long distance	. . .	1 + area code + 7 digits

Speed-dial numbers

Customer service	611
Emergency	911

Receiving calls

Caller dials the roamer access number below,
hears tone, dials cellular area code and number.

Green Bay, Sturgeon Bay	. .	414-428-7626
Manitowoc		414-254-7626

NationLink & RoamingAmerica (see p. 1157)
Dial *31 send to receive calls, *30 to deactivate.
Dial *32 send to tell caller the roamer access no.
This company's subscribers have level 6 service.

Billing information

Visitor rates

Most visitors	. . . . .	$3 day + 99¢ minute

Visitors without roaming agreements (p. 1159)
Call co. to use Amex, Discover, Mastercard, Visa

Identification numbers

City	Market	System	Billing	RSA
Green Bay	186	00217	30039	MSA
Manitowoc	717	00217	00000	WI-10
Sturgeon Bay	717	00005	00000	WI-10

WI-10 recently acquired from GenCell
Management. A billing number may be added.

For full coverage area, see Appleton, WI.

Cellular company

Cellcom
1556 State Highway 41
De Pere, WI 54115

Customer service	. . . .	414-339-4000
Corporate office	. . . . .	800-236-0055
. . . .		414-339-4010

Sending calls

Visitor dialing instructions

Local calls	. . . . . .	area code + 7 digits
Long distance	. . . . .	area code + 7 digits

Speed-dial numbers

Customer service	611
Emergency	911

Receiving calls

Caller dials the roamer access number below,
hears tone, dials cellular area code and number.

Green Bay	. . . .	414-621-7626
Manitowoc	. . . .	414-323-7626
Sturgeon Bay	. . . .	414-493-7626

Follow Me Roaming forwarding (see p. 1156)
Activate, dial *18 send. Deactivate dial *19 send.

Billing information

Visitor rates

Most visitors	. . . . .	$3 day + 85¢ minute

Visitors without roaming agreements (p. 1159)
Cannot call since company takes no credit cards.

Identification numbers

City	Market	System	Billing	RSA
Green Bay	186	00362	00000	MSA
Manitowoc, S. Bay	717	01822	00000	WI-10

For full coverage area, see Marinette, WI.

- - - - System A
───── System B
25.0 miles per inch

Differences in coverage area are temporary or result from one company providing a more conservative estimate than another.

© Communications Publishing (206) 232-8800

La Crosse, Wisconsin

System A

Cellular company

United States Cellular
9360 U.S. Highway 16
Onalaska, WI 54650

Local office	608-781-2600
Corporate office	312-399-8900

Sending calls

Visitor dialing instructions

Local calls	area code + 7 digits
Long distance	area code + 7 digits

Speed-dial numbers

Customer service	*611
Emergency	911
Roaming information	*711

Receiving calls

Caller dials the roamer access number below, hears tone, dials cellular area code and number.
. . . . 608-792-7626

Billing information

Visitor rates

Most visitors	$3 day + 75¢ minute

Visitors without roaming agreements (p. 1159)
Call co. to use Mastercard, Visa

Identification numbers

City	Market	System	Billing	RSA
All served cities	290	00413	00000	MSA

For full coverage area, see Tomah, WI.

System B

Cellular company

Century Cellunet
1824 E Main Street
Onalaska, WI 54650

Local office	608-781-9000
Corporate office	318-388-9000

Sending calls

Visitor dialing instructions

Local calls	area code + 7 digits
Long distance . . .	1 + area code + 7 digits

Speed-dial numbers

Customer service	*611
Emergency	911

Receiving calls

Caller dials the roamer access number below, hears tone, dials cellular area code and number.
. . . . 608-780-7626

Follow Me Roaming forwarding (see p. 1156)
Activate, dial *18 send. Deactivate dial *19 send.

Billing information

Visitor rates

Most visitors	$3 day + 75¢ minute

Visitors without roaming agreements (p. 1159)
Cannot call since company takes no credit cards.

Identification numbers

City	Market	System	Billing	RSA
All served cities	290	00396	00000	MSA

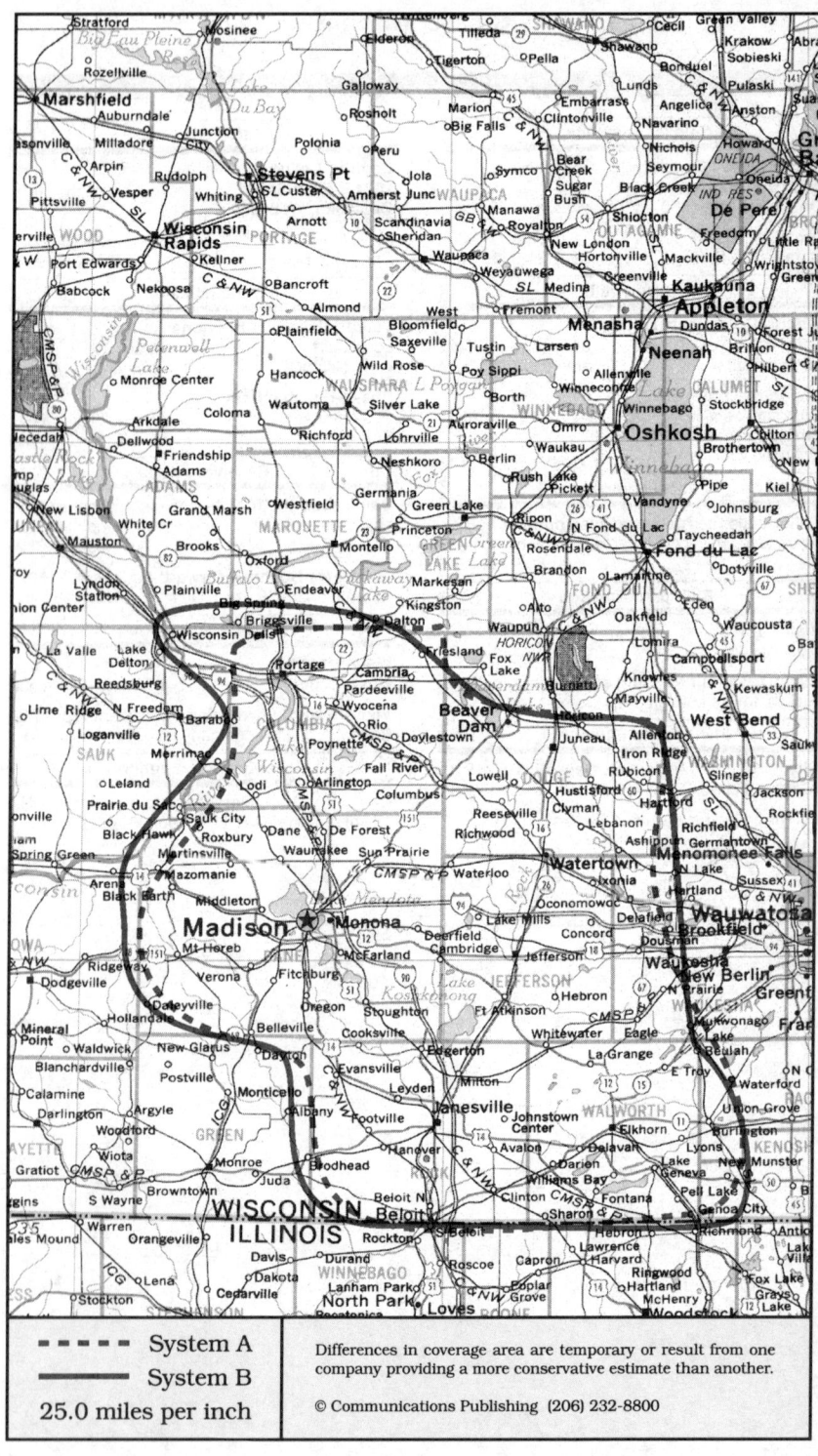

- - - - System A	Differences in coverage area are temporary or result from one
—— System B	company providing a more conservative estimate than another.
25.0 miles per inch	© Communications Publishing (206) 232-8800

Madison, Wisconsin
Beaver Dam, Beloit, Janesville, Lake Geneva, Madison, Portage, Watertown

System A	System B

Cellular company

Cellular One
BellSouth Cellular
3000 Milton Avenue, Unit 113
Janesville, WI 53545

1702 W Beltline Highway
Madison, WI 53713

Customer service	. . . . .		800-888-5611
	. . . .		414-948-2273
	. . . .		608-271-2273
Local office	. . .	Janesville	608-752-6699
	. . . .	Madison	608-283-0010
Corporate office	. . . .		404-604-6100

Sending calls

Visitor dialing instructions
Local calls . . . . area code + 7 digits
Long distance . . . 1 + area code + 7 digits

Speed-dial numbers
Customer service	. . . .	611
Emergency	. . . .	911
WIBA radio	. . . .	∗101
WTDY radio talk show	. . . . . .	∗123
WTSO radio	. . . .	∗6397

Receiving calls

Caller dials the roamer access number below,
hears tone, dials cellular area code and number.
Beloit, Janesville	. . . .	608-751-7626
Madison, Portage	. . . .	608-575-7626
Beaver Dam, Lake G., Watertown		414-948-7626

NationLink & RoamingAmerica (see p. 1157)
Dial ∗31 send to receive calls, ∗30 to deactivate.
Dial ∗32 send to tell caller the roamer access no.
This company's subscribers have level 6 service.

Billing information

Visitor rates
Most visitors . . . . . $3 day + 99¢ minute

Visitors without roaming agreements (p. 1159)
Call co. to use Amex, Discover, Mastercard, Visa

Identification numbers

City	Market	System	Billing	RSA
Beaver Dam	716	00217	30263	WI-9
Beloit, Janesville	216	00217	30041	MSA
Lake Geneva	716	00217	30221	WI-9
Madison	113	00217	00000	MSA
Portage	716	00217	30117	WI-9
Watertown	716	00217	30265	WI-9

For full coverage area, see Rockford, IL and
Appleton, WI and Monroe, WI.

Cellular company

Ameritech Mobile Communications
Administration
4781 Hayes Road, Suite 100
Madison, WI 53704

Sales
2501 W Beltline Highway
Madison, WI 53713

Customer service	. . . .		800-221-0994
	. . . .		708-706-7300
Local office	. .	administration	608-241-2733
	. . . .	sales	608-274-1439
Corporate office	. . . .		708-706-7600

Sending calls

Visitor dialing instructions
Local calls . . . . . . area code + 7 digits
Long distance . . . . area code + 7 digits

Speed-dial numbers
Customer service	. . . .	∗611
Directory assistance	. . . . . .	411
Emergency	. . . .	∗911
Make a traffic report WIBA	. . . . .	211
Recorded help message	. . . . .	∗90
Roaming information	. . . . .	∗711
Technical department for problems	. .	∗712

Receiving calls

Caller dials the roamer access number below,
hears tone, dials cellular area code and number.
. . . . 608-695-7626

Follow Me Roaming forwarding (see p. 1156)
Activate, dial ∗18 send. Deactivate dial ∗19 send.

Billing information

Visitor rates
Most visitors . . . . . $3 day + 85¢ minute

Visitors without roaming agreements (p. 1159)
Cannot call since company takes no credit cards.

Identification numbers

City	Market	System	Billing	RSA
Beloit, Janesville	216	00044	00000	MSA
Madison	113	00044	00000	MSA
Portage, Watertown	716	00044	00000	WI-9

For full coverage area, see Milwaukee, WI.

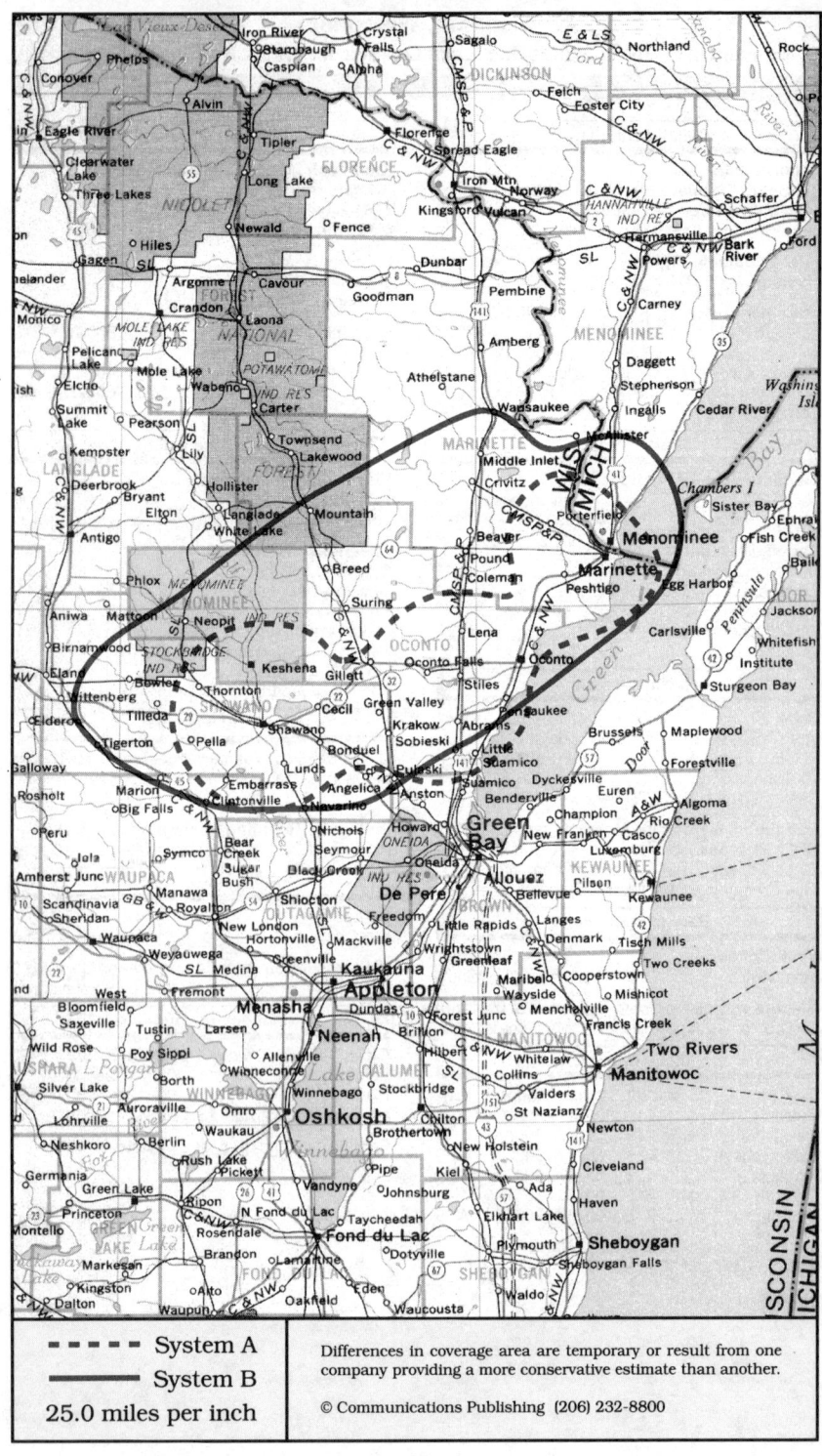

- - - - - System A	Differences in coverage area are temporary or result from one
———— System B	company providing a more conservative estimate than another.
25.0 miles per inch	© Communications Publishing (206) 232-8800

System A	System B

Cellular company

Cellular One
Quantum Communications
1738 Main Street
Marinette, WI 54143

Customer service	414-664-2273
Local office	715-732-2355
Corporate office	612-942-7650

Sending calls

Visitor dialing instructions

Local calls	area code + 7 digits
Long distance . . .	1 + area code + 7 digits

Speed-dial numbers

Customer service	611
Emergency	911

Receiving calls

Caller dials the roamer access number below,
hears tone, dials cellular area code and number.

Marinette	715-587-7626
Oconto	414-664-7626
Shawano	715-584-7626

NationLink & RoamingAmerica (see p. 1157)
Dial *31 send to receive calls, *30 to deactivate.
Dial *32 send to tell caller the roamer access no.
This company's subscribers have level 6 service.

Billing information

Visitor rates

Most visitors	$0 day + 99¢ minute

Visitors without roaming agreements (p. 1159)
Cannot call since company takes no credit cards.

Identification numbers

City	Market	System	Billing	RSA
All served cities	711	01809	00000	WI-4

Cellular company

Cellcom
1556 State Highway 41 S
De Pere, WI 54115

Customer service	414-339-4000
Corporate office	800-236-0055
. . . .	414-339-4010

Sending calls

Visitor dialing instructions

Local calls	area code + 7 digits
Long distance	area code + 7 digits

Speed-dial numbers

Customer service	611
Emergency	911

Receiving calls

Caller dials the roamer access number below,
hears tone, dials cellular area code and number.

Marinette	715-923-7626
Oconto	414-373-7626
Shawano	715-853-7626

Follow Me Roaming forwarding (see p. 1156)
Activate, dial *18 send. Deactivate dial *19 send.

Billing information

Visitor rates

Most visitors	$3 day + 85¢ minute

Visitors without roaming agreements (p. 1159)
Cannot call since company takes no credit cards.

Identification numbers

City	Market	System	Billing	RSA
All served cities	711	01810	00000	WI-4

For full coverage area, see Green Bay, WI and
Manitowoc, WI.

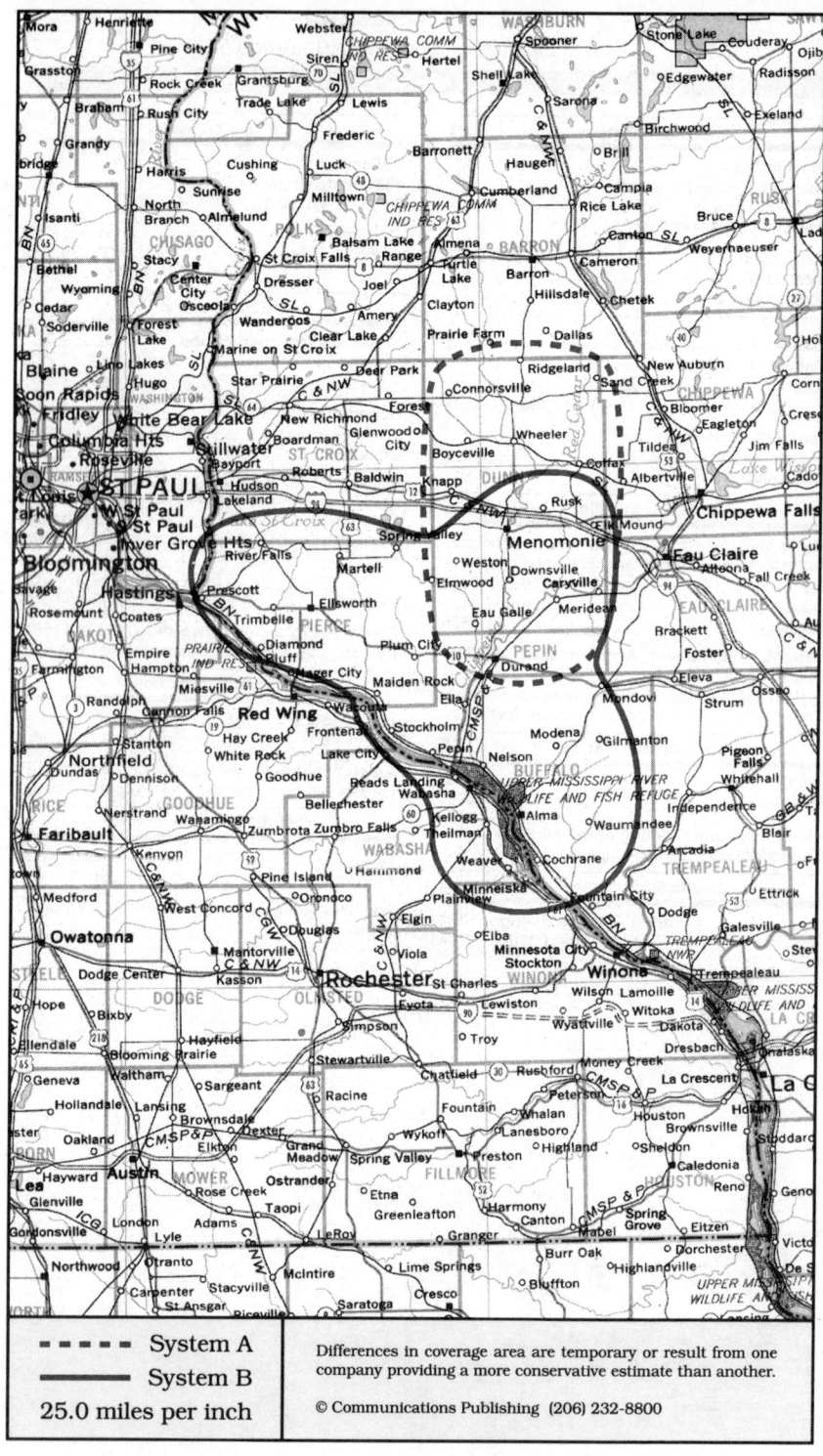

-----	System A
———	System B
25.0 miles per inch	

Differences in coverage area are temporary or result from one company providing a more conservative estimate than another.

© Communications Publishing (206) 232-8800

Menomonie, Wisconsin
Alma, Durand, Ellsworth, Menomonie

System A	System B

Cellular company

Cellular One
Cellular Information Systems
55 Holly Hill Lane #101, PO Box 4507
Greenwich, CT 06830

Corporate office 203-622-6317

Sending calls

Visitor dialing instructions
Local calls . . . . . area code + 7 digits
Long distance . . . 1 + area code + 7 digits

Speed-dial numbers
Customer service *611
Emergency 911

Receiving calls

Caller dials the roamer access number below,
hears tone, dials cellular area code and number.
Ashland 715-682-8726
Rice Lake 715-296-7626

Billing information

Visitor rates
Most visitors . . . . . $3 day + 85¢ minute

Visitors without roaming agreements (p. 1159)
Call co. to use Amex, Mastercard, Visa

Identification numbers

City	Market	System	Billing	RSA
All served cities	712	01811	00000	WI-5

For full coverage area, see Eau Claire, WI and
Menomonie, WI.

Cellular company

Cellcom
872 N Broadway
Menomonie, WI 54751

**Purchased by Cellcom but may be doing
business as Cellulink.**

Local office 715-235-2661
Corporate office 800-236-0055
. . . . 414-339-4010

Sending calls

Visitor dialing instructions
Local calls . . . . . . area code + 7 digits
Long distance . . . 1 + area code + 7 digits

Speed-dial numbers
Customer service 611
Emergency 911
Roaming information 711

Receiving calls

Caller dials the roamer access number below,
hears tone, dials cellular area code and number.
. . . . 715-495-7626

Billing information

Visitor rates
Most visitors . . . . . $3 day + 85¢ minute

Visitors without roaming agreements (p. 1159)
Cannot call since company takes no credit cards.

Identification numbers

City	Market	System	Billing	RSA
All served cities	712	01812	00000	WI-5

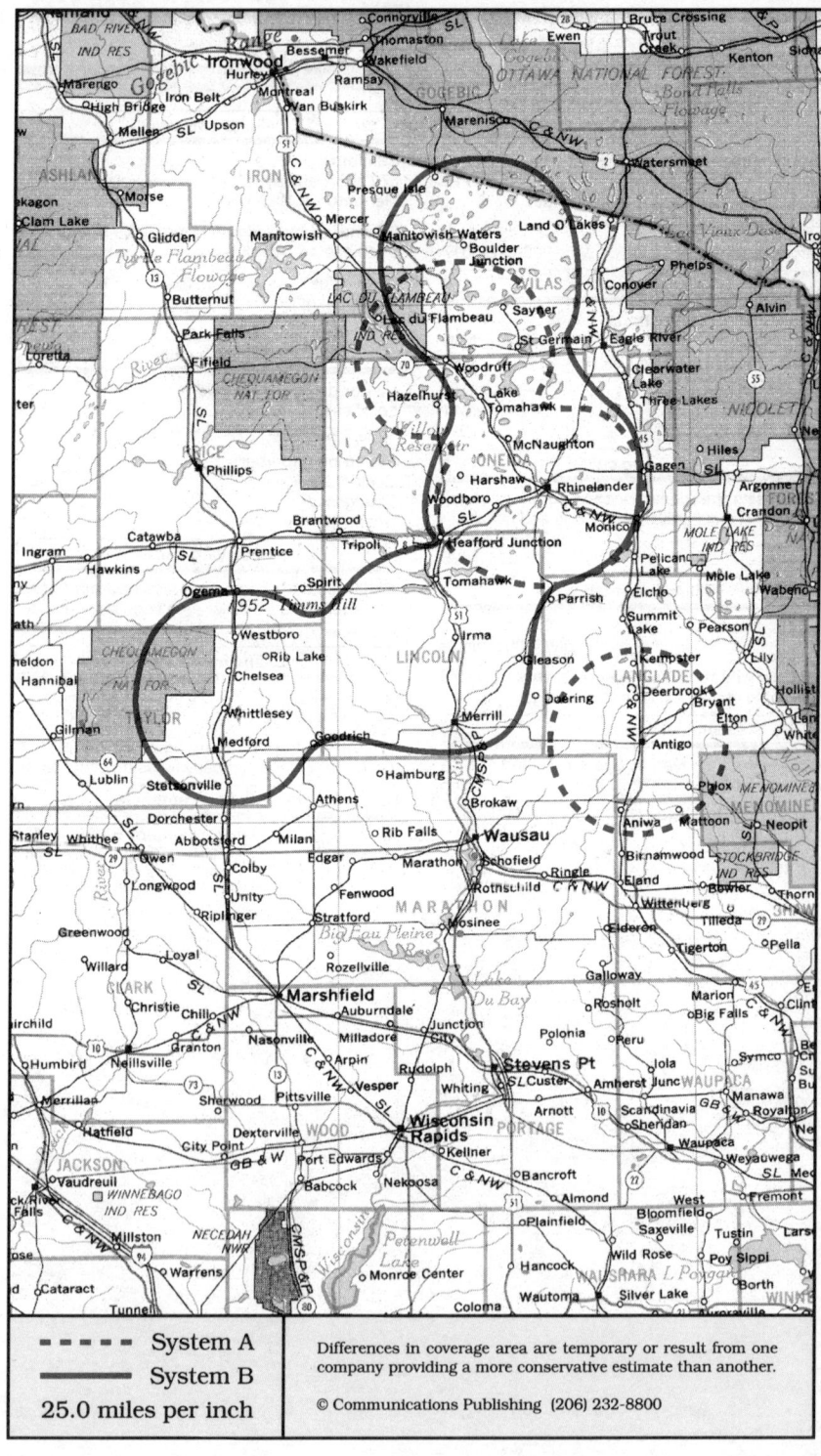

Legend	
- - - - - System A	
——— System B	
25.0 miles per inch	

Differences in coverage area are temporary or result from one company providing a more conservative estimate than another.

© Communications Publishing (206) 232-8800

Merrill, Wisconsin

Antigo, Medford, Merrill, Minocqua, Rhinelander

System A	System B

Cellular company

Cellular Information Systems
55 Holly Hill Lane #101, PO Box 4507
Greenwich, CT 06830

Corporate office 203-622-6317

Sending calls

Visitor dialing instructions
Local calls . . . . . area code + 7 digits
Long distance . . . 1 + area code + 7 digits

Speed-dial numbers
Customer service 611
Emergency 911

Receiving calls

Caller dials the roamer access number below, hears tone, dials cellular area code and number.
Antigo 715-623-1900
Minocqua 715-358-2026
Rhinelander 715-367-7626

NationLink & RoamingAmerica (see p. 1157)
Dial ∗31 send to receive calls, ∗30 to deactivate.
This company's subscribers have level 3 service.

Billing information

Visitor rates
Most visitors . . . . . $3 day + 85¢ minute

Visitors without roaming agreements (p. 1159)
Call co. to use Amex, Mastercard, Visa

Identification numbers

City	Market	System	Billing	RSA
All served cities	710	01807	00000	WI-3

Cellular company

Cellulink
Pacific Telecom Cellular
4600 W College Avenue
Appleton, WI 54915

Customer service 800-236-9700
. . . . 414-841-1111
Corporate office 414-841-1100

Sending calls

Visitor dialing instructions
Local calls . . . . . . area code + 7 digits
Long distance . . . . . area code + 7 digits

Speed-dial numbers
Customer service . . . . ∗611
Emergency 911
Roaming information . . . . ∗711

Receiving calls

Caller dials the roamer access number below, hears tone, dials cellular area code and number.
. . . . 715-493-7626

Follow Me Roaming forwarding (see p. 1156)
Activate, dial ∗18 send. Deactivate dial ∗19 send.

Billing information

Visitor rates
Most visitors . . . . . $3 day + 85¢ minute

Visitors without roaming agreements (p. 1159)
Cannot call since company takes no credit cards.

Identification numbers

City	Market	System	Billing	RSA
All served cities	710	01808	00000	WI-3

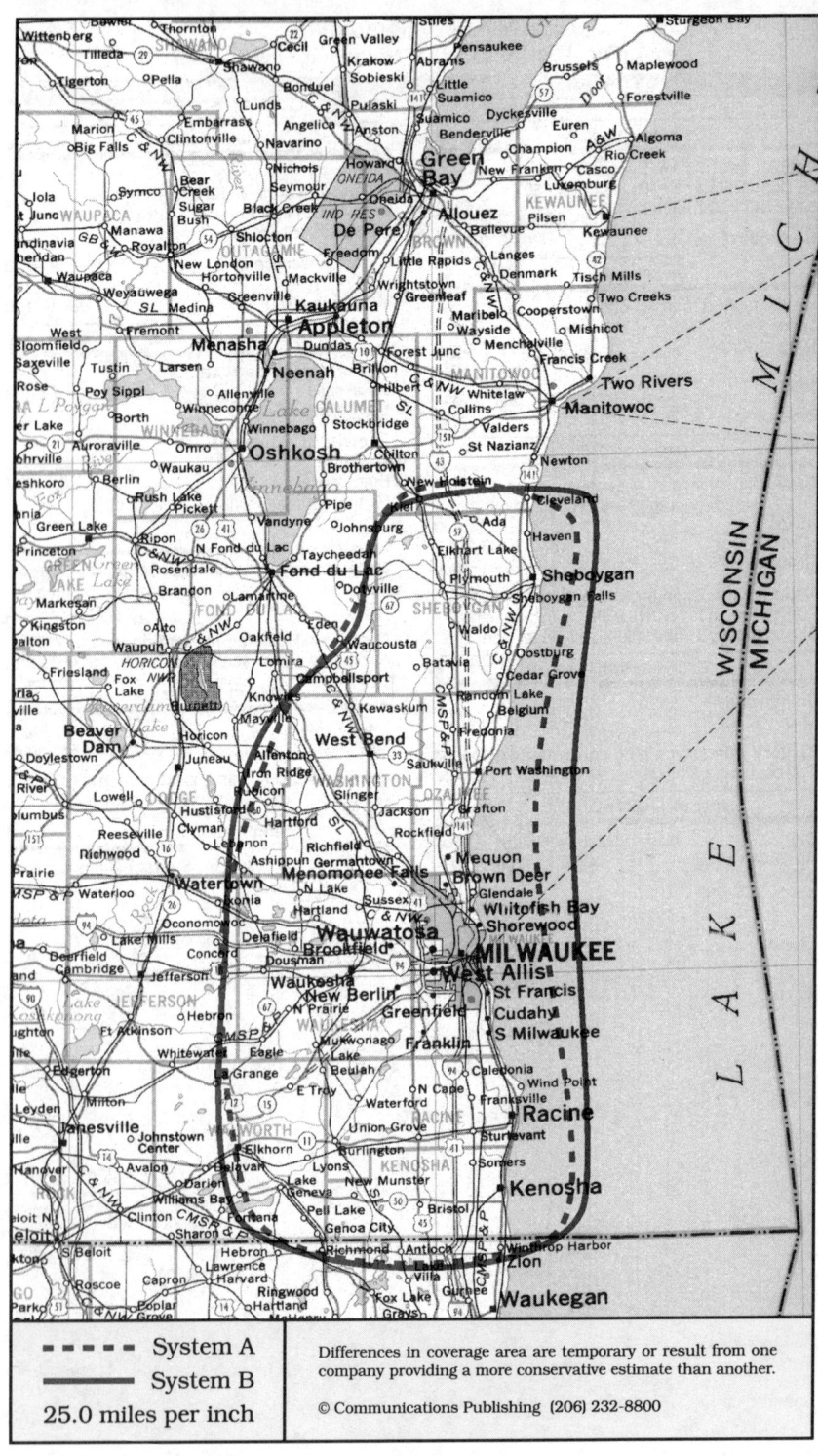

▪ ▪ ▪ ▪	System A
▬▬▬	System B
25.0 miles per inch	

Differences in coverage area are temporary or result from one company providing a more conservative estimate than another.

© Communications Publishing (206) 232-8800

Milwaukee, Wisconsin
Kenosha, Milwaukee, Racine, Sheboygan, Waukesha, West Bend

System A	System B

Cellular company

Cellular One
BellSouth Cellular
Main office
3545 N 124th Street, PO Box 806
Brookfield, WI 53008-0806

740 N Water Street
Milwaukee, WI 53202

6100 Washington Avenue
Racine, WI 53406

3124 Wilgus Road
Sheboygan, WI 53081

Customer service	some areas	800-894-7611
. . . .		414-783-5505
Local office . . .	Brookfield	414-783-5500
. . . .	Milwaukee	414-291-3140
. . . .	Racine	414-884-7000
. . . .	Sheboygan	414-459-4100
Corporate office .	BellSouth	404-604-6100

Sending calls

Visitor dialing instructions
Local calls	area code + 7 digits
Long distance . . .	1 + area code + 7 digits

Speed-dial numbers
Customer service	611
Emergency	911
WISN traffic	#ISN
WISN television news	#12
WKLH hotline	#KLH or #554
WLTQ traffic	#973
WQFM contest	#93
WOKY traffic	#920
WMIL traffic	#106
WKTI contest	#9494

Receiving calls

Caller dials the roamer access number below,
hears tone, dials cellular area code and number.
Milwaukee, Sheboygan . . .	414-254-7626
Racine	414-939-7626

NationLink & RoamingAmerica (see p. 1157)
Dial *31 send to receive calls, *30 to deactivate.
Dial *32 send to tell caller the roamer access no.
This company's subscribers have level 6 service.

Billing information

Visitor rates
Most visitors	$3 day + 95¢ minute

Visitors without roaming agreements (p. 1159)
Cannot call since company takes no credit cards.

Identification numbers

City	Market	System	Billing	RSA
Kenosha	244	00005	30035	MSA
Milwaukee	021	00005	00000	MSA
Racine	189	00005	30037	MSA
Sheboygan	277	00005	30121	MSA

For full coverage area, see Madison, WI.

Cellular company

Ameritech Mobile Communications
103 E Silver Spring Drive
Whitefish Bay, WI 53217

Customer service	800-221-0994
. . . .	708-706-7300
Local office	414-963-3140
Corporate office	708-706-7600

Sending calls

Visitor dialing instructions
Local calls	area code + 7 digits
Long distance	area code + 7 digits

Speed-dial numbers
Customer service	*611
Directory assistance	411
Emergency	*911
Recorded help message	*90
Roaming information	*711
Technical department for problems . .	*712
WISN television news	*12
WTMJ radio	211

Receiving calls

Caller dials the roamer access number below,
hears tone, dials cellular area code and number.
. . . .	414-791-7626

Follow Me Roaming forwarding (see p. 1156)
Activate, dial *18 send. Deactivate dial *19 send.

Billing information

Visitor rates
Most visitors	$3 day + 85¢ minute

Visitors without roaming agreements (p. 1159)
Cannot call since company takes no credit cards.

Identification numbers

City	Market	System	Billing	RSA
Kenosha	244	00044	00000	MSA
Milwaukee	021	00044	00000	MSA
Racine	189	00044	00000	MSA
Sheboygan	277	00044	00000	MSA

For full coverage area, see Madison, WI.

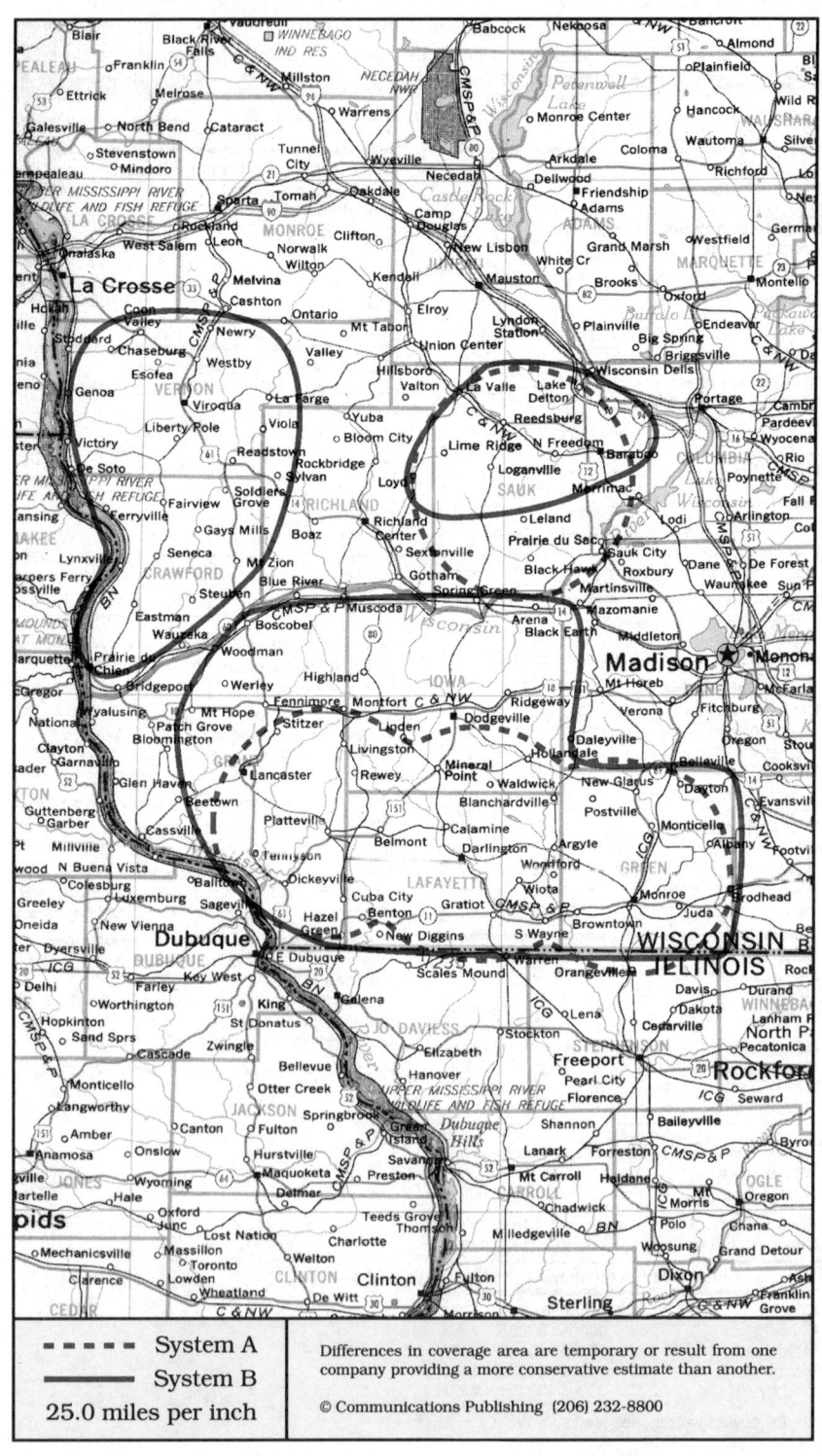

- - - - System A	Differences in coverage area are temporary or result from one company providing a more conservative estimate than another.
——— System B	
25.0 miles per inch	© Communications Publishing (206) 232-8800

Monroe, Wisconsin

Baraboo, Darlington, Dodgeville, Lancaster, Monroe, Platteville, Viroqua

System A	System B

Cellular company

Cellular One
BellSouth Cellular
185 E Pine
Platteville, WI 53818

Customer service	800-888-5611
. . . .	608-778-2273
Local office	608-348-2273
Corporate office	404-604-6100

Sending calls

Visitor dialing instructions

Local calls	area code + 7 digits
Long distance . . .	1 + area code + 7 digits

Speed-dial numbers

Customer service	611
Emergency	911

Receiving calls

Caller dials the roamer access number below,
hears tone, dials cellular area code and number.
. . . . 608-778-7626

NationLink & RoamingAmerica (see p. 1157)
Dial ✳31 send to receive calls, ✳30 to deactivate.
Dial ✳32 send to tell caller the roamer access no.
This company's subscribers have level 6 service.

Billing information

Visitor rates
Most visitors . . . . . $3 day + 99¢ minute

Visitors without roaming agreements (p. 1159)
Call co. to use Amex, Discover, Mastercard, Visa

Identification numbers

City	Market	System	Billing	RSA
All served cities	715	01817	00000	WI-8

For full coverage area, see Madison, WI.

Cellular company

United States Cellular
1700 N Broadway, Suite 120
Rochester, MN 55906

Local office . . . from Monroe	608-328-1234
Corporate office	312-399-8900

Sending calls

Visitor dialing instructions

Local calls	area code + 7 digits
Long distance	area code + 7 digits

Speed-dial numbers

Customer service	✳611
Emergency	✳911
Roaming information	✳711

Receiving calls

Caller dials the roamer access number below,
hears tone, dials cellular area code and number.
. . . . 608-642-7626

Billing information

Visitor rates
Most visitors . . . . . $3 day + 75¢ minute

Visitors without roaming agreements (p. 1159)
Call co. to use Mastercard, Visa

Identification numbers

City	Market	System	Billing	RSA
All served cities	715	01818	00000	WI-8

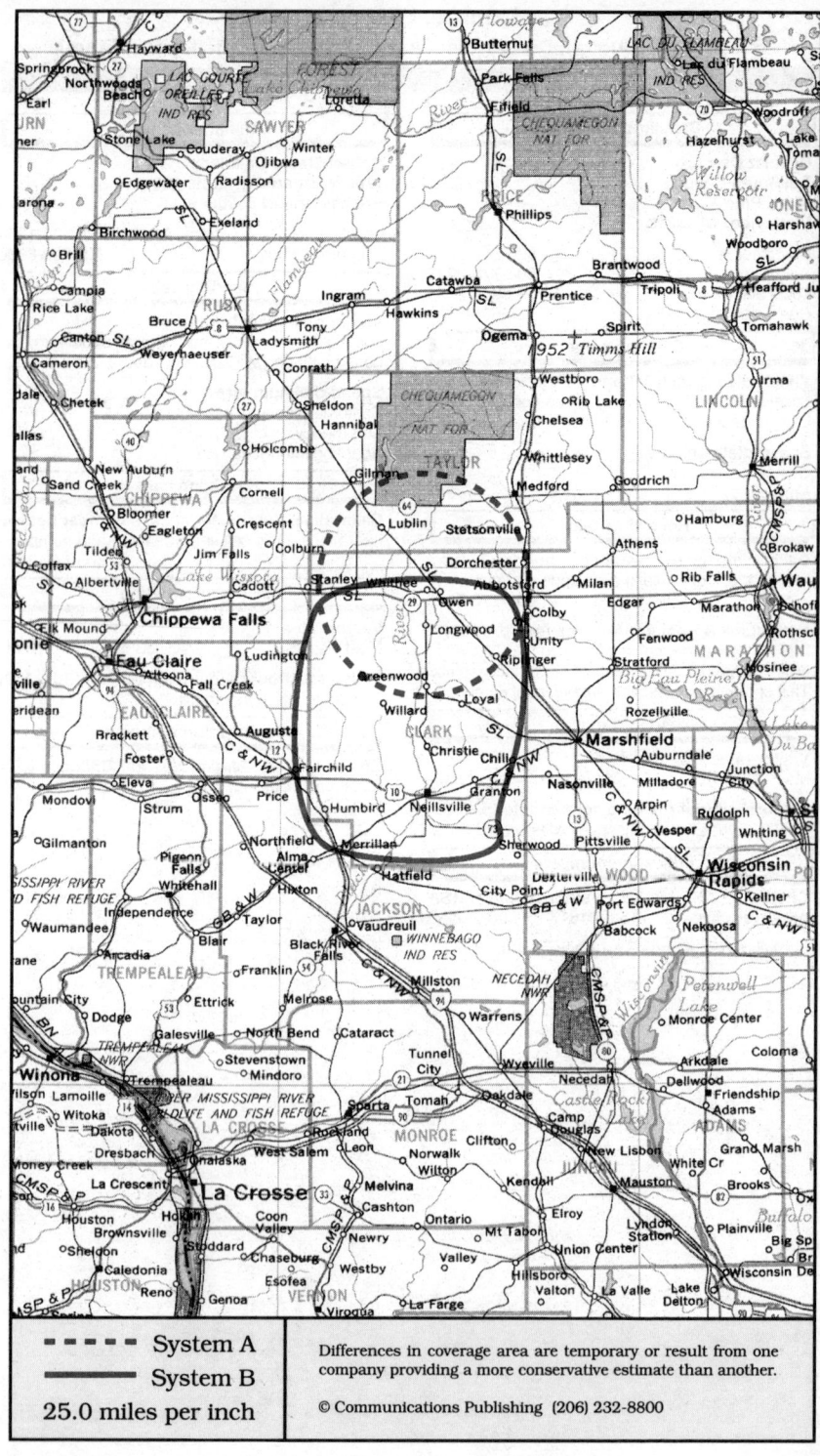

▪ ▪ ▪ ▪ ▪ System A	Differences in coverage area are temporary or result from one
────── System B	company providing a more conservative estimate than another.
25.0 miles per inch	© Communications Publishing (206) 232-8800

Neillsville, Wisconsin
Neillsville (Clark County)

System A	System B

Cellular company

Cellular One
Licensed to: Wausau Cellular
Managed by: Cellular Information Systems
1503 N 6th Street
Wausau, WI 54401

Local office 715-842-7900
Corporate office 203-622-6317

Sending calls

Visitor dialing instructions
Local calls . . . . . area code + 7 digits
Long distance . . . 1 + area code + 7 digits

Speed-dial numbers
Customer service 611
Emergency 911

Receiving calls

Caller dials the roamer access number below,
hears tone, dials cellular area code and number.
. . . . 715-571-7626

Billing information

Visitor rates
Most visitors . . . . . $3 day + 85¢ minute

Visitors without roaming agreements (p. 1159)
Call co. to use Amex, Discover, Mastercard, Visa

Identification numbers

City	Market	System	Billing	RSA
All served cities	713	00591	00000	WI-6

For full coverage area, see Wausau, WI.

Cellular company

Cellulink
Pacific Telecom Cellular
4600 W College Avenue
Appleton, WI 54915

Customer service 800-236-9700
. . . . 414-841-1111
Corporate office 414-841-1100

Sending calls

Visitor dialing instructions
Local calls . . . . . . area code + 7 digits
Long distance . . . . . area code + 7 digits

Speed-dial numbers
Customer service *611
Emergency 911
Roaming information *711

Receiving calls

Caller dials the roamer access number below,
hears tone, dials cellular area code and number.
. . . . 715-797-7626

Follow Me Roaming forwarding (see p. 1156)
Activate, dial *18 send. Deactivate dial *19 send.

Billing information

Visitor rates
Most visitors . . . . . $3 day + 85¢ minute

Visitors without roaming agreements (p. 1159)
Cannot call since company takes no credit cards.

Identification numbers

City	Market	System	Billing	RSA
All served cities	713	01814	00000	WI-6

For full coverage area, see Tomah, WI.

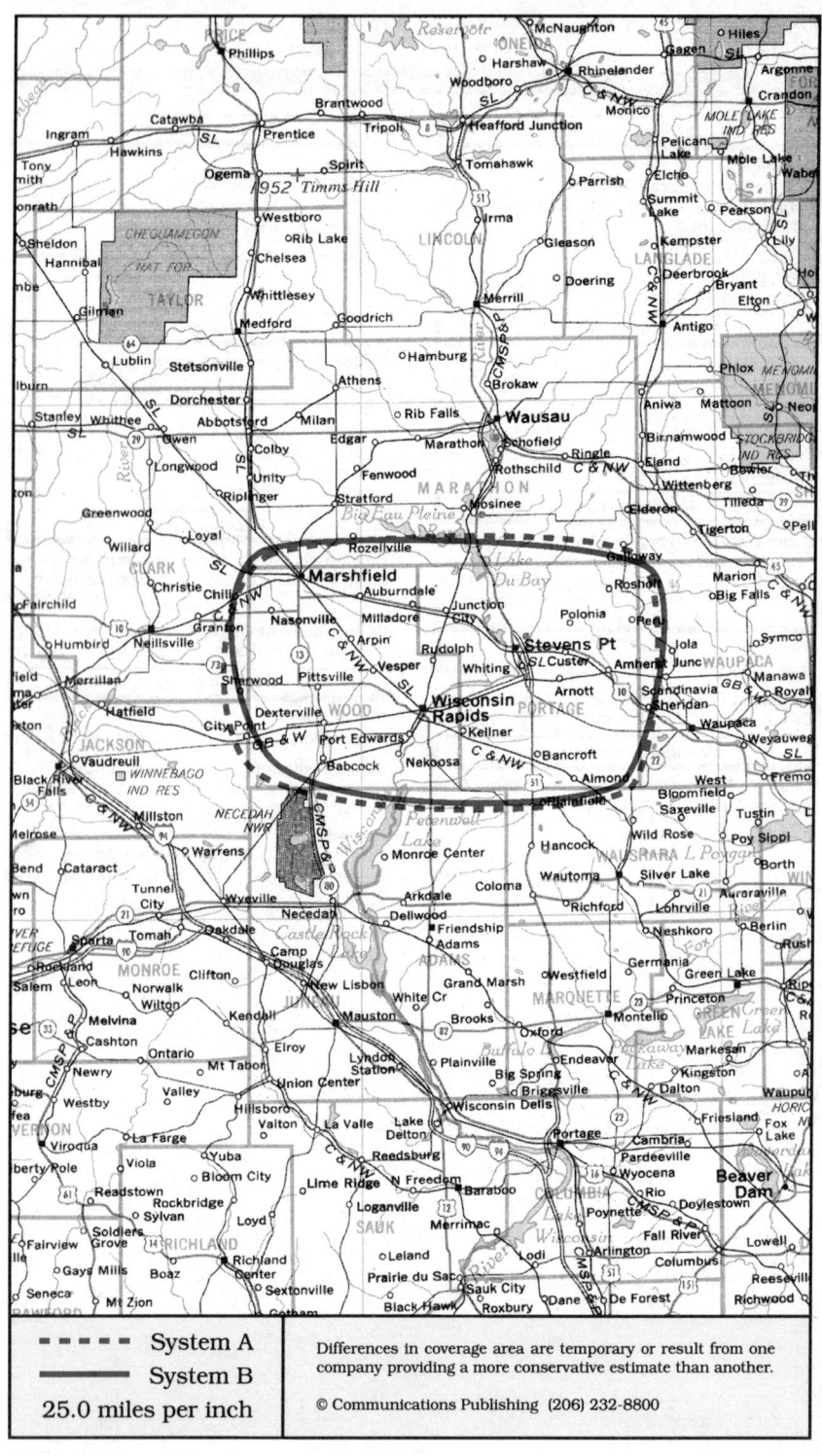

Stevens Point, Wisconsin
Marshfield, Stevens Point, Wisconsin Rapids

System A	System B

Cellular company

Cellular One
Licensed to: CJD Wisconsin Cellular
Managed by: Rural Cellular Management
1619 N Central Avenue
Marshfield, WI 54449

3409 Main Street
Stevens Point, WI 54481

4230 8th Street S
Wisconsin Rapids, WI 54494

Local office . . .	Marshfield	715-387-0700
. . . . . .	Stevens Point	715-345-0101
. . . .	Wisconsin Rapids	715-424-0400
Corporate office	Rural Cell. Mgt	707-422-2100

Sending calls

Visitor dialing instructions

Local calls	area code + 7 digits
Long distance . . .	1 + area code + 7 digits

Speed-dial numbers

Customer service	611
Emergency	911

Receiving calls

Caller dials the roamer access number below,
hears tone, dials cellular area code and number.
. . . . 715-498-7626

NationLink & RoamingAmerica (see p. 1157)
Dial *31 send to receive calls, *30 to deactivate.
Dial *32 send to tell caller the roamer access no.
This company's subscribers have level 9 service.

Billing information

Visitor rates
Most visitors . . . . . $0 day + 99¢ minute

Visitors without roaming agreements (p. 1159)
Call co. to use Amex, Discover, Mastercard, Visa
or Roamer Plus will intercept first call to bill you.

Identification numbers

City	Market	System	Billing	RSA
All served cities	714	01815	00000	WI-7

For full coverage area, see Waupaca, WI.

Cellular company

United States Cellular
35-A Park Ridge Drive
Stevens Point, WI 54481

Local office	715-345-1122
Corporate office	312-399-8900

Sending calls

Visitor dialing instructions

Local calls	area code + 7 digits
Long distance	area code + 7 digits

Speed-dial numbers

Customer service	611
Emergency	911
Roaming information	711

Receiving calls

Caller dials the roamer access number below,
hears tone, dials cellular area code and number.
. . . . 715-572-7626

Follow Me Roaming forwarding (see p. 1156)
Activate, dial *18 send. Deactivate dial *19 send.

Billing information

Visitor rates
Most visitors . . . . . $2 day + 75¢ minute

Visitors without roaming agreements (p. 1159)
Call co. to use Mastercard, Visa

Identification numbers

City	Market	System	Billing	RSA
All served cities	714	00570	30196	WI-7

For full coverage area, see Wausau, WI.

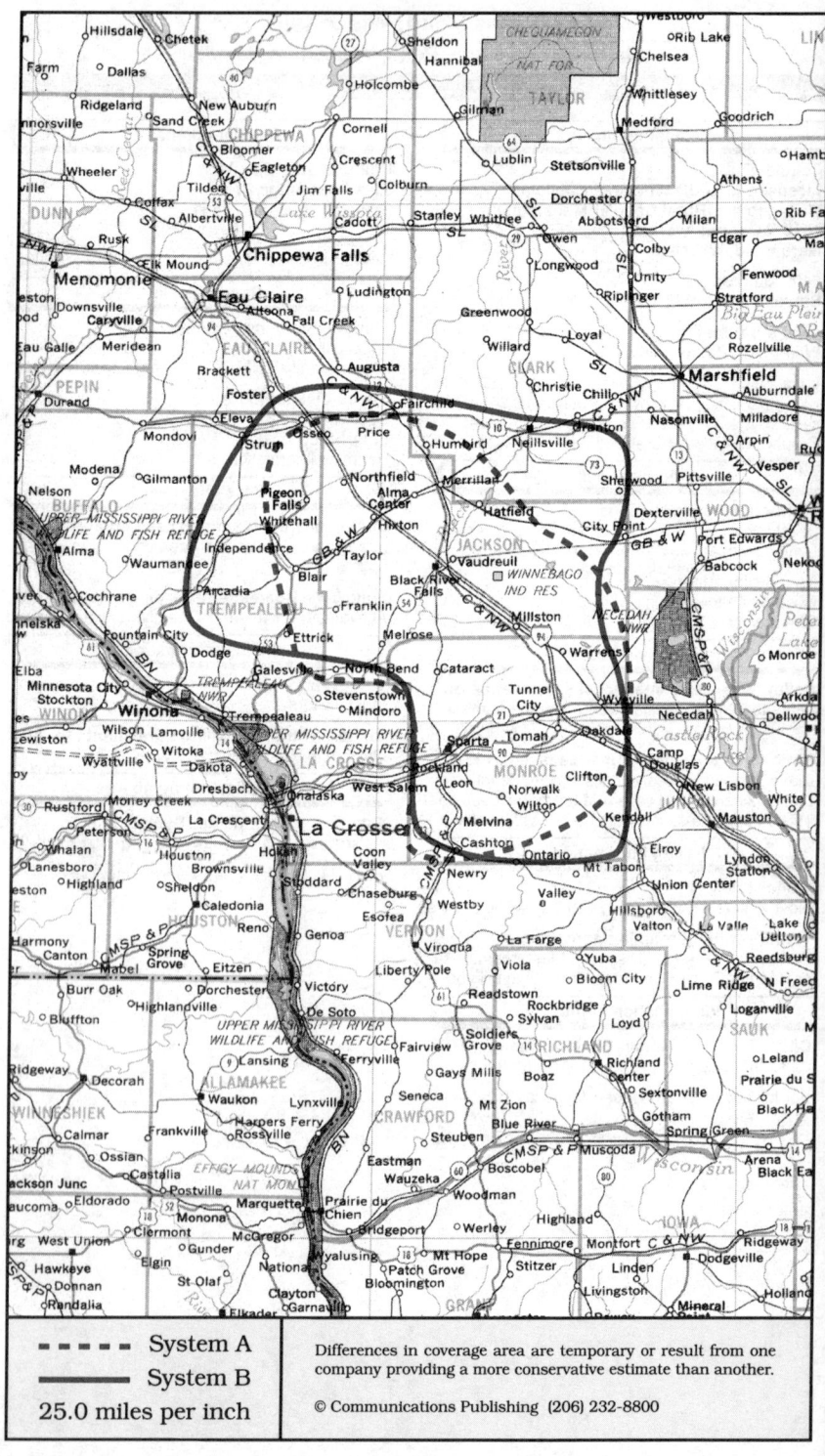

System A
System B
25.0 miles per inch

Differences in coverage area are temporary or result from one company providing a more conservative estimate than another.

© Communications Publishing (206) 232-8800

Tomah, Wisconsin
Black River Falls, Fort McCoy, Sparta, Tomah

System A	System B

Cellular company

United States Cellular
9360 U.S. Highway 16
Onalaska, WI 54650

Local office	608-781-2600
Corporate office	312-399-8900

Sending calls

Visitor dialing instructions

Local calls	area code + 7 digits
Long distance	area code + 7 digits

Speed-dial numbers

Customer service	*611
Emergency	911
Roaming information	*711

Receiving calls

Caller dials the roamer access number below,
hears tone, dials cellular area code and number.
. . . . 608-792-7626

Billing information

Visitor rates

Most visitors	$3 day + 75¢ minute

Visitors without roaming agreements (p. 1159)
Call co. to use Mastercard, Visa

Identification numbers

City	Market	System	Billing	RSA
All served cities	713	01813	00000	WI-6

For full coverage area, see La Crosse, WI.

Cellular company

Cellulink
Pacific Telecom Cellular
4600 W College Avenue
Appleton, WI 54915

Customer service	800-236-9700
. . . .	414-841-1111
Corporate office	414-841-1100

Sending calls

Visitor dialing instructions

Local calls	area code + 7 digits
Long distance	area code + 7 digits

Speed-dial numbers

Customer service	*611
Emergency	911
Roaming information	*711

Receiving calls

Caller dials the roamer access number below,
hears tone, dials cellular area code and number.
. . . . 608-797-7626

Follow Me Roaming forwarding (see p. 1156)
Activate, dial *18 send. Deactivate dial *19 send.

Billing information

Visitor rates

Most visitors	$3 day + 85¢ minute

Visitors without roaming agreements (p. 1159)
Cannot call since company takes no credit cards.

Identification numbers

City	Market	System	Billing	RSA
All served cities	713	01814	00000	WI-6

For full coverage area, see Neillsville, WI.

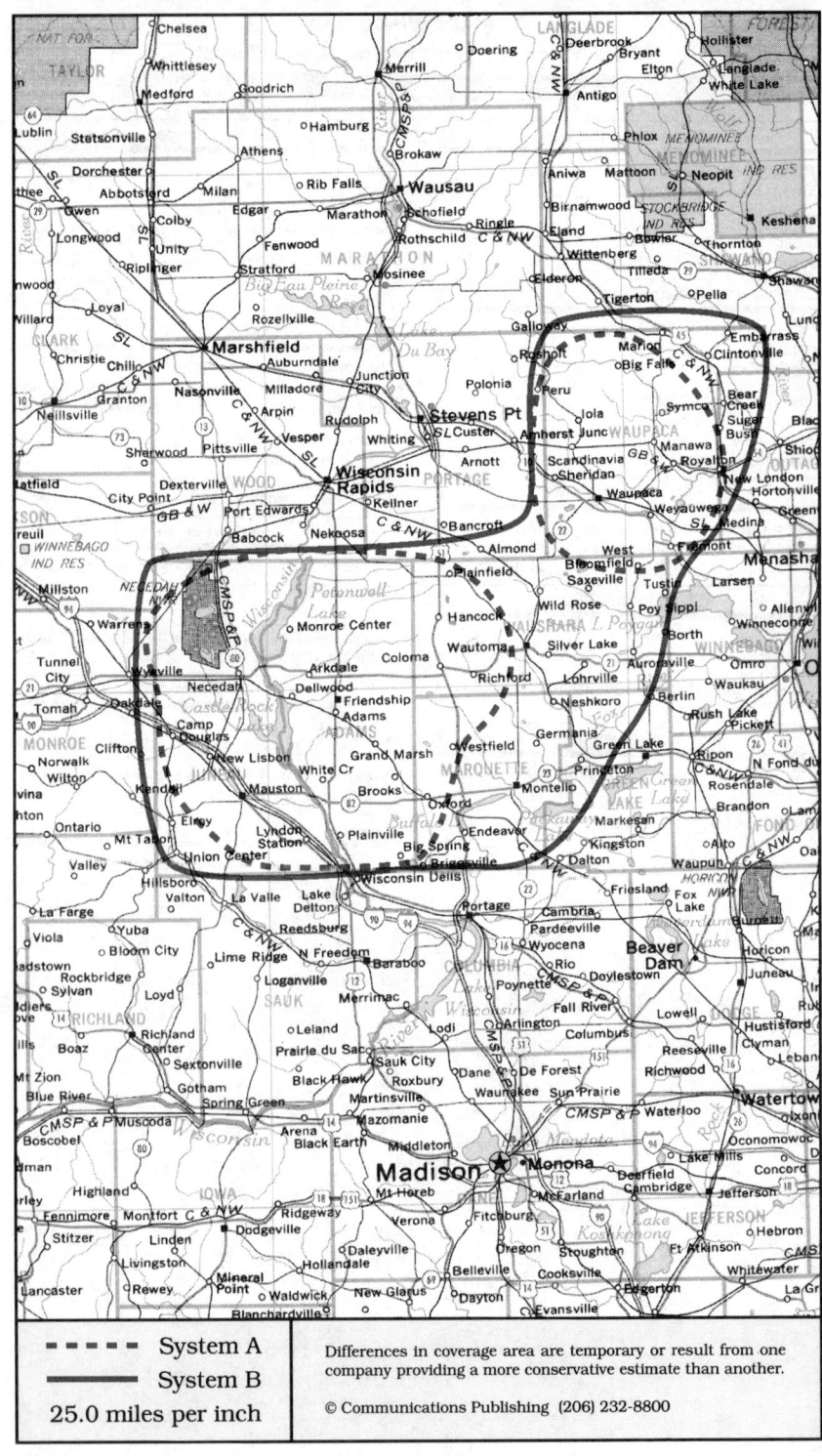

Legend	
▪ ▪ ▪ ▪	System A
——	System B
25.0 miles per inch	

Differences in coverage area are temporary or result from one company providing a more conservative estimate than another.

© Communications Publishing (206) 232-8800

Waupaca, Wisconsin
Mauston, Montello, Waupaca

|

System A

Cellular company

Cellular One
Licensed to: CJD Wisconsin Cellular
Managed by: Rural Cellular Management
4230 8th Street S
Wisconsin Rapids, WI 54494

Local office 715-424-0400
Corporate office Rural Cell. Mgt 707-422-2100

Sending calls

Visitor dialing instructions
Local calls . . . . . area code + 7 digits
Long distance . . . 1 + area code + 7 digits

Speed-dial numbers
Customer service 611
Emergency 911

Receiving calls

Caller dials the roamer access number below,
hears tone, dials cellular area code and number.
All cities except Mauston . . 715-498-7626
Mauston 608-547-7626

NationLink & RoamingAmerica (see p. 1157)
Dial *31 send to receive calls, *30 to deactivate.
Dial *32 send to tell caller the roamer access no.
This company's subscribers have level 9 service.

Billing information

Visitor rates
Most visitors . . . . . $0 day + 99¢ minute

Visitors without roaming agreements (p. 1159)
Call co. to use Amex, Discover, Mastercard, Visa
or Roamer Plus will intercept first call to bill you.

Identification numbers

City	Market	System	Billing	RSA
All served cities	714	01815	00000	WI-7

For full coverage area, see Stevens Point, WI.

System B

Cellular company

Cellulink
Pacific Telecom Cellular
4600 W College Avenue
Appleton, WI 54915

Customer service 800-236-9700
. . . . 414-841-1111
Corporate office 414-841-1100

Sending calls

Visitor dialing instructions
Local calls . . . . . . area code + 7 digits
Long distance . . . . . area code + 7 digits

Speed-dial numbers
Customer service *611
Emergency 911
Roaming information *711

Receiving calls

Caller dials the roamer access number below,
hears tone, dials cellular area code and number.
. . . . 414-572-7626

Billing information

Visitor rates
Most visitors . . . . . $3 day + 85¢ minute

Visitors without roaming agreements (p. 1159)
Cannot call since company takes no credit cards.

Identification numbers

City	Market	System	Billing	RSA
All served cities	714	01816	00000	WI-7

For full coverage area, see Appleton, WI.

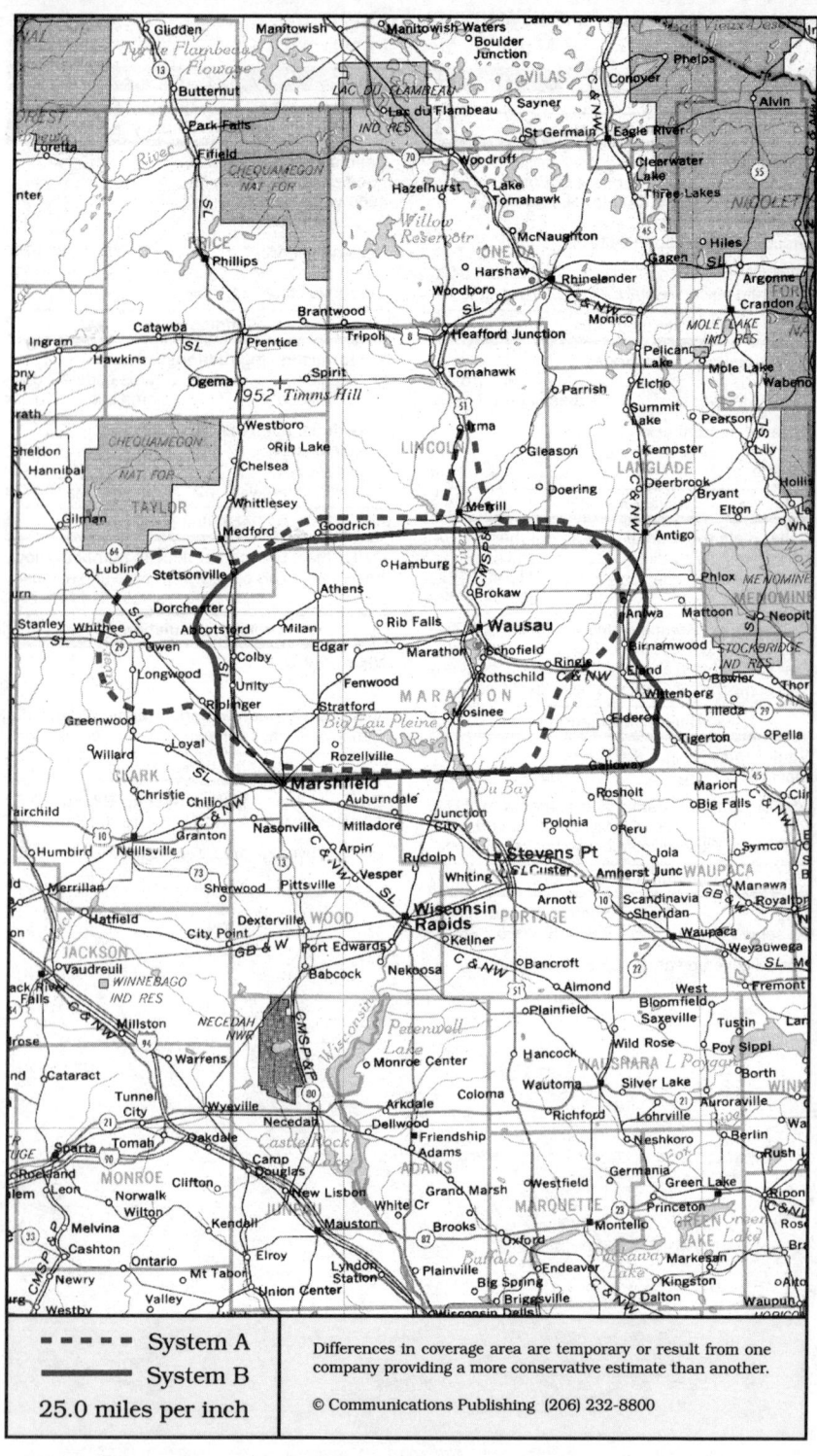

▪ ▪ ▪ ▪ System A ━━━━ System B 25.0 miles per inch	Differences in coverage area are temporary or result from one company providing a more conservative estimate than another. © Communications Publishing (206) 232-8800

Wausau, Wisconsin

System A

Cellular company

Cellular One
Licensed to: Wausau Cellular
Managed by: Cellular Information Systems
1503 N 6th Street
Wausau, WI 54401

Local office 715-842-7900
Corporate office 203-622-6317

Sending calls

Visitor dialing instructions
Local calls . . . . area code + 7 digits
Long distance . . . 1 + area code + 7 digits

Speed-dial numbers
Customer service 611
Emergency 911

Receiving calls

Caller dials the roamer access number below,
hears tone, dials cellular area code and number.
. . . . 715-571-7626

Billing information

Visitor rates
Most visitors . . . . . $3 day + 85¢ minute

Visitors without roaming agreements (p. 1159)
Call co. to use Amex, Discover, Mastercard, Visa

Identification numbers

City	Market	System	Billing	RSA
Wausau	263	00591	00000	MSA

For full coverage area, see Neillsville, WI.

System B

Cellular company

United States Cellular
2220 Grand Avenue
Wausau, WI 54401

Local office 715-842-4200
Corporate office 312-399-8900

Sending calls

Visitor dialing instructions
Local calls . . . . . area code + 7 digits
Long distance . . . . . area code + 7 digits

Speed-dial numbers
Channel 7 television . . . . ∗887
Channel 9 television . . . . ∗999
Customer service 611
Emergency 911
Roaming information 711
WRIG radio . . . . ∗1390

Receiving calls

Caller dials the roamer access number below,
hears tone, dials cellular area code and number.
. . . . 715-573-7626

Follow Me Roaming forwarding (see p. 1156)
Activate, dial ∗18 send. Deactivate dial ∗19 send.

Billing information

Visitor rates
Most visitors . . . . . $3 day + 75¢ minute

Visitors without roaming agreements (p. 1159)
Call co. to use Mastercard, Visa

Identification numbers

City	Market	System	Billing	RSA
All served cities	263	00570	00000	MSA

For full coverage area, see Stevens Point, WI.

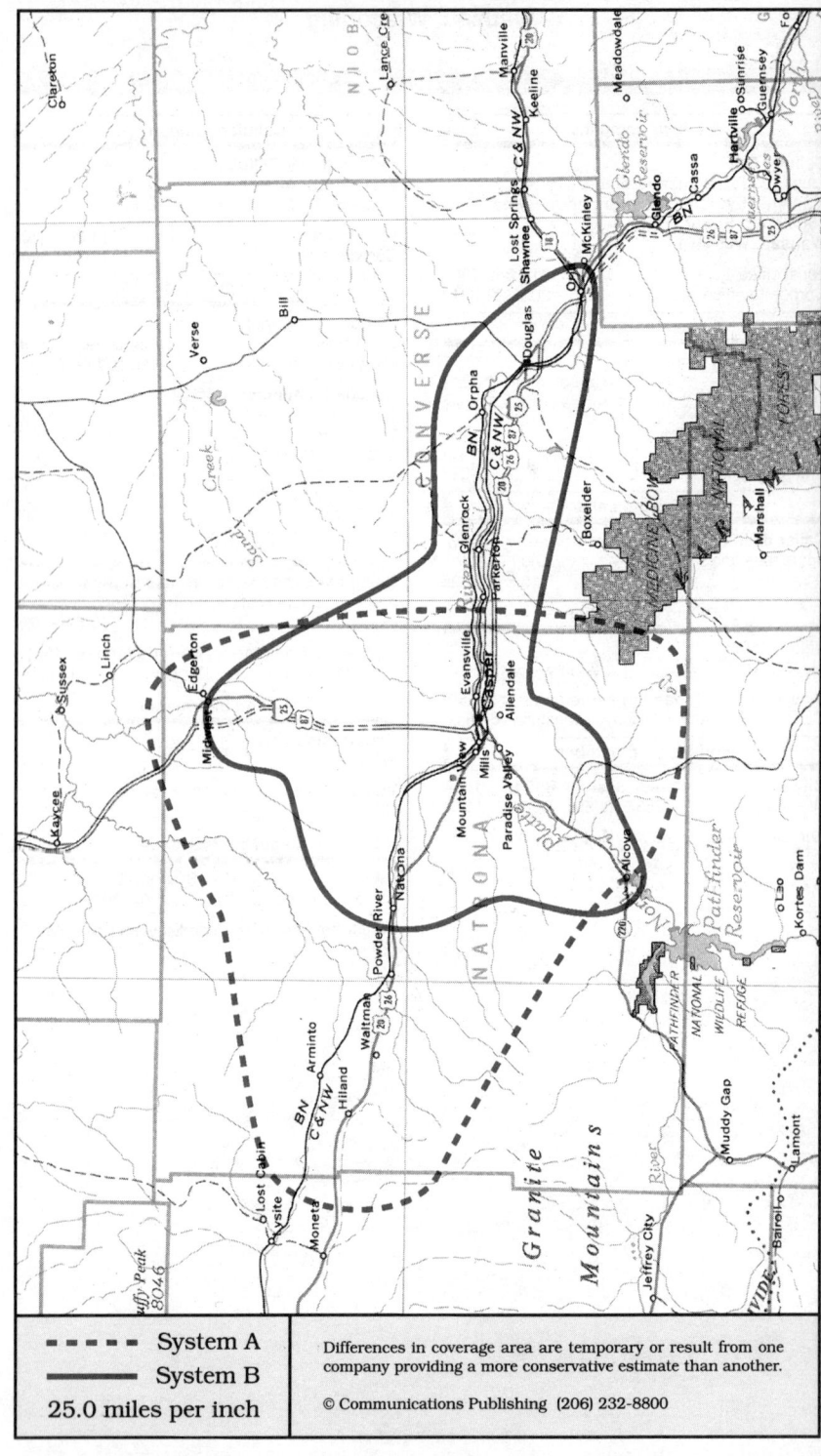

- - - - - System A	Differences in coverage area are temporary or result from one company providing a more conservative estimate than another.
———— System B	
25.0 miles per inch	© Communications Publishing (206) 232-8800

Casper, Wyoming
Casper, Douglas

System A	System B

Cellular company

Cellular One
Pacific Northwest Cellular
334 S Wolcott Street
Casper, WY 82601

Customer service	800-635-0304
. . . .	307-235-0110
Local office	307-235-5663
Corporate office	206-635-0300

Sending calls

Visitor dialing instructions

Local calls	area code + 7 digits
Long distance . . .	1 + area code + 7 digits

Speed-dial numbers

Customer service	611
Emergency	911

Receiving calls

Caller dials the roamer access number below,
hears tone, dials cellular area code and number.
. . . . 307-267-7626

Billing information

Visitor rates
Most visitors . . . . . $0 day + 99¢ minute

Visitors without roaming agreements (p. 1159)
Call co. to use Mastercard, Visa

Identification numbers

City	Market	System	Billing	RSA
All served cities	299	00301	00000	MSA

Cellular company

U S West Cellular
340 W "B" Street, Suite 204
Casper, WY 82601

Customer service		800-238-7848
. . . .		206-562-2895
. . . . .	local subscribers	800-626-6611
. . . .		206-747-1771
Local office		307-472-3829
Corporate office		206-747-4900

Sending calls

Visitor dialing instructions

Local calls	area code + 7 digits
Long distance . . .	1 + area code + 7 digits

Speed-dial numbers

Customer service	*611
Emergency	911
Roaming information	*711

Receiving calls

Caller dials the roamer access number below,
hears tone, dials cellular area code and number.

Casper	307-262-7626
Douglas	307-359-7626

Follow Me Roaming forwarding (see p. 1156)
Activate, dial *18 send. Deactivate dial *19 send.

Billing information

Visitor rates
Most visitors . . . . . $3 day + 75¢ minute

Visitors without roaming agreements (p. 1159)
Call co. to use Amex, Mastercard, Visa

Identification numbers

City	Market	System	Billing	RSA
Casper	299	00284	00000	MSA
Douglas	722	00284	00000	WY-5

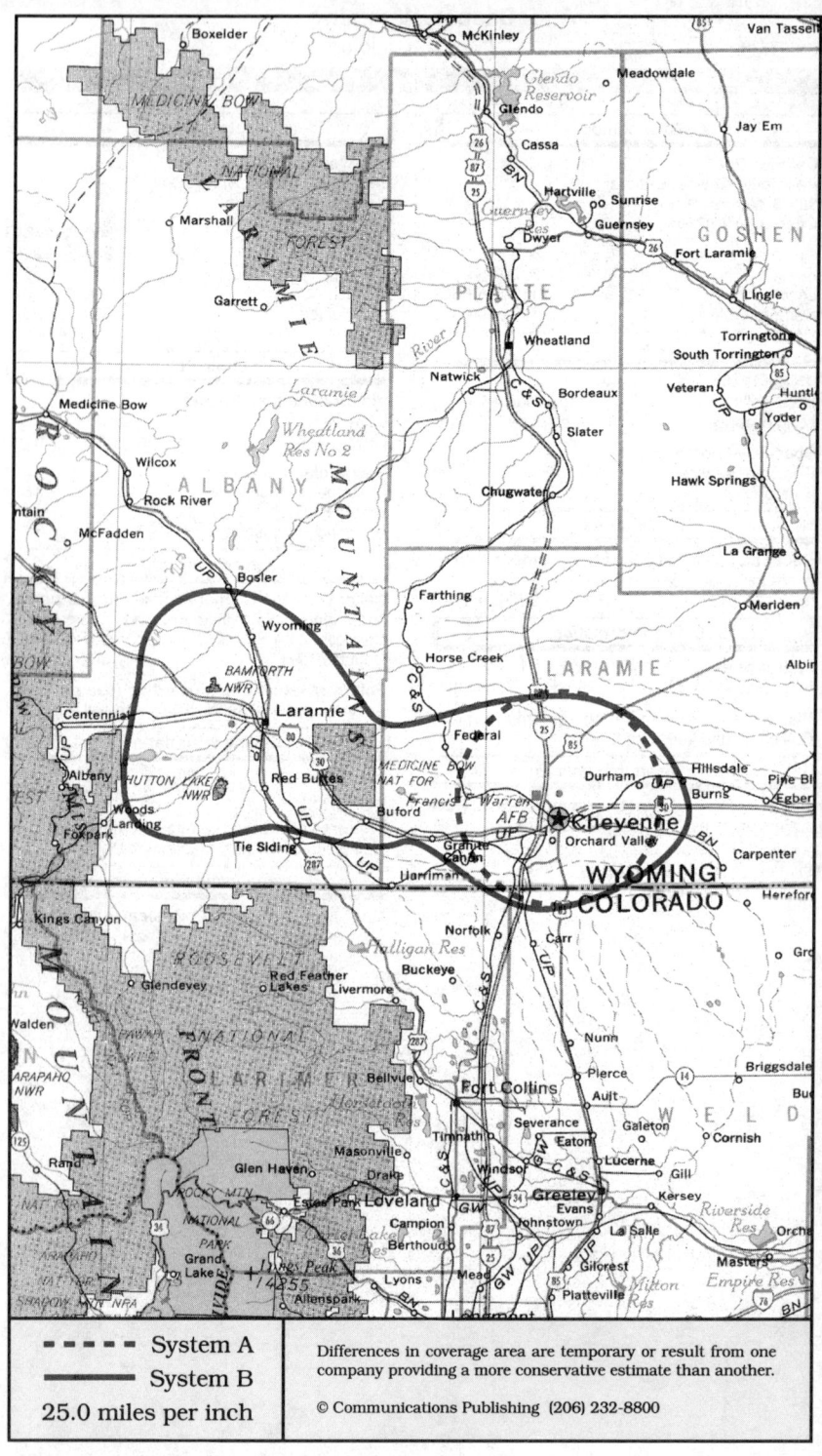

- - - - System A	Differences in coverage area are temporary or result from one company providing a more conservative estimate than another.
——— System B	
25.0 miles per inch	© Communications Publishing (206) 232-8800

Cheyenne, Wyoming
Cheyenne, Laramie

System A	System B

Cellular company

Cellular One
Pacific Northwest Cellular
11400 SE 8th Street, Suite 445
Bellevue, WA 98004-6431

Customer service	800-635-0304
. . . .	206-450-2914
Corporate office	206-635-0300

Sending calls

Visitor dialing instructions

Local calls	area code + 7 digits
Long distance . . .	1 + area code + 7 digits

Speed-dial numbers

Customer service	611
Emergency	911

Receiving calls

Caller dials the roamer access number below,
hears tone, dials cellular area code and number.
. . . . 307-775-8626

North American Cellular Network (see p. 1156)
All calls forward to you. Deactivate dial ✱35 send.

Billing information

Visitor rates
Most visitors . . . . . $0 day + 99¢ minute

Visitors without roaming agreements (p. 1159)
Cannot call since company takes no credit cards.

Identification numbers

City	Market	System	Billing	RSA
All served cities	721	01099	00000	WY-4

This company will also serve Laramie.

Cellular company

U S West Cellular
200 W 17th Street, Suite 20
Cheyenne, WY 82001

Customer service	800-238-7848
. . . .	206-562-2895
. . . . . local subscribers	800-626-6611
. . . .	206-747-1771
Local office	307-632-3705
. . . . . . from Laramie	307-745-8724
Corporate office	206-747-4900

Sending calls

Visitor dialing instructions

Local calls	area code + 7 digits
Long distance . . .	1 + area code + 7 digits

Speed-dial numbers

Customer service	✱611
Emergency	911
Roaming information	✱711

Receiving calls

Caller dials the roamer access number below,
hears tone, dials cellular area code and number.

Cheyenne	307-630-7626
Laramie	307-760-7626

Billing information

Visitor rates
Most visitors . . . . . $3 day + 75¢ minute

Visitors without roaming agreements (p. 1159)
Call co. to use Amex, Mastercard, Visa

Identification numbers

City	Market	System	Billing	RSA
All served cities	721	01830	00000	WY-4

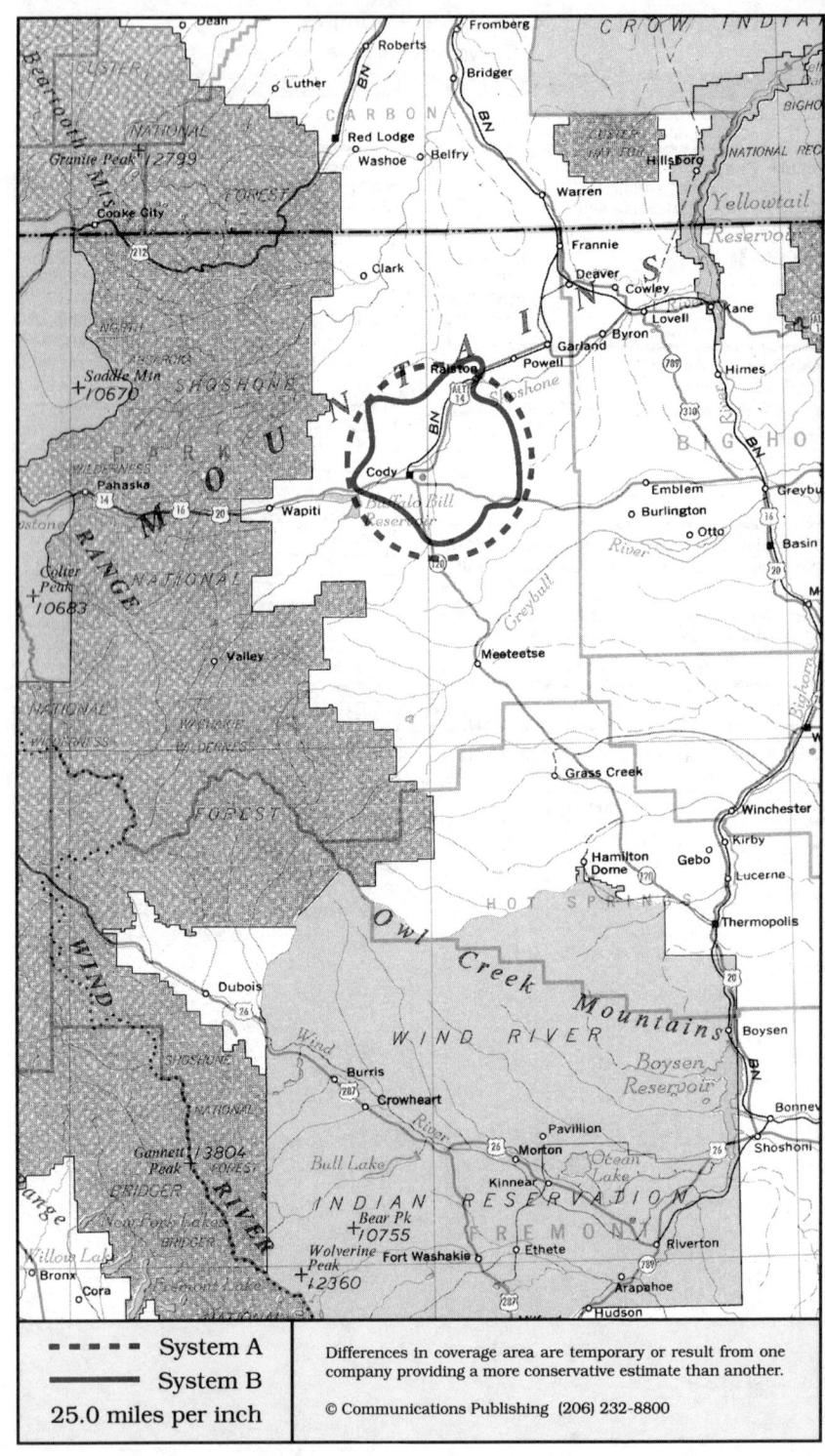

- - - - - System A	Differences in coverage area are temporary or result from one company providing a more conservative estimate than another.
——— System B	
25.0 miles per inch	© Communications Publishing (206) 232-8800

Cody, Wyoming

Cellular company

Cellular One
Metacomm Cellular
1307 Sheridan Avenue
Cody, WY 82414

Corporate office	some areas	800-540-1377
. . . .		307-527-7700

Sending calls

Visitor dialing instructions

Local calls	area code + 7 digits
Long distance . . .	1 + area code + 7 digits

Speed-dial numbers

Customer service	*611
Emergency	911

Receiving calls

Caller dials the roamer access number below, hears tone, dials cellular area code and number.
. . . . 307-578-1326

Billing information

Visitor rates

Most visitors	$3 day + 75¢ minute

Visitors without roaming agreements (p. 1159)
Call co. to use Mastercard, Visa

Identification numbers

City	Market	System	Billing	RSA
All served cities	718	01823	00000	WY-1

Cellular company

CommNet 2000
Cellular, Inc.
1701 Grand Avenue
Billings, MT 59102

Customer service	. . . .	800-597-5524
Local office	800-942-2060
. . . .	406-248-1990
Corporate office	303-694-3234

Sending calls

Visitor dialing instructions

Local calls	area code + 7 digits
Long distance . . .	1 + area code + 7 digits

Speed-dial numbers

Customer service	611
Emergency	911
Roaming information	711

Receiving calls

Caller dials the roamer access number below, hears tone, dials cellular area code and number.
. . . . 307-272-7626

Billing information

Visitor rates

Most visitors	$3 day + 95¢ minute

Visitors without roaming agreements (p. 1159)
Cannot call since company takes no credit cards.

Identification numbers

City	Market	System	Billing	RSA
All served cities	718	00262	30074	WY-1

- - - - System A ———— System B 30.1 miles per inch	Differences in coverage area are temporary or result from one company providing a more conservative estimate than another. © Communications Publishing (206) 232-8800

Rawlins, Wyoming
Lander, Rawlins

Cellular company

CommNet 2000
Cellular, Inc.
2052 E 17th Street
Idaho Falls, ID 83404

Service was not operating at press time. For nearby coverage area, see Rock Springs, WY.

Customer service	800-597-5523
Local office	800-767-2007
. . . .	208-525-8039
Corporate office	303-694-3234

Sending calls

Visitor dialing instructions
Local calls . . . . . area code + 7 digits
Long distance . . . 1 + area code + 7 digits

Speed-dial numbers
Customer service *611
Emergency 911
Roaming information 711

Receiving calls

Caller dials the roamer access number below, hears tone, dials cellular area code and number.
. . . . 307-352-2626

Billing information

Visitor rates
Most visitors . . . . . $3 day + 95¢ minute

Visitors without roaming agreements (p. 1159)
Cannot call since company takes no credit cards.

Identification numbers

City	Market	System	Billing	RSA
All served cities	720	01827	00000	WY-3

Cellular company

Union Cellular
Union Telephone
850 N State Highway 414, PO Box 160
Mountain View, WY 82939

Corporate office .	some areas	800-646-2355
. . . .		307-782-6131

Sending calls

Visitor dialing instructions
Local calls . . . . . . area code + 7 digits
Long distance . . . 1 + area code + 7 digits

Speed-dial numbers
Customer service 611
Emergency 911

Receiving calls

Caller dials the roamer access number below, hears tone, dials cellular area code and number.
. . . . 307-780-7626

Follow Me Roaming forwarding (see p. 1156)
Activate, dial *18 send. Deactivate dial *19 send.

Billing information

Visitor rates
Most visitors . . . . . $0 day + 70¢ minute

Visitors without roaming agreements (p. 1159)
Cannot call since company takes no credit cards.

Identification numbers

City	Market	System	Billing	RSA
All served cities	720	01828	00000	WY-3

For full coverage area, see Rock Springs, WY.

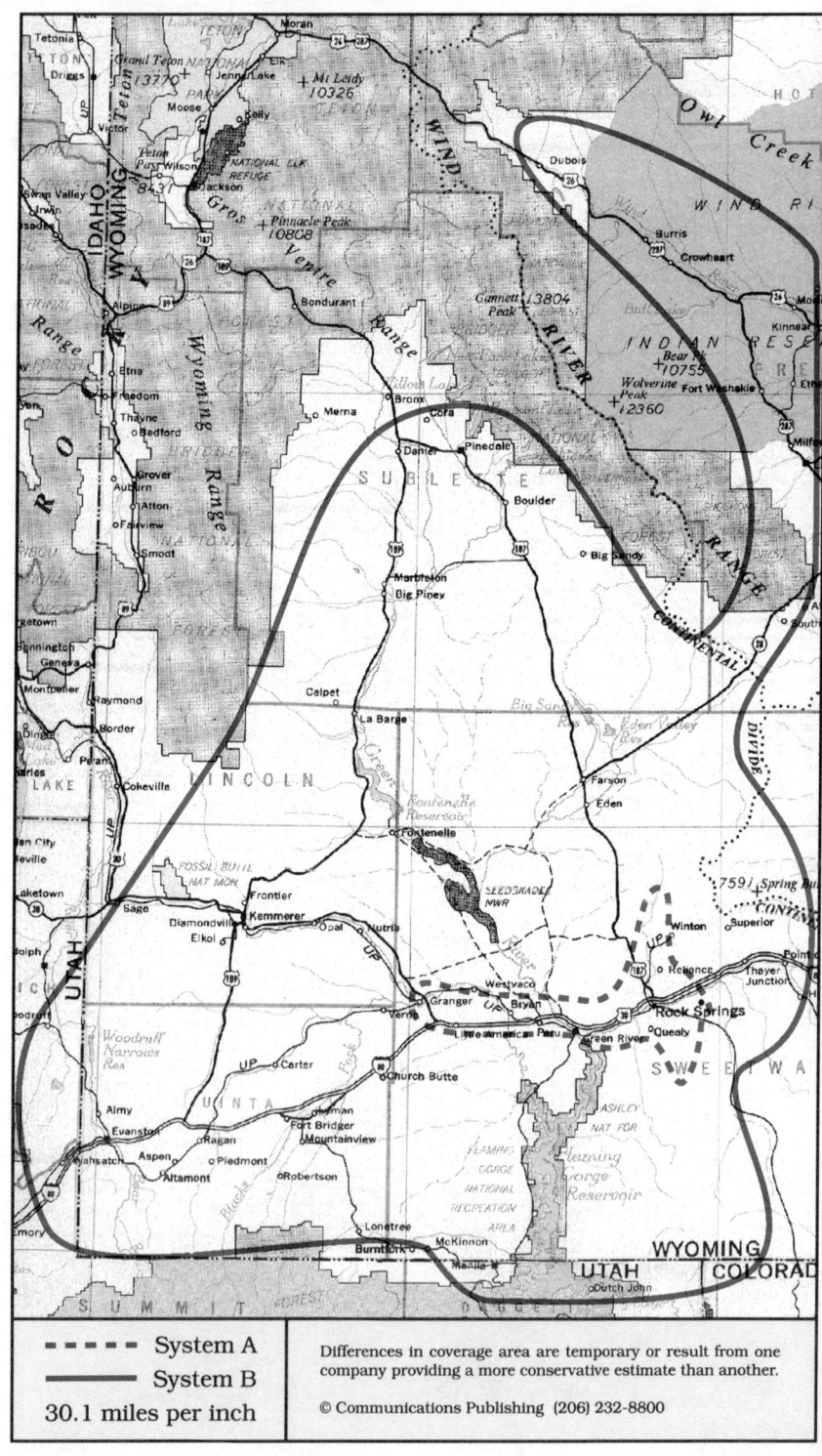

▪▪▪▪▪ System A	Differences in coverage area are temporary or result from one
———— System B	company providing a more conservative estimate than another.
30.1 miles per inch	© Communications Publishing (206) 232-8800

Rock Springs, Wyoming

Manila UT (Flaming Gorge), Rock Springs (Southwest Wyoming)

System A	System B

Cellular company

CommNet 2000
Cellular, Inc.
2052 E 17th Street
Idaho Falls, ID 83404

Customer service	800-597-5523
Local office	800-767-2007
. . . .	208-525-8039
Corporate office	303-694-3234

Sending calls

Visitor dialing instructions

Local calls	area code + 7 digits
Long distance . . .	1 + area code + 7 digits

Speed-dial numbers

Customer service	*611
Emergency	911
Roaming information	711

Receiving calls

Caller dials the roamer access number below,
hears tone, dials cellular area code and number.
. . . . 307-352-2626

Billing information

Visitor rates

Most visitors	$3 day + 95¢ minute

Visitors without roaming agreements (p. 1159)
Cannot call since company takes no credit cards.

Identification numbers

City	Market	System	Billing	RSA
All served cities	720	01827	00000	WY-3

Cellular company

Union Cellular
Union Telephone
850 N State Highway 414, PO Box 160
Mountain View, WY 82939

Corporate office . some areas	800-646-2355
. . . .	307-782-6131

Sending calls

Visitor dialing instructions

Local calls	area code + 7 digits
Long distance . . .	1 + area code + 7 digits

Speed-dial numbers

Customer service	611
Emergency	911

Receiving calls

Caller dials the roamer access number below,
hears tone, dials cellular area code and number.

Manila	801-880-7626
Rock Springs	307-780-7626

Follow Me Roaming forwarding (see p. 1156)
Activate, dial *18 send. Deactivate dial *19 send.

Billing information

Visitor rates

Most visitors	$0 day + 70¢ minute

Visitors without roaming agreements (p. 1159)
Cannot call since company takes no credit cards.

Identification numbers

City	Market	System	Billing	RSA
Manila (Utah area)	677	01742	00000	UT-5
Rock Springs	720	01828	00000	WY-3

For full coverage area, see Rawlins, WY.

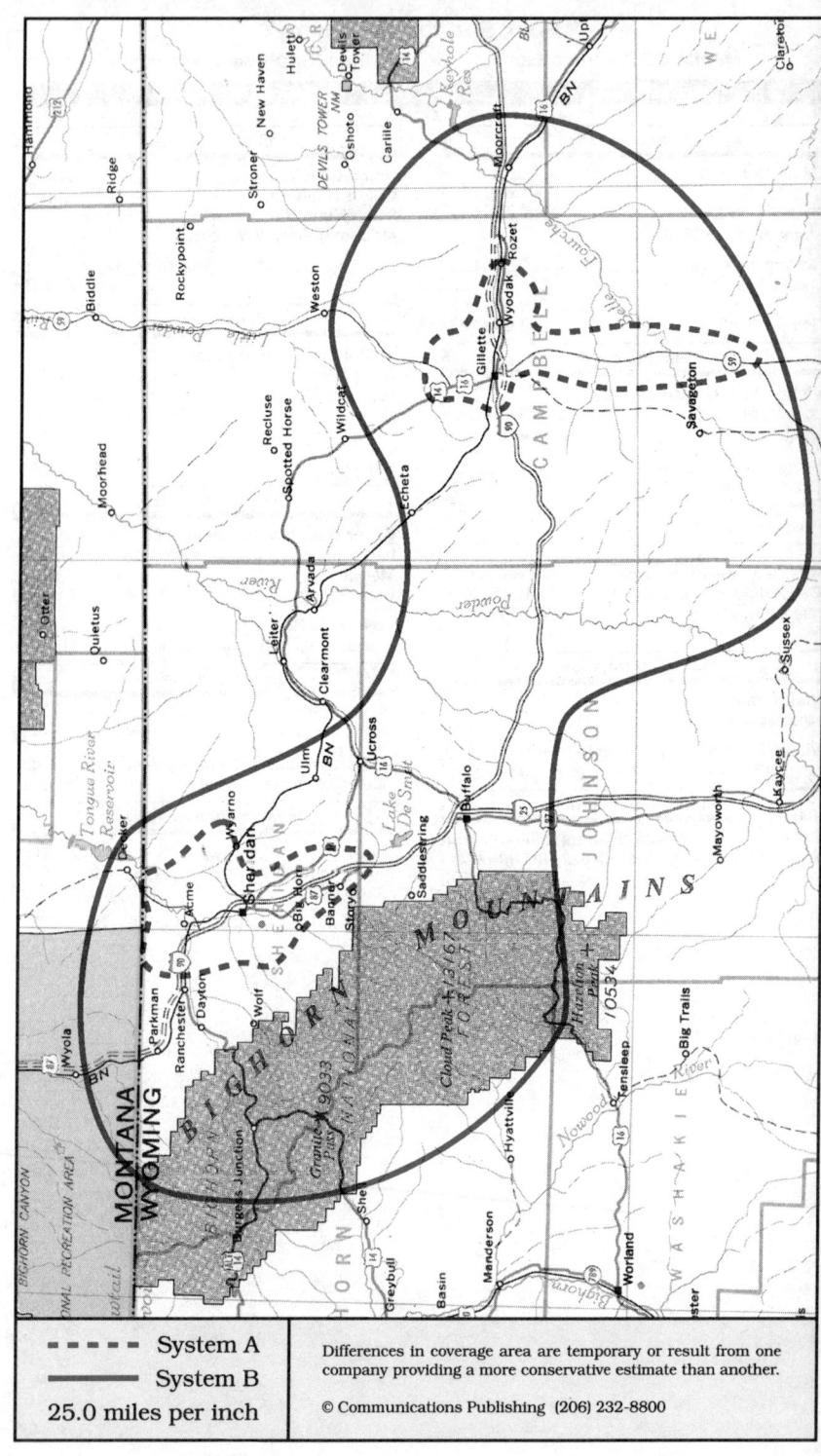

- - - - System A	Differences in coverage area are temporary or result from one company providing a more conservative estimate than another.
——— System B	
25.0 miles per inch	© Communications Publishing (206) 232-8800

Sheridan, Wyoming
Buffalo, Gillette, Sheridan

System A	System B

System A

Cellular company

Cellular One
GenCell Management
1891 Woolner Avenue
Fairfield, CA 94533

Customer service	800-888-7868
Corporate office	707-425-8000

Sending calls

Visitor dialing instructions

Local calls	area code + 7 digits
Long distance . . .	1 + area code + 7 digits

Speed-dial numbers

Customer service	611
Emergency	911

Receiving calls

Caller dials the roamer access number below,
hears tone, dials cellular area code and number.

Gillette	307-680-7626
Sheridan	307-751-7626

NationLink & RoamingAmerica (see p. 1157)
Caller hears your current roamer access number.
This company's subscribers have level 0 service.

Billing information

Visitor rates

Most visitors	$3 day + 99¢ minute

Visitors without roaming agreements (p. 1159)
American Roaming Network intercepts call to bill.

Identification numbers

City	Market	System	Billing	RSA
All served cities	719	01825	00000	WY-2

System B

Cellular company

CommNet 2000
Cellular, Inc.
1701 Grand Avenue
Billings, MT 59102

Customer service	800-597-5525
Local office	800-942-2060
. . . .	406-248-1990
Corporate office	303-694-3234

Sending calls

Visitor dialing instructions

Local calls	area code + 7 digits
Long distance . . .	1 + area code + 7 digits

Speed-dial numbers

Customer service	611
Emergency	911
Roaming information	711

Receiving calls

Caller dials the roamer access number below,
hears tone, dials cellular area code and number.

. . . .	307-752-7626

Billing information

Visitor rates

Most visitors	$3 day + 95¢ minute

Visitors without roaming agreements (p. 1159)
Cannot call since company takes no credit cards.

Identification numbers

City	Market	System	Billing	RSA
All served cities	719	00262	30076	WY-2

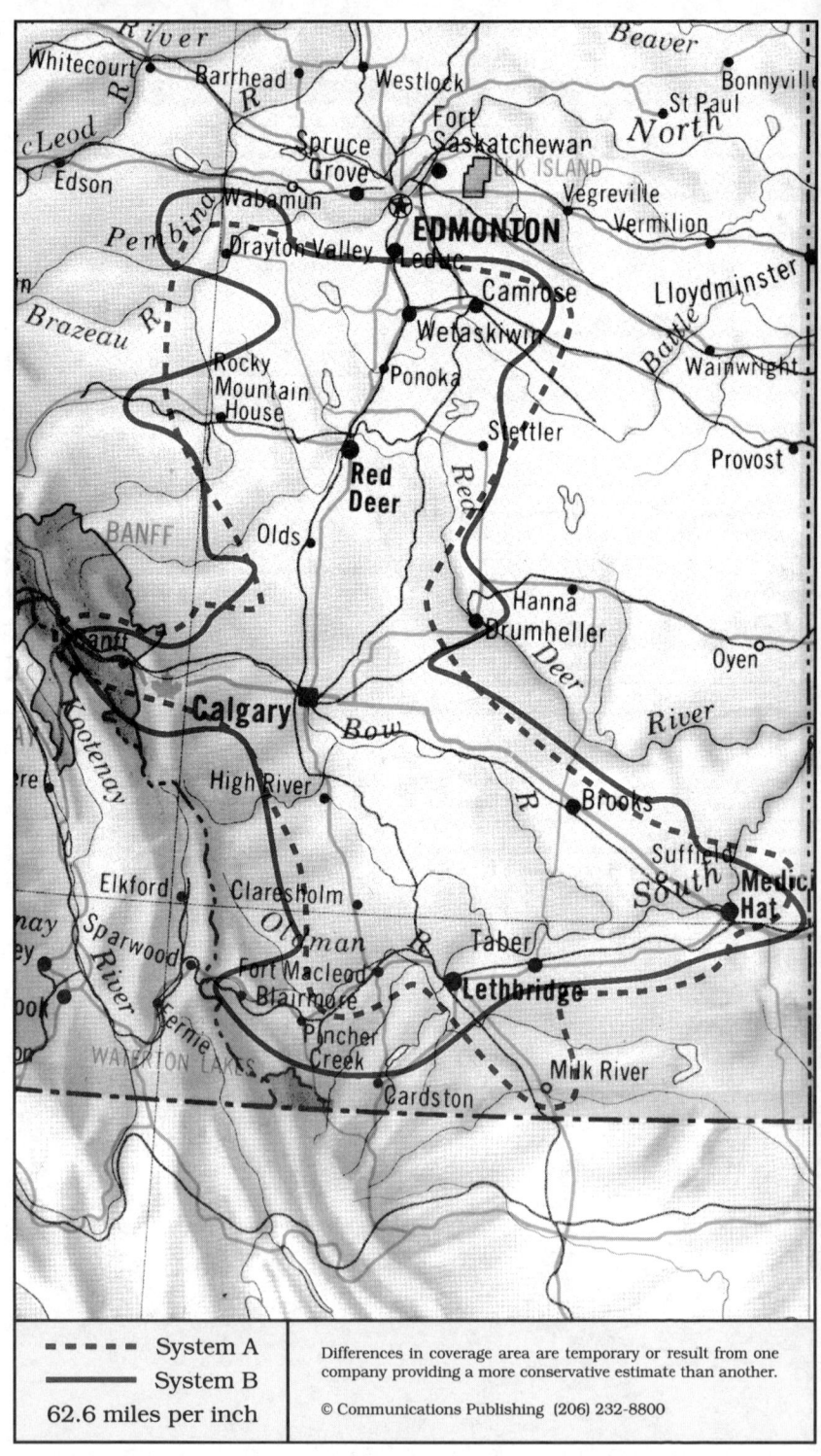

- - - - System A	Differences in coverage area are temporary or result from one
——— System B	company providing a more conservative estimate than another.
62.6 miles per inch	© Communications Publishing (206) 232-8800

Calgary, Alberta
Calgary, Camrose, Lethbridge, Medicine Hat, Red Deer

System A	System B

System A

Cellular company

Cantel
1810-112 4th Avenue SW
Calgary, Alberta T2P 0H3 Canada

Customer service	some areas	800-661-1874
. . . .		604-433-1440
Local office		403-265-1400
Corporate office		416-229-1400

Sending calls

Visitor dialing instructions
Local calls . . . . area code + 7 digits
Long distance . . . 1 + area code + 7 digits

Speed-dial numbers
Customer service 0
Directory assistance 411
Emergency 911
Landline telephone repair 611

Receiving calls

Caller dials the roamer access number below, hears tone, dials cellular area code and number.

City	Roamer access	System ID
Banff	none assigned	16757
Bassano	none assigned	16643
Brooks	403-362-9200	16645
Calgary	403-560-7626	16387
Calgary corridor	none assigned	16493
Camrose	403-679-9100	16769
Canmore	403-678-0100	16531
Gleichen	none assigned	16641
James River Bridge	none assigned	16537
Kininvie	none assigned	16647
Lethbridge	403-380-7626	16473
Medicine Hat	403-528-7626	16573
Red Deer	403-341-7626	16445
Stettler	403-740-7626	16763
Strathmore	none assigned	16639
Suffield	none assigned	16649
Vulcan	403-485-7626	16761
Warner	403-642-7626	16759

All Cantel subscribers receive calls automatically.

North American Cellular Network (see p. 1156)
All calls forward to you. Deactivate dial ∗35 send.

Billing information

Visitor rates
Most visitors ($ U.S.) . $2 day + $1.05 minute
All Cantel subscribers . .   billed at home rate

Visitors without roaming agreements (p. 1159)
Cannot call since company takes no credit cards.

Identification numbers

Listed above under the section **receiving calls**

For full coverage area, see Edmonton, Alberta.

System B

Cellular company

AGT Mobility
3030 2nd Avenue SE
Calgary, Alberta T2A 5N7 Canada

Customer service	some areas	800-661-3681
. . . .		403-248-2355
Corporate office		403-530-5300

Sending calls

Visitor dialing instructions
Local calls . . . . . . area code + 7 digits
Long distance . . . 1 + area code + 7 digits

Speed-dial numbers
CAA (Canadian Automobile Association) . 222
Customer service 811
Emergency 911
Make a traffic report . . . . 105 or 4665
Telephone repair 611

Receiving calls

Caller dials the roamer access number below, hears tone, dials cellular area code and number.

Baff, Canmore	403-678-7626
Brooks	403-362-1199
Calgary	403-540-7626
Camrose	403-679-7626
Drumheller	403-823-0099
Fort McMurray	403-799-0099
Lethbridge	403-382-7626
Medicine Hat	403-529-7626
Olds	403-556-4926
Pincher Creek	403-627-7626
Red Deer	403-340-4199
Rocky Mountain House . . .	403-845-8626
Stettler	403-742-7626
Sundre	403-638-6526
Wetaskiwin	403-352-1026

Billing information

Visitor rates
Most visitors ($Canadian) $2.75 day + 90¢ minute
Cellnet Canada visitors $0 day + 45¢ minute

Visitors without roaming agreements (p. 1159)
Call co. to use Mastercard, Visa

Identification numbers

City	System
All served cities	16384

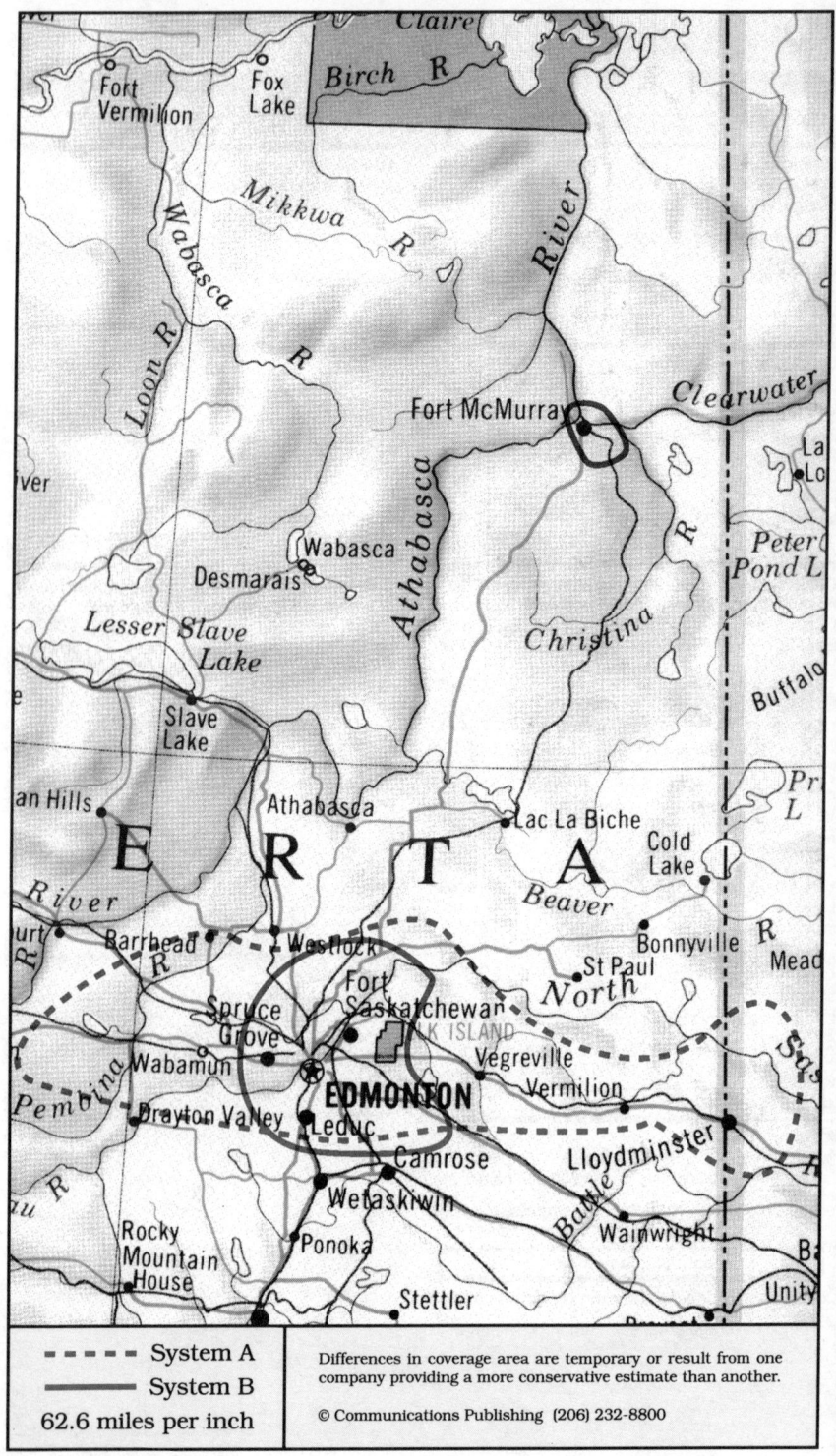

- - - - System A	Differences in coverage area are temporary or result from one company providing a more conservative estimate than another.
———— System B	
62.6 miles per inch	© Communications Publishing (206) 232-8800

Edmonton, Alberta
Edmonton, Fort McMurray, Lloydminster

|

Cellular company

Cantel
1950-10303 Jasper Avenue
Edmonton, Alberta T5J 3N6 Canada

Customer service	some areas	800-661-1874
. . . .		604-433-1440
Local office		403-429-1400
Corporate office		416-229-1400

Sending calls

Visitor dialing instructions
Local calls . . . . . area code + 7 digits
Long distance . . . 1 + area code + 7 digits

Speed-dial numbers
Customer service 0
Directory assistance 411
Emergency 911
Landline telephone repair 611

Receiving calls

Caller dials the roamer access number below,
hears tone, dials cellular area code and number.
Drayton Valley 403-542-9000
Edmonton 403-441-7626
Vegreville 403-632-0000

All Cantel subscribers receive calls automatically.

North American Cellular Network (see p. 1156)
All calls forward to you. Deactivate dial ∗35 send.

Billing information

Visitor rates
Most visitors ($ U.S.) . . $2 day + $1.05 minute
All Cantel subscribers . . billed at home rate

Visitors without roaming agreements (p. 1159)
Cannot call since company takes no credit cards.

Identification numbers

City	System
Drayton Valley	16767
Edmonton	16391
Edmonton corridor	16491
Entwistle	16765
Vegreville	16771

For full coverage area, see Calgary, Alberta.

Cellular company

Ed Tel Cellular
10044-108th Street, Room 856
Edmonton, Alberta T5J 1S7 Canada

Corporate office 403-441-2355

Sending calls

Visitor dialing instructions
Local calls . . . . 7 digits
Long distance . . . 1 + area code + 7 digits

Speed-dial numbers
Alberta Motor Association 222
Customer service 811
Directory assistance 411
Emergency 911
Landline telephone repair 611
Make a traffic report . . . .
. . 630 or 925 or 930 or 973 or 1200 or 1480

Receiving calls

Caller dials the roamer access number below,
hears tone, dials cellular area code and number.
. . . . 403-446-7626

Billing information

Visitor rates
Most visitors ($ Canadian) $0 day + $1 minute

Visitors without roaming agreements (p. 1159)
Cannot call since company takes no credit cards.

Identification numbers

City	System
All served cities	16388

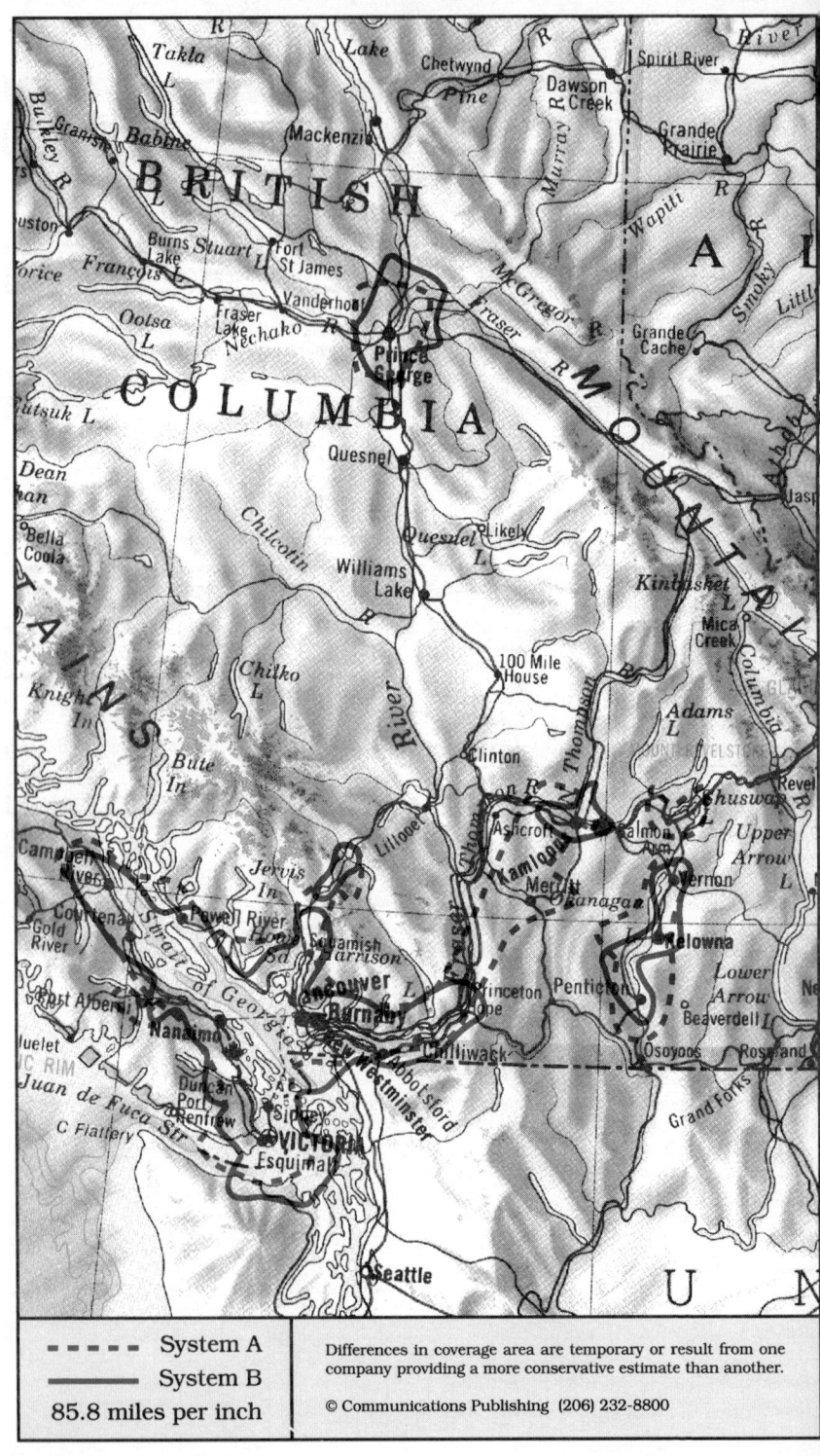

- - - - - System A	Differences in coverage area are temporary or result from one company providing a more conservative estimate than another.
───── System B	
85.8 miles per inch	© Communications Publishing (206) 232-8800

Vancouver, British Columbia
Campbell River, Kamloops, Kelowna, Prince George, Vancouver, Victoria

System A	System B

Cellular company

Cantel
1600-4710 Kingsway
Burnaby, British Columbia V5H 4M5 Canada

Customer service	some areas	800-663-0670
. . . .		604-433-1440
Local office		604-431-1400
Corporate office		416-229-1400

Sending calls

Visitor dialing instructions

| Local calls . . . . | area code + 7 digits |
| Long distance . . . | 1 + area code + 7 digits |

Speed-dial numbers

Customer service	0
Directory assistance	411
Emergency	911

Receiving calls

Caller dials the roamer access number below,
hears tone, dials cellular area code and number.

City	Roamer access	System ID
Abbotsford	604-852-7626	16425
Bowser	none assigned	16685
Campbell River	604-287-1000	16657
Chemainus, Duncan	604-246-8000	16427
Chilliwack	604-795-1400	16461
Comox, Courtenay	604-338-3000	16661
Hope	none assigned	16569
Kamloops	604-371-2000	16527
Kelowna	604-862-1000	16521
Merritt	604-378-0000	16817
Nanaimo	604-756-7626	16447
Okanagan Mountain	none assigned	16523
Oliver	none assigned	16803
Osoyoos	none assigned	16807
Parksville	604-248-1300	16687
Penticton	604-490-1000	16525
Port Alberni	604-720-3000	16735
Port Hardy	none assigned	16801
Prince George	604-565-1000	16737
Salmon Arm	604-832-0700	16733
Squamish	604-892-8000	16693
Texada Island	none assigned	16805
Vancouver	604-644-7626	16425
Vancouver corridor	none assigned	16489
Vernon	604-549-9000	16519
Victoria	604-727-8888	16427
Whistler	604-932-1300	16463

All Cantel subscribers receive calls automatically.

North American Cellular Network (see p. 1156)
All calls forward to you. Deactivate dial *35 send.

Billing information

Visitor rates

| Most visitors ($ U.S.) . | $2 day + $1.05 minute |
| All Cantel subscribers . . | billed at home rate |

Visitors without roaming agreements (p. 1159)
Cannot call since company takes no credit cards.

Identification numbers are listed above

Cellular company

B.C. Tel Mobility Cellular
4519 Canada Way
Burnaby, British Columbia V5G 4S4 Canada

Customer service	some areas	800-242-2355
. . . .		604-291-2355
Corporate office		604-294-1974

Sending calls

Visitor dialing instructions

| Local calls . . . . . | area code + 7 digits |
| Long distance . . . | 1 + area code + 7 digits |

Speed-dial numbers

Customer service	*811
Directory assistance	411
Emergency	*911
Telephone repair	*611

Receiving calls

Caller dials the roamer access number below,
hears tone, dials cellular area code and number.

Abbotsford	604-854-4811
Campbell River	604-287-6211
Chilliwack	604-795-6711
Courtenay	604-334-7626
Duncan	604-246-7626
Gibsons (sunshine coast) . .	604-886-6011
Hope	604-869-1011
Kamloops	604-371-7626
Kelowna	604-862-7626
Mayne Island	604-539-7626
Merritt	604-378-7626
Nanaimo	604-755-9211
Oliver, Osoyoos	604-498-7626
Parksville	604-248-7626
Penticton	.604-490-7626
Port Alberni	604-720-7626
Prince George	604-565-7626
Port Hardy	604-949-1111
Powell River	604-483-8011
Salmon Arm	604-832-5811
Saltspring Island	604-537-1811
Squamish	604-892-7626
Vancouver	604-290-7626
Vernon	604-549-0088
Victoria	604-380-7626
Whistler	604-932-7626

Billing information

Visitor rates

| Most visitors ($ Canadian) | $3 day + $1.25 minute |
| Cellnet Canada visitors | $0 day + 45¢ minute |

Visitors without roaming agreements (p. 1159)
Call co. to use Amex, Mastercard, Visa

Identification numbers

City	System
All served cities	16422

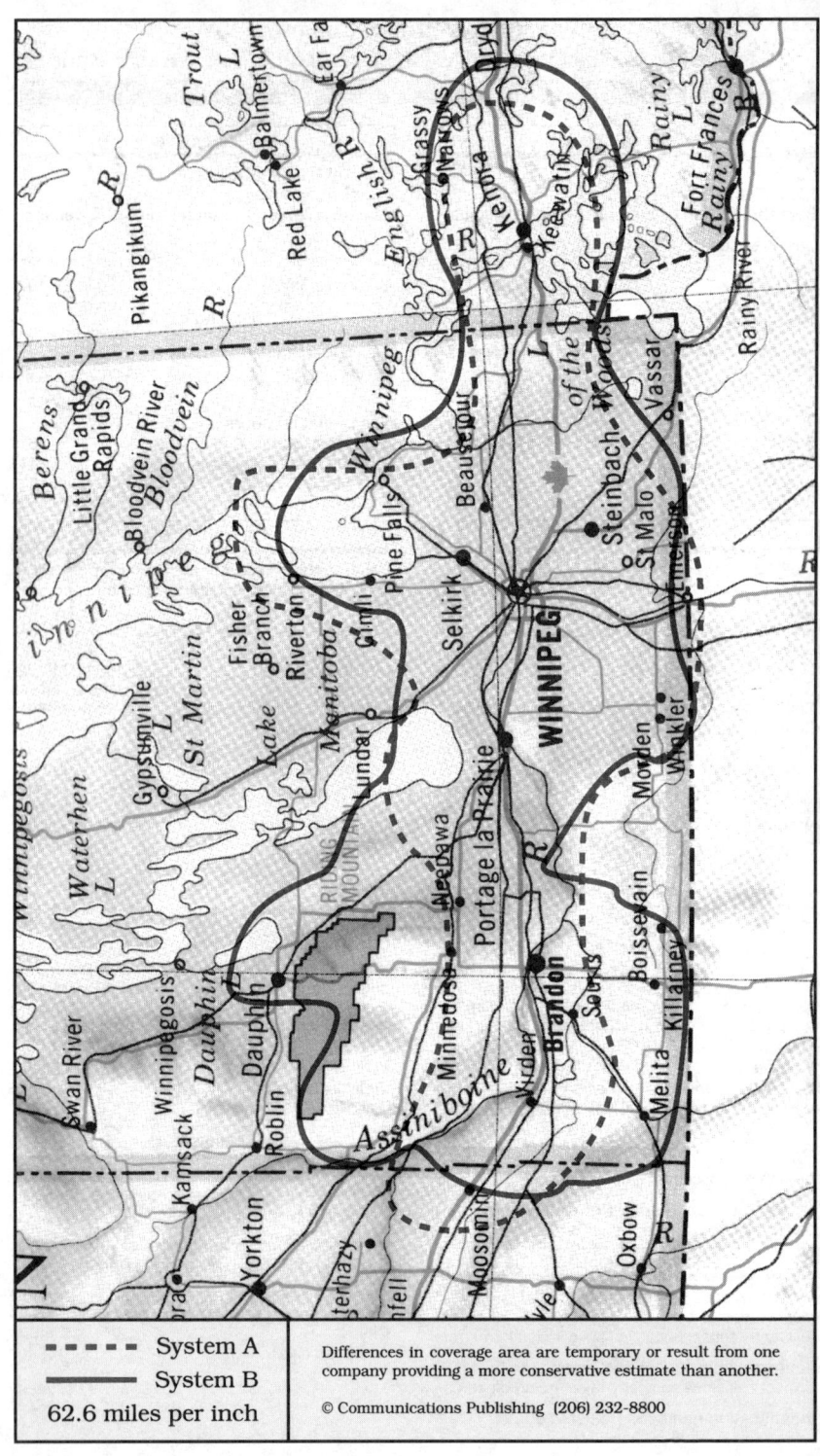

- - - - - System A	
——— System B	
62.6 miles per inch	

Differences in coverage area are temporary or result from one company providing a more conservative estimate than another.

© Communications Publishing (206) 232-8800

Winnipeg, Manitoba

Brandon, Dauphin, Kenora (Ontario), Riverton, Steinbach, Winnipeg

System A	System B

Cellular company

Cantel
330 Portage Avenue, Suite 1506
Winnipeg, Manitoba R3C 0C4 Canada

Customer service	some areas	800-661-1874
. . . .		604-433-1440
Local office		204-942-1400
Corporate office		416-229-1400

Sending calls

Visitor dialing instructions

Local calls	area code + 7 digits
Long distance . . .	1 + area code + 7 digits

Speed-dial numbers

Customer service	0
Directory assistance	411
Emergency	911
Landline telephone repair	611

Receiving calls

Caller dials the roamer access number below,
hears tone, dials cellular area code and number.

City	Roamer access	System ID
Brandon	204-726-3900	16471
Carberry	none assigned	16783
Kenora	807-468-0600	16513
Morden, Winkler	204-325-0100	16755
Morris	204-746-7626	16511
Oak Lake	none assigned	16785
Portage la Prairie	204-239-7626	16453
Selkirk	204-482-0900	16455
Steinbach	204-355-7626	16465
Virden	204-748-8100	16663
Winnipeg	204-955-7626	16431
Winnipeg corridor	none assigned	16497

All Cantel subscribers receive calls automatically.

North American Cellular Network (see p. 1156)
All calls forward to you. Deactivate dial ∗35 send.

Billing information

Visitor rates

Most visitors ($ U.S.) .	$2 day + $1.05 minute
All Cantel subscribers . .	billed at home rate

Visitors without roaming agreements (p. 1159)
Cannot call since company takes no credit cards.

Identification numbers

Listed above under the section **receiving calls**

Cellular company

MTS Mobility
555 Madison Street
Winnipeg, Manitoba R3H 0L6 Canada

Customer service	some areas	800-362-3347
. . . .		204-987-6666
Corporate office		204-987-6500

Sending calls

Visitor dialing instructions

Local calls	7 digits
Long distance . . .	1 + area code + 7 digits

Speed-dial numbers

Customer service	811
Directory assistance	411
Emergency	911
Landline telephone repair	611
Operator	0

Receiving calls

Caller dials the roamer access number below,
hears tone, dials cellular area code and number.

Manitoba	204-771-7626
Ontario	807-468-9411

Billing information

Visitor rates

Most visitors ($ Canadian)	$3.25 day + $1 minute
Cellnet Canada visitors	$0 day + 45¢ minute

Visitors without roaming agreements (p. 1159)
Call co. to use Amex, Mastercard, Visa
Cannot call since company takes no credit cards.

Identification numbers

City	System
All served cities	16428

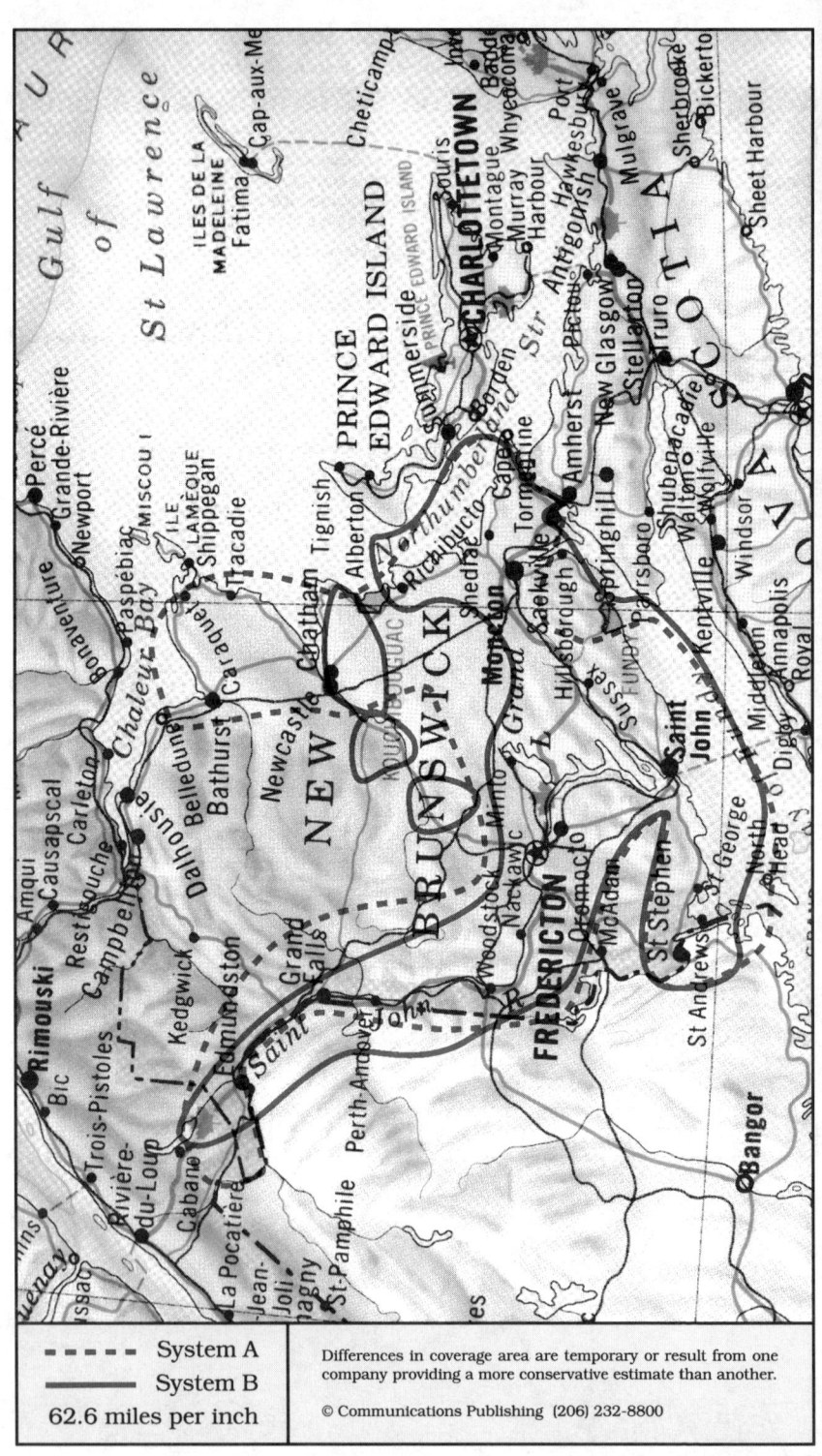

System A

System B

62.6 miles per inch

Differences in coverage area are temporary or result from one company providing a more conservative estimate than another.

© Communications Publishing (206) 232-8800

Saint John, New Brunswick
Chatham, Fredericton, Moncton, Saint John

System A	System B

Cellular company

Cantel
357 St. George Street
Moncton, New Brunswick E1C 1X3 Canada

418 Rothesay Avenue
Saint John, New Brunswick E2J 2C4 Canada

Customer service	some areas	800-361-5410
. . . .		514-340-9220
Local office	Moncton	506-853-5055
. . . .	St. John	506-658-6150
Corporate office		416-229-1400

Sending calls

Visitor dialing instructions

Local calls	area code + 7 digits
Long distance . . .	1 + area code + 7 digits

Speed-dial numbers

Customer service	0
Directory assistance	411
Emergency	911
Landline telephone repair	611

Receiving calls

Caller dials the roamer access number below,
hears tone, dials cellular area code and number.

City	Roamer access	System ID
Bathurst	506-547-4000	16607
Bristol, Florenceville	506-392-0000	16675
Chatham, Newcastle	506-627-8000	16609
Edmunston	506-737-0000	16619
Fairfield	none assigned	16483
Fredericton	506-450-5126	16469
Grand Falls	506-473-8000	16819
Moncton	506-856-0626	16467
Rothesay	none assigned	16565
Saint John	506-658-8626	16411
Saint John corridor	none assigned	16517
Sussex	506-432-1276	16501
Woodstock	506-325-0000	16603

All Cantel subscribers receive calls automatically.

North American Cellular Network (see p. 1156)
All calls forward to you. Deactivate dial ✱35 send.

Billing information

Visitor rates

Most visitors ($ U.S.) .	$2 day + $1.05 minute
All Cantel subscribers . .	billed at home rate

Visitors without roaming agreements (p. 1159)
Cannot call since company takes no credit cards.

Identification numbers

Listed above under the section **receiving calls**

Cellular company

NBTel Cellular
New Brunswick Telephone
One Brunswick Square, PO Box 1430
Saint John, New Brunswick E2L 4K2 Canada

Customer service .	inside NB	800-561-2355
. . . .		506-648-2355
Corporate office		506-694-2317

Sending calls

Visitor dialing instructions

Local calls	7 digits
Long distance . .	1 + area code + 7 digits

Speed-dial numbers

Coast Guard Alert	✱16
Customer service	811
Directory assistance	411
Emergency	911
Maritime Auto Association	✱222
Repair service . . .	611 or 1-800-566-0057
Roamer access number nearest caller . .	511

Receiving calls

Caller dials the roamer access number below,
hears tone, dials cellular area code and number.

Bathurst	.	506-547-7626
Buctouche	.	506-743-7626
Edmundston	.	506-737-7626
Florenceville	.	506-392-7626
Fredericton	.	506-457-7626
Moncton	.	506-852-7626
Newcastle	.	506-627-7626
Richibucto	.	506-523-8600
Sackville	.	506-364-7626
Saint John	.	506-636-0999
St. George	.	506-755-7626
St. Leonard	.	506-423-8600
St. Stephen	.	506-466-8600
Shippegan	.	506-336-7626
Sussex	.	506-432-7626
Tracadie	.	506-395-8600
Woodstock	.	506-325-7626

All Cellnet subscribers receive calls automatically

Follow Me Roaming forwarding (see p. 1156)
Activate, dial ✱18 send. Deactivate dial ✱19 send.

Billing information

Visitor rates
All rates are in Canadian dollars

Cellnet Canada visitors	$0 day + 45¢ minute
Most visitors	$3 day + $1.25 minute
Unicel of Maine visitors	$0 day + 65¢ minute

Visitors without roaming agreements (p. 1159)
Cannot call since company takes no credit cards.

Identification numbers

City	System
All served cities	16408

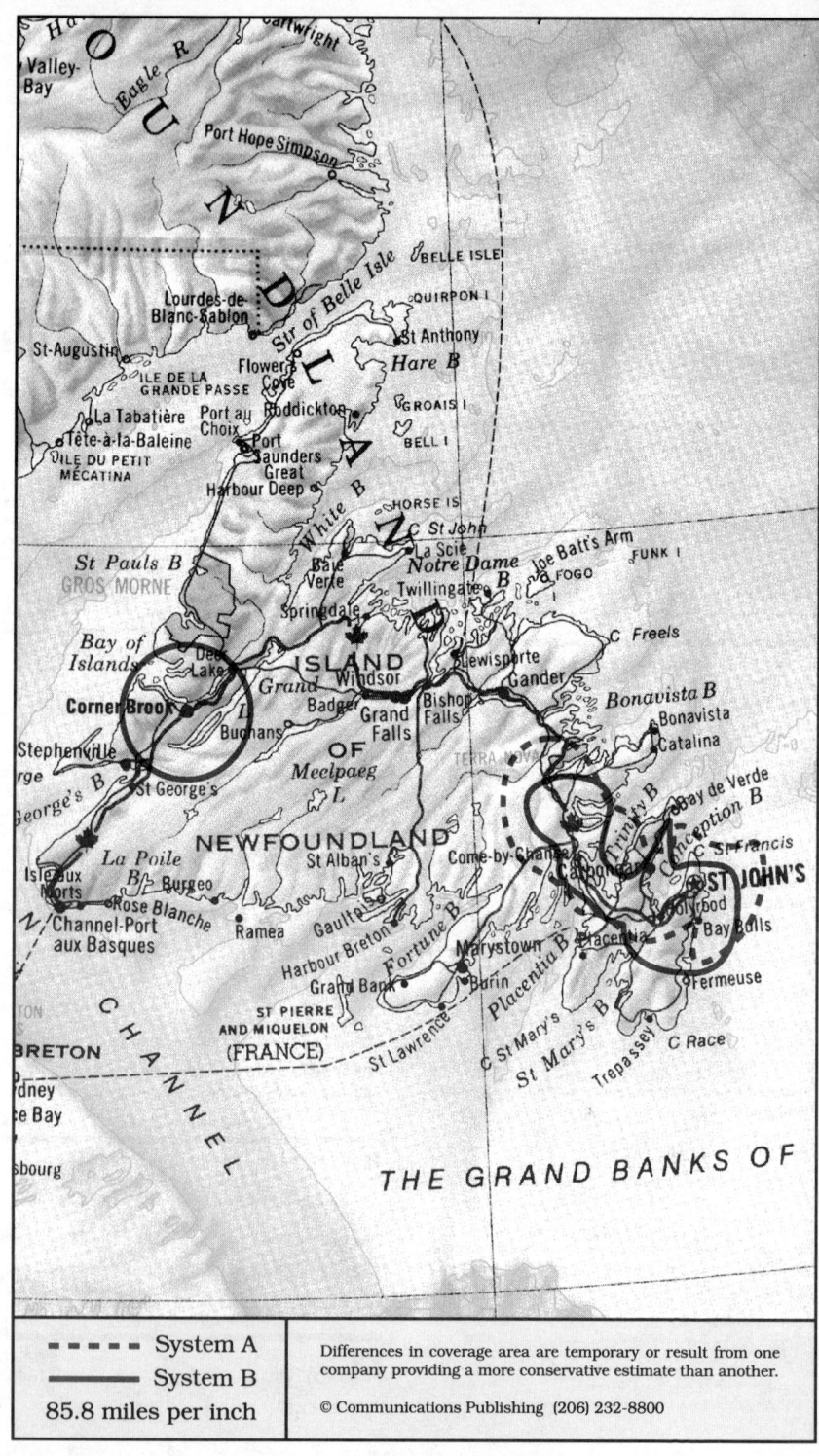

– – – – System A	
——— System B	
85.8 miles per inch	

Differences in coverage area are temporary or result from one company providing a more conservative estimate than another.

© Communications Publishing (206) 232-8800

St. John's, Newfoundland
Centerville, Corner Brook, St. John's

System A	System B
Cellular company	**Cellular company**

Cantel
58 Kenmount Road, Woodgate Plaza
St. John's, Newfoundland A1B 1W2 Canada

Customer service	some areas	800-361-5410
. . . .		514-340-9220
Local office	. . . . . .	709-738-1600
Corporate office	. . . .	416-229-1400

Sending calls

Visitor dialing instructions

Local calls	. . . .	area code + 7 digits
Long distance	. . .	1 + area code + 7 digits

Speed-dial numbers

Customer service	. . . .	0
Directory assistance	. . . . . .	411
Emergency	. . . .	911
Landline telephone repair	. . . . . .	611

Receiving calls

Caller dials the roamer access number below, hears tone, dials cellular area code and number.

Clarenville	. . . .	709-466-8626
St. John's	. . . .	709-749-7000

All Cantel subscribers receive calls automatically.

North American Cellular Network (see p. 1156)
All calls forward to you. Deactivate dial *35 send.

Billing information

Visitor rates

Most visitors ($ U.S.)	.	$2 day + $1.05 minute
All Cantel subscribers	. .	billed at home rate

Visitors without roaming agreements (p. 1159)
Cannot call since company takes no credit cards.

Identification numbers

City	System
Clarenville	 16691
Portugal Cove	 16625
St. John's	 16417
Newfoundland corridor	 16623

NewTel Cellular
Newfoundland Cellular
51 O'Leary Avenue, PO Box 2110
St. John's, Newfoundland A1C 5H6 Canada

Local office	. . . .	709-739-2355
Corporate office	. . . . .	709-739-2000

Sending calls

Visitor dialing instructions

Local calls	. . . .	7 digits
Long distance	. . .	1 + area code + 7 digits

Speed-dial numbers

Customer service	. . . .	811
Directory assistance	. . . . . .	411
Emergency	. . . .	911
Landline telephone repair	. . . . .	611
Technical support	. . . .	*11

Receiving calls

Caller dials the roamer access number below, hears tone, dials cellular area code and number.

Clarenville	. . . . . .	709-427-7626
Corner Brook	. . . . . .	709-632-7626
St. John's	. . . .	709-682-7626

Follow Me Roaming forwarding (see p. 1156)
Activate, dial *18 send. Deactivate dial *19 send.

Billing information

Visitor rates

Most visitors ($ Canadian)	$1 day + 60¢ minute
Cellnet Canada visitors	$0 day + 45¢ minute

Visitors without roaming agreements (p. 1159)
Cannot call since company takes no credit cards.

Identification numbers

City	System
All served cities	 16414

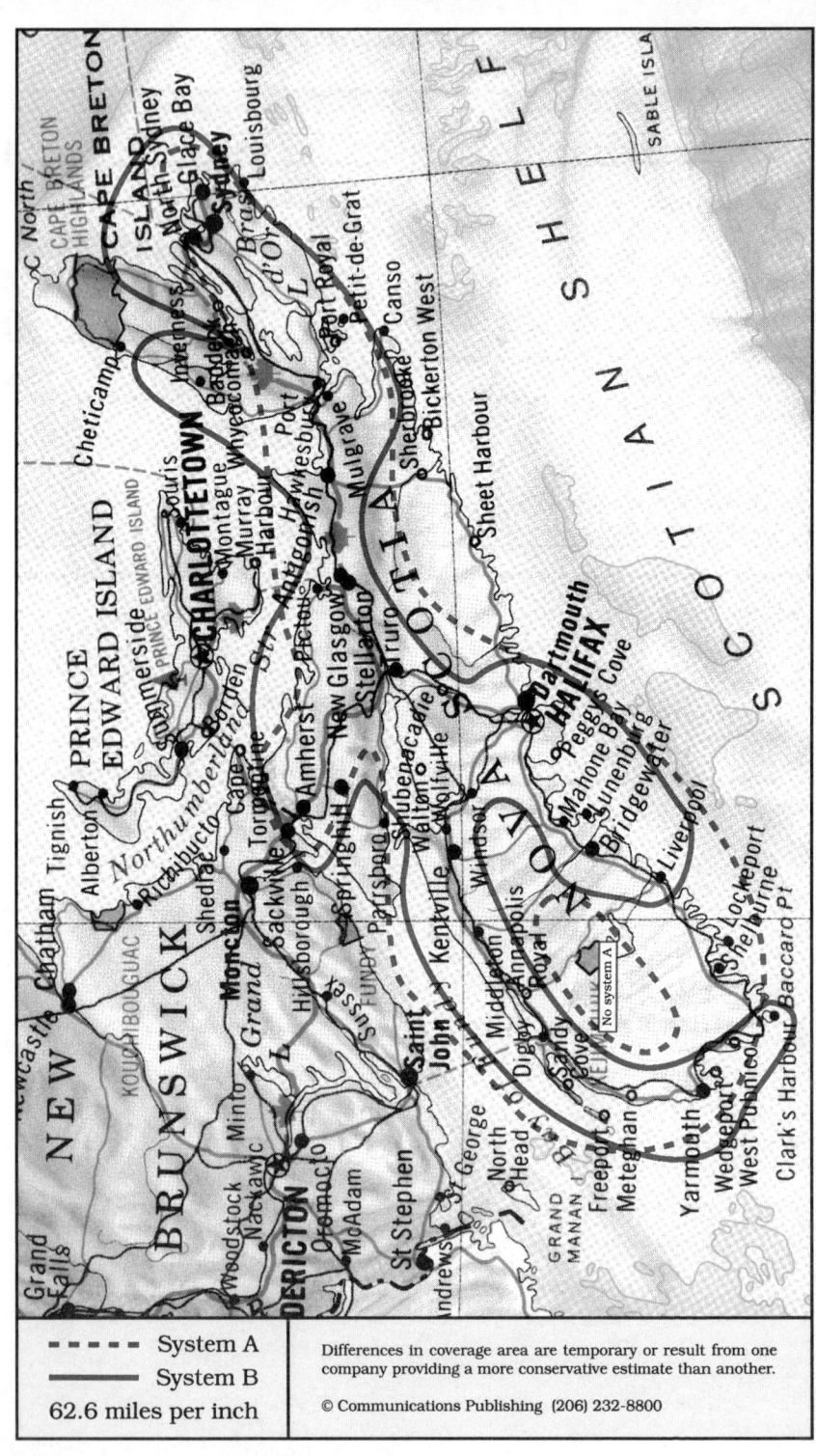

- - - - - System A	Differences in coverage area are temporary or result from one company providing a more conservative estimate than another.
———— System B	
62.6 miles per inch	© Communications Publishing (206) 232-8800

Halifax, Nova Scotia
Amherst, Halifax, Sydney, Yarmouth

System A	System B

System A

Cellular company

Cantel
6080 Yonge Street, Suite 905
Halifax, Nova Scotia B3K 5L2 Canada

Customer service	some areas	800-361-5410
. . . .		514-340-9220
Local office . .	some areas	800-265-6565
. . . .		902-453-1400
Corporate office		416-229-1400

Sending calls

Visitor dialing instructions

Local calls	area code + 7 digits
Long distance . . .	1 + area code + 7 digits

Speed-dial numbers

Customer service	0
Directory assistance	411
Emergency	911
Landline telephone repair	611

Receiving calls

Caller dials the roamer access number below,
hears tone, dials cellular area code and number.

Amherst	902-664-2500
Berwick, Kentville	902-679-0300
Bridgewater	902-527-3000
Chester	902-454-5393
Digby	902-532-4300
Halifax	902-452-7626
Liverpool	902-646-1300
New Glasgow, Westville . .	902-928-8226
Sydney	902-567-4000
Truro	902-893-1200
Yarmouth	902-749-4000

All Cantel subscribers receive calls automatically.

North American Cellular Network (see p. 1156)
All calls forward to you. Deactivate dial ∗35 send.

Billing information

Visitor rates

Most visitors ($ U.S.) .	$2 day + $1.05 minute
All Cantel subscribers . .	billed at home rate

Visitors without roaming agreements (p. 1159)
Cannot call since company takes no credit cards.

Identification numbers

City	System
Amherst	16479
Berwick, Kentville	16617
Bridgewater, Chester	16599
Digby	16791
Halifax	16393
Liverpool	16795
New Glasgow, Westville . .	16535
Nova Scotia corridor	16799
Sydney	16777
Truro	16475
Yarmouth	16793

System B

Cellular company

MT&T Cellular
Maritime Telephone
238 Brownlow Avenue, Suite 200, Tower 1
Dartmouth, Nova Scotia B3B 1Y2 Canada

Corporate office	902-421-2355

Sending calls

Visitor dialing instructions

Local calls	7 digits
Long distance . . .	1 + area code + 7 digits

Speed-dial numbers

After-hours repair service . . .	902-454-1300
Customer service	∗611
Directory assistance	411
Emergency	911
Police	4105

Receiving calls

Caller dials the roamer access number below,
hears tone, dials cellular area code and number.

Amherst	902-667-6800
Annapolis Royal	902-532-8100
Antigonish	902-863-7700
Aylesford	902-847-1200
Baddeck	902-295-7626
Barrington	902-637-7626
Bridgewater	902-527-7626
Chester	902-275-7626
Digby	902-245-8100
Grand Narrows	902-622-7626
Halifax	902-456-7626
Kentville	902-679-7626
Liverpool	902-354-8000
Meteghan	902-769-7626
Middleton	902-825-7200
New Glasgow	902-396-6800
Port Hawkesbury	902-227-7626
Pubnico	902-648-7626
Sydney	902-565-7626
Truro	902-893-0600
Weymouth	902-837-8000
Windsor	902-798-7626
Yarmouth	902-749-6200

Follow Me Roaming forwarding (see p. 1156)
Activate, dial ∗18 send. Deactivate dial ∗19 send.

Billing information

Visitor rates

Most visitors ($ Canadian)	$3 day + $1.25 minute
Cellnet Canada visitors	$0 day + 45¢ minute

Visitors without roaming agreements (p. 1159)
Cannot call since company takes no credit cards.

Identification numbers

City	System
All served cities	16390

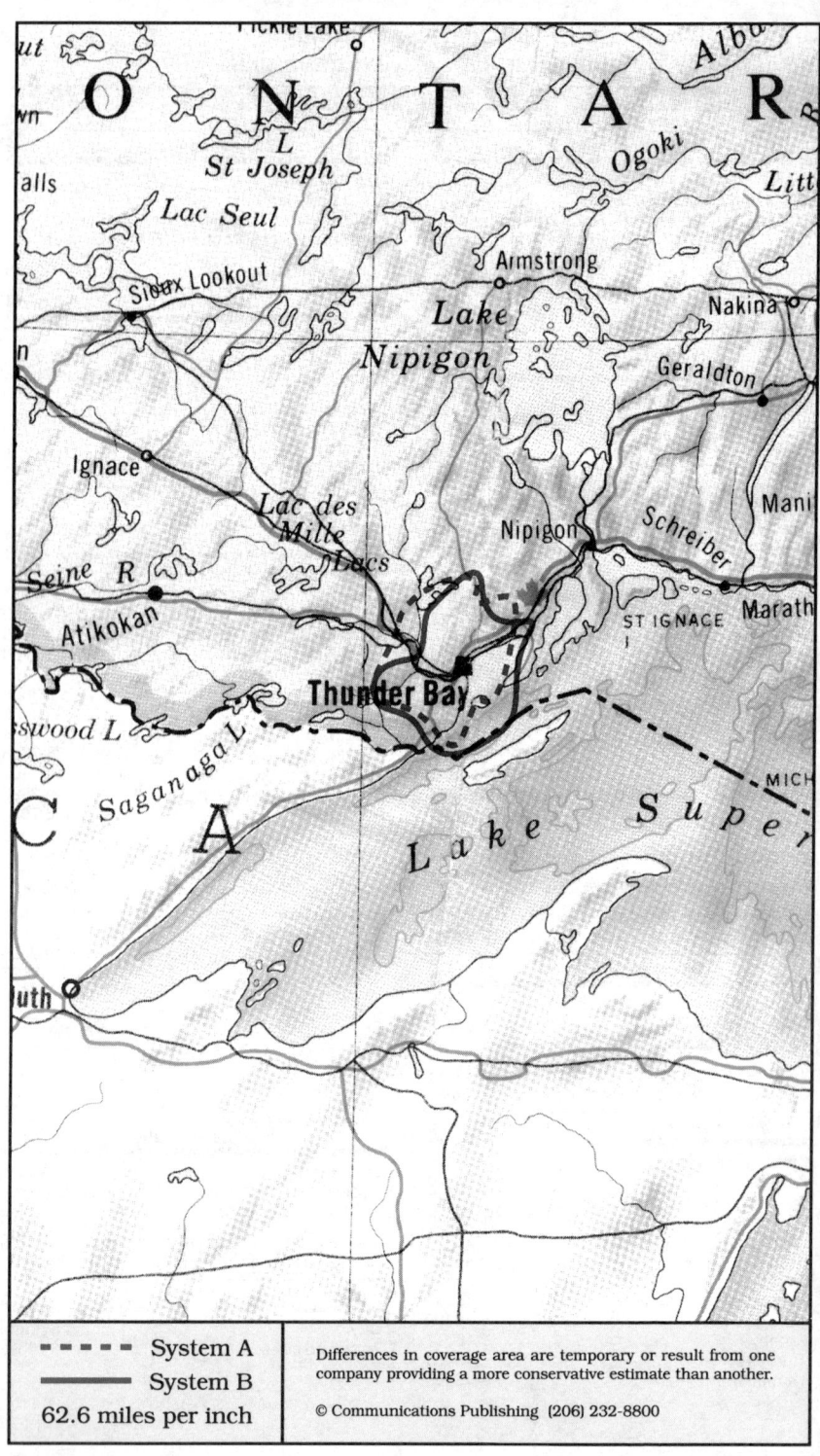

ONTAR
Albu
Pickle Lake
L
St Joseph
Ogoki
Litt
Lac Seul
Armstrong
Nakina
Sioux Lookout
Lake
Geraldton
Nipigon
Ignace
Lac des
Mille
Lacs
Nipigon
Schreiber
Mani
Seine R
ST IGNACE
Marath
Atikokan
I
Thunder Bay
Saganaga L
sswood L
MICH
CA
Lake Super
uth

Thunder Bay, Ontario

System A

Cellular company

Cantel
10 York Mills Road
North York, Ontario M2P 2C9 Canada

Customer service	some areas	800-268-7347
. . . .		604-433-1440
Corporate office	. . . . .	416-229-1400

Sending calls

Visitor dialing instructions

Local calls	area code + 7 digits
Long distance . . .	1 + area code + 7 digits

Speed-dial numbers

Customer service	0
Directory assistance	411
Emergency	911
Landline telephone repair	611

Receiving calls

Caller dials the roamer access number below,
hears tone, dials cellular area code and number.
. . . . 807-473-7626

All Cantel subscribers receive calls automatically.

North American Cellular Network (see p. 1156)
All calls forward to you. Deactivate dial *35 send.

Billing information

Visitor rates

Most visitors ($ U.S.) .	$2 day + $1.05 minute
All Cantel subscribers . .	billed at home rate

Visitors without roaming agreements (p. 1159)
Cannot call since company takes no credit cards.

Identification numbers

City	System
All served cities	16421

System B

Cellular company

Thunder Bay Cellular
1046 Lithium Drive
Thunder Bay, Ontario P7B 6G3 Canada

Corporate office	807-626-2355

Sending calls

Visitor dialing instructions

Local calls	7 digits
Long distance . . .	1 + area code + 7 digits

Speed-dial numbers

Canadian Automobile Association	*CCA or *222
Coast Guard	*16
Customer service	811
Emergency	911
Ontario Provincial Police . .	*OPP or *677
Repair service	807-622-5000 or 611

Receiving calls

Caller dials the roamer access number below,
hears tone, dials cellular area code and number.
. . . . 807-626-7626

Follow Me Roaming forwarding (see p. 1156)
Activate, dial *18 send. Deactivate dial *19 send.

Billing information

Visitor rates

Most visitors ($ Canadian)	$3 day + $1.25 minute
Cellnet Canada visitors	$0 day + 45¢ minute

Visitors without roaming agreements (p. 1159)
Cannot call since company takes no credit cards.

Identification numbers

City	System
All served cities	16418

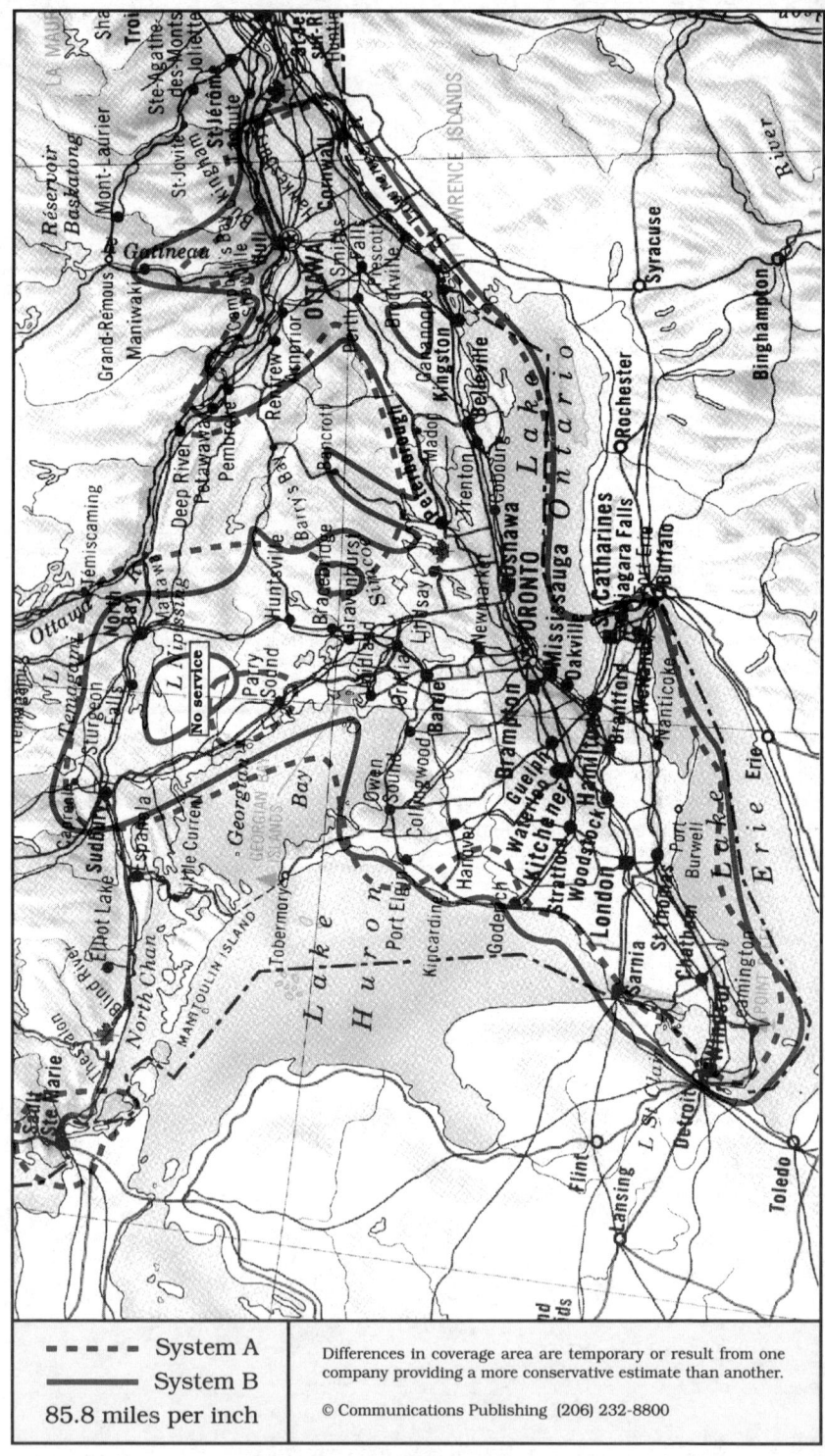

System A
System B
85.8 miles per inch

Differences in coverage area are temporary or result from one company providing a more conservative estimate than another.

© Communications Publishing (206) 232-8800

Toronto, Ontario
London, North Bay, Ottawa, Sault Ste. Marie, Toronto

System A	System B

Cellular company

Cantel
10 York Mills Road
North York, Ontario M2P 2C9 Canada

Customer service	some areas	800-268-7347
. . . . . .	from Ottawa	800-361-5410
. . . .		416-449-4811
Corporate office		416-229-1400

Sending calls

Visitor dialing instructions
Local calls . . . . . area code + 7 digits
Long distance . . . 1 + area code + 7 digits

Speed-dial numbers
Customer service 0
Directory assistance 411
Emergency 911
Landline telephone repair 611

Receiving calls

Caller dials the roamer access number below,
hears tone, dials cellular area code and number.
Ottawa* 613-769-7626
Toronto* 416-258-7626
*For a list of all Ontario roamer access numbers,
see the last Canadian page.

All Cantel subscribers receive calls automatically.

North American Cellular Network (see p. 1156)
All calls forward to you. Deactivate dial *35 send.

Billing information

Visitor rates
Most visitors ($ U.S.) . $2 day + $1.05 minute
All Cantel subscribers . . billed at home rate

Visitors without roaming agreements (p. 1159)
Cannot call since company takes no credit cards.

Identification numbers

For a list of all Ontario system identification
numbers, see the last Canadian page.

For full coverage area, see Montreal, Québec.

Cellular company

Bell Mobility Cellular
20 Carlson Court, Suite 100
Etobicoke, Ontario M9W 6V4 Canada

Customer service	in Canada	800-387-4869
. . . .	in U.S.	800-667-7626
. . . .		416-674-2233
Corporate office		416-674-2220

Sending calls

Visitor dialing instructions
Local calls . . . . . . area code + 7 digits
Long distance . . . 1 + area code + 7 digits

Speed-dial numbers
Cellular system repair 611
Cellular tutorials . . . . . *HELP or *4357
Customer service 811
Directory assistance 411
Emergency 911
News, sports, weather . . . #INFO or #4636
Ontario Provincial Police . . *OPP or *677
Roamer access number 511
Roaming information *711

Receiving calls

Caller dials the roamer access number below,
hears tone, dials cellular area code and number.
Ottawa* 613-725-7626
Toronto* 416-460-7626
*For a list of all Ontario roamer access numbers,
see the last Canadian page.

Follow Me Roaming forwarding (see p. 1156)
Activate, dial *18 send. Deactivate dial *19 send.

Billing information

Visitor rates
Most visitors ($ Canadian) $3 day + $1.25 minute
Cellnet Canada visitors $0 day + 45¢ minute

Visitors without roaming agreements (p. 1159)
Cannot call since company takes no credit cards.

Identification numbers

City	System
All served cities	16420

For full coverage area, see Montreal, Québec.

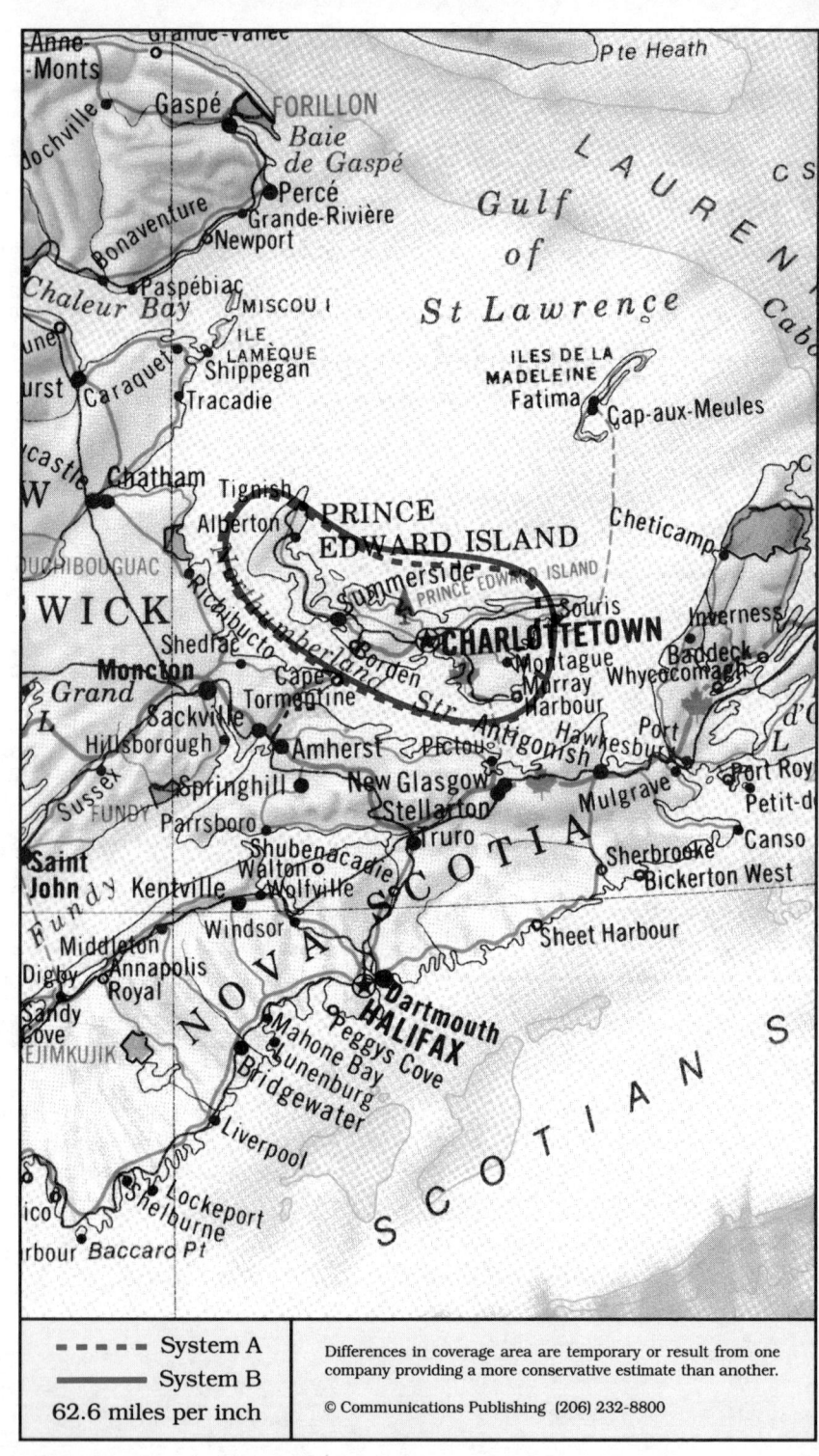

System A

System B

62.6 miles per inch

Differences in coverage area are temporary or result from one company providing a more conservative estimate than another.

© Communications Publishing (206) 232-8800

Charlottetown, Prince Edward Island
Charlottetown, Summerside

System A	System B

Cellular company

Cantel
330 University Avenue
Charlottetown, Prince Edward Island C1A 4M4

Customer service	some areas	800-361-5410
. . . .		514-340-9220
Local office		902-628-6228
Corporate office		416-229-1400

Sending calls

Visitor dialing instructions

Local calls	area code + 7 digits
Long distance . . .	1 + area code + 7 digits

Speed-dial numbers

Customer service	0
Directory assistance	411
Emergency	*RCMP or *7267
Landline telephone repair	611

Receiving calls

Caller dials the roamer access number below,
hears tone, dials cellular area code and number.

Charlottetown	902-628-3000
Summerside	902-888-1200

All Cantel subscribers receive calls automatically.

North American Cellular Network (see p. 1156)
All calls forward to you. Deactivate dial *35 send.

Billing information

Visitor rates

Most visitors ($ U.S.) .	$2 day + $1.05 minute
All Cantel subscribers . .	billed at home rate

Visitors without roaming agreements (p. 1159)
Cannot call since company takes no credit cards.

Identification numbers

City	System
Charlottetown	16595
Summerside	16667

Cellular company

Island Tel Cellular
69 Belvedere Avenue, PO Box 820
Charlottetown, Prince Edward Island C1A 7M1

Corporate office	902-368-2355

Sending calls

Visitor dialing instructions

Local calls	7 digits
Long distance . . .	1 + area code + 7 digits

Speed-dial numbers

Customer service	811
Emergency	0
Repair service	902-566-0155

Receiving calls

Caller dials the roamer access number below,
hears tone, dials cellular area code and number.

Alberton	902-853-7626
Charlottetown	902-628-7626
Montague	902-969-7626
Summerside	902-888-7626

Follow Me Roaming forwarding (see p. 1156)
Activate, dial *18 send. Deactivate dial *19 send.

Billing information

Visitor rates

Most visitors ($ Canadian)	$3 day + $1.25 minute
Cellnet Canada visitors	$0 day + 45¢ minute

Visitors without roaming agreements (p. 1159)
Cannot call since company takes no credit cards.

Identification numbers

City	System
All served cities	16430

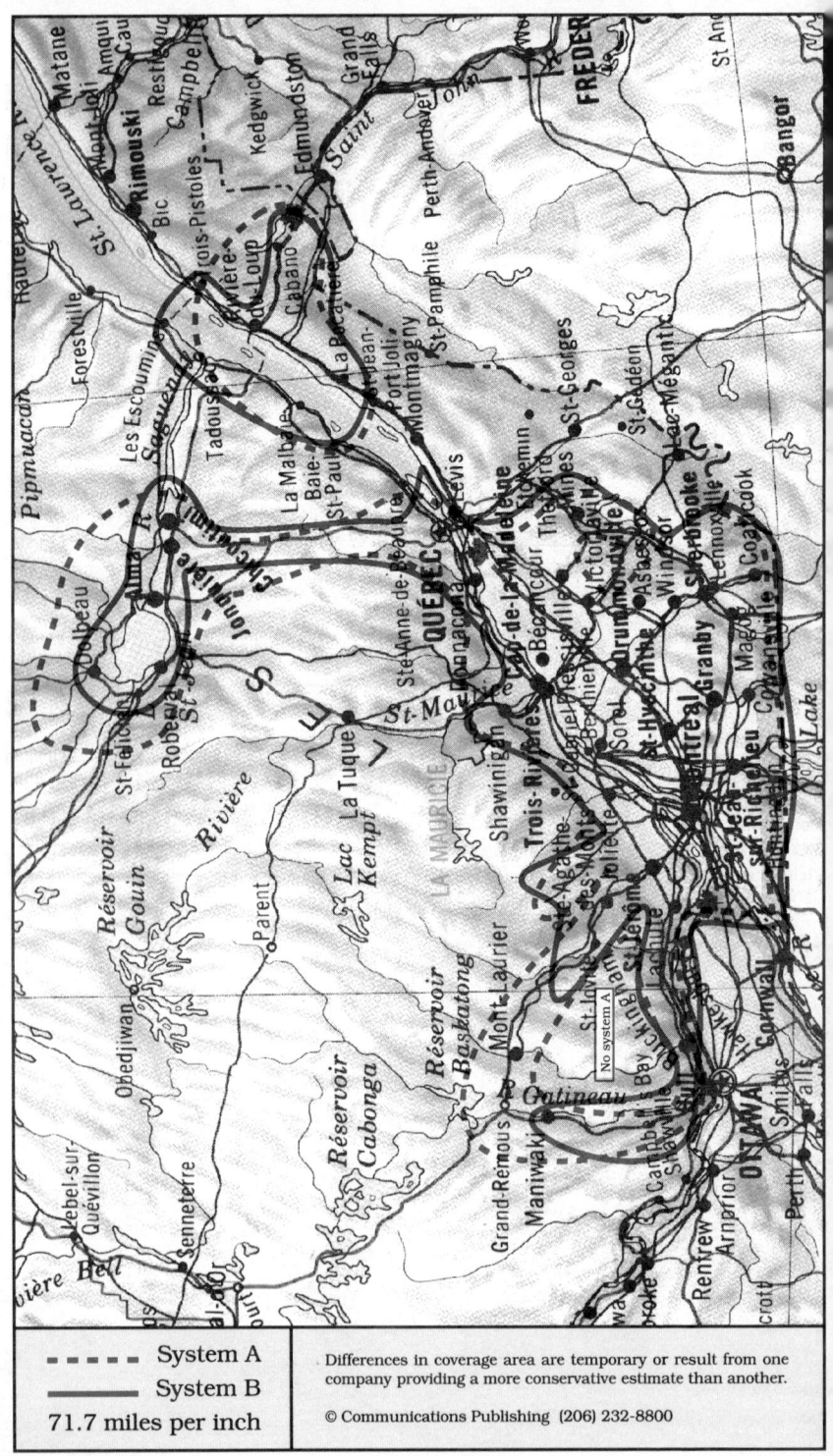

System A
System B
71.7 miles per inch

Differences in coverage area are temporary or result from one company providing a more conservative estimate than another.

© Communications Publishing (206) 232-8800

No system A

Montréal, Québec
Alma, Montréal, Mont-Laurier, Rivière du Loup, Thetford Mines, Sherbrooke

System A	System B

Cellular company

Cantel
6315 Cote de Liesse
St. Laurent, Québec H4T 1E5 Canada

Customer service	some areas	800-361-5410
. . . .		514-340-9220
Corporate office	. . . .	416-229-1400

Sending calls

Visitor dialing instructions

Local calls	. . . .	area code + 7 digits
Long distance	. . .	1 + area code + 7 digits

Speed-dial numbers

Customer service	0
Directory assistance	411
Emergency	911
Landline telephone repair	611

Receiving calls

Caller dials the roamer access number below,
hears tone, dials cellular area code and number.

City	Roamer access	System ID
Actonvale	none assigned	16787
Alma	418-662-0000	16677
Buckingham	none assigned	16809
Chicoutimi	418-693-3000	16389
Drummondville	819-472-9026	16631
Granby	514-777-9026	16651
Joliette	514-752-3000	16669
Montabello	none assigned	16811
Montréal	514-497-7626	16401
Montréal to Ottawa corridor		16487
Montréal to Québec City corridor		16543
Mont Laurier	819-623-8000	16731
Québec City	418-655-7626	16407
Rivière-du-Loup	418-867-7000	16587
Ste. Agathe	819-326-9626	16703
Ste. Hyacinthe	514-771-9326	16563
St. Jean-Sur-Richel.	514-358-0826	16559
St. Jerome	514-431-6126	16561
St. Polycarpe	none assigned	16547
Shawinigan	none assigned	16813
Sherbrooke	819-822-7626	16439
Sorel	514-746-6626	16671
Thetford Mines	418-335-1500	16515
Trois Rivières	819-372-6200	16433
Valleyfield	514-377-7526	16743
Victoriaville	819-752-1300	16679

All Cantel subscribers receive calls automatically.

North American Cellular Network (see p. 1156)
All calls forward to you. Deactivate dial *35 send.

Billing information

Visitor rates

Most visitors ($ U.S.) .	$2 day + $1.05 minute
All Cantel subscribers . .	billed at home rate

Visitors without roaming agreements (p. 1159)
Cannot call since company takes no credit cards.

Identification numbers

Listed above under the section **receiving calls**

Also see Toronto, Ontario and Rimouski, Québec.

Cellular company

Bell Mobility Cellular
1425 Transcanadienne, Suite 110
Dorval, Québec H9P 2W9 Canada

Customer service	in Canada	800-361-5551
. . . .	in U.S.	800-667-7626
. . . .		514-685-0040
Corporate office	. . . .	416-674-2220

Sending calls

Visitor dialing instructions

Local calls	. . . . .	area code + 7 digits
Long distance	. . .	1 + area code + 7 digits

Speed-dial numbers

Cellular system repair	611
Cellular tutorials	*HELP or *4357
Customer service	811
Directory assistance	411
Emergency	911
News, sports, weather . . .	#INFO or #4636
Ontario Provincial Police . .	*OPP or *677
Roaming information	*711

Receiving calls

Caller dials the roamer access number below,
hears tone, dials cellular area code and number.

Alma	418-662-4343
Chicoutimi	418-690-7626
Drummondville	819-475-7171
Durham	519-369-7111
Goderich	519-524-0101
Granby	514-777-6100
Huntsville	705-788-4141
Hull	613-725-7626
Joliette	514-752-7000
Montréal	514-386-7626
Mont Laurier	819-440-7626
Québec City	418-654-7626
Riviére du Loup	418-868-4026
Sherbrooke	819-821-0808
Sorel	514-746-9393
St. Hyacinthe	514-771-8000
St. Jean	514-357-0202
St. Jerome	514-565-7171
Ste. Agathe	819-324-7626
Thetford Mines	418-334-2026
Trois Rivières	819-372-7626
Valleyfield	514-377-6500
Victoriaville	819-357-0202

Follow Me Roaming forwarding (see p. 1156)
Activate, dial *18 send. Deactivate dial *19 send.

Billing information

Visitor rates

Most visitors ($ Canadian)	$3 day + $1.25 minute
Cellnet Canada visitors	$0 day + 45¢ minute

Visitors without roaming agreements (p. 1159)
Cannot call since company takes no credit cards.

Identification numbers

All served cities	16420

For full coverage area, see Toronto, Ontario.

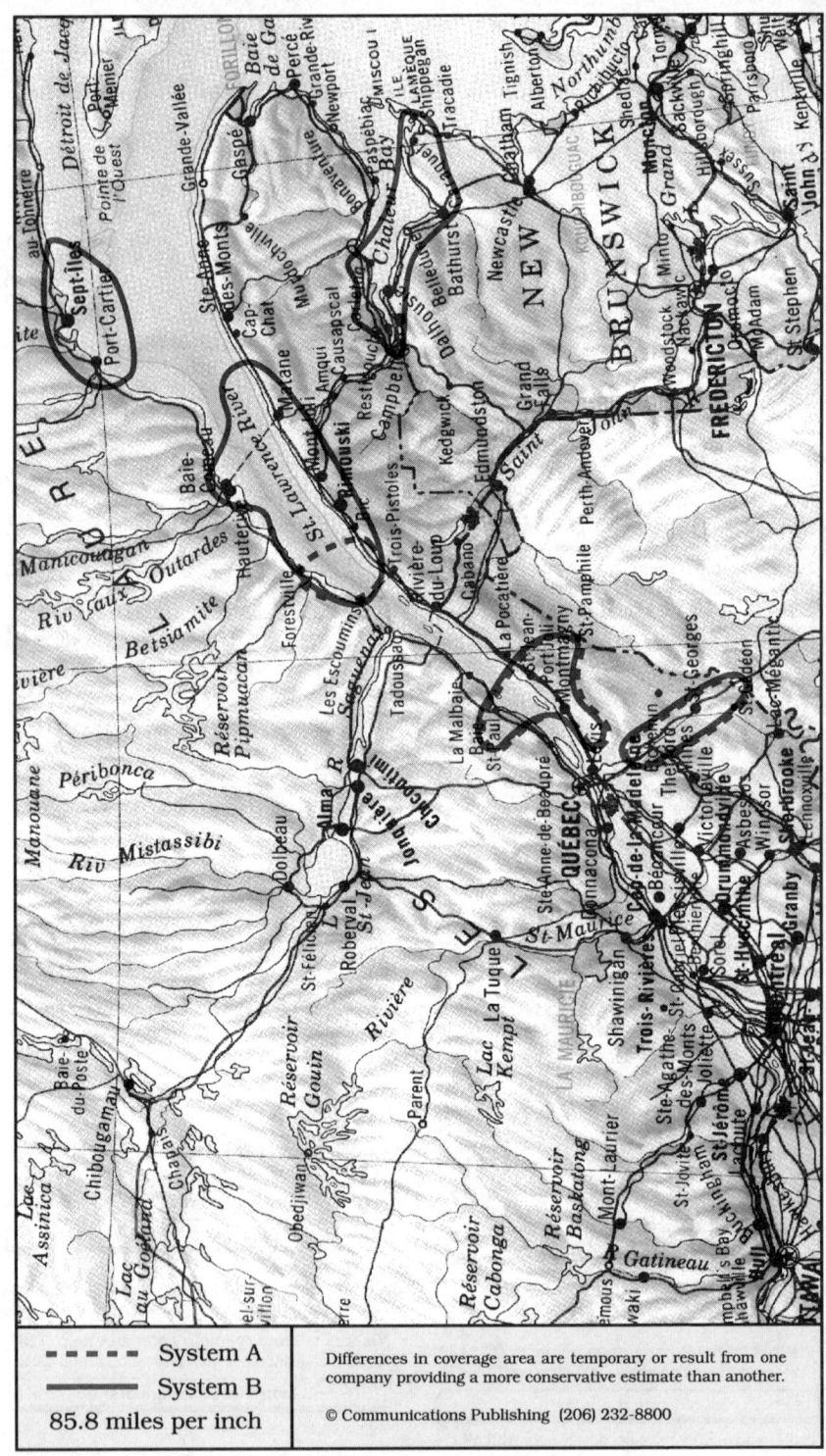

Rimouski, Québec

Bathurst (New Brunswick), Montmagny, Rimouski, St. Georges, Sept. Iles

System A	System B

Cellular company

Cantel
6315 Cote de Liesse
St. Laurent, Québec H4T 1E5 Canada

Customer service	some areas	800-361-5410
. . . .		514-340-9220
Corporate office	. . . . .	416-229-1400

Sending calls

Visitor dialing instructions

Local calls	area code + 7 digits
Long distance . . .	1 + area code + 7 digits

Speed-dial numbers

Customer service	0
Directory assistance	411
Emergency	911
Landline telephone repair	611

Receiving calls

Caller dials the roamer access number below,
hears tone, dials cellular area code and number.

City	Roamer access	System ID
Donnacona	418-285-6226	16589
Montmagny	418-248-8500	16673
Rimouski	418-725-6400	16815
St. Georges	418-774-7126	16549
Ste. Marie	418-387-0000	16481

All Cantel subscribers receive calls automatically.

North American Cellular Network (see p. 1156)
All calls forward to you. Deactivate dial ✱35 send.

Billing information

Visitor rates

Most visitors ($ U.S.) .	$2 day + $1.05 minute
All Cantel subscribers . .	billed at home rate

Visitors without roaming agreements (p. 1159)
Cannot call since company takes no credit cards.

Identification numbers

Listed above under the section **receiving calls**

For full coverage area, see Montreal, Québec.

Cellular company

Québec-Téléphone Cellulaire
9, Rue Jules A. Brillant, Departement 705
Rimouski, Québec G5L 7E4 Canada

Customer service	some areas	800-463-8988
Corporate office	. . . . .	418-722-4282

Sending calls

Visitor dialing instructions

Local calls	7 digits
Long distance . . .	1 + area code + 7 digits

Speed-dial numbers

Cellular system repair	611
Customer service	811
Emergency	0 or 411
Roaming information	✱711

Receiving calls

Caller dials the roamer access number below,
hears tone, dials cellular area code and number.

Baie-Comeau	418-295-6000
Donnacona	418-285-9191
Matane	418-566-5000
Montmagny	418-248-6700
Rimouski	418-725-9191
Sept-Iles	418-964-6000
St. Georges	418-226-7626
Ste. Marie	418-387-9191

Follow Me Roaming forwarding (see p. 1156)
Activate, dial ✱18 send. Deactivate dial ✱19 send.

Billing information

Visitor rates

Most visitors ($ Canadian)	$3 day + $1.25 minute
Cellnet Canada visitors	$0 day + 45¢ minute

Visitors without roaming agreements (p. 1159)
Cannot call since company takes no credit cards.

Identification numbers

City	System
All served cities	16458

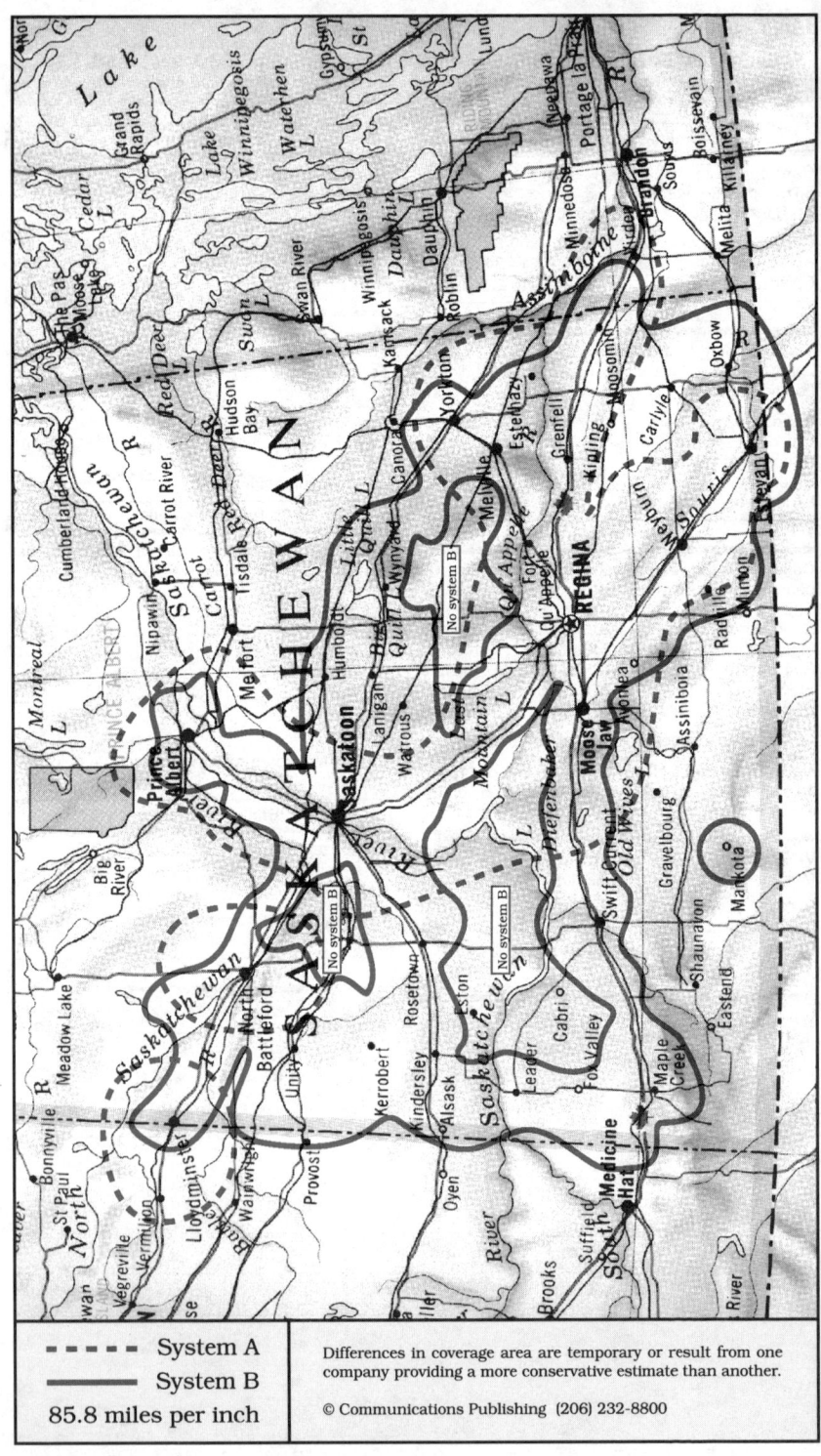

- - - - - System A	Differences in coverage area are temporary or result from one company providing a more conservative estimate than another.
—— System B	
85.8 miles per inch	© Communications Publishing (206) 232-8800

Regina, Saskatchewan

Estevan, Lloydminster, Maple Creek, Prince Albert, Regina, Saskatoon

System A	System B

Cellular company

Cantel
Suite 1400, 1920 Broad Street
Regina, Saskatchewan S4P 3V2 Canada

Suite 950, 410 22nd Street E
Saskatoon, Saskatchewan S7K 5T6 Canada

Customer service	some areas	800-661-1874
. . . .		604-433-1440
Local office	Regina	306-522-1400
. . . .	Saskatoon	306-665-1400
Corporate office	. . . .	416-229-1400

Sending calls

Visitor dialing instructions

Local calls	area code + 7 digits
Long distance . . .	1 + area code + 7 digits

Speed-dial numbers

Customer service	0
Directory assistance	1-306-555-1212
Emergency	911
Landline telephone repair	611

Receiving calls

Caller dials the roamer access number below,
hears tone, dials cellular area code and number.

City	Roamer access	System ID
Balgonie	none assigned	16611
Davidson	306-567-0100	16533
Estevan	306-461-7626	16749
Lloydminster,Alberta	403-871-7626	16567
Melville	306-728-1200	16753
Moose Jaw	306-681-7626	16503
Moosomin	306-435-0100	16781
North Battleford	306-481-7626	16577
Prince Albert	306-981-7626	16575
Regina	306-596-7626	16409
Regina corridor	none assigned	16499
Rosthern	306-232-3200	16751
Saskatoon	306-241-7626	16413
Weyburn	306-891-7626	16747
Yorkton	306-641-7626	16745

All Cantel subscribers receive calls automatically.

North American Cellular Network (see p. 1156)
All calls forward to you. Deactivate dial *35 send.

Billing information

Visitor rates

Most visitors ($ U.S.) .	$2 day + $1.05 minute
All Cantel subscribers . .	billed at home rate

Visitors without roaming agreements (p. 1159)
Cannot call since company takes no credit cards.

Identification numbers

Listed above under the section **receiving calls**

Cellular company

SaskTel Mobility Cellular
2106 First Avenue
Regina, Saskatchewan S4P 3Y2 Canada

403-A 50th Street E
Saskatoon, Saskatchewan S7K 6K1 Canada

Customer service	some areas	800-667-6840
. . . .		306-931-5123
Local office	Regina	306-777-2822
. . . .	Saskatoon	306-931-6029
Corporate office	. . . .	306-777-5545

Sending calls

Visitor dialing instructions

Local calls	7 digits
Long distance . . .	1 + area code + 7 digits

Speed-dial numbers

Customer service	811
Directory assistance	411
Emergency	911
Roaming information	711
Trouble reporting	611

Receiving calls

Caller dials the roamer access number below,
hears tone, dials cellular area code and number.*
*For A - K cities, see the last Canadian page

Lanigan	306-365-7626
Lloydminster	306-821-7626
Lumsden	306-731-7626
Macklin, Whitewood . . .	306-753-7626
Maple Creek	306-662-7626
Melville	306-728-7626
Moose Jaw	306-631-7626
Moosomin	306-435-7626
North Battleford	306-441-7626
Oxbow	306-483-7626
Prince Albert	306-961-7626
Qu'Appelle	306-699-7626
Regina	306-536-7626
Rosetown	306-831-7626
Rosthern	306-232-7626
Saskatoon	306-221-7626
Stoughton	306-457-7626
Swift Current	306-741-7626
Weyburn	306-861-7626
Wilkie	306-843-7626
Wynyard	306-554-7626
Yorkton	306-621-7626

Follow Me Roaming forwarding (see p. 1156)
Activate, dial *18 send. Deactivate dial *19 send.

Billing information

Visitor rates

Most visitors ($ Canadian)	$3 day + $1 minute
Cellnet Canada visitors	$0 day + 45¢ minute

Visitors without roaming agreements (p. 1159)
Cannot call since company takes no credit cards.

Identification numbers

City	System
All served cities	16410

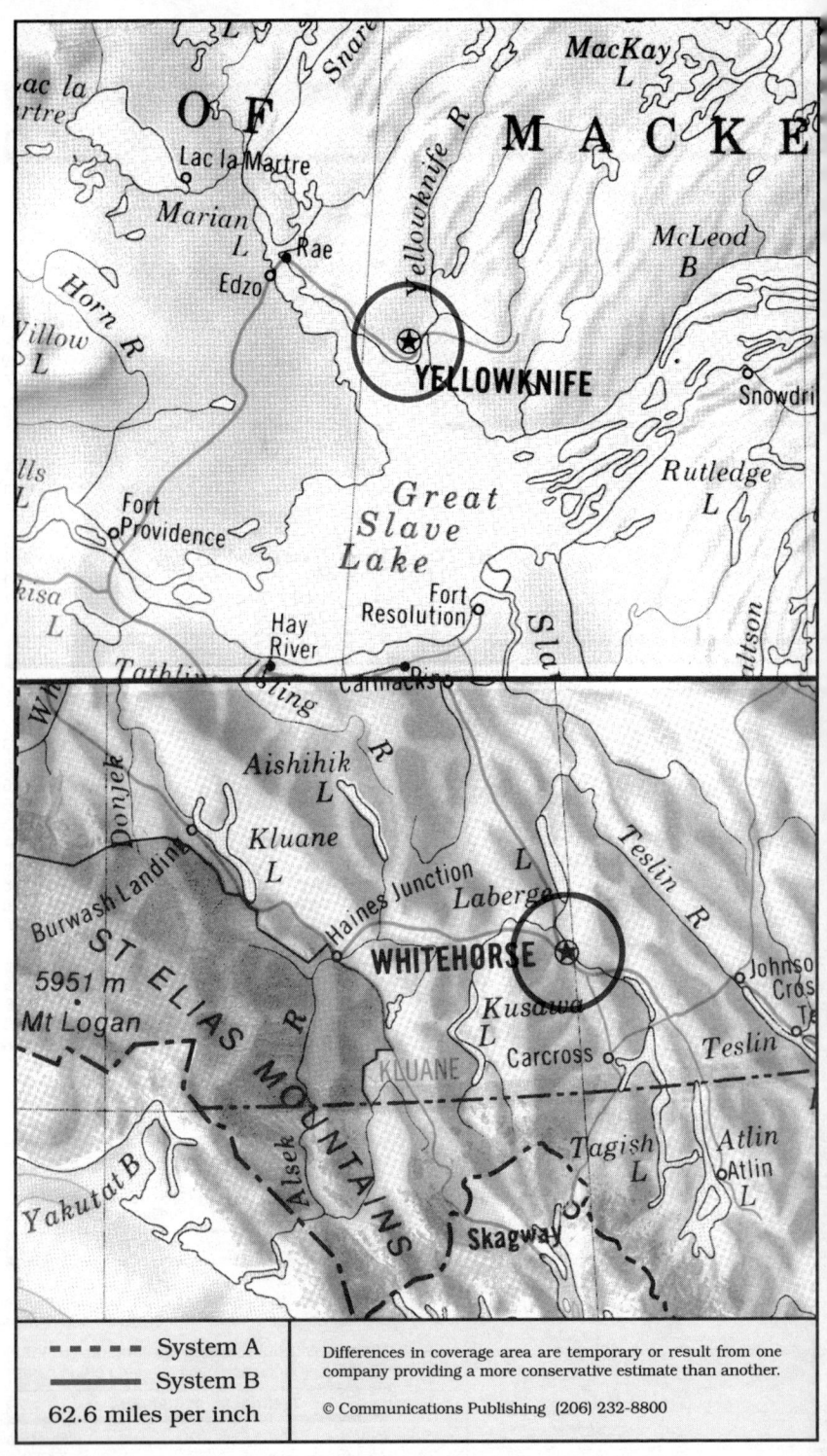

- - - - - System A	
——— System B	
62.6 miles per inch	

Differences in coverage area are temporary or result from one company providing a more conservative estimate than another.

© Communications Publishing (206) 232-8800

System A	System B

Cellular company

NorthwesTel
PO Bag 2727
Whitehorse, Yukon Territory Y1A 4Y4 Canada

Important notice:
This company does not plan to provide service to most short-term visitors until 1994. You may become a temporary subscriber by paying an activation fee and a monthly fee and having your cellular telephone reprogrammed with a Yukon telephone number.

Corporate office   403-668-5300

Sending calls

Visitor dialing instructions
Local calls . . . . . .   area code + 7 digits
Long distance . . .   1 + area code + 7 digits

Speed-dial numbers
Customer service   403-668-5300
Emergency 0

Receiving calls

Caller dials the roamer access number below,
hears tone, dials cellular area code and number.
For future use   403-668-1999

Billing information

Visitor rates
. . . .   No service for visitors

Visitors without roaming agreements (p. 1159)
Cannot call since company takes no credit cards.

Identification numbers

City	System
All served cities	Call company

Canadian Roamer Access Numbers

These was no room for these roamer access numbers on previous pages.

System A	System B

Province of Ontario (system A)

City	Roamer Access	System ID
Aurora	416-841-2900	16423
Barrie	705-721-2500	16435
Beeton	416-729-5000	16699
Belleville	613-969-2000	16449
Bracebridge	705-645-7000	16615
Brantford	519-758-7626	16397
Brockville	613-498-7000	16681
Carleton Place	613-253-9100	16707
Chatham	519-436-2000	16457
Cobourg	416-373-5000	16621
Collingwood	705-444-3100	16477
Cornwall	613-936-3000	16451
Dunville	416-774-0100	16789
Fenelton Galls	705-887-8500	16725
Fort Erie (planned)	416-994-2000	16797
Georgetown	416-873-3200	16585
Goderich	519-524-3500	16773
Guelph	519-837-7626	16689
Haliburton	705-754-5000	16739
Hamilton	416-541-7626	16395
Hull	819-773-0000	16405
Huntsville	705-788-6000	16695
Kingston	613-541-9100	16443
Kitchener	519-748-7626	16397
Leamington	519-322-6000	16579
Lindsay	705-878-7100	16583
London	519-668-4100	16399
Midland	705-526-1800	16553
Morrisburg	none assigned	16715
Newmarket	416-853-3800	16423
North Bay	705-495-5000	16507
Orangeville	519-942-5000	16529
Orillia	705-327-4000	16509
Oshawa	416-434-0100	16403
Ottawa	613-769-7626	16405
Ottawa to Kingston corridor		16545
Owen Sound	519-372-8000	16505
Parry Sound	705-746-3000	16705
Pembroke	613-732-6100	16697
Perth	613-264-3000	16713
Peterborough	705-749-3737	16441
Port Elgin	519-832-7000	16775
Port Perry	416-446-0658	16403
Renfrew	none assigned	16717
St. Catharines	416-641-7626	16415
Sarnia	519-339-7400	16459
Sault Ste. Marie	705-945-1800	16539
Simcoe	519-428-7800	16555
Smith Falls	613-284-3000	16711
Stratford	519-272-7000	16551
Sudbury	705-688-2199	16419
Tillsonburg	519-688-8000	16557
Toronto	416-258-7626	16423
Toronto suburbs	416-587-7626	16423
Toronto (N of Hwy41)	416-617-7626	16485
Uxbridge	416-852-8000	16701
Vankleck Hill	613-678-7000	16709
Welland	416-734-0000	16581
Windsor	519-977-3000	16429
Woodstock	519-421-8000	16541

Province of Saskatchewan (system A)

For all cities, see Regina, Saskatchewan page.

Province of Ontario (system B)

City	Roamer Access	System ID
Alma	418-662-4343	16420
Aurora	416-460-7626	16420
Barrie	705-722-1818	16420
Belleville	613-969-7626	16420
Bracebridge	705-645-0909	16420
Brantford	519-758-4550	16420
Brockville	613-498-5026	16420
Carleton Place	613-253-6426	16420
Chatham	519-436-4776	16420
Cobourg	416-373-2020	16420
Collingwood	705-444-4044	16420
Cornwall	613-937-7626	16420
Fort Erie	416-994-3030	16420
Guelph	519-766-2111	16420
Haliburton	705-457-5454	16420
Hamilton	416-572-7626	16420
Hawkesbury	613-678-0026	16420
Hespler	519-658-7626	16420
Kingston	613-541-7626	16420
Kitchener	519-741-7626	16420
Leamington	519-322-8181	16420
Lindsay	705-878-2121	16420
London	519-657-5858	16420
Midland	705-527-3131	16420
Milton	416-875-7626	16420
Newmarket	416-853-2355	16420
Niagara Falls	416-357-8111	16420
North Bay	705-495-9111	16420
Orangeville	519-942-6464	16420
Orillia	705-327-3626	16420
Oshawa	416-433-7626	16420
Ottawa	613-725-7626	16420
Owen Sound	519-372-5050	16420
Parry Sound	705-746-1616	16420
Pembroke	613-732-0303	16420
Peterborough	705-741-8888	16420
Port Elgin	519-389-7111	16420
St. Catharines	416-641-5858	16420
Sarnia	519-339-6868	16420
Sault Ste. Marie	705-946-8989	16420
Simcoe	519-428-5757	16420
Smith Falls	613-284-6026	16420
Stradford	519-272-5151	16420
Sudbury	705-688-6111	16420
Toronto	416-460-7626	16420
Welland	416-732-8111	16420
Windsor	519-977-4848	16420
Woodstock	519-421-6363	16420

Province of Saskatchewan (system B)*

City	Roamer Access	System ID
Burstall	306-679-7626	16410
Davidson	306-567-7626	16410
Delisle	306-493-7626	16410
Esterhazy	306-745-7626	16410
Estevan	306-421-7626	16410
Foam Lake	306-272-7626	16410
Fort Qu'Appelle	306-332-7626	16410
Grenfell	306-697-7626	16410
Gull Lake	306-672-7626	16410
Kenosee Lake	306-577-7626	16410
Kerrobert	306-834-7626	16410
Kindersley	306-463-7626	16410

*For L - Z cities, see Regina, Saskatchewan page

International Cellular

✈ Your cellular telephone operates only on AMPS systems. When traveling, you should declare it at customs, and bring proof of ownership to avoid paying duty upon return. Some countries prohibit the importation of cellular telephones; contact their consulates for more information or consider renting a cellular telephone upon arrival. The letters PTT indicate that cellular service is managed by a postal and telecommunications government agency.

Systems

System	MHz frequency	Description
ACS	400 & 800	Comvik Advanced Cellular System available in Sweden and Hong Kong.
AMPS	800	Advanced Mobile Phone Service. Analog cellular currently operating in the U.S., Canada, Mexico, Caribbean, Central and South America, etc.
CDMA	800	Code Division Multiple Access. A digital technology proposed for the U.S. by Qualcomm, San Diego, CA that offers 10 to 20 times the user capacity of AMPS.
DCS1800	1.8 GHz	Digital Cellular System. European digital. May be available in 1995.
ETACS	900	Enhanced Totally Accessed Communications System.
E-TDMA	800	Extended Time Division Multiple Access. A digital technology proposed for the U.S. by Hughes Network Systems/IMM, Washington DC that offers 8 to 10 times the user capacity of AMPS.
GSM	900	Groupe Special Mobile. European digital using TACS and NMT 900 frequences for international roaming throughout Europe.
IMTS	150 & 450	Improved Mobile Telephone Service. Radiotelephone service operating in the U.S. and Canada. Does not provide true cellular handoff.
JTACS	900	Japanese Totally Accessed Communications System.
N-AMPS	800	Narrowband Advanced Mobile Phone Service. A channel-splitting technology designed by Motorola, Arlington Heights IL to that offers 3 times the user capacity of AMPS. Currently in use in Japan.
Netz-B	160	IMTS style system with no handoff.
Netz-C	450	German digital cellular system.
NMT	450 & 900	Nordic Mobile Telephone.
PCN	1.7 & 1.8 GHz	Future Groupe Speciale Mobile system in England. Proposed for U.S.
RC2000	200 & 450 & 800	Radiocom 2000. A hybrid cellular system with limited handoff capability.
RTMS	450	Italian cellular system.
TACS	900	Totally Accessed Communications System. AMPS channels used for UHF TV in Europe. Digital TACS will use remaining AMPS channels.
TDMA	800	Time Division Multiple Access. A digital technology selected by CTIA, Washington DC for the U.S. that offers 3.7 times the user capacity of AMPS. Currently in use in Europe.

Service

* Service in countries with this mark are described in more detail beginning on page 1063.

Country	System	Company
Algeria	NMT-900	Algerian PTT
Andorra	NMT-450	Telefonica of Spain
* Anguilla	AMPS	Anguilla Boatphone
* Antigua	AMPS	Boatphone Antigua
* Argentina	AMPS	Companie de Radiocomunicaciones Moviles (CRM)
* Australia	AMPS	Telecom MobileNet
Austria	Netz-B, C	Austrian PTT
* Bahamas	AMPS	Bahamas Telecommunications (Batelco)
Bahrain	TACS	Bahrain Telecommunications Company
Bangladesh	AMPS	Hutchison Bangladesh Telecom Pvt.
* Barbados	AMPS	Barbados Communications Services
* Barbuda	AMPS	Boatphone Antigua
Belgium	NMT-450	RTT
* Bermuda	AMPS	Bermuda Cellular
* Bolivia	AMPS	Telefonica Celular de Bolivia
* Brazil	AMPS	Tele R.S.
Brunei	AMPS	Jabatan Telecoms Brunei
* Canada	AMPS	(see the main index that begins on page 1)
	IMTS	numerous companies
* Cayman Islands	AMPS	Cable and Wireless

Chile	AMPS	in Santiago A) Chile Telephone B) CIDCOM
	AMPS	except Santiago A) VTR Cellular/Millicom B) Telecom
China	TACS	PTT Beijing (A and B)
	NMT-450	PTT Chongqing
	TACS	PTT Guangzhou
	TACS	PTT Qinhuangdao
	TACS	PTT Shanghai (A and B)
	TACS	PTT Shenyang
	TACS	PTT Shenzhen
	TACS	PTT Tianjin
	TACS	PTT Zhuhai (Cities below will have service in the future:)
		Chengdu, Dalian, Fuzhou, MMEI, Shangdong, Xiamen
	NMT-450	PTT Zhi Hua
Colombia	AMPS	ETB
Costa Rica	AMPS	Millicom (described in this book) and Comvik
Cyprus	NMT-900	Cyprus Telecommunication Authority (CYTA)
Czechoslovakia	NMT-450	Eurotel Prague, Bratislavia (Bell Atlantic and U S West)
Denmark	NMT-450, 900	1) Statens Teletjeneste (PTT) 2) Dansk Mobiltelefon
Dominican Republic	AMPS	Codetel Comunicaciones Moviles
Egypt	AMPS, TACS	Arab Republic Egypt Telecom Org. (ARENTO)
El Salvador	AMPS	Telemovil El Salvador
England	not applicable	Please see United Kingdom
Estonia	NMT-450	Eesti Mobil Telefon
Faeroe Islands	NMT-450	Telefonwerk Faroya Logting (The Faeroe Telephone Co.)
Facrocse	NMT-450	Facrocse Telephone Company
Finland	NMT-450, 900	Telecom Finland (TELE)
France	NMT-450	Ligne Societe Francaise du Radiotelephone (SFR)
	NMT900, RC2000	France Telecom
Gabon	AMPS	OPT
Germany	Netz-B, C	Deutsche Bundespost Telekom
Ghana	TACS	Millicom Ghana
Great Britain	not applicable	Please see United Kingdom
✳ Grenada	AMPS	Grentel Boatphone
✳ The Grenadines	AMPS	St. Vincent & the Grenadines Boatphone (see St. Vincent)
✳ Guadeloupe	AMPS	France Antilles Boatphone
✳ Guam	AMPS	A) Guam Cellular Telephone B) Guam Tel. Authority
✳ Guatemala	AMPS	Comunicaciones Celulares (Millicom)
✳ Hong Kong	AMPS	1) Hutchison Radio (described in this book)
	TACS	1) Communications Services Ltd. 2) Hutchison Radio
	ETACS	Pacific Link
Hungary	NMT-450	Hungary PTT and WesTel Radiotelephone (US West)
	TACS	Contel Hungaria
Iceland	NMT-450	Post and Telecommunications Administration (PTT)
India	TACS	ONCG, DOT
Indonesia	AMPS	Perumtel and PI INTI
	NMT-450	Perumtel and PTT
Ireland	TACS	Telecom Eireann (PTT)
Ireland (northern)	not applicable	Please see United Kingdom.
Israel	AMPS	Motorola Tadiran (MTCC)
Italy	RTMS, TACS	Societe Italiana per l'Esercizio Telefonico (SIP)
✳ Jamaica	AMPS	Jamaican Boatphone
Japan	NTT	1) Nippon Telephone & Telegraph (NTT) 2) IDO
	JTACS	Daini Denden (DDI)
	NTACS	Nipon Idou Tsushin (IDO)
Jordan	AMPS	(not available)
Kenya	TACS	Kenya Post and Telecommunications (KPTC)
Korea	AMPS	Korea Mobile Telecommunications
Kuwait	TACS, ETACS	MTSC Kuwait Mobile Telephone Systems Company (PTT)
Laos	AMPS	EPTL Lao PTT
Latvia	NMT-450	Post and Telecommunications Administration (PTT)
Lithuania	NMT-450	PTT and Comliet (Millicom)
Luxembourg	NMT-450	Post and Telecommunications Administration (PTT)
Macao	TACS	Cable & Wireless (PTT)
Malaysia	AMPS	Syarikat Telekom Malaysia Berhad (STM) and PTT
	ETACS	Celcom
	NMT-450	Jabatan Telekom Malaysia

Malaysia (cont.)	NMT-900	Celcom of Sdn Berhad	
Malta	ETACS	Telemalta/Racal	
✳ Martinique	AMPS	France Antilles Boatphone	
Mauritius	ETACS	Emtel (Currimjee Jeewanjee, Comvik, Millicom)	
✳ Mexico Region 1	AMPS	Mexicali, Tijuana A) Baja Celular Méxicana	B) Telcel
✳ Region 2	AMPS	Hermosillo, Nogales A) Movitel del Noroeste	B) Telcel
✳ Region 3	AMPS	Chihuahua, Torréon A) Norcel	B) Telcel
✳ Region 4	AMPS	Matamoros,Monterrey A) Cedetel	B) Telcel
✳ Region 5	AMPS	Guadalajara, Morella A) Comcel	B) Telcel
✳ Region 6	AMPS	Irapuato, Querétaro A) Portacel	B) Telcel
✳ Region 7	AMPS	Acapulco, Puebla A) Telcom	B) Telcel
✳ Region 8	AMPS	Cancun, Mérida A) Portatel del Sureste	B) Telcel
✳ Region 9	AMPS	México City, Toluca A) IUSA Cell	B) Telcel
✳ Montserrat	AMPS	Montserrat Boatphone	
Morocco	NMT-450, TACS	Office National P & T (ONPT)	
Netherlands	NMT-450, 900	Niederlanden Postenijen Telegraphic en Telefonie	
✳ Netherlands Antilles	AMPS	St. Maarten Boatphone (described on St. Maarten page)	
✳ New Zealand	AMPS	Telecom Cellular	
	TACS	BellSouth New Zealand	
Nigeria	TACS	Nigerian Telecommunications (NITEL)	
Norway	NMT-450, 900	Norwegian Telecommunications Authority (NTA)	
Oman	NMT-450	General Telecom. Org. (GTO)	
Pakistan	AMPS	1) Paktel 2) Pakcom (Millicom)	
Panama	AMPS	INTEL	
Paraguay	AMPS	Telecel	
Peru	AMPS	Cellular International SA andTelemovil	
Philippines	AMPS	1) (A) Philippines LDT 2) (B) Extelcom 3) (B) Isla	
Poland	NMT-450	Polish PTT	
Portugal	Netz-C	Telecom Portugal	
✳ Puerto Rico (U.S.)	AMPS	A) Cellular One B) Servicios Celulares	
✳ St. Kitts & Nevis	AMPS	St. Kitts & Nevis Boatphone	
✳ St. Lucia & St. Vincent	AMPS	St. Lucia Boatphone	
✳ St. Maarten (Dutch)	AMPS	St. Maarten Boatphone	
✳ St. Martin (French)	AMPS	St. Martin Mobiles	
✳ St. Vincent	AMPS	St. Vincent and the Grenadines Boatphone	
Samoa (U.S.)	AMPS	American Samoa Government (011-684-633-1121)	
Saudi Arabia	NMT-450	Saudi Telecom (A) and (B) (PTT)	
Scotland	not applicable	Please see United Kingdom.	
✳ Singapore	AMPS	Singapore Telecom	
	ETACS	Singapore Telecommunications Authority	
South Africa	Netz-C	South African Post Office (SAPO)	
South Korea	AMPS	Korea Mobile Telecom (KTMC) (no roaming)	
Soviet Union	NMT-450	Ministry of Communications in Leningrad and Moscow	
Spain	NMT-450, 900	La Compania Telefonica Nacional de Espana (CTNE)	
	TACS	La Compania Telefonica Nacional de Espana (CTNE)	
Sri Lanka	ETACS	Celltel Lanka (Millicom)	
Sweden	NMT-450 & 900	1) Televerket 2) Comvik	
Switzerland	NMT-900	PTT	
Taiwan	AMPS	Taiwan Director General of Telephone	
Thailand	AMPS	1) Comm. Authority of Thailand (CAT) 2) United Comm.	
	NMT-450	Telephone Organization of Thailand (TOT)	
	NMT-900	Advanced Information Services (Shinawatra)	
Trinidad & Tobago	AMPS	Telecommunication Services of Trinidad & Tobago	
Tunisia	NMT-450	Ministry of Posts & Telecommunications (PTT)	
Turkey	NMT-450	PTT	
United Arab Emirates	TACS	Etisalat (Emirates Telecom)	
United Kingdom	TACS, ETACS	1) Cellnet 2)Vodafone	
United Kingdom	DCS-1800	1) Cellnet 2)Vodafone	
✳ United States	AMPS	(see the main index that begins on page 1)	
Uruguay	AMPS	Abiatar (BellSouth)	
✳ Venezuela	AMPS	1) Telcel (BellSouth) 2) Comp. Anonomia Nacional Tele.	
✳ Virgin Islands (British)	AMPS	CCT Boatphone (Caribbean Cellular Telephone)	
✳ Virgin Islands (U.S.)	AMPS	A) Cellular One B) Vitel Cellular	
Wales	not applicable	Please see United Kingdom.	
Yugoslavia	NMT-450	Zagreb PTT (in city of Zagreb only)	
Zaire	AMPS	1) TELCEL 2)Express Communications	

Anguilla

System A

Cellular company

Anguilla Boatphone
PO Box 269
The Valley, Anguilla, West Indies

Customer service	from U.S.	800-262-8366
. . . . .	from Canada	800-567-8366
. . . . .	from elsewhere	809-462-5051
Local office	from Anguilla	809-497-2100
. . . . .	St. Maarten	011-5995-22100
Fax machine	from Anguilla	809-497-3770
. . . . .	St. Maarten	011-5995-25678

Comments

Visitor registration
You must call this company and register for credit card billing on Amex, Discover, Mastercard, or Visa. Or you may register upon arrival by pressing *0 send* on your cellular telephone. An operator is available from 7 am to midnight.

Coverage area
Southern side of Anguilla facing St. Maarten.

Sending calls

Visitor dialing instructions

Anguilla	4 digits
Dutch St. Maarten	5 digits
French St. Martin	6 digits
Local calls	area code + 7 digits
Long distance	1 + area code + 7 digits
Worldwide calls	011 + country code + number

Speed-dial numbers

Ambulance	22111
AT&T U.S. operator	872
Boatphone cellular operator	0
Directory assistance	411 or 150
East Caribbean Cellular business office	611
Saba VHF marine operator (free)	155

Receiving calls

Caller dials the roamer access number below, after it stops ringing (a tone may not be audible), dials your home cellular area code and number.
. . . . 011-5995-7-7626

Caller dials your temporary number directly:
You will be assigned a temporary telephone number that a caller may dial to reach you. Caller may call the operator at 011-599-5-24100, tell her your name, and be told your cellular number.

Billing information

Visitor rates ($ U.S.)

Daily activation fee (only on days used)	$5.00
Cellular phone rental per day (optional)	$10.00

Per minute fee including long distance charges:

AT&T or MCI operator	$1.60
Bahamas, Bermuda, Western Europe	$7.00
Canada	$4.50
Directory assistance	$1.00
Eastern Europe, South America	$8.00
Incoming collect; other international	$9.00
Local or non-collect incoming calls	$1.60
U.S. and other Caribbean islands	$4.00

System identification number . . . **00037**

Antigua & Barbuda

System A

Cellular company

Boatphone Antigua
Lower Newgate Street, PO Box 1516
St. John's, Antigua, West Indies

Customer service	from U.S.	800-262-8366
. . . . .	from Canada	800-567-8366
. . . . .	from elsewhere	809-462-5051
Local office		809-462-5051
Fax machine		809-462-5052

Comments

Visitor registration
You must call this company and register for credit card billing on Amex, Discover, Mastercard, or Visa. Or you may register upon arrival by pressing *0 send* on your cellular telephone. An operator is available from 7 am to midnight. Cellular phones rent for $10 per day

Coverage area
Antigua, Barbuda, and the surrounding waters

Sending calls

Visitor dialing instructions

Local calls	7 digits
To other 809 area code	1 + 7 digits
To U.S. and Canada	1 + area code + 7 digits
Worldwide calls	011 + country code + number

Speed-dial numbers

AT&T U.S. operator	872
Boatphone cellular operator	0

Receiving calls

Caller dials the roamer access number below, after it stops ringing (a tone may not be audible), dials your home cellular area code and number.
. . . . 809-464-7626

Caller dials your temporary number directly:
You will be assigned a temporary telephone number that a caller may dial to reach you.

Billing information

Visitor rates ($ U.S.)

Daily activation fee (only on days used) $5.00
Per minute including long distance (land-based):
(Add 20% tax to all international calls)

AT&T or MCI operator or collect	$2.20
Boatphone operator or emergency no.	$0.00
Canada, United States	$4.00
Directory assistance	$1.00
800 numbers	$2.00
Europe, South America	$5.30
Incoming collect	$9.00
Local or incoming calls	$1.60
Nearby islands	$2.20
Other Caribbean islands	$4.00
Other international	$7.30

System identification number . . . **08145**

Argentina

Cellular company

Compania de Radiocomunicaciones Moviles
Tucuman 752
1425 Buenos Aires, C.F. Argentina

Information 011-54-1-325-5220
Fax machine 011-54-1-326-5300

Comments

Visitor registration
If your home company has a roaming agreement
with this company, it will program your telephone
with an Argentinean number before you leave.
Your telephone cannot be programmed in
Argentina. Visitors whose home companies do
not have roaming agreements may rent cellular
telephones.

Sending calls

Visitor dialing instructions
Local calls . . . . . area code + 7 digits
Long distance . . . 1 + area code + 7 digits

Speed-dial numbers
Customer service 325-5220

Receiving calls

This company has no roamer access number.
You will be assigned a temporary telephone
number that others may dial to reach you.

Billing information

Visitor rates
Cellular telephone rental per day . . . $15
Per minute fee . . . . 70¢

System identification number not available

Australia

Cellular company

Telecom MobileNet (headquarters)
Telecom Australia
181-189 Victoria Parade
Melbourne, Victoria 3066
Australia

Customer service . . . 011-61-3-285-0320
. . . . fax 011-61-3-349-1832
Corporate office 011-61-3-412-1802
. . . . fax 011-61-3-412-1890

Telecom MobileNet (local office)
145 Franklin Street
Adelaide, South Australia 5343
Australia

Roaming department . . 011-61-8-230-2333
. . . . free in Australia 018-018-468
. . . . fax 011-61-8-231-9305

Austel (government agency)
Assistant Manager, Industry Development
PO Box 7443, St. Kilda Road
Melbourne, Victoria 3004

Information 011-61-3-828-7335
. . . . fax 011-61-3-828-7438

Comments

Visitor registration
1) Send the following items to Austel:
• Copy of receipt for your telephone purchase or
  four months cellular telephone service bills.
• Letter from the Australian supplier of your brand
  of telephone stating that your model is identical
  to a model approved for sale in Australia.
2) Austel will provide an electronic serial number
  and certificate for three months connection.
3) Give the information from Austel and credit
  card number to the roaming dept. in Adelaide.

Coverage area
This company reports that it covers 90% of the
populated areas of Australia

Sending calls

Visitor dialing instructions
Local calls . . . . . . area code + 7 digits
Worldwide . . 011 + country code + number

Speed-dial numbers
Customer service 1104
Emergency 000

Receiving calls

This company has no roamer access number.
You will be assigned a temporary telephone
number that others may dial to reach you.

Billing information

Visitor rates ($ Australian)
Daily activation fee $3
Local calls (first 30 seconds) . . . . 29¢
Local calls (additional 30 seconds) . . . 19¢
Calls over 165 km (first 30 seconds) . . 40¢
Calls over 165 km (additional 30 seconds) 30¢

System identification number varies by area

Bahamas

System B

Cellular company

Batelco
Bahamas Telecommunications
The Mall at Marathon Road, PO Box N-3048
Nassau, Bahamas

Information	. . . . .	809-394-4000
Fax machine	. . . . .	809-393-4798

Comments

Visitor registration
You may use this system automatically if your
home company has a roaming agreement with
this company. (See *roaming agreements* in the
index.) Without a roaming agreement, you must
go to the office and pay the fees described below.
Credit cards are not accepted.

Coverage area
30 miles in all directions from Nassau

Sending calls

Visitor dialing instructions
Local calls	7 digits
To nearby islands	1 + 7 digits
To U.S. and Canada .	1 + area code + 7 digits
Worldwide . .	011 + country code + number

Speed-dial numbers
Customer service	394-4000
Directory assistance	916
Emergency	919

Receiving calls

Caller dials the roamer access number below,
hears tone, dials cellular area code and number,
if your home company has a roaming agreement
that allows you to use your phone automatically.
. . . . 809-359-7626

Caller dials your temporary cellular number
to reach you directly if you went to the office and
your telephone was programmed with a new
number because you had no roaming agreement.

Billing information

Visitor rates
With a roaming agreement (automatic service)
Daily activation fee	$3
Per minute fee	75¢

Without a roaming agreement (go to office)
Deposit (refundable)	$300
Activation fee	$70
Programming fee	$50
Per minute 7 am - 7 pm	45¢
Per minute other times	30¢

Group identification number	. . .	**00007**
Billing identification number	. . .	**32752**

Barbados

System B

Cellular company

Barbados Communications Services
Building #2, Chelston Park, Culloden Road
St. Michael, Barbados

Customer service	. from U.S.	800-262-8366
. . . . .	from Canada	800-567-8366
. . . . .	from elsewhere	809-462-5051
Local office		809-431-4000
Fax machine		809-431-4050

Comments

Visitor registration
You must call this company and register for credit
card billing on Amex, Discover, Mastercard, or
Visa. Or you may register upon arrival by
pressing *0 send* on your cellular telephone. An
operator is available from 7 am to midnight.
Cellular phones rent for $7.50 per day

Coverage area
Barbados

Sending calls

Visitor dialing instructions
Local calls	7 digits
To other 809 area code	1 + 7 digits
To U.S. and Canada . .	011 + 1 + number
Worldwide calls	011 + country code + number

Speed-dial numbers
AT&T U.S. operator	872
Boatphone cellular operator	0

Receiving calls

This company has no roamer access number.
You will be assigned a temporary telephone
number that others may dial to reach you.

Billing information

Visitor rates ($ U.S.)
Daily activation fee (only on days used)	$5.00
Per minute	60¢

System identification number	. . .	**08160**

Bermuda

Cellular company

Bermuda Cellular
Bermuda Telephone Company
30 Victoria Street, PO Box HM 1021
Hamilton, Bermuda HM DX

Visitor registration	809-292-6032
Customer service	809-292-5272
Corporate office	809-295-1001
Fax machine	809-292-8841

Comments

Visitor registration
You must call this company 24 hours a day and register to have your calls charged to an Amex, Mastercard, or Visa. Registered visitors receive free call forwarding, call transfer, call waiting, and three-way conference calling.

Coverage area
Five cell sites cover a 50-mile radius around the island of Bermuda.

Sending calls

Visitor dialing instructions
Local calls	7 digits
To U.S. and Canada .	1 + area code + 7 digits
To other countries	
011 + country code + area code + local number	

Speed-dial numbers
Customer service	292-5272
DIrectory assistance	411
Emergency	911
Operator	0

Receiving calls

This company has no roamer access number. You will be assigned a temporary telephone number that others may dial to reach you, and this number will be reserved for you as long as you continue to pay the $5 monthly fee. Or others may instead dial the operator at 809-292-6032 and request you by name.

Billing information

Visitor rates ($ U.S.)
Initial activation fee	$10
Daily activation fee	$5
Per minute fee (7 am - 7 pm)	60¢
Per minute fee (other times)	35¢

System identification number . . . 08198

Bolivia

Cellular company

Telefonica Celular de Bolivia
Belisario Salinas 470
La Paz, Bolivia

Information	011-591-239-2141
Fax machine	011-591-239-2150

Comments

Visitor registration
This company accepts American Express, Mastercard, and Visa. You must go to the office with your telephone so it can be programmed with a Bolivian telephone number. Some models of telephones may take one day to program.

Sending calls

Visitor dialing instructions
Please contact the company for information.

Receiving calls

This company has no roamer access number. You will be assigned a temporary telephone number that others may dial to reach you.

Billing information

Visitor rates ($ U.S.)
Daily activation fee	$0
Per minute fee including long distance approx.	$4

System identification number is not available

Brazil

System B

Cellular company

Tele R.G.
Ave. Chile, 500-Loja
C.P. 20031, Rio de Janeiro

Information   011-55-21-262-5518
Fax machine   011-55-21-262-5637

Comments

Visitor registration
You must go to the office with your cellular
telephone so it can be programmed with a
Brazilian telephone number. Further information
was not available when this book went to press.

Cayman Islands

System B

Cellular company

Cable and Wireless
Anderson Square, PO Box 293
Grand Cayman, Cayman Islands
British West Indies

Information   809-949-7800
Fax machine   809-949-0039

Comments

Visitor registration
You must bring your telephone and and American
Express, Mastercard, or Visa to this company's
sales department in the Anderson Square
Building, Georgetown and register for credit card
billing. A deposit will be held temporarily and then
refunded when service is discontinued, less
calling charges.

Coverage area
Coverage includes the Cayman Islands and 30
miles out to sea.

Sending calls

Visitor dialing instructions
Local calls . . . .   5 digits
To North America   .   01 + area code + 7 digits

Speed-dial numbers
Customer service   97800
Emergency   911
Fire   500
Hospital   555
Operator   0
Police   999
Repair service   112
Telephone problems   611
Time   844

Receiving calls

This company has no roamer access number.
You will be assigned a temporary telephone
number that a others may dial to reach you.

Billing information

Visitor rates ($ U.S.)
Temporary deposit $427
Initial activation fee $25
Daily activation fee (14 day minimum of $28) $2
Per minute fee . . . .   25¢
Handheld phone rental (14-day min.)   . $3/day
Transportable phone rental (14-day min.) $2/day

System identification number   . . .   **00700**

Costa Rica

Cellular company

Millicom Costa Rica
Centro Colon Building, Floor 1
Apartado Postal 89-1007
San Jose, Costa Rica

Information	011-50-6-572-527
Fax machine	011-50-6-338-551

Comments

Visitor registration
You must bring your cellular telephone to the office to be programmed with a Costa Rica number.

Sending calls

Visitor dialing instructions
Local calls . . . . 6 digits

Speed-dial numbers
Customer service 572-527

Receiving calls

This company has no roamer access number. You will be assigned a temporary telephone number that others may dial to reach you.

Billing information

Visitor rates ($ U.S.)
Deposit	$500
Daily activation fee	$10
Per minute fee	$1

System identification number is not available

Dominican Republic

Cellular company

Codetel Comunicaciones Moviles
Winston Churchill Avenue
Santo Domingo, Dominican Republic

Information . . . . from U.S.	800-235-4081
. . . .	809-220-7017
. . . .	809-543-2337
Fax machine	809-220-7025

Comments

Visitor registration
This company plans to begin making roaming agreements with other companies in April 1993. Until that time, you must bring your telephone (preferably a dual NAM phone that accepts two numbers) to the Winston Churchill Ave. office. You need to pay the deposit (credit cards are not accepted) and need a local co-signer. If you prefer, cellular telephones are available for rent.

Coverage area
Dominican Republic and 40 miles out to sea

Sending calls

Visitor dialing instructions
Local calls	7 digits
Long distance . . .	1 + area code + 7 digits

Speed-dial numbers
Emergency	711
Directory assistance	1411

Receiving calls

This company has no roamer access number. You will be assigned a temporary telephone number that others may dial to reach you.

Billing information

Visitor rates (in pesos)
Deposit	3000
Programming fee	300
Monthly activation fee (25 minutes free) .	300
Per minute fee (peak hours)	10
Per minute fee (non-peak hours)	2

System identification number . . . 08832

Grenada

Cellular company

Grentel Boatphone
The Carenage, PO Box 119
St. Georges, Grenada, West Indies

Customer service	from U.S.	800-262-8366
	from Canada	800-567-8366
	from elsewhere	809-462-5051
Local office	Grenada	809-440-1111
	St. Lucia	809-452-0361
Fax machine	Grenada	809-440-4100
	St. Lucia	809-452-0394

Comments

Visitor registration
You must call this company and register for credit card billing on Amex, Discover, Mastercard, or Visa. Or you may register upon arrival by pressing *0 send* on your cellular telephone. An operator is available from 7 am to midnight. Cellular phones rent for $10 per day

Coverage area
Grenada

Sending calls

Visitor dialing instructions
Local calls	7 digits
To other 809 area code	1 + 7 digits
To U.S. and Canada	1 + area code + 7 digits
Worldwide calls	011 + country code + number

Speed-dial numbers
AT&T U.S. operator	872
Boatphone cellular operator	0
Emergency	911

Receiving calls

Caller dials the roamer access number below, after it stops ringing (a tone may not be audible), dials your home cellular area code and number.
. . . . 809-493-7626

Caller dials your temporary number directly:
You will be assigned a temporary telephone number that a caller may dial to reach you.

Billing information

Visitor rates ($ U.S.)
Daily activation fee (only on days used) . $5.00
Per minute including long distance (land-based):

AT&T or MCI operator or collect	$2.20
Boatphone operator or emergency no.	$0.00
Canada, United States	$4.00
Directory assistance	$1.00
800 numbers	$2.00
Europe, South America	$5.30
Incoming collect	$9.00
Local or incoming calls	$1.60
Nearby islands	$2.20
Other international	$7.30
Other Caribbean islands	$4.00

System identification no. (north cell)	**08003**
System identification no. (south cell)	**08005**

Guadeloupe

Cellular company

France Antilles Boatphone
27 Imm. Connexion, Rue H. Berquerel
97122 Baie-Mahault, Guadeloupe
French West Indies

Customer service	from U.S.	800-262-8366
	from Canada	800-567-8366
	from elsewhere	809-462-5051
Local office		011-590-266-640
Fax machine		011-590-267-382

Comments

Visitor registration
You must call this company and register for credit card billing on Amex, Discover, Mastercard, or Visa. Or you may register upon arrival by pressing *0 send* on your cellular telephone. An operator is available from 7 am to midnight. Cellular phones rent for $10 per day

Coverage area
Guadeloupe

Sending calls

Visitor dialing instructions
Local calls	6 digits
To other 809 area code	1 + 7 digits
To U.S. and Canada	1 + area code + 7 digits
Worldwide calls	011 + country code + number

Speed-dial numbers
AT&T U.S. operator	872
Boatphone cellular operator	0
Emergency	911

Receiving calls

Caller dials the roamer access number below, after it stops ringing or you hear a warbling tone (a tone may not be audible), dials your home cellular area code and number.
. . . . 011-590-37-7626

Caller dials your temporary number directly:
You will be assigned a temporary telephone number that a caller may dial to reach you.

Billing information

Visitor rates (French francs)
Daily activation fee (only on days used) 27,00 F
Per minute including long distance (land-based):
(Excludes taxes; value added tax 7.5%)

All other destinations	43,80 F
Boatphone operator or emergency	0,00 F
Canada, United States	24,00 F
Europe except France	31,80 F
France	24,00 F
Incoming collect	9,60 F
Local or incoming calls	9,60 F
Martinique, St. Martin, St. Barthelemy	13,20 F
Other Caribbean islands	24,00 F

System identification number	**10810**

Guam

System A

Cellular company

Guam Cellular Telephone
PO Box 21989, Guam Main Facility
420 S Route 10, Ste 102-B, University Square
Mangilao, GU 96921

Information	011-671-734-0039
Fax machine	011-671-734-0043

Comments

Visitor registration
This company will begin providing credit card roaming in March 1993 using American Express, Mastercard, and Visa. Until then, you may sign up for service as a regular customer and pay the fees listed below. Calls under 30 seconds in length are not billed. Office is open 9 am to 6 pm Mon - Fri and 9 am to 2 pm Saturday.

Coverage area
There are two cells, one on Mount Barrigada and one on Mount Aluton. Cells are planned for the Tamuning/Tumon and Cabras Island/Orete Point areas.

Sending calls

Visitor dialing instructions
Local calls	area code + 7 digits
Long distance . . .	1 + area code + 7 digits

Speed-dial numbers
Emergency	911

Receiving calls

This company has no roamer access number. You will be assigned a telephone number that others may dial to reach you.

Billing information

Visitor rates ($ U.S.)
	plan 1	plan 2	plan 3
Deposit			$150
Initial activation fee			$25
Monthly activation fee .	$15	$35	$175
Per minute fee	80¢	40¢	38¢
Free minutes	0	0	385

Identification numbers

City	Market	System	Billing	RSA
All served cities	732	01851	00000	GU-1

Guam

System B

Cellular company

Guam Telephone Authority
PO Box 22109, S Marine Drive
Dededo, GU 96921

Information	011-671-649-3160
Fax machine	011-671-637-4822

Comments

Visitor registration
You must call this company and register for credit card billing on Mastercard or Visa.

Coverage area
Three cells cover most of Guam except the south

Sending calls

Visitor dialing instructions
Local calls	7 digits
Long distance (MCI)	001 + area code + number
Long distance (ITNE)	011 + area code + number

Speed-dial numbers
Customer service	*482
Emergency	911

Receiving calls

This company has no roamer access number. You will be assigned a temporary telephone number that others may dial to reach you.

Billing information

Visitor rates
Initial activation fee	$25
Daily activation fee	$5
Weekly activation fee (instead of daily) . .	$25
Two-week activation fee (includes 125 min.)	$150
Per minute fee	85¢

Identification numbers

City	Market	System	Billing	RSA
All served cities	732	01852	00000	GU-1

Guatemala

Cellular company

Comunicaciones Celulares
5th Avenue A 13-28
Zone 9, Guatemala City, Guatemala

Information   011-502-2-343-260
Fax machine   011-502-2-343-266

Comments

Visitor registration
You must bring your telephone to the office to be
programmed with a Guatemalan telephone
number. You may pay by American Express,
Diner's Club, Discover, Mastercard, or Visa.

Sending calls

Visitor dialing instructions
Local calls . . . . .   area code + 7 digits
Long distance . . .   1 + area code + 7 digits

Speed-dial numbers
Customer service   2-343-260
Long distance in Guat.   0 + area code + 6 digits
To U.S. and Canada   001 + area code + number

Receiving calls

This company has no roamer access number.
You will be assigned a temporary telephone
number that others may dial to reach you. They
will dial 011-502-982 then a 4 digit number.

Billing information

Visitor rates ($ U.S. approximately)
Daily activation fee $4
Per minute fee . . . .   40¢

System identification number is not available

Hong Kong

Cellular company

Hutchison Telephone
Ground Floor of Citicorp Center
18 Whitefield Road
North Point, Hong Kong

Visitor information . . .   011-852-807-9033
Corporate office   011-852-807-9000
Fax machine   011-852-887-9195

Comments

Visitor registration
With a roaming agreement—Most Canadians and
customers of GTE Mobilnet, NYNEX, and PacTel
should call their home customer service two
weeks prior to arrival and provide a credit card
number and dates of travel to advance register.
Other visitors—You must register to use this
system by bringing your telephone, passport, and
an American Express, Discover, Mastercard, or
Visa to the office during business hours.
Comments—To find Hong Kong time, add 19
hours to Eastern standard time or 16 hours to
Pacific standard time. Visitors receive free call
forwarding, call waiting, no answer transfer, and
three-way calling.

Coverage area
Territory-wide including offshore areas

Sending calls

Visitor dialing instructions
To Hong Kong island and new territories   7 digits
To Kowloon (6 digits) . . . . .   3 + 6 digits
To Kowloon (7 digits) . . . . . .   7 digits
To another cellular telephone   05 + the number
To Canada & U.S. 011 + 1 + area code + 7 digits

Speed-dial numbers
Emergency   999
General information   807-9000
Visitor information   807-9033

Receiving calls

This company has no roamer access number.
You will be assigned a temporary telephone
number that others may dial to reach you.

Billing information

Visitor rates ($ Hong Kong)
Daily activation fee $30
Per minute fee . . . . $3..50

System identification number is not available.

Jamaica

Cellular company

Jamaica Boatphone
17 Belmont Road
Kingston 5, Jamaica

Customer service .	from U.S.	800-262-8366
. . . . .	from Canada	800-567-8366
. . . . .	from elsewhere	809-462-5051
Local office		809-968-4000
Fax machine		809-968-3999

Comments

Visitor registration
You must call this company and register for credit card billing on Amex, Discover, Mastercard, or Visa. Or you may register upon arrival by pressing *0 send* on your cellular telephone. An operator is available from 7 am to midnight. Cellular phones rent for $10 per day

Sending calls

Visitor dialing instructions

Local calls	0 + 7 digits
To other 809 area code	1 + 7 digits
To U.S. and Canada .	1 + area code + 7 digits
Worldwide calls .	011 + country code + number

Speed-dial numbers

AT&T U.S. operator	872
Boatphone cellular operator	*0
Emergency	199

Receiving calls

Caller dials your temporary number directly:
You will be assigned a temporary telephone number that a caller may dial to reach you.

Billing information

Visitor rates ($ U.S.)

Daily activation fee (only on days used) .	$8.00
Per minute including long distance (land-based):	
AT&T or MCI operator or collect . .	$2.20
Boatphone operator or emergency no.	$0.00
Canada, United States	$4.00
Directory assistance	$1.00
800 numbers	$2.00
Europe, South America	$5.30
Incoming collect	$9.00
Local or incoming calls	$1.60
Nearby islands	$2.20
Other international	$7.30
Other Caribbean islands	$4.00

System identification number . . . 08186

Martinique

Cellular company

France Antilles Boatphone
Centre d'Affaires Californie II
Entre Elodie, 3 ème étage
97232 Lamentin, Martinique
French West Indies

Customer service	from U.S.	800-262-8366
. . . . .	from Canada	800-567-8366
. . . . .	from elsewhere	809-462-5051
Local office		011-596-504-888
Fax machine		011-596-504-890

Comments

Visitor registration
You must call this company and register for credit card billing on Amex, Discover, Mastercard, or Visa. Or you may register upon arrival by pressing *0 send* on your cellular telephone. An operator is available from 7 am to midnight. Cellular phones rent for $10 per day

Coverage area
Guadeloupe

Sending calls

Visitor dialing instructions

Local calls	6 digits
To other 809 area code	1 + 7 digits
To U.S. and Canada	1 + area code + 7 digits
Worldwide calls	011 + country code + number

Speed-dial numbers

AT&T U.S. operator	872
Boatphone cellular operator	0
Emergency	911

Receiving calls

Caller dials the roamer access number below, after it stops ringing or you hear a beep (a tone may not be audible), dials your home cellular area code and number.

. . . .	011-596-47-7626

Caller dials your temporary number directly:
You will be assigned a temporary telephone number that a caller may dial to reach you.

Billing information

Visitor rates (French francs)

Daily activation fee (only on days used)	27,00 F
Per minute including long distance (land-based):	
(Excludes taxes; value added tax 7.5%)	
All other destinations	43,80 F
Boatphone operator or emergency	0,00 F
Canada, United States	24,00 F
Europe except France	31,80 F
France	24,00 F
Incoming collect	9,60 F
Local or incoming calls	9,60 F
Martinique, St. Martin, St. Barthelemy	13,20 F
Other Caribbean islands	24,00 F

System identification number . . . 10800

System A

Cellular company

Baja Celular Méxicana
Ave. Obregon #674
21110 Zona Centro, Mexicali, Baja California

Paseo de los Héroes 9116
22320 Zona del Rio, Tijuana, Baja California

Local office	. .	Mexicali 011-52-6-552-9901
	. . . . . .	Mexicali 011-52-6-552-9902
	. . . . . .	Tijuana 011-52-6-684-8520
	. . . . . .	Tijuana 011-52-6-620-0611
Corporate office	. .	011-52-5-539-0565
Fax machine	. .	Mexicali 011-52-6-552-8781
	. . . . . .	Tijuana 011-52-6-684-8523

Comments

Visitor registration
You may use this system automatically if your home company has a roaming agreement with this company. (See *roaming agreements* in the index.) Without a roaming agreement, you must go to the office and register for credit card billing.

Coverage area
Ensenada, Mexicali, Tijuana

Sending calls

Visitor dialing instructions
Local calls . . . . 6 digits
Long distance within Mexico . . 91 + 8 digits
To Canada or United States 95 + 10 digit number

Speed-dial numbers
Customer service *711

Receiving calls

Caller dials the roamer access number below,
hears tone, dials cellular area code and number.
. . . . 011-52-6-620-7626

Caller dials the temporary telephone number
that was assigned to you if you used a credit card to register for service.

Billing information

Visitor rates ($ U.S.)
Daily activation fee (with roaming agreement) $0
Daily activation fee (credit card roamers) . $2
Per minute fee . . . . 75¢

System identification number . . . 24579

System B

Cellular company

Telcel
Radiomóvil Dipsa
Blvd. Agua Caliente 4558
22420 Col. Aviacion, Tijuana, Baja California

Local office	. . . .	voice 011-52-6-681-8466
	. . . .	fax 011-52-6-681-8151
Corporate office	. . . .	011-52-5-625-3700
	. .	roaming manager 011-52-5-625-3814
	. .	roaming manager 011-52-5-400-0051
	. . . .	fax 011-52-5-625-3817

Comments

Visitor registration
You may use this system automatically if Telcel has a roaming agreement with your company. (See *roaming agreements* in the index.) Without a roaming agreement, you must go to the office with a credit card to register for billing.

Coverage area
Ensenada, Mexicali, Tijuana

Sending calls

Visitor dialing instructions
Local calls . . . . 8 digits
Long distance within Mexico . . 91 + 8 digits
To Canada or United States 95 + 10 digit number

Speed-dial numbers
Customer service 111
Fire department 115
Police 112
Red Cross 113

Receiving calls

Caller dials the roamer access number below,
hears tone, dials cellular area code and number.
*These numbers were not yet on at press time.
Ensenada* 011-52-6-671-7626
Mexicali* 011-52-6-569-7626
San Luis Rio Colorado* . 011-52-6-538-7626
Tijuana 011-52-6-628-7626

Billing information

Visitor rates (N$ new pesos)
Daily activation fee . . . . N$6.60
Per minute fee . . . . N$2.20
Tax 10%

System identification number . . . 24578

International Cellular

México, Hermosillo & Nogales ➋

Cellular company

Movitel del Noroeste
Veracruz 304
83190 Col. San Benito, Hermosillo, Sonora

Local office	011-52-6-218-5050
. . . . fax	011-52-6-218-5009
México City office . . .	011-52-5-566-6613
. . . . fax	011-52-5-566-3554

Comments

Visitor registration
This company reports that it does not provide service to visitors. There are no roaming agreements with U.S. or Canadian companies and they do not have credit card roaming.

Coverage area
Hermosillo and Nogales

Sending calls

Visitor dialing instructions

Local calls	area code + 7 digits
Long distance . . .	1 + area code + 7 digits

Speed-dial numbers

Customer service	*611
Fire	*112
Police	*115
Red Cross	*111

Receiving calls

Visitors cannot send or receive calls.

Billing information

Visitor rates

. . . . not applicable

System identification number is not available

México, Hermosillo & Nogales ➋

Cellular company

Telcel
Radiomóvil Dipsa
Blvd. Navarrete #228
83200 Col. Villa Satelite, Hermosillo, Sonora

Local office . . . . voice	011-52-6-218-9080
. . . . fax	011-52-6-218-7514
Corporate office	011-52-5-625-3700
. . . roaming manager	011-52-5-625-3814
. . . roaming manager	011-52-5-400-0051
. . . . fax	011-52-5-625-3817

Comments

Visitor registration
You may use this system automatically if Telcel has a roaming agreement with your company. (See *roaming agreements* in the index.) Without a roaming agreement, you must go to the office with a credit card to register for billing.

Coverage area
Hermosillo, Nogales, Obregón

Sending calls

Visitor dialing instructions

Local calls	8 digits
Long distance within Mexico . .	91 + 8 digits
To Canada or United States	95 + 10 digit number

Speed-dial numbers

Customer service	111
Fire department	115
Police	112
Red Cross	113

Receiving calls

Caller dials the roamer access number below, hears tone, dials cellular area code and number.
*These numbers were not yet on at press time.

Ciudad Obregón* . . .	011-52-6-447-7626
Hermosillo	011-52-6-256-7626
Nogales*	011-52-6-318-7626

Billing information

Visitor rates (N$ new pesos)

Daily activation fee	N$6.60
Per minute fee	N$2.20
Tax	10%

System identification number . . . 24594

System A

Cellular company

Norcel
Telefonia Celular del Norte
Blvd. Manuel Gomez Morin 7045, Piso 4
Col. Rincones de Santa Rita
32509 Ciudad Juárez, Chihuahua

Information	011-52-1-618-1830
Fax machine	011-52-1-618-0480

Comments

Visitor registration
You may use this system automatically if your home company has a roaming agreement with this company. (See *roaming agreements* in the index.) Without a roaming agreement, you must go to the office and register for credit card billing.

Coverage area
Ciudad Cuahutemoc, Ciudad Juárez, Delicias, Durango, Torreón

Sending calls

Visitor dialing instructions
Local calls	6 digits
Long distance within Mexico . .	91 + 8 digits
To Canada or United States	95 + 10 digit number

Speed-dial numbers
Customer service	611
Police	114 or 115
Red cross	111

Receiving calls

This company has no roamer access number.
You will be assigned a temporary telephone number that others may dial to reach you. Visitors from El Paso receive their calls automatically.

Billing information

Visitor rates (N$ new pesos)
Daily activation fee	N$6.00
Per minute fee	N$1.20

System identification number is not available

System B

Cellular company

Telcel
Radiomóvil Dipsa
Av. Victor Hugo #301
Complejo Industrial Chihuahua
31109 Chihuahua, Chihuahua

Local office . . . . voice	011-52-1-481-0444
. . . . fax	011-52-1-481-0277
Corporate office	011-52-5-625-3700
. . . roaming manager	011-52-5-625-3814
. . . roaming manager	011-52-5-400-0051
. . . . fax	011-52-5-625-3817

Comments

Visitor registration
You may use this system automatically if Telcel has a roaming agreement with your company. (See *roaming agreements* in the index.) Without a roaming agreement, you must go to the office with a credit card to register for billing.

Coverage area
Chihuahua, Ciudad Juárez, Delicias, Torreón

Sending calls

Visitor dialing instructions
Local calls	8 digits
Long distance within Mexico . .	91 + 8 digits
To Canada or United States	95 + 10 digit number

Speed-dial numbers
Customer service	111
Fire department	115
Police	112
Red Cross	113

Receiving calls

Caller dials the roamer access number below,
hears tone, dials cellular area code and number.
*These numbers were not yet on at press time.
Chihuahua; Cd. Cuahutemoc	011-52-1-427-7626
Ciudad Juárez	011-52-1-626-7626
Delicias*	011-52-1-465-7626
Torreón	011-52-1-727-7626

Billing information

Visitor rates (N$ new pesos)
Daily activation fee	N$6.60
Per minute fee	N$2.20
Tax	10%

System identification number . . . 24582

México, Matamoros/Monterrey ❹

Cellular company

Cedetel
Celular de Telefonia
Angela Peralta 180
64049 Col. Ex-Seminario, Monterrey, N.L.

Customer service	011-52-8-345-6255
Switchboard	011-52-8-343-2020
President's office	011-52-8-342-2758
Fax	011-52-8-343-1089

Rio Orinoco 106 Ote
Col. del Valle, Garza Garcia, Nuevo Leon

Information	011-52-8-356-5680
Fax	dial main number, request transfer

Comments

Visitor registration
You may use this system automatically if this company has a roaming agreement with your company. (See *roaming agreements* in the index.) Without a roaming agreement, you cannot use the service.

Coverage area
Matamoros, Monterrey

Sending calls

Visitor dialing instructions

Local calls	6 digits
Long distance	1 + area code + 7 digits

Speed-dial numbers

Customer service	*611
Police	*114
Red cross	*111

Receiving calls

Caller dials the roamer access number below, hears tone, dials cellular area code and number.

011-52-8-325-7626

Billing information

Visitor rates ($ U.S.)

Daily activation fee	$2.85
Per minute fee	75¢

System identification number is not available

México, Matamoros/Monterrey ❹

Cellular company

Telcel
Radiomóvil Dipsa
Calzada del Valle #275 Ote
66220 Col. del Valle, Garza Garcia, N.L.

Local office	011-52-8-319-4001
	fax 011-52-8-356-5499
Corporate office	011-52-5-625-3700
roaming manager	011-52-5-625-3814
roaming manager	011-52-5-400-0051
	fax 011-52-5-625-3817

Comments

Visitor registration
You may use this system automatically if Telcel has a roaming agreement with your company. (See *roaming agreements* in the index.) Without a roaming agreement, you must go to the office with a credit card to register for billing.

Coverage area
Matamoros, Monterrey, Nuevo Laredo

Sending calls

Visitor dialing instructions

Local calls	8 digits
Long distance within Mexico	91 + 8 digits
To Canada or United States	95 + 10 digit number

Speed-dial numbers

Customer service	111
Fire department	115
Police	112
Red Cross	113

Receiving calls

Caller dials the roamer access number below, hears tone, dials cellular area code and number.
*These numbers were not yet on at press time.

Altamira*	011-52-1-267-7626
Cadereyta*	011-52-8-281-7626
Matamoros	011-52-8-918-7626
Monterrey	011-52-8-366-7626
Monterrey* (additional no.)	011-52-8-362-7626
Nuevo Laredo	011-52-8-735-7626
Reynosa	011-52-8-936-7626
Saltillo	011-52-8-427-7626
Tampico*	011-52-1-218-7626

Billing information

Visitor rates (N$ new pesos)

Daily activation fee	N$6.60
Per minute fee	N$2.20
Tax	10%

System identification number . . . 24584

México, Guadalajara & Morelia ⑤

System A

Cellular company

Comcel
Comunicaciones Celulares de Occidente
Av. Lázaro Cárdinas 3438
45000 Col. Chapalita, Guadalajara, Jalisco

Local office	011-52-3-679-6666
. . . .	fax 011-52-3-647-7800
Roaming department . .	011-52-3-679-1111
. . . .	fax 011-52-3-647-7400

Comments

Visitor registration
You must go to the office with your telephone to
have it reprogrammed with a Guadalajara
telephone number. American Express, Discover,
Mastercard, and Visa are accepted. The office is
open from 8:30 am to 6:30 pm Mon - Fri and from
9 am to 2 pm Saturday.

Coverage area
Colima, Morelia, Guadalajara

Sending calls

Visitor dialing instructions
Local calls	7 digits
To Canada or United States	95 + 10 digit number

Speed-dial numbers
Customer service	*611
Red Cross	111

Receiving calls

This company has no roamer access number.
You will be assigned a temporary telephone
number that others may dial to reach you.

Billing information

Visitor rates (N$ new pesos)
Daily activation fee	N$7.48
Per minute fee	N$2.00

System identification number . . . 24587

México, Guadalajara & Morelia ⑤

System B

Cellular company

Telcel
Radiomóvil Dipsa
Av. Chapultepec 408
44150 Sector Juarez, Guadalajara, Jalisco

Local office	voice 011-52-3-669-1095
. . . .	fax 011-52-3-669-1000
Corporate office	011-52-5-625-3700
. .	roaming manager 011-52-5-625-3814
. .	roaming manager 011-52-5-400-0051
. . . .	fax 011-52-5-625-3817

Comments

Visitor registration
You may use this system automatically if Telcel
has a roaming agreement with your company.
(See *roaming agreements* in the index.) Without a
roaming agreement, you must go to the office
with a credit card to register for billing.

Coverage area
Guadalajara, Morelia, Uruapán, Zamora

Sending calls

Visitor dialing instructions
Local calls	8 digits
Long distance within Mexico . .	91 + 8 digits
To Canada or United States	95 + 10 digit number

Speed-dial numbers
Customer service	111
Fire department	115
Police	112
Red Cross	113

Receiving calls

Caller dials the roamer access number below,
hears tone, dials cellular area code and number.
*These numbers were not yet on at press time.
Guadalajara*	011-52-3-662-7626
Morelia*	011-52-4-318-7626
Uruapán*	011-52-4-526-7626
Zamora*	011-52-3-548-7626

Billing information

Visitor rates (N$ new pesos)
Daily activation fee	N$6.60
Per minute fee	N$2.20
Tax	10%

System identification number . . . 24586

System A

Cellular company

Portacel
Sistemas Telefonicos Portatiles
Prol. Blvd. del Campestre 1616
Col. Lomas del Campestre, León
31705 Guanajuato

Local office	011-52-4-718-7401
. . . . fax	011-52-4-718-7459
México City legal office .	011-52-5-523-2812
. . . . fax	011-52-5-523-3739

Comments

Visitor registration
You may use this system automatically if this company has a roaming agreement with your company. (See *roaming agreements* in the index.) Without a roaming agreement, you cannot use the service.

Coverage area
All areas listing roamer access numbers below.

Sending calls

Visitor dialing instructions

Cellular to cellular	8 digits
Local calls	6 digits
Long distance within Mexico . .	91 + 8 digits
To Canada or United States . .	95 + 10 digits

Speed-dial numbers

Customer service for subscribers . . .	*611
Customer service for visitors	*711
Fire department	*112
Highway assistance	*116
Police (federal)	*114
Police (highway)	*113
Police (state)	*115
Red Cross	*111
Roame access number	*511

Receiving calls

Caller dials the roamer access number below, hears tone, dials cellular area code and number.
*This number was not yet on at press time.

Aguascalientes	011-52-4-919-7626
Celaya	011-52-4-619-7626
Irapuato	011-52-4-628-7626
Lebu	011-52-4-729-7626
Querétaro	011-52-4-197-7626
Salamanca*	011-52-4-646-7626
San Luis Potosi	011-52-4-828-7626
Zacatecas	011-52-4-946-7626

This company has no roamer access number.
You will be assigned a temporary telephone number that others may dial to reach you. XXXX

Billing information

Visitor rates (N$ new pesos)

Daily activation fee (approx $2.40 U.S.)	N$6.80
Per minute fee (approx $.80 U.S.) . .	N$2.60

System identification number . . . 24589

México, Irapuato & Querétaro ⑥

System B

Cellular company

Telcel
Radiomóvil Dipsa
Constituyentes 15 Oriente
76040 Col. San Francisquito, Querétaro, Qro.

Local office . . . . voice	011-52-4-214-1262
. . . . fax	011-52-4-214-1304
Corporate office	011-52-5-625-3700
. . . roaming manager	011-52-5-625-3814
. . . roaming manager	011-52-5-400-0051
. . . . fax	011-52-5-625-3817

Comments

Visitor registration
You may use this system automatically if Telcel has a roaming agreement with your company. (See *roaming agreements* in the index.) Without a roaming agreement, you must go to the office with a credit card to register for billing.

Coverage area
Aguascalientes, Irapuato, Querétaro

Sending calls

Visitor dialing instructions

Local calls	8 digits
Long distance within Mexico . .	91 + 8 digits
To Canada or United States . .	95 + 10 digits

Speed-dial numbers

Customer service	111
Fire department	115
Police	112
Red Cross	113

Receiving calls

Caller dials the roamer access number below, hears tone, dials cellular area code and number.
*These numbers were not yet on at press time.

Aguascalientes*	011-52-4-911-7626
Celaya*	011-52-4-117-7626
Irapuato*	011-52-4-621-7626
Leon*	011-52-4-724-7626
Querétaro*	011-52-4-226-7626

Billing information

Visitor rates (N$ new pesos)

Daily activation fee	N$6.60
Per minute fee	N$2.20
Tax	10%

System identification number . . . 24588

System A

Cellular company

Telcom
Telecomunicaciones del Golfo
Blvd. Norte #2210, Local 24
Fracc. Las Hadas Plaza
72070 San Pedro, Puebla, Puebla

Local office . . . voice	011-52-2-239-0000
. . . . fax	011-52-2-230-1874
México City office . . .	011-52-5-251-4433
. . . . . . accounting	011-52-5-251-5590
. . . roaming manager	011-52-5-500-9674
. . . . fax	011-52-5-251-5753

Comments

Visitor registration
This company reports that it will not provide service to visitors until it begins to sign roaming agreements with other cellular companies.

Coverage area
Acapulco and Puebla

Sending calls

Visitor dialing instructions
Local calls	8 digits
Long distance within Mexico . .	91 + 8 digits
To Canada or United States	95 + 10 digit number

Speed-dial numbers
Customer service	not available

Receiving calls

Visitors cannot send or receive calls at this time.

Billing information

Visitor rates (N$ new pesos)
. . . .	not applicable

System identification number is not available

System B

Cellular company

Telcel
Radiomóvil Dipsa
39 Poniente 2923
76050 Col. Amanecer las Animas, Puebla, Pue

Local office . . . . voice	011-52-2-237-9141
. . fax (request transfer)	011-52-2-237-9141
Corporate office	011-52-5-625-3700
. . . roaming manager	011-52-5-625-3814
. . . roaming manager	011-52-5-400-0051
. . . . fax	011-52-5-625-3817

Comments

Visitor registration
You may use this system automatically if Telcel has a roaming agreement with your company. (See *roaming agreements* in the index.) Without a roaming agreement, you must go to the office with a credit card to register for billing.

Coverage area
Acapulco, Puebla, Veracruz

Sending calls

Visitor dialing instructions
Local calls	8 digits
Long distance within Mexico . .	91 + 8 digits
To Canada or United States	95 + 10 digit number

Speed-dial numbers
Customer service	111
Fire department	115
Police	112
Red Cross	113

Receiving calls

Caller dials the roamer access number below, hears tone, dials cellular area code and number.
*These numbers were not yet on at press time.
Acapulco*	011-52-7-449-7626
Puebla*	011-52-2-238-7626
Veracruz*	011-52-2-929-7626

Billing information

Visitor rates (N$ new pesos)
Daily activation fee	N$6.60
Per minute fee	N$2.20
Tax	10%

System identification number . . . 24590

System A

Cellular company

Portatel del Sureste
Administration
Call 35 No. 366
Col. Emiliano Zapata, Mérida, Yucatán

. . . . 011-52-9-944-1222
. . . . fax 011-52-9-944-1229

Sales
Prolongacion Paseo del Montejo
97120 Fracc. Campestre, Mérida, Yucatán

. . . . 011-52-9-926-9011
. . . . fax 011-52-9-926-9321

Visitor registration
This company reports that it does not have any roaming agreements with other companies. It does not provide credit card roaming. Instead, you may rent a telephone from a local dealer.

System B

Cellular company

Telcel
Radiomóvil Dipsa
Calle 31 #92
97125 Colonia México, Mérida, Yucatan

Local office . . . . voice 011-52-9-927-8748
. . fax (request transfer) 011-52-9-927-8748
Corporate office 011-52-5-625-3700
. . . roaming manager 011-52-5-625-3814
. . . roaming manager 011-52-5-400-0051
. . . . fax 011-52-5-625-3817

Comments

Visitor registration
You may use this system automatically if Telcel has a roaming agreement with your company. (See *roaming agreements* in the index.) Without a roaming agreement, you must go to the office with a credit card to register for billing.

Coverage area
Cancun, Merida

Sending calls

Visitor dialing instructions
Local calls . . . . 8 digits
Long distance within Mexico . . 91 + 8 digits
To Canada or United States 95 + 10 digit number

Speed-dial numbers
Customer service 111
Fire department 115
Police 112
Red Cross 113

Receiving calls

Caller dials the roamer access number below, hears tone, dials cellular area code and number.
*These numbers were not yet on at press time.
Cancun* 011-52-9-845-7626
Mérida* 011-52-9-947-7626

Billing information

Visitor rates (N$ new pesos)
Daily activation fee . . . . N$6.60
Per minute fee . . . . N$2.20
Tax 10%

System identification number . . . 24592

System A

Cellular company

USA Cell
Industrias Unidas S.A.
Presidente Masaryk 490, Piso 8
11560 Col. Polanco, México D.F.

Customer service .	voice	011-52-5-500-0611
. . . .	fax	011-52-5-504-4160
Corporate office .	main	011-52-5-280-9021
. . . .	fax	011-52-5-500-0405
. . . .	roaming	011-52-5-505-2314

Comments

Visitor registration
You may use this system automatically if this company has a roaming agreement with your company. (See *roaming agreements* in the index.) Without a roaming agreement, you cannot use the service.

Coverage area
Cuernavaca, México City, Toluca, Valle de Bravo

Sending calls

Visitor dialing instructions

Local calls	7 digits
Long distance in Mexico . . .	91 + 8 digits
To Canada and U.S.	95 + area code + 7 digits
To other countries	98 + country code + number

Speed-dial numbers

Customer service	811
English-speaking customer service . .	*610
Red cross	*911

Receiving calls

Caller dials the roamer access number below, hears tone, dials cellular area code and number.
. . . . 011-52-5-502-7626

Or caller dials number below from rotary-dial phone and tells operator the cellular number.
. . . . 011-52-5-502-7627

Billing information

Visitor rates ($ U.S. approximate)

Daily activation fee	$3
Per minute fee	99¢

System identification number . . . 24581

México, México City & Toluca ❾

System B

Cellular company

Telcel
Radiomóvil Dipsa
Avenida Ejercito Nacional 488
11570 Colonia Polanco, Mexico, D.F.

Corporate office		011-52-5-625-3700
. . .	roaming manager	011-52-5-625-3814
. . .	roaming manager	011-52-5-400-0051
. . . .	fax	011-52-5-625-3817

Comments

Visitor registration
You may use this system automatically if Telcel has a roaming agreement with your company. (See *roaming agreements* in the index.) Without a roaming agreement, you must go to the office with a credit card to register for billing.

Coverage area
Cuernavaca, México City, Toluca

Sending calls

Visitor dialing instructions

Local calls	8 digits
Long distance within Mexico . .	91 + 8 digits
To Canada or United States	95 + 10 digit number

Speed-dial numbers

Customer service	111
Fire department	115
Police	112
Red Cross	113

Receiving calls

Caller dials the roamer access number below, hears tone, dials cellular area code and number.
*These numbers were not yet on at press time.

Cuernavaca*	011-52-7-327-7626
México City*	011-52-5-410-7626
México City* (alternate no.)	011-52-5-411-7626
México City* (alternate no.)	011-52-5-412-7626
Toluca*	011-52-7-264-7626

Billing information

Visitor rates (N$ new pesos)

Daily activation fee	N$6.60
Per minute fee	N$2.20
Tax	10%

System identification number . . . 24580

Montserrat (West Indies)

Cellular company

Montserrat Boatphone
Cable and Wireless
PO Box 219
Molyneaux Building, Houston Street
Plymouth, Montserrat, West Indies

Customer service	from U.S.	800-262-8366
. . . . .	from Canada	800-567-8366
. . . . .	from elsewhere	809-462-5051
Local office		809-491-2112
Fax machine		809-491-3599

Comments

Visitor registration
You must call this company and register for credit card billing on Amex, Discover, Mastercard, or Visa. Or you may register upon arrival by pressing *0 send* on your cellular telephone. An operator is available from 7 am to midnight. Cellular phones rent for $10 per day

Coverage area
Montserrat

Sending calls

Visitor dialing instructions

Local calls	7 digits
To other 809 area code	1 + 7 digits
To U.S. and Canada .	1 + area code + 7 digits
Worldwide calls .	011 + country code + number

Speed-dial numbers

AT&T U.S. operator	872
Boatphone cellular operator	0
Emergency	911

Receiving calls

Caller dials the roamer access number below, after it stops ringing (a tone may not be audible), dials your home cellular area code and number.
. . . . 809-464-7626

Caller dials your temporary number directly:
You will be assigned a temporary telephone number that a caller may dial to reach you.

Billing information

Visitor rates ($ U.S.)

Daily activation fee (only on days used) .	$5.00

Per minute including long distance (land-based):

AT&T or MCI operator or collect . .	$2.20
Boatphone operator or emergency no.	$0.00
Canada, United States	$4.00
Directory assistance	$1.00
800 numbers	$2.00
Europe, South America	$5.30
Incoming collect	$9.00
Local or incoming calls	$1.60
Nearby islands	$2.20
Other international	$7.30
Other Caribbean islands	$4.00

System identification number . . . 08145

New Zealand

Cellular company

Telecom Cellular
PO Box 550
Level 6, Med City Tower, 141 Willis Street
Wellington, New Zealand

Corporate office . . .	011-64-4-801-7400
. . . . in New Zealand	0800-651-000
. . . roaming manager	011-64-4-382-3299
Fax machine	011-64-4-382-8086

Comments

Visitor registration
Contact this company to request an application form. Return it with your American Express, Discover, Mastercard, or Visa number, plus your electronic serial number and planned dates of use. Otherwise your first call in New Zealand may be intercepted for credit card billing. Customers of B.C. Tel Mobility Cellular , Cantel, and GTE Mobilnet may register with their home companies at least five days prior to arrival.

Coverage area
Most populated areas of New Zealand

Sending calls

Visitor dialing instructions

Cellular to cellular	6 digits
To Auckland	09 + 7 digits
To Christchurch	03 + 7 digits
To Hamilton	07 + 7 digits
To Wellington	04 + 7 digits
Worldwide . . .	00 + country code + number

Speed-dial numbers

Customer service	*123
Traffic safety line	*555

Receiving calls

This company has no roamer access number.
You will be assigned a temporary telephone number that others may dial to reach you. Callers inside New Zealand will dial 025 + the number. Callers outside New Zealand dial 011 + 64 + 25 + the number.

Billing information

Visitor rates ($ New Zealand)

Daily activation fee	3.00
Per minute fee outgoing calls	2.00
Per minute fee incoming calls	0.00

System ID numbers . 08576, 08580, 08582

Puerto Rico

Cellular company

Cellular One
Cellular Communications of Puerto Rico
Metro Office Park, Building 6, Third Floor
Guaynabo, PR 00968-1705

Customer service	809-397-5000
Corporate office	809-397-1000
Fax machine	809-397-5765

Comments

Visitor registration
You may use this system automatically if this company has a roaming agreement with your company. (See *roaming agreements* in the index.) Without a roaming agreement, you must call to receive a temporary number and charge calls to American Express, Mastercard, or Visa.

Coverage area
Aguadilla, Culebra, Humaco, Mayaguez, Ponce, San Juan, Vieques

Sending calls

Visitor dialing instructions
Local calls	7 digits
Long distance	area code + 7 digits
Worldwide	011 + country code + number

Speed-dial numbers
Customer service	611
Emergency	911

Receiving calls

Caller dials the roamer access number below, hears tone, dials cellular area code and number.
809-397-7626

Billing information

Visitor rates ($ U.S.)
Daily activation fee	$2
Per minute fee	99¢

System identification number   00227

Puerto Rico

Cellular company

Servicios Celulares
Puerto Rico Telephone
PO Box 360998
Rio Pedras, San Juan, PR 00936-0998

Physical address
(Old Lottery Building) Munoz Rivera Ave. 1130
Rio Piedras, San Juan, PR 00926

Roaming department (for visitors)		809-753-1830
		809-754-7116
	fax	809-758-6104
Customer service		809-763-3333
	fax	809-763-8100
Sales department		809-766-2355

Comments

Visitor registration
You may use this system automatically if this company has a roaming agreement with your company. (See *roaming agreements* in the index.) Without a roaming agreement, you may rent a telephone at this location or in the American Airlines terminal at the airport. If your stay is prolonged, you may wish to contact the sales department and become a subscriber.

Coverage area
Puerto Rico, including islands of Culebra and Vieques.

Sending calls

Visitor dialing instructions
Local calls	7 digits
To U.S.and Canada	1 + area code + 7 digits
Worldwide	011 + country code + number

Speed-dial numbers
Customer service	611
Emergency	911
Operator	123 or 411

Receiving calls

Caller dials the roamer access number below, hears tone, dials cellular area code and number.
809-384-7626

Billing information

Visitor rates
Daily activation fee	$3
Per minute fee	$95

System identification number   00218

St. Kitts & Nevis

Cellular company

St. Kitts & Nevis Boatphone
Cayon & Fort Street, PO Box 437
Basseterre, St. Kitts, West Indies

Customer service .	from U.S.	800-262-8366
. . . . . .	from Canada	800-567-8366
. . . . .	from elsewhere	809-462-5051
Local office		809-465-3003
Fax machine		809-465-3033

Comments

Visitor registration
You must call this company and register for credit card billing on Amex, Discover, Mastercard, or Visa. Or you may register upon arrival by pressing *0 send* on your cellular telephone. An operator is available from 7 am to midnight. Cellular phones rent for $10 per day.

Coverage area
St. Kitts and Nevis

Sending calls

Visitor dialing instructions

Local calls	7 digits
To other 809 area code	1 + 7 digits
To U.S. and Canada	1 + area code + 7 digits
Worldwide calls .	011 + country code + number

Speed-dial numbers

AT&T U.S. operator	872
Boatphone cellular operator	0

Receiving calls

Caller dials the roamer access number below, after it stops ringing (a tone may not be audible), dials your home cellular area code and number.
. . . . 809-467-7626

Caller dials your temporary number directly:
You will be assigned a temporary telephone number that a caller may dial to reach you.

Billing information

Visitor rates ($ U.S.)

Daily activation fee (only on days used) .	$5.00
Per minute including long distance (land-based):	
AT&T or MCI operator or collect . .	$2.20
Boatphone operator or emergency no.	$0.00
Canada, United States	$4.00
Directory assistance	$1.00
800 numbers	$2.00
Europe, South America	$5.30
Incoming collect	$9.00
Local or incoming calls	$1.60
Nearby islands	$2.20
Other international	$7.30
Other Caribbean islands	$4.00

System identification number . . . **08033**

St. Lucia

Cellular company

St. Lucia Boatphone
Rodney Bay Marina, PO Box 2136
Gros Islet, St. Lucia, West Indies

Customer service .	from U.S.	800-262-8366
. . . . . .	from Canada	800-567-8366
. . . . .	from elsewhere	809-462-5051
Local office		809-452-0361
Fax machine		809-452-0394

Comments

Visitor registration
You must call this company and register for credit card billing on Amex, Discover, Mastercard, or Visa. Or you may register upon arrival by pressing *0 send* on your cellular telephone. An operator is available from 7 am to midnight. Cellular phones rent for $10 per day

Coverage area
St. Lucia

Sending calls

Visitor dialing instructions

Local calls	7 digits
To other 809 area code	1 + 7 digits
To U.S. and Canada	1 + area code + 7 digits
Worldwide calls	011 + country code + number

Speed-dial numbers

AT&T U.S. operator	872
Boatphone cellular operator	0

Receiving calls

Caller dials the roamer access number below, after it stops ringing (a tone may not be audible), dials your home cellular area code and number.
. . . . 809-493-7626

Caller dials your temporary number directly:
You will be assigned a temporary telephone number that a caller may dial to reach you.

Billing information

Visitor rates ($ U.S.)

Daily activation fee (only on days used)	$5.00
Per minute including long distance (land-based):	
AT&T or MCI operator or collect . .	$2.20
Boatphone operator or emergency no.	$0.00
Canada, United States	$4.00
Directory assistance	$1.00
800 numbers	$2.00
Europe, South America	$5.30
Incoming collect	$9.00
Local or incoming calls	$1.60
Nearby islands	$2.20
Other international	$7.30
Other Caribbean islands	$4.00

System identification number . . . **08011**

St. Maarten (Netherlands Antilles)

System B

Cellular company

St. Maarten Boatphone
East Caribbean Cellular Telephone
13 Camille Richardson Street
Phillipsburg, St. Maarten
Netherlands Antilles

Customer service	from Canada	800-567-8366
	from U.S.	800-262-8366
	from elsewhere	809-462-5051
Local office		011-599-522-100
Fax machine		011-599-525-678

Comments

Visitor registration
You must call this company and register for credit card billing. Or you may register upon arrival by pressing *0 send* on your cellular telephone. An operator is available from 7 am to midnight.

Coverage area
Anguilla, Dutch and French St. Maarten, St. Barthelemy, Saba, St. Eustatius, and vicinity.

Sending calls

Visitor dialing instructions

Dutch St. Maarten	5 digits
French St. Maarten, St. Barthelemy	6 digits
To 809 area code	1 + 809 + 7 digits
To U.S. and Canada	1 + area code + 7 digits
Worldwide calls	011 + country code + number

Speed-dial numbers

Ambulance	22111
AT&T U.S. operator	872
Boatphone cellular operator	0
Directory assistance	411 or 150
Dutch police	911
East Caribbean Cellular business office	611
MCI operator	624
Phillipsburg Boatphone office	22211
Saba VHF marine operator (free)	155
Weather	199

Receiving calls

Caller dials the roamer access number below, after it stops ringing (a tone may not be audible), dials your home cellular area code and number.
011-5995-7-7626

Caller dials your temporary number directly:
You will be assigned a temporary telephone number that callers may dial to reach you.

Billing information

Visitor rates ($ U.S.)

Daily activation fee (only on days used)	$5.00
Per minute including long distance (land-based):	
AT&T or MCI operator	$1.60
Bahamas, Bermuda, Western Europe	$7.00
Canada	$4.50
Directory assistance	$1.00
Eastern Europe, South America	$8.00
Incoming collect; other international	$9.00
Local or incoming calls	$1.60
U.S. and other Caribbean islands	$4.00

System identification number 00136

St. Martin (French W. Indies)

System A

Cellular company

St. Martin Mobiles
Box 240
Marigot, St. Martin 97150
French West Indies

Customer service	from Canada	800-567-8366
	from U.S.	800-262-8366
	from elsewhere	809-462-5051
Local office		011-590-273-000
Fax machine	Phillipsburg	011-599-525-678

Comments

Visitor registration
You must call this company and register for credit card billing. Or register upon arrival by pressing *0 send* on your telephone. Operator on duty from 7 am to midnight. You may request a local French cellular number and routing by France Telecom.

Coverage area
Anguilla, Dutch and French St. Martin, St. Barthelemy, Saba, St. Eustatius, and vicinity.

Sending calls

Visitor dialing instructions

Boatphone cellular operator	0
Dutch St. Maarten	19 + 5995 + 5 digits
French St. Martin, St. Barthelemy	6 digits
To France	0033 + number
To 809 area code	1 + 809 + 7 digits
To U.S. and Canada	1 + area code + 7 digits
Worldwide calls	011 + country code + number

Speed-dial numbers

AT&T U.S. operator	872
Boatphone cellular operator	0
Directory assistance	411 or 150
East Caribbean Cellular business office	611
French police	911
MCI operator	624
Saba VHF marine operator (free)	155

Receiving calls

Caller dials the roamer access number below, after it stops ringing (a tone may not be audible), dials your home cellular area code and number.
011-5995-7-7626

Caller dials your temporary number directly:
You are assigned a temporary telephone number callers may dial to reach you. From another country, they dial 011-590-27 + a 4 digit number.

Billing information

Visitor rates ($ U.S.)

Daily activation fee (only on days used)	$5.00
Per minute including long distance (land-based):	
AT&T or MCI operator	$1.60
Bahamas, Bermuda, Western Europe	$7.00
Canada	$4.50
Directory assistance	$1.00
Eastern Europe, South America	$8.00
Incoming collect; other international	$9.00
Local or incoming calls	$1.60
U.S. and other Caribbean islands	$4.00

System identification number 00135

St. Vincent & The Grenadines

Cellular company

St. Vincent and the Grenadines Boatphone
Villa, PO Box 103
Kingstown, St. Vincent, West Indies

Customer service .	from U.S.	800-262-8366
. . . .	from Canada	800-567-8366
. . . .	from elsewhere	809-462-5051
Local office . . .	St. Lucia	809-452-0361
. . . .	St. Vincent	809-457-4600
Fax machine . .	St. Lucia	809-452-0394
. . . .	St. Vincent	809-457-4940

Comments

Visitor registration
You must call this company and register for credit card billing on Amex, Discover, Mastercard, or Visa. Or you may register upon arrival by pressing *0 send* on your cellular telephone. An operator is available from 7 am to midnight. Cellular phones rent for $10 per day

Coverage area
St. Vincent and the Grenadines

Sending calls

Visitor dialing instructions

Local calls	7 digits
To other 809 area code	1 + 7 digits
To U.S. and Canada .	1 + area code + 7 digits
Worldwide calls .	011 + country code + number

Speed-dial numbers

AT&T U.S. operator	872
Boatphone cellular operator	0

Receiving calls

Caller dials the roamer access number below, after it stops ringing (a tone may not be audible), dials your home cellular area code and number.
. . . . 809-493-7626

Caller dials your temporary number directly:
You will be assigned a temporary telephone number that a caller may dial to reach you.

Billing information

Visitor rates ($ U.S.)

Daily activation fee (only on days used) .	$5.00

Per minute including long distance (land-based):

AT&T or MCI operator or collect . .	$2.20
Boatphone operator or emergency no.	$0.00
Canada, United States	$4.00
Directory assistance	$1.00
800 numbers	$2.00
Europe, South America	$5.30
Incoming collect	$9.00
Local or incoming calls	$1.60
Nearby islands	$2.20
Other international	$7.30
Other Caribbean islands	$4.00

System identification number . . .   08007

Singapore

Cellular company

Singapore Telecom
31 Exeter Road, Comcentre
Singapore 0923

Information	011-65-738-0123
Fax machine	011-65-737-4989

Comments

Visitor registration
Request a roamer application, complete it with your Mastercard or Visa number, and return to the company.

Coverage area
Singapore

Sending calls

Visitor dialing instructions

International calls	005 + area code + number

Speed-dial numbers

Customer service	738-0123
Emergency	999

Receiving calls

This company has no roamer access number.
You will be assigned a temporary telephone number that others may dial to reach you.

Billing information

Visitor rates (S$ Singapore)

Daily activation fee	S$5.00
Per minute fee	S$.50

System identification number is not available

Virgin Islands (United States)

System A

Cellular company

Cellular One
U.S. Virgin Islands Telephone
Crown Bay Marina, Suite 312
St. Thomas, VI 00802

Information	809-774-0005
Fax machine	809-774-6711

Comments

Visitor registration
You must call this company and register for credit card billing on American Express, Mastercard, or Visa. Visitors receive free call forwarding, call waiting, and conference calling.

Coverage area
St. Croix, St. John, St. Thomas

Sending calls

Visitor dialing instructions
Local calls	area code + 7 digits
Long distance	1 + area code + 7 digits

Speed-dial numbers
Ambulance	922
Coast Guard	999
Emergency	911
Virgin Islands police	915

Receiving calls

Caller dials the roamer access number below, hears tone, dials cellular area code and number. Or caller may reach you directly by dialing the temporary telephone number assigned to you.
. . . . 809-690-7626

Billing information

Visitor rates
Daily activation fee	$5
Per minute fee	90¢

Identification numbers

City	Market	System	Billing	RSA
St. Croix	731	01847	00000	VI-2
St. John, St. Thomas	730	01847	00000	VI-1

Virgin Islands (United States)

System B

Cellular company

Vitel Cellular
Vitelco
PO Box 7610
St. Thomas, VI 00801

Information	809-777-8581
Corporate office	809-777-8899
Fax machine	809-774-6323

Comments

Visitor registration
You must call this company and register for credit card billing.

Coverage area
St. Croix, St. John, St. Thomas, and the surrounding cruising waters.

Sending calls

Visitor dialing instructions
Local calls	7 digits
Long distance	1 + area code + 7 digits

Speed-dial numbers
Customer service	611 or 775-2922
Emergency	0

Receiving calls

This company has no roamer access number.
You will be assigned a temporary telephone number that others may dial to reach you.

Billing information

Visitor rates ($ U.S.)
Daily activation fee	$5
Per minute fee	99¢

Identification numbers

City	Market	System	Billing	RSA
St. Croix	731	01848	00000	VI-2
St. John, St. Thomas	730	01848	00000	VI-1

Venezuela

Cellular company

Telcel
Avenue Diego Cisneros
Centro Colgate ala Norte, Piso 3
1071 Los Ruices, Caracas, Venezuela

Information	011-58-2-201-8000
Fax	011-58-2-238-7779

Additional information not available at press time.

Virgin Islands (British)

Cellular company

CCT Boatphone
6 Mill Mall, PO Box 267
Road Town, Tortola, British Virgin Islands

Customer service .	from U.S.	800-262-8366
. . . .	from Canada	800-567-8366
. . . .	from elsewhere	809-462-5051
Local office		809-494-3825
Fax machine		809-494-4933

Comments

Visitor registration
You must call this company and register for credit card billing on Amex, Discover, Mastercard, or Visa. Or you may register upon arrival by pressing *0 send* on your cellular telephone. An operator is available from 7 am to midnight. Cellular phones rent for $10 per day

Coverage area
British Virgin Islands and surrounding waters

Sending calls

Visitor dialing instructions

Local calls	49 + 5 digits
To other 809 area code . . .	1 + 7 digits
To U.S. and Canada	1 + area code + 7 digits
Worldwide calls	011 + country code + number

Speed-dial numbers

AT&T U.S. operator	872
Boatphone cellular operator	0
Directory assistance	119
Emergency	911

Receiving calls

Caller dials the roamer access number below, after it stops ringing (a tone may not be audible), dials your home cellular area code and number.
. . . . 809-496-7626

Caller dials your temporary number directly:
You will be assigned a temporary telephone number that a caller may dial to reach you.

Billing information

Visitor rates ($ U.S.)

Daily activation fee (only on days used)	$5.00
Per minute including long distance (land-based):	
AT&T or MCI operator or collect . .	$2.20
Boatphone operator or emergency no.	$0.00
Canada, United States	$4.00
Directory assistance	$1.00
800 numbers	$2.00
Europe, South America	$5.30
Incoming collect	$9.00
Local or incoming calls	$1.60
Nearby islands (U.S. Virgin Islands) .	$2.20
Other international	$7.30
Other Caribbean islands	$4.00
Puerto Rico	$3.15

System identification number . . . 06191

Index of Roaming Agreements

The following pages list roaming agreements for the cellular companies in this book. First find your home cellular company or its parent company in the index below. When you turn to the page shown, you will find a list of companies that allow you to use their services automatically. If you use a system operated by one of these host companies, your home company is billed, and it adds a roaming charge to your monthly bill. Roaming agreements can change at the discretion of your home company. For more information about roaming agreements, please refer to page 1159.

Roaming Agreements

Index of Roaming Agreements

Roaming Agreements

Index of Roaming Agreements

AGT Cellular

ALLTEL Mobile Communications
Ameritech Mobile Communications
Atlantic Cellular
B.C. Tel Mobility Cellular
Baja Cellular Mexicana
Bell Atlantic Mobile
Bell Mobility Cellular
BellSouth Mobility
Blue Mountain Cellular
Bluegrass Cellular
C. C. Cellular
CTIA
Cal-One Cellular
Cantel
CellCom
Cellular Connection
Cellular Holdings
Cellular, Inc.
Cellular Information Systems
Cellular One of Chicago
Cellular One of Manchester
Cellular Plus
Centel Cellular
Century Cellunet
Coastel Communications
Contel Cellular
Cybertel Cellular Telephone
Dobson Cellular
Ed Tel Cellular
Enid Cellular
Gaia
GTE Mobilnet ("B" systems only)
Hutchinson Telephone
Independent Cellular Network
Iowa East Cellular
IUSA Cell
Kansas Independent Network
Lincoln Telephone
MACTel Cellular
Mercury Communications
Missouri Cellular
Mobiletel
Mountaineer Cellular
MT&T Cellular
MTS Cellular
NewTel Cellular
NYNEX Mobile Communications
Oneonta Telephone
Pacific Telecom Cellular
PacTel Cellular
Pacific Northwest Cellular
Petroleum Communications
Poka Lambro Telecommunications
Quantum Communications
Québec Téléphone Cellulaire
Radiomovil Dipsa, S.A.
RFB Cellular
Rochester Telephone Mobile
Comms.
SaskTel Cellular
Springwich Cell. Ltd Partner. (SNET)
Sotheastern Cellular
Southwestern Bell Mobile Systems
StarCellular
Thunder Bay Cellular
U S West Cellular
Union Telephone
United States Cellular
West Central Cellular

Advantage Cellular

Albany Telephone
ALLTEL Mobile Communications
Ally
Ameritech Mobile Communications
Appalachian Cellular
Auburn Telephone

Bahamas Telephono
Bell Atlantic Metromobile
Bell Atlantic Mobile
Bell Mobility Cellular
BellSouth Mobility
Big Horn Cellular
Bledsoe Telephone
Bluegrass Cellular
Blue Mountain Cellular
Blueridge Cellular
C. C. Cellular
Cal-One Cellular
Cantel
Canton Cellular
Cellcom
Cellular Connection
Cellular Information Systems
Cellular One of Albany
Cellular One of Beckley
Cellular One of Bismarck
Cellular One of Chicago
Cellular One of Dothan
Cellular One of Indiana
Cellular One of Meridian
Cellular One of Sioux Falls
Cellular One of Southwest Georgia
Cellular One of Upstate New York
Cellular One of Watertown
Cellular One of Wausau
Cellular One of Wilmington
Cellular Phone of Kentucky
Cellular Plus
Cellular Ventures
Cellulink
Cellwave
Centel Cellular
Century Cellunet
Coastel Communications
Contel Cellular
Crowley Cellular Telecommunications
Dicomm Cellular
Digital Cellular of Texas
Enid Cellular
Farmers Cellular
FGI Cellular Management
First Fayette Cellular
Gaia
GTE Mobilnet ("B" systems only)
Genesee Telephone
Hudson Valley Cellular
IUSA Cell
Independent Cellular Network
Iowa East Cellular
Kansas Cellular
Lake Huron Cellular
Lincoln Telephone
Litchfield County Cellular
MT&T Mobile
Mackinac Cellular
MACTel Cellular Systems
Maine Cellular
McCaw Cellular Communications
Mercury Cellular
MidSouth Cellular
Miscellco Cellular
Missouri Cellular
Montgomery Cellular
Mountaineer Cellular
National Cellular Limited Partnership
Nebraska Cellular
North Carolina Cellular
NYNEX Mobile Communications
Oneota Telephone
PacTel Cellular
Palmer Communications
Petroleum Communications
Pioneer Cellular
Poka Lambro Cellular
Québec Téléphone Cellulaire
Richmond Cellular Telephone

Rochester Telephone Mobile
Comms.
Savannah Cellular
Southeastern Cellular
Southwestern Bell Mobile Systems
Springwich Cell. Ltd Partner. (SNET)
StarCellular
Thunder Bay Telephone
Tsaconas Cellular
U S West Cellular
Union Telephone
United States Cellular
United Telespectrum
Unity Cellular Systems (Unicel)
Vanguard Cellular
Virginia RSA 10 Limited Partnership
Vitel Cellular
West Alabama Cellular
Westex Cellular
US Virgin Islands
XIT Cellular

ALLTEL Mobile Communications

AGT Cellular
Ameritech Mobile Communications
Atlantic Cellular
Bell Atlantic Mobile
Bell Mobility Cellular
BellSouth Cellular
Cantel
Cellcom
Cellular, Inc.
Cellular Information Systems
Cellular One of Baltimore
Cellular One of Chicago
Cellular One of Cincinnati
Cellular One of Cleveland
Cellular One of Columbus
Cellular One of Indianapolis
Cellular One of Milwaukee
Cellular One of Rome
Cellular One of Utica
Cellular One of Washington
Cellular One of Wilmington
Cellular Plus
Cellular South
Cellulink
Celutel
Centel Cellular
Centennial Cellular
Century Cellunet
CFW Cellular
Coastel Communications
Comcast
Contel Cellular
Crowley Cellular Telecomm
Cybertel Cellular
Ed Tel Cellular
GTE Mobilnet
Houston Cellular Telephone
Independent Cellular Network
Kansas Cellular
L.A. Cellular Telephone
Lincoln Telephone
Mactel Cellular
Maine Cellular
McCaw Cellular Communications
Mercury Cellular
Metrocel Cellular
Mobiletel
MT&T Cellular
MTS Cellular
NBTel Cellular
Nebraska Cellular
NYNEX Mobile Communications
PacTel Cellular
Palmer Communications
Petroleum Communications
Pioneer Cellular

Roaming Agreements

Public Service Cellular
Québec Téléphone Cellulaire
Radiofone
Richmond Cellular Telelephone
Rochester Telephone
Sasktel Cellular
Savannah Cellular
Southwestern Bell Mobile System
Springwich Cell. Ltd Partner. (SNET)
StarCellular
U S West Cellular
Union Telephone
United States Cellular
Unity Cellular Systems (Unicel)
Vanguard Cellular
West Central Cellular

Ameritech Mobile Communications

Advantage Cellular
AGT Cellular
ALLTEL Mobile Communications
Alton Cellular
Appalachian Cellular
Associated Communications
Atlantic Cellular
B.C. Tel Mobility Cellular
Bell Atlantic Mobile
Bell Mobility Cellular
BellSouth Mobility
Bledsoe Telephone
Bluegrass Cellular
Cal-One Cellular
Cantel Cellular
Carolina West Cellular
Cellcom of Green Bay
Cellular, Inc.
Cellular Information Systems
Cellular One of Benton Harbor
Cellular One of Chicago, IL
Cellular One of Erie, PA
Cellular One of Indianapolis, IN
Cellular One of Milwaukee, WI
Cellular One of Montgomery, AL
Cellular One of Roanoke, VA
Cellular One of Sharon, PA
Cellular One of Sioux Falls, SD
Cellular One of Upstate New York
Cellular One of Youngstown, OH
Cellular Plus
Cellular South
Cellular Three
Cellular Ventures
Cellulares Telefonica
Cellulink
Centel Cellular
Centennial Cellular
Century Cellular
Coastel Communications
Cone Enterprises
Contel Cellular
Cybertel Cellular Telephone
Danbury Cellular Telephone
Dobson Cellular
Ed Tel Cellular
Enid Cellular
Farmers Cellular
GenCell Management
GTE Mobilnet
Highland Cellular
Honolulu Cellular
Independent Cellular Network
Interstate Cellular (Interstate Cellular (Intercel))
Kansas Cellular
Kaplan Telephone
Lake Huron Cellular
Lincoln Telephone Cellular
MAC-TEL
Maine Cellular

MCTA
Mercury Cellular
Mercury Communications
Metrophone
Mid-Missouri Cellular
Mid-Texas Cellular
Mississippi Cellular
Missouri Cellular
MobileTel
Mountaineer Cellular
MT&T Cellular
MTS Cellular
National Cellular
Nebraska Cellular
NewTel Cellular
NYNEX Mobile Communications
PacTel Cellular
Palmer Communications
Petroleum Communications
Pine Cellular
PTSI
Public Service Cellular
Québec Téléphone Cellulaire
Rochester Telephone
Sasktel Cellular
Savannah Cellular
Shenandoah Cellular
Southeastern Cellular
Southwestern Bell Mobile Systems
Springwich Cell. Ltd Partner. (SNET)
StarCellular
Sterling Cellular
Thunder Bay Cellular
U S West Cellular
Union Telephone
United Bluegrass Cellular
United States Cellular
Unitel
Unity Cellular Systems (Unicel)
Valley Cellular
Vanguard Cellular Systems
West Central Cellular
Western Maine Cellular
Westex Cellular
XIT Cellular
Youngstown Cellular

Associated Communications

Advantage Cellular
Bay Area Cellular
Bell Atlantic Mobile
BellSouth Mobility
Cantel
Cellular One of Amarillo, TX
Cellular One of Baltimore/Washington
Cellular One of Berrien County
Cellular One of Boston, MA
Cellular One of Chicago, IL
Cellular One of Cincinnati, OH
Cellular One of Cleveland, OH
Cellular One of Columbus, OH
Cellular One of Detroit, MI
Cellular One of Grand Rapids, MI
Cellular One of Indianapolis, IN
Cellular One of Lansing, MI
Cellular One of La Salle, IL
Cellular One of Milwaukee, WI
Cellular One of Richmond, IN
Cellular One of Roanoke, VA
Cellular One of Syracuse, NY
Cellular One of Toledo, OH
Cellular One of Utica, NY
Cellular One of Warren, OH
Cellular One of Youngstown, OH
Cellular Plus
Centel Cellular
Centennial Cellular
Comcast

Contel Cellular
Crowley Cellular Telecommunications
Cybertel Cellular Telephone
GTE Mobilnet
Gulf Coast Cellular
Houston Cellular Telephone
Los Angeles Cellular Telephone
McCaw Communications
MetroCel Cellular
Metrophone
PacTel Cellular
Palmer Communications
Radiofone
Rochester Telephone Mobile Comms. (Genesee Tel. subscribers cannot use Rochester Tel. Mobile Comms.)
Springwich Cell. Ltd Partner. (SNET)
U S West Cellular
United Bluegrass Cellular
United States Cellular
Vanguard

Atlantic Cellular

Atlantic Cellular
Advantage Cellular
ALLTEL Mobile Communications
American Rural Cellular
Ameritech Mobile Communications
Ameritech Rural Cellular
Appalachian Cellular
Associated Communications
Atlas Cellular
Bakersfield Cellular Telephone
Bay Area Cellular
BCTel Cellular
Bell Atlantic Mobile
Bell Mobility Cellular
Bell South Cellular
Bell South Mobility
Blackwater Cellular
Bledsoe Telephone
Blue Grass Cellular
Blue Mountain Cellular
Blue Ridge Cellular
Cal-One Cellular
Cantel
Carolina West Cellular
C.C. Cellular
Cellcom of Green Bay
Cellcom of Hickory
Cellular 2000
Cellular Connection
Cellular, Inc.
Cellular Information Systems
Cellular of Indiana
Cellular One in Ohio
Cellular One of Baltimore
Cellular One of Beckley
Cellular One of Berrien County
Cellular One of Boston
Cellular One of Chicago
Cellular One of Danville
Cellular One of Detriot
Cellular One of East Central Penn
Cellular One of Grand Rapids
Cellular One of Indianapolis
Cellular One of Lake Charles
Cellular One of Lansing
Cellular One of Milwaukee
Cellular One of Muskegon
Cellular One of New York
Cellular One of North/Central PA
Cellular One of Ocean County
Cellular One of Ohio
Cellular One of Richmond
Cellular One of Syracues
Cellular One of Toledo
Cellular One of Upstate NY

Roaming Agreements

Cellular One of Utica
Cellular One of Victoria
Cellular One of Washington
Cellular One of Wichita Falls
Cellular Plus
Cellular South
Cellulink
Centel Cellular
Centennial Cellular
Century Cellunet
Chariton Valley Cellular
Coastel Communications
Comcast
Cone Enterprises
Contel Cellular
Cooper Cellular
Crowley Cellular Telecomm
Cybertel Cellular
Danbury Cellular Telephone
Dicomm Cellular
Digital Cellular
Dobson Cellular Systems
Eastex Cellular
Ed Tel Cellular
Enid Cellular
Farmers Cellular Telephone
First Kentucky Cellular
Five Star Cellular
Galveston Cellular
GenCell Management
GMD Cellular
Greater South Dakota Cellular
GTE Mobilnet
Honolulu Cellular Telephone
Houston Cellular Telephone
Hudson Valley Cellular
Independent Cellular Network
Interstate Cellular
Iowa East Cellular Telephone
Kansas Cellular
Lamar County Cellular
Lincoln Telephone
Litchfield County Cellular
Mactel Cellular
Maine Cellular
Maritime Cellular
McCaw Cellular Communications
Mercury Cellular
Mercury Communications
Metrocel Cellular
Metrophone
Mid-Missouri Cellular
Minnesota Southern Cellular
Miscellco Communications
Mississippi Cellular
Missouri Cellular
Mobiletel
Montana Cellular Telephone
Mountaineer Cellular
Mountaineer Mobile
MT&T Cellular
MTS Cellular
NewTel Cellular
North Carolina Cellular
North-West Cellular
NYNEX Mobile Communications
Pace Communications
PacTel Cellular
Palmer Communications
PC Cellular of Vermont
Pegasus Cellular
Petroleum Communications
PTSI Cellular
Public Service Cellular
Quantum Communications
Québec Téléphone Cellulaire
Radiofone
Richmond Cellular Telephone
Rochester Telephone

Santa Cruz Cellular
Savannah Cellular
Shenandoa Cellular
Sooner Cellular
Southeastern Cellular
Southwestern Bell Mobile Systems
Springwich Cell. Ltd Partner. (SNET)
StarCellular
Sterling Cellular
Texas Cellular
Thunder Bay Cellular
U S West Cellular
Union Telephone
United States Cellular
Unitel
Unity Cellular Systems (Unicel)
Vanguard Cellular
West Central Cellular
Westex

B.C. Tel Mobility Cellular

AGT
Ameritech Mobile
Bell Atlantic
Bell Mobility Cellular
BellSouth Mobility
Cellulink
Centel
Contel
Ed Tel Cellular
GTE Mobilnet ("B" systems only)
Independent Cellular
IUSA Cell
Lincoln Telephone Cellular
MT&T
MTS
NewTel Cellular
NYNEX Mobile Communications
PacTel Cellular
Petroleum Communications
SaskTel
Southwestern Bell Mobile
StarCellular
United States Cellular

Bahamas Telecommunications

ALLTEL Mobile Communications
Ally
Associated Communications
BellSouth Mobility
Blueridge Cellular
Cantel
Cellular One of Chicago
Cellular One of Danville
Cellular One of LaSalle
Centel Cellular
Century Cellunet
Mactel Cellular
Maritime Cellular
McCaw Cellular Communications
Metrophone
Miscellco Communications
MT&T Cellular
Nebraska Cellular
Pace Communications
PacTel Cellular
Petroleum Communications
Richmond Cellular Telephone
United States Cellular
Vanguard Cellular

Bakersfield Cellular

ALLTEL Mobile Communications
Baton Rouge Cellular
Bay Area Cellular
BellSouth Mobility
Cantel
Cellular Information Systems

Cellular One of Albany
Cellular One of Amarillo
Cellular One of Billings
Cellular One of Chicago
Cellular One of Dothan
Cellular One of Jacksonville
Centel Cellular
Crowley Cellular
GenCell Management
Honolulu Cellular
Houston Cellular Telephone
Los Angeles Cellular
McCaw Communications
Metro Mobile CTS
Metro One
MetroCel Cellular
NYNEX Mobile Communications
Cellular One of Ohio
PacTel Cellular
Palmer Communications
Petroleum Communications
Radiofone
Richmond Cellular
U S West Cellular
United States Cellular
Vanguard Cellular Systems
West Central Cellular
Western Cellular

Bay Area Cellular

Allegan Cellular
Atlantic Cellular
Bahamas Telephone
Bakersfield Cellular
Bell Atlantic Mobile
BellSouth Cellular
Big Horn Cellular
Blue Mountain Cellular
Blue Ridge Cellular
Bluegrass Cellular
Cantel
Cellcom of Hickory
Cellular Information Systems
Cellular One of Alabama
Cellular One of Amarillo, TX
Cellular One of Baltimore/Washington
Cellular One of Beaumont, TX
Cellular One of Beckley, WV
Cellular One of Berrien County, MI
Cellular One of Boston
Cellular One of Chicago
Cellular One of Clanton, AL
Cellular One of Clinton, IA
Cellular One of Columbia, MO
Cellular One of Columbia, TN
Cellular One of Columbus, IN
Cellular One of Danville, IL
Cellular One of Detroit
Cellular One of East/Central PA
Cellular One of Galveston, TX
Cellular One of Greenwood, SC
Cellular One of Hattiesburg, MS
Cellular One of Huntsville, TX
Cellular One of Indiana
Cellular One of Indianapolis
Cellular One of Lake Charles, LA
Cellular One of La Salle, IL
Cellular One of Lenoir, NC
Cellular One of Marianna, FL
Cellular One of Meridian/Waycross, GA
Cellular One of Mid-South Texas
Cellular One of Middle Tennessee
Cellular One of New York
Cellular One of Norfolk, NE
Cellular One of Northeast Colorado
Cellular One of Northeast Georgia
Cellular One of Sioux Falls, SD

Cellular One of Southwest Florida
Cellular One of Southwest Georgia
Cellular One of the Great Lakes of
  Iowa
Cellular One of Upstate New York
Cellular One of Victoria
Cellular One of Wichita Falls, TX
Cellular Plus
Cellular Telephone of Kentucky
Cellular, Inc.
Celutel
Centel
Centennial Cellular
Century Cellunet
Clear Communications
Comcast
CompComm
Contel
Cooper Cellular Management
Crowley Cellular
Cybertel Cellular
Dial Two
Dicomm
Dobson Cellular
Finger Lakes Telephone
GenCell Management
Genessee
GMD Partnership
Greater South Dakota Cellular
GTE (Southeastern Region)
Honolulu Cellular
Horizon Cellular
Houston Cellular
Lake Huron Cellular
Litchfield County Cellular
Los Angeles Cellular
Mackinac Cellular
McCaw Cellular Communications
Mercury Communications
Metrocell
Metrophone of Philadelphia
Mid-South Cellular
Miscellco
Missouri Cellular
Mountaineer Mobile
New Par
Pacific Northwest Cellular
PacTel Cellular
Palmer Communications
Petroleum Communications
PriCellular
Quantum Communications
Radiofone
Richmond Cellular
Rural Cellular Management
Sagir
Santa Cruz Cellular
Southeast Indiana Cellular
Sterling Cellular
Sussex Cellular
Texoma Cellular Management
U S West Cellular
United Bluegrass Cellular
United States Cellular
Unitel
Vanguard
Youngstown Cellular

Bell Atlantic Mobile

Advantage Cellular
ALLTEL Mobile Communications
Ally
Alpha Cellular
AMC Cellular
Atlantic Cellular
Atlas Cellular
B.C. Tel Mobility Cellular
Bell Mobility
Blue Mountain Cellular

Blue Ridge Cellular
Bluegrass Cellular
C. C. Cellular
Cal-One Cellular
Cantel
Carolina West Cellular
Cellular South
Cellular Holding
Cellular Information Systems
Cellular One of
  Baltimore/Washington
Cellular One of Brookings
Cellular One of Chicago
Cellular One of Columbia, MO
Cellular One of Frederick
Cellular One of Galesburg, IL
Cellular One of Great Lakes of Iowa
Cellular One of Indiana
Cellular One of Iowa East
Cellular One of Lake Charles, LA
Cellular One of Meridian
Cellular One of Northeast Mississippi
Cellular One of Sioux Falls
Cellular One of Southwest Georgia
Cellular One of Upstate New York
Cellular One of Waycross
Cellular Phone of Kentucky
Cellular Plus
Cellular Plus of Georgia
Cellular, Inc.
Celutel
Centel Cellular
Century Cellunet
Chariton Valley Cellular
Coastel Communications
Crowley Cellular
Dominion Resources
Easterbrooke
EdTel Cellular
FGI Cellular
First Fayette Cellular
First Kentucky Cellular
General Cellular
GMD Partnership
Highland Cellular
Illinois Valley Cellular
Interstate Cellular
Kansas Cellular
Kentucky Cellular
Lake Huron Cellular
Lincoln Telephone
Maine Cellular
Mercury Cellular and Paging
Mercury Communications
Mid-Missouri Cellular
Missouri Cellular
Mobiletel
MTS Cellular
National Cellular
Nebraska Cellular
NYNEX Mobile Communications
Ocean County Cellular
Ohio State Cellular
Palmer Communications
Panhandle Teleocmmunication
Peoples Cellular
Petroleum Communications
Prime Cellular Management
Public Service Cellular
Radiofone
RFB Cellular
Rochester Telephone
SaskTel Cellular
Shenandoah Mobile
Sterling Cellular
Sussex Cellular
Texas 16 Cellular
U S West Cellular
United Bluegrass Cellular
Unity Cellular

Valley Telecommunications
Vanguard Cellular
Virginia Cellular
Vitel Cellular
West Alabama Cellular
Westex Cellular

Bell Mobility Cellular

AGT Cellular
Advantage Cellular
Albany Telephone
ALLTEL Mobile Communications
American Rural Cellular
Ameritech Mobile Communications
Atlantic Cellular
B.C. Tel Mobility Cellular
Bell Atlantic Mobile
BellSouth Mobility
Bledsoe Cellular
Bluegrass Cellular
Buffalo Telephone
C. C. Cellular
C. V. Cellular
Cal-One Cellular
Cantel
Canton Cellular
Cellcom
Cellular Connection
Cellular Holdings
Cellular, Inc.
Cellular One of Aiken/Augusta
Cellular One of the Bay Area
Cellular One of Indiana
Cellular One of Sioux Falls
Cellular One of Syracuse, NY
Cellular One of Utica, NY
Cellular Plus
Cellular South
Cellulink
Centel Cellular
Century Cellunet
Chariton Valley Cellular
Contel Cellular
Dobson Cellular
Ed Tel Cellular
Enid Cellular
Farmers Cellular
Finger Lakes Cellular
Gaia
Genesee Telephone
Graceba Cellular
GTE Mobilnet ("B" systems only)
Highland Cellular
Hutchinson Telephone
Illinois Valley Cellular
Independent Cellular Telephone
Interstate Cellular (Interstate Cellular
  (Intercel))
Kansas Cellular
Kaplan Industries
Lake Huron Cellular
Lincoln Telephone Cellular
MACtel
Maine Cellular
Mega Comm
Mercury Cellular
MT&T Cellular
MTS Cellular
Nebraska Cellular
NewTel Cellular
NYNEX Mobile Communications
PacTel Cellular
PC Cellular
Petroleum Communications
Poka Lambro
PriCellular
Québec Téléphone Cellulaire
Richmond Cellular
Rochester Telephone

Sasktel
Savannah Cellular
Shenandoah Cellular
Southeastern Cellular
Southeast Indiana
Southwestern Bell Mobile Systems
Springwich Cell. Ltd Partner. (SNET)
StarCellular
Thunder Bay Cellular
Tsaconas Cellular
U S West Cellular
Union Telephone
United States Cellular
United Telespectrum
Unity Cellular Systems (Unicel)
XIT Cellular

BellSouth Cellular

ALLTELL Mobile Communications
Associated Communications
Atlantic Cellular
Bell Atlantic
BellSouth Mobility
Blue Mountain Cellular
Blue Ridge Cellular
C. C. Cellular
Cellcom of Hickory
Cellular Information Systems
Cellular One of Beaumont
Cellular One of Berrien County
Cellular One of Columbia, MO
Cellular One of Chicago/Gary
Cellular One of Galveston, TX
Cellular One of Milwaukee, WI
Cellular One of North Texas
Cellular One of Northeast Mississippi
Cellular One of Richmond, VA
Cellular One of Southwest Florida
Cellular One of Syracuse/Utica
Cellular One of Victoria, TX
Cellular Plus
Cellular South
Cellular Ventures
Cellular XL
Celutel
Century Cellunet
CFW Cellular
Contel
Cooper Cellular Management
Crowley Cellular Telecommunications
Dial Two
Houston Cellular Telephone
Independent Cellular Networks
Intercel
Iowa East Cellular
Kansas Independent Networks
McCaw Cellular Communications
Mercury Cellular and Paging
Mercury Communications
Mississippi Cellular
Missouri Cellular
MobileTel
New Par
NYNEX Mobile Communications
Pacific Telecom Cellular
Palmer Communications
Petroleum Communications
PriCellular
Quantum Communications
Shenandoah Cellular
Sooner Cellular
Sterling Cellular
Texoma Cellular
Union Cellular
United Bluegrass
United States Cellular
Vanguard Cellular Systems

BellSouth Mobility

Advantage Cellular
AGT Cellular
ALLTEL Mobile Communications
Ally
Alpha Cellular
American Rural Cellular
Ameritech Cellular
Appalachian Cellular
Associated Communications
Atlantic Cellular
Auburn Television
Bakersfield Cellular
Batelco
B.C. Tel Mobility Cellular
Bell Atlantic
Bell Mobility Cellular
Bell South Cellular
BellSouth Mobility
Big Horn Cellular
Bledsoe Telephone
Blue Mountain Cellular
Utica Telephone
Blue Ridge Cellular
Bluegrass Cellular
Brazos Cellular
C. C. Cellular
Cal-One Cellular
Cantel
Canton Cellular
Carolina West Cellular
Cellcom of Green Bay
Cellcom of Hickory
Cellular Communication of Puerto
  Rico
Cellular Connection
Cellular Information Systems
Cellular One of Altoona, PA
Cellular One of
  Baltimore/Washington
Cellular One of Beckley, WV
Cellular One of Buffalo, NY
Cellular One of Chicago, IL
Cellular One of Collier/Hendry
Cellular One of Columbia, MO
Cellular One of Danville
Cellular One of East/Central PA
Cellular One of Enid, OK
Cellular One of Galveston, TX
Cellular One of Greenwood, SC
Cellular One of Huntsville
Cellular One of Huntsville, TX
Cellular One of Indiana
Cellular One of Indianapolis
Cellular One of La Salle, IL
Cellular One of Lake Charles, LA
Cellular One of Meridian
Cellular One of Middle Tennessee
Cellular One of Nebraska RSA 3
Cellular One of North/Central PA
Cellular One of Northeast Georgia
Cellular One of Northeast Mississippi
Cellular One of Sioux Falls
Cellular One of Southwest Georgia
Cellular One of Syracuse, NY
Cellular One of Upstate New York
Cellular One of Utica, NY
Cellular One of Western Illinois
Cellular One of Wichita Falls, TX
Cellular Phone of Kentucky
Cellular Plus
Cellular Plus of Georgia
Cellular South
Cellular Systems International
Cellular Three
Cellular Ventures
Cellular XL
Cellular, Inc.
Cellulink

Cellwave
Centel Cellular
Century Cellular
Century Cellunet
Chariton Valley Cellular
Clear Communications
Coastel Communications
Contel Cellular
Cooper Cellular
Cross-Valliant Cellular
Crowley Cellular
Cybertel
Danbury Cellular
Dekalb Telephone
Digital Cellular
Dobson Cellular
Dominion Cellular
Easterbrooke Cellular
Eastex Cellular
EdTel Cellular
Enid Cellular
Farmers Cellular
FGI Cellular
Finger Lakes Telephone
First Fayette Cellular
Gaia
GenCell Management
GMD Partnership
Greater South Dakota Cellular
GTE Mobilnet
Highland Cellular
Honolulu Cellular
Houston Cellular
Hudson Valley RSA Cellular
Illinois Valley Cellular
Independent Cellular
Interstate Cellular
Iowa East Cellular
IUSA Cell
Kansas Independent
Leaco Rural Telephone
Lincoln Telephone
Litchfield Cellular
Los Angeles Cellular
Louisiana Cellular
Mackinac Cellular
MacTel Cellular
Maine Cellular
McCaw Cellular Communications
Mega Comm
Mercury Cellular
Mercury Communications
Metro Mobile
Metro One
Mid-Missouri Cellular
Milwaukee Telephone
Missouri Cellular
Mobiltel
Mountaineer Cellular
Mountaineer Mobile
MT&T Cellular
MTS Cellular
Nebraska Cellular
New Par
NewTel Cellular
NYNEX Mobile Communications
Ohio State Cellular
Oneonta Telephone
Pace Communications
PacTel Cellular
Palmer Communication
Panhandle Telecommunications
PC Cellular of Kentucky
PC Cellular of Vermont
Pegasus Cellular
Peoples Cellular
Petroleum Communications
Pine Cellular
Poka Lambro Telephone

PriCellular
Prime Cellular Management
Public Service Telephone
Puerto Rico Telephone
Quantum Communication
Québéc Téléphone Cellulaire
Radiofone
RFB Cellular
Richmond Cellular
Rochester Telephone
SaskTel Cellular
Shenandoah Mobile
Southeast Indiana Cellular
Southeastern Cellular
Southern Cellular
Southwestern Bell
Springwich Cell. Ltd Partner. (SNET)
StarCellular
Sterling Cellular
Texas 16 Cellular
Thunder Bay Cellular
U S West Cellular
UBET Cellular
Union Telephone
United Bluegrass Cellular
United States Cellular
Unitel
Unity Cellular
Valley Telecom
Vanguard Cellular
Virgin Islands Telephone
Vitel Cellular
West Central Cellular
Westex Cellular
Wisconsin Cellular
XIT Cellular
Youngstown Cellular

Blue Mountain Cellular

Adirondack Cellular
Advantage Cellular
Albany Telephone
ALLTEL Mobile Communications
Ally
Alpha Cellular
Appalachian Cellular
Atlantic Cellular
B.C. Tel Mobility Cellular
Bakersfield Cellular
Bay Area Cellular
Bell Atlantic
BellSouth Cellular
Bell South Mobility
Big Horn Cellular
Blackwater Cellular
Bledsoe
Blue Ridge Cellular
Bluegrass Cellular
Buckhead Telephone
Buffalo Telephone
Cal-One Cellular
Cantel
Canton Cellular
Cellcom of Green Bay
Cellcom of Hickory
Cellular Connection
Cellular of Sioux Falls
Cellular Holding
Cellular, Inc.
Cellular Information Systems
Cellular One of Baton Rouge
Cellular One of Berrien County
Cellular One of Columbia
Cellular One of Central Illinois
Cellular One of Chicago
Cellular One of Danville
Cellular One of Fort Meyers
Cellular One of Galesburg
Cellular One of Huntsville

Cellular One of Indiana
Cellular One of La Salle, IL
Cellular One of Meridian
Cellular One of Middle Tennessee
Cellular One of New York
Cellular One of Northeast Georgia
Cellular One of North Texas
Cellular One of Santa Cruz
Cellular One of Sioux Falls
Cellular One of Southwest Georgia
Cellular One of Southwest Oklahoma
Cellular One of Upstate New York
Cellular One of West Oklahoma
Cellular One of Wichita Falls
Cellular South
Cellular Ventures
Cellular XL
Celutel
Centel
Centennial Cellular
Century Cellunet
CJD Wisconsin Cellular
Coastel Communications
CompComm
Contel Cooper Cellular
Crowley Communications
Cybertel
Danbury Cellular
Detroit Cellular
Digital Cellular of Texas
Dominion Cellular
Easterbrooke Cellular
Farmers Cellular
Finger Lakes Cellular
First Kentucy Cellular
GTE Mobilenet
Galveston Cellular
General Cellular
Genessee Telephone
Greater South Dakota Cellular
Highland Cellular
Honolulu Cellular
Houston Cellular
Independent Cellular
Indianapolis Telephone
Iowa East Cellular
Kansas Independent
Kull Cellular
Lake Charles Cellular
Lake Huron Cellular
Lincoln Cellular
Lone Star Cellular
Los Angeles Cellular
McCaw Cellular Communications
Mega Communications
Mercury Cellular
Mercury Communications
Metro Mobile
Metrocel
Metrophone
Mid-South Cellular
Mid-Missouri Cellular
Milwaukee Telephone
Miscellco
Missouri Cellular
MobileTel
Mountaineer Mobile
Nebraska Cellular
Ocean County Cellular
PacTel Cellular
Palmer Communications
Panhandle Telecommunications
Petroleum Communications
PriCellular
Public Service
Quantum Communications
Radiofone
Richmond Cellular
Rural Cellular
Springwich Cell. Ltd Partner. (SNET)

Sterling Cellular
Sunshine Cellular
Sussex Cellular
Texas 16 Cellular
U S West
Union Telephone
United State Cellular
Unitel
Vanguard
West Alabama Cellular
Westex
X-Cell Cellular
Youngstown Cellular

Blue Ridge Cellular (VA)

Advantage Cellular
ALLTEL Mobile Communications
Ally
Appalachian Cellular
Associated Communications
Bay Area Cellular
BC Tel Cellular
Bell Atlantic Mobile
BellSouth Mobility
Bledsoe Telephone
Blue Mountain Cellular
Bluegrass Cellular
Cantel Cellular
Carolina West Cellular
Cellcom of Green Bay
Cellcom of Hickory
Cellular Connection
Cellular One of
  Baltimore/Washington
Cellular One of Chicago
Cellular One of Columbia, MO
Cellular One of Detroit, MI
Cellular One of East/Central PA
Cellular One of Gainesville, GA
Cellular One of La Salle, IL
Cellular One of Lake Charles, LA
Cellular One of Milwaukee, WI
Cellular One of Ohio
Cellular One of Southwest Florida
Cellular XL
Cellular, Inc.
Celutel
Centel Cellular
Centennial Cellular
Century Cellunet
Comcast Cellular
Cone Enterprises
Contel Cellular
Crowley Cellular
Cybertel Cellular
Enid Cellular
Gaia
Galveston Cellular
GenCell Management
GMD Cellular
GTE Mobilenet
Highland Cellular
Houston Cellular Telephone
Independent Cellular Network
Kaplan Telephone
Liberty Cellular
Mercury Cellular
Metrocel Cellular
Minerich Cellular
Missouri Cellular
Mountaineer Cellular
MTS Cellular
Northwest Cellular
NYNEX Mobile Communications
OK Cellular
Oneonta Cellular
PacTel Cellular
Palmer Communications
Pegasus

Roaming Agreements

Petroleum Communications
Pioneer Cellular
PTSI
Public Service Cellular
Radiofone
Richmond Cellular
Rural Cellular Management
Shenandoah Mobile
Southeastern Cellular
Sterling Cellular
Tsacona Cellular
U S West Cellular
Union Telephone
United States Cellular
Unity Cellular Systems (Unicel)
Vanguard Cellular
Virginia Cellular
Walnut Hill Cellular
Yorkville Telephone

Blue Ridge Cellular Telephone (NC)

Advantage Cellular
ALLTEL Mobile Communications
Associated Communications
Atlas Cellular
Batelco
Bay Area Cellular
BCTel Cellular
Bell Atlantic Mobile
BellSouth Cellular
BellSouth Mobility
Bluegrass Cellular
Cantel
Canton Cellular
Cellcom of Green Bay
Cellcom of Hickory
Cellular, Inc.
Cellular Information Systems
Cellular of Indiana
Cellular One of Amarillo
Cellular One of
  Baltimore/Washington
Cellular One of Berrien County
Cellular One of Chicago
Cellular One of Cincinnati
Cellular One of Cleveland
Cellular One of Columbus
Cellular One of Detroit
Cellular One of Fredericksburg
Cellular One of Gainesville, GA
Cellular One of Gary, IN
Cellular One of Grand Rapids
Cellular One of Greenwood
Cellular One of Indianapolis
Cellular One of Lake Charles, LA
Cellular One of Lansing, MI
Cellular One of Milwaukee
Cellular One of Norfolk
Cellular One of Richmond
Cellular One of Sioux Falls
Cellular One of Syracuse
Cellular One of Utica
Cellular One of Wichita Falls
Cellular Plus
Cellular South
Coastel Communications
Contel Cellular
Crowley Cellular Telecomm
Cybertel Cellular
Enid Cellular
Gencell
GTE Mobilnet
Honolulu Cellular Telephone
Houston Cellular Telephone
Independent Cellular Network
Iowa East
Island Tel Cellular
Lincoln Telephone
Mactel

McCaw Cellular Communications
MCTA
Mercury Cellular
Metrocel Cellular
MTS Cellular
NewTel Cellular
OK Cellular
Palmer Communications
Petroleum Communications
Pioneer Cellular
Radiofone
Richmond Cellular Telephone
Savannah Cellular
U S West Cellular
United States Cellular
Unity Cellular Systems (Unicel)
Vanguard Cellular

Brazos Cellular Communications

ALLTEL Mobile Communications
Ameritech Rural Cellular
Ameritech Mobile Communications
Atlas Cellular
BellSouth Mobility
Bledsoe
Bluegrass Cellular
C. C. Cellular
Cal-One Cellular
Cellcom
Cellular, Inc.
Cellular Information Systems
Cellular One of Amarillo
Cellular One of Bowling Green
Cellular One of Columbus, IN
Cellular One of Enid
Cellular One of Gainesville
Cellular One of Greenville
Cellular One of Marianna
Cellular One of North/Central PA
Cellular One of Northeastern
  Mississippi
Cellular One of the Ozarks
Cellular One of Southwestern Florida
Cellular One of Wichita Falls
Cellular Three
Centel Cellular
Century Cellunet
Contel
Crowley Communications
Digital Cellular
Dobson Cellular
Enid Cellular
Farmer Cellular
First Kentucky Cellular
GTE Mobilenet
HBF Cellular
Independent Cellular
Illinois Valley Cellular
Interstate Cellular (Interstate Cellular
  (Intercel))
Kansas Cellular
Liberty Cellular
Lincoln Telephone
McCaw Cellular Communications
Mercury Cellular
Metrocel
Mid-South Cellular
Mid-Texas Cellular
Miscellcom
Mobiletel
NYNEX Mobile Communications
Pace Communications
PacTel Cellular
Panhandle Telecommunications
Peoples Cellular
Pioneer Cellular
PriCellular
Panhandle Telecommunication
Radiofone

Southeastern Cellular
Southwestern Bell Mobile
Texas 16 Cellular
Tsaconas Cellular
U S West
United States Cellular
Unitel
Valley Telecommunications
West Central Cellular
Westex
XIT Cellular
Yorkville Telephone

C. C. Cellular

Advantage Cellular
AGT Cellular
ALLTEL Mobile Communications
Ally
Alpha Cellular
Ameritech Mobile Communications
Appalachian Cellular
Atlantic Cellular
Bakersfield Cellular
Bell Atlantic Mobile
Bell Mobility Cellular
BellSouth Cellular
Bell South Mobility
Big Horn Cellular
Bledsoe Telephone
Bluegrass Cellular
Blue Ridge Cellular
BMCT
Brazos Cellular Communications
BCTel Mobility Cellular
Cal-One Cellular
Cantel Cellular
Canton Cellular
Cellcom
Cellcom of Hickory
Cellular Connection
Cellular Holdings
Cellular, Inc.
Cellular Information Systems
Cellular One of Amarillo
Cellular One of Berrien County
Cellular One of Chicago
Cellular One of Columbia
Cellular One of Columbia, TN
Cellular One of Danville, IL
Cellular One of Farmington Hills, MI
Cellular One of La Salle, IL
Cellular One of Marion
Cellular One of Meridian
Cellular One of Middle Tennessee
Cellular One of North Texas
Cellular One of Sioux Falls
Cellular One of Southwest Oklahoma
Cellular One of Stauton, VA
Cellular One of Western Illinois
Cellular One of Western Oklahoma
Cellular One of Worthington, OH
Cellular Phones of Kentucky
Cellular Plus
Cellular South
Cellular XL
Cellulink
Celutel
Centel Cellular
Century Cellunet
Coastel Communications
Contel Cellular
Cooper Cellular Management
C. V. Cellular
Danbury Cellular
Dicomm Cellular
Digital Cellular of Texas
Dobson Cellular
Dominion Cellular
Ed Tel Cellular

1102

Enid Cellular
Farmers Cellular
FGI Cellular
First Fayette Cellular
First Kentucky Cellular
Gaia
GMD
Greater South Dakota Cellular
Great Lakes of Iowa
GTE Mobilenet
Highland Cellular
Houston Cellular
Illinois Valley Cellular
Independent Cellular Network
Indianapolis Telephone
Interstate Cellular
Kansas Cellular
Kentucky Cellular
Lincoln Telephone
Litchfield County Cellular
Mac-Tel Cellular
McCaw Cellular Communications
Maritime Telephone & Telegraph
Mercury Cellular & Paging
Mercury Communications
Metro Mobile CTS
Mid-Missouri Cellular
Mid-South Cellular
Mississippi Cellular Telephone
Missouri Cellular
Mobile Communications Systems
Mobiletel
Montana Cellular Telephone
Mountaineer Mobile
MTS Cellular
National Cellular Limited Partnership
Nebrask Cellular
NYNEX Mobile Communications
Ocean Coutny Cellular
Oneonta Telephone
Pace Communications
Pacific Northwest Cellular
Pacific Telecom Cellular
Panhandle Telecommunications
People Cellular
Petroleum Communications
Pine Cellular
PriCellular
Public Service Cellular
Quantum Communications
Rochester Telephone
RSA II Partnership
Sasktel Cellular
Shenandoah Mobile
Springwich Cell. Ltd Partner. (SNET)
Southeastern Cellular
Southwestern Bell Mobile
StarCellular
Sunshine Cellular
Sussex Cellular
Texas 16 Cellular
Tsaconas Cellular
Thunder Bay Cellular
U S West Cellular
Ubet Cellular
Union Telephone
United State Cellular
Unity Cellular Systems (Unicel)
Universal Telecell
Valley Telecommunications
Vanguard Cellular
West Central Cellular
Westex Telecommunications
XIT Cellular

Cal-One Cellular

AGT Cellular
Advantage Cellular
Ameritech Cellular

Atlantic Cellular
B.C. Tel Mobility Cellular
Bell Atlantic Mobile
Bel Cellular
BellSouth Mobility
Blue Moutain Cellular
Brazos Cellular
C.C. Cellular
Cellcom of Greenbay
Cellular Connection
Cellular One of Alpena, MI
Cellular One of Cheboygon, MI
Cellular Holding
Cellular, Inc.
Cellular 3
Centel Cellular
Coastal Communications
Dobson Cellular Systems
EdTell Cellular
GenCell Management
GMD Limited Partnership
Illinois Valey Cellular
Independent Cellular Network
Inland Cellular
Kansas Cellular
Mactel Cellular Systems
Maine Cellular
McCaw Cellular Communications
Mercury Cellular & Paging
Mid-Missouri Cellular
Mississippi Cellular
Nebraska Cellular
Pacific Telecommunications
PacTel Cellular
Panhandle Telecommunications
Peoples Cellular
Petroleum Communications
PriCellular
Public Service Cellular
Quantum Communications
Rochester Telephone Mobile
  Communications
Sasktel Communications
Shenandoah Cellular
Springwich Cell. Ltd Partner. (SNET)
U S West Cellular
Union Cellular Telephone
United States Cellular
Unity Cellular Systems (Unicel)
Wes-Tex Telecommunications
XIT Cellular

Cantel

Advantage Cellular
Albany Telephone
Allcell
Allegan Cellular
ALLTEL Mobile Communications
Ally
Ameritech Mobile Communications
Associated Communications
Atlantic Cellular
Auburn Television
Bahamas Telecommunications
Bakersfield Cellular
Bay Area Cellular
Bell Atlantic Mobile
Bell South Mobility
Bledsoe Telephone
Blue Mountain Cellular
Blue Ridge Cellular
Bluegrass Cellular
Boston Cellular Telephone
Buckhead Telephone
Buffalo Telephone
C. C. Cellular
Canton Cellular
Cedetel
Cellcom of Green Bay

Cellcom of Hickory
Cellular 2000
Cellular De Telefonia, S.A. DE C.V.
Cellular Holding
Cellular Information Systems
Cellular One of Amarillo, TX
Cellular One of
  Baltimore/Washington
Cellular One of Baton Rouge, LA
Cellular One of Beckley, WV
Cellular One of Berrien County
Cellular One of Brookings, SD
Cellular One of Chicago, IL
Cellular One of Columbia, TN
Cellular One of Detroit, MI
Cellular One of East/Central, PA
Cellular One of Frederick
Cellular One of Galesburg, IL
Cellular One of Huntsville, TX
Cellular One of Indiana
Cellular One of Indianapolis, IN
Cellular One of Jackson, MS
Cellular One of La Salle
Cellular One of Missoula, MT
Cellular One of New Jersey
Cellular One of New York, NY
Cellular One of North Texas
Cellular One of Puerto Rico
Cellular One of Richmond
Cellular One of Roanoke
Cellular One of Santa Cruz, CA
Cellular One of Sioux Falls
Cellular One of Southwest Florida
Cellular One of Southwest Oklahoma
Cellular One of the Ozarks
Cellular One of the Rio Grande Valley
Cellular One of Upstate New York
Cellular One of Virginia
Cellular One of Western Oklahoma
Cellular One of Wichita Falls
Cellular One of Wilmington
Cellular Plus
Cellular South
Cellular Systems International
Cellular, Inc.
Cellulink
Cellwave
Celutel
Centel Cellular
Centennial Cellular
Century Telephone
Clear Communications
Comcast
Comcast Cellular
Contel Cellular
Cooper Cellular
Crowley Cellular
Cybertel Cellular
Danbury Cellular
Danville Cellular
Digital Cellular of Texas
Dominion Cellular
Enid Cellular
Farmers Cellular
FGI Cellular
Finger Lakes Telephone
General Cellular
GMD Partnership
GTE Mobile Communications
Honolulu Cellular Telephone
Houston Cellular
Illinois Valley Cellular
Iowa East Cellular
IUSA Cell
Kansas Cellular
Lake Huron Cellular
Lincoln Telephone
Los Angeles Cellular
Mackinac Cellular

Maine Cellular
McCaw Communications
MCMG
Mega Comm
Mercury Cellular and Paging
Mercury Communications
Metrocel Cellular
Metrophone
Midsouth Cellular
Milwaukee Telephone
Miscellco Communications
Missouri Cellular
Mountaineer Mobile
National Cellular Limited Partnership
New Par
NYNEX Mobile Communications
Ocean County Cellular
Ohio State Cellular
P. C. Cellular of Vermont
Pace Communications
Pacific Northwest Cellular
PacTel Cellular
Palmer Cellular
Petroleum Communications
Pioneer Telephone
PriCellular
Public Service Cellular
Quantum Communications
Radiofone
Radiomovile Dipsa, S.A. DE C.V.
RFB Cellular
Richmond Cellular
Rochester Telephone
Rural Cellular Management
Springwich Cell. Ltd Partner. (SNET)
Southeast Indiana Cellular
Southeastern Cellular
Southwestern Bell Mobile Systems
StarCellular
Sterling Cellular
Sussex Cellular
Syracuse Cellular Telephone
Texoma Cellular
Tsaconas Cellular
U S West Cellular
Union Telephone
United State Cellular
Unitel
Utica Cellular Telephone
Unity Cellular Systems (Unicel)
Vanguard Cellular
Virginia 10 RSA Limited Partnership
Yorkville Telephone
Youngstown Cellular Telephone

Carolina West Cellular

ALLTEL Mobile Communications
Ameritech Mobile
Atlantic Cellular
Bell Atlantic
BellSouth Mobility
Bledsoe Telephone
Blue Ridge Cellular
Bluegrass Cellular
Cellcom of Green Bay
Cellular Connection
Cellular Phone of Kentucky
Centel Cellular
Contel Cellular
Farmers Cellular Telephone
GTE Mobilnet
Independent Cellular Network
Kaplan Telephone
Kentucky Cellular
Lincoln Telephone
Mercury Cellular
NYNEX Mobile Communications
Sterling Cellular
Union Telephone

United States Cellular

Cellcom (of Green Bay)

AGT Cellular
Albany Telephone
ALLTEL Mobile Communications
Ally
Alpha Cellular
Ameritech Mobile Communications
American Rural Cellular
Appalachian Cellular
Atlantic Cellular
Atlas Cellular
Auburn Television
Bahamas Telecommunications
Bakersfield Cellular
Baton Rouge Cellular
BCTel Mobility Cellular
Bell Atlantic Mobile
Bell Mobility Cellular
Bell Mobility Cellular
BellSouth Mobility
Benton Harbor Cellular
Bledsoe Cellular
Bluegrass Cellular
Blue Mountain Cellular
Blue Ridge Cellular
Boat Phone
Brazos Cellular Communications
Buckhead Telephone
Buffalo Telephone
C. C. Cellular
Cal-One Cellular
Cantel
Canton Cellular
Carolina West Cellular
Cellcom
Cellcom of Hickory
Cellular Connection
Cellular Holdings
Cellular, Inc.
Cellular Information Systems
Cellular of Indiana
Cellular One of Amarillo
Cellular One of Beaumont
Cellular One of Central Illinois
Cellular One of Chicago
Cellular One of Dobson, OK
Cellular One of Galesburg, IL
Cellular One of Gainesville,GA
Cellular One of Grand Forks, WY
Cellular One of Huntsville, TX
Cellular One of Lake Charles, LA
Cellular One of La Salle, IL
Cellular One of Northeast Colorado
Cellular One of Northeast Mississippi
Cellular One of Marion
Cellular One of Meridian
Cellular One of Ohio
Cellular One of the Rio Grande Valley
Cellular One of Sioux Falls, SD
Cellular One of Upstate New York
Cellular Phone of Kentucky
Cellular Plus
Cellular South
Cellular Three
Cellular XL
Cellutel
Centel Cellular
Centennial Cellular
Century Cellunet
Coastel Communications
Contel Cellular
Cross-Valliant Cellular
Crowley Cellular
Cybertel Cellular
Dominion Cellular
Ed Tel Cellular
Enid Cellular

Farmers Collular
FGI Cellular
Five Star Cellular
Galveston Cellular
General Cellular Genessee
  Telephone
GM Limited Partnership
Great Lakes of Iowa
Greater South Dakota Cellular
GTE Mobilnet ("B" systems only)
Highland Cellular
Honolulu Cellular
Horizon Cellular
Houston Cellular
Hudson Valley Cellular
Independent Cellular Network
Illinois Valley Cellular
Indianapolis Telephone
Interstate Cellular (Intercel)
Iowa East Cellular
Kansas Independent
Kaplan Telephone
Kuai Cellular One, MO
Lake Huron Cellular
Lincoln Telephone
Litchfield County Cellular
Lone Star Cellular
Mackinac Cellular
MacTel Cellular
Maine Cellular
Maritime Telephone
McCaw Communications
Mercury Communications
Mercury Cellular
MetaComm Cellular
Metrocel Cellular
Metrophone
Mid-Missouri Cellular
Mid-South Cellular
Milwaukee Telephone
Minnesota Cellular
Miscellco Communications
Missouri Cellular
Mobile Communications Systems
Mobiletel
Montana Cellular
Mountaineer Cellular
MTS Cellular
NCPT Cellular
Nebraska Cellular
New Par
NYNEX Mobile Communications
Oneonta Telephone
Pacific Northwest Cellular
Pacific Telecom Cellular
PacTel Cellular
Palmer Communications
Panhandle Telecommunications
PC Cellular of Kentucky
PC Cellular of Vermont
Petroleum Communications
Poka Lambro Telecommunications
PriCellular
Public Service Cellular
Québéc Téléphone Cellulaire
Quantum Communications
Quest West Alabama Cellular
Radiofone
RFB Cellular
Richmond Cellular
Rochester Telephone Mobile
  Comms.
Rural Cellular Management
Sagir
Santa Canta Cruz
SaskTel Cellular
Savannah Cellular
Springwich Cell. Ltd Partner. (SNET)
Southeastern Cellular
Southwest Alabama

Southwestern Bell Mobile Systems
StarCellular
Texas 16 Cellular
Texas Cellular
Thunder Bay Cellular
Tsacona Cellular
U S West Cellular
Union Telephone
Unitel
Uintah Basin
United States Cellular
Utica / Rome Telephone
Unity Cellular Systems (Unicel)
Valley Telecommunications
Vanguard Cellular
Virginia RSA 10
WesTex Cellular
Wichita Falls Cellular
Yorkville Telephone
XIT Cellular

Cellcom of Hickory

Albany Telephone
Atlantic Cellular
Bell Atlantic Mobile Systems
BellSouth Mobility
Buffalo Telephone
Cantel
Cellular Communications
Cellular Information Systems
Cellular One of Atlantic City, NJ
Cellular One of
  Baltimore/Washington
Cellular One of Chicago, IL
Cellular One of Indianapolis
Cellular One of Lake Charles, LA
Cellular One of La Salle, IL
Cellular One of Milwaukee
Cellular One of Wilmington, DE
Century Cellular
Comcast Cellular
Contel Cellular
Crowley Cellular Telecommunications
Cybertel Cellular
GTE Mobilnet
GenCell Management
Genesee Telephone
Gulf Coast Cellular
McCaw Communications
MetroCel Cellular
Metro One
Metro Mobile CTS
Metrophone
NYNEX Mobile Communications
PacTel Cellular
Palmer Communications
Providence Journal Cellular
Radiofone
Richmond Cellular
United States Cellular
Vanguard Cellular

Cellular Connection

Advantage Cellular
ALLTEL Mobile Communications
Ameritech Mobile Communications
Appalachian Cellular
Atlantic Cellular
B.C. Tel Mobility Cellular
Bell Mobility Cellular
BellSouth Mobility
Bledsoe Telephone
Blue Ridge Cellular
Blue Ridge Cellular Telephone
Bluegrass Cellular
Cal-One Cellular
Canton Cellular
Carolina West Cellular

C.C. Cellular
Cellcom
Cellular
Cellular One of Enid
Cellular One of Indiana
Cellular One of Indianapolis
Cellular One of Iowa
Cellular One of Meridian
Cellular One of New Jersey
Cellular One of New York
Cellular One of Norhteast Colorado
Cellular One of North Texas
Cellular One of Northeast Georgia
Cellular One of Ohio
Cellular One of Ozarks
Cellular One of Richmond KY
Cellular One of SC
Cellular One of West Oklahoma
Cellular One of West Virginia
Cellular Plus Georgia
Cellular Plus Iowa
Cellular South
Cellular XL
Centel Cellular
Century Cellunet
Coastel
Contel Cellular
Crowley Cellular
Cybertel Cellular
Dial Two
Dicomm Cellular
Digital Cellular Texas
Dobson Cellular Systems
Farmers Cellular Telephone
First Fayette Cellular
Gaia
Greater South Dakota Cellular
GTE Mobile Communications
Illinois Valley Cellular
InterCel Cellular
Iowa East Cellular Telephone
Kansas Cellular
Lincoln Telephone Cellular
Mercury Cellular & Paging
Metro-Cel Cellular
Mid-Missouri Cellular
Milwaukee Telephone
Mobile Communications
MobileTel
Mountaineer Celluar
MTS Cellular
Nebraska Cellular
NYNEX Mobile Communications
OK Cellular
Pace Communications
Pacific Telecom Cellular
Panhandle Telecommunications
Pegasus Cellular
Petroleum Communications Offshore
  Cellular
Pine Cellular Phones
PriCellular
Public Service Cellular
Quantum Communications
Québéc Téléphone Cellulaire
R & D Cellular
Richmnond Cellular
Sasktel Cellular
Savannah Cellular
Southeast Cellular
Southwestern Bell Mobile
StarCellular
Sterling Cellular
Texoma Cellular
Tsaconas Cellular
U S West Cellular
Union Telephone
United States Cellular
Unity Cellular Systems (Unicel)
Vanguard Cellular Systems

Virginia 10
Westex Cellular
XIT Cellular
Yorkville Telephone

Cellular Holding

AGT Cellular
ALLTEL Mobile Communications
Ally
American Rural Cellular
Ameritech Mobile Communications
Appalachian Cellular
Atlantic Cellular
Auburn Television
Bahamas Telecommunications
Bell Atlantic Mobile
Bell Mobility Cellular
Bell South Cellular
Bell South Mobility
Big Horn Cellular
Bledsoe
Blue Mountain Cellular
Blue Ridge Cellular of Virginia
Bluegrass Cellular
C. C. Cellular
Cal-One Cellular
Cantel Cellular
Canton Cellular
Carolina West Cellular
Cellcom
Cellular Connection
Cellular Information Systems
Cellular One of Baton Rouge
Cellular One of Central Illinois
Cellular One of Chicago
Cellular One of Columbia, TN
Cellular One of Danville, IL
Cellular One of East/Central PA
Cellular One of Enid
Cellular One of Galveston
Cellular One of Great Lakes of Iowa
Cellular One of Huntsville, TX
Cellular One of Lake Charles, LA
Cellular One of Marion, IN
Cellular One of Meridian
Cellular One of Middle Tennessee
Cellular One of Northeast Arizona
Cellular One of Northeast Mississippi
Cellular One of the Rio Grande Valley
Cellular One of Southwest Florida
Cellular One of Southwest Georgia
Cellular One of Western Illinois
Cellular Phone of Kentucky
Cellular Plus
Cellular Plus of Georgia
Cellular Properties
Cellular Ventures
Cellular, Inc.
Celutel
Centel Cellular
Century Cellunet
Chariton Valley Cellular
Clear Communications
Coastel Communications
Columbia Cellular Partnership
Comp Comm
Contel Cellular
Cross-Valliant Cellular
Crowley Cellular
Cybertel Cellular
Digital Cellular of Texas
Dobson Cellular
Dominion Cellular
Eastex
Enid Cellular
Farmers Cellular
FGI Cellular
First Kentucky Cellular
Five Star Cellular

Roaming Agreements

Gaia
GenCell Management
GMD Partnership
GTE Mobilnet
Highland Cellular
Houma Cellular
Houston Cellular Telephone
Illinois Valley Cellular
Independent Cellular Network
Interstate Cellular (Intercel)
Iowa East Cellular Telephone
Kansas Cellular
Kaplan Cellular
Lincoln Telephone
Litchfield County Cellular
Mackinac Cellular
MacTel
McCaw Cellular Communications
Mercury Cellular
Mercury Communications
Metrocel Cellular
Mid-Missouri Cellular
Miscellco Communications
Mississippi Cellular
Missouri Cellular
Mobile Communications
Mobiletel
Mountaineer Cellular
MT&T Cellular
MTS Cellular
National Cellular
Northwest Cellular
NYNEX Mobile Communications
Oneonta Cellular
Pace Communications
Pacific Northwest Cellular
PacTel Cellular
Palmer Communications
Panhandle Cellular
PC Cellular of Vermont
People Cellular
Petroleum Communications
Pine Cellular
PriCellular
Public Service Cellular
Quantum Communications
Québec Téléphone Cellulaire
Radiofone
Rochester Telephone
Shenandoah Cellular
Southeast Indiana Cellular
Southeastern Cellular
Southwestern Bell
Springwich Cell. Ltd Partner. (SNET)
StarCellular
Stillwater Cellular
Sunshine Cellular
Texas 16 Cellular
Thunder Bay Cellular
Tsaconas Cellular
U S West Cellular
Union Telephone
United States Cellular
Valley Telephone
Vanguard Cellular
West Alabama Cellular
West Central Cellular
XIT Cellular
Yorkville Telephone

Cellular, Inc. (Commnet 2000)

AGT Cellular
ALLTEL Mobile Communications
Ameritech Mobile Communications
Atlantic Cellular
Bell Atlantic Mobile
Bell Mobility Cellular
BellSouth Mobility
Bledsoe

Cal-One Cellular
Cantel
C.C. Cellular
Cellcom
Cellular Information Systems
Cellular One Danville
Cellular One of Chicago
Cellular One of Columbus, MO
Cellular One of Indianapolis
Cellular One of LaSalle
Cellular One of Milwaukee
Cellular One of Sioux Falls
Cellular One of Upstate New York
Cellular One of Wichita Falls
Cellular South
Cellulink
Centel Cellular
Centennial Cellular
Century Cellunet
Contel Cellular
Digital Cellular
Ed Tel Cellular
Enid Cellular
Five Star Cellular
GenCell Management
Greater South Dakota Cellular
GTE Mobilenet
Independent Cellular Network
Kansas Cellular
Lincoln Telephone
Mactel Cellular
McCaw Cellular Communications
Mercury Cellular
Mid-Missouri Cellular
Miscellco Communications
Mobilesouth Cellular
MTS Cellular
Nebraska Cellular
NYNEX Mobiel Communications
Pace Communications
PacTel Cellular
Petroleum Communication
PTSI Cellular
Richmond Cellular Telephone
Rochester Telephone
Sasktel Cellular
Savannah Cellular
Southwestern Bell Mobile Systems
Springwich Cell. Ltd Partner. (SNET)
StarCellular
Sterling Cellular
Texas Cellular
U S West Cellular
UBET Cellular
Union Telephone
United States Cellular
Unitel
Vanguard Cellular
Westex
XIT Cellular

Cellular Information Systems

Advantage Cellular
AGT Cellular
Albany Telephone
Alley
ALLTEL Mobile Communications
Alpha Cellular
AMC Cellular
Ameritech Mobile Communications
Appalachian Cellular
Associated Communications
Atlantic Cellular
Atlas Cellular
Auburn Television
Bakersfield Cellular
Bay Area Cellular
Bell Atlantic Mobile
Bell Mobility Cellular

BellSouth Mobility
Benton Harbor Cellular
Bledsoe Telephone
Bluegrass Cellular
Blue Mountain
Blue Ridge Cellular
Brazos Cellular
B.C. Tel Mobility Cellular
C. C. Cellular
Cantel
Canton
Cellcom of Hickory
Cellcom of Greenbay
Cellular Holdings
Cellular, Inc.
Cellular of Indiana
Cellular One Central
Cellular One of Allegan
Cellular One of Amarillo
Cellular One of Atlantic City
Cellular One of
  Baltimore/Washington
Cellular One of Baton Rouge
Cellular One of Boston
Cellular One of Brookings
Cellular One of Brownsville
Cellular One of Buffalo
Cellular One of Chicago
Cellular One of Collier County
Cellular One of Columbia
Cellular One of Danbury
Cellular One of Danville
Cellular One of Dobson
Cellular One of Dumas
Cellular One of Erie, PA
Cellular One of Frederick
Cellular One of Gainesville
Cellular One of Galesburg
Cellular One of Galveston
Cellular One of Greenville
Cellular One of Grandforks
Cellular One of Huntsville
Cellular One of Indianapolis
Cellular One of Lake Charles
Cellular One of La Salle
Cellular One of Marion
Cellular One of Meridian
Cellular One of Middle Tennessee
Cellular One of Milwaukee
Cellular One of New Brunswick
Cellular One of New Jersey
Cellular One of North Texas
Cellular One of Northeast Colorado
Cellular One of Northeast Texas
Cellular One of Ocean County
Cellular One of Ohio
Cellular One of Richmond
Cellular One of the Rio Grande Valley
Cellular One of Scottsbluff
Cellular One of Sioux Falls
Cellular One of Springfield
Cellular One of Stauton
Cellular One of Syracuse
Cellular One of Upstate New York
Cellular One of Watertown
Cellular One of Western Oklahoma
Cellular One of Wilmington
Cellular One of Utica
Cellular One of Victoria
Cellular One of Wichita Falls, TX
Cellular One of Williamsburg
Cellular One of Youngstown
Cellular Plus
Cellular South
Cellular Systems International
Cellular Telephone of Kentucky
Cellular Three
Cellular Ventures
Cellular XL

Roaming Agreements

Cellulink
Centel Cellular
Centennial
Century Cellular
Century Cellunet
Clear Communications
Coastel Communications
Codetel
Comcast Cellular
Contel Cellular
Crowley Cellular Telecommunications
Cybertel Cellular Telephone
Detroit Cellular
Dial Two
Dicomm Cellular
Digital Cellular
Dobson Cellular
Dominion Cellular
Ed Tel Cellular
Enid Cellular
Farmers Cellular
FGI Cellular
Five Star Cellular
GenCell Management
General Cellular
Genessee Telephone
Greater South Dakota Cellular
GTE Mobilnet ("B" systems only)
Gulf Coast Cellular
Highland Cellular
Honolulu Cellular
Horizon Cellular
Houston Cellular Telephone
Hudson Cellular
Independent Cellular
Indianapolis Telephone
Interstate Cellular (Intercel)
Iowa East Cellular
Kansas Cellular
Kansas Independent Networks
Kaplan Telephone
Los Angeles Cellular
Lake Huron Cellular
Lincoln Telephone
Litchfield County Cellular
Louisiana Cellular
MacTel Cellular
Maritime Telephone and Telegraph
McCaw Communications
Mercury Cellular
Mercury Communications
Metacomm Cellular
Metro Mobile CTS
Metro One
MetroCel Cellular
Metrophone
Mid-South Cellular
Milwaukee Telephone
Minnesota Southern Cellular
Miscellco
Missouri Cellular
Mobile Communications
MobileTel
Montana Cellular
Mountaineer Mobile
MTS Cellular
National Cellular
Nebraska Cellular
New Par
NYNEX Mobile Communications
Oneonta Telephone
Pacific Telecom Cellular
PacTel Cellular
Palmer Communications
Panhandle Communications
Petroleum Communications
Pine Cellular
Pioneer Cellular
Poka Lambro Telecommunications
Public Service Cellular

Quantum Communications
Québéc Téléphone Cellulaire
R & D Cellular
Radiofone
RFB Cellular
Richmond Cellular
Rochester Telephone
Rural Cellular
Sagir
Santa Cruz Cellular
Savannah Cellular
Springwich Cell. Ltd Partner. (SNET)
Sooner
Southeastern Arizona Cellular
Southeastern Cellular
Southwestern Bell Mobile Systems
StarCellular
Sterling Cellular
Sussex Cellular
Texas 16 Cellular
Texas Cellular
Thunderbay Telephone
Tsaconas Cellular
U S West Cellular
Union Telephone
Unitel
United States Cellular
Valley Telecommunications
Vanguard Cellular Systems
Victoria Cellular
Virginia 10 RSA
Virginia Cellular
West Central Cellular
Wes-Tex
Wichita Falls Cellular
XIT Cellular ("B" side)
Yorkville Telephone
Youngstown Cellular

Cellular One of Alpena, MI (RFB Cellular)

Adirondack Cellular
Albany Telephone
Allegan Cellular
ALLTEL Mobile Communications
Alpine Cellular
Auburn Television
Ameritech Mobile
Bakersfield Cellular
Bell Atlantic Mobile
Bell South Cellular
BellSouth Mobility
Blackwater Cellular
Bledsoe Telephone
Bluegrass Cellular
Buckhead Telephone
C. C. Cellular
Cal-One Cellular
Cantel
Canton Cellular
Cellcom
Cellular 2000
Cellular Connection
Cellular, Inc.
Cellular Information Systems
Cellular One of Berrien County
Cellular One of Boston, MA
Cellular One of Buffalo, NY
Cellular One of Chicago
Cellular One of Columbia, TN
Cellular One of Columbus, IN
Cellular One of Danville
Cellular One of East/Centel PA
Cellular One of Indiana
Cellular One of Indianapolis
Cellular One of La Salle
Cellular One of Middle Tennessee
Cellular One of Sioux Falls, SD
Cellular One of Southwest Florida

Cellular One of West Illinois
Cellular Phone of Kentucky
Centel Cellular
Centennial Cellular
Century Cellunet
Contel Cellular
Crowley Cellular
CJD Wisconsin Cellular
Cybertel
Dicomm Cellular
Easterbrooke Cellular
Farmers Cellular
Finger Lakes Telephone
GenCell Management
Genessee Telephone
Greater South Dakota Cellular
GTE Mobilnet
HLD Cellular
Horizon Cellular
Houston Cellular
Independent Cellular
Kentucky Telephone
Lake Huron Cellular
Mackinac Cellular
McCaw Cellular Communications
Mega Comm
Mercury Communications
Metrophone
Mid-South Cellular
Milwaukee Telephone
MT&T
NYNEX Mobile Communications
Pace Communications
Pacific Telecom Cellular
PacTel Cellular
Palmer Cellular
PC Cellular of Vermont
Quantum Communications
Southeastern Cellular
Southeast Indiana Cellular
Springwich Cellular
Sterling Cellular
Syracuse Telephone
U S West Cellular
Unitel
United States Cellular
Utica Telephone
Vanguard Cellular
Virginia Cellular
Yorkville Telephone

Cellular One of Amarillo, TX

Allcell
Associated Communications
Bakersfield Cellular
Bay Area Cellular
Cellular Information Systems
Cellular One of Baltimore/Washington
Cellular One of Beaumont
Cellular One of Berrien County
Cellular One of Chicago/Gary
Cellular One of Fort Myers
Cellular One of Galveston
Cellular One of Lake Charles
Cellular One of Milwaukee
Cellular One of Syracuse
Cellular One of Utica
Cellular One of Victoria
Cellular One of Wichita Falls, TX
Cellular One of Wilmington
Celutel
Centel Cellular
Contel Cellular
Crowley Cellular Telecommunications
Cybertel Cellular Telephone
FGI Cellular Management
GTE Mobilnet ("B" side)
GenCell Management

Roaming Agreements

Houston Cellular Telephone
Los Angeles Cellular
MTS Cellular
McCaw Communications
Metro Mobile CTS
Metro One
MetroCel Cellular
Mobiltel
PacTel Cellular
Palmer Communications
Petroleum Communications
U S West Cellular
United States Cellular
Vanguard Cellular Systems

Cellular One of Baltimore/ Washington

Allcell
ALLTEL Mobile Communications
Bay Area Cellular
Bell Atlantic Metro Mobile
Cantel
CCI
Cellcom of Hickory
Cellular Information Systems
Cellular One of Amarillo
Cellular One of Atlantic City
Cellular One of Berrien County
Cellular One of Boston
Cellular One of Chicago
Cellular One of Detroit
Cellular One of Indianapolis
Cellular One of Jacksonville
Cellular One of La Salle, IL
Cellular One of Lake Charles
Cellular One of New Brunswick
Cellular One of Richmond
Cellular One of Sioux Falls
Cellular One of Syracuse/Utica
Cellular One of Wichita Falls, TX
Cellular One of Wilmington
Cellular One of Wilmington N.C.
Cellular One of Youngstown
Centel
Century Cellular
Commonwealth Mobile Services
Contel
Crowley Cellular
Cybertel
GTE Mobilnet ("B" side)
GenCell Management
Genesee Telephone
Gulf Coast Cellular
Honolulu Cellular Telephone
Houston Cellular
Independent Cellular network
Los Angeles Cellular
McCaw Cellular Communications
Metro One
MetroCel
Metro Mobile
Metrophone
Milwaukee Telephone
New Par
NYNEX Mobile Communications
PacTel Cellular
Palmer Communications
Petroleum Communications
Radiofone
Southwestern Bell Mobile
Springwich Cell. Ltd Partner. (SNET)
U S West Cellular
United States Cellular
Vanguard Cellular

Cellular One of Berrien County

Albany Telephone
Allcell

Ameritech Mobile Communications
Associated Communications
Buffalo Telephone
Cellcom
Cellular Information Systems
Cellular One of Akron
Cellular One of Amarillo
Cellular One of Bakersfield
Cellular One of Baltimore/Washington
Cellular One of Beaumont
Cellular One of Canton
Cellular One of Chicago
Cellular One of Cleveland
Cellular One of Detroit
Cellular One of Elyria
Cellular One of Flint
Cellular One of Grand Rapids
Cellular One of Indianapolis
Cellular One of Lake Charles
Cellular One of Lansing
Cellular One of La Salle, IL
Cellular One of Lima, OH
Cellular One of Lorain, OH
Cellular One of Milwaukee, WI
Cellular One of of the Rio Grande Valley
Cellular One of Roanoke
Cellular One of Rome
Cellular One of Syracuse
Cellular One of Utica
Cellular One of Victoria
Cellular One of Wichita Falls, TX
Cellular One of Youngstown
Centel Cellular
Century Cellular
Comcast
Contel Cellular
Crowley Cellular Telecommunications
Cybertel Cellular Telephone
GTE Mobilnet
GenCell Management
Genesee Telephone
Gulf Coast Cellular
Houston Cellular Telephone
Jackson Cellular
Los Angeles Cellular
McCaw Communications
Metro Mobile CTS
Metro One
Metrocel Cellular
Metrophone
NYNEX Mobile Communications
PacTel Cellular
Palmer Communications
Petroleum Communications
Radiofone
U S West Cellular
United States Cellular
Vanguard Cellular Systems

Cellular One of Boston, MA

Associated Communications
Bay Area Cellular
Cantel
Cellular One of Baltimore/WashingtonCellular One of Baton Rouge
Cellular One of Chicago
Cellular One of Detroit
Cellular One of Grand Rapids
Cellular One of Lansing
Cellular One Santa Cruz
Cellular One of Syracuse
Comcast
Contel
Crowley Cellular Telecommunications
Honolulu Cellular
Houston Cellular

Indianapolis Telephone
McCaw Communications
Metro Mobile CTS
Metro One
MetroCel
Metrophone
Milwaukee Telephone
New Par
Palmer Communications
Radiofone
Southwestern Bell Mobile Systems
Springwich Cell. Ltd Partner (SNET)
U S West Cellular
United States Cellular
Vanguard Cellular Systems

Cellular One of Chicago/Gary

Allcell
ALLTEL Mobile Communications
Associated Communications
Bay Area Cellular
Bell Atlantic Mobile
BellSouth Mobility
Cantel
Cellcom
Cellular Information Systems
Cellular One of Akron
Cellular One of Amarillo
Cellular One of Augusta
Cellular One of Baltimore/Washington
Cellular One of Beaumont
Cellular One of Boston
Cellular One of Canton
Cellular One of Cedar Rapids
Cellular One of Cincinnati
Cellular One of Cleveland
Cellular One of Columbus
Cellular One of Dayton
Cellular One of Detroit
Cellular One of Elmira
Cellular One of Flint
Cellular One of Grand Rapids
Cellular One of Hamilton
Cellular One of Indianapolis
Cellular One of Jacksonville
Cellular One of Lansing
Cellular One of Mansfield
Cellular One of Milwaukee
Cellular One of Muskegon
Cellular One of Syracuse
Cellular One of Toledo
Cellular One of Utica
Cellular One of Wichita Falls, TX
Cellular One of Wilmington
Cellular One of Worcester
Cellular One of Youngstown
Cellular Plus
Cellulink
Centel Cellular
Century Cellular
Century Cellunet
Comcast
Contel Cellular
Crowley Cellular Telecommunications
Cybertel Cellular Telephone
Enid Cellular
GTE Mobilnet ("B" systems only)
Gulf Coast Cellular
Houston Cellular Telephone
Los Angeles Cellular
MCTA
MTS Cellular
Maine Cellular
McCaw Communications
Metro Mobile CTS
Metro One
MetroCel Cellular
Metrophone

Roaming Agreements

PacTel Cellular
Palmer Communications
Radiofone
Southwestern Bell Mobile Systems
Springwich Cell. Ltd Partner. (SNET)
United States Cellular
Vanguard Cellular Systems

Cellular One of Columbia, MO

Albany Telephone Company
Allegan Cellular
ALLTEL Mobile Communications
Alpha Cellular
Ameritech Mobile Communications
Appalachian Cellular
Atlantic Cellular
Bakersfield Cellular
Baton Rouge Cellular
Bell Atlantic Mobile
BellSouth Cellular
BellSouth Mobility
Big Horn Cellular
Blackwater Cellular
Bledsoe Telephone Company
Blue Mountain Cellular
Blue Ridge Cellular
Bluegrass Cellular
Buffalo Telephone Company
C. C. Cellular
Cantel
Canton Cellular
Carolina WST Cellular
Clear Communities
Cellcom of Green Bay
Cellcom of Hickory
Cellcom of Kentucky
Cellular 2000
Cellular Connection
Cellular Holdings
Cellular Inc.
Cellular Information Systems
Cellular One of Amarillo
Cellular One of
  Baltimore/Washington
Cellular One of Bay Area
Cellular One of Berrien County
Cellular One of Casper
Cellular One of Chicago
Cellular One of Columbia, TN
Cellular One of Danbury
Cellular One of Danville
Cellular One of Dobson
Cellular One of Dominion
Cellular One of East/Central PA
Cellular One of Galesburg
Cellular One of Galveston
Cellular One of Grand Forks
Cellular One of Huntsville
Cellular One of Indianapolis
Cellular One of Lake Charles
Cellular One of La Salle, IL
Cellular One of Meridian
Cellular One of Mid-Tennessee
Cellular One of Missoula
Cellular One of Northeast Georgia
Cellular One of North Central PN
Cellular One of Sioux Falls
Cellular One of South Dakota
Cellular One of Southern Minnesota
Cellular One of Southwest Florida
Cellular One of Southwest Oklahoma
Cellular One of Texas 17
Cellular One of The Rio Grand Valley
Cellular One of Upstate New York
Cellular One of Virginia
Cellular One of Waycross/Vidalia
Cellular One of Western Oklahoma
Cellular One of Wichita Falls
Cellular One of/GMD

Cellular Plus
Cellular Ventures
Cellular XL
Celutel
Celutel of Biloxi
Centel
Centennial Cellular
Century Cellunet
Chariton Valley Cellular
Clear Communications
Coastel Communications
Comcast Cellular Communications
Contel Cellular
Cooper Cellular Management
Crowley Cellular
Cybertel Cellular
Detroit Cellular
Dobson Cellular
Eastex Cellular
Farmers
Finger Lakes Telephone
First Kentuck Cellular
Gaia
GenCell Management
Genessee Telephone
GMD Limited Partnership
GTE Mobile Communications
Highland Cellular
Honolulu Cellular
Houston Cellular
Independent Cellular
Illinois Valley Cellular
Interstate Cellular (Intercel)
Iowa East Cellular
Jackson Cellular
Kansas Cellular
Kentucky Cellular
Lake Huron Cellular
Lincoln Telephone
Litchfield County Cellular
McCaw Cellulr
Mercury Communications
Metro Mobile
Metrocel
Mid-Missouri Cellular
Mid-South Cellular
Milwaukee Telephone
Miscellco Communications
Mississippi Cellular
Missouri Cellular
Mobiletel
National Cellular Limited
Nebraska Cellular
New Par
NYNEX Mobile Communications
Pace Communications
Pacific Telecom Cellular
PacTel Cellular
Palmer Cellular
Panhandle Cellular
Petroleum Communications
Pine Cellular
Pricecellular
Quantum Communications
Radiofone
Richmond Cellular
Rural Management
Savannah Cellular
Southern Cellular
Southwestern Bell
Sterling Cellular
Sussex Cellular
Syracuse Cellular
Texas 16 Celllular
Texoma Cellular
Tsaconas Cellular
U S West
Union Telephone
United States Cellular

Unitel
Utica Telephone
Vanguard Cellular
Victoria Cellular
X-Cell Cellular
Yorkville Telephone

Cellular One of Demopolis, AL

Advantage Cellular
ALLTEL Mobile Communications
Ally
Associated Communications
Bakersfield Cellular
Batelco
Bay Area Cellular
Bell Atlantic Mobile
BellSouth Mobility
Big Horn Cellular
Blue Mountain Cellular
Blue Ridge Cellular Telephone
Bluegrass Cellular
Cellcom of Green Bay
Cellcom of Hickory
Cellular of Indiana
Cellular One of Amarillo
Cellular One of Berrien County
Cellular One of Brookings
Cellular One of Chicago
Cellular One of Columbia, TN
Cellular One of East/Central PA
Cellular One of Gainesville, GA
Cellular One of Galesburg
Cellular One of Galveston, TX
Cellular One of Lake Charles, LA
Cellular One of LaSalle
Cellular One of Marianna, FL
Cellular One of Milwaukee
Cellular One of Santa Cruz
Cellular One of Scottsbluff
Cellular One of Sioux Falls
Cellular One of Upstate New York
Cellular Plus
Cellular Properties
Cellular South
Cellular Systems International
Cellular Ventures
Cellular XL
Celutel
Centel Cellular
Centennial Cellular
Century Celunet
Coastel Communications
Cone Enterprises
Crowley Cellular Telecommunications
Dominion Cellular
Enid Cellular
Farmers Cellular Telephone
GenCell Management
GMD
GTE Mobilnet
Honolulu Cellular Telephone
Houston Cellular Telephone
Independent Cellular Network
Interstate Cellular (Interstate Cellular
  (Intercel))
Lake Huron Cellular
Los Angeles Cellular
Mactel Cellular
McCaw Cellular Communications
Mercury Cellular
Mercury Communications
Metrocel Cellular
Metrophone
Miscellco Communications
Mississippi Cellular
Mobiletel
MT&T
Northwest Cellular
PacTel Cellular

Roaming Agreements

Palmer Communications
Petroleum Communicationsm
PriCellular
Prime Cellular
Public Service Cellular
Quantum Communications
Radiofone
Rural Cellular
Southeastern Cellular
Sunsshine Cellular
Sussex Cellular
United States Cellular
Unitel
Vanguard Cellular

Cellular One of Galveston, TX

Ally
Alpine Cellular
Appalachian Cellular
Atlas Cellular
Bakersfield Cellular Telephone
BellSouth Cellular
Big Horn Cellular
Blue Mountain Cellular
Blue Ridge Cellular
Blue Ridge Cellular Telephone
Cellcom
Cellcom of Hickory
Cellular of Indiana
Cellular One of Columbia TN
Cellular One of Danville
Cellular One of Lake Charles
Cellular One of LaSalle
Cellular One of Marianna
Cellular One of Middle TN
Cellular One of Southwest OK
Cellular One of Upstate NY
Cellular One of Western OK
Cellular Ventures
Crowley Cellular Telecomm
Danbury Cellular
Dobson Cellular
Dominion Cellular
GMD Partnership
Greater South Dakota Cellular
Highland Cellular
LA Cellular
Litchfield County Cellular
Mercury Communications
Miscellco Communications
Mountaineer Cellular
National Cellular
North Texas Cellular
Palmer Communications
Rural Cellular
Sterling Cellular
Texas 16 Cellular Telephone
U S West Cellular
Union Telephone
United States Cellular
Vanguard Cellular Telephone
West Alabama Cellular
Youngstown Cellular Telephone

Cellular One of Indianapolis, IN

ALLTEL Mobile Communications
Ameritech Mobile Communications
Associated Communications
Bay Area Cellular
Bell Atlantic Mobile
BellSouth Mobility
Cantel
Cellular, Inc.
Cellular Information Systems
Cellular One of Akron
Cellular One of Albany, GA
Cellular One of
  Baltimore/Washington
Cellular One of Boston

Cellular One of Canton
Cellular One of Chicago
Cellular One of Cincinnati
Cellular One of Cleveland
Cellular One of Columbus
Cellular One of Dayton
Cellular One of Detroit
Cellular One of Flint
Cellular One of Gary
Cellular One of Grand Rapids
Cellular One of Hamilton
Cellular One of Lansing
Cellular One of Middletown
Cellular One of Milwaukee
Cellular One of Muskegon
Cellular One of South Bend
Cellular One of Syracuse
Cellular One of Toledo
Cellular One of Utica
Cellular One of Worcester
Cellular One of Youngstown
Centel Cellular
Century Cellunet
Comcast
Crowley Cellular Telecommunications
Cybertel Cellular Telephone
Gulf Coast Cellular
Houston Cellular Telephone
Los Angeles Cellular
MCTA
McCaw Communications
Metro Mobile CTS
Metro One
MetroCel Cellular
Metrophone
PacTel Cellular
Radiofone
Springwich Cell. Ltd Partner. (SNET)
United States Cellular
Vanguard Cellular Systems

Cellular One of La Salle, IL

Allcell
ALLTEL Mobile Communications
Bahamas Telephone
Bay Area Cellular
Cantel
Cellcom of Hickory
Cellular One of Albany
Cellular One of Atlantic City
Cellular One of
  Baltimore/Washington
Cellular One of Beaumont
Cellular One of Berrien County
Cellular One of Buffalo
Cellular One of Chicago
Cellular One of Indianapolis
Cellular One of Lake Charles
Cellular One of Milwaukee
Cellular One of Sioux Falls
Cellular Information Systems
Celutel
Comcast
Contel
Crowley Cellular Telecommunications
Cybertel
Genesee Telephone
Houston Cellular
Lincoln Telephone Cellular
McCaw Communications
Metrocel
New Par
PacTel Cellular
Palmer Communications
Radiofone
U S West Cellular
United States Cellular

Cellular One of Lake Charles, LA

ALLTEL Mobile Communications
Ally
Associated Communications
Batelco
Bay Area Cellular
Bell Atlantic Mobile
BellSouth Mobility
Bledsoe Telephone
Blue Ridge Cellular
Cantel
C.C. Cellular
Cellcom of Green Bay
Cellcom of Hickory
Cellular 2000
Cellular, Inc.
Cellular Information Systems
Cellular One Amarillo
Cellular One of Richmond
Cellular One of Baltimore
Cellular One of Beaumont
Cellular One of Berrien County
Cellular One of Chicago
Cellular One of Cincinnati
Cellular One of Cleveland
Cellular One of Columbia MO
Cellular One of Columbus OH
Cellular One of Danville
Cellular One of Detroit
Cellular One of Fredericksburg
Cellular One of Grand Rapids
Cellular One of Huntsville
Cellular One of Indianapolis
Cellular One of Lake Charles
Cellular One of Lansing
Cellular One of Meridian
Cellular One of Milwaukee
Cellular One of Northeast PA
Cellular One of Ocean County
Cellular One of Santa Cruz
Cellular One of Sioux Falls, SD
Cellular One of Syracuse
Cellular One of Toledo
Cellular One of Upstate NY
Cellular One of Utica
Cellular One of Washington
Cellular One of Wichita Falls
Cellular Plus
Cellular XL
Cellulink
Celutel
Centel Cellular
Centennial Cellular
Comcast
Contel Cellular
Crowley Cellular Telecommunications
Cybertel Cellular
Danbury Cellular Telephone
Gencell
GTE Mobilnet
Houston Cellular Telephone
Iowa East Cellular
Kansas Cellular
Lincoln Telephone
McCaw Cellular Communications
Mercury Communications
Metrocel Cellular
Metrophone
Miscellco Communications
Missouri Cellular
Nebraska Cellular
NYNEX Mobile Communications
Pace Communications
PacTel Cellular
Palmer Communications
Petroleum Communications
Radiofone
Richmond Cellular Telephone
Rural Cellular
Southwestern Bell Mobile

Sterling Cellular
U S West Cellular
Union Telephone
United States Cellular
Vanguard Cellular

Cellular One of Milwaukee, WI

Albany Telephone
ALLTEL Mobile Communications
Ally
American Rural Cellular
Ameritech Mobile Comm
Appalachian Cellular
Atlantic Cellular
Auburn Television
Bakersfield Cellular Telephone
Baton Rouge Cellular
Bell Atlantic Mobile
BellSouth Cellular
BellSouth Mobility
Big Horn Cellular
Bledsoe Telephone
Blue Mountain Cellular
Blue Ridge Cellular
Bluegrass Cellular
Buffalo Telephone
Cantel
Canton Cellular
Cellcom
Cellcom Hickory
Cellular
Cellular Connection
Cellular Indiana
Cellular Information Systems
Cellular One
Cellular One of Amarillo
Cellular One of Atlantic City
Cellular One of
  Baltimore/Washington
Cellular One of Bay Area
Cellular One of Berrien County
Cellular One of Boston
Cellular One of Brookings
Cellular One of Chicago
Cellular One of Collier
Cellular One of Columbia
Cellular One of Columbia TN
Cellular One of Columbus
Cellular One of Danville
Cellular One of Delaware
Cellular One of East Central PA
Cellular One of Hunterdon
Cellular One of Huntsville TX
Cellular One of Indianapolis
Cellular One of Jackson
Cellular One of Lake Charles
Cellular One of Lasalle
Cellular One of Mackinac
Cellular One of Meridian
Cellular One of Middle TN
Cellular One of New Brunswick
Cellular One of North Texas
Cellular One of Northeast Colorado
Cellular One of Northeast Georgia
Cellular One of Ocean County
Cellular One of Santa Cruz
Cellular One of Sioux Falls
Cellular One of Southwest Florida
Cellular One of Trenton
Cellular One of Upstate New York
Cellular One of West Illinois
Cellular One of West Oklahoma
Cellular One of Wichita Falls
Cellular One of Wilmington
Cellular One of Youngstown
Cellular Plus
Cellular Upstate New York
Cellular Ventures
Cellular XL

Celutel
Centel Cellular
Century Cellular
Century Cellunet
Cibernet
CJD Wisconsin Cellular
Comcast Cellular Comm
Comp Communications
Contel Cellular
Cooper Cellular
Crowley Cellular
Cybertel Cellular
Danbury Cellular Telephone
Detroit Cellular Telephone
Dicomm Cellular
Dominion Cellular
Easterbrooke Cellular
Farmers Cellular
FGI Cellular
Gencell
Genesee Telephone
GMD
GTE Mobilnet
Gulf Coast Cellular
Highland Cellular
Houston Cellular
Independent Cellular Network
Iowa East Cellular Telephone
Kaplan Telephone
Lake Huron Cellular
Litchfield County Cellular
McCaw Cellular
Mega Communications
Mercury Communications
Metro Mobile
Metro One
Metrocel Cellular Telephone
Mid-Missouri Cellular
Minnesota Southern Cellular
  Telephone
Miscellco Communications
Missouri Cellular
Mobile Communications
Mobiletel
Montana Cellular Telephone
Mountaineer Mobile
MTS Cellular
National Cellular
Nebraska Cellular
New Par
Northwest Cellular
Ohio State Cellular
PacTel Cellular
Palmer Cellular
Petroleum Communications
PriCellular
Quantum Communications
Radiofone
RFG Cellular
Richmond Cellular Telephone
Rural Cellular
Shenandoah Mobile
Southeast Cellular
Southeast Indiana Cellular
  Telephone
Springwich Cellular
Sussex Cellular
Syracuse Telephone
Texas 16 Cellular
The Buckhead Telephone
Tsaconas Cellular
U S West Cellular
Union Telephone
United Bluegrass Cellular
United States Cellular
Unitel
Utica Telephone
Vanguard Cellular
Virginia Cellular

West Alabama Cellular
X-Cell Cellular
Yorkville Telephone

Cellular One of North Texas

ALLTEL Mobile Communications
Alpine
Amarillo Cellular
Ameritech Mobile Communications
Appalachian Cellular
Atlantic Cellular
Auburn Television
Bakersfield Cellular
Bay Area Cellular
BellSouth Cellular
Blackwater Cellular
Blue Mountain Cellular
Bluegrass Cellular
C. C. Cellular
Cellcom of Hickory
Cellular 2000 ("B" side)
Cellular Information Systems
Cellular One of Berrien County
Cellular One of Columbia
Cellular One of Columbia, TN
Cellular One of Danville
Cellular One of Dumas
Cellular One of East/Central PA
Cellular One of Huntsville
Cellular One of La Salle, IL
Cellular One of Sioux Falls
Cellular One of Southwest Florida
Cellular One of Upstate New York
Cellular One of Western Illinois
Cellular One of Wichita Falls
Cellular One of Worthington
Cellular XL
Celutel
Centennial Cellular
CJD Wisconsin Cellular
Clear Communications
Coastel Communications
Cone Enterprises
Contel Cellular
Crowley Cellular
Danbury Cellular
Detroit Cellular Telephone
Dobson Cellular ("A side")
Easterbrooke Cellular
First Kentucky Cellular
Galveston Cellular
General Cellular
Greater South Dakota Cellular
GTE Mobilenet
Highland Cellular
HLD Cellular
Houston Cellular
Indianapolis Telephone
Lake Charles Cellular
Lake Huron Cellular
Los Angeles Cellular
McCaw Communications
Mega Communications
Mercury Communications
Metrocel Cellular
Metrophone
Mid-Tennessee Partners
Mid-South Cellular
Milwaukee Telephone
Miscellco Communications
Missouri Cellular
Mobile Telenet
Nebraska Cellular
Ocean County Cellular
Oneonta Telephone
PacTal
Palmer Cellular
PC Cellular
Pegasus Cellular

Roaming Agreements

Petroleum Communications
Pine Cellular
PriCellular
Prime Cellular
R & D Cellular
Radiofone
Santa Cruz Cellular
Southeast Indiana Cellular
Southeastern Cellular
Sterling Cellular
Sussex Cellular
Syracuse Telephone
Texas 16 Cellular
Texoma Cellular Management
Tsaconas Cellular
U S West Cellular
Union Telephone
United States Cellular
Unitel
Utica Telephone
Vanguard Cellular
Westex
Wisconsin Cellular
X-Cell Cellular
Youngstown Cellular

Cellular One of Northeast Colorado

Bay Area Cellular
Bell Atlantic Mobile
Cellular One of Amarillo
Cellular One of Bakersfield
Cellular One of Chicago
Cellular One of Galveston
Centel
Centennial Cellular
Crowley Cellular
Dial Two
Dobson Cellular
Gencell
GTE Mobilenet
Highland Cellular
Houston Cellular
Kansas Cellular
LA Cellular
Metrocel
Nebraska Cellular
Sterling Cellular
U S West Cellular
United State Cellular
Unitel
X-Cell Cellular

Cellular One of Northeast Mississippi

Adirondack Cellular
ALLTEL Mobile Communications
Alpine Cellular
Bay Area Cellular
Bell Atlantic Mobile
BellSouth Cellular
BellSouth Mobility
Blackwater
Blue Mountain Cellular
Blue Ridge
Brazos Cellular
Canton
C. C. Cellular
Cellcom of Hickory
Cellular Holdings
Cellular Information Systems
Cellular One of Chicago
Cellular One of Columbia
Cellular One of Danville
Cellular One of East/Central PA
Cellular One of Indianapolis
Cellular One of Meridian
Cellular One of Middle Tennessee
Cellular One of Sioux Falls
Cellular One of Southeastern Indiana

Cellular One of Southwest Georgia
Cellular One of Southwestern OK
Cellular One of Western Illinois
Cellular One of Western Oklahoma
Cellular XL
Celutel
Centel
Century Cellunet
CJD Wisconsin
Clear Communications
Contel
Crowley Cellular
Cybertel
Detroit Cellular
Dobson Cellular
Dominion Cellular
Easterbrooke
Eastex
Farmers Cellular
Galveston Cellular
General Cellular
Greater South Dakota Cellular
GTE Mobilnet
HLD Cellular
Houston Cellular
Illinois Valley Cellular
Mackinac
McCaw Cellular Communications
MegaComm
Mercury Cellular
Metrocel
Metro Mobile CTS
Mid-South Cellular
Mississippi Cellular
New Par
North Texas Cellular
Pace Communications
PacTel Cellular
Palmer Communications
PC Cellular Kentucky
PC Cellular Vermont
PriCellular
Radiofone
Southeastern Cellular
Sterling Cellular
Stillwater Cellular
Sussex
Texas 16
Tsaconas
Union
U S West Cellular
United Bluegrass Cellular
United States Cellular
Vanguard
West Alabama

Cellular One of Puerto Rico

Cantel
PacTel Cellular
United States Cellular

Cellular One of Richmond, VA

ALLTEL Mobile Communications
Associated Communications
Bay Area Cellular
Bell Atlantic Mobile Systems
BellSouth Mobility
Cantel
Cellular Information Systems
Cellular One of
 Baltimore/Washington
Cellular One of Chicago
Cellular One of Cincinnati
Cellular One of Cleveland
Cellular One of Columbus
Cellular One of Detroit
Cellular One of Flint
Cellular One of Grand Rapids

Cellular One of Lansing
Cellular One of Milwaukee
Cellular One of Muskegon
Cellular One of Roanoke
Cellular One of Saginaw
Cellular One of Syracuse
Cellular One of Toledo
Cellular One of utica
Cellular One of Youngstown
Centel Cellular
Century Cellular
Crowley Cellular Telecommunications
Cybertel Cellular Telephone
Gulf Coast Cellular
Houston Cellular Telephone
MCTA
Metro Mobile CTS
Metro One
MetroCel Cellular
Radiofone
United States Cellular
Vanguard Cellular Systems

Cellular One of Sioux Falls, SD

Advantage
Allcell
Allegan Cellular
ALLTEL Mobile Communications
Ameritech Mobile Communications
Atlantic Cellular
Baton Rouge Cellular
Bay Area Cellular
Bell Atlantic Mobile
Bell Mobility Cellular
BellSouth Cellular
Bledsoe
Blue Mountain Cellular
Blue Ridge Cellular
Bluegrass Cellular
C. C. Cellular
Cantel
Canton Cellular
Cellcom
Cellcom of Hickory
Cellular Information Systems
Cellular of Indiana
Cellular One of Amarillo
Cellular One of Beckley, WV
Cellular One of Berrien County
Cellular One of Boston, MA
Cellular One of Brookings
Cellular One of Chicago
Cellular One of Clanton, AL
Cellular One of Cody, WY
Cellular One of Columbia, MO
Cellular One of Columbus, IN
Cellular One of Enid, OK
Cellular One of Galveston, TX
Cellular One of Greenwood, SC
Cellular One of Huntsville, TX
Cellular One of Indianapolis, IN
Cellular One of Lake Charles, LA
Cellular One of La Salle, IL
Cellular One of Meridian
Cellular One of Mid-South Texas
Celllular One of North Carolina 2
Cellular One of North Texas
Cellular One of Richmond, KY
Cellular One of Santa Cruz
Cellular One of South Mississippi
Cellular One of Southwest Florida
Cellular One of Southwest Oklahoma
Cellular One of the Rio Grande Valley
Cellular One of Upstate New York
Cellular One of Victoria, TX
Cellular One of West Alabama
Cellular One of Western Oklahoma
Cellular One of Worthington, IH
Cellular Phone of Kentucky

Cellular Plus
Cellular Systems International
Cellular Telephone of Kentucky
Cellular, Inc.
Celutel of Biloxi
Celutel
Centel Cellular
Centennial Cellular
Century Telephone
Chariton Valley Cellular
Clear Communications
Contel Cellular
Cooper Cellular
Crowley Cellular Telecommunications
Dicomm Cellular
Digital Cellular of Texas
Farmers Cellular Telephone
Gaia
GenCell Management
GMD Partnership
Great Lakes of Iowa
GTE Mobile Communications
Honolulu Cellular
Houston Cellular
Independent Cellular Network
Inland Cellular
Iowa East Cellular
Jackson Cellular
Kansas Cellular
Kentucky Cellular
Lincoln Telephone Cellular
Litchfield County Cellular
Los Angeles Cellular
McCaw Communications
MCMG
Mercury Cellular
Mercury Communications
MetroCel Cellular
Metrophone
MidSouth Cellular
Milwaukee Telephone
Miscellco Communications
Mississippi Cellular Telephone
Missouri Cellular
Mobile Communications Systems
MobileTel
Mountaineer Cellular
MTS Cellular
National Cellular
Nebraska Cellular
NYNEX Mobile Communications
Oneonta Telephone
Pace Communications
Pacific Northwest Cellular
Pacific Telecom Cellular
PacTel Cellular
Palmer Communications
Petroleum Communications
Pricelluler
Public Service Cellular
Quantum Communications
R & D Cellular
Radiofone
RFB Cellular
Richard L. Vega Group
Rural Cellular Management
Southeastern Cellular
Southwestern Bell Mobile Systems
Sterling Cellular
The Buckhead Telephone Company
U S West Cellular
Union Telephone
United States Cellular
Unitel
United States Cellular
Virginia Cellular
Washington/Baltimore Cellular
  Telephone
Westex
Yorkville Telephone Cooperative

Youngstown Cellular Telephone

Cellular One of Southwest Florida

Albany Telephone
ALLTEL Mobile Communications
Ally
Alpha Cellular
Ameritech Mobile Communication
Appalachian
Atlantic Cellular
Bay Area Cellular
Bell Atlantic Mobile
BellSouth Cellular
Blackwater Cellular
Blue Mountain Cellular
Bluegrass Cellular
Blue Ridge Cellular
Buckhead Telephone
Cantel
Canton Cellular
Cedetel
Cellcom
Cellcom of Hickory
Cellular Holding
Cellular, Inc.
Cellular Information Systems
Cellular One of
  Baltimore/Washington
Cellular One of Berrien County
Cellular One of Boston
Cellular One of Chicago
Cellular One of Columbia, MO
Cellular One of Columbia, TN
Cellular One of Columbus, IN
Cellular One of Danville
Cellular One of East Central PA
Cellular One of Hattiesburg
Cellular One of Indianapolis
Cellular One of Lake Charles, LA
Cellular One of La Salle, IL
Cellular One of Meridian
Cellular One of Mid-Tennessee
Cellular One of Northeast Georgia
Cellular One of North Texas
Cellular One of Panama City, FL
Cellular One of Sioux Falls
Cellular One of Southwest Georgia
Cellular One of Southwest Oklahoma
Cellular One of Upstate New York
Cellular One of Western Illinois
Cellular One of Western Oklahoma
Cellular One of Wichita Falls, TX
Cellular One of Worthington
Cellular Plus
Cellular South
Cellular Systems International
Cellular Ventures
Celutel
Centel Cellular
Century Cellular
Century Celunet
Comcast Cellular Communications
Contel Cellular
Crowley Cellular
Cybertel Cellular
Detroit Cellular
Dicomm Cellular Telephone
Dominion Cellular Telephone
Finger Lakes Telephone
GenCell Management
Genesee Telephone
GMD
Greater South Dakota Cellular
GTE Mobilenet
Highland Cellular
Honolulu Cellular
Houston Cellular
Interstate Cellular
Iowa East Cellular

First Kentucky Cellular
Lake Huron Cellular
Litchfield Cellular
MCMG
McCaw Cellular Communications
Mercury Communications
MetaComm Cellular Partners
Metro Mobile CTS
MetroCel Cellular
Metrophone
Milwaukee Telephone
Miscellco Communications
Missouri Cellular
MTS
National Cellular Limited Partnership
Nebraska Cellular
NYNEX Mobile Communications
PacTel Cellular
Palmer Cellular
Panama City Cellular
Pegasus Cellular
Petroleum Communications
Quantum Communications
Radiofone
Richmond Cellular
Rural Cellular Management
Santa Cruz Cellular
Sterling Cellular
Texas 16 Cellular
U S West Cellular
United States Cellular
Virginia 6 RSA
Virginia 10 RSA
Vanguard Cellular
West Alabama Cellular
Youngstown Cellular

**Cellular One of Syracuse, NY
(Cellular One of Utica, NY)**

Allcell
Associated Communications
Atlantic Cellular
Bay Area Cellular
Bell Atlantic Mobile Systems
Bell Mobility Cellular
Cantel
Cellular One of Akron
Cellular One of Amarillo
Cellular One of
  Baltimore/Washington
Cellular One of Berrien County
Cellular One of Boston
Cellular One of Canton
Cellular One of Chicago
Cellular One of Cincinnati
Cellular One of Cleveland
Cellular One of Columbus
Cellular One of Dayton
Cellular One of Detroit
Cellular One of Flint
Cellular One of Grand Rapids
Cellular One of Indianapolis
Cellular One of Lansing
Cellular One of Mansfield
Cellular One of Milwaukee
Cellular One of Richmond
Cellular One of Sioux Falls
Cellular One of Toledo
Cellular One of Warren
Cellular One of Wichita Falls, TX
Cellular One of Wilmington
Cellular One of Youngstown
Cellular Information Systems
Centel
Century Cellular
Comcast
Contel Cellular
Crowley Cellular
Cybertel

1113

Roaming Agreements

Gulf Coast Cellular
Honolulu Cellular
Houston Cellular
McCaw Communications
Metro Mobile CTS
Metro One
MetroCel Cellular
Metrophone
NYNEX Mobile Communications
Palmer Communications
Petroleum Communications
Radiofone
Rochester Telephone Mobile Comms.
Springwich Cell. Ltd Partner. (SNET)
United States Cellular
U S West Cellular
Vanguard Cellular Systems

Cellular One of Upstate New York

Adirondack Cellular
Advantage Cellular
Albany Telephone
ALLTEL Mobile Communications
AMC Cellular
Ameritech Mobile Communications
Appalachian Cellular
Atlantic Cellular
Bakersfield Cellular
Bay Area Cellular
B. C. Cellular
Bell Atlantic Mobile
Bell South Mobility
Big Horn Cellular
Bluegrass Cellular
Blue Mountain Cellular
Buffalo Telephone
Cantel
Canton Cellular
CellCom of Greenbay
CellCom of Hickory, NC
Cellular Communications of Puerto Rico
Cellular Inc.
Cellular Information Systems
Cellular of Indiana
Cellular One Great Lakes of Iowa
Cellular One of Amarillo
Cellular One of Atlantic City
Cellular One of Baltimore/Washington
Cellular One of Berrien County
Cellular One of Boston
Cellular One of Chicago
Cellular One of Columbia, MO
Cellular One of Columbia, TN
Cellular One of Danville, IL
Cellular One of East/Central PA
Cellular One of Fredrick, MD
Cellular One of Greater South Dakota
Cellular One of Indianapolis
Cellular One of La Salle, IL
Cellular One of Middle Tenessee
Cellular One of New Jersey
Cellular One of North Texas
Cellular One of Northeast Georgia
Cellular One of Ocean County
Cellular One of Santa Cruz
Cellular One of Sioux Fall
Cellular One of Southwest Florida
Cellular One of Southwest Oklahoma
Cellular One of Syracuse
Cellular One of Utica
Cellular One of Western Illinois
Cellular One of Western Oklahoma
Cellular One of Wichita Falls, TX
Cellular One of Wilmington, DE
Cellular Plus

Cellular South
Cellular Systems International
Cellular Telephone of Kentucky
Celutel
Centel Cellular
Centennial Cellular
Century Cellunet
Clear Communications
Coastel Communications
Contel Cellular
Crowley Cellular
CyberTel
Detroit Cellular
Dicomm Cellular
Dobson Cellular
Dominion Resources
Fingers Lake Telephone
Gaia
Galveston Cellular
GenCel Management
Genesse Telephone
GMD Partnership
GTE Mobilnet
Gulf Coast Cellular
Highland Cellular
Honolulu Cellular
Houston Cellular
Hudson Valley RSA
Independent Cellular
Interstate Cellular (Intercel)
Iowa East Cellular Telephone
Kansas Cellular
Lake Huron Cellular
Litchfield County Cellular
Los Angeles Cellular
Mackinac Cellular
Maine Cellular
McCaw Communications
Mega, Inc
Mercury Cellular
MetroCel
Metro Mobile
Metro One
Metrophone
Midsouth Cellular
Milwaukee Telephone
Miscellco Communications
Missouri Cellular
Mobile Communications
MTS Cellular
MT&T
National Cellular Limited
Nebraska Cellular
New Par
NYNEX Mobile Communications
Pacific Telecom Cellular
PacTel Cellular
Palmer Cellular
PC Cellular of Vermont
Petroleum Communications
PriCellular
Radiofone
Richmond Cellular
Rochester Telephone
Shenandoah Mobile Company
Southeast Indiana
Springwich Cellular
Sterling Cellular
Sussex Cellular
Texas 16 Cellular
U S West Cellular
Union Telephone
United States Cellular
Unity Cellular Systems (Unicel)
Vanguard
Victoria Cellular
West Alabama Cellular

Cellular One of Victoria, TX

Bell Atlantic Mobile
Cellular Information Systems
Cellular One of Akron
Cellular One of Amarillo
Cellular One of Brownsville
Cellular One of Canton
Cellular One of Chicago
Cellular One of Cincinnati
Cellular One of Dayton
Cellular One of Detroit
Cellular One of McAllen
Cellular One of Toledo
Cellular One of Wichita Falls, TX
Cellular One of Youngstown
Centel Cellular
Crowley Cellular Telecommunications
Houston Cellular Telephone
McCaw Communications
Metro Mobile CTS
MetroCel Cellular
Petroleum Communications
Radiofone
Southwestern Bell Mobile Systems
West Central Cellular

Cellular One of Western Oklahoma

Adirondack Cellular
ALLTEL Mobile Communications
Alpine Cellular
Amarillo Cellular
AMC Cellular
Ameritech Mobile Communications
Appalachian Cellular
Bakersfield Cellular
Bay Area Cellular
BellSouth Cellular
Blackwater Cellular
Blue Mountain Cellular
Bluegrass Cellular
C. C. Cellular
Cellcom of Hickory
Cellular 2000 ("B" side only)
Cellular One Huntsville
Cellular One of Berrien County
Cellular One of Chicago
Cellular One of Columbia
Cellular One of Columbia, TN
Cellular One of Danville
Cellular One of Detroit, MI
Cellular One of Dumas, TX
Cellular One of East/Central PA
Cellular One of Indianapolis
Cellular One of Lake Charles, LA
Cellular One of La Salle, LA
Cellular One of Milwaukee, WI
Cellular One of Sioux Falls
Cellular One of Southwest Florida
Cellular One of Upstate New York
Cellular One of Western Illinois
Cellular One of Wichita Falls, TX
Cellular One of Worthington
Cellular XL
Celutel
Centennial Cellular
Century Cellunet
CIS
CJD Wisconsin Cellular
Clear Communications
Coastel Communications
Cone Enterprises
Contel Cellular
Crowley Cellular
Danbury Cellular
Dobson Cellular ("A" side only)
Easterbrooke Cellular
First Kentucky Cellular
Galveston Cellular
General Cellular
Greater South Dakota

GTE Mobilnet
Highland Cellular
HLD Cellular
Houston Cellular
Lake Huron Cellular
Lawton Cellular
Los Angeles Cellular
McCaw Cellular Communications
Mega Comm
Mercury Communications
Metrophone
Mid-Tennessee Partners
Mid-South Cellulr
milwaukee Telephone
Miscellco Communications
Missouri Cellular
Nebraska Cellular
Ocean County Cellular
Oneonta Telephone
PacTel Cellular
Panhandle Telecommunications
PC Cellular
Pegasus Cellular
Petroleum Communications
Pine Cellular
PriCellular
Prime Cellular
R & D Cellular
Radiofone
Santa Cruz Cellular
Sooner Cellular
Southeast Indiana Cellular
Southeastern Cellular
Sterling Cellular
Sussex Cellular
Syracuse Telephone
Texas 16 Cellular
Texas 5
Tsaconas Cellular
U S West Cellular
Union Telephone
United State Cellular
Unitel
Utica Telephone
Vanguard Cellular
Westex
X-Cell Cellular
Youngstown Cellular

Cellular Plus (A Side)

Bakersfield Cellular
Baton Rouge Cellular
Bay Area Cellular
BellSouth Cellular
Blackwater FCellular
Blue Ridge Cellular
Buffalo Telephone
Cantel
Cellular One of Danville
Cellular One of Washington
Cellular Holding
Cellular Information System
Cellular One of Amarillo
Cellular One of Berrien City
Cellular One of Chicago
Cellular One of Collier
Cellular One of Columbia
Cellular One of Detroit
Cellular One of Galesburg
Cellular One of Huntsville TX
Cellular One of Indiana
Cellular One of Indianapolis
Cellular One of Lasalle IL
Cellular One of New Jersey
Cellular One of GA, FL, AL
Cellular One of Santa Cruz
Cellular One of Sioux Falls
Cellular One of Syracuse
Cellular Properties

Cellular South
Celutel
Centel
Century Cellunet
Contel Cellular
Crowley Cellular
Cybertel
FGI Cellular
First Kentucky Cellular
GTE Mobilnet
Highland Cellular
Houston Cellular
Iowa East Cellular
Lincoln Telephone Cellular
MacKinac Celluar
McCaw Cellular Communications
Megacomm Cellular
Metro Cell
Mid-Missouri Cellular
Mid-Tenn Cellular
National Cellular
Petroleum Communications
PriCellular
Radiofone
Richmond Cellular
Sunshine Cellular
Ten Woodland Road
Union Telephone
Vanguard Cellular
West Alabama Cellular
Youngstown Cellular

Cellular Plus (B Side)

Advantage Cellular
ALLTEL Mobile Communications
Ameritech Mobile Communications
Associated Communications
Atlantic Cellular
Bell Atlantic Metro Mobile
Bell Mobility Cellular
BellSouth Mobility
Bluegrass Cellular
Cantel
Cellular One of
  Baltimore/Washington
Cellular One of New York
Centel Cellular
Century Cellunet
Contel Cellular
Crowley Cellular Telecommunications
GTE Mobilnet
Highland Cellular
Hudson Valley Cellular
Independent Cellular Network
Maine Cellular
Metrophone
MT&T Cellular
NYNEX Mobile Communications
PacTel Cellular
Rochester Telephone
Shenandoah Mobile
Southwestern Bell Mobile Systems
Springwich Cell. Ltd Partner. (SNET)
StarCellular
U S West Cellular
United States Cellular
Unity Cellular

Cellular Plus of Georgia

Bell Atlantic Mobile
BellSouth Mobility
Cellular Holding
Independent Cellular Network
McCaw Communications
Mississippi Cellular
Pacific Telecom Cellular
United States Cellular

Cellular Three

Altel
Ameritech
Bell Atlantic Mobile
Bell South Mobility
Bledsoe Telephone
Brazos Cellular
Cal-One Cellular
Cellcom
Cellular, Inc.
Cellular Information
Cellular One of Amarillo
Centel Cellular
Contel Cellular
Digital Celluar Texas
Dobson Cellular
Enid Cellular
Farmers Cellular Telephone
FGI Cellular Management
Five Star Cellular
GTE Mobilenet
Kansas Cellular
MacTel
McCaw Cellular Communications
Metrocel Cellular Telephone
Mid-Tex Cellular
Miscellco Communications
OK Cellular
Pace Communications Rider
PacTel Cellular
Pioneer Cellular
PTSI Cellular
Southwestern Bell Mobile Systems
Sterling Cellular
Texas Cellular
U S West Cellular
Union Telephone
United States Cellular
Unitel
Valley Telecommunications
West Central Cellular
Westex Cellular
XIT Cellular

Cellular Ventures

Advantage Cellular
ALLTEL Mobile Communications
Amercell
Ameritech Mobile
Atlas Cellular
Auburn Television
Bakersfield Cellular Telephone
Batelco
Bay Area Cellular
BCTel Cellular
Bell Atlantic Mobile
BellSouth Cellular
BellSouth Mobility
Big Horn Cellular
Blackwater Cellular
Blue Mountain Cellular
Blue Ridge Cellular
Bluegrass Cellular
Canton Cellular
C.C. Cellular
Cellcom of Green Bay
Cellcom of Hickory
Cellular 2000
Cellular Connection
Cellular Information Systems
Cellular of Indiana
Cellular One of Aiken
Cellular One of Amarillo
Cellular One of Charleston
Cellular One of Chicago
Cellular One of Cincinnati
Cellular One of Cleveland
Cellular One of Columbia MO
Cellular One of Columbia TN

Cellular One of Columbus OH
Cellular One of Danville
Cellular One of Demopolis
Cellular One of Detroit
Cellular One of Florence
Cellular One of Galesburg
Cellular One of Galveston
Cellular One of Grand Rapids
Cellular One of Huntsville
Cellular One of Indianapolis
Cellular One of Lake Charles
Cellular One of Lansing
Cellular One of LaSalle
Cellular One of Milwaukee
Cellular One of Richmond
Cellular One of Santa Cruz
Cellular One of Sioux Falls
Cellular One of the Great Lakes of Iowa
Cellular One of the Rio Grande Valley
Cellular One of Toledo
Cellular One of Upstate NY
Cellular One of Wichita Falls
Cellular Phone of Kentucky
Cellular Plus
Cellular South
Cellular Systems International
Cellular XL
Cellutel
Centel Cellular
Centennial Cellular
Century Cellunet
Coastal Communications
Contel Cellular
Cooper Cellular
Crowley Cellular Telecommunications
Cybertel Cellular
Dominion Cellular
Enid Cellular
Gencell
GMD Limited
Greater South Dakota Cellular
GTE Mobilnet
Honolulu Cellular
Houston Cellular Telephone
Independent Cellular Network
Iowa East Cellular Telephone
Kansas Cellular
Lake Huron Cellular
Lincoln Telephone
Litchfield County Cellular
Mackinac Cellular
McCaw Cellular Communications
Mercury Cellular
Mercury Communications
Metrocel Cellular
Minerich Cellular
Miscellco Communications
Mississippi Cellular
Mobile Communications Systems
Mobiletel
MT&T Cellular
MTS Cellular
National Cellular
Nebraska Cellular
Northwest Cellular
PacTel Cellular
Palmer Communications
Petroleum Communications
Public Service Cellular
Quantum Communications
Radiofone
Rural Cellular
Sagir
Southwestern Bell Mobile Systems
Springwich Cell. Ltd Partner. (SNET)
Sterling Cellular
Texas 16 Cellular
U S West Cellular

United States Cellular
Unitel
Unity Cellular Systems (Unicel)
Vanguard Cellular
X-Cell Cellular

Cellular XL

ALLTEL Mobile Communications
Appalachian Cellular
Atlas Cellular
Auburn Television
Bay Area Cellular
Bell Atlantic Mobile
BellSouth Cellular
BellSouth Mobility
Blue Grass Cellular
Blue Mountain Cellular
Canton Cellular
Cellcom of Green Bay
Cellcom of Hickory
Cellular Connection
Cellular of Indiana
Cellular One of Berrien County
Cellular One of Chicago
Cellular One of Columbia MO
Cellular One of Danville
Cellular One of Detroit
Cellular One of Gainesville GA
Cellular One of Huntsville
Cellular One of Indianapolis
Cellular One of LaSalle
Cellular One of Meridian
Cellular One of Milwaukee
Cellular One of Sioux Falls
Cellular One of Southwest Florida
Cellular One of Wichita Falls
Cellular Systems International
Celutel
Centennial Cellular
Century Cellunet
Coastal Communications
Contel Cellular
Crowley Cellular Telecommunications
Cybertel Cellular
Danbury Cellular Telephone
Dominion Cellular
Farmers Cellular Telephone
GenCell Management
GTE Mobilnet
Highland Cellular
Honolulu Cellular
Houston Cellular Telephone
Independent Cellular Network
Lake Charles Cellular Telephone
Lake Huron Cellular
Litchfield County Cellular
Los Angeles Cellular Telephone
McCaw Cellular Communications
Megacomm
Mercury Cellular
Mercury Communications
Mountaineer Cellular
National Cellular
Nebraska Cellular
PacTel Cellular
Palmer Communications
PC Cellular of Kentucky
PC Cellular of Vermont
Petroleum Communications
Quantum Communcications
Radiofone
Richmond Cellular Telephone
Southeast Indiana Cellular
Southeastern Cellular
Springwich Cell. Ltd Partner. (SNET)
Sterling Cellular
Stillwater Cellular Telephone
Syracuse Telephone
Texas 16 Cellular Telephone

U S West Cellular
Union Telephone
United States Cellular
Utica Telephone
Unity Cellular Systems (Unicel)
Vanguard Cellular Systems
Victoria Cellular
West Alabama Cellular
Youngstown Cellular

Celutel

Adirondack Cellular Telephone
Albany Cellular
Ally
American Rural Cellular
Appalachian Cellular
Atlantic Cellular
Auburn Television
Bakersfield Cellular
Baton Rouge Cellular
Bay Area Cellular
Bell Atlantic Mobile
Bell South Cellular
Big Horn Cellular
Blue Mountain Cellular
Blue Ridge Cellular
Bluegrass Cellular
Buffalo Telephone
C. C. Cellular
Cantel Cellular
Canton Cellular
Cedetal Cellular de Telefonia
Cellcom of Green Bay
Cellular One of Wichita Falls, TX
Cellular Holding
Cellular Information Systems
Cellular One of Amarillo, TX
Cellular One of Baltimore/Washington
Cellular One of Berrien County, MI
Cellular One of Boston, MA
Cellular One of Chicago, IL
Cellular One of Columbia, MO
Cellular One of Columbia, TN
Cellular One of Danville
Cellular One of Dumas, TX
Cellular One of East/Central PA
Cellular One of Grand Fork, ND
Cellular One of Greater South Dakota
Cellular One of Huntsville, TX
Cellular One of Indiana
Cellular One of Indiana
Cellular One of Indianapolis
Cellular One of La Salle, IL
Cellular One of Lake Charles, LA
Cellular One of Meridian, MS
Cellular One of Middle Tennessee
Cellular One of Natchitoches, MS
Cellular One of North Texas
Cellular One of Northeast Georgia
Cellular One of Sioux Falls, SD
Cellular One of Southwest Florida
Cellular One of Southwest Oklahoma
Cellular One of Upstate New York
Cellular One of Victoria, TX
Cellular One of Waycross/Vidalia
Cellular One of West Alabama
Cellular One of Western Illinois
Cellular One of Natchez, MS
Cellular Phone of Kentucky
Cellular Plus
Cellular XL
Centel Cellular
Centennial Cellular
Century Cellunet
Clear Communications
Comcast Cellular
Contel Cellular
Crowley Cellular

Cybertel Cellular
Danbury Cellular
Detroit Cellular
Dicomm Cellular
Dominion Cellular
Finger Lakes Telephone
First Kentucky Cellular
Gaia
GenCell Management
Genessee Telephone
GMD Partnership
Great Lakes of Iowa
GTE Mobile Communications
Highland Cellular
Horizon Cellular
Houston Cellular
Illinois Valley Cellular
Iowa East Cellular
Lake Huron Cellular
Litchfield County Cellular
Los Angeles Cellular
Mackinac Cellular
McCaw Communications
MCMG
Mega Comm
Mercury Communications
Metrophone
Mid-South Cellular
Milwaukke Telephone
Miscellco
Mobile Communications
Montana Cellular Telephone
National Cellular
Oneonta Telephone
PC Cellular of Kentucky
PC Cellular of Vermont
Pacific Northwest Cellular
Pacific Telecom Cellular
PacTel Cellular
Palmer Cellular
Pine Cellular
PriCellular
Quantum Communications
R & D Cellular
Radiofone
Richmond Cellular
Rural Cellular Management
Southeast Indiana Cellular
Southeastern Cellular
Sterling Cellular
Stillwater Cellular
Sussex Cellular
Syracuse Cellular
Texas 16 Cellular
Tsaconas Cellular
U S West Cellular
Union Telephone
United States Cellular
Unitel
Utica Telephone
Vanguard Cellular
Virginia Cellular
Youngstown Cellular

Centel Cellular

Advantage Cellular
Advantage Cellular Systems
AGT Cellular
ALLTEL Mobile Communications
Alpha Cellular
Alton Cellular
American Rural Cellular
Ameritech Mobile Communications
Appalachian Cellular
Atlantic Cellular
Atlas Cellular
Bahamas Telephone
Baton Rouge Cellular
Bay Area Cellular

BCTel Mobility Cellular
Bell Atlantic Mobile
Bell Mobility Cellular
BellSouth Mobility
Blackwater Cellular
Bledsoe Telephone Cooperative
Bluegrass Cellular
Blue Mountain Cellular
Blueridge Cellular
Brazos Cellular
Buffalo Telephone
C. C. Cellular
Cal-One Cellular
Cantel
Canton Cellular
Carolina West Cellular
Cass Telephone Cellular
Cedetel
Cellcom
Cellcom of Hickory
Cellular 2000
Cellular Connection
Cellular, Inc.
Cellular Information Systems
Cellular One of Amarillo
Cellular One of Atlantic City
Cellular One of
   Baltimore/Washington
Cellular One of Berrien County
Cellular One of Chicago/Central IL
Cellular One of Clinton, IA
Cellular One of Collier/Hendry
Cellular One of Columbia
Cellular One of Detroit
Cellular One of Galesburg
Cellular One of Huntsville
Cellular One of Indiana
Cellular One of Indianapolis
Cellular One of Lake Charles
Cellular One of La Salle, IL
Cellular One of Meridian
Cellular One of New Brunswick
Cellular One of Northeast Colorado
Cellular One of Northeast Georgia
Cellular One of Sioux Falls
Cellular One of Southwest Georgia
Cellular One of Upstate New York
Cellular One of Victoria
Cellular One of Western Illinois
Cellular One of Wichita Falls, TX
Cellular One of Wilmington, DE
Cellular One of Wilmington, NC
Cellular Phone of Kentucky
Cellular Plus
Cellular Properties
Cellular South
Cellular Telephone Company of
   Kentucky
Cellular Three
Cellular Ventures
Cellular XL
Celutel
Century Cellular
Century Cellunet
CJD Wisconsin Cellular
Coastel Communications
Contel Cellular
Cross-Valliant Cellular
Crowley Cellular Telecommunications
Cybertel Cellular telephone
Danbury Cellular
Detroit Cellular
Digital Cellular of Texas
Dobson Cellular
Dominion Resources
Eastbrooke Cellular
Ed Tel Cellular
Enid Cellular
Farmers Cellular
FGI Cellular Management

Finger Lakes Telephone
First Fayette Cellular
Gaia
GenCell Management
Genessee Telephone
GMD Limited Partnership
GTE Mobilnet ("B" systems only)
Gulf Coast Cellular
HBF Cellular
Highland Cellular
HLD Cellular
Horizon Cellular
Houston Cellular Telephone
Illinois Valley Cellular
Independent Cellular Network
Interstate Cellular
IUSA Cell
Kansas Independent Networks
Lake Huron Cellular
Lamar County Cellular
Lincoln Telephone Cellular
Los Angeles Cellular Telephone
Mackinac Cellular
MACtel Cellular System
Maine Cellular
McCaw Communications
MegaComm
Mercury Cellular and Paging
Mercury Communications
Metrophone
Miid-Missouri Cellular
Mid-South Cellular
Mid-Tennesse Cellular
Milwaukee Telephone
Miscellco Communications
Missouri Cellular
MobileTel
Mountaineer Cellular
Mountaineer Mobile
MT&T Cellular
MTS Cellular
National Cellular
Nebraska Cellular
NYNEX Mobile Communications
OK Cellular
Pace Communications
Pacific Telecom Cellular
PacTel Cellular
Palmer Communications
Panhandle Telecommunications
Pegasus Cellular
Peoples Cellular
Petroleum Communications
Pine Cellular Phones
Public Service Cellular
Puerto Rice Telephone
Québec Téléphone Cellulaire
R & D Cellular
RFB Cellular
Radiofone
Richmond Cellular
Rochester Telephone Mobile
   Systems
Rural Cellular Management
Sasktel Cellular
Shenandoah Mobile
Southeastern Cellular
Southeast Indiana Cellular
Southwestern Bell Mobile Systems
Springwich Cell. Ltd. Partner. (SNET)
StarCellular
Sunshine Cellular
Syracuse Telephone
Telefonia Cellular del Norte
Texas 16 Cellular
Texas RSA South Limited
   Partnership
Thunder Bay Telephone
U S West Cellular

UBET Cellular
Union Telephone
United States Cellular
Unitel
Utica/Rome Cellular
Unity Cellular Systems (Unicel)
Valley Telecommunications
Vanguard Cellular Systems
Virginia Cellular
West Alabama Cellular
West Central Cellular
Westex Cellular
X-Cell
XIT Cellular
Yorkville Telephone
Youngstown Cellular

Centennial Cellular

ALLTEL Mobile Communications
Associated Communications
Baja Cellular
Bay Area Cellular
Bell Atlantic Mobile
Cantel
Cellcom of Hickory
Cellular Information Systems
Cellular One of Amarillo
Cellular One of
  Baltimore/Washington
Cellular One of Berrien County
Cellular One of Boston
Cellular One of Champaign
Cellular One of Chicago/Gary
Cellular One of Cincinnati
Cellular One of Cleveland
Cellular One of Columbus, OH
Cellular One of Decatur
Cellular One of Detroit
Cellular One of Grand Rapids
Cellular One of Indianapolis
Cellular One of Lansing
Cellular One of Milwaukee
Cellular One of Muskegon
Cellular One of New York
Cellular One of Richmond
Cellular One of Saginow
Cellular One of Springfield
Cellular One of Syracuse
Cellular One of Toledo
Cellular One of Victoria
Cellular One of Warren, OH
Cellular One of Wichita Falls
Cellular One of Wilmington
Cellular One of Youngstown
Cellular Plus
Celutel
Centel Cellular
Comcast
Contel Cellular
Crowley Cellular
Cybertel Cellular
GTE Mobilnet
Houston Cellular Telephone
Lincoln Telephone
Los Angeles Cellular Telephone
MACtel
McCaw Cellular Communications
Metrocel Cellular
Metrophone
PacTel Cellular
Petroleum Communications
Richmond Cellular Telephone
U S West Cellular
United States Cellular
Unitel
Vanguard Cellular

Century Cellunet

Advantage Cellular

AGT Cellular
ALLTEL Mobile Communications
Ameritech Mobile Communications
Batelco
Bell Atlantic Mobile
Bell Mobility Cellular
BellSouth Cellular
BellSouth Mobility
Cantel
Cellcom of Green Bay
Cellular, Inc.
Cellular Information Systems
Cellular One of Chicago
Cellular One of Cincinnati
Cellular One of Cleveland
Cellular One of Columbus OH
Cellular One of Detroit
Cellular One of Indianapolis
Cellular One of Lafayette
Cellular One of Lake Charles
Cellular One of Richmond
Cellular One of Sioux Falls
Cellular One of Wilmington
Cellular Plus
Cellular South
Cellulink
Celutel
Centel Cellular
Coastel Communications
Contel Cellular
Crowley Cellular Telecommunication
Cybertel Cellular
Ed Tel Cellular
Enid Cellular
Gencel
GTE Mobilnet
Houston Cellular Telephone
Independent Cellular Network
Kansas Cellular
Lincoln Telephone
Mactel Cellular
Maine Cellular
MCTA
Mercury Cellular
Metrocel Cellular
Mobiletel
Mobile South Cellular
MT&T Cellular
MTS Cellular
NYNEX Mobile Communications
PacTel Cellular
Petroleum Communications
Public Service Cellular
Radiofone
Rochester Telephone
Sasktel Cellular
Savannah Cellular
Southwestern Bell Mobile System
Springwich Cell. Ltd Partner. (SNET)
StarCellular
Thumb Cellular
U S West Cellular
Unity Cellular Systems (Unicel)
Vanguard Cellular
West Central Cellular

CFW Cellular

Advantage Cellular
AGT Cellular
ALLTEL Mobile Communications
Ameritech Mobile Communications
Associated Communications
Bakersfield Cellular
Bay Area Cellular
Bell Atlantic Mobile
Bell Mobility Cellular
BellSouth Cellular
BellSouth Mobility
Cantel

Cellcom
Cellular, Inc.
Cellular Information Systems
Cellular One of Amarillo
Cellular One of
  Baltimore/Washington
Cellular One of Berrien County
Cellular One of Chicago
Cellular One of Cincinnati
Cellular One of Cleveland
Cellular One of Columbus, OH
Cellular One of Detroit
Cellular One of Indianapolis
Cellular One of Jacksonville
Cellular One of Lake Charles
Cellular One of Milwaukee
Cellular One of New York
Cellular One of Richmond
Cellular One of Roanoke
Cellular One of Sioux Falls
Cellular One of Springfield
Cellular One of Syracuse
Cellular One of Utica
Cellular One of Victoria
Cellular One of Wichita Falls
Cellular One of Wilmington
Cellular One of Youngstown
Cellular Plus
Cellular South
Cellulink
Centel Cellular
Centennial Cellular
Century Cellunet
Coastel Communications
Comcast
Crowley Cellular Telecommunication
Cybertel Cellular
Ed Tel Cellular
Enid Cellular
GenCell Management
GTE Mobilnet
Honolulu Cellular Telephone
Houston Cellular Telephone
Independent Cellular Network
Lincoln Telephone
Maine Cellular
McCaw Cellular Communications
MCTA
Mercury Cellular
Metrocel Cellular
Metrophone
Mobiltel
MT&T Cellular
MTS Cellular
NYNEX Mobile Communications
PacTel Cellular
Palmer Communications
Petroleum Communication
Public Service Cellular
Radiofone
Richmond Cellular Telephone
Sasktel Cellular
Savannah Cellular
Southwestern Bell Mobile Systems
Springwich Cell. Ltd Partner. (SNET)
StarCellular
U S West Cellular
United States Cellular
Unity Cellular Systems (Unicel)
Vanguard Cellular
West Central Cellular

Chariton Valley Cellular

ALLTEL Mobile Communications
Ameritech Mobile Communication
Appalacian Cellular
Atlantic Cellular
Atlas Cellular
Bell Atlantic Mobile

BellSouth Mobility
Bluegrass Cellular
Cantel
C.C. Cellular
Cellcom
Cellular Connection
Cellular, Inc.
Cellular One of Beckley, WV
Cellular One of Columbia, MO
Cellular One of Hays
Cellular One of Indianapolis
Cellular One of Northeast Georgia
Cellular One of Sioux Falls
Cellular One of Wichita Falls
Cellular Plus
Centennial Cellular
Contel Cellular
Cybertel Cellular
Digital Cellular
Enid Cellular
GTE Mobilnet
Independent Cellular Network
Kansas Cellular
Lincoln Telephone
McCaw Cellular Communications
Missouri Cellular
Nebraska Cellular
Pace Communications
PacTel Cellular
Petroleum Communications
Shenandoah Cellular
Southwestern Bell Mobile Systems
U S West Cellular
Union Telephone
United States Cellular

Coastel Communications

Advantage Cellular
AGT Cellular
ALLTEL Mobile Communications
Ally
American Rural Cellular
Ameritech Mobile Communications
Atlantic Cellular
Atlas Cellular
Bell Atlantic Mobile
BellSouth Cellular
BellSouth Mobility
Bledsoe Telephone
Bluegrass Cellular
Blue Ridge Cellular
Cal-One Cellular
Canton Cellular
C.C. Cellular
Cellcom
Cellular Connection
Cellular One of Baton Rouge
Cellular One of Chicago
Cellular One of Cody
Cellular One of Columbia, MO
Cellular One of Huntsville
Cellular One of Lake Charles, LA
Cellular One of Meridian
Cellular One of Northeast Georgia
Cellular One of Upstate NY
Cellular One of Wilmington
Cellular Phone of Kentucky
Cellular South
Cellular XL
Celutel
Centel Cellular
Century Cellunet
Contel Cellular
Digital Cellular
Dobson Cellular Systems
Ed Tel Cellular
Enid Cellular
Farmers Cellular
GenCell Management

Greater South Dakota Cellular
GTE Mobilnet
Houston Cellular Telephone
Independent Cellular Network
Interstate Cellular
Iowa East Cellular
IUSA Cell
Kansas Cellular
Mactel Cellular
Maine Cellular
Mercury Cellular
Miscellco Communications
Missouri Cellular
Mobile Communications
Mobiletel
Mobilsouth Cellular
MT&T Cellular
Nebraska Cellular
NewTel
NYNEX Mobile Communication
Pace Communications
PacTel Cellular
Petroleum Communications
PTSI Cellular
Public Service Cellular
Quantum Communications
Radiofone
Richmond Cellular Telephone
Rochester Telephone
Sasktel Cellular
Savannah Cellular
Shenandoah Cellular
Sooner Cellular
Southeastern Cellular
Southwestern Bell Mobile Systems
Springwich Cell. Ltd Partner. (SNET)
StarCellular
Sterling Cellular
U S West Cellular
Union Telephone
Unitel
Unity Cellular Systems (Unicel)
Vanguard Cellular
Westex
XIT Cellular

Comcast Cellular One (Metrophone)

Adirondack Cellular Telephone
Albany Telephone
Alpine Cellular
Atlantic Cellular
Atlas Cellular One
Bakersfield Cellular
Bay Area Cellular
Bell Atlantic Mobile
Blackwater Cellular
Blue Mountain Cellular
Cantel
Cellcom of Green Bay
Cellcom of Hickory
Cellular, Inc.
Cellular Information Systems
Cellular One of Amarillo
Cellular One of Berrien County
Cellular One of Boston
Cellular One of Buffalo
Cellular One of Chicago
Cellular One of Columbia, MO
Cellular One of Danville
Cellular One of Detroit
Cellular One of Galveston
Cellular One of Greenville
Cellular One of Indiana
Cellular One of Indianapolis
Cellular One of Lake Charles
Cellular One of Lake Huron
Cellular One of LaSalle
Cellular One of Mid-Tenn
Cellular One of New Brunswick

Cellular One of North Texas
Cellular One of Northeast Georgia
Cellular One of Puerto Rico
Cellular One of Santa Cruz
Cellular One of Sioux Falls
Cellular One of Southwest Florida
Cellular One of Southwest Georgia
Cellular One of Southwest Oklahoma
Cellular One of Syracuse/Utica
Cellular One of
   Washington/Baltimore
Cellular One of West Illinois
Cellular One of West Oklahoma
Cellular One of Youngstown
Cellular Plus
Cellwave Celutel
Centel
Centennial Cellular
CJD Wisconsin
Contel
Crowley Cellular
Cybertel
Danbury Cellular
Dicomm Cellular
Easterbrooke Cellular
Eastex Cellular
FGI Cellular
First Kentucky
General Cellular
Genesee Cellular
GTE Mobilnet
Highland Cellular
HLD Cellular
Honolulu Cellular
Horizon Cellular
Houston Cellular
Los Angeles Cellular
McCaw Cellular Communications
Midland Communications
Missouri Cellular
Mountaineer Mobile
New Par
NYNEX Mobile Communications
Pacific Northwest
PacTel
Palmer Communications
Pegasus Cellular
Petroleum Communications
Radiofone
RFB Cellular
Richmond Cellular
Shenedoah Mobile
Sunshine Celllar
Sussex Cellular
U S West Cellular
Union Telephone
United States Cellular
Vanguard Cellular
Virginia Cellular

Cone Enterprises

Ameritech Mobile Communications
Atlantic Cellular
Blue Ridge Cellular
Cellular One of El Campo TX
Cellular One of North Texas
Cellular One of Syracuse/Utica NY
Cellular One of Western Oklahoma
Dial Two
Dobson Cellular (A Systems)
GenCell Management
GMD Limited Partnership
Interstate Cellular
Mercury Cellular & Paging
Shenandoah Cellular
Sooner Cellular
Texoma Cellular
West Alabama Cellular

Roaming Agreements

Contel Cellular ("A" systems)

ALLTEL Mobile Communications
Bay Area Cellular
Cantel
Cellcom of Green Bay
Cellcom Hickory
Cellular, Inc.
Cellular Information Systems
Cellular One of Amarillo
Cellular One of Bakersfield
Cellular One of
  Baltimore/Washington
Cellular One of Berrien County
Cellular One of Baton Rouge
Cellular One of Beaumont
Cellular One of Billings
Cellular One of Boston
Cellular One of Casper
Cellular One of Chicago
Cellular One of Indianapolis
Cellular One of La Salle, IL
Cellular One of Lake Charles
Cellular One of Sioux Falls
Cellular One of Wichita Falls, TX
Cellular Plus
Cellular South
Cellulink
Celutel
Centel
Centennial Cellular
Century Cellunet
Century Roanoke Cellular
Comcast Cellular
Crowley Cellular
Cybertel
Detroit Cellular
Ellis Thompson
FGI Cellular
GenCell Management
Genesee Telephone
Gulf Coast Cellular
Honolulu Cellular
Houston Cellular
Independent Cellular Network
Los Angeles Cellular
Lincoln Telephone Cellular
McCaw Communications
Mercury Cellular
Metro Mobile
Metro One
MetroCel
Metrophone
Milwaukee Telephone
New Par
PacTel Cellular
Palmer Communications
Petroleum Communications
Public Service Cellular
Radiofone
Richmond Cellular
Springwich Cell. Ltd Partner. (SNET)
Syracuse Telephone
U S West Cellular
United States Cellular
Utica Telephone
Vanguard Cellular
West Central Cellular
Youngstown Cellular

Contel Cellular ("B" systems)

AGT Cellular
Advantage Cellular
ALLTEL Mobile Communications
Ameritech Mobile Communications
Atlantic Cellular
B.C. Tel Mobility Cellular
Bell Atlantic Mobile
Bell Mobility Cellular

BellSouth Mobility

Cantel
Cellcom of Green Bay
Cellcom of Hickory
Cellular, Inc.
Cellular Information Systems
Cellular One of Amarillo
Cellular One of
  Baltimore/Washington
Cellular One of Baton Rouge
Cellular One of Beaumont
Cellular One of Billings
Cellular One of Casper
Cellular One of Chicago
Cellular One of Indianapolis
Cellular One of La Salle, IL
Cellular One of Lake Charles
Cellular One of Sioux Falls
Cellular One of Wichita Falls, TX
Cellular Plus
Cellular South
Cellulink
Centel
Century Cellunet
Coastel Communications
Crowley Cellular
Cybertel
Ed Tel Cellular
Enid Cellular
FGI Cellular
GTE Mobilnet ("B" systems only)
GenCell Management
Genesee Telephone
Gulf Coast Cellular
Honolulu Cellular
Houston Cellular
Independent Cellular
Kansas Independent Networks
Lincoln Telephone Cellular
MACtel
MT&T Cellular
MTS Cellular
Maine Cellular
McCaw Communications
Mercury Cellular
Metro Mobile
Metro One
MetroCel
Metrophone
Milwaukee Telephone
MobileTel
Nebraska Cellular
New Par
NYNEX Mobile Communications
PacTel Cellular
Palmer Communications
Petroleum Communications
Public Service Cellular
Radiofone
Richmond Cellular
Rochester Telephone
SaskTel Cellular
Shoals Cellular
Springwich Cell. Ltd Partner. (SNET)
Southwestern Bell
StarCellular
Syracuse Telephone
U S West Cellular
United States Cellular
Unitel
Unity Cellular
Utica Telephone
Vanguard Cellular
Victoria Cellular
West Central Cellular
Youngstown Cellular

Cooper Cellular Management
(Minnesota Southern Cellular)
(Montana Southern Cellular)

ALLTEL Mobile Communications

Alpine Cellular
Atlantic Cellular
Atlas Cellular
Bay Area Cellular
Bell Atlantic
BellSouth Cellular
BellSouth Mobility
Blackwater Cellular Telephone
Blue Mountain Cellular
Bluegrass Cellular
Cantel
C.C. Cellular
Cellular Information Systems
Cellular One of Chicago
Cellular One of Galveston
Cellular One of LaSalle
Cellular One of Milwaukee
Cellular One of Sioux Falls
Cellular One of Wichita Falls
Cellular Properties
Cellular Ventures
Cellulink
Century Cellunet
Contel Cellular
Cybertel
Gencell
Greater South Dakota Cellular
GTE Mobilenet
HLD Cellular
Honolulu Cellular
Houston Cellular
Kaplan Telephone
Lake Huron Cellular
McCaw Cellular Communications
Mercury Communications
Metrocell
Mid-Tenn Cellular
Missouri Cellular
National Cellular
Pacific Telecom Cellular
PacTel Cellular
Quantum Communications
Radiofone
Rural Cellular
Shenandoah Cellular
Sunshine Cellular
Ten Woodland
U S West Cellular
United States Cellular

Crowley Cellular Telecommuni-cations

Advantage Cellular
ALLTEL Mobile Communications
Associated Communications
Atlantic Cellular
Bakersfield Cellular
Bay Area Cellular
Bell Atlantic Mobile
Bellsout Mobility
BellSouth Cellular
Cantel
Cellcom of Hickory
Cellular Information Systems
Cellular One of Amarillo
Cellular One of Berrien County
Cellular One of Boston
Cellular One of Chicago
Cellular One of Cincinnati
Cellular One of Cleveland
Cellular One of Columbia MO
Cellular One of Columbus OH
Cellular One of Detroit
Cellular One of Galveston
Cellular One of INdianapolis
Cellular One of LaSalle
Cellular One of Milwaukee

Cellular One of New York
Cellular One of Richmond
Cellular One of Sioux Falls
Cellular One of Syracuse
Cellular One of Upstate New York
Cellular One of Utica
Cellular One of Victoria
Cellular One of Washington
Cellular One of Wilmington
Cellular One of Witchita Falls
Cellular One of Worcester
Cellular One of Youngstown
Cellular Plus (A only)
Cellular South
Celutel
Centel Cellular
Centennial Cellular
Century Cellunet
Comcast
Commonwealth Mobile Systems
Contel Cellular
Cybertel Cellular
GenCell Management
GTE Mobilnet
Houston Cellular Telephone
Independent Cellular Network
Los Angeles Cellular Telephone
McCaw Cellular Communications
MCTA
Mercury Cellular
Metrocel Cellular
Metrophone
Miscellco Communications
Missouri Cellular
Mobiletel
Pace Communications
PacTel Cellular
Palmer Communications
Petroleum Communications
Public Service Cellular
Radiofone
Richmond Cellular Telephone
Rochester Telephone
Southwestern Bell Mobile System
Springwich Cell. Ltd Partner. (SNET)
Sterling Cellular
U S West Cellular
United States Cellular
Vanguard Cellular
West Central Cellular

Cybertel Cellular Telephone

AGT Cellular
Albany Telephone
Allcell
ALLTEL Mobile Communications
Ameritech Mobile Communications
Bakersfield Cellular Telephone
Baton Rouge Cellular Telephone
Bell Atlantic Mobile
BellSouth Mobility
Buffalo Telephone
Cantel
Cellcom of Green Bay
Cellular Information Systems
Cellular One of Akron
Cellular One of Amarillo
Cellular One of Atlantic City
Cellular One of
  Baltimore/Washington
Cellular One of Bay Area
Cellular One of Berrien County
Cellular One of Casper
Cellular One of Chicago
Cellular One of Cincinnati
Cellular One of Cleveland
Cellular One of Columbus
Cellular One of Dayton
Cellular One of Detroit

Cellular One of Fort Myers
Cellular One of Fort Smith
Cellular One of Fort Wayne
Cellular One of Kalamazoo
Cellular One of Lafayette
Cellular One of Lake Charles
Cellular One of La Salle, IL
Cellular One of Mansfield
Cellular One of Milwaukee
Cellular One of New Brunswick
Cellular One of Roanoke
Cellular One of Sioux Falls
Cellular One of South Bend
Cellular One of Syracuse
Cellular One of Trenton
Cellular One of Wilmington
Cellulink
Centel Cellular
Century Cellunet
Contel Cellular
Crowley Cellular
GTE Mobilnet ("B" systems only)
GenCell Management
Genesee Telephone
Honolulu Cellular
Houston Cellular Telephone
Independent Cellular Network
Indianapolis Telephone
Lincoln Telephone Cellular
MTS Cellular
McCaw Communications
Metro Mobile CTS
MetroCel Cellular
Metrophone
Milwaukee Telephone
MobileTel
Montgomery Cellular Telephone
NYNEX Mobile Communications
PacTel Mobile Access
Petroleum Communications
Radiofone
Richmond Cellular Telephone
Springwich Cell. Ltd Partner. (SNET)
U S West Cellular
United States Cellular

Dial Two

Adirondack Cellular
Allegan Cellular
ALLTEL Mobile Communications
Ally
Ameritech Mobile
Appalachian Cellular
Associatied Communications
Atlantic Cellular
Atlas Cellular
Bakersfield Cellular
Batelco
Bay Area Cellular
Bell Atlantic Mobile
BellSouth Cellular
BellSouth Mobility
Bledsoe Telephone
Bluegrass Cellular
Blue Mountain Cellular
Blue Ridge Cellular
Buckhead Telephone
Cantel
Canton Cellular
C. C. Cellular
Cedetel
Cellcom of Greenbay
Cellcom of Hickory
Cellular Connetion
Cellular, Inc.
Cellular Information Systems
Cellular One of Amarillo, TX
Cellular One of
  Baltimore/Washington

Cellular One of Berrien County, MI
Cellular One of Bloomington, IL
Cellular One of Boston, MA
Cellular One of Chicago, IL
Cellular One of Columbia, MO
Cellular One of Columbia, TN
Cellular One of Columbus, IN
Cellular One of Danville, IL
Cellular One of Detroit
Cellular One of East/Central PA
Cellular One of El Campo, TX
Cellular One of Gainesville, GA
Cellular One of Galveston, TX
Cellular One of Huntsville, TX
Cellular One of Indiana
Cellular One of Indianapolis
Cellular One of Lake Charles, LA
Cellular One of La Salle, IL
Cellular One of Milwaukee, WI
Cellular One of Northeast Colorado
Cellular One of Northeast, MS
Cellular One of Norfolk, NE
Cellular One of Ohio
Cellular One of the Rio Grande Valley
Cellular One of Santa Cruz
Cellular One of Southwest Florida
Cellular One of Southwest Virginia
Cellular One of Syracuse, NY
Cellular One of Upstate New York
Cellular One of Utica, NY
Cellular One of Victoria, TX
Cellular One of Wichita Falls, TX
Cellular One of Youngstown, OH
Cellular Plus
Cellular South
Cellular Ventures
Cellular XL
Cellulink
Celutel
Centel Cellular
Centennial Cellular
Century Cellunet
Chariton Valley Cellular
Clear Communications
Compcomm
Cone Enterprises
Contel Cellular
Crowley Cellular
Cybertel Cellular
Danbury Cellular
Dicomm Cellular
Digital Cellular of Texas
Dobson Cellular
Dominion Cellular
Farmers Cellular
First Kentucky Cellular
Gaia
GenCell Management
GMD Cellular
Greater South Dakota Cellular
GTE Mobilenet
Highland Cellular
Honolulu Cellular
Horizon Cellular
Houston Cellular
Illinois Valley Cellular
IUSA Cell
Kansas Cellular
Los Angeles Cellular
Lake Huron Cellular
Lincoln Telephone
Litchfield County Cellular
Mackinac Cellular
McCaw Cellular Communications
Mercury Communications
Metrocel Cellular
Metro Mobile CTS
Metrophone
Mega Comm
Midsouth Cellular

Miscellco Communication
Mocell
Mountaineer Cellular
Mountaineer Mobile
Nebraska Cellular
New Par
NYNEX Mobile Communications
Ocean County Cellular
Pace Communications
Pacific Telecom Cellular
PacTel Cellular
Palmer Communications
Pegasus Cellular
Petroleum Communications
PriCellular
Quantum Communications
Radiofone
RFB Cellular
Richard I. Vega Group
Rural Cellular Management
Santa Cruz Cellular
Southeast Indiana Cellular
Sterling Cellular
Sussex Cellular
Texoma Cellular
U S West Cellular
Union Telephone
United States Cellular
Unitel
Vanguard Cellular
Virginia Cellular
West Alabama Cellular
X-Cell Cellular
Yorkville Telephone
Youngstown Cellular

Dicomm Cellular

Advantage Cellular
Allegan Cellular
ALLTEL Mobile Communications
AMC Cellular
Associated Communications
Atlantic Cellular
Bell Atlantic Mobile
Bell Atlantic Mobile
Bell South Cellular
Bluegrass Cellular
Blue Mountain Cellular
Blue Ridge Cellular
C. C. Cellular
Cantel
Canton Cellular
Cellcom of Hickory
Cellular of Indiana
Cellular Information Systems
Cellular One of
  Baltimore/Washington
Cellular One of Berrien County
Cellular One of Boston
Cellular One of Chicago
Cellular One of Detroit
Cellular One of Galesburg
Cellular One of West Virginia
  Highlands
Cellular One of Indianapolis
Cellular One of Northeast Colorado
Cellular One of Ohio
Cellular One of Southwest Florida
Cellular One of Syracuse/Utica
Cellular One of Upstate New York
Celutel
Centennial Cellular
Century Cellunet
Comcast Cellular Communications
Crowley
Cybertel
Gaia
GMD Partnership
General Cellular

GTE Mobile Communications
Greater South Dakota Cellular
Honolul Cellular
Horizon Cellular
Houston Cellular
Independent Cellular Network
First Kentucky Cellular
Litchfield County Cellular
McCaw Cellular Communications
MCMG
Mercury Communications
Metrophone
Mid-South Cellular
Milwaukee Cellular
Mobile Communications Systems
Mountaineer Mobile
N C Cellular
New Par
NYNEX Mobile Communications
Ocean County Cellular
Ohio State Cellular
PC Cellular
Pacific Telecom Cellular
PacTel Cellular
Palmer Cellular
Pegasus Cellular
Petroleum Communications
RFB Cellular
Radiofone
Richmond Cellular
Rural Cellular Management
Sioux Falls Cellular
Springwich Cell. Ltd Partner. (SNET)
Southeastern Cellular
Sterling Cellular Union Telephone
U S West Cellular
United Bluegrass
Unitel
United State Cellular
Vanguard Cellular
Virginia Cellular
Youngstown Cellular

Digital Cellular of Texas

Advantage Cellular
AGT Cellular
ALLTEL Mobile Communications
Ameritech mobile Communications
Atlantic Cellular
Bell Mobility Cellular
Blue Grass Cellular
Brazos Cellular Communications
Cantel
Cellular, Inc.
Cellular Information Systems
Cellular One of Lake Charles
Cellular One of wilmington
Cellular Plus
Cellular Systems International
Cellular Three
Cellulink
Centel Cellular
Century Cellunet
Coastel Communications
Contel Cellular
Crowley Cellular Telecommunictions
Dobson Cellular Systems
Enid Cellular
GenCell Management
GTE Mobilnet
Houston Cellular
Idependent Cellular Network
Kansas Cellular
Leaco Rural Telephone
Lincoln Telephone
McCaw Cellular Communications
Metrocel Cellular
Mid-Tex Cellular
Miscellco Communications

Mobiletel
MT&T Cellular
MTS Cellular
Nebraska Cellular
NYNEX Mobile COmmunications
PacTel Cellular
Petroleum Communications
PTSI Cellular
Radiofone
Rochester Telephone
Sasktel Cellular
Savannah Cellular
Southwestern Bell Mobile Systems
StarCellular
U S West Cellular
Union Telephone
United States Cellular
Unitel
Unity Cellular Systems (Unicel)
West Central Cellular
Westex
XIT Cellular

Dobson Cellular Systems

Advantage Cellular
ALLTEL Mobile Communications
Ameritech Mobile Communicatons
Appalachian Cellular
Atlantic Cellular
BCTel Cellular
Bell Atlantic Mobile
BellSouth Mobility
Bledsoe Telephone
Blue Grass Cellular
Blue Mountain Cellular
Brazos Cellular Communications
Cal-One Cellular
C.C. Cellular
Cellcom
Cellular Connection
Cellular, Inc.
Cellular Information Systems
Cellular of Indiana
Cellular One of Amarillo
Cellular One of Berrein Count
Cellular One of Columia MO
Cellular One of Gainesville GA
Cellular One of Highlands
Cellular One of Indianapolis
Cellular One of Lake Charles
Cellular One of Ozarks
Cellular One of Wichita Falls
Cellular South
Cellular Systems International
Cellular Three
Cellular XL
Centel Cellular
Century Cellunet
Coastel Communications
Contel Cellular
Crowley
Digital Cellular
Eastex Cellular
Farmers CellularTelephone
GenCell Management
GTE Mobilnet
Houston Cellular Telephone
Independent Cellular Network
Interstate
Kansas Cellular
Leaco Rural Telephone
Lincoln Telephone
McCaw Cellular Communications
Mercury Cellular
Metrocel Cellular
Mid-Tex Cellular
Miscellco Communications
Missouri Cellular
Mobiltel

Mountaineer Mobile
Nebraska Cellular
NYNEX Mobile Communications
Pace Communications
Pacific Telecom
PacTel Cellular
Petroleum Communications
Pine Cellular Telephone
Pioneer Cellular
PTSI Cellular
Rochester Telephone
Shenandoah Cellular
Southeastern Cellular
Southwestern Bell Mobile Systems
Springwich Cell. Ltd Partner. (SNET)
StarCellular
U S West Cellular
Union Telephone
United States Cellular
Unitel
Unity Cellular Systems (Unicel)
West Central Cellular
Westex
Wilkes Telephone
XIT Cellular

Dominion Cellular Systems

Boatphone
Cellular Communications of Puerto
  Rico
Cellular One of Columbia, TN
Crowley Cellular
Lone Star Cellular

Ed Tel Cellular

AGT Cellular
ALLTEL Mobile Communications
Ameritech Mobile Communications
Atlantic Cellular
B.C. Tel Mobility Cellular
Bell Atlantic Mobile
Bell Mobility Cellular
BellSouth Mobility
Bledsoe
Bluegrass Cellular
C. C. Cellular
Cal-One Cellular
Cellcom
Cellular, Inc.
Centel Cellular
Century Cellular
Century Cellunet
Coastel Communications
Contel Cellular ("B" systems)
Enid Cellular
Five Star Cellular
GTE Mobilnet ("B" systems)
Independent Cellular
Indiana RSA 7
Island Telephone Cellular
Kansas Independent
Liberty
Lincoln Telephone
MACtel Cellular
MT&T Mobile
MTS Cellular
NBTel Cellular
NewTel Cellular
NYNEX Mobile Communications
Panhandle Telephone System
Pace
Pacific Telecom Cellular
PacTel Cellular
Poka Lambro
Québéc Téléphone Cellulaire
Rochester Telephone Mobile
  Communications
Sasktel Cellular

Springwich Cell. Ltd Partner. (SNET)
Southwestern Bell Mobile Systems
StarCellular
Thunder Bay Cellular
U S West Cellular
Union Telephone Cellular
United States Cellular
Virginia RSA 10
XIT Cellular

Enid Cellular

Alberta Government Telephone
ALLTEL Mobile Communications
Ameritech Mobile Communications
Bell Atlantic Mobile
BellSouth Mobility
Buffalo Telephone
Cantel
Cellcom
Cellular, Inc.
Cellular One of Akron
Cellular One of
  Baltimore/Washington
Cellular One of Cincinnati
Cellular One of Cleveland
Cellular One of Columbus
Cellular One of Dayton
Cellular One of Ft. Smith
Cellular One of Youngstown
Cellular Plus
Cellular South
Cellular Technology
Cellulink
Centel Cellular
Century Cellunet
Coastel Communications
Contel Cellular
Ed Tel Cellular
GTE Mobilnet ("B" systems only)
IUSA Cell
Independent Cellular Network
Kansas Independent Networks
MACtel Cellular
MT&T Cellular
MTS Cellular
Maine Cellular
Mercury Cellular & Paging
Nebraska Cellular
NYNEX Mobile Communications
PacTel Cellular
Petroleum Communications
Public Service Cellular
Rochester Telephone Mobile
  Comms.
Sasktel Cellular
Savannah Cellular
Shoals Cellular
Southwestern Bell Mobile Systems
Springwich Cell. Ltd Partner. (SNET)
StarCellular
U S West Cellular
United States Cellular
Utica/Rome Cellular Telephone
Unity Cellular Systems (Unicel)
Wes-Tex Telecommunications
West Central Cellular
XIT Cellular

Finger Lakes Telephone

Advantage Cellular
ALLTEL Mobile Communications
Ally
AMC Cellular
Appalachian Cellular
Associated Comm.
Atlantic Cellular
Auburn Television Group
Bay Area Cellular
Bell Atlantic

Bell Mobility Cellular
Bell South Mobility
Blue Mountain Cellular
Blue Ridge Cellular
Bluegrass Cellular
Cantel
Canton Cellular
Cellcom Hickory
Cellular Connection
Cellular Information Systems
Cellular One of Amarillo
Cellular One of
  Baltimore/Washington
Cellular One of Berrien County
Cellular One of Boston
Cellular One of Chicago
Cellular One of Columbia
Cellular One of East Central PA
Cellular One of Indiana
Cellular One of Lake Charles
Cellular One of Merridian
Cellular One of Natchitoches
Cellular One of Northeast GA
Cellular One of Santa Cruz
Cellular One of Southeast Florida
Cellular One of Upstate New York
Cellular One of Waycross
Cellular One of West Illinois
Cellular XL
Cellutel
Centel Cellular
Centennial Cellular
Clear Communications Group
Comcast Cellular Communications
Contel Cellular
Crowley Cellular
Danbury Cellular Telephone
DIcomm Cellular Telephone
Eastex Cellular One
Ellis Thompson
Farers Cellular Telephone
First Fayette Cellular
First Kentucky Cellular
Gaia
GenCell Management
GMD/ Cellular One
Greater South Dakota Cellular
GTE Mobilnet
Highland Cellular
Honolulu Cellular Telephone
Horizon Cellular
Houston Cellular Telephone
Independent Cellular Network
Indianapolis Cellular Telephone
Iowa East Cellular
Kansas Cellular
Kentucky Cellular
Kull Cellular
Lake Huron Cellular
Litchfield Count Cellular
McCaw Communications
MCMG
Mercury Communications
Metrocel Cellular Telephone
Metrophone
Minerich Cellular
Miscellco Communications
Missouri Cellular
Mountaineer Mobile
National Cellular
NYNEX Mobile Communications
Ocean County Cellular
Ohio State Cellular
Pace Communications
PacTel Cellular
Palmer Communications
Petroleum Communications
PriCellular
Quantum Communications

Roaming Agreements

RFB Cellular
Rochester Telephone
Rural Cellular
Springwich Cellular
Sterling Cellular
Sussex cellular
Syracuse Telephone
Texoma Cellular
Tsaconas
U S West Cellular
Union Telephone
United States Cellular
Unitel
Utica Telephone
Vanguard Cellular Systems
Virginia 10
Virginia Cellular
Wichita Falls Cellular Telephone
Yorkville Telephone
Youngstown Cellular Telephone

Five Star Cellular

AGT Cellular
ALLTEL Mobile Communications
Ameritech Mobile Communications
Atlantic Cellular
BCTel Cellular
Cellular Information Systems
Cellular One of Amarillo
Centel Cellular
Century Cellunet
Coastel Communications
Contel Cellular
Crowley CellularTelecommunications
GenCell Management
GTE Mobilnet
Houston Cellular Telephone
Independent Cellular Network
Kansas Cellular
Lincoln Teleopone
McCaw Cellular Communications
Metrocel Cellular
Miscellco Communicaton
Mobiletel
MT&T Cellular
MTS Cellular
Nebraska Cellular
Petroleum Communications
PTSI Cellular
Sasktel Cellular
Southwestern Bell Mobile Systems
StarCellular
Westex

GenCell Management

Albany Telephone
Allcell
Ally
Alpha Cellular
Ameritech
Appalachian Cellular
Atlantic Cellular
Bakersfield Cellular
Bay Area Cellular Telephone
Bell Atlantic
Bell South Cellular
BellSouth Mobility
Big Horn Cellular
Bledsoe Cellular
Blue Mountain Cellular
Blue Ridge Cellular
Bluegrass Cellular
Buckhead Telephone
Buffalo Telephone
Cal-One Cellular
Cantel
Canton Cellular
Cedetal
Cellcom of Green Bay

Cellcom of Hickory
Cellular Phone of Kentucky
Cellular Holding
Cellular Information Systems
Cellular One of Amarillo
Cellular One of
  Baltimore/Washington
Cellular One of Berrien County
Cellular One of Boston
Cellular One of Chicago
Cellular One of Collier/Hendry
Cellular One of Detroit
Cellular One of Indianapolis
Cellular One of Jacksonville
Cellular One of Lake Charles
Cellular One of Northeast Colorado
Cellular One of Northeast Georgia
Cellular One of Sioux Falls
Cellular One of Southwest Florida
Cellular One of the Rio Grande Valley
Cellular One of Upstate New York
Cellular One of Victoria
Cellular One of Western Illinois
Cellular One of Wichita Falls, TX
Cellular One of Youngstown
Cellular Plus
Cellular Properties
Cellular Technology
Cellular XL
Cellular, Inc.
Celutel
Centel Cellular
Centennial Cellular
Century Cellular
Century Cellunet
Clear Communications
Coastel Communications
Comcast Cellular
Cone Enterprises
Contel
Contel Cellular
Crowley Cellular
Cybertel Cellular
Danbury Cellular
Dicomm Cellular
Dobson Cellular
Dominion Cellular
Eastex Cellular
Enid Cellular
Farmers Cellular
FGI Cellular
Finger Lakes Telephone
First Fayette Cellular
Gaia
Galveston Cellular
Genesee Telephone
Great Lakes of Iowa
Greater South Dakota Cellular
GTE Mobilnet ("B" systems only)
Horizon Cellular
Houston Cellular
Illinois Valley Cellular
Independent Cellular
Iowa East Cellular
Kansas Cellular
Los Angeles Cellular
Mackinac Cellular
Maine Cellular
McCaw Communications
Mercury Cellular
Mercury Communications
Metro Mobile
Metro One
Metro Phone Cellular
Metrocel Cellular
Midtex Cellular
Milwaukee Telephone
Minnesota Southern Cellular
Miscellco Communications

Missouri Cellular
Mobile Communications Systems
MobileTel
Mountaineer Cellular
NCPT Cellular
New Par
North Carolina Cellular
NYNEX Mobile Communications
PC Cellular of Vermont
Pacific Northwest Cellular
Pacific Telecom Cellular
PacTel Cellular
Palmer Communications
Petroleum Communications
PriCellular
Prime Cellular Management
Quantum Communications
Radiofone
RFB Cellular
Richmond Cellular
Rural Cellular Management
Santa Barbara Cellular
Santa Cruz Cellular
Shenandoah Cellular
Sooner Celluar
Southwestern Bell
StarCellular
Sterling Cellular
Sunshine Cellular
Sussex Cellular
Syracuse Telephone
Tsacona Cellular
U S West Cellular
Union Telephone
United States Cellular
Unitel
Utica Telephone
Vanguard Cellular Systems
Virginia Cellular
West Alabama Cellular
Westex
X-Cell Cellular
Yorkville Telephone

GMD Limited Partnership

Albany Telephone
ALLTEL Mobile Communications
Ally
Alpha Cellular
Alton Cellular
Amarillo Cellular
AMC Cellular
American Rural Cellular
Appalachian Cellular
Associated Communications
Atlantic Cellular
Atlas Cellular
Auburn Television
Bachtel Cellular
Bahamas Telecommunications
Bakersfield Cellular
Bay Area Cellular
B.C. Tel Mobility Cellular
Bell Atlantic Mobile
Bell Mobility Cellular
Bell South Cellular
Bell South Mobility
Bledsoe Telephone
Bluegrass Cellular
Blue Mountain Cellular
Blue Ridge Cellular
Boat Phone
Brazos Cellular
C. C. Cellular
Cable and Wireless
Cal-One Cellular
Cantel
Canton Cellular

Roaming Agreements

Carolina West Cellular
CCPR Services of Puerto Rico
Cellcom of Green Bay
Cellcom of Hickory
Celludyne
Cellular Holdings
Cellular Information Systems
Cellular of Sioux Falls
Cellular One of
  Baltimore/Washington
Cellular One of Berrien County, MI
Cellular One of Chicago
Cellular One of Indiana
Cellular One of Mid-South Texas
Cellular One of Northeast Colorado
Cellular One of Scottsbluff, NE
Cellular One of Southwest Florida
Cellular One of Upstate New York
Cellular Phone of Kentucky
Cellular Three
Cellular XL
Cellular, Inc.
Celutel
Centel Cellular
Centennial Cellular
Century Cellunet
Clear Communications
Colmbia Cellular Partnership
Comcast Cellular
Cone Enterprises
Contel Cellular
Cross-Valliant Cellular
Crowley Cellular
CTEC
Dakota Cellular
Danbury Cellular
Dobson Cellular
Dominion Cellular
East Caribbean Cellular
Eastex Cellular
EdTel Cellular
Finger Lakes Telephone
First Fayette Cellular
Galveston Cellular
GenCell Management
Georgia Independent RSA 7&10
GMD Limited
Greater South Dakota Cellular
GTE Mobilnet
Highland Cellular
Honolulu Cellular
Horizon Cellular
Hutchinson Telephone of Hong Kong
Independent Cellular Network
Inland Cellular
Interstate Cellular (Intercel)
Iowa East Cellular
IUSA Cell, S.A. de C.V.
JHP Cellular
Kansas Independent
Kaplan Telephone
Kull Cellular
Lake Charles Cellular
Lake Huron Cellular
Lamar County Cellular
Lincoln Telephone
Litchfield County Cellular
Lone Star Cellular
MacTel Cellular
Maine Cellular
Matanuska-Kenai Cellular
  Partnership
McCaw Cellular Communications
MCMG
Mega Comm
Mercury Cellular & Paging
Mercury Communications
Midland Communications
Minnesota Southern Cellular
Miscellco Communications

Missouri 7 Limited Partnership
Missouri Cellular
Missouri RSA 5 Partnership
Mobile Communications
Mobile Telenet
Montana Cellular
Mountaineer Cellular Telephone
Mountaineer Mobile
MT&T Cellular
MTS Mobility
National Cellular Limited Partnership
Nationwide RSA Service
NBTel Cellular
Nebraska Cellular
NewPar
NewTel Cellular
NYNEX Mobile Communications
PC Cellular of Kentucky
Pacific Telecom Cellular
PacTel Cellular
Palmer Communications
Panhandle Telecommunications
Peoples Cellular
Petroleum Communications
Pioneer Telephone
Point Communications
Poka Lambro Telecommunications
Public Service Cellular
Puerto Rico Telephone
Quantum Communications
Québec Téléphone Cellulaire
R & D Cellular
Radiofone Cellular
RFB Cellular
Rochester Telephone
RSA II Partnership
Rural Cellular Management
RVC Services
Saco River Cellular
Santa Cruz Cellular
SaskTel Cellular
Singapore Telecommunications
Sooner Cellular
Southeast Indiana Cellular
Southern Cellular
Southwestern Bell Mobile
Springwich Cell. Ltd Partner. (SNET)
Sterling Cellular
Syracuse Telephone
Taylor Telecommunications
Tele RJ of Rio De Janeiro
Texas 5
Texas RSA 1 Limited Partnership
Thunder Bay Cellular
U S West New Vector
Uintah Basin Telephone
Union Telephone
United State Cellular
Unitel
Unity Cellular
US Virgin Island Telephone
Utica Telephone
Vanguard Cellular
Victoria Cellular
Virginia 10 RSA
Virginia Cellular
West Alabama Cellular
West Central Cellular
X-Cell Cellular
Yorkville Telephone
Youngstown Cellular

GTE Mobilnet

AGT Cellular
Advantage Cellular
Albany Telephone
ALLTEL Mobile Communications
Ameritech
Atlantic

B.C. Tel Mobility Cellular
Bakersfield
Baton Rouge Cellular
Bell Atlantic Mobile
Bell Mobility Cellular
BellSouth Mobility
Buffalo Telephone
Cantel
Cellcom
Cellcom of Hickory, NC
Cellular, Inc.
Cellular Information Systems
Cellular One of Akron
Cellular One of Albany, GA
Cellular One of Amarillo
Cellular One of
  Baltimore/Washington
Cellular One of Billings
Cellular One of Chicago
Cellular One of Cincinnati
Cellular One of Cleveland
Cellular One of Columbus
Cellular One of Dayton
Cellular One of Dothan, AL
Cellular One of Indianapolis
Cellular One of Jacksonville
Cellular One of Lake Charles
Cellular One of Panama City
Cellular One of Sioux Falls
Cellular One of Wichita Falls, TX
Cellular One of Wilmington
Cellular One of Youngstown
Cellular South
Cellulink
Cellulink of Eau Claire
Centel
Century Cellular
Century Cellunet
Coastel
Columbus Cellular
Commonwealth
Contel
Crowley
Cybertel
Ed Tel Cellular
Enid
Genesee Telephone
Gulf Coast
Honolulu Cellular
Houston Cellular
Independent Cellular
Kansas Cellular
Lincoln Telephone Cellular
MACtel
MT&T
MTS
Maine Cellular
Mercury
Metro One
MetroCel
Milwaukee
MobileTel
Montgomery
Nebraska Cellular
NYNEX Mobile Communications
PacTel Cellular
Petroleum Communications
Public Service Cellular
Radiofone
Rochester
Southwestern Bell Mobile Systems
Springwich Cell. Ltd Partner. (SNET)
StarCellular
U S West Cellular
United States Cellular
Unity Cellular Systems (Unicel)
Vanguard
West Central Cellular

Honolulu Cellular Telephone

ALLTEL Mobile Communications
Ally
Ameritech Mobile Communications
Appalachian Cellular
Atlantic Cellular
Atlas Cellular
Bakersfield Cellular
Bay Area Cellular
Bell Atlantic Mobile
BellSouth Mobility
Big Horn Cellular
Blackwater Cellular
Bledsoe Telephone
Blue Mountain Cellular
Blue Ridge Cellular Telephone
Cantel
Cellcom of Green Bay
Cellular 2000
Cellular Communications of Puerto Rico
Cellular Information Systems
Cellular of Indiana
Cellular One of Beaumont
Cellular One of Boston
Cellular One of Chicago
Cellular One of Cincinnati
Cellular One of Cleveland
Cellular One of Columbia MO
Cellular One of Columbus OH
Cellular One of Detroit
Cellular One of Gainesville
Cellular One of Galesburg
Cellular One of Greenwood
Cellular One of Huntsville
Cellular One of Indianapolis
Cellular One of Jacksonville NC
Cellular One of Marianna
Cellular One of McCaw
Cellular One of Milwaukee
Cellular One of New York
Cellular One of Richmond
Cellular One of Santa Cruz
Cellular One of Sioux Falls
Cellular One of Syracuse
Cellular One of Upstate New York
Cellular One of Utica
Cellular One of Washington
Cellular One of Youngstown
Cellular Plus
Cellular Ventures
Cellular XL
Comcast
Contel Cellular
Cooper Cellular
Cybertel Cellular
Danbury Cellular
Dominion Cellular
Farmers Cellular Telephone
GenCell Management
GTE Mobilnet
Houston Cellular
Interstate Cellular (Intercel)
Iowa East Cellular
Lake Huron Cellular
Lincoln Telephone
Los Angeles Cellular Telephone
Metrocel Cellular
Metrophone
Miscellco Communications
Mobiletel
Mountaineer Mobile
National Cellular
Ocean County Cellular
Pace Communications
Petroleum Communications
Radiofone
Rural Cellular
Texas 16 Cellular Telephone

U S West
Union Telephone
United States Cellular
Vanguard Cellular
West Alabama Cellular

Horizon Cellular

Albany Telephone
AMC Cellular
American Rural Cellular
Atlantic Cellular
Bakersfield Cellular
Bay Area Cellular
Bell Atlantic Mobile
Bluegrass Cellular
Buffalo Telephone
Cantel Cellular
Canton Cellular
Cellcom
Cellcom of Hickory
Cellular Information Systems
Cellular One of
  Baltimore/Washington
Cellular One of Berrien County
Cellular One of Boston
Cellular One of Chicago
Cellular One of Columbia, TN
Cellular One of Danville
Cellular One of Detroit
Cellular One of Indianapolis
Cellular One of Lake Charles
Cellular One of La Salle, IL
Cellular One of Middle Tennessee
Cellular One of Upstate New York
Cellular Phone of Kentucky
Celutel
Centel Cellular
Contel Cellular
Comcast Cellular
Crowley Cellular
Dial Two
Dicomm Cellular
Dobson Cellular
Gaia
General Cellular
Genesee Telephone
GMD Partnership
GTE Mobilnet
Houston Cellular
Iowa East Cellular
McCaw Communications
Metrocel Cellular
Metrophone
Mountaineer Mobile
Mid-South Cellular
New Par
NYNEX Mobile Communications
Ocean County Cellular
PC Cellular of Vermont
PTI Cellular
PacTel Cellular
Palmer Cellular
Pegasus Cellular
PriCellular
Radiofone
RFB Cellular
Richmond Cellular
Shenadoah Mobile
Southwestern Bell
Sterling Cellular
Sunshine Cellular
Sussex Cellular
Syracuse Telephone
United State Cellular
U S West Cellular
Utica Telephone
Vanguard Cellular
Youngstown Cellular

Houston Cellular Telephone

Allcell
ALLTEL Mobile Communications
Associated Communications
Atlantic Cellular
Bakersfield Cellular
Bay Area Cellular
Bell Atlantic Mobile
BellSouth Mobility
Cantel
Cellular Information Systems
Cellular One of Amarillo
Cellular One of
  Baltimore/Washington
Cellular One of Berrien County
Cellular One of Boston
Cellular One of Chicago/Gary
Cellular One of Detroit
Cellular One of Ft. Myers
Cellular One of Indianapolis
Cellular One of La Salle, IL
Cellular One of Lake Charles
Cellular One of Milwaukee
Cellular One of Richmond
Cellular One of Roanoke
Cellular One of Syracuse
Cellular One of Utica
Cellular One of Victoria
Cellular One of Wilmington
Cellular One of Youngstown
Cellular South
Celutel
Centel Cellular
Century Cellunet
Coastel Communications
Comcast
Contel Cellular
Crowley Cellular TeleCommunications
Cybertel Cellular Telephone
FGI Cellular Management
GTE Mobilnet ("B" systems only)
GenCell Management
Gulf Coast Cellular
Los Angeles Cellular Telephone
MTS Cellular
McCaw Communications
Mercury Cellular and Paging
Metro Mobile CTS
Metro One
MetroCel Cellular
Metrophone
MobileTel
Pace Communications
PacTel Cellular
Palmer Communications
Petroleum Communications
Radiofone
U S West Cellular
United States Cellular
Vanguard Cellular Systems

Illinois Valley Cellular

Advantage Cellular
ALLTEL Mobile Communications
Ally
Ameritech Mobile Communications
Atlantic Cellular
Bakersfield Cellular
Bell Atlantic
Bell Mobility Cellular
Bell South Mobility
Bledsoe
Blue Mountain Cellular
Blue Ridge Cellular
Bluegrass Cellular
Brazos Cellular
Cal-One Cellular
Cantel
Carolina West Cellular

Cellcom of Green Bay
Cellular Connection
Cellular Holdings
Cellular Information Systems
Cellular One of Columbia, MO
Cellular One of Galesburg
Cellular One of Indiana
Cellular One of Lake Charles
Cellular One of Mid-South Texas
PriCellular
Cellular One of West Alabama
Cellular Plus
Cellular Properties
Cellular, Inc.
Celutel
Centel Cellular
Century Cellunet
Churchill Cellular
Coastel Communications
Consolidated Communications
  Mobile
Contel
Cooper Cellular
Cybertel
Digital Cellular of Texas
Dobson Cellular
Enid Cellular
Farmers Cellular
First Kentucky Cellular
GenCell Management
GMD Limited
GTE Mobilnet
Highland Cellular
Independent Cellular Network
Interstate Cellular (Intercel)
Kansas Cellular
Lake Huron Cellular
Lincoln Telephone
Litchfield County Cellular
MacTel Cellular
Maine Cellular
Mega Comm
Mercury Cellular
Mid-Missouri Cellular
Mid-Tennessee Cellular
Mississippi Cellular
Missouri Cellular
Mobile Telephone
Mountainer Cellular
MT&T Cellular
NYNEX Mobile Communications
Pacific Northwest Cellular
PacTel Cellular
Panhandle Telephone
PC Cellular of Kentucky
PC Cellular of Vermont
Peoples Cellular
Pine Cellular
Public Service Cellular
R & D Cellular
Radiofone
Rochester Telephone
Southeast Indiana Cellular
Southwestern Bell Mobile
Springwich Cell. Ltd Partner. (SNET)
StarCellular
Sterling Cellular
Stillwater Cellular
Sunshine Cellular
Sussex Cellular
Ten Woodland Road
Thunder Bay Cellular
U S West Cellular
UBET Cellular
Union Telephone
United States Cellular
Unity Cellular Systems (Unicel)
Vanguard Cellular
Virginia 10 RSA Limited

West Central Cellular
Westex Cellular
XIT Cellular
Yorkville Telephone

Independent Cellular Network

Advantage Cellular
AGT Cellular
ALLTEL Mobile Communications
Ally
Alpha Cellular
Ameritech Mobile Communications
Appalachian Cellular
Atlantic Cellular
Atlas Cellular
Baton Rouge Cellular Telephone
B.C. Tel Mobility Cellular
Bell Atlantic Mobile
Bell Mobility Cellular
BellSouth Cellular
BellSouth Mobility
Blue Mountain Cellular
Blue Ridge Cellular Telephone
Blue Ridge Cellular Lenoir, NC
Brazos Cellular
Buffalo Telephone
C. C. Cellular
Cal-One Cellular
Cantel
Canton Cellular
Carolina West Cellular
Cellcom
Cellcom of Hickory
Cellular Holdings
Cellular Inc.
Cellular Information Systems
Cellular One of
  Baltimore/Washington
Cellular One of Chicago
Cellular One of Columbia, MO
Cellular One of Indiana
Cellular One of Indianapolis
Cellular One of Meridian
Cellular One of Milwaukee
Cellular One of Northeast Georgia
Cellular One of Sioux Falls
Cellular One of Southwest Georgia
Cellular One of Syracuse/Utica
Cellular One of Upstate New York
Cellular Phone of Kentucky
Cellular Plus of Georgia
Cellular Plus of Pennsylvania
Cellular Systems International
Cellular Ventures
Cellular XL
Centel Cellular
Century Cellular
Century Cellunet
Coastel Communications
Columbia Cellular Partnership
Contel Cellular
Crowley Cellular
Cybertel Cellular
Dicomm Cellular
Dobson Cellular
Dominion Cellular
Easterbrook Cellular
Ed Tel Cellular
Enid Cellular
Finger Lakes Telephone
First Kentucky Cellular
GenCell Management
GTE Mobilnet ("B" systems only)
Highland Cellular
houston Cellular
Illinois Valley Cellular
Interstate Cellular (Intercel)
Iowa East Cellular
Kansas Cellular

Lincoln Telephone
Litchfield County Cellular
MACTel Cellular System
Maine Cellular
McCaw Cellular Communications
MCMG
Mercury Cellular Telephone & Paging
Mercury Communications
Metro One
Metrophone
Miscellco
Mississippi Cellular
Missouri Cellular
Mobile Tel
Mountaineer Mobile
MTS Cellular
MT&T Cellular
National Cellular
Nebraska Cellular
NYNEX Mobile Communications
Pace Communications
Pacific Telecom Cellular
PacTel Cellular
Palmer Cellular
Panama City Telephone
Panhandle Telecommunication
Petroleum Communications
Public Service Cellular
Quantum Communications
Québéc Téléphone Cellulaire
Radiofone
RFB Cellular
Richmond Cellular
Rochester Telephone Mobile
  Comms.
SaskTel Cellular
Savannah Cellular
Shenandoah Mobile
Southwestern Bell Mobile Systems
Springwich Cell. Ltd Partner. (SNET)
Southwestern Bell Mobile
StarCellular
Sterling Cellular
TelCel
Texas 16 Cellular
Thunder Bay Cellular
U S West Cellular
Union Telephone
United States Cellular
Unity Cellular Systems (Unicel)
Universal Telecell
Virginia 10 RSA Limited
Vanguard Cellular
West Alabama Cellular
Wichita Falls Cellular
Westex Telecommunications
Youngstown Cellular Telephone

Inland Cellular

B.C. Tel Mobility Cellular
Bluegrass Cellular
Cal-One Cellular
Cellular, Inc.
Cellular One of Sioux Falls
Contel Cellular
GTE Mobilenet
MACtel Cellular Systems
Missouri Cellular
Pacific Telecom Cellular
PacTel Cellular
U S West Cellular
United States Cellular

Interstate Cellular (Intercel)

ALLTEL Mobile Communications
Ameritech Mobile Communications
Appalachian Cellular
Atlantic Cellular
Atlas Cellular

Roaming Agreements

Bell Atlantic Mobile
BellSouth Cellular
BellSouth Mobility
Bledsoe Telephone
Blue Grass Cellular
Brazos Cellular
C.C. Cellular
Cellcom of Hickory
Cellular Information Systems
Cellular One in Ohio
Cellular One of Gainesville
Cellular One of Indianapolis
Cellular One of Lenior
Cellular Plus
Cellular South
Centel Cellular
Century Cellunet
Coastel Communications
Cone Enterprises
Contel Cellular
Cybertel Cellular
Dobson Cellular
Farmers Cellular Telephone
Five Star Cellular
Greater South Dakota Cellular
GTE Mobilnet
Highland Cellular
Houston Cellular
Independent Cellular Network
Lake Huron Cellular
Litchfield County Cellular
McCaw Cellular Communications
Mercury Cellular
Mercury Communicatons
Metrocel Cellular
Mobiletel
MT&T Cellular
NYNEX Mobile Communications
Pace Communications
PacTel Cellular
Palmer Communications
Petroleum Communications
Public Service Cellular
Radiofone
Savannah Cellular
Shenandoah Mobile
Southeastern Cellular
Southwestern Bell Mobile Systems
Springwich Cell. Ltd Partner. (SNET)
Sterling Cellular Management
Texas Cellular
U S West Cellular
Union Telephone
United States Cellular
Unity Cellular Systems (Unicel)
Vanguard Cellular
Yorkville Telephone

Iowa East Cellular Telephone

Advantage Cellular
Allegan Cellular
ALLTEL Mobile Communications
Ally
Associated Communications
Atlas Cellular
Bakersfield Cellular
Bay Area Cellular
Bell Atlantic Mobile
BellSouth Cellular
BellSouth Mobility
Bledsoe Telephone
Blue Ridge Cellular Telephone
Cellcom of Green Bay
Cellcom of Hickory
Cellular Connection
Cellular Information Systems
Cellular of Indiana
Cellular One of Amarillo
Cellular One of Berrein County

Cellular One of Chicago
Cellular One of Cincinnati
Cellular One of Cleveland
Cellular One of Columbia, MO
Cellular One of Columbus, OH
Cellular One of Danville
Cellular One of Detroit
Cellular One of Gainesville, GA
Cellular One of Galesburg
Cellular One of Galveston
Cellular One of Indianapolis
Cellular One of LaSalle
Cellular One of Marianna
Cellular One of Milwaukee
Cellular One of Richmond
Cellular One of Santa Cruz
Cellular One of Syracuse
Cellular One of Upstate New York
Cellular One of Wichita Falls
Cellular Plus
Cellular South
Cellular Systems International
Cellular Ventures
Cellulink
Centel Cellular
Centennial Cellular
Century Cellunet
Coastel Communications
Crowley Cellular Telecommunications
Cybertel Cellular
Digital Cellular
Dominion Cellular
Farmers Cellular Telephone
GenCell Management
Greater South Dakota Cellular
GTE Mobilnet
Honolulu Cellular Telephone
Houston Cellular Telephone
Kansas Cellular
Lake Huron Cellular
Lincoln Telephone
Litchfield County Cellular
Metrocel Cellular
Mississippi Cellular
Mobiletel
MTS Cellular
National Cellular
Nebraska Cellular
NYNEX Mobile Communications
Oneonta Telephone
Pace Communications
Palmer Communications
Petroleum Communications
PTSI Cellular
Public Service Cellular
Quantum Communications
Radiofone
Rural Cellular
Sanannah Cellular
Southwestern Bell Mobile Systems
Texas 16 Cellular Telephone
U S West Cellular
Union Telephone
United States Cellular
Unity Cellular Systems (Unicel)
Vanguard Cellular
West Alabama Cellular

IUSA Cell (Mexico)

B.C. Tel Mobility Cellular
Cantel
Cellular Information Systems
Cellular One of Chicago
Coastel Communications
Comcel (Guadalajara, Mexico)
Enid Cellular
Petroleum Communications
U S West Cellular
United States Cellular

Kansas Independent Cellular Networks

Advantage Cellular
AGT Cellular
ALLTEL Mobile Communications
American Rural Cellular
Ameritech Mobile Communications
Appalachian Cellular
Atlantic Cellular
Atlas Cellular
Auburn Television
BCTel Cellular
Bell Atlantic Mobile
BellSouth Cellular
Bledsoe Telephone
Blue Grass Cellular
Blue Mountain Cellular
Blue Ridge Cellular
Brazos Cellular
Cal-One Cellular
Cantel
C.C. Cellular
Cellcom of Green Bay
Cellular Connection
Cellular, Inc.
Cellular Information Systems
Cellular of Indiana
Cellular One of Berrien County
Cellular One of Chicago
Cellular One of Columbia, MO
Cellular One of Columbia, TN
Cellular One of Danville
Cellular One of Gainesville, GA
Cellular One of Greenville
Cellular One of Hattiesburg
Cellular One of Indianapolis
Cellular One of Jefferson
Cellular One of Lake Charles
Cellular One of LaSalle
Cellular One of Lenior
Cellular One of Marianna
Cellular One of Middle Tennessee
Cellular One of Syracuse
Cellular One of Union
Cellular One of Upstate New York
Cellular One of Western Illinois
Cellular One of Wichita Falls
Cellular Phone of Kentucky
Cellular Plus
Cellular South
Cellular Systems International
Cellular Three
Cellular Ventures
Cellulink
Centel Cellular
Centennial Cellular
Century Cellunet
Coastel Communications
Contel Cellular
Crowley Cellular
Cybertel Cellular
Digital Cellular
Dobson Cellular
Eastex Cellular
Ed Tel Cellular
Enid Cellular
Farmers Cellular Telephone
First Kentucky Cellular
Gaia
Gencell
Greater South Dakota Cellular
GTE Mobilnet
Highland Cellular
Houston Cellular
Independent Cellular Network
Iowa East Cellular Telephone
Kentucky Cellular
Kull Cellular

Lake Huron Cellular
Lincoln Telephone
Litchfield County Cellular
Mactel Cellular
Maine Cellular
McCaw Cellular Communications
Mercury Cellular
Mercury Communications
Mid-Missouri Cellular
Miscellco Communications
Missouri Cellular
Mobiletel
MT&T Cellular
MTS Cellular
Nebraska Cellular
NYNEX Mobile Communications
Oneonta Telephone
Pace Communications
PacTel Cellular
Petroleum Communications
PTSI Cellular
Public Service Cellular
Quantum Communications Group
Radiofone
Richmond Cellular Telephone
Rochester Telephone
Sasktel Cellular
Savannah Cellular
Shenandoah Cellular
Southeastern Cellular
Southwestern Bell Mobile Systems
Springwich Cell. Ltd Partner. (SNET)
StarCellular
Sterling Cellular
U S West Cellular
Union Telephone
United States Cellular
Unitel
Unity Cellular Systems (Unicel)
Valley Telecommunications
Vanguard Cellular
West Central Cellular
Westex Cellular
X-Cell Cellular
Xit Cellular
Yorkville Telephone

Lincoln Telephone Cellular

Advantage Cellular
ALLTEL Mobile Communications
Ally
American Rural Cellular
Ameritech Mobile Communications
Appalachian Cellular
Atlantic Cellular
BCTel Cellular
Bell Atlantic Mobile
Bell Mobility Cellular
BellSouth Mobility
Bighorn Cellular Telephone
Bledsoe Telephone
Blue Ridge Cellular
Blue Ridge Cellular Telephone
Bluegrass Cellular
Brazos Cellular
Cantel
Canton Cellular
Carolina West Cellular
C.C. Cellular
Cellcom
Cellular Connection
Cellular, Inc.
Cellular of Indiana
Cellular One of Brookings
Cellular One of Chicago
Cellular One of Danville
Cellular One of LaSalle
Cellular One of Sioux Falls

Cellular Plus
Cellular Telephone of Kentucky
Cellular Ventures
Cellulink
Centel Cellular
Century Cellunet
Chariton Valley Cellular
Contel Cellular
Cybertel Cellular
Digital Cellular
Dobson Cellular
Ed Tel Cellular
Enid Cellular
Farmers Cellular
Gaia
GTE Mobilnet
Highland Cellular
Honolulu Cellular Telephone
Independent Cellular Network
Iowa East Cellular
Kansas Cellular
Kentucky Cellular
Mactel Cellular
McCaw Cellular Communications
MCTA
Mercury Cellular
Mercury Communications
Metrocel Cellular
Mid-Mid Missouri Cellular
Minerich Cellular
Missouri Cellular
MTS Cellular
Nebraska Cellular
Newtel Cellular
NYNEX Mobile Communications
Pace Communications
PacTel Cellular
Petroleum Communications
PTSI
Québéc Téléphone Cellulaire
Radiofone
Richmond Cellular Telephone
Rochester Telephone
Shenandoah Mobile
Southwestern Bell Mobile Systems
Springwich Cell. Ltd Partner. (SNET)
Sterling Cellular
Thunder Bay Cellular
U S West Cellular
Union Cellular
United States Cellular
Unitel
Unity Cellular Systems (Unicel)
Valley Cellular
Vanguard Cellular
Westex Cellular
X-Cell Cellular
Xit Cellular

Los Angeles Cellular Telephone

Allcell
ALLTEL Mobile Communications
Associated Communications
Bakersfield Cellular
Bay Area Cellular
Cantel
Cellular One of Amarillo
Cellular One of Bakersfield
Cellular One of
  Baltimore/Washington
Cellular One of Baton Rouge
Cellular One of Beaumont
Cellular One of Berrien County
Cellular One of Chicago
Cellular One of Indianapolis
Cellular One of Kalamazoo
Cellular One of Lake Charles
Cellular One of Milwaukee
Cellular One of Santa Cruz

Cellular One of Sioux Falls
Cellular One of South Bend
Centennial Cellular
Comcast
Contel Cellular
Crowley Cellular Telecommunications
General Cellular
Honolulu Cellular Telephone
Houston Cellular Telephone
McCaw Communications
Metro Mobile CTS
Metro One
MetroCel Cellular
Metrophone
New Par
Palmer Communications
Radiofone
U S West Cellular

MACtel Cellular System
(Anchorage, AK)

Advantage Cellular
AGT Cellular
ALLTEL Mobile Communications
Ally
Ameritech Mobile Communications
Atlantic Cellular
Bahamas Cellular
BCTel Mobility Cellular
Bell Atlantic Mobile
Bell Mobility Cellular
BellSouth Mobility
Big Horn Cellular
Bluegrass Cellular
Blue Ridge Cellular
C. C. Cellular
C. V. Cellular
Cal-One Cellular
Canton
Cellcom of Greenbay
Cellcom of Hickory
Cellular 3
Cellular, Inc.
Cellular Information Systems
Cellular One of South Dakota
Cellular Phone of Kentucky
Cellular Plus
Cellular South
Cellulink
Centel Cellular
Century Cellunet
Coastel Communications
Contel Cellular
Dobson Cellular Systems
EdTel Cellular
Enid Cellular
GTE Mobilnet ("B" systems only)
Hudson Valley Cellular
Illinois Valley Cellular
Inland Cellular
Kansas Cellular
Lincoln Telephone Cellular
Maine Cellular
Mercury Cellular and Paging
Missouri Cellular
Mobile South Cellular
Mountaineer Cellular
MTS
MT&T Cellular
Nebraska Cellular
NewTel Cellular
NYNEX Mobile Communications
PacTel Cellular
Public Service Cellular
Québéc Téléphone Cellulaire
Saskatchewan Telecom
Springwich Cell. Ltd Partner. (SNET)
Shenandoah Mobile
Southwestern Bell Mobile Systems

StarCellular
Sterling Cellular
Thunder Bay Cellular
U S West Cellular
Union Telephone
United States Cellular
Unity Cellular
Vanguard Cellular
West Alabama Cellular
Yorkville Cellular

Maine Cellular

Advantage Cellular
ALLTEL Mobile Communications
Ameritech Mobile Communications
Atlantic Cellular
Bell Atlantic Mobile
Bell Mobility Cellular
BellSouth Mobility
Bledsoe Telephone
Blue Grass Cellular
Blue Ridge Cellular
Cal-One Cellular
Cantel
Cellular One of Upstate New York
Cellular Plus
Cellulink
Centel Cellular
Century Cellunet
Coastal Communications
Contel Cellular
CPS Cellular Telephone
Digital Cellular
Enid Cellular
GenCell Management
GTE Mobilnet
Independent Cellular Network
Kansas Cellular
Mactel Cellular
MT&T Cellular
MTS Cellular
Nebraska Cellular
New Brunswick Cellular
NewTel Cellular
Northwest Cellular
NYNEX Mobile Communications
Pace Communications
PacTel Cellular
Palmer Communications
Petroleum Communications
PTSI Cellular
Public Service Cellular
Québec Téléphone Cellulaire
Richmond Cellular Telephone
Rochester Telephone
Savannah Cellular
Shenandoah Mobile
Southwestern Bell Mobile
Springwich Cell. Ltd Partner. (SNET)
StarCellular
Sterling Cellular
U S West Cellular
United States Cellular
Unity Cellular Systems (Unicel)
XIT Cellular

McCaw Communications

Adirondack Cellular
Advantage Cellular
Albany Telephone
Allegan Cellular
ALLTEL Mobile Communications
Ally
Alpine Cellular
Amarillo Cellular Telephone
AMC Cellular
American Rural Cellular
Ameritech
Appalachian Cellular

Atlantic Cellular
Auburn Television
BacTel
Baja Cellular
Bakersfield Cellular
Bay Area Cellular
B.C. Tel Mobility Cellular
Bell Atlantic Mobile
Bell South Cellular
BellSouth Mobility
Blackwater Cellular
Blue Mountain Cellular
Blue Ridge Cellular
Bluegrass Cellular
Boston Cellular
Brazos Cellular
Buckhead Telephone
Buffalo Telephone
C. C. Cellular
Cal-One Cellular
Cantel
Canton
Cedetel
Cellcom of Green Bay
Cellcom of Hickory
Celltelco
Celluar One of Collier/Hendry
Cellular 2000
Cellular CL
Cellular Communications of Puerto
  Rico
Cellular Connections
Cellular Holding
Cellular Information Systems
Cellular One of
  Baltimore/Washington
Cellular One of Baton Rouge, LA
Cellular One of Berrien County
Cellular One of Chicago
Cellular One of Columbia, MO
Cellular One of Columbia, TN
Cellular One of Columbus, IN
Cellular One of Danville
Cellular One of Detroit
Cellular One of East/Central PA
Cellular One of Enid, OK
Cellular One of Finger Lakes
Cellular One of Frederick, MD
Cellular One of Galveston, TX
Cellular One of Grand Forks, ND
Cellular One of Honolulu, HI
Cellular One of Huntsville, TX
Cellular One of Indiana
Cellular One of Indianapolis, IN
Cellular One of Lake Charles, LA
Cellular One of Lenoir, NC
Cellular One of Meridian, MS
Cellular One of Middle Tennessee
Cellular One of North Texas
Cellular One of Northeast Colorado
Cellular One of Northeast Georgia
Cellular One of Richmond, KY
Cellular One of Southwest Oklahoma
Cellular One of the Great Lakes of
  Iowa
Cellular One of Upstate New York
Cellular One of Victoria, TX
Cellular One of Waycross/Vidalia
Cellular One of West Okalhoma
Cellular One of Western Illinois
Cellular One of Wilkesboro, NC
Cellular Plus
Cellular Plus of Georgia
Cellular South
Cellular Three
Cellular, Inc.
Cellulink
Cellwave
Celutel

Centel Cellular
Centennial Cellular
Century Cellunet
Chariton Cellular
CJD Cellular
Clear Communications
Coastel Communications
Comcast
Contel Cellular
Crowley Communications
Cybertel
Dicomm
Digital Cellular of Texas
Dobson Cellular
Dominion Cellular
Easterbrooke
Enid Cellular
First Fayette Cellular
First Kentucky Cellular
Gaia
GenCell Management
Genessee Telephone
GMD Cellular
Greater South Dakota Cellular
GTE Mobilnet
HBF Cellular
Highland Cellular
HLD Cellular
Houston Cellular
Independent Cellular Network
Interstate Cellular (Interstate Cellular
  (Intercel))
Iowa East Cellular
IUSA Cell
Kansas Independent Network
Lake Huron Cellular
Lincoln Cellular
Litchfield Cellular
Los Angeles Cellular
Mackinac Cellular
Maine Cellular
Mega Comm
Mercury Cellular
Mercury Cellular
Mercury Communications
Metacomm
Metrophone
Mid-Missouri Cellular
Mid-South Cellular
Mid-Texas Cellular
Midland Communications
Milwaukee Telephone
Minnesota Cellular
Miscellco
Missouri Cellular
Mobile Communications Limited
  Partnership
Montana Cellular
Mountaineer Cellular
Mountaineer Mobile
National Cellular
Nebraska Cellular
New Par
North Carolina Cellular
NYNEX Mobile Communications
Oklahoma West Cellular
PC Cellular of Kentucky
PC Cellular of Vermont
Pace Communications
Pacific Northwest Cellular
PacTel Cellular
Palmer Communications
Peoples Cellular
Petroleum Communications
Pine Cellular
Poka Lambro Cellular
PriCellular
Prime Cellular
Pro-Max
PTSI Cellular

Public Service Cellular
Quantum Communications
R & D Cellular
Radiofone
RFB Cellular
Richmond Cellular
Rochester Telephone
Santa Cruz Cellular
Sioux Falls Cellular
Springwich Cell. Ltd Partner. (SNET)
Southeast Indiana Cellular
Southwestern Bell Mobile Systems
Sterling Cellular
Stillwater Cellular
Sussex Cellular
Texas 16 Cellular
Tsaconas Cellular
U S West Cellular
UBET Cellular
Union Telephone
United States Cellular
Unitel
Valley Telecommunications
Vanguard Cellular
Virgin Island Telecommunications
Virginia 10 Limited Partnership
Virginia Cellular
West Alabama Cellular
Westex Cellular
X-Cell Cellular
XIT Cellular
Youngstown Cellular

MCMG

Advantage Cellular
Allegan Cellular
ALLTEL Mobile Communications
Ally
Alpha Cellular
AMC Cellular
Ameritech
Appalachian Cellular
Atlantic Cellular
Auburn Television
Bahamas Cellular Telephone
Bakersfield Cellular
Bell Atlantic Mobile
Bay Area Cellular
Bell South Cellular
BellSouth Mobility
Bledsoe Telephone
Blue Mountain Cellular
Blue Ridge Cellular
Bluegrass Cellular
Buckhead Telephone
C. C. Cellular
Cantel
Canton Cellular
Cellcom of Green Bay
Cellcom of Hickory
Cellular Connection
Cellular, Inc.
Cellular Information Systems
Cellular One of Albany, NY
Cellular One of Amarillo, TX
Cellular One of
   Baltimore/Washington
Cellular One of Berrien County, MI
Cellular One of Boston, MA
Cellular One of Buffalo, NY
Cellular One of Chicago, IL
Cellular One of Columbia, MO
Cellular One of Galesburg, Il
Cellular One of Huntsville, TX
Cellular One of Indiana
Cellular One of Indianapolis, IN
Cellular One of Lake Charles, LA
Cellular One of La Salle, IL
Cellular One of Meridian, MS

Cellular One of New York City
Cellular One of Northeast Colorado
Cellular One of Northeast Georgia
Cellular One of Rochester, NY
Cellular One of Santa Cruz, CA
Cellular One of South Dakota
Cellular One of Sioux Falls, SD
Cellular One of Southwest Florida
Cellular One of Southwest Texas
Cellular One of Upstate New York
Cellular One of Victoria, TX
Cellular One of West Virginia
   Highlands
Cellular One of Wichita Falls, TX
Cellular Phone of Kentucky
Cellular Plus
Cellular XL
Celutel
Centel
Centennial Cellular
Clear Communications
Comcast Cellular
CompComm
Contel
Cooper Cellular Management
Crowley Cellular
CyberTel
Danbury Cellular
Detroit Cellular
Dicomm Cellular
Dobson Cellular
Dominion Resources
Farmers Cellular
FGI Cellular
Finger Lakes Telephone
First Faytette Cellular
First Kentucky Cellular
Gaia
Galveston Cellular
GenCell Management
GMD
Great Lakes of Iowa
Greater South Dakota Cellular
GTE Mobilnet
HBF Cellular
Horizon
Houma/Thibodaux Cellular
Houston Cellular
Illinois Valley Cellular
Independent Cellular Network
Iowa East Cellular
Jackson Cellular
Kansas Cellular
Lake Huron Cellular
Lincoln Cellular
Litchfield County Cellular
Mackinac
McCaw Cellular Communications
MegaComm
Mercury Cellular
Mercury Communications
MetaComm Cellular
MetroCel
Metrophone
Mid-South Cellular
Milwaukee Telephone
Miscellco
Missouri Cellular
Mobile Communications Systems
Mobile Telenet
Mountaineer Cellular
National Cellular
Nebraska Cellular
New Par
NYNEX Mobile Communications
Ocean County Cellular
Pacific Northwest Cellular
Pacific Telecom Cellular
PacTel Cellular

Palmer Cellular
Panhandle Telepphone
PC Cellular of Kentucky
PC Cellular of Vermont
Pegasus Cellular
Petroleum Communications
PriCellular
Public Service Cellular
Quantum
R & D Cellular
Radiofone
RFB Cellular
Richard Vega Group
Richmond Cellular
Rural Cellular Management
Southeast Indiana Cellular
Shenandoah Mobile
Springwich Cell. Ltd Partner. (SNET)
Southeastern Cellular
Southwestern Bell
Stillwater Cellular
Sussex Cellular
Syracuse Telephone
Tsaconas Cellular
Texas 16 Cellular
Texas 5
U S West Cellular
Union Telephone
Unitel
United States Cellular
Unity Cellular Systems (Unicel)
Utica Telephone
Vanguard Cellular
Virginia Cellular
West Alabama Cellular Telephone
Yorkville Telephone

Mercury Cellular and Paging

Advantage Cellular
ALLTEL Mobile Communications
Ally
American Rural Cellular
Ameritech Mobile Communications
Atlantic Cellular
Auburn Television Grop
Bell Atlantic Mobile
Bell Mobility Cellular
BellSouth Cellular
BellSouth Mobility
Big Horn Cellular
Bledsoe Telephone
Blue Ridge Cellular
Blue Ridge Cellular & Telephone
Bluegrass Cellular
Brazos Cellular
Cal-One Cellular
Cantel
Carolina West Cellular
C.C. Cellular
Cellcom
Cellular Connection
Cellular, Inc.
Cellular Information Systems
Cellular of Indiana
Cellular One of Beaumont
Cellular One of Chicago
Cellular One of Gainesville, GA
Cellular One of LaSalle
Cellular One of Marianna
Cellular One of Syracuse
Cellular One of Upstate New York
Cellular South
Cellular Ventures
Cellular XL
Centel Cellular
Century Cellunet
Coastel Communications
Cone Enterprises
Contel Cellular

Crowley Cellular
Cybertel Cellular
Digital Cellular
Dobson Cellular Systems
Dominion Cellular
Enid Cellular
Farmers Cellular Telephone
GenCell Management
Georgia Cellular Plus
GMD
Greater South Dakota Cellular
GTE Mobilnet
Highland Cellular
Houston Cellular Telephone
Independent Cellular Network
Interstate Cellular
Iowa East Cellular
Kansas Cellular
Lincoln Telephone
Mactel Cellular
McCaw Cellular Communications
Mercury Communications
Metrocel Cellular
Mid-Missouri Cellular
Miscellco Communications
Mississippi Cellular
Mobiletel
Mocell
Mountaineer Cellular
National Cellular
Nebraska Cellular
Northwest Cellular
NYNEX Mobile Communications
OK Cellular
PacTel Cellular
Palmer Communications
Petroleum Communications
Pioneer Cellular
PTSI Cellular
Quantum Communication
Radiofone
Richmond Cellular Telephone
Rochester Telephone
Shenandoah Cellular
Southeastern Cellular
Southwestern Bell Mobile Systems
Springwich Cell. Ltd Partner. (SNET)
StarCellular
Sterling Cellular
Texas 16 Cellular
U S West Cellular
UBET Cellular
Union Telephone
United States Cellular
Unity Cellular Systems (Unicel)
Valley Telecommunications
Vanguard Cellular
West Alabama Cellular
West Central Cellular
West Virginia Cellular
Westex
XIT Cellular

Mercury Communications

Advantage Cellular
Allegan Cellular
ALLTEL Mobile Communications
Alpha Cellular
American Rural Cellular
Ameritech Mobile Communications
Appalachian Cellular
Atlantic Cellular
Auburn Television
Bahamas Telecommunications
Bay Area Cellular
BCTel Mobility Cellular
Bell Atlantic Mobile
BellSouth Cellular
BellSouth Mobility

Big Horn Cellular
Blackwater Cellular
Bledsoe Telephone
Blue Mountain Cellular
Blue Ridge Cellular
Bluegrass Cellular
Cantel
Canton Cellular
Cellcom
Cellcom of Hickory
Cellular Communications of Puerto Rico
Cellular Inc.
Cellular Information Systems
Cellular of Indiana
Cellular One of Amarillo
Cellular One of Berrien County
Cellular One of Boston
Cellular One of Chicago
Cellular One of Collier/Hendry
Cellular One of Columbia, MO
Cellular One of Danville, IL
Cellular One of East/Central PA
Cellular One of Indianapolis
Cellular One of Lake Charles, LA
Cellular One La Salle, IL
Cellular One of Meridian, MS
Cellular One of Northeast Georgia
Cellular One of Northeast Mississippi
Cellular One of Sioux Falls
Cellular One of Western Illinois
Cellular Phone of Kentucky
Cellular South
Cellular Telephone of Kentucky
Cellular X L
Celutel
Centel Cellular
Centennial Cellular
Century Cellunet
CJD Wisconsin Cellular
Clear Communications
Contel Cellular
Cooper Cellular
Crowley Cellular
Cybertel Cellular
C. C. Cellular
Danbury Cellular
Detroit Cellular
Dicomm Cellular Telephone
Dobson Cellular Systems
Dominion Cellular
Easterbrooke Cellular
Eastex
Enid Cellular
Farmers Cellular Telephone
First Fayette Cellular
Gaia, Inc.
Galveston Cellular Telephone
General Cellular
GMD Partnership
Great Lakes of Iowa
GTE Mobilenet
Highland Cellular
Houston Cellular
Independent Cellular
Interstate Cellular
Iowa East Cellular
Kansas Cellular
Lake Huron Cellular
Lincoln Telephone
Litchfield County Cellular
Lone Star Cellular
Mackinac Cellular
McCaw Cellular Communications
MCMG
Mercury Cellular & Paging
MetaComm Cellular
Metro Mobile CTS
MetroCel Cellular
Mid-South Cellular

Milwaukee Telephone
Miscellco Communications
Missouri Cellular
Mobile Communications Systems
Nountaineer Mobile
National Cellular Limited Partnership
New Par
NYNEX Mobile Communications
Ohio State Cellular
Pace Communications
Pacific Telecom Cellular
PacTel Cellular
Palmer Communications
PC Cellular of Kentucky
Pegasus Cellular
Petroleum Communications
Pine Cellular Phones
PriCellular
Quantum Communications
Radiofone
RFB Cellular
Santa Cruz Cellular Telephone
Southwestern Bell Mobile Systems
Sterling Cellular
Sussex Cellular
Syracuse Telephone
Texas 16 Cellular Telephone
Texoma Cellular Management
Union Telephone
U S West Cellular
Utica Telephone
United States Cellular
Vanguard Cellular Systems
Virginia 10 RSA Limited Partnership
West Alabama Cellular
Yorkville Telephone

Mid-Missouri Cellular

ALLTEL Mobile Communications
American Rural Cellular
Ameritech Mobile
Appalachian Cellular
Atlantic Cellular
Atlas Cellular
Bachtel Cellular
Bell Atlantic Cellular
BellSouth Mobility
Bledsoe Telephone
Bluegrass Cellular
Blue Mountain Cellular
C. C. Cellular
Cal-One Cellular
Cellcom of Green Bay
Cellcom of Hickory
Cellular Connection
Cellular Holdings
Cellular, Inc.
Cellular One of Chicago
Cellular One of Columbia
Cellular One of Delmington, DE
Cellular One of Dobson
Cellular One of Galveston
Cellular One of Greater South Dakota
Cellular One of Indiana
Cellular One of Iowa
Cellular One of Northeast Georgia
Cellular Connection
Cellular Information Systems
Cellular Plus
Cellular South
Centel Cellular
C. V. Cellular
Cybertel
Digital Cellular of Texas
Dobson Cellular
Enid Cellular
Farmers Cellular
Five Star Cellular
GTE Mobile Communications

Roaming Agreements

Highland Cellular
Hudson Valley Cellular
Illinois Valley Cellular
Kansas Cellular
Lake Huron Cellular
Lincoln Telephone
Litchfield County Cellular
Mercury Cellular
Milwauke Telephone
Missouri Cellular
Missouri 1 Atchinson RSA Limited
Nebraska Cellular
NYNEX Mobile Communications
Pacific Telecom Cellular
Peoples Cellular
Pine Cellular
Pioneer Cellular
Sasktel Cellular
Shenandoah Mobile
Southwestern Bell Mobile Systems
Sterling Cellular
XIT Cellular
Yorkville Telephone

Miscellco Communications

Advantage Cellular
ALLTEL Mobile Communications
Appalachian Cellular
Associated Communications
Atlantic Cellular
Atlas Cellular
Auburn Television Group
Bakersfield Cellular Telephone
Batelco
Bay Area Cellular
Bell Atlantic Mobile
BellSouth Cellular
Bluegrass Cellular
Brazos Cellular
Cantel
Canton Cellular
Cellcom
Cellular Connection
Cellular Information Systems
Cellular of Indiana
Cellular One of Amarillo
Cellular One of Berrien County
Cellular One of Chicago/Gary
Cellular One of Cincinnati
Cellular One of Cleveland
Cellular One of Columbia, MO
Cellular One of Columbia, TN
Cellular One of Columbus, OH
Cellular One of Danville
Cellular One of Detroit
Cellular One of Gainesville, GA
Cellular One of Galveston
Cellular One of Grand Rapids
Cellular One of Huntsville
Cellular One of Indianapolis
Cellular One of Lake Charles
Cellular One of Lansing
Cellular One of LaSalle
Cellular One of Lima
Cellular One of Meridian
Cellular One of Middle Tennessee
Cellular One of Milwaukee
Cellular One of Muskegon
Cellular One of Richmond
Cellular One of Santa Cruz
Cellular One of Southwest Oklahoma
Cellular One of Syracuse
Cellular One of Upstate New York
Cellular One of Utica
Cellular One of Victoria
Cellular One of Wichita Falls
Cellular One of Youngstown
Cellular Plus (A only)
Cellular South

Cellular Systems International
Cellular Three
Cellular Ventures
Celutel
Centel Cellular
Centennial Cellular
Century Cellular
Chariton Valley Cellular
Coastel Communications
Contel Cellular
Crowley Cellular Telecommunications
Cybertel Cellular
Danbury Cellular Telephone
Digital Cellular
Dobson Cellular
Dominion Cellular
Enid Cellular
GenCell Management
Greater South Dakota Cellular
GTE Mobilnet
Honolulu Cellular Telephone
Houston Cellular Telephone
Independent Cellular Network
Kansas Cellular
LA Cellular
Lake Huron Cellular
Litchfield County Cellular
Mactel Cellular
McCaw Cellular Communications
Mercury Cellular
Mercury Communications
Metacomm Cellular
Metrocel Cellular
Mobile Communications Systems
Mobiletel
MT&T Cellular
National Ventures
Nebraska Cellular
North West Cellular
PacTel Cellular
Petroleum Communications
PTSI Cellular
Radiofone
Richmond Cellular Telephone
Rural Cellular
Savannah Cellular
Shenandoah Mobile
Southwestern Bell Mobile Systems
Sterling Cellular
Texoma Cellular
U S West Cellular
Union Telephone
United States Cellular
Unity Cellular Systems (Unicel)
Vanguard Cellular

Mississippi Cellular Telephone

Advantage Cellular
ALLTEL Mobile Communications
Ally
American Rural Communications
Ameritech Mobile Communications
Appalachian Cellular
Atlantic Cellular
Auburn Telephone
Bahamas Telephone
Bell Atlantic Mobile
Bell Mobility Cellular
BellSouth Cellular
Bell South Mobility
Bledsoe
Bluegrass Cellular
Blue Mountain Cellular
Blue Ridge Cellular
C. C. Cellular
Cal-One Cellular
Cantel Cellular
Canton Cellular
Cellcom

Cellular Connections
Cellular Holdings
Cellular Inc.
Cellular Information Systems
Cellular One of Baton Rouge
Cellular One of Chicago
Cellular One of Southwest Florida
Cellular One of Enid
Cellular One of Lake Charles
Cellular One of Meridian
Cellular One of Northeast Georgia
Cellular One of Northeast Mississippi
Cellular One of Sioux Falls
Cellular One of Upstate New York
Cellular Plus of Georgia
Cellular Plus of Iowa
Cellular Phone of Kentucky
Cellular Ventures
Centel Cellular
Century Cellunet
Chariton Valley Cellular
Clear Communications
Coastel Communications
Columbia Cellular
Contel
Cellular Crowley Cellular
C. V. Cellular
Cybertel Cellular
Digital Cellular of Texas
Dobson Cellular
Dominion Resources
Eastex
Enid Cellular
Farmers
First Kentucky Cellular
Gaia
GenCell Management
GMD Partnership
Greater South Dakota Cellular
GTE Mobilnet
Highland Cellular
Houston Cellular
Illinois Valley Cellular
Independent Cellular
Interstate Cellular
Iowa East Cellular
Kansas Independent
Lincoln Telephone
McCaw Cellular Communications
Mackinac Cellular
MacTel Cellular
MCTA
Mercury Cellular
Mercury Communications
Metrocel Cellular
Mid-Missouri Cellular
Miscellco
Missouri Cellular
Mobile Communications
Mobiletel
Mountaineer Cellular
MTS Cellular
MT&T
National Cellular Limited
NYNEX Mobile Communications
Oneota Telephone
Pace Communications
Pacific Northwest Cellular
Pacific Telecom Cellular
PacTel Cellular
Palmer Communications
Panhandle Cellular
People Cellular
Petroleum Communications
Pine Cellular
PriCellular
Public Service Cellular
Quantum Communications
Québéc Téléphone Cellulaire

Radiofone
Richmond
Rochester Telephone
Shenandoah Mobile Company
Southeastern Cellular
Southwestern Bell
Springwich Cell. Ltd Partner. (SNET)
StarCellular
Sterling Cellular
Stillwater Cellular
Texas 16 Cellular
Thunder Bay Cellular
Tsaconas Cellular
U S West Cellular
Union Telephone
Unitel
United State Cellular
Valley Telecommunications
Vanguard Cellular
West Alabama Cellular
West Central Cellular
Westex Cellular
XIT Telecommunications
Yorkville Telephone

Missouri Cellular

Advantage Cellular
Albany Telephone
Allegan
ALLTEL Mobile Communications
Alpha Cellular
Alpine Cellular
American Cellular
American Rural
Ameritech Mobile
Appalachian Cellular
Atlantic Cellular
Auburn Television
Bahama Cellular
Batelco
B.C. Tel Mobility Cellular
BellSouth Cellular
BellSouth Mobility
Blackwater
Bledsoe Cellular
Bluegrass Cellular
Blueridge Cellular
BMCT
Buffalo Cellular
Cantel
C. C. Cellular
Cellcom of Greenbay
Cellcom of Hickory
Cellular 2000
Cellular Holdings
Cellular, Inc.
Cellular Information Systems
Cellular One Central
Cellular One of Amarillo
Cellular One of Bay Area
Cellular One of
  Baltimore/Washington
Cellular One of Berrien County
Cellular One of Chicago
Cellular One of Columbia
Cellular One of Columbia, TN
Cellular One of Danbury
Cellular One of Danville
Cellular One of East/Central PA
Cellular One of Gainesville
Cellular One of Galesburg
Cellular One of the Great Lakes
Cellular One of Indiana
Cellular One of Indianapolis
Cellular One of Lake Charles
Cellular One of La Salle
Cellular One of Meridian
Cellular One of Mid-Tennessee
Cellular One of North Carolina

Cellular One of Northeast Georgia
Cellular One of Ohio
Cellular One of Santa Cruz
Cellular One of South Dakota
Cellular One of Southwest Florida
Cellular One of Upstate New York
Cellular One of Waycross
Cellular One of Western Illinois
Cellular One of Wichita Falls
Cellular Phone of Kentucky
Cellular Plus
Cellular South
Cellular Systems International
Cellular Telephone of Kentucky
Cellular XL
Centel
Century Cellular
Century Cellunet
CFW Communications
Chariton Valley
Coastel Communications
Contel
Cooper
Crowley
Cellular Systems International
Cybertel
Detroit Cellular
Dobson Cellular
Easterbrook Cellular
Enid
Farmers Cellular
FGI Cellular
General Cellular
Genesee Telephone
GTE Mobilenet
Gulf Coast Cellular
Gulf Telephone
HBF Cellular
Highland Cellular
Illinois Valley Cellular
Independent Cellular
Inland Cellular
Interstate Cellular (Interstate Cellular
  (Intercel))
Iowa East Cellular
Kansas Cellular
Los Angeles Cellular
Lake Huron Cellular
Lincoln Telephone
Litchfield County Cellular
MacTel
McCaw Communications
MegaComm
Mercury Cellular
Mercury Communications
Metro Mobile
Metrocel
Metrophone
Mid-Missouri Cellular
Milwaukee Telephone
Miscellco
Mississippi Cellular
Mobile South
Mobiletel
Mountaineer Telephone
MT&T
MTS Cellular
NYNEX Mobile Communications
Oneonta Cellular
Pace Communications
Pacific Telecom Cellular
PacTel Cellular
Palmer Cellular
Panhandle Telecommunications
PC Cellular of Kentucky
PC Cellular of Vermont
Pegasus Cellular
Petroleum Communications
Pine Cellular

PriCellular
PTSI Cellular
Public Service Cellular
Quantum Cellular
Radiofone
Richland Cellular
Richmond Cellular
Rural Cellular Management
SaskTel
South East Cellular
South Eastern Cellular
Springwich Cell. Ltd Partner. (SNET)
Stillwater Cellular
Sunshine Cellular
Sussex Cellular
Syracuse Cellular
Texoma Cellular
Thumb Cellular
Tsaconas Cellular
Union Telephone
Unitel
U S West Cellular
United States Cellular
Utica Cellular
Valley Telecommunications
Vanguard
Victoria Cellular
Virginia Cellular
Walnut Hill Cellular
Westex
Wisconsin Cellular
Yorkville Cellular
Youngstown Cellular

MobileTel

AGT Cellular
ALLTEL Mobile Communications
Ally
American Rural Cellular
Ameritech Mobile Communications
Appalachian Cellular
Atlantic Cellular
Auburn Television
Bell Atlantic Mobile
BellSouth Cellular
BellSouth Mobility
Big Horn Cellular
Bledsoe Telephone
Blue Mountain Cellular
Bluegrass Cellular
Brazos Cellular
Buffalo Telephone
C. C. Cellular
Cellcom
Cellular Connection
Cellular Holding
Cellular Information Systems
Cellular of Indiana
Cellular One of Amarillo
Cellular One of Baton Rouge
Cellular One of Chicago
Cellular One of Columbia
Cellular One of Indianapolis
Cellular One of Meridian
Cellular One of Ohio
Cellular One of Southwest Georgia
Cellular One of Sioux Falls
Cellular One of Wilmington
Cellular South
Cellular Telephone Company of
  Kentucky
Cellular Ventures
Cellulink
Centel Cellular
Century Cellunet
Coastel Communications
Contel Cellular
Crowley Cellular Telecommunications
C V Cellular

Cybertel Cellular
Digital Cellular of Texas
Enid Cellular
Farmers Cellular
FGI Cellular Management
Gaia
GenCell Management
Greater South Dakota Cellular
GTE Mobilnet ("B" systems only)
Highland Cellular
Honolulu Cellular Telephone
Houston Cellular Telephone
Illinois Valley Cellullar
Independent Cellular Network
Interstate Cellular (Interstate Cellular
  (Intercel))
Iowa East Cellular
Kansas Independent Networks
Lincoln Telephone
Mackinac Cellular
Mercury Cellular and Paging
Mid-South Cellular
Milwaukee Cellular
Miscellco
Missouri Cellular
Mountaineer Cellular
National Cellular
Nebraska Cellular
NewTel Cellular
NYNEX Mobile Communications
OK Cellular
Oneonta Telephone
Pace Communications
Pacific Northwest Cellular
PacTel Cellular
Panhandle Telecommunications
Petroleum Communications
Peoples Cellular
Pine Cellular
Pioneer Cellular
PTSI Cellular
Radiofone
Richmond Cellular
Shenandoah Mobile
Southwestern Bell Mobile Systems
Southeastern Cellular
Springwich Cell. Ltd Partner. (SNET)
StarCellular
Texas 16 Cellular
Texas Cellular
Tsaconas Cellular
U S West Cellular
Union Telephone
United States Cellular
Unitel
Vanguard Cellular Systems
West Alabama Cellular
Westex Cellular
XIT Cellular
Yorkville Telephone

MT&T Cellular

AGT Cellular
Advantage Cellular
ALLTEL Mobile Communications
Ameritech Mobile Communications
Associated Communications
B.C. Tel Mobility Cellular
Bakersfield Cellular
Bay Area Cellular
Bell Atlantic Mobile
Bell Mobility Cellular
BellSouth Mobility
Buffalo Telephone
Cellular Information Systems
Cellular Plus
Centel Cellular
Century Cellunet
Coastel Communications

Contel Cellular
Ed Tel Cellular
Enid Cellular
GTE Mobilnet ("B" systems only)
Island Tel Cellular
MCTA
MTS Cellular
Maine Cellular
NB Tel Cellular
NewTel Cellular
NYNEX Mobile Communications
OK Cellular
PacTel Cellular
Petroleum Communications
Pioneer Cellular
Québéc Téléphone Cellulaire
Rochester Telephone
Sasktel Cellular
Southwestern Bell Mobile Systems
Springwich Cell. Ltd Partner. (SNET)
StarCellular
U S West Cellular
United States Cellular
Unity Cellular Systems (Unicel)

MTS Cellular

AGT Cellular
ALLTEL Mobile Communications
Ally
Ameritech Mobile Communication
Appalachian Cellular
Associated Communications
Atlantic Cellular
Bachtel Cellular
B.C. Mobility Cellular
Bell Atlantic Mobile
Bell Mobility Cellular
BellSouth Mobility
Bledsoe Telephone
Blue Ridge Cellular
Bluegrass Cellular
C.C. Cellular
Cellcom
Cellular 2000
Cellular Connection
Cellular, Inc.
Cellular Information Systems
Cellular One of Amarillo
Cellular One of Chicago
Cellular One of Destrehan
Cellular One of Gainesville GA
Cellular One of Indianapolis
Cellular One of Milwaukee
Cellular One of New York
Cellular One of Roanoke
Cellular One of Sioux Falls
Cellular One of Springfield
Cellular One of Upstate New York
Cellular One of Wichita Falls
Cellular South
Cellular Ventures
Centel Cellular
Century Cellunet
Contel Cellular
Cybertel Cellular
Digital Cellular
Ed Tel Cellular
Enid Cellular
Five Star Cellular
Greater South Dakota Cellular
GTE Mobilnet
Hagerstown Cellular
Independent Cellular Network
Kansas Cellular
Lamar County Cellular
Lincoln Telephone
Litchfield County Cellular
MacTel Cellular
Maine Cellular

Mississippi Cellular
Missouri Cellular
MT&T Cellular
Nebraska Cellular
New Brunswick Cellular
Newtel Cellular North West Cellular
NYNEX Mobile Communications
OK Cellular
Pace Communications
PacTel Cellular
Palmer Communications
Petroleum Communications
Pioneer Cellular
Public Serive Cellular
Québéc Téléphone Cellulaire
Richmond Cellular Telephone
Rochester Telephone Mobile
  Communications
Sasktel Cellular
Shenandoah Cellular
Southwestern Bell Mobile Systems
StarCellular
Texas Cellular
Thumb Cellular
Thunder Bay Cellular
U S West Cellular
United States Cellular
Unitel
Unon Telephone
Unity Cellular Systems (Unicel)
Vanguard Cellular
Wilkes Telephone
XIT Cellular

NB Tel Cellular

Advantage Cellular
AGT Cellular
ALLTEL Mobile Communications
Ally
Ameritech Mobile Communications
Atlantic Cellular
Bakersfield Cellular
Bay Area Cellular
B.C. Tel Mobility Cellular
Bell Atlantic Mobile
Bell Mobility Cellular
BellSouth Mobility
Blueridge Cellular
Buffalo Telephone
C. C. Cellular
Canton Cellular
Cellular Information Systems
Cellular One of Chicago, IL
Cellular One of Indiana
Cellular One of New York
Cellular One of Upstate New York
Cellular Plus
Cellular South
Centel
Century Cellunet
Cibernet
Coastel Communications
Commonwealth Mobile Services
Contel Cellular
Cybertel Cellular
Digital Cellular of Texas
EdTel Cellular
Enid Cellular
GTE Mobilnet
Independent Cellular Network
Interstate Cellular
IUSA Cell, S.A. de C.V.
MacTel Cellular
Maine Cellular
Misselco Communications
Mobile Communications
Mobilesouth Cellular
MT&T Cellular
MTS Cellular

Nebraska Cellular
NewTel Cellular
Northwest Cellular
NYNEX Mobile Communications
OK Cellular
Pace Communications
PacTel Cellular
Petroleum Communication
Pioneer Cellular
PriCellular
Québec Téléphone Cellulaire
Richmond Cellular
Rochester Telephone Mobile
  Communications
SaksTel Cellular
Shenandoah Mobile
Southwestern Bell Mobile Systems
Springwich Cell. Ltd Partner. (SNET)
StarCellular
Thunder Bay Cellular
U S West New Vector
Union Telephone
United States Cellular
Unity Cellular
Virgin Islands Telephone (VITEL)
XIT Cellular

Nebraska Cellular

Advantage Cellular
ALLTEL Mobile Communications
American Rural Cellular
Ameritech Mobile
Appalachian Cellular
Atlantic Cellular
Bahamas Telecommunications
B.C. Tel Mobility Cellular
Bell Atlantic Mobile
Bell Mobility Cellular
BellSouth Mobility
Blue Mountain Cellular
Bluegrass Cellular
C. C. Cellular
C. V. Cellular
Cal-One Cellular
Canton
Cellcom of Green Bay
Cellcom of Hickory
Cellular Connection
Cellular, Inc.
Cellular Information Systems
Cellular One of Berrien County
Cellular One of Chicago
Cellular One of Columbia, MO
Cellular One of Danville
Cellular One of Destreham
Cellular One of East/Central PA
Cellular One of Galesburg
Cellular One of Hattiesburg
Cellular One of Huntsville
Cellular One of La Salle, IL
Cellular One of Lake Charles
Cellular One of Lake Huron
Cellular One of Indiana
Cellular One of Middle Tennessee
Cellular One of Norfolk
Cellular One of North Texas
Cellular One of Northeast Colorado
Cellular One of Northeast Georgia
Cellular One of Scottsbluff, NE
Cellular One of Southwest Florida
Cellular One of Sioux Falls
Cellular One of Upstate New York
Cellular One of West Virginia
Cellular One of Western Oklahoma
Cellular One of Wichita Falls
Centel
Centennial Cellular
Century Cellunet
Chariton Valley Cellular

Coastel Communications
Contel
Digital Cellular of Texas
Dobson Cellular
Enid Cellular
First Cellular of Omaha, NE
First Kentucky Cellular
General Cellular
GMD Partnership
Great Lakes of Iowa
Greater South Dakota Cellular
GTE Mobilenet
Houston Cellular
Hudson Valley Cellular
Illinois Valley Cellular
Independent Cellular
Kansas Independent Networks
Kull Cellular
Lincoln Telephone Cellular
Litchfield County Cellular
MacTel Cellular Systems
Maine Cellular
McCaw Communications
Mercury Cellular
Mid-Missouri Cellular
Mid-Texas Cellular
Milwaukee Telephone
Miscellco Communications
Missouri Cellular
MobileTel
Mountaineer Cellular
MT&T
MTS Cellular
Northwest Cellular
NYNEX Mobile Communications
Pacific Northwest Cellular
PacTel Cellular
Palmer Communications
Petroleum Communications
Pine Cellular Phones
PTSI Cellular
Quantum Communications
Radiofone
Richmond Cellular
Rochester Telephone
SaskTel Cellular
Shenandoah Mobile
Southwestern Bell Mobile
Springwich Cell. Ltd Partner. (SNET)
StarCellular
Sterling Cellular
U S West Cellular
UBET Cellular
Union Telephone
United States Cellular
Unitel
Unity Cellular Systems (Unicel)
Valley Telecommunications
Vanguard
West Central Cellular
Westex
X-Cell Cellular
XIT Cellular

New Par

Albany Telephone
AllCell
Allegan Cellular
ALLTEL Mobile Communications
Ally
Alpha Cellular
Alpine
Altoona Cellular
Associated Communications
Atlantic Cellular
Bakersfield Cellular
Baton Rouge Cellular
Bay Area Cellular
Bell Atlantic Mobile

BellSouth Cellular
Bell South Mobility
Blackwater Cellular
Bluegrass Cellular
Blue Mountain Cellular
Blue Ridge Cellular
Buffalo Telephone
C. C. Cellular
C. T. Cube
Cantel
Cellcom of Hickory
Cellular 2000
Cellular Communications of Puerto
  Rico
Cellular Information Systems
Cellular of Indiana
Cellular One of Amarillo, TX
Cellular One of
  Baltimore/Washington
Cellular One of Berrien County
Cellular One of Billings, MT
Cellular One of Collier/Hendry
Cellular One of Columbia, MO
Cellular One of Erie
Cellular One of Huntsville, TX
Cellular One of Jacksonville
Cellular One of Lake Charles
Cellular One of La Salle, IL
Cellular One of Marquette
Cellular One of Massachucetes/NH
Cellular One of Meridian
Cellular One of Middle Tennessee
Cellular One of North Texas
Cellular One of Panama City
Cellular One of the Rio Grande Valley
Cellular One of Sioux Falls
Cellular One of Southwest Florida
Cellular One of Toledo, OH
Cellular One of Upstate New York
Cellular One of Utica, NY
Cellular One of Wichita Falls
Cellular One of Wilmington
Cellular Phone of Kentucky
Cellular Plus
Cellular Proporties
Cellular Technology
Cellular Telephone Company
Cellular Ventures
Cellular XL
Celutel
Centel Cellular
Centennial Cellular
Century Cellunet
Contel Cellular
Crowley Cellular
Cybertel Cellular
Danbury Cellular
Dial Two
Dicomm Cellular
Cominion Resources
Fayetteville Cellular
First Cellular Group
First Fayette Cellular
First Kentucky Cellular
GenCell Management
GMD Limited
Graceba Cellular
Greater South Dakota Cellular
GTE Mobilnet
Highland Cellular
Houston Cellular
IFC Cellular
Independent Cellular Network
Indianapolis Telephone
Iowa East Cellular
Lake Huron Cellular
Longbranch Cellular
Los Angeles Cellular
McCaw Communications
Mega Comm

Mercury Communications
Metacomm Cellular
MetroCel Cellular
Metrophone
Mid South Cellular
Midland Communications
Milwaukee Telephone
Miscellco Communications
Missouri Cellular
Mobile Communications
MObile Telenet
Montgomery Cellular
National Cellular
New Brunswick Cellular
PC Cellular
Pacific Telecom Cellular
PacTel Mobile
Palmer Cellular
Panama City Cellular
Petroleum Communications
PriCellular
Quantum Communications
R & D Cellular
Radiofone
Richmond Cellular
Roanoke Valley Cellular
Rochester Telephone
Rogers Radiocall
Rural Cellular Management
Santa Barbara Cellular
Santa Cruz Cellular
Springwich Cell. Ltd Partner. (SNET)
Sooner Cellular
Southeast Indiana Cellular
Southern Cellular
Sterling Cellular
Sunshine Cellular
Sussex Cellular
Syracuse Telephone
Texoma Cellular
The Cellular Telephone Company
United States Cellular
United TeleSpectrum
Unitel
Vanguard Cellular
Victoria Cellular
Virginia Cellular
Yankee Cellular
Youngstown Cellular

NewTel Cellular

AGT Cellular
Ameritech Mobile
Atlantic Cellular
B.C. Tel Mobility Cellular
Bell Mobility Cellular
Bell South Mobility
Cellular One of Upstate New York
Contel Cellular
GTE Mobility
Island Tel Cellular
MT&T Cellular
MTS Cellular
NB Tel Cellular
NYNEX Mobile Communications
Oneonta Telephone
PACE Communications
Petroleum Communications
Québec Téléphone Cellulaire
SaskTel Cellular
South East Cellular
Thunder Bay Cellular
Tsaconas Cellular
United State Cellular
Unity Cellular
Yorkville Telephone

NYNEX Mobile Communications

AGT Cellular
ALLTEL Mobile Communications
Ameritech Mobile Communications
Associated Communications
Bakersfield Cellular
Bay Area Cellular
BCTel Cellular
Bell Atlantic Mobile
Bell Mobility Cellular
BellSouth Cellular
BellSouth Mobility
Cantel
Cellcom of Hickory
Cellular, Inc.
Cellular Information Systems
Cellular One of Amarillo
Cellular One of
  Baltimore/Washington
Cellular One of Benton Harbor
Cellular One of Boston
Cellular One of Brownsville
Cellular One of Chicago/Gary
Cellular One of Cincinnati
Cellular One of Cleveland
Cellular One of Columbus, OH
Cellular One of Detroit
Cellular One of Indianapolis
Cellular One of Milwaukee
Cellular One of Richmond
Cellular One of Sioux Falls
Cellular One of Springfield
Cellular One of Syracuse/Utica
Cellular One of Victoria
Cellular One of Wichita Falls
Cellular One of Youngstown
Cellular Plus ("B" side)
Cellular South
Cellulink
Centel Cellular
Centennial Cellular
Century Cellunet
Coastel Communications
Comcast
Contel Cellular
Crowley Cellular Telecommunications
Cybertel Cellular
Ed Tel Cellular
Enid Cellular
GenCell Management
GTE Mobilnet
Honolulu Cellular Telephone
Houston Cellular Telephone
Independent Cellular Network
JHP
Los Angeles Cellular Telephone
Mactel Cellular
Maine Cellular
McCaw Cellular Communications
MCTA
Mercury Cellular
Metrocel Cellular
Metrophone
Mobiletel
MT&T Cellular
MTS Cellular
NBTel Cellular
PacTel Cellular
Palmer Communications
Petroleum Communications
Public Service Cellular
Radiofone
Richmond Cellular Telephone
Sasktel Cellular
Savannah Cellular
Southwestern Bell Mobile Systems
Springwich Cell. Ltd Partner. (SNET)
StarCellular
U S West Cellular
United States Cellular
Unity Cellular Systems (Unicel)

Vanguard Cellular

OK Cellular

Blue Ridge Cellular
Blue Ridge Cellular Telephone
Cellular Connection
Cellular Three
Centel
Mercury Cellular and Paging
MobileTel
MTS Mobility
MT&T
NB Tel Cellular
PacTel Cellular
Peoples Cellular
Unity Cellular Systems (Unicel)
Unitel
United States Cellular

Pace Communications

BellSouth Mobility
Cellular Information Systems
Cellular One of Lake Charles
Century Cellunet
Coastel Communications
Contel Cellular
Houston Cellular
MobileTel
Petroleum Communications
Radiofone

Pacific Telecom Cellular (Cellulink)

Advantage Cellular
AGT Cellular
Akron Cellular Telephone
Allegan
ALLTEL Mobile Communications
Ally
Alpha Cellular
Alpine Cellular
AMC Cellular
Ameritech Mobile Communications
Appalachian Cellular
Atlantic Cellular
Bahamas Telecommunications
Bakersfield Cellular Telephone
BCTel Mobility Cellular
Bell Atlantic Mobile
Bell Mobility Cellular
BellSouth Cellular
Big Horn Cellular
Blackwater Cellular
Bledsoe Telephone
Blue Mountain Cellular
Blue Ridge Cellular
Bluegrass Cellular
Brazos Cellular
C. C. Cellular
Cal-One Cellular
Cantel
Canton Celular
Carolina West Cellular
CEDETEL
Cellcom of Green Bay
Cellcom of Hickory
Cellular Information Systems
Cellular One of Amarillo
Cellular One of Berrien County
Cellular One of Chiacgo
Cellular One of Columbia, MO
Cellular One of Columbia, TN
Cellular One of Danville, IL
Cellular One of East/Central PA
Cellular One of Frederick, MD
Cellular One of Gainesville, GA
Cellular One of Galesburg, IL
Cellular One of Indiana
Cellular One of Indianapolis

Roaming Agreements

Cellular One of La Salle, IL
Cellular One of Lakes Charles, LA
Cellular One of Mid-South Texas
Cellular One of Middle Tennessee
Cellular One of of Northeast
Cellular One of Sioux Falls, SD
Cellular One of South Dakota
Cellular One of the Great Lakes of Iowa
Cellular One of Upstate New York
Cellular One of Victoria, TX
Cellular Phone of Kentucky
Cellular Plus
Cellular Plus of Georgia
Cellular Three
Cellular Ventures
Cellular XL
Cellular, Inc.
Celutel
Centel
Century Cellunet
Chariton Valley Cellular
Clear Communications
Coastel Communications
Contel Cellular
Cooper Cellular
Cross-Valliant Cellular
Crowley Cellular
CyberTel Cellular
Danbury Cellular
Dial Two
DiComm Cellular
Digital Cellular of Texas
Dobson Cellular
Dominion Resources
Eastex Cellular
EdTel Cellular
Enid Cellular
Farmers Cellular
FGI Cellular Management
First Kentucky Cellular
Gaia
General Cellular
Georgia Independent RSA 7 &10
GMD Limited Partnership
Great Lakes of Iowa
GTE Mobilnet
Honolulu Cellular
Houston Cellular
Illinois Valley Cellular
Independent Cellular
Inland Cellular Telephone Company
Iowa East Cellular Telephone Company
IUSA Cell
Kansas Independent Network
Kaplan Telephone
Kull Cellular
Lake Huron Cellular
Lincoln Telephone
Litchfield County Cellular
Mackinac Cellular
MacTel Cellular
Maine Cellular Telephone
McCaw Cellular Communications
MCMG
Mercury Cellular Telephone
Mercury Communications
Metacomm Cellular
MetaComm Cellular
Metrocel Cellular Telephone
Mid-Missouri Cellular
Mid-South Cellular
Milwaukee Telephone Company
Miscellco Communications
Missouri Cellular
Mobile Communications
MobilTel
Mountaineer Cellular

Mountaineer Mobile
MT&T Mobile
MTS Cellular
National Cellular
Nebraska Cellular Telephone
New Par
NYNEX Mobile Communications
Pace Communications
Pacific Northwest Cellular
PacTel Cellular
Palmer Cellular
Panhandle Telecommunications
Peoples Cellular
Petroleum Communications
PriCellular
Québéc Téléphone Cellulaire
Qunatum Communications
Radiofone
RFB Cellular
Richmond Cellular
Rochester Telephone Mobile Communications
Rural Cellular Management
SaskTel Cellular
Savannah Cellular
Springwich Cell. Ltd Partner. (SNET)
Southwestern Bell Mobile Systems
Springwich Cellular
Sterling Cellular
Sunshine Cellular
Texas-16 Cellular
Thunder Bay Cellular
U S West Cellular
Union Telephone
United States Cellular
Unitel
Unity Cellular
Vanguard Cellular
Virginia 10 Limited Partnership
West Alabama Cellular
West Central Cellular
Westex Cellular
X-Cell Cellular
XIT Cellular
Yorkville Telephone

PacTel Cellular

Advantage Cellular
AGT Cellular
Allcell
Allcell Laredo
Allegan Cellular
ALLTEL Mobile Communications
Ally
Alpha Cellular
Alpine Cellular
AMC Cellular
Ameritech Mobile
Appalachian Cellular
Atlantic Cellular
Atlas Cellular
Auburn Television
Bakersfield Cellular
BCTel Mobility Cellular
Bell Atlantic Mobile
Bell Mobility Cellular
Bell South Cellular
Bell South Mobility
Blackwater Cellular
Bledsoe Telephone
Blue Grass Cellular
Blue Mountain Cellular
Blue Ridge Cellular
Brazos Cellular
Buckhead Telephone
Buffalo Telephone
Cal-One Cellular
Cantel Cellular Canada
Canton Cellular

Carolina West Cellular
Celfon
Cellcom
Cellcom Hickory
Cellular 2000
Cellular Communications Puerto Rico
Cellular Holding
Cellular One of Central Illinois
Cellular One of Chicago
Cellular One of Columbia
Cellular One of Columbia Tenn
Cellular One of Danville
Cellular One of Dumas
Cellular One of East/Central PA
Cellular One of Enid
Cellular One of Frederick
Cellular One of Gainesville GA
Cellular One of Grand Forks
Cellular One of Huntsville
Cellular One of Indiana
Cellular One of Indianapolis
Cellular One of Kansas
Cellular One of Lake Iowa
Cellular One of Lasalle
Cellular One of Marion & Huntington
Cellular One of Meridian
Cellular One of Middle TN
Cellular One of New Jersey
Cellular One of New York
Cellular One of Northeast Mississippi
Cellular One of NorthTexas
Cellular One of Ocean County
Cellular One of Ozarks
Cellular One of Panama City
Cellular One of Puerto Rico
Cellular One of Santa Cruz
Cellular One of Sioux Falls
Cellular One of Southwest Florida
Cellular One of Southwest Oklahoma
Cellular One of Washington
Cellular One of Waycross
Cellular One of West Illinois
Cellular One of West Oklahoma
Cellular One of Wilmington
Cellular One of Woodard OK
Cellular Phone Kentucky
Cellular Plus GA
Cellular Plus PA
Cellular South
Cellular Systems International
Cellular Telephone Kentucky
Cellular Three
Cellular Ventures
Cellular XL
Cellulink
Celutel
Centel Cellular
Centennial Cellular
Century Cellular
Century Cellunet
Chariton Valley Cellular
Churchill County Cellular
Clear Communications
Coastel Cellular
Commonwealth Mobile Services
Contel Cellular
Cross-Valliant Cellular
Crowley Cellular
Danbury Cellular Telephone
Detroit Cellular
Dial Two
Dicomm Cellular Telephone
Digital Cellular
Dobson Cellular
Dominion Resources
Easterbrooke Cellular
ED Tel Cellular
Enid Cellular
Farmers Cellular Telephone

Finger Lakes Cellular
First Fayette Cellular
First Kentucky Cellular
Five Star Cellular
Florida 9 Cellular
Gaia
Galveston Cellular
Gencell
Genesee Telephone
GMD
Greater S Dakota Cellular
GTE Mobile Communications
HFB Cellular
Highland Cellular
HLD Cellular
Honolulu Cellular Telephone
Houston Cellular
Illinois Valley Cellular
Independent Cellular Network
Indianapolis Telephone
Inland Cellular Telephone
Interstate Cellular
Iowa E Cellular Telephone
Kansas Independent Networks
Kaplan
Kentucky Cellular
Lake Huron Cellular
Leaco Rural Telephone
Liberty Cellular
Lincoln Telephone
Litchfield County Cellular
Lone Star Cellular
Louisiana Cellular
Mac-Tel Cellular Systems
Maine Cellular
Maritime Tel & Tel
McCaw Cellular Communications
Mega Communications
Mercury Cellular Telephone
Mercury Communications
Metacomm Cellular
Metro Mobile
Metrocel Cellular
Metrophone
Mid-Missouri Cellular
Mid-South Cellular
Mid-Tenn Cellular
Mid-Tex Cellular
Midland Communications
Milwaukee Telephone
Minerick Cellular
Miscellco
Missouri 1 Atchison
Missouri Cellular
Mo-Tel Cellular
Mobile Communications
Mobiletel
Montana Cellular Telephone
Mountaineer Cellular
Mountaineer Mobile
MTS Cellular
National Cellular
Nebraska Cellular
New Par
Northwest Cellular
NYNEX Mobile Communications
OK Cellular
Oneonta Telephone
Pace Communications
Palmer Communications
Panhandle Telecommunications
PC Cellular Vermont
Pegasus Cellular Telephone
Peoples Cellular
Petroleum Communications
Pine Cellular Phones
Pioneer Telephone
Poka-Lambro Telecommunications
Pri-Cellular
Public Service Cellular

Puerto Rico Telephone
Quantum Communications
Québéc Téléphone Cellulaire
R & D Cellular
Radiofone
RFB Cellular
Richmond Cellular Telephone
Rochester Telephone
Rural Cellular
Sasktel Cellular
Shenandoah Mobile
Springwich Cell. Ltd Partner. (SNET)
Sooner Cellular
Southeast Cellular
Southeast Indiana Cellular
  Telephone
Southern Cellular
Southwestern Bell Mobile
StarCellular
Sterling Cellular
Sunshine Cellular
Sussex Cellular
Syracuse Telephone
Telcel
Texas 16 Cellular
Texas Cellular
Texoma Cellular
Thumb Cellular
Thunder Bay Cellular
Tsaconas Cellular
U S West Cellular
UBET Cellular
Union Telephone
United States Cellular
Universal Telecell
Unity Cellular Systems (Unicel)
Utica Telephone
Valley Telecommunications
Vanguard Cellular
Virginia Cellular
Vitel Cellular
Wanut Hill Cellular
West Alabama Cellular
West Central Cellular
West-Tex Telecomm
Wichita Falls Celltelco
Wisonsin Cellular
XIT Cellular
Yorkville Telephone
Youngstown Cellular

Palmer Communications

Advantage Cellular
Allcell
ALLTEL Mobile Communications
Ameritech Mobile Communication
Associated Communication
Bay Area Cellular
Bell Atlantic Mobile
BellSouth Mobility
Cantel
Cellcom of Hickory, NC
Cellular Information Systems
Cellular One of Amarillo, TX
Cellular One of Cincinnati, OH
Cellular One of Cleveland
Cellular One of Bakersfield
Cellular One of Baltimore
Cellular One of Beaumont
Cellular One of Berrien County
Cellular One of Casper
Cellular One of Chicago/Gary
Cellular One of Dayton
Cellular One of Detroit
Cellular One of Flint, MI
Cellular One of Grand Rapids
Cellular One of Indianapolis, IN
Cellular One of Lake Charles, LA
Cellular One of Lansing, MI

Cellular One of Lima, OH
Cellular One of Milwaukee
Cellular One of Muskegon
Cellular One of Richmond, VA
Cellular One of Roanoke
Cellular One of Springfield
Cellular One of Syracuse, NY
Cellular One of the Rio Grande Valley
Cellular One of Toledo, OH
Cellular One of Youngstown, OH
Cellular One of Utica, NY
Cellular South
Centel Cellular
Comcast
Contel Cellular
Crowley Cellular
Cybertel
GTE Mobilnet ("B" systems only)
GenCell Management
Gulf Coast Cellular
Houston Cellular Telephone
Los Angeles Cellular
MCTA
MTS Cellular
McCaw Communications
Mercury Cellular and Paging
MetroCel Cellular
Metro Mobile CTS
Metro One
Metrophone
MobileTel
PacTel Cellular
Petroleum Communications
Radiofone
Santa Cruz Cellular
U S West Cellular
United States Cellular
Vanguard Cellular

Panhandle Telecommunication

Advantage Cellular
ALLTEL Mobile Communications
Ameritech Mobile Communications
Appalachian Cellular
Atlantic Cellular
BCTel Mobility Cellular
Bell Atlantic Mobile
BellSouth Mobility
Bledsoe
Blue Mountain Cellular
Blue Ridge Cellular
Bluegrass Cellular
Brazos Cellular
C. C. Cellular
C.I.S.
Cal-One Cellular
Cantel
Canton
Cellcom
Cellcom of Hickory
Cellular Connection
Cellular Holdings
Cellular, Inc
Cellular One of
  Baltimore/Washington
Cellular One of Boston
Cellular One of Chicago
Cellular One of Greater South Dakota
Cellular One of Lake Charles
Cellular One of Amarillo
Cellular One of Columbia
Cellular One of Columbia, TN
Cellular One of Columbus, IN
Cellular One of Danville
Cellular One of Dobson
Cellular One of Dumas
Cellular One of East/Central PA
Cellular One of Indiana
Cellular One of Middle Tennesee

Cellular One of the Ozarks
Cellular One of Southwest Oklahoma
Cellular One of Western Oklahoma
Cellular One of Wichita Falls
Cellular Phone of Kentucky
Cellular Plus
Cellular Systems
Cellular Three
Centel Cellular
Century Cellular
Chariton Valley Cellular
Coastel Communications
Contel Cellular
Crowley Cellular
Cybertel
Dobson Cellular
Eastex Cellular
Ed Tel Cellular
Enid Cellular
Farmers Cellular
GenCell Management
GMD Partnership
GTE Mobilnet
Highland Cellular
Illinois Valley Cellular
Independent Cellular Network
Kansas Cellular
Lake Huron Cellular
Leaco Rural Telephone
Lincoln Telephone
Mackinac Cellular
MacTel Cellular
Maine Cellular
McCaw Cellular Communications
Mercury Cellular
Mercury Communications
Metrocel
Mid-Missouri Cellular
Mid-Texas Cellular
Miscellco
Mississippi Cellular
Missouri Cellular
Mobiletel
Mountaineer Cellular
Mountaineer Mobile
MT&T Cellular
MTS Cellular
Nebraska Cellular
NYNEX Mobile Communications
Oklahoma Cellular RSA 6
Oklahoma Western Telephone
Pace Communications
Pacific Telecom Cellular
PacTel Cellular
Peoples Cellular
Pine Cellular
Poka Lambro Telecommunications
Public Service Cellular
Quantum Communications
Québéc Téléphone Cellulaire
Rochester Telephone
Sagir
Sasktel Cellular
Southeastern Cellular
Southwestern Bell Mobile
Springwich Cellular
StarCellular
Sussex Cellular
Taylor Telecommunications
Tsaconas Cellular
U S West Cellular
UBET cellular
Union Telephone
United State Cellular
Unitel
Unity Cellular Systems (Unicel)
Valley Telecommunications
Virginia 10 RSA Limited
West Central Cellular

XIT Cellular
Yorkville Telephone

Peoples Cellular

ALLTEL Mobile Communications
Bell Atlantic
Bell South Mobility
Blue Grass Cellular
Cal-One Cellular
C.C. Cellular
Cell Communications
Cellular 2000
Cellular 29 Plus
Cellular Holding
Cellular One of Chicago
Cellular Phone Kentucky
Cellular Three
Cellulink
Century Cellunet
Chariton Valley Cellular
Cybertel Cellular
Dobson Cellular
Enid Cellular
First Cellular Omaha
Five Star Cellular
GTE Mobilnet
Kansas Cellular
Kentucky Cellular
Lamar Cellular
Leaco Cellular
Liberty Cellular
Lincoln Cellular
McCaws Cellular
Mercury Cellular
Metrocell
Mid Missouri Cellular
Mississippi Cellular
Mobiletel
Mountaineer Cellular
Northwest Cellular
OK Cellular
Pace Communications
PacTel Cellular
Pine Cellular
Pioneer Cellular
PTSI Panhandle Telecomm
Public Service Cellular
Southwest Bell Mobile
Texas Cellular
Thumb Cellular
U S West Cellular
Union Telephone
United States Cellular
Walnut Hill Cellular
West Central Cellular
Westex Cellular
XIT Rural Telephone
Yorkville Telephone

Petroleum Communications

Advantage Cellular
AGT Cellular
Albany Telephone
ALLTEL Mobile Communications
Ally
American Rural Cellular
Ameritech Mobile
Appalachian Cellular
Associated Communications
Atlantic Cellular
Atlas Cellular
Auburn Television
Bahama Cellular
Bakersfield Cellular
Batelco
Baton Rouge Cellular Telephone
Bay Area Cellular
Bell Atlantic Mobile
B.C. Tel Mobility Cellular

Bell South Cellular
BellSouth Mobility
Big Horn Cellular
Bluegrass Cellular
Blue Mountain Cellular
Blueridge Cellular
Buffalo-Erie Telephone
C. C. Cellular
Cal-One Cellular
Cantel Cellular
Canton Cellular
Celfon
Cellcom of Green Bay
Cellcom of Hickory
Cellular 2000
Cellular Communications of Puerto
Rico
Cellular Connection
Cellular, Inc.
Cellular Information Systems
Cellular of Indiana
Cellular One of Amarillo, TX
Cellular One of Atlantic City, NJ
Cellular One of
Baltimore/Washington
Cellular One of Berrien County, MI
Cellular One of Boston, MA
Cellular One of Chicago, IL
Cellular One of Cincinnati, OH
Cellular One of Columbia, MO
Cellular One of Columbia, TN
Cellular One of Columbus, OH
Cellular One of Danville, IL
Cellular One of Delaware
Cellular One of Detroit, MI
Cellular One of Dumas, TX
Cellular One of East/Central PA
Cellular One of Gainesville, GA
Cellular One of Hattiesburg, MS
Cellular One of Hickory, NC
Cellular One of Huntsville, TX
Cellular One of Indianapolis, IN
Cellular One of Lafayette, LA
Cellular One of La Salle, IL
Cellular One of Lake Charles, LA
Cellular One of Lenoir, NC
Cellular One of Meridian, MS
Cellular One of Middle Tennessee
Cellular One of Milwaukee, WI
Cellular One of New Jersey
Cellular One of New York, NY
Cellular One of North Texas
Cellular One of Ocean County, NJ
Cellular One of Richmond, IN
Cellular One of Santa Cruz, CA
Cellular One of Shenandoah, PA
Cellular One of Sioux Falls, SD
Cellular One of Southwest Oklahoma
Cellular One of Syracuse, NY
Cellular One of Upstate New York
Cellular One of Utica, NY
Cellular One of Victoria, TX
Cellular One of Walla Walla, WA
Cellular One of Western Illinois
Cellular One of Western Oklahoma
Cellular One of Wilmington, NC
Cellular Phone
Cellular Phone Kentucky
Cellular Plus ("A" side)
Cellular South
Cellular Telephone of Kentucky
Cellular Ventures
Cellular XL
Cellulink
Centel
Centennial Cellular
Century Cellular
Century Cellunet
Chariton Valley Cellular

Coastel Communications
Comcast
Contel Cellular
Crowley Cellular
Cybertel Cellular
Detroit Cellular Telephone
Dicomm Cellular
Digital Cellular of Texas
Enid Cellular
FGI Cellular
Five Star Cellular
Finger Lakes Telephone
Gaia
GenCell Management
Genesee Telephone
GMD Cellular
GTE Mobilnet ("B" systems only)
Gulf Coast Cellular
HDL Rural Cellular
Highland Cellular
Honolulu Cellular
Houston Cellular
Independent Cellular Network
Interstate Cellular
Iowa East Cellular
IUSA Cell
Kansas Cellular
Kaplan Telephone
Lake Huron Cellular
Lincoln Telephone Cellular
Litchfield County Cellular
Mackinac Cellular
Maine Cellular Telephone
Maritime Cellular
McCaw Communications
MCMG
Mercury Cellular
Mercury Communications
MetroCel
Metro Mobile CTS
Metro One
Metrophone
Mid-South Cellular
Miscellco
Missouri Cellular
Mobile Communications
MobileTel
MobileSouth Cellular
Mountaineer Cellular
Motel Cellular
MTS Cellular
National Cellular
NCPT Cellular
Nebraska Cellular
NewTel Cellular
Northwest Cellular Link
NYNEX Mobile Communications
Oneonta Telephone
Pace Communications
PacTel Cellular
Palmer Communications
Pegasus Cellular
Pine Cellular Phones
PriCellular
Public Service Cellular
Quantum Communications
Québéc Téléphone Cellulaire
Radiofone
Richmond Cellular
Rural Cellular Management
Savannah Cellular
Shenandoah Mobile
Southeastern Cellular
Southwestern Bell Mobile
Springwich Cell. Ltd Partner. (SNET)
StarCellular
Sterling Cellular
Syracuse Cellular Telephone
Texas 16 Cellular
Texas Cellular

Thunder Bay Cellular
Tsaconas Cellular
U S West Cellular
United States Cellular
Unitel
Unity Cellular Systems (Unicel)
Vanguard Cellular
Virginia Cellular
West Alabama Cellular
West Central Cellular
Westex Cellular
Wichita Falls Cellular Telephone
Youngstown Cellular

Pioneer Cellular

Advantage Cellular
ALLTEL Mobile Communications
B.C. Tel Mobility Cellular
Blue Ridge Cellular
Blue Ridge Cellular Telephone
Brazos Cellular
Cellular Information Systems
Cellular 3
Dobson Cellular Systems (B systems)
MT&T Cellular
MTS Cellular
Mercury Cellular & Paging
Mid-Missouri Cellular
MobileTel
MTS Mobility
NB Tel Cellular
Peoples Cellular
Quantum Communications
Unity Cellular Systems (Unicel)
United States Cellular

PriCellular

Adirondack Cellular
AllTel Mobile Communications
Ally
Alpine
Appalachian Cellular
Atlantic Cellular
Bakersfield Cellular
Bay Area Cellular
B.C. Tel Mobility Cellular
Bell Atlantic Mobile
Bell Mobility Cellular
BellSouth Cellular
BellSouth Mobility
Big Horn Cellular
Bledsoe Telephone
Blue Mountain Cellular
Bluegrass Cellular
Blue Ridge Cellular
Brazos Cellular
Cal-One Cellular
Cantel
Canton Cellular
Carolina West Cellular
C. C. Cellular
Cellcom of Greenbay
Cellcom of Hickory
Cellular 2000
Cellular Connection Alaska
Cellular Holdings
Cellular, Inc.
Cellular Information Systems
Cellular One of Albany
Cellular One of Ameraillo
Cellular One of
  Baltimore/Washington
Cellular One of Berrien County
Cellular One of Boston
Cellular One of Brookings
Cellular One of Buffalo
Cellular One of Chicago
Cellular One of Collier / Hendry

Cellular One of Columbia, MO
Cellular One of Columbus, TN
Cellular One of Danville
Cellular One of Enid
Cellular One of Finger Lakes
Cellular One of Frederick
Cellular One of Haralson
Cellular One of Huntsville
Cellular One of Indiana
Cellular One of Indianapolis
Cellular One of Lake Charles
Cellular One of La Salle, IL
Cellular One of Meridian
Cellular One of Middle Tennessee
Cellular One of Norfolk
Cellular One of Northeast Georgia
Cellular One of Northeast Mississippi
Cellular One of the Ozarks
Cellular One of Paramus
Cellular One of Santa Cruz
Cellular One of Sioux Falls
Cellular One of Southwest Alabama
Cellular One of Southwest Oklahoma
Cellular One of Syracuse, NY
Cellular One of Upstate New York
Cellular One of Utica
Cellular One of Wayside
Cellular One of West Virginia
Cellular One of Western Oklahoma
Cellular One of Wichita Falls, TX
Cellular Plus
Cellular South
Cellular Systems International
Cellular Telephone Company of
  Kentucky
Cellular Ventures
Cellular XL
Celutel
Centel Cellular
Centennial Cellular
Century Cellunet
Chariton valley Cellular
CJD Wisconsin Cellular
Clear Communications
Coastel Communications
Contel Cellular
Cooper Cellular
Crowley Cellular
C.V. Cellular
Cybertel Cellular
Danbury Cellular
Detroit Cellular
Dial Two
Digital Cellular
Dobson Cellular
Dominion Cellular
Easterbrooke Cellular
Enid Cellular
Farmers Cellular
Gaia
Galveston Cellular
General Cellular
Genesee Telephone
GMD Limited
GTE Mobile Communications
Honolulu Cellular
Houston Cellular
Illinois Valley Cellular
Independent Cellular
Interstate Cellular
Jackson Cellular
Kansas Cellular
Kull Cellular
Lake Huron Cellular
Litchfield County Cellular
Los Angeles Cellular
Mackinac Cellular
McCaw Communications
Mercury Cellular
MetroCel

Metrophone
Mid-Missouri Cellular
Mid-Texas Cellular
Mid-South Cellular
Milwaukee Telephone
Miscellco Communications
Missouri Cellular
Mobile Communications
Mountaineer Cellular
MTS Cellular
MT&T Cellular
Nebraska Cellular
NYNEX Mobile Communications
Ocean County Cellular
Oneonta Telephone
Pace
Pacific Northwest Cellular
Pacific Telecom Cellular
PacTel Cellular
Palmer Communications
PC Cellular of Kentucky
Petroleum Communications
Public Service Cellular
Quantum Communications
R & D Cellular
Radiofone
Richmond Cellular
Rochester Telephone
Rural Cellular Management
SaskTel
Shenandoah Mobile
Southeastern Cellular
Southwestern Bell Mobile
Springwich Cell. Ltd Partner. (SNET)
Sterling Cellular
Stillwater Cellular
Sunshine Cellular
Sussex Cellular
Texoma
Thunder Bay Cellular
Texas 16
Tsacanas
U S West Cellular
UBET Cellular
Union Telephone
United States Cellular
Unitel
Unity
Vanguard Cellular
Virginia Cellular
West Alabama Cellular
West Central Cellular
Westex
X-Cell Cellular
Yorkville Telephone Cooperative

Prime Cellular Management

Adirondack Cellular
ALLTEL Mobile Communications
Alpine
Atlantic Cellular
Bay Area Cellular
Bell Atlantic Mobile
BellSouth Cellular
BellSouth Mobility
Blackwater
Blue Mountain Cellular
Brazos Cellular
Canton
C. C. Cellular
Cellcom of Greenbay
Cellcom of Hickory
Cellular 2000
Cellular Holdings
Cellular Information Systems
Cellular One of Amarillo
Cellular One of
  Baltimore/Washington
Cellular One of Berrien County

Cellular One of Chicago
Cellular One of Columbia
Cellular One of Danville
Cellular One of East/Central PA
Cellular One of Indiana
Cellular One of Indianapolis
Cellular One of Lake Huron
Cellular One of Meridian
Cellular One of Middle Tennessee
Cellular One of Sioux Falls
Cellular One of Southwest Florida
Cellular One of Southwest Oklahoma
Cellular One of Southeast Indiana
Cellular One of Western Oklahoma
Cellular Phone of Kentucky
Cellular XL
Celutel
Centel
CJD Wisconsin
Clear Communications
Contel
Crowley
Cybertel
Detroit Cellular
Digital Cellular
Dobson Cellular
Dominion Cellular
Easterbrooke
Eastex
Galveston
General Cellular
Greater South Dakota
GTE Mobile Communications
HLD Cellular
Houston Cellular
Illinois Valley Cellular
Lake Charles Cellular
Los Angeles Cellular
Mackinac
McCaw Cellular Communications
MegaComm
Mercury
Metrocel
Metromobile
Mid-South Cellular
Mississippi Cellular Telephone
Mobile Telenet
New Par
North Texas Cellular
PacTel Cellular
Palmer Communications
PC Cellular of Kentucky
PC Cellular of Vermont
PriCellular
R & D Cellular
Radiofone
Sterling
Stillwater
Sussex
Texas 16 Cellular
U S West Cellular
Union Cellular
United Bluegrass Cellular
Unitel
United States Cellular
Vanguard Cellular
Virginia 10
Victoria Cellular
West Alabama

Public Service Cellular

Advantage
ALLTEL Mobile Communications
Ally
American Rural Cellular
Ameritech Mobile Communications
Appalacian Cellular
Atlantic Cellular
Atlas Cellular

Auburn Television
Bahamas Telecommunications
Bell Atlantic Mobile
Bell South Cellular
Bledsoe Telephone
Blue Mountain Cellular
Blue Ridge Cellular
Bluegrass Cellular
Cal-One Cellular
Cantel
Canton Cellular
Cellcom
Cellular Connection
Cellular Holding
Cellular, Inc.
Cellular Information Systems
Cellular One of Chicago
Cellular One of Meridian
Cellular One of Northeast Georgia
Cellular One of Sioux Falls
Cellular Phone of Kentucky
Cellular Plus
Cellular Properties
Centel Cellular
Century Cellunet
Coastel Communications
Contel Cellular
Cross-Valliant Cellular
Crowley Cellular Telecommunications
Dobson Cellular
Dominion Cellular
Enid Cellular
Farmers Cellular
First Fayette Cellular
First Kentucky Cellular
Gaia
GMD Limited Partnership
Great Lakes of Iowa
GTE Mobilnet ("B" systems only)
Highland Cellular
Illinois Valley Cellular
Independent Cellular Network
Interstate Cellular (Interstate Cellular
  (Intercel))
Iowa East Cellular
Kansas Cellular
Kaplan Telephone
Lake Huron Cellular
Litchfield County Cellular
Mackinac Cellular
MacTel
Maine Cellular
McCaw Communications
Mercury Cellular
MetaComm Cellular
Mid-South Cellular
Mid-Tennessee Cellular
Mississippi Cellular
Missouri Cellular
Mobile Communications
Mountaineer Cellular
MTS Cellular
National Cellular
NYNEX Mobile Communications
Oneonta Telephone
Pacific Telecom Cellular
PacTel Cellular
Palmer Cellular
Panhandle Telecommunications
Peoples Cellular
Petroleum Communications
Radiofone
Richmond Cellular
Southeastern Cellular
Southwestern Bell Mobile Systems
Springwich Cell. Ltd Partner. (SNET)
Sterling Cellular
Sunshine Cellular
Ten Woodland

Texas 16 Cellular
U S West Cellular
Union Telephone
United States Cellular
Vanguard Cellular Systems
Virigina 10 RSA Limited Partnership
West Alabama Cellular Telephone
Westex Cellular
XIT

Quantum Communications

Advantage Cellular
AGT Cellular
ALLTEL Mobile Communications
Ally
Associated Communications
Atlantic Cellular
Atlas Cellular
Bakersfield Cellular
BCTel Cellular
Bell Atlantic Mobile
Bell Mobility Cellular
BellSouth Cellular
BellSouth Mobility
Bledsoe Telephone
Blue Ridge Cellular
Bluegrass Cellular
Brazos Cellular Communications
Cal-One Cellular
Cantel
C.C. Cellular
Cellcom
Cellcom of Hickory
Cellular, Inc.
Cellular Information Systems
Cellular of 2000
Cellular One of Amarillo
Cellular One of Appleton
Cellular One of Baton Rouge
Cellular One of Berrien County
Cellular One of Burlington
Cellular One of Cincinnati
Cellular One of Cleveland
Cellular One of Columbus OH
Cellular One of Danville
Cellular One of Detroit
Cellular One of Frederick
Cellular One of Galesburg
Cellular One of Galveston
Cellular One of Grand Forks
Cellular One of Huntsville
Cellular One of Indianapolis
Cellular One of Jacksonville
Cellular One of LaSalle
Cellular One of Lima
Cellular One of Meridian
Cellular One of Milwaukee
Cellular One of Northeast Georgia
Cellular One of Richmond
Cellular One of Rome
Cellular One of Sioux Falls
Cellular One of Syracuse
Cellular One of the Rio Grande Valley
Cellular One of Toledo
Cellular One of Upstate New York
Cellular One of Utica
Cellular One of Victoria
Cellular One of Wilmington
Cellular One of Youngstown
Cellular Plus
Cellular South
Cellular Three
Cellular XL
Cellulink
Celutel
Centel Cellular
Centennial Cellular
Century Cellunet
Coastel Communications

Codetel
Comcast
Contel Cellular
Crowley Cellular Telecommunications
Cybertel Cellular
Digital Cellular
Dobson Cellular Systems
Ed Tel Cellular
First Fayette Cellular
Five Star Cellular
GenCell Management
Greater South Dakota Cellular
GTE Mobilnet
Honolulu Cellular Telephone
Houston Cellular Telephone
Independent Cellular Network
Interstate Cellular
Iowa East Cellular
IUSA Cell
Kansas Cellular
LA Cellular Telephone
Lincoln Telephone
Mactel Cellular
Maine Cellular
Maritime Tel & Tel
McCaw Cellular Communications
Mercury Cellular
Mercury Communications
Metrocel Cellular
Mid-Missouri Cellular
Miscellco Communications
Mountaineer Mobile
Mobiletel
Mobilsouth Cellular
Mountaineer Mobile
MT&T Cellular
MTS Cellular
Nebraska Cellular
NB Tel Cellular
New Par
NYNEX Mobile Communications
Pace Communications
PacTel Cellular
Palmer Communications
Petroleum Communications
Pioneer Cellular
PTSI Cellular
Public Service Cellular
Radiofone
Richmond Cellular Telephone
Rochester Telephone Mobile
 Communications
Sasktel Cellular
Savannah Cellular
Southwestern Bell Mobile Systems
Springwich Cell. Ltd Partner. (SNET)
StarCellular
Sterling Cellular
Telcel
Telefonia Cellular
Texas Cellular
Thunder Bay Cellular
UBET Cellular
U S West Cellular
Union Telephone
United States Cellular
Unity Cellular Systems (Unicel)
Vanguard Cellular
West Central Cellular
Westex Cellular
XIT Cellular

Québéc Téléphone Cellulaire

Advantage Cellular
AGT Cellular
ALLTEL Mobile Communications
Ameritech Mobile Communications
Atlantic Cellular
BCTel Cellular

Bell Atlantic Mobile
Bell Mobility Cellular
BellSouth Mobility
Cellcom
Cellular Plus
Cellular South
Cellulink
Centel Cellular
Century Cellunet
Contel Cellular
CPS Cellular Telephone
Ed Tel Cellular
Enid Cellular
GTE Mobilnet
Independent Cellular Network
Lincoln Telephone Cellular
Maine Cellular
MT&T Cellular
MTS Cellular
NewTel Cellular
NYNEX Mobile Communications
PacTel Cellular
PTSI
Rochester Telephone Mobile
 Communications
Sasktel Cellular
Shenadoah Mobile
Southwestern Bell Mobile Systems
Springwich Cell. Ltd Partner. (SNET)
StarCellular
Thunder Bay Cellular
U S West Cellular
United States Cellular
Unity Cellular Systems (Unicel)
Western Maine Cellular

Radiofone

ALLTEL Mobile Communications
Associated Communications
Bakersfield Cellular
Bay Area Cellular
Bell Atlantic Mobile
BellSouth Mobility
Cantel
Cellcom of Hickory
Cellular Information Systems
Cellular One of
 Baltimore/Washington
Cellular One of Beaumont
Cellular One of Berrien County
Cellular One of Boston
Cellular One of Chicago/Gary
Cellular One of Detroit
Cellular One of Flint
Cellular One of Fort Myers, FL
Cellular One of Galveston, TX
Cellular One of Grand Rapids
Cellular One of Indianapolis
Cellular One of La Salle, IL
Cellular One of Lake Charles, LA
Cellular One of Milwaukee, WI
Cellular One of Richmond
Cellular One of Roanoke
Cellular One of Sioux Falls, SD
Cellular One of Syracuse, NY
Cellular One of Utica/Rome
Cellular One of Victoria, TX
Cellular One of Wilmington, DE
Cellular One of Youngstown, OH
Cellular South
Celutel
Centel Cellular
Century Cellunet
Coastel Communications
Comcast
Contel Cellular
Crowley Cellular Telecommunications
Cybertel Cellular Telephone
GTE Mobilnet ("B" systems only)

GenCell Management
Houston Cellular Telephone
Lincoln Telephone Cellular
Los Angeles Cellular Telephone
McCaw Communications
Mercury Cellular and Paging
Metro Mobile CTS
Metro One
MetroCel Cellular
Metrophone
Mobiltel
PacTel Cellular
Palmer Communications
Petroleum Communications
Public Service Cellular
Southwestern Bell Mobile Systems
U S West Cellular
Vanguard Cellular Systems
West Central Cellular

Rochester Telephone Mobile Communications

AGT Cellular
Advantage Cellular
Albany Telephone
ALLTEL Mobile Communications
Ameritech Mobile Communications
Appalachian Cellular
Atlantic Cellular
Atlas
Bell Atlantic Mobile
Bell Mobility Cellular
BellSouth Mobility
Bluegrass Cellular
Blue Ridge Cellular
C. C. Cellular
Cal-One Cellular
Cantel
Cellcom
Cellular Holdings
Cellular, Inc.
Cellular Information Systems
Cellular One of Chicago
Cellular One of Syracuse
Cellular One of Upstate New York
Cellular One of Utica/Rome
Cellular Plus
Cellulink
Centel Cellular
Century Cellunet
Clear Communications
Coastel Communications
Contel Cellular
Cross-Valliant Cellular
Crowley Cellular Telecommunications
Dobson Cellular
Ed Tel Cellular
Enid Cellular
GTE Mobilnet ("B" systems only)
Honolulu Cellular
Hudson Valley RSA 6
Illinois Valley Cellular
Independent Cellular Network
Kansas Cellular
Lincoln Telephone Cellular
Litchfield County Cellular
Maine Cellular
Mercury Cellular
Mid-South Cellular
Mississippi Cellular
Mountaineer Cellular
MT&T Cellular
MTS Cellular
Nebraska Cellular
NYNEX Mobile Communications
Pace Communications
Pacific Northwest Cellular
PacTel Cellular
Pegasus Cellular

PRTC Celulares Telefonica
Québéc Téléphone Cellulaire
Shenandoah Cellular
Southwestern Bell Mobile Systems
Springwich Cell. Ltd Partner. (SNET)
StarCellular
Sussex Cellular
Thunder Bay Cellular
U S West Cellular
Union Telephone
United States Cellular
Unity Cellular Systems (Unicel)

Rural Cellular Management

AMC Cellular Association
Ally
Alpine
Albany Telephone
ALLTEL Mobile Communications
Amarillo Cellular
Ameritech Mobile ("B" side)
Atlantic Cellular
Bakersfield Cellular
Bay Area Cellular
Bell Atlantic Mobile
Bell South Cellular
BellSouth Mobility ("B" side)
Blackwater Cellular
Bluegrass Cellular ("B" side)
Blue Mountain Cellular
Blue Ridge Cellular
Buckhead Telephone
Buffalo Telephone
CJD Wisconsin
Cantel Cellular
Canton Cellular
Cellcom
Cellcom of Hickory
Cellular 2000
Cellular One of Berrien County
Cellular One of Boston
Cellular One of Chicago
Cellular One of Danville
Cellular One of Detroit
Cellular One of Galesburg
Cellular One of Galveston
Cellular One of Huntsville
Cellular One of Indianapolis
Cellular One of Indiana
Cellular One of La Salle, IL
Cellular One of Meridian
Cellular One of Northeast Georgia
Cellular One of Santa Cruz
Cellular One of Sioux Falls
Cellular One of Southwest Florida
Cellular One of Southwest Georgia
Cellular One of South Indiana
Cellular One of Virginia RSA 6
Cellular One of Western Illinois
Cellular Inc.
Cellular Information Systems
Cellular Plus of Iowa
Cellular Ventures
Cellular XL
Cellulink
Celutel
Centel Cellular
Century Cellular
Clear Communications
Columbia Cellular Partnership
Contel Cellular
Cooper Cellular Management
Crowley Cellular
Cybertel
Danbury Cellular
Dicomm Cellular
Dobson Cellular
Dominion Resources
Easterbrooke Cellular

FGI Cellular
Finger Lakes Telephone
First Fayette Cellular
First Kentucky Cellular
GMD Limited Partnership
Gaia
General Cellular
Genessee Telephone
GTE Mobilnet
Great South Dakota Cellular
Highland Cellular
HLD Cellular
Honolulu Cellular
Houston Cellular
Independent Cellular Network
Indiana Cellular
Iowa East Cellular
Kull Cellular
Los Angeles Cellular
Lake Charles Cellular
Lake Huron Cellular
Litchfield County Cellular
MacKinac Cellular
McCaw Cellular Communications
Mega Comm
Metacomm
Mercury Communications
Metro One
Metro Mobile
Metrophone
Metrocel
Mid-Tennessee Cellular
Mid-South Cellular
Milwaukee Telephone
Miscellco
Missouri Cellular
Mobile Communications
Mobile Telenet
Mountaineer Mobile
National Cellular
New Par
Ocean Coutny Cellular
Ohio State Cellular
PC Cellular of Kentucky
PC Cellular of Vermont
PacTel Cellular
Palmer Communications
Panama City Cellular
Pegasus Cellular
Petroleum Communications
PriCellular
Prime Cellular
Quantum Communications
R & D Cellular
RFB Cellular
Radiofone
Richmond Cellular
Southeastern Indiana Cellular
Sooner Cellular
Stillwater Cellular
Sterling Cellular
Sunshine Cellular
Sussex Cellular
Syracuse Telephone
Ten Woodland
Texas 5
U S West Cellular
Union Cellular
United States Cellular
Unitel
Utica Telephone
Vanguard Cellular
Victoria Cellular
Washington/Baltimore Cellular
Wichita Falls Telephone
Youngstown Cellular

SaskTel Mobility Cellular

AGT Cellular

ALLTEL Mobile Communications
Ameritech Mobile Communications
Appalachian Cellular
Atlantic Cellular
BCTel Mobility Cellular
Bell Atlantic Mobile
Bell Mobility Cellular
BellSouth Mobility
Bledsoe
Bluegrass Cellular
Cal-One Cellular
C. C. Cellular
Cellcom
Cellular, Inc.
Cellular One of Indiana
Cellular Phone of Kentucky
Cellulink
Centel
Century Cellunet
Coastel Communications
Contel Cellular ("B" systems only)
Cross-Valliant Cellular
Digital Cellular of Texas
Dobson Cellular
Ed Tel Cellular
Enid Cellular
Farmer's Cellular
Five Star Cellular
Greater South Dakota Cellular
GTE Mobilnet ("B" systems only)
Highland Cellular
Illinois Valley Cellular
Independent Cellular
Kaplan Cellular
Liberty Cellular
MacTel
MT&T Cellular
MTS Cellular
Mid-Missouri Cellular
Miscellco
Missouri Cellular
NB Tel Cellular
Nebraska Cellular
NewTel Cellular
NYNEX Mobile Communications
PacTel Cellular
Panhandle Telecommunications
PriCellular
Quantum Communications
Québéc Téléphone Cellulaire
Shenandoah Mobile
Southeastern Bell Mobile Systems
Southwestern Bell Mobile Systems
Springwich Cell. Ltd Partner. (SNET)
Thumb Cellular
U S West Cellular
Union Telephone
United States Cellular
XIT Cellular
Yorkville Telephone

Shenandoah Cellular

Advantage Cellular
ALLTEL Mobile Communications
Ally
Ameritech Mobile Communications
Associated Communications
Atlantic Cellular
Atlas Cellular
Auburn Television Group
BCTel Cellular
Bell Atlantic Mobile
Bell Mobility Cellular
BellSouth Cellular
BellSouth Mobility
Blue Ridge Cellular
Blue Ridge Cellular Telephone
Bluegrass Cellular
Cal-One Cellular

C.C. Cellular
Cellcom of Green Bay
Cellcom of Hickory
Cellular Connection
Cellular, Inc.
Cellular Information Systems
Cellular of Indiana
Cellular One of Chicago, IL
Cellular One of Gainesville, GA
Cellular One of Indianapolis
Cellular One of Lake Charles, LA
Cellular One of Milwaukee, WI
Cellular One of Richmond
Cellular One of Syracuse/Utica
Cellular One of Upstate New York
Cellular One of Washington
Cellular One of Wichita Falls
Cellular Plus
Cellular South
Cellulink
Centel Cellular
Centennial Cellular
Century Cellunet
CGH Partnership
Coastel Communications
Comcast
Cone Enterprises
Cooper Cellular
Crowley Cellular Telecommunications
Cybertel Cellular
Dobson Cellular
Eastex Cellular
Ed Tel Cellular
Enid Cellular
GenCell Management
Greater South Dakota Cellular
GTE Mobilnet
Highland Cellular
Horizon Cellular
Houston Cellular Telephone
Independent Cellular Network
Interstate Cellular
Kansas Cellular
Lake Huron Cellular
Lamar County Cellular
Lincoln Telephone
Mactel Mercury Cellular
McCaw Cellular Communications
Metrophone
Mid-Missouri Cellular
Minerich Cellular
Miscellco Communications
Mobiletel
Mountaineer Cellular
Mountaineer Mobile
MT&T Cellular
MTS Cellular
Nebraska Cellular
NYNEX Mobile Communications
Ocean County Cellular
Pace Communications
PacTel Cellular
Palmer Communications
Petroleum Communications
PTSI Cellular
Public Service Cellular
Quantum Communications
Radiofone
Rochester Telephone Mobile
  Communications
Sasktel Cellular
Savannah Cellular
Southwestern Bell Mobile Systems
Springwich Cell. Ltd Partner. (SNET)
StarCellular
Sterling Cellular
Thumb Cellular
Union Telephone
United States Cellular
Unitel

Unity Cellular Systems (Unicel)
Vanguard Cellular
Virginia Cellular
West Central Cellular
Westex Cellular

Sooner Cellular

Adirondack Cellular
ALLTEL Mobile Communications
Alpine
Amarillo Cellular
AMC Cellular
Ameritech Mobile Communications
Appalachian Cellular
Atlantic Cellular
Bakersfield Cellular
Bay Area Cellular
BellSouth Cellular
Blackwater Cellular
Blue Mountain Cellular
Bluegrass Cellular
C. C. Cellular Cellcom of Hickory
Cellular 2000 ("B" side)
Cellular One of Berrien County
Cellular One of Chicago
Cellular One of Columbia, GA
Cellular One of Columbia, TN
Cellular One of Danville
Cellular One of Dumas
Cellular One of East/Central PA
Cellular One of Huntsville
Cellular One of La Salle, IL
Cellular One of Sioux Falls
Cellular One of Southwest Florida
Cellular One of Upstate New York
Cellular One of Western Illinois
Cellular One of Wichita Falls
Cellular One of Worthington
Cellular XL
Celutel
Centennial Cellular
Century Cellunet
CIS
CJD Wisconsin Cellular
Clear Communications
Coastel Communications
Cone Enterprises
Contel Cellular
Crowley Cellular
Danbury Cellular
Detroit Cellular
Dobson Cellular ("A" side)
Easterbrooke Cellular
First Kentucky Cellular
Galveston Cellular
General Cellular
Greater South Dakota Cellular
GTE Mobilenet
Highland Cellular
HLD Cellular
Houston Cellular
Indianapolis Telephone
Lake Charles Cellular
Lake Huron Cellular
Lawton Cellular
Los Angeles Cellular
McCaw Cellular Communications
Mega Communications
Mercury Communications
Metrocel
Metrophone
Mid-Tennessee Partners
Mid-South Cellular
Milwaukee Telephone
Miscellco Communications
Missouri Cellular
Mobile Telenet
Nebraska Cellular
Ocean County Cellular

Oneonta Telephone
PacTel Cellular
Palmer Cellular
Panhandle Telecommunications
PC Cellular
Pegasus Cellular
Petroleum Communications
Pine Cellular
PriCellular
Prime Cellular Management
R & D Cellular
Radiofone
Santa Cruz Cellular
Southeast Indiana Cellular
Sterling Cellular
Sussex Cellular
Syracuse Telephone
Texas 16 Cellular
Texas 5
Tsaconas Cellular
U S West
Union Telephone
United States Cellular
Unitel
Utica Telephone
Vanguard Cellular
Westex
Wisconsin Cellular
X-Cell Cellular
Youngstown Cellular

Southwestern Bell Mobile Systems

Advantage Cellular
AGT Cellular
ALLTEL Mobile Communications
Ameritech Mobile Communications
Atlantic Cellular
B.C. Tel Mobility Cellular
Bell Atlantic Mobile
Bell Mobility Cellular
BellSouth Mobility
Buffalo Telephone Co.
Cantel Cellular
Cellcom
Cellular, Inc.
Cellular Information Systems
Cellular One of Akron, OH
Cellular One of
  Baltimore/Washington
Cellular One of Boston
Cellular One of Canton
Cellular One of Chicago
Cellular One of Cincinnati
Cellular One of Cleveland
Cellular One of Columbus
Cellular One of Dayton
Cellular One of Hamilton
Cellular One of Mansfield
Cellular One of Northern Ohio
Cellular One of Springfield
Cellular One of Youngstown
Cellular Plus
Cellular South
Cellulink
Centel Cellular
Century Cellunet
Coastel Communications
Contel Cellular
Ed Tel Cellular
GTE Mobile Communications
Independent Cellular Network
Kansas Independent Networks
Lincoln Telephone Cellular
MT&T Cellular
MTS Cellular
Maine Cellular
Mercury Cellular and Paging
Mobiltel
Nebraska Cellular

NYNEX Mobile Communications
PacTel Cellular
Petroleum Communications
Pioneer Telephone Cooperative
Public Service Cellular
Radiofone
Rochester Telephone Mobile
  Comms.
Sasktel Cellular
Springwich Cell. Ltd Partner. (SNET)
StarCellular
U S West Cellular
United States Cellular
Unity Cellular Systems
West Central Cellular
Westex Cellular
XIT Cellular

Springwich Cellular Limited
Partnership (SNET)

Advantage Cellular
ALLTEL Mobile Communications
Alpha Cellular
Ameritech Mobile Communications
Appalachian Cellular
Atlantic Cellular
Atlas Cellular
Bell Atlantic Mobile
Bell Mobility Cellular
BellSouth Mobility
Blue Mountain Cellular
Buffalo Telephone
C. C. Cellular
Cal-One Cellular
Canton Cellular
Carolina West Cellular
Cellcom
Cellular Holdings
Cellular, Inc.
Cellular of Indiana
Cellular One of Boston
Cellular One of Casper
Cellular One of Columbia, TN
Cellular One of Danville
Cellular One of East/Central PA
Cellular One of Gainesville, GA
Cellular One of Grand Forks
Cellular One of Great Lakes
Cellular One of Indianapolis
Cellular One of Middle Tennessee
Cellular One of Missoula
Cellular One of Ohio
Cellular One of Syracuse
Cellular One of Upstate New York
Cellular One of
  Wsashington/Baltimore
Cellular Plus
Cellular South
Cellular XL
Cellulink
Cellwave
Centel
Century Cellunet
Coastel Communications
Contel
Cybertel Cellular
Dicomm Cellular
Dobson Cellular
EdTel Cellular
Enid Cellular
GMD Partnership
GTE Mobilnet ("B" systems only)
Independent Cellular
Interstate Cellular
Kansas Cellular
Lake Huron Cellular
Lincoln Telephone Cellular
Maine Cellular
McCaw Communications

Mercury Cellular
Milwaukee Telephone
Missouri Cellular
MobileTel
MT&T Cellular
Nebraska Cellular
NYNEX Mobile Communications
PC Cellular of Vermont
Pacific Northwest Cellular
PacTel Cellular
Pegasus Cellular
Petroleum Communications
PriCellular
Public Service Cellular
Québec Téléphone Cellulaire
RFB Cellular
Rochester Telephone
SaskTel
Savannah Cellular
Shenandoah Cellular
Southwestern Bell
StarCellular
Sterling Cellular
U S West Cellular
United States Cellular
Union Cellular
Unity Cellular Systems (Unicel)
Vanguard Cellular

StarCellular

AGT Cellular
Advantage Cellular
Ameritech Mobile Communications
Atlantic Cellular
B.C. Tel Mobility Cellular
Bell Atlantic Mobile
Bell Mobility Cellular
BellSouth Mobility
Cellcom
Cellular, Inc.
Cellular Information Systems
Cellular One of Lake Charles, LA
Cellular Plus
Cellular South
Cellulink
Centel Cellular
Century Cellunet
Coastel Communications
Contel Cellular
Ed Tel Cellular
Enid Cellular
GenCell Management
GTE Mobilnet ("B" systems only)
Independent Cellular Network
Kansas Independent Networks
MACtel Cellular System
MT&T Cellular
MTS Cellular
Maine Cellular
Mercury Cellular and Paging
MobileTel
PacTel Cellular
Rochester Telephone Mobile
  Communications
Southwestern Bell Mobile Systems
Springwich Cell. Ltd Parter. (SNET)
U S West Cellular
United States Cellular
Unity Cellular Systems (Unicel)
XIT Cellular

Sterling Cellular Management (FGI)

Advantage Cellular
Albany Cellular
ALLTEL Mobile Communications
Alpha Cellular
American Rural Cellular
Ameritech Mobile

Roaming Agreements

Appalachian Cellular
Atlantic Cellular
Auburn Television
Bahamas Telecommunications
Bakersfield Cellular
Bay Area Cellular
Bell Atlantic Metro Mobile
BellSouth Mobility
Bledsoe Telephone
Bluegrass Cellular
Blue Mountain Cellular
Blue Ridge Cellular
Buffalo Telephone
Cantel Cellular
Canton Cellular
Carolina West Cellular
Cellcom of Green Bay
Cellcom of Hickory
Cellular Communications
Cellular Connection
Cellular Holding
Cellular, Inc.
Cellular Information Systems
Cellular of Indiana
Cellular One of Atlantic City
Cellular One of Amarillo
Cellular One of
  Baltimore/Washington
Cellular One of Baton Rouge
Cellular One of Berrien County
Cellular One of Boston
Cellular One of Brookings
Cellular One of Chicago
Cellular One of Clayton
Cellular One of Collier/Hendry
Cellular One of Columbia
Cellular One of Columbia, TN
Cellular One of Detroit
Cellular One of Frederick, MD
Cellular One of Galesburg
Cellular One of Galveston
Cellular One of Huntsville, TX
Cellular One of Indianapolis
Cellular One of La Salle
Cellular One of Lake Charles, LA
Cellular One of Meridian
Cellular One of Middle Tennessee
Cellular One of Northeast Georgia
Cellular One of Northeast Colorado
Cellular One of Santa Cruz
Cellular One of Sioux Falls
Cellular One of Syracuse/Utica
Cellular One of Southwest Georgia
Cellular One of Upstate New York
Cellular One of Western Oklahoma
Cellular One of Wichita Falls
Cellular Phone of Kentucky
Cellular Plus
Cellular South
Cellular Systems International
Cellular Three
Cellular Ventures
Cellular XL
Cellwave
Celutel of Biloxi
Centel Cellular
Century Cellular
Century Cellunet
Clear Communications
Coastel Cellular
Compcomm
Contel Cellular
Crowley Cellular
Cybertel Cellular
C. C. Cellular
Danbury Cellular
Dicomm Cellular
Dobson Cellular
Ellis Thompson Cellular
Farmers Cellular

Gaia
General Cellular Managment
Genessee Telephone
GTE Mobilnet
Gulf Coast Cellular
Highland Cellular
Houston Cellular
Independent Cellular
Interstate Cellular
Kansas Cellular
Kull Cellular
Los Angeles Cellular
Lake Huron Cellular
Lincoln Telephone
Litchfield County Cellular
Mackinac Cellular
MacTel Cellular
Maine Cellular
McCaw Cellular Communications
Mega Comm
Mercury Cellular
Mercury Communications
Metro Mobile
Metro One
Metrocel Cellular
Metrophone
Midland Communications
Mid-Missouri Cellular
Mid-South Cellular
Miscellco Communications
Missouri Cellular
Mobiletel
Mountaineer Mobile
MT&T Cellular
National Cellular
Nebraska Cellular
North-West Cellular
NYNEX Mobile
Ocean County Cellular
Ohio State Cellular
Pace
Pacific Northwest Cellular
PacTel Cellular
Palmer Cellular
Petroleum Communications
PriCellular
Prime Cellular Management
Public Service Cellular
Puerto Rice Telephone
Quantum Communications
Radiofone
RFB Cellular
Richmond Cellular
Rural Cellular Management
Savannah Cellular
Shenandoah Mobile
Sooner Cellular
Springwich Cellular
StarCellular
Sussex Cellular
Texas 16 Cellular Telephone
Texoma Cellular
U S West Cellular
UBET Cellular
Union Telephone
United States Cellular
Unitel
Unity Cellular Systems (Unicel)
Valley Telecommunications
Vanguard Cellular
Virginia Cellular
Victoria Cellular
West Alabama Cellular
X-Cell Cellular

Telcel

AGT Cellular
Ameritech Mobile Communications
BC Tel Cellular

Bell Atlantic
Bell South Cellular
Cantel
Cellular One of Great Lakes of Iowa
Cellular One of Meridian
Cellular One of Washington
Cellular Phone of Kentucky
Coastel Communications
Contel Cellular
Independent Cellular Network
NYNEX Mobile Communications
PacTel Cellular
PriCellular
Southwestern Bell Mobile
U S West Cellular
United States Cellular

Texas Cellular

Advangage Cellular
AGT Cellular
ALLTEL Mobile Communications
Ameritech mobile Communications
Associated Communications
Atlantic Cellular
BCTel Cellular
Bell Atlantic Mobile
Bell Mobility Cellular
BellSouth Mobility
Brazos Cellular Communications
Cantel
Cellcom of Green Bay
Cellular, Inc.
Cellular Information Systems
Cellular One of Boston
Cellular One of Chicago
Cellular One of Columbia, MO
Cellular One of Lake Charles, LA
Cellular One of LaSalle
Cellular One of Washington
Cellular One of Wichita Falls
Cellular One of Wilmington
Cellular One of Youngstown
Cellular Plus
Cellular South
Cellular Systems International
Cellulink
Centel Cellular
Century Cellunet
Coastel Communications
Contel Cellular
Crowley Cellular Telecommunications
Digital Cellular
Ed Tel Cellular
Enid Cellular
Five Star Cellular
Gencell
GTE Mobilenet
Houston Cellular
Independent Cellular Network
Kansas Cellular
Lincoln Telephone
Maine Cellular
McCaw Cellular Communications
Mercury Cellular
Metrocel Cellular
Miscellco Communications
Mobiletel
MT&T Cellular
MTS Cellular
Nebraska Cellular
NYNEX Mobile Communications
Petroleum Communications
PTSI Cellular
Public Service Cellular
Radofone
Rochester Telephone Mobile
  Telecommunications
Sasktel Cellular
Savannah Cellular

Southwestern Bell Mobile Systems
Springwich Cell. Ltd Partner. (SNET)
StarCellular
U S West Cellular
United States Cellular
Unitel
Unity Cellular Systems (Unicel)
West Central Cellular
Wetex
XIT Cellular

Texoma Cellular Management

Adirondack Cellular
ALLTEL Mobile Communications
Alpine
Amarillo Cellular
AMC Cellular
Ameritech Mobile Communications
Appalachian Cellular
Atlantic Cellular
Auburn Television
Bakersfield Cellular
Bay Area Cellular
BellSouth Cellular
Blackwater Cellular
Blue Mountain Cellular
Bluegrass Cellular
C. C. Cellular
Cellcom of Hickory
Cellular 2000 ("B" side)
Cellular One of Berrien County
Cellular One of Chicago
Cellular One of Columbia, GA
Cellular One of Columbia, TN
Cellular One of Danville
Cellular One of Dumas
Cellular One of East/Central PA
Cellular One of Huntsville
Cellular One of La Salle, IL
Cellular One of Sioux Falls
Cellular One of Southwest Florida
Cellular One of Upstate New York
Cellular One of Western Illinois
Cellular One of Wichita Falls
Cellular One of Worthington
Cellular XL
Celutel
Centennial Cellular
Century Cellunet
CIS
CJD Wisconsin Cellular
Clear Communications
Coastel Communications
Cone Enterprises
Contel Cellular
Crowley Cellular
Danbury Cellular
Detroit Cellular
Dobson Cellular ("A" side)
Easterbrooke Cellular
First Kentucky Cellular
Galveston Cellular
General Cellular
Greater South Dakota Cellular
GTE Mobilnet
Highland Cellular
HLD Cellular
Houston Cellular
Indianapolis Telephone
Lake Charles Cellular
Lake Huron Cellular
Lawton Cellular
Los Angeles Cellular
McCaw Cellular Communications
Mega Communications
Mercury Communications
Metrophone
Mid-Tennessee Partners
Mid-South Cellular

Milwaukee Telephone
Miscellco Communications
Missouri Cellular
Nebraska Cellular
Ocean County Cellular
Oneonta Telephone
PacTel Cellular
Panhandle Telecommunications
PC Cellular
Pegasus Cellular
Petroleum Communications
Pine Cellular
PriCellular
Prime Cellular Management
R & D Cellular
Radiofone
Santa Cruz Cellular
Sooner Cellular
Southeast Indiana Cellular
Sterling Cellular
Sussex Cellular
Syracuse Telephone
Texas 16 Cellular
Texas 5
Tsaconas Cellular
U S West
Union Telephone
United States Cellular
Unitel
Utica Telephone
Vanguard Cellular
Westex
Wisconsin Cellular
X-Cell Cellular
Youngstown Cellular

Thunder Bay Cellular

Advantage Cellular
AGT Cellular
ALLTEL Mobile Communications
American Rural Cellular
Amertech Mobile Communications
Associated Communications
Atlantic Cellular
BCTel Cellular
Bell Atlantic Mobile
BellSouth Mobility
Cantel
Cellcom of Green Bay
Cellular, Inc.
Cellular One of Sioux Falls
Cellular One of Syracuse
Cellular One of Utica
Cellular South
Cellulink
Centel Cellular
Century Cellunet
Ed Tel Cellular
GTE Mobilnet
Hutchinson Telephone
Independent Cellular Network
Lincoln Telephone
Mactel Cellular
Maine Cellular
MT&T Cellular
MTS Cellular
New Brunswick Cellular
Newtel Cellular
NYNEX Mobile Communications
Pace Communications
PacTel Cellular
Petroleum Communications
Québéc Téléphone Cellulaire
Richmond Cellular Telephone
Rochester Telephone Mobile
   Communications
Sasktel Cellular
Shenandoah Cellular
Southwestern Bell Mobile Systems

Springwich Cell. Ltd Partner. (SNET)
StarCellular
U S West Cellular
United States Cellular
Unity Cellular Systems (Unicel)

U S West Cellular

Advantage Cellular
AGT Cellular
Albany Telephone
ALLTEL Mobile Communications
Ally
Alpha Cellular
Alpine
Ameritech Mobile Communications
Appalachian Cellular
Atlantic Cellular
Baja Cellular
Bakersfield Cellular
BAMS - Metro Mobile
Bay Area Cellular
BCTel Mobility Cellular
Bell Atlantic Mobile
BellSouth Mobility
Big Horn Cellular
Blackwater Cellular
Bledsoe
Blue Mountain Cellular
Blue Ridge Cellular
Blue Ridge Cellular Virginia
Bluegrass Cellular
Brazos Cellular Communication
Cal-One Cellular
Cantel
Canton Cellular
Carolina West Cellular
CBIS
C. C. Cellular
Cellcom Green Bay
Cellcom of Hickory
Cell Comm of Puerto Rico
Cellular 2000
Cellular de Telefonia, SA Da CV
Cellular, Inc.
Cellular Information Systems
Cellular One of Amarillo
Cellular One of
   Baltimore/Washington
Cellular One of Berrien County
Cellular One of Boston
Cellular One of Casper
Cellular One of Chicago
Cellular One of Columbia
Cellular One of Columbia, TN
Cellular One of Danville
Cellular One of Frederick
Cellular One of Grand Forks
Cellular One of Huntsville
Cellular One of Indiana
Cellular One of Indianapolis
Cellular One of Lake Charles
Cellular One of La Salle, IL
Cellular One of Meridian
Cellular One of Northeast Colorado
Cellular One of Northeast Georgia
Cellular One of North Texas
Cellular One of Middle Tennessee
Cellular One of Sioux Falls
Cellular One of Southwest Florida
Cellular One of Upstate New York
Cellular One of Western Illinois
Cellular One of Western Oklahoma
Cellular One of Wilmington
Cellular One of Youngstown
Cellular Phone of Kentucky
Cellular Plus
Cellular Plus of Iowa
Cellular South
Cellular Technical Services

Cellular Three
Cellular XL
Celulares Telefonica
Celutel
Centel Cellular
Centennial Cellular
Century Cellunet
Chariton Valley Cellular
CJD Wisconsin Cellular
Clear Communications
Coastel
Comcast
Contel Cellular
Cross-Valliant Cellular
Crowley Cellular Telecommunications
CTIA - DC
Cybertel Cellular Telephone
Danbury Cellular
Dicomm Cellular
Digital Cellular
Dobson Cellular
Dominion Cellular
Easterbrooke Cellular
Ed Tel Cellular
EDS
Enid Cellular
Farmers Cellular
First Kentucky Cellular
Gaia
Galveston Cellular
GenCell Management
Genesee Cellular One of Buffalo
Genesee Telephone
GMD Limited Partnership
Great Lakes of Iowa
Greater South Dakota Cellular
GTE Mobilnet ("B" systems only)
Highland Cellular
HLD Cellular
Honolulu Cellular
Houston Cellular Telephone
Hudson Valley RSA
Illinois Valley Cellular
Independent Cellular Network
Inland Cellular Telephone
Interstate Cellular
Iowa East Cellular
IUSA Cell, SA de CV
Kansas Cellular
Lake Huron Cellular
Land Line
Leaco Rural Telephone
Los Angeles Cellular
Lincoln Telephone Cellular
Litchfield County Cellular
Mackinac Cellular
MACtel Cellular
Maine Cellular
McCaw Communications
Mega Comm
Mercury Cellular and Paging
Mercury Communications
MetroCel Cellular
Metrophone
Mid South Cellular
Mid-Missouri Cellular
Milwaukee Telephone
Miscellco
Missouri Cellular
Mobile Communications
MobileTel
Montana Cellular
Mountaineer Cellular
Mountaineer Mobile
MT&T Cellular
MTS Cellular
National Cellular
Nebraska Cellular
New Par

NYNEX Mobile Communications
Ocean County Cellular
Pacific Telecom Cellular
PacTel Cellular
Palmer Communications
PC Cellular of Vermont
Pegasus Cellular
Peoples Cellular
Petroleum Communications
Pine Cellular
PriCellular
Public Service Cellular
Quantum Communications
Québéc Téléphone Cellulaire
R & D Cellular
Radiofone
RFB Cellular
Richmond Cellular
Rochester Telephone Mobile
  Comms.
Santa Cruz Cellular
Sasktel Cellular
Savannah Cellular
Shenandoah Mobile Company
Southeast Indiana Cellular
Southwestern Bell Mobile
Springwich Cell. Ltd Partner (SNET)
StarCellular
Sterling Cellular
Sunshine Cellular
Sussex Cellular
Syracuse/Utica Telephone
Texas 16 Cellular
Texoma
Thunder Bay Cellular
Tsconas
Uintah Basin Telephone Association
Union Telephone
United States Cellular
Unitel
Unity Cellular Systems (Unicel)
Valley Telecommunications
Vanguard Cellular Systems
Virginia Cellular
West Alabama Cellular
West Central Cellular
XIT Cellular
X-Cell Cellular
Youngstown Cellular

UBET Cellular

BellSouth Mobility
C.C. Cellular
Cellular, Inc.
Centel Cellular
FGI Cellular
Illinois Valley Cellular
McCaw Communications
Mercury Cellular & Paging
Nebraska Cellular
PacTel Cellular
Panhandle Telecommunications
PriCellular
Quantum Cellular
U S West Cellular
Union Cellular Telephone

Union Cellular Telephone

Advantage Cellular
ALLTEL Mobile Communications
Ameritech Mobile Communications
Associated Communications
Atlantic Cellular
BCTel Cellular
BellSouth Cellular
BellSouth Mobility
Bluegrass Cellular
Cal-One Cellular
Cantel

C.C. Cellular
Cellcom of Green Bay
Cellular 2000
Cellular, Inc.
Cellular Information Systems
Cellular One of Berrien County
Cellular One of Columbia, MO
Cellular One of Danville
Cellular One of Fairfield
Cellular One of Galesburg
Cellular One of Huntsville
Cellular One of Indianapolis
Cellular One of LaSalle
Cellular One of Milwaukee
Cellular One of the Rio Grande Valley
Cellular One of Santa Cruz
Cellular One of Sioux Falls
Cellular One of Syracuse/Utica
Cellular One of Upstate New York
Cellular One of Wichita Falls, TX
Cellular Plus
Cellular Three
Cellular XL
Cellulink
Celutel
Century Cellunet
Coastel Communications
Contel Cellular
Crowley Cellular Telecommunications
Danbury Cellular Telephone
Digital Cellular of Texas
Enid Cellular
GenCell Management
Greater South Dakota Cellular
GTE Mobilnet
Honolulu Cellular Telephone
Houston Cellular Telephone
Independent Cellular Network
Interstate Cellular
Kansas Cellular
Lincoln Telephone
Mactel Cellular
McCaw Cellular Communications
Mercury Cellular
Metrocel
Miscellco Communications
Missouri Cellular
Mobiletel
MT&T Cellular
MTS Cellular
Nebraska Cellular
NBTel Cellular
PacTel Cellular
Petroleum Communications
Public Service Cellular
Quantum Communications
Radiofone
Richmond Cellular Telephone
Sasktel Cellular
Southwestern Bell Mobile Systems
StarCellular
Sterling Cellular
U S West Cellular
UBET Cellular
United States Cellular
Unity Cellular Systems (Unicel)
Westex
XIT Cellular

United Bluegrass Cellular

Albany Telephone
ALLTEL Mobile Communications
Alpine
Ameritech Mobile
Appalachian Cellular
Atlantic Cellular
Atlas Cellular
Bakersfield Cellular
Baton Rouge Cellular

Bay Area Cellular
Bell Atlantic Mobile
BellSouth Cellular
Blackwater Cellular
Blue Mountain Cellular
Blue Ridge Cellular
Buffalo Telephone
C. C. Cellular
Cantel
Cellcom of Green Bay
Cellcom of Hickory
Cellular 2000
Cellular Connection
Cellular Information Systems
Cellular of Indiana
Cellular One of
 Baltimore/Washington
Cellular One of Berrien County
Cellular One of Boston
Cellular One of Chicago
Cellular One of Columbia, MO
Cellular One of Columbia, TN
Cellular One of Columbus, IN
Cellular One of Danville
Cellular One of Detroit
Cellular One of East/Central PA
Cellular One of Huntsville
Cellular One of Indianapolis
Cellular One of Lake Charles, LA
Cellular One of La Salle, IL
Cellular One of Middle Tennessee
Cellular One of Northeast Georgia
Cellular One of North Texas
Cellular One of Santa Cruz
Cellular One of Sioux Falls
Cellular One of Southwest Oklahoma
Cellular One of Syracuse
Cellular One of the Ozarks
Cellular One of Western Illinois
Cellular One of Western Oklahoma
Cellular One of Wichita Falls
Cellular Phone of Kentucky
Cellular XL
Celutel
Centel Cellular
Centennial Cellular
Century Cellunet
CJD Wisconsin Cellular
Clear Communications
Contel Cellular
Crowley Cellular
Cybertel Cellular
Dicomm Cellular
Easterbrooke Cellular
Eastex Cellular
FGI Cellular
Finger Lakes Telephone
First Fayette Cellular
First Kentucky Cellular
Gaia
Galveston Cellular
GenCell Management
Genesee Telephone
GMD Partnership
Greater South Dakota Cellular
GTE Mobilnet
Highland Cellular
HLD Cellular
Honolulu Cellular
Houston Cellular
Kull Cellular
Lake Huron Cellular
Lincoln Telephone
Litchfield County Cellular
Los Angeles Cellular
McCaw Cellular Communications
Mercury Communications
Metro Mobile
Metrocel Cellular

Metrophone
Mid-south Cellular
Milwaukee Telephone
Miscellco Communications
Missouri Cellular
Mountaineer Cellular
Mountaineer Mobile
New Par
NYNEX Mobile Communications
Pacific Telecom Cellular
PacTel Cellular
Palmer Cellular
Petroleum Communications
PriCellular
Radiofone
Richmond Cellular
Rural Cellular Management
Savannah Cellular
South Eastern Cellular
Tsaconas Cellular
U S West Cellular
Union Telephone
United States Cellular
Utica Telephone
Vanguard Cellular
Virginia Cellular
Youngstown Cellular

United States Cellular

Advantage Cellular
AGT Cellular
Albany Telephone Co.
ALLTEL Mobile Communications
Ally
Alpha Cellular
Alpine
American Rural Cellular
Ameritech Mobile Communications
Appalachian Cellular
Atlantic Cellular
Atlas Cellular
Auburn Cellular
Auburn Television
B.C. Tel Mobility Cellular
Bachtel Cellular
Bahamas Telecommunications
Bakersfield Cellular
Bell Atlantic Mobile
Bell Mobility Cellular
BellSouth Mobility
Big Horn Cellular
Blackwater Cellular
Bledsoe
Blue Mountain Cellular
Blue Ridge Cellular
Bluegrass Cellular
Brazos Cellular
Buckhead Telephone
Buffalo Telephone
C. C. Cellular
C. V. Cellular
CGH Cellular
CJD Wisconsin Cellular
Cal-One Cellular
Cantel
Canton
Carolina West Cellular
Cedetel
Cellcom
Cellcom of Hickory
Cellular 2000
Cellular Connections
Cellular Holdings
Cellular, Inc.
Cellular Information Systems
Cellular One of Akron
Cellular One of Amarillo
Cellular One of Atlantic City, NJ
Cellular One of

Baltimore/Washington
Cellular One of Baton Rouge
Cellular One of the Bay Area
Cellular One of Beaumont
Cellular One of Berrien County
Cellular One of Billings
Cellular One of Boston
Cellular One of Chicago
Cellular One of Collier/Hendry
Cellular One of Columbia
Cellular One of Columbia, TN
Cellular One of Cumberland, MD
Cellular One of Danville
Cellular One of Dothan
Cellular One of Erie
Cellular One of Frederick
Cellular One of Fort Myers
Cellular One of Fort Wayne
Cellular One of Galesburg
Cellular One of Galveston
Cellular One of Huntsville
Cellular One of Indiana
Cellular One of Indianapolis
Cellular One of Jacksonville, NC
Cellular One of Kalamazoo
Cellular One of Lake Charles
Cellular One of La Salle, IL
Cellular One of Macon
Cellular One of Mansfield
Cellular One of Marion
Cellular One of Meridian
Cellular One of Middle Tennessee
Cellular One of Milwaukee
Cellular One of Montgomery
Cellular One of North Texas
Cellular One of Northeast Colorado
Cellular One of Northeast Georgia
Cellular One of Northern Ohio
Cellular One of Ocean County
Cellular One of Panama City
Cellular One of Puerto Rico
Cellular One of Roanoke
Cellular One of Santa Cruz
Cellular One of Sioux Falls
Cellular One of South Bend
Cellular One of South Dakota
Cellular One of Southwest Alabama
Cellular One of Southwest Georgia
Cellular One of Syracuse
Cellular One of the Rio Grande Valley
Cellular One of Upstate New York
Cellular One of Utica/Rome
Cellular One of Victoria, TX
Cellular One of Virginia
Cellular One of Waycross
Cellular One of Western Illinois
Cellular One of Western Oklahoma
Cellular One of Wichita Falls, TX
Cellular One of Wilmington
Cellular Phones of Kentucky
Cellular Phones of Aiken-Augusta
Cellular Plus of Georgia
Cellular Plus of Iowa City
Cellular South
Cellular Systems International
Cellular Telephone of Kentucky
Cellular Three
Cellular Ventures
Cellular XL
Cellulink
Cellwave
Celutel
Centel Cellular
Centennial Cellular
Century Cellunet
Century Telephone
Clear Communications
Coastel Communications
Codetel
Columbus Cellular

Comcast
Commonwealth Mobile
Contel Cellular
Cooper Cellular
Crowley Cellular Telecommunications
Cybertel Cellular Telephone
Danbury Cellular
Detroit Cellular Telephone Co.
Dicomm Cellular
Digital Cellular of Texas
Dobson Cellular
Dominion Resources
Easterbrooke Cellular
Eastex Cellular
Ed Tel Cellular
Enid Cellular
Farmers Cellular
FGI Cellular Management
Finger Lakes Telephone
First Fayette Cellular
Five Star Cellular
Florida 9 Cellular
Gaia
GenCell Management
Genesee Telephone Co.
GMD Limited
Graceba Cellular
Greater South Dakota Cellular
GTE Mobilnet ("B" systems only)
Gulf Coast Cellular
HBF Cellular
HLD Cellular
Highland Cellular
Honolulu Cellular
Houston Cellular Telephone
Hudson Valley RSA Cellular
Illinois Valley Cellular
Independent Cellular
Inland Cellular
Interstate Cellular
Iowa East Cellular
IUSA Cell
Jackson Cellular
Kansas Independent Networks
Kaplan Telephone
Kull Cellular
Lake Huron Cellular
Lamar County Cellular
Leaco Rural Telephone
Liberty Cellular
Lincoln Telephone Cellular
Litchfield Cellular
Mackinac Cellular
Mactel Cellular
Maine Cellular
McCaw Communications
Mercury Cellular and Paging
Mercury Communications
Metro Mobile CTS
Metro One
MetroCel Cellular
Metrophone
Mid-Missouri Cellular
Mid-South Cellular
Mid-Tex Cellular
Milwaukee Telephone Co.
Miscellco
Missouri Cellular
Mobile Communications
Mobiltel
Montana Cellular
Mountaineer Cellular
Mountaineer Mobile
MT&T Cellular
MTS Cellular
National Cellular
Nebraska Cellular
NYNEX Mobile Communications
OK Cellular

Oklahoma Western Telephone
Oneonta Telephone
Pacific Telecom Cellular
PacTel Cellular
Palmer Communications
Panhandle Telecommunications
PC Cellular of Kentucky
PC Cellular of Vermont
Peoples Cellular
Petroleum Communications
Pine Cellular
Pioneer Cellular
Portacel Telefonia Cellular
Public Service Cellular
Quantum Communications
R & D Cellular
Radiofone
RFB Cellular
Richmond Cellular Telephone Co.
Rochester Telephone Mobile
   Comms.
Rural Cellular Management
Sasktel Cellular
Savannah Cellular
Shenandoah Mobile
Springwich Cell. Ltd Partner. (SNET)
Sooner Cellular
Southeast Cellular
Southeast Indiana Cellular
Southeastern Cellular
Southwestern Bell Mobile Systems
StarCellular
Stillwater Cellular
Sussex Cellular
Syracuse/Utica Telephone Co.
Texas 16 Cellular
Texas Cellular
Thumb Cellular
Thunder Bay Cellular
Tsaconas Cellular
U S West Cellular
UBET Cellular
Union Telephone
Unitel
Unity Cellular
Valley Telecommunications
Vanguard Cellular Systems
Virginia Cellular
Walnut Hill Cellular
West Alabama Cellular
West Central Cellular
Western Maine
Westex Cellular
X-Cell Cellular
XIT Cellular
Yorkville Telephone
Youngstown Cellular

Unitel (Liberty Cellular)

Adirondack Cellular
ALLTEL Mobile Communications
Alpha Cellular
Atlantic Cellular
Bakersfield Cellular
B.C. Tel Mobility Cellular
Bell Atlantic
BellSouth Mobility
Blackwater Cellular
Blue Mountain Cellular
Blue Ridge Cellular
Bluegrass Cellular
Brazos Cellular
Cantel
C.C. Cellular
Cellcom
Cellular
Cellular 2000
Cellular Information
Cellular Information Systems

Cellular One of Alabama
Cellular One of Boston
Cellular One of Boston
Cellular One of Chicago
Cellular One of Grand Forks
Cellular One of Indiana
Cellular South
Cellular Three
Centel
Centennial Cellular
Century Cellunet
Century Communications
CJD Wisconsin Cellular
Coastel
Contel Cellular
Crowley
Cybertel Cellular
Dentel Cellular
Detroit Cellular
Dicomm Cellular
Digital Cellular
Dobson Cellular
Easterbrooke Celluler
Enid Cellular
FGI Cellular
Finger Lake Telephone
Five Star Cellular
General Cellular
GTE Mobilnet
HLD Cellular
Houston Cellular
ICN
Independent Cellular
Kansas Cellular
Lake Huron Cellular
Liberty Cellular
Lincoln Telecommunications
Lincoln Telephone
Mackinac Cellular
McCaw Cellular Communications
Metro Mobile
Metrocel Cellular
Mid Tex Cellular
Mid-South Cellular
Mobiletel
Mountain
MTS Cellular
MTX Cellular
National Cellular
Nebraska Cellular
New Par
Northwest Bell
Northwest Cellular
NYNEX Mobile
Ohio State Cellular
OK Cellular
PacTel Cellular
Palmer Cellular
Petroleum Communications
Pioneer Cellular
PTSI Cellular
Radiofone
RFB Cellular
Richmond Cellular
Rural Cellular
Shenandoah Mobility
Southeast Indiana
Southwestern Bell
Sterling Cellular
Sussex Celluarl
Tsaconas Cellular
U S West Cellular
Union Telephone
United States Cellular
Valley Telecommunications
Vanguard
Westex Cellular
Yorkville Telephone

Unity Cellular Systems (Unicel)

Advantage Cellular
ALLTEL Mobile Communications
Ally
American Rural Cellular
Ameritech Mobile Communications
Atlantic Cellular
Batelco
Bell Atlantic Mobile
Bell Atlantic Mobile
Bell Mobility Cellular
BellSouth Mobility
Bledsoe Telephone
Bluegrass Cellular
Blueridge Cellular
Cal-One Cellular
Cantel
Canton
C.C. Cellular
Cellcom
Cellular 2000
Cellular Connection
Cellular Indiana
Cellular One
Cellular One of Boston
Cellular One of Columbia TN
Cellular One of Danville
Cellular One of East Central Penn
Cellular One of Gainsville GA
Cellular One of Iowa
Cellular One of Marianna FL
Cellular One of New York
Cellular One of Northeast Georgia
Cellular One of Oklahoma City
Cellular One of Sioux Falls
Cellular Phone Kentucky
Cellular Plus
Cellular XL
Cellulink
Centel Cellular
Century Cellunet
Coastel Communications
Contel Cellular
Cross Valliant Cellular
Digital Cellular
Dobson Cellular
Enid Cellular
Farmers Cellular Telephone
First Kentucky Cellular
Gaia
GTE Mobilnet
Highland Cellular
Hudson Valley Cellular
Illinois Valley Cellular
Indpendent Cellular Network
Interstate Cellular
Iowa East Cellular Telephone
Kansas Cellular
Lake Huron Cellular
Lincoln Telephone
Litchfield County Cellular
Mactel Cellular
Maine Cellular
Mercury Cellular
Miscellco Communications
Missouri Cellular
MT&T Mobility
MTS Mobility
Nebraska Cellular
Newtel Cellular
NYNEX Mobile Communications
OK Cellular
Pace Communications
PacTel Cellular
Petroleum Communications
Pioneer Cellular
PriCellular
PTSI Cellular
Quantum Communications
Québéc Téléphone Cellulaire

Richmond Cellular Telephone
Rochester Telephone Mobile
  Communications
Savannah Cellular
Shenandoah Mobile
Southeast Cellular
Southwestern Bell Mobile
Springwich Cell. Ltd Partner. (SNET)
StarCellular
Sterling Cellular
Thunder Bay Cellular
Union Telephone
Utica Rome
Vanguard Cellular
Western Maine Cellular
XIT Cellular
Yorkville Telephone

Vanguard Cellular Systems

Adirondack Cellular
Advantage Cellular
Albany Telephone
Ameritech Mobile
Alpha Cellular
Alpine
Allegan Cellular
ALLTEL Mobile Communications
Ally
Amarillo Cellular
American Rural Cellular
AMC Cellular
Appalachian Cellular
Atlantic Cellular
Bahamas Telecommunications
Bakersfield Cellular
Bay Area Cellular
Bell Atlantic Mobile
BellSouth Mobility
Blackwater Cellular
Bluegrass Cellular
Blue Mountain Cellular
Blue Ridge Cellular
Buckhead Telephone
Buffalo Telephone
C. C. Cellular
Cantel
Canton Cellular
CEDETEL
Cellcom of Hickory
Cellular 2000
Cellular Communications of Puerto
  Rico
Cellular Holding
Cellular, Inc.
Cellular Information Systems
Cellular One of Altoona
Cellular One of
  Baltimore/Washington
Cellular One of Berrien County
Cellular One of Boston
Cellular One of Chicago
Cellular One of Collier County
Cellular One of Columbia
Cellular One of Indiana
Cellular One of Indianapolis
Cellular One of Lake Charles
Cellular One of La Salle, IL
Cellular One of Meridian
Cellular One of Middle Tennessee
Cellular One of North Colorado
Cellular One of Northeast Georgia
Cellular One of Northeast Mississippi
Cellular One of North Texas
Cellular One of Panama City
Cellular One of Portsmouth
Cellular One of Richmond
Cellular One of Santa Cruz
Cellular One of Southwest Oklahoma
Cellular One of Tennessee

Cellular One of Upstate New York
Cellular One of Western Oklahoma
Cellular One of Western Illinois
Cellular Phone of Aiken-Augusta
Cellular Phone of Kentucky
Cellular Plus
Cellular South
Cellular Systems International
Cellular Ventures
Cellular XL
Celutel
Centennial Cellular
Centel Cellular
Century Cellunet
CGH Partnership
CJD Wisconsin Cellular
Clear Communications
Coastel Communications
Contel Cellular
Cooper Cellular
CPS Cellular Telephone
Crowley Cellular Development
Cybertel
Danbury Cellular
Detroit Cellular
Dicomm Cellular
Dobson Cellular of Enid
Dominion Resources
Easterbrooke Cellular
Eastex Cellular
FGI Cellular
First Fayette Cellular
First Kentucky Cellular
GTE Mobilnet ("B" systems only)
Gaia
Galveston Cellular
GenCell Management
Genesee Telephone
GMD Limited Partnership
Graceba Cellular
Greater South Dakota Cellular
Gulf Coast Cellular
Highland Cellular
Honolulu Cellular
Horizon Cellular
Houston Cellular
Independent Cellular Network
Interstate Cellular
Iowa Cellular
Iowa East Cellular
IUSA Cell
Jacksonville Cellular
Kansas Cellular
Lincoln Telephone
Litchfield County Cellular
Louisiana Cellular
MTS Cellular
MacTel Cellular
Mackinac Cellular
McCaw Communications
MegaCommunications
Mercury Communications
Mercury Cellular and Paging
Metacom
Metro Mobile
Metro One
MetroCel Cellular
Metrophone
Midland Communications
Milwaukee Telephone
Miscellco
Missouri Cellular
Mobile Communications
MobileTel
Mountaineer Cellular
Mountaineer Mobile
National Cellular
Nebraska Cellular
New Par

NYNEX Mobile Communications
Ohio State Cellular
PacTel Cellular
Palmer Communications
PC Cellular of Kentucky
Petroleum Communications
PriCellular
Public Service Cellular
Quantum Communications
Radiofone
RFB Cellular
Roanoke Valley Cellular
Savannah Cellular
Shenandoah Mobile
South Bend Metronet
Southeast Indiana Cellular
Springwich Cell. Ltd Partner. (SNET)
Sunshine Cellular
Syracuse Telephone
Texas 16 Cellular
U S West Cellular
Union Telephone
United States Cellular
Unitel
Unity Cellular Systems (Unicel)
Virginia Cellular
Vitel Cellular
West Alabama Cellular
Wilmington Cellular
Youngstown Cellular

West Central Cellular

ALLTEL Mobile Communications
Ameritech Mobile Communications
Associated Communications
Bell Atlantic Mobile
Bellsouth Mobilty
Cellular Information Systems
Cellular One of Boston
Cellular One of Chicago
Cellular One of Victoria
Cellular One of Washington
Cellulink
Centel Cellular
Century Cellunet
Contel Cellular
Crowley Cellular Telecommunications
Cybertel Cellular
Enid Cellular
GTE Mobilnet
Mid-Tex Cellular
PacTel Cellular
Radiofone
Southwestern Bell Mobile System

Wes-Tex Cellular

Advantage Cellular
ALLTEL Mobile Communications
American Rural Cellular
Ameritech
Appalachian Cellular
Atlas Cellular
Bell Atlantic
Bell Atlantic Mobile
Bell South
Bledsoe Telephone
Blue Grass Cellular
Blue Mountain Cellular
Brazos Cellular
Cal-One Cellular
C.C. Cellular
Cellcom
Cellcom of Hickory
Celllular One of North Texas
Cellular
Cellular 2000
Cellular Connection
Cellular Information Systems
Cellular One of Brookings, SD

Cellular One of Columbia TN
Cellular One of Danville IL
Cellular One of East/Central PA
Cellular One of Enid
Cellular One of Gainesville
Cellular One of Indiana
Cellular One of Middle Tennessee
Cellular One of Natcghitoches
Cellular One of Natchez, MS
Cellular One of Ozarks
Cellular One of Santa Cruz
Cellular One of Sioux Falls
Cellular One of Southwest Oklahoma
Cellular One of Western Oklahoma
Cellular One of Wichita Falls
Cellular South
Cellular System International
Cellular Three
Cellular XL
Cellulink
Centel
Century Cellunet
Coastal Cellular
Contel
Crowley
C.V. Cellular
Digital Cellular
Dobson Cellular
Enid Cellular
Farmers Cellular
Five Star Cellular
GenCell Management
GTE Mobilnet
Illinois Valley Cellular
Independent Cellular Network
Kansas Cellular
Kaplan Telephone
Lake Huron Cellular
Leaco Cellular
Lincoln Cellular
Litchfield Cellular
McCaw Cellular Communications
Mercury Cellular
Metro Cel
Missouri Cellular
Mobile Tel
Nebraska Cellular
Oneonta Telephone
Pace Communications
PacTel Cellular
Peoples Cellular
Petroleum Communications
Pine Cellular
PriCellular
PTSI Cellular
Public Service Cellular
Richmond Cellular
Shenandoah Mobile
Southeast Cellular
Southwestern Bell Mobile Systems
Springwich Cell. Ltd Partner. (SNET)
StarCellular
Texas Cellular
U S West Cellular
UBET Cellular
Union Telephone
United States Cellular
Unity Cellular
Universal Telecell
Valley Telecom
Virginia 10
West Central Cellular
XIT Cellular
Yorkville Telephone

Wilcom Cellular

Ameritech Mobile Communication
Associated Communications
Bell Atlantic Mobile

BellSouth Mobility
Cantel
Cellular Information Systems
Cellular One of Chicago/Gary
Cellular One of Cincinnati
Cellular One of Cleveland
Cellular One of Columbus, OH
Cellular One of Detroit
Cellular One of Grand Rapids
Cellular One of Indianapolis
Cellular One of Lansing
Cellular One of Milwaukee
Cellular One of Muskegon
Cellular One of Richmond
Cellular One of Roanoke
Cellular One of Sioux Falls
Cellular One of Syracuse
Cellular One of Toledo
Cellular One of Utica
Cellular One of Washington
Cellular One of Wilmington
Centel Cellular
Centennial Cellular
Comcast
Contel Cellular
Crowley Cellular Telecommunications
Cybertel Cellular
GTE Mobilnet
Houston Cellular Telephone
Independent Cellular Network
McCaw Cellular Communications
Metrocel Cellular
Metrophone
Mobiletel
NYNEX Mobile Communications
Palmer Communications
Petroleum Communications
Richmond Cellular Telephone
Southwestern Bell Mobile Systems
U S West Cellular
United States Cellular
Vanguard Cellular

X-CELL Cellular

ALLTEL Mobile Communications
Auburn Television Group
Bay Area Cellular
Blue Mountain Cellular
Cellular One of Berrien County
Cellular One of Chicago, IL
Cellular One of Columbia, MO
Cellular One of Milwaukee, WI
Cellular One of Norfolk, NE
Cellular One of Northeast Colorado
Cellular One of the Great Lakes of Iowa
Cellular One of Western Illinois
Cellular Phone of Kentucky
Centel Cellular
Centennial Cellular
Contel Cellular
Crowley Cellular
Cybertel
GenCell Management
GMD Partnership
Kansas Cellular
Lake Huron Cellular
Lincoln Cellular
Los Angeles Cellular
McCaw Cellular Communications
Metrocel Cellular
Mid South Cellular
Miscellco
Nebraska Cellular
PriCellular
Quantum Communications
Sagir
Sterling Cellular
Texoma Cellular

U S West Cellular
United States Cellular

Wilkes Cellular Telephone
Yorkville Cellular

XIT Cellular

Advantage Cellular
ALLTEL Mobile Communications
Ameritech Mobile Communications
Appalachian Cellular
Atlantic Cellular
B. C. Tel Mobility Cellular
Bell Atlantic Mobile
Bell Mobility Cellular
BellSouth Mobility
Bledsoe Telephone
Cal-One Cellular
C.C. Cellular
Cellcom
Cellular Connection
Cellular, Inc.
Cellular Information Systems
Cellular One of Amarillo
Cellular One of Boston
Cellular One of Chicago
Cellular South
Cellular Three
Cellulink
Centel Cellular
Century Cellunet
Chariton Valley Cellular
Coastel Communications
Contel Cellular
CT Cube
Digital Cellular
Dobson Cellular Systems
Ed Tel Cellular
Enid Cellular
Farmers Cellular Telephone
Five Star Cellular
GTE Mobilnet
Independent Cellular Network
Kansas Cellular
Kaplan Telephone
Lamar County Cellular
Leaco Cellular
Lincoln Telephone
Maine Cellular
McCaw Cellular Communications
Mercury Cellular
Mid-Missouri Cellular
Mid-Tex Cellular
Mobiletel
Mountaineer Cellular Telephone
MT&T Cellular
MTS Cellular
Nebraska Cellular
Northwest Cellular
NYNEX Mobile Communications
PacTel Cellular
PTSI Cellular
Public Service Cellular
Rochester Telephone Mobile
  Communications
Sasktel Cellular
Savannah Cellular
Southeastern Cellular
Southwestern Bell Mobile Systems
Springwich Cell. Ltd Partner. (SNET)
StarCellular
Texas Cellular
Thumb Cellular
U S West Cellular
Union Telephone
United Cellular
United States Cellular
Unity Cellular Systems (Unicel)
Valley Cellular
Walnut Hill Cellular
West Maine Cellular
Westex

Yorkville Telephone Cooperative

Bell Mobility Cellular
Blue Ridge Cellular
Brazos Cellular Communications
Cantel
Cellcom of Green Bay
Cellular Connection
Cellular Holdings
Cellular Information Systems
Cellular One of Columbia, MO
Cellular One of Sioux Falls, SD
Dial Two
Fingers Lake
GenCell Management
GMD
Illinois Valley Cellular
Interstate Cellular (Interstate Cellular
  (Intercel))
Kansas Cellular
MACtel Cellular
MCMG
Mercury Communications
Mid-Missouri Cellular
Milwaukee Telephone
Missouri Cellular
Mobiletel
Mountaineer Cellular
Mountaineer Mobile
MT&T Cellular
MTS
Nebraska Cellular
NewTel Cellular
NYNEX Mobile Communications
Pace Communications
PC Cellular of Kentucky
PC Cellular of Vermont
PacTel Cellular
Palmer Communications
Panhandle Telecommunications
Petroleum Communications
Pine Cellular
Prime Cellular Management
Public Service Telephone
Quantum Communications
Québéc Téléphone Cellulaire
Radiofone
RFB Cellular
Richmond Cellular
Rochester Telephone
Rural Cellular Management
SaskTel Cellular
Shenandoah Mobile
Springwich Cell. Ltd Partner. (SNET)
Southeastern Cellular
Southern Cellular
Southwestern Bell
StarCellular
Sterling Cellular
Syracuse Telephone
Texas 16 Cellular
Thunder Bay Telephone
U S West Cellular
UBET Cellular
Union Telephone
United States Cellular
Unitel
Unity Cellular Systems (Unicel)
Valley Telecommunications
Vanguard Cellular
Vitel Cellular (U.S. Virgin Islands)
West Central Cellular
Westex Cellular
XIT Cellular
Youngstown Cellular

Answering calls

When your cellular telephone rings, simply pick up the handset to answer the call. If you have a speakerphone, press *send* if you wish to answer the call on the speakerphone.

Call waiting

If you have selected a call waiting option as part of your cellular service, you may hear a tone while in the middle of a conversation. The tone notifies you that a second call is coming in. To place the first call on hold and answer the second call, simply press *send* on most systems. You can alternate between the calls as often as you like by pressing *send*.

Charges for incoming calls

You are charged both for calls you send and calls you receive. This is because both types of calls use cellular radio channels, which are a limited resource. To avoid charges for incoming calls, you may choose to keep your cellular telephone number private, not to answer incoming calls, or to turn off your cellular telephone when you are not making calls.

--------------------------------- Traveling ---------------------------------

All cities by roamer access number

When you travel outside your home service area, people call you by first dialing a roamer access number. The number varies depending on which city you are visiting and which system you are using (A or B). The number is found in each listing's *receiving calls* section.

The caller dials the roamer access number and hears a tone or recorded message, then dials the 10-digit area code and number of your cellular telephone using a Touch-Tone telephone. This rings your cellular telephone. The caller pays any long distance charge to the roamer access number for each attempt to call you, and you pay the visitor rate for each call you answer, including any daily fees and per minute fees.

Auto-Access (some B systems in California and Nevada)

If your home company offers Auto-Access (some California and Nevada B systems), you can use it to receive calls in other Auto-Access cities. For you to use it, both your home company and the host company must list Auto-Access in their *receiving calls* sections.

To activate Auto-Access on a host system, dial *28 send* and wait to hear the confirmation tone. After approximately 30 minutes, all calls made to your cellular telephone number will be forwarded to you indefinitely whenever you travel to another Auto-Access area. You are charged long distance from your home area to the area you are visiting for calls you answer.

To turn off Auto-Access until you reactivate it, simply dial *29 send* in an Auto-Access area. Please contact your home company if you have questions about Auto-Access. The service is provided by Pacific Telesis Group.

Follow Me Roaming & Phone Me Anywhere (some A and B systems)

If your home company offers Follow Me Roaming, you can use it to receive calls in all cities where it is available. For you to use the service, both your home company and the host company must list Follow Me Roaming in their *receiving calls* sections.

To activate Follow Me Roaming on a host system, dial *18 send* and wait to hear a confirmation message. After a short activation delay, all calls made to your cellular telephone number will be forwarded to you at your expense until midnight or until you deactivate the service. To deactivate it, dial *19 send* on the host system, or activate Follow Me Roaming on another host system, or dial the code for *cancel call forwarding* on your home system. Please contact your home company if you have questions about Follow Me Roaming. The service is provided by GTE Telecommunication Services, at (813) 273-3000.

NationLink & RoamingAmerica (some A systems)

NationLink, also known as RoamingAmerica, helps you receive calls when you travel. How it operates depends on the service level of your home company and of the host company. Please turn to your home city and find your home company, then look in its *receiving calls* section. If NationLink is not listed, you cannot use NationLink. Otherwise, note which level of service your company offers its subscribers, and read the description of that level below:

Level 0   You cannot use NationLink, but some people who visit your home city can.

Level 1   Caller notification in cities that show *32 available, if you dial *32 send.

Level 2   Caller notification is automatic after you place your first call in cities that show *32 available or if the city you visit has a home subscriber level 0 (such as Mobile AL).

Level 3   Call forwarding in cities that show *31 available, if you dial *31 send.

Level 4   Call forwarding in cities that show *31 available, if you dial *31 send. In Florida, you receive calls automatically after you place your first call without dialing *31.

Level 5   Call forwarding in cities that show *31 available, and forwarding happens automatically in these cities after you place your first call.

Level 6   Call forwarding in cities that show *31 available, if you dial *31 send. Caller notification in cities that show *32 available, if you dial *32 send.

Level 7   Call forwarding in cities that show *31 available, if you dial *31 send (automatic in FL after 1st call). Caller notification in cities that show *32, if you dial *32 send.

Level 8   Caller notification happens automatically after your first call if the city you visit shows *32 available or if it has a home subscriber level of 0 (such as Mobile, AL). If the city lists *31, you may dial *31 send to activate call forwarding instead.

Level 9   Call forwarding happens automatically after your first call in cities that show *31. If the city lists *32, you may dial *32 send to activate caller notification instead.

Level 10   Please read the description for level 6 above.

Level 11   Please read the description for level 0 above.

Level 12   Call forwarding in cities that show *31 available, if you dial *31 send. Caller notification in cities that show *32 available, if you dial *32 send. (Cities that list a home subscriber level of 0 give auto call forwarding after you place the first call.)

Level 13   Caller notification happens automatically after your place your first call in cities that show *32 or in cities you visit that have a home subscriber level of 0 (such as Mobile, AL). In some nearby cities, you must manually choose call forwarding by dialing *31 send or caller notification by dialing *32 send. Ask your home co.

Level 14   Please read the description for level 6 above.

If call forwarding is listed for your level above, calls made to your cellular telephone number will be forwarded to you at your expense until midnight Eastern time unless reactivated.

If caller notification is listed for your level above, anyone who dials your home cellular number will hear a message telling how to call you using the correct roamer access number.

To deactivate NationLink, dial *30 send on the host system, or activate the service on a different host system, or dial the code for *cancel call forwarding* on your home system. If you have questions about NationLink, please contact your home company. NationLink is provided by EDS Personal Communications, at (617) 890-1000.

North American Cellular Network (NACN) (some A systems)

If your home company offers NACN, you can use it to receive calls in all cities where it is available. Look to see if your home company and the host company both list NACN. in their *receiving calls* sections. To activate the service, simply turn on your telephone. After a short delay, calls made to your home cellular number will forward to you. For each call that you answer, you will be billed for long distance from your home number to your location and any per day and per minute fees in the city you are visiting. NACN. turns off automatically when you return home, or you may turn it off by dialing *35 send (*633* send in some cities).

General

Cellular telephone service is provided by two competing companies in each city. One operates system A and the other system B. The person who activated your cellular telephone set it to use one of these systems, and the company operating that system in your home city assigned your cellular telephone number and bills you each month.

How you are billed

You are billed for each call you send or receive based on how many minutes the call lasts. Most companies round up to the nearest minute, but some bill in six-second increments for the fraction of a minute used. Your home company may offer calling plans that include a set number of free minutes each month based on the amount you pay as a fixed monthly fee.

Your first monthly bill may be higher than usual because it includes a fee to activate your telephone, a normal monthly fee, pre-payment for the following month, and possibly a security deposit. Friends who have cellular telephones can give you a better idea what to expect for an average monthly bill.

Selecting a system when you travel

When you use your cellular telephone outside your home area, you are known as a *roamer*. A *roam* light may appear on your telephone handset, indicating that you have left the area served by your home company and will be using the service of a host company.

You probably do not have to adjust your cellular telephone to use it when you travel. It will recognize that you have left your home area and will attempt to send a call on the same type of system that you use in your home area (A or B). If no company provides service on that system, it will attempt to use the other system. This feature was set by the person who activated your cellular telephone.

You may wish to select the system you use when you travel. One of the two companies listed for each city may charge lower rates, or cover the area where you are traveling better, or allow you to send calls more easily because it has a roaming agreement with your home cellular company, as described in the *roaming agreements* section that follows.

To select between system A and system B on your handset, press the function button, then the A/B button—or refer to your owner's manual. The dealer who sold you the telephone also can assist you. If you change the system setting on your telephone when you travel, be sure to set it back when you return home so you can place calls on your home system.

Sending calls when you travel

Follow the *visitor dialing instructions* listed in this book for each city you visit. Most host companies require you to dial the area code and seven-digit number for local calls. To call long distance, some host companies require you to precede your call with a *1*, while others require you to precede your call with a *0* so an operator can bill the toll portion of the call.

The host company bills your home company for each direct-dial call you make when you travel—charging the rates shown in this book—unless the two companies have negotiated reduced rates with one another. Your home company then may bill you more or less than the amount it paid when it charges the call to your monthly bill, depending on its policy.

Roaming agreements

Roaming agreements are contracts made between cellular companies to provide service to each other's customers. For example, if your home company has a roaming agreement with a company in Chicago, you can use your cellular telephone in Chicago, and any daily fees, per minute fees, and long distance fees will appear on your home bill.

Cellular companies have been increasing the number of their roaming agreements so that in most cities, you can use your cellular telephone automatically as soon as you arrive. Often the host company that offers you automatic roaming operates on the same system that you use at home (A or B).

To find out which companies have roaming agreements that allow you automatic service, please find your home company in the index that begins on page 1090 of the red section. Then turn to the page indicated for your home company to find the list of other companies that offer you automatic service. Since new roaming agreements are made every day, you may wish to contact your home cellular company to obtain the latest information.

Without a roaming agreement

If your home company does not have a roaming agreement in a city that you visit, you will not be able to send and receive calls automatically. Instead, you should refer to the page of the city you are visiting and look at the *billing information* sections of the two companies that operate there. A paragraph called *visitors without roaming agreements* tells you what steps you should take to begin service.

Often visitors without roaming agreements are allowed to charge their calls to major credit cards, such as American Express, Diner's Club, Discover, MasterCard, and Visa. We have listed the credit cards that each company accepts. You may call the host company during normal office hours to provide your credit card number, or you may be transferred to the billing department automatically when you attempt to place the first call.

If the company does not accept credit cards, you either will be denied service, or the first call you attempt to make will be intercepted and automatically routed to a private company that accepts credit cards on behalf of the cellular company. Two of these private companies are listed below. To see if a cellular company intercepts calls and forwards them to one of these two companies, look in its paragraph called *visitors without roaming agreements*.

American Roaming Network

If you attempt to place a call in a city where your home company does not have a roaming agreement, your call may be intercepted and forwarded to an operator at this company who will request your American Express, Mastercard, Visa, or calling card, or will bill your call collect or to a third party. This service is provided by National Communications, with offices at (305) 938-0300, for cellular companies that do not accept credit cards directly.

Roamer Plus

If you attempt to place a call in a city where your home company does not have a roaming agreement, your call may be intercepted and forwarded to a Roamer Plus operator who will request your American Express, Mastercard, Visa, or calling card, or will bill your call collect or to a third party. This service is provided by Boston Communications Group, with offices at (617) 439-4141, for cellular companies that do not accept credit cards directly.

Airplanes

Your cellular telephone should be turned off before you board an airplane. Using it onboard is prohibited both on the ground and in the air.

Antennas

You receive the best performance from a cellular antenna by mounting it in the center of your roof. The second best choice is to install a glass-mount antenna at the top of your rear windshield. Wherever it is mounted, your cellular antenna performs best when it is pointed straight up, not tilted back at an angle. Remove your antenna before driving into a carwash.

Digital cellular

Digital technology allows more telephone conversations on the limited radio channels assigned to cellular service. Experiments with digital cellular service have begun in some cities, and most metropolitan areas will provide it within five years. Because of the large number of analog telephones already in use, analog and digital service will be offered in tandem for the foreseeable future. The new transitional *dual-mode* cellular telephone is able to place calls on both types of systems.

Emergencies

Dial *911* or *0* to reach an emergency operator, and state your location and cellular telephone number. Cellular callers have saved lives by reporting accidents and fires, and in some cases even their own heart attacks. Battery backup and wireless transmission help cellular telephone systems to operate during many natural disasters that leave land-line telephones dead. To protect your life, please wear your safety belt whenever you drive.

International

Cellular telephone service in Canada, Mexico, and the United States is known as Advanced Mobile Phone Service (AMPS). AMPS is widely used in North America, Central America, South America, the Caribbean, Southeast Asia, and the South Pacific. Countries in other areas have cellular systems that are not compatible with AMPS. For more information, please refer to the international section that begins on page 1059.

Operators

In the United States, dialing an operator varies from one cellular company to another. In some cases, dialing *0* reaches a long distance operator. In others, dialing *0* reaches a local operator and *00* reaches a long distance operator. Some companies route all long distance operator calls to one company, but others allow you to choose the company by dialing 102880 for AT&T, 102220 for MCI, and 103330 for Sprint.

Privacy

It is not a good idea to discuss confidential information on a cellular telephone because your call is broadcast over public airwaves. It is unlikely that anyone will hear your calls, since it is both illegal and difficult to monitor the 832 channels used by most cellular telephones. However, if you are concerned about privacy, your dealer can help you to purchase scrambling devices that mount on the incoming and outgoing telephones.

Speakerphones

To use a speakerphone to make a call, turn on your telephone, dial the number without picking up the handset, and press send. When your telephone rings, you can answer a call on a speakerphone by pressing send. To hang up, press end.

These pages list the corporate offices of the largest cellular operating companies, which manage cellular systems in multiple cities. Single offices that are included in the main listings are not repeated here.

ALLTEL Mobile Communications
10825 Financial Parkway, Suite 401
Little Rock, AR 72211
501-661-8500

Ameritech Mobile Communications
2000 W Ameritech Drive
Hoffman Estates, IL 60195-5000
708-706-7600

Associated Communications
200 Gateway Towers
Pittsburgh, PA 15222
412-281-1907

Atlantic Cellular
15 Westminster Street, Suite 830
Providence, RI 02903
401-421-7090

Atlas Cellular
(subsidiary of PriCellular)
45 Rockefeller Plaza, Suite 3201
New York, NY 10020
212-459-0800

Bachow and Associates
(affiliated with the Bachtel Companies)
1600 Market Street, Suite 2020
Philadelphia, PA 19103
215-972-7550

Bell Atlantic Mobile
180 Washington Valley Road
Bedminster, NJ 07921
908-306-7000

Bell Mobility Cellular
20 Carlson Court
Etobicoke, Ontario M9W 6V4 Canada
416-674-2220

BellSouth Cellular
(subsidiary of BellSouth Enterprises)
1100 Abernathy Road, Suite 500
Atlanta, GA 30328
404-604-6100

BellSouth Mobility
(subsidiary of BellSouth Enterprises)
5600 Glenridge Drive
Atlanta, GA 30342
404-847-3600

Boatphone
Lower Newgate Street
PO Box 1516
St. Johns, Antigua, British West Indies
809-462-5051

Boston Communications Group
265 Franklin Street, Suite 1102
Boston, MA 02110
617-439-4141

Cal-One Cellular
PO Box 627
Fort Jones, CA 96032
916-468-5222

Cantel
(official name is Rogers Cantel)
10 York Mills Road E
North York, Ontario M2P 2C9 Canada
416-229-1400

Cellular Holding
(subsidiary of Potosi)
PO Box 739
Meadville, MS 39653
601-384-3211

Cellular, Inc.
(Same address as CommNet 2000)
5990 Greenwood Plaza Boulevard, Suite 300
Englewood, CO 80111
303-694-3234

Cellular Information Systems
55 Holly Hill Lane, Suite 101
PO Box 4507
Greenwich, CT 06830
203-622-6317

Cellular Plus
(subsidiary of CTEC)
1400 Spruce Street
Avoca, PA 18641-2294
717-883-8832

Cellular South
(also known as Cellular Holding)
PO Box 739
Meadville, MS 39653
601-384-3211

Celutel
(also known as Symphony Management)
900 Bestgate Road, Suite 400
Annapolis, MD 21401
410-573-5200

Centel Cellular
O'Hare Plaza, 8725 Higgins Road, Suite 650
Chicago, IL 60631
312-399-2644

Centennial Cellular
(subsidiary of Century Communications)
50 Locust Avenue
New Canaan, CT 06840
203-972-2000

Century Cellunet
520 Riverside Drive
PO Box 4065
Monroe, LA 71211-4065
318-388-9000

Century Communications
50 Locust Avenue
New Canaan, CT 06840
203-972-2000

Clear Communications
135 Mount Vernon Highway
Atlanta, GA 30328
404-843-9100

Comcast Cellular Communications
480 E Swedesford Road
Wayne, PA 19087
215-975-5000

CommNet 2000
(same address as Cellular Inc.)
5990 Greenwood Plaza Boulevard, Suite 300
Englewood, CO 80111
303-694-3234

CompComm
900 Haddon Avenue, Suite 412
Collingswood, NJ 08108
609-854-1000

Cone Enterprises
PO Box 10321
Lubbock, TX 79408
806-744-1661

Contel Cellular
(affiliated with GTE Mobile Communications)
245 Perimeter Center Parkway
Atlanta, GA 30346
404-804-3400

Cooper Cellular Management
3372 Edgewater Drive
Orlando, FL 32804
407-425-9074

Crowley Cellular Telecommunications
2800 W Higgins Road, Suite 805
Hoffman Estates, IL 60195
708-843-9081

Dobson Cellular Systems
(subsidiary of Dobson Communications)
13439 N Broadway Extension, Suite 100
Oklahoma City, OK 73114
405-749-9744

GenCell Management
(also known as General Cellular)
1891 Woolner Avenue
Fairfield, CA 94533
707-425-8000

GMD Limited Partnership
333 E Arlington Boulevard
Greenville, NC 27858
919-321-0066

GTE Mobile Communications
245 Perimeter Center Parkway
Atlanta, GA 30346
404-391-8000

GTE Mobilnet
(affiliated with GTE Mobile Communications)
245 Perimeter Center Parkway
Atlanta, GA 30346
404-391-8000

Horizon Cellular
101 Lindenwood Drive, Suite 125
Malvern, PA 19355
215-651-5900

Independent Cellular Network
2100 Electronics Lane
Ft. Myers, FL 33912
813-489-1600

McCaw Cellular Communications
5400 Carillon Point
PO Box 97060
Kirkland, WA 98033
206-827-4500

Mercury Communications
200 E Capitol Street, Suite 1647
Jackson, MS 39201-2202
601-948-4800

MCMG
(Minerich Cellular Management Group)
1262 Old Hillsboro Road
Franklin, TN 37064
615-791-0202

Miscellco Communications
120 N Congress Street, Suite 500
PO Box 24328
Jackson, MS 39225-4328
601-948-1212

Nationwide Cellular Service
20 E Sunrise Highway
PO Box CS-2320
Valley Stream, NY 11581-1252
516-568-2000

New Par
350 E Wilson Bridge Road
Worthington, OH 43085
614-436-4331

NYNEX Mobile Communications
2000 Corporate Drive
Orangeburg, NY 10962
914-365-7200

Pacific Northwest Cellular
11400 SE 8th Street, Suite 445
Bellevue, WA 98004-6431
206-635-0300

Pacific Telecom Cellular
(also known as PTI)
4600 W College Avenue
Appleton, WI 54915
414-841-1200

PacTel Cellular
2999 Oak Road
Walnut Creek, CA 94596
510-210-3600

Palmer Communications
12800 University Drive, Suite 500
Fort Myers, FL 33907-5333
813-433-4350

PriCellular
(affiliated with Atlas Cellular)
45 Rockefeller Plaza, Suite 3201
New York, NY 10020
212-459-0800

Prime Cellular
134 Clinton Road
Fairfield, NJ 07004
201-227-1434

Quantum Communications
7901 Flying Cloud Drive, Suite 250
Eden Prairie, MN 55344
612-942-7650

Radiofone
3131 N I-10 Service Road
PO Box 8760
Metairie, LA 70011-8760
504-837-8330

Rural Cellular
(affiliated with Cellular 2000)
2819 Highway 29 S Midway Mall
PO Box 127
Alexandria, MN 56308
612-762-2000

Rural Cellular Management
2573 Clay Bank Road, Suite 9
Fairfield, CA 94533
707-422-2100

Southeastern Cellular
PO Box 500
Millry, AL 36558
205-846-2090

Southwestern Bell Mobile Systems
17330 Preston Road, Suite 100-A
Dallas, TX 75252
214-733-2000

Springwich Cellular Limited Partnership
(subsidiary of SNET)
555 Long Wharf Drive, Room 750
New Haven, CT 06511
203-553-7600

Sterling Cellular Management
1080 Holcomb Bridge Road, Bldg 100, Ste 200
Roswell, GA 30076
404-552-5030

Symphony Management
(also known as Celutel)
900 Bestgate Road, Suite 400
Annapolis, MD 21401
410-573-5200

Texoma Cellular Management
4500 Seymour Highway, Suite 105
Wichita Falls, TX 76309
817-691-9566

United States Cellular
8410 W Bryn Mawr, Suite 700
Chicago, IL 60631
312-399-8900

Unitel
Bayport 1, Suite 400
West Atlantic City, NJ 08232
609-646-9400

U S West Cellular
(subsidiary of U S West New Vector Group)
3350—161st Avenue SE
PO Box 7329
Bellevue, WA 98008-1329
206-747-4900

Vanguard Cellular Systems
2002 Pisgah Church Road, Suite 300
Greensboro, NC 27455-3314
919-282-3690

X-Cell Cellular
2401 15th Street, Suite 230
Denver, CO 80202
303-433-3700

The following reservation numbers for travel services are provided for your convenience. To obtain additional toll-free numbers, you may dial 1-800-555-1212. Normal airtime charges apply when calling from a cellular telephone.

Numbers generally may be dialed from throughout the continental United States, unless the letters *exc* (except) precede a state name. Often the same number may be dialed from Alaska, Bermuda, Canada, Hawaii, Puerto Rico, and the Virgin Islands. In some areas, your call must be preceded by a "1" or a "0." Some branch locations of these companies also have 800 reservation numbers.

Airlines

Airline	Notes	Number
Aerlingus	exc NY	800-223-6537
Aero Peru		800-777-7717
Aeromexico		800-237-6639
Air Canada	exc Canada	800-776-3000
Air France		800-237-2747
Air India		800-223-7776
Air Jamaica		800-523-5585
Air New Zealand		800-262-1234
Alaska Airlines		800-426-0333
Alitalia		800-223-5730
Aloha Airlines		800-367-5250
America West	some states	800-247-5692
American Airlines		800-433-7300
Australian Airlines		800-922-5122
British Airways		800-247-9297
Canadian Airlines	exc Canada	800-426-7000
Continental Airlines		800-525-0280
Delta Airlines		800-221-1212
El Al		800-223-6700
Finnair		800-950-5000
Gulf Air		800-553-2824
Hawaiian Airlines		800-367-5320
Horizon Air		800-547-9308
Iberia Airlines		800-772-4642
Icelandair		800-223-5500
Japan Airlines	exc HI	800-525-3663
	from HI	808-521-1441
KLM Royal Dutch Airlines		800-777-5553
Korean Airlines		800-421-8200
Lufthansa Airlines		800-645-3880
MarkAir	exc AK	800-426-6784
	from AK	800-478-0800
Mesa Airlines		800-637-2247
Mexicana Airlines		800-531-7921
Northwest Airlines	domestic	800-225-2525
	international	800-447-4747
Olympic Airways		800-223-1226
Philippine Airlines		800-435-9725

Airline	Notes	Number
Qantas Airways		800-227-4500
Sabena Belgian Airlines		800-955-2000
Saudi Arabian Airlines	exc NY	800-472-8342
Scandinavian Airlines		800-221-2350
Singapore Airlines		800-742-3333
Southwest Airlines		800-531-5601
Swissair		800-221-4750
Tower Air		800-221-2500
TWA		800-221-2000
United Airlines		800-241-6522
US Air		800-428-4322
Varig Brazilian		800-468-2744
Viasa Venezuelan Airlines		800-426-6784

Hotels

Hotel	Notes	Number
Admiral Benbow Inns		800-451-1986
Amberley Suite Hotels		800-227-7229
Americana Hotels		varies by hotel
Aristocrat Inns of America		varies by hotel
Atlas Hotels	exc CA	800-854-2608
	CA only	800-542-6082
Bally's–Las Vegas	exc NV	800-634-3434
Bally's–Park Place	exc NV	800-225-5977
Bally's–Reno	exc NV	800-648-5080
Best Western		800-528-1234
Caesar's Palace		800-634-6661
Canadian Pacific Hotels	exc Canada	800-828-7447
	Canada	800-268-9411
	Ont. & Quebec	800-268-9420
Circus Circus,Las Vegas	exc NV	800-634-3450
Circus Circus, Reno	exc NV	800-648-5010
Clarion Hotels/Resorts		800-252-7466
Club Med		800-258-2633
Comfort Inns		800-228-5150
Compri Hotels		800-426-6774
Courtyard by Marriott		800-321-2211
Days Inns of America		800-325-2525
Dillon Inns		800-253-7503
Doubletree Hotels		800-528-0444
Downtowner/Passport		800-238-6161
Drury Inns		800-325-8300
Econo Lodge/Travel M. H.		800-446-6900
Embassy Suites		800-362-2779
Exel Inns		800-356-8013
Fairfield Inn by Marriot		800-228-2800
Fairmont Hotels		800-527-4727
Family Inns		800-251-9752
Four Seasons		800-332-3442
Friendship Inns	exc Canada	800-453-4511
	Canada only	800-828-0013
Glacier National Park	(summer)	406-226-5551
Guest Quarters Hotels	exc DC	800-424-2900
Hampton Inn Hotels		800-426-7866
Harley Hotels		800-321-2323
Harrah's Hotels-Casino	exc NV	800-648-3773

Hotel	Notes	Number
Helmsley Hotels		800-221-4982
Hilton Hotels		800-445-8667
Holiday Inn		800-465-4329
Howard Johnson		800-654-2000
Hyatt Hotels		800-233-1234
Intercontinental Hotels		800-327-0200
Journey's End Motels		800-668-4200
L-K Motels		800-282-5711
La Quinta Hotels		800-531-5900
Marriott Hotels		800-228-9290
Master Host		800-251-1962
Meridien Hotels		800-543-4300
Motel 6		no 800 number
Nendels Inns & Motor Inns		800-547-0106
Omni Intl. Hotels		800-843-6664
Preferred Hotels		800-323-7500
Princess Hotels International		800-223-1818
Quality Inns	(all rooms)	800-228-5151
	(non-smoking)	800-228-5864
Quality Royale		800-228-5152
Radisson Hotels		800-333-3333
Ramada Inn	exc Canada	800-228-2828
	from Canada	800-268-8998
Ramada Renaissance		800-228-9898
Red Carpet Inns		800-251-1962
Red Lion Inns		800-547-8010
Red Roof Inns		800-843-7663
Residence Inns		800-331-3131
Rodeway Inn		800-228-2000
Scottish Inns		800-251-1962
Sheraton Hotels & Motor Inns		800-325-3535
Shilo Inns	exc Canada	800-222-2244
	from Canada	800-228-4489
Shoney's Inns		800-222-2222
Sonesta International		800-766-3782
Stouffer Hotels		800-468-3571
Suisse Chalet		800-524-2538
Super 8		800-843-1991
Travel Host Motels	exc Canada	800-346-4974
	from Canada	800-624-7574
Travelodge		800-255-3050
Treadway Inns & Resorts		800-631-0182
Trusthouse Forte		800-225-5843
Utell International		800-223-9868
Vagabond Inns	exc Canada	800-522-1555
	from Canada	800-468-2251
Westin Hotels		800-228-3000
Yellowstone National Park, WY		307-344-7311
Yosemite National Park, CA		209-252-4848

Rental Cars

Rental car company	Notes	Number
Agency Rent-a-Car		800-321-1972
Airways Rent-a-Car		800-952-9200
Alamo Rent-a-Car		800-327-9633
Allstar Rent-A-Car		800-426-5243

Rental car company	Notes	Number
American International Rent-a-Car		800-527-0202
Auto Europe		800-223-5555
Avis Rent-a-Car	domestic	800-331-1212
	from Canada	800-879-2847
	international	800-331-1084
Budget Rent-a-Car	domestic	800-527-0700
	international	800-472-3325
Dollar Rent-a-Car	for mainland	800-421-6868
	for Hawaii	800-367-7006
European Car Reservation		800-535-3303
Eurorent Rent-a-Car		800-521-2235
Exchange Car Rental		800-777-2836
Foremost Euro-Car		800-272-3299
General Rent-a-Car		800-327-7607
Hertz Rent-a-Car	exc Canada	800-654-3131
	from Canada	800-263-0600
	intl rentals	800-654-3001
Lloyd's International Rent-a-Car		800-654-7037
National Car Rental	domestic	800-227-7368
	international	800-227-3876
Payless Rent-a-Car		800-729-5377
Rent-a-Wreck		800-535-1391
Sears Rent-a-Car	exc Canada	800-527-0770
	from Canada	800-268-8900
Showcase Rental Car		800-421-6808
Superior Rent A Car		800-237-8106
Thrifty Rent-a-Car	exc Tulsa	800-367-2277
Ugly Duckling Rent.		800-843-3825
USA Rent-a-Car		800-872-2277
Value Rent-a-Car		800-327-2501

Travel Services

Travel service	Number
American Automobile Association	800-222-4357
American Express lost traveler's cheques	800-221-7282
AT&T operator (from inside U.S.)	10288-0
DHL Worldwide Express Delivery	800-225-5345
Directory assistance for toll-free nos.	800-555-1212
Federal Express	800-238-5355
Hotelecopy FaxMail	800-322-4448
MCI Operator (from inside U.S.)	10222-0
To purchase this *Cellular Travel Guide*	800-927-8800
U S Sprint operator (from inside U.S.)	10333-0
Visa lost traveler's cheques	800-227-6811

User's Guide

A Block and B Block–The two sets of channels assigned by the government to the two cellular telephone companies operating in each service area.

A/B Switch–A switch or button on a cellular telephone that allows the user to choose between systems operating on the A Block and B Block channels.

Airtime–The amount of billable talking time you have accumulated, usually in minutes.

American Roaming Network–A service that intercepts calls made by roamers who lack roaming agreements to be billed normally. Calls are billed to a credit card (also see page 1159).

AMPS–Advanced mobile phone service; the cellular telephone system used in the U.S. and Canada. Other countries that use AMPS, including Mexico, are listed in the international section of this book.

Authorized agent–A company chosen by the licensed cellular telephone company to market its services and equipment, such as a cellular telephone dealer.

Billing identification number (BID)–A unique identifying number assigned by CTIA to smaller areas within a cellular system that share the same system identification number.

Call diversion–This service feature transfers incoming calls to a third number after a specified number of rings or when your cellular telephone is busy or turned off.

Carrier–A cellular telephone company licensed to provide cellular telephone service.

Cell site–The area immediately surrounding one broadcasting and receiving antenna in a cellular telephone system. The system in a city may be made up of dozens of individual cell sites, and your call is transferred from one to the next as you drive through each one.

Cellular Geographic Service Area (CGSA)–The service area composed of one or more counties in which a cellular telephone company is authorized to operate.

Cellular operating system–The network of cell sites connected to a single cellular telephone switching center and operated by an individual cellular telephone company.

Cellular Telecommunications Industry Association (CTIA)–A cellular industry association that sponsors publications, conventions, and other services (202)785-0081.

Channel–One of the radio frequencies on which your cellular telephone and the cellular operating system communicate automatically. Most systems use 666 or 832 channels.

CIBER (Cellular Intercarrier Billing Exchange Record)–A standardized reporting format used by cellular telephone companies that allows your home company to bill you for calls you make in another company's service area.

Clearinghouse–A company that sorts the cellular calls you make outside your home service area so they can be billed back to your home company and appear on your bill.

Coverage area–The geographic area reliably served by a cellular operating system. Because topographic features such as mountains and lakes affect the area where a cellular telephone will operate, a system's coverage area is never identical to its CGSA.

Crosstalk–Technical problem causing two simultaneous conversations to be heard on a telephone.

Deadspot–An area within a cellular system where service is not available due to steep terrain, a tunnel, heavy foliage, or electronic interference.

Dropped call–A telephone call that is disconnected accidentally by the cellular system.

Electronic serial number (ESN)–The unique identifying number of a cellular telephone, which has been embedded in its circuits by the manufacturer.

Follow Me Roaming–A service provided by GTE Telecommunications Services that helps a cellular user receive calls made to his home cellular number while visiting another city (see page 1156).

Handoff–The transfer of a cellular conversation from one voice channel to another. Handoff should occur imperceptibly as you travel from one cell site to an adjacent one.

Home carrier–The cellular telephone company with which your telephone is registered and to which you pay monthly access charges.

Host carrier–The cellular telephone company that provides you service when you are outside the area served by your home carrier.

Improved Mobile Telephone Service (IMTS)–The mobile telephone service that has been operating in the U.S. and Canada since the 1960s and is not compatible with cellular.

Incoming call–A call received by a cellular telephone.

Incomplete call–A call attempt that results in a busy signal or no answer. Usually it is not charged.

Landline–Pertaining to a telephone in a fixed location, such as your home or office.

Mobile Telephone Switching Office (MTSO)–The control center of a cellular operating system that connects all the cell sites, computers, and switching equipment to the local and long distance telephone companies.

Metropolitan statistical area (MSA)–One of 306 urban areas licensed for cellular telephone service in the U.S. by the Federal Communications Commission (also see page 1168).

National Cellular Resellers Association–A cellular industry association that lobbies on behalf of cellular resellers (202)429-2014.

NationLink–(Also known as RoamingAmerica) A service provided by EDS that helps a cellular user to receive calls made to his cellular number when in another city (also see page 1157).

Non-wireline–Description of a company operating an A block of cellular channels in a city.

North American Cellular Network–An automatic service that helps a cellular user receive calls made to his home cellular number while visiting another city (also see page 1156).

NPA–The telephone industry's abbreviation for area code.

NXX–The telephone industry's abbreviation for the first three digits of a 7-digit telephone number.

Off-peak rate–A discounted *per minute* billing rate usually offered on evenings and weekends.

Optional features–Custom calling features available from a cellular telephone company that may include call forwarding, call waiting, conference calling, incoming call restrictions, outgoing call restrictions, operator access restrictions, and voice mail.

Outgoing call–A call sent from a cellular telephone.

Peak rate–The *per minute* billing rate for calls completed during weekday business hours.

Reciprocal roaming agreement–Please refer to *roaming agreement* below.

Reseller–A company that purchases cellular airtime at a wholesale price from a licensed cellular operating company and resells it, generally at a discounted price.

Roam–An message displayed on a cellular telephone when it is operated outside the area served by the home carrier with which it is registered.

Roamer–A cellular telephone user who is outside the service area of his home carrier.

Roamer Plus–A service that intercepts calls made by roamers who lack roaming agreements to be billed normally. Calls are billed to a credit card (also see page 1159).

Roamer access number–A telephone number that allows cellular users to receive calls when they are outside their home service areas. A caller dials the roamer access number listed in this directory for the system serving the area that the cellular user is visiting, waits to hear a tone or tones, and then dials the 10-digit area code and number of the cellular telephone.

Roaming agreement–An agreement between two cellular telephone companies allowing one company's customers to make calls using the other's system. A customer is billed by his home carrier for calls he makes on the host carrier's system plus any daily fees (also see page 1159).

Rural service area (RSA)–One of 428 rural areas licensed for cellular telephone service in the United States by the Federal Communications Commission (also see page 1170).

Seatbelt–Safety equipment designed to save your life.

System identification number (SID)–A unique identifying code number assigned to each cellular telephone system. Also see *billing identification number* on the previous page.

Telocator Network of America (Telocator)–A cellular industry association that sponsors publications, conventions, and other services (202)467-4770.

Wireline–Description of a company operating a B block of cellular channels in a city.

Metropolitan Cellular Markets

The following pages list all cellular markets licensed by the U.S. Federal Communications Commission. The first 306 markets are numbered in order of population, and each is known as a Metropolitan Statistical Area (MSA). Cellular service is available in every MSA. For information about service in a particular MSA, please refer to the city index that begins on page one.

MSA	City	State
1	New York	New York
2	Los Angeles	California
3	Chicago	Illinois
4	Philadelphia	Pennsylvania
5	Detroit	Michigan
6	Boston	Massachusetts
7	San Francisco	California
8	Washington	District of Columbia
9	Dallas	Texas
10	Houston	Texas
11	St. Louis	Missouri
12	Miami	Florida
13	Pittsburgh	Pennsylvania
14	Baltimore	Maryland
15	Minneapolis	Minnesota
16	Cleveland	Ohio
17	Atlanta	Georgia
18	San Diego	California
19	Denver	Colorado
20	Seattle	Washington
21	Milwaukee	Wisconsin
22	Tampa	Florida
23	Cincinnati	Ohio
24	Kansas City	Missouri
25	Buffalo	New York
26	Phoenix	Arizona
27	San Jose	California
28	Indianapolis	Indiana
29	New Orleans	Louisiana
30	Portland	Oregon
31	Columbus	Ohio
32	Hartford	Connecticut
33	San Antonio	Texas
34	Rochester	New York
35	Sacramento	California
36	Memphis	Tennessee
37	Louisville	Kentucky
38	Providence	Rhode Island
39	Salt Lake City	Utah
40	Dayton	Ohio
41	Birmingham	Alabama
42	Bridgeport	Connecticut
43	Norfolk	Virginia
44	Albany	New York
45	Oklahoma City	Oklahoma
46	Nashville	Tennessee
47	Greensboro	North Carolina
48	Toledo	Ohio
49	New Haven	Connecticut
50	Honolulu	Hawaii
51	Jacksonville	Florida
52	Akron	Ohio
53	Syracuse	New York
54	Gary	Indiana
55	Worcester	Massachusetts
56	NE PA (Scranton)	Pennsylvania
57	Tulsa	Oklahoma
58	Allentown	Pennsylvania
59	Richmond	Virginia
60	Orlando	Florida
61	Charlotte	North Carolina
62	New Brunswick	New Jersey
63	Springfield	Massachusetts
64	Grand Rapids	Michigan
65	Omaha	Nebraska
66	Youngstown	Ohio
67	Greenville	South Carolina
68	Flint	Michigan
69	Wilmington	Delaware
70	Long Branch	New Jersey
71	Raleigh/Durham	North Carolina
72	West Palm Beach	Florida
73	Oxnard	California
74	Fresno	California
75	Austin	Texas
76	New Bedford	Massachusetts
77	Tucson	Arizona
78	Lansing	Michigan
79	Knoxville	Tennessee
80	Baton Rouge	Louisiana
81	El Paso	Texas
82	Tacoma	Washington
83	Mobile	Alabama
84	Harrisburg	Pennsylvania
85	Johnson City	Tennessee
86	Albuquerque	New Mexico
87	Canton	Ohio
88	Chattanooga	Tennessee
89	Wichita	Kansas
90	Charleston	South Carolina
91	San Juan	Puerto Rico
92	Little Rock	Arkansas
93	Las Vegas	Nevada
94	Saginaw/Bay/Midland	Michigan
95	Columbia	South Carolina
96	Fort Wayne	Indiana
97	Bakersfield	California
98	Davenport/Moline	Iowa & Illinois
99	York	Pennsylvania
100	Shreveport	Louisiana
101	Beaumont	Texas
102	Des Moines	Iowa
103	Peoria	Illinois
104	Newport News	Virginia
105	Lancaster	Pennsylvania
106	Jackson	Mississippi
107	Stockton	California
108	Augusta	Georgia
109	Spokane	Washington
110	Huntington/Ashland	West Virginia
111	Vallejo	California
112	Corpus Christi	Texas
113	Madison	Wisconsin
114	Lakeland	Florida
115	Utica/Rome	New York
116	Lexington/Fayette	Kentucky
117	Colorado Springs	Colorado
118	Reading	Pennsylvania
119	Evansville	Indiana
120	Huntsville	Alabama
121	Trenton	New Jersey
122	Binghamton	New York
123	Santa Rosa/Petaluma	California
124	Santa Barbara	California
125	Appleton	Wisconsin
126	Salinas	California
127	Pensacola	Florida
128	McAllen	Texas
129	South Bend	Indiana
130	Erie	Pennsylvania
131	Rockford	Illinois
132	Kalamazoo	Michigan
133	Manchester/Nashua	New Hampshire
134	Atlantic City	New Jersey
135	Eugene/Springfield	Oregon
136	Lorain/Elyria	Ohio
137	Melbourne	Florida
138	Macon	Georgia
139	Montgomery	Alabama
140	Charleston	West Virginia
141	Duluth	Minnesota
142	Modesto	California
143	Johnstown	Pennsylvania
144	Orange County	California
145	Hamilton	Ohio
146	Daytona Beach	Florida
147	Ponce	Puerto Rico

148	Salem	Oregon	228	Vineland/Millville	New Jersey
149	Fayetteville	North Carolina	229	Medford	Oregon
150	Visalia/Tulare	California	230	Decatur	Illinois
151	Poughkeepsie	New York	231	Mansfield	Ohio
152	Portland	Maine	232	Eau Claire	Wisconsin
153	Columbus	Georgia	233	Wichita Falls	Texas
154	New London/Norwich	Connecticut	234	Athens	Georgia
155	Savannah	Georgia	235	Petersburg	Virginia
156	Portsmouth	New Hampshire	236	Muncie	Indiana
157	Roanoke	Virginia	237	Tyler	Texas
158	Lima	Ohio	238	Sharon	Pennsylvania
159	Provo/Orem	Utah	239	Joplin	Missouri
160	Killeen/Temple	Texas	240	Texarkana	Texas & Arkansas
161	Lubbock	Texas	241	Pueblo	Colorado
162	Brownsville	Texas	242	Olympia	Washington
163	Springfield	Missouri	243	Greeley	Colorado
164	Fort Myers	Florida	244	Kenosha	Wisconsin
165	Fort Smith	Arkansas & Oklahoma	245	Ocala	Florida
166	Hickory	North Carolina	246	Dothan	Alabama
167	Sarasota	Florida	247	Lafayette	Indiana
168	Tallahassee	Florida	248	Burlington	Vermont
169	Mayaguez	Puerto Rico	249	Anniston	Alabama
170	Galveston	Texas	250	Bloomington/Normal	Illinois
171	Reno	Nevada	251	Williamsport	Pennsylvania
172	Lincoln	Nebraska	252	Pascagoula	Mississippi
173	Biloxi/Gulfport	Mississippi	253	Sioux City	Iowa & Nebraska
174	Lafayette	Louisiana	254	Redding	California
175	Santa Cruz	California	255	Odessa	Texas
176	Springfield	Illinois	256	Charlottesville	Virginia
177	Battle Creek	Michigan	257	Hagerstown	Maryland
178	Wheeling	West Virginia & Ohio	258	Jacksonville	North Carolina
179	Topeka	Kansas	259	State College	Pennsylvania
180	Springfield	Ohio	260	Lawton	Oklahoma
181	Muskegon	Michigan	261	Albany	Georgia
182	Fayetteville	Arkansas	262	Danville	Virginia
183	Asheville	North Carolina	263	Wausau	Wisconsin
184	Houma	Louisiana	264	Florence	South Carolina
185	Terre Haute	Indiana	265	Fort Walton Beach	Florida
186	Green Bay	Wisconsin	266	Glens Falls	New York
187	Anchorage	Alaska	267	Sioux Falls	South Dakota
188	Amarillo	Texas	268	Billings	Montana
189	Racine	Wisconsin	269	Cumberland	Maryland & West Virginia
190	Boise	Idaho	270	Bellingham	Washington
191	Yakima	Washington	271	Kokomo	Indiana
192	Gainesville	Florida	272	Gadsden	Alabama
193	Benton Harbor	Michigan	273	Kankakee	Illinois
194	Waco	Texas	274	Yuba City	California
195	Cedar Rapids	Iowa	275	St. Joseph	Missouri
196	Champaign	Illinois	276	Grand Forks	North Dakota & Minnesota
197	Lake Charles	Louisiana	277	Sheboygan	Wisconsin
198	St. Cloud	Minnesota	278	Columbia	Missouri
199	Steubenville	Ohio	279	Lewiston/Auburn	Maine
200	Parkersburg	West Virginia	280	Burlington	North Carolina
201	Waterloo/Cedar Falls	Iowa	281	Laredo	Texas
202	Arecibo	Puerto Rico	282	Bloomington	Indiana
203	Lynchburg	Virginia	283	Panama City	Florida
204	Aguadilla	Puerto Rico	284	Elmira	New York
205	Alexandria	Louisiana	285	Las Cruces	New Mexico
206	Longview/Marshall	Texas	286	Dubuque	Iowa
207	Jackson	Michigan	287	Bryan/College Station	Texas
208	Fort Pierce	Florida	288	Rochester	Minnesota
209	Clarksville	Tennessee & Kentucky	289	Rapid City	South Dakota
210	Fort Collins/Loveland	Colorado	290	La Crosse	Wisconsin
211	Bradenton	Florida	291	Pine Bluff	Arkansas
212	Bremerton	Washington	292	Sherman/Denison	Texas
213	Pittsfield	Massachusetts	293	Owensboro	Kentucky
214	Richland	Washington	294	San Angelo	Texas
215	Chico	California	295	Midland	Texas
216	Janesville/Beloit	Wisconsin	296	Iowa City	Iowa
217	Anderson	Indiana	297	Great Falls	Montana
218	Wilmington	North Carolina	298	Bismarck	North Dakota
219	Monroe	Louisiana	299	Casper	Wyoming
220	Abilene	Texas	300	Victoria	Texas
221	Fargo/Moorhead	North Dakota & Minnesota	301	Lawrence	Kansas
222	Tuscaloosa	Alabama	302	Enid	Oklahoma
223	Elkhart/Goshen	Indiana	303	Aurora/Elgin	Illinois
224	Bangor	Maine	304	Joliet	Illinois
225	Altoona	Pennsylvania	305	Alton/Granite City	Illinois
226	Florence	Alabama	306	Gulf Of Mexico	Louisiana & Texas
227	Anderson	South Carolina			

User's Guide

The following pages list cellular markets 307 to 734 licensed by the U.S. Federal Communications Commission. They are numbered in alphabetical order by state, and each is known as a Rural Service Area (RSA). An RSA includes one or more counties and often is named by the one in bold.

All RSAs had at least one cellular system operating when this book went to press. To locate more information about service in an RSA, turn to the main index that begins on page one, and under the appropriate state, find the page number for the RSA. When you turn to the page listed, you can identify which company serves the RSA by looking at the identification numbers section at the bottom of that page. For example, the main entry for Miami, Florida indicates at the bottom of its page that both companies are licensed to serve an MSA (Miami) and an RSA (FL-11 or Key West).

RSA Counties

307 Alabama 1
- Blount
- Cullman
- **Franklin**
- Lawrence
- Marion
- Morgan
- Winston

308 Alabama 2
- Cherokee
- De Kalb
- **Jackson**

309 Alabama 3
- Choctaw
- Fayette
- Greene
- Hale
- **Lamar**
- Marengo
- Pickens
- Sumter

310 Alabama 4
- **Bibb**
- Chilton
- Dallas
- Lowndes
- Perry
- Wilcox

311 Alabama 5
- Chambers
- Clay
- **Cleburne**
- Coosa
- Randolph
- Talladega
- Tallapoosa

312 Alabama 6
- Clarke
- Conecuh
- Escambia
- Monroe
- **Washington**

313 Alabama 7
- **Butler**
- Coffee
- Covington
- Crenshaw
- Geneva
- Pike

314 Alabama 8
- Barbour
- Bulloch
- Henry
- **Lee**
- Macon

315 Alaska 1
- Fairbanks
- Nome
- North Slope
- Northwest Arctic
- Southwest Fairbanks
- **Wade-Hampton**
- Yukon-Koyukuk

316 Alaska 2
- Aleutian Islands
- **Bethel**
- Bristol Bay

- Dillingham
- Kenai Peninsula
- Kodiak
- Matanuska-Susitna
- Valdez-Cordova

317 Alaska 3
- Angoon
- **Haines**
- Juneau
- Ketchikan Gateway
- Petersburg
- Prince of Wales-Outer Ketchikan
- Sitka
- Skagway
- Wrangell
- Yakutat

318 Arizona 1
- **Mohave**

319 Arizona 2
- **Coconino**
- Yavapai

320 Arizona 3
- Apache
- **Navajo**

321 Arizona 4
- **Yuma**

322 Arizona 5
- **Gila**
- Pinai

323 Arizona 6
- Cochise
- **Graham**
- Greeniee
- Santa Cruz

324 Arkansas 1
- Boone
- Carroll
- **Madison**
- Newton

325 Arkansas 2
- Baxter
- Fulton
- Izard
- **Marion**
- Searcy
- Stone

326 Arkansas 3
- Independence
- Jackson
- Lawrence
- Randolph
- **Sharp**

327 Arkansas 4
- **Clay**
- Craighead
- Greene
- Mississippi
- Poinsett

328 Arkansas 5
- Arkansas
- **Cross**
- Lee
- Monroe
- Phillips
- St. Francis

329 Arkansas 6
- **Cleburne**

- Prairie
- White
- Woodruff

330 Arkansas 7
- Conway
- Perry
- **Pope**
- Van Buren
- Yell

331 Arkansas 8
- **Franklin**
- Johnson
- Logan
- Scott

332 Arkansas 9
- Howard
- Montgomery
- Pike
- **Polk**
- Sevier

333 Arkansas 10
- Clark
- Dallas
- **Garland**
- Grant
- Hot Spring

334 Arkansas 11
- Columbia
- **Hempstead**
- Lafayette
- Nevada

335 Arkansas 12
- Ashley
- Bradley
- Calhoun
- Chicot
- Cleveland
- Desha
- Drew
- Lincoln
- **Ouachita**
- Union

336 California 1
- **Del Norte**
- Humboldt
- Siskiyou
- Trinity

337 California 2
- Lassen
- **Modoc**
- Plumas

338 California 3
- **Alpine**
- Amador
- Calaveras
- Mariposa
- Tuolumne

339 California 4
- **Madera**
- Merced
- San Benito

340 California 5
- **San Luis Obispo**

341 California 6
- Inyo
- **Mono**

342 California 7
- **Imperial**

343 California 8
- Colusa
- Glenn
- **Tehama**

344 California 9
- Lake
- **Mendocino**

345 California 10
- Nevada
- **Sierra**

346 California 11
- **El Dorado**

347 California 12
- **Kings**

348 Colorado 1
- Grand
- Jackson
- **Moffat**
- Rio Blanco
- Routt

349 Colorado 2
- **Logan**
- Morgan
- Phillips
- Sedgwick
- Washington
- Yuma

350 Colorado 3
- Clear Creek
- Delta
- Eagle
- **Garfield**
- Gunnison
- Mesa
- Montrose
- Pitkin
- Summit

351 Colorado 4
- Chafee
- Custer
- Fremont
- Lake
- **Park**

352 Colorado 5
- Cheyenne
- **Elbert**
- Kit Carson
- Lincoln

353 Colorado 6
- Dolores
- Hinsdale
- La Plata
- Montezuma
- Ouray
- San Juan
- **San Miguel**

354 Colorado 7
- Alamosa
- Archuleta
- Conejos
- Mineral
- Rio Grande
- **Saguache**

355 Colorado 8
- Bent
- Crowley
- **Kiowa**
- Otero

Prowers
356 Colorado 9
Baca
Costilla
Huerfano
Las Animas
357 Connecticut 1
Litchfield
358 Connecticut 2
Windham
359 Delaware 1
Kent
Sussex
360 Florida 1
Collier
Hendry
361 Florida 2
Glades
Highlands
Indian River
Okeechobee
362 Florida 3
Charlotte
De Soto
Hardee
363 Florida 4
Citrus
Hernando
Lake
Sumter
364 Florida 5
Flagler
Putnam
365 Florida 6
Dixie
Gilchrist
Levy
366 Florida 7
Columbia
Hamilton
Sewannee
Union
367 Florida 8
Jefferson
Lafayette
Madison
Taylor
368 Florida 9
Calhoun
Franklin
Gulf
Liberty
369 Florida 10
Holmes
Jackson
Walton
Washington
370 Florida 11
Monroe
371 Georgia 1
Fannin
Gilmer
Gordon
Murray
Pickens
Towns
Union
Whitfield
372 Georgia 2
Banks
Barrows
Dawson
Franklin
Habersham
Hall
Lumpkin
Rabun
Stephens
White
373 Georgia 3
Bartow
Chattooga

Floyd
Polk
374 Georgia 4
Elbert
Hart
Jasper
Lincoln
Morgan
Ogelthorpe
Putnam
Taliaferro
Wilkes
375 Georgia 5
Carroll
Coweta
Haralson
Heard
Troup
376 Georgia 6
Crawford
Harris
Lamar
Meriweather
Monroe
Pike
Spalding
Talbot
Taylor
Upson
377 Georgia 7
Baldwin
Hancock
Johnson
Laurens
Washington
Wilkinson
378 Georgia 8
Bullock
Burke
Candler
Emanuel
Glascock
Jefferson
Jenkins
Screven
Treutlen
Warren
379 Georgia 9
Clay
Crisp
Dooly
Macon
Marion
Quintman
Randolph
Schley
Stewart
Sumter
Terrell
Webster
380 Georgia 10
Ben Hill
Bleckley
Coffee
Dodge
Irwin
Jeff Davis
Montgomery
Pulaski
Telfair
Turner
Wheeler
Wilcox
381 Georgia 11
Appling
Bacon
Brantley
Charlton
Evans
Pierce
Tattnall
Toombs
Ware

382 Georgia 12
Camden
Glynn
Liberty
Long
McIntosh
Wayne
383 Georgia 13
Baker
Calhoun
Decatur
Early
Grady
Miller
Mitchell
Seminole
Thomas
384 Georgia 14
Atkinson
Berrien
Brooks
Charlton
Clinch
Colquitt
Cook
Echols
Lanier
Lowndes
Tift
Worth
385 Hawaii 1
Kauai
386 Hawaii 2
Kalwao
Maui
387 Hawaii 3
Hawaii
388 Idaho 1
Benewah
Bonner
Boundary
Clearwater
Kootenai
Latah
Lewis
Nez Perce
Shosone
389 Idaho 2
Adams
Gem
Idaho
Payette
Valley
Washington
390 Idaho 3
Boise
Custer
Lehmi
391 Idaho 4
Canyon
Elmore
Owyhee
392 Idaho 5
Blaine
Butte
Camas
Cassia
Gooding
Jerome
Lincoln
Minidoka
Twin Falls
393 Idaho 6
Bannock
Bear Lake
Bingham
Bonneville
Caribou
Clark
Franklin
Fremont
Jefferson
Madison

Oneida
Power
Teton
394 Illinois 1
Carroll
De Kalb
Jo Daviess
Lee
Ogle
Stephenson
Whiteside
395 Illinois 2
Bureau
Ford
Iroquois
La Salle
Livingston
Marshall
Putnam
Stark
396 Illinois 3
Fulton
Hancock
Henderson
Knox
McDonough
Mercer
Schuyler
Warren
397 Illinois 4
Adams
Brown
Calhoun
Cass
Greene
Macoupin
Morgan
Pike
Scott
398 Illinois 5
De Witt
Logan
Mason
Moultrie
Piatt
399 Illinois 6
Bond
Christian
Effingham
Fayette
Marion
Montgomery
Shelby
400 Illinois 7
Clark
Coles
Crawford
Cumberland
Douglas
Edgar
Jasper
Vermillion
401 Illinois 8
Alexander
Franklin
Jackson
Jefferson
Johnson
Massac
Perry
Pulaski
Randolph
Union
Washington
Williamson
402 Illinois 9
Clay
Edwards
Gallatin
Hamilton
Hardin
Lawrence

Pope
Richland
Saline
Wabash
Wayne
White
403 Indiana 1
Jasper
La Porte
Newton
Pulaski
Starke
White
404 Indiana 2
Kosciusko
Lagrange
Noble
Steuben
405 Indiana 3
Blackford
Grant
Huntington
Jay
406 Indiana 4
Carroll
Cass
Clinton
Fulton
Miami
Wabash
407 Indiana 5
Benton
Fountain
Montgomery
Parke
Putnam
Warren
408 Indiana 6
Fayette
Franklin
Henry
Randolph
Rush
Union
Wayne
409 Indiana 7
Davies
Dubois
Greene
Knox
Martin
Owen
Perry
Pike
Spencer
410 Indiana 8
Bartholomew
Brown
Crawford
Harrison
Jackson
Lawrence
Orange
Washington
411 Indiana 9
Decatur
Jefferson
Jennings
Ohio
Ripley
Scott
Switzerland
412 Iowa 1
Adams
Fremont
Mills
Montgomery
Page
Taylor
413 Iowa 2
Clarke
Decatur
Lucas

Ringgold
Union
Wayne
414 Iowa 3
Appanoose
Davis
Jefferson
Monroe
Van Buren
Wapello
415 Iowa 4
Des Moines
Henry
Lee
Louisa
Muscatine
416 Iowa 5
Cedar
Clinton
Jackson
Jones
417 Iowa 6
Iowa
Jasper
Keokuk
Mahaska
Marion
Poweshiek
Washington
418 Iowa 7
Adair
Audubon
Cass
Guthrie
Madison
419 Iowa 8
Crawford
Harrison
Monona
Shelby
420 Iowa 9
Calhoun
Carroll
Greene
Ida
Sac
421 Iowa 10
Boone
Hamilton
Humboldt
Story
Webster
Wright
422 Iowa 11
Benton
Grundy
Hardin
Marshall
Tama
423 Iowa 12
Allamakee
Buchanan
Clayton
Delaware
Fayette
Winneshiek
424 Iowa 13
Butler
Chickasaw
Floyd
Howard
Mitchell
425 Iowa 14
Cerro Gordo
Franklin
Hancock
Kossuth
Winnebago
Worth
426 Iowa 15
Buena Vista
Clay
Dickinson

Emmet
Palo Alto
Pochahontas
427 Iowa 16
Cherokee
Lyon
O'Brien
Osceola
Plymouth
Sioux
428 Kansas 1
Cheyenne
Decatur
Rawlins
Sheridan
Sherman
Thomas
429 Kansas 2
Graham
Norton
Osborne
Phillips
Rooks
Smith
430 Kansas 3
Clay
Cloud
Jewell
Lincoln
Mitchell
Ottawa
Republic
Washington
431 Kansas 4
Geary
Marshall
Nemaha
Pottawatomie
Riley
432 Kansas 5
Atchison
Brown
Doniphan
Leavenworth
Jackson
433 Kansas 6
Gove
Greeley
Lane
Logan
Scott
Wallace
Wichita
434 Kansas 7
Barton
Ellis
Ness
Pawnee
Rush
Russell
Trego
435 Kansas 8
Dickinson
Ellsworth
Marion
McPherson
Rice
Saline
436 Kansas 9
Chase
Greenwood
Lyon
Morris
Wabaunsee
437 Kansas 10
Allen
Anderson
Bourbon
Coffey
Franklin
Linn
Miami

Woodson
438 Kansas 11
Finney
Grant
Hamilton
Haskell
Kearney
Morton
Seward
Stanton
Stevens
439 Kansas 12
Clark
Ford
Gray
Hodgeman
Meade
440 Kansas 13
Barber
Comanche
Edwards
Kiowa
Pratt
Stafford
441 Kansas 14
Cowley
Harper
Harvey
Kingman
Reno
Sumner
442 Kansas 15
Chautauqua
Cherokee
Crawford
Elk
Labette
Montgomery
Neosho
Wilson
443 Kentucky 1
Ballard
Calloway
Carlisle
Fulton
Graves
Hickman
Marshall
McCracken
444 Kentucky 2
Caldwell
Crittenden
Hopkins
Livingston
Lyon
Trigg
Union
Webster
445 Kentucky 3
Allen
Breckenridge
Butler
Edmondson
Grayson
Hancock
Logan
McLean
Meade
Muhlenberg
Ohio
Simpson
Todd
Warren
446 Kentucky 4
Anderson
Green
Hardin
Larue
Marion
Mercer
Nelson
Spencer
Taylor

Washington
447 Kentucky 5
  Adair
  Barren
  Clinton
  Cumberland
  Hart
  McCreary
  Metcalfe
  Monroe
  Russell
  Wayne
448 Kentucky 6
  Boyle
  Casey
  Garrard
  Laurel
  Lincoln
  Madison
  Pulaski
  Rockcastle
449 Kentucky 7
  Carroll
  Franklin
  Gallatin
  Grant
  Harrison
  Henry
  Owen
  Pendleton
  Shelby
  Trimble
450 Kentucky 8
  Bath
  Bracken
  Fleming
  Lewis
  Mason
  Menifee
  Montgomery
  Nicholas
  Robertson
  Rowan
451 Kentucky 9
  Elliott
  Floyd
  Johnson
  Lawrence
  Magoffin
  Martin
  Morgan
  Pike
452 Kentucky 10
  Breathitt
  Estill
  Jackson
  Knott
  Lee
  Letcher
  Owsley
  Perry
  Powell
  Wolfe
453 Kentucky 11
  Bell
  Clay
  Harlan
  Knox
  Leslie
  Whitley
454 Louisiana 1
  Bienville
  Claiborne
  Jackson
  Lincoln
  Union
455 Louisiana 2
  East Carroll
  Franklin
  Madison
  Morehouse
  Richland
  Tensas

West Carroll
456 Louisiana 3
  De Soto
  Natchitoches
  Red River
  Sabine
  Vernon
457 Louisiana 4
  Caldwell
  Catahoula
  Concordia
  La Salle
  Winn
458 Louisiana 5
  Acadia
  Allen
  Avoyelles
  Beauregard
  Cameron
  Evangeline
  Jefferson Davis
  Pointe Coupee
  St. Landry
  Vermilion
459 Louisiana 6
  Assumption
  Iberia
  Iberville
  St. Mary
460 Louisiana 7
  East Feliciana
  St. Helena
  Tangipahoa
  Washington
  West Feliciana
461 Louisiana 8
  St. Charles
  St. James
  St. John The Baptist
462 Louisiana 9
  Plaquemines
463 Maine 1
  Franklin
  Oxford
464 Maine 2
  Aroostook
  Piscataquis
  Somerset
465 Maine 3
  Kennebec
  Knox
  Lincoln
  Waldo
466 Maine 4
  Hancock
  Washington
467 Maryland 1
  Garrett
468 Maryland 2
  Calvert
  Caroline
  Dorchester
  Kent
  Queen Anne's
  St. Mary's
  Somerset
  Talbot
  Wicomico
  Worcester
469 Maryland 3
  Frederick
470 Massachusetts 1
  Franklin
471 Massachusetts 2
  Barnstable
  Dukes
  Nantucket
472 Michigan 1
  Baraga
  Dickinson
  Gogebic
  Houghton
  Iron

Keweenaw
  Marquette
  Menominee
  Ontonagon
473 Michigan 2
  Alger
  Chippewa
  Delta
  Luce
  Mackinac
  Schoolcraft
474 Michigan 3
  Antrim
  Charlevoix
  Emmet
  Grand Traverse
  Kalkaska
475 Michigan 4
  Alcona
  Alpena
  Cheboygan
  Crawford
  Montmorency
  Oscoda
  Otsego
  Presque Isle
476 Michigan 5
  Benzie
  Lake
  Leelanau
  Manistee
  Mason
  Missaukee
  Osceola
  Wexford
477 Michigan 6
  Arenac
  Clare
  Gladwin
  Iosco
  Ogemaw
  Roscommon
478 Michigan 7
  Gratiot
  Isabella
  Mecosta
  Montcalm
  Newaygo
479 Michigan 8
  Allegan
480 Michigan 9
  Branch
  Cass
  Hillsdale
  Lenawee
  St. Joseph
481 Michigan 10
  Huron
  Sanilac
  Tuscola
482 Minnesota 1
  Kittson
  Marshall
  Pennington
  Red Lake
  Roseau
483 Minnesota 2
  Beltrami
  Clearwater
  Lake of t. Woods
  Mahnomen
  Norman
484 Minnesota 3
  Itasca
  Koochiching
485 Minnesota 4
  Cook
  Lake
486 Minnesota 5
  Becker
  Big Stone
  Douglas

Grant
  Otter Trail
  Pope
  Stevens
  Swift
  Todd
  Traverse
  Wadena
  Wilkin
487 Minnesota 6
  Aitkin
  Carlton
  Cass
  Crow Wing
  Hubbard
  Isanti
  Kanabec
  Mille Lacs
  Morrison
  Pine
488 Minnesota 7
  Chippewa
  Kandiyohi
  McLeod
  Meeker
  Nicollet
  Renville
  Sibley
489 Minnesota 8
  Lac qui Parle
  Lincoln
  Lyon
  Redwood
  Yellow Medicine
490 Minnesota 9
  Brown
  Cottonwood
  Jackson
  Martin
  Murray
  Nobles
  Pipestone
  Rock
  Watonwan
491 Minnesota 10
  Blue Earth
  Faribault
  Freeborn
  Le Sueur
  Rice
  Steele
  Waseca
492 Minnesota 11
  Dodge
  Fillmore
  Goodhue
  Houston
  Mower
  Wabasha
  Winona
493 Mississippi 1
  Coahoma
  Lafayette
  Marshall
  Panola
  Quitman
  Tate
  Tunica
494 Mississippi 2
  Alcorn
  Benton
  Lee
  Pontotoc
  Prentiss
  Tippah
  Tishomingo
  Union
495 Mississippi 3
  Bolivar
  Carroll
  Holmes
  Leflore
  Sunflower

	Tallahatchie		Clinton		Bollinger		Garden
496	Mississippi 4		Daviess		Cape Girardeau		Kimball
	Calhoun		**De Kalb**		Madison		Morrill
	Clay		Livingston		**Perry**		Scotts Bluff
	Chickasaw	508	Missouri 5		Wayne		Sheridan
	Grenada		Chariton	522	Missouri 19		**Sioux**
	Monroe		**Linn**		Butler	534	Nebraska 2
	Yalobusha		Macon		Dunklin		Boyd
497	Mississippi 5		Randolph		Mississippi		Brown
	Humphreys		Shelby		New Madrid		**Cherry**
	Issaquena	509	Missouri 6		Pemiscot		Garfield
	Sharkey		Audrain		Scott		Holt
	Warren		**Marion**		**Stoddard**		Keya Paha
	Washington		Monroe	523	Montana 1		Rock
	Yazoo		Pike		Flathead		Wheeler
498	Mississippi 6		Ralls		Glacier	535	Nebraska 3
	Attala	510	Missouri 7		Lake		Antelope
	Choctaw		Cooper		**Lincoln**		Burt
	Lowndes		Howard		Pondera		Cedar
	Montgomery		Johnson		Sanders		Cuming
	Noxubee		Lafayette		Teton		Dixon
	Oktibbeha		Pettis	524	Montana 2		**Knox**
	Webster		**Saline**		Blaine		Madison
	Winston	511	Missouri 8		Chouteau		Pierce
499	Mississippi 7		**Callaway**		Hill		Stanton
	Kemper		Lincoln		Liberty		Thurston
	Lauderdale		Montgomery		**Toole**		Wayne
	Leake		Warren	525	Montana 3	536	Nebraska 4
	Neshoba	512	Missouri 9		Garfield		Arthur
	Newton		**Bates**		**Phillips**		Blaine
	Scott		Cedar		Valley		Custer
500	Mississippi 8		Henry	526	Montana 4		**Grant**
	Adams		St. Clair		**Daniels**		Greenley
	Amite		Vernon		Dawson		Hooker
	Claiborne	513	Missouri 10		McCone		Howard
	Franklin		**Benton**		Roosevelt		Logan
	Jefferson		Camden		Richland		Loup
	Lincoln		Dallas		Sheridan		McPherson
	Pike		Hickory		Wibaux		Sherman
	Wilkinson		Polk	527	Montana 5		Thomas
501	Mississippi 9	514	Missouri 11		Granite		Valley
	Copiah		Cole		Lewis and Clark	537	Nebraska 5
	Jefferson Davis		Gasconade		**Mineral**		**Boone**
	Lawrence		Miller		Missoula		Butler
	Marion		**Moniteau**		Powell		Colfax
	Simpson		Morgan		Ravalli		Dodge
	Walthall		Osage	528	Montana 6		Merrick
502	Mississippi 10	515	Missouri 12		Broadwater		Nance
	Clarke		Crawford		**Deer Lodge**		Platte
	Covington		Dent		Jefferson		Polk
	Jasper		**Maries**		Judith Basin		Saunders
	Jones		Phelps		Meagher		Washington
	Smith		Pulaski		Silver Bow	538	Nebraska 6
	Wayne	516	Missouri 13		Wheatland		Buffalo
503	Mississippi 11		St. Francois	529	Montana 7		Dawson
	Forrest		Ste. Genevieve		**Fergus**		**Keith**
	George		**Washington**		Golden Valley		Lincoln
	Greene	517	Missouri 14		Musselshell		Perkins
	Lamar		Barry		Petroleum	539	Nebraska 7
	Pearl River		**Barton**		Stillwater		**Hall**
	Perry		Dade		Sweet Grass		Hamilton
504	Missouri 1		Lawrence	530	Montana 8		Seward
	Atchison		McDonald		**Beaverhead**		York
	Gentry	518	Missouri 15		Gallatin	540	Nebraska 8
	Holt		Douglas		Madison		**Chase**
	Nodaway		Howell		Park		Dundy
	Worth		Ozark	531	Montana 9		Franklin
505	Missouri 2		**Stone**		Big Horn		Frontier
	Grundy		Taney		**Carbon**		Furnas
	Harrison	519	Missouri 16		Rosebud		Gosper
	Mercer		**Laclede**		Treasure		Harlen
	Putnam		Texas	532	Montana 10		Hayes
	Sullivan		Webster		Carter		Hitchcock
506	Missouri 3		Wright		Custer		Kearney
	Adair	520	Missouri 17		Fallon		Phelps
	Clark		Carter		Powder River		Red Willow
	Knox		Iron		**Prairie**	541	Nebraska 9
	Lewis		Oregon	533	Nebraska 1		**Adams**
	Schuyler		Reynolds		Banner		Clay
	Scotland		Ripley		Box Butte		Fillmore
507	Missouri 4		**Shannon**		Cheyenne		Jefferson
	Caldwell	521	Missouri 18		Dawes		Nuckolis
	Carroll				Deuel		Saline

Rural Cellular Markets

Thayer
Webster
542 Nebraska 10
Cass
Gage
Johnson
Nemaha
Otoe
Pawnee
Richardson
543 Nevada 1
Churchill
Humboldt
Pershing
544 Nevada 2
Elko
Eureka
Lander
545 Nevada 3
Carson City
Douglas
Lyon
Storey
546 Nevada 4
Esmerelda
Mineral
Nye
547 Nevada 5
Lincoln
White Pine
548 New Hampshire 1
Cheshire
Coos
Grafton
Sullivan
549 New Hampshire 2
Belknap
Carroll
Merrimack
550 New Jersey 1
Hunterdon
551 New Jersey 2
Ocean
552 New Jersey 3
Sussex
553 New Mexico 1
Cibola
McKinley
Rio Arriba
San Juan
Taos
554 New Mexico 2
Colfax
Harding
Mora
Union
555 New Mexico 3
Catron
Sierra
Socorro
Valencia
556 New Mexico 4
Curry
De Baca
Guadalupe
Los Alamos
Quay
Roosevelt
San Miguel
Santa Fe
Torrance
557 New Mexico 5
Grant
Hidalgo
Luna
558 New Mexico 6
Chavez
Eddy
Lea
Lincoln
Otero
559 New York 1

Jefferson
Lewis
St. Lawrence
560 New York 2
Clinton
Essex
Franklin
Fulton
Hamilton
561 New York 3
Allegany
Cattaraugus
Chautauqua
Genesee
Steuben
Wyoming
562 New York 4
Cayuga
Chenango
Cortland
Schuyler
Seneca
Tompkins
Yates
563 New York 5
Delaware
Otsego
Schoharie
Sullivan
Ulster
564 New York 6
Columbia
Greene
565 North Carolina 1
Cherokee
Clay
Graham
Haywood
Jackson
Macon
Swain
Transylvania
566 North Carolina 2
Avery
Caldwell
Mitchell
Watauga
Yancey
567 North Carolina 3
Alleghany
Ashe
Surry
Wilkes
568 North Carolina 4
Cleveland
Henderson
Lincoln
McDowell
Polk
Rutherford
569 North Carolina 5
Anson
Montgomery
Richmond
Scotland
570 North Carolina 6
Chatham
Moore
Lee
571 North Carolina 7
Caswell
Franklin
Granville
Person
Rockingham
Vance
Warren
572 North Carolina 8
Edgecombe
Halifax
Nash
Northampton
Wilson

573 North Carolina 9
Bertie
Camden
Chowan
Gates
Hertford
Pasquotank
Perquimans
574 North Carolina 10
Harnett
Johnston
Wayne
575 North Carolina 11
Bladen
Columbus
Hoke
Robeson
576 North Carolina 12
Duplin
Pender
Sampson
577 North Carolina 13
Carteret
Craven
Greene
Jones
Lenoir
Pamlico
578 North Carolina 14
Beaufort
Dare
Hyde
Martin
Pitt
Tyrell
Washington
579 North Carolina 15
Cabarrus
Davie
Iredell
Rowan
Stanly
580 North Dakota 1
Burke
Divide
McLean
Mountrail
Renville
Ward
Williams
581 North Dakota 2
Benson
Bottineau
Cavalier
McHenry
Pierce
Ramsey
Rolette
Towner
582 North Dakota 3
Barnes
Dickey
Griggs
La Moure
Nelson
Pembina
Ransom
Richland
Sargent
Steele
Trail
Walsh
583 North Dakota 4
Adams
Billings
Bowman
Dunn
Golden Valley
Grant
Hettinger
McKenzie
Mercer
Oliver

Sioux
Slope
Stark
584 North Dakota 5
Eddy
Emmons
Foster
Kidder
Logan
McIntosh
Sheridan
Stutsman
Wells
585 Ohio 1
Defiance
Henry
Paulding
Williams
586 Ohio 2
Erie
Huron
Sandusky
Seneca
587 Ohio 3
Ashtabula
588 Ohio 4
Darke
Logan
Mercer
Shelby
Union
589 Ohio 5
Crawford
Hancock
Hardin
Marion
Wyandot
590 Ohio 6
Ashland
Coshocton
Holmes
Knox
Licking
Morrow
Wayne
591 Ohio 7
Guensey
Harrison
Monroe
Muskingum
Noble
Tuscarawes
592 Ohio 8
Adams
Brown
Clinton
Fayette
Highland
593 Ohio 9
Gallia
Jackson
Pike
Ross
Scioto
594 Ohio 10
Athens
Hocking
Meigs
Morgan
Perry
Vinton
595 Ohio 11
Columbiana
596 Oklahoma 1
Beaver
Cimarron
Texas
597 Oklahoma 2
Alfalfa
Ellis
Harper
Major
Woods

Marshall
Moore
649 Tennessee 7
 Bledsoe
 Bradley
 Louden
 McMinn
 Meigs
 Monroe
 Polk
 Rhea
650 Tennessee 8
 Johnson
651 Tennessee 9
 Maury
652 Texas 1
 Dallam
 Deaf Smith
 Hartley
 Moore
 Oldham
 Sherman
653 Texas 2
 Armstrong
 Carson
 Collingsworth
 Donley
 Gray
 Hansford
 Hemphill
 Hutchinson
 Lipscomb
 Ochitree
 Roberts
 Wheeler
654 Texas 3
 Bailey
 Castro
 Cochran
 Hale
 Hockley
 Lamb
 Lynn
 Parmer
 Swisher
 Terry
 Yoakum
655 Texas 4
 Briscoe
 Childress
 Cottle
 Crosby
 Dickens
 Floyd
 Garza
 Hall
 Kent
 King
 Motley
 Stonewall
656 Texas 5
 Archer
 Baylor
 Foard
 Hardeman
 Haskell
 Knox
 Shackelford
 Stephens
 Throckmorton
 Wilbarger
 Young
657 Texas 6
 Cooke
 Jack
 Montague
 Palo Pinto
658 Texas 7
 Camp
 Cass
 Delta
 Fannin
 Franklin

Hopkins
Hunt
Lamar
Marion
Morris
Rains
Red River
Titus
Upshur
Wood
659 Texas 8
 Andrews
 Borden
 Coke
 Dawson
 Fisher
 Gaines
 Glasscock
 Howard
 Martin
 Mitchell
 Nolan
 Scurry
 Sterling
660 Texas 9
 Bosque
 Brown
 Coleman
 Comanche
 Eastland
 Erath
 Hamilton
 Hill
 Mills
 Runnels
 Somervell
661 Texas 10
 Anderson
 Falls
 Freestone
 Henderson
 Leon
 Limestone
 Milam
 Navarro
 Robertson
 Van Zandt
662 Texas 11
 Angelina
 Cherokee
 Nacogdoches
 Panola
 Rusk
 Sabine
 San Augustine
 Shelby
663 Texas 12
 Brewster
 Culbertson
 Hudspeth
 Jeff Davis
 Presidic
664 Texas 13
 Pecos
 Reeves
 Terrell
665 Texas 14
 Crane
 Crockett
 Irion
 Loving
 Reagan
 Schleicher
 Sutton
 Upton
 Ward
 Winkler
666 Texas 15
 Blanco
 Burnet
 Concho
 Gillespie
 Kendall

Kerr
Kimble
Lampasas
Llano
Mason
McCulloch
Menard
San Saba
667 Texas 16
 Austin
 Bastrop
 Burleson
 Caldwell
 Colorado
 Fayette
 Gonzales
 Jackson
 Lavaca
 Lee
 Matagorda
 Washington
 Wharton
668 Texas 17
 Grimes
 Houston
 Jasper
 Madison
 Newton
 Polk
 San Jacinto
 Trinity
 Tyler
 Walker
669 Texas 18
 Bandera
 Dimmit
 Edwards
 Frio
 Kinney
 La Salle
 Maverick
 Medina
 Real
 Uvalde
 Val Verde
 Zavala
670 Texas 19
 Atascosa
 Brooks
 Duval
 Jim Hogg
 Jim Wells
 Kenedy
 Kleburg
 Live Oak
 McMullen
 Starr
 Willacy
 Zapata
671 Texas 20
 Aransas
 Bee
 Calhoun
 De Witt
 Goliad
 Karnes
 Refugio
 Wilson
672 Texas 21
 Chambers
673 Utah 1
 Box Elder
 Cache
 Rich
674 Utah 2
 Morgan
 Summit
 Wasatch
675 Utah 3
 Juab
 Millard
 Sanpete

Sevier
676 Utah 4
 Beaver
 Iron
 Washington
677 Utah 5
 Carbon
 Dagget
 Duchesne
 Emery
 Grand
 Uintah
678 Utah 6
 Garfield
 Kane
 Piute
 San Juan
 Wayne
679 Vermont 1
 Caledonia
 Essex
 Franklin
 Lamoille
 Orange
 Orleans
 Washington
680 Vermont 2
 Addison
 Bennington
 Rutland
 Windham
 Windsor
681 Virginia 1
 Buchanan
 Dickenson
 Lee
 Norton City
 Russell
 Wise
682 Virginia 2
 Bland
 Galax City
 Grayson
 Smyth
 Tazewell
 Wythe
683 Virginia 3
 Carroll
 Floyd
 Giles
 Montgomery
 Patrick
 Pulaski
 Radford City
684 Virginia 4
 Bedford
 Bedford City
 Franklin
 Henry
 Martinsville City
685 Virginia 5
 Alleghany
 Bath
 Buena Vista City
 Clifton Forge City
 Covington City
 Lexington City
 Rockbridge
686 Virginia 6
 Augusta
 Harrisonburg City
 Highland
 Nelson
 Rockingham
 Staunton City
 Waynesboro City
687 Virginia 7
 Buckingham
 Charlotte
 Cumberland
 Halifax
 Prince Edward
 South Boston City

688 Virginia 8
  Amelia
  Brunswick
  Lunenburg
  Mecklenburg
  Nottoway
689 Virginia 9
  Emporia City
  Franklin City
  Greensville
  Isle of Wight
  Southampton
  Surrey
  Sussex
690 Virginia 10
  Clarke
  Fauquier
  Frederick
  Page
  Rappahannock
  Shenandoah
  Warren
  Winchester City
691 Virginia 11
  Culpeper
  Fredericksburg City
  Louisa
  Madison
  Orange
  Spotsylvania
  Stafford
692 Virginia 12
  Accomack
  Caroline
  Essex
  King and Queen
  King George
  King William
  Lancaster
  Mathews
  Middlesex
  Northampton
  Northumberland
  Richmond
  Westmoreland
693 Washington 1
  Clallam
  Island
  Jefferson
  San Juan
  Skagit
694 Washington 2
  Chelan
  Douglas
  Okanogan
695 Washington 3
  Ferry
  Pend Oreille
  Stevens
696 Washington 4
  Grays Harbor
  Mason
697 Washington 5
  Adams
  Grant
  Kittitas
  Lincoln
698 Washington 6
  Cowlitz
  Lewis
  Pacific
  Wahkiakum
699 Washington 7
  Klickitat
  Skamania
700 Washington 8
  Asotin
  Columbia
  Garfield
  Walla Walla
  Whitman
701 West Virginia 1
  Calhoun

Jackson
  Mason
  Roane
702 West Virginia 2
  Doddridge
  Gilmer
  Lewis
  Pleasants
  Ritchie
  Tyler
  Wetzel
703 West Virginia 3
  Barbour
  Harrison
  Marion
  Monongalia
  Preston
  Taylor
704 West Virginia 4
  Berkeley
  Grant
  Hampshire
  Hardy
  Jefferson
  Morgan
  Pendleton
705 West Virginia 5
  Braxton
  Clay
  Nicholas
  Pocahantas
  Randolph
  Tucker
  Upshur
  Webster
706 West Virginia 6
  Boone
  Lincoln
  Logan
  McDowell
  Mingo
  Wyoming
707 West Virginia 7
  Fayette
  Greenbriar
  Mercer
  Monroe
  Raleigh
  Summers
708 Wisconsin 1
  Barron
  Burnett
  Polk
  Washburn
709 Wisconsin 2
  Ashland
  Bayfield
  Iron
  Price
  Rusk
  Sawyer
710 Wisconsin 3
  Florence
  Forest
  Langlade
  Lincoln
  Oneida
  Taylor
  Vilas
711 Wisconsin 4
  Marinette
  Menominee
  Oconto
  Shawano
712 Wisconsin 5
  Buffalo
  Dunn
  Pepin
  Pierce
713 Wisconsin 6
  Clark
  Jackson
  Monroe

Trempealeau
714 Wisconsin 7
  Adams
  Green Lake
  Juneau
  Marquette
  Portage
  Waupaca
  Waushara
  Wood
715 Wisconsin 8
  Crawford
  Grant
  Green
  Iowa
  Lafayette
  Richland
  Sauk
  Vernon
716 Wisconsin 9
  Columbia
  Dodge
  Fond du Lac
  Jefferson
  Walworth
717 Wisconsin 10
  Door
  Kewaunee
  Manitowoc
718 Wyoming 1
  Big Horn
  Hot Springs
  Park
  Washakie
719 Wyoming 2
  Campbell
  Crook
  Johnson
  Sheridan
  Weston
720 Wyoming 3
  Carbon
  Fremont
  Lincoln
  Sublette
  Sweetwater
  Teton
  Uinta
721 Wyoming 4
  Albany
  Goshen
  Laramie
  Niobrara
  Platte
722 Wyoming 5
  Converse
723 Puerto Rico 1
  Rincon
724 Puerto Rico 2
  Adjuntas
  Guanica
  Guanyanilla
  Lajas
  Lares
  Las Marias
  Maricao
  Penuelas
  Sabana Grande
  San Sebastian
  Yauco
725 Puerto Rico 3
  Ciales
  Jayuya
  Morovis
  Orocovis
  Utuado
726 Puerto Rico 4
  Aibonito
  Arroyo
  Barranquitas
  Coamo
  Comerio

Guayama
  Maunabo
  Patillas
  Salinas
  Santa Isabel
  Yabucoa
727 Puerto Rico 5
  Cieba
  Naguabo
728 Puerto Rico 6
  Vieques
729 Puerto Rico 7
  Culebra
730 Virgin Islands, U.S. 1
  Alleg
  St. John & environs
  St. Thomas
731 Virgin Islands, U.S. 2
  St. Croix & environs
732 Guam and environs
733 American Samoa
  Eastern District
  Manu's District
  Rose Island
  Swains Island
  Western District
734 Northern Mariana Isls.
  Rota
  Saipan
  Tinian

United States

ALABAMA
all locations	205

ALASKA
all locations	907

ARIZONA
all locations	602

ARKANSAS
all locations	501

CALIFORNIA
Anaheim	714
Bakersfield	805
Beverly Hills	310
Concord	510
Eureka	707
Fresno	209
Glendale	818
Hayward	510
Long Beach	310
Los Angeles	213
Modesto	209
Oakland	510
Orange	714
Riverside	909
Sacramento	916
San Diego	619
San Bernardino	909
San Fernando	818
San Francisco	415
San Jose	408
Santa Barbara	805
Santa Monica	310
Stockton	209
Torrance	310
Walnut Creek	510

COLORADO
Aspen	303
Boulder	303
Colorado Springs	719
Denver	303
Fort Collins	303
Grand Junction	303
Leadville	719
Pueblo	719
Steamboat Springs	303

CONNECTICUT
all locations	203

DELAWARE
all locations	302

DIST. OF COLUMBIA
Washington	202

FLORIDA
Apalachicola	904
Atlantic Beach	904
Avon Park	813
Bowling Green	813
Cape Canaveral	305
Cape Kennedy	305
Cocoa Beach	407
Coral Gables	305
Daytona Beach	904
De Land	904
Edgewater	904
Everglades	813
Fort Lauderdale	305
Fort Myers	813
Fort Pierce	407
Gulf Port	813
Hialeah	305
Hollywood	305
Inverness	904
Jacksonville	904
Key Largo	305
Key West	305
Melbourne	407
Miami	305
Miami Beach	305
Orlando	407
Palm Beach	305
Pensacola	904
St. Augustine	904
St. Cloud	305
St. Petersburg	813
Sarasota	813
South Miami	305
Tallahassee	904
Tampa	813
West Palm Beach	407
Winter Haven	813

GEORGIA
Athens	404
Atlanta	404
Augusta	404
Columbus	404
Decatur	404
La Grange	404
Macon	912
Norcross	404
Rome	404
Savannah	912
Stockbridge	404
Valdosta	912
Waycross	912

HAWAII
all locations	808

IDAHO
all locations	208

ILLINOIS
Alton	618
Arlington Heights	708
Aurora	708
Bloomington	309
Broadview	708
Calumet City	708
Casey	217
Champaign	217
Chicago	312
Chillicothe	309
Columbia	618
Danville	217
Decatur	217
Des Plaines	708
East St. Louis	618
Elgin	708
Elmwood Park	708
Evanston	708
Glen Carbon	618
Harrisburg	618
Havana	309
Highland Park	708
Joliet	815
LaGrange	708
Lake Forest	708
Lansing	708
La Salle	815
Madison	618
Morris	815
Mt. Vernon	618
North Chicago	708
Oak Park	708
Park Forest	708
Park Ridge	708
Peoria	309
Plainfield	815
Quincy	217
Rockfalls	815
Rock Island	309
Rockford	815
Schaumburg	708
Seneca	815
Springfield	217
Sterling	815
Waukegan	708
Winnetka	708

INDIANA
Evansville	812
Ft. Wayne	219
Gary	219
Hammond	219
Indianapolis	317
Kokomo	317
Michigan City	219
South Bend	219
Warsaw	219

IOWA
Cedar Rapids	319
Council Bluffs	712
Des Moines	515
Dubuque	319
Mason City	515
Sioux City	712

KANSAS
Abilene	913
Arkansas City	316
Atchison	913
Augusta	316
Bloomington	913
Chanute	316
Coffeyville	316
Dodge City	316
El Dorado	316
Garden City	316
Hutchinson	316
Kansas City	913
Lawrence	913
Leavenworth	913
Manhattan	913
Ottawa	913
Salina	913
Topeka	913
Wichita	316

KENTUCKY
Ashland	606
Covington	606
Lexington	606
Louisville	502
Paducah	502
Shelbyville	502
Winchester	606

LOUISIANA
Baton Rouge	504
Lake Charles	318
New Orleans	504
Shreveport	318
Springfield	504

MAINE
all locations	207

MARYLAND
Annapolis	410
Baltimore	410
Cumberland	301
Frederick	301
Hagerstown	301
Rockville	301
Salisbury	410

MASSACHUSETTS
Arlington	617
Boston	617
Brockton	508
Brookline	617
Cambridge	617
Chicopee	413
Concord	508
Dorchester	617
Dover	508
East Boston	617
Fall River	508
Falmouth	508
Framingham	508
Holyoke	413
Hyannis	508
Hyde Park	617
Ipswich	508
Jamaica Plain	617
Lawrence	508
Lexington	617
Lowell	508
Milford	508
Milton	617
New Bedford	508
Newton	617
Northhampton	413
Peabody	508
Plymouth	508
Quincy	617
Randolph	617
Reading	617
Rockland	508
Salem	508
Southwick	413
Springfield	413
Stoneham	617
Stoughton	617
Tauton	508
Wakefield	617
Waltham	617
Warren	413
Wellesley	617
Westfield	413
Westminster	508
Whitman	617
Wilmington	508
Winchester	617
Worcester	508

MICHIGAN
Ann Arbor	313
Battle Creek	616
Bay City	517
Benton Harbor	616
Cheboygan	616
Dearborn	313
Detroit	313
East Lansing	517
Flint	313
Grand Rapids	616
Jackson	517
Kalamazoo	616
Lansing	517
Lincoln Park	313
Marquette	906
Milford	313
Monroe	313
Muskegon	616
New Haven	313
Oak Park	313
Saginaw	517

MINNESOTA
Austin	507
Duluth	218
Lake Minnetonka	612
Minneapolis	612
New Prague	612
Northfield	507
Plainview	507
Proctor	218
Red Wing	612
Rochester	507
St. Cloud	612
St. Louis Park	612
St. Paul	612

Area Codes

Sauk Rapids	612
Warren	218
MISSISSIPPI	
all locations	601
MISSOURI	
Bolivar	417
Branson	417
Chillicothe	816
Columbia	314
Ft. Leonard Wood	314
Grandview	816
Independence	816
Joplin	417
Kansas City	816
Oran	314
Poplar Bluff	314
St. Joseph	816
St. Louis	314
Springfield	417
University City	314
MONTANA	
all locations	406
NEBRASKA	
Grand Island	308
Hastings	402
Lincoln	402
North Platte	308
Omaha	402
Sidney	308
NEVADA	
all locations	702
NEW HAMPSHIRE	
all locations	603
NEW JERSEY	
Allentown	609
Asbury Park	201
Atlantic City	609
Bayonne	201
Caldwell	201
Camden	609
Delaware	201
Dover	201
East Orange	201
Englewood	201
Garfield	201
Glen Ridge	201
Hanover	201
Hoboken	201
Jersey City	201
Kenilworth	201
Lakewood	201
Linwood	609
Long Branch	201
Lyndhurst	201
Medford	609
Middletown	201
Montclair	201
Mount Holly	609
Newark	201
New Brunswick	201
Northfield	609
Oakhurst	201
Orange	201
Paterson	201
Perth Amboy	201
Plainfield	201
Princeton	609
Ridgewood	201
Salem	609
Springfield	201
Trenton	609
Union	201
Vineland	609
Wayne	201
Westfield	201
West New York	201
Wharton	201
Wyckoff	201
NEW MEXICO	
all locations	505
NEW YORK	
Albany	518
Allegany	716
Amsterdam	518
Apalachin	607
Babylon	516
Bronx	212
Brooklyn	718
Buffalo	716
Carthage	315
(Cellular phones)	917
Corning	607
Elmira	607
Far Rockaway	718
Flanders	516
Flushing	718
Fulton	315
Garden City	516
Goshen	914
Great Neck	516
Grand Island	716
Greenwich	518
Hudson	518
Hyde Park	914
Ithaca	607
Kennedy Int'l Airp.	718
Kingston	914
Lackawanna	716
La Guardia Airport	718
Lake George	518
Lake Success	516
Levittown	516
Long Beach	516
Manhattan	212
Mount Vernon	914
New Rochelle	914
New York City	212
Niagara Falls	716
Ossining	914
Oswego	315
Pelham	914
Potsdam	315
Poughkeepsie	914
Queens	718
Rochester	716
Rockville Centre	516
Roosevelt	516
Rye	914
Scarsdale	914
Schenectady	518
Staten Island	718
Syracuse	315
Ticonderoga	518
Troy	518
Utica	315
West Point	914
White Plains	914
Yonkers	914
NORTH CAROLINA	
Asheboro	919
Asheville	704
Burlington	919
Carthage	919
Charlotte	704
Durham	919
Fayetteville	919
Goldsboro	919
Greensboro	919
Greenville	919
Havelock	919
Kannapolis	704
Newton	704
Raleigh	919
Rocky Mount	919
Sparta	919
Valdese	704
Wendell	919
Wilson	919
Winston-Salem	919
NORTH DAKOTA	
all locations	701
OHIO	
Akron	216
Amherst	216
Ashtabula	216
Brunswick	216
Camden	513
Canton	216
Cincinnati	513
Cleveland	216
Columbus	614
Dayton	513
East Cleveland	216
Euclid	216
Garfield Heights	216
Germantown	513
Hamilton	513
Hudson	216
Lakewood	216
Lancaster	614
Mansfield	419
Marion	614
Massillon	216
McArthur	614
Middleton	216
Newtown	513
Niles	216
Plymouth	419
Reading	419
Salem	216
Sandusky	419
Solon	216
South Euclid	216
Springfield	513
Steubenville	614
Toledo	419
Warren	216
Youngstown	216
Zanesville	614
OKLAHOMA	
Ardmore	405
Bartlesville	918
Bethany	405
Claremore	918
El Reno	405
Guthrie	405
Muskogee	918
Norman	405
Oklahoma City	405
Shawnee	405
Stillwater	405
Tulsa	918
Wynona	918
OREGON	
all locations	503
PENNSYLVANIA	
Allegheny	412
Allentown	215
Altoona	814
Bellevue	412
Bethlehem	215
Bristol	215
Brook Haven	215
Chester	215
Clarion	814
Duquesne	412
Easton	215
Erie	814
Germantown	215
Gettysburg	717
Harrisburg	717
Hershey	717
Lancaster	717
Lebanon	717
Levittown	215
McKeesport	412
Media	215
Milton	717
Penn Hills	412
Philadelphia	215
Pittsburgh	412
Pottstown	215
Reading	215
St. Marys	814
Scranton	717
Sharon	412
Strasburg	717
Wilkes-Barre	717
York	717
PUERTO RICO	
all locations	809
RHODE ISLAND	
all locations	401
SOUTH CAROLINA	
all locations	803
SOUTH DAKOTA	
all locations	605
TENNESSEE	
Athens	615
Bethesda	615
Carthage	615
Chattanooga	615
Covington	901
Gainesville	901
Jackson	901
Johnson City	615
Knoxville	615
Madison	615
Memphis	901
Nashville	615
Oak Ridge	615
TEXAS	
Abilene	915
Alamo	512
Amarillo	806
Arlington	817
Austin	512
Beaumont	409
Bonham	903
Bowie	817
Brownsville	210
Carthage	903
Corpus Christi	512
Crockett	713
Dallas	214
Denison	903
Denton	817
Eldorado	915
El Paso	915
Fort Bliss	915
Fort Worth	817
Galveston	409
Grand Prairie	214
Granger	512
Harlingen	210
Hereford	806
Houston	713
Killeen	817
Kingsville	512
Kyle	512
Laredo	210
Liberty	409
Longview	903
Lubbock	806

McAllen	210
Mesquite	214
Mineola	903
Nixon	512
Odessa	915
Pecos	915
Plainview	806
Port Arthur	409
Ranger	817
Rio Hondo	512
San Antonio	210
Sanger	817
Sulphur Springs	903
Sweetwater	915
Temple	817
Texarkana 214	903
Texas City	409
Tyler	903
Waco	817
Wichita Falls	817
UTAH	
all locations	801
VERMONT	
all locations	802
VIRGINIA	
Charlottesville	804
Fredericksburg	703
Harrisonburg	703
Lynchburg	804
Newport News	804
Norfolk	804
Richmond	804
Roanoke	703
Winchester	703
VIRGIN ISLANDS	
all locations	809
WASHINGTON	
Bellevue	206
Pullman	509
Everett	206
Mercer Island	206
Seattle	206
Spokane	509
Tacoma	206
Vancouver	206
Walla Walla	509
Yakima	509
WEST VIRGINIA	
all locations	304
WISCONSIN	
Appleton	414
Beloit	608
Eau Claire	715
Green Bay	414
La Crosse	608
Madison	608
Milwaukee	414
Racine	414
Wausau	715
WYOMING	
all locations	307

Canada

ALBERTA	
all locations	403
BRITISH COLUMBIA	
all locations	604
MANITOBA	
all locations	204
NEW BRUNSWICK	
all locations	506
NEWFOUNDLAND	
all locations	709
N.W. TERRITORIES	
all locations	403
NOVA SCOTIA	
all locations	902
ONTARIO	
Barrie	705
Belleville	613
Bracebridge	705
Brantford	519
Chatham	519
Cobourg	416
Collingwood	705
Cornwall	613
Hamilton	416
Kingston	613
Kitchener	519
London	519
Midland	705
Newmarket	416
Niagara Falls	416
North Bay	705
Orangeville	519
Orillia	705
Oshawa	416
Ottawa	613
Peterborough	705
St. Catherines	416
Sarnia	519
Simcoe	519
Sudbury	705
Toronto	416
Welland	416
Windsor	519
Woodstock	519
PRINCE EDWARD I.	
all locations	902
QUEBEC	
Alma	418
Baie-Comeau	418
Chicoutimi	418
Drummondville	819
Granby	514
Montreal	514
Noranda	819
Québec City	418
Rivière-du-Loup	418
Rouyn	819
Sherbrooke	819
St. Hyacinthe	514
St. Jean-sur-Richel.	514
St. Jerome	514
Thetford Mines	418
Trois Rivières	819
Val-d'or	819
Valleyfield	514
SASKATCHEWAN	
all locations	306
YUKON TERRITORY	
all locations	403

Numerical Order

201	New Jersey
202	District of Columbia
203	Connecticut
204	Manitoba
205	Alabama
206	Washington
207	Maine
208	Idaho
209	California
210	Texas
212	New York
213	California
214	Texas
215	Pennsylvania
216	Ohio
217	Illinois
218	Minnesota
219	Indiana
301	Maryland
302	Delaware
303	Colorado
304	West Virginia
305	Florida
306	Saskatchewan
307	Wyoming
308	Nebraska
309	Illinois
310	California
312	Illinois
313	Michigan
314	Missouri
315	New York
316	Kansas
317	Indiana
318	Louisiana
319	Iowa
401	Rhode Island
402	Nebraska
403	Alberta/NWT/Yukon
404	Georgia
405	Oklahoma
406	Montana
407	Florida
408	California
409	Texas
410	Maryland
412	Pennsylvania
413	Massachusetts
414	Wisconsin
415	California
416	Ontario
417	Missouri
418	Quebec
419	Ohio
501	Arkansas
502	Kentucky
503	Oregon
504	Louisiana
505	New Mexico
506	New Brunswick
507	Minnesota
508	Massachusetts
509	Washington
510	California
512	Texas
513	Ohio
514	Quebec
515	Iowa
516	New York
517	Michigan
518	New York
519	Ontario
601	Mississippi
602	Arizona
603	New Hampshire
604	British Columbia
605	South Dakota
606	Kentucky
607	New York
608	Wisconsin
609	New Jersey
612	Minnesota
613	Ontario
614	Ohio
615	Tennessee
616	Michigan
617	Massachusetts
618	Illinois
619	California
700	varies
701	North Dakota
702	Nevada
703	Virginia
704	North Carolina
705	Ontario
706	Georgia
707	California
708	Illinois
709	Newfoundland
712	Iowa
713	Texas
714	California
715	Wisconsin
716	New York
717	Pennsylvania
718	New York
719	Colorado
800	varies
801	Utah
802	Vermont
803	South Carolina
804	Virginia
805	California
806	Texas
807	Ontario
808	Hawaii
809	Caribbean Islands
812	Indiana
813	Florida
814	Pennsylvania
815	Illinois
816	Missouri
817	Texas
818	California
819	Quebec
900	varies
901	Tennessee
902	Nova Scotia/P.E.I.
903	Texas
904	Florida
906	Michigan
907	Alaska
908	New Jersey
909	California
912	Georgia
913	Kansas
914	New York
915	Texas
916	California
917	New York
918	Oklahoma
919	North Carolina

User's Guide

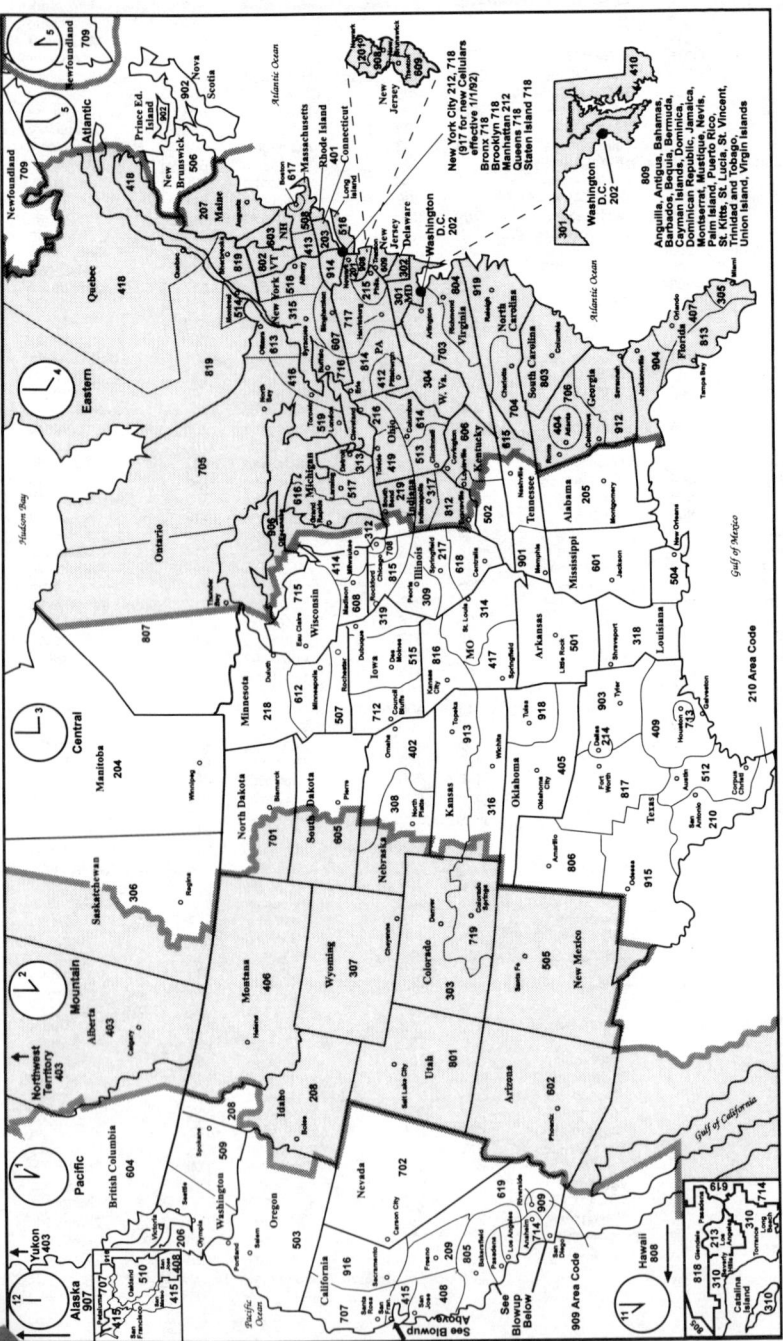